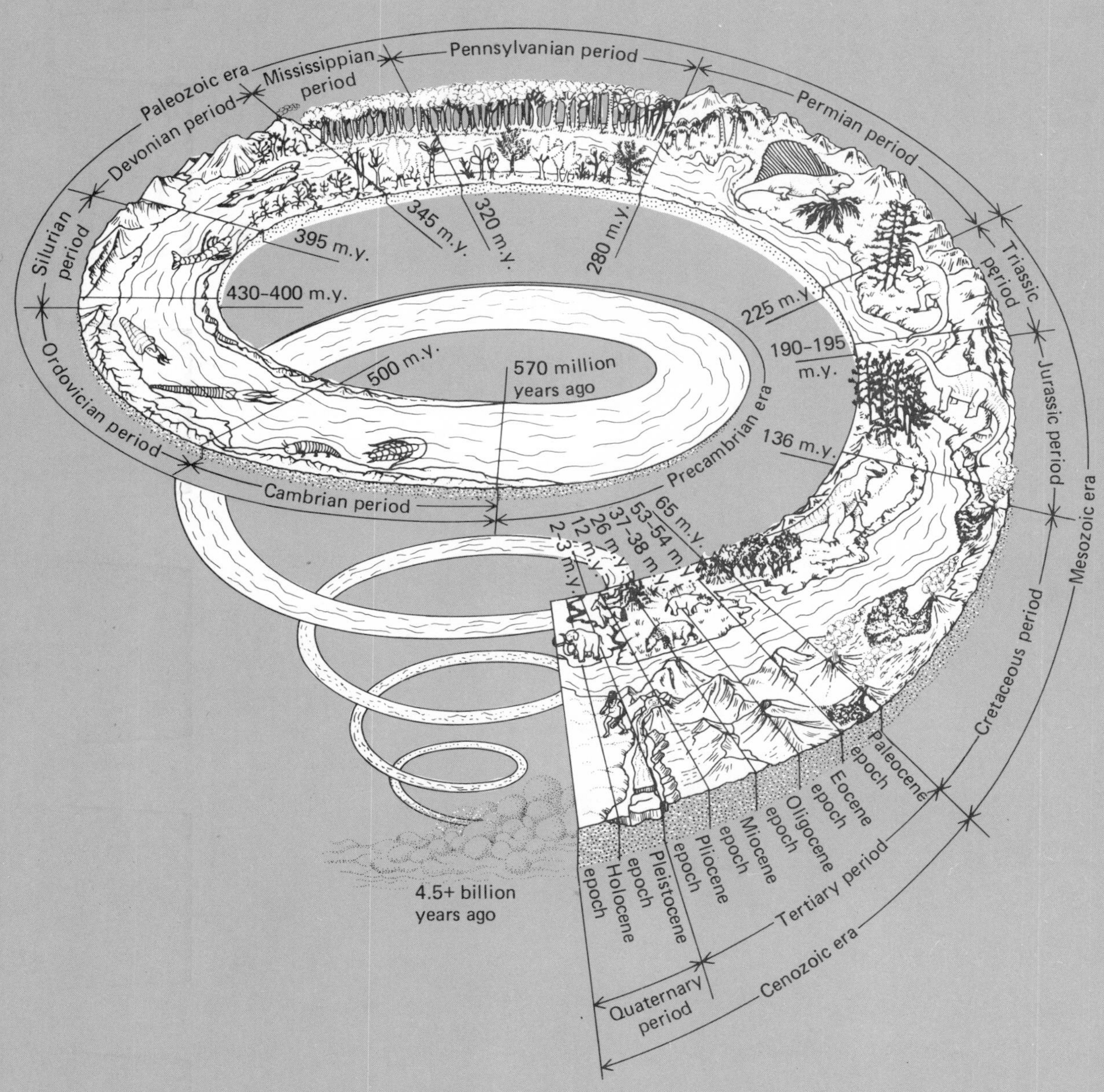

A geologic time scale and a history of the earth since the beginning of the Paleozoic Era.

General Zoology

Sixth Edition

General Zoology

Sixth Edition

Claude A. Villee
Harvard University

Warren F. Walker, Jr.
Oberlin College

Robert D. Barnes
Gettysburg College

SAUNDERS COLLEGE PUBLISHING
Philadelphia New York Chicago San Francisco Montreal Toronto London Sydney Tokyo Mexico City Rio de Janeiro Madrid

Address orders to:
383 Madison Avenue
New York, NY 10017

Address editorial correspondence to:
West Washington Square
Philadelphia, PA 19105

Text Typeface: Zapf Book
Compositor: Clarinda Company
Acquisitions Editor: Michael Brown
Developmental Assistant: Leesa Massey
Project Editor: Patrice L. Smith
Copyeditor: Rebecca Gruliow
Managing Editor & Art Director: Richard L. Moore
Art/Design Assistant: Virginia A. Bollard
Text Design: Caliber Design
Cover Design: Lawrence R. Didona
New Text Artwork: J & R Technical Services
Production Manager: Tim Frelick
Assistant Production Manager: Maureen Iannuzzi
Page Layout: Carol C. Bleistine

Cover credit: A male and female shrimp *(Pereclemenes imperator)* on a sea cucumber *(Bohadschiz graffei)*. Courtesy of Charles R. Seaborn.

Library of Congress Cataloging in Publication Data

Villee, Claude Alvin, 1917-
General zoology

Includes bibliographies and index.

1. Zoology. I. Walker, Warren Franklin.
II. Barnes, Robert D. III. Title.
QL47.2.V5 1984 591 83-20253

ISBN 0-03-062451-7

GENERAL ZOOLOGY/6e ISBN 0-03-062451-7

Library of Congress catalog card number 83-20253

4567 032 98765432

CBS COLLEGE PUBLISHING
Saunders College Publishing
Holt, Rinehart and Winston
The Dryden Press

Preface

A text book is remarkably like a living organism; it can grow (or decrease in size), adapt, evolve, and relate to its environment. In text books this process, which tends to occur cyclically, is termed a new edition. Our text of GENERAL ZOOLOGY is a prime example of one that changes greatly in form and substance from one edition to the next, yet remains unchanged in its goal of presenting clearly and understandably the theories and facts that constitute modern zoological science.

The sixth edition represents another major revision as we continue our effort to provide students and teachers with an accurate, up-to-date, and readable text. Many chapters have been rewritten and reorganized so that the sequence of chapters seems to be optimal. In Part I, the text begins with animal cells, their properties and the chemical basis of modern cell biology. This is followed in Part II by a presentation of the structure and functions of animals and their organ systems. We have used a comparative approach to the problems faced by all animals in their attempts to survive, and to the variety of solutions that the various groups have evolved. Symmetry, life style, skeletons, muscles, nutrition, gas exchange, circulatory systems, homeostasis, nervous systems, sense organs and hormonal controls are discussed in turn. A consideration of reproduction at the end of this section sets the stage for the discussion of the continuity of life in Part III. In this we consider both Mendelian and molecular aspects of genetics and follow with the basics of population genetics. The discussion of evolution, which has been extensively rewritten, develops from the discussion of population genetics and proceeds with discussions of evolutionary concepts and evidence.

Part IV deals with the immense diversity of animal forms, ranging from the protozoa (which most biologists now regard as protists, classified with algae and other unicellular organisms) and sponges (which lack most of the features typical of animals) to the primates, including human beings. Part IV is introduced by a chapter on the origin of life and a discussion of the methods used to classify animals and to determine their evolutionary relationships. A summary of the major phyla concludes this chapter. Chapters are devoted to each of the major invertebrate phyla and to each of the classes of vertebrates. The vertebrates are considered in detail because of the comparisons possible with the human condition and because of their importance in tracing the evolution of human beings. Notable changes have been made in the discussion of flight, the evolution of endothermy, and reproductive patterns in mammals and in the discussion of human evolution.

Finally, in Part V we examine the relations of animals to their environment as evidenced by their behavior and by their interrelations with their environment both living (biotic) and nonliving (abiotic). The chapter on behavior has been rewritten to focus on concepts of comparative ethology and the evolution of behavior.

We have added many features that will make it easier for students to use the book. Each chapter begins with an Orientation, a brief statement of the chapter objectives and the main points on which the student should focus his or her attention. This is followed, in the chapters dealing with the animal groups, by a summary of the important features of the group. In addition to the information and concepts considered in the text, we have placed some less essential but illustrative material in separate boxes. The derivation of most important terms is given in the text where the term is first mentioned, and should assist the student in reaching an understanding of these new terms. The glossary and index have been combined so that the student need consult only one list to find the meaning of a term and where in the text it is defined and used.

This edition has many new illustrations, both color and black and white. Illustrations in color are used where this will help clarify animal structure and depict the great beauty of many animal forms.

Acknowledgments

Many new illustrations have been prepared for this edition and we are deeply grateful for the care and artistry with which Mary Ann Nelson, Caroline Herbert, Britt Griswold, and Young Sohn prepared the many new line drawings. The contributions of other scientists and artists who have permitted us to use certain of their drawings and photographs are acknowledged with each figure. We also thank Jan Creidenberg for his help with various tasks in the preparation of the manuscript.

Our special thanks are due to Michael Brown, Lee Walters, and Leesa Massey for their help in developing this edition and to Patrice Smith for guiding it through the production phase. We are indebted to the following persons for the contribution of their time and experience to the sixth edition: Jerry M. Baskin, University of Kentucky; William A. Brueske, Humboldt State University; John A. Byers, University of Idaho; David Cook, Wayne State University; Fred Coyle, Western Carolina University; Jack A. Cranford, Virginia Polytechnic Institute; Gayle Davis, University of Alabama; Lee C. Drickamer, Williams College; William G. Dyer, University of Southern Illinois; Frank Fisher, Rice University; Lemuel Fraser, University of Wisconsin, Madison; Peter Goldman, Northeast Missouri State University; Judith Goodenough, University of Massachusetts, Amherst; Mark S. Hafner, Louisiana State University; W. Holt Harner, Broward Community College; Richard L. Hurley, Humboldt State University; Gerard F. Iwantsch, Fordham University; Gwilym S. Jones, Northeastern University; Byron H. Knapp, East Stroudsburg State College; Paul K. Lago, University of Mississippi; Harris J. Linder, University of Maryland; Margaret T. McCann; Elizabeth A. McMahan, University of North Carolina, Chapel Hill; Dale M. Madison, State University of New York, Binghamton; Donna Maglott, Howard University; Dean E. Metter, University of Missouri, Columbia; Dorothy B. Mooren, University of Wisconsin, Milwaukee; Willard C. Myser, The Ohio State University; David O. Norris, University of Colorado, Boulder; John A. Novak, Colgate University; Paul Peck, Montgomery College, Rockville; Robert Reed, Shippensburg State College; Daniel J. Simons, Montgomery College, Rockville; Curtis J. Swanson, Wayne State University; Barry D. Valentine, The Ohio State University; Frederick vom Saal, University of Missouri, Columbia; Harry Weller, Miami University of Ohio; F. M. Williams, Pennsylvania State University. We are grateful to them for their comments and suggestions, but we, of course, assume responsibility for any errors and shortcomings that remain.

Claude A. Villee
Warren F. Walker, Jr.
Robert D. Barnes

Contents Overview

PART ONE Animals Cells 1

1 Introduction: The Physical and Chemical Basis of Life 3
2 Cells and Tissues 30
3 Cell Metabolism 62

PART TWO The Organism: Animal Form and Function 85

4 Symmetry, Form, and Life Style 87
5 Body Covering, Support, and Movement 96
6 Animal Nutrition 125
7 Gas Exchange 148
8 Internal Transport 165
9 Regulation of Internal Body Fluids 193
10 Nervous Systems and Neural Integration 212
11 Receptors and Sense Organs 238
12 Endocrine Systems and Hormonal Integration 260
13 Reproduction I: Gamete Formation and Fertilization 289
14 Reproduction II: Embryonic Development 308

PART THREE Continuity of Animal Life 335

15 Heredity 337
16 Molecular Aspects of Genetics 362
17 The Concept of Evolution 387
18 The Mechanisms of Evolution 407

PART FOUR The Diversity of Animals 429

19 Origin of Life and Diversity of Life 431

20 Protozoa 443

21 Sponges 462

22 Cnidarians 470

23 Flatworms 491

24 Pseudocoelomates 505

25 Mollusks 517

26 Annelids 550

27 Arthropods 571

28 Bryozoans 617

29 Echinoderms 622

30 Protochordates 640

31 Vertebrates: Fishes 651

32 Vertebrates: Amphibians and Reptiles 673

33 Vertebrates: Birds 699

34 Vertebrates: Mammals 724

35 Vertebrates: Primates 746

PART FIVE Animals and Their Environment 759

36 Behavior 761

37 Dynamic Processes in Ecology 785

38 Ecology of Populations and Communities 803

Contents

PART ONE Animal Cells 1

1 Introduction: The Physical and Chemical Basis of Life 3

Zoology and Its Subsciences 4
The Scientific Method 5
History of Zoology 6
Applications of Zoology 11
Characteristics of Living Things 11
The Organization of Matter: Atoms and Molecules 14
Chemical Compounds 19
Electrolytes: Acids, Bases, and Salts 20
Organic Compounds of Biological Importance 21

2 Cells and Tissues 30

The Cell Theory 31
The Plasma Membrane 32
The Nucleus and Its Functions 33
Chromosomes 36
Cytoplasmic Organelles 36
The Cell Cycle 41
Mitosis 44
Regulation of Mitosis 48
The Study of Cellular Activities 48
Energy 50
Molecular Motion 51
Diffusion 51
Exchanges of Material Between Cell and Environment 52
Tissues 53

3 Cell Metabolism 62

Entropy and Energy 64
Chemical Reactions 64
Enzymes 65
Respiration and Cellular Energy 68
The Tricarboxylic Acid (TCA) Cycle 71

The Chemiosmotic Theory of Oxidative Phosphorylation 76
The Molecular Organization of Mitochondria 77
The Dynamic State of Cellular Constituents 80
Biosynthetic Processes 81

PART TWO The Organism: Animal Form and Function 85

4 Symmetry, Form, and Life Style 87

Motility 88
Animal Architecture 91
Size 92
Colonial Organization 93
Predictability in Animal Design 93

5 Body Covering, Support, and Movement 96

Body Covering 97
Support 99
Movement 105

6 Animal Nutrition 125

Digestion and Absorption 126
Evolution of the Animal Gut 127
Diets and Feeding Mechanisms 128
The Vertebrate Pattern 132
Regulation of Food Supply 140
Nonfoods in the Diet 141
Fuel Utilization: Metabolic Rates and Energy 143

7 Gas Exchange 148

Gases 149
Steps in Gas Exchange 149
Environmental Gas Exchange in Aquatic Animals 150
Environmental Gas Exchange in Terrestrial Animals 152
Gas Transport 160

8 Internal Transport 165

Functions of Transport Systems 166
Methods of Internal Transport 166
Blood and Interstitial Fluid 167
Red Blood Cells 168
Hemostasis 168
White Blood Cells 170
Immunity 171
Blood Group Factors 175
Invertebrate Circulatory Patterns 177
Vertebrate Circulatory Patterns 177
Fetal and Neonatal Circulations 182
The Propulsion of Blood and Hemolymph 183
The Peripheral Flow of Blood and Hemolymph 187

9 Regulation of Internal Body Fluids 193

Nitrogenous Wastes 194
Excretory Organs 195
The Vertebrate Kidney 198
Osmotic Regulation in Marine Animals 205
Osmoregulation in Freshwater Organisms 208
Osmoregulation in Terrestrial Animals 208

10 Nervous Systems and Neural Integration 212

Irritability and Response 213
The Neuron 213
The Nerve Impulse 216
Transmission at the Synapse 218
Evolution and Organization of the Nervous System 221
Organization of the Vertebrate Nervous System 223
The Peripheral Nervous System 224
The Central Nervous System: The Spinal Cord 228
The Central Nervous System: The Brain 229

11 Receptors and Sense Organs 238

Receptor Mechanisms 239
Sensory Coding and Sensation 241
Mechanoreceptors 242
Chemoreceptors 248
Photoreceptors 249
The Vertebrate Eye 251
The Compound Eye of Arthropods 256
Thermoreceptors 257

12 Endocrine Systems and Hormonal Integration 260

Endocrine Glands 261
Molecular Mechanisms of Hormone Action 261
The Hormones of Vertebrates 264
The Thyroid Gland 264
The Parathyroid Glands 268
The Islet Cells of the Pancreas 269
The Adrenal (Suprarenal) Glands 270
The Pituitary Gland 272
Hypothalamic Releasing Hormones: Neurosecretion 276
The Pineal 277
The Testes 277
The Ovaries 278
The Estrous and Menstrual Cycles 279
The Hormones of Pregnancy 281
Hormonal Control of Lactation 281
Prostaglandins 282
The Hormones of Arthropods 283

13 Reproduction I: Gamete Formation and Fertilization 289

Asexual Versus Sexual Reproduction 290
Meiosis 290

Spermatogenesis 293
Oögenesis 295
Fertilization 297
Parthenogenesis 298
Adaptations for Fertilization 298
Egg Deposition 302
Predictability in Gonoduct Design 302
Vertebrate Reproductive Patterns 302

14 Reproduction II: Embryonic Development 308

Cleavage 309
Gastrulation and Mesoderm Formation 311
Morphogenetic Movements and Differentiation 314
Organogenesis 317
Controls in Embryonic Development 319
Adaptations for Development 322
Mammalian Development 326

PART THREE Continuity of Animal Life 335

15 Heredity 337

History of Genetics 338
Chromosomal Basis of the Laws of Heredity 339
A Monohybrid Cross 339
The Mathematical Basis of Genetics: The Laws of Probability 341
Population Genetics 342
Gene Pools and Genotypes 344
Incomplete Dominance 345
Carriers of Genetic Diseases 345
Mendel's Laws of Segregation and Independent Assortment 346
Deducing Genotypes 346
The Genetic Determination of Sex 347
Linkage and Crossing Over 350
Genic Interactions 352
Polygenic Inheritance 355
Multiple Alleles 357
Inbreeding, Outbreeding, and Hybrid Vigor 359
Problems in Genetics 359

16 Molecular Aspects of Genetics 362

The Chemistry of Chromosomes 363
The Genetic Code 367
The Synthesis of DNA: Replication 369
Transcription of the Code: The Synthesis of RNA 371
The Synthesis of a Specific Polypeptide Chain 373
Changes in Genes: Mutations 375
Human Cytogenetics 377
Gene–Enzyme Relations 380
Genes and Differentiation 380
Lethal Genes 383
Penetrance and Expressivity 384

17 The Concept of Evolution 387

Evidence for Evolution 388
The Theory of Natural Selection 402
Development of Evolutionary Ideas 402

18 The Mechanisms of Evolution 407

Microevolution: Evolutionary Changes Within Populations 408
Speciation 417
Macroevolution or Transpecific Evolution 421

PART FOUR The Diversity of Animals 429

19 The Origin and Diversity of Life 431

The Origin of Life 432
The Grouping of Species 433
Taxonomic Nomenclature 436
Adaptive Diversity 437
How to Study Animal Groups 437
Synopsis of the Phyla of Metazoan Animals 439

20 Protozoa 443

Phylum Ciliophora 444
Phylum Sarcomastigophora: Flagellates 449
Phylum Sarcomastigophora: Sarcodines 454
Sporozoans 457
Classification of Protozoa 460

21 Sponges 462

Structure and Function of Sponges 463
Regeneration and Reproduction 467
Classification of Phylum Porifera 468

22 Cnidarians 470

Cnidarian Structure and Function 471
Class Hydrozoa 474
Class Scyphozoa 479
Class Anthozoa 481
Classification of Cnidarians 488

23 The Flatworms 491

Class Turbellaria 492
Symbiosis and Parasitism 497
Flukes 498
Class Cestoda 500
Classification of Flatworms 503

24 Pseudocoelomates 505

Phylum Rotifera 507
Phylum Gastrotricha 509
Phylum Nematoda 509
Classification of Pseudocoelomates 515

25 Mollusks 517

General Features of the Phylum Mollusca 518
The Ancestral Mollusks 518
Class Gastropoda 521
Polyplacophora and Monoplacophora 528
Class Bivalvia 531
Class Cephalopoda 539
Classification of the Mollusca 547

26 Annelids 550

Metamerism and Locomotion 551
Classification 552
Class Polychaeta 552
Class Oligochaeta 561
Class Hirudinea 565
Classification of the Phylum Annelida 569

27 Arthropods 571

Arthropod Classification 575
Subphylum Trilobitomorpha 575
Subphylum Chelicerata 576
Class Merostomata 576
Class Arachnida 577
Subphylum Crustacea 586
Subphylum Uniramia 597
Classification of the Phylum Arthropoda 614

28 Bryozoans 617

Structure of a Bryozoan Individual 618
Organization of Colonies 619
Reproduction 620

29 Echinoderms 622

Class Stelleroidea: Asteroids 624
Class Stelleroidea: Ophiuroids 628
Class Echinoidea 631
Class Holothuroidea 633
Class Crinoidea 635
Fossil Echinoderms 636
Classification of the Phylum Echinodermata 638

30 Protochordates 640

Subphylum Urochordata 641
Chordate Metamerism 644

Subphylum Cephalochordata 644
Subphylum Vertebrata 646
Phylum Hemichordata 646
Deuterostome Relationships and Chordate Origins 647
Classification of the Phylum Hemichordata 649
Classification of the Phylum Chordata 649

31 Vertebrates: Fishes 651

Aquatic Adaptations 652
Class Agnatha 653
Jaws and Paired Appendages 658
Class Chondrichthyes 658
Class Osteichthyes 660
Classification of Fishes 669

32 Vertebrates: Amphibians and Reptiles 673

The Transition from Water to Land 674
Class Amphibia 674
Class Reptilia 683
Classification of Amphibians and Reptiles 696

33 Vertebrates: Birds 699

Principles of Flight 700
The Adaptive Features of Birds 701
The Origin and Evolution of Birds 712
Migration and Navigation 717
Classification of Birds 720

34 Vertebrates: Mammals 724

Major Adaptations of Mammals 725
Primitive Mammals 730
Therians 730
Classification of Mammals 743

35 Vertebrates: Primates 746

Primate Adaptations 747
The Groups of Primates 747
Human Characteristics 751
Early Evolution of Apes and Hominids 752
The Ape-Men 752
The Protohominids 753
Early *Homo* 754
Homo Sapiens 756

PART FIVE Animals and Their Environment 759

36 Behavior 761

Causation 763
Development 768

Genetics and Evolution of Behavior 772
Behavioral Ecology 773

37 Dynamic Processes in Ecology 785

The Concepts of Ranges and Limits 786
Habitat and Ecologic Niche 787
Competitive Exclusion Principle 789
The Concept of the Ecosystem 790
The Physical Environment 792
The Cyclic Use of Matter 792
Solar Radiation 794
Energy Flow and Food Chains 798

38 Ecology of Populations and Communities 803

Populations and Their Characteristics 804
Population Cycles 808
Population Dispersion and Territoriality 809
Biotic Communities 810
Community Succession 811
Geographical Ecology 812
The Tundra Biome 814
The Forest Biomes 815
The Grassland Biome 817
The Chaparral Biome 818
The Desert Biome 818
The Edge of the Sea: Marshes and Estuaries 819
Marine Life Zones 820
Freshwater Life Zones 824
The Dynamic Balance of Nature 827
Human Ecology 828

Glossary/Index 833

PART ONE **Animal Cells**

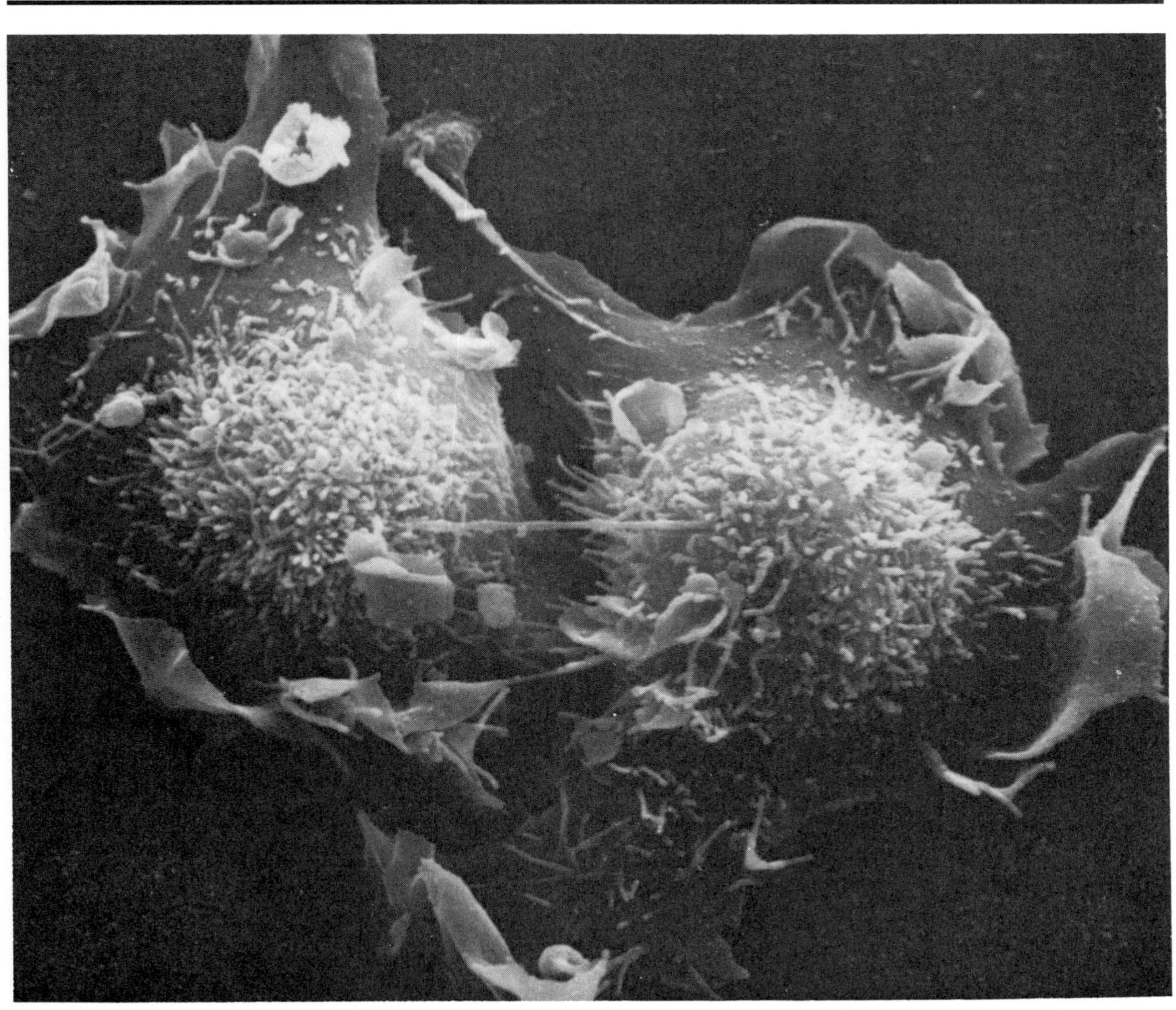

1 Introduction: The Physical and Chemical Basis of Life

ORIENTATION

In this chapter we will discuss the characteristics that distinguish living from nonliving systems and the surprising difficulties that occur in making a clear differentiation between the two. We will discuss the inorganic and organic chemical components of living systems and the physical principles that regulate their activity. We will focus on the structure and biological functions of the several classes of organic molecules. These molecules, containing carbon, make up almost all of the compounds found in living systems.

Zoology and its Subsciences

Zoology, one of the biological sciences, deals with animals and the many different aspects of animal life. A "zoo" is a collection of animals, but a visit to a zoo, interesting though it is, can barely begin to suggest the enormous variety of animals living today. More than one million different kinds of animals have been described and classified—nearly 800,000 of these are insects! Each of these kinds has become adapted, through a long evolutionary process, to the particular set of environmental circumstances under which it survives best. In addition to the ones living at present, a host of other kinds of animals have lived in past ages but are now extinct.

Modern zoology concerns itself with much more than the simple recognition and classification of the many kinds of animals. It includes studies of the structure, function, and embryonic development of each part of an animal's body; of the nutrition, health, and behavior of animals; of their heredity and evolution; and of their relations to the physical environment and to the plants and other animals of that region.

Enough facts about animals and their ways are known to fill a whole library of books, and more information appears every year from the intensive researches of zoologists in the field and in the laboratory. No zoologist today can know more than a small fraction of this enormous body of knowledge. Zoology is now much too broad a subject to be encompassed by a single scientist or to be treated thoroughly in a single textbook. Most zoologists are specialists in some limited phase of the subject—in one of the subdivisions of zoology. The sciences of **anatomy, physiology,** and **embryology** deal with the structure, function, and development, respectively, of an animal. Each of these may be further subdivided according to the kind of animal investigated, e.g., invertebrate physiology, arthropod physiology, insect physiology, or comparative physiology. **Parasitology** deals with those forms of life that live in or on and at the expense of other organisms. **Cytology** is concerned with the structure, composition, and function of cells and their parts, and **histology** is the science of the structure, function, and composition of tissues. The science of **genetics** investigates the mode of transmission of characteristics from one generation to the next and is closely related to the science of **evolution,** which studies the way in which new species of animals arise and how the present kinds of animals are related by descent to previous animals. The study of the classification of organisms, both animals and plants, is called **taxonomy.** The science of **ecology** is concerned with the relations of a group of organisms to their environment, including both the physical factors and the other forms of life that provide food or shelter for them, compete with them in some way, or prey upon them.

Some zoologists specialize in the study of one group of animals. There are **mammalogists, ornithologists, herpetologists,** and **ichthyologists,** who study mammals, birds, reptiles and amphibians, and fishes, respectively; **entomologists,** who investigate insects; **protozoologists,** who study the single-celled animals, and so on.

Advances in chemistry and physics have made possible quantitative studies of the molecular structures and events underlying biological processes. The term **molecular biology** has been applied to analyses of gene structure and function and genic control of the synthesis of enzymes and other proteins, studies of subcellular structures and their roles in regulatory processes within the cell, investigations of the mechanisms underlying cellular differentiation, and analyses of the molecular basis of evolution by comparative studies of the molecular structure of specific proteins—enzymes, hormones, cytochromes, hemoglobins—in different species.

The science of zoology thus includes both a tremendous body of facts and theories about animals and the means for learning more. The ultimate source of each fact is in some carefully controlled observation or experiment. In earlier times, some scientists kept their discoveries to themselves, but there is now a strong tradition that scientific discoveries are public property and should be freely published. In a scientific publication the investigator must give all of the relevant details of the means by which the discovery was made so that others can repeat the observation. It is this criterion of *repeatability* that leads us to accept a certain observation or experiment as representing a true fact; observations that cannot be repeated by competent investigators are discarded.

When a scientist has made some new observation or carried out a series of experiments that add to our knowledge in a field, he writes a report, called a "paper," in which he describes his methods in sufficient detail so that another worker can repeat them, gives the results of his observations, discusses the conclusions to be drawn from them, perhaps formulates a theory to explain them or discusses how they are explained by a previous theory, and finally indicates the place of these new facts in their particular field of science. The knowledge that his discovery will be subjected to keen scrutiny by his colleagues is a strong stimulus for repeating the observations or experiments carefully before publishing them. He then submits his paper to one of the professional journals in the particular field of his discovery. Several thousand zoological journals are published currently all over the world. Some of the more important American ones are the *Journal of Experimental Zoology, Journal of Cellular and Comparative Physi-*

ology, Biological Bulletin, Physiological Zoology, American Journal of Physiology, Anatomical Record, Ecology, and the journals devoted to research on a particular group of animals, such as the *Journal of Mammalogy.* The paper is read by one or more of the board of editors of the journal, all of whom are experts in the field. If it is approved, it is published and becomes part of "the literature" of the subject.

At one time, when there were far fewer journals, it might have been possible to read them all each month as they appeared, but this is obviously impossible now. Journals such as *Biological Abstracts* assist the hard-pressed zoologist by publishing, classified by fields, short summaries or abstracts of each paper published, giving the facts found, the conclusion reached, and an exact reference to the journal in which the full report appears. The ultimate in brevity is *Current Contents,* which simply lists the authors and titles of the papers published in several hundred journals, together with the name, volume, and pages of the journal in which the paper was published. A considerable number of journals devoted solely to reviewing the newer developments in particular fields of science have sprung up in the past 50 years; some of these are *Physiological Reviews, Quarterly Review of Biology, Nutrition Reviews, Annual Review of Physiology,* and *Vitamins and Hormones.* The new fact or theory thus becomes widely known through publication in the appropriate professional journal and by reference in abstract and review journals and eventually may become a sentence or two in a textbook.

The professional societies of zoologists and the various special branches of zoology have annual meetings at which new discoveries may be reported. Two of the largest annual meetings are those of the American Institute of Biological Sciences and the Federation of American Societies for Experimental Biology. There are, in addition, national and international gatherings, called **symposia,** of specialists in a given field to discuss the newer findings and the present status of the knowledge in that field. For example, the discussions of the Cold Spring Harbor Symposia in Quantitative Biology, held each June at the Long Island Biological Laboratory in Cold Spring Harbor, are published and provide an excellent review of some particular field. A different subject is discussed each year.

The Scientific Method

The ultimate aim of each science is to reduce the apparent complexity of natural phenomena to simple, fundamental ideas and relations, to discover all the facts and the relationships among them. The essence of the scientific method is the posing of questions and the search for answers, but they must be "scientific" questions, arising from observations and experiments, and "scientific" answers, ones that are testable by further observation and experiment. The Danish physicist Niels Bohr put it this way, "The task of science is both to extend the range of our experience and to reduce it to order."

There is, however, no single "scientific method," no regular, infallible sequence of events that will reveal scientific truths. Different scientists go about their work in different ways. The ultimate source of all the facts of science is careful, close observation and experiment, free of bias and done as quantitatively as possible. The observations or experiments may then be analyzed, or simplified into their constituent parts, so that some sort of order can be brought into the observed phenomena. Then the parts can be reassembled and their interactions made clear. On the basis of these observations, the scientist constructs a **hypothesis,** a trial idea about the nature of the observation, or about the connections between a chain of events, or even about cause and effect relationships between different events. It is in this ability to see through a mass of data and construct a reasonable hypothesis to explain their relationships that scientists differ most and that true genius shows itself.

The role of a hypothesis is to penetrate beyond the immediate data and place them into a new, larger context, so that we can interpret the unknown in terms of the known. There is no sharp distinction between the usage of the words "hypothesis" and "theory," but the latter has, in general, the connotation of greater certainty than a hypothesis. A **theory** is a conceptual scheme that tries to explain the observed phenomena and the relationships between them, so as to bring into one structure the observations and hypotheses of several different fields. The theory of evolution, for example, provides a conceptual scheme into which fit a host of observations and hypotheses from paleontology, anatomy, physiology, biochemistry, genetics, and other allied sciences.

A good theory correlates many previously separate facts into a logical, easily understood framework. The theory, by arranging the facts properly, suggests new relationships between the individual facts and suggests further experiments or observations that might be made to test these relationships. A good theory should be simple and should not require a separate proviso to explain each fact; it should be flexible, able to grow and to undergo modifications in the light of new data. A theory is not discarded because of the existence of some isolated fact that contradicts it, but only because some other theory is better able to explain all of the known data. A hypothesis must be subject to some sort of experimental test—i.e., it must make a prediction that can be verified in some way—or it is mere speculation. Conversely, unless a prediction follows as the logical outgrowth of some theory, it is no more than a guess.

The finding of results contrary to those predicted by the hypothesis causes the investigator, after he has assured himself of the validity of his observation, either to discard the hypothesis or to change it to account for both the original data and the new data. Hypotheses are constantly being refined and elaborated. Few scientists would regard any hypothesis, no matter how many times it may have been tested, as a statement of absolute and universal truth. It is rather regarded as the best available approximation to the truth for some finite range of circumstances.

The history of science shows that, although many scientists have made their discoveries by following the precepts of the ideal scientific method, there have been occasions on which important and far-reaching theories have resulted from making incorrect conclusions from erroneous postulates or from the misinterpretation of an improperly controlled experiment! There are instances in which, in retrospect, it seems clear that all the evidence for the formulation of the correct theory was known, yet no scientist put the proper two and two together. And there are other instances in which scientists have been able to establish the correct theory despite an abundance of seemingly contradictory evidence.

The proper design of experiments is a science in itself and one for which only general rules can be made. In all experiments, the scientist must ever be on his guard against bias in himself, bias in the subject, bias in his instruments, and bias in the way the experiment is designed.

Each experiment must include the proper **control group** (indeed, some experiments require several kinds of control groups). The control group is one treated exactly like the experimental group in all respects but one, the factor whose effect is being tested. The use of controls in medical experiments raises the difficult question of the ethical justification of withholding treatment from a patient who might be benefited by it. If there is sufficient evidence that one treatment is indeed better than another, a physician would hardly be justified in further experimentation. However, the medical literature is full of treatments now known to be useless or even detrimental, which were used for many years, only to be abandoned finally as experience showed that they were ineffective and that the evidence that had originally suggested their use was improperly controlled. There is a time in the development of any new treatment when the medical profession is not only morally justified, but really morally required, to do carefully controlled tests on human beings to be sure that the new treatment is better than the former one.

In medical testing, it is not sufficient simply to give a treatment to one group of patients and not to give it to another, for it is widely known that there is a strong psychological effect in simply giving a treatment of any sort. For example, a group of students in a large western university served as subjects for a test of the hypothesis that daily doses of extra amounts of vitamin C might help to prevent colds. This idea grew out of the observation that people who drink lots of fruit juices seem to have fewer colds. The group receiving the vitamin C showed a 65 percent reduction in the number of colds contracted during the winter in which they received treatment as compared with the previous winter when they had no treatment. There were enough students in the group (208) to make this result statistically significant. In the absence of controls, one would have been led to the conclusion that vitamin C does help to prevent colds. A second group of students were given "placebos," pills identical in size, shape, color, and taste to the vitamin C pills but without any vitamin C. The students were not told who was getting vitamin C and who was not; they only knew they were getting pills that might help to prevent colds. The group getting the placebos reported that they had a 63 percent reduction in the number of colds! This controlled experiment thus showed that vitamin C had nothing to do with the decrease in the number of colds and that the reductions reported in both groups were either psychological effects or simply the result of a lesser amount of cold virus on the campus that year. In a "double blind" study the physician does not know which subject is receiving the experimental compound and which the placebo. Each patient is given pills with a different code number, and only after the experiment has been completed is the code revealed.

History of Zoology

Our interest in animals is probably somewhat older than the human race, for the apemen and man-apes that preceded us in evolution undoubtedly learned at an early time which animals were dangerous, which could be hunted for food, clothing, or shelter, where these were to be found, and so on. Some of our prehistoric ancestors' impressions of the contemporary animals have survived in the cave paintings of France and Spain (Fig. 1.1). Some animals were regarded as good or as evil spirits. Later man decorated pottery, tools, cloth, and other objects with animal figures.

FIGURE 1.1 Paintings by Upper Paleolithic man from the wall of the cavern at Lascaux, Dordogne, France. (Photo by Windels Montignac.)

The early Egyptians had a wealth of knowledge about animals and had domesticated cattle, sheep, pigs, cats, geese, and ducks. The Greek philosophers of the fifth and sixth centuries B.C., Anaximander, Xenophanes, Empedocles, and others, speculated on the origin of the animals of the earth. One of the earliest classifications of animals is found in a Greek medical book of that time which classifies animals primarily as to whether or not they are edible. Aristotle (384–322 B.C.) was one of the greatest Greek philosophers and wrote on many topics. His *Historia animalium* contains a great deal of information about the animals of Greece and the nearby regions of Asia Minor. Aristotle's descriptions are quite good and are recognizable as those of particular animals living today. The breadth and depth of his zoological interests are impressive—he made a careful study of the development of the chick and of the breeding of sharks and bees, and he had notions about the functions of the human organs, some of which, not too surprisingly, were quite wrong. He believed that nature strives to change from the simple and imperfect to the more complex and perfect and arranged all organisms in a "ladder of nature," going from the simple to the complex. His contributions to logic, such as the development of the system of inductive reasoning from specific observations to a generalization that explains them all, have been of inestimable value to all branches of science.

The Greek physician Galen (131–201 A.D.) was one of the first to do experiments and dissections of animals to determine structure and functions (Fig. 1.2). The first experimental physiologist, he made some notable discoveries on the functions of the brain and nerves and demonstrated that arteries carry blood and not air. His descriptions of the human body were the unquestioned authority for some 1300 years, even though they contained some remarkable errors, being based on dissections of pigs and monkeys rather than of human bodies. Pliny (23–79 A.D.) and others in succeeding centuries compiled encyclopedias (Pliny's *Natural History* was a 37-volume work) regarding the kinds of animals and

FIGURE 1.2 Galen (131–201 A.D.). The founder of experimental physiology. (Courtesy of *J.A.M.A.*, from the series of Medical Greats.)

FIGURE 1.3 Andreas Vesalius (1514–1564). In his *De Humani corporis fabrica,* he established the basis for modern anatomy. (From Garrison: *History of Medicine.* 4th ed. Philadelphia, W. B. Saunders Co., 1929.)

where they lived that are remarkable mixtures of fact and fiction. Some of the ones written in the Middle Ages were called "bestiaries." The zoological books written in the Middle Ages are, almost without exception, copied from Aristotle, Galen, and Pliny; no original observations were made to corroborate or refute the accuracy of these authorities.

The Renaissance in science began slowly with scholars such as Roger Bacon (1214–1294) and Albertus Magnus (1206–1280), who were interested in all branches of natural science and philosophy. The genius Leonardo da Vinci (1452–1519) was an anatomist and physiologist as well as a painter, engineer, and inventor. He made many original observations in zoology, some of which came to light only much later when his notebooks were deciphered.

One of the first to question the authority of Galen's descriptions of human anatomy was the Belgian Andreas Vesalius (1514–1564), who was professor at the University of Padua in Italy (Fig. 1.3). By dissecting human bodies and making detailed, clear drawings of what he saw, Vesalius revealed many of the inaccuracies in Galen's descriptions of the human body. He published his observations and illustrations in *De humani corporis fabrica (On the Structure of the Human Body)* in 1543. Since Vesalius dared to reject the authority of Galen, he was the object of much adverse criticism and was finally forced to leave his professorial post.

Just as Vesalius had emphasized the importance of relying on original observation rather than on authority in anatomy, so did William Harvey (1578–1657) in physiology (Fig. 1.4). Harvey, an English physician, received his medical training at the University of Padua, where Vesalius had taught. He returned to England and investigated the circulation of the blood. In 1628 he published *Exercitatio anatomica de motu cordis et sanguinis in animalibus (Anatomical Studies on the Motion of the Heart and Blood in Animals).* At that time blood was believed to be generated in the liver from food and to pass just once to the organs of the body where it was used up. The heart was believed to be nonmuscular and to be expanded passively by the inflowing blood. Harvey described, from direct observations on animals, how first the atria and then the ventricles fill and empty by muscular contraction. He showed by experiment that when an artery is cut blood spurts from it in rhythm with the beating of the heart and that when a vein is clamped it becomes full of blood on the side away from the heart and empty on the side toward the heart. He showed that valves in the veins permit blood to flow

FIGURE 1.4 William Harvey (1578–1657). He proved the circulation of the blood by the first quantitative physiologic demonstration. (From Garrison: *History of Medicine.* 4th ed. Philadelphia, W. B. Saunders Co., 1929.)

toward the heart but not in the reverse direction (Fig. 1.5). From these experiments he concluded that blood is carried away from the heart in arteries and back to the heart in veins. Furthermore, by measuring how much blood is delivered by each beat of the heart, and by measuring the number of heartbeats per minute, he could calculate the total flow of blood through the heart per minute or hour. He found this flow to be so great that it could not be generated anew in the liver but must be recirculated, used over and over again. This was the first quantitative physiological argument. He inferred that there must be small vessels connecting arteries and veins to complete the circular path of the blood but, lacking a microscope, he was unable to see them. In later years he made a careful study of the development of the chick, published in 1651 as *Exercitationes de generatione animalium.* In this work he postulated that mammals, like the chick, develop from an egg.

The development of the compound microscope by the Janssens in 1590 and by Galileo in 1610 provided the means for attacking many problems in zoology and botany. Robert Hooke, (1635–1703), Marcello Malpighi (1628–1694), Antony van Leeuwenhoek (1632–1723), and Jan Swammerdam (1637–1680) were some of the first microscopists. They studied the fine structure of plant and animal tissues. Hooke was the first to describe the presence of "cells" in plant tissue, Leeuwenhoek was the first to describe bacteria, protozoa, and sperm, and Malpighi was the first to describe the capillaries connecting arteries with veins. The light microscope has been modified and improved greatly in the past century, and our ability to see the fine structure of cells has been greatly extended by the invention of the phase microscope and of the transmission and scanning electron microscopes. These, with good resolution at magnifications as great as 100,000 to 500,000 diameters, have revealed a whole new level of complexity in the structure of all kinds of cells.

John Ray (1627–1705) and Linnaeus (Karl von Linné) (1707–1778) brought order into the classification of animals and plants and devised the binomial system (two names, genus and species) for the scientific naming of the kinds of animals and plants. Linnaeus first used this binomial system consistently in the tenth edition of his *Systema naturae* (1758).

Contributions to our understanding of the embryonic development of animals were made by Fabricius, the professor of Anatomy at Padua who taught William Harvey, and by Harvey, Malpighi, and Kaspar Wolff (1759). Wolff proposed the theory of epigenesis, the concept that there is a gradual differentiation of structure during development from a relatively structureless egg. Karl Ernst von Baer (1792–1876) established the theory of germ layers and emphasized the need for comparative studies of development in different animals.

Following William Harvey, physiology was advanced by René Descartes (1596–1650), who was a philosopher rather than an experimenter. He believed that "animal spirits" are generated in the heart, stored in the brain, and pass through the nerves to the muscles, causing contraction or relaxation, according to their quantity. Charles Bell (1774–1842) and François Magendie (1783–1855) made notable contributions to our understanding of the function of the brain and spinal nerves. Johannes Müller (1801– 1858) studied the properties of nerves and capillaries; his textbook of physiology stimulated a great deal of interest and research in the field. Claude Bernard (1813–1878) was one of the great advocates of experimental physiology and contributed significantly to our understanding of the role of the liver, heart, brain, and placenta. Henry Bowditch (1840–1911) discovered the "all-or-none" principle of the contraction of heart muscle and established the first laboratory for teaching physiology in the United States. Ernest Starling (1866–1927) made many contributions to the physiology of circulation and the nature of lymph and with William Bayliss (1866–1924) elucidated the hormonal control of the secretion of pancreatic juice.

The Scottish anatomist John Hunter (1728–1793) and the French anatomist Georges Cuvier (1769–1832) were

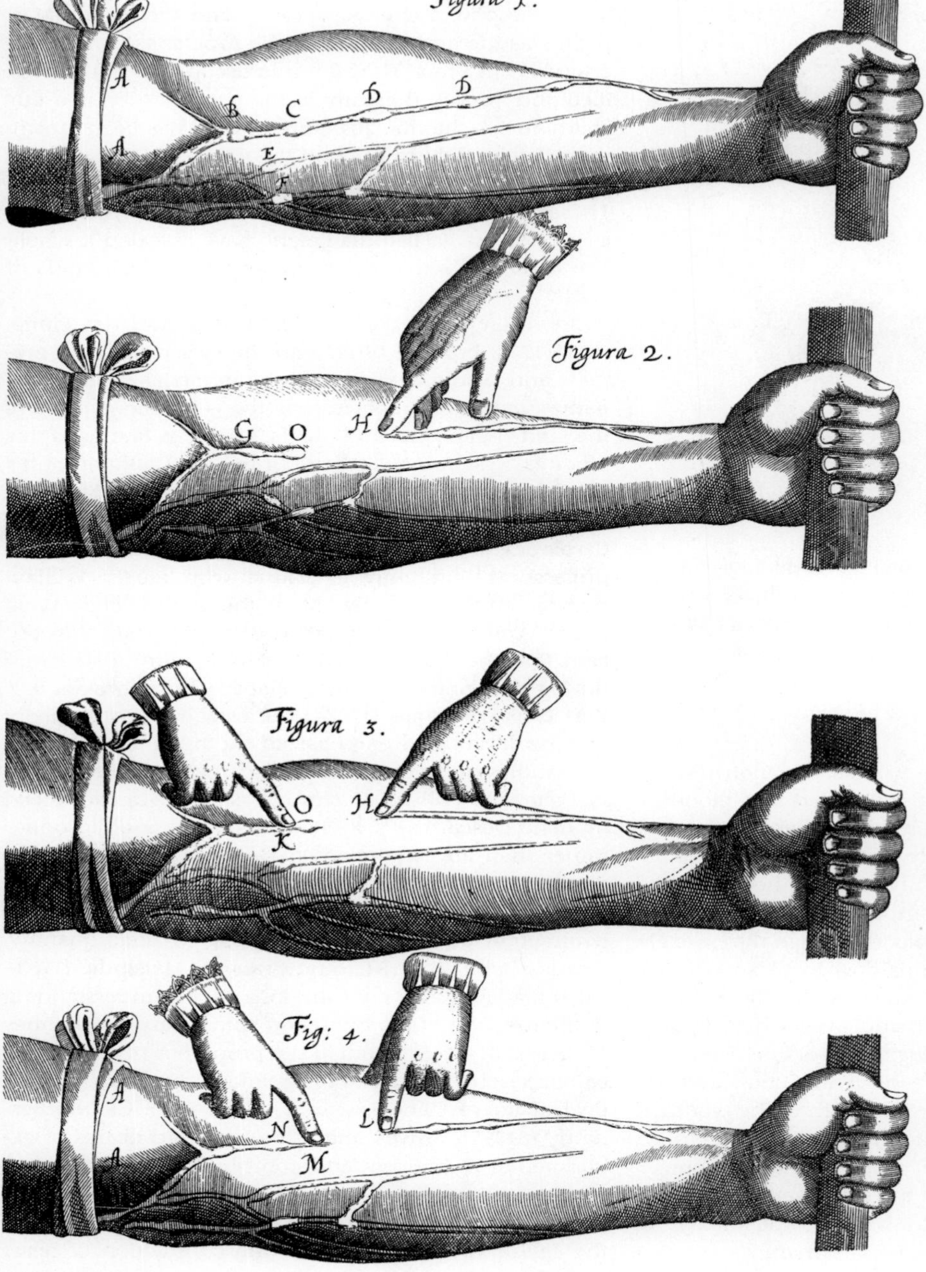

FIGURE 1.5 Harvey's illustrations to demonstrate the direction of blood flow. *1,* The formation of "knots" at valves. *2,* Stripping a portion of a vessel to show that there is no back flow. *3* and *4,* Demonstration of valvular blocking of blood flow. (From Harvey: *Exercitatio Anatomica de Motu Cordis et Sanguinis in Animalibus.* Translation by C. D. Leake: *Anatomical Studies on the Motion of the Heart and Blood in Animals.* Springfield, Illinois, Charles C Thomas, 1931.)

pioneers in the field of comparative anatomy, studying the same structure in different animals. Richard Owen (1804–1892) developed the concepts of homology and analogy. Cuvier was one of the first to study the structure of fossils as well as of living animals and is credited with founding the science of paleontology. Cuvier believed strongly in the unchanging nature of species and carried on bitter debates with Lamarck, who in 1809 proposed a theory of evolution based on the idea of the inheritance of acquired characters.

One of the most important and fruitful concepts in biology is the **cell theory,** which has gradually grown since Robert Hooke first saw, with the newly invented microscope, the dead cell walls in a piece of cork. The French biologist René Dutrochet clearly stated in 1824 that "all organic tissues are actually globular cells of exceeding smallness, which appear to be united only by simple adhesive forces; thus all tissues, all animal organs are actually only a cellular tissue variously modified." Dutrochet recognized that growth is the result of the increase in the volume of individual cells and of the addition of new cells. Two Germans, botanist M. J. Schleiden and zoologist Theodor Schwann, studied many different plant and animal tissues and are generally credited with formulating the cell theory (1838), for they showed that cells are the units of structure in plants and animals and that organisms are aggregates of cells arranged according to definite laws. The presence of a nucleus within the cell, now recognized as an almost universal feature of cells, was first described by Robert Brown in 1831.

Zoology, along with the other biological sciences, has expanded at a tremendous rate in the twentieth century, with the establishment of the subsciences of cytology, embryology, genetics, evolution, biochemistry, biophysics, endocrinology, and ecology. The discoveries and new techniques of chemistry and physics have made possible new approaches to the biological sciences that have attracted the attention of many biologists. So many men have contributed to the growth of zoology that only a few in each field can be mentioned: Mendel, deVries, Morgan, Bridges, and Müller in genetics; Darwin, Dobzhansky, Wright, and Goldschmidt in evolution; Harrison and Spemann in embryology; and Crick, Watson, Kornberg, and Nirenberg in molecular biology. Many others will be mentioned as these subjects are discussed in detail in the text.

The establishment and growth of the marine biological laboratories, such as the ones at Naples (Italy), Woods Hole (MA), Pacific Grove (CA), Friday Harbor (WA), and elsewhere, have played an important role in fostering research in zoological sciences. There are comparable stations for the study of freshwater biology, such as the one at Douglas Lake, Michigan.

Applications of Zoology

Some of the practical uses of a knowledge of zoology will become apparent as the student proceeds through this text. Zoology is basic in many ways to the fields of human and veterinary medicine and public health, agriculture, conservation, and to certain of the social sciences. There are esthetic values in the study of zoology, for a knowledge of the structure and functions of the major types of animals will greatly increase the pleasure of a stroll in the woods or an excursion along the seashore. Trips to zoos, aquariums, and museums are also rewarding in the glimpses they give of the host of different kinds of animals. Many of these are beautifully colored and shaped, graceful or amusing to watch, but all will mean more to a person equipped with the basic knowledge of zoology that enables him to recognize them and to understand the ways in which they are adapted to survive in their native habitats.

Characteristics of Living Things

Most biologists are agreed that all the varied phenomena of life are ultimately explainable in terms of the same physical and chemical principles that define nonliving systems. A corollary of this is that when enough is known of the chemistry and physics of vital phenomena it may be possible to synthesize living matter.

To differentiate the living from the nonliving and then to separate the living into plants and animals are difficult to do sharply and clearly. Organisms such as cats, clams, and cicadas are clearly recognizable as animals, but sponges, for example, were considered to be plants until well into the nineteenth century, and there are single-celled organisms which, even today, are called animals by zoologists, plants by botanists, and protists by others. Even the line between living and nonliving is rather tenuous, for the viruses, too small to be seen with an ordinary light microscope, can be considered either the simplest living things or very complex, but nonliving, organic chemicals.

All living things have, to a greater or lesser degree, the properties of specific organization, irritability, movement, metabolism, growth, reproduction, and adaptation.

Organization. Each kind of living organism is recognized by its characteristic form and appearance; the adult organism usually has a characteristic size. Nonliving things generally have much more variable shapes and sizes. Living things are not homogeneous but are made of different parts, each with special functions; thus the bodies of animals and plants are characterized by a specific, complex organization.

The structural and functional unit of both animals and plants is the **cell.** It is the simplest bit of living matter than can exist independently and exhibit all the characteristics of life. The processes of the entire organism are the sum of the coordinated functions of its constituent cells. These cellular units vary considerably in size, shape, and function. Some of the smallest animals have bodies made of a single cell; the human body, in contrast, is made of countless billions of cells fitted together.

The cell itself has a specific organization, and each kind of cell has a characteristic size and shape by means of which it can be recognized. A typical cell, such as a liver cell (Fig. 1.6), is polygonal in shape, with a **plasma membrane** separating the living substance from the surroundings. Almost without exception, each cell has a **nucleus,** typically spherical or ovoid in shape, which is separated from the rest of the cell by a **nuclear membrane.** The nucleus, as we shall see later, has a major role in controlling and regulating the cell's activities. It contains the hereditary units or **genes.** A cell experimentally deprived of its nucleus usually dies in a short time; even if it survives for several days, it is unable to reproduce.

The bodies of higher animals are organized in a hierarchy of increasingly complex levels: Cells are organized into **tissues,** tissues into **organs,** and organs into **organ systems.**

Irritability. Living things are irritable; they respond to stimuli, to physical or chemical changes in their immediate surroundings. Stimuli that are effective in evoking a response in most animals and plants are changes in light (either in its color, intensity or direction), temperature, pressure, sound, and in the chemical composition of the earth, water, or air surrounding the animal. In humans and other complex animals, certain cells of the body are highly specialized to respond to certain types of stimuli: The rods and cones in the retina of the eye respond to light, certain cells in the nose and in the taste buds of the tongue respond to chemical stimuli, and special groups of cells in the skin respond to changes in temperature or pressure. In lower animals, such specialized cells may be absent, but the whole organism responds to any one of a variety of stimuli. Single-celled amoebas will respond by moving toward or away from heat or cold, certain chemical substances, or the touch of a microneedle. Indeed, many of the cells of higher animals have a similar generalized sensitivity.

Movement. Living things are characterized further by their ability to move. The movement of most animals is quite obvious—they wiggle, swim, run, or fly. The movements of plants are much slower and less obvious but are present nonetheless. A few animals—sponges, corals, hydroids, oysters, certain parasites—do not move from place to place, but most of these have microscopic,

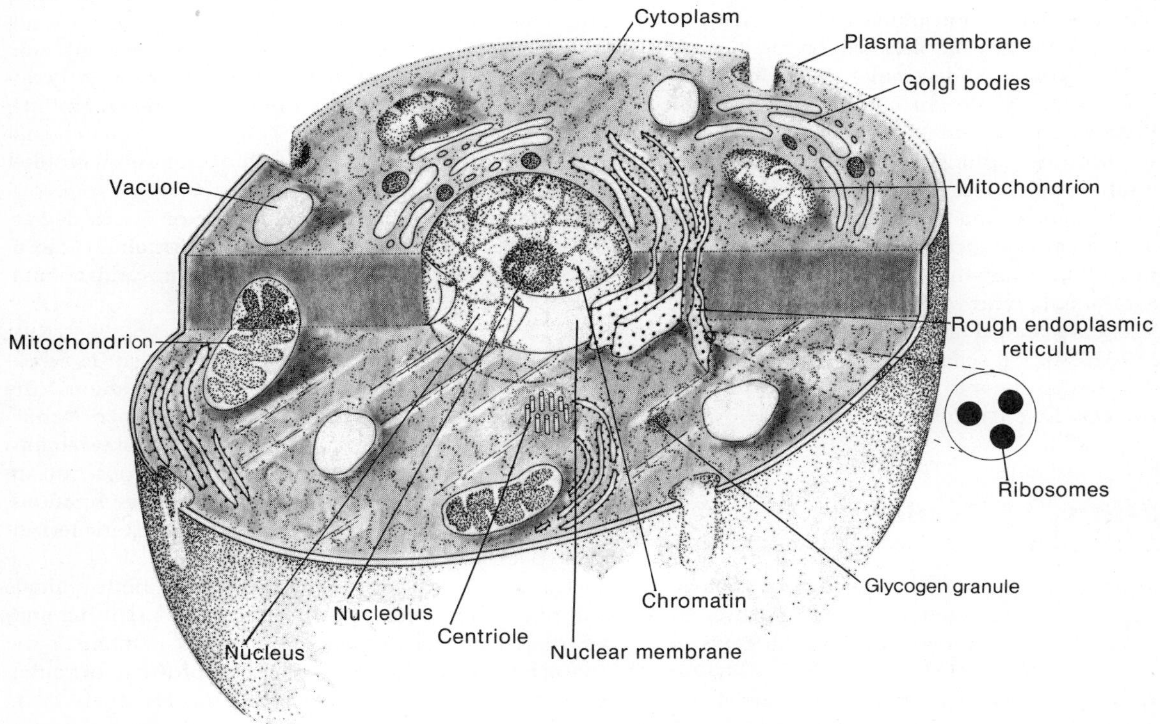

FIGURE 1.6 Schematic drawing of a generalized animal cell.

hairlike projections from the cells, called **cilia** or **flagella,** to move their surroundings past their bodies and thus bring food and other necessities of life to themselves. The movement of an animal body may be the result of muscular contraction, of the beating of cilia or flagella, or of the slow oozing of a mass of cell substance (known as amoeboid motion).

Metabolism. All living things carry on a wide variety of chemical reactions, the sum of which we call **metabolism** (Gr. *metabolē,* change). All cells are constantly taking in new substances, altering them chemically in a multitude of ways, building new cell components, and transforming the potential energy contained in large molecules of carbohydrates, fats, and proteins into kinetic energy and heat as these substances are converted into simpler ones. The never-ending flow of energy within a cell, from one cell to another and from one organism to the next, is the very essence of life, one of the unique and characteristic attributes of living things. The rate of metabolism is affected by temperature, age, sex, general health and nutrition, hormones, and many other factors, even the time of day. The study of energy transformations in living organisms is termed **bioenergetics.**

Those metabolic processes in which simpler substances are combined to form more complex substances and that result in the storage of energy and the production of new cellular materials are termed **anabolic** (Gr. *ana,* up + *bolē,* throw). The opposite processes, in which complex substances are broken down to release energy, are called **catabolic** (Gr. *kata,* down + *bolē,* throw). Both types of metabolism occur continuously and are intricately interdependent so that they become, in practice, difficult to distinguish. Furthermore, the synthesis of most molecules requires energy, so that some catabolic processes must occur to supply the energy to drive the anabolic reactions of these syntheses.

Growth. Both plants and animals grow; nonliving things do not. The increase in mass may be brought about by an increase in the size of the individual cells, by an increase in the *number* of cells, or by both. An increase in cell size may occur by the simple uptake of water, but this swelling is not generally considered to be growth. The term **growth** is restricted to those processes that increase the amount of living substance of the body, measured by the amount of nitrogen, protein, or nucleic acid present. Growth may be uniform in the several parts of an organism, or, perhaps more commonly, growth is differential, greater in some parts than in others, so that the body proportions change as growth occurs.

Reproduction. If there is any one characteristic that can be said to be the *sine qua non* of life, it is the ability to reproduce. Although at one time worms were believed to arise from horse hairs in a trough of water, maggots from decaying meat, and frogs from the mud of the Nile, we now know that each can come only from previously existing ones. One of the fundamental tenets of biology is that "all life comes only from living things."

The classic experiment disproving the **spontaneous generation of life** was performed by an Italian, Francesco Redi, about 1680, Redi proved that maggots do not come from decaying meat by this simple experiment: He placed a piece of meat in each of three jars, leaving one uncovered, covering the second with a piece of fine gauze, and covering the third with parchment. All three pieces of meat decayed, but maggots appeared on only the meat in the uncovered jar. A few maggots appeared on the gauze of the second jar, but not on the meat, and no maggots were found on the meat covered by parchment. Redi thus demonstrated that the maggots did not come from the decaying meat, but hatched from eggs laid by blowflies attracted by the smell of the decaying meat. Further observations showed that the maggots develop into flies which in turn lay more eggs. Louis Pasteur, about 200 years later, showed that bacteria do not arise by spontaneous generation but only from previously existing bacteria. The submicroscopic filtrable viruses do not arise from nonviral material by spontaneous generation; the multiplication of viruses requires the presence of previously existing viruses.

The problem of the original source of life will be discussed later (p. 432), but it is likely that billions of years ago, when physical and chemical conditions on the earth's surface were quite different from those at present, the first living things *did* actually arise from nonliving material.

The process of reproduction may be as simple as the splitting of one individual into two. In most animals, however, it involves the production of specialized eggs and sperm that unite to form the zygote or fertilized egg, from which the new organism develops. In some animals, the liver flukes for example, reproduction involves several quite different forms, each of which gives rise to the next in succession, until the cycle is completed and the adult reappears.

Adaptation. The ability of an animal or plant to adapt to its environment is the characteristic that enables it to survive the exigencies of a changing world. Each particular species can achieve adaptation either by seeking a suitable environment or by undergoing modifications to make it better fitted to its present surroundings. Adaptation may involve immediate changes that depend upon the irritability of cells or the responses of the enzyme systems to inducers or repressors, or it may be the result of a long-term process of mutation and selection. It is obvious that no single kind of organism can adapt to all the conceivable kinds of environment; hence there will be certain areas where it cannot survive. The list of factors that may limit the distribution of a species is almost endless: water, light, temperature, food, predators, competitors, parasites, and so on.

The Organization of Matter: Atoms and Molecules

To get a more complete idea of what living matter is and what it can do requires some understanding of certain basic principles of physics and chemistry.

Matter and Energy. The universe consists of **matter** and **energy,** which are related by the famous Einstein equation, $E = mc^2$, where E = energy, m = mass, and c = the velocity of light, which is a constant. Although matter can be converted into energy in a nuclear reactor, in the familiar, everyday world matter and energy are separate and distinguishable. Matter occupies space and has mass, and energy is the ability to produce a change or motion in matter—the ability to do work. Energy may take the form of heat, light, electricity, motion, or chemical energy. We shall return (p. 63) to the kinds of energy transformations of importance in biological phenomena.

Atoms. Regardless of the form—gaseous, liquid, or solid—that matter may assume, it is always composed of units called **atoms.** In nature there are 92 different kinds of atoms, ranging from hydrogen, the smallest, to uranium, the largest.

At one time the atom was believed to be the ultimate, smallest unit of matter. However, atoms are divisible into even smaller particles organized around a central core, as our solar system of planets is organized around the sun. The exact number and kind of these particles and their arrangement in the atom are the subject of continuing research. For our purposes we need consider only three types: **electrons,** which have a negative electric charge and an extremely small mass or weight; **protons,** which have a positive electric charge and are about 1800 times as heavy as electrons; and **neutrons,** which have no electric charge, but have essentially the same mass as protons.

BOX 1.1

Atomic Structure

The classic Rutherford-Bohr concept of atomic structure stated that the positively charged protons are concentrated in the center of the atom, the nucleus, and the negatively charged electrons circle the nucleus in precisely defined orbits or "shells," just as the planets circle the sun in precise orbits. This concept is still useful in interpreting chemical phenomena, but it is now clear that electrons do not move in precise orbits. Rather, the electrons surrounding the atomic nucleus are present in discrete energy levels, termed quantum levels. The motion of an electron is described by quantum numbers and orbitals. The term "orbital" describes the motion of an electron in space corresponding to a particular energy level of the atom. The concept of the behavior of matter, termed "quantum mechanics," describes how an electron moves around the nucleus. Quantum theory defines the probability that an electron will be found a certain distance out from the nucleus, but it does not predict the exact path along which the electron moves. By considering the probability of finding an electron at any given point in space over a period of time, we obtain an averaged picture of the motion of an electron. An orbital is such an averaged picture of the motion of an electron. Quantum theory also provides a prediction of how the electron orbitals change as the energy state of the atom, described by the principal quantum number, n, increases. Electrons can move from one orbital to another as they gain or lose energy. No more than two electrons may occupy an orbital. Electron orbitals have different sizes and shapes and may be spherical, elliptical, or may be represented by more complex three-dimensional coordinates. Elliptical orbitals have various orientations in space and are assigned to one of three axes, each at right angles to the other two. The orbitals are arranged in a series of increasing energy levels: The innermost orbital, with the least energy, is 1s (spherical). At the second energy level there is a spherical orbital, 2s, larger than the first, since the more energetic electrons can wander through a larger space, plus three elliptical orbitals, each at right angles to the other two. These are called $2p_x$, $2p_y$, and $2p_z$. Each can hold two electrons, so there are a total of six 2p electrons. The third energy level contains one 3s (spherical) orbital, three 3p elliptical orbitals, and five more complex-shaped orbitals termed 3d.

The electrons in the several orbital series can be diagramed with a horizontal line representing the orbital and a vertical arrow representing each electron. Thus the 1s orbital of the hydrogen atom in its lowest energy state is represented by $\underline{\uparrow}$.

The six electrons of the carbon atom are assigned:

Increasing energy ↑			
2p	↑	↑	—
2s	↑↓		
1s	↑↓		

Note that when an orbital series has fewer electrons (e.g., 6) than the number required to fill all of the orbitals (in this case, 10), the electrons do not pair but occupy separate orbitals. The orbital series of three other atoms of great importance in biological phenomena, nitrogen, oxygen, and phosphorus, are diagramed in the figure below.

Orbital series	Nitrogen	Oxygen	Phosphorus
3p			↑ ↑ ↑
3s			↑↓
2p	↑ ↑ ↑	↑↓ ↑ ↑	↑↓ ↑↓ ↑↓
2s	↑↓	↑↓	↑↓
1s	↑↓	↑↓	↑↓

In using the older, Bohr planetary model of atomic structure, electrons are assigned to specific shells, or paths, around the nucleus. The first shell next to the atomic nucleus can hold two electrons, the second shell eight, and each shell further out also has a maximum number of electrons it may contain. However, no atom may have more than eight electrons in its outermost shell. The inner shells are filled first, and if there are not enough electrons to fill each shell, the outermost is left incomplete. Hydrogen, with one electron, has an incomplete innermost shell. Helium has two electrons, which just fill the innermost shell. Because helium's outermost shell is filled, the atom is chemically inert, i.e., does not form compounds with other elements. The element neon is another inert gas; with 10 electrons two are placed in the inner shell, filling it, and eight are placed in the second shell, completing it and precluding the formation of compounds with other elements. The oxygen atom, with two electrons in its inner shell and six electrons in the second shell, is chemically active and forms compounds with almost all of the elements except inert gases such as helium and neon. Diagrams of the Bohr planetary models of the atoms of hydrogen, carbon, nitrogen, and oxygen are given in the figure below.

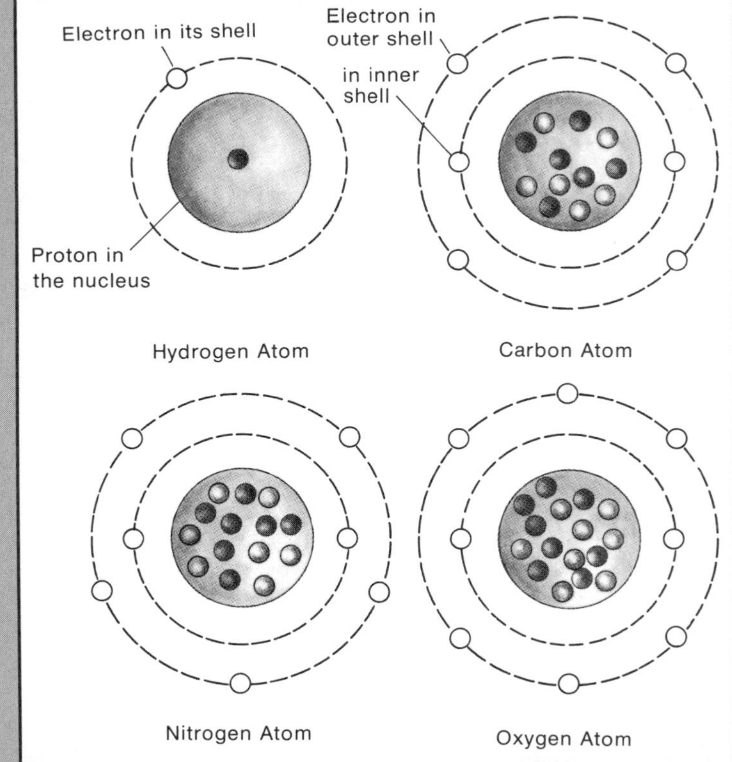

Diagrams of the structure of the atoms of the four chief elements of living matter: hydrogen, carbon, nitrogen, and oxygen. The symbols used are ⊖, neutron; ◯, proton; ○, electron. The electron shells represent spheres around the nucleus within which the electrons are located 90 percent of the time.

The center of the atom, corresponding in position to the sun in our solar system, is the **nucleus,** which contains protons and neutrons; it composes almost the total mass of the atom. Just as the solar system is mostly empty space, so is the atom, with electrons moving in the empty space around the nucleus. The electrons do not circle the nucleus in specific orbits, as the planets circle the sun; instead they whirl around the nucleus, now close to it, then further away. Each type of atom has a characteristic number of electrons whirling around the nucleus and characteristic numbers of protons and neutrons in the nucleus. The number of electrons is the atomic number of the element and determines its place in the periodic table (Box 1.1). In all atoms, the number of protons in the nucleus equals the number of electrons circling around it, so that the atom as a whole is in a state of electrical neutrality. The different kinds of matter reflect differences in the number and arrangement of these basic particles. Living systems are composed of exactly the same kinds of atoms, with the same kind of atomic structure, as nonliving systems.

Elements. An **element** is a substance composed of atoms, all of which have the same number of protons in the atomic nucleus and therefore the same number of electrons circling in orbitals.

BOX 1.2

Elements Important to Organisms

The elements can be arranged in a periodic table (see the figure below), ranging from the lightest element, hydrogen, to the heaviest natural element, uranium. Elements in the same vertical column have very similar chemical properties, reflecting basic similarities in the structure of their atoms. Although most of these elements could be found in a sample of tissue from some animal, a much lesser number of elements are really necessary for life. The elements carbon, hydrogen, oxygen, and nitrogen make up about 96 percent of the substances in the human body and in most living things. Potassium, calcium, sulfur, and phosphorus are usually present to the extent of about 1 percent each. Since bone is largely composed of calcium and phosphorus, the amounts of these elements present in a bony animal are much greater than in a completely soft-bodied animal. Lesser amounts of sodium, chlorine, iron, iodine, magnesium, copper, manganese, cobalt, zinc, and a few others complete the list of elements essential for life. Many of these trace elements are required as components of specific enzyme systems. Some elements, such as lead and mercury, are toxic themselves or are metabolized into toxic compounds by living things.

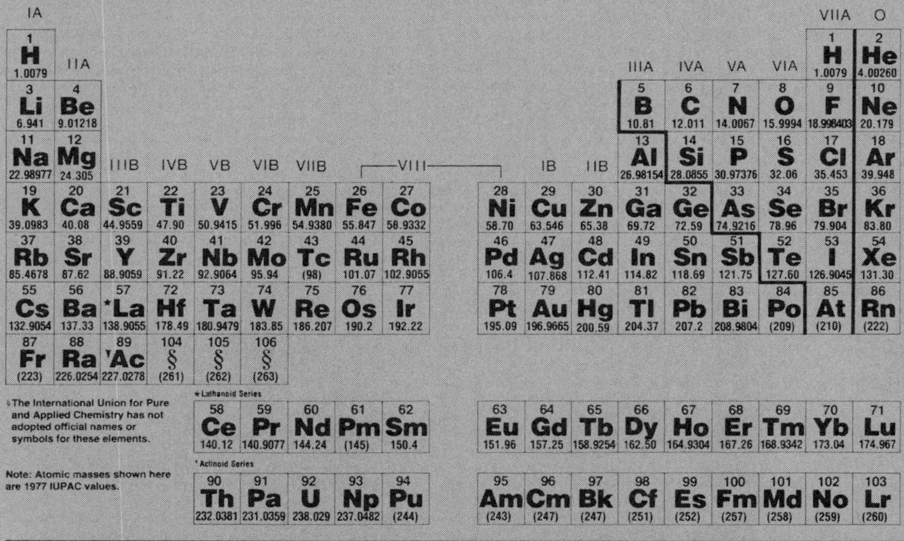

Periodic table of the elements.

For convenience in writing chemical formulas and reactions, chemists have assigned to each of the elements a symbol, usually the first letter of the name of the element: O, oxygen; H, hydrogen; C, carbon; N, nitrogen. A second letter is added to the symbol of those elements with the same initial letter: Ca, calcium; Co, cobalt; Cl, chlorine; Cu, copper; Na, sodium (L. *natrium*).

Ions. The chemical properties of an element are determined primarily by the number and arrangement of electrons revolving in the outermost energy shell around the atomic nucleus and to a lesser extent by the number of electrons in the inner shells. These, in turn, depend upon the number and kind of particles, protons and neutrons, in the nucleus.

The number of electrons in the outermost shell varies from zero to eight in different kinds of atoms (see the figure in Box 1.1). If there are zero or eight electrons in the outer shell, the element is chemically inert and will not readily combine with other elements (e.g., the "noble" gases such as helium or argon). When there are fewer than eight electrons in the outermost shell, the atom tends to lose or gain electrons in order to achieve an outer shell of eight electrons. Since the number of protons in the atomic nucleus is not changed, this loss or gain of electrons produces an atom with a net positive or negative charge. Such electrically charged atoms are known as **ions.** Atoms with one, two, or three electrons in the outer shell tend to lose them to other atoms and become positively charged ions because of the excess protons in the nucleus (e.g., Na^+, sodium ion; Ca^{++}, calcium ion). These are termed **cations** because they migrate to the cathode (negative electrode) of an electrolytic cell. Atoms with five, six, or seven electrons in the outer shell tend to gain electrons from other atoms and become negatively charged ions or **anions** (e.g., Cl^-, chloride ion). Anions migrate to the anode or positively charged electrode of an electrolytic cell. Because they bear opposite electric charges, anions and cations are attracted to each other and unite forming an ionic or electrostatic bond. Atoms such as carbon, which have four electrons in the outer orbit, neither lose nor gain electrons but share them with adjacent atoms.

Isotopes. Most elements are composed of two or more kinds of atoms, which differ in the number of neutrons in the atomic nucleus. The different kinds of atoms of an element are called **isotopes** (Gr. *isos,* equal + *topes,* place) because they occupy the same place in the periodic table of the elements. All the isotopes of a given element have the same number of electrons circling the atomic nucleus.

Although the isotopes of a given element have the same chemical properties, they can be differentiated physically. Some are radioactive and can be detected and measured by the kind and amount of radiation they emit. Others can be differentiated by the slight difference in the mass of the atoms caused by an extra neutron in the nucleus. Substances containing ^{15}N (heavy nitrogen) instead of ^{14}N, the usual isotope, or ^{2}H (heavy hydrogen, deuterium) instead of ^{1}H, will have a greater mass detectable with a **mass spectrometer.**

A tremendous insight has been gained into the details of the metabolic activities of cells by preparing some substance—sugar, for example—labeled with radioactive carbon (^{11}C or ^{14}C) or heavy carbon (^{13}C) in place of ordinary carbon (^{12}C). The labeled substances are fed or injected into an animal or plant, or cells are incubated in a solution containing the tracer, and the labeled products resulting from the organism's or cell's normal metabolic processes are then isolated and identified. Such experiments have traced the sequence of reactions undergone by a given compound and determined the form or forms in which the labeled atoms finally leave the cell or organism.

BOX 1.3

Chemical Bonds

The constituent atoms of a molecule are joined together by forces called chemical bonds (see the figure below). Ionic, covalent, and hydrogen bonds are important in the molecules present in biological materials. Ionic and hydrogen bonds are relatively weak and easily broken. Covalent bonds are strong, and their formation is endergonic, requiring the input of energy. Both the formation and cleavage of covalent bonds are carried out by enzymatic reactions within the cell.

Ionic bonds result from the attraction of particles with unlike charges, that is, between a positively charged atom or molecule and a negatively charged atom or molecule.

Hydrogen bonds are formed when a hydrogen atom is shared between two atoms, one of which is usually oxygen. They tend to form between a hydrogen atom covalently bonded to oxygen or nitrogen and a strongly electronegative atom, usually oxygen or ni-

Box continued on following pages.

$Na^+ + Cl^- \longrightarrow Na^+Cl^-$ Ionic bond

Hydrogen bond:

$>C{=}O + HO{-}C< \longrightarrow >C{=}O \cdots\cdots HO{-}C<$

or

$>C{=}O + H_2N{-} \longrightarrow >C{=}O \cdots\cdots H{-}N(H){-}$

Glycoside bond: $R{-}C(H){=}O + HO{-}C{-}R' \longrightarrow R{-}C{-}O{-}CR' + H_2O$

Peptide bond: $R{-}C(=O){-}OH + H_2N{-}C{-}R' \longrightarrow R{-}C(=O){-}N(H){-}R' + H_2O$

Ester bond: $R{-}C(=O){-}OH + HO{-}C{-}R' \longrightarrow R{-}C(=O){-}O{-}C{-}R' + H_2O$

Phosphate ester bond: $R{-}C{-}OH + HO{-}P{-}O_3H_2 \longrightarrow R{-}C{-}O{-}P{-}O_3H_2 + H_2O$

Thioester bond: $R{-}C(=O){-}OH + HS{-}C{-}R' \longrightarrow R{-}C(=O){-}O{-}S{-}C{-}R' + H_2O$

Types of chemical bonds. Glycoside, peptide, ester, phosphate ester, and thioester bonds are all covalent bonds.

trogen, in another molecule or in another part of the same molecule. Hydrogen bonds are readily formed and broken; they are geometrically quite precise. They have a specific length and a specific direction, which is important in their role in determining the structure of macromolecules, such as proteins and nucleic acids. The water molecules in liquid water are held together in part by hydrogen bonds. The oxygen atom and the two hydrogen atoms in a water molecule form a triangle. The electrons belonging to the hydrogen atoms are more strongly attracted to the oxygen nucleus than to the hydrogen nuclei and tend to be located nearer the oxygen atom. Because of this, the two hydrogen atoms have a small local positive charge and the oxygen atom has a small local negative charge, although the water molecule as a whole is electrically neutral. Molecules that are positively charged at one end and negatively charged at the other are said to be polar. Such molecules are usually soluble in water since the electrostatic attraction of the negative and positive charges tends to align the water molecules around them. When the positively charged hydrogen atom of one water molecule is next to an atom carrying an electronegative charge, such as the oxygen atom in another water molecule, the attraction between them forms a hydrogen bond.

In a covalent bond the electrons associated with two different atoms are shared, and each is subject to the attractive forces of both atomic nuclei. The covalent bond of the hydrogen molecule, H_2, that holds the two atoms together results from the fact that each of the two electrons is attracted simultaneously to the two protons. The two atoms are also separated by repulsive forces; the two protons repel each other, and the two electrons do the same. The H—H bond is stable and has a length determined by the balance between the forces of attraction and the forces of repulsion. Similar factors determine the length of other covalent bonds, such as —C—C—, —C—N—, or C=O.

Two of the carbon atom's six electrons are in the 1s orbital, and there is one in each of the others, 2s, $2p_x$, $2p_y$, and $2p_z$. This arrangement is most stable because it keeps the electrons as far apart as possible. The four electrons in the 2s and 2p orbitals are "valence" electrons, available for covalent bonding. A simple way of showing this relationship is the "electron dot" method, in which each electron in the outer shell is shown as a dot around the letter representing the atom: $\cdot\dot{\underset{\cdot}{C}}\cdot$. When one carbon and four hydrogens share electron pairs, a molecule of methane, CH_4, is formed: $\begin{matrix} & H & \\ & \ddot{} & \\ H: & C & :H \\ & \ddot{} & \\ & H & \end{matrix}$. The electrons whirl around both the carbon nucleus and the four hydrogen nuclei; each atom shares its outer shell electrons with the other, thereby completing the 1s orbital of each hydrogen and the 2s and 2p orbitals of the carbon. Such bonds are termed nonpolar covalent bonds.

The oxygen atom has eight electrons, two in the 1s orbital and six occupying the 2s and 2p orbitals. The state of the oxygen atom in which the electrons are farthest apart is that in which there are two electrons in the 2s orbital, two electrons in one 2p orbital, and one in each of the other orbitals. These six electrons in the outer shell are available for electron sharing to form covalent bonds: $:\dot{\underset{\cdot}{O}}:$. When oxygen shares electrons with two hydrogen atoms, H_2O is formed: $H:\ddot{\underset{\cdot\cdot}{O}}:H$. In this covalent bond the shared electrons tend to be pulled more strongly to one element than to the other. Such a bond is called a polar covalent bond. In a water molecule the electrons tend to lie closer to the oxygen nucleus than to the hydrogen nuclei, giving the oxygen atom a partial negative charge and the hydrogen atoms a partial positive charge. Covalent bonds may have all degrees of polarity, from ones in which the electrons are exactly shared, as in the hydrogen molecule, to ones in which the electrons are much closer to one atom than to the other and in which the bond is quite polar. In a sense, an ionic bond is simply one extreme of this situation, in which the electrons are pulled completely from one atom to the other.

Hydrophobic bonds and van der Waal's bonds are weak bonds of importance in the structure of protein molecules. Hydrophobic interactions occur when nonpolar molecules or groups tend to cluster together in an aqueous environment. The nonspecific attractive force between any two atoms when they are about 3 to 4 Å apart is termed a van der Waal's bond. This bond is weaker and less specific than a hydrogen bond but is important in biological molecules. The basis of the van der Waal's bonds is the changing distribution of electronic charge around atoms with time. Although a single van der Waal's bond is very weak, a great many of them may occur in the interactions between protein molecules, and their sum is large and biologically important.

The covalent bonds important in joining biological molecules are ones formed by removing an OH group from one molecule and an H from the other, that is, by the removal of the constituents of a water molecule. In biosynthetic reactions the bond is usually formed not by the removal of a water molecule but by substituting a phosphate group or some other group for the OH group on one molecule and then removing the phosphate and an H from the second molecule, thus liberating inorganic phosphate and forming the bond.

The glycosidic bonds joining sugar molecules are formed by removing the H from an alcohol group of one sugar and an OH from an aldehyde group of the other. The peptide bonds of proteins are formed by removing an OH from the carboxyl group of one amino acid and an H from the amino group of another. The ester bonds of fats are formed by removing the OH from the carboxyl group of a fatty acid and an H from the alcohol group of glycerol. Other ester bonds of great biological importance are phosphate esters, formed by removing an H from phosphoric acid and an OH from a sugar, and thioesters, formed by removing an OH from the carboxyl group of an acid and an H from an SH group in another molecule.

Chemical Compounds

Most elements are present in living material not as free elements but as **chemical compounds,** substances composed of two or more different kinds of atoms or ions joined together. Compounds can be split apart into two or more simpler substances. The amounts of the elements in a given compound are always present in a definite proportion by weight. This reflects the fact that the atoms are joined together by chemical bonds in a precise way to form the compound. The assemblage of atoms held together by chemical bonds is called a **molecule.** A molecule is the smallest particle of a compound that has the composition and properties of a larger portion of it. A molecule is made of two or more atoms—which may be the same, as in a molecule of oxygen, O_2, or nitrogen, N_2—or they may be atoms of different elements. The properties of a chemical compound are usually quite different from the properties of its constituent elements. Each water molecule, for example, contains two atoms of hydrogen and one atom of oxygen, but the chemical properties of water are quite different from those of either hydrogen or oxygen. The chemical formula of water, H_2O, states the kinds of atoms that are present in the molecule and their relative proportions.

A molecule of the simple sugar, glucose, is composed of three kinds of atoms: 6 carbon atoms, 12 hydrogen atoms, and 6 oxygen atoms; its formula in chemical shorthand is $C_6H_{12}O_6$. Any larger portion of glucose—a gram or kilogram— will also contain carbon, hydrogen, and oxygen in these same proportions.

The weight of any single atom or molecule is exceedingly small, much too small to be expressed in terms of grams or micrograms. **Atomic** weights are expressed in terms of the atomic weight unit, the **dalton,** approximately the weight of a proton or neutron. An atom of the lightest element, hydrogen, weighs one dalton, for its nucleus contains a single proton and no neutrons. A carbon atom weighs 12 daltons; its nucleus contains 6 protons and 6 neutrons. Oxygen has an atomic weight of 16 daltons. The **molecular weight** of a compound is the sum of the atomic weights of its constituent atoms. Thus the molecular weight of water, H_2O, is $(2 \times 1) + 16$, or 18 daltons. The molecular weight of glucose, $C_6H_{12}O_6$, is $(6 \times 12) + (12 \times 1) + (6 \times 16)$, or 180 daltons. The amount of a compound whose mass in grams is equivalent to its molecular weight is termed 1 **mole** (1 mole of glucose is 180 g.). A one-molar (1 M) solution contains 1 mole of a substance in 1 liter (L) of water.

A large part of each cell is simply **water.** The percentage of water in human tissues ranges from about 20 percent in bone to 85 percent in brain cells. The water content is greater in embryonic and young cells and decreases as aging occurs. About 70 percent of our total body weight is water; as much as 95 percent of a jellyfish is water. Water has a number of important functions in living systems. Most of the other chemicals present are dissolved in it; they must be dissolved in water in order to react. Water also dissolves the waste products of metabolism and assists in their removal. Water has a great capacity for absorbing heat with a minimal change in its own temperature; thus, it protects the living material against sudden thermal changes. Since water absorbs a large amount of heat as it changes from a liquid to a gas, the mammalian body can dissipate excess heat by the evaporation of sweat. Water's high heat conductivity makes possible the even distribution of heat throughout the body. Finally, water has an important function as a lubricant. It is present in body fluids wherever one organ rubs against another and in joints where one bone moves on another.

A **mixture** contains two or more kinds of atoms or molecules which may be present in varying proportions. Air is a mixture of oxygen, nitrogen, carbon dioxide, and water vapor, plus certain rare gases such as argon. The proportions of these constituents may vary widely. Thus, in contrast to a pure compound, which has a fixed ratio of its constituents and definite chemical and physical properties, a mixture has properties which vary with the relative abundance of its constituents.

The composition of every cell is constantly changing; the cell constituents are in a "dynamic state." There is a continuous synthesis of large, energy-rich molecules and continual decomposition of these into smaller, energy-poor ones. Some of the most important compounds are present in cells only in extremely minute amounts at any given time, although the total amount formed and used in a 24-hour period may be quite large. An appreciation of this may be gained from the following consideration: When substances undergo chemical reactions in sequence (and almost all the reactions of importance biologically are sequences or "cycles"), such as $A \rightarrow B \rightarrow C \rightarrow D$, the rate of the whole process is controlled by the rate of the slowest reaction in the chain. For example, if reaction $A \rightarrow B$ is 10 times as fast as $B \rightarrow C$, and if $C \rightarrow D$ is 100 times as fast as $B \rightarrow C$, then the least reactive substance, B, will tend to accumulate, and the most reactive one, C, will be present in the smallest amount. For this reason many of the most active and important substances are present in cells in extremely minute amounts. This, coupled with their chemical instability, has made their detection and isolation difficult.

The compounds present in cells are of two main types: **inorganic** and **organic.** The latter include all the compounds (other than carbonates) that contain the element carbon. The element carbon is able to form a much wider variety of compounds than any other element because the outer shell of the carbon atom contains four electrons, which can be shared in a number of different ways with adjacent atoms.

Electrolytes: Acids, Bases, and Salts

Among the inorganic compounds present in living systems are water, carbon dioxide, acids, bases and salts. An **acid** is a compound that releases hydrogen ions (H^+)

when dissolved in water.* Acids turn blue litmus paper to red and have a sour taste. Hydrochloric (HCl) and sulfuric (H_2SO_4) are inorganic acids; lactic ($C_3H_6O_3$) from sour milk and acetic (CH_3COOH) from vinegar are two common organic acids. A **base** is a compound that releases hydroxyl ions (OH^-) when dissolved in water. Bases turn red litmus paper blue. Common inorganic bases include sodium hydroxide (NaOH) and ammonium hydroxide (NH_4OH).

For convenience in stating the degree of acidity or alkalinity of a fluid, the hydrogen ion concentration may be expressed in terms of pH, the logarithm of the reciprocal of the hydrogen ion concentration, log $1/[H^+]$. On this scale, a neutral solution has a pH of 7 (its hydrogen ion concentration is 0.0000001 or 10^{-7} molar), alkaline solutions have pH's ranging from 7 to 14 (the pH of 1 M NaOH), and acids have pH's from 7 to 0 (the pH of 1 M HCl). Most animal cells are neither strongly acid nor alkaline but contain a mixture of acidic and basic substances; their pH is about 7. Any considerable change in the pH of a cell is inconsistent with life. Since the scale is a logarithmic one, a solution with a pH of 6 has a hydrogen ion concentration 10 times as great as that of one with a pH of 7.

When an acid and a base are mixed, the hydrogen ion of the acid unites with the hydroxyl ion of the base to form a molecule of water (H_2O). The remainder of the acid (anion) combines with the rest of the base (cation) to form a **salt.** For example, hydrochloric acid (HCl) reacts with sodium hydroxide (NaOH) to form water and sodium chloride (NaCl) or common table salt:

$$H^+Cl^- + Na^+OH^- \rightarrow H_2O + Na^+Cl^-$$

A salt may be defined as a compound in which the hydrogen atom of an acid is replaced by some metal. One property of a metal is its ability to form cations.

When a salt, an acid, or a base is dissolved in water, it separates into its constituent ions. These charged particles can conduct an electric current; hence these substances are known as **electrolytes.** Sugars, alcohols, and the many other substances that do not separate into charged particles when dissolved, and therefore do not conduct an electric current, are called **nonelectrolytes.**

Cells and extracellular fluids contain a variety of mineral salts, of which sodium, potassium, calcium, and magnesium are the chief cations (positively charged ions) and chloride, bicarbonate, phosphate, and sulfate are the important anions (negatively charged ions).

Although the concentration of salts in cells and in body fluids is small, this amount is of great importance for normal cell functioning. The concentrations of the respective cations and anions are kept remarkably constant under normal conditions; any marked change results in impaired function and finally in death. A great many of the enzymes that mediate the chemical reactions occurring in the body require one or another of these ions—for example, magnesium, manganese, cobalt, potassium—as cofactors. These enzymes are unable to function in the absence of the ion. Normal nerve function requires a certain concentration of calcium in the body fluids; a decrease in this results in convulsions and death. Normal muscle contraction requires certain amounts of calcium, potassium, and sodium. In addition to these several specific effects of particular cations, mineral salts serve an important function in maintaining the osmotic relationships between each cell and its environment.

*Other definitions of acid and base, such as those by Brønsted or Lewis, may be useful in understanding certain more complex reactions. Brønsted defines an acid as a proton donor and a base as a proton acceptor.

Organic Compounds of Biological Importance

The major types of organic substances present in cells are the carbohydrates, proteins, fats, nucleic acids, and steroids. Some of these are required for the structural integrity of the cell, others to supply energy for its functioning, and still others are of prime importance in regulating metabolism within the cell. **Carbohydrates** and **lipids** are the chief sources of chemical energy in almost every form of life; **proteins** are structural elements but are of even greater importance as catalysts (enzymes) and regulators of cellular processes. **Nucleic acids** are of prime importance in the storage and transfer of information used in the synthesis of specific proteins and other molecules.

The basic pattern of the types of substances, and even their relative proportions, is remarkably similar for cells from the various parts of the body and for cells from different animals. A bit of human liver and the substance of an amoeba each contain about 80 percent water, 12 percent protein, 2 percent nucleic acid, 5 percent fat, 1 percent carbohydrate, and a fraction of 1 percent of steroids and other substances. Certain specialized cells, of course, have unique patterns of chemical constituents; the mammalian brain, for example, is rich in certain kinds of fats.

Carbohydrates. The simplest of the organic substances are the carbohydrates—the sugars, starches, and celluloses—which contain carbon, hydrogen, and oxygen in a ratio of approximately 1 C:2 H:1 O. Carbohydrates are found in all living cells, usually in relatively small amounts, and are important as readily available sources of energy. Both **glucose** and **fructose** are single sugars with the formula $C_6H_{12}O_6$. However, the arrangement of the atoms within the two molecules is different,

```
       H                     H
       |                     |
       C=O               H—C—O—H
       |                     |
   H—C—O—H                 C=O
       |                     |
  HO—C—H                HO—C—H
       |                     |
   H—C—O—H               H—C—O—H
       |                     |
   H—C—O—H               H—C—O—H
       |                     |
   H—C—O—H               H—C—O—H
       |                     |
       H                     H
    Glucose              Fructose
```

FIGURE 1.7 Structural formulas of two simple sugars.

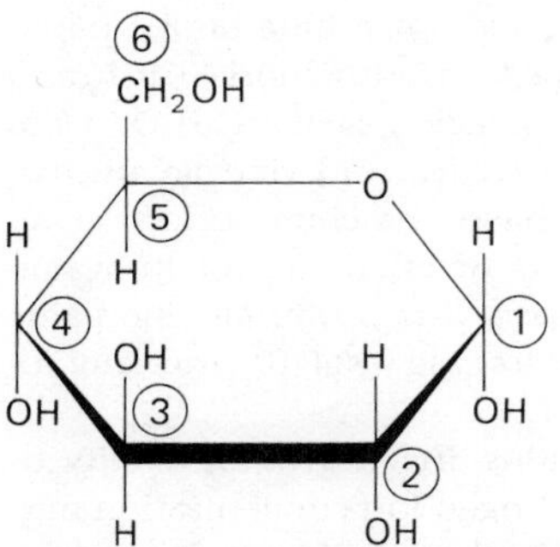

FIGURE 1.8 Structural formula of glucose in the ring form. The molecule is a six-membered ring containing five carbon atoms and one oxygen atom. This is the usual form of glucose in solution.

and the two sugars have somewhat different chemical properties and quite different physiological roles. Such differences in the molecular structures of substances with the same chemical formula are frequently found in organic chemistry. Chemists indicate the arrangement of atoms in a molecule by a **structural formula** in which the atoms are represented by their symbols—C, H, O, and so on—and the chemical bonds or forces that hold the atoms together are indicated by connecting lines. Hydrogen has one such bond; oxygen, two; nitrogen, three; and carbon, four. The structural formulas of glucose and fructose are compared in Figure 1.7. Note that the lower four carbon atoms in the two sugars have identical groups; only the upper two show differences. The molecules of glucose and other single sugars in solution are not extended straight chains as shown in Figure 1.7 but are present as flattened, boat-shaped, or chair-shaped rings formed when a chemical bond connects carbon 1 to the oxygen attached to carbon 5 (or carbon 4). Glucose in solution typically consists of a ring of five carbon atoms and one oxygen atom (Fig. 1.8).

Carbon atoms can unite with each other as well as with many other kinds of atoms and form an almost infinite variety of compounds. Carbon atoms linked together may form long chains (as in fatty acids), branched chains (certain amino acids), rings (purines and pyrimidines), and complex rings (steroids). Molecules are in fact three-dimensional structures, not simple two-dimensional ones, as these formulas suggest. There are more complex ways of representing the third dimension of the molecule. Since the properties of the compound depend in part on the exact nature of its three-dimensional structure, its **conformation,** such three-dimensional formulas are helpful in understanding the intimate relations between molecular structure and function.

Glucose is the only hexose that occurs in any quantity in the cells and body fluids of both vertebrates and invertebrates. The other carbohydrates eaten by vertebrates are converted to glucose in the liver. Glucose is an indispensable component of mammalian blood and is normally present in a concentration of about 0.1 percent. A prolonged increase in the concentration of glucose in the blood, as in untreated or poorly controlled diabetes mellitus, can cause metabolic alterations and extensive damage to certain tissues, such as the eye and the kidney. A reduced concentration of glucose in the blood leads to an increased irritability of certain brain cells, so that they respond to slight stimuli. They discharge nerve impulses that result in muscular twitches, convulsions, and finally unconsciousness and death. Brain cells use glucose as their prime metabolic fuel, and a certain minimum concentration of glucose in the blood is required to supply this. A complex physiological control mechanism, which operates like the "feedback" controls of electronic devices and which involves the liver, pancreas, pituitary and adrenal glands, maintains the proper concentration of glucose in the blood.

The double sugars, with the formula $C_{12}H_{22}O_{11}$, consist of two molecules of single sugar joined by the removal of a molecule of water. **Sucrose,** or table sugar, is a combination of glucose and fructose (Fig. 1.9). Other common double sugars are **maltose,** composed of two molecules of glucose, and **lactose,** composed of glucose and galactose. Lactose, found in the milk of all mammals, is an important item in the diet of the young of these forms. Fructose, the sweetest of the common sugars, is more than 10 times sweeter than lactose; sucrose is intermediate.

Most animal cells contain some **glycogen** or animal starch, the molecules of which are made of a very large number—thousands—of molecules of glucose joined

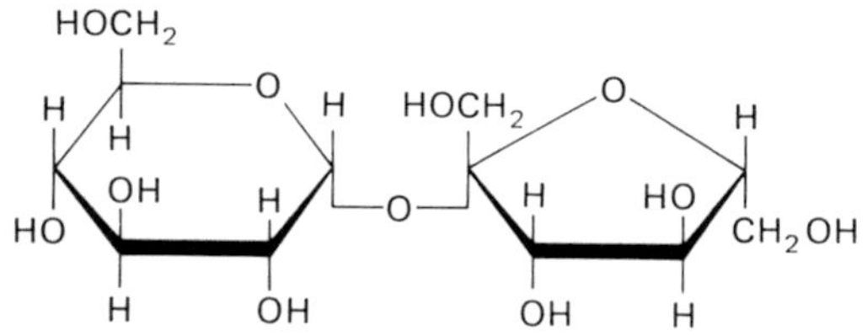

FIGURE 1.9 Structural formula of sucrose, a disaccharide (double sugar) composed of a molecule of glucose and a molecule of fructose joined by a glucosidic linkage.

together by the removal of an H from one and an OH from the next. Glycogen is the form in which animal cells typically store carbohydrate for use as an energy source in cell metabolism. The glycogen molecules within a living cell are constantly being built up and broken down. Glucose and other single sugars are not a suitable storage form of carbohydrate for, being soluble, they readily pass out of the cells. The molecules of glycogen, much larger and less soluble than glucose, cannot pass through the plasma membrane. Glycogen is typically stored intracellularly as microscopic granules, which can be made visible by special stains. Glycogen is readily converted into small molecules such as glucose-phosphate (p. 73) to be metabolized within the cell.

Cellulose, also composed of hundreds of molecules of glucose, is an insoluble carbohydrate which is a major constituent of the tough outer wall of plant cells. The bonds joining the glucose molecules in cellulose are different from those joining the glucoses in glycogen and are not split by the amylases that digest glycogen and starch.

Glucosamine and galactosamine, nitrogen-containing derivatives of the sugars glucose and galactose, are important constituents of supporting substances such as connective tissue fibers, cartilage and chitin, a constituent of the hard outer shell of insects, spiders, and crabs.

Carbohydrates serve as a readily available fuel to supply energy for metabolic processes. Glucose is metabolized to carbon dioxide and water with the release of energy. A few carbohydrates combine with proteins (**glycoproteins**) or lipids (**glycolipids**) to serve as structural components of certain cells. **Ribose** and **deoxyribose** are five-carbon sugars that are components of ribonucleic acid (RNA) and deoxyribonucleic acid (DNA) (Fig. 1.14, p. 27).

Fats. The term **fat,** or lipid, refers to a heterogeneous group of compounds that share the property of being soluble in chloroform, ether, or benzene but are only very sparingly soluble in water. True fats are composed of carbon, hydrogen, and oxygen but have much less oxygen in proportion to the carbon and hydrogen than carbohydrates have. Fats have a greasy or oily consistency; some, such as beef tallow fat or bacon fat, are solid at ordinary temperatures; others, such as olive oil or cod liver oil, are liquid. Each molecule of fat is composed of one molecule of **glycerol** and three molecules of **fatty acid.** All such neutral fats, termed triacylglycerols, contain glycerol but may differ in the kinds of fatty acids present. Fatty acids are long chains of carbon atoms with a carboxyl group (—COOH) at one end. All fatty acids in nature have an even number of carbon atoms—palmitic has 16 and stearic, 18. Fatty acids with one or more double bonds between carbon atoms are called **unsaturated.** Oleic acid has 18 carbons and one double bond (and hence has two less hydrogen atoms than stearic). A fat common in beef tallow, tristearin, $C_{57}H_{110}O_6$, has three molecules of stearic acid and one of glycerol (Fig. 1.10). Fats containing unsaturated fatty acids are usually liquid at room temperature, whereas **saturated** fats, such as tristearin, are solids.

Fats are important as biological fuels and as structural components of cells, especially cell membranes. Glycogen, or starch, is readily converted to glucose and metabolized to release energy quickly; the carbohydrates serve as short-term sources of energy. Fats yield more than twice as much energy per gram as do carbohydrates and thus are a more economical form for the storage of food reserves. Carbohydrates can be transformed by the body into fats and stored in this form—a restatement of the generally known fact that starches and sugars are "fattening."

Certain kinds of fats, phospholipids and cholesterol, are important structural elements of the body. The plasma membrane around each cell and the nuclear membrane contain phospholipids as important constituents, and the myelin sheath around the nerve fibers (p. 214) has a high cholesterol content. Animals store fat as globules within the cells of adipose tissue. The layer of adipose tissue just under the skin serves as an insulator against the loss of body heat. Whales, which live in cold water and have no insulating hair, have an especially thick layer of fat (blubber) just under the skin for this purpose. The subcutaneous fat in humans keeps the skin firm, in addition to restricting the loss of body heat.

The fat deposits are not simply long-term stores of foodstuff used only in starvation but are constantly being used and re-formed. Studies with labeled fatty acids showed that mice replace half of their stored fats in seven days.

Besides the true fats, composed of glycerol and fatty acids, lipids include several related substances that contain components such as phosphorus, choline, and sugars in addition to fatty acids. The **phospholipids** are important constituents of the membranes of plant and

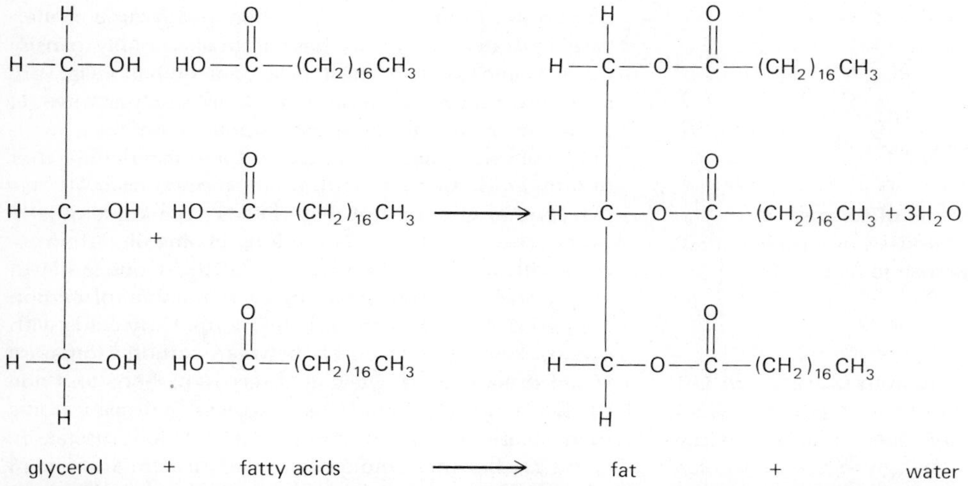

$$\begin{array}{l} H-\overset{\displaystyle H}{\underset{|}{\overset{|}{C}}}-OH \quad HO-\overset{\displaystyle O}{\overset{\|}{C}}-(CH_2)_{16}CH_3 \\ H-\overset{|}{\underset{|}{C}}-OH \quad HO-\overset{\displaystyle O}{\overset{\|}{C}}-(CH_2)_{16}CH_3 \\ H-\overset{|}{\underset{\displaystyle H}{\underset{|}{C}}}-OH \quad HO-\overset{\displaystyle O}{\overset{\|}{C}}-(CH_2)_{16}CH_3 \end{array} \longrightarrow \begin{array}{l} H-\overset{\displaystyle H}{\overset{|}{C}}-O-\overset{\displaystyle O}{\overset{\|}{C}}-(CH_2)_{16}CH_3 \\ H-\overset{|}{\underset{|}{C}}-O-\overset{\displaystyle O}{\overset{\|}{C}}-(CH_2)_{16}CH_3 + 3H_2O \\ H-\underset{\displaystyle H}{\underset{|}{C}}-O-\overset{\displaystyle O}{\overset{\|}{C}}-(CH_2)_{16}CH_3 \end{array}$$

$$\text{glycerol} + \text{fatty acids} \longrightarrow \text{fat} + \text{water}$$

FIGURE 1.10 The synthesis of a fat, tristearin, from a molecule of glycerol and three molecules of stearic acid. In the formula, $(CH_2)_{16}$ represents a chain of sixteen carbon atoms joined in a line, $-\overset{H}{\underset{H}{C}}-\overset{H}{\underset{H}{C}}-$. . ., to each of which are attached two hydrogen atoms.

animal cells in general and of nerve cells in particular. The fatty acid portion of the phospholipid molecule is **hydrophobic,** not soluble in water. The other portion, composed of glycerol, phosphate, and a nitrogenous base such as choline, is ionized and readily water-soluble. For this reason, phospholipid molecules in a film tend to be oriented with the polar, water-soluble portion pointing one way and the nonpolar, fatty-acid portion pointing the other. The plasma membrane is a lipid bilayer composed of two layers of phospholipid molecules with the nonpolar parts very close together and the polar, water-soluble portions facing outward on each side of the membrane.

Beeswax, lanolin, and other **waxes** contain a fatty acid plus an alcohol other than glycerol. **Cerebrosides,** as their name indicates, are fatty substances found especially in nerve tissue. They contain galactose, long-chain fatty acids, and a long-chain amino alcohol, sphingosine.

Steroids. Steroids are complex molecules containing carbon atoms arranged in four interlocking rings, three of which contain six carbon atoms each and the fourth of which contains five. Vitamin D, male and female sex hormones, the adrenal cortical hormones, bile salts, and cholesterol are examples of steroids. **Cholesterol** (Fig. 1.11) is an important structural component of nervous tissue and other tissues, and the steroid hormones are of great importance in regulating certain aspects of metabolism.

Proteins. Proteins differ from carbohydrates and true fats in that they contain nitrogen in addition to carbon, hydrogen, and oxygen. Proteins typically contain sulfur and phosphorus also. Proteins are among the largest molecules present in cells and share with nucleic acids the distinction of great complexity and variety. Hemoglobin, the red pigment found in the blood of all vertebrates and many invertebrates, has the formula $C_{3032}H_{4816}O_{872}N_{780}S_8Fe_4$. Although the hemoglobin molecule is enormous compared with a glucose or triolein molecule, it is only a small- to medium-sized protein.

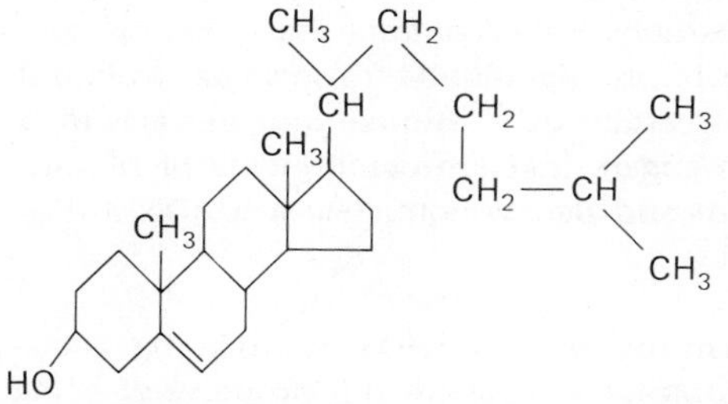

FIGURE 1.11 Structural formula of a sterol, cholesterol.

Many, indeed most, of the proteins within a cell are **enzymes,** biological catalysts that control the rates of the many chemical processes of the cell.

Protein molecules are made of simpler components, the **amino acids,** some 30 or more of which are known. Since each protein contains hundreds of amino acids, present in a certain proportion and in a particular order, an almost infinite variety of protein molecules is possible. Analytical tools, such as the amino acid analyzer, permit one to determine the sequence of the amino acids in a given protein molecule. **Insulin,** the hormone secreted by the pancreas and used in the treatment of diabetes, was the first protein whose structure was elucidated. Work culminating in 1957 revealed the sequence of the 124 amino acids that compose the molecules of the enzyme ribonuclease, secreted by the pancreas.

It is possible to distinguish several different levels of organization in the protein molecule. The first level is the so-called **primary structure,** which depends upon the sequence of amino acids in the **polypeptide chain.** This sequence, as we shall see, is determined in turn by the sequence of nucleotides in the RNA and DNA of the nucleus of the cell. A second level of organization of protein molecules involves the coiling of the polypeptide chain into a helix or into some other regular configuration. The polypeptide chains ordinarily do not lie out flat in a protein molecule, but they undergo coiling to yield a three-dimensional structure. One of the common secondary structures in protein molecules is the **α-helix,** which involves a spiral formation of the basic polypeptide chain. The α-helix is a very uniform geometric structure with 3.6 amino acids occupying each turn of the helix. The helical structure is determined and maintained by the formation of hydrogen bonds between amino acid residues in successive turns of the spiral.

A third level of structure of protein molecules involves the folding of the peptide chain upon itself to form globular proteins. Again, weak bonds such as hydrogen, ionic and hydrophobic bonds form between one part of the peptide chain and another part, so that the chain is folded in a specific fashion to give a specific overall structure of the protein molecule. Covalent bonds such as disulfide bonds (—S—S—) are important in the tertiary structure of many proteins. The biological activity of a protein depends in large part on the specific tertiary structure that is held together by these bonds. When a protein is heated or treated with any of a variety of chemicals, the tertiary structure is lost. The coiled peptide chains unfold to give a random configuration accompanied by a loss of the biological activity of the protein. This change is termed **denaturation.**

Proteins with two or more subunits have a quaternary structure. This refers to the combination of two or more like or unlike peptide chain subunits, each with its own primary, secondary, and tertiary structures, to form the biologically active protein molecule.

Each cell contains hundreds of different proteins, and each kind of cell contains some proteins that are unique to it. There is evidence that each species of animal and plant has certain proteins that are different from those of all other species. The degree of similarity of the proteins of two species is a measure of their evolutionary relationship. Each species has a characteristic pattern of constituent proteins. This pattern differs at least slightly from that of related species and more markedly from those of more distantly related species. Because of the interactions of unlike proteins, grafts of tissue removed from one animal will usually not grow when implanted on a host of a different species but degenerate and are sloughed off by the host. Indeed, even grafts made between members of the same species will usually not grow; only grafts between genetically identical donors and hosts—identical twins or members of a closely inbred strain—will grow.

Amino acids differ in the number and arrangement of their constituent atoms, but all contain an **amino group** (NH_2) and a **carboxyl group** (COOH). The amino group enables the amino acid to act as a base and combine with acids; the carboxyl group enables it to combine with bases. For this reason, amino acids and proteins are important biological "buffers" and resist changes in acidity or alkalinity. Amino acids are linked together to form proteins by **peptide bonds** between the amino group of one and the carboxyl group of the adjacent one (Fig. 1.12). The proteins eaten by an animal

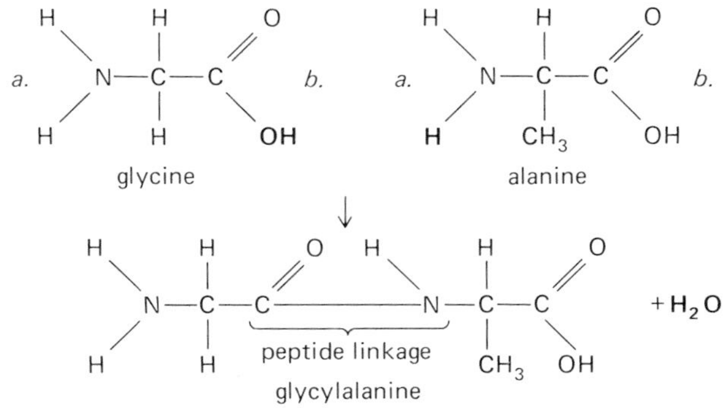

a. b. glycine a. b. alanine

peptide linkage

glycylalanine $+ H_2O$

FIGURE 1.12 Structural formulas of the amino acids glycine and alanine, showing *(a)* the amino group and *(b)* the acid (carboxyl) group. These are joined in a peptide linkage to form glycylalanine by the removal of water.

are not incorporated directly into its cells but are first digested to the constituent amino acids to enter the cell. Subsequently each cell combines the amino acids into the proteins that are characteristic of that cell. Thus, a human eats beef proteins in a steak but breaks them down to amino acids in the process of digestion, then rebuilds them as human proteins—human liver proteins, human muscle proteins, and so on.

Proteins and amino acids may serve as energy sources in addition to their structural and enzymatic roles. The amino group is removed by an enzymatic reaction, **deamination,** and then the remaining carbon skeleton enters the same metabolic paths as glucose and fatty acids and eventually is converted to carbon dioxide and water by the Krebs tricarboxylic acid cycle (p. 71) and associated paths. The amino group is excreted as ammonia, urea, uric acid, or some other nitrogenous compound, depending on the kind of animal. In prolonged fasting, after the supply of carbohydrates and fats has been exhausted, the cellular proteins may be used as a source of energy.

Animal cells can synthesize some, but not all, of the different kinds of amino acids; different species differ in their synthetic abilities. Humans, for example, are apparently unable to synthesize eight of these; they must be either supplied in the food eaten or perhaps synthesized by the bacteria present in the intestine. Plant cells apparently can synthesize all the amino acids. The ones that an animal cannot synthesize, but must obtain in its diet, are called **essential amino acids.** It must be kept in mind that these are no more essential for protein synthesis than any other amino acid but are simply essential constituents of *the diet,* without which the animal fails to grow and eventually dies. Animals differ in their biosynthetic capabilities; what is an essential amino acid for one species may not be essential in the diet of another.

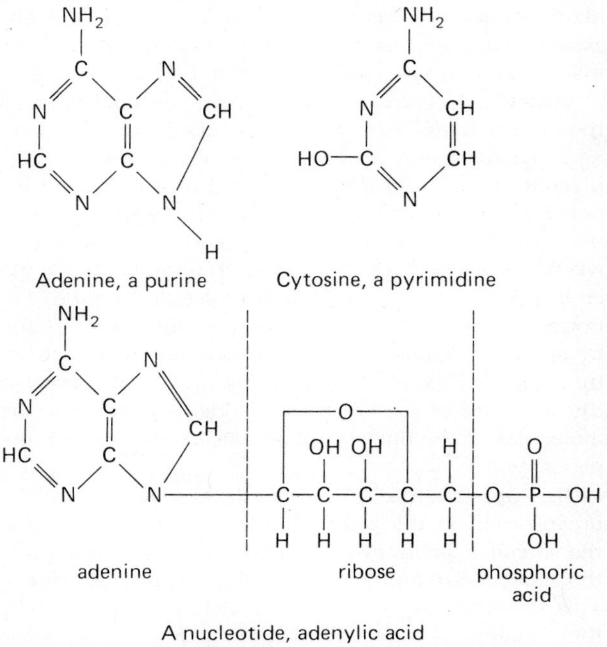

Nucleic Acids. Nucleic acids are complex molecules, larger than most proteins, and contain carbon, oxygen, hydrogen, nitrogen, and phosphorus. They were first isolated by Miescher in 1870 from the nuclei of pus cells and gained their name from the fact that they are acidic and were first identified in nuclei. There are two classes of nucleic acid—one containing the five-carbon sugar, ribose, and called **ribose nucleic acid** or **RNA,** and one containing a five-carbon sugar with one less oxygen atom, deoxyribose, and called **deoxyribonucleic acid** or **DNA.** There are many different kinds of RNA and DNA that differ in their structural details and in their metabolic functions. DNA occurs in the chromosomes in the nucleus of the cell and, in much smaller amounts, in mitochondria and chloroplasts. It is the primary repository for biological information. RNA is present in the nucleus, especially in the nucleolus, in the ribosomes, and in lesser amounts in other parts of the cell.

Adenine, a purine

Cytosine, a pyrimidine

adenine ribose phosphoric acid

A nucleotide, adenylic acid

FIGURE 1.13 Structural formulas of a purine, adenine; a pyrimidine, cytosine; and a nucleotide, adenylic acid.

Nucleic acids are composed of units, called **nucleotides,** each of which contains a nitrogenous base, a five-carbon sugar, and phosphoric acid. Two types of nitrogenous bases, purines and pyrimidines, are present in nucleic acids (Fig. 1.13). Purines have a double-ring structure, in contrast to the single ring of the pyrimidines. RNA contains the purines **adenine** and **guanine** and the pyrimidines **cytosine** and **uracil,** together with the pentose, ribose, and phosphoric acid. DNA contains adenine and guanine, cytosine and the pyrimidine **thymine,** together with deoxyribose and phosphoric acid. The molecules of nucleic acids are made of linear chains of nucleotides, each of which is attached to the next by bonds between the sugar part of one and the phosphoric acid of the next. These are called phosphodiester bonds because each phosphate group is joined by ester linkages to *two* sugar molecules, forming a bridge between the two (Fig. 1.14). The specificity of the nucleic acid resides in the specific sequence of the four kinds of nucleotides present in the chain. For example, CCGATTA might represent a segment of a DNA molecule, with C = cytosine, G = guanine, A = adenine, and T = thymine.

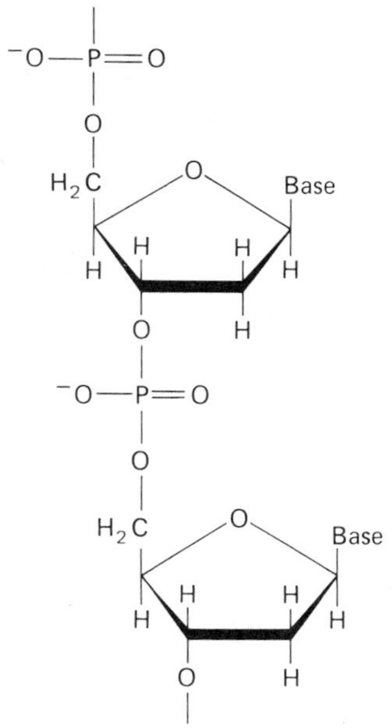

FIGURE 1.14 A portion of a molecule of DNA showing a phosphodiester bond between two adjacent deoxyribose molecules.

An enormous mass of evidence now indicates that DNA is responsible for the specificity and chemical properties of the **genes,** the units of heredity. There are several kinds of RNA, each of which plays a specific role in the biosynthesis of specific proteins by the cell (p. 373).

Nucleotides and Coenzymes. Related structurally to nucleic acids but with quite different roles in cellular function are several mono- and dinucleotides. Each is composed of phosphoric acid, ribose, and a purine or pyrimidine base, like the units composing the nucleic acids. Each of the bases may form a nucleoside triphosphate, with base, sugar, and three phosphate groups linked in a row. **Adenosine triphosphate,** abbreviated **ATP,** composed of adenine, ribose, and three phosphates, is of major importance as the "energy currency" of all cells. The two terminal phosphate groups are joined to the nucleotide by energy-rich bonds, indicated by the symbol ~P. The biologically useful energy of these bonds can be transferred to other molecules; most of the chemical energy of the cell is stored in these ~P bonds of ATP, ready to be released when the phosphate group is transferred to another molecule. Either the terminal or the two terminal phosphate groups are transferred in biological reactions, leaving adenosine diphosphate, ADP, or adenosine monophosphate, AMP. **Guanosine triphosphate, GTP,** is specifically required in certain steps in the synthesis of proteins; **uridine triphosphate, UTP,** is specifically required in certain steps in carbohydrate metabolism, e.g., glycogen synthesis; and **cytidine triphosphate, CTP,** is specifically required for certain steps in the synthesis of fats and phospholipids. All four nucleoside triphosphates are necessary for the synthesis of RNA, and the four deoxyribose nucleoside triphosphates—dATP, dGTP, dCTP, and dTTP—are required for the synthesis of DNA.

Completing the list of nucleotides important in metabolic processes are the dinucleotides, NAD, NADP, and FAD. **Nicotinamide adenine dinucleotide,** abbreviated **NAD,** consists of nicotinamide, ribose, and phosphate attached to an adenine nucleotide, phosphate, ribose, and adenine (Fig. 1.15). NAD is of major importance as a primary electron and hydrogen acceptor in cellular oxidation reactions. Enzymes called **dehydrogenases** remove electrons and hydrogen from molecules such as lactic acid and transfer them to NAD, which in turn passes them on to other electron acceptors. **Nicotinamide adenine dinucleotide phosphate,** abbreviated **NADP,** serves as electron and hydrogen acceptor in certain other reactions. NADP is exactly like NAD except that it has a third phosphate group attached to the ribose of the adenine nucleotide. **Flavin adenine dinucleotide, FAD,** consists of riboflavin-ribitol-phosphate-phosphate-ribose-adenine and serves as a hydrogen and electron acceptor for certain other dehydrogenases. Notice that these dinucleotides have vitamins—nicotinamide or riboflavin—as component parts. These molecules, NAD, NADP, and FAD, are termed **coenzymes;** they are cofactors required for the functioning of certain enzyme systems but are only loosely bound to the enzyme molecule and are readily removed. When they have accepted electrons and hydrogens, they are changed from their oxidized form,

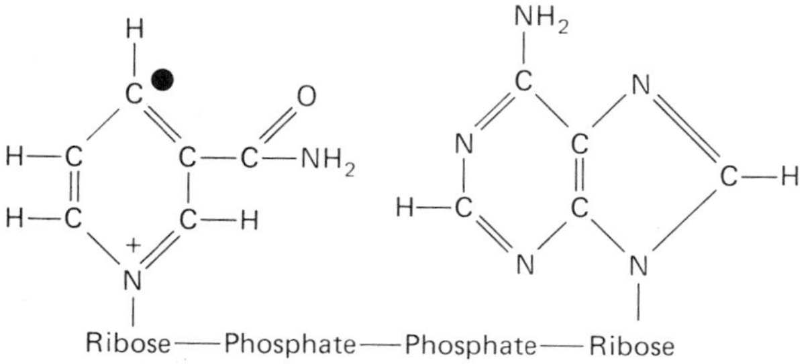

FIGURE 1.15 Nicotinamide adenine dinucleotide (NAD). Nicotinamide, on the left, undergoes reversible oxidation and reduction, accepting a hydrogen at the carbon marked ●.

e.g., NAD, to their reduced form, NADH. They are converted back to the oxidized form when they transfer their electrons to the next acceptor in the chain of respiratory enzymes (p. 69).

Processes in which electrons (e^-) are removed from an atom or molecule are termed **oxidations.** The reverse process, the addition of electrons to an atom or molecule, is termed **reduction.** A simple example of oxidation and reduction is the reversible reaction

$$Fe^{++} \rightleftharpoons Fe^{+++} + e^-$$

The reaction towards the right is an oxidation, the removal of an electron, and the reaction towards the left is a reduction, the addition of an electron. Every oxidation reaction, in which an electron is given off, must be accompanied by a reduction, in which the electron is accepted by another molecule; electrons do not exist in the free state. Oxidation-reduction reactions are common in biological systems (p. 70).

In summary, the chemical components of the cells of animals and humans include water, inorganic salts, and organic substances—nucleic acids, proteins, carbohydrates, and fats, as well as many others. Carbohydrates and fats play some role in the structure of cells but are especially important as fuels; carbohydrates are readily available fuel; fats serve as long-term stores of energy. Nucleic acids have a primary role in storing and transmitting information. Proteins are structural and functional constituents of cells but may serve as fuel after deamination. The body can convert each of these substances into the others to some extent. The cell's contents compose a very complex, multicompartmental system. Many of the properties of cells—muscle contraction, amoeboid motion, and so on—involve changes in the biophysical state of this complex system.

Summary

1. Zoology is an ancient science, with its roots in prehistoric times, but it is also a very modern science, using a wide assortment of sophisticated physical and chemical methods to increase our understanding of animals and their ways of life. Like other scientists, zoologists use the scientific method of posing questions and answering them according to the results of carefully designed experiments or observations with appropriate control groups.

2. All living things have the properties of specific organization, irritability, movement, metabolism, growth, reproduction, and adaptation. The structural and functional unit of all living things is the cell. A typical cell has an outer plasma membrane, a nucleus containing one or more nucleoli, and various organelles within the cytoplasm.

3. All matter, whether living or nonliving, is composed of atoms joined by bonds to form molecules of chemical compounds. Each atom has a central nucleus containing protons and neutrons and one or more electrons whirling around the nucleus. The atoms of the 92 naturally occurring elements differ in the number of electrons circling the nucleus. The lightest element, hydrogen, has one electron, and uranium, the heaviest, has 92. Four elements, carbon, oxygen, nitrogen, and hydrogen, make up the bulk of all living matter.

4. Water makes up a large fraction of every cell and has a number of important functions in living systems. It is an excellent solvent; it has a great capacity for absorbing heat with a minimal change in temperature, and absorbs a large amount of heat as it changes from a liquid to a gas; it has a high heat conductivity and is an excellent lubricant where one organ rubs on another.

5. Living systems contain a variety of acids (compounds that release hydrogen ions, H^+, when dissolved in water), bases (compounds that release OH^- ions when dissolved in water), and salts (compounds in which the hydrogen ion of an acid is replaced with a metal ion). Acids, bases, and salts are charged particles and can conduct an electric current and hence are termed electrolytes. Cells and extracellular fluids contain a variety of mineral salts, of which sodium, potassium, calcium, and magnesium are the principal cations (positively charged ions) and chloride, bicarbonate, phosphate, and sulfate are the important anions (negatively charged ions).

6. The organic compounds in cells include carbohydrates, proteins, fats, nucleic acids, and steroids. Carbohydrates and lipids are the chief sources of energy; proteins serve as structural elements, as enzymes, and as regulators of cellular processes. Nucleic acids store and transfer genetic information, and certain nucleotides, such as ATP and NAD, are important in energy transfer and as coenzymes for many reactions. Combinations of carbohydrates with proteins (glycoproteins) and with lipids (glycolipids) serve as structural components of cells. In addition to serving as fuels, fats play an important role as structural constituents of cells, especially of cell membranes. Steroids, complex molecules composed of four rings of carbon atoms, include cholesterol, an important structural component of cells and cell membranes, and the steroid hormones such as the male and female sex hormones.

7. Proteins are composed of amino acids joined by peptide bonds between the amino group of one and the carboxyl group of the adjacent one. The sequence of amino acids in the peptide chain determines the primary structure of the protein. The coiling of the peptide chain into a helix or some other regular configuration and the folding of the coiled chain determines the secondary and tertiary structure of the protein. Proteins composed of two or more subunits have a quaternary structure determined by the specific combination of the subunits. The biological activity of a protein depends in large part on its specific tertiary and quaternary structure; heating the protein or treating it with certain chemicals causes the coiled peptide chain to unfold, and then biological activity is lost Amino acids may serve as energy sources. The amino group is removed enzymatically (deamination), and the carbon skeleton is metabolized to carbon dioxide and water. Animal cells can synthesize some, but not all, amino acids. The ones animals cannot synthesize but must obtain in their diet are termed essential amino acids.

8. There are two classes of nucleic acids, DNA and RNA, each composed of nucleotides—units containing a nitrogenous base, a five-carbon sugar, and a phosphate. The molecules of nucleic acids are linear chains of nucleotides attached to each other by phosphodiester bonds between the sugar of one and the phosphate of the next. The specificity of the nucleic acids resides in the specific sequence of the four kinds of nucleotides.

References and Selected Readings

Baker, J. J. W., and G. E. Allen: *Matter, Energy and Life.* Reading, Mass., Addison Wesley, 1981. A presentation of thermodynamic principles and their application to studies of living systems.

Baum, S. J., and C. W. Scaife: *Chemistry: A Life Science Approach.* 2nd ed. New York, Macmillan, 1980. A chemistry text for students of biology.

Becker, R., and W. Wentworth. *General Chemistry.* 2nd ed. Boston, Houghton Mifflin, 1980. A general text with study guide.

Condon, E. U., and H. Odabesi: *Atomic Structure.* New York, Cambridge University Press, 1980. A detailed discussion of atomic structure and its effect on chemical properties.

Fessenden, R. J., and Joan S. Fessenden: *Chemical Principles for the Life Sciences.* 2nd ed. New York, Allyn, 1979. A chemistry text written primarily for students of biology.

Jones, M. M., J. T. Netterville, D. O. Johnston, and J. L. Wood: *Chemistry, Man and Society.* Philadelphia, W. B. Saunders Co., 1972. A standard text of college chemistry; good reference for presentation of chemical principles of life processes.

Lee, G. L., H. O. Van Orden, and R. O. Ragsdale: *General and Organic Chemistry.* Philadelphia, W. B. Saunders Co., 1971. A standard text of college chemistry, including organic chemistry; good source for information regarding principles of chemistry.

White, E. H.: *Chemical Background for the Biological Sciences.* 2nd ed. Englewood Cliffs, Prentice Hall, Inc., 1970. Extensive presentation of the structure of atoms and molecules and the nature of chemical reactions.

The *Scientific American* publishes excellent discussions of many topics related to biology. Several hundred of these have been reprinted as "offprints" by Wm. Freeman Co., San Francisco. They are too numerous to cite individually but compose a rich source of collateral reading.

2 Cells and Tissues

ORIENTATION

The cell theory describes the cell as the basic structural and functional unit of life. We will, in this chapter, discuss the parts of a typical animal cell and their activities. We will see how the functions of a cell are regulated by its nucleus and will discuss the cell cycle, mitosis, and the growth of organisms. In other parts of the chapter we will explain how energy is produced and used to maintain the chemical composition of cells and to move, repair damage, grow, and reproduce. We will describe how exchanges occur between the cell and its environment and how these maintain the chemical composition of the cell within the narrow limits necessary for life (homeostasis). Last, we will discuss how cells are organized into tissues and how they perform specialized functions within a complex organism.

The living substance of all animals is organized into units called cells. Each cell contains a nucleus and is surrounded by a plasma membrane. Mammalian red blood cells lose their nucleus in the process of maturation, and a few types of cells such as those of skeletal muscles have several nuclei per cell, but these are rare exceptions to the general rule of one nucleus per cell. In the **protozoa** all of the living material is found within a single plasma membrane. These creatures may be considered to be unicellular, i.e., single-celled, or acellular, with bodies not divided into cells. Many protozoa have a high degree of specialization of form and function within this single cell (Fig. 2.1), and the single cell may be quite large, larger than certain multicellular organisms. Thus, it would be wrong to infer that a single-celled organism is necessarily smaller or less complex than a many-celled one.

The Cell Theory

The term "cell" was applied by Robert Hooke, some 300 years ago, to the small, boxlike cavities he saw when he examined cork and other plant material under the newly invented compound microscope. The important part of the cell, we now realize, is not the cellulose wall seen by Hooke but the cell contents. In 1839 the Bohemian physiologist Purkinje introduced the term "protoplasm" for the living material of the cell. As our knowledge of cell structure and function has increased, it has become clear that the living contents of the cell compose an incredibly complex system of heterogeneous parts. Purkinje's term "protoplasm" has no clear meaning in a chemical or physical sense, but it may be used to refer to all the organized constituents of a cell.

In this same year, 1839, a German botanist, Matthias Schleiden, and Theodor Schwann, his fellow countryman and a zoologist, formulated the generalization that has since developed into the **cell theory:** The bodies of all plants and animals are composed of cells, the fundamental units of life. The cell is both the structural and the functional unit in all organisms, the fundamental unit possessing all the characteristics of living things. A further generalization, first clearly stated by Virchow in 1855, is that new cells can come into existence only by the division of previously existing cells. The corollary of this, that all cells living today can trace their ancestry back to the earliest living things, was stated by August Weismann about 1880.

The bodies of higher animals are made of many cells, differing in size, shape, and functions. A group of cells that are similar in form and specialized to perform one or more particular functions is called a **tissue.** A tissue may contain nonliving cell products in addition to the cells themselves. A group of tissues may be associated into an **organ,** and organs into **organ systems.** For example, the vertebrate digestive system is composed of a number of organs: esophagus, stomach, intestine, liver, pancreas, and so on. Each organ, such as the stomach, contains several kinds of tissue—epithelium, muscle, connective tissue, nerves—and each tissue is made of many, perhaps millions of cells.

A single-celled animal placed in the proper environment will survive, grow, and eventually divide. For most single-celled animals, a drop of sea water or pond water will provide the environment required. It is more difficult to culture cells removed from a multicellular animal—a human, chick, or frog. This was first accomplished in 1907 by Ross Harrison of Yale, who was able to grow cells from a salamander in a drop of nutrient medium containing blood plasma. Since then, many different kinds of cells from animals and plants have been

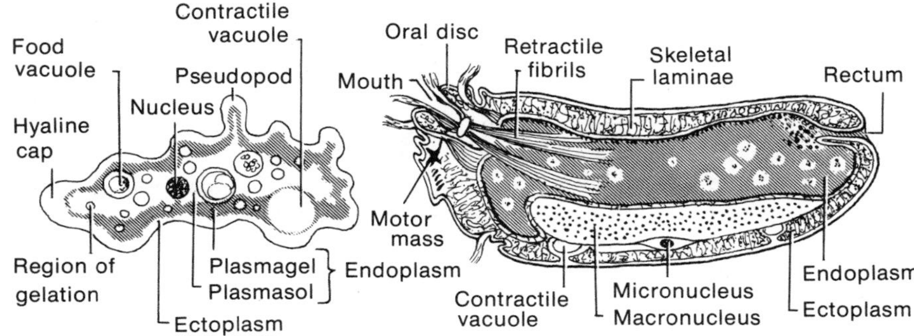

FIGURE 2.1 Diagrams of an amoeba *(left)* and *Epidinium (right)* to illustrate the range in complexity of the protozoa.

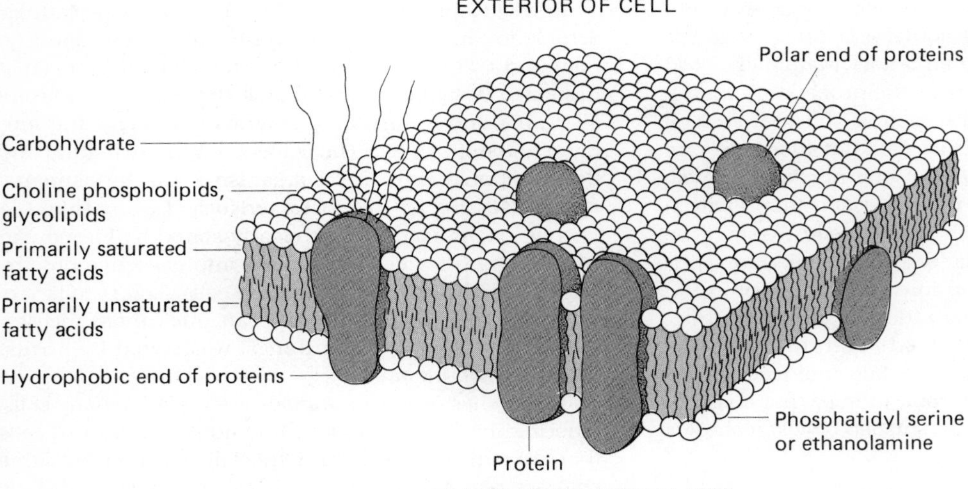

FIGURE 2.2 Diagram of the molecular architecture of biological membranes such as the plasma membrane.

cultured *in vitro,** and many important facts about cell physiology have been revealed in this way.

The Plasma Membrane

The outer surface of each cell is bounded by a delicate, elastic covering termed the **plasma membrane.** This is of prime importance in regulating the contents of the cell, for all materials entering or leaving it must pass through this membrane. It hinders the entrance of certain substances and facilitates the entrance of others.

The plasma membrane behaves as though it has ultramicroscopic pores through which certain substances pass. The size of these pores determines the maximal size of the molecules that can pass through the membrane. Factors other than molecular size, such as the electrical charge, if any, carried by the particles, the number of water molecules, if any, bound to the surface of the particle, and the solubility of the particle in lipids, may also be important in determining whether or not the substance will pass through the membrane. Infoldings of the plasma membrane may be continuous with channels that extend deep into the interior of the cell, providing pathways for the entrance of some materials and for the removal of secretory and excretory products.

**In vitro,* Latin *in glass*. The cells are removed from the animal body and incubated in glass vessels.

All biological membranes appear to have a basically similar **lipid bilayer** structure. The central region of the membrane consists of two layers of lipids, mostly phospholipids but including some glycolipids and cholesterol as well. Each layer is just one molecule thick, with the hydrophobic (water-repelling) tails of the lipids aligned toward each other and with their polar head groups on the outside. The lipid bilayer is a mixture of lipid molecules in the fluid state. Special **membrane proteins** are associated with the lipid bilayer, some present only on the outer or the inner surface. Some are found only *within* the membrane and still others extend completely *through* the lipid bilayer. Some of these membrane proteins are enzymes; others are receptors for hormones or other specific compounds. Certain of the proteins can move laterally within the membrane but are unable to rotate (Fig. 2.2).

The plasma membrane is much more than a simple cell envelope that prevents the ready movement of dissolved materials in either direction. It is an active, functional structure with enzymatic mechanisms that move specific molecules in or out of the cell against a concentration gradient. Multiple small evaginations of the plasma membrane, termed **microvilli,** are present in certain cells, such as those lining the kidney tubules (p. 200).

High-resolution electron micrographs of the plasma membrane (Fig. 2.3) show a dense–light–dense three-layered structure. The plasma membranes of animal, plant, and bacterial cells and the membranes of many

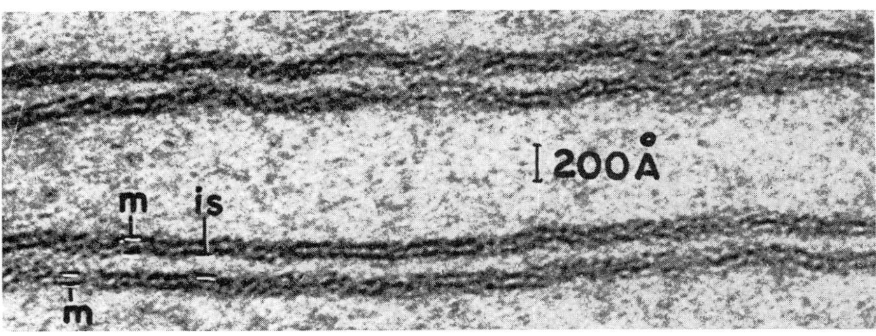

FIGURE 2.3 Electron micrograph of the cell membranes of intestinal cells showing the three-layered structure. Membrane, *m;* intercellular space, *is.* Magnification ×240,000.

subcellular organelles all appear to have a similar three-layered structure.

Nearly all plant cells (but not most animal cells) have a thick **cell wall** made of cellulose that lies outside the plasma membrane. This cell wall is nonliving, secreted by the cell substance. It is pierced in many places with tiny holes through which the contents of one cell connect with the contents of the adjacent cell and through which materials can pass from one cell to the next. These tough, firm cell walls provide support for the plant body.

The Nucleus and its Functions

The **nucleus** of the cell is usually spherical or ovoid. It may have a fixed position in the center of the cell or at one side, or it may be moved around as the cell moves and changes shape. The nucleus is separated from the cytoplasm by a nuclear membrane which controls the movement of materials into and out of the nucleus (Fig. 2.4). The electron microscope reveals that the nuclear membrane is double-layered and that there are ex-

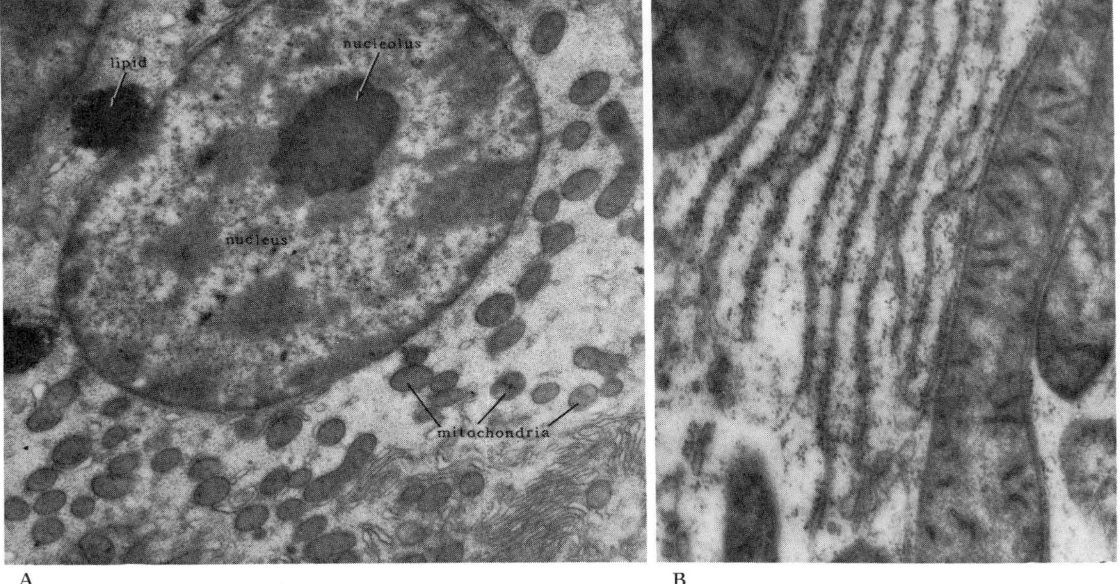

FIGURE 2.4 *A,* Electron micrograph of the nucleus and surrounding cytoplasm of a frog liver cell. The spaghetti-like strands of the endoplasmic reticulum are visible in the lower right corner. Magnified 16,500 times. *B,* High-power electron micrograph of mitochondria and endoplasmic reticulum within a rat liver cell. Granules of ribonucleoprotein (ribosomes) are seen on the strands of endoplasmic reticulum, and structures with double membranes are evident within the mitochondria in the upper left corner and on the right. Magnified 65,000 times. (Electron micrographs courtesy of Dr. Don Fawcett.)

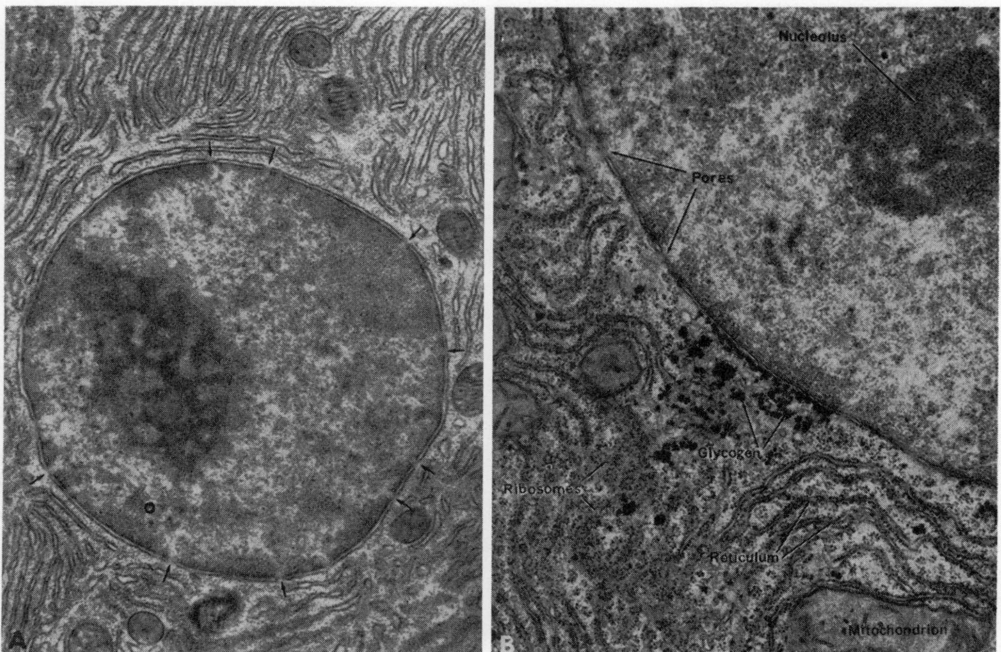

FIGURE 2.5 Electron micrographs showing pores in the nuclear membrane. The endoplasmic reticulum is evident in both pictures; *B,* shows the ribosomes on the endoplasmic reticulum. *A,* Magnified 20,000 times; *B,* magnified 50,000 times. (Electron micrographs courtesy of Drs. Don W. Fawcett and Keith R. Porter.)

tremely fine channels through the nuclear membrane through which the nuclear contents and cytoplasm are continuous (Fig. 2.5) and through which even large molecules of RNA may pass.

The nucleus is an important center for the control of cellular processes and is required for growth and for cell division. An amoeba can survive for some days after the nucleus has been removed by a microsurgical operation, but it cannot grow and eventually dies. To demonstrate that it is the absence of the nucleus, not the operation itself, that causes the ensuing death, one can perform a **sham operation.** A microneedle is inserted into an amoeba and moved around inside the cell to simulate the operation of removing the nucleus, but the needle is withdrawn without actually removing the nucleus. An amoeba subjected to this sham operation will recover, grow, and divide.

A classic series of experiments demonstrating the role of the nucleus in the control of cell growth was performed by Hämmerling with the single-celled *Acetabularia mediterranea.* This marine alga, 4 to 5 cm. long, is mushroom-shaped, with "roots" and a stalk surmounted by a flattened, disc-shaped umbrella. The single nucleus is located near the base of the stalk. Hämmerling cut across the stalk (Fig. 2.6) and found that the lower part containing the nucleus could live and regenerate an umbrella. The upper part eventually died without regenerating a stalk and roots. In further experiments, Hämmerling first severed the stalk just above the nucleus, then made a second cut just below the umbrella. The section of stalk thus isolated, when replaced in sea water, was able to grow a partial or complete umbrella. This might seem to show that a nucleus is not necessary for regeneration. However, when Hämmerling cut off this second umbrella the stalk was unable to form a new one. From experiments such as these, Hämmerling concluded that the nucleus supplies some substance necessary for umbrella formation. This substance passes up the stalk and instigates umbrella growth. In the experiments described here, some of this substance remained in the stalk after the initial cuts, enough to produce one new umbrella. After that amount of "umbrella substance" was exhausted by the regeneration of an umbrella, no second regeneration was possible in the

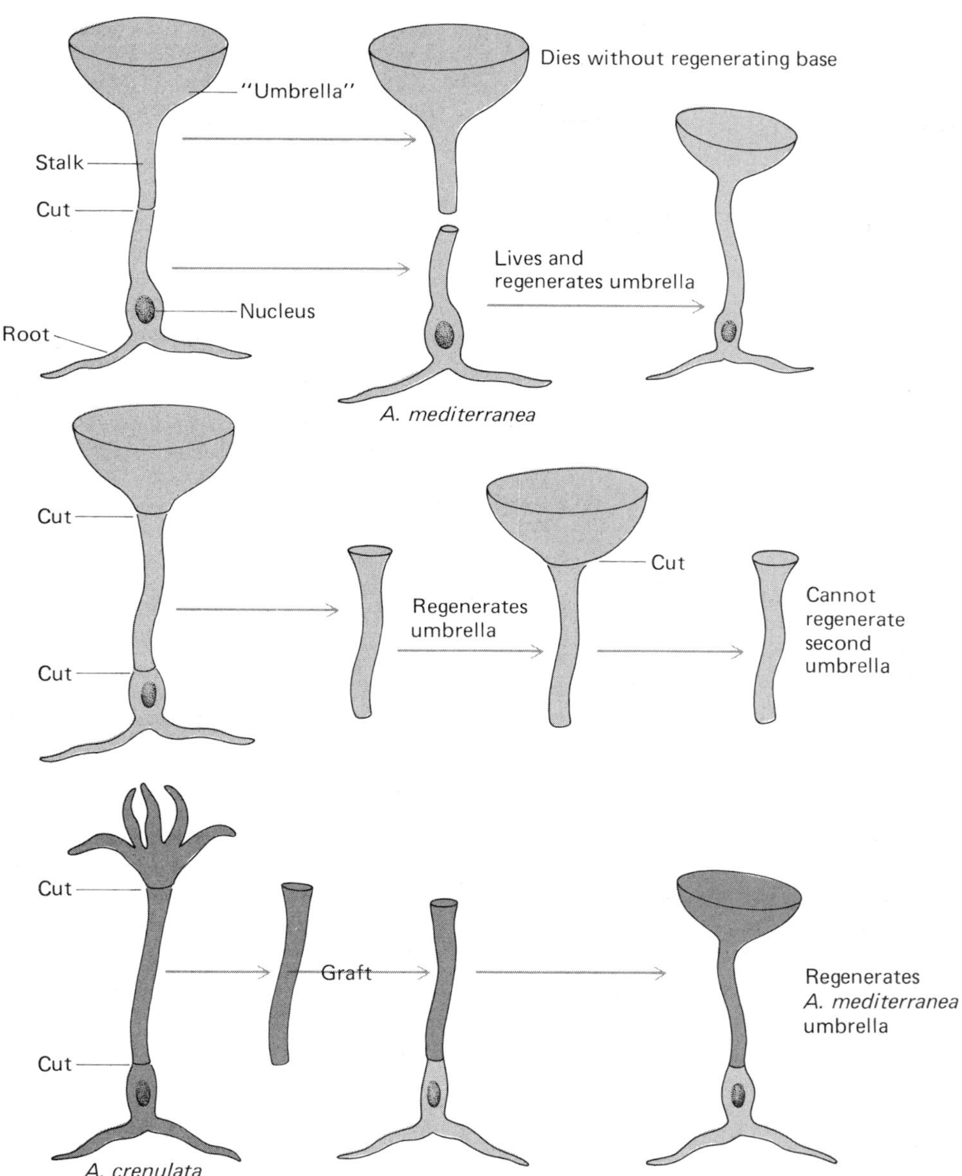

FIGURE 2.6 Hämmerling's experimental demonstration of the production of an umbrella-regenerating substance by the nucleus of the alga, *Acetabularia.* Lower line, when a stalk from *A. crenulata* is grafted onto the base (light gray) of an *A. mediterranea* plant, the stalk regenerates an umbrella whose shape is that characteristic of *A. mediterranea* plants.

absence of a nucleus. What hypothesis would you make about the nature of "umbrella substance"? How would you go about testing your hypothesis?

A second species, *Acetabularia crenulata,* has a branched instead of a disc-shaped umbrella. When a piece of *crenulata* stalk (without a nucleus) was grafted onto the base of a *mediterranea* plant (containing a *mediterranea* nucleus) a new umbrella developed at the top of the stalk. The shape of the umbrella was determined not by the species of the stalk but by the species of the base (Fig. 2.6, lower part). The nucleus, through the action of its genes, provided the specific information that controlled the type of umbrella that was regenerated and could override the tendency of the stalk to form an umbrella characteristic of its own species.

Chromosomes

When a cell has been killed by fixation with the proper chemicals and then stained with the appropriate dyes, several structures are visible within the nucleus (Fig. 2.7). Within the semifluid ground substance are suspended a fixed number of extended, linear, threadlike bodies called **chromosomes,** composed of DNA and proteins and containing the units of heredity, the **genes.** In a stained section of a nondividing cell the chromosomes typically appear as an irregular network of dark-staining threads and granules termed **chromatin.** Just prior to nuclear division these strands condense into compact, rod-shaped chromosomes which are subsequently distributed to the two daughter cells in exactly equal numbers. Each type of organism has a characteristic number of chromosomes present in each of its constituent cells. The fruit fly has 8 chromosomes, the toad, 22, the rat, 42, humans, 46, the goat, 60, and the duck, 80. The somatic cells of higher plants and animals each contain two of each kind of chromosome. The 46 chromosomes in each human cell include 2 of each of 23 different kinds. They differ in length, shape, and the presence of knobs or constrictions along their length.

A cell with two complete sets of chromosomes is said to be **diploid.** Sperm and egg cells, which have only one of each kind of chromosome, one full set of chromosomes, are said to be **haploid.** They have just half as many chromosomes as the somatic cells of that same species. When the egg is fertilized by the sperm the two haploid sets of chromosomes are joined, and the diploid number is restored.

The nucleus also contains one or more small spherical bodies, **nucleoli,** which are evident by phase microscopy. The cells of any particular animal have the same number of nucleoli. The nucleolus disappears when a cell is about to divide and reappears after division is complete. If the nucleolus is destroyed by carefully localized ultraviolet or X-irradiation, cell division is inhibited. This does not occur in control experiments in which regions of the nucleus other than the nucleolus are irradiated. The nucleolus plays a key role in the synthesis of the ribonucleic acid constituents of ribosomes.

Cytoplasmic Organelles

Two small, dark-staining cylindrical bodies, called **centrioles,** are found in the cytoplasm near the nucleus of animal cells. With the electron microscope each cen-

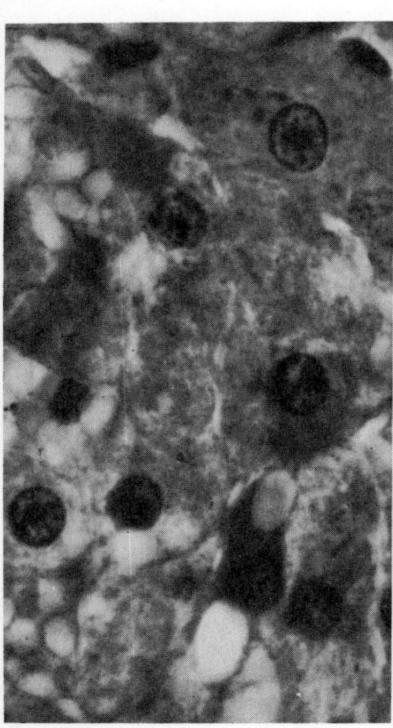

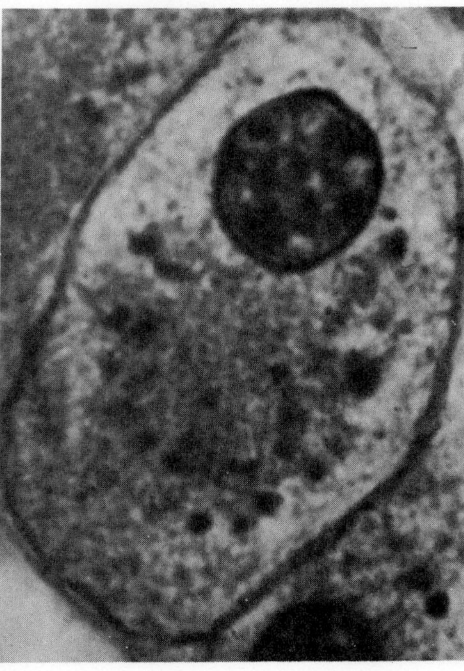

FIGURE 2.7 Tissue sections of human adrenal gland stained to show cellular details; *left,* magnified 600 times; *right,* magnified 1500 times. (Courtesy of Dr. Kurt Benirschke.)

triole is revealed to be a hollow cylinder with a wall in which are embedded nine parallel, longitudinally oriented groups of microtubules, with three tubules in each group. The cylinders of the two centrioles are typically oriented with their long axes perpendicular to each other.

When cell division begins, the centrioles move to opposite sides of the cell. From each centriole there extends a cluster of raylike filaments called an **aster,** and between the separating centrioles a **spindle** forms, composed of threads of protein with properties similar to those of the contractile proteins in muscle, actin and myosin.

Those cells that bear cilia on their exposed surfaces have a structure at the base of each cilium, the **basal body,** that resembles the centriole in the presence of nine parallel tubules. Each cilium contains nine peripherally located longitudinal filaments and two centrally located ones. Like centrioles, basal bodies can duplicate themselves.

The cytoplasm may contain droplets of fat and crystals or granules of protein or glycogen which are simply stored for future use. In addition, it contains the metabolically active cell organelles, **mitochondria, endoplasmic reticulum,** and **Golgi bodies.**

Mitochondria range in size from 0.2 to 5 micrometers (μm.) and in shape from spheres to rods and threads (Figs. 2.4 and 2.18). Their number may range from just a few to more than 1000 per cell. When living cells are examined, their mitochondria appear to move, change shape and size, fuse with other mitochondria to form longer structures, or cleave to form smaller ones. Each mitochondrion is bounded by a double membrane, an outer smooth membrane and an inner one folded into parallel plates that extend into the central cavity and may fuse with folds from the opposite side. Each of these inner and outer membranes is composed of a bilayer of phospholipid molecules with associated membrane proteins. The shelflike inner folds, termed **cristae,** contain the enzymes of the electron transmitter system, of prime importance in converting the potential energy of foodstuffs into biologically useful energy for cellular activities. The semifluid material within the inner compartment, the matrix, contains certain enzymes of the Krebs tricarboxylic acid cycle (p. 71). The mitochondria, whose prime function is the release of biologically useful energy, have been dubbed the "powerhouses" of the cell.

Cells, such as those of the pancreas that are especially active in protein synthesis, are crowded with the membranous labyrinth of the endoplasmic reticulum (Figs. 2.4 and 2.5); other cells may have only a scanty supply of such membranes. Two types are found, **granular** or **rough endoplasmic reticulum,** to which are bound many **ribosomes,** and **agranular** or **smooth endoplasmic reticulum,** consisting of membranes alone. Both smooth and rough endoplasmic reticulum may be found in the same cell. The agranular endoplasmic reticulum may play some role in the process of cellular secretion. The tightly packed sheets of endoplasmic reticulum may form tubules some 50 to 100 nm. in diameter. In other regions of the cell, the cavities of the endoplasmic reticulum may be expanded, forming flattened sacs called **cisternae.** The membranes of the endoplasmic reticulum divide the cytoplasm into a multitude of compartments in which different groups of enzymatic reactions may occur. The endoplasmic reticulum serves a further function as a system for the transport of substrates and products through the cytoplasm to the exterior of the cell and to the nucleus.

Ribosomes are ubiquitous, occurring in all kinds of cells from bacteria to higher plants and animals. Ribosomes contain RNA and protein and are composed of two nearly spherical subunits, which are combined to form the active protein-synthesizing unit (Fig. 2.8). Ribosomes are synthesized in the nucleus and pass to the

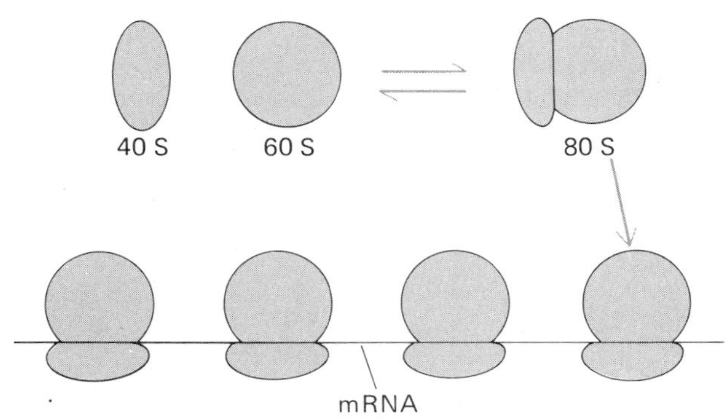

FIGURE 2.8 Diagram illustrating the assembly of a small subunit (40 S) and a large subunit (60 S) to form a ribosome (80 S), the subcellular organelle on which proteins are synthesized. The 80 S ribosomes attach to an RNA strand to form a polysome. The sedimentation coefficients of biological macromolecules, such as proteins and nucleic acids, are expressed in Svedberg units, abbreviated as S (named after The Svedberg, the inventor of the ultracentrifuge). Although larger molecules have larger sedimentation coefficients, the relationship between molecular weight and sedimentation coefficient is not a simple arithmetical one. Hence adding a 40 S and a 60 S subunit results in a molecule with an 80 S, not a 100 S, sedimentation coefficient.

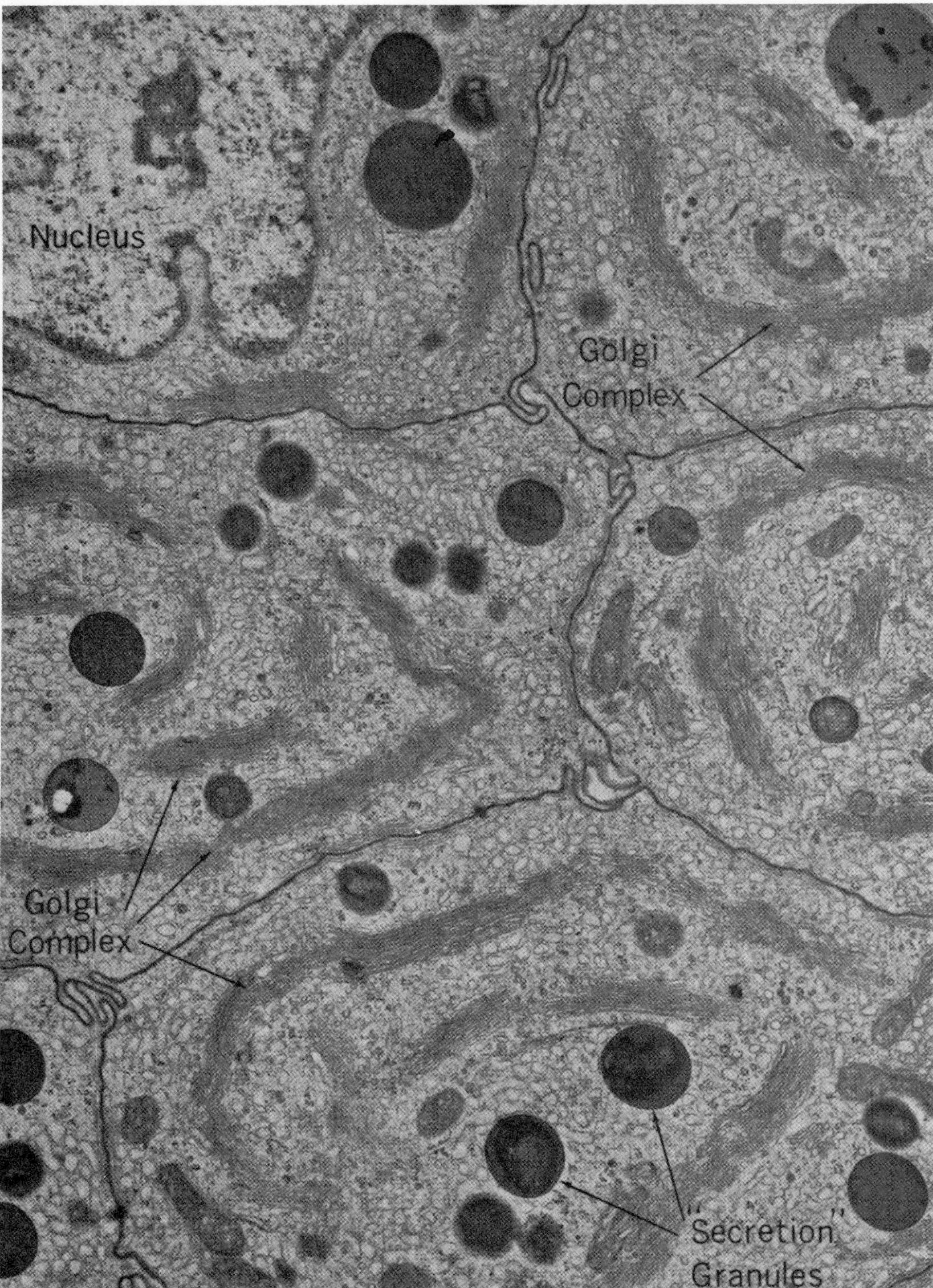

FIGURE 2.9 Electron micrograph of a rabbit epididymis showing the extensive Golgi complex, evident in the parallel arrays of the membranes. Magnification ×9500. (From Fawcett, D. W.: *The Cell.* Philadelphia, W. B. Saunders Co., 1966.)

cytoplasm where they are active in synthesizing proteins. Ribosomes may be bound to the membranes of the endoplasmic reticulum, or they may be free in the matrix of the cytoplasm. In many cells, clusters of five or six ribosomes, termed **polysomes,** appear to be the functional unit that is effective in protein synthesis. It is estimated that a bacterial cell, such as *Escherichia coli,* contains some 6000 ribosomes and that the rabbit reticulocyte (precursor of the red blood cell) contains some 100,000 ribosomes. The protein components of ribosomes from different cells are remarkably similar in their amino acid composition; however, the nucleotide composition of the RNA of ribosomes from different species varies considerably.

Golgi bodies, found in all cells except mature sperm and red blood cells, consist of an irregular network of canals. In the electron microscope (Fig. 2.9) Golgi bodies appear as parallel arrays of membranes without granules. The canals may be distended in certain regions to form small vesicles or vacuoles filled with material. The Golgi bodies appear to serve as temporary storage places for proteins and other compounds synthesized in the endoplasmic reticulum. These materials are repackaged within the Golgi bodies in large sacks made of membranes from the Golgi bodies. In these packets they move to the plasma membrane, which fuses with the membrane of the vesicle, opening the vesicle and releasing its contents to the exterior of the cell.

Many, if not most, cells have hollow cylindrical cytoplasmic subunits termed **microtubules** (Fig. 2.10), which appear to be important in maintaining or controlling the shape of the cell. Microtubules play a role in such cellular movements as that of chromosomes for they form the mitotic spindle (p. 45). They also serve as channels for the oriented flow of cytoplasmic constituents within the cell. Microtubules are also the major structural components of cilia and flagella.

Microtubules are composed of several types of proteins; **tubulin** is present in the largest amount. Tubulin is made of unlike α and β subunits that differ in their

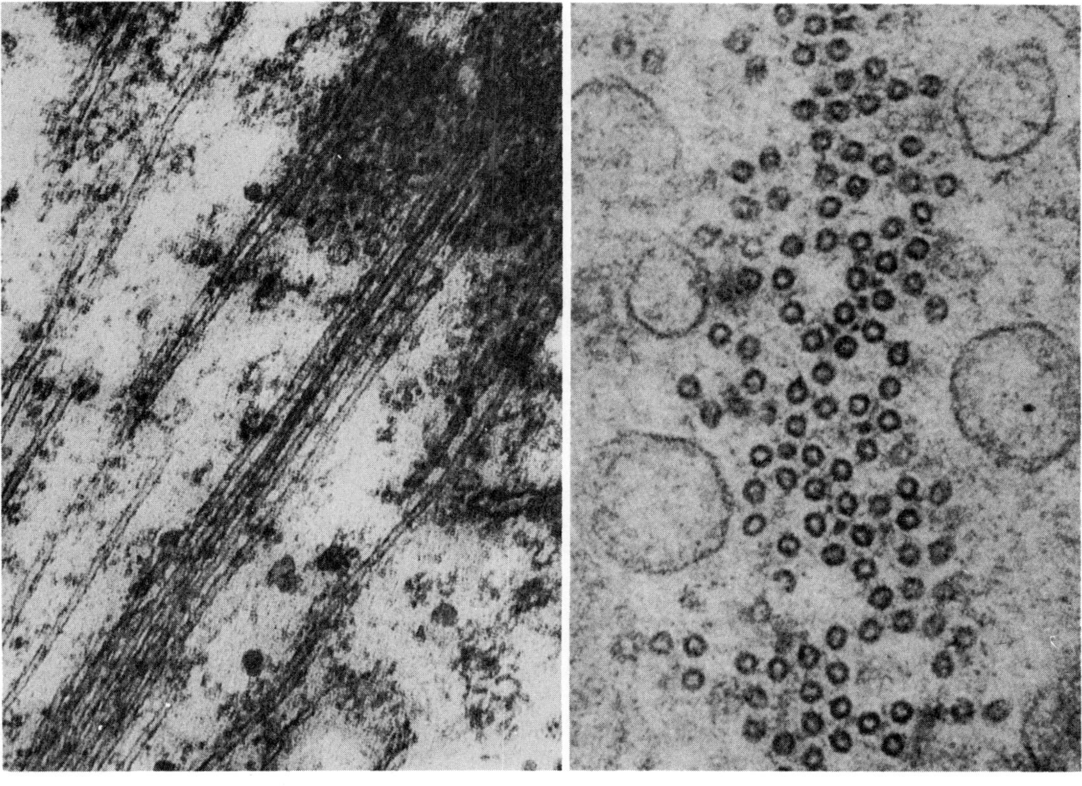

FIGURE 2.10 Electron micrographs of cytoplasmic microtubules. *A,* Microtubules of the mitotic spindle seen here in longitudinal section. The chromosomes are at the lower right. Magnification ×70,000. (Courtesy E. Roth.) *B,* The microtubules in transverse section present circular profiles. Those shown here are from the manchette of a mammalian spermatid. Magnification ×140,000.

amino acid composition. Microtubules are long, hollow cylinders with a wall thickness of 4.5 to 7 nm. and diameters ranging from 20 to 30 nm. The thickness of the wall corresponds to the dimensions of the subunits and suggests that the walls are one molecule thick. Microtubules can grow by the addition of more α and β subunits or can shorten by the disassembly of subunits. The microtubules of nerve axons play a role in the rapid transport of proteins and other molecules such as the hormones of the posterior pituitary and the hypothalamic releasing factors down the axon to the tip.

Solid cytoplasmic **filaments** are present in some cells in addition to the hollow microtubules. These protein filaments play additional roles in cell structure and motion. The cytoplasm of the skeletal muscle fibers contains many long myofibrils, protein filaments that participate in muscle contraction.

Lysosomes are intracellular organelles about the size of mitochondria but less dense. They are membrane-bounded structures that contain a variety of enzymes capable of hydrolyzing the macromolecular constituents of the cell. In the intact cell these enzymes are segregated within the lysosome, presumably to prevent their digesting the contents of the cell. Rupture of the lysosome membrane releases the enzymes and accounts, at least in part, for the lysis of dead cells and the resorption of cells such as those in the tail of a tadpole during metamorphosis. Since they contain enzymes that can hydrolyze the major cellular constituents when they rupture and release them, the lysosomes have been termed "suicide bags" by the Belgian biochemist Christian De Duve.

The cytoplasm of certain cells, chiefly those of lower animals, contains **vacuoles,** bubble-like cavities filled with fluid and separated from the rest of the cytoplasm by a vacuolar membrane. Food is digested within food vacuoles of many protozoa. Other vacuoles play a role in osmoregulation.

The Microtrabecular Lattice. Even the ground substance of the cell, which appears to be entirely homogeneous and structureless by light microscopy or by conventional electron microscopy, has been shown to have an internal structure when examined by high-voltage (one million volt) electron microscopy. The cytoplasmic ground substance contains an irregular three-dimensional lattice of very slender protein threads that extend throughout the cytoplasm and are attached to the cell membrane. The interlinked filaments form a three-dimensional spider web in which are suspended the various intracellular organelles. The microtubules and microfilaments are coated with a material very similar to that composing the individual lattice filaments. Thus the microtubules, microfilaments, endoplasmic reticulum, and mitochondria are integrated with the lattice and are suspended in it. This network has been named the **microtrabecular lattice** by Keith Porter, the American cell biologist who discovered it. The microtrabecular lattice extends throughout the cytoplasm and links the subcellular organelles, microtubules, and microfilaments into a structural and functional unit termed the **cytoplast.** The individual fibers of the lattice are 6 to 16 nm. thick. Administration of the drug cytochalasin B inhibits many kinds of cellular movements. Cells exposed to this drug have microtrabecular lattices that become thickened and coarse, and the spaces between their lattice fibers become enlarged.

Many of the enzymes that have been believed to be "soluble" in the cytoplasm may be bound to this microtrabecular lattice. The lattice may serve as a sort of intracellular musculature, undergoing local contractions and changes in shape that continually redistribute and reorient the intracellular organelles as the cell engages in its various functions. Cells can undergo a variety of internal movements such as the movement of chromosomes in cell division and the movement of vesicles and granules within the cell. The movement of pigment granules within the pigment cells in the skin of fishes, frogs, and reptiles is under hormonal or nervous control. These pigment granules can be moved quite rapidly by the microtrabecular lattice. The movement of the pigment granules is accompanied by marked changes in the structure of the lattice. The aggregation of the pigment granules into the center of the cell is accompanied by a shortening and thickening of the microtrabecular threads. This draws the granules into clusters and moves them along paths defined by microtubules to the central region of the cell (Fig. 2.11). In the reverse process, by which the pigment granules are dispersed through the cell, the microtrabecular fibers elongate and move the pigment granules back to the peripheral region of the cell along the microtubules. If the microtubules are disrupted by treating the cell with colchicine, the granules cannot move in an organized fashion. These movements enable the animal to lighten or darken its skin color rapidly to match the color of its surroundings.

Where, you may ask, is life localized—in the mitochondria? in the ribosomes? in the ground substance? The answer, of course, is that life is not a function of any single one of these parts but of the whole integrated system of many component parts, organized in the proper spatial relationship and interdependent on one another in a great variety of ways.

Most animal cells are quite small, too small to be seen with the naked eye. The diameter of the human red blood cell is about 7.5 μm., but most animal cells have diameters ranging from 10 to 50 μm. There are a few species of giant amoebas with cells about 1 mm. in diameter. The largest cells are the yolk-filled eggs of

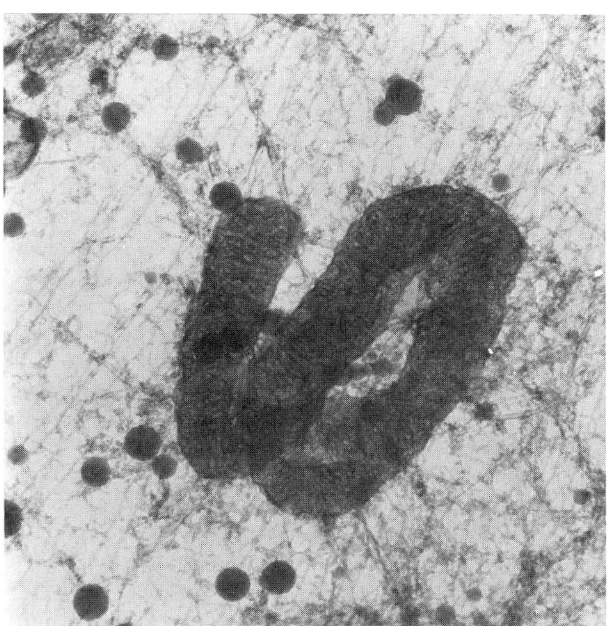

FIGURE 2.11 A large, S-shaped mitochondrion suspended in microfilaments, the very thin, wirelike structures. The dark circular structures are pigment granules in this chromatophore (pigment) cell. (Courtesy of Drs. Keigi Fujiwara, Hugh Randolph Byers, and Elena McBeath, Harvard Medical School.)

birds and sharks. The egg cell of a large bird such as a turkey or goose may be several centimeters across. Only the yolk of a bird's egg is the true egg cell; the egg white and shell are noncellular material secreted by the bird's oviduct as the egg passes through it.

The limit of the size of a cell is set by the physical fact that, as a sphere gets larger, its surface increases as the square of the radius but its volume increases as the cube of the radius. The metabolic activities of the cell are roughly proportional to cell volume. These activities require nutrients and oxygen and release carbon dioxide and other wastes which must enter and leave the cell through its surface. The upper limit of cell size is reached when the surface area can no longer provide for the entrance of enough raw materials and the exit of enough waste products for cell metabolism to proceed normally. The limiting size of the cell will depend on its shape and its rate of metabolism. When this limit is reached the cell must either stop growing or divide.

Much has been learned in recent years about the role each of these particles plays in the economy of the cell. Cells are homogenized in special glass grinding tubes to break the cell membrane and release the intracellular structures. Then, by subjecting the homogenate to increasing amounts of centrifugal force in an ultracentrifuge, first the nuclei, then the mitochondria, and finally the microsomes can be sedimented separately. Microsomes are ribosomes together with fragments of the endoplasmic reticulum. When these sedimented particles are examined in the electron microscope, they are found to have the same structure exhibited by comparable structures in the intact cell. The separated particles can then be suspended in suitable incubation media and their metabolism can be studied. Such separated mitochondria and microsomes will carry out many biochemical reactions, and much is now known about the functions of each of these particles. The liquid left after the homogenate has been subjected to high centrifugal force to sediment the microsomes contains many other enzymes which apparently exist in the cell more or less freely in the ground substance.

The Cell Cycle

The doubling of all the constituents of the cell, followed by its division into two daughter cells, is generally termed the **cell cycle.** A cell is "born" when its parent cell divides; it undergoes a cycle of growth and division and gives rise to two daughter cells. The cell cycle in a typical plant or animal cell requires about 20 hours for completion. Only an hour or so is devoted to cell division, or mitosis; the rest of the time is required for interphase growth. Under optimal conditions of nutrition and temperature, the length of the cell cycle for any given kind of cell is constant. Under less favorable conditions it may be slowed, but it has not been possible to speed up the cell cycle and make cells grow faster. From this we infer that the duration of the cell cycle is the time required for carrying out some precise program that has been built into each cell. This program appears to include two parts: one having to do with the replication of the genetic material in the chromosomes and the other involving the doubling of all of the other constituents of the cell involved in growth.

The period of DNA synthesis, during which there is an exact replication of each chromosome, is termed the **S phase** (Fig. 2.12). The time, or gap, between mitotic division and DNA replication is termed the **G_1 phase,** or gap 1 phase. During the G_1 phase certain key processes occur that make it possible for the cell to enter the S phase and become committed to a future cell division. Following the completion of the S phase, the cell is usually not ready to divide immediately but undergoes a second gap phase, the **G_2 phase,** during which there is an increase in protein synthesis and final preparation for cell division. The completion of G_2 is marked by the beginning of mitotic division, the **M phase.**

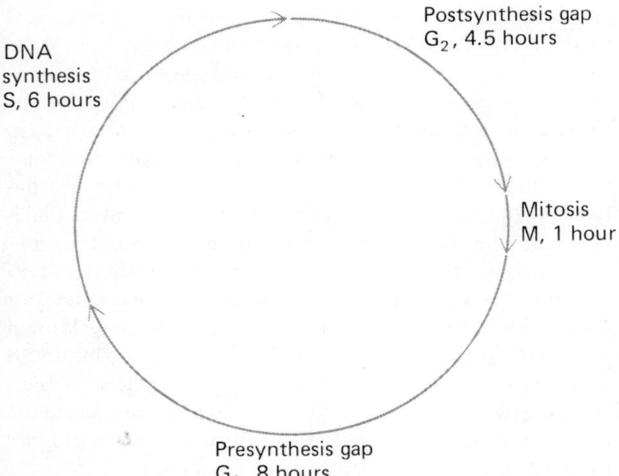

FIGURE 2.12 Diagram of the cell cycle showing durations of the four phases in a typical cell.

In a typical cell cycle, the $\mathbf{G_1}$ phase occupies about eight hours, the S phase six hours, the G_2 phase four and one-half hours, and the M phase one hour. Although the duration of each phase may vary somewhat, the greatest variation is found in the G_1 phase, which may be as short as a few hours or minutes or as long as several days or weeks. Each human cell contains 46 threads of DNA with a total length of 2 m. or more, all of which are stuffed into a nucleus about 5 μm. in diameter. In the complex process of replication an exact copy is made of each of these 46 threads. Replication in the S phase does not simply begin at one end of each thread and travel along to the other end; rather, each thread undergoes replication in many segments according to a definite program. The segments do not replicate in tandem in any one chromosome, nor does any one chromosome complete its replication before the next chromosome begins. When the last segments have been replicated, DNA synthesis shuts down and does not resume until the next cycle.

Scanning electron microscopy reveals that cells grown in culture have quite different external appearances during the different phases (Fig. 2.13). During the M phase the cells are spherical and not strongly attached to the substrate. As they enter the G_1 phase they begin to flatten, the cell surface shows bubble-like blisters, and microvilli appear. Later in the G_1 phase the margins of the cells become thin and active and appear ruffled. During the S phase the cells flatten further, and their surfaces become smooth. As the cell enters the G_2 phase it once again thickens; the surface shows ruffles and microvilli but not as many surface blisters as during the G_1 phase. Two important transition points occur in the cell cycle: one between the G_1 phase and the S phase, when the replication of the chromosomes starts, and the other between the G_2 phase and the M phase, when the chromosomes condense and mitosis begins.

Certain procedures utilizing Sendai viruses permit an investigator to fuse or hybridize two or more cells. By producing hybrids between cells in different phases of the cell cycle or between cells of different species with different cycle lengths, it has been possible to investigate the signals that move a cell from one phase to the next (Fig. 2.14). Something in the S phase pervades the cell and is responsible for initiating the replication of chromosomes. Moreover, it can induce any nucleus to enter the S phase whether or not the nucleus is ready. When cells in the G_1 phase are fused with ones in the S phase, nuclei of the G_1 cells begin to make DNA long before they normally would. The signal to begin chromosome replication might be some specific molecule, or it might be some specific alteration in the internal environment of the cell. Similarly, something in the M phase forces chromosomes to condense, whether or not they have replicated. Fusion of an M phase cell with one in any other phase causes chromosomes in the nuclei of the other phase to condense.

An organism, such as a higher animal, that can be viewed as a society of cells, includes some cells that will reproduce and others that will not. The cells of tissues that perform certain special services—those of the nervous system and the muscular system—do not in general reproduce at all. In tissues such as the skin, the blood-forming system, and epithelial linings, new cells are produced at a rate that just compensates for the continued loss of old cells. The cells involved in immune reactions and the healing of wounds multiply at a rapid rate as a bodily response to some external provocation. Finally, in malignant cancer the cells no longer abide by the rules of governance of the organism and go through their cycles in an anarchic fashion. There is nothing special about the cell cycle in cancer cells compared with the cell cycle in normal cells. The tempo, the cycle, and its phases are the same, but the difference is that cancer cells repeat the cycle without restraint.

A very simple rule emerges from these studies. Cells that do not divide never enter the S phase, and conversely, cells that enter the S phase almost always complete that phase and go on to divide. Thus replication of the chromosomes is a strong commitment ultimately to undergo division, and the major control of cell division lies in whether or not a cell enters into replication. However, cells that are not going to divide are not cells that have become stuck somehow in the transition between the G_1 and the S phases. Nondividing cells are probably best considered as ones that have not entered

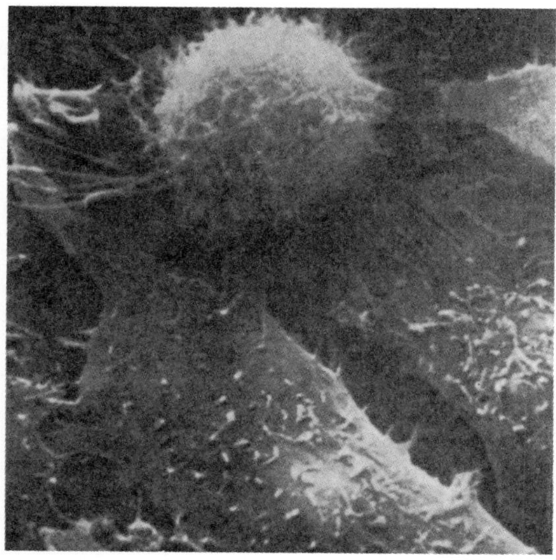

A

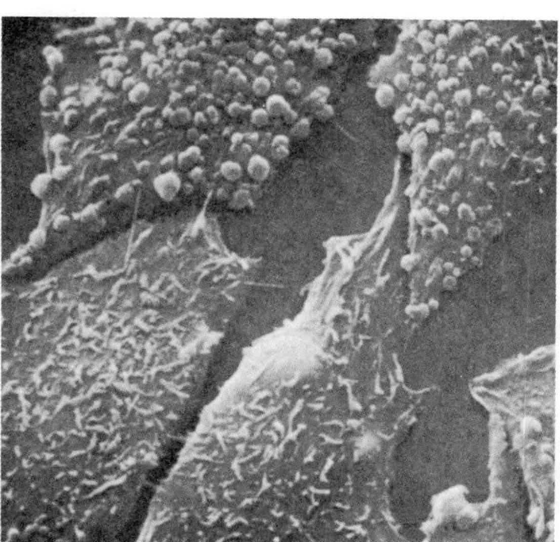

B

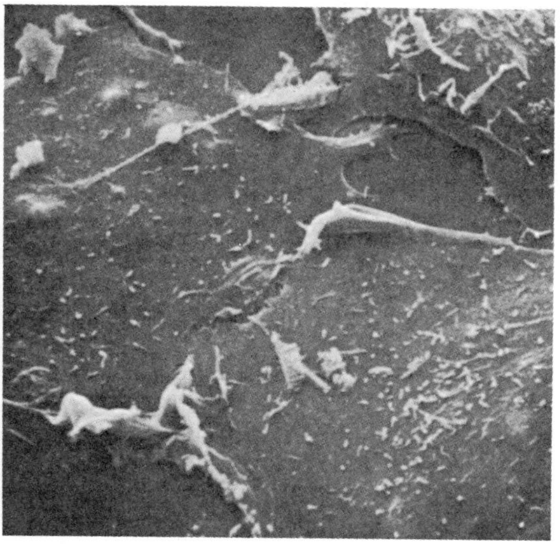

C

FIGURE 2.13 Hamster ovary cells in mitosis. *A,* The shape of the cell is changing from flat to round. A number of long, thin projections covering the cell surface help to secure the cell to the substrate. *B,* Flattened cells in late G_1 phase; the surfaces are covered with microvilli and blisterlike spheres (blebs). Several cells are establishing contact with each other. *C,* Increased flattening of cells having entered the S phase; no blebs now remain, and microvilli are less abundant. "Ruffles" (light areas) appear on the edges of the cells. (Scanning electron micrographs by Keith R. Porter, David M. Prescott, and Jearl F. Frye. From Mazia, D.: The cell cycle. *Scientific American, 230*:54, 1974. © 1974 by Scientific American, Inc. All rights reserved.)

the cycle at all. The many kinds of specialized cells in the body are noncycling cells, making no progress at all toward cell division. Many of the cells that have left the cell cycle can be made to reenter it by placing them under spcific culture conditions in the laboratory. Presumably agents that cause cancer must in some way cause noncycling cells to enter the cycle. The conversion of lymphocytes from noncycling to cycling cells is of importance in the immune response. The cells begin to grow and divide, producing cells that contribute to the formation of antibodies. Noncycling lymphocytes can be transformed into cycling ones in the laboratory by exposing them to **lectins,** specific kinds of plant proteins. When noncycling lymphocytes are artificially stimulated by exposure to lectins, nearly 24 hours pass before they enter the S phase and begin to replicate the chromosomes. The delay apparently is due to the need of these stimulated lymphocytes to produce the enzymes required for the replication process as a preliminary to entering the S phase. Thus they are not stuck in

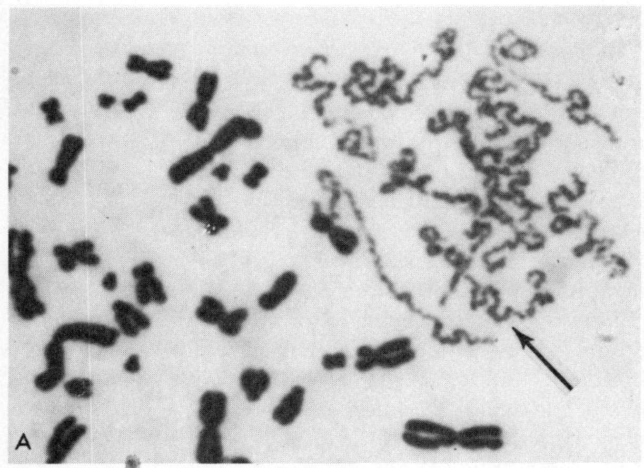

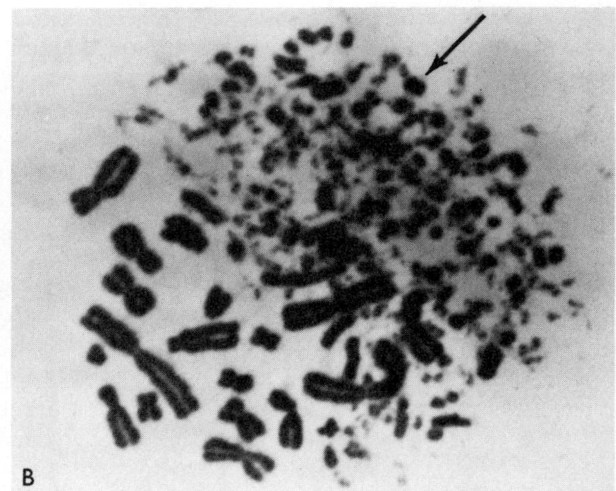

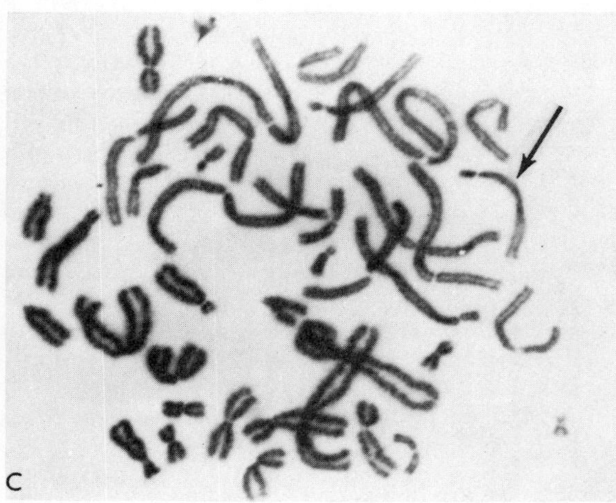

FIGURE 2.14 Prematurely condensed chromosomes (PCC) of Chinese hamster ovary cells. Cells synchronized in mitosis were fused with interphase cells with the help of U.V. inactivated Sendai virus. Metaphase chromosomes are darkly stained and well condensed; PCC are lightly stained with variable morphology. *A,* G_1 cell was fused with a mitotic cell. G_1-PCC *(arrow)* have a single chromatid and are greatly spiralized. *B,* Heterophasic binucleate cell formed by fusion of a cell in S phase with one in mitosis. S-PCC appear to be fragmented because of uneven condensation of S chromatin, dependent upon the state of DNA replication. Some chromosome regions have replicated; others have not yet started replication or are in the active process of replication. Replicated elements consist of double chromatids *(arrow);* unreplicated regions are seen as single chromatids. Actual site of replication (replicon) in the chromatin does not undergo condensation; hence it appears as a gap between the condensed regions. *C,* Product of fusion between G_2 and mitotic cells. G_2-PCC *(arrow)* are long and slender, with the two chromatids still attached to each other. (Courtesy of Dr. Potu N. Rao, The University of Texas System Cancer Center, M. D. Anderson Hospital and Tumor Institute, Houston, Texas.)

the G_1 phase, ready to go into the S phase when they are signaled. When stimulated, a noncycling cell must first do the things that a cycling cell does in the G_1 phase before it can enter the S phase and replicate its chromosomes.

Mitosis

Because of the limitation on the size of individual cells, growth is accomplished largely by an increase in the *number* of cells. The process of cell division, called **mitosis** (Gr. *mites,* a thread + *osis,* a process), is extremely regular and ensures the qualitatively and quantitatively equal distribution of the hereditary factors between the two resulting daughter cells. Mitotic divisions occur during embryonic development and growth; in the replacement of cells that wear out, such as blood cells, skin, the intestinal lining, and so on; and in the repair of injuries.

When a dividing cell is stained and examined under the microscope, dark-staining **chromosomes** are visible within the nucleus. Each consists of a central thread, the **chromonema,** along which lie the **chromomeres**—small, beadlike, dark-staining swellings. It has been suggested that the chromomeres are, or contain, the genes, for breeding experiments have shown clearly that these hereditary units lie within the chromosome in a linear order. However, the correlation between chromomeres and genes is not regular; some chromomeres contain several genes and some genes have been located between chromomeres. Each chromosome has, at a fixed point along its length, a small clear circular zone called a **kinetochore** that controls the movement of the chromosome during cell division. As the chromosome becomes shorter and thicker just before cell division occurs, the kinetochore region becomes accentuated and appears as a constriction.

If we tamper experimentally with the mechanism of cell division, and the resulting cells receive more or less than the proper number of chromosomes, marked abnormalities of growth, and perhaps the death of these cells, will follow. Mitosis may be defined as the regular process of cell division by which each of the two daughter cells receives exactly the same number and the same kind of chromosomes that the parent cell contained. Although each chromosome appears to split longitudinally into two halves, each original chromosome has brought about the synthesis of an exact replica of itself immediately adjacent to itself during the preceding S phase. The old and new chromosomes are identical in structure and function and at first lie so close to one another that they appear to be one. As mitosis proceeds and the chromosomes contract, the line of cleavage between them becomes visible. When the process is complete, the original and the new chromosomes separate and become incorporated into different daughter cells. The role of the complicated mitotic machinery is to separate the "original" and "replica" chromosomes and to deliver them to opposite ends of the dividing cell so they will become incorporated into different daughter cells.

The term mitosis in a strict sense refers to the division of the nucleus into two daughter nuclei, and the term **cytokinesis** is applied to the division of the cytoplasm to form two daughter cells, each containing a daughter nucleus. Nuclear division and cytoplasmic division, although almost invariably well synchronized and coordinated, are separate and distinct processes.

Each mitotic division is a continuous process, with each stage merging imperceptibly into the next one. For descriptive purposes mitosis may be divided into four stages: **prophase, metaphase, anaphase,** and **telophase** (Fig. 2.15). Between mitoses a cell is said to be in the resting stage. The nucleus is "resting" only with respect to division, however, for during this time it may be very active metabolically. It is difficult to visualize from a description or diagram of mitosis, or from examining a fixed and stained slide of cells, just how active a process cell division is. Motion pictures made by phase microscopy reveal that a cell undergoing division bulges and changes shape like a gunny sack filled with a dozen unfriendly cats.

Prophase. The chromatin threads condense, and the chromosomes appear as a tangled mass of coiled threads within the nucleus. Early in prophase the threads are stretched maximally so that the individual chromomeres are visible. Later in prophase the chromosomes shorten and thicken, and the chromomeres lie so close together that individual ones cannot be distinguished. Each part of the doubled chromosome is called a chromatid; the two chromatids are held together at the kinetochore, which remains single until the metaphase.

Early in prophase the centrioles migrate to opposite sides of the cell. Between the separating centrioles a **spindle** forms. The protein threads of the spindle are arranged like two cones base to base, broad at the center or equator of the cell and narrowing to a point at either end or pole. With a microneedle attached to a micromanipulator, the spindle can be moved as a unit from one part of the cell to another. By appropriate techniques spindles may be isolated from dividing cells (Fig. 2.16). At the end of prophase, the centrioles have gone to the opposite poles of the cell, the spindle has formed between them, and the chromosomes have become short and thick.

Metaphase. When the chromosomes are fully contracted and appear as short, dark-staining rods, the nuclear membrane disappears and the chromosomes line up in the equatorial plane of the spindle. The short period during which the chromosomes are in this equatorial plane is known as the metaphase. This is much shorter than the prophase; although times for different cells vary considerably, the prophase lasts from 30 to 60 minutes or more, and the metaphase lasts only 2 to 6 minutes.

During the metaphase the kinetochore of each chromosome divides, and the two chromatids become completely separate daughter chromosomes. The division of the kinetochores occurs simultaneously in all the chromosomes, under the control of some as yet unknown mechanism. The daughter kinetochores begin to move apart, marking the beginning of the anaphase.

Anaphase. The chromosomes separate (Fig. 2.17), and one of the daughter chromosomes goes to each pole. The period during which the separating chromosomes move from the equatorial plate to the poles, known as the anaphase, lasts some 3 to 15 minutes. The spindle fibers apparently act as guide rails along which the

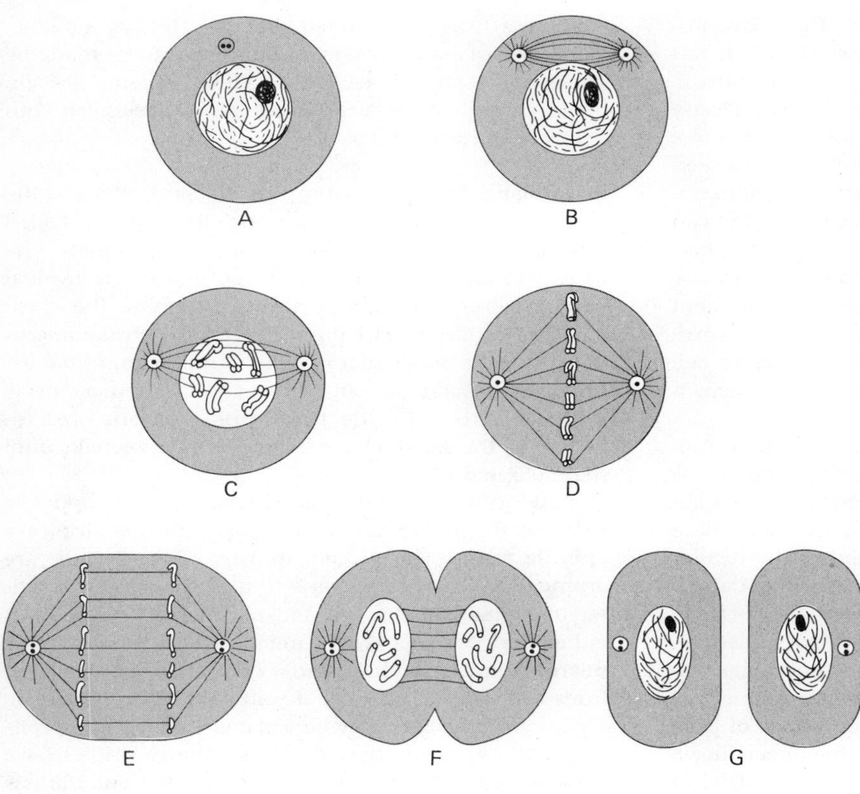

FIGURE 2.15 Mitosis in a cell of a hypothetical animal with a diploid number of six (haploid number of three); one pair of chromosomes is short, one pair is long and hooked, and one pair is long and knobbed. *A,* Resting stage. *B,* Early prophase; centriole divided and chromosomes appearing. *C,* Later prophase: centrioles at poles, chromosomes shortened and visibly double. *D,* Metaphase: chromosomes arranged on the equator of the spindle. *E,* Anaphase: chromosomes migrating toward the poles. *F,* Telophase: nuclear membranes formed, chromosomes elongating, cytoplasmic division beginning. *G,* Daughter cells: resting phase.

chromosomes move toward the poles. Without such guide rails the chromosomes would merely be pushed randomly apart, and many would fail to be incorporated into the proper daughter nucleus. The mechanism moving the chromosomes apart is not clear. Experiments indicate that some of the spindle fibers are contractile and can pull the chromosomes toward the poles. Spindles isolated from cells about to divide can be induced to contract when ATP is added. The chromosomes moving toward the poles usually assume a V shape with the kinetochore at the apex pointing toward the pole. It appears that whatever force moves the chromosome to the pole is applied at the kinetochore. A chromosome that lacks a kinetochore, perhaps as a result of exposure to X-radiation, does not move at all in mitosis.

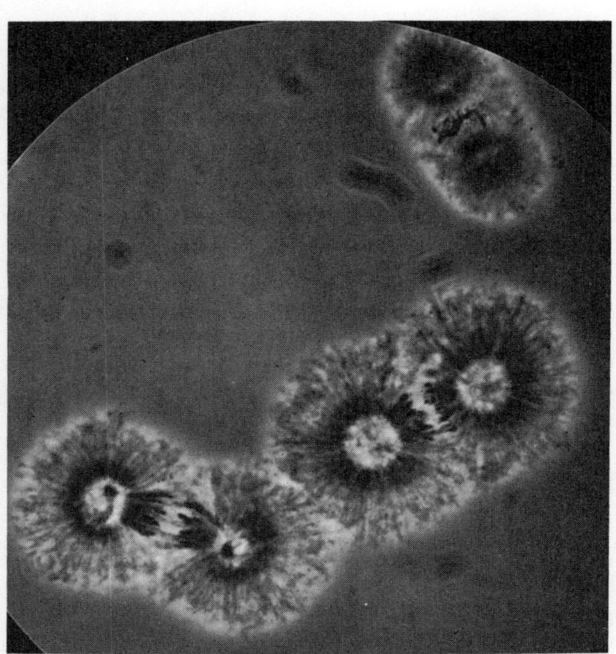

FIGURE 2.16 Photomicrograph of the mitotic apparatus isolated from dividing cells of a sea urchin embryo. Each mitotic apparatus includes spindle fibers, asters, and chromosomes. A metaphase figure appears in the upper right and two anaphase figures below. (Courtesy of Daniel Mazia.)

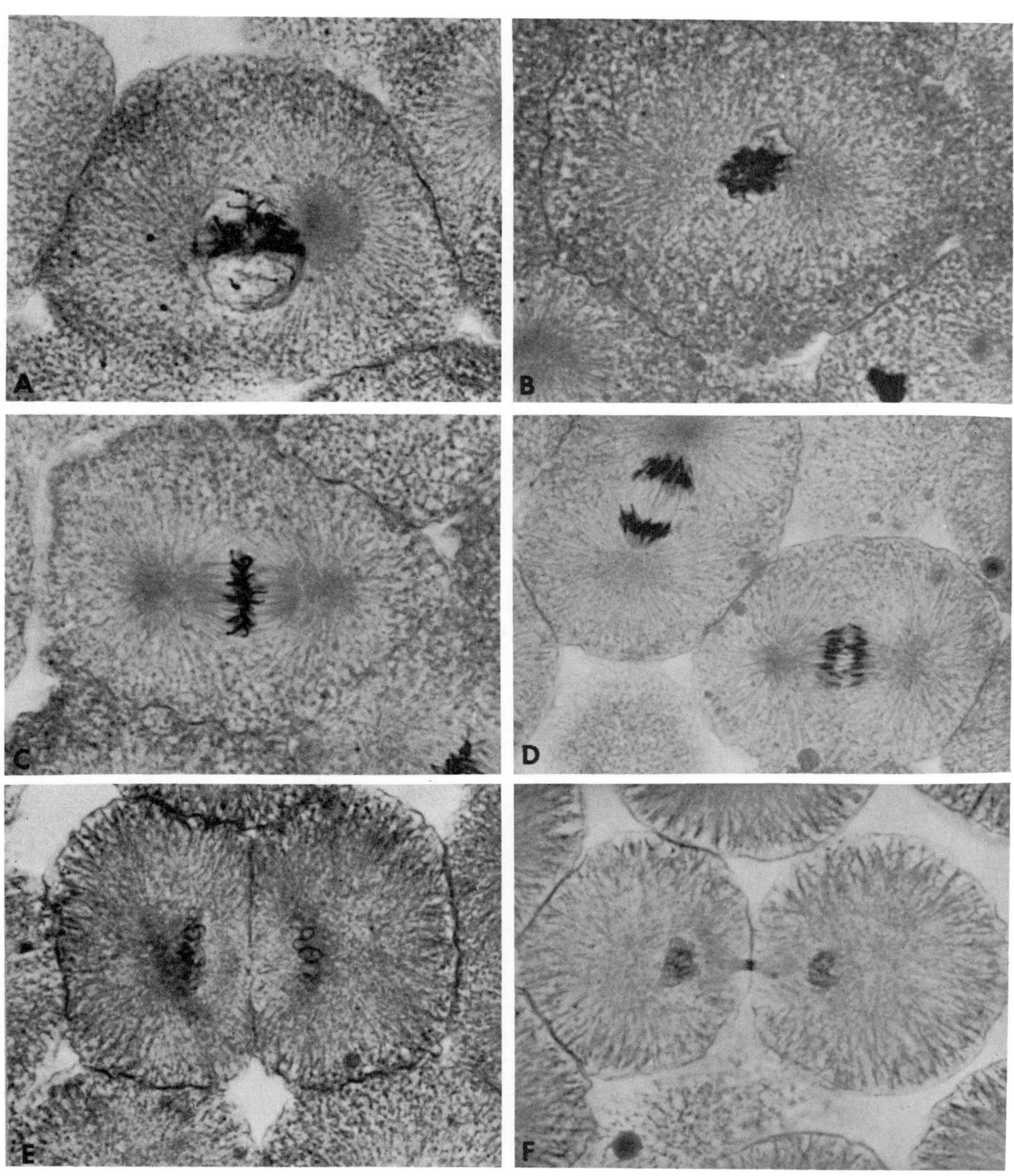

FIGURE 2.17 Photomicrographs of stages in mitosis in the white fish blastula. *A,* Early prophase; *B,* later prophase; *C,* metaphase; *D,* two cells in early and late anaphase, respectively; *E,* early telophase; and *F,* late telophase. The dark spot connecting the two cells in *E* is the remainder of the spindle. (Photographs courtesy of Dr. Susumu Ito.)

Telophase. When the chromosomes have reached the poles of the cell, the last phase of mitosis, telophase, begins. Several processes occur simultaneously in this period: A nuclear membrane forms around the group of chromosomes at each pole, the chromosomes elongate, stain less darkly, and return to the resting condition in which only irregular chromatin threads are visible, and the cytoplasm of the cell begins to divide. Division of the cytoplasm (**cytokinesis**) is accomplished in animal cells by the formation of a furrow that circles the cell at the equatorial plate and gradually deepens until the two halves of the cell are separated as independent daughter cells. The events of telophase require some 30 to 60 minutes for their completion.

The mitotic process results in the formation of two daughter cells from a single parent cell. Since all the cells of the body are formed by mitosis from a single fertilized egg, each cell has the same number and kind of chromosomes and the same number and kind of genes as every other cell.

The speed and frequency of cell division vary greatly from tissue to tissue and from one animal to another. In the early stages of embryonic development, there may be only 30 minutes or so between successive cell divisions. In certain adult tissues, notably the nervous system, mitoses are extremely rare. In other adult tissues, such as the red bone marrow, where red blood cells are produced, mitotic divisions must occur frequently to supply the 10,000,000 red blood cells each human being produces every second of the day and night.

Regulation of Mitosis

The factors that initiate and control cell division are not known exactly. The possible role of the ratio of cell surface to cell volume was discussed previously (p. 41). The ratio of *nuclear* surface to *nuclear* volume may also be important. Since normal cell function requires the transport of substances back and forth through the nuclear membrane, growth will eventually result in a state in which the area of the nuclear membrane is insufficient to meet the demands of the volume of cytoplasm. Cell division, by splitting the volume of cytoplasm into two parts and increasing the area of nuclear membrane, will restore optimal conditions.

The nucleus-cytoplasmic mass theory, proposed by von Hertwig in 1908, requires some changes, for factors other than cell mass are involved in controlling mitosis. Cells can be made to divide before they have doubled in size, and the eggs of many organisms grow to very large size before dividing. There is some evidence that the two cyclic nucleotides, cyclic AMP and cyclic GMP, play roles in regulating many cellular functions, including cell division. The two have opposite, antagonistic effects on the growth and division of certain cells grown in culture, with cGMP stimulating growth and cAMP inhibiting it.

The Study of Cellular Activities

Despite great differences in size, shape, and location in the body, all cells have many metabolic activities in common. The myriad enzymes in each cell enable it to release energy by converting sugars, fats, and proteins to carbon dioxide and water. Each cell synthesizes its own structural proteins and enzymes. Superimposed on this basic pattern of metabolism common to all cells may be other activities peculiar to each type of cell. For example, muscle cells have special contractile proteins, **myosin** and **actin;** particular digestive enzymes are produced by the cells lining the stomach and intestine; and the cells of the pituitary, adrenal, and thyroid glands manufacture characteristic hormones.

Each of the many ways of studying cellular activity provides useful information about cell morphology and physiology. Living cells suspended in a drop of fluid can be examined under an ordinary microscope or with one equipped with **phase contrast lenses.** In this way one can study the movement of an amoeba or a white blood cell or the beating of the cilia on a paramecium. Cells from a many-celled animal—frog, chick, or man—can be grown by "**tissue culture**" for observation over a long period of time. A complex nutritive medium, made of blood plasma, an extract of embryonic tissues, and a mixture of salts, glucose, amino acids, and vitamins, is prepared and sterilized. A drop of this is placed in a cavity on a special microslide, the cells to be cultured are added aseptically, and the cavity is sealed with a glass cover slip. After a few days the cells have exhausted one or more of the nutritive materials and must be transferred again to a fresh drop of medium. Connective tissue cells (fibroblasts) will survive in culture and undergo a relatively fixed number of mitoses (cell generations)—about 50—after which they can no longer divide. In contrast, cancer cells can survive in culture and undergo divisions indefinitely. Muscle cells in culture divide a few times and then fuse to form a muscle fiber, becoming contractile.

Cell morphology may be studied by using a bit of tissue that has been killed quickly with a special "fixative," then sliced with a machine called a microtome, and stained with special dyes. The thin stained slices, mounted on a glass slide and covered with a glass cover slip, are then ready for examination under the micro-

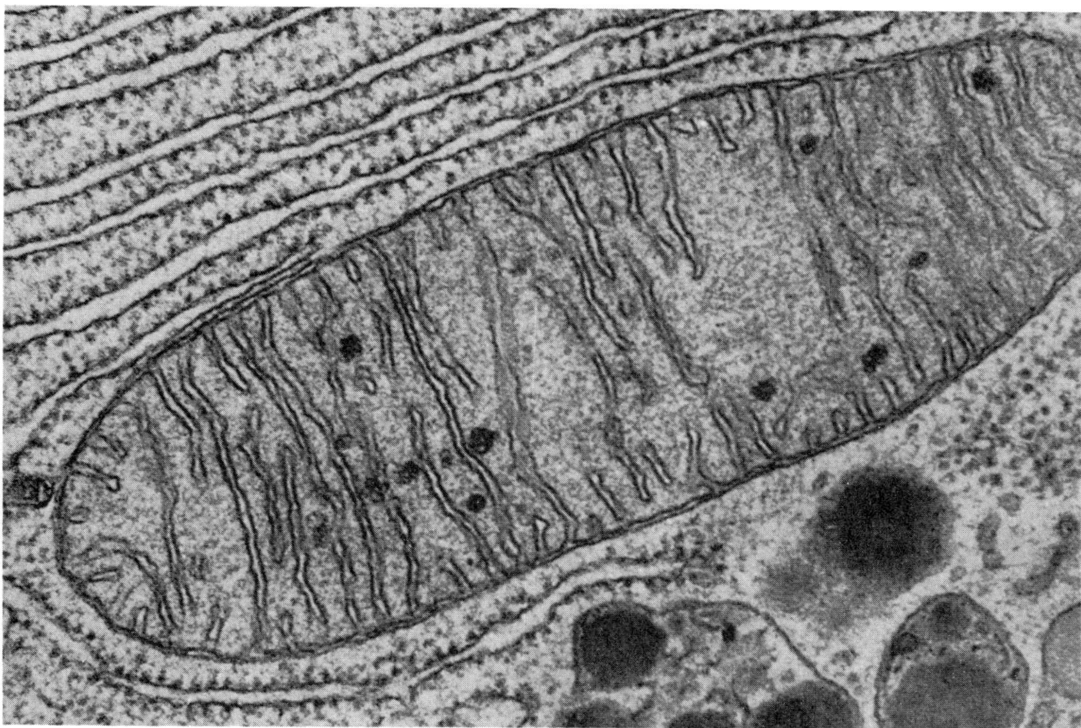

FIGURE 2.18 Electron micrograph of a typical mitochondrion from the pancreas of a bat, showing the cristae, matrix, and matrix granules. Endoplasmic reticulum is seen at the upper left and some lysosomes at the lower right. Magnification ×79,000. (Courtesy of K. R. Porter.)

scope. The nucleus, mitochondria, and other specialized parts of the cell differ chemically and will combine with different dyes and be stained characteristic colors (Fig. 2.7). For observation in the electron microscope a bit of tissue is fixed with osmic acid, mounted in acrylic plastic for cutting in extremely thin sections, and then placed on a fine grid to be inserted into the path of the electron beam. Both light microscopy and electron microscopy have revealed many details about cell structure (Fig. 2.18).

Some clue as to the location and functioning of enzymes within cells can be obtained by **histochemical** studies, in which a tissue is fixed and sliced thin by methods that do not destroy enzymic activity. Then the proper chemical substrate for the enzyme is provided and, after a specified period of incubation, some substance is added that will form a colored compound with one of the products of the reaction mediated by the enzyme. The regions of the cell that have the greatest enzymic activity will have the largest amount of the colored substance (Fig. 2.19). Methods have been devised that permit the demonstration and localization of a wide variety of enzymes. Such studies have given interesting insights into the details of cell function.

Another method of investigating cell function is to measure, by special microchemical analyses, the amounts of chemical used up or produced as a bit of tissue is incubated in an appropriate medium. In such experiments much has been learned of the roles in cell metabolism of vitamins, hormones, and other chemicals by adding these substances one by one and observing the resulting effects.

Every living cell, whether it is an individual unicellular animal or a single component of a multicellular one, must be supplied constantly with nutrients and oxygen. These materials are constantly being metabolized—used up—as the cell goes about its business of releasing energy from the nutrients to provide for its myriad activities. Some of the substances required by the cell are brought to it and taken in by complex active processes that require the expenditure of energy by the cell. Other substances are brought to the cell by the simpler, more

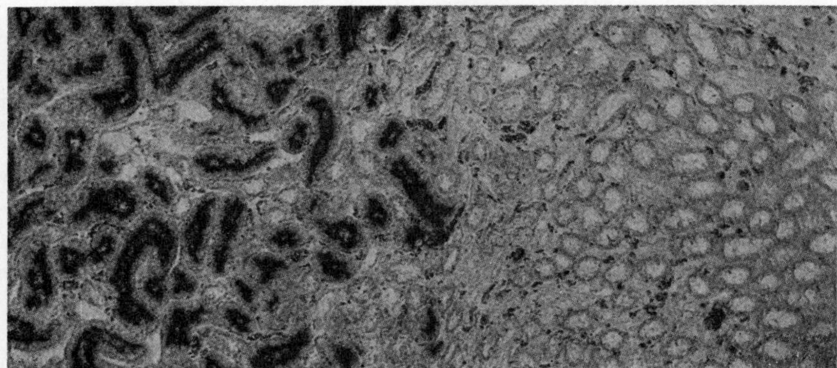

FIGURE 2.19 Histochemical demonstration of the location of the enzyme alkaline phosphatase within the cells of the rat's kidney. The tissue is carefully fixed and sectioned by methods that do not destroy the enzyme's activity. The tissue section is incubated at the proper pH with a naphthyl phosphate. Some hydrolysis of the naphthyl phosphate occurs wherever the phosphatase enzyme is located. The naphthol released by the action of the enzyme couples with a diazonium salt to form an intensely blue, insoluble azo dye, which remains at the site of the enzymatic activity. The photomicrograph thus reveals the sites of phosphatase activity; i.e., the sites at which the azo dye is deposited. The cells of the proximal convoluted tubules *(left)* have a lot of enzyme; those of the loop of Henle *(right)* have little or no activity. (Courtesy of R. J. Barnett.)

easily understood physical process of **diffusion.** To understand this process, so important in many biological phenomena, we must first consider some of the basic physical concepts of energy and molecular motion.

Energy

Energy may be defined as the ability to do work, to produce a change in matter. It may take the form of heat, light, electricity, motion, or chemical energy. Physicists recognize two kinds of energy: **potential energy,** the capacity to do work owing to the position or state of a body, and **kinetic energy** (Gr. *kinētikos,* moving), the capacity to do work owing to the motion of a body. A rock at the top of a hill has potential energy; as it rolls downhill the potential energy is converted to kinetic energy.

Energy derived ultimately from solar radiation, trapped by photosynthetic processes in plants, is stored in the molecules of foodstuffs as the chemical energy of the bonds connecting their constituent atoms. This chemical energy is a kind of potential energy. When these food molecules are taken within a cell, chemical reactions occur that change this potential energy into heat, light, motion, or some other kind of kinetic energy. Light is a kind of kinetic energy that may be thought of as the movement of photons or light quanta. All forms of energy are at least partially interconvertible, and living cells constantly transform potential energy into kinetic energy or the reverse (Table 2.1). If the conditions are suitably controlled, the amount of energy entering and leaving any given system can be measured and compared. Such experiments have shown that energy is neither created nor destroyed but simply transformed from one form to another. This is an expression of one of the fundamental laws of physics, the law of the con-

TABLE 2.1 Energy Transformations in Cells

Transformation	Type of Cell
Chemical energy to electrical energy	Nerve, brain
Sound to electrical energy	Inner ear
Light to chemical energy	Chloroplast
Light to electrical energy	Retina of eye
Chemical energy to osmotic energy	Kidney
Chemical energy to mechanical energy	Muscle cell, ciliated epithelium
Chemical energy to radiant energy	Luminescent organ of firefly
Chemical energy to electrical energy	Sense organs of taste and smell

servation of energy. Living things as well as nonliving systems obey this law.

Molecular Motion

The constituent molecules of all substances are constantly in motion. The prime difference between solids, liquids, and gases is the freedom of movement of the molecules present. The molecules of a solid are very closely packed, and the forces of attraction between the molecules permit them to vibrate but not to move around. In the liquid state the molecules are somewhat farther apart and the intermolecular forces are weaker, so that the molecules can move about with considerable freedom. The molecules in the gaseous state are so far apart that the intermolecular forces are negligible and molecular movement is restricted only by external barriers. Molecular movement in all three states of matter is the result of the inherent heat energy of the molecules, the kinetic energy determined by the temperature of the system. By increasing this **molecular kinetic energy,** one can change matter from one state to another. When ice is heated it becomes water, and when water is heated it is converted to water vapor.

If a drop of water is examined under the microscope, the motion of its molecules is not evident. If a drop of India ink is added, the carbon particles move continually in aimless zigzag paths because they are constantly being bumped by water molecules and the recoil from this bump imparts the motion to the carbon particle. The motion of such small particles is called **brownian movement,** after Robert Brown, an English botanist, who in 1827 first observed the motion of pollen grains in a drop of water.

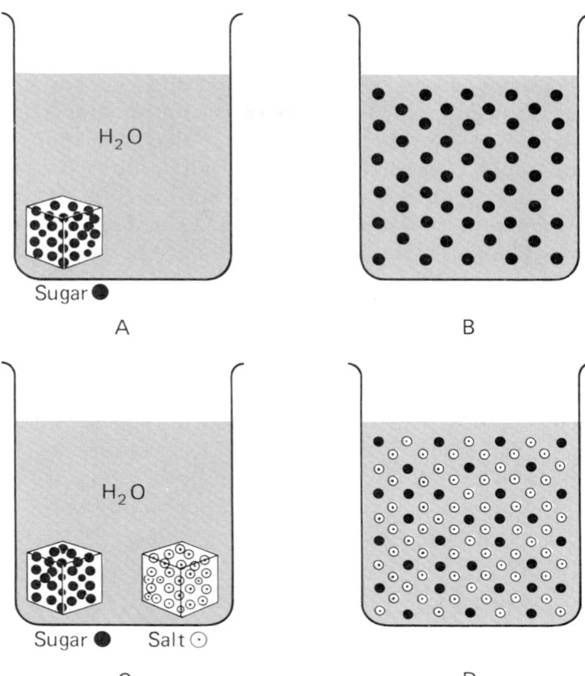

FIGURE 2.20 Diffusion. When a cube of sugar is placed in water *(A)* it dissolves and its molecules become uniformly distributed throughout the water as a result of the molecular motion of both sugar and water molecules *(B)*. When lumps of sugar and salt are placed in water *(C)*, each type of molecule diffuses independently of the other, and both salt and sugar become uniformly distributed in the water *(D)*.

Diffusion

Molecules in a liquid or gaseous state will diffuse, that is, move in all directions, until they are spread evenly throughout the space available. **Diffusion** may be defined as the movement of molecules from a region of high concentration to one of lower concentration, brought about by their kinetic energy. The rate of diffusion is a function of the concentration difference, the size of the molecule, and the temperature. If a bit of sugar is placed in a beaker of water, the sugar will dissolve, and the individual sugar molecules will diffuse and come to be distributed evenly throughout the liquid (Fig. 2.20). Each molecule tends to move in a straight line until it collides with another molecule or the side of the container; then it rebounds and moves in another direction. By this random movement of molecules, the sugar eventually becomes evenly distributed throughout the water in the beaker. This process could be demonstrated by tasting drops of liquid taken from different parts of the beaker. If a colored dye is used in place of sugar, the process of diffusion can be observed directly. The molecules of sugar or dye continue to move after they have become evenly distributed throughout the liquid in the container; however, as fast as some molecules move from left to right, others move from right to left, so that an equilibrium is maintained.

Any number of substances will diffuse independently of each other. If a lump of salt is placed in one part of a beaker of water and a lump of sugar in another, the molecules of each will diffuse independently of the other, and each drop of water in the beaker will eventually have some salt and some sugar molecules.

The rate of movement of a single molecule is several hundred meters per second, but each molecule can go only a fraction of a nanometer before it bumps into another molecule and rebounds. Thus the progress of a molecule in a straight line is quite slow. Diffusion is quite rapid over short distances, but it takes a long time—days and even weeks—for a substance to diffuse a distance measured in centimeters. This fact has important biological implications for it places a sharp limit on the number of molecules of oxygen and nutrients that can reach an organism by diffusion alone. Only a very small organism that requires relatively few molecules per second can survive if it remains in one place and allows molecules to come to it by diffusion. A larger organism must have some means of moving to a new region or some means of stirring its environment to bring molecules to it. As a third alternative it may live in some spot where the environment is constantly moving past it—in a river, for example, or in the intertidal region at the seashore.

Exchanges of Material Between Cell and Environment

All nutrients and waste products must pass through the plasma membrane to enter or leave the cell. Cells are almost invariably surrounded by a watery medium—the fresh or salt water in which an organism lives, the tissue sap of a higher plant, or the plasma or extracellular fluid of a higher animal. In general, only dissolved substances can pass through the plasma membrane, but not all dissolved substances penetrate the plasma membrane with equal facility. The membrane behaves as though it had ultramicroscopic pores through which substances pass, and these pores, like the holes in a sieve, determine the maximal size of molecule that can pass. Factors other than simple molecular size, such as the electric charge, if any, of the diffusing particle, the number of water molecules bound to the diffusing particle, and its solubility in fatty substances, may also be important in determining whether or not the substance can pass through the plasma membrane.

A membrane is said to be **permeable** if it will permit any substance to pass through, **impermeable** if it will allow no substance to pass, and **differentially permeable** if it will allow some but not all substances to diffuse through. The nuclear and plasma membranes of all cells and the membranes surrounding food and contractile vacuoles are differentially permeable membranes. Permeability is a property of the *membrane,* not of the diffusing substance.

The diffusion of a dissolved substance through a differentially permeable membrane is known as **dialysis.** If a pouch made of collodion, cellophane, or parchment is filled with a sugar solution and placed in a beaker of water, the sugar molecules will dialyze through the membrane (if the pores are large enough), and eventually the concentration of sugar molecules in the water outside the pouch will equal that within the pouch. The molecules then continue to diffuse, but there is no net change in concentration, for the rates in the two directions are equal.

A different type of diffusion is observed if a membrane is prepared with smaller pores so that it is permeable to the small water molecules but not to the larger sugar molecules. A pouch may be prepared of a membrane with these properties and filled with a sugar solution. The pouch is then fitted with a cork and glass tube and placed in a beaker of water so that the levels of fluid inside and outside the pouch are the same. The sugar molecules cannot pass through the membrane and so must remain inside the pouch. The water molecules diffuse through the membrane and mix with the sugar solution so that the level of fluid within the pouch rises. The liquid within the pouch is 5 percent sugar and therefore only 95 percent water; the liquid outside the membrane is 100 percent water. The water molecules are moving in both directions through the membrane, but there is a greater movement from the region of higher concentration (100 percent, outside the pouch) to the region of lower concentration (95 percent, within the pouch). This diffusion of water or solvent molecules through a membrane is called **osmosis** (Gr. *ōsmos,* action of pushing) and is illustrated diagrammatically in Figure 2.21.

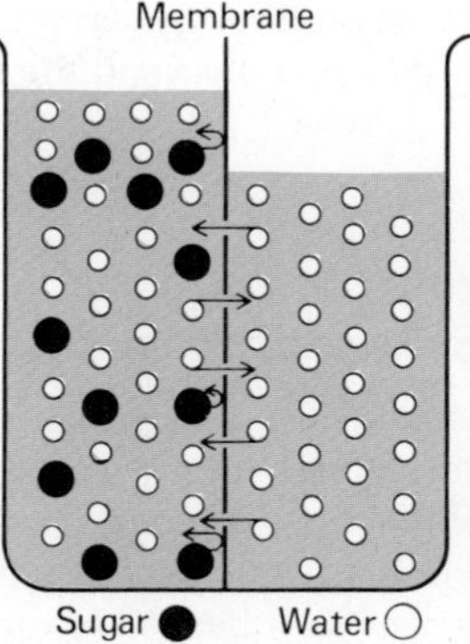

FIGURE 2.21 Diagram illustrating osmosis. When a solution of sugar in water is separated from pure water by a semipermeable membrane that allows water but not the larger sugar molecules to pass through, there is a net movement of water molecules through the membrane to the sugar solution. The water molecules are diffusing from a region of higher concentration (pure water) to a region of lower concentration (the sugar solution).

If an amount of water equal to that originally present in the pouch enters, the solution in the pouch will be diluted to 2.5 percent sugar and 97.5 percent water, but the concentration of water outside the pouch will still exceed that inside, and osmosis will continue. Equilibrium is reached when the water in the glass tube rises to a height such that the weight of the water in the tube exerts a hydrostatic pressure just equal to the tendency of the water to enter the pouch. Osmosis then occurs with equal speed in both directions through the differentially permeable membrane, and there will be no net change in the amount of water in the pouch. The pressure of the column of water is called the **osmotic pressure** of the sugar solution. The osmotic pressure results from the tendency of the water molecules to pass through the differentially permeable membrane and to equalize the concentration of water molecules on its two sides. A more concentrated sugar solution would have a greater osmotic pressure and would "draw" water to a higher level in the tube. A 10 percent sugar solution would cause water to rise approximately twice as high in the tube as a 5 percent solution. Only molecules in solution have an osmotic effect. Particles in suspension have no osmotic effect.

It is evident from this discussion that dialysis and osmosis are simply two special forms of diffusion. Diffusion is the general term for the movement of molecules from a region of high concentration to a region of lower concentration, brought about by their kinetic energy. Dialysis is the diffusion of dissolved molecules (solutes) through a differentially permeable membrane, and osmosis is the diffusion of solvent molecules through a differentially permeable membrane. In biological systems the solvent molecules are almost universally water.

The salts, sugars, and other substances dissolved in the fluid within each cell give the intracellular fluid a certain osmotic pressure. When the cell is placed in a fluid with the same osmotic pressure as that of its intracellular fluid, there is no net entrance or exit of water, and the cell neither swells nor shrinks. Such a fluid is said to be **isosmotic** with the intracellular fluid of the cell. Normally, the blood plasma and body fluids are isosmotic with the intracellular fluids of the body cells. If the surrounding fluid contains more dissolved substances than the fluid within the cell, water will tend to pass out of the cell, and the cell will shrink. Such a fluid is said to be **hyperosmotic** to the cell. If the surrounding fluid has a lower concentration of dissolved substances than the fluid in the cell, water will tend to pass into the cell and the cell will swell. This fluid is said to be **hyposmotic** to the cell. Many cells have the ability to pump water or certain solute molecules into or out of the cell and in this way can maintain an osmotic pressure that differs from that of the surrounding medium. Amoebas, paramecia, and other protozoa that live in pond water, which is very hyposmotic to their intracellular fluid, have evolved **contractile vacuoles** (see Fig. 2.1) that collect water from the interior of the cell and pump it to the outside. Without such a mechanism the cells would quickly burst from the water entering the cell.

The power of certain cells to accumulate selectively certain kinds of molecules from the environmental fluid is truly phenomenal. Human cells (and those of vertebrates in general) can accumulate amino acids so that the concentration within the cell is 2 to 50 times that in the extracellular fluid. Cells also have a much higher concentration of potassium and magnesium, and a lower concentration of sodium, than the environmental fluids. Certain primitive chordates, the tunicates, (Chap. 30), can accumulate vanadium so that the concentration inside the cell is some 2,000,000 times that in the surrounding sea water. The transfer of water or of solutes in or out of the cell against a concentration gradient is physical work and requires the expenditure of energy. Some active physiological process is required to perform these transfers; hence a cell can move molecules against a gradient only as long as it is alive. If a cell is treated with some metabolic poison, such as cyanide, it quickly loses its ability to maintain concentration differences on the two sides of its plasma membrane.

Tissues

In the evolution of both plants and animals, one of the major trends has been toward the structural and functional specialization of cells. The cells composing the body of one of the higher animals are not all alike but are differentiated and specialized to perform certain functions more efficiently than an unspecialized animal cell could. This specialization has also had the effect of making the several parts of the body interdependent, so that an injury to, or the destruction of, cells in one part of the body may result in the death of the whole organism. The advantages of specialization are so great that they more than outweigh the disadvantages. The cells of the body that are similarly specialized are known as a **tissue** (Fr. *tissu*, woven). A tissue may be defined as a group or layer of similarly specialized cells that together perform certain special functions. Each tissue is composed of cells that have a characteristic shape, size, and arrangement; the different types of tissue of the vertebrate body are readily recognized when examined microscopically. Certain tissues are composed of nonliving cell products in addition to the cells; connective tissue contains many extracellular protein fibers in addition to the fibroblasts or connective tissue cells, and bone and cartilage are made largely of proteins and salts secreted by the bone or cartilage cells.

The tissues of a multicellular animal may be classified in six major groups, each of which has several subgroups. These are epithelial, connective, muscular, blood, nervous, and reproductive tissues.

Epithelial Tissues. Epithelial tissues are composed of cells that form a continuous layer or sheet covering the surface of the body or lining cavities within the body. There is usually a noncellular **basement membrane** underlying the sheet of epithelial cells. This membrane is secreted by cells and consists of minute protein fibrils imbedded in a mucopolysaccharide matrix. The epithelial cells in the skin of vertebrates are usually connected by small cytoplasmic processes or bridges. The epithelia of the body protect the underlying cells from mechanical injury, from harmful chemicals and bacteria, and from desiccation. The epithelial lining of the digestive tract absorbs water and nutrients for use in the body. The lining of the digestive tract and a variety of other epithelia produce and give off a wide spectrum of substances. Some of these are used elsewhere in the body and others are waste products that must be eliminated. Since the entire body is covered by an epithelium, all of the sensory stimuli pass through some epithelium to reach the specific receptors for those stimuli. The functions of epithelia are thus protection, absorption, secretion, and sensation.

The cells in epithelial tissues may be flat, cuboidal, or columnar in shape, they may be arranged in a single layer or in many layers, and they may have fine hairs or cilia on the free surface. On the basis of these structural characteristics epithelia are subdivided into the following groups.

Squamous epithelium is made of thin, flattened cells the shape of flagstones or tiles (Fig. 2.22). It is found on the surface of the skin and the lining of the mouth, esophagus, and vagina. The endothelium lining the cavity of blood vessels and the mesothelium lining the coelom are squamous epithelia. In the lower animals the skin is usually covered with a single layer of squamous epithelium, but in the higher animals the outer layer of the skin consists of stratified squamous epithelium, made of several layers of these flat cells.

The kidney tubules are lined with **cuboidal epithelium,** made of cells that are cube-shaped and look like dice (Fig. 2.22). Many other parts of the body, such as the stomach and intestines, are lined by cells that are taller than they are wide. An epithelium composed of such elongated, pillar-like cells is known as **columnar epithelium** (Fig. 2.22). Columnar epithelium may be simple, consisting of a single layer of cells, or stratified, composed of several layers of cells.

Either cuboidal or columnar epithelial cells may have cilia on their free surface. Ciliated cuboidal epithelium is found in the sperm ducts of earthworms and other animals, and ciliated columnar epithelium lines the ducts of the respiratory system of terrestrial vertebrates. The rhythmic, concerted beating of the cilia moves solid particles in one direction through the ducts. Epithelial cells, usually columnar ones, may be specialized to re-

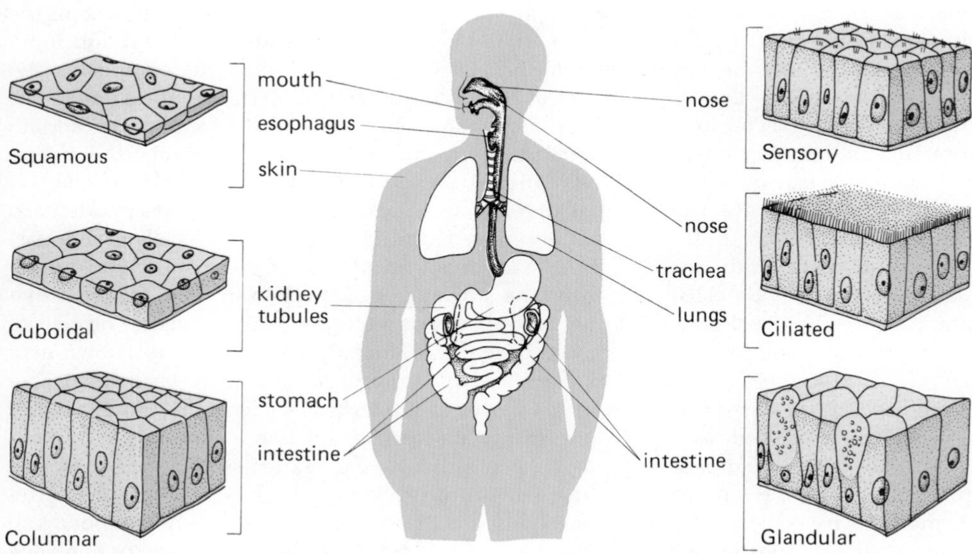

FIGURE 2.22 Diagram of the types of epithelial tissue and their location in the body.

ceive stimuli. The groups of cells in the taste buds of the tongue or the olfactory epithelium in the nose are examples of sensory epithelium. Columnar or cuboidal epithelia may also be specialized for secreting certain products, such as milk, wax, saliva, perspiration, or mucus. The outer epithelium of many invertebrates secretes a thin, continuous, noncellular protective layer. This may be a thin cuticle, as in the worms, or a hard shell or exoskeleton, as in crabs and mollusks.

Connective Tissues. The connective tissues—bone, cartilage, tendons, ligaments, fibrous connective tissue, and adipose tissue—support and bind together the other tissues and organs. Connective tissue cells characteristically secrete a nonliving material called the **matrix,** and the nature and function of each connective tissue is determined primarily by the nature of this intercellular matrix. The actual connective tissue cells may form only a small and inconspicuous part of the tissue. The matrix, rather than the connective tissue cells themselves, does the actual connecting and supporting.

Fibrous connective tissue consists of a thick, interlacing, matted network of fibers in which are distributed the fibroblast cells that secrete the matrix (Fig. 2.23). The matrix consists of fibers and of an amorphous ground substance composed of polysaccharides and polysaccharide–protein complexes. The most abundant fibers

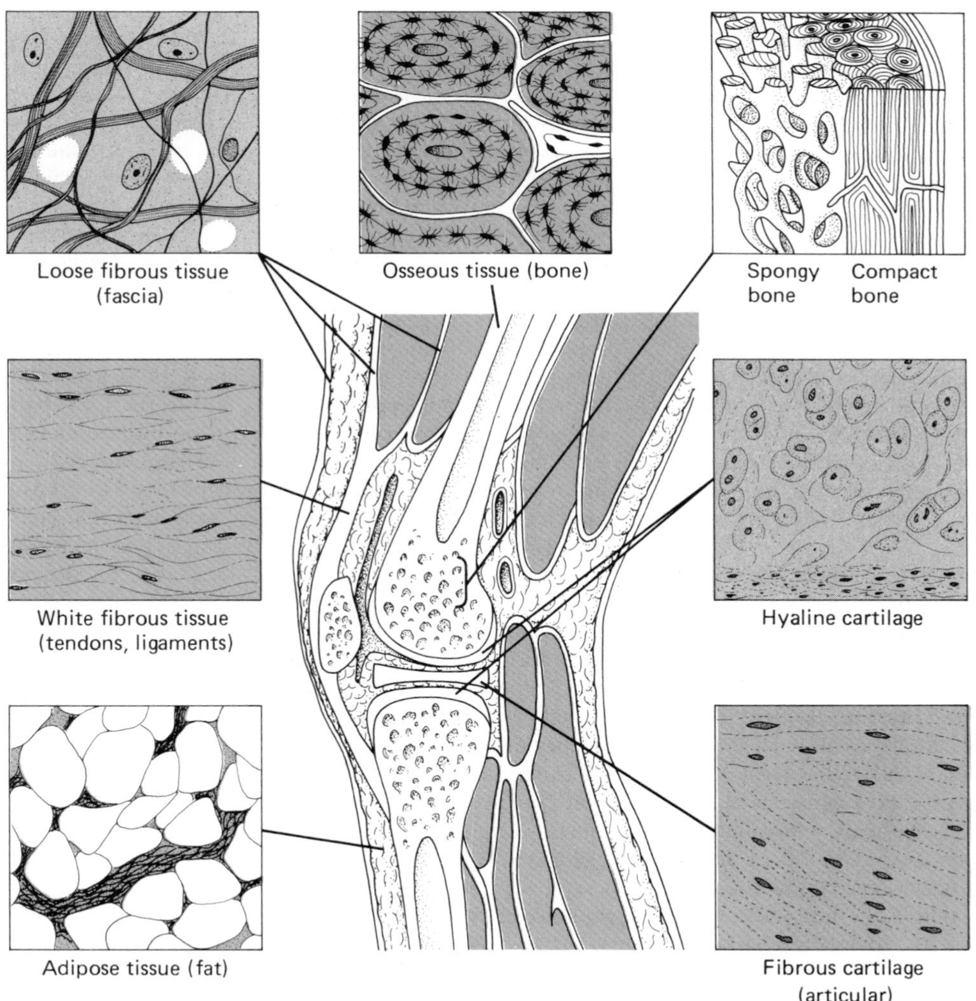

FIGURE 2.23 Diagram of the types of connective tissue and their location in the knee joint.

are composed of a protein called **collagen,** which is rich in the amino acids glycine, proline, and hydroxyproline. Treating the fibers with hot water converts some of the collagen into the soluble protein **gelatin.** The amino acid compositions of gelatin and collagen are nearly identical. Because so much connective tissue is present, nearly one third of all the protein in the human body is collagen. The collagen units composing the fibers consist of a helix made of the peptide chains wound around each other in a supercable and joined by hydrogen bonds. **Collagen fibers** are whitish in color in living preparations, seldom branch, are very flexible, and are relatively inelastic after twists and turns in them have been pulled straight. Thinner elastic fibers, composed of the protein **elastin,** are yellowish in color, branch and rejoin extensively, and are more elastic than collagen. Very delicate **reticular fibers,** which are similar in many ways to the subunits of which collagen is composed, form delicate latticeworks around many gland cells, in the lymph nodes, and in some other organs. Many white blood cells, phagocytic macrophages, and other cells are also found in fibrous connective tissue.

There are many specialized types of fibrous connective tissue. **Adipose tissue** is rich in fat cells, specialized fibroblasts that store large quantities of fat in a vacuole in the cytoplasm. Ligaments and tendons are specialized fibrous connective tissues. **Tendons** are composed of thick, closely packed bundles of collagen fibers, which form flexible cables that connect a muscle to a bone or to another muscle. A **ligament** is fundamentally similar in constitution to a tendon and connects one bone to another. An especially thick mat of fibrous connective tissue is located in the lower layer of the skin of most vertebrates; when this is chemically treated—"tanned"—it becomes leather.

The supporting skeleton of vertebrates is composed of cartilage or bone. **Cartilage** appears as the supporting skeleton in the embryonic stages of all vertebrates but in the adult is largely replaced by bone in all but sharks and rays. Cartilage can be felt in our bodies as the supporting framework of the external ear flap or the tip of the nose. Its matrix is similar to that of fibrous connective tissue but is firmer and more elastic. Its ground substance contains a great deal of a special carbohydrate–protein complex called **chondroitin** that obscures the abundant collagen fibers. Modified fibroblasts, the **cartilage cells,** that secrete the matrix become entrapped within it (Fig. 2.23), but these cells are still living, continue to divide, separate, and secrete more matrix around themselves. Cartilage can grow from within, rather than by the complex remodeling process seen in bone, so it is an ideal embryonic skeletal material. Although it does not resist compression and tension stresses as well as bone, it is more elastic and smoother. Cartilage occurs in adult vertebrates where elasticity is needed, as with the costal cartilages attaching the ribs to the breastbone, or where both elasticity and low friction are needed, as on the ends of bones in moveable joints.

Bone consists of a dense matrix composed of proteins, principally collagen, and calcium salts identical with the mineral hydroxyapatite, $Ca_3(PO_4)_2 \cdot CaCO_3$. Approximately 65 percent of the bone is made of this mineral. Bone cells (osteoblasts) remain alive and secrete a bony matrix, both protein and calcium salts, throughout life. The osteoblasts become surrounded and trapped by their own secretion and remain in microscopic cavities (lacunae) in the bone as living osteocytes (Fig. 2.23). The protein is laid down as minute fibers that contribute strength and resiliency, and the mineral salts contribute hardness to bone.

At the surface of each bone is a thin fibrous layer called the **periosteum** (Gr. *peri,* around + *osteum,* bone) to which the muscles are attached by tendons. The periosteum contains cells, some of which differentiate into osteoblasts and secrete protein and salts to bring about growth and repair. Most bones are not solid but have a marrow cavity in the center. The apparently solid matrix of the bone is pierced by many microscopic channels (haversian canals) in which lie blood vessels and nerves to supply the bone cells. The bony matrix is deposited, usually in concentric rings or lamellae, around these haversian canals. Each bone cell is connected to the adjacent bone cells and to the haversian canals by cellular extensions occupying minute canals (canaliculi) in the matrix. The bone cells obtain oxygen and raw materials and eliminate wastes by way of these canaliculi. Bone contains not only bone-secreting cells but also bone-destroying cells. By the action of these two types of cells, the shape of a bone may be altered to resist changing stresses and strains. Bone formation and destruction is regulated by the availability of calcium and phosphate, by the presence of vitamin D, and by calcitonin and parathyroid hormone, secreted by the thyroid and parathyroid glands. The marrow cavity of the bone may contain yellow marrow (largely a fat depot) or red marrow, the tissue in which red and certain white blood cells are formed.

The organization of cartilage and bone into skeletal systems is considered in Chapter 5.

Muscular Tissues. The movements of most animals result from the contraction of elongate, cylindrical, or spindle-shaped cells, each of which contains many microscopic, elongate, parallel, contractile fibers called **myofibrils,** composed of the proteins myosin and actin (p. 48). Muscle cells perform mechanical work by con-

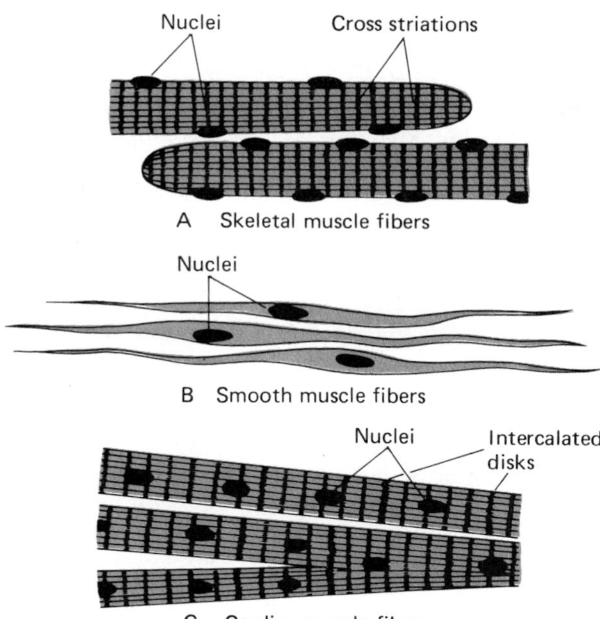

FIGURE 2.24 Types of muscle tissue.

tracting, by getting shorter and thicker; they are unable to do work by pushing. Three types of muscle tissue are found in vertebrates: skeletal, cardiac, and smooth (Fig. 2.24). **Cardiac muscle** is found only in the walls of the heart; **smooth muscle** in the walls of the digestive tract, the urinary and genital tracts, and the walls of arteries and veins; and **skeletal muscle** makes up the muscle masses that are attached to and move the bones of the body.

Skeletal muscle cells are among the exceptions to the rule that cells have but one nucleus; each of these cells has many nuclei because the cells develop embryologically by the end-to-end fusion of smaller cells. The nuclei of skeletal muscle cells have an unusual position, at the periphery of the cell, just below the plasma membrane. Skeletal muscle cells are extremely long, 2 or more centimeters in length; indeed, investigators believe that some muscle cells extend from one end of the muscle to the other, so that their length is equal to that of the muscle. Muscle fibers range in thickness from 10 to 100 μm.; continued strenuous muscle activity increases the thickness of the fiber. The myofibrils of skeletal muscle have alternate dark and light cross bands or **striations.** These striations have a fundamental role in contraction, during which the dark band remains constant but the light band shortens (p. 109). The contraction of skeletal muscles is generally voluntary, under the control of the will. Striated muscles can contract very rapidly but cannot remain contracted; a striated muscle fiber must relax and rest before it is able to contract again. The muscles of insects, spiders, crabs, and other arthropods have cross striations and contract very rapidly.

Cardiac muscle cells are striated but have centrally located nuclei. Cardiac muscle is unicellular, not multicellular, as it appears by light microscopy. The "intercalated discs" visible in the fibers have been identified by electron microscopy as cell membranes tightly bound together. Contraction of cardiac muscle is involuntary.

Smooth muscle cells are not striated, have pointed ends, have centrally located nuclei, and their contraction is involuntary. Smooth muscle contracts slowly but can remain contracted for long periods of time. The voluntary muscles of some invertebrates, such as the ones that close the shell of an oyster, are smooth muscles. The distinguishing features of the three types of muscle are summarized in Table 2.2.

Vascular Tissues. The **blood,** composed of a liquid part—**plasma**—and of several types of **formed elements**—red cells, white cells, and platelets—may be

TABLE 2.2 Comparison of Vertebrate Muscle Tissues

	Skeletal	**Smooth**	**Cardiac**
Location	Attached to skeleton	Walls of viscera: stomach, intestines, etc.	Wall of heart
Shape of fiber	Elongate, cylindrical, blunt ends	Elongate, spindle-shaped, pointed ends	Elongate, cylindrical; fibers branch and fuse
Number of nuclei per fiber	Many	One	One
Position of nuclei	Peripheral	Central	Central
Cross striations	Present	Absent	Present
Speed of contraction	Most rapid	Slowest	Intermediate
Ability to remain contracted	Least	Greatest	Intermediate
Type of control	Voluntary	Involuntary	Involuntary

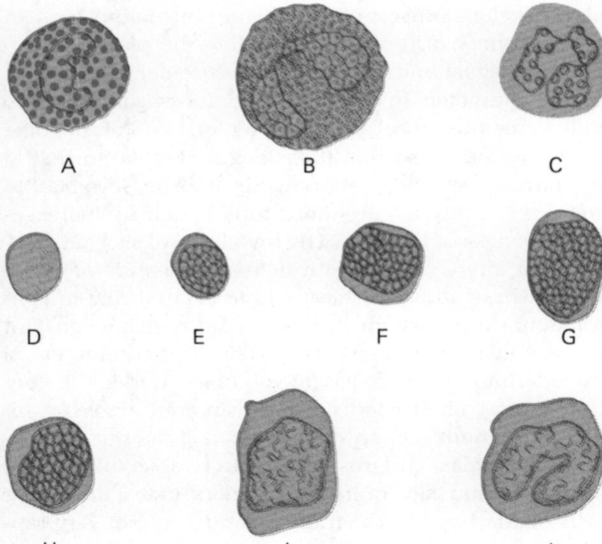

FIGURE 2.25 Types of white blood cells. *A*, Basophil; *B*, eosinophil; *C*, neutrophil; *E–H*, a variety of lymphocytes; *I* and *J*, monocytes; *D*, a red blood cell drawn to the same scale.

classified as a separate type of tissue or as one kind of connective tissue. The latter classification is based on the fact that blood cells and connective tissue cells originate from similar cells; however, the adult cells are quite different in structure and function. The **red cells** (erythrocytes) of vertebrates contain the red pigment hemoglobin, which has the property of combining easily and reversibly with oxygen. Oxygen, combined as oxyhemoglobin, is transported to the cells of the body in the red cells. Mammalian red cells are flattened, biconcave discs without a nucleus; those of other vertebrates are more typical cells with an oval shape and a nucleus.

The five different kinds of **white blood cells**—lymphocytes, monocytes, neutrophils, eosinophils, and basophils (Fig. 2.25)—lack hemoglobin but move around and engulf bacteria. They can slip through the walls of blood vessels and enter the tissues of the body to engulf bacteria there.

The fluid plasma transports a great variety of substances from one part of the body to another. Some of the substances transported are in solution; others are bound to one or another of the plasma proteins. The plasma of vertebrates is a light yellow color; in certain invertebrates the oxygen-carrying pigment is not localized in cells but is dissolved in the plasma and colors it red or blue. **Platelets** are small fragments broken off from cells in the bone marrow; they play a role in the clotting of blood (Chap. 8).

Nervous Tissues. Cells specialized for the reception of stimuli and the transmission of impulses are called **neurons.** A neuron typically has an enlarged cell body, containing the nucleus, and two or more cytoplasmic processes, the nerve fibers, along which the nerve impulse travels (Fig. 2.26). Nerve fibers vary in width from a few micrometers to 30 or 40 and in length from a millimeter or two to a meter or more. Two types of nerve fibers are distinguished: **axons,** which transmit impulses away from the cell body, and **dendrites,** which transmit them to the cell body. The junction between the axon of one neuron and the dendrite of the next neuron in the chain is called a **synapse.** At the synapse the axon and dendrite do not actually touch; there is a small gap between the two. Transmission of an impulse across the synapse involves a different mechanism from that moving an impulse along the nerve fiber. An impulse can travel across the synapse only from an axon

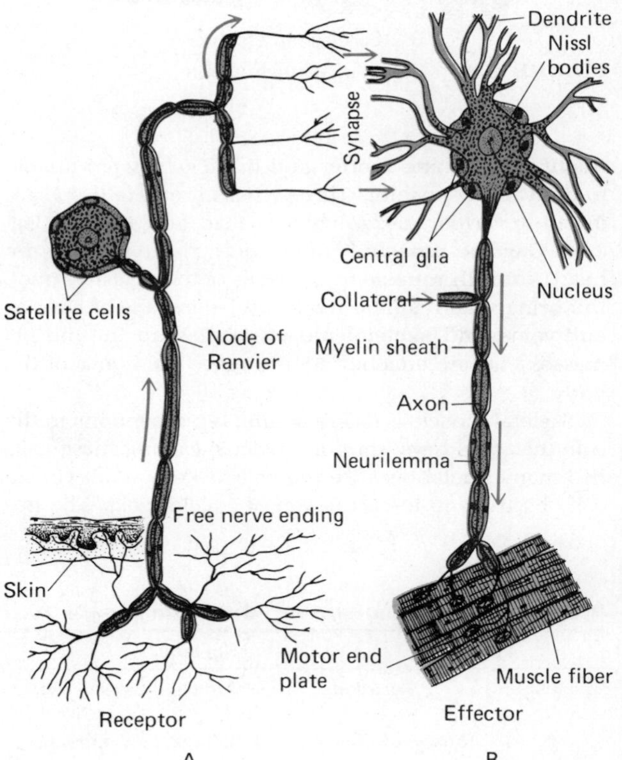

FIGURE 2.26 Diagrams of *(A)* an afferent neuron and *(B)* an efferent neuron. The arrows indicate the direction of the normal nerve impulse. (From King, B. G., and M. J. Showers: *Human Anatomy and Physiology.* 6th ed. Philadelphia, W. B. Saunders Co., 1969.)

to a dendrite; thus, the synapse serves as a valve to prevent the backflow of impulses (p. 219).

The cell bodies of neurons commonly occur in groups; there are columns of cell bodies in the spinal cord, sheets of cell bodies over the surface of parts of the brain, nodules of cell bodies ("nuclei") within the brain, and the ganglia of the cranial and spinal nerves. A **ganglion** is a group of nerve cell bodies located in vertebrates outside the central nervous system. A nerve consists of a group of axons and dendrites bound together by connective tissue. Each nerve fiber—axon or dendrite—is surrounded by a **neurilemma** or a **myelin sheath,** or both. The neurilemma is a delicate, transparent, tubelike membrane made of cells that envelop the fiber. The myelin sheath is made of noncellular, fatty material that forms a glistening white coat between the fiber and neurilemma. The myelin sheath is interrupted at fairly regular intervals along the nerve by constrictions called the nodes of Ranvier. Nerve fibers are either "medullated" and have a thick myelin sheath or "nonmedullated" and have an extremely thin myelin sheath.

Nervous tissue within the brain and spinal cord contains, in addition to neurons, several different kinds of supporting cells called **neuroglia.** These have many cytoplasmic processes, and the cells and their processes form an extremely dense supporting framework in which the neurons are suspended. The neuroglia are believed to separate and insulate adjacent neurons, so that nerve impulses can pass from one neuron to the next only over the synapse, where the neuroglial barrier is incomplete.

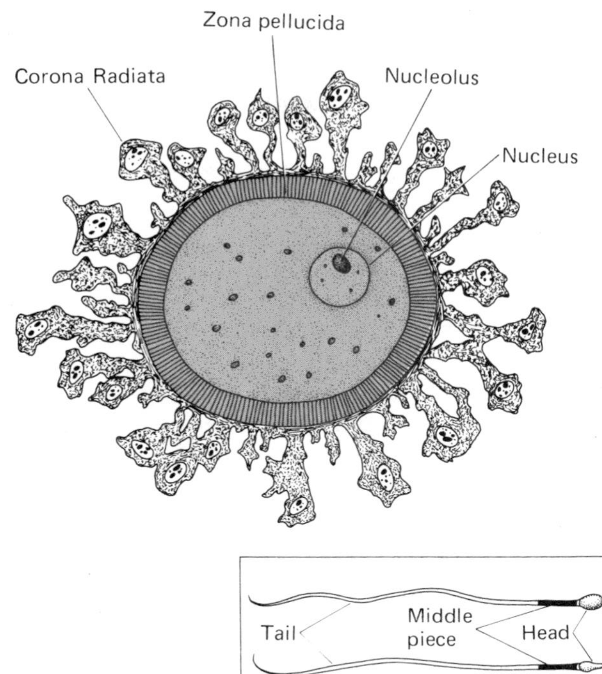

FIGURE 2.27 Human egg and sperm, magnified 400 times. *Inset.* Side and top views of a sperm, magnified about 2000 times. The egg is surrounded by other cells which form the corona radiata.

Reproductive Tissues. The **egg cells** (ova) formed in the ovary of the female and the **sperm cells** produced by the testes of the male constitute the reproductive tissues—cells specially modified for the production of offspring (Fig. 2.27). Egg cells are generally spherical or oval and are nonmotile.

Summary

1. The living substance of all animals is organized into structural and functional units called cells. The bodies of higher animals are composed of many cells differing in size, shape, and function. A group of cells that are similar in form and specialized to perform one or more specific functions is termed a tissue. A group of tissues may be associated into an organ and organs into organ systems.

2. The outer surface of each cell is bounded by a delicate elastic plasma membrane composed of a lipid bilayer in which are embedded a variety of special membrane proteins. The plasma membrane is an active functional structure that regulates the materials entering and leaving the cell and contains enzymes that move specific molecules in or out of the cell against a concentration gradient. Microvilli are small evaginations of the plasma membrane that play a role in the entrance of certain substances.

3. The nucleus is separated from the rest of the cell by a double-layered nuclear membrane, each sheet of which is a lipid bilayer. There are pores in the nuclear membrane, fine channels through which the nuclear contents and cytoplasm are in direct contact. The nucleus is required for growth and for cell division and is an important center for the control of cellular processes. Within the nuclear matrix are suspended a fixed number of long, linear threadlike chromosomes, composed of DNA and proteins and containing the units of heredity, the genes. Each kind of organism has a characteristic number of pairs of chromosomes in each cell; that is, two of each kind of chromosome. A cell with two complete sets of chromosomes is said to be diploid. Sperm and egg cells have only one of each kind of chromosome and are termed haploid.

4. The cytoplasm contains a number of organelles: centrioles, mitochondria, endoplasmic reticulum, and

Golgi bodies, each with specific functions. The centrioles play a role in cell division. Mitochondria are bounded by a double membrane, an outer smooth one and an inner one folded into parallel plates, termed cristae, extending into the central cavity. Each membrane is composed of a lipid bilayer and specific membrane proteins. The mitochondria contain a variety of enzymes involved in generating biologically useful energy in the form of ATP. The membranes of the endoplasmic reticulum are lipid bilayers with membrane proteins and organized into a maze of tubes extending through much of the cell. These, with their associated ribosomes, play a role in the synthesis of proteins and in the secretion of cell products. Golgi bodies are other membranous organelles that serve as the sites for temporary storage and repackaging of proteins and other compounds synthesized in the endoplasmic reticulum. Other subcellular organelles are hollow cylindrical microtubules and solid microfilaments.

5. The doubling of the constituents of a cell, followed by its division into two daughter cells, is termed the cell cycle, composed of G_1, S, G_2, and M phases. The period of DNA synthesis, during which there is an exact replication of each chromosome, is termed the S phase. During the preceding G_1 phase specific processes occur that enable the cell to enter the S phase. A G_2 phase follows the S phase in which there is increased protein-synthesis and preparation for mitosis, for cell division, termed the M phase.

6. The process of mitosis involves the doubling of the number of chromosomes and the distribution of the chromosomes to the two daughter cells so that each daughter cell receives exactly the same number and kind of chromosomes as the other, each receiving two complete sets of chromosomes. Although mitosis is a continuous process, for descriptive purposes it is divided into prophase, when the chromosomes shorten and thicken, metaphase, when they line up on the equatorial plane, anaphase when they migrate to the poles of the cell, and telophase, when they reach the poles and the nuclear membranes reform.

7. All molecules are constantly in motion as a result of their inherent heat energy. Molecules in a liquid or gaseous state diffuse, move in all directions until they are spread evenly throughout the space available. Diffusion is the movement of molecules from a region of high concentration to one of lower concentration owing to their kinetic energy. Dialysis and osmosis are two special forms of diffusion. Dialysis is the diffusion of dissolved molecules (solutes) through a differentially permeable membrane and osmosis is the diffusion of solvent molecules through a differentially permeable membrane. The fluid within each cell has a certain osmotic pressure due to the presence of the salts, sugars, and other substances dissolved in it.

8. Vertebrates and other multicellular animals are composed of epithelial, connective, muscular, blood, nervous, and reproductive tissues, each composed of similarly specialized cells that perform certain special functions. Epithelial tissues form a layer or sheet of cells covering the surface of the body or lining a cavity. Connective tissues include bone, cartilage, tendons, ligaments, fibrous connective tissue, and adipose tissue and support and bind together the other tissues. The nature and function of connective tissues are determined by the kind of nonliving matrix secreted by the cells. The movements of most animals result from the contraction of elongate, cylindrical, or spindle-shaped cells containing microscopic parallel contractile fibers, myofibrils, composed of actin and myosin. The skeletal, cardiac, and smooth muscles of vertebrates differ in their shape, the number and location of nuclei, and the presence or absence of striations, cross bands that play a role in contraction. Blood vascular tissues include red cells, several kinds of white cells, platelets, and the liquid plasma in which the formed elements float. Erythrocytes (red cells) contain hemoglobin and transport oxygen and carbon dioxide. The white cells, lymphocytes, monocytes, neutrophils, eosinophils, and basophils move around and engulf bacteria. Nervous tissues are composed of neurons, cells specialized for the reception of stimuli and the transmission of nerve impulses. Each contains a nucleus and two or more cytoplasmic fibers, axons and dendrites, extending from the nucleus. Reproductive tissues include eggs—spherical, nonmotile cells usually containing a store of nutrients (yolk)—and sperm—small motile cells with a head containing the nucleus, a middle piece, and a long tail, used in propelling the sperm. A typical egg has a large nucleus, called the germinal vesicle, and a variable amount of yolk in the cytoplasm. Shark and bird eggs have enormous amounts of yolk that provides nourishment for the developing embryo until it hatches from the shell. Sperm cells are small and modified for motility. A typical sperm has a long **tail,** the beating of which propels the sperm to its meeting and union with the egg. The **head** of the sperm contains the nucleus surrounded by a thin film of cytoplasm. The tail is connected to the head by a short **middle piece.** An **axial filament,** formed by the centriole in the middle piece, extends to the tip of the tail. Most of the cytoplasm is sloughed off as the sperm matures; this decreases the weight of the sperm and perhaps renders it more motile.

References and Selected Readings

Ambrose, E. J.: *Cell Biology.* 2nd. ed. Baltimore, University Park Publishers, 1978. A fine presentation of the details of cell structure and function.

Avers, C. J.: *Cell Biology.* 2nd ed. New York, D. Van Nostrand, 1981. A briefer presentation of the principles of cell biology.

Bloom, W., and D. Fawcett: *Textbook of Histology.* 10th ed. Philadelphia, W. B. Saunders Co., 1975. One of the standard texts of histology, with many superb illustrations of the structure of tissues.

Clemens, M. M.: *Biochemistry of Cell Regulation.* Boca Raton, CRC Press, 1980. A two-volume source book of gene expression and regulation of metabolism.

DeWitt, W.: *Biology of the Cell.* Philadelphia, Saunders College Publishing, 1977. An evolutionary approach to the principles of cell biology.

Giese, A. C.: *Cell Physiology.* 5th ed. Philadelphia, Saunders College Publishing, 1979. A classic presentation of the functions of cells and their substructures.

Hoagland, M.: *The Roots of Life: A Layman's Guide to Genes, Evolution and the Ways of Cells.* New York, Avon, 1979. A paperback written for the general public by a working scientist.

Jordan, E. G., and C. A. Cullis: *The Nucleolus.* New York, Cambridge University Press, 1982. Papers presented at a symposium on the nucleolus that summarize many aspects of research.

Margulis, L.: *Symbiosis in Cell Evolution.* San Francisco, W. H. Freeman & Co., 1981. A fascinating summary of the hypothesis that eukaryotic cell organelles evolved from prokaryotes.

Sato, G.: *Functionally Differentiated Cell Lines.* New York, A. R. Liss, 1981. A report of research on cells cultured *in vitro.*

Toner, P. G., and K. Carr: *Cell Structure.* 2nd ed. London, Churchill, 1979. A brief text of cytology.

Tribe, M. A.: *Metabolism and Mitochondria.* New York, Cambridge University Press, 1976. A brief text of intermediary metabolism and the role of mitochondria.

3 Cell Metabolism

ORIENTATION

Obtaining biologically useful energy from the foods animals eat is a primary function of animals of all kinds. This function is achieved by the activities of a host of biological catalysts, called enzymes, that carry out sequences of chemical reactions by which adenosine triphosphate, ATP, and other energy-rich compounds are produced. We will examine the structure of the mitochondria, within which many of these reactions occur, and the nature of the mitochondrial membranes that permit the reactions to occur. We will discuss the tricarboxylic acid (TCA) cycle, the final common pathway by which sugars, amino acids, and fatty acids are metabolized. We will determine how these reactions lead to the production of ATP that can then be used to drive all the reactions of the cell, the synthesis of new materials, the transmission of nerve impulses, movement, and the constant repair of cell parts that occurs in normal, healthy cells.

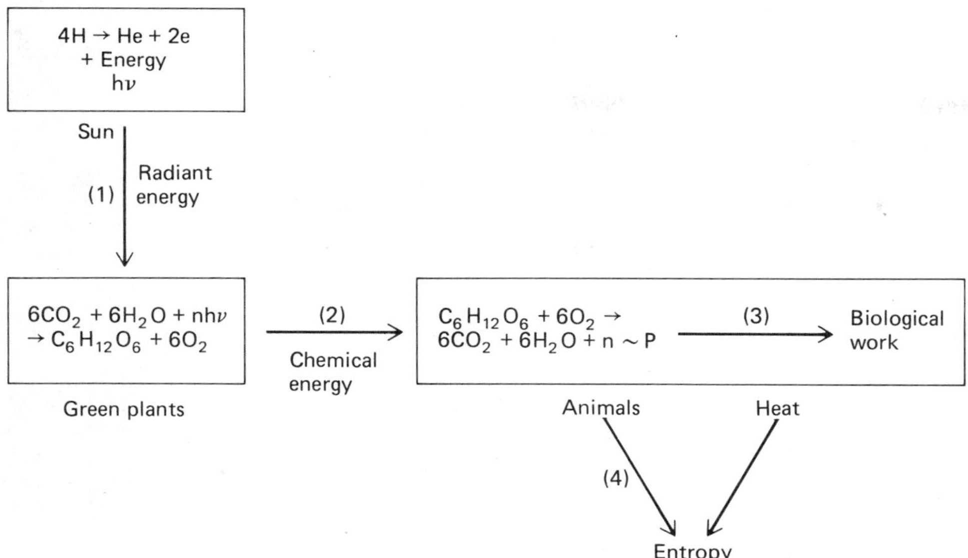

FIGURE 3.1 Energy transformations in the biological world. (1) The radiant energy of sunlight is transformed in photosynthesis into chemical energy in the bonds of organic compounds. The symbol v refers to the frequency of the light and the symbol h refers to Planck's constant, which relates energy and frequency. (2) The chemical energy of organic compounds is transformed during cellular respiration into biologically useful energy, the energy-rich phosphate bonds of ATP and other compounds. (3) The chemical energy of energy-rich phosphate bonds is utilized in cells to do mechanical, electrical, osmotic, or chemical work. (4) Finally energy flows to the environment as heat in the "entropy sink."

The never-ending flow of energy within a cell, from one cell to another and from one organism to another organism, is the essence of life itself. Three major types of energy transformations can be distinguished in the biological world (Fig. 3.1). In the first, the radiant energy of sunlight is captured by the green pigment **chlorophyll,** present in green plants, and is transformed by the process of **photosynthesis** into chemical energy. This is used to synthesize carbohydrates and other complex molecules from carbon dioxide and water. The radiant energy of sunlight, a form of **kinetic energy,** is transformed into a type of **potential energy.** The chemical energy is stored in the molecules of carbohydrates and other foodstuffs as the energy of the bonds that connect their constituent atoms.

In a second type of energy transformation, the chemical energy of carbohydrates and other molecules is transformed by the process termed **cellular respiration** into the biologically useful energy of energy-rich phosphate bonds. This kind of energy transformation occurs in the mitochondrion. A third type of energy transformation occurs when the chemical energy of these energy-rich phosphate bonds is utilized by the cell to do work—the mechanical work of muscular contraction, the electrical work of conducting a nerve impulse, the osmotic work of moving molecules against a gradient, or the chemical work of synthesizing molecules for growth. As these transformations occur (Table 3.1), energy finally flows to the environment and is dissipated as heat.

TABLE 3.1 Energy Transformations in Cells

Transformation	Type of Cell
Chemical energy to electrical energy	Nerve, brain
Sound to electrical energy	Inner ear
Light to chemical energy	Chloroplast
Light to electrical energy	Retina of eye
Chemical energy to osmotic energy	Kidney
Chemical energy to mechanical energy	Muscle cell, ciliated epithelium
Chemical energy to radiant energy	Luminescent organ of firefly
Chemical energy to electrical energy	Sense organs of taste and smell

Entropy and Energy

The branch of physics that deals with energy and its transformations, thermodynamics, is based on certain relatively simple principles that are universally applicable to chemical processes, whether these occur in living or nonliving systems.

The amount of energy entering and leaving any system may be measured and compared. It is always found that energy is neither created nor destroyed but is only transformed from one form to another. This is an expression of the first law of thermodynamics, sometimes called the **law of the conservation of energy:** The total energy of any object and its surroundings (i.e., a system) remains constant. As any given object undergoes a change from its initial state to its final state it may absorb energy from the surroundings or deliver energy to the surroundings. The difference in the energy content of the object in its initial and final state must be just equalled by a corresponding change in the energy content of the surroundings. Heat is a convenient form in which energy may be measured, which is why the study of energy has been called thermodynamics, i.e., heat dynamics. Heat, however, is not a useful way of transferring energy in biological systems, for living organisms are basically **isothermal** (equal temperature). There is no significant temperature difference between the different parts of the cell or between the different cells in a tissue. Stated in another way, cells do not act as heat engines. They have no means of allowing heat to flow from a warmer to a cooler body.

The **second law of thermodynamics** may be stated briefly as "the entropy of the universe increases." **Entropy** may be defined as a randomized state of energy that is unavailable to do work. The second law may be phrased as "Physical and chemical processes proceed in such a way that the entropy of the system becomes maximal." Entropy then is a measure of randomness or disorder. In almost all energy transformations there is a loss of some heat to the surroundings, and since heat involves the random motion of molecules, such heat losses increase the entropy of the surroundings. Living organisms and their component cells are highly organized and thus have little entropy. They preserve this low entropy state by increasing the entropy of their surroundings. You increase the entropy of your surroundings when you eat a candy bar and convert its glucose to carbon dioxide and water and return them to the surroundings.

The force that drives all processes is the tendency of the system to reach the condition of maximum entropy. Heat is either given up or absorbed by the object to allow the system to reach the state of maximum entropy. The changes of heat and entropy are related by a third dimension of energy termed **free energy.** Free energy may be visualized as that component of the total energy of a system that is available to do work under isothermal conditions; it is thus the thermodynamic parameter of greatest interest in biology. Entropy and free energy are related inversely; as entropy increases during an irreversible process, the amount of free energy decreases. All physical and chemical processes proceed with a decline in free energy until they reach an equilibrium in which the free energy of the system is at a minimum and the entropy is at a maximum. Free energy is useful energy; entropy is degraded, useless energy.

BOX 3.1

You and a Light Bulb

Some interesting calculations of the overall energy changes involved in metabolism in the human body have been made by E. G. Ball of Harvard University. Since the conversion of oxygen to water involves the participation of hydrogen atoms and electrons, the total flow of electrons in the human body can be calculated and expressed in amperes. From the average oxygen consumption of an adult 70-kg. man at rest—264 mL per minute—and the fact that each oxygen atom requires two hydrogen atoms and two electrons to form a molecule of water, Dr. Ball calculated that 2.86×10^{22} electrons flow from foodstuff, via dehydrogenases, primary acceptors, and the cytochromes, to oxygen each minute in all the cells of the body. Since an ampere equals 3.76×10^{20} electrons per minute, this flow of electrons amounts to 76 amperes. This is quite a bit of current, for an ordinary 100-watt light bulb uses just a little less than 1 ampere.

The flow of electrons from substrate to oxygen involves a potential difference of 1.13 volts (from -0.32 to $+0.81$ volt). Since volts $\times$ amperes $=$ watts, $1.13 \times 76 = 85.9$ watts. The human body, then, uses energy at about the same rate as a 100-watt light bulb but differs from it in having a much larger flow of electrons passing through a much smaller voltage change.

Chemical Reactions

A chemical reaction is a change involving the molecular structure of one or more substances; matter is changed from one substance, with its characteristic properties, to another, with new properties, and energy is released or absorbed. Hydrochloric acid, for example, reacts with the base, sodium hydroxide, to yield water and the salt, sodium chloride; in the process energy is released as heat:

$HCl + NaOH \rightarrow NaCl + H_2O$ + energy (heat)

Most chemical reactions are reversible and this reversibility is indicated by a double arrow: $\rightleftharpoons$.

The unit of energy most widely used in biological systems is the **Calorie,** which is the amount of heat required to raise 1 kg. of water 1°C. (strictly speaking, from 14.5° to 15.5°C.). This unit, more correctly called a kilogram calorie or kilocalorie, is the unit commonly used in calculating the energy content of foods. Other forms of energy—radiant, chemical, electrical, the energy of motion or position—can be converted to heat and measured by their effect in raising the temperature of water.

Atoms are neither destroyed nor created in the course of a chemical reaction; thus the sum of each kind of atom on one side of the arrow must equal the sum of that kind of atom on the other side. This is an expression of one of the basic laws of physics, the **law of the conservation of matter.** The direction of a reversible reaction is determined by the energy relations of the several chemicals involved , their relative concentrations, and their solubility.

The equilibrium point for any given reaction is determined by the tendency of the reaction components to reach maximum entropy, or minimum free energy, for the system. Reactions such as the union of HCl and NaOH to form NaCl and H_2O proceed with a decrease in free energy; i.e., they release energy to the system and increase its entropy. Such reactions are said to be **exergonic.** Exergonic reactions tend to go to completion; all of the reactants are converted into the products. Other reactions, termed **endergonic,** require an input of energy; entropy is decreased and free energy is increased. Such reactions do not go far in the direction of completion under standard conditions. In reactions that are freely reversible, the free energy change is zero.

Catalysis. Many of the substances that are rapidly metabolized by living cells are remarkably inert outside the body. A glucose solution, for example, will keep indefinitely in a bottle if it is kept free of bacteria and molds. It must be subjected to high temperature or to the action of strong acids or bases before it will decompose. Living cells cannot utilize conditions as extreme as these, for the cell itself would be destroyed long before the glucose, yet glucose is rapidly decomposed within cytoplasm at ordinary temperatures and pressures and in a solution that is neither acidic nor basic. The reactions within the cell are brought about by special agents known as **enzymes** that belong to the class of substances known as catalysts.

A **catalyst** is an agent that affects the velocity of a chemical reaction without altering its end point and without being used up in the course of the reaction. The list of substances that may serve as a catalyst in one or more reactions is long indeed. Metals, such as iron, nickel, platinum and palladium, when ground into fine powders are widely used as catalysts in industrial processes, such as the hydrogenation of cottonseed and other vegetable oils to make margarine or the cracking of petroleum to make gasoline. A minute amount of catalyst will speed up the reaction of vast quantities of reactants, for the molecules of catalyst are not exhausted in the reaction but are used again and again.

Enzymes

Enzymes are protein catalysts produced by cells. They regulate the speed and specificity of the thousands of chemical reactions that occur within cells. Although enzymes are synthesized within cells, they do not have to be inside a cell to act as a catalyst. Many enzymes have been extracted from cells with their activity unimpaired. They can then be purified and crystallized, and their catalytic abilities can be studied. Enzyme-controlled reactions are basic to all the phenomena of life: respiration, growth, muscle contraction, nerve conduction, photosynthesis, nitrogen fixation, deamination, digestion, and so on.

Properties of Enzymes. Enzymes are usually named by adding the suffix *-ase* to the name of the substance acted upon; e.g., sucrose is split by the enzyme **sucrase** to give glucose and fructose. Most enzymes are soluble in water or dilute salt solution, but some, for example the enzymes present in the mitochondria, are bound together by lipoprotein (a phospholipid–protein complex) and are insoluble in water.

The catalytic ability of some enzymes is truly phenomenal. For example, one molecule of the iron-containing enzyme **catalase,** extracted from beef liver, will bring about the decomposition of 5,000,000 molecules of hydrogen peroxide (H_2O_2) per minute at 0° C. The substance acted upon by an enzyme is known as its **substrate;** thus, hydrogen peroxide is the substrate of the enzyme catalase.

The number of molecules of substrate acted upon by a molecule of enzyme per minute is called the **turnover number** of the enzyme. The turnover number of catalase is thus 5,000,000. Most enzymes have high turnover numbers, thus they can be very effective although present in the cell in relatively minute amounts. Hydrogen peroxide is a poisonous substance produced as a by-product in a number of enzyme reactions. Catalase protects the cell by destroying the peroxide.

Hydrogen peroxide can be split by iron atoms alone, but it would take 300 *years* for an iron atom to split the same number of molecules of H_2O_2 that a molecule of catalase, which contains one iron atom, splits in one *second.* This example of the evolution of a catalyst emphasizes one of the prime characteristics of enzymes—they are very efficient catalysts.

Enzymes differ in the number of kinds of substrates they will attack. **Urease** is an example of an enzyme that is absolutely specific. Urease decomposes urea to ammonia and carbon dioxide and will attack no substance other than urea. Most enzymes are not quite so specific and will attack several closely related substances. **Peroxidase,** for example, will decompose several different peroxides in addition to hydrogen peroxide. A few enzymes are specific only in requiring that the substrate have a certain kind of chemical bond. The **lipase** secreted by the pancreas willl split the ester bonds connecting the glycerol and fatty acids of a wide variety of fats.

In theory, enzyme-controlled reactions are reversible; the enzyme does not determine the *direction* of the reaction but simply accelerates the *rate* at which the reaction reaches equilibrium. The classic example of this is the action of the enzyme lipase on the splitting of fat or the union of glycerol and fatty acids. If one begins with a fat, the enzyme catalyzes the splitting of this to give some glycerol and fatty acids. If one begins with a mixture of fatty acids and glycerol, the enzyme catalyzes the synthesis of some fat. When either system has operated long enough, the same equilibrium mixture of fat, glycerol, and fatty acid is reached:

$$\text{fat} \rightleftharpoons \text{glycerol} + 3 \text{ fatty acids}$$

Since reactions give off energy when going in one direction, it is obvious that an equivalent amount of energy in the proper form must be supplied to drive the reaction in the opposite direction.

To drive an energy-requiring reaction, some energy-yielding reaction must occur at about the same time. In most biological systems, energy-yielding reactions result in the synthesis of "energy-rich" phosphate esters, such as the terminal bonds of **adenosine triphosphate** (abbreviated as ATP). The energy of these energy-rich bonds is then available for the conduction of an impulse, the contraction of a muscle, the synthesis of complex molecules, and so on. Biochemists use the term "coupled reactions" for two reactions that must occur together so that one can furnish the energy, or one of the reactants, needed by the other.

Enzymes generally work in teams in the cell, with the product of one enzyme-controlled reaction serving as the substrate for the next. We can picture the inside of a cell as a factory with many different assembly lines (and disassembly lines) operating simultaneously. Each of these assembly lines is composed of a number of enzymes, each of which catalyzes the reaction by which one substance is converted into a second. This second substance is passed along to the next enzyme, which converts it into a third, and so on along the line. Eleven different enzymes, working consecutively, are required to convert glucose to lactic acid. The same series of 11 enzymes is found in human cells, in green leaves, and in bacteria.

Some enzymes, such as pepsin and urease, consist solely of protein. Many others, however, consist of a protein (called the **apoenzyme**) and some smaller organic molecule (called a **coenzyme**), usually containing phosphate. Coenzymes can usually be separated from their enzymes and, when analyzed, have proved to contain some vitamin—thiamine, niacin, riboflavin, pyridoxine, and so on—as part of the molecule. This finding has led to the generalization that *all vitamins function as parts of coenzymes in the cell.*

Neither the apoenzyme nor the coenzyme alone has catalytic properties; only when the two are combined is activity evident. Certain enzymes require for activity, in addition to a coenzyme, the presence of one or more ions. Magnesium (Mg^{++}) is required for the activity of several of the enzymes in the chain that converts glucose to lactic acid. **Salivary amylase,** the starch-splitting enzyme of saliva, requires chloride ion as an activator. Most, if not all, of the elements required by plants and animals in very small amounts—the so-called **trace elements,** manganese, copper, cobalt, zinc, iron, and others—serve as enzyme activators.

Enzymes may be present in the cell either dissolved in the liquid phase or bound to, and presumably an integral part of, one of the subcellular organelles. The respiratory enzymes, which catalyze the metabolism of lactic acid and the carbon chains of fatty acids and amino acids to carbon dioxide and water, are integral parts of the mitochondria.

The Mechanism of Enzyme Catalysis. Many years ago Emil Fischer, the German organic chemist, suggested that the specificity of the relationship of an enzyme to its substrate indicated that the two must fit together like a lock and key (Fig. 3.2). The idea that an enzyme combines with its substrate to form a reactive intermediate enzyme–substrate complex that subsequently decomposes to release the free enzyme and the reaction products was formulated mathematically by Leonor Michaelis more than 50 years ago. By brilliant

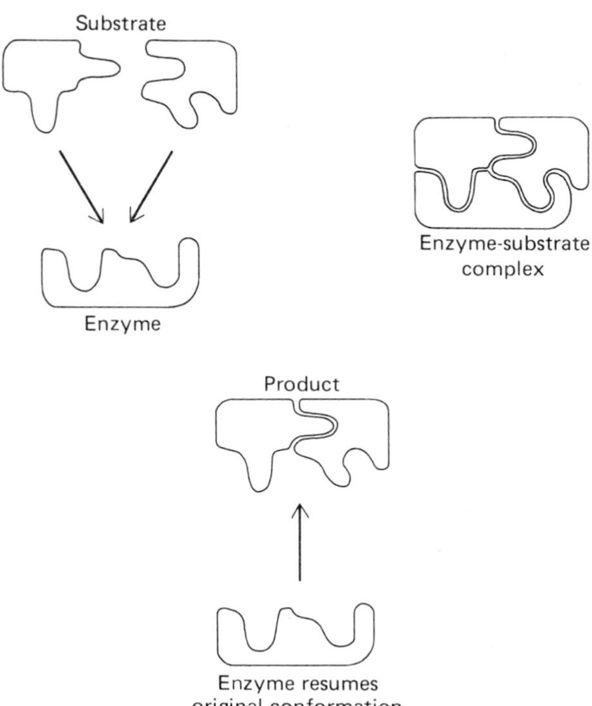

FIGURE 3.2 The "induced-fit" theory of enzyme action. It is postulated that the enzyme has an active site to which the substrate(s) fit (top left), forming an enzyme-substrate complex (top right). The binding of the substrate induces changes in the shape of the enzyme, making it more reactive. When the product is released (bottom), the enzyme regains its original shape.

inductive reasoning, he assumed that such a complex does form and then calculated what the relationships between enzyme concentration, substrate concentration, and the velocity of the reaction should be. Exactly these relationships are observed experimentally, giving strong evidence of the validity of Michaelis' assumption that an enzyme–substrate complex forms as an intermediate.

Direct evidence of the existence of enzyme–substrate complexes was obtained by David Keilin of Cambridge University and Britton Chance of the University of Pennsylvania. Chance isolated a brown-colored peroxidase from horseradish and found that when this was mixed with the substrate, hydrogen peroxide, a green-colored enzyme–substrate complex, formed. This, in turn, changed to a second, pale red complex that finally split to give the original brown enzyme and the products of the reaction. By observing the rates of changes of color, Chance was able to calculate the rates of formation and dissociation of these complexes.

It is clear that when the substrate is part of an enzyme–substrate complex, it is much more reactive than it is when free. It is not clear, however, *why* this should be true. One explanation postulates that the enzyme unites with the substrate at two or more places, and the substrate molecule is held in a position that strains its molecular bonds and renders them more likely to break. It is probable that only a relatively small part of the enzyme molecule (termed the "active site") is involved in combining with the substrate.

One approach to the study of enzymic action is to investigate the detailed structure of the enzyme molecule itself. It is now possible to determine the sequence of amino acids in the peptide chain, or chains, composing the enzyme protein. Ribonuclease, which consists of 124 amino acids in a single chain that is looped on itself like a pretzel, was the first enzyme for which it was possible to give the exact sequence of amino acids in the entire molecule. Studies of the sequence of amino acids and the three-dimensional structure of the active site of the enzyme have been fruitful in throwing light on the mechanism of enzyme action. Studies of the constituents and topography of certain active sites have revealed

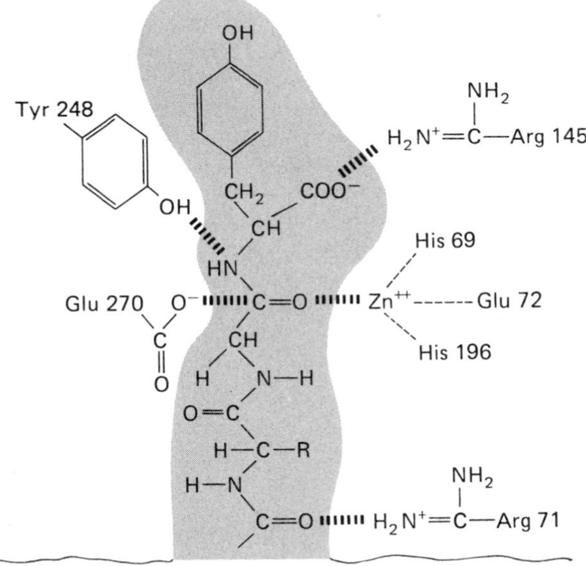

FIGURE 3.3 Model of an active site of an enzyme, carboxypeptidase. The gray represents a cleft in the surface of the enzyme into which a part of the substrate fits. The substrate is bound to the enzyme by five weak bonds, shown by the dashed lines. Seven of the amino acids of the enzyme's active site are indicated by their abbreviations. The number beside each amino acid refers to its position in the polypeptide chain of the enzyme.

the presence of certain amino acids (e.g., serine) or metal ions that actually participate in the enzyme-catalyzed reaction as proton donors or acceptors or as nucleophilic (electron-donating) or electrophilic (electron-attracting) agents (Fig. 3.3). A metal ion present at the active site may bind with one or more groups on the substrate molecule and, by forming a chelated intermediate, participate in the strain-producing process.

Factors Affecting Enzymic Activity

Temperature. The velocity of most chemical reactions is approximately doubled by each 10° increase in temperature, and, over a moderate range of temperature, this is true of enzyme-catalyzed reactions as well. Enzymes, and proteins in general, are inactivated by high temperatures. Native protein molecules exist at least in part as **spiral coils,** or helices, and the denaturation process appears to involve the unwinding of the helix. Most organisms are killed by exposure to heat because their cellular enzymes are inactivated. The enzymes of humans and other warm-blooded animals operate most efficiently at a temperature of about 37°C.—body temperature—whereas those of cold-blooded animals work optimally at about 25°C. Enzymes are generally not inactivated by freezing; their reactions continue slowly, or perhaps cease altogether at low temperatures, but their catalytic activity reappears when the temperature is again raised to normal.

Acidity. All enzymes are sensitive to changes in the acidity and alkalinity—the pH—of their environment and will be inactivated if subjected to strong acids or bases. Most enzymes exert their greatest catalytic effect only when the pH of their environment is within a certain rather narrow range. On either side of this optimum pH, as the pH is raised or lowered, enzymic activity rapidly decreases. The protein-digesting enzyme secreted by the stomach, pepsin, is remarkable in that its pH optimum is 2.0; it will work only in an extremely acid medium. The protein-digesting enzyme secreted by the pancreas, trypsin, in contrast, has a pH optimum of 8.5, well on the alkaline side of neutrality. Most intracellular enzymes have pH optima near neutrality, pH 7.0. This marked influence of pH on the activity of an enzyme is what would be predicted from the fact that enzymes are proteins. The number of positive and negative charges associated with a protein molecule, and the shape of the molecular surface, are determined in part by the pH. Probably only one particular state of the enzyme molecule, with a particular number of negative and positive charges, is active as a catalyst.

Concentration of Enzyme, Substrate, and Cofactors. If the pH and temperature of an enzyme system are kept constant and if an excess of substrate is present, the rate of the reaction is directly proportional to the amount of enzyme present. If the pH, temperature, and enzyme concentration of a reaction system are held constant, the initial reaction rate is proportional to the amount of *substrate* present, up to a limiting value. If the enzyme system requires a coenzyme or specific activator ion, the concentration of this substance may, under certain circumstances, determine the overall rate of the enzyme system.

Enzyme Inhibitors. Enzymes can be inhibited by a variety of chemicals, some of which inhibit reversibly, others irreversibly. **Cytochrome oxidase,** one of the "respiratory enzymes," is inhibited by cyanide, which forms a complex with the atom of iron present in the enzyme molecule and prevents it from participating in the catalytic process. Cyanide is poisonous to humans and other animals because of its action on the cytochrome enzymes.

Enzymes themselves may act as poisons if they get into the wrong place. As little as 1 mg. of crystalline trypsin injected intravenously will kill a rat. Certain snake, bee, and scorpion venoms contain enzymes that destroy blood cells or other tissues.

Respiration and Cellular Energy

All the phenomena of life—growth, movement, irritability, reproduction, and others—require the expenditure of energy by the cell. Living cells are not heat engines; they cannot use heat energy to drive these reactions but must use chemical energy, chiefly in the form of energy-rich phosphate bonds, abbreviated ~P. These bonds have a relatively high free energy of hydrolysis (i.e., the difference in the energy content of the reactants and the products after the bond has been split is relatively high). The free energy of hydrolysis is not localized in the covalent bond joining the phosphorus atoms to the oxygen or nitrogen atom. Thus the term "energy-rich phosphate bond" is actually a misnomer, but it is so deeply ingrained by long usage that it is not likely to be changed.

ATP is the "energy currency" of the cell. All the energy-requiring reactions of cellular metabolism utilize ATP to drive the reaction. ATP is an energy-rich molecule because its triphosphate unit contains two **phosphoanhydride bonds,** ones formed by removing the elements of water, H_2O, from adjacent molecules of phosphoric acid. A large amount of free energy is released when ATP is hydrolyzed to adenosine diphosphate (ADP) and inorganic phosphate (P_i) or when ATP is hydrolyzed to adenosine monophosphate (AMP) and pyrophosphate (PP_i). Energy-rich molecules do not pass freely from one cell to another but are made at the site in which they are to be utilized. The energy-rich bonds of ATP that will drive the reactions of muscle contraction, for example, are produced in the muscle cells.

The term **cellular respiration** refers to the enzymic processes within each cell by which molecules of carbohydrates, fatty acids, and amino acids are metabolized, ultimately to carbon dioxide and water, with the conservation of biologically useful energy. Many of the enzymes catalyzing these reactions are located in the cristae and walls of the mitochondria (see Fig. 3.10).

All living cells obtain biologically useful energy by enzymic reactions in which electrons flow from one energy level to another. For most organisms, oxygen is the ultimate electron acceptor; oxygen reacts with the electrons and with hydrogen ions to form a molecule of water. Electrons are transferred to oxygen by a system of enzymes, localized within the mitochondria, called the **electron transmitter system.**

Electrons are removed from a molecule of some foodstuff and transferred by the action of a specific enzyme to some primary electron acceptor. Other enzymes transfer the electrons from the primary acceptor through the several components of the electron transmitter system and eventually combine them with oxygen (Fig. 3.4). The chief source of energy-rich phosphate bonds, ~P, in the cell is from the flow of electrons through the acceptors and the electron transmitter system. This flow of electrons has been termed the "electron cascade," and we might picture a series of waterfalls over which electrons flow, each fall driving a water wheel, an enzymic reaction by which the energy of the electron is captured in a biologically useful form, that of the energy-rich phosphate bonds of ATP.

Processes in which electrons (e^-) are removed from an atom or molecule are termed **oxidations;** the reverse

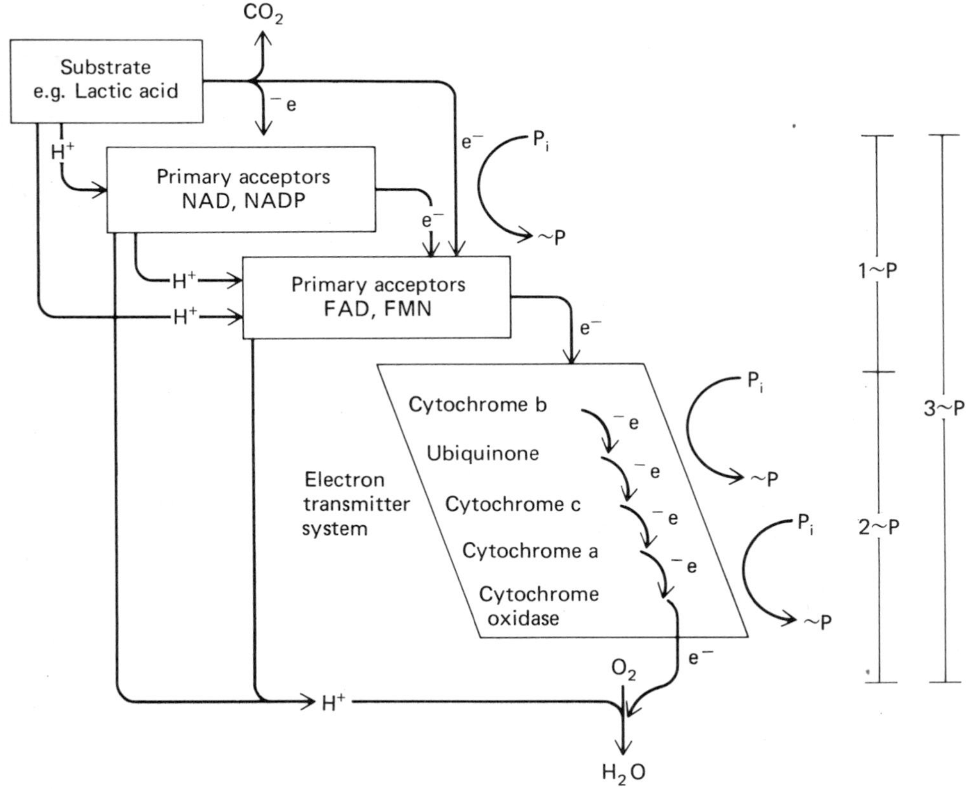

FIGURE 3.4 Diagram of the reactions of the "electron cascade," the succession of metabolic steps by which electrons are transferred from substrate to oxygen and the energy is trapped in a biologically useful form as energy-rich phosphate bonds, ~P. The passage of electrons from NAD to oxygen generates 3 ~P (vertical line to the right of the diagram), whereas the passage of electrons from FAD to oxygen generates 2 ~P.

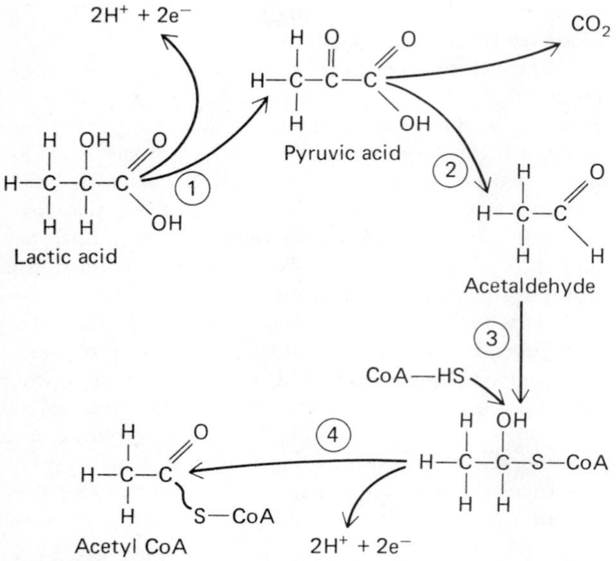

FIGURE 3.5 The enzymic conversion of lactic acid to acetyl coenzyme A.

process, the addition of electrons to an atom or molecule, is termed **reduction.** An example of oxidation and reduction is the reversible reaction

$$Fe^{++} \underset{\text{reduction}}{\overset{\text{oxidation}}{\rightleftharpoons}} Fe^{+++} + e^{-}$$

The reaction toward the right is an oxidation (the removal of an electron) and the reaction toward the left is a reduction (the addition of an electron). Each oxidation reaction, in which an electron is given off, must be accompanied by a reduction, a reaction in which the electron is accepted by another molecule, for electrons do not exist in the free state.

The passage of electrons in the electron transmitter system is a series of oxidation and reduction reactions termed **biological oxidation.** When the energy of this flow of electrons is captured in the form of ~P, the overall process is termed **oxidative phosphorylation.** In most biological systems, two electrons and two protons (that is, two hydrogen atoms) are removed together, and the process is known as **dehydrogenation.**

The specific compounds of the electron transmitter system that are alternately oxidized and reduced are proteins known as **cytochromes.** Each cytochrome contains a heme group similar to the one present in hemoglobin. In the center of the heme group is an iron atom that is alternately oxidized and reduced—converted from Fe^{++} to Fe^{+++} and back—as it gives off and takes up an electron. Another component of the electron transmitter system is **ubiquinone** (also called coenzyme Q).

Lactic acid, the acid of sour milk, is an important intermediate in metabolism. Its molecular structure permits it to undergo a dehydrogenation (Fig. 3.5, step 1), in which two hydrogen ions and two electrons are removed enzymatically by lactic dehydrogenase, and pyruvic acid is produced. The molecular structure of pyruvic acid does not permit further dehydrogenation. It undergoes decarboxylation, the loss of carbon dioxide (step 2), to yield acetaldehyde. Acetaldehyde, in turn, combines with coenzyme A (step 3) in a "make-ready" reaction to yield a product that can be dehydrogenated (step 4) to yield two hydrogen ions, two electrons, and acetyl coenzyme A. The electrons in both step 1 and step 4 are transferred to a primary electron acceptor, nicotinamide adenine dinucleotide, NAD (Fig. 3.4). The dehydrogenation and decarboxylation of pyruvate require lipoic acid and thiamine pyrophosphate as coenzymes.

The oxidation of succinic acid, another important intermediate in metabolism, (Fig. 3.6) involves a **flavin** as the hydrogen and electron acceptor. The product, fumaric acid, cannot be dehydrogenated directly but undergoes a make-ready reaction in which a molecule of water is added to yield malic acid. The oxidation of malic acid requires NAD as primary acceptor and yields oxaloacetic acid. Oxaloacetic acid may undergo several reactions. It usually unites with acetyl CoA to form citric acid, as discussed later, but may undergo other reactions, one of which is decarboxylation, to yield pyruvic acid. The further metabolism of pyruvic acid to acetyl coenzyme A could occur by the reactions just discussed (Fig. 3.5, step 4).

In the reactions by which carbohydrates, fats, and proteins are oxidized, the cell utilizes these three simple types of reactions: dehydrogenation, decarboxylation, and make-ready reactions. They may occur in different orders in different chains of reactions. All the dehydrogenation reactions are, by definition, oxidations—reactions in which electrons are removed from a molecule. The electrons cannot exist in the free state for any finite period of time and must be taken up immediately by other compounds, electron acceptors. Two of the primary electron acceptors of the cell are pyridine nucleotides NAD and NADP. The functional end of both pyridine nucleotides is the vitamin **nicotinamide.** Nicotinamide accepts one hydrogen ion and two electrons from a molecule undergoing dehydrogenation (e.g., lactic acid) and becomes reduced nicotinamide adenine dinucleotide, NADH, releasing one proton (see Fig. 3.4).

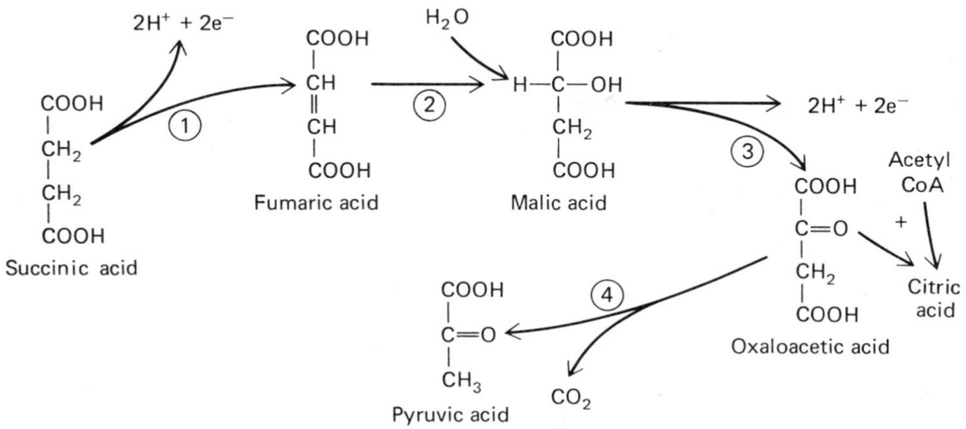

FIGURE 3.6 The oxidation of succinic acid. The condensation of oxaloacetic acid with acetyl CoA is shown in Figure 3.7.

The FAD of succinic dehydrogenase is bound very tightly to the protein part of the enzyme and cannot be removed easily. Such tightly bound cofactors are termed **prosthetic groups** of the enzyme. The pyridine nucleotide effective in the lactic dehydrogenase system, in contrast, is very loosely bound and is readily removed. Such loosely bound cofactors are termed **coenzymes.**

The reduced pyridine nucleotides, NADH or NADPH, cannot react with oxygen. Their electrons must first be passed to FAD or FMN and then be passed through the intermediate acceptors of the electron transmitter system before they can react with oxygen. The flavin primary acceptors usually pass their electrons to the electron transmitter system, but some flavoproteins can react directly with oxygen. When this occurs, hydrogen peroxide, H_2O_2, is produced and no ~P is formed. An enzyme that can mediate the transfer of electrons directly to oxygen is termed an **oxidase;** one that mediates the removal of electrons from a substrate to a primary or intermediate acceptor is termed a **dehydrogenase.**

The Tricarboxylic Acid (TCA) Cycle

The acetyl coenzyme A formed by the oxidation of lactic acid, or formed by the oxidation of fatty acids, undergoes a series of reactions involving dehydrogenation, decarboxylation, and make-ready reactions that have been termed the **Krebs tricarboxylic acid (TCA) cycle.** Acetyl coenzyme A combines with oxaloacetic acid to yield citric acid (Fig. 3.7, step 1). Citric acid cannot undergo dehydrogenation. Two additional make-ready reactions involving the removal and addition of a molecule of water (step 2) yield isocitric acid, which can undergo dehydrogenation (step 3). The hydrogen acceptor is a pyridine nucleotide, usually NAD, and the product is oxalosuccinic acid, which undergoes decarboxylation to yield α-ketoglutaric acid (step 4).

Just as pyruvic acid is converted to the coenzyme A derivative of an acid with one less carbon atom, α-ketoglutaric acid (five carbons) is converted to succinyl coenzyme A (four carbons) by reactions requiring NAD, coenzyme A, thiamine pyrophosphate, and lipoic acid as coenzymes. α-Ketoglutaric acid is metabolized by a dehydrogenation, a decarboxylation, and a make-ready reaction using coenzyme A to yield succinyl coenzyme A (step 5).

The bond joining coenzyme A to succinic acid is an energy-rich one, ~S. The reaction of succinyl coenzyme A with inorganic phosphate yields succinyl phosphate and coenzyme A (step 6). The phosphate group is then transferred to ADP to form ATP and free succinic acid (step 7). This is an example of an energy-rich bond synthesized at the **substrate level** by reactions not involving the electron transmitter system. Only a small fraction of the total energy-rich bonds made in metabolism are formed by reactions such as this. Normal cells metabolizing in a medium containing oxygen synthesize most of their ATP by oxidative phosphorylation in the electron transmitter system.

The further oxidation of succinic acid to fumaric, malic, and oxaloacetic acid (see Fig. 3.6), completes the cycle, for the oxaloacetic acid is then ready to combine with another molecule of acetyl coenzyme A to form a

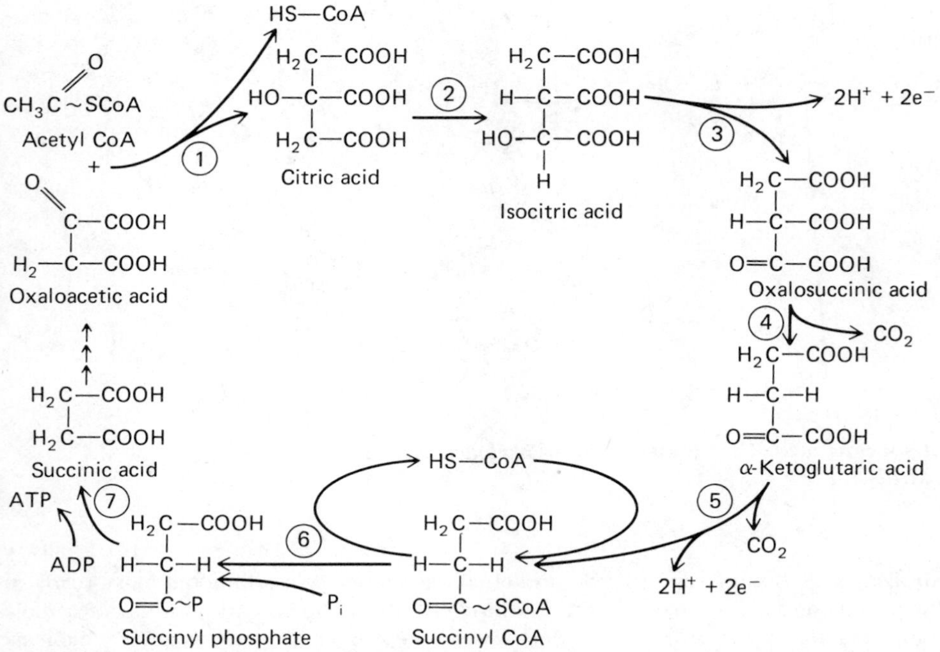

FIGURE 3.7 The tricarboxylic acid cycle (Krebs citric acid cycle). Coenzyme A (HS–CoA) is added in reaction 5 and released in reaction 6, being replaced by phosphate.

molecule of citric acid. In the course of this cycle, two molecules of CO_2 and eight hydrogen atoms are removed, and one molecule of ~P is synthesized at the substrate level. The Krebs tricarboxylic acid (TCA) cycle, also called the citric acid cycle, is the final common pathway by which the carbon chains of carbohydrates, fatty acids, and amino acids are metabolized.

Fatty Acid Oxidation. The metabolism of the carbon chain of a fatty acid first requires the activation of the fatty acid by a reaction with ATP and then with coenzyme A to yield the fatty acyl coenzyme A. For example, palmitic acid, which contains 16 carbons, reacts to form palmityl coenzyme A (Fig. 3.8, step 1). Palmityl coenzyme A undergoes a dehydrogenation between the second and third carbons of the chain (Fig. 3.8, step 2). The hydrogen acceptor for this dehydrogenation is a flavin, for the group undergoing dehydrogenation is —CH_2—CH_2—. The product of this dehydrogenation undergoes a make-ready reaction, the addition of a molecule of water, to yield a molecule that has an H—C—OH group

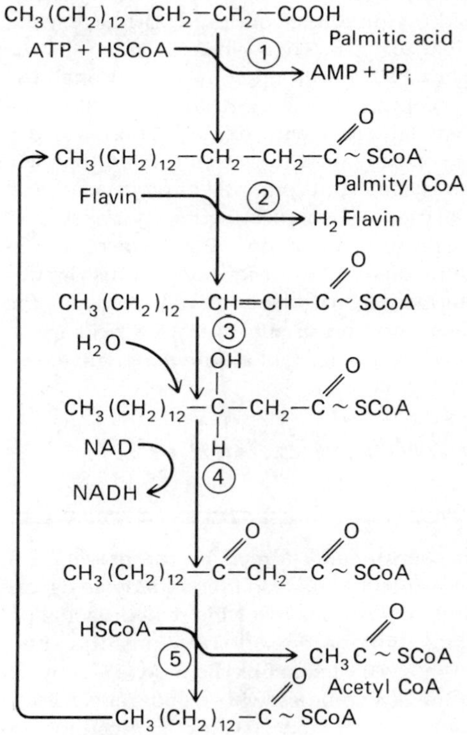

FIGURE 3.8 Reactions in the oxidation of fatty acids. Reactions 2 through 5 cleave a fatty acid chain two carbons at a time.

(Fig. 3.8, step 3). This, in turn, undergoes a dehydrogenation utilizing NAD as the hydrogen acceptor (Fig. 3.8, step 4). The resulting molecule can undergo a make-ready reaction with coenzyme A to clip off a two-carbon unit, acetyl coenzyme A, and leave a carbon chain that is two carbons shorter (Fig. 3.8, step 5). This molecule is in the activated state—it contains coenzyme A on its carboxyl group—and is ready to be dehydrogenated by the enzyme that uses flavin as a hydrogen acceptor. This repeating series of reactions, which includes dehydrogenations and make-ready reactions but not decarboxylations, cleaves a fatty acid chain two carbons at a time (Fig. 3.8). Seven such series of reactions will split palmitic acid to eight molecules of acetyl coenzyme A, which then enter the TCA cycle (Fig. 3.7).

Glycolysis. Glucose and other sugars are also converted ultimately to acetyl coenzyme A. The series of glycolytic reactions begins, as with fatty acids, by one in which glucose is "activated." Glucose reacts with ATP to yield glucose-6-phosphate and ADP, a reaction catalyzed by the enzyme **hexokinase.** After this make-ready reaction, other make-ready reactions establish a configuration that can undergo dehydrogenation (see Fig. 3.9). A rearrangement yields fructose-6-phosphate (Fig. 3.9, step 2), and the transfer of a second phosphate from ATP forms fructose-1, 6-diphosphate (step 3). Fructose-1, 6-diphosphate is split by an enzyme, **aldolase,** into two three-carbon sugars, glyceraldehyde-3-phosphate and dihydroxyacetone phosphate (step 4). These two sugars are interconverted by a separate enzyme.

Glyceraldehyde-3-phosphate reacts with a compound containing an —SH group that is part of the enzyme molecule to yield an H—C—OH group (with vertical bonds above and below the C) that can undergo dehydrogenation with NAD as the hydrogen acceptor. The product of the reaction, phosphoglyceric acid bound to the SH group of the enzyme, then reacts with inorganic phosphate to yield 1-3-diphosphoglyceric acid and free enzyme —SH (Fig. 3.9, step 5). The phosphate in carbon 1 is an energy-rich group that can react with ADP to form ATP. This, like the energy-rich phosphate of succinyl phosphate, is one made at the substrate level. The resulting 3-phosphoglyceric acid undergoes a make-ready rearrangement to yield 2-phosphoglyceric acid. Then, in an unusual reaction, an energy-rich phosphate is generated by the removal of water by a **dehydration,** rather than by the removal of two hydrogens, a **dehydrogenation** (step 6). The product, phosphoenolpyruvic acid, can transfer its phosphate group to ADP to yield ATP and free pyruvic acid. This is the second energy-rich phosphate group generated at the substrate level in the metabolism of glucose to pyruvic acid.

Each glucose molecule yields two molecules of glyceraldehyde-3-phosphate, and hence a total of four energy-rich phosphates is produced as glucose is metabolized to pyruvic acid. However, two energy-rich phosphates are utilized in the process, one to convert glucose to glucose-6-phosphate and the second to convert fructose-6-phosphate to fructose-1, 6-diphosphate. The net yield in the process is 2~P (4~P produced minus 2~P used up in the reactions). Pyruvic acid is then metabolized to acetyl coenzyme A by the reactions described previously (see Fig. 3.5).

Amino Acid Oxidation. Amino acids are oxidized by reactions in which the amino group is first removed, a process called **deamination;** then the carbon chain is metabolized and eventually enters the tricarboxylic acid cycle. The amino acid alanine, for example, yields pyruvic acid when deaminated, glutamic acid yields α-ketoglutaric acid, and aspartic acid yields oxaloacetic acid. These three amino acids can enter the TCA cycle directly. Other amino acids may require several reactions in addition to deamination to yield a substance that is a member of the TCA cycle, but ultimately the carbon chains of all the amino acids are metabolized in just this way.

The Electron Transmitter System. The major reactions by which biologically useful energy is produced occur when the electrons are transferred from the primary acceptors, such as the pyridine nucleotides, through the electron transmitter system and provide the energy to drive the reactions of oxidative phosphorylation. The electrons entering the electron transmitter system from NADH have a relatively high energy content. As they pass along the chain of enzymes, they lose much of their energy, some of which is conserved in the form of ATP.

The enzymes of the electron transmitter system are located within the substance of the mitochondrion in the mitochondrial membranes (Fig. 3.10). It is not known whether a particular electron, in passing from pyridine nucleotide to oxygen, must go through each and every one of the intermediates or whether it might skip some steps. It probably has to pass through at least three different steps to account for the 3 ~P that are made for each pair of electrons that pass from pyridine nucleotide to oxygen. The flow of electrons is tightly coupled to the phosphorylation process and will not occur unless phosphorylation can occur also. This, in a sense, prevents waste, for electrons will not flow unless energy-rich phosphate can be formed.

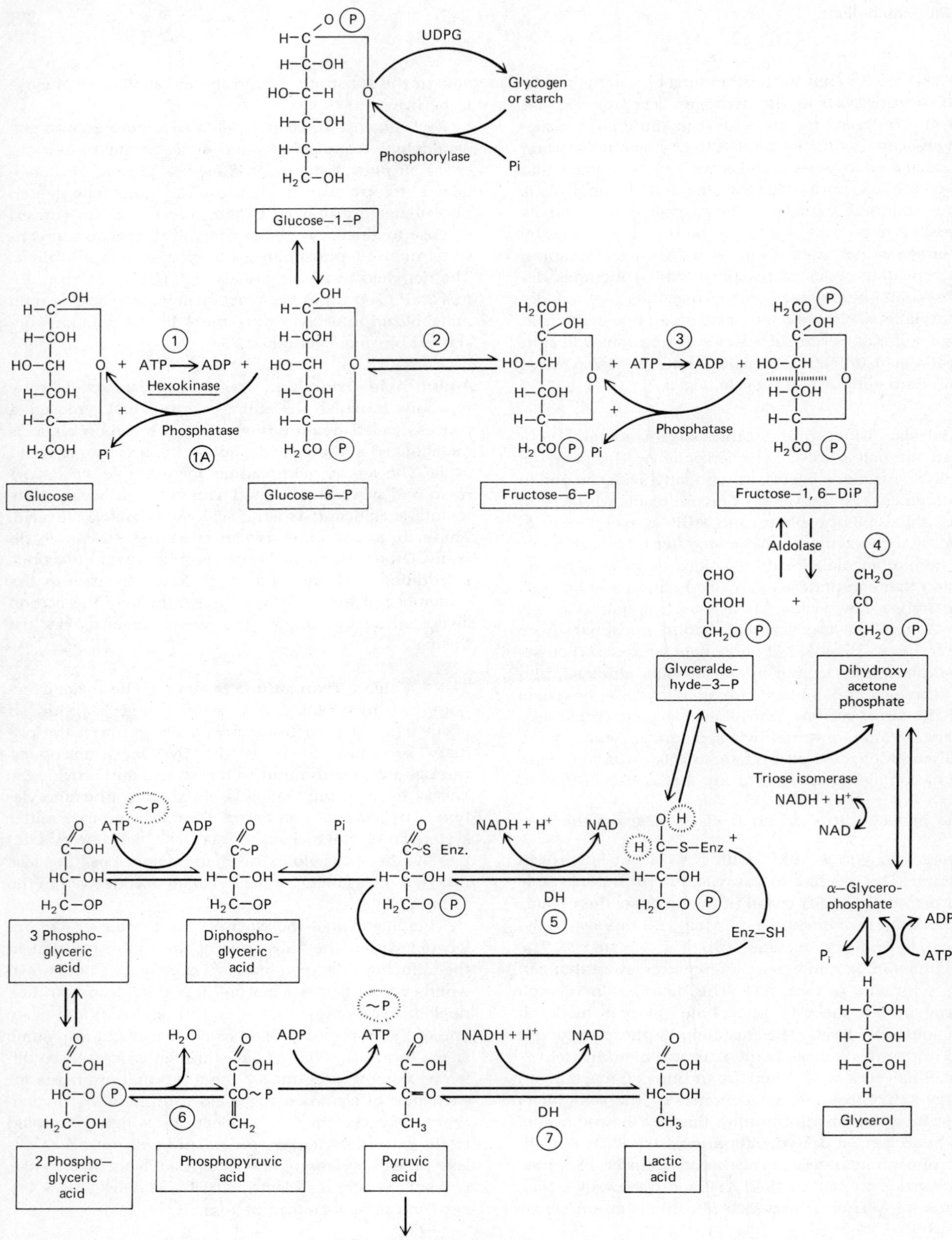

UDPG
Glycogen or starch
Phosphorylase
Pi
Glucose–1–P
Glucose
Glucose–6–P
Fructose–6–P
Fructose–1, 6–DiP
ATP → ADP
Hexokinase
Phosphatase
1
1A
2
3
4
Aldolase
Glyceralde-hyde–3–P
Dihydroxy acetone phosphate
Triose isomerase
NADH + H+
NAD
α–Glycero-phosphate
ADP
ATP
Glycerol
3 Phospho-glyceric acid
Diphospho-glyceric acid
Enz–SH
DH
5
2 Phospho-glyceric acid
Phosphopyruvic acid
Pyruvic acid
Lactic acid
6
7
To acetyl coenzyme A

◀ **FIGURE 3.9** Diagram of the series of reactions by which glucose and other sugars are metabolized to pyruvic acid. Note that most of the steps are reversible, as indicated by double arrows. Other steps are essentially irreversible; glucose-6-phosphate is converted to glucose by a separate enzyme, a glucose-6-phosphatase (reaction 1A), and not by the hexokinase that catalyzes the conversion of glucose to glucose-6-phosphate (reaction 1).

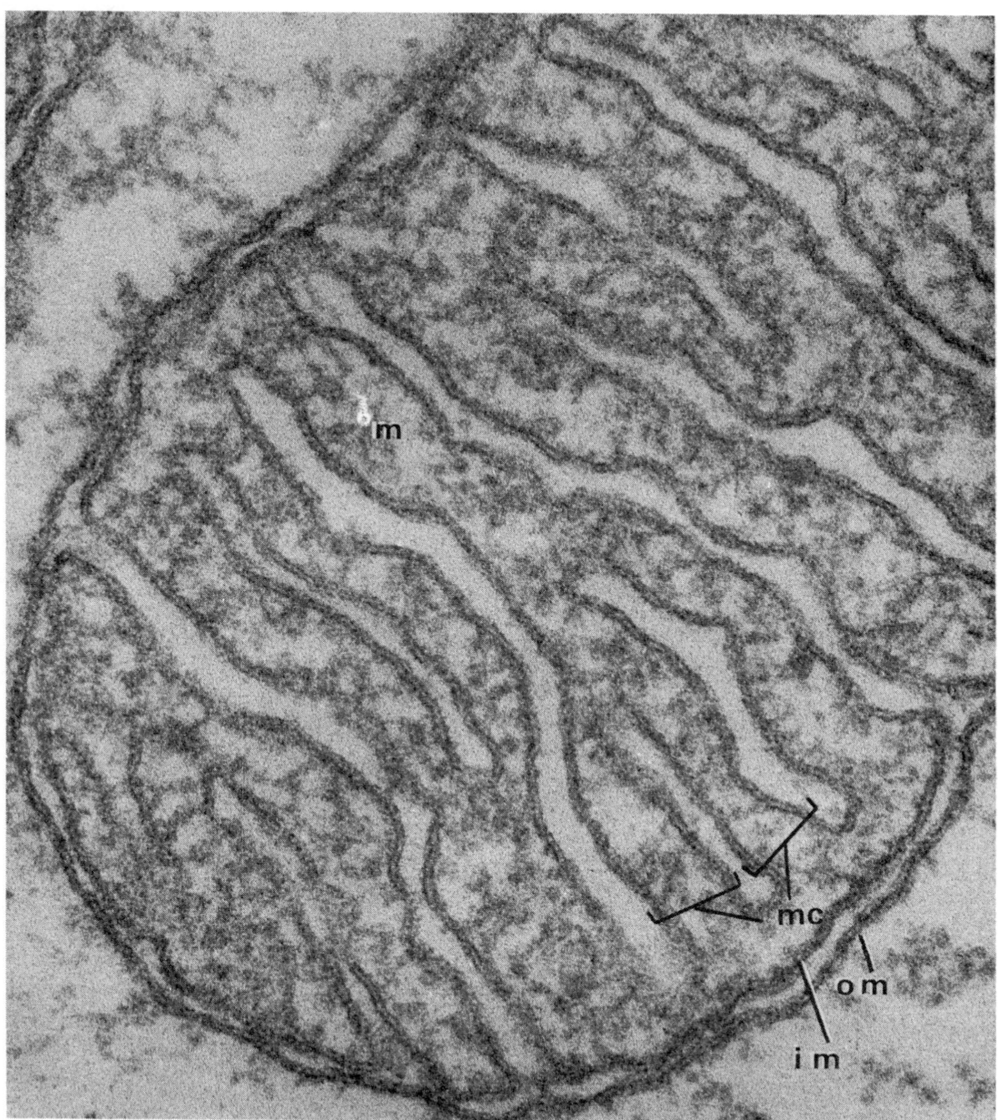

FIGURE 3.10 Electron micrograph of a mitochondrion from a pancreatic centroacinar cell. The double-layered unit membrane is evident in the smooth outer membrane *(om)* and in the inner membrane *(im)*, which folds to form the cristae *(mc)*. Magnification ×207,000. (Courtesy of G. E. Palade; from DeRobertis, E. D., and DeRobertis, E. M.: *Cell and Molecular Biology*. 7th ed. Philadelphia, Saunders College Publishing, 1980.)

The Chemiosmotic Theory of Oxidative Phosphorylation

Although it had long been known that oxidative phosphorylation occurs in the mitochondria, and many experiments showed that the transfer of electrons from NADH to oxygen resulted in the production of 3~P, just how the ~P were synthesized remained a mystery. From the oxidation–reduction potentials of the various members of the electron transmitter system it was possible to suggest where in the system the phosphorylation was coupled and the enzyme that converts ADP and P_i to ATP was characterized. Several theories suggested that some sort of energy-rich intermediate was formed and transferred its energy to drive the synthesis of ATP, but no such intermediate could be discovered despite an intensive search in many laboratories.

A very different mechanism, the **chemiosmotic hypothesis,** was suggested by Peter Mitchell in 1961. This has been supported by experimental evidence from many laboratories, and Mitchell was rewarded with a Nobel Prize in 1978 for his breakthrough discovery. Mitchell proposed that electron transport and ATP synthesis are coupled by a proton gradient across the inner mitochondrial membrane. According to this model, the stepwise transfer of electrons from NADH or $FADH_2$ through the electron carriers to oxygen results in the pumping of protons across the inner mitochondrial membrane and the generation of a membrane potential with the cytoplasmic side being positive (Fig. 3.11). Protons are pumped out of the matrix by three kinds of electron transfer complexes, each associated with a particular step in the electron transport system. The protons subsequently flow back into the matrix of the mitochondria through special sites in the inner membrane where the enzyme that synthesizes ATP from ADP and P_i is located. The protons are moving down an energy gradient, and energy is released to drive the synthesis of ATP. Thus the proton gradient across the inner mitochondrial membrane couples phosphorylation with oxidation.

The model requires that the various electron transmitter systems be spatially oriented with respect to the inner and outer faces of the inner mitochondrial membrane. This enables the carriers to pump protons out of the matrix as they transport electrons. The spatial organization of the ATP-synthesizing enzyme permits it to utilize the energy of the proton gradient to drive the synthesis of ATP. The model also requires that the inner mitochondrial membrane be impermeable to the flow of electrons (except through the special sites where the ATP-synthesizing enzyme is located), otherwise the system would be short-circuited by the leakage of protons. A wealth of experimental evidence confirms these hypotheses and demonstrates that a proton gradient is generated across the inner mitochondrial membrane as electrons are transported. Certain of the electron carriers are arranged in the membrane in such a way that they pick up protons from the matrix side and release them on the outer side of the inner membrane, thereby establishing the proton gradient across the membrane.

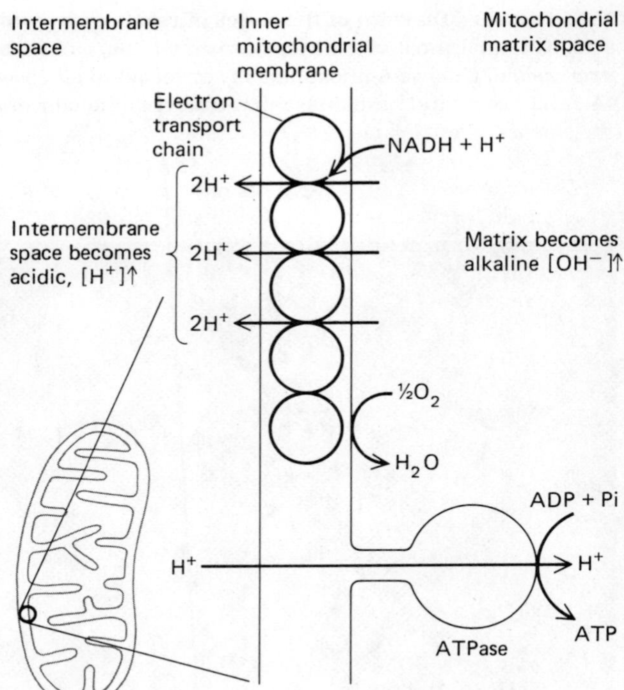

FIGURE 3.11 The chemiosmotic theory of the formation of a proton gradient across the inner mitochondrial membrane and its role in generating ATP. The inner mitochondrial membrane is labeled *im* in Figure 3.10.(inset) Diagram of a mitochondrion indicating the area of the inner membrane enlarged in the diagram.

Control of Oxidative Phosphorylation

The flow of electrons from pyridine nucleotides to oxygen, a total drop of 1.13 volts (from −0.32 to +0.81 volt), would yield 52 kilocalories (kcal.) per pair of electrons if the process were 100 percent efficient.* Under

*This may be calculated from the formula $\Delta G' = -nF\Delta E$, where $\Delta G'$ is the change in free energy, n is the number of electrons involved (two), F is the faraday (23.04 kcal.), and ΔE is the difference in the oxidation–reduction potentials of the reactants (1.13 volts). $\Delta G' = -2 \times 23.04 \times 1.13 = 52.06$ kcal.

experimental conditions most cells will produce at most 3 ~P per pair of electrons as the electrons pass from pyridine nucleotide to oxygen. Each ~P is equivalent to about 10 kcal. Hence, the 3~P amount to about 30 kcal. The efficiency of the electron-transmitter system may then be calculated as 30/52 or about 57 percent.

The amount of ATP present in any cell is usually rather small. In muscle cells, in which large amounts of energy may be expended in a short time during the contraction process, an additional substance, creatine phosphate, serves as a reservoir of ~P. The terminal phosphate of ATP is transferred by the enzyme creatine kinase to creatine to yield creatine phosphate and ADP. The phosphate bond of creatine phosphate is an energy-rich bond. The ~P of creatine phosphate must be transferred back to ADP, converting it to ATP, to be utilized in muscle contraction and other energy-requiring reactions.

The fact that phosphorylation is tightly coupled to oxidation or electron flow provides a system of control that can regulate the rate of energy production and adjust it to the momentary rate of energy utilization. In a resting muscle cell, for example, oxidative phosphorylation will occur until all of the ADP has been converted to ATP. Then, since there are no more acceptors of ~P, phosphorylation must stop. Since oxidation is tightly coupled to phosphorylation, the flow of electrons will cease.

When the muscle contracts, the energy required is obtained from the splitting of the energy-rich terminal phosphate of ATP to yield ADP and inorganic phosphate plus energy. The ADP formed can then serve as an acceptor of ~P, phosphorylation begins, and the flow of electrons to oxygen occurs. Oxidative phosphorylation continues until all of the ADP has been converted to ATP. Electric generating systems have an analogous control device that adjusts the rate of production of electricity to the rate of utilization of electricity.

The conversion of glucose to carbon dioxide and water yields about 4 kcal./gm. Let us now dissect the overall equation

$$C_6H_{12}O_6 + 6\ O_2 \rightarrow 6\ CO_2 + 6\ H_2O + \text{energy}$$

and review where the energy is released (Table 3.2 and Fig. 3.12).

In glycolysis (reaction 1, Table 3.2 and Fig. 3.12), glucose is activated by the addition of 2 ~P and converted to 2 pyruvate + 2 NADH + 4 ~P. Then the two pyruvates are metabolized (reaction 2) to 2 CO_2 + 2 acetyl CoA + 2 NADH. Finally, in the TCA cycle (reaction 3), the two acetyl CoA are metabolized to 4 CO_2 + 6 NADH + 2 H_2FP + 2 ~P.

These reactions can be added together (reaction 4) by eliminating items that are present on both sides of the arrows. Then, since the oxidation of NADH in the electron-transmitter system yields 3 ~P per mole, the 10 NADH = 30 ~P. The oxidation of H_2FP yields 2 ~P per mole, and 2 H_2FP = 4 ~P. Summing these we see that the complete aerobic metabolism of 1 mole (180 gm.) of glucose yields 38 ~P (reaction 5). Each ~P is equivalent to about 10 kcal. and the 38 ~P = 380 kcal.

When a mole of glucose is burned in a calorimeter, some 690 kcal. are released as heat. The metabolism of glucose in an animal cell releases 380/690, or about 55 percent, of the total energy as biologically useful energy, ~P. The remainder of the energy is dissipated as heat.

The Molecular Organization of Mitochondria

Mitochondrial shapes range from spherical to elongated, sausage-like structures, but an average mitochondrion is an ellipsoid about 3 μm. long and a little less than 1 μm. in diameter. Some very generously endowed large cells, such as the giant amoeba, *Chaos chaos*, have several hundred thousand mitochondria, and an average mammalian liver cell has about 1000. The mitochondrial protein accounts for about 20 percent of the total protein of the liver cell.

The inner membrane of the mitochondrion is folded repeatedly (Fig. 3.10) to form the shelflike cristae extending into, or all the way across, the central cavity. Both the inner and outer mitochondrial membranes are lipid bilayers containing protein molecules.

The enzymes of the TCA cycle have been found in the soluble matrix within the mitochondrion. The enzymes of the electron-transmitter system are tightly bound to the inner membrane. Each group of these

TABLE 3.2 The Reactions by Which Biologically Useful Energy is Released

(1) $C_6H_{12}O_6$ + 2 ~ P	→	2 pyruvate + 2 NADH + 4 ~ P
(2) 2 pyruvate	→	2 CO_2 + 2 acetyl CoA + 2 NADH
(3) 2 acetyl CoA	→	4 CO_2 + 6 NADH + 2 FPH_2 + 2 ~ P
(4) Sum $C_6H_{12}O_6$	→	6 CO_2 + 10 NADH + 2 FPH_2 + 4 ~ P
(5) $C_6H_{12}O_6$ + 6 O_2	→	6 CO_2 + 6 H_2O + 30 ~ P + 4 ~ P + 4 ~ P

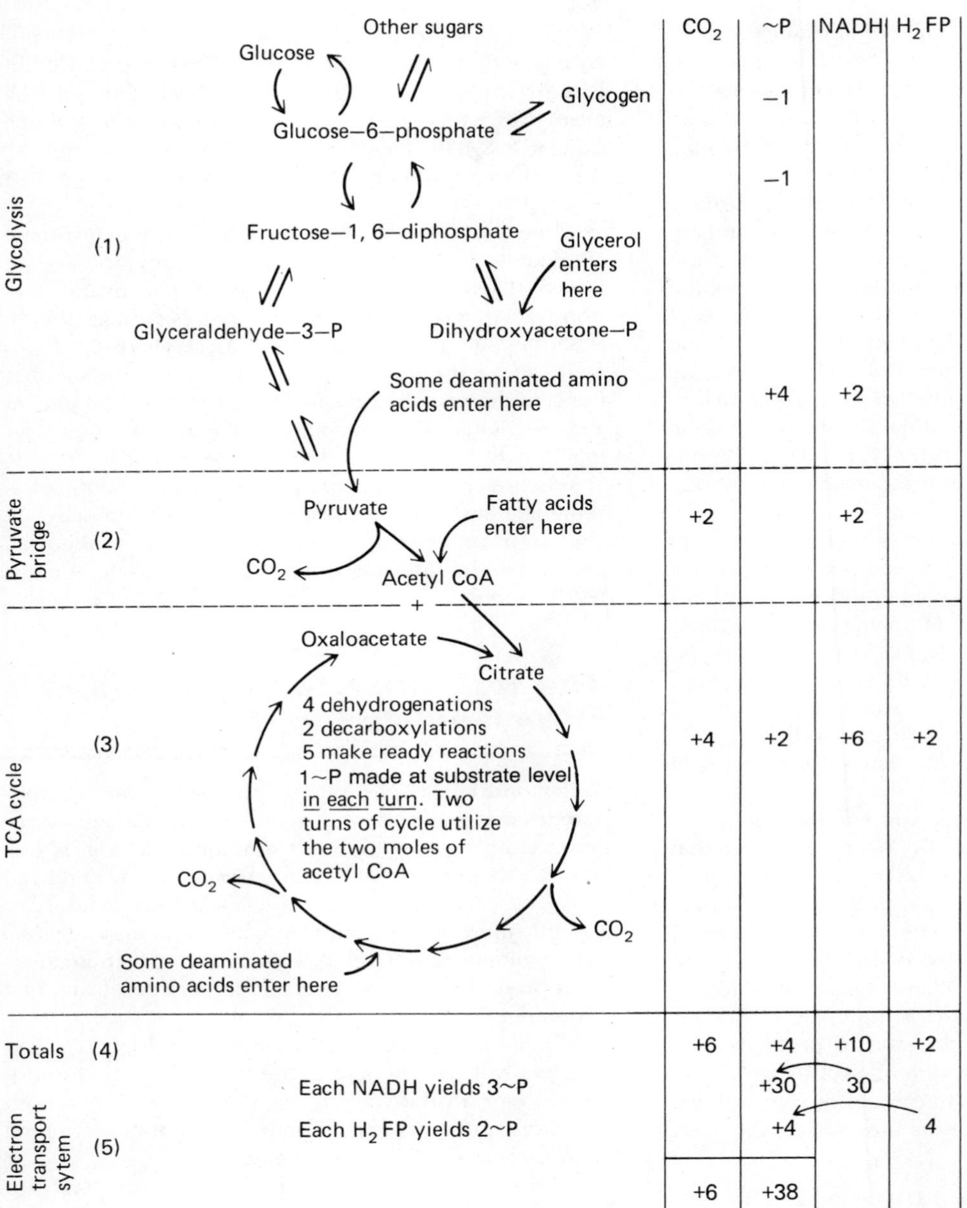

	CO_2	~P	NADH	H_2FP
Glycolysis (1)		−1		
		−1		
		+4	+2	
Pyruvate bridge (2)	+2		+2	
TCA cycle (3)	+4	+2	+6	+2
Totals (4)	+6	+4	+10	+2
Each NADH yields 3~P		+30	30	
Each H_2FP yields 2~P		+4		4
Electron transport sytem (5)	+6	+38		

FIGURE 3.12 Summary of the reactions by which glucose is metabolized to carbon dioxide with the release of biologically useful energy, ~P.

electron-transmitter enzymes, termed a **respiratory assembly,** is one of the fundamental units of cellular activity. It has been estimated that the mitochondrion of the liver cell contains some 15,000 respiratory assemblies and that they make up about one quarter of the mass of the mitochondrial membranes. Thus the mitochondrial membrane is not just a protective skin but an important functional part of the mitochondrion.

High-resolution electron micrographs of mitochondria, made by Humberto Fernández-Moran at the University of Chicago, have demonstrated the presence of small particles on the outer surface of the outer membrane and the inner surface of the inner membrane (Fig. 3.13). The particles on the inner membrane typically have a spherical knoblike head, a cylindrical stalk, and a base plate. These particles may be sites of the enzyme reactions carried out by mitochondria. The particles of the inner membrane are not the respiratory assemblies, as was suggested when they were first found, but they do contain one or more "coupling factors" necessary for oxidative phosphorylation.

Much of the cell's biologically useful energy, ATP, is generated by enzyme systems located in the *inner* membrane of the mitochondrion, yet most of the energy

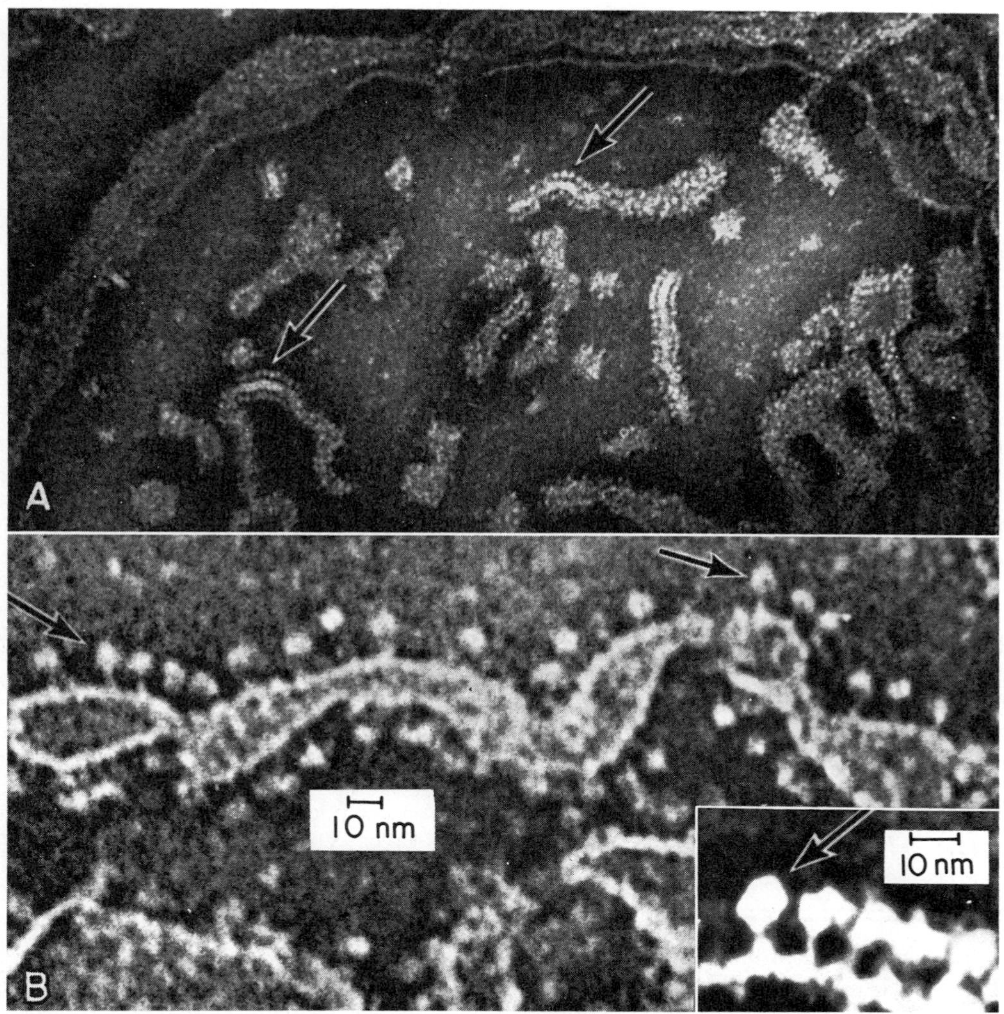

FIGURE 3.13 Electron micrograph of a mitochondrion swollen in a hypotonic solution and negatively stained with phosphotungstate. *A,* Isolated cristae; magnification ×85,000. *B,* At higher magnification, ×500,000, the elementary particles attached by a stalk to the surface of the cristae are evident. *Inset:* Magnified ×650,000, the polygonal shape of the elementary particle and its slender stalk are clearly visible. (From DeRobertis, E. D., and E. M. DeRobertis, Jr.: *Cell and Molecular Biology.* 7th ed. Philadelphia, Saunders College Publishing, 1980.)

utilized by the cell is required for processes that take place *outside of* the mitochondrion. ATP is used in the synthesis of proteins, fats, carbohydrates, nucleic acids, and other complex molecules, in the transport of substances across the plasma membrane, in the conduction of nerve impulses, and in the contraction of muscle fibers, all of which are reactions occurring largely or completely outside of the mitochondrion, in other parts of the cell. Several mechanisms, too complex to be considered here, have evolved to mediate the transfer of ~P across the mitochondrial membrane.

A single cell may have 1,000 or more mitochondria, and the function of each one must be controlled appropriately to generate the amount of energy required by the cell at any given moment. The rate at which ~P is utilized by a cell may vary over a remarkably wide range as the cell becomes active or quiescent. This relationship has been measured in the muscle cells of a frog. One hundred grams of frog muscle utilizes 1.6μ moles of ~P per minute when quiescent and 3300μ moles of ~P per minute in a state of tetanus (continuous contraction). The rate of production of ATP in the cell is controlled in large part by the rate of utilization of ATP by the cell. The flow of electrons in the electron-transmitter system is tightly coupled to phosphorylation, and oxidative phosphorylation can occur only when there is ADP to be converted to ATP. These facts provide the basis for the system by which a cell's utilization of ATP, which produces ADP, is used to regulate the rate at which ATP is produced. In addition, the structure and biological activity of some of the enzymes involved in glucose oxidation are affected by the concentration of ADP present. In this way, an increased concentration of ADP can lead to increased activity of these enzymes and an increased production of ATP. The problem of the nature and interrelations of these various biological control systems is very much in the forefront in biology today.

The Dynamic State of Cellular Constituents

The body of an animal or human appears to be unchanging as days and weeks go by, and it would seem reasonable to infer that the component cells of the body, and even the component molecules of the cells, are equally unchanging. In the absence of any evidence to the contrary, it was generally held, until about 1937, that the constituent molecules of animal and plant cells were relatively static and that, once formed, they remained intact for a long period of time. A corollary of this concept is that the molecules of food that are not used to increase the total cellular mass are rapidly metabolized to provide a source of energy. It followed from this that one could distinguish two kinds of molecules: relatively static ones that made up the cellular "machinery" and ones that were rapidly metabolized and thus correspond to cellular "fuel."

However, in 1937 Rudolf Schoenheimer and his colleagues at Columbia University began a series of experiments in which amino acids, fats, carbohydrates, and water, each suitably labeled with some "heavy" or radioactive isotope, were fed to rats. Schoenheimer's experiments, which have been confirmed many times since, showed that the labeled amino acids fed to the rats were rapidly incorporated into body proteins. Similarly, labeled fatty acids were rapidly incorporated into the fat deposits of the body, even though in each case there was no increase in the *total* amount of protein or fat. Such experiments have demonstrated that the fats and proteins of the body cells—and even the substance of the bones—are constantly and rapidly being synthesized and broken down. In the adult the rates of synthesis and degradation are essentially equal so that there is little or no change in the total mass of the animal's body. The distinction between "machinery" molecules and "fuel" molecules becomes much less sharp. The machinery is constantly being overhauled, and some of the machinery molecules are broken down and used as fuel.

The one exception to the rule of molecular flux is provided by the molecules of DNA that constitute the genes within the nucleus of the cell. Experiments with labeled atoms have shown that the molecules of DNA are remarkably stable and are broken down and resynthesized only very slowly, if at all. The amount of DNA per nucleus is constant, and new molecules of DNA must be synthesized before a cell can divide. The stability of the DNA molecules may be important in ensuring that hereditary characters are transmitted to succeeding generations with as few chemical errors as possible. In contrast, the molecules of RNA undergo constant synthesis and degradation; indeed, the amount of RNA per cell may vary within wide limits.

From the rate at which labeled atoms are incorporated it has been calculated that one half of all the tissue proteins of the human body are broken down and rebuilt every 80 days. The proteins of the liver and blood serum are replaced very rapidly, one half of them being synthesized every 10 days. Some enzymes in the liver have half-lives as short as two to four hours. The muscle proteins, in contrast, are replaced much more slowly, one half of the total number of molecules being replaced every 180 days. The celebrated aphorism of Sir

Frederick Gowland Hopkins, the late English biochemist, sums up this concept very succinctly: "Life is a dynamic equilibrium in a polyphasic system."

Biosynthetic Processes

Our discussion thus far has dealt with processes that break down molecules of foodstuffs and conserve their energy in the biologically useful form of ~P. Cells have the ability to carry out an extensive array of biosynthetic processes utilizing the energy of ATP and, as raw materials, some of the five-, four-, three-, two-, and one-carbon compounds that are intermediates in the metabolism of glucose, fatty acids, amino acids, and other compounds.

The whole subject of intermediary metabolism, by which an enormous variety of compounds is synthesized, is much too complicated to be discussed in detail here. The enzyme controlling each reaction is genetically determined, and the overall maze of enzyme reactions by which any compound is synthesized includes a variety of self-adjusting control mechanisms to regulate and integrate the reactions. Several basic principles of cellular biosynthesis can be distinguished:

1. Each cell, in general, synthesizes its own proteins, nucleic acids, lipids, polysaccharides, and other complex molecules and does not receive them preformed from other cells. Muscle glycogen, for example, is synthesized within the muscle cell and is not derived from liver glycogen.
2. Each step in the biosynthetic process is catalyzed by a separate enzyme.
3. Although certain steps in a biosynthetic sequence will proceed without the use of energy-rich phosphate, the overall synthesis of these complex molecules requires chemical energy. Why should this be true? Can you relate this to your understanding of the concept of entropy?
4. The synthetic processes utilize as raw materials relatively few substances, among which are acetyl coenzyme A, glycine, succinyl coenzyme A, ribose, pyruvate, and glycerol.
5. These synthetic processes are in general not simply the reverse of the processes by which the molecule is degraded but include one or more separate steps that differ from any step in the degradative process. These steps are controlled by different enzymes, and this permits separate control mechanisms to govern the synthesis and the degradation of the complex molecule.
6. The biosynthetic process includes not only the formation of the macromolecular components from simple precursors but their assembly into the several kinds of membranes that compose the outer boundary of the cell and the intracellular organelles. Each cell's constituent molecules are in a dynamic state and are constantly being degraded and synthesized. Thus even a cell that is not growing, not increasing in mass, uses a considerable portion of its total energy for the chemical work of biosynthesis. A cell that is growing rapidly must allocate a correspondingly larger fraction of its total energy output to biosynthetic processes, especially the biosynthesis of protein. A rapidly growing bacterial cell may use as much as 90 percent of its total biosynthetic energy for the synthesis of proteins.

Many of the steps in biosynthetic processes involve the formation of peptide bonds, glycosidic bonds, and ester bonds. Although these bonds are broken by hydrolysis, by the addition of water, they are *not* formed by reactions in which water is removed. The biosynthesis of lactose in the mammary gland, for example, does not proceed via

$$\text{glucose} + \text{galactose} \rightleftharpoons \text{lactose} + H_2O$$

This reaction would require energy, some 5.5 kcal. per mole, to go to the right if all reactants were present in the concentration of 1 mole per L. However, the concentrations of glucose and galactose in the mammary gland are probably less than 0.01 mole per L, whereas the concentration of water is very high, about 55 moles per L. Thus, the equilibrium point of the reaction under these conditions would be very far to the left.

Instead, one or more of the reactants is activated by a reaction with ATP in which the terminal phosphate is enzymatically transferred to glucose, with the conservation of some of the energy of ATP. The glucose phosphate, with a higher energy content than free glucose, can react with galactose via another enzyme-catalyzed reaction to yield lactose and inorganic phosphate.

$$\begin{aligned} &\text{ATP} + \text{glucose} \rightarrow \text{ADP} + \text{glucose-1-phosphate} \\ &\text{glucose-1-phosphate} + \text{galactose} \rightarrow \\ &\qquad\qquad \text{lactose} + \text{phosphate} \\ \text{Sum:}\ &\text{ATP} + \text{glucose} + \text{galactose} \rightarrow \\ &\qquad\qquad \text{lactose} + \text{ADP} + P_i \end{aligned}$$

This reaction proceeds to the right because there is net decrease in free energy. The 10 kcal. of the ~P bond are used to supply the 5.5 kcal. needed to assemble the glucose and galactose into lactose; the overall decrease in free energy is 4.5 kcal. per mole. Since water is not a product of this reaction, the high concentration of water in the cell does not inhibit it.

The same principle applies to the synthesis of the peptide bonds of proteins, the ester bonds of lipids and nucleic acids, and the glycosidic bonds of polysaccharides, but in many of these not one but *two* terminal phosphates of ATP are transferred as a unit in the reaction by which one of the molecules is activated. The activated molecule may be X ~PP or X ~AMP, depending on which part of the ATP is transferred to the substrate molecule. In either case the reaction ultimately involves the hydrolysis of 2 ~P with an energy utilization of 20 kcal. per mole.

$$X + ATP \rightarrow X \sim PP + AMP$$
$$X \sim PP + Y \rightarrow XY + PP_i$$
$$PP_i + H_2O \rightarrow P_i + P_i$$
$$\text{Sum: } X + Y + ATP \rightarrow XY + AMP + 2P_i$$

BOX 3.2

Bioluminescence

A number of animals, and some molds and bacteria as well, have an enzymic mechanism for the production of light. Luminescent animals are found among the protozoa, sponges, cnidaria, ctenophores, nemerteans, annelids, crustaceans, centipedes, millipedes, beetles, flies, echinoderms, mollusks, hemichordates, tunicates, and fishes. From this wide and irregular distribution of the light-emitting ability it is clear that the enzymes for luminescence have appeared independently in a number of different evolutionary lines. It is sometimes difficult to establish that a given organism is itself luminescent; in a number of instances animals once believed to be luminescent have been shown instead to contain luminescent bacteria.

Several different exotic East Indian fish known as lantern eyes have **light organs** under their eyes in which live luminous bacteria. The light organs contain long cylindrical cells that are well provided with blood vessels to supply an adequate amount of oxygen to the bacteria. The bacteria emit light continuously, and the fish have a black membrane, somewhat similar to an eyelid, that can be drawn up over the light organ to turn off the light. How the bacteria come to collect in the fish's light organ, as they must in each newly hatched fish, is a complete mystery.

Some animals have accessory lenses, reflectors, and color filters with the light-producing organ, and the whole complex assembly is like a lantern. Certain shrimp have such complicated light-emitting organs.

The production of light is an enzyme-controlled reaction, the details of which differ in different organisms. Bacteria and fungi produce light continuously if oxygen is available. Most luminescent animals, in contrast, give out flashes of light only when their luminescent organs are stimulated. The name **luciferin** has been given to the material that is oxidized to produce light and **luciferase** to the enzyme that catalyzes the reaction. The luciferins from the crustacean *Cypridina* and from the firefly *Photinus* have been isolated and crystallized and shown to be quite different chemically.

The oxidation of luciferin by luciferase can occur only in the presence of oxygen. It is possible to extract luciferin and luciferase from a firefly, mix the two substances in a test tube with added Mg^{2+} and adenosine triphosphate, and demonstrate the emission of light in the test tube. The energy for the reaction is supplied by the ATP, and under certain conditions the amount of light emitted is proportional to the amount of ATP present. This system can be used to measure the amount of ATP in a tissue extract.

The amount of light produced by certain luminescent animals is amazing. Many fireflies produce as much light, in terms of lumens per square centime-

Anomalops katoptron, a luminescent fish from the waters of the Malay Archipelago. The crescent-shaped luminescent organs below the eyes are equipped with reflectors. (After Steche)

ter, as do modern fluorescent lamps. Different kinds of animals may emit lights of different colors, red, green, yellow, or blue. One of the more spectacular luminescent creatures is the "railroad worm" of Uruguay, the larva of a beetle, that has a row of green lights along each side of its body and a pair of red lights on its head. The light produced by luminescent organisms is entirely in the visible part of the spectrum; no ultraviolet or infrared light is produced. Since very little heat is given off in the process, bioluminescence has been called "cold light."

The advantage an animal derives from the emission of light is not certain in every case. For deep-sea animals, which live in perpetual darkness, light organs might be useful to enable members of a species to recognize one another, to serve as a lure for prey or as a warning for would-be predators. Experiments have shown that the light emitted by fireflies serves as a signal to bring the two sexes together for mating. The light emitted by bacteria and fungi probably serves no useful purpose to the organisms but is simply a by-product of oxidative metabolism, just as heat is a by-product of metabolism in other plants and animals.

Summary

1. A basically similar pattern of cellular metabolism is present in all animals and, indeed, in all forms of life. Although the number of enzymatic reactions of cellular metabolism is large, the number of kinds of reactions is relatively small.

2. The living world is composed of autotrophs that utilize the radiant energy of sunlight to synthesize organic compounds, such as glucose, from CO_2 and H_2O and of heterotrophs that convert glucose and other organic molecules to CO_2 and H_2O, while conserving some of the energy as ATP.

3. Entropy may be defined as a randomized state of energy that is unavailable to do work. The second law of thermodynamics states that "physical and chemical processes proceed in such a way that the entropy of the system becomes maximal." Living organisms and their cells are highly organized and nonrandom; therefore they have little entropy. They preserve this low state of entropy by increasing the entropy of their surroundings.

4. Free energy may be visualized as that component of the total energy of a system that is available to do work under isothermal conditions. Entropy and free energy are inversely related: As entropy increases during an irreversible process, the amount of free energy decreases.

5. In a chemical reaction matter is changed from one substance, with its characteristic properties, to another, with new properties, and energy is released or absorbed. Reactions that proceed with a decrease in free energy, i.e., that release energy to the system and increase its entropy, are said to be exergonic. Endergonic reactions require an input of energy to proceed; entropy is decreased and free energy is increased. In cells endergonic and exergonic reactions may be coupled, with one providing the energy to drive the other.

6. A catalyst is an agent that affects the velocity of a chemical reaction without altering its end point and without being used up in the course of the reaction. Enzymes are protein catalysts produced by cells; they have a high degree of specificity for the reaction(s) they catalyze. The substance acted upon by an enzyme is termed its substrate.

7. Exergonic reactions are coupled to endergonic reactions, usually via the synthesis of ATP from ADP and P_i; this occurs largely in the electron-transmitter system present in the mitochondria. As electrons are passed through the electron-transmitter system from substrate to primary acceptors to cytochromes and finally to oxygen, some of the energy is conserved as ATP. The spatially oriented flow of electrons generates a proton gradient across the inner mitochondrial membrane, and the flow of protons back through the membrane is coupled to the generation of $\sim$P.

8. Enzymes combine with their substrates to form reactive, intermediate enzyme–substrate complexes that decompose to release the free enzyme and the reaction products. The substrate is much more reactive when it is part of an enzyme–substrate complex than when it is free. The enzyme may bind with the substrate at two or more sites on the molecule, holding it in a position that strains its bonds and makes them more likely to break.

9. The activity of an enzyme is affected by the temperature, the pH of the surroundings, and the concentrations of enzyme, substrate, and cofactors.

10. The oxidation of glucose occurs by a sequence of reactions organized into three systems: glycolysis, the Krebs tricarboxylic acid (TCA) cycle, and the electron-transmitter system. The conversion of pyruvate to acetyl CoA is a metabolic one-way bridge between glycolysis and the TCA cycle. Both glucose and fatty acids must be activated by reacting with ATP before they can be metabolized. Glucose is converted to glucose-6-phosphate, and fatty acids are converted to their coenzyme A derivatives.

11. In dehydrogenations electrons and protons are removed from a substrate and transferred to the primary acceptors NAD and FAD. Decarboxylations are reactions in which a carboxyl group is removed as carbon dioxide. Cellular metabolism also includes a large number of "make-ready" reactions in which molecules undergo rearrangements so they can undergo further dehydrogenations and decarboxylations.

12. Four dehydrogenations, two decarboxylations, and five make-ready reactions compose one turn of the TCA cycle. This metabolizes the two-carbon unit of acetyl coenzyme A to carbon dioxide, generating 1 ~P at the substrate level and 11 ~P in the electron-transmitter system.

13. Fatty acids are oxidized and cleaved to two-carbon acetyl coenzyme A units by a cyclic series of four reactions. These include two dehydrogenations, one reaction in which a molecule of water is added across a double bond, and a reaction in which a carbon–carbon bond is split by the addition of coenzyme A.

14. The sequence of reactions that convert glucose to pyruvic acid, with the net production of two ~P, is termed glycolysis. The glycolytic enzymes are not bound to mitochondria but are dissolved in the cytoplasm.

15. The tight coupling of phosphorylation to electron flow in the electron-transmitter system regulates the rate of energy production to the rate of energy utilization.

16. The cells of living things exist in a dynamic state and are constantly building up and breaking down the many different cell constituents. Only the molecules of DNA that constitute the genes are stable and are broken down and resynthesized very slowly, if at all.

17. Each cell in general synthesizes its own complex macromolecules, and each step in the process is catalyzed by a separate enzyme system. These synthetic reactions are strongly endergonic and require ATP to drive them.

References and Selected Readings

Becker, W. M.: *Energy and the Living Cell: An Introduction to Bioenergetics.* New York, Harper & Row, 1977. A paperback giving a brief discussion of bioenergetics.

Carter, L. C.: *Guide to Cellular Energetics.* San Francisco, W. H. Freeman & Co., 1973. A paperback giving a brief discussion of bioenergetics.

Lehninger, A. L.: *Bioenergetics.* 2nd ed. New York, Benjamin, 1971. A fine presentation of energy metabolism in living systems.

Lehninger, A. L.: *Principles of Biochemistry.* New York, Worth, 1982. A standard detailed text of biochemistry; an excellent reference.

Peusner, L.: *Concepts in Bioenergetics.* Englewood Cliffs, Prentice-Hall, 1974. A brief overview of bioenergetics.

Racker, E.: *A New Look at Mechanisms in Bioenergetics.* New York, Academic Press, 1976. A well-written exposition of bioenergetics and the mechanisms involved by one of the major contributors to the field.

Smith, E. L., R. L. Hill, I. R. Lehman, R. J. Lefkowitz, P. Handler, and A. White: *Principles of Biochemistry.* 7th ed. New York, McGraw Hill, 1983. A standard text of biochemistry; an excellent reference.

Stryer, L.: *Biochemistry.* 2nd ed. San Francisco, W. H. Freeman & Co., 1981. A well-illustrated text written in simpler style than other biochemistry texts.

PART TWO The Organism: Animal Form and Function

4 Symmetry, Form, and Life Style

ORIENTATION

Despite the enormous diversity within the Animal Kingdom, most animals share a similar basic body plan. In this chapter we will examine the common general features of animal architecture and explore the ways the design of animals is influenced by their ability to move and by their size.

There is a tremendous diversity of animal life with an almost endless variety of structural and functional modifications. The million or more species described so far are distributed almost everywhere over the earth's surface and are adapted for many different habitats and life styles. Yet through all of this diversity a number of common ground plans can be distinguished. This is true because most animals ultimately share a common evolutionary origin; different groups have diverged at different points along evolutionary lines and thus share many features that developed in earlier ancestral forms. In addition, many unrelated groups have invaded and occupied similar environments and habitats and have thus been subjected to similar selection pressures. Not surprisingly, similar solutions to common problems (convergent evolution) appear frequently in animal evolution.

The following chapters will explore some of the problems of animal existence and some of the solutions that animals have evolved during the some one billion years they have occupied this planet. We will largely be concerned with multicellular animals **(Metazoans)** (Gr. *meta,* after + *zoon,* animals), but from time to time reference will be made to the unicellular animal-like protozoans (Gr. *protos,* first + *zoon,* animals). This chapter will serve to introduce the general architecture of animals and its relationship to animal life styles.

BOX 4.1

The plane dividing the body into two identical halves passes from anterior to posterior and from dorsal to ventral; it is called the **sagittal** plane (or section). Any plane lateral to the midline is said to be **parasagittal.** A **cross,** or **transverse,** plane passes between lateral surfaces from dorsal to ventral. The **frontal** plane passes from anterior to posterior but between lateral surfaces and is perpendicular to the other two planes. "Proximal" and "distal" are also commonly used anatomical terms of reference. **Proximal** means close to the base or to some point of reference; **distal** means away from the base or from some point of reference.

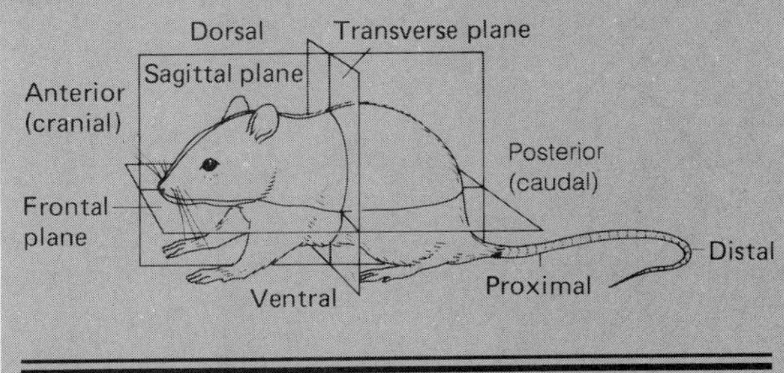

Motility

Certainly the most distinguishing feature of animals is their ability to move. Some forms live attached to the substratum, but such attached species have evolved from motile ancestors, typically have motile larval stages, and can still move parts of the body.

Most animals are bilaterally symmetrical, a type of symmetry in which the body can be divided through only one plane to produce two mirror image halves (Box 4.1). **Bilateral symmetry** is a natural consequence of animal motility. The front, or **anterior,** end of the body is directed forward and meets the environment first. It is thus different from the rear, or **posterior,** end. The **ventral** body surface facing the ground, or bottom **(substratum),** is different from the opposite, or **dorsal,** surface directed upward. The two **lateral** body surfaces contact the environment in much the same way and are similar.

Correlated with motility and bilateral symmetry is the development of a head, termed **cephalization** (Gr. *kephale,* head). Sense organs and associated nervous tissues tend to be more highly concentrated at the forward-moving anterior end, and the mouth is also most frequently located in this region of the body.

Most attached, or **sessile,** organisms are radially symmetrical or exhibit a tendency toward **radial symmetry.** Radial symmetry is an arrangement of similar parts around a central body axis, as in a pie, wheel, or column (Fig. 4.1). Such a symmetry is adaptive for a sessile life style, for the organism can meet its environment equally from all directions. Plants (which are usually attached to the substratum) typically exhibit a radial body plan (think of the perfect "Christmas tree").

The ability of most animals to move from place to place has made possible the evolution of an enormous range of adaptations. Those involving diets, feeding modes, and locomotion are only some of the most conspicuous. Sessile animals, on the other hand, are greatly limited in their life styles. They must depend upon food that floats or swims close by or must create a water current to bring in food. Sessile animals also cannot seek mates. Most rely upon the chance meeting of sperm and eggs shed freely into the water or upon sperm brought to the body in water currents.

Many zoologists have concluded that the first multicellular animals probably were radial even though they were motile. They probably moved through the water

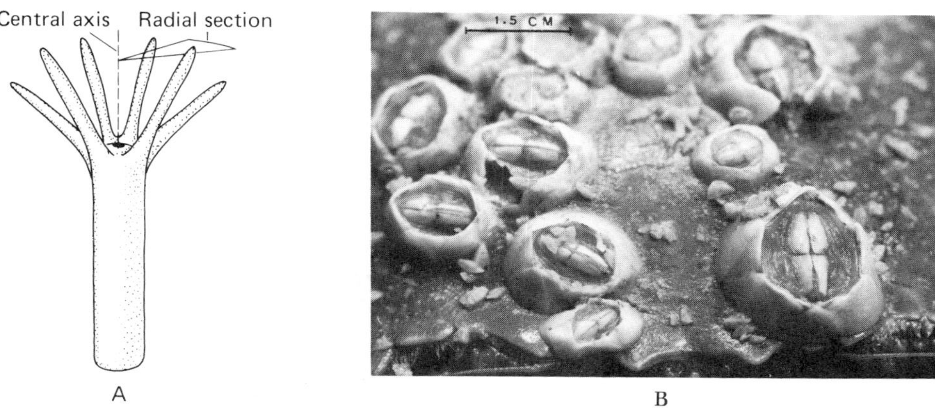

FIGURE 4.1 Radial symmetry. *A*, Hydra, an animal with primary radial symmetry. *B*, Sessile barnacles, bilaterial animals with some secondary radial features. (By Betty M. Barnes.)

BOX 4.2

Some radially symmetrical animals are not perfectly so but are actually biradial. The body can be cut into two planes that will produce identical sections or halves. The slitlike mouth and pharynx imposes a biradial symmetry on the sea anemone illustrated in this figure. If the sea anemone were bilaterally symmetrical, it could be cut through only one plane to produce two identical halves. If it were perfectly radial, a number of pie-shaped sections could be cut from around the central axis.

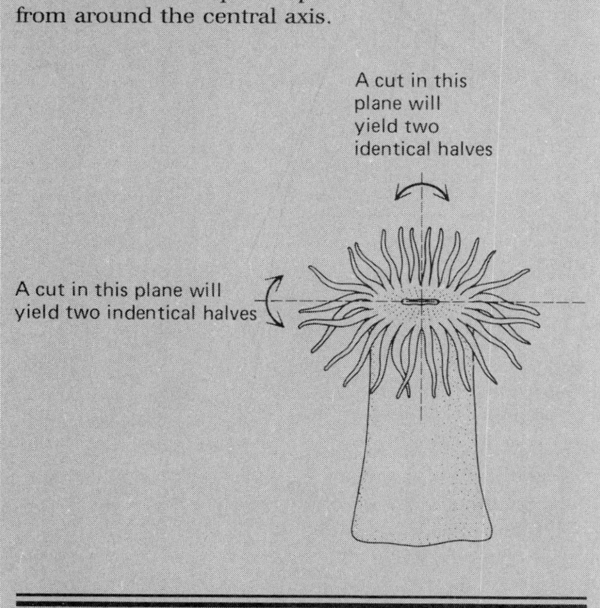

with one pole forward (Fig. 4.2*A*) and would thus have possessed anterior and posterior ends but no dorsal and ventral surfaces. From such ancestors the sessile and radial hydras, sea anemones, and corals may have evolved. If so, their radial symmetry would be *primary*, i.e., would be like that of their immediate ancestors. The sea stars and sea urchins living today are radial but motile (Fig. 4.3). Moreover, their radial symmetry appears to be *secondary;* i.e., the ancestors of these animals were bilateral and free-swimming, for the free-swimming larvae of sea stars and sea urchins are bilateral (Fig. 4.3). In the evolutionary history of the group, some ancestral form perhaps became attached, and the more adaptive radial symmetry evolved (Fig. 4.4). Later echinoderms, such as sea stars and sea urchins, became motile again but retained the radially symmetrical body plan.*

Aside from echinoderms, most attached animals with bilateral ancestors show only a slight tendency toward radial symmetry. Barnacles, for example, are basically bilateral, but the outer protective plates are arranged in a ring around the body (Fig. 4.1*B*).

If the first animals were radial even though motile, what led to the evolution of bilaterality? Perhaps some early population of the ancestral radial form took up an existence close to the bottom, creeping or swimming

*The concepts of primary and secondary origins is not limited to symmetry. They may be applied to any structures or conditions. For example, the anal opening has disappeared in some animals, and the absence of an anus in such animals is said to be secondary. The absence of an anus in hydras and flatworms, however, is said to be primary because we believe they evolved from ancestors that never possessed an anus.

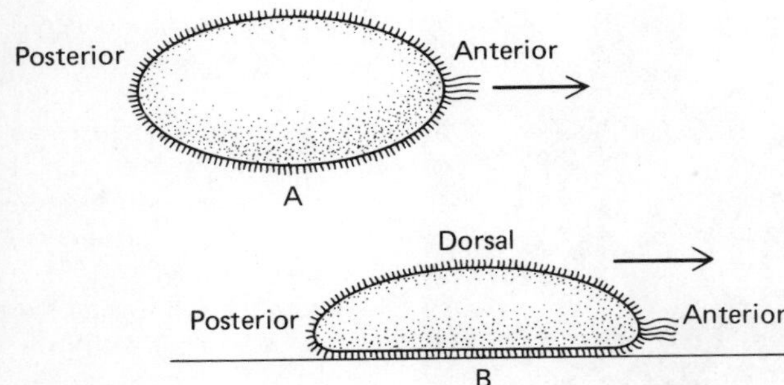

FIGURE 4.2 Hypothetical, ancestral, multicellular animal. *A,* Free-swimming, radially symmetrical ancestor. *B,* Creeping, bilaterally symmetrical ancestor.

Attached, radially symmetrical echinoderms. Many fossil species and living sea lilies

Motile, radially symmetrical echinoderms

Hypothetical, motile, bilaterally symmetrical echinoderm ancestor

Gut

Mouth

Anus

Sea star

Sea urchin

Mouth

Anus

Arms

Free-swimming bilateral larva

Larva attached and developing radial symmetry

Radially symmetrical young sea star

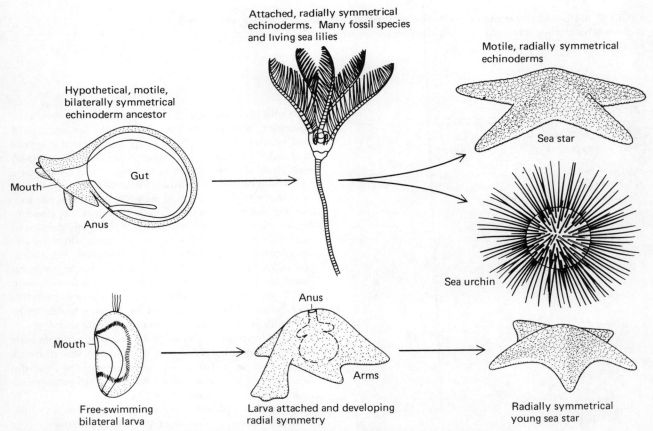

FIGURE 4.3 Evolution of symmetry in echinoderms. The ancestors of echinoderms were probably free-moving, bilaterally symmetrical animals. Attachment led to a shift to the more adaptive radial symmetry. Most early fossil echinoderms and the living sea lilies are attached and radial. Many modern echinoderms, such as sea stars and sea urchins, have become detached and motile but have retained the secondary radial symmetry of their attached ancestors. This sequence is paralleled in the embryonic development of many living sea stars. The free-swimming larva that develops from the egg is bilaterally symmetrical. It settles to the bottom, becomes attached, and then changes to the radially symmetrical adult form. On completion of the transformation, the little sea star becomes free of its attachment and crawls away.

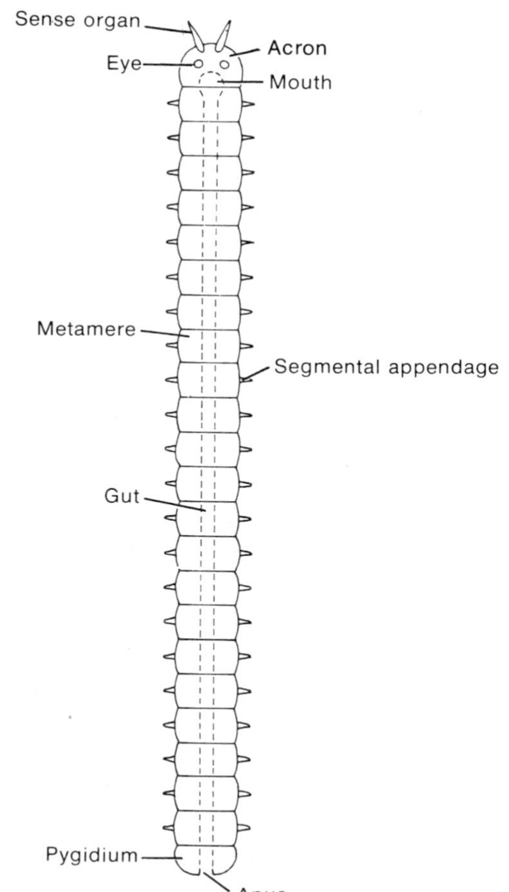

FIGURE 4.4 Diagram of a metameric animal, such as an annelid worm. In this animal the acron composes most of the head, but in many metameric animals a varying number of anterior segments combine with the acron to form the head.

over the surface (Fig. 4.2*B*). They would already have been elongate with an anterior and posterior end. With the assumption of a bottom habit, one surface became directed toward the bottom. Ventral and dorsal sides could then have differentiated, and the animal would have become bilateral.

Animal Architecture

Most bilateral animals have a relatively uniform basic design. The mouth is typically located at the anterior end, and the gut tube runs through the body, parallel to the anterior–posterior axis.

The animal's outer casing, the body wall, is covered on the exterior by the **integument,** or skin. The integument is composed of one or more layers of epithelial cells that may also form such specialized structures as hairs, spines, feathers, and nails. In a number of animal groups, such as crabs and clams, all or part of the integumentary epithelium secretes an outer, protective, nonliving shell or a cuticular covering.

The integument is the principal barrier between the external environment and the inner environment of the animal. In addition to an obviously important role in protection, it may also be involved in various regulatory functions that will be discussed in later chapters.

Beneath the integument the body wall is composed of one to several muscle layers. An outer layer of circular muscles and an inner layer of longitudinal muscle, as in earthworms, is a common arrangement, but other arrangements occur. The vertebrate body wall is composed of distinct muscles that are oriented in various ways in different regions of the body (Fig. 4.5).

In many animals there is a fluid-filled space, or **body cavity,** located between the body wall and the central gut tube and other internal organs. The body cavity serves different functions in different animals. The fluid may function as a hydraulic skeleton, a medium for internal transport, a reservoir for waste deposition, or a site for sperm and egg maturation, and the space may facilitate growth and movement of internal organs.

Body cavities are defined on the basis of the way they form in embryonic development. The most widespread type of body cavity is a **coelom** (Gr. *koiloma,* hollow cavity) that arises as a space within embryonic mesoderm (Fig. 14.8) (see also page 315). The cavity comes to be lined by a layer of epithelial cells, called **peritoneum** (Gr. *peri,* around + *teinein,* stretch). Various internal organs may bulge part way into the coelom, pushing the peritoneum ahead. Such organs thus always lie behind the peritoneum and are said to be **retroperitoneal** (Gr. *retro,* behind). Other organs may push so far into the coelom that they are suspended by a fold of peritoneum that is then called a **mesentery** (Gr. *mesos,* middle + *enteron,* gut).

Many animals possess only a small body cavity or lack one altogether. Those having such a solid type of body structure include insects and crabs, which are believed to have lost the body cavity in the course of their evolutionary history, and flatworms, which may never have possessed one.

The structural design of many animals involves the division of the body into a linear series of similar parts, or *segments* (Fig. 4.4). The segmental repetition of parts, called **metamerism** (Gr. *meta,* after + *meros,* part), includes both internal and external structures. The anterior part of the body bearing the sense organs and brain,

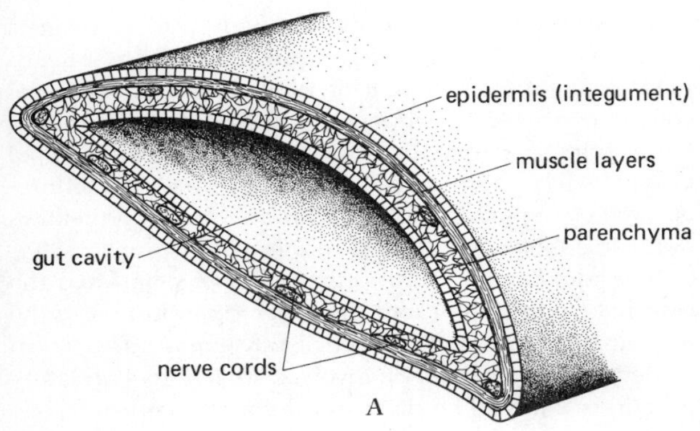

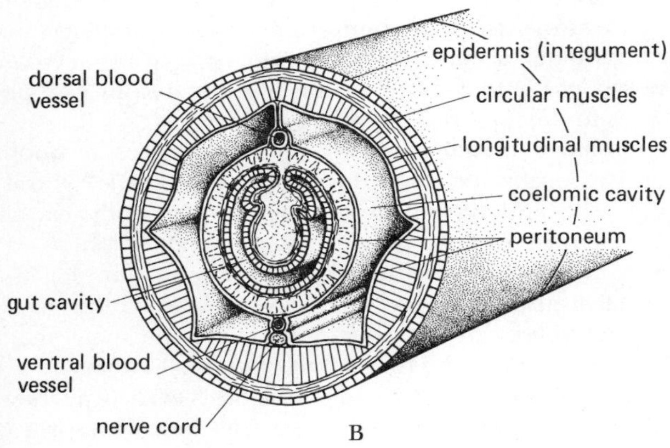

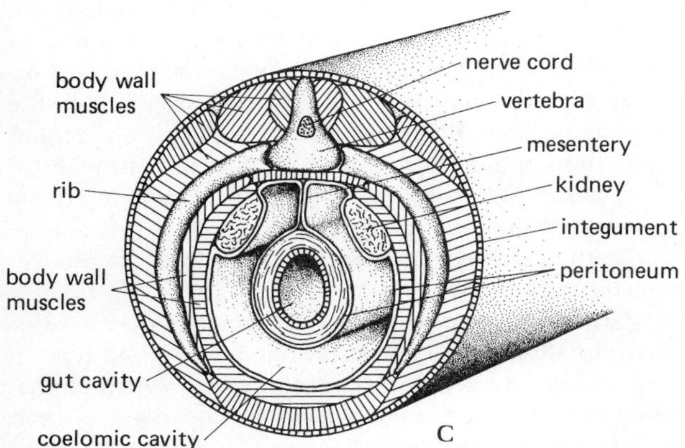

FIGURE 4.5 Three-dimensional cross sections through the middle of the body of three bilateral metazoans. *A,* Flatworm, which has a solid type of body structure. *B,* Earthworm and *C,* vertebrate, both of which have a coelomate type of structure. A fluid-filled coelom, or body cavity, fills the interior of the body.

the **acron** (Gr. *akron,* top), is not a segment, nor is the terminal section, the **pygidium** (Gr. *pyge,* rump + *idiom,* narrow), which bears the anus. New segments are formed at the rear, in front of the pygidium. Metamerism is believed to have evolved twice in the Animal Kingdom, once in the ancestors of the segmented worms (annelids) and arthropods and once in the ancestors of the vertebrates. In both it is believed to have represented an adaptation for locomotion (p. 120).

In all groups of metameric animals, a varying number of anterior trunk segments join the acron, forming a more complex "head." In vertebrates, the acron is vestigial, and the "head" is actually composed of a large number of anterior segments.

Size

The size of multicellular animals varies over a wide range. At one extreme are microscopic forms, such as rotifers and certain flatworms; at the other are animals of vast bulk, such as elephants and whales. Many structural and functional features of animals correlate with their size. As size increases, *volume or weight increases by the cube, whereas surface area increases by the square of the linear dimension* (Fig. 4.6). This means that a significantly larger or smaller size of a particular animal may not be functional. For example, a sparrow the size of a vulture would not have enough wing surface area to carry the increased body weight. A microscopic roundworm that lives between soil particles and lacks gills or internal transport system would have insufficient surface area for gas exchange if it were the size of an earthworm. *Therefore change in size usually demands some change in design.* The design change may be a disproportionate increase in the size of different parts of the body. For example, the cross-sectional area of the legs of elephants must be disproportionately greater than that of other mammals in order to hold up the elephant's vast bulk. Alternatively, design change may involve the evolution of new structures to compensate for altered functional demands. Gills, for example, make their appearance when the general body surface area of aquatic animals is inadequate for the exchange of gases with the internal volume of tissue.

There is a common misconception that small animals are always primitive and large animals are specialized. Small size, or miniaturization, has been an important part of the adaptive evolution of a number of animal groups, such as mites, bryozoans, hummingbirds, and mice. The large ratio of surface to volume of very small animals may also pose problems. They are

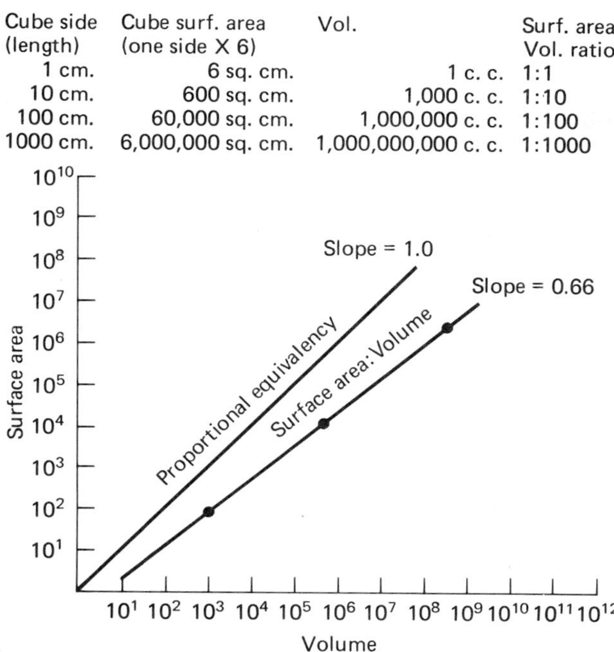

Cube side (length)	Cube surf. area (one side X 6)	Vol.	Surf. area: Vol. ratio
1 cm.	6 sq. cm.	1 c. c.	1:1
10 cm.	600 sq. cm.	1,000 c. c.	1:10
100 cm.	60,000 sq. cm.	1,000,000 c. c.	1:100
1000 cm.	6,000,000 sq. cm.	1,000,000,000 c. c.	1:1000

FIGURE 4.6 Relationship of surface area to volume in four cubes of increasing size. When plotted logarithmically the slope is 0.66, as opposed to a slope of 1.0 if their increase were equivalent.

subject to much greater heat and water loss than are large animals.

The world of large animals is dominated by gravitational forces. Falling is a much greater danger than sticking. For very small animals surface adhesive and frictional forces are the controlling ones, and gravitational forces are relatively insignificant.

Colonial Organization

Most animals are solitary, i.e., individuals of a given species live independently of each other except for sexual pairing and sometimes the rearing of young. Other animals cannot exist except in close association; such forms are said to be **colonial.** In many colonial animals, including the familiar corals, the individuals within the colony are connected anatomically (Fig. 4.7). All individuals arise by budding except the first, which develops from the fertilized egg. In the colonies of ants, bees, termites, baboons, and humans, the individuals are not anatomically connected but live together and are functionally or behaviorally interdependent (Fig. 4.7). Such groups of animals are usually said to be *social* colonies.

Polymorphism (Gr. *polys,* multi + *morphe,* form), or the specialization of forms for the division of labor, characterizes many colonial species. Individuals have become specialized for functions such as feeding, defense, or reproduction. The worker and soldier castes of ants are familiar examples of polymorphism in social insects (Fig. 27.38), but structurally united colonies may also exhibit relatively complex polymorphism. Not all colonial animals are polymorphic, but polymorphism occurs only in species with colonial organization. Sexual dimorphism, which refers to various differences of size, color, and structure between males and females of a given species, is not polymorphism. Many noncolonial animals are sexually dimorphic.

Predictability in Animal Design

The sea is the ancestral environment of animals. From the sea there have been numerous invasions to fresh water and land. Of these three major environments the sea is in many ways the most uniform and least stressful. Oxygen is usually adequate, there is no danger of desiccation (drying up), and the salt content of the sea is somewhat similar to that of extracellular fluids.

The salt content of fresh water is much less than that of tissue fluids. Pools, ponds, and even streams may periodically dry up, streams fluctuate in velocity and turbidity, and the oxygen content may be very low in swamps and at the bottom of some lakes. On land oxygen is plentiful, but desiccation is an ever present danger. Each of these three environments poses different problems of existence, and a variety of structural and functional adaptations have evolved in the animals inhabiting them.

Many structural and functional adaptive themes appear over and over again in animal design, for they are often related to size, degree of motility, or environment. A knowledge of these relationships can have great predictive value. Thus if the size of an animal, whether it is sessile or sedentary, and something of its life style are known, it is possible to predict, or make educated guesses, about many aspects of its structure and physiology. The following chapters will provide a basis of understanding for making many such predictions.

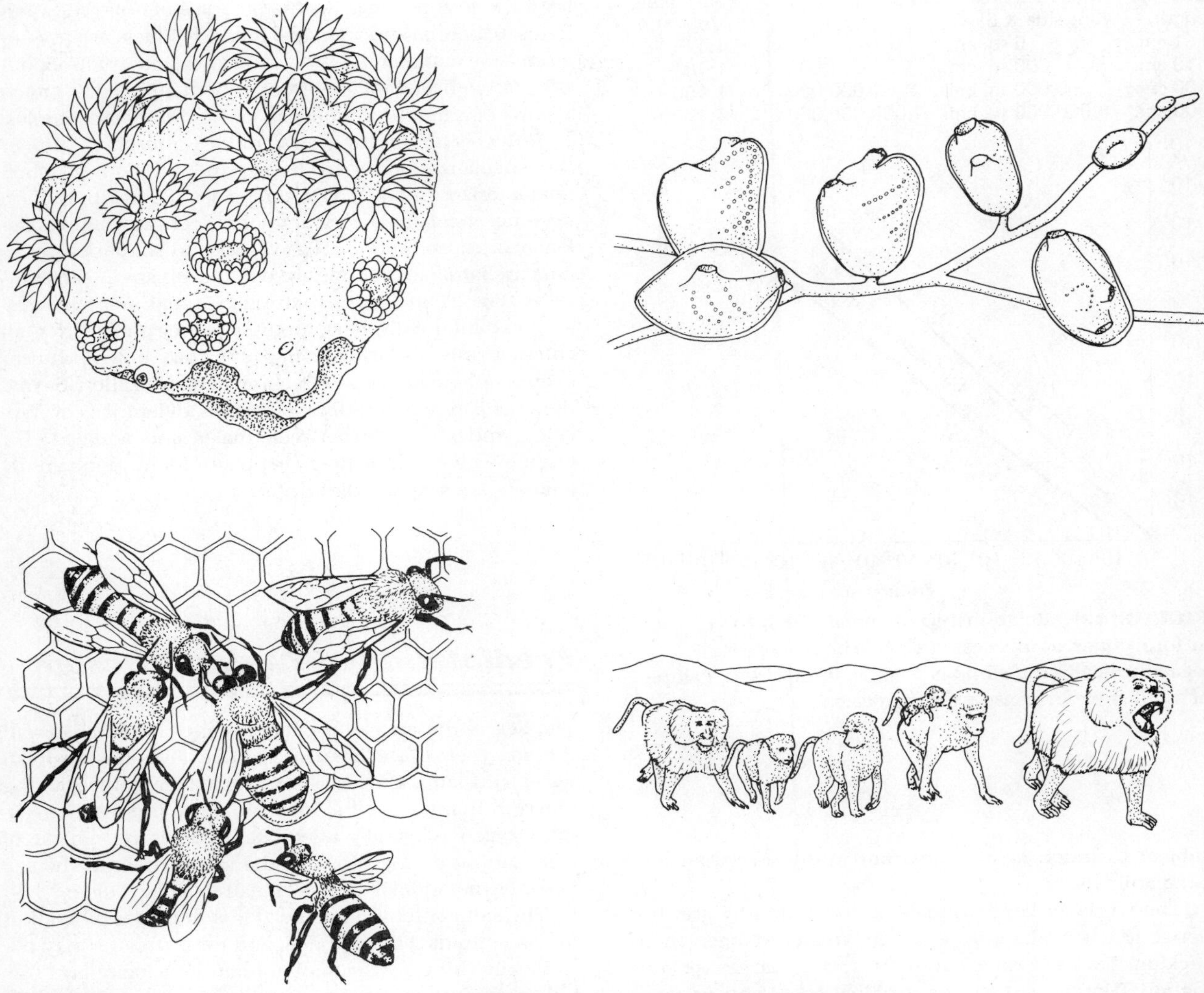

FIGURE 4.7 Four examples of colonial animals. The individuals of the coral and sea squirt colonies are attached together. Individuals of the bee and baboon colonies live together and are dependent upon each other through their social organization. (*C* and *D* based on a drawing by S. Landry in Wilson, E. O.: *The Insect Societies.* Cambridge, Massachusetts, Harvard University Press, 1971 and in Wilson, E. O.: *Sociobiology.* Cambridge, Massachusetts, Harvard University Press, 1975.)

Summary

1. Most animals are bilaterally symmetrical; the body can be cut along only one plane to produce two identical halves. Most animals also exhibit some degree of cephalization, the concentration of sense organs and nervous tissue at the anterior end of the body. Bilateral symmetry and cephalization are correlated with animal motility, whereby one end of the animal's body meets the environment first.

2. Sessile animals are radially symmetrical or exhibit some tendency toward radial symmetry; the body is composed of similar parts arranged around a central axis. Radial symmetry is of advantage in sessility, for the animal can meet the environment equally from all directions.

3. Most animals possess a body design in which there is an outer casing, called the body wall, composed of the skin, or integument, and one or more layers of muscle. A gut tube generally extends through the middle of the body from the mouth near the anterior end and the anus toward the posterior end.

4. In many animals a fluid-filled space, called a body cavity, occupies much of the interior of the body and partially or completely surrounds many internal organs. The most common type of body cavity is a coelom, which is lined by peritoneum. Body cavities serve a variety of functions depending upon the group of animals in which they occur.

5. Three large groups of animals—annelids, arthropods, and vertebrates—exhibit a segmental organization, called metamerism, in which the body is divided into a linear series of similar parts (segments). Metamerism probably evolved as a locomotor adaptation.

6. The structural evolution of animals has been greatly influenced by the limitations of size, for as animals become larger or smaller the volume increases or decreases by the cube, but the surface area increases or decreases by the square of the linear dimension. Thus the ratio of surface area to volume changes as the size of the animal changes.

7. Most animals are solitary in that they are not dependent upon other individuals of the same species except for purposes of reproduction. Some, however, exhibit colonial organization, in which the individuals of a colony are structurally united together (e.g., corals) or live in behaviorally dependent associations, called social colonies (e.g., bees and ants). The colonies of many species are composed of structurally and functionally different members, a condition known as polymorphism.

References and Selected Readings

The following general references contain a great deal of information on the anatomy and physiology of organ systems.

Alexander, R. M.: *Size and Shape. Studies in Biology.* No. 20. London, Edward Arnold, 1971. A brief account of some of the relationships between size and shape and modes of existence in organisms.

Alexander, R. M.: *The Chordates.* Cambridge, Cambridge University Press, 1975. An excellent analysis in functional terms of the structure of selected chordates.

Alexander, R. M.: *The Invertebrates.* Cambridge, Cambridge University Press, 1979. A somewhat quantitative approach to the study of invertebrates.

Barnes, R. D.: *Invertebrate Zoology.* 4th ed. Philadelphia, W. B. Saunders Company, 1980. A textbook and reference on the anatomy, physiology, ecology, and classification of invertebrates.

Bloom, W., and D. W. Fawcett: *A Textbook of Histology.* 10th ed. Philadelphia, W. B. Saunders Co., 1975. The microscopic and ultrastructure of the cells and tissues of mammalian organ systems are thoroughly described.

Fretter, V., and Graham, A.: *A Functional Anatomy of Invertebrates.* New York, Academic Press, 1976. A review of major invertebrate groups with an emphasis upon the function of their major anatomical features.

Guyton, A. C.: *Textbook of Medical Physiology.* 6th ed. Philadelphia, W. B. Saunders Co., 1981. A detailed consideration of mammalian physiology is presented in this standard textbook.

Kluge, A. G.: *Chordate Structure and Function.* 2nd ed. New York, Macmillan Publishing Co., 1977. A comparative anatomy of chordates, with an introduction to their evolution and development.

Prosser, C. L. (Ed.): *Comparative Animal Physiology,* 3rd ed. Philadelphia, W. B. Saunders Co., 1973. A very valuable source book on the physiology of both invertebrates and vertebrates.

Romer, A. S. and T. S. Parsons: *The Shorter Version of the Vertebrate Body.* 5th ed. Philadelphia, W. B. Saunders Co., 1978. The morphological aspects of the evolution of vertebrates are thoroughly considered in this widely used textbook of comparative anatomy.

Schmidt-Nielsen, K.: *Animal Physiology,* 2nd ed. Cambridge, Cambridge University Press, 1978. A physiological account of the interrelationships between animals and their environments, with emphasis upon oxygen, food, temperature, water, movement, and integration.

Young, J. Z., and M. J. Hobbs: *The Life of Mammals.* 2nd ed. New York, Oxford University Press, 1975. Structure and function are carefully considered in this thorough account of the gross and microscopic anatomy and the physiology of mammals.

5 Body Covering, Support, and Movement

ORIENTATION

Since the internal environment of an animal is never identical to its external environment, a body covering (integument: skin, hair, scales, feathers, horns) must provide some degree of protection yet permit exchanges between external and internal environments. Animals, especially large ones and terrestrial ones, also need some type of support (principally an exoskeleton or an endoskeleton). Finally, animals must move materials within their bodies, and either must move themselves or move fluids and materials over their surfaces, to obtain food and other materials and to reproduce. In this section we will examine the three principal kinds of animal movement: amoeboid, ciliary, and muscle. Mechanisms of protection, support, and movement often are interrelated and will be explored in this chapter.

Body Covering

Functions of the Body Covering

Protozoa and animals have a body covering separating their internal environment from the outside world. This may be as simple as the plasma membrane of some single-celled protozoans or as complex as the skin or **integument** (L., *integumentum,* a covering) of vertebrates. Its complexity is related to the degree of chemical and physical difference between internal and external environments and to the size and mode of life of the animal. A well-developed integument protects against abrasion, bacterial penetration, ultraviolet radiation, and other assaults of the external world. In meeting these needs, the body covering may become hard enough to serve as the main support or skeleton of the animal, e.g., the shell of a snail or the exoskeleton of a crayfish. Although the body covering protects the organism, it does not isolate it from the external world. It has a regulatory role in many animals, and exchanges of gases, ions, and water take place across it. It also contains many receptor cells that help the animal detect changes in its external environment. Glands, scales, claws, feathers, and hair develop from the integument and serve a variety of functions. Pigment cells come to lie within it. In birds and mammals, the skin helps to control the loss of body heat and is important in thermoregulation. The body covering is truly a "Jack-of-all-trades."

The Integument of Invertebrates

The integument of the metazoan invertebrates (Fig. 5.1) usually consists of a single layer of columnar epithelial cells known as the **epidermis** (Gr., *epi,* upon + *derma,* skin) that rests upon a basement membrane. When the epidermis is exposed to the surface, many of its cells may be ciliated, and others are glandular. In some cases the glands secrete an overlaying, noncellular cuticle or shell, in which case the epidermal layer is nonciliated and usually is called a **hypodermis.** Cuticles and shells are discussed further in connection with the skeleton.

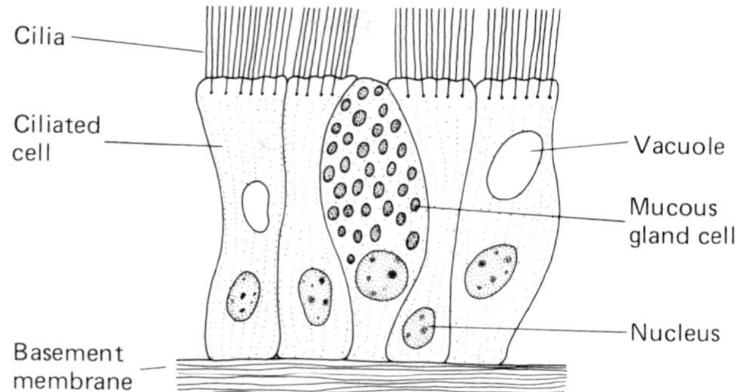

FIGURE 5.1 The integument of free-living flatworms and many other invertebrates is a simple epidermis consisting of a single layer of columnar epithelial cells resting on a basement membrane.

The Vertebrate Integument

The integument of all vertebrates consists of an **epidermis,** of stratified squamous epithelium, overlaying a **dermis** composed of a thick layer of dense connective tissue (Fig. 5.2) made up mostly of collagen fibers. When tanned, we know this as leather. The basal cells of the epidermis, the **stratum germinativum,** are columnar in shape and divide mitotically to form new cells, many of which move to the surface where they finally slough off. As they shift outward, they change in shape, becoming cuboidal and eventually flattened or squamous. In some vertebrates, the epithelial cells synthesize and accumulate a horny, water-insoluble protein, **keratin** (Gr., *keras,* horn), and die in the process. The dermis contains blood vessels and sense organs that may come close to the surface in dermal papillae and ridges that protrude into the base of the epidermis. Bone may form in the dermis, and glands may bulge into it from the epidermis. Often fat is deposited in the deeper part of the dermis and just beneath it.

The epidermis of fishes is relatively thin and contains no keratin. Mucous cells and multicellular glands within it cover the surface with mucus that reduces friction and prevents exchanges of water with the environment and the attachment of many ectoparasites. The skin often forms overlapping folds, and **bony scales** develop in the dermis of these folds (Fig. 5.3). The superficial parts of the scales may become exposed as the epidermis wears off. Spines are modified scales, and poison glands are associated with them in some species.

Amphibian skin is similar to fish skin except for the absence of bony scales and the synthesis of some keratin. In addition to mucous glands, the skin of some amphibians contains poison glands that secrete distasteful and sometimes highly toxic materials (see Fig. 32.10). Certain South American Indians use the toxin of frog skin (curare) to poison their arrow tips.

As vertebrates adapted to terrestrial conditions, the amount of keratin in the epidermis increased, and the outer layers of cells formed a well-defined horny layer, the **stratum corneum** (L., *corneus,* horny), that greatly reduces the loss of body water (Fig. 5.2). This layer is particularly thick in reptiles and forms **horny scales**

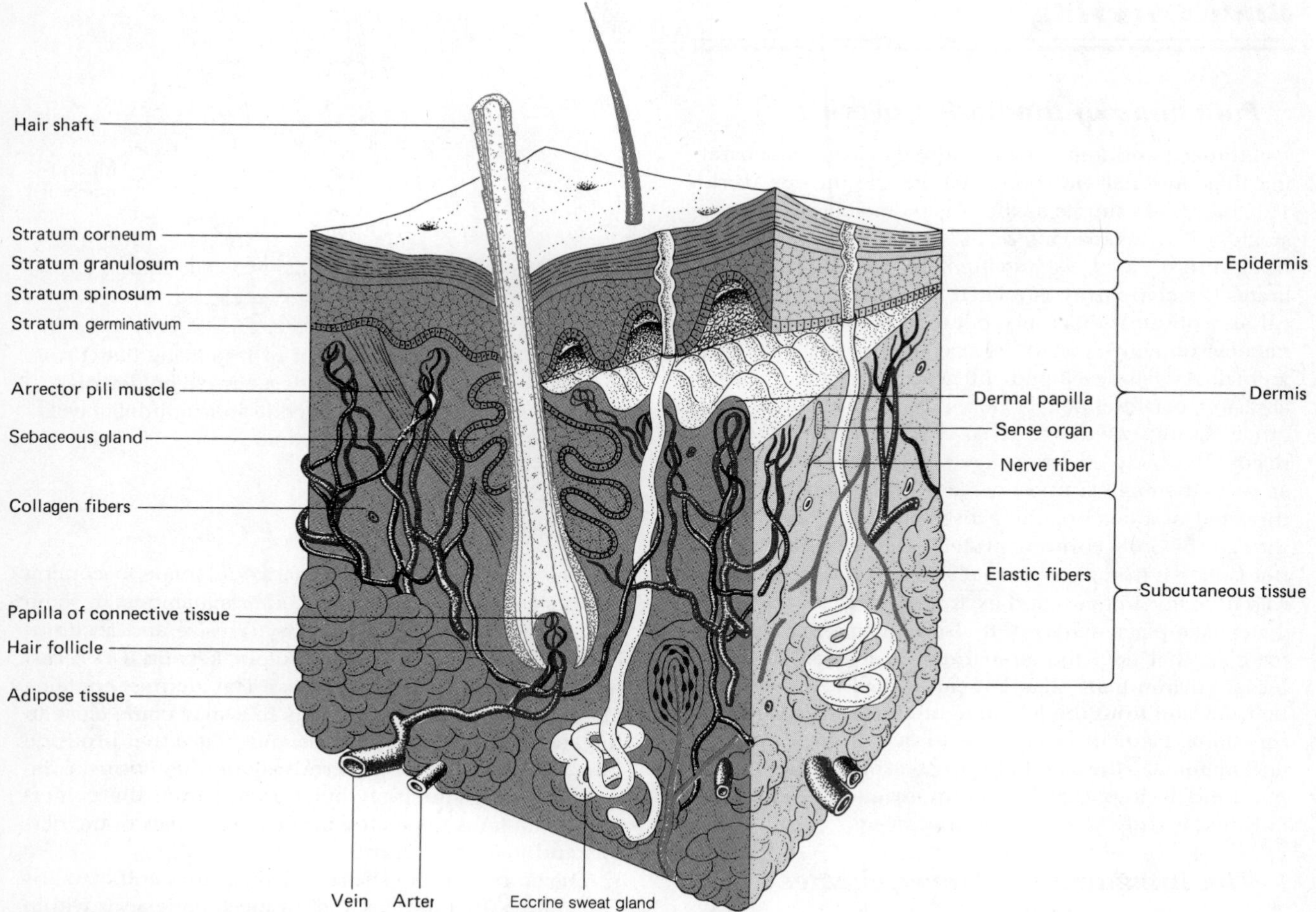

FIGURE 5.2 Diagrammatic vertical section through human skin. (Modified from Pillsbury, Shelley, and Kligman: *Manual of Cutaneous Medicine*. Philadelphia, W. B. Saunders Co., 1961.)

and **plates** (Fig. 5.3). In the shell of turtles, bone formed in the dermis underlies the plates. The **feathers** of birds are essentially elongated and frayed horny scales. Mammalian **hair** is a different type of epidermal derivative, one that emerges from a follicle of epithelial cells that has grown down into the dermis (Fig. 5.2). Feathers and hair entrap a layer of still air and provide insulation against heat loss.

Glands are virtually absent from the dry, horny skin of reptiles. Contrary to popular opinion, snakes are not the least bit slimy. Only a few **scent glands** producing odoriferous secretions used in sexual recognition remain. Bird skin too is nearly aglandular, but many glands have reevolved in mammalian skin. All develop as ingrowths from the epidermis into the dermis (Fig. 5.2). **Sebaceous glands** (L., *sebum,* tallow) are sac-shaped structures whose cells break down and release an oily secretion into the hair follicles. They are particularly abundant in our scalp. **Eccrine sweat glands** are coiled, tubular glands that secrete a watery solution onto the surface, and the evaporation of this solution helps cool the body. Similar **apocrine sweat glands,** found in our arm pits and pubic area, produce a more odoriferous secretion. In many mammals these secretions are chemical signals or pheromones (p. 777) involved in courtship, marking territories, and so forth.

Skin Coloration

Pigment within the skin, the degree of vascularity and, in some animals, the presence of refractive gran-

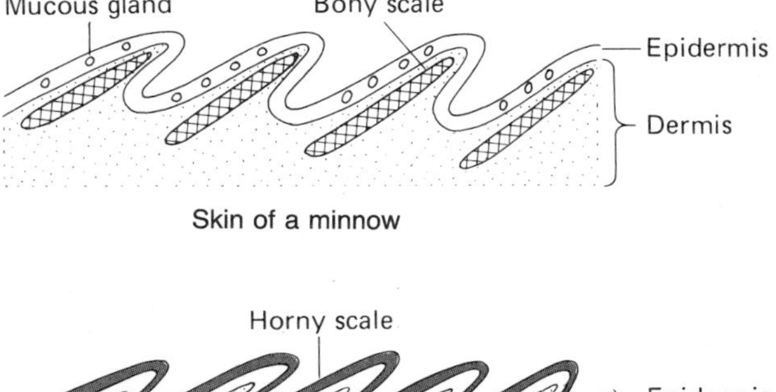

FIGURE 5.3 Vertical sections through the skin of a bony fish and a reptile to show the differences in bony and horny scales.

ules or surface striations interact to produce brilliant coloration. Some color patterns conceal the animal from predators or enable an animal to lie unseen as it awaits its prey, but others are used for species recognition or as warning signals to frighten another animal or establish territory. Squids, shrimp, crabs, and lower vertebrates, such as the frog, may have several kinds of pigment contained in stellate cells known as **chromatophores** (Gr., *khrōma,* color + *phoros,* bearing). Changes in general color tone are effected by the migration of pigment within these cells. The skin of a frog darkens when **melanin** (Gr., *melas,* black) pigment streams into the processes of the cells; it lightens when the pigment concentrates near the center of the cell. Pigment migration is controlled by nerves in some vertebrates and by hormones in others (p. 273). In mammals, melanin is present both within and between basal cells of the epidermis. Some melanin is present in the skin of all human beings, except albinos, but it is especially abundant in the skin of blacks.

Support

Functions of Skeletons

Most animals have some sort of skeleton, within or on the outside of their bodies, that maintains body shape and provides support. Support is particularly important in large organisms and those that live out of water. Some aquatic animals, especially small ones, can survive without a skeleton because water provides a great deal of buoyancy. Only a few terrestrial animals, certain worms and slugs, lack a skeleton; air is much less dense than water and affords little support.

Support is not the only skeletal function, for a skeleton may be present in small animals in which support is not needed. Fish only a few millimeters long and insects less than 1 mm. in length have skeletons. Protection of the organism or certain of its internal parts frequently is an important skeletal function. Sessile animals in particular need protection, for they cannot retreat to shelter if the need arises. Skeletons may also store calcium, phosphorus, and other important minerals, provide a site for the manufacture of blood cells, and transmit forces from muscles to the point of force application. Skeletal parts are used in many movements, including biting and locomotion.

Skeletal Materials and Types

Most skeletons are composed of hard parts and are relatively rigid except at joints. Calcium compounds are the most common component of mineral skeletons and are present in groups ranging from protozoans to vertebrates. Calcium is present as calcium carbonate in most invertebrate skeletons but as calcium phosphate in vertebrate skeletons. Salts of magnesium, strontium, and other substances are sometimes used. Siliceous compounds form skeletons in some protozoans and sponges. Skeletons may be composed of specialized organic compounds, such as chitin or collagen, or complexes of organic and inorganic materials. The latter are particularly strong, for the inorganic crystals resist compression, and the organic component resists tension and provides elasticity.

In most invertebrates the hard materials are secreted on the outside of the body by the underlying integument; hence they form **exoskeletons.** In contrast, vertebrates (together with some invertebrates) have **endoskeletons** located within the body wall or deeper body tissues. Not all skeletons are composed of hard materials. A fluid **hydroskeleton,** in the body cavity, blood vessels, or other body spaces, can provide support or induce movement. Such fluid skeletons are very important in the support and movement of worms and many other animals.

A few representative skeletons are considered here to illustrate their variety, advantages, and limitations; others will be described in the survey of the Animal Kingdom.

Some Exoskeletons. Many amoebae living in fresh water and damp soil have a protective shell that partly covers their bodies. Cytoplasmic processes, **pseudo-**

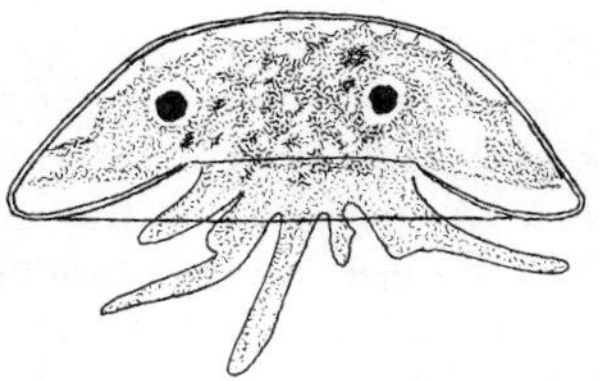

FIGURE 5.4 Lateral view of *Arcella vulgaris*, one of the fresh water, shelled amoebae. (From R. D. Barnes, after Deflandre.)

podia (Gr., *pseudes,* false + *podos,* foot), protrude through an opening at one end. *Arcella* (Fig. 5.4) has a chitinoid shell that is secreted by the cytoplasm. Some other species secrete a cementing matrix in which minute sand grains or other soil particles become embedded.

Stony corals are made up of thousands of interconnected individuals, **polyps,** that secrete a calcium carbonate skeleton beneath themselves and around their sides, so that the lower part of the polyp typically is located in a skeletal cup (Fig. 5.5). Radiating calcareous septa in the bottom of the cups project up into folds at the base of the polyps. As polyps grow and multiply, new material is secreted and added to material already present. Some species of coral are low and encrusting; others are upright and branching. Coral reefs are composed largely of the exoskeletons of innumerable generations of stony corals.

Chitons, snails, clams, and nearly all other mollusks are covered by a shell secreted by a modified body wall known as the **mantle.** Molluscan shells have evolved into many shapes: a single spiral in snails, bivalved in clams, and chambered in the nautilus. The shell is composed of layers of calcium carbonate crystals deposited within an organic framework. **Conchiolin,** a horny protein, covers the outer surface of the shell, and its inner surface is very smooth and often pearllike.

FIGURE 5.5 Skeleton of a scleratinian, or stony, coral. The cups in which the polyps are located are widely separated. (By Betty M. Barnes.)

Arthropods, including crabs, insects, and spiders, have an exoskeleton, or **cuticle** (L., *cuticula,* diminutive of *cutis,* skin), composed of the polysaccharide **chitin** and protein bound together to form a complex glycoprotein. Calcium carbonate is incorporated into the exoskeletons of crabs and other crustaceans. The exoskeleton covers the entire body surface and extends inward to line much of the digestive tract, as well as the air-transporting tubes (tracheae) of insects.

The exoskeleton is secreted by the underlying **hypodermis** (Fig. 5.6*A*). It is composed of a thin, surface **epicuticle** and a much thicker **procuticle,** that is subdivided into an **exocuticle** and **endocuticle.** The epicuticle, especially of terrestrial species, contains waxes; when these are absent, the exoskeleton is relatively permeable to water and gases. The exocuticle is stronger than other parts, for it has been tanned; its molecular structure has been stabilized by reactions with phenols and by the formation of additional cross linkages in the protein molecules. Sensory processes are borne on the exoskeleton, and it frequently is perforated by a few pores through which secretions are discharged.

The exoskeleton of the arthropod is divided into plates on the thorax and abdomen and into a series of strong, tubelike segments on the appendages. These are connected to each other by thinner, folded **articular membranes** where movement occurs (Fig. 5.6*B*). An exoskeleton provides excellent protection, good attachment for the muscles that lie inside of it, and good support if the animal is small, as is true of most arthopods. A small animal has a large surface area relative to its mass (p. 92), so a surface skeleton is very effective. As size increases, the surface area decreases relative to the mass to be supported, so a surface skeleton would have to become progressively thicker and thicker to provide adequate support.

A major disadvantage of the arthropod exoskeleton is its restriction upon growth. This has been solved by periodically shedding the old skeleton. During the process of moulting, or **ecdysis** (Gr., *ekdysai,* to strip), the hypodermis detaches from the exoskeleton and secretes enzymes—chitinase and proteinases—into a **molting fluid** that accumulates beneath the skeleton (Fig. 5.7). The untanned endocuticle is gradually digested, and a new skeleton is secreted beneath the old one. Waxes and other lipids in its epicuticle prevent it from being digested. Eventually, the old exoskeleton ruptures along lines where the exocuticle is very thin, and the animal pulls out of its old encasement. The new skeleton is soft

Epicuticle
Exocuticle
Endocuticle
Procuticle
Hypodermis
Tracheole
Muscle
Muscle
Tonofibrils (insertion fibrils)
A
Articular membrane
B

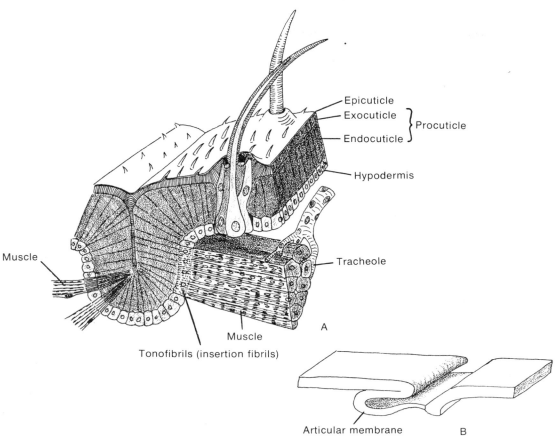

FIGURE 5.6 *A,* Three dimensional section of an arthropod integument (insect). *B,* An intersegmental joint. Articular membrane is folded beneath segmental plate. (*A* after Weber from Kaestner; *B* after Weber from Vandel.)

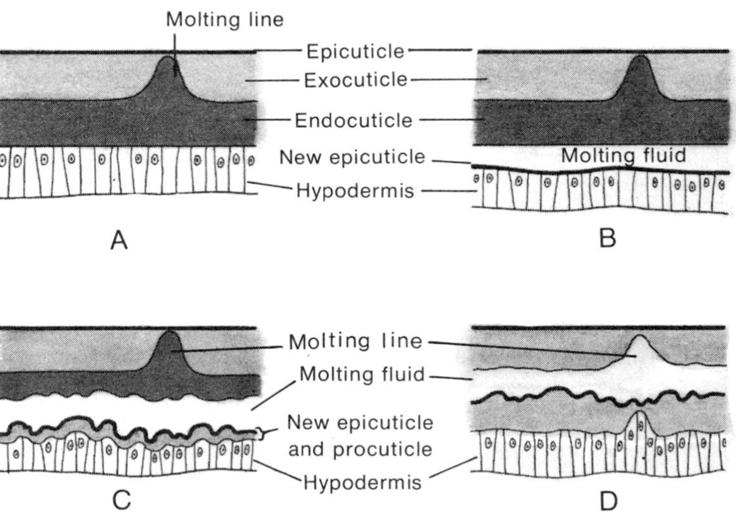

FIGURE 5.7 Molting in an arthropod. *A,* The fully formed exoskeleton and underlying hypodermis between molts. *B,* Separation of the hypodermis and secretion of molting fluid and the new epicuticle. *C,* Digestion of the old endocuticle and secretion of the new procuticle. *D,* The animal just before molting, encased within both new and old skeleton.

and pliable for a few hours or days and is stretched to accommodate the increased size of the animal. Animals that have recently molted are very vulnerable to predators, for they are soft, and their weak skeletons do not transmit muscle forces well.

Some Endoskeletons. An endoskeleton does not protect the entire organism as effectively as an exoskeleton, but it transmits muscle forces well, and the animal need not undergo molting in order to grow. Although not limited to large animals, endoskeletons are particularly effective for support in large animals such as vertebrates. A series of small internal cylinders and rods can provide the support needed with a minimum of weight.

Endoskeletons are found in many different phyla. Most radiolarians, marine protozoa closely allied to amoebas, have very delicate internal skeletons composed of elaborate silica lattices (Fig. 5.8). Radiolarians are relatively large planktonic organisms; some reach several millimeters in diameter. Their skeletons provide a large area with minimal weight and help to keep these organisms afloat.

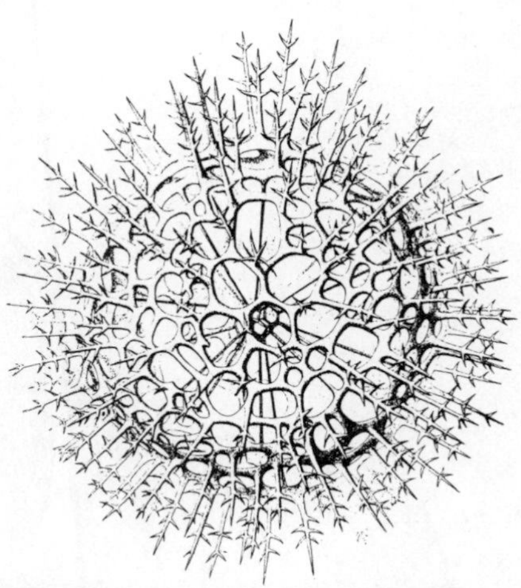

FIGURE 5.8 The internal, siliceous skeleton of a radiolarian. (Courtesy of E. Giltsch, Jena.)

Sponges are supported and their shape maintained by internal needle-like mineral spicules or a network of amoeboid organic spongin fibers, or both, that are secreted by wandering cells called **sclerocytes.** Depending upon the class of sponge, the spicules are composed of calcium carbonate or of silicon dioxide. It is the spongin skeleton that we use in natural bath sponges.

The body walls of starfish, sea urchins, and most other echinoderms are supported by small internal calcareous pieces, **ossicles,** that are interconnected in a lattice-like array in the wall of starfish but form a nearly solid internal shell, or **test,** in sea urchins. The test is perforated by many small openings for various organs. Spines develop beneath the epidermis and attach to many ossicles. They are long and mobile in sea urchins and serve as organs of locomotion.

The skeleton of fishes, frogs, cats, and other vertebrates is an endoskeleton composed of many individual bones and cartilages joined to each other to provide a strong but moveable framework. Some parts—for example, the bony scales of fishes and the plates of a skull—may lie in a very superficial position, yet they develop within or just beneath the skin.

Hydroskeletons. Hydroskeletons may occur in conjunction with other skeletons, or they may provide the primary skeletal support. The coelomic fluid of an earthworm is such a hydroskeleton. Contraction of circular or longitudinal muscles in the body wall increases the pressure of the fluid forcing the body to elongate or widen (p. 120).

Blood flowing into the foot of a clam acts as a hydroskeleton and pushes the foot into the sand or mud. Jumping spiders have only flexor muscles that can bend their legs. The forceful extension of the legs during a jump comes from the rush of body fluids into the leg.

Amphioxus, one of the lower chordates, has a **notochord** (Gr., *nōtos,* back + L., *chorda,* string), a longitudinal cord of cells filled with liquid; its turgidity is maintained by a firm, surrounding connective tissue sheath. The notochord is a hydroskeleton and is analogous to a rubber tube filled with water under pressure. It can bend yet is firm enough to provide support and resist shortening of the body. Contraction of the segmented, longitudinal muscles on either side of the body wall results, therefore, in side to side undulatory movements that sweep posteriorly along the body. All chordates, including vertebrates, have a notochord as larvae or embryos. It is replaced in most adult vertebrates by the vertebral column.

The Vertebrate Skeleton

Parts of the Skeleton. During embryonic development most of the vertebrate skeleton is composed of cartilage, but the cartilage is replaced by bone in most adults. This bone is called **cartilage replacement bone** to distinguish it from the **dermal bone** that develops within or just beneath the skin without any cartilaginous precursor. These two types of bone differ only in their mode of development; they are similar histologically (p. 56).

The skeleton can be divided into a **somatic skeleton** (Gr., *soma,* body), located in the body wall and appendages, and a deeper **visceral skeleton** (L., *viscera,* bowels), associated primitively with the pharyngeal wall and gills. The somatic skeleton contains both dermal and cartilage replacement bones, but only cartilage replacement bones are in the visceral skeleton. The somatic skeleton can be further subdivided into an **axial skeleton** (vertebral column, ribs, sternum, and most of the skull) and an **appendicular skeleton** (girdles and limb bones).

The Fish Skeleton. The parts of the skeleton can be seen more clearly in a dogfish (a small shark) than in terrestrial vertebrates (Fig. 5.9). The vertebral column is composed of vertebrae, each of which has a biconcave **centrum** that develops around and largely replaces the embryonic notochord. Dorsal to each centrum is a **vertebral arch** surrounding the **vertebral canal** in which the spinal cord lies. Short **ribs** attach to the vertebrae, a sternum is absent, and the individual vertebrae are rather loosely held together. Strong vertebral support is not required in the aquatic environment.

The skull of the dogfish is an odd-shaped box of cartilage, the **chondrocranium** (Gr., *khondros,* cartilage + *kranion,* brain case), encasing the brain, nose, eye, and inner ear. It forms the core of the skull of all vertebrates. Other components of a vertebrate skull include the anterior arches of the visceral skeleton and dermal bones that encase the chondrocranium and anterior visceral arches. These dermal bones have been lost during the evolution of cartilaginous fishes but were present in the fishes ancestral to terrestrial vertebrates, or tetrapods.

The visceral skeleton consists of seven pairs of >-shaped visceral arches hinged at the apex of the >. They are interconnected ventrally but are free dorsally. Each arch lies in the wall of the pharynx and supports gills in very primitive vertebrates. In jawed vertebrates the first or **mandibular arch** becomes enlarged and, together with associated dermal bones, forms the upper and lower jaws. The entire jaw in the dogfish is derived from the mandibular arch; there are no surrounding dermal bones. The second or **hyoid arch** of the dogfish helps to support the jaws. Its dorsal portion, known as the **hyomandibular,** extends as a prop from the **otic capsule** (Gr., *ōtikos,* ear), the part of the chondrocranium housing the inner ear, to the angle of the jaw. The third to seventh visceral arches, the **branchial arches** (Gr., *brankia,* gills), support the gills; gill slits lie between them.

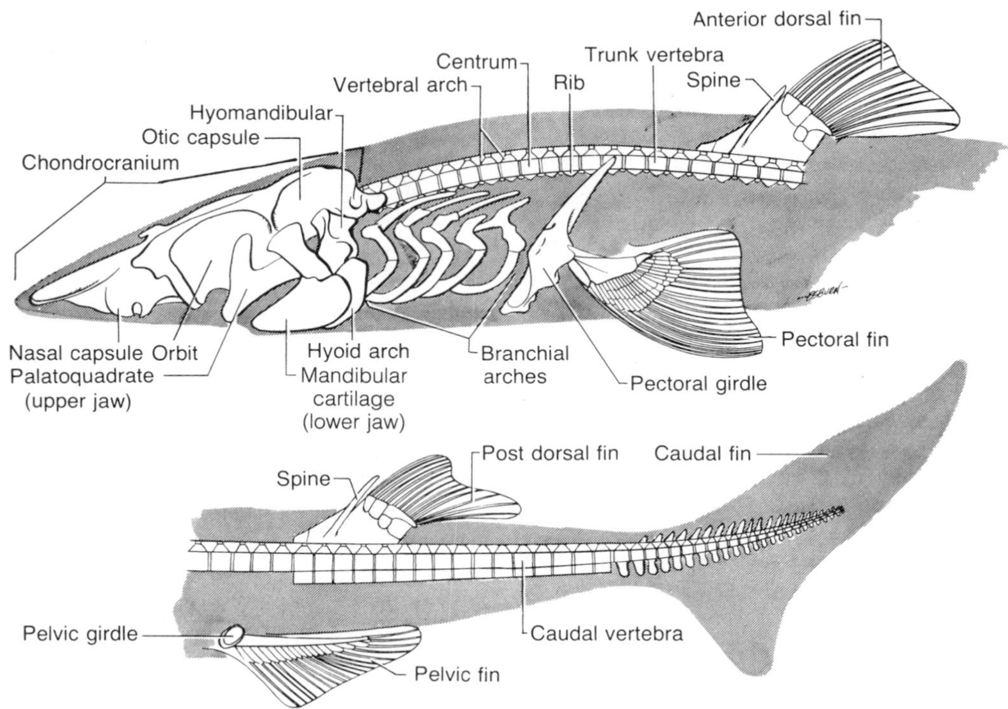

FIGURE 5.9 A lateral view of the skeleton of a dogfish.

The appendicular skeleton is very simple in the dogfish. A U-shaped bar of cartilage, the **pectoral girdle,** in the body wall posterior to the gill region supports the **pectoral fins.** The **pelvic girdle** is a transverse bar of cartilage in the ventral body wall anterior to the termination of the digestive tract. It supports the **pelvic fins** but is not connected with the vertebral column.

The Tetrapod Skeleton. The numerous changes that have occurred in the evolution of the skeleton from primitive fishes are correlated with the shift from water to land and with the greater activity of the tetrapod. The human skeleton, although adapted to our upright posture, illustrates the main features (Figs. 5.10 and 5.11). The vertebral column in all tetrapods is thoroughly ossified, and the individual vertebrae are strongly united by overlapping **articular processes** borne on the vertebral arches. A well-developed **vertebral spine,** extending dorsally from the arch, and lateral **transverse processes** serve for the attachment of ligaments and muscles. Ribs articulate on the centrum and transverse processes.

Correlated with changes in the method of locomotion and the independent movement of various parts of the body, there is more regional differentiation of the vertebral column.

Mammals have seven neck or **cervical vertebrae,** of which the first two, the **atlas** (Gr. mythology, *Atlas,* a god who supported the heavens upon his shoulders) and **axis,** are modified to permit extensive movement of the head. The skull rocks up and down at the joint between the skull and atlas; turning movements occur between the atlas and axis. The vertebral column of the mammalian trunk is differentiated into thoracic and lumbar regions. We have 12 **thoracic** and 6 **lumbar vertebrae.** Only the thoracic vertebrae bear distinct **ribs,** most of which connect via the costal cartilages with the ventral breast bone, or **sternum.** Rudimentary ribs, present in the other regions during embryonic development, fuse onto the transverse processes. Five **sacral vertebrae,** which are fused to form the **sacrum** (L., *sacer,* sacred, because this region was used in sacrifices), transfer weight from the trunk to the pelvic girdle and hind legs. We have no external tail, but four **caudal vertebrae** remain. They are fused together to form an internal **coccyx** (Gr., *kukkux,* bone shaped like a cuckoo's bill) to which certain anal muscles attach. The basic pattern of vertebrae is much the same in other terrestrial vertebrates, although there are variations in the number of vertebrae and in their regional differentiation, depending upon methods of locomotion and support.

The skull (Figs. 5.10 and Box 5.1) can be divided into a portion housing the brain and encasing the inner ear, the **cranium,** and the bones forming the jaws and encasing the eyes and nose, the **facial skeleton.** The

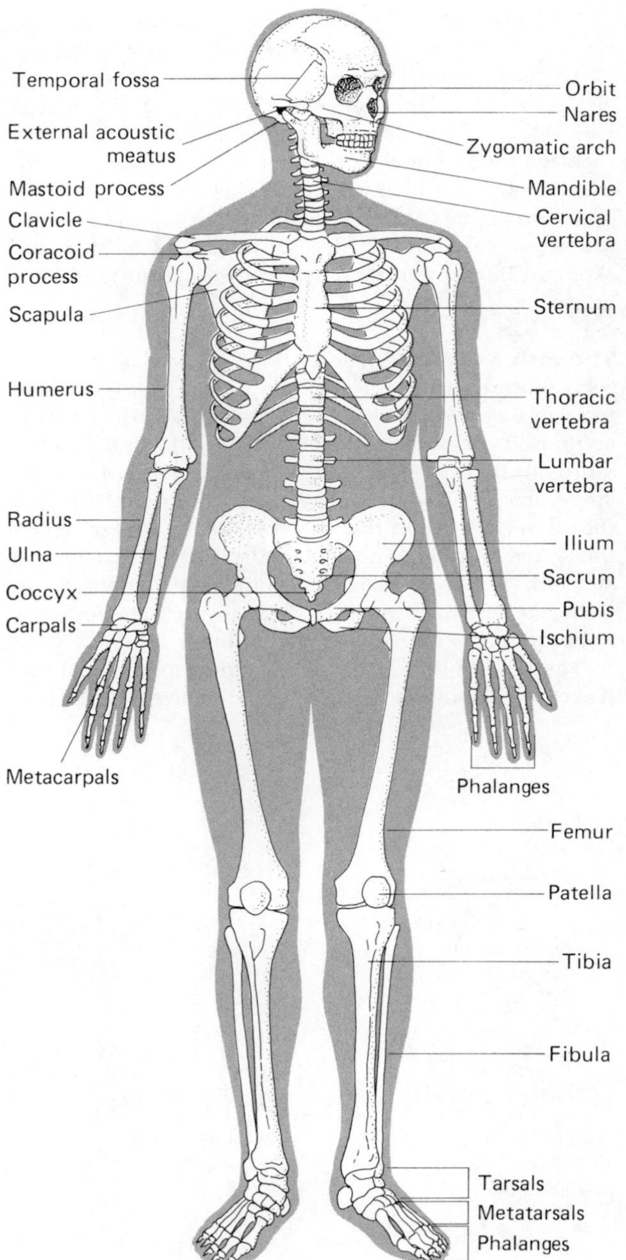

FIGURE 5.10 A view of the human skeleton.

mammalian brain and cranium are very large. The eyes are lodged in sockets known as **orbits.** A **temporal fossa** lies posterior to each orbit in mammals and contains certain powerful jaw muscles. It is bounded laterally and ventrally by a handle-like bar of bone, the **zygomatic arch** (Gr., *zugōma,* yoke), from which other

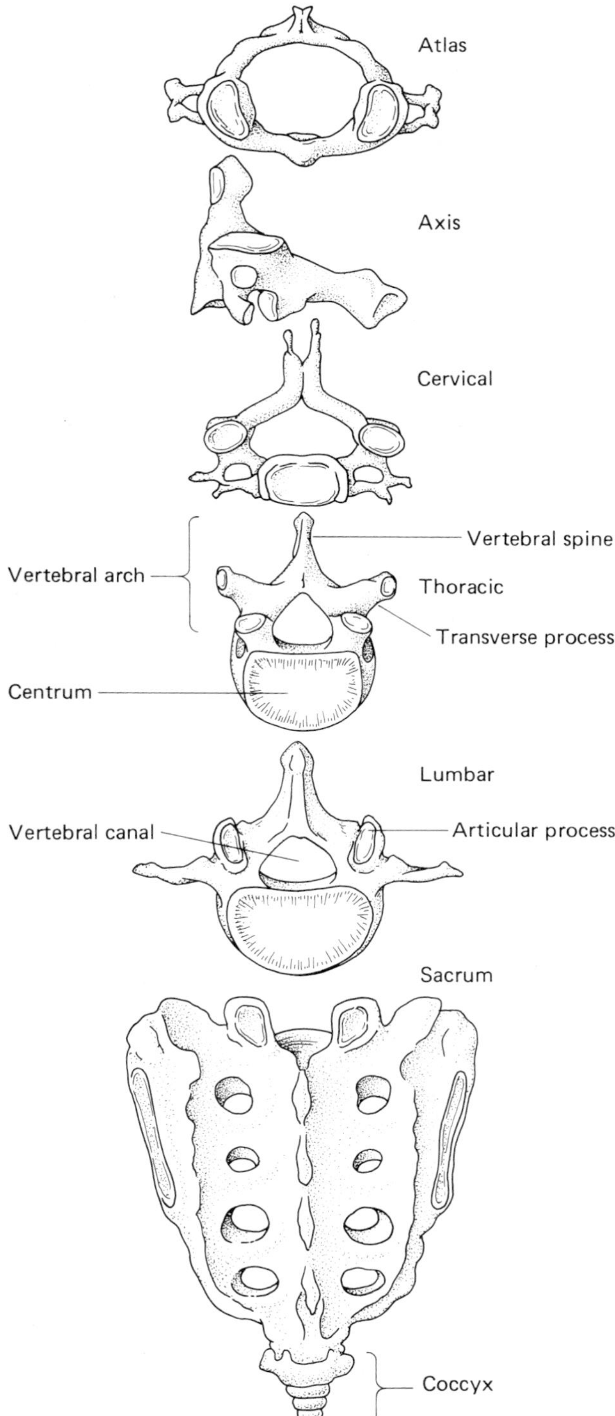

FIGURE 5.11 Parts and types of mammalian vertebrae as represented by those of a human being. The axis is seen in lateral view; the sacrum and coccyx in dorsal view; the others are viewed from the anterior end.

muscles extend to the lower jaw. External nostrils, or **nares,** lead to the paired nasal cavities, and internal nostrils, or **choanae,** lead from these cavities. The choanae of primitive terrestrial vertebrates open into the roof of the front part of the mouth cavity, but in mammals a bony **hard palate** extends the nasal cavities more posteriorly so the choanae open into the pharynx. This permits a mammal to chew and manipulate food in the mouth while continuing to breathe; breathing need be stopped only momentarily when the food is swallowed. A warm-blooded mammal needs more food and oxygen than a cold-blooded frog, since the mammal must be able to sustain a high rate of metabolism. An **external acoustic meatus** leads to the eardrum, which is sunk beneath the body surface in mammals. There are many foramina for blood vessels and nerves; most conspicuous is the large **foramen magnum,** through which the spinal cord enters the skull. Many of the individual bones forming the skull of the human are shown in Box 5.1.

Although the appendicular skeleton of the dogfish is quite different from that of terrestrial vertebrates, there is a close resemblance between the appendicular skeleton of the terrestrial vertebrates and their ancestral crossopterygian fishes (Fig. 5.12). The humerus of the arm and the femur of the leg (Fig. 5.12) represent the single proximal bone of the crossopterygian fin; the **radius** and **ulna,** or **tibia** and **fibula,** correspond to the next two bones. The **carpals** or **tarsals, metatarsals** or **metacarpals,** and **phalanges** of the hand or foot are structurally comparable to the more peripheral elements of the crossopterygian fin.

The girdles of tetrapods need to be stronger than those of fish because body weight must be transferred through them to the limbs and ground. The pectoral girdle is bound onto the body by muscles, but the pelvic girdle extends dorsally and is firmly attached to the vertebral column. A **pubis, ischium,** and **ilium** are present on each side of the pelvic girdle, though all have fused together in the adult. Our pectoral girdle includes a **scapula,** a **coracoid process** that is a distinct bone in most lower tetrapods, and a **clavicle.** The clavicle is the only remnant of a series of dermal bones that are primitively associated with the girdle. All other girdle bones are cartilage replacement bones.

Movement

Most movement in animals depends upon changes in cell shape (amoeboid movement), the beating of cytoplasmic processes (cilia and flagella), or the contraction of muscle cells. Amoeboid and ciliary action are effective for individual cells and small animals, and ciliary action

BOX 5.1

As the brain grew larger during the course of vertebrate evolution, the cartilage replacement bones of the chondrocranium no longer completely encased it. They form a ring of bone around the foramen magnum (the **occipital bone**), encase the inner ear (part of the **temporal bone**), and form the floor of the cranium. The sides and roof of the cranium are completed by dermal bones, such as the **frontal** and **parietals,** and by a portion of the mandibular arch known as the **alisphenoid.** The last is a cartilage replacement bone.

The jaws of most vertebrates are formed of dermal bones that encase the cartilaginous mandibular arch, but the jaw joint lies between the posterior ends of the rods forming the arch, that is, between the dorsal palatoquadrate and ventral mandibular cartilage. Usually a **quadrate bone** ossifies from the posterior end of the palatoquadrate and an **articular bone** from the end of the mandibular cartilage. In most tetrapods the hyomandibular has been transformed into the **stapes,** (L., *stapes,* stirrup), a slender rod of bone extending from the eardrum to the part of the skull housing the inner ear, but it still retains a connection with the quadrate. During the evolution of mammals the jaw mechanism became stronger and the bite more powerful. A new joint evolved between the dentary and squamosal, both dermal bones, and the original jaw joint, located just posterior to the new one, became redundant. The articular and quadrate, which had become quite small, were overspread by the eardrum and incorporated in the middle ear as the outer two auditory ossicles, **malleus** (L., *malleus,* hammer) and **incus** (L., *incus,* anvil), respectively. The incus of mammals, like the quadrate from which it evolved, connects with the stapes. Mammals have three auditory ossicles that form a pressure-amplifying leverage system (Chapter 11).

The ventral part of the hyoid arch, together with the remains of the third visceral arch, forms the **hyoid bone** (a sling for the support of the tongue) and the **styloid process** of the skull, to which the hyoid is connected by a ligament. With the loss of gills in tetrapods, the remaining visceral arches have become greatly reduced, but parts of them form the cartilages of the larynx.

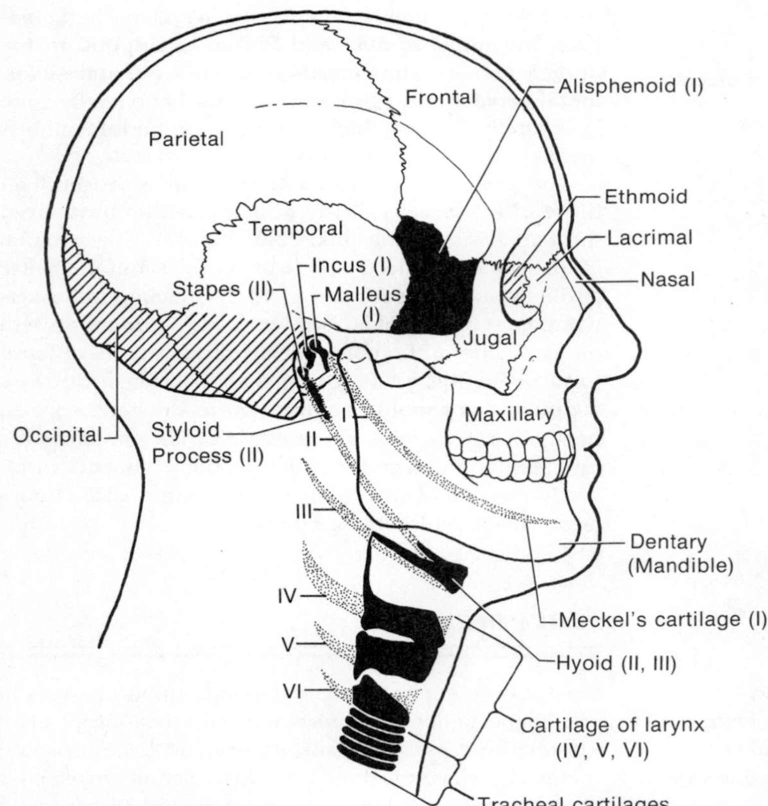

Components of the human skull. Dermal bones have been left plain, chondrocranial derivatives are hatched, those parts of the embryonic visceral skeleton that disappear are stippled, and parts of the visceral skeleton that persist are shown in black. Roman numerals refer to visceral arches and their derivatives. (Modified after Neal and Rand.)

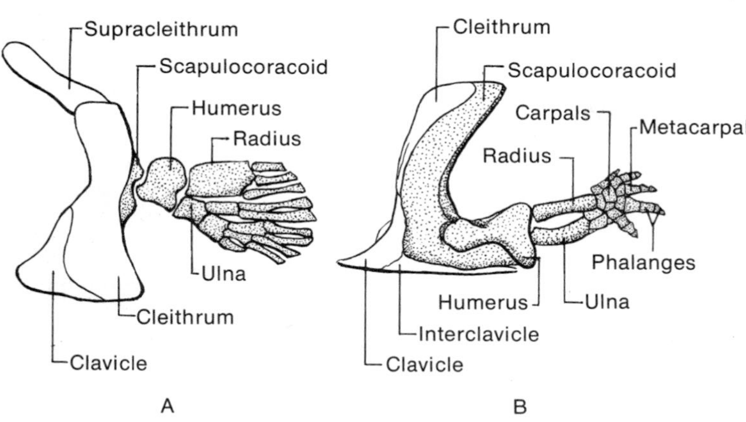

FIGURE 5.12 Lateral views of the appendicular skeleton of a crossopterygian *(A)* and ancestral amphibian *(B)* to show the changes that occurred in the transition from water to land. Dermal bones have been left plain; cartilage replacement bones are stippled. (*A* modified after Gregory; *B* after Romer.)

sometimes plays a role in the movement of larger animals, as in moving a sheet of mucus on mammalian respiratory epithelium. However, most movement in large animals and the internal transport of food, gases, blood, and other materials results from the contraction of muscles.

Muscle

Muscle is a biocontractile system in which cells, or parts of them, are elongated and specialized to develop tension along their axis of elongation. Contractile stalks are found in some protozoans, and relatively unspecialized contractile cells occur in sponges around the opening through which water is discharged from the central cavity. As animals and their movements become more complex, simple contractile cells are replaced by well-differentiated muscular tissue. Simple contractile cells persist in certain higher animals; the myoepithelial cells that aid in the discharge of secretions from mammalian sweat and mammary glands are an example. But most vertebrate muscle cells are part of highly organized smooth, cardiac, and skeletal muscular tissues (p. 60). The elongated cells, or fibers, of these tissues contain contractile fibrils, and the contractile elements of the last two types are so arranged that the muscle fibers have a cross-striated appearance when seen under a light microscope. Smooth and striated muscle fibers are also found among invertebrates. As in vertebrates, invertebrate striated fibers occur where brief, rapid actions are required. Rapidly swimming jellyfish have bands of striated fibers, and nearly all of the musculature of arthropods is striated.

In most animals, groups of fibers are bound together and invested by connective tissue through which nerves and blood vessels run. However, some of the very small muscles of invertebrates consist of only a few muscle fibers, with little connective tissue between them. Collagen, a major constituent of vertebrate connective tissue, is absent in arthropods.

Muscles usually attach to skeletal elements that are moved as the muscles shorten and lengthen. In arthropods, muscles attach directly to the exoskeleton (Fig. 5.6 *A*), or to inward extensions of the skeleton, but in vertebrates the attachments to bones are by the investing connective tissue fibers, some of which become continuous with the connective tissue **periosteum** (Gr., *peri,* around + *osteon,* bone) covering bones and some of which penetrate the bones. The connective tissue between muscle and bone may form a cordlike **tendon** (Fig. 5.13 *A*).

Muscles perform work only by contracting and must work against antagonistic forces. The adductor muscle that holds the shell of a clam closed works against the force of an elastic spring ligament in the shell hinge. When the muscle relaxes, the ligament opens the valves of the shell and stretches the muscle to its original length. Most muscles are arranged in antagonistic sets, such as the biceps and triceps of the arm (Fig. 5.13 *A*). Contraction of the biceps causes a decrease of the angle between upper arm and forearm, an action called **flexion,** and contraction of the triceps increases this angle, or **extends** the forearm.

Not all muscles attach to bones. Many form a mass of interlacing muscle fibers, as in the foot of a snail, or form layers in the walls of hollow organs (Fig. 5.13 *B*). These fibers do not have definable origins and insertions but pull upon each other and squeeze upon fluids within the organ so as to change the size and shape of the organ of which they are a part. The original length of the fibers is restored by the pull of fibers having a different direction, or by forces generated in the fluids.

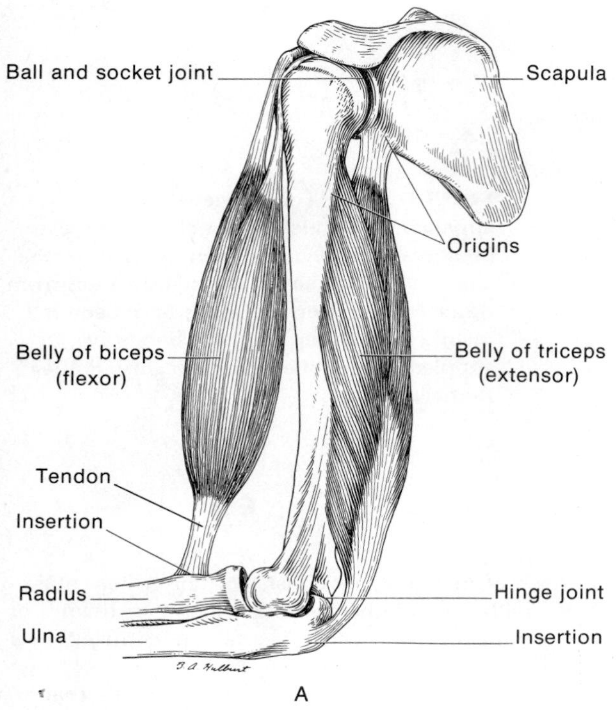

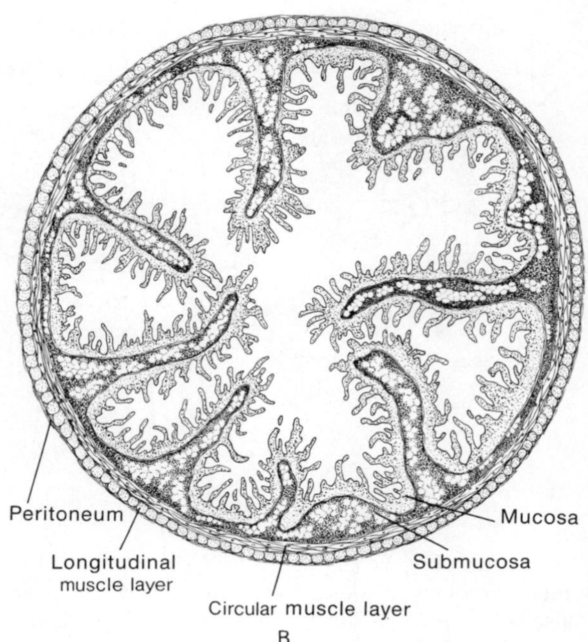

FIGURE 5.13 Arrangement of muscles. *A,* Antagonistic skeletal muscles attached to bones in the human arm. *B,* Antagonistic layers of smooth muscles in a segment of the wall of a mammalian intestine.

Biochemistry of Muscle Contraction

Striated Muscle Contraction. The mechanism of muscle contraction is best understood from studies of vertebrate skeletal muscle (Fig. 5.14). It is also essentially the same in cardiac muscle. The cytoplasm of a striated muscle fiber is packed with many longitudinal **myofibrils** (Gr., *myos,* muscle + L., *fibrilla,* small fiber), that can be seen with a light microscope (Fig. 5.14 *C* and *D*). Electron microscopy and biochemical analysis show that each myofibril is composed, in turn, of many longitudinal protein myofilaments of two types, thick and thin. The **thin filaments** are composed of globular **actin** monomers with which **tropomyosin** and **troponin** complexes are associated (Fig. 5.14 *E*). The **thick filaments** are composed of the tails of elongated **myosin** molecules. Heads of the myosin molecules protrude toward the thin filaments as potential **cross-bridges** that can link the filaments. Each head and tail are joined by a compliant hinge. Thick and thin filaments are packed in such a way that each thick filament is surrounded by six thin ones. Since the myosin heads spiral around the thick filament, cross-bridges can be formed with all of the surrounding thin filaments (Fig. 5.14 *E* and *G*). The denser, thick filaments are in register with each other and account for the darker, anisotropic (in polarized light) **A bands.** The thin filaments, also in register, account for the lighter, isotropic **I bands.** The two types of filament partially overlap, which explains the somewhat denser and less dense (Hensen's disk, or **H zone**) portions of the A bands. **Z lines,** to which the thin filaments attach, cross the entire myofibril in the center of the I bands. The portion of a myofibril between Z lines constitutes a **sarcomere** (Gr., *sarkos,* flesh + *meros,* part).

When striated muscle contracts, the I bands narrow, and the H zones may disappear as the Z lines move closer together (Fig. 5.14 *F*). The degree of narrowing of the I bands is a function of the extent of the contraction. Such observations in the late 1950s led H. E. Huxley and Jean Hanson of Cambridge University and King's College, London, to propose the sliding filament hypothesis of muscle contraction, a theory that has since been elaborated by many investigators.

No interaction occurs between the filaments at rest because the active sites on the actin filaments to which myosin heads can attach are blocked by tropomyosin. When a muscle fiber is stimulated, changes occur in its plasma membrane, or **sarcolemma,** that lead to the unblocking of the active sites. At rest, the plasmalemma has an electrical charge, known as the **resting potential,** that derives from a large concentration of Na^+ ions on the outside. When a nerve impulse reaches the termination, or **motor end plate,** of a neuron, a transmitter substance, usually **acetylcholine,** is released by the neuronal ending, and causes a change in the permea-

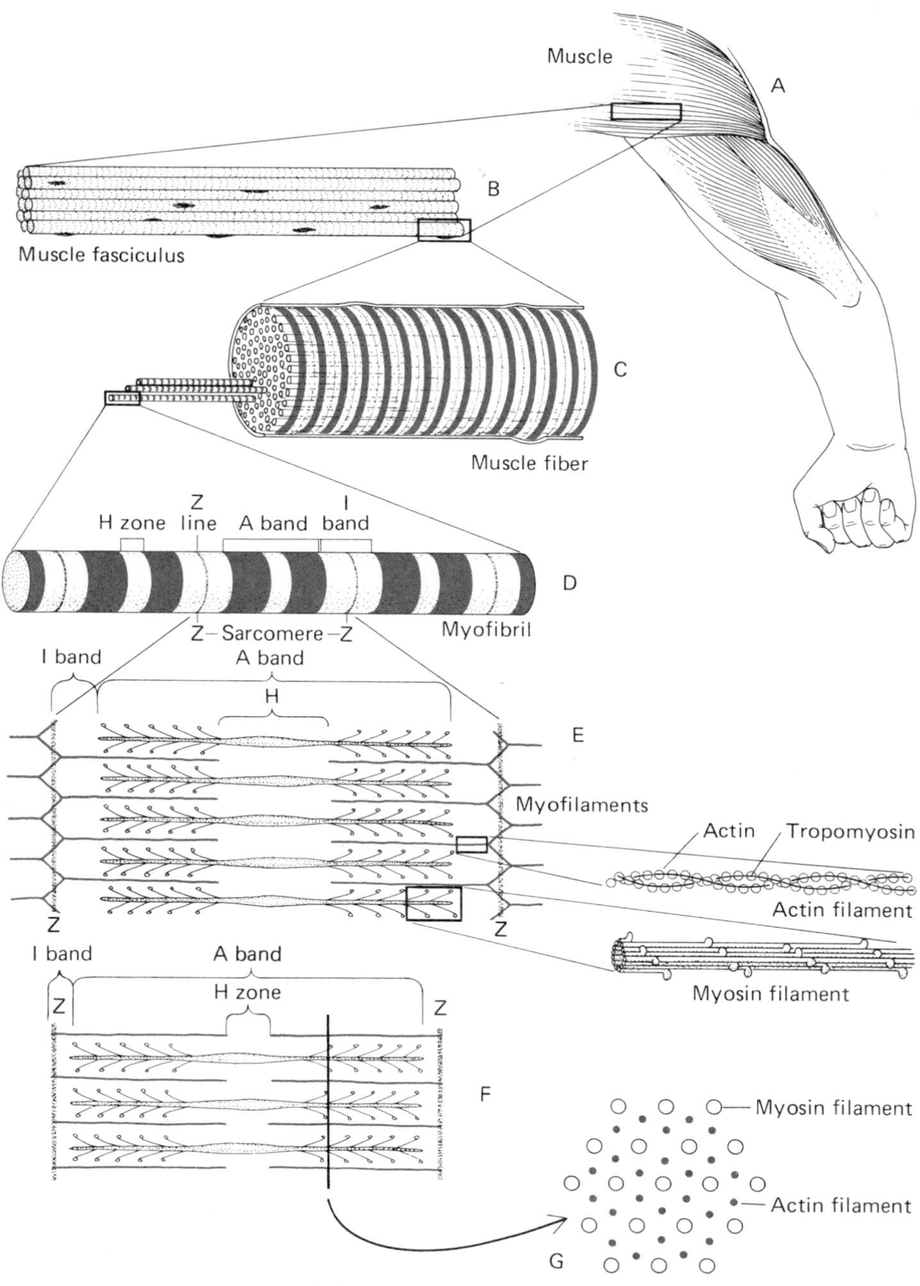

FIGURE 5.14 Diagrams of the structure and contraction of skeletal muscle. *A,* The deltoid muscle of the human shoulder; *B,* a group of muscle fibers; *C,* a portion of a muscle fiber showing the myofibrils of which it is composed; *D,* a portion of a myofibril; *E,* enlargement of one resting sarcomere from a myofibril showing the actin and myosin myofilaments of which it is composed; *F,* the same sarcomere in the contracted state; *G,* a cross section through a sarcomere showing that each myosin filament is surrounded by six actin filaments to which its heads attach. (Modified from Bloom, W., and D. W. Fawcett: *A Textbook of Histology.* 10th ed. Philadelphia, W. B. Saunders Co., 1975.)

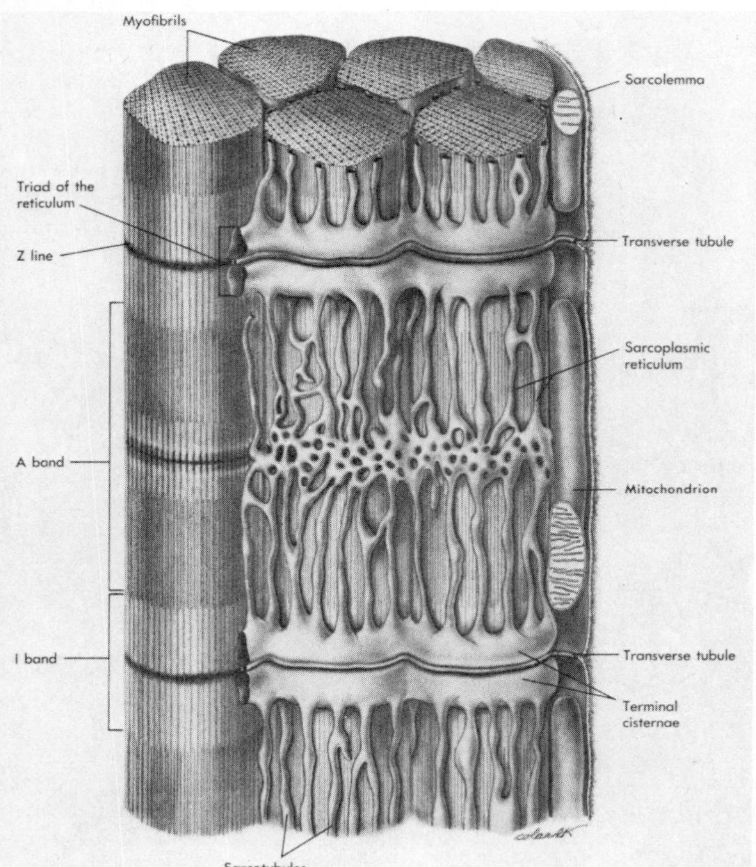

FIGURE 5.15 A drawing of a portion of a sarcomere of frog striated muscle to show the interrelations of the membrane of the cell, sarcoplasmic reticulum, and myofibrils. (From Bloom W., and D. W. Fawcett: *A Textbook of Histology.* 10th ed. Philadelphia, W. B. Saunders Co., 1975.)

bility of the sarcolemma such that Na^+ ions quickly flow inward. This momentarily reverses the electrical charge and gives rise to the **action potential.** The action potential is self-propagating and spreads quickly along the sarcolemma and deep into the fiber, where the sarcolemma is infolded as the **transverse tubules** (Fig. 5.15). In many muscles these tubules are adjacent to the Z lines. Ca^{++} ions in adjacent storage sacs of the sarcoplasmic reticulum, the **cisternae,** are then released and combine with troponin, changing its configuration in such a way that tropomyosin is pulled away from the active sites on the actin. Myosin heads now bind to these sites.

How this results in the sliding of the myofilaments past each other is not entirely clear. Probably the myosin heads, which previously have been energetically charged, or cocked, use their stored energy to bend slightly at their hinge regions (Fig. 5.16). Since the myosin heads at the two ends of the myosin filament face in opposite directions, this would pull the actin filaments, to which the heads are attached, toward the center of the sarcomere. As the heads bend, they expose a region that has ATPase activity. ATP is broken down, and the released energy straightens and recocks the myosin heads and causes them to detach from the actin sites to which they are currently attached and to reattach to adjacent ones. The numerous myosin heads attaching and reattaching act as a ratchet pulling the ac-

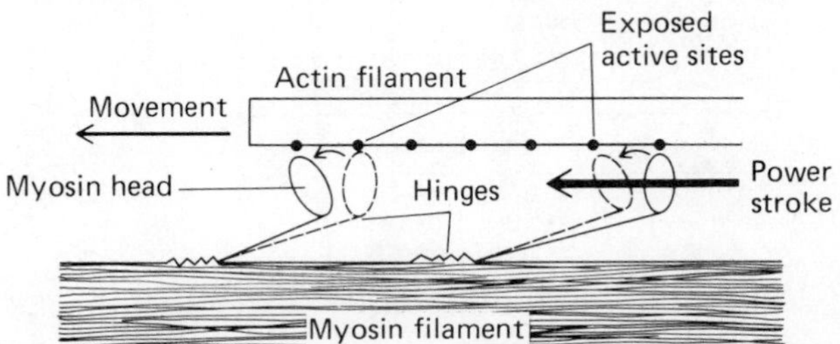

FIGURE 5.16 Diagram of the ratchet mechanism for muscle contraction. As shown here, some myosin heads are attached (on *right* side) and undergoing a configurational change that exerts a power stroke while other ones (on *left* side) are detached and preparing to reattach. (From Guyton, A. C.: *Textbook of Medical Physiology.* 5th ed. Philadelphia, W. B. Saunders Co., 1976.)

tin filaments deeper and deeper into the sarcomere. As this is going on, a calcium pump is driving free Ca^{++} ions back into the sacroplasmic cisternae, and the enzyme **acetylcholinase** breaks down acetylcholine at the motor end plate. If muscle stimulation is not continued, the level of free calcium becomes so low that tropomyosin again moves over the active sites on the actin filaments, blocks them, and contraction stops. In this model, ATP is needed to energetically charge the myosin heads and to detach them from the actin sites prior to the next cycle of attachment, pull, and detachment. ATP also drives the calcium pump. Energy is needed for relaxation, that is, to detach the myosin heads and to remove free Ca^{++} ions. A muscle deprived of a source of ATP remains contracted. We know this as muscle cramps, or, after death, as **rigor mortis.**

The attachment and detachment of each myosin head utilizes one molecule of ATP. When the small amount of ATP stored in muscle is used up, ATP is resynthesized from ADP by the transfer of a phosphate group from other energy-rich phosphates: **creatine phosphate** in vertebrate and some invertebrate muscles and **arginine phosphate** in most invertebrates. Ultimately these energy-rich phosphates are restored by the metabolism of food products in the glycolytic and citric acid cycles (p. 71).

Muscle fibers with relatively slow rates of contraction do not use energy faster than it can be restored by the metabolism of food products and oxygen delivered to them. However, the fibers used in very rapid bursts of activity store substantial amounts of glycogen. If oxygen is not delivered fast enough for oxidative phosphorylation, considerable energy is provided by the anaerobic glycolysis of the glycogen. Since the product, pyruvic acid, cannot enter the citric acid cycle in the absence of oxygen, it is converted to lactic acid, which diffuses from the cells. When sufficient oxygen becomes available, most of the lactic acid is reconverted (primarily in liver cells) to glucose, which is carried back to the muscle fibers and restores muscle glycogen. The amount of oxygen needed to remove the accumulated lactic acid represents an **oxygen debt.** An animal continues to breath deeply after vigorous exercise until this debt is repaid. The existence of this mechanism is of great importance to animals, for it permits them to contract muscles forcibly and repeatedly in emergencies without having the contractions limited by the rate at which oxygen can be delivered to the tissues.

Smooth Muscle Contraction. The contraction of smooth muscle utilizes the same basic mechanism as that of striated muscle. Its fibers contain actin and myosin, but these myofilaments are scattered rather than being in register (Fig. 5.17*A*). Since smooth muscle fibers are much smaller than striated fibers, Ca^{++} ions are not stored within the sarcoplasmic reticulum but are concentrated in the extracellular fluid. Activation involves the influx of Ca^{++} ions and then the formation of cross-bridges between actin and myosin. As bridges form, the actin filaments pull upon "dense bodies" that are held in a fixed position within the cytoplasm, and the fiber shortens.

Although smooth muscle fibers in different organs look very much alike, they have quite different properties. The smooth muscles of vertebrates can be grouped into two major types, multiunit and visceral. **Multiunit smooth muscle** fibers are found in the iris and ciliary body of the eye, form the arrector pili muscles of the hairs, and are present in the walls of many blood vessels. They are stimulated to contract by nerve impulses that reach each fiber at a motor end plate (Fig 5.17*B*). Certain fibers, such as the radial fibers that dilate the pupil of the iris, are activated by neurons of the sympathetic nervous system (p. 227); others, including the circular fibers in the iris that constrict the pupil, by neurons of the parasympathetic nervous system. **Visceral smooth muscle** fibers that contribute to the wall of the digestive organs, uterus, and many other visceral organs are more tightly packed together (Fig. 5.17*C* and *D*). Only a few of the fibers have motor end plates, but the action potential generated in one fiber can flow to adjacent ones through **gap junctions** that tightly unite their plasma membranes at many points. Groups of visceral smooth muscle fibers usually receive both a sympathetic and parasympathetic innvervation. These have antagonistic effects upon the organs, one promoting contraction and the other inhibiting it. Visceral smooth muscle also has a degree of spontaneous activity and will contract automatically when stretched, as occurs when an organ fills with food or liquid.

Physiology of Muscle Contraction

Motor Units. By muscle contraction physiologists mean the state of activity by which tension is developed within muscle. Contraction is initiated normally by a nerve impulse. In a typical vertebrate skeletal muscle, one **neuron** will branch in such a way as to have one or more motor end plates on each of a number of different muscle fibers. The neuron and the fibers it supplies constitute a **motor unit.** The degree of fine control over the activity of a muscle is inversely proportional to the number of muscle fibers in its motor units. Ocular muscles that move our eyeball may have as few as two or three fibers to a unit, whereas some of the leg muscles that hold our body erect contain several hundred per unit.

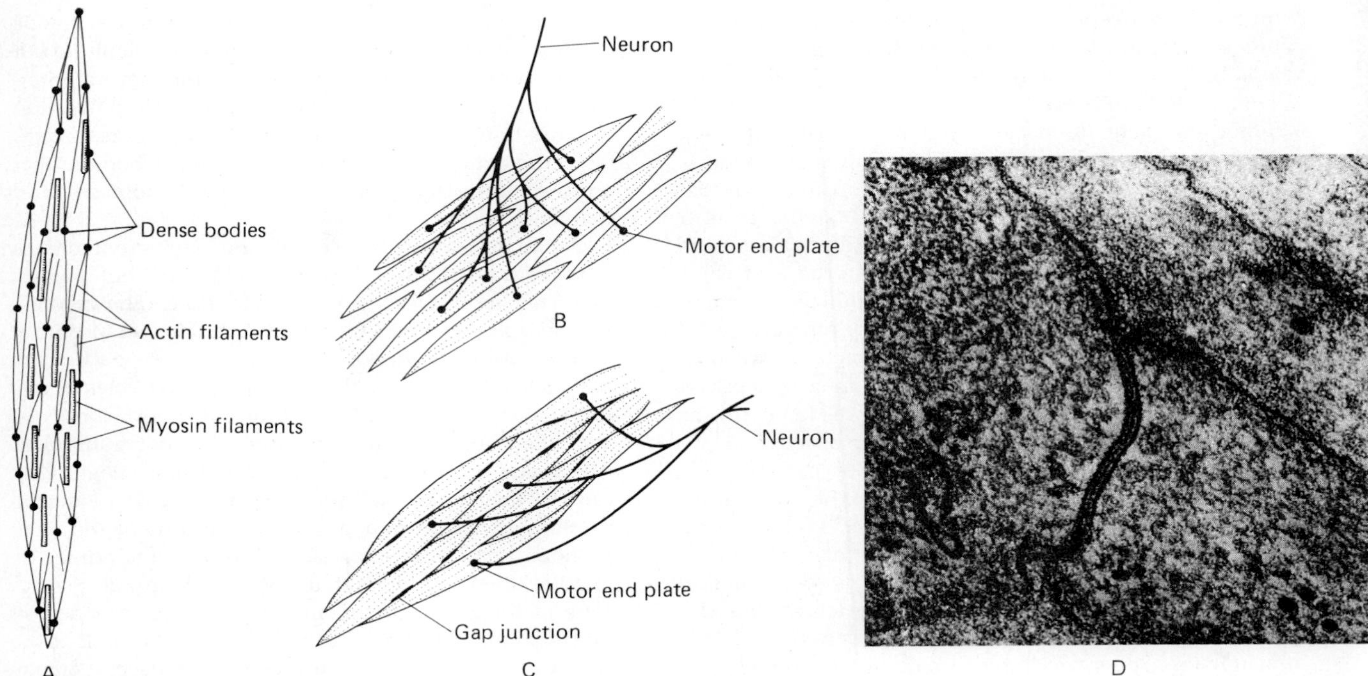

FIGURE 5.17 Smooth muscle structure and innervation. *A,* A diagram showing the arrangement of actin and myosin filaments and dense bodies in a smooth muscle fiber. *B,* The innervation of multiunit smooth muscle. *C,* Innervation of visceral smooth muscle. *D,* Electron micrograph of a gap junction between two visceral smooth muscle fibers. (*A* from Guyton, A. C.: *Medical Physiology.* 6th ed. Philadelphia, W. B. Saunders Co., 1980; *D* from Williams, P. L., and R. Warwick (Eds.): *Gray's Anatomy.* 36th British ed. Philadelphia, W. B. Saunders Co., 1980.)

Stages in Contraction. Muscle fibers may be activated by a single nerve impulse, but two or more in rapid succession may be needed. The first impulse initiates changes that are built upon by subsequent ones, if they arrive before the effects of the first have worn off. The early impulse is said to **facilitate** activation.

Activation requires a few milliseconds, so there is a delay, or **latent period,** between the application of a stimulus and the onset of cross-bridge formation and the shortening of the contractile elements (Fig. 5.18). The first effect is to stretch the elastic components in the muscle, including the hinge region of the myosin molecules, connective tissue, and tendon. Tension develops and the muscle may shorten during this **contraction period.** Tension decreases during the longer **relaxation period.**

If the muscle is working against a constant load, the muscle as a whole will shorten as tension develops, and the contraction is described as **isotonic** (Gr., *isos,* equal + *tonos,* strain) or **positive work.** If the ends of the muscle are attached to bones that are not free to move, internal shortening and tension develop as the elastic elements are stretched, but there is little or no overall shortening of the muscle. This is **isometric contraction** (Gr., *isos,* equal + *metron,* measure). In **negative work contraction,** the length of the muscle increases as tension develops. As a human stands up from a sitting position, the hamstring muscles on the back of the thigh contract as they pull the thigh under the hips, yet, at the same time, they increase in length as the lower leg, to which they also attach, straightens out.

Types of Muscle. If muscles are to be efficient in movement, the tension they develop and their speed of contraction must match the conditions under which they are used. Not all muscle fibers are alike in these properties. Most skeletal muscles of the higher vertebrates are **twitch,** or **phasic, muscles.** Each fiber has a

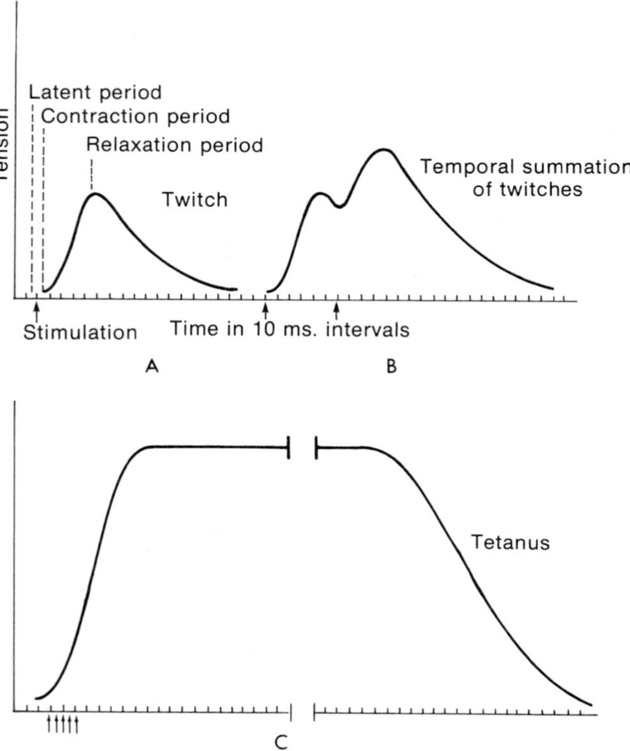

FIGURE 5.18 **Isometric contraction in a motor unit.** ***A,*** **A twitch;** ***B,*** **temporal summation of two twitches;** ***C,*** **tetanus. Vertical arrows indicate points of stimulation.**

single motor end plate. When a motor unit is stimulated electrically through its neuron, a single stimulus of threshold intensity will elicit a simple **twitch** (Fig. 5.18). This is an **all-or-none** phenomenon because the action potential spreads through the entire fiber, and the twitch is maximal; a stronger stimulus will not cause a stronger twitch. Gradation of force of contraction, however, can develop in two ways. If a second stimulus reaches the motor unit before the effects of the first have worn off, a second twitch will be superimposed upon the first. This is possible because the first twitch does not last long enough to fully engage all the myosin heads or to stretch all of the elastic components. This phenomenon is called **temporal summation.** If impulses reach the motor unit so rapidly that no relaxation occurs between twitches, tension develops very rapidly to a plateau beyond which it will not go even when the rate of stimulation is increased. The muscle is in **tetanus,** and this state will be maintained until the impulses cease or the muscle tires (Fig. 5.18). Graded contractile responses can also be elicited by recruiting more and more motor units—a phenomenon called **spatial summation.**

Seldom are all motor units in a muscle active at the same time; some are relaxing as others contract. In this way a muscle can sustain a contraction for a long time without fatigue. Most muscles in our body are under a state of partial contraction, known as **tonus,** all of the time. This maintains organs in proper spatial relationship to each other.

Phasic muscle fibers differ in their speed of contraction and fatigue. **Slow phasic fibers,** used in sustained postural contractions and slow movements, are often red in color because they contain muscle hemoglobin (**myoglobin**). Since myoglobin has a higher affinity for oxygen than the hemoglobin in the blood (p. 163), it accelerates the transfer of oxygen from blood to muscle. Mitochondria are abundant since most of the energy of slow phasic fibers is derived from oxidative phosphorylation. These fibers contract slowly and no faster than oxygen and food are delivered to them; hence they do not go into oxygen debt or fatigue easily. **Fast phasic** or **glycolytic fibers,** used when quick, short bursts of activity are required, lack myoglobin and are whitish in color. Mitochondria are few in number since most energy is derived quickly by the anaerobic glycolysis of stored glycogen. These fibers contract rapidly but go into oxygen debt and soon fatigue. Energy is replaced after the burst of activity. Slow and fast phasic fibers are sometimes segregated in different muscles—dark and light meat of a chicken is a familiar example—but often there will be a mixture of both types in a single muscle. This two-geared system (slow and fast) is efficient but not economical of space or weight. This is not a problem in most vertebrates, which are relatively large animals.

A few vertebrate muscles are tonic. **Tonic muscles** resemble slow phasic fibers in contracting and fatiguing slowly, but differ in having multiple motor end plates on each fiber (Fig. 5.19C). They are not all-or-none. Activation by a single nerve impulse is limited to a few sarcomeres because the action potential does not propagate far. Gradation in force is derived from more frequent impulses that cause a wider spread of the action potential. In mammals, tonic fibers occur in situations where a graded force is needed in a small muscle, as in the ocular muscles and the intrafusal fibers of muscle spindles (p. 243).

Force, Work, and Power. The **force** that a muscle can generate is a function of the number of cross-bridges that can be formed at one time, and this, in turn, is determined by the number of fibers that are packed into the muscle. Since the number of fibers can

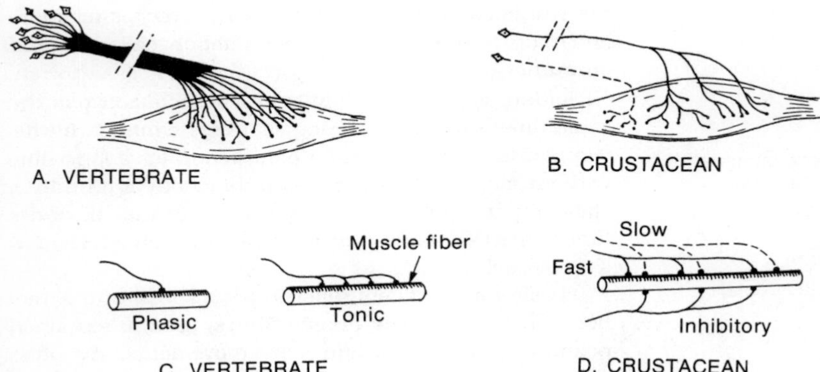

FIGURE 5.19 Patterns of nerve supply in vertebrate and crustacean muscles. *A*, Vertebrate muscles are supplied by large nerves composed of many neurons; *B*, crustacean muscles are supplied by small nerves that may contain as few as two neurons. Pattern of neuron endings on individual muscle fibers is shown in *C*, vertebrate phasic and tonic muscle, and *D*, in crustacean muscle. (Modified from Schmidt-Neilsen, K.: *Animal Physiology*. Cambridge University Press, 1975.)

be estimated by determining the cross-sectional area, force is proportional to the cross section of a muscle. The maximum force of contraction per unit of area is surprisingly similar for many muscles, ranging from about 4 to 6 kg cm^{-2} for muscles as diverse as the adductor of a clam, the jumping muscles in a grasshopper, and the leg muscles of a human being. The **work** that muscles can do, which is the product of the force generated and the distance through which the force works, also is quite similar per unit of muscle, e.g., per gram, but obviously is different for entire muscles. Large muscles, which have a large cross-sectional area and force and are long enough to contract a considerable distance (usually about one third of their resting length), do a great deal more work than do small muscles. Total work is proportional to the mass of a muscle. One quality that is quite different on a unit basis is power output because **power** is the rate of doing work, that is, work/time. The small muscles of a shrew can contract one third of their length much faster than can the large muscles of an elephant, so their power, per gram of muscle, is much higher. Because they contract faster, the cycle of cross-bridge formation and release also is much faster in the shrew. Since a molecule of ATP is used for each cross-bridge, more ATP per gram of muscle is used. This is an important factor, along with surface area–volume considerations (p. 145), in accounting for the high level of metabolism of small animals.

Some Specializations of Invertebrate Muscles

Molluscan Catch Muscle. The adductor muscle that closes the shell of a clam contains two kinds of fibers that in some species, such as scallops, are segregated into two parts. One is striated, acts as a phasic muscle, and quickly closes an open shell. The other, which is smooth, is tonic and holds the shell against the force of the elastic hinge that tends to open it. An unusual attribute of the tonic muscle is its ability to hold the shell closed for many hours, or even days, and sometimes against a strong external force, such as a starfish might exert in trying to pull the shell open. It is as though there were some sort of "catch" mechanism that maintains the contracted state, but how it works is not fully understood. It is known that the catch muscle contains relatively large filaments of paramyosin. Some investigators have postulated that paramyosin stabilizes the actomyosin cross-bridges so that they are not released. Others hypothesize that continued nerve impulses are necessary to maintain contraction. Discovery that oxygen consumption continues during catch, albeit at very low levels, adds support to this theory.

Polyneuronal Innervation of Crustacean Muscle. Arthropod muscles differ greatly from vertebrate muscles in their innervation (Fig. 5.19). The muscles in the legs and claws of crabs and other crustaceans, for example, are often supplied by only two (or a few) neurons, and one neuron may extend to two or more muscles. Each muscle fiber receives multiple terminations from a neuron, and two or more different neurons terminate on each fiber. As many as five may innervate a single fiber. At least one is inhibitory; the others induce varying degrees of fast or slow contraction. The final response of a muscle depends on the rate of stimulation of the different neurons and the interaction between them. These complexities enable a few neurons and muscle fibers to act as a many-geared system and to generate as wide a range of speed, force, and duration of muscle contraction as is brought about by the much larger nervous and muscular systems of vertebrates, in which speed and force of contraction are controlled by the type and number of muscle fibers activated.

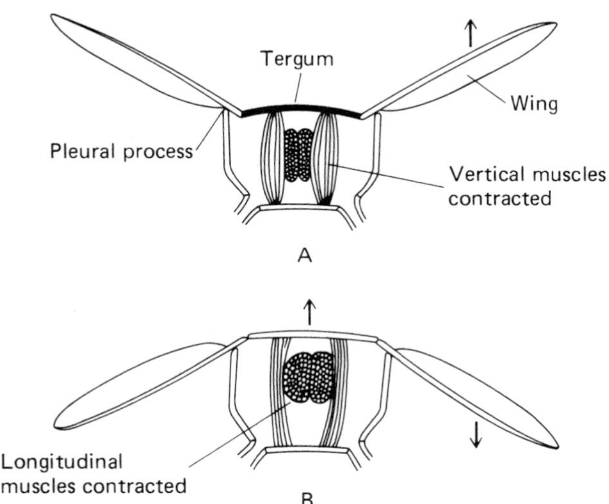

FIGURE 5.20 Diagrams of the indirect flight muscles of an insect as seen in cross sections of the thorax. *A,* Contraction of vertical muscles lowers the tergum and raises the wings. *B,* Contraction of longitudinal muscles raises the tergum and lowers the wings. (After Ross.)

Insect Flight and Tymbal Muscles. The flight and tymbal (sound-producing) muscles of some insects are of interest, for they contract at frequencies far faster than can be induced by nerve impulses. A midge may beat its wings 2000 times per second, and some tymbal membranes oscillate as fast as 7000 times per second! Muscle activation, recovery, and reactivation by neuronal stimulation require more time than this. The flight muscles of flies, wasps, and beetles do not attach directly to the wings, as muscles do in slower-flying insects, but rather to the walls of the thorax (Fig. 5.20). For this reason, these muscles are called indirect flight muscles. Vertical fibers depress the thorax roof, or **tergum,** against a fulcrum formed by the lateral thoracic walls and hence raise the wings. Longitudinal fibers shorten the thorax in an anteroposterior direction, which increases its height and lowers the wings. These two sets of muscles are antagonistic to each other and to the elasticity of the thoracic wall. A nearly isometric contraction of one set develops a tension that causes a sudden change in thorax shape, and this stretches the other set of muscles. The stretching of this set, rather than another nerve impulse, induces its contraction. Recovery time is very short because the muscles contract as little as 2 percent of their resting length. Much of the activity is **myogenic;** that is, it is inherent within the muscles themselves as compared with the usual **neurogenic** mechanism in which each contraction requires a nerve impulse. Nerve impulses in indirect flight muscles are necessary to keep the muscles in an active state, but there may be as many as 40 wing beats for each nerve impulse.

The Vertebrate Muscular System

In describing the vertebrate muscular system and tracing its evolution, it is convenient to group the muscles into **somatic muscles,** associated with the body wall and appendages, and **visceral muscles,** associated with the pharynx and other parts of the gut tube. This grouping parallels the major divisions of the skeletal system. Somatic muscles are striated and under voluntary control. Most of the visceral muscles are smooth and involuntary; however, the visceral muscles associated with the visceral arches, called branchial muscles, are striated, and some are under voluntary control.

Evolution of Somatic Muscles. Most of the somatic musculature of fishes consists of segmental **myomeres** (Fig. 5.21). This is an effective arrangement for bringing

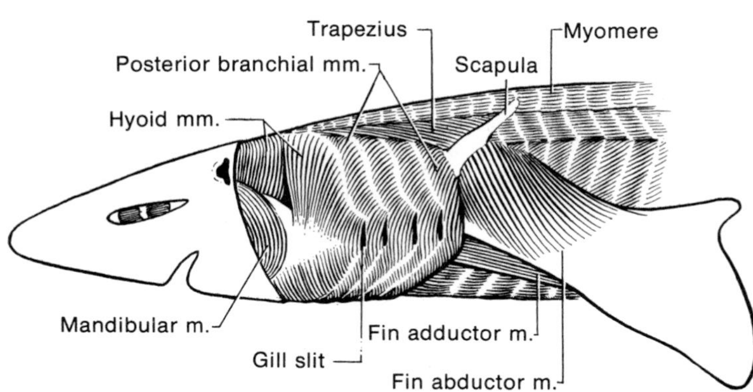

FIGURE 5.21 A lateral view of the anterior muscles of a dogfish. (Modified after Howell.)

about the lateral undulations of the trunk and tail that are responsible for locomotion. The muscles of the paired fins are very simple and in many fishes consist of little more than a single dorsal **abductor** that pulls the fin up and caudally and a ventral **adductor** that pulls the fin down and anteriorly.

The transition from water to land entailed major changes in the somatic muscles. The appendages became increasingly important in locomotion, and movements of the trunk and tail became less important. The segmental nature of the trunk muscles was largely lost, although traces can be seen in the **rectus abdominis** that extends longitudinally on each side of the midventral line (Fig. 5.22). Back muscles remain powerful, for they play an important role in supporting the vertebral column and body, but trunk muscles on the flanks form thin sheets, such as the **external oblique,** that help to support the abdominal viscera and assist in breathing movements. Some trunk muscles attach onto the pectoral girdle and in a quadruped help transfer body weight to the girdle and appendage. The primitive single fin abductor and adductor became divided into many components, and these became larger and more powerful. Our **latissimus dorsi** and **triceps** (Fig. 5.22), for example, are appendicular muscles that evolved from the fish abductor, whereas the **pectoralis** and **biceps** evolved from the adductor.

Evolution of Branchial Muscles. Branchial muscles are well developed in fishes and are grouped according to the visceral arches with which they are associated (Fig. 5.21): **mandibular muscles** and certain **hyoid muscles** of the first two arches are concerned with jaw movements; most of the rest, with respiratory movements of the gill apparatus. Branchial muscles obviously became less important in tetrapods, for the gills were lost and the visceral arches were reduced. Those of the mandibular arch remain as the **temporalis, masseter,** and other jaw muscles (Fig. 5.22). Most of those of the hyoid arch moved to a superficial position and became the **facial muscles** that are responsible for movements of the scalp, nose, cheeks, and lips. Those of the remaining arches are associated with the pharynx and larynx, and some, e.g., the **sternocleidomastoid** and **trapezius,** are important muscles associated with the pectoral girdle, (Fig. 5.22).

Nonmuscular Movement

Nearly all cells have some capacity to move and change shape because their cytoplasmic ground substance contains a microtrabecular lattice (p. 40) of very delicate protein fibers, which, being composed partly of actin and myosin, are contractile. Additional contractile myofilaments or microtubules often are associated with the

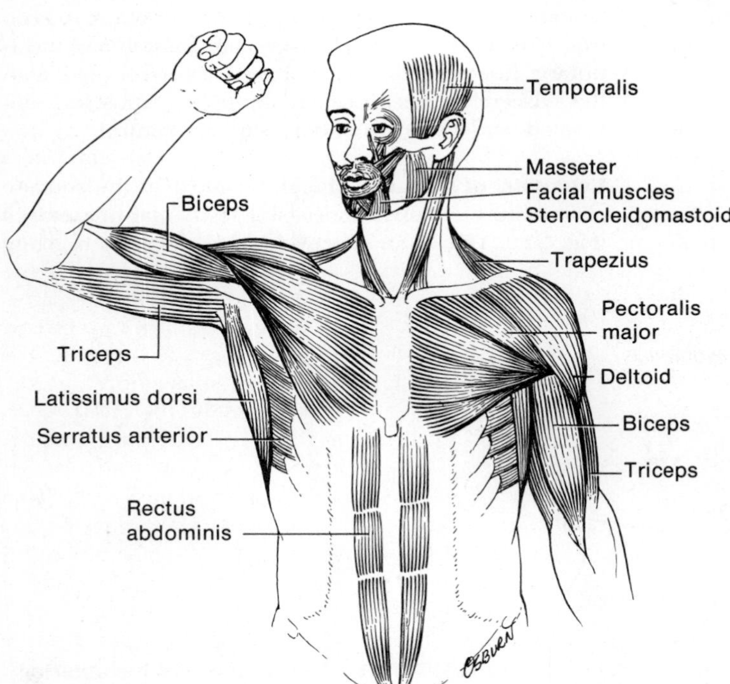

FIGURE 5.22 An anterior view of certain of the superficial muscles of a human being.

lattice. Specialized contractile mechanisms emerge from this basic framework of the cell.

Amoeboid Movement. In many cells one or more processes, **pseudopodia,** develop on the cell surface, and the rest of the cell flows in their direction. This is known as amoeboid movement because it occurs in all amoebas and related protozoa, but the phenomenon is widespread and is observed in the amoebocytes of sponges, many of the coelomic and blood corpuscles of other invertebrates, and in certain white blood cells of vertebrates. It also occurs in embryonic tissues, in wound healing, and in many cell types in tissue culture. The pseudopodia may be used to surround and engulf a food particle, or they may attach to the substratum and be used in locomotion.

Several pseudopodia may start to form on different parts of the cell, but usually one becomes dominant, and the cell advances in that direction. There is no permanent front end; the dominant pseudopod may appear on any surface. The cytoplasm of an amoeba can be divided into a semirigid **ectoplasm** beneath the cell membrane and a deeper, more fluid **endoplasm** (Fig. 5.23). According to the prevalent theory of pseudopod formation, the deepest, axial endoplasm flows forward, separated from the ectoplasm by a sheer zone of endoplasm. It then spreads laterally in the "fountain zone" near the front of the pseudopod, becomes semirigid, and adds to the advancing sleeve of ectoplasm. Concurrently, at the posterior end of the cell, in the "recruitment zone," ectoplasm is converted to the forward-moving stream of axial endoplasm.

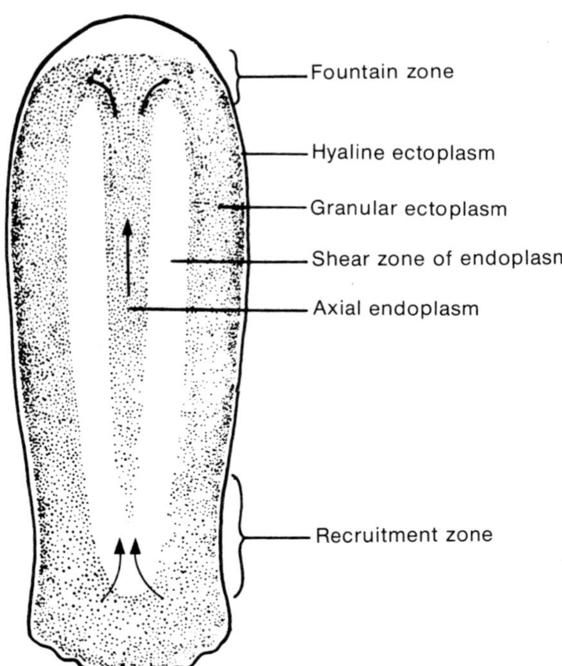

FIGURE 5.23 Amoeboid movement, endoplasm flowing into an advancing pseudopod. (Modified from Allen, R. D.: "Ameboid Movement," in Brachet, J. and A. E. Mirsky (Eds.): *The Cell.* New York, Academic Press, 1961.)

The basis for amoeboid movement, as well as for other types of cell movement, is essentially the same as that in muscle contraction, for it involves contractile proteins, such as **actin** and **myosin.** Actin molecules are in a diffuse state in the endoplasm, but they polymerize and become more filamentous in the fountain zone. As they become filamentous, they interact with myosin molecules, and this leads to a contraction. The contraction probably involves a sliding of actin and myosin filaments past each other as it does in muscle. Endoplasm is pulled forward and added to the ectoplasm, which represents contracted or still contracting actin and myosin. This contraction exerts additional force on the endoplasm. Actin and myosin become dissociated in the recruitment zone. Actin is converted into a less structured state and moves forward again in the endoplasm. Local changes in the availability of calcium ions appear to be responsible for changes in the interaction of actin and myosin. As in muscle cells, the concentration of calcium must attain a certain threshold level for actin and myosin to interact and contract. When the concentration is less than this critical threshold, actin and myosin dissociate.

Ciliary and Flagellar Movement. Another type of motion involves the beating of slender, hairlike, cytoplasmic processes projecting from the cell surface. These are called **cilia** (L., *cilium,* eyelid) when each cell has one to many short processes and **flagella** (L., *flagellum,* whip) when each has one or a few longer processes, but the fundamental structure of cilia and flagella is the same. Cilia and flagella are nearly ubiquitous among protozoa and metazoa, where they are used to move a cell through liquid or liquid along the surface of cells. They occur in ciliates and flagellates among the protozoa, in the collar cells of sponges, on many epithelial surfaces, ranging from the gastrovascular cavity of cnidarians to the lining of vertebrate reproductive ducts, and in the tails of spermatozoa of most animals. Only roundworms and arthropods lack motile cilia or flagella, but ultrastructural comparisons reveal that processes of many of the sensory cells of arthropods are modified cilia.

Ciliary and flagellar movement has been analyzed by means of very high speed cinephotography. It turns out to be very complex and also variable among different species. Sometimes undulatory waves move along the flagellum, but successive waves are oriented 90 degrees to each other. This imparts a helical twist or rotary action to the flagellum, so that it acts as a propellor moving water parallel to the axis of the flagellum (Fig. 5.24 *A*). A flagellum may pull a cell forward by a tip-to-base

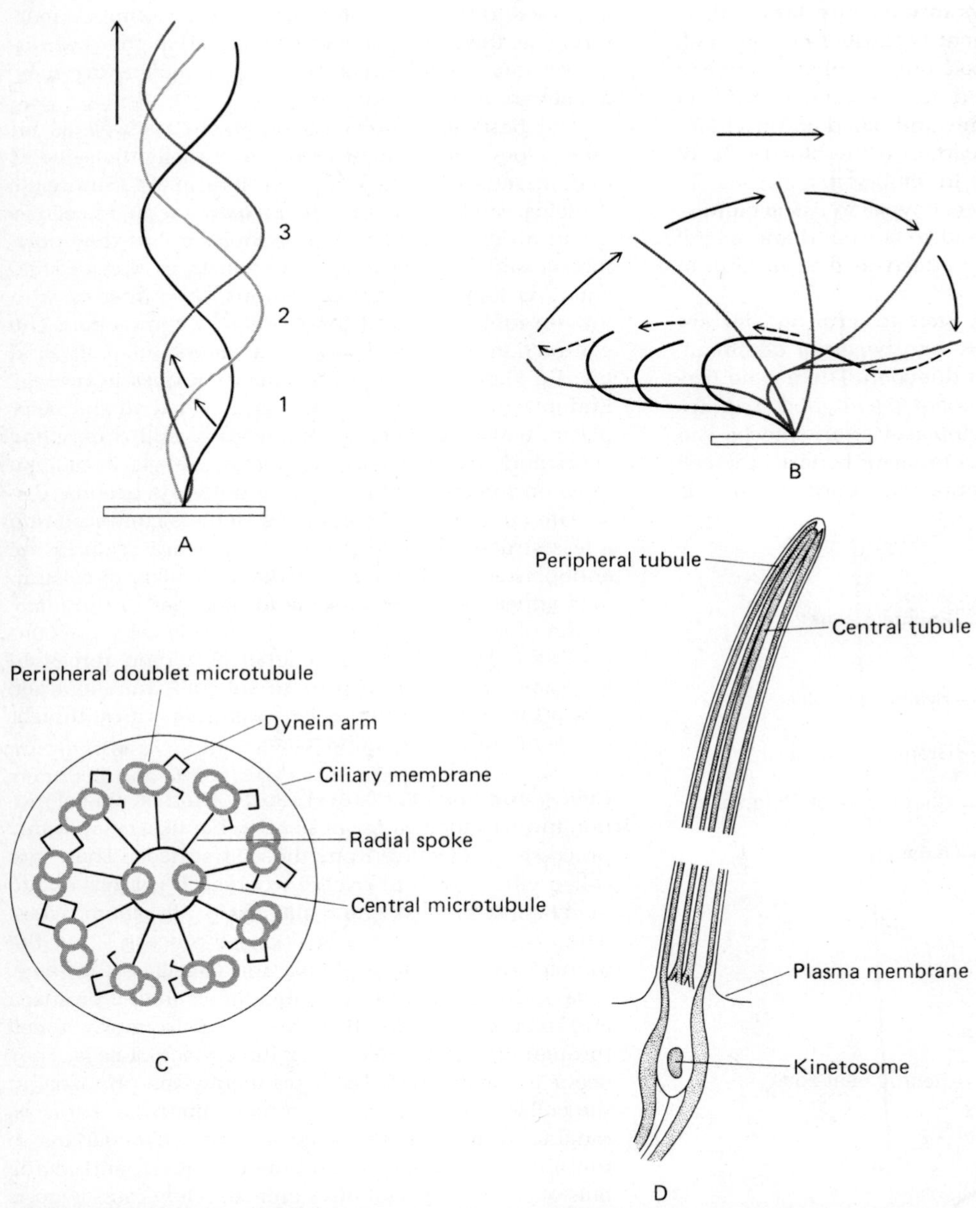

FIGURE 5.24 Ciliary and flagellar movement. *A,* Three successive stages in the undulatory movement of a flagellum. *B,* Successive stages in the oarlike action of a cilium; effective stroke shown in white and recovery stroke in black. In both *A* and *B,* water is moved in the direction of the large arrow, and the cell moves in the opposite direction. *C* and *D,* Diagrams of transverse and longitudinal sections of a cilium. (*A* and *B* from Satir, P.: How cilia move. *Scientific American, 231*(Oct.):44–52, 1974. *C* and *D,* modified from the work of Gibbons, I. R.)

undulation or push it by a base-to-tip undulation (p. 450). The beat of a cilium is more planar; the action is oarlike, with a rather "stiff-armed" effective stroke and a "curling," or feathered, recovery stroke moving water parallel to the cell surface (Fig. 5.24 *B*).

Each cilium contains a sheaf of nine double microtubules, situated just beneath the cell membrane, and a core of two single microtubules (Fig. 5.24 *C* and *D*). Radial spokes extend between the outer doublets and the two central microtubules. The walls of the microtubules are made up of protein filaments composed of **tubulin** (p. 39). Arms composed of the protein **dynein** extend between the outer doublets and form temporary cross-bridges between them. The arrangement is similar in many ways to that in muscle filaments, although the contractile proteins are different. According to the sliding microtubule hypothesis, the release of energy from ATP and the successive forming and breaking of dynein bridges cause one or more microtubules to slide past others. In a sense, the dynein bridges of one double microtubule "walk along" an adjacent doublet. The resulting shearing forces are resisted by the radial spokes, so that some of the sliding action is converted to a bending action.

If there are many cilia upon a cell, their action must be integrated if movement is to be effective. Often traveling, or **metachronal** (Gr., *metachronas,* done afterwards), **waves** pass along the surface (Fig. 5.25). If a paramecium encounters an obstacle, ciliary action is temporarily reversed; the animal backs away from the obstacle, turns, and again moves forward.

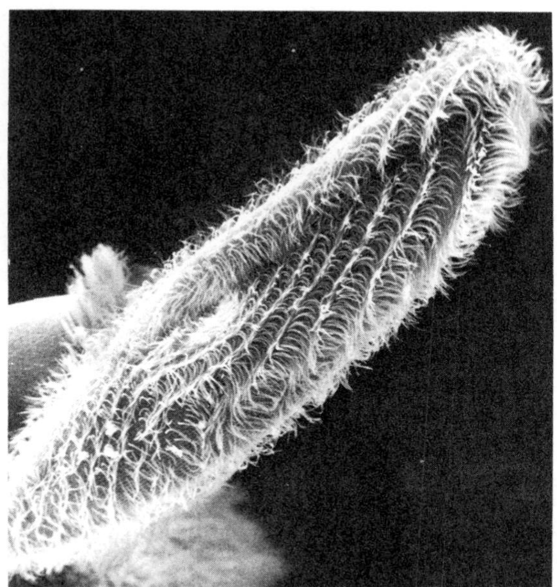

FIGURE 5.25 Scanning electron microscope photograph of coordinated waves of ciliary beating passing over the surface of *Paramecium caudatum.* (From Schmidt-Nielsen, K.: *Animal Physiology.* 2nd ed. Cambridge University Press, 1979; photograph by R. L. Hammersmith.)

Many small, multicellular animals, such as flatworms, and even larger forms, such as snails, use cilia to glide over the substratum. Mucus secreted by epidermal glands is essential for this type of locomotion, although not for ciliary swimming. The mucus functions as a blanket adhering to the substratum, and it is over this blanket that the animal crawls. The cilia gain traction by gripping the mucus with the cilium tip during the effective stroke. Only the cilium tip makes contact with the mucus; a watery fluid is believed to lie between the blanket and the animal's surface at the cilium base. The mucus is probably secreted as granules that then pass out to the blanket. A similar arrangement operates over many nonlocomotor ciliated surfaces, such as the gill of clams and the trachea of vertebrates. Here, however, the mucus blanket moves, carrying with it trapped particles.

Cilia and flagella are associated with basal bodies (**kinetosomes**) (Fig. 5.24 *D*) that are essentially similar to the centrioles associated with the mitotic spindles (p. 45). Basal bodies are related to the development and growth of the cilia and also appear to be involved in integrating the activities of different cilia. Basal bodies are interconnected in paramecium.

Locomotion

To move, an animal must support itself, develop a thrust against the surrounding medium (water or air) or the substratum (ocean bottom or ground) and maintain stability. Most terrestrial animals support themselves with a skeleton. Much of the support for aquatic animals, even for those with a skeleton, comes from the water. If the **density** (mass per unit volume) of the animal is the same as that of the surrounding water, the animal has **neutral buoyancy** and neither rises nor sinks. Many aquatic organisms have a density nearly the same as that of water, and little energy need be expended to keep afloat. Others have tissues denser than water, but the tendency to sink can be mitigated by including within their mass some material less dense than water. Examples are the numerous chambers of gas in the shell of *Nautilus,* a molluscan relative of the octopus, and the **swim bladder** of most bony fish. Gas usually is secreted into or reabsorbed from the fish's swim bladder by means of special glands associated with the circulatory system as the animal respectively sinks or rises in the water, so the fish maintains a neutral buoyancy over a wide range of depth. The major gas secreted by the gland is oxygen. The glandular tissue has a very active lactic dehydrogenase that secretes lactic acid. The acidity of the lactic acid causes the release of oxygen from oxyhemoglobin that carries the oxygen in the blood (p. 161).

Other aquatic organisms may have a great deal of oil in their tissues. This is true for many small planktonic organisms. Sharks do not have swim bladders but have very large livers with a high oil content. The oil, of course, serves as an important energy reserve as well as reducing the density of the animal to some extent. Still other marine animals, such as many jellyfish and some deep-water squids, achieve near neutral buoyancy by regulating their ion content, increasing the lighter ions, and reducing certain heavy ones.

The motion of an animal is derived from pushing against the surrounding medium or substratum. This, in accordance with Newton's third law of motion, generates an equal and opposite thrust against the animal. Part of this force moves the animal forward. For example, when the segmented muscles of the trunk and tail of a fish contract, they cause the trunk and tail to push back against the water with a particular force (Fig. 5.26); the water pushes against the trunk and tail with an equal but opposite force that can be resolved into lateral and forward components. As the tail moves from side to side, the lateral components cancel each other, and the forward components propel the animal. Segmentation, or metamerism, of the muscles of primitive chordates and fishes evolved in connection with this particular type of locomotion.

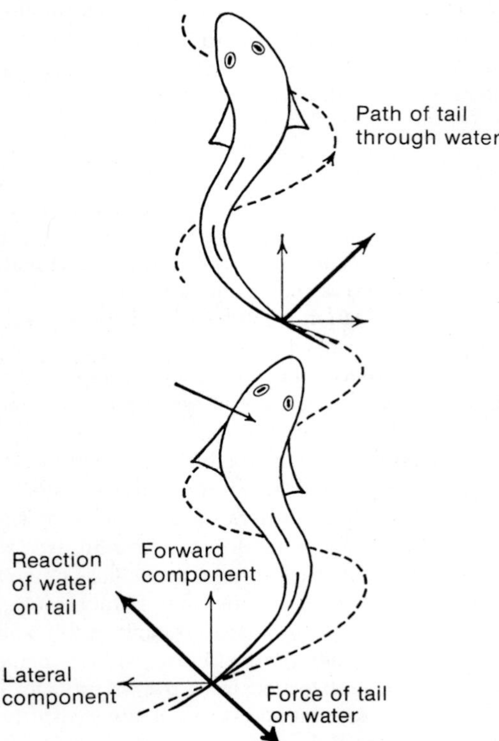

FIGURE 5.26 Diagram of a dogfish swimming to illustrate the thrust of the tail against the water and the reaction of the water on the tail. (Modified from Marshall and Hughes.)

The cilia of paramecium or the flagellum of a sperm tail push against the water. The up and down undulations of a leech's body or the lateral undulations of a snake's trunk push against the water or ground. A squid moves by emitting a powerful jet of water from its funnel. The thrusts of a bird's or insect's wings generate both an upward force that supports the animal and a forward one that moves the animal forward.

The thrust may come from a change in body shape, as when an amoeba forms pseudopodia. Thrusts are developed by the creeping foot of a snail or the burrowing foot of a clam by changes in foot shape. An earthworm develops thrust by the contraction of muscles in its body wall that develop forces in the coelomic fluid of its hydroskeleton. Contraction of circular muscles in the body wall increases the pressure of the fluid, stiffens the body and forces it to elongate, stretching longitudinal muscles in the body wall in the process (Fig. 5.27 *A*). Contraction of longitudinal muscles also increases coelomic pressure but causes the body to shorten and widen as circular muscles are stretched (Fig. 5.27 *B*). The partitioning of the coelom and muscles into discrete segments in the earthworm and most other annelids enables body lengthening and shortening to be confined to a few segments at a time. Alternate waves of elongation and contraction move posteriorly along the length of the animal (Fig. 5.27 *C*). First the front of the body is elongated and pushed forward. Then this part shortens and widens, bristles in the body wall anchor it to the substratum, and more posterior parts of the body are pulled forward. The segmentation of annelids evolved independently of that of chordates as a burrowing mechanism. It is basically a modification of the body wall muscles, but blood vessels, nerves, and other internal organs are also segmentally arranged.

The legs of arthropods and terrestrial vertebrates act as lever arms that transform a muscle pull to a thrust upon the ground (Fig. 5.28). The point of rotation, or **fulcrum,** of many of these lever systems is at or near the proximal end of the appendage. In lever systems that flex an appendage or pull it posteriorly relative to the body, the point of application of the **in-force** by muscles and the point of delivery of **out-force** by the lever system lie on the same side of the fulcrum, but the in-force is applied closer. The forces are **vector quantities,** for they have both magnitude and direction. The

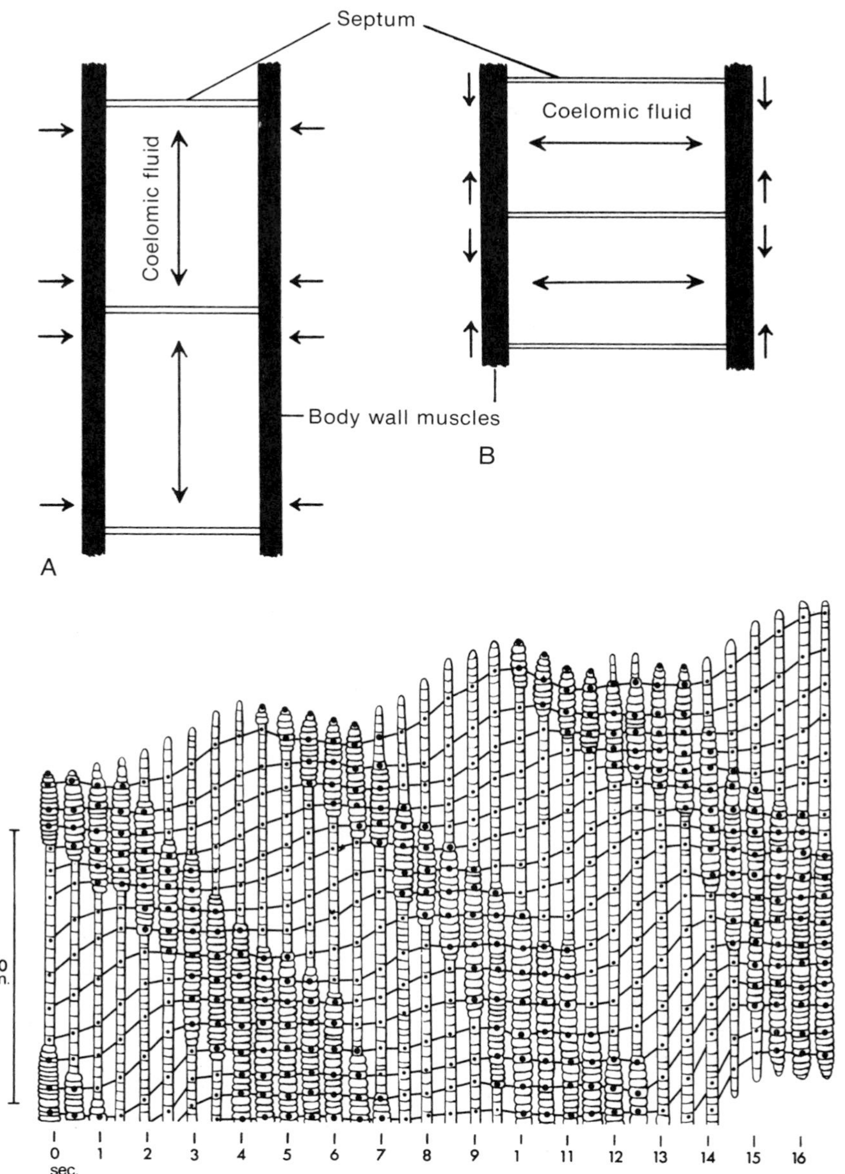

FIGURE 5.27 *A* and *B*, Diagrammatic frontal section through two annelid segments. External arrows indicate direction of force exerted by body wall muscles. Internal arrows indicate direction of force exerted by coelomic fluid pressure: (*A*) during circular muscle contraction and (*B*) during longitudinal muscle contraction. *C*, Diagram showing mode of locomotion of an earthworm. Segments undergoing longitudinal muscle contraction are marked with larger dots and drawn twice as wide as those undergoing circular muscle contraction. The forward progression of a segment during the course of several waves of circular muscle contraction is indicated by the horizontal lines connecting the same segments. (After Gray and Lissmann.)

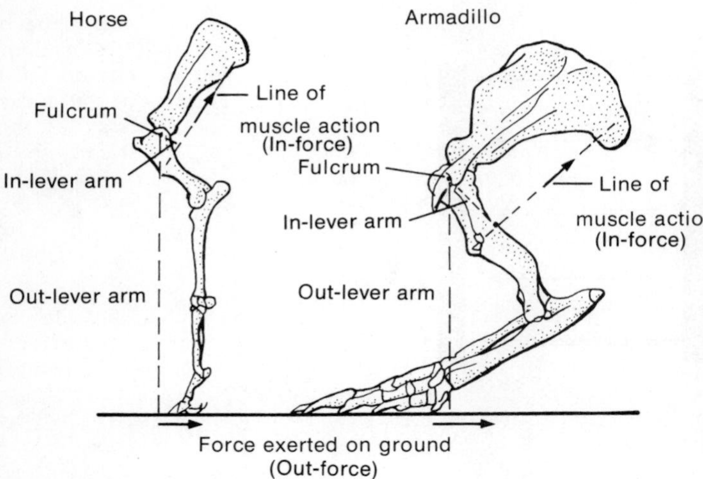

FIGURE 5.28 Forelimbs of a horse and an armadillo drawn to the same size to show how changes in the in-lever and out-lever arm relationships adapt the limb lever system for velocity or force. (After J. Maynard Smith, from Young, J. Z.: *Life of Mammals.* 2nd ed. New York, Oxford University Press, 1975.)

in-lever arm (the perpendicular distance between the fulcrum and the line of muscle action) is shorter than the **out-lever arm** (the perpendicular distance from the fulcrum to the distal end of the lever system). When the lever system is in equilibrium, the following products are equal:

in-force × in-lever arm =
out-force × out-lever arm

Since the in-lever arm is the shorter, it is evident that the in-force developed by the muscles must exceed the out-force delivered. The system is mechanically inefficient from the point of view of in-force and out-force relationships, but it provides for compactness and relatively high velocity. **Velocity,** also a vector quantity with both magnitude and direction, is also related to the length of the lever arms. A point on a lever farther from the fulcrum moves faster than one near the fulcrum:

in-velocity × out-lever arm =
out-velocity × in-lever arm

It follows from these force and velocity relationships that force and velocity can be altered by changing the length of the in- and out-lever arms relative to each other, but force and velocity are inversely related. In a horse, the gear ratio (length of out-lever/length of in-lever) is much higher than in an armadillo (Fig. 5.28). Consequently, the distal end of a horse's limb, which is adapted for running, can move faster but with relatively less force than an armadillo's limb; whereas the armadillo's limb, adapted for burrowing, has a lower gear ratio and can deliver relatively more force but at the expense of velocity.

In most lever systems that extend an appendage or pull it forward relative to the body, for example the extension of the human arm by the triceps muscle (Fig. 5.13), the in-lever (perpendicular distance from the elbow joint to the line of action of the triceps) and the out-lever (perpendicular distance from the elbow joint to the hand) are on opposite sides of the fulcrum rather than on the same side, but the principles of force and velocity discussed before still apply.

In addition to providing support and propulsion, the locomotor mechanisms of aquatic and flying animals must provide for stability against roll, pitch, and yaw and for reduced friction. The number and combination of feet placed on the ground at one time by a terrestrial animal represents an interaction between the amount of support and thrust optimum for a particular body form and velocity.

The amount of energy required for locomotion tends to be minimized by various devices. For example, as a part of the body moves in one direction, some of the energy expended may be stored as potential **strain energy** by the stretching of a tendon or the deforming of elastic material in a joint. This energy is then released and contributes to the recovery movement. Wing movements of an insect utilize the elasticity of the thoracic wall. When running mammals land upon their toes, tendons in the foot and leg are stretched because body weight, continuing to push downward, flexes the leg and foot. The elastic recoil of this strain energy helps raise the body and swing the leg forward.

An interesting special use of strain energy in fleas was discovered. Calculations showed that the power of the limb muscles, that is, the rate at which they do work, is inadequate for fleas to jump the distances that they do. Apparently the slow contraction of their leg muscles before the jump stores strain energy in special pads of elastic material, **resilin,** associated with the joints. The sudden release of this energy acts as a catapult and makes the jump possible.

Summary

1. A body covering protects and separates the body from the outside world. The body covering of most invertebrates is an epidermis of simple columnar epithelium; that of vertebrates, a many-layered epithelium overlying a dense connective tissue. Pigment is present in the skin of many animals and usually is contained in chromatophores. Bony scales characterize the skin of fishes. The skin becomes highly cornified in terrestrial vertebrates, and reptiles are clothed with horny scales. Skin glands are abundant in all vertebrates except for reptiles and birds. Feathers develop from the skin of birds; hair, from that of mammals.

2. Skeletons provide support, protection, mineral storage, transfer of muscle forces, and sometimes have other functions. They are usually composed of calcium or other mineral compounds that often are bound to protein fibers. Chitin and other materials are sometimes used.

3. Most invertebrates have exoskeletons secreted on the body surface by the epidermis. They are very effective in small animals, but would become very cumbersome in a large animal. The arthropod exoskeleton does not grow with the animal and must be molted periodically.

4. Endoskeletons occur in a few invertebrates and in all vertebrates. Although not limited to large animals, an endoskeleton is more efficient for support in a large animal than is an exoskeleton.

5. Hydroskeletons depend on the turgidity of a liquid within some body organ or space. They are found in worms and some other invertebrates.

6. The vertebrate skeleton can be divided into a somatic skeleton in the body wall and appendages and a deeper visceral skeleton associated with the gills and jaws. During evolution from fish to terrestrial vertebrates, the somatic skeleton becomes larger and stronger in connection with support and movement on land, and the visceral skeleton becomes reduced, although parts of it remain.

7. Elongated muscle cells usually are bound together by connective tissue and organized into muscles. Muscles only perform work by contracting and must act against an antagonistic force.

8. When a muscle fiber is stimulated, calcium ions are released, and heads on the myosin myofilaments bind to the now unblocked active sites on actin myofilaments. The heads bend slightly, detach, and reattach to adjacent active sites. One molecule of ATP is used for a cycle of attachment, bending, and detachment of a myosin head. Pumping out calcium blocks the active sites and prevents further contraction. Energy for the resynthesis of ATP comes first from the transfer of a phosphate group to ADP from creatine or arginine phosphate and ultimately from the metabolism of food. Muscles that contract more rapidly than oxygen can be delivered store glycogen and break it down anaerobically into lactic acid.

9. Most muscles of higher vertebrates are twitch or phasic muscles whose motor units respond in an all-or-none fashion. Gradation of tension comes from both temporal and spatial summation. Many muscles contain both slowly contracting phasic fibers that derive energy from oxidative phosphorylation and fast phasic fibers that can derive energy from anaerobic glycolysis. Tonic muscle fibers do not respond in an all-or-none fashion. Gradation in force comes from the rate of stimulation and the distance the action potential spreads along the fiber.

10. Some invertebrate muscles have unique characteristics. Molluscan catch muscles remain contracted for long periods of time. Patterns of innervation enable arthropods to develop a wide range of force and speed of contraction in systems that have only a few neurons and muscle fibers. Indirect flight muscles of insects contract more rapidly than nerve impulses can reach them.

11. In the evolution from fish to terrestrial vertebrates, the segmentation of trunk muscles is reduced, appendicular muscles increase in size and complexity, and branchiomeric muscles become reduced.

12. Pseudopod formation during amoeboid motion depends on the interaction of actin and myosin molecules. In ciliary and flagellar movement, microtubules of tubulin are pulled by dynein bridges.

13. Effective locomotion requires a means for support, generating a thrust, and maintaining stability. Aquatic animals are supported by water, but air sacs, oil, or other light materials help them achieve neutral buoyancy. Terrestrial animals are supported by the skeleton. Thrust is derived from forces generated by undulations of the body, a jet of water, air moving over wings, changes in body shape, or the pushing of legs. The number and combination of feet upon the ground provides optimum thrust and stability. The storage of some expended energy as elastic storage energy in tendons and ligaments minimizes the total energy needs.

References and Selected Readings

Additional information on the skeleton and movement can be found in the general references cited at the end of Chapter 4.

Cohen, C.: The protein switch of muscle contraction. *Scientific American, 233* (Nov.): 36, 1975. A discussion of the role of calcium and regulatory proteins in unblocking active sites on actin filaments.

Elder, H. Y., and Trueman, E. R.: *Aspects of Animal Movements.* Cambridge, Cambridge University Press, 1980. A recent review by the Society for Experimental Biology of muscle contraction, skeletal factors in locomotion, swimming, flying, walking, and other types of movement.

Goldman, R., T. Pollard, and J. Roseblum: *Cell Motility.* Cold Spring Harbor Laboratories, Conferences in Cell Proliferation, 1976. A three-volume collection of papers reviewing the state of our knowledge on muscle contraction, amoeboid and ciliary action, and other types of cell motility.

Gray, J.: *Animal Locomotion.* New York, W. W. Norton & Company, 1968. A thorough analysis of the principles and types of animal locomotion.

Lazarides, E., and J. P. Revel: The molecular basis of cell movement. *Scientific American, 240* (May): 100, 1979. A review of muscle contraction and the role of contractile proteins in other types of cell motility.

Pritchard, J. J.: *Bones.* London, Oxford University Press, 1974. The microscopic structure, growth, remodeling, and architecture of bone are summarized in this Oxford Biology Reader.

Satir, P.: How cilia move. *Scientific American, 231* (Oct.): 44, 1974. The beating of cilia and flagella and mechanisms underlying their action are described.

Trueman, E. R.: *The Locomotion of Soft-Bodied Animals.* Bristol, England, Edward Arnold Press, 1975. An excellent review of crawling, swimming, burrowing, and their evolution among soft-bodied invertebrates.

Wilkie, D. R.: *Studies in Biology.* 11. Muscle. London, The Camelot Press, 1968. An excellent summary of muscle structure and action from the biochemical level to muscles as organs working within the body.

6 Animal Nutrition

ORIENTATION

The production of energy that drives living systems and the formation of new cells or cellular components necessary for maintenance and growth in organisms demand a continual supply of organic molecules. The processes and problems related to that supply are termed nutrition and are important to all organisms. This chapter will explore the ways animals obtain organic nutrients, called feeding, and the ways in which such nutrients are handled within the animal body in order to make them available for cellular synthesis and combustion. In the course of that exploration we will see some of the steps by which the animal digestive system may have evolved. Finally, we will look at rates of energy utilization that nutrient procurement make possible.

Foods are organic compounds used in the synthesis of new biomolecules and as fuels in the production of cellular energy. Most foods are fatty acids, glycerols, sugars, and amino acids. Foods, or organic nutrients, are the same for all organisms, whether they are bats, oak trees, or fungi.

Organisms differ not in the nature of their foods but in the way they acquire them. **Autotrophs** (Gr. *autos,* self + *trophe,* nourish), which include all green plants, algae, some protozoans, and some bacteria, synthesize their foods from inorganic compounds.

Heterotrophs (Gr. *heteros,* other + *trophe,* nourish) acquire their foods from the cells or the organic products of other organisms. Excepting certain protozoa, all animals are heterotrophs. Those that consume the cells or cellular products of other organisms are said to be **holozoic** (Gr. *holos,* whole + *zoon,* animal). Some animal parasites, as well as many bacteria and fungi, although heterotrophic, are not holozoic. They absorb food from the body of their host or digest organic materials externally and absorb the products (e.g., mold on bread). They are said to be **saprozoic** (Gr. *sapros,* rotten + *zoon,* animal) or **saprophytic.**

Digestion and Absorption

Digestion. Digestion is a necessary part of heterotrophic nutrition. The large molecules of carbohydrates, fats, and proteins consumed as parts of cells and tissues must be broken down into smaller constituent units, such as sugars and amino acids, in order to be transported across cell membranes in the process of absorption. Even if membrane transport of large molecules were not a problem, the organic compounds synthesized by a heterotroph are often not the same as those consumed as food. Digestion is therefore necessary before reassembly can occur.

Digestion involves the addition of water to the molecule being fragmented, a reaction termed **hydrolysis:**

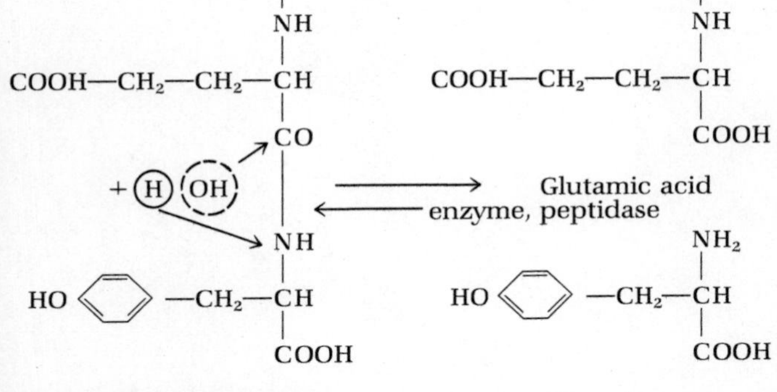

```
                      |                                         |
                      NH                                        NH
                      |                                         |
  COOH—CH2—CH2—CH                          COOH—CH2—CH2—CH
                      |                                         |
                      CO                                       COOH
 + (H) (OH)           |     ————————→         Glutamic acid
                      |     ←———————— enzyme, peptidase
                      NH                                       NH2
                      |                                         |
  HO-C6H4 —CH2—CH                         HO-C6H4 —CH2—CH
                      |                                         |
                     COOH                                      COOH

Two terminal amino acids                   Tyrosine
     of a protein
```

Although hydrolysis is an exothermic reaction, only a small amount of energy is released, and enzymes must catalyze the reaction in order for it to take place rapidly.

Enzymes are more or less specific in cleaving certain types of bonds, and thus different enzymes attack different types of food molecules or different parts of the molecule. However, many digestive enzymes are not as substrate-specific as are intracellular enzymes. This is an advantage in digestion, for food will generally include a wide variety of similar but not identical compounds.

Most animals from sponges to humans possess the same general types of enzymes. The major classes are as follows: **Proteases** cleave the peptide bonds of proteins. **Exopeptidases** separate the terminal amino acid from the rest of the protein molecule. **Endopeptidases** attack the center of the molecule, splitting it into two smaller fractions. The point of cleavage depends upon the enzyme. Trypsin, for example, can cleave only those peptide bonds adjacent to the amino acids arginine and lysine.

Most protein-digesting enzymes are released in an inactive form, and conditions at the site of digestion bring about their activation. For example, pepsin, an endopeptidase, is secreted in the stomach of humans and other vertebrates as inactive pepsinogen. The acidity of the stomach, produced by secretions of hydrochloric acid, converts pepsinogen to pepsin.

The secretion of protein-splitting enzymes in an inactive form and the presence of a mucous lining of the digestive tract are important adaptations for preventing these enzymes from digesting the tissues that secrete them.

Carbohydrases digest carbohydrates. **Polysaccharidases** digest high molecular weight carbohydrates. **Amylase,** which breaks down starch, is the principal animal polysaccharidase, but some animals also possess **cellulase,** which breaks down cellulose. **Oligosaccharidases** split low molecular weight trisaccharides and disaccharides into simple sugars. Each requires a different enzyme; the disaccharides sucrose and maltose are hydrolyzed by **sucrase** and **maltase.**

Lipases split the ester bonds of fats, separating the fatty acids from glycerol.

Absorption. After digestion, the smaller food compounds are absorbed by cells of the digestive tract by diffusion or active transport. Different molecules, even those of similar size and structure, such as the simple sugars fructose, galactose, and glucose, have different routes of passage. Fats are unusual in that there is considerable absorption of certain fatty acids and glycerides as minute aggregate droplets. Some details will be provided later in the section on the vertebrate digestive system, but the subject of the absorption of organic nutrients by animals is still incompletely understood.

Evolution of the Animal Gut

We do not know what the first multicellular animals, or metazoans, were like. Many zoologists speculate that they were small, radially symmetrical, slightly elongate animals (see Fig. 4.2*A*). A layer of monociliated cells covered the outside of the body, and a solid mass of cells filled the interior. It is quite probable that there was neither mouth nor gut. Small food particles, such as bacteria, algae, protozoans, and cell fragments, were engulfed by the exterior cells. The food particle became enclosed within a food vacuole located in the cell, and digestive enzymes passed from the cytoplasm into the vacuole, where digestion occurred **intracellularly** (Fig. 20.2). The products of digestion were absorbed from the vacuole. Any undigested wastes were discharged to the exterior. Absorbed food material reached interior cells by diffusion. There are organisms living today that feed in this manner. Among multicellular animals, sponges have neither mouth nor gut, and the small particles on which they feed are digested intracellularly. Intracellular digestion is also characteristic of many protozoans, such as *Paramecium* and amoebas.

All other animals possess a mouth and gut. If a gut is not present, the gutless condition is secondary; i.e., the gut has been lost. The early gut was probably a simple sac with only one opening, the mouth. A digestive system of this type is found in many of the more primitive animals, such as the little flatworm *Macrostomum* (Fig. 6.1). These little freshwater animals, less than 1 mm. in length, swim or crawl about over debris in pools and ponds. The ventral mouth leads into a simple ciliated tube, the pharynx, that in turn opens into a large simple sac, the **intestine** or **enteron.** These flatworms feed on other small animals that they swallow whole. When the prey reaches the intestinal sac, certain cells produce enzymes that pass into the cavity and digest proteins **extracellularly.** The prey soon fragments, and the resulting particles are phagocytized by other intestinal cells, where digestion is completed intracellularly. Indigestible wastes are **egested** back out of the mouth.

Note the difference in the size of the food of these flatworms and that of sponges and our hypothetical metazoan. The evolution of a gut cavity makes possible the utilization of larger food masses. This in turn demands that digestion be extracellular, at least in part.

The more familiar planarians are larger flatworms, in which the gut sac is highly branched, facilitating the diffusion of absorbed food material to various parts of the body (Fig. 23.2). Planarians also have a muscular tubular pharynx that can be extended out of the mouth into the body of the prey or into dead animal remains. Special pharyngeal enzymes aid in penetration by the pharynx, and the contents of the prey are sucked up into the intestine.

Hydras, sea anemones, and corals also have a digestive cavity with only one opening (Fig. 23.1) They feed on other small animals that, after being swallowed whole, are first digested extracellularly and then intracellularly.

Most animals possess a digestive tract with separate intake and exit openings, mouth and anus. Such an arrangement permits simultaneous and continuous processing of food as it passes along the gut tube. One process need not be halted to permit another to take place. Parts of the gut tube may be restricted to specific functions, and this has led to regional specialization. The following are the most commonly encountered specializations of the animal digestive tract:

Buccal, or Mouth, Cavity The anterior region of the gut into which the mouth opens. The buccal cavity commonly contains teeth, jaws, salivary glands and

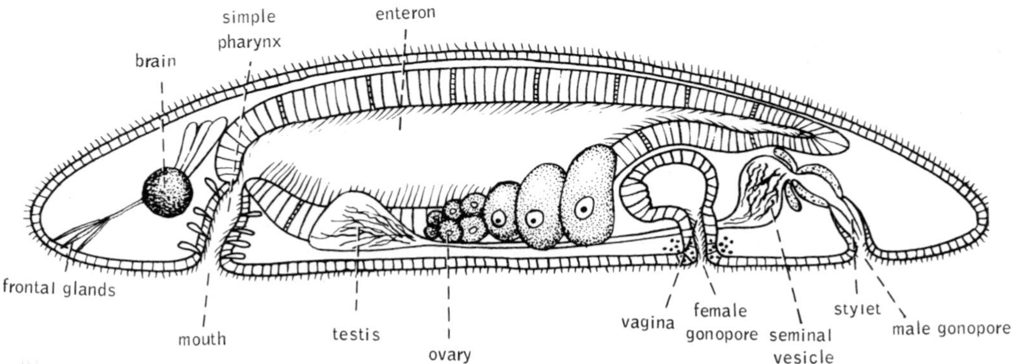

FIGURE 6.1 Semidiagrammatic sagittal section of the freshwater flatworm *Macrostomum.* (After Ax, P. in Dougherty, E. C. (Ed.): *The Lower Metazoa.* University of California Press, 1963.)

other structures concerned with feeding and food intake, or **ingestion.**

Pharynx An anterior region of the gut that is usually muscular and frequently specialized for the ingestion of food. The vertebrate pharynx can be defined by position but not by function.

Esophagus A tubular section that transports food to more posterior regions of the gut.

Crop An area specialized for temporary food storage.

Gizzard A region of the gut specialized for the mechanical breakdown, or **trituration,** of food into smaller particles prior to digestion. The walls are muscular, and the surface facing the lumen often bears hard plates, teeth, or ridges.

Stomach A dilated, sometimes muscular, region of the gut, within which storage and digestion take place. In some animals the stomach may also be a site of absorption.

Intestine A tubular region involved in digestion and absorption. Feces are usually formed in the posterior part of the intestine.

Rectum A terminal part of the gut in which feces are formed and stored prior to elimination.

Two other kinds of specialization of the gut deserve mention. Parts of the gut tube, especially the intestine, commonly contain structural modifications that increase the surface area. The walls may be infolded or bear finger-like projections; or there may be outpocketings called ceca present.

Secretory cells are usually a part of the gut lining, but they may be concentrated in special glandular regions or organs. The vertebrate liver and pancreas are two examples. Although in the adult such organs may be connected to the gut by only a small duct, they arise as outpocketings of the gut tube during embryonic development.

Diets and Feeding Mechanisms

No digestive tract possesses all of the specializations described above. Whatever specializations are present depend largely upon (1) the diet and (2) the feeding habits of the animals.

The feeding habits and diet are related in turn to the animal's life style; i.e., whether it is free moving or attached, lives in burrows or in tubes and so forth.

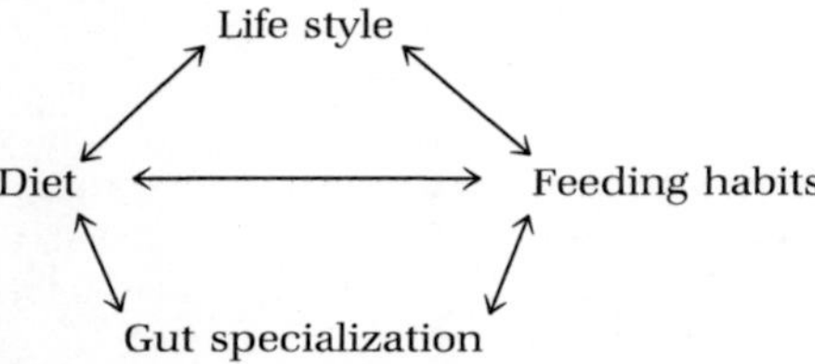

Kinds of Diets. Animals have exploited virtually all types of organic food sources. **Herbivorous** animals utilize plants as food. Most herbivores feed on only a few, or in some cases on only one, plant species. Moreover, the diet may be restricted to certain parts of the plant, such as roots, leaves, sap, nectar, and seeds. There are also many species of animals that feed on algae or fungi. All of these restrictions account for the great diversity of herbivorous diets.

Carnivorous animals feed on the bodies of other animals. Like herbivores, carnivores are usually restricted to the kinds of food found in their habitat and geographical range. Although most carnivores usually consume all or most of the bodies of their prey, some eat only certain parts, such as blood, skin, and so on. Animal **parasites** are actually carnivores, but the injury they inflict on the prey, or **host,** is usually not sufficient to cause death.

Many animals utilize nonliving organic substances as food sources. Such specialized diets may be limited to hair, feathers, shed skin, feces (coprophagy), or decomposing animal bodies (carrion feeders). Nonliving plant products, such as cellulose, and other organic compounds of decomposing vegetation are important food sources.

Many animals do not use plant or animal remains until they have been broken up into small fragments, or **detritus,** and for some the bacterial decomposers rather than the detritus itself may serve as food. In terrestrial habitats, such as a forest floor, detritus is called **humus** and consists mostly of plant remains. In the sea, detritus derived from various sources is at first suspended in the sea water but gradually sinks to the bottom and becomes mixed with sand grains. The deposited material is an important food source for many animals.

The diet of some animals is augmented by or dependent upon other organisms, or **symbionts,** that live within them. Algal symbionts in hydras and corals provide their hosts with glucose or glycerol. The organic compounds obtained from these algae only supplement the diets of their hosts that also feed in more conventional ways.

Although large numbers of animals are herbivorous or feed on plant products, such as wood, relatively few animals possess the enzymes to digest cellulose. Cellulose is most commonly digested by bacterial and protozoan symbionts. In cattle and other ruminants, for example, grass or hay undergoes microbial fermentation in the large rumen and small reticulum, two of the four chambers of the stomach (Fig. 6.2). Food is regurgitated from time to time, and the animal ruminates, or chews, its cud. This reworking provides further mechanical breakdown of the fermenting cellulose substrate. The rumen, which has a capacity of up to approximately 200 L, contains a large colony of bacteria and other microorganisms that play a twofold role in the animal's nutri-

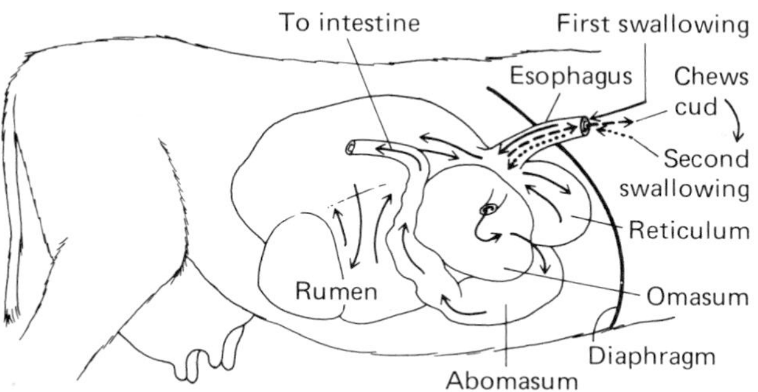

FIGURE 6.2 Passage of food through the compartmented stomach of a cow, a ruminant. Food first enters the large rumen and small reticulum, where fermentation, mixing, and sorting of plant materials take place. The rumen contents are regurgitated and chewed again, providing additional mechanical breakdown of plant fibers. Periodically chewed and fermented material is released from the reticulorumen into the omasum, where water, salts, and organic acids are absorbed. The residue then passes into the glandular abomasum, which is equivalent to the stomach of other mammals. (Modified from Houpt, T. R., in Goldstein, L. (Ed.): *Introduction to Comparative Physiology*. New York, Holt, Rinehart & Winston, 1977.)

tion. They produce **cellulases** that split the cellulose into disaccharide units (cellobiose), and other enzymes that convert the sugars to smaller units, mostly organic acids. The fermenting material and its products are buffered by the alkaline saliva that is secreted in great quantity by the salivary glands—a daily amount roughly equal to one third of the body weight of the cow! As the microorganisms multiply, they synthesize amino acids and proteins. Remarkably, the nitrogen source includes some of the cow's urea, which diffuses into the rumen. Much of the fermentation product is absorbed in the reticulorumen, which slowly releases liquified material through the valvelike omasum into the final region, the abomasum. The abomasum is the only one which contains gastric glands, and it is here that the digestion of microorganisms and other materials begins.

There are many herbivorous mammals, such as horses and rabbits, that are not ruminants but are also dependent upon microorganisms for cellulose digestion. The cecum at the beginning of the large intestine is an important site of fermentation in many of these species. Termites and wood roaches feed upon wood but harbor symbiotic protozoan flagellates rather than bacteria for cellulose breakdown.

Feeding Mechanisms. The food of animals is obtained through a variety of feeding mechanisms. Although similar diets and feeding habits have evolved many times in different groups, the feeding processes are not identical. Each mechanism utilizes the special features of the body plan of the particular group of animals involved.

Cropping and Sucking Herbivores. The buccal cavity of many herbivores is equipped with a variety of structures, such as jaws, teeth, blades, and scrapers, that crop and ingest fragments of plant tissue. The ingestive organs of other herbivores are designed for piercing plant tissue and sucking out sap or cell contents. The herbivore gut, which may contain colonies of cellulose-digesting organisms, is often longer than that of carnivores. The difference in gut length associated with diet is especially striking in the development of frogs. Algae-feeding herbivorous tadpoles have a long, greatly coiled gut, but during the metamorphosis of the tadpole to the insect-feeding adult form, the gut becomes much shorter.

Raptorial Feeding. Raptorial feeding (capture of prey) is generally associated with a carnivorous diet. Buccal structures are often involved in seizing and holding prey, but many animals utilize other parts of the body. The prey is either swallowed whole or is ingested in fragments after being bitten and torn. The digestive secretions contain a high percentage of proteases, and digestion is usually rapid.

Parasitism. Parasites are specialized carnivores that do not immediately kill their prey. Those that attach to the outside of the host (**ectoparasites**) are adapted for clinging and generally have ingestive organs for sucking blood or tissue fluids that are obtained by piercing or biting the integument. Those that live within the body of the host (**endoparasites**) may also obtain food by sucking, or they may absorb simple food compounds directly from the host.

Suspension Feeding. Many aquatic animals feed on minute plants, animals, and detritus suspended in water. The suspended plants and animals are collectively known as **plankton.** The most common type of suspension feeding is **filter feeding,** in which suspended food is filtered or screened from the water. Filter feeding typically involves three processes: (1) creation of a water current; (2) filtering by some specialized part of the body; and (3) transport of the collected food from the filter to the mouth. Fanworms, a group of marine tube-dwelling worms, are good examples of filter feeders (Fig. 6.3). In these worms, a circle of feather-like structures, **radioles,** projects from the head in the form of a funnel. Cilia on the radioles produce a water current that passes down into the funnel, and as water passes between the radioles, food particles are trapped in mucus on the radiole surface (the filter). Tracts of cilia, different from those producing the water current, transport food particles down the radioles to the mouth.

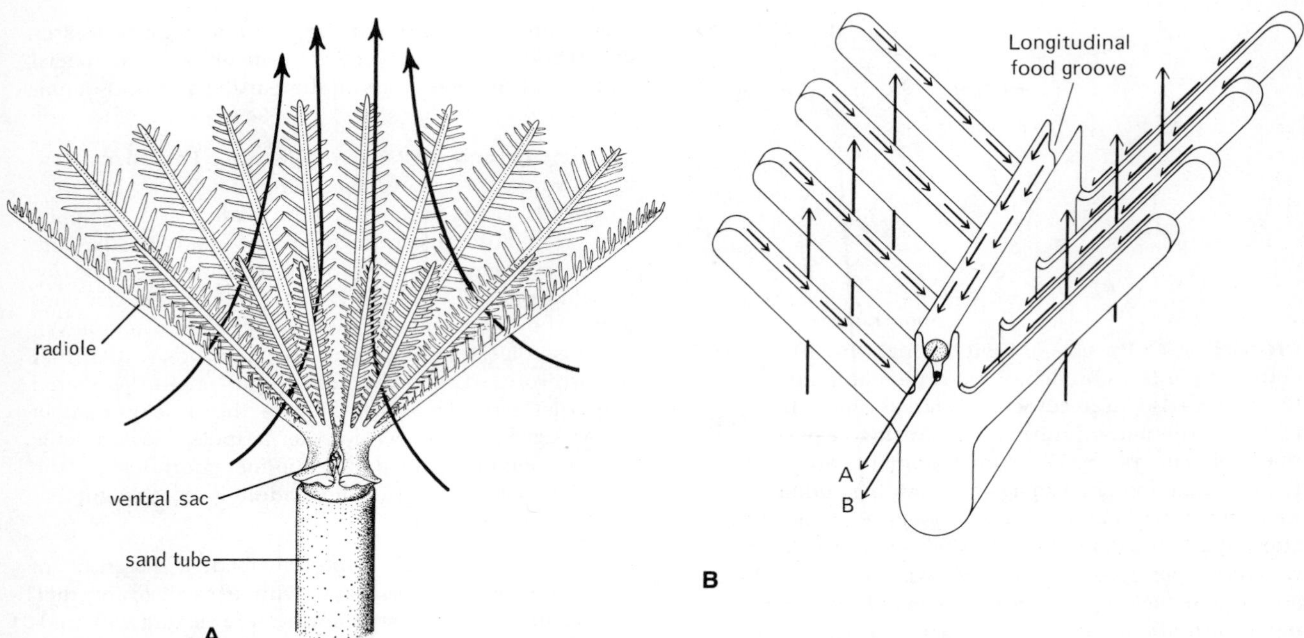

FIGURE 6.3 Filter feeding in the annelid fanworm *Sabella. A,* Water current passing through radioles projecting from opening of tube. *B,* Water current *(large arrows)* and ciliary tracts *(small arrows)* over a section of one radiole. The spheres *A* and *B* indicate the different sizes of particles sorted. (Modified from Nicol, E. A. T.: The feeding mechanism, formation of the tube, and physiology of digestion in *Sabella pavonina. Trans. R. Soc. Edinb., 56*(23):537–596, 1930.)

Deposit Feeding. The organic detritus deposited on the bottom of aquatic habitats and mixed with sand grains is utilized as a food source by many animals, called deposit feeders. Some, such as the marine lugworms, are **nonselective deposit feeders.** Lugworms live in burrows and ingest the substratum, both sand grains and deposited detritus, at the end of the burrow. During passage through the gut, the organic material is digested away from the mineral component. Like many other nonselective deposit feeders, lugworms periodically back up to the surface and defecate the mineral matter in conspicuous piles, called **castings.** Although terrestrial, earthworms are in part nonselective deposit feeders.

Some deposit-feeding animals ingest only organic material. The marine terebellid worms are such **selective deposit feeders** (Fig. 6.4). The terebellids possess a large number of long, delicate head tentacles that project out over the substratum from the tubes or burrows in which worms live. Light organic particles adhere to mucus on the tentacles and are transported back to the mouth by cilia, as well as by contraction of the tentacles.

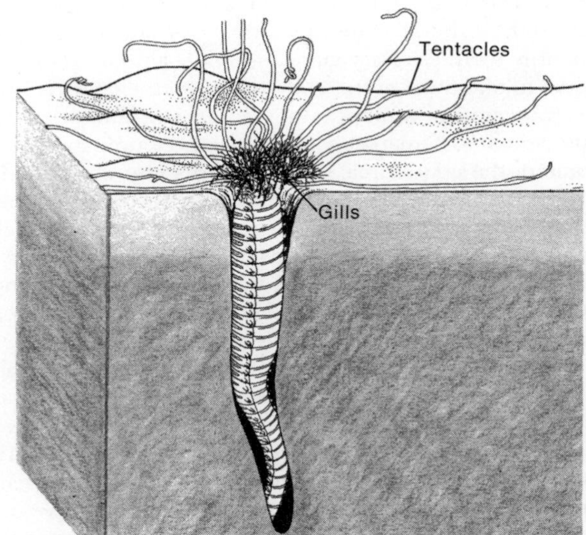

FIGURE 6.4 The annelid worm *Amphitrite* at aperture of burrow with tentacles outstretched over substratum.

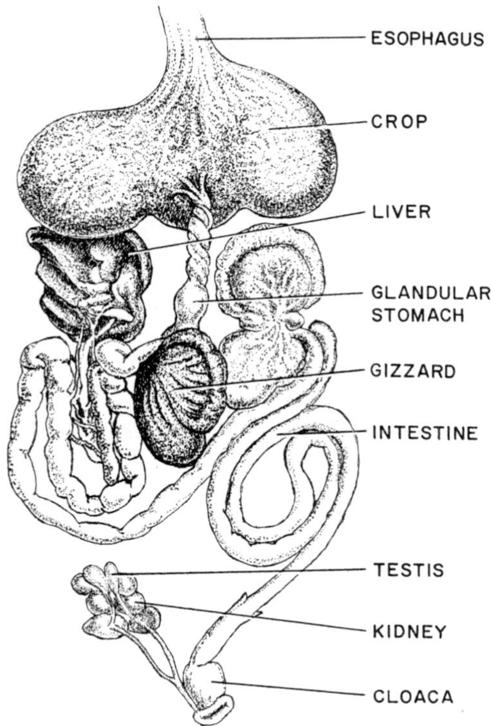

FIGURE 6.5 Digestive system of a pigeon; that of a finch would be similar. (After Schimkewitsch and Stresemann, from Welty J. C.: *The Life of Birds.* 3rd ed. Philadelphia, Saunders College Publishing, 1982.)

Feeding habits and diets greatly affect the gut modifications discussed earlier. The close interrelationship between all of the factors can be seen by comparing two related species with different feeding habits. A finch and an owl are related closely enough so that gut modifications correlate more with differences in diet than with differences in phylogenetic position. A seed-eating finch possesses a heavy beak especially adapted for cracking seeds. The edges are sharp and the interior surface is ridged. During feeding, seeds first pass down to the stomach, but as the stomach fills, the crop, which in birds is a side pocket of the esophagus, opens and receives the seeds (Fig. 6.5). Later, contractions of the crop send seeds to the stomach. The upper part of the stomach is glandular and produces proteases in an acid medium, as in most other vertebrates. The lower portion of the stomach is modified as a gizzard with heavy muscular walls. The interior keratinaceous surface is ridged. Ingested mineral particles, or grit, lodge between the ridges and aid in crushing and grinding seeds.

In contrast to finches, owls are raptorial and feed on mice and other small rodents that are swallowed entire. The prey is caught and killed with the powerful talons and short curved beak (Fig. 6.6). There is no crop, and the glandular part of the stomach is well developed. The gizzard is reduced to a valve that prevents hair and bones from entering the intestine. These indigestible items are regurgitated. The intestine is shorter than that of the finch.

The insects provide many similar illustrations. Grasshoppers, for example, have a pair of jawlike mandibles

A

B

FIGURE 6.6 *A,* Owl eating a mouse. (Only owls and parrots can lift food to the beak with the foot.) *B,* Pellets of undigestible material regurgitated by an owl. (Photos by C. R. Austing, from Welty, J. C.: *The Life of Birds.* 3rd ed. Philadelphia, Saunders College Publishing, 1982.)

on either side of the mouth. In grass-eating species, the mandibles have opposing chisel-like cutting surfaces and more basally flattened grinding molar surfaces (Fig. 6.7*A*). The grass is manipulated and aligned by a pair of maxillae and a labium located behind the mandibles.

Pieces of food removed by the mandibles are moistened with salivary secretions. They pass through a pharynx and an esophagus into a crop, where they are temporarily stored (Fig. 6.7*B*). The contents of the crop slowly pass into a proventriculus, a short gizzard-like region with projecting cuticular teeth. Digestion and absorption take place in the ventriculus, or midgut. Wastes are carried posteriorly by a tubular intestine to the terminal rectum that opens to the exterior through the anus.

Leafhoppers and squash bugs have the same feeding appendages as grasshoppers, but they are highly modified for piercing and sucking (Fig. 6.8*A*). A needle-like stylet formed by the mandibles and maxillae contains two channels, one for conveying salivary secretions outward and one for the passage of plant juices into the gut. The elongated labium functions as a guide for the stylet.

The pharynx is modified as a pump for drawing in fluid that then passes through an esophagus to the midgut. The proventriculus is reduced to a valve.

Fluid feeders take in large amounts of water. In leafhoppers, the intestine loops forward over the anterior part of the midgut, and much of the ingested water passes directly across the midgut walls into the intestinal lumen, bypassing the remainder of the midgut (Fig. 6.8*B*).

The Vertebrate Pattern

Mouth. The basic pattern of the digestive system is similar in all vertebrates. In very primitive vertebrates the mouth is unsupported by jaws, but most vertebrates

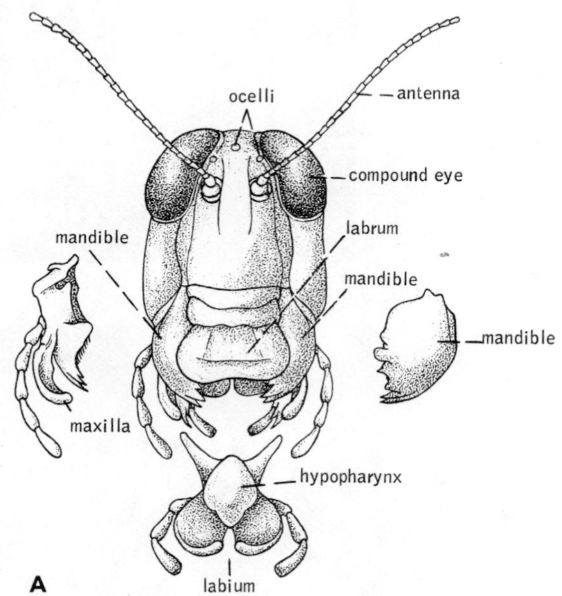

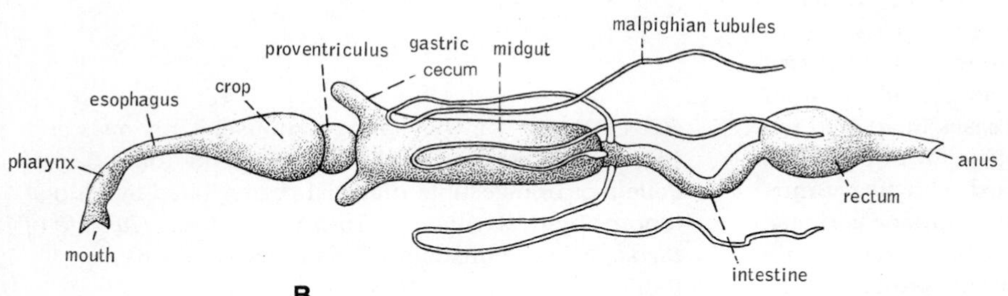

FIGURE 6.7 *A,* Front view of a grasshopper, showing mouthparts. *B,* Lateral view of digestive tract of a chewing insect, such as a grasshopper. (After Snodgrass.)

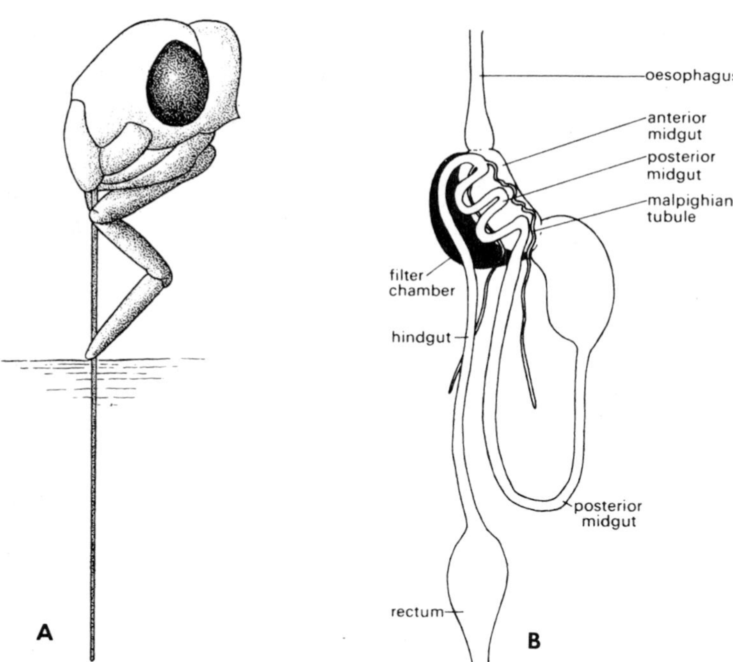

FIGURE 6.8 *A,* head of a hemipteran insect penetrating a plant surface with its beak. *B,* Digestive tract of a spittlebug. (*A* after Kullenberg; *B* after Snodgrass, from Chapman, R. F.: *The Insects: Structure and Function.* Copyright 1969 by Elsevier Science Publishing Co., Inc.)

have jaws and a good complement of teeth that aid in obtaining food.

Teeth are similar in structure to the scales of sharks, which are composed of enamel- and dentin-like materials and are believed to have evolved from bony scales. A representative mammalian tooth (Fig. 6.9) consists of a **crown** projecting above the gum, a **neck** surrounded by the gum, and one or more **roots** embedded in sockets in the jaws. The crown is covered by a layer of **enamel.** Enamel, which is the hardest substance in the body, consists almost entirely of crystals of calcium salts. Calcium, phosphate, and fluoride are important

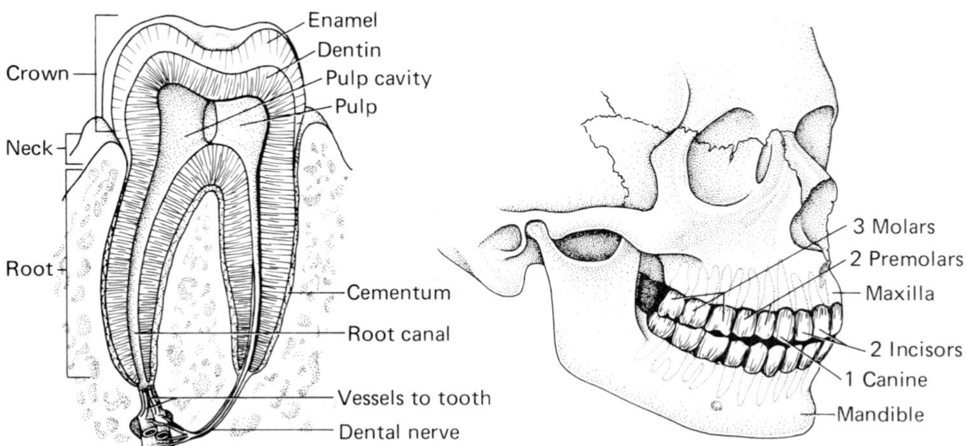

FIGURE 6.9 The teeth of humans. *Left,* A vertical section through a molar tooth; *right,* types of teeth. (From Jacob, Francone and Lossow: *Structure and Function in Man.* 4th ed. Philadelphia, W. B. Saunders Publishing Co., 1978.)

constituents of enamel, and all must be present in the diet in suitable amounts for proper tooth development and maintenance. The rest of the tooth is composed of **dentin,** a substance similar in composition to bone. In the center of the tooth is a **pulp cavity,** containing blood vessels and nerves. A layer of **cement** covers much of the root and holds the tooth firmly in place in the jaw.

In mammals, the teeth are differentiated into several types that are used for the seizing and mechanical breakdown of food, and there is a close correlation between the number and shape of different tooth types and the diet and feeding habit of the species (p. 738). Mammalian teeth, unlike those of lower vertebrates, are not continuously replaced by new sets. Humans first develop a set of deciduous, or **milk, teeth**—two incisors, one canine, and two premolars on each side of each jaw. These are later replaced by **permanent teeth;** in addition, three molars develop on each side of each jaw behind the premolars. The molars remain throughout life and are not replaced (Fig. 6.9).

A fish can easily manipulate and swallow food because the flow of water aids in carrying it back into the pharynx. Oral glands and a muscular tongue are poorly developed in fishes. The evolution of these structures accompanied the transition from water to land, and they became more elaborate in the higher terrestrial vertebrates. The **tongue** of the frog and the anteater is a specialized food-gathering device (Fig. 6.10), but in most mammals its chief function is to manipulate food in the mouth and to aid in swallowing. In many mammals, the tongue pushes the food between the teeth so that it is thoroughly masticated and mixed with saliva. The food is then shaped into a ball, a **bolus,** and moved into the pharynx by raising the tongue (Fig. 6.11). The tongue also has numerous microscopic taste buds; the human tongue is of great importance in speech.

In addition to a liberal sprinkling of simple glands in the lining of the mouth cavity, mammals have evolved several pairs of conspicuous **salivary glands** that are connected to the mouth by ducts. The location of the human **parotid, mandibular,** and **sublingual glands** is shown in Figure 6.11. In primitive terrestrial vertebrates, oral glands simply secrete a mucous and watery fluid that lubricates the food, and this is still the major function of saliva. The saliva of most mammals and of a few other terrestrial vertebrates contains **salivary amylase** that splits starch and glycogen into the disaccharide maltose. Chloride ions present in the saliva are necessary to activate amylase. Its pH optimum is close to neutrality, so its action is eventually stopped by the acidic gastric juice of the stomach. But, since it takes one half hour or longer for the food and gastric juice to become thoroughly mixed, 40 percent or more of the starches are split before the amylase is inactivated.

FIGURE 6.10 Tongue action of a frog. Notice that the tongue is stretched about one third of the length of the body. (From Van Riper in *Natural History,* Vol. LXVI, No. 6.)

Pharynx and Esophagus. The pharynx of vertebrates extends from the back of the mouth cavity to the beginning of the esophagus. Part of the human pharynx and that of other mammals lies above the **soft palate** (Fig. 6.12) and receives the internal nostrils, or **choanae,** and the openings of the pair of eustachian, or **auditory, tubes** from the middle ear cavities. Another part lies beneath the soft palate, and a third part lies just posterior to these parts and leads to the esophagus and larynx. The larynx is the first part of the passageway to the lungs, and the opening into the larynx is the glottis. Passage of food into the pharynx initiates a series of reflexes: The muscular soft palate rises and prevents food from entering the nasal cavities; breathing momentarily stops; the larynx is elevated, and the flaplike epiglottis swings over the glottis, preventing food from entering the larynx; the tongue prevents food from returning to the mouth; and muscular contractions of the pharynx move the bolus into the esophagus.

The pharynx of the terrestrial vertebrate is a rather short region in which the food and air passages cross, but in fishes it is a more extensive area associated with the gill slits. Successive waves of contraction and relaxation of the muscles, known as **peristalsis,** propel the bolus down the esophagus to the stomach. The muscles relax in front of the food and contract behind it. When the food reaches the end of the esophagus, the cardiac sphincter that closes off the entrance to the stomach relaxes and allows it to enter. The esophagus is generally a simple conducting tube, but in some animals its structure has been modified for storage. The crop of the pigeon, for example, is a modified part of the esophagus (Fig. 6.5).

Stomach. The **stomach** is usually a J-shaped pouch (Fig. 6.11) whose chief functions are the storage and mechanical churning of food and the initiation of the enzymic hydrolysis of proteins. Lampreys, lungfishes, and

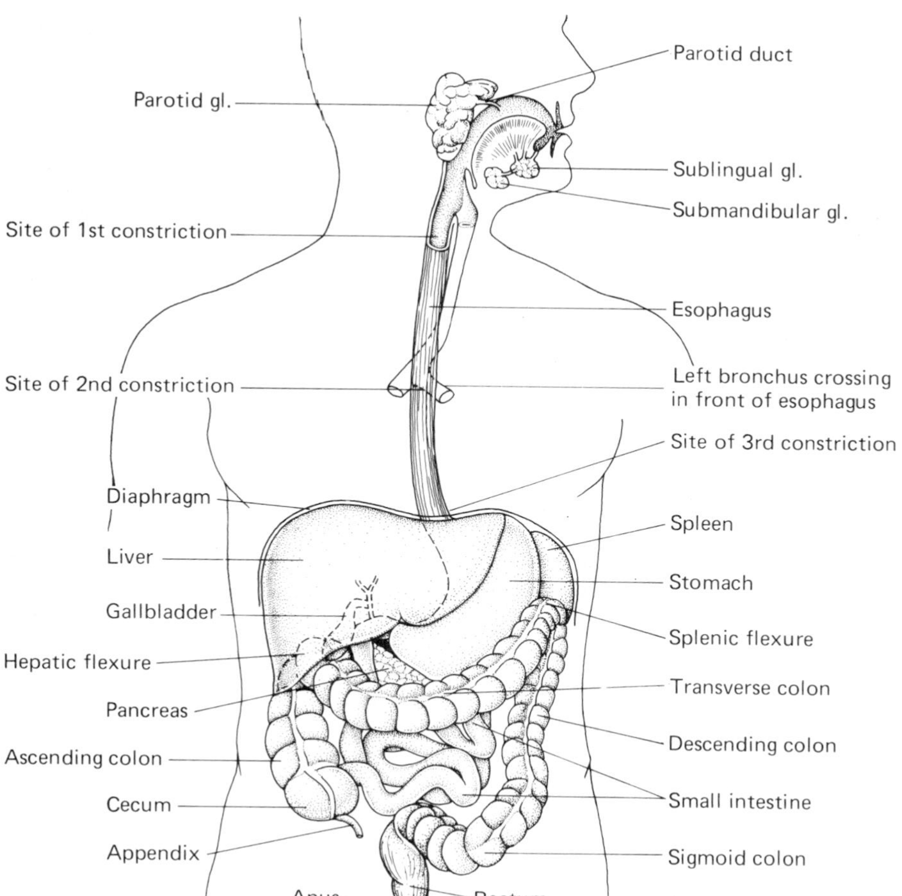

FIGURE 6.11 The human digestive system. (From Gardner, W. D., and W. A. Osburn: *Structure of the Human Body.* 2nd ed. Philadelphia, Saunders College Publishing, 1973.)

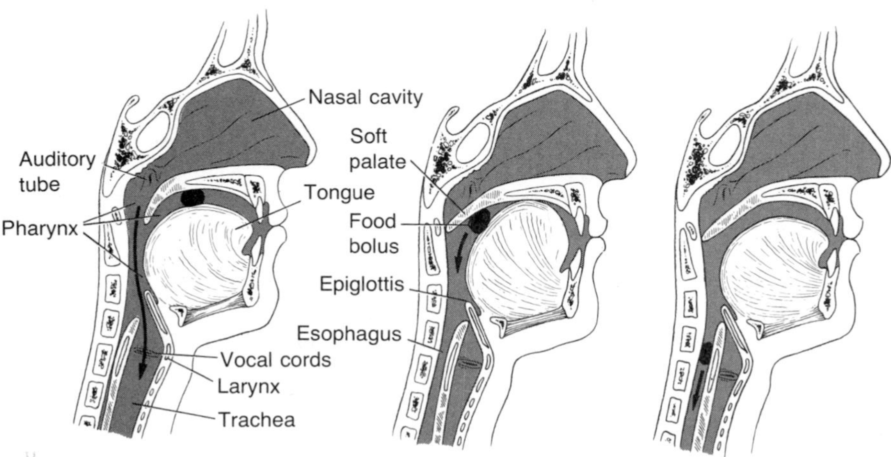

FIGURE 6.12 Diagrams showing the shifts in position of the tongue, soft palate, and epiglottis when a bolus of food is swallowed.

some other primitive fishes do not have a stomach, and the absence of this organ is thought to have been a characteristic of the ancestral vertebrates. The early vertebrates were probably filter feeders that fed more or less continuously on minute food particles that could be digested by the intestine alone. Presumably, the evolution of jaws and the habit of feeding less frequently and on larger pieces of food required that there be an organ for storage and initial conversion of this food into a state in which it could be digested further in the intestine.

After food enters the stomach, the **cardiac sphincter** at its anterior end and the **pyloric sphincter** at the posterior end close. Muscular contractions of the stomach churn the food, breaking it up mechanically and mixing it with the gastric juice secreted by tube-shaped **gastric glands.** The gastric juice contains hydrochloric acid and the proteolytic enzyme **pepsin.** In addition, **rennin** is particularly abundant in the stomach of young mammals and causes the milk protein casein to coagulate and remain in the stomach long enough to be acted on by pepsin.

As the food is reduced to a creamy material known as **chyme,** churning waves sweeping down the stomach briefly open the pyloric sphincter. The acidic food enters the intestine in 1- to 3-mL spurts (in humans) and is quickly neutralized by the alkaline secretions flowing into the intestine from the liver and pancreas.

Liver and Pancreas. The liver and pancreas are large, glandular outgrowths from the anterior part of the intestine (Fig. 6.11). **Liver** cells continually secrete **bile** that passes through hepatic ducts into the **common bile duct** and then up the cystic duct into the **gallbladder.** Bile does not enter the intestine immediately, for a sphincter at the intestinal end of the bile duct is closed until food enters the intestine. Contraction of the wall of the gallbladder forces bile out. The bile that is finally poured into the intestine is concentrated, for a considerable amount of water and some salts are absorbed from the bile in the gallbladder.

Although bile contains no digestive enzymes, it nevertheless has a twofold digestive role. Its alkalinity, along with that of the pancreatic secretions, neutralizes the acid food entering the intestine and creates a pH favorable for the action of pancreatic and intestinal enzymes. Its bile salts emulsify fats, breaking them up into smaller globules and thereby providing more surfaces on which fat-splitting enzymes can act. These salts are also essential for the absorption of fats and fat-soluble vitamins (A, D, E, K). Most of the bile salts are not eliminated with the feces but are absorbed from the intestine with the fats and return to the liver by the blood stream to be used again.

The color of bile results from the presence of **bile pigments** derived from the breakdown in the liver of hemoglobin from red blood cells. The bile pigments are converted by enzymes of the intestinal bacteria to the brown pigments responsible for the color of the feces. If their excretion is prevented by a gallstone or some other obstruction of the bile duct, they are reabsorbed by the liver and gallbladder, the feces are pale, and the pigments accumulate in the skin, giving it the yellowish tinge characteristic of jaundice.

The **pancreas** is an important digestive gland, producing quantities of enzymes that act upon carbohydrates, proteins, and fats. Enzymes that attack nucleic acids are produced by the pancreas of cows and other ruminants, but the secretion of nucleases is very slight in most other mammals. These enzymes enter the intestine by way of a pancreatic duct that joins the common bile duct. In some vertebrates an accessory pancreatic duct empties directly into the intestine. The pancreas contains patches of hormone-producing tissue, the **islets of Langerhans,** which will be considered in a later section.

Intestine. Most of the digestive processes, and virtually all of the absorption of the end products of digestion, occur in the intestine. Enzymes acting in the intestine are produced by the pancreas and by many of the epithelial cells lining the intestine. Adequate surface area for absorption is made available by the length of the intestine and by outgrowths and foldings of the lining.

The structural details of the intestine vary considerably among vertebrates. Primitive fishes have a short, straight intestine extending posteriorly from the stomach. Since its internal surface is increased by a helical fold known as the spiral valve, it is called a **valvular intestine.** Terrestrial vertebrates have lost the spiral valve and make up for this by an increase in the length of the intestine, which becomes more or less coiled. The tetrapod intestine has become further differentiated into an anterior **small intestine** and a posterior **large intestine** (Fig. 6.11). The first part of the small intestine is the **duodenum** and, in mammals, the two succeeding parts are the **jejunum** and **ileum.** The large intestine, or **colon,** of the frog and most vertebrates leads to a posterior chamber known as the **cloaca** (L. *cloaca,* sewer). The cloaca, which also receives the products of the urinary and reproductive systems, opens on the body surface by the **cloacal aperture.** In mammals, the cloaca has become divided into a ventral part that receives the urogenital products, and a dorsal **rectum** that opens on the body surface at the **anus.** A blind pouch called the **cecum** is present at the junction of small and large intestines of mammals. This is very long in herbivores,

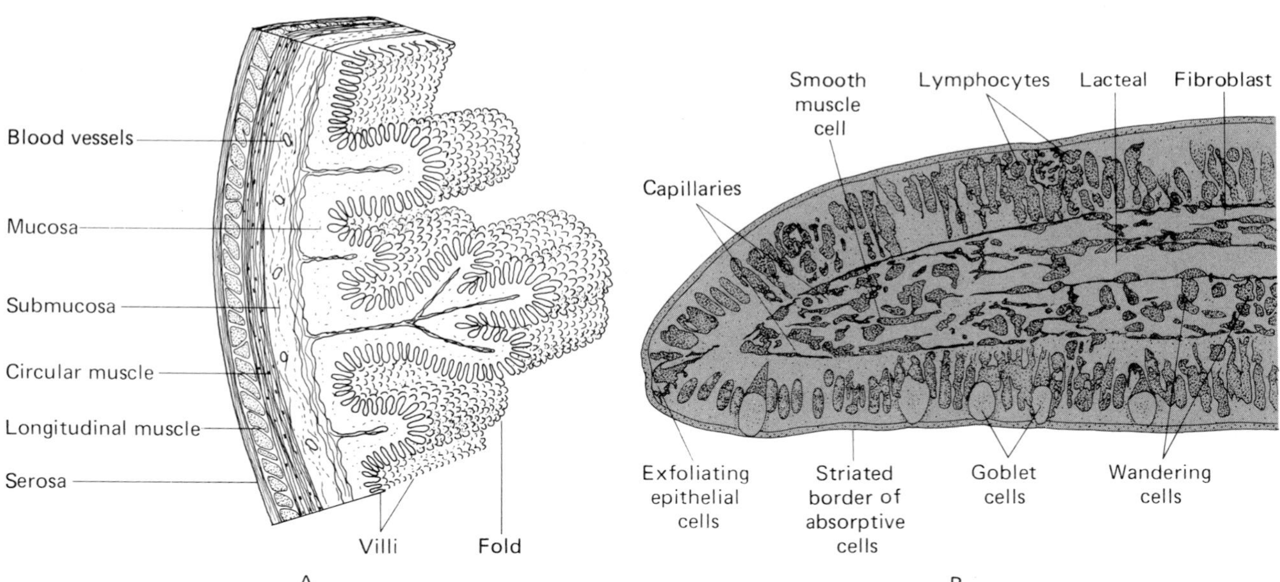

FIGURE 6.13 *A,* **Section of the small intestine viewed from the interior.** *B,* **Longitudinal section of the distal end of an intestinal villus. Microvilli compose the striated border of the absorptive cells.** (From Bloom and Fawcett: *A Textbook of Histology.* 10th ed. Philadelphia, W. B. Saunders Co., 1975.)

such as the rabbit and horse, and contains colonies of bacteria that digest cellulose. Humans have a small cecum with a **vermiform appendix** on its end, a structure found in many other primates and that is well developed in some rodents. The appendix is probably not vestigial as is often stated. An **ileocecal** valve located between the small and large intestines prevents bacteria in the colon from moving back up into the small intestine.

A transverse section of the small intestine of a mammal illustrates the microscopic structure of the digestive tract (Figs. 6.13 and 6.14). There is an outer covering of **visceral peritoneum** (the serous coat), a layer of **smooth muscle,** a layer of vascular connective tissue, the **submucosa,** and, finally, the innermost layer, the **mucosa** (or mucous membrane). The outer fibers of the muscular coat are usually described as longitudinal; the inner, as circular. Actually, both layers are spiral; the outer is an open spiral, and the inner, a tight spiral. When a section of the intestine is distended by **chyme** (contents), the stretched muscle responds by contracting. Some of the contraction results in localized constrictions, called **segmentation,** that aid in mixing of the contents. Other contractions, called **peristalsis,** are more wavelike. One muscle layer of the intestinal wall contracts over a distance of several centimeters, followed by the other layer. Such contraction moves the food down the intestine a short distance. The mucosa consists of a layer of smooth muscle, connective tissue, and, finally, the simple columnar epithelium next to the lumen. Many mucus-producing **goblet cells** are present in the lining epithelium, and their secretion helps to lubricate the food and protect the lining of the intestine from abrasion and chemical injury. In the small intestine of mammals and birds, the mucosa bears numerous minute, finger-shaped **villi** containing blood capillaries and small lymphatic vessels. Circular folds in the intestinal mucosa, villi, and microvilli form an enormous surface area for absorption; the surface area of the human small intestine may be as much as 550 m.^{2}. The villi are moved about by the muscle layer in the mucosa, the **muscularis mucosae,** and by strands of smooth muscle that extend into them.

At the base of the villi are tubelike areas known as the **crypts of Lieberkühn.** Mitotic division of epithelial cells in the bottom of the crypts continually produces new cells that migrate outward over the villi and are sloughed off. In the course of the outward migration the cells differentiate into the mucus-producing goblet cells and **absorptive cells.** There is a rapid turnover of epi-

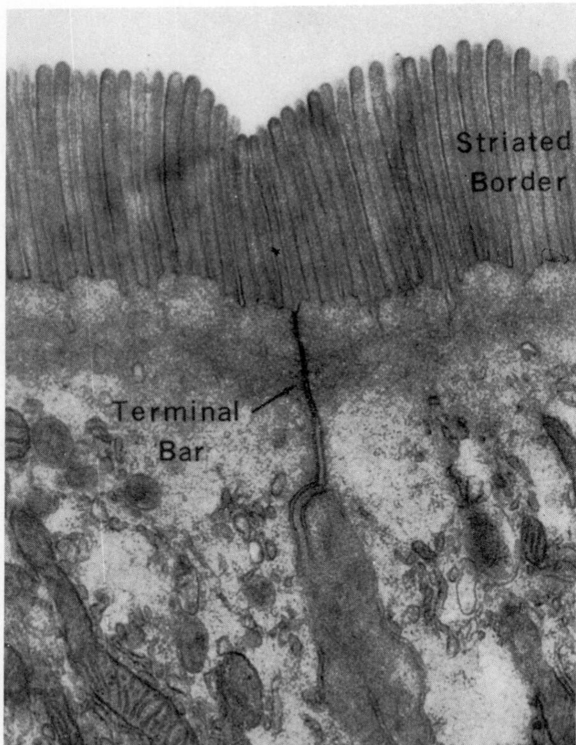

FIGURE 6.14 Electron micrograph of part of the surface of two epithelial cells from a villus to show the striated border composed of numerous microvilli. It is on the surfaces of the microvilli that digestion and absorption occur. The terminal bar is a specialization of the plasma membrane of adjacent cells that unites them tightly and closes the intercellular space. Original photograph, ×30,000. (From Bloom and Fawcett: *A Textbook of Histology*. 10th ed. Philadelphia, W. B. Saunders Co., 1975.)

thelial cells, for an individual cell lasts only about 48 hours. The contents of the shed cells are the source of the watery intestinal fluid.

Starches not broken down in the stomach by salivary amylase are soon converted to maltose in the neutral environment of the small intestine by **pancreatic amylase.** There may also be traces of amylase in the intestinal juice. The disaccharides maltose, sucrose, and lactose are acted upon by specific enzymes: **maltase, sucrase,** and **lactase.** These and other enzymes produced by the intestinal cells are not completely free in the main part of the intestinal lumen, for cell-free extracts of intestinal juice contain very little enzyme. Rather, the enzymes are located in the microvilli surfaces that compose the striated border of the absorptive cells of the villi (Figs. 6.13 and 6.14). The disaccharides are hydrolyzed as they enter this border; all yield glucose; sucrose and lactose also yield fructose and galactose, respectively. These monosaccharides are also absorbed by the absorptive cells, but the mechanism of active transport is still uncertain.

The endopeptidases **trypsin** and **chymotrypsin** are secreted by the pancreas as inactive precursors, **trypsinogen** and **chymotrypsinogen.** Their activation is initiated by **enterokinase,** an intestinal enzyme that splits off a portion of the trypsinogen molecule and converts it to trypsin. Once formed, trypsin helps to continue the activation of trypsinogen, and it alone activates chymotrypsinogen. The various endopeptidases—pepsin, trypsin, chymotrypsin—split bonds adjacent to specific amino acids; their concerted action is necessary for the fragmentation of protein molecules. Pepsin only splits bonds adjacent to phenylalanine or tyrosine; trypsin bonds next to arginine or lysine; and chymotrypsin bonds next to leucine, methionine, phenylalanine, tyrosine, or tryptophan. Exopeptidases secreted by the pancreas and small intestine attack the terminal peptide bonds, stripping off one amino acid after another. Exopeptidases known as **carboxypeptidases** split off the terminal amino acid with the free carboxyl group (p. 25); **aminopeptidases** separate the terminal amino acid with the free amino group (NH_2).

The intestinal exopeptidases are also located in the surface membrane of the absorptive cells of the villus, and these same cells are the site of amino acid absorption. There are different carriers for different groups of amino acids, and there is now evidence that tri- and dipeptides can be absorbed and then digested within the absorptive cells.

Fats emulsified by bile salts are attacked by esterases that cleave the ester bonds joining fatty acids to glycerol. **Lipase,** most of which is secreted by the pancreas, although some is produced in the small intestine, is the principal esterase of vertebrates. The mixture of bile salts, fatty acids, and partly digested fats collectively emulsifies fats further into particles, many of which are small enough to be absorbed directly.

The absorption of fats and fatty acids presents a special problem, for unlike the other end products of digestion, they are not water-soluble. Their uptake is facilitated by their combination with bile salts, for this makes a soluble complex. Bile salts are freed within the mucosal cells and recycled, and the fatty acids and glycerol are combined with phosphate to form **phospholipids.** These phospholipids are further stabilized with protein and released into the lymphatic system as small globules called **chylomicrons,** that are then transported to the blood stream. Other absorbed materials enter the capillaries of the blood vessels.

The total volume of secretions by the digestive tract is very large, about 8 L a day in the human. Although most of the water ingested and present in saliva and gastrointestinal secretions is absorbed in the small intestine, the material that enters the large intestine is still quite fluid. Most of the remaining water and many salts are absorbed as the residue passes through the colon. If the residue passes through very slowly and too much water is absorbed, the feces become very dry and hard, and **constipation** may result; if it goes through very rapidly and little water is absorbed, **diarrhea** results.

The large intestine of mammals harbors a great population of symbiotic bacteria, much like that of the rumen of ruminants. Some herbivorous species, such as horses, are dependent upon these bacteria for the breakdown of plant material; others, such as humans, are not. Nevertheless, the bacterial actions within the human large intestine are quite comparable to those in a horse. By the time food reaches the large intestine of humans most digestible compounds have been removed. Depending upon the diet, the residue includes a large amount of plant cellulose and pectins, large numbers of sloughed intestinal cells, and the polysaccharides of mucus, which is produced in great quantities along the entire length of the gut. The intestinal bacteria act on this substrate, and as in the rumen of cows, the fermentation products are organic acids and the gases methane and hydrogen.

In herbivorous mammals, such as horses, the organic acids are an important food material. Although humans are not dependent upon these compounds, at least some must be absorbed, for the organic acid content of the feces is less than expected. Most of the gases are reabsorbed and eventually eliminated through the lungs. The well-known flatulence produced by beans results from the rapid fermentation of a small carbohydrate molecule, stachyose, stored in bean seeds. Not all humans can digest milk sugar, an inability especially prevalent among some black populations, and the lactose is also fermented by the colon bacteria. As in ruminants, the bacteria obtain some nitrogen from the host's urea, which diffuses into the large intestine. As much as 25 percent of the human feces consists of eliminated bacteria, attesting to the large size and rapid reproduction of the gut flora.

The Control of Digestive Functions. Each of the secretions of the digestive tract is secreted at an appropriate time. We salivate when we eat, and gastric juice is produced when food reaches the stomach. The control of these digestive secretions is partly nervous and partly endocrine. The smell of food or its presence in the mouth stimulates sensory nerves that carry impulses to a salivation center in the medulla of the brain. From there the impulses are relayed along motor nerves to the salivary glands, which then secrete.

The control of gastric secretion is more complex. Years ago the famous Russian physiologist Pavlov performed an experiment in which he brought the esophagus of a dog to the surface of the neck and severed it. When the dog ate, the food did not reach the stomach, yet some gastric juice was secreted provided that the **vagus** (L. *vagus*, wandering) **nerve,** which carries motor fibers to the stomach and other internal organs, was left intact. If the vagus nerve was cut, this secretion did not occur. This experiment proved that the control of gastric secretion was at least partly nervous. Subsequently, it was discovered that if the vagus was cut but food was permitted to reach the stomach, a considerable flow of gastric juice was produced. Obviously, the vagus nerve is not the only means of stimulating the gastric glands. Further investigation revealed that when partly digested food reaches the pyloric region of the stomach, certain of the mucosal cells produce the hormone **gastrin** that is absorbed into the blood through the stomach wall and ultimately reaches the gastric glands, stimulating them to secrete.

Gastric secretion is reduced and finally stopped as the stomach empties and food enters the duodenum. When food, especially fats, enters the duodenum, the duodenal mucosa produces the hormone **enterogastrone,** or gastric inhibitory peptide, which, on reaching the stomach, inhibits the secretion of the gastric glands and slows down the churning action of the stomach. This not only helps to prevent the stomach from digesting its own lining but also enables fatty foods to stay for a longer period in the duodenum where they can be acted on by bile salts and lipase.

One of the first hormones to be discovered was **secretin,** which initiates pancreatic secretion. In 1902 Bayliss and Starling were investigating the current belief that the secretion of pancreatic juice was under nervous control. They found that the pancreas secreted its juice when acid food entered the small intestine even though the nerves to and from the intestine were cut. A stimulant of some sort apparently traveled in the blood. The injection of acids into the blood stream had no effect, so they reasoned that some stimulating principle must be produced by the intestinal mucosa upon exposure to acid foods. When they injected extracts of such a mucosa into the circulatory system, the pancreas secreted. It is now recognized that secretin itself only stimulates the pancreas to produce a copious flow of a bicarbonate-rich alkaline liquid that neutralizes the acid material entering from the stomach. A second hormone produced by the intestinal mucosa, **cholecystokinin-pancreozymin,** is necessary to stimulate the release of pancreatic enzymes. These hormones also cause the gallbladder to contract and release the bile. Vagal stimulation also plays a role in the release of bile. Figure 6.15 summarizes the control systems of digestive functions.

Regulation of Food Supply

Autotrophic organisms can regulate the acquisition of different food compounds by regulating their rates of synthesis. Regulation by controlling diet is much more difficult for heterotrophic organisms. Most animals feed discontinuously, which results in more food being acquired than can be utilized at one time, and the food ingested at a particular feeding may contain just adequate amounts of some substances and excess amounts of others.

These problems are overcome by storage and interconversion of the several types of food. Glycogen, the typical animal carbohydrate, functions as a short-term storage product, one that is called up rapidly when available sugars in the blood or elsewhere are depleted. Glycogen is stored in most vertebrate cells but is especially abundant in liver and muscle. Significantly, all of the blood draining the intestine passes through the liver by way of the **hepatic portal** venous system. In the

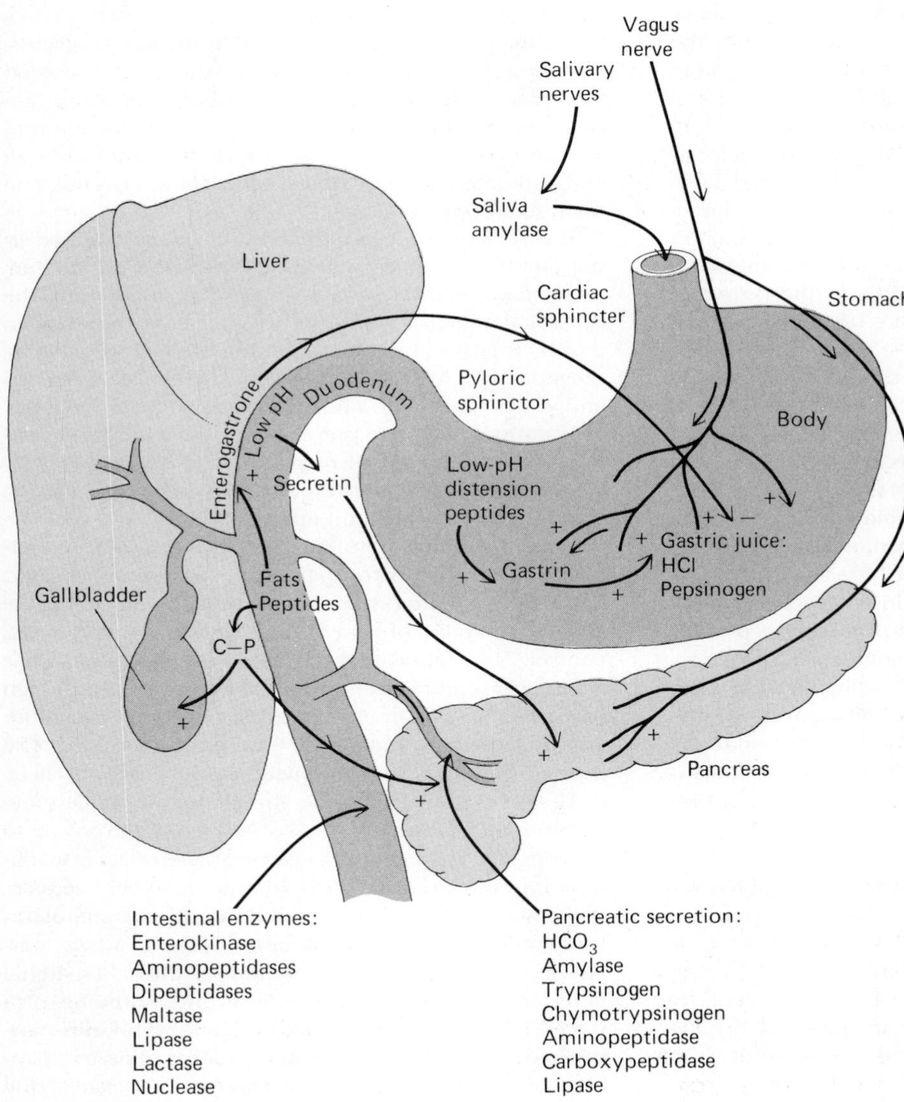

FIGURE 6.15 Control of the digestive functions of the stomach, intestine, gallbladder, and pancreas. Plus and minus symbols indicate excitatory or inhibitory action. C–P is the hormone cholecystokinin-pancreozymin. (Modified from Houpt, T. R., in Goldstein, L. (Ed.): *Introduction to Comparative Physiology*. New York, Holt, Rinehart & Winston, 1977.)

course of passage through the liver, some of the excess sugars absorbed from the intestine are taken up by the liver cells and converted to glycogen. The hormone **insulin,** produced by clusters of pancreatic endocrine cells, the **islets of Langerhans,** controls the cellular uptake of glucose and glycogen synthesis. A rise in blood sugar stimulates the pancreatic cells to produce insulin. Insulin is transported via the blood stream throughout the body where it induces glycogen synthesis in muscle cells and in the liver. The reverse reaction, the conversion of glycogen to glucose, is regulated by another pancreatic hormone, **glucagon,** and by **epinephrine.** However, muscle cells lack the enzyme to convert glucose-6-phosphate to glucose, so that muscle glycogen can only be used as an energy store for muscle cells.

Food taken in as carbohydrate or protein can be converted to fats for storage and then utilized subsequently as a source of energy when the animal is not feeding. Within two or three hours after absorbing a fatty meal, the chylomicrons disappear from the blood; some are taken up by liver cells; others are digested within the blood stream by lipoprotein lipase. Lipoprotein lipase is produced in great quantities by the fat depots of the body, and it is believed that most of the hydrolyzed fat is quickly absorbed and resynthesized by these tissues. Fats stored in the liver or adipose tissue undergo constant breakdown and resynthesis, even though the total amount stored may change very little over long periods of time.

Proteins are not usually stored in animals, but there is continual turnover of tissue proteins; i.e., proteins are cleaved to amino acids, which may be transported elsewhere, and new proteins are constructed. In the human the protein turnover amounts to about 400 g. a day, which is about four times the usual daily protein intake. During prolonged starvation many animals will utilize their tissue proteins, sacrificing muscles, gonads, and other organs.

The ability of animals to convert food compounds from one form to another is of great importance in balancing the uncertainties of diet with internal food requirements. Excess sugars can be converted and stored as glycogen or can be converted to fatty acids and stored as fats. However, most animals cannot synthesize the polyunsaturated fatty acids they require; these, called **essential fatty acids,** must be obtained in the diet.

Like fatty acids, some but not all of the 20 amino acids used in protein synthesis can be synthesized by animals. Most animals must obtain about half or more of the 20 from their diet (**essential amino acids**). The large amount of excess amino acids that may be absorbed from the diet, especially in carnivorous animals, cannot be stored but can be used as fuels or converted to other compounds, such as sugar or fatty acids, via pathways discussed earlier (p. 74) and summarized in Figure 6.16. Protein metabolism will be discussed further in Chapter 9.

Nonfoods in the Diet

An animal's diet not only provides food but also is the principal source of other substances essential to cellular constituents. The diet is a major source of water, both free water and water from metabolism; indeed, there are many animals that never drink water, obtaining it entirely from food. The diet is also the primary source of inorganic ions, such as calcium, phosphate, potassium, sodium, chloride, iodide, and others involved in various cell functions and in the formation of skeletons and shells.

Traces of copper are needed as a component of certain enzyme systems and for the proper utilization of iron in hemoglobin. Traces of manganese, molybdenum, cobalt, and zinc are required as components of certain enzyme systems; zinc, for example, is a component of carbonic anhydrase, alcohol dehydrogenase, and a number of other enzymes.

Still another provision of the animal diet is vitamins. Vitamins are relatively simple organic compounds required in small amounts in the diet. They differ widely in their chemical structure but have in common the fact that they cannot be synthesized in adequate amounts by the animal and hence must be present in the diet. All plants and animals require many of these same substances to carry out specific metabolic functions, but organisms differ in their ability to synthesize them; thus, what is a "vitamin" for one animal or plant is not necessarily one for another. Only humans, monkeys, and guinea pigs require vitamin C, **ascorbic acid,** in their diets; other animals can synthesize ascorbic acid from glucose. The mold *Neurospora* requires the vitamin biotin. Insects cannot synthesize cholesterol, and it might be argued that for insects cholesterol is a vitamin.

A diet deficient in a vitamin results in a pathological condition with characteristic symptoms. A deficiency in ascorbic acid, for example, causes the disease known as **scurvy.** Since ascorbic acid is involved in the maintenance of intercellular materials, such as collagen fibers, scurvy is characterized by such symptoms as slow wound healing, bleeding gums, and loosening teeth. Studies of the deficiency diseases led to the discovery of vitamins. Scurvy was a common disease on the long voyages of sailing ships and was soon recognized to be the consequence of poor diet. In 1747 an English naval

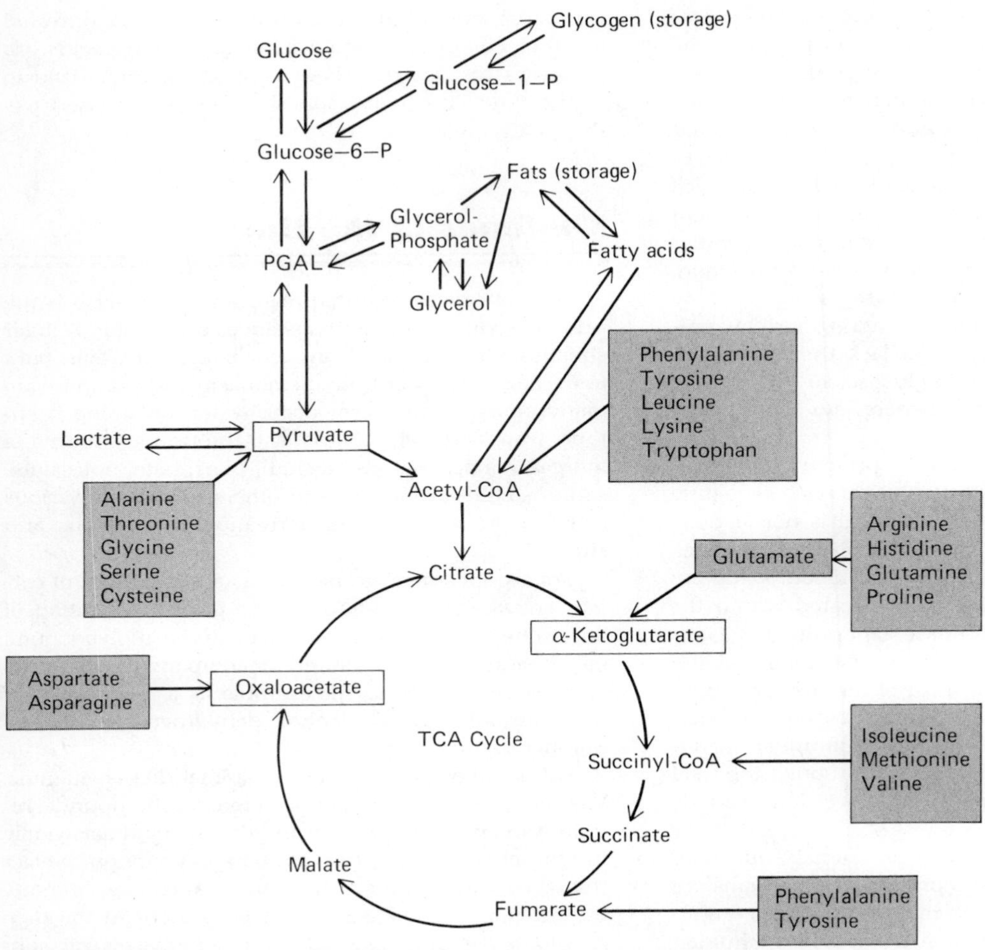

FIGURE 6.16 **Diagram of metabolic pathways by which food compounds are interconverted, stored, or used as fuels. Keto acids in unshaded boxes; amino acids in shaded boxes. (Modified in part from Lehninger.)**

physician, John Lind, discovered that scurvy could be quickly cured with citrus fruit. Subsequent use of citrus fruits to treat scurvy led some 50 years later to the British navy requiring that its sailors eat limes as a preventive measure, even though there was no understanding of why limes prevented scurvy.

Christiaan Eijkman, a Dutch medical officer working in Indonesia in the late 1800s, was the first to recognize that a deficiency disease is caused by the absence of some substance that is not a food itself. The disease beriberi results from a deficiency of thiamine, one of the B vitamins. Eijkman discovered that chickens fed polished rice developed a disease rather like beriberi in humans. He found that both the chickens and beriberi victims could be cured by feeding them seed coats removed from the rice grains in polishing.

The name "vitamine" was coined because it was thought that these substances were vital amines. When this was found not to be the case, the name was shortened to "vitamin." Our understanding of vitamins soon outstripped the simple system of letter designation. The B vitamins are differentiated by number, but are related only in being found in similar foods. Some vitamins have names only. All of the human vitamins are probably now known.

Vitamins can be divided into two groups, those soluble in water and those soluble in fats. The water-soluble vitamins include ascorbic acid (C) and the B complex. All function as coenzymes and are thus important in various metabolic reactions. Their precise roles, especially those of the B complex, are fairly well understood. The fat-soluble vitamins, A, D, E, and K, have diverse functions and appear to be vitamins only for vertebrates.

Despite American preoccupation with vitamins, an ordinary balanced diet will readily supply the daily minimum vitamin requirements. Certain deficiency diseases are actually rare because the vitamin is never lacking. Riboflavin, for example, occurs widely in food and is also synthesized by the intestinal bacteria. Vitamin K is also synthesized by intestinal bacteria, and a deficiency of vitamin K is more often associated with some problem in absorption rather than with diet. Although vitamin D can be obtained from a variety of foods, such as egg yolks, fish, and liver, this vitamin can be synthesized when the skin is exposed to ultraviolet light. Its deficiency disease, rickets, is chiefly a consequence of inadequate diet during winter in northern climates.

The fat-soluble vitamins A and D can be toxic at high levels, but toxic concentrations can never be obtained through diet, only by overdoses of vitamin pills or with fortified food.

A summary of the common vitamins needed by humans, their characteristics, and their associated deficiency disease is presented in Table 6.1.

Fuel Utilization: Metabolic Rates and Energy

Foods that are not stored or utilized in the synthesis of new cell components are utilized as fuel in the production of cellular energy, a process termed cellular respiration. As described in Chapter 3, cell respiration is a process of energy transfer. Substrates, such as glucose, are metabolized to carbon dioxide and water, and some of the energy is conserved in the energy-rich phosphates of ATP. These energy transfers are only about 40 percent efficient, and most of the energy loss is in the form of heat. Since the heat conversion factor is constant and since energy production is almost entirely an aerobic process, the amount of heat produced and the amount of oxygen utilized are relatively precise reflections of the amount of fuel consumed, or the animal's **metabolic rate.**

These relationships are further influenced by activity. Muscular activity obviously demands more energy than that required in the resting state or for what is termed the **standard metabolic rate.** But the standard metabolic rates of different animals are not the same. In general, more active animals have higher metabolic rates even in the resting state. We might compare them to different automobiles that have their carburetors adjusted so that one idles faster than another.

The heat produced in cell metabolism is eventually dissipated across the body surface, but the rate of loss varies in different animals. Most animals are **ectothermic** (Gr. *ektos*, outside + *therme*, heat); their rate of heat loss is so high and their rate of heat production so low that the body temperature is set by that of the external environment rather than by internal metabolism. The body temperatures of such animals fluctuate with the temperature of the environment; the animals are said to be **poikilothermic** (Gr. *poikilos*, various + *therme*, heat). Ectothermic and poikilothermic are actually different ways of describing the same condition and can be used synonymously.

Many aquatic ectothermic animals, such as intertidal clams and worms, have metabolic rates that also vary with the environmental oxygen supply. For example, when the tide goes out, an intertidal razor clam will be restricted to a pool of water in the bottom of its burrow. The amount of oxygen in the burrow water steadily falls as it is used by the clam and by microorganisms in the surrounding sand, and the oxygen may eventually be entirely depleted. However, as the environmental oxygen falls the clam's metabolic rate and oxygen demands also fall, and when the oxygen is used up, the clam may utilize anaerobic respiration entirely. Of course, the clam is inactive during such periods of low environmental oxygen. Animals whose oxygen demands adjust to the oxygen supply in the environment are said to be **oxyconformers.** Not all poikilotherms have this ability. A snake will die whenever the environmental oxygen supply cannot meet the oxygen demands of the metabolic rate characteristic of a particular temperature.

Birds and mammals are **endothermic.** Their body temperature is derived from internal heat production. Heat loss is less than in ectotherms, and the rate of internal heat production is greater. Body temperature does not fluctuate and is independent of environmental temperatures. Animals with constant body temperature are said to be **homeothermic.**

Birds and mammals are the only living continuously endothermic animals. Many other animals, such as some lizards, certain active fish, and some insects, are intermittently endothermic.

Homeothermism is maintained in birds and mammals by regulating heat production and by controlling heat loss across the body surface. Fur and feathers act as insulating barriers, and in aquatic mammals, such as seals, porpoises, and whales, there is a thick layer of fat associated with the dermis of the skin. The sweat glands of mammals serve to cool the skin. Sweat is poured out onto the surface of the skin, and some of the heat re-

TABLE 6.1 Common Vitamins

Vitamins	Common Sources	Function	Diseases and Symptoms if Deficient in Diet
Water Soluble			
B Complex			
B_1, Thiamine	Yeast, meat, whole grain, eggs, milk, green vegetables	Thiamine pyrophosphate coenzyme in decarboxylation of α-keto acids (carbohydrate metabolism)	Beriberi (polyneuritis): spasms of rigidity of legs, nerve and muscle degeneration
B_2, Riboflavin	Same as B_1: colon bacteria	Flavin mononucleotide and flavin adenine dinucleotide are coenzymes for dehydrogenase reactions—electron transport in mitochondria and certain oxidations in the endoplasmic reticulum	Similar to niacin deficiency, but only mild deficiency probably ever occurs
B_6, Pyridoxine	Same as B_1	Pyridoxal-phosphate is a coenzyme for many reactions involving amino acid metabolism: transamination, decarboxylation, and others	Dermatitis, gastrointestinal disturbances, but deficiency rare
Niacin, or Nicotinic Acid	Same as B_1	Nicotinamide adenine dinucleotide (NAD) and nicotinamide adenine dinucleotide phosphate (NADP) are coenzymes for many dehydrogenase reactions in cellular oxidation	Pellagra: cracked, scaly skin, irritated mucous membranes, nervous disorders
B_{12}, Cobalamin	Synthesized by gut bacteria, found in animal foods only: meat, fish, eggs, milk	Cobalt-containing coenzymes involved in amino acid conversions and for DNA synthesis; cell division	Anemia (red cell precursors are most rapidly dividing cells)
Folic Acid	Same as B_1	Coenzyme tetrahydrofolic acid involved in conversion of glycine to serine and in DNA synthesis; cell division	Anemia; probably most common vitamin deficiency worldwide
C, Ascorbic Acid	Citrus fruits and fresh vegetables	Maintenance of intercellular substances: collagen fibers of connective tissue, capillary walls	Scurvy: bleeding gums, loosening teeth, slow wound healing
Fat Soluble			
A, Retinol	Butter, eggs, fish liver oils; carotene in plants can be converted to A; large amounts of A are stored in the liver	Component of the light-sensitive pigment, visual purple in the retina; maintenance and growth of epithelial cells	Night blindness, inflammation of the eyes, scaly skin, easy infection
D, Cholecalciferol	Butter, eggs, fish oils, liver; cholesta 5,7-dienol in skin converted to D on exposure to ultraviolet light	Absorption and utilization of calcium and phosphorus	Rickets: weak bones, and defective teeth
E, Alpha-tocopherol	Vegetable oils, vegetables, egg yolks, milk fat, liver; widely distributed in foods and stored in body	Prevents oxidation of polyunsaturated fats	Human deficiency detected only in premature babies: red blood cells rupture
K, Several Naphthoquinone compounds	Green vegetables, colon bacteria	Synthesis of prothrombin in liver, hence normal blood clotting	Bleeding

*Two other B complex vitamins, pantothenic acid and biotin, are rarely deficient in human diets.

quired for evaporation is removed from the body. Panting in dogs functions in the same way. It is significant that sweat glands are largely restricted to certain large mammals, such as primates, horses, and camels. Very small mammals, such as rodents, moles, and shrews, lack sweat glands but possess a very great surface area for heat loss relative to their size.

Various controls come into operation at temperature extremes. On a hot day, loss of excess internal heat must be facilitated. Blood flow into the skin is increased through the opening of arterioles. Sweating may occur. Activity may be reduced, and the animal may remain in its burrow or den.

At low temperatures heat loss must be curtailed. Blood flow to the skin is reduced. Shivering may occur, the muscle contraction thereby increasing internal heat production. The insulation of many temperate and arctic animals is very effective against heat loss.

Behavioral modifications are also important at lower temperatures. Many animals avoid exposure by remaining in burrows or retreats. Bears and some other mammals will sleep for long periods during the winter, but this is not hibernation. True hibernators are small mammals with high metabolic rates, such as some rodents, hamsters, bats, and hedgehogs, that undergo a marked state of torpor at certain lower critical temperatures. The metabolic rates decline greatly, heart beat slows, and the body temperature drops to a low level. The energy saving is very great, and the animal slowly utilizes a store of fat acquired before hibernation. Some birds also hibernate, and there are a few hummingbirds and bats (mammals) that exhibit a daily state of torpor. Adaptations for thermoregulation will be described in more detail in Chapters 33, 34, and 35.

If the oxygen consumptions per gram of body weight (a measure of fuel utilization and metabolic rate) of different species within the same group of animals are compared, striking differences are evident. Note that by using oxygen consumption per gram of body weight it is possible to compare animals that are greatly different in size, such as rats and horses. When the rates of different mammals are plotted against the mass of the animal, as in Figure 6.17, it can immediately be seen that metabolic rate and size are inversely related. For example, the metabolic rate of a shrew is many times greater than that of an elephant.

Why is the metabolic rate of small animals greater than that of large ones? Among mammals, homeothermism is certainly a factor. More internal heat must be produced to compensate for the greater heat loss from the larger surface area relative to volume in small species. But this is not the complete explanation. When metabolic rates are plotted against body size using a logarithmic scale, an equivalent relationship will have a slope of 1.0; i.e., an increase in one will result in an equivalent increase in the other (Fig. 6.18). If the increase in metabolic rate in small mammals were simply related to heat loss through increase in surface area, then the slope should parallel the slope for the relationship of surface area to volume, a slope of 0.67 (volume increases by the cube and surface area by the square). Instead, the actual slope is about 0.75. Moreover, the

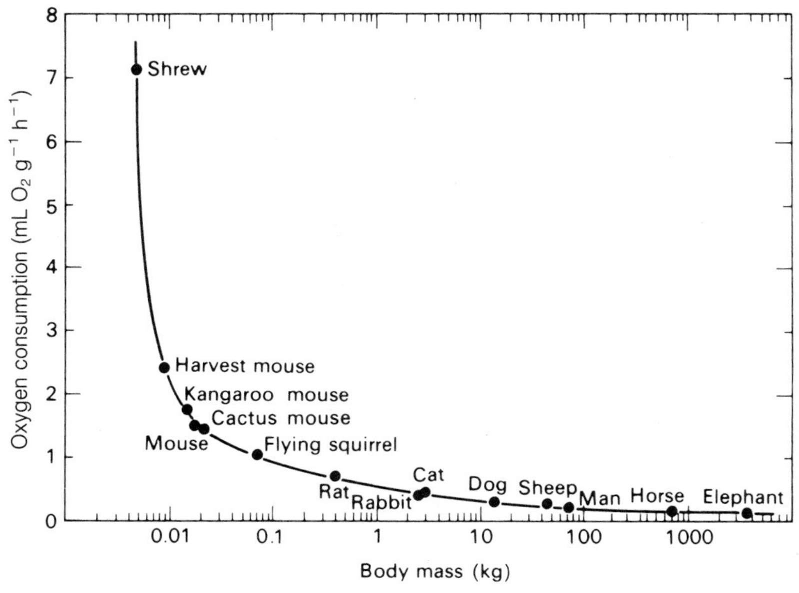

FIGURE 6.17 Rate of oxygen consumption of various mammals plotted against body weight. Ordinate has an arithmetic scale, but abscissa is logarithmic. (From Schmidt-Nielsen, K.: *Animal Physiology: Adaptation and Environment.* Cambridge University Press, 1975.)

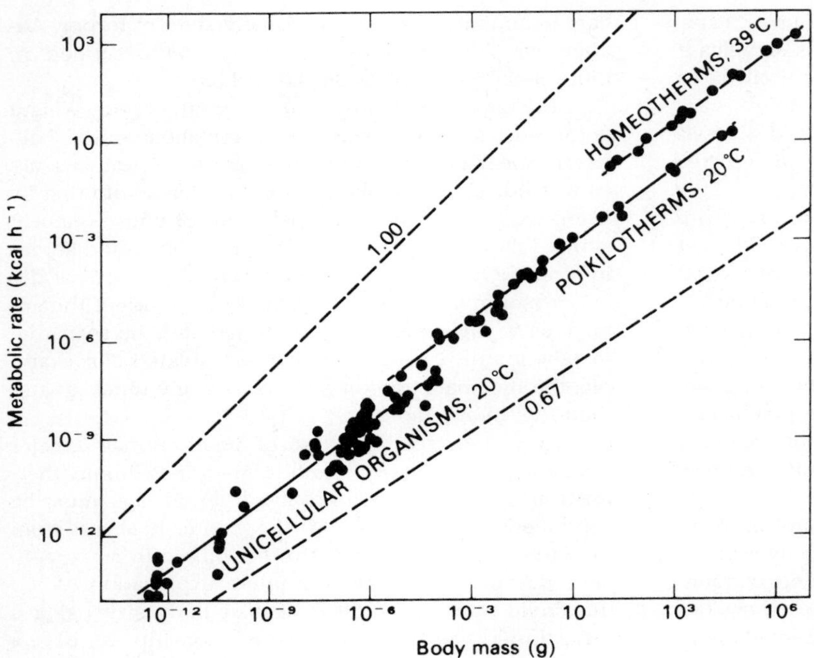

FIGURE 6.18 Metabolic rates of different organisms measured in the number of kilocalories consumed per hour, plotted against body weight, using log coordinates. The regression lines (slope) are about 0.75. (After Hemmingsen, from Schmidt-Nielsen, K: *Animal Physiology: Adaptation and Environment.* Cambridge University Press, 1975.)

slope is the same for animals that do not maintain a constant body temperature, poikilotherms, as it is for homeotherms. The reasons for this are not fully understood, but obviously the amount of surface area available for heat loss cannot be the only factor involved. The fact that the power output of the muscles of small animals is greater than that of large animals may be another factor (p. 114).

Summary

1. Food consists of organic nutrients (sugars, amino acids, fatty acids, and glycerol) that are used for the synthesis of new biomolecules and as fuel in the production of cellular energy. Organisms differ not in the nature of their foods but in the way in which they acquire them. Animals are heterotrophic, obtaining their food substances by the consumption of the bodies or products of other organisms.

2. Digestion, the reduction of large molecules to small ones, is a necessary adjunct to heterotrophic nutrition. Only small molecules can pass through cell membranes and be reassembled as new and different large molecules within cells. The hydrolytic reactions of digestion are catalyzed by enzymes; different enzymes attack different types and sizes of food molecules. Most animals possess the same general types of digestive enzymes.

3. The earliest metazoans were probably gutless, like living sponges. Minute food particles were engulfed by surface cells and digested intracellularly. Digestion in most animals occurs within a gut cavity, that permits extracellular digestion of large food masses. Primitively, as in hydras, sea anemones, and flatworms, the mouth is the only opening into the gut cavity, and digestion is initially extracellular and then intracellular. Wastes are egested back out of the mouth. Most animals possess a gut with separate entrance and exit openings that permit a one-way passage and simultaneous processing of food in different parts of the gut tube. The gut thus becomes regionally specialized along its length. Digestion is usually entirely extracellular. Specializations of the gut are correlated with the animal's diet and mode of feeding.

4. Symbiotic microorganisms contribute to the nutrition of many animals, either by sharing their products of photosynthesis (symbiotic algae in corals and some other animals) or by sharing their products of digestion (cellulose-digesting bacteria in ruminants and other animals).

5. Many animals feed on small food particles, that may consist of living organisms (plankton, fine algae, bacteria) or the fragmented remains of organisms (detritus or deposit material). Such particulate food may be obtained by filter feeding or by deposit feeding.

6. The vertebrate mouth cavity contributes to ingestion and may provide some mechanical breakdown of food through chewing. Salivary secretions moisten food and may initiate some digestion of starch. Food is delivered by the esophagus to the stomach, where the endopeptidase, pepsin, in an acid medium attacks proteins. In the duodenum the acidic chyme, containing partially digested food, is neutralized by alkaline secretions from the pancreas and liver. A battery of pancreatic enzymes attack all classes of large food molecules. Fats are emulsified by bile salts secreted by the liver.

7. In the small intestine the products of gastric and pancreatic digestion are reduced to their basic units (simple sugars, amino acids, and so on) by enzymes located on the surfaces of absorptive cells of the villi. These same cells are the site of absorption by active transport. Much of the fat emulsified by the bile salts is absorbed into lymph vessels as small droplets. Mucus secreted along much of the gut lubricates food and protects the gut lining.

8. The large intestine, or colon, is a site of water reabsorption and feces formation. In mammals the large intestine supports a bacterial flora that ferments undigestible food residues, largely plant material, if this is part of the diet. The fatty acid fermentation products are an important food source for many herbivorous animals.

9. The functions of the digestive tract are regulated by nervous and hormonal controls. Animals adjust differences between food supply and food utilization through short- and long-term food storage and by converting one type of food material into another. These processes are under hormonal control.

10. The diet of animals provides a number of necessary substances, largely inorganic ions and vitamins. Vitamins are small organic molecules that cannot be synthesized by the animal. They are required for a wide range of metabolic reactions.

11. The metabolic rate refers to the amount of energy utilized by an organism over a given period of time. In animals the rate of energy utilization, excluding that resulting from muscle action, is called the standard metabolic rate. Metabolic rates of different animals can be determined and compared by measuring oxygen consumption and heat production.

12. Animals that have relatively constant body temperatures are said to be homeothermic; those with fluctuating body temperatures are said to be poikilothermic. Animals in which the body temperature is largely determined by internal heat production are said to be endothermic; those in which the body heat level is largely determined by the environment, because of low heat production and high heat loss, are said to be ectothermic. Many ectothermic animals are oxyconformers; their oxygen consumption is adjusted to the amount of oxygen available in the environment. There is a relationship between metabolic rate, size, and surface area of animals; smaller animals, especially those that are endothermic, exhibit higher rates.

References and Suggested Readings

In addition to the titles listed below, many of the general references listed at the end of Chapter 4 cover the topic of nutrition and related subjects.

Davenport, H. W.: Why the stomach does not digest itself. *Scientific American,* 226(Jan.):86–93, 1972. An exploration of reasons pepsin in an acid medium does not damage the stomach walls.

Hamilton, E. M., and E. N. Whitney: *Nutrition, Concepts and Controversies.* 2nd ed. St. Paul, West Publishing Co., 1982. An interesting account of most aspects of human nutrition, with an analysis of various current dietary fads and predjudices.

Jennings, J. B.: *Feeding, Digestion, and Assimilation in Animals,* 2nd ed. New York, Pergamon Press, 1973. A brief account of all aspects of animal nutrition.

Jørgensen, C. B.: *The Biology of Suspension Feeding,* New York, Pergamon Press, 1966. A detailed review of filter feeding and other types of suspension feeding.

Moog, F.: The lining of the small intestine. *Scientific American, 245*(Nov.):154–176, 1981. An account of recent studies on digestion and absorption in the small intestine of mammals.

7 Gas Exchange

ORIENTATION

In order to sustain cellular respiration, the cells of animals must be continually supplied with oxygen, and carbon dioxide must be removed. The amount of oxygen available is quite different in aquatic and terrestrial environments, and animals differ in how much they need. These factors affect the type and location of the surfaces of the body where gas exchanges occur and the ways water or air are moved across them. Gases move between the gas exchange surfaces and the cells and tissues of the body by some combination of diffusion, tracheal systems, and transport by respiratory pigments in blood and other fluids.

The energy required for the myriad activities of cells is derived primarily from the reactions of biological oxidation (p. 68). The essential feature of these reactions is the transfer of hydrogen atoms or their electrons from donors to acceptors, accompanied by the transfer of energy to phosphate ester bonds. The ultimate electron acceptor in the metabolism of most animal cells is oxygen, which is converted to water. For these processes to continue, the supply of oxygen must be renewed constantly, because only small amounts of oxygen can be stored in blood or in tissues, and the carbon dioxide produced must be removed. Thus, the survival of each cell requires the continuous exchange of gases with its environment.

TABLE 7.1 Amount of Oxygen Dissolved in Water at Sea Level*

Temperature °C	Fresh Water	Sea Water
0	10.29	7.97
20	6.59	5.31

*Oxygen content is expressed in milliliters of oxygen per liter of water.

Gases

Availability of Oxygen. The availability of oxygen varies in different habitats. Dry air contains about 21 percent oxygen and 0.04 percent carbon dioxide; the remainder is nitrogen with traces of other gases. The availability of a gas is expressed as a **partial pressure,** that is, the pressure of that gas in a mixture of gases. This pressure is calculated by multiplying the total pressure of the mixture of gases by the percentage of the particular gas in the mixture. Dry air at sea level has a pressure of 760 mm. Hg and is 21 percent oxygen; hence the partial pressure of oxygen at sea level is $760 \times 0.21 = 159.60$ mm. Hg. The partial pressure of oxygen is decreased when water vapor is a gaseous component of air, as it usually is. In human lungs, for example, where the air is saturated with water vapor, the partial pressure of oxygen is reduced to 104 mm. Hg. It is also lower at high altitudes, where the total atmospheric pressure is reduced. At 6000 m. (19,685 feet), the partial pressure of oxygen is reduced to 80 mm. Hg, although the air at 6000 m. is still 21 percent oxygen.

Partial pressure determines the availability of oxygen because it affects the amount that will go into solution. Oxygen must be in solution before it can be used by an organism or its cells. Aquatic organisms derive their oxygen from oxygen dissolved in water. In terrestrial animals, oxygen first dissolves in the moist surfaces of the body or respiratory organ and then enters the blood or body fluids. When a gas and liquid are in contact, an equilibrium is reached when the rate at which molecules pass from the gas to the liquid equals the rate at which they pass from the liquid to the gas. The higher the partial pressure of oxygen in air, the more oxygen will go into solution and be available to the cells. The tendency of a gas to escape from solution, its **tension,** is numerically equal to the partial pressure of the gas in the air with which it is in equilibrium.

The tension of a gas is an important force in determining its movement between liquids and between liquid and air; but it does not measure the *quantity* of the gas in solution. Quantity is affected not only by tension but also by the solubility of the particular gas and by factors that affect solubility, such as temperature and salinity (Table 7.1). It is important to note that aquatic animals have far less oxygen available to them than do terrestrial ones who are living in an environment that is 21 percent oxygen (i.e., 210 mL of oxygen per L).

Oxygen Conformity and Regulation. Some animals, known as **oxygen conformers,** have the ability to adjust their level of metabolism and oxygen consumption according to the availability of oxygen. The annelid worm *Glycera,* for example, lowers its metabolism and oxygen consumption gradually as the water in its burrow becomes stagnant during low tide. As the tide comes in, oxygen consumption goes up.

Most animals are **oxygen regulators.** Their oxygen consumption, although varying with such factors as temperature, body size, and level of activity, is independent of the availability of environmental oxygen unless the oxygen level drops below a certain partial pressure known as the critical level. Below this, the animal's oxygen consumption necessarily declines rapidly, and death may follow.

Steps in Gas Exchange

To obtain oxygen and eliminate carbon dioxide, an animal must have a **respiratory membrane,** a moist, thin, and permeable surface exposed to the environment, through which gases can move. The membrane may simply be the body surface, but often it is limited to parts of a **respiratory organ,** such as a gill lamella, the end of a tracheal tubule, or a lung alveolus (Fig. 7.1). There must also be a way of getting gases to and from the respiratory membrane, that is, of ventilating its surface, and a way of transporting gases between the membrane and the cells of the body. The location of the respiratory membrane and modes of ventilation and transport vary greatly among animals, depending on their size, oxygen needs, and the environment in which they live.

Passage of gases across the surface of the respiratory membrane and in and out of the cells of the body is always by **diffusion** (p. 51). If the gases are not already in water, they will go into solution on the moist surface of the respiratory membrane and pass through it, following their concentration gradients (Fig. 7.1). Since oxygen is being used by the cells, there is always less in the cells and in the body than in the medium, water or air, in which the animal is living. Conversely, the cells are producing carbon dioxide so there is always more in the cells and in the body than in the surrounding medium.

Ventilation of the surface of the membrane and transport of gases within the body can also take place by diffusion if the animal is small enough. Since diffusion rate is slow (its rate is inversely proportional to distance), cells must be within 0.5 to 1.0 mm. of an oxygen source unless their metabolic rate is exceptionally low. Because of their low metabolism, small size, and shape, diffusion is adequate in flatworms, for example. Diffusion is inadequate in large and active animals so they need some system of **bulk flow,** whereby gases are carried passively in a moving fluid, both to ventilate the surface of the respiratory membrane, and to transport gases in the body. We will first examine the location and nature of respiratory membranes and the ways they are ventilated, and then we will consider the transport of gases in the body.

Environmental Gas Exchange in Aquatic Animals

Most small aquatic organisms, such as protozoas, hydras, flatworms, and rotifers, have no special respiratory organs. Their body surface, in some cases supplemented by the lining of the digestive tract or other body spaces, is the only respiratory membrane that they need since they possess a very favorable ratio of surface area to volume. As oxygen diffuses into the body and carbon dioxide out, the oxygen is rapidly depleted in the

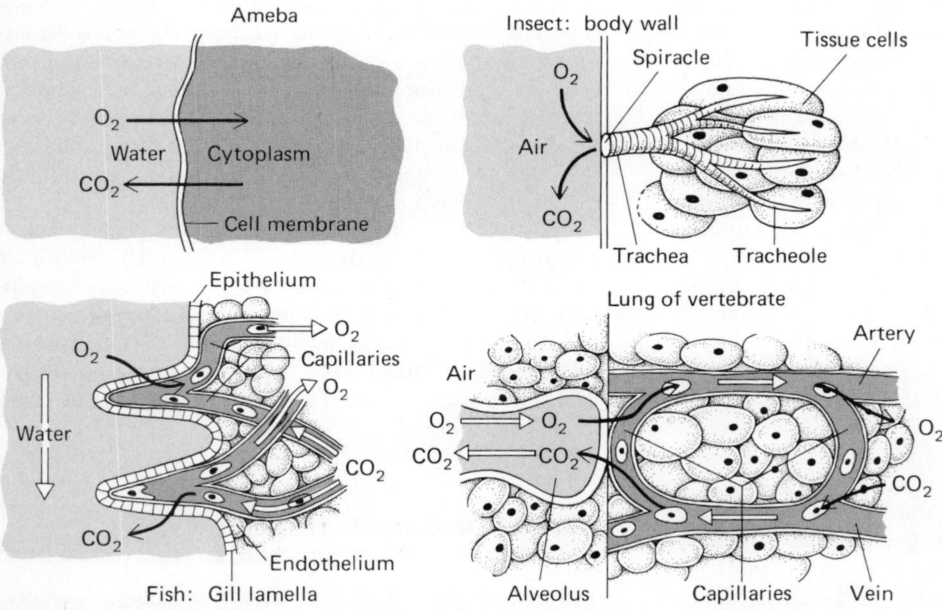

FIGURE 7.1 A diagram of some of the types of respiratory organs present in animals. Diffusion movements are shown by solid arrows; bulk flow movements by open arrows.

boundary layer of water next to the body surface, and carbon dioxide rapidly accumulates. If an animal is stationary and not living in a habitat where the boundary layer is replaced by moving currents, then ciliated or flagellated cells move water across the organism's surface or through its cavities so that there is a continual supply of oxygen.

Gills. Larger and more active aquatic animals have respiratory organs, usually gills, to which the respiratory membranes are confined or that supplement other surfaces. Gills are organs composed of many delicate lamellae, or filaments, that extend outward from an exposed surface. Their existence is possible in an aquatic environment because the density of water provides adequate support. Gills would tend to dry out, collapse, and clump together in air. Although primarily for gas exchange, gills may also be used for other purposes, including filter feeding, excretion, ion exchange, and osmoregulation.

Gills must have a large surface area, and a large volume of water must be moved across them because water contains so little dissolved oxygen. The weight of the water that must be moved is 100,000 times the weight of the available oxygen, whereas in air the weight of the inert medium, mostly nitrogen, is only 3.5 times the weight of the available oxygen. Usually water is moved in one direction across the gills by water currents, the action of cilia, the locomotion of the animal, or a muscular pump that carries water through a gill chamber. Rarely are gills moved back and forth or is water pumped in and out because continually accelerating and decelerating this dense medium is energetically expensive.

Gases diffuse rapidly between the water crossing the gills and body fluids circulating through them. The cuticle, if one is present, and the epithelial layers separating the water from the body fluids are very thin. The effectiveness of the gills is increased in some mollusks and crustaceans and in fishes by the evolution of a **counter current exchange** mechanism in which water flows across the gills in a direction opposite to the flow of blood through them. A difference in oxygen tension between the water and blood is maintained throughout the system (Fig. 7.2). Blood leaving a gill has become nearly fully saturated with oxygen. It is exposed to water just starting across it and flowing in the opposite direction. This water is fully saturated with oxygen, hence oxygen diffuses into the blood. At the other end of a gill, the water has given up most of its oxygen, but it is exposed to blood just entering the gill that contains even less oxygen. Counter current exchange mechanisms are extraordinarily efficient, and as much as 90 percent of the oxygen dissolved in the water may be removed. This is far more than the approximately 25 percent of oxygen that mammals extract from the air in their lungs.

Invertebrate Gills. Many sorts of gills and gill-like structures, some simple, others quite elaborate, are found among invertebrates. Polychaete annelid worms have paired lateral projections, the **parapodia,** that are used in locomotion, but parts may also be modified as gills. Molluscan gills are composed of many filaments arranged in different ways. Most aquatic arthropods have gills that are modified appendages or outgrowths closely associated with appendages. In crayfishes and crabs, the gills arise from the base of the appendages and adjacent body wall and extend up into a **branchial chamber** (L. *branchiae,* gills) that lies on each side of the animal between the body wall and a part of the carapace that has grown laterally and ventrally over the gills (Fig. 7.3*A*). Each gill consists of a central axis to

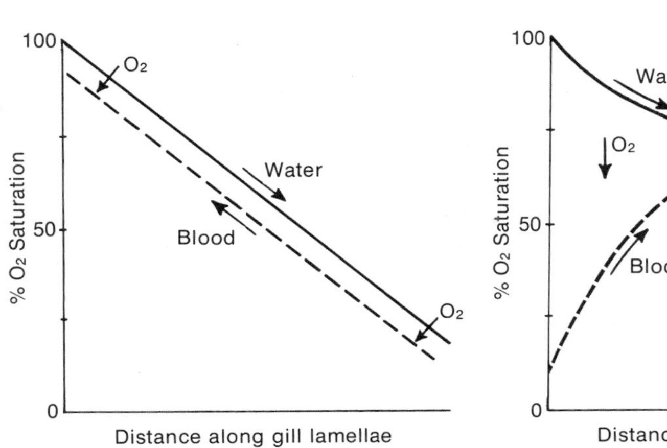

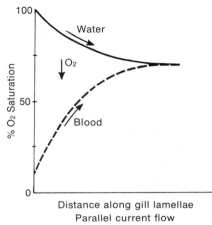

FIGURE 7.2 Theoretical diagrams to illustrate the effects of countercurrent or parallel flow on the exchange of oxygen between water and blood. The equilibrium attained in parallel flow would be somewhere above 50 percent saturation of the blood because of the presence of hemoglobin, yet the blood becomes less fully saturated than it would during countercurrent flow. (Modified after Hughes.)

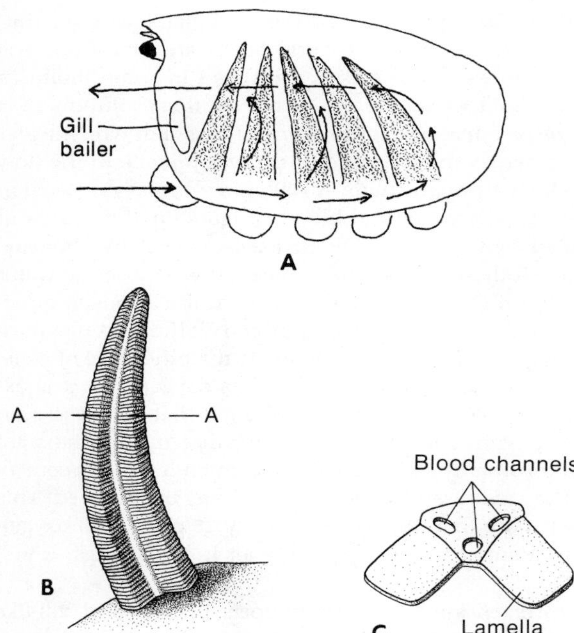

FIGURE 7.3 Gills of decapod crustacea. Lateral view of a crab *(A)* to show the location of the gills and the ventilating current of water through the branchial chamber. The carapace is drawn as though it were transparent. *B,* One gill of a crab consists of a stack of lamellae attached to a central axis. *C,* Cross section of a crab gill. (*B* and *C* modified from Calman.)

which are attached either lamellae or filaments, depending on the species (Fig. 7.3*B* and *C*). The ventilating current is produced by the rapid sculling action of the **gill bailer,** a semilunar-shaped process of one of the oral appendages (the second maxilla). The gill bailers drive two exhalant water streams out of the branchial chambers to each side of the mouth. Water enters the branchial chambers of the crayfish at the back of the carapace and between the legs. In crabs, the carapace is tightly sealed along its ventral margin, and the inhalant apertures are at the base of the great claws (chelipeds). Channels in each gill axis carry blood to and from spaces in the lamellae; gas exchange occurs across the thin walls of the lamellae.

Vertebrate Gills. Fishes and many amphibians exchange gases with the environment by means of gills. Some fish larvae and larval amphibians have **external gills,** filamentous processes that extend outward from the side of the head near the openings of the gill slits. Adult fishes have **internal gills** situated within the gill pouches. In most species the pouches are covered by a fold of the body, the **operculum,** (L. *operculum,* lid) that forms an **opercular chamber** lateral to them. The gills themselves consist of lamellae attached to a visceral arch at the base of each gill (Fig. 7.4). Secondary folds perpendicular to the lamellae contain capillary beds, and it is here that gas exchange with the water occurs. Exchange is particularly effective because of a countercurrent flow between blood and water (Fig. 7.4). As you should expect, active fish have larger gill areas than slower-moving ones. The gill area per gram of body weight of the fast-swimming mackerel is 50 times that of the goosefish, which spends most of its time lying on the bottom waiting for food to come to it.

A pumping action of the mouth and pharynx causes a ventilating current of water to flow across the gills (Fig. 7.4*C* and *D*). During inspiration, both the pharynx and opercular chambers expand, causing a decrease in pressure within them relative to the surrounding water. These chambers are acting as suction pumps. Water is drawn in only through the mouth, for a thin membrane on the free edge of the operculum acts as a valve, preventing entry in this direction. Once in the pharynx, the water passes across the gills into the opercular chamber as a result of the lower pressure there. Expiration begins with the closure of the mouth, or of oral valves, and a contraction of the pharynx and opercular chamber, so that the pressure in them exceeds that of the surrounding water. These chambers are now acting as force pumps. The membrane at the edge of the operculum is pushed open, and water is discharged. During both inspiration and expiration there is a pressure gradient moving water across the gills. Water does not enter the esophagus, for this is collapsed except when swallowing occurs. Food and other particles are prevented from clogging the gills by gill rakers that act as strainers.

Some pelagic fish keep their mouths open and use their forward motion to drive water across the gills, a process called **ram ventilation.** When a mackerel's speed exceeds 0.4 m. per second, pumping movements of the operculum slow down. When speed exceeds 0.6 m. per second, they stop and the fish depends on ram ventilation.

Environmental Gas Exchange In Terrestrial Animals

A particular advantage of terrestrial life is that oxygen is much more abundant than in water. A problem is that body water can be lost easily through any exposed surface that is moist, thin, permeable, and vascular enough to serve as a respiratory membrane. Most terrestrial animals have adapted to these conflicting conditions by

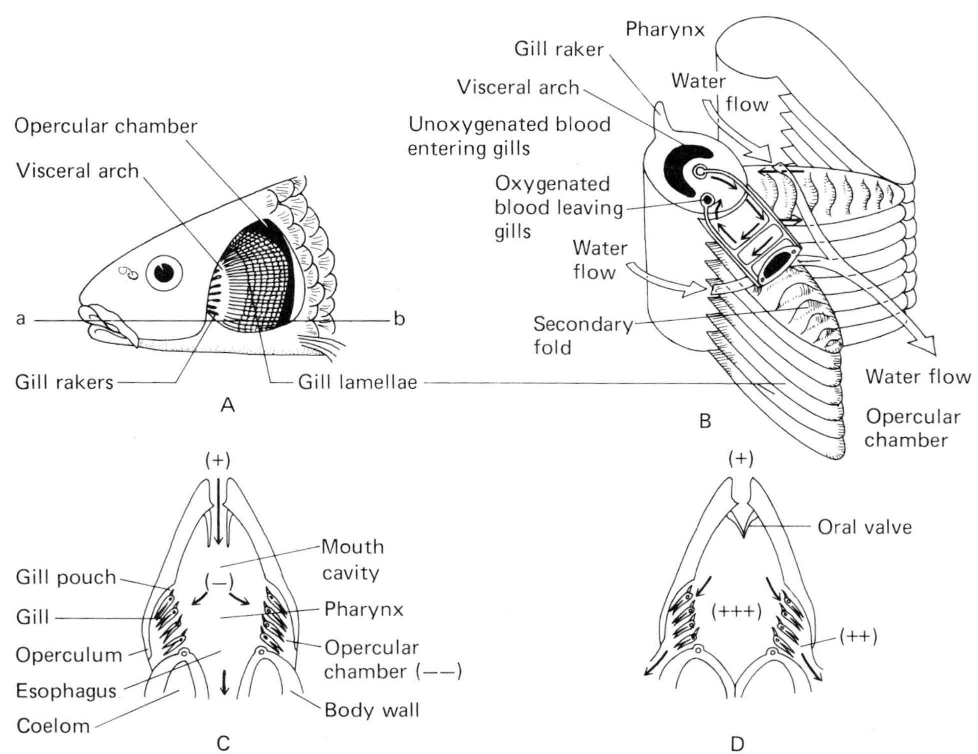

FIGURE 7.4 Gills of fishes. *A,* The operculum has been cut away to show the gills in the opercular chamber. *B,* An enlargement of a portion of one gill. *C* and *D,* Frontal sections through the mouth and pharynx in the plane of line a–b in *A*. Inspiration occurs in *C* and expiration in *D*. Relative water pressures in the various parts of the system are shown by + and −.

evolving respiratory organs, either trachea or lungs, in which the respiratory membranes lie deep within the body in spaces with a relative humidity close to 100 percent (saturation at the animal's body temperature). Moreover, only a little of this air need be exchanged with each breath because the oxygen content of air is so high. All of these factors minimize water loss. No more air is moved in and out of these spaces than is needed to meet the animal's metabolic needs at the particular time and circumstances. The rate of ventilation is more carefully regulated than in animals utilizing gill ventilation. Air is much less dense than water and can be moved with little energy expenditure. In some animals diffusion is adequate. Oxygen, for example, diffuses 300,000 times faster in air than in water.

A few terrestrial animals, those whose metabolism is not high and who live in damp habitats, such as earthworms and certain salamanders, can exchange all or some of their gases through the general body surface and have no special respiratory organs. Gills can be used for gas exchange in air, provided that they lie in chambers in which a high humidity can be maintained or that the animals live in damp habitats. They are usually confined to animals, such as crabs, crayfish, and pillbugs, that venture onto the land only temporarily or live in damp places.

Tracheal Systems. Tracheal systems are the commonest gas exchange organs of terrestrial arthropods. Most body segments of insects have paired lateral apertures, the **spiracles** (L. *spiraculum,* air hole), that lead into a system of **tracheal tubules** (L. *trachia,* windpipe) (Fig. 7.5*A* and *B*). Filtering devices prevent small particles from clogging the system. Valves are present and can open and close as the animal's needs vary. The tracheae have a more or less ladder-like pattern with interconnecting transverse and longitudinal trunks. They terminate in minute tubules, the **tracheoles,** that are generally less than 1 μm. in diameter (Figure 7.5*C*). The tracheoles are formed by special tracheole cells. They contain the respiratory membrane, and they permeate all of the tissues of the body. A small amount of liquid

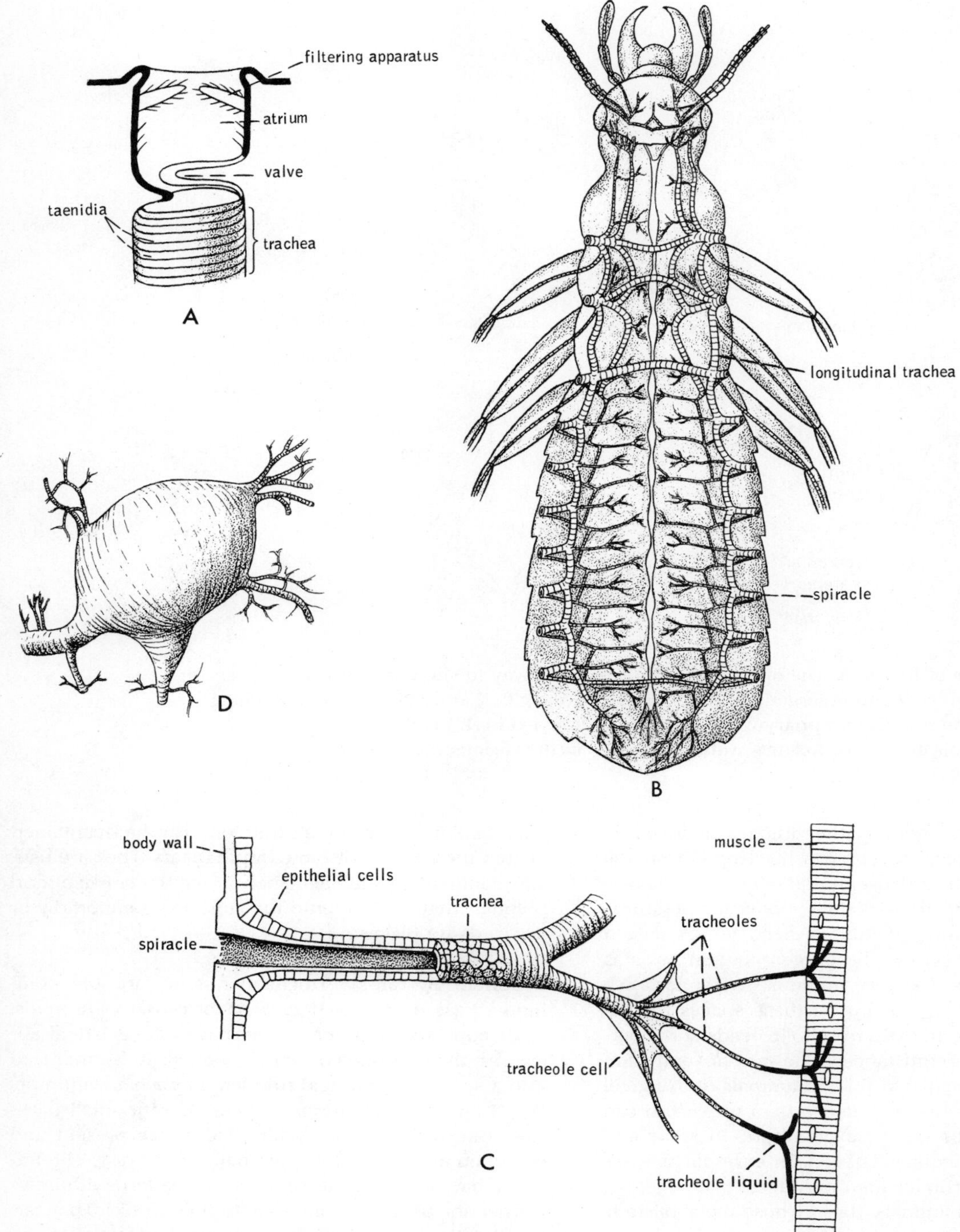

FIGURE 7.5 *A,* A spiracle with atrium, filtering apparatus, and valve. (After Snodgrass.) *B,* A tracheal system of an insect. (After Ross.) *C,* Diagram showing relationship of spiracle and tracheoles with tracheae. (Modified from Ross.) *D,* An air sac. (After Snodgrass.)

is present in the ends of the tracheoles, and gases dissolve in this. As metabolites increase in active tissues, the osmotic pressure of the tissues increases, some of the fluid leaves the tracheoles, and air comes closer to the tissues. In insect flight muscles, where oxygen needs are very high, tracheoles penetrate the muscle cells and lie as close as 0.07 μm. to the mitochondria. A cuticle lines the entire system, but only that in the tracheae is shed during molting.

Cuticular rings, **taenidia** (L. *taenia,* ribbon), strengthen and hold the tracheae open. In small insects, diffusion is the only force needed to exchange gases because it is so rapid through the air-filled tubules. Larger and more active insects have a ventilating system that includes **air sacs** that can be compressed and expanded by the action of body muscles (Fig. 7.5*D*). The opening and closing of spiracles is carefully regulated to permit adequate gas exchange yet to prevent water loss. A synchronous pattern of the opening and closing of successive spiracles keeps a unidirectional flow of air moving through the larger tracheae.

A few insects have readapted to the aquatic environment and usually come to the surface to get air. Diving beetles have a dense mat of hairlike cuticular processes that are nonwettable and entrap a permanent bubble of air. The spiracles open only into this bubble, and gases diffuse between it and the water. The air bubble acts as a gill!

A tracheal system usually combines ventilation of the respiratory membrane with distribution of gases to and from the tissues. Blood is not used for gas transport, but does, of course, provide the cells with food and removes nitrogenous wastes. Because of the large role that diffusion plays, tracheal systems are suitable only for small animals: insects, centipedes, millipedes, and small spiders. Large spiders have lungs, or a combination of lungs and tracheae, and gases are transported by blood.

Diffusion Lungs. Lungs, defined as invaginations from some body surface, are the respiratory organs of some terrestrial invertebrates and of nearly all terrestrial vertebrates. Invertebrate lungs are known as **diffusion lungs,** for gases move in and out of them primarily by diffusion, although some body movements often help in ventilation. One group of mollusks, the pulmonate snails, have converted the mantle cavity into a lung, and scorpions and the larger spiders have book lungs that have evolved from invaginations of the abdomen. Each opens to the surface by a spiracle and contains many leaflike lamellae held apart by bars that enable air to circulate freely (Fig. 7.6).

Ventilation Lungs of Lower Vertebrates. The lungs of vertebrates are called **ventilation lungs** because air is moved in and out of them by a pumping action of adjacent body parts. Secretions of phospholipids function as surfactants and reduce the surface tension of the film of liquid inside the lungs and hence the amount of energy needed to expand and fill them. Lungs usually develop as a bilobed outgrowth from the floor of the pharynx, extend posteriorly, grow around the digestive tract, and come to lie in the dorsal part of the pleuroperitoneal cavity. This cavity is the part of the lower vertebrate coelom that contains the lungs and digestive organs; the pericardial cavity is the part that surrounds the heart.

Lungs probably evolved as an accessory respiratory organ in swamp-dwelling fish that lived in ponds where the oxygen tension was low or that even dried up occasionally. Although they have gills, African and South

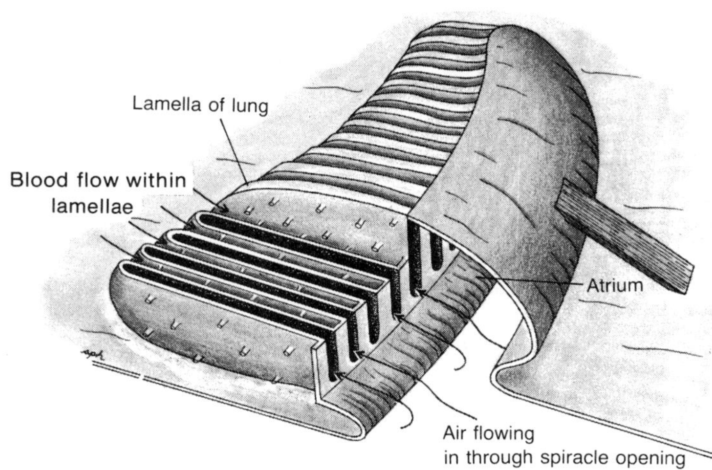

FIGURE 7.6 Section through a book lung of a spider.

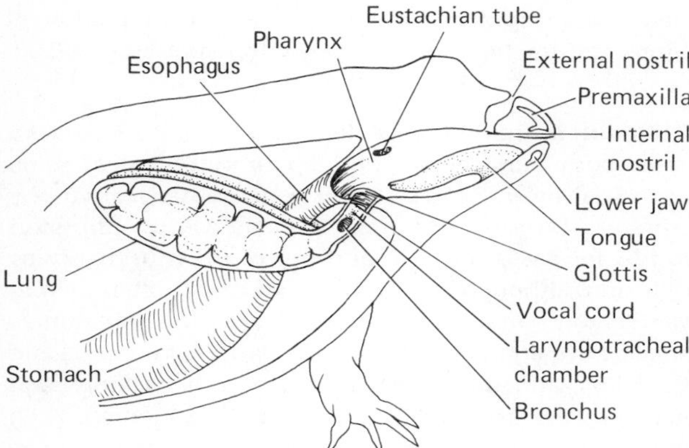

FIGURE 7.7 A diagrammatic longitudinal section of the respiratory system of the frog.

American lungfish obtain most of their oxygen through their lungs.

The lungs have been converted to a **swim bladder** in most fishes descended from ancestral lunged fish. The swim bladder is primarily a hydrostatic organ (see p. 119); however, the gas within it is mostly oxygen, so it is also a reserve that can be used when the fish cannot obtain enough oxygen through its gills.

Amphibians evolved from primitive fishes with lungs. Their aquatic larvae exchange gases through gills, but lungs appear and begin to function in late larvae that surface to gulp air. The lungs are saclike, and their internal surface area is increased somewhat by pocket-like folds (Fig. 7.7), but the surface area is far smaller than in higher terrestrial vertebrates. The lungs are ventilated by up and down movements of the pharynx floor that first draw air through the nostrils into the pharynx and then force it into the lungs. The lungs are ventilated by a **force pump.** The elastic recoil of the lungs is primarily responsible for expelling air (Box 7.1). Contemporary amphibians supplement pulmonary respiration by gas exchange through their thin, moist, and vascular skin. Indeed, this is the primary method of gas exchange in many amphibians and in all, except gilled species, when they are under water. Most of the carbon dioxide is eliminated through the skin because the rate of lung ventilation is not sufficient to flush it out. Enough oxygen also enters through the skin to meet the animals' needs except during periods of increased activity in the spring and summer. A considerable amount of water is also exchanged through the skin, and this loss has been a factor that has prevented amphibians from fully exploiting the terrestrial environment. Cutaneous respiration also places a limit on the maximum size of amphibians; small organisms have more body surface relative to mass than large ones. Higher terrestrial vertebrates have evolved lungs with greater internal surface areas and different ways of ventilating the lungs. They have thus been able to dispense with cutaneous respiration and the constraints that it imposes.

BOX 7.1

Breathing Movements of the Frog

(*A*) The glottis is closed and air from a previous cycle is held in the stretched lungs at greater than atmospheric pressure. The floor of the pharynx is lowered. Fresh air enters through the nostrils and accumulates in the pharynx floor. (*B*) The glottis opens, and the elastic recoil of the lungs, sometimes assisted by the contraction of flank muscles, expels the stale air through the dorsal part of the pharynx. Chemical analysis has shown that there is little mixing of fresh and stale air. (*C*) The nostrils close, the pharynx is compressed, and air is forced into the lungs. (*D*) By continued lowering and raising of the pharynx floor, additional fresh air is pumped into the lungs until they become well inflated. (Figure modified from C. Gans, H. J. DeJongh, and J. Farber. Bullfrog ventilation; How does the frog breathe? *Science,* 163:1223–1225, 1969.)

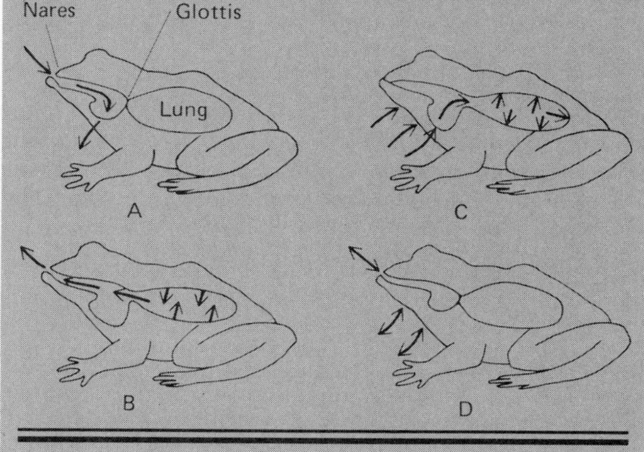

Mammalian Respiratory System. In mammals (Fig. 7.8), the air is drawn into the paired **nasal cavities** through the **external nostrils,** or **nares.** These cavities are separated from the mouth cavity by a hard palate, and the animal can breathe while food is in its mouth. The surface area of the cavities is increased by a series of ridges known as **conchae** (Gr. *konkhē,* shell), and the nasal mucosa (in addition to having receptors for smell) is vascular and ciliated and contains many mucous glands. In the nasal cavities the air is warmed and

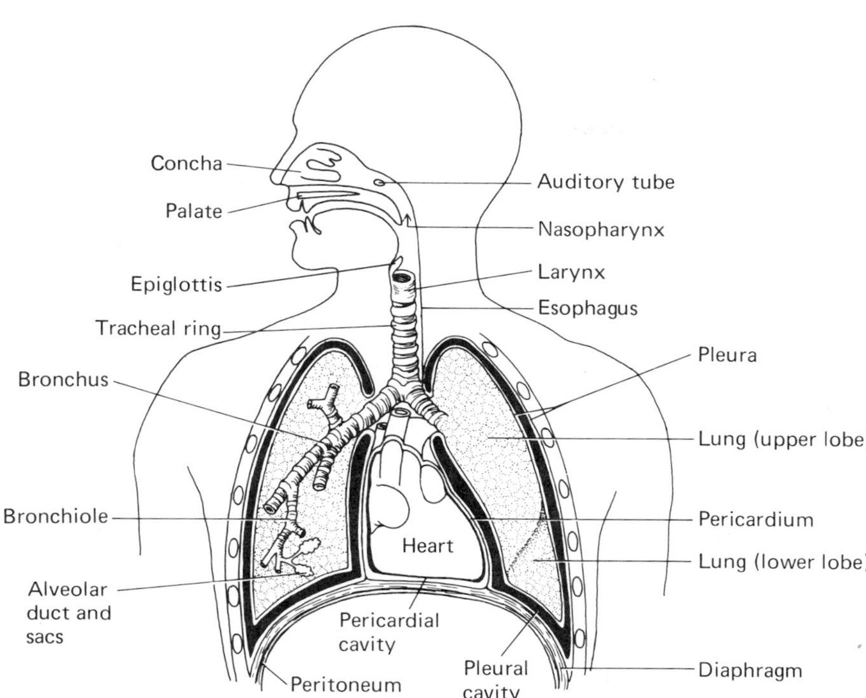

FIGURE 7.8 The human respiratory system. Details of the alveolar sacs, drawn enlarged here, are shown in Figure 7.11.

moistened, and minute foreign particles are entrapped in a sheet of mucus that is carried by ciliary action into the pharynx, where it is swallowed or expectorated. Inspired air is moistened in primitive terrestrial vertebrates, such as the frog, but cold-blooded tetrapods do not need so much conditioning of the air as birds and mammals.

Air continues through the **internal nostrils,** or **choanae,** passes through the **pharynx** and enters the **larynx,** which is open except when food is swallowed. The raising of the larynx during swallowing can be demonstrated by placing your hand on the Adam's apple, the external protrusion of the larynx. The **epiglottis** flips back over the entrance of the larynx when it is raised.

The larynx is composed of cartilages derived from certain of the visceral arches and serves both to guard the entrance to the windpipe, or **trachea,** and to house the **vocal cords** (Fig. 7.9). The vocal cords are a pair of folds in the lateral walls of the larynx. They can be brought closer together, or be moved apart, by the pivoting of laryngeal cartilages connected to their dorsal ends. The **glottis** (Gr. *glotta,* tongue) of mammals is the opening between the vocal cords. When the cords are close together, air expelled from the lungs vibrates them, and they in turn vibrate the column of air in the upper respiratory passages, just as the reed in an organ pipe vibrates the air in the pipe. This phenomenon is called **phonation,** or the production of a vocal sound. **Speech** is the shaping of the vocal sounds into patterns that have meaning for us; it is accomplished by the pharynx, mouth, tongue, and lips. Muscle fibers in the vocal cords and larynx control the tension of the cords and the pitch of our voice. We normally speak when air

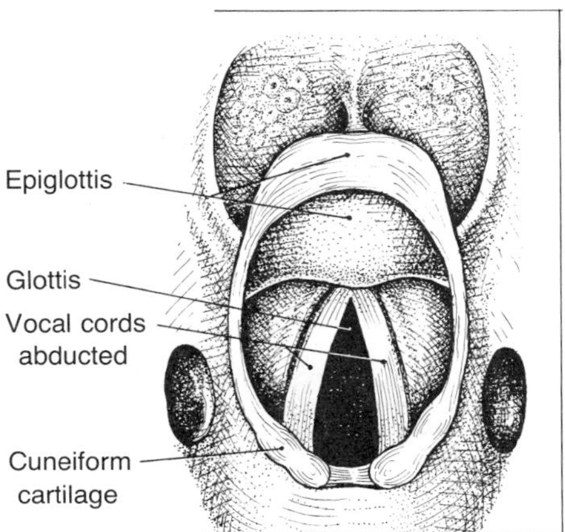

FIGURE 7.9 A view of the vocal cords from looking into the larynx from above. Normal position of vocal cords. (From Jacob, Francone and Lossow: *Structure and Function in Man.* 4th ed., Philadelphia, W. B. Saunders Co., 1978.)

is expired. Ventriloquists have trained themselves to speak during inspiration.

The trachea extends down the neck and finally divides into **bronchi** (Gr. *bronkhos,* windpipe) that lead to the pair of **lungs** (see Fig. 7.8). Unlike the esophagus, which is collapsed except when a ball of food is passing through, the trachea is held open by **C**-shaped cartilaginous rings, and air can move freely through it. Its mucosa continues to condition the air.

The lungs of amphibians lie in the anterodorsal part of the pleuroperitoneal cavity. This cavity has become divided in mammals. The pleural cavities lie within the chest, or **thorax,** and are separated from the peritoneal cavity by a muscular **diaphragm.** A coelomic epithelium, the **pleura,** lines the pleural cavities and covers the lungs. Each bronchus enters a lung, accompanied by blood vessels and nerves, in a mesentery-like fold of pleura (see Fig. 7.8).

The bronchi branch profusely within the lungs, and the walls of the respiratory passages become progressively thinner (see Fig. 7.8). Each passage eventually terminates in an **alveolar sac** whose walls are so puckered by pocket-shaped **alveoli** (L. *alveolus,* small cavity) that it resembles a cluster of miniature grapes (Fig. 7.10). The alveolar walls are extremely thin and relationships between them and the surrounding capillaries were not clear until they were studied by electron microscopy (Fig. 7.11). Alveoli are so densely packed that they abut against each other, with a dense capillary bed lying between them. Gases diffuse between the capillaries and the alveoli on each side of them. Occasionally interalveolar pores extend through the septa separating alveoli. Some cells lining the alveoli have secretory properties and probably release the surfactant phospholipids. Large phagocytic cells, called "dust cells" because they ingest and sometimes accumulate minute foreign particles, are abundant.

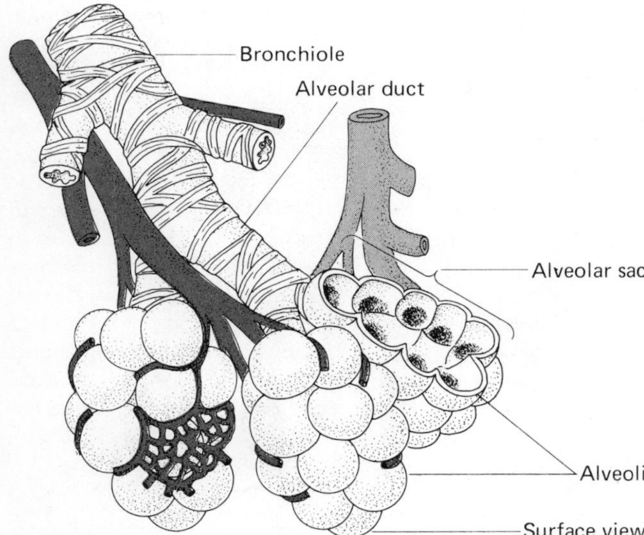

FIGURE 7.10 Termination of the respiratory passages in the mammalian lungs. Alveoli have a diameter of 0.2 to 0.3 mm.

Endotherms, as one would expect, have larger respiratory exchange surfaces than ectotherms to sustain their higher levels of metabolism. The alveolar surface of a normal, adult human male, for example, is estimated to be 70 $m.^2$. One might also expect that small mammals would have relatively more respiratory surface than large ones because their metabolism is much higher, but this is not the case. A shrew, a human being, and a whale all have lungs that are about 6 percent of body volume. The increased gas exchange needs of small mammals are met primarily by increased rates of ventilation and by changes in some characteristics of the blood.

Mammalian Lung Ventilation. Mammalian lungs are ventilated by changing the size of the thoracic cavity and, consequently, the pressure within the lungs. The lungs follow the movements of the chest wall, for they are at once separated from it and united to it by the adhesive force of a thin layer of fluid that lies within the pleural cavities. During normal, quiet **inspiration,** the size of the thorax is increased slightly, intrapulmonary pressure falls to about 1 mm. Hg below atmospheric pressure, and air passes into the lungs until intrapulmonary and atmospheric pressures are equal. Mammalian lungs are therefore **suction lungs** because air is not forced directly into them but enters passively as a consequence of the expansion of the thorax. During normal **expiration,** the size of the thorax is decreased, intrapulmonary pressure is raised to about 1 mm. Hg above atmospheric pressure, and air is driven out of the lungs until equilibrium is again reached.

During inspiration, the thorax is enlarged by the contraction of the dome-shaped **diaphragm** and the **external intercostal muscles.** The diaphragm pushes the abdominal viscera posteriorly and increases the length of the chest cavity (Fig. 7.12); the external intercostals raise the sternal ends of the ribs, push the sternum forward, and expand the dorsoventral diameter of the chest. Expiration results primarily from the relaxation of the inspiratory muscles and the elastic recoil of the lungs and chest wall, which are stretched during inspiration. But during heavy breathing, antagonistic expiratory muscles can decrease the size of the thoracic cavity. Contraction of **abdominal muscles** forces the abdominal viscera against the diaphragm and pushes it

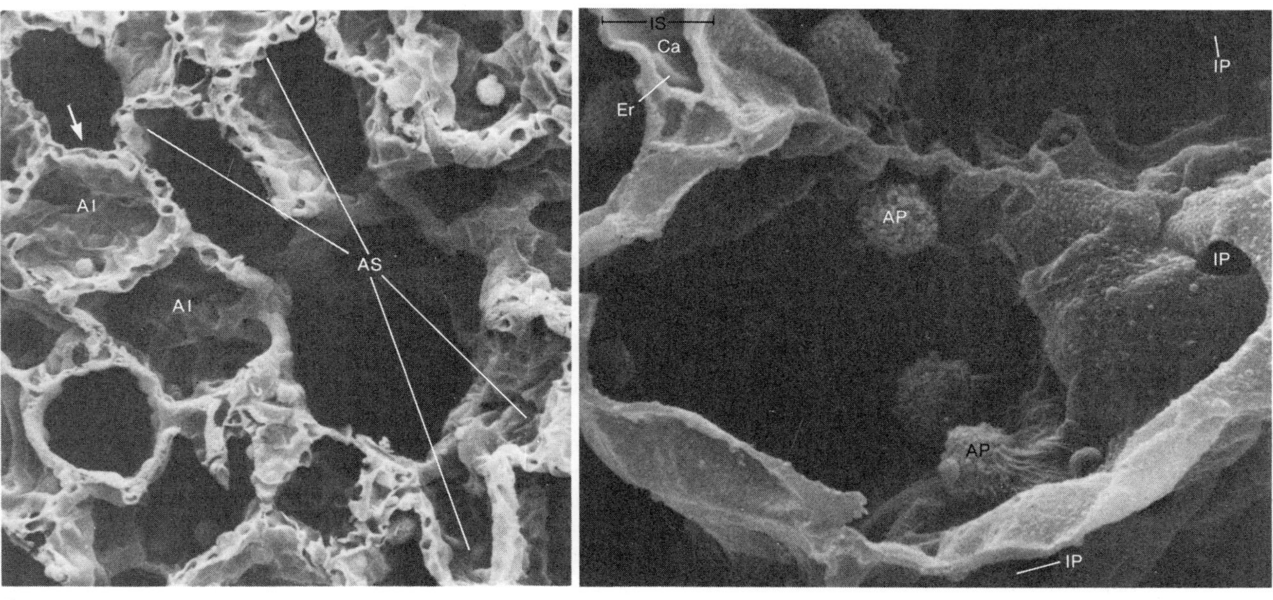

FIGURE 7.11 *A,* Scanning electron micrograph of a group of alveoli *(Al)* branching from an alveolar sac *(AS).* Numerous capillaries, some indicated by arrows, lie between alveoli. *B,* An enlarged view showing pores *(IP)* between alveoli, a capillary *(Ca)* and an erythrocyte *(Er)* in the interalveolar septum *(Is),* and several alveolar phagocytes *(AP).* (From Kessel, R. G., and R. H. Kardon: *Tissues and Organs: A Text-Atlas of Scanning Electron Microscopy.* San Francisco, W. H. Freeman & Co., 1979.)

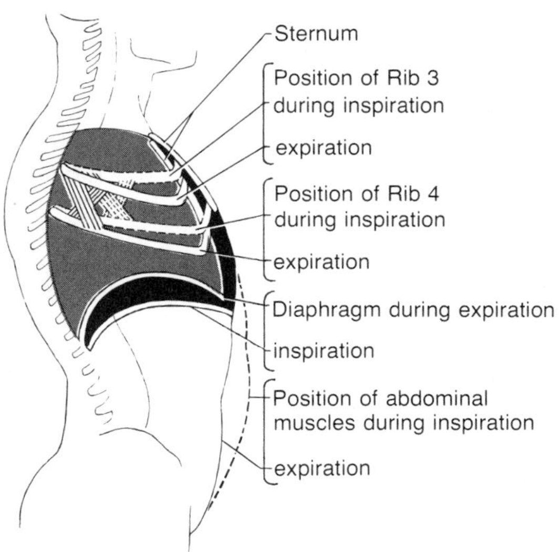

FIGURE 7.12 Mechanics and control of breathing. The elevation of the ribs and depression of the diaphragm during inspriation increases the size of the chest cavity, indicated by the black area. See text for explanation.

anteriorly; **internal intercostals** pull the sternal ends of the ribs posteriorly.

The lungs of an adult human male can hold about 6 L of air, but in quiet breathing they contain only about half this amount, of which 0.5 L is exchanged in any one cycle of inspiration and expiration. This half liter of **tidal air** is mixed with the 2.5 L of air already in the lungs. Vigorous respiratory movements can lower and raise the intrapulmonary pressure 80 to 100 mm. Hg below and above atmospheric pressure, and under these conditions 4 to 5 L of air can be exchanged. This maximum is know as the **vital capacity.** There is always, however, at least a liter of **residual air** left in the lungs to mix with the tidal air, for the strongest respiratory movements cannot collapse all the alveoli and respiratory passages. Alveolar air is always a mixture of tidal and residual air, and its gas content is in equilibrium with gases in the blood. If we exercise, more tidal air is exchanged, but more oxygen diffuses into the blood, and more carbon dioxide leaves the blood so the composition of the alveolar air remains about the same as at rest. The composition of alveolar air is quite different from that of atmospheric air; it contains more water vapor, more carbon dioxide, and less oxygen (Table 7.2).

TABLE 7.2 A Comparison of Atmospheric Air On a Cool, Dry Day at Sea Level with Alveolar Air*

	Atmospheric Air		Alveolar Air	
	pp (mm. Hg)	*Percent*	*pp (mm. Hg)*	*Percent*
Nitrogen	597.0	78.62	569.0	74.9
Oxygen	159.0	20.84	104.0	13.6
Carbon dioxide	0.3	0.04	40.0	5.3
Water vapor	3.7	0.50	47.0	6.2
Total	760.0	100.00	760.0	100.0

*From A. C. Guyton. *A Textbook of Medical Physiology.* 5th ed. Philadelphia, W. B. Saunders Co., 1976.

Respiratory movements are cyclic and are controlled by the respiratory center in the brain. Inspiratory neurons in this center have an inherent rhythm, becoming active every few seconds and sending nerve impulses out to the inspiratory muscles. After a few seconds of inactivity they again become active. The oscillatory activity of these neurons is all that is required for quiet breathing, but many factors can affect their activity. Of greatest importance is the level of carbon dioxide and hydrogen ions in the blood. Their increase when an animal becomes more active affects a chemosensitive area in the brain beside the respiratory center and this, in turn, affects the center. More impulses are sent to the inspiratory muscles and, if needed, to expiratory muscles as well, and the frequency and force of breathing movements increase. Sensory input from peripheral receptors also has an effect. If oxygen levels of the blood fall too low, chemoreceptors (the **aortic** and **carotid** bodies) associated with large arteries near the heart are affected. Impulses from them lead to an increase in the activity of the respiratory center. If an animal breathes very deeply, stretch receptors in the bronchi will be stimulated and they, in turn, will dampen the activity of the respiratory center.

Gas Transport

The tracheal system of arthropods brings oxygen from the body surface directly to the tissues and removes carbon dioxide. Other animals that are too large to exchange gases by diffusion alone have a transport system that carries gases between the respiratory membrane and the tissues. A small amount of oxygen and carbon dioxide can be carried in physical solution in coelomic fluid, blood or hemolymph. Most of these gases, however, are transported in other ways. Carbon dioxide reacts chemically with water and most ends up being carried as bicarbonate ion and salts as described below. Oxygen and some carbon dioxide bind loosely with colored compounds known as **respiratory pigments.** Hemoglobin, which gives the red color to our blood, is the most familiar one. If these gases were only in physical solution, human blood could carry only about 0.2 ml. of oxygen and 0.3 ml. of carbon dioxide in each 100 ml. of blood. The reactions of carbon dioxide with water and the properties of hemoglobin enable blood to carry 100 or more times as much gas: 20 ml. of oxygen and about 50 ml. of carbon dioxide per 100 ml. The respiratory pigments enable the blood to transport far more oxygen than could be transported in simple solution. Carriers are important for most animals. Transport in simple solution is adequate only for a few animals such as clams and some worms which have low oxygen needs or large surface areas for gas exchange. Oxygen in most terrestrial arthropods is transported in the tracheal system and not by the blood.

Respiratory Pigments. The four types of respiratory pigments found in animals are all metal-containing proteins. Hemoglobin and hemocyanin are the most common.

The protein portion of a **hemoglobin** molecule is a **globin** consisting, in most vertebrates, of four peptide chains, typically two α chains and two β chains. Each chain has attached to it a molecule of **heme,** a porphyrin composed of four pyrroles with an iron atom at the center (Fig. 7.13). The heme is the same in all hemoglobins and is also very similar to the heme in the cytochromes of the electron transport system (p. 70). The globins differ greatly in the number of peptide chains and their amino acid composition, and these differences impart characteristic properties to the numerous varieties of hemoglobin. Hemoglobin, the most widespread of the respiratory pigments, is found in nearly all vertebrates and in many different invertebrates, especially annelid worms. It is contained within the red blood cells, or **erythrocytes,** of vertebrates but is in solution in the circulating fluids in most of the invertebrates. Having hemoglobin within cells prevents the undesirable osmotic effects that a molecule of this size would have in free solution, and the chemical environment with which hemoglobin molecules react can be more carefully regulated. In those animals in which hemoglobin is dissolved in the plasma, it is always a giant molecule. The hemoglobin molecule of *Arenicola,* a marine polychaete, for example, is a polymer containing 96 heme units, whereas vertebrate hemoglobin, which is contained within cells, has no more than four.

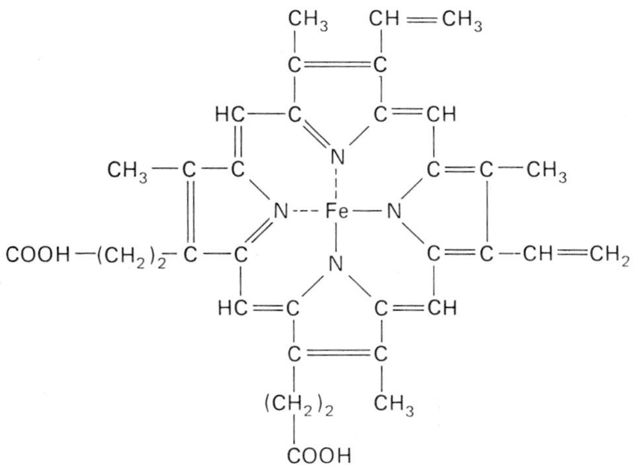

FIGURE 7.13 The structure of a heme. (From Prosser, C. L.: *Comparative Animal Physiology.* 3rd ed. Philadelphia, Saunders College Publishing, 1973.)

Hemocyanin is next in abundance to hemoglobin. It is found in mollusks, apart from clams and their allies, in the larger crustaceans, and in a few arachnids, including the horseshoe crab, *Limulus.* It is a large molecule carried in solution and gives the blood a bluish tinge. Unlike hemoglobin, hemocyanin does not contain a porphyrin, and oxygen is carried between two copper atoms that are a part of the protein chain. Both hemoglobin and hemocyanin have evolved independently numerous times.

Gas Reactions in the Blood. As oxygen diffuses from the respiratory organ into the blood, it combines with hemoglobin (Hb) to form **oxyhemoglobin** (HbO_2):

$$Hb + O_2 \underset{\text{Tissues}}{\overset{\text{Lung}}{\rightleftharpoons}} HbO_2 \qquad (1)$$

During the formation of oxyhemoglobin in the lungs, where oxygen tension is high, one molecule of oxygen forms a loose bond with one of the four iron atoms in the molecule of hemoglobin. Subsequently, additional molecules of oxygen bind to the other three iron atoms, and the hemoglobin is fully saturated with oxygen. Since a single mammalian erythrocyte contains as many as 265,000,000 molecules of hemoglobin, a great deal of oxygen can be carried. The reaction is reversible, and hemoglobin releases much of its oxygen when blood reaches the tissues where the oxygen tension is low.

As carbon dioxide diffuses into the blood from tissue cells, most of it combines with water to form carbonic acid which, in turn, dissociates into hydrogen and bicarbonate ions:

$$CO_2 + H_2O \underset{\text{Tissues}}{\overset{\text{Lung}}{\leftrightharpoons}} H_2CO_3 \leftrightharpoons H^+ + HCO_3^- \qquad (2)$$

In the lungs, the reactions are reversed and carbon dioxide is released. The ionizing reactions proceed very rapidly, but the reactions between carbon dioxide and water are much slower, taking several seconds. This is too long for much carbonic acid to be formed in the second or less that blood travels through a particular section of a capillary. Fortunately, the red blood cells, or erythrocytes, contain the enzyme **carbonic anhydrase** that increases the rate of reaction between carbon dioxide and water by nearly 5000-fold. In addition to these reactions, 15 to 20 percent of the carbon dioxide entering the blood combines with hemoglobin to form **carbaminohemoglobin.** The binding site is not with the iron but with amino groups of the globulin. This reaction also is reversed in the lungs.

The reaction of carbon dioxide with water in the tissue capillaries could greatly increase the acidity of the blood (decrease its pH) because carbonic acid is a fairly strong acid, i.e., it dissociates easily and releases its hydrogen ions. This is ameliorated by the buffering action of blood proteins, of which hemoglobin is the most important because it is the most abundant (p. 168). Oxyhemoglobin (HbO_2) combines with excess hydrogen ions to form reduced, or **acid, hemoglobin** (HHb). Since acid hemoglobin is a weaker acid than carbonic acid (it holds on to hydrogen ions to a greater extent), the pH of the blood does not fall very much (arterial blood has a pH of 7.6; venous blood, one of 7.2). As oxyhemoglobin takes up hydrogen ions, changes occur in its molecular configuration that cause more oxygen molecules to be released. More oxygen is driven off than would leave simply because of the low partial pressure of oxygen in the tissues. The extra oxygen driven off because of the entrance of carbon dioxide is called the **Bohr effect.** The reverse reactions occur in the lungs. As carbon dioxide leaves the lungs, the blood becomes less acid because hydrogen ions are used to form carbon dioxide (equation 2, in preceding paragraph). Acid hemoglobin gives up hydrogen ions, which permits more carbon dioxide to be formed and given off, which in turn stabilizes the pH of the blood and enables more oxygen to be taken up than would be the case from the increased partial pressure of oxygen alone. These reactions along with ones that involve the bicarbonate ions are shown in Box 7.2.

The reactions of gases in the blood enable the blood to carry a great deal more oxygen and carbon dioxide than could be carried in physical solution. They minimize any change in the pH of blood as oxygen and carbon dioxide are transported. They facilitate the unloading of oxygen and the uptake of carbon dioxide in the

BOX 7.2

Reactions in the Blood That Enable it to Take Up, Transport, and Give Up Oxygen and Carbon Dioxide

1. Oxygen leaves oxyhemoglobin because of the low partial pressure of oxygen in the tissues.
2. Carbon dioxide enters the blood because of its high partial pressure, reacts with water to form carbonic acid and hydrogen and bicarbonate ions.
3. Much oxyhemoglobin is present as a potassium salt, $KHbO_2$. Many hydrogen ions replace potassium, drive off more oxygen, and form acid hemoglobin, HHb, and free potassium ions.
4. Many potassium ions, at relatively high concentration within cells, combine with some bicarbonate ions to form potassium bicarbonate, $KHCO_3$.
5. Other bicarbonate ions diffuse into the plasma where they combine with sodium, which is at relatively high concentration outside of cells, to form sodium bicarbonate, $NaHCO_3$.
6. Electrical balances are maintained by a shift of chloride ions from the plasma into the erythrocytes.
7. All of these reactions move in the opposite directions in the lungs.

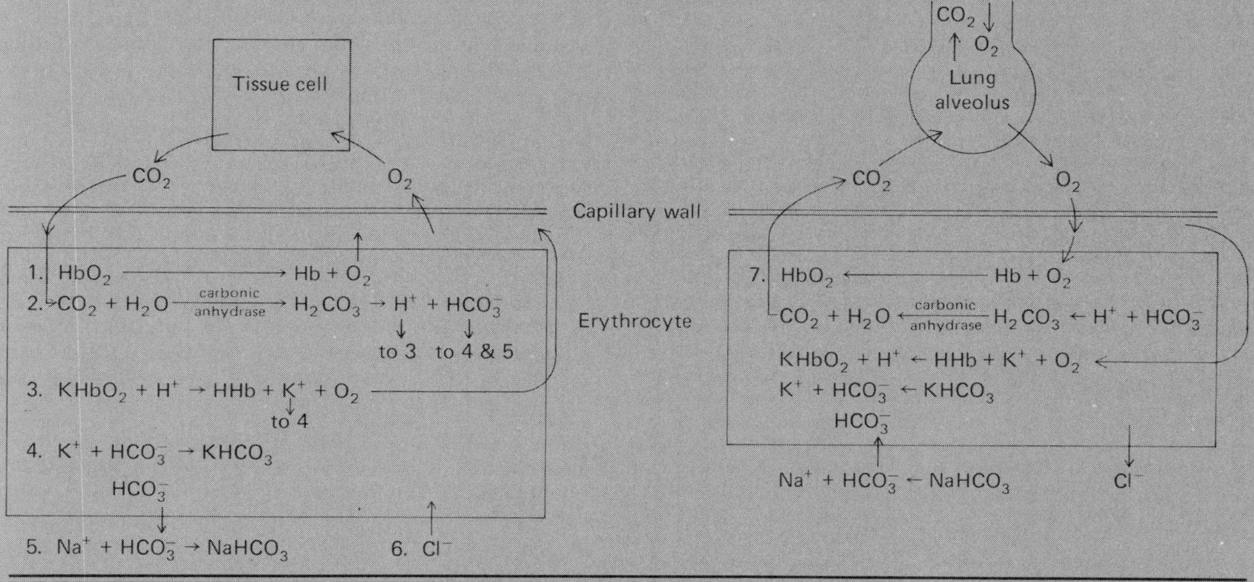

tissue capillaries and the reverse movements of gas in lung capillaries.

Oxygen Dissociation Curves. Hemoglobin is an allosteric molecule, and changes in its shape affect the amount of oxygen it can carry. When one molecule of oxygen binds to hemoglobin, intramolecular changes occur that affect the way one heme interacts with the other hemes. These changes facilitate the uptake of the second molecule of oxygen, which, in turn, facilitates the binding of the remaining molecules of oxygen. Reverse changes occur in the tissues. The release of one molecule of oxygen facilitates the release of the second, and so on. Because of these properties, the amount of oxygen bound to hemoglobin is not directly proportional to the oxygen tension. By experimentally determining what percent of the hemoglobin is saturated with oxygen at different tensions, physiologists have determined oxygen dissociation curves for different hemoglobins. Because these curves are somewhat S-shaped, or sigmoid, more oxygen can be delivered to the tissues than if the relationship were linear. Hemoglobin becomes nearly fully saturated in the lungs at an oxygen tension of about 100 mm. Hg (Fig. 7.14). Since the top of the curve is somewhat flat, nearly full saturation can be achieved over a broad range of tensions. In a sense, there is a wide margin of safety. Saturated human arterial blood holds 20 mL of oxygen per 100 mL of blood

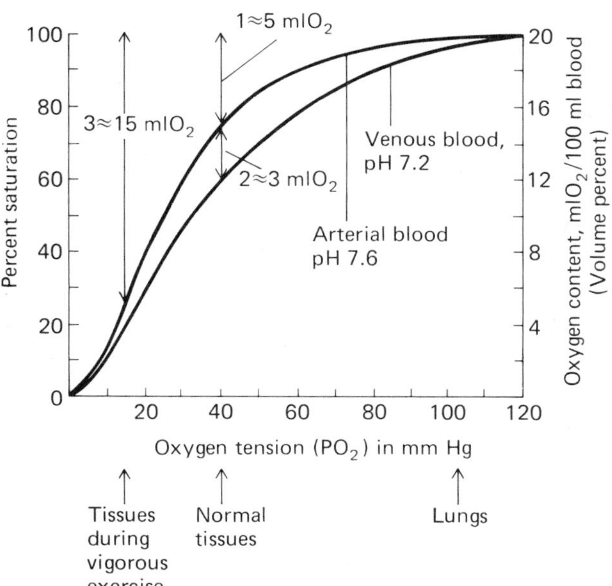

FIGURE 7.14 Oxygen dissociation curves for human arterial and venous blood. *Arrow 1* indicates the amount of oxygen delivered to normal tissues because of the reduction in oxygen tension between the lungs and tissues. *Arrow 2* indicates the extra amount of oxygen delivered because of the entry of carbon dioxide into the blood (Bohr effect). *Arrow 3* indicates the amount of oxygen delivered by arterial blood to tissues during exercise. A small amount would be delivered because of the Bohr effect.

(20 volume percent). No oxygen leaves the blood as it passes through the arteries to the tissues. But in the thin-walled tissue capillaries, the blood is in an environment where oxygen tension is about 40 mm. Hg under normal circumstances. Because the dissociation curve is sigmoid, this is at a point on the curve where the blood can only be 75 percent saturated, so about 5 mL of oxygen is given off per 100 mL of blood. Oxygen dissociation curves are also affected by the pH of the blood. As carbon dioxide enters and the blood becomes more acid, the curve is shifted to the right, and an extra 3 mL of oxygen is given up at a tension of 40 mm. Hg (the Bohr effect). During vigorous exercise, the tissue tension of oxygen falls to nearly 15 mm. Hg. Under this condition, arterial blood can only be about 25 percent saturated, so it must give up nearly 15 mL of oxygen per 100 mL of blood.

Hemoglobins differ in the composition of the globin part of the molecule, and this difference imparts different oxygen carrying properties. A shift of the oxygen dissociation curve to the left facilitates oxygen uptake because the blood can become fully saturated at lower oxygen tensions. Mammalian fetal hemoglobin has a

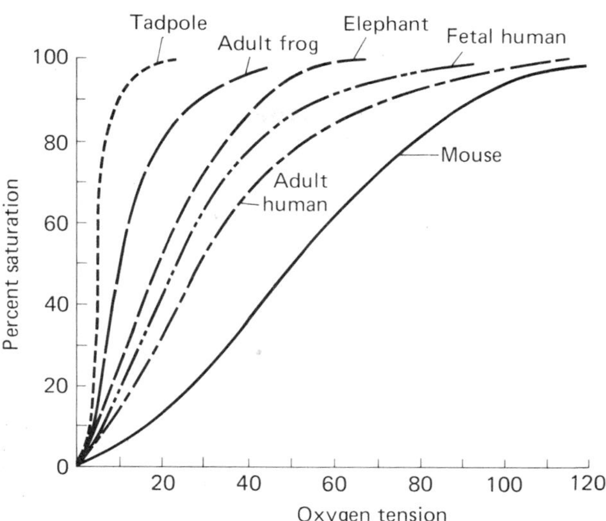

FIGURE 7.15 Oxygen dissociation curves of representative animals that show the effect of left or right shifts in the curves on the uptake and delivery of oxygen. The curves are equilibrium curves that lie between venous and arterial blood.

curve of this type, and this enables the fetus to take oxygen from the blood of the mother (Fig. 7.15). Aquatic animals have curves shifted to the left because oxygen tension is less in water than its partial pressure in air. An interesting example is seen in frogs. The curve for the hemoglobin of the aquatic tadpole is far to the left of the curve for the adult frog. Although a shift of the curve to the left makes it easier to take up oxygen, there is a trade-off because less oxygen can be unloaded in the tissues at a given oxygen tension. This is not a problem for a fetus or aquatic animals because their metabolic rates are low relative to those of an adult mammal or many terrestrial animals. Saturating the blood with oxygen is not a problem for most terrestrial animals because air contains so much oxygen, but terrestrial animals with high levels of metabolism need to unload a great deal of oxygen in their tissues. Small mammals, for example, with their relatively higher levels of metabolism, need more oxygen. They have evolved hemoglobins that shift the oxygen dissociation curves to the right. At a tissue oxygen tension of 40 mm. Hg, an elephant's blood is about 85 percent saturated, whereas a mouse's blood is only 30 percent saturated, and far more oxygen is delivered to the mouse's tissues (Fig. 7.15).

Summary

1. The air from which animals obtain the oxygen they need to sustain cellular respiration is a mixture of gases containing about 21 percent oxygen and 0.04 percent carbon dioxide. The partial pressure of oxygen is the product of its percentage in the air and the pressure of the air. The amount of oxygen that will diffuse into water or the liquid environment of the body depends on its solubility and partial pressure. Gas in a liquid has a tension that is in equilibrium with the partial pressure of the gas in the air.

2. A few invertebrates are oxygen conformers and can adjust their level of metabolism to available oxygen supplies, but most animals are oxygen regulators and need a constant supply of oxygen.

3. Animals must have a moist and permeable respiratory membrane for gas exchange. This may be the body surface or may be limited to a part of a respiratory organ. Passage of gases across this membrane, and between body cells and the liquid bathing them, occurs by diffusion. In addition, there must be a way to bring gases to and from the respiratory membrane (ventilation) and to carry them to or away from the vicinity of the cells (gas transport). Ventilation and transport are by diffusion in small animals, but some system of bulk flow is needed in most animals.

4. The respiratory organs of large aquatic invertebrates and all aquatic vertebrates are gills. Since water contains less oxygen than air, gills must have a large surface area, and a large volume of water must move across them. Gas exchange is often facilitated by moving water across the gills in one direction and blood through them in the opposite direction (counter current exchange). Gills are ventilated by water currents, ciliary action, movements of the gills, or a muscular pump that sends water through a branchial chamber.

5. Loss of body water is minimized in terrestrial animals by having the respiratory membrane located deep in a body chamber where the relative humidity is high and by moving no more air through the chamber than is needed to meet the animal's requirements.

6. A few small terrestrial animals living in moist microhabitats utilize the body surface or gills for gas exchange. Most terrestrial arthropods have a system of tracheal tubes that lead from surface spiracles to all of the tissues of the body. The terminal tracheoles are ventilated, and gas is transported through the system, primarily by diffusion. Other terrestrial animals have lungs. Invertebrate lungs are ventilated primarily by diffusion; vertebrate lungs, by a muscular pumping action of the floor of the pharynx (amphibians) or ribs and diaphragm (mammals).

7. Vertebrate endotherms have a larger gas exchange surface than ectotherms, but small mammals do not have relatively larger lungs than do large ones. Their relatively higher metabolism is sustained by higher rates of ventilation and by adaptations of the blood and circulatory system. Rates of ventilation are controlled by a respiratory center in the brain that responds primarily to levels of carbon dioxide and hydrogen ions in the blood.

8. Transport of gases between the respiratory membrane and the cellular environment is by diffusion in small animals, by the tracheal system in most terrestrial arthropods, and by the blood in other animals. Respiratory pigments, which bind loosely and reversibly with oxygen, enable the blood to carry more oxygen than could be carried in simple solution. Hemocyanin and hemoglobin are the most common respiratory pigments, the latter are found in the erythrocytes of vertebrates.

9. Most carbon dioxide is transported as bicarbonate ions. Hydrogen ions are buffered by hemoglobin and, as oxyhemoglobin takes on hydrogen ions, additional oxygen is driven off (Bohr effect).

10. The properties of hemoglobin are such that plots of the amount of oxygen it binds relative to oxygen tension form a sigmoid curve. This facilitates binding oxygen at the respiratory surface, where tension is relatively high, and unloading much of it in the tissues, where tension is low. Carbon dioxide shifts the dissociation curve to the right and causes more oxygen to be unloaded. During exercise, tissue tensions of oxygen fall more than normally, and more oxygen can be released. Hemoglobins are adapted to an animal's mode of life and the environment in which it lives.

References and Selected Readings

Additional information on gas exchange can be found in the general references on vertebrate organ systems cited at the end of Chapter 4.

Avery, M. E., N. Wang, and H. W. Taeusch: The lung of the newborn infant. *Scientific American, 228* (April):75,1973. An account of the discovery and importance of the surfactant that lowers surface tension in the lungs.

Comroe, J. H., Jr.: The lung. *Scientific American, 214* (Feb.):57, 1966. A fascinating summary of pulmonary anatomy and physiology.

Harrison, R. J., and G. L. Kooyman: *Diving in Marine Mammals.* London, Oxford University Press, 1971. An Oxford Biology Reader summarizing anatomical and physiological adaptations for diving.

Hock, R. J.: The physiology of high altitude. *Scientific American, 222* (Feb.):52, 1970. Respiratory and circulatory adaptations of people to the low oxygen levels at high altitudes are explored.

Hughes, G. M.: *Comparative Physiology of Vertebrate Respiration.* Cambridge, Harvard University Press, 1963.

Hughes, G. M.: *The Vertebrate Lung.* London, Oxford University Press, 1973. The structure and physiology of the lungs of terrestrial vertebrates are summarized in this Oxford Biology Reader.

Randall, D. J., W. W. Burggren, A. P. Farrell, and M. S. Haswell: *The Evolution of Air Breathing in Vertebrates.* New York, Cambridge University Press, 1981. A thorough account of the physiological problems encountered in the evolution of air breathing in many vertebrate groups.

Zuckerkandle, D.: The evolution of hemoglobin. *Scientific American, 212* (May):110, 1965. An analysis of the various types of mammalian hemoglobin and the evolutionary information that they can provide.

8 Internal Transport

ORIENTATION

The cells of metazoans are bathed by an interstitial fluid that is kept nearly constant in its composition. Transport systems bring oxygen, food, and other needed materials from sites of intake in the body to the interstitial fluids and carry away nitrogenous wastes and other products present in excess amounts. The composition of the blood enables it to perform these functions, as well as to prevent undue loss of blood after an injury, and to protect the body against the invasion of microorganisms. Various types of transport systems will be examined along with methods of moving fluids through the systems and of exchanging materials with the interstitial fluid.

Functions of Transport Systems

Materials are transported within cells primarily by diffusion, but materials also must be carried to and from the liquid environment in which cells live. For protozoa and small aquatic animals the surrounding fluid is simply pond water or a puddle; in larger animals it is the **interstitial fluid** (L. *inter,* between + *sistere,* to set) between the cells. Interstitial fluid contains the largest component of body water outside of the cells (Fig. 8.1). Other extracellular water lies within blood vessels (**plasma water**) or has been secreted by cells into various epithelial spaces, such as the coelom or lumen of the digestive tract (**transcellular water**). All of these fluids are involved in the transport process. As cells metabolize they obtain oxygen, glucose, minerals, and other needed substances from the interstitial fluid and discharge carbon dioxide, nitrogenous wastes, and other by-products of metabolism into it. Addition and removal of materials from the blood and interstitial fluid by the digestive tract, lungs, kidneys, and other organs take place in such a way that these fluids remain in a steady state. Most cells cannot tolerate wide changes in their immediate environment.

The problem of internal transport is one of moving materials between the interstitial fluid bathing the cells and the sites where needed materials enter the body and wastes are removed. **Diffusion** alone is adequate in very small animals but is supplemented in most larger animals by the flow of liquids through body cavities or a circulatory system that carries gases, food, and other materials. This is known as **bulk flow.** The degree of development and complexity of the transport system is very much related to the size of the animal and its rate of metabolism. An animal cannot be large or sustain a high rate of metabolism without a very efficient circulatory system that can quickly supply the interstitial fluid with the multitude of materials the cell needs and rapidly remove waste products. By utilizing bulk flow, transport systems are essentially decreasing diffusion distance. Oxygen does not have to diffuse from the lungs to the tissue. It diffuses into the blood, which carries it quickly by bulk flow to the tissues where diffusion again operates.

Transport systems carry things, but they do so in such a way that they help maintain a constant internal environment, or **homeostasis** (Gr. *homios,* alike + *stasis,* standing). In addition to carrying gases and the expected nutrients and waste products, the transport systems of some animals distribute heat from active tissue through the body and to or away from sites where heat can be lost or gained, according to the animal's needs. It transports **hormones,** chemical messengers, from the glands that secrete them to target organs. Certain components of the blood act as buffers, sopping up and holding excess acids and/or bases until they can be removed from the body. Many of the blood's components also play a vital role in defending the body against invading microorganisms.

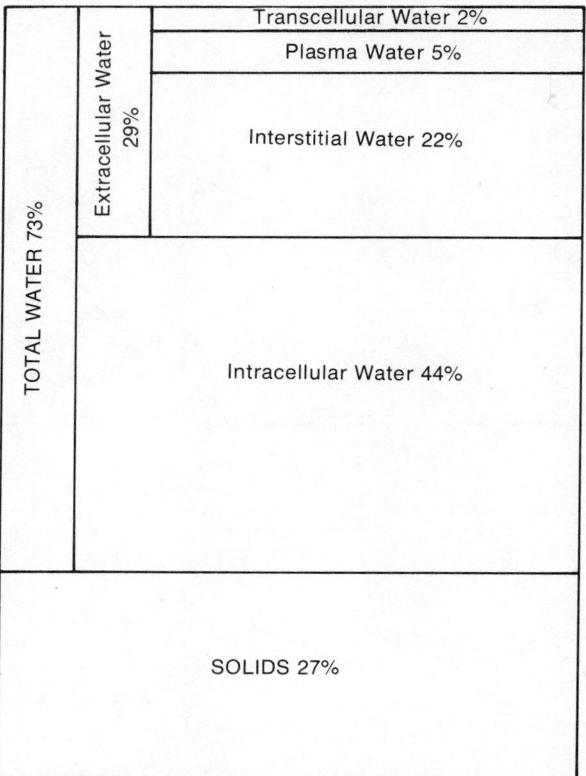

FIGURE 8.1 The distribution of solids and water in the adult mammalian body, expressed as a percentage of the fat-free total body weight.

Methods of Internal Transport

Materials are distributed in many ways between the interstitial fluid and the sites where they enter and leave the body. Contractions of the body wall stir fluids in the body spaces of primitive worms (Fig. 8.2). Ciliary currents move fluid through the body cavities of sponges, sea anemones, and starfish. Most other animals have a distinct circulatory system through which blood or hemolymph is moved by the contractions of certain blood vessels, specialized hearts or adjacent body mus-

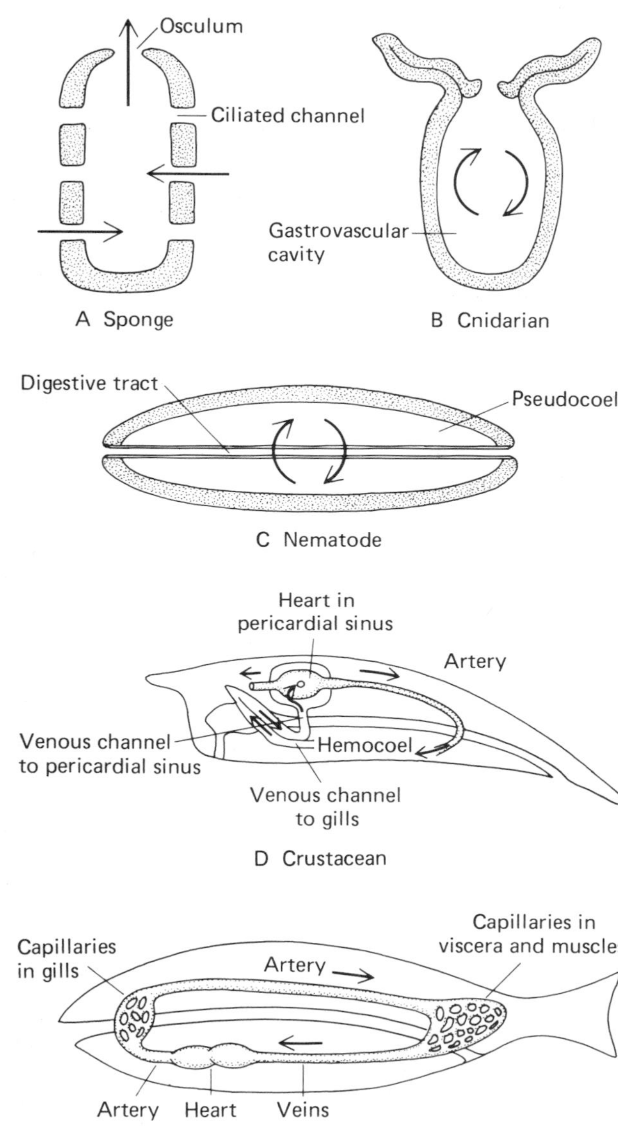

FIGURE 8.2 Methods of internal transport. Water circulation through the cavities of a sponge *(A)* and a cnidarian *(B)*. *C,* Movement of body fluids within the body cavity of a primitive worm. *D,* Open circulatory system of a crustacean. *E,* Closed circulatory system of a vertebrate. (*A, B,* and *C* from Prosser, C. L.: *Comparative Animal Physiology.* 3rd ed. Philadelphia, Saunders College Publishing, 1973.)

cles. In the octopus and other cephalopod mollusks, and in annelid worms, vertebrates and other animals with **closed circulatory systems,** blood is always confined to vessels or vessel-like channels. In the closed system of vertebrates, for example, blood moves from the heart through arteries to extensive networks of capillaries in the body tissues and returns through veins. The capillaries are very small, thin-walled vessels, and small molecules diffuse easily between the blood and the interstitial fluid. Most mollusks and arthropods and some of the lower chordates (tunicates) have **open circulatory systems.** Liquid moves from the heart through arteries directly to spaces among the cells. Collectively these spaces form a **hemocoel,** and the liquid moving through them is most appropriately called **hemolymph,** for it combines properties of blood and lymphlike interstitial fluid. Many of the tissue spaces are quite diffuse, whereas others are organized into special channels, but the channels do not have the lining of endothelial cells that characterizes blood vessels of vertebrates. Hemolymph in them is in direct contact with body cells.

A closed circulatory system is not necessarily more advanced than an open system, but a higher pressure and more rapid flow of blood can develop easily because the blood is confined to vessels. Materials are quickly distributed between organs. The octopus and squid can be very active mollusks partly because they have closed systems, whereas clams, snails, and other mollusks with open systems are sluggish, slow-moving creatures. High pressure can develop in open systems for special purposes. The legs of jumping spiders are extended not by the action of muscles but by the sudden flow of hemolymph into them. High pressure can develop in the spider's leg because its volume is limited by the exoskeleton. Animals with closed systems are able to regulate the flow of materials to the tissues more precisely than animals with open systems. By dilating and constricting appropriate vessels, blood can be shunted to particularly active tissues, for example, to muscles during rapid locomotion, whereas less active tissues will have a reduced flow through them. Circulating fluids and materials are distributed more evenly to all of the tissues in animals with open systems.

Blood and Interstitial Fluid

Adult human beings have about 5.7 L of **blood.** Our blood, as well as that of other animals, is composed of a liquid component, the **plasma,** and various formed elements that are carried in the plasma, among them red blood cells **(erythrocytes),** white blood cells **(leukocytes),** and **platelets.** The plasma itself is about 90 percent water, 7 to 8 percent soluble plasma proteins, 1 percent electrolytes, and the remaining 1 to 2 percent is composed of a variety of substances: (1) nutrients such as glucose, amino acids, lipids, and vitamins; (2) metabolic intermediates such as pyruvate and lactate; (3) nitrogenous wastes, including urea and uric acid; (4) dissolved gases; and (5) hormones. Many of these molecules are free in solution; others, including some of the trace metals, are bound to specific transport proteins.

The primary plasma proteins are albumins, globulins, and fibrinogen. Albumin constitutes 60 percent of all plasma proteins and functions primarily to maintain the osmotic pressure of the blood. This is essential for normal fluid exchanges between the blood and interstitial fluid, as explained later. The immunoglobulins are antibodies important in the body's defense. Other globulins bind with and transport iron and other substances, are blood clotting factors, and perform many other functions. Fibrinogen is the source of the fibrin of blood clots. Most of the plasma proteins are synthesized in the liver, but antibodies are made by lymphoid tissues and tissue plasma cells.

The major electrolytes in the blood and interstitial fluid are sodium, chloride, and bicarbonate ions. Only small amounts of other ions, including potassium, magnesium, and phosphate, are present. This is in sharp contrast to the intracellular fluids, in which the most abundant ions are potassium, phosphate, and magnesium.

The plasma is a complex liquid in dynamic equilibrium with other body fluids. Many substances move in and out of it, but its composition and properties remain remarkably constant.

Plasma and interstitial fluid are separated by the endothelial wall of the capillaries, and they differ in composition only in the materials that do not pass through this wall or that do not pass through readily. Most of the blood cells remain in the plasma, but certain leukocytes that are capable of amoeboid movement squeeze between the cells of capillary walls and enter the tissue spaces. Most of the large protein molecules remain in the plasma, but some enter the interstitial fluid. Other components of the plasma and interstitial fluid diffuse readily through the capillary wall, so their concentration in the two compartments tends toward equilibrium.

Red Blood Cells

The respiratory pigments of invertebrates are carried free in solution or within corpuscles of various kinds that circulate in the blood, hemolymph, or coelomic fluid (see p. 160). The vertebrate respiratory pigment, hemoglobin, is always located within red blood cells, or **erythrocytes.** Most vertebrates have oval-shaped, nucleated erythrocytes, but those of mammals lose their nuclei as they develop and become biconcave discs. Such a shape provides more surface area than a sphere of equal volume, and the increased surface area, in turn, facilitates the passage of gases and other materials through their plasma membrane. Red cells are the most numerous of the cellular elements of the blood, there being about 5,000,000 per cubic millimeter in adult human blood.

Mature mammalian erythrocytes do not survive indefinitely. Experiments that involve tagging them with radioactive iron show that they have an average life span of 120 days. Cells lining the blood spaces of the spleen and liver eventually engulf or **phagocytize** (Gr. *phagein,* to eat + *cyte,* cell) the red cells and digest them. The iron of the heme is salvaged by the liver and is reused, but the rest of the heme is metabolized to **bile pigments** and excreted. Under normal circumstances, the erythrocyte population is in a steady state, with new ones being synthesized as rapidly as the old are destroyed. But if delivery of oxygen to the tissues is reduced, as during hemorrhage or when ascending to a high altitude, more cells are made available so that delivery of oxygen to the tissues is restored to normal levels. Additional erythrocytes can be released immediately into the circulating blood from reserve stores in the **spleen,** but the reduced oxygen tension also triggers a set of reactions that leads to an increased rate of synthesis of erythrocytes. Cells in the kidney in particular respond to a lower oxygen tension by synthesizing a hormone known as **erythropoietin.** This passes in the blood to the red bone marrow where it stimulates the first stage in red cell production, the formation of hemocytoblasts from primordial stem cells. In lower vertebrates, red cells are produced in vascularized connective tissues of the kidney, liver, and spleen. These sites are important during the embryonic development of mammals, but the red bone marrow is the primary source of erythrocytes in the adult.

Hemostasis

The loss of blood or hemolymph through an injury can severely impair the distribution of materials to and from the body cells, and animals with circulatory systems have evolved mechanisms for **hemostasis;** that is, for stopping the flow of blood after an injury. Because of its great medical importance, hemostasis has been studied most intensively in vertebrates, and especially in mammals, but the phenomenon occurs in invertebrates as well.

Hemostasis in all animals involves a short-term closure of the blood vessels, followed by the formation of a clot, that by its contraction holds the injured tissues together until healing occurs. In animals with closed circulatory systems, nerve reflexes cause small broken vessels to constrict, thereby reducing blood flow. Many crustaceans with open systems can reflexly discard a limb that has been caught by a predator. The limb is severed at a distinct **autotomy plane,** and a specialized membrane here constricts around the perforation so there is very little loss of hemolymph.

Temporary closure of vessels in most animals is also achieved by a clumping of blood cells in the injured vessel or hemolymph space. In lower vertebrates the cells that clump are **thrombocytes** (Gr. *thrombos,* clot + *cyte,* cell), but in mammals **platelets** play this role. When they contact damaged vascular surfaces they swell and become sticky, adhering to each other and to the walls of the blood vessels. Platelets are nonnucleated fragments of cytoplasm that bud off from giant cells in the bone marrow. They number about 300,000 per cubic millimeter in human blood and survive for 8 to 10 days.

A clot forms in vertebrates when two small peptides are removed from the soluble plasma protein, **fibrinogen,** by the action of the enzyme **thrombin** forming **fibrin monomer.** These monomers spontaneously polymerize, forming at first loosely bound fibrin threads. Another enzyme, **fibrin stabilizing factor,** soon establishes stronger, covalent bonds between the monomers (Box 8.1). The fibrin threads form a delicate network within which the formed elements of the blood are trapped (Fig. 8.3). A **clot** has formed. The clot subsequently contracts and squeezes out the liquid phase, called **serum.** Serum differs from plasma in that it lacks fibrinogen and cannot clot.

Thrombin, of course, is not present in the blood in its active form, otherwise the blood would spontaneously coagulate. When blood vessels are injured and blood comes into contact with exposed collagen, clotting factor XII in the blood is activated, and the platelets release phospholipids. These events trigger an **intrinsic pathway** in which a series of proenzyme-enzyme transformations occurs that has been compared to an "enzyme cascade." (See Box 8.1) The conversion of one proenzyme into an active enzyme triggers the conversion of the second one, and so on through many steps, finally leading to the conversion of prothrombin to thrombin. Most of these steps require calcium ions, and some require additional blood clotting factors, many of which are plasma globulins. An advantage of this complexity is the amplification that results. The few molecules activated in the first step affect more and more at each successive step until millions are involved. Although this process leads to the formation of a great deal of fibrin, it takes a few minutes. A second extrinsic pathway that is triggered by phospholipids and other factors released by the injured tissues themselves can act more quickly by directly activating one of the later steps. This initiates clotting but does not produce as much fibrin.

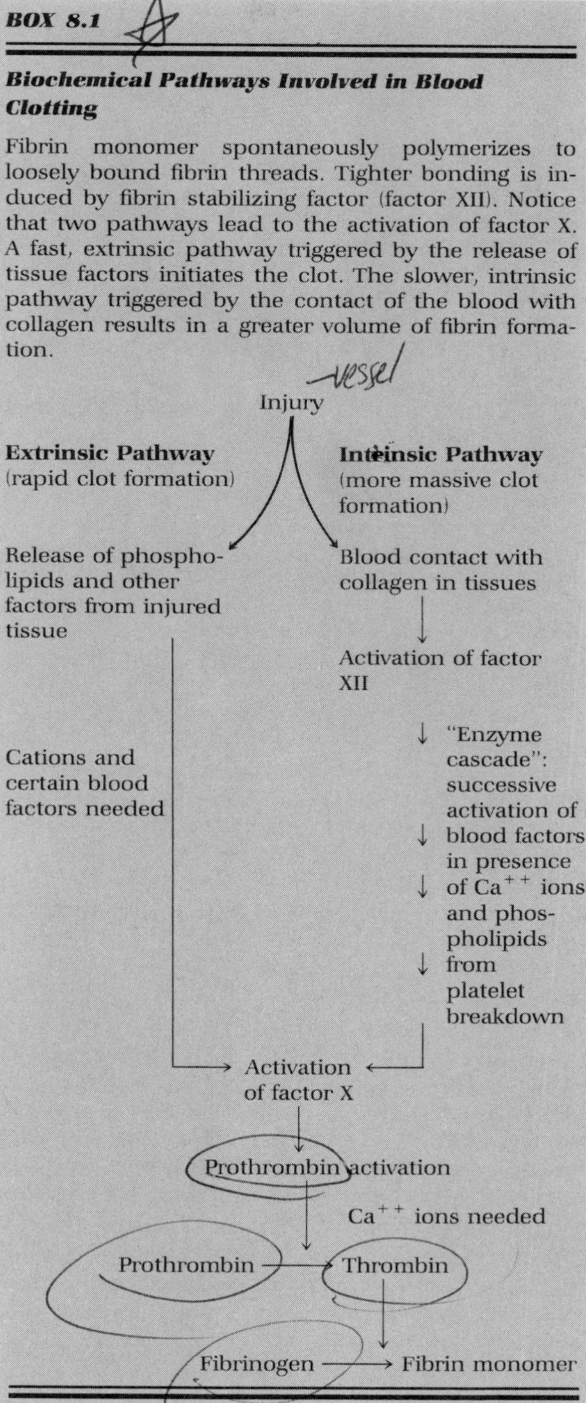

BOX 8.1

Biochemical Pathways Involved in Blood Clotting

Fibrin monomer spontaneously polymerizes to loosely bound fibrin threads. Tighter bonding is induced by fibrin stabilizing factor (factor XII). Notice that two pathways lead to the activation of factor X. A fast, extrinsic pathway triggered by the release of tissue factors initiates the clot. The slower, intrinsic pathway triggered by the contact of the blood with collagen results in a greater volume of fibrin formation.

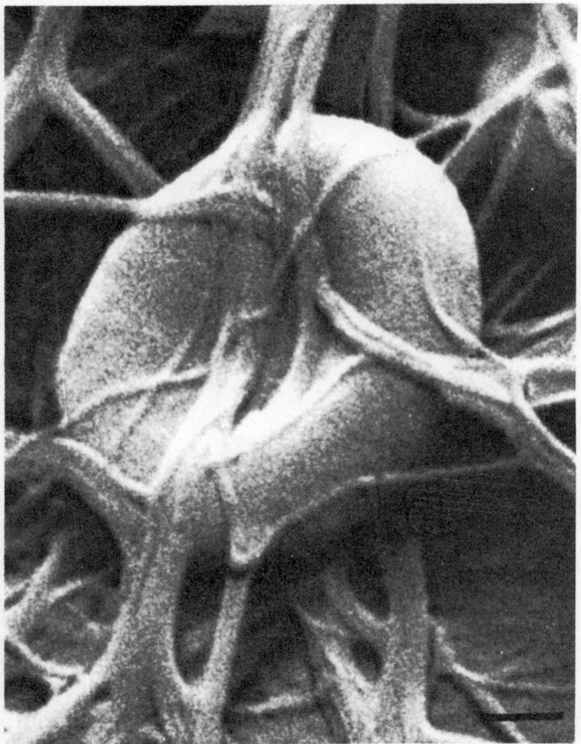

FIGURE 8.3 Scanning electron micrograph of a portion of a human blood clot showing an erythrocyte enmeshed in a network of fibrin. (From *Science, 173* (3993): 1971. Cover picture. Photograph by Emil Bernstein, the Gillette Company Research Institute.)

After a few days, when healing is well underway, enzymes from the injured tissue and factors in the blood convert **profibrinolysin** (Gr. *lysis,* loosing), a blood factor that has been incorporated in the clot, into its active form. **Fibrinolysin** is a proteolytic enzyme that cleaves the fibrin molecules and breaks up the clot. This same mechanism can destroy clots that start to form in vessels. Intravascular clots do not form easily, however, because of the smoothness of the vessel linings that prevents the release of the various activating factors and the presence of inhibitors and anticoagulants, such as **heparin.** Some heparin is produced by basophilic leukocytes in the blood, and more diffuses into the capillaries from tissue mast cells. Despite these mechanisms, a clot, known as a **thrombus,** may develop in a vessel and can be very serious if it plugs a vessel that supplies a vital area.

In the hereditary disease **hemophilia** there is a deficiency of one of the blood factors in the enzyme cascade so that clots do not form well and a slight scratch may lead to fatal bleeding.

White Blood Cells

In addition to cells containing respiratory pigments and ones involved with hemostasis, the circulating fluids contain a variety of colorless cells that perform many different functions. The circulatory system transports them to the particular part of the body where they may be needed. Many are important in body defense and wound healing; a few among invertebrates have an excretory role. In vertebrates, these cells are known as white blood cells, or **leukocytes** (Gr. *leukos,* white + *cyte,* cell). Five distinct types can be recognized. **Neutrophils, basophils,** and **eosinophils** have irregular-shaped nuclei (hence they are often called **polymorphonucleocytes**) and distinctive cytoplasmic granules (Fig. 2.25). **Monocytes** have a large nucleus indented on one side and considerable cytoplasm; **lymphocytes,** a nearly spherical nucleus and little cytoplasm. Lymphocytes originate in and are stored in the lymph nodes and other lymphoid tissues; the other leukocytes, in red bone marrow. Collectively they number about 7000 per cubic millimeter in human blood under normal circumstances. Large numbers are released from storage and synthesized during infections. All have short transit times of a day or less through the circulatory system, then squeeze between cells in capillary walls and enter the tissues where their functions are performed. The monocytes enlarge, develop granules, and transform into tissue **macrophages** that reside in many body tissues, sometimes for months. The life span of others is only a few days. Many leukocytes are destroyed within the tissues; others are lost from the body through the lungs, digestive tract, kidneys, and reproductive tract.

Bacteria and viruses that have invaded the body produce toxins that, together with **histamine** and other factors released by injured tissues, cause blood vessels in an infected area to dilate and to become more permeable to plasma and leukocytes. The area becomes **inflamed.** Polymorphonucleocytes and macrophages migrate by amoeboid action toward the infection to which they are attracted by the chemotactic effect of the gradient of toxins. Neutrophils and macrophages act as scavengers and phagocytize, that is, engulf and digest, microbes and destroyed tissues. Many of these cells themselves break down and contribute, along with the microbes and tissue debris, to **pus** that may form.

Basophils and the tissue **mast cells** into which they transform release more histamine and heparin. Both of these substances increase the inflammatory reaction and the migration of other leukocytes into the area.

Eosinophils are attracted by foreign proteins, which they probably help to detoxify. They are particularly abundant in certain invertebrate parasitic infections, such as trichinosis caused by pork roundworms (p. 514). By releasing enzymes that help activate profibrinolysin, they break down blood clots. They also secrete materials that break down histamine and digest the antigen-antibody complex after an immune reaction (see later), so they are important in cleaning up the debris of an infection.

Immunity

Antibodies and Immunity. The phagocytosis of invading bacteria and viruses is a primary body defense that is developed in many invertebrates and all vertebrates. Birds, mammals, and, to a lesser extent, lower vertebrates have evolved a second line of defense, for they have the capacity to synthesize special proteins, known as **antibodies,** that combine with and destroy foreign substances **(antigens)** (Gr. *anti,* against + *genos,* birth) that enter the body (Fig. 8.4). Antigens are usually

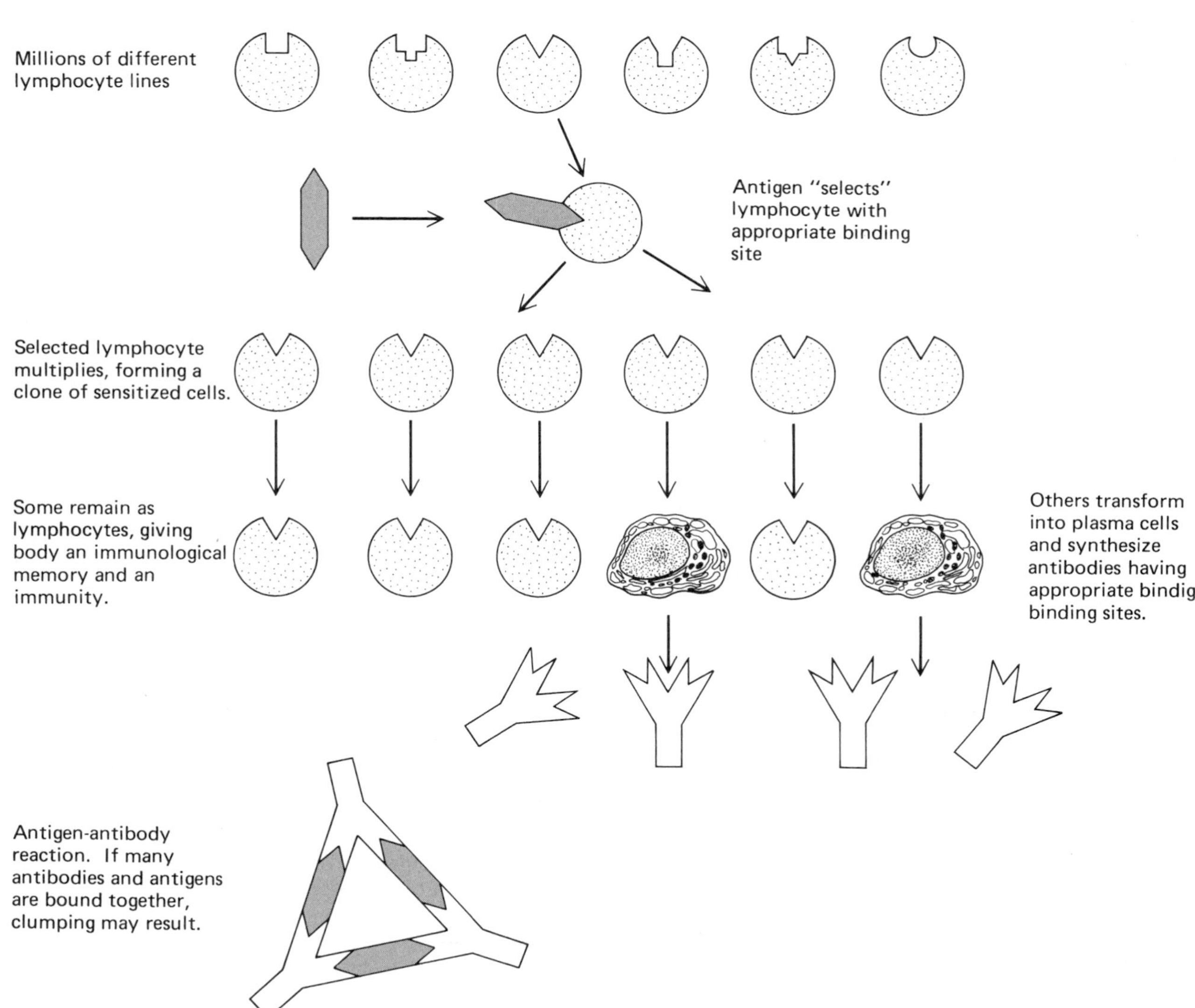
Millions of different lymphocyte lines
Antigen "selects" lymphocyte with appropriate binding site
Selected lymphocyte multiplies, forming a clone of sensitized cells.
Some remain as lymphocytes, giving body an immunological memory and an immunity.
Others transform into plasma cells and synthesize antibodies having appropriate bindig binding sites.
Antigen-antibody reaction. If many antibodies and antigens are bound together, clumping may result.

FIGURE 8.4 Activity of B lymphocytes in an antigen-antibody reaction.

proteins associated with invading bacteria or viruses, but any foreign protein and certain foreign polysaccharides and DNA may induce antibody formation. Often the antigen itself is first phagocytized by tissue macrophages, and the antibody response is not to the original antigen but to antigenic substances released by the macrophages. Only a specific part of the antigenic material, the **immunological specific determinant group,** causes an antibody to be produced. Some antigens have several determinant groups and induce the synthesis of several different antibodies. The antigen-antibody reaction is generally very specific. Antibodies that have developed in response to mumps viruses, for example, will not combine with other antigens. The configuration of the antigen and antibody molecules is such that only antibodies that have developed in response to a given antigen can fit on the surface of the antigen and react with it.

Once the body has produced antibodies in response to an antigen, it retains the capacity to produce similar antibodies for many years. If a subsequent invasion of the same type of antigen occurs during this period, antibodies specific for it may be present, and more can be synthesized very quickly by the multiplication of a clone of sensitized cells remaining from the first infection. The body has developed an **immunity** to this particular antigen. The immunity that is acquired as a result of having once had mumps, smallpox, and many other infectious diseases lasts a very long time, generally for life. The immunity to certain other diseases lasts for a much shorter time and, after it is lost, one can get the disease again.

However, it is not necessary to contract a disease to develop an immunity to it. By **vaccination,** antibodies are produced in response to the introduction of a similar but less virulent microorganism than the one against which protection is offered, or to the introduction of the disease organism in a harmless state.

The production of antibodies in response to an antigen that enters the body, either through disease or vaccination, is known as an **acquired immunity.** Acquired immunities are typically long-lasting. Temporary protection can be given to a person who has been exposed to a new and dangerous antigen by injecting an antiserum containing antibodies from an organism that has already produced them. A person bitten by a poisonous snake, for example, can be given antiserum obtained from a horse that has developed antibodies specific to this snake venom. The horse has developed an acquired immunity in response to successive injections of small amounts of the snake venom, but the immunity conferred on the person receiving the antiserum is a **passive immunity.** A newborn infant receives a passive immunity to certain diseases by the passage of maternal antibodies through the placenta and in the milk secreted during the first few days of lactation. Passive immunities last only a few days or weeks.

Lymphocytes. The key cells in the antigen-antibody response are the lymphocytes, of which there are two distinct populations. Although both are derived from common stem cells in bone marrow, the two undergo maturation processes in different tissues before entering lymph nodes and the general circulation. **T-lymphocytes** are processed in the **thymus,** an organ of pharyngeal pouch origin located at the base of the neck, and in the adjacent parts of the thoracic cavity in fetal and young mammals. The thymus atrophies after maturity. **B-lymphocytes** were first identified in chickens where they undergo their processing in the **bursa of Fabricius,** a pouchlike diverticulum of the cloaca. The site of processing of mammalian B-lymphocytes is not entirely clear, for mammals have no bursa, but it appears to be the fetal liver and probably aggregations of lymphocytes associated with the digestive tract: tonsils, appendix, and islands of lymphoid tissue in the intestinal wall (Fig. 8.5).

On exposure to a particular antigen, B- or T-lymphocytes that are competent to respond to this particular antigenic stimulus begin to multiply rapidly, but it may require nearly a week for them to reach a peak in numbers and activity. This is known as the **primary immune response** (see Fig. 8.4). Some of the progeny cells combat the foreign antigen in ways described later, but others remain as lymphocytes with the capacity to react again with this particular antigen. They give the body an **immunological memory.** Should the body be exposed again to the same antigen, a clone of sensitized cells is present. They can multiply and respond more quickly, in a day or two, in what is called the **secondary immune response.** The secondary response is not only more rapid, but it is also more potent and lasts far longer.

In other respects the B- and T-lymphocytes differ in the way they respond. Some of the progeny cells of the stimulated B-lymphocytes develop an elaborate granular endoplasmic reticulum capable of protein synthesis (see Fig. 8.4), lodge in the tissues, and, now known as **plasma cells,** synthesize appropriate antibodies. The antibodies are released into the blood and carried as part of the gamma globulin fraction. This **humoral antibody response** combats bacteria and viruses in the blood.

T-lymphocytes, on exposure to appropriate antigens, are transformed into **lymphoblasts** that produce antibody-like substances, known as **lymphokines,** that are not released into the circulation. Rather, the lymphoblasts themselves react directly with the antigen, usually

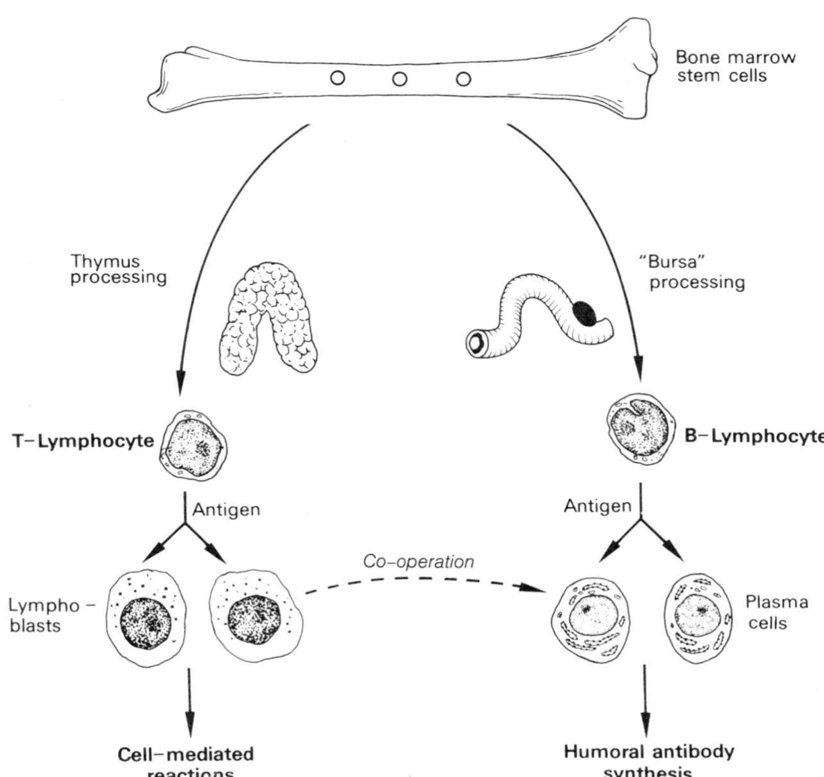

FIGURE 8.5 Diagram of the development of the two types of mammalian immune systems from stem cells in the bone marrow. (From Frobisher, M., R. Hinsdill, K. Crabtree, and C. Goodheart: *Fundamentals of Microbiology.* 9th ed., Philadelphia, Saunders College Publishing, 1974.)

fungi, intracellular viruses, cancer cells, invertebrate parasites, and foreign tissues. This **cell-mediated response** is responsible for the rejection of skin and other tissue grafts. Most adult vertebrates will not accept a homograft from another individual that is genetically very different but will accept autografts from other parts of the body or isografts from an identical twin. When performing an organ transplant, the surgeon tries to obtain an organ from an individual as genetically similar as possible to the recipient, and even then steps must be taken to suppress the activity of the T-lymphocytes.

Antibody Structure, Action, and Diversity. Antibodies consist of one or more Y-shaped molecules composed of four protein chains bound together by disulfide bonds (Fig. 8.6). Two heavy chains make up the stem and part of the arms of the Y, and two light chains complete the arms. The basal segment of each chain is fairly constant in its amino acid composition, but the parts at the ends of the arms are highly variable. It is the variable segments that contain the specific antigen

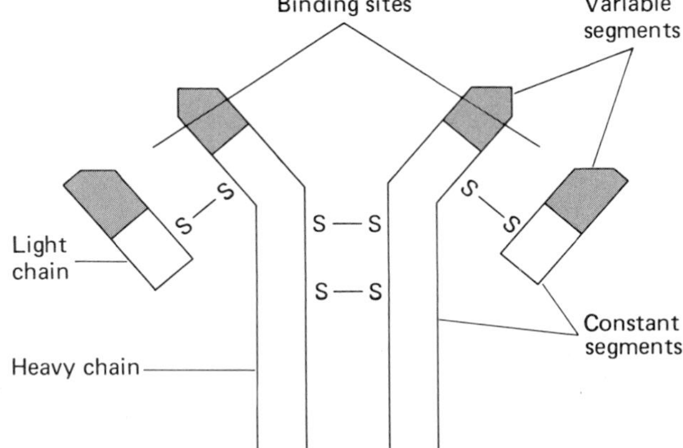

FIGURE 8.6 Diagram of the arrangement of the four protein chains in an antibody molecule of the immunoglobulin gamma (IgG) type. The specific antigen binding sites are located at the variable ends of the molecule. Since there are two variable ends, each molecule can bind with two antigen sites of the same type.

binding sites. The ends of the arms have been compared with the notches on a key, and the many configurations here fit the many antigens. Antibodies are all **immunoglobulins** (abbreviated Ig), and the composition of their heavy chains determines to which of five general classes they belong and the way they will perform their immunological functions. For example, if the heavy chains are of the alpha type, the antibodies (IgA) will pass through epithelial surfaces with saliva, tears, sweat, and mucus and react with microorganisms at their points of entry into the body. The most abundant antibodies, IgG, circulate in the gamma globulin fraction of the plasma.

Antibodies attack antigens in many different ways. Their combination may increase the permeability of capillaries, thereby increasing the invasion of white blood cells into the infected area. They may bind with the antigens so as to cause the antigens to clump and precipitate out of the system (See Fig. 8.4). They may promote phagocytosis by other cells. Finally, they may activate the plasma **complement system** that will break down the plasma membrane of the invading microorganisms and lead to their lysis. The complement system consists of at least nine enzyme precursors. Union of an antigen with an antibody activates other sites on the antibody which, in turn, activates the first of the complement precursors. This activates the second precursor, and so on through an amplifying enzyme cascade similar to the one leading to blood clotting. The large amount of final complement formed lyses the invaders and sometimes also increases the inflammatory reaction.

In the late 1950s, Sir Macfarlane Burnet proposed the **clonal selection theory** to explain how the appropriate antibodies are produced in response to a particular antigen (see Fig. 8.4). According to this view, which is now generally accepted, lymphocytes differentiate genetically during embryonic development through somatic mutation or other mechanisms. Millions of different lines of lymphocytes are generated, each programmed genetically with the capacity to make a specific antibody in response to a specific antigen. When an antigen enters the body certain lymphocytes already exist that have the competence to respond to it, for there is a veritable dictionary of different lymphocytes within the body (see Fig. 8.4). The appropriate lymphocytes are stimulated (or selected) by the antigen, multiply rapidly, and form a clone of identical cells that synthesize the antibodies needed to combat the antigen. The time required for this accounts for the latent period between an infection and the body's response.

The clonal selection theory helps us to understand how the body distinguishes between "self" and "nonself." Antibodies are synthesized in response to foreign proteins ("**nonself**") but normally not in response to the body's own proteins ("**self**"). During embryonic development, when the immune system is not completely formed, lymphocytes encounter the various body tissues. If a given lymphocyte has developed the capacity to react with one of the body's own proteins, a mild reaction will occur, and the lymphocyte will be eliminated before it has had a chance to multiply and produce other cells of its type. By birth the individual will have a wide assortment of lymphocytes capable of reacting only with foreign antigens.

Considerable evidence supports this notion. If foreign antigens are injected into an individual during fetal or neonatal life before the immune system has matured, the body assumes they are "self," and lymphocytes that start to react against them are eliminated. It is possible, for example, to inject a newborn mouse of strain X with cells from strain Y. The Y cells are incorporated as "self," and at a later time the X strain mouse will accept a skin graft from a Y mouse, something it would not normally do. The X mouse has developed an **acquired immunological tolerance.** A converse situation sometimes develops in which certain of the body's own tissues are no longer recognized as "self." Certain types of arthritis and multiple sclerosis are **autoimmune diseases** in which antibodies are synthesized that destroy some of the body's own proteins.

A major problem is how so many genetic lines of lymphocytes can develop with the capacity to code for perhaps billions of different antibodies. Somatic mutations of lymphocyte genes doubtless play a role, but mutation rates are not high enough to generate this much diversity. Recent discoveries in molecular genetics and microbiology are providing some insights. The genes in the early embryonic cells that ultimately will control the synthesis of antibody proteins are not continuous sequences of DNA nucleotides; rather nearly 140 different coding sequences are separated by noncoding spacing sequences. Embryonic cells have in effect a "gene kit," components of which can be put together in different combinations to make the definitive genes that code for specific antibodies. As embryonic development proceeds, certain segments of the "gene kit" are cut out, and eventually four different segments are spliced together to make the definitive gene for a particular cell. Different combinations of segments will be cut out and spliced together in different cells. It has been estimated that over 20 billion combinations are possible. This process has been called "somatic recombination." Somatic recombination appears to be random, so some functionless genes probably are formed, but it appears that the cell can try several times to make a functional gene. Once a particular lymphocyte has made an active gene that synthesizes a protein, inhibitory mechanisms become effective, and this cell continues, with appropriate stimulation, to make only this kind of antibody.

Allergy. **Allergic reactions** sometimes develop as side effects of immunity. In one type of allergy reexposure to antigens greatly stimulates the IgE antibodies that have previously attached to mast cells and basophils, leading to the release of large amounts of histamine. In **hay fever,** released histamine is confined to the upper respiratory tract, causing vasodilation, increased capillary permeability, and swelling in these areas. In **anaphylaxis** (New L. *ana,* intensification + Gr. **phylaxis,** a guarding), the release of histamine is widespread. Respiratory passages contract, plasma loss is extensive, and death by circulatory shock may occur within minutes. Epinephrine and other antihistamines can alleviate the symptoms.

Blood Group Factors

In addition to having the capacity to make antibodies in response to foreign antigens, and in certain diseases in response to some of our own tissues, some antigens and antibodies normally are present in our blood at all times. These are the blood group factors. The **ABO factors** were discovered when transfusions were frequently made in World War I. Sometimes the transfusions were successful, but often they were not, and erythrocytes in the blood of the recipient would clump (agglutinate), with fatal results. Careful analysis by Landsteiner showed that specific antigenic proteins, called A and B, might be present in the plasma membranes of the erythrocytes. These antigens are called **agglutinogens** since they may cause agglutination of the red cells. Some individuals have antigen A, some B, some both A and B, and some neither. Naturally occurring antibodies **(agglutinins)** specific for these agglutinogens, and designated *a* and *b*, may be present in the plasma. An individual with red cells containing a certain agglutinogen does not, of course, have the agglutinin specific for it in his plasma. If he did, his own red cells would be agglutinated.

Four groups of persons, O, A, B, and AB, can be recognized according to the presence or absence of these agglutinogens and agglutinins (Fig. 8.7). Before a transfusion is made, both donor's and recipient's cells are typed by mixing them with serum containing an *a* agglutinin (anti-A serum) and with serum containing a *b* agglutinin (anti-B serum), and observing which, if either, causes the cells to clump. Transfusions between mem-

Blood group	Agglutinogen in erythrocyte	Agglutinins in plasma	Reaction with	
			Anti-A serum	Anti-B serum
O Universal donor	none	a and b		
A	A	b	agglutination	
B	B	a		agglutination
AB Universal recipient	A and B	None	agglutination	agglutination

FIGURE 8.7 The ABO blood groups. The two right-hand columns show the appearance of the blood of each group when exposed to anti-A and anti-B serum. The blood group to which an individual belongs is identified by the antiserum in which its red cells agglutinate; if there is no agglutination with either antiserum, the individual belongs to group O.

bers of the same group are perfectly safe, and transfusions between different groups are also safe provided that the donor's erythrocytes do not contain an agglutinogen that will react with the recipient's agglutinins. The agglutinins in the donor's plasma become so diluted in the recipient that they have no effect, and they may be disregarded unless an unusually large transfusion is given. Members of Group O, who have neither of the agglutinogens, can give blood to members of any group and are "universal donors." But since their plasma contains both of the agglutinins, they can receive blood only from members of their own group. Members of Group AB, in contrast, have neither agglutinin, and can receive blood from members of any group. Since they have both agglutinogens, they can give blood only to members of their own group. They are "universal recipients." Members of Groups A and B can give blood to members of Group AB and receive from members of Group O.

Another blood factor of importance in transfusions is the **Rh factor,** particularly because of complications that can result when incompatibilities develop between mother and fetus (p. 358).

Many other naturally occurring factors have been discovered, but they are of less significance in transfusions. Since the inheritance of most of them has been determined (p. 358), they are important in legal medi-

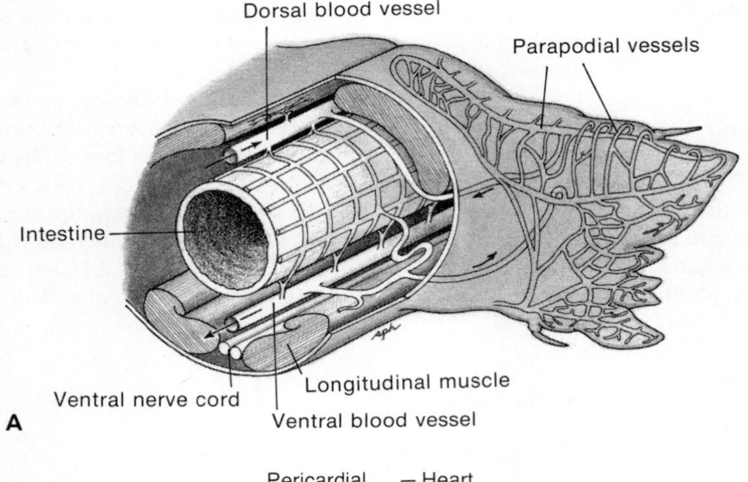

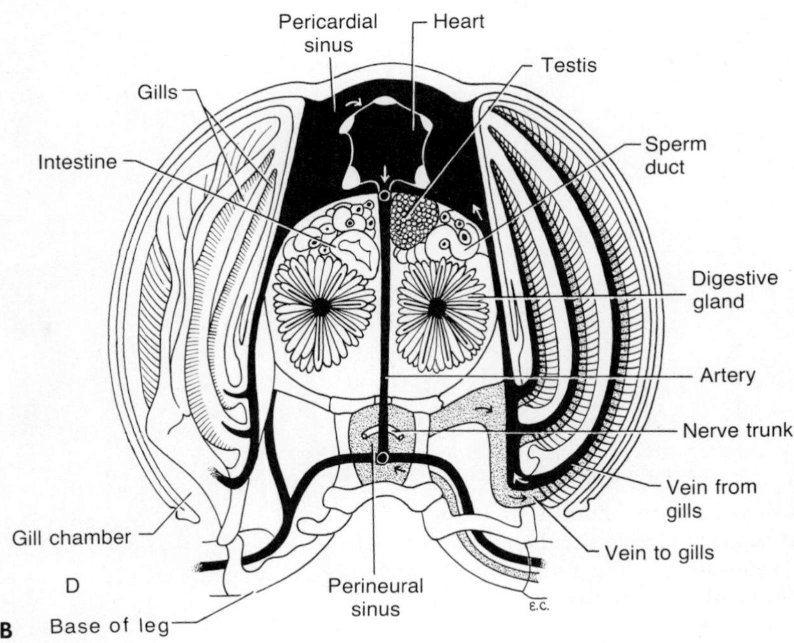

FIGURE 8.8 Representative invertebrate circulatory systems. Arrows indicate the direction of blood or hemolymph flow. *A,* The closed system of an annelid as represented by a portion of the polychaete *Nereis.* Anterior is toward the right. The paired circumesophageal vessels that carry blood from the dorsal to the ventral vessel are not shown. *B,* The open system of an arthropod as seen in a cross section through the thorax of a crayfish. (A Modified from Nicoll; *B* after Howes.)

cine in determining from whom a trace of blood may have come, identifying parents in cases of disputed paternity, and so forth. Different racial groups are characterized by different frequencies of these factors, but their biological significance has received little study.

Invertebrate Circulatory Patterns

Animals, such as roundworms and flatworms, that are small or have a shape that provides a large surface area and a small internal distance, lack a circulatory system, but one is present in other invertebrate phyla: mollusks, annelids, arthropods, echinoderms. Circulation in most echinoderms is provided by the coelom.

Annelids. Annelid worms are the major invertebrate group with a closed circulatory system. A pulsating **dorsal vessel** conveys blood anteriorly (Fig. 8.8*A*). Near the front of the body one or more pairs of **circumesophageal vessels** carry the blood around the gut to a **ventral vessel.** In some annelids, the circumesophageal vessels pulsate and act as accessory hearts. Blood flows posteriorly in the ventral vessel and is distributed by branches in each segment to the gut and body wall, thence back into the dorsal vessel. Typically there is a branch from the ventral vessel to the gut and a pair of branches to the nephridia, body wall, and parapodia in each body segment. Although the blood usually is confined to small, thin-walled vessels, the vessels are not lined with endothelium as are the small capillaries of vertebrates.

Arthropods. Arthropods, although related to annelids, have evolved an open circulatory system. The **heart,** which may be long and tubular or compact and limited to the thorax, is situated dorsally within an expanded portion of the hemocoel known as the **pericardial sinus** (Fig. 8.8*B*). Hemolymph returning from tissue spaces enters the pericardial sinus rather than going directly into the heart. Hemolymph enters the heart through paired lateral openings, the **ostia,** that contain valves that prevent it from leaving. Major arteries leave the front, back, and often the underside of the heart and lead into branches that carry hemolymph to tissue spaces in the various organs. In crustaceans the tissue spaces drain into sinuses, whence the hemolymph passes through channels in the gills and back to the pericardial sinus. Insects have a circulatory system, but its functions do not include the transport of gases, which are carried instead by the tracheal system (see p. 153).

Vertebrate Circulatory Patterns

Fishes. The circulatory system is closed in all vertebrates. It has undergone many changes during the evolution of vertebrates, most of which are correlated not only with the shift from gills to lungs as the site of external gas exchange that occurred during the transition from water to land but also with the development of the efficient, high-pressure circulatory system necessary for an active terrestrial vertebrate.

In most fish (Fig. 8.9), all of the blood entering the heart from the veins has a low oxygen and a high carbon dioxide content; i.e., it is venous blood. The heart consists of a **sinus venosus,** a single **atrium,** a single **ventricle,** and a **conus arteriosus** arranged in linear sequence. The muscular contraction of the heart in-

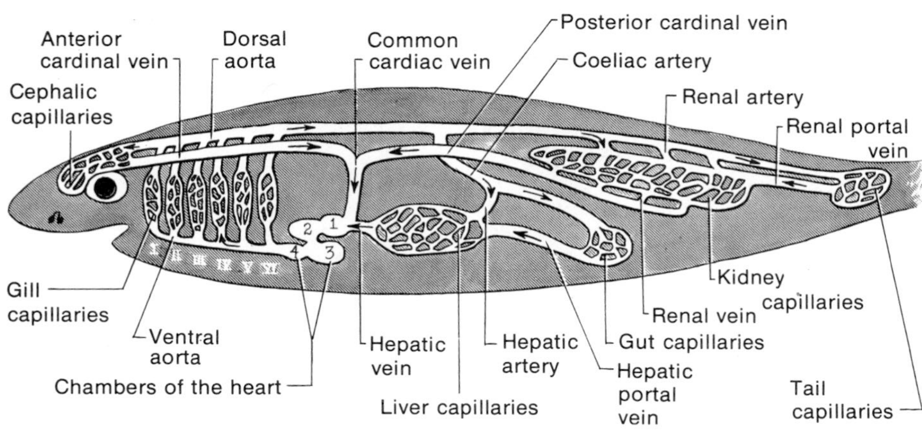

FIGURE 8.9 The major parts of the circulatory system of a primitive fish. *1,* Sinus venosus; *2,* atrium; *3,* ventricle; *4,* conus arteriosus of the heart. The aortic arches are numbered with Roman numerals. Only traces of the first aortic arch remain in the adults of most fishes.

creases the blood pressure, which is very low in the veins, and sends the blood out through an artery, the **ventral aorta,** to five or six pairs of **aortic arches** that extend dorsally through capillaries in the gills to the **dorsal aorta.** Carbon dioxide is removed and oxygen is added as the blood flows through the gills; i.e., it changes to arterial blood. The dorsal aorta distributes this through its various branches to all parts of the body.

Blood pressure decreases as blood flows along because of the viscosity of the blood and the friction within it and between it and the lining of the vessels.

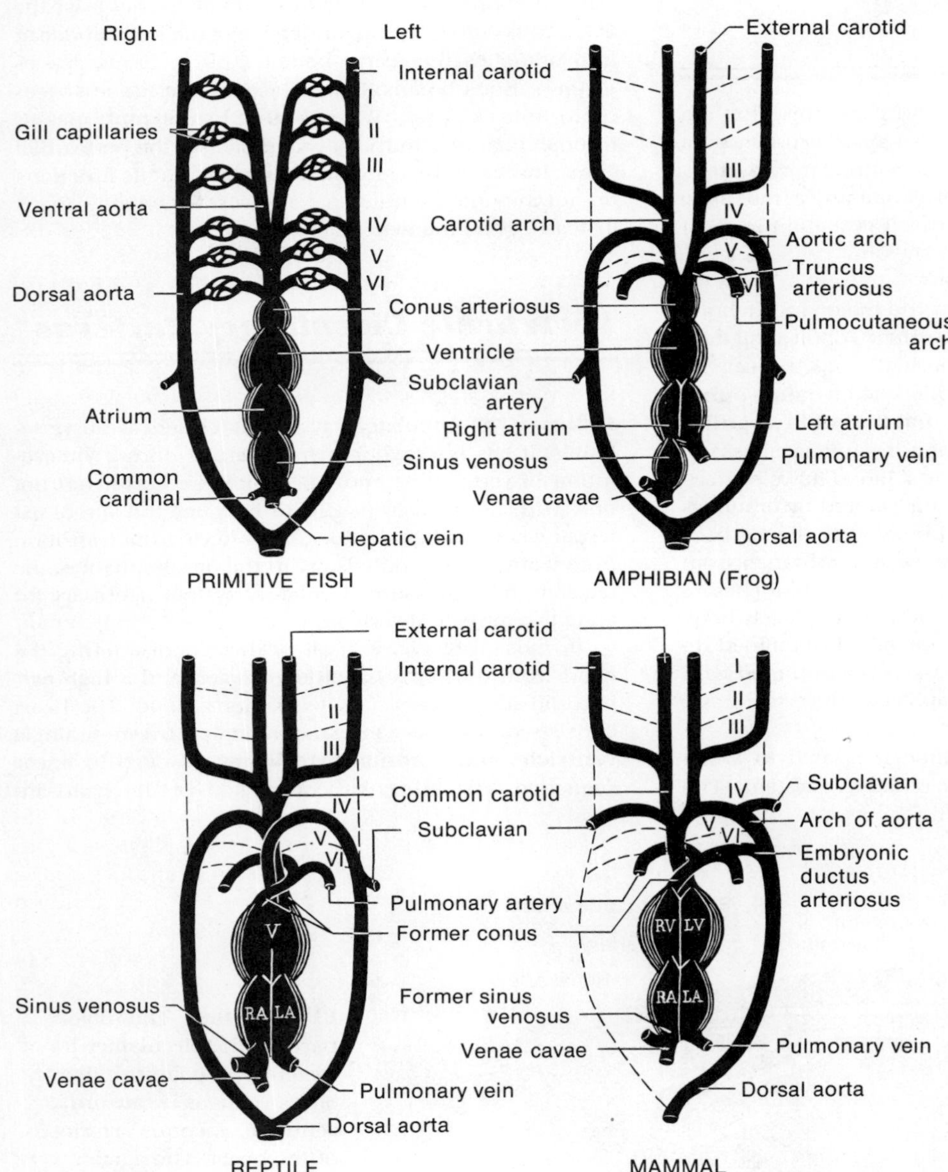

FIGURE 8.10 Diagrams of the heart and aortic arches to show the changes that occurred in the evolution from primitive fishes to mammals. All are ventral views. The heart tube has been straightened so that the atrium lies posterior to the ventricle.

Pressure is reduced considerably as the blood passes through the capillaries of the gills, for friction is greatest in vessels of small diameter. Mean blood pressure in the ventral aorta of a dogfish during heart contraction, for example, is about 30 mm. Hg; that in the dorsal aorta is about 20 mm. Hg. This relatively low blood pressure in the aorta becomes even lower when the blood reaches the capillaries in the tissues. Circulation in most fishes is rather sluggish.

Veins drain the capillaries of the body (where blood pressure is further reduced) and lead to the heart, but not all veins go directly to the heart. In fish, blood returning from the tail first passes through capillaries in the kidneys before entering veins leading to the heart. Veins that drain one capillary bed and lead to another are called portal veins, and these particular veins are known as the **renal portal system.** Other veins, known as the **hepatic portal system,** drain the digestive tract and lead to capillary-like passages in the liver. Since much of the blood returning to the heart has passed through one or the other of these portal systems in addition to the capillaries in the gills and tissues, its pressure is near 0 mm. Hg.

Amphibians. Correlated with the shift from gills to lungs, many changes occurred in the heart and aortic arches (Fig. 8.10). The aortic arches were reduced in number, the first two and the fifth being lost. Those that remain are no longer interrupted by gill capillaries. In a primitive tetrapod, such as the frog, the third pair of aortic arches forms part of the **internal carotid arteries** supplying the head; the fourth, the **systemic arches** leading to the dorsal aorta; and the sixth, the **pulmocutaneous arches** leading to the lungs and skin. New veins, the **pulmonary veins,** return oxygenated blood from the lungs to the heart. The heart now receives blood from both the body and the lungs. Though blood streams from the body and lungs are separated in the frog by a divided atrium, the two converge in a single ventricle (Fig. 8.11). Experiments using indicators to identify the streams have shown that, although there is some mixing, a high degree of separation is achieved. The slight differences in the times the streams enter the ventricles, a spongy ventricular muscle that helps hold them apart, the deflective action of a spiral valve within the conus arteriosus, and the internal partitioning of the arterial trunks that lead from the conus to the vessels supplying the body and lungs all play a role in separating the two streams. The pulmocutaneous arch receives primarily right atrial blood relatively low in oxygen content, the systemic arch receives a rather complete mixture of blood, and the carotid arch receives primarily left atrial blood with a high oxygen content. The mixture of left and right atrial blood going to the body is not as detrimental as it may at first appear because some of the right atrial blood comes from the skin where oxygenation also occurs. When the frog is under water the lungs are collapsed, and the resistance to the flow of blood through them increases. The lack of a division within the ventricle permits oxygenated blood returning to the right atrium from the skin to be sent directly to the body. Energy need not be expended to drive blood through the lungs.

The loss of gills and the gill capillaries, together with the evolution of a double circulation through the heart, result in a much higher blood pressure in the arteries of a primitive tetrapod than in a fish.

Reptiles, Birds, and Mammals. Higher tetrapods depend upon their lungs for external gas exchange. Since no gas exchange occurs in the skin, there is no mixing of aerated blood from the skin with blood from the body. The mixing in the heart of arterial blood from the lungs with venous blood from the body is lessened in reptiles by a partial division of the ventricle and by complete division in birds and mammals (Fig. 8.10).

In mammals, venous blood from the body enters the **right atrium,** into which the primitive sinus venosus has become incorporated (Figs. 8.10 and 8.12). Arterial blood from the lungs enters the **left atrium.** The atria pass the blood on to the **right** and **left ventricles,** respectively. The primitive conus arteriosus has become completely divided, part contributing to the pulmonary artery leading from the right ventricle to the lungs and the rest to the arch of the aorta leading from the left ventricle to the body.

The sixth pair of aortic arches form the major part of the mammalian **pulmonary arteries,** and the third pair contribute to the **internal carotid arteries.** But it will be observed in Fig. 8.10 that only the left side of the fourth arch, known as the **arch of the aorta,** leads to the dorsal aorta. The right fourth arch contributes to the right **subclavian artery** to the shoulder and arm and does not connect posteriorly with the aorta.

The major change in the veins is the complete loss of a renal portal system. Blood from the tail and posterior appendages enters a **posterior vena cava** (Fig. 8.12) that continues forward to the heart. It receives blood from the kidneys but does not carry blood to them. An **anterior vena cava** drains the head and arms. The hepatic portal system is still present.

These evolutionary changes have resulted in a very efficient circulatory system. Mammals have relatively more blood than lower vertebrates; it is distributed under greater pressure, and there is no mixing of arterial and venous blood. A human being, for example, has 7.6

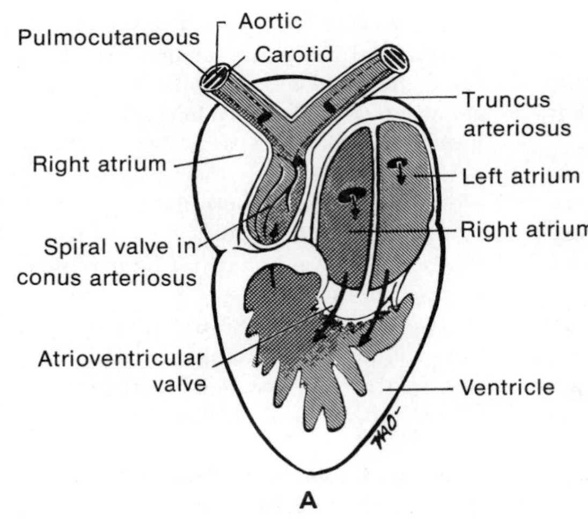

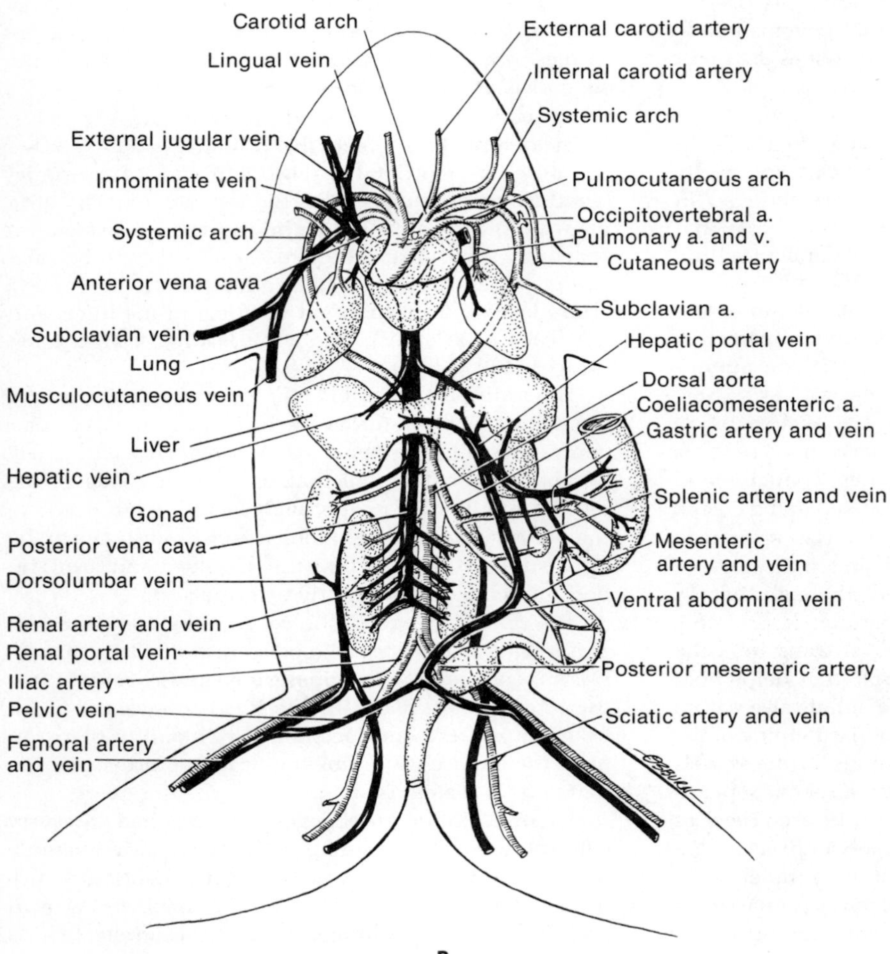

FIGURE 8.11 *A,* Ventral view of a dissection of the heart of a frog showing the partial separation that occurs in the ventricle and conus arteriosus between blood high in oxygen content, entering from the lungs and left atrium, and less highly oxygenated blood entering from the skin, body, and right atrium. *B,* Ventral view of the major arteries and veins of the frog. Veins are shown in black; arteries are in white. Certain of the anterior veins have been omitted from the right side of the drawing, and certain of the anterior arteries from the left side.

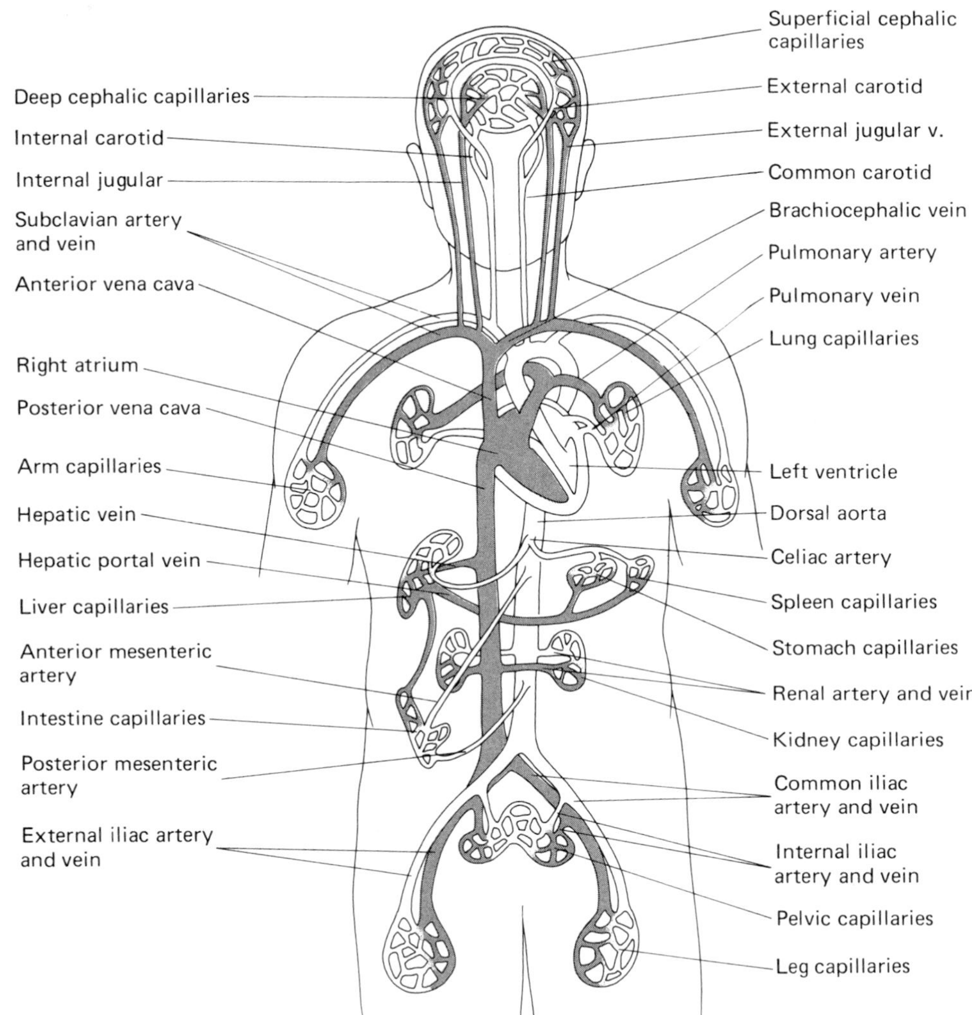

FIGURE 8.12 The major arteries and veins of a human being as seen in an anterior view.

mL. of blood per 100 g. of body weight compared with 2 mL. per 100 g. in a fish. The mean pressure in our dorsal aorta is about 100 mm. Hg.

As blood pressures have increased during the evolution of the circulatory system, more liquids and plasma proteins have escaped from the capillaries into the interstitial fluid than have been returned by the veins. A separate **lymphatic system** evolved; this plays a role in returning fluid and plasma proteins from the tissues to the main part of the circulatory system. Lymphatic vessels arise as outgrowths from the veins and, in general, tend to parallel the veins and ultimately empty into them.

The system reaches its greatest development in mammals. **Lymphatic capillaries** occur in most of the tissues of the body. They are more permeable than the capillaries between the arteries and veins, and pressures within them are exceedingly low. Their high permeability makes them the most likely route for the spread of microorganisms or cancer cells within the body. **Lymph nodes** lie at many points where small lymphatic vessels converge. They are major sites for the production of ymphocytes, and cells within them can phagocytize invading bacteria or respond to them by initiating antibody production.

Fetal and Neonatal Circulations

The placenta of the mammalian fetus, rather than the digestive tract, lungs, and kidneys, is the site for exchange of materials. This, together with the fact that the vessels in the unexpanded lungs of the fetus are not developed enough to handle the total volume of blood that is circulating through the body, requires certain differences in the fetal circulation (Fig. 8.13). Blood rich in oxygen returns from the placenta in an **umbilical vein,** passes rather directly through the liver via the **ductus venosus,** and enters the posterior vena cava, where it is mixed with blood returning from the posterior half of the fetus. The posterior vena cava empties into the right atrium, which also receives venous blood from the head by way of the anterior vena cava.

The lungs cannot accommodate all of this blood early in development, yet a large volume of blood must pass through all chambers of the heart to ensure their normal development. The right side must pump blood even though little can go through the lungs, and the left side must pump blood even though little returns from the lungs. The entrance of the posterior vena cava is directed toward an opening, the **foramen ovale,** in the partition separating the two atria. Most of the blood from the posterior vena cava tends to go through this into the left atrium, thence to the left ventricle and out to the body through the arch of the aorta. This blood bypasses the lungs yet permits the left side of the heart, which otherwise would receive little blood from the collapsed lungs, to function and develop normally. The rest

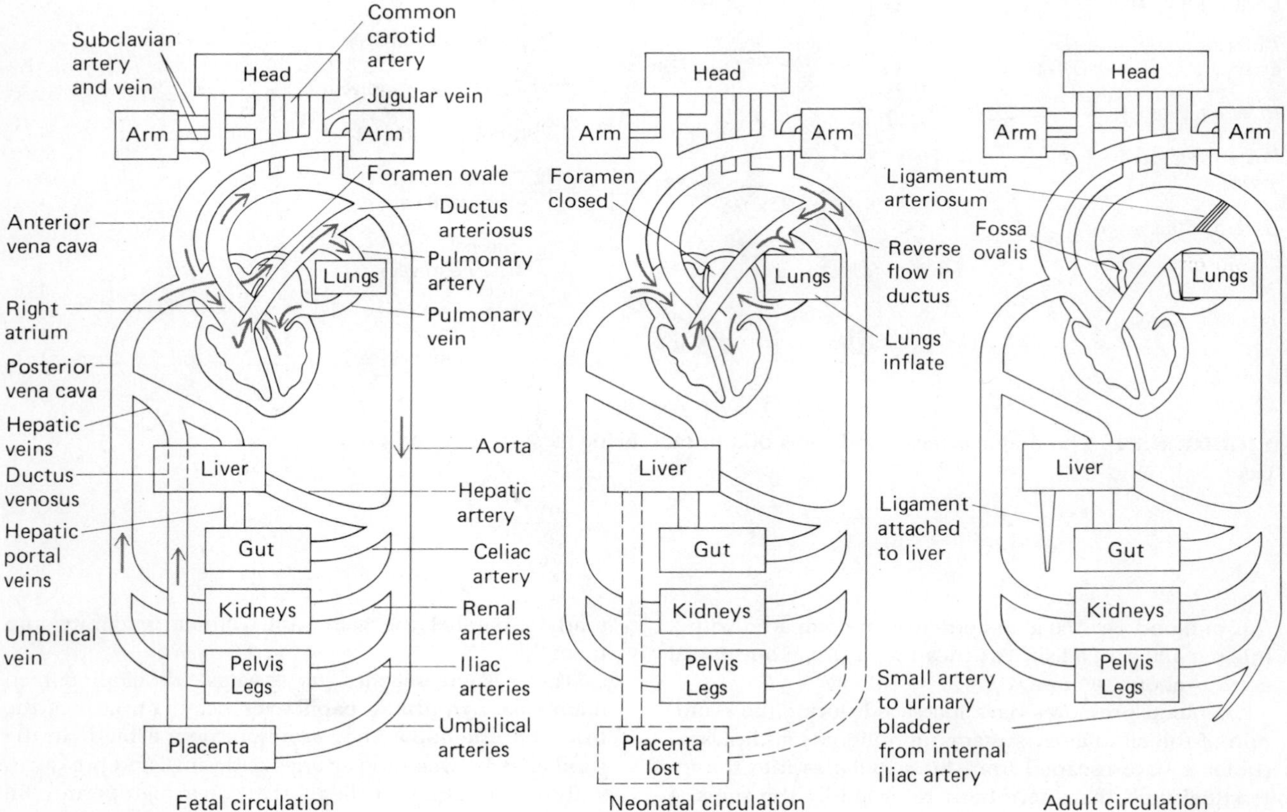

FIGURE 8.13 Diagrams in ventral views of the mammalian fetal circulation and the changes at birth that give rise to the neonatal circulation and soon to the adult pattern.

of the blood from the posterior vena cava enters the right ventricle along with the blood from the anterior vena cava and starts out of the pulmonary artery toward the lungs. However, only a fraction of this blood passes through the lungs to return to the left atrium and mix with blood from the posterior vena cava. Most of the blood in the pulmonary artery goes through another bypass, the **ductus arteriosus,** to the dorsal aorta. The ductus arteriosus represents the dorsal part of the left sixth aortic arch (Fig. 8.10). Since the ductus arteriosus enters the aorta after the arteries to the head and arms have been given off, these parts of the body receive the blood with the highest oxygen content. After the entrance of the ductus arteriosus, the blood in the aorta is highly mixed. This is the blood that is distributed to the rest of the body and, by way of **umbilical arteries,** to the placenta.

Throughout fetal life the lungs and most of the vessels within them are collapsed. The resistance to blood flow through the lungs (pulmonary resistance) is greater than the resistance to flow through the body and placenta (systemic resistance). As a consequence there is a large flow of blood through the placenta and a small flow through the lungs. The volume of blood returning to the left atrium is less than the volume entering the right atrium. This facilitates the opening of the valve in the foramen ovale and the continued bypassing of the lungs.

These conditions are immediately reversed at birth. The placenta is expelled, and blood volume in the systemic circuit is reduced (Fig. 8.13). Carbon dioxide accumulates in the fetal blood, activating the respiratory system; the lungs fill with air, and the pulmonary vessels that were collapsed open up. Resistance to blood flow in the pulmonary circulation is now less than that in the systemic circulation. More blood flows through the lungs and returns to the left atrium than during fetal life. The valve in the foramen ovale is pushed against the interatrial septum and soon adheres to it by the growth of tissue. This bypass of the lungs is cut off. Its location continues to be represented by a depression, the **fossa ovalis.**

The circulatory pattern is now very close to the adult condition except that the ductus arteriosus remains open. Since pulmonary resistance is less than systemic resistance, the direction of flow in the ductus arteriosus is reversed. Some of the blood that has already been through the lungs and is leaving the heart in the arch of the aorta flows back to the lungs through the ductus arteriosus. This pattern of circulation, which lasts from several hours to a day or two in the human infant, is known as the **neonatal circulation** (Gr, *neo,* new + *natus,* born). Experiments on newborn lambs show that this reversal of flow, and the consequent double aeration of some of the blood, is of great significance because a changeover is taking place between the synthesis of fetal and adult types of hemoglobin. Although fetal hemoglobin has a high affinity for oxygen (p. 163), it also holds onto it to a greater extent than does adult hemoglobin. If the ductus arteriosus is experimentally tied off during this period, 10 to 20 percent less oxygen is delivered to the tissues.

Muscles in the wall of the ductus arteriosus eventually contract and stop all blood flow through it, and the adult pattern is established (Fig. 8.13). Eventually, the duct becomes permanently occluded by the growth of fibrous tissue into its lumen and is converted into the **ligamentum arteriosum.** The stimulus for these changes is unknown, but if the duct remains open, an undue strain is placed upon the heart, for it must pump this extra amount of blood that is recirculating through the lungs in addition to the normal amount of blood to the tissues.

So-called "blue-babies" are infants born with circulatory defects that produce cyanosis, that is, insufficient oxygenation of the blood. Various congenital defects can cause cyanosis, but a common one is pulmonary stenosis, an abnormal constriction of the base of the pulmonary arteries. When this occurs, a persistent open ductus arteriosus can be life-saving. Can you determine why?

The base of each umbilical artery persists as a small artery supplying the urinary bladder; the anterior end of the umbilical vein, as a ligament attached to the liver.

The Propulsion of Blood and Hemolymph

Hearts. All animals with circulatory systems have a muscular pump of some type that propels the blood or hemolymph. In its simplest form this is a pulsating vessel in which blood is pushed along by waves of peristaltic contraction (Fig. 8.14*A*). Valves are not present. **Peristaltic tubular hearts** of this type are found in some primitive worms.

Some invertebrates and all vertebrates have **chambered hearts** with muscular walls and valves between chambers, permitting blood to move in one direction only. The heart may lie within a cavity, such as the pericardial sinus of arthropods or the pericardial coelom of mollusks and vertebrates, that facilitates its contraction and expansion.

Usually, the residual pressure in fluid returning to a chambered heart is sufficient for the blood to enter. The first chamber of the heart is thin-walled so that it easily expands as fluid enters. It contracts with force sufficient

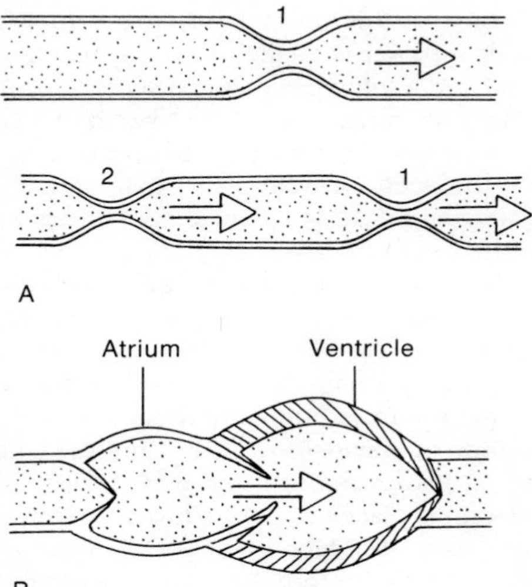

FIGURE 8.14 Circulatory pumping mechanisms. *A,* A wave of peristaltic contraction *(1)* travels along the peristaltic tubular heart and is soon followed by another *(2). B,* The chambered heart of a mollusk or vertebrate. The atrium is contracting, and the ventricle is filling. Arrows indicate the direction of blood flow.

only to drive the fluid into succeeding chambers, where enough pressure is developed to deliver the blood or hemolymph to the tissues and back to the heart. Sometimes, however, the pressure is so low in vessels or spaces just before the heart that the heart, or its first chamber, must actively expand and literally suck fluid in. Arthropods have a **suction heart.** Thin muscle strands extend from the outside of the heart to the wall of the hemolymph-filled pericardial sinus in which it lies. Their contraction enlarges the single-chambered heart, and blood is sucked in through the ostia. Valves at the bases of the aortae leaving the heart close so that hemolymph is not drawn back into the heart from them. When the heart contracts, they open and valves in the ostia close.

The Mammalian Heart. The heart of mammals and most other vertebrates is not a suction heart, for it receives blood under low pressure from the veins. The heart lies within the pericardial cavity and is covered with a smooth coelomic epithelium, the **visceral pericardium.** It is lined by the simple squamous epithelium, the **endothelium,** which lines all parts of the circulatory system. The rest of its wall is composed of dense connective tissue that forms a fibrous skeleton and **cardiac muscle.** Although striated, cardiac muscle differs from skeletal muscle in that its fibers branch and anastomose profusely (see Fig. 2.24). Atrial muscles are separated from ventricular muscles, but within each group the individual muscle fibers are firmly united by special junctions that offer little impediment to the spread of action potentials from one fiber to another. Functionally the atrial and ventricular muscles each act as a unit, or syncytium. When one cell becomes active, the activity spreads rapidly to the others. Thus, the atria and ventricles follow the "all-or-none" law that applies to individual motor units of phasic skeletal muscles (see p. 112).

During a heart cycle, the atria and ventricles contract and relax in succession. Contraction of these chambers is known as **systole** (Gr. *systolē,* drawing together); relaxation, as **diastole** (Gr. *diastolē,* dilation). Ventricular systole is very powerful and drives the blood out into the pulmonary artery and arch of the aorta under high pressure. Since the muscle fibers of the ventricles are arranged in a spiral, the blood is not just pushed out but is virtually wrung out of them. When the ventricles relax, their elastic recoil reduces the pressure within them, and blood enters from the atria. Atrial contraction does not occur until the ventricles are nearly filled with blood. The atria are primarily antechambers that accumulate blood during ventricular systole.

Blood being pumped by the heart is prevented from moving backward by the closure of a system of valves (Fig. 8.15*A*). One with three cusps, known as the **tricuspid valve,** lies between the right atrium and ventricle; one with two cusps, the **bicuspid valve,** between the left chambers. These valves operate automatically as pressures change, opening when atrial pressure is greater than ventricular, closing when ventricular pressure is greater. **Tendinous cords** extend from the free margins of the cusps to the ventricular wall and prevent them from turning into the atria during the powerful ventricular contractions. When the ventricles relax, blood in the pulmonary artery and aorta, which is under pressure, tends to back up into them. This closes the **pulmonary** and **aortic valves** at the base of each of these vessels and prevents blood from returning to the ventricles. The pulmonary and aortic valves are each composed of three semilunar-shaped pockets that fill with blood and press against each other. Abnormalities in the structure of the valves, occurring congenitally or produced by disease, may prevent their closing properly. Blood then leaks back during diastole; the leaking blood is heard as a "heart murmur."

Heart Beat and Its Integration. Heart musculature has an inherent capacity for beating, and a heart, if properly cultured, will continue to beat rhythmically when excised from the body. An integrating system within the heart initiates the beat and stimulates suc-

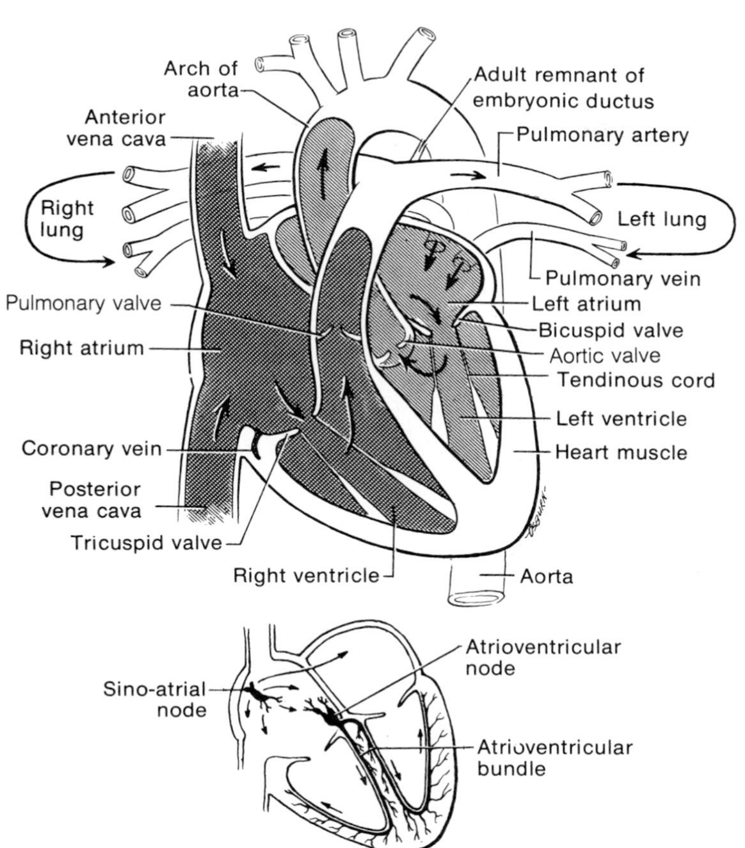

FIGURE 8.15 The adult mammalian heart. *Upper,* Course of blood through the heart; *lower,* distribution of the specialized cardiac muscle that forms the conducting system of the heart.

cessive chambers in turn. In mammals, each contraction is initiated in the **sinoatrial node,** or "pacemaker"—a node of specialized cardiac muscle **(Purkinje fibers)** located in that part of the wall of the right atrium into which the primitive sinus venosus is incorporated (Fig. 8.15*B*). The impulse spreads to all parts of the atria and to another node of Purkinje fibers, the **atrioventricular node,** from which it continues along pathways of Purkinje fibers to all parts of the ventricles. Since there is no muscular connection between atria and ventricles, an impulse can reach the ventricular muscles only through the Purkinje fibers. It does so very rapidly so that ventricular contraction begins at the apex of the heart and spreads quickly toward the origin of the great arteries leaving the heart.

Cardiac Output. Small animals, with higher oxygen consumption per unit of volume (p. 92), need more blood sent to each unit of tissue than do large animals. During exercise animals need an increased blood flow. Cardiac output, defined as the volume of blood or hemolymph pumped per unit of time, must be adapted to these different needs. Cardiac output can be increased by (1) having a large heart, (2) increasing the rate of heart beat (pulse rate), or (3) increasing the volume of output per beat or stroke (stroke volume). Accordingly one might expect small animals to have relatively larger hearts than larger ones, but this is not the case. The heart of a shrew, a human, or an elephant each constitutes about 0.6 percent of body weight. Changing the rate of heart beat is the major factor in adjusting output to metabolic needs. The pulse rate of a shrew at rest is 600 beats per minute, that of humans is about 70 beats per minute, and that of an elephant is only 25 beats per minute. Rate also increases during exercise, but the rate of increase is not directly proportional to the increased oxygen needs. The blood itself delivers more oxygen because the oxygen tension in active tissues drops, and a greater proportion of the oxygen carried by the blood is unloaded (p. 163). Stroke volume also increases slightly during exercise because the increased pressure and more rapid return of venous blood stretches the heart musculature. This causes the heart to contract with greater force and to send out the greater volume of blood received during each period of atrial diastole. Within physiological limits, the greater the tension on cardiac (or any other) muscle, the more powerful will be its contraction. This capacity of the heart to adjust its output per stroke to the volume of blood delivered to it is known as *Starling's law of the heart.*

The heart of a normal adult human being who is not exercising sends about 70 mL. of blood per beat out into the aorta. At the normal rate of 72 beats per minute, this is a total output of 5 L. per minute, which is approximately equivalent to the total amount of blood in the body. A similar observation made in 1628 by William Harvey led him to conclude that the blood recirculates (p. 8). Until that time it was believed that blood was continually produced in the liver, pumped to the tissues and consumed. Harvey's calculations showed that the amount of blood pumped by the heart each hour was much more than could possibly be produced and consumed. He made the correct inference that the blood must recirculate, even though he could not see the microscopic capillaries that connect arteries and veins.

Although a large volume of blood flows through the cavities of the heart, this blood may not provide for the metabolic needs of the heart musculature. In mammals, a pair of **coronary arteries** arise from the base of the arch of the aorta and supply capillaries in the heart wall. This capillary bed is drained ultimately by a **coronary vein** that empties into the right atrium. Obviously, any damage to the coronary vessels, the plugging of one of the larger arteries by a thrombus, for example, could have serious consequences, for the heart muscles cannot function without a continuing supply of oxygen and food.

Cardiac Control. Various control mechanisms adjust cardiac output to an animal's needs. The hearts of annelids and mollusks are primarily **myogenic.** That is, heart beat initiates in the heart musculature itself, and rate of beat is influenced by extrinsic factors that affect the musculature directly. Increases in temperature or of the pressure of the fluid entering the heart cause an acceleration. Inhibiting and accelerating neurons terminate on the molluskan heart but are of less importance in regulating beat.

The hearts of arthropods and vertebrates, although having an underlying myogenic rhythm, are **neurogenic;** nerve impulses greatly influence the activity of the heart. In mammals, motor neurons originate in the **vasomotor center** in the medulla of the brain and end primarily on the sinoatrial node (Fig. 8.16). The activation of **accelerator neurons,** which are sympathetic in nature (p. 227), increases the heart rate and also the force of contraction because many of these fibers also extend to the cardiac muscles. The activation of **decelerator neurons,** which are parasympathetic fibers in the vagus nerve (p. 227), decreases the heart rate.

Many influences impinge upon the vasomotor center. **Baroceptors** (Gr. *baros,* pressure + L. *capere,* to take) in the atrial wall, and especially in the walls of large arteries near the heart, detect changes in blood pressure and send signals on sensory neurons to the vasomotor center (Fig. 8.16). By a combination of excitation and in-

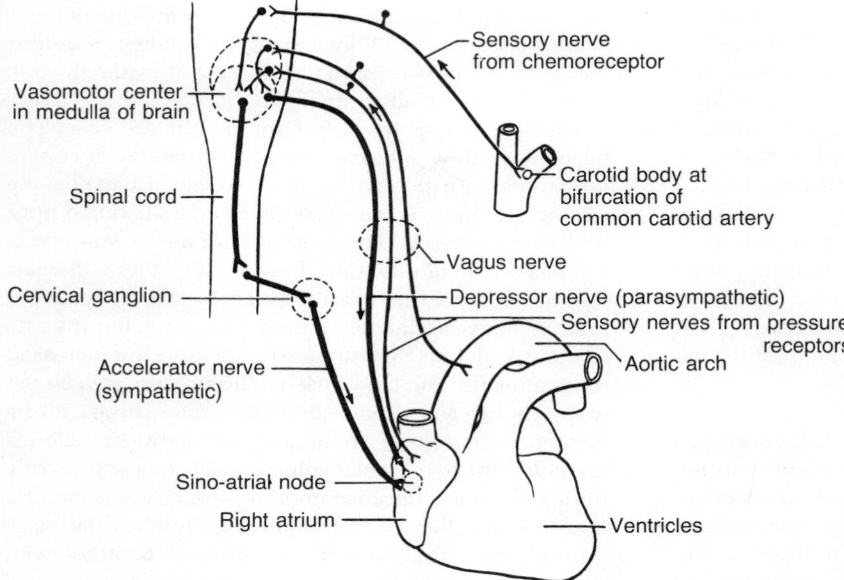

FIGURE 8.16 Diagram of major nerve reflexes controlling heart output and blood pressure. Motor nerve fibers have been drawn more heavily than sensory fibers. The motor nerves to the heart are parts of the autonomic nervous system, which is discussed more fully in Chapter 10.

hibition of appropriate motor neurons, heart rate will decrease if pressure is too high or increase if pressure is too low. These pressure receptors also activate motor neurons innervating muscles in the walls of peripheral arteries and cause their relaxation (if pressures are too high) or contraction. **Chemoreceptors** in the **carotid bodies** at the junctions of the internal and external carotid arteries monitor oxygen and carbon dioxide content in the blood. In addition to affecting the rate of breathing (p. 160), they send impulses to the vasomotor center. If oxygen levels are too low, heart rate and blood pressure increase, thereby delivering more blood to the tissues. Many other factors, including emotional state, can also affect the vasomotor center.

The Peripheral Flow of Blood and Hemolymph

Arteries. Arteries are vessels leading from the heart to capillaries or to hemocoel spaces. Usually the fluid within them is rich in oxygen and low in carbon dioxide, but the gas content of vessels depends on the site of external gas exchange. The pulmonary artery of an adult mammal and the ventral aorta of a fish contain blood low in oxygen, for the blood is on the way to the respiratory membranes. The arteries of vertebrates are lined with endothelium and have a relatively thick wall containing elastic connective tissue and smooth muscle. Invertebrate arteries are similar but lack muscle unless they are pulsating vessels. The walls of the large arteries leaving the heart of vertebrates are richly supplied with elastic tissue. The force of each ventricular systole forces blood into the arteries and stretches them to accommodate it. During diastole, the elastic recoil of the first part of the artery to expand helps to push the blood into the adjacent part of the artery, which in turn expands. If the arteries were rigid pipes, they would deliver blood to the tissues in spurts that coincided with ventricular systole. The blood would pound like steam rushing into empty radiator pipes. The elasticity of the larger arteries transforms what would otherwise be an intermittent flow into a steady flow. The alternate stretching and contracting of the arteries travels peripherally very rapidly (7.5 m. per second) and can be detected as the **pulse,** but the blood itself does not move as fast.

As the arteries extend to the tissues, they branch and rebranch. At each branch, the lumen becomes smaller, but the total cross-sectional area of all these branches increases greatly. The velocity of blood flow, therefore, decreases, for the blood, like a river widening out and flowing into a lake, is moving into an area that grows larger and larger. The mean blood pressure is also decreased continually because of the friction of the blood moving in the vessels (Fig. 8.17). Blood pressure continues to decrease as the blood flows through the capillaries and veins. The rate of flow, however, increases as the blood passes from the capillaries to the venules, and as these smaller veins lead into fewer larger ones. The blood is now moving into a smaller and smaller area and, like water flowing out of a lake into a narrowing river, moves faster and faster.

Capillary Exchange. In animals with open circulatory systems, the hemolymph flows from arteries into the hemocoel and directly bathes all of the tissues and cells. Cells are separated from the hemolymph only by their plasma membranes. In animals with closed systems, blood flows from arteries into capillaries, and exchanges must occur between the blood and interstitial fluid through the capillary walls (Fig. 8.18A). Capillaries are small, numerous, and exceedingly thin-walled vessels. Their diameter is about that of blood cells, and they frequently are spaced very close together. Capillary density

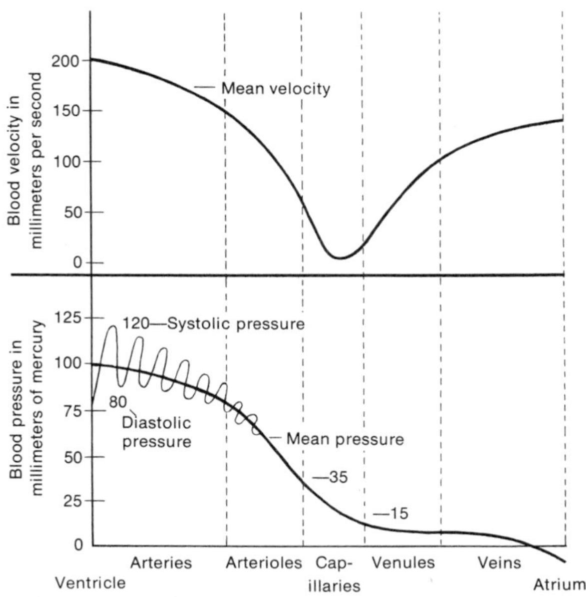

FIGURE 8.17 Variation in blood velocity and pressure in different parts of the cardiovascular system. The velocity does not return to its original value, for the cross-sectional area of the veins is greater than the cross-sectional area of the arteries. The blood pressure in the veins near the heart is less than atmospheric pressure because of the negative pressure within the thorax.

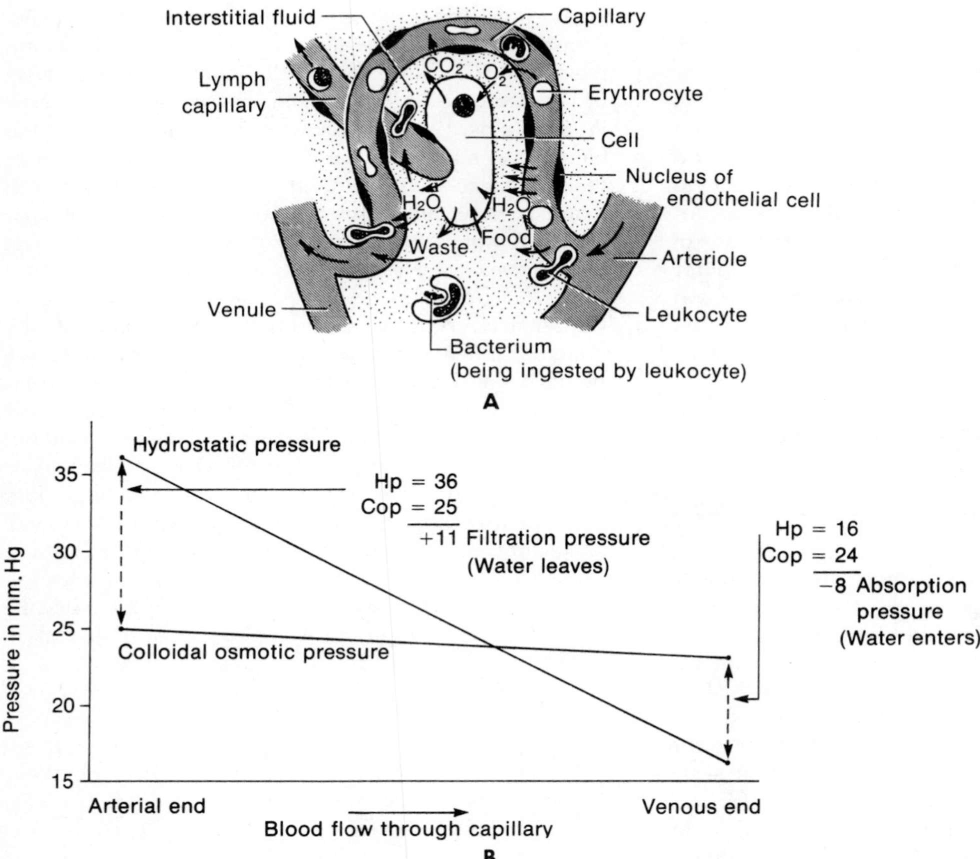

FIGURE 8.18 Capillary exchange. *A,* Diagram of the exchange of materials between the blood and interstitial fluid in a capillary bed. *B,* Graph of the forces within a capillary responsible for the movement of water in and out. The forces graphed are net forces, i.e., the difference between the true hydrostatic (Hp) and colloidal osmotic (Cop) pressures in the plasma and comparable forces (which are rather low) in the interstitial fluid.

is particularly high in the active tissues of small mammals. Schmidt-Nielsen and Pennycuik found that a cross section of mouse muscle contains about 2000 capillaries per square millimeter, whereas a comparable section of horse muscle has only about one half as many.

Capillary walls are more permeable than the plasma membrane of cells. Oxygen, glucose, amino acids, various ions, and other needed substances easily diffuse through them into the interstitial fluid following their concentration gradients. Conversely, carbon dioxide, nitrogenous wastes, and other by-products of metabolism easily diffuse into the blood. Many of these molecules undoubtedly diffuse through the membranes of the endothelial cells forming the capillary walls, but electron microscope studies have shown that the increased permeability also results from more direct passages through the walls. Some of these passages are present where endothelial cells come together and overlap, for the cells are not firmly united with one another over the full extent of these surfaces (Fig. 8.19). In addition, small pinocytotic vesicles with diameters of about 50 nm. frequently invaginate on either surface of endothelial cells, bud off, move through the cell, and open on the opposite surface. In the glomerulus of the kidney, the intestinal villi, and in many glands, the capillaries are fenestrated with small pores up to 100 nm. in diameter. These are closed only by the delicate protein fibers and glycoproteins that comprise the basement membrane of an epithelial surface.

Water moves each way through the walls, and maintaining the proper amount in the interstitial fluid involves a combination of forces. As we have seen, most of

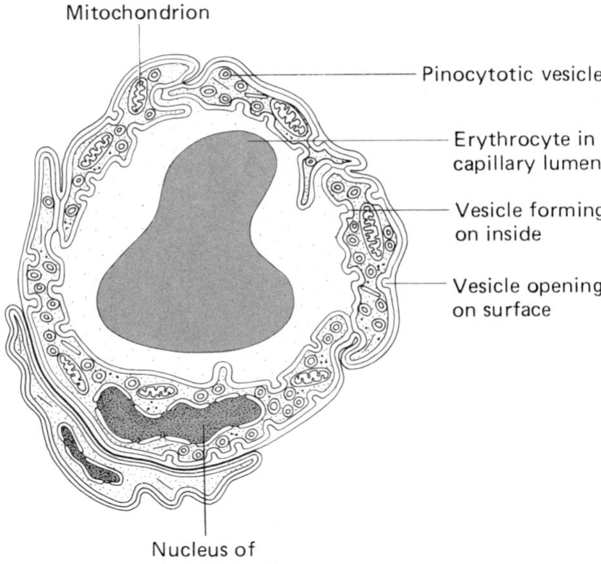

FIGURE 8.19 Drawing of a cross section of a capillary based on an electronmicrograph. Note the pinocytotic vesicles passing across the wall cells. (Modified from Williams, P. L., and R. Warwick (eds.): *Gray's Anatomy.* 36th British edition. Philadelphia, W. B. Saunders, 1980.)

the large plasma proteins remain in the blood, where they exert an osmotic force called the **colloidal osmotic pressure** that draws water into the capillaries. An opposite force, the blood or **hydrostatic pressure,** forces water out. At the arterial end of the capillary bed, hydrostatic pressure exceeds the colloidal osmotic pressure, so that there is a net force (called the **filtration pressure**) that moves water out of the capillaries (Fig. 8.18*B*). As blood moves through the capillary bed, friction causes a great reduction of the hydrostatic pressure. A smaller reduction in the colloidal osmotic pressure also occurs because of the loss of some plasma proteins. At the venous end of a capillary bed, colloidal osmotic pressure exceeds hydrostatic pressure so that there is a net force (the **absorption pressure**) that draws water into the capillaries. The absorption pressure is frequently a bit lower than the filtration pressure, but the tissues do not flood. Capillaries have somewhat more absorption surface at their venous end through which more water can enter. Terrestrial vertebrates, whose blood pressure is somewhat higher than that of fishes, also have a system of **lymphatic capillaries** (p. 181).

Regulation of Capillary Flow and Exchange. Blood does not flow evenly through all of the capillaries of the body, or even through the capillary beds of individual organs, but varies according to the needs of the tissues. Some may be very active and require a large blood flow, whereas others may be relatively inactive. The degree of contraction of smooth muscle in the walls of the small arteries, or **arterioles,** preceding the capillaries regulates blood flow into a capillary bed. Within a capillary bed there are some **thoroughfare channels** through which some blood flows all of the time. Tiny, muscular **precapillary sphincters** surround the beginning of capillaries that branch from these channels so that parts of a bed may be open or closed according to needs.

Many factors affect the degree of contraction of the muscles in the arterioles and precapillary sphincters. If the tissues are very active and the availability of oxygen and other nutrients falls, these muscles relax, and more blood enters the capillaries. It is not certain whether this is due to a deficiency in the nutrients themselves or to the production of a vasodilator substance under these circumstances. Whatever the reason, there is a high degree of autoregulation of blood flow through the capillaries.

Nervous regulation is superimposed upon the autoregulation. Muscles in the blood vessel walls have a certain tonus, that is, they are partially contracted because of a steady outflow of nerve impulses along sympathetic **vasoconstrictor nerves** from a **vasomotor center** in the brain stem. This tonus maintains normal blood pressure and rate of blood flow. Impulses impinging upon the vasomotor center can increase or decrease the rate of outflow of vasoconstrictor impulses. An increase causes a greater degree of contraction of the muscles in the small arteries. This increases the general level of blood pressure in the body and rate of blood flow but decreases pressure and blood flow through the capillaries peripheral to the constricting arteries. A decrease in the number of vasoconstrictor impulses has the opposite effect, i.e., it lowers general blood pressure and rate of flow but increases pressure and flow in the peripheral capillaries. In this way the small arteries can regulate the amount of blood going to particular organs (Fig. 8.20).

Variations in capillary pressure and blood flow naturally affect the filtration rate and the exchange of materials between capillaries and the interstitial fluid. If an exceptionally large volume of fluid leaves the capillaries, a tissue swelling, or **edema,** occurs. Many factors can cause edema. Among them are a poor return of venous blood and lymph (see later discussion) which causes a backing up of blood in the capillary beds and an increase in capillary hydrostatic pressure, the loss of plasma proteins in burns or other severe injuries, which lowers the blood's colloidal osmotic pressure, and the release of histamine in allergic reactions (p. 175), which both relaxes the smooth muscles in the arterioles and increases the permeability of capillary walls.

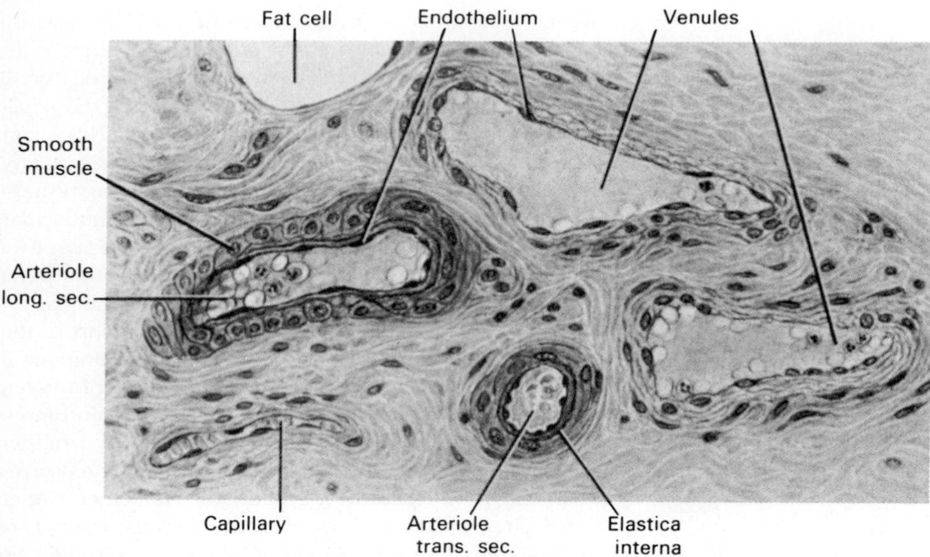

FIGURE 8.20 Arterioles and accompanying venules from the submucosa of a human rectum, magnification ×250. (From Copenhaver, W. M., R. P. Bunge, and M. B. Bunge: *Bailey's Textbook of Histology.* 16th ed. Baltimore, Williams & Wilkins, 1971.)

Venous and Lymphatic Return. Veins are vessels that return blood to the heart. In animals with open systems the venous channels do not have an endothelial lining, although they may be lined with connective tissue. Their structure in vertebrates is fundamentally similar to that of arteries, though a vein is larger and has a much thinner and more flaccid wall than its companion artery (Fig. 8.20). Since they are larger, the veins hold more blood than the arteries and are an important reservoir for blood. Lymphatic vessels have even thinner walls. Valves present in both veins and lymphatics permit the blood and lymph to flow only toward the heart.

Though blood pressure is low in the veins (see Fig. 8.17) and lowest in the large veins near the heart, it is still the major factor in the return of blood to the heart in a mammal. Two other factors assist it. One is the fact that the elastic lungs are always stretched to some extent and tend to contract and pull away from the walls of the pleural cavities. This creates a slight subatmospheric or negative pressure within the thoracic cavity that is greatest during inspiration. The larger veins, of course, pass through the thorax, and the reduction of pressure around them decreases the pressure within them and increases the pressure gradient. The other factor is that the contraction and relaxation of body muscles exert a "milking" action on the veins and force the blood toward the heart. All these factors increase during exercise, which makes for a more rapid return of blood and an increased cardiac output.

The return of lymph is dependent upon similar forces. The interstitial fluid itself has a certain pressure derived from the flow of liquid out of the capillaries. This establishes a pressure gradient in the lymphatics that is made steeper by the negative intrathoracic pressure. The "milking" action of surrounding muscles and, for lymphatics returning from the intestine, the movement of the villi, help considerably. Some lower vertebrates have lymph "hearts"—specialized pulsating segments of lymphatic vessels.

Summary

1. Transport systems provide a means for the bulk flow of materials between the interstitial fluid bathing the cells and the sites where materials enter and leave the body. They also help to maintain a constant internal environment (homeostasis).

2. Although any body cavity can be used in transport, larger animals have well-developed circulatory systems. In open systems, hemolymph passes from the vessels into the interstitial spaces; in closed systems, the blood is confined to the vessels. In general, the more active animals have closed systems.

3. The blood of all animals consists of some cells carried in a liquid plasma composed of water, plasma proteins, electrolytes, and materials in transit. Vertebrate blood carries platelets, leukocytes, and the red blood cells that transport gases.

4. Animals have hemostatic mechanisms that reduce loss of blood after injury: muscles in the injured vessels contract, a plug forms, and the blood clots. Platelets are necessary for the last two events.

5. Leukocytes protect the body by promoting an inflammation of the infected area, phagocytosing microbes, destroying toxins, and producing antibodies.

6. Invading foreign proteins and some other materials act as antigens. Lymphocytes respond by synthesizing antibodies with the specific molecular configuration needed to react with each type of antigen. B-lymphocytes are transformed into plasma cells that release antibodies into the circulation. T-lymphocytes are transformed into lymphoblasts that react directly with an antigen. During embryonic development lymphocyte genes undergo complex somatic recombinations resulting in innumerable different lines of lymphocytes, each programmed genetically to synthesize a different antibody. When a particular antigen enters the body, lymphocytes competent to respond to it are stimulated and produce a clone of responding cells. Some of these cells remain after the initial infection, giving the body an acquired immunity and the ability to respond more rapidly to a subsequent invasion. Antibodies are Y-shaped immunoglobulins that cause the antigen to precipitate, promote its phagocytosis, or, by activating the plasma complement system, cause its lysis.

7. Naturally occurring antigens and antibodies form the basis for the ABO, Rh, and other blood groups. Many of these are of medical importance in transfusions and of legal importance in questions of paternity or identifying the source of blood, but their biological significance is not clear.

8. The heart of most fishes consists of a linear series of chambers that receive only venous blood and pump it to the gills. Vessels leaving the branchial region distribute aerated blood to the body. With the loss of gills and the evolution of lungs in the ancestors of terrestrial vertebrates, a double circulation evolved. In mammals, the right atrium and ventricle receive venous blood from the body and pump it to the lungs. The left chambers of the heart receive oxygenated blood from the lungs and pump it to the body. The double circulation is incomplete in amphibians, but this is an adaptation that permits blood to bypass the functionless lungs when the animals are under water.

9. The fetal circulatory system of mammals enables the fetus to receive and eliminate materials through the placenta, bypass the functionless lungs, and supply both sides of the heart with enough blood to pump so that they can develop normally. Some blood is recirculated through the lungs during the brief neonatal period.

10. A chambered heart with valves preventing the backflow of blood is the most common type. The interconnection of the cardiac muscle fibers allows those of the atrium and the ventricle to respond as units in an all-or-none fashion. The inherent rhythm of contraction of cardiac muscle originates in the sinoatrial node and spreads on cardiac muscle and special Purkinje fibers to other parts. Nerves terminating on the sinoatrial node and on cardiac muscle fibers increase or decrease their rate of contraction and the force of contraction. The heart beats faster in small animals with high metabolic rates than in larger animals and in all animals during exercise.

11. Because of their elastic expansion and contraction, arteries play an important role in delivering blood to the tissues. The arterioles regulate the volume of blood going to particular tissues and, together with controls on the rate of heart beat, regulate the blood's pressure and rate of flow.

12. Exchanges of solutes in the capillary beds occur by diffusion, by the passage of materials through pores between the cells of the capillary walls, and by the passage of pinocytotic vesicles across the cells. Water movements are regulated by the interaction of the hydrostatic pressure of the blood and its osmotic pressure.

13. Materials return from the capillaries and interstitial fluid through the veins and lymphatic vessels. The pressure in these vessels is low, and the contraction of surrounding body muscles is the major force in the return of blood and lymph. Valves in the vessels prevent back flow.

References and Selected Readings

Attention is again directed to the general references on vertebrate organ systems cited at the end of Chapter 4.

Cooper, M. D., and A. R. Lawton III: The development of the immune system. *Scientific American, 234* (Nov.): 58, 1974. The differentiation of T- and B-lymphocytes and their vital roles in immune responses are discussed.

Hamilton, W. F., and P. Dow (Eds.): *Handbook of Physiology.* Section 2. Circulation. Vol. 1, 2, 3. Washington, D. C., American Physiological Society, 1962, 1963, 1965. A standard source book for serious students in physiology.

Leder, P.: The genetics of antibody diversity. *Scientific American, 246* (May): 102, 1982. A summary of the fascinating research that led to the theory of somatic recombination of nucleic acid sequences during the development of lymphocytes.

Martin, A. W.: Circulation in invertebrates. *Annual Reviews of Physiology, 36*:171–186, 1974. A scholarly review of the invertebrate circulatory system.

Mayerson, H. S.: The lymphatic system. *Scientific American, 208* (June):80, 1963. The importance of this second drainage system of the tissues is thoroughly discussed.

Schmidt-Nielsen, K.: Countercurrent systems in animals. *Scientific American, 244* (May): 118, 1981. A summary of the biological uses in the circulatory system and respiratory passages of the counter current principle.

Wiggers, C. J.: The heart. *Scientific American, 196* (May):87, 1957. A fine discussion of the activities of the heart and the safety factors that enable it to continue operating even though partially impaired by coronary disease.

Zweifach, B. W.: The microcirculation of the blood. *Scientific American, 200* (Jan.):54, 1959. A discussion of the factors that control circulation in capillary beds.

9 Regulation of Internal Body Fluids

ORIENTATION

The internal body fluids of animals—coelomic fluid, blood, fluid between and within cells—is kept relatively constant both in volume and in content. This chapter will explore the mechanisms by which this constancy is maintained, or regulated. One important mechanism is excretion, and thus excretory processes will be studied in some detail. Since the problems of regulation are different for marine animals, as compared with those living in freshwater, and the problems of terrestrial animals are very different from those of aquatic species, regulation in each of the three major environments—sea, fresh water, and land—must be examined.

Living organisms are composed mostly of water. A 70-kg. adult human contains about 49 L of water; about 35 L are within cells, and the rest is extracellular body fluids (see Fig. 8.1). Some of these extracellular fluids lie in the minute spaces between cells; others may be located in special cavities, such as the coelom; and a large amount may be present in the blood vascular system. These extracellular fluids are a dynamic part of the internal environment of the animal. They contain salts and proteins (Fig. 9.1); they are in osmotic balance with the intracellular environment; and they are routes for the passage of food, gases, and wastes. The composition of both intracellular and extracellular fluids must be maintained within relatively narrow limits, regardless of the external environment.

Balancing water lost against water gained is a more complicated process than it might seem. Humans adrift in a lifeboat on a tropical sea, for example, are surrounded by water but cannot drink it. Their bodies are cooled by the evaporation of sweat, but in the process more water is lost. Without provisions, they might catch and eat fish, but the use of proteins as cellular fuel produces nitrogenous wastes that require water for removal. However, they could squeeze out the fish fluids and use them as a source of water. The dilemma of these people in a lifeboat illustrates four important factors that may complicate the regulation of internal body fluids: (1) the salt content of the environment in which an animal lives; (2) the water demands of special physiological processes; (3) the need to maintain internal salt balance; and (4) the elimination of nitrogenous wastes. These complications are interrelated, but since excretory organs are involved with other aspects of regulation, in addition to the elimination of wastes, we will begin with excretion.

Nitrogenous Wastes

Protein Metabolism. The continual breakdown of organic compounds within living systems results in **metabolic wastes.** A major waste is carbon dioxide, which is derived from the decarboxylation of metabolites at various points in cell respiration. The degradation of nucleic acids and proteins produces nitrogenous wastes. Amino acids are deaminated in reactions that replace the amino group with an oxygen atom, converting the amino acid to a keto acid.

$$\underset{\text{Alanine}}{\begin{matrix} CH_3 \\ | \\ HC\text{—}NH_2 \\ | \\ COOH \end{matrix}} \longrightarrow \underset{\text{Pyruvic acid}}{\begin{matrix} CH_3 \\ | \\ C{=}O \\ | \\ COOH \end{matrix}} \quad + \underset{\text{Ammonia}}{NH_3}$$

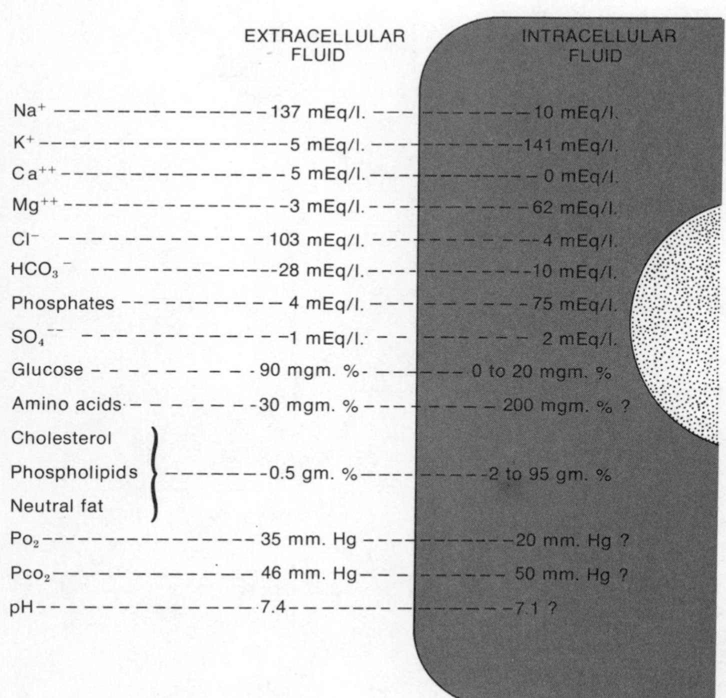

FIGURE 9.1 Diagram illustrating the chemical composition of extracellular fluid and intracellular fluid. (Redrawn after Guyton, A. C.: *Physiology of the Human Body*. 5th ed. Philadelphia, Saunders College Publishing, 1979.)

The keto acid is then metabolized in the TCA cycle as a source of biologically useful energy or is converted to other compounds. The processes of deamination and transamination convert the carbon chain of each amino acid into the corresponding α-keto acid. Some α-keto acids, such as pyruvate, α-ketoglutarate, and oxaloacetate, feed directly into the TCA cycle (Fig. 6.16). Other α-keto acids may require a series of reactions with many intermediate steps before resulting in a compound that enters the respiratory cycle.

Most of the pathways for oxidation of amino acids generally differ from those involved in the synthesis of amino acids. There are, however, some shared reversible reactions. The addition of an amino group to α-ketoglutaric acid yields the amino acid glutamic acid.

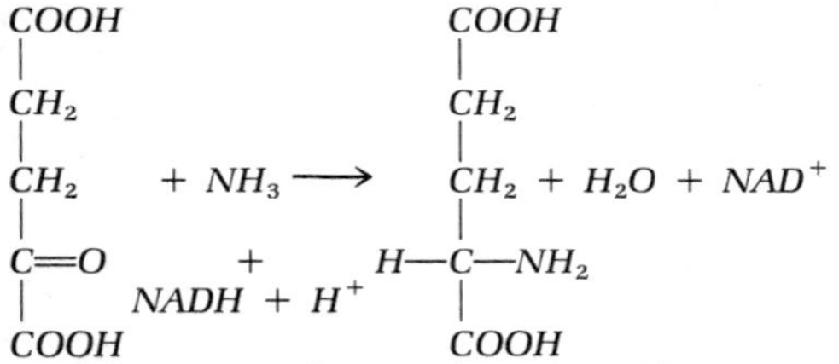

$$\begin{array}{lcl} COOH & & COOH \\ | & & | \\ CH_2 & & CH_2 \\ | & & | \\ CH_2 \quad + NH_3 & \longrightarrow & CH_2 + H_2O + NAD^+ \\ | & & | \\ C{=}O \quad + & & H{-}C{-}NH_2 \\ | \quad NADH + H^+ & & | \\ COOH & & COOH \\ \alpha\text{-Ketoglutaric acid} & & \text{Glutamic acid} \end{array}$$

The liver and kidney are major sites of amino acid metabolism in vertebrates. (Remember that the liver first receives the products of digestion from the intestine.)

Forms of Nitrogenous Wastes. The amino group removed in deamination appears as a molecule of ammonia, a toxic compound that cannot be permitted to accumulate and must be eliminated. Most aquatic animals, including most fish, excrete ammonia itself. Other animals eliminate more complex but less toxic nitrogenous wastes synthesized from ammonia by processes that require energy and thus have an ATP cost. Urea, a simple compound synthesized from carbon dioxide, two amino groups, and three ATPs, is the waste product of mammals, amphibians, and some fish. Urea is synthesized within the liver by a cyclic series of reactions (Box 9.1).

Uric acid, the principal waste product of birds, terrestrial reptiles, insects, and land snails, is also produced in small amounts by humans and other mammals from the degradation of nucleic acids. It is a relatively insoluble purine (Fig 9.2). Ammonia, urea, and uric acid are the most common forms of nitrogenous wastes eliminated by animals; spiders, however, excrete guanine.

Excretory Organs

Very small aquatic animals have enough surface area in relation to volume to eliminate ammonia by simple diffusion to the exterior. Even large aquatic animals may lose some ammonia by way of the gill surface. Animals with a volume greater than a few cubic centimeters, however, require excretory organs for the removal of their nitrogenous wastes.

Excretory organs are typically tubular or saccular structures adapted for concentrating wastes. Most belong to one of two general types. In one type, the inner

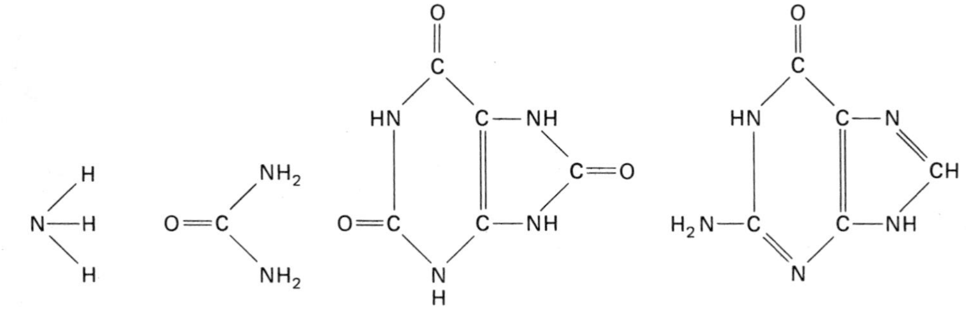

Ammonia
Most aquatic animals

Urea
Sharks and rays
Some bony fish
Amphibians
Mammals

Uric acid
Land snails
Insects
Turtles
Lizards
Snakes
Birds

Guanine
Spiders

FIGURE 9.2 The principal nitrogenous waste products of various animal groups.

BOX 9.1

Cycle of reactions involved in the liver's synthesis of urea. Two amino groups (shown in bold) are required. One is provided by a molecule of ammonia, which combines with CO_2 to form carbamoyl phosphate. The other amino group is provided by aspartate. In the utilization of amino acids as fuels, few are deaminated directly. Most lose their amino group by transamination with ketoglutaric acid, which becomes glutamic acid. The amino acid residues, now keto acids, enter the respiratory cycle at different points, as shown in Fig. 6.16. Glutamic acid is then deaminated to form ketoglutaric acid, and the amino group removed is the source of ammonia. The ornithine cycle is closely related to the citric acid cycle in that fumaric can be converted to malic acid, which can be converted to oxaloacetic acid. Oxaloacetic acid can, in turn, undergo transamination with an amino acid into aspartic acid, which then eventually delivers its amino group to urea.

Ornithine is simply a carrier for the various contributions (indicated by boxes) to the final molecule of urea. The energy cost for the production of one molecule of urea is four high energy phosphate bonds: two ATPs in the formation of carbamoyl phosphate and two high-energy bonds from one ATP in the formation of argininosuccinate from aspartate and citrulline. The synthesis of carbamoyl phosphate and its addition to ornithine to form citrulline occurs in the mitochondrion. Ornithine enters the mitochondrion, and citrulline leaves it. The other reactions occur outside of the mitochondrion. (The *-ate* endings simply indicate the ionized form of the acid—sulfuric acid and sulfate; glutamic acid and glutamate.)

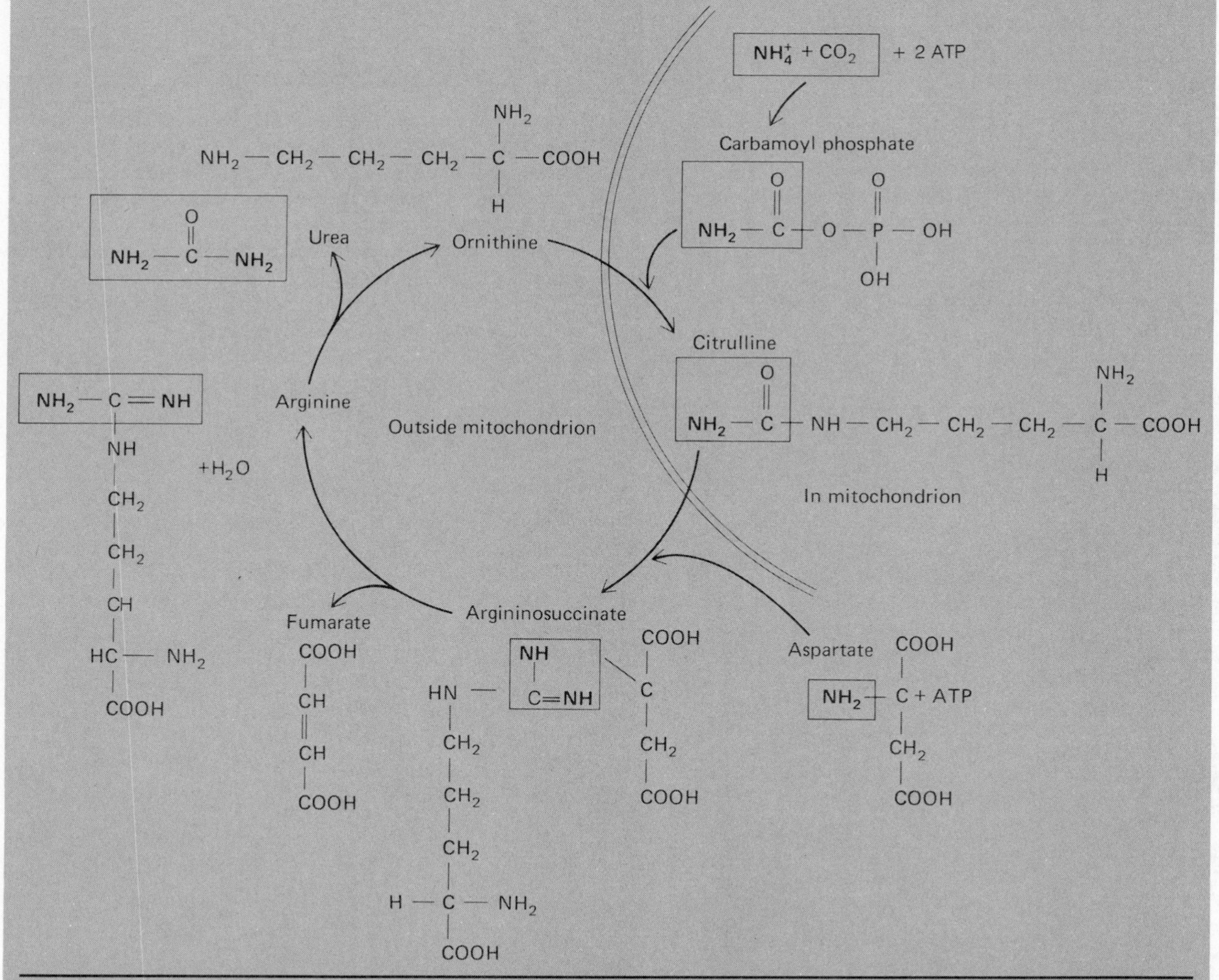

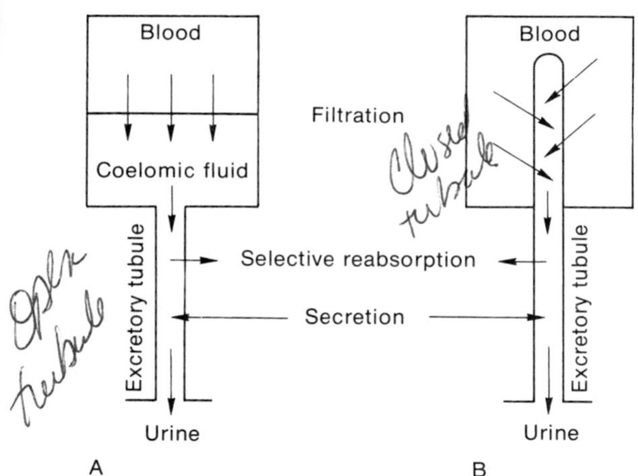

FIGURE 9.3 Models of the two principal types of excretory organs in animals, based on their function. In type A, the excretory tubule opens into the coelom, and wastes are concentrated from coelomic fluid, which receives filtrate from the blood. In type B, the excretory tubule is closed, and wastes are concentrated from blood filtrate.

end of the tubule opens to the coelom (Fig. 9.3*A*). Small molecules filtered from the blood into the coelomic fluid pass in turn into the excretory tubule. As the coelomic fluid passes down the tubule, it is subjected to varying degrees of **selective reabsorption.** Some organic substances, such as sugars and amino acids, may be reabsorbed, and, depending on the environment of the animal, there may be some reabsorption of salts and water. In addition to selective reabsorption, **secretion** of wastes by the tubule wall may occur. The secretion and reabsorption of materials from the lumen of the tubule are facilitated by the blood or blood vessels surrounding the tubule. As a result of selective reabsorption and secretion, the wastes within the tubular fluid become more concentrated, and the final mixture of water, wastes, and salts is expelled to the exterior as **urine.**

In a second type of excretory organ, the tubule does not open into the coelom (Fig. 9.3*B*). Rather, the blind end of the tubule receives filtrate directly from the blood. The blood filtrate is modified and concentrated by selective reabsorption and secretion during its passage through the tubule. In neither type of excretory organ are wastes "screened out" from the blood as is popularly believed. Remember that blood filtrate is not waste; it is simply the fluid portion of blood (minus large molecules and cells) that can pass through the capillary wall. *Excretory organs do not screen wastes; they concentrate them.*

Various modifications of these two general plans have evolved in a number of different animal groups. A common change has been the reduction or loss of filtration and an increase in the importance of secretion.

The excretory organs of earthworms consist of two tubules, or **nephridia** (Gr., *nephros*, kidneys) per segment. Each arises in a ciliated funnel, or **nephrostome** (Gr. *stoma*, mouth), that opens into the coelom of the preceding segment (Fig 9.4). The tubule perforates the septum and extends back into the following segment, where it may loop or coil before passing through the body wall. The external opening is called a **nephridiopore.**

Snails and clams possess similar nephridia, but these animals are not metameric, and usually only one or two are present.

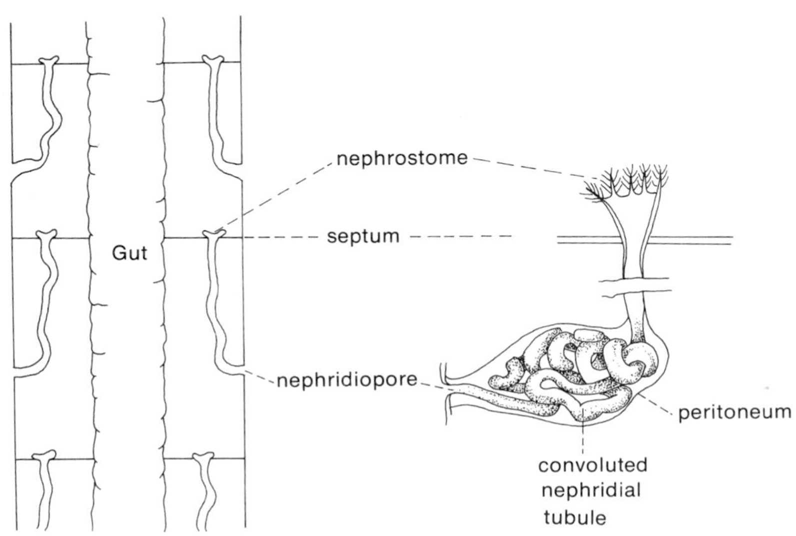

FIGURE 9.4 Annelid metanephridia. *A*, Diagrammatic dorsal view of two segments showing position of nephridia. *B*, One nephridium of *Nereis vexillosa*, a marine annelid in which the nephridial tubule is greatly coiled. (Modified from Jones.)

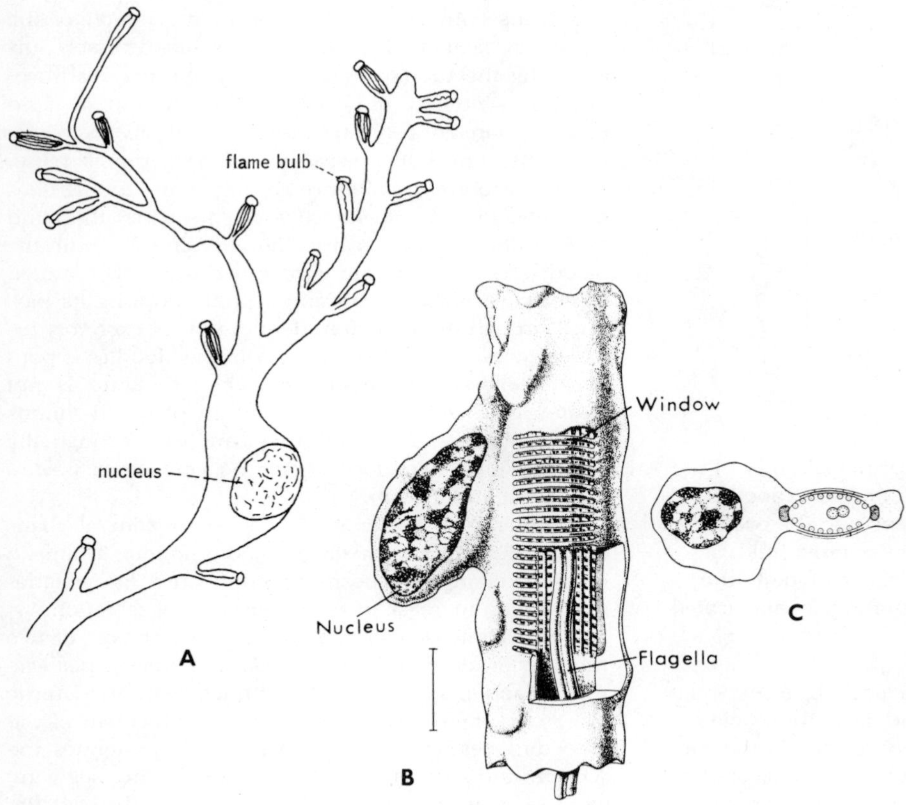

FIGURE 9.5 Protonephridia. *A*, Protonephridium of the flatworm *Mesostoma*, in which a number of flame bulbs are a part of one cell. *B*, Flame cell of the freshwater flatworm *Stenostomum*. *C*, Cross section at level of nucleus. (*A* After Reisinger from Hyman; *B* and *C* from Kummel, G.: *Z. Zellforsch* *57*:172–201, 1962.)

Some small animals, including flatworms and rotifers, possess blind tubules called **protonephridia** (Gr., *protos*, first + *nephros*, kidney). The blind end bears a tuft of cilia or a flagellum and is called a flame bulb or flame cell, depending upon its structure (Fig. 9.5). The wall of this terminal part of the tubule is partially perforated by slits. A cell membrane covers each slit, much like a pane of glass over a window. The beating of the cilia or flagella drives water down the tubule and creates a negative pressure within the flame bulb or flame cell. Surrounding water is then pulled inward, passing through the membranous panes covering the slits. The protonephridial tubule opens to the exterior by way of a nephridiopore. Although often considered to be excretory organs, these tubules are usually not involved in the excretion of nitrogenous wastes but may function as pumps to remove excess water.

The excretory organs of crayfish, shrimps, and crabs, the paired green glands, consist of a saccule located in the head and bathed in blood. Filtrates pass from the blood through the wall of the sac and then down a tubule. Selective reabsorption occurs in certain parts of the tubule and bladder, and the resulting urine is expelled to the outside through an opening at the base of each antenna (Fig. 27.21*A*).

Malpighian tubules of insects and spiders are also blind-ending tubules bathed in blood (Fig. 6.7). Unlike crustacean green glands, they can number from two to many and open into the intestine rather than to the exterior. Wastes concentrated from the blood thus pass from the tubules into the intestine and rectum. During passage through the rectum, much of the water is absorbed, and the pasty urine is eliminated through the anus with the fecal material.

The Vertebrate Kidney

The kidneys of vertebrates are paired organs that lie dorsal to the coelom on each side of the dorsal aorta. All vertebrate kidneys are composed of units called kidney tubules, or **nephrons,** that end blindly and receive a filtrate from the blood. The number and arrangement of the nephrons differ among the various groups of vertebrates. In ancestral vertebrates each kidney contained one nephron for each body segment that lay between the anterior and posterior ends of the coelom (Fig. 9.6). These nephrons drained into an **archinephric duct** that continued posteriorly to the cloaca. Such a kidney is called a **holonephros,** for it extends the entire length of the coelom. A holonephros is found today in the larvae of certain cyclostomes but not in any adult vertebrate.

In the kidney of adult fish and amphibians (Fig. 9.6*C*), the most anterior tubules have been lost, some of the middle tubules are associated with the testis, and there

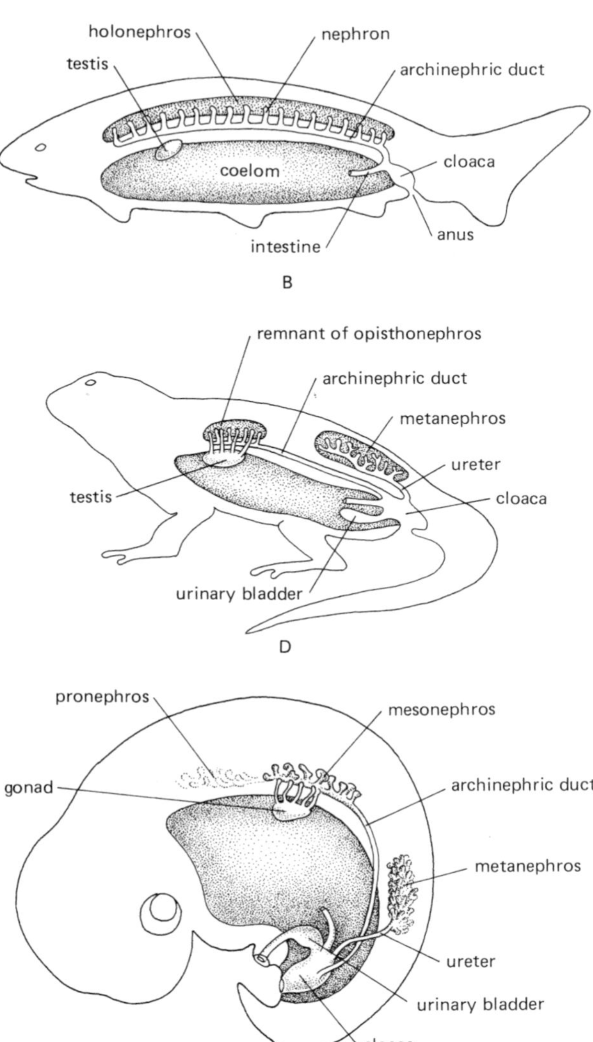

FIGURE 9.6 A comparison of the evolution and embryonic development of the vertebrate kidney and its ducts. *A–D,* The evolutionary sequence of kidneys. *A* and *B,* Dorsal and lateral views of hypothetical ancestral vertebrate with a holonephros. *C,* A fish with an opisthonephros. *D,* A reptile with a metanephros. *E* and *F,* The developmental sequence of kidneys in a reptile. *G,* Mammalian embryo in which the cloaca has become divided by a fold into rectum and urogenital sinus.

is a concentration and multiplication of tubules posteriorly. Such a posterior kidney is known as an **opisthonephros** (Gr., *opisthe*, behind + *nephros*, kidney). The original archinephric duct functions as both excretory and in males as a sperm duct.

The kidney of reptiles, birds, and mammals is known as a **metanephros** (Gr., *meta*, after + *nephros*, kidney) (Fig. 9.6*D*). All of the middle tubules not associated with the testis have been lost, and there is an even greater multiplication and posterior concentration of tubules. The number of nephrons is particularly large in birds and mammals; their high rate of metabolism yields a large amount of wastes. It is estimated that humans have about 1,000,000 nephrons per kidney, whereas certain salamanders have less than 100. The tubules producing urine drain into a **ureter,** that evolved as an outgrowth from the old archinephric duct. The archinephric duct has been taken over completely by the male genital system as the sperm duct.

The evolutionary sequence of kidneys is holonephros, opisthonephros, and metanephros. In the embryonic development of higher vertebrates—reptiles, birds, and mammals—there is a somewhat parallel sequence. Nephrogenic mesoderm (kidney mesoderm) arises dorsally along the entire length of the embryo, but only the most posterior part develops into the adult metanephros (Fig. 9.6*E*, *F*, and *G*).

Most tetrapods have a **urinary bladder** that develops as a ventral outgrowth from the cloaca. Generally, the excretory ducts from the kidneys lead to the dorsal part of the cloaca, and urine must flow across it to enter the bladder. However, in mammals (Fig. 9.6*G* and 9.7) the ureters lead directly to the bladder, and the bladder opens to the body surface through a short tube, the **urethra.** The cloaca becomes divided and disappears as such in all but the most primitive mammals. The dorsal part of the cloaca forms the rectum, and the ventral part contributes to the urethra of higher mammals (Fig. 9.6*G*).

Urine is produced continually by the kidneys and is carried down the ureters by peristaltic contractions. It accumulates in the bladder, for a smooth muscle sphincter at the entrance of the urethra and a striated muscle sphincter located more distally along the urethra are closed. Urine is prevented from backing up into the ureters by valvelike folds of mucous membrane within the bladder. When the bladder becomes filled, stretch receptors are stimulated, and a reflex is initiated that leads to the contraction of the smooth muscles in the bladder wall and the relaxation of the smooth muscle sphincter. Relaxation of the striated muscle sphincter is a voluntary act.

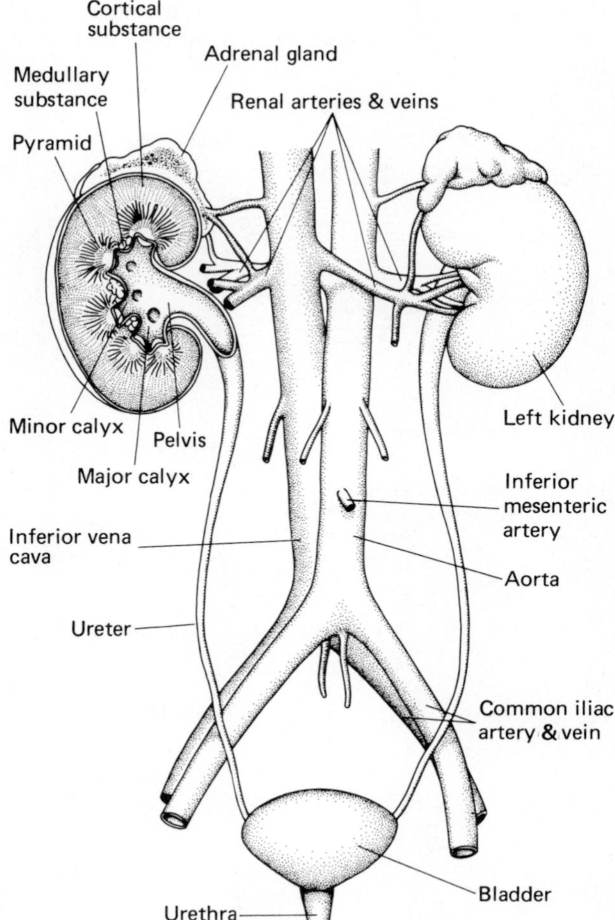

FIGURE 9.7 The human excretory system as seen in a ventral view. The right kidney has been sectioned to show the internal structures.

Nephron Structure. The proximal end of each nephron (Fig. 9.8), known as **Bowman's capsule,** is a hollow ball of squamous epithelial cells, one end of which has been pushed in by a knot of capillaries called a **glomerulus** (L., *glomus*, ball). Bowman's capsule and the glomerulus constitute a **renal corpuscle.** The rest of the nephron is a tubule, largely composed of cuboidal epithelial cells and subdivided in mammals into a **proximal convoluted tubule**, a **loop of Henle**, and a **distal convoluted tubule**. A **collecting tubule** receives the drainage of several nephrons and leads to the **renal pelvis**, an expansion within the kidney of the proximal end of the ureter (Fig. 9.7). The location of the different parts of a nephron within the kidney and their relationship to blood vessels have important functional consequences. As shown in Figure 9.8, the renal corpuscles and convoluted tubules lie in the outer **cortex** of the kidney, and a dense capillary network surrounds the convoluted tubules; the loops of Henle extend toward

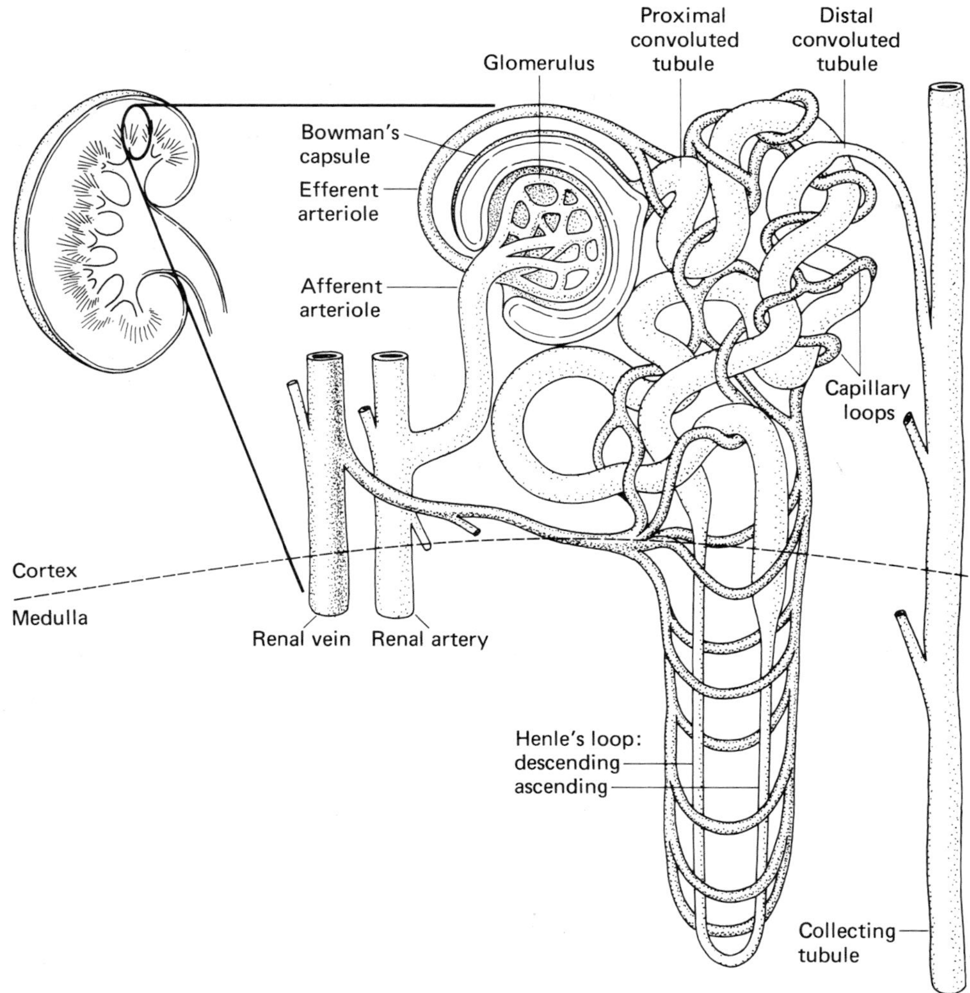

FIGURE 9.8 A diagram of a mammalian nephron and associated blood vessels.

the center, or **medulla**, of the kidney. Most of the human nephrons extend only a short distance into the medulla, but about one fifth of them (the **juxtamedullary nephrons**) have long loops of Henle that extend, along with the collecting tubules and capillary loops, far into the medulla. It is the convergence of these structures into subdivisions of the renal pelvis (the **calyces**) that forms the **renal pyramids** (Fig. 9.7).

Urine Formation. The kidney tubules produce **urine**, a watery solution containing waste products of metabolism removed from the blood but not substances needed by the organism. The most abundant nitrogenous waste in humans and other mammals is urea, but lesser amounts of ammonia, uric acid, and creatinine are present. The yellowish color of urine is due to **urochrome**, a pigment derived from the breakdown of hemoglobin and, hence, related to the bile pigments.

That the first step in urine formation is **glomerular filtration** was demonstrated in the 1920s by Dr. A. N. Richards, who developed a micropipette technique for removing and analyzing minute samples of fluid from the lumen of Bowman's capsule. The wall, or **filtration barrier**, through which fluid must pass to reach the lumen of Bowman's capsule is composed of the inner capillary endothelium, which contains many large pores, a middle basement membrane, and an outer layer of cells called **podocytes** (Gr., *pous*, foot + Fr., *cyte*, cell) that are derived from the infolded wall of Bowman's capsule. The podocytes are large cells with primary and

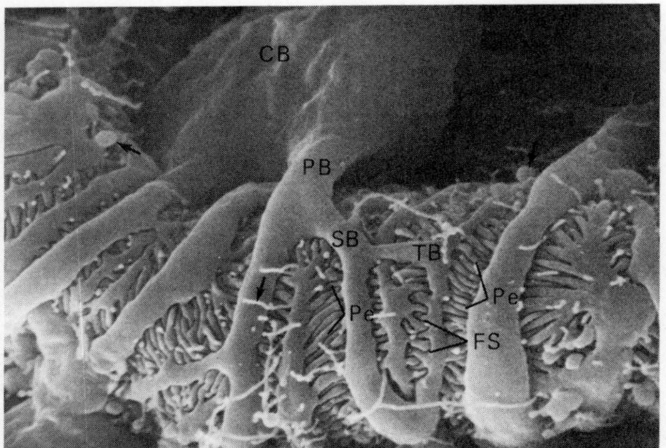

FIGURE 9.9 Scanning electron photomicrograph of part of a podocyte of a Bowman's capsule. These highly modified cells represent that part of the capsule wall that envelops the glomerulus and contributes to the filtration barrier through which fluid from the glomerulus must pass. From the central part of the cell *(CB)* containing the nucleus extend primary *(PB)*, secondary *(SB)*, and tertiary branches *(TB)*, which terminate in finger-like projections, called pedicels *(Pe)*. The fine slits *(FS)* between adjacent interdigitating pedicels of the podocyte are the primary avenues of filtrate passage. (From Kessel, R. G., and R. H. Kardon: *Tissues and Organs: A Text-Atlas of Scanning Electron Microscopy*. San Francisco, W. H. Freeman & Co., p. 231, 1979.)

secondary finger-like processes that interdigitate with each other (Fig. 9.9). The intervening slits are the passageways for the filtrating fluid. The endothelial pores and podocyte slits make the filtration barrier many times more permeable than other capillary beds but at the same time restrict passage to molecules of 40,000 mol. wt. or less.

Water, various ions, and small organic molecules, including simple sugars and amino acids as well as nitrogenous wastes, pass easily through the filtration barrier. Blood cells and large molecules, including fats and plasma proteins, remain in the blood. The only small molecules to be held in the blood are those bound to plasma proteins; among these are iron and other trace minerals and certain vitamins. The resulting filtrate in Bowman's capsule is thus very much like plasma and is essentially the same as the interstitial fluid of other capillary beds.

Estimates of the filtration rate are obtained by the use of **inulin**, a polysaccharide from artichokes that is filtered easily and is not removed or added to the filtrate as the fluid continues down the kidney tubules. For example, when inulin is injected intravenously until the plasma concentration reaches 1 mg. per mL, 125 mg. of inulin appear in the urine per minute. For this to occur, 125 mL of plasma must be filtered each minute. This amounts to 180 L of filtrate per day! The kidneys have a rich blood supply, receiving about one quarter of the cardiac output in a mammal. This enables them to process the total blood volume every four or five minutes.

A glomerulus lies between an **afferent arteriole**, a branch of a renal artery, and an **efferent arteriole** that leads to capillaries on other parts of the tubule (Fig. 9.8). The efferent arteriole is the smaller of the two, and this ensures a high **filtration pressure** (about 60 mm. Hg) in the glomerular capillaries to drive water and small solute molecules from the blood. The filtration pressure exceeds the pressures within the kidney tubule and the osmotic pressure of the blood (both of which promote the return of fluid to the blood) by 20 mm. Hg. As a consequence, there is a substantial net production of glomerular filtrate.

The volume of urine in humans is only about 1 percent of the glomerular filtrate, and nearly all of the glucose, amino acids, and ions present in the glomerular filtrate are taken back into the blood by **tubular reabsorption** as the filtrate passes down the tubules. Other substances may be added to the filtrate by **tubular secretion.** Each minute human kidneys excrete an amount of urea equal to that present in 75 mL of plasma. It can be said that 75 mL of plasma has been cleared of urea, or that urea has a **clearance rate** of 75 mL per minute. Other substances have different clearance rates. From inulin studies we know that the filtration rate is 125 mL per minute. If the clearance rate of a freely filterable substance (such as urea) is less than this, some of this substance must be reabsorbed in the tubules; if the clearance rate exceeds 125 mL per minute, then some of the substance must be added by tubular secretion. In mammals, virtually all of the glucose and amino acids that pass into the filtrate, and some urea, are reabsorbed in the proximal convoluted tubule (Fig. 9.10). Sodium, chloride, and bicarbonate ions are reabsorbed in both the proximal and distal convoluted tubules. More than 99 percent of the water in the glomerular filtrate is reabsorbed; this occurs at many levels in the tubules. Creatinine, ammonia, hydrogen, and potassium ions and various drugs (penicillin) are among the few substances added to the filtrate by tubular secretion in mammals, and most of this occurs in distal parts of the tubule.

Reabsorption, which plays such an important role in urine formation, involves both the passive diffusion of materials back into the capillaries surrounding the tubules and the active uptake of materials by the tubular cells and their secretion into the blood against a concentration gradient. This, of course, requires the expen-

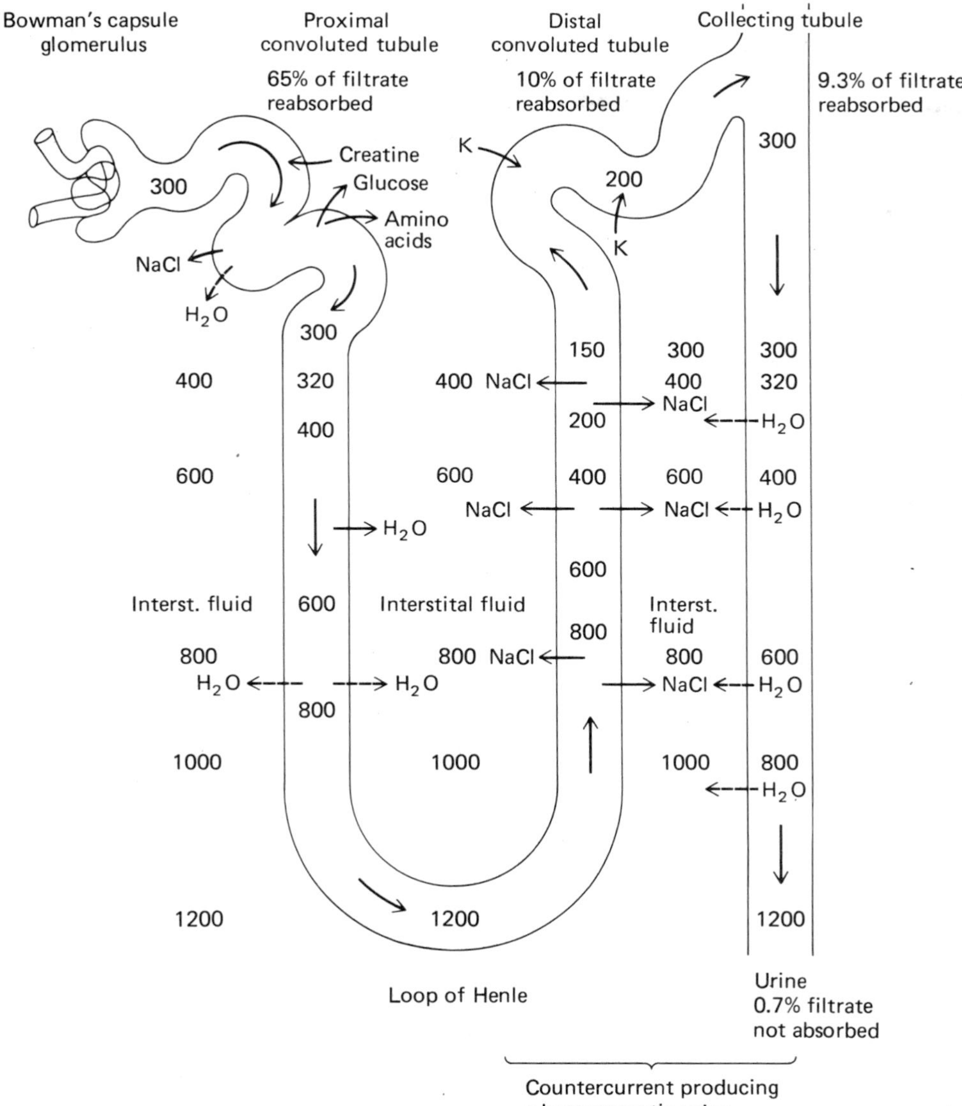

FIGURE 9.10 Diagram of the processes, including the countercurrent multiplying mechanism, that transform the glomerular filtrate into urine as it passes through a nephron and collecting tubule. Solid arrows into and out of the tubules indicate active transport; broken arrows indicate passive diffusions. The section of the ascending limb of the loop of Henle, shown with a heavy wall, is impermeable to water. The numbers indicate the relative concentrations of osmotically active solutes in milliosmols per liter. (Modified from Guyton, A. C.: *Human Physiology and Mechanisms of Disease*. 3rd ed. Philadelphia. W. B. Saunders Co., 1982.)

diture of energy by the tubular cells. Apart from urea, most of the reabsorption of solutes that takes place in the proximal convoluted tubule is an active process because the concentration of these substances in the filtrate in this region is the same as their concentration in the blood. The active reabsorption of one ion can bring about the passive reabsorption of another with the opposite charge. For example, when sodium, which is positively charged, is removed from the filtrate, the remaining fluid will exhibit an excess of negative charges, i.e., a charge imbalance. As a result chloride ions passively follow the reabsorbed sodium ions.

Those materials that can be actively reabsorbed are taken back in varying amounts, depending upon their concentration in the blood. If the concentration of one of these materials in the blood and glomerular filtrate rises above a certain level, known as the **renal threshold**, not all of it will be reabsorbed into the blood from the tubule, and the amount present in excess of the renal threshold will be excreted. The quantitative value of the renal threshold differs for different substances. In **diabetes mellitus** the impaired cellular utilization of glucose leads to a high concentration of glucose in the blood; the renal threshold for glucose (about 150 mg. of glucose per 100 mL blood) is exceeded, and sugar appears in the urine in large amounts. The osmotic pressure of the body fluids is controlled by the amount of salts that are returned to the blood from the glomerular filtrate; the pH of the fluids is regulated by the elimination or retention of basic and acidic substances.

Water is reabsorbed passively. As solutes, especially sodium, are actively taken out of the filtrate in the proximal tubule, the water in the filtrate tends to become more concentrated than it is in the blood. However, water molecules passively diffuse out as fast as solutes are pumped out; hence the concentration of the filtrate remains the same as that of the blood. About 65 percent of all the water taken back is reabsorbed in this way in the proximal tubule, but the tubular filtrate cannot be made more concentrated (i.e., to contain less water) than the blood by this mechanism.

In most of the lower vertebrates the urine is not more concentrated than the blood; however, birds and mammals do produce a hyperosmotic urine. The unique feature of the mammalian nephron is the loop of Henle. The descending and ascending limbs of the loop of Henle lie parallel to each other, so that the direction of flow of fluid in one is opposite to that in the other (Fig. 9.10). The active transport of chloride ions out of the ascending limb into the interstitial fluid (with sodium ions following passively by charge attraction*) and their passive diffusion back into the descending limb result in a countercurrent multiplying mechanism. Sodium and chloride ions moved out of the ascending limb go right back into the descending limb. The recycling of these ions in the loop of Henle, together with additional sodium and chloride ions being continually brought to the loop in the glomerular filtrate, results in an accumulation of sodium and chloride in the loop of Henle and the surrounding interstitial fluid of the medulla. A concentration gradient from cortex to medulla is established as shown in Figure 9.10. The degree of accumulation depends upon the length of the loop; there is more opportunity to pump out chloride in a long loop. The capillary loops associated with the juxtamedullary nephrons provide a countercurrent mechanism that permits most of the sodium and chloride that starts out of the medulla in the blood to diffuse back into the blood entering the medulla. This, combined with the rather sluggish rate of blood flow through these vascular loops, means that little sodium and chloride is carried away from the medulla by this route.

*Note that this is the reverse of the situation that occurs in the proximal convoluted tubule but that the result is the same—sodium chloride is moved out of the tubule.

The osmotic gradient established by these mechanisms makes it possible for additional water to be reabsorbed passively and for a hyperosmotic urine to be produced. Water simply follows the osmotic gradient, moving from an area of low osmotic pressure (high concentration of water) to one of high osmotic pressure (low concentration of water). The glomerular filtrate, which was isosmotic to the blood in the proximal tubule, loses water as it passes down into the loop of Henle. It does not regain this water as it ascends the loop of Henle, for the cells of the ascending limb of the loop have a low permeability to water. However, the filtrate becomes more dilute because of the large amount of sodium and chloride pumped out. By the time the filtrate reaches the distal tubule it is again isosmotic, or in some cases hypo-osmotic, to the blood. The filtrate now descends through the medulla again, this time in the collecting tubule. It passes as a countercurrent to the adjacent gradient of sodium-chloride in the filtrate of the ascending loop of Henle and loses water to the surrounding interstitial spaces. The filtrate now becomes very hyperosmotic.

Kidney Regulation. The amount of material reabsorbed or excreted by the kidney depends to some extent on the rate of glomerular filtration. If this is too high, certain essential substances are flushed through the tubules before reabsorption is completed; if the filtration rate is too low, some products normally excreted may be reabsorbed. Filtration pressure, and hence rate, can be varied within limits by a feedback mechanism that controls the constriction or relaxation of the afferent and efferent arterioles leading to the glomeruli. These vessels are regulated by the solute level in a section of the distal convoluted tubule that passes between them. Together the three constitute the **juxtaglomerular** apparatus (Box 9.2).

The amount of water excreted and hence the volume of body fluids are also very much affected by **vasopressin (antidiuretic hormone)**, released by the posterior

BOX 9.2

The juxtaglomerular apparatus is a feedback mechanism permitting solute levels in the distal convoluted tubule to regulate the constriction or dilation of the afferent and efferent arterioles leading to and from the glomerulus. This in turn affects filtration pressure. Shortly before it empties into the collecting tubule, a distal convoluted tubule passes between the afferent and efferent arterioles and comes into an intimate relationship with modified muscular cells (juxtaglomerular cells) around the afferent and efferent arterioles. These cells contain an enzyme that on release into the bloodstream acts on a plasma protein to form a vasoconstrictor, called angiotensin. The apparatus is sensitive to chloride levels. If glomerular filtration is too low, there will be an overabsorption of chloride ions before the filtrate reaches the distal tubule in the juxtaglomerular apparatus. Special cells in the apparatus will have a low chloride content, that will cause activation of the vasoconstrictor and constriction of the efferent arteriole, as well as of arterioles elsewhere in the body. In some manner the low chloride content of the distal tubule also causes the afferent arteriole to the glomerulus to dilate. The combination of the dilation of the afferent and constriction of the efferent arterioles increases filtration pressure and the fluid volume passing into Bowman's capsule.

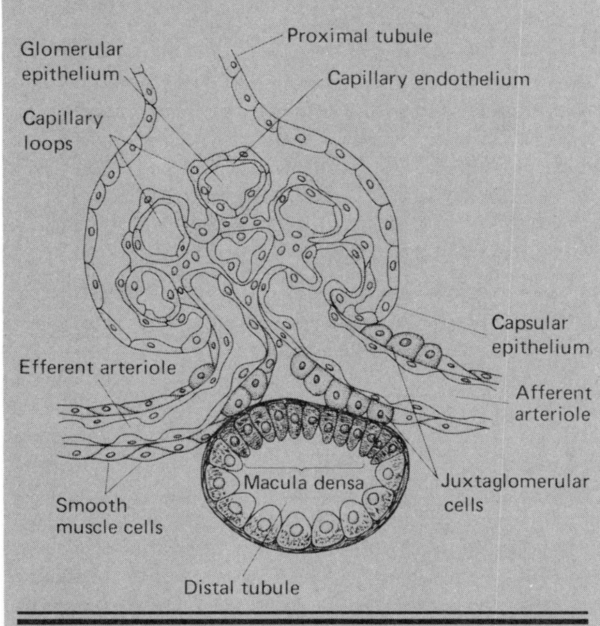

lobe of the pituitary (p. 273). This increases the permeability of the cells of the collecting tubule to water so that more water is reabsorbed into the blood. If an excess of water is present in the body fluids, the blood volume and pressure increase. This raises the glomerular filtration pressure, and more filtrate is produced. An increase in the amount of water in the tissue fluid inhibits the release of vasopressin, the permeability of collecting tubule cells is lowered, and less water is reabsorbed. Increased production of filtrate and decreased reabsorption of water rapidly bring the volume of body fluids down to normal. If the volume of body fluids falls below normal, as in a severe hemorrhage, these factors work in the opposite direction: Less glomerular filtrate is produced, more water is reabsorbed, and the volume of body fluid is soon raised to normal.

The osmotic pressure of the tubular contents also affects the amount of water removed. If a large amount of salts or sugars is being eliminated, the osmotic pressure of the tubular contents is increased, and less water can be reabsorbed. The urine volume is greater when there is a large amount of osmotically active substances in the urine, as after a large intake of salt or in diabetes mellitus.

Vasopressin regulates the volume of water in the urine. Sodium and potassium concentrations in the urine, and thus in the blood, are directly regulated by another hormone, **aldosterone**, which is secreted by the adrenal glands attached to the kidneys. Aldosterone stimulates reabsorption of sodium from, and secretion of potassium into, the distal tubules and collecting ducts. Elimination of potassium is an important function of the kidney since much more potassium is consumed in food than can be utilized in cellular functions. Note that all of these regulations of kidney function depend upon feedback mechanisms involving solute concentrations. The pituitary (vasopressin), via the hypothalamus, and the adrenal gland (aldosterone) monitor solute levels in the blood. The juxtaglomerular apparatus monitors solute levels in the kidney tubules. (Box 9.2).

Osmotic Regulation in Marine Animals

The amount of salt in the open ocean is about 35 parts per thousand (35‰), or about 3.5 percent. In estuaries and bays the salinity may be considerably lower, depending upon the inflow of fresh water from rivers and streams. The term **brackish** refers to estuarine waters that are more saline than fresh but less saline than the open ocean.

With the exception of such vertebrates as fishes, porpoises, and whales, the internal body fluids of most marine animals are isosmotic with sea water. The salt content of their blood, coelomic fluid, and intercellular fluids is about the same as that of their marine environment, although the content of specific ions may be somewhat different. Their intracellular fluids are isosmotic with extracellular fluids, but their osmotic pressure is influenced by the large amount of organic compounds present. Intracellular osmotic pressures are commonly adjusted with free amino acids.

If the salinity of the external environment changes slightly, the concentration of salts in the body fluids of the animal also changes. Such marine animals are said to be **osmoconformers**; i.e., the salt concentration of their internal environment conforms to that of the surrounding external environment.

Since a minimal internal salt content is necessary for life, osmoconformers require environments of relatively high salinity. They are **stenohaline** (Gr., *stenos,* narrow + *halinos,* saline), restricted to a narrow range of salinities, usually near that of the open ocean.

Osmoconformers, like aquatic animals in general, eliminate their nitrogenous wastes as ammonia. Ammonia is highly soluble and, since it is toxic, a large amount of water must be available so that it can be kept at low levels in the urine. Water supply is not a problem for most aquatic animals. As might be expected, the urine of osmoconformers is isosmotic with both their internal body fluids and the surrounding sea water.

Not all marine animals are osmoconformers and stenohaline. If a spider crab and a blue crab are placed in an aquarium in which the salinity is slowly lowered, their blood salts will gradually fall; i.e., both act as osmoconformers (Fig. 9.11). With furthur decline in environmental salt concentrations, the spider crab dies, but the blue crab does not. If the blue crab's blood salts are analyzed, they are found to be at higher levels than the salts in the environment; it holds onto its blood salts; it does not conform to the level of environmental salts. It is **osmoregulating.**

Marine osmoregulators can invade or inhabit estuarine waters, where salinities are low. They can usually tolerate a wide range of salinities and are said to be **euryhaline** (Gr., *eurys,* wide + *halinos,* saline).

When osmoregulating crabs are placed in dilute sea water, the urine flow may greatly increase as the animal eliminates the excess of water diffusing inward. But in the process salts are lost, for most crabs can only produce a urine isosmotic or only slightly hypo-osmotic with their blood. For example, in normal sea water the European crab *Carcinus* excretes a volume of urine equivalent to 3.6 percent of its body weight each day; in sea water diluted to 1.4 percent salinity it eliminates a urine equivalent to 33.0 percent. Replacement of salt lost in the urine is provided by the gills, which actively pick up salt from the ventilating current. Thus, in osmoregulating crabs the green glands are the water pumps, and the gills are centers of salt replacement.

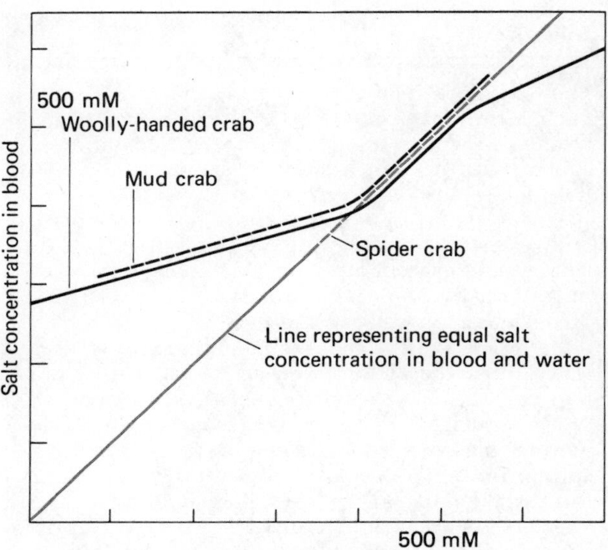

FIGURE 9.11 The osmoregulatory ability of three crabs. The osmotic concentration of blood changes with the concentration of salts in the surrounding sea water. The spider crab cannot osmoregulate, and the osmotic concentration of its blood conforms to the salt concentration of the sea water. Below and above certain concentrations, the crab dies. The blood of both the mud crab and the woolly-handed crab conforms to an upper range of external salt concentration, but at the lower concentrations the crabs maintain higher osmotic concentrations in the blood by absorbing salts through the gills. These two crabs can live in brackish estuarine environments where the spider crab cannot. The woolly-handed crab can excrete salt at high blood concentrations. The osmoregulatory ability of the blue crab is similar to that of the mud crab. (Adapted from Schmidt-Nielsen.)

The Vertebrate Problem. The salt content of the internal body fluids of most vertebrates is about 1 percent and may reflect the invasion of fresh water early in their evolutionary history. Marine vertebrates are thus believed to be immigrants to the sea. The salt content of their body fluids is only about one third that of sea water, and water tends to diffuse outward, especially across surfaces such as the gills. They are therefore living in a physiological desert, and water conservation is a major problem.

Marine bony fish cannot produce urine more concentrated than their blood, for the kidney tubules lack

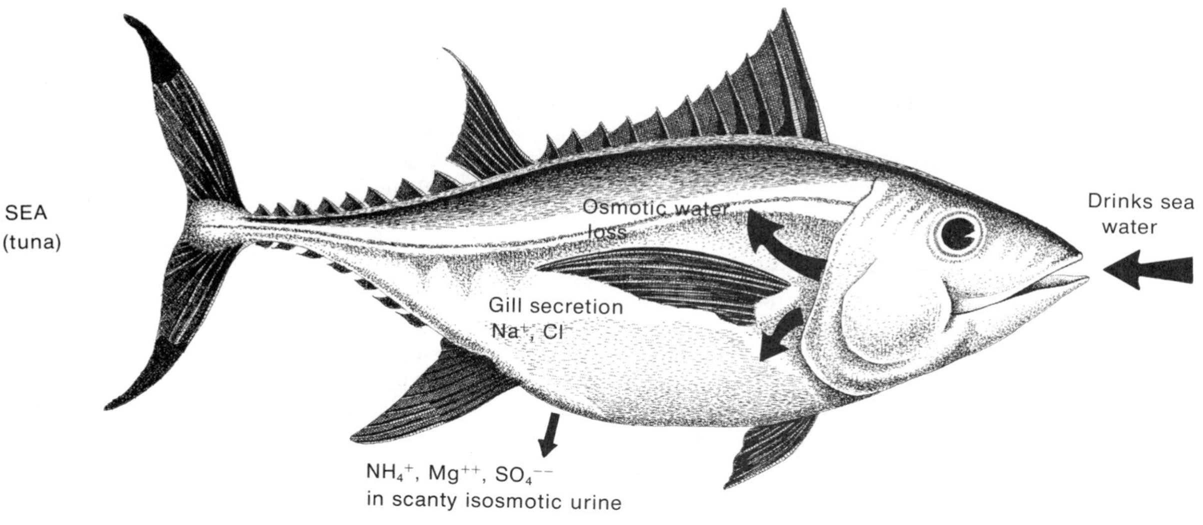

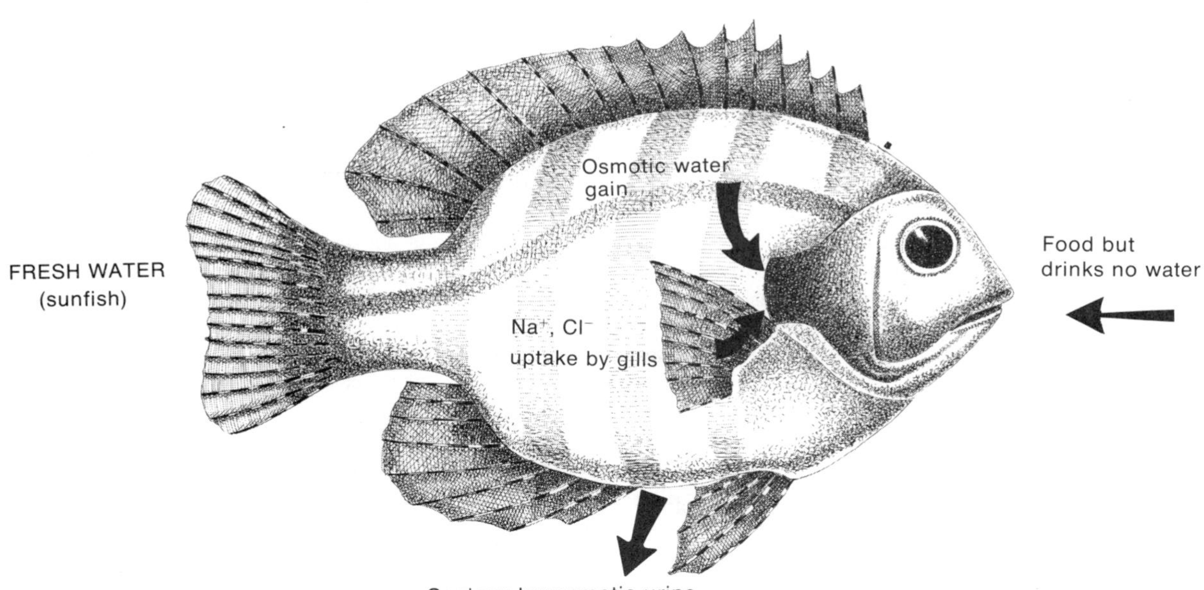

FIGURE 9.12 **Osmoregulation in bony fish. The body fluids of fish have a salt content of about one percent. Fish are thus hypo-osmotic to sea water and hyperosmotic to fresh water.**

loops of Henle. Marine fish obtain their water with food and by drinking sea water. The excess salt (Na^+ and Cl^-) taken into the blood is excreted by active transport across the gill epithelium. The kidneys remove divalent magnesium and sulfate ions, in addition to ammonium ion (NH_4^+), which is the principal nitrogenous waste of most fish (Fig. 9.12). However, most ammonia is eliminated through the gills rather than by the kidneys. The glomeruli of marine fish are usually smaller in size and in number, and some species have lost the glomeruli altogether. What would be the adaptive significance of such reduction in glomeruli?

Other marine vertebrates solve the problem of living in a hyperosmotic medium differently. Sharks produce urea as their nitrogenous waste and retain it in their internal body fluids in such high concentration that the osmotic pull of water inward is in equilibrium with the outward diffusion resulting from the higher external salt concentration. The gills of sharks are not permeable to urea as are those in most fishes. The concentration of urea in the body fluids of sharks may reach 100 times that in mammals, a level far exceeding the tolerance of other vertebrates. The salts that accumulate through both food intake and inward diffusion across surfaces such as the gills are removed by the kidneys and by a special salt-excreting gland attached to the caudal end of the shark intestine.

Sea birds and sea turtles excrete excess salt from a pair of salt glands in their heads. In sea birds, these glands open into the nasal passageways; in sea turtles, they open into the eyes. Turtles literally weep salty tears. The glands excrete only in response to an internal salt load, and the concentration of salt in the excreted fluid can reach a level almost twice that of sea water. Only birds and mammals can excrete a urine saltier than the blood, but the urine of seabirds is only twice as concentrated as the blood, which is still less than the concentration of salt in sea water.

Seals, porpoises, and whales obtain much of their water from their food; some food (such as mollusks and crustaceans) is isosmotic with sea water, and some (such as fish) is less concentrated than sea water. However, marine mammals can utilize the kidney for osmoregulation. They have very long loops of Henle that make it possible to produce a urine that has a salt concentration greater than that of sea water.

Humans cannot drink sea water because they cannot produce a urine with a salt content greater than approximately 2.2 percent. If we were to drink a liter of sea water, it would require one and one third liters of urine water to eliminate the excess salt. Moreover, the magnesium and sulfate ions in sea water cause diarrhea, and additional water is lost.

Osmoregulation in Freshwater Organisms

The salt concentration in fresh water varies with the water source but is always very low. The external environment is thus very hypo-osmotic to the internal body fluids of freshwater animals, and they must cope with the tendency of water to diffuse into the body, especially into thinly covered areas, such as the gills. Salts tend to diffuse outward, and the internal body fluids lose salts through excretion.

Given these conditions, the following generalizations are not unexpected: (1) All freshwater animals are osmoregulators. (2) Excretory organs function as water pumps, and the urine is usually hypo-osmotic to the internal body fluids as a result of selective reabsorption of salts. (3) Salt concentration of internal body fluids is maintained at the lowest level compatible with the animal's metabolic needs. Thus, less energy is required to maintain internal salt balance. (4) Nitrogenous wastes are usually eliminated as ammonia, since plenty of water is available to remove it in dilute concentrations. However, freshwater fish and adult amphibians more commonly excrete nitrogenous wastes as urea.

Freshwater fish excrete a hypo-osmotic urine but still suffer some salt loss. Replacement of salt is provided by gills that, in contrast to those of marine fish, pick up salts from the ventilating current (Fig. 9.12). Crayfish osmoregulate similarly. The green glands produce a hypo-osmotic urine, and gills are also the site of salt replacement.

The exoskeleton of crayfish and of some other crustaceans living in fresh or brackish water is less permeable than that of marine forms. This reduces, but does not eliminate, the inward diffusion of water.

The internal fluids of freshwater clams and snails have extremely low salt concentrations, only about 0.16 percent, in comparison with about 1 percent in freshwater fish. Their nephridia produce a copious hypo-osmotic urine amounting to approximately 50 percent of the body weight per day, and the concentration of salts in the urine is about one half of that of the blood. Salts are actively absorbed across the body surface.

Most very small freshwater animals have no specialized excretory organs. The surface area is adequate to permit elimination of ammonium ions by diffusion. Nevertheless, they must osmoregulate. Freshwater sponges and protozoans, such as amoebas and *Paramecium*, possess water-pumping organelles called **contractile vacuoles** or **water expulsion vesicles**. The vacuole slowly accumulates water and, when full, collapses to the outside through a temporary opening. *Paramecium* possesses a contractile vacuole at each end of the body. The vacuoles fill through tiny radiating canals and empty through a canal to the outside. The rate of pulsation is governed by the rate at which water is taken in by diffusion and feeding.

Osmoregulation in Terrestrial Animals

The danger of desiccation is a major threat to animals living on land. Water is lost in urine and in feces, but evaporation is the principal route of water loss. At any given temperature, a given volume of air is capable of holding a certain amount of water vapor. The difference between the amount of water vapor actually present and the amount that would be if the air were saturated is known as the **saturation deficit.** The saturation deficit determines the steepness of the evaporation gradient between the animal and the external environment and varies with season, climate, habitat, and time of day, all of which affect the distribution and behavior of land animals. However, of any of these variables, the movement of air across an evaporating surface, sweeping away the more saturated overlying layers of air, is the single most important factor increasing desiccation.

The general body surface and the gas exchange surface are the two principal sites of evaporation. In many snakes, for example, 64 percent of the water loss by evaporation may occur across the skin and 36 percent through gas exchange in the lungs.

Water lost must be replaced with water gained by drinking and eating and, in some animals, by direct uptake of water vapor in the surrounding air.

Adaptive Strategies. The many different groups of land animals, such as vertebrates, spiders, insects, and snails, are representatives of separate invasions of the terrestrial environment. Within each of these groups various adaptations have evolved that increase the likelihood of maintaining a balance between water lost and gained. The principal adaptive strategies are

1. Reduction of evaporation by:
   A. Internal gas exchange organs. The gas exchange organs of most land animals are located internally or at least within a protective covering where the thin, moist exchange surface is less likely to dry up, and there is less water lost by evaporation.
   B. Modification of the integumental barrier. Loss of water by evaporation across the general body surface is reduced by various modifications of the skin that make the integumental barrier more impervious to the outward passage of water.
   C. Occupation of humid habitats. Subterranean burrows and spaces beneath stones and logs and in leaf mold, where saturation deficits are lower than in exposed locations, are less stressful environments for maintaining water balance.
   D. Nocturnal activity. Many species move out of protective retreats only at night, when the danger of desiccation is greatly reduced.
2. Reduction of water loss from excretion by:
   A. Production of insoluble nitrogenous wastes. Some land animals excrete their nitrogenous wastes in the form of uric acid or guanine, which have very low solubility and thus require little water for removal.
   B. Production of a hyperosmotic urine. Mammals, birds, and insects are able to excrete urine with a salt concentration greater than that of the blood; such a concentrated urine conserves water.
   C. Low urine output. Some terrestrial animals, such as many land crabs and pill bugs, conserve excretory water by greatly reducing urine flow and utilizing other avenues for the elimination of nitrogenous wastes.
3. Reduction of water loss from egestion. Some animals, especially those living in deserts, conserve water by defecating relatively dry feces.
4. Toleration of internal water loss. Most animals can tolerate relatively limited fluctuations in the levels of internal water. For example, if a human loses more than 12 percent of his body water, death may result. But a few animals can tolerate the loss of more than half of their body water. This can be replaced later when water is again available.
5. Utilizing the water of oxidation. Significant amounts of water are liberated in various reactions of cell respiration. The oxidation of 100 g. of glucose, for example, yields 60 g. of water. A few desert animals survive entirely on this water source.

Earthworms are poorly adapted land animals and are only active within their burrows in moist soil or at the surface at night. When the soil becomes dry, they move to deeper levels, where they may lose water and become dormant. Earthworms excrete ammonia and urea, and the nephridia can, in the event of excess water, produce a hypo-osmotic urine.

Land snails and slugs are also subject to great evaporative water loss across the skin. Moreover, additional water is lost in the secretion of the mucous trail over which they crawl. However, these animals are able to tolerate considerable desiccation. The European land snail *Helix* can survive a water loss equivalent to 50 percent of its body weight, and the slug *Limax* has been claimed to survive an 80 percent loss. They excrete uric acid, which conserves some water. Nevertheless, snails and slugs are confined to humid habitats or to exposure only at night, when saturation deficits are low.

Spiders, insects, and pill bugs are all small terrestrial arthropods, each representative of a different past invasion of land. Spiders and insects are highly successful terrestrial animals, as their great numbers attest. Contributing to their success has been the evolution of a waxy layer on the surface of the exoskeleton. When exposed to the same saturation deficit, their evaporation per square centimeter is comparable to that of mammals and is only a fraction of that of snails and slugs. The internal tracheal tubules, which are the gas exchange organs of insects as well as of many spiders, reduce respiratory evaporation. Insects also conserve water by excreting uric acid and producing a hyperosmotic urine via the malpighian tubules and rectum. In fact, insects are the only animals other than birds and mammals that produce a hyperosmotic urine. Spiders excrete guanine, which has relatively low solubility in water. Nevertheless, very small spiders, possessing a large surface area in relation to volume, are confined to leaf mold and other more humid habitats.

Pill bugs are not insects but crustaceans, relatives of crabs and shrimps (p. 594). They are mostly nocturnal animals, living beneath stones and wood and in leaf mold. In contrast to spiders and insects, they lack a waxy outer covering and are subject to much greater water loss. They are ammonotelic, like most aquatic animals, but are remarkable in that they excrete gaseous ammonia rather than ammonium ions. The gills are the principal sites of excretion, and urine output from the green glands is very small.

Frogs and toads suffer the highest rate of evaporative water loss of all terrestrial vertebrates, a rate far higher than that of insects and spiders (Table 9.1). In spite of the warty skin of the toad, it loses water about as rapidly as the frog. Nocturnal habits, especially of toads, and protective environments are the principal defences of these animals against desiccation. Both excrete urea, which is less toxic than ammonia but very soluble.

Most reptiles are much better adapted for life on land than amphibians. The skin of most snakes, for example, is more resistant to water loss from evaporation, and its nitrogenous wastes are excreted as uric acid. Uricotelism probably first evolved in reptiles as an adaptation for reproduction on land, since it permits accumulation within the egg of nitrogenous wastes produced by the developing embryo.

Birds excrete uric acid. As in reptiles, uric acid is important in conserving water. The integument of the bird, with its covering of feathers, reduces water loss, and the avian kidneys can produce a hyperosmotic urine.

Mammals are also well adapted for life on land. The skin is an effective barrier against evaporation. Mammals excrete urea, but most mammals live where sources of drinking water and water in food are adequate. Moreover, the mammalian kidney can produce a hyperosmotic urine. Human beings can produce a urine that contains as much as 6 percent urea and 2.2 percent salts, providing considerable conservation of water. On the other hand, mammals can also produce a dilute (hypo-osmotic) urine when it is beneficial.

Kangaroo rats, mammals abundant in southwestern North America, have been studied more extensively than any other desert animal. These small rodents live in subterranean burrows and come to the surface only at night. They rarely drink since standing water is almost never available. They eat only dry food, mostly seeds, and rely almost entirely on the water produced in the oxidation of food. There are no sweat glands. The feces contain little moisture, and their very long loops of Henle can produce a highly concentrated urine, containing as much as 23 percent urea and 7 percent salt. Kangaroo rats could drink sea water!

TABLE 9.1 Comparison of Water Loss by Evaporation from the Body Surface of Different Animals

	Evaporation* (μg.)
Earthworm	400
Garden snail, active	870
Garden snail, inactive	39
Salamander	600
Frog	300
Human (not sweating)	48
Rat	46
Water snake	41
Pond turtle	24
Box turtle	11
Iguana lizard	10
Gopher snake	9
Desert tortoise	3
Cockroach	49
Flour mite	2
Tick	0.8

*Micrograms per cm.[2] body surface per hour per mm. Hg saturation deficit. (After Schmidt-Nielsen, K.: The neglected interface: the biology of water as a liquid gas system. *Q. Rev. Biophys.*, 2:283, 1969.)

Camels are unable to store water (the hump contains stored fat) but can tolerate great water loss. They can drink an enormous amount of water at one time, as much as one third the body weight in 10 minutes. There then may follow a gradual loss that can take the animal through a prolonged period without water. Camels have sweat glands, but sweating commences at a higher internal body temperature than in other mammals. Moreover, camels begin the day with a lower early morning temperature than do other mammals.

Summary

1. The extracellular body fluids (water plus solutes) of an animal are in osmotic equilibrium with intracellular fluids and are maintained, or regulated, at a relatively constant level, regardless of external environmental conditions.

2. The regulation of internal body fluids may be complicated by the excretion of nitrogenous wastes that require some water for their removal. Most nitrogenous waste products of animals result from the deamination of amino acids utilized as cellular fuels. Ammonia, the highly soluble and toxic waste product of deamination, is eliminated directly where sufficient water for dilution is available. Many animals convert ammonia to less toxic urea or uric acid, but this conversion requires energy.

3. Small aquatic animals eliminate their nitrogenous waste (ammonia) by diffusion across the body surface. Larger animals possess tubular or saccular excretory organs that concentrate wastes. The concentration process typically involves modification of coelomic fluid or blood filtrate through selective reabsorption and secretion. The principal excretory organs are nephridia (mollusks and annelids), green glands (shrimps, crayfish, crabs), malpighian tubules (insects, spiders), and kidneys (vertebrates).

4. In the mammalian kidney each tubular nephron, the structural and functional unit of the kidney, receives filtrate from a knot of capillaries, the glomerulus. The filtrate, essentially like plasma minus the large proteins, is received in a capsule (Bowman's capsule) and then passes through the tubule, where glucose and other organic nutrients, some salts, and over 90 percent of the water are reabsorbed. The kidneys of mammals can produce a urine that is hypo- or hyperosmotic to the blood, depending upon internal fluid levels. A hyperosmotic urine is produced by a hairpin loop of the tubule (loop of Henle) that makes possible a gradient of solute concentration and the extraction of a large amount of water from the fluid in the counterflowing collecting tubule. Loops of Henle are found only in the kidneys of birds and mammals; other vertebrates (fish, amphibians, and reptiles) cannot produce a urine with a salt concentration greater than that of the blood.

5. The internal body fluids of most marine animals are in osmotic equilibrium with the surrounding seawater, which in the open ocean contains about 3.5 percent salt. They are osmoconformers, for any decrease in environmental salts results in an equivalent decrease in the salts of their internal body fluids. Marine vertebrates have internal body fluids with a salt concentration of only about one third that of seawater; they therefore live in a desiccating environment and must be osmoregulators. They drink seawater or eat salty food but can eliminate excess salt by secretion through the gills (fish) or head glands (sea turtles and birds) or by excreting a urine saltier than seawater (whales, porpoises, and seals).

6. Freshwater animals and marine animals living in brackish estuaries are osmoregulators. They live in an environment with a salt content less than that in their internal body fluids and thus must hold on to salts and get rid of excess water. They excrete a hypo-osmotic urine and may replace lost salts by absorption through certain body surfaces, such as the gills (fish).

7. Desiccation, especially evaporation from the skin and gas exchange organs, is the principal problem in the regulation of internal body fluids in terrestrial animals. A variety of adaptive strategies have evolved that aid in balancing the limitations of water gained with water lost. Many species avoid desiccating environmental conditions by living in humid habitats or by nocturnal activity. Evaporative water loss is reduced in arachnids (spiders, scorpions, mites), insects, reptiles, birds, and mammals by modifications of the integumental barrier and in all terrestrial animals by internal gas exchange organs.

8. A urine hyperosmotic to the blood can be produced only by insects, birds, and mammals. Terrestrial snails, insects, reptiles, and birds excrete uric acid, which is relatively insoluble and requires little water for elimination. Mammals excrete urea, which although very soluble, is less toxic than ammonia. A few animals, such as some land snails, slugs, and camels, have become adapted to survive great water loss for a short time.

References and Selected Readings

Many of the references listed at the end of Chapter 4 provide excellent coverage of the topic of regulation of internal body fluids and related subjects.

Bentley, P. J.: *Endocrines and Osmoregulation: A Comparative Account of the Regulation of Water and Salt in Vertebrates.* New York, Springer-Verlag, 1971.

Fertig, D.S., and V. W. Edmounds: The physiology of the house mouse. *Scientific American, 221*(Oct.):103–110, 1969. An account of adaptations of the common house mouse, many of which are shared with desert rodents.

Schmidt-Nielsen, K.: The physiology of the camel. *Scientific American, 201* (Dec.): 140–151, 1959. An interesting account of the adaptations of camels for life in deserts.

10 Nervous Systems and Neural Integration

ORIENTATION

Nerve cells, or neurons, are highly adapted to transmit waves of excitation along their length, a process that enables animals to integrate their various internal functions and to adjust to changes in the external environment. In this chapter we will describe the structural features of neurons and how they function in the transmission of nerve impulses. We will discuss current theories of the nature of the nerve impulse and how it is transmitted along a neuron. Each neuron is connected to other neurons at synapses, which serve as valves; impulses cross the synapse by specific chemicals, neurotransmitters, secreted at the tip of one axon and taken up by receptors in the adjacent neuron. We will then consider how neurons are assembled into the remarkably complex structures of the central and peripheral nervous systems of the vertebrates and how this system evolved.

Irritability and Response

A fundamental property of all cells is response to stimuli. Cells are said to be irritable, and in response to a stimulus, waves of excitation are conducted along their surfaces. Such waves of excitation are even conducted, although very slowly, by eggs and plant cells. The nerve cell, or **neuron**, represents an evolutionary adaptation for the rapid transmission of a wave of excitation. Neurons are found in all multicellular animals except sponges and collectively constitute the **nervous system**. This system makes possible the rapid integration of internal functions and adjustments to changes in the external environment. The evolution of the nervous system is closely connected with the evolution of animal locomotion. Indeed, heterotrophic nutrition, muscular movement, and nervous integration are three distinct and interrelated features of animal design.

The Neuron

The structural and functional unit of the nervous system of all multicellular animals is the neuron (Fig. 10.1). The average neuron is slightly less than 0.1 mm. in diameter, but it may be several meters long. Traditionally, it is pictured as having three parts—the axon, the cell body, and the dendrite; a long **axon** emerges from one end of the cell body, and bushy **dendrites** emerge from the other. However, there are so many exceptions to this generalization that the usefulness of this anatomical

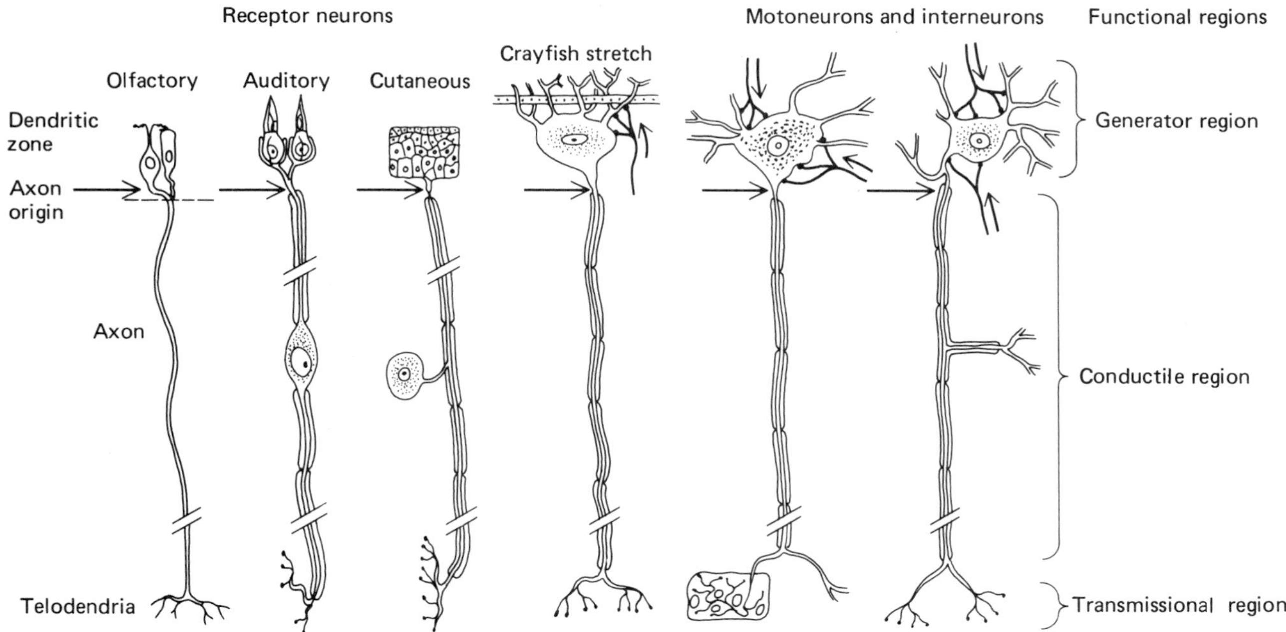

FIGURE 10.1 Diagram of a variety of receptor and effector neurons, arranged to illustrate the idea that impulse origin, rather than cell-body position, is the most reasonable focal point for the analysis of neuron structure in functional terms. Thus, the impulse conductor, or axon, may arise from any response-generator structure, whether transducing receptor terminals or synapse-bearing surfaces (dendrites, cell body surface, or axon hillock). The interior of the cell body (chromidial neuroplasm, or perikaryon) is conceived of as related primarily to the outgrowth of axon and dendrites and to metabolic functions other than membrane activity. Thus, the position of the perikaryon in the neuron is not critical with respect to the "neural aspects" of neuron function—namely, response generation, conduction, and synaptic transmission. Except for the stretch neuron of the crayfish, the neurons shown are those of vertebrates. (Redrawn from Bodian, D.: The generalized vertebrate neuron. *Science, 137*:323–326, 1972.)

subdivision is rather limited. Functional distinctions are more accurate, for there are three functional parts of the neuron. The dendrites constitute the part of the neuron specialized for receiving excitation, whether from environmental stimuli or from another cell. The axon is specialized to distribute or conduct excitation away from the dendritic zone. It is generally long and smooth but may give off an occasional collateral. Each axon ends in a distribution or emissive apparatus termed the **telodendria.** The **cell body**, which contains the nucleus, is concerned with metabolic maintenance and growth of the cell and may be situated anywhere with respect to the other parts.

The neurons of higher animals are specialized into three functionally distinct regions. The **generator region** is usually restricted to the dendrites and is characterized by specializations of the neuronal membrane that permit it to be excited by neurotransmitters from an adjacent neuron. The **conductile region** makes up most of the axon of the neuron. The **transmissional region** is composed of the tips of the axon and is specialized for the secretion of neurotransmitters, such as acetylcholine and norepinephrine. The cell body, or soma, usually does not participate in electrical activity. When a nerve is stimulated artificially by an electrical stimulus, the resulting impulse is propagated along the axon (the conductile region) in both directions. The action potential propagating toward the transmissional region is termed **orthodromic**, and the impulse transmitted towards the dendrite is termed the **antidromic** action potential. Under normal conditions a nerve impulse always arises at the portion of the conductile region adjacent to the generator region and is conducted orthodromically to the transmissional region.

The generator regions of receptor neurons (p. 240) are activated by a variety of mechanical, chemical, thermal, and other stimuli, and the generator regions of interneurons and motor neurons are stimulated through synaptic transmission by a neurotransmitter secreted by the adjacent neuron. In response to the neurotransmitter, the generator region may set up one or more propagated action potentials or nerve impulses in the conductile region.

The generator region of the neuron acts as an amplifier, generating an action potential greater than the stimulus. The generator region may also act as a transducer, converting one type of energy state or flow into a different type.

The idea that a nerve fiber is simply a sort of cytoplasmic telephone wire has rapidly changed; we have come to appreciate that neurons are very dynamic and metabolically active cells. These cells carry out the intracellular transport of materials over long distances by the active cytoplasmic streaming and undulations of the axons. Neurons may be true secretory cells, producing and secreting neuro-hormones, such as vasopressin, oxytocin, and the hypothalamic-releasing factors. The cell body contains an abundant accumulation of endoplasmic reticulum and ribosomes, detected by light microsopy in the nineteenth century and termed **Nissl substance.** The Golgi apparatus of neurons is also abundant and well developed. The cytoplasm of the axon contains longitudinally oriented **neurofibrils** and **neurotubules** that are associated with the intracellular transport of molecules.

It is the axon that is usually responsible for the tremendous length of a neuron. The axon of a sensory cell, barely 0.1 mm. in diameter, located in the toe of a giraffe traverses a distance of several meters before ending in the spinal cord. The bundling together of many axons makes up the nerves and nerve trunks seen in an anatomical dissection. A common connective tissue sheath surrounds the nerves.

Neurons are classified as **sensory** neurons, **motor** neurons, or **interneurons** on the basis of their function. Sensory (**afferent**) neurons conduct information from the receptor endings. Motor (**efferent**) neurons conduct information to the effectors, such as muscles, glands, electric organs, and light organs. Interneurons connect two or more neurons and usually lie entirely within the central nervous system. In contrast, sensory and motor neurons typically have one of their endings in the central nervous system and the other close to the animal's external or internal environment.

The cell bodies of neurons are typically grouped together in masses called **ganglia**. In the simplest sense, a ganglion is any aggregation of neural cell bodies. The **dorsal root ganglia** of vertebrates (Fig. 10.2) are collections of cell bodies of sensory neurons, and the vertebrate **autonomic ganglia** are groups of motor neuron cell bodies. However, the aggregations of nerve cell bodies *within* the central nervous system of vertebrates are termed **nuclei**.

Enveloping the axon outside the central nervous system is a cellular sheath, the **neurilemma**, composed of Schwann cells. These cells, migrating in from the mesenchyme, line up along axons and wrap around them. On some axons the Schwann cell lays down within its folds a spiral wrapping of insulating fatty material called **myelin** (Gr., *myelos*, marrow) (Figs. 10.3 and 10.4). At the gaps, or **nodes of Ranvier**, between adjacent Schwann cells the axon is free of myelin. Axons lying within the brain and spinal cord have no neurilemma sheaths, and their myelin is provided by satellite cells (oligodendrocytes) rather than by Schwann cells. Groups of neurons consisting of heavily myelinated fibers, such as those in

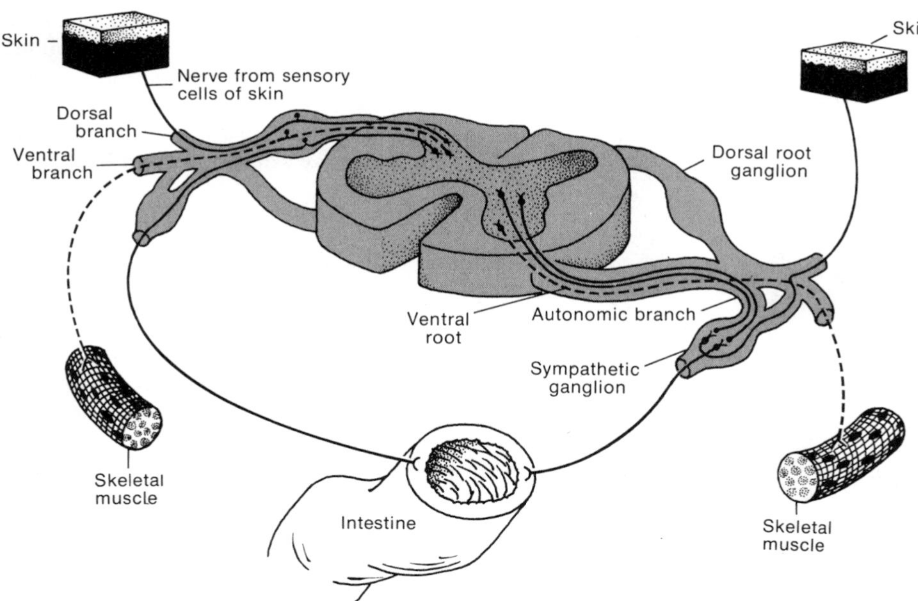

FIGURE 10.2 Diagram of the primary types of sensory and motor neurons of the spinal nerves and their connections with the spinal cord. For convenience, the sensory neurons are shown on the left and the motor neurons on the right, though both kinds are found on each side of the body.

the brain and spinal cord and those to skin and skeletal muscle, are white in appearance. Those with little or no myelin are gray.

The principal function of the Schwann cells and myelin sheath is to provide for saltatory conduction (p. 218), but the myelin may have other functions as well. It is affected by the neuron it surrounds, for when the distant nerve cell body is cut off from its axon, the myelin surrounding the axon begins degenerating within a few minutes. The neuron thus has some trophic function that is necessary for the well-being of the Schwann cell.

There is also an interdependence between the neuron and its sheath cells. An axon separated from its cell body by a cut soon degenerates. A hollow tube of

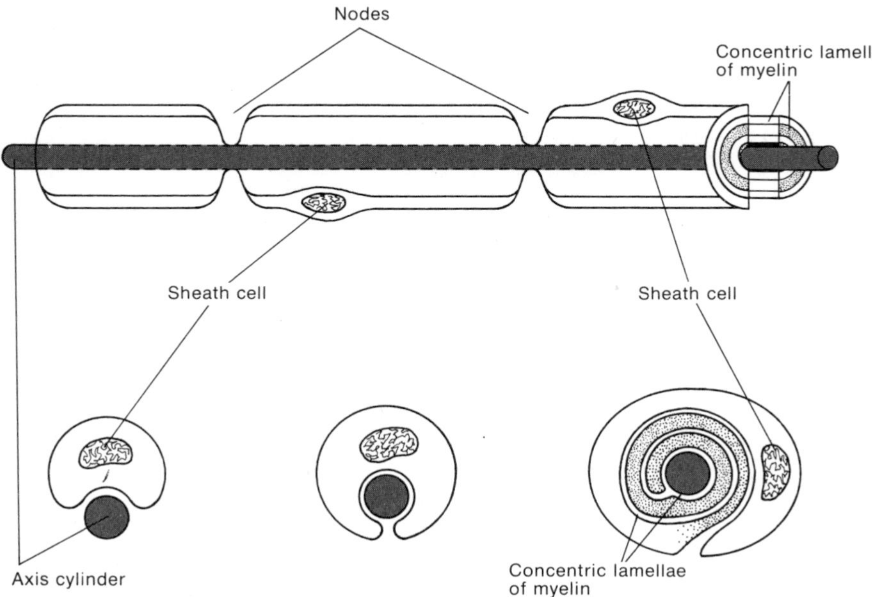

FIGURE 10.3 Sheath cells on neuron. *Upper,* Dissection of a myelinated nerve fiber. *Lower,* Envelopment of axis cylinder by a sheath cell. (After Geren, B. B., 1954; from Ballard, W. W.: *Comparative Anatomy and Embryology.* New York, The Ronald Press Co., 1967.)

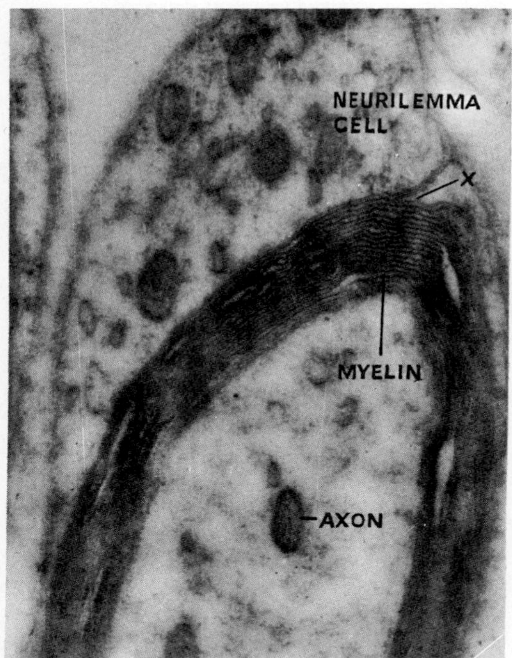

FIGURE 10.4 Electron micrograph of sciatic nerve fiber of a seven-day-old mouse showing the development of the myelin sheath by the folding of the cell walls of the neurilemma. The spiral infolding of the cell membrane of the neurilemma cell is visible as it is forming the thick, compact, many-layered myelin sheath. Note that at *X* the layers of myelin are continuous with the cell membrane of the neurilemma. Magnification ×83,000. (Courtesy of Dr. Betty G. Uzman.)

Schwann cells remains, but the myelin eventually disappears. As long as the cell body of the neuron has not been injured, it is capable of regenerating a new axon. Sprouting begins within a few days after the cut. The growing axon enters the old sheath tube and proceeds along it to its final destination in the central nervous system or periphery. The length of time required for regeneration depends on how far the nerve has to grow and may require as much as two years. Regeneration following a cut within the spinal cord or brain is very feeble and may be totally absent.

It is a remarkable fact that each regenerating axon of a cut nerve finds its way back to its former point of connection, whether this be a specific muscle or sense organ in the periphery.

It is now clear that the nervous system has, in addition to its role of transmitting impulses, important trophic relations with all the organs it innervates. The presence of nerves is essential for the regeneration of amputated amphibian limbs, for the normal maintenance of taste buds, and for the continued functional integrity of muscles. Something other than impulses is transported along the long cellular extensions of the neurons, and it has been demonstrated that there is active flow of cytoplasm away from the cell body.

The Nerve Impulse

The study of the nature of the nerve impulse has been fraught with special difficulties because nothing visible occurs when an impulse passes along a nerve. Only with the development of microchemical techniques was it possible to show that the nerve fiber expends more energy, consumes more oxygen, and gives off more carbon dioxide and heat when an impulse is transmitted than it does in the resting state. This indicates that metabolic, energy-requiring processes are involved in the conduction of an impulse or in the recovery of a nerve after conduction, or both. The transmission of a nerve impulse obeys the "all-or-none" law: The conduction of the impulse is independent of the nature or strength of the stimulus starting it, provided that the stimulus is strong enough to start an impulse. The energy for the conduction of the impulse comes from the nerve, not from the stimulus, so that although the speed of the conducted impulse is independent of the strength of the stimulus, it is affected by the state of the nerve fiber. Drugs or low temperature can retard or prevent the transmission of an impulse. The impulses transmitted by all types of neurons are essentially the same. That one impulse results in a sensation of light, another in a sensation of pain, and a third in the contraction of a muscle is a function of the way the nerve fibers are connected and not of any special property of the impulses.

According to the present **membrane theory** of nerve conduction, the electrical events in the nerve fiber are governed by the **differential permeability** of the neuronal membrane to sodium and potassium ions, and these permeabilities are regulated by the **electric field** across the membrane. The interaction of these two factors, differential permeability and electric field, leads to a requirement for a critical threshold of change in order for excitation to occur. Excitation is a regenerative release of electrical energy from the nerve membrane, and the propagation of this change along the fiber is the **action potential**, the brief all-or-none electrochemical depolarization of the membrane.

The resting neuron is a long, cylindrical tube whose plasma membrane separates two solutions of different chemical composition with the same total number of ions. In the external medium, sodium and chloride ions predominate; within the cell, potassium and various organic ions predominate. Potassium and chloride ions diffuse relatively freely across this membrane, but the permeability to sodium is low. Potassium tends to leak out of the neuron and sodium tends to leak in, but because of the selective permeability of the membrane, potassium tends to leak out faster than sodium leaks in. This, plus the fact that the negatively charged organic ions within the cell cannot get out, creates an increasingly negative charge on the inside of the membrane. As the inside becomes more negative in relation to the outside, the exit of potassium ions is impeded. Ionic conditions would eventually change and come to a new equilibrium if something were not done to counteract this leakage of ions. The steady state is maintained by the **sodium pump,** which actively transports sodium ions from the inside to the outside against a concentration and electrochemical gradient. The sodium pump requires energy, utilizing ATP derived from metabolic processes within the nerve cell. The differential distribution of ions on the two sides of the membrane results in a potential difference of 60 to 90 millivolts across the membrane, the **resting membrane potential,** with the interior being negatively charged with respect to the exterior (Fig. 10.5).

The extrusion of sodium ions is accompanied by the entrance of potassium ions; there appears to be an exchange of cations at the cell surface, with a potassium ion entering for each sodium ion extruded. The relatively low ionic permeability of the membrane is such that even when the pump is poisoned with cyanide, many hours elapse before the concentration gradients of sodium and potassium across the membrane disappear.

Electrical studies of the **cable properties** of the nerve fiber show that the axon could hardly serve as a passive transmission line because its cable losses are enormous. When a weak signal is applied to the fiber, one too small to excite its usual relay mechanism, the signal fades out within a few millimeters of its origin. The nerve impulse could not be propagated over the long distances in the nerve unless there were some process to boost the signal repeatedly.

Although the permeability of the membrane to sodium is very low at the usual resting membrane potential, it increases as the membrane potential decreases. This permits the leakage of sodium ions down an electrochemical gradient into the interior of the nerve. This further decreases the membrane potential and further increases the permeability to sodium. The process thus is self-reinforcing and progressive, resulting in the upward deflection of the action potential. The entering sodium ions drop the transmembrane potential to zero and beyond, to about −40 or −50 millivolts. After one or

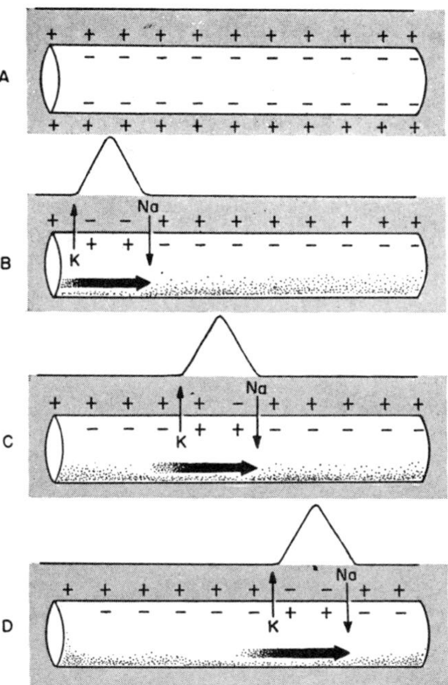

FIGURE 10.5 Diagram illustrating the membrane theory of nerve transmission. *A,* Resting nerve, showing the polarization of the membrane with positive charges on the outside and negative charges on the inside. *B–D,* Successive stages in the conduction of a nerve impulse, showing the wave of depolarization of the membrane and the accompanying action potential propagated along the nerve.

two milliseconds, the permeability of sodium decreases and potassium begins to move out. This movement leads to the restoration of the resting potential; that is, to the repolarization in the membrane. When the membrane is completely repolarized, the permeability to potassium becomes normal, and the excess sodium that entered during the process is slowly removed by the sodium pump.

The actual quantities of ions that move in and out during the passage of an action potential are so small that there is no detectable change in the concentration of either ion in the fiber during an impulse.

Because of the changes in permeability that accompany the depolarization of the nerve membrane, the fiber cannot immediately transmit a second impulse. This period of inexcitability, termed the **absolute refractory period,** is brief and lasts until normal permeability relations have been restored. *A nerve impulse, then, is a wave of depolarization that passes along the nerve fiber. The change in membrane potential in one region renders the adjacent region more permeable, and the wave of de-*

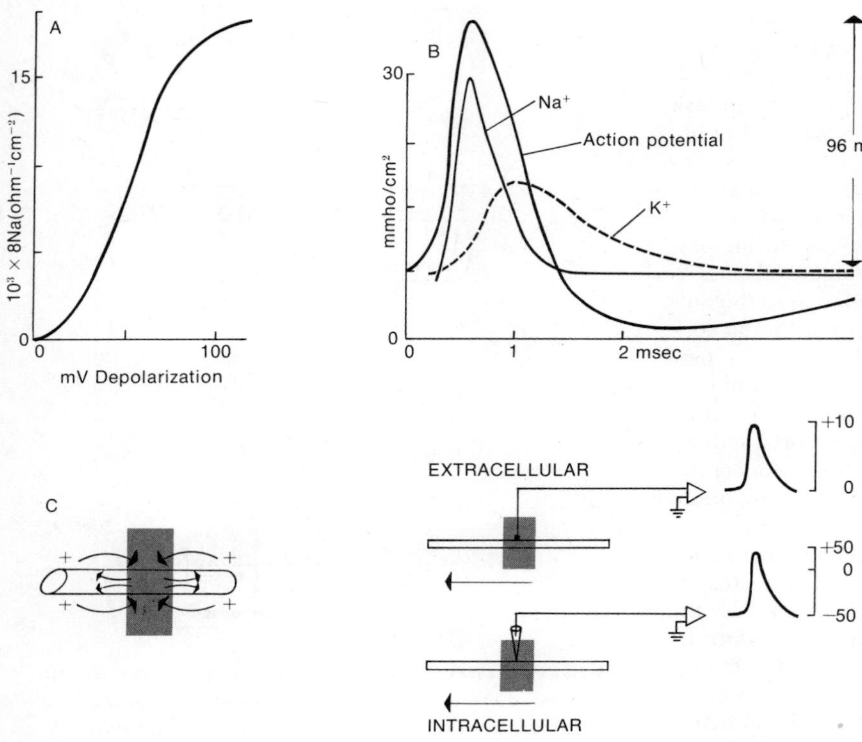

FIGURE 10.6 Generation of the nerve impulse. *A,* Increase in Na^+ permeability as membrane depolarizes. *B,* Na^+ and K^+ movement across membrane during action potential. *C,* Diagram illustrating flow of electronic current in vicinity of action potential (shaded) and examples of extracellular and intracellular recording of action potentials. With a penetrating microelectrode the entire action potential is recorded, whereas with an extracellular electrode only the positive overshoot is detected. (*A* and *B* adapted from Katz, B.: The Croonian lecture: the transmission of impulses from nerve to muscle, and the subcellular unit of synaptic action. *Proceedings of the Royal Society of Biology, 155:*455, 1962.)

polarization is transmitted along the fiber. The entire cycle of depolarization and repolarization requires only a few milliseconds.

At any point where an action potential has been generated, **an electrotonic** current flows ahead within the neuron, through the membrane, and returns to the outer surface (Fig. 10.6C). The circuit is completed by current flowing in the solution that bathes the nerve. The effectiveness of electrotonic currents in propagating the impulse depends upon the magnitude of the current and the resistance of the neuronal membranes, the cytoplasm, and the surrounding medium. These factors determine at what distance from the active sites the membrane permeability will be sufficiently increased to start the regenerative sodium entry. The concept that the action potential is propagated by electrotonic currents moving ahead of it is termed the **local circuit theory of propagation.**

The rate of conduction of impulses increases as the diameter of the axon increases because the internal resistance declines. As a result, large nerve fibers conduct impulses more rapidly than do small ones. Giant nerve fibers have evolved in various members of the Animal Kingdom, although not in the higher vertebrates. The giant axons of the squid, crayfish, and earthworm conduct impulses rapidly and have been studied intensively. They generally serve the purpose of conducting danger signals, for which speed rather than detailed information is critical.

In the vertebrates, a high rate of conduction is achieved by a different evolutionary development. The myelin sheath surrounding the neuron insulates and prevents the flow of current between the fluid external to the sheath and the fluid within the axon. At the **nodes of Ranvier** there are gaps in the sheath where there is no insulation. At these points free ionic communication between the inside and outside of the membrane is possible. Impulses are generated only at the nodes, and nerve impulses leap from one node to the next (Fig. 10.7). This type of transmission, termed **saltatory conduction,** enables a myelinated nerve that is only a few micrometers in diameter to conduct impulses at velocities of up to 100 m. per second, whereas the largest unmyelinated axons, nearly 1 mm. in diameter, conduct with velocities of 20 to 50 m. per second.

Transmission at the Synapse

The nervous systems of most animals are composed of individual discontinuous neurons. This requires some mechanism for the transfer of the neural message from the axon of one neuron to the dendrite or cell body of

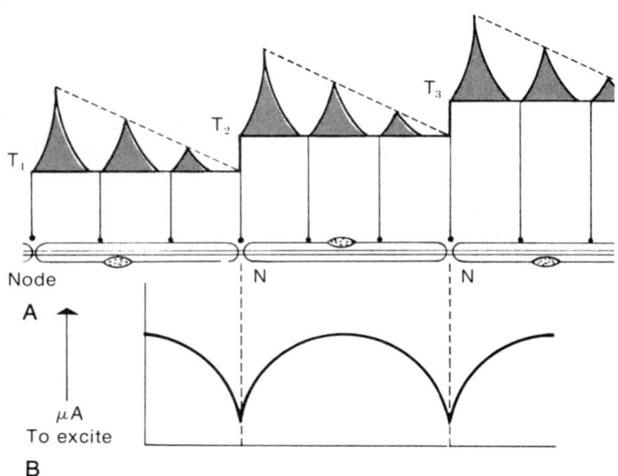

FIGURE 10.7 Saltatory conduction in a myelinated axon. *A,* Recording at and between successive nodes of Ranvier demonstrates electrotonic transmission between nodes, as impulse appears instantaneously at all points within a node while diminishing in magnitude. At the node, a time delay occurs, and the impulse regains initial magnitude, showing the node to be the site of the active impulse-generating process. *B,* Support for the idea that electrotonic current flowing from a previous active site exits and excites at the next node comes from a demonstration that current (μA = microamperes) required to excite the axon is least at the nodes. (From Case, J.: *Sensory Mechanisms.* New York, The Macmillan Co., 1966.)

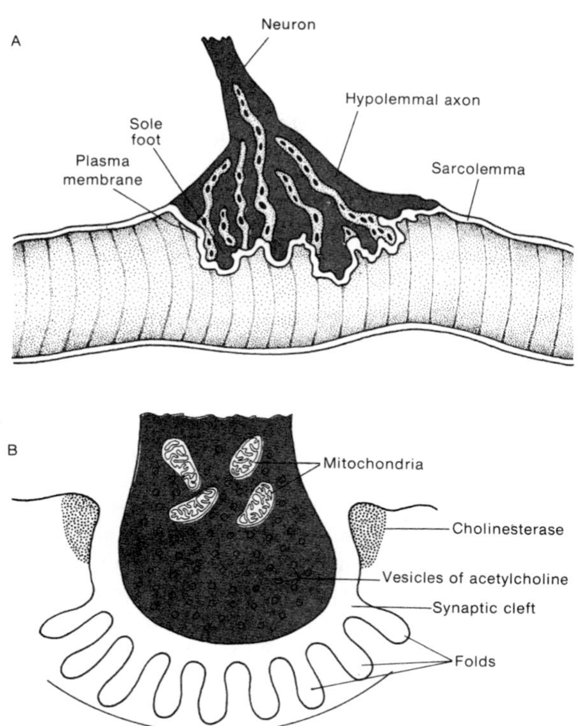

FIGURE 10.8 Diagram of a neuromuscular junction showing *(below)* the vesicles of acetylcholine in the sole foot of the neuron that penetrates into the membrane of the muscle fiber. This membrane becomes extensively folded to form the "synaptic gutter" in the plasma membrane of the muscle cell. The transmission of a nerve impulse from one neuron to a second neuron across a synapse is believed to involve a similar structural and functional junction.

the next or, at the neuromuscular junction, to the muscle. The junction between the axon of one neuron and the dendrite of the next is termed a *synapse* (Gr., *synapsis* conjunction). The generator region of most neurons has a great many synapses—as many as several thousand—with many different neurons.

At some specialized synapses, transmission is accomplished electrically. The arrival of the action potential at the end of the axon of the presynaptic neuron sets up electric currents in the external fluid of the synaptic gap. These currents, in turn, stimulate the dendrite of the postsynaptic cell to generate an action potential. Thus the transmission and the generation of the action potential take place in the postsynaptic cell in basically the same way as in a single nerve fiber. Electrical synapses of this sort are found in parts of the nervous system of the crayfish and the fish.

At most synapses, however, a gap of some 20 nm. separates the two plasma membranes, and the impulse is transmitted across this gap by special chemical transmitters. A specific chemical is synthesized by the neuron and released from the tip of the axon (**neurosecretion**) when a nerve impulse reaches it. This diffuses across the synaptic gap and attaches to specific molecular sites in the dendrite (an example of **chemoreception**), producing a change in the membrane potential of the dendrite and cell body and leading to the initiation of a new nerve impulse. The chemical transmitted passes from axon to dendrite by simple diffusion. Over the short distance involved, diffusion is rapid enough to account for the speed of transmission observed at the synapse. Transmission at the neuromuscular junction and certain other synapses involves the secretion and chemoreception of acetylcholine. This potent stimulant causes a local depolarization of the membrane of the muscle cell, which sets up propagated impulses in the membrane and causes a contraction of the muscle fiber (Fig. 10.8). **Curare** prevents the transmission of impulses from nerve to muscle, specifically at this type of synapse, by combining with the receptors for acetylcholine and preventing their normal reaction with it.

The sympathetic postganglionic fibers accelerate the heart rate by releasing **norepinephrine.** Such fibers are termed **adrenergic,** whereas those that secrete acetylcholine are termed **cholinergic.** In the synaptic area are potent enzymes: (1) acetylcholinesterase, which specifically hydrolyzes and inactivates acetylcholine, and (2) monoamine oxidase, which oxidizes and inactivates norepinephrine. These enzymes prevent the continuous stimulation of a dendrite or muscle by the neurotransmitter material.

Acetylcholine is released by motor nerves in discrete tiny packets, each of which contains about 1000 molecules. The mechanism that releases acetylcholine requires calcium ions and is inhibited by magnesium ions. The transmitter substance is stored within the nerve endings in small intracellular structures that discharge their entire contents to the surface. Electron micrographs of the tips of the neurons at the synapse reveal masses of **synaptic vesicles** that appear to be the sites of storage of the neurotransmitters (Fig. 10.9). Thus, the arrival of the nerve impulse leads to the liberation of the contents of one or more of these vesicles into the synaptic space. How an action potential induces the release of a packet of acetylcholine molecules from its vesicle is still unknown, but the existence of these vesicles provides a satisfactory explanation for the polarity of the synapse; that is, for the fact that a nerve impulse will travel in one direction but not in the reverse between axon and dendrite. Norepinephrine is concentrated in the synaptic vesicles of adrenergic fibers and is released by the arrival of an action potential. Other neurons within the central nervous system may have other neurochemical transmitters, such as **serotonin** or **dopamine.**

Synapses are of great importance functionally because they are points at which the flow of impulses through the nervous system is regulated. Not every impulse reaching a synapse is transmitted to the next neuron. The synapses, by regulating the route of nerve impulses through the nervous system, determine the response of the organism to specific stimuli. Synapses are the "switches" of the nervous system.

Synaptic regulation occurs in four principal ways: by spatial summation, temporal summation, facilitation, and inhibition. The dendrites of most postsynaptic neurons receive hundreds, even thousands, of axon terminals. Since there is spread and overlap of depolarization produced by the transmitter substances delivered by different terminals, that delivered simultaneously by a number of different axon terminals can have a combined effect, i.e., can be summed. This spatial summation may be sufficient to reach the firing threshold of the postsynaptic neuron.

Neurons typically transmit impulses in rapid volleys. When such a volley reaches the synapse, the axon terminals make successive releases of transmitter sub-

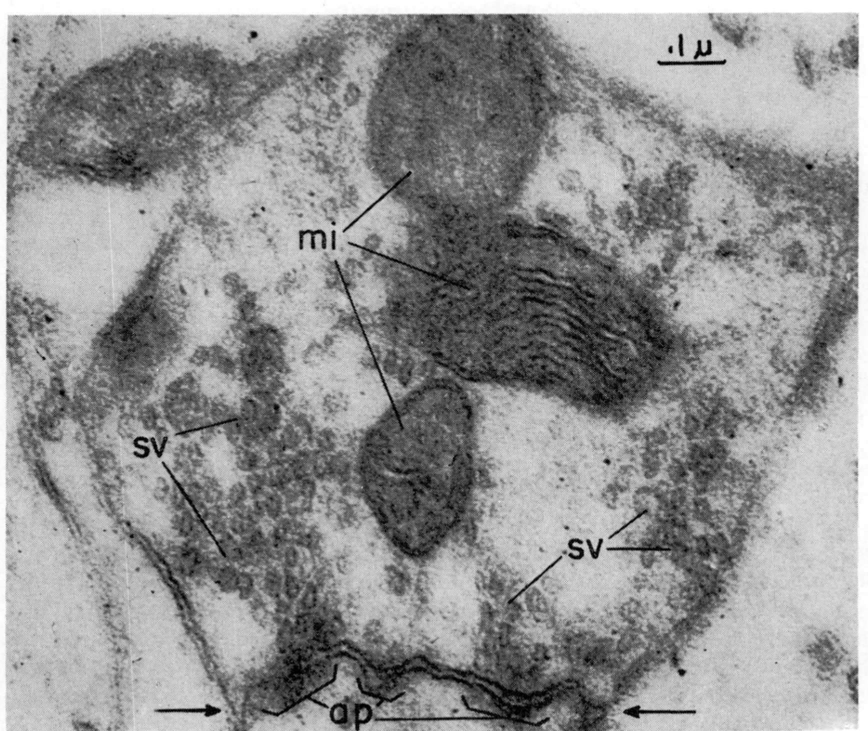

FIGURE 10.9 Electron micrograph of a synaptic ending in the stimulated olfactory bulb of the rat. The synaptic ending contains three mitochondria *(mi)* and several synaptic vesicles *(sv)*. The zone of contact between the two neurons is indicated by the two arrows. The nerve membranes appear to be thickened at "active points" *(ap)* in this zone of contact. (From DeRobertis, E., and A. Pellegrino De Iraldi: *Anatomical Record, 139:*299, 1961.)

stances. This occurs so rapidly that the second or third release occurs before the first has been erased. The accumulated transmitter substance (a summation in time) may be sufficient to reach the firing threshold of the postsynaptic neuron.

Transmitter substances, and each type of neuron produces only one type, bind with either excitatory or inhibitory receptor sites on the postsynaptic neuron. When an excitatory site is activated, there is a partial depolarization of the postsynaptic membrane. A single excitatory impulse crossing a synapse usually does not cause enough depolarization to affect the axon hillock and initiate an impulse in the postsynaptic neuron, but it does make it easier for a second or third impulse arriving soon after, or arriving on different presynaptic neurons, to cause enough depolarization of the postsynaptic neuron to trigger a nerve impulse. The first impulse **facilitates** the effect of later ones. If a transmitter substance binds with an inhibiting receptor site on the postsynaptic neuron, the postsynaptic membrane becomes hyperpolarized. This **inhibits** the activity of the neuron because it makes it more difficult for excitatory sites to depolarize the membrane. Changes in the membrane potential in the generator region of a neuron, like those in a receptor, are graded. A certain threshold of depolarization must be reached before an all-or-none wave of depolarization is initiated in the axon. Facilitation and inhibition are of prime importance in effecting the integration of body activities. Whether a postsynaptic neuron becomes active or not depends upon the interaction of the many excitatory or inhibitory impulses that may be impinging upon it.

Inhibition and facilitation can occur only in the synapse, since once an impulse starts along a neuron it can be neither stopped nor accelerated. Information transmitted along the axon as a spike potential is coded and decoded at the synapse by processes involving neurosecretion and chemoreception, graded thresholds, and the facilitation and summation of inhibitory and excitatory impulses. These are the substrates of animal behavior and the basis of the complex human processes of learning, memory, and intelligence.

Evolution and Organization of the Nervous System

The presence of a nervous system provides an animal with the means for receiving a variety of information from the outside environment and about its internal environment, for coordinating these bits of information and for responding appropriately. A well-developed nervous system is associated with motility and the increased contact with a changing environment that motility brings about. All animals but sponges have a nervous system, and the range of behavior of an animal is roughly proportional to the complexity of its nervous system.

Nervous systems function as transducers and amplifiers; they convert one form of energy into another; they can take in a small signal and put out a much larger one. The central nervous system of the higher animal operates as an analytical computer.

Nerve Net Systems. In the primitive nerve net systems of cnidarians and some flatworms, the tips of the neurons are undifferentiated as to axon and dendrite, and excitation can be transmitted across the synapse in either direction. Thus each end of the neuron can both secrete a neurotransmitter and receive excitation from the adjacent neuron by its chemoreceptors.

The cnidarian system includes sensory neurons, motor neurons, and interneurons; there are some connections from sensory to motor neurons, but usually the connections are via interneurons. As a result of this nerve net arrangement, the conduction of impulses is generally diffuse, and excitation tends to spread in all directions from the initial point. However, simple cnidarian nerve nets permit response patterns and can give rise to some coordinated behavior.

Higher Nervous Systems. A nervous system of the nerve net type with diffuse conduction is sufficient to meet the needs of sessile, radially symmetrical animals, such as the cnidarians, but not to meet the needs of motile, bilaterally symmetrical animals. Several trends in the evolution of the nervous system are evident: the appearance of synapses that permit transmission of impulses in one direction only to minimize or eliminate the diffuse conduction of the nerve net system; the aggregation of nerve processes to form one or more longitudinal bundles of nerve fibers; a concentration of cell bodies into groups (ganglia); and the further concentration of ganglia at the anterior end of the body (cephalization), correlated with the appearance of sense organs, to form a brain. The brain and longitudinal cords make up the central nervous system characteristic of the higher animals. Smaller bundles of nerves from the central system to the distal regions of the body constitute the peripheral nervous system. Thus nerves are composed of bundles of fibers of sensory or motor neurons, and ganglia are clusters of nerve cell bodies.

The most primitive flatworms have an epidermal nerve net system similar to that of the cnidarians. Most flatworms have a nervous system that is sunk in beneath the muscle layer and organized in four or five pairs of cords arranged radially (Fig. 10.10). An aggregation of nervous tissue surrounds the statocyst at the anterior end of the body, the first hint, perhaps, of a "brain." In

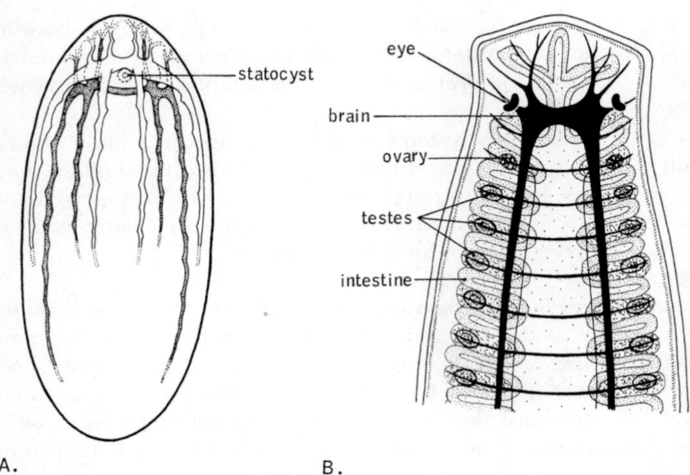

FIGURE 10.10 Turbellarian nervous systems. *A,* Radial arrangement of nerves in the acoel *Anaperus. B,* Anterior end of the nervous system of the planarian *Procerodes.* (*A* After Westblad; *B* modified after Lang.)

other flatworms and certain related groups of animals, there is a tendency for the number of pairs of nerve cords to be reduced. The flatworm nervous system lacks well-developed ganglia.

A brain can be defined functionally as a site of sensory integration and motor command. The brain of animals is generally anterior because this is the part of the body that meets the environment first and contains the greatest concentration of receptors and sense organs. Among higher animals two types of nervous systems are common: the dorsal, hollow, single nerve cord found in chordates and the ventral, solid, usually paired nerve cords with segmental ganglia characteristic of annelids and arthropods.

The anterior brain and ventral nerve cord of an earthworm, for example, are composed of neurons, with axons and dendrites arranged in definite nerve cords and fibers (Fig. 26.9). Its system is differentiated into central and peripheral nervous systems with sensory neurons, motor neurons, and interneurons joined by synapses, so that nerve impulses travel in one direction only. This enables the central nervous system to act as an integrator, selecting certain incoming sensory impulses and passing them to effectors while inhibiting or suppressing others. The ventral nerve cord extending the entire length of the body enables separate segments to move in a coordinated fashion. Each segment has a pair of ganglia, a collection of nerve cell bodies.

Within the ganglia the cell bodies are located peripherally, making up the **rind,** and the processes forming synaptic connections are located centrally, forming the **neuropile**. There are also axons of neurons that extend through the ganglia to more anterior or posterior levels. The cords of many invertebrate animals contain giant axons, that provide for very rapid conduction of impulses involved in escape reflexes, such as pulling back into a tube (some worms) or darting backwards (crayfish). Earthworms (*Lumbricus*) have three giant axons running through the upper middle part of the cord that are involved in rapid contraction for escape back into the burrow. The most anterior part of the cord is an enlarged ganglion, a "brain" that sends impulses (motor commands) down the cord to coordinate movements. After the brain has been removed, the earthworm can move almost as well as before, but it persists in futile efforts to go ahead instead of turning aside when it comes to an obstacle. The brain, therefore, appears to be necessary for adaptive movements. The nervous systems of arthropods and mollusks are generally similar to that of the earthworm. In all of these, the nerve cord is ventral to the digestive system and is solid.

Despite the small size of many invertebrates, their nervous systems are enormously complex. The crayfish system, for example, contains nearly 100,000 neurons, and the housefly has more than one million neurons.

The nervous system and sense organs of cephalopods are among the most highly developed of all invertebrates. The ganglia, which in other mollusks are located at different points along the nerve cords, are concentrated at the anterior end to form a complex brain housed within a capsule. Functional centers of the brain have been identified, and the behavior of the octopus and other cephalopods has been studied intensively.

In addition to generating and conducting nerve impulses, the neurons of many types of animals have neurosecretory functions. Some nerves secrete what appears to be a carrier protein, termed **neurophysin,** to which certain hormones are bound. Specific functions of neurohormones have been demonstrated in annelids, crustaceans, insects, and vertebrates. The neurosecretory neurons typically have axons that terminate on or near a blood vessel or blood sinus. Neurohormones in certain annelids regulate the conversion of young worms into the sexually active reproductive forms. In crustaceans, neurohormones regulate carbohydrate metabolism, molting and metamorphosis, water regulation, sexual maturation, and the movement of pigment in chromatophores.

Organization of the Vertebrate Nervous System

The neurons of the vertebrate nervous system can be assigned either to the **central nervous system**, which includes the brain and spinal cord, or to the **peripheral nervous system**, which includes the nerves that extend between the central nervous system and the receptors or effectors. The functional organization of the nervous system can be introduced by considering a specific example. When you touch a hot object (Fig. 10.11), receptors in the skin are stimulated and initiate impulses in afferent neurons. These neurons are part of a spinal nerve and extend into the spinal cord, where they synapse with interneurons. The interneurons in turn transmit impulses to efferent neurons that extend from the cord and carry impulses back through the spinal nerve to a group of extensor muscles in your hand. Their contraction withdraws your hand from the stove. For the movement to be effective, the antagonistic flexor muscles should relax, which involves the inhibition of impulses going to these muscles. Normally, some impulses pass to all of the muscles of the body continually and cause a partial contraction, called muscular **tonus**. Such a stimulus and response is a simple **spinal reflex**, and the neuronal pathway along which the impulse travels is called a **reflex arc**.

A reflex (L., *reflexus*, bent back) is a simple, automatic response to a given stimulus that involves only a few neurons, all connecting within the same general level of the central nervous system. A spinal reflex, for example, does not depend upon connections with higher levels of the cord or brain. An impulse may be carried to the cerebral cortex by other interneurons, and you then become aware of the stimulus and may decide to do something about it—perhaps withdraw the entire arm or turn off the stove. Reflexes are the functional units of the nervous system, and many of our activities are the result of them. They are important in controlling heart rate, blood pressure, breathing, salivation, movements of

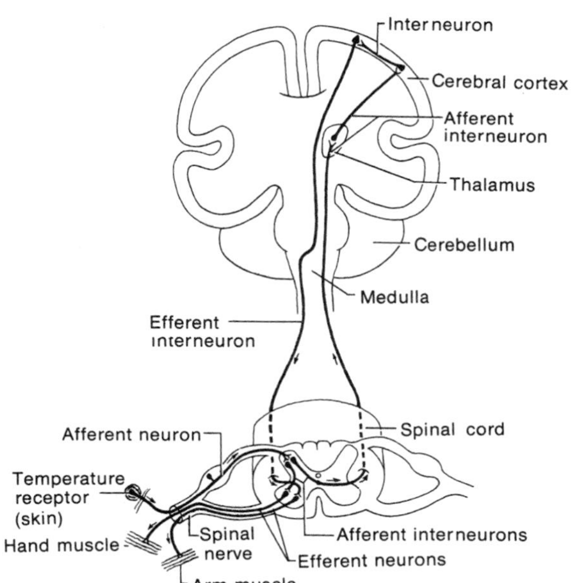

FIGURE 10.11 Diagram of the neurons involved in the passage of impulses to and from the brain and the relationship of these to the neurons in a reflex arc (shown simplified).

the digestive tract, and so on. When we step on something sharp or come in contact with something hot we do not wait until the pain is experienced by the brain and then, after deliberation, decide what to do. Our responses are immediate and automatic. The foot or hand is being withdrawn by reflex action *before* the pain is experienced. Many of the more complicated activities of our daily lives, such as walking, are regulated to a large extent by reflexes.

The reflexes present at birth and common to all human beings are called **inherited reflexes**. Others, acquired later as the result of experience, are called **conditioned reflexes.** The minimum anatomical requirements for reflex behavior are a sensory neuron with a receptor to detect the stimulus, connected by a synapse to a motor neuron that is attached to a muscle or some other effector, such as the extensor stretch reflex. This simplest type of reflex arc is termed **monosynaptic** because there is only one synapse between the sensory and motor neurons. More typically reflex arcs include one or more internerurons between the sensory and motor neurons. A simple reflex in which stimulation of a receptor produces contraction of a single muscle is typified by the **knee jerk**. When the tendon of the knee cap is tapped, and thereby stretched, receptors in the tendon are stimulated; an impulse travels over the reflex arc, up to the spinal cord, and down again; and

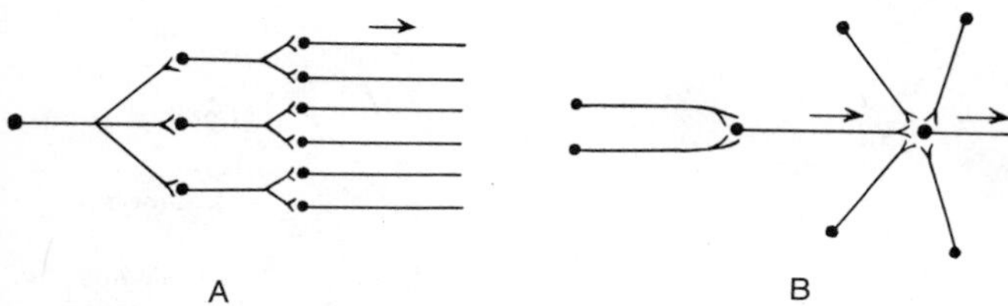

FIGURE 10.12 Diagrams of important types of neuronal interrelationships. *A,* A divergent pathway; *B,* a convergent pathway. A given neuron may be involved in more than one of these pathways.

the muscle attached to the tendon contracts, resulting in a sudden straightening or extending of the leg.

A number of reflexes involve the connections of many interneurons in the spinal cord. Not only does the spinal cord function in the conduction of impulses to and from the brain but it also plays an important role in the integration of reflex behavior. Its importance in this respect is demonstrated in a "spinal" animal, one whose brain has been destroyed or removed. If a piece of acid-soaked paper is applied to the back of a spinal frog, one leg will come up and flick the paper away. No matter how many times the piece of paper is replaced on the skin, the leg will come up and flick it away. This response, involving many muscles working in a coordinated fashion, is purely reflex and demonstrates one of the chief characteristics of a reflex; fidelity of repetition. A frog with a brain might make the response two or three times, but eventually it would do something else, perhaps hop away. Most reflexes have some survival value for the animal. The anatomical configuration responsible for the reflex was selected during evolution because of this survival value.

The integration and correlation of the body's activities involve pathways with many neurons, not just two or three, as in the simpler reflexes. Some pathways are **divergent** (Fig. 10.12*A*); that is, the axon of one neuron may branch many times, each branch synapses with a different neuron, and these in turn may branch further. Such an arrangement permits a single impulse to exert an effect over a wide area. Indeed, a single impulse may ultimately activate a thousand or more neurons. In **convergent** pathways (Fig. 10.12*B*) neurons from many different areas converge upon a single neuron or group of neurons. Complex integrated responses in animals depend upon divergent and convergent neuronal circuits coupled with synaptic regulation.

The Peripheral Nervous System

Emerging from the brain and spinal cord and connecting them with every receptor and effector in the body are the paired cranial and spinal nerves that make up the **peripheral nervous system**. The only nerve cell bodies present in the peripheral nervous system are those of the sensory neurons aggregated into ganglia near the brain or spinal cord and of certain motor neurons of the autonomic system, discussed further on.

Spinal Nerves

There is a pair of peripheral nerves for each body segment of the vertebrate body. Afferent and efferent neurons lie together in most of a spinal nerve, but near the cord the nerve splits into a dorsal and ventral root and the neurons are segregated (Fig. 10.13). The **dorsal root** contains the afferent neurons and bears an en-

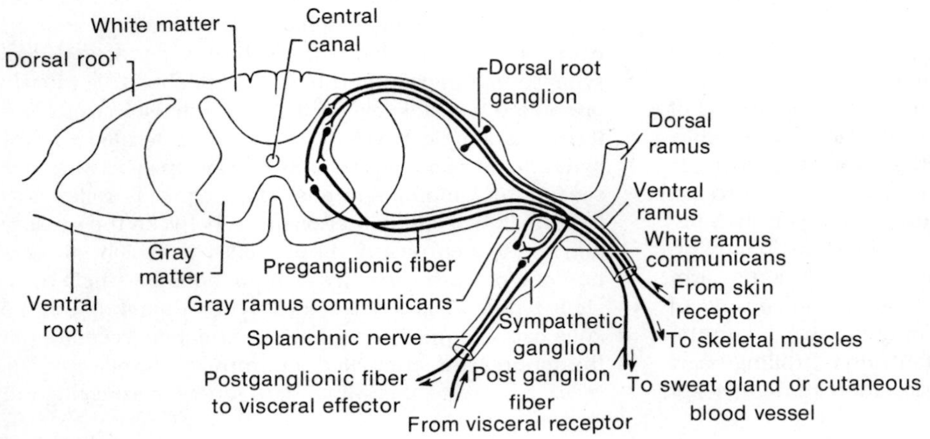

FIGURE 10.13 A diagrammatic cross section through the spinal cord and a spinal nerve. Each spinal nerve is formed by the union of dorsal and ventral roots and divides laterally into several branches (rami) going to different parts of the body. The dorsal ramus contains the same types of neurons as the ventral ramus.

largement, the **dorsal root ganglion**, that contains their cell bodies. The cell bodies of afferent neurons are nearly always located in ganglia on both spinal and cranial nerves. The afferent neurons enter the spinal cord and terminate in synapses with the dendrites or cell bodies of interneurons. Most of these cell bodies are located in the dorsal portion of the gray matter of the cord. The **ventral root** contains the efferent neurons; their cell bodies nearly always lie in the ventral portion of the gray matter of the cord.

The spinal nerves of all vertebrates are essentially alike, although in the most primitive vertebrates the roots do not unite peripherally, and some of the efferent neurons leave the cord in the dorsal root. In most vertebrates the roots unite to form a spinal nerve that subsequently divides into a **dorsal ramus** that supplies the skin and muscles in the dorsal part of the body, a **ventral ramus** that innervates the lateroventral parts of the body, and usually one or more **communicating rami** to the visceral organs. Afferent and efferent neurons are found in each ramus. Humans have 31 pairs of spinal nerves; those supplying the receptors and effectors of the limbs are larger than the others, and their ventral rami are interlaced to form a complex network, or **plexus**, from which nerves extend to the limbs (Fig. 10.14).

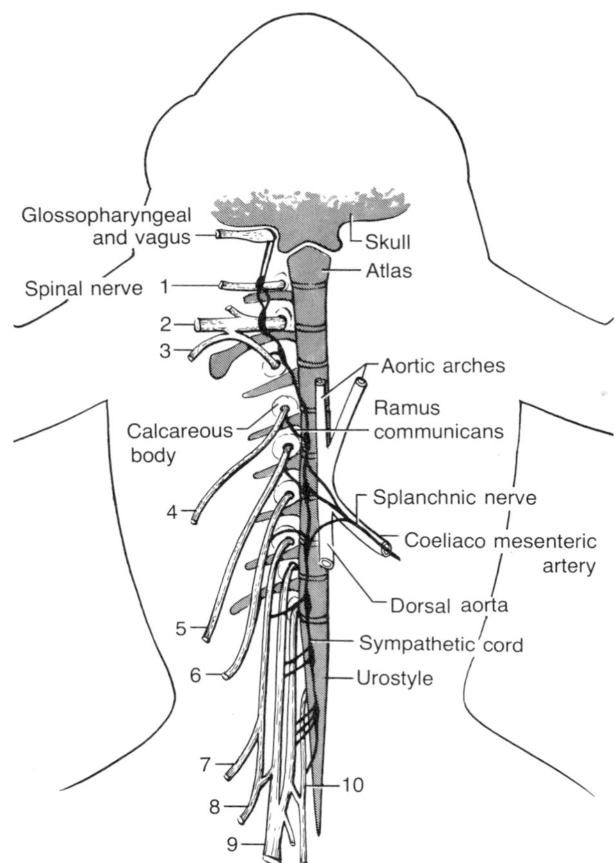

FIGURE 10.14 A ventral view of the ventral rami of the spinal nerves and the sympathetic cord lying on the right side of the vertebral column of a frog. (Modified after Gaupp.)

Cranial Nerves

The nerves from the nose, eye, and ear have evolved with the organs of special sense. They are composed entirely of afferent fibers, except for a few efferent neurons in the optic and vestibulocochlear (auditory) nerves that extend to the sense organs and may modulate their activity. The remaining cranial nerves contain large numbers of both afferent and efferent fibers, and they are considered to be serially homologous with the separate roots of the spinal nerves of primitive vertebrates. Some of them are essentially the cephalic counterparts of dorsal roots; others are the counterparts of ventral roots. The location of the cell bodies of the neurons of cranial nerves and of the endings within the brain follows the pattern described for spinal neurons.

Reptiles, birds, and mammals have 12 pairs of cranial nerves, omitting the minute and poorly understood nervus terminalis. Though distributed to the nasal mucosa, this nerve is not olfactory. The other cranial nerves and their distribution are shown in Table 10.1, and their stumps can be seen in an illustration of a sheep brain (Fig. 10.15).

Fishes and amphibians have only the first 10 pairs of cranial nerves and lack spinal accessory and hypoglossal nerves. Neurons comparable to those in the spinal accessory and hypoglossal are incorporated in other nerves, however. The trigeminal, facial, glossopharyngeal, and vagus nerves of fishes are primarily associated with the muscles of the visceral arches, and, as shown in Table 10.1, they supply the derivatives of this musculature in the higher vertebrates. Muscles may change shape and function tremendously during the course of evolution, but their innervation remains remarkably constant. One of the most important cranial nerves, the tenth, or **vagus**, forms part of the autonomic system and innervates the internal organs of the chest and upper abdomen.

The Autonomic Nervous System

The heart, lungs, digestive tract, and other internal organs, together with the ciliary and iris muscles of the eye, the small muscles associated with hairs, and many of the glands of the body, are innervated by a special set

TABLE 10.1 Cranial Nerves of Humans

Nerve	Origin of Afferent Neurons	Distribution of Efferent Neurons
I, Olfactory	Olfactory portion of nasal mucosa (smell)	
II, Optic	Retina (sight)	A few to retina
III, Oculomotor	A few fibers from proprioceptors in extrinsic muscles of eyeball (muscle sense)	Most fibers to four of the six extrinsic muscles of eyeball, a few to muscles in ciliary body and pupil
IV, Trochlear	Proprioceptors in extrinsic muscles of eyeball	Another extrinsic muscle of eyeball
V, Trigeminal	Teeth, and skin receptors of the head (touch, pressure, temperature, pain); proprioceptors in jaw muscles	Muscles derived from musculature of first visceral arch, i.e., jaw muscles
VI, Abducens	Proprioceptors in an extrinsic muscle of eyeball	One other extrinsic muscle of eyeball
VII, Facial	Taste buds of anterior two thirds of tongue (taste)	Muscles derived from musculature of second visceral arch, i.e., facial muscles; salivary glands; tear glands
VIII, Vestibulocochlear	Semicircular canals, utriculus, sacculus (sense of balance); cochlea (hearing)	A few to cochlea
IX, Glossopharyngeal	Taste buds of posterior third of tongue; lining of pharynx	Muscles derived from musculature of third visceral arch, i.e., pharyngeal muscles concerned in swallowing; salivary glands
X, Vagus	Receptors in many internal organs: larynx, lungs, heart, aorta, stomach	Muscles derived from musculature of remaining visceral arches (excepting those of pectoral girdle), i.e., muscles of pharynx (swallowing) and larynx (speech); muscles of gut, heart; gastric glands
XI, Spinal accessory	Proprioceptors in certain shoulder muscles	Visceral arch muscles associated with pectoral girdle, i.e., sternocleidomastoid and trapezius
XII, Hypoglossal	Proprioceptors in tongue	Muscles of tongue

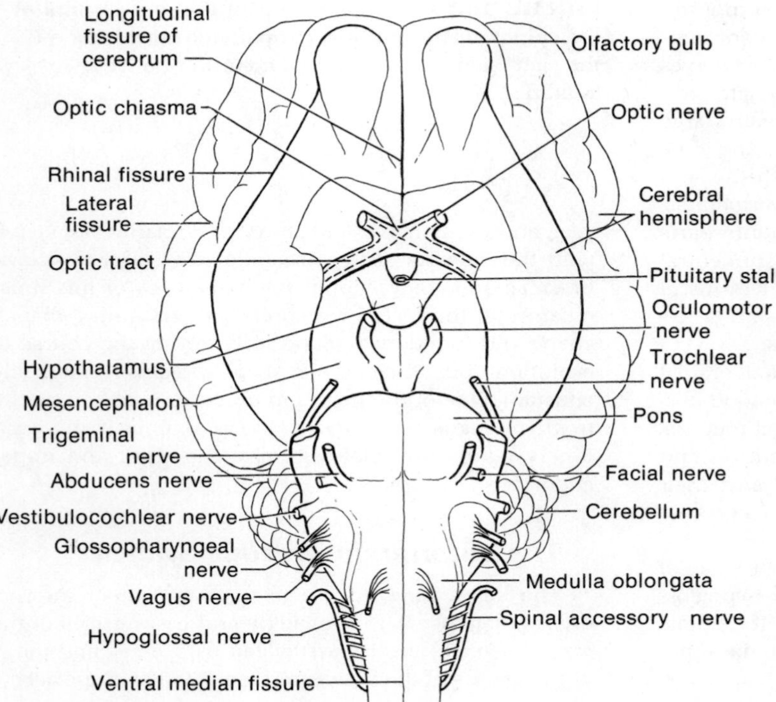

FIGURE 10.15 A ventral view of the brain of a sheep. The stumps of all but the first pair of cranial nerves are visible. The olfactory nerves consist of the processes of olfactory cells, which enter the olfactory bulbs in many small groups that cannot be seen with the unaided eye. The rhinal fissure separates the ventral olfactory portion of each cerebral hemisphere from the rest of the hemisphere. The paths of the optic fibers in the optic chiasms have been indicated by broken lines.

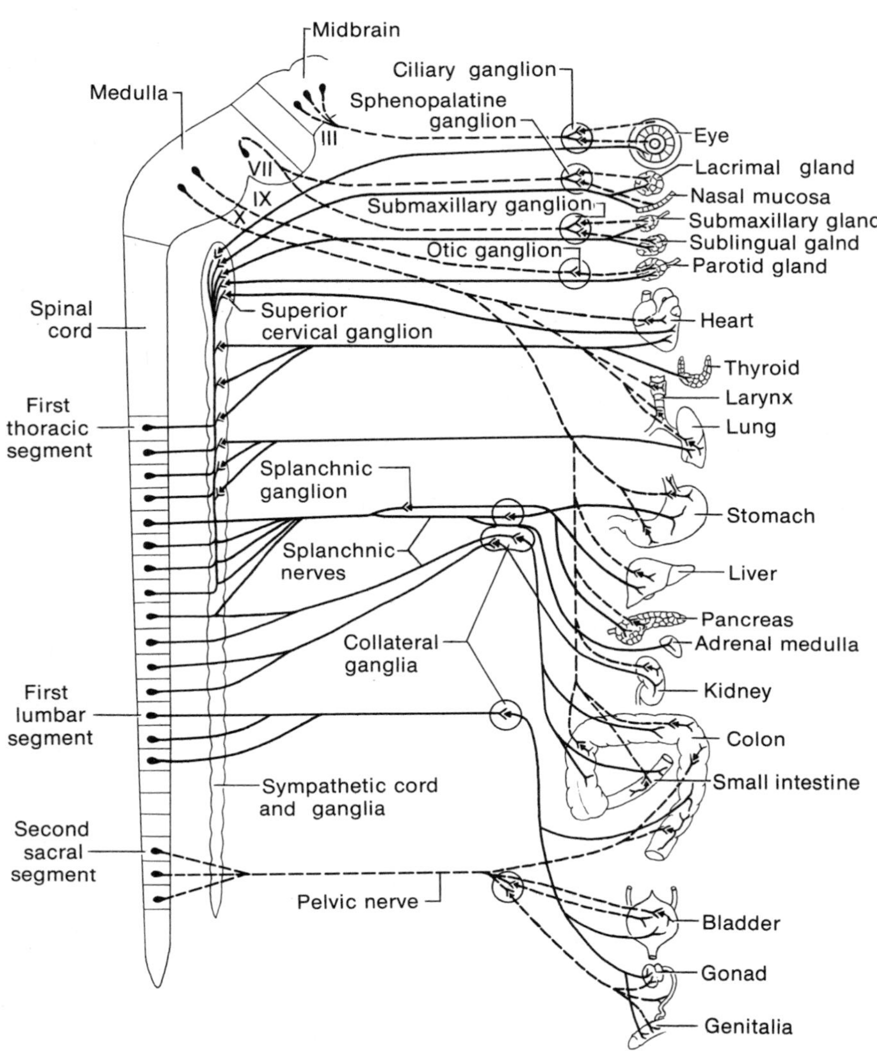

FIGURE 10.16 The human autonomic nervous system. Sympathetic fibers are drawn in solid lines; parasympathetic fibers in broken lines. The sympathetic fibers that go to the skin are not shown. (After Howell.)

of peripheral nerves collectively called the **autonomic nervous system**. This system is composed of two parts: the **sympathetic** and **parasympathetic** nerves.

The organs innervated by the autonomic system function automatically, requiring no thought on our part; indeed, they cannot be controlled voluntarily. The autonomic nervous system is usually defined as a motor system, and the afferent fibers that return from internal organs are not part of this system, even though they may be in a nerve composed largely of autonomic fibers. The autonomic system is distinguished from the rest of the nervous system by several features. There is no voluntary control by the cerebrum over these nerves. Most organs receive a *double* set of fibers—one set via the sympathetic nerves and one via the parasympathetic nerves, and the axon endings of each set secrete a different transmitter substance to the effectors. The fibers of the sympathetic system secrete norepinephrine, and those of the parasympathetic system secrete acetylcholine. Thus it is the neurosecretion that accounts for the differences in response. The nature of the nerve impulse is the same in both systems.

Another peculiarity of the autonomic system is that motor impulses reach the effector organ from the brain or spinal cord not by a single neuron, as do those in all other parts of the body, but by a relay of two neurons. The cell body of the first neuron in the chain, the **preganglionic neuron**, is located in the brain or spinal cord; that of the second neuron, the **postganglionic neuron**, is located in a ganglion somewhere outside the central nervous system (Fig. 10.16). The cell bodies of the postganglionic neurons of the sympathetic nerves are close to the spinal cord; those of the parasympathetic nerves are close to or actually within the walls of the organs they innervate.

Sympathetic and parasympathetic systems usually have opposite effects upon the organs innervated. These effects are summarized in Table 10.2. Sympathetic stimulation speeds up the rate and increases the force of the heart beat, causes most peripheral arteries to constrict (thereby increasing blood pressure), increases the glucose content of the blood, and, in general, has effects that enable the body to adjust to conditions of stress. It inhibits the activity of the digestive tract. Parasympathetic stimulation speeds peristalsis of the digestive tract and similar vegetative processes, but it slows down the heart rate and decreases blood pressure. In general, the effects of the sympathetic system adapt the animal to respond to the environmental stress by fighting or running away. The parasympathetic system adapts it for more passive responses.

The Central Nervous System: The Spinal Cord

The tubular spinal cord, surrounded and protected by the neural arches of the vertebrae, has two important functions; to transmit impulses to and from the brain and to act as a reflex center. In cross section, two regions are evident; an inner, butterfly-shaped mass of **gray matter** made up of nerve cell bodies and an outer mass of **white matter** made up bundles of axons and dendrites (Fig. 10.17). The "wings" of the gray matter are divided into two dorsal and two ventral horns. The latter contain the cell bodies of motor neurons whose axons pass out through the spinal nerves to the muscles. All of the other neurons in the spinal cord are interneurons.

The axons of the white matter are segregated into bundles with similar functions: the **ascending tracts** transmit impulses to the brain, and the **descending tracts** carry impulses from the brain to the effectors. Neurologists have carefully noted the symptoms of patients with injured spinal cords and later correlated these with the particular tracts found to be damaged when the patient's nervous system was examined after death. From these observations they have been able to map out the location and functions of the various tracts (Fig. 10.17). For example, the dorsal columns of the white matter transmit impulses originating in the sense organs of muscles, tendons, and joints through which we are aware of the position of the parts of the body. In advanced syphilis, these columns may be destroyed, so that the patient cannot tell where his arms and legs are unless he looks at them. He must watch his feet in order to walk.

One curious fact, still not satisfactorily explained, emerged from these studies of the location and function

TABLE 10.2 Selected Effects of Autonomic Stimulation*

Organ	Sympathetic Stimulation	Parasympathetic Stimulation
Skin		
Hair muscles	Contraction	....
Sweat glands	Secretion	....
Eye		
Iris sphincter	....	Contraction
Iris dilator	Contraction	....
Circulatory system		
Heart (rate and force)	Increased	Decreased
Coronary arteries	Dilation	Constriction
Most other arteries	Constriction	Dilation
Lung bronchi	Dilation	Constriction
Digestive organs		
Muscles of stomach and intestine	Decreased peristalsis	Increased peristalsis
Salivary glands	Some mucus secretion	Secretion
Gastric and intestinal glands	....	Secretion
Pancreas	....	Secretion
Liver	Bile flow inhibited Glucose released	Bile flow stimulated
Urinary organs		
Urinary bladder muscles	Relaxation	Contraction
Bladder sphincter	Contraction	Relaxation
Adrenal medulla	Secretion	....

*Dotted line indicates the absence of innervation.

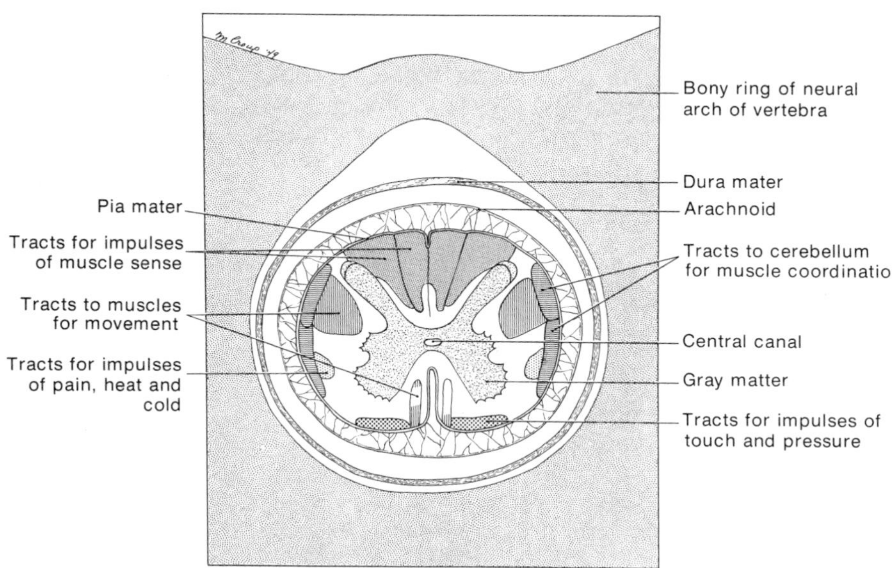

FIGURE 10.17 Cross section of the spinal cord surrounded by the bony vertebra, showing the meninges *(dura mater, pia mater,* and *arachnoid),* the gray matter, and some of the important nerve tracts of the white matter.

of the fiber tracts. All the fibers in the spinal cord *cross over* from one side of the body to the other somewhere along their path from sense organ to brain or from brain to muscle. Thus the right side of the brain controls the left side of the body and receives impressions from the sense organs of the left side. Some fibers cross in the spinal cord itself, and others cross in the brain.

In the center of the gray matter is a small canal that extends the entire length of the neural tube and is filled with **cerebrospinal fluid**. This fluid is similar to plasma but contains much less protein. The spinal cord and brain are wrapped in three sheets of connective tissue known as **meninges**. Meningitis is a disease in which these wrappings become infected and inflamed. One of these sheaths (**dura mater**) is fastened against the bone of the cranium; another (**pia mater**) is located on the surface of the brain and spinal cord; and a third, the **arachnoid**, lies between. The spaces between the arachnoid and the pia are filled with more cerebrospinal fluid, so that the spinal cord (and the brain) floats in this liquid and is protected from bouncing against the bone of the vertebrae or the skull with every movement. An **extradural space**, in which lie blood vessels, connective tissue, and fat, may exist in the vertebral column between the dura mater and the bone.

The Central Nervous System: The Brain

The brain is the enlarged anterior end of the neural tube. In the human, the enlargement is so great that the resemblance to the spinal cord is obscured, but in lower vertebrates, the relationship of brain to spinal cord is clearer. The brain develops as a series of enlargements of the anterior end of the embryonic neural tube (Fig. 10.18). In an early embryo there are three swellings, the forebrain, midbrain, and hindbrain, but both the forebrain and the hindbrain are later subdivided so that five regions are present in the adult. The forebrain divides into **telencephalon** (Gr., *tele*, end + *enkephalos*, brain) and **diencephalon**. The hindbrain divides into a **metencephalon**, the dorsal portion of which forms the **cerebellum**, and a **myelencephalon** that becomes the **medulla oblongata** (Table 10.3).

The central canal of the spinal cord extends into the brain and is continuous with several large interconnected chambers known as **ventricles** (Fig. 10.19). A large ventricle lies in each cerebral hemisphere and is connected with the third ventricle in the thalamus by an **interventricular foramen of Monro.** The **cerebral aqueduct of Sylvius** extends from the third ventricle through the mesencephalon to a fourth ventricle in the metencephalon and medulla oblongata. All of these cavities are filled with cerebrospinal fluid, much of which is produced by vascular **choroid plexuses** that develop in the thin roof of the medulla and diencephalon and in the lateral ventricles of mammals. Cerebrospinal fluid escapes from the brain through foramina in the roof of the medulla and slowly circulates in the spaces between the layers of connective tissue, the meninges, that encase the brain and spinal cord. The cerebrospinal fluid lies in the subarachnoid space, between the arachnoid and pia mater, and extends far into the nervous tissue in **perivascular spaces** that sur-

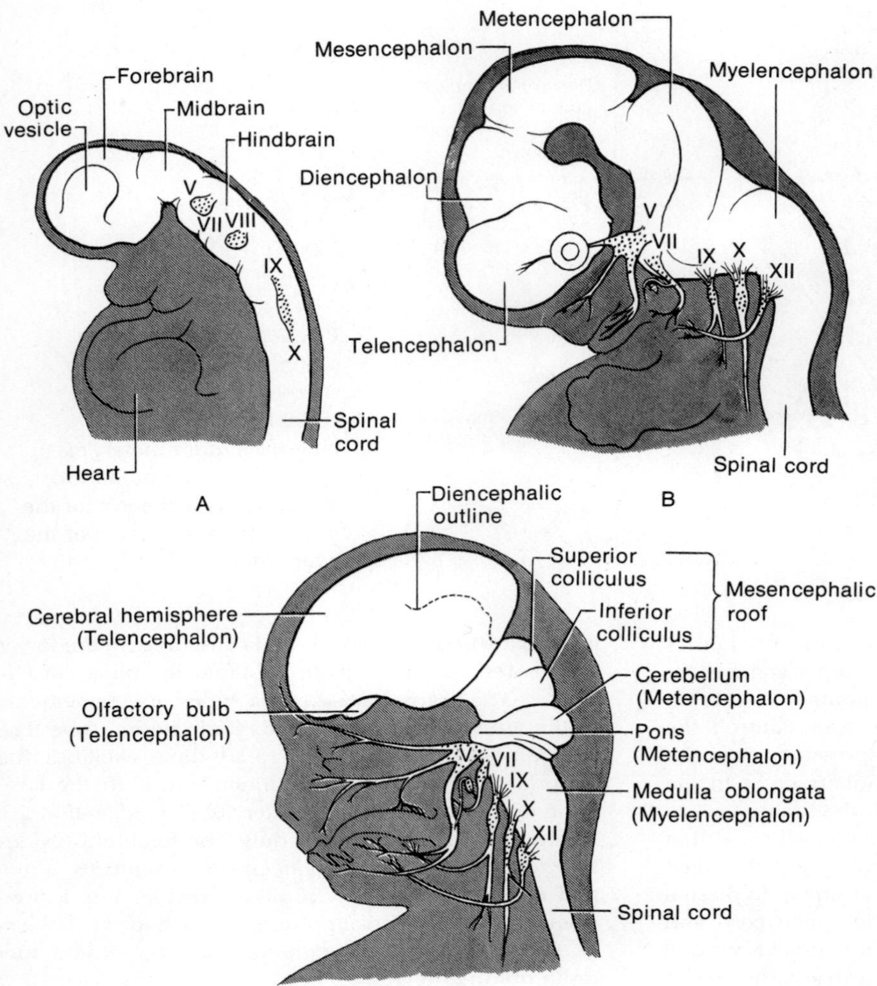

FIGURE 10.18 Three stages in the development of the human brain. *A,* The three primary brain regions can be recognized in an embryo that is about 3.5 weeks old. *B,* All five brain regions are evident in an embryo 7 weeks old. *C,* The various structures found in a fully developed brain are beginning to differentiate in an embryo 11 weeks old. (After Patten.)

TABLE 10.3 Brain Regions and Derivatives

Brain Region	Major Derivatives
Forebrain (prosencephalon)	
Telencephalon	Olfactory bulbs
	Cerebral hemispheres
Diencephalon	Epithalamus, pineal body
	Thalamus
	Hypothalamus, pituitary gland (part)
Midbrain (mesencephalon)	Superior colliculi
	Inferior colliculi
Hindbrain (rhombencephalon)	
Metencephalon	Cerebellum
	Pons
Myelencephalon	Medulla oblongata

round the blood vessels penetrating the brain. It is produced continuously and reenters the circulatory system by filtering into certain venous sinuses located in the dura mater covering the brain. The cerebrospinal fluid forms a protective liquid cushion for the brain and spinal cord and may help nourish the tissues of the central nervous system. The perivascular spaces serve some of the functions of a lymphatic system by returning plasma protein that may have escaped from capillaries in the brain. Nervous tissue does not have a lymphatic drainage.

Medulla Oblongata

The medulla oblongata (Figs. 10.20 and 10.21) lies between the spinal cord and the rest of the brain and is

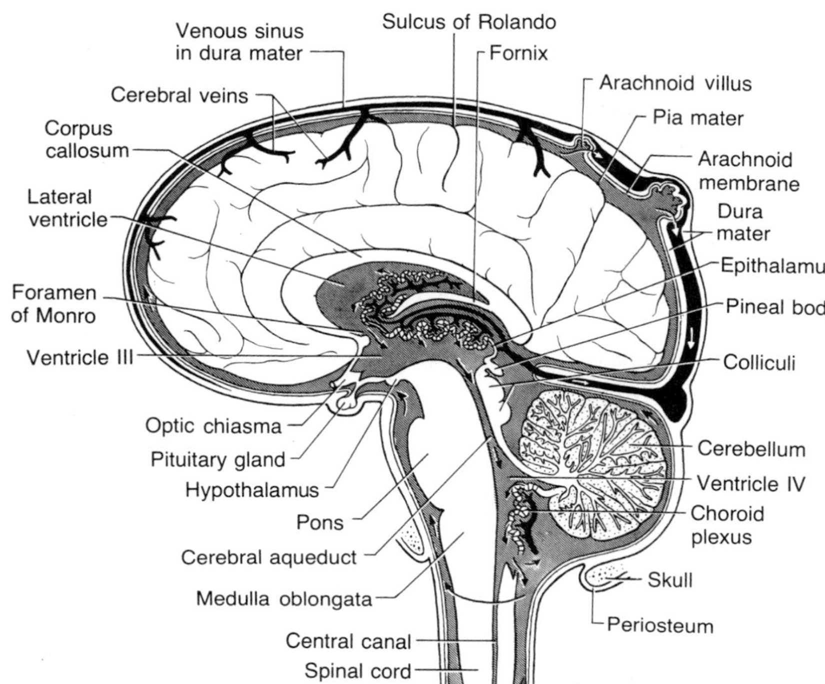

FIGURE 10.19 A sagittal section of the human brain and its surrounding meninges. Cerebrospinal fluid is produced by the choroid plexuses, circulates as indicated by the arrows, and finally enters a venous sinus in the dura matter. (Modifed after Rasmussen.)

fundamentally the same in all vertebrates. The gray columns of the spinal cord extend into the medulla, but within the brain they become discontinuous, breaking up into discrete islands of cell bodies known as **nuclei.** The dorsal nuclei receive the afferent neurons from cranial nerves that are attached to this region and contain the cell bodies of afferent interneurons. These are sensory nuclei, just as the dorsal columns of the cord are sensory columns. The ventral nuclei contain the cell bodies of the efferent neurons of the cranial nerves and hence are motor nuclei. In mammals, the reflexes that regulate the rate of heart beat, blood pressure, respira-

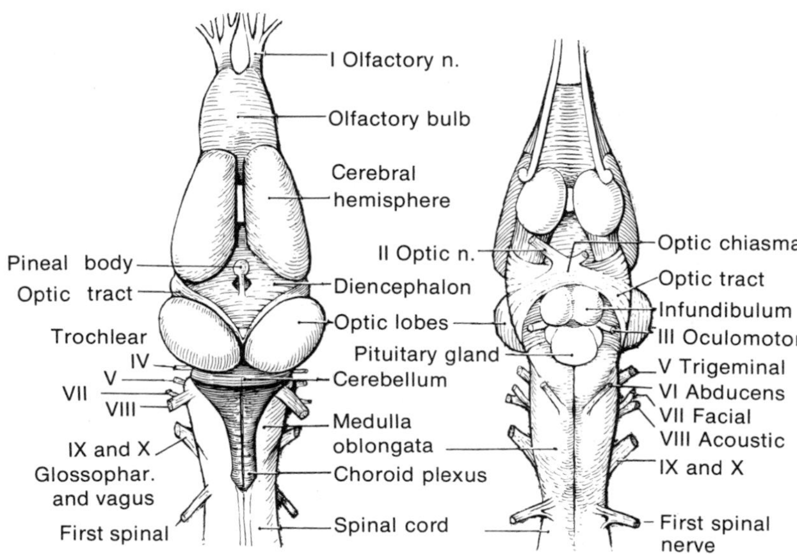

FIGURE 10.20 *Left,* A dorsal view and *right,* a ventral view of the brain of the frog. (Modified after Gaupp.)

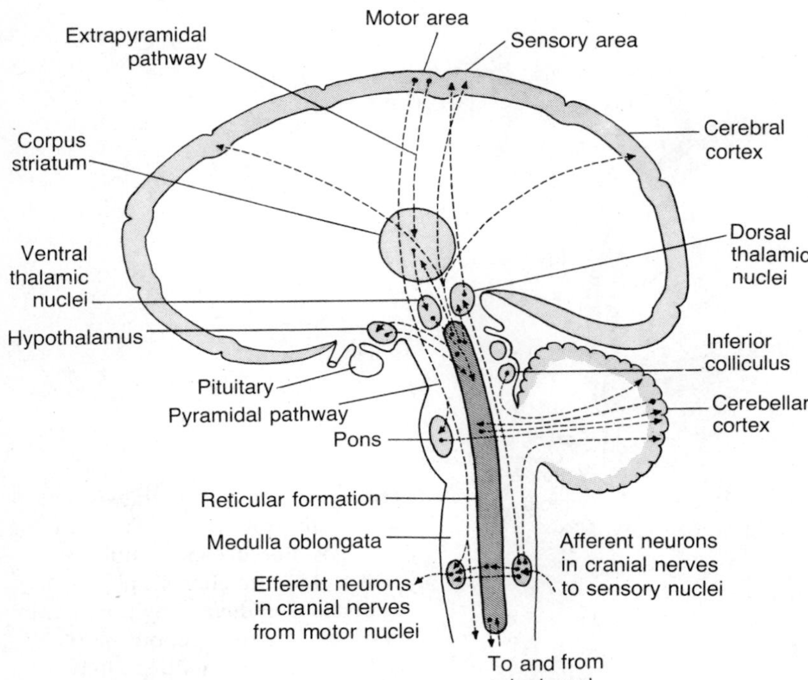

FIGURE 10.21 Diagrammatic lateral view of the human brain on which some of the major masses of gray matter, the reticular formation, and certain important pathways have been projected.

tory movements, salivary secretion, swallowing, and many other processes are mediated by these nuclei. Afferent impulses come into the sensory nuclei and are relayed by the interneurons to the motor nuclei, and efferent impulses go out to the effectors. Motor and sensory nuclei associated with other cranial nerves are also found in the metencephalon and the mesencephalon.

Reticular Formation

Between the motor and sensory nuclei throughout the brain stem is a network of thousands of cell bodies and their processes known as the **reticular formation.** Dendrites and axons of these neurons branch profusely and have numerous connections with each other and with many of the sensory and motor pathways passing to and from the brain (Fig. 10.21). Certain of these neurons receive input from as many as 4000 other neurons and feed out to 25,000 other neurons. The reticular formation is believed to be a very primitive part of the brain, from which the long ascending and descending fiber tracts and the various centers have evolved. Convergence and interaction of neurons from many of the sense organs, the cerebellum, the cerebrum, the thalamus, the hypothalamus, and other areas occur in the reticular formation, and the region appears to be able to suppress, enhance, or otherwise modify impulses passing through it. For example, branches from cells in the sensory nuclei, in addition to synapsing with motor cells and ascending to higher centers, feed into this system. If the sensory input is important to the animal (an unusual noise, the cry of a baby at night, and so on), the reticular formation, in turn, sends nonspecific impulses to the cerebral cortex. These fan out widely in the cortex and arouse the animal if asleep or help to keep it alert if awake. Without this source of stimulation, the cortex does not function and cannot interpret the specific sensory information brought to it. In short, the reticular formation appears to be involved in what we call consciousness; that is, an awareness of oneself and one's environment.

The Cerebellum and Pons

All vertebrates have a **cerebellum** (L. dim. of *cerebrum,* brain) that develops in the dorsal part of the metencephalon and is a center for balance and motor coordination. It is small in many of the lower vertebrates, such as the frog (Fig. 10.20), in which muscle move ments are not complex but is very large in birds and mammals. Impulses enter it from most of the sense organs, but those from the proprioceptive organs in the muscles and the parts of the ear concerned with equilibrium play a particularly prominent role (Fig. 10.21).

The cerebellum keeps track of the orientation of the body in space and the degree of contraction of the skeletal muscle. In mammals this information is sent by way of part of the reticular formation and thalamus to the cerebrum, where voluntary movements are initiated. Copies, so to speak, of the motor commands sent out by the cerebrum to the muscles are sent to the cerebellum, which monitors the responses to the body and returns corrective signals to the cerebrum or, in some cases, directly to the muscle. Much of the gray matter of the mammalian cerebellum lies on the surface, where there is more room for the increased number of cell bodies. The surface is also complexly folded, which further increases the surface area available for cell bodies.

The floor of the metencephalon is unspecialized in lower vertebrates and simply contributes to the medulla oblongata. In mammals, this region has evolved into a **pons** (L. bridge), a thick bundle of fibers extending crossways from one cerebellar hemisphere to the other and carrying impulses from one to the other, thus coordinating muscle movements on the two sides of the body. The pons also contains nuclei that relay impulses from the cerebrum to the cerebellum.

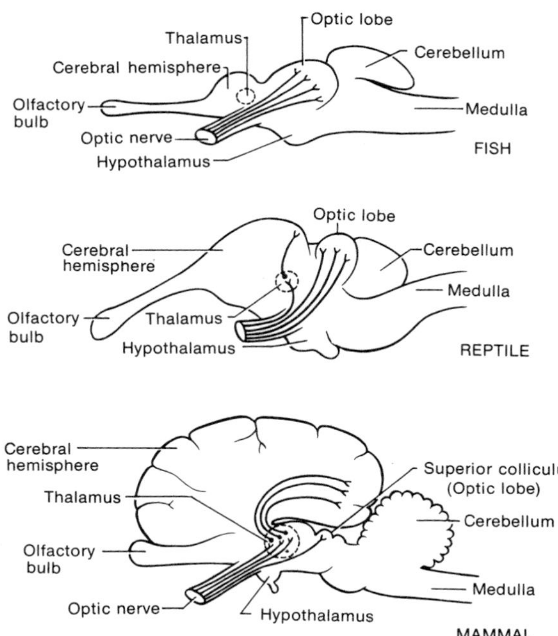

FIGURE 10.22 Diagrammatic lateral view of the brain of a fish, a reptile, and a mammal to show the increasing importance of the cerebral hemispheres and the decreasing importance of the optic lobes as integration centers. The shift in optic pathways is shown; a similar shift occurs for other sensory modalities.

The Optic Lobes

The optic lobes of primitive fishes and amphibians (Figs. 10.20 and 10.22) receive the termination of all of the fibers in the optic nerve and the projections from most other sense organs. This sensory information is integrated, and motor impulses are sent to the appropriate efferent neurons. The optic lobes of fishes and amphibians are the master integrating center of the brain. The cerebral hemispheres of the lower vertebrate are small and concerned primarily with integrating olfactory impulses. In reptiles, some of the visual and other types of sensory data are sent to the cerebral hemispheres, and the cerebrum begins to assume certain functions of the optic lobes (Fig. 10.22). Still more sensory information is sent to the cerebral hemispheres of birds and mammals, and the hemispheres of mammals have taken over most of the functions of the optic lobes. In mammals the optic lobes (superior colliculi) remain relatively small centers that regulate the movements of the eyeballs and pupillary and accommodation reflexes. A pair of inferior colliculi are present posterior to the superior colliculi in mammals. They are a center for certain auditory reflexes. The reflex constriction of the pupil when light shines on the eye and the pricking up of a dog's ears in response to sound are controlled by reflex centers in the superior and inferior colliculi. The midbrain also contains a cluster of nerve cells regulating muscle tonus and posture.

Thalamus and Hypothalamus

In front of the midbrain the central canal widens and becomes the third ventricle, the roof of which contains a cluster of blood vessels (**choroid plexus**) secreting some of the cerebrospinal fluid. The thick walls of the third ventricle, the **thalamus** (Gr., *thalamos,* inner chamber) are relay centers for sensory impulses. Fibers from the spinal cord and lower parts of the brain synapse here with other neurons going to the various sensory areas of the cerebrum. The thalamus appears to regulate and coordinate the external manifestations of emotions. Thus, by stimulating the thalamus a sham rage can be produced in a cat—the hair stands on end, the claws protrude, the back becomes humped, and other signs of anger are evinced. However, as soon as the stimulation stops, the appearance of rage ceases.

The thalamus is more than just a relay station, for considerable processing of the sensory input occurs here. The thalamic nuclei have many connections with each other, with the hypothalamus, with the reticular formation, and with many parts of the cerebral hemispheres. Pain, temperature, and certain other sensory modalities reach the level of conscious awareness in the thalamus. Other sensations are to some extent sorted out in this region, and a determination is made as to which of these we will concentrate on among the hundreds of stimuli continually impinging upon us. The thalamus influences the manner in which we view many stimuli; that is, the different degree of agreeableness or disagreeableness that we may place upon the same type and intensity of stimuli at different times. An experimental rat with an electrode implanted in the "pleasure center" in the thalamus will push a lever that stimulates the center thousands of times per hour. Other areas appear to be pain centers, and an experimental animal will avoid activating an electrode implanted in one of them.

The hypothalamus is an important center for the control of many autonomic functions, with centers regulating body temperature, appetite, water balance, carbohydrate and fat metabolism, blood pressure, and sleep. It is curious that the anterior part of the hypothalamus prevents a rise in body temperature, and the posterior part prevents a fall. The hypothalamus controls certain functions of the anterior lobe of the pituitary, such as the secretion of gonadotropins, by producing **releasing factors.** It also produces the hormones **oxytocin** and **vasopressin,** which are released in the posterior lobe of the pituitary. The parts of the hypothalamus that produce neurosecretions are neuroendocrine transducers, for they convert nerve signals into endocrine signals.

Cerebral Hemispheres

The parts of the brain considered so far have to do with unlearned automatic behavior determined by the fundamental structure of these parts, which is essentially the same from fish to human. The **cerebral hemispheres,** the largest and most anterior parts of the human brain, have a basically different function—controlling learned behavior. The complex psychological phenomena of consciousness, intelligence, memory, insight, and the interpretation of sensations have their physiological basis in the activities of the neurons of the cerebral hemispheres.

As the cerebral hemispheres assumed the dominant role in nervous integration during the course of evolution, they enlarged and grew posteriorly over the diencephalon and mesencephalon (Fig. 10.22). A layer of gray matter has developed on the surface of the cerebrum and forms a gray cortex that provides more area for the increased number of cell bodies. The cortex is also complexly folded, having numerous ridges **(gyri)** with furrows **(sulci)** between them, and this further increases the surface area. Over 12 billion neurons and even more neuroglial cells are present.

Parts of the cerebral hemispheres are still concerned with their primitive function of olfactory integration, but their great enlargement is correlated with the evolution of other integrative centers (Fig. 10.23). Impulses from the eyes, ears, skin, and many other parts of the body are carried to the cerebral cortex by afferent interneurons after being relayed in the thalamus. The impulses terminate in specific parts of the cerebral cortex; their locations have been determined by correlating brain injuries with loss of sensation and also by electrical stimulation during brain operations. Many human brain operations are performed under local anesthesia, and the patient can describe the sensations that are felt when particular regions of the cerebral cortex are stimulated. Impulses from the skin terminate in the gyrus located just posterior to the central **sulcus of Rolando,** a prominent fissure extending down the side of each hemisphere and dividing the hemisphere into an anterior **frontal** and posterior **parietal lobe.** The extent of the area receiving impulses from any part of the body is proportional to the number of sense organs in that part

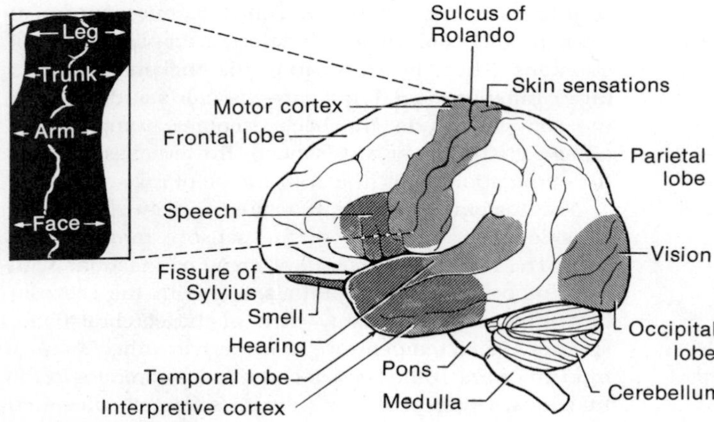

FIGURE 10.23 Important cortical areas of the human brain as seen in a lateral view. Most association areas of the cortex have not been hatched.

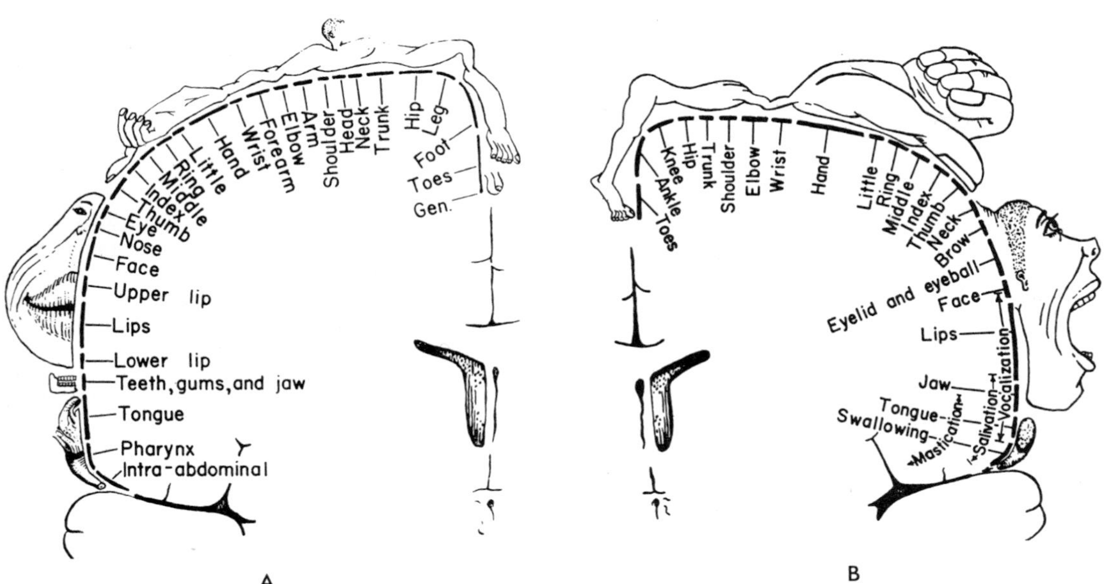

FIGURE 10.24 Vertical sections through *(A)* the sensory cortex receiving from the skin and *(B)* the motor cortex. The proportions of the cortex related to the various body regions have been emphasized by drawing body parts in similar proportions. (From Penfield and Rasmussen: *The Cerebral Cortex.* New York, Macmillan.)

of the body, not to the surface area of that part. Thus, the area receiving impulses from the hands is as extensive as the area receiving impulses from the trunk (Fig. 10.24A).

Impulses from the ear are carried to part of the **temporal lobe** that is separated from the frontal and parietal lobes by the **lateral fissure of Sylvius.** Impulses from the eye are received in the **occipital lobe,** which lies just posterior to the parietal lobe. The path of the optic fibers of mammals is an exception to the generalization that most afferent impulses cross at some point during their ascent to the brain. Half of the fibers in each optic nerve cross in the optic chiasma and end on the opposite side of the brain, but the other half do not. Thus, destruction of one occipital lobe results in inability to perceive images that fall on half of each retina rather than complete loss of vision in one eye.

Appropriate motor impulses to the skeletal muscles are initiated in response to all the sensory data that enter the cerebrum. The cell bodies of the efferent interneurons are contained in the motor cortex, which lies just anterior to the sulcus of Rolando. The motor cortex is subdivided, in the manner of the adjacent sensory cortex, into areas associated with different parts of the body. Fibers to the hand occupy a large portion, for the muscles that control finger movements contain more motor units than most (Fig. 10.24B). This is correlated with the intricacy of our finger movements. Many efferent interneurons pass directly to the motor nuclei of the brain and to the motor columns of the spinal cord, crossing to the opposite side along the way (Fig. 10.21). This direct pathway to the lower motor neurons is called the **pyramidal pathway.** It is phylogenetically a new pathway, being present only in mammals and reaching its greatest development in primates. Skilled and learned movements are, for the most part, mediated by this pathway. Many other motor impulses leave the cortex by way of the **extrapyramidal pathway,** which involves the relay of motor impulses in a mass of gray matter (the **corpus striatum**) situated deep within each cerebral hemisphere and additional relays in the thalamus and reticular formation of the brain stem. This is phylogenetically an older system and is associated with grosser movements, automatic postural adjustments, and stereotyped responses.

When all the areas of known function are plotted, they cover almost all of the rat's cerebral cortex, a large part of the dog's, and a moderate amount of the monkey's, but only a small part of the total surface of the human cortex. The remainder, known as **association areas,** is made up of neurons that are not directly connected to sense organs or muscles but supply interconnections between the other areas. In the human, billions of association neurons are arranged in most parts of the cortex in six vertical layers, with numerous interconnections and with the afferent neurons bringing impulses into the cortex and the efferent ones carrying them out. Such an organization makes possible the exceedingly complex functional interrelationships between different parts of the cortex and between the cortex and subcortical regions. Information in one hemisphere can also affect the other by impulses crossing on commissural fibers that extend between them. A particularly large commissure, known as the **corpus callosum** (L. *corpus*, body + *callosus*, hard) and found only in placental mammals, can be seen in a sagittal section of the brain (Fig. 10.19). All of these interconnections permit the integration of many different sensory modalities, the comparison of current sensory input with information stored in the brain, and the composition of a motor response appropriate to the present requirements of the organism. In some way, the association areas integrate into a meaningful unit all the diverse impulses constantly reaching the brain so that the proper response is made. They interpret and manipulate the symbols and words by which our thought processes are carried on. When disease or accident destroys the functioning of one or more association areas, the condition known as **aphasia** (Gr. *a*, neg + *phasis*, speech) may result, in which the ability to recognize certain kinds of symbols is lost. For example, the names of objects may be forgotten, although their functions are remembered and understood.

Summary

1. Nerve cells, or neurons, adapted for the rapid transmission of a wave of excitation, are found in all multicellular animals except sponges. The nervous system evolved along with animal locomotion and heterotrophic nutrition. The prototype neuron consists of a long, slender axon emerging from one end of the cell body containing the nucleus and bushy dendrites emerging from the other. Functionally, a generator region, restricted to the dendrites and characterized by specializations that permit it to be excited by neurotransmitters from an adjacent neuron, is connected to the conductile region that makes up most of the axon, and this ends in the transmissional region, the tips of the axon specialized for the secretion of neurotransmitters. The generator regions of sensory neurons are activated by a variety of mechanical, chemical, thermal, and other stimuli; the generator regions of motor neurons and interneurons are stimulated through synaptic transmission by a neurotransmitter secreted by an adjacent neuron. Neurons may be true secretory cells, producing and secreting neurohormones, such as vasopressin, oxytocin, and the hypothalamic releasing hormones. The cell body contains endoplasmic reticulum, ribosomes, and Golgi apparatus; the axons contain longitudinally oriented neurofibrils and neurotubules.

3. Sensory neurons are either receptors or connectors of receptors that conduct impulses from receptor endings. Motor neurons conduct impulses to muscles and glands, and interneurons connect two or more neurons. The cell bodies of neurons are typically grouped together in masses, called ganglia in the peripheral nervous system and nuclei in the central nervous system.

4. Enveloping the axon outside the central nervous system is a cellular sheath, the neurilemma, composed of Schwann cells. These lay down on certain axons a special wrapping of insulating fatty material called myelin. Between adjacent Schwann cells are gaps, the nodes of Ranvier, where the axon is free of myelin.

5. The conduction of an impulse along a neuron is independent of the nature or strength of the stimulus, provided that the stimulus is strong enough to start it. The energy for the conduction of the impulse comes from the nerve, not the stimulus. Electrical events in the neuron are controlled by the differential permeability of the neuronal membrane to sodium and potassium ions, regulated in turn by the electric field across the membrane. In the resting state the inside of the neuronal membrane is negatively charged, maintained by the sodium pump, which actively transports sodium ions from the inside to the outside, resulting in a resting membrane potential of 60 to 90 millivolts. The conduction of an impulse involves the depolarization of the membrane, an increased permeability of the membrane, and an inflow of sodium ions. The membrane potential drops to zero or below, then the permeability to sodium decreases, and the resting potential is restored; the membrane is repolarized.

6. Transmission across the synapse is achieved in a few instances by electric currents that pass across the gap. At most synapses, however, the tip of the axon secretes a special neurotransmitter that diffuses across the gap and attaches to specific molecular sites in the adjacent dendrite, leading to the initiation of a new nerve impulse. Synapses are the points at which the flow of impulses through the nervous system is regulated.

7. Cnidarians and some flatworms have simple nerve net systems with synapses that will transmit excitation in either direction. The evolution of the nervous system has been characterized by the development of synapses that permit transmission of impulses in one direction only, the aggregation of nerve fibers into longitudinal bundles, the concentration of nerve cell bodies into ganglia, and the concentration of ganglia at the anterior end, coupled with the appearance of sense organs, to form a brain. The brain and spinal cord make up the central nervous system, and bundles of nerves from the central system to distal parts of the body constitute the peripheral nervous system. Chordates have evolved a dorsal, hollow single nerve cord, and annelids and arthropods have evolved ventral, solid, paired nerve cords with segmental ganglia.

8. A reflex is an innate, stereotyped automatic response to a given stimulus that depends only on the anatomical relationships of the neurons involved. The neuronal pathway along which the impulse travels is called a reflex arc. Inherited reflexes are present at birth, whereas conditioned reflexes are acquired later as the result of experience. Reflexes account for only a small part of the behavior of higher animals but play an important role in the behavior of many lower animals.

9. A pair of peripheral nerves composed of afferent and efferent neurons is present in each segment of the vertebrate body. Near the spinal cord each nerve splits into a dorsal root containing afferent (sensory) neurons and a ventral root containing efferent (motor) neurons. Peripherally the spinal nerve splits into dorsal, ventral, and communicating rami. Fish and amphibia have 10, and reptiles, birds, and mammals have 12 cranial nerves. The spinal cord has an inner mass of gray matter, composed of nerve cell bodies, and an outer mass of white matter made up of axons and dendrites and their myelin sheaths.

10. The autonomic nervous system, composed of sympathetic and parasympathetic nerves, regulates the heart, lungs, digestive tract, and other internal organs. Most organs receive one set of fibers from the sympathetic and another set from the parasympathetic system; these typically have opposite effects on the organ innervated. Impulses reach these organs by two neurons in series, a preganglionic and a postganglionic neuron.

11. The spinal cord and brain are wrapped in protective sheets of connective tissue, the meninges.

12. The brain develops as a series of enlargements at the enlarged anterior end of the embryonic neural tube that results in five regions: telencephalon (olfactory lobes and cerebral hemispheres), diencephalon (thalamus), mesencephalon (midbrain), metencephalon (cerebellum), and myelencephalon (medulla oblongata). Each cerebral hemisphere contains a chamber, the lateral ventricle, connected with the third ventricle in the thalamus and with a fourth ventricle in the metencephalon and medulla oblongata. These cavities are filled with cerebrospinal fluid produced by vascular choroid plexuses. Nuclei in the medulla contain the cell bodies of neurons that reflexly regulate heart rate, respiratory movements, salivation, swallowing, and many other processes. The cerebellum is a center for balance and motor coordination. The pons is a band of transverse fibers that transmit impulses from one side of the cerebellum to the other. The midbrain and its optic lobes are centers that integrate certain functions, such as the movements of the eyes. The thalamus is a center for relaying and processing sensory input. The hypothalamus controls many autonomic functions, such as the regulation of body temperature, appetite, water balance, metabolism, blood pressure, and sleep. The cerebral hemispheres control learned behavior and integrate many sensory modalities, comparing them with information stored in the brain and composing an appropriate motor response.

References and Selected Readings

Aidley, D. J.: *The Physiology of Excitable Cells.* New York, Cambridge University Press, 1974. Presents discussions of the nature of the nerve impulse and of some of the experiments that have led to our present theories regarding the nerve impulse.

Brazier, M. A.: *The Electrical Activity of the Nervous System.* 4th ed. London, Krieger, 1977. A relatively brief, very readable discussion of the nervous system.

Bullock, T. H., and G. A. Horridge: *Structure and Function of the Nervous System of Invertebrates.* San Francisco, W. H. Freeman & Co., 1969. A two-volume, comprehensive treatise covering the nervous systems of a variety of invertebrate animals.

Freedman, R., and J. E. Morris: *The Brains of Animals and Man.* New York, Holiday House, Inc., 1972. A well-illustrated discussion of modern research on brain function.

Gardner, E.: *Fundamentals of Neurology.* 7th ed. Philadelphia, W. B. Saunders Co., 1980. A fine, concise account of the morphology and physiology of the human nervous system.

Garrod, D. R., and J. D. Feldman (Eds.): *Development in the Nervous System.* New York, Cambridge University Press, 1982. The papers presented at a symposium on the embryology of the nervous system.

Guyton, A. C.: *Organ Physiology: Structure and Function of the Nervous System.* 3rd ed. Philadelphia, W. B. Saunders Co., 1981. A masterful presentation of the structure and function of the human nervous system.

House, E. L: *A Systematic Approach to Neuroscience.* 3rd ed. New York, McGraw-Hill Book Co., 1979. A standard text of neurobiology.

Levi-Montalcini, R.: *Nerve Cells, Transmitters and Behavior.* New York, Elsevier, 1980. A brief but interesting discussion of nerve cells and their properties.

Millar, R. P. (Ed.): *Neuropeptides: Biochemical and Physiological Studies.* London, Churchill, 1981. A multiauthor book covering research on the chemical structure and functional role of peptide hormones and neurotransmitters.

Noback, C., and R. Demarest: *The Human Nervous System: Basic Principles of Neurobiology.* New York, McGraw Hill Book Co., 1980. An excellent, superbly illustrated discussion of the nervous system and sense organs.

Rueben, J. P. (Ed.): *Electrobiology of Nerve Synapse and Muscle.* New York, Raven Press, 1976. A multiauthor book covering research on transmission at the synapse.

Sherrington, C. S.: *Integrative Action of the Nervous System.* New Haven, Yale University Press, 1948. A true classic in neurobiology, written by an outstanding pioneer in the field.

11 Receptors and Sense Organs

ORIENTATION

The various kinds of sense organs and receptors enable animals to detect changes in the environment so that they can make appropriate adaptive responses to those changes. We will examine in this chapter the nature of sense organs and their receptor cells and how each of these functions in detecting changes in the environment. We will consider the mechanisms by which a receptor cell can detect a very small change in the environment and how its receptor potential generates a nerve impulse in the sensory neuron. We will then discuss the means by which impulses from the sense organ are coded and decoded in the central nervous system. Receptors for each kind of environmental stimulus have distinct and different structures and function in specific ways. We will describe the structures and functions of mechanoreceptors, chemoreceptors, photoreceptors, thermoreceptors, and electroreceptors. We will explore the physics of hearing and the chemistry of vision.

In order to survive each organism has evolved some means for making appropriate, adaptive reponses to changes in the environment of importance to it and for avoiding making responses to unimportant signals. The organism maintains a homeostatic balance with its environment. This requires that the organism have receptors or sense organs to detect the changes in the environment, systems of nerves and endocrine organs to integrate and coordinate the information received and to trigger the response, and effectors, such as muscles, glands, melanocytes, nematocysts, electric organs, or luminescent structures, to carry out the responses. Sense organs enable their possessors to obtain the information needed as they search for food, find and attract a mate, and escape enemies; they are of great importance in the survival of the individual and of the species.

The receptor cells within sense organs can be remarkably sensitive to the appropriate stimulus. The human eye can be stimulated by an extremely weak beam of light, the human ear can detect very faint sounds, and the human taste buds can detect very dilute solutions of compounds such as vinegar. The sense organs of some animals are sensitive to stimuli that are completely ineffective in humans. Dogs and cats can hear high-pitched whistles inaudible to us. Bats emit very high-pitched noises and are guided in flight by the echoes that rebound from objects in their path. They can catch insects, guided by the echoes from their small prey; some insects can also hear these high-pitched noises and take appropriate evasive action.

Simply put, the function of any receptor cell is to detect a very small amount of energy in a specific form and to transmit the information to the nervous system. The nerve impulse initiated in a sensory neuron by a specific receptor is not specific and is the same no matter where it originates. Differences in the intensity of the stimulus are coded in the number of nerve fibers activated and in the number of impulses passing along a given nerve fiber. Awareness of the sensation depends on the precise part of the brain to which the impulse passes.

The various types of receptors can be classified in several ways: whether they are dispersed (the receptors for touch in the skin) or concentrated (the photoreceptors in the retina of the eye), on the basis of their structure or on the basis of the kind of energy change they monitor (mechanoreceptors, chemoreceptors, photoreceptors, and so on).

Receptor Mechanisms

Receptor cells have the dual function of detecting change and transmitting information concerning the nature of the change to the central nervous system. In the course of evolution, sense organs have become highly specific for one kind of environmental stimulus: One organ detects light, another detects mechanical pressure, a third detects certain chemicals, and so on. No sense organ would be very useful if it responded only to gross changes in the environment. However, if it were so sensitive that it responded to every moving molecule or electron, it would transmit much more noise than signal. Each sense organ has thus evolved specificity and an optimal, not necessarily maximal, sensitivity, so that it maintains an optimal ratio of signal to noise. Sense organs should have the capacity for recording not only "on" and "off" but also rate, magnitude, and direction of change.

Each sense organ has a specialized structure that may consist not only of receptor cells but also of a variety of accessory tissues that support and protect the receptor cells, control the intensity of the signal reaching the receptors, help determine signal direction, and so forth. The receptor cells of the human eye are the rods and cones; the accessory structures are the cornea, lens, iris, and ciliary muscles (see Fig. 11.14). Some of the capacities of a sense organ are intrinsic to the receptor cells; others are conferred upon the sense organ by the accessory structures. Among invertebrates, for example, one type of mechanoreceptor may be sensitive only to gross touch because it is associated with a stout, rigid spine, whereas another type may be an auditory receptor because it is connected to a long filamentous hair that is moved by sound waves.

Receptors are often parts of nerve cells, the axons of which extend directly into the central nervous system or connect synaptically with one or more interneurons connected with the central nervous system. Other receptors, such as the human taste buds, are modified epithelial cells connected to one or more sensory nerve cells (see Fig. 11.11). In those sense organs in which the receptor is a primary neuron, this one cell both detects and transmits the information to the central nervous system. In organs in which the receptor cell is not a modified neuron, it detects, but the information is transmitted by the associated neuron. Each receptor is specialized to receive one particular form of energy more efficiently than others. Rods and cones absorb the energy of photons of certain specific energies and wave lengths. Temperature receptors respond to radiant energy transferred by radiation, conduction, or convection. Taste and smell are detected by the potential energy in the mutual attraction and repulsion of molecules.

The various kinds of environmental energy cause the receptors to perform biological work. This work transforms metabolic energy into electrical energy. These relationships are exemplified by a very simple sense organ, the tactile hair of an insect. This hair, plus its

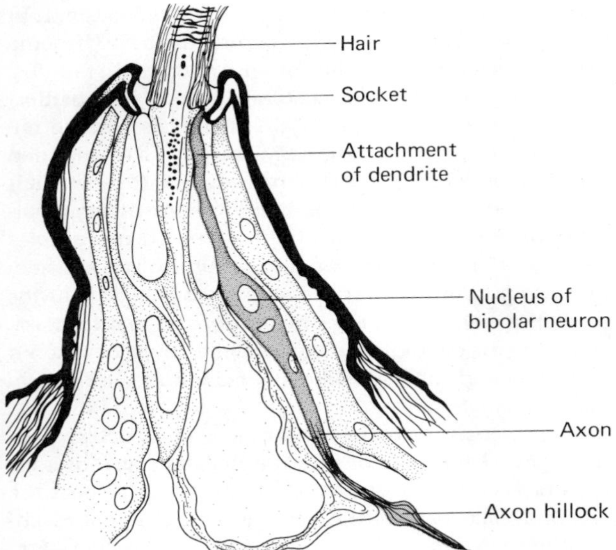

FIGURE 11.1 A tactile hair from a caterpillar, showing the attachment of the dendrite of the bipolar neuron (the mechanoreceptor) at the point where the shaft of the hair enters the socket. (From Hsu, F.: Étude cytologique et comparée sur les sensilla des insectes. *La Cellule, 47*:1–60, 1938.)

associated cells, is a complete sense organ (Fig. 11.1). The **bipolar neuron** at its base is the receptor; the dendrite is attached to the base of the hair, near the socket; the axon passes directly to the central nervous system without synapsing. In its unstimulated state this neuron maintains a steady resting potential; i.e., there is a potential difference between the inside and outside of the neuron. As in muscles (p. 108), this potential difference exists because the ionic compositions of the fluids on each side of the semipermeable membrane are different. The difference is maintained by the sodium pump and by metabolic work performed by the cell. When the hair is touched its shaft moves in the socket and mechanically deforms the dendrite. This stimulus, the deformation of the cell, changes the permeability of the neuronal membrane to ions, with the result that the potential difference between the two sides of the membrane decreases, disappears, or increases. If it decreases or disappears, the cell is said to be **depolarized.** If it increases, the cell is said to be **hyperpolarized.** The state of depolarization caused by the stimulus is termed the **receptor potential.** The receptor potential spreads relatively slowly down the dendrite, decaying exponentially. When a special area of the cell near the axon, the **axon hillock,** becomes depolarized, **action potentials** are generated. The action potentials then travel along the axon to the central nervous system. The primary receptor detects an event in the environment (movement of the hair), generates electrical energy at the expense of its metabolic energy (the receptor potential), and transmits information (action potentials) to the central nervous system.

The amplitude and duration of the receptor potential are related to the strength and duration of the stimulus; a strong stimulus causes a greater depolarization of the receptor membrane than does a weak one. The action potentials are repetitive, and the frequency at which they are generated is related to the magnitude of the receptor potential. The strength of the stimulus is reflected in the frequency of the action potentials. The amplitude of each action potential bears no relation to the stimulus. It is characteristic of the particular neuron under the usual recording conditions.

The action potential (p. 214) is an "all-or-none" phenomenon. The receptor potential, in contrast, is a **graded response.** Once a stimulus has triggered a receptor to generate action potentials, the stimulus has no further control over them. Even though the stimulus may continue unabated, neither the receptor potential nor the action potentials will continue unchanged (Fig. 11.2). The receptor potential gradually falls, and the frequency of the action potentials decreases. This phenomenon is termed **adaptation.** Some receptors adapt very rapidly and completely; others do so more slowly (Fig. 11.3).

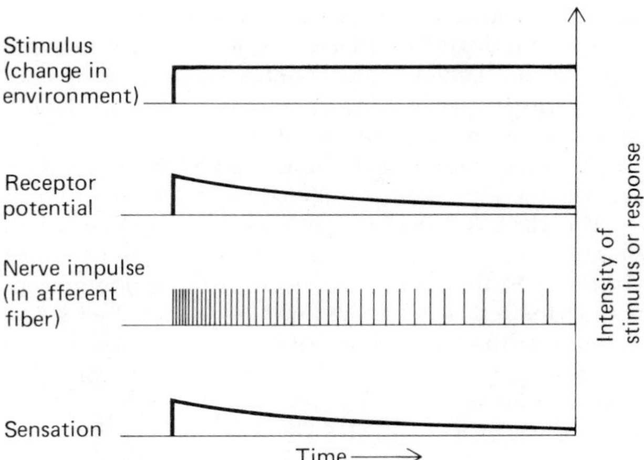

FIGURE 11.2 A diagram showing the relations among the stimulus, the receptor potential, the action potential, and sensation. (From Adrian, E. D.: *The Basis of Sensation.* London, Chatto & Windus, Ltd., 1949.)

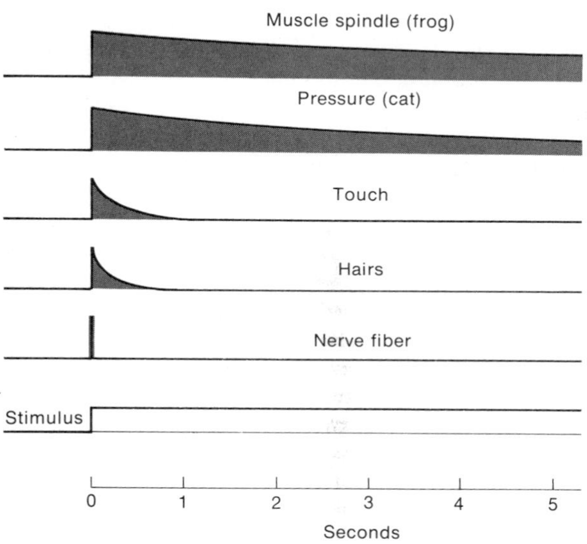

FIGURE 11.3 A diagram showing the relation between the stimulus and the different rates of adaptation for different receptors and a nerve fiber. The heights of the curves indicate the rates of discharge of action potentials. (From Adrian, E. D.: *The Basis of Sensation*. London, Chatto & Windus Ltd., 1949.)

Sensory Coding and Sensation

The stimulation of any sense organ or sensory receptor initiates a type of coded message composed of action potentials transmitted by the nerve fibers and decoded in the central nervous system. Impulses from the sense organs may differ in (1) the total number of fibers transmitting; (2) the specific fibers carrying action potentials; (3) the total number of action potentials passing over a given fiber; (4) the frequency of the action potentials passing over a given fiber; or (5) the time relations between action potentials in different specific fibers. How the sense organs initiate different codes and how the brain analyzes and interprets them to produce the various sensations are not yet understood. It is important to remember that all action potentials are qualitatively the same. Light of wave length 400 nm. (blue), sugar molecules (sweet), and sound waves of 440 hertz (A above middle C) all initiate action potentials to the brain via the appropriate nerves; these action potentials are identical. How can the organism then assess its environment accurately? The qualitative differentiation of stimuli must depend on the sense organ itself or the brain, or both. In fact it depends upon both. Our ability to discriminate red from green, hot from cold, or red from cold is due to the fact that particular sense organs and their individual receptor cells are connected to specific cells in particular parts of the brain.

The frequency of the repetitive action potential along sensory nerve fibers codes the intensity of the stimulus. Since each receptor normally responds to but one category of stimuli, a message arriving in the central nervous system along this nerve fiber is interpreted as meaning that a particular stimulus has occurred. Interpretation of the message and, in the case of human beings, the quality of sensation depend on which central neurons receive the message. Sensation, when it occurs, occurs in the brain. Rods and cones do not see—only the combination of rods, cones, and centers in the brain can see. Furthermore, many sensory messages never give rise to sensations. For example, when chemoreceptors in the carotid sinus and the hypothalamus sense internal changes in the body, appropriate physiological adjustments are made, but our consciousness is not stirred.

Since only those nerve impulses that reach the brain can result in sensations, any blocking of the passage of the impulse along the nerve fiber by an anaesthetic has the same effect as removing the original stimulus entirely. The sense organs, of course, will continue to initiate impulses that can be detected by the proper electrical apparatus, but the anaesthetic prevents them from reaching their destination.

Another method of coding information, termed **cross-fiber patterning,** is used in olfactory organs. It is unlikely that the olfactory organ contains a specific receptor for each of the thousands of individual odors that can be recognized. There is evidence that only a limited number of categories of receptors exist, each of which responds to a spectrum of odors. There is no rigid specificity because the spectra overlap. Perception of different characteristic odors probably depends upon the pattern of response of all the fibers responding together.

The temporal pattern of action potentials generated in a single neuron may serve as a code for different stimuli. Single taste receptors in flies, for example, generate action potentials at an even, regular frequency when the stimulus is salt but generate irregular frequencies when the stimulus is acid.

The axon of a sensory neuron in invertebrates typically extends to the central nervous system without synapsing. In these circumstances the message generated at the periphery arrives unaltered. In the compound eye, as in most vertebrate sense organs, many interneurons are interposed between the receptor and the brain centers. The vertebrate retina or olfactory bulb has an exceptionally complicated neural circuitry. In many animals, especially vertebrates, the messages generated by the peripheral sense organs pass through complex neuronal pathways to reach the brain. As a consequence of all of these synaptic connections, the original message is altered and may lose or gain some of its information. The message that finally arrives at the brain has been well censored by interneurons and bears even less resemblance to the original stimulus than did the action potentials from the receptor.

Mechanoreceptors

Receptors are customarily classified according to the nature of their effective stimuli. Thus there are mechanoreceptors, chemoreceptors, photoreceptors, thermoreceptors, and electroreceptors. Each of these receptors consists of one or more specialized cells or free nerve endings that respond as miniature transducers, for, like the piezoelectric crystal in a phonograph pickup, they convert one form of energy into another. Each receptor is affected by the type of energy to which it is attuned and converts it into an electric current, the receptor potential, which in turn initiates the action potentials of the nerve impulse. Unlike nerves or many muscles the activity of a receptor is not an "all-or-none" phenomenon but varies with the strength of the stimulus. The receptor potential must attain a certain threshold to initiate an action potential. If the intensity of the receptor potential exceeds the threshold, it will initiate additional action potentials, additional nerve impulses.

Mechanoreceptors are sensitive to stretch, compression, or torque imparted to tissues by the weight of the body, the relative movement of parts, the gyroscopic effects of moving parts, and the impact of the substratum or the surrounding medium (air or water). Mechanoreceptors are concerned with enabling an organism to maintain its primary body attitude with respect to gravity and with maintaining postural relations (information that is essential for all forms of locomotion and for all coordinated and skilled movements). Mechanoreceptors provide information about the shape, texture, weight, and topographical relations of objects in the external environment; they are necessary for the operation of certain internal organs; and they supply information about the presence of food in the stomach, feces in the rectum, urine in the bladder, and fetus in the uterus. Mechanoreceptors may be categorized as tactile, proprioceptive, and auditory.

The Tactile Sense

Among the simplest tactile receptors are the **tactile hairs** of invertebrates. The tactile hair of an insect is a **phasic receptor** (i.e., it responds only when the hair is moving). When the hair is displaced, a receptor potential develops and a few action potentials are generated, but all activity ceases when motion ceases, even though the hair is maintained in the displaced position.

The remarkable tactile sensitivity of human beings, especially in the fingertips and lips, is due to a large and diverse number of dispersed receptors in the skin (Fig. 11.4). By comparing the distribution of the different types of receptors and the types of sensations produced, it has been found that free nerve endings are responsible for pain perception, that basket nerve end-

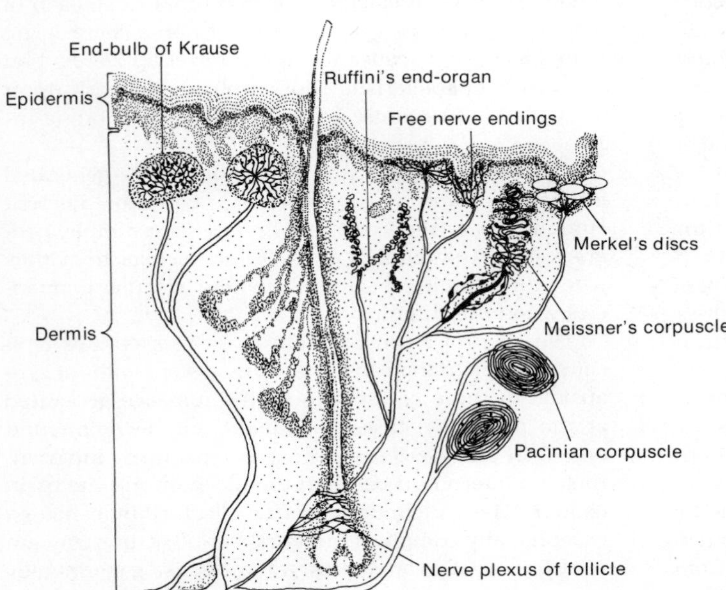

FIGURE 11.4 Diagrammatic section through the skin showing the types of sense organs present. The sense organs respond to the following stimuli: cold, end-bulbs of Krause; warmth, Ruffini's end-organs; touch, Meissner's corpuscles and Merkel's discs; deep pressure, pacinian corpuscles; and pain, free nerve endings.

ings around hair bulbs (Meissner's corpuscles and Merkel's discs) are responsible for touch, that the end bulbs of Krause and Ruffini's endings are responsible for sensations of cold and warmth, and that pacinian corpuscles mediate the sensation of deep pressure.

The pacinian corpuscle has been studied extensively. The bare axon is surrounded by lamellae interspersed with fluid. Compression causes displacement of the lamellae, which provides the deformation stimulating the axon. Even though the displacement is maintained under steady compression, the receptor potential rapidly falls to zero and action potentials cease. This is a phasic receptor responding to velocity.

Proprioception

Among many invertebrates, such as insects, the sense organs most commonly concerned with relaying postural information are hairs, plates (campaniform organs), and other modified cuticular structures. These are **tonic** (static) sense organs. Unlike phasic receptors, the receptor potential is maintained (though not at a constant magnitude) as long as the stimulus is present and action potentials continue to be generated. Thus there is continued information about the position of the organ concerned.

Each skeletal muscle, tendon, and joint is equipped with **proprioceptors** sensitive to muscle tension or stretch or to the rate of change in these qualities. Impulses from the proprioceptors are extremely important in ensuring the harmonious contraction of the different muscles involved in a single movement. Without them, complicated skillful acts, such as tying knots even when our eyes are closed, would be impossible. Impulses from these organs are also important in the maintenance of balance. Proprioceptors are probably more numerous and more continuously active than any of the other sense organs, although we are less aware of them than we are of the others. We can get some idea of what life without proprioceptors would be like when an arm or leg "goes to sleep"—a feeling of numbness results from the lack of proprioceptive impulses.

Muscle Spindles

The mammalian muscle spindle is a versatile stretch receptor and illustrates how sensory performance may be modified by the consequences of its own action. In the muscles of higher vertebrates there are, in addition to the usual striated muscle fibers termed **extrafusal fibers,** special **intrafusal fibers** associated with sensory nerve endings. A bundle of intrafusal fibers, together with their sensory endings, is termed a **muscle spindle.** Intrafusal fibers are striated except in the region of the nucleus. In the region of attachment of the extrafusal muscle fibers to the tendon there is another sense organ, the **tendon** or Golgi **organ.** There are two sets of motor neurons to the muscle: The alpha efferents innervate the ordinary, or extrafusal, muscle fibers, and the gamma efferents innervate the intrafusal fibers. For violent muscular contraction, impulses from the central nervous system come mostly by way of the alpha efferents. If, as a consequence, the muscle is stretched excessively, the Golgi organ and muscle spindle are stimulated. Messages from the Golgi organ pass up the sensory nerves to a point within the spinal cord where they synapse with the alpha efferents. They inhibit the alpha efferents and the muscle stops contracting; thus tension is kept within bearable limits, and the muscle is kept at a constant length under a specific load. In the production of slow voluntary movements, impulses from the central nervous system pass down the gamma efferents to the intrafusal fibers, which begin a slow, graded contraction. As a consequence, the muscle spindle is stimulated and sends impulses to the synapse with the alpha efferent, which is excited to cause the extrafusal fibers to contract. Since the intrafusal fibers are connected in parallel with the other fibers, rather than in series, as are the Golgi organs, they become slack (Fig. 11.5). The spindle then stops exciting the alpha efferents, and the muscle ceases contracting. The net result is to cause the muscle to come to a new state of tension; thus muscle tone is maintained, and precise voluntary movements can be made.

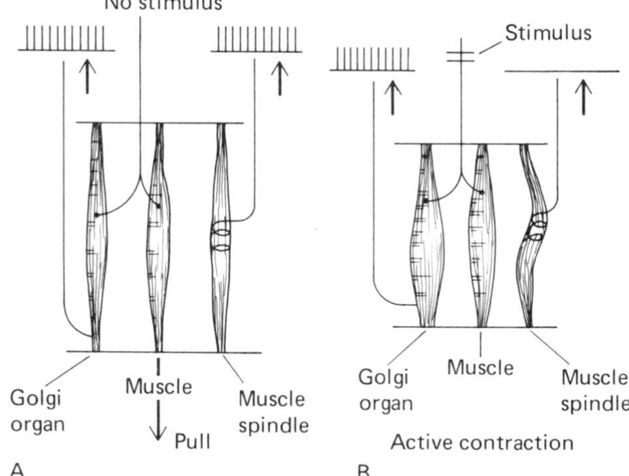

FIGURE 11.5 **Golgi organ receptors have a series relationship to the muscle fibers. Spindles are in parallel. Pull on the muscle *(A)* increases the rate of firing of both receptors. Active contraction of the muscles *(B)* will cause an increase in discharge of the Golgi tendon organ and a decrease in rate of discharge from the spindle. (From Ochs, S.: *Elements of Neurophysiology,* New York, John Wiley & Sons, Inc., 1965.)**

While the muscle has been contracting, it has been stretching its antagonist. Naturally, the spindle of the antagonist muscle excites it to contract. If it were to continue to do so in the face of the pull being exerted upon it, the excessive strain would stimulate its Golgi organ. This inhibits its contraction. Thus antagonistic muscles are prevented from fighting each other.

The Importance of Knowing Which Way Is Up

It is clearly important for every animal to know at all times which way is up, so that it can contract its muscles or beat its cilia to move appropriately. When the organ in our inner ear responsible for equilibrium is malfunctioning, we become dizzy and are unable to stand upright. Organs of balance, called **statocysts,** (Gr. *statos,* standing + *kystis,* sac), are found in most phyla of animals. These are usually hollow spheres of sense cells, in the middle of which is a **statolith,** a particle of sand or calcium carbonate pressed by gravity against certain mechanoreceptors. As the animal's body changes position, the statolith is pressed against different sense cells, and the animal is then stimulated to regain its orientation with respect to gravity. The statocysts of jellyfish are associated at specific sites between the scallop-like folds of the margin of the bell to form a **rhopalium.** The jellyfish carry out swimming movements that are primarily vertical; if the bell becomes tilted, stimuli from the statocysts bring about asymmetrical contractions, and the bell is righted. The horizontal movements of the jellyfish depend largely upon water currents. Decapods, such as crabs and lobsters, have a statocyst at the base of each first antenna that opens by a pore to the exterior. The animals use sand grains as statoliths and must replace them after each molt. A newly molted crayfish given iron filings instead of sand grains will place them within the statocyst chamber and then will orient itself to a magnetic field rather than to gravity.

Membranous Labyrinths. All vertebrates have the ability to perceive differences in the orientation of their bodies with respect to their surroundings and to maintain their equilibrium. Although vision and proprioceptive impulses from the muscles and joints play a part, this ability is primarily a function of the inner ear. The inner ear contains a complex of membranous walls, sacs, and ducts called the membranous labyrinth, filled with a liquid **endolymph** and surrounded by a protective liquid cushion, the **perilymph.** The membranous labyrinths of nearly all vertebrates include the saccule, utricle, and three semicircular ducts.

The **utricle** (L. *utriculus,* a bag) and **saccule** are small hollow sacs containing patches of sensitive hair cells and small ear stones, or **otoliths,** made of calcium carbonate. Normally the pull of gravity causes the otoliths to press against particular hair cells, stimulating them to initiate impulses to the brain via sensory nerve fibers at their bases. When the head is tipped, the otoliths press upon the hairs of other cells and stimulate them.

Each of the three **semicircular ducts** consists of a semicircular tube connected at both ends to the utricle (Fig. 11.6). The ducts are so arranged that each is at right angles to the other two. At one of the openings of each canal into the utricle is a small bulblike enlargement, the **ampulla,** containing a clump of hair cells similar to those in the saccule and utricle but lacking otoliths. These cells are stimulated by movements of the endolymph within the ducts. When the head is turned, there is a lag in the movement of the fluid within the ducts; the hair cells move in relation to the fluid and are stimulated by its flow. This stimulation produces not only the awareness of rotation but also certain reflex movements in response to it, movements of the eyes and head in a direction opposite to the original rotation. The three ducts are located in three different planes; hence movement in any direction will stimulate the movement of fluid in at least one of the ducts.

Differences in the position of the head (**static equilibrium**) affect the way gravity pulls the otoliths on the underlying hair cells in the utricle and saccule (and we constantly know which way is up). Rapid forward movement (**linear acceleration**) causes the otoliths, which have more inertia than the fluid, to push back upon certain hair cells. During sudden turns of the head in various planes (**angular acceleration**), the endolymph of the semicircular ducts, because of its inertia, does not move as fast as the hair cells in the ampulla, and their

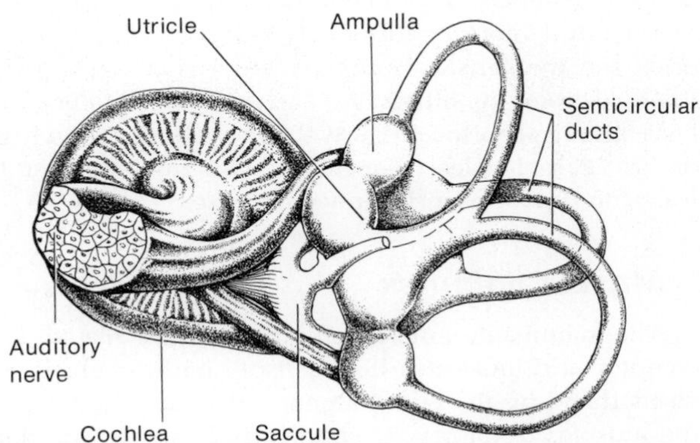

FIGURE 11.6 The right semicircular ducts and cochlea of an adult human, shown dissected free of surrounding bone and enlarged about five times, seen from the inner and posterior side. Note that the plane of each semicircular duct is perpendicular to those of the other two.

differential rate of movement stimulates the hair cells. We cannot distinguish between static equilibrium and linear acceleration on the basis of signals from the ear alone; inputs from other sensory organs are also necessary.

The Halteres of Flies. Millions of years before humans invented the gyroscope, flies had evolved a balancing organ to stabilize flight. Any flying machine must maintain stability if it is to be controllable in the air. Flies must be able to control lift and to stabilize in all three planes of rotation; that is, they must correct for pitch, roll, and yaw. Flies accomplish this with information derived from the **halteres,** a pair of marvelously modified hind wings. Each haltere is a heavy mass of tissue on a thin stalk (Fig. 11.7) and resembles an Indian club. The base is folded and articulated in a complicated fashion and is equipped with some 418 mechanoreceptors. These respond to strains produced in the cuticle by gyroscopic torque generated by the beating of the halteres. These oscillating masses generate forces at the base of the stalk as the whole fly rotates. They probably do not act as stabilizing gyroscopes of the sort placed in ships to offset the movement of the waves. Their action is indirect in that their mechanoreceptors signal the central nervous system to make the necessary corrections in flight.

Phonoreception

Phonoreception, or hearing, refers to the detection of pressure waves resulting from a mechanical disturbance some distance away. The human ear can detect sound waves with frequencies of from 20 to about 20,000 hertz, and some animals—bats, for example—can detect frequencies of over 100,000 hertz. Frequencies of less than 20 hertz are usually perceived as vibrations and not as sound. Mammals, birds, and some reptiles have a **cochlear duct,** an elongated cul-de-sac extending from the sacculus concerned with phonoreception. Fishes have a homologous, but very much smaller, diverticulum known as the **lagena** (Fig. 11.8). The sound waves reach-

A

B

FIGURE 11.7 *A,* Ventral view and *B,* dorsal view of a model of the left haltere of the blowfly *Lucilla sericata.* Rows of sense organs are visible at the base. Magnification ×130. (Courtesy of J. W. S. Pringle.)

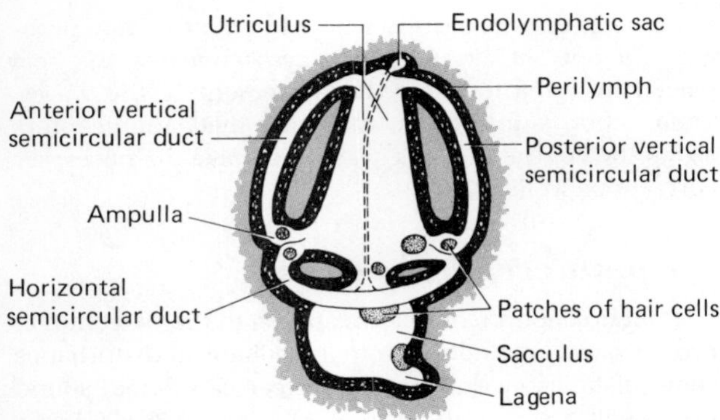

FIGURE 11.8 The left ear of a fish seen in a lateral view. Only an inner ear is present, embedded within spaces in the otic capsule of the skull. (Modified after Kingsley.)

ing a fish are in water, and the tissues of the fish are mostly water. Thus sound waves readily penetrate to the inner ear. The fish's problem is to avoid being "transparent" to sound. Many trap or interrupt the passage of a sound wave with a large otolith in the sacculus; others use the swim bladder as the initial receptor, or hydrophone.

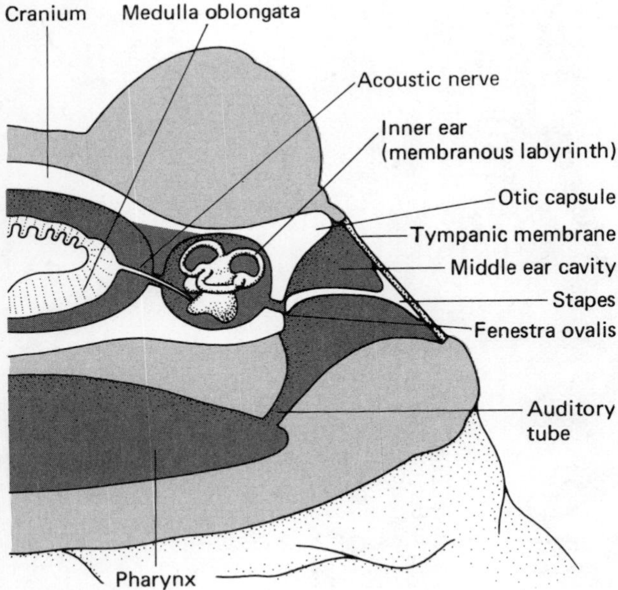

FIGURE 11.9 A diagrammatic cross section through the head of a frog to show the ear and its relation to surrounding parts.

Phonoreception in Tetrapods. In the tetrapods, a part of the membranous labyrinth of the ear, usually the lagena or cochlear duct, is specialized for phonoreception. A problem of terrestrial life is to amplify the force of the sound waves in air (a very light compressible medium) enough to induce a pressure wave in the denser incompressible liquid of the inner ear. A variety of devices have evolved to amplify and transmit either ground or airborne vibrations to the phonoreceptor. These constitute the external and middle ears. Frogs detect higher-frequency vibrations by an external **tympanic membrane** (Fig. 11.9) that reponds to vibrations in the air and a bone, the stapes, that transmits the vibrations across the middle ear cavity from the tympanic membrane to the oval window (fenestra ovalis) in the otic capsule. The oval window communicates with the inner ear.

The hearing apparatus of mammals is similar but more elaborate (Fig. 11.10). The outer ear consists of two parts: the skin-covered cartilaginous pinna, or auricle, and the auditory canal (external auditory meatus) leading from it to the middle ear. In humans, the pinna, or visible ear, is of some slight use for directing sound waves into the canal, but in other animals, such as the cat, the larger, movable pinna is very important. It acts as an ear trumpet that concentrates and increases slightly the pressure of the sound waves that vibrate the tympanic membrane, or eardrum, which is situated in a protected site at the internal end of the auditory canal. The middle ear is a small chamber containing three tiny bones: the hammer-shaped **malleus,** the anvil-shaped **incus,** and the stirrup-shaped **stapes,** arranged in sequence. These transmit sound waves across the middle ear cavity to the oval window and increase their amplitude. The major pressure amplification results from the fact that the area of the eardrum is nearly 20 times greater than the area of the membrane in the oval window. Virtually all of the force that impinges on the tympanic membrane reaches the membrane in the oval window, and since this membrane is smaller, the force per unit area is increased.

The narrow **auditory,** or **eustachian, tube** connects the middle ear to the pharynx and serves to equalize the pressure on the two sides of the eardrum. If the middle ear were completely closed, any variation in atmospheric pressure would cause a pronounced and painful bulging or caving in of the eardrum. The pharyngeal end of the eustachian tube normally is collapsed and closed so that we do not become unpleasantly aware of our own voice. The auditory tube is opened during yawning or swallowing, and during an abrupt ascent or descent in an elevator or airplane. Such actions help to prevent a cracking sensation of the eardrums produced by the changes in the atmospheric pressure.

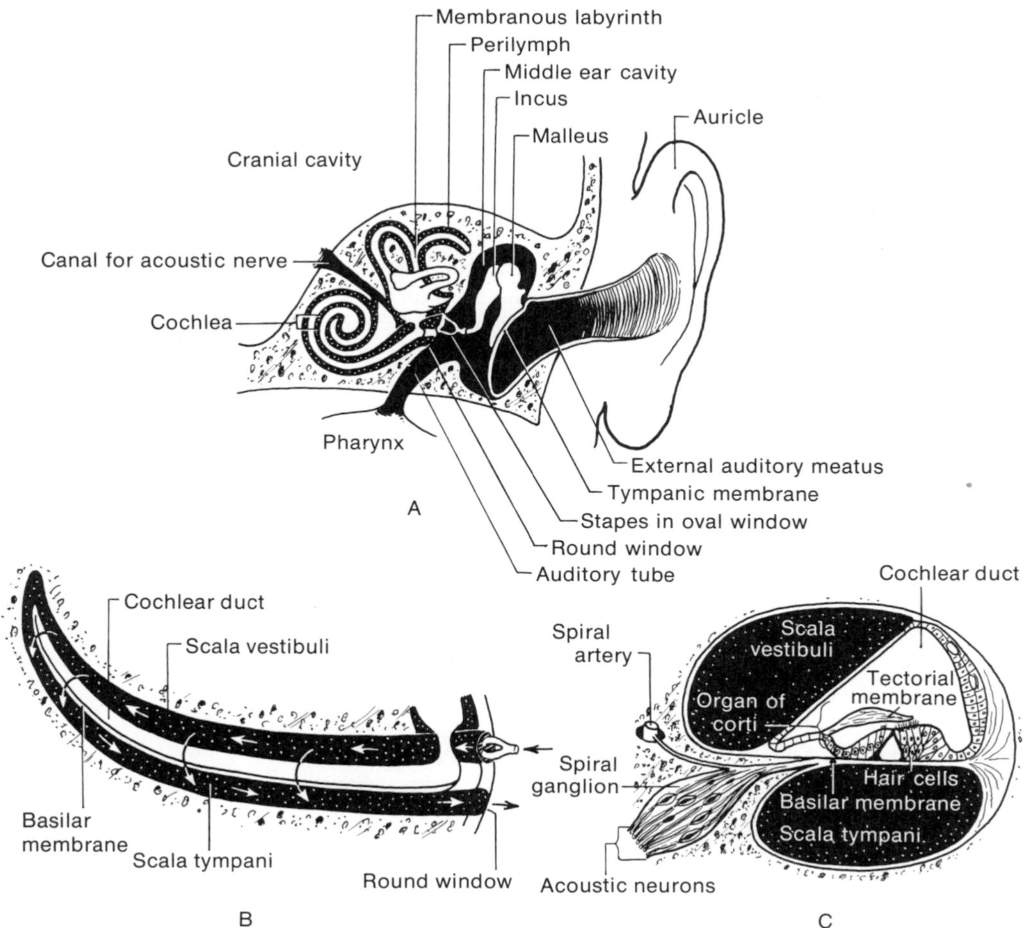

FIGURE 11.10 The mammalian ear. *A,* Schematic drawing of the outer, middle, and inner ear of a human being. *B,* Diagram of the cochlea as though it were uncoiled. *C,* An enlarged cross section through the cochlea. The cochlear duct and other parts of the membranous labyrinth are filled with endolymph and are shown in white; perilymph is black with white stipples.

The inner ear consists of a complicated group of interconnected canals and sacs often referred to, most appropriately, as the **membranous labyrinth.** The part of the labyrinth concerned with hearing is a spirally coiled tube of two and one half turns, resembling a snail's shell, called the **cochlea** (Gr. *kochlias,* snail). The cochlea consists of three canals separated from each other by thin membranes and coming almost to a point at the apex. The oval window is connected to the base of one of these tubes, the **vestibular canal.** At the base of the **tympanic canal** is another opening, the round window, covered by a membrane that also leads to the middle ear. These two canals are interconnected at the apex of the cochlea and are filled with a fluid known as **perilymph.** Between the two lies a third canal, the **cochlear duct,** filled with **endolymph** and containing the actual organ of hearing, the **organ of Corti.** This structure consists of five rows of cells extending the entire length of the coiled cochlea; each cell is equipped with hairlike projections extending into the cochlear canal. Each organ of Corti contains about 24,000 hair cells. The hair cells rest upon the **basilar membrane** separating the cochlear duct from the tympanic canal (Fig. 11.10). Overhanging these cells is the **tectorial membrane** (L. *tectum,* roof) attached along one edge to the membrane on which the hair cells rest and with its other edge free.

The hair cells initiate impulses in the fibers of the auditory nerve.

Pressure waves produced by the stapes at the oval window pass through the vestibular canal, cross the cochlear duct, and escape through the tympanic canal and the round window. The basilar membrane, which supports the organ of Corti, is set in vibration. The basilar and the tectorial membranes are hinged at different places; hence, a pressure wave causes a slight differential movement between them and develops a shearing force that stimulates the intervening hair cells, which are mechanoreceptors. The ear is extremely sensitive: At certain frequencies it can detect any displacement of these membranes that is as slight as the diameter of a hydrogen atom. Sensory neurons of the acoustic nerve extend from the hair cells into the brain.

How sounds are analyzed by the cochlea is a complex process not completely understood. It is well established that pitches or tones of different frequencies are detected in different regions of the cochlea. According to the current hypothesis, a movement of the stapes against the liquid in the vestibular canal causes the proximal part of the basilar membrane to move toward the tympanic canal. This sets up a tension in the basilar membrane and initiates a wave that travels along the membrane. The mechanism is analogous to a wave sent along a whip by a jerking movement at the proximal end. The wave travels along the entire basilar membrane, but the physical properties of the membrane, including its width and elasticity, change progressively, so that the length and amplitude of the wave vary. Waves of different frequency reach maximal amplitude at different points along the cochlea. Long wave lengths, or low notes, cause a maximal displacement near the apex of the cochlea, whereas shorter wave lengths, or high notes, cause a maximal displacement at the proximal end of the basilar membrane. More intense sounds, in addition to stimulating hair cells to move vigorously, cause a longer length of the basilar membrane to vibrate, but the displacement peak remains the same for any given frequency. Two musical instruments playing the same note have dissimilar qualities because of differences in the number and kinds of overtones, or harmonics, that are present. Both instruments stimulate the same part of the basilar membrane, but they vary with respect to the other parts of the membranes that are stimulated simultaneously. Loud sounds cause resonance waves of greater amplitude and lead to a more intense stimulation of the hair cells and the initiation of a greater number of impulses per second passing over the auditory nerve to the brain.

Careful histologic work has shown that the nerve fibers from each particular part of the cochlea are connected to particular parts of the auditory area of the brain, so that certain brain cells are responsible for the perception of high tones, whereas other cells are responsible for the perception of low tones.

Impaired hearing or deafness may be caused by injuries or malformations of either the sound-transmitting mechanisms of the outer, middle, or inner ears or of the sound-perceiving mechanisms of the latter. The external ear may become obstructed by wax secreted by the glands in its wall, the middle ear bones may become fused after an infection, or the inner ear or auditory nerve may be injured by a local inflammation or by the fever accompanying some disease. Conduction deafness can be corrected by a hearing aid that amplifies vibrations enough to be transmitted directly through the skull bones to the cochlea. If the auditory nerve or the cochlea is damaged, the resulting deafness usually cannot be corrected. Continued exposure to loud, high-pitched noises, as with a boilermaker or a member of a rock band, may destroy certain hair cells in the cochlea, leading to partial deafness.

Relatively few animals have a sense of hearing. The vertebrate ear began as an organ of equilibrium, the cochlea being a later evolutionary outgrowth of the saccule that reached full development only in mammals.

Chemoreceptors

Throughout the Animal Kingdom, many sexual, reproductive, social, and feeding activities are regulated or influenced by specific chemical aspects of the environment. Insects, for example, use a great many chemicals in communication, for defense from predators, and for the recognition of specific foods. Many vertebrates employ chemical secretions to mark territory, to attract their sexual partners, or to defend themselves. Chemoreception is also involved in the tracking of prey by carnivores and in the detection of carnivores by their intended prey.

Among mammals specific and highly sensitive chemoreceptive systems compose the senses of taste and smell. Human beings depend primarily on visual or auditory cues. We use our chemical senses much less than other mammals do and tend to minimize their importance.

The organs of taste are budlike structures located in mammals on the tongue and soft palate, but in lower vertebrates they are found in many parts of the mouth and pharynx, and even in some tissues on the skin of the head. Each taste cell, which is an epithelial cell and a receptor, has at its free surface a border of microvilli, many of which project into a tiny pore connecting with the fluids bathing the surface of the tongue. The con-

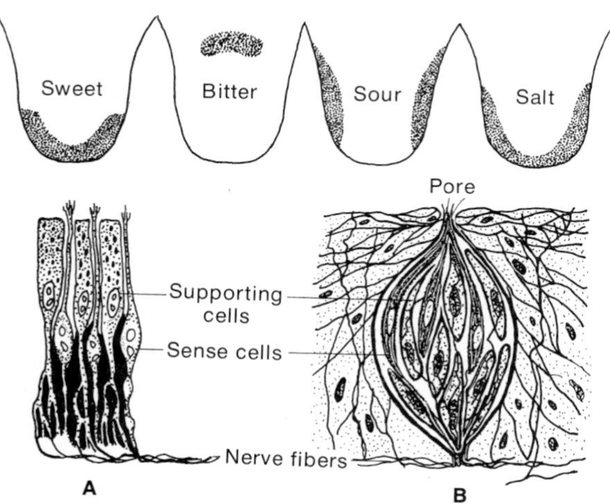

FIGURE 11.11 *Above,* The distribution on the surface of the tongue of taste buds sensitive to sweet, bitter, sour, and salt. *Below: A,* Cells of the olfactory epithelium of the human nose. *B,* Cells of a taste bud in the epithelium of the tongue.

nections with the nerve cells are complex, for each taste cell is innervated by more than one neuron. Some neurons may connect with one taste cell and others with many.

Traditionally there are four basic tastes: sweet, sour, salt, and bitter. To this must now be added water. It is true that the greatest sensitivity to each of these tastes is restricted to a certain area of the human tongue (Fig. 11.11). Some taste buds are specific to salt, acid, or sugar, but most respond to two or more categories of taste solutions. Thus the detection and processing of information in the taste organs of the tongue are very complicated. Taste discrimination probably depends on a code that consists of cross-fiber patterning; that is, each receptor responds to more than one kind of chemical, but no two respond exactly alike, so that the total pattern of messages going to the brain is different for different solutions.

Flavor does not depend on the perception of taste alone. It is compounded of taste, smell, texture, and temperature. Smell affects flavor because odors pass from the mouth to the nasal chamber by way of the choanae.

The Sense of Taste in Insects

One of the most thoroughly studied organs of taste is the taste hair of the fly (Fig. 11.12). The terminal segments of the legs and the mouth parts of flies, moths, butterflies, and certain other insects are equipped with very sensitive hairs and pegs. In the fly each one of these contains four taste receptors and a tactile receptor. All are primary sensory neurons both receiving and transmitting stimuli. One taste receptor is more or less specific to sugars, one to water, and two to salts. If water is placed on one hair of a thirsty fly, action potentials generated by the water cell pass directly to the central nervous system and cause the fly to respond by extending its retractible proboscis and drinking. Similarly, sugar on one hair stimulates the sugar receptor and causes the fly to feed. Salts cause the fly to reject the solution.

The Sense of Smell (Olfaction)

The sense of smell of terrestrial vertebrates is served by primary neurons located in the nasal epithelium in the upper part of the nasal cavity (Fig. 11.13). Each of these neurons has a short axon that passes through the cribriform plate of the skull and immediately synapses with other neurons in the brain. In the rabbit, for example, there are some 10^8 receptors that make millions of synaptic connections within the olfactory bulb of the brain. The possibilities for processing olfactory data generated by the receptors even before they reach the cerebrum are enormous. In contrast to the sensations of taste, the various odors cannot be assigned to specific classes. Each substance has its own distinctive smell. The olfactory organs respond to remarkably small amounts of a substance. The synthetic substitute for the odor of violets, ionone, can be detected by most people when it is present to the extent of one part in more than 30 billion parts of air. The sense of smell is rapidly fatigued, and air originally having a powerful stimulus may seem odorless after a few minutes. This fatigue is specific for the particular substance producing it. Thus, receptors that have become insensitive to one substance will react quite normally to another. This suggests that there are many different kinds of sense cells, each specific for a particular chemical. Some people either completely lack a sense of smell or are able to smell some substances but not others.

Photoreceptors

Light-sensitive cells exist in almost all organisms. Even protozoa respond to changes in light intensity, usually moving away from the source of light. Some protozoa have eye spots that are more sensitive to light than is the rest of the cell. In some animals, such as earthworms, isolated light-sensitive cells, or photoreceptors,

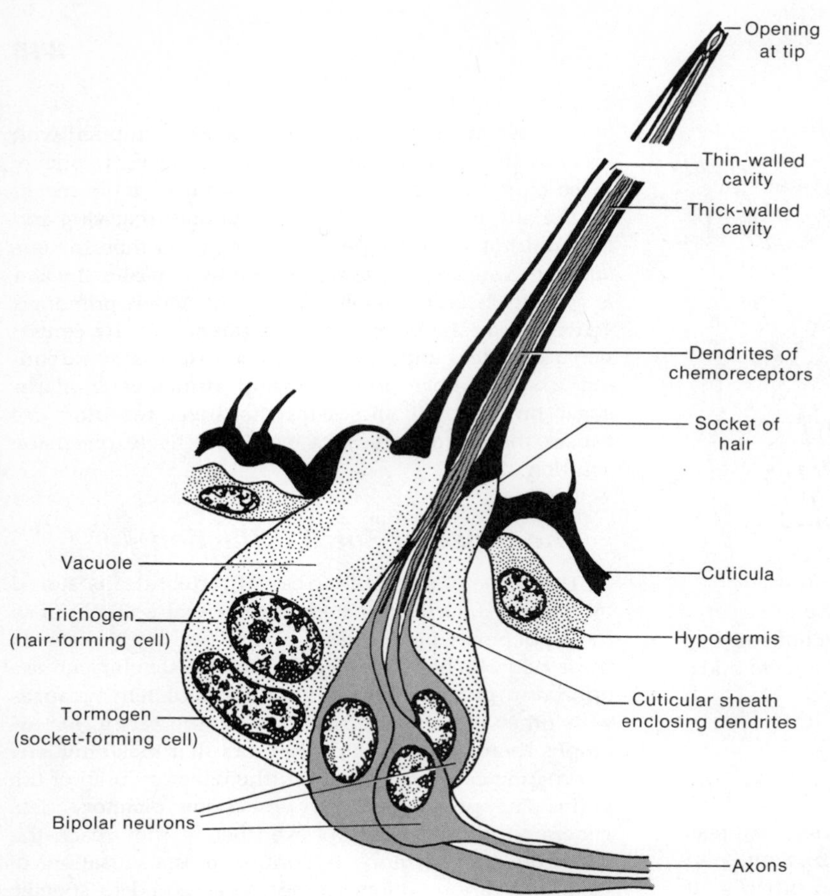

FIGURE 11.12 Diagram of a chemoreceptive hair of the blowfly. Three of the five neurons are shown. (From Dethier, V. G.: "Insects and the concept of motivation," in Levine, D. (Ed.): *Nebraska Symposium on Motivation.* Lincoln, University of Nebraska Press, 1966.)

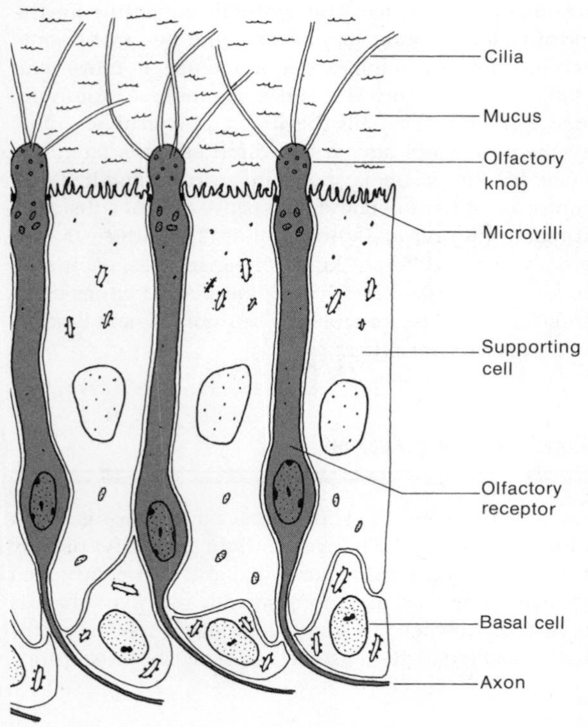

are dispersed over the body surface; in most animals they are concentrated to form eyes. The circumference of the bell of the jellyfish contains clusters of simple eyes, or **ocelli** (L. dim. of *oculus,* eye), that can detect general light intensity. Flatworms, such as planaria, also have ocelli, bowl-shaped structures containing black pigment at the bottom of which are clusters of light-sensitive photoreceptor cells. These are shaded by the pigment from light coming from all directions except from above and slightly to the front. This arrangement enables the planaria to detect the direction of the source of light, but the planarian eye cannot form images. The eyes of invertebrates are usually specializations of the skin, and only in vertebrates is part of the eye, the retina, derived partly from the embryonic brain and partly from the skin.

The function of the eye of most animals, and undoubtedly its original function, was to provide

FIGURE 11.13 Simplified diagram of olfactory epithelium indicating the various cellular components. (From Moulton, D. G., and L. Beilder: Structure and function in the peripheral olfactory system. *Physiological Review,* 47:4, 1967.)

information about the general light intensity of the environment. Such information may have important implications for an animal. A low light intensity might mean that the animal is safely under cover or that it is dusk or night, at which time activity is increased or decreased. A point of high light intensity, such as that produced by the sun or at the opening of a burrow or at the water surface, might provide a cue for orientation. A sudden change in light intensity might indicate the passing of a predator and initiate a withdrawal or escape reflex. An important step in the evolution of eyes was the development of a lens to concentrate light. With lens systems and greatly increased numbers of photoreceptor cells, eyes became able to form images. The most highly developed eyes are found in certain spiders, insects, shrimps, and crabs, in squids, cuttlefish, and octopods, and in the vertebrates. Two fundamentally different types of eyes have evolved: the **camera eye** of the vertebrates and cephalopods and the **compound,** or **mosaic, eye** of the arthropods.

The Vertebrate Eye

The ancestral vertebrates had eyes of two types: a median eye on top of the head that probably distinguished only between light and dark and a pair of image-forming eyes on the sides of the head. Cyclostomes and a few reptiles retain functional median eyes, but in most groups it has become a small pineal body attached to the top of the brain. The mammalian pineal gland is a small organ with an endocrine role.

The eyes of different groups of vertebrates vary in their adaptation for seeing beneath water, in the air, and under varying light intensities, but all are similar in their major features. The analogy between a vertebrate eye and a camera is complete. The eye (Fig. 11.14) has a **lens** that can be focused for different distances, a *diaphragm* (the **iris**) that regulates the size of the light opening (the **pupil**), and a light-sensitive **retina** that is located at the rear of the eye and corresponds to the film of the camera. Next to the retina is a sheet of cells filled with black pigment that absorbs extraneous light and prevents internally reflected light from blurring the image (cameras are also black on the inside). This sheet, the **choroid,** contains blood vessels that nourish the retina. Most of the outer coat of the eyeball is a tough, opaque, curved sheet of connective tissue, the **sclera,** that helps protect the inner structures and maintain the rigidity of the eyeball. On the front surface of the eye, this sheet becomes the transparent **cornea** through which light enters. The surface of the cornea is covered with a layer of stratified epithelium, the **conjunctiva,** that is continuous with the epidermis.

A transparent elastic lens located just behind the iris bends the light rays coming in, bringing them to a focus on the retina. The lens is aided by the curved surface of the cornea and by the refractile properties of the liquids inside the eyeball. The cavity between the cornea and the lens is filled with a watery substance, the **aqueous humor;** the larger chamber between the lens and the

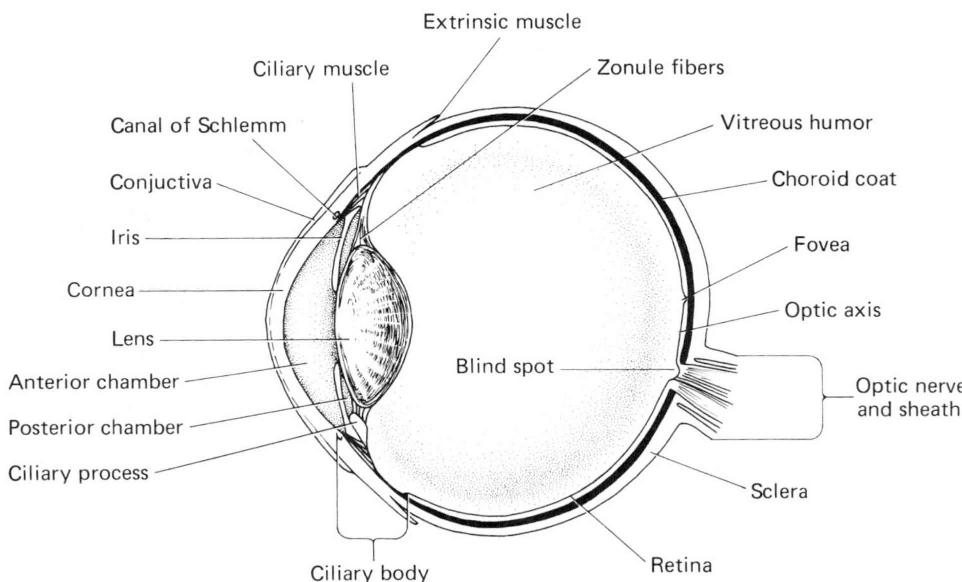

FIGURE 11.14 Diagram of a section through a mammalian eye.

retina is filled with a more viscous, gelatinous fluid, the **vitreous** (L. *vitreus,* glassy) **humor.** Both fluids are important in maintaining the shape of the eyeball. The aqueous humor is secreted by the **ciliary body,** a doughnut-shaped structure that attaches the zonule fibers holding the lens to the eyeball. It is drained through the canal of Schlemm at the base of the cornea. The aqueous humor helps to nourish the cornea and lens, which lack blood vessels, and by maintaining the intraocular pressure, it helps to maintain the shape of the eyeball. An imbalance between the rates of production and drainage of the aqueous humor can lead to increased intraocular pressure and the disease **glaucoma,** in which the pressure flattens and eventually injures the retina. In mammals a number of zonule fibers extend from the ciliary body to the lens and help to hold the lens in place. Muscles within the ciliary body affect the lens' shape through the zonule fibers and focus the image sharply on the retina as considered subsequently.

The amount of light entering the eye is regulated by the iris, a ring of muscle that appears in the human as blue, green, or brown, depending on the amount and nature of the pigment present. The iris is composed of two sets of muscle fibers; the one arranged circularly contracts to decrease the size of the pupil, and the one arranged radially contracts to increase the size of the pupil. The response of these muscles to changes in light intensity is not instantaneous but requires some 10 to 30 seconds to adapt either to a very dimly lighted area or to the high light intensity of full sunlight.

Each eye has six muscles stretching from the surface of the eyeball to various points in the bony socket of the skull. These enable the entire eye to be moved and oriented in a given direction. In the human and other animals with binocular vision these muscles are innervated in such a way that the eyes normally move together and focus on the same area. In terrestrial vertebrates a pair of movable eyelids cover the eyeballs, and the cornea is kept moist, cleansed, and perhaps nourished by the secretions from tear glands. Tears are drained from the median corner of the eye by a lacrimal duct that leads into the nasal cavity. Many mammals, such as cats, have a **nictitating membrane,** a third eyelid in the median corner of the eye. It is moved passively over the cornea when the eyeball is retracted slightly and aids in cleaning and protecting the eye. This membrane is reduced to a vestigial **semilunar fold** in the human.

The light-sensitive part of the vertebrate eye is the **retina,** a hemisphere made up of an abundance of receptor cells called, according to their shape, **rods** and **cones.** In the human eye there are some 125,000,000 rods and 6,500,000 cones. In addition, the retina contains many sensory and connector neurons and their axons. Curiously, the light-sensitive cells are located at the *back* of the retina and directed away from the source of light. To reach them, light must pass through several layers of neurons. This seemingly inappropriate arrangement is explained by the fact that the eye develops as an outgrowth of the brain and folds in such a way that the sensitive cells eventually lie on the farthermost side of the retina (Fig. 11.15). The polarity of the light-sensitive cells is retained during their various developmental maneuvers. The fact that the retina and optic nerves are developmentally parts of the brain also

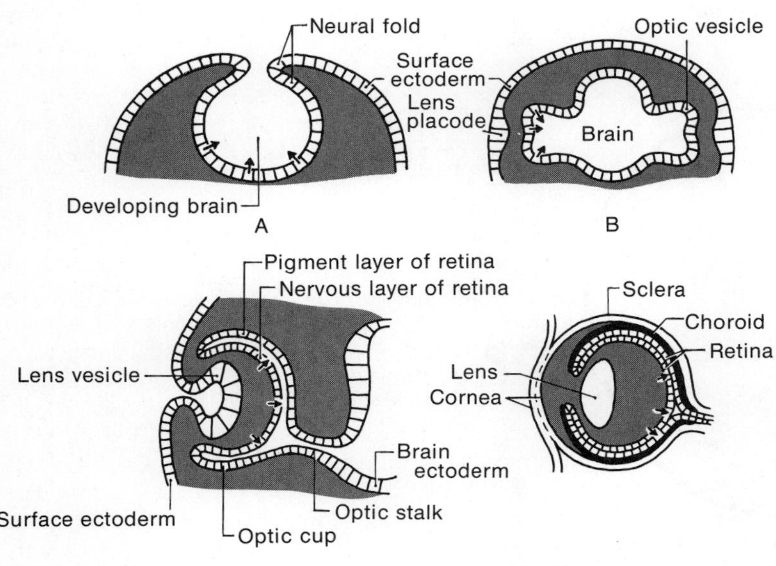

FIGURE 11.15 The development of the eye. *A,* Cross section through an embryo in which the anterior portions of the neural folds are closing to form the brain; *B,* the optic vesicles evaginate from the sides of the forebrain; *C,* an optic cup develops from each optic vesicle, and the lens forms from adjacent surface ectoderm; *D,* the choroid, sclera, and part of the cornea develop from surrounding mesoderm. Arrows indicate the original polarity of the ectoderm cells. (*D* from Romer.)

explains why at least two types of afferent neurons, bipolar and ganglion cells, are involved in transmitting impulses from the rods and cones. Chains of neurons are common in brain tracts, but in most peripheral nerves only one neuron extends from a receptor cell to the brain or spinal cord.

In the center of the retina of the human eye, directly in line with the center of the cornea and lens (the optical axis), is the region of keenest vision, a small depressed area called the **fovea** (L. a small pit). Here are concentrated the light-sensitive cones responsible for vision in bright light, for the perception of detail, and for color vision. The other light-sensitive cells, the rods, are more numerous in the periphery of the retina, away from the fovea. These function in twilight or dim light and are insensitive to colors.

Refraction of Light

Light that enters the eye is bent toward the optic axis in such a way that it forms a sharp inverted image upon the retina (Fig. 11.16). The lens is important in bending light rays, but the cornea and aqueous and vitreous humors are also involved. The cornea is the major refractive agent in terrestrial vertebrates, for the difference between the refractive index of air and the cornea is greater than that between any of the other refractive media. The action of the cornea places the image approximately on the retina; the lens brings it into sharp focus.

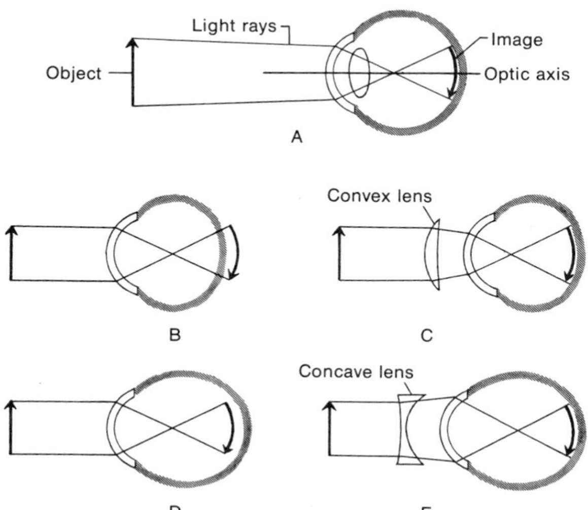

FIGURE 11.16 Image formation by the eye. *A,* Normal eye; *B,* far-sighted eye; *C,* far-sighted eye corrected by a convex lens; *D,* near-sighted eye; *E,* near-sighted eye corrected by a concave lens.

When the eye is at rest, distant objects are in focus. The refractive power of the eye must be increased in viewing a near object, or its image will be blurred, for the image will come into sharp focus at a point behind the retina. Accommodation for near vision is accomplished in mammals by the contraction of muscles within the ciliary body. This brings the point of origin of the zonule fibers a bit closer to the lens and releases their tension. The front of the elastic lens bulges out slightly, and its refractive powers are increased accordingly. When the ciliary muscles are relaxed, intraocular pressure pushes the wall of the eyeball outward and increases the tension of the zonule fibers, and the lens is flattened a bit. The lens becomes less elastic with age, and our ability to focus on near objects decreases.

The refractive parts of the eye form a sharp image of an object on the retina only if the eyeball is of an appropriate length. In theory, if the eyeball is shorter than normal, as in farsighted people, the image of an object falls behind the retina. Accommodation is necessary to bring the image into focus, and the power of accommodation may not be great enough to focus on a near object. This can be corrected by placing a convex lens in front of the eye (Fig. 11.16*C*). Nearsighted people have eyeballs that are longer than normal, and the image falls short of the retina. This can be corrected by placing a concave lens in front of the eye (Fig. 11.16*E*).

The Chemistry of Vision

The light strikes the rods and cones and activates them; they, in turn, initiate nerve impulses. The outer segment of each rod contains an elaboration of the membrane system of the cell, and a great deal of the pigment **rhodopsin** (Gr. *rhodan,* rose + *opsis,* sight) is associated with these membranes. All visual pigments have a common plan, a chromophore (retinal) bound to a protein (opsin). The combination is known as rhodopsin, or **visual purple.** Rhodopsin is the visual pigment in the rods of most vertebrates. The cones contain iodopsin, a visual pigment consisting of the same chromophore (retinal) but a different protein. Iodopsin is found in humans, chickens, cats, snakes, frogs, and crayfish. Freshwater fish have different visual pigments. Their rod pigment, **porphyropsin,** consists of $retinal_2$ plus opsin. The cone pigment, **cyanopsin,** consists of $retinal_2$ plus a different protein. Cyanopsin is found in the eyes of freshwater fish, tortoises, and tadpoles. It is interesting that euryhaline eels, salmon, and trout have both rhodopsin and porphyropsin, but the pigment commonly associated with spawning (in fresh water) predominates. The sea lamprey has mostly rhodopsin on its downstream migration to the sea and mostly porphyropsin when swimming upstream as a sexually mature adult. Amphibians change from porphyropsin to rhodopsin when they develop from tadpole to adult.

Retinal is the aldehyde of vitamin A (retinol) and is formed by an oxidation catalyzed by alcohol dehydrogenase. Quanta of light striking rods or cones trigger a chemical change in retinal that results in the emission of a nerve impulse of the receptor cell. These structures are ready to discharge, having been charged with the requisite energy by internal chemical reactions.

The eye is called upon to respond to an enormous range of light intensities, and the rhodopsin system is peculiarly adapted to provide for a wide range of responses. The eyes of mollusks, arthropods, and vertebrates, which arose quite independently in the course of evolution, all utilize the same basic chemical reaction—the conversion by light of the *cis* form of retinal to the *trans* form. This involves a rearrangement of the molecular structure that can occur very rapidly. Light energy converts rhodopsin (with the *cis* form of retinal) into **lumirhodopsin,** an unstable compound containing the *trans* form of retinal. It decays first into **metarhodopsin** and then into free *trans* retinal and releases opsin (Fig. 11.17). The light energy is needed only for the cis–trans isomerization; the other changes do not require light. Some of the changes affect, in a way not yet clear, the permeability of the membrane with which rhodopsin is associated and lead to the development of the receptor potential. The resynthesis of rhodopsin begins with the conversion of the all-*trans* retinal into 11-*cis* retinal, an energy-requiring reaction catalyzed by the enzyme retinal isomerase. The 11-*cis* retinal then combines with opsin in a reaction that does not require the input of energy to form rhodopsin. Photoisomerization itself may be the activating process, partly because of its great speed and the speed with which the light affects the rods. It can be shown that a single quantum of light can be absorbed by a single molecule of rhodopsin and lead to the excitation of a single rod. When the eye is exposed to a flash of light lasting only one millionth of a second, the eye sees an image of light that persists for nearly one tenth of a second. This is the length of time that the retina remains stimulated following a flash and presumably reflects the length of time that lumirhodopsin persists in the rods. This persistence of images in the retina enables your eye to fuse the successive flickering images on a cinema or a television screen, and you have the impression of seeing a continuous moving picture.

A cycle of breakdown and resynthesis of rhodopsin goes on continuously if the eyes are exposed to any light, but the equilibrium point shifts to adapt for seeing in very dim or very bright light. When the eye is suitably shielded from light, the breakdown of rhodopsin is prevented, and its concentration gradually builds up over some minutes until essentially all of the opsin has been converted to rhodopsin. The sensitivity of the eye to light, a function of the amount of rhodopsin present, can increase a thousandfold if the eye is dark-adapted for a few minutes and about a hundred thousandfold if the eye is dark-adapted for as long as an hour.

The chemistry of the cones and of color vision itself is less understood, but it is known that the cones con-

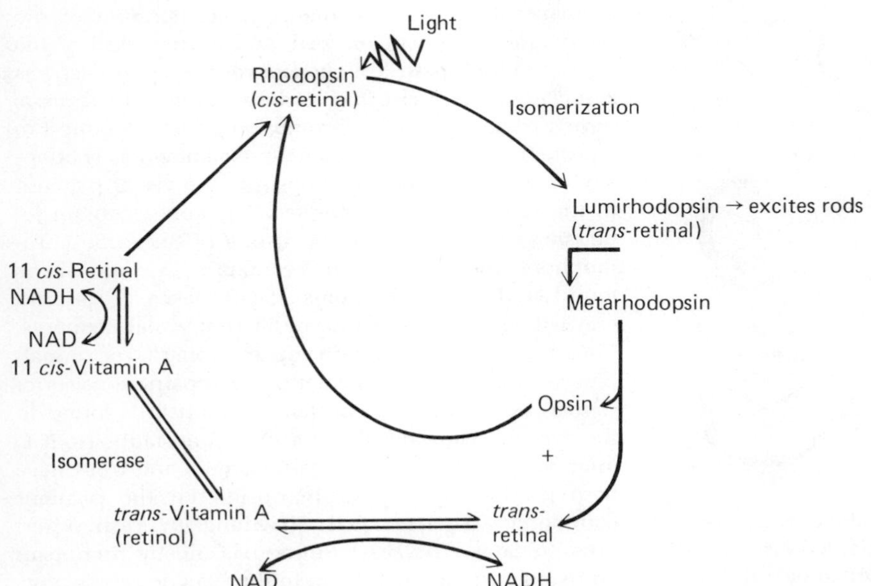

FIGURE 11.17 The rhodopsin-retinal system underlying the sensitivity of rods to light.

tain a light-sensitive pigment, iodopsin, composed of retinal and a different opsin. The cones are considerably less sensitive to light than are rods and cannot provide vision in dim light. The prime function of the cones is to perceive colors. The evidence from psychological tests is consistent with the hypothesis that there are three different types of cones that respond respectively to blue, green, and red light. Each can respond to light with a considerable range of wave lengths; the green cones, for example, can respond to light of any wave length from 450 to 675 nm. (this includes blue, green, yellow, orange, and red light), but they respond to green light more strongly than to any of the others. Intermediate colors, other than blue, green, and red, are perceived by the simultaneous stimulation of two or more types of cones. According to this theory, yellow light (i.e., light with a wave length of 550 nm.) stimulates green and red cones to an approximately equal extent, and this is interpreted by the brain as yellow color. Colorblindness results when one or more of the three types of cones is absent because of the lack of the gene necessary for the formation of that kind of cone or cone pigment.

Organization of the Retina

The rods and cones of the retina synapse with bipolar cells (Fig. 11.18), which in turn synapse with ganglion cells. The axons of the ganglion cells cross the retina, come together at the blind spot, and pass to the brain as the **optic nerve.**

The synaptic relationships of rods and cones differ in an important way. Many rods—as many as several hundred in the peripheral part of the retina—converge on a single bipolar cell, and a number of bipolar cells may converge on a single ganglion cell. Far fewer cones converge on a single bipolar cell, and in a fovea many of the cones synapse with a single bipolar and ganglion cell. The difference in neuronal pathways, together with a difference in the threshold of illumination, explains why rods are more effective than cones in receiving light of low intensity. Stimuli from a number of rods, each of which is below the threshold of a bipolar cell, converge upon a single cell, have an additive effect upon it, and may activate it. Much less summation, or none at all, occurs in the more "private line" system of the cones. Since the greatest density of rods and the greatest degree of convergence occur in the periphery of the retina, one can see best in dim light by looking out of the side of the eye, so that the image falls to the side of the fovea. Rods are particularly abundant in the eyes of nocturnal vertebrates.

Although rod vision is more sensitive to dim light than cone vision, it also follows from their different neuronal pathways that rod vision is less acute. The eye cannot distinguish which rods are receiving light among a number of rods that converge on a single bipolar cell. In contrast, cone vision in the fovea is extremely acute, because most of the cones have a direct line to the brain. Distinct pathways to the brain from cones sensitive to different wave lengths of light are also necessary for an animal to distinguish and interpret colors. Rod vision may be compared to a high-speed, coarse-grain, black-and-white photographic film and cone vision to a slower, fine-grained, color film.

Some integration or processing of visual images occurs in the retina by the lateral connections of the horizontal and amocrine cells (Fig. 11.18). For example, light

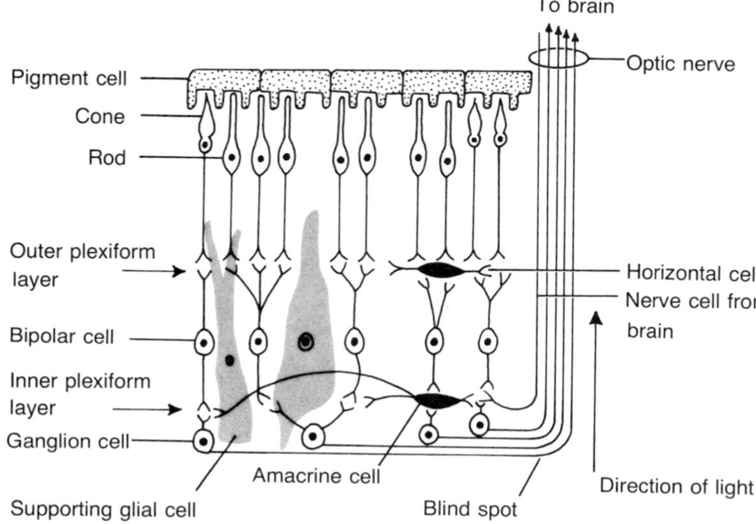

FIGURE 11.18 A diagrammatic vertical section through the retina to illustrate major types of interconnections among the component cells.

falling on a group of rods and cones will stimulate the bipolar cells with which they are in synaptic contact, but concurrent impulses travelling on horizontal cells will inhibit the activity of adjacent bipolar cells. This intensifies the boundary line between light and dark and also is important in helping to detect the movement of an image across the retina.

Eyes of Other Vertebrates

The eyes of all vertebrates are essentially alike, but those of primitive vertebrates differ from mammalian eyes in several respects, for the problems associated with sight beneath water are not identical with those in the air. The water itself cleans and moistens the eye, and fishes have not evolved movable eyelids or tear glands. Secondly, the refractive index of water is nearly the same as that of the cornea, so the cornea of a fish's eye does not bend light rays very much. Most refraction is accomplished by the lens, which is nearly spherical and hence has a greater refractive power than the oval lens of tetrapods. It is interesting in this connection that the lens of a frog's eye flattens a bit during metamorphosis when a change in environment occurs. Finally, the method of accommodating for near and far vision is different, for in the fishes and amphibians the lens is moved back and forth in a camera fashion and does not change shape.

The Cephalopod Eye

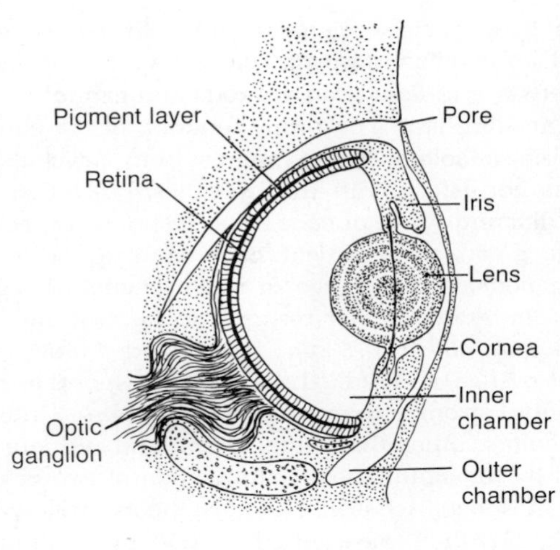

FIGURE 11.19 The camera type of cephalopod eye, complete with shutter (iris); the eye of the squid, *Loligo*. (After Williams.)

The eye of a cephalopod, such as a squid, octopus, or nautilus, is highly developed and parallels the eye of vertebrates to a striking degree (Fig. 11.19). The cephalopod eye has a retina, a movable lens, an iris diaphragm, and a cornea, but the photoreceptors in the retina are directed toward the source of light instead of away, as in vertebrates. The cephalopod eye has a large number of photoreceptor cells, which suggests that object discrimination is possible. To discriminate an object requires that there be enough photoreceptors present so that the neuron light points can form an object when put together. Behavioral studies suggest that cephalopods are very nearsighted animals.

The Compound Eye of Arthropods

Most arthropods have eyes. Some are simple, having only a few photoreceptors; others are large, with thousands of retinal cells. A transparent lens-cornea is the usual integumentary contribution to the eye, and the focus is usually fixed since the immovable lens is continuous with the surrounding exoskeleton. Insects and many crustaceans, such as crabs and shrimp, have a **compound eye,** so called because it is composed of many long cylindrical units, each of which possesses all of the elements for light reception. Each unit, called an **ommatidium** (Gr. dim. of *omma,* eye), contains an outer cornea, a middle crystalline cone, and an inner **retinula** (Fig. 11.20). On the external surface, the cornea of each ommatidium is distinct and forms one facet of the compound eye. The cornea functions as the lens, and the crystalline cone funnels light down to the retinula, which is equivalent to the retina and is composed of a rosette of seven monopolar neurons. Their inner photosensitive surfaces (rhabdomeres) together form a central **rhabdome.** Movable screening pigment is commonly present between adjacent ommatidia.

The term "compound eye" tends to imply that each ommatidium forms a separate image, but this is untrue. The seven retinula cells function as a single photoreceptor unit and together transmit a single signal. The image formed by the entire eye, although sometimes called mosaic, depends upon all of the signals transmitted by the ommatidia and thus is really no different from that of other types of eyes. Compound eyes may function differently in bright light than they do in weak light. In bright light the screening pigment is extended between the ommatidium, and each retinula responds only to the light received by its facet and crystalline cone (Fig. 11.20*A*). Under these conditions the eye is believed to be

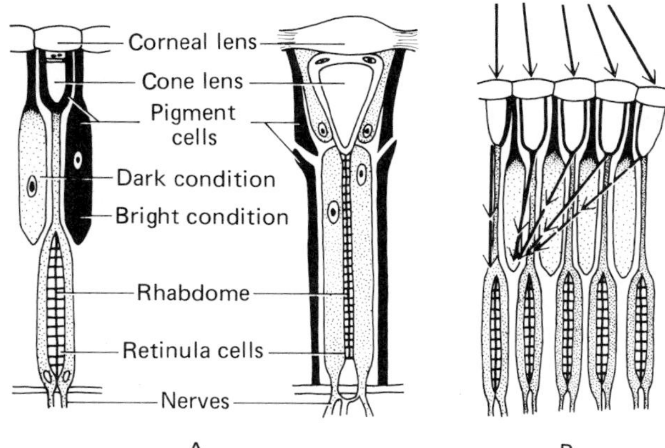

FIGURE 11.20 *A,* Insect ommatidia, showing a diurnal type *(left)* and a nocturnal type *(right).* In the diurnal type, pigment is shown in two positions, adapted for very dark conditions on the left side and for relatively bright conditions on the right. *B,* Nocturnal type of eye adapted for dark conditions, showing how light can be concentrated upon one rhabdome from several lenses. If the pigment moved downward, light from peripheral lenses would be screened out.

especially effective in detecting movement, for slight changes in the position of the moving object will stimulate different ommatidia.

In weak light the screening pigment is contracted, and the light received by an ommatidium can cross over to adjacent units; thus, a retinula may be fired by the light received through several ommatidia (Fig. 11.20*B*). Under these conditions the eye probably functions to detect changes in general light intensity or the position of a bright light source or shadow. Although the eyes of some arthropods can adapt to bright or dim light, the compound eyes of most species are usually adapted for functioning in either bright or weak light but not in both. They are either diurnal or nocturnal or live in habitats in which there is reduced light; hence, the eyes usually function under only limited light conditions.

Visual acuity varies greatly and depends upon the number of photoreceptors present, regardless of the type of eye. The eyes of some hunting spiders and the compound eyes of many insects and some crabs and shrimps appear to be capable of detecting objects. In all of these species the number of photoreceptors is very great. The eye of the dragonfly, for example, has 10,000 ommatidia, and the eye of a wolf spider has 4500 photoreceptors. However, when one compares these numbers with the 130 million photoreceptors in the human eye, it is evident that the images formed in even the most highly developed arthropod eye must be very crude.

Although the compound eye of the arthropod forms only coarse images, it compensates for this by being able to follow flicker to high frequencies. Flies are able to detect flickers of up to approximately 265 per second, whereas the human eye can detect flickers of only 45 to 53 per second. Because flickering lights fuse above these values, we see motion pictures as smooth movement and the ordinary 60-cycle light in a room as a steady light. To an insect, both motion pictures and light must flicker horribly. Because the insect has such a high critical flicker fusion rate, any movement of prey or enemy is immediately detected by one of the eye units. Hence, the compound eye is peculiarly well suited to the arthropod's way of life.

Compound eyes are superior to our eyes in two other respects. They are sensitive to different wave lengths of light from the red into the ultraviolet, and they are able to analyze the plane of polarization of light. An insect can see well into the ultraviolet, and its world of color is much different from ours. Since different flowers reflect ultraviolet to different degrees, two flowers that appear identically white to us may appear strikingly different to insects. How the world might appear to an insect with ultraviolet vision can be appreciated by viewing a landscape through a television camera with an ultraviolet transmitting lens. The sky that appears equally blue to us in all quadrants reveals quite different patterns to an insect because the plane of polarization of the light is not the same in all parts of the sky, and hence the insect's eye can detect the difference. Honeybees and some other arthropods employ this ability as a navigational aid.

Thermoreceptors

Cells sensitive to variations in temperature are found in a wide variety of animals. Paramecia will avoid warm or cold water and will collect in a region where the temperature is moderate. Some insects have thermoreceptors either in the antennae or all over the body. Insects that suck blood from warm-blooded animals are attracted to their prey by the temperature gradients nearby. Blood-sucking bugs, for example, are much less able to find their prey after their antennae, which contain their thermoreceptors, have been removed. Fish have fairly sensitive thermoreceptors, for a change of only 0.5° C. will change the behavior of sharks and bony fish. Certain snakes have sensitive thermoreceptors, usually located in pits on the sides of their heads, by which they can detect the presence and location of warm-blooded prey.

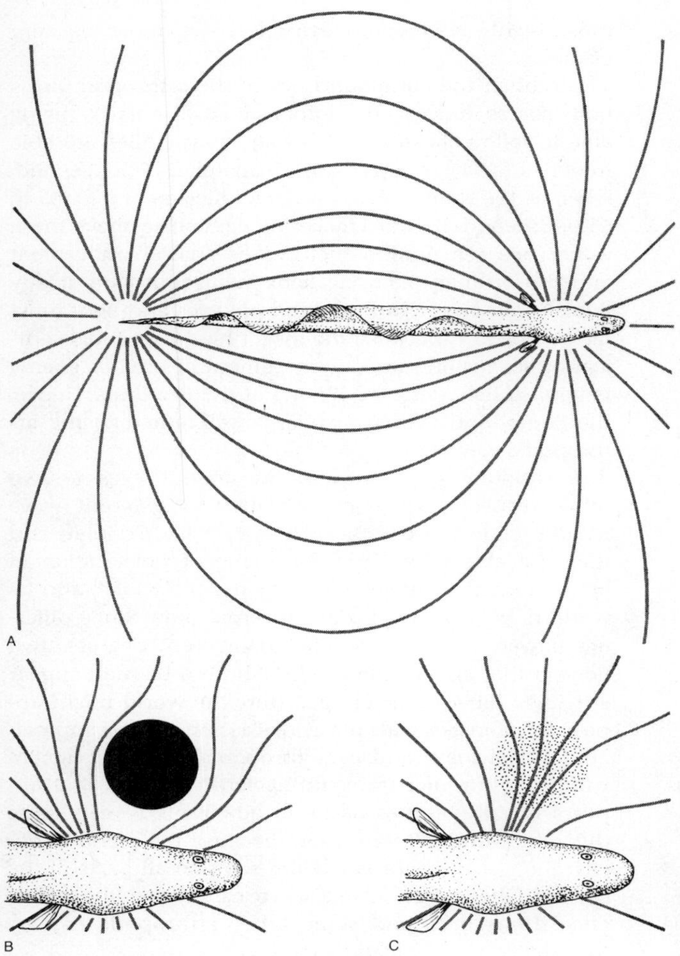

FIGURE 11.21 Electrolocation in the fish *Gymnarchas*. The fish generates an electric field, its tail negative to its head *(A)*. Objects that conduct less *(B)* or more *(C)* electricity than the surrounding water distort the electric field, and this is detected by special sense organs along the sides of the fish.

Electroreceptors

Certain skeletal muscles of fish, such as the electric ray or electric eel, have in the course of evolution developed into electric organs capable of discharging pulses of electric energy. The electric eel, *Electrophorus*, can generate pulses in excess of 500 volts. In other fishes the pulses of electricity given off are very weak. Researchers have found that these weak pulses are part of an electric guidance system. Certain species of tropical fish live in water that is so turbid that their eyes are essentially useless. As a means of orienting themselves, they set up a pulsating electric field around their bodies by means of their special electric organs. The electric field generated has the fish's tail negative with respect to its head. Any object in the environment that has an electrical conductivity different from that of the water will distort the field, and these changes can be detected by the fish by special sense organs in its lateral line system (Fig. 11.21). In most fish this system detects low-frequency vibrations and movements in the water.

Summary

1. The function of each receptor cell is to detect a very small amount of energy in a specific form, such as light or sound waves, and to transmit the information to the nervous system. Receptors may be classified by the kind of energy change they monitor.

2. Each sense organ has a specialized structure consisting of one or more receptor cells plus a variety of accessory tissues. Receptors may be parts of nerve cells or modified epithelial cells connected to sensory nerve cells.

3. The stimulus causes a change in the permeability of the neuronal membrane to ions so that the resting potential is decreased, the cell is depolarized, and the depolarization is termed the receptor potential. This receptor potential spreads down the dendrites, depolarizes the axon hillock, and generates an action potential that travels along the axon to the central nervous system. The amplitude and duration of the receptor potential are related to the strength and duration of the stimulus.

4. The stimulation of any sense organ or sensory receptor initiates a coded response composed of action potentials transmitted by nerves and decoded in the central nervous system.

5. Mechanoreceptors are sensitive to stretch, compression, or torque imparted to tissues by the movement of parts. The tactile sense is due to a large number of several types of sense organs in the skin, each specialized for one kind of sensation. Each muscle and joint is equipped with proprioceptors sensitive to muscle tension and stretch. These permit the harmonious contraction of the different muscles involved in a single movement. Muscle spindles are stretch receptors that serve to keep the tension of the muscle within bearable limits and relatively constant.

6. The position of an animal with respect to gravity is sensed by organs of balance, such as statocysts or otocysts. In vertebrates this is the function of the membranous labyrinth of the inner ear, the utricle, saccule, and semicircular ducts.

7. Phonoreception in vertebrates is achieved by the saccule and lagena in fishes and by the cochlea in birds and mammals. Sound waves are transmitted from the tympanic membrane to the cochlea by the malleus, incus, and stapes. The organ of Corti within the cochlea is the actual organ of hearing in which sound waves stimulate specific mechanoreceptors that initiate impulses in the sensory neurons of the acoustic nerve.

8. Chemoreceptors, organs of taste and smell, are stimulated by specific chemicals in the environment. Taste cells of mammals are epithelial cells, each connected with more than one sensory neuron. Primary neurons in the nasal epithelium of terrestrial vertebrates serve the sense of smell.

9. Light-sensitive cells, photoreceptors, are found in nearly all animals. Eyes able to form images evolved with the development of lenses and large numbers of photoreceptor cells. The eyes of vertebrates and cephalopods are camera eyes, with a lens, iris, and light-sensitive retina plus other accessory tissues. The rods and cones in the retina are light-sensitive cells containing visual pigments composed of retinal bound to a protein. Light striking a rod or cone triggers a chemical change in retinal that results in the emission of a nerve impulse. Many arthropods have compound eyes, composed of many photoreceptors, each with its own lens-cornea. Each unit of the eye, called an ommatidium, responds to incident light and transmits a single signal to the brain.

10. Many animals have thermoreceptors, sensitive to variations in temperature. A few fish have electroreceptors that can detect changes in the electric fields surrounding them.

References and Selected Readings

Alpern, M., M. Lawrence, and B. Wolsk: *Sensory Processes.* Belmont, California, Brooks-Cole Publishing Co., 1967. A general discussion of the perception of sensory stimuli.

Barlow, H. B., and J. D. Mollon: *The Senses.* Cambridge, Cambridge University Press, 1983. Part of a series covering general physiology; well written.

Brown, E. L., and K. A. Diffenbacher: *Perception and the Senses.* Oxford, Oxford University Press, 1979. A discussion of the interpretation of sensory data in the brain.

Burt, E. T., and A. Pringle: *The Senses of Animals.* London, Crane Russak Co., 1974. A fascinating description of the special senses of a variety of animals.

Davson, H.: *Physiology of the Eye.* 4th ed. New York, Academic Press, 1980. A complete discussion of the function of the eye; excellent reference.

Dethier, V. G.: *The Hungry Fly.* Cambridge, Harvard University Press, 1976. An account of the experiments dealing with the sense of taste in the fly.

Goldsmith, T. H., and T. D. Bernard: *The Visual System of Insects. In* Rockstein, M. (Ed.): *The Physiology of Insecta.* Vol. 2. New York, Academic Press, 1965.

Grant, P. T., and A. M. Mackie: *Chemoreception in Marine Organisms.* New York, Academic Press, 1974. An extensive discussion of chemoreceptors in a variety of marine animals.

Gregory, R. L.: *Eye and Brain: The Psychology of Vision.* 2nd ed. New York, McGraw-Hill Book Company, 1973. A fine discussion of the role of the brain in the perception of sensory stimuli.

Hara, T. J.: *Chemoreception in Fishes.* New York, Elsevier, 1982. One of a series on developments in the science of fisheries.

Katsuki, Y.: *Receptive Mechanisms of Sound in the Ear.* Cambridge, Cambridge University Press, 1982. A good source book for information on the sense of hearing.

McDevitt, D.: *Cell Biology of the Eye.* New York, Academic Press, 1982. A detailed source book on the eye and its functions.

Quarton, G. C., T. Melnechuk, and F. O. Schmitt (Eds.): *The Neurosciences: A Study Program.* New York, Rockefeller University Press, 1967. A fine series of chapters, each written by a different expert, covering a wide range of topics related to neurobiology. An excellent reference book.

Stevens, S. S., and F. Warshovsky: *Sound and Hearing.* New York, Time-Life Science Library, 1965. A well-illustrated account of the sense of hearing.

Thompson, R. F.: *Introduction to Physiological Psychology.* New York, Harper & Row, 1975. A text on the biological foundations of psychology written for the undergraduate student.

12 Endocrine Systems and Hormonal Integration

ORIENTATION

Chemical messengers, hormones, constitute the endocrine system which, together with the nervous system, integrate the activities of the many organs and tissues of complex, multicellular animals. Each kind of hormone is typically secreted by specific cells that constitute the endocrine gland. The hormone passes into the blood stream and is carried around the body to specific target organs, with cells containing specific receptor proteins that take up and bind the hormone. The term hormone is an operational term and defines a compound's function, not its chemical nature. Among the substances with hormonal activity are proteins, amino acids, amines, fatty acids, and steroids. We will examine how hormones are synthesized and secreted, how they are transported through the blood stream, and how they interact with their specific receptors in their target cells and alter their metabolism. We will go on to consider how the amount of hormone secreted by a gland is regulated, what happens when an endocrine gland produces too much or too little of its hormone, and how several hormones may interact to regulate such important functions as the estrous and menstrual cycles, the production and secretion of milk, the molting of insects and crustaceans, and many aspects of behavior.

Nerves and sense organs enable an animal to adapt very rapidly—with responses measured in milliseconds—to changes in the environment. The swift responses of muscles and glands are typically under nervous control. The **hormones** (Gr. *hormaein,* to set in motion) secreted by the glands of the endocrine system diffuse or are transported by the blood stream to other parts of the body and coordinate their activities. The responses under endocrine control are in general somewhat slower—measured in minutes, hours, or weeks—but longer lasting than those under nervous control. The long-range adjustments of metabolism, growth, and reproduction are typically under endocrine control. Hormones play a key role in maintaining within narrow limits the concentrations of glucose, sodium, potassium, calcium, phosphate, and water in the blood and extracellular fluids of vertebrates. Hormonal regulation and integration of cellular processes also occur in many invertebrates.

Endocrine Glands

Endocrine glands secrete their products into the blood stream rather than into a duct leading to the exterior of the body or to one of the internal organs, as do exocrine glands. For this reason they are sometimes referred to as glands of internal secretion. The thyroid, parathyroids, pituitary, and adrenals function only in the secretion of hormones and are truly ductless glands. The pancreas secretes digestive enzymes via ducts and hormones carried by the blood stream.

In current usage, the term **hormone** refers to a special chemical messenger that is produced by some restricted region of an organism and diffuses or is transported by the blood stream to another region of the organism, where it is effective in very low concentrations in regulating and coordinating cellular activities. Most hormones are carried in the blood from the site of production to the site of action, but neurohormones may pass down an axon, and prostaglandins may be transferred in the seminal fluid. This definition of hormones includes some remarkably diverse chemical compounds: amino acids and amines, peptides, proteins, fatty acids, purines, steroids, and gibberellins. It seems unlikely that all of these diverse compounds would affect cell function by the same or similar mechanisms. Indeed, it is becoming apparent that hormones may have several independent mechanisms of action by which they regulate cellular activities. All of the hormones are required for normal body function, and each must be present in a certain optimal amount. Either a hyposecretion (deficiency) or hypersecretion (excess) of any one may result in a characteristic pathological condition.

In any endocrine system we can distinguish three parts—the **secreting cell,** the **transport mechanism,** and the **target cell**—each characterized by a greater or lesser degree of specificity (Fig. 12.1). In general, each type of hormone is synthesized and secreted by a specific kind of cell. Some hormones are transported in the blood in solution, but most are bound to some protein component of the serum. Some are bound nonspecifically to albumin; others are selectively bound to specific high-affinity proteins. The hormones are taken up and bound by specific **receptors** in their target cells. Neurohormones are produced in neurons, pass down the axon, and are secreted at the axonal tip (Fig. 12.1*B*), a process termed **neurosecretion**.

Molecular Mechanisms of Hormone Action

Hormones, such as thyroxine and growth hormone, affect the metabolic condition of nearly every cell of the body. Most hormones, however, elicit a detectable response only in certain cells of the body. With some hormones only a few types of cells respond; with others a broader spectrum of response can be observed. An engrossing question at present is "how does the target cell recognize its appropriate hormone?" Each type of target cell contains a specific protein, a **receptor**, that forms a tight and specific combination with one type of hormone. The receptors for several hormones have been at least partially purified. The blood and interstitial fluid bathing each cell in the body contain the entire spectrum of hormones, but the specific receptor enables the cell to pick out one specific hormone and ignore all the others. The hormone-receptor interaction is characterized by high affinity, high specificity, and low capacity; that is, the receptors are readily saturated by low concentrations of hormone.

Hormones are remarkably effective substances; a very small quantity produces a marked effect in the structure and function of one or another part of the body. Only small amounts of a hormone are secreted by the endocrine gland at any one time, and the amount circulating in the blood is very small. Even the amount excreted each day in the urine is not very large. Because of this, the isolation of a pure hormone can be a difficult job indeed. To obtain a few milligrams of pure estradiol, one of the female sex hormones, more than two tons of pig ovaries must be extracted.

Any theory of the molecular mechanisms by which a given hormone produces its effects in specific tissues

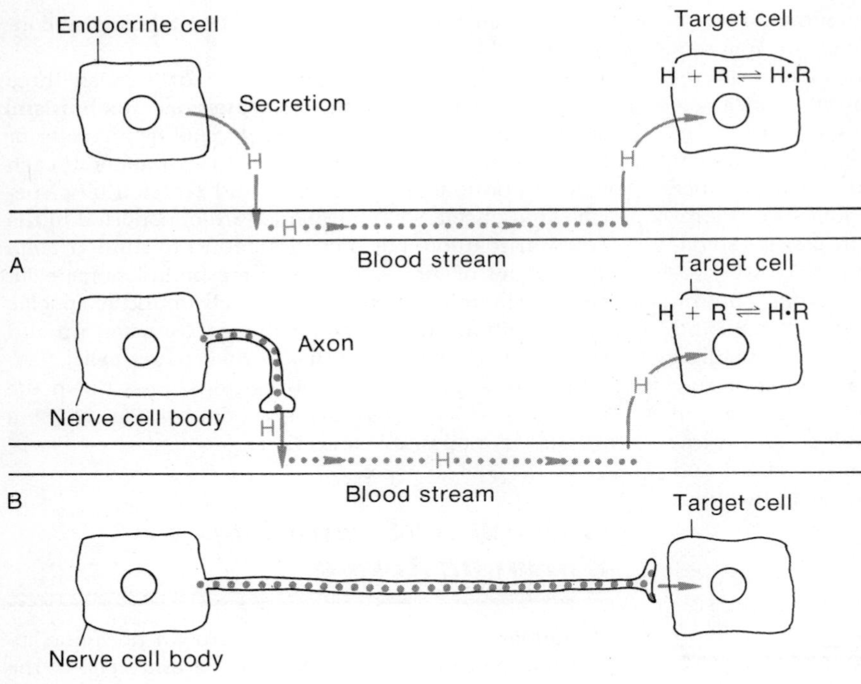

FIGURE 12.1 Diagram illustrating the differences between endocrine secretion *(A)*, neurosecretion *(B)*, and neurotransmission *(C)*. In *(A)* and *(B)*, *H* represents hormone and *R* represents receptor.

must account for the high degree of specificity of many hormones and for the remarkable degree of biological amplification inherent in hormonal processes. Hormones circulate in the blood stream in very low concentrations—steroid hormones at concentrations of about 10^{-9} M and peptide and protein hormones at concentrations of about 10^{-12} M. The several current theories regarding the mechanism of hormone action suggest that the hormone first combines with a specific receptor protein. The effects of hormones in facilitating the entrance of certain substrates into the cell, as for example in the uptake of glucose by muscle cells stimulated by insulin, have suggested that the hormone combines with some protein or other component of the cell membrane. This leads to a change in the molecular architecture of the membrane and hence in its permeability to specific substrates.

The receptors for steroid hormones are soluble proteins located in the cytoplasm that combine with the steroid and transport it into the nucleus where it activates certain previously repressed genes (Fig. 12.2). This leads to the production of additional or new kinds of messenger RNA that code for the synthesis of specific new proteins. This theory accounts for the marked amplification of hormonal effects because a small amount of hormone, by turning on the transcription of a specific gene, could result in the production of many molecules of messenger RNA and many, many more molecules of protein. Evidence supporting this hypothesis is derived from studies of the effects of estradiol on the uterus, of estradiol and progesterone on the oviduct, and of dihydrotestosterone on the seminal vesicle and prostate.

In contrast, the receptors for peptide and protein hormones are located on the plasma membranes of the target cell. The peptide hormone is bound to the receptor on the surface of the cell, and it is believed that this combination stimulates in some fashion the adenyl cyclase on the inner side of the plasma membrane, resulting in an increased production of cyclic AMP.

The cyclic AMP is regarded as an intracellular "second messenger" that mediates the effect of the hormone. Glucagon, for example, stimulates the adenyl cyclase of liver cells (Fig. 12.3). The resulting cyclic AMP activates a protein kinase that transfers a phosphate group from ATP to a third enzyme, phosphorylase kinase, and activates it so that it can in turn convert an inactive fourth enzyme, phosphorylase b, to active phosphorylase a. The latter then catalyzes the production of glucose-1-phosphate from glycogen. This is the first step in glycolysis (p. 73). At each of these successive steps in the enzymic cascade, there is an amplification of tenfold to a hundredfold, just as in the sequence of enzymes involved in blood clotting. The cascade effect permits a very small amount of glucagon to lead to the production of a very large amount of glucose-1-phosphate.

It is not clear whether hormones are typically used up as they carry out their regulatory action in the target tissue. Estradiol, however, is not used up or changed

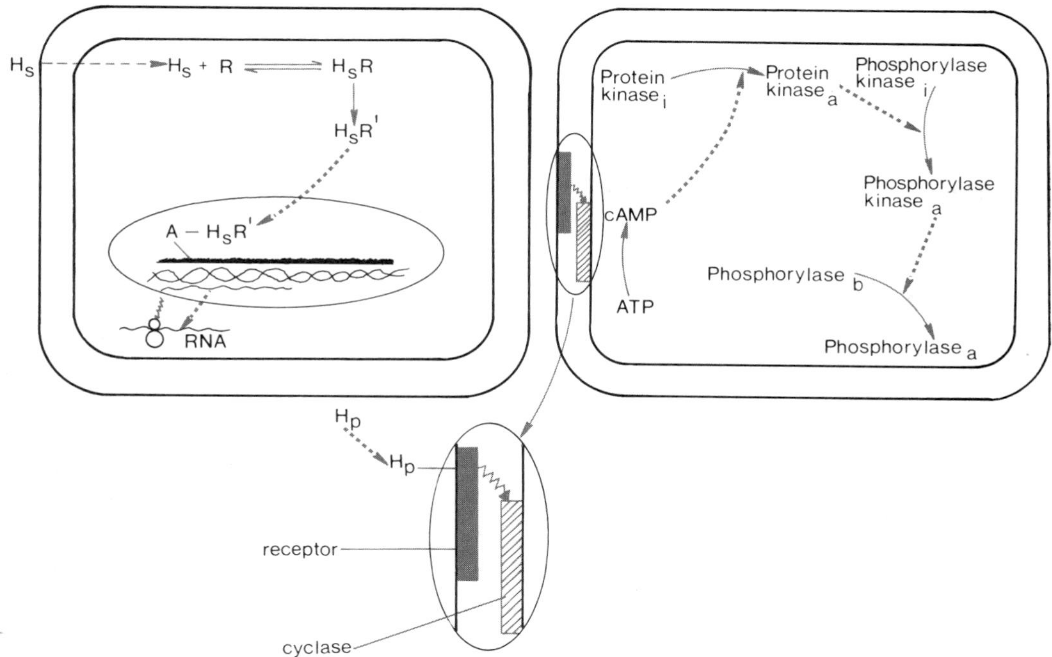

FIGURE 12.2 A comparison of current concepts of the receptors for steroid hormones (H_s) and peptide hormones (H_p). Note that the steroid hormone enters the cell, binds with a soluble receptor, passes into the nucleus, and is attached to the chromatin, stimulating RNA synthesis. In contrast, the peptide hormone remains outside of the cell, binding to a receptor on the plasma membrane. This activates a membrane-bound adenyl cyclase, producing cyclic AMP. The latter stimulates a cascade of protein kinases, which phosphorylate an enzyme, thereby activating (or inactivating) it.

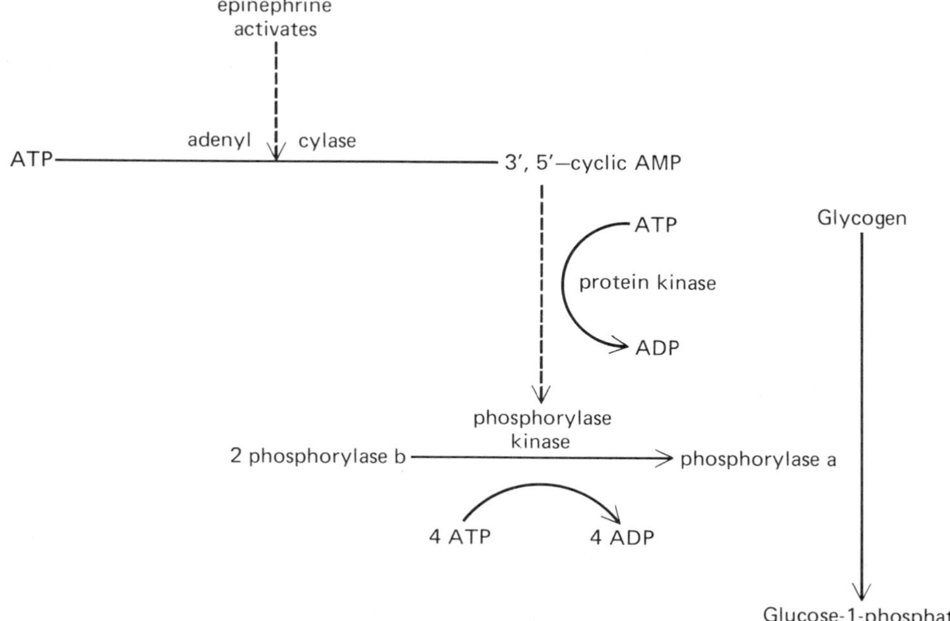

FIGURE 12.3 The sequence of enzymic events by which epinephrine or glucagon stimulates adenyl cyclase and brings about the synthesis of 3′, 5′-adenosine monophosphate (cyclic AMP). This in turn activates a protein kinase that phosphorylates phosphorylase kinase, which in turn phosphorylates and activates phosphorylase. This finally brings about the cleavage of glycogen and the secretion of glucose.

chemically as it stimulates the growth of the uterus. Hormones bound to their receptors appear to be relatively stable, but hormones circulating in the blood have relatively short biological half-lives. They are inactivated and eliminated from the body and must be replaced by new hormone molecules synthesized in the endocrine glands.

It seems unlikely that all hormones have a common molecular mechanism by which their effects are produced. Indeed, there is evidence that certain hormones produce their effects not by a single mechanism but by several different mechanisms acting in parallel within a single target cell.

The Hormones of Vertebrates

The location of the human endocrine glands is shown in Figure 12.4. Their relative position in the body is much the same in all of the vertebrates. The source and physiological effects of the principal mammalian hormones are listed in Table 12.1.

The Thyroid Gland

All vertebrates have a **thyroid gland** located in the neck. In mammals two lobes lie on either side of the larynx and are joined by a narrow isthmus of tissue extending across the ventral surface of the trachea near its junction with the larynx. The thyroid develops as a ventral outgrowth of the pharynx, but the connection is usually lost early in development. The gland is composed of cuboidal epithelial cells arranged in hollow spheres one cell thick. The cavity of each follicle is filled with a gelatinous colloid secreted by the epithelial cells lining it (Fig. 12.5).

Follicle cells have the remarkable ability to accumulate iodide from the blood. This is used in the synthesis of **thyroglobulin,** a protein secreted into the colloid and stored. Proteolytic enzymes secreted in the lysosomes of the thyroid cells hydrolyze thyroglobulin to its constituent amino acids, one of which is **thyroxine,** a derivative of the amino acid tyrosine containing 65 percent iodine. Thyroxine passes into the blood stream, where it is transported loosely bound to certain plasma proteins. In tissues, thyroxine, which contains four atoms of iodine per molecule, may be converted to triiodothyronine, which contains one less atom of iodine and is several times more active than thyroxine.

The first clue to thyroid function came from observations made by a British physician, Sir William Gull, in

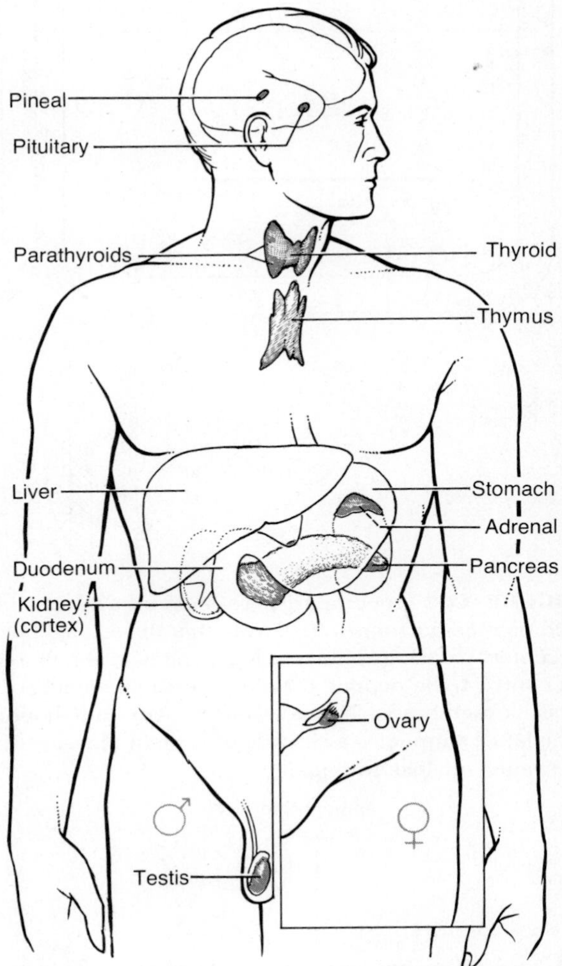

FIGURE 12.4 The approximate locations of the major endocrine glands in men and women.

1874, who noted the association of spontaneous decreased function of the thyroid with puffy dry skin, dry brittle hair, and mental and physical lassitude. In 1895, using a calorimeter to measure the rate of metabolism in patients by the amount of heat they produced, Magnus-Levy found that patients with **myxedema** (Gr. *myxa,* mucus + *oidema,* swelling) had metabolic rates notably lower than normal. When these patients were fed thyroid tissue, their metabolic rate was raised toward normal. This led to the idea that the thyroid secretes a hormone that regulates the metabolic activity of all body cells.

The role of the thyroid hormone in all vertebrates is to increase the rate of oxidative, energy-releasing pro-

TABLE 12.1 Vertebrate Hormones and their Physiologic Effects

Hormone	Source	Physiologic Effect
Thyroxine	Thyroid gland	Increases basal metabolic rate
Parathyroid hormone (PTH)	Parathyroid glands	Regulates calcium and phosphorus metabolism
Calcitonin	Parafollicular cells of thyroid	Antagonist of PTH
Insulin	Beta cells of islets in pancreas	Increases glucose utilization by muscle and other cells, decreases blood sugar concentration, increases glycogen storage and metabolism of glucose
Glucagon	Alpha cells of islets in pancreas	Stimulates conversion of liver glycogen to blood glucose
Secretin	Duodenal mucosa	Stimulates secretion of pancreatic juice
Pancreozymin	Duodenal mucosa	Stimulates release of bile by gallbladder and release of enzymes by pancreas
Epinephrine	Adrenal medulla	Reinforces action of sympathetic nerves; stimulates breakdown of liver and muscle glycogen
Norepinephrine	Adrenal medulla	Constricts blood vessels
Cortisol	Adrenal cortex	Stimulates conversion of proteins to carbohydrates
Aldosterone	Adrenal cortex	Regulates metabolism of sodium and potassium
Dehydroepiandrosterone	Adrenal cortex	Androgen; stimulates development of male sex characters
Growth hormone	Anterior pituitary	Controls bone growth and general body growth; affects protein, fat, and carbohydrate metabolism
Thyrotropin (TSH)	Anterior pituitary	Stimulates growth of thyroid and production of thyroxine
Adrenocorticotropin (ACTH)	Anterior pituitary	Stimulates adrenal cortex to grow and produce cortical hormones
Follicle-stimulating hormone (FSH)	Anterior pituitary	Stimulates growth of graafian follicles in ovary and of seminiferous tubules in testis
Luteinizing hormone (LH)	Anterior pituitary	Controls production and release of estrogens and progesterone by ovary and of testosterone by testis
Prolactin (LTH)	Anterior pituitary	Maintains secretion of estrogens and progesterone by ovary; stimulates milk production by breast; controls "maternal instinct"
Oxytocin	Hypothalamus, via posterior pituitary	Stimulates contraction of uterine muscles and secretion of milk
Vasopressin	Hypothalamus, via posterior pituitary	Stimulates contraction of smooth muscles; antidiuretic action on kidney tubules
Melanocyte-stimulating hormone (MSH)	Anterior lobe of pituitary	Stimulates dispersal of pigment in chromatophores
Testosterone	Interstitial cells of testis	Androgen; stimulates development and maintenance of male sex characters
Estradiol	Cells lining follicle of ovary	Estrogen; stimulates development and maintenance of female sex characters
Progesterone	Corpus luteum of ovary	Acts with estradiol to regulate estrous and menstrual cycles
Prostaglandins	Seminal vesicle and other tissues	Stimulate uterine contractions
Chorionic gonadotropin	Placenta	Acts together with other hormones to maintain pregnancy
Placental lactogen	Placenta	Has effects like prolactin and growth hormone
Relaxin	Ovary and placenta	Relaxes pelvic ligaments
Melatonin	Pineal gland	Inhibits ovarian function

cesses in all body cells. The amount of energy released by an organism at rest under standard conditions, measured in a calorimeter by the amount of heat given off or calculated from the oxygen consumed, is decreased in thyroid deficiency and increased when thyroxine is administered or when the thyroid gland is overactive. Complete removal of the thyroid gland from the mammal reduces the animal's metabolic rate to half the normal value, and the body temperature decreases slightly. Individual tissue slices from an animal with thyroid de-

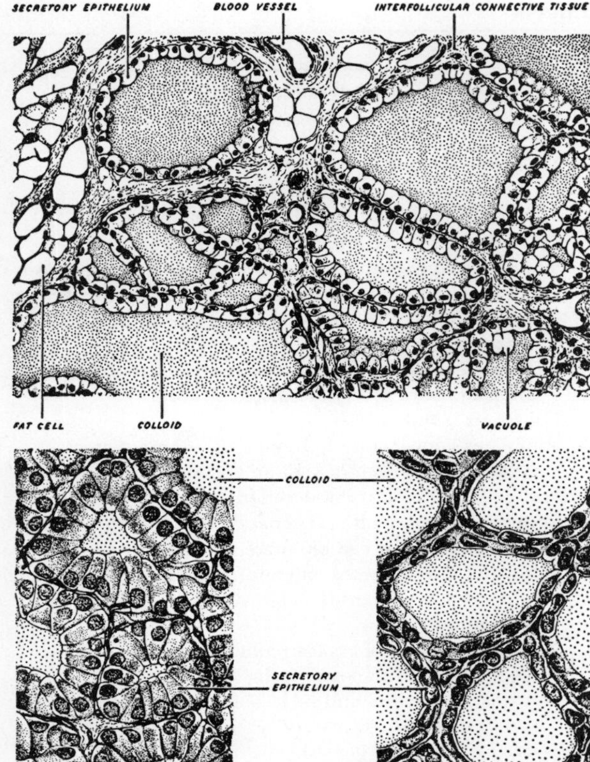

FIGURE 12.5 *Upper,* Cells of the normal thyroid gland of the rat. *Lower left,* Thyroid from a normal rat that had received 10 daily injections of thyroid stimulating hormone. *Lower right,* Thyroid from a rat six months after complete removal of the pituitary gland. (From Turner, C. D., and J. T. Bagnara: *General Endocrinology.* 6th ed. Philadelphia, Saunders College Publishing, 1976.)

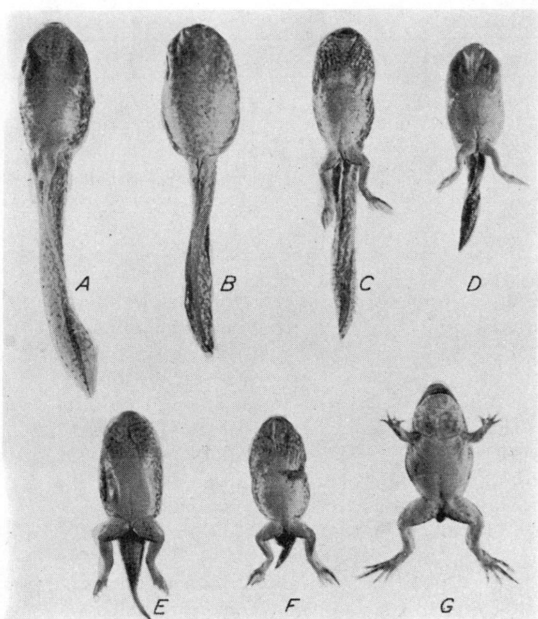

FIGURE 12.6 The effect of thyroid feeding upon the tadpoles of *Rana catesbeiana. A,* The untreated control, which was killed at the end of the experiment. The metamorphosed animal at the lower right *(G)* was killed two weeks after starting the feeding of thyroid gland. The remaining animals *(B* to *F)* were removed from the experiment at intervals during this period. Note the effect of thyroid substances on the metamorphosis of the mouth, tail, and paired appendages. (From Turner, C. D.: *General Endocrinology.* 3rd ed. Philadelphia, W. B. Saunders Co., 1960.)

ficiency show a reduced metabolic rate, decreased oxygen consumption, and decreased utilization of carbohydrates, fats, and proteins when incubated *in vitro.* Since nutrients are metabolized at a lower rate, they tend to be stored, and the hypothyroid animal tends to become obese.

By its action on metabolic processes, thyroxine has a marked influence on growth and differentiation. Removing the thyroid of a young animal causes decreased body growth, retarded mental development, and delayed or decreased differentiation of gonads and external genitalia. All of these changes are reversed by the administration of thyroxine. The metamorphosis of frog and salamander tadpoles into adults is controlled by the thyroid. The removal of the larval thyroid completely prevents metamorphosis, and administering thyroxine to tadpoles causes them to metamorphose prematurely into miniscule adults (Fig. 12.6). The effect of thyroxine on amphibian metamorphosis appears to be not simply a secondary result of its effect on metabolism, for tadpole metabolism can be increased by dinitrophenol, but premature metamorphosis does not occur.

The production and discharge of thyroxine is regulated by **thyroid stimulating hormone** (TSH), secreted by the anterior lobe of the pituitary. In 1916, P. E. Smith found that the removal of the pituitary of frog tadpoles produces deterioration of the thyroid and prevents metamorphosis. Similar pituitary control of thyroid function has been demonstrated in rats, humans, and other mammals. The secretion of TSH by the pituitary is regulated in part by the amount of thyroxine in the blood. Decreased production of thyroxine leads to lowered concentration of thyroxine in the blood stream. Thyroxine inhibits the production of TSH by the pituitary and

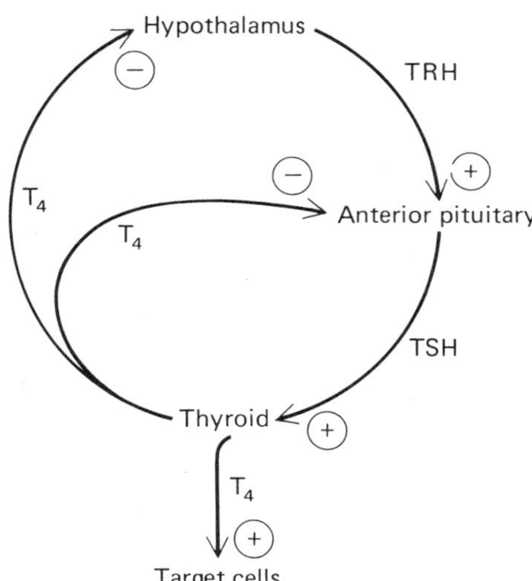

FIGURE 12.7 Control of the thyroid gland by feedback inhibition. Thyrotropic releasing hormone, TRH, secreted by the hypothalamus, stimulates certain cells in the anterior pituitary to produce and secrete thyroid stimulating hormone, TSH. This in turn stimulates the thyroid gland to secrete thyroxine (abbreviated T_4 because it contains four atoms of iodine). Thyroxine (and another hormone secreted by the thyroid, triiodothyronine, abbreviated T_3 because it has three atoms of iodine) stimulates its target cells but inhibits the secretion of TRH by the hypothalamus and the secretion of TSH by the pituitary. Stimulations are indicated by ⊕ and inhibitions by ⊖.

the production of thyrotropin releasing hormone (TRH, see p. 277) by the hypothalamus. Thus a decreased concentration of thyroxine in the blood releases those in hibitions and leads to an increased production of TRH by the hypothalamus and of TSH by the pituitary (Fig. 12.7). The TSH passes to the thyroid and raises its output of thyroxine. When the concentration of thyroxine in the blood is returned to normal, the release of TSH is decreased. By this feedback mechanism, the output of thyroxine is kept relatively constant, and the basal metabolic rate is kept within the normal range.

A deficiency in the amount of thyroxine secreted by the thyroids in an adult results in **myxedema,** characterized by a low metabolic rate and decreased heat production. The body temperature may drop to several degrees below normal so that the patient constantly feels cold. His pulse is slow and he is physically and mentally lethargic. His appetite usually remains normal, and since the food consumed is not used up at the normal rate, there is a tendency toward obesity. The skin becomes waxy and puffy, owing to the deposition of mucous fluid in the subcutaneous tissues, and the hair usually falls out. Myxedema responds well to the administration of thyroxine or dried thyroid gland. Since thyroxine is not digested appreciably by the digestive juices or the bacteria in the gut, it can be given by mouth.

Hypothyroidism results when the diet does not contain enough iodine for the synthesis of thyroxine. The gland compensates for the insufficiency by increasing in size under stimulation by TSH, and the resulting goitrous enlargement may become a large disfiguring mass. The symptoms accompanying the goiter resemble those of myxedema but are milder. The incidence of such simple goiters has been greatly reduced by the addition of iodine as potassium iodide to table salt.

Hypothyroidism present from birth is known as **cretinism.** Children suffering from the disease are dwarfs of low intelligence who never mature sexually (Fig. 12.8*A*). If treatment with thyroxine is begun early, normal physical growth and mental development can be effected.

The overproduction of thyroid hormone produces Graves' disease or **exophthalmic goiter** (Fig. 12.8*C*). The thyroid may be enlarged or may be of nearly normal size, but it produces an excessive amount of hormone. The patient's basal metabolic rate increases to as much as twice the normal amount. The excessively rapid heat production causes the hyperthyroid person to feel uncomfortably warm and to perspire profusely. Because the food he eats is used up quickly, he tends to lose weight even on a high caloric diet. High blood pressure, nervous tension, irritability, muscular weakness, and tremors are symptomatic of the condition, but probably the most characteristic symptom is the protrusion of the eyeballs, called exophthalmos, which gives the patient a wild, staring expression (Fig. 12.8*C*). The swelling of the gland as the result of hyperthyroidism is known as exophthalmic goiter in order to distinguish it from the simple goiter caused by insufficient dietary iodine.

Hyperthyroidism can be treated by surgically removing some of the thyroid gland, by killing the cells with X-rays, by administering the drug thiouracil, which inhibits the synthesis of thyroxine, or by injecting radioactive iodine, ^{131}I. The radioactive iodine is accumulated by the thyroid and its radiation destroys the cells.

In addition to the follicular cells that secrete thyroxine, the thyroid contains parafollicular cells that secrete **calcitonin,** a peptide containing 32 amino acids. This hormone acts with parathyroid hormone to regulate the concentration of calcium in the blood. Its effects oppose those of parathyroid hormone; it inhibits bone resorption and leads to a decrease in the concentration of calcium in the blood and body fluids.

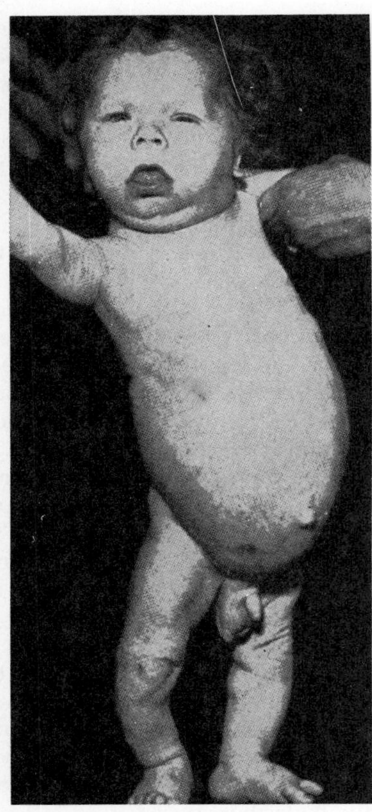

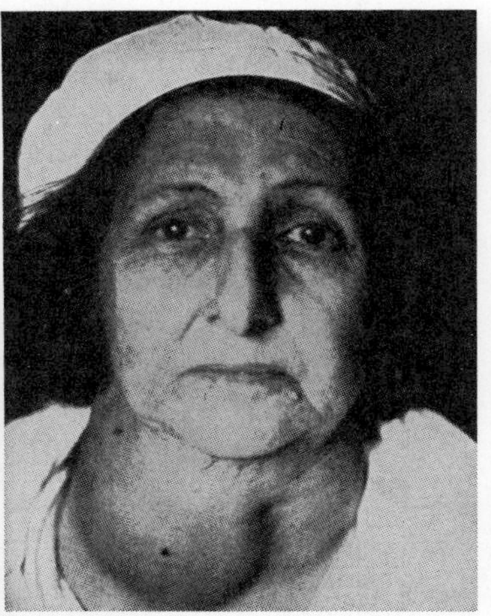

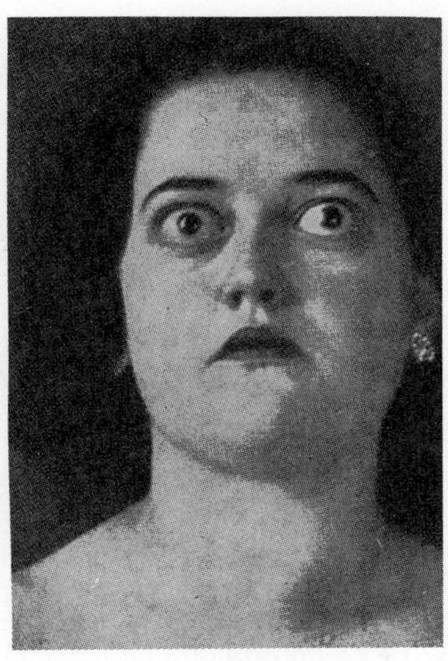

FIGURE 12.8 *A,* A cretin. *B,* Simple goiter. *C,* Exophthalmic goiter. (*A* and *B* from Selye: *Textbook of Endocrinology.* Acta Endocrinologia, Inc.; *C* from Houssay: *Human Physiology.* New York, McGraw-Hill Book Co., Inc.)

The Parathyroid Glands

Imbedded in the thyroid glands of terrestrial vertebrates are small masses of tissue called **parathyroid glands.** There are usually two pairs of parathyroids that develop as outgrowths of the third and fourth pairs of pharyngeal pouches. Each gland consists of solid masses and cords of epithelial cells. The hormone secreted by these cells, **parathyroid hormone** (PTH), regulates the concentrations of calcium and phosphorus in the blood and body fluids. It is essential for life; the complete removal of the parathyroid results in death in a few days. PTH is a single peptide chain containing 84 amino acids. Parathyroid hormone promotes the absorption of calcium from the lumen of the intestine, the release of calcium from the bones, and the reabsorption of calcium from the glomerular filtrate in the kidney tubules. Parathyroidectomy produces a decreased concentration of calcium in the serum, a decreased excretion of phosphorus, and a resulting increased amount of phosphorus in the serum. The parathyroidectomized animal is subject to muscular tremors, cramps, and convulsions—a condition known as **tetany** (Gr. *tetanos,* stretched), that results from the decreased concentration of calcium in the body fluids. Injecting a solution of calcium stops the tetanic convulsions, and further convulsions can be prevented by repeated administration of calcium.

Parathyroid deficiencies are rare, occurring occasionally when the glands are removed inadvertently during an operation on the thyroid or when degeneration results from an infection. The administration of PTH cannot be used for the long-term treatment of parathyroid deficiencies, for the patient becomes refractory to repeated injections of the hormone. The deficiency can be treated successfully by a diet rich in calcium and vitamin D and low in phosphorus.

Hyperfunction of the parathyroid, induced by a tumor of the gland, is characterized by high calcium and low phosphorus content of the blood and by increased urinary excretion of both calcium and phosphorus. The calcium comes, at least in part, from the bones, and as a result, the bones become demineralized, soft, and easily broken. The increased concentration of calcium in the body fluids results in deposits of calcium in abnormal places—the kidneys, intestinal wall, heart, and lungs. The disease can be treated by removing the excess parathyroid tissue surgically or by destroying it with X-rays.

The Islet Cells of the Pancreas

Scattered among the pancreatic acinar cells that secrete the digestive enzymes are clusters of endocrine cells, the **islets of Langerhans,** quite different in appearance. They have a richer supply of blood vessels than the acinar cells and no associated ducts. The islet cells can be differentiated into several types by the staining reactions of their cytoplasmic granules. They develop as buds from the pancreatic ducts and eventually lose all connection with the ducts.

In 1886, two German investigators, Minkowski and Von Mering, were studying the role of the pancreas in digestion by removing the gland surgically from dogs and noting the subsequent effects on digestive functions. The caretaker for the animals noticed that their urine attracted swarms of flies to the cages. Large amounts of sugar were found in the dogs' urine, and the resemblance to human diabetes was recognized. Diabetes had been known since the first century A.D., but its cause was unknown, and no treatment was effective. After 1892, when the discoveries were published, many scientists tried to prepare effective extracts of pancreas, but none of the preparations was very effective and many were toxic. The digestive enzymes of the pancreas destroyed the hormone before it could be extracted and purified. Finally, in 1922, two Canadians, Banting and Best, obtained an active substance by tying the pancreatic ducts, waiting several weeks for the acinar cells to degenerate, and then making an extract from the remaining islet cells. Since the islet cells develop before the enzyme-producing cells in embryonic animals, they were also able to obtain active extracts from fetal pancreas. The first preparation of pure crystalline insulin was made in 1927 by Abel. Commercial insulin has been extracted from beef, sheep, or pig pancreas by an acid alcohol method that rapidly inactivates the proteolytic enzymes. The human insulin gene has been isolated, transferred to the colon bacillus, *Escherichia coli,* and by these recombinant DNA techniques, human insulin has been produced.

Insulin, secreted by the β cells of the pancreas, consists of two peptide chains. Fred Sanger and his colleagues at the University of Cambridge determined the sequence of the 21 amino acids in one chain and the 30 amino acids in the other. Insulin was the first protein whose amino acid sequence was determined, and these studies earned Sanger a Nobel prize, for they not only clarified the structure of insulin but also established methods by which the amino acid sequence of other proteins could be determined. Insulin is synthesized in the β cells of the pancreas as a single peptide composed of 84 amino acids. This compound, **proinsulin,** undergoes folding; three disulfide bonds are formed; and a C peptide containing 33 amino acids is removed from the center of the original chain by hydrolytic cleavage. This leaves two peptide chains joined by the disulfide bridges. Proinsulin is an example of a **prohormone;** other peptide hormones are also synthesized within cells as prohormones that undergo partial cleavage to yield smaller peptides with full hormonal activity. Steiner showed in 1976 that the peptide synthesized on the ribosomes of the β cells is an even longer chain, **preproinsulin,** that loses a portion of the amino terminal part of the chain to become proinsulin.

Most commercial preparations of insulin contain a second pancreatic hormone that *increases* blood sugar concentration instead of decreasing it as insulin does. This hormone, **glucagon,** has been separated from insulin, crystallized, and found to be a single peptide chain composed of 27 amino acids. Glucagon is secreted by the α cells of the pancreatic islets.

Insulin and glucagon take part in the regulation of carbohydrate metabolism, along with certain hormones secreted by the pituitary, adrenal medulla, and adrenal cortex. By way of cyclic AMP, glucagon activates the enzyme **phosphorylase** that is involved in the conversion of liver glycogen to blood glucose and thus raises the concentration of glucose in the blood (Fig. 12.3). Insulin increases the rate of conversion of blood glucose to intracellular glucose-6-phosphate, thereby decreasing the concentration of glucose in the blood, increasing the storage of glycogen in skeletal muscle, and increasing the metabolism of glucose to carbon dioxide and water. A deficiency of insulin decreases the utilization of sugar, and the alterations in carbohydrate metabolism that result secondarily produce many other changes in the metabolism of proteins, fats, and other substances.

The surgical removal of the pancreas or its hypofunction in **diabetes mellitus** impairs glucose utilization and results in elevated concentrations of glucose in the blood (**hyperglycemia**). The patient excretes large amounts of glucose in the urine (**glycosuria**) because the concentration of sugar in the blood exceeds the renal threshold (p. 204). Extra water is required to excrete this sugar, and the urine volume increases; as a result the patient tends to become dehydrated and thirsty. The tissues, unable to obtain enough glucose from the blood, convert protein into carbohydrate. Much of this is excreted and there is a steady loss of weight. Fat deposits are mobilized and the lipids are metabolized. The fatty acids are not metabolized completely but tend to accumulate as partially oxidized **ketone bodies,** such as acetoacetic acid. The ketone bodies are volatile and have a sweetish smell that gives a peculiar and characteristic odor to the breath of diabetics. Ketone bodies are acidic and must be excreted in urine, causing an acidosis (decrease in the alkaline reserves of the body fluids).

Untreated diabetes is ultimately fatal because of the acidosis, the toxicity of the accumulated ketone bodies, and the continuous loss of weight. The injection of insulin alleviates all of the diabetic symptoms. (The patient is enabled to utilize carbohydrates normally, and the other symptoms disappear.) Insulin injections do not "cure" diabetes, since the pancreas does not begin secreting its hormone again. The effect of an injection of insulin lasts for only a short time, a day at most, for the injected insulin is gradually destroyed in the tissues, as is the insulin secreted by the patient's pancreas.

The administration of a large dose of insulin to a normal or diabetic person causes a marked decrease in the concentration of blood sugar. The nerve cells, which require a certain amount of glucose for normal function, become hyperirritable and then fail to respond as the glucose level decreases. The patient becomes bewildered, incoherent, and comatose and may die unless glucose is administered.

The secretion of insulin and glucagon is controlled by the concentration of glucose in the blood. When the concentration of glucose in the blood increases (after a meal, for example), the secretion of insulin is stimulated and it acts to restore the glucose level to normal. A rise in the concentration of insulin in the blood can be detected within two minutes after an experimentally induced rise in blood glucose. A major effect of insulin is a dramatic increase in the transport rate of glucose into skeletal muscle and adipose tissue. This leads to a decrease in the concentration of glucose in the blood and removes the stimulus for the secretion of insulin. As the glucose content of the blood falls below the optimal range, the release of glucagon from the α cells of the pancreas is stimulated, the glycogen phosphorylase system of the liver is stimulated by the glucagon, and the glucose released by the liver returns the concentration of glucose in the blood to normal.

The Adrenal (Suprarenal) Glands

The small paired adrenal glands of mammals are located at the anterior end of each kidney. Each consists of a pale outer **cortex** and a darker inner **medulla.** In cyclostomes and fishes the two parts are spatially separate. In amphibians, reptiles, and birds their anatomic relations are quite variable, and the two parts are interspersed. Cortical tissue develops from coelomic mesoderm near the mesonephric kidneys, whereas the medullary tissue is ectodermal, derived from the neural crest cells that also form the sympathetic ganglia (p. 317).

The cells of the adrenal medulla are arranged in irregular cords and masses around the blood vessels (Fig. 12.9). The medulla secretes two closely related hormones: **epinephrine** (also called adrenalin) and **norepinephrine.** These are comparatively simple amines derived from the amino acid tyrosine.

The secretions of the adrenal medulla function during emergencies to reinforce and prolong the action of

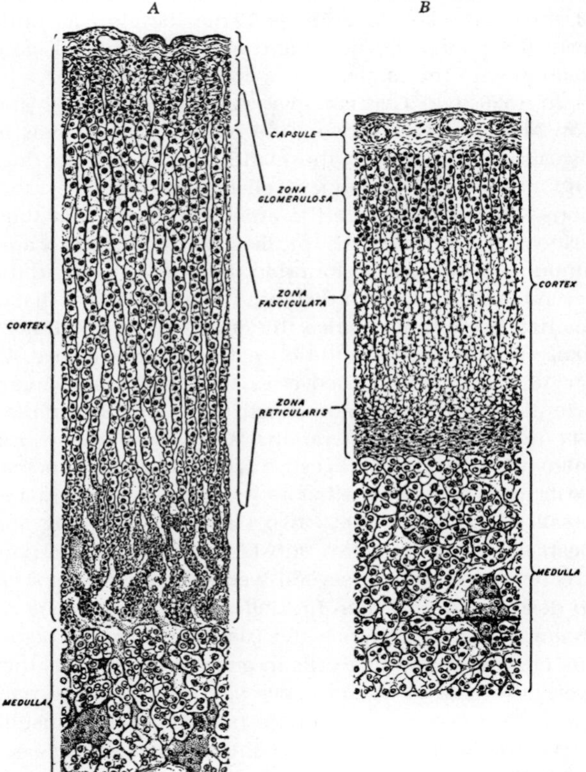

FIGURE 12.9 Sections through the adrenal cortex and medulla of *(A)* normal and *(B)* hypophysectomized rats. Since the functional capacity of the adrenal cortex is conditioned by the release of ACTH, hypophysectomy results in tremendous shrinkage of the cortex. The medulla is not influenced by hypophysectomy. Both sections are drawn to scale. (From Turner, C. D., and J. T. Bagnara: *General Endocrinology.* 6th ed. Philadelphia, W. B. Saunders Co., 1976.)

the sympathetic nervous system. The responses evoked by the two systems are very similar. Epinephrine secretion is increased by stresses such as cold, pain, trauma, emotional states, and certain drugs. The following changes resulting from the action of the sympathetic nerves and epinephrine could prepare an animal to attack its prey, defend itself against enemies, or run away: (1) The efficiency of the circulatory system is increased by increased blood pressure and heart rate and the dilation of the large blood vessels; (2) the increase in the ability of blood to coagulate and the constriction of the vessels in the skin tend to minimize the loss of blood if the animal is wounded; (3) the intake of oxygen is increased by the increased rate of breathing and the dilation of the respiratory passages; (4) the mobilization of glycogen stores of the liver and muscle makes glucose available for energy; (5) the release of ACTH from the pituitary is stimulated. ACTH in turn stimulates the release from the adrenal cortex of glucocorticoids that increase the conversion of proteins to carbohydrate and make further carbohydrate available for quick energy. Epinephrine is used clinically in treating asthma (it dilates respiratory passages), in increasing blood pressure, and in stimulating a heart that has stopped beating. Norepinephrine has much weaker effects on blood sugar and heart rate but is a more powerful vasoconstrictor.

The adrenal cortex is composed of three layers of cells surrounding the medulla (Fig. 12.9) and secretes a number of steroid hormones: (1) **glucocorticoids,** which promote the conversion of proteins to carbohydrates; (2) **mineralocorticoids,** which regulate sodium and potassium metabolism; and (3) **androgens,** which have male sex hormone activity. The most potent glucocorticoid is **cortisol,** and the most potent mineralocorticoid is **aldosterone.** Deoxycorticosterone is also an effective regulator of salt and water metabolism and is widely used clinically.

Steroids are synthesized from cholesterol not only in the adrenal cortex but also in the testis, ovary, and placenta (Table 12.2). A sequence of enzymic reactions converts cholesterol to progesterone. Progesterone is not only one of the primary female sex hormones but also an important intermediate in the synthesis of adrenal cortical hormones, androgens, and estrogens.

The human disease resulting from decreased secretion of adrenocortical hormones was first described in 1855 by the English physician Thomas Addison. **Addison's disease** is usually caused by a tubercular or syphilitic infection of the cortex that destroys its cells. The hypofunction of the adrenal in Addison's disease or its complete surgical removal is characterized by low blood pressure, muscular weakness, digestive upsets, increased excretion of sodium and chloride in the urine, increased retention of potassium in body fluids and cells, and a peculiar bronzing of the skin caused by the deposition of melanin. There is a marked decline in the concentration of blood sugar and in the glycogen content of liver, muscle, and other tissues. The animal's ability to produce carbohydrates from proteins is greatly impaired. The loss of sodium produces an acidosis, and the loss of body fluid leads to lower blood pressure and decreased rate of blood flow. The appetite for food and water decreases and there is loss of weight. The upsets in the digestive tract include diarrhea, vomiting, and pain. Muscles are more readily fatigued and less able to do work. The basal metabolic rate decreases, and the animal is less able to withstand exposure to cold and other stresses. Death ensues within a few days after complete adrenalectomy. Patients with Addison's disease are treated by the oral or intramuscular administration of a natural adrenal steroid, such as cortisol, or of a synthetic steroid.

The development and function of the adrenal cortex is regulated by **adrenocorticotropic hormone,** ACTH, secreted by the anterior lobe of the pituitary. ACTH stimulates the production of corticoids by increasing the activity of one or more of the enzymes involved in the conversion of cholesterol to cortisol. Stimulation of the adrenal cortex by ACTH leads to an increased production of cortisol and an increased concentration of that hormone in the blood. This increased concentration of cortisol inhibits the secretion of ACTH by the pituitary. Cortisol may act directly by inhibiting the syn-

TABLE 12.2 Glands Producing Steroid Hormones

Adrenal Cortex	Ovary	Testis	Placenta
Cortisol	Estradiol	Testosterone	Progesterone
Deoxycorticosterone	Progesterone	Androstenedione	Estradiol
Aldosterone	Androgens	Estradiol	
Androsterone		Estrone	
Dehydroepiandrosterone		Corticoids	
Estradiol			

thesis of ACTH in the pituitary or indirectly by decreasing the production of corticotropin-releasing hormone, CRH, by the hypothalamus.

The primary stimulus that elicits the secretion of cortisol is some sort of physical stress—an injury, a painful disease, exposure to heat or cold—that generates impulses that are forwarded to the hypothalamus. The hypothalamus secretes CRH, which passes along the hypothalamo-hypophyseal portal system of blood vessels directly to the anterior lobe of the pituitary. The CRH stimulates the appropriate cells in the pituitary to secrete ACTH, and this is carried in the blood to the adrenal cortex where it stimulates the production and release of cortisol. The mobilization of amino acids and lipids from peripheral tissues and the process of gluconeogenesis in the liver provide substrates for the repair of the damage and decrease the stimulus that led to the production of the releasing factor. The increased concentration of cortisol in the blood acts in the hypothalamus or pituitary, or both, to decrease the production and release of ACTH by negative feedback control.

An inherited defect of any one of the several enzymes involved in the synthesis of cortisol from pregnenolone and cholesterol may lead to enlargement of the adrenal cortex. The commonest defect is a deficiency of the enzyme catalyzing the insertion of a hydroxyl group at carbon 21 of the steroid. Such a defect leads to an accumulation of intermediates that, since they cannot be converted to cortisol, are converted to androgens, such as androstenedione (Fig. 12.10). This may be converted either in the adrenal or elsewhere in the body into testosterone, the most potent androgen. The lack of cortisol to inhibit ACTH production in the pituitary results in the oversecretion of ACTH. The negative feedback control cannot operate, and the adrenal grows larger and secretes even more androgens. Eventually, the individual becomes virilized. Female fetuses lacking the enzyme have external genitalia that are masculinized to varying degrees, but male fetuses with the enzymic defect may show little or no abnormality at birth. After birth, virilization in both males and females is manifested by enlargement of the phallus, early development of pubic and axillary hair, lowering of the voice and other effects of androgens. Patients with this condition, **adrenal cortical hyperplasia,** can be treated by injecting cortisol to turn off the production of ACTH by the pituitary.

The Pituitary Gland

The pituitary gland, or **hypophysis cerebri,** is an endocrine gland lying in a small depression on the floor of the skull just below the hypothalamus, to which it is attached by a narrow stalk. Its only known function is the secretion of hormones. The pituitary has a double origin: A dorsal outgrowth (**Rathke's pouch**) from the roof of the mouth grows up and surrounds a ventral evagination (the **infundibulum**) from the hypothalamus (Fig. 12.11). Both parts are of ectodermal origin. Rathke's pouch soon loses its connection to the mouth, but the connection to the brain (the **infundibular stalk**) remains. The hypophysis has three lobes: the anterior and intermediate lobes derived from Rathke's pouch and a posterior lobe from the infundibulum. The anterior lobe has no nerve fibers and is stimulated to release its hormones by hormonal factors reaching it through its blood vessels. The anterior lobe receives a double blood supply—arterial and portal. Some branches of the inter-

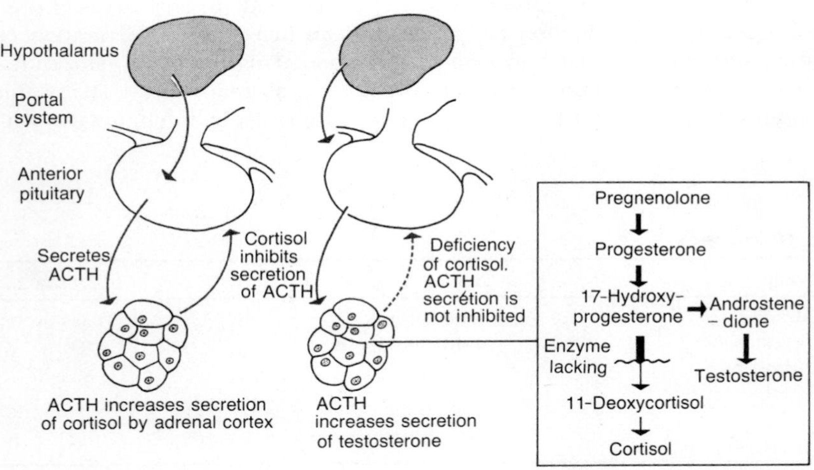

FIGURE 12.10 *Left,* The normal control of adrenal function by ACTH. A releasing factor from the hypothalamus stimulates the pituitary to secrete ACTH, which increases the synthesis of cortisol in the adrenal cortex. The cortisol then inhibits the secretion of ACTH by the pituitary. *Right,* In one type of adrenal cortical virilism, the adrenals lack an enzyme, and the conversion of progesterone to cortisol is greatly reduced. ACTH secretion is not inhibited because of the lack of cortisol. The hypersecretion of ACTH results in increased adrenal size and the secretion of androgens, such as testosterone, instead of cortisol.

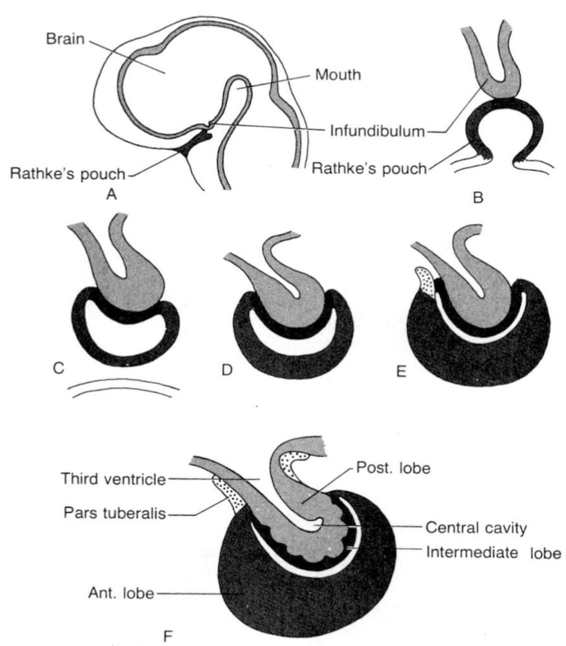

FIGURE 12.11 The development of the pituitary gland. *A,* Sagittal section through head of young embryo. *B–F,* Sagittal sections of successive stages of developing pituitary gland.

nal corotid artery pass directly to the pituitary; others serve a capillary bed around the infundibular stalk and the median eminence of the hypothalamus (Fig. 12.12). Portal veins from these capillaries pass down the infundibular stalk and empty into those surrounding the secretory cells of the anterior lobe. They provide a direct route by which releasing factors secreted by the hypothalamus pass directly to the anterior lobe and control the secretion of its hormones.

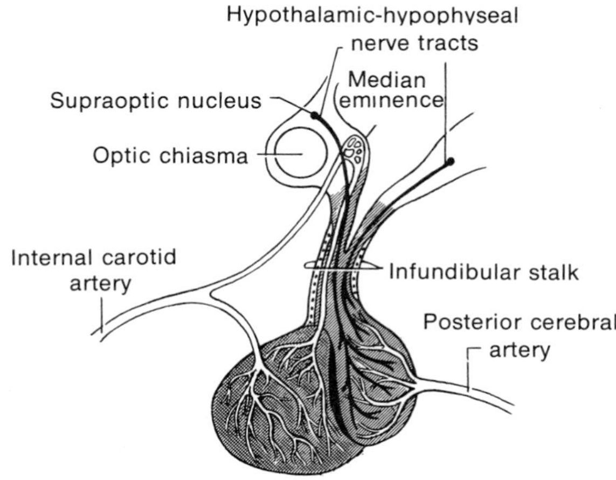

FIGURE 12.12 Blood supply of the pituitary gland.

The posterior lobe is composed of nonmyelinated nerve fibers and branching cells that contain brownish cytoplasmic granules. Its two hormones, **oxytocin** and **vasopressin** (also known as **antidiuretic hormone,** ADH), are not produced in the posterior lobe but are secreted by neurosecretory cells in the supraoptic and paraventricular nuclei of the brain, pass down the axons of the hypothalamic-hypophyseal tract, and are stored and released by the posterior lobe. There are two types of neurons in these nuclei; one type produces only oxytocin and the other only vasopressin. Both the supraoptic and paraventricular nuclei have about equal numbers of the two kinds of neurons.

The brilliant work of Vincent du Vigneaud, for which he was awarded the Nobel Prize in 1955, led to the isolation of these two hormones, the determination of their molecular structure, and their synthesis. Each is a peptide containing nine amino acids, seven of which are the same for both hormones. These two substances, with quite different physiologic properties, differ in only two amino acids. Oxytocin stimulates the contraction of uterine muscles and sometimes is injected during childbirth to contract the uterus. Vasopressin causes a contraction of smooth muscles; its contraction of the muscles in the walls of arterioles causes a general increase in blood pressure. It also regulates the reabsorption of the water by the kidney tubules (see p. 205).

An injury of the brain nuclei, the posterior lobe, or the connecting nerve tracts may lead to a deficiency of antidiuretic hormone (vasopressin) and the condition known as **diabetes insipidus.** This disease, characterized by the failure of the kidney to concentrate urine, results in the patient's excreting as much as 30 or 40 L of urine daily and hence suffering from excessive thirst. Injection of vasopressin does not cure the disease, just as injection of insulin does not cure diabetes mellitus, but it does relieve all the symptoms, and by repeated injections of vasopressin the patient can live a normal life. In one form of diabetes insipidus the receptors in the kidney for ADH are missing, and hence the injection of antidiuretic hormone does not alleviate the disorder.

The intermediate lobe of the pituitary secretes a **melanocyte-stimulating hormone** (MSH) that darkens the skin of fishes, amphibians, and reptiles by dispersing the pigment granules in the chromatophores. The skin of a frog becomes darkened in a cool dark environment and light-colored in a warm light place (Fig. 12.13). Hypophysectomy produces a permanent blanching of the skin, and injection of MSH causes darkening. The loca-

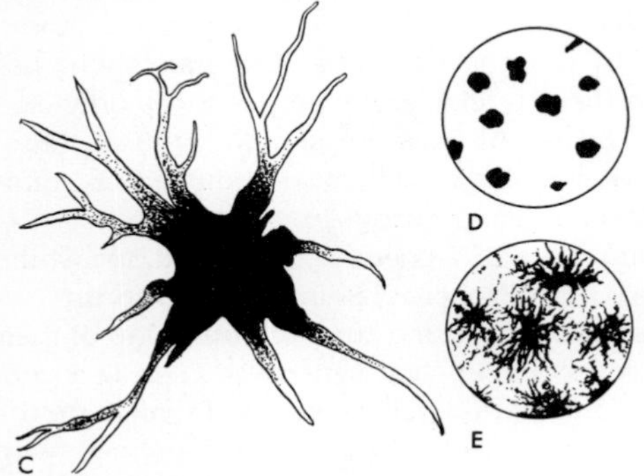

FIGURE 12.13 Integumentary adaptations in normal frogs *(Rana pipiens)*. *A,* Light-adapted animal; *B,* dark-adapted animal. *C,* A chromatophore, greatly magnified, showing the pigment. *D,* A section of skin of frog adapted to a warm light environment. *E,* Skin adapted to a cool dark environment. (*A* and *B* from Turner, C. D.: *General Endocrinology.* 2nd ed. Philadelphia, W. B. Saunders Co., 1955.)

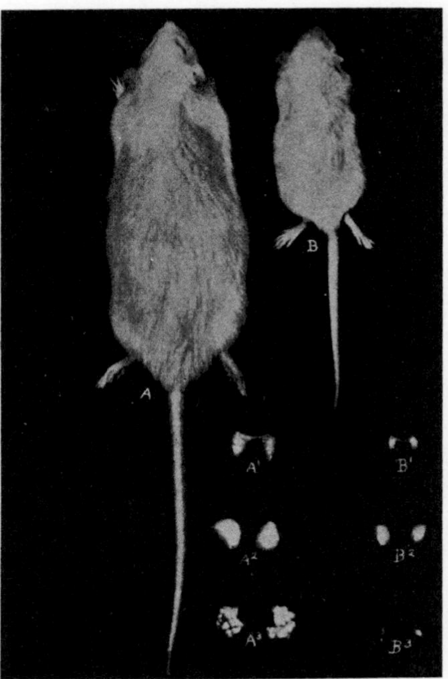

FIGURE 12.14 The effects of hypophysectomy in the rat. *A,* Normal littermate control; *B,* Littermate hypophysectomized when 36 days of age. These photographs were made at 144 days of age, when the control animal weighed 80 g. A^1, A^2, and A^3 are thyroids, adrenals, and ovaries from hypophysectomized animal. Note marked differences in size. (From Turner, C. D.: *General Endocrinology.* 2nd ed. Philadelphia, W. B. Saunders Co., 1955.)

tion of the pigment in the chromatophores is controlled directly by the amount of MSH present, not by nerves. The pituitaries of birds and mammals are rich in MSH, but MSH has little or no effect on their pigmentation.

The anterior lobe contains at least five or six types of cells, each of which probably produces and secretes a different kind of hormone: growth hormone (**somatotropin**), thyroid-stimulating hormone (**TSH**), adrenocorticotropic hormone (**ACTH**), follicle-stimulating hormone (**FSH**), luteinizing hormone (**LH**), and prolactin (luteotrophic hormone, **LTH**).

Growth hormone was the first pituitary hormone to be described. As early as 1860 it was recognized that gigantism was correlated with an enlargement of the pituitary. A growth-promoting extract of beef pituitary was made by Evans and Long in 1921, and pure growth hormone was isolated by C. H. Li in 1944. Human growth hormone is a single peptide chain of 191 amino acids. It controls general body and bone growth and leads to an increase in the amount of cellular protein (Figs. 12.14 and 12.15). Overactivity of the pituitary during the growth period leads to very tall but well proportioned persons, and underactivity leads to small persons of normal body proportions, called **midgets.** After normal growth has been completed, hypersecretion of growth hormone produces **acromegaly,** characterized by the thickening of the skin, tongue, lips, nose, and ears and by growth of the bones of the hands, feet, jaw, and skull. The other bones have apparently lost their ability to respond to growth hormone.

Growth hormone affects the rates of a number of metabolic processes leading to increased protein synthesis, the mobilization of lipid from tissues, an increased lipid concentration in the blood, an increased deposition of glycogen in liver and muscle, and an increased concentration of glucose in the blood. It is curious that a man in his eighties has about as much growth hormone in his pituitary as does a rapidly grow-

FIGURE 12.15 The effect of growth hormone on the dachshund. *Top,* Normal dog. *Bottom,* Dog that received injections of growth extract for a period of six months. (From Evans, Simpson, Meyer, and Reichert.)

ing child. Growth hormone is rapidly degraded after it has been secreted into the blood stream; its biological half-life is about 25 minutes.

Adrenocorticotropic hormone, ACTH, is a peptide containing 39 amino acids in a single chain. It is synthesized rapidly, little is stored in the pituitary, and it is rapidly removed from the plasma. Its biological half-life is less than 20 minutes. This hormone stimulates the growth of the adrenal cortex and its production of adrenocorticosteroids.

The atrophy of the thyroid that follows extirpation of the pituitary can be prevented by the administration of an extract containing **thyroid-stimulating hormone** (TSH). TSH is a basic glycoprotein with a molecular weight of about 25,000.

Follicle-stimulating hormone (FSH) and **luteinizing hormone** (LH) are both glycoproteins. FSH, with a molecular weight of about 32,000, contains about 8 percent of carbohydrate, including some sialic acid, which apparently is required for biological activity, for FSH is no longer hormonally active after the sialic acid has been removed by treating the hormone with the enzyme neuraminidase. LH is a slightly smaller protein with a molecular weight of about 26,000. A third gonadotropin, **prolactin** or luteotrophic hormone, is a protein with a single peptide chain of 198 amino acids but no carbohydrate components.

The three glycoprotein hormones of the pituitary—TSH, FSH, and LH—are each composed of two subunits, termed α and β. The α subunits of all three are very similar peptides with 96 amino acids in the chain, whereas the β subunits are distinctive and appear to be responsible for the biological specificity of the hormone. The hormones can be separated into their subunits, which have little or no biological activity, and then recombined to give full activity. The combination of a TSH α subunit with an FSH β subunit results in a protein with FSH activity.

The ovaries or testes of a hypophysectomized young animal never become mature, and they neither produce gametes nor secrete enough sex hormones to develop the secondary sex characters. Hypophysectomy of an adult results in involution and atrophy of the gonads. Both FSH and LH are necessary for achieving sexual maturity and for the regulation of the estrous or menstrual cycles. The effect of FSH is primarily on the development of graafian follicles in the ovaries; it does not cause any significant production of estradiol. Luteinizing hormone is also necessary for the development of follicles, and it controls the release of ripe eggs from the follicle, the formation of corpora lutea, and the production and release of estrogens and progesterone (Fig. 12.16). Prolactin, or LTH, maintains the secretion of estrogens and progesterone by the ovary and stimulates the secretion

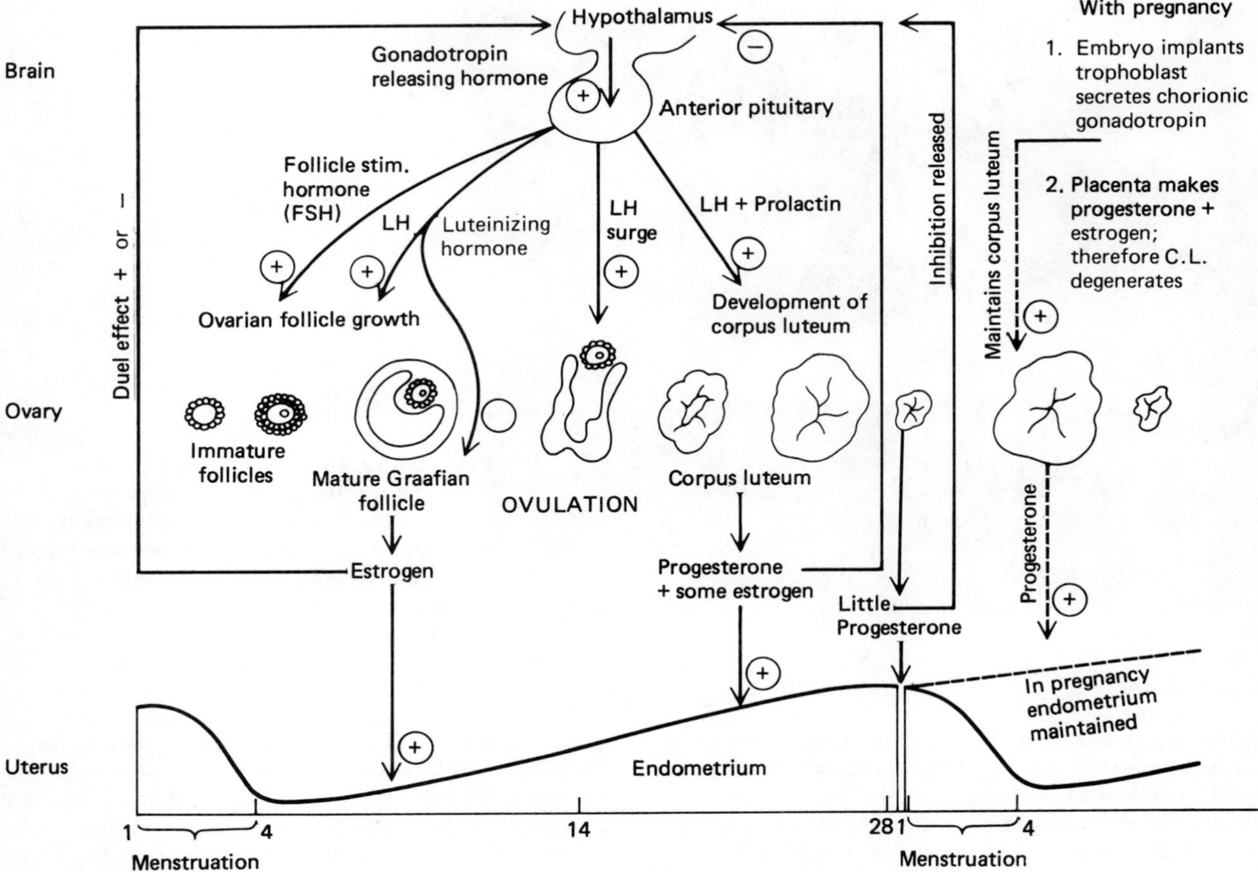

FIGURE 12.16 The menstrual cycle in the human female. The solid lines indicate the course of events if the egg is not fertilized; the dashed lines indicate the course of events when pregnancy occurs. The actions of the hormones of the pituitary and ovary in regulating the cycle are indicated by arrows.

of milk by the breast after parturition. It is effective only after the breast has been appropriately stimulated by estrogen and progesterone. Prolactin induces behavior patterns leading to the care of the young (the "maternal instinct") in mammals and other vertebrates. Roosters treated with prolactin will take care of chicks, taking them to food and water, sheltering them under their wings, and protecting them from predators.

FSH and LH control the development and functions of the testis. FSH increases the size of the seminiferous tubules, and LH stimulates the interstitial cells of the testis to produce the male sex hormone.

Hypothalamic Releasing Hormones: Neurosecretion

The control of pituitary function, which ensures that the proper amount of each of these hormones will be released at the proper moment in response to the requirements of the organism, is a complex process indeed. The release of each trophic hormone is controlled in part by the concentration of the respective target hormone in the circulating blood. For example, the release of ACTH is inhibited by cortisol. This mechanism ensures that in the normal animal the secretions of the pituitary and its target organ are kept in balance.

The hypothalamus provides an important control of pituitary function. Axons from certain centers in the hypothalamus end in the median eminence in the floor of the third ventricle. The tips of these axons secrete specific releasing or inhibiting factors that are carried by the portal veins to the hypophysis, where they stimulate or inhibit the release of specific pituitary hormones. The secretion of hormones and other substances by neurons is termed **neurosecretion.**

The first of the hypothalamic substances to be identified as a distinct entity was CRH, or **corticotropin releasing hormone,** which evokes the release of ACTH; CRH was shown in 1982 to be a peptide containing 41 amino acids. The release of CRH from the hypothalamus is inhibited by glucocorticoids and increased by adrenalectomy. The first releasing hormone to be isolated and synthesized was TRH, or **thyrotropin releasing hormone,** a tripeptide. The administration of TRH causes the release of prolactin, as well as TSH, and thus TRH may be a physiologically important regulator of the release of both.

Another substance isolated from the median eminence, **gonadotropin releasing hormone** (GnRH), induces the release of both FSH and LH from the pituitary tissue. GnRH has been isolated and identified as a decapeptide.

The release of growth hormone in the pituitary is stimulated by a 44–amino acid peptide isolated in 1983, termed **growth hormone releasing hormone,** or a GHRH. A growth-hormone-inhibiting hormone, or **somatostatin,** has been identified and purified from the hypothalamus. Somatostatin is a peptide containing 14 amino acids.

Thus the concept that the pituitary is the master gland of the endocrine system has had to be changed in view of the finding that the master pituitary is, in turn, the slave of the hypothalamus. However, both the hypothalamus and pituitary may be regulated by the hormone secreted by the target glands—the thyroid, adrenal, gonads, and so on (Fig. 12.7). Thus, in the endocrine system there is no master–slave relationship to be discerned; instead, the system is an egalitarian society composed of a federated group of autonomous organs, each of which can provide feedback controls for certain other glands.

The Pineal

The **pineal gland,** a small round structure on the upper surface of the thalamus lying between the cerebral hemispheres, develops as an outgrowth of the brain (see Fig. 10.19). It secretes **melatonin,** a methoxy indol synthesized from serotonin. A specific enzyme involved in the synthesis of melatonin is found only in the pineal. The enzyme is stimulated by norepinephrine released by the tips of the sympathetic nerves that extend to the pineal from a cervical sympathetic ganglion. Light falling on the retina of the eye increases the synthesis of melatonin by the pineal. A small nerve, the **inferior accessory optic tract,** passes from the optic nerve through the medial forebrain and connects with the sympathetic nervous system. Melatonin inhibits ovarian functions either directly or by way of an effect on the pituitary.

The Testes

Although Berthold concluded in 1849 from his experiments with roosters that the testis produces a bloodborne substance needed for the development of male sex characteristics, the major androgen, testosterone, was not identified until 1935. Testosterone, secreted by the interstitial cells of the testis, induces growth by stimulating the formation of cell proteins. The administration of androgens leads to an increase in body weight, owing to the synthesis of protein in muscle and, to a lesser extent, in the liver and kidney.

Testosterone and other androgens stimulate the development and maintenance of the secondary male characters; the enlargement of the external genitals; the growth of the accessory glands, such as the prostate and seminal vesicles; the growth of the beard and body hair, and the deepening of the voice. The secondary sex characters of other animals—the antlers of deer and the combs, wattles, and plumage of birds—are also controlled by androgens. Male sex hormones are responsible, in part, for increased libido in both sexes and for the development of mating behavior.

The removal of the testis (castration) of an immature male prevents the development of the secondary sex characters. A castrated man, a **eunuch,** has a high-pitched voice, beardless face, and small genitals and accessory glands. Castration was practiced in the past to provide guardians for harems and sopranos for choirs. Many domestic animals are castrated to make them more placid. The injection of testosterone into a castrated animal restores all the sex characters to normal. The anal fin of the male mosquito fish, *Gambusia,* is differentiated into a penis-like organ used to transfer sperm to the female. It fails to develop if the fish is castrated but appears if the castrate male (or the female) is treated with testosterone.

It should be clearly understood that males produce female sex hormones, estrogens, and that females produce androgens. One of the richest sources of female sex hormones is the urine of stallions. The normal differentiation of the sex characters is a function of a balance between the two.

The removal of the pituitary causes regression of both interstitial cells and seminiferous tubules of the testes. Androgen secretion is decreased, and the secondary sex characters regress. Normal development of the seminiferous tubules in spermatogenesis apparently requires the combined action of FSH, LH, and testosterone. The administration of excessive amounts of testosterone or estrogen may produce regression of the testes, presumably by inhibiting the release of FSH and LH from the pituitary.

The cyclic growth and regression of the testes in animals with periodic breeding seasons appear to be mediated by the pituitary. Such animals have very low amounts of circulating gonadotropin in the nonbreeding season. Changes in the ambient temperature or in the amount of daily illumination produce stimuli that are mediated by the brain and hypothalamus to induce gonadotropin secretion by the pituitary and the consequent growth and funtional state of the testes.

The Ovaries

The ovaries of vertebrates are endocrine organs as well as the source of eggs; they produce the steroid hormones **estradiol** and **progesterone.** Some mammalian ovaries produce a third hormone, the protein **relaxin,** that softens the ligaments of the pubic symphysis at the time of parturition.

In the human, the major sources of female sex hormones are the cells lining the ovarian follicles and those of the corpus luteum formed after ovulation has occurred (Fig. 12.17). The follicle cells secrete primarily estradiol, and the luteal cells secrete primarily progesterone. Estradiol regulates and controls the body changes in the female at the time of puberty or sexual maturity: the broadening of the pelvis, development of the breasts, growth of the uterus, vagina, and external genitalia, change in voice quality, and onset of the menstrual cy-

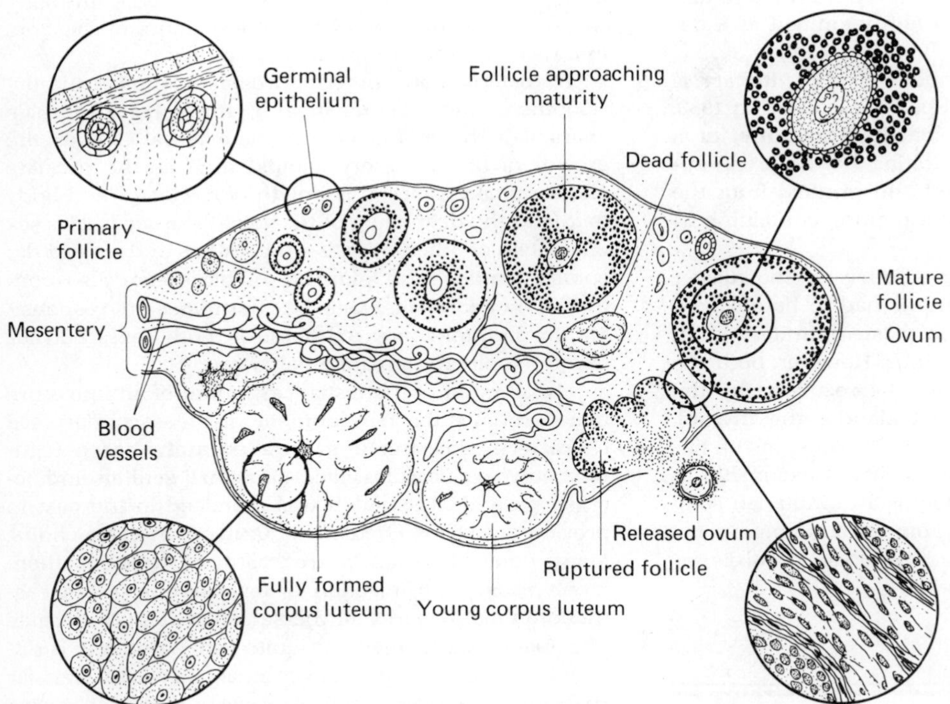

FIGURE 12.17 Stages in the development of an egg, follicle, and corpus luteum in a mammalian ovary. Successive stages are depicted clockwise, beginning at the mesentery. Insets show the cellular structure of the successive stages.

cle. Progesterone is required for the completion of each menstrual cycle, for the implantation of the fertilized egg in the uterus, and for the development of the breasts during pregnancy.

The Estrous and Menstrual Cycles

The females of most mammalian species show cyclic periods of the sex urge and will permit copulation only at certain times, known as periods of **estrus** or "heat," when conditions are optimal for the union of egg and sperm. Most wild animals have one estrous period per year; the dog and cat have two or three; and rats and mice have estrous periods every four or five days. Estrus is characterized by heightened sex urge, ovulation, and changes in the lining of the uterus and vagina. After estrus, the uterine lining thickens, and its glands and blood vessels develop to provide an optimal environment for the implanting embryos. In contrast, the cycle of the primates is characterized by periods of vaginal bleeding called **menstruation** that result from the degeneration and sloughing of the endometrial lining of the uterus. Primates, unlike other mammals, show little or no cyclic change in the sex urge and permit copulation at any time in the menstrual cycle.

A key event in the estrous cycles of lower mammals and the menstrual cycles of primates is **ovulation,** the release of a mature egg from a follicle in the ovary (Fig. 12.16). This must be released when sperm are likely to be present in the oviduct and when the lining of the uterus, the **endometrium,** is in the proper condition to permit implantation of the fertilized egg. The coordination of these events involves some half dozen hormones in addition to neural mechanisms. Ovulation is triggered by a surge of LH, secreted by the pituitary in response to GnRH, secreted by the hypothalamus (Fig. 12.18). This in turn is triggered by a surge of estradiol that operates by positive feedback control to induce the release of GnRH and LH. Prolonged treatment of the female with a constant amount of either estradiol or progesterone, as in the oral contraceptive pills, can block the LH surge and hence block ovulation. A sharp rise in estradiol must occur to affect the positive feedback center in the hypothalamus so that the pituitary is induced to secrete LH. This theory is supported by experiments in which a single dose of estrogen induced ovulation in animals whose ovaries had been primed with suitable doses of FSH and LH.

The pattern of gonadotropin release is cyclic in the female mammal but noncyclic in the male. In the rat, the male–female patterns of gonadotropin release are

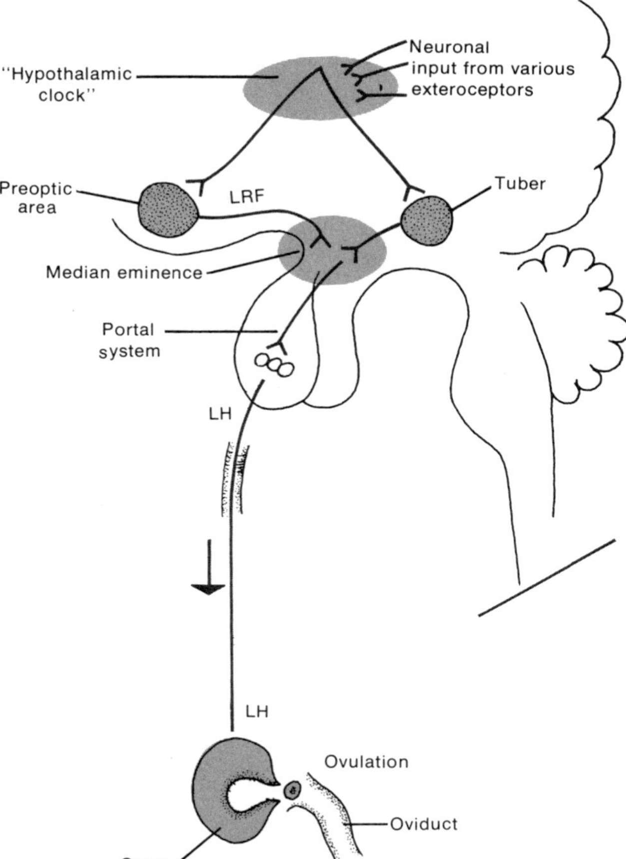

FIGURE 12.18 Spontaneous ovulation in humans and other primates. The secretion of releasing factors from the hypothalamus is stimulated by the surge of estradiol secreted by the ovary and is modulated by a variety of neuronal inputs.

determined in early postnatal life. The hypothalami of both genetic male, XY, and female, XX, rats will develop the female cyclic pattern of release unless exposed to testosterone early in postnatal life. However, human females with congenital adrenal cortical hyperplasia who have been exposed to large amounts of androgen during fetal life subsequently develop normal ovulatory cycles if treated with cortisol.

The lining of the uterus is almost completely sloughed at each menstrual period and thus is thinnest just after the menstrual flow. At that time, under the influence of FSH, one or more of the follicles in the ovary begins to grow rapidly (Fig. 12.16). The follicular cells produce estradiol, which stimulates the growth of the endometrium and some growth of the uterine glands and blood vessels. Ovulation occurs in response to the LH surge from the pituitary, and in the human this occurs about 14 days before the beginning of the next menstrual period. The corpus luteum develops and, under the stimulus of LH and prolactin, secretes progesterone and estradiol, which promote the further growth of the endometrium. The endometrial glands become secretory, and the blood vessels become long and coiled. Progesterone decreases the activity of the uterine muscles and brings the uterus into a condition such that the developing embryo formed from the fertilized egg can undergo implantation and development. Progesterone inhibits the development of other follicles. If fertilization and implantation do not occur, the corpus luteum begins to regress and secretes less progesterone; the endometrium, no longer provided with enough progesterone to be maintained, begins to slough off, resulting in menstruation.

In the rabbit, cat, ferret, mink, and certain other mammals, ovulation occurs reflexly in response to the stimulus of the act of copulation (Fig. 12.19). Nerves in the lining of the vagina send impulses up the spinal cord to the brain and stimulate the hypothalamus to secrete GnRH. The stimulus for ovulation may be a single copulation, as in the rabbit, or a minimum of 19 per day, as required by the short-tailed shrew. Direct electrical stimulation of the appropriate regions of the hypothalamus, the **tuber cinereum** and **preoptic region,** can cause ovulation in the rabbit, cat, or monkey. In the rabbit the neural stimulation of coitus causes the hypothalamus to produce GnRH, which passes via the portal system to the anterior pituitary and causes it to release a surge of LH. This passes to the ovary and causes the follicle to rupture, thereby releasing the ovum. Such animals are termed **reflex ovulators.**

A second and larger group of mammals are **spontaneous ovulators;** their ovulation is stimulated not by coitus but by the series of feedback hormonal controls described for the human. In these mammals, the timing and frequency of ovulation may be influenced by environmental factors. A female rat kept under normal conditions of day and night lighting will ovulate early in the morning, between 1:00 and 2:30 A.M. If rats are kept for two or more weeks under artificial conditions in which the period of light and dark are reversed, the time of ovulation will be shifted 12 hours. Rats exposed to continuous light eventually stop ovulating and go into constant estrus, with persistent vaginal cornification. Primates are spontaneous ovulators, and there is ample evidence that the rhythm of the human menstrual cycle and ovulation can be influenced by environmental factors. Nurses on night duty and airline stewardesses who travel long distances to different time zones frequently note perturbations in their menstrual cycles.

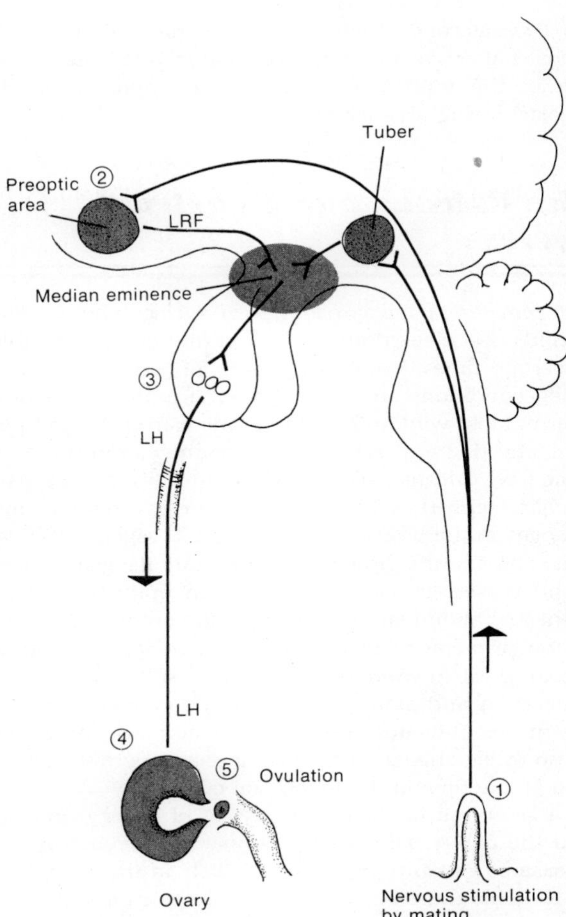

FIGURE 12.19 Diagram of the nervous pathways involved in the reflex ovulation in an animal such as the rabbit. The stimulation of receptors in the vagina by the mating act generates impulses that pass to the hypothalamus and bring about the secretion of releasing factors. These pass to the pituitary and cause the release of luteinizing hormone, which goes via the bloodstream to the ovary and initiates ovulation, the release of the egg.

Oral Contraceptives

Estrogen and progesterone block ovulation not by a direct effect on the ovary but by preventing the secretion of gonadotropin-releasing hormone by the hypothalamus. If a woman were to ovulate after she was pregnant and if this second egg were also fertilized, she would then have two fetuses in her uterus—one several months younger than the other. At parturition both would be ejected, and the younger fetus would be at a severe disadvantage. This has been prevented by the evolution of a hormonal system that prevents ovulation once conception has occurred. The high levels of progesterone and estrogen that are maintained throughout pregnancy prevent an LH surge.

The oral contraceptives contain synthetic estrogens and progestins that prevent ovulation in a similar manner. Natural estradiol and progesterone are rapidly metabolized in the body, but synthetic hormones with slightly altered molecular structures are metabolized more slowly and hence persist for a longer time. The oral contraceptives, like the natural hormones, inhibit the secretion of releasing hormones and LH and thus prevent ovulation. A woman taking "the pill" has no midcycle surge of LH and FSH and does not ovulate.

The oral contraceptives in general use are combination pills, containing both an estrogen and a progestin. These are taken daily for a period of 21 days followed by a break of seven days, during which menstruation occurs. Gonadotropin secretion is suppressed, preventing ovulation. In addition, the oral contraceptive prevents the normal maturation of the endometrium and produces an altered cervical mucus hostile to penetration by sperm. Another type of pill, containing only progestin, is taken every day throughout the cycle and, although ovulation may occur, fertilization is prevented by the changes in the nature of the cervical mucus and the endometrium.

The Hormones of Pregnancy

If the egg has been fertilized and implants in the endometrium, the cells of the trophoblast in the developing placenta secrete **chorionic gonadotropin.** Its strong luteinizing and luteotrophic activities maintain the maternal corpus luteum and stimulate its continued secretion of progesterone. One of the earliest signs of pregnancy is the appearance of chorionic gonadotropin in the blood and urine. The peak of its production is reached in the second month of pregnancy, after which the amount in blood and urine decreases to low levels. Several pregnancy tests involve the detection of this gonadotropin in a sample of the urine from a woman suspected of being pregnant. A very sensitive radioimmunoassay test for chorionic gonadotropin can diagnose pregnancy just a few days after implantation. By the 16th week or so of pregnancy in the human, the placenta itself produces enough progesterone so that the corpus luteum is no longer needed and undergoes involution. The placenta also produces estrogens. The human placenta, and probably the placentas of other mammals, secretes another protein hormone, **placental lactogen,** with properties somewhat similar to those of pituitary growth hormone and prolactin.

Thus the placenta is an endocrine organ that produces hormones similar to those of the ovary and pituitary, in addition to being an organ for the support and nourishment of the fetus. These placental hormones, together with those of the maternal and fetal endocrine glands, control the many adaptations necessary for the continuation and successful termination of pregnancy.

The production of estrogens and progesterone increases gradually throughout pregnancy and reaches a peak just before or at the time of parturition. The factors that determine the onset of labor, the expulsion of the fetus from the uterus, remain a mystery. There are many hormonal changes that occur close to the time of parturition—decreases in estrogen and progesterone, increases in certain hormones secreted by the fetal adrenal, and changes in prostaglandin production in the uterus—but whether these are causes, effects, or unrelated phenomena remains to be determined.

Hormonal Control of Lactation

The growth and development of the mammary glands after puberty and the production and secretion of milk after parturition are controlled by a complex sequence of hormonal events. The initial development of the breast and the proliferation of glandular elements are stimulated by estradiol and require the presence of insulin and growth hormone. Progesterone stimulates the further development of the glands at and after puberty. During pregnancy the additional development of the mammary glands is stimulated by estradiol and by progesterone produced primarily in the placenta. The alveoli and ducts of the mammary glands continue to develop and become secretory; however, progesterone inhibits the production of milk, and it is only after parturition and the sudden decrease in progesterone in the blood that occurs with the loss of placenta that milk production can begin. Throughout pregnancy the production of pituitary gonadotropins is inhibited by the placental gonadotropins, but with the expulsion of the placenta at birth, the pituitary begins to secrete large quantities of prolactin that stimulate the breast to produce milk (Fig. 12.20). If the breast is regularly emptied of milk, the pituitary will continue to secrete prolactin, but if milk is not removed, secretion will stop. Nerve impulses from the nipple to the hypothalamus stimulate the secretion of prolactin-releasing hormone, which passes to the pituitary and increases the production of prolactin. ACTH, TSH, and growth hormone may also

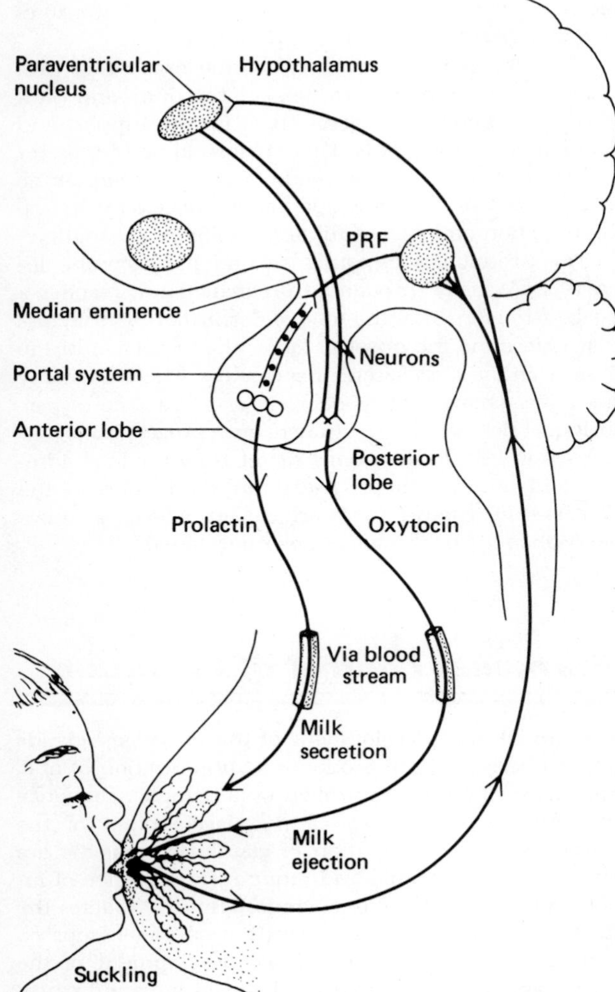

FIGURE 12.20 A diagram of the nervous and hormonal controls that stimulate the production, secretion, and release of milk in the mammary gland.

play a role in controlling both the growth of the mammary glands and their production of milk.

The *secretion* of milk from the alveolar glands into the milk duct is under the control of prolactin. The *transport* of milk from the alveolus to the nipple, where it can be removed by the suckling infant, is triggered by the **milk ejection reflex,** a neurohormonal reflex involving oxytocin. Shortly after the infant is put to breast, the mammary gland seems suddenly to fill with milk, which comes under pressure and may spurt from the nipple. Milk flow can occur in anticipation of the suckling reflex, and, in contrast, stress or discomfort can inhibit this flow so that less milk is obtained by the suckling child. The rapid increase in pressure within the mammary gland is due to the sudden expulsion from the alveoli of milk that has been synthesized previously. The movement of milk results from the contraction of myoepithelial cells that squeeze the alveoli and expel their contents. The act of suckling, or the milking of a cow or mare, triggers nerve impulses from the receptors in the nipple that pass up the spinal cord to the hypothalamus and cause the release of oxytocin from the posterior lobe of the pituitary. In the mammary gland, oxytocin stimulates the myoepithelial cells to contract, increasing intramammary pressure and bringing about the ejection of milk. Thus the milk ejection response is a reflex, but unlike ordinary reflexes the afferent arc is nervous but the efferent arc is hormonal. Like ordinary neural reflexes, the response can be conditioned, which explains ejection before milking in response to stimuli associated with nursing.

During lactation the release of gonadotropins is suppressed, with the result that ovulation occurs only occasionally and menstrual periods are usually irregular.

Prostaglandins

There is great interest at present in the prostaglandins, derivatives of 20-carbon polyunsaturated fatty acids. Many different prostaglandins have been discovered or synthesized, and their chemistry is quite complex. Although prostaglandins were first found in semen and are produced by the seminal vesicle, it is now clear that they are produced by a great many types of tissues and released into the blood stream. The production rate in the human, the amount synthesized per day, was measured by Samuelson in Sweden. In four normal adult males, the production rate ranged from 109 to 226 μg. per 24 hours, and in two normal adult females, it ranged from 23 to 48 μg. per 24 hours. These experiments indicate that there is a sex difference in the rate of production of prostaglandins and that only a small fraction of the polyunsaturated fatty acids consumed in the diet (about 10 g. per day) is converted to prostaglandins. The high concentration of prostaglandin in human semen appears to be essential for normal fertility for men with low concentrations of prostaglandin in their seminal fluid have greatly reduced fertility, even though their semen may contain the normal number of sperm.

Prostaglandins have a variety of effects on smooth muscles, the nervous system, and blood pressure and have been implicated in the control of many different types of biological events—perhaps by regulating the production of cyclic AMP by adenyl cyclase. Prostaglandins decrease arterial blood pressure, but they increase the motility of uterine muscles and are used in increasing uterine contractions at the time of childbirth. They are also used to induce abortions earlier in pregnancy. Prostaglandins also increase the motility of intestinal muscles and may cause severe cramps, nausea, vomiting, and diarrhea. In a number of species certain prostaglandins have a sedative, tranquilizing effect on the animal.

The Hormones of Arthropods

The hormones and neurosecretions of crustaceans and insects are probably better known than those of other invertebrates. In these forms there is experimental evidence for a battery of hormones that regulate chromatophore expansion, molting, growth, sexual reproduction, and development. A number of the crustacean hormones, such as those controlling chromatophore expansion, are released from a gland in the eye stalk. These hormones have been studied extensively, since the eye stalk can be removed readily, and extracts of the eye stalk can be prepared and tested.

All arthropods—crabs, spiders, insects—pass through a series of successive molts during development. Because the exoskeleton is firm and rigid, these animals can grow only by periodically shedding the chitinous exoskeleton and growing a new, larger one. Before the old exoskeleton is shed a new one develops underneath (p. 100). Many insects, such as moths, butterflies, flies, and many others pass through successive stages that are quite unlike one another (Fig. 12.21). From the egg hatches a wormlike **larva**—called a caterpillar (moths) or maggot (flies)—that crawls around, eats voraciously, and molts several times, each time becoming a larger larva. The last larval molt forms a **pupa** (L. a doll), that neither moves nor feeds. The moth or butterfly larva spins a cocoon and pupates (molts to form a pupa) within it. During pupation the structures of the larva are broken down and used as raw materials for the formation of parts of the adult animal. Each part—legs, wings, eyes, and antennae—develops from a group of cells, called a **disc,** that develops directly from the egg. They have never been a functional part of the larva but remain quiescent during the larval period. During the pupal stage these discs grow and differentiate into the adult structures, but they remain collapsed and folded. When the adult hatches out of the pupal case, blood is pumped into these collapsed structures, they unfold and inflate, and the chitinous exoskeleton hardens. This striking change in appearance from larva to adult is termed **metamorphosis.**

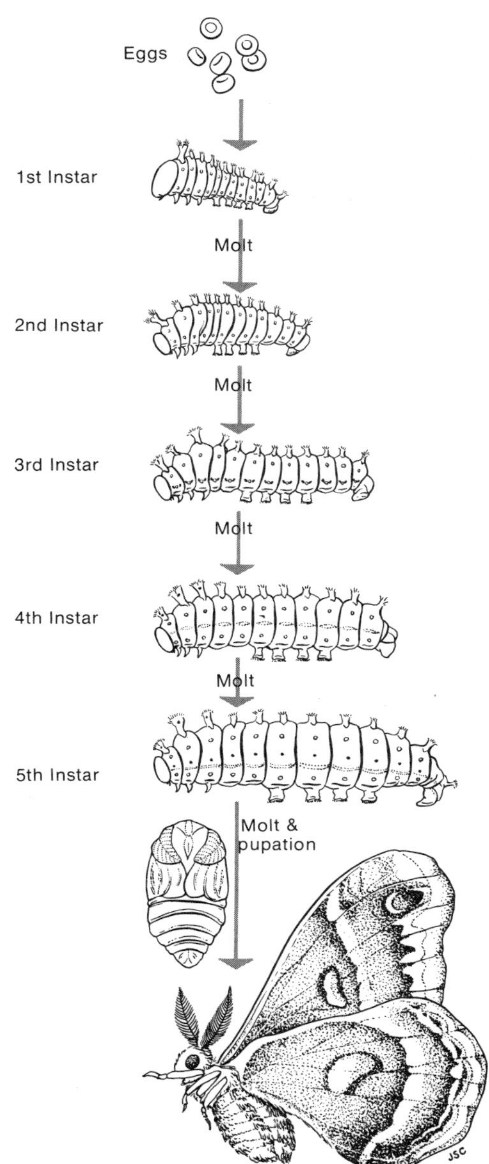

FIGURE 12.21 Stages in the complete metamorphosis of a cecropia moth.

The molting of insects and other arthropods is under hormonal control. The major organ initiating molting in insects is the **intercerebral gland** on the surface of the brain (Fig. 12.22). Axons from these neurosecretory cells pass posteriorly to the **corpus cardiacum,** composed of the expanded tips of these axons. The intercerebral

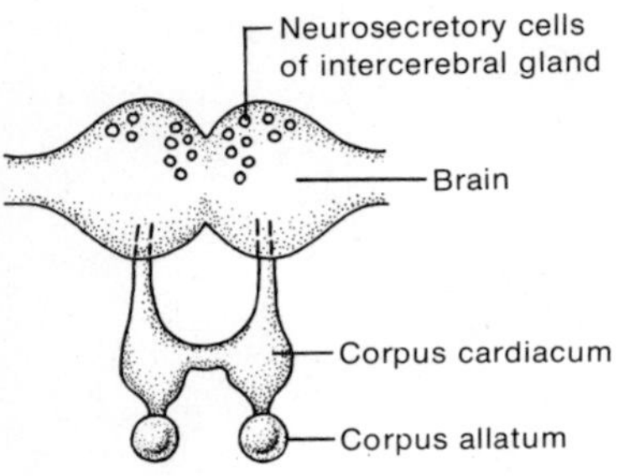

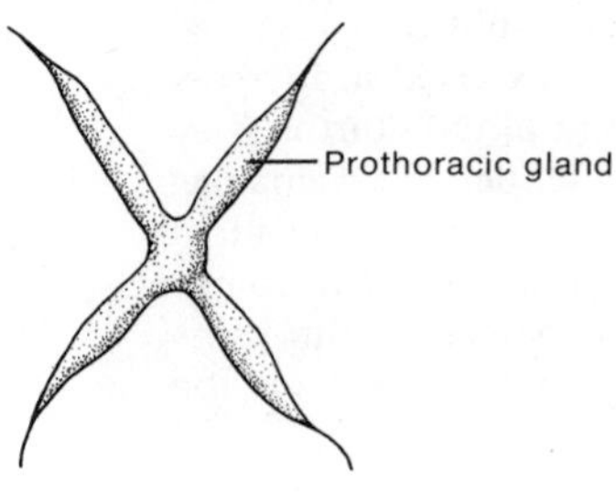

FIGURE 12.22 The endocrine glands of the cockroach. Those on the left lie in the head dorsal to the esophagus; the prothoracic gland (right) is in the ventral part of the prothorax, among the muscle cells. (After Bodenstein, D.: *Recent Progress in Hormone Research.* Vol. 10. New York, Academic Press, Inc., 1954.)

gland secretes several hormones that regulate various aspects of body activity. One of these, **prothoracicotropic hormone,** is released from the corpora cardiaca and initiates the molting process by stimulating the prothoracic gland. The prothoracic gland, a diffuse set of strands of large ectodermal cells in the ventral part of the prothorax, secretes **ecdysone** (Gr. *ekdysis,* a getting out), a steroid synthesized from cholesterol. The conditions that stimulate the activity of the intercerebral glands have been studied in the blood-sucking bug *Rhodnius prolixus.* Stretch receptors in its abdomen are stimulated when the bug has a large meal of blood, and these stimulate the intercerebral glands to activity. *Rhodnius* feeds only infrequently, but one large meal is

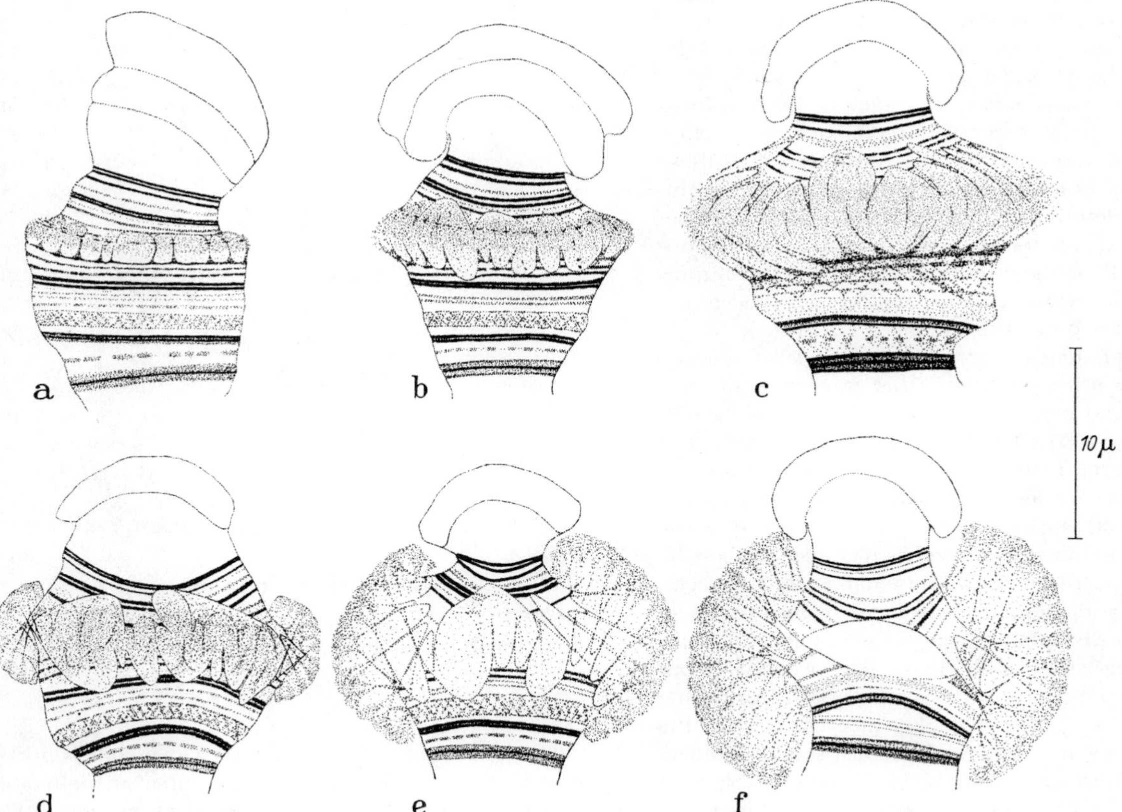

FIGURE 12.23 Diagram illustrating the changing appearance of a many-stranded (polytene) chromosome in a salivary gland of *Chironomus tentans* as a chromosome puff gradually appears. The material that makes up the puff has been shown by histochemical tests and by autoradiography with tritium-labeled uridine to consist largely of RNA. (From Beermann: *Chromosoma, 5*:1390198, 1952.)

enough to support the metabolic activity associated with a molt. Ecdysone stimulates the hypodermis to secrete molting fluid, thus leading to molt and metamorphosis. The same system of prothoracicotropic hormone and ecdysone initiates the molt from larval to pupal stage.

In 1960 Clever and Karlson found that within 15 minutes of injecting minute amounts of ecdysone into the larva of the midge *Chironomus* the puffing or swelling of a specific region of a particular chromosome occurred. Dipteran insects may have in certain tissues giant chromosomes composed of many chromatids. In these giant chromosomes, individual bands, corresponding to specific genetic loci, are visible under the microscope and can be identified. The puffing of certain bands in the chromosomes occurs at specific times in the course of normal development and is known to represent the synthesis of RNA at those sites (Fig. 12.23). Injecting ecdysone causes puffing at band I-18C and the subsequent production of the enzyme **dopa decarboxylase** in epidermal cells. Dopa decarboxylase catalyzes the conversion of DOPA to N-acetyl dihydroxyphenylethylamine, an agent involved in the hardening of the cuticle.

The control of metamorphosis involves a third set of endocrine organs, the **corpora allata** (L. *corpus*, body + *adlatus*, added), small glands located in the head, just behind the corpora cardiaca. If these glands are removed from a young larva, it will undergo metamorphosis at the next molt, even though this may be one or more molts too soon. Conversely, if the corpora allata from young larvae are transplanted into older larvae about to undergo metamorphosis at the next molt, the juvenile form is retained at the molt. The corpora allata secrete **juvenile hormone,** a derivative of an 18-carbon fatty acid. A substance extracted from certain American paper products (e.g., *The New York Times*) prepared from the woods of certain trees has marked juvenile hormone activity. It is conjectured that the trees protect themselves from some types of insects by secreting a juvenile hormone-like material that prevents the larvae from becoming adults and reproducing. It has also been suggested that we use juvenile hormone as an insecticide to prevent the multiplication of insects. Under test conditions, spraying caterpillars, or the foliage on which they are feeding, with solutions containing juvenile hormone has prevented their pupating normally. Instead of molting into giant larvae, they die.

Juvenile hormone inhibits metamorphosis but permits molting to occur, thus ensuring that the larva will molt several times and reach a large size before pupating. Juvenile hormone is not produced during the last larval stage; hence pupation can occur at the ensuing molt.

The corpora allata secrete hormones that affect other phenomena in certain groups of insects. The deposition of yolk in eggs and the secretions of the accessory sex glands are under the control of hormones from the corpora allata. In some cockroaches the male is attracted to the female by a pheromone that she produces and secretes under the control of a hormone from the corpora allata.

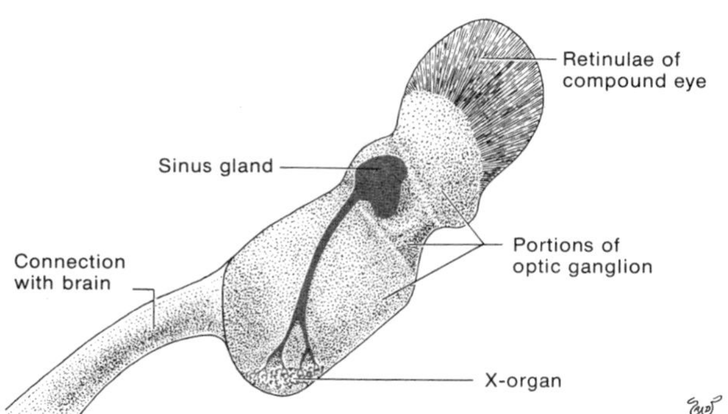

FIGURE 12.24 A diagram showing the location of the sinus gland and X organ in the eyestalk of the crab.

Some pupae, such as those of the giant silkworm *Platysamia cecropia,* remain in a state of arrested development or dormancy (called **diapause**) over the winter. If a newly formed pupa is kept at 24°C., it will remain inactive for five or six months but eventually will begin to develop. If a newly formed pupa is chilled to 4°C. for six weeks and then kept at 24°C. it begins at once to develop. Thus, chilling initiates further development by stimulating the release of prothoracicotropic hormone from the intercerebral gland. If intercerebral glands are removed and chilled and then transplanted into a diapausing pupa that has not been chilled, the pupa begins development.

The molting process in crustaceans is under the control of a hormone that is accumulated in the **sinus glands** in the eyestalk (Fig. 12.24). These glands secrete several hormones, one of which initiates molting. The sinus glands are composed of the expanded tips of axons surrounding a blood sinus. The cell bodies of these axons are located some distance away, in the eyestalk in the **X organ.** The hormone is produced in the cell bodies of the X organ and passes along the axons to the sinus gland, where it is stored and released. When the eyestalk of a crab or other crustacean is removed, a molt usually occurs, indicating that the hormone made in the X organ and secreted by the sinus gland tends to *inhibit* molting. The **Y organ,** composed of diffuse strands of ectodermal cells at the base of the large muscles of the mandibles, is an endocrine gland that produces ecdysone that induces molting (Fig. 12.25). Ecdysone is now believed to be involved in regulating the molting of all arthropods. The hormone of the X organ inhibits molting by preventing the secretion of molting hormone (ecdysone) by the Y organ. The X organ secretes, in addi-

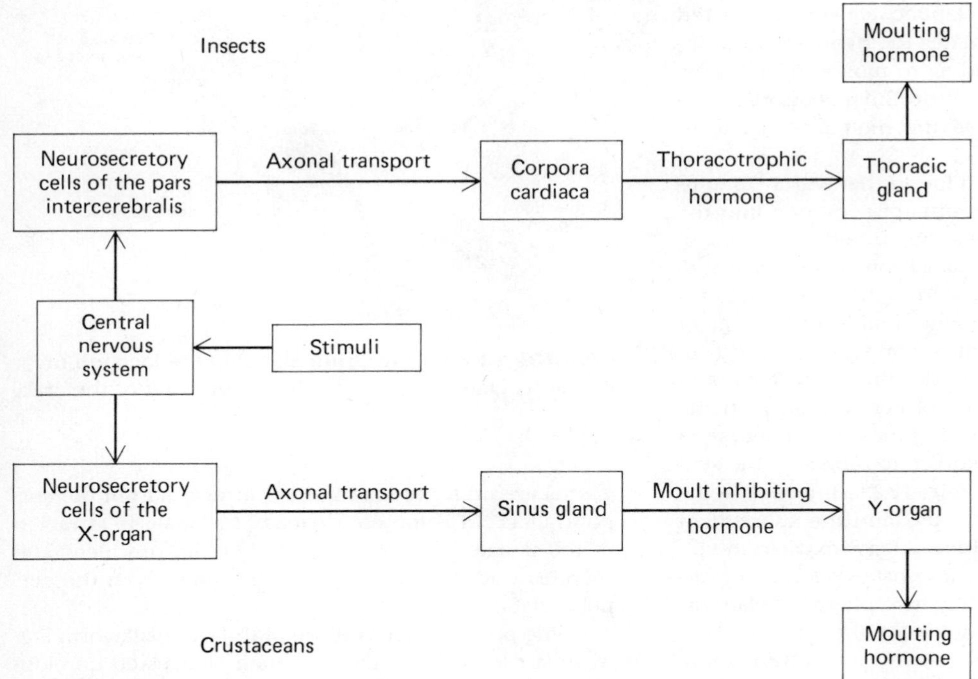

FIGURE 12.25 Comparison of the hormonal control of molting in crustaceans with that in insects. The insect thoracic glands and the crustacean Y-organs are analogous in both producing ecdysone. But in insects the hormone from the corpora cardiaca stimulates secretion of ecdysone, while in crustaceans the sinus gland hormone inhibits secretion. (From Highnam, K.C., and L. Hill: *The Comparative Endocrinology of the Invertebrates.* 2nd ed. London, Edward Arnold (Publishers), Ltd., p. 215.)

tion to the antimolting hormone, hormones that affect the distribution of pigment in the compound eyes and the pigmentation of the body and several that influence metabolism and reproduction.

Summary

1. Hormones are chemical messengers produced by cells in one part of the body that diffuse or are transported in the blood to another region where they are effective in very low concentrations in regulating and coordinating cellular activities. Hormones are remarkably diverse chemical compounds. In general, each type of hormone is produced and secreted by a specific kind of cell. The target cells of each hormone have specific receptors, proteins that bind the hormone with high affinity, high specificity, and low capacity. The steroid hormone combines with its specific receptor and may be transported to the nucleus where it activates specific genes, leading to the production of messenger RNA coding for the synthesis of specific proteins. Peptide hormones bind to their receptors on the surface of the target cell and activate adenylate cyclase, generating cyclic AMP, which activates protein kinase and leads to the phosphorylation of specific proteins and activates or inactivates specific enzymes.

2. The vertebrate thyroid gland produces thyroxine and triiodothyronine, which increase the rate of oxidative, energy-releasing processes in body cells. Lack of thyroxine results in retarded body growth and mental development. The production and secretion of thyroxine is regulated by thyroid stimulating hormone (TSH) produced by the pituitary and by TRH secreted by the hypothalamus.

3. The parathyroid glands produce parathyroid hormone (PTH), which, along with calcitonin secreted by the thyroid, regulates the metabolism of calcium and phosphorus.

4. The islet cells of the pancreas produce insulin and glucagon, hormones that regulate the metabolism of glucose and other carbohydrates. Insulin increases the uptake of glucose from the blood by muscles and adipose tissue. Glucagon, by way of cyclic AMP, activates phosphorylase, an enzyme that catalyzes the breakdown of glycogen in the liver; this leads to an increased content of glucose in the blood.

5. The adrenal glands consist of an outer cortex and an inner medulla. The medulla secretes epinephrine and norepinephrine, which reinforce and prolong the action of the sympathetic nervous system. The adrenal cortex produces three types of steroid hormones: glucocorticoids, mineralocorticoids, and androgens with male sex hormone activity. Glucocorticoids stimulate the conversion of amino acids to carbohydrates; mineralocorticoids regulate the metabolism and excretion of sodium and potassium; and the adrenal androgens normally control the growth of axillary and pubic hair. The development and function of the adrenal cortex is regulated by adrenocorticotrophic hormone (ACTH) secreted by the pituitary.

6. The pituitary gland lying on the floor of the skull just beneath the brain has a double origin: A dorsal outgrowth of the roof of the mouth (Rathke's pouch) meets and surrounds a ventral evagination of the hypothalamus, the infundibulum. The anterior and intermediate lobes of the pituitary are derived from Rathke's pouch and the posterior lobe from the infundibulum. The anterior lobe has a double blood supply, arteries and a portal system that supplies blood from the hypothalamus, providing a means to transfer releasing hormones from the hypothalamus to the pituitary. The posterior lobe secretes oxytocin and vasopressin produced in the hypothalamus. Oxytocin stimulates uterine contraction, and vasopressin increases blood pressure. The intermediate lobe produces melanocyte stimulating hormone (MSH), which darkens the skins of fishes, amphibians, and reptiles by dispersing the pigment granules in the chromatophores. The anterior lobe produces, in addition to TSH and ACTH, growth hormone, follicle stimulating hormone (FSH), luteinizing hormone (LH), and prolactin (Prl). Growth hormone controls general body and bone growth by increasing the rates of metabolic processes involved in protein synthesis. FSH and LH (and TSH as well) are glycoproteins composed of α and β subunits. The α subunits of all three are identical; the β subunits are unique and determine the specificity of the hormone. FSH stimulates the growth of follicles in the ovary and of seminiferous tubules in the testis. LH stimulates the release of ripe eggs from the follicle and the formation of corpora lutea; in the male it regulates the production of androgens in the interstitial cells of the testis.

7. Specific releasing hormones, produced in the hypothalamus and secreted into the pituitary portal system, control pituitary function. Corticotropin releasing hormone evokes the release of ACTH; thyrotropin releasing hormone induces the release of TSH and of prolactin; gonadotropin releasing hormone causes the release of both FSH and LH. These releasing hormones control the pituitary, the pituitary hormones control the target glands, and the hormones of the target glands in turn control the hypothalamus.

8. The pineal gland, a small round structure on the upper surface of the thalamus, secretes melatonin, a hormone that inhibits ovarian function.

9. The interstitial cells of the testes produce androgens, such as testosterone; these stimulate the development and maintenance of the secondary male sex characteristics and the accessory male glands, such as the prostate and seminal vesicle. The antlers of deer and the combs, wattles, and plumage of birds are controlled by androgens. Androgens are also responsible, at least in part, for increased libido in both sexes and for the development of mating behavior. The removal of the pituitary causes regression of both interstitial cells and seminiferous tubules.

10. The ovaries produce the female sex hormones, progesterone and estradiol. Estradiol controls the body changes in the female at the time of puberty or sexual maturity: broadening of the pelvis, development of breasts, growth of uterus, vagina, and external genitalia. Progesterone is required for the completion of each menstrual cycle, for the implantation of the fertilized egg, and for the maintenance of pregnancy.

11. The menstrual cycles of primates and the estrous cycles of other mammals are regulated by the complex interactions of FSH, LH, prolactin, estradiol, and progesterone. In some animals, such as the rabbit and ferret, ovulation is induced reflexly by stimulation of the vagina in copulation. In humans and many other mammals ovulation is stimulated not by copulation but by an intricate sequence of feedback controls involving gonadotropin releasing hormone, LH, estradiol, and perhaps FSH and progesterone as well. Oral contraceptives contain synthetic analogues of estradiol and progesterone and act by preventing the secretion of gonadotropin releasing hormone.

12. The placenta produces the protein hormones chorionic gonadotropin and placental lactogen and the steroid hormones progesterone and estradiol.

13. Lactation is under a very complex hormonal control that involves estradiol and progesterone plus prolactin and, in some species, growth hormone, insulin, and ACTH as well. The secretion of milk from the alveolar glands is regulated by prolactin, but the transport of milk from the alveolus to the nipple is controlled by oxytoxin, which stimulates contraction of myoepithelial cells that squeeze the alveoli.

14. Prostaglandins, derived from 20-carbon polyunsaturated fatty acids, have a variety of effects on smooth muscles, the nervous system, and blood pressure and play a role in the inflammatory process.

15. Arthropods secrete a variety of hormones that regulate many bodily processes, such as chromatophore expansion, molting, growth, sexual reproduction, and development. Molting is regulated by the steroid ecdysone and metamorphosis by juvenile hormone, a derivative of an 18-carbon fatty acid. Ecdysone is regulated in turn by prothoracicotropic hormone, secreted by the intercerebral glands. The sinus glands in the eyestalk of crustacea produce several hormones, one of which inhibits molting; molting is induced by another hormone produced by the Y organ.

References and Selected Readings

Bentley, D. J.: *Comparative Vertebrate Endocrinology.* 2nd ed. Cambridge, Cambridge University Press, 1983. An interesting overview of endocrine phenomena in a variety of animals.

Burnette, W. J.: *Invertebrate Endocrinology.* Berlin, Springer-Verlag, 1973. A source book for information regarding the production and roles of hormones in invertebrate animals.

Kistner, R.: *The Pill.* New York, Delacorte Press, 1969. An in-depth survey of steroid oral contraceptives and their side effects.

McKerns, K. W. (Ed.): *Hormones and Cancer.* New York, Plenum Press, 1983. A series of research papers on the role of endocrines in producing and treating cancer.

Norris, D.: *Vertebrate Endocrinology.* Philadelphia, Lea & Febiger, 1980. A good text on hormones produced by vertebrates and their functions.

Stanbury, J. B., J. B. Wyngaarden, D. S. Fredrickson, J. Goldstein, and M. Brown: *The Metabolic Basis of Inherited Disease.* 5th ed. New York, McGraw-Hill, 1983. A large source book with an enormous amount of information, mostly concerning the human.

Tepperman, J.: *Metabolic and Endocrine Physiology.* 2nd ed. Chicago, Year Book Medical Publishers, 1968. A clear presentation of the metabolic effects of the mammalian hormones.

Turner, C., and J. Bagnara: *General Endocrinology.* 4th ed. Philadelphia, W. B. Saunders Co., 1976. A standard text of general endocrinology.

Tulchinsky, D., and K. J. Ryan: *Maternal and Fetal Endocrinology.* Philadelphia, W. B. Saunders Co., 1980. A detailed presentation of the roles of hormones in pregnancy and development.

Yen, S. S., and R. B. Jaffe: *Reproductive Endocrinology.* Philadelphia, W. B. Saunders Co., 1978. A detailed presentation of the roles of hormones in reproduction.

13 Reproduction I: Gamete Formation and Fertilization

ORIENTATION

Reproduction is the formation of new individuals from existing ones and is characteristic of all living organisms. This and the next chapter will explore the reproductive processes of animals. We will begin by considering the differences between asexual and sexual processes, but most of the two chapters will be devoted to sexual reproduction. This first chapter covers egg and sperm formation, including the mechanism by which the number of chromosomes is reduced to one half the number of nonreproductive cells. We will then study the process of fertilization and some of the many adaptations in animals that increase the chances that this event will take place.

If there is one feature of an organism that qualifies as the essence of life it is the ability to reproduce—to perpetuate the species. The survival of the species as a whole requires that its individual members multiply, that each generation produce new individuals to replace ones killed by predators, parasites, or aging. This can be contrasted with those processes needed for the day-to-day survival of the individual organism, processes such as nutrition, gas exchange, integration, and excretion. Reproduction is not necessary for the survival of the individual organism, but without reproduction the species becomes extinct. Thus the survival of the individual is directed in part toward fulfilling a reproductive capacity that is critical for the continued existence of the species.

At the molecular level reproduction involves the unique capacity of the nucleic acids for self-replication, which depends on the specificity of the relatively weak hydrogen bonds between pairs of nucleotides (Chapter 16). Reproduction at the level of the whole organism ranges from the simple fission of unicellular organisms—a process that does not involve sex at all—to the incredibly complicated morphological, physiological, biochemical, and behavioral processes involved in the reproduction of higher animals. The primary events of reproduction in all animals (and in plants as well) are the formation of gametes, fertilization, and the transformation of the fertilized egg into a new individual. Many adaptations have evolved in animals that make these events possible. This chapter will deal briefly with asexual reproduction and then with sexual reproduction from gamete formation through fertilization. The following chapter will describe the fate of the fertilized egg.

Asexual Versus Sexual Reproduction

In **asexual reproduction** a single parent splits, buds, or fragments to give rise to two or more offspring that have hereditary traits identical with those of the parent. It is thus a kind of natural cloning. Although not all animals can reproduce asexually, the process is widespread among both the higher and lower groups; indeed, the production of identical human twins by the splitting of a single fertilized egg is a kind of asexual reproduction.

Perhaps the simplest form of asexual reproduction is the splitting of the body of the parent into two more or less equal parts, each of which becomes a new, independent, whole organism. This form of reproduction, termed **fission,** occurs chiefly among the protists (Fig. 13.1*A*), the single-celled animals and plants, but some animals, such as planarian flatworms (Fig. 13.1*B*) divide transversely by fission, each half regenerating the missing part.

Hydras and certain other cnidarians reproduce by **budding**; a small part of the parent's body becomes differentiated and separates from the rest (Fig. 13.1*C*). This part develops into a complete new individual and may take up independent existence, or the buds from a single parent may remain attached as a colony of many individuals.

Salamanders, lizards, starfish, and crabs can grow a new tail, leg, or other organ if the original one is lost. When this ability to regenerate the whole from a part is extremely marked, it becomes a method of reproduction.

Certain starfish have the ability to regenerate an entire new starfish from a single arm (Fig. 13.1*D*), and many sea anemones can regenerate new individuals from fragments of tissue torn from the basal disc as it slowly moves across the bottom.

Despite the large number of individuals that can be produced by asexual reproduction, the process has the serious limitation that every offspring is genetically identical to its parent. Asexual reproduction does not increase the genetic variability of the species, which might lead to new adaptations or improve old ones. Thus very few animals can rely solely on asexual reproduction; it almost always occurs in addition to sexual reproduction.

Sexual reproduction typically involves two parents; each contributes a specialized reproductive cell, a gamete, and these fuse to form the fertilized egg. The **egg** is typically large and nonmotile, with a store of nutrients to support the development of the embryo that results when the egg is fertilized. The **sperm** is typically small and motile, adapted to swimming actively to the egg by beating its long whiplike tail. Sexual reproduction is biologically advantageous, for it makes possible the combination of the best inherited characteristics of the parents and provides for the possibility that some of the offspring may be better adapted to survive than either parent was. Evolution can proceed much more rapidly and effectively with sexual, than with asexual, reproduction.

Meiosis

The development of a new animal begins with **gametogenesis,** the formation of eggs and sperm in members of the parental generation. During gametogenesis a pair of cell divisions called **meiosis** (Gr. *meioun,* diminish) reduces the number of chromosomes from the diploid to the haploid condition. When two gametes unite in fertilization the fusion of their nuclei reconstitutes the

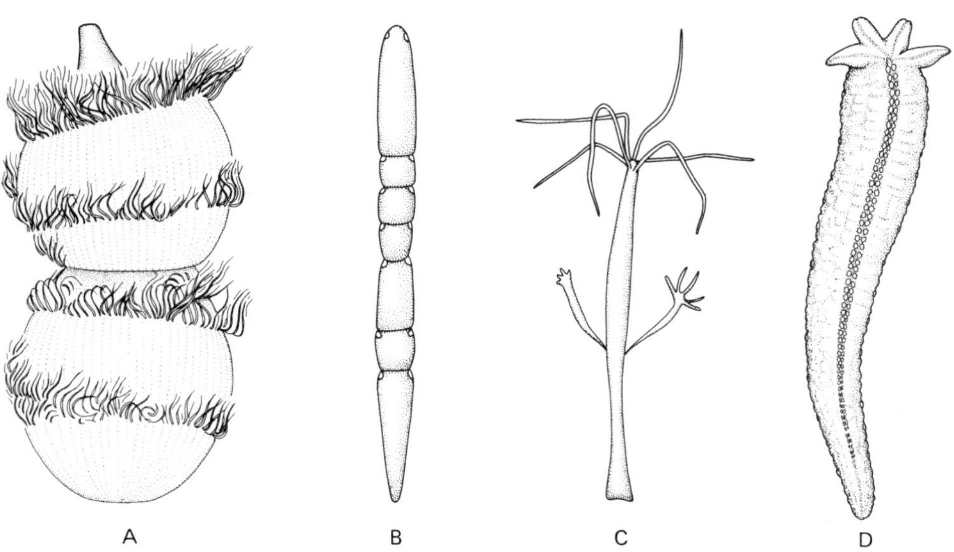

FIGURE 13.1 Types of asexual reproduction in animals. *A,* An electron scanning micrograph of an early stage of binary fission in the ciliate protozoan, *Didinium nasutum.* This type of asexual reproduction is common among protozoans. *B,* Transverse fission in the flatworm *Stenostomum.* In these worms formation of new fission planes begins before the old ones are complete, producing a chain of regenerating individuals. *C,* Budding of new individuals from the body wall of hydra. *D,* Regeneration of a new disc and arms from a single arm in the starfish *Linckia.* Most other starfish require a part of the disc for regeneration to occur. (*A* courtesy of Eugene B. Small; *B,* after Child; *C,* modified from Richters.)

2N number of chromosomes. In meiosis the members of each pair of chromosomes separate and pass to different daughter cells. As a result of this, each gamete contains one and only one of each kind of chromosome; in other words, it contains one complete set of chromosomes. This is accomplished by the pairing, or **synapsis,** of the like chromosomes and a separation of the members of the pair, with one going to each pole of the dividing cell. The like chromosomes that pair during meiosis, called **homologous chromosomes,** are identical in size and shape, have identical chromomeres along their length, and contain similar hereditary factors, or genes. For example, one pair of homologous chromosomes may contain genes that affect eye color, hair texture, and so forth. Since each member of the pair originally came from different parents, one from the mother and one from the father, the ways in which the genes affect these traits (brown eyes or blue eyes, etc.) may or may not be the same. A set of one of each kind of chromosome is called the 1N, or **haploid, number.** A set of two of each kind is called the 2N, or **diploid, number.** The haploid number for the human is 23 and the diploid number is 46. Gametes—eggs and sperm—have the haploid number. Fertilized eggs (zygotes) and all the cells of the body developing from the zygote have the diploid number. A fertilized egg gets exactly half of its chromosomes and half of its genes from its mother and the other half from its father. Only the last two cell divisions, which result in mature functional eggs or sperm, are meiotic; all other cell divisions are mitotic.

The process of meiosis consists of two successive cell divisions (Fig. 13.2). Each includes prophase, metaphase, anaphase, and telophase stages, but there are important differences between mitosis and meiosis, especially in the prophase of the first meiotic division. In this the chromosomes first appear as long, thin threads, becoming shorter and thicker, as in mitosis. The homologous chromosomes undergo synapsis while they are still elongate and thin. The homologous chromosomes come to lie close together side by side along their entire length and are twisted around each other. After synapsis has occurred, the chromosomes continue to shorten and thicken. Each one becomes visibly double, consisting of two chromatids, as in mitosis. This doubling has occurred in the S phase before meiosis begins. At the end of the first meiotic prophase the chromosomes have doubled and undergone synapsis to yield a bundle of four homologous chromatids called a **tetrad.** Each pair of chromosomes gives rise to a bundle of four so the number of tetrads equals the haploid number of chro-

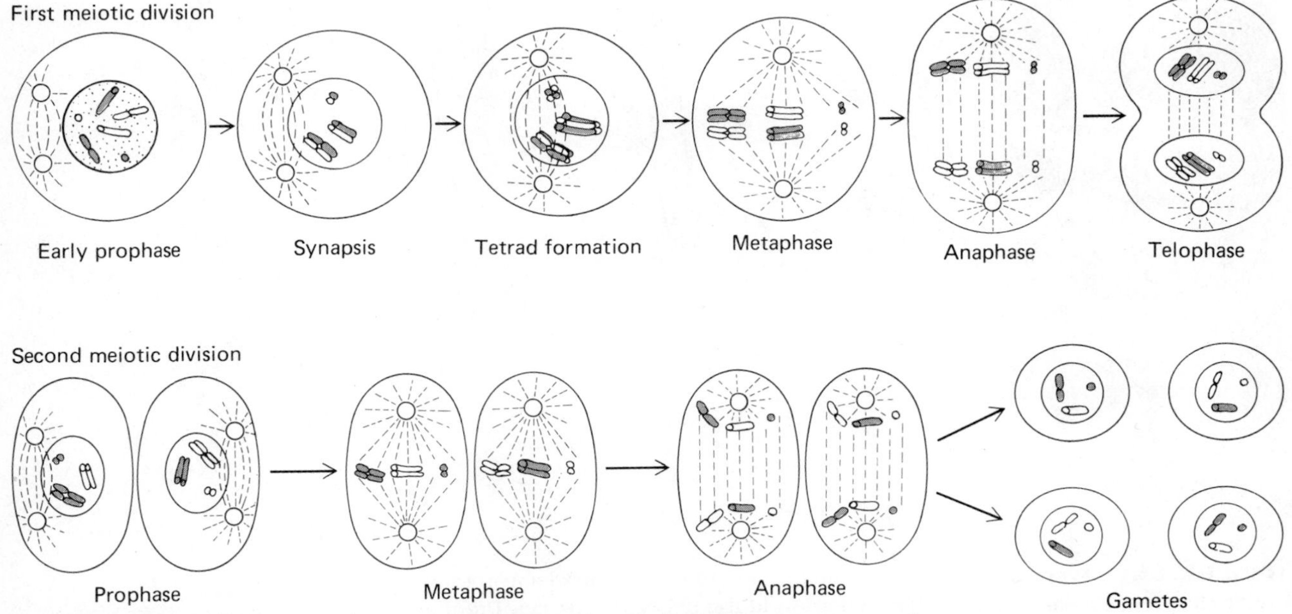

FIGURE 13.2 Meiosis in a hypothetical animal with a diploid chromosome number of six.

mosomes. In human cells there are 23 tetrads (and a total of 92 chromatids) at this stage. The centromeres have not divided and there are only two centromeres for the four chromatids of each tetrad.

While these events are occurring, the two centrioles pass to opposite poles, a spindle forms between the centrioles, and the nuclear membrane dissolves. The tetrads line up around the equator of the spindle, and the cell is said to be in metaphase. In the anaphase of the first meiotic division the daughter chromatids formed from each chromosome, still united by the centromere, separate and move toward opposite poles. Thus the homologous chromosomes of each pair, but not the daughter chromatids of each chromosome, are separated in anaphase 1. This differs from mitotic anaphase in which the centromeres do divide, and the daughter chromatids pass to opposite poles. At telophase the haploid number of double chromosomes is present at each pole. The orientation of one tetrad on the metaphase plate and the direction its homologous chromosomes follow as they separate are independent of the orientation of other tetrads. Although the resulting cells in the first telophase will have one double chromosome of each type, they probably will not be identical genetically, for each cell will have a mixture of maternal and paternal chromosomes. For example, one cell might have maternal chromosomes A and C and paternal chromosome B, along with their replicates, whereas the other cell would have paternal A and C and maternal B. At the end of a mitotic division, the daughter cells are identical genetically.

In most animals there is no clear interphase between the two meiotic divisions. The chromosomes do not divide into daughter chromatids, and there is no synthesis of DNA as there is in the S phase between mitotic divisions. The chromosomes do not form chromatin threads; instead the centriole divides, a new spindle forms in each cell at right angles to the spindle of the first division, and the haploid number of double chromosomes lines up on the equator of each spindle. The telophase of the first meiotic division and the prophase of the second meiotic division are usually of rather short duration. The lining up of the double chromosomes on the equator of the spindle constitutes the metaphase of the second meiotic division. The metaphases of the first and second meiotic divisions can be distinguished because in the first the chromosomes are arranged in bundles of four and in the second the chromosomes are arranged in bundles of two. The centromeres divide and the daughter chromatids, now chromosomes, separate and move to opposite poles. Thus in the telophase of the second meiotic division in humans, 23 chromosomes, one of each kind, arrive at each pole. The cytoplasm then divides, nuclear membranes form, and the chromosomes gradually elongate and become chromatin threads.

The two successive meiotic divisions yield four nuclei, each of which has one and only one of each kind of chromosome, a haploid set. The members of the homologous pairs of chromosomes are segregated into separate daughter cells. The four cells resulting from the two meiotic divisions become gametes and do not undergo any further mitotic or meiotic divisions.

Fundamentally the same process occurs in the meiotic divisions in the testis, which result in sperm, and in the meiotic divisions in the ovary, which result in eggs, but there are some important differences in detail.

Spermatogenesis

Gametogenesis involves either the formation of sperm, called **spermatogenesis,** or the formation of eggs, called **oogenesis.** In all animals except sponges, gametogenesis is restricted to certain areas of the body, which are usually developed as gonadal organs, the **testes** or **ovaries.** The gametes exit from the body by way of **gonoducts, oviducts** from the ovaries, and **sperm ducts** from the testes.

The testes of animals vary greatly in structure. They may be saccular, tubular, or composed of a large number of little chambers. The testes of vertebrates consist of thousands of cylindrical sperm tubules in each of which develop billions of sperm. The walls of the sperm tubules are lined with primitive, unspecialized germ cells called **spermatogonia** (Gr. *sperma,* seed + *gonos,* offspring). Throughout embryonic development and during early postnatal development the spermatogonia divide mitotically, giving rise to additional spermatogonia to provide for the growth of the testis. After sexual maturity some spermatogonia begin to undergo spermatogenesis, the formation of mature sperm, while others continue to divide mitotically and produce more spermatogonia for later spermatogenesis. In most wild

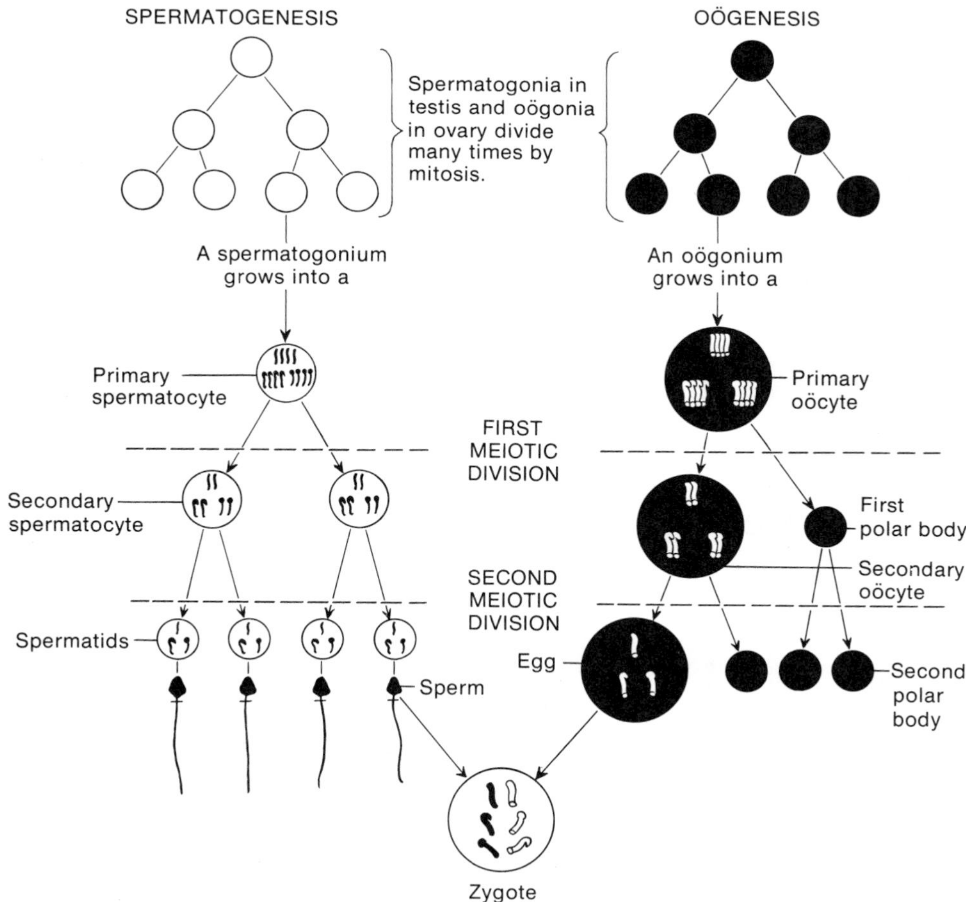

FIGURE 13.3 Comparison of the formation of sperm and eggs.

animals there is a definite breeding season, either in spring or fall, during which the testis increases in size and spermatogenesis occurs. Between breeding seasons the testis is small and contains only spermatogonia. In man and most domestic animals spermatogenesis occurs throughout the year once sexual maturity is reached.

Spermatogenesis begins with the growth of the spermatogonia into larger cells known as **primary spermatocytes** (Fig. 13.3). These divide (first meiotic division) into two **secondary spermatocytes** of equal size that in turn undergo the second meiotic division to form four **spermatids** of equal size. The spermatid, a spherical cell with a generous amount of cytoplasm, is a mature gamete with the haploid number of chromosomes. A complicated process of growth and differentiation, though not cell division, converts the spermatid into a functional sperm. The nucleus shrinks in size and becomes the head of the sperm (Fig. 13.4), while the sperm sheds most of its cytoplasm. Secretory granules from the Golgi bodies congregate at the front end of the sperm and form a cap, the **acrosome** (Gr. *akros,* tip + *soma,* body). This contains enzymes that may play a role in penetrating the cell membrane of the egg.

The two centrioles of the spermatid move to a position just in back of the nucleus. A small depression appears on the surface of the nucleus, and one of the centrioles, the proximal centriole, takes up a position in the depression at right angles to the axis of the sperm. The second, or distal, centriole, just behind the proximal centriole, gives rise to the **axial filament** of the sperm tail (Fig. 13.5). Like the axial filament of flagella, it consists of two longitudinal fibers in the middle and a ring of nine pairs, or doublets, of longitudinal fibers surrounding the central two.

The mitochondria move to the point at which head and tail meet and form a small **middle piece** that provides energy for the beating of the tail. Most of the cytoplasm of the spermatid is discarded as **residual bodies** that are taken up by phagocytosis by the **Sertoli cells** in the seminiferous tubules. These cells protect, support, and nourish the developing sperm. The mature

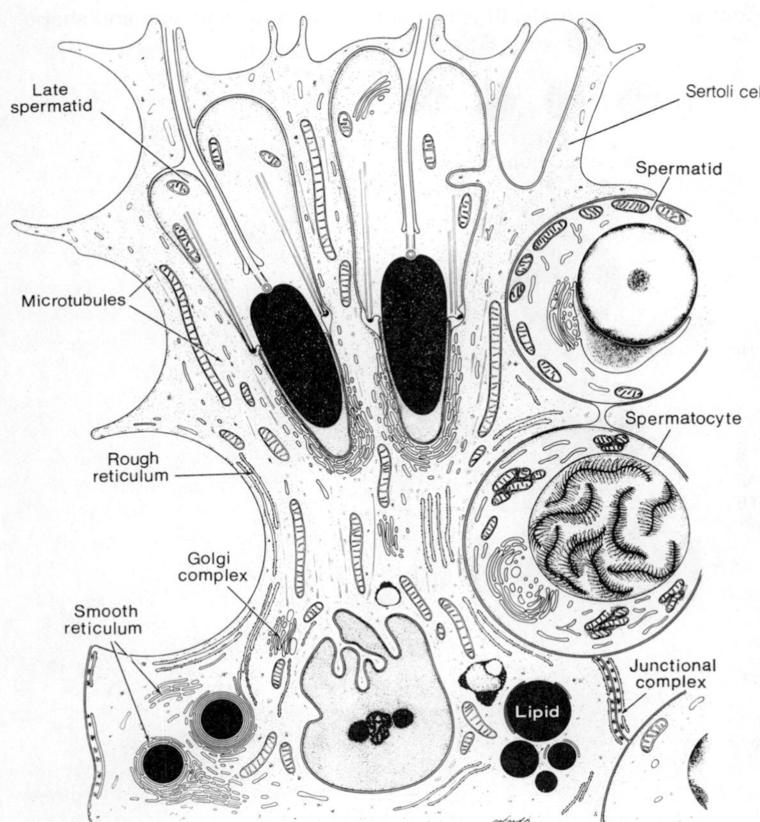

FIGURE 13.4 Diagram of a section of a human seminiferous tubule to show the stages in spermatogenesis and in the transformation of a spermatid into a mature sperm. Stages are embedded within a large, supportive Sertoli cell during the course of their development. (From Bloom, W. and D. W. Fawcett: *A Textbook of Histology,* 10th ed. Philadelphia, W. B. Saunders Co., 1975.)

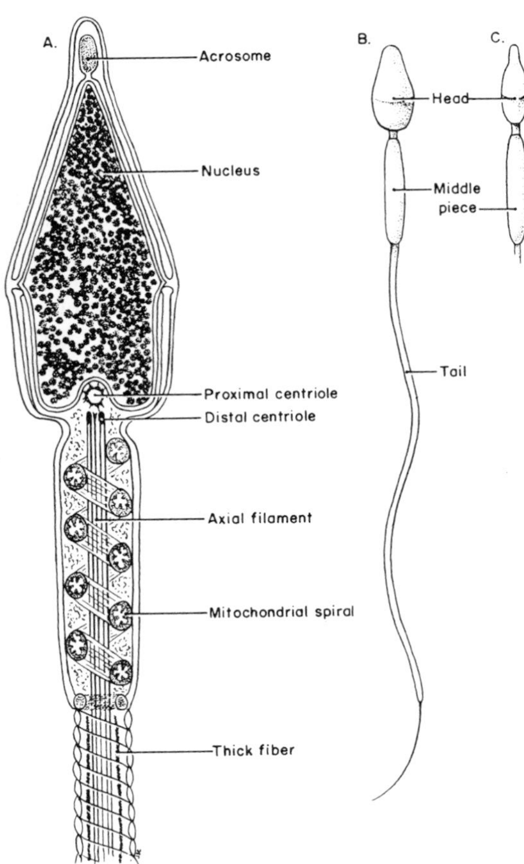

FIGURE 13.5 *A,* Diagram of the head and middle piece of a mammalian sperm, greatly enlarged, as seen in the electron microscope. *B* and *C,* Top and side views of a sperm seen by light microscopy.

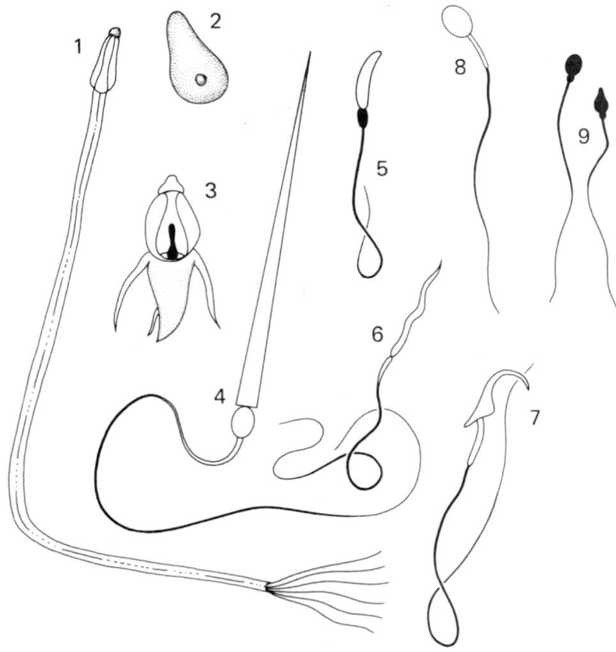

FIGURE 13.6 Spermatozoa from different species of animals, illustrating the differences in size and shape. *1,* Gastropod. *2,* Ascaris. *3,* Hermit crab. *4,* Salamander. *5,* Frog. *6,* Chicken. *7,* Rat. *8,* Sheep. *9,* Man.

sperm retains only a thin sheath of cytoplasm surrounding the mitochondria in the middle piece and the axial filament of the tail.

Spermatogenesis occurs in a cyclic fashion along all parts of a seminiferous tubule. In any one section of the tubule, the various stages will be passed through in succession. In adjacent sections of the tubule the cells will tend to be a stage behind or ahead. In human beings it takes 16 days for a mature sperm to develop from a spermatogonium.

The spermatozoa of various animal species may be quite different. There are great variations in the size and shape of the tail and in the characteristics of the head and middle piece (Fig. 13.6). The sperm of a few animals, such as crabs, lobsters, and the parasitic roundworm *Ascaris* have no tail and move instead by amoeboid motion.

Oögenesis

The ova, or eggs, develop in the ovary from immature sex cells, **oögonia.** Early in development the oögonia undergo many successive mitotic divisions to form additional oögonia, all of which have the diploid number of chromosomes. Some or all of the oögonia develop into **primary oöcytes** and begin the first meiotic division. The events occurring in the nucleus—synapsis, the formation of tetrads, and the separation of the homologous chromosomes—are similar to those occurring in spermatogenesis, but the division of the cytoplasm is unequal, resulting in one large cell, the **secondary oöcyte,** that contains the yolk and nearly all the cytoplasm and one small cell, the first **polar body,** that consists of practically nothing but a nucleus (Fig. 13.3). It was named a polar body before its significance was understood because it appeared as a small speck at the animal pole of the egg.

In the second meiotic division the secondary oöcyte divides unequally to yield a large **oötid** with essentially all the yolk and cytoplasm, and a small second polar body, both of which have the haploid number of chromosomes. The first polar body may divide at about the same time into two additional polar bodies. The oötid undergoes further changes, but no further cell division, to become a mature ovum. The three small polar bodies soon disintegrate so that each primary oöcyte gives rise to just one ovum in contrast to the four sperm formed from each primary spermatocyte. The unequal cytoplasmic division ensures that the mature egg will have enough cytoplasm and stored yolk to survive if it is fertilized. The primary oöcyte in a sense puts all of its yolk in one ovum; the egg has neatly solved the problem of reducing its chromosome number without losing the cytoplasm and yolk needed for development after fertilization.

Like the testes, the structure of ovaries varies greatly. In many animals, notably the vertebrates, the oögonia and oöcytes are surrounded by a layer of follicle cells. In the human this occurs early in fetal development, and by the third month the oögonia begin to develop into **primary oöcytes.** When a human female is born, her two ovaries contain some 400,000 primary oöcytes that have attained the prophase of the first meiotic division. These primary oöcytes remain in prophase until the woman reaches sexual maturity. During each monthly reproductive cycle one or more follicles begins to enlarge. Not all reach maturity because many atrophy, but usually one follicle will. At the time of ovulation (15 to 45 years after meiosis began!) the egg will be in the secondary oöcyte stage. In most vertebrates sperm

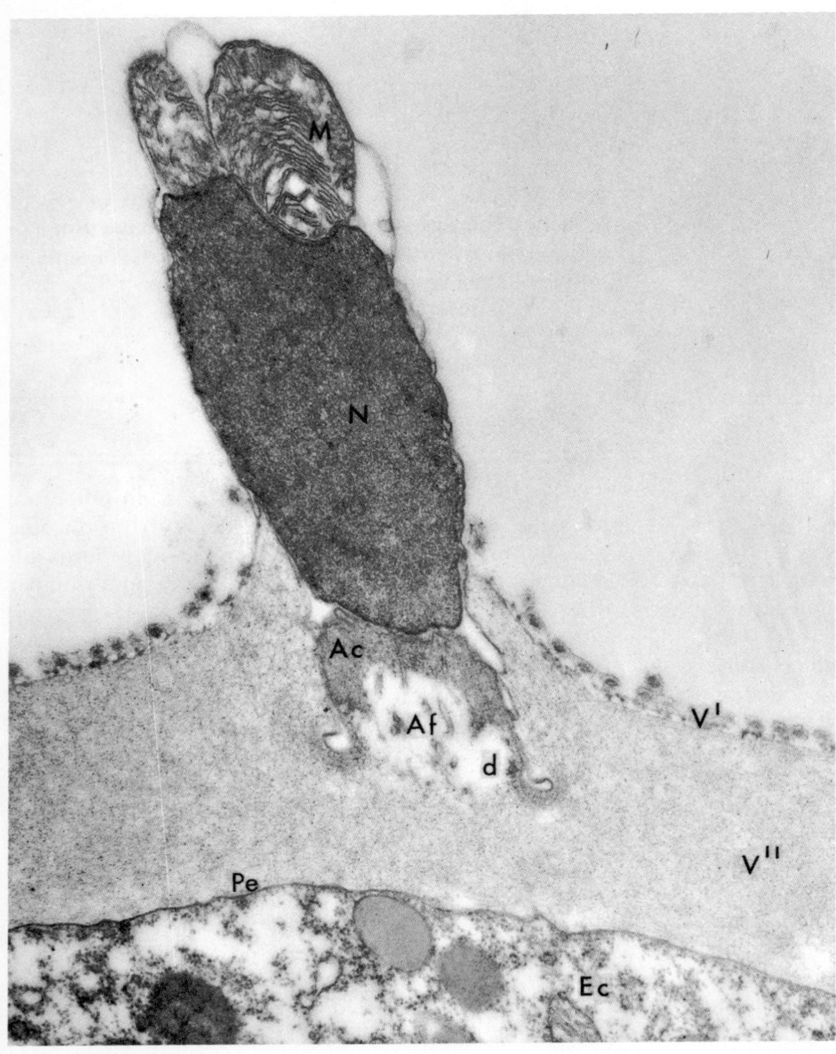

FIGURE 13.7 Fertilization. *A,* Sperm of an annelid worm penetrating the vitelline membrane of the egg. Abbreviations: *Ac,* acrosome; *Af,* acrosomal filaments; *d,* area where substance of vitelline membrane has been dissolved away; *Ec,* egg cytoplasm; *N,* nucleus of sperm; *Pe,* plasma membrane of egg; *V′*, external layer of vitelline membrane; *V″*, internal layer of vitelline membrane. (*A,* Courtesy of L. H. Colwin and A. L. Colwin, 1961.) *B,* Stages in the fertilization of a sea urchin egg. Sperm size, thickness of vitelline envelope, and perivitelline space have all been exaggerated for purposes of clarity. The fertilization membrane, which lifts from the egg surface, is formed from the vitelline envelope plus cortical material released from the egg.

penetration is needed as a stimulus for the second meiotic division.

The composition of the stored nutrients in the yolk varies from one species to another but usually includes proteins, phospholipids, and neutral fats organized in granules. The amount and distribution of yolk in an egg plays an important role in determining the pattern of development that ensues following fertilization. Eggs of animals, such as sea urchins and mammals, with small amounts of yolk evenly distributed within the cytoplasm, are called **isolecithal** (Gr. *isos,* equal + *lekithos,* yolk) or **homolecithal.**

Many animals, including flatworms, snails, clams, and most vertebrates, have eggs with the yolk concentrated at the lower, or **vegetal, pole.** Such eggs are termed **telolecithal** (Gr. *telos,* end + *lekithos,* yolk). The upper, or **animal, pole** contains the nucleus and less yolky cytoplasm. Frogs have moderately telolecithal eggs, but those of reptiles and birds are extreme, with as much as 90 percent of the egg composed of yolk. A cap of nonyolky cytoplasm containing the nucleus rests on the animal pole. The largest eggs are of this type, and the record is held by the ostrich, which has a yolk (the actual egg cell) over 6 cm. in diameter.

The eggs of arthropods, especially insects, have a different pattern of yolk distribution and are termed **centrolecithal.** The yolk is concentrated in the center of the egg, and the cytoplasm is present as a thin layer on the entire surface of the egg. In addition, an island of cytoplasm in the center of the egg contains the nucleus.

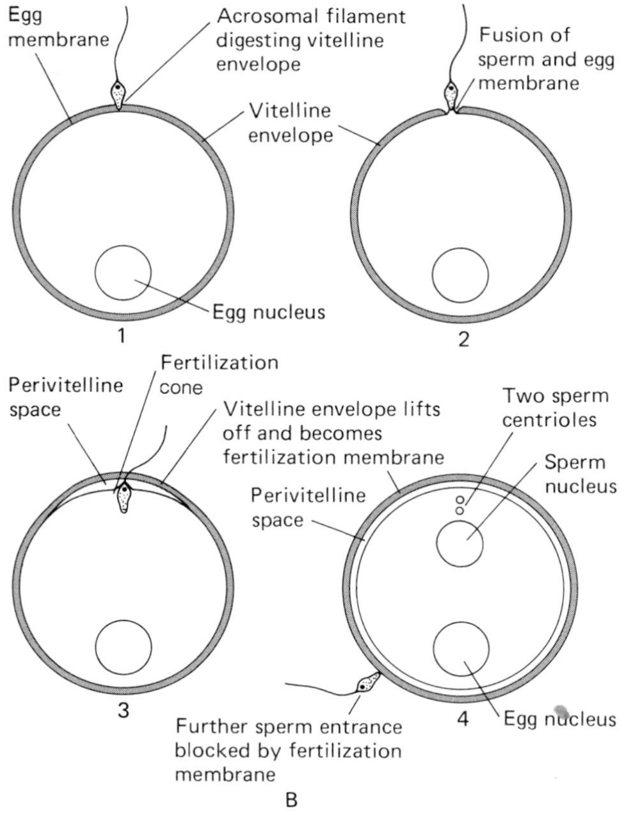

Fertilization

Developing eggs eventually reach a state of maturation at which they are capable of fusion with a sperm, a process known as **fertilization.** By various means to be described later, sperm and eggs of the same species are brought into proximity with each other. The swimming movements of the sperm bring it into contact with the egg. Each gamete contributes to a complex series of cellular changes that results in the entrance of the sperm and the activation of the egg. Egg cells, unlike other cells, have a **vitelline membrane,** or envelope, lying over the plasma membrane (Fig. 13.7*A*). As a sperm approaches the egg surface an **acrosomal reaction** occurs. In many species one or more acrosomal filaments develops and penetrates the vitelline membrane. Concurrently enzymes released from the acrosome dissolve a pathway through this membrane. When the acrosomal material reaches the egg's plasma membrane, the surface of the egg bulges outward as a **fertilization cone** (Fig. 13.7*B*). The plasma membranes of sperm and egg then come together and break down in the region of contact, forming a passage through which the sperm nucleus is drawn into the egg cytoplasm. Its tail may remain behind. As these events are taking place, the surface of the egg undergoes a **cortical reaction** that spreads outward from the point of sperm contact. Mucopolysaccharides that have been stored in cortical granules are released to the surface. As they imbibe water and swell, the vitelline membrane is raised from the egg surface as a **fertilization membrane.** This prevents other sperm from penetrating the egg.

Within the egg cytoplasm, the compact nuclear material of the sperm head swells to form a **sperm,** or **male pronucleus,** that moves toward the **egg,** or **female pronucleus.** The two pronuclei either fuse together to form a zygote nucleus or each contributes its chromosomes to the first cleavage spindle.

The cortical reaction activates the egg, throwing its metabolic machinery, so to speak, from neutral into forward gear. The egg may now complete its meiotic divisions. The egg then prepares for the cleavage divisions that lead to the formation of the embryo.

The union of one haploid set of chromosomes from the sperm with another haploid set from the egg, which occurs in fertilization, reestablishes the diploid chromosome number. Thus the fertilized egg, or **zygote** (Gr. *zygotos,* yolked or united), and all of the body cells developing from it by mitosis have the diploid number of chromosomes. In each individual exactly half of the chromosomes and half of the genes come from the mother and the other half from the father. All of the phenomena of mendelian genetics depend upon these simple facts. Because of the nature of gene interactions, the offspring may resemble one parent more than the other, but the two parents make essentially equal contributions to its inheritance.

Parthenogenesis

The eggs of many species can be stimulated to cleave and develop without fertilization. The development of an unfertilized egg into an adult is known as **parthenogenesis** (Gr. *parthenos,* virgin + *gene,* birth). Changes in temperature, pH, or the salt content of the surrounding water or chemical or mechanical stimulation of the egg itself will stimulate the eggs of many species to parthenogenetic development. A variety of marine invertebrates, frogs, salamanders, and even rabbits have been produced parthenogenetically. The resulting adult animals are generally weaker and smaller than normal and are infertile.

Parthenogenesis occurs naturally among many groups of animals, including vertebrates. In fact, there are some species of animals, such as certain microscopic, freshwater animals called rotifers and certain lizards, in which males are unknown. Parthenogenetic species commonly have eggs produced by mitosis rather than by meiosis, so that the population is diploid. But this is not always true, and some species of rotifers produce eggs by both mitotic and meiotic divisions. In some water fleas (crustaceans) and some aphids (insects), parthenogenesis occurs for several generations, and then some males are produced that develop and mate with the females.

Parthenogenesis is related to sex determination in some animals. The queen honey bee is fertilized by a male just once during her lifetime, in her "nuptial flight." The sperm are stored in a pouch connected with the genital tract and enclosed by a muscular valve. If sperm are released from the pouch as she lays eggs, fertilization occurs and the eggs develop into females—queens and workers. If the eggs are not fertilized, they develop into males—drones.

Adaptations for Fertilization

Most sperm can swim, but the motile powers of sperm are sufficient for transport over only microscopic distances. Moreover, the life span of at least one of the two gametes is usually brief. Thus in order for fertilization to occur there must be some means of bringing sperm and eggs together in time and in space.

Reproductive Synchrony. The synchronization of reproductive function in male and female requires first that gametes are *produced at the same time* in most of the population of a given species inhabiting a certain geographical area, and second that they be *released at the same time.* Some species of animals have some part of the population always reproductively active, but in the great majority of animals production of gametes is seasonal. Such synchronous production is generally dependent upon some environmental cues, such as change in photoperiod, ambient temperature, food supply, and lunar and tidal cycles, that activate internal neural and hormonal controls. For example, in birds increasing day length in spring stimulates via the nervous system the secretion of hormones that initiate migration, nesting, and egg production. This is why commercial poultry farms equip their houses with artificial lighting.

Gamete release can also be triggered by external cues. They may be environmental signals, such as tidal or lunar cycles. Some species of marine worms, for example, emerge from their burrows and swarm near the surface at certain specific phases of the moon (p. 561). Similarly, the grunion, a small coastal Pacific fish, is famous for its predictable spawning time on California beaches.

A common stimulus for gamete release is that provided by the opposite sex. For example, in many sessile or sedentary marine animals using external fertilization, the release of gametes by one sex will stimulate the release of the gametes of nearby members of the opposite sex, sometimes leading to epidemic gamete release. In most mammals the female comes into heat, i.e., is attractive and receptive to the male, only near the time of ovulation (egg release from the ovary).

The specific reproductive synchronization of a particular male for a particular female frequently involves some sort of courtship behavior (Box 36.2). The courtship, initiated by the male or female, may be a very brief ceremony, or in certain species of birds may last for many days. The courtship behavior serves two additional roles: it tends to decrease aggressive tendencies, and it establishes species and sexual identification; i.e.,

A

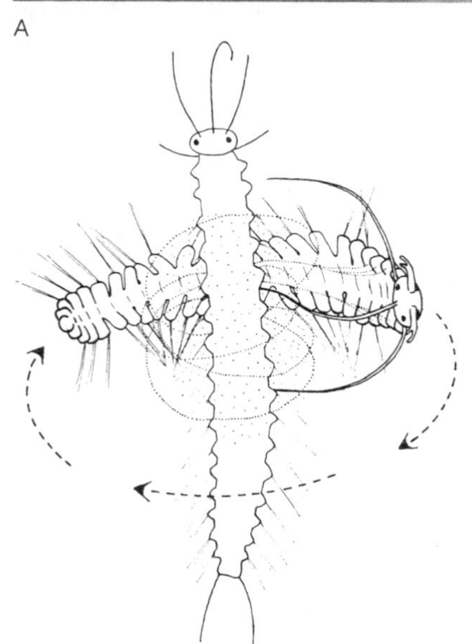
B

FIGURE 13.8 *A,* Leopard frogs in amplexis, a position assumed at the time of egg and sperm release. The male is above and grips the female with his forelimbs. *B,* Syllid polychaete worms during swarming. Males and females leave the bottom and come to the surface of the water. The male swims rapidly around the female while he releases sperm. (After Girdholm.)

it identifies a member of the same species but of the opposite sex. Special structural and functional adaptations have evolved in some species for which control of aggression seems to be especially difficult.

External Fertilization. In many marine and freshwater animals the eggs and sperm are liberated into the surrounding water, and fertilization occurs outside of the body. This was certainly the primitive site of fertilization. Yet given the limited locomotor powers of sperm, how is this possible? You could probably predict a number of the following conditions that ensure that at least some sperm and eggs meet. More than one of the following may be operative for a particular species:

1. Large numbers of gametes are produced and released to compensate for the small percentage that actually undergo fertilization. For example, the American

oyster releases from 15 to 115 million eggs at one spawning, of which only a small part will be fertilized and an even smaller part will survive to settlement.

2. Many sessile animals depend upon natural water currents to carry gametes to neighboring animals of the same species.

3. The sperm of some sessile animals are brought to other individuals in the currents they produce in ventilating and feeding.

4. A favorable habitat will usually be occupied by more than one individual of the same species. There may be hundreds or thousands of individuals, and they may occur in dense aggregations, as on an oyster reef. Thus sperm and eggs do not need to be carried far to increase the chances of contact.

5. The individual members of some animal species live relatively separated and solitary lives in holes or burrows but emerge and come together as a swarm during reproductive periods. Marine annelids, called palolo worms (p. 561) are a good example.

6. Many animals, frogs, fish, and certain worms, for example, come into close physical contact with each other at the time of gamete release, causing sperm and eggs to be discharged together. In fact, the contact, initiated by chemical signals (pheromones) or other factors, may trigger the gamete release.

7. Obviously, most of these factors are highly dependent upon synchronous release of gametes to be effective.

Internal Fertilization. Fertilization within the body of the female has evolved in many marine and freshwater animals and in all terrestrial species. Desiccating conditions on land make external fertilization impossible. The likelihood of fertilization is increased by internal fertilization because sperm are placed in close proximity to the eggs. This greatly increases the percentage of eggs that are fertilized and thus permits reduction in the number of eggs and sperm produced. The energy spent by animals using external fertilization in producing wasted sperm and eggs is conserved. Internal fertilization offers still another advantage that has been exploited in many animals. Sperm can be stored in the female after being received from the male. Eggs can then be produced and fertilized continually or in batches without further need of the male.

Internal fertilization requires copulation, the deposition of sperm from the male system into the female system. Complex neural and hormonal mechanisms have usually evolved to bring about the necessary attraction and precopulatory behavior this requires. Sperm are usually transferred within a fluid medium, called **semen,** but there are many animals, such as leeches, octopods, lobsters, and some salamanders, in which great numbers of sperm are transferred together as a package, called a **spermatophore.** The spermatophore disintegrates after being received by the female system. Spermatophores are usually formed in the terminal parts of the male system, and their shape varies from one group of animals to another (Fig. 13.9).

Various modifications of the gonoducts have evolved in association with internal fertilization. In the male a part of the sperm duct, or areas adjacent to it, may be modified for specific reproductive functions. A part of the duct may be given over for sperm storage; this part is often called a **seminal vesicle,** but the human organ termed the seminal vesicle secretes part of the seminal fluid rather than stores sperm. There may be glandular areas for the production of seminal fluid, which serves as a vehicle for the sperm and may also activate, nourish, and protect them. The terminal part of the sperm duct may open onto or into a copulatory organ, a **penis,** that provides for the transfer of sperm to the female (Fig. 13.10).

In the female the terminal portion of the oviduct may be modified as a **vagina** for receiving the male copula-

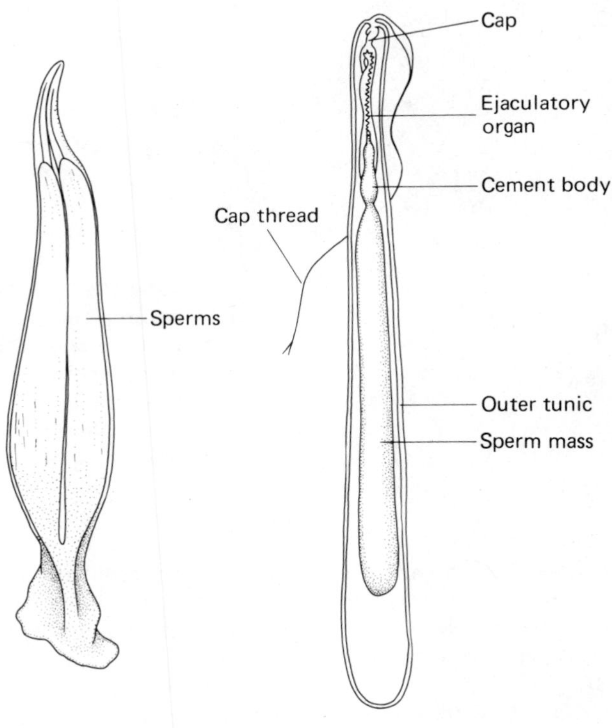

FIGURE 13.9 Sperm packages, or spermatophores. *A,* Of a leech. *B,* Of a squid. (*A,* After Pavlovsky; *B,* after Watase.)

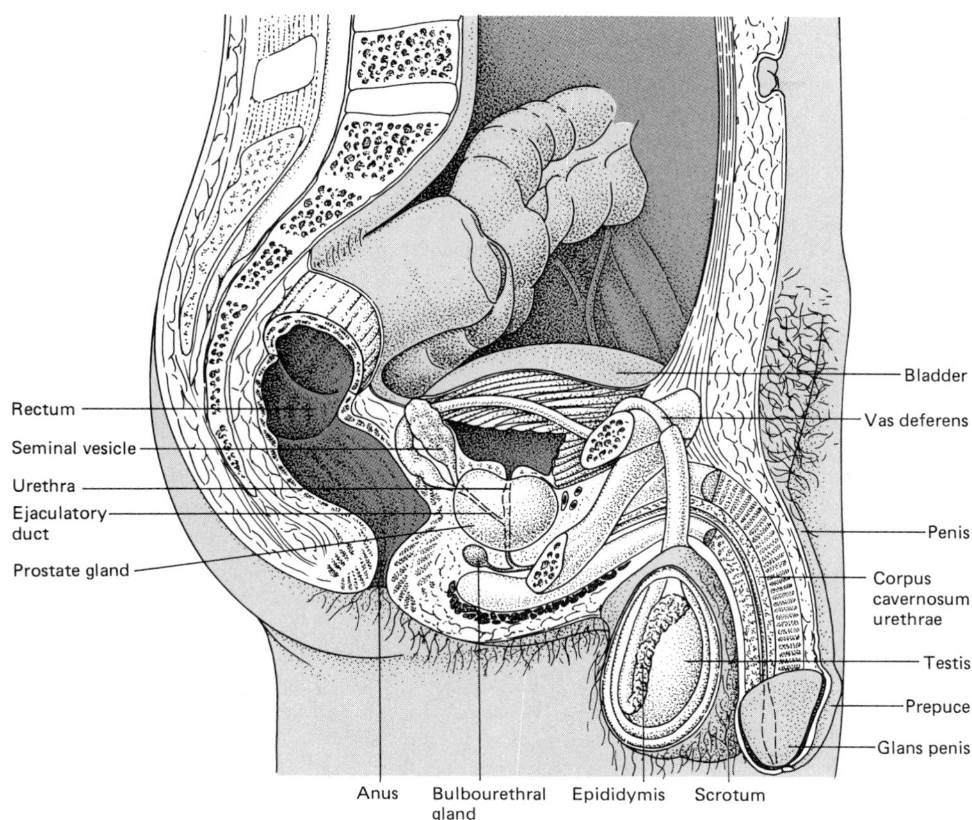

FIGURE 13.10 A diagrammatic sagittal section through the pelvic region of a man to show the genital organs. The prostatic utricle is a vestige of the oviduct that is present in the sexually indifferent stage of the embryo.

tory organ, and a section of the oviduct, the **seminal receptacle,** may be modified for the storage of sperm after their transfer from the male.

Certain flatworms and leeches and a few other animals have an unusual mode of sperm transfer, called **hypodermic impregnation,** that bypasses much of the female reproductive tract. The sperm are injected by the penis into the body wall of the copulatory partner and make their way through the tissues to the eggs. Those flatworms with this mode of transfer have a dagger-like stylet associated with the penis, with which the partner is stabbed!

Hermaphroditism. The great majority of animals have separate sexes, i.e., an individual is either male or female, and this probably represents the primitive condition. Animals with separate sexes are sometimes described as **gonochoristic** (Gr. *gonos,* offspring + *chòrismos,* separated) or **dioecious,** although the latter term is more frequently used by botanists than by zoologists.

In many animals both sperm and eggs can be produced in the same individual, a condition called **hermaphroditism.*** Where both male and female systems are present at the same time, as in earthworms and flatworms, the animals are said to be **simultaneous hermaphrodites,** and there is typically reciprocal transfer of sperm, i.e., each receives sperm from the opposite partner in copulation. However, some species of hermaphroditic animals are first one sex and then the other. Most commonly testes develop first, and the individual functions as a male; later in time the male gonad atrophies, and a female gonad appears. This phe-

*Gr. mythology *Hermaphroditus,* the son of Hermes and Aphrodite, who became united in one body with the nymph Salmacis.

nomenon is termed **protandry** (Gr. *protos,* first + *anèr,* male). Some oysters and slipper shells among mollusks are protandric. Individual slipper shells tend to live in clusters stacked one upon another, permitting the penis of the upper individual to reach the gonopore of the individual below. The young specimens are always male, but after a period of transition, the male reproductive system degenerates, and the animal then develops into a female. The timing of the sex change appears to be influenced by the sex ratio of the entire association.

The less common condition of individuals becoming first female and then male, which occurs, for example, in some fish, is called **protogyny** (Gr. *protos,* first + *gynē,* woman).

Some hermaphroditic animals, such as the parasitic tapeworms, are capable of self-fertilization. (Does this violate the generalization that sexual reproduction involves two individuals?) Since a particular host animal may be infected with but a few parasites, self-fertilization can be an important adaptation for the survival of certain parasitic species. But self-fertilization has the serious disadvantage of restricting the mixing of genetic material within the population, and most hermaphroditic animals have cross fertilization as do those with separate sexes.

The adaptive significance of hermaphroditism in groups of animals that are sessile is clear, for when animals are widely dispersed any individual that settles nearby is a potential mate. Thus, although most crustaceans are motile and gonochoristic (have separate sexes), the sessile barnacles are hermaphroditic. However, there are many exceptions to this rule, and there are motile hermaphroditic forms and sessile forms with separate sexes. We have no good explanation as to why such animals as free-living flatworms and land snails are hermaphroditic.

Egg Deposition

The eggs of some marine animals, such as certain worms, starfish, and some fish, are simply dispersed into the sea water after external fertilization, and the developing eggs become part of the plankton. In many of these animals there is no shell or case, and the egg is covered only by a membrane. Such eggs do not "hatch" because there is nothing to hatch from.

The eggs of most animals are enclosed within some type of envelope, usually secreted by the oviduct, and deposited on the bottom, often attached to some object. The envelope may be gelatinous, leathery, or horny. Each egg may be covered separately; or a few to many eggs may be enclosed within one case; or the eggs may be embedded within a common mass, with or without separate envelopes. If the egg covering involves more than a mucous covering, fertilization must take place before the envelope is added. Thus many animals with egg cases have internal fertilization. In insects the egg shell is produced before fertilization, but a hole, the micropyle, is present to permit entrance of the sperm. Because of the protection afforded the developing embryo, egg deposition can also be a factor accounting for a reduction in the total number of eggs produced.

Predictability in Gonoduct Design

The reproductive tracts, or gonoducts, of animals reflect the reproductive habits characteristic of the species. If the reproductive habits are known, it is possible to predict a great deal about the design of the reproductive system. Let us examine two common animals, a starfish and a whelk, that have very different reproductive habits.

Starfish have external fertilization, and the eggs, which lack shells or cases, are dispersed into the sea water. What would you predict about the reproductive systems of such a starfish? Their gonoducts are adapted only for transport and are simple tubes extending from the gonads to the external gonopore.

Whelks, which are large predatory marine snails, have internal fertilization and deposit their eggs in horny cases of varying shape, depending upon the species (Figs. 25.9 and 25.14). The coils of the long sperm duct can store sperm prior to discharge, and the more distal part of the duct contains glands for the production of semen. The duct terminates in a tentacle-like penis located on the right side of the head. The female system contains a terminal part of the oviduct adapted for the reception of the male penis at copulation, a seminal receptacle for the storage of sperm, and a large glandular distal region of the oviduct, that secretes the material that forms the egg cases and an albumin surrounding the egg within the case. The soft case is passed to a glandular region of the foot, where the case is molded and attached to the bottom.

Vertebrate Reproductive Patterns

Gonads. The sexes are separate in all but a few bony fishes. The **testes** (Figs. 13.10 and 13.11) are paired organs of modest size, each consisting of numerous highly coiled **seminiferous tubules** whose total length in man has been estimated at 250 meters! This provides an area

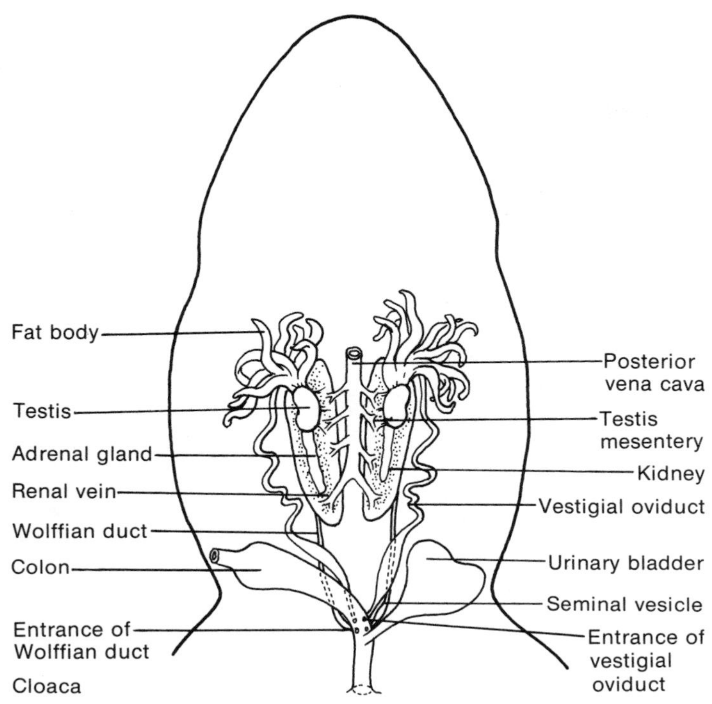

FIGURE 13.11 Ventral view of the urogenital system of a male frog. The vestigial oviduct shown in this figure tends to be absent in many male frogs.

large enough for the production of billions of sperm. As the sperm mature, they enter the lumen of the tubule and move toward the genital ducts.

The ovaries of fishes or amphibians, which produce thousands or hundreds of eggs, fill much of the body cavity (Fig. 13.12). Higher vertebrates produce fewer eggs,

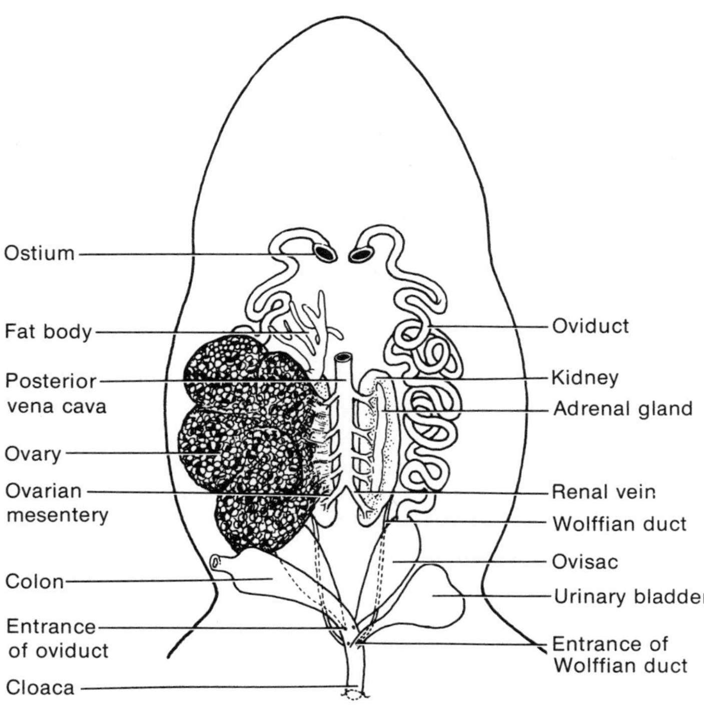

FIGURE 13.12 Ventral view of the urogenital system of a female frog. The left ovary has been removed.

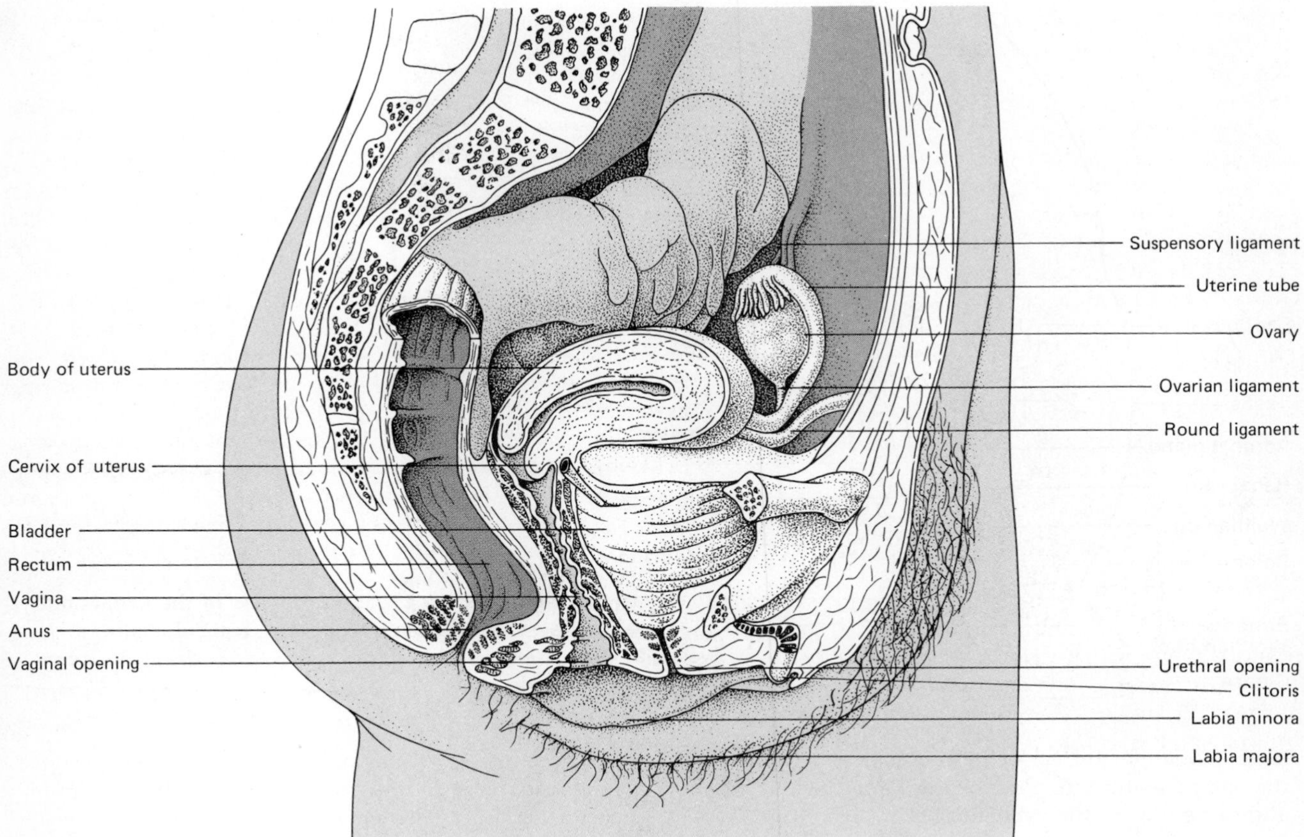

FIGURE 13.13 A diagrammatic sagittal section through the pelvic regions of a woman to show the genital organs.

for fertilization is internal and the eggs are deposited in situations in which there is a greater chance for survival of the developing embryos. The ovaries of reptiles and birds are still large, and the eggs contain much yolk. Mammalian eggs contain very little yolk, and the ovaries are quite small, 2.5 cm. in length in humans (Fig. 13.13).

In nearly all vertebrates, the gonads are suspended by mesenteries in the abdominal cavity, and they remain there throughout life. But in the males of most mammals, the testes undergo a posterior migration, or descent, and move out of the main part of the abdominal cavity into a sac of skin and associated layers of the body wall known as the **scrotum** (Fig. 13.10).

In most mammals, spermatogenesis does not go to completion unless the testes are descended. They remain descended in the majority of species, but in rabbits and rodents they are migratory—descending into the scrotum during the breeding season, withdrawing into the abdominal cavity at other times. Spermatogenesis, like other vital processes, can only occur within a limited temperature range. This range is exceeded by the temperature in the abdominal cavity but not by the temperature in the scrotum, which is approximately 4°C. lower.

Male Reproductive Tracts. In the males of most fishes and amphibians (frog, Fig. 13.11), microscopic tubules, called **vasa efferentia,** carry sperm from the seminiferous tubules through the mesentery supporting the testis to the anterior kidney tubules (the kidney is an opisthonephros, p. 200). Sperm pass through these tubules to the archinephric duct, which carries both sperm and urine to the cloaca. In the evolution of reptiles, birds, and mammals, the kidney changes from an opisthonephros drained by the archinephric duct to a metanephros drained by a ureter (p. 200). The pattern of sperm discharge remains the same as in amphibians, but the terminology is different because terms were given to the mammalian structures before their evolutionary relationship to primitive kidney structures was known. The anterior end of the primitive opisthonephros lies against the surface of the testis as a band known as the **epididymis** (Gr. *epi,* upon + *didymos,* testicle). Since the epididymis is so close to the testis, the primitive vasa efferentia are very short and are incorporated into a network of small tubules, the **rete testis**, situated between testis and epididymis. Passages within the epididymis represent primitive kidney tubules together with a highly coiled portion of the archinephric duct. The rest of the primitive archinephric duct passes as the sperm duct, the **vas deferens,** to join the urethra, much of which evolved from the ventral part of the cloaca.

Sperm are stored in the epididymis and vas deferens. They also undergo a maturation process in the epididymis that gives them the ability to move when they are ejaculated. Sperm removed from the seminiferous tubules are nonmotile. Humans thus utilize passages homologous to those of a frog.

Other differences between the male reproductive organs of lower and higher vertebrates are correlated with differences in mode of reproduction. Since frogs mate in the water, fertilization is external. Fertilization is internal in reptiles, birds, and mammals, and, except for most birds, the male has a penis. Accessory sex glands produce the seminal fluid, or semen, that transports the sperm. In mammals, the penis develops around the urethra and contains three cavernous bodies composed of spongy **erectile tissue.** Arterial dilatation coupled with a restriction of venous return causes the vascular spaces within the erectile tissue to become filled with blood during sexual excitement, making the penis turgid and effective as a copulatory organ. The accessory sex glands are a pair of **seminal vesicles** that connect with the distal end of the vasa deferentia, a **prostate gland** surrounding the urethra at the point of entrance of the vasa deferentia and a pair of **Cowper's glands** located more distally along the urethra. These secrete the various components of the seminal fluid.

Female Reproductive Tracts. Eggs are removed from the coelom in most female vertebrates by a pair of oviducts, but the oviducts are modified for various modes of reproduction. Fishes and nearly all amphibians reproduce in the water. Most are oviparous, fertilization is external, and the eggs develop into larvae that can care for themselves. In the frog (Fig. 13.12), each oviduct is a simple coiled tube that extends from the anterior end of the coelom to the cloaca. The oviducts contain glandular cells that secrete layers of jelly about the eggs, and their lower ends are expanded for temporary storage of the eggs, but they are not otherwise specialized.

Fertilization is internal in reptiles, birds, and mammals. They reproduce on the land, and the free larval stage has been replaced by the evolution of a cleidoic egg (p. 324). Most reptiles and all birds are oviparous, and the eggs develop externally. The oviducal glands, which secrete the albumin and a shell around the egg, are more complex in the oviducts of reptiles than in those of amphibians and fishes, but in other respects the oviducts of reptiles have not changed greatly.

Most mammals and a few fishes and reptiles retain the fertilized egg within the reproductive tract until embryonic development is complete, an adaptation that will be described more fully in the next chapter. The embryo receives its nutrients from the mother. The oviducts are modified accordingly. In the human female (Figs. 13.13 and 13.14), the **ostium** lies adjacent to the ovary and may even partially surround it. When ovulation occurs, the discharged eggs are close enough to the ostium to be easily carried into it by ciliary currents. The anterior portion of each oviduct is a narrow tube known as the **fallopian tube,** and eggs are carried down it by ciliary action and muscular contractions. The remainder of the primitive oviducts have fused with each other to form a thick-walled muscular **uterus** and part of the **vagina.** The terminal portions of the vagina and urethra develop from a further subdivision of the ventral part of the cloaca. The vagina is a tube specialized for the reception of the penis and is lined with stratified squamous epithelium. It is separated from the main part of the uterus, in which the embryo develops, by the sphincter-like neck of the uterus known as the **cervix.** The orifices of the vagina and urethra are flanked by

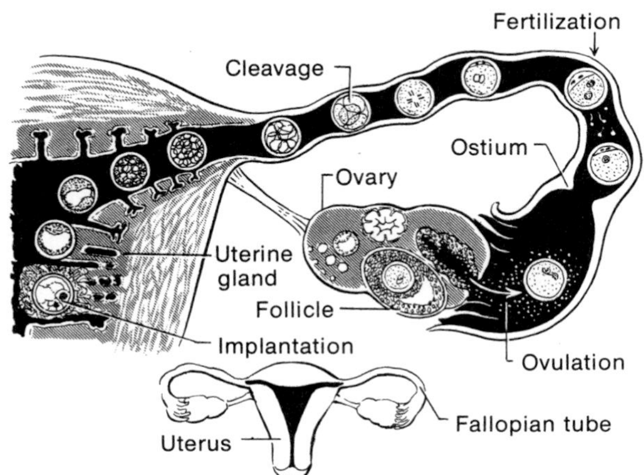

FIGURE 13.14 A diagram to show the path of an egg from the ovary to the uterus and the changes that occur en route. The last stage is about 1½ weeks long.

paired folds of skin, the **labia minora** and the **labia majora.** A small bundle of sensitive erectile tissue, the **clitoris** (Gr. *kleitoris,* a small hill), lies just in front of the labia minora. Structures comparable to these are present in the sexually indifferent stage of the embryo and develop into more conspicuous organs in the male. The labia majora are comparable to the scrotum; the labia minora and clitoris, to the penis. A pair of glands, homologous to Cowper's glands in the male, discharge a mucous secretion near the orifice of the vagina. A fold of skin, the **hymen,** partially occludes the opening of the vagina but is ruptured during the first intercourse.

Mammalian Fertilization. During copulation, the sperm that have been stored primarily in the epididymis are ejaculated by the sudden contraction of muscles in and around the male ducts, and the accessory sex glands concurrently discharge their secretions. The seminal fluid that is deposited may contain as many as 400,000,000 sperm. Mucus in the seminal fluid serves as a conveyance for the sperm; proteolytic enzymes break it down into a more watery fluid after the semen has been deposited in the vagina and permit the sperm to become highly motile. Fructose provides a source of energy, alkaline materials prevent the sperm from being killed by acids normally in the vagina, and certain fatty acids (prostaglandins) promote the contraction of the smooth muscle in the walls of the uterus and fallopian tubes.

Sperm move from the vagina through the uterus and up the fallopian tube in a little over one hour. How they do this is not entirely understood. They can swim, tadpole fashion, by the beating of the tail, but muscular contraction of the uterus and fallopian tubes and ciliary currents in the tubes are primarily responsible for moving the sperm to the upper part of the oviduct. Fertilization occurs in the upper part of the fallopian tube (Fig. 13.14), but the arrival of an egg and the sperm in this region need not coincide exactly. Sperm retain their fertilizing powers for a day or two, and the egg moves slowly down the fallopian tube, retaining its ability to be fertilized for about a day. The chance of fertilization is further increased in many species of mammals (but not in human beings) by the female coming into "heat" (estrus) and receiving the male only near the time of ovulation. Ovulation, "heat," and changes in the uterine lining in preparation for the reception of a fertilized egg are controlled by an intricate endocrine mechanism, considered in Chapter 12.

Only one sperm fertilizes each egg, yet unless millions are discharged fertilization does not occur. One reason for this is that only a fraction of the sperm deposited in the vagina reach the upper part of the fallopian tube. The others are lost or destroyed along the way. When the egg enters the fallopian tube, it is still surrounded by a few of the follicle cells that encased the egg within the ovary (see Fig. 12.16), and a sperm cannot penetrate the egg until these are dispersed. This requires the enzyme **hyaluronidase,** which can break down **hyaluronic acid,** a component of the intercellular cement. Hyaluronidase is produced by the sperm themselves, and large numbers are apparently necessary to produce enough of it.

Summary

1. Sexual reproduction involves the union of haploid sperm and eggs and makes possible the mixing of the genetic characteristics of the species population, but some animals are also able to reproduce asexually, producing new individuals from fragments or divisions of the parent's body.

2. In the formation of gametes the chromosome number is reduced to one half by two meiotic divisions. These divisions produced four spermatids from one spermatogonium, and each spermatid is then transformed into a small, compact cell adapted for transporting genetic material to the egg. In oogenesis the cytoplasm is unequally divided among the four daughter cells so that one, the egg, obtains all of the yolk material. The amount and distribution of yolk material varies greatly in the eggs of different species of animals.

3. Fertilization involves all of the events from the penetration of the egg membrane by the sperm acrosome to the union of sperm and egg chromosomes within a single nucleus, restoring the diploid number. Parthenogenesis, the development of the egg without fertilization, occurs naturally in many different animal groups.

4. The most important adaptation that increases the likelihood of fertilization is synchrony in the production and release of gametes. Many aquatic animals have external fertilization, which is possible where individuals of the species come together during the reproductive period or live close together, and sperm can be transported to the eggs in water currents.

5. Internal fertilization within the body of the female is characteristic of many aquatic animals and of all terrestrial species. It requires copulation and various modifications of the reproductive tracts of both sexes, such as a copulatory organ (usually a penis), semen-producing glands, seminal vesicle, vagina, and seminal receptacle.

6. Primitively, animals are gonochoristic, i.e., the sexes are separate, but many species are simultaneous or protandric hermaphrodites. However, cross rather than self-fertilization is generally the rule. In simultaneous hermaphrodites cross fertilization is reciprocal. Hermaphroditism is clearly adaptive for many parasitic and sessile animals, but its origin and significance in other groups is uncertain.

7. The eggs of some marine animals are planktonic, but most marine species and all freshwater species deposit their eggs within envelopes or cases that are attached to the bottom or to the parent. The female reproductive tract is modified for the secretion of egg cases, and the number of eggs produced is less than when the eggs are planktonic.

8. The vertebrate reproductive tract varies greatly, reflecting different adaptations for fertilization and egg deposition. In mammals the male's penis deposits sperm in the vagina, and fertilization occurs in the upper end of the fallopian tube. The large number of sperm released increases the likelihood that some will make the passage up the uterus and fallopian tube and collectively contribute to the enzymatic dispersal of follicle cells retained around the ovulated egg.

References and Selected Readings

For convenience, all references on reproduction have been placed at the end of the second chapter on this subject, Chapter 14, p. 334.

14 Reproduction II: Embryonic Development

ORIENTATION

In this chapter we continue our study of animal reproduction by following the history of the fertilized egg through embryo formation and the attainment of the adult body form. We begin by examining patterns of egg division, called cleavage, and then the way this mass of embryonic cells undergoes rearrangement and differentiation to lay down the ground plan of the adult animal. Upon that ground plan the more detailed development of the organ systems takes place. Finally, we will look at adaptations that have evolved in animals to protect and support the developing embryo.

The division, growth, and differentiation of a fertilized egg into the remarkably complex and interdependent system of organs that is the adult animal is certainly one of the most fascinating of all biological phenomena. Not only are the organs complicated and reproduced in each new individual with extreme fidelity of pattern, but also many of these organs begin to function while they are still developing. The human fetal heart begins to beat during the fourth week of gestation, long before its development is complete.

The pattern of cleavage, blastula formation, and gastrulation is seen with various modifications in all multicellular animals. The details of later development are quite different in animals of different phyla but are similar in more closely related forms. The main outlines of human development can be discerned by studying the embryos of rats, pigs, chickens, and even frogs.

Cleavage

The entrance of the sperm into the egg (Fig. 13.7) initiates a rapid series of changes—the completion of the meiotic divisions, the fusion of male and female pronuclei, and the complex movements of the egg's cytoplasmic constituents; the rates of oxygen consumption and protein synthesis in the eggs of certain species rise sharply. Some evidence supports the hypothesis that the RNA present in the fertilized egg, which directs protein synthesis during the early stages of cleavage, was synthesized in the oöcyte before fertilization and hence is a product of the maternal genotype rather than of the embryo's genotype. The RNA is masked in the unfertilized egg, perhaps by combination with protein, and is unavailable for translation into a peptide chain. The ribosomes of the unfertilized egg are generally single, and polyribosomes appear only after fertilization has occurred. One of the effects of fertilization appears to be an unmasking of the maternal messenger RNA so that it can be translated in protein synthesis. Later, at about the time of gastrulation, the genome of the embryo is transcribed to form embryonic messenger RNA, which codes for proteins that play a role in the gastrulation process and in later development.

The final event triggered by fertilization is the repeated mitotic division of the egg, termed cleavage. The plane of the first cleavage division of an isolecithal egg passes through the animal and vegetal poles of the egg, yielding two equal cells, or **blastomeres** (Fig. 14.1). The second cleavage division also passes through animal and vegetal poles, but at right angles to the first, and divides the two cells into four. The third cleavage division is horizontal—its plane is at right angles to the planes of the first two cell divisions, and the embryo is split into four cells above and four cells below this line of cleavage. Cleavage patterns of this type are said to be **radial,** for the cleavage planes are always either parallel or at right angles to the polar axis of the egg. Further divisions result in embryos containing 16, 32, 64, 128 cells, and so on, until a hollow ball of cells, the **blastula** (Gr. *blastos,* bud), is formed. The size of the embryo does not increase during cleavage; indeed, the embryo is using materials stored in the yolk and its mass decreases. Since the cells become oriented to, and anchored together at, the surface, a cavity develops in the center. This is the **blastocoel.** The wall of the blastula consists of a single layer of cells in embryos with little yolk.

Yolk impedes cleavage so that cleavage planes pass more easily and rapidly through nonyolky than yolky cytoplasm. Thus the amount and distribution of yolk material within the egg markedly influences the cleavage pattern and can result in blastomeres of different sizes, large **macromeres** and small **micromeres.** The eggs of frogs are moderately telolecithal, and the cleavage divisions of the cells originating from the lower part of the frog's egg are slowed by the inert yolk so that the blastula consists of many small cells at the animal pole and a few large cells at the vegetal pole (Fig. 14.1). The lower wall of the blastula is much thicker than the upper one, and the blastocoel is flattened and displaced upward. In the bird's egg only a small disc of cytoplasm at the animal pole undergoes cleavage divisions (Figs. 14.1 and 14.2). The lower, yolk-filled part of the egg never cleaves. Cleavage is said to be **incomplete.**

Animals with centrolecithal eggs, such as insects, also have incomplete cleavage, but the cleavage pattern, called **superficial,** differs from that in frogs and birds because the yolk distribution is different. Cleavage begins with the division of the nucleus in the central island of cytoplasm (Fig. 14.3). After several nuclear divisions without cytoplasmic division the nuclei migrate out from the center of the egg. Each nucleus is surrounded by a bit of the original central cytoplasm. When the nuclei reach the surface of the egg the cytoplasm surrounding them fuses with the superficial layer of cytoplasm, resulting in a syncytium covering the surface of the egg. The cytoplasm subsequently becomes subdivided by membranes that extend in from the surface. These blastomeres are connected to the yolk mass for a time but eventually become separated from it; the constituents of the yolk are gradually used up as nutrients for the developing embryo. This stage can be compared to the formation of the blastula even though there is no cavity comparable to the blastocoel. The blastoderm surrounds a mass of uncleaved yolk rather than a cavity.

Many invertebrate animals, including flatworms, mollusks, and annelids, share a pattern of early embryonic

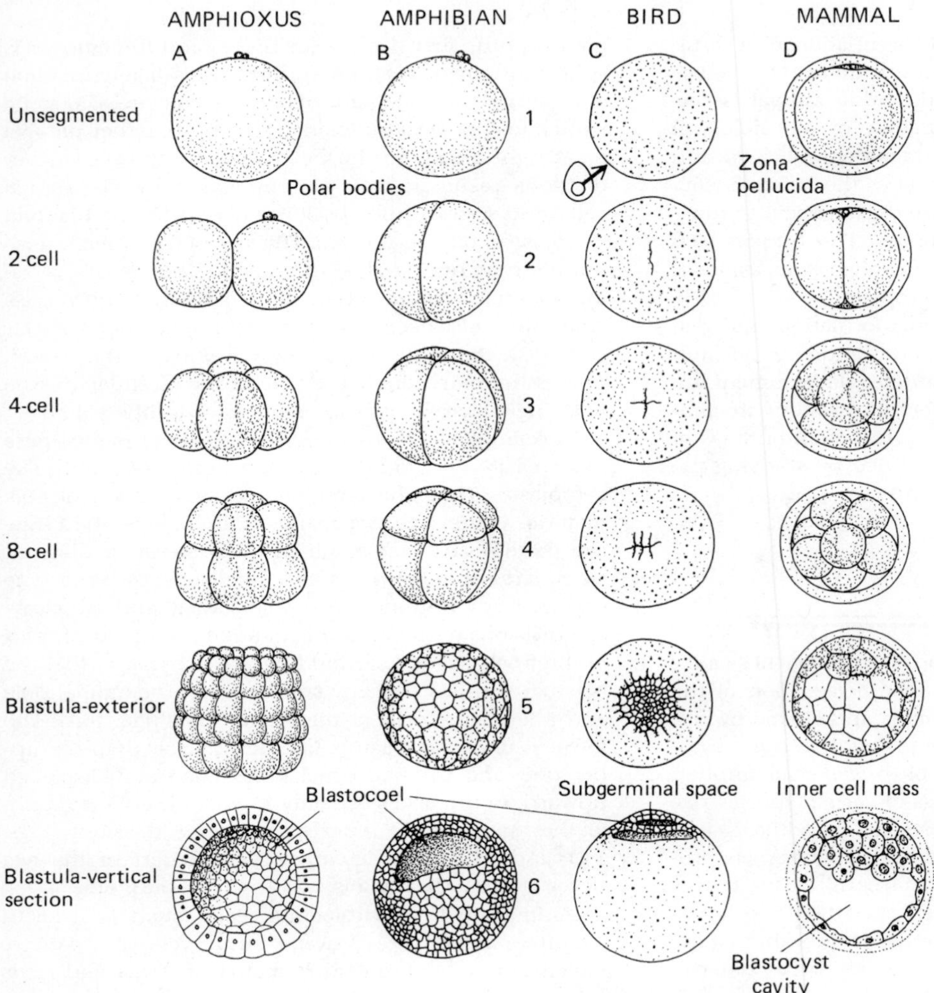

FIGURE 14.1 Stages in cleavage and blastula formation in eggs of chordates. All are lateral views except for that of the bird, which is viewed from the animal pole. *A, Amphioxus* (holoblastic cleavage, isolecithal egg with little yolk). *B,* Frog (holoblastic cleavage, moderately telolecithal egg with much yolk). *C,* Bird (meroblastic discoidal cleavage, telolecithal egg with much yolk). *D,* Mammal (holoblastic cleavage, isolecithal egg with essentially no yolk). (Modified from Storer and Usinger: *General Zoology.* 3rd ed. Copyright 1957, by McGraw-Hill Book Co., Inc.)

development called **spiral cleavage,** in contrast to the radial pattern of cleavage in most of the animals we have thus far discussed. The first and second cell divisions are meridional and at right angles to each other, forming four cells of nearly equal size (Fig. 14.4). However, beginning with the third division the cleavage planes are oblique. The mitotic spindles are inclined to one side (one of them is indicated by the solid line in the eight-cell stage, Fig. 14.4). As a result, the upper four cells are displaced circularly so that each upper cell touches *two* lower cells. The third division is also *unequal,* separating four upper small micromeres from four lower large macromeres. The fourth division is also oblique but always in the opposite direction from the third (Fig. 14.4).

The fifth division continues the pattern, and the axes of division are oblique in the direction of those of the third division. As viewed from the animal pole, the cleavage pattern appears spiral.

Associated with the spiral and radial patterns of cleavage are differences in the timing of the fixation of the fates of the blastomeres. The fate of blastomeres in spiral cleavage tends to be fixed very early. If at the four-cell stage the blastomeres are separated and allowed to

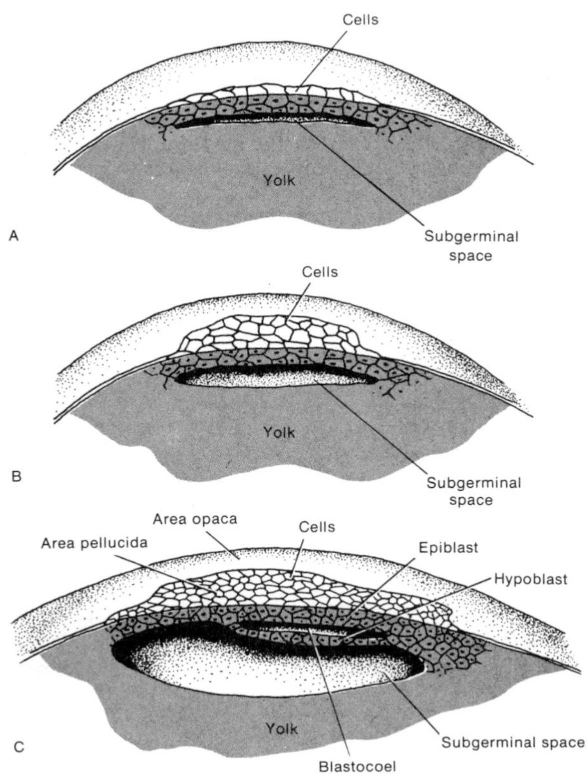

FIGURE 14.2 Successive stages in the cleavage of a hen's egg. *A,* Cleavage is restricted to a small disc of cytoplasm on the upper surface of the egg yolk, called the blastodermic disc. *B,* A subgerminal space appears beneath the blastodermic disc, separating it from the unsegmented yolk. *C,* The blastodermic disc cleaves into an upper epiblast and a lower hypoblast separated by the blastocoel.

develop independently, each will form only one quarter of the embryo. Cleavage is thus said to be not only spiral but also **determinate.** The fate of blastomeres in animals with radial cleavage is usually determined much later, and cleavage is said to be **indeterminate.** Separated blastomeres at the four-cell stage in many animals with radial indeterminate cleavage form complete embryos.

There is nothing about the spiral pattern that causes early fixation of blastomere fate or vice versa. The two conditions just happened to become linked in the evolution of cleavage patterns.

Gastrulation and Mesoderm Formation

Cleavage converts the zygote into a multicellular body, but there is no change of shape. The egg mass is simply divided into smaller cellular units. **Gastrulation,** which follows cleavage, lays down the ground plan of the adult body form. This is achieved by morphogenetic movements of cells and by changes in the shape of the embryo.

In embryos with little yolk, such as the primitive chordate amphioxus, the single-layered blastula is converted into a double-layered sphere, a **gastrula** (Gr. *gaster,* stomach), by the inpushing, or **invagination**, of a section of one wall of the blastula (Fig. 14.5). This eventually meets the opposite wall and obliterates the original blastocoel. The new cavity of the gastrula is the **archenteron** (Gr. *arche,* beginning + *enteron,* gut), and the opening of the archenteron to the exterior, the **blastopore,** marks the site of the invagination that produced the gastrula. The outer of the two walls of the gastrula is the **ectoderm,** which eventually forms the epidermis and nervous system. The inner layer of cells lining the archenteron is mainly the presumptive **endoderm,** which will form the lining of the digestive tract and its outgrowths, such as the liver and pancreas, plus the presumptive **mesoderm,** which will form the remaining organs of the body. As the gastrula elongates in the an-

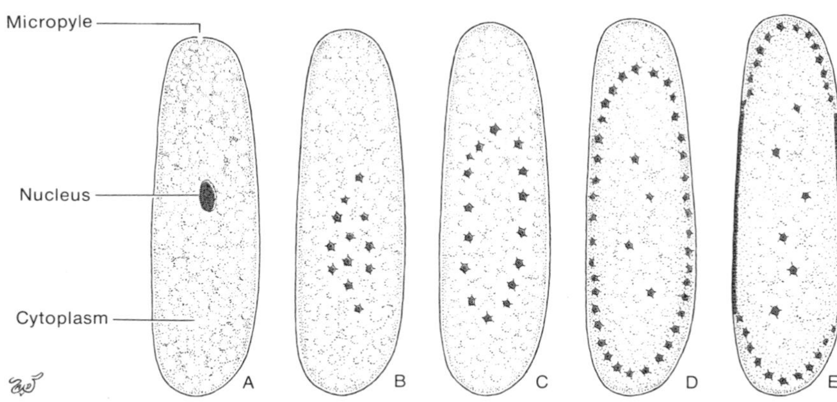

FIGURE 14.3 *A,* The centrolecithal egg of a beetle. *B–E,* Stages in the cleavage of the centrolecithal egg.

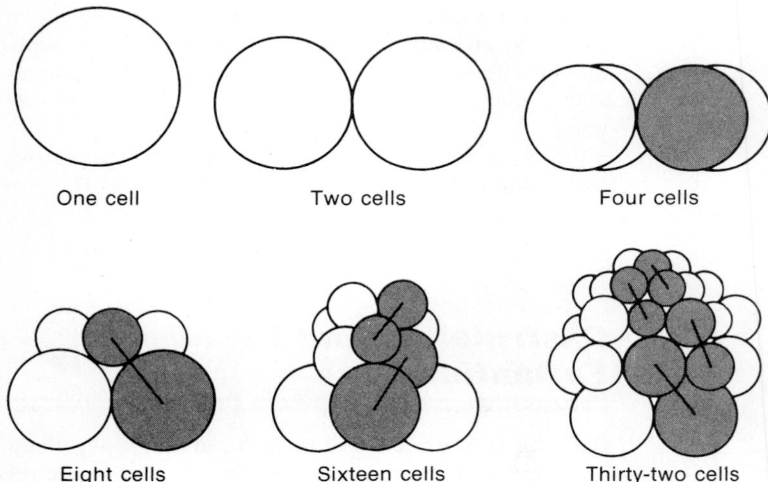

FIGURE 14.4 Spiral cleavage. One quadrant (the progeny of one cell of the four-cell stage) is shaded. Lines indicate the axes of the preceding mitoses.

terior-posterior axis the presumptive notochord forms as a longitudinal band of cells occupying the middorsal part of the inner layer. The presumptive mesoderm forms two longitudinal bands of cells, one on each side of the presumptive notochord. The remainder of the lateral, ventral, and anterior parts of the inner layer are presumptive endoderm cells.

Gastrulation in both frog and chick eggs is greatly modified by the presence of yolk, although an archenteron is formed. Gastrulation in the frog involves, first of

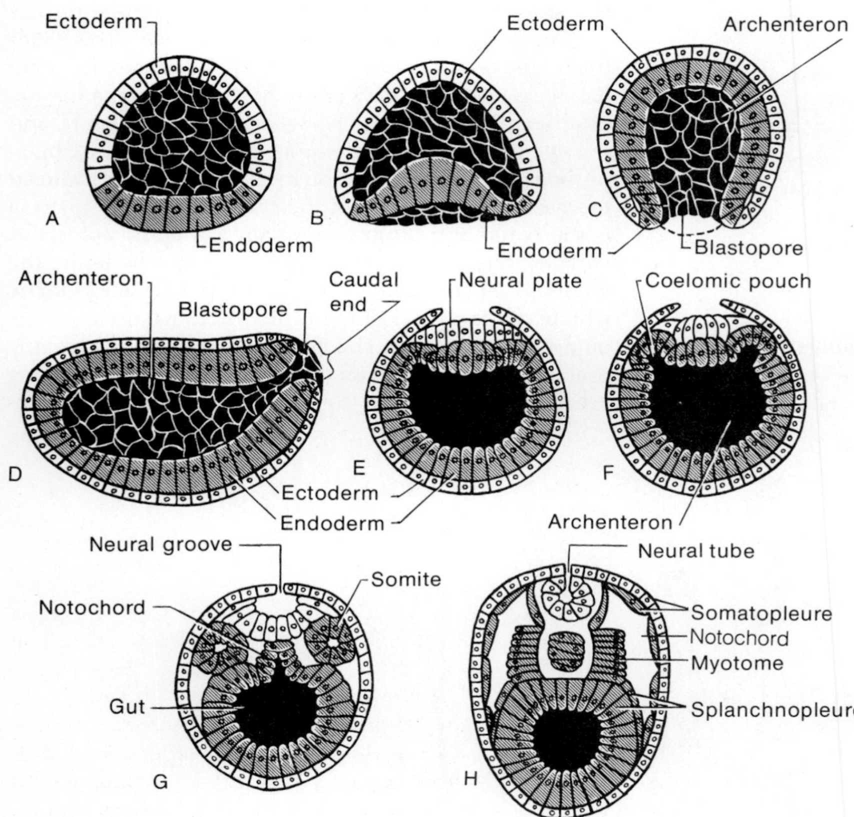

FIGURE 14.5 Stages in gastrulation *(A–C)* and mesoderm formation *(F–H)* in *Amphioxus*. Note that the mesoderm forms by the evagination of pouches from the archenteron.

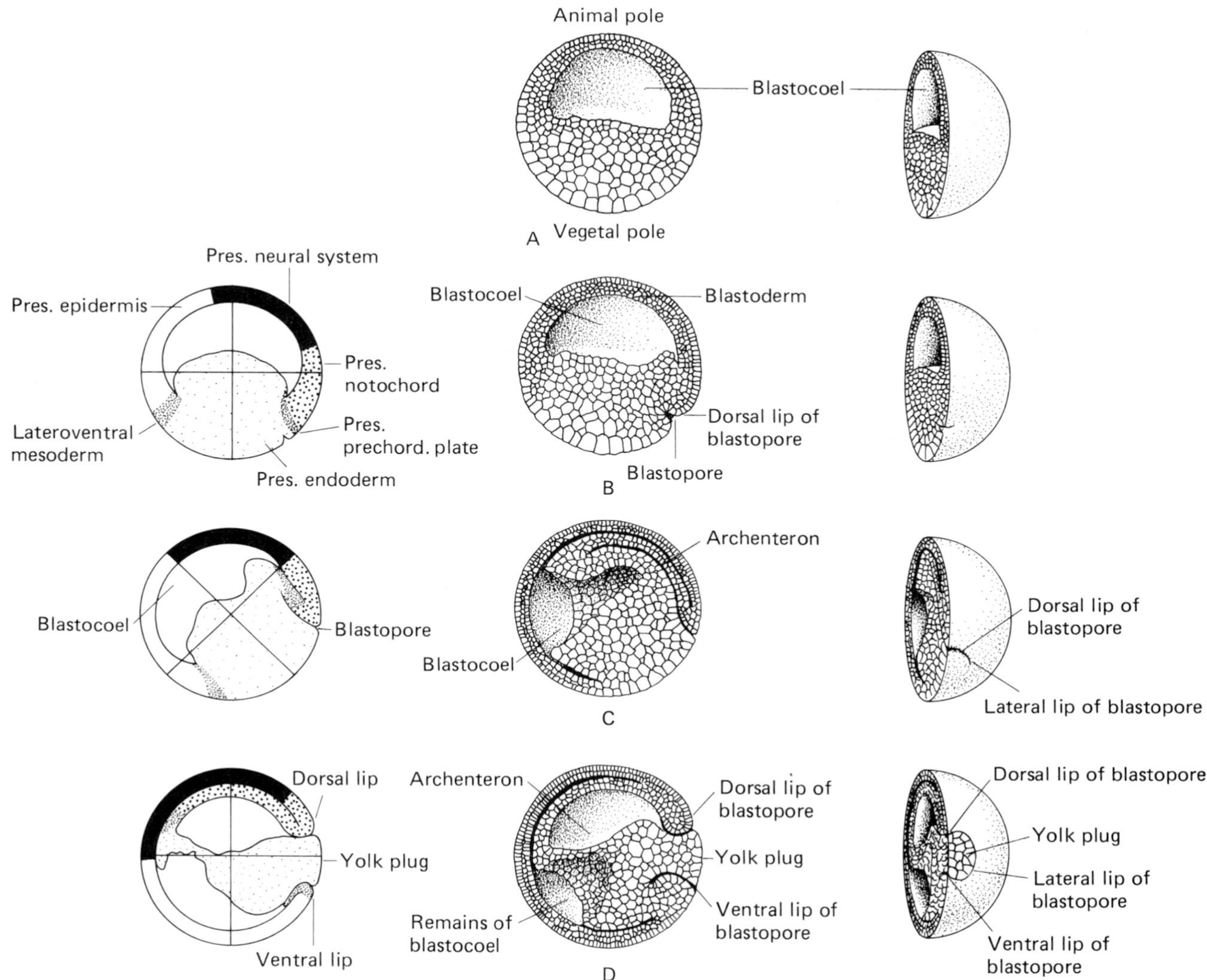

FIGURE 14.6 Stages in gastrulation in a frog. *A* is a late blastula stage, and *D* is a late gastrula. The diagrams in the left column show the positions of presumptive, or future, gastrula cells in earlier stages of gastrulation. (From Balinsky, B.I.: *An Introduction to Embryology.* 5th ed. Philadelphia, Saunders College Publishing, 1981.)

all, **pregastrulatory movements** of the yolk-filled cells. They tend to move inward slightly, bulging up the floor of the blastocoel. Concurrently, cells of the animal hemisphere are pulled toward the ventral hemisphere. A slight indentation begins to appear on what is to become the posterior end of the embryo (Figs. 14.6 and 14.7). The indentation represents the beginning of the **blastopore,** and the cells directly above it constitute the **dorsal lip** of the blastopore. This indentation results from a change in cell shape (see the next section) and is the amphibian equivalent of invagination. As a result of continued multiplication and changes in cell shape, animal hemisphere cells move toward the dorsal lip of the blastopore, roll over it, and move inward. This process is called **involution.** But cells are moving toward the blastopore faster than they are involuting, so the lip of theblastopore overgrows the mass of yolk-filled cells, a process called **epiboly.** All of these processes begin at

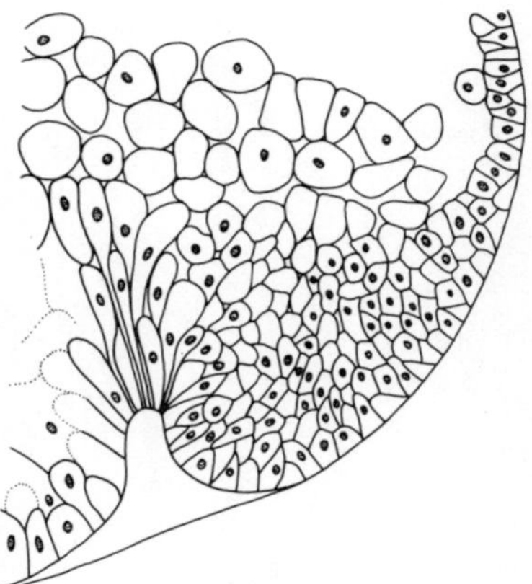

FIGURE 14.7 Early formation of the blastopore in gastrulation in a salamander. Note that the cells at the bottom of the pit are bottle-shaped, which contributes to their movement into the interior. (After Vogt from Balinsky, B. I.: *An Introduction To Embryology.* 5th ed. Philadelphia, Saunders College Publishing, 1981.)

the dorsal lip, but the lip spreads laterally and ventrally until a complete blastopore is formed that surrounds a mass of yolk-filled cells, the **yolk plug** (Fig. 14.6). As the blastopore lip spreads, processes of involution and epiboly follow, and the yolk plug decreases in size.

The continued mitotic divisions during gastrulation lead to more cells and more nuclear material but little or no change in the total volume or mass of the embryo. The rate of metabolism, as measured by the rate of oxygen consumption, increases two- or threefold over the rate during cleavage, presumably to supply the biologically useful energy for the morphogenetic movements and for the sharply increased rate of synthesis of messenger RNA and of proteins.

In most animals a third layer of cells, the mesoderm, develops between ectoderm and endoderm. In annelids, mollusks, and certain other invertebrates the mesoderm develops from special cells that are differentiated early in cleavage (p. 311). These migrate to the interior and come to lie between the ectoderm and endoderm. They then multiply to form two longitudinal cords of cells that develop into sheets of mesoderm between the ectoderm and endoderm. The coelomic cavity originates by the splitting of the sheets to form pockets and hence is called a **schizocoel** (Fig. 14.8).

The mesoderm arises in primitive chordates as a series of bilateral pouches from the endoderm (Figs. 14.5 and 14.8). These lose their connection with the gut and fuse one with another to form a connected layer. The cavity of the pouch is retained as the coelom, called an **enterocoel** (Gr. *enteron,* gut + *koilos,* hollow) because it is derived indirectly from the archenteron. The mesoderm in amphibia is formed from cells that roll in at the lips of the blastopore, then separate from the roof of the archenteron, and move anteriorly as a sheet between ectoderm and endoderm (Fig. 14.6). In birds and mammals a thickened band of ectoderm and endoderm cells, the primitive streak, develops on the surface of the developing embryo, marking the longitudinal axis of the embryo (Fig. 14.9). At the primitive streak cells migrate in from the surface, proliferate, and spread as a sheet of mesoderm between the ectoderm and endoderm. The primitive streak is a dynamic structure and persists even though the cells composing it are constantly changing as they migrate. It is comparable in some respects to the blastopore of more primitive vertebrates.

However the mesoderm may originate, it eventually forms two sheets that come to lie between the ectoderm and endoderm; one sheet becomes attached to the inner endoderm and the other to the outer ectoderm. The cavity between the two becomes the coelom, or body cavity. This splitting of the mesoderm permits the development of two independent sets of muscles—one in the body wall adapted for locomotion and the other in the gut wall adapted for churning and moving the contents of the gut (Fig. 14.8).

The sheets of mesoderm in vertebrates grow ventrally, and the ones from either side meet in the ventral midline; the coelomic cavities on the two sides then fuse into one. The mesoderm grows dorsally along each side of the notochord and neural tube and becomes differentiated into paired segmental blocks of tissue, the **somites** (Gr. *soma,* body), from which the main muscles of the trunk develop (Figs. 14.11 and 14.12*B*).

The primitive skeleton of the chordate, the **notochord** (Gr. *nōton,* back + *chordē,* cord), is a flexible, unsegmented, longitudinal rod found in the dorsal midline of all chordate embryos (Figs. 14.5, 14.10, and 14.11). It is formed at the same time and in a similar way as the mesoderm, as an outgrowth of the roof of the archenteron, from the dorsal lip of the blastopore or from the primitive streak. Later in the development of vertebrates the notochord is replaced by the vertebral column derived from part of the mesoderm.

Morphogenetic Movements and Differentiation

Gastrulation in all of the types of embryos involves the movement, or migration, of cells, which occurs in specific ways and leads to specific arrangements of cells. These morphogenetic movements involve considerable

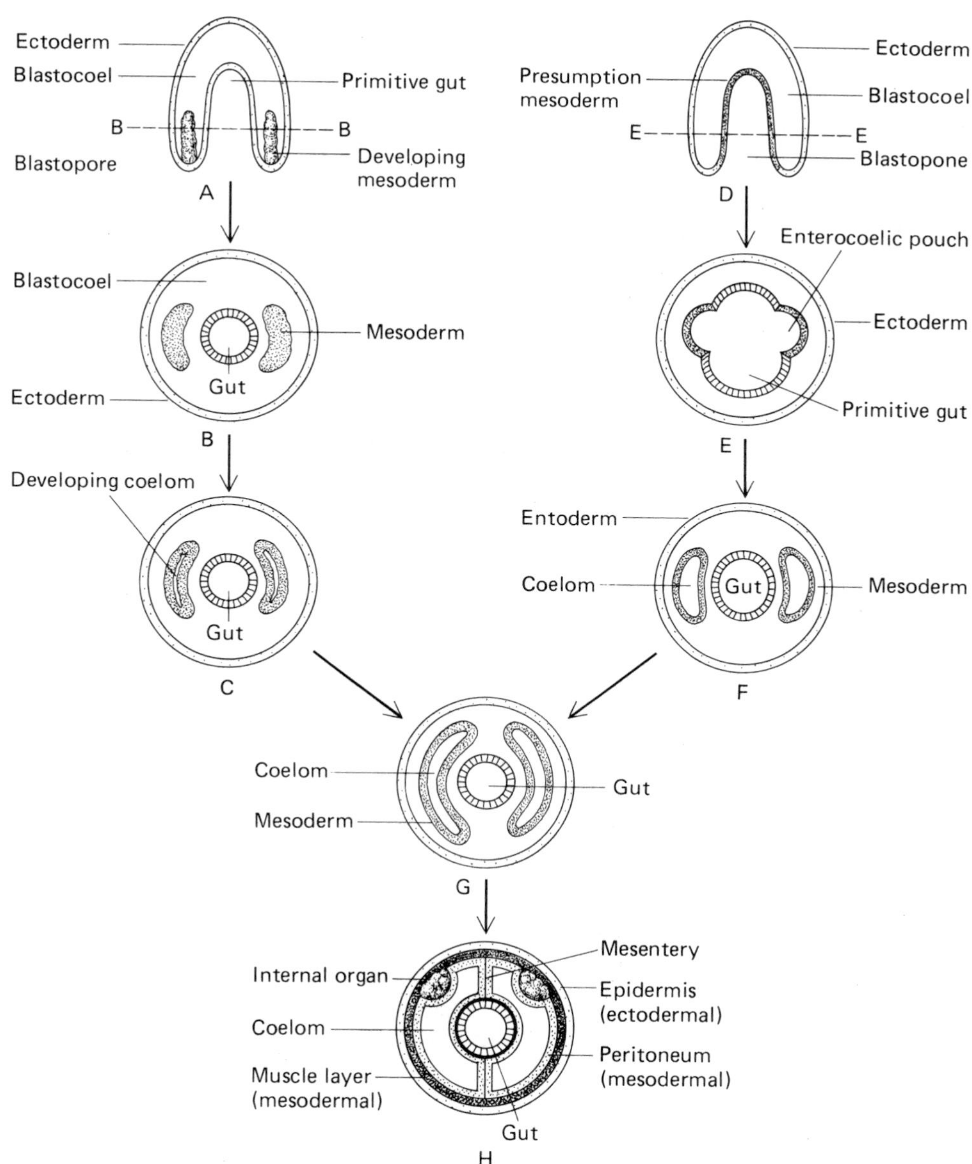

FIGURE 14.8 Comparison of the two principal modes of mesoderm and coelom formation in animals. *A–C,* Coelom formation by schizocoely as occurs in mollusks and annelids. *A,* Frontal section through gastrula, showing developing mesoderm derived from cells set aside very early in cleavage. *B,* Cross section of *A* taken at level *B–B. C,* Mesodermal mass splits to form coelomic cavity. *D–F,* Coelom formation by enterocoely as occurs in echinoderms and primitive chordates. *D,* Frontal section through gastrula showing presumptive (future) mesoderm as part of primitive gut wall. *E,* Cross section of *D* taken at level *E–E.* Presumptive mesoderm forms wall of outpocketing of primitive gut, the enterocoelic pouches. *F,* Pouches separate as coelomic vesicles, the walls forming the mesoderm and the cavity of the coelom. *G–H,* The fate of the coelomic vesicles is more or less the same regardless of their mode of origin. The outer wall of the vesicle becomes associated with ectoderm to form body wall muscles; the inner wall of the vesicle becomes associated with the gut to form the muscles and blood vessels of the gut wall.

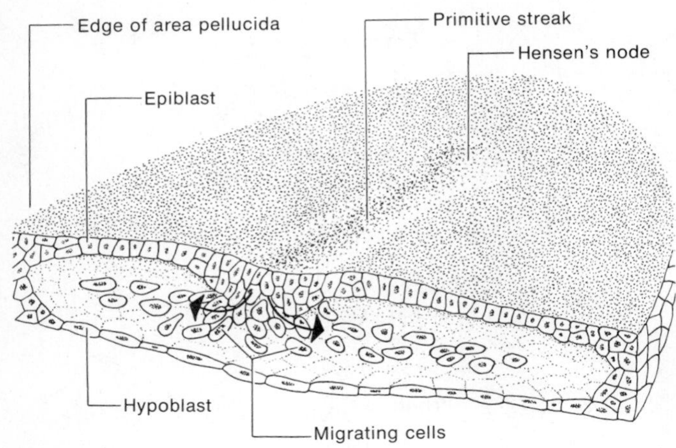

FIGURE 14.9 Gastrulation in the bird. The anterior half of the area pellucida of a chick embryo is cut transversely to show the migration of mesodermal cells from the primitive streak.

parts of the embryo, which stretch, fold, contract, or expand.

Some of the movements result from changes in the shapes of cells. Cells composing expanding presumptive ectoderm in frogs, for example, shift from several layers of cuboidal cells into a single layer of flattened cells. If removed from the embryo and cultured, a section of the presumptive ectoderm from the early gastrula of a frog will expand actively, just as it does *in situ* when contributing to the movements of gastrulation. Invaginating cells are bottle-shaped with the narrowed end at the outer surface (Fig. 14.7). A bit of the blastopore lip transplanted from one embryo to another will invaginate and form an archenteron cavity independent of the archenteron on the host embryo.

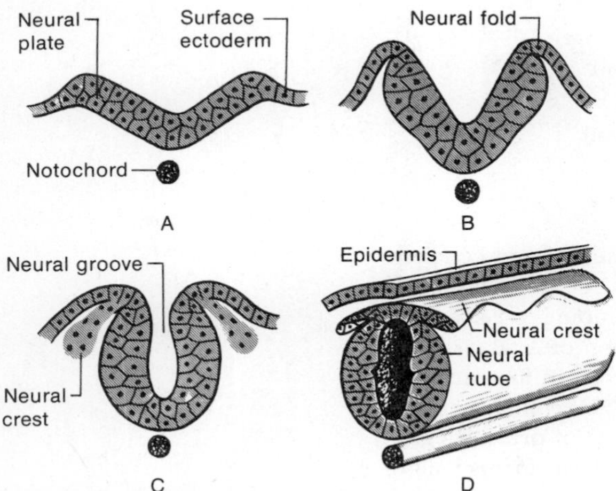

FIGURE 14.10 A series of cross-sectional diagrams through the surface ectoderm to show the formation of the neural tube and neural crest. (After Arey.)

The movements of the cells and the positions they take up are guided, at least in part, by **selective affinities** of certain cells that can be demonstrated if the cells of an early embryo are disaggregated and then incubated in various combinations. Epidermal cells become concentrated on the exterior of the cell mass, and mesodermal cells take up a position between the epidermis and endoderm. When cells touch as a result of random movement, they may remain in contact if held together or they may move apart if not bound strongly. The specific affinities of different cells are related to the specific kinds of glycoproteins present in the cell membrane. These substances enable the cells to recognize "like" and "unlike" cells. The cells stick to like cells but not to unlike ones.

With gastrulation, cells begin to differentiate, a process that continues throughout organogenesis. Since all cells have the same chromosomes, and thus the same genetic information, why do some develop into muscle cells and others into nerve cells? It is probable that most genes are inactive most of the time and that an important mechanism in development is the turning on, or activation, of appropriate genes in the right cells at the right time.

Substances produced by groups of similar differentiating cells may diffuse out and bring about the differentiation of adjacent cells. This process, known as **induction,** is an important developmental mechanism and has been demonstrated many times in the development of vertebrate organization. For example, the roof of the vertebrate archenteron induces the overlying ectoderm to form the neural tube, the forerunner of the brain and nerve cord. Flank ectoderm in a frog gastrula transplanted to a position over the archenteron roof will

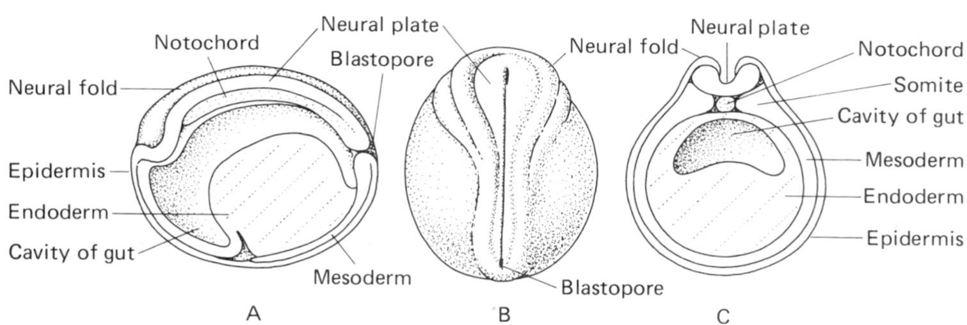

FIGURE 14.11 Neural fold stage in a frog embryo. Neural plate is rolling inward to form neural tube, the precursor of the brain and spinal cord. *A* is a midsagittal section; *B*, a dorsal view; *C*, a cross section. (From Balinsky, B. I.: *An Introduction to Embryology*. 5th ed. Philadelphia, Saunders College Publishing, 1981.)

participate in the formation of neural tube instead of skin.

Organogenesis

By the end of gastrulation the adult ground plan has been laid down, organ rudiments have been delineated, and the stage is set for the formation of organ systems (organogenesis) (Fig. 14.12).

In vertebrates the ectodermal cells over the notochord along the middorsal anterior-posterior axis become a thickened **neural plate.** The center of this becomes depressed and forms the **neural groove** (Figs. 14.10 and 14.11), and the outer edges of the plate rise in two longitudinal **neural folds** that meet at the anterior end and appear, when viewed from above, like a horseshoe. These folds gradually come together at the top, forming a hollow **neural tube.** The cavity at the anterior part of the neural tube becomes the ventricles of the brain, and the cavity in the posterior part becomes the neural canal extending the length of the spinal cord. The brain region is the first to appear, and a long spinal cord develops slightly later.

The various motor nerves grow out of the brain or spinal cord, but the sensory nerves have a separate origin. When the neural folds fuse to form the neural tube, bits of nervous tissue, the **neural crest,** are left over on either side of the tube (Fig. 14.10). These migrate downward from their original position and form the dorsal root ganglia of the spinal nerves and the postganglionic sympathetic neurons. From sensory cells in the dorsal root ganglia, dendrites grow out to the sense organs, and axons grow into the spinal cord. Other neural crest cells migrate and become the cells of the adrenal medulla and of the neurilemma sheath of peripheral neurons. Neural crest cells form still other structures, including the meninges (pia mater and arachnoid), pigment cells, papillae of the teeth, and visceral arches of the head skeleton.

A pair of saclike protrusions, the **optic vesicles,** appear on the lateral walls of the forebrain and grow laterally (Fig. 14.12). The base of the vesicle becomes constricted as the optic nerve. The optic vesicle comes in contact with the inner surface of the overlying epidermis, then flattens out and invaginates to form a double-walled **optic cup.** The inner, thicker layer of the cup becomes the sensory **retina** of the eye, and the outer, thinner layer of the cup becomes the pigment layer of the retina. When the optic cup touches the overlying epidermis it induces the latter to develop into a **lens** rudiment.

The mesoderm of vertebrates extends laterally from the somites (p. 314) on either side of the notochord and then downward between the inner endoderm and the exterior ectoderm. The mesodermal somites form all of the vertebral skeleton and most of the skeletal muscles. The latter grow out to the site of their ultimate location, such as the abdominal wall or the digits of the developing limbs.

Mesoderm lateral to the somites forms the dermis of the skin, the musculature of the gut, all of the blood vessels and connective tissue, the kidneys, and the gonads.

Many organs develop in the embryo without having to function at the same time, but the heart and circulatory system must function while undergoing development. The heart forms first as a simple tube from the fusion of two thin-walled blood vessels beneath the developing head (Figs. 14.12 and 14.13). In this early condition it is essentially like a fish heart, consisting of four

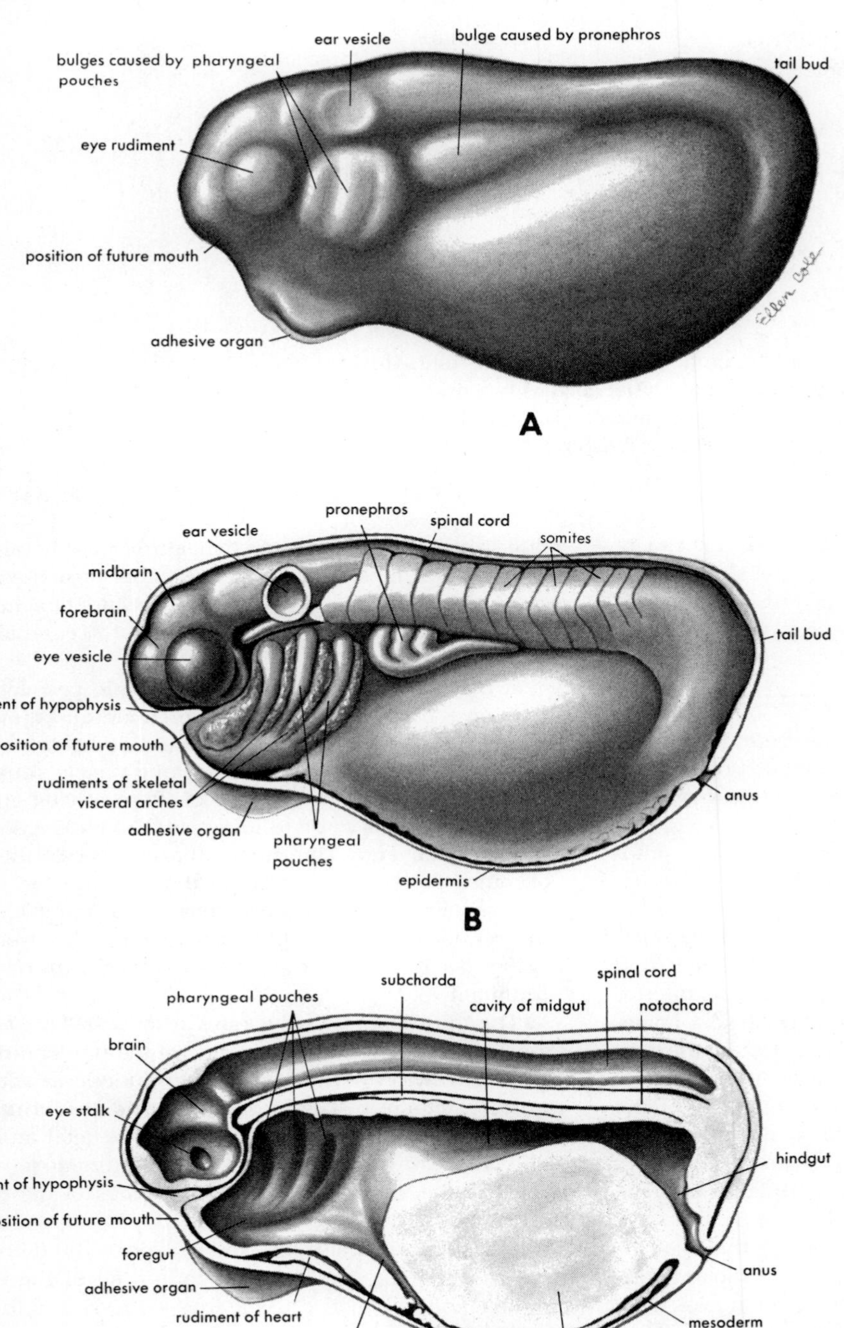

FIGURE 14.12 A frog embryo in an early tail bud stage. *A,* External view; *B,* same embryo with the skin of the left side removed; *C,* same embryo cut in the median plane. (From Balinsky, B. I.: *An Introduction to Embryology.* 5th ed. Philadelphia, Saunders College Publishing, 1981.)

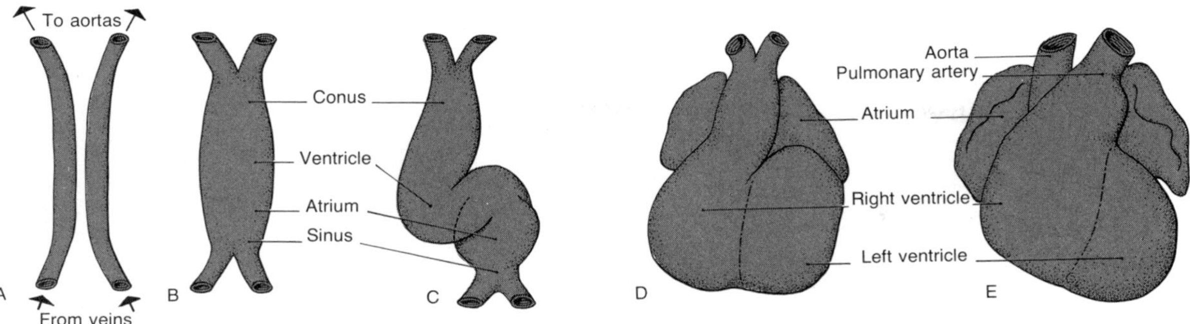

FIGURE 14.13 Ventral views of successive stages in the development of the heart. See text for discussion.

chambers arranged in a series: the **sinus venosus,** which receives blood from the veins; the single **atrium;** the single **ventricle;** and the **arterial cone,** which leads to the aortic arches.

Initially the heart is a fairly straight tube, with the atrium lying posterior to the ventricle, but since the tube grows faster than the points to which its front and rear ends are attached, it bulges out to one side (Fig. 14.13). The ventricle then twists in an S-shaped curve down and in front of the atrium, coming to lie posterior and ventral to it as it does in the adult. The sinus venosus gradually becomes incorporated into the atrium as the latter grows around it.

When it first appears, the embryonic heart is a single structure with only one of each chamber, whereas the adult heart of birds and mammals is a double pump, with separate right and left atria and ventricles. This separation prevents the mixing of aerated blood from the lungs with nonaerated blood from the rest of the body. The mammalian heart begins separating into four chambers at an early stage. The changes are remarkable in their adaptation for blood flow during fetal life while at the same time preparing for life after birth (p. 182).

The digestive tract is first formed from the archenteron of the gastrula (Figs. 14.6 and 14.12) and elongates with the growth of the embryo. The lungs, liver, and pancreas originate as hollow, tubular outgrowths of the original foregut and hence are composed of endoderm, but these outgrowths are always associated with some mesodermal tissue, which forms the blood and lymph vessels, connective tissue, and muscles of these organs. The endoderm forms only the internal epithelium of the digestive tract and lungs and the actual secretory cells of the pancreas and liver.

The most anterior part of the foregut flattens out to become, in cross section, a flattened oval rather than a circle and develops into the **pharynx.** In the pharyngeal region a series of four or five paired **pharyngeal pouches** bud out laterally from the endoderm and meet a corresponding set of inpocketings from the overlying ectoderm (Fig. 14.12). In lower vertebrates, such as fishes, the two sets of pockets fuse to make a continuous passage from the pharynx to the outside—the **gill slits** that function as respiratory organs. In the human and other higher vertebrates this normally does not occur; the pouches develop but give rise to important new structures, with the earlier pattern then disappearing. The first pair of pouches becomes the cavities of the middle ear and their connections with the pharynx, the eustachian tube. The thymus and the parathyroid glands develop from the walls of other pouches.

The mouth cavity arises as a shallow pocket of ectoderm, the **stomodeum** (Gr. *stoma,* mouth + *hodos,* way), that grows in to meet the anterior end of the foregut; the membrane between the two ruptures and disappears during the fifth week of development. Similarly, the anus is formed from an ectodermal pocket, the **proctodeum** (Gr. *proktos,* anus + *hodos,* way) that grows in to meet the hindgut; the membrane separating these two disappears early in the third month of development.

Controls in Embryonic Development

At about 72 hours of incubation two ridges appear on either side of a chick embryo, foreshadowing the formation of the limbs (Fig. 14.14*B*). Similar precursors appear in frogs and mammals. Further development leads to more distinct limb buds, then gradually the differentiation of fore- and hindlimb features, cartilage, bone, and muscle. What determines the precise positioning of

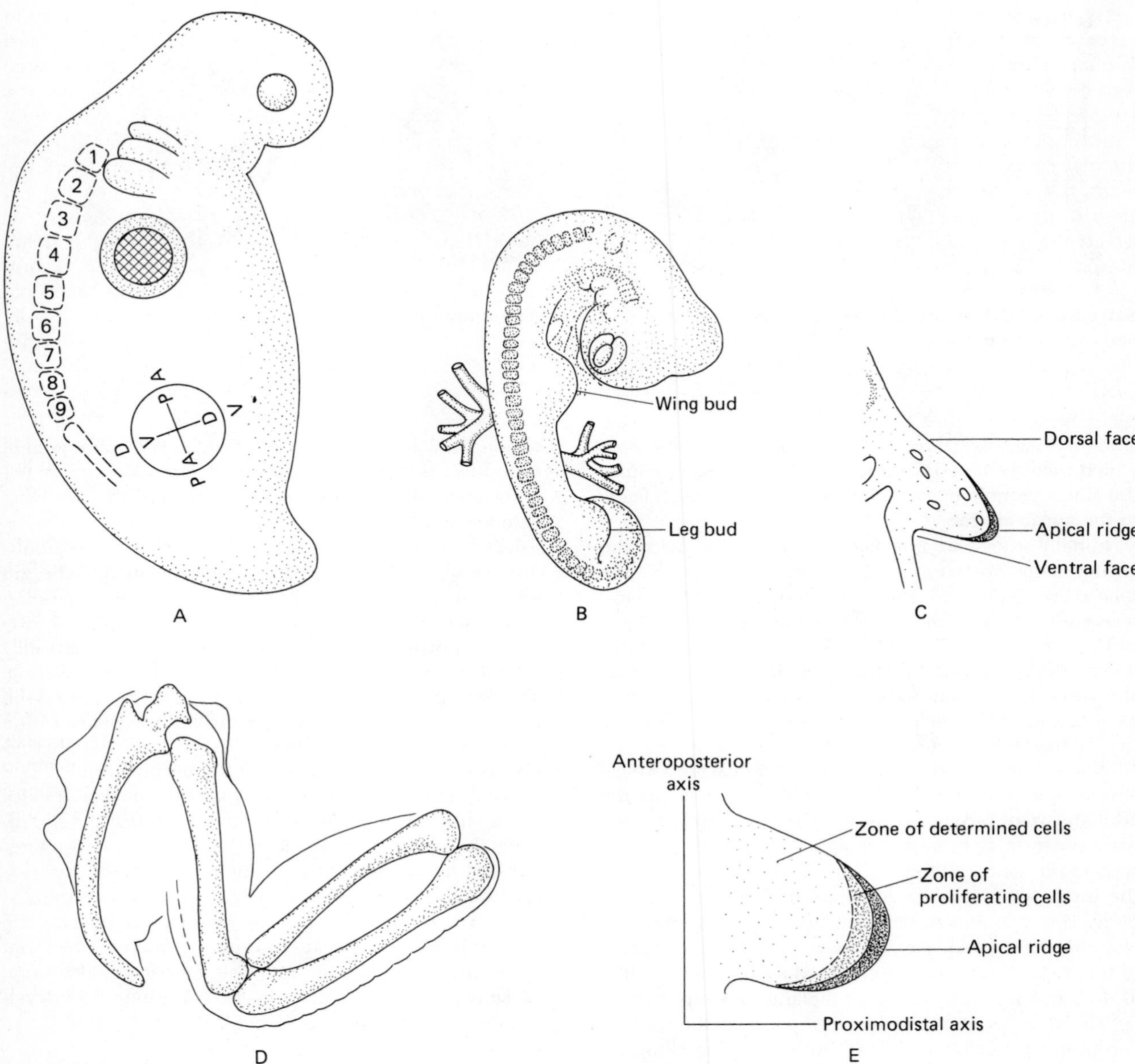

FIGURE 14.14 Experimental studies on vertebrate limb development. *A,* Limbs of an amphibian embryo. In forelimb field, if central area of field is removed, it is replaced by cells in peripheral part of field. If hindlimb field is rotated at a very early stage, limb development is normal; if rotated at this stage (tail bud), the limb will be inverted. Letters indicate dorsal, ventral, anterior, posterior. *B,* Lateral view of chick embryo showing limb buds. *C,* Section of a limb bud. *D,* Photograph of a chick that had the apical ridge removed from the wing bud early in development. Note that the last bones to form were the radius and ulna. *E,* Frontal view of a limb bud showing its orientation along two axes and the zone of proliferating cells (progressive zone) beneath the apical ridge. (*A,* Modified from Swett; *B* and *C,* after Amprino from Berrill, N.J., and G. Karp: *Development.* New York, McGraw-Hill Book Co., 1976; *D* based on a photo from Summerbell and J. H. Lewis: *Journal of Embryology and Experimental Morphology,* 1976.)

a limb bud, the differentiation of some limb bud cells as cartilage and others as muscle, the organization of the forelimb bud as a wing and of the hindlimb bud as a leg, and the mirror image development of all of these structures on the opposite sides of the body?

All of these questions focus on the regulation of morphogenesis and differentiation in embryonic development, fundamental problems that are slowly being unraveled by experimental embryology. The ease with which the developing vertebrate limbs, especially those of frogs and chicks, can be manipulated—removed, rotated, transplanted, their components separated—has made them favorite experimental subjects. As a result, we have accumulated a considerable amount of information about their developmental control, much of which is applicable to the development of other structures and to other organisms.

The mesodermal and overlying ectodermal cells compose the limb fields, which can early be plotted as circular areas on either side of the embryo (Fig. 14.14*A*). The limb bud arises within the field, and if the center of the field is removed, surrounding field cells will fill in and replace it (Fig. 14.14*A*). If the early limb bud is split and its two parts kept separate, two limbs instead of one will form. If the limb field mesoderm is shifted to another location, it will form, together with the new overlying ectoderm, an out-of-place, or ectopic, limb. Wing field mesoderm grafted as a replacement for leg mesoderm will produce wing parts where the leg should be. If the limb field of a late frog gastrula is rotated 180°, that is, turned upside down, the limb will develop normally, but if the same operation is performed later at the tail bud stage, the limb will develop in an inverted position (Fig. 14.14*A*). The three axes of the limb—dorsal to ventral, anterior to posterior (thumb to little finger), and proximal to distal—are not all determined at the same time. First the anterior-posterior axis is fixed, then the dorsoventral, and finally the proximodistal axis. Thus by rotating the limb field at just the right stage, it is possible to produce a limb that has the palm down but the thumb to the rear!

These experiments tell us a number of things: (1) Mesoderm rather than ectoderm is the embryonic tissue that is first determined within the limb field. (2) Embryonic cells have a greater initial potential than their final fate. Thus peripheral cells in the limb field are initially capable of limb formation although they never contribute to actual limb formation, and the cells of the early limb bud have the potential to form all parts of two complete limbs if the limb bud is split. (3) There is progressive determination, with cells becoming steadily more restricted in their developmental potential. (4) Position is critically important in the determination of cell fate.

We might think of the embryo at the limb field stage as embracing three theoretical gradients, one extending from posterior to anterior, one from dorsal to ventral, and one from the midline to the sides of the body (Fig. 14.15). The gradients are established at gastrulation, perhaps by the dorsal lip of the blastopore. Whether the gradients reflect the distribution of a few or many cytoplasmic constituents changing along the gradient axis is uncertain, but the latter is more likely. In either case, the fate of any cell at this stage would be determined first by the germ layer to which it belongs and second by its particular position with regard to the three em-

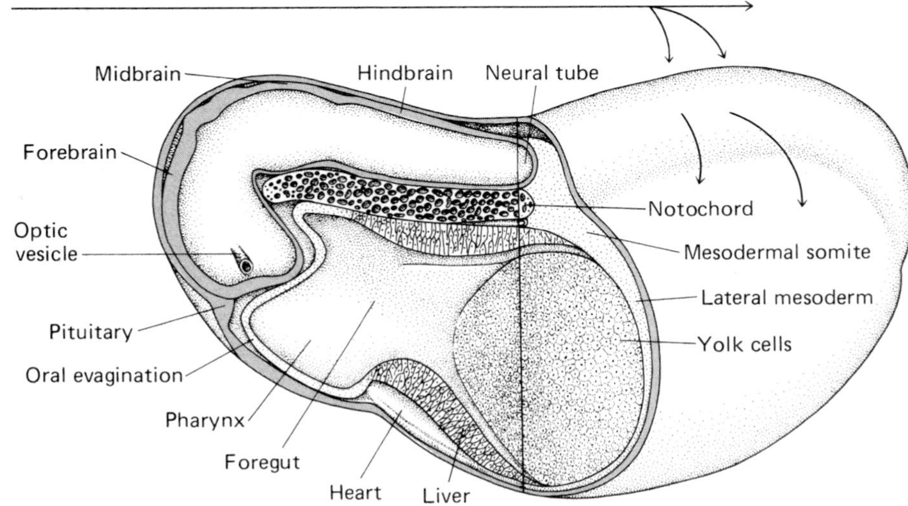

FIGURE 14.15 Lateral stereodiagrammatic view of a 3.00-mm frog embryo. Arrows indicate three theoretical gradients in relation to which any cell in the embryo might be characterized. (Modified from Huettner, A. F.: *Fundamentals of Comparative Embryology of Vertebrates*, rev. ed. Copyright 1949 by Macmillan Publishing Co., renewed 1977 by M. R. Huettner, R. A. Huettner, and R. J. Huettner.)

bryonic gradients. That position would be characterized by a distinct cellular environment that would activate certain genes. Those genes active in controlling early embryonic processes would be shared by neighboring cells. Thus, embryonic fields are broad and many overlap; cells are capable of fates beyond those for which they are normally destined. But the activation of the genes would further modify the cellular environment and perhaps even that of adjacent cells (induction). The new environment would cause some genes to be switched off and others to be turned on. The cellular environment would be further changed, and the cells would be directed steadily down the road of differentiation.

Let us return to the development of the limb and follow it further. The ectoderm covers the distal end of the bud as a caplike apical ridge extending along the anterior-posterior margin (Fig. 14.14*C*). The mesoderm of the limb bud is derived from two different parts of the somites opposite the neural tube (Fig. 14.14*B*), and although they are similar in appearance, cells from one part are already determined to form cartilage and bone, and those from the other part are determined to form muscle. If ventral leg muscle mesoderm is transplanted to the dorsal side of the wing, the graft forms leg muscle in the wing, but *dorsal* leg muscle. The graft has thus been directed by signals both from its original position (muscle instead of cartilage) and also from its new position (dorsal instead of ventral).

If the apical ridge is removed from the limb bud during the course of development, the proximal part of the limb will develop but distal parts will be missing. The wing, for instance, will contain the humerus, radius, and ulna but will be missing bones of the wing tip (Fig. 14.14*D*). If the hind limb apical ridge is substituted for the wing apical ridge, a perfectly normal wing will be formed.

What then is the role of the apical ridge ectoderm? The area beneath the apical ridge is a site of cell proliferation (progress zone), where new cells are continually being added to the distal end of the mesodermal mass of the growing limb (Fig. 14.14*E*). The apical ridge apparently functions to suppress the differentiation of cells undergoing mitosis beneath it. Those cells that move out of this zone and join the growing mass behind escape the suppression and start differentiation. Some limb gradients—proximal-distal, anterior-posterior—must be in operation. The gradients would inform a cell where it was within the mesodermal mass. A precartilage cell, for example, would therefore "know" it was on the radius side of the forearm rather than in the humerus position. Indeed, there is some experimental evidence for the existence of such gradients.

There is still another sort of question about limb development that might be asked. The processes we have described, incomplete as our understanding is, would be operative for any group of vertebrates. Why then are the forelimbs of a horse, dog, rat, and human each different? Developmentally, how does this difference come about?

Pattern and form in organisms are, in part, a consequence of unequal growth rates of various parts of the body. Thus, the human thumb differs from the little finger because the rates of deposition of embryonic tissue in these two parts are not exactly the same. They undergo allometric growth with regard to each other. The allometry affects not only length but thickness, nail size, and so on. Further, the difference in growth rate between these two digits varies from one mammal to another, and it is such differences that account for forelimb shape and form characteristic of various mammals.

There obviously must be some sort of controlling mechanism in allometric growth that signals this part or that part of the limb to speed up or slow down mitosis and differentiation of one embryonic tissue or another. It is a marvelous regulatory system, but at the present time we know very little about how it works. One of the important unanswered questions of developmental biology is how the genotype of an animal is translated into the phenotype. There is some sort of genetic "blueprint" for the human hand, for example, as opposed to that for a dog's paw, but how does it direct construction? The direction for hand development is probably in the form of a very precise sequence of chemical signals, coded by DNA, that controls the stop, start, slow, and accelerate "switches" directing allometric growth. One complement of DNA, the regulatory genes, may be controlling another complement, the genes that are actually synthesizing the protein materials necessary for growth.

Adaptations for Development

Planktonic Development and Larvae. The sea is the ancestral home of animals, and among some of the many living marine animals we find what appear to be primitive life cycles. The eggs are fertilized externally in the sea water and begin development in the plankton (Fig. 14.16). Continued development leads to a **larva** that is an independent motile and feeding stage of development (Fig. 14.17). The larva provides for dispersion of the species and requires only enough yolk in the egg to support prelarval development. Following planktonic life, the larva settles to the bottom and undergoes metamorphosis to the adult form.

A life cycle including a larval stage is called **indirect** and is characteristic of more than half of the species of temperate and tropical marine bottom-dwelling inver-

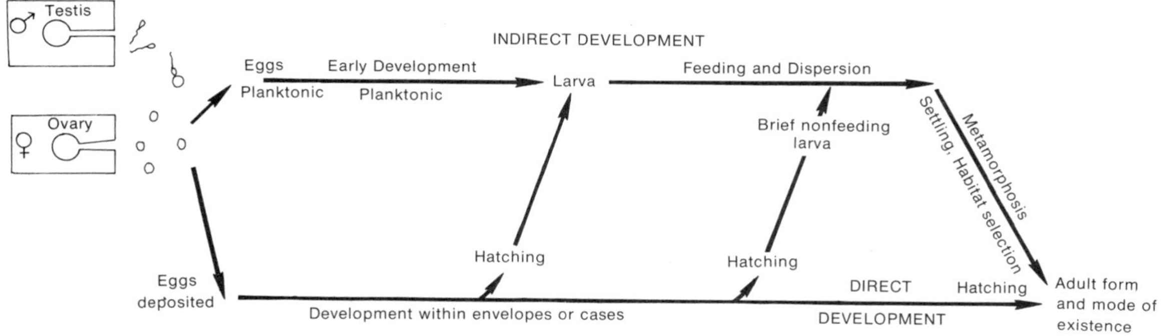

FIGURE 14.16 A comparison of direct and indirect development.

tebrates. Sessile organisms, such as barnacles and oysters, are dependent upon larvae for dispersion. The principal disadvantage of indirect development is the high mortality to which the planktonic eggs, embryos, and larvae are subjected. Many marine animals have retained larvae but eliminated early planktonic development by having the eggs deposited on the bottom within some sort of protective envelope. Hatching occurs at the larval stage. Some have long-lived larvae that feed on diatoms and other minute organisms (planktotrophic larvae); others have evolved brief nonfeeding, yolk-laden larvae (lecithotrophic larvae) that are planktonic only long enough to permit adequate dispersion of the species (Fig. 14.16).

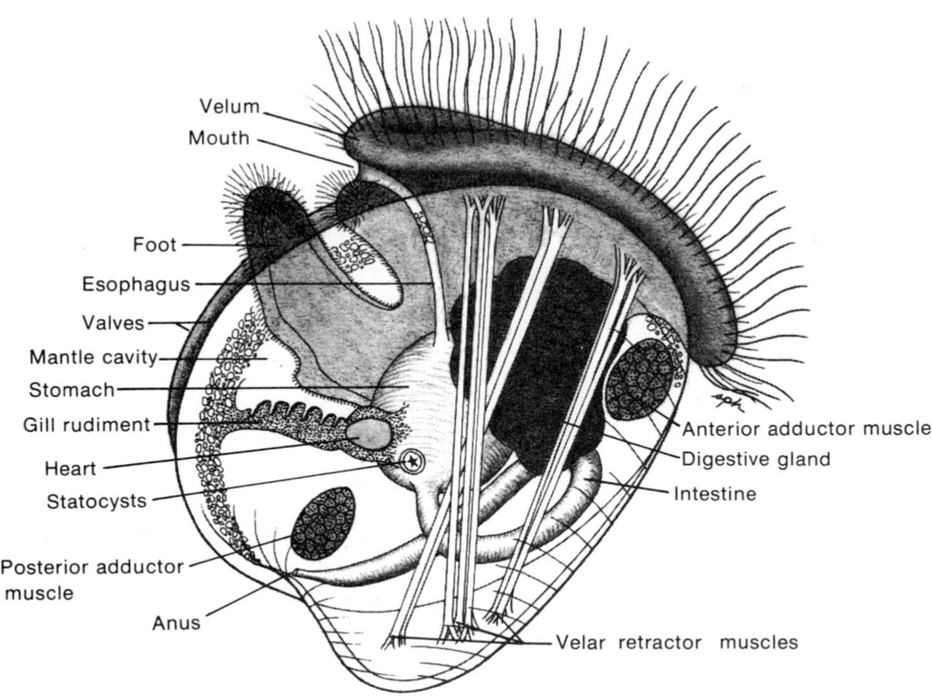

FIGURE 14.17 The veliger larva of an oyster. The large ciliated organ called the velum is used in swimming and feeding. A bivalve shell is present, and although the foot disappears in the attached adult, it is present in the larva. (After Galtsoff.)

Other species have dispensed with larvae altogether. The entire sequence of developmental events up to the adult body form occurs within a protective egg envelope or case, and the young on hatching take up the adult mode of existence. The eggs of such species are relatively small in number and contain large amounts of yolk. Such a developmental pattern with no larval stage is said to be **direct** and is characteristic of many marine and most freshwater and terrestrial animals.

The familiar larvae of amphibians and insects are exceptions to these generalizations. Tadpoles do not represent the retention of the larval stage of a marine ancestor but are freshwater developmental stages of terrestrial adults and reflect the evolution of amphibians. Insect larvae represent a specialization that has evolved in higher members of this large group of terrestrial animals, permitting the exploitation of food sources and habitats different from those of the adults. Primitive insects lack larvae.

Cleidoic Eggs and Extraembryonic Membranes. Many terrestrial animals that have direct development tend to have eggs that are to some degree self-contained systems. Such eggs are called **cleidoic eggs** (Gr. *kleis,* bar + *ōon,* egg). The egg is provided with all of the necessary food material and is enclosed within a protective covering, or shell. It also contains water to prevent desiccation of the embryo and may even store embryonic waste materials. However, eggs can never be completely self-contained, for there must be exchange of gases with the outside environment. The best developed cleidoic eggs are found in reptiles, birds (Fig. 14.18), and insects, all of which are highly successful terrestrial animals.

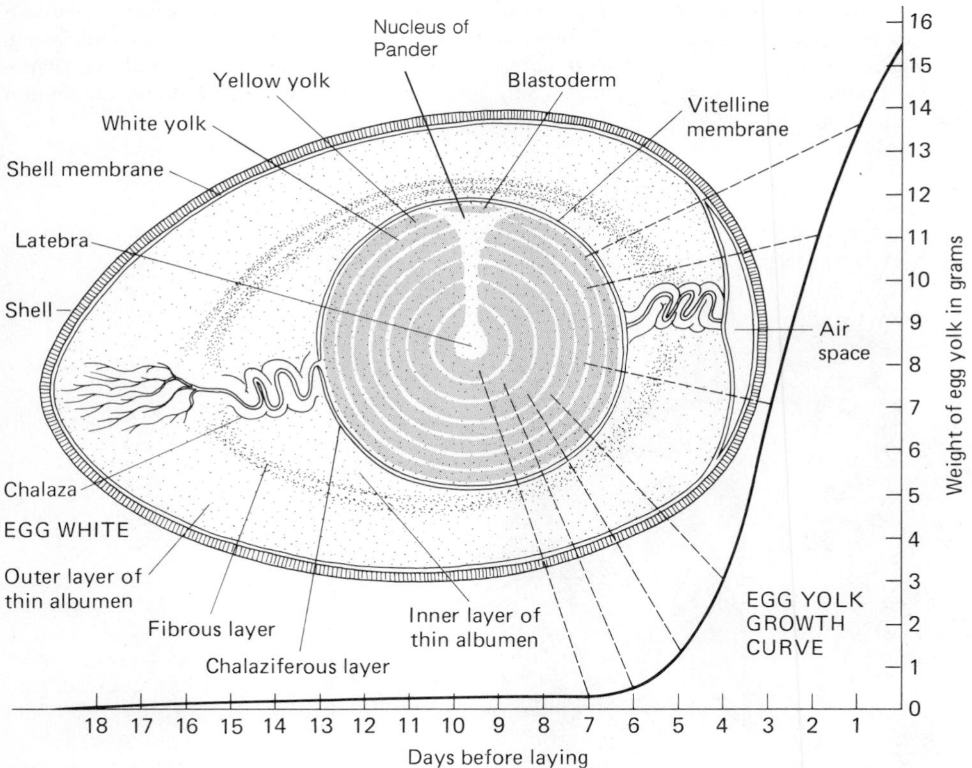

FIGURE 14.18 Structure of the cleidoic egg of a hen at the time of laying. The "yolk" is the actual ovum (one cell) which develops within the ovary. When the ovum passes down the oviduct it becomes surrounded by albumin, the fibrous shell membranes, and the porous calcareous shell. The chalaza keeps the yolk suspended within the albumin, which provides fluid and an accessory source of nutrition for the embryo. The nucleus of Pander and latebra are concentrations of white yolk within the ovum. The graph shows the development of the yolk within the ovary during the 18 days prior to laying. The spherical egg of reptiles is very similar except that the shell is not as heavily calcified. (After Witschi from Patten, B. M.: *Early Embryology of the Chick*. 5th ed. New York, McGraw-Hill Book Co., 1971.)

Reptiles were the first vertebrates to reproduce on land. This involved the evolution of an egg with a large amount of yolk surrounded by a tough outer protective shell. In addition, four unique **extraembryonic membranes** evolved; these sheets of embryonic cells come to lie outside of the embryo proper (hence the name extraembryonic). They play an important role in maintaining the embryo during the course of development.

One extraembryonic membrane, the **yolk sac,** surrounds the yolk mass (Fig. 14.19*A*). Yolk is digested by enzymes produced by the yolk sac, and the products of digestion are transported to the embryo by yolk sac blood vessels. The **amnion** (Gr. *amnion,* fetal membrane) completely encloses the embryo, and its fluid-filled interior provides a protective aquatic environment in which the embryo develops. The **chorion** (Gr. *chorion,* skin) lies beneath the shell and surrounds both the yolk sac and the amnion. The fourth and last extraembryonic membrane to form, the **allantois,** grows out from the hindgut and comes to lie against the inner side of the chorion like a great flattened balloon. Blood from the embryo is circulated out into the allantois by allantoic vessels. Here gas exchange occurs, oxygen diffusing inward through the shell and chorion and carbon dioxide diffusing outward. The cavity of the allantois also functions as a "septic tank" for the deposition of nitrogenous wastes in the form of highly insoluble uric acid.

Yolk sac, allantois, and amnion all connect to the center of the belly region of the embryo, and the connecting stalk is homologous with the umbilical cord of mammals. The yolk sac is largely resorbed during the course of development, but the connection with the other extraembryonic membranes is broken when the young reptile hatches from the egg and the membranes are left behind with the shell.

Birds and mammals possess the same extraembryonic membranes as do reptiles, from which they evolved. The three groups are often referred to as **amniotes** because of this common possession of an amnion. Reproduction in birds is very much like that of reptiles, except that birds brood their eggs. Excepting the primitive egg-laying monotremes, mammals do not possess cleidoic eggs, and the extraembryonic membranes contribute to the formation of the placenta described in a later section.

Brooding. Parental care of the eggs, termed **brooding,** greatly reduces the mortality rate and occurs in a wide variety of animals. Brooding may be external, as in birds and octopods. The latter attach their eggs to the rocky wall of their lair, cleaning and guarding them until they hatch. The number of eggs produced by brooding species is always small, but the survival rate is much greater than in nonbrooding forms.

Internal brooding, in which the fertilized eggs develop within the reproductive tract of the female, occurs

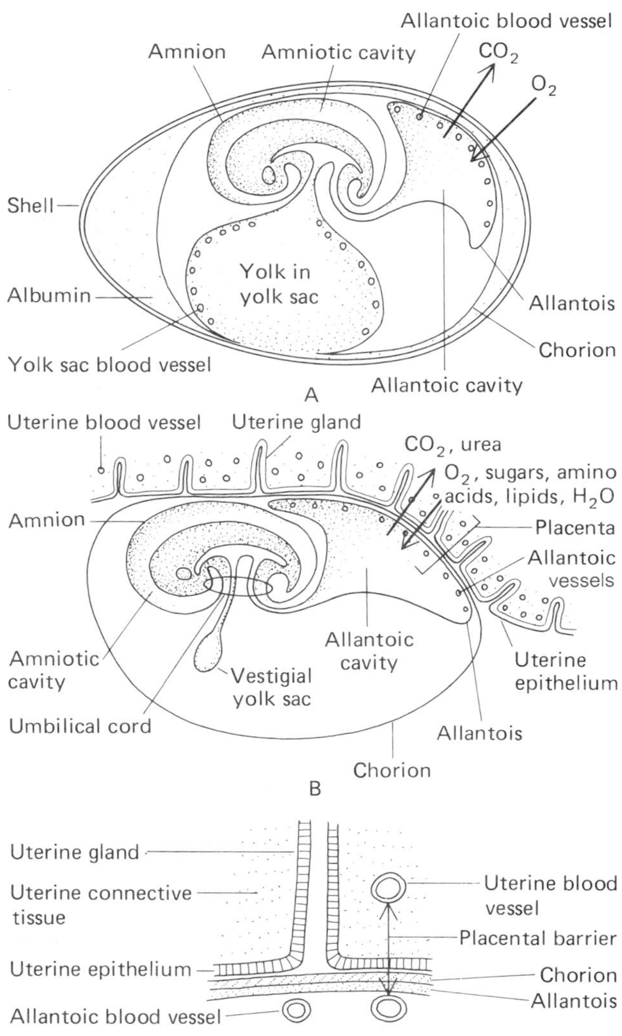

FIGURE 14.19 Comparison of the extraembryonic membranes of reptiles, birds, and mammals. *A,* Reptiles and birds. Embryo and extraembryonic membranes lie within the shell. Food is derived from digestion of yolk in yolk sac, and gas exchange occurs between allantoic blood vessels and air outside of shell. *B,* Primitive placental mammal having no placental villi. Chorionic vesicle containing embryo rests against glandular epithelium of inner surface of uterine wall. Exchange of gases, wastes, and food occurs between allantoic and uterine blood vessels, and the intervening tissues constitute the placenta and placental barrier. The yolk sac is vestigial. *C,* Diagrammatic representation of placental barrier where no erosion has occurred. The barrier includes all of the tissues between the uterine and allantoic blood vessels through which a molecule or ion must diffuse or be actively transported.

in almost every group of animals. When development is completed, the young pass out of the female; she is said to have given birth to her young.

Development of the egg outside of the body of the female, whether or not it is brooded, is termed **oviparous** (L. *ovum*, egg + *parere*, to bring forth) development. Internal brooding is now usually called **viviparous** (L. *vivus*, living + *parere*, to bring forth) whether the embryo obtains its food from yolk within the egg (**aplacental viviparity** or an older term **ovoviviparity**) or whether the food is supplied directly by the mother (**placental viviparity**). Aplacental viviparity is most common and is found in many invertebrates and in many sharks, fish, and snakes. Placental viviparity is much less common; scorpions and mammals are the most notable but not the only examples.

Internal brooding usually occurs within a modified part of the oviduct called the *uterus*. Gas exchange takes place between the embryo and the uterine wall, and in viviparous development there are adaptations for the transfer of food materials from the mother to the embryo.

Mammalian Development

As we have seen, the ancestors of mammals were reptiles that had cleidoic eggs and extraembryonic membranes. However, from an ancestral oviparous habit, mammals evolved a viviparous mode of development, in which the embryo is brooded internally and the mother, rather than yolk, provides food material. The mammalian egg has therefore shifted from the reptilian extreme telolecithal type to one in which there is little yolk. Cleavage has again become complete. The reptilian extraembryonic membranes, however, have been retained and put to new uses in viviparous development.

Establishment of the Embryo in the Uterus. The fertilized egg passes down the fallopian tube, undergoing cleavage along the way, and arrives in the uterus as a **blastocyst,** a hollow sphere that is the equivalent of the blastula of lower vertebrates (Fig. 14.20*A*). The embryo will develop from the inner cell mass on one side, and the periphery of the sphere, the **trophoblast** (Gr. *trophé*, nourishment + *blastos*, bud), is equivalent to the chorionic ectoderm. The trophoblast forms very early because it plays a very important role in the implantation of the embryo in the uterus and the establishment of the placenta.

Energy for early development is supplied by the small amount of food within the egg and by secretions from glands in the uterine lining. Gastrulation begins about a week following fertilization, and a primitive streak like that of reptiles and birds forms on the inner cell mass of the blastocyst. Considering the reptilian ancestry of mammals, this is not as peculiar an arrangement as it may seem. Imagine greatly increasing the blastocyst in size and filling the interior with yolk. You would then have something like the reptilian extreme telolecithal egg with the primitive streak developing on one side.

Extraembryonic membrane formation is very precocious, for the embryo must be prepared quickly for intrauterine life. The trophoblast, as we have seen, forms before the embryo begins to differentiate, and the rest of extraembryonic membrane development takes place rapidly (Fig. 14.20*B–D*). In reptiles and birds the allantois and chorion lie against the shell and provide for the exchange of gases with the outside environment, with allantoic blood vessels transporting oxygen and carbon dioxide to and from the embryo. The mammalian condition is very similar, but the allantois and chorion are in contact with the uterine wall instead of a shell (Fig. 14.19*B*). Now the uterus is the outside environment, supplying not only oxygen but also food material and removing wastes. The allantoic blood vessels are called the **umbilical arteries** and **veins,** and the connecting cord of membranes through which the blood vessels pass to the embryo is the **umbilical cord.** The yolk sac in most mammals has become very small and nonfunctional.

Those tissues involved in the exchange between mother and embryo—the allantois, chorion, and adjacent uterine lining—constitute the **placenta.** Note that the placenta has a dual origin, part from the embryo and part from the mother. Exchange occurs between the uterine circulation (maternal) on one side and the umbilical circulation (embryonic) on the other. The two streams never mix; they are kept separate by a barrier of placental tissues (Fig. 14.19*C*). Most gases and waste products simply diffuse across the membrane. However, active transport may play a role in the transfer of certain substances; the fetal blood contains a higher concentration of amino acids, calcium, and ascorbic acid than the maternal blood.

In the evolution of mammals two modifications occurred that facilitate exchange across the placental barrier. First, there has been a tendency to increase the placental surface area by branching evaginations of the chorioallantoic membranes, called **villi,** that become embedded in the uterine wall (Figs. 14.20 and 14.21). The development and distribution of villi vary from one group of mammals to another. The second modification has been to reduce the number of tissues separating maternal and fetal blood streams. Thus when the embryonic villi push into the uterine wall, they digest away some of the uterine epithelium and underlying connec-

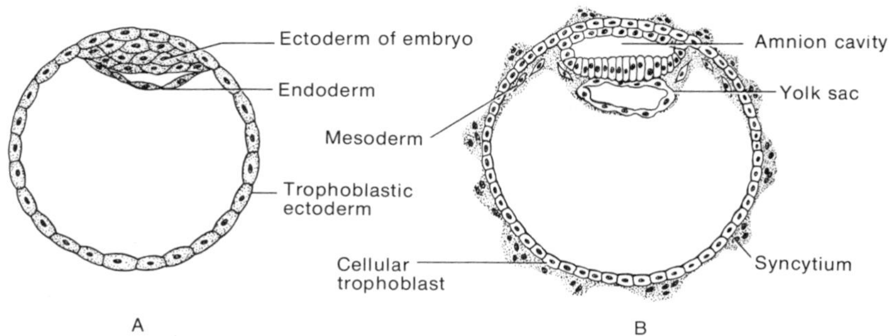

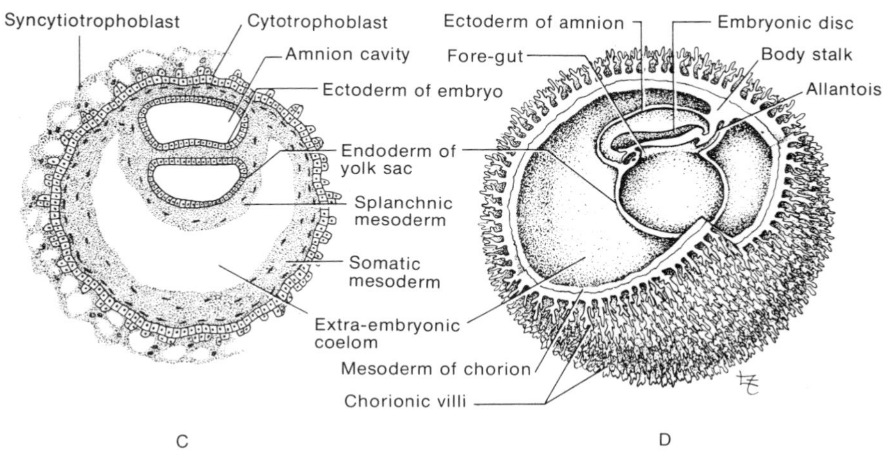

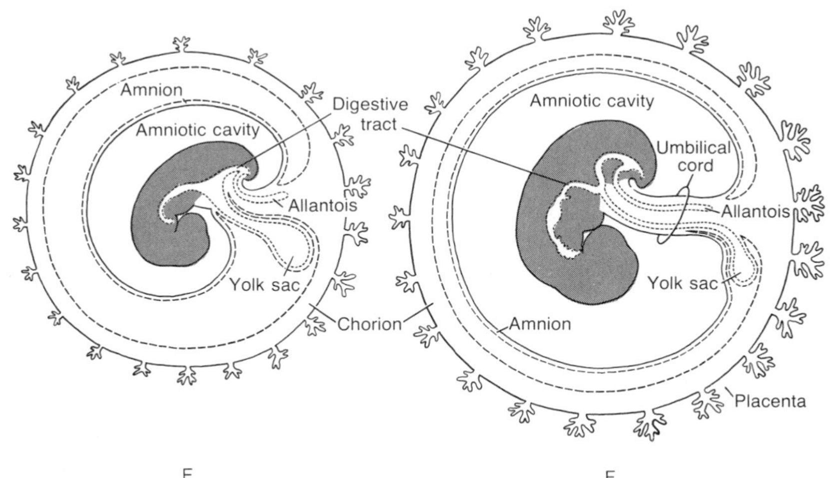

FIGURE 14.20 Diagrams of human embryos 10 (*A*) to 20 (*D*) days old, showing the formation of the amniotic and yolk sac cavities and the origin of the embryonic disc. *E* and *F* show later stages in the development of the umbilical cord and body form in the human embryo. In *E* and *F* the solid lines represent layers of ectoderm; the dashed lines, mesoderm; and the dotted lines, endoderm.

tive tissue and blood vessels. The degree of erosion varies in different groups of mammals. In the pig, for example, there are no villi, and the chorion simply rests against the uterine lining. It can be easily lifted away. In the human, on the other hand, the uterine tissue is so eroded that the highly branched villi are bathed in pools of maternal blood. The human placental barrier is thus reduced to the embryonic chorion and allantois. In no

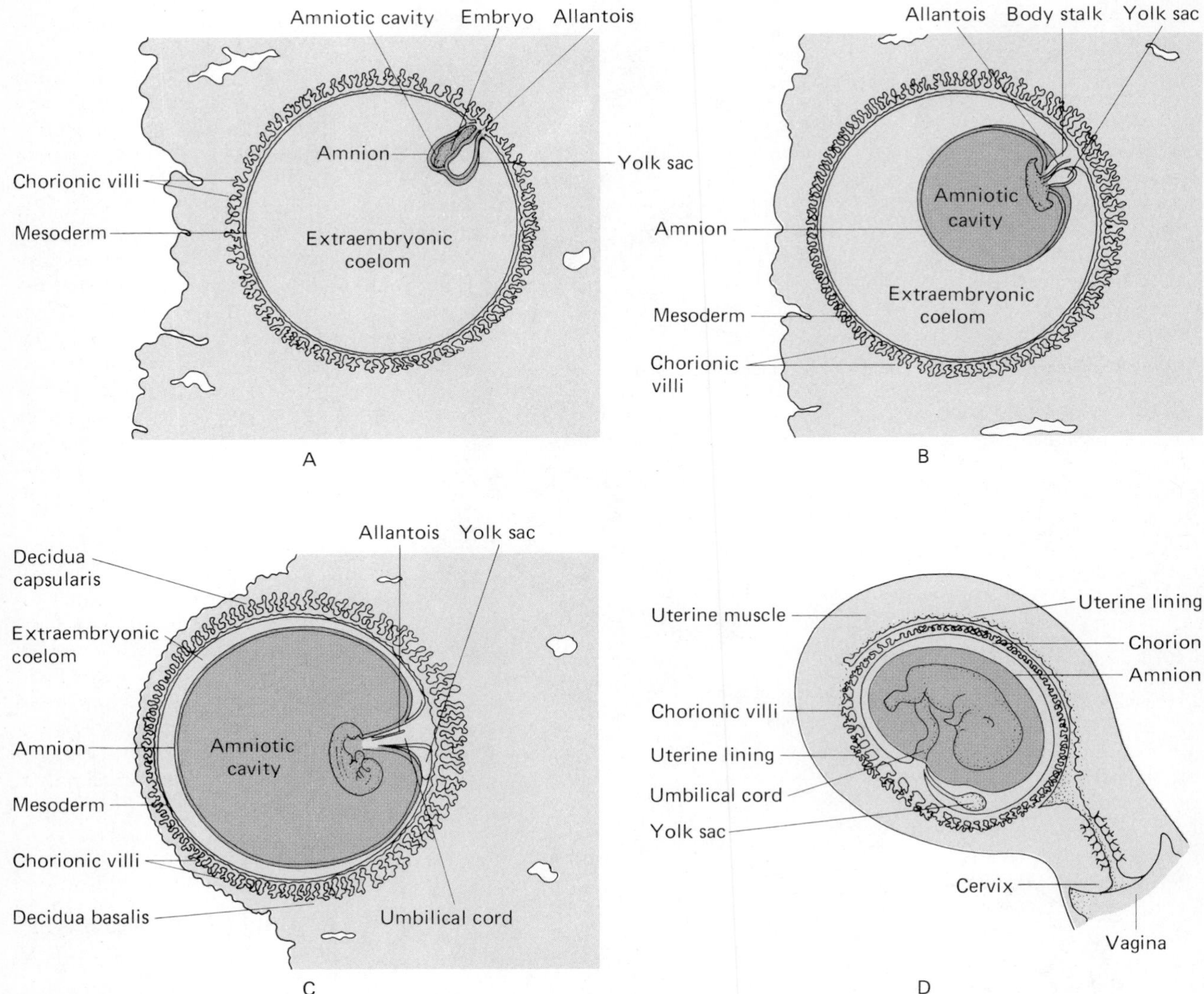

FIGURE 14.21 Development of the human placenta. *A,* Diagram of a 25-day-old human embryo (2.4 mm. long). The blastocyst is completely buried within the uterine wall, and villi are growing outward from the entire chorion surface. *B,* Embryo at 29 days (4.3 mm.). Note that the villi toward the inner side of the uterine wall are already a little larger than those on the outer side. *C,* Embryo at 33 days (5.00 mm. long). Chorionic vesicle has begun to bulge into the uterine lumen, pushing with it the overlying layer of uterine wall. Functional villi are now restricted to the inner side of the uterine wall. *D,* A later human embryo and its position within the entire uterus. (*A–C* from Huettner, A. F.: *Fundamentals of Comparative Embryology of the Vertebrates.* Revised ed. New York, The Macmillan Co., 1949; *D* modified after Patten.)

mammals have the intervening tissues completely disappeared; some barrier must be present. Why? Occasionally breaks occur in the placenta, and this can lead to problems (see p. 176).

The invasion of the uterine wall by the trophoblast cells of the developing fetal membranes should bring about an immunological response from the mother, and she would be expected to reject the embryo as she would a graft of foreign tissue. But this does not happen. Of the various factors that contribute to the protection of the embryo from this danger, two seem to be of particular importance. In some manner, perhaps because of a special fibroid coating, the antigens of the invading trophoblast cells are not made manifest, at least not immediately. Thus for some time the mother's immune system does not recognize that the embryo is foreign. Then the production of the hormones of pregnancy tend to suppress the maternal lymphocytes that would be involved in the rejection process.

In most mammals the embryo and surrounding membranes develop within the lumen of the uterus, and only the chorionic villi invade the uterine wall. However, in humans the entire blastocyst burrows into the uterine wall during gastrulation and becomes completely covered over (Fig. 14.21*A*). The spherical chorionic vesicle containing the embryo develops villi over its entire surface, and the villi invade the surrounding maternal tissue in all directions. But as the embryo, and later the fetus, gets larger and larger, it bulges into the uterine cavity and stretches the overlying uterine tissue so that villi growth is halted (Fig. 14.21*C*). All of the functional villi become restricted to the opposite side away from the lumen. The human placenta thus comes to be disc-shaped, with a diameter about the size of a small dinner plate when fully formed.

Twinning. In primates, whales, horses, and many other species of mammals, offspring are usually produced singly. In other animals, more than one (up to 25 in the pig) are produced in a single litter. About once in every 88 human births, two individuals are delivered at the same time. More rarely, three, four, five, and even six children are born simultaneously. About three fourths of multiple births are the result of the simultaneous release of two eggs, both of which are fertilized and develop. Such fraternal (**dizygotic,** "two egg") twins may be of the same or different sex and have only the same degree of family resemblance that brothers and sisters born at different times have. They are entirely independent individuals with different hereditary characteristics.

In contrast, identical (**monozygotic,** "one egg") twins are formed from a single fertilized egg that at some early stage of development divides into two (or more) independent parts, each of which develops into a separate fetus. Such twins, of course, are of the same sex, have identical hereditary traits, and are so similar that it is difficult to tell them apart. Monozygotic twinning may occur in any of several ways. The two blastomeres produced by the first cleavage may separate and each become an embryo, the blastocyst may contain two inner cell masses, or two primitive streaks may form on a single embryonic disc (Fig. 14.22). Twins have been produced experimentally by the first method in lower vertebrates. This method is less likely to occur in mammals than in others, for the mammalian egg and cleavage stages are surrounded by a strong membrane, the zona pellucida, that should prevent the blastomeres from separating.

Where two primitive streaks form on the same embryonic disc, the identical twins sometimes develop without separating completely and are born joined together (Siamese twins). All grades of union have been known to occur, from almost complete separation to fusion throughout most of the body, so that only the head or the legs are double. Sometimes the two twins are of different sizes and degrees of development, one being quite normal, while the second is only a partially formed parasite on the first. Such errors of development usually die during or shortly after birth.

Development of the Human Body Form. The two-week-old human embryo is a flat disc located on one side of the blastocyst (Figs. 14.20 and 14.21). The cells of this disc undergo gastrulation movements in the same manner as described for the chick (p. 314), but the resulting gastrula, like that of all other mammals, birds, and reptiles, is still flat with no delineation of the body outline. The conversion of the disc-shaped gastrula into a roughly cylindrical embryo is accomplished by three

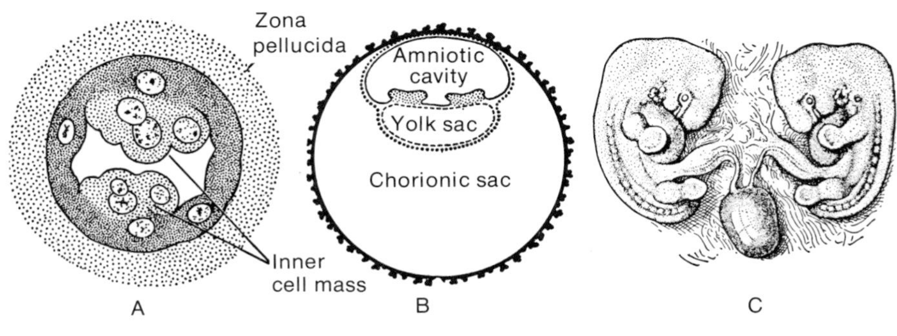

FIGURE 14.22 Two methods of monozygotic twinning: *A*, by division of inner cell mass; *B*, by the formation of two primitive streaks. In the latter case, the twins would be connected to a single yolk sac, as in *C*. (After Arey.)

processes: (1) the growth of the embryonic disc, which is more rapid than the growth of the surrounding tissue (Fig. 14.20*A*); (2) the underfolding of the embryonic disc, especially at the front and rear ends (Fig. 14.20*D*); and (3) the constriction of the ventral body wall to form the future umbilical cord and to separate the embryo proper from the extraembryonic parts (Fig. 14.20*E*). In addition, the body begins to separate into head and trunk, and the pectoral and pelvic appendages appear.

Growth is rapid at the anterior end of the embryonic disc, and soon the head region bulges forward from the original embryonic area. The tail, which even human embryos have at this stage, bulges to a lesser extent over the posterior end. The sides of the disc grow downward, eventually to form the sides of the body. The embryo becomes elongated because growth is more rapid at the head and tail ends than laterally. The enlarging of the embryo has been compared to the increase in size of a soap bubble blown from a pipe, which, as it grows, swells out in all directions above the mouth of the pipe (the yolk sac). What is to become the mouth and heart originally lies in front of the embryonic disc, and as the disc grows and bulges over the tissues in front, the mouth and heart swing underneath to the ventral side. A similar underfolding occurs at the posterior end. By such growth and underfolding the lateral and eventually the ventral walls of the body are formed, and the embryo becomes more or less cylindrical in shape.

When the embryo is still a simple disc its entire undersurface is open to the yolk cavity. As the body walls fold, a foregut and hindgut (which form, respectively, the anterior and posterior parts of the digestive tract) are cut off from the yolk sac but remain connected to it by the yolk stalk. As the embryo grows and folds, it becomes more and more separated from those embryonic tissues that contribute only to the placenta and surrounding membranes. These extraembryonic tissues are connected to the embryo by a cylindrical tube, the **umbilical cord.** This takes place about four weeks after development has commenced and allows the embryo to float free in the liquid-filled amniotic cavity, connected to the placenta only by the umbilical cord (Figs. 14.20 and 14.21).

The month-old embryo, which is about 5 mm. long, is now recognizable as a vertebrate of some kind. It has become cylindrical, with a relatively large head region and with prominent gill arches and a tail. Meanwhile, somites are forming rapidly in the mesoderm on either side of the notochord and the beating heart is present as a large bulge on the ventral surface behind the gills. The arms and legs are still mere buds on the sides of the body.

By the end of six weeks the embryo is about 12 mm. long, the head begins to be differentiated, the arms and legs have grown out, but the tail and gill arches are still present.

At the end of two months of growth, when it is 25 mm. long, the embryo begins to look definitely human. The face has begun to develop, showing the rudiments of eyes, ears, and nose. The arms and legs have developed, at first resembling tiny paddles, but by this stage the beginnings of fingers and toes are evident (Fig. 14.23). Some of the bones are beginning to ossify, and the organ systems have differentiated. The tail, which was prominent during the fifth week of development, has begun to shorten and be concealed by the growing buttocks. As the heart moves posteriorly on the ventral side and the gill pouches become less conspicuous, a neck region appears. Now most of the internal organs are well laid out so that development in the remaining seven months consists mostly of an increase in size and completion of some of the details of organ formation (Fig. 14.24). At this time, about three months, human form has been more or less attained, and the developing human is called a **fetus** instead of an embryo.

The embryo is about 75 mm. long after three months of development, 250 mm. long after five months, and 50 cm. long after nine months. During the third month the nails begin to form, and the sex of the fetus can be distinguished; by four months the face looks quite human; and by five months, hair appears on the body and head. During the sixth month, eyebrows and eyelashes ap-

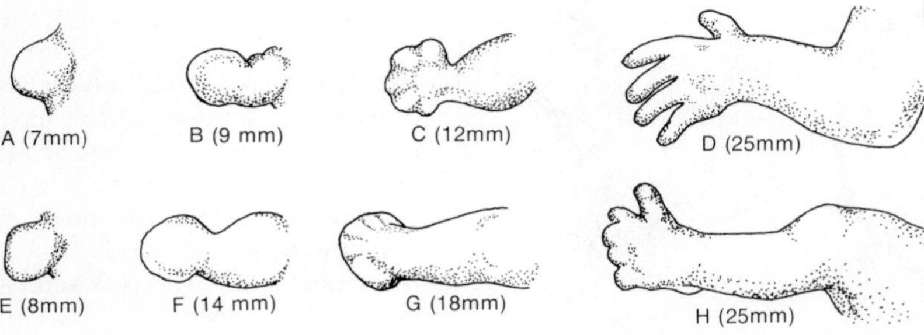

FIGURE 14.23 Stages in the development of the human arm (*upper row*) and leg (*lower row*) between the fifth and eighth weeks.

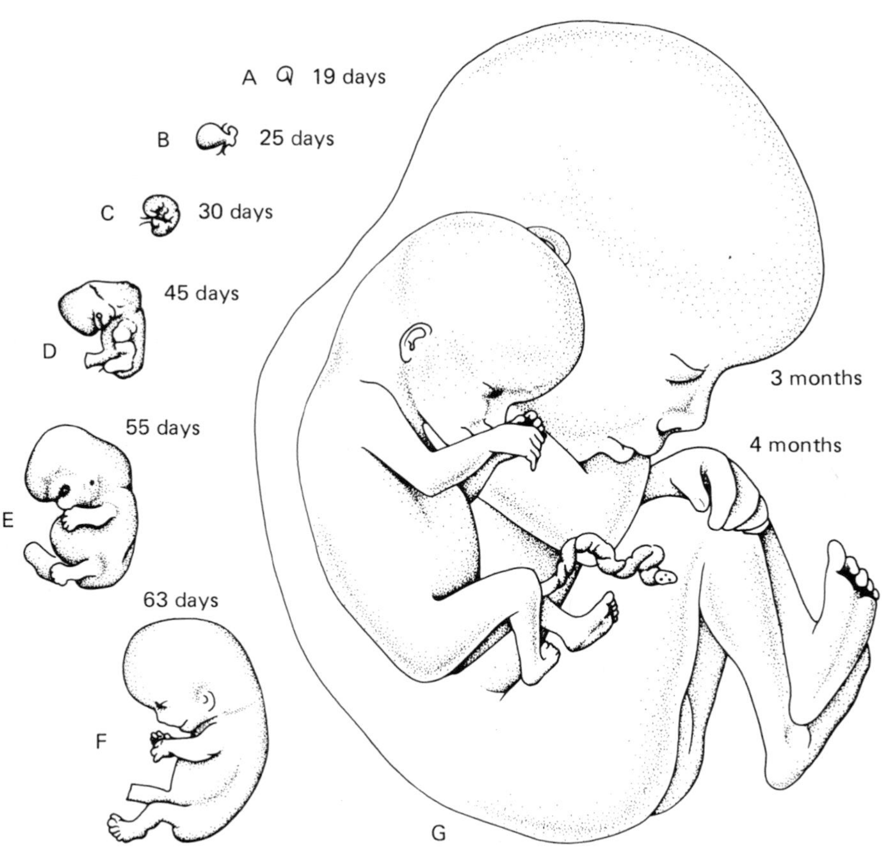

FIGURE 14.24 A life-size, graded series of human embryos. Note the characteristic position of the arms and legs in the four-month-fetus. (From Arey, L. B.: *Developmental Anatomy.* 7th ed. Philadelphia, W. B. Saunders Co., 1974.)

pear. After seven months, the fetus resembles an old person with red and wrinkled skin. During the eighth and ninth months, fat is deposited under the skin, causing the wrinkles partially to smooth out; the limbs become rounded, the nails project at the finger tips, the original coat of hair is shed, and the fetus is "at full term," ready to be born. The total gestation period, or time of development, for human beings is about 280 days from the beginning of the last menstrual period before conception until the time of birth.

Malformations. In view of the extreme complexity of the developmental process it is indeed remarkable that it occurs so regularly and that so few malformations occur. Nevertheless, about one child in a hundred is born with some major defect, such as a cleft palate, club foot, or spina bifida. Some of these congenital defects are inherited; others result from environmental factors. Experiments with fruit flies, frogs, and mice have shown that X-rays, ultraviolet rays, temperature changes, and a variety of chemical substances may induce alterations in development. The kind of defect produced depends on the time in development at which the environmental agent is applied and does not depend to any great extent on the kind of agent used. For example, X-rays, the administration of cortisone, and the lack of oxygen will all produce similar defects in mice—harelip and cleft palate—if applied at comparable times in development during which particular organs are differentiating and growing most rapidly and are most susceptible to interference.

Birth. As the embryo develops, the uterus enlarges considerably to accommodate it. At the time of conception, the human uterus does not protrude far above the pubic symphysis (Fig. 13.13), but nine months later, when embryonic development has been completed, it extends up in the abdominal cavity nearly to the level of the breasts. During this enlargement, the individual muscle fibers in its wall increase in size, and additional muscle develops from undifferentiated cells in the uterine wall. The uterus becomes a powerful muscular organ ready to assume its role in childbirth, or **parturition.**

The factors that initiate birth are not entirely clear, but hormones secreted by the ovary and pituitary and perhaps by the fetal adrenal gland may play a role. Hormones produced by the pituitary, ovary, and placenta have prepared the mother's body for the birth. The mammary glands have enlarged and are ready for milk production, the uterine musculature has increased, and the pubic and other pelvic ligaments have relaxed so that the pelvic canal can enlarge slightly. Birth begins by a series of involuntary uterine contractions, "labor," that gradually increase in intensity and push the fetus, generally head first, against the cervix (Fig. 14.25). The cervix gradually dilates, but in human beings as much as 18 hours or more may be required to open the cervical canal completely at the first birth. The sac of amniotic fluid that surrounds the fetus acts as a wedge and also helps to open the cervix. The amnion normally ruptures during this process, and the amniotic fluid is discharged. When the head begins to move down the vagina, particularly strong uterine contractions set in, and the baby is born within a few minutes. A few more contractions of the uterus force most of the fetal blood from the placenta to the baby, and the umbilical cord can be cut and tied, although tying is unnecessary since contraction of the umbilical arteries would prevent excessive bleeding of the infant. Other mammals simply bite through the cord. Within a week the stump of the cord shrivels, drops off and leaves a scar known as the **navel.**

In many small mammals, such as dogs and cats, the birth process is very rapid, and the young are passed to the outside surrounded by the placenta and all of the extraembryonic membranes, which the mother removes and eats. Where birth is slow, as in humans, the placenta remains in place until the infant is outside of the mother's body. Then uterine contractions continue for a while after birth, and the placenta and remaining extraembryonic membranes are expelled as the "afterbirth." Much of the uterine lining is lost at birth, for the human placenta is an intimate union of fetal membranes and maternal tissue. Uterine contractions prevent excessive bleeding at this time. Following the birth, the uterine lining is gradually reconstituted, and the uterus decreases in size, though it does not become as small as it was originally.

Postnatal Development. Development does not, of course, cease at birth. At birth the teeth and genital organs of the human infant are only partly formed, and the body proportions are quite different from those of the adult. The head composes about half of the length of the two-month-old human fetus, but its growth terminates early in childhood so that the head of the adult is proportionately smaller than that of the newborn. The arms attain their proportionate size shortly after birth, but the legs attain theirs only after some 10 years of growth. The last organs to mature in humans are the genitals, which do not begin to grow rapidly until 12 to 14 years after the infant is born.

The degree of maturity and self-sufficiency of the newly hatched bird or newly born mammal is adaptive for the life-style of the species and varies widely from one group to another. The young of chickens and ducks, which nest on the ground, can run around and eat solid food just after hatching, but the young of robins, which nest in trees and shrubs, are blind, have very few feathers, and cannot stand. Newborn colts and calves have fur and can run with the herd. Newborn rats and mice, which live in nests, or humans, which are carried about,

FIGURE 14.25 Photographs of two models from the Dickinson-Belski series on human birth. (From *The Birth Atlas.* New York, Maternity Center Association.)

are quite helpless and require a lot of parental care to survive. The developmental processes that occur after birth involve some multiplication and differentiation of cells, but in large part they involve the growth of cells formed earlier. The weight or size of an animal or plant as a function of time usually yields an S-shaped growth curve, one remarkably like the growth curve of a population of individuals (p. 804). Although biologists can identify some of the factors ("environmental resistance") that stabilize the size of a population at a certain level (p. 806), very little is known of the factors that lead to the cessation of cell multiplication and growth. Some plants and animals do not, in fact, stop growing but continue to grow, though perhaps at a slower pace, all through their lives.

The human egg is about 100 μm. in diameter, just barely visible to the naked eye. A baby at birth is about 50 cm. long, roughly 5000 times as long as the egg. In developing from an infant to an adult the height increases only an additional three and one half times, to about 175 cm. The maximal rate of linear growth occurs before birth, in the fourth month of fetal life. There is a final growth spurt at the time of adolescence that reaches its peak at about age 12 years in girls and about age 14 in boys.

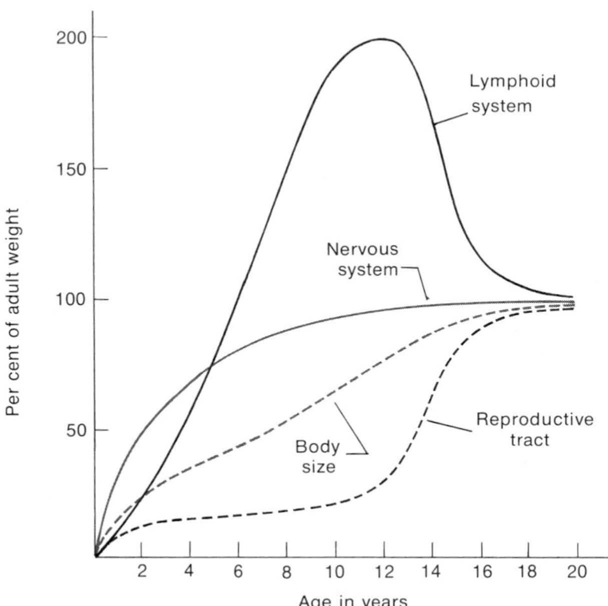

FIGURE 14.26 Diagram showing the relative rates of growth of the several different organ systems during human development.

Each structure and organ has its characteristic rate of growth. The growth rates of the various organs can be assigned to one of four types (Fig. 14.26). The growth curve of the skeleton follows that of the body as a whole. The brain and spinal cord grow relatively rapidly early in childhood and nearly reach their adult size by the age of nine. Lymphoid tissue, including the thymus, has a third type of growth curve, reaching a maximum at age 12 that exceeds the adult value. It subsequently undergoes involution until about age 20, when it attains adult values. The fourth pattern of growth is shown by the reproductive system, which grows very slowly until age 12 or thereabouts and then undergoes rapid growth at puberty.

Summary

1. Activation of the egg by fertilization initiates mitotic divisions, called cleavage. The three most common types of cleavage are radial cleavage (echinoderms and vertebrates), in which the cleavage planes are parallel or at right angles to the polar axis, spiral cleavage (annelids and mollusks), in which the cleavage planes are oblique to the polar axis, and superficial cleavage (arthropods), in which there are nuclear but not cytoplasmic divisions. The amount and distribution of yolk, which impedes cleavage, greatly affects the cleavage pattern. Cleavage frequently leads to a spherical multicellular stage called a blastula, containing an interior cavity, the blastocoel. The total mass of the blastula is less than that of the egg.

2. Gastrulation converts the blastula into a bilateral embryo (gastrula) possessing the ground plan of the adult. The conversion occurs by morphogenetic movements of embryonic cells. As in cleavage, the pattern of gastrulation is greatly affected by the amount and distribution of yolk. The germ layers—ectoderm, mesoderm, and endoderm—become evident at gastrulation.

3. Following gastrulation, organ rudiments derived from one or more germ layers are soon laid down. In all animals the nervous system, the epidermal layer of the skin, and the mouth and anal regions are derived from ectoderm; the lining of the gut and various organs associated with the gut, such as the liver and pancreas, are derived from endoderm; muscle layers, blood vessels, and connective tissue are derived from mesoderm.

4. Position is a primary factor in determining the fate of embryonic cells and in regulating the course of development. Position determines the nature of the cytoplasmic environment and the surrounding cellular environment, which, interacting with the nucleus, regulate the sequential activation of genes and thus the ultimate fate of the cell.

5. Primitively, as in many marine animals, development includes a motile and feeding larval stage (indirect development) that provides for dispersal and an early source of nutrition outside of the egg. However, larvae are subject to high mortality or are incompatible with certain environmental conditions and have therefore become suppressed in many marine species and in most freshwater species (direct development).

6. Cleidoic eggs, which are more or less self-contained systems enclosed within a protective shell, have evolved in some animal groups, especially terrestrial ones. Extraembryonic membranes—yolk sac, amnion, chorion, and allantois—provide protection and maintenance for the developing embryo within the cleidoic eggs of reptiles and birds.

7. Parental care, or brooding of eggs, either outside or inside of the body of the female, is a widespread adaptation enhancing the survival of the embryo. Brooding permits reduction in the number of eggs produced.

8. The mammalian embryo is brooded within the uterus, where it arrives as a blastula (blastocyst) following fertilization in the upper end of the fallopian tube. The chorion and allantois of their reptilian ancestors have become adapted for exchange of gases, food, and waste between the embryonic and uterine blood streams. Those parts of the chorioallantois and uterine wall involved in the exchange constitute the placenta.

9. Twinning, or multiple births, in mammals results from the ovulation of more than one egg from the ovaries, from the early separation of the blastomeres in the cleavage of the egg, or from the formation of more than one embryonic center within the blastocyst.

References and Selected Readings

Attention is again called to the general references cited at the end of Chapter 4.

Arey, L. B.: *Developmental Anatomy.* 7th ed. Philadelphia, W. B. Saunders Co., 1974. A standard text on human development, emphasizing morphology.

Asdell, S. A.: *Patterns of Mammalian Reproduction.* 2nd ed. Ithaca, Comstock Publishing Co., 1964. An important source book on differences in reproduction and reproductive cycles that occur in the various kinds of mammals.

Austin, C. R., and R. V. Short (eds.): *Reproduction in Mammals.* Vols. 1–8. Cambridge University Press, 1972–1980. (Vol. 1, 2nd ed., 1982.) These eight volumes introduce a wide range of topics in mammalian reproduction.

Balinsky, B. I.: *An Introduction to Embryology.* 5th ed. Philadelphia, Saunders College Publishing, 1981. A well-written and well-illustrated account of animal development.

Beaconsfield, P., G. Birdwood, and R. Beaconsfield: The placenta. *Scientific American, 243* (Aug.):94–102, 1980. An account of the structure and physiology of the human placenta.

Epel, D.: The program of fertilization. *Scientific American, 237* (Nov.):125–138, 1977. An exploration of the cytological events of fertilization.

Karp, G., and N. J. Berrill: *Development.* 2nd ed. New York, McGraw-Hill Book Co., 1981. An excellent introductory text of animal development, with particular emphasis on the results of experimental studies.

Metz, C. B., and A. Monroy (Eds.): *Fertilization.* New York, Academic Press, 1967. Fine discussions of sperm motility and metabolism and of the process of fertilization.

Nijhout, H. F.: The color patterns of butterflies and moths. *Scientific American, 245* (Nov.):140–151, 1981. An interesting study of the developmental mechanisms controlling pattern formation.

Page, E. W., C. A. Villee, and D. B. Villee: *Human Reproduction.* 3rd ed. Philadelphia, W. B. Saunders Co., 1981. A general account of human reproduction ranging from the behavioral to the endocrine, genetic, and biochemical aspects of the subject.

Rugh, R., and L. B. Shettles: *From Conception to Birth: The Drama of Life's Beginnings.* New York, Harper & Row, 1971. A popularly written book, with superb illustrations.

Turner, C. D., and J. T. Bagnara: *General Endocrinology.* 6th ed. Philadelphia, W. B. Saunders Co., 1976. Contains an excellent chapter on the biology of sex and reproduction.

PART THREE Continuity of Animal Life

15 Heredity

ORIENTATION

That offspring tend to resemble their parents is true whether we are considering plants, animals, or humans. In this chapter we will explore the processes at the cellular level that account for this resemblance. We will reach an understanding of Mendel's laws and of how they are applied to the solution of problems in genetics. We will discuss the laws of probability and their application to genetics so that we can calculate the probability of specific genetic events. We will describe the Hardy-Weinberg principle and see how it is used in analyzing gene pools and in estimating the frequency of heterozygous genetic carriers in a population. We will reach an understanding of the genetic mechanisms by which sex is determined and by which sex-linked traits are inherited. Since there are many more genes than there are chromosomes, each chromosome must contain many genes. The genes within a given chromosome tend to be inherited together and are said to be linked. The specific associations of the genes within a chromosome change from time to time by the process of crossing over. Pairs of genes may interact in a variety of ways in producing a given trait, and we will have a look at these. The crossing of two unrelated organisms sometimes produces offspring that are stronger and better adapted for survival than either parent. We will explore the genetic basis for this phenomenon of "hybrid vigor."

History of Genetics

It must have been thousands of years ago when man first made one of the fundamental observations of heredity—that "like tends to beget like." But his curiosity as to why this is true and how it is brought about remained unsatisfied until the beginning of the present century. Mendel's careful work with peas revealed the fundamental principles of heredity, but the report of his work, published in 1866, was far ahead of its time. It is clear that his work was known to a number of the leading contemporary biologists, but in the absence of our present knowledge of chromosomes and their behavior, its significance was unappreciated.

In 1900, three different biologists, working independently—de Vries in Holland, Correns in Germany, and von Tschermak in Austria—rediscovered the phenomenon of regular, predictable ratios of the types of offspring produced by mating pure-bred parents. They then found Mendel's published report and, realizing his priority in these discoveries, gave him credit for his work by naming two of the fundamental principles of heredity **Mendel's laws.**

With the genetic and cytologic facts at hand, W. S. Sutton and C. E. McClung independently came to the conclusion (1902) that the hereditary factors are located in the chromosomes. They also pointed out that, since there are many more hereditary factors than chromosomes, there must be more than one hereditary factor per chromosome. By 1911, T. H. Morgan was able to postulate, from the regularity with which certain characters tended to be inherited together, that the hereditary factors (which he named "genes") were located in the chromosomes in linear order, "like the beads on a string."

Gregor Johann Mendel (1822–1884), an Austrian abbot, spent some eight years breeding peas in the garden of his monastery at Brünn, now part of Czechoslovakia. He succeeded in reaching an understanding of the basic principles of heredity because (1) he studied the inheritance of single contrasting characters (such as green versus yellow seed color, wrinkled versus smooth seed coat), instead of attempting to study the complete inheritance of each organism; (2) his studies were quantitative; he counted the number of each type of offspring and kept accurate records of his crosses and results; and (3) by design or by good fortune, he chose a plant, and particular characters of that plant, that gave him clear ratios. If he had worked with other plants or with certain other traits of peas, he might have been unable to get these ratios.

Mendel found that the offspring of a cross of yellow and green all had yellow seed coats; the result was the same whether the male or the female parent had been the yellow one. Thus, the character of one parent can "dominate" over that of the other, but which of the contrasting characters is dominant depends upon the specific trait involved, not upon which parent contributes it. This observation, repeated for several different strains of peas, led Mendel to the generalization—the "law of dominance"—that when two factors for the alternative expression of a character are brought together in one individual, one may be expressed completely and the other not at all. The character that appears in the first generation is said to be **dominant;** the contrasting character is said to be **recessive.**

Mendel then took the seeds produced by this first generation of the cross (called the **first filial generation,** abbreviated F_1), planted them, and had the resulting plants fertilize themselves to produce the second filial generation, the F_2. He found that both the dominant and the recessive characters appeared in this generation, and upon counting the number of each type (Table 15.1) he found that, whatever set of characters he used, the ratio of plants with the dominant character to those with the recessive character was very close to 3:1. From such experiments Mendel concluded that (1) there must be discrete unit factors that determine the inherited characters; (2) these unit factors must exist in pairs; and

TABLE 15.1 An Abstract of the Data Obtained by Mendel From His Breeding Experiments With Garden Peas

Parental Characters	First Generation	Second Generation	Ratios
Yellow seeds × green seeds	All yellow	6022 yellow : 2001 green	3.01 : 1
Round seeds × wrinkled seeds	All round	5474 round : 1850 wrinkled	2.96 : 1
Green pods × yellow pods	All green	428 green : 152 yellow	2.82 : 1
Long stems × short stems	All long	787 long : 277 short	2.84 : 1
Axial flowers × terminal flowers	All axial	651 axial : 207 terminal	3.14 : 1
Inflated pods × constricted pods	All inflated	882 inflated : 299 constricted	2.95 : 1
Red flowers × white flowers	All red	705 red : 224 white	3.15 : 1

(3) in the formation of gametes the members of these pairs separate from each other, with the result that each gamete receives only one member of the pair. The unit factor for green seed color is not affected by existing for a generation within a yellow seeded plant (e.g., the F_1 individuals). The two separate during gamete formation and, if a gamete bearing this factor for green seed coat unites with another gamete with this factor, the resulting seed has a green color. The generalization known as Mendel's first law, the **law of segregation,** may now be stated as follows: Genes exist in pairs in individuals, and in the formation of gametes each gene separates or segregates from the other member of the pair and passes into a different gamete, so that each gamete has one, and only one, of each kind of gene.

In other experiments Mendel observed the inheritance of two pairs of contrasting characters in a single cross. He mated a pure-breeding strain with round yellow seeds and one with wrinkled green seeds. The first filial generation all had round yellow seeds, but when these were self-fertilized he found in the F_2 generation all four possible combinations of seed color and shape. When he counted these he found 315 round yellow seeds, 108 round green seeds, 101 wrinkled yellow seeds, and 32 wrinkled green seeds. There is a close approximation of a 3:1 ratio for seed color (416 yellow to 140 green) and for seed shape (423 round to 133 wrinkled). Thus, the inheritance of seed color is independent of the inheritance of seed shape; neither one affects the other. When the two types of traits are considered together, it is clear that there is a ratio of 9 with two dominant traits (yellow and round):3 with one dominant and one recessive (round and green):3 with the other dominant and recessive (yellow and wrinkled):1 with the two recessive traits (green and wrinkled). Mendel's second law, the **law of independent assortment,** may now be given as follows: The distribution of each pair of genes into gametes is independent of the distribution of any other pair.

Chromosomal Basis of the Laws of Heredity

The laws of heredity follow directly from the behavior of the chromosomes in mitosis, meiosis, and fertilization. Within each chromosome are numerous hereditary factors, the **genes,** each of which controls the inheritance of one or more characteristics. Each gene is located at a particular point, called a **locus** (plural, loci), along the chromosome. Since the genes are located in the chromosomes, and each cell has two of each kind of chromosome, it follows that each cell has two of each kind of gene. The chromosomes separate in meiosis and recombine in fertilization and so, of course, do the genes within them. Genes are arranged in a linear order within the chromosomes; **homologous chromosomes** have similar genes arranged in a similar order. When chromosomes undergo synapsis during meiosis (p. 290) the homologous chromosomes become attached point by point and, presumably, gene by gene.

Studies of inheritance are possible only when there are two alternate, contrasting conditions, such as Mendel's yellow and green or round and wrinkled peas. These contrasting conditions, inherited in such a way that an individual may have one or the other but not both, were originally termed allelomorphic traits. At present the terms allele and gene are used more or less interchangeably; both refer to the hereditary factor responsible for a given trait. The term **allele** emphasizes that there are two or more alternative kinds of genes at a specific locus in a specific chromosome.

A Monohybrid Cross

Brown and black coat color are allelomorphic traits in guinea pigs. Each body cell of the guinea pig has a pair of chromosomes that contain genes for coat color; since there are two chromosomes, there are two genes per cell. A "pure" black guinea pig (one of a pedigreed strain of black guinea pigs) has two genes for black coat, one in each chromosome, and a "pure" brown guinea pig has two genes for brown coat. The brown gene controls the formation of an enzyme involved in one of the steps in the production of a brown pigment in the hair cells, whereas the black gene produces a different enzyme that alters the chemical reactions in the synthesis of pigment and results in the production of black pigment. In working genetic problems, letters are conventionally used as symbols for the genes. A pair of genes for black pigment is represented as **BB,** and a pair of genes for brown pigment by **bb.** A capital letter is used for one gene and the corresponding lower case letter is used to represent the gene for the contrasting trait, the allele.

A monohybrid cross is a cross between two individuals that differ in a single pair of genes. The mating of a "pure" brown male guinea pig with a "pure" black female guinea pig is illustrated in Figure 15.1. During meiosis in the testes of the male the two **bb** genes separate so that each sperm has only one **b** gene. In the formation of ova in the female the **BB** genes separate so that each ovum has only one **B** gene. The fertilization of this egg by a **b** sperm results in an animal with the genetic formula **Bb.** These guinea pigs contain one gene for brown coat and one for black coat. What color would

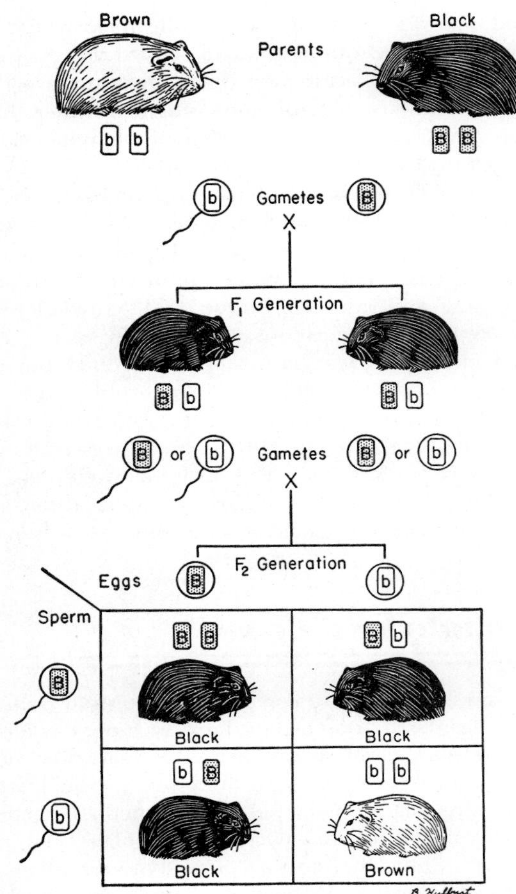

FIGURE 15.1 An example of a monohybrid cross: the mating of a brown with a black guinea pig.

you expect them to be—dark brown, gray, or perhaps spotted? In this instance they are just as black as the mother. The gene for black coat color is said to be **dominant** to the gene for brown coat color. It will produce black body color even when only one dose of the black gene is present. The brown gene is said to be **recessive** to the black one; it will produce brown color only when present in double dose (**bb**). The phenomenon of dominance supplies part of the explanation as to why an individual may resemble one of his parents more than the other despite the fact that both make equal contributions to his genetic constitution.

Homozygous and Heterozygous Organisms

An animal with two genes exactly alike, two blacks (**BB**) or two browns (**bb**), is said to be **homozygous** for the character. An organism with one dominant and one recessive gene (**Bb**) is said to be **heterozygous** or "hybrid." Thus, in the mating under consideration the black and brown parents were homozygous, **BB** and **bb**, respectively, and the offspring in the F_1 were all heterozygous, **Bb. Recessive genes** are those that will produce their effect only when homozygous; a **dominant gene** is one that will produce its effect whether it is homozygous or heterozygous.

In the process of gamete formation in these heterozygous black F_1 guinea pigs, the chromosome containing the **B** gene undergoes synapsis with, and then separates from, the homologous chromosome containing the **b** gene, so that each sperm or egg has a **B** gene or a **b** gene. It follows that sperm (or eggs) containing **B** genes and ones with **b** genes are produced in equal numbers by heterozygous **Bb** individuals. Since there are two kinds of eggs and two kinds of sperm, the mating of two of these heterozygous black guinea pigs permits four different combinations of eggs and sperm. The four possible combinations occur with equal frequency, which can be determined by algebraic multiplication:

$$(\tfrac{1}{2}\mathbf{B} + \tfrac{1}{2}\mathbf{b}) \text{ eggs} \times (\tfrac{1}{2}\mathbf{B} + \tfrac{1}{2}\mathbf{b}) \text{ sperm} = \tfrac{1}{4}\mathbf{BB} + \tfrac{1}{2}\mathbf{Bb} \times \tfrac{1}{4}\mathbf{bb} \text{ zygotes.}$$

The possible combinations of eggs and sperm may be represented in a "Punnett square" (Fig. 15.1), devised by the English geneticist R. C. Punnett. The possible types of eggs are written across the top of the Punnett square and the possible types of sperm are arranged down its left side, then the squares are filled in with the resulting zygote combinations. Three fourths of the offspring are either **BB** or **Bb** and consequently have a black coat color, and one fourth are **bb,** with a brown coat color. This 3:1 ratio is characteristically obtained in the second generation of a **monohybrid cross,** i.e., a mating of two individuals that differ in a single trait governed by a single pair of genes. The genetic mechanism responsible for the 3:1 ratios obtained by Mendel in his pea breeding experiments is now evident.

Phenotype and Genotype

The appearance of an individual with respect to a certain inherited trait is known as its **phenotype.** The organism's genetic constitution, usually expressed in symbols, is called its **genotype.** In the cross we have been considering, the phenotypic ratio of the F_2 generation is 3 black coated guinea pigs:1 brown coated guinea pig, and the genotypic ratio is 1 **BB**:2**Bb**:1**bb.**The phenotype may be any observable or detectable characteristic—shape, size, color—or the presence or absence of a specific enzyme required for the metabolism of a specific substrate. Genetic changes in enzymes are probably the basis for most, if not all, changes in color and morphological features. Guinea pigs with the genotypes **BB** and **Bb** are alike phenotypically; they both have black coats. One third of the black guinea pigs in the F_2 generation of the mating of black × brown are homozygous, **BB,** and the other two thirds are heterozygous, **Bb.** The two types can be differentiated by making a **test cross,** by mating them with homozygous brown (**bb**) guinea pigs. If all of the offspring are black, what inference would you make about the genotype of the black parent? If any of the offspring are brown, what conclusion would you draw regarding the genotype of the black parent? Can you be more certain about one of these two inferences?

This sort of testing is of great importance in the commercial breeding of animals or plants where the breeder is trying to establish a strain that will breed true for a certain characteristic. When the individuals in a breeding stock are selected on the basis of their own phenotypes, the breeding program is not maximally effective because it does not differentiate between homozygous and heterozygous individuals. The method of **progeny selection** is one in which a breeder tests the genotypes of his breeding stock by making test matings and observing the offspring. If the offspring are superior with respect to the desired trait, the parents are thereafter used regularly for breeding. Two bulls, for example, may look equally healthy and vigorous, yet one will have daughters with qualities of milk production that are distinctly superior to those of the daughters of the other bull.

The Mathematical Basis of Genetics: The Laws of Probability

The discussions of heredity earlier in this chapter were concerned with inheritance in individuals. Geneticists may also be concerned with the inheritance of characteristics in a population as a whole. All genetic events are governed by the laws of probability and, although the outcome of any single event is highly uncertain, in a large number of events the laws of probability provide a reasonable prediction of the fraction of those events that will be of one type or the other. In tossing a coin, where the probability, p, of obtaining a "heads" is one chance in two, or ½, one cannot predict the outcome of any *single* toss of the coin. But in 100 tosses about 50 will come up "heads" and 50 will come up "tails." Probabilities are usually expressed as the fraction obtained by dividing the number of "favorable" events by the total number of possible events.

When expressed in this fashion, the limits of probability are from zero to one. A probability of zero indicates that the event is impossible; there is no favorable possibility. A probability of one represents a certainty; that is, all the possible events are favorable ones. If you are engaged in a game of chance in which the favorable event involves turning up a "three" on a die, the probability of this favorable event is one in six. If a bag contains 10 red, 40 black, and 50 white marbles the chance of picking a single red marble out of the bag is 10 out of 100 or ⅒.

The Product Law

If two events are independent the probability of their coinciding is the *product* of their individual probabilities. For example, the probability of obtaining a head on the first toss of a coin is ½, and the probability of obtaining a head on the second toss of a coin (an independent event) is ½. The probability of obtaining two heads on successive tosses of the coin is the product of their probabilities, ½ × ½, or ¼. There is one chance in four of obtaining two heads on two successive tosses of a coin. This "product rule" of probability also holds for three or more independent events. For example, the probability of choosing at random an individual who is male, has blood group A, and was born in June is 0.5 × 0.4 × 0.084 = 0.0168.

The Sum Law

The probability that one or another of two mutually exclusive events will occur is the sum of their separate probabilities. For example, in rolling a die the probability that the die will come up *either* two or five is ⅙ + ⅙ = ⅓.

It is important to realize that all genetic ratios are expressions of probability, based on the laws of chance; they do not express certainties. The offspring of the mating of two individuals heterozygous for the same gene pair (i.e., Aa × Aa) will appear in the ratio of three with the dominant trait to one with the recessive trait. If the number of offspring is large enough this ratio will be very closely approximated, as Mendel's experiments demonstrated (Table 15.1). However, if the number of offspring is small, the ratio of the two types may be quite different from the expected 3:1. Why should this be? If there are only four offspring, any distribution from all four with the dominant trait to all four with the recessive trait might be found, although the latter would occur only very rarely. A better statement is that there are three chances in four (¾) that any particular offspring of two heterozygous individuals will show the dominant trait and one chance in four (¼) that it will show the recessive trait.

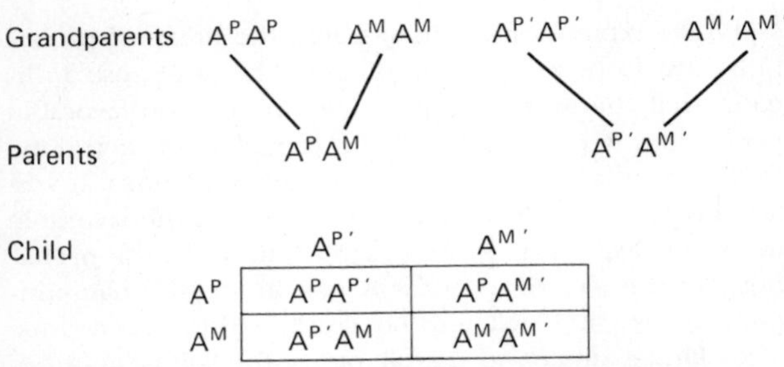

FIGURE 15.2 An example of the application of the laws of probability to genetics, illustrating both the "product law" of independent events and the "sum law" of nutually exclusive events. See text for discussion.

What is the probability that a child will inherit from his father the particular allele, A^P, that the father inherited from the child's grandfather? The father will pass on to his son one of his two alleles, A^P or A^M. The probability that the son will receive from his father the same allele that the father received from his grandfather is one in two (Fig. 15.2). The probability that the child will receive from his father the allele that the father obtained from the grandfather (A^P) and also will obtain from his mother the allele that she received from the grandmother ($A^{M'}$) is the product of their independent occurrences, $\frac{1}{2} \times \frac{1}{2} = \frac{1}{4}$. The probability that the child will obtain either the two alleles from the two grandfathers, $A^PA^{P'}$, or the two alleles from the two grandmothers, $A^MA^{M'}$, is the sum of their independent probabilities, $\frac{1}{4} + \frac{1}{4} = \frac{1}{2}$.

How often in the mating of two heterozygous black guinea pigs, **Bb,** would one expect to get a litter of four brown guinea pigs (we are assuming litter size is always four)? The probability that the first one will be brown is $\frac{1}{4}$. The probability that the second one will be brown is also $\frac{1}{4}$. The fertilization of each egg by a sperm is an independent event, and we use the product law to calculate the combined probability of two (or more) events occurring together. Thus the probability that all four of the offspring of any given mating will be brown is $\frac{1}{4} \times \frac{1}{4} \times \frac{1}{4} \times \frac{1}{4}$, or $\frac{1}{256}$.* In other words, there is one chance in 256 that all four guinea pigs will have brown coat color!

The probability that any given offspring will show the dominant trait, black coat color, is $\frac{3}{4}$. We can, in a similar fashion, calculate the probability that all four guinea pigs in the litter will be black. This is $\frac{3}{4} \times \frac{3}{4} \times \frac{3}{4} \times \frac{3}{4}$, or 81 chances in 256 that all four guinea pigs will have black coat color.

How many different ways are there of getting three black and one brown guinea pig in a given mating? A look at Figure 15.3 shows that there are four ways: The first to be born could be brown, and the next three would be black, or the second one born could be brown and the first, third, and fourth black. The other two possibilities are that the brown one would be the third or the fourth to be born. To calculate the probability that three of the four offspring will be black and one brown we must multiply the number of possible combinations by the probability of each type. There are $4 \times \frac{3}{4} \times \frac{3}{4} \times \frac{3}{4} \times \frac{1}{4}$, or $\frac{108}{256}$ chances that there will be three black guinea pigs and one brown guinea pig in a litter of four. It may be surprising at first that the 3:1 ratio is actually obtained in less than half of the total number of litters of four. However, when we add up the total *numbers* of black and brown offspring from a large number of matings, the 3:1 ratio is more and more closely approximated.

We have calculated the probability that all four would be black or that three would be black and one brown. The probability for other combinations, e.g., two black and two brown, can be calculated in a similar way (Fig. 15.3).

Probabilities can be multiplied or added just like any other fractions. In this example there are no other posibilities; the four guinea pigs must be either four black, or three black and one brown, or two black and two brown, or one black and three brown, or four brown. The sum of the five probabilities will add up to one: $\frac{81}{256} + \frac{108}{256} + \frac{54}{256} + \frac{12}{256} + \frac{1}{256} = \frac{256}{256} = 1$.

Population Genetics

The question that sometimes puzzles beginning geneticists is: Why, if brown eye genes are dominant to blue eye genes, haven't all the blue eye genes disappeared? The answer lies partly in the fact that a recessive gene, such as the one for blue eyes, is not changed by having existed for a generation next to a brown eye gene in a heterozygous individual, **Bb.** The remainder of the explanation lies in the fact that as long as there is no selection for either eye color, that is, as long as people with blue eyes are just as likely to marry and have as many children as people with brown eyes, successive generations will have the same proportion of blue- and brown-eyed people as the initial one.

A brief excursion in mathematics will show why this is true. If we consider the distribution of a single pair of genes, **A** and **a,** any member of the population will have

*For ease in multiplication, the fractions may be converted to decimals, but the fraction $\frac{1}{256}$ emphasizes that there is one chance in 256 of obtaining the desired result.

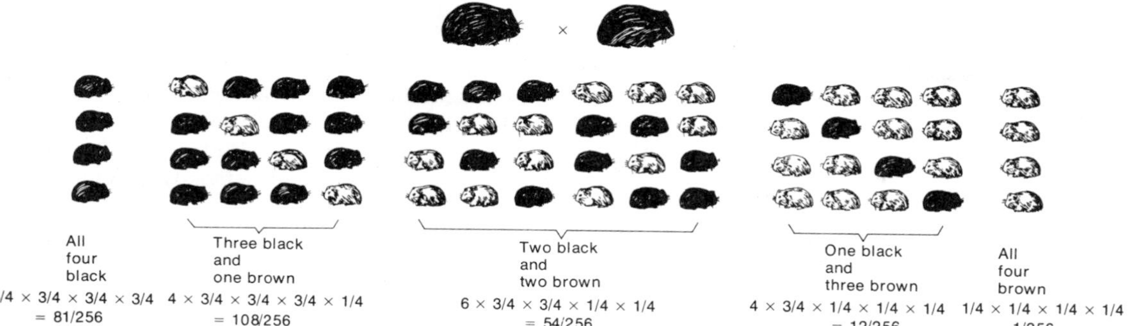

FIGURE 15.3 Possible combinations of black and brown offspring in litters of four, resulting from mating two heterozygous black guinea pigs. Each vertical column represents one possible litter.

the genotype **AA, Aa,** or **aa.** No other possibilities exist. Now let us suppose that these genotypes are present in the population in the ratio of ¼**AA**:½**Aa**:¼**aa.** If all the members of the population select their mates at random without regard to whether they are **AA, Aa,** or **aa** and if all of the types of pairs produce, on the average, comparable numbers of offspring, the succeeding generations will also have genotypes in the ratio ¼**AA**:½**Aa**:¼**aa.** This can be demonstrated by putting down all the possible types of matings, the frequency of their occurrence at random, and the kinds and proportions of offspring produced by each type of mating. When all the types of offspring are summed, it will be found that the next generation will also have genotypes in the ratio ¼**AA**:½**Aa**:¼**aa** (Table 15.2).

G. H. Hardy, an English mathematician, and G. Weinberg, a German physician, independently observed in 1908 that the frequencies of the members of a pair of allelic genes in a population are described by the expansion of a binomial equation. If we let p be the frequency of **A** genes in the population and q be the frequency of **a** genes in the population, since a gene must be either **A** or **a**, $p + q = 1$. Thus, if we know the value of either p or q, we can calculate the value of the other.

When we consider all of the matings in any given generation, a p number of **A**-containing eggs and a q number of **a**-containing eggs are fertilized by a p number of **A**-containing sperm and a q number of **a**-containing sperm: (p**A** + q**a**) × (p**A** + q**a**). The proportion of the types of offspring of all these matings is described by the algebraic product: p^2**AA** + $2pq$**Aa** + q^2**aa.**

We can also use a Punnett Square to demonstrate these results:

eggs / sperm	**A** (frequency p)	**a** (frequency q)
A (frequency p)	**AA** (frequency p^2)	**Aa** (frequency pq)
a (frequency q)	**Aa** (frequency pq)	**aa** (frequency q^2)

Sum: p^2**AA** + $2pq$**Aa** + q^2**aa.**

If p, the frequency of gene **A**, equals ½, then q, the frequency of gene **a**, equals $1 - p$ or $1 - ½$, or ½. From the formula, the frequency of genotype **AA**, p^2, equals $(½)^2$ or ¼; the frequency of **Aa**, $2pq$, equals 2 × ½ × ½, or ½; and the frequency of **aa**, q^2, equals $(½)^2$, or ¼. Any population in which the distribution of alleles **A** and **a**

TABLE 15.2 The Offspring of the Random Mating of a Population Composed of ¼AA, ½Aa, and ¼aa Individuals

Mating Male Female	Frequency	Offspring
AA × AA	1/4 × 1/4	1/16 AA
AA × Aa	1/4 × 1/2	1/16 AA + 1/16 Aa
AA × aa	1/4 × 1/4	1/16 Aa
Aa × AA	1/2 × 1/4	1/16 AA + 1/16 Aa
Aa × Aa	1/2 × 1/2	1/16 AA + 1/8 Aa + 1/16 aa
Aa × aa	1/2 × 1/4	1/16 Aa + 1/16 aa
aa × AA	1/4 × 1/4	1/16 Aa
aa × Aa	1/4 × 1/2	1/16 Aa + 1/16 aa
aa × aa	1/4 × 1/4	1/16 aa
		Sum: 4/16 AA + 8/16 Aa + 4/16 aa

*From Villee, C. A.: *Biology*. 7th ed. Philadelphia, W. B. Saunders Co., 1977.

conforms to the relation p^2**AA** + $2pq$**Aa** + q^2**aa** is in **genetic equilibrium.** The proportions of these alleles in successive generations will be the same (unless altered by selection or mutation).

Gene Pools and Genotypes

The genetic constitution of a population of a given organism is termed the **gene pool.** Stated differently, all the genes of all the individuals in a population make up the gene pool. This may be contrasted to the **genotype,** which is the genetic constitution of a single *individual.* Any individual may have only two alleles of any given gene. In contrast, the gene pool of the population may contain any number of different alleles of a specific gene. The A, B, and O blood groups (p. 175) are inherited by three alleles, $\mathbf{a}^A$, $\mathbf{a}^B$, and **a.** In the population there are three alleles, but any given individual can have no more than two of the three.

The gene pools of different populations may differ in the proportions of the specific alleles. One population may have the alleles **A** and **a** in a ratio of 0.5 to 0.5. Another population of the same species may have the two alleles in the ratio 0.7**A**:0.3**a**. If all the individuals in this second population have equal chances of surviving to adulthood and equal chances of producing gametes, then 70 percent of the sperm produced by the entire population of males would have gene **A,** and 30 percent would have gene **a.** Similarly, 70 percent of the eggs produced by the entire population of females would contain gene **A,** and 30 percent would have gene **a.** The random union of the eggs and sperm would result in offspring in the ratio of 0.49**AA** + 0.42**Aa** + 0.09**aa.** Notice that the gene pool of the offspring is identical to the gene pool of the parents!

		sperm 0.7**A**	sperm 0.3**a**
eggs	0.7**A**	0.49**AA**	0.21**Aa**
eggs	0.3**a**	0.21**Aa**	0.09**aa**

By similar calculations you can show that the next generation, and each succeeding generation, will contain an identical gene pool, 0.7**A** and 0.3**a**. The three kinds of genotypes will be present in the ratio 0.49**AA**:0.42**Aa**:0.09**aa** in succeeding generations, provided that (1) there are no mutations for **A** or **a**; (2) the three kinds of genotypic individuals have equal probabilities of surviving, mating, and producing offspring, and there is no selection of mates according to these genotypes; and (3) the population of individuals is large enough.

This principle that a population is genetically stable in succeeding generations is termed the **Hardy-Weinberg principle.** The process of evolution, stated in the simplest terms, represents departure from the Hardy-Weinberg principle of genetic stability. Evolution involves changes in the gene pool of a population that result from mutations and selection. Thus an understanding of the Hardy-Weinberg principle is of prime importance in understanding the mechanism of evolutionary change.

Estimating the Frequency of Genetic Carriers

The values of p and q, which are **gene frequencies,** cannot be measured directly. However, since the recessive phenotype can be distinguished you can determine q^2, the frequency of genotype **aa.** From this you can calculate the gene frequencies q (which is the square root of q^2) and p (which is $1 - q$). Finally you can calculate the frequencies of the other genotypes, p^2**AA** and $2pq$**Aa.**

Albinos are individuals with no pigment at all in their skin or hair. **Albinism** is an inherited trait in which the individual lacks a specific enzyme, **tyrosinase,** that catalyzes one of the reactions involved in the production of the dark pigment **melanin.** Albinism is inherited by a single pair of genes, and albinos, the homozygous recessive individuals, occur about once in 20,000 births. From this fact you can calculate that the frequency of **aa** individuals (q^2) is $\frac{1}{20,000}$. From the value of q^2 you can determine q by taking its square root. The square root of $\frac{1}{20,000}$ is about $\frac{1}{141}$. Since $p = 1 - q$ or $1 - \frac{1}{141}$, $p = \frac{140}{141}$. You now have values for both p and q and can calculate the value of $2pq$, which represents the frequency of the genetic "carrier" **Aa** individuals: $2 \times \frac{140}{141} \times \frac{1}{141} = \frac{1}{70}$. Surprising as it may seem, one person in 70 is a **carrier** of albinism, although only one person in 20,000 is homozygous and displays the trait.

As another example of the use of the Hardy-Weinberg principle, let us consider the inheritance of the blood groups M, MN, and N. In the United States, among the white population 29.16 percent have blood type M, 49.58 percent have blood type MN, and 21.26 percent have blood type N. Applying the Hardy-Weinberg principle to these data, $q^2 = 0.2126$, from which we calculate that $q = 0.46$. To estimate the value for p, we subtract 0.46 from 1 and arrive at 0.54. The square of this estimate of p, $(0.54)^2$, is equal to 0.29, and $2pq = 2 \times 0.54 \times 0.46$, or 0.49. The excellent agreement between the values observed and the theoretical values is evidence that M and N blood types are inherited by a single pair of genes, with neither gene being dominant to the other (they may be termed **codominants**) so that the heterozygote shows the two kinds of blood antigens.

If the gene pool contains three alleles, as in the ABO blood groups, their gene frequencies can be symbolized as $\mathbf{a}^A = p$, $\mathbf{a}^B = q$, and $\mathbf{a} = r$. The frequencies of the genotypes can then be calculated by expanding the trinomial $(p + q + r)^2$.

Gene	Frequency	$\mathbf{a}^A$ p	$\mathbf{a}^B$ q	$\mathbf{a}$ r
$\mathbf{a}^A$	p	$p^2\mathbf{a}^A\mathbf{a}^A$	$pq\ \mathbf{a}^A\mathbf{a}^B$	$pr\ \mathbf{a}^A\mathbf{a}$
$\mathbf{a}^B$	q	$pq\ \mathbf{a}^A\mathbf{a}^B$	$q^2\mathbf{a}^B\mathbf{a}^B$	$qr\ \mathbf{a}^B\mathbf{a}$
$\mathbf{a}$	r	$pr\ \mathbf{a}^A\mathbf{a}$	$qr\ \mathbf{a}^B\mathbf{a}$	$r^2\ \mathbf{aa}$

Sum: $p^2\mathbf{a}^A\mathbf{a}^A + 2pq\ \mathbf{a}^A\mathbf{a}^B + q^2\mathbf{a}^B\mathbf{a}^B + 2pr\ \mathbf{a}^A\mathbf{a} + 2qr\ \mathbf{a}^B\mathbf{a} + r^2\ \mathbf{aa}$.

The frequencies of the individual alleles can then be calculated from the frequencies of those phenotypes with distinctive genotypes (Table 15.3). The frequencies of the O, A, B, and AB blood groups can be quite different in different populations. The frequency of the $\mathbf{a}^B$ allele varies most strikingly, with a 20-fold difference between the frequency of the allele in the African blacks and the American Indians (Table 15.3).

Incomplete Dominance

In many different species and for a variety of traits it has been found that one gene is not completely dominant to the other. Heterozygous individuals have a phenotype that can be distinguished from that of the homozygous dominant; it may be intermediate between the phenotypes of the two parental strains. Such pairs of genes are said to be **codominant** or to show incomplete dominance. The mating of red short-horn cattle with white ones yields offspring that have an intermediate, roan coat. The mating of two roan cattle yields offspring in the ratio of 1 red:2 roan:1 white; thus the genotypic and phenotypic ratios are the same; each genotype has a recognizably different phenotype. If you saw a white cow nursing a roan calf, what would you guess was the coat color of the calf's father? Is there more than one possible answer?

This phenomenon of **incomplete dominance** is found with a number of traits in different animals and with some human characteristics. Studies of a number of human diseases inherited by recessive genes—sickle cell anemia, Mediterranean anemia, gout, epilepsy, and many others—have shown that the individuals who are heterozygous for the trait have slight but detectable differences from the homozygous normal individual.

TABLE 15.3 The Frequency of the ABO Blood Groups in Different Populations*

Population	Frequency of Blood Groups				Gene Frequencies		
	O	*A*	*B*	*AB*	$\mathbf{a}^A$	$\mathbf{a}^B$	$\mathbf{a}$
Northern Europeans	.40	.45	.10	.05	.29	.08	.63
African blacks	.42	.24	.28	.06	.13	.23	.64
American Indians	.67	.29	.03	.01	.17	.01	.82

*From Villee, C. A.: *Biology*. 7th ed. Philadelphia, W. B. Saunders Co., 1977.

Carriers of Genetic Diseases

In the human many inherited diseases are transmitted by recessive genes, and it is important to be able to distinguish the homozygous normal individual from the heterozygous individual who may superficially appear to be normal but who is a **carrier** for the trait. The mating of two carriers, two heterozygous individuals, would provide one chance in four for the appearance of a homozygous recessive individual showing the inherited disease. The human trait called **Tay-Sachs disease** is inherited by a single pair of genes, but an individual must have two such genes to show the trait. Thus the gene for Tay-Sachs disease may be said to be recessive to the normal gene. Individuals with this disease have an abnormally high content in the brain of a complex glycolipid called a ganglioside. This is a normal constituent of cell membranes, especially at nerve endings, and may play some role in synaptic transmission. Individuals with Tay-Sachs disease lack the gene that codes for one of the enzymes, hexoseaminidase A, required for the degradation of this ganglioside. Because of the lack of this degradative enzyme, the ganglioside accumulates in the brain and causes retarded mental development, paralysis, dementia, and blindness; the patients usually die as infants, and only a few survive as long as four years. This disease results from the lack of an enzyme required to break down a substance rather than the more usual situation where the enzyme needed to synthesize a substance is missing. The gene coding for this disease has a high incidence in Eastern European Jews, and in this population as many as 1 person in 30 may carry the gene for Tay-Sachs disease.

The human disease **phenylketonuria** is also inherited by a single pair of genes. The homozygous recessive individual lacks the enzyme needed to convert one amino acid, phenylalanine, to another, tyrosine. The phenylalanine accumulates in the tissues and causes retarded mental development. The heterozygote appears perfectly normal, but when given a standard amount of phenylalanine in his diet, may accumulate more in his blood and excrete more in his urine. A very simple test, adding ferric chloride to the urine in the diapers of the newborn, distinguishes the homozygous phenylketonuric, who can then be kept on a diet containing a minimal amount of phenylalanine. This is effective in minimizing the amount of mental retardation as the infant develops.

Mendel's Laws of Segregation and Independent Assortment

Mendel's first law, the law of segregation, is illustrated by the mating of the black and brown guinea pigs described previously. It may be stated as follows: Genes exist in individuals in pairs, and in the formation of gametes each gene separates or segregates from the other member of the pair and passes into a different gamete, so that each gamete has one and only one of each kind of gene.

A mating that involves individuals differing in two traits is called a **dihybrid cross.** The principles involved and the procedure of solving problems are exactly the same in monohybrid and in dihybrid or trihybrid crosses. In the latter the number of types of gametes is greater, and the number of types of zygotes is correspondingly larger. When two pairs of genes are located in different (nonhomologous) chromosomes, each pair is inherited independently of the other; that is, each pair separates during meiosis independently of the other. When a black, short-haired guinea pig (**BBSS,** short hair is dominant to long hair) and a brown, long-haired guinea pig (**bbss**) are mated, the **BBSS** individual produces gametes all of which are **BS.** The **bbss** guinea pig produces only **bs** gametes. Each gamete contains one and only one of each kind of gene. The union of **BS** gametes and **bs** gametes yields only individuals with the genotype **BbSs.** All of the offspring are heterozygous for hair color and for hair length, and all are phenotypically black and short-haired.

When two of the F_1 individuals are mated, each produces four kinds of gametes in equal numbers, **BS, Bs, bS, bs;** thus 16 combinations are possible among the zygotes (Fig. 15.4). There are 9 chances in 16 of obtaining a black, short-haired individual, 3 chances in 16 of obtaining a black, long-haired individual, 3 chances in 16 of obtaining a brown, short-haired individual, and 1 chance in 16 of obtaining a brown, long-haired individual.

Mendel's second law, the law of independent assortment, states that the members of one pair of genes separate, or segregate, from each other in meiosis independently of other pairs of genes and come to be assorted at random in the resulting gamete. The segregation of the **Bb** genes is independent of the segregation of the **Ss** genes.

In a similar fashion problems involving three pairs of genes may be solved. An individual heterozygous for three pairs of genes will yield eight types of gametes in equal numbers. The union of these 8 types of eggs and 8 types of sperm will yield 64 possible zygotes in the F_2 generation.

Deducing Genotypes

The science of genetics resembles mathematics in that when the student has mastered the few basic principles involved he can solve a wide variety of problems. These basic principles include the following: (1) Inheritance is biparental; both parents contribute to the genetic constitution of the offspring. (2) Genes are not altered by existing together in a heterozygote. (3) Each individual has two of each kind of gene, but each gamete has only one of each kind. (4) Two pairs of genes located at different chromosomes are inherited independently. (5) Gametes unite at random; there is neither attraction nor repulsion between an egg and a sperm containing identical genes.

In working genetics problems, it is helpful to use the following procedure:

1. Write down the symbols used for each pair of genes.
2. Determine the genotypes of the parents by deducing them from the phenotypes of the parents and, if necessary, from the phenotypes of the offspring.
3. Derive all of the possible types of gametes each parent would produce.
4. Prepare the appropriate Punnett Square, and write the possible types of eggs across its top and the types of sperm along its side.
5. Fill in the squares with the appropriate genotypes and read off the genotypic and phenotypic ratios of the offspring.

As an example of the method of solving a problem in genetics, let us consider the following: the length of fur in cats is an inherited trait; the gene for long hair (**l**), as in Persian cats, is recessive to the gene for short hair (**L**) of the common tabby cat. Let us suppose that a short-haired male is bred to three different females, two of which, A and C, are short-haired and one, B, long-haired (Fig. 15.5). Cat A gives birth to a short-haired kitten, but

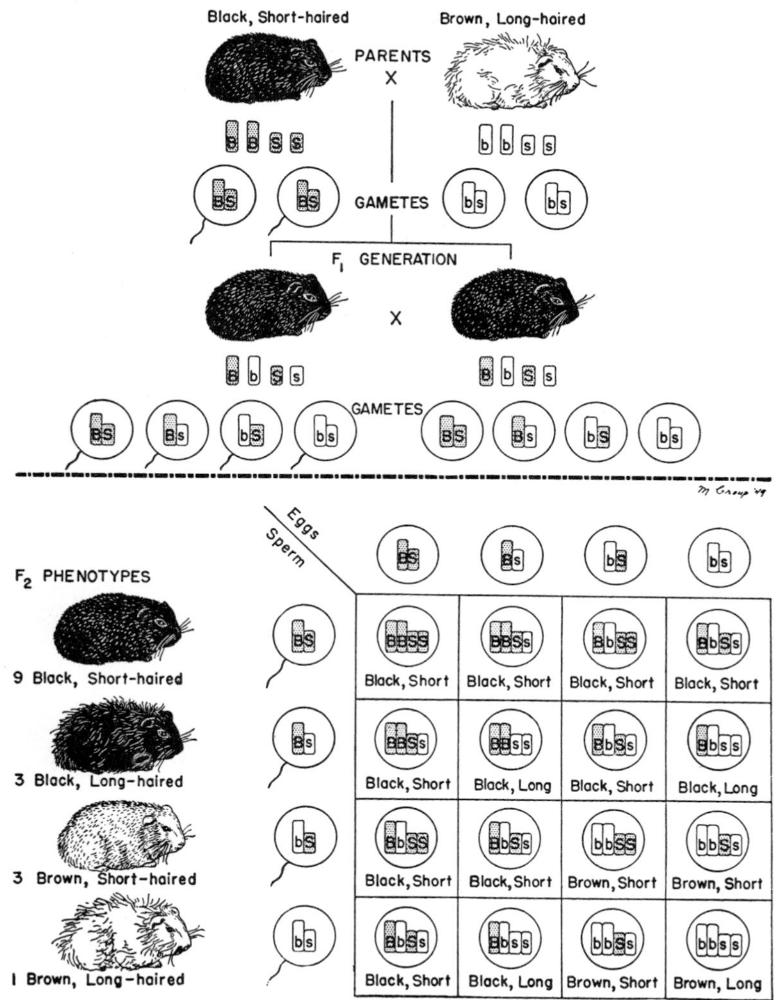

FIGURE 15.4 An example of a dihybrid cross: the mating of a black, short-haired guinea pig and a brown, long-haired one, illustrating independent assortment.

cats B and C each produce a long-haired kitten. What offspring could be expected from further mating of this male with these three females?

Since the long-haired trait is recessive we know that all the long-haired cats must be homozygous. We can deduce, then, that cat B and the kittens produced by cats B and C have the genotype **ll.** All the short-haired cats have at least one **L** gene. The fact that any of the offspring of the male cat has long hair proves that he is heterozygous, with the genotype **Ll.** The kitten produced by cat B received one **l** gene from its mother but must have received the other from its father. The fact that cat C gave birth to a long-haired kitten proves that she, too, is heterozygous and has the genotype **Ll.** It is impossible to decide, from the data at hand, whether the short-haired cat A is homozygous **LL** or heterozygous **Ll.** A test cross with a long-haired male would be helpful in deciding this. Further mating of the short-haired male with cat B would give half long-haired and half short-haired kittens, whereas further mating of the short-haired male with cat C would give three times as many short-haired kittens as long-haired ones.

The Genetic Determination of Sex

The sex of an organism is a genetically determined trait. There is an exception to the general rule that all homologous pairs of chromosomes are identical in size and shape: the so-called **sex chromosomes.** In one sex of each species of animals there is either an unpaired chromosome or an odd pair of chromosomes, the two

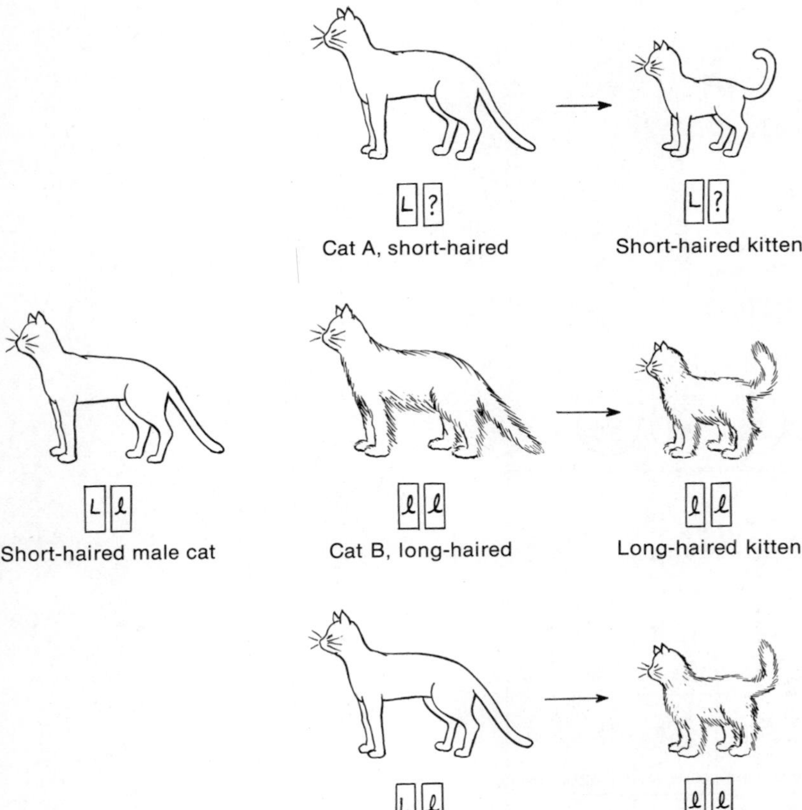

FIGURE 15.5 An example of problem-solving in genetics: deducing parental genotypes from the phenotypes of the offspring. See text for discussion.

members of which differ in size and shape. In most species the females have two identical chromosomes, called X chromosomes, and males have either a single X chromosome or one X plus a generally somewhat smaller one called the Y chromosome. In butterflies and birds the system is reversed; the male has two X chromosomes and the female one X and one Y. The Y chromosome usually contains few genes, and in most species the X and Y chromosomes are distinguished by their different size and shape. Yet in meiosis the X and Y chromosomes act like homologous chromosomes; they undergo synapsis, separate, pass to opposite poles, and become incorporated into different gametes (Fig. 15.6). Human males have 22 pairs of ordinary chromosomes, called **autosomes,** 1 X and 1 Y chromosome, whereas females have 22 pairs of autosomes and 2 X chromosomes.

The experiments of C. B. Bridges revealed that the sex of fruit flies, *Drosophila,* is determined by the ratio of the number of X chromosomes to the number of haploid sets of autosomes. Normal males have one X and two haploid sets of autosomes, a ratio of 1:2, or 0.5. Normal females have two X and two haploid sets of autosomes, a ratio of 2:2, or 1.0.

The Y Chromosome and Maleness

In humans, and perhaps in other mammals, maleness is determined in large part by the presence of the Y chromosome. An individual with an XXY constitution is a nearly normal male in external appearance, though with underdeveloped gonads (Klinefelter's syndrome). An individual with one X but no Y chromosome has the appearance of an immature female (Turner's syndrome). It is possible to determine the "nuclear sex" of an individual by careful microscopic examination of some of his cells. Individuals with two X chromosomes have a "chromatin spot" at the edge of the nucleus that is evident in cells from the skin or from the mucosal lining of the mouth (Fig. 15.7). Cells from male individuals with only one X chromosome do not show a chromatin spot.

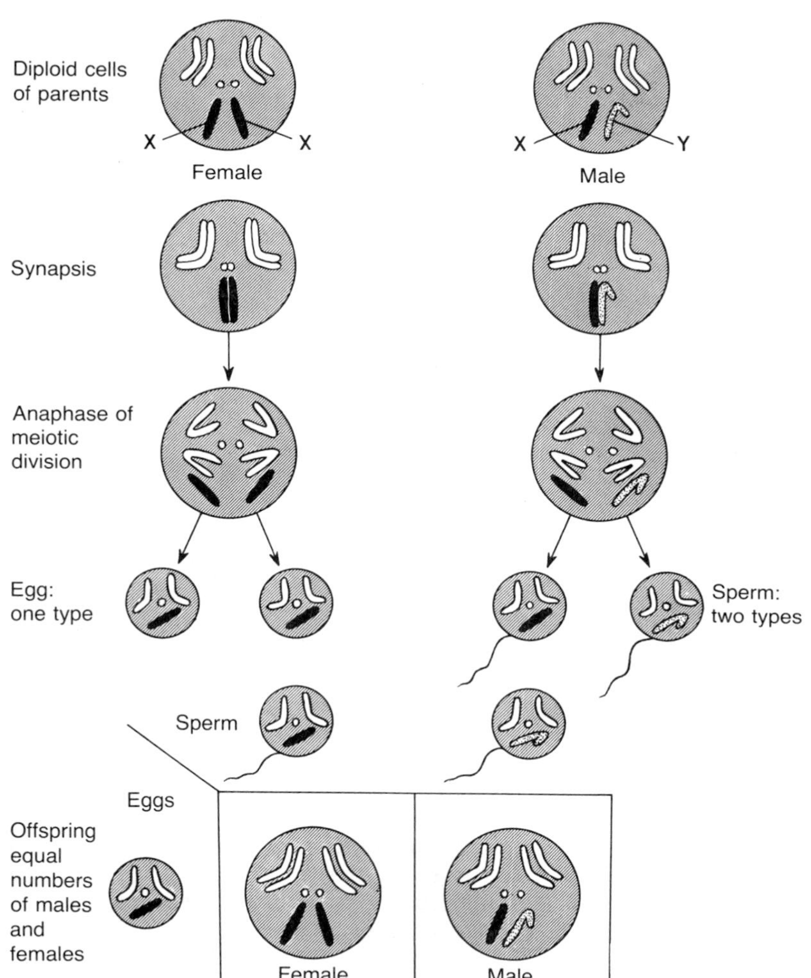

FIGURE 15.6 Diagram illustrating the transmission of the sex chromosomes of the fruit fly.

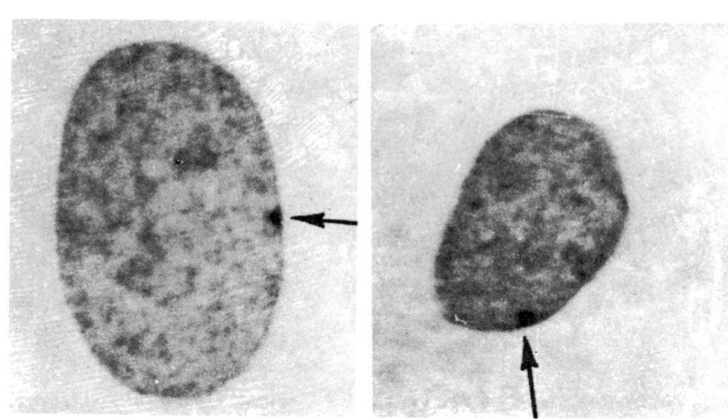

FIGURE 15.7 Sex chromatin in human fibroblasts cultured from the skin of a female. The chromatin spot at the periphery of each nucleus is indicated by the arrow. (Feulgen, ×2200. Courtesy of Dr. Ursula Mittwoch, Galton Laboratory, University College, London.)

Nuclear Sexing

The chromatin spot represents one of the two X chromosomes that becomes condensed and dark-staining. The other X chromosome, like the autosomes, is a fully extended thread not evident by light microscopy. Mary Lyon has suggested that only one of the two X chromosomes in the female is active; the condensed chromatin spot represents the inactive X chromosome. Which of the two becomes inactive in any given cell is a matter of chance; thus the cells in the body of a female are of two types, according to which X chromosome is inactive. Since the two X chromosomes may have different genetic complements, the cells may differ in the effective genes present. Mice have several genes for coat color in the X chromosome, and females heterozygous for two such genes may show patches of one coat color in the midst of areas of the other, a phenomenon termed variegation. The inactivation of one X chromosome apparently occurs early in embryonic development, and thereafter all the progeny of that cell have the same inactive X chromosome. Although one X chromosome appears to be inactive, there are marked abnormalities of development when one X chromosome is completely missing from the chromosomal complement of the cell, as in the XO condition of Turner's syndrome.

All the eggs produced by XX females have one X chromosome. Half of the sperm produced by XY males contain an X chromosome, and half contain a Y chromosome. The fertilization of an X-bearing egg by an X-bearing sperm results in an XX, or female, zygote, and the fertilization of an X-bearing egg by a Y-bearing sperm results in an XY, or male, zygote. Since there are equal numbers of X- and Y-bearing sperm, there are equal numbers of male and female offspring. In human beings, there are approximately 107 males born for every 100 females, and the ratio at conception is said to be even higher, about 114 males to 100 females. One possible explanation of the numerical discrepancy is that the Y chromosome is smaller than the X chromosome, and a sperm containing a Y chromosome, being a little lighter and perhaps able to swim a little faster than a sperm containing an X chromosome, would win the race to the egg slightly more than half of the time. Both during the period of intrauterine development and after birth, the death rate among males is slightly greater than that among females, so that by the age of 10 or 12 there are equal numbers of males and females. In later life there are more females than males in each age group.

Sex-Linked Characteristics

The X chromosome contains many genes, and the traits controlled by these genes are said to be **sex-linked,** because their inheritance is linked with the inheritance of sex. The Y chromosome contains very few genes, so that the somatic cells of an XY male contain only one of each kind of gene in the X chromosome instead of two of each kind as in XX females. A male receives his single X chromosome, and thus all of his genes for sex-linked traits, from his mother. Females receive one X from the mother and one from the father. In writing the genotype of a sex-linked trait it is customary to write that of the male with the letter for the gene in the X chromosome plus the letter Y for the Y chromosome. Thus AY would represent the genotype of a male with a dominant gene for trait "A" in his X chromosome.

The phenomenon of sex-linked traits was discovered by T. H. Morgan and C. B. Bridges in the fruit fly, *Drosophila.* These flies normally have eyes with a dark red color, but Morgan and Bridges discovered a strain with white eyes. The gene for white eye, **w,** proved to be recessive to the gene for red eye, **W,** but in certain types of crosses the male offspring had eyes of one color and the female offspring had eyes of the other color. Morgan reasoned that the peculiarities of inheritance could be explained if the genes for eye color were located in the X chromosome; later work proved the correctness of this guess. Crossing a homozygous red-eyed female with a white-eyed male (**WW** × **wY**) produces offspring all of which have red eyes (**Ww** females and **WY** males). But crossing a homozygous white-eyed female with a red-eyed male (**ww** × **WY**) yields red-eyed females and white-eyed males (**Ww** and **wY**) (Fig. 15.8).

In humans, **hemophilia** (bleeder's disease) and **color blindness** are sex-linked traits. About four men in every hundred are color-blind, but somewhat less than 1 percent of all women are color-blind. Only one gene for color-blindness produces the trait in males, but two such genes (the trait is recessive) are necessary to produce a color-blind female.

Not all the characters that differ in the two sexes are sex-linked. Some, the **sex-influenced traits,** are inherited by genes located in autosomes rather than X chromosomes, but the expression of the trait, the action of the gene that produces the phenotype, is altered by the sex of that animal, presumably by the action of one of the sex hormones. The presence or absence of horns in sheep, mahogany-and-white spotted coat versus red-and-white spotted coat in Ayrshire cattle, and pattern baldness in man are examples of such sex-influenced traits.

Linkage and Crossing Over

Each species of animal has many more pairs of genes than it has pairs of chromosomes. Obviously there must be many genes per chromosome. Humans have 23 pairs

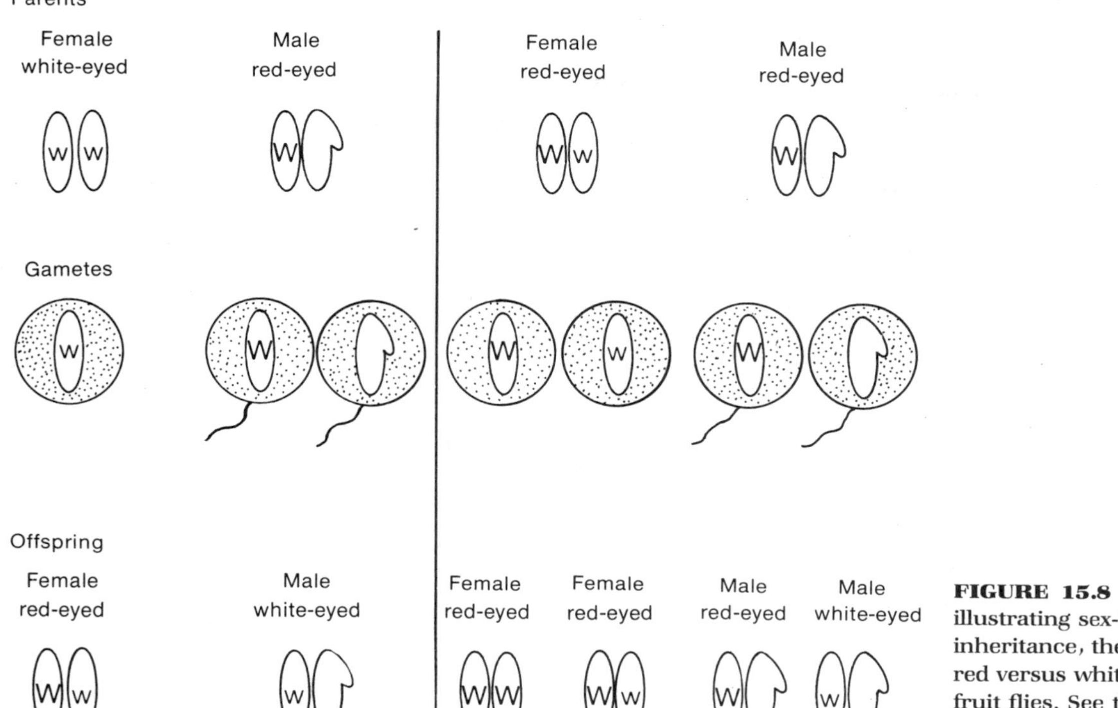

FIGURE 15.8 Diagram illustrating sex-linked inheritance, the inheritance of red versus white eye color in fruit flies. See text for discussion.

of chromosomes, but thousands of pairs of genes. The chromosomes pair and separate during meiosis as units; thus, all the genes in any given chromosome tend to be inherited together and are said to be linked. If the chromosomal units never changed, the traits would always be inherited together, and linkage would be absolute. However, during meiosis when the chromosomes are pairing and undergoing synapsis, homologous chromosomes may exchange entire segments of chromosomal material, a process called **crossing over** (Fig. 15.9). This exchanging of segments occurs at random along the length of the chromosome. Several exchanges may occur at different points along the same chromosome at a single meiotic division. It follows that the greater the distance is between any two genes in the chromosome, the greater will be the likelihood that an exchange of segments between them will occur.

In fruit flies the pair of genes **V** for normal wings and **v** for vestigial wings and the pair of genes **B** for gray body color and **b** for black body color are located in the same pair of chromosomes. They tend to be inherited together and are said to be linked. What would you predict the offspring would be like from a cross of a homozygous **VVBB** fly with a homozygous **vvbb** fly? They are all flies with gray bodies and normal wings and have the genotype **VvBb.** When one of these F_1 heterozygotes is crossed with a homozygous **vvbb** fly (Fig. 15.10), the offspring appear in a ratio that differs from that of the ordinary test cross for a dihybrid. If the two pairs of genes were not linked but were in different chromosomes, the offspring would appear in the ratio of ¼ gray-bodied, normal-winged : ¼ black-bodied, normal-winged : ¼ gray-bodied, vestigial-winged : ¼ black-bodied, vestigial-winged flies. If the genes were completely linked and no exchange of chromosomal segments occurred, then only the parental types—flies with gray bodies and normal wings and flies with black bodies and vestigial wings—would appear among the offspring, and these would be present in equal numbers (Fig. 15.10). However, there is an exchange of segments between the locus of gene **V** and the locus of gene **B.** Because of this crossing over of part of the chromosomes, some gray-bodied, vestigial-winged flies and some black-bodied, normal-winged flies (the cross-over types) appear among the offspring (Fig. 15.10). Most of the offspring, of course, resemble the parents and are gray, normal or black, vestigial. In this particular instance, crossing over occurs between these two points in this chromosome in about one cell in every five or in 20 percent of the total undergoing meiosis. In such crosses, about 40 percent of the offspring are gray flies with normal wings. Another 40 percent are black flies with ves-

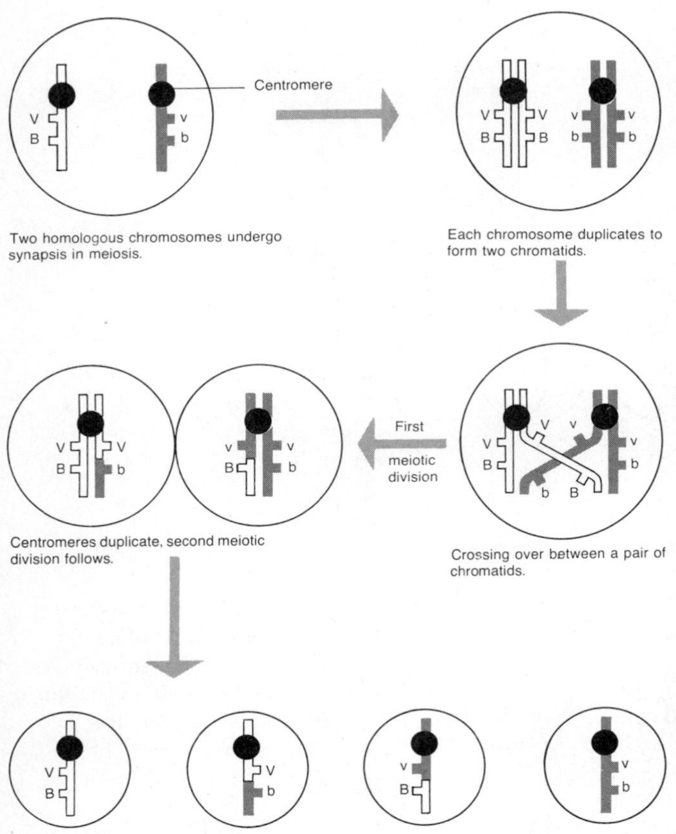
Centromere
V
B
v
b
Two homologous chromosomes undergo synapsis in meiosis.
Each chromosome duplicates to form two chromatids.
First
meiotic
division
Crossing over between a pair of chromatids.
Centromeres duplicate, second meiotic division follows.
Four haploid gametes produced; here two crossover and two noncrossover gametes.

FIGURE 15.9 Diagram illustrating crossing over, the exchange of segments between chromatids of homologous chromosomes. Crossing over permits recombination of genes (e.g., vB and Vb); the farther apart genes are located on a chromosome, the greater is the probability that crossing over between them will occur.

tigial wings. Ten percent are gray flies with vestigial wings, and 10 percent are black flies with normal wings. The distance between two genes in a chromosome is measured in "cross-over units" or "centimorgans" (named in honor of T. H. Morgan, the American geneticist who studied crossing over in the fruit fly) that represent the percentage of crossing over that occurs between them. Thus, **V** and **B** are said to be 20 centimorgans apart.

In a number of species the frequency of crossing over between specific genes has been measured. All of the experimental results are consistent with the hypothesis that genes are present in a linear order in the chromosomes. Thus, if the three genes A, B, and C occur in a single chromosome, the amount of crossing over between A and C is either the sum of, or the difference between, the amounts of crossing over between A and B and B and C. For example, if the crossing over between A and B is five units and that between B and C is three units, the crossing over between A and C will be found to be either eight units (if C lies to the right of B) or two units (if C lies between A and B) (Fig. 15.11). By putting together the results of a great many such crosses, detailed maps of the location of specific genes on specific chromosomes have been made (Fig. 15.12).

Since crossing over occurs at random, more than one cross-over may occur in a single chromosome at a given time. One can observe among the offspring only the frequency of *recombination* of characters, not the actual frequency of cross-overs. The frequency of crossing over will be somewhat larger than the observed frequency of recombination because the simultaneous occurrence of two cross-overs between two particular genes leads to the reconstitution of the original combination of genes (Fig. 15.13).

Genic Interactions

Several pairs of genes may interact to affect the production of a single trait; one pair of genes may inhibit or reverse the effect of another pair; or a given gene may

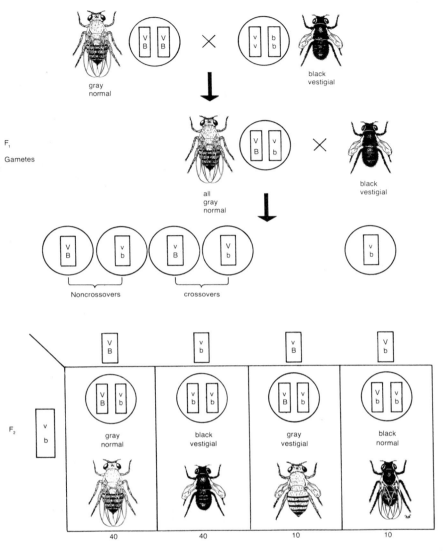

FIGURE 15.10 Diagram of a cross involving linkage and crossing over. The genes for vestigial versus normal wings and black versus gray body in fruit flies are linked; they are located in the same chromosome.

produce different effects when the environment is altered in some way. The genes are inherited as units but may interact with one another in some complex fashion to produce the trait.

Complementary Genes

Two independent pairs of genes that interact to produce a trait in such a way that neither dominant can produce its effect unless the other is present too are called **complementary genes.** The presence of at least one dominant gene from each pair produces one character; alternative conditions result from the absence of either or both dominants.

In poultry a number of genes interact to control the shape and size of the comb on the top of the head (Fig. 15.14). The gene for rose comb (**R**) is dominant to that for single comb (**r**). Another pair of genes governs the

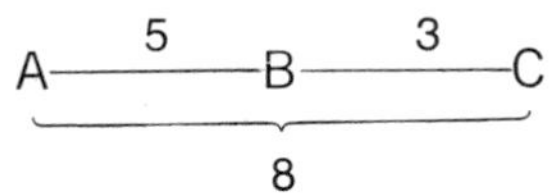

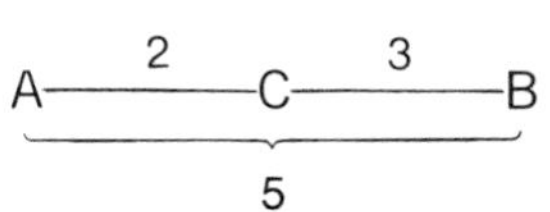

FIGURE 15.11 Diagram illustrating the principle of inferring the order of genes on a chromosome from their crossover values.

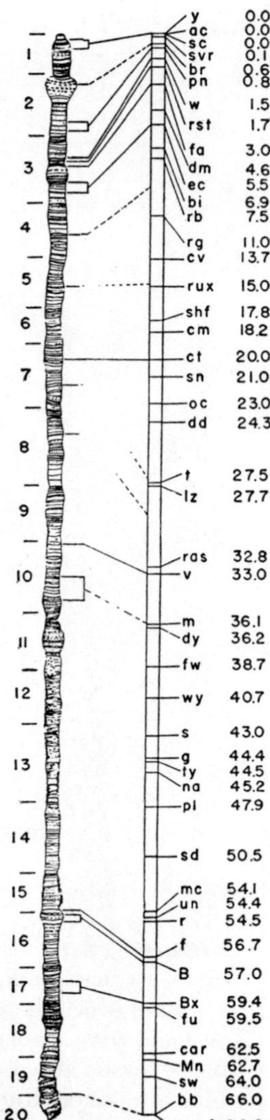

FIGURE 15.12 Diagram of the X chromosome of a fruit fly as seen in a cell of the salivary gland together with a map of the loci of the genes located in the X chromosome, with the distances between them as determined by frequency of crossing over. (From Hunter and Hunter: *College Zoology*.)

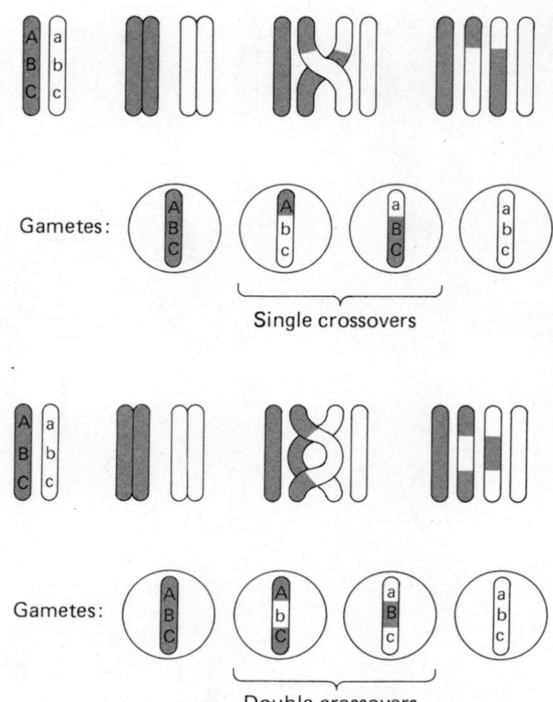

FIGURE 15.13 Diagram illustrating the results of single and double crossing over in a pair of homologous chromosomes. Note that the simultaneous occurrence of two crossovers in a given chromosome leads to a reconstitution of the original combination of genes, e.g., A and C.

inheritance of pea comb (**P**) and single comb (**p**). Thus a chicken with a single comb must have the genotype **pprr.** The genotype of a pea-combed fowl is either **PPrr** or **Pprr** and for a rose-combed chicken is either **ppRr** or **ppRR.** When a homozygous pea-combed fowl is mated with a homozygous rose-combed fowl, the offspring have neither pea or rose combs, but a completely different type called a walnut comb. The phenotype of walnut comb is produced whenever a chicken has one or more **R** gene plus one or more **P** gene. Thus **P** and **R** may be termed complementary genes. What would you predict about the types of combs among the offspring of the mating of two heterozygous walnut-combed chickens, **PpRr**?

Supplementary Genes

The term **supplementary genes** is applied to two independent pairs of genes that interact in the production of a trait in such a way that one dominant will produce its effect whether or not the second is present, but the second gene can produce its effect only in the presence of the first. In guinea pigs, in addition to the pair of genes for black versus brown coat color (**B** and **b**), the gene **C** controls the production of an enzyme that converts a colorless precursor into the pigment melanin and, hence, is required for the production of any pigment at all in the coat. The homozygous recessive, **cc,**

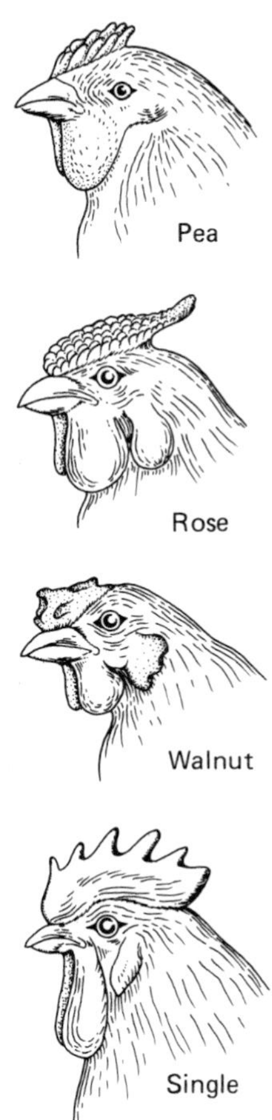

FIGURE 15.14 Diagram of some of the types of combs found in fowl.

lacks the enzyme, no melanin is produced, and the animal is a white-coated, pink-eyed **albino,** no matter what combination of **B** and **b** genes may be present. The eyes have no pigment in the iris, and the pink color results from the color of the blood in the tissues of the eye. The mating of an albino, **ccBB,** with a brown guinea pig, **CCbb,** produces offspring all of which are genotypically **CcBb** and have black-colored coats! When two of these F_1 black guinea pigs are mated, offspring appear in the F_2 in the ratio of 9 black:3 brown:4 albino. Make a Punnett Square to prove this.

The coat color of Duroc-Jersey pigs represents a slightly different type of gene interaction. Two independent pairs of genes (**R-r** and **S-s**) regulate coat color; at least one dominant of each pair must be present to give the full red-colored coat. Partial color, sandy, results when only one type of dominant is present, and an animal that is homozygous for both recessive traits (**rrss**) has a white-colored coat. The mating of two different strains of sandy-colored pigs, **RRss** × **rrSS,** yields offspring all of which are red, and the mating of two of these red F_1 individuals produces an F_2 generation in the ratio of 9 red:6 sandy:1 white (Fig. 15.15). Genes that interact in this fashion have been termed "mutually supplementary."

Polygenic Inheritance

Many human characteristics, height, body form, intelligence, and skin color, and many commercially important characters of animals and plants, such as milk production in cows, egg production in hens, the size of fruits, and the like, are not separable into distinct alternate classes and are not inherited by single pairs of genes. Nonetheless these traits are governed by genetic factors; there are several, perhaps many, different pairs of genes that affect the same characteristic. The term **polygenic inheritance** is applied to two or more independent pairs of genes that affect the same character in the same way and in an additive fashion. When two varieties that differ in some trait controlled by polygenes are crossed, the F_1 are very similar to one another and are usually intermediate in the expression of this character between the two parental types. Crossing two F_1 individuals yields a widely variable F_2 generation, with a few members resembling one grandparent, a few resembling the other grandparent, and the rest showing a range of conditions intermediate between the two.

The inheritance of human skin color was carefully investigated by C. B. Davenport in Jamaica. He concluded that the inheritance of skin color is controlled by two pairs of genes, **A-a** and **B-b,** inherited independently. The genes for dark pigmentation, **A** and **B,** are incompletely dominant, and the darkness of the skin color is proportional to the sum of the dominant genes present. Thus, a full black person has four dominant genes, **AABB,** and a white person has four recessive genes, **aabb.** The F_1 offspring of a mating of white and black are all **AaBb,** with two dominant genes and a skin color (mulatto) intermediate between white and black. The mating of two such mulattoes produces offspring with skin colors ranging from full black to white (Table 15.4). A mulatto with the genotype **AaBb** produces four kinds of eggs or sperm with respect to the genes for skin color: **AB, aB, Ab,** and **ab.** From a Punnett Square for the mating of two doubly heterozygous mulattoes (**AaBb**) it will be evident that there are 16 possible zygote combinations: 1 with four dominants (black), 4 with three dominants (dark brown skin), 6 with two domi-

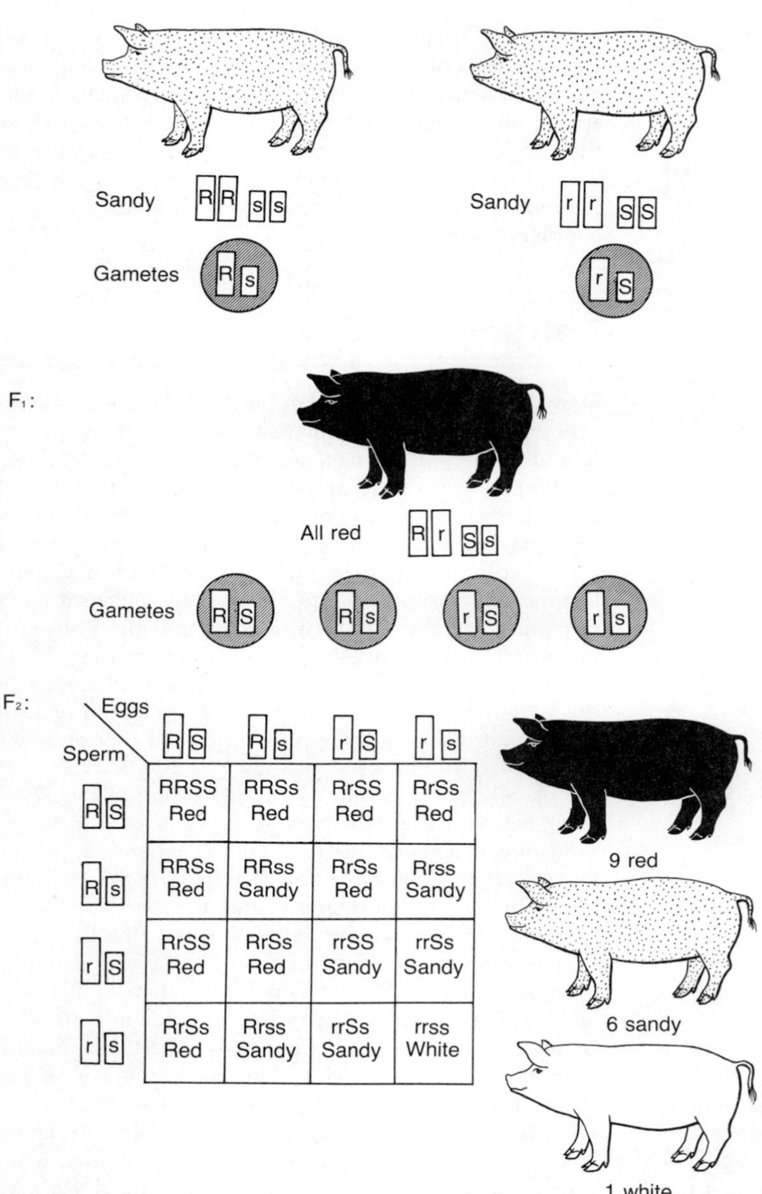

FIGURE 15.15 Diagram of the mode of inheritance of coat color in Duroc-Jersey pigs, illustrating inheritance by "mutually supplementary" genes.

nants (mulatto), 4 with one dominant (light brown skin), and 1 with no dominants (white skin). The genes **A** and **B** produce about the same amount of pigmentation, and the genotypes **AaBb, AAbb,** and **aaBB** produce the same phenotype, mulatto skin color.

This example of polygenic inheritance is fairly simple, for only two pairs of genes appear to be involved. With a larger number of pairs of genes, perhaps 10 or more, there are so many classes and the differences between them are so slight that the classes are not distinguishable; a continuous series is obtained. The inheritance of human stature is governed by a large number of pairs of genes, with shortness dominant to tallness. Since height is affected not only by these genes but also by a variety of environmental agents, there are adults of every height from perhaps 140 cm. (55 inches) up to 203 cm. (84 inches). If we measure the height of 1083 adult men selected at random and draw a graph of the number having each height, we will obtain a bell-shaped normal curve, or **curve of normal distribution** (Fig.

TABLE 15.4 Polygenic Inheritance of Skin Color in Humans

Parents	AaBb (Mulatto)	AaBb (Mulatto)
Gametes	AB Ab aB ab	AB Ab aB ab
Offspring:		
1 with 4 dominants—AABB—phenotypically black		
4 with 3 dominants—2 AaBB and 2 AABb—phenotypically "dark"		
6 with 2 dominants—4 AaBb, 1 AAbb, 1 aaBB—phenotypically mulatto		
4 with 1 dominant—2 Aabb, 2 aaBb—phenotypically "light"		
1 with no dominants—aabb—phenotypically white		

15.16). It is evident that there are few extremely tall or extremely short men, but many of intermediate height. If we measured the heights of 1000 adult American women and plotted the measurements in the same way, the data would generate a curve with a similar shape, but the mean would be less; women, on the average, are shorter than men.

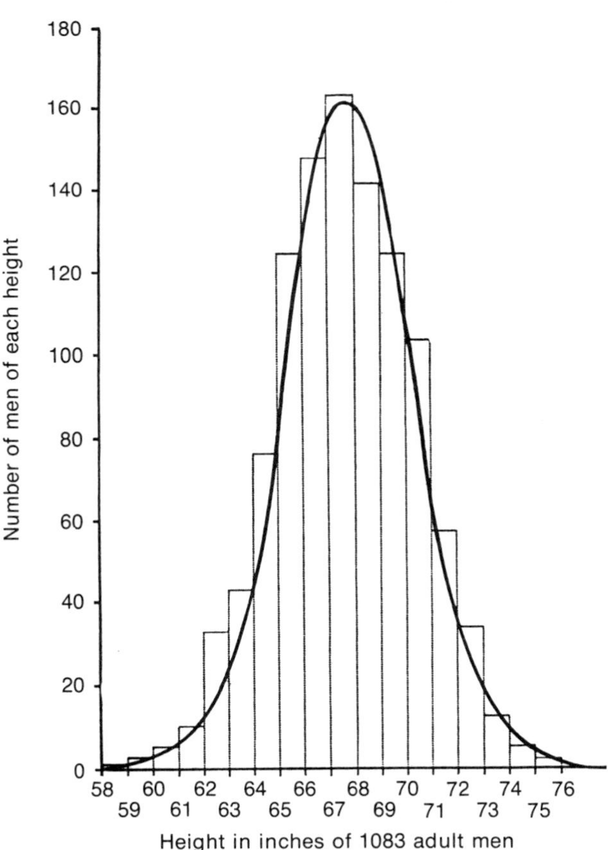

FIGURE 15.16 An example of a "normal curve," or curve of normal distribution: the heights of 1083 white men. The blocks indicate the actual number of men whose heights were within the unit range. For example, there were 163 men between 67 and 68 inches in height. The smooth curve is a normal curve based on the mean and standard deviation of the data.

All living things show comparable variations in certain of their characteristics. If one were to measure the length of 1000 shells from the same species of clam, or the weight of 1000 hens' eggs, or the amount of milk produced per year by 1000 dairy cows, or the intelligence quotient (I.Q.) of 1000 grade school children and make graphs of the number of individuals in each subclass, one would obtain a normal curve of distribution in each instance. The variation is due in part to the action of polygenes and in part to the effects of a variety of environmental agents.

When a geneticist attempts to establish a new strain of hens that will lay more eggs per year, or a strain of turkeys with more breast meat, or a strain of sheep with longer, finer wool, he selects individuals that show the desired trait in greatest amount for further breeding. There is a limit, of course, to the effectiveness of selective breeding in increasing some desirable trait or in decreasing some undesirable one. When the strain becomes homozygous for all the genetic factors involved, further selective breeding will be ineffective.

The inheritance of certain traits depends not only on a single pair of genes that determines the presence or absence of the trait but also on a number of polygenes that determine the extent of the trait. For example, the presence or absence of spots in the coat of most mammals is determined by a single pair of genes. The size and distribution of the spots, however, are determined by a series of polygenes. The term **modifying factors** has been suggested for polygenes that affect the degree of expression of another gene.

Multiple Alleles

In all the types of inheritance discussed so far, there have been only two possible alleles, one dominant and one recessive gene, which could be represented by capital and lower case letters, respectively. In addition to a dominant and a recessive gene, there may be one or more additional kinds of gene found at the same location in the chromosome that affect the same trait in an alternative fashion. The term **multiple alleles** is applied to three or more different kinds of gene, three or more alternative conditions at a single locus in the chromosome, each of which produces a distinctive phenotype. The population as a whole has three or more different alleles, but each individual has any two, and no more than two, of the possible types of alleles, and any gamete has only one. The members of an allelic series are indicated by the same letter, with suitable distinguishing superscripts.

Multiple alleles govern the inheritance of the human blood groups O, A, B, and AB, that are distinguished by the kinds of agglutinogens in the red blood cells. The importance of these in transfusions was discussed earlier (p. 175). The three alleles of the series, $\mathbf{a^A}$, $\mathbf{a^B}$, and **a**, regulate the kind of agglutinogen in the red blood cells (Table 15.5). Gene $\mathbf{a^A}$ produces agglutinogen A, gene $\mathbf{a^B}$ produces agglutinogen B, and gene **a** produces no agglutinogens. Gene **a** is recessive to the other two, but neither $\mathbf{a^A}$ nor $\mathbf{a^B}$ is dominant to the other; each produces its characteristic agglutinogen independently of the other. The agglutinins, anti-A and anti-B, are immunoglobulins that appear in the plasma of those individuals lacking the corresponding antigens on their red cells.

Since blood types are inherited and do not change in a person's lifetime, they are useful indicators of parentage. In cases of disputed parentage, genetic evidence can show only that a certain man or woman *could be* the parent of a particular child and never that he or she *is* the parent. In certain circumstances, however, the genetic evidence can definitely exclude a particular man or woman as the parent of a given child. Thus, if a child of blood group A is born to a type O woman, no man with type O or type B blood could be its father (Table 15.6). Could an AB man be the father of an O child? Could an O man be the father of an AB child? Could a type B child with a type A mother have a type A father? a type O father?

TABLE 15.5 The Inheritance of the Human Blood Groups

Blood Group	Genotypes	Agglutinogen in Red Cells	Agglutinin in Plasma
O	aa	None	a and b
A	a^Aa^A, a^Aa	A	b
B	a^Ba^B, a^Ba	B	a
AB	a^Aa^B	A and B	None

TABLE 15.6 Exclusion of Paternity Based on Blood Types

Child	Mother	Father Must Be of Type	Father Cannot Be of Type
O	O	O, A, or B	AB
O	A	O, A, or B	AB
O	B	O, A, or B	AB
A	O	A or AB	O or B
A	A	A, B, AB, or O	———
B	B	A, B, AB, or O	———
A	B	A or AB	O or B
B	A	B or AB	O or A
B	O	B or AB	O or A
AB	A	B or AB	O or A
AB	B	A or AB	O or B
AB	AB	A, B, or AB	O

Approximately 41 percent of native white Americans are type O, 45 percent are type A, 10 percent are type B, and 4 percent are type AB. The frequency of the blood groups in other races may be quite different; American Indians, for example, have a low frequency of group A and a high frequency of group B. No one blood type is characteristic of a single race; the racial differences lie in the *relative frequency* of the several blood types. Studies of the relative frequencies of the blood groups found in different races living today and in mummies and prehistoric skeletons have provided valuable evidence as to the relationships of the present races of man.

We now know of 11 additional sets of blood groups, inherited by different pairs of genes, all of which are helpful in establishing paternity. The most important of these are the Rh alleles, which determine the presence or absence of a different agglutinogen, the **Rh factor,** first found in the blood of rhesus monkeys. There are actually several alleles at the **Rh** locus, but to simplify matters we shall consider just two: **Rh,** which produces the Rh positive antigen, and the recessive **rh,** which does not produce the antigen. Genotypes **RhRh** and **Rhrh** are phenotypically Rh positive, and genotype **rhrh** is phenotypically Rh negative. An Rh negative woman married to an Rh positive man may have an Rh positive child. If some blood manages to pass across the placenta from the fetus to the mother it will stimulate the formation, in her blood, of antibodies to the Rh factor. Then, in a subsequent pregnancy, some of these Rh antibodies may pass through the placenta to the child's blood, and react with the Rh antigen in the child's red blood cells. The red cells are agglutinated and destroyed, and a serious, often fatal, anemia, called **erythroblastosis fetalis,** ensues. The introduction of fetal red cells into the maternal blood stream occurs during parturition or, to a lesser degree, during spontaneous or induced abortion. The production of maternal anti-Rh antibodies to these fetal antigens can be blocked by injecting small amounts of anti-D gamma globulin into Rh negative women just after childbirth or abortion.

Inbreeding, Outbreeding, and Hybrid Vigor

It is commonly believed that the mating of two closely related individuals—brother and sister or father and daughter—is harmful and leads to the production of monstrosities. The marriage of first cousins is forbidden by law in some states. Carefully controlled experiments with many different kinds of animals and plants have shown that there is nothing harmful in the process of inbreeding itself. It is, in fact, one of the standard procedures used by commercial breeders to improve strains of cattle, corn, cats, and cantaloupes. In all animals or plants it simply tends to make the strain homozygous.

All natural populations of individuals are heterozygous for many traits; some of the hidden recessive genes are for desirable traits, others are for undesirable ones. **Inbreeding** will simply permit these genes to become homozygous and lead to the unmasking of the good or bad traits. If a stock is good, inbreeding will improve it; but if a stock has many undesirable recessive traits, inbreeding will lead to their phenotypic expression.

The crossing of two completely unrelated strains, called **outbreeding,** is another widely used genetic maneuver. It is frequently found that the offspring of such a mating are much larger, stronger, and healthier than either parent. Much of the corn grown in the United States is a special hybrid variety developed by the United States Department of Agriculture from a mating of four different inbred strains. Each year, the seed to grow this uniformly fine hybrid corn is obtained by mating the original inbred lines. If the hybrid corn were used in mating, it would give rise to many different kinds of corn since it is heterozygous for many different traits. The mule, the hybrid offspring of the mating of a horse and donkey, is a strong, sturdy animal, better adapted for many kinds of work than either of its parents. This phenomenon of **hybrid vigor,** or **heterosis,** does not result from the act of outbreeding itself, but from the heterozygous nature of the F_1 organisms that result from outbreeding. Each of the parental strains is homozygous for certain undesirable recessive traits, but the two strains are homozygous for *different* traits, and each one has dominant genes to mask the undesirable recessive genes of the other. As a concrete example, let us suppose that there are four pairs of genes, **A, B, C,** and **D;** the capital letters represent the dominant gene for some desirable trait, and the lower case letters represent the recessive gene for its undesirable allele. If one parental strain is then **AAbbCCdd** and the other **aaBBccDD,** the offspring will all be **AaBbCcDd** and have all of the desirable and none of the undesirable traits. The actual situation in any given cross is undoubtedly much more complex and involves many pairs of genes.

Problems in Genetics

Since many students find that solving problems is helpful in learning the principles of genetics we have provided the following:

1. In peas, the gene for smooth seed coat is dominant to the one for wrinkled seeds. What would be the result of the following matings: heterozygous smooth × heterozygous smooth? Heterozygous smooth × wrinkled? Heterozygous smooth × homozygous smooth? Wrinkled × wrinkled?
2. In peas, the gene for red flowers is dominant to the one for white flowers. What would be the result of mating heterozygous red-flowered, smooth-seeded plants with white-flowered, wrinkled-seeded plants?
3. The mating of two black short-haired guinea pigs produced a litter that included some black long-haired and some white short-haired offspring. What are the genotypes of the parents and what is the probability of their having black short-haired offspring in subsequent matings?
4. Human color blindness is a sex-linked, recessive trait. What is the probability that a woman with normal vision whose husband is color-blind will have a color-blind son? a color-blind daughter? What is the probability that a woman with normal vision whose father was color-blind but whose husband has normal vision will have a color-blind son? a color-blind daughter?
5. A blue-eyed man, both of whose parents were brown-eyed, marries a brown-eyed woman whose father was blue-eyed and whose mother was brown-eyed. Their first child has blue eyes. Give the genotypes of all the individuals mentioned and give the probability that the second child will also have blue eyes.
6. Outline a breeding procedure whereby a true-breeding strain of red cattle could be established from a roan bull and a white cow.
7. Suppose you learned that shmoos may have long, oval, or round bodies and that matings of shmoos gave the following results:

long × oval gave 52 long and 48 oval

long × round gave 99 oval
oval × round gave 51 oval and 50 round
oval × oval gave 24 long, 53 oval, and 27 round

What hypothesis about the inheritance of shmoo shape would be consistent with these results?

8. A mating of an albino guinea pig and a black one gave six white (albino), three black, and three brown offspring. What are the genotypes of the parents? What kinds of offspring, and in what proportions, would result from the mating of the black parent with another animal that has exactly the same genotype as it has?

9. Mating a red Duroc-Jersey hog to sow A (white) gave pigs in the ratio of 1 red:2 sandy:1 white. Mating this same hog to sow B (sandy) gave 3 red:4 sandy:1 white. When this hog was mated to sow C (sandy) the litter had equal numbers of red and sandy piglets. Give the genotypes of the hog and the three sows.

10. The size of the egg laid by one variety of hens is determined by three pairs of genes; hens with the genotype **AABBCC** lays eggs weighing 90 g. and hens with the genotype **aabbcc** lay eggs weighing 30 g. Each dominant gene adds 10 g. to the weight of the egg. When a hen from the 90-g. strain is mated with a rooster from the 30-g. strain, the hens in the F_1 generation lay eggs weighing 60 g. If a hen and rooster from this F_1 generation are mated, what will be the weights of the eggs laid by the hens of the F_2 generation?

11. Mrs. Doe and Mrs. Roe had babies at the same hospital and at the same time. Mrs. Doe took home a girl and named her Nancy. Mrs. Roe received a boy and named him Harry. However, she was sure that she had had a girl and brought suit against the hospital. Blood tests showed that Mr. Roe was type O, Mrs. Roe was type AB, Mr. and Mrs. Doe were both type B, Nancy was type A, and Harry was type O. Had an exchange occurred?

12. A woman who is type O and Rh negative is married to a man who is type AB and Rh positive. The man's father was type AB and Rh negative. What are the genotypes of the man and woman and what blood types may occur among their offspring? Is there any danger that any of their offspring may have erythroblastosis fetalis?

Summary

1. Mendel's inferences from his experiments with the breeding of garden peas have been tested repeatedly in all kinds of organisms and found to be true. These are (a) the units of heredity exist in pairs in individuals, but gametes have only one of each kind of gene, and (b) during the formation of eggs and sperm each pair of genes separates independently of the members of other pairs of genes.

2. Each chromosome behaves genetically as though it were composed of genes arranged in a linear order; the members of a homologous pair of chromosomes have genes arranged in a similar order.

3. In a cross of two individuals that differ in a single pair of genes, a monohybrid cross, the offspring are heterozygous, having one of each kind of gene. Because one of the alleles may be dominant, the offspring usually resemble one of the parents more than the other. In the mating of two of these offspring there are three chances in four that one of the second filial generation will resemble that parent and one chance in four that it will resemble the other parent.

4. Some important basic genetic terms are locus, allele, dominant, recessive, monohybrid, homozygous, heterozygous, phenotype, and genotype.

5. All genetic ratios are expressed in terms of probabilities. Thus in mating of two individuals heterozygous for a single pair of genes, there are three chances in four that any given offspring will show the dominant trait and one chance in four that it will show the recessive trait. The probability of two independent events occurring together is the product of the probabilities of each occurring separately. Probabilities are expressed as a fraction, the number of favorable events divided by the total number of events. This can range from zero (an impossible event) to one (a certain event).

6. In situations in which the investigator cannot make specific crosses of pure strains of organisms he must rely on the laws of probability and the principles of population genetics to discover methods of inheritance by the analysis of gene pools.

7. If two events are independent the probability of their coinciding is the product of their individual probabilities. The probability that one or the other of two mutually exclusive events will occur is the sum of their separate probabilities.

8. The Hardy-Weinberg principle states that the frequency of the members of a pair of alleles in a population is described by the expansion of a binomial equation, $(p\mathbf{A} + q\mathbf{a})^2$. It follows that any population in which the distribution of alleles **A** and **a** conforms to the relation $p^2\mathbf{AA} + 2pq\mathbf{Aa} + q^2\mathbf{aa}$ is in genetic equilibrium. The proportions of these alleles will be the same in successive generations. Using the Hardy-Weinberg principle we can calculate the frequency of heterozygous genetic carriers from the frequency of the recessive phenotype.

9. The relationship between a gene and its phenotype may be quite complex. Several pairs of genes may interact to affect the appearance of a single trait. One pair of genes may inhibit, reverse, or augment the effect of another pair of genes. Interacting pairs of genes may be either complementary or supplementary.

10. One member of a pair of genes may not be completely dominant to the other, and the offspring may be intermediate between the two parental types. The genes are said to be codominant, to show incomplete dominance.

11. The sex of humans and other vertebrates is determined by special X and Y chromosomes. Females have two X chromosomes, and males have one X and one Y. Maleness is determined in large part by the presence of the Y chromosome. The male produces two kinds of sperm, some containing an X and some containing a Y chromosome. Females produce only one kind of egg, containing an X chromosome. The fertilization of an X-bearing egg by an X-bearing sperm results in an XX, female, zygote and the fertilization of an X-bearing egg by a Y-bearing sperm results in an XY, male, zygote. Males produce equal numbers of X-bearing and Y-bearing sperm.

12. In the cells of females one of the X chromosomes becomes dense and dark-staining, forming a chromatin spot evident in the nucleus of female cells. Male cells have no chromatin spot, and hence this difference permits the nuclear sexing of individuals.

13. Traits controlled by genes located in the X chromosome are called sex-linked because their inheritance is linked with that of sex. A male receives his single X chromosome and all of his genes for sex-linked traits from his mother. A female receives one X chromosome from her mother and the other X chromosome from her father. Hemophilia and color blindness are two human sex-linked traits.

14. Since each individual has many more genes than chromosomes, each chromosome must contain many genes. All the genes in any given chromosome tend to be inherited together—they are said to be linked. At the time of synapsis in meiosis, homologous chromosomes may exchange entire segments of chromosomal material, a process called crossing over. By measuring the frequency of crossing over between various genes it is possible to construct a genetic map of the chromosome.

15. Two or more independent pairs of genes may have similar and additive effects on the phenotype. This type of inheritance, polygenic inheritance, is involved in many human characteristics, such as height, body form, intelligence, and skin color, as well as a host of characteristics in plants and animals. In polygenic inheritance the F_1 generation is intermediate between the two parental types and shows little variation, whereas the F_2 generation shows wide variation between the two parental types.

16. The term multiple alleles is applied to three or more genes that can occupy a single locus on the chromosome, each of which produces a specific phenotype. Any individual has any two of the genes, and any gamete has only one. An example is the inheritance of the OAB human blood types, governed by three alleles.

17. Inbreeding, the mating of two closely related individuals, greatly increases the probability for an individual to become homozygous for recessive genes, whereas outbreeding, the mating of individuals of totally unrelated strains, increases the probability that the offspring will be heterozygous for many traits. These heterozygous individuals may be stronger and better able to survive than either parent, a phenomenon termed hybrid vigor.

References and Selected Readings

Avers, C.: *Genetics.* New York, D. Van Nostrand, 1980. A standard text of general genetics.

Burns, G. W.: *The Science of Genetics.* 5th ed. New York, Macmillan, 1983. A standard text of general genetics.

Edwards, J. H.: *Human Genetics.* New York, Methuen, 1978. A good text of the genetics of the human.

Farnsworth, M. W.: *Genetics.* New York, Harper & Row, 1978. A standard text of general genetics.

Gardner, E. J., and D. P. Snustad: *Principles of Genetics.* 6th ed. New York, John Wiley & Sons, 1981. A standard text of general genetics.

Goodenough, U.: *Genetics.* 2nd ed. New York, Holt, Rinehart & Winston, 1978. A standard text of general genetics.

Green, E. L.: *Genetics and Probability.* New York, Oxford University Press, 1980. A good presentation of a difficult subject, the importance of probability in genetics.

Thompson, J. S., and Thompson, M. W.: *Genetics in Medicine.* 3rd ed. Philadelphia, W. B. Saunders Co., 1979. A text of human genetics and its role in medical practice.

16 Molecular Aspects of Genetics

ORIENTATION

There has been a tremendous explosion of information regarding the molecular mechanisms by which genes undergo replication and transcription and the means by which the gene transcript is "read out" or translated to yield the specific protein for which the gene codes. The genetic code is composed of triplet codons composed of three adjacent nucleotides that specify the location of a single amino acid in the polypeptide chain. DNA molecules consist of two complementary chains of polynucleotides twisted together in a double helix. Genetic information flows from the genic DNA to messenger RNA and then to the specific sequence of amino acids in the peptide chain. In this chapter we will describe the many complex molecular mechanisms by which this flow of information is achieved, the nature of messenger RNA and the processes by which it is synthesized and translated, and the role of the ribosomes in the synthesis of proteins. Finally we will discuss the nature of the changes in genes, termed mutations, and the relationship between a specific gene and the enzyme or other protein produced as a result of the transcription and translation processes.

Since the rediscovery of Mendel's laws in 1900, geneticists have been attempting to determine the physical structure and chemical composition of the hereditary units, the genes. A further goal was to discover the mechanisms by which they transfer biological information from one generation to another and control the development and maintenance of the organism. During the first half of this century, much was learned about the complexity of the molecular structure of proteins. As a result, nearly all biochemists assumed that any complex biological unit with such marked specificity as the gene must be protein. There was great difficulty, however, in explaining how protein molecules could be duplicated precisely, as genes must be with each cell division.

Our present belief that DNA and RNA, rather than proteins, are the primary agents for the transfer of biological information came about gradually, culminating in 1953 in the proposal by James Watson and Francis Crick of a model of the DNA molecule that explained how it could transfer information and undergo replication. This proposal stimulated an enormous flood of research and has led to the present **"central dogma"** of biology: Genes are composed of DNA and are located within the chromosomes. Each gene contains information coded in the form of a specific sequence of purine and pyrimidine nucleotides within its DNA molecule that specifies amino acid sequences in a polypeptide chain.

The Chemistry of Chromosomes

Chromosomes contain DNA, RNA, histones, and other proteins. There are about 6×10^{-9} mg. of DNA per nucleus in the somatic cells of a wide range of vertebrates, and 3×10^{-9} mg. of DNA per nucleus in egg or sperm cells. That the amount of DNA, like the number of genes, is constant in all the cells of the body, and the amount of DNA in germ cells is only half the amount in somatic cells, is evidence that DNA is an essential part of the gene. From the amount of DNA per cell, one can estimate the number of nucleotide pairs per cell and thus the amount of genetic information present. Most plant and animal cells have between 1 to 6×10^9 nucleotide pairs. The number is smaller in bacteria (2×10^6) and yet smaller in bacteriophages (7×10^4).

The chromosomes of higher animals and plants (but not bacteria) contain five different types of histones that are bound by salt linkages to the phosphate groups of the DNA molecules. The DNA-histone complex is called **chromatin.** It now appears that the basic unit of a chromosome, termed a **nucleosome,** consists of a globular set of eight histones, two of each of types II to IV, to which is attached a segment of DNA about 200 nucleotides long. This globule is a solid structure with no holes, and the DNA is believed to be wound around the outside of the globule. The location and the role of histone I is still uncertain. Evidence indicates that all of the DNA of any given chromosome is present in a single long polynucleotide molecule. In between successive nucleosomes a portion of the DNA is free and connects one nucleosome with the adjacent one. In addition to histones, the chromosome contains a variety of acidic proteins that are believed to function in controlling gene activity by determining which genes will be expressed at any given time in any given cell.

Genetic Information Is Transmitted by DNA

The first direct evidence that DNA can transmit genetic information came from the experiments of Avery and his coworkers with the "transforming agents" isolated from pneumococci and certain other bacteria. A transforming agent isolated from strain III of pneumococcus (the bacteria causing pneumonia) will, when added to culture medium, transform strain II organisms into strain III pneumococci. This is a "permanent" transformation; the strain III organisms so produced multiply to give only strain III offspring. When transforming agents have been isolated and characterized, they have been found to be pure DNA.

DNA has also been shown to be the carrier of genetic information in bacterial viruses that consist of a "head," made up of a protein membrane around a DNA "core," and a tail composed of protein. The DNA present in the head of the virus is transferred inside the bacterial cell, but most of the tail and the head membrane remain outside. Within the bacterial cell the nucleic acid leads to the production both of additional nucleic acid cores and of protein coats, and many virus particles are formed and released when the infected bacterial cell bursts.

Bacterial genes may be transferred passively from one bacterium to another by a bacterial virus, or bacteriophage, a process termed **transduction.** As a virus particle forms within the host bacterium, it may enclose and come to contain a small segment of the bacterial genetic material along with the 'phage DNA. When the 'phage is subsequently released, it becomes attached to a new bacterium and injects the segment of bacterial chromosome from the previous host into the new host along with its own 'phage DNA. The segment of bacterial DNA may undergo crossing over with the new host's chromosome and thus incorporate genes from the previous host strain. Since only DNA is transferred in this way, this is further evidence confirming the hypothesis that genes are DNA.

The Watson-Crick Model of DNA

Highly purified DNA has been prepared from a wide variety of animals, plants, and bacteria and found in each case to consist of a sugar—**deoxyribose**—phosphoric acid, and nitrogenous bases. Four major kinds of bases are found in DNA: two purines, **adenine** and **guanine,** and two pyrimidines, **cytosine** and **thymine.** Although the amount of DNA varies in different organisms, certain consistent ratios of bases have been found. The total amount of purines equals the total amount of pyrimidines ($A + G = T + C$), the amount of adenine equals the amount of thymine ($A = T$), and the amount of guanine equals the amount of cytosine ($G = C$). The pattern of X-ray diffraction provides a number of clues about the structure of the DNA molecule. From such X-ray diffraction pictures, Franklin and Wilkins inferred that the nucleotide bases, which are flat molecules, are stacked one on top of the other like a group of saucers. These X-ray diffraction patterns showed three major periodicities in crystalline DNA, one of 0.34 nm., one of 2.0 nm., and one of 3.4 nm.

On the basis of the analytical results and the X-ray diffraction patterns, Watson and Crick proposed in 1953 a model of the DNA molecule (Fig. 16.1). It had been known that the adjacent nucleotides in DNA are joined in a chain by phosphodiester bridges. It seemed clear to Watson and Crick that the 0.34-nm. periodicity corresponded to the distance between successive nucleotides in the chain. It was a reasonable guess that the 2.0-nm. periodicity corresponded to the width of the chain. To explain the 3.4-nm. periodicity, they postulated that the chain was coiled in a helix. A **helix** is formed by winding a chain around a cylinder; in contrast, a **spiral** is formed by winding a chain around a cone. This 3.4-nm. periodicity corresponded to the distance between successive turns of the helix. Since 3.4 nm. is just 10 times the 0.34-nm. distance between the successive nucleotides, it was clear that each full turn of the helix contained 10 nucleotides. From these data Watson and Crick could calculate the density of a chain of nucleotides coiled in a helix 2 nm. wide, with turns that were 3.4 nm. long. Such a chain would have a density only half as great as the known density of DNA. Consequently, they postulated that there were two chains—a **double helix** of nucleotides—that made up the DNA molecule.

The next problem was to determine the spatial relationships between the two chains that make up the double helix. Having tried a number of arrangements with their scale models, they found that the best fit with all the data was with one in which the two nucleotide helices were wound in opposite directions (Figs. 16.1 and 16.2), with the sugar-phosphate chains on the outside of the helix and the purines and pyrimidines on the inside, held together by hydrogen bonds between bases on the opposite chains. These hydrogen bonds hold the chains together and maintain the helix. A double helix can be visualized by imagining the form that would result from taking a ladder and twisting it into a helical shape, keeping the rungs of the ladder perpendicular. The sugar and phosphate molecules of the nucleotide chains make up the railings of the ladder, and the rungs are formed by the nitrogenous bases held together by hydrogen bonds.

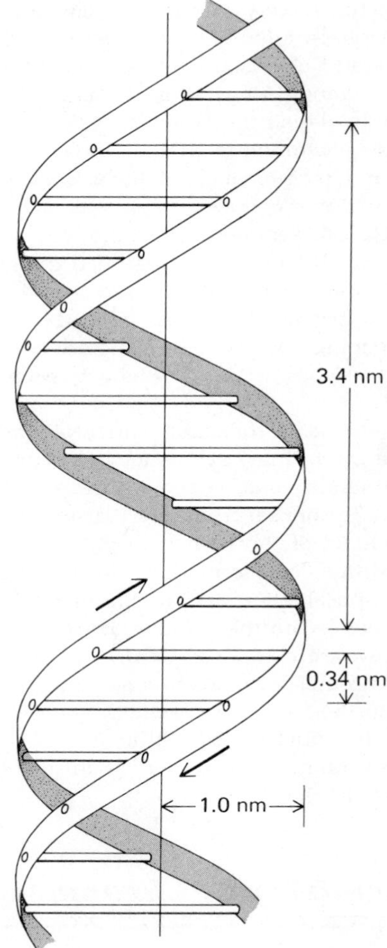

FIGURE 16.1 A schematic drawing of the double helix model of DNA, together with certain of its dimensions in nanometers. (From Anfinsen: *Molecular Basis of Evolution.* New York, John Wiley & Sons.)

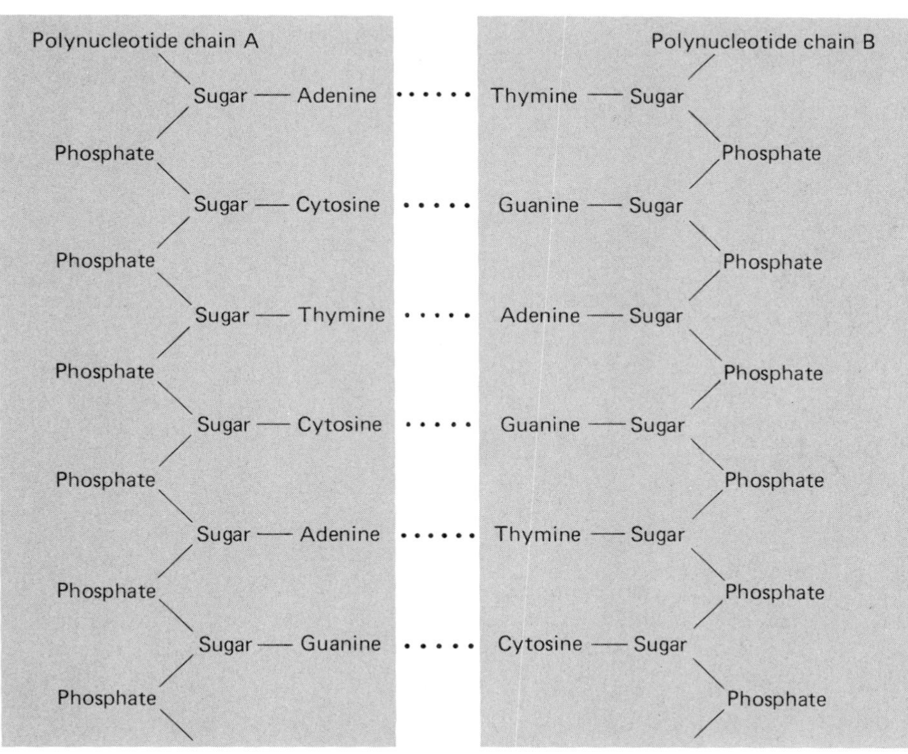

FIGURE 16.2 Schematic diagram of a portion of a DNA molecule, showing the two polynucleotide chains joined by hydrogen bonds (. . . . . .). The chains are not flat, as represented here, but are coiled around each other in helices (see Fig. 16.1).

Base Pairing

Further studies of the possible models made it clear to Watson and Crick that each crossrung must contain one purine and one pyrimidine. The space available with the 2.0-nm. periodicity would accommodate one purine and one pyrimidine but not two purines, which would be too large, and not two pyrimidines, which would not come close enough together to form proper hydrogen bonds. Further examination of the detailed molecule showed that although a combination of adenine and cytosine was the proper size to fit as a rung on the ladder, they could not be arranged in such a way that they would form proper hydrogen bonds. A similar consideration ruled out the pairing of guanine and thymine; however, adenine and thymine would form hydrogen bonds, and guanine and cytosine could form hydrogen bonds. The nature of the hydrogen bonds requires that adenine pair with thymine and that guanine pair with cytosine. This concept of **specific base pairing** explains why the amounts of adenine and thymine in any DNA molecule are always equal and the amounts of guanine and cytosine are always equal. For every adenine in one chain there will be a thymine in the other chain. Similarly for every guanine in the first chain there will be a cytosine in the second chain. Thus the two chains are complementary to each other; i.e., the sequence of nucleotides in one chain dictates the sequence of nucleotides in the other. The two strands are **antiparallel;** they extend in opposite directions and have their terminal phosphate groups at opposite ends of the double helix.

Nucleotides in DNA and RNA are linked by 3′, 5′ phosphodiester bridges (Fig. 16.3) between the adjacent ribose (in RNA) or deoxyribose (in DNA) nucleotides.

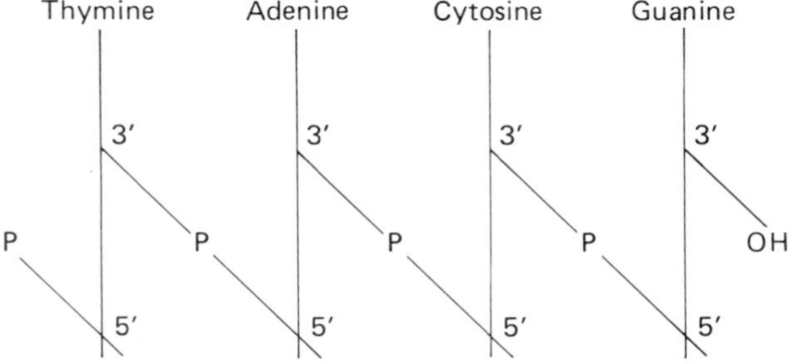

FIGURE 16.3 Diagram of the 3′, 5′ phosphodiester bridges in a sequence of nucleotides in a portion of a DNA molecule. The 5′ end of the chain, with a phosphate group attached to the 5′ position of the deoxyribose, is at the left.

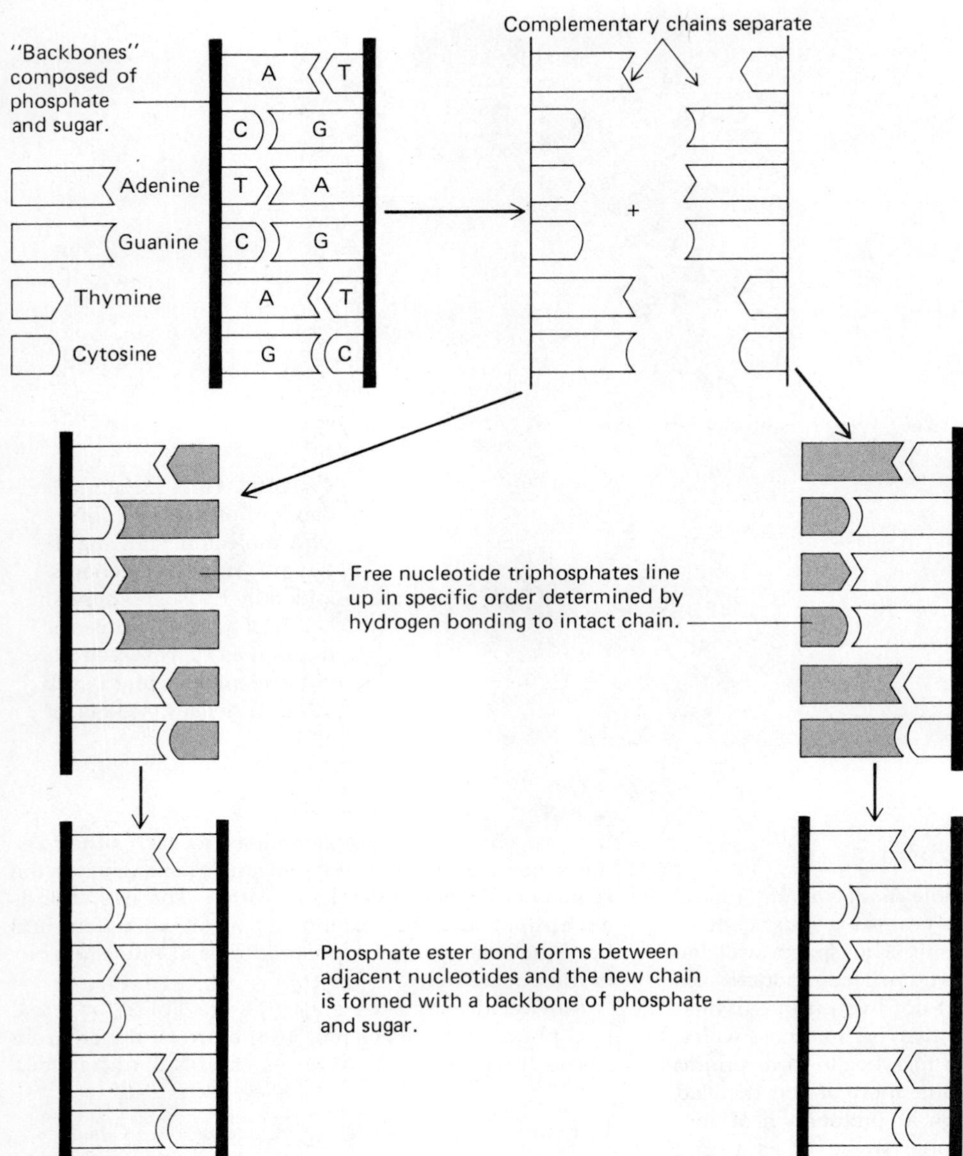

FIGURE 16.4 Diagrammatic scheme of how DNA molecules may undergo replication in the reaction catalyzed by the enzyme DNA polymerase.

Each chain has a free phosphate group attached to the 5′ position and a free OH group at the 3′ end. DNA and RNA sequences are written with the 5′ end of the chain at the left and the 3′ end at the right of the sequence.

The most distinctive properties of the genetic material are that it carries information and undergoes replication. The Watson-Crick model explains how DNA molecules may carry out these two functions. When a DNA molecule undergoes replication, the two chains separate and each one brings about the formation of a new chain that is complementary to it. Thus two new chains are established (Fig. 16.4). The nucleotides in the new chain are assembled in a specific order because each purine or pyrimidine in the original chain forms hydrogen bonds with the complementary pyrimidine or purine nucleotide triphosphate from the surrounding medium and lines them up in a complementary order. Phosphate ester bonds are formed by the reaction catalyzed by **DNA polymerase** to join the adjacent nucleotides in the chain, and a new polynucleotide chain results. The new and original chains then wind around each other, and two new DNA molecules are formed. Each chain, in other words, serves as a template or a mold against which a new partner chain is synthesized. The end result is two complete double-chain molecules, each identical to the original double-chain molecule.

A second prime function of DNA, in addition to its role in replication, is the transcription of the information contained in its specific sequence of nucleotides sometime between cell division. The product of the transcription process, messenger RNA (mRNA), then combines with ribosomes in the cytoplasm to carry out the synthesis of enzymes and other specific proteins. Thus we can visualize how each gene can lead to the production of a specific enzyme. We shall return to this subject in another section.

The Genetic Code

The Watson-Crick model of the DNA molecule implied that genetic information is transmitted by some specific sequence of its constituent nucleotides. In 1954, George Gamow, an imaginative physicist, was one of the first to suggest that the minimum coding relation between nucleotides and amino acids would be three nucleotides per amino acid. Four nucleotides taken two at a time provide for only 16 combinations ($4^2 = 16$), whereas four nucleotides taken three at a time provide for 64 combinations ($4^3 = 64$). At first glance, this would seem to provide many more code symbols than are needed, since there are only 20 different amino acids. It was believed at one time that some of these 64 combinations were simply "nonsense" codes that did not specify any amino acid. However, there is now strong evidence that all but three of the 64 combinations do, in fact, code for one or another amino acid and that as many as six different nucleotide triplets may specify the same amino acid. Because more than one triplet may code for the same amino acid, the code is said to be "degenerate."

The fundamental characteristics of the genetic code are now well established; it is a triplet, nonoverlapping code with three adjacent nucleotide bases, termed a **codon,** specifying each amino acid (Table 16.1).

The code is commaless; no "punctuation" is necessary, since the code is read out beginning at a fixed point, and the entire strand is read, three nucleotides at a time, until the readout mechanism comes to a specific "termination" code that signals the end of the message. The nature of the signal, "begin reading here," at least in bacteria, also appears to be specified by a specific sequence of bases.

TABLE 16.1 The Genetic Code: The Sequence of Nucleotides in the Triplet Codons of Messenger RNA that Specify a Given Amino Acid

First Position (5′ end)*	Second Position	Third Position (3′ end)* *U*	*C*	*A*	*G*
U	U	Phe	Phe	Leu	Leu
	C	Ser	Ser	Ser	Ser
	A	Tyr	Tyr	Terminator	Terminator
	G	Cys	Cys	Terminator	Try
C	U	Leu	Leu	Leu	Leu
	C	Pro	Pro	Pro	Pro
	A	His	His	Glu·NH_2	Glu·NH_2
	G	Arg	Arg	Arg	Arg
A	U	Ileu	Ileu	Ileu	Met
	C	Thr	Thr	Thr	Thr
	A	Asp·NH_2	Asp·NH_2	Lys	Lys
	G	Ser	Ser	Arg	Arg
G	U	Val	Val	Val	Val
	C	Ala	Ala	Ala	Ala
	A	Asp	Asp	Glu	Glu
	G	Gly	Gly	Gly	Gly

*The 5′ end has a phosphate group attached to the sugar, and the 3′ end has a deoxyribose with a free OH group.

Experimental Evidence for a Triplet Code

From a mathematical analysis of the coding problem, Crick concluded early in 1961 that three consecutive nucleotides in a strand of mRNA provide the code that determines the position of a single amino acid in a polypeptide chain and suggested the term "codon" for this unit of information. Experimental evidence to support this was quickly forthcoming from the laboratory of Nirenberg and Matthaei, who prepared artificial RNAs of known composition. When the artificial mRNA polyuridylic acid (UUUUU) was added to a system of purified enzymes for the synthesis of proteins, phenylalanine, and no other amino acid, was incorporated into protein; the polypeptide that resulted contained only phenylalanine. The inference that UUU is the code for phenylalanine was inescapable. Further similar experiments by Nirenberg and by Ochoa showed that polyadenylic acid provided the code for lysine and that polycytidylic acid coded for proline. Making mixed nucleotide polymers (such as poly AC) and using them as artificial messengers made possible the assignment of many other nucleotide combinations to specific amino acids.

These experiments did not reveal the order of the nucleotides within the triplets, but this has been inferred from other kinds of experiments. Nirenberg and Leder discovered that even when no protein synthesis is occurring, specific amino acyl transfer RNA (tRNA) molecules will be attached to ribosomes when mRNA is present. Synthetic mRNA molecules as short as trinucleotides will suffice to promote the binding of specific amino acyl tRNAs to the ribosomes. It is possible to synthesize trinucleotides of known sequence; using these, the coding assignment of all 64 possible triplets has been determined. For example, GUU, but not UGU or UUG, induces the binding of valine tRNA to ribosomes. UUG induces the binding of leucine tRNA. Since GUU and UUG code for different amino acids, it follows that the reading of the code in the mRNA strand makes sense only in one direction.

Codon–Amino Acid Specificity

Careful examination of the coding relationships (Table 16.1) shows that there is a pattern to the "degeneracy" in the code. For a number of amino acids the first two nucleotides of the codon are specific, but any of the four nucleotides may be present in the 3′ position. All of the four possible combinations will code for the same amino acid. In these, although the code may be read three nucleotides at a time, only the first two nucleotides appear to contain specific information. Only methionine and tryptophan have single triplet codes; all the other amino acids are specified by from two to as many as six different nucleotide triplets. Codons that specify the same amino acid are termed synonyms. As shown in Table 16.1, UCU and UCC are synonyms for serine and CAU and CAG are synonyms for histidine.

The genetic code is said to be "degenerate" because there is more than one code word for most amino acids. The significance of this fact is not fully understood. However, the code is not ambiguous; each codon specifies only one amino acid.

Universality of the Code. There is good reason to believe that the genetic code is universal; that is, a given codon specifies the same amino acid in the protein-synthesizing systems of all organisms from viruses to the human. RNA from the chick oviduct used as the template in a protein-synthesizing system of ribosomes and tRNA from the bacterium *Escherichia coli,* for example, results in the synthesis of **ovalbumin,** the characteristic protein of egg white.

Indirect but quite persuasive evidence for the universality of the code comes from an analysis of the amino acid sequences in protein. **Cytochrome c,** a protein constituent of the electron transmitter system, has been isolated from several organisms, and the amino acid sequence of each peptide has been determined. The differences from one species to another are very small, and the amino acid substitutions are those that would be expected if a single nucleotide were substituted for another in that codon. Similar analyses of a number of other proteins have given similar results.

Colinearity. DNA is a linear polynucleotide chain, and a protein is a linear polypeptide chain. The sequence of amino acids in the peptide chain is dictated by the order of the corresponding nucleotide bases in codons in the mRNA, and this in turn is determined by the sequence of nucleotides in one of the two polynucleotide chains of the DNA molecule. Changing the sequence of nucleotides in the DNA molecule produces a corresponding change in the sequence of amino acids in peptides. The DNA molecule and the resulting polypeptide chain are said to be **colinear.**

This concept of colinearity was implicit in the original Watson-Crick model of the DNA molecule. Direct evidence of colinearity came from analyses by Charles Yanofsky of the genetic control of the enzyme tryptophan synthetase in bacteria. **Tryptophan synthetase** is composed of four subunits, two A chains and two B chains. The A chain is a polypeptide containing 267 amino acids. Yanofsky carefully mapped the genetic location of each of a large number of mutants with altered A chains. He then collected A-chain polypeptides from each of the mutant strains and analyzed them to determine which amino acid had been changed. His analyses showed that the relative position of the changed nucleotide within a gene, as determined by genetic analysis, corresponds to the relative position of the altered amino acid in the peptide chain of the enzyme molecule, as determined by direct chemical analysis of the peptide. Each amino acid substitution could be accounted for by a change in a single nucleotide in a codon.

It is now clear that DNA does not control the production of a polypeptide by any direct interaction with the amino acid. Instead, it forms an intermediate template, an RNA molecule, that in turn directs the synthesis of the peptide chain. The DNA has been compared to a master model that is carefully preserved in the nucleus and used only to synthesize secondary working models that pass out to the ribosomal mechanism in the cytoplasm and are utilized for the actual synthesis of proteins.

The coding relationships between DNA, RNA, and protein involve (1) the **replication** of DNA to form new DNA; (2) the **transcription** of DNA to form an mRNA template; and (3) the **translation** of the code of the mRNA template into the specific sequence of amino acids in a protein. The arrows in the formula indicate the direction of transfer of genetic information:

$$\text{DNA} \xrightarrow{\text{transcription}} \text{RNA} \xrightarrow{\text{translation}} \textbf{Protein}$$

$$\text{DNA} \xrightarrow{\text{replication}} \text{DNA}$$

The Synthesis of DNA: Replication

The synthesis of DNA is catalyzed by an enzyme system, **DNA polymerase,** first isolated from the cells of the common colon bacillus, *Escherichia coli,* by Kornberg and colleagues in 1957. DNA polymerase requires the triphosphates of all four deoxyribonucleosides (abbreviated dATP, dGTP, dCTP, and dTTP) as substrates, magnesium ions, and a small amount of high molecular weight DNA polymer to serve as a primer and template for the reaction. The product of the reaction is more DNA polymer identical to that used as the primer and template and a molecule of pyrophosphate (PP_i) for each molecule of deoxyribonucleotide incorporated.

$$\left.\begin{matrix}\text{dATP}\\ \text{dGTP}\\ \text{dCTP}\\ \text{dTTP}\end{matrix}\right\}_n \xrightarrow[\text{DNA polymerase}]{\text{DNA polymer, } Mg^{++}} \text{DNA} + \text{nPP}_i$$

The DNA polymerase from *Escherichia coli* can use template DNA prepared from a wide variety of sources—bacteria, viruses, mammals, and plants—and will produce DNA with a nucleotide ratio comparable to that of the template used. Thus the sequence of nucleotides in the product is dictated by the sequence in the primer and not by the properties of the polymerase or the ratio of the substrate molecules present in the reaction mixture.

In the cells of higher organisms, the synthesis of DNA occurs only during the S phase of the cell cycle (p. 41), when chromosomes are in their extended form and are not readily visible. Thus, if an enzyme similar to the Kornberg enzyme catalyzes the synthesis of DNA *in vivo,* there must be some sort of biological signal that initiates DNA synthesis at this time and turns it off at other times. It appears that both the enzyme DNA polymerase and the substrates, dATP, dGTP, dCTP, and dTTP, are present all the time, and hence there must be some change in the DNA template that initiates DNA synthesis and then turns it off.

Semiconservative Replication

During DNA replication two strands are formed, each complementary to one of the existing DNA strands in the double-stranded helix. The double-stranded helix unwinds; one strand provides a template for one new one, and the other original strand also provides a template for a second new strand. This is called a **semiconservative** mechanism: The two original strands of DNA are retained in the product, one in each of the two daughter helices.

Strong experimental evidence supports the concept that each eukaryotic chromosome is a single DNA molecule. The DNA molecules of the fruit fly *Drosophila* have been shown to be as long as 2.1 cm., just the length of the longest chromosome. A question that immediately comes to mind is how can a eukaryotic cell manage to replicate such an enormous molecule in the allotted time during the cell cycle? The answer is that replication does not simply begin at one end of the molecule and proceed to the other. Instead replication begins at many sites along the chromosome (some 2 μm. apart on the average in the rapidly dividing fertilized egg) and proceeds in both directions from the origin at about the same rate, 1 μm. per minute.

The experiment of Meselson and Stahl (Fig. 16.5) provided evidence that DNA replication is carried out by a semiconservative mechanism, at least in bacteria. Bacteria were grown for several generations in a medium containing heavy nitrogen, ^{15}N. Thus, all of the nitrogen in the DNA of those bacteria was ^{15}N. When a sample of the DNA was isolated and centrifuged in a cesium chloride density gradient, the DNA collected at a level that

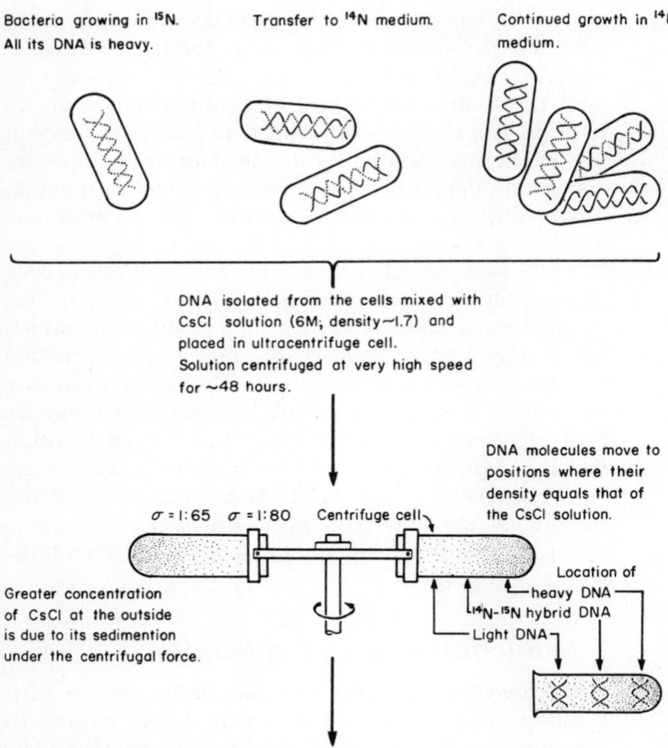
Bacteria growing in ¹⁵N.
All its DNA is heavy.
Transfer to ¹⁴N medium.
Continued growth in ¹⁴N medium.
DNA isolated from the cells mixed with CsCl solution (6M; density~1.7) and placed in ultracentrifuge cell.
Solution centrifuged at very high speed for ~48 hours.
DNA molecules move to positions where their density equals that of the CsCl solution.
σ = 1:65 σ = 1:80 Centrifuge cell
Location of heavy DNA
¹⁴N-¹⁵N hybrid DNA
Light DNA
Greater concentration of CsCl at the outside is due to its sedimention under the centrifugal force.
The location of DNA molecules within the centrifuge cell can be determined by ultraviolet optics. DNA solutions absorb strongly at 2600 A.

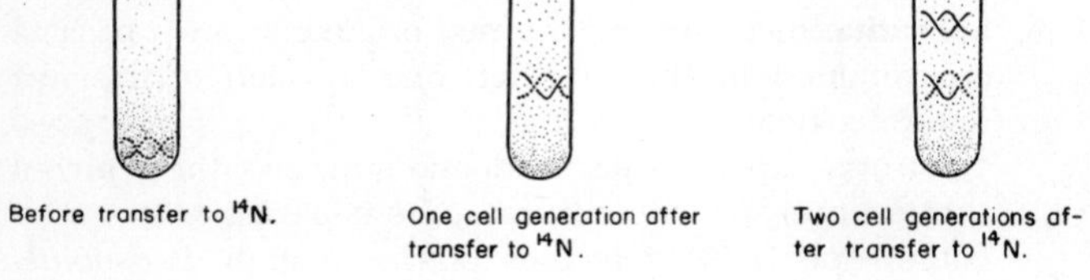
Before transfer to ¹⁴N.
One cell generation after transfer to ¹⁴N.
Two cell generations after transfer to ¹⁴N.

FIGURE 16.5 Diagram of the experiment of Meselson and Stahl which indicated that DNA is replicated by a semiconservative mechanism: The two original strands of DNA are retained in the product, one in each daughter helix.

reflected the presence of the heavy nitrogen in the DNA molecules.

The bacteria were then removed from the ^{15}N medium, placed in a medium containing ordinary nitrogen, ^{14}N, and allowed to divide once in this medium. When some of the DNA from this generation was isolated and centrifuged, all the DNA was lighter and had a density corresponding to its being half labeled with ^{15}N and half labeled with ^{14}N. If the Watson-Crick theory is correct and replication is semiconservative, this result would be expected because one strand of the double-stranded DNA in each organism is ^{15}N-labeled and the other is ^{14}N-labeled.

When the organisms were allowed to divide again in ^{14}N medium, each molecule of progeny DNA again received one parental strand and made one new strand containing ^{14}N. As a result, some double-stranded DNA containing only ^{14}N was formed and appeared as a light DNA on centrifugation. The parental ^{15}N-containing strands made complementary strands containing ^{14}N and appeared on centrifugation with a density characteristic of the half ^{15}N, half ^{14}N double-stranded state. Thus the original parental strands of DNA are not dispersed or split apart during the replication process but are conserved and passed on to the next generation of cells. Each strand of the parental double helix is conserved in a different daughter cell; hence the process is termed semiconservative.

Transcription of the Code: The Synthesis of RNA

Unlike DNA, molecules of RNA do not usually have complementary base ratios, and the inference is that RNA is not a double helix like DNA but is single-stranded. RNA contains ribose instead of deoxyribose and uracil rather than thymine.

Three kinds of RNA molecules are required for protein synthesis: **messenger RNA (mRNA),** which transmits genetic information from the DNA molecule in the nucleus to the cytoplasm; **ribosomal RNA (rRNA),** which makes up a large portion of the cytoplasmic particles called **ribosomes** on which protein synthesis occurs; and **transfer RNA (tRNA),** which acts as an adaptor to bring amino acids into line in the growing polypeptide chain in the appropriate place.

Messenger RNA is synthesized by a **DNA-dependent RNA polymerase.** It requires DNA as a template and uses as substrate the four triphosphates of the ribonucleosides commonly found in RNA. The products are RNA and inorganic pyrophospate. Transfer RNA and rRNA are also produced in the nucleus by DNA-dependent, RNA-synthesizing systems and are transcribed from complementary deoxynucleotide sequences in DNA.

The RNA that is produced by the use of these carefully defined templates is exactly that predicted by the kinds of base pairing permitted in the Watson-Crick model. Although the DNA template is double-stranded and contains two different but complementary template sequences (which would have quite different genetic information), it appears that only one DNA strand is selected for transcription and only one kind of mRNA is produced. The molecular basis for this distinction between the two strands is unknown.

Initiation and Termination of Transcription

The transcription process begins at specific sites, termed promoters, on the DNA. The recognition signal at the promoter, which brings about the binding of RNA polymerase so that transcription can begin, appears to consist of a sequence of seven base pairs. One of the subunits of the polymerase, the sigma subunit, activates the polymerase to recognize the promoter sites and also decreases the binding of the polymerase to other regions of the DNA. RNA chains are synthesized beginning at the 5′ end and proceeding in the 5′→3′ direction. The 5′ end of a new RNA chain always starts with either pppG or pppA; that is, it has a triphosphate group at its 5′ terminus and a free OH group at its 3′ terminus.

The termination of transcription, like its initiation, is under the precise control of specific sequences of nucleotides in the DNA. Thus the DNA template contains nucleotide sequences that are interpreted by RNA polymerase as "stop" signals. A specific rho protein assists the polymerase in recognizing certain base pairs as the signal for the termination of transcription.

Electron microscopy has established that most cells contain an extensive system of tubules with thin membranes termed the endoplasmic reticulum. Associated with the endoplasmic reticulum or floating freely in the cytoplasm are small particles termed ribosomes composed of RNA and protein. The RNA of ribosomes appears to consist of two components with molecular weights of about 600,000 and 1,300,000.

Transfer RNA

Molecules of tRNA are considerably smaller than the molecules of mRNA or rRNA. Each functions as a specific adaptor in protein synthesis, binding to and identifying one specific amino acid; i.e., each of the 20 amino acids is attached to one or more specific kinds of tRNA.

Transfer RNAs are polynucleotide chains of some 70 nucleotides. Each of the several kinds of tRNA has an identical sequence of nucleotides, CCA, at the 3′ end to which the amino acid is attached. In addition each has guanylic acid at the opposite, 5′, end of the nucleotide chain. The chain is doubled back on itself, forming three or more loops of unpaired nucleotides; the folding is stabilized by hydrogen bonds between complementary bases in the intervening portions of the chain to form a cloverleaf configuration (Fig. 16.6). The loop nearest the amino acid acceptor (CCA) has seven nucleotides. The seven-membered middle loop contains an anticodon, a nucleotide triplet complementary to the codon in mRNA that specifies the amino acid attached to the 3′ end. The remaining loop near the 5′ end is the largest and contains 8 to 12 nucleotides. The folding results in a constant distance between the anticodon and the amino acid in all of the tRNAs examined so far.

Reverse Transcription

The central dogma states that biological information flows from DNA to RNA to protein. An important exception to this rule was discovered by Howard Temin in 1964 when he found that infection with certain RNA tumor viruses, such as the Rous sarcoma virus, is blocked

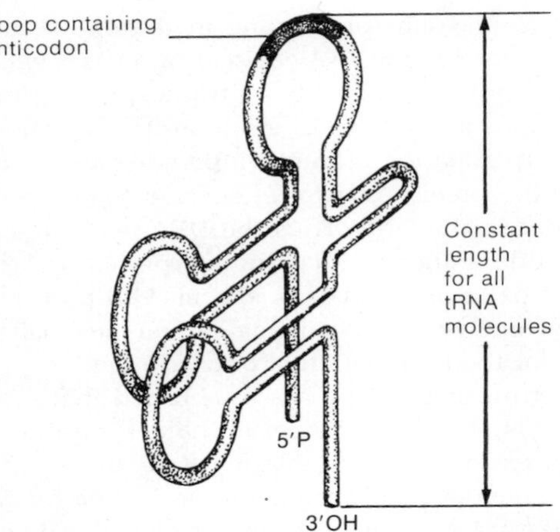

FIGURE 16.6 A diagram of the three-dimensional cloverleaf structure of transfer RNA. One loop contains the triplet anticodon which forms specific base pairs with the mRNA codon. The amino acid is attached to the terminal ribose at the 3′ OH end, which has the sequence CCA of nucleotides. Each transfer RNA also has guanylic acid, G, at the 5′ end (P). The pattern of folding permits a constant distance between anticodon and amino acid in all transfer RNAs examined.

by inhibitors of DNA synthesis and of DNA transcription. This suggested that DNA synthesis and transcription are required for the multiplication of RNA tumor viruses and that information flows in the reverse direction, that is, from RNA to DNA. Temin proposed that a DNA provirus is formed as an intermediate in the replication of these tumor viruses and in their cancer-producing effect. Temin's hypothesis required a new kind of enzyme—one that would synthesize DNA using RNA as a template. Just such an enzyme was discovered by Temin and by David Baltimore in 1970—discoveries for which they received the Nobel Prize in 1975. This RNA-directed DNA polymerase (also called reverse transcriptase) has been found to be present in all RNA tumor viruses.

The question of whether viruses cause human cancer is being investigated intensively at present. Suggestive evidence that viruses probably do play a role in certain human cancers is now at hand. First, human leukemias, sarcomas, lymphomas, and breast adenocarcinomas have been shown to contain large RNA molecules similar to those of tumor viruses that cause cancer in mice. Second, these human cancer cells contain particles with reverse transcriptase activity. Third, the DNAs of some human cancer cells have virus-like sequences of nucleotides in their DNA, sequences not found in the DNA of comparable normal cells.

The Processing of RNA

In prokaryotes, such as bacteria, the RNA molecules formed by transcription are cleaved and chemically modified in several ways. Transfer RNA and rRNA molecules are formed by the splitting out of segments of the RNA chain, and these are chemically modified by the addition of methyl groups to certain bases. Certain bases are modified to form unusual bases, such as pseudouridylate and ribothymidylate. The mRNA molecules of prokaryotes undergo very little, if any, modification. Some of them begin the translation process, the formation of a polypeptide chain, at one end of the mRNA while the other end is still being transcribed.

In the eukaryotes the rRNA and tRNA precursors undergo extensive cleavage and chemical modification, as they do in prokaryotes. The mRNA of eukaryotes, unlike that of prokaryotes, undergoes extensive cleavage and splicing of sequences from the large initial product of translation, termed heterogeneous nuclear RNA, hnRNA. In eukaryotes, the primary transcripts, hnRNA, are not used directly as mRNAs but undergo extensive processing before being transported out of the nucleus to the ribosomes in the cytoplasm.

Eukaryotic mRNAs have a "cap" at the 5′ end that consists of 7-methyl guanosine joined to the mRNA by a 5′-5′ pyrophosphate bond. The cap protects the mRNA from attack by enzymes, such as phosphatases and nucleases. The 3′ end of the eukaryotic mRNAs has a large tail of 150 to 200 adenylic acids—a "poly A tail." This is added by a special enzyme, a poly A polymerase. The poly A tail also enhances the stability of the mRNA.

A surprising finding was that most eukaryotic genes contain sequences that may be quite long but that are not present in the corresponding mRNA. In the processing of hnRNA to form mRNA, the initial transcript is cleaved, certain sequences called intervening sequences, or introns, are excised, and the split ends are spliced together enzymatically. The sequences that are retained in the mRNA are called exons. Thus the hnRNA consists of alternating sequences of introns and exons. Introns may be more than 1000 base pairs in length. The β-globin gene has 2 introns, the ovalbumin gene has 7, and the gene for conalbumin has 16 introns (Fig. 16.7). The gene for ovalbumin contains about 7700 base pairs, but its mRNA is only 1895 base pairs long. The enzymes that cleave the hnRNA, remove the introns, and splice the exons back together are very precise. If they were not, and slipped by a single nucleotide, the reading frame would be changed and an entirely different protein would be synthesized. There are several hypotheses regarding the origin and role of these exon and intron sequences in eukaryotic genes, but none has been fully established. They may play some role in controlling the flow of information from nucleus to cytoplasm and in controlling cellular differentiation.

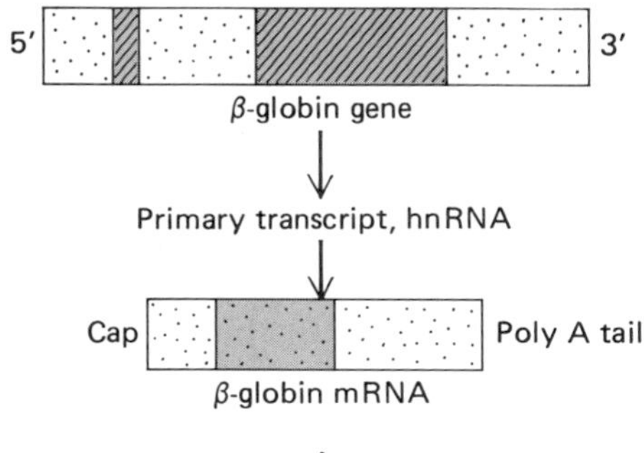

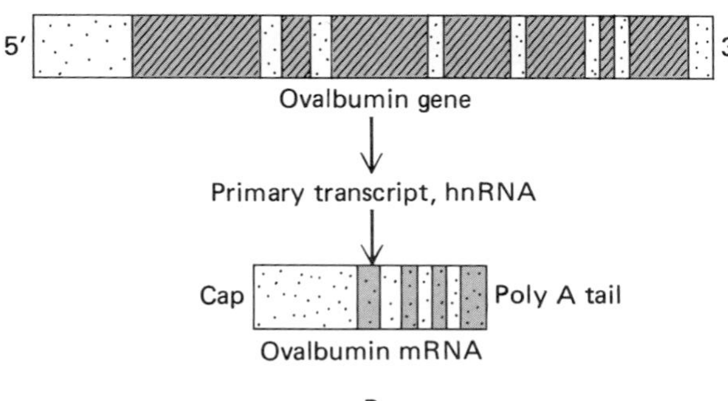

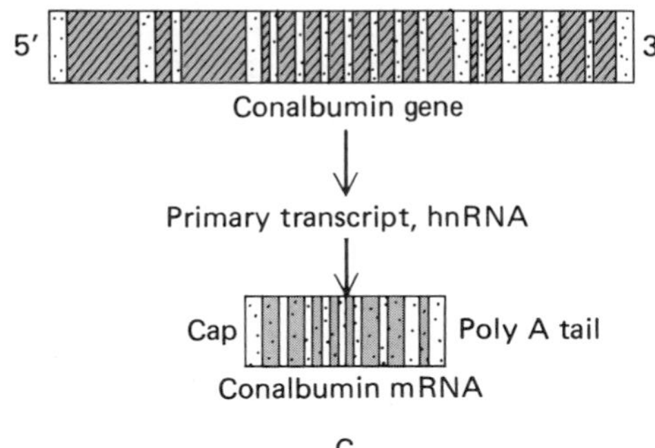

FIGURE 16.7 Maps of the β-globin, ovalbumin, and conalbumin genes showing the exons (the regions incorporated into mRNA and subsequently translated) and the intervening sequences or introns (the noncoding regions that are removed as hnRNA is processed and converted into mRNA). Introns are indicated in diagonal lines and exons by stippling. During processing of the primary transcript, in addition to the removal of the introns, a cap is added at the 5′ end and a polyadenylate tail is added at the 3′ end.

The Synthesis of a Specific Polypeptide Chain

Activation of Amino Acids

The first step in the synthesis of a peptide requires the activation of the amino acid by an enzyme-mediated reaction with ATP. There is a separate specific amino acid–activating enzyme for each amino acid. After activation the same enzyme then catalyzes the transfer of the amino acid to the specific tRNA for the amino acid. The amino acid is attached at the 3′ end of the tRNA, which contains cytidylic, cytidylic, and adenylic acid; the amino acid is attached to the ribose of the terminal adenylic acid. If these three nucleotides are removed the tRNA is unable to function. The enzymes that activate amino acids and transfer them to tRNA are highly specific and play a key role in ensuring the correct translation of the genetic message. Attaching the amino acid to tRNA not only activates its carboxyl group so it can form a peptide bond but ensures that it will be inserted at the correct place in the peptide because the specific tRNAs recognize specific codons on the mRNA.

Ribosomal Functions

The amino acid bound to its specific tRNA is transferred to the ribosomes. The role of the ribosome is to provide the proper orientation of the amino acid–tRNA, the mRNA, and the growing polypeptide chain so that the genetic code on the mRNA can be read accurately. There are some 15,000 ribosomes in a rapidly growing cell of *E. coli.* These ribosomes account for nearly one third of the total mass of the cell. Only one polypeptide chain can be formed at a time on any given ribosome.

The template for the synthesis of a specific protein is supplied by the mRNA formed on one strand of the double helix of DNA. The mRNA undergoes processing in the nucleus, and the poly A–rich RNA then passes out of the nucleus and becomes associated with the ribosomes. The poly A tail may play some role in transport through the nuclear membrane or in protecting the mRNA from destruction by ribonuclease.

Ribosomes from different kinds of cells may differ somewhat in details, but there is a general similarity in their structures. Ribosomes can be separated into two subunits whose structure is apparent in electron micrographs; the smaller subunit seems to sit like a cap on the flat surface of the larger subunit (Fig. 16.8). Ribosomes have GTPase activity, which plays an important role in the transfer of amino acids from tRNA to the forming peptide chain. Protein-synthesizing particles somewhat similar to ribosomes are also found in the

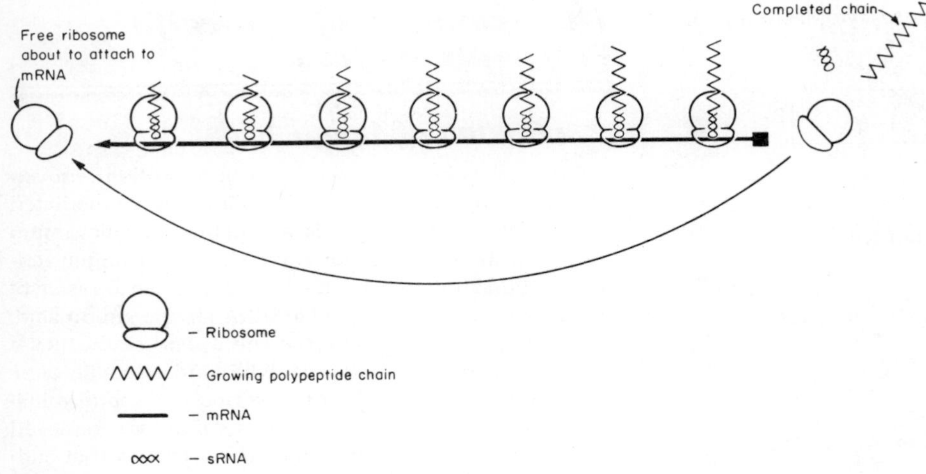

FIGURE 16.8 The postulated mechanism by which a polypeptide chain is synthesized. Each ribosome "rides" along the RNA, reading and translating the genetic message. In this diagram the ribosomes are pictured as moving from left to right. Amino acids are added as dictated by the specific base-pairing of the mRNA codon and the transfer RNA anticodon. (After Watson, J. D.: *The Molecular Biology of the Gene.* New York, W. A. Benjamin, 1965.)

nucleus, in the chloroplasts of plant cells, and in mitochondria.

Protein synthesis has been studied intensively in preparations of rabbit reticulocytes, which are engaged primarily in making just one protein—hemoglobin. Alexander Rich and his coworkers showed that the ribosomes most active in protein synthesis are those that interact in clusters of five or more (Fig. 16.9). These clusters, termed **polysomes,** are held together by the strand of mRNA. Peptide chains are formed by the sequential addition of amino acids beginning at the N-terminal end, the end having a free amino group. This N-terminal amino acid (at the left end of the peptide chain as written) is coded for by the sequence of three nucleotides at (or near) the 5′ end of the nucleic acid (at the left end of the polynucleic chain as written).

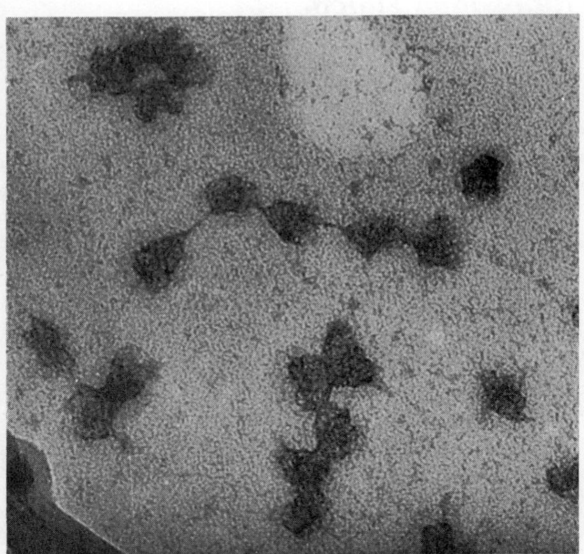

FIGURE 16.9 Electron micrograph of polyribosomes isolated from the reticulocytes of a rabbit. The preparation was stained with uranyl nitrate. The electron micrograph shows that polyribosomes tend to occur in clusters of four, five, or six and that the clusters are connected by a thin strand of mRNA. (Photograph courtesy of Dr. Alexander Rich.)

Electron micrographs suggest that individual ribosomes become attached to one end of a polysome cluster and gradually move along the messenger RNA strand as the polypeptide chain attached to it increases in length by the sequential addition of amino acids (Figs. 16.8 and 16.10). Thus each ribosome appears to ride along the extended mRNA molecule, "reading the message" as it goes and in some way bringing the tRNA molecule charged with its specific amino acid into line at the right position. After it completes the reading of one molecule of mRNA and releases the polypeptide that it has synthesized, the ribosome appears to jump off the end of one mRNA chain, move to a new mRNA, and begin reading it (Figs. 16.8 and 16.10). The growing peptide chain always remains attached to its original ribosome; there is no transfer of a peptide chain from one ribosome to another. Several ribosomes may be working simultaneously on a single strand of mRNA, each reading a different part of the message.

Three stages—initiation, elongation, and termination—can be distinguished in the process of protein synthesis. The overall process requires the coordinated action of more than 100 different macromolecules, including mRNA, all the specific tRNAs, activation enzymes, initiation factors, elongation factors, and termi-

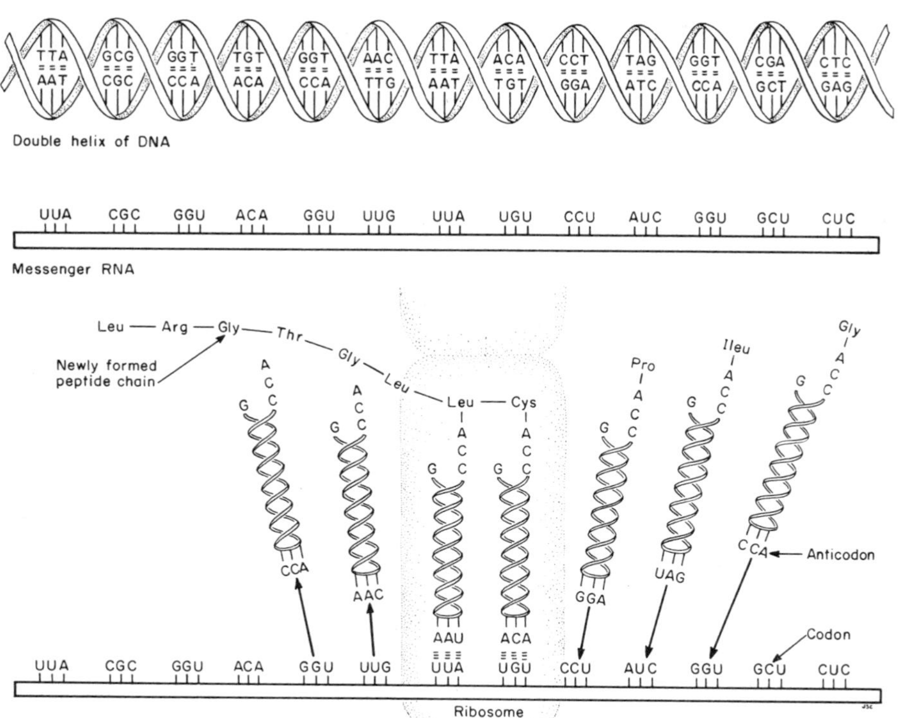

FIGURE 16.10 Diagram of the postulated mechanism of protein synthesis on the ribosome, illustrating the relationship between the triplet code of the DNA helix, the complementary triplet code of mRNA, and the complementary triplet code (anticodon) of transfer RNA. Molecules of transfer RNA charged with specific amino acids are depicted coming from the right, assuming their proper place on mRNA at the ribosome, transferring the amino acid to the growing peptide chain, and then (*left*) leaving the ribosome to be recharged with amino acids for further reactions. The growing polypeptide chain remains attached to its original ribosome.

nation factors, in addition to the ribosomes themselves, each made of subunits and many kinds of protein and RNA. In *Escherichia coli*, in which the process has been studied in greatest detail, mRNA, formylmethionyl tRNA, and the smaller ribosomal subunit come together to form an "initiation complex." The formylmethionyl tRNA recognizes the special initiation sequence on the mRNA. The larger ribosomal subunit then joins the complex to form a complete ribosome, and this is ready for the next phase.

The elongation process begins with the binding to the complex of the next succeeding amino acyl tRNA, which is recognized by the specific codon next in line. The elongation cycle continues with the enzymatic formation of a peptide bond, the elimination of the uncharged tRNA, and ends with the transfer of the tRNA containing the peptide from one site within the ribosome to the other. This leaves the second site ready to accept the next amino acyl tRNA molecule. Both the binding of the amino acyl tRNA and the translocation from one site to the other require GTP. A specific protein elongation factor is required for the binding of the amino acyl tRNA to the ribosome.

The synthesis of the peptide chain is terminated by "release factors" that recognize the terminator codons UAA, UGA, and UAG. This leads to the hydrolysis of the bond between the polypeptide and the tRNA. When the peptide chain is completed and released from the ribosome, the ribosome dissociates into its two subunits, the larger and smaller subunits.

Two different enzymes are required to carry out the transfer of the amino acids from tRNA to the peptide linkage in the peptide chain. One enzyme binds the amino acid tRNA, and the other is a synthetase that forms the peptide bond. The reactions that require GTP are the binding of the tRNA to the ribosome and the translocation of the amino acid–tRNA complex from one site in the ribosome to the other, rather than the synthesis of the peptide bond. GTP is used, and the products are GDP and inorganic orthophosphate.

An overview of the several steps involved in the synthesis of a specific polypeptide chain is provided in Figure 16.11. Clearly much has yet to be learned about the biosynthesis of proteins. In the cell-free system the synthesis of the complete α chain of hemoglobin, which is 141 amino acids long, requires 1.5 minutes; i.e., about two amino acids are added each second. Even the best cell-free protein-synthesizing system operates at a rate only 0.01 as rapid as that within an intact, living cell.

Changes in Genes: Mutations

Although genes are remarkably stable and are transmitted to succeeding generations with great fidelity, they

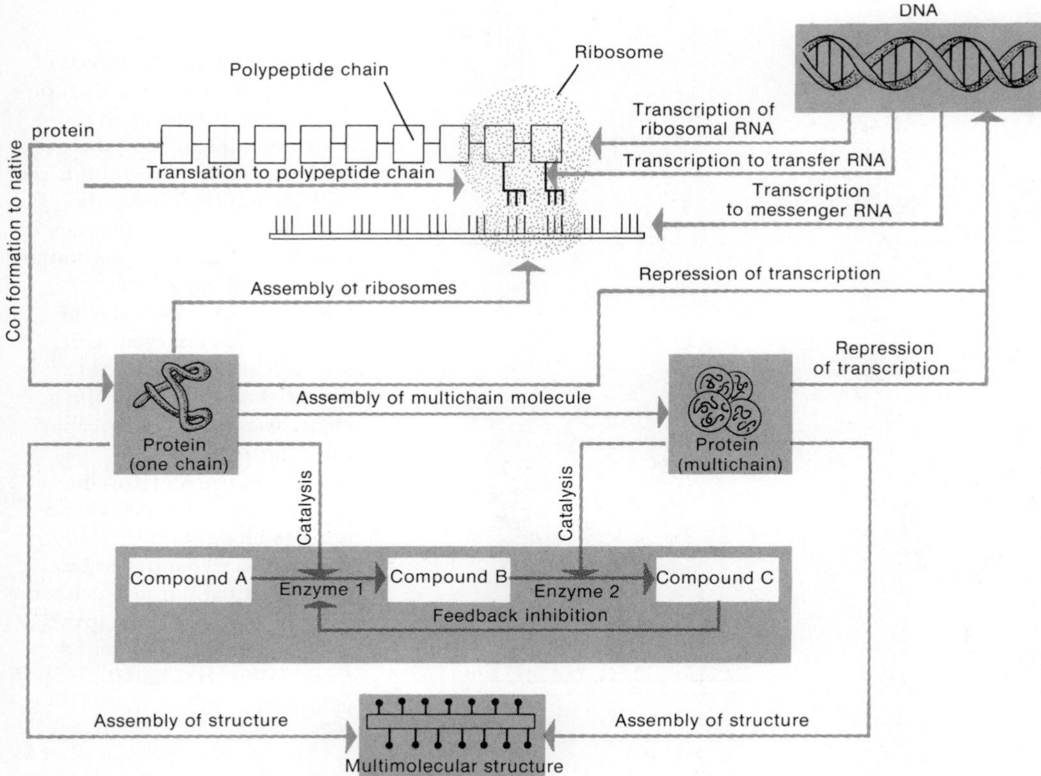

FIGURE 16.11 Overview of the process by which biological information is transferred from DNA via RNA to specific polypeptides. The peptide subunits are then assembled into multichain proteins.

do from time to time undergo changes called **mutations.** After a gene has mutated to a new form this new form is stable and usually has no greater tendency to mutate again than the original gene. A mutation has been defined as any inherited change not due to segregation or to the normal recombination of unchanged genetic material. Mutations provide the diversity of genetic material that characterizes the chromosomes of most organisms and makes possible a study of the process of inheritance.

Some mutations, termed **chromosomal mutations,** are accompanied by a visible change in the structure of the chromosome. A small segment of the chromosome may be missing (a **deletion**) or be represented twice in the chromosome (a **duplication**) (Fig. 16.12). A segment of one chromosome may be transferred to a new position on a new chromosome (a **translocation**), or a segment may be turned end for end and attached to its usual chromosome (an **inversion**). All of these changes have important genetic effects because they alter the amount of genetic material or its arrangement, and hence the way one gene may interact with other genes.

Another type of mutation is characterized by a change in the number of individual chromosomes or in the number of *sets* of chromosomes. Chromosome number is indicated by the term **ploidy.** Most animals are diploid, having two of each kind of chromosome, but some species may have more than two complete sets and be **polyploid.** An individual may have one or two extra chromosomes or may be lacking one chromosome.

Other mutations, termed point, or **gene, mutations,** involve small changes in molecular structure that are not evident under the microscope. These gene mutations involve some change in the sequence of nucleotides within a particular section of the DNA molecule, usually the substitution of one nucleotide for another in a given codon.

From your knowledge of the DNA molecule, you might predict that replacing one of the purine or pyrimidine nucleotides by an analogue, such as azaguanine

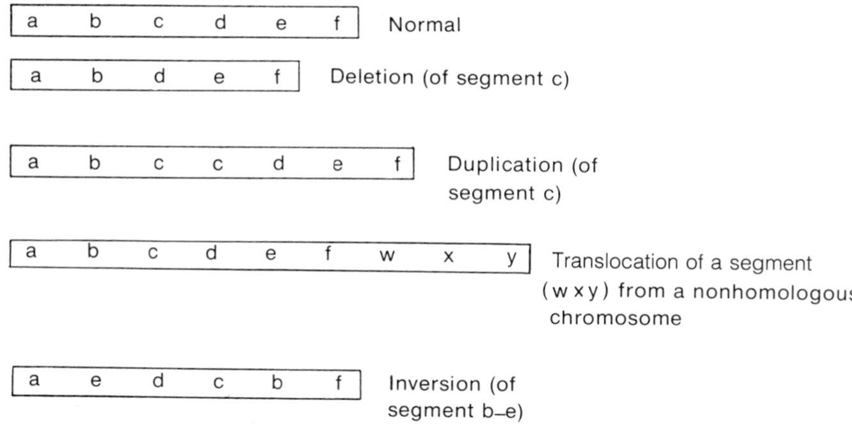

FIGURE 16.12 Diagram illustrating the types of mutations that involve changes in the structure of the chromosome.

or bromouracil, would result in mutation. In several experiments in which such analogues were incorporated into bacteriophage DNA, no mutations were evident. Because of the degeneracy of the genetic code (Table 16.1), a number of changes in base pairs could occur without changing the amino acid specified. In other experiments, the incorporation of bromouracil into DNA did lead to an increased rate of mutation. Other chemicals known to be mutagenic include nitrogen mustards, epoxides, nitrous acid, and alkylating agents. These are all chemicals that can react with specific nucleotide bases in the DNA and change their nature. When an analogue is incorporated into DNA it may lead to mistakes in the pairing of nucleotides during subsequent replication processes. For example, when bromouracil is incorporated into DNA in place of thymine it will pair with guanine rather than with adenine, the normal pairing partner of thymine. This would lead to the substitution of a G-C pair of nucleotides at the point in the double helix previously occupied by an A-T pair of nucleotides (Fig. 16.13).

Many "spontaneous" gene mutations may result from errors in base pairing during the replication process; thus the A-T normally present at a given site may be replaced by G-C, C-G, or T-A. The altered DNA will be transcribed to give an altered mRNA, and this will be translated into a peptide chain with one amino acid different from the normal sequence. If the altered amino acid is located at or near the active site of the enzyme, the altered protein may have markedly decreased or altered catalytic properties. If the altered amino acid is elsewhere in the protein, it may have little or no effect on the properties of the enzyme and may thus go undetected. The true number of gene mutations may be much greater than the number observed.

Other mutations may result from the insertion or deletion of a single base pair in the DNA molecule. This would shift the reading of the genetic message, alter all the codons lying beyond that point, and change completely the nature of the resulting peptide chain and its biological activity. Thus if the normal sequence is (CAG) (TTC) (ATG), the insertion of a G between the two T's results in (CAG) (TGT) (CAT) (G . . .).

Gene mutations can be induced not only by exposing the cell to certain chemicals but also by a variety of types of radiation—X-rays, gamma rays, cosmic rays, ultraviolet rays, and the several types of radiation that are by-products of atomic power. How radiation may lead to changes in base pairs is not clear, but the radiant energy may react with water molecules to release short-lived, highly reactive free radicals that attack and react with specific bases.

Mutations occur spontaneously at low, but measurable, rates that are characteristic of the species and of the gene; some genes are much more prone to undergo mutation than others. The rates of spontaneous mutation of different human genes range from 10^{-3} to 10^{-5} mutations per gene per generation. Since we have a total of some 2.3×10^4 genes, this means that the total mutation rate is on the order of one mutation per person per generation. Each one of us, in other words, has some mutant gene that was not present in either of our parents.

Human Cytogenetics

Although the normal human chromosome number is 46, some rare instances of abnormal chromosome numbers have been reported. These usually are associated with some change in the phenotype of the individual. A severely defective male child was found to be a *triploid* individual with a total of 69 chromosomes, 66 auto-

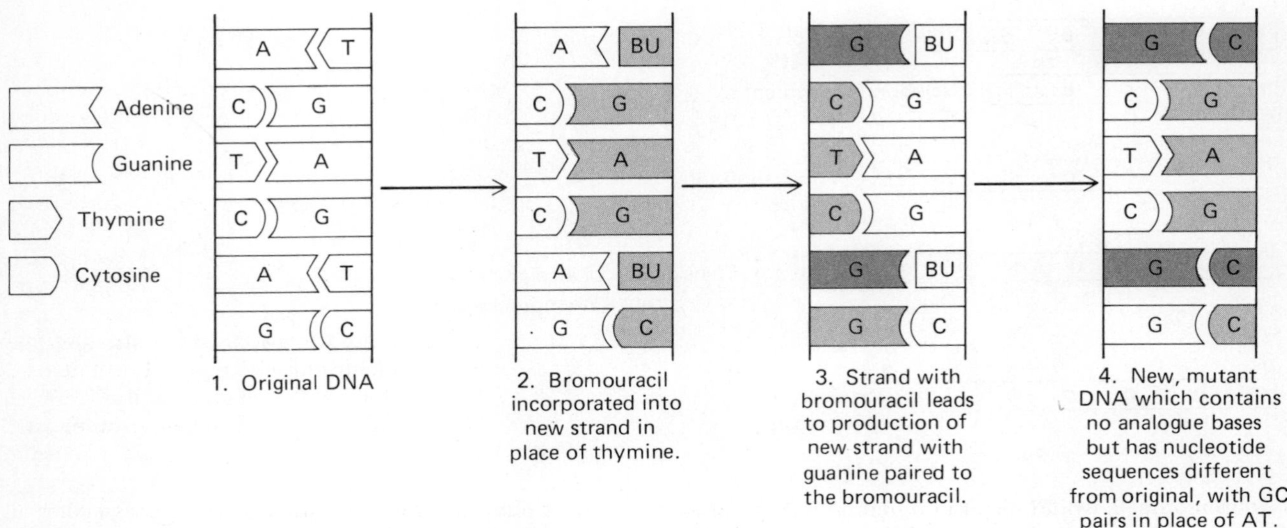
Adenine
Guanine
Thymine
Cytosine
A T
C G
T A
C G
A T
G C
BU
1. Original DNA
2. Bromouracil incorporated into new strand in place of thymine.
3. Strand with bromouracil leads to production of new strand with guanine paired to the bromouracil.
4. New, mutant DNA which contains no analogue bases but has nucleotide sequences different from original, with GC pairs in place of AT.

FIGURE 16.13 Diagrammatic scheme of how an analogue of a purine or pyrimidine might interfere with the replication process and cause a mutation, an altered sequence of nucleotides in the DNA, indicated in black. In this instance, two new GC pairs are indicated. Probably a single substitution of a GC pair for an AT pair would be sufficient to cause a mutation if it occurred in the triplet code at a point that changed the amino acid specified. The nucleotides of the new chain at each replication are indicated by the dotted blocks.

somes, 2 X chromosomes, and 1 Y chromosome. It seems likely that this zygote was formed from a normal haploid egg that was fertilized by an unusual diploid sperm or from an exceptional diploid egg fertilized by a normal haploid sperm.

Nondisjunction refers to the failure of a pair of homologous chromosomes to separate normally during the reduction division. Two X chromosomes, for example, might fail to separate and both might enter the egg nucleus, leaving the polar bodies with no X chromosome. Alternatively, the two joined X chromosomes might go into the polar body, leaving the female pronucleus with no X chromosome. Nondisjunction of the XY chromosomes in the male might lead to the formation of sperm that have both an X and a Y chromosome or to sperm with neither an X nor a Y chromosome. Chromosomal nondisjunction may occur during either the first or second meiotic division; it may also occur during mitotic divisions and lead to the establishing of a group of abnormal cells in an otherwise normal individual.

Cytogenetic studies have clarified the origin of one of the more distressing abnormal human conditions, that of **Down's syndrome,** or "mongolism." Individuals suffering from this have abnormalities of the face, eyelids, tongue, and other parts of the body and are greatly retarded in both their physical and mental development. The term "mongolism" was originally applied to this condition because affected individuals often show a fold of the eyelid similar to that typical of members of the Mongolian race. Down's syndrome is a not uncommon congenital malformation, occurring in 0.15 percent of all births. It had been known for some time that the appearance of Down's syndrome is related to the age of the mother and that it increases greatly with maternal age. For example, Down's syndrome is a hundredfold more likely in the offspring of women 45 years or older than in the offspring of mothers under 19. The occurrence of Down's syndrome, however, is independent of the age of the father, and it is also independent of the number of preceding pregnancies in the woman. Cytogenetic studies revealed that individuals with Down's syndrome have one extra small chromosome 21, a total of 47. The presence of this extra small chromosome is believed to arise by nondisjunction in the maternal oöcyte.

It is not clear why the DNA transcribing system does not simply ignore the redundant bit of genetic information and produce cells identical to those of the normal individual, but the presence of this extra chromosome leads to the complex physical and mental abnormalities that characterize Down's syndrome. Whether the extra genes in the third chromosome 21 lead to the production of an extra amount of certain enzymes and whether this is the basis for the abnormal physical and mental development is not known. When a certain chromosome or part of a chromosome has been added or deleted ("genetic imbalance") comparable defects are observed in all types of organisms.

Another condition caused by an upset in chromosome number is that of individuals who are outwardly nearly normal males but have small testes. They produce few or no sperm, they have seminiferous tubules that are very aberrant in appearance, and they usually have gynecomastia (a tendency for formation of female-like breasts). This condition, called **Klinefelter's syndrome,** usually becomes apparent only after puberty, when the small testes and gynecomastia may bring the individual to the attention of his physician. The cells of these individuals show a chromatin spot, and at one time they were thought to be XX individuals, i.e., genetic females. However, when their chromosomes were counted it was found that they have 47 chromosomes; their cells have *two* X and one Y chromosome. The fact that they are nearly normal males in their external appearance emphasizes the male-determining effect of the Y chromosome in man.

Another condition resulting from changes in chromosome number is **Turner's syndrome,** in which the external genitalia, though feminine, are those of an immature female. The internal reproductive tract is present and resembles that of an immature but perfectly formed female. The uterus is present but small, and the gonads may be absent. The cells of these individuals are "chromatin negative," which suggests that they are males. However, they have only one X chromosome but no Y chromosome. This type of disorder again emphasizes the importance of the Y chromosome in determining the male characteristics.

An individual with an extra chromosome, with three of one kind, is said to be **trisomic,** and an individual lacking one of a pair is said to be **monosomic.** Thus individuals with Down's syndrome are trisomic for chromosome 21, and individuals with Turner's syndrome are monosomic for the X chromosome. The features of certain human chromosomal abnormalities are summarized in Table 16.2.

Studies of human genetics have advanced rapidly in recent years with the development of the process of amniocentesis. This involves inserting a long needle through the abdominal wall and uterine wall of a pregnant woman and withdrawing a sample of amniotic fluid for analysis. The amniotic fluid contains cells derived from the fetus. These can be grown in culture and analyzed for their karyotype or for the presence of a specific enzyme. This was first used in 1956 to diagnose the sex of the fetus by determining the presence or absence of sex chromatin in the cells recovered from the amniotic fluid. It has been shown that amniocentesis can be carried out successfully with little or no risk to either mother or fetus not only near term but fairly early in gestation. The amniotic fluid is centrifuged to sediment the cells, which can be transferred to slides, fixed, and stained. Or the cells can be incubated for a suitable

TABLE 16.2 Some Human Chromosomal Abnormalities*

Abnormality	Genetic Features	Clinical Aspects
Turner's syndrome (gonadal dysgenesis)	XO	Short stature, streak ovary, juvenile female genitalia, poorly developed breasts
Klinefelter's syndrome	XXY	Gynecomastia, small testes
Triple X females	XXX	Two "Barr bodies" present, fairly normal females, but secondary sex characteristics may be poorly developed
Down's syndrome	Trisomy 21	Epicanthal folds, protruding tongue, hypotonia, mental retardation
Trisomy 18	Trisomy 18	Mental retardation, multiple congenital malformations
D trisomy	Trisomy 15	Mental retardation, severe multiple anomalies, cleft palate, polydactyly, central nervous system defects, eye defects
Translocation mongolism	15/21, 21/22, or 21/21 translocation	Mongolism, clinically similar to trisomy 21
Philadelphia chromosome	Deletion of one arm of chromosome 21	Chronic granulocytic leukemia
Orofaciodigital syndrome	Translocation of part of chromosome 6 to 1	Defects of upper lip, palate, and mouth, stubby toes with short nails
Cri du chat syndrome	Deletion of short arm of chromosome 5	Mental retardation, facial anomalies

*From Page, E. W., Villee, C. A., and Villee, D. B.: *Human Reproduction.* 3rd ed. Philadelphia, W. B. Saunders Co., 1981.

period, treated with colchicine, and stained so the complete karyotype can be visualized. Certain enzyme deficiencies can be detected by incubating these amniotic fluid cells with the appropriate substrate and measuring the product. An increased concentration of α-fetoprotein in amniotic fluid has been found in pregnancies with infants having defects in the development of the neural tube.

Gene–Enzyme Relations

If we assume that a specific gene leads to the production of a specific enzyme by the method outlined above, we must next inquire how the presence or absence of a specific enzyme may affect the development of a specific trait. The expression of any structural or functional trait is the result of a number, perhaps a large number, of chemical reactions that occur in series, with the product of each reaction serving as the substrate for the next: A→B→C→D. The dark color of most mammalian skin or hair is due to the pigment **melanin** (D), produced from dihydroxyphenylalanine (dopa) (C), produced in turn from tyrosine (B) and phenylalanine (A). Each of these reactions is controlled by a particular enzyme; the conversion of dopa to melanin is mediated by **tyrosinase. Albinism,** characterized by the absence of melanin, results from the absence of tyrosinase. The gene for albinism (a) does not produce the enzyme tyrosinase, but its normal allele (A) does.

Similar one-to-one relationships of gene, enzyme, and biochemical reaction have been described for humans. **Alkaptonuria** is a trait, inherited by a recessive gene, in which the patient's urine turns black on exposure to air. The urine contains homogentisic acid, a normal intermediate in the metabolism of the amino acids phenylalanine and tyrosine. The tissues of normal people have an enzyme that oxidizes homogentisic acid so that it is eventually excreted as carbon dioxide and water. Alkaptonuric patients lack this enzyme because they lack the gene that controls its production. As a result, homogentisic acid accumulates in the tissues and blood and spills over into the urine.

The relationship between gene, enzyme, and phenotypic trait in these examples is clear and readily understood. The relationship between gene, enzyme, and such traits as size, shape, and behavior is far from understood but probably involves complex interactions between a number of gene products

The question of whether the genes normally operate so as to lead to the production of the *maximum* number of enzymes all the time has been given consideration in recent years. From a variety of experimental evidence it would appear that this is not the case. Each gene is probably "repressed" to a greater or lesser extent under normal conditions and then, in response to some sort of environmental demand for that particular enzyme, the gene becomes "derepressed" and leads to an increased production of the enzyme. When a single gene is fully derepressed it can lead to the synthesis of fantastically large amounts of enzymes—one enzyme may compose 5 to 8 percent of the total protein of the cell! If all enzymes were produced at this same fantastically high rate, metabolic chaos would result. Thus the phenomena of gene repression and derepression would appear to be necessary to prevent this chaos and to provide a means for increasing or decreasing the rate of synthesis of one particular enzyme in response to variations in environmental requirements.

Genes and Differentiation

One of the important unsolved problems of modern biology is the nature of the mechanisms that regulate developmental processes. How can a single fertilized egg give rise to the many different types of cells that differ so widely in their structure, functions, and chemical properties?

We now have a detailed working hypothesis as to how biological information is transferred from one generation of cells to the next and how this information may be transcribed and translated in each cell so that specific enzymes and other proteins are synthesized. The operation of this system would produce a multicellular organism in which each cell would have the same assortment of enzymes as every other cell. Additional hypotheses are needed to account for (1) the means by which the *amount* of any given enzyme produced in a cell is regulated; (2) the control of the *time* in the course of development when each kind of enzyme appears; and (3) the mechanism by which *unique patterns* of proteins are established in each of the several kinds of cells in a multicellular organism despite the fact that they all contain identical quotas of genetic information.

The Preformation Theory and Epigenesis

A theory widely held by early embryologists was that the egg or sperm contained a completely formed but minute germ that simply grew and expanded to form the adult. This **preformation theory** was gradually displaced by the contrasting theory of **epigenesis,** which stated that the unfertilized egg is structureless, not organized, and that development proceeds by the progressive differentiation of parts. However, development is not simply epigenetic. Certain potentialities, though not structures, may be localized in certain regions of the egg and early embryo; this restricts the development possibilities of that part. When the embryos of echinoderms or chordates are separated experimentally at the two- or four-cell stage, each of the separated cells will form an embryo complete in all details although smaller than normal. However, when embryos of annelids or mollusks are separated at the two-cell stage, neither cell can develop into a whole embryo. Each cell develops only into those structures it would have formed normally—half an embryo, perhaps, or some part of one. This localization of potentialities eventually occurs in the development of all eggs; it simply occurs earlier in some than in others.

Some sort of chemical or physiological differentiation must be present before any structural differentiation is visible, but the basic problem of how the chemical differentiation arises remains unsolved. By appropriate experiments it has been possible to map out the location of these potentialities in frog, chick, and other embryos (Fig. 16.14).

Differential Nuclear Division

Cellular differentiation might be explained if genetic material were parceled out differentially at cell division, and the daughter cells received different kinds of genetic information. However, experiments of Briggs and King showed that even the nucleus from a differentiated cell taken from an advanced stage of embryonic development can, when placed in an enucleated egg, lead to the development of an normal embryo. Thus it clearly has retained a full set of genetic information. Thus the differences in the kinds of enzymes and other proteins found in different cells of the same organism must arise by differences in the activity of the same set of genes in different cells.

Differential Gene Activity

It seems likely that cellular differentiation, which is accompanied by the synthesis of different proteins by different cells, is controlled by differential gene activity. Differential gene activity may be controlled at the transcription of DNA into RNA, at the translation of RNA into protein, or at both of these levels. We know that the continued synthesis of any protein requires the continued synthesis of its corresponding mRNA. Each kind of mRNA has a half-life ranging from a few minutes in certain microorganisms to 12 or 24 hours or longer in mammals. Although each molecule of RNA template can serve to direct the synthesis of many molecules of its protein, the RNA is eventually degraded and must be replaced. This provides a mechanism by which a cell can alter the kind of protein being synthesized as new types of mRNA replace the previous ones. Thus the cell can respond to exogenous stimuli with the production of new types of enzymes.

A mechanism that controls the transcription of DNA to regulate the production of messenger RNA would probably be the most economical one biologically, for it would clearly be to the cell's advantage not to have its ribosomes encumbered with nonfunctional molecules of mRNA. However, some biologists have postulated that in the unfertilized egg DNA may be transcribed to form messenger RNA, but the mRNA is "masked" and inactive as a template for protein synthesis until it is subsequently unmasked by a separate process.

Some striking evidence regarding the differential activity of genes comes from cytologic studies of insect tissues. In certain insect tissues the chromosomes undergo repeated duplication, and the daughter strands line up exactly in register, locus by locus, so that characteristic bands appear along the length of the **giant chromosomes.** When these bands are examined carefully, either in the same tissue at different times or in different tissues at the same time, certain differences in

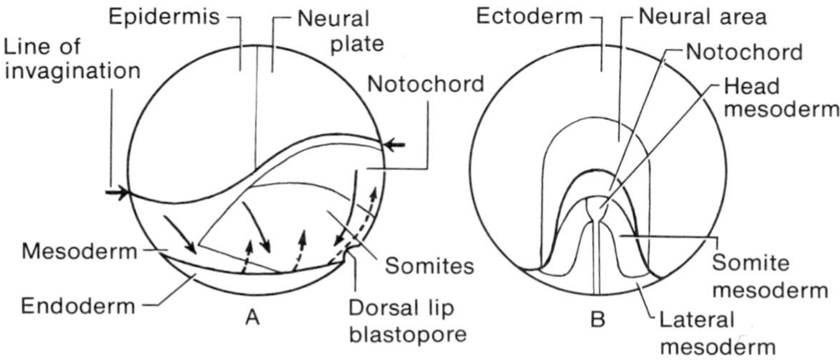

FIGURE 16.14 Embryo maps. *A,* Lateral view of a frog gastrula showing the presumptive fates of its several regions. *B,* Top view of chick embryo showing location in the primitive streak stage of the cells that will form particular structures of the adult.

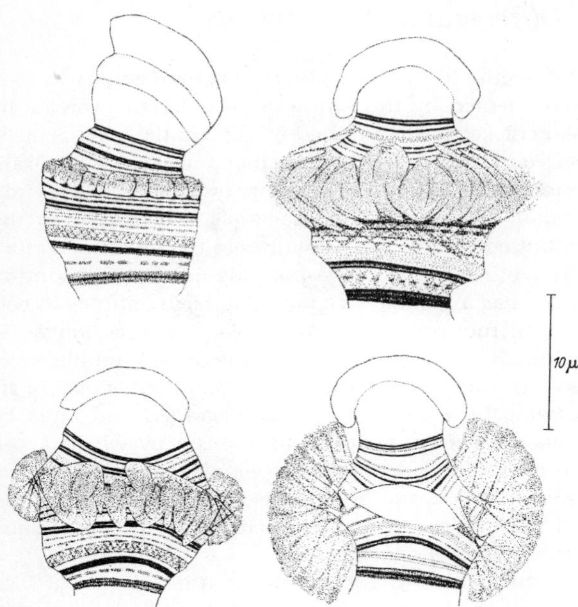

FIGURE 16.15 Diagrams illustrating changing appearance of a polytene chromosome of a salivary gland of *Chironomus tentans* as a chromosome puff gradually appears. The material comprising the puff has been shown by histochemical tests and by autoradiography with tritium-labeled uridine to be largely ribonucleic acid. (From Beermann: *Chromosoma, 5*:139–198, 1952.)

appearance become evident. A particular section of a chromosome may have the appearance of a diffuse puff (Fig. 16.15). Histochemical tests and other evidence have shown that the puff consists of RNA. It has been inferred that genes show this puffing phenomenon when they become active and that the puff represents the mRNA produced by the active gene in that band. The appearance of puffs at certain regions in the chromosome can be correlated with specific cellular events, such as the initiation of molting and pupation.

Enzyme Induction

The induction of enzymes by environmental stimuli has been cited as a model for embryonic differentiation. Bacteria, and to some extent animal cells, respond to the presence of certain substrate molecules by forming enzymes to metabolize them. As the embryo develops, the gradients established as a result of growth and cell multiplication could result in quantitative and even qualitative differences in enzymes. The induction or inhibition of one enzyme could lead to the accumulation of a chemical product that would induce the synthesis of a new enzyme and confer a new functional activity on these cells. Enzymes can be induced in an embryo by the injection of an appropriate substance. Adenosine deaminase, for example, has been induced in the chick embryo by the injection of adenosine; however, no enzyme has been induced that is not normally present to some extent in the embryo. Adaptive changes in enzymes are temporary and reversible, whereas differentiation is a permanent, essentially irreversible process. Morphogenesis appears to be too complex a phenomenon to be explained in terms of any single process, such as enzyme induction.

The DNA of the genes not being transcribed at any given moment may be bound to a histone or to an acidic nuclear protein that makes the DNA unavailable for the transcription system.

Morphogenetic Substances

Experiments by Grobstein have reemphasized that extrinsic factors as well as nuclear factors may play a role in the process of differentiation. For example, embryonic pancreatic epithelium will continue to differentiate in organ culture only in the presence of mesenchyme cells. The mesenchyme can be replaced by a chick embryo juice, and the active principle of the embryo juice appears to be a protein, for it is inactivated by trypsin but not by ribonuclease or deoxyribonuclease.

Suspensions of dissociated, individual healthy cells can be prepared by treating a tissue briefly with a dilute solution of trypsin. When the suspension of cells is placed in tissue culture medium the cells may reaggregate and continue to differentiate in conformity with their previous pattern. The cells in tissue culture reaggregate not in a chaotic mass but in an ordered fashion, forming recognizable morphogenetic units. The cells appear to have specific affinities, for epidermal cells join with each other to form a sheet, disaggregated kidney cells join to form kidney tubules, and so on (Fig. 16.16). The mechanism by which one cell recognizes another cell and joins with it to form a tubule, sheet, or other structural unit appears to involve characteristic tissue-specific glycoproteins on the surface of the cell.

Embryonic tissues growing *in vivo* have differential sensitivities to changes in nutrients, to the presence of inhibitors and antimetabolites, and to various environmental agents. Any of these factors, applied during the appropriate critical period in development, may change the course of development and differentiation and mimic the phenotype of a mutant gene, producing what is termed a **phenocopy.**

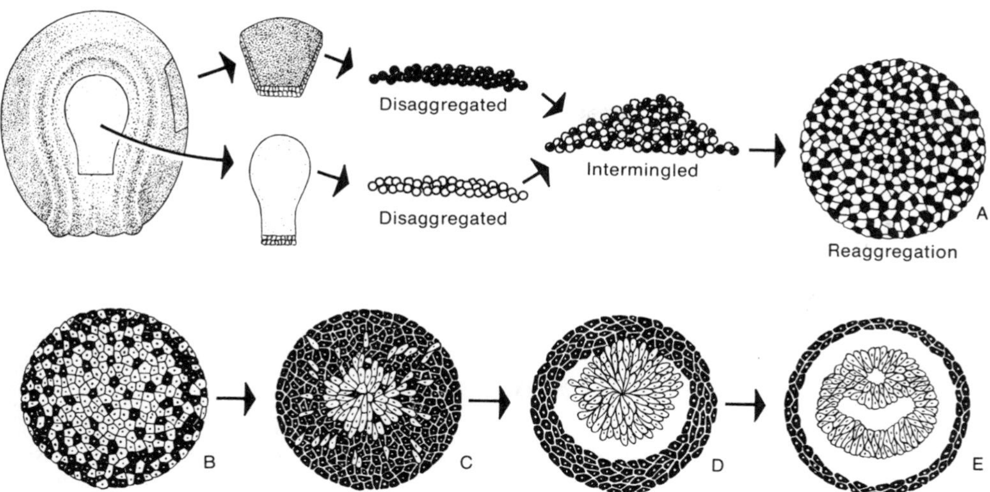

FIGURE 16.16 A mass of neural plate cells and epidermal cells from the ventral surface was removed from an amphibian embryo and disaggregated. The two types of cells subsequently reaggregated in tissue culture to form a heterogeneous mass. Finally each type of cell migrated in an appropriate fashion so that the medullary plate cells rounded up and formed a neural tubelike structure in the center, whereas the epidermal cells migrated to the periphery of the mass and joined to form a new epidermis.

A striking demonstration that the same set of genes operating in dissimilar environments may have different morphological effects was provided by experiments with three races of frogs found in Florida, Pennsylvania, and Vermont. Each of these races normally develops at a speed that is adapted to the length of the usual spring and summer season in its locale.

When an egg is fertilized with a sperm from a different race and the original egg nucleus is removed before the sperm nucleus can unite with it, it is possible to establish a cell with "northern" genes operating in "southern" cytoplasm or the reverse. Northern genes in southern cytoplasm resulted in poorly regulated development; the animal's head grew more rapidly than the posterior region and became disproportionately large. Southern genes introduced into northern cytoplasm led again to poorly regulated development, but the head rather than the posterior region was retarded and disproportionately small. Genes from the Pennsylvania race of frog acted as "northern" with Florida cytoplasm but as "southern" with Vermont cytoplasm. The same set of genes had diverse morphological effects when they operated in different cytoplasmic environments.

Differentiation may be controlled at least in part by influences originating outside the cell, by "organizers" from neighboring cells in early differentiation, by the mesenchymal proteins studied by Grobstein, or by hormones from distant cells. Such systemic influences participate in the integration of the differentiation of individual cells into the larger pattern of differentiation of the tissues of the whole organism. Eventually it should be possible for developmental biologists to bridge the gap between studies of development at the level of the whole organism and studies at the molecular level and to trace in detail the sequence of events from the initial action of the gene to the final expression of the phenotype. This is one of the most exciting areas of biological research at present.

Lethal Genes

Certain genes produce such a tremendous deviation from the normal development of an organism that it is unable to survive. The presence of these **lethal genes** can be detected by certain upsets in the expected genetic ratios. For example, some mice in a certain strain had yellow coat color, but experimenters found it impossible to establish a true-breeding strain with yellow coat. Instead, when two yellow mice were bred, offspring were produced in the ratio of 2 yellow : 1 nonyellow. A yellow mouse bred to a black mouse gave half yellow mice and half black mice among the offspring. Then investigators noticed that the litters of yellow × yellow matings were somewhat smaller than other litters of mice, being only about three-quarters as large. They reasoned that one quarter of the embryos, those homozygous for yellow, did not develop. When the uterus of the mother was opened early in pregnancy the abnormal embryos, those homozygous for the yellow trait, were found. Embryos homozygous for yellow color begin development, then cease developing, die, and are resorbed.

Some lethal genes, such as yellow, produce a phenotypic effect when heterozygous and hence are said to be dominant. Many, perhaps most, of the lethal genes appear to have no effect when heterozygous but cause death when homozygous and are called recessive lethal genes. These can be detected only by special genetic techniques. When wild populations of fruit flies and other organisms are analyzed, the presence of many recessive lethals is revealed. In the light of our present theory about the relations between genes and development we can suppose that a lethal gene is a mutant that causes the absence of some enzyme of primary importance in intermediary metabolism. The absence of this enzyme prevents the proper development of the organism.

Penetrance and Expressivity

Each recessive gene described so far produces its trait when it is homozygous, and each dominant gene produces its effect when it is homozygous or heterozygous, but other genes are known that do not always produce their expected phenotypes. Genes that always produce the expected phenotype are said to have complete penetrance. If only 70 percent of the individuals of a stock homozygous for a certain recessive gene show the character phenotypically, the gene is said to have 70 percent penetrance. The term **penetrance** refers to the statistical regularity with which a gene produces its effect when present in the requisite homozygous (or heterozygous) state. The percentage of penetrance of a given gene may be altered by changing the conditions of temperature, moisture, nutrition, and so forth under which the organism develops.

Some stocks that are homozygous for a recessive gene may show wide variations in the appearance of the character. Fruit flies homozygous for a recessive gene that produces shortening and scalloping of the wings exhibit wide variations in the *degree* of shortening and scalloping. Such differences are known as variations in the **expressivity** or expression of the gene. The expressivity of the gene may also be altered by changing the environmental conditions during the organism's development. In view of the long and sometimes tenuous connection between the gene in the nucleus of the cell and the final production of the trait, it is easy to understand why the expression of the trait might vary or why the mutant trait might be completely absent.

Summary

1. Genes are composed of DNA and are located in the chromosomes. Each gene contains information coded in the form of a specific sequence of purine and pyrimidine nucleotides. The genetic code is a triplet code; each codon is a group of three adjacent nucleotides that specify the location of a single amino acid in the polypeptide chain.

2. The DNA molecule consists of two complementary chains of polynucleotides twisted about each other in a double helix; the two chains are joined by hydrogen bonds between specific pairs of purine and pyrimidine bases. Genetic information flows from the DNA of the gene to mRNA and then to the specific sequence of amino acids in peptide chains synthesized on the ribosomes. Messenger RNA contains a sequence of ribonucleotides complementary to the sequence of deoxyribonucleotides in the gene.

3. To be incorporated into a peptide chain amino acids are first activated by reacting with ATP and then are transferred to a specific adaptor molecule termed transfer RNA. Each kind of tRNA has an anticodon, a sequence of three nucleotides complementary to the specific codon in messenger RNA. The amino acid–tRNA complexes are arranged on the mRNA in an order dictated by the complementarity of the nucleotide triplets in the mRNA codon and the tRNA anticodon. Thus the information coded as a specific sequence of deoxyribonucleotides in DNA is transcribed as a specific sequence of ribonucleotides in mRNA and ultimately is translated as a specific sequence of amino acids in the protein molecule.

4. The Watson-Crick model of the DNA molecule proposed that it consists of two very long helical chains of deoxyribonucleotides wound around a common central axis and extending in opposite directions. The deoxyribose-phosphate backbone of each chain is on the outside of the double helix, and the purines and pyrimidines are on the inside. The two chains are joined by hydrogen bonds between specific pairs of bases. Adenine (A) pairs with thymine (T), and guanine (G) pairs with cytosine (C) so that one strand of the double helix is the complement of the other.

5. The two strands of the double helix unwind and separate to permit the replication of the DNA. Each strand acts as a template for the formation of a new strand complementary to the parent. The replication of the DNA is semiconservative; each daughter molecule contains one strand from the parent molecule and one new molecule. Replication is a complex process catalyzed by several enzymes, primarily DNA polymerase. The substrates in the reaction are the four deoxyribonucleoside triphosphates. DNA polymerase catalyzes the formation of a phosphodiester bond only if the base on the nucleotide being added is complementary to the base in the template strand. DNA synthesis occurs in the cells of higher organisms only during the S phase of the cell cycle.

6. The genetic code is said to be "degenerate"—there is more than one code word for most amino acids—but the code is not ambiguous, for each codon specifies only one amino acid. The genetic code is universal; a given codon specifies the same amino acid in the protein-synthesizing systems of all organisms from viruses to humans. The linear, unbranched polynucleotide chain of DNA is colinear with the linear unbranched peptide chain. In the binding of the anticodon to the codon the pairing of the first two bases is quite precise, but the factors recognizing the third base of the codon appear to be less stringent, and the anticodon may bind to any of three codons. Crick described this imprecision of base pairing at the third base as a "wobble."

7. Messenger RNA is synthesized by a DNA-dependent RNA polymerase that uses DNA as the template and the triphosphates of the four ribonucleotides as the substrate. The process yields as products an RNA chain and inorganic pyrophosphate, PP_i. Transfer and ribosomal RNA are also produced by DNA-dependent RNA synthesizing systems and are transcribed from complementary deoxyribonucleotide sequences in DNA. After RNA has been synthesized it may undergo reactions that cleave it or chemically modify it. The RNA of eukaryotes may undergo extensive cleavage and splicing, with the removal of certain sequences called introns and the retention of others termed exons. A "cap" of 7-methyl guanosine is added at the 5′ end, and a polyadenylic acid tail of 150 to 200 adenylic acids is added at the 3′ end. They serve to protect the RNA from attack by enzymes and to facilitate its transport across the nuclear membrane.

8. Peptide chains are synthesized, beginning at the amino terminal end, by the sequential addition of amino acids to the carboxyl end of the growing chain. Each amino acid is bound to a specific tRNA, with the carboxyl group of the amino acid attached to the 3′OH group of the ribose of the adenosine at the CCA (3′) end of the tRNA. The binding of the amino acid to its respective tRNA is catalyzed by a specific enzyme.

9. Protein synthesis occurs on the ribosomes in the cytoplasm of the cells and involves three separate processes: initiation, elongation, and termination.

10. Initiation is marked by the binding of the first amino acid to the start signal on the mRNA. The first amino acid, typically formyl methionyl tRNA, unites with mRNA, two initiation factors, GTP and the smaller ribosomal subunit, to form an initiation complex. The formyl methionyl tRNA recognizes the special initiation sequence on the mRNA. The larger ribosomal subunit then joins the complex to form a complete ribosome.

11. The elongation process begins with the binding of the second amino acyl tRNA to the complex. The elongation cycle continues with the enzymatic formation of a peptide bond, the release of the uncharged tRNA, and the movement of the tRNA containing the growing peptide chain from one site within the ribosome to another, leaving the second site empty and ready to accept the next amino acyl tRNA. Specific elongation factors are required to bind the amino acyl tRNA. GTP is used in binding the tRNA to the ribosome and in moving the amino acyl tRNA complex from one site in the ribosome to the other.

12. The synthesis of the peptide chain is terminated by release factors that recognize the terminator codons on the mRNA.

13. The ribosome is formed from its subunits each time a single peptide is synthesized. This permits the mRNA to become attached by complementary base pairing to the smaller ribosomal subunit at just the right point so that the message is read out correctly.

14. Genes may undergo changes, called mutations, that result in changes in the gene products, the kind of protein synthesized. Chromosomal mutations, involving a visible change in the chromosome, include deletions, duplications, inversions, and translocations of segments of the chromosome. Point mutations involve the substitution of one nucleotide for another in a specific codon so that it codes for a different amino acid.

15. Each gene codes for a specific single enzyme that catalyzes a single step in the biosynthesis of a particular chemical compound. The mutant allele of the gene does not produce the enzyme, and hence the biosynthesis of the chemical is prevented. Conditions such as phenylketonuria, alkaptonuria, and albinism in humans result from the presence of the mutant allele instead of the normal gene and the lack of the enzyme produced by the normal gene. In these "inborn errors of metabolism" the lack of the enzyme interferes with the normal sequence of reactions, and either a normal intermediate accumulates (alkaptonuria) or the normal product cannot be made (albinism).

16. Studies of the karyotype, the number and kind of chromosomes present in a nucleus, permit us to detect individuals who are polyploid, having one or more complete extra sets of chromosomes. We can also detect individuals who are trisomic and have one extra chromosome, or who are monosomic, lacking one member of a pair. Down's syndrome, Klinefelter's syndrome, and Turner's syndrome are examples of trisomics or monosomics.

References and Selected Readings

Cooke, R.: *Improving on Nature: The Brave New World of Genetic Engineering.* Chicago, Time-Life Books, 1978. An interesting survey of the methods and promise of genetic engineering.

Facklam, H., and M. Facklam: *From Cell To Clone: The Story of Genetic Engineering.* New York, Harcourt Brace Jovanovich, 1979. Genetic engineering for the layman.

Glover, D. M.: *Genetic Engineering.* New York, Methuen, 1980. A paperback, simplified presentation of genetic engineering.

Goulian, M., and P. Hanawalt (Eds.): *DNA Synthesis and Its Regulation.* Menlo Park, California, Benjamin, 1975. A carefully edited series of research papers dealing with DNA synthesis.

Hanawalt, P. E., C. Friedberg, and C. F. Fox (Eds.): *DNA Repair Mechanisms.* New York, Academic Press, 1979. Carefully edited research papers dealing with DNA repair.

Kornberg, A.: *DNA Replication.* San Francisco, W. H. Freeman & Co., 1980. Discussion of DNA by a masterful scientist and writer.

Stahl, F. W.: *Genetic Recombination.* San Francisco, W. H. Freeman & Co., 1979. Discussion of genetic recombinations, primarily in microorganisms.

Stent, G., and R. Calendar: *Molecular Genetics.* 2nd ed. New York, Academic Press, 1978. A well-written survey of molecular genetics by scientists in the field.

Watson, J. D., and J. Tooze: *The DNA Story: A Documentary History of Gene Cloning.* San Francisco, W. H. Freeman & Co., 1981. An interesting history of a very new science.

Watson, J. D.: *Molecular Biology of the Gene.* 3rd ed. New York, Benjamin-Cummings, 1976. A fine text on molecular genetics and how information flows from DNA to RNA to protein.

Watson, J. D.: *The Double Helix.* New York, Atheneum Publishing, 1968. An interesting insight into the motivations and thought processes of one of the classic molecular biologists; "how to become a Nobel Laureate."

17 The Concept of Evolution

ORIENTATION

Although biologists are not in complete agreement as to all aspects of the mechanisms of evolution, the notion that organic evolution has occurred, and is the explanation for the great diversity of organisms, is supported by a large body of evidence from all fields of biology. It is as well documented as any scientific hypothesis and is regarded as one of the important unifying principles of biology. In this chapter we will examine the evidence that bears upon the concept of evolution, explain Darwin's theory of natural selection as a possible mechanism for evolution, and consider how evolutionary ideas developed and have changed over the years.

An immense variety of animals inhabit every conceivable place on land and in the water. They exhibit tremendous variations in size, shape, and degree of complexity and in methods of obtaining food, of evading predators, and of reproducing their kind. How all these species came into existence, how they came to have the particular adaptations that make them peculiarly fitted for survival in a particular environment, and why there are orderly degrees of resemblance between forms that permit their classification in genera, orders, classes, and phyla are fundamental problems of zoology. From detailed comparisons of the structures of living and fossil forms, from the sequence of the appearance and extinction of species in times past, from the physiological and biochemical similarities and differences between species, and from analyses of heredity and variation in many different animals and plants has come one of the great unifying concepts of biology, that of **evolution.** It provides a framework into which all fields of biology are woven, and it gives us an awareness of our place in nature and time.

The term evolution means an unfolding, or unrolling, a gradual, orderly change from one state to the next. The principle of **organic evolution** applies this concept to living things: All the various plants and animals living today have descended from simpler organisms by modifications that have accumulated in successive generations.

Evidence for Evolution

The concept of organic evolution is supported by an enormous body of evidence, much of which was assembled by Charles Darwin and published in 1859 in his classic book, *On the Origin of Species by Means of Natural Selection.* Darwin's evidence was largely based on taxonomy, morphology, paleontology, geographic distribution, and different varieties developed by plant and animal breeders. Since then much more information in support of the concept has come from genetics, biogeography, physiology, and biochemistry. Indeed, all fields of biology and geology have made their contributions. Although biologists still are not in complete agreement on some aspects of the *mechanisms* for evolutionary change (Chapter 18), the concept that evolution has taken place is now as well documented as any scientific hypothesis. It is consistent with all of the information that can be brought to bear upon it. A scientific theory should be falsifiable in the sense that it should lead to observations or testable predictions that, if not true, would require that the theory be modified or rejected. A theory that cannot generate testable observations or predictions is simply not a scientific theory. Evolution is testable in the sense that one would predict that mammalian fossils would not be found in rocks of the same age as those containing the ancestral fishes, or that a newly discovered insect species would have many structural, biochemical, and genetic features in common with its presumed close allies. If such predictions were not true, the theory of evolution would be falsified and would need replacement or modification.

Taxonomy

The science of taxonomy began long before the doctrine of evolution was accepted; indeed the founders of scientific taxonomy, Ray and Linnaeus, were firm believers in the fixity, the unchangingness, of species. Present-day taxonomists are concerned with the naming and describing of species primarily as a means of discovering or clarifying evolutionary relationships. This is based upon the assumption that their degree of resemblance is a measure of their degree of relationship. The characteristics of living things are such that they can be fitted into a hierarchical scheme of categories, each more inclusive than the previous one—species, genera, families, orders, classes, and phyla. This can best be interpreted as proof of evolutionary relationship. If the kinds of animals and plants were not related by evolutionary descent, their characteristics would be present in a confused, random pattern, and no such hierarchy of forms could be established.

Adult Morphology

A study of the details of the structure of any particular organ system in the diverse members of a given phylum reveals a basic similarity of form that is varied to some extent from one class to another. A bird's wing, a bat's wing, a dolphin's front flipper, and a human arm and hand, though superficially dissimilar, are composed of very similar bones, muscles, and nerves (Fig. 17.1). Each has a single bone, the humerus, in the proximal part of the limb, followed by two bones (the radius and ulna) in the forearm, a group of carpals, and a variable number of fingers. This is particularly striking because wings, flippers, and the human arm are used in different ways, and there is no mechanical need for them to be so similar. Darwin pointed out that such basic structural similarities in organs used in different ways is the expected outcome if evolution, or "descent with modification," to use his phrase, has taken place. We call structures of this type **homologous** (Gr. *homologos,* agreeing). Criteria for homology are similarity of relationships (i.e., placement in the body), similarity in connections to nerves and other parts, and similarity in

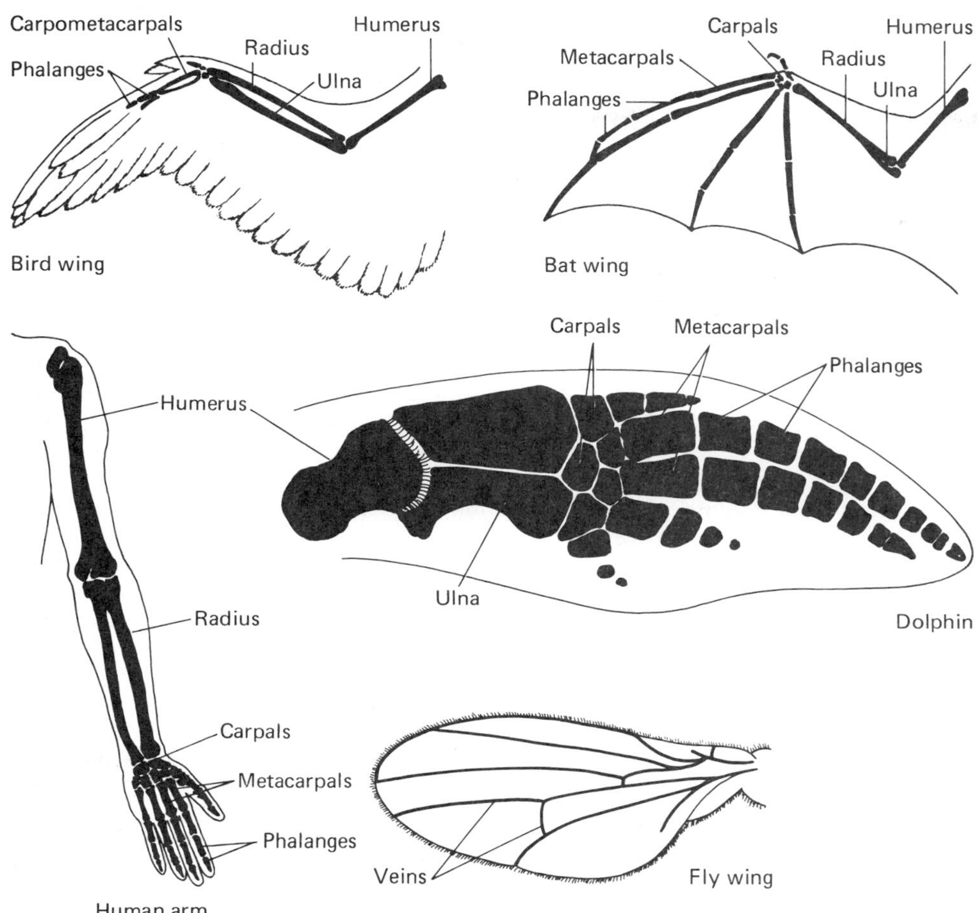

FIGURE 17.1 Structural similarities. The bones of the bird wing, bat wing, dolphin flipper, and human arm are homologous for they have a basic, underlying similarity. Although superficially similar to the vertebrate wings, the fly wing is analogous to them because it is constructed quite differently.

mode of embryonic development. As the theory of evolution became accepted and we learned more about inheritance, biologists realized that homologous organs share a basic resemblance because they share a common evolutionary origin and an underlying gene complex. Biologists are sometimes accused of circular reasoning, defining homologous organs as having a common origin and then using them as evidence for evolution, but this is not the way the concept was used historically. Homologous organs were recognized and criteria for their identification were established long before the acceptance of evolution.

Another type of resemblance between organs is **analogy,** or similarity of function. Many biologists, as we do, limit the term analogy to functional similarity among nonhomologous organs so that the two concepts do not overlap. There may or may not be some superficial structural similarity among analogous organs, depending upon the nature of their function. The trachea of insects and the lungs of mammals are analogous organs that have evolved to meet the common problem of breathing air. There are many ways of breathing air, and the morphological resemblances between them need not be great. An airfoil is one of the few solutions to flight, and unrelated flying animals have evolved wings that, being airfoils, resemble one another at least superficially, as do the wings of insects and vertebrates (Fig. 17.1). In more fundamental ways the wings are quite different. Vertebrate wings are modified pectoral appendages supported by bones; insect wings, outgrowths of the dorsal thoracic wall supported by chitinous "veins." In making judgments as to whether similar organs are homologous or analogous, an investigator must be mindful of the level of comparison being made. The skeletons of the bird's and bat's wings are homologous. They meet the criteria of homology, and we know that they evolved from the forelimb of a common vertebrate ancestor. The flying surfaces of their wings are quite different: superficial feathers growing out from the posterior margin of the wings in the bird and essentially a webbed hand in the bat. At this level of comparison bird and bat wings are analogous. Flight evolved independently in the two groups. Although the pectoral appendage was utilized as a wing in each group, it was modified in different ways.

Homologous organs are important evidences of evolution and are used to help determine the interrelationships of animals (Chapter 19). Analogous organs are not evidence of "descent with modification" but are of evolutionary interest because they show how unrelated groups adapt to common problems as their evolution leads to their convergence in a similar habitat.

Vestiges

Many species of animals have organs, or parts of organs, that are small and are either functionless or retain only some of the functions of the organ's well-developed homologue in another species. We call such structures **vestigial organs.** There are many examples: the pelvic girdle of a whale, auricular muscles attaching to our external ear flaps, our reduced tail vertebrae (coccyx), the semilunar fold in our eye (Fig. 17.2). Some of these may be without any function, although it is logically impossible to prove a negative because there may be some subtle, unrecognized function. Whatever it may or may not do, the semilunar fold does not sweep across the surface of the eye and help to cleanse it as its homologue, the nictitating membrane, does in most other mammals. Our coccyx and the pelvic girdle of the whale still serve as points of attachment for certain pelvic muscles, but, again, this is only one of the functions that a well-developed pelvis and tail have. The presence of occasional vestigial organs is to be expected as an ancestral species evolves and adapts to different modes of life. Some organs will become more important; others, less important.

Embryologic Resemblances

As Darwin pointed out, there is a closer resemblance between embryos of different animals than between their adults, and it is not easy to distinguish the early embryo of a human, fish, frog, chick, or pig (Fig. 17.3). Segmented muscles, gill pouches, a tubular heart undivided into left and right sides, a system of aortic arches in the gill region (Fig. 17.4), and many other features are found in the embryos of all vertebrates, but none of them persists in the adults of reptiles, birds, or mammals. Why are they present? Again, their presence is consistent with evolutionary theory. All of these structures are necessary stages in the development of a fish. The small, segmented, premuscular blocks of the embryo give rise to the segmented muscles of the adult fish, which are used in its particular type of swimming. The gill pouches break through to the surface as gill slits. The heart remains undivided for it pumps only venous blood forward to the gills that develop in association with the aortic arches. If higher vertebrates evolved from fish, they would inherit a fish's pattern of development, but accumulating genetic changes would modify the pattern to some extent to meet new needs. Some developmental steps would be abbreviated, and new characters would evolve that are adaptive and enable the embryo to survive to later stages. For example, most

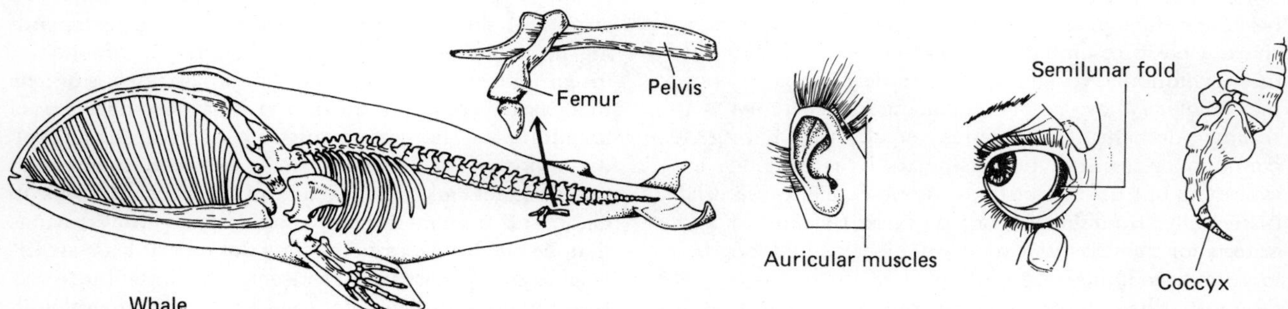

FIGURE 17.2 Examples of vestigial organs. The pelvis and femur are from a whale; the others, from the human body.

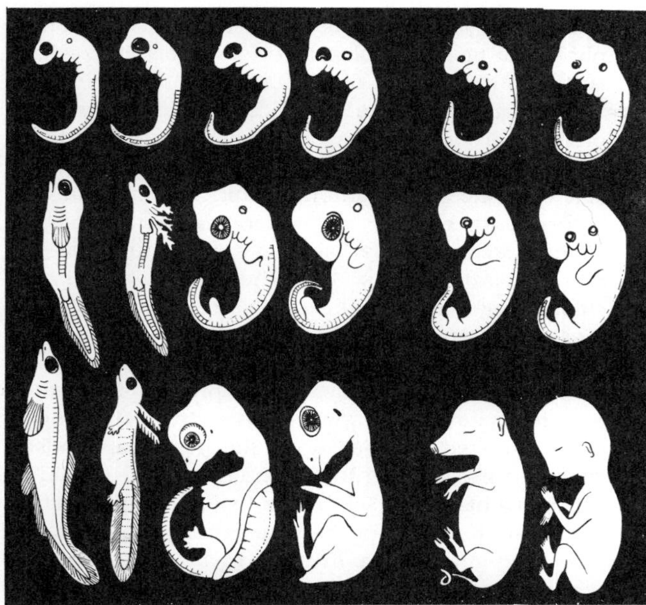

FIGURE 17.3 Comparison of early and later stages in the development of vertebrate embryos. Note the similarity of the earliest stages of each.

mammalian embryos have other structures, such as a placenta, that enable them to survive and to develop within the mother's uterus. Such secondary traits may alter the original characters common to all vertebrates so that the basic resemblances are blurred. The production by mammals of eggs with little yolk and modifications of cleavage and blastulation such that a trophoblast develops very early are correlated with the formation of a placenta. They are unique mammal features not characteristic of primitive vertebrates.

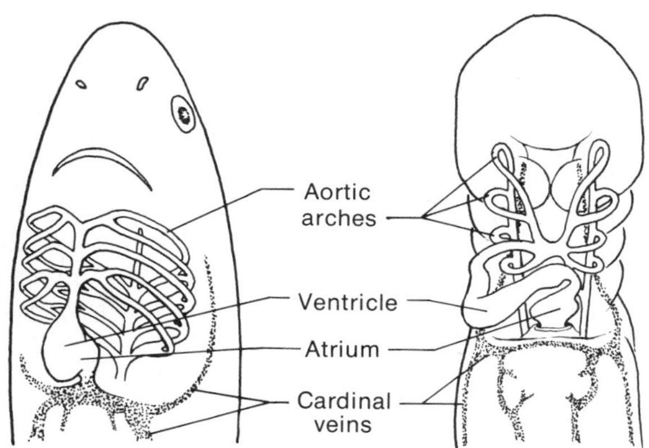

FIGURE 17.4 Ventral views of the heart and aortic arches of a human embryo *(right)* and an adult shark *(left)*. Both have a single atrium and single ventricle, several aortic arches, and anterior and posterior cardinal veins emptying in the heart. Gills, not shown, are associated with the shark's aortic arches.

All chordate embryos develop a dorsal hollow nerve cord, a notochord, and pharyngeal pouches. The early human embryo at this stage resembles a fish embryo, with gill pouches, pairs of aortic arches, a fishlike heart with a single atrium and ventricle, and a well-differentiated tail complete with muscles for wagging it. At a slightly later stage the human embryo resembles a reptilian embryo. Its gill pouches regress, and the atrium becomes divided into right and left chambers. Still later in development, the human embryo develops a mammalian, four-chambered heart. During the seventh month of intrauterine development the human embryo, with its coat of embryonic hair and in the relative size of body and limbs, resembles a baby ape more than it resembles an adult human.

Our increasing understanding of physiologic genetics provides us with an explanation of these phenomena. All chordates have in common a certain number of genes that regulate the processes of early development. As our ancestors evolved from fish, through amphibian and reptilian stages, they accumulated mutations for new characteristics but kept some of the original "fish" genes, which still control early development. Later in development the genes that the human shares with amphibians influence the course of development so that the embryo resembles a frog embryo. Subsequently, some of the genes that we have in common with reptiles come into control. Only after this do most of the peculiarly mammalian genes exert their influence, and these are followed by the action of genes we have in common with other primates. It is not surprising, therefore, to see a close resemblance among the early embryos of vertebrates or those of other groups of animals.

Ernst Haeckel in 1866 argued that the resemblance between embryos was of even greater significance. He believed that embryos of every species represent adult stages in the evolution of that species, that the embryo during its embryonic development (ontogeny) literally reclimbs its evolutionary tree (phylogeny). By carefully working out embryonic sequences, we could reconstruct the course of evolution. This notion stimulated a great deal of descriptive embryological research, but Haeckel's view that "ontogeny recapitulates phylogeny" became untenable as the fossil record became more complete and as we gained an increased understanding of embryonic processes. No *adult* ancestors of mammals ever had the external gills present in larval amphibians or the extraembryonic membranes of embryonic reptiles. What recapitulation there is, is at the embryonic level. Animals possess embryos that may resemble the embryos of their ancestors. Thus embryonic stages of different animals may resemble each other more closely than do the adults.

Physiological and Biochemical Resemblances

As studies of the physiology and biochemical traits of organisms have been made using a wide variety of animal types, it has become clear that there are functional similarities and differences that parallel closely the morphological ones. Indeed, if one were to establish taxonomic relationships based on physiological and biochemical characters instead of on the usual structural ones, the end result would be much the same.

The blood serum of each species of animal contains certain specific proteins. The degree of similarity of these serum proteins can be determined by **antigen-antibody** reactions (p. 171). When serum proteins are compared by this method, our closest "blood relations" have been found to be the great apes and then, in order, the Old World monkeys, the New World monkeys, and finally the tarsioids and lemurs. The biochemical relationships of a variety of forms tested in this way correlate with and complement the relationships determined by other means. Cats, dogs, and bears are closely related; cows, sheep, goats, deer, and antelopes constitute another closely related group. Similar tests of the sera of crustaceans, insects, and mollusks have shown that those forms regarded as being closely related from morphological or paleontological evidence also show similarities in their serum proteins.

Investigations of the sequence of amino acids in the protein portion of the cytochromes from different species have revealed great similarities, of course, and specific differences, the pattern of which demonstrates the nature and number of underlying mutations, the changes in nucleotide base pairs, that must have occurred in evolution (Box 17.1). The evolutionary relationships inferred from these studies agree completely with those based on morphological studies.

It might seem unlikely that an analysis of the urinary wastes of different species would provide evidence of evolutionary relationship, yet this is true. The kind of waste excreted depends upon the particular kinds of enzymes present, and the enzymes are determined by genes that have been selected in the course of evolution. The waste products of the metabolism of purines, adenine and guanine, are excreted by human beings and other primates as uric acid, by other mammals as allantoin, by amphibians and most fishes as urea, and by most invertebrates as ammonia.

Genetics and Cytology

For the past several thousand years human beings have been selecting and breeding animals and plants for their own uses, and a great many varieties, adapted for different purposes, have been established. These results of artificial selection provide striking models of what may be accomplished by evolution. All of our breeds of dogs have descended from one, or perhaps a very few, species of wild dog or wolf. A comparable range of varieties has been produced by artificial selection in cats, chickens, sheep, cattle, and horses.

Geneticists have been able to trace the ancestry of certain modern plants by a combination of cytological techniques in which the morphology of the chromosomes is compared and by breeding techniques that compare the kinds of genes and their order in particular chromosomes in a series of plants. In this way, the present cultivated tobacco plant, *Nicotiana tobacum,* was shown to have arisen from two species of wild tobacco, and corn was traced to teosinte, a grasslike plant that grows wild in the Andes and Mexico.

Geology and Paleontology

All of the lines of evidence we have been considering are most logically interpreted as the result of evolution. The most direct evidence for evolution comes from the

BOX 17.1

The diagram illustrates the difference in amino acid sequences in the cytochrome c's from a variety of different species of animals and plants. The numbers refer to the number of different amino acids in the cytochrome c's of the species compared. For example, there is 1 difference between a human being (abscissa) and a monkey (ordinate), 10 between a human and a kangaroo, and 48 between a human and the mold *Neurospora*. Since differences in the amino acids arose by mutations in the genetic code, the amino acid differences reflect the degree of genetic similarity and difference between the organisms compared. Not surprisingly, we are closer genetically to monkeys than to the others listed. (From Dayoff, M. O., and R. V. Eck: *Atlas of Protein Sequence and Structure*. Silver Springs, Maryland, National Biomedical Research Foundation, 1968.)

	Human	Monkey	Pig, Bovine, sheep	Horse	Dog	Rabbit	Kangaroo	Chicken, turkey	Duck	Rattlesnake	Turtle	Tuna fish	Moth	Neurospora	Candida	Yeast
Human	0															
Monkey	1	0														
Pig, bovine, sheep	10	9	0													
Horse	12	11	3	0												
Dog	11	10	3	6	0											
Rabbit	9	8	4	6	5	0										
Kangaroo	10	11	6	7	7	6	0									
Chicken, turkey	13	12	9	11	10	8	12	0								
Duck	11	10	8	10	8	6	10	3	0							
Rattlesnake	14	15	20	22	21	18	21	19	17	0						
Turtle	15	14	9	11	9	9	11	8	7	22	0					
Tuna fish	21	21	17	19	18	17	18	17	17	26	18	0				
Moth	31	30	27	29	25	26	28	28	27	31	28	32	0			
Neurospora	48	47	46	46	46	46	49	47	46	47	49	48	47	0		
Candida	51	51	50	51	49	50	51	51	51	51	53	48	47	42	0	
Yeast	45	45	45	46	45	45	46	46	46	47	49	47	47	41	27	0

sciences of geology and paleontology which deal, respectively, with the history of the earth and with the finding, cataloging, and interpretation of fossils.

The Fossil Record. The term **fossil** (L. *fossilis*, something dug up) refers not only to the bones, shells, teeth, and other parts of an animal's body that may survive but

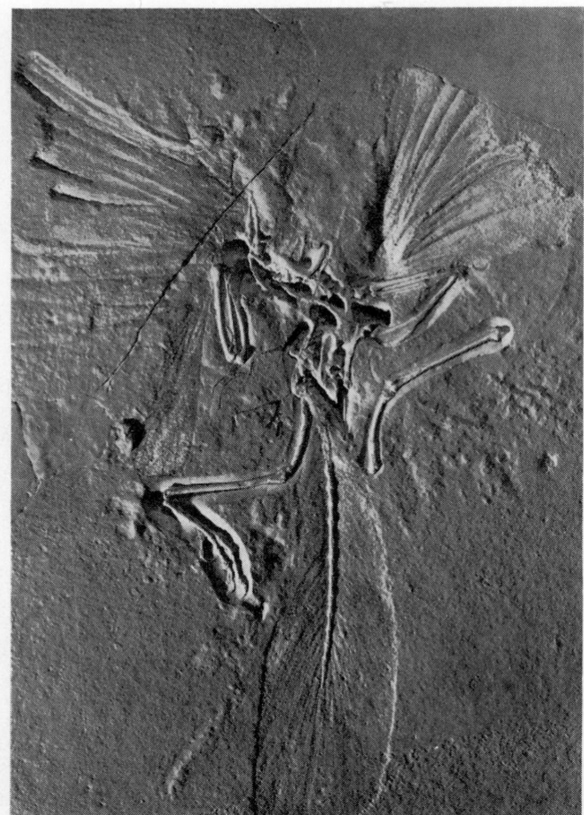

FIGURE 17.5 An example of a fossil, the remains of *Archaeopteryx*, a tailed, toothed bird from the Jurassic period. (Courtesy of the American Museum of Natural History.)

also to any impression or trace left by previous organisms (Fig. 17.5). Footprints or trails made in soft mud that subsequently hardened are a common type of fossil. From such remains one can infer something of the structure and locomotion of the animals that made them.

The commonest vertebrate fossils are skeletal parts. From the shapes of bones and the positions of the bone scars that indicate points of muscle attachment, paleontologists can make inferences about an animal's posture and style of walking, the position and size of its muscles, and hence the contours of its body. Careful study of fossil remains has enabled paleontologists to make reconstructions of what the animal must have looked like in life.

In one type of fossil, the original hard parts, or more rarely the soft tissues of the body, have been replaced by minerals, a process called **petrifaction.** Iron pyrites, silica, and calcium carbonate are some of the common petrifying minerals. The muscle of a shark more than 300,000,000 years old was so well preserved by petrifaction that not only the individual muscle fibers but also their cross striations could be observed in thin sections under the microscope.

Molds and casts are superficially similar to petrified fossils but are produced in a different way. **Molds** are formed by the hardening of the material surrounding a buried organism, followed by the decay and removal of the body of the organism. The mold may subsequently be filled by minerals that harden to form **casts** that are exact replicas of the original structures.

Some animal remains have been exceptionally well preserved by being embedded in bogs, tar, amber, or ice. The remains of woolly mammoths, deep frozen in Siberian ice for more than 25,000 years, were so well preserved that the meat was edible!

The formation and preservation of a fossil require that some structure be buried under conditions that will slow decay. This is most likely to happen if an animal's remains are covered quickly by water-borne sediments. Remains of aquatic organisms may be trapped in bogs, mud flats, sand bars, and deltas. Those of terrestrial animals that lived on a flood plain may also be covered by water-borne sediments, or, if they lived in arid regions, by wind-blown sand. Occasionally terrestrial animals may be trapped in a tar pit, as in La Brea near Los Angeles, or be covered by volcanic ash, as in Pompeii following the eruption of Mt. Vesuvius. It follows from the conditions of preservation that the fossil record is not a random sample of past life. There is a bias toward aquatic animals and those living in a few terrestrial habitats. Decay is very rapid on the forest floor, and few fossils of forest animals are known. These facts must be considered in interpreting the fossil record.

Through the accumulation of sediments and movements of the earth's crust, sediments containing fossils are compressed, and some sedimentary rocks are uplifted. Doubtless many fossils are distorted or destroyed in the process. Only a few of the sedimentary rocks that have been formed can be studied where they are exposed in mountain ranges or where river valleys or our road building efforts have cut deeply into them.

In order to be interpreted, the sedimentary layers containing fossils must be arranged in chronological order. The layers of sedimentary rock should occur in the sequence of their deposition, with the newer, later strata on top of the older, earlier ones, but subsequent geologic events may have changed the relationship of the layers. Not all of the expected strata may occur in some particular region, for that land may have been exposed rather than submerged during one or more geologic ages. In some regions the strata formed previously have subsequently emerged, been washed away, and then

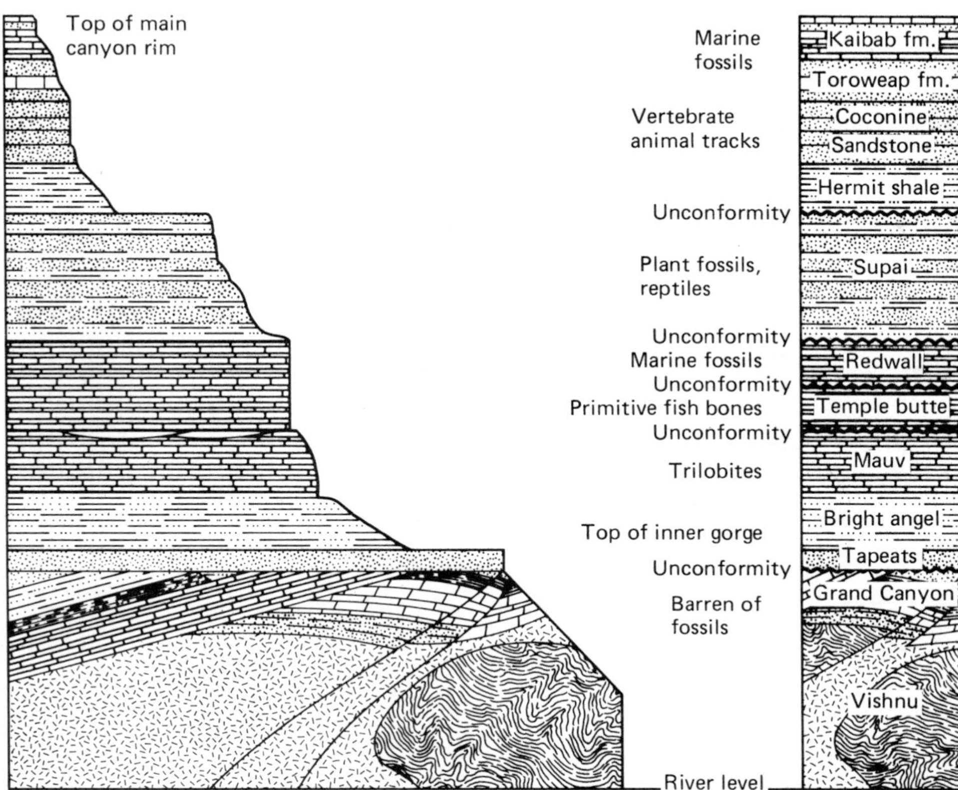

FIGURE 17.6 *Left,* Drawing of a vertical section through the Grand Canyon showing the sequence of rock layers. *Right,* Diagram showing the names of the layers, the major fossils they contain, and the positions of unconformities. (From Press, F., and R. Siever: *Earth.* 2nd ed. San Francisco, W. H. Freeman & Co., 1978.)

relatively recent strata have been deposited directly on very ancient ones. Geologists call an old erosion surface an **unconformity** (Fig. 17.6). Certain sections of the earth's crust, in addition, have undergone massive foldings and splittings so that the original relationship of layers to one another has been distorted. Fortunately geologists can identify specific layers of rock by such features as the type of rock (limestone, sandstone, shale, and so forth), grain in the rock, what layers lie above or below a given layer, and by certain key invertebrate fossils, known as **index fossils,** that characterize a specific layer. Utilizing these tools, geologists can arrange fossils in chronological order, identify layers in widely separated localities, and fill in some of the gaps in one locality by layers present in a different locality.

Earth history has been very complex. Sea basins have appeared and disappeared, and extensive uplifts and submergences of portions of the earth's crust have occurred. Geologists now attribute these changes to plate tectonics (Gr. *tektōn,* a builder), that is, to the slow movements of enormous crustal plates across a more plastic underlying mantle. An upwelling of molten material, as is now taking place in the midocean ridges, may push plates apart and form ocean basins between them. Plates may collide, leading to the upthrust of mountain ranges or the formation of deep ocean trenches where one plate subducts beneath another. The most recent separation of North and South America from Europe and Africa began about 100 million years ago (Fig. 17.7), and the Atlantic ocean is continuing to widen. The subduction of Pacific plates beneath North and South America has been responsible for the recent (geologically speaking) uplift of the Rocky Mountains and the Andes, and the collision of India with Asia has caused the uplift of the Himalayas.

Geologic Time. In Darwin's day, geologists could make only rough guesses as to the age of the rocks by

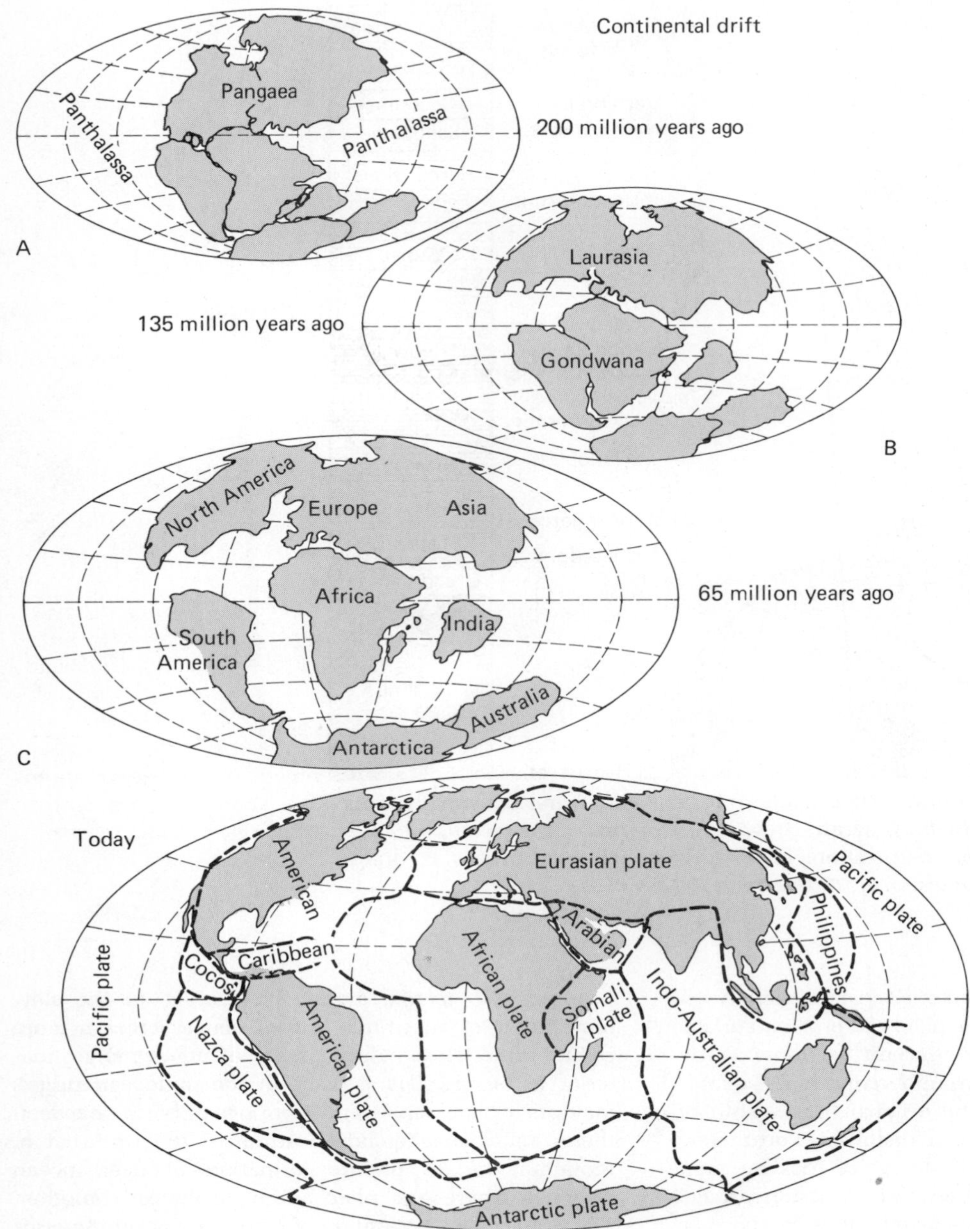

FIGURE 17.7 Continental drift. *A,* The supercontinent Pangaea of the Triassic period, about 200 million years ago. *B,* Breakup of Pangaea into Laurasia (Northern Hemisphere) and Gondwana (Southern Hemisphere) 135 million years ago in the Cretaceous period. *C,* Further separation of land masses, which occurred in the Tertiary period, 65 million years ago. Note that Europe and North America are still joined and that India is a separate land mass. *D,* The continents today. (From Norstog, K., and R. W. Long: *Plant Biology.* Philadelphia, Saunders College Publishing, 1976.)

measuring the thickness of layers and estimating rates of sedimentation. Owing to the discovery and study of radioactivity in the late nineteenth and early twentieth centuries, geologists now have several geochronometers to use. All are based on the fact that certain isotopes of radioactive elements emit or capture atomic particles and change into different isotopes, or isotopes of a different element, at measurable rates that are not affected by the temperatures or pressures to which the rocks have been subjected. For example, uranium-238 (^{238}U) emits alpha particles (two protons and two neutrons) and is transformed into lead-206 (^{206}Pb) with a half-life of 4.5 billion years. That is, one half of the ^{238}U is transformed into ^{206}Pb in this period. It will take another 4.5 billion years for half of what is left to be transformed, and so on. Radioactive potassium (^{40}K) captures an orbital electron in its nucleus and is transformed into argon-40 (^{40}Ar) with a half-life of 1.3 billion years. By measuring the proportions of the remaining original element and the derived element in a sample, one can calculate the time that has elapsed since the rock crystallized and the elements remained associated. These geochronometers can be used only on igneous rocks that crystallized from molten intrusions. Sedimentary rocks are dated by their location relative to the igneous intrusion.

The uranium and potassium clocks are useful for dating rocks formed millions or billions of years ago. The half-lives of these elements are so long that the changes that would occur during periods of thousands of years would be very small and close to the limits that our instruments can detect. Events that have occurred in the last 40,000 years or so can be dated by the amount of carbon-14 and carbon-12 in the remains of organisms, for ^{14}C has a half-life of only 5570 years. Carbon in organic remains is derived from the fixation of atmospheric carbon dioxide so the ratio of ^{12}C to ^{14}C in living organisms is the same as in the atmosphere. After death, ^{12}C does not change, but the amount of ^{14}C decreases because its nucleus emits a beta particle, which becomes an orbiting electron, and it is transformed into nitrogen-14 (^{14}N). The change in the ratio ^{12}C:^{14}C in the remains is a function of time.

The oldest dated rocks on earth are about four billion years old, dating from the time when the earth's crust began to form. Astronomical evidence indicates that the earth and the other planets began to condense about 4.7 billion years ago. The oldest fossils of cells are over three billion years old, but fossil organisms do not become abundant until the beginning of the Cambrian period, 570 million years ago. Geologists divide time from the Cambrian on into three eras: Paleozoic, Mesozoic, and Cenozoic (Fig. 17.8). These, in turn, are divided into periods, which are usually named for a region where outcrops of the type of rock formed at that time are found (Cambrian, from Cambria, an old name for Wales, Devonian from Devon in England, Jurassic from the Jura Mountains, and so forth). The two most recent periods are further divided into epochs.

The Succession of Floras and Faunas

A study of the fossil record reveals a succession of floras and faunas that are direct evidence for evolution. From simple beginnings, organisms become more complex and diverse as time goes on. From their origins in the seas, they spread into fresh water, onto the land, and even into the air. Progression is not continuous but is punctuated by episodes of rapid species formation as ecological opportunities are exploited, periods of stasis as habitats become filled, and extinction when environmental conditions change more rapidly than organisms can adjust to the changes (Fig. 17.8).

Precambrian Time. The oldest deposits, which are from South African rocks 3.4 billion years old, contain evidence of bacteria and bacterial activity. From then until nearly the end of Precambrian time, a period of some 2.5 billion years, only fossils of prokaryotes are found. Marine multicellular organisms appeared late in the Precambrian: algae, jellyfish, sponges, some worms, and a few arthropods. Most were soft-bodied creatures.

The Paleozoic Era. A great expansion in the number and diversity of organisms occurred late in the Precambrian and early in the Cambrian period. All of the major groups of invertebrates appeared, including those with shells and exoskeletons. At one time paleontologists believed that this explosion of life was an artifact of a good Cambrian and a poor Precambrian fossil record, but new discoveries indicate that many primitive forms of life were present during the Precambrian. The sudden increase in diversity probably resulted from the evolution of some environmental characteristic or feature of organisms that promoted diversity. Possibly the accumulation of oxygen from the photosynthetic activity of bacteria and algae reached a threshold that permitted the expansion of life. The most numerous Cambrian animals were brachiopods (p. 621) and trilobites (p. 575). One of the present-day brachiopods, *Lingula,* is the oldest known genus of animals and is almost identical with its Cambrian ancestors.

Evolution since the Cambrian has been characterized by the development and ramification of the lines already present rather than by the establishment of entirely new

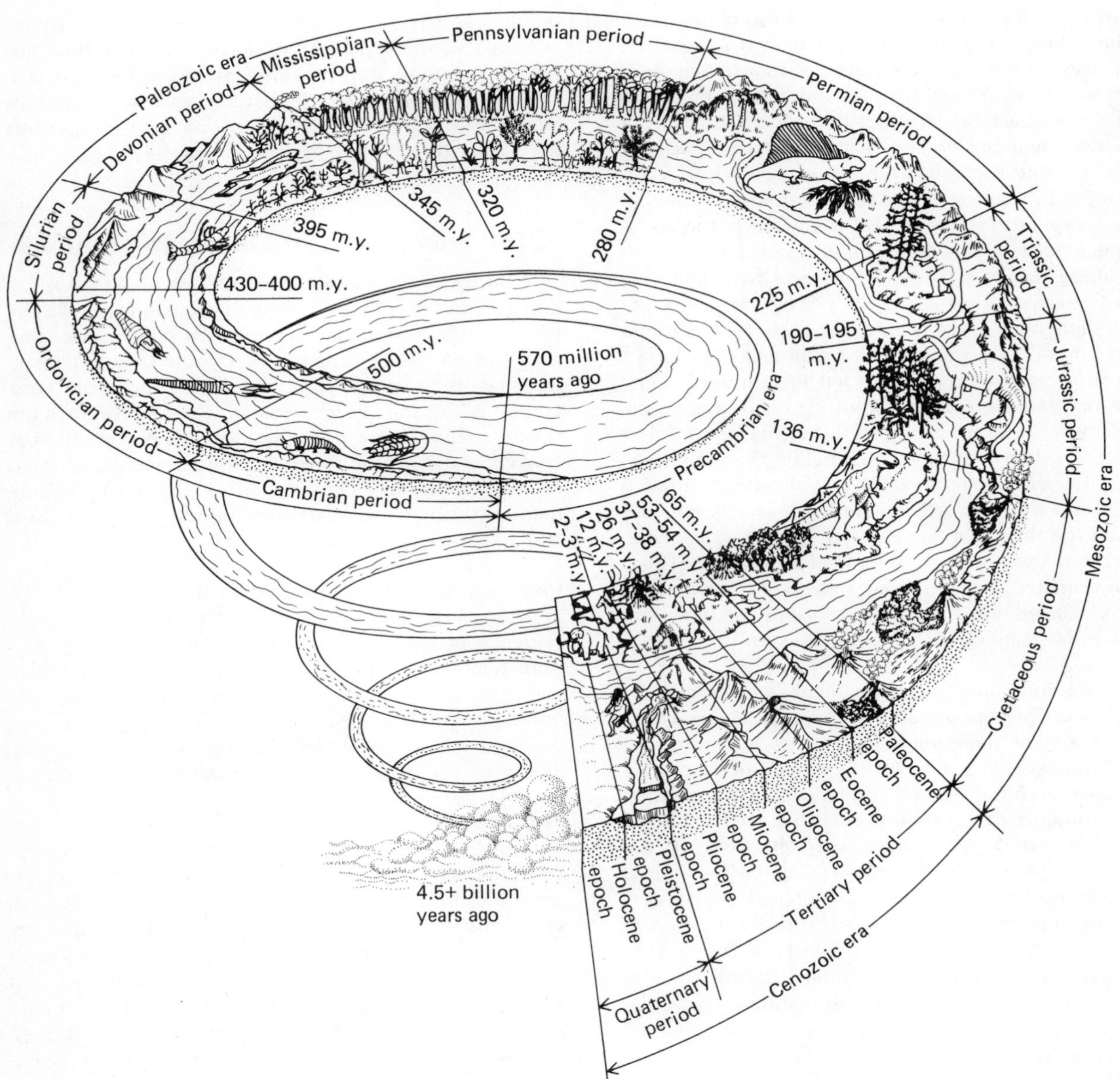

FIGURE 17.8 The geologic time scale and the history of the earth since the beginning of the Paleozoic Era. (From Press, F., and R. Siever: *Earth.* W. H. Freeman and Co., 1978; after Geological Survey publication.)

groups. For example, primitive mollusks were present in the early Cambrian, but *Nautilus*-like forms do not appear in the fossil record until the upper Cambrian; bivalve mollusks, not until the Ordovician; and the ancestors of the modern squids, not until later in the Paleozoic. The first vertebrates, the jawless, limbless, armored, bottom-dwelling fishes called **ostracoderms** (p. 653), appeared in the late Cambrian. Two important events of the Silurian were the evolution of land plants and of the first air-breathing animals, primitive scorpions.

The Devonian seas contained corals, sea lilies, and brachiopods in addition to a great variety of fishes. Trilobites were still present but were declining in numbers and importance. The first land vertebrates, the amphibians called **labyrinthodonts,** appeared in the latter part of the Devonian; this period also saw the first true forests of ferns, "seed ferns," club mosses, and horsetails and the first wingless insects and millipedes.

The Mississippian and Pennsylvanian periods are frequently grouped together as the Carboniferous, for during this time there flourished the great swamp forests whose remains gave rise to the major coal deposits of the world. The earliest stem reptiles appeared in the Pennsylvanian, and from these there evolved in the succeeding Permian period a group of early, mammal-like reptiles, the **pelycosaurs,** from which the mammals eventually evolved. Two important groups of winged insects, the ancestors of the cockroaches and the ancestors of the dragonflies, evolved during the Carboniferous.

The Permian period was characterized by widespread changes in topography and climate associated with the coming together of all of the continents (Fig. 17.7) into the single supercontinent of Pangea (Gr. *pan,* all + *gaia,* earth). The climate probably was harsh. There are indications of glaciation in the southern part of Pangea, in regions now forming parts of South America, Africa, India, Australia, and Antarctica. Shallow seas and swamps were drained as land masses came together. Originally separate faunas came into competition. Nearly half of the families of marine invertebrates became extinct. The trilobites finally disappeared, and the brachiopods, stalked echinoderms, cephalopods, and many other kinds of invertebrates were reduced to small, unimportant, relic groups.

Mesozoic Era. The Mesozoic era began some 225,000,000 years ago and lasted 160,000,000 years. It is subdivided into the Triassic, Jurassic, and Cretaceous periods. During the Triassic and Jurassic most of the continental area was above water, warm, and fairly dry. During the Cretaceous the sea once again overspread large parts of the continents. In the latter part of the Cretaceous the interior of the North American continent was submerged and cut in two by the union of a bay from the Gulf of Mexico and one from the Arctic Sea. Upheaval of the Rockies, Alps, Himalayas, and Andes mountains occurred at the end of the Cretaceous. The Mesozoic was characterized by the tremendous evolution, diversification, and specialization of the reptiles and is commonly called the Age of Reptiles. Mammals originated in the Triassic and birds in the Jurassic. Most of the modern orders of insects appeared in the Triassic. Conifers and other gymnosperms, which were the dominant terrestrial plants near the beginning of the Mesozoic Era, began to be replaced by the flowering plants, or angiosperms.

Massive extinctions of marine, freshwater, and terrestrial organisms occurred during the last part of the Cretaceous. The dinosaurs and most of the dominant reptilian groups disappeared. The reasons are not entirely clear. Climatic changes certainly occurred as new mountain ranges began to form. There is also evidence that the earth collided with a giant asteroid that would have sent dust clouds circulating around the earth for years.

Cenozoic Era. The Cenozoic era is subdivided into the earlier Tertiary period, which lasted some 63,000,000 years, and the present Quaternary period, which includes the last one or two million years.

The Tertiary is subdivided into five epochs, the Paleocene, Eocene, Oligocene, Miocene, and Pliocene. The Rockies, formed at the beginning of the Tertiary, were considerably eroded by the Oligocene. Another series of uplifts in the Miocene raised the Sierra Nevadas and a new set of Rockies and resulted in the formation of the western deserts. The climate of the Oligocene was rather mild, and palm trees grew as far north as Wyoming. The uplifts of the Miocene and Pliocene and the successive ice ages of the Pleistocene killed off many of the mammals that had evolved.

Four periods of glaciation occurred in the Pleistocene, between which the sheets of ice retreated. At their greatest extent, these ice sheets extended as far south as the Missouri and Ohio rivers and covered 4,000,000 square miles of North America. The Great Lakes, which were carved out by the advancing glaciers, changed their outlines and connections several times. During the Pleistocene glaciations enough water was removed from the oceans and locked in the vast sheets of ice to lower the level of the oceans as much as 100 meters. This created land connections, highways for the dispersal of many land forms, between Siberia and Alaska at Bering Strait and between England and the continent of Europe. Many mammals, including the saber-toothed tiger, the mammoth, and the giant ground sloth, became extinct in the Pleistocene after primitive human beings had appeared.

Geographic Distribution

In the course of his travels around the world in *H.M.S. Beagle,* Darwin was greatly impressed by his observations that plants and animals were *not* found everywhere that they could exist if climate and topography were the only factors determining their distribution. Central Africa, for example, has elephants, gorillas, chimpanzees, lions, and antelopes, while Brazil, with a similar climate and other environmental conditions, has none of these but does have prehensile-tailed monkeys, sloths, and tapirs, animals not found in Africa. The present distribution of organisms, and the sites at which their fossil remains are found, are understandable only on the basis of the evolutionary history of each species and the tectonic movements of the plates of the earth's crust.

Each new species arises from the transformation of a former species, or more commonly from the divergence of a population of a previous species (Chapter 18). This, of course, occurs in a limited geographic area, the **center of origin** of the species. If the new species is successful, its population grows and it expands its **range,** or the territory that it occupies, until it reaches **barriers** that it cannot cross. A barrier may be a physical one, such as an ocean, mountain or desert, an environmental one, such as unfavorable climate, or a biological barrier, such as the absence of food or the presence of other species that prey upon it or compete with it for food or shelter.

Oceans are particularly effective barriers to the dispersal of nonflying terrestrial species, yet islands are populated by such species. How did they get there? Some islands, such as the British Isles, have recently (geologically speaking) been connected to continents and share many species with them. But oceanic islands, such as the Galapagos Islands that lie in the Pacific 600 miles west of Ecuador, have been formed by volcanic activity and have never been attached to continents. It is not surprising, therefore, that most of the species on these islands are found only in this limited geographic area. Such species are called **endemic species.** Indeed, many of the species in the Galapagos archipelago are endemic to individual islands. Yet all show some remote affinity to the South and Central American faunas. Darwin was much impressed by the differences in the nature of species between continental and oceanic islands.

Oceanic islands are colonized by chance. Mats of trees and other vegetation, for example, are carried down continental rivers into the ocean. Occasionally some of these are picked up by ocean currents and carried far to sea. The Galapagos Islands are close to the intersection of the Humboldt current flowing north from South America and the Panama current flowing south from Central America. Presumably ancestral species reached these islands by riding on these mats. No frogs, toads, or other animals are found that could not have survived exposure to salt water on a long ocean voyage. Once on the islands and isolated from the mainland, the new arrivals underwent further speciation.

Biogeographic Realms. Careful studies of the distribution of plants and animals have revealed a distinctive pattern that was pointed out by Alfred Russel Wallace in the late 1800s. We can recognize six major **biogeographic realms,** each characterized by the presence of many unique species (Fig. 17.9). These realms are the direct outcome of the centers of origin of species and their past migrations and the barriers they encountered, including those resulting from past connections and separations of continents.

North America, Europe, and Asia were long interconnected as Laurasia and were separated from the southern continents (Fig. 17.7*B*). It is not surprising, therefore, that they share many species, such as foxes, wolves, and elk. They are sometimes grouped into a single **Holarctic realm.** Eventually North America separated, and its fauna diverged to some extent. The **Palearctic realm** (Europe and Asia) is characterized by oxen, sheep, goats, robins, and magpies. Unique organisms of the **Nearctic realm** (North America) include the mountain goat, prairie dog, skunk, raccoon, bison, bluejay, and turkey.

The southern continents, originally part of Gondwana, separated from each other, but South America and Australia remained interconnected by way of Antarctica for a long time (Fig. 17.7*C*). Unique species that evolved in the **Ethiopian realm** (Africa) include the gorilla, chimpanzee, zebra, rhinoceros, hippopotamus, giraffe, and aardvark. Although Africa eventually became connected with Europe and Asia, the Mediterranean Sea and North African deserts were effective barriers to much faunal interchange.

The **Oriental realm** (India and Southeastern Asia) is characterized by the Indian elephant, black panther, many primitive primates, and gibbon. Many of these species originally evolved in India but spread into Southeast Asia when the movement of plate tectonics led to the collision of India and Asia. But the resulting elevation of the Himalaya mountains prevented a further faunal interchange.

Throughout much of the Tertiary period South America was isolated and contained many unique species of animals, including many marsupials and primitive eutherians. When South America became connected with North America late in the Tertiary, there was considerable faunal interchange. The armadillo and opossum moved north. Many cats and other carnivores moved south, and many South American animals became extinct. Climatic conditions in Central America preclude a complete faunal exchange, and South America remains a distinctive **Neotropical realm** characterized by such species as llamas, prehensile-tailed monkeys, sloths, tapirs, anteaters, toucans, and certain parrots found nowhere else.

Australia and its neighboring islands, which constitute the **Australian realm,** have remained isolated until historic times and have retained a unique fauna, including the duck-billed platypus, echidna, kangaroo, wombat, koala bear, and other marsupials. Its assortment of curious birds includes the cassowary and emu, the lyrebird, cockatoo, and bird-of-paradise.

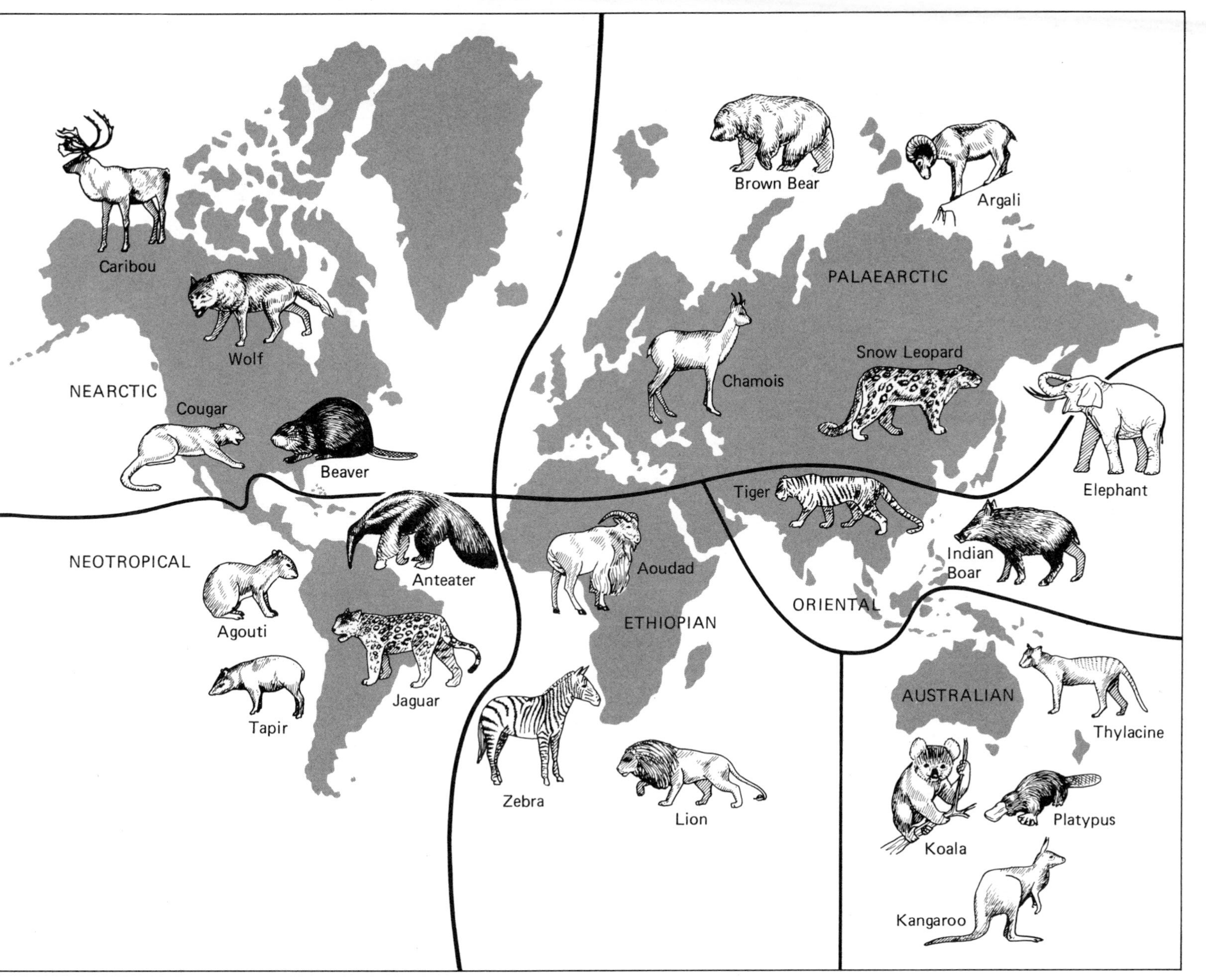

FIGURE 17.9 The biogeographic realms of the world with some of their characteristic animals.

Why certain animals appear in one region yet are excluded from another in which they are well adapted to survive (and in which they flourish when introduced by humans) can be explained only by their evolutionary history.

The Theory of Natural Selection

It was Darwin's genius not only to compile the evidence for evolution but to propose a plausible hypothesis, the theory of **natural selection,** to account for evolutionary change. The theory of natural selection, as developed by Darwin and Wallace, is based on a combination of observations well known at the time and certain inferences drawn from them.

1. It could be observed that organisms have a very high reproductive potential. More organisms of each kind are born than can possibly obtain food and survive. If all of the offspring of any species remained alive and reproduced, they would soon crowd all others from the earth.
2. This high reproductive potential is not achieved, since the population size of each species remains fairly constant or fluctuates within limits.
3. It follows from these observations that, despite some chance elimination of offspring, those that are better fitted to the particular conditions would survive in greater number than those that are less fit.
4. Darwin was aware of the existence of numerous inherited variations, although he and his contemporaries did not understand the basis for variation or inheritance.
5. Given variation and "survival of the fittest" (a phrase coined by Herbert Spencer, a contemporary philosopher, and not used by Darwin), Darwin and Wallace further deduced that the surviving individuals would transmit their superior qualities to the next generation. Superior features would accumulate generation after generation, lead to adaptive change, and result in the "origin of species."

Natural selection operates in many ways. Nineteenth-century Darwinians emphasized the direct combat between individuals of the same or different species, which they saw as leading to the survival of the fittest. Herbert Spencer viewed nature as "red in tooth and claw." Although this type of selection certainly occurs, we now recognize that the essence of natural selection is not so much survival of the fittest as it is greater reproductive success of the fitter. The basis of natural selection is *differential reproduction;* those individuals best adapted to existing conditions produce more viable offspring.

Development of Evolutionary Ideas

Early Evolutionary Ideas

Although Darwin made the concept of evolution credible, evolutionary ideas are very old and antedate Darwin by many years. Ever since the evolution of our unique mental capacities, our ancestors have probably wondered about their origins, their place in nature, and their destiny. Burial sites of stone age people include tools and other artifacts that the deceased might need for a trip into the hereafter. As different races and cultures developed, various ideas about creation emerged. Plato (428–348 B.C.) visualized a creator making the world from chaos and then creating the gods who, in turn, made men. Women and other animals arose through the reincarnation of men's souls. The more flawed the soul, the more lowly the reincarnate. Aristotle (384–322 B.C.) was a keen observer of nature and saw much evidence of design and purpose. He arranged all organisms in one "scale of nature" extending from the simple to the complex. Existing organisms were seen to be imperfect but moving toward a more perfect state. This is sometimes interpreted as an evolutionary idea, but Aristotle is very vague on the nature of the movement. It may have been a closer and closer approach to the creator's ideal for each particular species. Aristotle certainly did not articulate a notion of the transmutation of species.

As these ideas were emerging in ancient Greece, ancient myths were being assimilated and developed by the Hebrews. According to these, God created the world and all therein in seven days, and later a great deluge was survived only by Noah and the inhabitants of his ark. These and other familiar stories became incorporated in the Book of Genesis. These Hebrew ideas of origin were adopted by the early Christians and spread with Christianity through the Roman world. Judeo-Christian ethics, and religious and biblical authorities thus came to dominate early western culture.

During the Renaissance there were increasing challenges to the authority of the church of Rome that culminated in the Reformation. These were accompanied by an increased interest in the study of nature and a movement away from reliance on interpretations of Aristotle and early authorities. In 1543 Copernicus proposed, and soon thereafter Galileo showed quite convincingly, that the sun and not the earth was the center for the rotation of the planets. Modern scientific thought tied to observations, experiments, and rigorous inductive and deductive logic emerged in the seventeenth century with the work of such luminaries as Francis Bacon, William Harvey, Isaac Newton, and René Descartes.

It was not, however, until the 18th century that the new mechanistic science began to have much effect upon interpretations of the biological world. The wonderous adaptations of organisms appeared to be far removed from materialism and to be indisputable evidence of divine design and purpose. But the discovery of many new species as new continents were explored, the discovery of many more fossils, which previously had been interpreted as poor sinners drowned in the great deluge, and increasing awareness of the numerous structural similarities between organisms gradually led many to believe that the organic, as well as physical, world might be guided by natural laws rather than by direct divine intervention.

The Frenchman Pierre-Louis de Maupertius suggested in 1745 that some races might begin as chance departures from natural design. Cautious evolutionary ideas were later advanced by Denis Diderot (1746), Georges Louis LeClerc, comte de Buffon (1779), Erasmus Darwin (1794), who was Charles Darwin's grandfather, and others.

The most thoroughly considered preDarwinian view of evolution was proposed by another Frenchman, Jean Baptiste de Lamarck, in his *Philosophie Zoologique* (1809). Like most biologists of his time, Lamarck believed that all living things are endowed with a vital force that controls the development and functioning of their parts and enables them to overcome handicaps in the environment. He believed that any trait acquired by an organism during its lifetime was passed on to succeeding generations—that acquired characters are inherited. Developing the notion that new organs arise in response to the demands of the environment, he postulated that the size of the organ is proportional to its use or disuse. The changes produced by the use or disuse of an organ are transmitted to the offspring, and this process, repeated for many generations, would result in marked alterations of form and function. Lamarck explained the evolution of the giraffe's long neck by suggesting that some short-necked ancestor of the giraffe took to browsing on the leaves of trees, instead of on grass, and that, in reaching up, it stretched and elongated its neck. The offspring, inheriting the longer neck, stretched still farther, and the process was repeated until the present long neck was achieved.

Lamarck's ideas did not take root in his time partly because he did not provide supporting evidence of evolutionary change and partly because they were vigorously opposed by another and more influential French biologist, Baron Georges Cuvier. Cuvier, regarded by many as the father of paleontology, was a keen student of fossils. He was very much aware of the extinction of species and their replacement by new ones. He saw no evidence in the fossil record of transitions between species. He attributed the succession of faunas to a series of catastrophes, of which Noah's flood was the most recent, followed by new divine creations of species.

Although there have been periodic resurrections of Lamarck's theory, even in our century, it has not been taken seriously by most biologists because it is not consistent with our knowledge of genetics. Environmentally induced somatic modifications occur. A few may be inherited via the cytoplasm (certain cell organelles), but there is no way that somatic modifications can alter the nuclear genetic material in the germ cells, the eggs and sperm, in a direction appropriate to the environmental stimulus. Moreover, many evolutionary phenomena cannot be explained by use or disuse. An example would be the evolution of worker castes in social insects for they are sterile and cannot perpetuate themselves.

During the early nineteenth century, Charles Lyell developed an earlier view of Hutton's into the geologic principle of uniformitarianism, which he published in his *Principles of Geology* (1830–1833). Lyell proposed that the mountains and valleys and other physical features of the earth's surface were not created in their present form, or were not formed by a succession of catastrophes, but were formed by the continuation over long periods of time of the processes of vulcanism, uplift, erosion, glaciation, and so on, that we see going on at the present time. Uniformitarianism was of great importance for the further development of the notion of organic evolution. First, organic evolution is in a sense an application of the principle of uniformitarianism to the organic world. Processes that we see going on today, continued over long periods of time, may account for the origin of species. Second, it followed from Lyell's ideas that the earth was far older than the creation date of 4004 B.C. calculated by Bishop Ussher in 1650 by adding up the geneologies in Genesis. An adequate amount of time was available for the slow organic changes involved in natural selection.

Darwin and Wallace

The time was clearly ripe for someone to "put it all together" and propose a well-documented theory of evolution—if not Charles Darwin, then someone else. Indeed, Alfred Russel Wallace independently proposed a theory nearly identical to Darwin's and at about the same time.

Darwin was born in 1809 and was sent at the age of 15 to study medicine at the University of Edinburgh. Finding the lectures intolerably dull, he transferred, after two years, to Christ's College, Cambridge University, to study theology. At Cambridge he joined a circle of friends interested in natural history. Through them he came to know Professor Henslow, the naturalist. Shortly after leaving college, and upon the recommendation of Professor Henslow, Darwin was appointed naturalist on the ship *Beagle,* which was to make a five-year cruise around the world to prepare navigation charts for the British Navy (Fig. 17.10). The *Beagle* left Plymouth in 1831 and cruised slowly down the east coast and up the west coast of South America. While the rest of the company mapped the coasts and harbors, Darwin studied the animals, plants, and geologic formations of both coastal and inland regions. He made extensive collections of specimens and copious notes of his observations. The *Beagle* then spent some time at the Galapagos Islands, west of Ecuador, where Darwin continued his observations of the flora and fauna, comparing them to those on the South American mainland. These observations convinced Darwin that the theory of special creation was inadequate and set him to thinking about alternative explanations.

Upon his return to England in 1836, Darwin spent his time assembling the notes of his observations for publication and searching for some reasonable explanation for the diversity of organisms and the peculiarities of their distribution. As Darwin wrote in his notebook:

> In October (1838), that is fifteen months after I had begun my systematic inquiry, I happened to read for amusement Malthus *On Population,* and being well prepared to appreciate the struggle for existence which everywhere goes on, from long-continued observation of the habits of animals and plants, it at once struck me that under these circumstances favorable variations would tend to be preserved, and unfavorable ones to be destroyed. The result of this would be the origin of new species. Here then I had at last got a theory by which to work.

Darwin spent the next 20 years building up a tremendous body of facts that demonstrated that evolution had occurred and formulating his arguments for natural selection. As Darwin was pondering his ideas, Alfred

H.M.S. BEAGLE IN STRAITS OF MAGELLAN. MT. SARMIENTO IN THE DISTANCE.
Frontispiece.

FIGURE 17.10 H.M.S. Beagle in the Straits of Magellan. (Courtesy of the American Museum of Natural History.)

Russel Wallace, who was studying the flora and fauna of Malaya and the East Indies, was similarly struck by the diversity of living things and the peculiarities of their distribution. Like Darwin, he happened to read Malthus' treatise and came independently to the same conclusion, that evolution occurred by natural selection. In 1858 Wallace sent a manuscript to Darwin and asked him, if he thought it of sufficient interest, to present it for publication. Darwin's friends persuaded him to present Wallace's paper along with an abstract of his own views, which he had prepared and circulated to a few friends several years earlier, to a meeting of the Linnaean Society in July 1858. Darwin's monumental *On the Origin of Species by Means of Natural Selection* was published in November 1859.

Evolution After Darwin

Evolution and Religion. As did preDarwinian ideas about evolution, so too Darwin's concept soon ran into conflict with the established church. Many people find comfort in the notion of supernatural design and purpose operating in the organic world and in the creation stories of Genesis, and they are disturbed by the notion that the organic world as well as the physical world operate under definable "laws of nature." However, Darwin's overwhelming evidence for organic change combined with an increasing understanding of the origins and purposes of religion led most people to realize that science and religion examine different phenomena, have different assumptions and methodologies, and different goals. Each is important in its own sphere. Science and religion have been reconciled in the minds of most, but not among some Protestant fundamentalist groups in the United States who believe that science as well as religion has its origins in a literal interpretation of Genesis.

Occasionally these groups have been influential enough to limit the teaching of evolution. The state of Tennessee passed a law prohibiting the teaching of evolution in public schools. Thomas Scopes, a high school teacher, was brought to trial in 1925 for violating the law. The trial attracted a great deal of attention, largely because of the colorful protagonists, Clarence Darrow for the defense and William Jennings Bryan for the prosecution. The case was intended to be a constitutional test of the first amendment, but because of certain technicalities it never went beyond the state court. Scopes was given a modest fine, and the law lapsed into limbo. Eventually a similar Arkansas law was found unconstitutional by the Supreme Court in 1968.

Having failed in attempts to prohibit the teaching of evolution, some fundamentalists have tried to require the concurrent teaching of Genesis in biology texts and classrooms not as religion, which would violate the constitutional mandate for the separation of church and state, but as "creation science." A law requiring the teaching of Genesis as science was adopted by the state of Arkansas in 1981, but it too was ruled unconstitutional by federal judge William Overton in January 1982. Judge Overton devoted 13 pages of his decision to demonstrate that creation science is not science and concluded that "the only inference which can be drawn . . . is that the act was passed . . . with the specific purpose of advancing religions."

Natural Selection. Although the concept that evolution has occurred was generally accepted by the late nineteenth century, Darwin's notion of natural selection as the guiding agent came under attack in the late nineteenth and early twentieth centuries. It depends on a continuing source of inherited variation. Experiments made by Johannsen in 1909 on beans showed that natural selection will not affect an inbreeding population after the first few generations for inherited variation quickly is exhausted. As mutations and the sources of variation became better understood, it appeared to many geneticists that more knowledge of the causes of mutation would itself explain evolution. Natural selection appeared to be unnecessary. T. H. Morgan, one of the pioneers in explaining inheritance, espoused this view.

Beginning in the 1930s a synthesis of evolutionary views from different biological disciplines began to emerge. Among the leaders in promoting an integration were the late Theodosius Dobzhansky, a pioneer student of population genetics, George Gaylord Simpson, a noted vertebrate paleontologist, and Ernst Mayr, a systematist and field naturalist who contributed an understanding of species as they exist in nature. The studies of these and others have gradually led to our contemporary theory of evolution, which is known as the modern synthesis or **NeoDarwinism** because natural selection is a key but by no means the only ingredient.

Summary

1. The concept of organic evolution provides a framework of understanding into which all fields of biology are woven.

2. Concepts of our place in nature are probably as old as humankind. The Hebrew ideas as set forth in the Book of Genesis were taken over by early Christians and dominated western culture until the Renaissance.

3. The notion that the organic world as well as the physical world operates under definable laws of nature began to be advanced in the eighteenth century by several writers but particularly by Lamarck who proposed a theory of evolution based on the inheritance of characters acquired by use and disuse.

4. These ideas, together with advances in geology, prepared the intellectual climate for the acceptance of organic evolution by natural selection as proposed nearly simultaneously by Darwin and Wallace in the middle of the nineteenth century.

5. Natural selection is basically differential reproduction. Those individuals with inherited characters that are best adapted to the present environment will, on average, leave more offspring than those with less suited characters. There is therefore a differential transmission from generation to generation of the more favorable inherited traits.

6. Evidence for organic evolution comes from all fields of biology and geology: taxonomy (the ability to group organisms into a hierarchy of forms), resemblances in fundamental aspects of structure among organisms (homology), the presence of vestigial organs, similarities of development among organisms, physiological and biochemical resemblances (in serum proteins, cytochromes, and other characters) that parallel morphological resemblances, the numerous breeds of domestic plants and animals, the fossil record, which shows an increasing complexity and diversity with time, the present-day distribution of organisms, and the biogeographic realms, which are a direct result of the history of organisms and the shifting of continents by plate tectonics.

7. Although evolutionary notions were initially opposed by the church, the overwhelming evidence for organic change coupled with new understandings of the different methodologies and purposes of science and religion have led to a reconciliation in the minds of most people.

8. The idea of natural selection fell into disrepute as more was learned about the nature of inheritance and as mutations were discovered early in the twentieth century. But in the 1930s advances in population genetics, paleontology, taxonomy, and other fields led to a synthetic neoDarwinian view that incorporates natural selection as one of the factors involved in evolutionary change.

References and Selected Readings

Baldwin, E. B.: *An Introduction to Comparative Biochemistry.* 4th ed. Cambridge University Press, 1964. A classic exposition of some of the biochemical similarities among animals that aid in defining evolutionary relationships.

Ben-Avraham, Z.: The movements of continents. *American Scientist 69* (May–June):291, 1981. A review of the theory of plate tectonics and the factors involved in crustal movements and mountain building.

Bingham, R.: On the life of Mr. Darwin. *Science 82* (Apr.):1982. An imaginative interview with an elderly Darwin in which he reviews his life and contributions.

Colbert, E. H.: *Evolution of the Vertebrates.* 3rd ed. New York, John Wiley & Sons, Inc., 1980. A text recommended for its presentation of information regarding vertebrate fossils.

Harris, C. L.: *Evolution, Genesis and Revelations.* Albany, State University of New York Press, 1981. A history of the development of evolutionary ideas. Selected readings are included from the writings of the ancient Greeks to contemporary authors.

Irvine, W.: *Apes, Angels, and Victorians.* Cleveland, World Publishing Co., 1959. An account of the lives of Charles Darwin and his public defender, Thomas Henry Huxley, and the impact of evolutionary ideas on Victorian society.

Moore, R. C. (Ed.): *Treatise on Invertebrate Paleontology.* Geological Society of America and University of Kansas Press, 1952–1983. Many volumes providing a detailed systematic treatment of fossil invertebrates.

Raup, D. M., and S. M. Stanley: *Principles of Paleontology.* 2nd ed. A textbook of paleontology that emphasizes approaches to the study of fossils and the utilization of data derived from such studies.

Romer, A. S.: *Vertebrate Paleontology.* 3rd ed. Chicago, University of Chicago Press, 1966. A comprehensive, well-written text, highly recommended as a source book for further reading.

Thenius, E.: *Fossils and the Life of the Past.* New York, Springer-Verlag, 1973. A translation from the German of an excellent little book that explains the nature of the fossil record and the information it provides of the structure and mode of life of past organisms and the environments in which they lived.

18 The Mechanisms of Evolution

ORIENTATION

The gene pools of populations and species contain large reservoirs of genetic variation that have been introduced through mutation. Allele frequencies in populations may vary in a random way because mutations may occur more frequently in some directions than in others (mutation pressure), because of an unequal migration of individuals between populations (gene flow), and because the individuals that happen to breed may not be representative of the population as a whole, especially in small populations (genetic drift). Natural selection interacts with these forces, brings the variation of gene pools into harmony with environmental changes, and leads to adaptation. New species arise when a part of a species becomes reproductively isolated from the main part of the species and natural selection and other forces act in a different direction. The accumulation of many speciation events over long periods of time leads to macroevolutionary changes.

Darwin made a very significant contribution in his theory of natural selection to our understanding of the ways evolutionary changes take place. The rediscovery of the principles of inheritance early in this century, the discovery of mutations, and the development of the Hardy-Weinberg equilibrium (p. 344) and the other principles of population genetics provided us with an understanding of the mechanisms by which natural selection operates. Studies in systematics and ecology have given us insights into the nature of variation in natural populations and how populations interact. Discoveries in paleontology have shown us something about the nature of extinct populations and how they change over long periods of time. Advances in molecular genetics and embryology have greatly increased our knowledge of the nature of genes, the ways they can change, how they regulate development, and how changes in development affect adult organisms. Gradually the background needed to understand the many factors influencing evolution has emerged, and a **synthetic theory of evolution,** sometimes called **neoDarwinism,** has developed. Although there is a consensus as to the major factors at work, biologists are not in full agreement about the relative importance of particular ones. The synthetic theory continues to change and develop as we learn more about the very complex factors involved.

Microevolution: Evolutionary Changes Within Populations

Species, Populations, and Gene Pools. It is quite clear that adult individuals do not evolve. What changes are patterns of embryonic development among groups of individuals. The group of prime interest to biologists is the species (L. *species,* particular kind). Many definitions of a species have been proposed, but the most productive, from the perspective of evolutionary theory, is the concept of a **biological species.** In sexually reproducing organisms, a species is a group of individuals that have the ability to reproduce with one another and that are reproductively isolated from other such groups in nature. Corollaries of this concept are that members of the same species share morphological, physiological, and ecological features, have a common geographic range, and have common isolating mechanisms that prevent them from crossing with other species under natural conditions. Exceptions occur. Occasionally, different species cross, or individuals of the same species from widely separated parts of the range will not cross successfully even when brought together. If species are evolving groups, some exceptions are to be expected.

Species are realities of nature, not abstractions. Living matter is not a continuum but is broken up into distinct units (lions, tigers, house cats) that are reproductively isolated from each other under normal conditions. There are limits to the amount of variability that can exist within a species without resulting in many inviable recombinations. Reproductive isolation ensures that these limits are not exceeded.

Members of a species are not distributed uniformly throughout their range but rather occur in clusters spatially separated to some extent from other clusters. The bullfrogs of one pond form a cluster separated from those in an adjacent pond. Some exchanges occur by migrations between ponds, but the frogs in one pond are more likely to reproduce with those in the same pond. Members of a species tend to be distributed in local interbreeding populations, called **demes** (Gr. *demōs,* people). The territorial limits of any particular deme may be vague and difficult to define, yet the concept is useful for purposes of theoretical analysis.

Members of a deme share a common **gene pool** (Fig. 18.1) that is the aggregation of all of the genes and their alleles of all of the individuals of the population. Each individual is genetically unique and has a specific genotype. In simple terms, evolution is a change in the types and frequencies of specific alleles in the gene pools of populations and species. The gene pool of a population is the smallest evolutionary unit. You and I as individuals are not evolving, but the gene pool from which we obtained our genes and to which we contribute genes may be changing because of differential reproduction, among other reasons.

Variation in Local Populations. Changes in the types and frequency of genes in gene pools, whether by natural selection or other means, are not possible unless there is a source of inherited variation. It is axiomatic that variation is the raw material for evolutionary change.

New variations are introduced into a species gene pool only through **mutation,** the ultimate source of variation on which evolutionary changes depend. Mutations result from a change in the nucleotide base pairs of the genetic code, from a rearrangement of genes within chromosomes so their interactions produce different effects (p. 375), or from a change in the number of chromosomes. Most mutations occur spontaneously. Rates are relatively stable for a particular locus but vary thousands of times among loci and between different species. In higher animals a gene mutation rate of 3×10^{-5} (1 gamete in 300,000) is quite representative. This may seem low, but the effect is significant when multiplied by the number of loci. Assuming conservatively

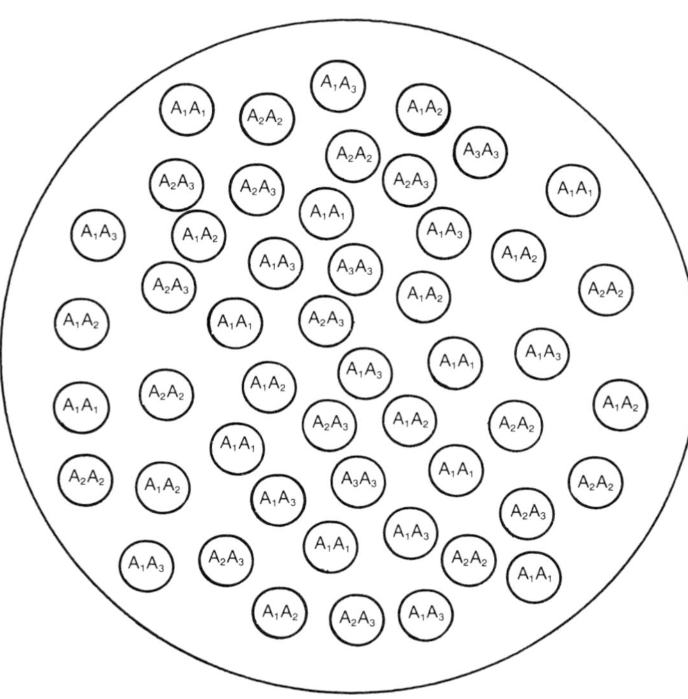

FIGURE 18.1 A diagram illustrating the concept of the gene pool. Diploid individuals are represented by small circles. Each has only two members of any set of alleles, but the pool may contain many ($\mathbf{A_1}$, $\mathbf{A_2}$, $\mathbf{A_3}$, . . .). The allele frequencies in this case are $\mathbf{A_1} = 0.40$, $\mathbf{A_2} = 0.35$, $\mathbf{A_3} = 0.25$.

that human beings have 100,000 paired loci (2×10^5) and an average mutation rate of 3×10^{-5}, then each of us on average carries six mutant genes: $(2 \times 10^5) \times (3 \times 10^{-5}) = 6$. Multiply this, in turn, by population size, and it is evident that mutation maintains a large supply of variability.

A given gene may mutate to give rise to a series of multiple alleles ($A \rightarrow a_1 \rightarrow a_2 \rightarrow a_3$ . . .). A gene also may mutate back to its original form ($a_1 \rightarrow A$).

A mutation is a random change with respect to the direction of evolution. If a population is adapting to a dry environment, mutations appropriate for dry conditions are no more apt to occur than ones for wet conditions or ones that have no relationship to the environmental problem. The effect of a random change upon the members of a population well adapted to its current environment is more likely to be deleterious than beneficial. If mutations are random and frequently deleterious, how can they be an important source of variation for adaptive evolutionary change? The answer is complex.

First of all, not all mutations are deleterious; some are neutral and a few are beneficial. Mutations that bring about significant departures from the phenotypic norm and are most likely to be deleterious are rather rare. Most mutations produce small changes in the phenotype that often are detectable only by biochemical techniques. By acting against seriously abnormal phenotypes, natural selection quickly eliminates, or reduces to low frequencies, major deleterious mutations. Small mutations, even ones with slightly deleterious phenotypic effects, have a better chance of being incorporated into the gene pool where at some later time they may be helpful to the population.

Whether a mutant is deleterious or beneficial must be evaluated in the context of all of its phenotypic effects in the particular environment in which it acts. Many genes are **pleiotropic** (Gr. *plion,* more + *trope,* turn), affecting more than one trait. The mutant allele for sickle cell anemia (p. 416) produces an altered hemoglobin, HbS, that is less soluble than normal hemoglobin, HbA, especially at low oxygen tensions. Most individuals homozygous for the allele die. Individuals heterozygous for the sickle cell allele are more resistant to a particularly virulent type of malaria caused by the protozoan *Plasmodium falciparum* than are individuals homozygous for the allele for normal hemoglobin. In certain parts of Africa, India, and Southern Asia, where this disease is prevalent, heterozygous individuals survive in greater numbers. In a heterozygous individual each allele produces its specific kind of hemoglobin, and the red blood cells contain the two kinds in roughly equivalent amounts. Such cells do not undergo sickling except at very low oxygen tensions, but the red cells containing the HbS are resistant to infection with the malarial organism; this renders carriers partially immune to this type of malaria. Each of the two types of homozygous individuals is at a disadvantage. Those homozygous for the sickling allele are likely to die of anemia; those homozygous for the normal allele may die of malaria. The frequency of the sickle cell allele reaches 40 percent in the gene pools of black populations in parts of Africa. Resistance to *falciparum* malaria is of no advantage in North America, and the frequency of the allele in the gene pools of American blacks has been reduced to 4 or 5 percent, partly because of dilution of the pool through migration from other populations and partly by selection against it.

Pleiotropic genes and those whose effects vary with the environmental context are very common. This suggests that a mutant that is deleterious in a well-adapted organism living in a stable environment may become beneficial if the environment changes. In the perspective of hundreds or millions of years, most environments change greatly.

Identical mutations recur time and time again in evolving populations. Beneficial ones have been incorporated into the gene pool; very deleterious ones have been eliminated; neutral and mildly deleterious ones may be carried. There are genetic mechanisms, involving incompletely understood interactions between different genes, that permit certain attributes of a mutant to be expressed, whereas its more deleterious effects may be suppressed. Some dominant-recessive interactions of a set of alleles are not dependent so much on the alleles themselves as on modifying genes at other loci with which they interact. Mutants whose deleterious effects have been suppressed can be carried in gene pools. When environmental conditions change, their expression may be changed.

Mechanisms of this type enable gene pools to carry tremendous reservoirs of variability, much of it at low frequency and much of it hidden. Until recently biologists could not estimate the total amount of variability in populations because they could only recognize those loci that had one or more mutations conspicuous enough to have effects in breeding experiments. Biologists now take a random sample of proteins from an organism and, using biochemical techniques, determine how many are **polymorphic,** i.e., exist in two or more forms as determined by different amino acid sequences or, in the case of enzymes, by different isozymes. Since each variety of a particular protein is coded by a different mutant, their number reflects the amount of mutation. It is estimated that 25 percent of vertebrate loci are polymorphic, i.e., exist in populations as two or more alleles. In any individual, some loci will be homozygous and some heterozygous. The average heterozygosity for an individual vertebrate is about 6 percent. If an individual had 100,000 paired loci, 6000 would be heterozygous.

The reservoir of variability is sufficient for populations to respond to selection pressures in many directions. This has been demonstrated in experiments with the fruit fly *Drosophila.* A population is established from offspring of a single female fly fertilized in the wild (Fig. 18.2). Her progeny has a certain variability, and by appropriate selection procedures it is possible to either increase or decrease certain traits, such as the mean number of bristles on a certain part of the thorax, and to extend the range of variability beyond that in the initial population. Since the change occurs in relatively few generations and can go in either direction, stored genetic variability rather than new mutations must be the source of variation.

Recombination of genetic traits already in a population through sexual reproduction further increases the variability available to a population for it produces new combinations of alleles that result in new genotypes and phenotypes. The effect of recombination can be surprisingly great. Nine different genotypes are generated in a dihybrid cross (AaBb × AaBb) involving only two genes on different chromosomes and each with only two alleles (p. 346). If we were dealing with five genes on different chromosomes and six alleles per gene (and these are not exceptionally high numbers), the number of different genotypes possible would be 4,084,101. Some of

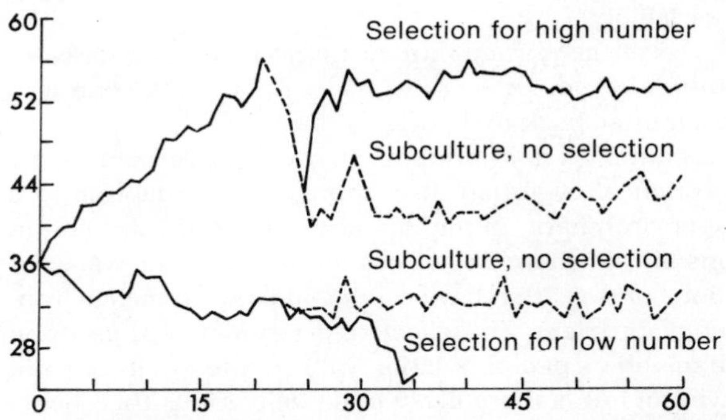

FIGURE 18.2 Mean bristle number on a part of the thorax in lines of *Drosophila melanogaster* derived from the progeny of a single female with 36 bristles. Generations are plotted on the abscissa; bristle number on the ordinate. Solid lines represent generations during which selection for high or low bristle number were made. Dashed lines represent subcultures that were continued without selection. (From Mayr, E.: *Populations, Species and Evolution: An Abridgment of Animal Species and Evolution.* Cambridge, Harvard University Press, 1970. Based on experiments by Mather and Harrison.)

the combinations that are generated may be adaptively superior, and natural selection could favor changes in linkage groups or other mechanisms that would perpetuate them.

Forces Acting on Variation. The reservoir of variability of populations remains in equilibrium as the Hardy-Weinberg principle shows (p. 344). If no new variation is introduced, the population is large, mating is random, and natural selection is not operating. The Hardy-Weinberg equilibrium is a theoretical norm that defines our expectations, but most natural populations are subjected to one or more of four forces that upset the equilibrium and shift it to new values: mutation pressure, migration, genetic drift, and natural selection.

Mutation Pressure. Because genes mutate in different directions (**A** $\rightleftharpoons$ **a**) and rates are often not the same, a **mutation pressure** may exist for a change in a certain direction. Given mutation in two directions the alleles should attain an equilibrium value, but this seldom happens because one allele may be opposed by natural selection. The value will then be determined by the relative strength of mutation introducing the allele and of selection removing it, as explained later under natural selection.

Migration. **Migration** of individuals between demes causes a **gene flow** that can have significant evolutionary consequences. The amount of migration is dependent on patterns of breeding and dispersal in a species. Although the males of many large animals may wander far in search of mates, other organisms are more apt to mate with their near neighbors. Consequently there is some degree of inbreeding. Inbreeding does not change allele frequencies directly but does so indirectly because it increases homozygosity (p. 359) and enhances the effects of natural selection by exposing recessive alleles.

If the amount of migration is large and populations differ in allele frequencies, significant changes result. The United States has been a melting pot for racial groups, each of which originally had many attributes of a deme. American blacks have a frequency of the Rh^O allele (one of the alleles of the Rh blood group, p. 358) of 0.45. In contrast the frequency in African blacks is 0.63. The reduced frequency in American blacks has been attributed to an influx over many generations of alleles from the white population in which the frequency of Rh^O is very low, 0.03. Gene frequencies have changed in the white population too, for migration has occurred in both directions, but the effects would not be so evident in the larger white population.

Genetic Drift. Population size has a very important effect on allele frequencies because the probability of a departure from the initial frequency by the chance mating of individuals (a special case of statistical sampling errors) is inversely related to population size. Assume you have a gene pool in which the alleles **A** and **a** each have a frequency of 0.5. This can be simulated by using a bowl of 1000 beads, one half red and one half blue. In a random sample of 100 beads, it is unlikely that exactly 50 will be one color and 50 the other, but if the sample is as large as 100 it is also unlikely that the deviation from 50:50 will be great. In a small sample the probability of a significant departure from the true value is increased. If the sample size is two, the probability that both will be red is $0.5 \times 0.5 = 0.25$, the probability that both will be blue is $0.5 \times 0.5 = 0.25$, and the probability that one will be blue and one red is only 0.5. The effect is the same when a breeding population is very small. The probability of two individuals with the same traits mating is increased, and there is a reasonable probability that the variability of a descendent generation will deviate by chance from the parent population. If the population remains small for many generations, sampling errors accumulate, and the population's variability drifts in a random way. One allele may be eliminated by chance from the population and another may become fixed (frequency of one). Sewall Wright termed this phenomenon **genetic drift.**

Drift affects allele frequencies in a random way. Its direction may or may not coincide with the direction of mutation pressure, migration, or selection, so it may reinforce or oppose these forces. It has been shown that if the product of any one of these directional forces and population size is much less than one, then drift will be of primary importance in determining gene frequencies. Consequently drift is important only in small populations, but there are times when populations are small enough for drift to be an important evolutionary factor.

Because of fluctuations in the environment, depletion of food supply, disease, and other factors, some populations periodically experience rapid and marked decreases in the number of individuals (p. 419). The population goes through a **bottleneck,** and drift can affect the variability of the few survivors. As the population again increases in size, the frequencies of many alleles may be quite different from those preceding the decline. The unusual distribution of ABO blood group allele frequencies in human populations (Fig. 18.3) may have resulted partly from population bottlenecks encountered by early bands of human beings. The I^B allele responsible for the B antigen has frequencies as high as 0.30 in parts of Asia and India, but is 0.10 or less in Europe and much of Africa. This allele is absent in native Americans although the ancestors of these people migrated from Asia.

Drift is also important when a few individuals fortuitously cross a barrier at a species border and establish a colony on the other side, as in the colonization of

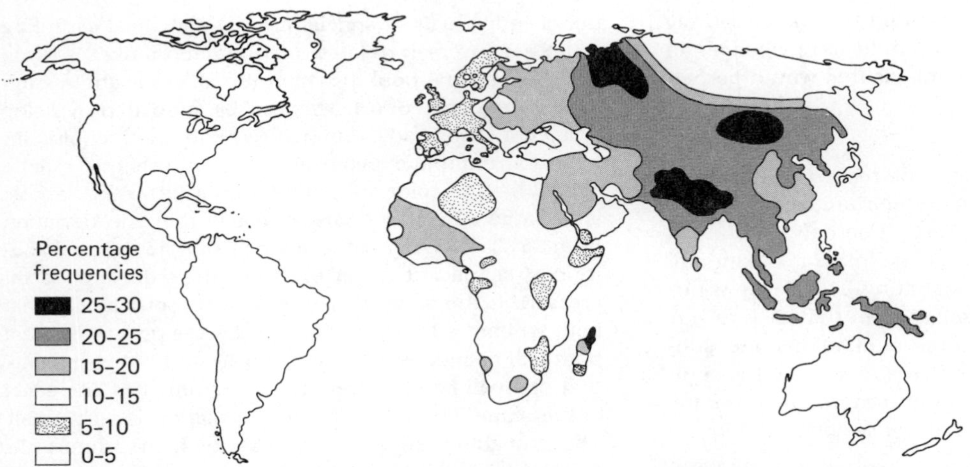

FIGURE 18.3 Geographic distribution of the I^B allele responsible for the B blood group in human beings. (From Ayala, F. J., and J. W. Valentine: *Evolving.* Menlo Park, The Benjamin/Cummings Publishing Co., 1979.)

oceanic islands. The colonizers carry with them a small and random sample of the variability of the gene pool from which they came, and their number is small enough initially for drift to operate. This phenomenon has been called the **founder effect** by Ernst Mayr.

Natural Selection. Drift is a random change. Mutation and migration may have directions, but the directions are unrelated to the environment. **Natural selection** is the only force that brings the variability in gene pools into harmony with the environment, checks the disorganizing effects of the other forces, and leads to adaptation. Natural selection, which in essence is simply differential reproduction, works on phenotypes and hence only indirectly on genetic loci and their alleles. Many factors may lie between the genotype and the final phenotype. We remind you that one gene locus may affect many structures and functions, one character may be controlled by many loci, loci interact in complex ways, and the expression of the genes may be influenced by environmental factors.

Natural selection may occur at any of a number of different times in the life cycle of an organism. There may be nonrandom mating (which Darwin called sexual selection), nonrandom fecundity (differences in the number of offspring produced), or nonrandom survival to reproductive age. The last is particularly common and frequently involves subtle interactions between organisms and the environment in which they live.

The outcome of selection depends upon the nature of the environment-population interaction. If an environment is homogeneous and stable, **stabilizing selection** works against the extreme variants of a population (Fig. 18.4). If an environment is changing, **directional selection** will shift the mean and range of variation of a population toward adaptation to the new conditions. If an environment is patchy and heterogeneous, that is, composed of two or more subenvironments, **disruptive selection,** or **balancing selection** will result in a **polymorphic** population in which several different phenotypes are maintained.

To study selection, biologists have developed measurements of the reproductive failure or success of one phenotype relative to other ones. The **selection coefficient** (s) is a measure on a scale of 0 to 1 of the elimination of one phenotype relative to the more successful one (0 = minimum elimination, 1 = maximum elimination). **Adaptive value,** or **fitness,** measures relative success and is the complement $(1 - s)$ of the selection coefficient. Each phenotype has both a selection coefficient and an adaptive value, and their sum must equal 1. If 40 percent of the homozygous recessive phenotype is eliminated relative to the dominant one, then its $s = 0.4$ and $1 - s = 0.6$.

Stabilizing Selection. Research on human birth weights and mortality done in 1951 by Karn and Penrose illustrates the effect of stabilizing selection. Body weight at birth is controlled primarily by a set of polygenes and varies along a normal curve (Fig. 18.5). Individuals with very low or very high birth weights have high selection coefficients and die more frequently than individuals closer to the mean. Selection operates to reduce the frequencies of alleles for extreme variants, and

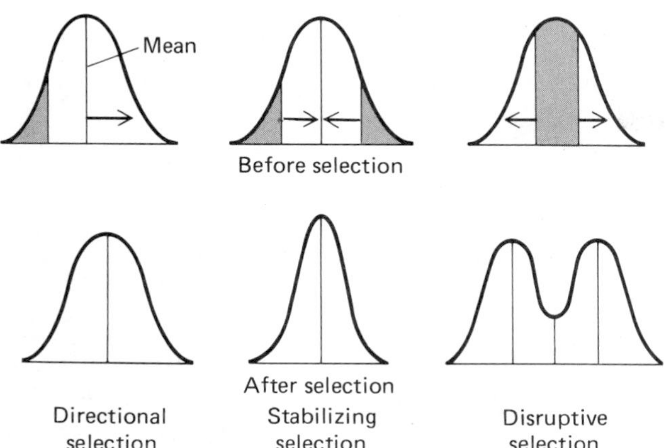

FIGURE 18.4 A diagram showing the effects of different types of natural selection on the distribution of variation within a deme. Abscissa represents variation; ordinate, the number of individuals. Hatched areas represent variants selected against; arrows, the direction of selection.

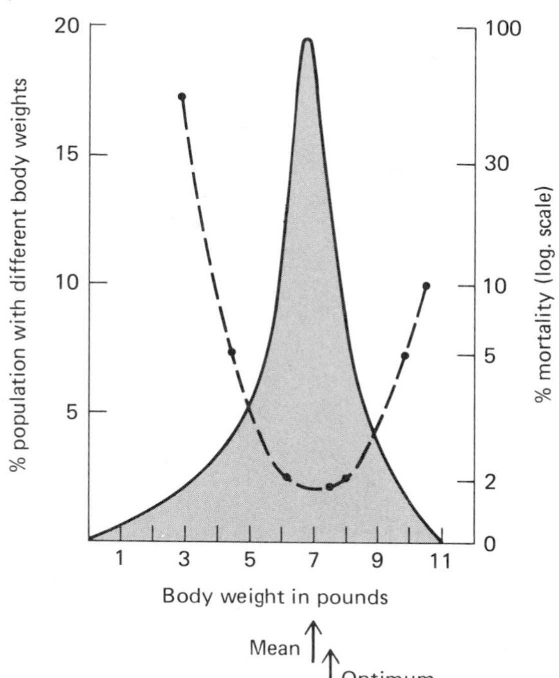

FIGURE 18.5 Stabilizing selection as seen in the distribution of human birth weights. Individuals with extremely low or high weights have higher mortality rates, shown by the broken line, than individuals close to the mean. (Modified from Cavalli-Sforza, L. L., and W. F. Bodmer: *The Genetics of Human Populations.* W. H. Freeman and Co., 1971.)

clusters the population's variability close to the size with the minimum mortality rate.

An equilibrium for a set of alleles (**A** and **a**) is reached when the rate at which a deleterious mutant is added to a population through new mutation equals the rate at which it is eliminated by selection. (A quantitative example is given in Box 18.1.) By keeping the level of deleterious mutations low, stabilizing selection promotes stability.

BOX 18.1

Equilibrium in Stabilizing Selection Against a Recessive Allele

If homozygous recessive individuals (**aa**, frequency = q^2) are selected against and have a selection coefficient of s, the rate of elimination of the recessive allele (**a**) from the population = q^2s. The rate of introduction of **a** will be the frequency of **A** ($=p$) available to mutate to **a** times the mutation rate of **A** → **a** ($=u$) = pu. Equilibrium is attained when $q^2s = pu$. If the frequency of **A** (p) is close to 1, as would be expected if selection has been working against **a** for many generations, p can be eliminated from the right-hand term, and the equation can be solved for q, which is the frequency of **a:** $q^2s \approx u$, $q^2 \approx u/s$, $q \approx \sqrt{u/s}$.

Assume $u = 10^{-5} = 0.00001$ and $s = 0.1$. At equilibrium **a** will have a frequency of approximately 0.01: $q \approx \sqrt{u/s} \approx \sqrt{0.00001/0.1} \approx \sqrt{0.0001} \approx 0.01$.

Directional Selection. Directional selection has been studied in many laboratory and field situations. Most of the peppered moths (*Biston betularia*) in rural parts of England have a black and white peppered wing color, and only a few are all black or melanic (Fig. 18.6). The situation is reversed in industrial regions. Why? Moths rest by day on tree trunks where some are eaten by birds. Prof. H. B. D. Kettlewell of Oxford University postulated in the 1950s that the peppered pattern is less conspicuous on the lichen-covered trees of rural areas, whereas the melanic form is less conspicuous on the soot-covered trees of industrial regions. To test this hypothesis, hundreds of male moths of each type were raised, marked with a spot of paint under their wings and released in both rural and industrial areas; the survivors were recaptured after a certain period of time by attracting them with light or virgin females. Significantly more melanic forms survived in the industrial areas and more peppered forms in the rural areas. Natural selection shaped the gene pool of each deme to the local conditions. In some localities, selection operated in one

FIGURE 18.6 *A,* Drawing from a photograph of a melanic and peppered form of the moth, *Biston betularia,* resting on a lichen-covered tree trunk in a rural area. *B,* Similar individuals on a soot-covered tree trunk in an industrial area. (From Ehrlich, P. R., R. W. Holm, and D. R. Parnell: *The Process of Evolution.* New York, McGraw-Hill Book Co., 1974, after H. B. D. Kettlewell.)

direction; in other localities, in the opposite direction. Seldom, however, does a population become entirely of one type. Male peppered moths fly considerable distances, so gene flow between populations helps to maintain some polymorphism. As air quality has improved in industrial regions in recent years, there has been a reversal of past trends and a slight increase in the percentage of light forms.

When selection works against an allele, the allele continues to decline generation by generation. The rate of decline depends on the frequency of the allele and the degree to which its phenotype is exposed to selection. It is clear from the Hardy-Weinberg equation (p. 344) that a recessive allele will decline rapidly when frequencies are near 0.5 because 0.25 of the population will be homozygous recessive (Fig. 18.7). When the frequency

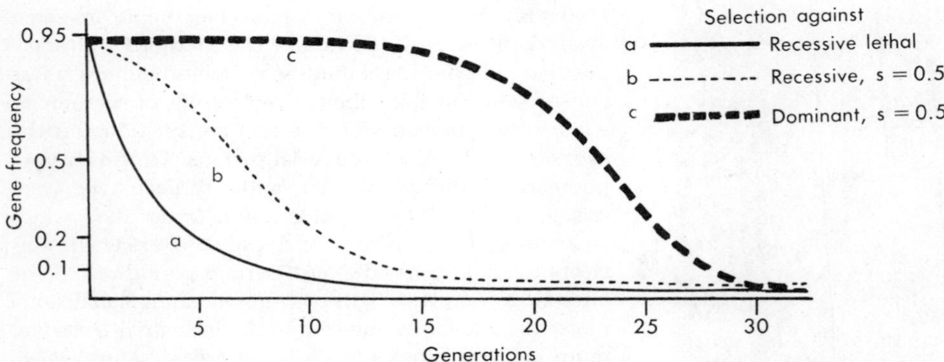

FIGURE 18.7 The outcome of directional selection: a = selection against a recessive lethal allele; b = selection against a recessive allele with a selection coefficient of 0.5; c = selection against a dominant allele with a selection coefficient of 0.5. (From Levine, L.: *Biology of the Gene.* St. Louis, C. V. Mosby Co., 1973.)

is very low, 0.1 for example, the rate of decline will be slow because only 0.01 will be homozygous recessive. Most of the recessive alleles in the population are now sheltered from selection in heterozygotes. As in stabilizing selection an equilibrium eventually is reached when the rate of elimination of the allele by selection equals its rate of introduction by mutation.

Somewhat surprisingly, the rate of elimination of a dominant deleterious (but not lethal) allele will also be slow at very high frequencies because the population is made up almost entirely of homozygous dominant individuals (Fig. 18.7). After enough of the recessive allele has accumulated in the population to be selected for the dominant one will begin to decline rapidly. An algebraic and quantitative model of directional selection is shown in Box 18.2.

BOX 18.2

Changes in Allele Frequency After One Generation of Selection Against Homozygous Recessive Individuals

Genotypes	AA	Aa	aa	Total
Initial frequency	p^2	$2pq$	q^2	
$p = 0.7, q = 0.3$	0.49	0.42	0.09	1
Selection coefficient	s 0.0	s 0.0	s 0.4	
Surviving to reproduce (initial frequency × adaptive value)	$p^2(1 - s)$ 0.49	$2pq(1 - s)$ 0.42	$q^2(1 - s)$ 0.054	$1 - sq^2$ 0.964

Value of q
in survivors $= pq(1 - s) + q^2(1 - s)/1 - sq^{2*}$
$= q(1 - sq)^{\dagger}/1 - sq^2$
$= 0.3 \times 0.88/0.964$
$= 0.274$

Value of p
in survivors $= 1 - q$

To compute the genotypes of the next generation, use the values of p and q in the survivors in the Hardy-Weinberg equation. To see how q changes over a series of generations, given the same selection coefficient, repeat the formula given here with changed values of q, generation by generation. This is an easy exercise to do if one has access to a computer or a programmable pocket calculator.

*This is one half the surviving **Aa** individuals plus all the surviving **aa** individuals divided by the surviving population's size.

†Since there is no selection against the heterozygotes, $pq(1 - s)$, $(1 - s)$ can be dropped from the first term in the original equation, which then becomes pq. The rest of the numerator simplifies: $pq + q^2(1 - s) = pq + q^2 - sq^2 = q(p + q - sq) = q(1 - sq)$.

Disruptive Selection. A particularly clear example of disruptive selection involves **Batesian mimicry,** a type of mimicry in which an edible insect species (the **mimic**) has a color pattern resembling that of a distasteful **model** species. Many experiments have demonstrated that bird predators learn to recognize, and tend to avoid, a conspicuously colored insect that is distasteful or dangerous. Selection favors conspicuous patterns in distasteful species because predators learn to avoid them quickly. An edible species that mimics a distasteful model shares in the immunity of the model, provided the models significantly outnumber the mimics. In some localities in Africa there are three different distasteful species of butterfly. Certain females of the edible swallowtail butterfly (*Papilio dardanus*) mimic one model; other females mimic other models. Selection has favored varieties of the swallowtail that resemble any one of the model species and has operated against intermediates. The single population has been disrupted into three different types.

Balancing Selection. When heterozygous individuals are superior in fitness to either homozygote, neither allele can be eliminated. The reproduction of the favored heterozygotes produces homozygotes of each type as well as more heterozygotes, and a polymorphic population is maintained. The reasons for heterozygotic superiority, known as **heterosis,** are not always clear. In the case of a single set of alleles, each produces a slightly different protein which may complement the other. In sickle cell anemia, the normal allele produces sufficient normal hemoglobin to provide the needed oxygen transport under most circumstances, and the abnormal allele produces a hemoglobin that increases resistance to falciparum malaria. In a polygenic situation, an unfavorable effect of each allele in the homozygote is not expressed in the presence of the other allele in the heterozygote (p. 359). An equilibrium between alleles at a locus that is dependent on the relative fitness of the two homozygotes eventually is established by balancing selection.

Since a population subject to balancing selection remains polymorphic, it maintains considerable evolutionary plasticity and can respond easily to changing conditions. The frequency of the Hb^s allele responsible for sickle cell anemia varies considerably between populations in Africa, being highest where the incidence of falciparum malaria is particularly high (Fig. 18.8).

Balancing selection maximizes the fitness of a population as a whole to its environment but at the cost of

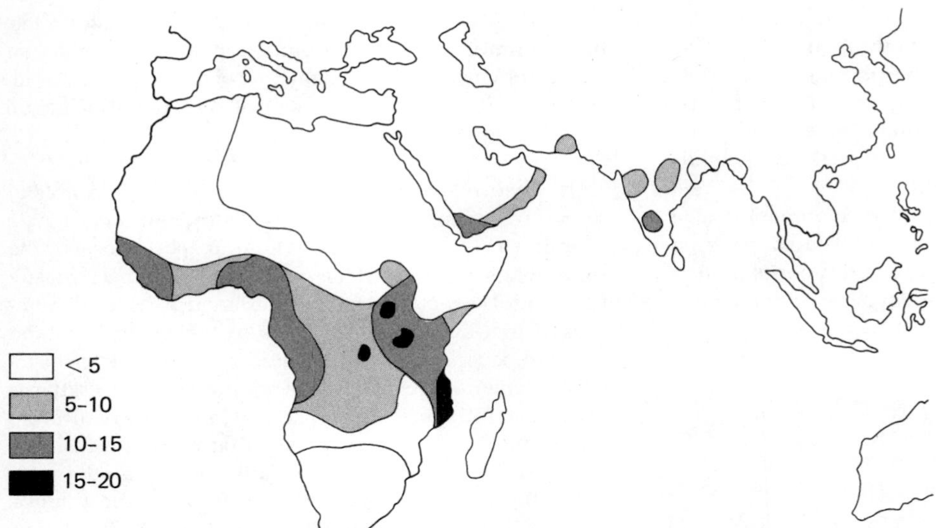

FIGURE 18.8 Geographic distribution of the allele that in homozygous condition is responsible for sickle-anemia. Its frequency is highest in regions where falciparum malaria is high. (From Ayala, F. J., and J. W. Valentine: *Evolving.* Menlo Park, The Benjamin/Cummings Publishing Co., 1979.)

generating homozygotes that are less fit. These constitute the **genetic load** of the population.

Kin Selection. The types of selection we have been considering affect the variability of a population by acting on the individuals within it. Selection may also act on a group of individuals as a whole in its competition with other groups. The groups that have been studied most intensively are kinship units, that is, siblings or other very closely related groups of individuals. Some biologists believe that kin selection has been responsible for the evolution of sterile worker castes in social insects. Since the workers are sterile females, they cannot perpetuate the alleles responsible for their characters through their own reproduction. But they are sisters of the new queen so the many alleles they share with her will be passed on through her reproduction. To the extent that genetic traits leading to the improvement of a worker caste give the new colony an advantage over competing colonies, group selection will favor their perpetuation and spread.

Negative and Positive Selection. Natural selection has two facets, for it both eliminates unfit individuals and favors fit ones. Critics claim that without selection we would have today all of the existing varieties together with those that have been eliminated. In doing so critics overlook the creative aspects of evolution that are derived from selective elimination and survival. Selection is the only force known that can bring genetic variation into harmony with the environment and lead to adaptation. By eliminating less favorable genes and alleles, selection changes the composition of the gene pool in a favorable direction and increases the probability that the favorable genes and alleles responsible for an adaptation will come together in the same individuals. Sir Garvin de Beer, the noted British biologist, pointed out on the occasion of the centenary of *The Origin of Species* that "the effects of natural selection are the reverse of chance when considered *ex post facto*; . . . they channel random variation into adaptive directions and thereby simulate the appearance of purposive change . . . . Natural selection has been paradoxically defined as a mechanism for generating an exceedingly high degree of improbability."*

Integrity of the Gene Pool. As a consequence of natural selection, the entire gene pool of a population becomes a complex, interacting system. The genes and alleles in the pool have come together in different combinations in individuals through generations of sexual reproduction. Their interactions to produce well-adapted individuals and a well-adapted population have been tested by selection many times. Any genes or al-

*Sir Garvin de Beer: *The Darwin-Wallace Centenary. Endeavour,* April, 1958.

leles that interact very unfavorably with others have been eliminated or their effects modified. The remaining genes and alleles interact in harmony and are said to be **coadapted.**

The existence of coadaptive interactions can be demonstrated when something upsets them. In selecting strongly for high or low bristle number in *Drosophila* (Fig. 18.2), Mather and Harrison found that the line with low bristle numbers died after 35 generations. Sterility began to appear in the line with high bristle numbers after 20 generations. Selection was stopped for several generations, and bristle number decreased. When selection was resumed, bristle number increased to the low 50s and fluctuated around this level despite continued selection. Subcultures taken from both high and low lines before they have reached maximal or minimal numbers of bristles maintained their new values without further selection. In interpreting these data, Mather and Harrison pointed out that they were not working with one set of alleles for high or low bristle number, but rather they were selecting for interacting gene complexes. If selection moved a complex too far too fast, disharmonious interactions leading to death or sterility began to appear. If selection did not go too far (as in the subcultures) or was stopped for a while (the high line), new and viable interactions could be established. Plant and animal breeders have discovered similar phenomena. When one character (high yield, for example) is selected for strongly, undesirable "correlated responses," such as a reduction in disease resistance, frequently occur.

The interactions of genes and alleles give the gene pool of the population a certain cohesiveness and integrity. During the formation of a new species a population separates from the parental pool, loses its cohesiveness with it, and establishes a new and different pool of coadapted genes.

Speciation

Central to the question of how species are formed is an understanding of the isolating mechanisms that prevent closely related species living in the same geographic area from mating with each other. We call these **sympatric species** (Gr. *syn,* with + *patra,* native land) to distinguish them from **allopatric species** (Gr. *allos,* other + *patra,* native land), which are closely related species living in different geographical areas. Sympatric species are very common. Bullfrogs, green frogs, wood frogs, leopard frogs, and pickerel frogs, all belonging to the genus *Rana,* may be found in or near the same pond. They and other sympatric species are prevented from mating by many reproductive isolating mechanisms.

Isolating Mechanisms. Premating isolating mechanisms, which prevent potential mates of different species from mating, are of several types. (1) Different sympatric species commonly live or breed in different habitats. Wood frogs breed in temporary woodland ponds; bullfrogs, in larger, permanent bodies of water. They are separated by **habitat isolation.** (2) Sympatric species frequently breed at different times of the year or times of day, so are kept apart by **temporal isolation.** The fruit flies *Drosophila pseudoobscura* and *D. persimilis* are sympatric over much of their range, but *pseudoobscura* is more active in the afternoon and *persimilis* in the morning. (3) Most species of animals have a courtship behavior, so mating between species normally is prevented by **ethological isolation. Courtship** is an exchange of signals between a male and a female. A male approaches a female and gives a sign of some sort that may be visual, auditory, or chemical. If the female belongs to the same species, she recognizes the signal and returns one. The encouraged male gives another signal and further exchanges of signals eventually result in both partners being stimulated sufficiently to copulate. If courtship starts between members of two different species, one partner may not recognize one of the signals and may fail to respond, and courtship will stop. (4) Sometimes different species will court and attempt copulation, but the structure of their external genital organs is incompatible so successful copulation is prevented by **mechanical isolation.** This occurs between some insect species.

When premating isolating mechanisms fail, as they occasionally do, **postmating isolating mechanisms** may prevent effective crossing. (1) The sperm of one species may die in the reproductive tract of the other or be unable to fertilize the egg. Isolation is achieved by **gamete mortality** or **incompatibility.** (2) Eggs may be fertilized by foreign sperm but the hybrids may die during development because of disharmonic genic interactions, that is, the genes from parents belonging to different species may not interact well in regulating the precise mechanisms needed for normal embryonic development. Isolation is achieved by **hybrid mortality.** Nearly all of the hybrids die in the embryonic stage when eggs of a bullfrog are fertilized artificially with sperm of a leopard frog. (3) Hybrids of an interspecific cross may develop to maturity but may be adaptively inferior to either parental species because their mixture of characters is not well suited to either parental habitat, or the hybrids may have a mixture of courtship behavior so that they will not mate. In either case isolation is achieved by **hybrid inferiority.** (4) Occasionally, interspecific hybrids not only survive but also are unusually vigorous and exhibit heterosis. The mule, a hybrid of a male ass and a female horse, is a familiar example. The hybrids are typically **sterile** because of synaptic failure during meiosis (p. 418) and are prevented from either perpetuating themselves or back crossing with either parental species.

Most sympatric species are isolated effectively in their native habitats by the effect of some combination of isolating mechanisms. If the environment of sympatric species is altered by bringing them into the laboratory or a zoo, interspecific crosses take place more easily. Allopatric species normally are separated geographically, but their reproductive isolating mechanisms can be tested by bringing them together. Frequently the postmating mechanisms are more effective than the premating ones.

Geographic Speciation. A new species arises, of course, from one or more preexisting species. We can imagine several ways by which this can occur. One of the most important is **geographic,** or **allopatric, speciation.** This is such a slow process that none of us can see it from beginning to end in our lifetime, but we can find examples of the steps we believe take place. An essential element in geographic speciation is **geographic variation.** Most species do vary geographically in their structure, physiology, ecology, and behavior because selection works in different directions in different parts of the geographic range. The northern populations of a warm-blooded species tend to have a larger body size in the Northern Hemisphere than southern populations because of a more favorable surface–volume relationship. A large animal has more heat-producing mass relative to its heat-losing surface than a small one. This particular phenomenon is widespread and is known as **Bergmann's rule.** A different sort of geographic variation is seen in the white-crowned sparrow and some other birds. Northern populations are migratory, whereas southern ones are not. Such characteristics clearly are adaptive, but geographic variation in a feature such as the number of scales along the belly of a species of snake does not appear to be adaptive in itself. Many such characters may be pleiotropic manifestations of a gene complex that is adaptive in other ways. Others may be the result of genetic drift or other random changes. Although geographically different parts of a species tend to diverge, gene flow by migration between adjacent demes prevents the species from breaking into two reproductively isolated parts (Fig. 18.9*A*). The species as a whole has a unity and cohesiveness maintained by gene flow and the sharing of many interacting gene complexes.

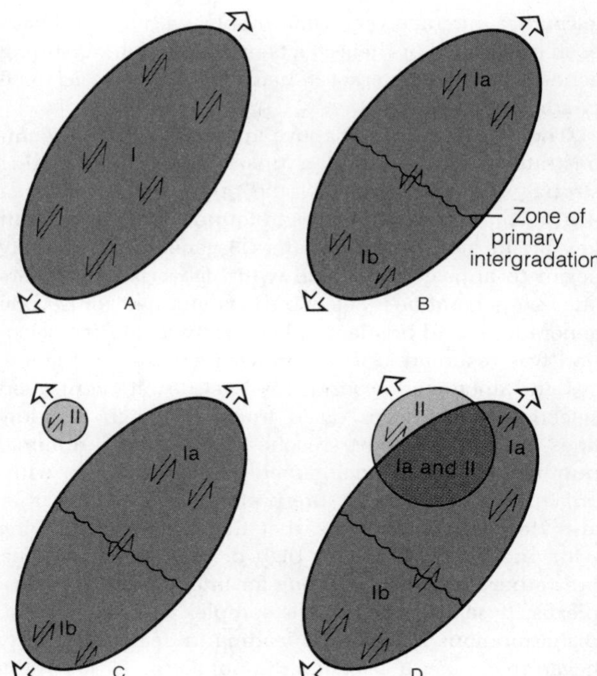

FIGURE 18.9 Steps in geographic speciation. The ovals and circles represent ranges of species I and II; large arrows, the direction of natural selection; half arrows, gene flow. *A,* A wide-ranging species showing geographic variation resulting from different selection pressures. *B,* The species has differentiated into subspecies. *C,* A geographic isolate has separated from one subspecies and differentiated into a different, allopatric species. *D,* The range of the new species expands so that this species becomes partly sympatric with the ancestral species.

Frequently, geographic variation is gradual over the range of the species, but the change may be particularly rapid in a relatively narrow geographic band, which usually correlates with abrupt environmental change or with a semibarrier. Regions of this type are used to subdivide a species into **subspecies** (Fig. 18.9*B*). At subspecies borders we find a zone of **primary intergration** where hybridization between the two subspecies takes place. The fact that a zone of intergradation is limited to a narrow geographic band means that there is some restriction of gene flow between the subspecies, probably due to slight hybrid inferiority or partial sterility. If the two subspecies can accumulate genetic differences faster than they share them through the remaining gene flow, they could become distinct species. The process would be very slow because the populations are large, and it would take many generations for favorable variations to spread and accumulate.

Most allopatric species evolve when a small population becomes physically isolated from the main part of the species and gene flow between them stops. Genetic differences spread and accumulate rapidly in the small, isolated group. This process is most likely to occur at the border of a species where environmental conditions exceed the range of tolerance of members of the species and there is a **barrier** to their spread. Depending on the species, this might be a mountain range, a desert, a forest, a large body of water, and so on. Barriers are seldom absolute, and a few members of a species somehow cross them and establish a colony on the other side. A few individuals may stray high into a mountain pass and reach a valley on the other side of the range. Oceanic islands have been colonized by individuals blown out from the mainland in a storm or carried out on mats of vegetation drifting in ocean currents. When the borders of a species are closely examined, it is not uncommon to find small colonies, called **geographic isolates,** beyond the species border that have little if any gene exchange with the parental species (Fig. 18.9*C*).

The few colonizers will carry with them a small and random sample of the variability of the parental species (founder effect). The population will be small for several generations, and genetic drift will take place. Probably the physical, and certainly the biological, environment (other species with whom the colonizers interact) will be different on the other side of the barrier, so selection will operate in different directions than in the parental species. Mutations occur at random and these will be different in the two groups. Conditions exist for the breakdown of the old coadaptive gene complex and the evolution of a new interacting genetic system. Mayr describes the isolates as passing through a "**genetic revolution.**" Doubtless many isolates become extinct in the process, but the few that succeed constitute new species. This model of speciation could take place relatively rapidly, in a few generations.

When closely related species are examined genetically, we find many differences between them in the genes they carry, in allele frequencies, and often in chromosome rearrangements (inversions, translocations). Richard Goldschmidt argued in the 1940s that the chromosome differences were the essence of speciation and that speciation could not take place unless macromutations of this type occurred. He felt that macromutations created a significant departure, a "hopeful monster," that would at once be reproductively isolated from the parental species. Although Goldschmidt's views were not at first accepted, biologists now recognize that major chromosome rearrangements frequently accompany speciation. These rearrangements would most likely be incorporated and spread in small, geographically isolated populations.

The development of postmating reproductive isolating mechanisms is believed to be a by-product of the genetic revolution that the geographic isolates pass through. The newly formed species is at first allopatric, living in a different geographic area than the parental species. It may remain so, and there are many examples of closely related species that replace one another geographically. But sometimes the new species invades the range of the parental species and becomes sympatric with it (Fig. 18.9*D*). If the new species mates with the parental species at all, the hybrids probably will be inferior in some way because of unfavorable interactions between different gene complexes. If this is the case, natural selection will intensify any slight reproductive differences between them. Those individuals whose behavior, ecology, or other attributes favor their mating with their own kind (**homogamy**) will have a greater reproductive success than those less discriminating individuals who waste their reproductive effort and gametes in producing inferior hybrids. In this way premating reproductive isolating mechanisms gradually will evolve.

When two species meet and compete for the same food or other similar resources, one of two outcomes will occur. If one species is far more successful than the other in exploiting a limited environmental resource, it will drive the less successful one to extinction. This was first demonstrated by Gause in laboratory populations of two species of the protozoa, *Paramecium,* and is known as Gause's principle of **competitive exclusion.** More commonly in field situations the competing species will partition or share the resources in different ways and so come to coexist. It is unlikely, for example, that two different bird species will be equally effective in feeding on the same range of seed size. If the bill size of one is slightly larger, this species will be more effective in feeding upon the larger seeds. Individuals of the species with the larger bills will be favored by selection because they are more likely to avoid competing directly with birds feeding on the smaller seeds. Over many generations, one species tends to restrict the other from feeding on the particular seed size to which it is becoming specialized. The accompanying intensification of morphological differences, in this case bill size, is called **character displacement.**

An excellent example of the results of geographic speciation and subsequent overlap of ranges is seen in a group of birds known as **Darwin's finches** that inhabit the Galapagos Islands (p. 400). There are now 14 species of finches in the archipelago, differing primarily in feeding habits and in bill size and shape. Several species occur on the same islands, and they are reproductively isolated by courtship differences in which bill display is an important signal. The Galapagos are volcanic islands that were colonized originally by chance by one or a few ancestral species of finch. Those on each island were geographically isolated and diverged as they adapted to the food and other resources of their island. Occasionally, different species, or incipient species, came together on the same island. Three species of *Geospiza* coexist on Abindon and Bindloe islands. They have quite different bill sizes and husk seeds of different size (Fig. 18.10). The most logical explanation for this is that competition led to character displacement with respect to bill size. This argument is particularly convincing when one sees that the nature of the character displacement is different on other islands where the competing species are different. *G. magnirostris* does not occur on Charles or Chatham islands, and many individuals of *fortis* have large enough bills to feed on the larger seeds. The very small islands of Daphne and Crossman each have a single species, *fortis* and *fulginosa,* respectively. On these islands their bill size is intermediate compared to those of birds on islands in which these two species are in competition.

Character displacement that accompanies geographic speciation promotes the divergence of an ancestral species into many different ones specialized for different foods and modes of life. We call this **adaptive radiation.** The species of *Geospiza* are an example of adaptive radiation on a small scale.

Sympatric Speciation. Many sympatric species arise by the secondary overlap of species that originally diverged geographically. It is also theoretically possible that sympatric species can diverge in one geographic area by strong selection to different ecological conditions. A very convincing example of sympatric evolution is found among cichlid fish, a widespread family of freshwater fish to which the familiar aquarium angelfish belong. The great tectonic lakes of the Rift Valley of Africa have been isolated for a long period of time and contain large numbers of endemic species of fish, i.e., species found only in a particular lake. Many of these fish are cichlids, which in their long isolated position have become highly specialized for all sorts of habitats and feeding modes. Among the cichlid lake faunas are "species flocks," clusters of numerous species that are more closely related to each other than to any other

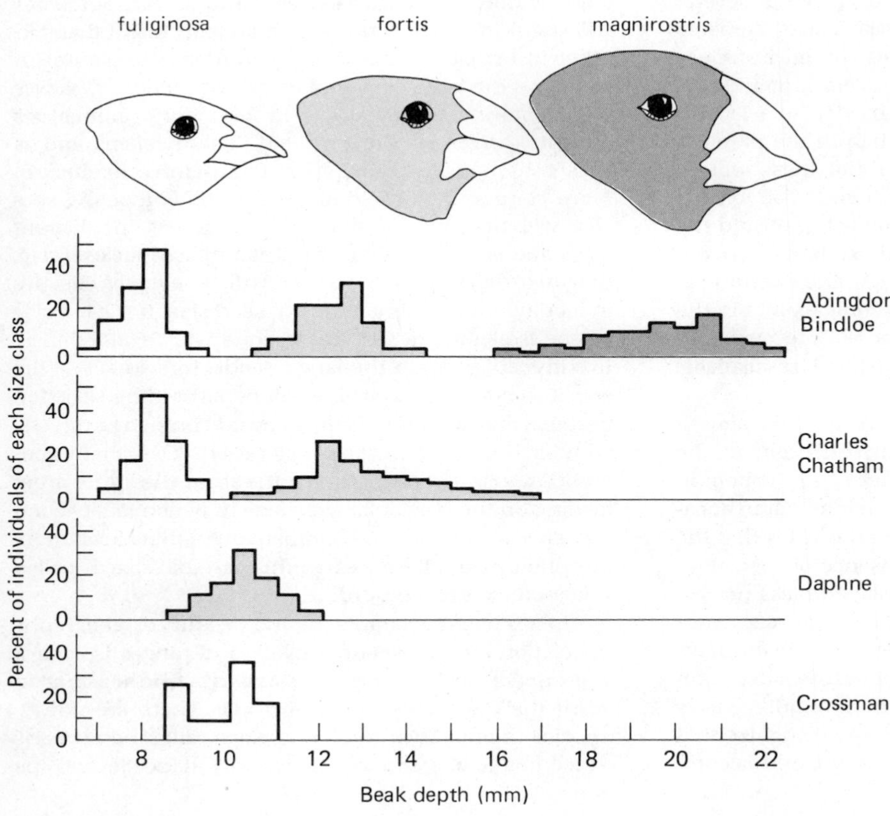

FIGURE 18.10 Character displacement in bill size in three species of *Geospiza* on several of the Galapagos Islands. (Modified from Lack, D.: *Darwin's Finches.* Cambridge University Press, 1947.)

species within or without the lake! It is difficult to imagine such speciation not having occurred sympatrically within the lake in which the flock now dwells, but the mechanism is uncertain. Perhaps selection leading to specialization within local populations (demes) scattered through the lake took place faster than gene flow between populations.

One very rapid and well-documented mode of sympatric speciation is by polyploidy, an abrupt change in the number of chromosome sets, for example from the diploid (2N) to the tetraploid (4N) condition. This may occur by **allopolyploidy,** the hybridization of two different species and the continuation of the hybrids as a third species. Let **AA** symbolize the entire chromosome complement of one diploid species; **BB,** the chromosomes of another diploid species (Fig. 18.11). A hybrid between the two species has the chromosome composition **AB.** It is sterile because the **A** and **B** chromosomes do not synapse normally in the meiotic divisions preceding gamete formation. However, in an occasional mitotic division, the replicated chromosomes fail to separate and a tetraploid cell (**AABB**) is formed. This cell gives rise by mitosis to a clone of tetraploid cells. When certain of these cells begin to differentiate into reproductive cells, normal meiosis occurs because one set of **A** chromosomes can synapse with the other set of **A** chromosomes, and likewise **B** with **B.** Diploid gametes (**AB**) are produced, and their union forms a tetraploid zygote that develops into a tetraploid individual. Tetraploid organisms can cross normally and perpetuate themselves, but they cannot cross easily with either diploid parental species.

In some cases polyploidy results from a doubling of chromosomes within one species rather than from hybridization (**AA** to **AAAA**). This is referred to as **autopolyploidy.** Polyploidy of either type involves not only a change in the number of chromosome sets but also an accompanying change in the amount of DNA. Chromosome numbers may also change by chromosome fissions and fusions, but these changes would not involve a change in the amount of DNA.

Botanists estimate that approximately one half of the species of flowering plants have evolved through allopolyploidy. As one example, there are species of violet with every multiple of 6 chromosomes (=n) from 12 (=2n) to 54 (=9n). Polyploidy is less common in animals than in plants, but the reasons for this are not completely understood. The evolution of allopolyploidy usually involves a period of partial sterility that lasts until a fortuitous doubling of chromosomes occurs. Perennial plants capable of vegetative and asexual reproduction are more likely to survive such a period than annual plants and most animals that are dependent on sexual reproduction. Because of the way they grow, plants also have more opportunities than animals for a doubling of chromosomes in tissues that will give rise to spores and gametes. Polyploids are also more common among organisms in which both sexes are combined in one individual (as in many plants) than in organisms where the sexes are separate. It only occurs in animals that are hermaphroditic, parthenogenetic, or in which sex is determined not by chromosome balances, which would be upset by changes in chromosome numbers, but by strong determining factors on the Y or Z chromosome.

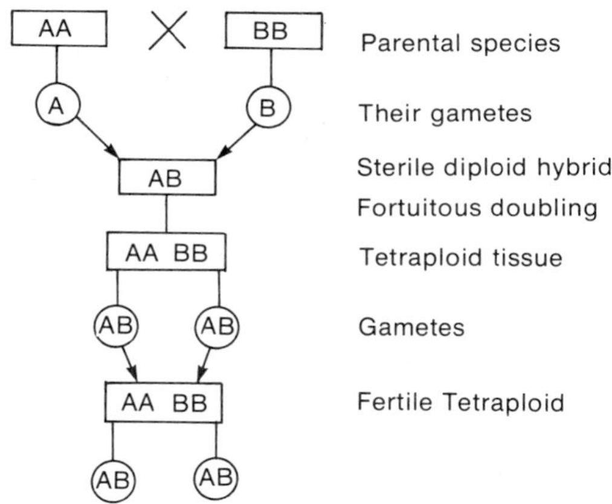

FIGURE 18.11 Diagram of one method by which allopolyploidy can occur. **AA** and **BB** symbolize the diploid chromosome complement of two different species.

Macroevolution or Transpecific Evolution

The species generated by the mechanisms of allopatric and sympatric speciation are reproductively isolated from the parent species but would usually be only a little different in structure, physiology, and behavior. How then can we account for the large differences that are so apparent between members of two different families or classes of animals, such as fish and amphibians? The changes leading to these larger, more striking differences are referred to as **macroevolution** or **transpecific evolution** (since the differences are more evident between taxonomic categories above the species level). They arise from a long series of speciation events, each moving the daughter species farther and farther away from the original ancestral parental form. Different rates of species formation and extinction along various branching lines will reveal, when looking backward at the fossil record or at the surviving members, major morphological changes, evolutionary trends, and adaptive shifts.

Adaptive Shifts. Speciation, as we have seen, usually involves an adaptive shift because the derived species has entered an environment where ecological conditions are not identical to those of the parent species. The success of shifts is not assured; individuals entering new environments frequently die. For a population to become established in a new situation two favorable circumstances must coincide. First of all no shift can be made unless the individuals entering the new environment are viable under the new conditions. This means that either the ecological differences between the ancestral and the new environment must not be great, or, if they are, as in the transition from water to land, the new entrants must already have evolved features, such as lungs in the vertebrate water–land transition, necessary in the new habitat. The new entrants need some degree of **preadaptation.** Obviously, preadaptive features are not evolved in anticipation of a shift. They may have evolved as neutral features in the ancestral habitat, but usually they confer some advantage in the ancestral habitat yet at the same time open up new evolutionary pathways. Many fishes live in aquatic habitats in which there is a variable supply of oxygen. At times the pool may be heavily shadowed so that photosynthetic activity is low, it may become stagnant, or it may be subjected to seasonal drought. Any variation under these circumstances toward an accessory respiratory organ that enables fish to extract oxygen from the air and supplement gill breathing would be favored. Lungs are simply one of several types of accessory respiratory organs found in fishes living in these habitats. Lungs are of value in the present context because they contribute to the survival of populations that find themselves temporarily in stagnant water or dried up pools. At the same time, lungs make possible a shift from water to land.

Second, successful shifts are not made, even in a preadapted species, unless the habitat a new species is entering has food and other resources that are not fully exploited by existing species. Resources will be under-exploited during periods when many species living in a habitat become extinct. Also the continual expansion and diversification of living organisms provide opportunities and resources that others may use. The colonization of the land by plants and invertebrates produced habitats and resources suitable for primitive vertebrates. The evolution of grasses created opportunities for the evolution of antelope and other grazing herbivores. The evolution of large herbivorous mammals created the opportunity for the evolution of lions and other large carnivores.

Adaptive Trends. When an environmental shift is made features evolve that adapt the population to the new conditions. Two questions arise. How can natural selection promote a feature in its very early stages of development when it may not be large enough or significant enough to be favored? How can natural selection favor the continuation of a trend over a long period of time and sometimes to the point where the structure may become rather burdensome?

An example of the first question is seen in electric eels and electric rays that have modified muscles forming electric organs capable of generating an electric current strong enough to ward off predators and stun prey. Clearly selection could not favor their present function early in their evolution when the electric current would have been very weak. African elephantfish and South American knifefish use slightly enhanced muscle action potentials, which all muscles have (p. 110), as part of an electrolocating system. Possibly the more powerful electric organs began in this way. Any variation toward an increase in the power of muscle action potentials would be favored in the evolution of electrolocating systems. As these systems continued to improve, they might reach a threshold of power that could discourage predators. Selection would then favor a continued power increase in this context. Finally the organs would be powerful enough to stun prey. We do not know, of course, precisely how powerful electric organs evolved. We hypothesize that they were adaptive at all stages in their evolution but adaptive in different ways.

The Irish elk (Fig. 18.12) that lived in Europe during the last glacial period of the Pleistocene is an example of the second question. Its antlers, which were used in combats with other males over mates, became exceptionally large and probably were a physiological burden for they were shed annually. Under some circumstances selection may carry a particular structure beyond the point of optimum usefulness because selection does not act on individual organs but on entire organisms. There is a growth relationship between body size and antler size in the Irish elk and other deer such that antlers grow at a faster rate than the body. Accordingly they become relatively larger as the body increases in size during the life of an individual and during the evolution of a group. The Irish elk lived during a cold period when selection would favor large body size as a temperature-conserving mechanism. (A large body has a favorable surface : mass ratio.) Many Pleistocene mammals were very large. As body size increased the antlers became relatively larger and larger. Since selection acts on the entire organism, this trend would be expected to stop at a point when body size was less than optimum and antler size greater than optimum.

Evolutionary trends sometimes continue for long periods of time because the direction of environmental

FIGURE 18.12 The male Irish elk of the late Pleistocene had huge antlers that were shed annually. (Neg. no. 35802. Courtesy American Museum of Natural History.)

change and natural selection may be the same for millions of years. Also as organisms begin to specialize for a particular mode of life they frequently evolve certain attributes that cannot be reversed easily. These aspects of trends are evident in the evolution of horses (Fig. 18.13). The ancestral horse, *Hyracotherium,* was a terrier-sized, forest-dwelling animal with three functional toes well suited for support on the soft forest floor and low-crowned molar teeth well adapted for browsing on soft forest vegetation. As the climate of North America became drier during much of the past 70 million years, grasslands replaced many forests. Horses shifted and adapted to the open, grassland conditions. They increased in size, and their legs and feet adapted for running at high speed on open, hard ground. The distal portion of their legs elongated, and the number of toes was reduced to one. Their molar teeth enlarged faster than other parts of the body, and cement spread from the roots over the surface of the teeth and into the valleys between ridges on the tooth surfaces. The teeth became high-crowned and well adapted for grinding hard grasses. The premolar teeth came to resemble the molar teeth. A commitment to a particular mode of life was made, and the more specialized the adaptation, the greater was the commitment. Lost toes and low-crowned teeth cannot be regained. Adaptations for running and grazing continued until they reached a high level of perfection. It became increasingly difficult for a major change in evolutionary direction to occur.

Diversification and Extinction. When a group is successful in making a major adaptive shift, from water to land, for example, evolutionary diversification is at first very rapid as descendent species take advantage of the numerous ecological opportunities and divide the resources of the environment among themselves. There is an extensive adaptive radiation. As opportunities are exploited, the rate of evolution of this group slows down. The new species frequently become highly specialized to narrow modes of life. They survive as long as their modes of life are possible. But climates and environments are continually in flux over long periods of time, and seldom does a particular habitat and potential mode of life remain unchanged. In the course of their adaptation to one mode of life, a few species evolve preadaptive features that enable them to make a shift as conditions change, but many cannot and become **extinct.** Extinction simply results from an imbalance between a species and the environment in which it is living.

The Nature of Macroevolution. Long-term evolutionary changes may occur in various ways. A species living in a changing environment may undergo selection that slowly shifts the mean and range of variation of the species in the direction of the environmental gradient. This is termed **phyletic speciation** (Fig. 18.14*A*). Populations at the beginning and end of this sequence are sufficiently different that biologists are justified in regarding them as distinct species even though drawing a dividing line between them presents a problem because the generations overlap morphologically and probably reproductively. Seldom is the fossil record complete, so paleontologists use these discontinuities as pragmatic species boundaries. Phyletic species then are not the same as species we have considered earlier where divergence occurs rather quickly in small, semi-isolated populations by the development of reproductive isolation. The fossil record also shows that organisms diversify in time, so there must be episodes of branching superimposed upon phyletic speciation.

Recently the concept of phyletic speciation has been questioned. Analysis of groups well documented in the fossil record suggests that new species arise rather quickly (in geologic terms) as they would by normal modes of speciation. Once a new species is formed it continues unchanged for several million years and then frequently becomes extinct. Before extinction, descendent species branch off in different directions. This pattern of species appearing and disappearing rather abruptly has been called **punctuated equilibria.** What

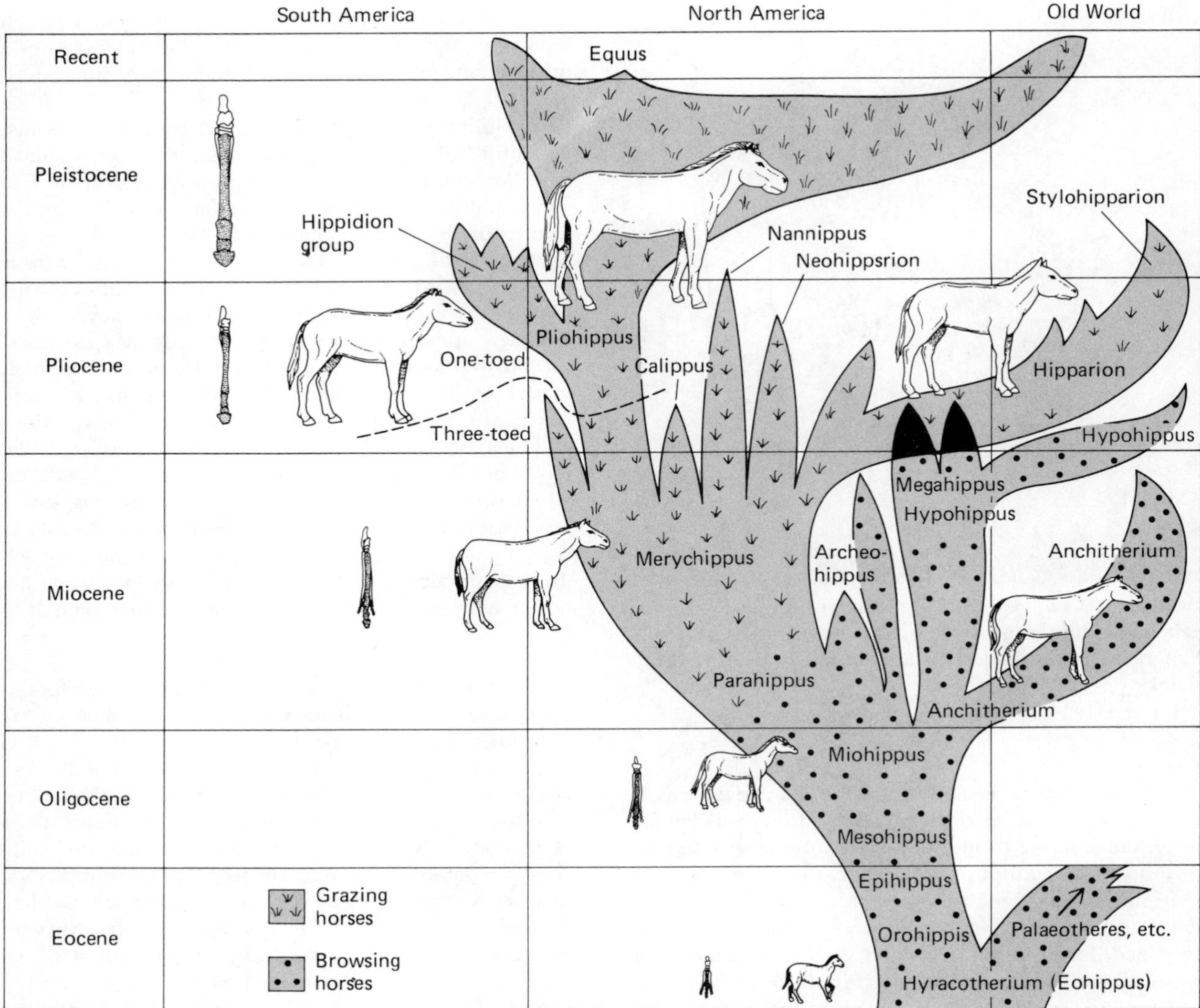

FIGURE 18.13 The evolution of horses. (Modified from Simpson, G. G.: *Horses.* London, Oxford University Press, 1951.)

appears to be a phyletic speciation trend may occur if descendent species branch off more frequently in one direction than in another or if extinctions are more common in one direction than in another (Fig. 18.14*B*). Directions in which branching is favored, or in which extinctions occur, are determined by success of adaptation to the environment, or simply by factors that influence the rate of speciation, not all of which are adaptive: mutation pressure; pattern of distribution and reproductive modes that affect the ease with which a species breaks up into small, semi-isolated groups; fecundity; and so on.

Both phyletic speciation and diversifying speciation by modes we have considered imply relatively slow (in geologic terms) and small genetic changes in isolating mechanisms and in morphology. We know, however, that rather quick and abrupt morphological changes sometimes occur in the history of life. Vertebrates evolved rather quickly from some lower chordate ancestor probably allied to the tunicates or sea squirts. One mechanism that can explain quick and major morphological shifts with a minimum of genetic change is a change in regulatory genes such that the rate of embryonic development of the body is shifted relative to the

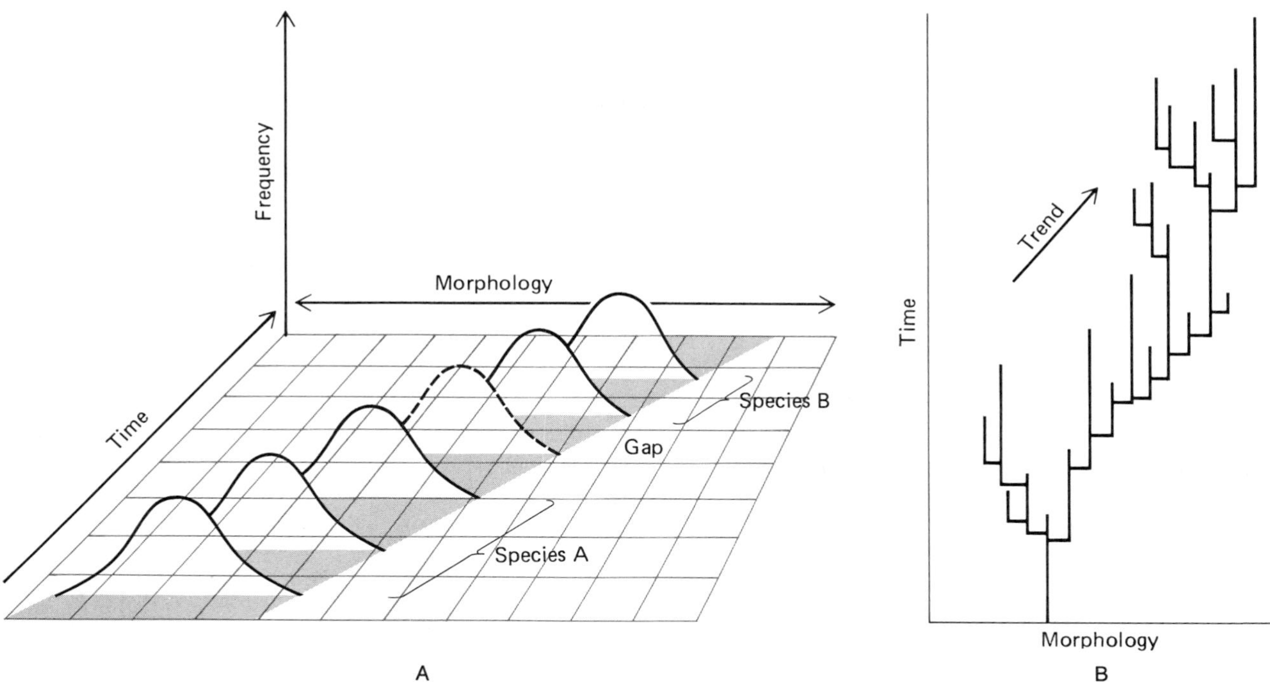

FIGURE 18.14 Two models of long-term evolutionary change. *A,* Phyletic speciation. The structure of a species changes slowly and continuously as it adapts to a changing environment. Gaps in the fossil record are convenient species boundaries. *B,* Punctuated equilibria. New species arise from former ones relatively quickly (horizontal lines) by normal processes of speciation and remain unchanged for millions of years. A trend to the right appears because there have been more speciation events, and less extinction, in this direction. (Modified from Luria, S. E., S. J. Gould, and S. Singer: *A View of Life.* Menlo Park, Benjamin/Cummings Publishing Co., 1981.)

rate of sexual development. An organism becomes sexually mature while morphologically still an embryo or larva and fails to complete its somatic development. This phenomenon is known as **paedomorphosis** (Gr. *pais,* child + *morphosis,* form). As Professor Stephen J. Gould of Harvard University has pointed out, paedomorphosis may result either from an acceleration of sexual development relative to somatic development (called **progenesis**) or by a slowing down of somatic development relative to sexual development (**neoteny**) (Gr. *neos,* young + *teinein,* to extend). All vertebrate embryos have a notochord in their back, a dorsal nerve cord, and muscles in the trunk and tail. These are essential for the undulatory swimming of ancestral vertebrates. Adult tunicates are sessile, saclike animals lacking these organs. Tunicate larvae, however, are tadpole-shaped, motile creatures with a well-developed tail containing a notochord, nerve cord, and muscle. At metamorphosis the larva attaches to the substratum and loses its tail. Vertebrates probably evolved from a paedomorphic tunicate larva capable of reproducing while a larva and never metamorphosing. Selection acting on such a larva and favoring increased mobility could lead quickly to the characteristic vertebrate morphology. Paedomorphosis is a way of eliminating specialized adult characters (in this case those associated with a sessile adult) and opening new avenues of evolution (increased mobility of the adult).

Summary

1. Biological species are groups of individuals capable of reproducing with one another but reproductively isolated from other such groups. They can be subdivided into small, local reproducing groups, the demes, which are the smallest units of evolutionary change.

2. Variation is introduced into populations by mutations. The overall effect of a mutation must be interpreted in the context of all of its pleiotropic effects and the environment in which it is acting. The reservoir of variation in most populations is sufficient for them to respond to selection in many directions.

3. The variability in a population can be shifted from its Hardy-Weinberg equilibrium value by mutation pressure, migration, genetic drift, and natural selection.

4. Natural selection may operate at any stage of the life cycle in many different ways. Stabilizing selection promotes stability; directional selection, a shift in the variability in the direction of the environmental change; disruptive selection, adjustment to a heterogeneous environment. When heterozygous individuals have the highest fitness, balancing selection establishes an equilibrium among all genotypes. Selection usually operates on individuals, but in group selection one group of individuals is favored as a whole over another group.

5. Selection is a creative force that changes the composition of the gene pool of a species in a favorable direction and thereby increases the probability that a group of favorable variations will be brought together.

6. Sympatric species are kept separate by some combination of premating reproductive isolating mechanism (habitat, temporal, ethological, mechanical) and postmating isolating mechanisms (gamete mortality, gamete incompatibility, hybrid mortality, hybrid inferiority, hybrid sterility). Allopatric species are separated geographically.

7. The most common way new species of animals arise is by allopatric speciation. Small populations become physically isolated from the parent species, usually around the periphery of the range. Gene flow is interrupted, selection is different in the isolated group, the founder effect and genetic drift operate, and the population may undergo a major genetic reorganization leading to some degree of postmating reproductive isolation. If the species that has developed in geographic isolation secondarily overlaps the range of the parent species, natural selection will promote the development of premating isolating mechanisms for they will reduce the futile waste of gametes.

8. When closely related species overlap, competition is reduced by resource partitioning which is facilitated by character displacement. This leads to an adaptive radiation.

9. Species may also arise sympatrically by allopolyploidy and autopolyploidy.

10. Long-term, or macroevolutionary, changes first involve an adaptive shift. If environmental differences are great the shifting population must have some degree of preadaptation. The habitat the new species is entering must also contain resources that are not fully exploited.

11. Natural selection can favor the evolution of structures early as well as late in their development, but selection may operate in a different way when a structure is small and just beginning to appear than when a structure is well formed. When different parts of an organism are interrelated by growth rates or other factors, natural selection, which operates on the entire organism, will strike a compromise in which one part may be below optimum in size and another part somewhat above optimum in its expression.

12. As organisms partition resources and become increasingly specialized to particular modes of life, it becomes increasingly difficult for a shift in evolutionary direction to occur.

13. Considerable diversification and adaptive radiation occur during the early evolution of a group of species, but as ecological opportunities are exploited the rate of new species formation decreases.

14. Evolutionary trends may involve phyletic speciation, that is, the gradual transformation of a species as a whole with time, or new species may arise from preexisting species by normal processes of speciation. After its formation a new species remains unchanged for millions of years and then usually becomes extinct (punctuated equilibria). Trends result from the pattern and direction of the formation and extinction of new species.

15. Major changes in evolutionary direction may result from slight changes in genes that control various embryonic developmental rates. In paedomorphosis, the rate of body development slows down relative to the rate of development of the reproductive organs, and juvenile features are retained by adults.

References and Selected Readings

Ayala, F. J.: The mechanisms of evolution. *Scientific American, 239*(Sept.):56, 1978. A summary of how the variability of populations is shaped into species and of the genetic differences between species.

Ayala, F. J., and J. W. Valentine: *Evolving, the Theory and Process of Organic Evolution.* Menlo Park, The Benjamin-Cummings Publishing Co. 1979. A contemporary text on the history and mechanisms of evolution. Many quantitative aspects of evolutionary theory are developed, and numerous suggestions are given for additional reading.

Bishop, J. A., and L. M. Cook: Moths, melanism and clean air. *Scientific American, 232*(Jan.):90, 1975. A study of the effect of air pollution control on industrial melanism in the moths of Britain.

Darwin, C.: *The Origin of Species.* 1859. Available in a number of recent reprint editions. This classic is well worth sampling for its clear, logical arguments and for its wealth of examples.

Dobzhansky, T.: *Genetics of the Evolutionary Process.* New York, Columbia University Press, 1970. A rewriting and extension of his classic book on *Genetics and the Origin of Species.* The original edition (1937) was instrumental in the development of neoDarwinism.

Friedman, M. J., and W. Trager: The biochemistry of resistance to malaria. *Scientific American, 244*(Mar.):154, 1981. How genes for sickle-anemia and thalassemia protect against malaria.

Goldschmidt, R. B.: *The Material Basis of Evolution.* New Haven, Yale University Press, 1940. Presents in detail Goldschmidt's views on the importance of large mutations in the evolution of species.

Gould, S. J.: *Ever Since Darwin.* New York, W. W. Norton & Co., 1977. A fascinating collection of essays, many originally published in *Natural History Magazine,* on many aspects of evolution.

Grant, P. R.: Speciation and the adaptive radiation of Darwin's finches. *American Scientist, 69*(Nov.–Dec.): 653, 1981. A recent study of speciation in Darwin's finches.

Lack, D.: *Darwin's Finches.* Cambridge, Cambridge University Press, 1947. A classic case study of speciation.

Lewontin, R. C.: Adaptation. *Scientific American, 239*(Sept.):212, 1978. The relationship between adaptation and natural selection is not always a simple one.

Mayr, E.: *Populations, Species, and Evolution.* Cambridge, Harvard University Press, 1970. An abridgment and revision of his earlier book on *Animal Species and Evolution.* A scholarly account of evolutionary theory with an emphasis upon systematic and field aspects.

Milkman, R. (Ed.): *Perspectives on Evolution.* Sunderland, Sinauer Associates, 1982. A collection of essays aimed at advanced students by leading investigators of evolutionary theory.

Rensberger, B.: Evolution since Darwin. *Science,* 82 (Apr.): 1982. A summary of the development of neoDarwinism and of contemporary controversies in evolutionary theory.

Simpson, G. G.: *The Meaning of Evolution.* Revised ed. New Haven, Yale University Press, 1967. An excellent, nontechnical presentation of evolutionary concepts with an emphasis on the paleontological aspects.

Williams-Ellis, A.: *Darwin's Moon.* London, Blackie & Son, 1966. An excellent biography of Alfred Russel Wallace, with a vivid portrayal of Wallace's life and times, which reemphasizes his solid contributions to the advancement of the theory of evolution.

PART FOUR The Diversity of Animals

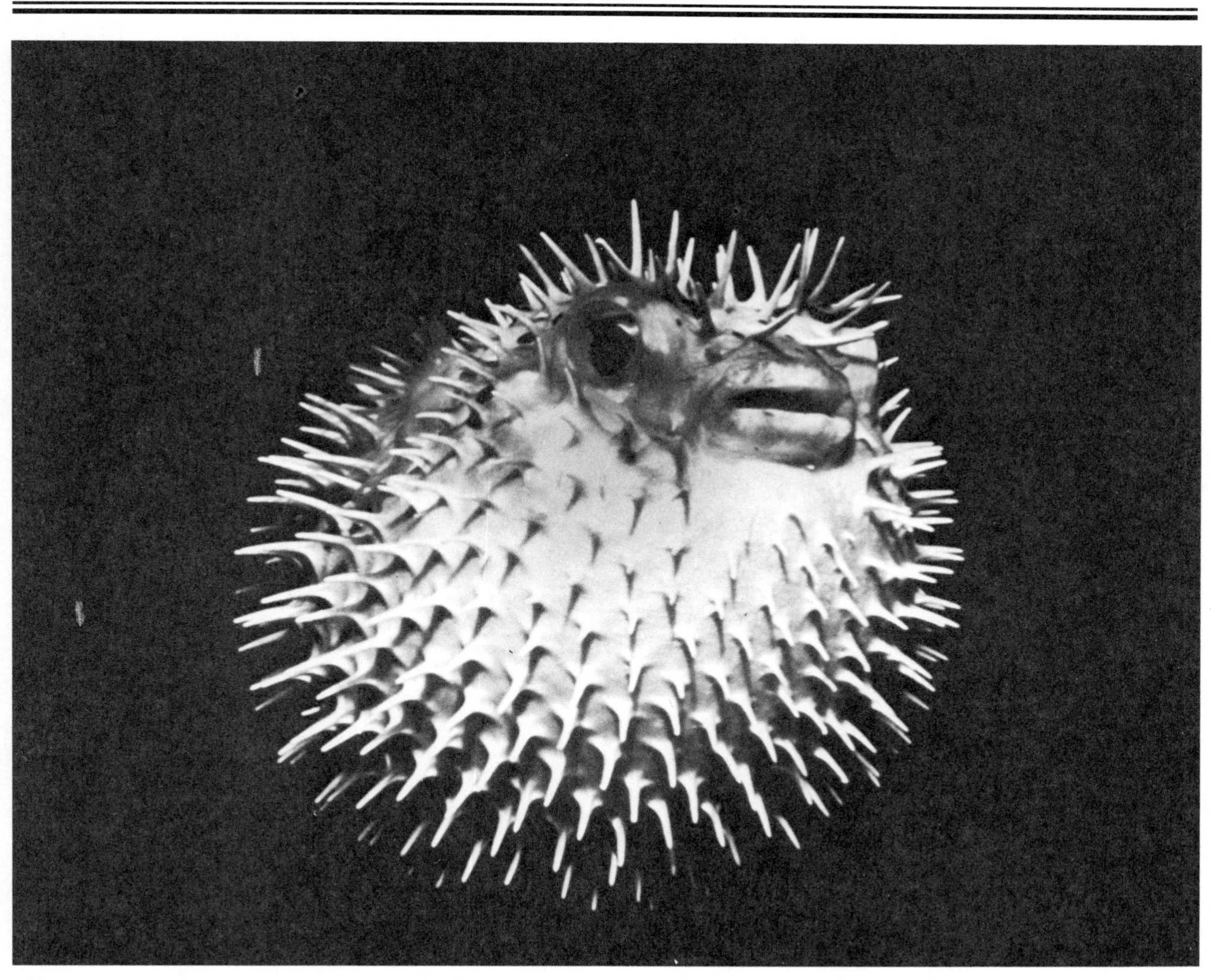

19 The Origin and Diversity of Life

ORIENTATION

Although spontaneous generation of life does not occur at the present time, life did evolve by a slow process of prebiotic chemical evolution about three and one half billion years ago when physical and chemical conditions on earth were quite different than they are today. Millions of organisms subsequently evolved, and over one million species of animals are living today. Most biologists strive to determine the evolutionary relationships among animals by examining the fossil record and by analyzing the degree of similarity among living species. We will explore the methods by which these analyses are made and summarize the approach we will take in studying the adaptive diversity of animals.

The Origin of Life

The question of the beginning of life on this planet has been investigated by many different biologists. Some have postulated that some kind of spores or microorganisms may have been carried through space from another planet to this one. This explanation is unsatisfactory, not only because it begs the question of the ultimate source of these spores but also because it is extremely unlikely that any sort of living thing could survive the extreme cold and intense irradiation of interplanetary travel.

The spontaneous origin of living things at the present time is believed to be extremely improbable. Francesco Redi's experiments showed, about 1680, that maggots do not arise *de novo* from decaying meat and laid to rest the old superstition that animals could appear by spontaneous generation (p. 13). Some 200 years later Louis Pasteur showed conclusively that microorganisms, such as bacteria, do not arise by spontaneous generation but come only from previously existing bacteria. Although the spontaneous generation of life at present is unlikely, it is most probable that billions of years ago, when chemical and physical conditions on the earth's surface were quite different from those at present, the first living things did arise from nonliving material. This concept, that the first things did evolve from nonliving things, was put forward particularly by the Russian biochemist A. I. Oparin in his book *The Origin of Life* (1938).

According to the currently accepted Kant-Laplace hypothesis, our solar system originated about four and one half billion years ago through the rotation and condensation of a vast cloud of interstellar dust and gases. The earth must have been molten at first because of the intense heat generated by meteorite impacts and radioactivity. The oldest rocks on earth are 3.9 billion years old, so a crust must have begun to form about that time. Life may have originated soon afterwards because paleontologists have found well-developed cells in the Fig Tree chert of South Africa that is 3.4 billion years old. This chert was formed by the precipitation in a shallow sea of jelly-like masses that included prokaryotic bacteria and blue-green algae (Fig. 19.1).

We can only speculate as to the sequence of events that led to these simple organisms. It seems certain that the early atmosphere, which was formed by volcanic gases, was strongly reducing and contained large quantities of methane, ammonia, nitrogen, and water vapor. Probably lesser amounts of hydrogen and carbon dioxide were present. Most carbon was in metallic carbides, and oxygen was in various oxides in the earth's crust as well as in water vapor. Conditions existed for the formation of many organic compounds. A number of reactions are known by which organic substances can be synthesized from inorganic ones. Metallic carbides could react with water to form acetylene, which would subsequently polymerize to form compounds with long chains of carbon atoms. High-energy radiation, such as cosmic rays, can catalyze the synthesis of organic compounds. Irradiating solutions of inorganic compounds with ultraviolet light or passing electric charges through the solution to simulate lightning also produce organic compounds. Harold Urey and Stanley Miller in 1953 exposed a mixture of water vapor, methane, ammonia, and hydrogen gases to electric discharges for a week and demonstrated the production of amino acids, such as glycine and alanine, together with other complex organic compounds. Similar reactions doubtless occurred in the earth's prebiotic atmosphere. Amino acids and other compounds could be produced in nature at the present time by lightning discharges or ultraviolet radiations; however, any organic compound produced in this way might undergo spontaneous oxidation or it would be taken up and degraded by molds, bacteria, and other organisms.

Most, if not all, of the reactions by which the more complex organic substances were formed probably occurred in the sea in which the inorganic precursors and the organic products of the reaction were dissolved and mixed. The sea became a sort of prebiotic, organic soup. This must have thickened in certain limited regions of the earth by evaporation, freezing, or adsorption on the surface of mineral particles. Molecules collided with increasing frequency, reacted, and aggregated to form new molecules of increasing size and complexity. As more has been learned of the role of specific hydrogen bonding and other weak intermolecular forces in the pairing

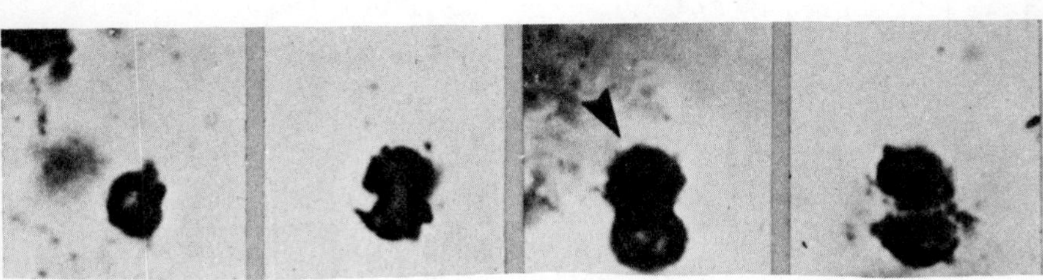

FIGURE 19.1 Prokaryotic cells undergoing division from a South African chert 3.4 billion years old. (From Knoll, A. H., and E. S. Barghorn: Archean microfossils showing cell division from the Swaziland system of South Africa. *Science, 198*:396, 1977.)

of specific nucleotide bases and the effectiveness of these processes in the transfer of biological information, it has become clear that similar forces could have operated early in evolution before "living" organisms appeared.

The known forces of intermolecular attraction and the tendency for certain molecules to form liquid crystals provide us with means by which large, complex, specific molecules can form spontaneously. Oparin suggested that natural selection can operate at the level of these complex molecules before anything recognizable as life is present. As the molecules came together to form colloidal aggregates, these aggregates began to compete with one another for raw materials. Some of the aggregates that had some particularly favorable internal arrangement would acquire new molecules more rapidly than others and would eventually become the dominant types.

Once some protein molecules had formed and had achieved the ability to catalyze reactions, the rate of formation of additional molecules would be greatly stepped up. Next, in combination with molecules of nucleic acid, these complex protein molecules probably acquired the ability to catalyze the synthesis of molecules like themselves. These hypothetic, autocatalytic particles composed of nucleic acids and proteins would have some of the properties of a virus, or perhaps of a free-living gene. The next step in the development of a living thing was the addition of the ability of the autocatalytic particle to undergo inherited changes—to mutate. Then, if a number of these free genes had joined to form a single larger unit, the resulting organism would have been similar to certain present-day viruses. A major step in early evolution was the development of a protein-lipid membrane around this aggregate that permitted the accumulation of some and the exclusion of other molecules. All the known viruses are parasites that can live only within the cells of higher animals and plants. However, a little reflection will suggest that free-living viruses, ones which do not produce a disease, would be very difficult to detect; such organisms may indeed exist.

The first living organisms, having arisen in a sea of organic molecules and in an atmosphere devoid of oxygen, presumably obtained energy by the fermentation of certain of these organic substances. These heterotrophs could survive only as long as the supply of organic molecules in the organic soup, accumulated from the past, lasted. Before the supply was exhausted, however, the heterotrophs evolved further and became autotrophs, able to make their own organic molecules by chemosynthesis or photosynthesis. One of the by-products of photosynthesis is gaseous oxygen, and it is likely that all the oxygen in the atmosphere was produced and is still produced in this way. It is estimated that all the oxygen of our atmosphere is renewed by photosynthesis every 2000 years, and all the carbon dioxide molecules pass through the photosynthetic process every 300 years. All the oxygen and carbon dioxide in the earth's atmosphere are the products of living organisms and have passed through living organisms over and over again during the course of evolution.

An explanation of how an autotroph may have evolved from one of these primitive, fermenting heterotrophs was presented by N. H. Horowitz in 1945. According to Horowitz' hypothesis, an organism might acquire, by successive mutations, the enzymes needed to synthesize complex from simple substances, in the *reverse* order to the sequence in which they are used in normal metabolism. Let us suppose that our first primitive heterotroph required organic compound Z for its growth. Substance Z, and a variety of other organic compounds, Y, X, W, V, U, and so on, were present in the organic sea broth that was the environment of this heterotroph. They had been synthesized previously by the action of nonliving factors of the environment. The heterotroph would survive nicely as long as the supply of compound Z lasted. If a mutation occurred that enabled the heterotroph to synthesize substance Z from substance Y, the strain of heterotroph with this mutation would be able to survive when the supply of substance Z was exhausted. A second mutation, one that established an enzyme catalyzing a reaction by which substance Y could be made from the simpler substance X, would again have great survival value when the supply of Y was exhausted. Similar mutations, setting up enzymes enabling the organism to use successively simpler substances, W, V, U, . . . and finally some inorganic substance, A, would eventually result in an organism able to make substance Z, which it needs for growth, out of substance A by way of all the intermediate compounds. When, by other series of mutations, the organism was able to synthesize all of its requirements from simple inorganic compounds, as the green plants can, it would have become an autotroph. Once the first simple autotrophs had evolved, the way was clear for the further evolution of the vast variety of green plants, bacteria, molds, and animals that inhabit the world today. These considerations lead us to the conclusion that the origin of life, as an orderly natural event on this planet, was not only possible but also was almost inevitable.

The Grouping of Species

The various members of the Animal Kingdom are probably all related through some degree of common ances-

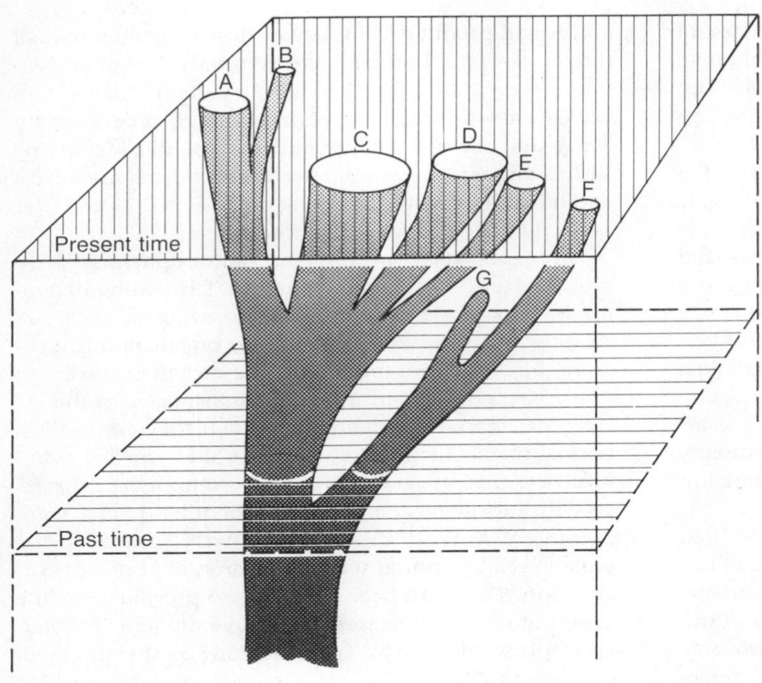

FIGURE 19.2 Diagrammatic representation of the evolutionary relationships of six (hypothetical) monophyletic species. Circular cross sections of the branches represent the species at the present time; junctions of branches represent points of common ancestry. Species G is extinct.

try and have varying degrees of similarity in structure and function.

Their great number and diversity require some framework of order within which they can be viewed. That framework is a system of classification, and the branch of zoology concerned with classification is called **taxonomy** or **systematic zoology.**

The basic unit in classifying organisms is the **species,** which in sexually reproducing forms is a group of individuals capable of reproducing with each other but reproductively isolated in nature from other groups (p. 417). In short, members of a species share a common gene pool with all that this implies with respect to a common geographic range, habitat, and similar characteristics, ranging from biochemical through morphological to behavioral ones. In sorting out species, biologists usually search for distinctive or **diagnostic** morphological characters that can be recognized easily. For example, it is easier to recognize a giraffe by its long neck than by its blood proteins. However, it is important to remember that species differ from one another in all the ways expressed by their genotypes and not just by a few conspicuous features.

A species population may be thought to have not only a dimension in space (range) but also a dimension in time. The population extends backward in time, merging with other species populations much like the branches of a tree (Fig. 19.2). Thus, species have varying degrees of evolutionary relationships with one another depending upon the distance to the point of population mergence or common ancestry. The evolutionary relationship of organisms is known as **phylogeny,** (Gr. *phylon,* race + *genesis,* descent), and this relationship provides a theoretical basis for the grouping, or classification, of organisms. Species that resemble each other closely, and therefore are presumed to be closely related, are grouped together within a **genus.** Similar genera are grouped together within a **family;** similar families, within **orders;** orders, within **classes;** and classes, within **phyla** (Fig. 19.3). Thus the taxonomic categories, or **taxa,** basically are hierarchically arranged, nested sets of resemblances. That is, a phylum contains a group of similar classes; each class, a group of similar orders; and so on down to species. Theoretically at least, any category in this hierarchy of classification should constitute a **monophyletic** grouping; i.e., all the species or subgroups within any group should have a common ancestry.

Although there is general agreement on what constitutes a species, the grouping of species into higher taxa is much more subjective and difficult to define. For example, in Figure 19.2, should species A–F be placed within a single genus or do they represent three genera, A–B, C–E, and F? (All of the groupings would be monophyletic.) The same type of problem exists in determining the limits of family, order, class, and phylum. In gen-

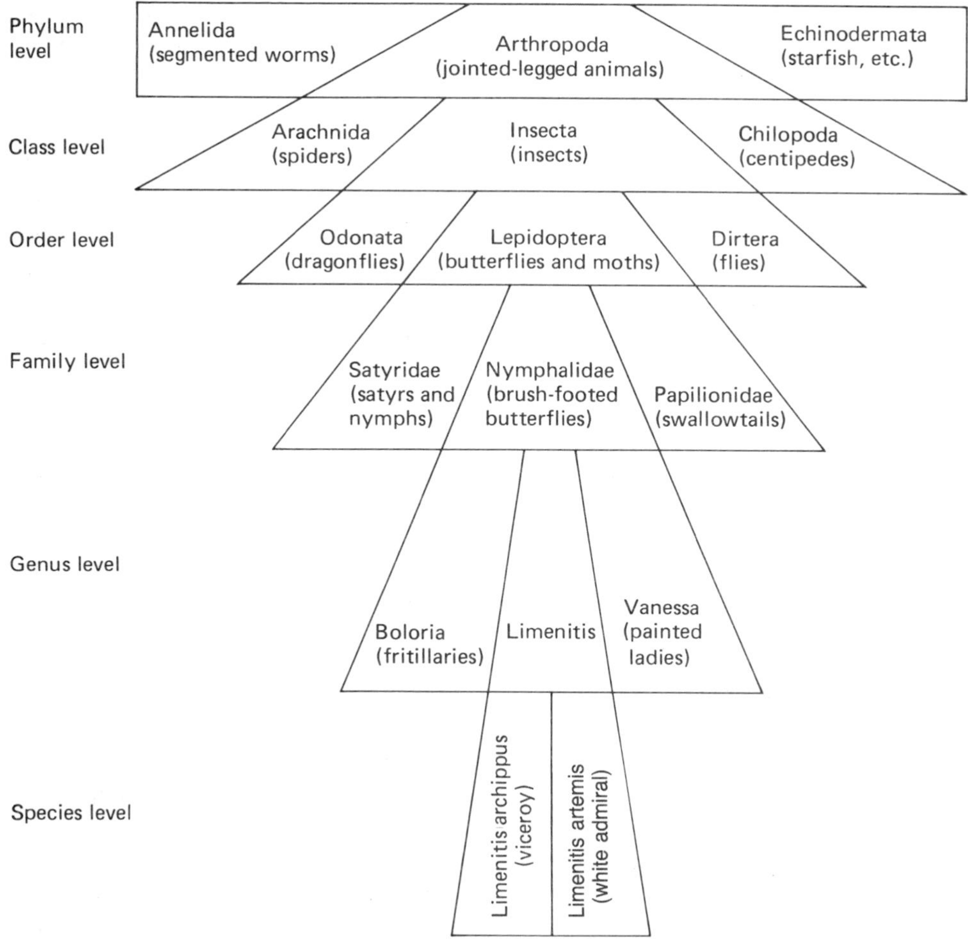

FIGURE 19.3 Part of the classification of arthropods showing the hierarchy of taxa and how lower ones are included within, or are nested in, the higher ones.

eral, the greater the gap (i.e., the extent of dissimilarity) between species, the more likely a separate grouping is justified.

Biologists base their judgements about degree of relationship on the fossil record, when available, and on the extent of similarity between living species. Any characteristics that reflect the degree of genetic similarity may be used. Morphological ones are generally favored because they are easier to work with. In assessing similarities, most biologists search for homologous structures (p. 388) in different animals, for the presence of homologous organs in different animals implies **divergent evolution** from a common ancestor (Fig. 19.4). Similarities among other organs, referred to as **analogous organs** (p. 389), result not from a shared ancestry but from **convergent evolution** (Fig. 19.4), i.e., adaptation by unrelated or very distantly related organisms to similar environmental conditions. The resemblance between analogous organs is related to their functions, and there may or may not be some superficial structural similarity.

In the course of evolution species branch from one another and as they continue along separate lines of evolution, diverge to different degrees. In reconstructing evolutionary trees, **classic phylogeneticists** first sort out homologous and analogous characters and use the shared homologies to determine probable branching points. Branching points can be recognized by the point where a lineage acquires a **derived character** found only in members of this lineage. The three auditory os-

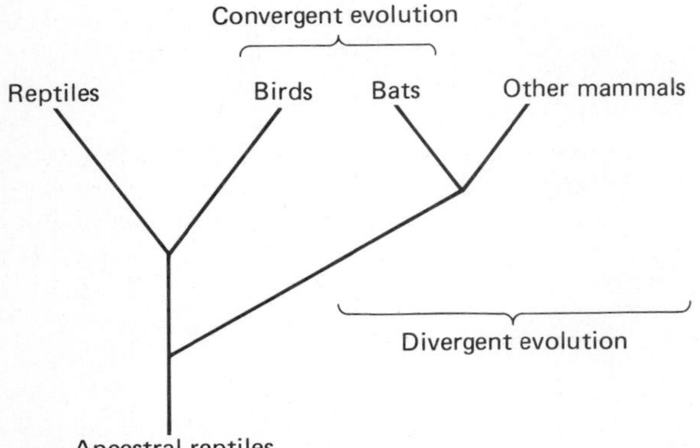

FIGURE 19.4 Diagram of convergent and divergent evolution. In divergence, an ancestral group breaks up into two or more lines of evolution which may lead far from the ancestral design. During this process distantly related groups (birds and bats) may converge and come to resemble one another in some ways as they adapt to similar modes of life (flying).

sicles in the middle ear are useful in identifying a branching point between mammals and reptiles because their evolution was a unique event, and all mammals and only mammals share this derived character. But three auditory ossicles are a **primitive character** when one mammal is compared with another since all have inherited them. They are of no value in sorting out mammalian groups. Biologists must search for other derived characters to find branching points among mammals. The pouch, or marsupium, in which the young undergo part of their development is a derived character that separates marsupials from placental mammals. Classic phylogeneticists also consider the amount of divergence of a lineage since it branched from a stem group. Consider, for example, the grouping of turtles, lizards, snakes, dinosaurs, and birds (Fig. 19.5). Birds evolved from early dinosaurs, but they have diverged far from their dinosaur ancestors as they have adapted to a flying mode of life. A classic phylogeneticist would emphasize the shared characters of turtles, lizards, snakes, crocodiles, and dinosaurs and place them in the class Reptilia. All have horny scales and are cold-blooded, among other shared features. Even though they separated from dinosaurs quite recently, birds are placed in the class Aves on the basis of their feathers, warm-bloodedness, and many other features indicating a great deal of divergence since branching from the dinosaur stock.

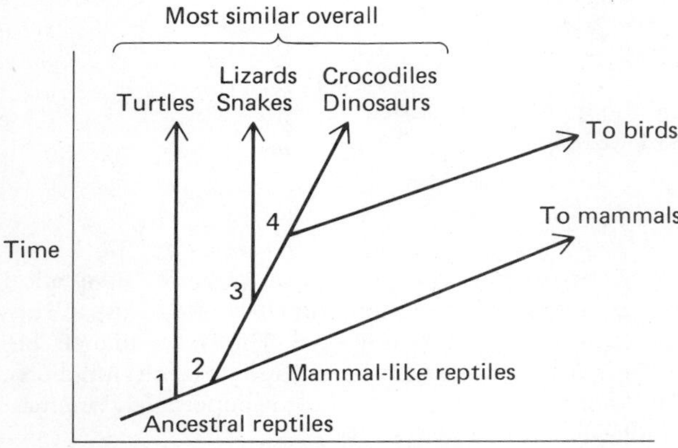

FIGURE 19.5 A diagram of branching points (numbered) and degrees of similarity in the evolution of major reptile groups and birds. Note that although turtles, lizards, snakes, and crocodiles are the most similar of these living forms, birds, dinosaurs, and crocodiles are most closely related in having branched most recently from a common ancestor.

Another school of biologists, **cladists** (Gr. *klados*, branch), believe that the evaluation of the overall degree of similarity is too subjective to be useful and focus only on branching points. A cladist would emphasize the recent common ancestry of crocodiles, dinosaurs, and birds and combine them in a group separate from lizards, snakes, and other reptiles.

A third school of biologists, the **pheneticists** or **numerical taxonomists** (Gr. *phainein*, to appear), attempt to be completely objective by examining all characteristics, computing averages of the overall amount of similarity between groups, and feeding these data into a computer to construct a classification. The weakness of this approach is that all characteristics are given equal weight. The pheneticists make no attempt to distinguish between homologous and analogous organs or between derived and primitive characters. They treat biological data as one would treat any other set of numerical data, such as measurements of a mixed assortment of nails, for example. They are trying to develop an easy-to-use classification and are less interested in reconstructing the course of evolution.

Taxonomic Nomenclature

The naming of species and higher taxonomic categories is another aspect of taxonomy. The scientific name of a species is composed of words usually derived from Latin or Greek, of which the first is the **generic** name and the second is the **species** name. Together the generic and specific names constitute the scientific name, which is therefore a binomial, i.e., composed of two words. All species in the same genus have the same generic name. As an illustration, the scientific names of the categories to which the familiar box turtles belong are listed below. Note that the family name ends in *idae* and that the generic and species names are always italicized.

Phylum Chordata
  Subphylum Vertebrata
    Class Reptilia
      Order Testudinata—Turtles and tortoises
        Family Emydidae—Freshwater and marsh turtles
          Genus *Terrapene*—Box turtles
          Terrapene carolina—Common box turtle
          Terrapene ornata—Ornate box turtle

Law of Priority. All taxonomic names must be unique; i.e., the same name cannot be applied to more than one species or higher taxonomic category. When duplication occurs, the first species or category to which the name was applied takes priority in the claim to the name. The application of the **law of priority** for animals begins with the tenth edition of the *Systema Naturae,* a catalog of the plants and animals known to the Swedish biologist Carolus Linnaeus in 1758. The *Systema Naturae* represents the first uniform application of the present system of binomial nomenclature, although the system was not invented by Linnaeus.

Adaptive Diversity

The Animal Kingdom displays a tremendous adaptive diversity. Every conceivable habitat within the ancestral oceanic environment has been exploited, and there have been numerous invasions of organisms into fresh water and onto land. The entry of animals into new habitats posed new problems of existence, which were met with modification of old structures and new systems for new functions. Since different animal groups invaded the same habitats, the evolutionary history of animals is filled with convergent adaptations.

Adaptation to new conditions rarely has resulted in the loss of all of a species' ancestral characteristics. Some are modified to the demands of the new environment; some remain unchanged and are said to be primitive. Thus every species possesses some primitive and some specialized features. A primitive species is one that has retained many ancestral characteristics. But the term primitive is useful only in a comparative sense. One can speak of a primitive animal, a primitive vertebrate, a primitive mammal, or a primitive human. But a primitive human is not a primitive mammal, nor is a primitive mammal a primitive vertebrate.

The evolutionary history of the Animal Kingdom can be depicted as having a treelike form (a **phylogenetic tree**), the diverging lines dividing trunks and branches representing diverging evolutionary lines (Fig. 19.6). Those groups of animals that diverged from the main line early in the history of the Animal Kingdom (i.e., from the lower part of the phylogenetic tree) are often described as being "lower animals." Others that arose later at a higher level of the tree are often said to be "higher animals." However, these terms are often carelessly used and seem to imply that all animals and animal structures fit into a hierarchy of simple to complex. Even the "lowest" animals, such as sponges, may display certain specialized and unique features in addition to their primitive characteristics.

How to Study Animal Groups

The study of animal groups can be a tedious and painful task if it amounts to nothing more than committing to memory a list of characteristics of one group after another. A much more useful and interesting approach is one that emphasizes relationships and adaptations. This is the approach that will be utilized in the following chapters. Two questions should be continually asked as these chapters are studied: (1) How is the structure and physiology of the group correlated with its mode of existence or the environmental demands of the habitat in which it lives? (2) How does the structure and physiology of the group reflect the group's evolutionary origin and its relationship to other groups? These questions can provide a meaningful framework into which facts about animal groups can be fitted. Moreover, with a little knowledge of the structural ground plan of animal phyla and an understanding of the problems posed by different modes of existence and different environments, it is possible to make intelligent guesses about possible adaptations that might be encountered.

There are about 30 animal phyla, more than we can possibly discuss in detail in this book. Our discussion of

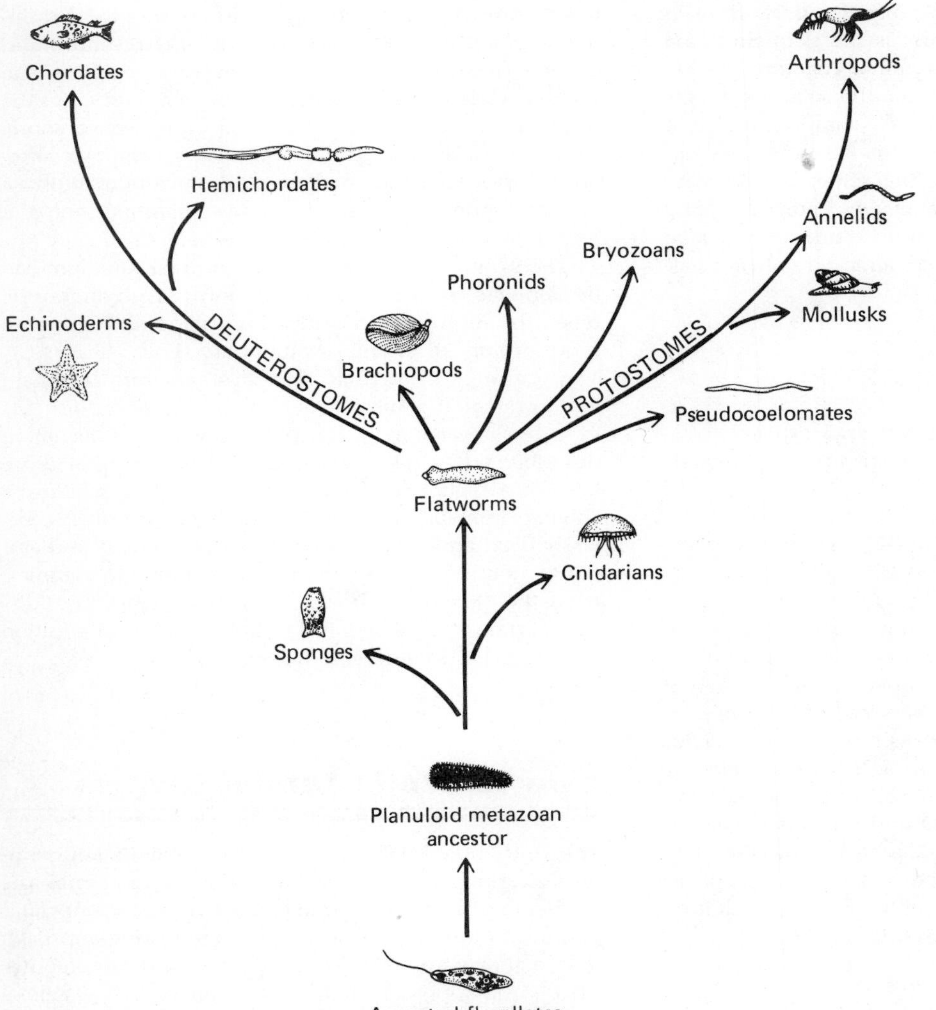

FIGURE 19.6 A phylogenetic tree of the Animal Kingdom. This is but one of many that might be constructed to reflect different ideas about the evolutionary relationships of different groups of animals.

animal groups will be limited to the major phyla. Minor phyla—those with relatively few species—are included in the following synopsis of the Animal Kingdom, and some will be briefly described at the end of certain of the later chapters. Detailed accounts of these groups can be found in the references cited at the end of this chapter.

Protozoa are often treated as animals, for many are able to move and are heterotrophic. But protozoa are unicellular organisms, excluded by definition from the Metazoa, or multicellular animals. They have had a separate evolutionary history, and their complexity has developed in a different way from that of metazoans. Protozoa will be discussed in the next chapter.

Synopsis of the Phyla of Metazoan Animals

The following diagnoses are limited to distinguishing characteristics. The approximate number of species described to date is indicated in parentheses.

PARAZOA Animals with poorly differentiated tissues and no organs.

Phylum Placozoa (1) Microscopic, flattened, marine animal *(Trichoplax adhaerens)* composed of ventral and dorsal epitheloid layers enclosing loose mesenchyme-like cells.

Phylum Pórifera (5000) Sponges. Sessile; no anterior end; some primitively radially symmetrical, but most are irregular. Mouth and digestive cavity absent; body organized about a system of water canals and chambers. Marine, but a few found in fresh water.

EUMETAZOA Animals with organs, mouth, and digestive cavity.

Radiata Tentaculate radiate animals with few organs. Digestive cavity, with mouth the principal opening to the exterior.

Phylum Cnidaria, or Coelenterata (9000) Hydras, hydroids, jellyfish, sea anemones, and corals. Free-swimming, or sessile, with tentacles surrounding mouth. Specialized cells bearing stinging organoids called nematocysts. Solitary or colonial. Marine, with a few in fresh water.

Phylum Ctenophora (90) Comb jellies. Free-swimming; biradiate, with two tentacles and eight longitudinal rows of ciliary combs (membranelles). Marine.

Bilateria Bilateral animals.

Protostomes Cleavage is determinate and commonly spiral; mouth arising from or near blastopore.

Acoelomates Area between body wall and internal organs filled with parenchyma.

Phylum Platyhelminthes (12,700) Flatworms. Body dorsoventrally flattened; digestive cavity (when present) with a single opening, the mouth. The turbellarians are free-living, a few terrestrial. Trematodes and cestodes are parasitic.

Phylum Mesozoa (50) An enigmatic group of minute parasites of marine invertebrates. No organs, and body composed of few cells.

Phylum Rhynchocoela, or Nemertina (900) Nemerteans. Long, dorsoventrally flattened body with a complex proboscis apparatus. Digestive cavity with mouth and anus. Marine, with a few terrestrial and in fresh water.

Phylum Gnathostomulida (80) Gnathostomulids. Minute wormlike animals. Body covered by a single layer of epithelial cells, each of which bears a single cilium. Anterior end with bristle-like sensory cilia. Mouth cavity with a pair of cuticular jaws. Marine.

Pseudocoelomates Animals in which the blastocoel sometimes persists, forming a body cavity. Digestive tract with mouth and anus. Body usually covered with a cuticle.

Phylum Rotifera (1500) Rotifers. Anterior end bearing a ciliated crown; posterior end tapering to a foot. Pharynx containing movable cuticular pieces. Microsopic. Largely in fresh water, some marine, some inhabitants of mosses.

Phylum Gastrotricha (430) Gastrotrichs. Elongated body with flattened ciliated ventral surface. Few to many adhesive tubes present; cuticle commonly ornamented. Microscopic. Marine and in fresh water.

Phylum Kinorhyncha (100) Kinorhynchs. Somewhat elongated body. Cuticle segmented and bearing posteriorly directed spines. Spiny, retractile anterior end. Less than 1 mm. in length. Marine.

Phylum Nematoda (12,000) Roundworms. Slender cylindrical worms with tapered anterior and posterior ends. Cuticle often ornamented. Free-living species usually only a few millimeters or less in length. Marine, freshwater, terrestrial, and parasitic forms.

Phylum Nematomorpha (230) Hairworms. Extremely long threadlike bodies. Adults free-living in damp soil, in fresh water, and a few marine. Juveniles parasitic.

Phylum Acanthocephala (1150) Acanthocephalans. Small wormlike endoparasites of vertebrates. Anterior retractile proboscis bearing recurved spines.

Schizocoelous Coelomates Body cavity a coelom, formed embryonically by a splitting of the mesoderm, or, if a body cavity is absent, the coelom has been lost. Digestive tract with mouth and anus.

Phylum Priapulida (9) Priapulids. Cucumber-shaped marine animals, with a large anterior proboscis. Body surface covered with spines and tubercles. Peritoneum of coelom greatly reduced.

Phylum Sipuncula (320) Sipunculans. Cylindrical marine worms. Retractable anterior end, bearing lobes or tentacles around mouth.

Phylum Mollusca (50,000) Snails, chitons, clams, squids, and octopods. Ventral surface modified in the form of a muscular foot, having various shapes; dorsal and lateral surfaces of body modified as a shell-secreting mantle, although shell may be reduced or absent. Marine, in fresh water, and terrestrial.

Phylum Echiura (140) Echiurans. Cylindrical marine worms, with a flattened nonretractile proboscis. Trunk with a large pair of ventral setae.

Phylum Annelida (11,400) Segmented worms—polychaetes, earthworms, and leeches. Body wormlike and metameric. A large longitudinal ventral nerve cord. Marine, in fresh water, and terrestrial.

Phylum Pogonophora (80) Marine, deepwater animals, with a long body housed within a chitinous tube. Anterior end of body bearing from one to many long tentacles. Digestive tract absent.

Phylum Tardigrada (400) Water bears. Microscopic segmented animals. Short cylindrical body bearing four pairs of stubby legs terminating in claws. Freshwater and terrestrial in lichens and mosses; few marine.

Phylum Onychophora (70) Onychophorans. Terrestrial, segmented, wormlike animals, with an anterior pair of antennae, and many pairs of short conical legs terminating in claws. Body covered by a thin cuticle.

Phylum Arthropoda (900,000) Crabs, shrimp, mites, ticks, scorpions, spiders, and insects. Body metameric with jointed appendages and encased within a chitinous exoskeleton. Vestigial coelom. Marine, freshwater, terrestrial, or parasitic.

Phylum Pentastomida (90) Wormlike endoparasites of vertebrates. Anterior end of body with two pairs of leglike projections terminating in claws and a median snoutlike projection bearing the mouth.

Lophophorate Coelomates Mouth surrounded or partially surrounded by a crown of hollow tentacles (a lophophore).

Phylum Phoronida (10) Phoronids. Marine, wormlike animals with the body housed within a chitinous tube.

Phylum Bryozoa (4000) Bryozoans. Colonial, sessile; the body usually housed within a chitinous or chitinous-calcareous exoskeleton. Mostly marine, a few in fresh water.

Phylum Brachiopoda (335) Lamp shells. Body often attached by a stalk and enclosed within two unequal dorsoventrally oriented calcareous shells. Marine.

Phylum Entoprocta (150) Entoprocts. Body attached by a stalk. Mouth and anus surrounded by a tentacular crown. Mostly marine.

Deuterostomes, or Enterocoelous Coelomates Cleavage radial and indeterminate; mouth arising some distance (anteriorly) from blastopore. Mesoderm and coelom develop primitively by enterocoelic pouching of the primitive gut.

Phylum Chaetognatha (70) Arrow worms. Marine planktonic animals with dart-shaped bodies bearing fins. Anterior end with grasping spines flanking a ventral preoral chamber.

Phylum Echinodermata (6000) Starfish, sea urchins, sand dollars, and sea cucumbers. Secondarily pentamerous radial symmetry. Most existing forms free-moving. Body wall containing calcareous ossicles usually bearing projecting spines. A part of the coelom modified into a system of water canals with external tubular projections used in feeding and locomotion. Marine.

Phylum Hemichordata (80) Acorn worms. Body divided into proboscis, collar, and trunk. Anterior part of trunk perforated with varying number of pairs of pharyngeal clefts. Marine.

Phylum Chordata (42,795) Pharyngeal pouches, notochord, and dorsal hollow nerve cord present at some time in life history. Marine, freshwater, and terrestrial.

Subphylum Urochordata (1250) Sea squirts, or tunicates. Sessile, or planktonic nonmetameric invertebrate chordates enclosed within a cellulose tunic. Notochord and nerve cord present only in larva. Solitary and colonial. Marine.

Subphylum Cephalochordata (45) Fishlike metameric invertebrate chordates, commonly called amphioxus.

Subphylum Vertebrata (41,500) The vertebrates—fish, amphibians, reptiles, birds, and mammals. Metameric. Trunk supported by a series of cartilaginous or bony skeletal pieces (vertebrae) surrounding or replacing notochord in the adult.

Summary

1. Life originated between 3.9 billion years ago, when the earth's crust formed, and 3.4 billion years ago, from which time fossils of prokaryotic cells have been found.

2. It is known that ultraviolet radiation, electric discharges, and other forms of energy can spontaneously synthesize amino acids and many other simple organic compounds in a reducing atmosphere, such as characterized the early earth.

3. The accumulation of these in ancient seas formed an "organic soup." Increasing concentrations of the soup in certain puddles and intermolecular attractions led to the formation of molecules of increasing size and complexity. Probably some of these combinations acquired the ability to catalyze the synthesis of molecules like themselves. The development of a membrane around groups of these molecules would set the stage for the emergence of life.

4. The first organisms must have been heterotrophic, deriving their energy by the anaerobic fermentation of other compounds in the soup. As the more complex compounds were consumed, natural selection would favor those organisms with the ability to catalyze the synthesis of the needed compounds from simpler precursors. In this way simple, autotrophic green plants probably evolved.

5. As plants continued to photosynthesize, oxygen and ozone accumulated in the atmosphere, and it changed from a reducing one to the type we have now. The evolution of oxidative metabolism became possible.

6. Life cannot now arise spontaneously because ozone reduces the penetration of certain types of radiation, and simple organic compounds, even if they did form, would be quickly oxidized or consumed by existing organisms.

7. Species are grouped into genera, orders, classes, and phyla. These taxa are hierarchically arranged, nested sets of resemblances.

8. Most biologists use homologous organs as a basis for grouping species because the presence of such organs in different species implies divergent evolution. Analogous organs in different species imply convergent evolution.

9. Evolutionary trees can be constructed by searching for points where one lineage separated from another. A diverging lineage acquires some unique derived character at a branch point that continues among descendent members of the group.

10. Cladists use only branching points in reconstructing evolutionary trees. Classic phylogeneticists also consider the degree to which a descendent lineage has diverged from the ancestral group. Pheneticists construct classifications by assessing the total degree of resemblance among all characters. These classifications may not reflect evolutionary relationship.

11. The scientific name for each species is a binomial that combines the genus and species names. Each taxon must have a unique name, and that first published since 1758 has priority in usage.

References and Selected Readings

Attenborough, D.: *Life on Earth.* Boston, Little, Brown & Co., 1979. A fascinating and beautifully illustrated book that parallels the BBC television series on the emergence and diversification of life.

Eldredge, N., and J. Cracraft: *Phylogenetic Patterns and the Evolutionary Process.* New York, Columbia University Press, 1980. A detailed consideration of cladistic methodology.

Gould, S. J.: The telltale wishbone. In Gould, S. J.: *The Panda's Thumb.* New York, W. W. Norton & Co., 1980. An essay comparing traditional and cladistic systems of classification with reference to dinosaurs and birds.

Mayr, E.: Biological classification: Toward a synthesis of opposing methodologies. *Science, 214*(Oct.):510, 1981. A comparison of the merits and deficiencies of traditional, cladistic, and phenetic systems of classification.

Mayr, E.: *Principles of Systematic Zoology.* New York, McGraw-Hill Book Co., 1969. A general account of zoological taxonomy including the text of the International Code of Zoological Nomenclature.

Ross, H. H.: *Biological Systematics.* Reading, Mass., Addison-Wesley Publishing Co., 1974. A short text on systematics with emphasis on evolution.

Schuster, P.: Prebiotic evolution. In Gutfreund, H. (Ed.): *Biochemical Evolution.* Cambridge, Cambridge University Press, 1981. An excellent review for advanced students of the origin of life.

Sneath, P. H. A., and R. R. Sokal: *Numerical Taxonomy.* San Francisco, W. H. Freeman & Co., 1973. An elaboration of the methods of numerical taxonomy.

The following invertebrate zoology texts and reference works contain general discussions of all of the invertebrate phyla, including the minor phyla omitted from this text. References that deal solely with particular groups will be listed at the end of each chapter in Part 4.

Barnes, R. D.: *Invertebrate Zoology.* 4th ed. Philadelphia, W. B. Saunders Co., 1980.

Gosner, K. L.: *A Field Guide to the Atlantic Seashore.* Peterson Field Guide Series. Boston, Houghton Mifflin Co., 1979. A good guide for the identification of commonly encountered coastal marine invertebrate animals.

Hickman, C. P.: *Biology of the Invertebrates.* 2nd ed. St. Louis, C. V. Mosby Co., 1973.

Hyman, L. H.: *The Invertebrates.* Vol. 1–6. New York, McGraw-Hill Book Co., 1940–1967. Vol. I (1940): Protozoa Through Ctenophora; Vol. II (1951): Platyhelminthes and Rhynchocoela; Vol. III (1951): Acanthocephala, Aschelminthes, and Entoprocta; Vol. IV (1955): Echinodermata; Vol. V (1959): Smaller Coelomate Groups; Vol. VI (1967): Mollusca, Part 1.

Kaestner, A.: *Invertebrate Zoology*. Vol. 1–3. New York, Wiley–Interscience, 1967–1979.

Meglitsch, P. A.: *Invertebrate Zoology*. 2nd ed. Oxford, Oxford University Press, 1972.

Morris, R. H., D. P. Abbott, and E. C. Haderlie: *Intertidal Invertebrates of California*. Stanford University Press, 1980. A superb work that not only provides for identification but also summarizes information on the biology of the species included.

Parker, S. P. (Ed.): *Synopsis and Classification of Living Organisms*. Vol. 1 and 2. New York, McGraw-Hill Book Co., 1982. Descriptions of the families and higher taxa of all living organisms. Information is not restricted to morphology.

Pennak, R. W.: *Fresh-Water Invertebrates of the United States*. 2nd ed. New York, John Wiley & Sons, 1978. An excellent guide for the identification of freshwater invertebrates. There is a summary of biological information for each group and a description of methods for collecting and preserving them.

Smith, D. L.: *A Guide to Marine Coastal Plankton and Marine Invertebrate Larvae*. Dubuque, Iowa, Kendall/Hunt Publishing Co., 1977.

20 Protozoa

ORIENTATION

Protozoa are unicellular organisms belonging to several different phyla. The great diversity of protozoa is reflected in the many kinds of habitats in which they are found and the many different modes of existence that they exhibit. In this chapter we will examine each of the protozoan phyla and their major subgroups. Since protozoa are unicells, we will see how organelles have become adapted in each of these groups to meet the needs of existence that are met by organs in multicellular animals.

Protozoa are unicellular, eukaryotic (nucleated) organisms. The multicellular **metazoans,** which are the subject of the greater part of this book, undoubtedly evolved from some group of protozoa, most probably the flagellated protozoa, and the metazoan characteristics of motility and heterotrophic nutrition are probably a protozoan inheritance. Despite these similarities, however, protozoa and metazoans are very different from each other, especially in the way they have become complex.

The evolution of complexity in multicellular plants and animals has occurred through a division of labor among the cells, certain cells becoming specialized for certain functions. Complexity in unicellular organisms has evolved through the specialization of different parts of the cell, for although protozoa are single cells, they are also complete organisms. The cell must perform not just certain functions but must retain the ability to perform all of the functions demanded of an organism.

We will consider the protozoa to be a group of closely related phyla of unicellular organisms. Most biologists place them together with algae in the Kingdom Protista, but we will include them in our survey because of their animal-like nature and traditional inclusion within the Animal Kingdom. The name *protozoa* will be used as a common name.

The flagellated protozoa are generally considered to be primitive and ancestral to the other protozoa, but let's break with tradition and start our examination of protozoan organisms with the ciliates. The ciliates are the most animal-like protozoa and will serve especially well to contrast with the metazoans we will be studying subsequently.

Phylum Ciliophora

Members of the phylum Ciliophora (L. *cilium,* eyelid + Gr. *pherein,* to bear) are characterized by possessing at some time in their life history ciliary organelles for locomotion and feeding. Ciliates are widespread in the sea and in fresh water, and a few species are commensals within the gut of vertebrates. Most are microscopic, but the largest (about 3 mm.) can be seen with the naked eye.

There is also always a distinct anterior end, and primitive species are radially symetrical (Fig. 20.1). However, most ciliates are asymmetrical. Their shape is maintained by a complex **pellicle,** a living outer layer of denser cytoplasm containing the peripheral and surface organelles.

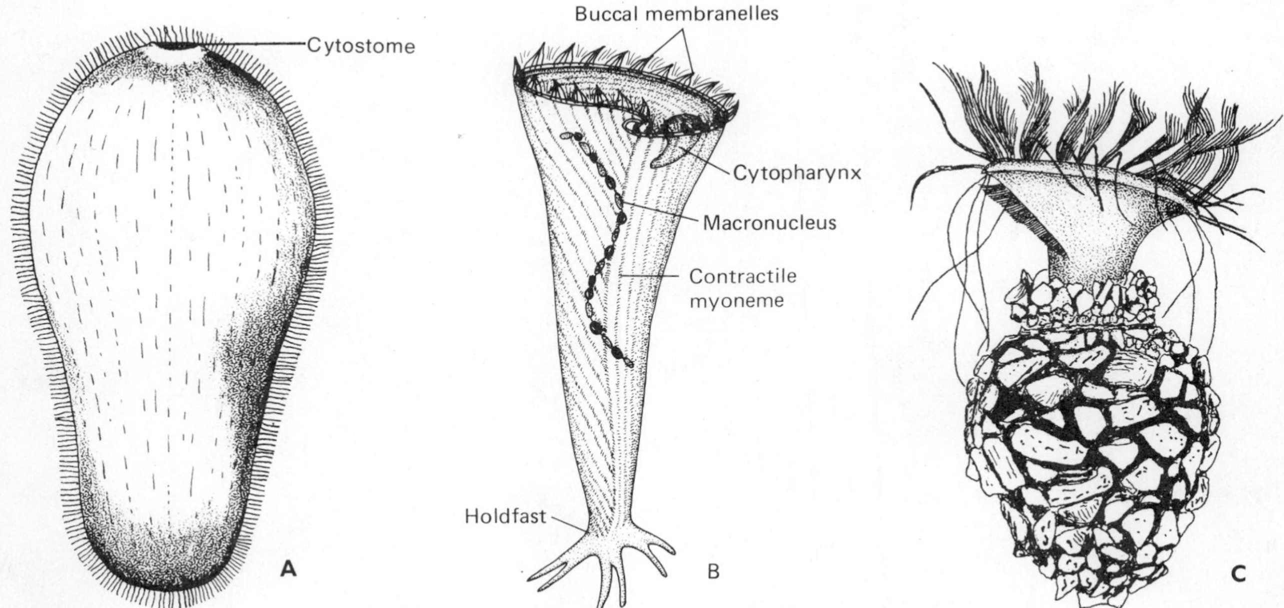

FIGURE 20.1 Ciliates. *A, Prorodon,* a primitive ciliate; *B, Stentor; C, Tintinnopsis,* a marine ciliate with a test composed of foreign particles. (*A* and *C* after Fauré-Fremiet from Corliss, J. O.: *The Ciliated Protozoa.* New York, Pergamon Press, 1961.)

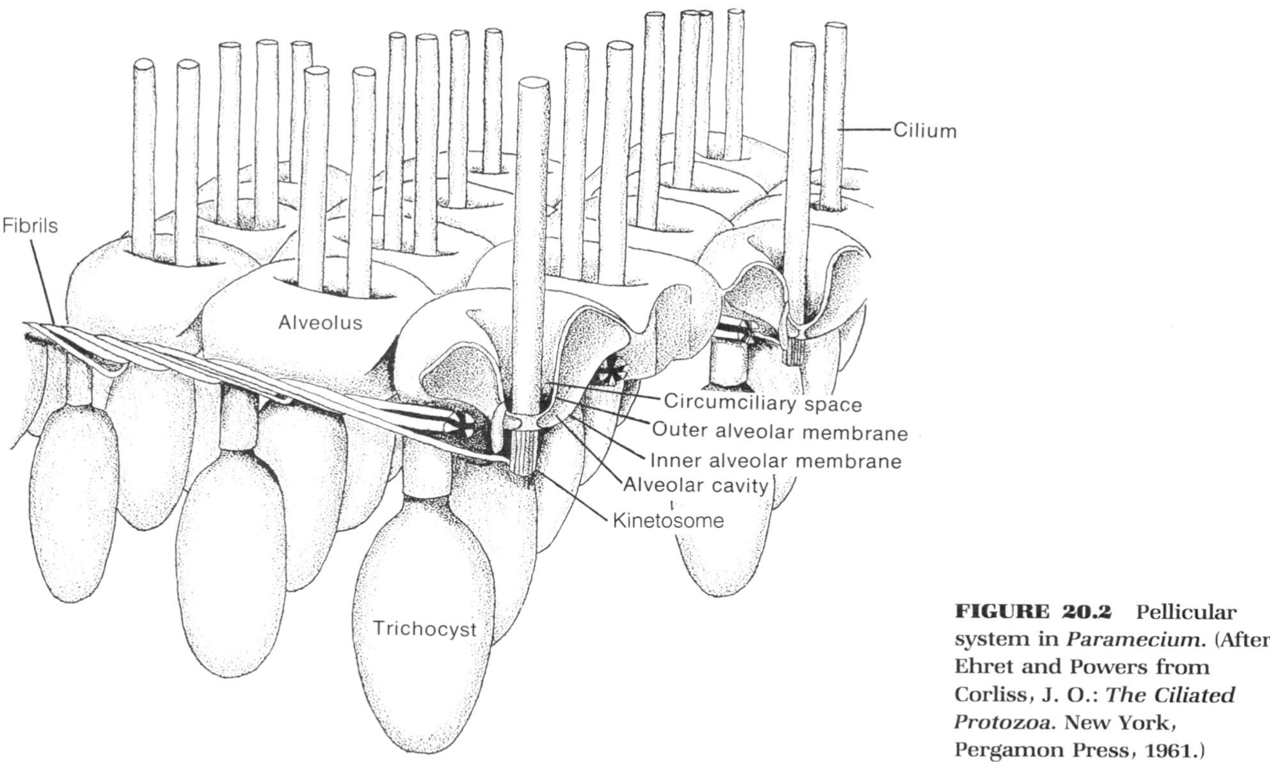

FIGURE 20.2 Pellicular system in *Paramecium.* (After Ehret and Powers from Corliss, J. O.: *The Ciliated Protozoa.* New York, Pergamon Press, 1961.)

The cilia arise from subsurface basal granules, or **kinetosomes** (Gr. *kinein,* to move + *soma,* body) (Fig. 20.2). The kinetosomes are connected together in longitudinal rows by fibrils, and all the fibrils and kinetosomes of a row make up a **kinety.** The kinetosomes and fibrils constitute the subsurface, or **infraciliature.**

The cilia covering the general body surface are called the **somatic ciliature.** Primitively, longitudinal rows of somatic cilia cover the entire surface of the body, but in many species the somatic ciliature is reduced to girdles, tufts, and bristles or is lacking altogether (Fig. 20.3). However, even those species with no somatic cilia possess an infraciliature persisting from cilia of earlier developmental stages.

The complex pellicle of ciliates is generally composed of three membranes, the outer one covering the body surface, including the cilia. The two inner membranes may be so folded, as in *Paramecium,* that they form large vesicles (alveoli) around the cilia (Fig. 20.2). Other common pellicular organelles of ciliates are bottle- or rod-shaped bodies, termed **trichocysts** (Gr. *thrix,* hair + *kystis,* bladder) (Fig. 20.2). These can be discharged and transformed into fine threads that may serve in anchoring the organism during feeding, in defense, or in prey capture.

Locomotion. The majority of ciliates swim by ciliary propulsion (see p. 117). In the forward swimming of such forms as *Paramecium,* the entire body of the organism spirals because the cilia beat somewhat obliquely to the long axis of the body. Beating occurs in synchronized waves down the length of the body. The function of the subsurface kinety system is still uncertain. It probably plays a role in the anchorage of the cilia, but there is no evidence that it is involved in coordinating ciliary beat.

Some of the most specialized ciliates, members of the class **Hypotricha** (Fig. 20.3*B*), have the somatic ciliature restricted to isolated tufts of fused cilia located in rows or groups on the side of the body that is kept against the substratum. All of the cilia of a tuft beat together.

Some ciliates are sessile. *Stentor,* a trumpet-shaped form, is attached by the tapered end and shortens by the contraction of pellicular microfilaments (myonemes) similar to muscle myofibrils (Fig. 20.1*B*). *Stentor* can also release itself and swim about. *Vorticella* has a bell-shaped body connected to the substratum by a long stalk that contains a bundle of contractile microfilaments (Fig. 20.3*D*). The ciliate retracts, coiling the stalk like a spring, and extends by a sudden release and popping movement. Some related forms are colonial, and

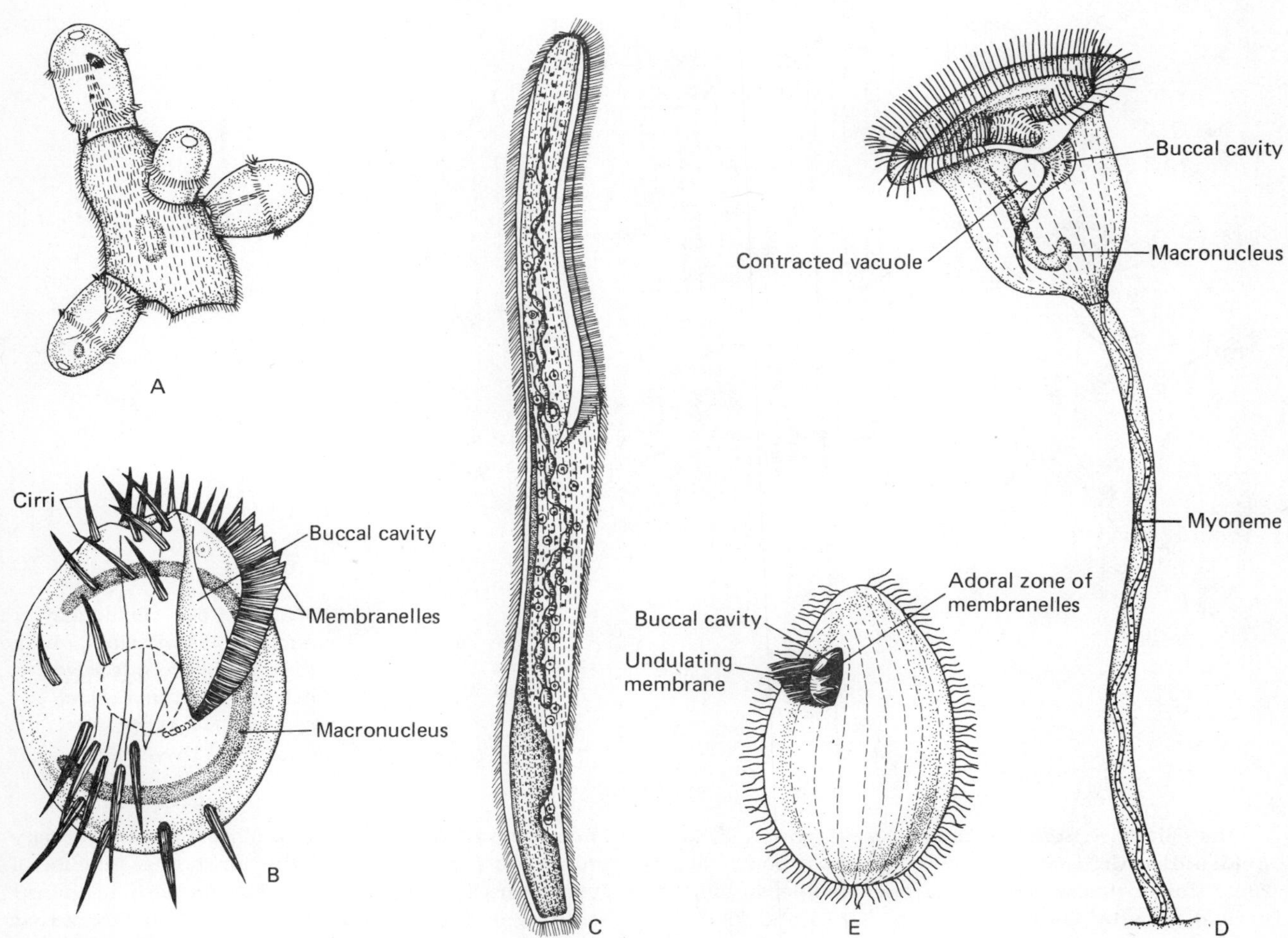

FIGURE 20.3 *A, Didinium,* a raptorial ciliate, attacking a *Paramecium. B, Euplotes,* a hypotrich ciliate that moves by means of ciliary tufts (cirri) located on the undersurface of the body; *C, Spirostomum;* the greatly elongated body of this group of ciliates can reach 3 mm. in length. The macronucleus is beadlike. *D, Vorticella,* a genus of common sessile ciliates having a contractile myoneme in the stalk. *E, Tetrahymena,* an easily cultured ciliate that has been the subject of many ultrastructural and physiological studies. (*A* after Mast from Dogiel; *B* after Pierson from Kudo; *C* after Stein from Grell, K. G.: *Protozoology.* New York, Springer-Verlag, 1973, p. 456; *D* after Sleigh, M. A.: *The Biology of Protozoa.* London, Edward Arnold, 1973, p. 206; *E* after Corliss.)

the individuals of the colony are connected together by a common stalk. Some other sessile ciliates live in tubes, which are either secreted or composed of foreign material cemented together (Fig. 20.1).

Nutrition. Ciliates possess a mouth, or **cytostome** (Fr. *cyte,* cell + Gr. *stoma,* mouth), which opens into a short canal, the **cytopharynx.** This in turn leads into the interior, more fluid cytoplasm. Primitively, the mouth is located at the anterior end (Fig. 20.1*A*), but in most ciliates it has been displaced posteriorly to varying degrees (Fig. 20.4). Species in which the oral apparatus is located at the anterior end are usually carnivores, and this is probably the primitive mode of feeding in ciliates. The mouth can be opened to a great diameter to ingest prey (Fig. 20.3*A*). The prey or contents of the prey pass into a food vacuole, which forms within the endoplasm at the end of the cytopharynx.

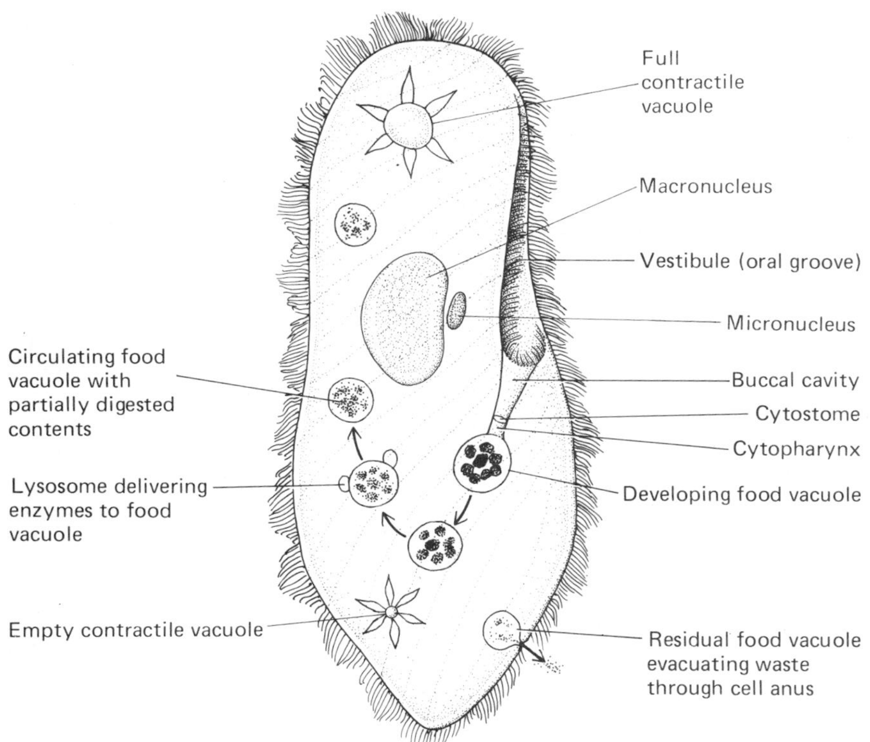

FIGURE 20.4 *Paramecium,* **a generalized drawing combining features of several species and showing formation and fate of food vacuoles.**

Many higher ciliates are suspension feeders and possess a more complex buccal apparatus. Typically, the mouth lies at the bottom of a buccal cavity, which contains compound ciliary organelles (Figs. 20.1 and 20.3). These organelles, which constitute the buccal, as opposed to the somatic, ciliature, consist of two types: **undulating membranes** and **membranelles.** An undulating membrane is a long row of fused cilia that forms a movable membrane (Fig. 20.3*E*). A membranelle is a short row of fused cilia that form a plate. Membranelles are typically arranged in fairly large numbers, one behind another. The function of the buccal ciliature is to produce a feeding current and drive suspended food particles—detritus, bacteria, and so forth—into the cytopharynx.

In the sessile *Vorticella* and the trumpet-shaped *Stentor,* the ciliary organelles wind around the distal end of the animal and spiral down into a pit on one side (Figs. 20.1 and 20.3*D*). In the familiar *Paramecium,* a buccal cavity, a cytostome, and a cytopharynx form a funnel located at the posterior end of a lateral oral groove, or vestibule. The somatic cilia within the oral groove produce a feeding current that sweeps from front to back, and the compound ciliary organelles of the funnel drive food particles down into the mouth (Fig. 20.4).

In all of these suspension feeders, the food particles collect at the end of the cytopharynx within a food vacuole, which gradually increases in size like a soap bubble at the end of a pipe (Fig. 20.4). When the vacuole reaches a certain size, it breaks free from the cytopharynx and circulates within the fluid cytoplasm. In those species that are carnivores, the prey, when swallowed, also becomes enclosed within a food vacuole. Like other organelles, the walls of food vacuoles are composed of lipid bilayer membranes.

Studies on digestion in ciliates have demonstrated that the vacuole shrinks slightly and its contents become at first increasingly acid. Digestive enzymes, which include all of the major classes—amylases, lipases, nucleases, and proteases—are delivered by lysosomes to the vacuole and initiate digestion of its contents. The vacuole becomes somewhat larger and less acid, and the products of digestion are absorbed from the vacuole into the surrounding cytoplasm. The residual undigestible waste eventually connects with the **cytoproct,** a fixed cell anus in the pellicle (Fig. 20.4). The contents of the vacuole are discharged, and the vacuole itself disappears.

Although only a few ciliates are parasitic, many are commensals. Among the most interesting commensals are the highly specialized species living within the stomach of some hoofed mammals. These are capable of digesting cellulose (p. 128).

Water Balance. Because of their microscopic size, ciliates have no special structures for gas exchange or excretion. They do have one or more organelles for water balance, called **contractile vacuoles** or **water expulsion vesicles,** which have fixed positions within the body. *Paramecium,* for example, has a contractile vacuole at each end (Fig. 20.4). In most ciliates the organelle is composed of a ring of radiating tubules that deliver water into a central vesicle, which gradually increases in size. On reaching a definitive size, the vacuole rapidly empties its contents through a canal in the pellicle. Then the vacuole fills again.

One might expect that only freshwater ciliates would have water expulsion vesicles, but they are also found in marine species, in which they serve to rid the body of water taken in during feeding. The water expulsion vesicles of marine ciliates pulsate at a slower rate than do those in fresh water.

Reproduction. Ciliates are distinguished from other protozoa in possessing two types of nuclei. One type, the **macronucleus,** is large and governs the nonreproductive functions of the cell. It is highly polyploid and is the principal source of cytoplasmic RNA. The shape and number vary greatly in different species.

The **micronuclei,** which range in number from one to 20, are small, round, and typically located in the vicinity of the macronucleus. They are diploid and function in reproduction.

Asexual. Ciliates reproduce asexually by means of transverse fission (Fig. 13.1). Each micronucleus divides mitotically, but the macronucleus usually divides by constriction. Regeneration of organelles lost in fission is a complex process and depends largely upon replication of existing structures. In some highly specialized ciliates, such as the hypotrichs, all of the organelles are resorbed during division, and new organelles are formed from a small number of persisting "germinal" kinetosomes.

Sexual. Sexual reproduction involves a process of **conjugation** and an exchange of nuclear material. Two individuals, called **conjugants,** meet, probably by random contact, adhere, and their cytoplasm fuses in the region of adhesion (Fig. 20.5*A*). The macronucleus is not involved in conjugation and is resorbed during the course of the process. All of the micronuclei undergo two meiotic divisions. Then all but one of these haploid nuclei disappear. The remaining nucleus divides mitotically to form two haploid nuclei: a stationary and a wandering nucleus (Fig. 20.5*D* and *E*). The wandering nucleus of each conjugant migrates to the opposite conjugant and fuses with the stationary nucleus to form a **zygote** nucleus. The conjugants separate, and the zygote nucleus undergoes a number of mitotic divisions to restore the nuclear condition characteristic of the species. The macronucleus develops from a micronucleus.

Restoration of nuclear number is commonly associated with cytosomal divisions. The process is highly variable and is best illustrated with an example. The adult *Paramecium caudatum* has one macronucleus and one micronucleus. Following conjugation the "zygote" nucleus divides three times to produce eight nuclei. Four become macronuclei and four micronuclei. However, three of the micronuclei are resorbed. The remaining micronucleus undergoes two mitotic divisions accompanying two cytosomal divisions. Each of the resulting four daughter individuals obtains one micronucleus from these divisions and inherits one of the four macronuclei formed prior to fission.

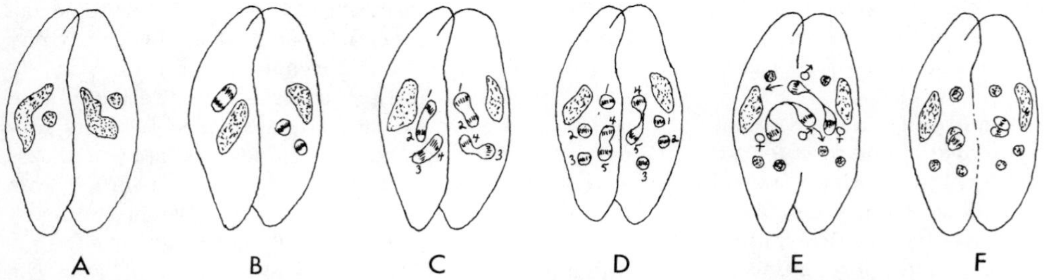

FIGURE 20.5 Sexual reproduction in *Paramecium caudatum. A* to *F,* Conjugation. *B* to *D,* Micronuclei undergo three divisions, the first two of which are meiotic. *E,* "Male" micronuclei are exchanged. *F,* They fuse with the stationary micronucleus of the opposite conjugant. (Modified after Clakins from Wichterman.)

In some species of ciliates, including *Paramecium,* the individuals of the population belong to genetically determined mating types. In *P. bursaria,* for example, there are a number of mating types within each variety, and conjugation can only occur between individuals of different mating types. Apparently adhesion of the cytoplasm will not take place between individuals of the same mating type.

Note that conjugation itself does not result in any increase in the number of individuals but does provide for an exchange of genetic material within the population. All of the individuals that result from the asexual reproduction of an original parent without intervening conjugation are genetically identical and constitute a **clone.** There have been numerous studies to determine how frequently conjugation must occur. Some species can undergo asexual divisions indefinitely without conjugation. In other species, there does seem to be an absolute requirement for conjugation after a given period of time or the clone will perish.

Most ciliates are capable of forming **cysts** under adverse environmental conditions. The body becomes encased within a protective secreted covering; there is some loss of water; and the metabolic rate is sharply reduced. Encystment is very important for the survival of ciliates when pools and ditches dry up and for dispersion by wind and on the muddy feet of aquatic birds and mammals.

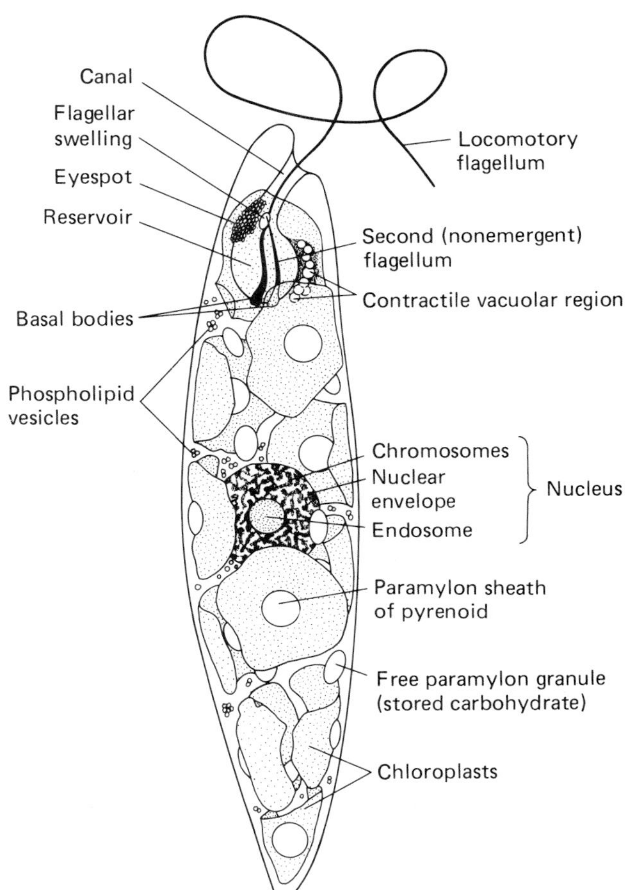

FIGURE 20.6 Structure of the phytoflagellate *Euglena gracilis.* (From Leedale, G. F.: *Euglenoid Flagellates.* Englewood Cliffs, N. J., Prentice-Hall, Inc., 1967.)

Phylum Sarcomastigophora: Flagellates

The phylum Sarcomastigophora (Gr. *sarx,* flesh + *mastix,* whip + *pherein,* to bear) contains the largest number of protozoa, some 18,000 species. They are a very diverse assemblage but are united in having only one type of nucleus and in possessing flagella or pseudopodia or both as locomotor or feeding organelles.

The subphylum Mastigophora includes the flagellated members of the Sarcomastigophora. They are believed to be the oldest of the eukaryotic organisms and the ancestors, directly or indirectly, of the other major groups of organisms—ciliates, algae, fungi, plants, metazoan animals, and others. The ciliates, for example, probably derive from some group of multinucleated mastigophorans that possessed many flagella.

A flagellum has an ultrastructure similar to that of a cilium but is longer (Fig. 20.6). In contrast to the lashing planar beat of a cilium, the beating of a flagellum commonly involves undulations in one or two planes (p. 117). When the undulations pass from the base to the tip of the flagellum, a pushing force is created, like that of a boat propeller (Fig. 20.7*A*). When the undulations pass from the tip to the base, a pulling force is produced, like that of an airplane propeller (Fig. 20.7*B*). Whether the flagellate body is pulled or pushed also depends upon where the flagellum is attached or how it is held in relation to the body (Fig. 20.7).

Phytoflagellates. The subphylum Mastigophora is composed of two groups: the phytoflagellates and the zooflagellates. The phytoflagellates are plantlike marine and freshwater forms. Most possess chlorophyll and exhibit autotrophic nutrition; the ten orders are classified with different groups of algae by algologists. The body is commonly asymmetrical, and the structure varies greatly. They possess one, or more commonly, two fla-

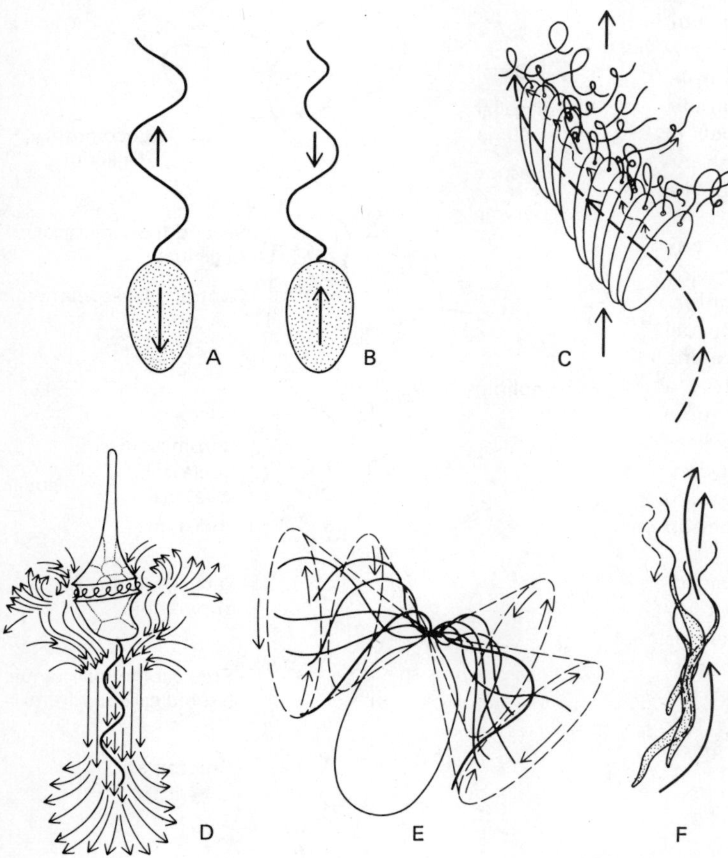

FIGURE 20.7 Flagellary locomotion. *A,* Pushing force (like a boat propeller) generated by base-to-tip undulations of the flagellum. *B,* Pulling force (like an airplane propeller) generated by tip-to-base undulations of the flagellum. *C,* Movement in *Euglena viridis.* Actual path indicated by dashed arrows. *D,* Locomotion in the dinoflagellate *Ceratium.* Arrows indicate the water currents generated by the transverse and posterior flagella. *E,* Locomotion in the phytoflagellate *Polytomella.* Arrows indicate the spiral pattern of the flagellar beat. *F,* Locomotion of the blood parasite *Trypanosoma.* Dotted arrow indicates movement of undulating membrane; solid arrow, the actual path of movement. (From Jahn, T. L., and E. C. Bovee: "Motile behavior of protozoa," in Chen, T. (Ed.): *Research in Protozoology.* New York, Pergamon Press, 1967, pp. 41–200.)

gella, which are generally carried at the anterior end. A nonliving cell wall or envelope is often present.

The mostly freshwater **euglenids** are quite representative of the phytoflagellates. They have rather spindle-shaped bodies covered by a living pellicle (Fig. 20.6). One or two flagella arise from a deep recess of the anterior end. In *Euglena,* for example, there is a long principal trailing flagellum and a short second flagellum, which does not emerge from the recess and cannot be seen with the light microscope. In the same vicinity there is an "eye" spot containing a red carotene pigment that apparently shields a light-sensitive area at the base of the flagellum (Fig. 20.6). One or two contractile vacuoles are also present.

Euglenids are typically green, but there are some colorless species, such as *Peranema.* This common little flagellate swims against the bottom, and its large, conspicuous, forward-projecting flagellum undergoes lateral undulations in propelling the organism forward. Another smaller flagellum trails. *Peranema* is predaceous and feeds on other protozoans, including *Euglena.* The prey is ingested through an anterior cytostome, which can be greatly distended in swallowing (Fig. 20.8*A*).

The members of the **Volvocida** are small green phytoflagellates possessing two flagella and a cellulose wall. Although some species, such as *Chlamydomonas,* are solitary, the group as a whole tends to be colonial, forming platelike or spherical aggregations. The most highly developed colonies are the large hollow spheres of *Volvox* (Fig. 20.8*D*). The entire colony moves, rotating through the water.

The brownish or yellowish **dinoflagellates** (Gr. *dinos,* rotation) are common members of marine plankton, although there are also freshwater species. The various species of dinoflagellates have a variety of shapes, but most are more or less oval or shaped like a top (Fig. 20.8*B* and *C*). The pellicle contains deposits of cellulose and may be thick or thin. When thick, it is often divided into plates or valves (Fig. 20.8*B*). One flagellum is located within a transverse groove that rings the body, and the other is located posteriorly in a longitudinal groove (Fig. 20.8*B* and *C*).

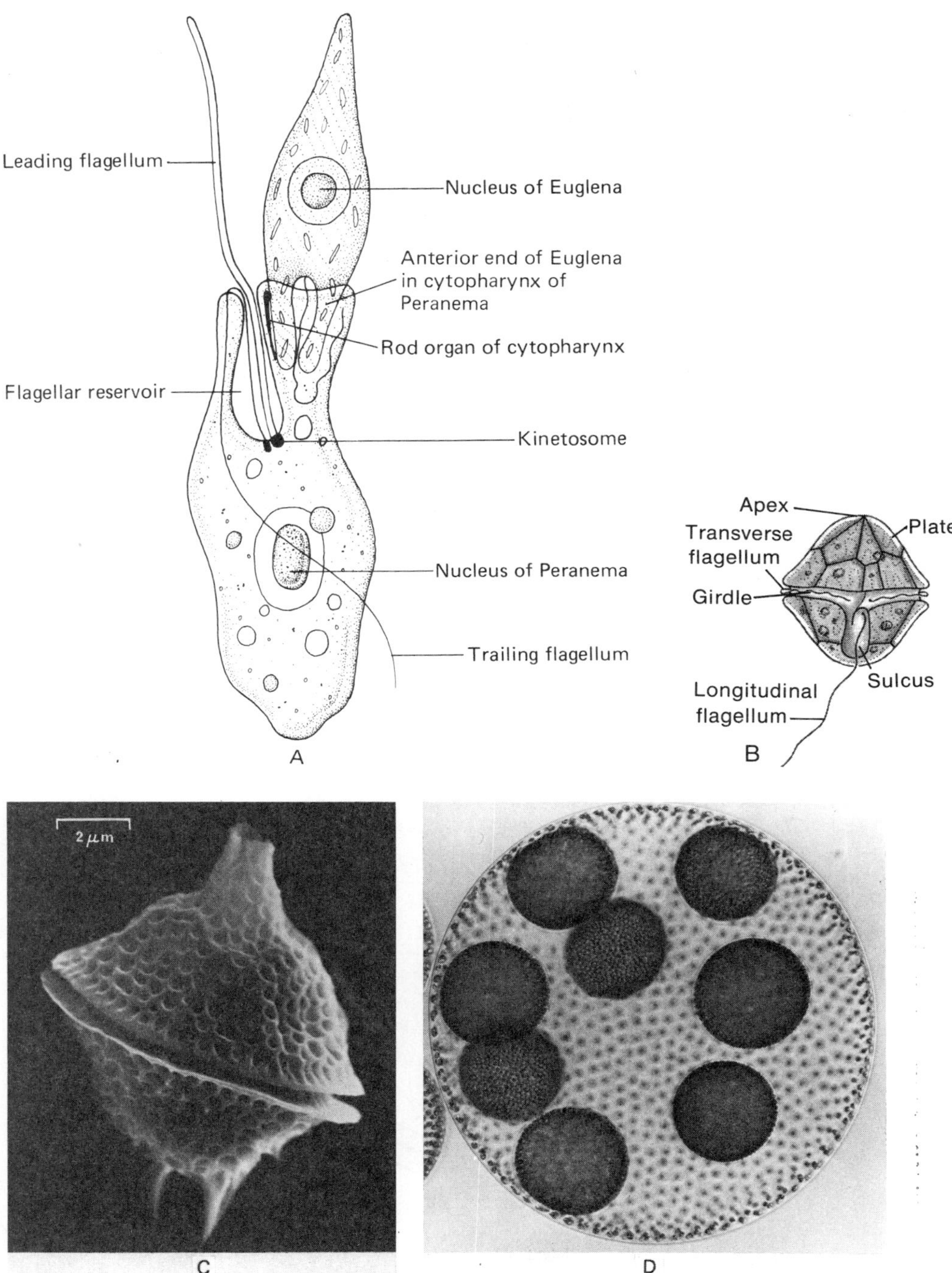

FIGURE 20.8 *A, Peranema* swallowing a *Euglena.* Associated with the cytopharynx are two hooked rods of uncertain function. Only one is shown here. *B,* A freshwater dinoflagellate, *Glenodinium cinctum. C,* Scanning electron micrograph of the marine dinoflagellate *Gonyaulax digitale.* Only the equatorial girdle shows in this view. *D,* A *Volvox* colony with daughter colonies inside. (*A* modified after Chen; *B* after Pennak; *C* by J. D. Dodge; *D* courtesy Carolina Biological Supply Co.)

Many species of dinoflagellates, as well as some other phytoflagellates, lose their flagella and pass into a nonmotile vegetative phase. The symbiotic algae of many corals, sea anemones, jellyfish, and other invertebrates are dinoflagellates in a nonmotile state.

The red tides that periodically plague the Gulf coast and other parts of the world are commonly caused by species of dinoflagellates. Optimum environmental conditions result in tremendous population growths, or blooms, of certain species. They occur in such great densities that their metabolic wastes poison fish and other marine animals.

Zooflagellates. The zooflagellates lack chlorophyll and are heterotrophic. Most species are either commensal or parasitic, and they constitute a much smaller part of the flagellate fauna than the phytoflagellates. They are undoubtedly a polyphyletic assemblage that evolved from different groups of phytoflagellates through loss of chlorophyll and autotrophic nutrition. The following examples will serve to illustrate their diversity.

The choanoflagellates (Gr. *choanē,* funnel) are a strange little group of free-living marine and freshwater forms with bodies resembling the collar cell of sponges (Fig. 20.9*A*), in which a palisade of microvilli surrounds the base of the flagellum. The water current produced by the beating of the flagellum is filtered by the collar microvilli. Most are colonial and attached, but one species has the collar cells embedded in a gelatinous mass. Some zoologists have suggested that the sponges may have evolved from the choanoflagellates.

The trypanosomid zooflagellates are parasites responsible for a large number of diseases of humans and domesticated mammals. They live in the bloodstream, white blood cells, and certain other tissues of the vertebrate host and are transmitted by blood-sucking insects, usually flies. During some stage in its life cycle the organism has an elongate, flattened body. A single flagellum extends anteriorly and laterally along the side of the body as an undulating membrane (Box 20.1).

The most complex flagellates, indeed among the most complex protozoans, are the gut symbionts of termites and wood roaches. The body is commonly saclike, and the anterior end bears a cap and rostrum complex. Numerous flagella may arise from the rostrum as well as from longitudinal grooves in the anterior half of the body (Fig. 20.9*B*). These flagellates engulf bits of wood ingested by the host and have enzymes that digest the cellulose to glucose within their food vacuoles. The product, glucose, is shared with the host, which is in-

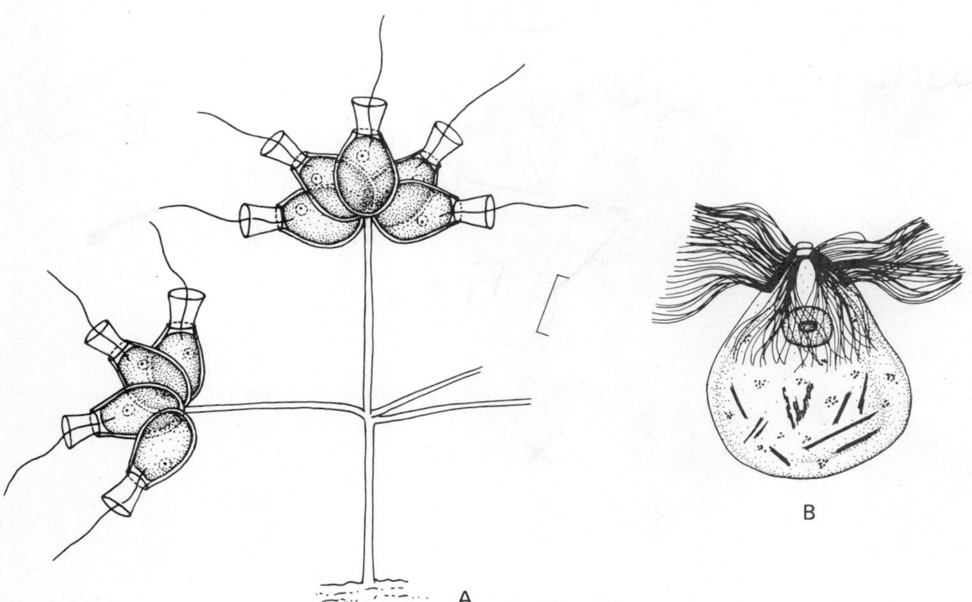

FIGURE 20.9 *A, Codosiga botrytis,* a colonial freshwater choanoflagellate. *B, Barbulanympha ufalula,* a complex zooflagellate that lives in the gut of wood roaches and digests the cellulose of the wood fragments eaten by the roach. (From Farmer, J. N.: *The Protozoa.* St. Louis, C. V. Mosby Co., 1980, p. 211; *B* from Sleigh, M. A.: *The Biology of Protozoa.* London, Edward Arnold, 1973, p. 143.)

BOX 20.1

Species of the trypanosomid genera *Leishmania* and *Trypanosoma* are agents of numerous diseases of humans and domesticated animals in subtropical and tropical regions of the world. Part of the life cycle is passed within or attached to gut cells of blood-sucking insects, mostly various kinds of flies, and part in the blood or in white blood cells and lymphoid cells of the vertebrate host, although other tissues may be invaded. Intracellular stages are aflagellate, but during the life cycle there are motile, extracellular flagellate stages (*C* in the figure) in the bloodstream or in the invertebrate host.

Leishmania is the agent of the widespread kala-azar and related diseases of Eurasia, Africa, and America, causing skin lesions (*A* in the figure, on a boy's wrist) and interference with immune responses, among other effects. Sandflies are the blood-sucking insect host.

Chagas disease of tropical America is caused by *Trypanosoma cruzi* and is transmitted by blood-sucking hemipteran bugs. Extensive damage may be caused in the human host when the parasite leaves the circulatory system and invades the liver, spleen, and heart muscles. *Trypanosoma brucei rhodesiense* and *T. b. gambiense* are the causal agents of African sleeping sickness and are transmitted by the tsetse fly (*B* and *C*). The parasite invades the cerebrospinal fluid and brain, producing the lethargy, drowsiness, and mental deterioration that mark the terminal phase of the disease.

There are also various trypanosome diseases of horses, cattle, and sheep that are of considerable economic importance. (*A* courtesy of S. S. Hendrix; *C* from Sleigh, M. A.: *The Biology of Protozoa.* London, Edward Arnold, 1973, p. 141.)

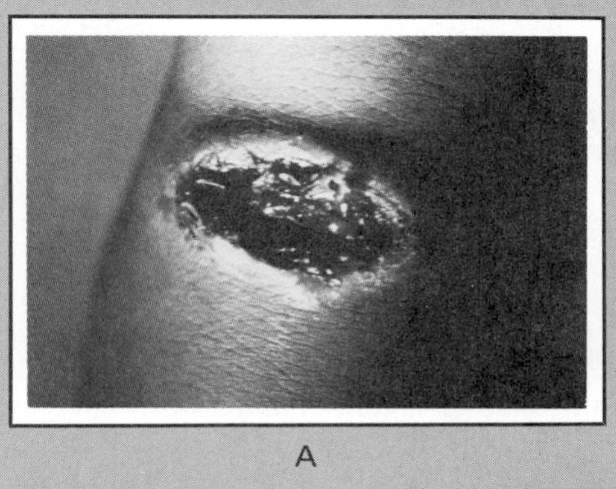

A

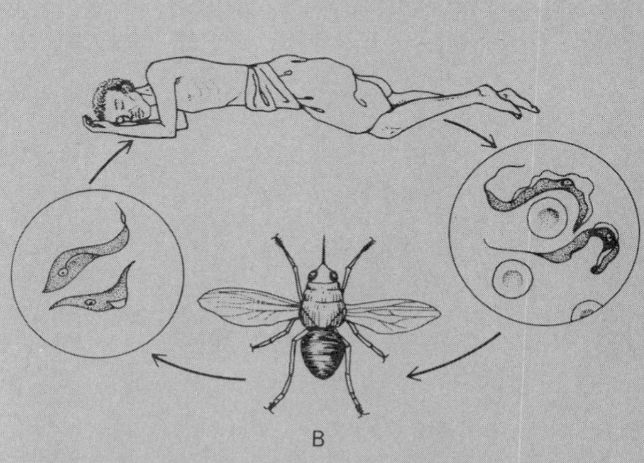

B

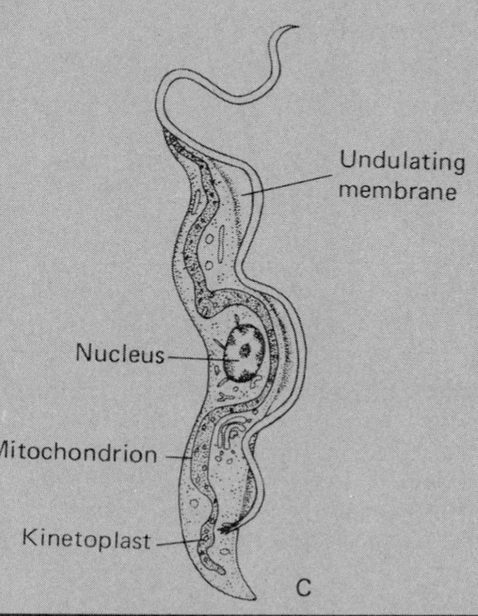

C

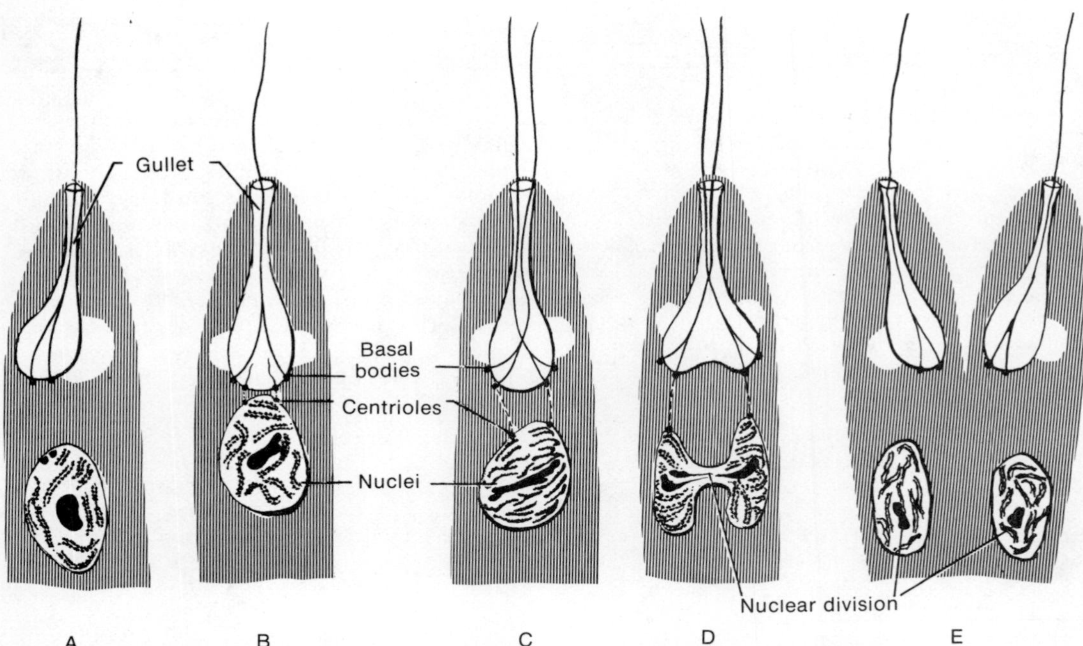

FIGURE 20.10 Details of longitudinal fission in *Euglena*. *A,* The centriole has already divided. *B,* Each centriole produces a new basal body and flagellum. The nucleus is in prophase, and the contractile vacuole is double. *C,* The old pair of flagellar roots separate and fuse with the new roots. *D,* Mitosis proceeds, and the gullet begins to divide. *E,* Anterior end dividing following duplication of organelles. (Redrawn from Ratcliffe, 1927.)

capable of digesting cellulose. The insect obtains a new gut fauna following each molt by licking other individuals, by rectal feeding, or by eating fecal cysts.

Reproduction. Asexual reproduction is typically by binary fission, but in contrast to ciliates, flagellates divide longitudinally to produce two more or less equal halves (Fig. 20.10). The kinetosomes and other organelles duplicate or re-form before or after actual division. Not much is known about sexual reproduction in most flagellates.

Phylum Sarcomastigophora: Sarcodines

The subphylum Sarcodina contains the amoebas and other protozoa that possess flowing extensions of the body known as **pseudopodia.** Sarcodines possess fewer organelles than ciliates and flagellates and are therefore relatively simple in structure. However, skeletal structures have reached a degree of development that is equalled by few other protozoa.

Amoebas

The most familiar sarcodines are the freshwater and marine amoebas, some of which are naked and some of which are enclosed within a shell (Fig. 20.12). Amoebas have straplike or large blunt pseudopodia, used in locomotion and in feeding (Fig. 20.11).

In the shelled species, the shell is secreted or composed of mineral particles cemented together. A large opening in the shell permits extension of the body and the pseudopodia (Fig. 20.12).

Locomotion. The pseudopodia and other parts of the body are bounded by a thick layer of gelatinous ectoplasm. Ectoplasm and the more fluid interior endoplasm are different molecular states of cytoplasm, and amoeboid flow, which is described on page 117, involves a rapid change of the molecular organization at the pseudopodial tip.

Nutrition. Amoebas feed on diatoms, algae, rotifers, and other protozoans. The food is surrounded by pseudopodia and eventually enclosed within a food vacuole (Fig. 20.11), where digestion proceeds as in ciliates. The

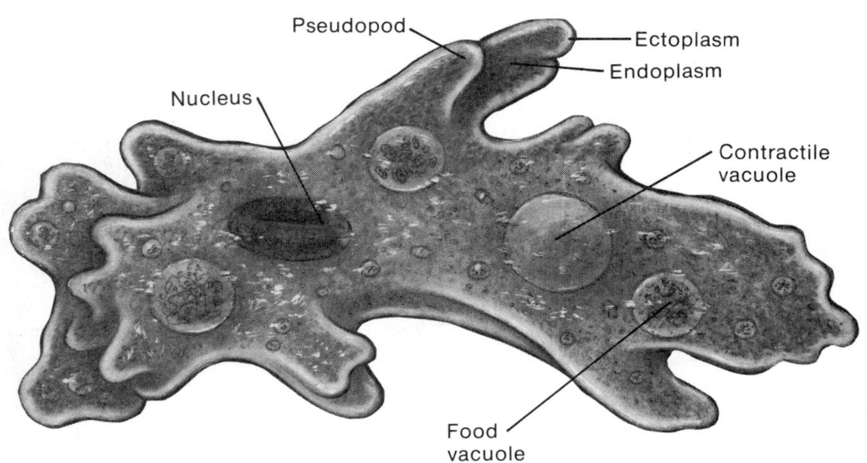

FIGURE 20.11 An amoeba. The animal is flowing to the right.

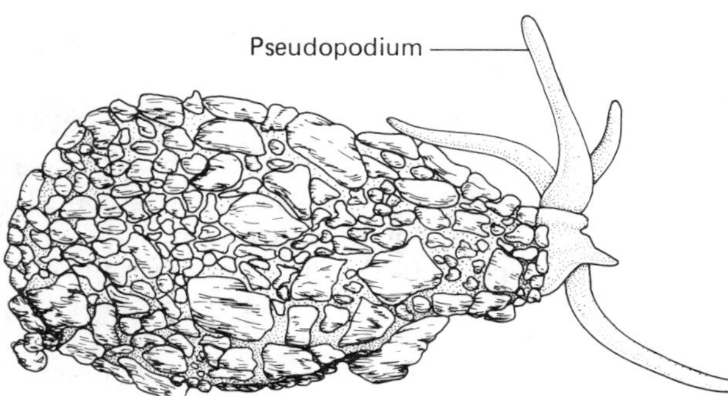

FIGURE 20.12 *Difflugia oblonga,* a shelled amoeba with a test of mineral particles. (Modified from Deflandre.)

vacuole containing undigestible residue ruptures at the posterior end.

A number of commensal and parasitic amoebas inhabit the gut of different animals, including humans. The commensal species, such as *Entamoeba coli* of humans, feed on bacteria and intestinal debris. The parasitic species invade the intestinal tissue. *Entamoeba histolytica* is the cause of amoebic dysentery. Both commensal and parasitic species leave the host as cysts in the feces, and reinfestation occurs through the mouth.

One or more contractile vacuoles are present in freshwater amoebas but are absent from marine forms.

Foraminiferans

Foraminiferans (L. *foramen,* opening + *pherein,* to bear) are marine Sarcodina that secrete a chitin-like shell with one chamber or, in most species, a multichambered calcareous shell (Fig. 20.13). They possess delicate anastomosing pseudopodia, which arise from cytoplasm that flows out of the large shell opening and back over the shell surface. The shell of many species is covered with tiny perforations, but the perforations are usually sealed by a membrane. The pseudopodial cytoplasm forms an adhesive net and can extend many times beyond the diameter of the shell. Any small organism swimming into the net is restrained and slowly surrounded by cytoplasm.

Although there are some planktonic forams, such as *Globigerina* (Fig. 20.13*A*), the majority are benthic and haul themselves slowly over the bottom by means of pseudopodia.

In most forams the shell is composed at first of a single chamber with an opening at one end. When the foram outgrows the first chamber, it secretes another one. This process continues, producing a multichambered shell that is occupied entirely by one individual.

Accumulations of foram shells are an important constituent of fine ocean bottom sediments. They have an extensive fossil record that begins in the Cambrian, and there are great limestone deposits, such as those forming the great cliffs overlooking the sea at Dover, England, composed largely of foram shells. The paleontology of these organisms is especially important in the search for oil deposits.

Heliozoans

Heliozoans are mostly freshwater Sarcodina. Their spherical bodies are free or attached by a stalk. The body is composed of a central cytoplasmic core, or **medulla,** that contains one to many nuclei, and an outer **cortex** of highly vacuolated cytoplasm (Fig. 20.14). Radiating from the cortex are many needle-like pseudopodia, called **axopods** (Gr. *axōn,* axle + *pous,* pod),

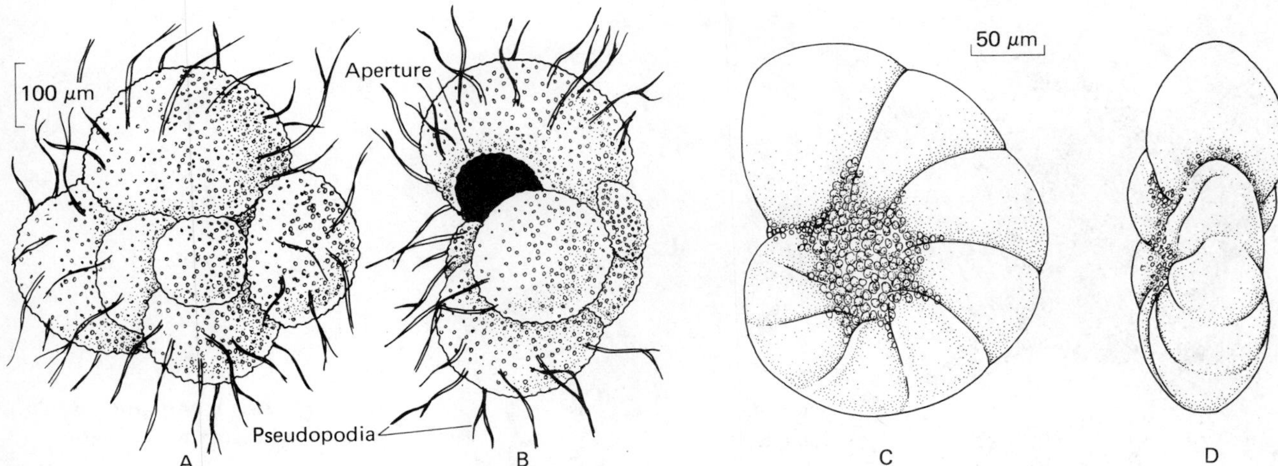

FIGURE 20.13 *A* and *B*, *Globigerina*, a planktonic foraminiferan. *A* shows the upper surfaces of the shell; *B* is a side view, showing the aperture and pseudopodia. *C* and *D*, Side and edge views of *Nonion*, a benthic foraminiferan. (All based on micrographs by J. W. Murray.)

from which the name Heliozoa, sun animals, is derived. The axopods contain a central cytoplasmic rod of microtubules that extends from the medulla (Fig. 20.14*B*). Many heliozoans possess a skeleton of siliceous scales, tubes, spheres, or needles embedded in the cortex. Where the skeleton is composed of needles, they radiate out of the cortex like the axopods. Some species even have sand grains or living diatoms embedded within the cortex.

The axopods function solely as food-trapping organelles. On contact, small organisms adhere to the axopods, which then withdraw. The prey is covered by cytoplasm and is gradually withdrawn into the cortex to be digested within a food vacuole.

Radiolarians

The radiolarians (L. *radiolus*, little ray) are a group of planktonic marine Sarcodina in which skeletal structures are highly developed. The body is somewhat like that of heliozoans in being more or less spherical with a central nucleated core of cytoplasm and a broad outer vacuolated cortex (Fig. 20.15). The cortical cytoplasm of many species contains symbiotic dinoflagellates. The pseudopodia are axopods or netlike as in forams. Radiolarians differ from heliozoans in that the central cytoplasm is encased within a membranous capsule that is perforated to permit communication with the outer cortex.

Some radiolarians have a skeleton of radiating strontium sulfate needles, but most have a skeleton of silicate arranged as concentric spherical lattices within and outside of the cortical cytoplasm (Fig. 20.15*B*).

The radiating pseudopodia project through the skeletal openings and function as food-trapping structures in the same way as in heliozoans and forams.

Plankton samples taken at different depths reveal a distinct vertical stratification of radiolarians, which are capable of some depth regulation through the retraction or extension of the radiating axopods or the cortical cytoplasm. Some are found only at great depths (4600 m.); some species undergo seasonal movements, dropping from the surface to lower depths in the summer. Radiolarian skeletons also accumulate on the ocean bottom and may predominate in the ooze. The fossil record of radiolarians extends back into the preCambrian, and they have contributed to large sedimentary deposits.

Reproduction in Sarcodina. Asexual reproduction is generally by binary fission. In shelled amoebas, heliozoans, and radiolarians, the skeleton is divided, or else one daughter cell gets the skeleton, and the other secretes a new one. Sexual reproduction usually involves the fusion of isogametes, which in radiolarians and foraminiferans are flagellated, but little is known about sexual reproduction in radiolarians. Reproduction in forams involves an alternation of asexual and sexual stages.

Considerable evidence suggests that the Sarcodina evolved from flagellate ancestors. Radiolarians and forams have flagellated developmental stages, and some flagellates lose their flagella and undergo amoeboid stages.

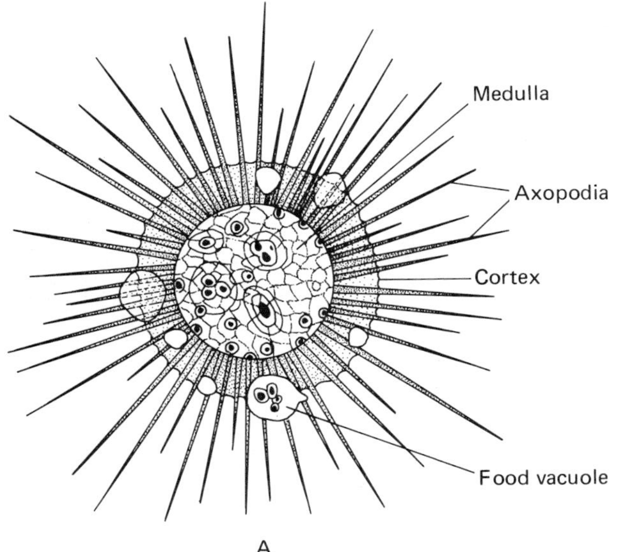

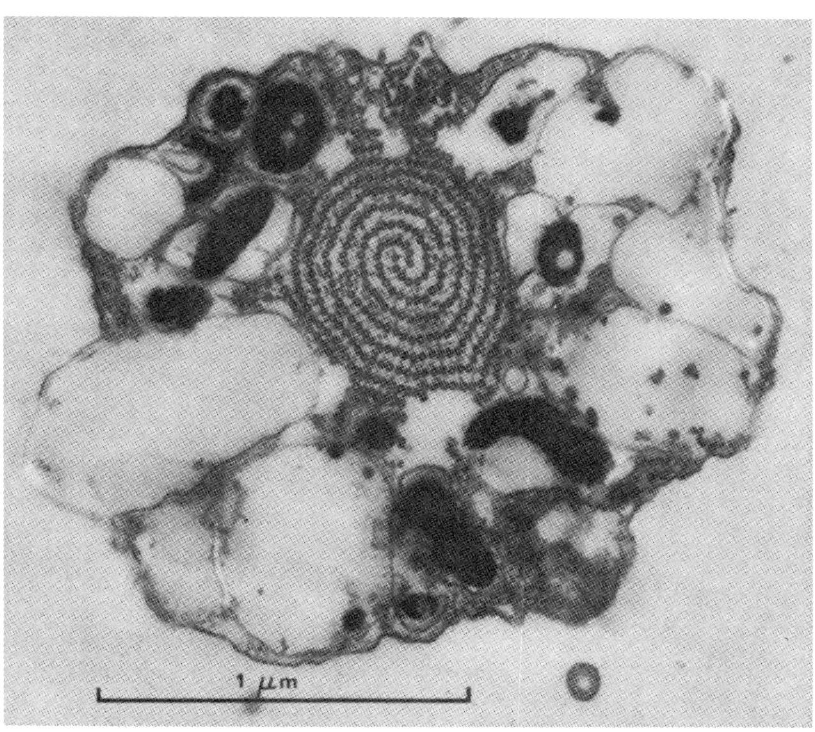

FIGURE 20.14 A multinucleate heliozoan, *Actinosphaerium eichorni. B,* Electron photomicrograph of a section through an axopod. Note that the axial rod is composed of a double spiral of microtubules. (*A* after Doflein from Tregouboff; Photomicrograph by A. C. Macdonald, in Sleigh M. A.: *The Biology of Protozoa.* London, Edward Arnold, 1973, p. 162.)

Sporozoans

Sporozoans are parasitic protozoa, living within or between cells of their invertebrate or vertebrate hosts. They belong to two phyla, the Apicomplexa and the Microspora, both formerly composing an old protozoan grouping, the Sporozoa. *Sporozoan* (Gr. *sporos,* seed + *zoon,* animal) continues to be used as a common name.

Most known sporozoans and all those of economic and medical importance belong to the phylum Apicomplexa, so named because they possess a complex of ringlike, tubular, and filamentous organelles at the api-

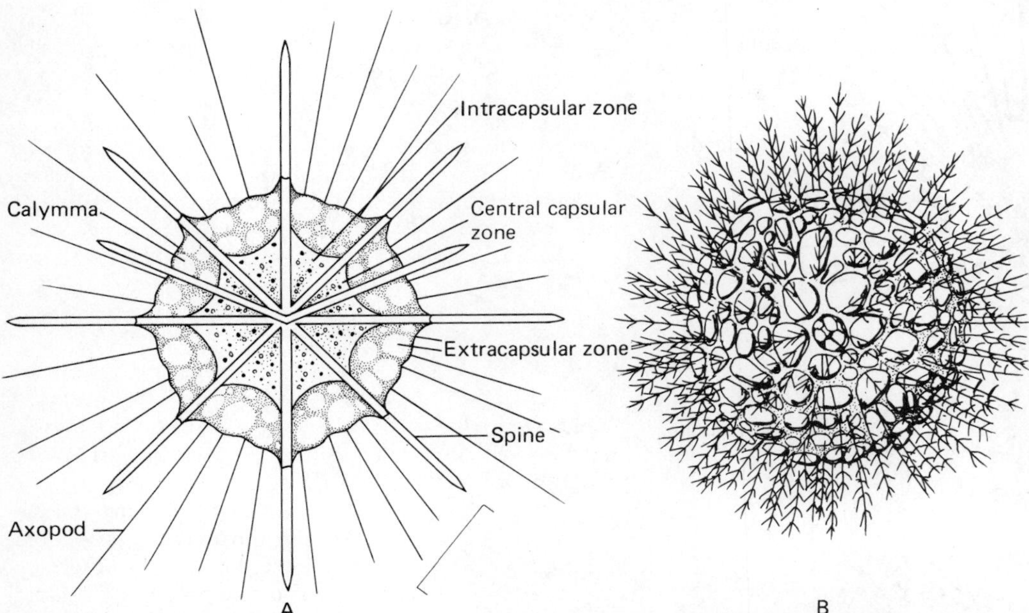

FIGURE 20.15 Radiolarians. *A, Acanthometra,* a radiolarian with a skeleton of radiating strontium sulfate rods. *B,* A siliceous skeleton of a radiolarian. (*A* from Farmer, J. N.: *The Protozoa.* St. Louis, C. V. Mosby Co., 1980, p. 353; *B* after Haeckel.)

cal end, visible only with the electron microscope (Fig. 20.16). The function of the apical complex is uncertain but may include entry into the host cell. One or more feeding pores are located on the side of the body.

The life cycle of apicomplexans typically involves an asexual and a sexual phase. An infective stage, called a sporozoite, invades the host and undergoes asexual multiplication by fission, producing individuals called merozoites. Merozoites can continue schizogamy but eventually form gametes that fuse to form a zygote. The zygote undergoes meiosis to form sporozoites.

The nature and life cycle of apicomplexan sporozoans can be illustrated by the coccidians, which include the parasites causing malaria in humans. Although in decline today, malaria was once widespread throughout the world and was one of the worst scourges of mankind. The untreated disease can be long-lasting and terribly debilitating. Malaria has played a major and often unrecognized role in directing the course of human history. The name means literally "bad air" because the disease was thought to be caused by the air of swamps and marshes. Although malaria had been recognized since ancient times, the causative agent was not recognized until 1880 when a physician with the French army in North Africa identified the coccidian parasite, *Plasmodium,* in the blood cells of a malarial patient. In 1897 the mosquito was recognized to be the vector.

The introduction of the parasite into a human host is brought about by the bite of certain species of mosquitoes, which inject the sporozoites along with their salivary secretions into the capillaries of the skin (Fig. 20.17). The parasite is carried by the blood stream to the liver, where it invades a liver cell. Here further development results in asexual reproduction through multiple fission. These daughter cells (cryptozoites) invade other liver cells and continue to reproduce. After a week or so there is an invasion of red blood cells by parasites produced in the liver. Within the red cell the parasite increases in size and undergoes multiple fission. These individuals (merozoites), produced by fission within the red cells, escape and invade other red cells. The liberation and reinvasion does not occur continually but occurs simultaneously from all infected red blood cells. The timing of the event depends upon the period of time required to complete the developmental cycle within the host's cells. The release causes chills and fever, the typical symptoms of malaria.

Eventually some of the parasites invading red cells do not undergo fission but become transformed into **gametocytes.** The gametocyte remains within the red blood cell. If such a cell is ingested by a mosquito, the

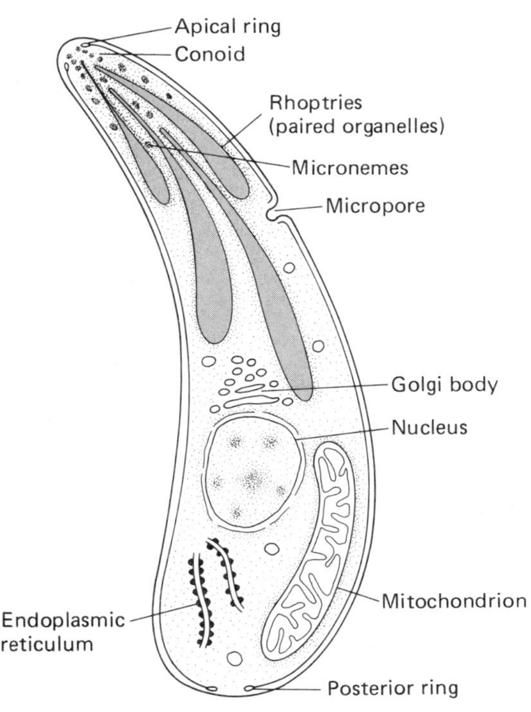

FIGURE 20.16 A lateral view of a generalized apicomplexan sporozoan. (From Farmer, J. N.: *The Protozoa*. St. Louis, C. V. Mosby Co., 1980, p. 360.)

gametocyte is liberated within the new host's gut. After some further development, a male gametocyte (microgametocyte) fuses with a female gametocyte (macrogametocyte) to form a zygote. The zygote enters the stomach wall and gives rise to a large number of spore stages (sporozoites). It is these stages, which migrate to the salivary glands, that are introduced into the human host by the bite of mosquitos.

The asexual stage of other coccidians occurs in blood cells or in gut cells. There are a number of diseases of domesticated animals caused by coccidians, such as *Eimeria* in chickens, turkeys, pigs, and sheep, and *Babesia* in cattle (red-water fever).

The phylum Microspora contains a smaller number of parasites but ones found in most animal groups, especially arthropods. They lack the apical complex of other sporozoans, and the sporelike stage is characterized by a polar filament that is extruded when this stage is taken into the host. The filament appears to be involved in some way with the invasion of the host's cell.

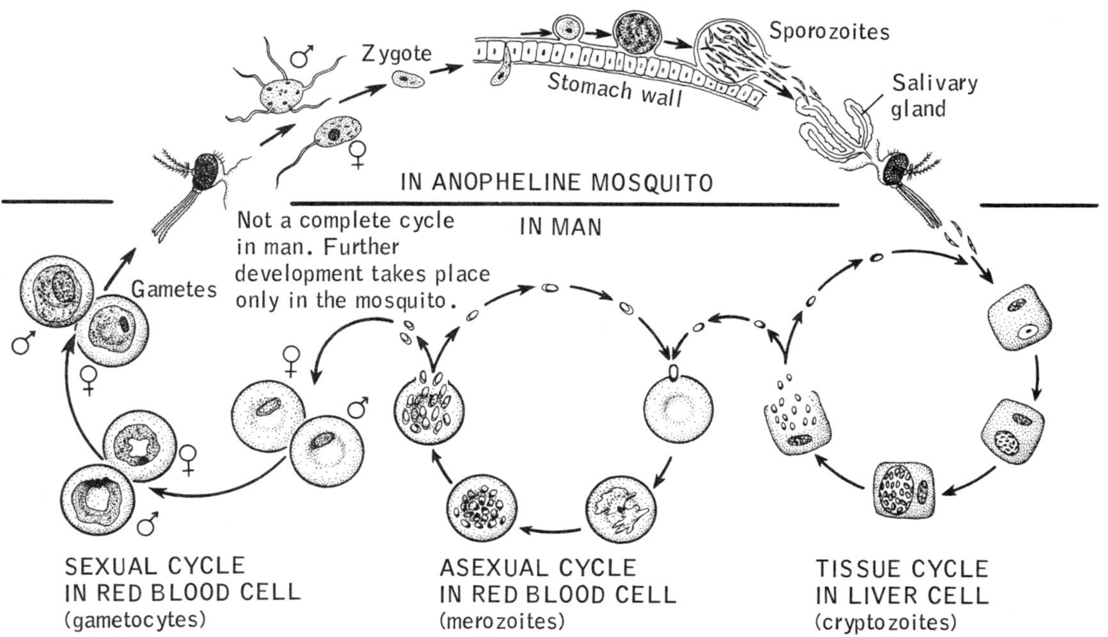

FIGURE 20.17 The life cycles of *Plasmodium* in a mosquito and in man. (Redrawn and modified from Blacklock and Southwell.)

Classification of Protozoa

Phylum Sarcomastigophora Some 18,000 unicellular organisms having one type of nucleus and possessing flagella or pseudopodia as locomotor or feeding organelles.

Subphylum Mastigophora Flagellates. Sarcomastigophorans possessing one or more flagella.

Class Phytomastigophora Plantlike flagellates. Mostly autotrophic and usually with one or two flagella. (These organisms are also placed within the algal phyla.)

Class Zoomastigophora Animal-like flagellates. Heterotrophic, with one to many flagella.

Subphylum Sarcodina Sarcodines. Sarcomastigophorans possessing pseudopodia.

Superclass Rhizopoda Pseudopodia used for both locomotion and feeding.

Class Lobosa Shelled and naked amoebas. Pseudopodia large or straplike.

Class Granuloreticulosa Foraminiferans. Chiefly marine species with mostly multichambered shells and strandlike anastomosing pseudopodia.

Superclass Actinopoda Floating or sessile Sarcodina with radiating rodlike pseudopodia (axopods) used in feeding only.

Classes Polycystina and Phaedaria Radiolarians with siliceous skeleton.

Class Acantharia Radiolarians with a skeleton of strontium sulfate.

Class Heliozoa Heliozoans. Mostly freshwater sarcodines lacking a central capsule.

Phylum Apicomplexa Sporozoans. Some 3900 parasitic protozoans having an apical complex of organelles.

Phylum Microspora Sporozoans. About 1000 parasitic protozoans lacking an apical complex of organelles but having a coiled polar filament within the infective sporelike stage.

Phylum Ciliophora Ciliates. About 7200 ciliated unicellular organisms possessing two types of nuclei.

Class Kinetofragminophora Ciliates lacking compound ciliary organelles in oral region. *Didinium,* suctorians.

Class Oligohymenophora Ciliates with a small number of compound ciliary organelles in the oral region. Such organelles often hidden. *Paramecium, Vorticella.*

Class Polyhymenophora Ciliates with a large number of conspicuous compound ciliary organelles in the oral region. *Spirostomum, Stentor,* hypotrichs.

Summary

1. The protozoa are unicellular or colonial organisms belonging to various protistan phyla. Most species are motile and heterotrophic, which accounts for their traditional placement in the Animal Kingdom. Protozoa are found in the sea and in fresh water, and there are many parasitic species.

2. Ciliates, members of the phylum Ciliophora, possess complex organelles, especially as part of the pellicle, the outer layer of the body. Cilia are used for swimming and in some organisms for feeding, and all ciliates possess an infraciliature of kinetosomes and connecting fibrils. Some ciliates are predatory, and others are suspension feeders. Food is ingested through a cytostome and cytopharynx; digestion occurs within a food vacuole. Contractile vacuoles provide for water balance. Ciliates possess two types of nuclei. The large polyploid macronucleus is variously shaped and functions in cellular regulation. The small one to many micronuclei are involved in reproduction. Following meiosis the haploid micronuclei function as gametes and are exchanged during conjugation of a pair of ciliates. The zygote micronucleus in each member of the pair then divides to form the macro- and micronuclei characteristic of the species.

3. The phylum Sarcomastigophora contains those protozoa that have only one type of nucleus and possess flagella or pseudopodia or both. The flagellate members, a diverse assemblage, are included in the subphylum Mastigophora and are divided into phytoflagellates and zooflagellates. The phytoflagellates are mostly biflagellated chlorophyll-bearing autotrophs and include many common and widespread marine and freshwater groups, such as euglenids and dinoflagellates. Zooflagellates bear one to many flagella and are heterotrophic. Although there are some free-living species, most are commensal or parasitic; the trypanosomids are of the greatest medical importance. Asexual reproduction is by longitudinal fission.

4. The subphylum Sarcodina includes those sarcomastigophorans that possess pseudopodia. Although pellicular organelles are not highly developed, skeletons are a characteristic feature of most groups.

5. Amoebas and foraminiferans use their pseudopodia for both feeding and locomotion. The amoebas, which are mostly inhabitants of fresh water, are either shell-less or possess shells of secreted or foreign materials and have straplike or large blunt pseudopodia. Foraminiferans are largely marine, and most possess chambered, calcareous shells. The delicate, anastomosing pseudopodia form a food-trapping net.

6. Radiolarians and heliozoans possess radiating, rod-like pseudopodia (axopods), which are used only in feeding. Radiolarians are planktonic marine sarcodines with well-developed skeletons of strontium sulfate or silicon dioxide. The mostly freshwater heliozoans are free or attached and may or may not possess skeletons.

7. Sarcodines reproduce asexually by binary fission; sexual reproduction is by fusion of isogametes but is poorly understood in some groups.

8. Sporozoans are parasitic protozoans composing the phyla Apicomplexa and Microspora. The former are characterized by certain electron-dense apical organelles; flagella are present only in the gametes of some species. The life cycles of sporozoans are complex, with alternating asexual and sexual stages.

9. Sporozoans parasitize both invertebrates and vertebrates, and some require two hosts. The coccidian apicomplexans, which live within blood or gut cells of their hosts, include most of the economically and medically important species, such as *Plasmodium,* the causative agent of malaria.

References and Selected Readings

Accounts of the protozoa can be found in the references cited at the end of Chapter 19. The parasitology texts listed at the end of Chapter 24 cover the parasitic protozoa. The references cited below are devoted exclusively to protozoa.

Bold, H. C., and M. J. Wynne: *Introduction to the Algae.* Englewood Cliffs, New Jersey, Prentice-Hall, 1978. A general account of the algal phyla, which include the phytoflagellates.

Corliss, J. O.: *The Ciliated Protozoa: Characterization, Classification, and Guide to the Literature.* 2nd ed. New York, Pergamon Press, 1979.

Dodson, E. O.: The kingdoms of organisms. *Systematic Zoology, 20*:265–281, 1971. This paper and Whittaker's review present various schemes for the classification of living organisms.

Farmer, J. N.: *The Protozoa: Introduction to Protozoology.* St. Louis, C. V. Mosby Co., 1980. A textbook of protozoology with more emphasis on classification than that of Sleigh.

Hawking, F.: The clock of the malaria parasite. *Scientific American, 222*(Jun.):123–131, 1970. The controlling mechanisms in the cyclical development of *Plasmodium.*

Jeon, K. W. (Ed.): *The Biology of Amoeba.* New York, Academic Press, 1973. Papers dealing with various aspects of the biology of *Amoeba.*

Jones, A. R.: *The Ciliates.* New York, St. Martin's Press, 1974. A general biology of the ciliated protozoans.

Margulis, L.: How many kingdoms? Current views of biological classification. *The American Biology Teacher, 43*(9):482–489, 1981.

Sleigh, M. A.: *The Biology of the Protozoa.* New York, American Elsevier Publishing Co., 1973. A good, well-balanced general account of the protozoa.

Whittaker, R. H.: New concepts of the kingdoms of organisms. *Science, 163:*150–159, 1971.

21 Sponges

ORIENTATION

This chapter will investigate sponges, which are primitive in lacking organs but specialized in being sessile, or attached. We will examine the structure of sponges, which is built around a system of water canals and chambers. We will then see how the physiology of sponges is dependent upon the flow of water through the body and how the water flow accommodates the lack of organs. We will also see how the evolution of sponges is related to flow dynamics. The system of water canals and chambers is the key to the understanding of sponges.

CHARACTERISTICS OF PORIFERA

1 Sponges are sessile marine or freshwater animals that are radially symmetrical or irregular in shape.
2 There are no organs, head, mouth, or gut cavity. The body structure is organized around a system of canals and chambers through which water flows.
3 Flagellated collar cells which line the chambers not only create the water current but also filter out fine food particles.
4 Support is provided by internal siliceous or calcareous spicules.
5 Sponges are either hermaphroditic, or the sexes are separate and development includes a free-swimming larva.

Sponges, which compose the phylum **Porifera,** lack most of the typical animal features. There is no head or anterior end; there is no mouth or gut cavity; and the body is immobile. In fact, some zoologists believe that sponges had an evolutionary origin separate from other animals. Certainly they departed early from the main line of animal evolution, and their structure reflects a combination of primitive and specialized characteristics.

With the exception of three freshwater families, the 5,000 species of sponges are marine. They are sessile animals, living attached to the bottom, most commonly to rock, shell, coral, pilings, and other hard surfaces.

Structure and Function of Sponges

The simplest sponges are shaped like little vases (Fig. 21.1). The interior cavity, called the **atrium** or **spongocoel** (L. *spongia,* sponge + *koilos,* cavity), opens to the outside through a large opening at the top, the **osculum.** The body of the sponge surrounding the spongocoel is perforated by pores, from which the phylum name porifera, or "pore-bearer" is derived. The outer surface of the sponge body is covered by flattened epidermal cells, **pinacocytes** (Gr. *pinax,* tablet + cell), and the pores are formed by **porocytes,** cells that are perforated like a ring. The spongocoel is lined with flagellated **choanocytes,** or **collar cells,** so called because of the collar-like arrangement of long microvilli around the base of the flagellum. Between the epidermis and the flagellated cells lies a layer called the **mesohyl** or mesenchyme, containing amoeboid cells of different types and skeletal pieces embedded within a gelatinous protein matrix.

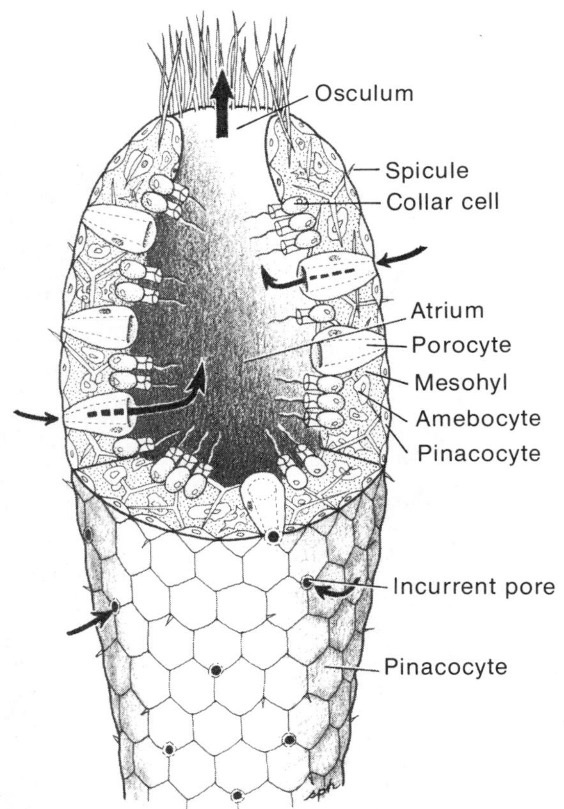

FIGURE 21.1 Structure of an asconoid sponge. (Modified from Buchsbaum.)

The skeleton of most sponges consists of **spicules** of calcium carbonate or silicon dioxide secreted by amoebocytes (Fig. 21.2). Each spicule may be a single needle or several needles fused at certain angles to each other. They are generally microscopic and unconnected, but in some sponges the spicules are fused together to form a complex skeleton (Fig. 21.3*B*). The size and shape of the spicule and the number of rays present are characteristic for each species of sponge, and spicule structure is an important characteristic in the classification of sponges. Many sponges possess an organic skeleton of coarse spongin fibers (Fig. 21.3*A*) instead of, or in addition to, spicules, and it is from species with only a spongin skeleton that commercial sponges are obtained. The skeleton, whether of different types of spicules or of spongin, is highly organized with regard to other features of sponge structure (Fig. 21.2*C*).

Many sponges are brightly colored red, orange, yellow, or purple. The color is derived from pigment located within the amoebocytes. Many of the freshwater species are green from symbiotic algae living within the amoebocytes.

The beating of flagella lining the atrium (spongocoel) causes water to be sucked in through the pores and driven up and out of the spongocoel through the osculum.

The flagellum of the collar cells describes a spiral in its beat. As a result, water is driven from a collar cell in the same manner as air is blown from a fan (Fig. 21.4).

Sponges are filter feeders, removing bacteria and other very fine suspended organic matter from the water. An initial screening is provided by the pores and passageways, which permit only very small particles to enter. When such particles pass over the collar cells, they may become trapped on the collar surface formed by the microvilli (0.1 μm. apart) between which water passes. The trapped particle passes down to the base of the collar where it is engulfed by the cell. The food par-

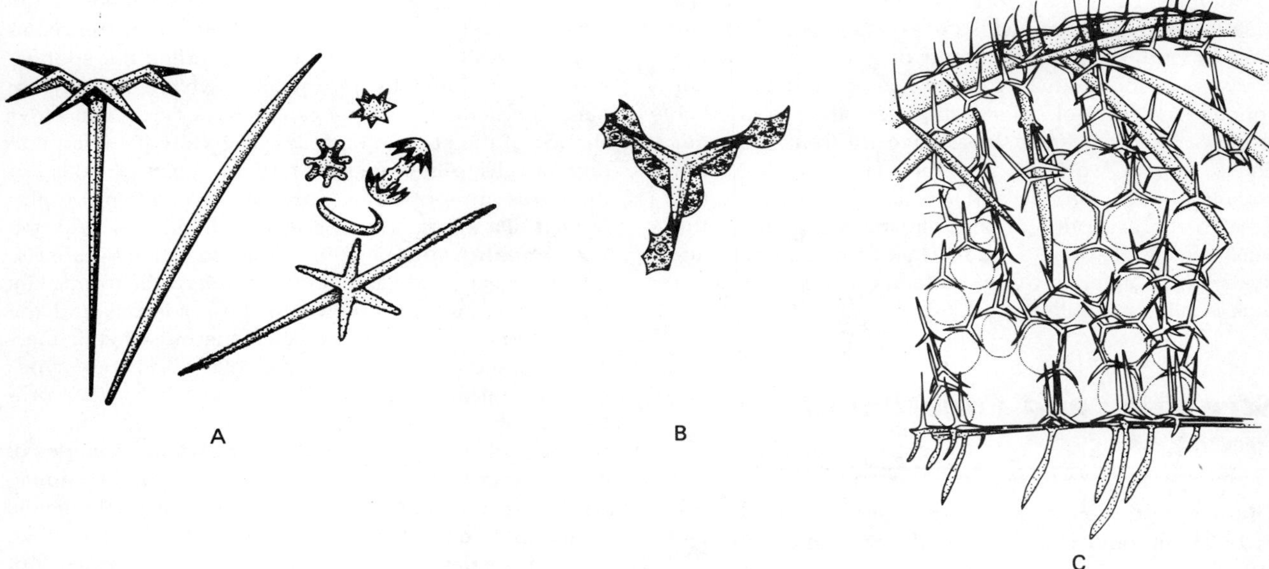

FIGURE 21.2 Sponge spicules. *A,* A few different types, showing variations in shape and size. *B,* Secretion of a calcareous triradiate spicule by amoebocytes. *C,* Spicules in their natural position within a calcareous leuconoid sponge. The dotted circles are the location of flagellated chambers. (*A* redrawn from Hyman and other sources; *B* after Minchin from Jones; *C* after Borojevic from Bergquist, P. R.: *Sponges.* London, Hutchison, p. 147, 1978.)

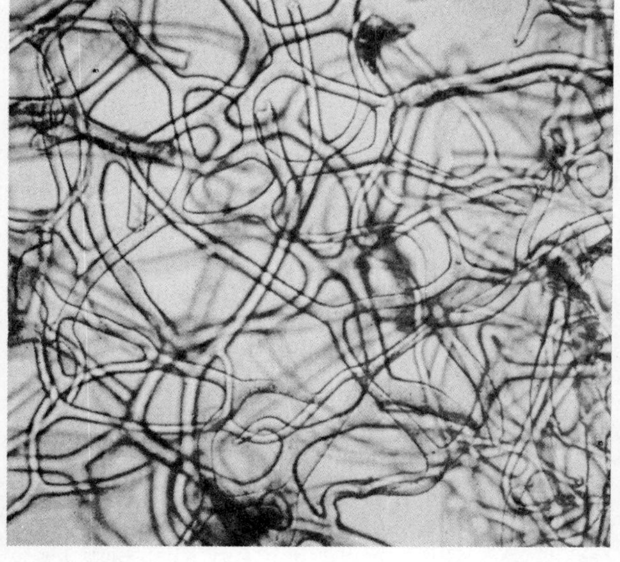

FIGURE 21.3 *A,* Spongin fibers (they appear translucent in the photograph). *B,* Photograph of the siliceous skeleton of the glass sponge (hexactinellid), *Euplectella,* Venus's flower-basket. The spicules are fused to form intersecting girders. (*A* Courtesy of the General Biological Supply House, Inc.)

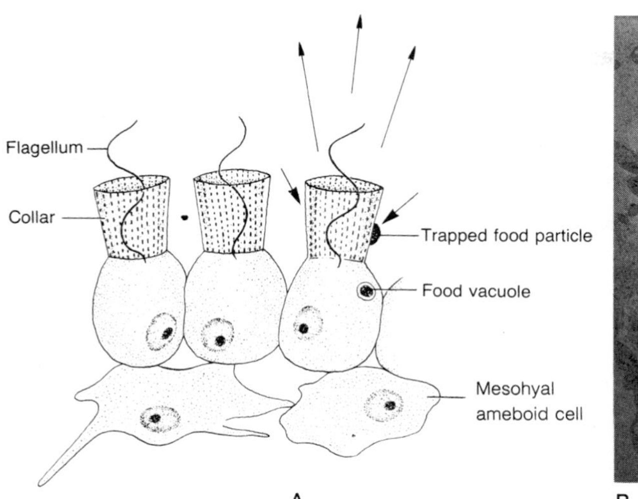

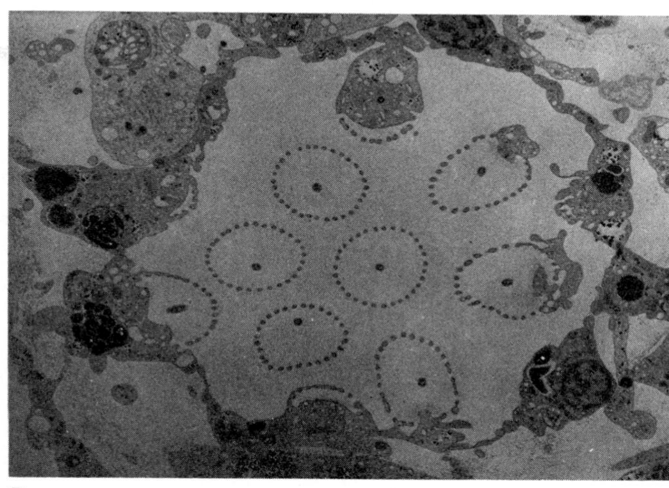

FIGURE 21.4 *A,* Three collar cells, or choanocytes, of the flagellated layer. Arrows indicate direction of water current. *B,* Electron photomicrograph of a section through a flagellated chamber of a leuconoid sponge. The circles are formed by the microvilli of the collar cells, and the dot in the center of each is the flagellum. (*B* from Rutzler, K., and G. Rieger: Sponge burrowing: fine structure of *Cliona lampa* penetrating calcareous substrata. *Marine Biology, 21*:144–162, 1973.)

ticle is then transferred to an amoebocyte for intracellular digestion. The products of digestion are passed by diffusion to the other cells of the body.

The current of water passing through the sponge body not only provides a source of food material but also serves for gas exchange, for the removal of wastes, and for the transfer of gametes.

Sponges with the simple vaselike structure are called **asconoid** sponges and are small, not more than a few centimeters high (Figs. 21.1 and 21.5). The asconoid structure imposes limitations in size for as the volume of the spongocoel increases, the flagellated surface area does not increase proportionally. Consequently, a large asconoid sponge would contain more water than its collar cells could efficiently move. In the evolution of sponges this problem was solved by repeated folding of the flagellated layer to increase its surface area. The first stage of folding is exhibited by **syconoid** sponges, in which the flagellated layer is evaginated outward into finger-like projections (Fig. 21.5). The evaginations are called **flagellated canals,** and the corresponding invaginations of the external surface are called **incurrent canals.** Pores are located between the incurrent and flagellated canals. The flagellated canals of syconoid sponges open into a central atrium devoid of collar cells.

In the great majority of sponges the surface area of the flagellated layer has been further increased by the formation of many small chambers within which the collar cells are located. In these so-called **leuconoid** sponges, water enters dermal pores on the body surface and passes through a system of incurrent canals, eventually reaching the flagellated chambers (Fig. 21.6). Many leuconoid sponges have no spongocoel; water leaves the body through converging excurrent canals opening to the exterior through an osculum. The development of flagellated chambers greatly increases the water-moving ability of a sponge. The number of flagellated chambers can range from 10,000 to 18,000 per mm.^{3}, and one leuconoid sponge 10 cm. high $\times$ 1 cm. in diameter was found to possess some 2,250,000 flagellated chambers and could pass 22.5 L of water through its body in 24 hours.

Within the system of canals and chambers water always flows in the direction of the larger opening. Moreover, the flagella of the choanocytes are directed toward the larger opening of the flagellated chamber. The speed of the water current is very slow through the chambers because the total cross-sectional area of all of the chambers served by an incurrent canal is very large. The flow speed greatly increases as the water is carried toward the osculum.

Most leuconoid sponges are irregular in shape, with many oscula located over the body surface (Fig. 21.6). Often the oscula are elevated like chimneys, which enhances the exhalant water flow. A far greater size is at-

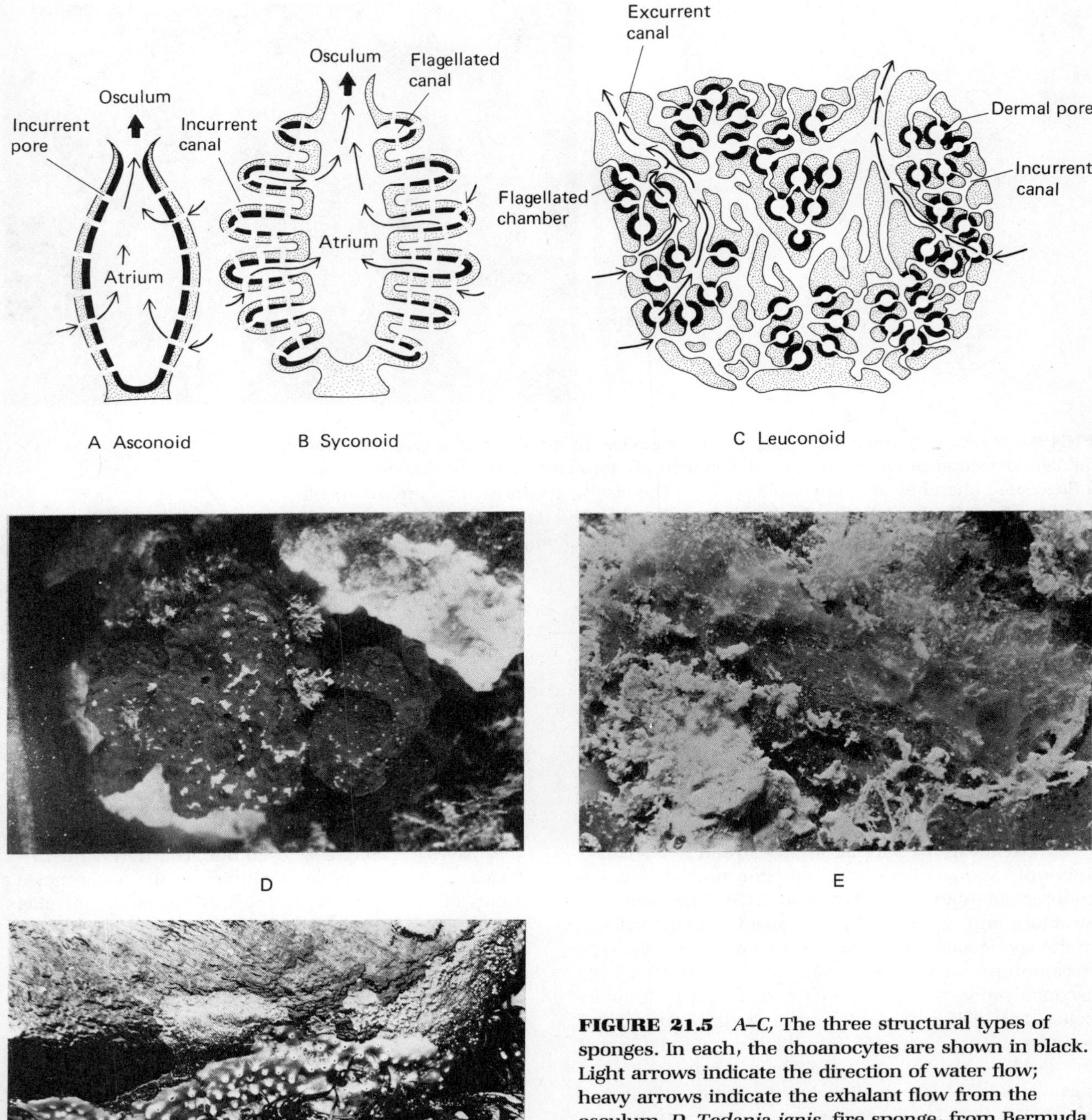

FIGURE 21.5 *A–C,* The three structural types of sponges. In each, the choanocytes are shown in black. Light arrows indicate the direction of water flow; heavy arrows indicate the exhalant flow from the osculum. *D, Tedania ignis,* fire sponge, from Bermuda. This species has a moundlike growth form. If handled, it produces a reaction like that from poison ivy in some persons. *E, Dysidea etheria,* a soft blue encrusting sponge from Bermuda. F, *Halichondria panicea*, an encrusting sponge from the low tide mark in Scotland. The tissue around the elevated oscula is pale green.

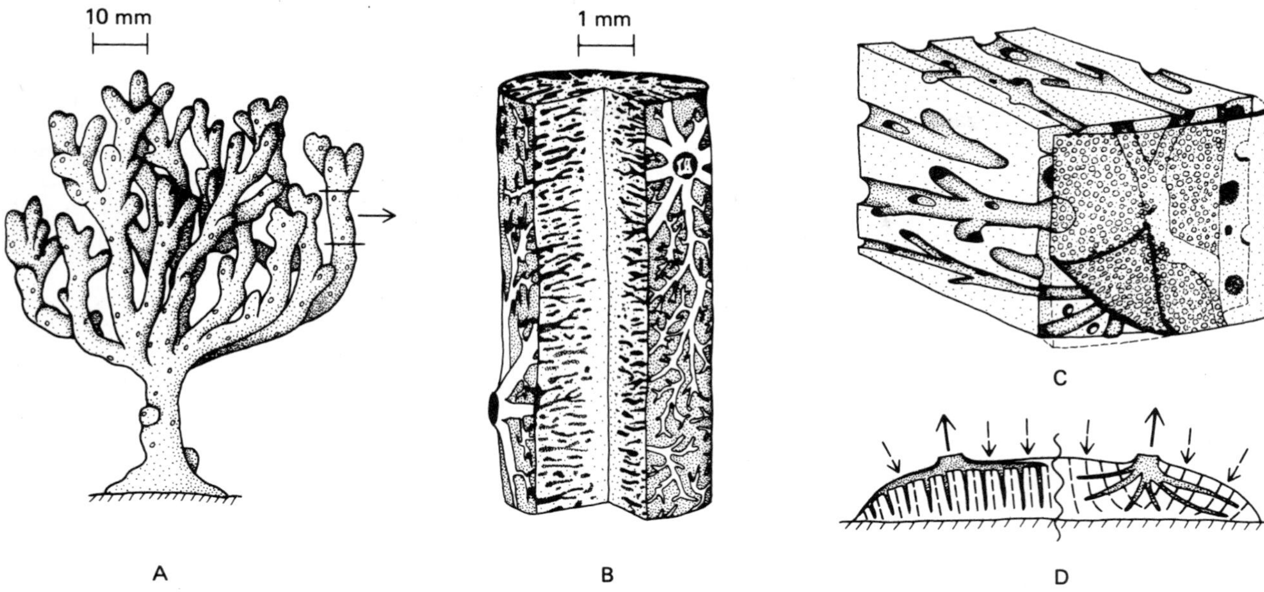

FIGURE 21.6 *A, B,* and *C,* Successive enlargements of a branching, solid leuconoid sponge, *Microciona prolifera,* with numerous oscula. Two oscula are shown in *B* and numerous dermal pores in *C.* In sponges with a solid structure the incurrent and excurrent canals may run parallel or at right angles to each other, as shown in the diagram *(D)* of two adjacent encrusting sponges. (From Reiswig, H. M.: The aquiferous systems of three Demospondiae. *Journal of Morphology, 145*:493–502, 1975.)

tained than in asconoid and in syconoid forms because any increase in bulk includes the addition of a large number of flagellated chambers necessary to drive the water current. The larger leuconoid sponges would more than fill a bushel basket, and some species reach 2 m. in height. Most of the sponges commonly encountered in shallow water are leuconoid.

Regeneration and Reproduction

The truly remarkable regenerative powers of sponges are well illustrated by the classic experiment of forcing a small bit of sponge through bolting silk and dissociating the cells. Within a short period of time, the dissociated cells reaggregate in the proper relationship. Ability to regenerate is closely correlated with asexual reproduction. A bud or small fragment of sponge breaks free of the parent and gives rise to a new sponge. Some sponges produce special asexual reproductive bodies, called **gemmules** (L. *gemmula,* little bud), consisting of an aggregate of essential cells, especially amoebocytes. Certain amoebocytes are **totipotent,** i.e., capable of giving rise to any other type of cell, and these play an important role in sponge regeneration and growth.

The majority of sponges are hermaphroditic, but eggs and sperm are usually produced at different times. Hermaphroditism is a common adaptation to a stationary existence in animals (p. 301). Sperm and eggs originate from amoebocytes and choanocytes and develop within the mesohyl; they are not located within a gonad. Mature sperm are shed into the water canals and exit through the osculum. Some are carried into the water canals of neighboring sponges. There sperm are trapped and engulfed by collar cells, which then move to an adjacent ripe egg. The carrier cell plasters itself against the egg and transfers the sperm. Fertilized eggs may be released into the water canals, carried out with the water stream, and undergo development in the sea water, but in most sponges the eggs are brooded and develop within the parental mesohyl.

Embryonic development leads to a free-swimming larva (Fig. 21.7), a stage that is important for species dispersion in sessile animals. In most species, the larval condition is reached at the blastula stage. After a brief free-swimming existence, the larva settles to the bottom and develops into an adult sponge.

Although sponges are primitive in that they lack organs, including a gut, and have only a small number of different kinds of cells, they are highly specialized ani-

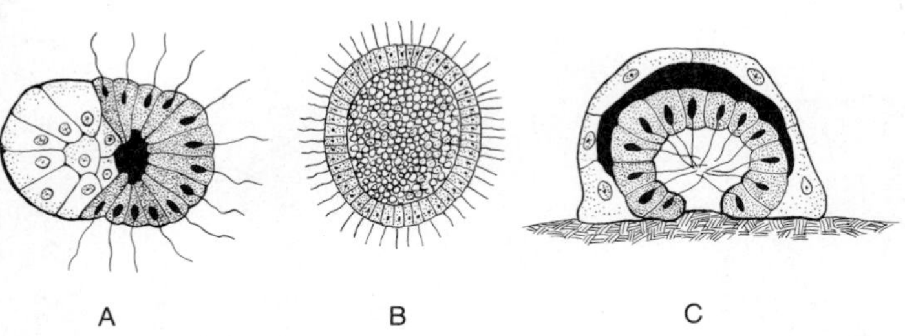

FIGURE 21.7 Sponge development. *A,* An amphiblastula larva. *B,* Parencymula larva. *C,* Gastrulation of an amphiblastula larva following settling. (*C* and *D* redrawn from Hyman, L. H.: *The Invertebrates,* Vol. I. New York, McGraw-Hill Book Co., 1940.)

mals in other respects. The specializations are largely adaptations to a stationary mode of existence—the absence of an anterior end, the circulation of water through the body for filter feeding, gas exchange, and water removal, and the condition of hermaphroditism. Certainly sponges evolved early in the evolution of the Animal Kingdom, and it seems highly unlikely that they gave rise to any other groups of animals.

Classification of Phylum Porifera

Class Calcarea Calcareous sponges. Spicules are usually separate and composed of calcium carbonate. Body form is asconoid, syconoid, or leuconoid. Calcareous sponges are mostly drab in color and small, not exceeding 10 cm. in height.

Class Hexactinellida Glass sponges. Spicules are siliceous and always include a six-pointed type (hexaxon). Some spicules are commonly fused together to form a highly organized skeleton. Although leuconoid, many species are cup-, vase-, or urnlike in shape, reaching a height of 10 to 100 cm. Most glass sponges occur in deeper water (200 to 2000 m.) than do other sponges and are thus less frequently encountered.

Class Demospongiae This class contains the greatest number of sponge species and includes most of the commonly encountered sponges. The skeleton of Demospongiae is composed of separate siliceous spicules. But some species, e.g., the commercial sponges, possess a skeleton of spongin fibers and some possess both spongin fibers and siliceous spicules. The body structure is always leuconoid, and an irregular symmetry with many oscula is common. Many Demospongiae are brightly colored. The small number of freshwater sponges belong to this class as well as the widespread marine boring sponges (Box 21.1).

Class Sclerospongiae A small group of tropical sponges with siliceous spicules and spongin fibers but encased within a solid external skeleton of calcium carbonate. Found within marine caves and tunnels associated with coral reefs.

BOX 21.1

Boring sponges, members of the family Clionidae, excavate tunnels in calcareous substrates—shell, coral, and coralline rock—which then become filled with the sponge body. Dermal pores and oscula are located at the surface. The boring is a chemical process, in which chips of calcium carbonate are etched out by single sponge cells. Boring sponges are found throughout the world and play an important role in breaking down coral and shell. Any shell-laden beach will contain many specimens perforated by boring sponges.

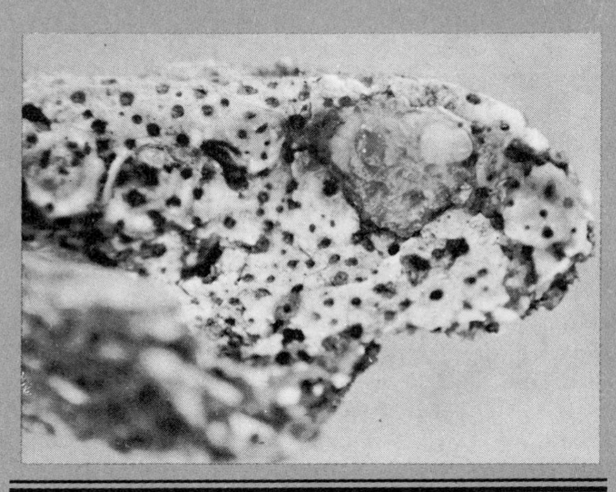

Summary

1. Sponges, members of the phylum Porifera, are sessile, aquatic animals, largely marine. The body structure is built around a system of water canals, a specialization correlated with their sessility. Sponges are primitive in being radially symmetrical, in lacking a mouth, gut, and other organs and in having a low level of cell differentiation.

2. The most primitive sponges are vase-shaped (asconoid form), with a central cavity or atrium. The body wall is composed of an outer layer of pinacocytes; a middle layer, or mesohyl, containing amoebocytes and the skeleton; and an inner layer of flagellated collar cells, or choanocytes. The beating of the flagella of the choanocytes sucks water into the atrium through pores in the body wall and expels it from the atrium through a distal opening, the osculum.

3. The skeleton of sponges is composed of coarse spongin fibers or needle-like spicules of calcium carbonate or silicon dioxide or of both spongin fibers and siliceous spicules.

4. The surface area of the flagellated layer has been increased in the evolution of sponges by folding of the body wall (syconoid form) or by the formation of many minute chambers to which the flagellated cells are confined (leuconoid form). Most sponges are leuconoid, and many have a solid structure and irregular shape.

5. Sponges are dependent upon the stream of water flowing through the body for food, gas exchange, and waste removal. Sponges are filter feeders on fine, suspended organic particles, the collar cell microvilli being the final filter.

6. Most sponges are hermaphroditic, and eggs are fertilized by sperm brought into the body by the water stream. The eggs are usually brooded to the larval stage within the parent body. The flagellated larval stage exits through the water canals and after a planktonic existence settles to form an adult sponge.

References and Selected Readings

Detailed accounts of sponges may also be found in the references listed at the end of Chapter 19.

Bergquist, P. R.: *Sponges.* London, Hutchinson & Co., 1978. Excellent coverage of all aspects of the biology of sponges.

Vogel, S.: Organisms that capture currents. *Scientific American,* 239(Aug.):128–135, 1978. A description of the way that sponges and some other animals utilize ambient water currents.

22 Cnidarians

ORIENTATION

The phylum Cnidaria contains the familiar hydras, sea anemones, corals, and jellyfish. They are radially symmetrical animals, with one end of the body bearing the mouth and tentacles. Some live attached with the mouth directed upward; others are free-swimming with the mouth directed downward. This chapter will first explain the general structure and function of cnidarians and will then examine the three cnidarian classes. We will see how the free-swimming and attached life styles have been exploited in each class and how the attached condition may have evolved.

CHARACTERISTICS OF CNIDARIA

1 Members of the phylum Cnidaria are mostly marine animals that are free-swimming (medusa, or jellyfish) or live attached (polyp).

2 The body is radially symmetrical, with the mouth and surrounding tentacles located at one end of the radial axis. The mouth is the only opening into the digestive cavity.

3 There are few organs, and the body wall is composed of two principal layers: an outer epidermis and an inner gastrodermis, separated by the mesoglea.

4 Explosive stinging or adhesive cells, called cnidocytes, are unique to the phylum.

5 The nervous system is commonly in the form of a net with receptor cells dispersed over the body surface.

6 The gonads are only aggregations of developing gametes, and there are no gonoducts. Fertilization is usually external, and development leads to a free-living planula larva.

Members of the phylum Cnidaria, or Coelenterata, are marine animals that include jellyfish, sea anemones, and corals. The hydras are among the few freshwater species.

Cnidarian Structure and Function

Cnidarians are radially symmetrical, and the oral end of the axis terminates in the mouth and a circle of tentacles (Fig. 22.1). The mouth opens into a blind **gastrovascular cavity.** A layer of cells, the **epidermis,** covers the outer surface of the body; another layer, the **gastrodermis,** lines the gastrovascular cavity (Fig. 22.1). The **mesoglea** (Gr. *mesos,* middle + *gloia,* glue), located between these two layers, varies from a thin noncellular membrane, as in hydras, to a thick jelly layer with or without cells, as in jellyfish. The epidermis and gastrodermis contain several kinds of cells. Cnidarians have few organs.

Two types of cnidarian body form can be distinguished. **Polypoid** cnidarians, such as hydras and sea anemones, have a cylindrical body with the oral end (bearing the mouth and tentacles) directed upward and the aboral end attached to the substratum (Figs. 22.1 and 22.8). **Medusoid** cnidarians, such as jellyfish (Figs. 22.1 and 22.9), have bell- or saucer-shaped bodies with the aboral end convex and directed upward and the oral end concave and directed downward. Medusae are usually free-swimming.

Movement. The body and tentacles of cnidarians can be extended or contracted and bent to one side or the other. Movement is brought about by the contraction of longitudinal and circular muscle fibers. However, the fibers are located not in true muscle cells but in basal extensions of epidermal and gastrodermal cells (Fig. 22.2). These **epitheliomuscle cells** and **nutritive muscle cells** have characteristics of the epithelial and muscle tissues of most other animals. The contractile layers are variously developed in different cnidarians. The gastrodermal fibrils of hydras, for example, are poorly developed, and movement is largely the result of contraction of the epidermal cells.

Nutrition and Nematocysts. Most cnidarians are carnivorous. Small forms, such as hydras and corals, feed upon **planktonic** animals, especially crustaceans, but the larger jellyfish and sea anemones can consume small fish. Gland cells lining the gut secrete proteolytic enzymes that rapidly digest the prey. Mixing of the gut contents is aided by the beating of the flagella of the gastrodermal cells. Small fragments of tissue are then engulfed by other gastrodermal cells, and digestion is completed intracellularly (Fig. 22.2). Undigestible waste materials are ejected through the mouth.

Cnidarians capture their prey with the aid of special stinging cells, called **cnidocytes,** located largely in the

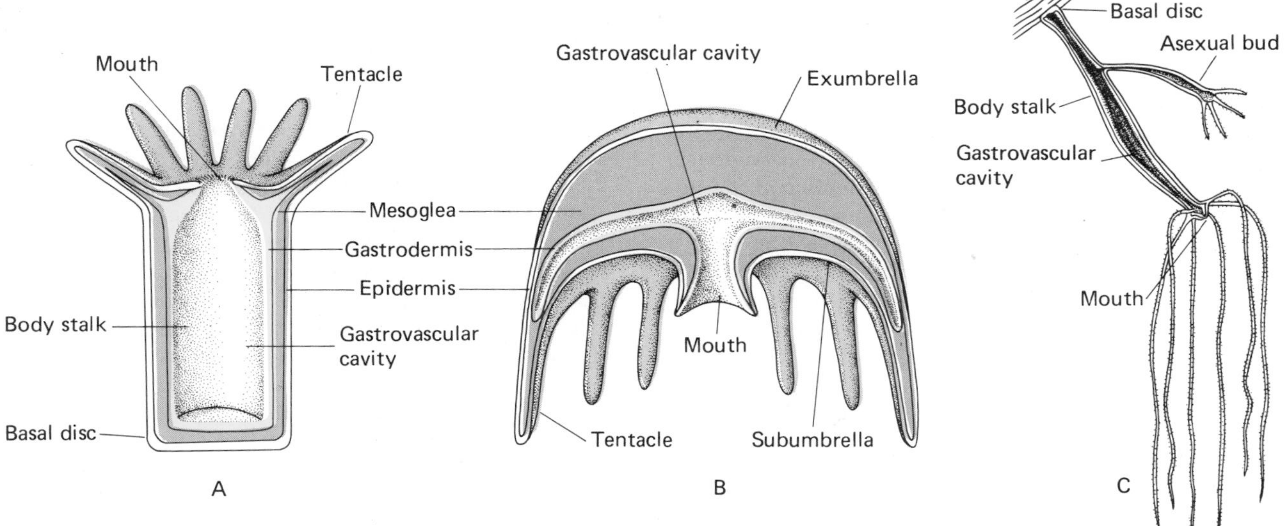

FIGURE 22.1 *A,* Polypoid body form; *B,* medusoid body form; *C,* an attached hydra with tentacles hanging in water.

BOX 22.1

Related Phylum

Members of the phylum Ctenophora, called comb jellies, sea walnuts, or more commonly ctenophores, are pelagic marine animals probably derived from some early line of medusoid cnidarians. Like medusae, the globose body, which is the size of a pea or golf ball, is transparent and contains a jelly-like mesoglea. The name *ctenophore,* which means "comb-bearer," comes from the presence of eight longitudinal ciliated bands, which can be seen in the figure. The ciliated bands propel the animal through the water, oral end forward. Ctenophores are carnivores on other planktonic animals, and most, including *Pleurobrachia* (shown here), possess long aboral tentacles with specialized adhesive cells used in prey-catching. All ctenophores are bioluminescent. (From Greve, W.: Southern North Sea comb jellies and their interspecific relations. Film C 1182 of the Inst. Wiss. Film., Göttingen. Publication from W. Greve, Publ. Wiss. Film., Sekt. Biol., Bd. 9, H.1 (1976) 53–62.

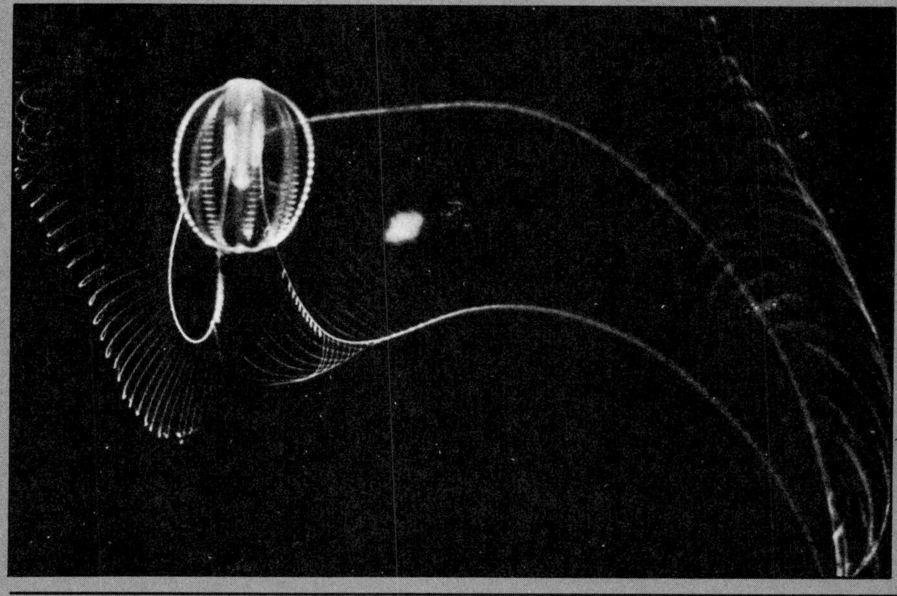

epidermal layer (Fig. 22.2). It is from these cells that the name *Cnidaria* (L. *cnide,* nettle) is derived. A cnidocyte contains a surface projection, the **cnidocil,** and the **nematocyst (cnida),** the actual stinging element. The undischarged nematocyst is composed of a bulb and a long thread coiled within the bulb (Fig. 22.3). When discharged, the nematocyst (Gr. *nēmatos,* thread + *kystis,* bladder) is expelled from the cnidocyte, and the thread is everted out of the bulb in the process. The mechanism of discharge is not completely understood, but it is believed that stimulation by the prey changes the permeability of the bulb wall so fluid rushes into the interior. The elevated fluid pressure both everts the thread and hurls the nematocyst from the cnidocyte. Some types of nematocysts function by entanglement; others are driven into the body of the prey and may inject a toxin (Fig. 22.3). It is these toxic penetrants that produce the sting of jellyfish and other cnidarians. Sea anemones have special adhesive cnidocytes in addition to those containing nematocysts.

Cnidocytes are typically embedded within the surface epidermal cells (Fig. 22.3C). Although they may be located throughout the epidermis, they are especially prevalent on the tentacles, where they may be concentrated in ringlike or wartlike batteries (Fig. 22.9). In hydras, contact with prey may result in the discharge of 25 percent of the tentacular nematocysts which are replaced in about 48 hours. The old cnidocytes are resorbed, and new cnidocytes differentiate from **interstitial cells** (Fig. 22.2), totipotent cells similar to certain amoebocytes of sponges.

After the prey has been quelled by the nematocysts, it is pulled by the tentacles toward the mouth, which often dilates enormously in swallowing. Mucus secreted by glands in the mouth region facilitates the process of swallowing.

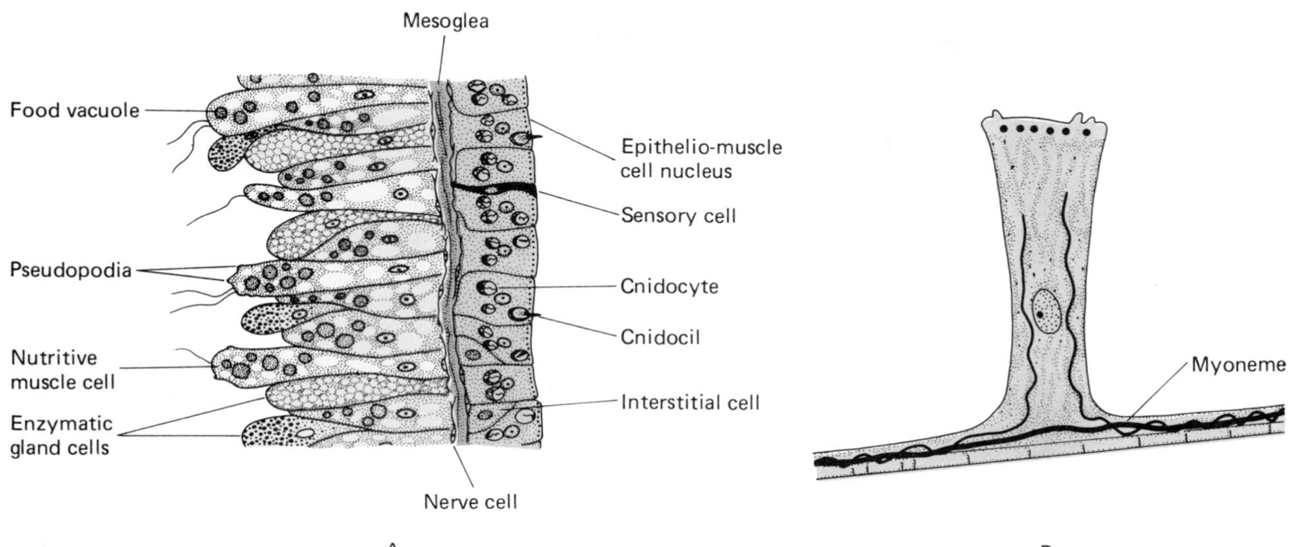

FIGURE 22.2 *A,* Body wall of hydra (longitudinal section); *B,* an epitheliomuscle cell. (After Gelei from Hyman, L. H.: *The Invertebrates,* Vol. I. New York, McGraw-Hill Book Co., 1940.)

There are no special systems for internal transport, gas exchange, or secretion. All of these processes take place by diffusion.

Nervous System. The cnidarian nervous system displays a number of primitive features. The neurons, located at the base of the epidermis and gastrodermis, are usually arranged as **nerve nets** rather than as nerve bundles (Fig. 22.4). The neurons may possess two or many processes, and the axons and dendrites of interneurons are not differentiated. Conduction commonly occurs in a radiating manner. Thus transmission in some neurons and across some synaptic junctions can occur in more than one direction.

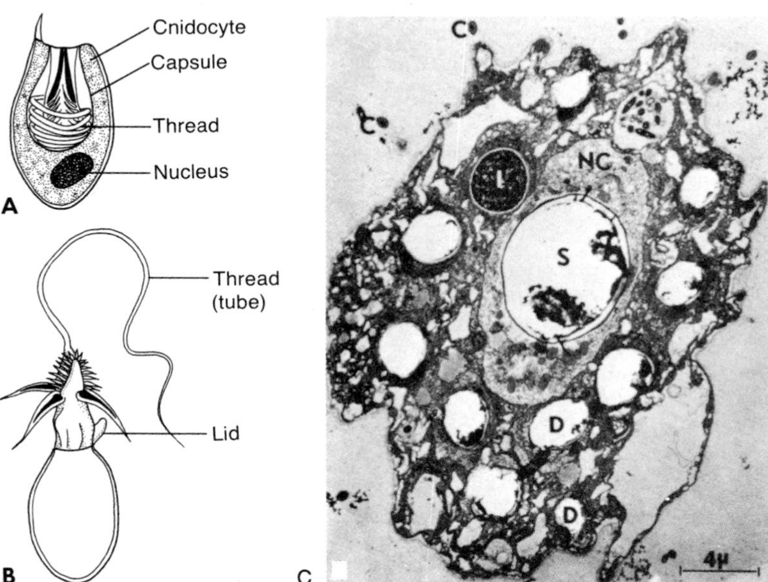

FIGURE 22.3 *A,* Undischarged penetrant nematocyst within cnidocyte of hydra; *B,* a discharged penetrant nematocyst; *C,* cross-section of an epitheliomuscle cell containing a central penetrant nematocyst (S) surrounded by smaller nematocysts (D and I). The letters C indicate cnidocils of adjacent cnidocytes, and NC is the central cnidocyte. The cnidocytes are enfolded by the epitheliomuscle cell rather than being intracellular. (By Westfall, J., S. Yamataka, and P. Enos: 20th Annual Proc. Electron Micro. Soc. Am., 1971.)

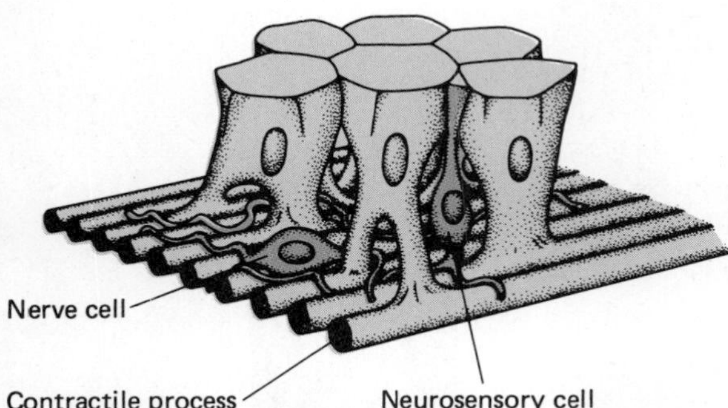

FIGURE 22.4 Diagram of the cnidarian epidermis, showing epitheliomuscle cells, sensory cell, and nerve net. (From Mackie, G. O., and L. M. Passano: Epithelial conduction in hydromedusae. *Journal of General Physiology, 52*:600, 1968.)

Growth and Reproduction. Polypoid cnidarians exhibit a high level of regenerative ability. For example, when a major part of the body, such as the oral end, is lost, the remaining part undergoes reorganization to form a new mouth region and tentacles.

Asexual reproduction is very common, especially in polypoid species. New individuals are usually formed by **budding.** A bud arises as an outpocketing of the body wall and thus contains an extension of the gastrovascular cavity and all of the body wall layers (Fig. 22.1). The bud separates from the parent, or in colonial species may remain attached as a new individual of the colony.

The sexes of most cnidarians are separate. The gametes develop from interstitial cells and form aggregations in specific locations in the epidermis or gastrodermis. Since there is no surrounding wall of sterile cells, the gonads are not comparable to the gonads of higher animals.

Fertilization is commonly external with development occurring in the plankton. At the completion of gastrulation, a characteristic larval stage, termed a **planula,** is attained. The planula is slightly elongate and radially symmetrical (Fig. 22.5), composed of a solid or hollow interior mass of cells surrounded by an outer layer of ciliated cells.

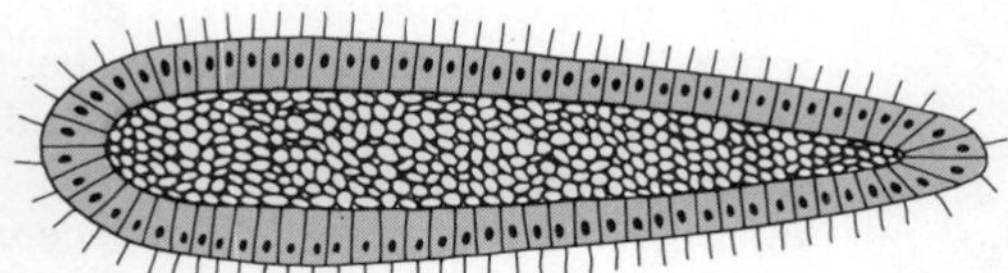

FIGURE 22.5 *Gonionemus*: planula larva that develops from the egg. (Based on a figure from Hyman, L. H.: *The Invertebrates,* Vol. I. New York, McGraw-Hill Book Co., 1940.)

Class Hydrozoa

The class Hydrozoa (Gr. *hydra,* water serpent + *zōon,* animal) includes the hydras and many colonial species called **hydroids.** While hydrozoan cnidarians are very abundant, they are small and not as conspicuous as the larger jellyfish, sea anemones, and corals. Most hydrozoans are marine, but the few species of freshwater cnidarians are members of this class.

Hydrozoans may exhibit a medusoid body form or a polypoid body form, or both, during the life history. The mesoglea is never cellular, cnidocytes are limited to the epidermis, and the gametes develop within the epidermis.

Colony Formation. The hydras and the small hydrozoan jellyfish are solitary, but many species are colonial (Figs. 22.6 and 22.8). If one can imagine hydras budding but the bud remaining attached to the parent, some notion of a hydroid colony can be attained. The individuals, or polyps, of a hydroid colony are usually attached to a main stalk, which is in turn anchored to the substratum (algae, rock, shell, or wharf piling) by a rootlike stolon. The arrangement of polyps on the stalk and the branching of the stalk vary with the species. In some hydroids, separate polyps spring directly from the stolon. The tissue layers of the stalks and stolons are continuous with the tissue layers of the polyps. Thus all of the polyps are interconnected, and there is a common gastrovascular cavity for the entire colony.

Skeleton Formation. The solitary hydrozoans have no skeletons, and some hydroid colonies in which the polyps arise directly from the substratum have only anchoring skeletons. Most hydroid colonies, however, are 3 to 10 cm. high, and support is provided by an external chitinous skeleton secreted by the epidermis. In some species, the skeletal envelope is limited to the stalk (Fig. 22.8*A*); in others, the skeleton extends around the body of the polyps but is open at the oral end for the emergence of the tentacles and mouth (Fig. 22.6).

Polymorphism. Many hydroid colonies have two or more structurally and functionally different kinds of individuals within the same species. This is termed polymorphism (Gr. *poly,* many + *morphē,* shape). The commonest type of individual is the feeding polyp (**gastrozooid**), resembling a hydra (Fig. 22.6). These polyps capture food and carry out the initial extracellular

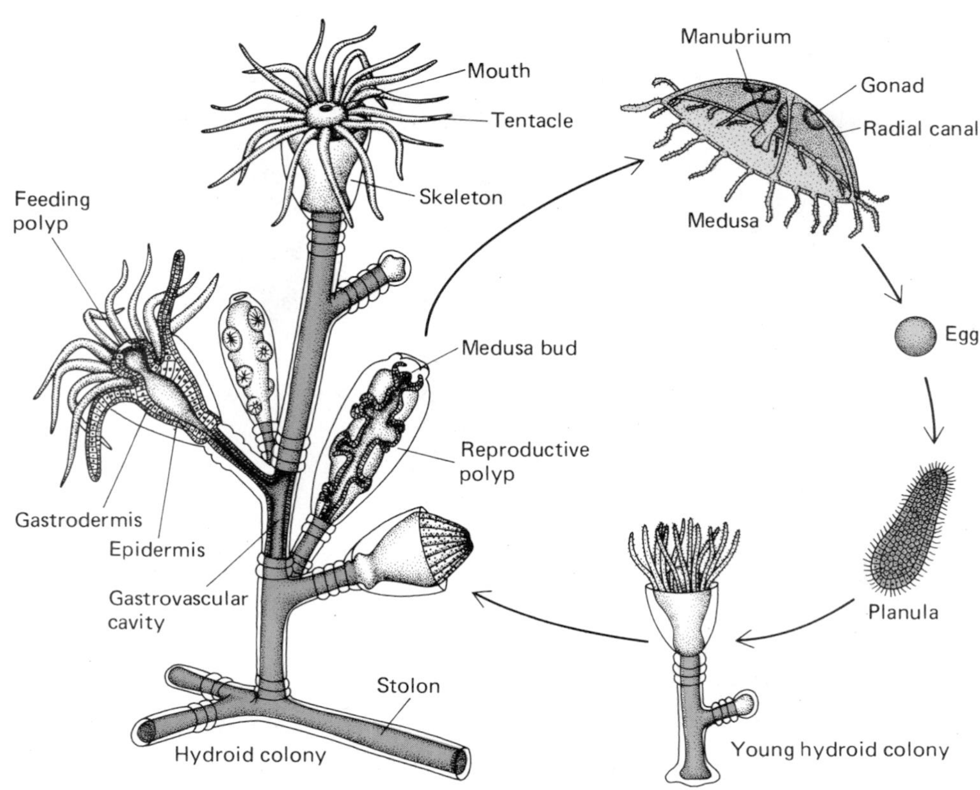

FIGURE 22.6 Life cycle of *Obelia,* showing structure of the hydroid colony.

phase of digestion. Intracellular digestion and absorption occur throughout the colony since the gastrovascular cavities are continuous.

Some hydroid colonies include defensive individuals, highly modified polyps with clublike bodies bearing great numbers of cnidocytes. Tentacles and mouth have disappeared (Fig. 22.8*B*). In some hydroids, sexually reproducing individuals known as medusae are budded from the body of feeding polyps. In many hydroids, however, there has been a further division of labor, and medusae are budded from special reproductive polyps (**gonozooids**—Gr. *gonos,* offspring + *zōon,* animal + *-oid,* like) no longer involved in feeding (Fig. 22.6).

Most colonial hydrozoans consist of groups of attached polyps. However, one group of colonial species, the **Siphonophora,** includes large pelagic colonies composed of both medusoid and polypoid individuals. There are individuals adapted as floats and pulsating swimming bells, from which hang feeding polypoid individuals with long tentacles (Fig. 22.7). The siphonophoran **Portuguese man-of-war** has a single gas-filled float from which the tentacles of the polyps may hang down several meters (Fig. 22.7). The nematocysts can produce painful stings, and entanglement with the tentacles can be a dangerous encounter for a swimmer in deep water.

Medusae. Sexual individuals are usually medusae.* Hydrozoan jellyfish (**hydromedusae**) are small, usually not more than a few centimeters in diameter (Figs. 22.6 and 22.9). The large jellyfish often encountered at beaches belong to another class of cnidarians. A varying number of tentacles hang from the bell margin of the hydromedusa, and the **manubrium** (L. *manubrium,* handle), a fold of body wall surrounding the mouth, hangs from the center of the convex undersurface. The contraction of a ring of muscle fibers within the bell margin reduces the diameter of the bell and drives water from beneath the animal. A shelflike inward projecting fold of the bell margin, the **velum** (L. *velum,* veil), increases the force of the water jet (Fig. 22.9). These pulsations propel the animal mostly in an upward direction and maintain the animal at a specific depth.

**Medusa* in Greek mythology was one of three gorgons, or monsters, who had snakes for hair.

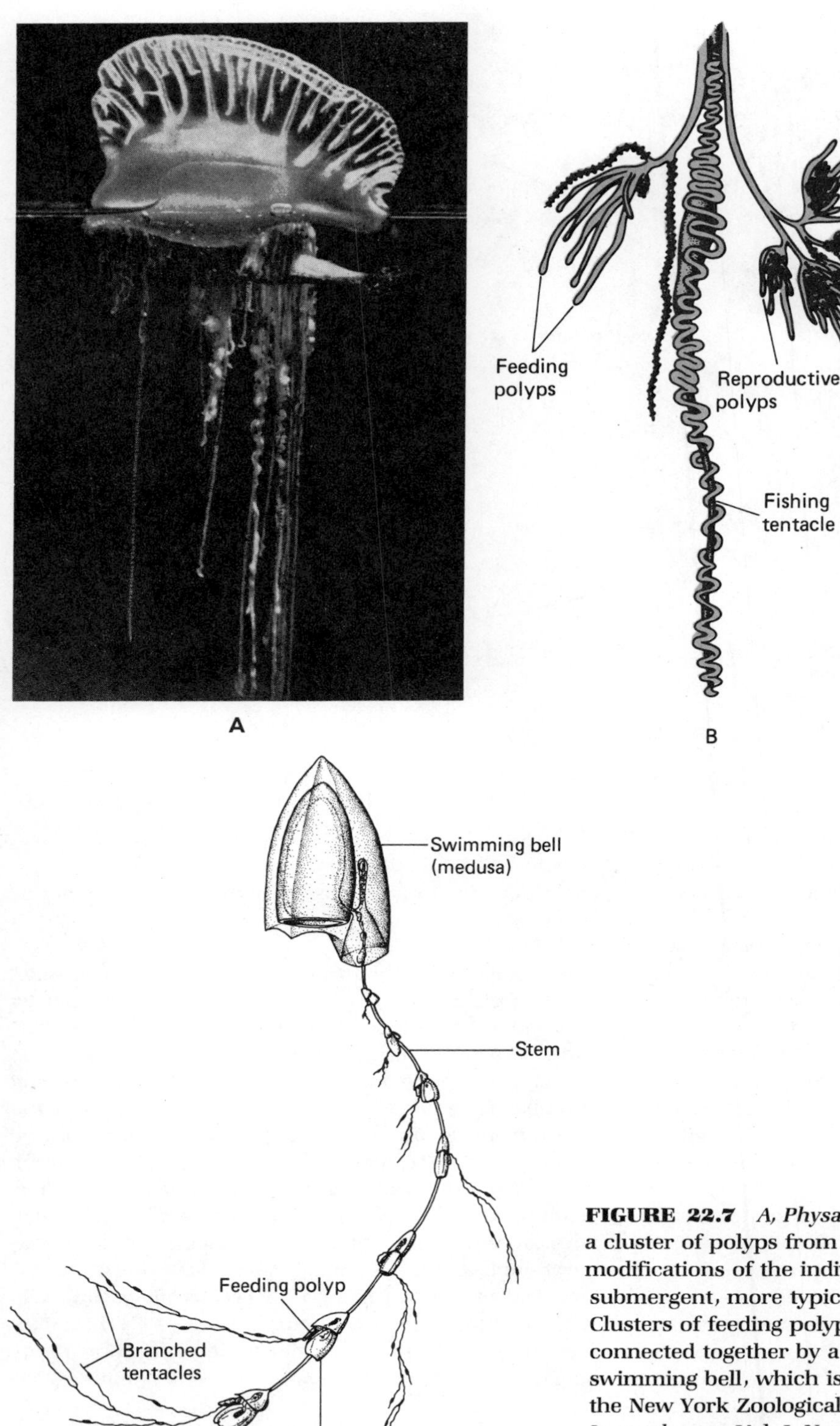

FIGURE 22.7 *A, Physalia,* the Portuguese man-of-war; *B,* a cluster of polyps from *Physalia,* showing the various modifications of the individual polyps; *C, Muggiaea,* a submergent, more typical siphonophore than *Physalia.* Clusters of feeding polyps and reproductive medusae are connected together by a long stem hanging beneath a swimming bell, which is about 1 cm. long. (*A* courtesy of the New York Zoological Society; *B* after Hyman, L. H.: *The Invertebrates,* Vol. I. New York, McGraw-Hill Book Co., 1940.)

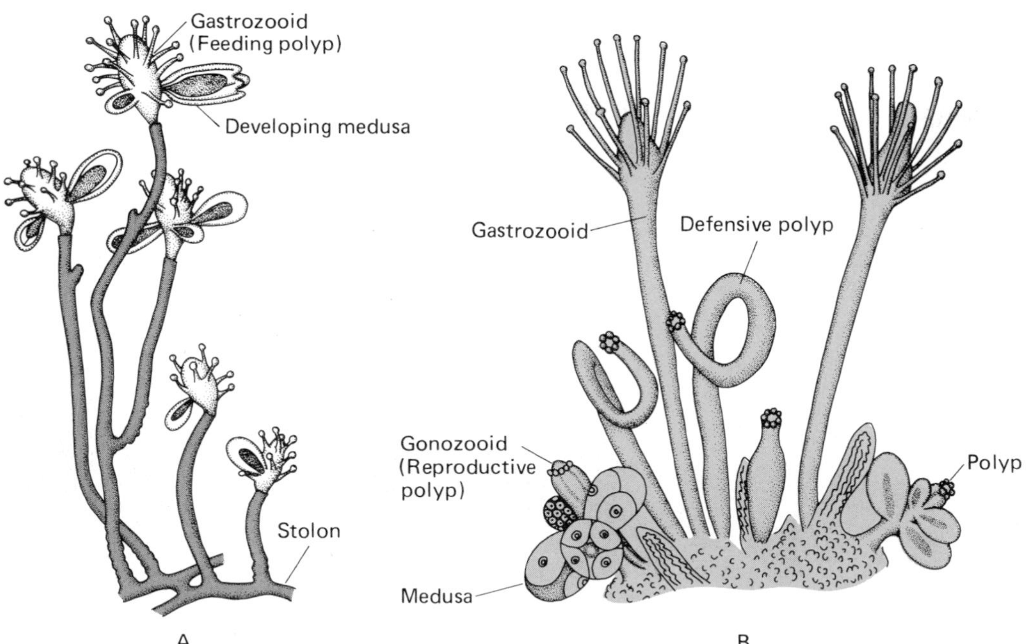

FIGURE 22.8 Hydroid colonies. *A, Coryne tubulosa,* a dimorphic form in which medusoids are formed directly on gastrozooids. *B, Hydractinia.* Polyps arise separately from mat of stolons, and skeleton is limited to stolons. (*A,* modified after Naumov; *B* after Hyman, L. H.: *The Invertebrates,* Vol. I. New York, McGraw-Hill Book Co., 1940.)

Located at the tentacle bases or between tentacles on the bell margin are **statocysts** (balancing organs) and **ocelli** (patches of photoreceptor cells), which orient the jellyfish to the pull of gravity from below and to light penetrating the water column from above (Fig. 22.9).

Hydromedusae feed on other small planktonic organisms. The gastrovascular cavity is not a single sac, as in polyps. The mouth opens into a central stomach from which extend four radial canals (Fig. 22.9). The radial canals connect in turn with a ring canal that extends around the bell margin.

The gametes of hydromedusae develop within the epidermis beneath the radial canals (Fig. 22.9). The eggs are fertilized when shed into the sea water, and development to the planula larva occurs in the plankton. In some species, fertilization and early development occur within the epidermis.

Life Cycles. Hydrozoans exhibit great variation in their life cycles. It is probable that the ancestral ones had no polyp stage. In similar species living today, the medusae produce planula larvae, which transform into free-swimming **actinula** larvae (Fig. 22.10). These, in turn, develop into new medusae. In some groups, the actinula larva becomes attached and gives rise by budding to a polypoid colony, which produces new medusae. In such a cycle, asexual reproduction by the polyps greatly increases the reproductive potential of the species. In species such as *Obelia* (Fig. 22.6), the actinula is bypassed, and the planula larva develops directly into the hydroid colony. With the evolution of the polyp, we see a progressive reduction of the medusa stage. In many hydrozoans, the medusae are formed, but they remain attached to the polyps. Finally, there are polypoid species, such as hydras (Fig. 22.11*B*) in which there are no medusa at all. Eggs and sperm are produced directly in the epidermis of the polyp.

Many zoologists believe that the medusa is the more primitive cnidarian body form. According to this view, the hydrozoan life cycle initially lacked a polypoid stage, and the planula developed first into an actinula and then into an adult medusa. Such a life cycle is exhibited by some living hydrozoans, such as *Aglaura* (Fig. 22.10). In other hydrozoans, the actinula stage perhaps became attached to the bottom by the aboral end and produced more actinulae by budding. The latter developed directly into medusae. This attached actinula became the first polyp, and secondary budding would represent the

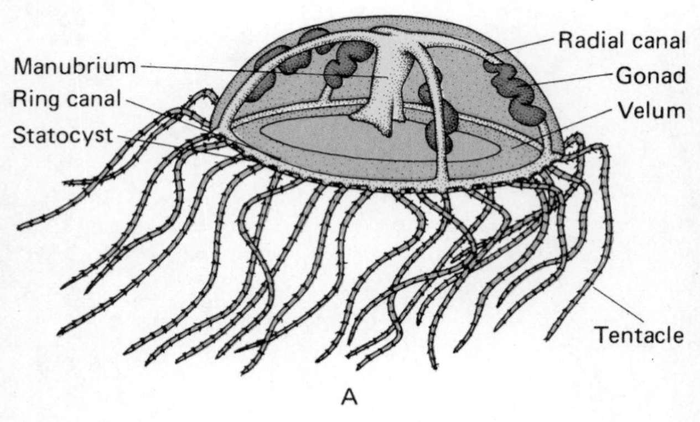

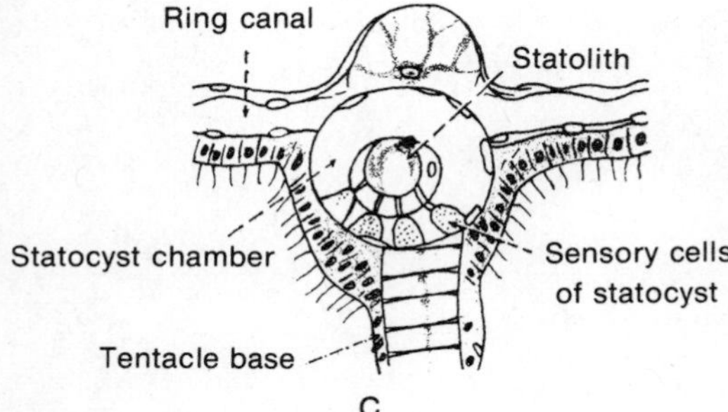

FIGURE 22.9 The hydromedusa, *Gonionemus vertens. A,* Lateral view. *B,* An individual actively swimming. *C,* Statocyst of a hydromedusa. (*B* courtesy of Charles Seaborn; C after the Hertwigs from Hyman, L. H.: *The Invertebrates,* Vol. I. New York, McGraw-Hill Book Co., 1940.)

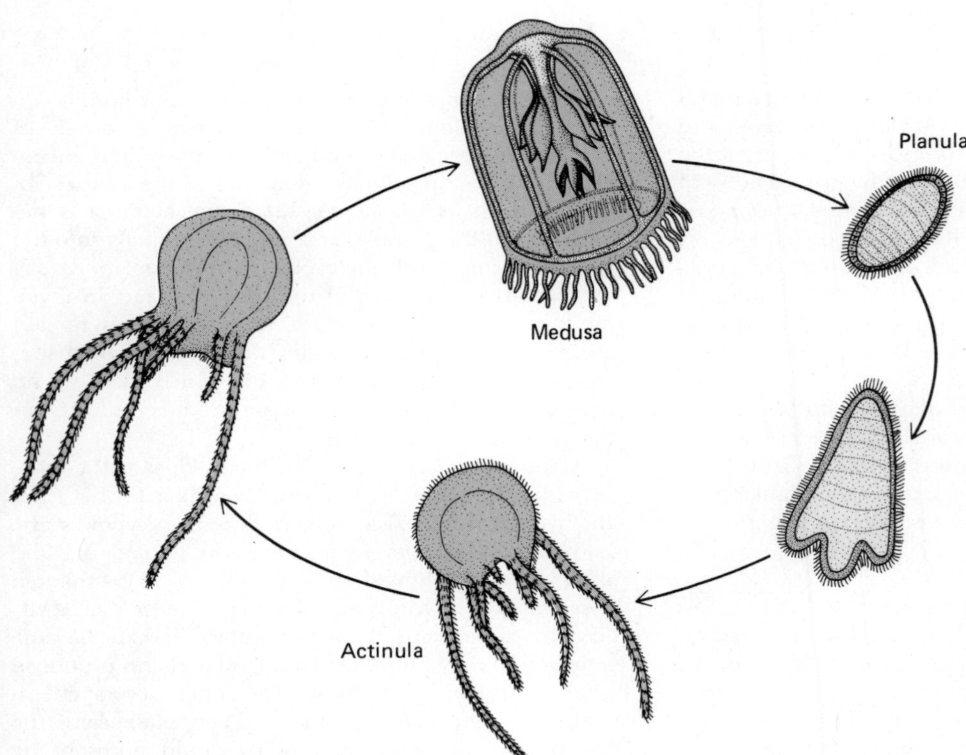

FIGURE 22.10 Life cycle of *Aglaura,* a hydrozoan that has no polypoid stage. Planula larva develops into an actinula, which develops directly into a medusa. (From Bayer and Owre.)

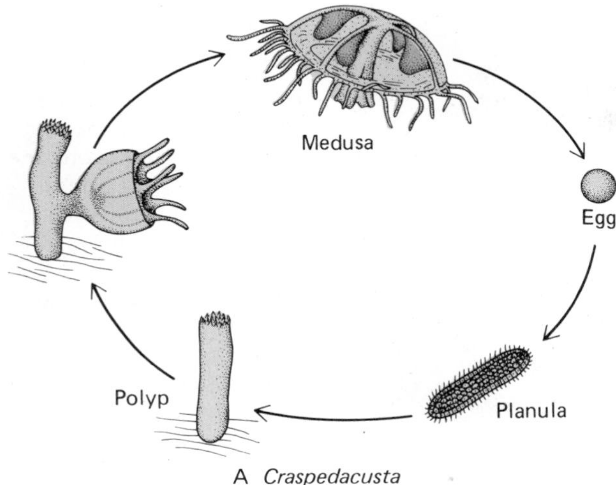

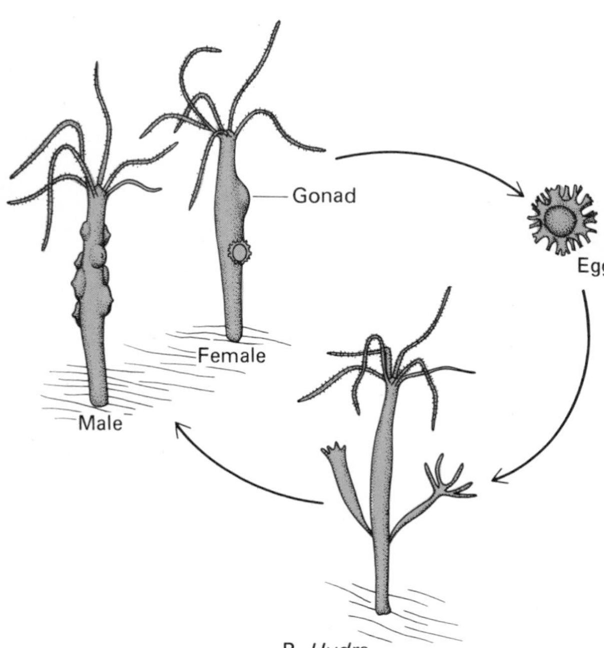

FIGURE 22.11 Life cycles of the freshwater *Craspedacusta* (*A*) and *Hydra* (*B*). Both have solitary polypoid stages; the medusoid stage is absent in *Hydra*.

formation of medusa buds. Such a life cycle has the advantage of permitting a long life for an actinula, which can give rise to not one but many medusae. This type of life cycle occurs in such hydrozoans as the marine *Gonionemus* and the little freshwater jellyfish *Craspedacusta* (Fig. 22.11).

With the evolution of the polyp, there now occurred a reduction of the medusa in different groups of polypoid species. The medusa developed by budding but was not free or detached. There was a gradual reduction of medusoid tissues until the polyp epidermis itself produced gametes.

Present evidence indicates that colonies evolved independently in different lines of hydrozoans, and this occurred independent of the evolutionary trend toward medusa reduction. A solitary polyp would produce more polyps by budding during certain times of the year and then produce medusa buds at another season. If the first type of buds remained attached to each other, this could have led to the evolution of a hydroid colony. Colony formation did not evolve in polyps of all hydrozoans. The hydras, for example, probably evolved from some line of solitary polyps that underwent reduction of the medusoid form.

Class Scyphozoa

The medusae of the class Scyphozoa (Gr. *skyphos,* cup + *zōon,* animal) are the cnidarians to which the name jellyfish is usually applied; they are considerably larger and more conspicuous than hydrozoan medusae (Fig. 22.12). The medusoid body form is dominant in the Scyphozoa, the polypoid form being strictly a larval stage. In contrast to hydromedusae, the mesoglea of scyphomedusae may be cellular; some cnidocytes are present in the gastrodermis, and the gametes develop within the gastrodermis rather than within the epidermis.

Many scyphomedusae are about the size of a saucer, and the translucent body is often tinted with orange, pink, purple, and other colors. Tentacles of varying numbers and length hang from the margin of the bell. Four long divisions of the manubrium, called oral arms, hang from the center of the underside of the bell and are often more conspicuous than the tentacles (Figs. 22.12 and 22.13). The nematocysts, especially those in the tentacles and oral arms, can produce painful stings. The sea nettles, such as the Chesapeake *Chrysaora,* can be a painful nuisance to swimmers at certain times of the year when they occur in large numbers (Fig. 22.12*A*). The sea wasps of the Indo-Pacific have such virulent nematocysts that their sting causes severe lesions and even death (Fig. 22.12*B*).

As in hydromedusae, statocysts and ocelli are present, but in the scyphomedusae they are associated at specific sites between the scallops of the bell margin to form a clublike **rhopalium** (Gr. *rhopalon,* club) (Fig. 22.13).

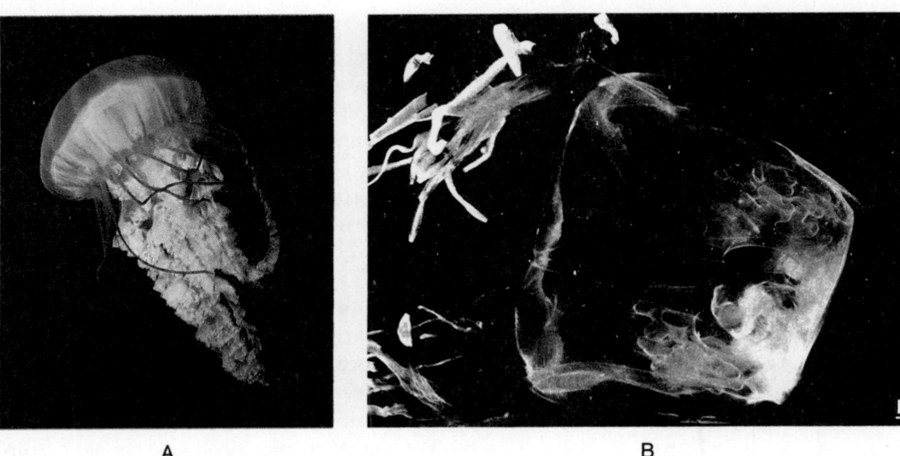

FIGURE 22.12 *A,* The sea nettle, *Chrysaora helvola,* a scyphozoan found along the Pacific coast. *B, Chironex fleckeri,* a dangerous sea wasp along the northern coast of Australia. (*A* courtesy of Charles Seaborn; *B* by J. H. Barnes, in Rees, W. J. (Ed.): *The Cnidaria and Their Evolution.* Zoological Society of London, 1966.)

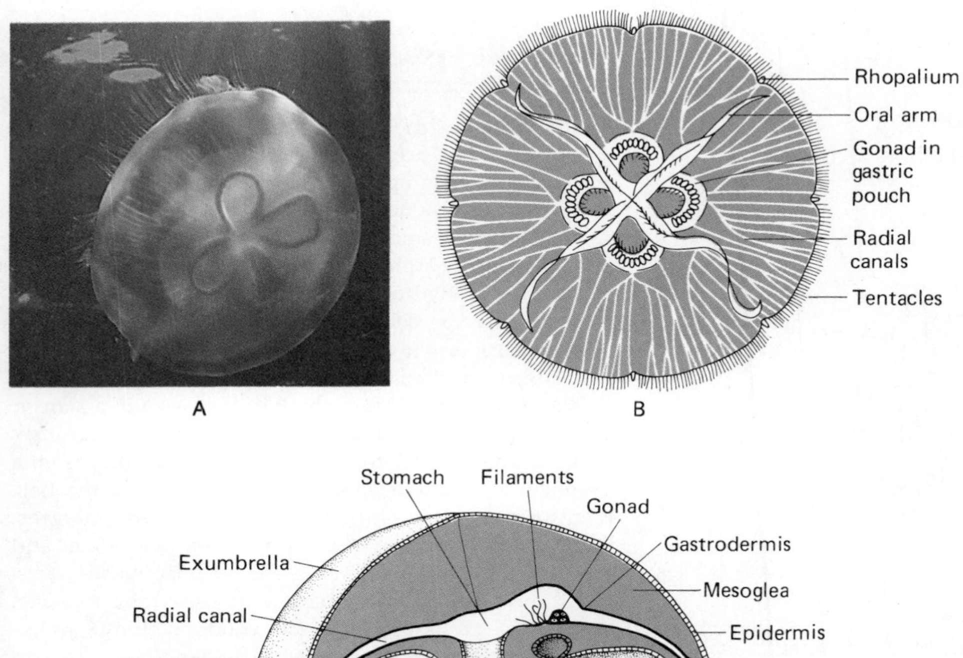

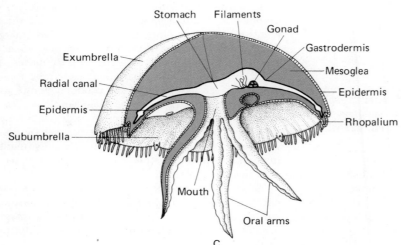

FIGURE 22.13 *A, Aurelia,* a scyphozoan medusa. *B,* Oral view. *C,* Side view in section. (*A* courtesy of Charles Seaborn.)

There is no velum in scyphomedusae, but swimming movements are rather like those of hydromedusae. However, some species can turn on their side and swim very rapidly in a horizontal direction.

Scyphozoans feed on animals of various sizes, including fish, that come in contact with the tentacles and oral arms. Some species, such as *Aurelia,* feed on plankton that adheres to the undersurface of the bell. The plankton is then swept by ciliated surface cells to the bell margin, which is wiped by the oral arms. There is a central stomach, as in the hydromedusae, but the stomach floor bears filaments containing cnidocytes (Fig. 22.13). These gastrodermal nematocysts perhaps function to quell any prey still alive when it enters the stomach. In most of the commonly encountered north temperate jellyfish, numerous canals extend radially to the bell margin. A ring canal may or may not be present.

Gametes develop in the gastrodermis of the stomach floor, and the "gonads" are often conspicuous within the transparent body (Fig. 22.13). The eggs and sperm are shed through the mouth. Fertilization and early development may occur in the sea water, or the eggs may be brooded on the oral arms. In either case, a free-swimming planula larva is attained. The planula develops into a polypoid larval stage, called a **scyphistoma,** that is about the size of a hydra and lives attached to the bottom (Fig. 22.14). The scyphistoma has a life span of one to several years, during which it feeds and produces more scyphistomae by budding. At certain seasons it ceases feeding and undergoes a special form of budding to produce young jellyfish. The buds are produced at the oral end of the body and in the course of formation in some species are stacked up like plates **(strobila).** When detached, the young jellyfish **(ephyra)** is tiny and displays only a rudimentary medusoid form (Fig. 22.14).

In describing the evolution of hydrozoans, we theorized that the polyps evolved from a larval stage. Note that the scyphozoan life cycle displays just such a polypoid larva, which by budding produces both more polyps and medusae. Thus this theory of the origin of the polypoid stage is supported by evidence from comparative development of scyphozoans as well as hydrozoans.

Class Anthozoa

The 6000 species of anthozoans (Gr. *anthos,* flower + *zōon,* animal) constitute the largest class of cnidarians and include the familiar sea anemones and various

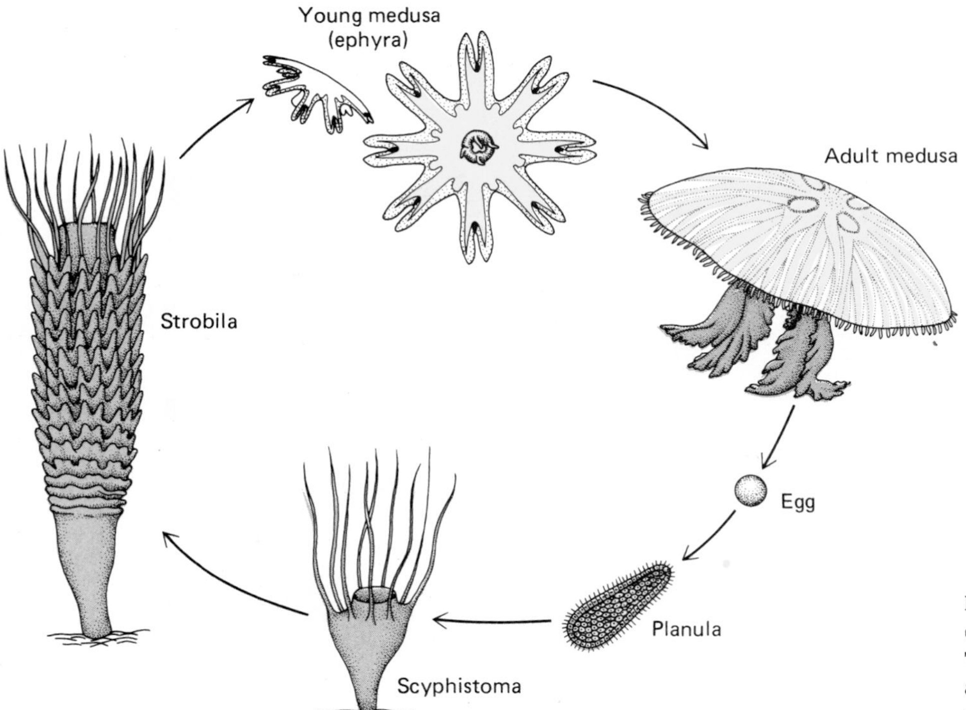

FIGURE 22.14 Life cycle of the scyphozoan *Aurelia.* The polypoid stage is a larva and produces medusae by transverse budding.

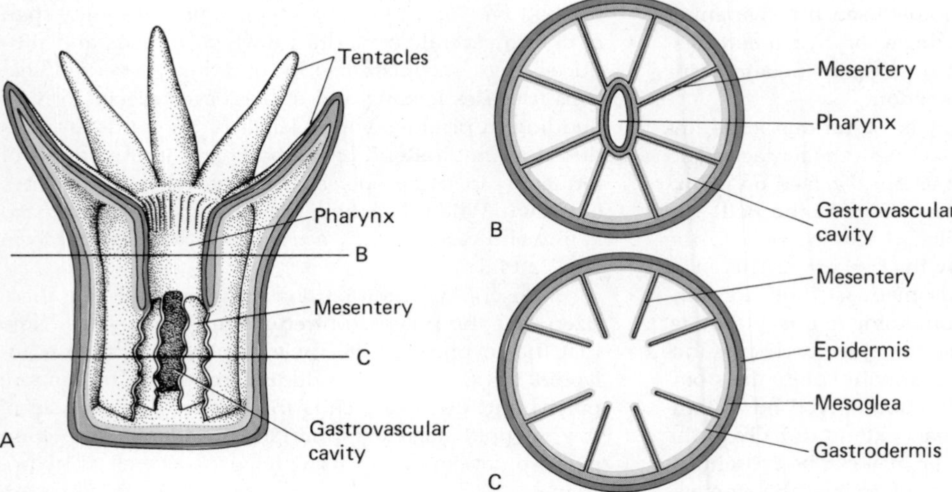

FIGURE 22.15 Structure of an anthozoan polyp. *A,* Longitudinal section; *B,* cross section taken at level of pharynx; *C,* cross section below level of pharynx.

types of corals. The class is entirely polypoid, and the cellular mesoglea, gastrodermal cnidocytes, and gastrodermal gametes suggest that the anthozoans may have evolved from the polypoid larva of some ancient group of scyphozoans.

The distinctive feature of anthozoans is the presence of a **pharynx** and **mesenteries.** The pharynx is a tube that hangs from the mouth into the gastrovascular cavity like a sleeve (Fig. 22.15). Since the pharynx is derived from an infolding of the body wall around the mouth, it possesses the same layers as the outer body. The mesenteries are sheetlike partitions that extend into the gastrovascular cavity from the outer body wall toward the pharynx. Each is composed of two layers of gastrodermis separated by a layer of mesoglea. At least eight of the mesenteries, called complete mesenteries, connect with the mesoglea and gastrodermis of the pharynx. In many anthozoans, there are additional incomplete mesenteries extending partway into the gastrovascular cavity.

FIGURE 22.16 Sea anemones. *A,* View of expanded oral end of *Aiptasia. B, Anthopleura,* a genus of common sea anemones found in shallow water along the Pacific coast of the United States. (*B* courtesy of Charles Seaborn.)

The functional significance of the mesenteries is difficult to understand. One might expect that they would provide greater surface area for digestion and absorption, but studies have shown that only the free margin is involved in digestion and absorption. The mesenteries may serve for internal gas exchange and may limit the diameter of the body.

Sea Anemones. The largest of the anthozoans are the solitary sea anemones (Fig. 22.16). The majority live attached to hard substrates. Most are one to several centimeters in diameter and are often brightly colored. A species on the Great Barrier Reef of Australia attains a diameter of 1 m.

The body of a sea anemone is heavier than that of a hydrozoan polyp, and the oral end, bearing tentacles peripherally and a central slit-shaped mouth, is disclike (Fig. 22.16). At one or both ends of the mouth there is a ciliated groove, the **siphonoglyph**, which drives water into the gastrovascular cavity. When sea anemones con-

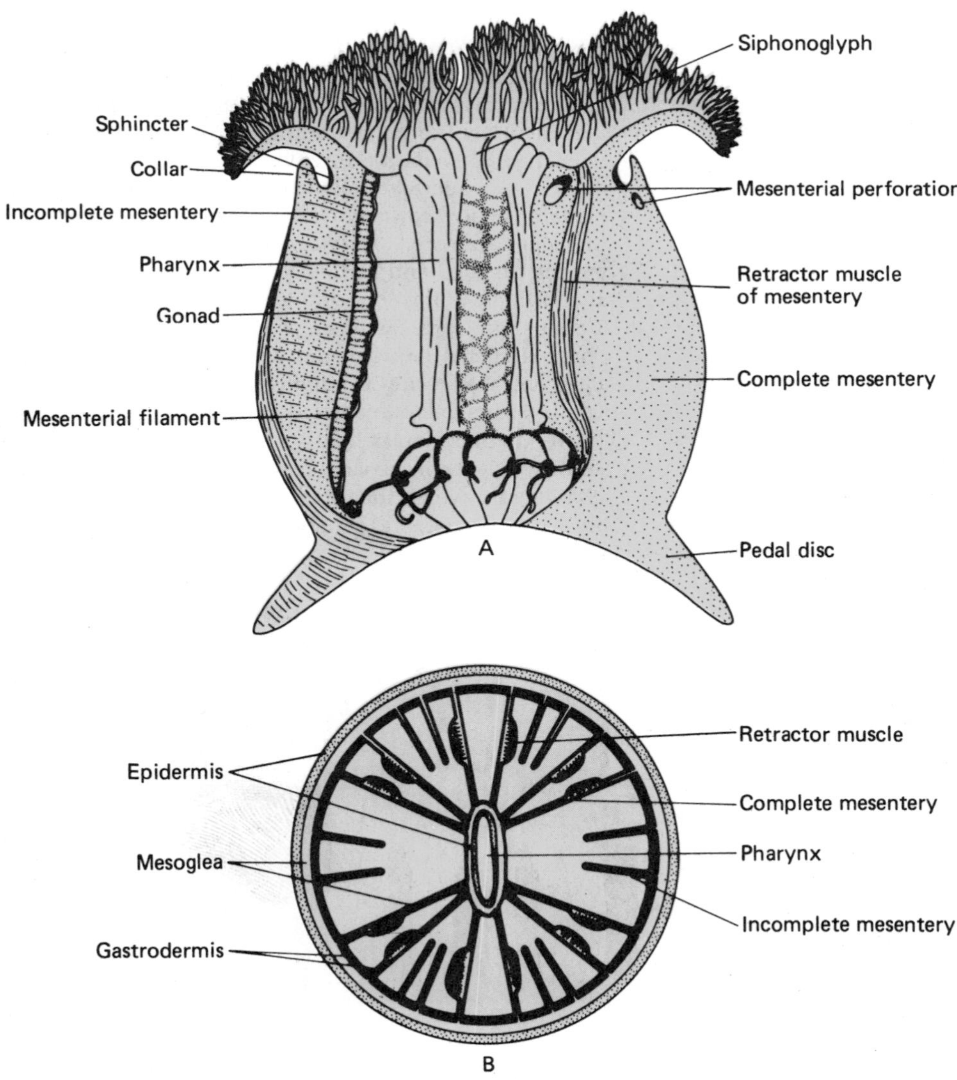

FIGURE 22.17 Structure of a sea anemone. *A,* Longitudinal section; *B,* cross section at the level of the pharynx. (*A* after Hyman, L. H.: *The Invertebrates,* Vol. I. New York, McGraw-Hill Book Co., 1940.)

tract, the upper, collar-like rim of the body is pulled over the oral disc collar (Fig. 22.17). The muscle fibers are entirely gastrodermal. Circular fibers are located in the gastrodermis of the body wall, and longitudinal retractor fibers are located in the mesenteries.

Sea anemones feed upon small fish and other invertebrates. The prey is passed down the pharynx and into the center of the gastrovascular cavity below the pharynx. The free edge of the mesenteries bears a glandular ciliated band called the **mesenterial filament.** This is the location of the gastrodermal cnidocytes, of enzyme production for extracellular digestion, and of intracellular digestion and absorption. A perforation in the upper part of each complete mesentery permits circulation of the gastrovascular contents.

Some sea anemones reproduce asexually by splitting of the body or by the regeneration of fragments of tissues separated from the base of the body. The sexes are usually separate, and the gametes develop in gastrodermal bands just behind the free edge of the mesenteries (Fig. 22.17). Fertilization and early development may occur externally in the sea water or within the gastrovascular cavity. The planula larva develops into a ciliated planktonic polypoid larva in which mesenteries and pharynx make their appearance. The polyp soon settles and becomes attached as a young sea anemone.

Corals. Most other anthozoans are various types of corals. The species with which the name coral is most generally associated are the **scleractinian,** or **stony, corals.** The polyps of scleractinian (Gr. *sklēros,* hard) corals are similar to those of sea anemones but are usually smaller and almost always are connected together in colonies (Fig. 22.18). The connection is by means of a lateral fold of the body wall, and the entire colony has the form of a sheet. A calçium carbonate skeleton is secreted by the epidermis of the undersurface of the connecting sheet and the lower part of the polyp. The living colony is thus sitting on top of a skeleton that is actually external. The lower part of the polyp is situated within a skeletal cup, the bottom of which contains radiating septa that project up into folds in the base of the polyp (Figs. 22.18 and 22.19).

Species of corals display various growth forms. Some are low and encrusting; others are upright and branching (Fig. 22.20). The surface configuration of the skeleton depends upon the relative positions of the polyps. The coral appears pockmarked when the polyps are well separated. In brain coral, which has a surface configuration of troughs and ridges, the polyps are joined together in rows (Fig. 22.19). No skeletal material is deposited between the polyps within a row, but skeleton is deposited between the rows.

The gastrodermal cells of most scleractinian corals contain yellow-brown symbiotic algae (**zooxanthellae**—Gr. *zōon,* animal + *xanthos,* yellow). The algae utilize the nitrogenous and carbon dioxide wastes of the coral and facilitate the deposition of calcium carbonate ($CaCO_3$). The coral obtains glycerol from the algae. Phosphate is cycled between the two. When either light or algae are absent, $CaCO_3$ deposition is greatly reduced. Corals with symbiotic algae do not occur at depths below the penetration of light.

Most other anthozoans, largely tropical, are octocorallian corals, which include the sea fans, sea whips, sea pens, sea pansies, and others. There are only eight mes-

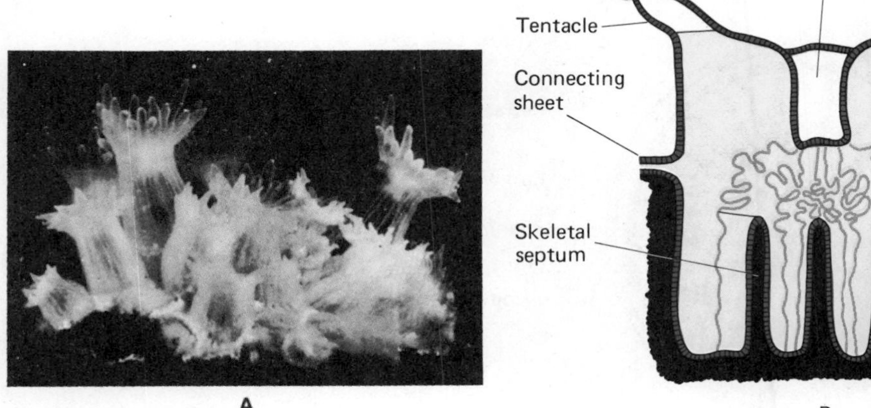

FIGURE 22.18 *A,* Polyps of the star coral. *B,* A coral polyp within its skeletal cup. (*A* Courtesy of the American Museum of Natural History; *B* after Hyman, L. H.: *The Invertebrates,* Vol. I. New York, McGraw-Hill Book Co., 1940.)

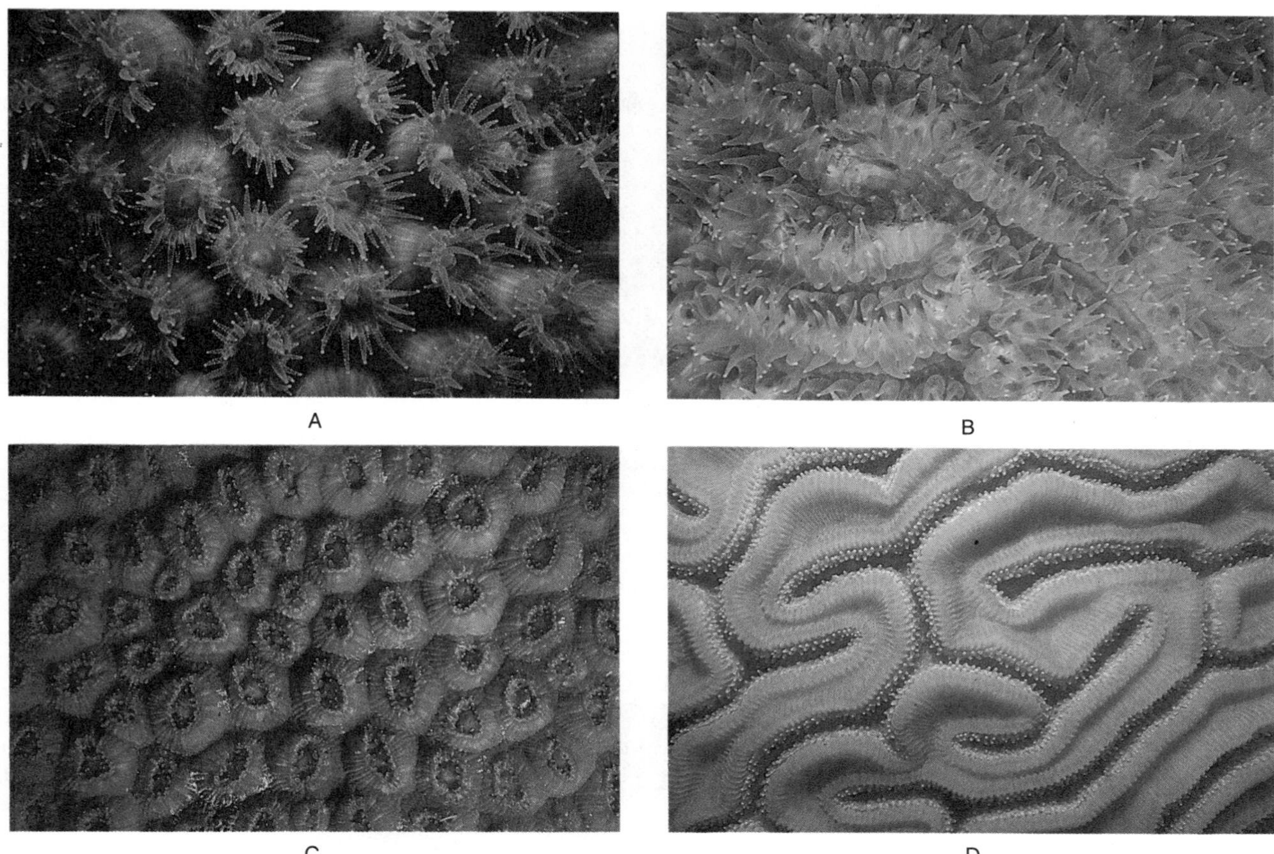

FIGURE 22.19 Photographs of two West-Indian scleractinian corals, *Montastrea cavernosa,* a platelike species, and *Diploria strigosa,* a brain coral, in which the polyps are arranged in rows. The polyps are expanded and feeding at night (*a* and *b*) and are contracted during the day (*c* and *d*). In the contracted states, the underlying skeletal cups and troughs in which the polyps are located can be seen through the body tissues. The yellow-brown color of the living colonies is due to the presence of symbiotic algae *(Zooxanthellae).* (Photographs courtesy of Charles Seaborn.)

enteries, all complete, within the body of the polyp and only eight tentacles, which are **pinnate;** i.e., they have little side branches (Fig. 22.21). Octocorallian polyps are usually very small but are organized into colonies that may attain considerable size. The interconnection of the polyps of a colony is quite different from that of scleractinian coral. A common mass of tissue (**coenenchyme**), composed of mesoglea and gastrodermal tubes, unites the lower half of the colony. A skeleton of separate microscopic pieces and sometimes also an organic horny material may be present *within* the mesoglea.

A skeleton supports the colony and is especially important in large upright forms. A central horny skeletal core extends through the long rodlike colonies of sea whips and sea rods. The sea fan is similar to the sea whip, but its branches are all in one plane, often with lattice-like cross connections.

Coral Reefs. Coral reefs are tropical, shallow-water, calcareous formations that support a great diversity of marine plants and animals. Moreover, certain of these plants and animals secrete the calcium carbonate that composes the underlying reef formation.

Contemporary coral reefs, located in the Caribbean and in the tropical parts of the Indo-Pacific oceans, have been built chiefly by scleractinian corals. Coral reefs can be classified into three major types, depending upon their location. **Fringing reefs,** which are the most com-

FIGURE 22.20 Coral reef in the Philippines. The brown "bush" in foreground on right is a gorgonian octocoral (sea rod). (Courtesy of Charles Seaborn.)

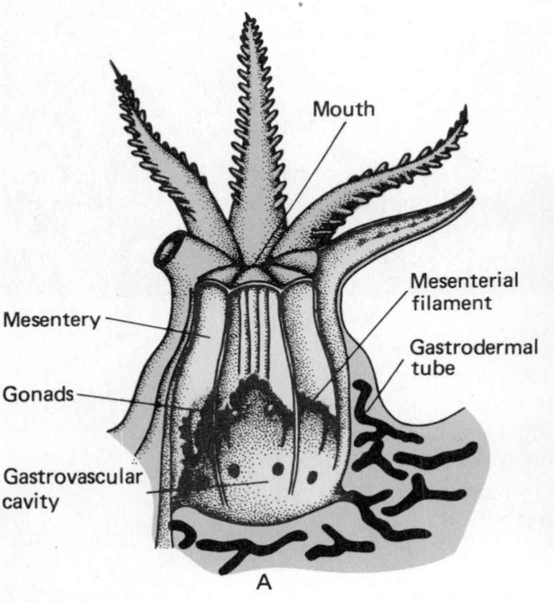

FIGURE 22.21 Octocorallian corals. *A,* Structure of a single polyp. Only four of the eight tentacles are shown. *B,* Gorgonian corals (sea rods) and brain coral on a Bermuda patch reef. The gorgonians are the erect plantlike colonies. *C,* A sea pansy. (*A* modified after F. M. Bayer in *The Treatise on Invertebrate Paleontology,* courtesy of the Geological Society of America and University of Kansas, 1956.)

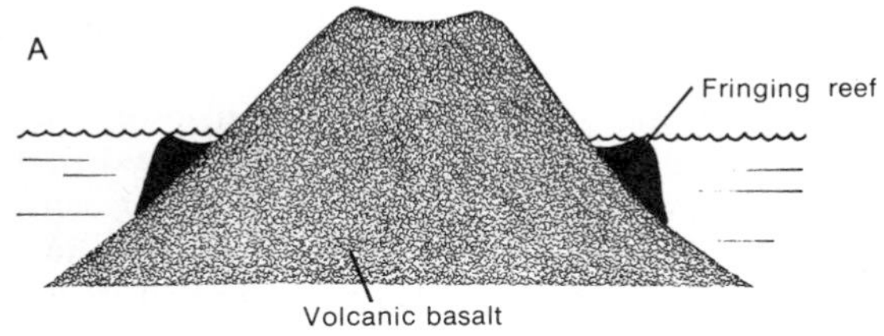

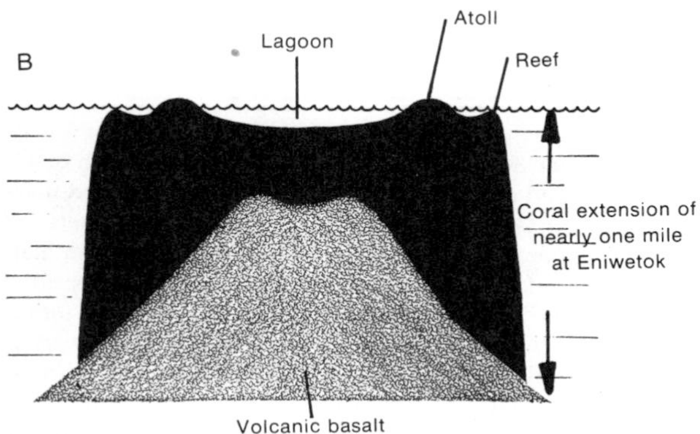

FIGURE 22.22 Formation of an atoll. *A,* Fringing reef around an emergent volcano. *B,* Continuous deposition of coral as volcanic cone subsides leads to the formation of a great coralline cap; emergent part of cap is atoll.

mon type of reef, are located adjacent to the shores of islands or continental coasts (Fig. 22.23). **Barrier reefs** parallel the coast but are separated from the shore by a lagoon of varying width. The best known is the Great Barrier Reef, which parallels the northeast coast of Australia for over 1000 miles. **Atolls** lie above old submerged volcanoes and are more or less circular reefs containing an interior lagoon. The South Pacific contains the greatest number of atolls. Small **patch reefs,** ranging from 3 to 50 m. across, commonly rise from the floor of lagoons of barrier reefs and atolls.

Most reefs have one side, the **reef front,** facing the open ocean (Fig. 22.23). Behind the reef front, which rises close to the surface, is a **reef flat** covered by only a meter or less of water. The greatest growth of large massive corals is on the reef front (Fig. 22.23). The shallow reef flat may contain pools, sand beds, and rocky surfaces and supports a very different fauna from that on the reef front.

The coralline rock that forms the underlying reef substratum is composed largely of chunks of dead coral, riddled and broken by boring animals and encrusted with tubes and skeletons of other organisms. These larger pieces of calcium carbonate become filled in with fine calcareous sediment, like mortar between bricks.

As the reef increases in thickness, there is a gradual compaction of the lower deposits. Core drillings of coral reefs have disclosed coralline deposits of great depths. On the Pacific atoll of Eniwetok, coral limestone extends downward for almost a mile before reaching basaltic rock. Since the deposition of coral by living colonies occurs only in the upper lighted zone, thick deposits of coral can only be explained by fluctuations in sea level and by subsidence of the bedrock upon which the reef is resting. Both have occurred at Eniwetok. The geologic history of Eniwetok began during the early Cenozoic Period, when a fringing reef developed around an emergent volcanic cone (Fig. 22.22). This oceanic peak gradually subsided to below sea level. Coral deposition occurred at a rate equalling subsidence, and the original fringing reef was transformed into an atoll resting on vertical walls almost one mile in height.

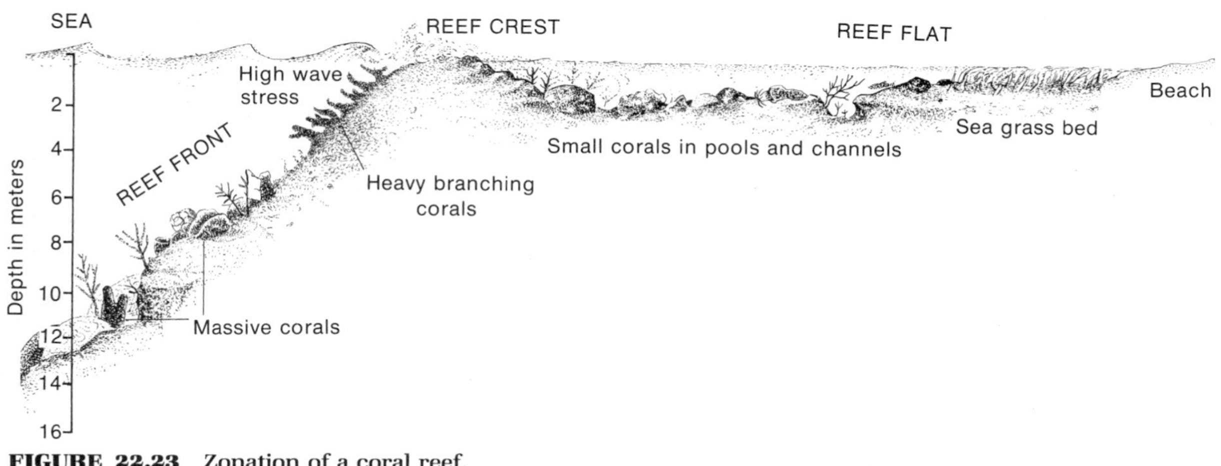

FIGURE 22.23 Zonation of a coral reef.

Fluctuating sea level periodically caused emergence and submergence. The rise in sea level from the last Pleistocene ice age low of 120 m. below the present level gave Eniwetok its present emergent form.

Subsidence of the Gulf Basin and fluctuating sea level have also contributed to reef formation in the Caribbean. Submarine caverns with stalactites in the Bahamas and elsewhere indicate a period of past emergence and subsequent submergence. Most modern reefs, such as those in the Caribbean, vary greatly in thickness and developed at various times during the approximately 120-m. rise in sea level following the last ice age 25,000 years ago.

Although existing coral reefs have been formed by scleractinian coral, reefs in the past have been produced by other organisms. Two extinct groups of anthozoans, the tabulate and rugose corals, formed great reefs during the Paleozoic. Reefs have also been formed in the past by algae that secrete calcium carbonate. Coralline algae are still prevalent today and are common members of reef communities, but there are no existing reefs composed primarily of living coralline algae.

Classificiation of Cnidarians

Phylum Cnidaria

Class Hydrozoa Common but inconspicuous forms, usually with both polypoid and medusoid stages. There are many colonial polypoid forms, called hydroids, that are surrounded by a chitinous skeleton.

Order Trachylina Medusoid hydrozoans lacking polypoid stage. This order contains perhaps the most primitive members of the class.

Order Hydroida Hydrozoans with a well-developed polypoid generation. Medusoid stage present or absent. The majority of hydrozoans belong to this order.

Order Siphonophora Pelagic hydrozoan colonies of polypoid and medusoid individuals. Colonies with float or swimming bells. Largely in warm seas. *Physalia* (Portuguese man-of-war). *Abyla, Agalma.*

Order Hydrocorallina Colonial polypoid hydrozoans that secrete a calcium carbonate skeleton. Mostly tropical. *Millepora,* stinging coral, or fire coral, a common member of coral reefs. *Stylaster.*

Class Scyphozoa Medusa stage is most conspicuous, with bell diameters commonly from 2 to 40 cm. Widespread marine forms, mostly free-swimming.

Order Stauromedusae, or **Lucernariida** Small sessile scyphozoans attached by a stalk on the aboral side of the trumpet-shaped body. Chiefly in cold littoral waters.

Order Cubomedusae* Sea wasps. Scyphomedusae with bells having four flattened sides and simple margins. Tropical and subtropical oceans. Rapid swimmers; many species are highly virulent and dangerous.

Order Coronatae Bell of medusa with a deep groove or constriction, the coronal groove, extending around the exumbrella. Many deep sea species.

Order Semaeostomae Scyphomedusae with bowl-shaped or saucer-shaped bells having scalloped margins. Gastrovascular cavity with radial canals or channels extending from central stomach to bell margin. Occur throughout the oceans of the world.

Order Rhizostomae Bell of medusa lacking tentacles. Oral arms of manubrium branched and bearing secondary mouth openings that lead into arm canals. Arm canals of manubrium pass into stomach. Original mouth lost through fusion of oral arms. Tropical and subtropical littoral scyphozoans.

Class Anthozoa Either solitary or colonial, with medusoid stage completely absent. Includes sea anemones and corals, plus many other forms.

Subclass Octocorallia Polyps with eight mesenteries and eight pinnate tentacles. All colonial.

Order Telestacea Lateral polyps on simple or branched stems. Skeleton of calcareous spicules.

Order Alcyonacea Soft corals. Coenenchyme forming a rubbery mass and colony having a massive, mushroom, or variously lobate growth form. Skeleton of separate calcareous spicules. Largely tropical.

Order Coenothecalia Contains only the Indo-Pacific blue coral, *Heliopora,* having a massive calcareous skeleton.

Order Gorgonacea Sea rods, sea whips, sea fans. Common tropical and subtropical octocorallians, having a largely upright growth form of branching rods. Skeleton consists of a central horny axial rod and calcareous spicules.

Order Pennatulacea Sea pens and sea pansies. Colony having a fleshy, flattened, or elongate body. Skeleton of calcareous spicules.

Subclass Zoantharia Polyps with more than eight tentacles and tentacles rarely pinnate. Solitary or colonial.

Order Zoanthidea Small mostly colonial anemone-like anthozoans having one siphonoglyph and no skeleton.

Order Actiniaria Sea anemones. Solitary anthozoans with no skeleton, with mesenteries in hexam-

*These medusoid cnidarians are now sometimes treated as a separate class.

erous cycles, and usually with two siphonoglyphs.

Order Scleractinia or **Madreporaria** Stony corals. Mostly colonial anthozoans secreting a heavy external calcareous skeleton. Sclerosepta arranged in hexamerous cycles. Many fossil species.

Order Rugosa, or **Tetracoralla** An extinct order of mostly solitary corals possessing a system of major and minor radiating sclerosepta. Cambrian to Permian.

Order Corallimorpharia Tentacles radially arranged. Resemble true corals but lack skeletons.

Order Ceriantharia Anemone-like anthozoans with greatly elongate bodies adapted for living in sand burrows. One siphonoglyph; mesenteries all complete.

Order Antipatharia Black or thorny corals. Gorgonian-like species with upright, plantlike colonies. Polyps arranged around an axial skeleton composed of a black thorny material and bearing thorns. Largely in deep water in tropics.

Subclass Tabulata Extinct colonial anthozoans with heavy calcareous skeletal tubes containing horizontal platforms, or tabulae, on which the polyps rested. Sclerosepta absent or poorly developed.

Summary

1. Members of the phylum Cnidaria are largely marine animals and include jellyfish, sea anemones, and corals. The familiar hydras are among the few freshwater species.

2. The body is radially symmetrical, with the mouth and surrounding tentacles located at one end of the radial axis. Some cnidarians are columnar in shape (polypoid form) and live attached to the bottom, with the oral end directed upward. Others are bowl-shaped (medusoid, or jellyfish, form) and are free-swimming, with the oral end directed downward.

3. The body wall is composed of two principal layers: an outer epidermis and an inner gastrodermis, separated by the mesoglea. The mesoglea may be an acellular membrane or a thick gelatinous layer with or without cells. The epidermis and gastrodermis are composed of different cell types; the explosive stinging cells called cnidocytes are unique to the phylum. Most cnidarians are carnivorous, and cnidocytes are important in prey capture. The mouth is the only opening into the digestive cavity, where digestion is both extracellular and intracellular.

4. The nervous system is commonly in the form of a net with receptor cells dispersed over the body surface. The statocysts and ocelli of jellyfish are the only cnidarian sense organs.

5. Many cnidarians have the sexes separate; some are hermaphroditic. The gonads are only aggregations of developing gametes, and there are no gonoducts. Fertilization is usually external, and development leads to a free-swimming planula larva.

6. Members of the class Hydrozoa are polypoid, medusoid, or exhibit both forms in their life history. The mesoglea is acellular in all members of the class. Most polypoid hydrozoans are colonial and polymorphic (hydroids) and usually possess an external chitinous skeleton.

7. Hydromedusae are very small and are the sexual, or gamete-producing, individuals. Some hydrozoans, considered to be primitive, have no polypoid stages, and the planula develops directly into a medusa; others, such as the hydras, are entirely polypoid and produce sperm and eggs. Hydroid species possess both polyps and medusae. The medusae may be freed or remain attached to the polyps from which they are budded. Attached medusae show all degrees of reduction.

8. The class Scyphozoa contains the large medusoid cnidarians commonly called jellyfish. The mesoglea is cellular, and the manubrium is divided into four long oral arms that hang down from the underside of the bell. The planula of scyphozoans develops into a small polypoid larva that buds off juvenile medusae.

9. Members of the large class Anthozoa are entirely polypoid. A tubular pharynx leads from the mouth into the gastrovascular cavity, which is radially partitioned by gastrodermal mesenteries. The mesoglea is cellular. Sea anemones are large solitary anthozoans lacking a secreted skeleton. Scleractinian corals are mostly colonial and secrete an external skeleton of calcium carbonate. Octocorals, which include the sea rods, sea fans, sea pens, and sea pansies, are colonial and possess only eight tentacles and mesenteries. The skeleton usually consists of internal calcareous spicules.

10. Coral reefs are tropical, calcareous platforms supporting an array of marine plants and animals, some of which produce the calcium carbonate that composes the platform. Reefs are only found in clear shallow water because reef corals contain symbiotic zooxanthellae that require light.

References and Selected Readings

Detailed accounts of cnidarians may be found in the references listed at the end of Chapter 19.

Colin, P. L.: *Caribbean Reef Invertebrates and Plants.* Neptune City, New Jersey, T. F. H. Publications, 1978. A field guide to the invertebrates and plants occurring on coral reefs of the Caribbean, the Bahamas, and Florida.

Goreau, T. F., N. I. Goreau, and T. J. Goreau: Corals and coral reefs. *Scientific American, 241*(Feb.):124 1979. A good brief account of the biology and geology of coral reefs.

Kaplan, E. H.: *A Field Guide to Coral Reefs of the Caribbean and Florida.* The Peterson Field Guide Series. Boston, Houghton Mifflin Co., 1982. An excellent guide to the common invertebrates and fishes of this region, including considerable ecological information about reefs.

Muscatine, L., and H. M. Lenhoff: *Coelenterate Biology.* New York, Academic Press, 1974. Review papers on many aspects of the biology of cnidarians.

23 The Flatworms

ORIENTATION

The flatworms and all the remaining members of the Animal Kingdom are either bilaterally symmetrical or, if radially symmetrical, have clearly evolved from bilateral ancestors. Bilaterality and cephalization (head development) are widespread animal features correlated with motility.

The chapter will begin by examining the free-living flatworms found in the sea and freshwater, and we will see how the dorsoventrally flattened shape of the body, which is a distinguishing characteristic of the phylum, is of advantage to these animals. We will then turn to the three classes of parasitic flatworms. We will study the ways in which their structure is correlated with their parasitic habit and the adaptations of their life cycle that increase the chances that some progeny of the parasite will eventually reach the host of the adult worm.

CHARACTERISTICS OF FLATWORMS

1. Members of the phylum Platyhelminthes include free-living marine or freshwater flatworms and parasitic flukes and tapeworms. The bodies are greatly flattened dorsoventrally.
2. The body surface of free living flatworms is covered by cilia, which are used in locomotion. Adult parasitic forms lack external cilia, and the body is covered by a specialized cuticle.
3. Free-living flatworms are carnivores or scavengers. The mouth is the only opening into the digestive tract. A digestive tract is absent in tapeworms.
4. There is no internal transport system, protonephridia are present, and the nervous system is composed of a varying number of longitudinal cords.
5. Flatworms are hermaphroditic. Sperm transfer is reciprocal, fertilization is internal, and in most species development is direct.

Many zoologists believe that members of the phylum **Platyhelminthes** are probably the most primitive of all the bilateral animals. The phylum is composed of three classes. The class **Turbellaria** contains the **free-living flatworms.** The class **Trematoda,** the **flukes,** and the class **Cestoda,** the **tapeworms,** are both entirely parasitic and probably evolved independently from turbellarians. The distinguishing characteristics of the phylum will be introduced with the following discussion of the free-living species.

Class Turbellaria

Members of the class Turbellaria, the free-living flatworms, occur both in the sea and in fresh water, and a few species are terrestrial in humid forests. The aquatic turbellarians are bottom dwellers, living within algal masses and beneath stones and other objects. The common familiar freshwater **planarians** inhabit the undersurface of stones and fallen leaves in springs, streams, and lakes. There are also marine flatworms that live with other animals. *Bdelloura,* for example, lives on the gills of horseshoe crabs.

Many tiny marine turbellarians live in the spaces between sand grains. Animals that live in these interstitial spaces compose the **interstitial fauna.** Virtually every phylum of animal has some representatives adapted for this habitat.

Turbellarians range in size from microscopic species to ones more than 60 cm. in length (land planarians), but most are less than 10 mm. long. The body tends to be dorsoventrally flattened, the condition to which the name Platyhelminthes (Gr. *platys,* flat + *helminthos,* worm) refers, and the larger the species, the more pronounced is the flattened shape (Fig. 23.1). The anterior end often bears eyes and in some species tentacles (Fig. 23.1*D*). Planarians commonly possess lateral projections that give the anterior end a triangular shape (Figs. 23.1 and 23.2).

A ciliated epidermal layer covers the body, although in the planarians only the ventral surface is ciliated. Beneath the epidermis is a muscle layer of circular, diagonal, and longitudinal fibers (Fig. 23.2). A network of loosely connected cells, called **parenchyma,** surrounds the gut and other organs and fills the interior of the body. Flatworms are therefore said to possess a solid, or **acoelomate** (*a-,* without + Gr. *koilos,* hollow), body structure, there being no cavity between the body wall and the internal organs.

Locomotion. Very small flatworms swim or crawl about bottom debris by ciliary propulsion. Contractions of the muscle layer permit turning, twisting, and folding of the body. The movement of larger turbellarians over 5 mm. in length also involves delicate undulatory waves of muscle contraction. The dorsoventral flattening of the body is probably in part an adaptation for locomotion. With increased size, a flattened shape provides a large surface area upon which the body can be carried. Glands in the epidermis and underlying tissue secrete a mucous film over which the animal glides.

Many flatworms possess special duogland systems that provide for temporary anchorage. One gland of the pair secretes an adhesive substance; the other provides a secretion that breaks the adhesive bond, releasing the animal from anchorage.

Nutrition. Turbellarians are carnivorous, feeding upon other small invertebrates. The mouth is located along the midventral line. In most species, the gut is differentiated into a **pharynx** adapted for ingestion and an **intestinal sac** that functions in digestion and absorption. No anus is present, and wastes are ejected through the mouth, as in cnidarians.

In small turbellarians the intestine is a simple unbranched sac (Fig. 23.3). Most of the larger flatworms have a highly branched intestinal sac and a tubular pharynx (Fig. 23.2). The tubular pharynx is muscular and, except for the end attached to the intestinal sac, lies free within a pharyngeal cavity. The cavity opens to the exterior through the mouth. When the animal feeds, the pharyngeal tube is projected out of the mouth (Fig. 23.2).

During feeding, flatworms crawl upon their prey, pinning them down and enveloping them with mucus. The prey is often swallowed whole. However, planarians, as well as many other species with a tubular pharynx, extend the pharynx and insert it into the body of the prey or into dead animal matter with the aid of proteolytic enzymes secreted by pharyngeal glands. Fragments of the body of the prey are then pumped into the intestine. The many divisions of the gut, which extend throughout much of the body, greatly increase the surface area available for digestion and absorption.

Digestion occurs in much the same way as in cnidarians. An initial extracellular digestion within the lumen of the intestinal sac is followed by intracellular digestion within the cells of the intestinal wall.

Gas Exchange, Internal Transport, and Water Balance. Gas exchange takes place across the general body surface, which, because of its flattened shape, is sufficiently great to meet oxygen demands. Moreover, the small vertical distances resulting from the flattened shape greatly facilitate internal transport by diffusion. Diffusion is also sufficient for the movement of food ma-

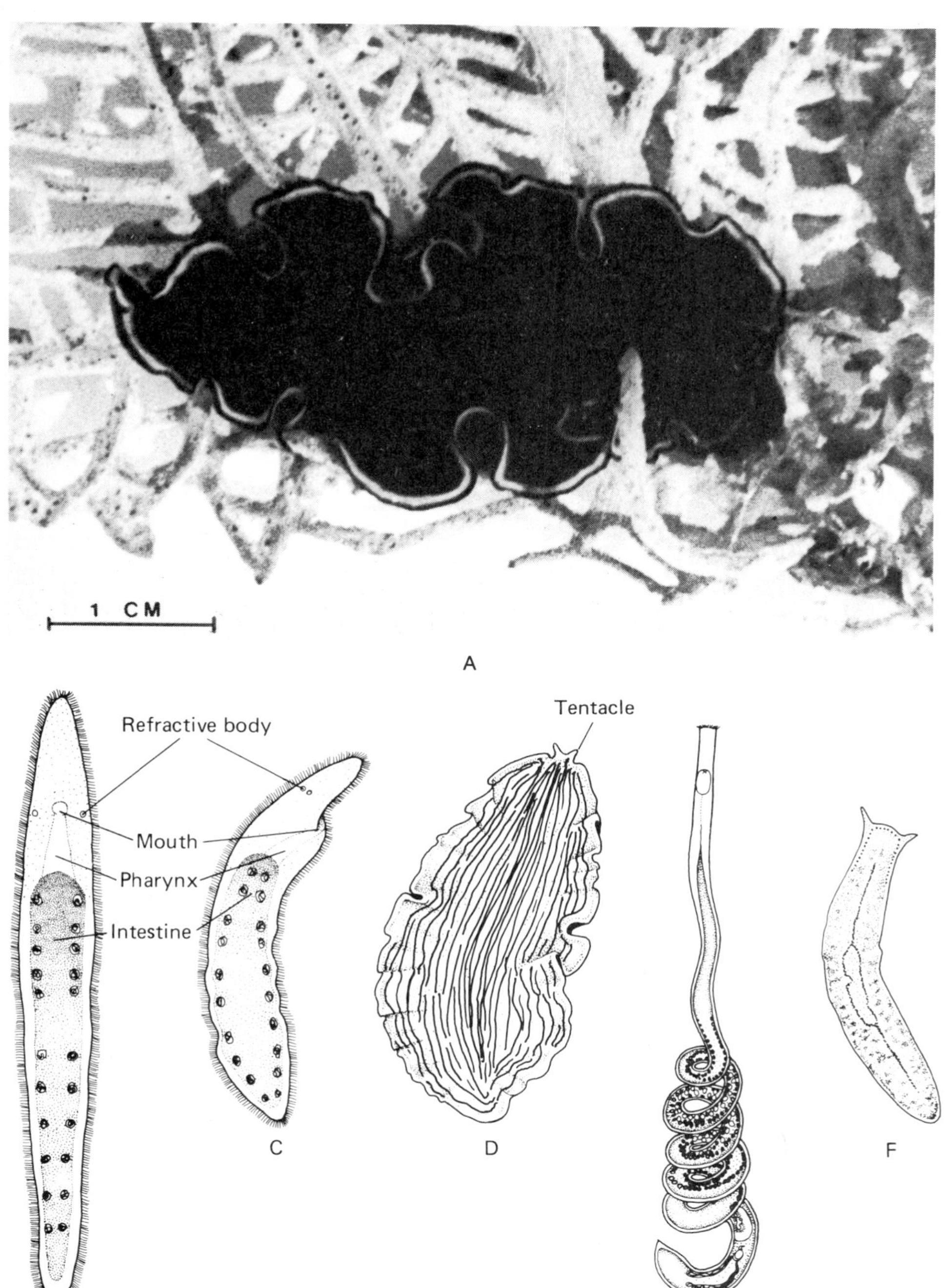

FIGURE 23.1 *A,* A large West Indian polyclad crawling on a sea fan. The light marginal band is orange. (By Betty M. Barnes.) *B* and *C,* The catenulid *Stenostomum,* a common microscopic freshwater turbellarian. *B,* Dorsal view. *C,* Contracted specimen with anterior end turned in lateral view. A simple ciliated pharynx connects the anterior ventral mouth and the intestine, which is an elongated sac. *D, Prostheceraeus,* a marine polyclad. (Drawn from a photograph by D. P. Wilson.) *E, Nematoplana,* a marine interstitial turbellarian. (From Rieger, R., and J. Ott: *Vie et Milieu,* Supplement 22, 1971.) *F, Polycelis,* a freshwater planarian. (After Steinmann and Bresslau.)

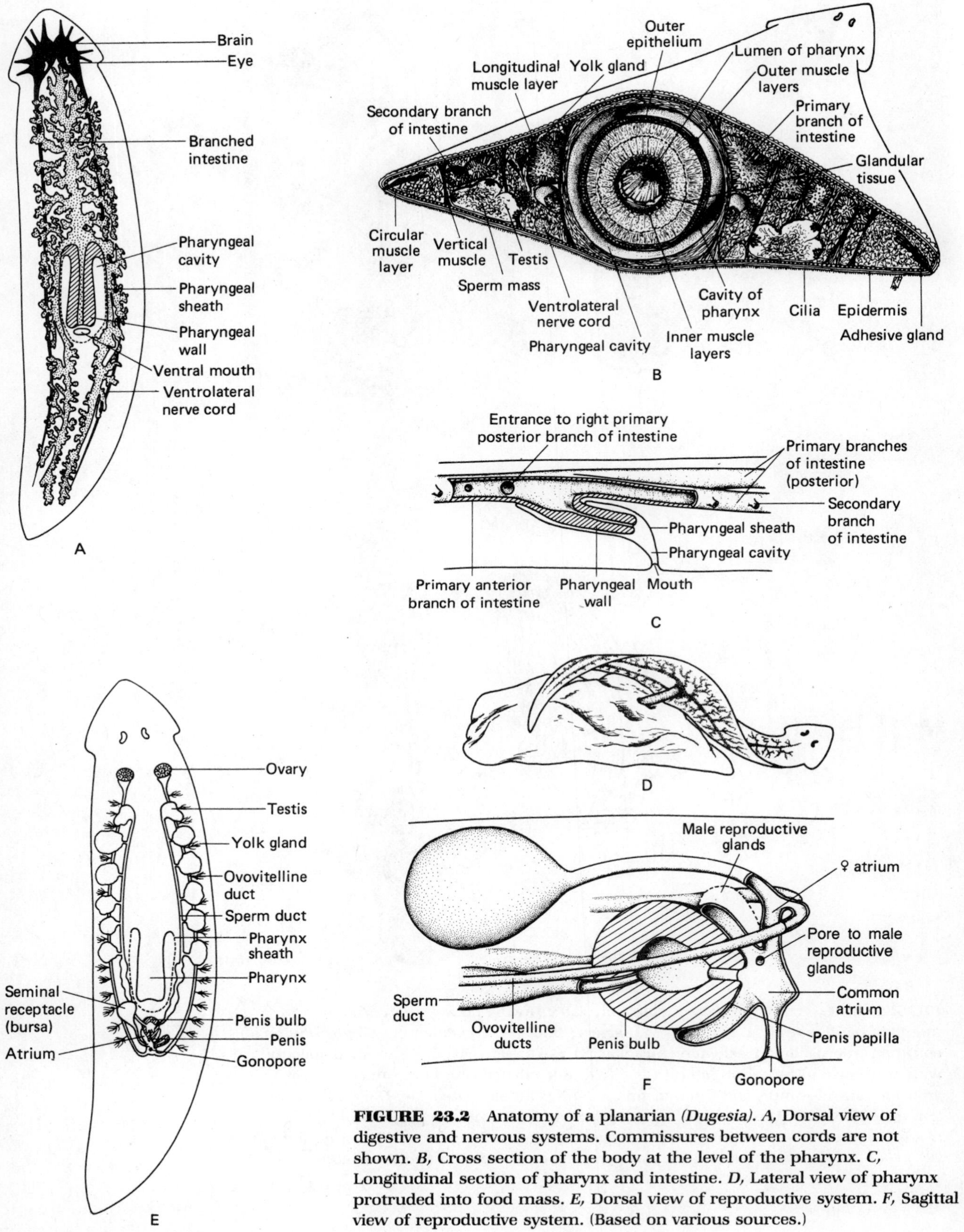

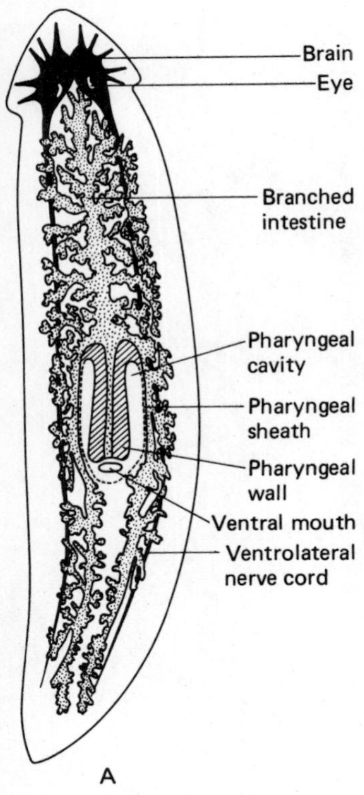

Outer epithelium
Lumen of pharynx
Longitudinal muscle layer
Yolk gland
Outer muscle layers
Secondary branch of intestine
Primary branch of intestine
Glandular tissue
Circular muscle layer
Vertical muscle
Testis
Sperm mass
Ventrolateral nerve cord
Pharyngeal cavity
Cavity of pharynx
Inner muscle layers
Cilia
Epidermis
Adhesive gland
B

Entrance to right primary posterior branch of intestine
Primary branches of intestine (posterior)
Secondary branch of intestine
Pharyngeal sheath
Pharyngeal cavity
Primary anterior branch of intestine
Pharyngeal wall
Mouth
C

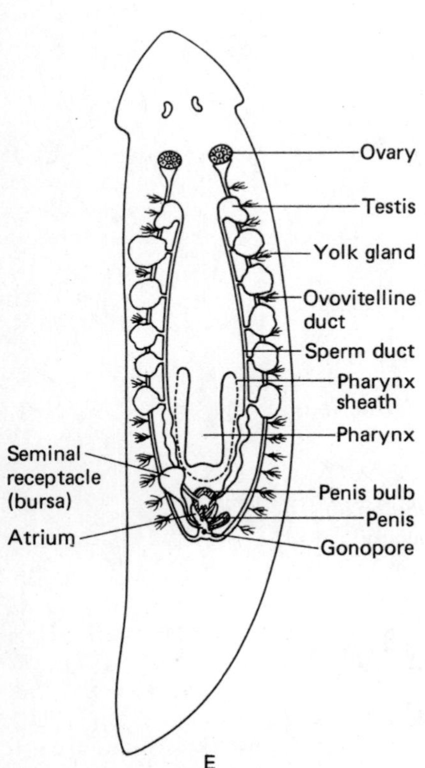

D

Male reproductive glands
♀ atrium
Pore to male reproductive glands
Common atrium
Sperm duct
Ovovitelline ducts
Penis bulb
Penis papilla
Gonopore
F

FIGURE 23.2 Anatomy of a planarian *(Dugesia)*. *A,* Dorsal view of digestive and nervous systems. Commissures between cords are not shown. *B,* Cross section of the body at the level of the pharynx. *C,* Longitudinal section of pharynx and intestine. *D,* Lateral view of pharynx protruded into food mass. *E,* Dorsal view of reproductive system. *F,* Sagittal view of reproductive system. (Based on various sources.)

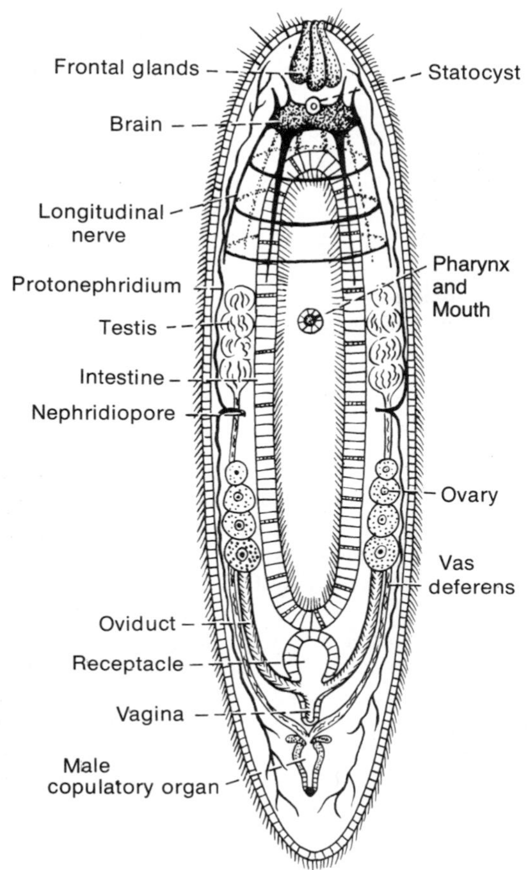

FIGURE 23.3 Dorsal view of a generalized rhabdocoel turbellarian. (After Ax, P. in Dougherty, E. C. (Ed.): *The Lower Metazoa.* University of California Press, 1963.)

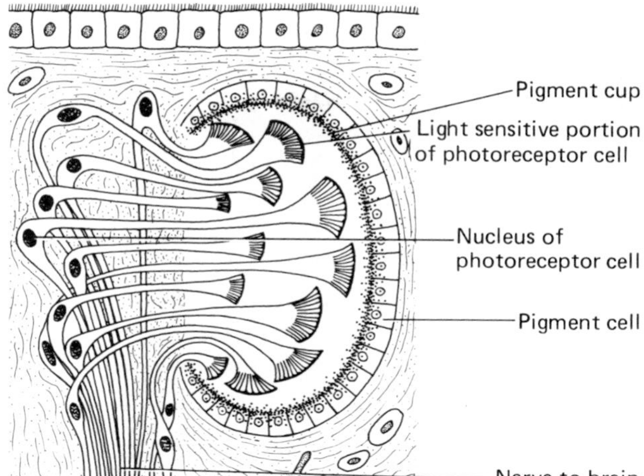

FIGURE 23.4 Diagrammatic section through the eye of a planarian.

terials, for in the large flatworms the gut branches are so extensive that no tissue is very far from some intestinal cells.

Most flatworms possess a system of tubules, called **protonephridia** (Gr. *prōto,* primitive + *nephros,* kidney), described on page 198. The nitrogenous wastes of flatworms, which are chiefly ammonia, are removed by general diffusion across the body surface, and the protonephridial system appears to function in the elimination of other kinds of metabolites and in water balance.

Nervous System and Sense Organs. In the majority of flatworms, neurons are organized in longitudinal bundles or nerve cords that lie just below the epidermis, but there are few ganglia. The most primitive arrangement of nerves in flatworms appears to be one in which there are four or five pairs of longitudinal cords radially arranged around the body (Fig. 23.3). The cords merge at the anterior end, and the slight enlargement is sometimes called a brain.

In the majority of turbellarians, the radial arrangement of nerve cords has been lost through the predominance of certain pairs and the loss of others. Planarians have a very large ventral pair and a reduced lateral pair (Fig. 23.2). The other pairs are absent.

A statocyst is present in many flatworms. Flatworms may have two or more simple eyes (Figs. 23.1 and 23.2) that function in detecting light intensity and direction. Usually, the eye has a cuplike form resulting from the invagination of the pigment cells (Fig. 23.4). In some animals the opening of the retinal cup becomes filled in and covered with a lens-cornea, but lenses are not present in most species.

Flatworms avoid bright light and the response is probably an adaptation to remain under cover.

Regeneration and Reproduction. Many flatworms are capable of asexual reproduction, and this ability is closely correlated with the ability to regenerate. The most common method of asexual reproduction is by **transverse fission.** In many freshwater planarians, for example, a fission plane forms behind the pharynx, and during movement the animal breaks in two. Following fission, the two halves regenerate the missing parts. In a number of small flatworms, such as *Stenostomum* and *Microstomum,* fission planes and regeneration proceed more rapidly than the separation of the parts, and chains of individuals are formed as a result (Fig. 13.1*B*).

Parenchyma is the principal source of new cells for regeneration. As in cnidarians, the regenerative parts retain their original polarity (Fig. 23.5). A piece taken from

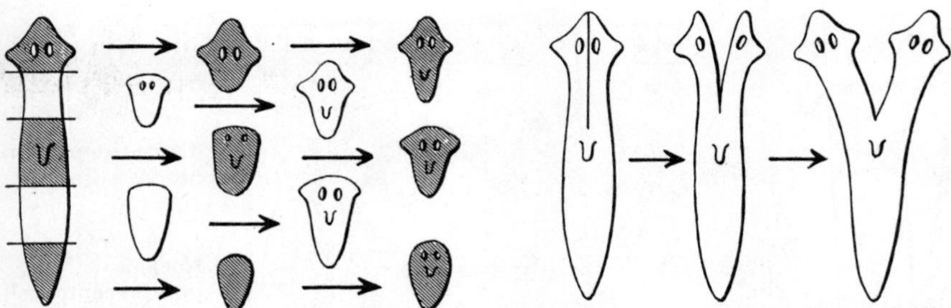

FIGURE 23.5 Polarity and regeneration in *Dugesia. Left,* Each of five pieces regenerates, but the rapidity with which the head develops depends upon the level of the piece. *Right,* A two-headed form produced by repeated splitting of the anterior end.

the middle of the body always regenerates a new anterior end at the severed anterior surface and a new posterior end at the posterior surface. There is also a lateral polarity; thus, a two-headed planarian can be produced by cutting the anterior end longitudinally along the midline (Fig. 23.5). The rate of regeneration of pieces taken at different levels reflects a distinct metabolic gradient along the anterior-posterior axis. Anterior pieces regenerate more rapidly than do posterior pieces of equal size.

Most flatworms are simultaneous hermaphrodites but usually exhibit cross-fertilization. There is typically simultaneous and mutual copulation, the penis of each animal being received by the female system of the other animal (Fig. 23.6), and the sperm are stored for a period of time. The eggs are fertilized internally and deposited on the bottom in jelly masses or in cocoons (Fig. 23.6).

A representative male system may contain one testis or many pairs of testes (Figs. 23.2 and 23.3). A small duct connects each testis with a main sperm duct that extends along each side of the body. The sperm ducts join together to form an ejaculation duct, which exits through a penis. Commonly, the penis is located within a chamber, the genital atrium, which may also contain the terminal part of the female system. The atrium opens to the outside through a gonopore.

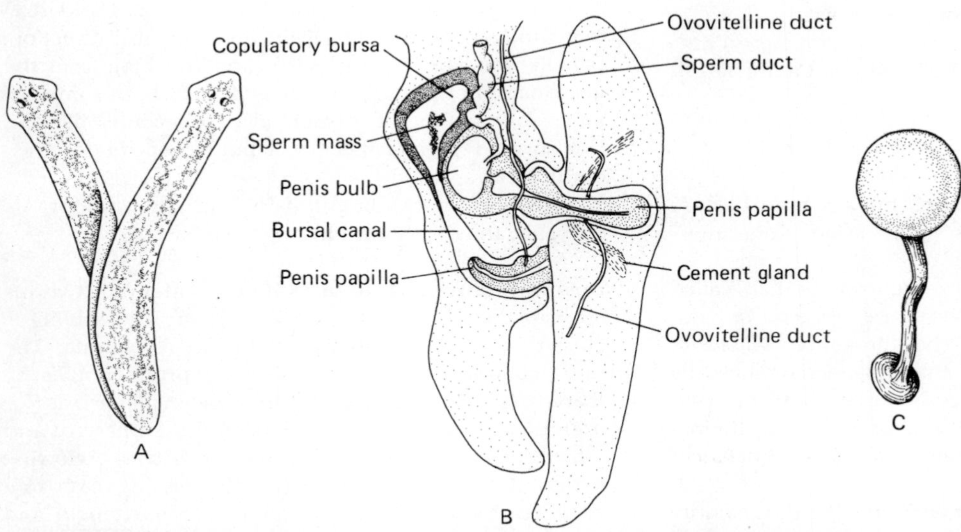

FIGURE 23.6 Reproduction in planarian flatworms. *A,* Copulatory position. *B,* Section through a copulating pair. *C,* An attached egg cocoon. (*B* after Burr from Hyman, L. H.: *The Invertebrates,* Vol. II. New York, McGraw-Hill Book Co., 1951.)

The female system (Figs. 23.2 and 23.3) contains either a single ovary or many pairs of ovaries, but only a single pair of oviducts is present. In many turbellarians yolk glands are located along the length of the oviduct. Yolk cells are released as the eggs travel down the oviduct, and the eggs, when deposited, are surrounded by yolk. The deposition of yolk outside an egg cell (**ectolecithal**) is an unusual condition. In most animals the yolk is contained within the egg cytoplasm (**entolecithal**). A copulatory sac that stores sperm, a vagina or atrium, and glands for the production of egg envelopes are typically present. There may be a separate female gonopore, or there may be a common genital atrium and gonopore for both systems.

Some freshwater turbellarians produce two kinds of eggs. Summer eggs have a thin shell and hatch within a short period of time; winter or dormant eggs have a thick resistant shell and are capable of withstanding cold and desiccation. Having two types of eggs is an adaptation shared with many other freshwater animals.

Spiral cleavage is present in turbellarians with entolecithal eggs, and some marine species have larvae.

Symbiosis and Parasitism

Many organisms depend for their existence upon an intimate physical relationship with other organisms. Such a relationship is known as **symbiosis** (Gr. *symbiosis,* a living together). The larger organism is termed the host and the smaller, the **symbiote.** The symbiote always derives some benefit from the relationship, but the consequence to the host varies. A symbiotic relationship is usually **obligate** for the symbiote, meaning that the symbiote cannot live without the host. Where the symbiote can exist either with and without the host, the relationship is said to be **facultative.** When both the symbiote and the host benefit, the relationship is called **mutualism.** Algae live within the amoebocytes of freshwater sponges and within the gastrodermal cells of scleractinian corals. The metabolism of each species is enhanced by the presence of the symbiotic partner.

In **commensalism,** the host is neither benefited nor harmed. The symbiote, or commensal, is protected by the host or utilizes excess food of the host. The clown fish living among the tentacles of sea anemones is a commensal. The sea anemones can exist without their symbiotes, but the fish cannot exist alone.

The most common symbiotic relationship is **parasitism.** Here the host is harmed by the presence of the symbiote, or parasite. Parasites that live on the outside of their host are said to be **ectoparasitic;** those on the inside, **endoparasitic.** Most parasites can utilize several closely related or ecologically similar species as hosts. Some parasites require two hosts to complete their life cycles, in which case the host for the larval or developmental stage is termed the **intermediate host;** the host for the adult stage is termed the **primary host.**

The usual benefit derived by a parasite from its relationship with the host is nutritive. The parasite feeds upon the host's tissues or body fluids or utilizes the food ingested by the host. When the parasite is enclosed by some part of the host's body, there may be other advantages, such as protection. The relationship, however, poses problems and is not without cost. The primary problem of parasites is that of reaching and penetrating new hosts.

From these generalizations, many of the adaptations that are encountered in parasites can be predicted. Ectoparasites that feed infrequently may have a part of the gut modified for storage. On the other hand, some endoparasites utilize digested food of the host and have lost the gut entirely. The problems of penetration and attachment to the host have resulted in the evolution of a variety of structures, such as suckers, hooks, and teeth. Enzymes may facilitate penetration in some species. The most common point of entrance for endoparasites is through the mouth of the host. The parasite may remain in the gut of the host or, in some species, break through the gut wall to reach other organs.

The problem of reaching new hosts has been met in most parasites through the production of enormous numbers of eggs or other developmental stages. The species is perpetuated if only a few individuals encounter the proper host and reach adulthood. The reproductive system of parasites is highly developed for the production of great numbers of gametes. There are some parasites in which most of the body is concerned with reproduction.

The structure of endoparasites usually reflects the less rigorous demands of the limited and uniform environment in which they live. Locomotor processes are reduced, and the sense organs found in free-living species are absent. The nervous system in turn is greatly reduced. Gas exchange organs are never present. Nevertheless, the internal environment of the host may be harsh. There may be no oxygen present, as in the gut, and the parasite must rely upon anaerobic respiration. The pH and osmotic pressure of the host organism may be very different from that of the parasite, and the parasite may be subjected to the digestive enzymes and the immune responses of the host.

A "good" parasite doesn't kill its host, for a dead host results in a dead parasite. A host may carry a small population of parasites without any serious consequences.

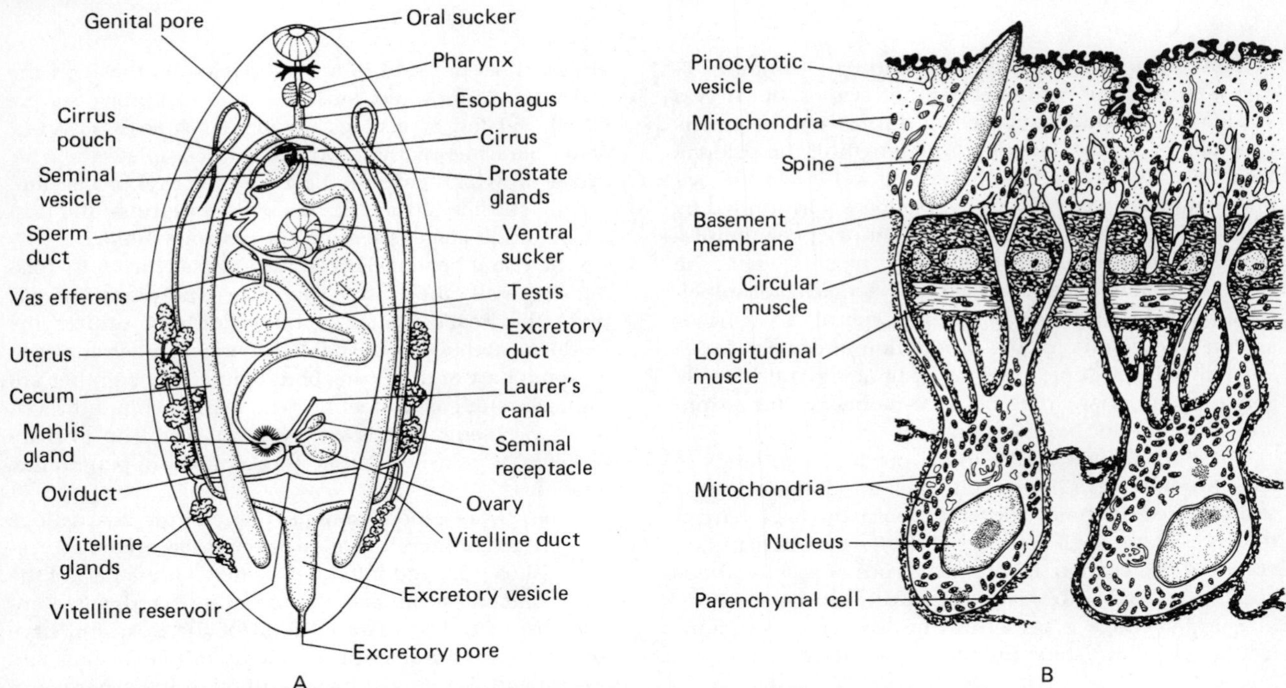

FIGURE 23.7 *A,* Structure of a generalized trematode. *B,* Section through the body wall of the sheep liver fluke, *Fasciola hepatica.* (*A,* after Chandler, A. C., and C. P. Read: *Introduction to Parasitology.* 10th ed. New York, John Wiley and Sons, 1961. *B* after Threadgold from Smyth, J. D.: *Introduction to Animal Parasitology.* New York, John Wiley and Sons, 1976.)

Where the stress of a parasitic infection is manifested in the host, the condition is recognized as a disease. Parasitic disease may be caused by a number of processes of the parasite. Cells and tissues may be destroyed. Blood vessels, ducts, or the gut may be clogged. The host may be robbed of food. The parasite may produce wastes or substances that have a toxic or allergic effect on the host.

The parasitic flatworms comprise three classes: the Monogenea and the Trematoda, which contain the flukes, and the Cestoda, which contains the tapeworms. These three groups of flatworm parasites, along with the parasitic roundworms, include the majority of parasitic worms of great economic and medical significance. The flukes and tapeworms evolved independently from the free-living turbellarians.

Flukes

The classes Monogenea and Trematoda, to which the flukes belong, contain over 8000 species of ectoparasites and endoparasites. The majority are parasites of vertebrates, especially fish, but immature stages are harbored by invertebrates.

Flukes are flattened and usually oval in shape (Figs. 23.7, 23.8, and 23.9), the majority being not more than a few centimeters long. The body is covered by a nonciliated cytoplasmic syncytium, the **tegument,** the nuclei of which are sunken into the parenchyma (Fig. 23.7). The mouth is located at the anterior end. Adhesive suckers are usually present around the mouth and may also be present midventrally. The monogenetic flukes possess a large posterior attachment organ, called an **opisthaptor,** provided with various structures, such as suckers and hooks. The digestive tract is composed of a muscular pumping pharynx behind the mouth, a short esophagus, and usually two long blind intestinal sacs (Fig. 23.7). Trematodes feed on cell fragments, mucus, tissue fluids, and blood of the host. Protonephridia are present, and the nervous system is similar to that of turbellarians, except that sensory structures are absent in endoparasitic forms.

Mutual copulation is the rule, and the reproductive system is adapted for the production of a tremendous number of eggs. Egg production has been estimated to be 10,000 to 100,000 times greater in trematodes than in the free-living turbellarians.

Life Cycles. Eggs are passed out of the host and hatch as mobile aquatic ciliated **miracidium larvae** (Gr.

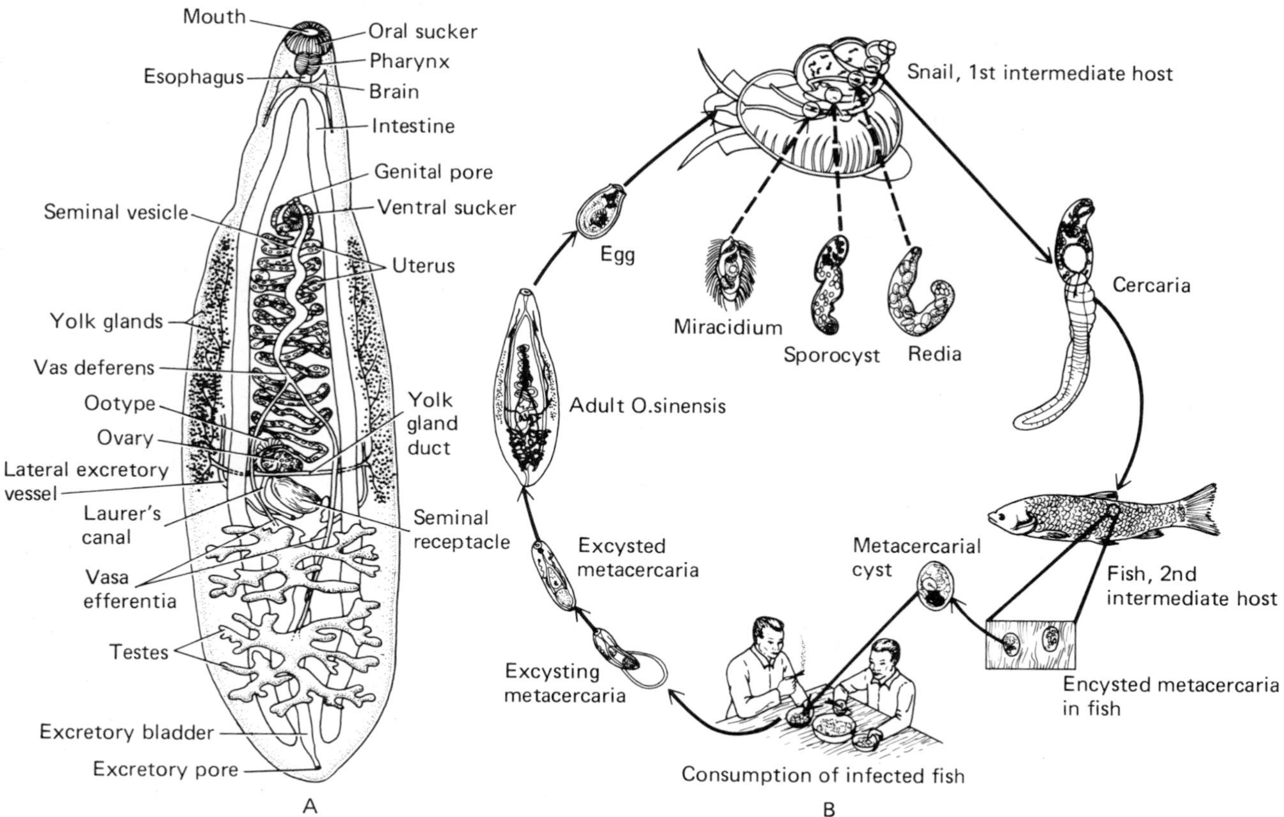

FIGURE 23.8 The Chinese liver fluke, *Opisthorchis sinensis. A,* Dorsal view of adult worm; *B,* life cycle. (*A* after Brown from Noble and Noble; *B* after Yoshimura from Noble and Noble.)

meirakion, stripling), which provide the means of reaching a new host, or the miracidium hatches from an egg ingested by the host (Fig. 23.8). In the class Monogenea (Gr. *monos,* single + *genea,* race), which are mostly ectoparasites of fish, the miracidium (called an **oncomiracidium** in this class) develops directly into the adult after contact with a host. In the class Trematoda, to which most flukes belong, development requires at least one intermediate host (digenean flukes). The free-swimming miracidium enters a molluscan host, usually a snail, and develops into a **sporocyst.** Within the host the sporocyst gives rise to a number of **redia,** which migrate to the host's digestive gland or the gonad. Redia may produce more redia or develop into a tailed **cercaria larva** (Fig. 23.8). The free-swimming cercaria (Gr. *kerkos,* tail) passes from the first intermediate to a second intermediate host or becomes attached to aquatic vegetation or sticks and stones, where it develops into a quiescent stage, called a **metacercaria.** If eaten by the primary host, the metacercaria is freed from the tissues of the intermediate host and develops into an adult. There are many variations of this generalized outline of the life cycle.

Neobenedenia melleni. This monogenean ectoparasite grazes on the skin and eyes of a number of different marine fish. The worm can cause blindness and serious damage to the host's integument. The eggs, when shed, fall to the bottom. A larva, an oncomiracidium having a precocious adhesive organ, escapes from the egg, and if a host fish is encountered, attaches to the skin. Following attachment and while feeding upon the host, the miracidium develops into an adult parasite.

Echinostoma revolutum. A great many trematodes infect the gut or gut derivatives of their primary host. Lungs, bile ducts, pancreatic ducts, and intestines are common sites. The life cycle of the common cosmopolitan *Echinostoma revolutum,* which infects the intestine of many aquatic birds and mammals, is a good illustration of the generalized life cycle described earlier. The eggs pass out with the host's feces and hatch in water. The free-swimming miracidium can penetrate and de-

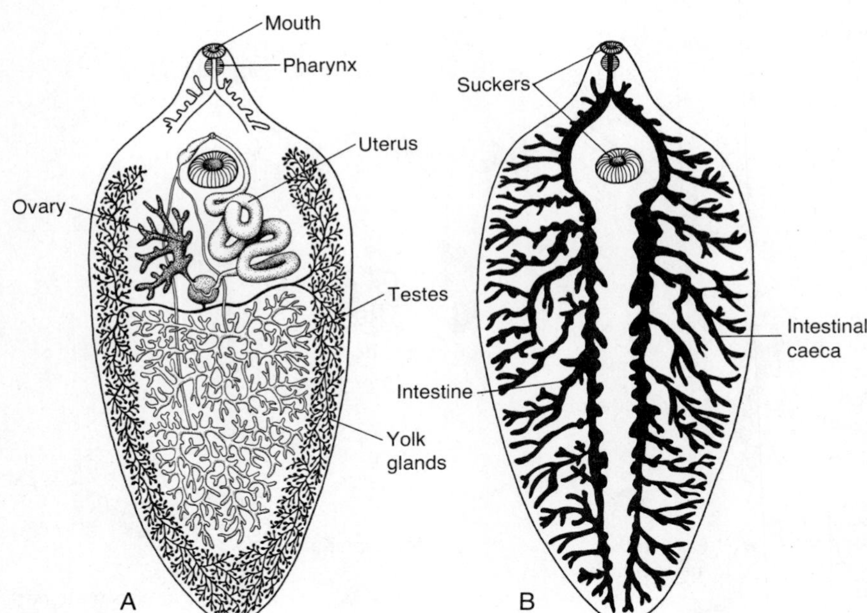

FIGURE 23.9 Structure of the sheep liver fluke, *Fasciola hepatica. A,* Reproductive system; *B,* digestive system.

velop within species of a number of genera of common snails. A cercaria larva leaves this snail and penetrates another snail, mussel, planarian, tadpole, or fish, where it undergoes further development. If the second intermediate host is ingested by the primary host, the immature worm is liberated and develops into an adult.

Opisthorchis sinensis. This species, commonly called the Chinese liver fluke, lives in the bile ducts of humans, dogs, cats, and pigs (Fig. 23.8). The intermediate hosts for the miracidium and cercaria are a snail and a fish, respectively. Human infestation has been common in the Orient because human feces are used to fertilize ponds, increasing the production of algae and water plants upon which the fish graze. The fish produced in these ponds are eaten raw, in part by preference and in part because of limited fuel. A few worms cause no disease symptoms, but several hundred can cause destruction of liver tissue, clogging of ducts, formation of bile stones, and hypertrophy of the liver. Over 20 million people, from Vietnam to Japan, are believed to be infected.

Fasciola hepatica. Known as the sheep liver fluke, this species reaches 3 cm. in length and is one of the largest trematodes. It is also an inhabitant of its host's bile ducts, and humans, cattle, pigs, rodents, and other mammals may be hosts in addition to sheep (Fig. 23.9). The life cycle is similar to that of the Chinese liver fluke, but a snail is the only intermediate host. The cercariae leave the snail and encyst as metacercariae on vegetation along the edges of ponds and streams, where they are eaten by grazing sheep or other mammalian hosts.

Class Cestoda

About 3400 species of cestodes, or tapeworms, have been described. All are endoparasites, and the majority are adapted for living in the gut of vertebrates.

The body form of adult tapeworms is quite different from that of turbellarians and flukes. The anterior end consists of a knoblike **scolex** containing suckers and, in some, hooks that anchor the worm to the host (Fig. 23.10). A narrow neck region connects the scolex to the **strobila,** which makes up the greater part of the tapeworm body. The strobila is composed of flattened sections, called **proglottids,** arranged in a linear series. New proglottids are continually being formed in the neck region and old ones detach at the end of the strobila. Tapeworms are generally long, and some specimens, with thousands of proglottids, reach lengths of over 75 feet.

The digestive system is completely absent in tapeworms, and digested food of the host is absorbed through the tegument, which is like that of trematodes but is covered with projections similar to microvilli. Longitudinal nerve cords and excretory ducts run the length of the strobila. Much of the body tissue is given over to the reproductive system, which is complete within each proglottid (Fig. 23.10). A common gonopore

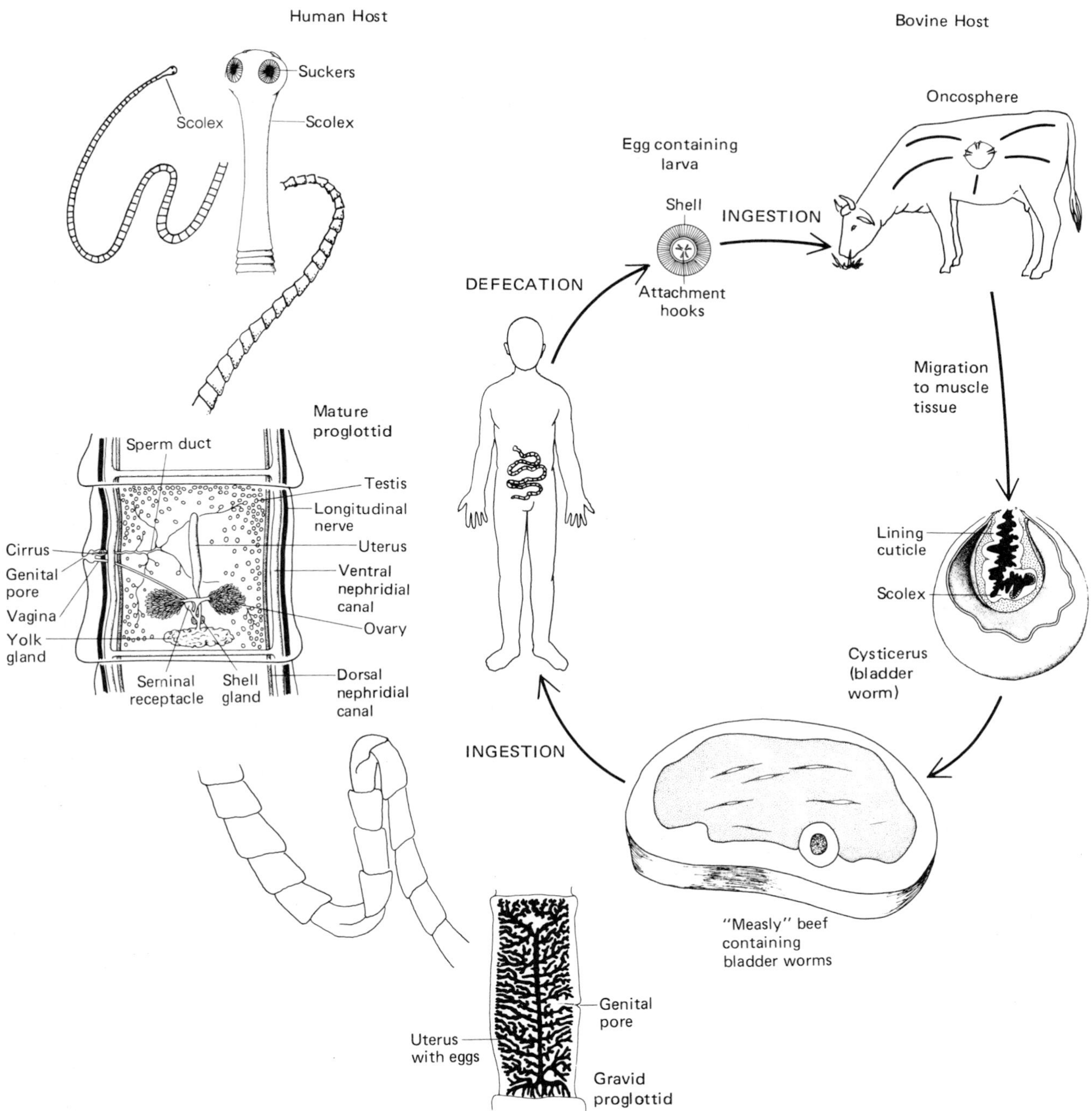

FIGURE 23.10 Life cycle of the beef tapeworm, *Taeniarhynchus saginatus.* (Adapted from various sources.)

is usually present on one edge of each proglottid. Mutual copulation between the proglottids of two different worms often occurs, but copulation between proglottids on the same strobila and self-fertilization within one proglottid is more usual.

Eggs containing embryos, which are surrounded by a shell, may be continually shed through the gonopore into the host's intestine, or they may be stored in a blind sac, called the uterus. In the latter case, terminal proglottids packed with eggs break away from the strobila and may rupture within the host's intestine, or they may leave with the feces and rupture later.

Life Cycles. One or more invertebrate or vertebrate intermediate hosts are required to complete the life cycle, which involves an **oncosphere** larva and one or more subsequent developmental stages.

Species of the family Taeniidae are among the best known tapeworms. *Taeniarhynchus saginatus,* the beef tapeworm, is one of the most common species in humans, where it lives in the intestine and frequently reaches a length of over 3 meters (Fig. 23.10). Proglottids containing embryonated eggs are eliminated through the anus, usually with feces. If an infected person defecates in a pasture, the eggs may be eaten by grazing cattle, sheep, or goats. On hatching within the intermediate host, a spherical oncosphere larva, bearing three pairs of hooks, bores into the intestinal wall, where it is picked up by the circulatory system and transported to striated muscle. Here the larva leaves the bloodstream and develops into a **cystocercus** stage (Fig. 23.10). The cystocercus, sometimes called a **bladder worm,** is an oval stage about 10 mm. in length, with the scolex invaginated into the interior. If raw or insufficiently cooked beef is ingested by humans, the cystocercus is freed, the scolex evaginates, and the larva develops into an adult worm within the gut.

A severe infection of adult tapeworms may cause diarrhea, weight loss, and reactions to the toxic wastes of the worm. The worms may be eliminated with drugs. Much more serious is cystocercus infection. Fortunately, the cystocercus stage of the beef tapeworm will not develop within humans, but this is not the case for the pork tapeworm, *Taenia solium,* and for the dog tapeworm, *Echinococcus granulosus.* The pork tapeworm has a life cycle rather like that of the beef tapeworm except that pigs rather than cattle are the intermediate hosts. The adult *Echinococcus,* which lives in the intestine of dogs, is minute, with only a few proglottids present at any one time. Many different mammals, including humans, can act as intermediate hosts, although herbivores are the most important in completing the life cycle. The cystocerci of the pork tapeworm develop in subcutaneous connective tissue and in the eye, brain, heart, and other organs. The bladder worm, or hydatid, of *Echinococcus* develops mostly in the lung or liver but can develop in many other sites as well. The bladder worms of both of these species can be very dangerous when growing in such places as the brain and can do much damage elsewhere. Hydatid cysts can reach a large size and contain a great volume of fluid (up to many liters!) which if released into the host can cause severe reactions. Bladder worm cysts can be removed only by surgery.

Copepods—tiny aquatic crustaceans—and aquatic annelids are the intermediate hosts for the tapeworms of many fish. Insects and mites are intermediate hosts for various tapeworms living in terrestrial vertebrates.

BOX 23.1

Related Phylum

The some 650 species comprising the phylum **Rhynchocoela** are called **nemerteans** or **ribbon worms.** They are mostly marine animals that burrow in sand or live in algae or beneath stones. Like flatworms, they are dorsoventrally flattened, covered by a ciliated epidermis and acoelomate. However, most nemerteans are longer and somewhat larger than flatworms. The burrowing ribbon-shaped species of *Cerebratulus* and *Lineus* may reach a length of 2 m., although most nemerteans are less than 20 cm. long. The distinguishing feature of nemerteans is the long proboscis used in capturing prey. The proboscis, which is usually unconnected to the gut, lies in a long, fluid-filled chamber called the rhynchocoel. When the fluid pressure of the rhynchocoel is elevated by the contractions of surrounding muscles, the proboscis is shot out, turning wrong side out in the process. The everted proboscis of many nemerteans bears a poisonous, dagger-like stylet at the tip. (Based on figures from Coe and Gibson.)

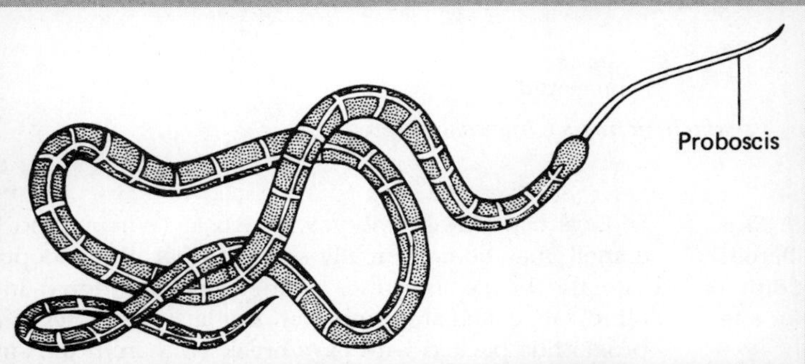

Classification of Flatworms

Phylum Platyhelminthes (Flatworms)

Class Turbellaria Free-living ciliated flatworms.

Of the nine orders of turbellarians, the following are commonly encountered and most easily described.

Order Acoela Small marine flatworms, usually measuring less than 2 mm. in length. Mouth and sometimes a simple pharynx present, but no digestive cavity. Protonephridia absent. No distinct gonads. Oviducts and yolk glands absent.

Order Macrostomida and Catenulida. Small marine and freshwater species having a nonmuscular pharynx and a simple saclike intestine.

Order Seriata, Suborder Tricladida Relatively large marine, freshwater, and terrestrial flatworms, known as planarians, ranging from 2 to more than 60 cm. in length in the terrestrial forms. Digestive system with a tubular muscular pharynx and a three-branched intestinal sac. Both mouth and pharynx located in the middle of the body. Eyes present in most species.

Order Polycladida Marine flatworms of moderate to large size, averaging from 2 to 20 mm. in length, with a greatly flattened and more or less oval shape. Intestinal sac elongate and centrally located with many highly branched diverticula. Eyes numerous. Yolk glands absent.

Class Monogenea Flukes, mostly ectoparasites of fish. Body elongate or oval and covered by a nonciliated tegument. Posterior end of worm bears a large adhesive organ (opisthaptor) provided with suckers and hooks. Digestive system present, with mouth at anterior end. Life cycle does not require an intermediate host.

Class Trematoda* Flukes, endoparasites, mostly of vertebrates. Body elongate or oval and covered by a nonciliated tegument. Anterior and more posteriorly located suckers present but no opisthaptor. Digestive system present, with mouth at anterior end. Life cycle requires one or more intermediate hosts.

Class Cestoda Tapeworms. Endoparasitic flatworms lacking a digestive system. Ciliated epidermis lacking; body covered by a cuticle.

*This class is commonly treated as an order (Digenea), and under this system the class Trematoda contains both groups of flukes.

Summary

1. Members of the phylum Platyhelminthes, called flatworms, are free-living or parasitic animals that are dorsoventrally flattened and have an acoelomate body construction. Most possess protonephridial tubules and are simultaneous hermaphrodites.

2. The class Turbellaria contains the free-living flatworms, found largely in the sea and in fresh water. The body is covered with cilia, which are used in locomotion. Turbellarians are carnivores or scavengers. The mouth is the only opening into the digestive tract, which consists of a pharynx and an intestine. The intestine, which is the site of extracellular and intracellular digestion, is a simple sac in small species and a highly branched sac in larger forms.

3. Gas exchange and elimination of nitrogenous waste occur across the general body surface. Diffusion also provides for internal transport. Some flatworms possess a nerve net type of nervous system, but most have a varying number of longitudinal nerve cords. The principal sense organs are a statocyst and two to many simple eyes.

4. Turbellarians exhibit copulation, reciprocal sperm transfer, and internal fertilization. Primitively, some marine species have spiral cleavage, and a few have a planktonic larva. Most exhibit ectolecithal and direct development.

5. The parasitic flukes belong to two classes, the Monogenea and the Trematoda, the latter containing the majority of species. The body of flukes is covered by a nonciliated syncytium. Suckers or other holdfast organs are typically present.

6. The monogenean flukes (Monogenea) are largely ectoparasites of fish, and the life cycle usually involves a miracidium larva and only one host. The digenean flukes (Trematoda) are endoparasites of many different vertebrates, and the life cycle requires at least one intermediate host, commonly a mollusk. Flukes are the cause of numerous parasitic diseases of humans and domesticated animals.

7. The class Cestoda contains the endoparasitic tapeworms, most of which live in the gut of vertebrates. The body is composed of an anterior section (scolex) modified for attachment and a long string of segment-like proglottids produced by budding from the scolex. Each proglottid contains complete male and female reproductive systems. Tapeworms are gutless, and food is absorbed from the host across a nonciliated tegument.

8. To complete the life cycle tapeworms require one or more intermediate hosts, which may be invertebrates or vertebrates. The intermediate host usually obtains the parasite orally, and the primary host becomes infected by eating the intermediate host. Beef and pork tapeworms are the principal species causing human parasitic disease.

References and Selected Readings

Detailed accounts of flatworms may be found in the references listed at the end of Chapter 19. The works listed below deal specifically with the biology of animal parasites and commensals and the general biology of symbiosis.

Kearn, G. C.: Experiments on host-finding and host-specificity in the monogenean skin parasite *Entobdella soleae. Parasitology,* 57:585–605, 1967. An interesting study of the way in which the oncomiracidium larva of a monogenean parasite finds its flatfish host.

Noble, E. R., and G. A. Noble: *Parasitology.* 4th ed. Philadelphia, Lea & Febiger, 1976.

Schell, S. C.: *How to Know the Trematodes.* Dubuque, W. C. Brown Co., 1970.

Schmidt, G. D., and L. S. Roberts: *Foundations of Parasitology.* 2nd ed. St. Louis, C. V. Mosby Co., 1981.

Smyth, J. D.: *Introduction to Animal Parasitology.* 2nd ed. New York, John Wiley & Sons, 1977.

Whitfield, P. J.: *The Biology of Parasitism: An Introduction to the Study of Associating Organisms.* Baltimore, University Park Press, 1979. This volume deals with various aspects of the biology of parasitism rather than with a survey of parasitic groups.

24 Pseudo-coelomates

ORIENTATION

Pseudocoelomates is the name given to a number of phyla of small animals that possess a type of body cavity called a pseudocoel. In this chapter we will study three of these pseudocoelomate phyla. We will see what characteristics they share other than a pseudocoel, what characteristics might be correlated with the small size of most species, and the reasons for placing each in a separate phylum. Finally, we will examine the parasitic pseudocoelomates called roundworms, some of which are of great economic and medical importance.

CHARACTERISTICS OF THE PSEUDOCOELOMATE PHYLA

1 The pseudocoelomate phyla are mostly minute animals, in which the body is covered by a cuticle. The body cavity, when present, is a pseudocoel. Adhesive glands for temporary anchorage are usually present.

Phylum Rotifera

2 Rotifers are mostly freshwater benthic or planktonic animals with somewhat elongate or saclike bodies.

3 An anterior ring of cilia (ciliated crown) functions in swimming and, in some, for feeding.

4 A posterior foot with adhesive toes provides for temporary attachment in benthic species.

5 Food is seized or ground with cuticular jaws located within a muscular pharynx (mastax).

Phylum Gastrotricha

6 Gastrotrichs are marine and freshwater animals living within bottom debris or between sand grains.

7 The bottle-shaped or straplike body bears cilia on the ventral surface, with which the animal moves.

8 Food is ingested by a muscular pharynx.

Phylum Nematoda

9 Nematodes, called roundworms, are marine, freshwater, or terrestrial animals, and many are parasitic.

10 The long cylindrical body, tapered at each end, is adapted in free-living species for living in interstitial spaces.

11 The interaction of the elastic cuticle, longitudinal body wall muscles, and fluid-filled pseudocoel produce body undulations that drive the animal through interstitial spaces.

12 Food is ingested by a muscular pumping pharynx.

Flatworms are **acoelomates.** Almost all other bilateral animals possess a body cavity or at least are derived from forms that had a body cavity. The body cavity of most animals arises as a cavity inside of mesoderm and is called a **coelom** (Fig. 14.8) but in some animals the body cavity, called a **pseudocoel,*** represents a persistent embryonic blastocoel that is never obliterated by gastrulation or by developing mesodermal structures (Fig. 24.1). A pseudocoel therefore does not have a peritoneal lining as does a coelom.

Pseudocoelomates are a diverse assemblage of free-living and parasitic animals, but they share a number of features in addition to the pseudocoel. They are all poorly cephalized, i.e., they lack a distinct head with well-developed sensory structures. Free-living species are all essentially aquatic. Even the nematodes that live in soil are dependent upon water films around soil particles. Pseudocoelomates usually are microscopic, and they possess neither a blood vascular system nor organs for gas exchange.

The organs, such as the intestine and gonads, of many pseudocoelomates have fixed numbers of cells, which are constant for a particular species. For example, the intestine of the roundworm *Rhabditis longicauda* is always composed of 18 cells. Nuclear divisions are completed at hatching, and subsequent growth involves an increase in the size of the cells, but no cell division occurs. This strange condition of cell constancy, called **eutely,** is possibly correlated with the small size of pseudocoelomates, for it is also found in certain other minute but unrelated animals. The cells of these tiny animals are perhaps near the lower limit of effective size for metazoans, and cell constancy may be an adaptation to preserve an adequate cell size.

*The name pseudocoel (Gr. *pseudo,* false + *koilos,* hollow) was an unfortunate choice in the history of zoological terminology because there is nothing false about this type of body cavity, nor is it really a "false coelom." It is simply a body cavity that has a different embryonic origin than a coelom.

The body of pseudocoelomates is covered by a flexible nonliving **collagen cuticle,** which is often sculptured (see Figs. 24.5 and 24.8). A varying number of adhesive glands is usually present, and these open through the cuticle individually as small tubes (see Fig. 24.5). The adhesive tubes are used for temporary anchorage to the substratum.

Water balance is achieved by a pair of **protonephridia** or by structures probably derived from protonephridia. The nervous system is relatively simple, composed of a varying number of longitudinal nerves.

The digestive tract of pseudocoelomates, like that of higher animals, is a tube with two openings: a **mouth** and an **anus.** A muscular pharynx and a stomach or intestine are the principal specializations of the gut tube.

Reproduction in most pseudocoelomates is entirely sexual, and with few exceptions, the sexes are separate.

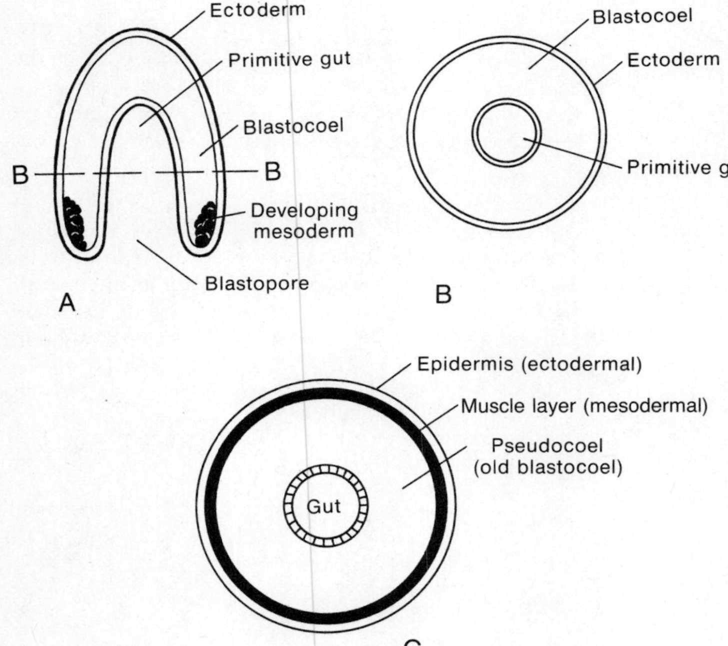

FIGURE 24.1 Embryonic origin of the pseudocoel. *A,* Frontal section of the gastrula, showing the remains of the blastocoel and the developing mesoderm; *B,* cross section of *A* taken at level *B–B; C,* diagrammatic cross section of adult pseudocoelomate.

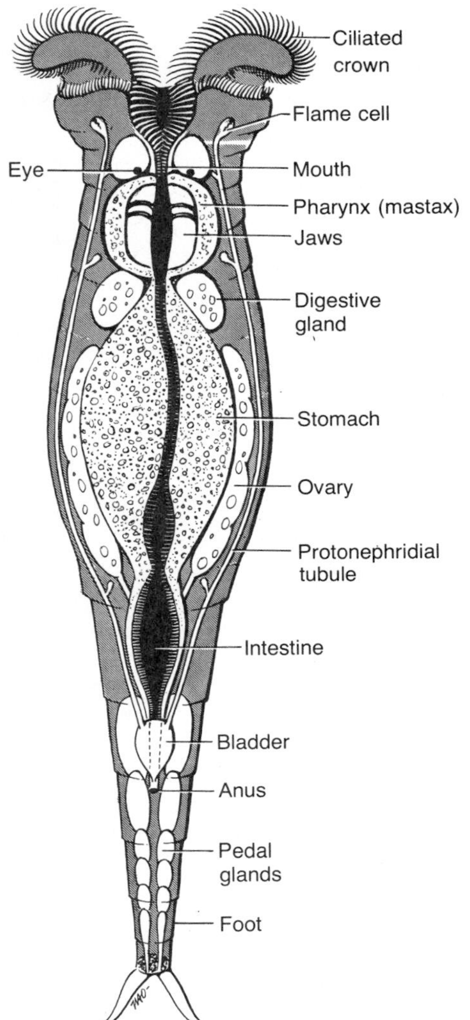

FIGURE 24.2 Ventral view of a rotifer, ***Philodina roseola,*** showing many of the internal structures. (Redrawn from Hyman, L. H.: *The Invertebrates,* Vol. III. New York, McGraw-Hill Book Co., 1951.)

Cleavage is determinate, but the spiral condition is often lost. Development is typically direct in free-living species.

There are seven groups of pseudocoelomate animals. Some zoologists believe that many of these groups are closely enough related to constitute classes of a single phylum, called the **Aschelminthes.** Most zoologists, however, object to the phylum Aschelminthes, pointing out that although pseudocoelomates share a number of features, there are no distinctive unifying characteristics. We will treat each of the pseudocoelomate groups as a separate phylum, but limitations of space permit us to describe only the three largest phyla; others are listed on p. 439.

Phylum Rotifera

The phylum Rotifera consists of some 1500 species, most of which live in fresh water.

The body of a rotifer is composed of a trunk and a posterior foot (Figs. 24.2 and 24.3). The anterior end of the trunk bears a crown of **cilia**, a distinguishing characteristic of the phylum. The distribution of crown cilia varies in different rotifers (Figs. 24.2 and 24.3). A common type of crown is one in which the cilia are arranged in the form of two discs (Fig. 24.2). The two discs of cilia beat in a circular manner, one clockwise and one counterclockwise, and look like two wheels spinning, from which the name *Rotifera* (L. *nota*, wheel + *ferre*, to carry) is derived.

The foot is a narrow posterior extension of the trunk and commonly terminates in a pair of adhesive glands, which open to the exterior through tubes, called toes (Fig. 24.3*A*). In some rotifers the foot is divided into ringlike sections, which can telescope into one another (Fig. 24.2).

Rotifers swim by means of the ciliated crown and crawl in a leechlike fashion, using the terminal adhesive glands as a means of attachment. Many species of rotifers live within algae and bottom debris and even in the capillary water of soil and mosses. They swim and crawl intermittently. Some rotifers are planktonic and are never found on the bottom (Fig. 24.3*B*), and a few are sessile (Fig. 24.3*C*).

A second distinguishing feature of rotifers is the pharyngeal apparatus, called the **mastax.** The mastax differs from the pharynx of the other pseudocoelomates in possessing seven cuticular teethlike pieces projecting into the lumen (Figs. 24.2 and 24.4). The teeth vary in size and shape depending upon the feeding habit. Predatory rotifers have a mastax that can project from the mouth and grasp and seize prey, mostly protozoans and other rotifers (Fig. 24.4*B*). Rotifers with disclike ciliated crowns are filter feeders. The beating cilia drive small particles into the mouth. The mastax of these rotifers contains broad flattened pieces adapted for grinding (Fig. 24.4*A*).

Digestion is probably extracellular within the large stomach. A short intestine passes from the stomach to the anus, which is located on the ventral side at the end of the trunk.

Reproduction is entirely sexual. The male, which is usually smaller than the female, stabs the female on any

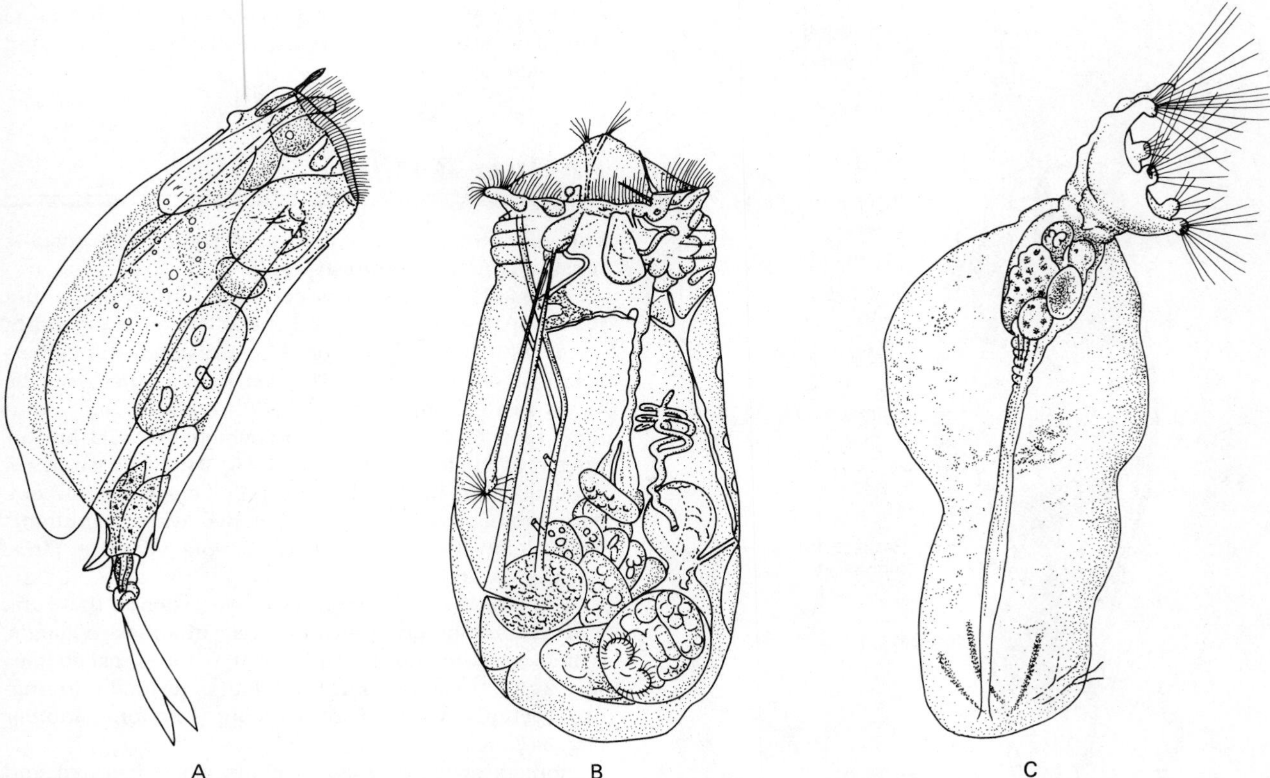

FIGURE 24.3 Three rotifers. *A,* Lateral view of a species of *Euchlanis.* The body is dorsoventrally flattened, and the foot possesses well-developed toes through which the adhesive glands open. This rotifer swims and anchors intermittently. *B,* A species of *Asplancha.* This predaceous planktonic rotifer has a transparent, sac-like body with no foot. *C,* A species of *Collotheca,* a sessile rotifer, in which the body is enclosed within a gelatinous case. The ciliated crown is reduced, and the anterior end, which has the form of a funnel, is adapted for trapping prey. (*A* redrawn from Myers in Hyman, L. H.: *The Invertebrates.* Vol. III. New York, McGraw-Hill Book Co., p. 120, 1951; *B* from Nogrady, T.: *Synopsis and Classification of Living Organisms,* Vol. I. New York, McGraw-Hill Book Co., 1982, p. 868; *C* after Hudson.)

part of her body with his penis. This type of copulation, called **hypodermic impregnation,** occurs in other animals, including some flatworms. The sperm then migrate to the eggs for fertilization. During the life of a female, 10 to 20 shelled eggs, one for each ovarian nucleus, are deposited singly on the bottom or are attached to the body of the female. When the females hatch, they have all of the adult features and attain sexual maturity after a growth period of a few days. The smaller males are sexually mature on hatching. Sessile rotifers are initially free-swimming.

Many freshwater species produce both rapidly hatching thin shelled eggs and dormant thick shelled eggs. **Parthenogenesis** is common. In some species, both parthenogenesis and development from fertilized eggs occur. But in other species, there are no known males, and all individuals are females produced parthenogenetically. Parthenogenesis is perhaps an adaptation for a rapid expansion in number of a population after it has been reduced in freshwater pools and streams by desiccation and other extreme environmental conditions.

As might be expected, the rotifers living on mosses or in soil are capable of withstanding extreme environmental conditions. During the short periods of time when these plants are filled with water, the animals

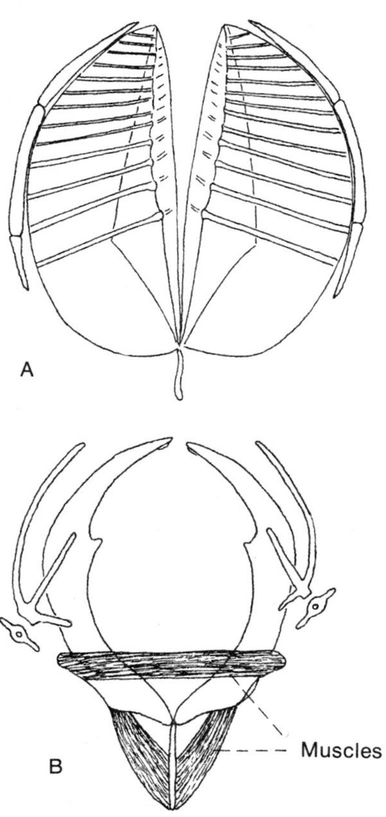

FIGURE 24.4 Two types of mastax teeth. *A,* Mastax with teeth adapted for grinding. Note the two large ridged plates that provide grinding surfaces. *B,* Mastax with grasping, forceps-like teeth. (*A* [after Beauchamp] and *B* from Hyman, L. H.: *The Invertebrates,* Vol. III. New York, McGraw-Hill Book Co., 1951.)

swim about in the water films on the leaves and stems. They are capable of withstanding desiccation, usually without the formation of protective covering, and can remain in a dormant state for as long as three to four years.

Phylum Gastrotricha

Gastrotricha constitutes one of the smaller phyla of pseudocoelomates, with about 425 species. Its members are found in the sea and in fresh water, and there are many more marine species than is true of rotifers. Many gastrotrichs live in the interstitial spaces between sand grains, others among surface debris and algae. Gastrotrichs are usually a little smaller than rotifers. The body is elongate, and the ventral surface is flattened and ciliated, from which the name *Gastrotricha*—stomach hairs—is derived. The anterior end may bear bristles or tufts of cilia. The posterior end is sometimes forked (Fig. 24.5). The **cuticle** of gastrotrichs is commonly in the form of scales, which are sometimes ornamented with spines. Adhesive tubes are present, often in rows along the sides of the body. The adhesive tubes of many gastrotrichs are now known to contain duogland systems as in flatworms (p. 492).

Gastrotrichs glide over the bottom propelled by the ventral cilia and may temporarily attach with the adhesive tubes. They feed on bacteria, small protozoa, algae, and detritus. Food is swept into the anterior mouth by cilia or is pumped in by the muscular pharynx.

The pseudocoel consists of only slitlike spaces between organs. Thus gastrotrichs do not really possess a body cavity.

In contrast to most other pseudocoelomates, gastrotrichs are **hermaphroditic.** However, in most freshwater species, the male system is degenerate, and all individuals are thus parthenogenetic females. As in freshwater rotifers, both dormant and nondormant eggs are produced.

Phylum Nematoda

The approximately 12,000 species of nematodes, or roundworms, constitute the largest phylum of pseudocoelomates and one of the most widespread and abundant groups of animals. They occur in enormous numbers in the interstitial spaces of marine and freshwater bottom sediments and in the water films around soil particles. One square meter of bottom mud off the Dutch coast has been reported to contain as many as 4,420,000 nematodes. Several hundred billion nematodes may be present in an acre of good farmland. Decomposing plant and animal bodies often contain large populations of nematodes. For example, examination of one decomposing apple revealed some 90,000 nematodes belonging to several different species.

Food crops, domestic animals and humans are parasitized by many species, and consequently roundworms are of tremendous economic and medical importance.

Most free-living nematodes are less than a millimeter in length, but species parasitic in the gut of vertebrates may attain much larger sizes; the horse nematode, *Parascaris equorum,* reaches a length of 35 cm.

The bodies of nematodes (Gr. *nēmatos,* thread) are usually long, cylindrical and tapered at both ends (Fig. 24.6). Lips, small sensory bristles and papillae encircle

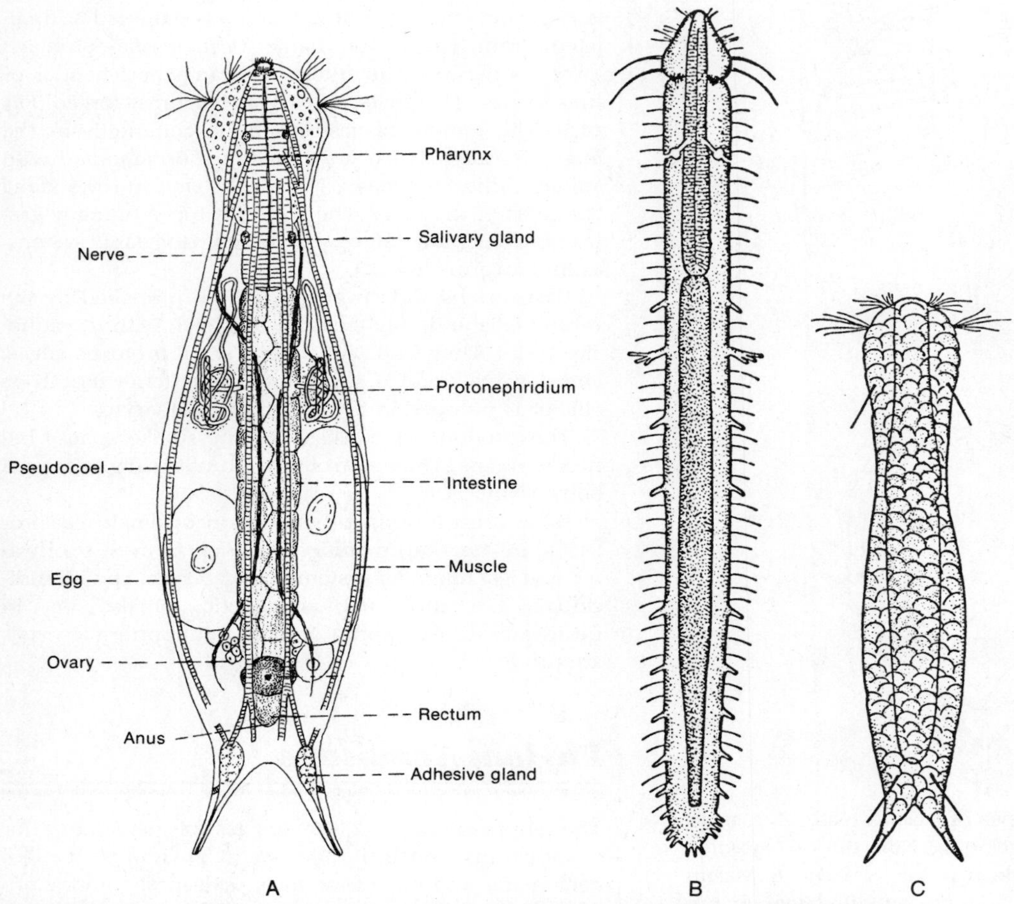

FIGURE 24.5 *A,* Internal structure of a gastrotrich, ventral view. *B, Pleurodasys,* a marine gastrotrich with lateral adhesive tubes (glands); *C, Lepidodermella,* a freshwater gastrotrich in which the cuticle has a scalelike ornamentation. (*A* modified after Zelinka from Pennak; *B* and *C* from Grasse, P.: *Traite de Zoologie,* 4:3, 1965.)

the mouth at the anterior end (Fig. 24.7). The anus is located a short distance in front of the posterior end, which in free-living species usually terminates in an adhesive gland (a duogland system). The collagen cuticle is composed of three layers, the outer one often sculptured. The cuticle is shed four times in the life of a nematode, but unlike the exoskeleton of insects and crustaceans, the nematode cuticle grows between molts. Beneath the cuticle is an epidermis composed of cells whose nuclei are restricted to middorsal, midventral, and midlateral lines (Fig. 24.8). The subepidermal muscle layer is composed only of longitudinal fibers. Contractions of muscle fibers, which are antagonistic to the elastic cuticle and the hydrostatic pressure of the pseudocoel, produce undulatory or thrashing locomotor movements that drive nematodes through the spaces between sand or soil particles.

The mouth opens into a **buccal cavity**, which may be provided with teeth or a **stylet** (Figs. 24.7*A* and *B*). The buccal cavity is connected to a tubular muscular **pharynx**. The remainder of the gut, a long straight intestine, is the site of digestion and absorption (Fig. 24.6).

Many free-living nematodes are predaceous; others feed on the contents of plant cells. Some consume dead organic matter. The stylet is used to puncture prey or plant cells, and the pharynx to pump out the contents.

Nematodes lack protonephridia, but some possess a system of glands and tubules that open through a midventral pore. The system may have an osmoregulatory function.

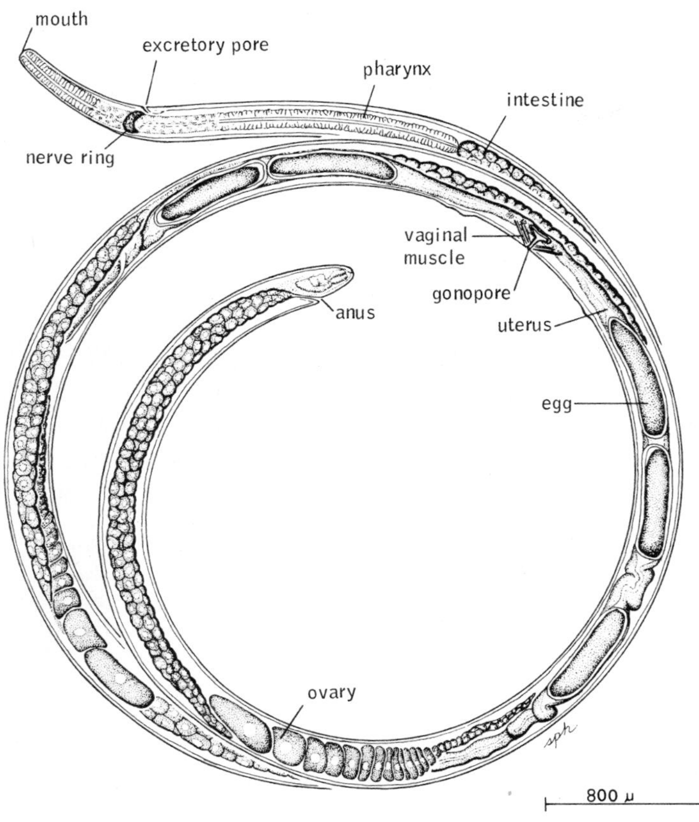

FIGURE 24.6 Female of the marine nematode *Pseudocella.* (After Hope.)

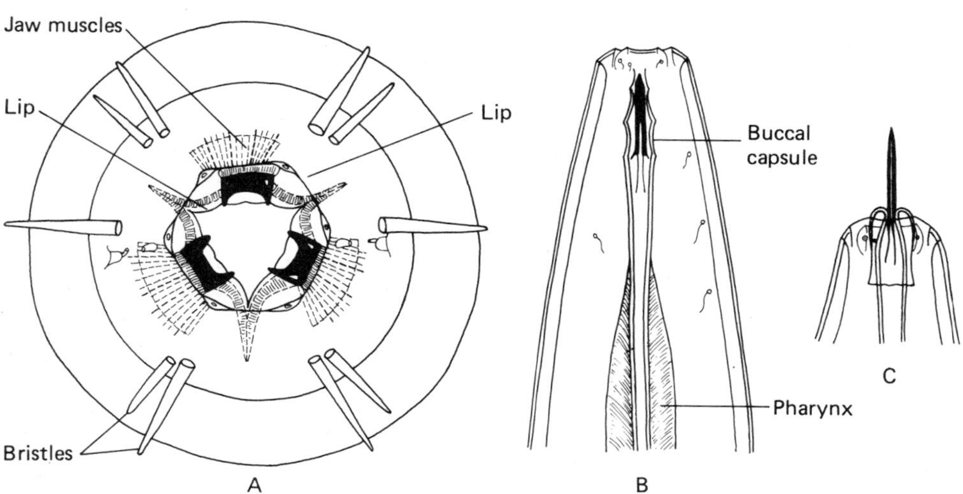

FIGURE 24.7 *A,* Anterior end of a marine nematode with jaws and bristles; *B,* anterior end of a predaceous nematode, with a retracted, spearlike stylet; *C,* stylet projected. (*A* after de Coninck, from Hyman; *B* and *C* after Thorne, from Hyman, L. H.: *The Invertebrates,* Vol. III. New York, McGraw-Hill Book Co., 1951.)

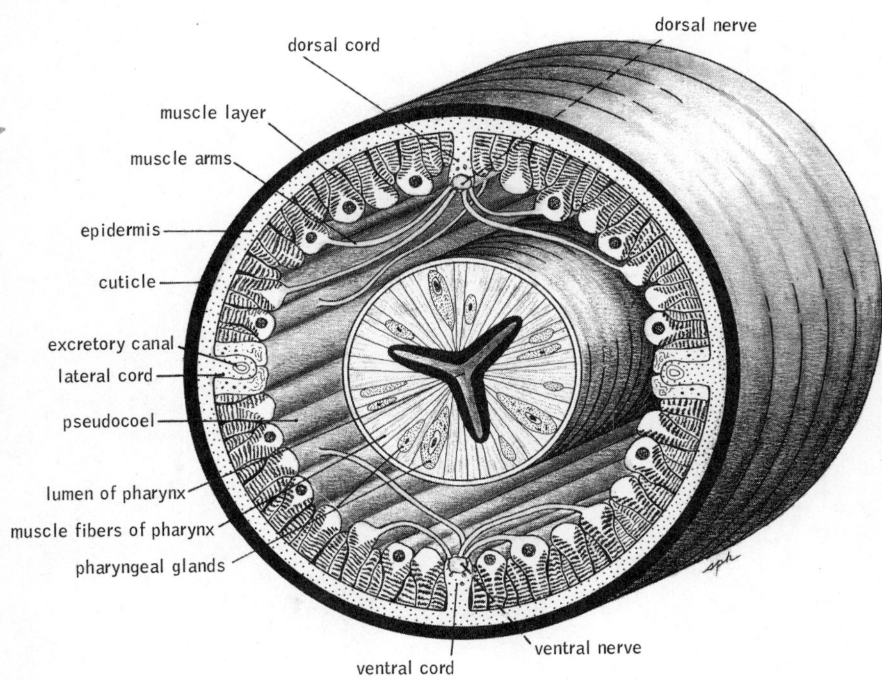

FIGURE 24.8 Diagrammatic cross section through the body of a nematode at the level of the pharynx. Note that the muscles send extensions, or arms, to the nerve cord, instead of the more usual reverse arrangement. Only a few of the many muscle arms are shown. A cuticle lines the pharynx as well as covering the surface of the body.

The female reproductive system is paired and tubular and includes two ovaries (Fig. 24.6). Each oviduct empties into a uterus and then into a common vagina that opens to the exterior in the midregion of the body. Male nematodes are usually smaller than females. The male reproductive system is a long coiled tube composed of a testis, sperm duct, seminal vesicle, and muscular ejaculatory duct. The latter opens into the rectum so that the anus functions as a gonopore as well as for egestion of wastes. The rectum contains two short curved spicules, which can be projected from the anus. During copulation, the posterior end of the male is curled around the body of the female in the region of the gonopore. The spicules are used to hold open the gonopore of the female during sperm transmission. In some nematodes, including the parasitic *Ascaris lumbricoides* of humans, the male forms a loop with its tail through which the female is attracted to enter. During passage the male is able to locate the female gonopore.

The sperm of nematodes are peculiar in lacking a flagellum and acrosome. The eggs possess a thick shell. Free-living nematodes deposit their eggs in the bottom debris and soil in which they live. The young have most of the adult features on hatching and, as already indicated, undergo four molts before attaining maturity. Adults do not molt but continue to grow.

Parasitic Nematodes. Parasitic nematodes attack both plants and animals and exhibit all degrees of complexity in their relationship with the host and in their life cycle.

Ascaroids. The ascaroid nematodes, which feed on the intestinal contents of humans, dogs, cats, pigs, cattle, horses, chickens, and other vertebrates, include the largest species of nematodes. They are entirely parasitic within a single host, and the life cycle typically involves transmission by the ingestion of eggs or larvae passed in the feces of another host. The juvenile stages usually penetrate the intestinal wall to enter the circulatory system, where they are carried to the lungs. Here they break into the alveoli and migrate back to the intestine via the trachea and esophagus.

The human ascaroid, *Ascaris lumbricoides,* reaches a length of 49 cm. and is one of the best known parasitic nematodes (Fig. 24.9). The species is widely distributed throughout the world, including the southeastern United States, particularly in children. The embryonated eggs are notoriously resistant to adverse environmental conditions and may remain viable in soil for up to 10 years. The very closely related species in pigs probably had a common evolutionary origin with the human species, the common ancestor being confined to one host or the other prior to human domestication of the pig.

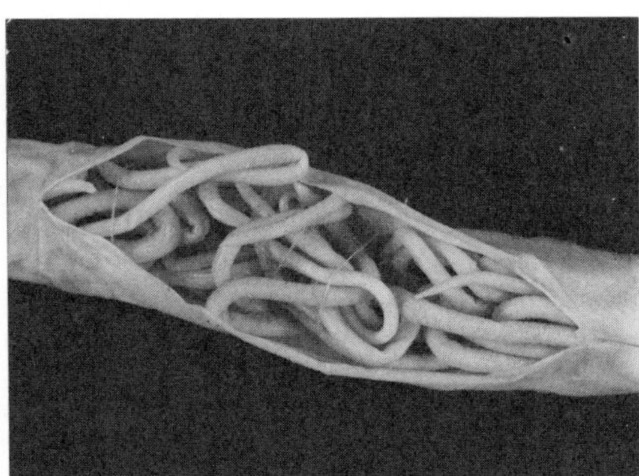

FIGURE 24.9 Adult specimens of *Ascaris suum* from the intestine of a pig. *Ascaris lumbricoides* of humans is very similar. (From Schmidt, G. D., and L. S. Roberts: *Foundations of Parasitology*. 2nd ed. St. Louis, C. V. Mosby Co., 1981.)

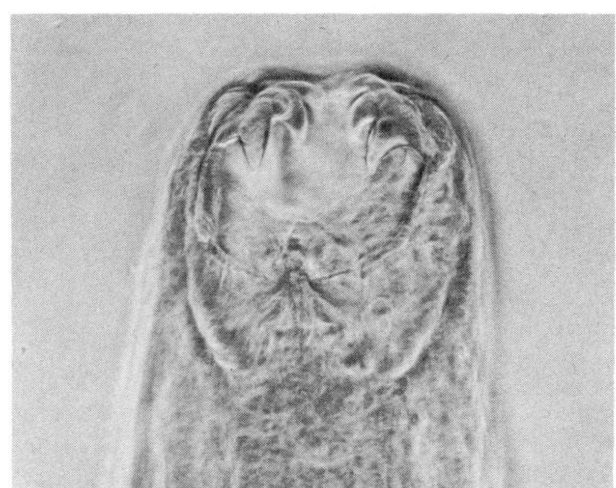

FIGURE 24.10 Anterior end of a dog hookworm, showing buccal region and teeth.

Toxicara canis and *cati* are two small ascaroid species common in dogs and cats. It is these species for which puppies and kittens are usually wormed.

Hookworms. The **hookworms** are another group of parasites of the digestive tract of vertebrates. Most members of this group feed on the host's blood. The mouth region is usually provided with cutting plates, hooks, teeth, or combinations of these structures for attaching and lacerating the gut wall (Fig. 24.10).

An infection of more than about 25 worms will produce symptoms of hookworm disease, and a heavy infection can produce serious danger to the host through loss of blood and tissue damage. An adult worm may live as long as two years in the intestine. Hookworms are one of the parasites causing serious infections in humans. It is estimated that over 380 million people are infected with *Necator americanis,* the most important species throughout the tropical regions of the world (despite the species name).

The life cycle of hookworms involves an indirect migratory pathway by the juveniles, as in ascaroids. The fertilized eggs leave the host in its feces and hatch outside the host's body on the ground. The larva gains reentry by penetrating the host's skin (feet in humans) and is carried in the blood to the lungs. From the lungs the larva migrates to the pharynx where it is swallowed and passes to the intestine. Not all species of this group gain reentry by skin penetration. Some, such as dog hookworms, have a pattern similar to the dog ascaroid.

Filarioids. The **filarioid** nematodes have life cycles requiring an arthropod intermediate host. The filarioids are threadlike worms inhabiting the lymphatic vessels and some other tissue sites in the vertebrate host, especially birds and mammals. The female is **ovoviviparous,** and the larvae are called **microfilariae.** Blood sucking insects, such as fleas, certain flies, and especially mosquitoes, are the intermediate hosts. A number of species parasitize humans, producing **filariasis.**

The chiefly African and Asian *Wuchereria bancrofti* illustrates the life cycle. The male is 40 mm. by 0.1 mm., and the female is about 90 mm. by 0.24 mm. (Fig. 24.11). Adults live in the ducts adjacent to the lymph glands of humans, especially in the lower part of the body. The microfilariae are found in the blood and are present in the peripheral bloodstream. When certain species of mosquitoes bite the host, the microfilariae enter the mosquito with the host's blood. Development within the intermediate host involves a migration through the gut to the thoracic muscles and after a certain period into the proboscis. From the proboscis the microfilariae are introduced back into the primary host when the mosquito feeds (Fig. 24.11). In severe filariasis, the blocking of the lymph vessels by large numbers of worms results in serious short-term lymphatic inflammation marked by pain and fever. Over a long period increase of connective tissue in affected areas may result in terrible enlargements of the legs, arms, and scrotum. Such enlargement is called **elephantiasis** (Fig. 24.11) and can only be corrected by surgery.

Some other important nematode parasites are *Dirofilaria immitis,* the heartworm of dogs; *Enterobius vermicularis,* intestinal pinworms of children; *Trichinella spiralis,* the juvenile stages of which encyst in the striated muscles, causing **trichinosis** (Box 24.1); *Dracunculus medinensis,* the 120 cm. long guinea worm of the mideast and Africa, which in the final part of its life cycle migrates to the subcutaneous tissue to release larvae and produces an ulcerated opening to the exterior.

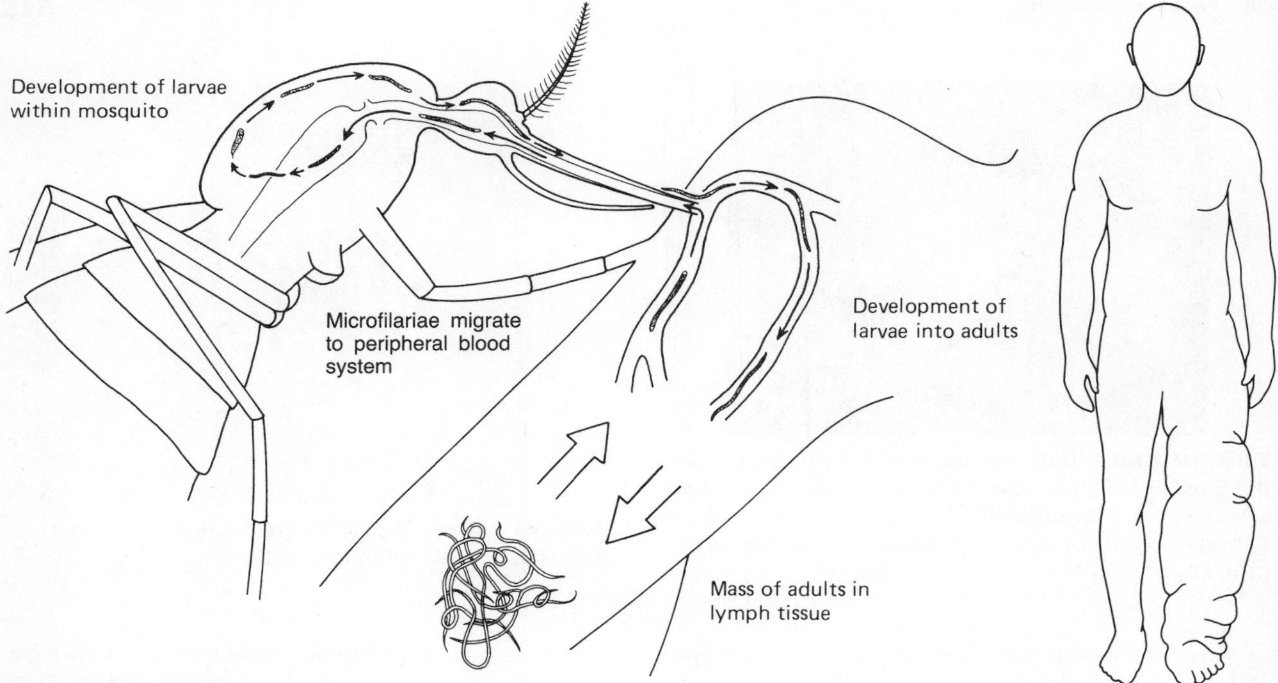

FIGURE 24.11 Life cycle of *Wuchereria bancrofti.*

BOX 24.1

Trichinella spiralis

The minute adults (female about 3 mm. long) live in the intestine of various carnivorous and omnivorous mammals. Following copulation, the female burrows into the submucosa of the intestinal wall of the host and gives birth to a large number of larvae. The larvae are carried by the bloodstream to striated muscles throughout the body, where they form cysts (*A*). The cysts must be eaten by another mammal to reestablish the adults in the intestine of a new host. Note that the same animal is acting as primary and intermediate host for this parasite. Humans usually obtain the nematode from eating insufficiently cooked pork (*E*). The cyst can be passed to pigs and rats in garbage (*F*). Pigs can also obtain the parasite by eating rats (*C*). The migration and invasion of tissues by the female and larvae can cause considerable tissue damage and can lead to death of the host. There is no cure for trichinosis at the present time.

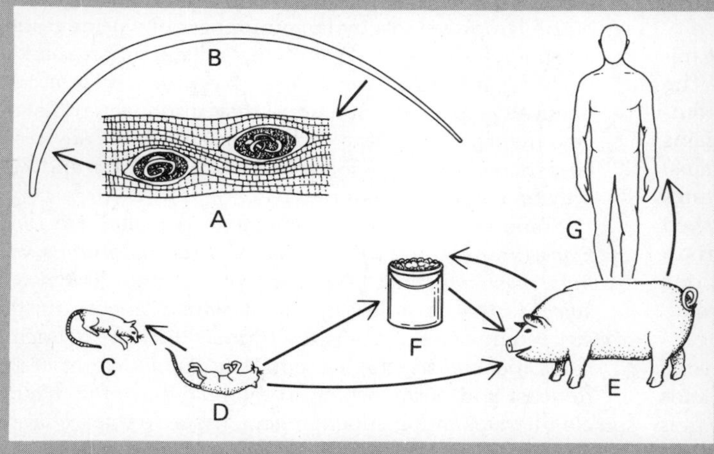

BOX 24.2

Related Phyla

Horsehair worms, members of the phylum Nematomorpha, have extremely long (up to 36 cm.) threadlike bodies (*A*). The nonfeeding adults live in freshwater or damp soil. The juveniles are parasitic in arthropods living around water. They inhabit the tissue spaces of the host, entering soon after hatching. The adult form emerges when the host is near water. (Courtesy of R. W. Pennak.)

The phylum Acanthocephala is a group of endoparasitic pseudocoelomates having an anterior retractable proboscis bearing recurved spines (*B*), hence the name *acantho cephala*—spiny head. The adults live-hooked to the gut wall of vertebrates. Embryonated eggs are passed out in the host's feces. If the eggs are eaten by certain crustaceans or insects, the juvenile acanthocephalan develops within the tissue spaces of the intermediate host. The primary vertebrate host becomes infected by eating the intermediate host. (After Yamaguti from Cheng, T. C.: *The Biology of Animal Parasites*. Philadelphia, W. B. Saunders Co., 1964.)

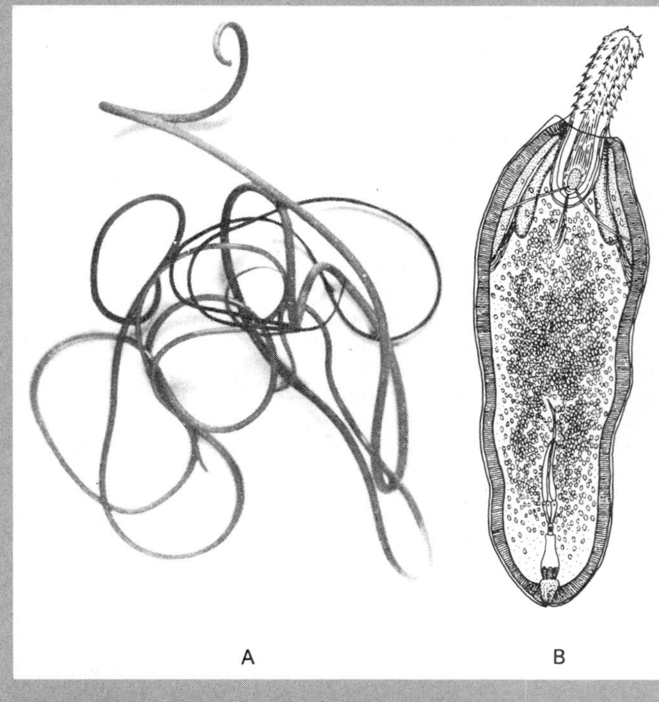

CLASSIFICATION OF PSEUDOCOELOMATES

(Classification of the pseudocoelomate phyla is technically difficult; therefore, we will supply no explicit listing of classification here. The interested student can consult the references below and at the end of Chapter 19 for further information.)

Summary

1. Pseudocoelomates are a group of phyla in which the body cavity, when present, is derived from a persistent blastocoel and is called a pseudocoel. The pseudocoelomate phyla are diverse in form, but most are poorly cephalized, have the body covered with a collagenous cuticle, and possess one or more adhesive glands. Free-living species are microscopic, which may be correlated with the tendency of many of these animals to have organs with fixed cell numbers (eutely) and to lack specialized systems for gas exchange and internal transport systems. Gastrotrichs and rotifers possess protonephridia.

2. Rotifers (phylum Rotifera) are largely freshwater pseudocoelomates that possess an anterior ciliated crown used in swimming. Some rotifers are planktonic; others swim about bottom debris and algae, temporarily anchoring by means of adhesive glands opening at the posterior end of the body.

3. Rotifers are predaceous or suspension feeders, the latter using the ciliated crown to drive food into the mouth. The muscular pharynx, called the mastax, contains a number of cuticular pieces used in prey catching or in grinding small particles.

4. Gastrotrichs (phylum Gastrotricha) are marine and freshwater animals with strap- or bottle-shaped bodies. They move by the beating of ventral cilia. Most species are interstitial but many live among surface debris and algae. They feed on fine particles pumped in by a muscular pharynx or driven in by cilia. Gastrotrichs are hermaphroditic.

5. The phylum Nematoda, containing the roundworms, is the largest group of pseudocoelomates. There are marine, freshwater, soil-inhabiting, and parasitic species. The long cylindrical and tapered body is adapted for living in interstitial spaces. The animal is propelled by whiplike movements produced by contractions of longitudinal muscles acting against the flexible cuticle and large fluid-filled pseudocoel.

6. Nematodes have varied feeding habits, and jaws and stylet are characteristic of many predaceous and plant-feeding species. Food is ingested with a muscular pharynx and digested within a long straight intestine. The sexes are separate, and copulation and internal fertilization are the rule. The cuticle is molted four times in the life cycle of a nematode.

7. There are many parasitic nematodes, which inhabit various parts of the bodies of a wide range of hosts. Some species utilize a single host (gut-inhabiting ascaroids, hookworms); others require two hosts (filaroids).

References and Selected Readings

Detailed accounts of the pseudocoelomate phyla may be found in the references listed at the end of Chapter 19. Parasitic forms are described in the parasitology texts at the end of Chapter 23. The following are a few works devoted to specific groups.

Croll, N. A., and B. E. Matthews: *Biology of Nematodes.* New York, John Wiley & Sons, 1977. A general account of the phylum.

Crowe, N. H., and K. A. Maden: Anhydrobiosis in tardigrades and nematodes. *Transactions of the American Microscopic Society, 93:*513–524, 1974. A study of the ability of nematodes to withstand extreme desiccating conditions.

D'Hondt, J. L.: Gastrotricha. *Oceanography and Marine Biology Annual Review, 9:*141–192, 1971. A review of biological knowledge of gastrotrichs.

Dropkin, V. H.: *Introduction to Plant Nematology.* New York, John Wiley & Sons, 1980. The biology of plant-feeding nematodes.

Dumont, H. J., and J. Green (Eds.): *Rotatoria.* The Hague, Junk, 1980. A collection of papers from a symposium on rotifers.

Gilbert, J. J.: Developmental polymorphosis in the rotifer *Asplanchna sieboldi. American Scientist, 68* (Nov.–Dec.):636–646, 1980. A study of the role of dietary vitamin E in adapting the body form to environmental changes.

Maggenti, A.: *General Nematology.* New York, Springer-Verlag, 1981. A general biology of nematodes.

Nicholas, W. L.: *The Biology of Free-living Nematodes.* Oxford, Clarendon Press, 1975. A general biology of the nonparasitic nematodes.

25 Mollusks

ORIENTATION

The mollusks constitute a very large phylum of many diverse animals, such as snails, chitons, clams, and octopods. This chapter begins with the ground plan of a generalized mollusk, which in at least some respects is certainly similar to the ancestral members of the phylum. Then the principal classes of mollusks, beginning with the one containing the snails, will be surveyed. In our survey we will see what characteristics of the living members of each class contribute to the reconstruction of the ancestral mollusks. We will also determine what major changes occurred in the evolution of the different classes. We will explore the adaptive significance of those changes and see how they might account for the great success and diversity of the phylum.

CHARACTERISTICS OF MOLLUSKS

1 The phylum **Mollusca** is the second largest phylum in the Animal Kingdom and contains such familiar forms as snails, slugs, clams, oysters, scallops, squids, and octopods. Although the greatest number of mollusks are marine, there are snails and clams that inhabit fresh water and many species of snails and slugs that are terrestrial.

2 The body of a mollusk is usually covered by a shell secreted by the underlying integument, called the mantle.

3 A ventral muscular foot is the locomotor organ of most mollusks.

4 The shell and mantle overhang the body, creating a mantle cavity that houses the gills of aquatic species and forms a lung in land snails.

5 Except for clams, most species employ as a feeding organ a unique beltlike structure bearing chitinous teeth called a radula. The radula can be protruded to scrape or tear and pull food into the mouth.

6 The circulatory system is usually open, and the heart, which lies within a pericardial coelom, consists of a pumping ventricle and paired auricles that receive blood from the gills.

7 The excretory organs are metanephridia, usually two, that drain the coelom and empty into the mantle cavity.

8 The nervous system consists of a pair of pedal cords to the foot and visceral cords supplying the organs of the mantle and visceral mass. The pedal and visceral cords unite anteriorly in a cerebral ganglion. Sense organs include one or two statocysts, one or two pairs of head tentacles, a pair of eyes, and a pair of osphradia, the last being sense organs that monitor the ventilating current passing through the mantle cavity.

9 Mollusks may have separate sexes or may be hermaphroditic, and paired gonads are adjacent to the pericardial coelom. Primitively, the gametes exit through the nephridia, but separate and often complex gonoducts are usually present. The earliest larval stage is a trochophore, but in many snails and clams there is only a later larval state, called a veliger.

General Features of the Phylum Mollusca

The phylum **Mollusca** includes some 50,000 living and 35,000 fossil species. Mollusks are about as well known taxonomically as mammals and birds. This reflects, at least in part, the long popularity of shell collecting.

Two distinguishing characteristics of the mollusks are the adaptation of the ventral surface of the body as a muscular foot for locomotion and the modification of the integument of the dorsal surface, the **mantle**, for the secretion of a protective calcareous shell. The shell has been lost in some mollusks, but the mantle almost always remains.

The Ancestral Mollusks

We have no direct evidence of the first mollusks. The phylum evolved in the Archeozoic seas, and by the time of the Paleozoic, when the fossil record first becomes clear, the different molluscan classes were already defined. Nevertheless, we can make some inferences about the nature of the ancestral mollusks on the basis of what can be observed in living forms. Such a hypothetical ancestor can serve as an introduction to the general features of the phylum.

Biologists postulate that the ancestral mollusk was small, perhaps a few centimeters in length, and adapted for living on hard rocky bottoms in ancient seas (Fig. 25.1). Probably, the animal's ventral surface formed a broad, flat, muscular creeping **foot.** The **head** was probably poorly developed, although in many living species the head bears tentacles, eyes, or other specialized sense organs. The dorsal surface of the body was covered by a low shield-shaped **shell** secreted by the underlying mantle and composed of one to several layers of calcium carbonate covered on the outside by an organic layer, the **periostracum** (Gr. *peri*, around + *ostrakon*, shell). Much of the shell of living mollusks is secreted by the mantle margin so that the shell grows in diameter as well as in thickness. The shell of the ancestral mollusks provided protection as long as the animal was attached to the rocky substratum; pairs of retractor muscles extending from the foot to the shell permitted the shell to be pulled down over the body.

The overhanging shell and mantle at the posterior end created a large mantle cavity containing two or several pairs of gills and two excretory openings (Fig. 25.1). The **anus** was located at the dorsal side of the cavity opening. The bipectinate **gill** structure postulated for the ancestral mollusk is similar to that found in many

BOX 25.1

Protostomes and Deuterostomes

All of the remaining members of the Animal Kingdom either have a type of body cavity called a coelom or, if a coelom is absent, have evolved from coelomate ancestors. The coelom appears during development as a cavity within the mesoderm (Fig. 14.8) and has a mesodermal lining, the **peritoneum.** The coeloms of living animals arise in a number of different ways and probably evolved independently several times within the Animal Kingdom.

Comparative studies of the development of coelomates led to the recognition of two major groups of metazoan phyla: the **protostomes** and the **deuterostomes.** These groups probably represent two principal lines of evolution within the Animal Kingdom. The protostome phyla—mollusks, annelids, arthropods, and certain lesser phyla—are characterized by **determinate spiral cleavage** (Fig. 14.4); the **mesoderm** can be traced back to a single cell of the **blastula** (p. 310). The **coelom** is a schizocoele, formed by an internal splitting of the mesodermal masses, a process called schizocoely (Fig. 14.8). The **mouth** originates from the blastopore, hence the name protostome (Gr. *protos*, first + *stoma*, mouth). Protostomes commonly possess a **trochophore** larva (Fig. 25.4), described on page 521. Although flatworms lack a coelom, their development by spiral cleavage justifies classifying them as acoelomate protostomes.

The features just described do not characterize the embryogeny of all protostomes; rather, they represent the primitive or basic plan of protostome development and are most commonly encountered among marine species. They have been partly or completely lost in the many species whose reproductive circumstances have led to marked changes in the amount and distribution of yolk material (in turn, the amount and distribution of yolk greatly influence the cleavage pattern) and to the suppression of larval stages, a characteristic of most freshwater and terrestrial animals and even of many marine species.

In contrast to protostomes, the deuterostome coelomates are characterized by **indeterminate radial cleavage.** The mouth appears during development as a new opening at the opposite end of the embryo from the blastopore, hence the name deuterostome (Gr. *deuteros*, second + *stoma*, mouth). The formation of the mesoderm and coelom is quite different from that of protostomes (p. 315).

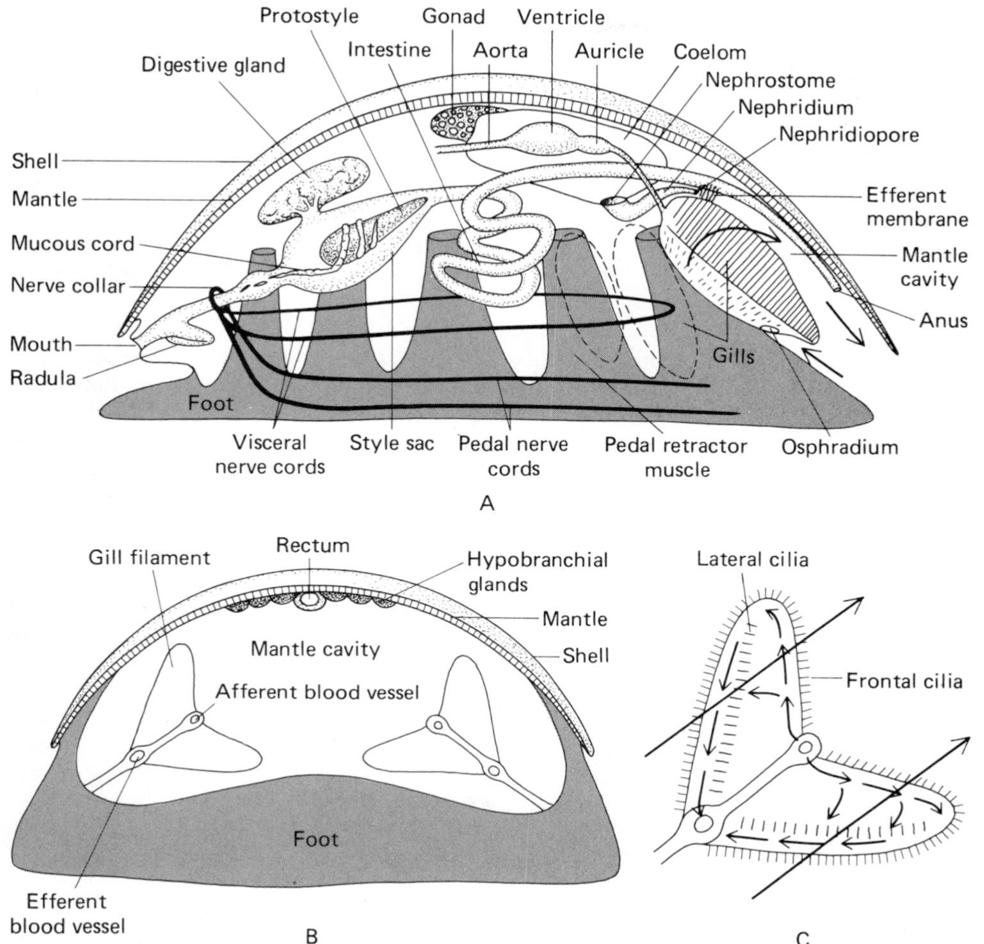

FIGURE 25.1 *A,* Hypothetical ancestral mollusk (lateral view). Arrows indicate path of water current through mantle cavity. *B,* Transverse section through body of ancestral mollusk at level of mantle cavity. *C,* Cross section through one gill, showing two filaments. Large arrows indicate direction of ventilating current produced by lateral gill cilia. Small arrows indicate flow of blood within gill. Frontal cilia on filament edge remove sediment.

living species. Each gill is composed of a longitudinal axis to which is attached on either side flattened gill filaments (Fig. 25.1). Blood circulates through the filaments, flowing from a supply to a drainage vessel in the gill axis. The filaments are ciliated and produce a ventilating current of water that enters the lower part of the mantle cavity, makes a U-turn across the gills, and then flows out of the cavity dorsally and posteriorly (Fig. 25.1). Wastes discharged by the **nephridia** and anus are removed in the exhalant stream of water.

As in most living mollusks, the mouth cavity of the ancestral mollusk probably contained a **radula** (L. a scraper), a unique rasping organ consisting of a beltlike membrane bearing a large number of teeth (Fig. 25.2). The radula rests upon a cartilaginous supporting skeleton, the **odontophore** (Gr. *odous*, tooth + *pherein*, to bear), supplied with protractor and retractor muscles. The ancestral mollusks were probably **microphagous**—they fed upon small particles of algae scraped with the radula from rocks.

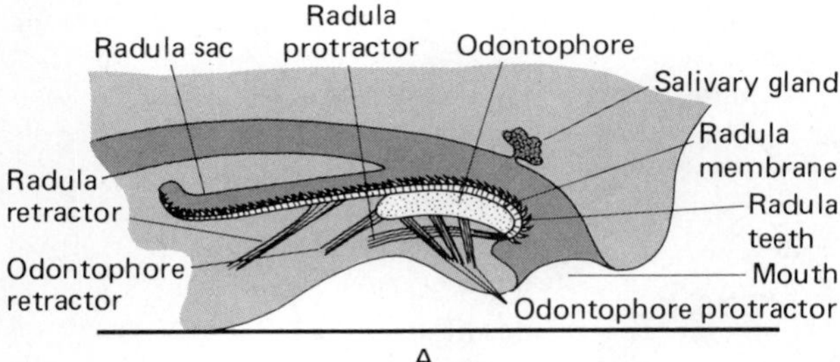

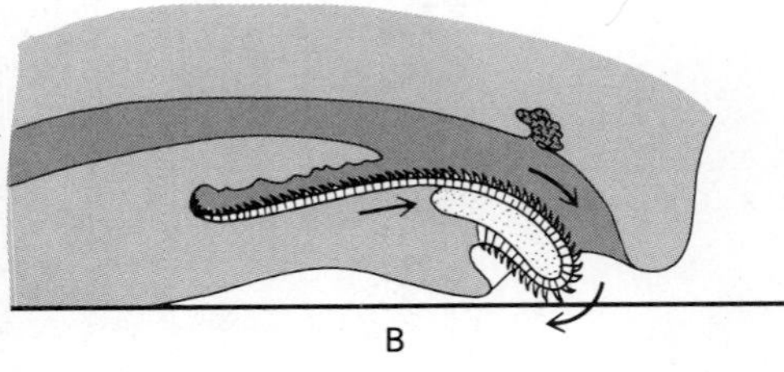

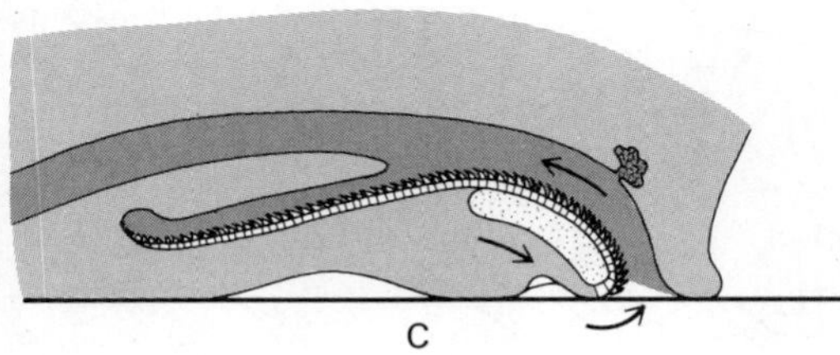

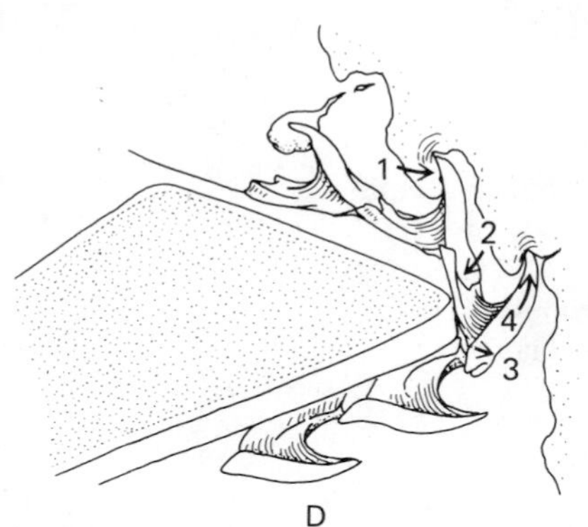

FIGURE 25.2 Molluscan radula. *A,* Mouth cavity, showing radula apparatus (lateral view). *B,* Protraction of the radula against the substratum. *C,* Forward substratum is scraped by radula teeth. *D,* The cutting action of the radula teeth when they are erected over the end of the odontophore (gray). (*D* from Solem, A.: *The Shell Makers: Introducing Mollusks.* Reprinted by permission of John Wiley & Sons, New York, 1974, p. 150.)

In living mollusks, the end of the radula and odontophore can project from the mouth and lick the adjacent surface, like the tongue of a mammal. New teeth are secreted at the posterior end of the radula as old teeth are worn away anteriorly. Particles of algae brought into the mouth are enveloped by mucus secreted by the **salivary glands.** A rope of mucus containing the algal particles then passes posteriorly through an **esophagus** to the **stomach.** The posterior end of the stomach, the **style sac,** is ciliated and rotates the mucous mass, winding in the string of mucus like a windlass winds in a rope (Fig. 25.1). Particles dislodged from the mucus are sorted by a ciliated area. Large undigestible particles are conducted posteriorly to the **intestine,** and fine digestible particles are conveyed to the ducts of a pair of **digestive glands** located to either side of the stomach. Digestion occurs intracellularly within the digestive gland. The long intestine functions only in the formation of feces.

An open blood vascular system provides for internal transport in most mollusks and was undoubtedly characteristic of the ancestral form. The heart, located within the coelomic cavity, is composed of a **ventricle** and two lateral **auricles,** which receive blood from each gill (Fig. 25.3). The contractile ventricle forces blood via an anterior aorta to the tissue spaces of the body—head, foot, visceral mass, and mantle. Blood then collects within larger sinuses from which it is returned to the heart by way of the gills (Fig. 25.3).

The pair of excretory organs, often called **kidneys,** are **metanephridia.** In contrast to the blind protone-

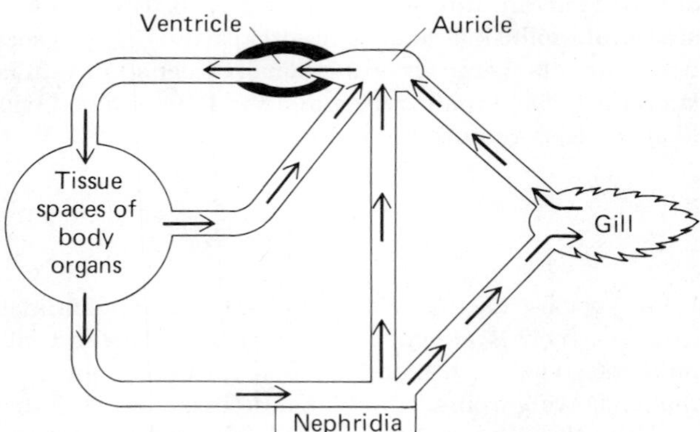

FIGURE 25.3 Diagram showing the principal features of the molluscan circulatory system. Auricles, gills, and nephridia are usually paired.

phridia of flatworms and pseudocoelomates, the inner ends of these tubes open into the coelom by way of a ciliated **nephrostome** (Fig. 25.1). Nephridia of this type were described on page 197. Special glandular tissue lining the coelom secretes wastes into the coelomic fluid. Fluid containing wastes collected by the blood from body tissues filters from the heart into the coelomic fluid. Coelomic fluid containing wastes passes into the inner end of the nephridium. The urine is expelled into the mantle cavity and flushed out by the exhalant ventilating current.

The nervous system of the ancestral mollusk was probably composed of a nerve ring around the esophagus, from the underside of which extended a ventral pair of pedal cords innervating the foot and a more dorsal pair of visceral cords innervating the mouth and the organs of the visceral mass (Fig. 25.1). Sense organs included a pair of **statocysts** in the foot and the **osphradia** (Gr. *osphradion,* strong scent). An osphradium is a patch of sensory epithelium near each gill, which monitors chemical substances in the water current passing through the mantle cavity (Fig. 25.1).

A pair of **gonads** was located to the front and sides of the coelom (Fig. 25.1). Eggs and sperm were released from the gonads into the coelom and were then carried to the mantle cavity by way of the nephridia. In the ancestral mollusks, fertilization was probably external either within the mantle cavity or in the surrounding sea water.

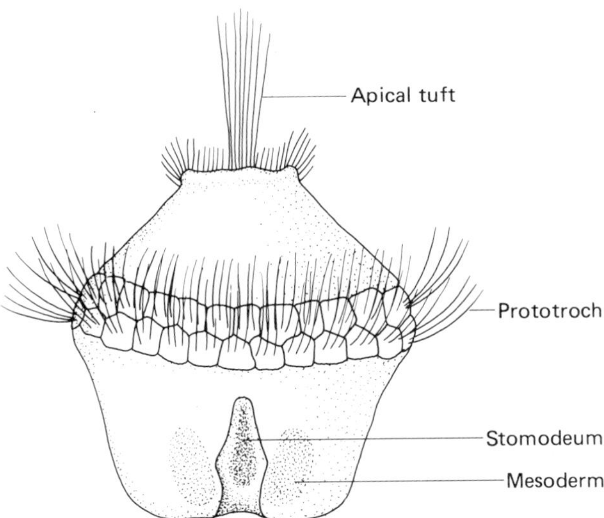

FIGURE 25.4 The trochophore larva of *Patella,* a marine snail. (After Patten.)

In living mollusks, cleavage is typically spiral, and the resulting gastrula develops into a free-swimming trochophore larva (Gr. *trochos,* wheel + *pherein,* to bear) (Fig. 25.4). The larval body is ringed about the middle with a girdle of **cilia,** the **prototroch** (Gr. *proto,* first + *trochos,* wheel), and the anterior pole bears an apical tuft of cilia. A digestive tract is present. The beating cilia of the prototroch provide for locomotion and may also serve to collect fine plankton for food. In the ancestral mollusks, the trochophore probably developed directly into the adult body.

Using this hypothetical ancestral form as a basis for comparison, we will examine five of the seven classes of mollusks.

Class Gastropoda

Among the 35,000 living species of gastropods, the largest and most diverse class of mollusks, are snails, whelks, conchs, limpets, and sea slugs. Most gastropods are marine, but some live in fresh water and others have become adapted for life on land.

Torsion and Shell Spiraling. Gastropods (Gr. *gastēr,* stomach + *pous,* foot) possess a well-developed **head** bearing two tentacles with an eye at the base of each and a broad, flat, creeping **foot** and a **shell** composed of a single piece. However, gastropods are distinguished from all other mollusks by the curious twisting of the visceral mass that occurs during development (Fig. 25.5).

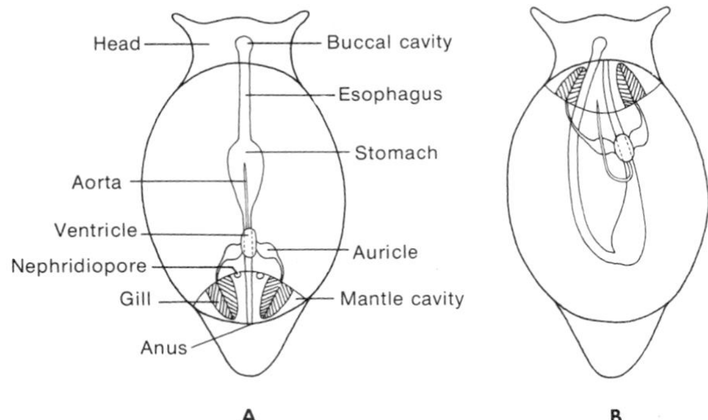

FIGURE 25.5 Dorsal view of hypothetical ancestral gastropod. *A,* Prior to torsion; *B,* after torsion. (Modified from Graham.)

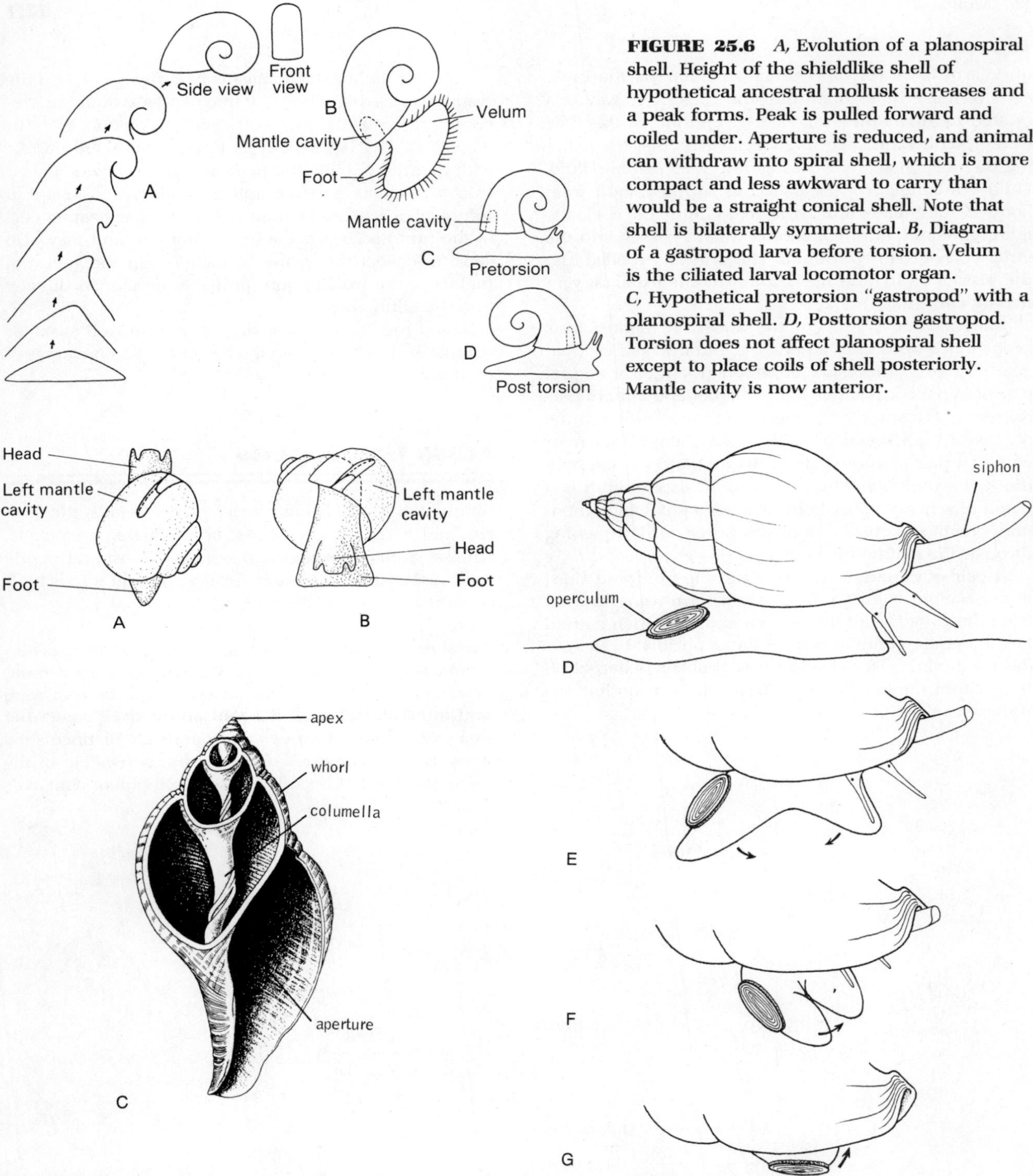

FIGURE 25.6 *A,* Evolution of a planospiral shell. Height of the shieldlike shell of hypothetical ancestral mollusk increases and a peak forms. Peak is pulled forward and coiled under. Aperture is reduced, and animal can withdraw into spiral shell, which is more compact and less awkward to carry than would be a straight conical shell. Note that shell is bilaterally symmetrical. *B,* Diagram of a gastropod larva before torsion. Velum is the ciliated larval locomotor organ. *C,* Hypothetical pretorsion "gastropod" with a planospiral shell. *D,* Posttorsion gastropod. Torsion does not affect planospiral shell except to place coils of shell posteriorly. Mantle cavity is now anterior.

FIGURE 25.7 *A* and *B,* Position in which shell is carried in gastropods. Slot that is found in some primitive gastropods indicates the original middorsal line of the mantle cavity. Note that the mantle cavity is restricted largely to the left side. *C,* Longitudinal section through a shell. *D–G,* Withdrawal into shell by folding of foot and closure with operculum. (*A* and *B* modified after Yonge.)

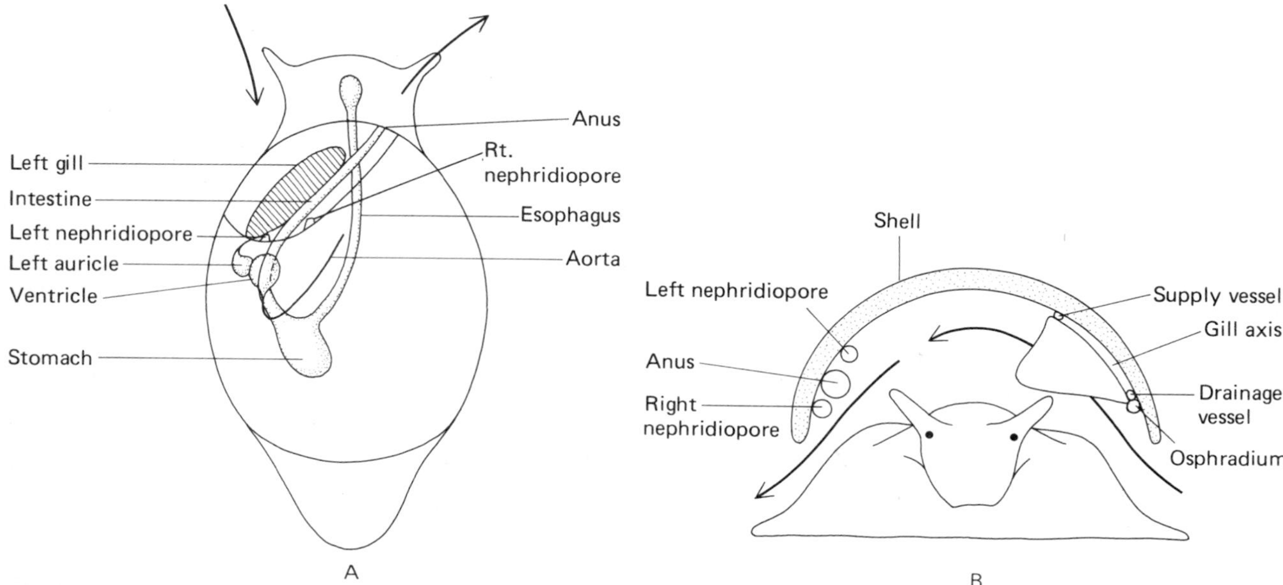

FIGURE 25.8 Typical prosobranch gastropod, with a single gill and mantle cavity on left side of animal. Arrows indicate path of water current through mantle cavity. *A,* Dorsal view; *B,* diagrammatic transverse section through head and mantle cavity. (Modified after Graham.)

This condition, called **torsion,** involves a 180° counterclockwise twist (viewed from above) that results in the **mantle cavity** and **anus** being located at the anterior end of the body. The gut, nervous system, and blood vascular system are correspondingly twisted. During embryonic development the mantle cavity forms first at the posterior end of the body but shifts to an anterior position.

The spiral shell of gastropods is *not* a consequence of torsion; indeed the spiral shell is formed during development before torsion occurs. The shieldlike shell of the ancestral mollusk protected only its dorsal surface, and the animal would have been vulnerable when dislodged from its substratum. In the course of evolution the gastropod shell became higher and more conelike, with a reduced aperture (Fig. 25.6). Such a shell provided a protective retreat into which the entire body could be withdrawn. The spiral condition appears to have been an adaptation for making the shell more compact and less awkward to carry than if it were a long cone. Fossils of the earliest known gastropods and the mollusks, which may have been ancestral to gastropods, had bilaterally coiled shells (Fig. 25.6). The asymmetry characteristic of the shell of all modern gastropods provided additional compaction and developed later.

The gastropod shell is balanced obliquely upon the body, with the spire directed posteriorly and upward on the right side (Fig. 25.7). The large first whorl of the shell bulges into the right side of the mantle cavity (Fig. 25.7 *B*). This partial occlusion of the mantle cavity has resulted in the reduction and loss of the right gill and in turn the reduction and loss of the right auricle and right kidney. Some primitive gastropods still have two gills, two auricles, and two kidneys, but most have only those on the left side.

Evolution of Water Circulation and Gas Exchange. Whatever the evolutionary significance of torsion may have been, the anterior position of the anus imposed a sanitation problem on the early gastropods, for waste would have been dumped on top of the head. The problem has been solved in most aquatic gastropods by the evolution of structures to generate a water current oblique to the body. Except in certain primitive bipectinate species, a single left gill lacks the filaments on one side and is anchored by the axis to the mantle wall. Water enters the mantle cavity on the left side of the head, makes a turn across the gills, and exits on the right side of the head (Fig. 25.8). The anus is located at the right edge of the mantle cavity, and wastes are removed by the exhalant water current. Commonly, the edge of the

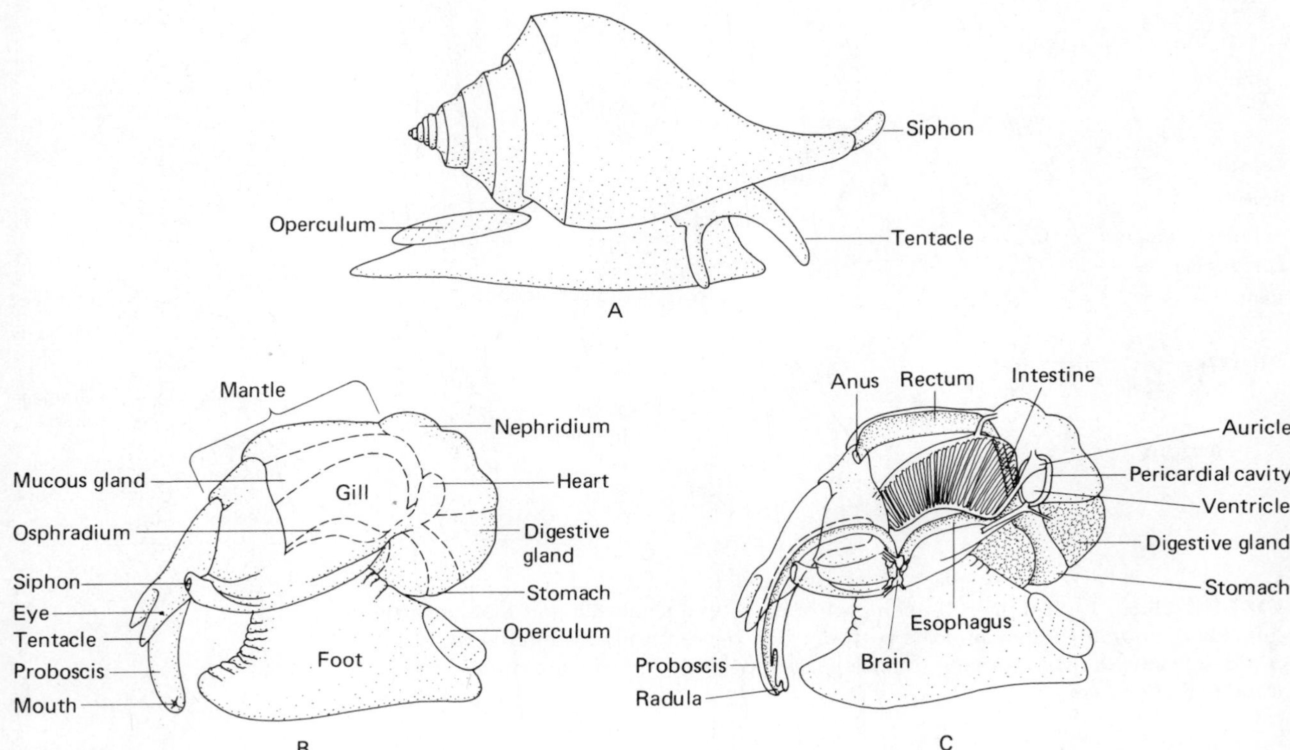

FIGURE 25.9 Anatomy of *Busycon canaliculatum*. *A*, Position of shell when foot is extended. *B*, Left side (shell removed), showing external organs and internal organs visible through the integument. *C*, Same view with digestive, respiratory, circulatory, and nervous systems indicated.

mantle on the left side is drawn out to form a siphon into which the inhalant water stream passes (Fig. 25.7*D*).

Most marine snails, including the common whelks, conchs, slipper shells, and limpets, compose the subclass **Prosobranchia.** The name Prosobranchia (Gr. *prosō*, forward + *brangchia*, gills) refers to the position of the gill in the anterior mantle cavity. From the prosobranchs two other subclasses of gastropods evolved. Members of the subclass **Opisthobranchia**—the bubble shells, sea butterflies, sea hares, and sea slugs—are characterized by a degree of detorsion, or untwisting of the visceral mass, from which the name Opisthobranchia (Gr. *opisthe*, behind + *brangchia*, gills) is derived. Detorsion has been accompanied by reduction of the shell and mantle cavity, as seen in sea slugs (Fig. 25.10G). Gas exchange occurs either through secondary gills located around the now posterior anus or across the external mantle surface, which may bear club-shaped projections (**cerata**).

Members of the third subclass, the **Pulmonata** (L. *pulmo*, lung), are adapted for living on land. The mantle cavity of the pulmonates evolved into a lung, and gills have been lost. Desiccation is minimized by reducing the opening into the mantle cavity to a very small pore (Fig. 25.11). The animal ventilates its mantle cavity by raising and lowering the floor of the mantle cavity, thereby moving air out of and into the chamber. Pulmonates probably evolved in very shallow estuarine habitats where they were often left out of water. Many pulmonates have remained aquatic, largely in fresh water. Some of these freshwater species must come to the surface to obtain air; others have abandoned the lung altogether and acquired secondary gills outside of the mantle cavity.

Shell. The shells of gastropods exhibit great variation in shape and coloring. Shell form and sculpture may contribute to protection, shell strengthening, or ease of carriage. The ventral margin of the aperture of the shell

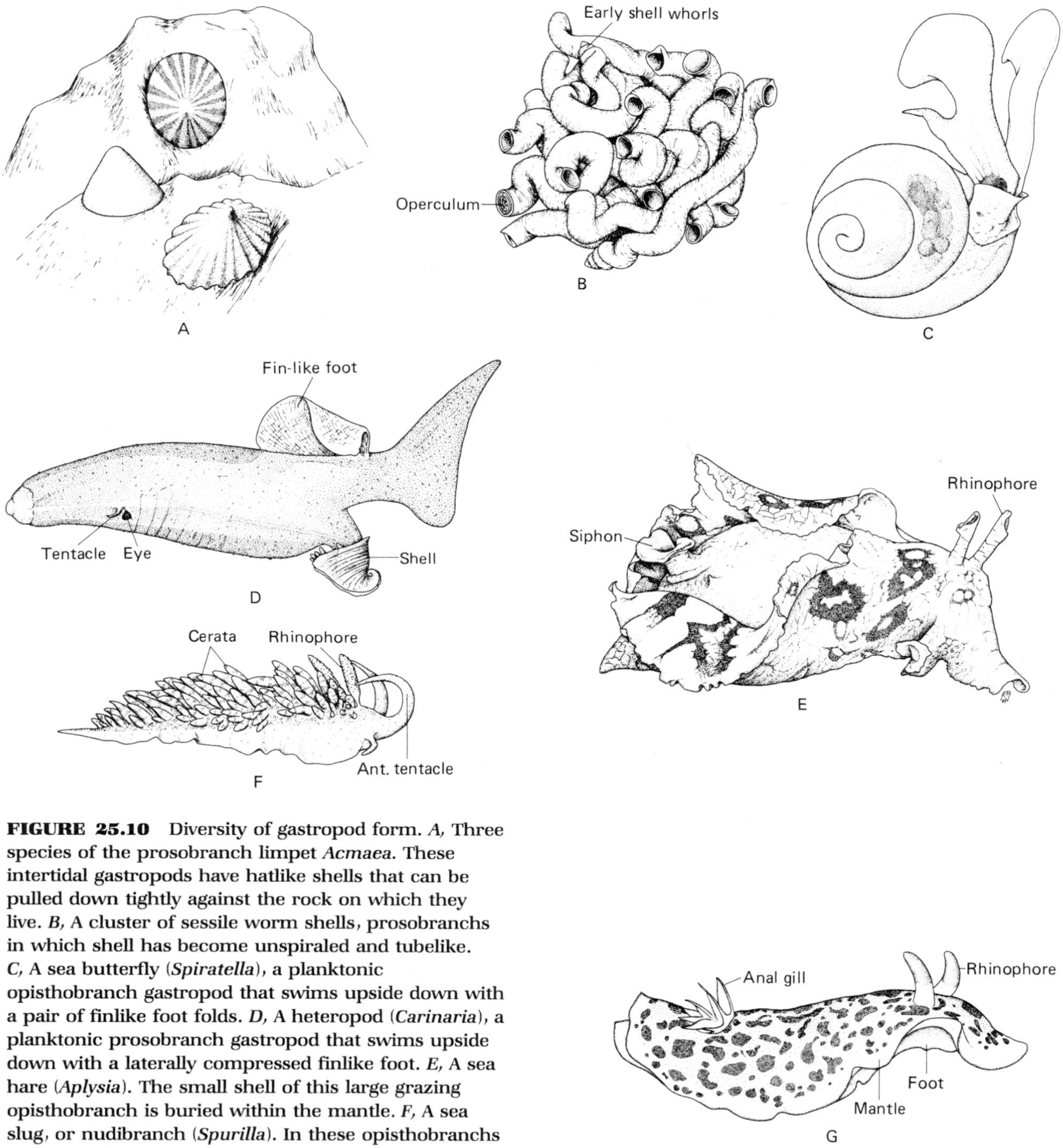

Early shell whorls
Operculum
A
B
C
Fin-like foot
Tentacle
Eye
Shell
D
Rhinophore
Siphon
E
Cerata
Rhinophore
Ant. tentacle
F
Anal gill
Rhinophore
Foot
Mantle
G

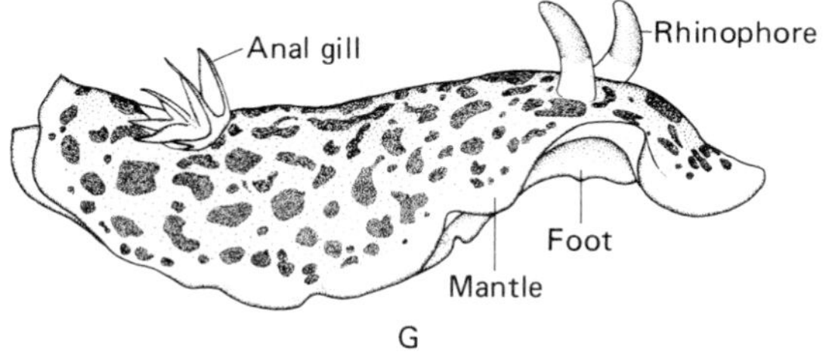

FIGURE 25.10 Diversity of gastropod form. *A,* Three species of the prosobranch limpet *Acmaea.* These intertidal gastropods have hatlike shells that can be pulled down tightly against the rock on which they live. *B,* A cluster of sessile worm shells, prosobranchs in which shell has become unspiraled and tubelike. *C,* A sea butterfly (*Spiratella*), a planktonic opisthobranch gastropod that swims upside down with a pair of finlike foot folds. *D,* A heteropod (*Carinaria*), a planktonic prosobranch gastropod that swims upside down with a laterally compressed finlike foot. *E,* A sea hare (*Aplysia*). The small shell of this large grazing opisthobranch is buried within the mantle. *F,* A sea slug, or nudibranch (*Spurilla*). In these opisthobranchs the shell has been lost and the body is secondarily bilaterally symmetrical. The dorsal surface of the body is covered with clublike projections (cerata) in many species, such as this one. *G,* A sea slug with anal gills instead of cerata. (*C* based on a photograph by D. P. Wilson; *D* based on a photograph by A. M. Keen; *F* based on a photograph by D. DeFreese; *G* based on a photograph by P. Zahl.)

A

B

FIGURE 25.11 Two terrestrial pulmonates. In the slug (*A*) the opening into lung can be seen at the lower edge of the saddle-like mantle. In the snail (*B*) the eyes can be seen at the top of the second pair of tentacles. The first pair of tentacles are small and do not show in this photograph. (Photographs by Betty M. Barnes.)

of many prosobranchs is drawn out to house the siphon (Fig. 25.9*A*). Another common modification is a greatly expanded last whorl. The spiral apex develops during juvenile states. Then the last whorl becomes extremely large and covers most of the body of the animal as a low convex plate. Abalones, slipper shells, and limpets have shells of this type. The shell of limpets has become secondarily bilateral (Fig. 25.10).

The central axis of the asymmetrical spiral shell is the **columella** (Fig. 25.7*C*), to which the single large retractor muscle of the body is anchored. The gastropod withdraws into the shell by folding the middle of the foot. The anterior half of the foot and the head are withdrawn first, followed by the posterior half of the foot. Many prosobranchs have a round plate, the **operculum,** on the back of the foot, which plugs the aperture when the animal is withdrawn (Fig. 25.7*D–G*).

Locomotion. Gastropods creep upon the broad, flat, ventral foot by a process of body deformation. Small waves of contraction sweep along the length of the foot, one wave closely following another wave (Fig. 25.12). The area of the foot in the contracted region is lifted; at relaxation it is replaced on the substratum a little in front of the point from which it was raised. During each wave of contraction a small section of the foot performs a little step; the summation of all of these little steps gives the appearance of a gliding motion. The substratum is lubricated by large amounts of mucus produced by a large gland in the foot. In some gastropods, the waves

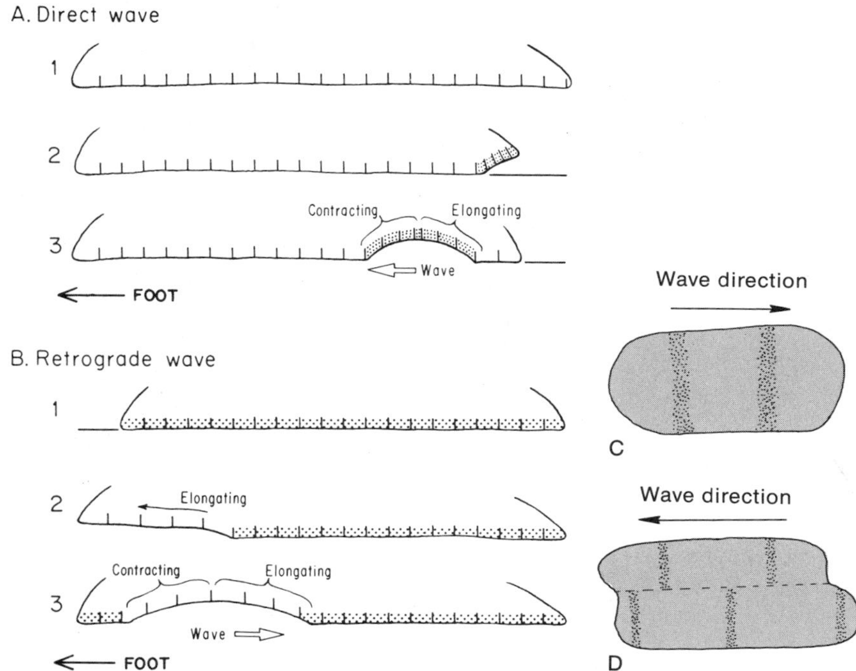

FIGURE 25.12 Pedal creeping in gastropods. *A,* Direct pedal muscular wave, in which the wave is moving in the same direction as the animal, and the muscle fibers in the front part of the wave are contracted. *B,* Retrograde wave, in which the wave is moving in the direction opposite to movement of the animal, and the muscle fibers in the front part of the wave are elongated. *C,* Ventral view of the foot showing monotaxic waves (stippled). *D,* Ditaxic waves. (*A* and *B* from Miller, S. L.: Adaptive design of locomotion and foot form in prosobranch gastropods. *Experimental Marine Biology and Ecology, 14*:99–165, 1974.)

sweep from back to front (direct waves) (Fig. 25.12*A*). In many other species, the waves move from front to back (retrograde waves), opposite to the direction in which the animal is moving (Fig. 25.12*B*). Waves may extend all the way across the foot (monotaxic), or those on one side of the foot may alternate with those on the other side (ditaxic) (Fig. 25.12*C* and *D*). The most common mode of muscular pedal creeping in gastropods is by ditaxic retrograde waves. Gastropods living on soft substrata, such as sand and mud, move by means of the cilia that cover the foot surface. Many species living on soft bottoms can burrow into the substratum.

The foot of the abalone, limpet, slipper shell, and other gastropods living on hard substrata also functions as an adhesive organ. The sucker-like foot and the low platelike or caplike shell enable the animal to live attached to rocks that are swept by currents (Fig. 25.10*A*). Some gastropods are semisessile or sessile. Slipper shells, for example, move about very little. Worm shells have lost the foot, and the shell, which has uncoiled and become tubelike, is attached to rocks and other objects (Fig. 25.10*B*).

The ability to swim has evolved in a number of different gastropod groups—prosobranch heteropods, opisthobranch sea butterflies, and sea hares. All have the foot modified in some way as a propulsive fin (Fig. 25.10*D*).

Nutrition. Gastropods are microphagous or macrophagous, feeding on food particles scraped up with the radula. Some species feed on algae or terrestrial vegetation; other gastropods are carnivorous. The radula of many carnivores have large heavy teeth and may be located at the end of an extensible proboscis (Fig. 25.9). The prey—other invertebrates, especially other mollusks—is held with the foot, and the tissues of the victim are torn and devoured with the radula. Some carnivores use the radula as a drill to penetrate the shells of bivalve prey that have been previously softened with secretions from a gland on the foot or proboscis. Many aquatic gastropods are scavengers or detritus feeders. A few filter-feeding gastropods, like the semisessile slipper shells, use the gills to filter plankton from the ventilating current.

Some gastropods are ectoparasites of bivalves (Fig. 25.13); others are endoparasites in the body wall of echinoderms, such as starfish, or in the coelom of sea cucumbers.

FIGURE 25.13 *Brachystomia,* an ectoparasitic snail, feeding on the body fluids of a clam. (After Abbott.)

The stomach of macrophagous gastropods has the form of a simple sac in which extracellular digestion occurs. Enzymes are provided by the salivary glands, glands along the esophagus, or the digestive glands. Absorption occurs within the digestive glands, and the intestine functions only in the formation of feces.

Circulation, Gas Exchange, Excretion, and Water Balance. Except for the asymmetry produced by torsion and the loss of the right gill, the circulatory system of gastropods is like that described for the ancestral mollusks. The blood of most gastropods contains **hemocyanin**, a respiratory pigment (p. 161).

Ammonia is the excretory waste of aquatic gastropods, but terrestrial pulmonates excrete uric acid. By the addition of a **ureter** formed from the mantle wall, the excretory opening of land pulmonates is located outside of the mantle cavity. The anus also opens outside the mantle cavity; thus, the lung is not fouled with waste. Pulmonates are not especially adapted for avoiding water loss through desiccation, and large amounts of water are lost in the mucus secreted when crawling. Most pulmonates are therefore restricted to humid environments or must be nocturnal. During the winter or during very dry periods the animals hide in leaf mold or beneath wood or stones or attach the shell to vegetation by mucous cords. The aperture of the shell is then covered with a mucous film, which dries to form a protective **epiphragm.** The animal is inactive during this time, and the metabolic rate drops to a very low level. Such a period of dormancy (termed **estivation**) is a common adaptation of semitropical and tropical animals living in regions with long dry seasons. Nocturnal activity and estivation permit some pulmonates to inhabit deserts and other arid regions.

Sense Organs. The head of a gastropod bears one or two pairs of sensory tentacles. The eyes, one located at the base of each tentacle (at the top of the tentacle in land snails), are not highly developed (Fig. 25.11); indeed, in most species they are not capable of object discrimination. The **osphradium** is an important sense organ. The mobile siphon moves about, drawing in water from different areas, and the chemoreceptors of the osphradium enable some species to determine the location of prey or other food.

Reproduction. Most species of prosobranchs have separate sexes. Pulmonates and opisthobranchs are hermaphroditic. In primitive prosobranchs, the gametes are discharged through the nephridia. In most gastropods, however, a complex reproductive tract has developed. Copulation followed by internal fertilization is the usual mode of reproduction. The eggs of aquatic species are deposited in strings or masses or within special cases molded by the foot (Fig. 25.14). The large yolky eggs of pulmonates are laid singly and deposited in small clusters in the soil and leaf mold or beneath bark, logs, or stones.

Cleavage is typically spiral. In primitive marine prosobranchs the initial trochophore larva develops further into a **veliger** larva that displays many gastropod features, such as a spiral shell and foot (Fig. 25.15). Like the trochophore, the veliger is planktonic and swims by means of the **velum**, a large ciliated organ derived from the prototroch of the trochophore. The velum also collects fine suspended particles upon which the larva feeds. In many marine and in all freshwater and terrestrial species, development is direct; i.e., no larval stage is present. Little snails emerge from the eggs or cases. In all of these forms with direct development, the eggs contain more yolk than the eggs of species with a feeding larval stage.

Polyplacophora and Monoplacophora

The **Polyplacophora** and the **Monoplacophora** have a broad, flat, creeping foot. The class Polyplacophora (Gr. *polys*, many + *plax*, plate + *pherein*, to bear) includes about 600 species of **chitons** ranging from a few millimeters to over 35 cm. in length and adapted for living on rocks and other hard substrata (Fig. 25.16). In many ways chitons parallel the prosobranch limpets. The head is reduced, and much of the ventral surface is occupied by the broad foot. Chitons, however, are distinguished from all other mollusks in possessing a shell

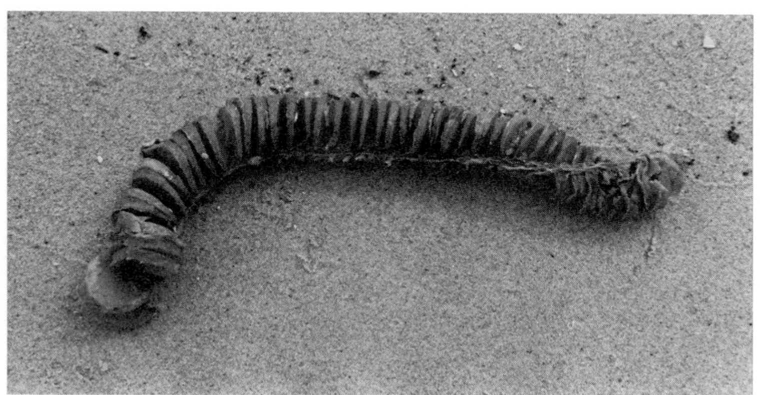

FIGURE 25.14 *A,* Egg cases of the welk, *Busycon carica.* The many cases, each with the shape of a pill box, are connected together by a cord. Development is direct, and the little welks emerge from a hole at the margin of the case. *B,* Egg case, or "sand collar" of the moon shell, *Polinices duplicatus.* (By Betty M. Barnes.)

composed of eight plates arranged linearly from anterior to posterior and overlapping one another. The lateral margins of the plates are overgrown by the mantle to varying degrees.

Chitons crawl about much like snails but are immobile when out of water at low tide. When the chiton is clinging to rocks and other objects, the mantle margin is held tightly against the substratum. The inner edge of

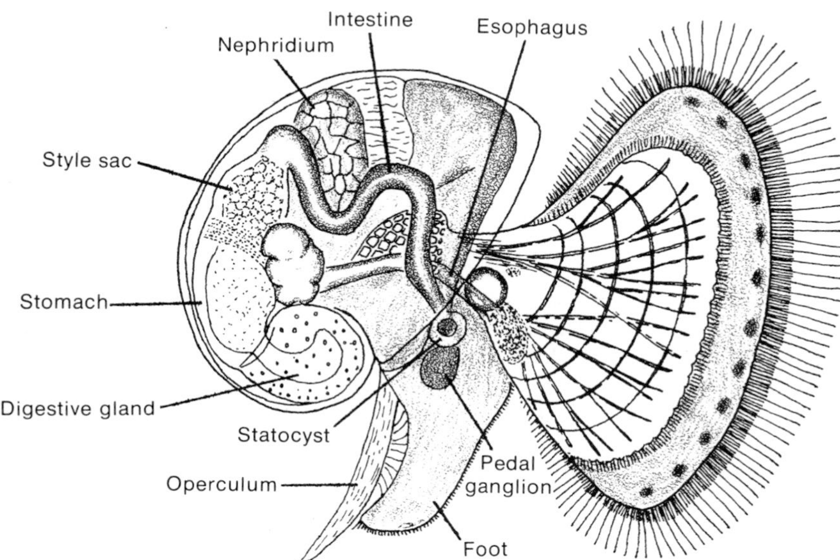

FIGURE 25.15 Lateral view of the veliger larva of gastropod (slipper shell). The ciliated velum is seen on the right. (After Werner from Raven.)

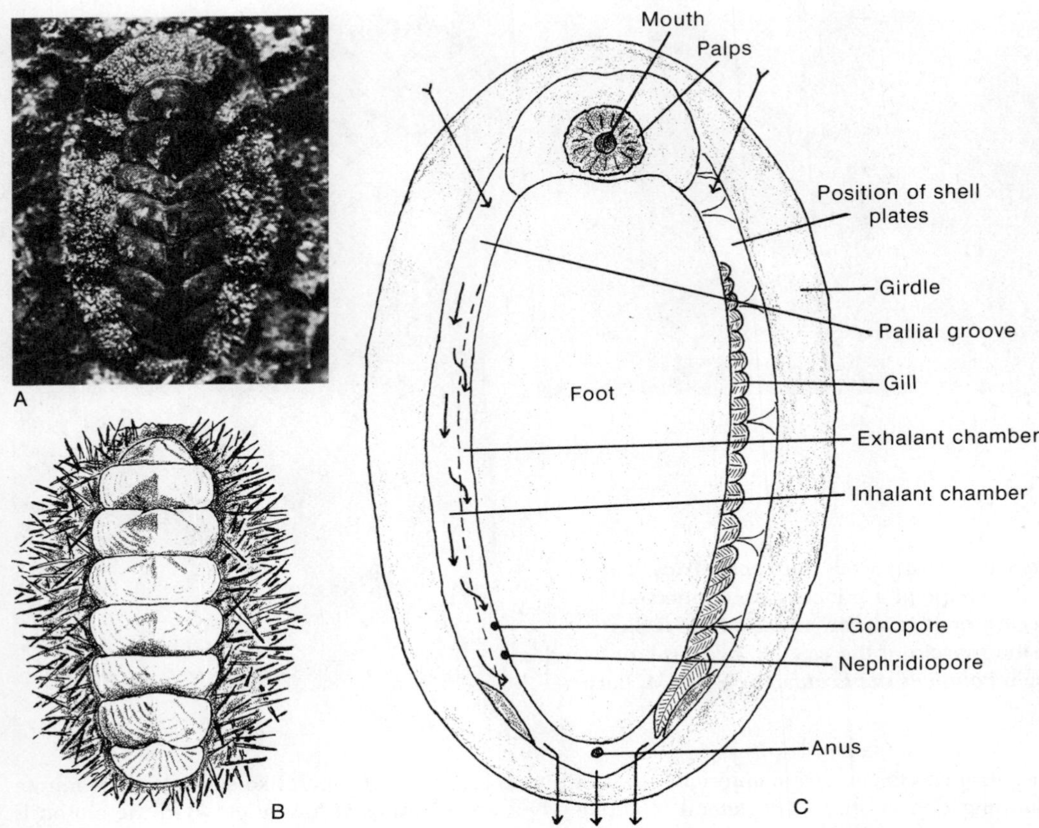

FIGURE 25.16 *A,* Photograph of an intertidal species of the genus *Chiton* from the Bahamas. *B,* A species of *Chiton* in which the mantle bears spines. *C,* Ventral view of a chiton, showing direction of water currents. (*B* after Borradaile and others; *C* after Lang and Haller.)

the mantle is then lifted to create a partial vacuum. This vacuum, combined with the adhesion of the foot, enables the chiton to hold tenaciously to the substratum.

The mantle cavity of chitons is limited to two lateral troughs, or grooves, one on each side of the body between the foot and mantle edge (Fig. 25.16). Within each groove are many gills.

The ventilating current, propelled by cilia on the gills, flows through the mantle grooves from anterior to posterior (Fig. 25.16). The current enters each groove through an opening created by the raised mantle margin. After passing the length of the groove and over the gills, the two currents exit as a single stream at the posterior end of the body, where the mantle edge is raised.

Chitons feed largely on algae scraped from rock surfaces with a radula.

Chitons have separate sexes, and each of the two gonads is provided with a gonoduct that opens near the nephridiopore in the mantle trough (Fig. 25.16C). The nephridia do not serve as gonoducts in chitons. The animals do not copulate, but sperm released by a male fertilize the eggs in the sea, or fertilization occurs in the mantle trough of the female. Eggs are shed singly or in strings, or they may be brooded within the mantle cavity. Spiral cleavage results in a trochophore larva, which develops directly into the adult without an intervening veliger.

The members of the class Monoplacophora were first known from fossils and were thought to be extinct until 10 living specimens were dredged up from a great depth in the Pacific, off the coast of Central America, in 1952. Since that time additional specimens have been collected from depths of 2000 to 7000 m. in other parts of the world. The specimens belong to several species of *Neopilina,* relics of a class that once contained more widely distributed species.

The dorsal surface of monoplacophorans (Gr. *monos,* single + *plax,* plate + *pherein,* to bear) is covered by a symmetrical shield-shaped shell, the apex of which is a little peaked and directed anteriorly (Fig 25.17). The ventral surface is similar to that of chitons, with the mantle cavity in the form of two grooves located to either side of the foot. The unusual feature of monoplacophorans, and the one that evoked special interest, is the replication of parts (Fig. 25.17). There are eight pairs of pedal retractor muscles and six pairs of nephridia. The mantle groove contains five or six pairs of unipectinate gills drained by two pairs of auricles. A radula is present, and the stomach is similar to that described for the ancestral mollusk. Two pairs of nephridia function as gonoducts for two pairs of gonads.

Class Bivalvia

The class **Bivalvia** contains 8000 species of marine and freshwater mollusks commonly called clams or bivalves; it includes the familiar mussels, cockles, oysters, and scallops. The distinguishing characteristics of the class represent adaptations for burrowing in a soft substratum. The body is greatly compressed laterally (Fig. 25.18). The head is very much reduced, the radula is lost, and the shell is composed of two lateral pieces, or valves, hinged together dorsally. Lateral compression has resulted in a great overhang of the mantle and shell, and the large mantle cavity extends to both sides of the body. The anteriorly directed foot is laterally compressed and somewhat bladelike, hence the older name of the class **Pelecypoda**—hatchet foot.

Mantle and Shell. The mantle margin possesses three folds (Fig. 25.19). The muscular inner fold of the mantle edge of one side is pressed against that of the other when the valves are closed. Where these muscles are anchored to the shell, they produce a scar, which appears as a long line (pallial line) a short distance back from the shell margin (Figs. 25.18*C* and 25.19*B*). The middle fold is sensory. The outer fold secretes the **periostracum** and the outer part of the calcareous material of the shell. The remainder of the calcareous material is secreted by the general mantle surface. The calcareous portion of the shell consists of two to four layers laid down as thin sheets (nacre) or prisms or in more complex ways (Fig. 25.19).

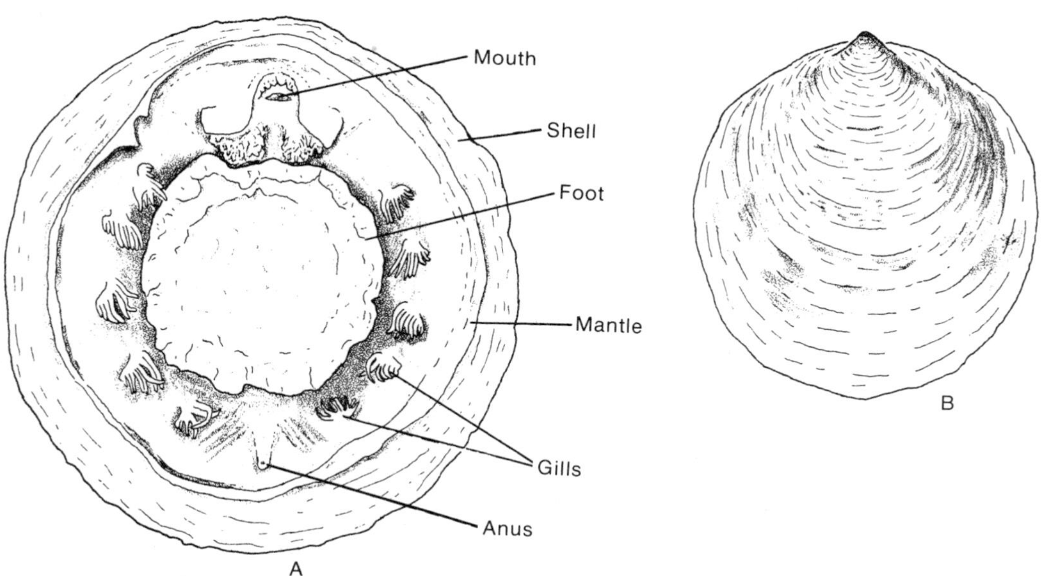

FIGURE 25.17 The monoplacophoran, *Neopilina. A,* Ventral view; *B,* dorsal view of shell. (Adapted from Lemche and Wingstrand.)

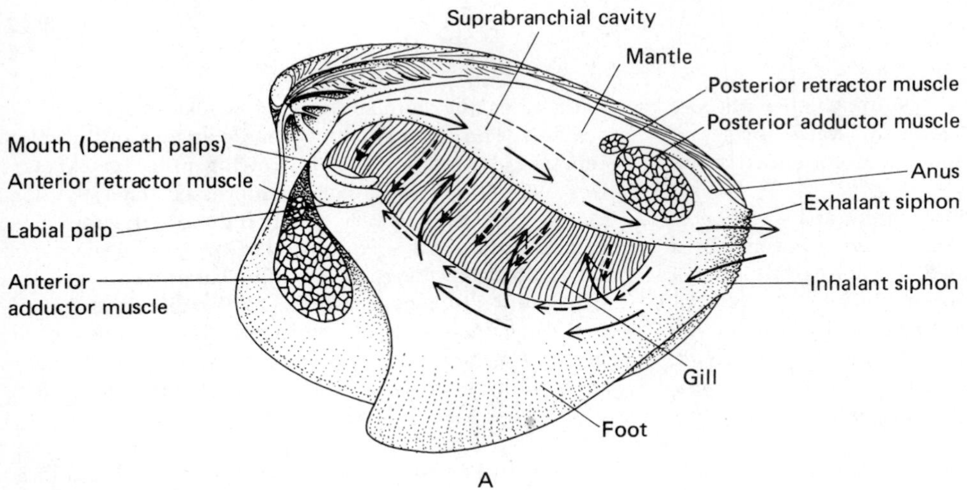

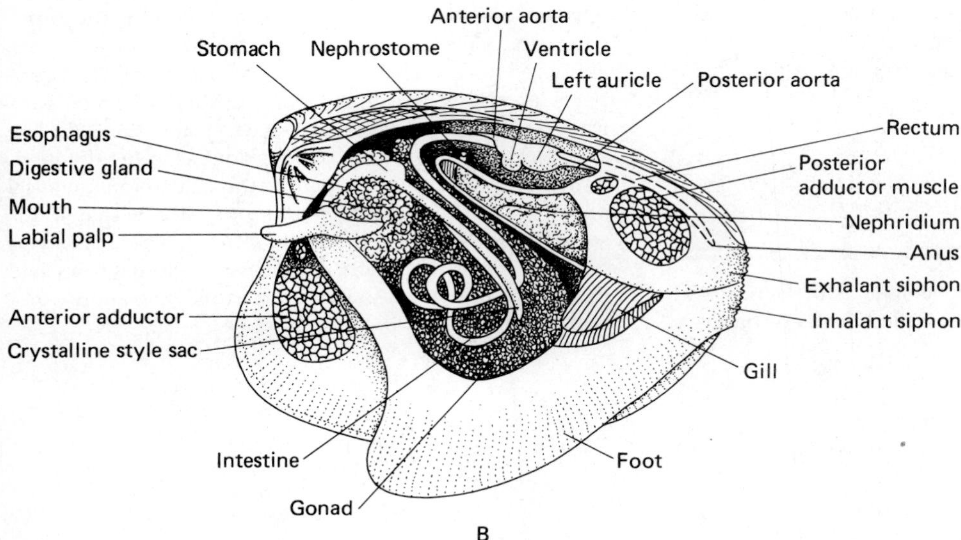

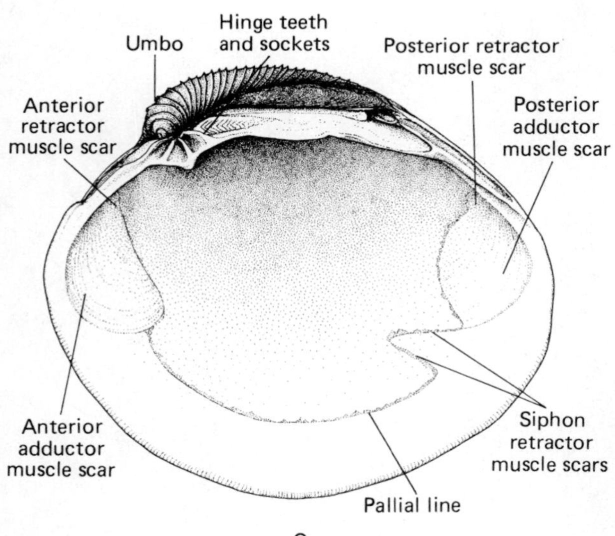

FIGURE 25.18 Anatomy of *Mercenaria mercenaria*. *A,* Animal with left valve and mantle removed. Heavy arrows indicate path of water current; dashed arrows, path of filtered particles. *B,* Section through visceral mass showing internal organs. *C,* Interior view of right valve, showing muscle scars.

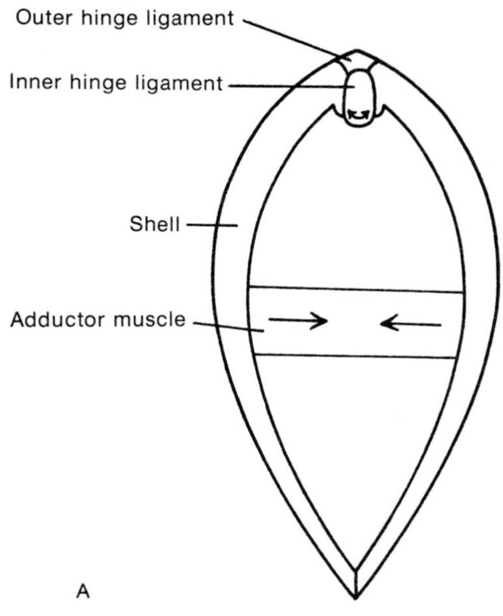

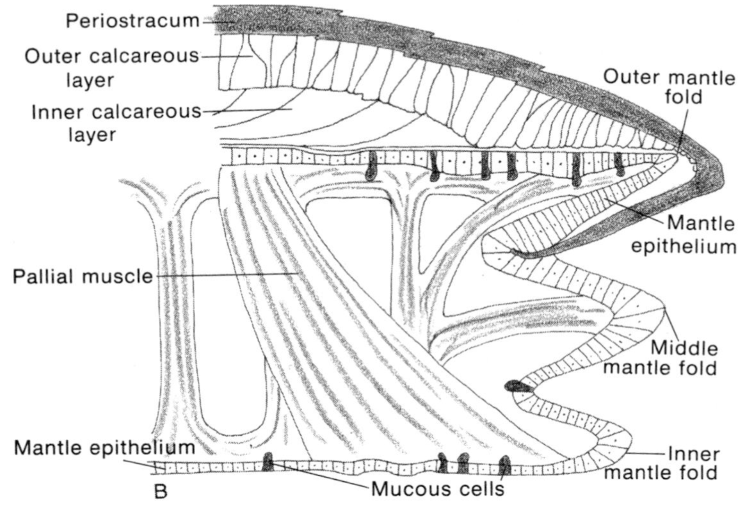

FIGURE 25.19 *A,* Diagrammatic transverse section through a bivalve to show hinge ligament and antagonistic adductor muscle. *B,* Transverse section through margin of shell and mantle of a bivalve, showing mantle lobes. (After Kennedy W. J., et al.: Environmental and biological controls on bivalve shell minerology. *Biological Review, 44*:499–530, 1969.)

Pearls are formed by the deposition of concentric layers of calcareous material around a parasite, a sand grain, or some other foreign object that becomes lodged between the mantle and shell and then within a pocket of mantle tissue.

The hinge ligament that connects the two valves dorsally is composed of elastic protein covered by a layer of periostracum. The valves are opened by tension resulting from compression and stretching of the elastic ligament (Fig. 25.19*A*) and closed by anterior and posterior adductor muscles extending transversely between the valves. The attachment points of the muscles are visible as scars on the inner surface of the valves (Fig. 25.18*C*). Commonly, this inner surface in the region of the hinge possesses teeth or ridges, which interlock with sockets and grooves on the opposite valve (Fig. 25.18*C*).

Protobranch Bivalves. Early bivalves belonged to the subclass Protobranchia, some species of which are still living today. The protobranchs (Gr. *protos,* first + *brangchia,* gills) possess a single pair of posterior-lateral bipectinate gills like those of primitive gastropods (Fig. 25.20*B*). The ventilating current enters the mantle cavity between the posterior and ventral gape of the valves, passes up through the gills, and exits posteriorly and dorsally.

The early protobranch bivalves, like many living species, were probably selective deposit feeders. Such Protobranchs utilize a pair of long tentacles for deposit feeding. Each tentacle is associated with two large flap-like folds, called **labial palps**, located to either side of the mouth (Fig. 25.20*A*). During feeding, the tentacles are extended into the bottom sediments. Deposit material

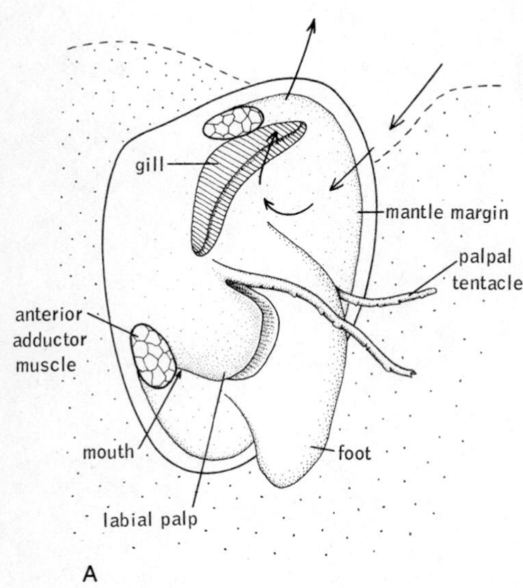

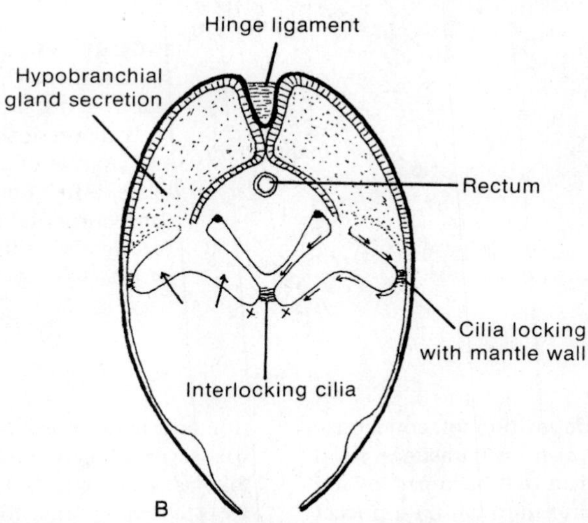

FIGURE 25.20 *A,* A generalized protobranch, showing its position in substratum and the path of water current (*arrows*). *B,* The protobranch, *Nucula,* showing the position of the gills in the mantle cavity (transverse section). (*B* after Yonge.)

adheres to the mucous-covered surface of the tentacle and then is transported by cilia back to the palps. Each pair of palps functions as a sorting device. Light particles are carried by certain cilia to the mouth; heavy particles are carried by other cilia in grooves to the palp margins, where they are ejected to the mantle cavity.

The evolution of selective deposit feeding in protobranch bivalves was correlated with the invasion of a soft bottom environment and the shift of the mouth away from the substratum as the body become laterally compressed for burrowing. The radula disappeared.

Evolution of Lamellibranchs. In some group of early protobranch bivalves, filter feeding evolved. An explosive evolution followed, and the filter feeders, called **lamellibranchs**, came to dominate the bivalve fauna. The gills and ventilating current of protobranchs preadapted them for filter feeding. As the lamellibranchs evolved, plankton in the ventilating current came to be utilized as a source of food, the gills became the filters, and the frontal gill cilia (which originally served for the transport of sediment) became adapted for the transport of the trapped plankton from the filter to the mouth.

The principal modification of the gills for filtering was the lengthening and folding of the gill filaments, which greatly increased their surface area (Fig. 25.21). The long folded filaments are supported by the development of cross connections between the two halves, by connections between adjacent filaments, and by connection of the tips of the filaments to the foot or mantle wall. The folding converted the original single gill into two gills, each new gill formed by one series of folded filaments. The lengthened filaments and their attachment to one another give the gills a sheet- or plate-like form, hence the name Lamellibranchia (L. *lamella*, small plate + Gr. *brangchia*, gills).

Although adjacent filaments are connected, openings (**ostia**) remain for the passage of water between the filaments (Fig. 25.21). The interior space between the two folded halves of the filaments forms water tubes, which connect with the **suprabranchial cavity**, the portion of the mantle cavity above the gills (Figs. 25.18 and 25.21). In lamellibranchs, the ventilating current, now also the feeding current, enters posteriorly and ventrally as in protobranchs. On reaching the gills, the water, propelled by the lateral gill cilia, enters openings between the filaments on all of the gill surfaces. Within the interior water tubes, the water stream flows upward to the suprabranchial cavity, where it turns posteriorly and flows outward through the shell gape (Fig. 25.21).

Suspended particles are filtered out by special laterofrontal cilia as the ventilating current passes between the filaments (Fig. 25.21*D*). The filtered particles are passed to short frontal cilia, become covered with mucus, and are transported downward or upward to food grooves. The food grooves are located ventrally along the gill margins and are dorsally adjacent to the points where the gill is attached (Fig. 25.22 *A* and *B*).

The food grooves carry the collected particles to the labial palps, which retain their original sorting function (Fig. 25.22). Fine particles, mostly phytoplankton, are conveyed to the mouth.

A cord of mucus filled with phytoplankton is carried down the esophagus and wound into the stomach (Fig. 25.23), which is much like that of protobranchs and the ancestral mollusk (they are all microphagous). However, the mass of mucus within the style sac has become compacted into a stiff rod, the **crystalline style**. The rod is secreted by the style sac and rotated by the style sac cilia. In addition to mucus, the style contains carbohydrate-splitting enzymes. The rotating end of the style is abraded against a chitinous piece in the stomach. The rotation of the crystalline style winds in the mucous rope from the esophagus and stirs the contents of the stomach; the abrasion of the end liberates enzymes and initiates the digestion of carbohydrates within the stomach. The churning of the stomach mass throws particles against the sorting region, which separates and conveys fine particles to the digestive glands. Here digestion is completed intracellularly. The rejected coarse particles are carried along a groove to the intestine, where they are compacted into fecal pellets and eventually ejected.

Like primitive gastropods, bivalves possess a single pair of nephridia. The blood lacks respiratory pigments, probably correlated with the large gill surface area and low metabolic rate.

Adaptive Groups of Bivalves. The evolution of filter feeding enabled lamellibranchs to exploit habitats other than shallow depths with soft bottoms. Several groups have evolved, each adapted for survival in a specific environment.

Soft-Bottom Burrowers. Most lamellibranchs have continued to inhabit sand and mud bottoms, but the ability to utilize the ventilating current in feeding has enabled many species to survive while burrowing deeper into the substratum.

Burrowing is accomplished with the foot, which is extended anteriorly between the gape of the valves into the substratum (Fig. 25.24). Initial extension of the foot is provided by a pair of protractor muscles, which extend from the foot to a point on the shell below the anterior adductor muscle. The substratum is simulta-

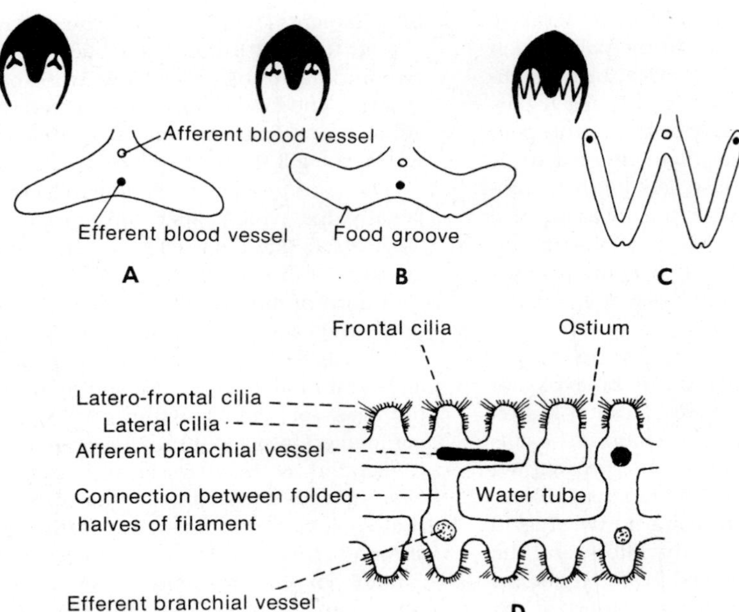

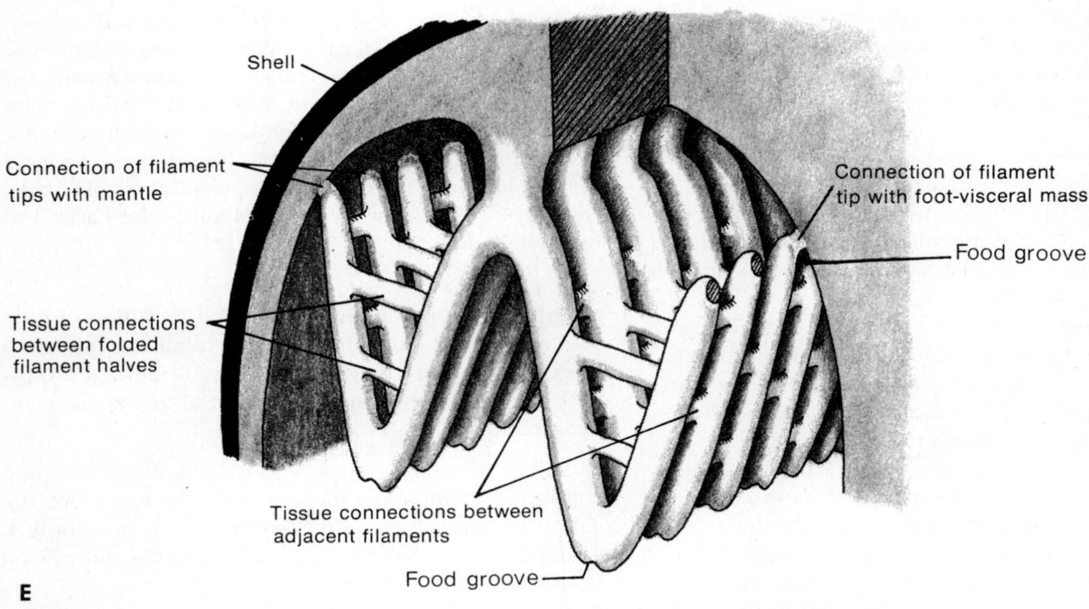

FIGURE 25.21 Evolution of lamellibranch gills. *A,* Primitive protobranch gill (position relative to foot-visceral mass and mantle indicated in cross section; see Fig. 25.20). *B,* Development of food groove in hypothetical intermediate condition. *C,* Folding of filaments at food groove to produce the lamellibranch condition. *D,* Frontal section of lamellibranch gill showing five fused adjacent filaments. *E,* Tissue connections that provide support for the folded lamellibranch filaments. (*A–C* after Yonge.)

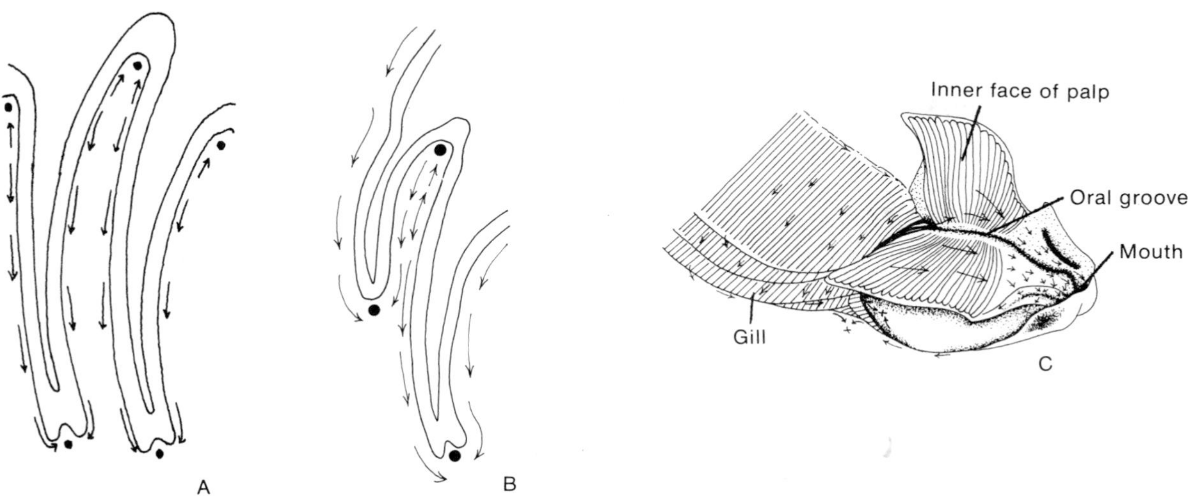

FIGURE 25.22 Transport and sorting of filtered particles by lamellibranch gills. *A* and *B*, Transverse sections of gills on one side (inner gill on right) showing direction of frontal cilia beat and position of anteriorly moving food tracts (black dots). *A* shows primitive condition with five tracts; *B*, condition in many lamellibranchs. *C*, One pair of palps spread apart and anterior section of gills of oyster. Arrows indicate ciliary tracts. (*A* and *B* after Atkins; *C* after Yonge.)

neously softened by water ejected from the mantle cavity as the valves are closed. The closing valves exert pressure on the water remaining within the mantle cavity, which in turn exerts pressure on the visceral mass, driving blood into the foot (Fig. 25.24). The elevated blood pressure further extends the foot and dilates the distal end, anchoring it in the substratum. Two pairs of pedal retractors pull the valves down upon the anchored foot. The pedal retractors may contract somewhat alternately, causing the valves to rock and facilitating their movement through the substratum.

Burrowers in soft bottoms must cope with the tendency of the surrounding sediment to enter with the ventilating and feeding water stream. The problem becomes greater as the animal burrows deeper. Relatively permanent burrows with mucus-compacted walls are formed by many species, especially those that burrow deeply. There has also been an evolutionary tendency for the opposing mantle edges to fuse together. The most common fusion points are around the apertures for inhalant and exhalant water currents, but fusion may also occur ventrally, leaving only an opening for the protrusion of the foot (Fig. 25.24). Ventral mantle fusion undoubtedly aids in the elevation of hydrostatic pressure within the mantle cavity during burrowing. **Siphons,** fused tubelike extensions of the mantle around

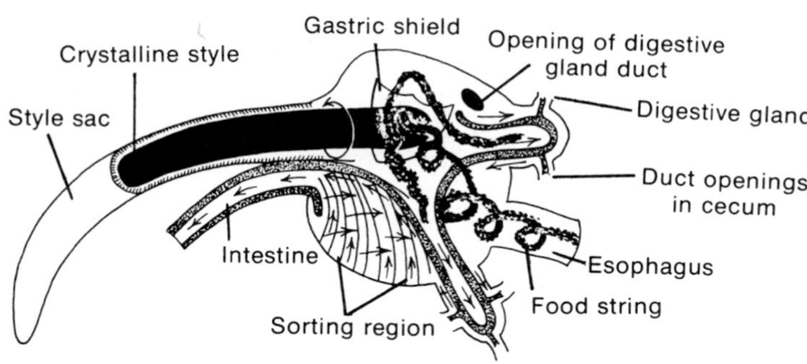

FIGURE 25.23 Diagram of the stomach and digestive gland duct of a lamellibranch, showing rotation of crystalline style and winding of mucous food string. Arrows indicate ciliary pathways. (After Morton.)

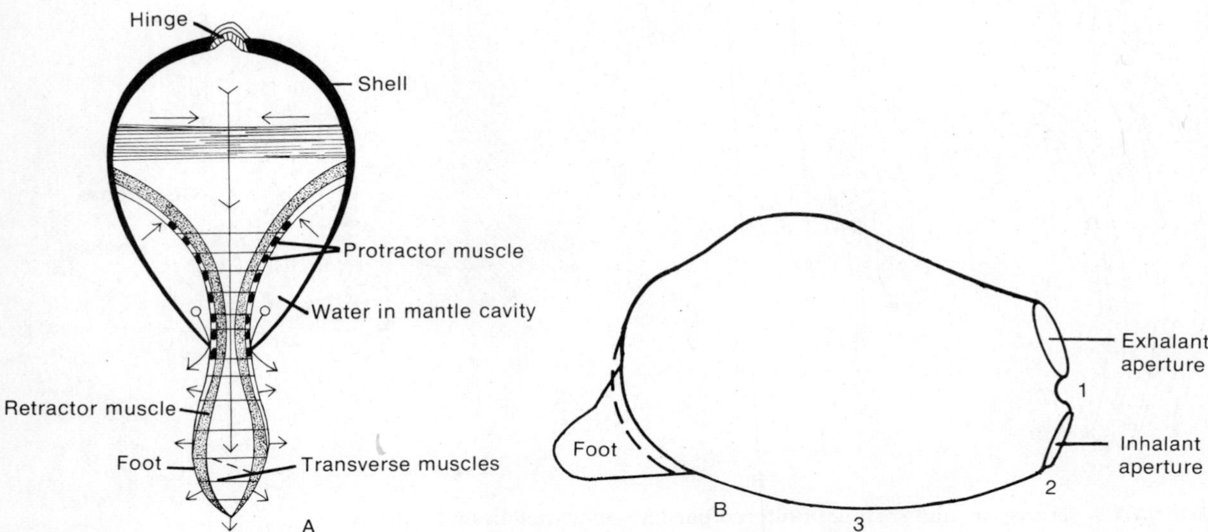

FIGURE 25.24 *A,* Diagrammatic cross section of a bivalve, showing hydrostatic forces that produce dilation of foot. Central vertical arrow indicates flow of blood into foot. *B,* Areas of mantle fusion in bivalves: (*1*) between inhalant and exhalant apertures or siphons, the most common point of fusion; (*2*) below inhalant aperture or siphon; and (*3*) between inhalant aperture and foot aperture. (*A* modified after Trueman, E. R.: *Science, 152*:523. Copyright 1966 by the American Association for the Advancement of Science.)

the inhalant and exhalant openings, are yet another adaptation to reduce the intake of sediment. The siphons are extended upward to the surface and permit the animal to obtain water relatively free of sediment (Fig. 25.25).

Attached Surface Dwellers. Many bivalves have abandoned burrowing and live on the surface of the sea bottom, especially on hard surfaces, such as rock and coral. The animal attaches itself in one of two ways. It may, like the oyster, lie on its side with one valve fused to the substratum (Fig. 25.26*A* and *C*). The foot is absent, and the anterior adductor muscle is reduced or absent. Other bivalves, such as the common mussels, are attached by means of horny **byssal threads,** secreted by a gland in the reduced foot (Fig. 25.26*A* and *B*).

Sediment is much less a problem for bivalves living attached to the surface of hard substrata. It is not surprising therefore that these species exhibit no mantle fusion and do not possess siphons. Some surface-dwelling bivalves, including both mussels and oysters, inhabit the intertidal zone and may be found attached to rocks, pilings, and sea walls that are exposed at low tide. During this period the animals are inactive and keep the valves closed to reduce desiccation.

Unattached Surface Dwellers. A few bivalves, such as scallops and file shells, rest unattached on the bottom or are attached only temporarily. These bivalves can swim in a jerky manner for short distances by rapidly clapping the valves and driving a jet of water from the mantle cavity. Correlated with the mobility is the fact that the sensory lobe of the mantle margin is highly developed and may bear eyes and tentacles (Fig. 25.26*G*).

Hard-Bottom Burrowers. Several groups of bivalves have evolved the ability to burrow into peat, clay, sandstone, coral, and even limestone rock. They use the anterior margins of the valves to drill. Cutting occurs when the two valves are opened, pulled down, or rocked against the head of the burrow. The drilling force is provided by pulling the body against the attachment of the suckerlike foot or against the attachment of byssal threads (Fig. 25.26*D* and *E*).

Shipworms are highly modified to drill into wood. The valves, which function as the drill, are very tiny and

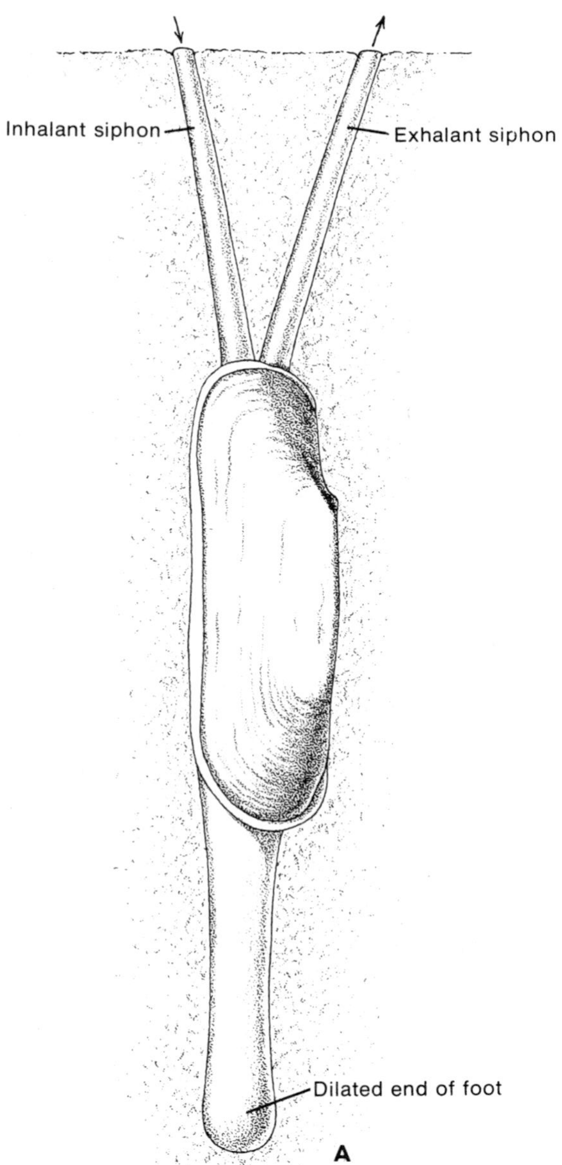

FIGURE 25.25 The razor clam *Tagelus,* a common deep-burrowing bivalve in shallow water along the east coast of the United States.

no longer cover the greatly elongated body and siphons (Fig. 25.26*F*). The animal occupies the entire burrow, which is lined with calcium carbonate secreted by the mantle, and the ends of the siphons are located at the burrow opening. Shipworms ingest the sawdust that they excavate in drilling and digest it with cellulase or with the aid of symbiotic bacteria. Shipworms can do extensive damage to marine timbers, and before the use of metal and fiberglass hulls and antifouling paints, shipworms greatly shortened the life of ships with wooden bottoms.

Reproduction and Development in Bivalves. Most bivalves have separate sexes. The two nephridia serve as gonoducts in the protobranchs, but separate gonoducts are present in most lamellibranchs. The gametes are shed in the exhalant water current, and fertilization occurs in the surrounding sea water. Some bivalves, including most freshwater species, brood their eggs within the water tubes of the gills.

Successive trochophore and veliger larvae are typical of the development of marine bivalves. The veliger is somewhat laterally compressed and covered by a bivalved shell (Fig. 25.27*A*). Many freshwater bivalves have a highly specialized developmental history. There is no veliger, but a modified larval stage, a **glochidium,** is released from the brood chambers of the female (Fig. 25.27*B*). The glochidium settles to the bottom of the stream or lake. When certain species of fish swim over the bottom, the glochidia become attached to the fins or gills by means of an attachment thread or a hook on the ventral margin of each valve. The tissues of the host then overgrow the glochidium, which now becomes a parasite for the remainder of larval development. The fish serves for the dispersion of the bivalves. When development is complete, the young clam breaks free of the host, drops to the bottom, and becomes a free-living adult.

Class Cephalopoda

The class **Cephalopoda** includes *Nautilus,* **squids, cuttlefish,** and **octopods.** There are only some 200 living species, but the rich fossil record indicates that in past geologic periods the class contained thousands of species.

The cephalopod characteristics represent adaptations for a swimming, predatory mode of existence, although there are many species that have secondarily adopted other life styles. The dorsoventral axis has become greatly lengthened (Fig. 25.28). The foot has become divided into **tentacles**, or arms, and has shifted somewhat anteriorly around the mouth, hence the name cephalopod (Gr. *kephalē*, head + *pous*, foot). Cephalopods swim by means of jet propulsion. The force is generated by the contraction of the **mantle,** which becomes locked about the **head**, and water is expelled

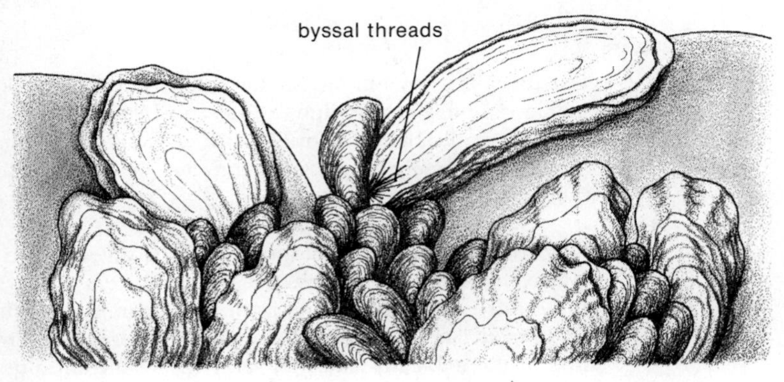

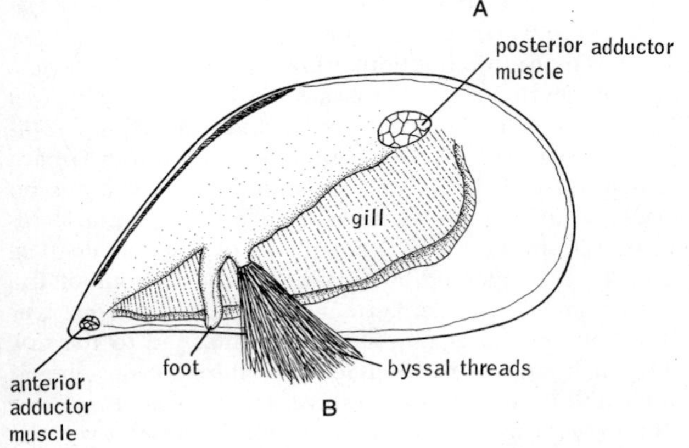

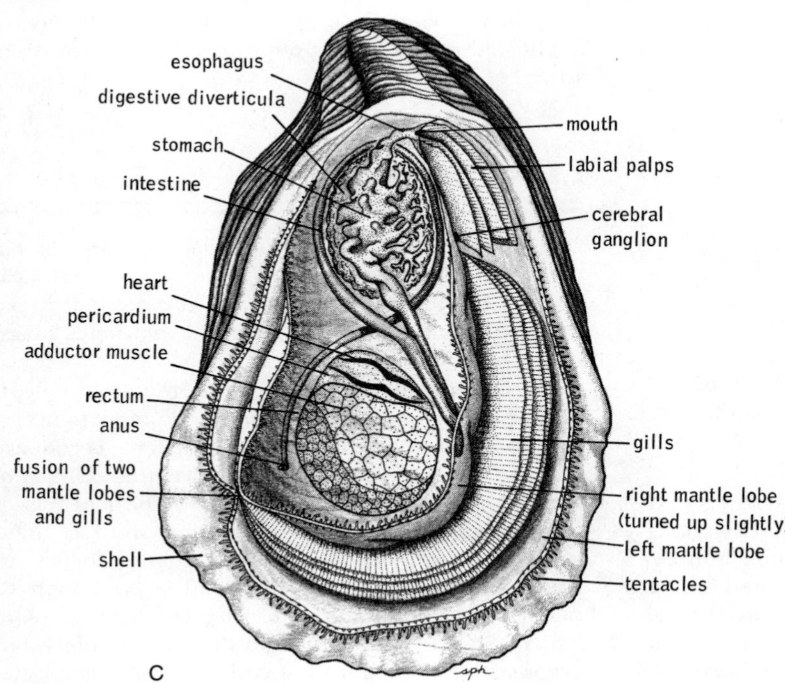

FIGURE 25.26 Bivalve adaptive diversity. *A,* Mussels (*Mytilus*) anchored with byssal threads among oysters (*Crassostrea*). *B,* Lateral view of a mytilid with left valve removed. *C,* Anatomy of the American oyster *Crassostrea virginica.* (Modified after Galtsoff.) *D,* Rock-boring bivalve (*Pholas*) in burrow with siphon projecting to surface. *E,* Ventral view of a similar rock-boring species, showing sucker-like foot. (By Katherine E. Barnes.) *F,* A wood-boring shipworm. Body is greatly elongated. Shell reduced to small drilling pieces at anterior end, but mantle secretes a calcareous lining to burrow. *G,* Gape view of a scallop, showing eyes and tentacles on sensory fold of mantle margin.

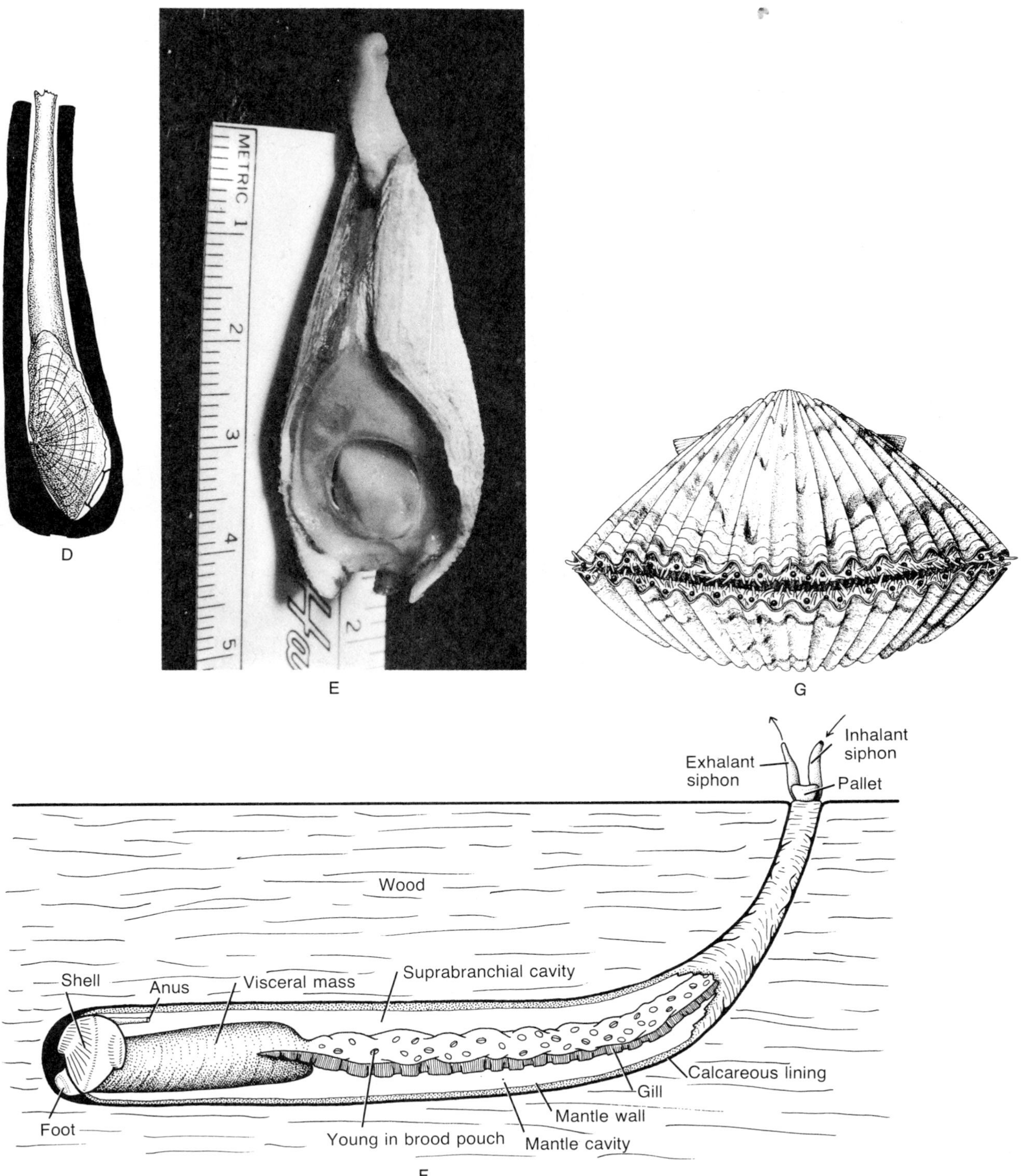

FIGURE 25.26 *Continued*

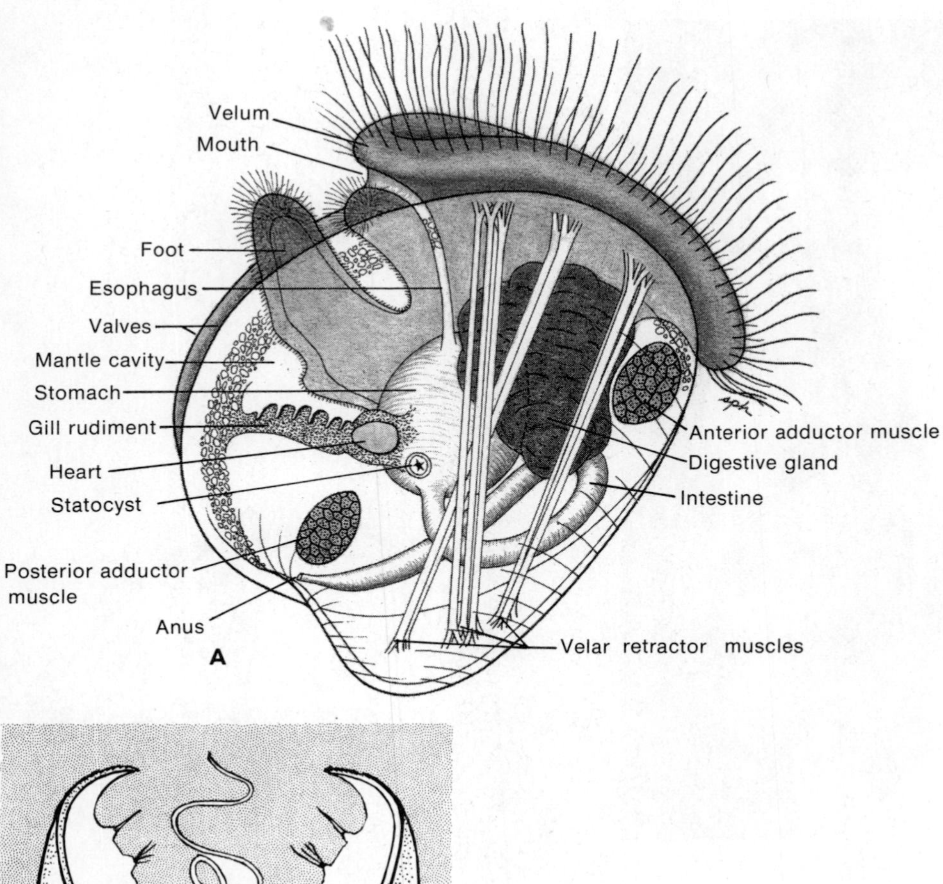

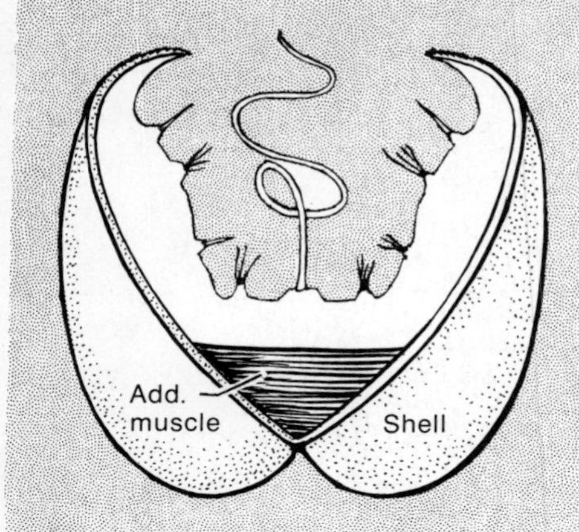

FIGURE 25.27 *A,* A fully developed veliger larva of an oyster. *B,* Glochidium, the larva of a freshwater bivalve. (*A* after Galtsoff.)

from the mantle cavity through a short funnel derived from a part of the foot.

The great increase in the dorsoventral axis and the assumption of a swimming habit have led to a shift in the orientation of the body. The tentacles, which represent the original ventral surface, are directed forward, and the **visceral mass**, which represents the original dorsal surface, is now directed posteriorly.

Cephalopod Shells. Only *Nautilus* among living cephalopods possesses a well-developed **shell** (Fig. 25.29), but a shell was characteristic of thousands of fossil spe-

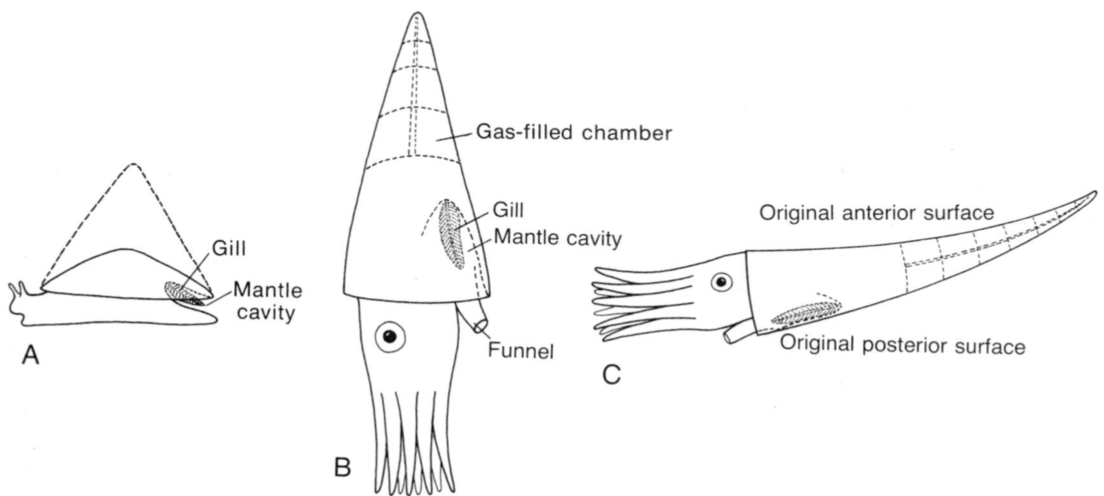

FIGURE 25.28 Evolution of a cephalopod. *A,* Lateral view of hypothetical ancestral mollusk. Dashed line indicates increase in length of dorsoventral axis and a change in the shape of the shell from a shield to a cone. *B,* An early cephalopod oriented in a position that is comparable to the ancestral mollusk. *C,* Actual swimming position of an early cephalopod.

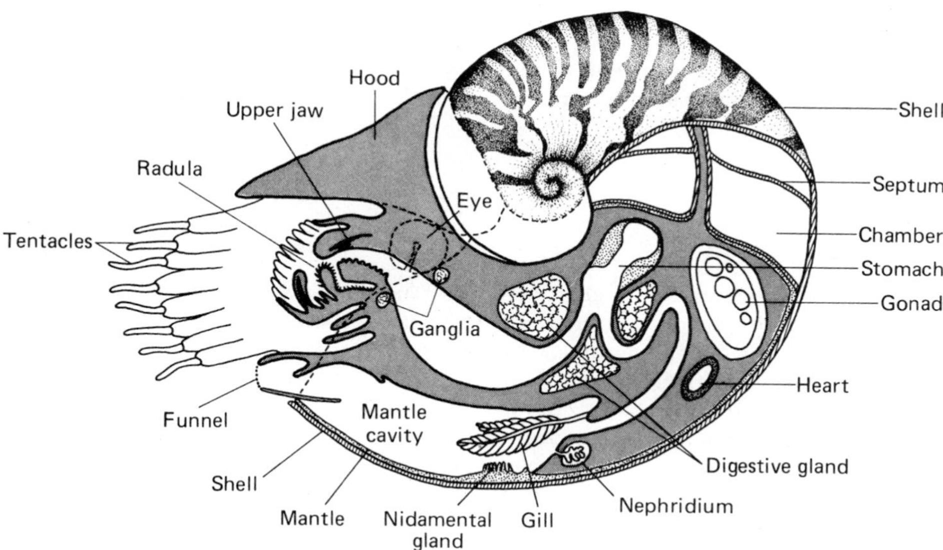

FIGURE 25.29 Lateral view of *Nautilus.* A diagrammatic section of the chambered shell is shown. The mantle of the left side is cut away to show the mantle cavity and two of the gills. When the animal retracts, the leathery hood protects the shell opening.

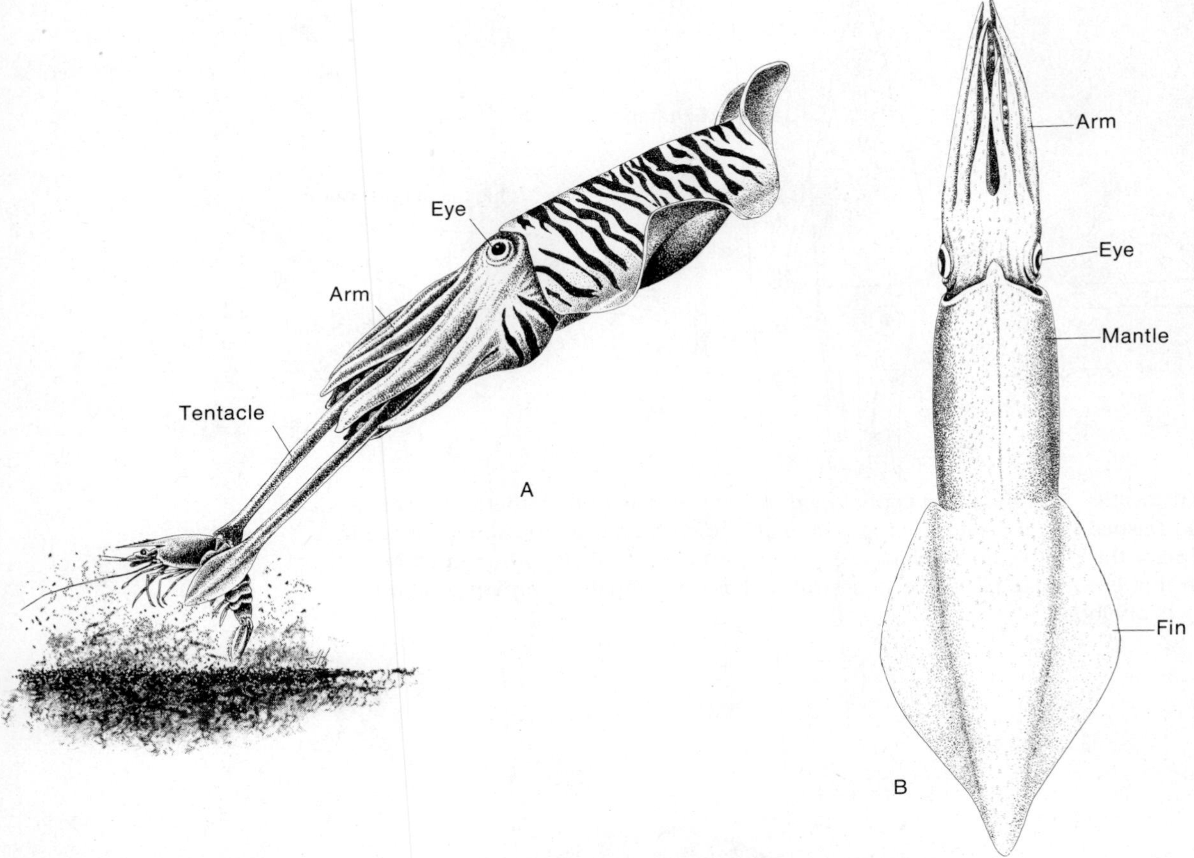

FIGURE 25.30 *A,* The cuttle fish *Sepia* seizing a shrimp with its tentacles. *B,* Dorsal view of the squid *Loligo* in swimming position. Tentacles are retracted and arms are held together, functioning as a rudder.

cies. The cephalopod shell, unlike that of other mollusks, is divided by transverse septa into interior chambers. The living animal occupies only the outer chamber, which opens to the exterior. The posterior chambers are filled with gas, which provides buoyancy for swimming. The gas is secreted by a cord of mantle tissue that extends back through the septa (Fig. 25.28).

In squids and cuttlefish, the shell has become greatly reduced and completely overgrown by the mantle. In octopods, the shell has disappeared completely.

The largest living cephalopods are giant squids of the genus *Architeuthis,* which may reach 16 m. in length and are bottom dwellers at 200 to 400 m. There are no giant octopods. The largest shelled fossil species were ones with straight conical shells up to 5 m. long and ones with coiled shells that were 2 m. in diameter.

Locomotion. The squids are the most powerful cephalopod swimmers. The body of a squid is torpedo-shaped (Fig. 25.30*B*); its posterior lateral fins are used as stabilizers. The arms are held together and function as a rudder. The siphon can be directed anteriorly or posteriorly to permit backward or forward swimming.

In very rapid swimming, contractions of the mantle muscles are synchronized by a system of **giant neu-**

FIGURE 25.31 An octopus. (Courtesy of G. I. Bernard, Oxford Films/Animals Animals, Inc.)

rons. The greater the diameter of a neuronal process, the faster it will conduct an impulse. The motor neurons radiating out of the mantle ganglia are of different diameters depending upon the distance that the impulse must be transmitted. Long neurons have larger diameters than do short neurons. As a result of these structural differences, an impulse originating in the mantle ganglion reaches all of the muscle fibers at the same time, ensuring simultaneous contraction.

Cuttlefish have shorter bodies than squids and are agile but not powerful swimmers (Fig. 25.30*A*). Their fins undulate and contribute to propulsion. Many species live in shallow water, hovering and darting over the bottom. Some lie on the bottom covered with sand during the day or while waiting for prey.

The body of octopods is rather globular (Fig. 25.31). They are largely bottom dwellers and crawl about with the arms, swimming only to escape. They are most commonly found on rocky and coralline bottoms. Some deep-sea octopods and squids have become pelagic. The arms are connected together by webbing and the animals swim like jellyfish.

Nutrition. Cephalopods are highly adapted for raptorial feeding and a carnivorous diet. Fish, shrimp, crabs, and other mollusks serve as food. The prey is seized by the many tentacles. Squids and cuttlefish have eight arms and two long prehensile tentacles (Fig. 25.30*A*); octopods possess only the eight arms. Except in *Nautilus,* the arms and tentacles are provided with suckers.

The principal ingestive organ is a pair of powerful, highly mobile, beaklike, horny **jaws,** which can cut and tear prey (Fig. 25.32*B*). The radula functions as a tongue for pulling in pieces of flesh bitten off by the jaws. As a further aid to dispatching the victim, a pair of salivary glands have been modified as poison glands.

Internal Structure and Physiology. Many features of their internal structure and physiology reflect the active, raptorial mode of life of cephalopods. The stomach is quite different from that of other mollusks. Digestion is extracellular, and the digestive glands produce large amounts of powerful proteolytic enzymes. Cilia are no longer needed on the gills and mantle since muscle contraction provides for the propulsion of the ventilating current, now also the locomotor current. A pair of **secondary hearts** elevates the blood pressure on entrance into the gills (Fig. 25.32*A*) and blood passes through the gill filaments within **capillaries,** increasing the rate of transport and reflecting a higher metabolic rate. Within the blood, oxygen is carried by **hemocyanin.** A single pair of **nephridia** provides for excretion.

The nervous system and sense organs of cephalopods are among the most highly developed of invertebrates. The ganglia, which in many other mollusks are located at different points along the nerve cords, are concentrated in cephalopods at the anterior end to form a complex **brain** housed within a cartilaginous capsule. Functional centers of the brain have been identified, and the behavior of the octopus and other cephalopods has been studied extensively.

The **eye** is highly developed and parallels the eye of vertebrates to a striking degree. A **retina,** a movable **lens,** an **iris diaphragm**, and a **cornea** are present (Fig. 11.19), but the photoreceptors in the retina are directed *toward* the source of light (*direct* eye) instead of *away* from the source of light as in vertebrates (*indirect* eye).

The arms are provided with many tactile cells and chemoreceptors and play an important role in the structural and chemical discrimination of surfaces. A **statocyst** is located on either side of the brain.

An **ink gland** opens into the intestine and is provided with an ejector mechanism that enables the animal to discharge a cloud of dark ink. It has been suggested that the cloud functions not as a smoke screen but as a dummy, confusing the predator and enabling the animal to escape. The alkaloids in the ink may narcotize the chemoreceptors of predators such as fish.

Chromatophores are large, pigment-laden cells to which muscle fibers are attached. Yellow, orange, red, blue, and black chromatophores may be present, depending upon the species. The contraction of the muscle cells expands the chromatophores, and the color changes produced appear to function as background

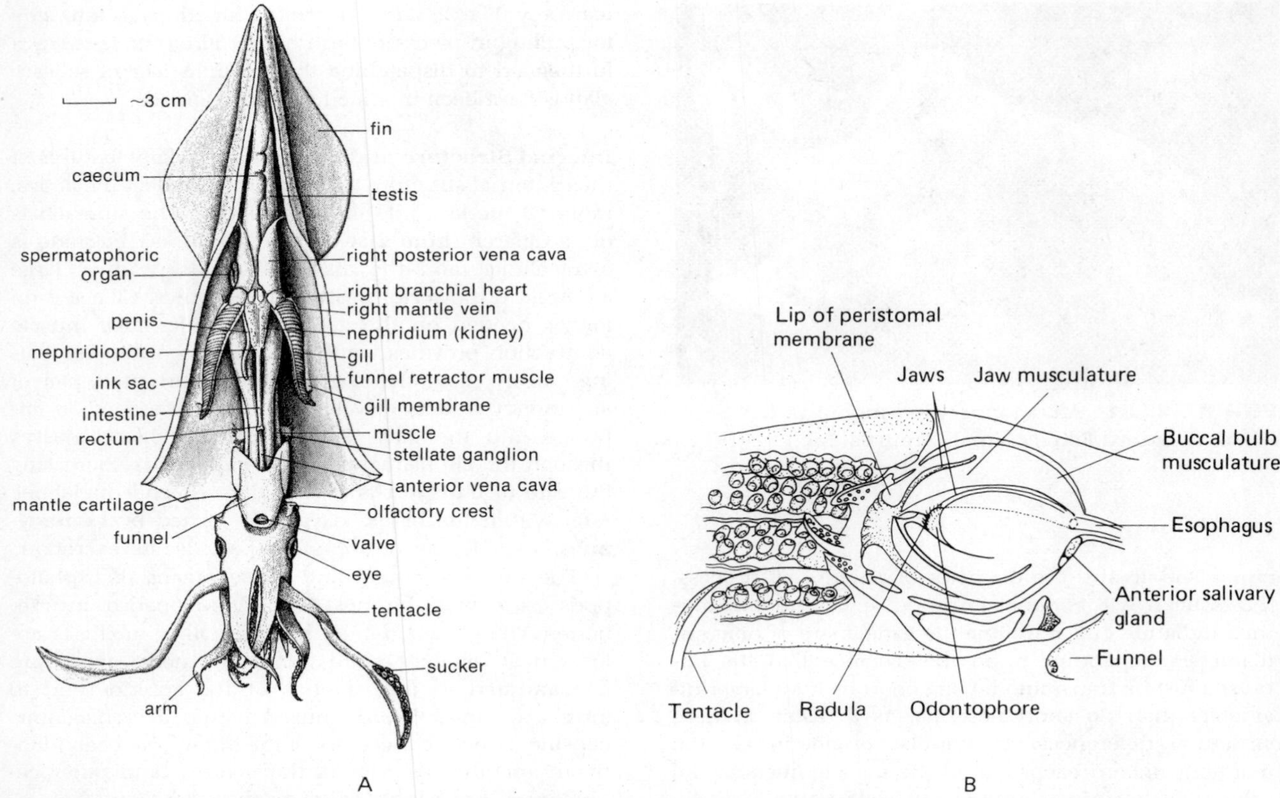

FIGURE 25.32 *A,* Ventral view of the squid *Loligo* with mantle cut and opened to show structures within mantle cavity. *B,* Longitudinal view through buccal mass, showing base of arms and tentacles on right side.

simulations, in the courting behavior of males, in aggressive responses between individuals, and in alarm reactions to intruders.

Deep-sea cephalopods do not have chromatophores but many are **bioluminescent.** The light is produced in special organs called **photophores** located on different parts of the body, often in patterns and composed of different colors. Photophores are found in other deep-sea animals, such as shrimps and fish, and probably play some role in species and sex recognition and in countershading.

Reproduction and Development. The sexes are separate in cephalopods. The gonad in each sex is provided with a relatively complex gonoduct. Copulation involves the transfer of sperm in packets termed **spermatophores.** Masses of sperm are cemented to one another or encased within special secretions in certain regions of the sperm duct.

In copulation, the male seizes the female head on and interlocks his arms with hers. One especially modified arm of the male reaches into his mantle cavity and plucks a mass of spermatophores from the gonoduct. The copulatory arm of the male then deposits the spermatophores in the mantle cavity of the female. The spermatophores discharge their sperm shortly afterwards, and the eggs are usually fertilized on leaving the female gonoduct.

In some pelagic cephalopods, the eggs are planktonic. In most, however, the eggs are deposited in strings on the bottom or are attached to stones or other objects. Octopods remain with their eggs, guarding and cleaning them of sediment. Development is direct, and the young have the adult form on hatching.

Classification of the Mollusca

Class Gastropoda Snails. Mantle and visceral mass exhibit some degree of torsion. Shell typically spiraled. Foot flattened and head well developed.

Subclass Prosobranchia Marine and freshwater, gill-bearing species in which the mantle cavity and contained organs are located anteriorly. Shell is present and operculum is common. Most marine snails with well-developed shells belong to this subclass.

Subclass Opisthobranchia Visceral mass has undergone detorsion, and some degree of reduction of shell and mantle cavity is common. Hermaphroditic. Entirely marine. Bubble shells, sea hares, sea butterflies, and sea slugs.

Subclass Pulmonata Gills absent and mantle cavity converted into a lung. Shell usually present but never an operculum. Hermaphroditic. Terrestrial and freshwater. Most land snails of temperate regions are pulmonates.

Class Polyplacophora Chitons. Body greatly flattened dorsoventrally. Head reduced. Shell composed of eight linearly arranged, overlapping plates. Marine.

Class Aplacophora Small group of aberrant marine mollusks with wormlike bodies and no shells.

Class Monoplacophora Small group of marine deepwater species. Body flattened dorsoventrally. Shieldlike shell. Various organs replicated—eight pairs of retractor muscles, five or six pairs of gills, six pairs of nephridia, and two pairs of auricles.

Class Bivalvia Clams, or bivalves. Body greatly flattened laterally. Shell composed of two lateral valves hinged dorsally. Head reduced.

Subclass Protobranchia Primitive marine bivalves with one pair of unfolded gills.

Subclass Lamellibranchia Marine and freshwater filter feeding bivalves with folded gills. Most bivalves belong to this subclass.

Subclass Septibranchia Small group of marine species having the gills modified as a pumping septum.

Class Scaphopoda Tusk or tooth shells. Burrowing marine mollusks having a tusklike shell open at each end.

Class Cephalopoda *Nautilus*, cuttlefish, squids, and octopods. Marine, mostly swimming mollusks, having the foot divided into tentacles, or arms. Shell, when present, usually divided into chambers, of which only the most recent is occupied by the living animal.

Subclass Nautiloidea Mostly fossil cephalopods with straight or coiled chambered shells. *Nautilus* is only living genus.

Subclass Ammonoidea Fossil cephalopods with coiled shells having wrinkled septa.

Subclass Coleoidea Shell internal or absent. Cuttlefish, squids, and octopods.

Summary

1. Members of the phylum Mollusca constitute one of the largest phyla of animals and are found in the sea, in fresh water, and on land. They are distinguished by the presence of a muscular foot, a calcareous shell secreted by the underlying integument, called the mantle, and the feeding organ, the radula.

2. The earliest mollusks are thought to have possessed a flat creeping foot, a dorsal shield-shaped shell, and a poorly developed head. Primitively, mollusks possess gills housed within a mantle cavity created by the overhanging mantle and shell. The gills are composed of flattened filaments projecting to either side of a supporting axis. Lateral cilia on the filaments create the ventilating current.

3. Primitively, the radula, a belt of recurved chitinous teeth stretched over a cartilage base, functions as a scraper, and the stomach is adapted for processing fine algae and other particles scraped up by the radula.

4. The blood vascular system is open, and the excretory organs are metanephridia. The usual sense organs are statocyst, eyes, tentacles, and osphradia. Primitively, the nephridia function as gonoducts, fertilization is external, cleavage is spiral, and a trochophore is the first larval stage.

5. The Gastropoda, including the snails and slugs, is the largest and most diverse class of mollusks. The distinguishing feature is torsion, a twisting of the visceral mass during the development of the snail. The head is well developed, the foot is broad and flat, and the shell is asymmetrically coiled.

6. Most living gastropods possess a single unipectinate gill located within the left side of the mantle cavity. The ventilating current enters the mantle cavity on the left and leaves on the right. Most marine gill-bearing gastropods are members of the subclass Prosobranchia. The subclass Opisthobranchia contains a smaller number of marine forms that have undergone partial detorsion and includes the sea butterflies, sea hares, and sea slugs. Members of the subclass Pulmonata are freshwater and terrestrial snails in which the gills have been lost and the mantle cavity converted into a lung.

7. Most gastropods creep by means of the foot, propelled by cilia or by waves of muscle contraction. They exhibit a great diversity of feeding habits, for which the radula is variously modified. The stomach is usually a simple sac and the site of extracellular digestion.

8. Most prosobranchs are gonochoristic; pulmonates and opisthobranchs are hermaphroditic. In either case there is usually copulation and internal fertilization. Development is indirect in many marine species, but in most of these a veliger is the hatching stage.

9. Chitons, members of the class Polyplacophora, are adapted for living on hard substrates, especially in the intertidal zone. The body is dorsoventrally flattened; the broad foot and mantle provide for gripping; and the shell is composed of eight linearly arranged overlapping plates. Food, especially algae, is scraped from rock surfaces by the radula.

10. The small number of living species of the class Monoplacophora are deep sea relics of a much larger and more widespread group of ancient mollusks. The repetition of both external and internal structures—gills, retractor muscles, auricles, and nephridia—is a distinctive feature of living monoplacophorans.

11. Members of the class Bivalvia are adapted for burrowing in soft sediments, although many species have secondarily become adapted for other life styles. The body is laterally compressed and covered by two lateral shell valves. The valves are forced open by dorsal elastic hinge ligaments; the valves are closed by the contraction of two (one in some) adductor muscles that extend between the valves. The head is greatly reduced. Fertilization is external, and development is usually indirect with both a trochophore and veliger larva.

12. Many primitive bivalves (protobranchs) are selective deposit feeders, but most members of the class are filter feeders. In the adaptation of the gills for filter feeding, the filaments became long and folded, all of the filaments on one side of the original gill forming two filtering surfaces. The gills are strengthened by tissue junctions between various parts of the gill. Trapped particles (primarily phytoplankton) are transported up or down the gill filament by frontal cilia and then anteriorly to the mouth in food grooves. Particles are sorted by the labial palps before ingestion. Bivalves have lost the radula. The stomach contains a crystalline style, which functions in mixing and extracellular digestion. Much digestion occurs intracellularly within the digestive glands.

13. Bivalves burrow by extending the anteriorly directed and bladelike foot out into the sediment. The foot becomes anchored by engorgement with blood, and the body is then pulled down behind the foot. Most burrowing species have the mantle margins drawn out as siphons, permitting more direct access to surface waters.

14. Members of the class Cephalopoda are designed for a pelagic raptorial existence. The foot has become modified as arms or tentacles arranged around the mouth. The many species of fossil cephalopods had cone-shaped or bilaterally symmetrical coiled shells. Only *Nautilus* among living species possesses an external shell; in all others the shell is reduced or lost. Cephalopods swim by ejecting water from the mantle cavity through a funnel. The gas content, which can be regulated, within the compartments of the shells of *Nautilus* and cuttlefish provides for buoyancy. Octopods are benthic and exhibit only escape swimming.

15. Prey is seized with the arms and tentacles, which are provided with suckers, and then bitten and torn with a pair of powerful beaklike jaws. The radula is used for ingestion. Many features of cephalopods are correlated with their very active life and higher metabolic rate: absence of gill cilia, closed blood vascular system, branchial hearts, presence of hemocyanin, highly developed eyes, complex nervous system and behavior, chromatophores, and ink gland. Cephalopods exhibit head-on copulation; one of the arms of the male transfers spermatophores to the female. Development is direct.

References and Selected Readings

The references listed below are devoted solely to mollusks. Accounts of mollusks may also be found in the general references listed at the end of Chapter 19.

Abbott, R. T.: *American Seashells.* 2nd ed. Princeton, New Jersey, D. Van Nostrand Co., 1974. An authoritative guide to the marine mollusks of the Atlantic and Pacific coasts of the Western Hemisphere.

Burch, J. B.: *How to Know the Eastern Land Snails.* Dubuque, Iowa, W. C. Brown Co., 1962. A field guide to the pulmonates.

Fretter, V., and J. Peake (Eds.): *Pulmonates*. Vols. I–III. London, Academic Press, Vol. I 1975, Vol. IIA 1978, Vol. IIB 1979. These three volumes cover all aspects of the biology of pulmonate gastropods.

Linsley, R. M.: Shell form and the evolution of gastropods. *American Scientist, 66*(Ju.–Aug.):432–441, 1978. How shell coiling is related to problems of ventilating water flow.

Purchon, R. D.: *The Biology of the Mollusca.* 2nd ed. Oxford, Pergamon Press, 1977. A general account of the mollusks.

Rehder, H. A.: *The Audubon Society Field Guide to North American Seashells.* New York, Knopf, 1981.

Roper, D. F. E., and K. J. Boss: The giant squid. *Scientific American, 246*(Apr.):96–105, 1982. A review of our knowledge of giant squids from beached specimens and other sources.

Runham, N. W., and P. J. Hunter: *Terrestrial Slugs.* London, Hutchinson University Library, 1970. A general biology of pulmonate slugs.

Solem, G. A.: *The Shell Makers.* New York, John Wiley & Sons, 1974. This little book is largely concerned with gastropods, especially land snails. A good complement to the volume of Yonge and Thompson.

Wells, M. J. : *Octopus: Physiology and Behavior of an Advanced Invertebrate.* London, Chapman & Hall, 1978.

Yonge, C. M., and J. E. Thompson: *Living Marine Molluscs.* London, Wm. Collins & Sons, 1976. A general biology of the mollusks.

Yonge, C. M.: Giant clams. *Scientific American, 232*(Apr.): 96–105, 1975. A general account of the biology of these unusual bivalves.

26 Annelids

ORIENTATION

Members of the phylum Annelida, called segmented worms, are common and widespread animals in the sea, fresh water, and soil. Their structural design is closely related to the patterns of locomotion that have developed within the phylum. In this chapter we will begin by examining the metameric ground plan of annelids and see how it reflects the ancestral adaptations for peristaltic burrowing in soft marine sediments. Following this introduction, we will survey the very diverse living marine annelids in which new modes of locomotion and life styles have evolved. Then we will see how the freshwater annelids and earthworms have retained a peristaltic mode of movement although most have left the sea. Finally, we will examine leeches, which are related to the freshwater annelids but have evolved yet another pattern of locomotion correlated with a predatory or bloodsucking life style.

CHARACTERISTICS OF THE ANNELIDS

1 The phylum Annelida contains the segmented worms, which inhabit the sea, fresh water, and land.

2 The long body is metameric, and externally segments are usually demarcated by grooves and serial repetition of appendages. Internally, metamerism is reflected in the compartmentation of the coelom by a septum between each segment. Annelidan metamerism is believed to have evolved as an adaptation for peristaltic burrowing.

3 A pharynx and intestine are usually the most conspicuous parts of the generally straight gut tube.

4 The excretory organs are paired segmental nephridia.

5 The circulatory system is at least partially closed and is composed of a dorsal vessel that pumps blood anteriorly and a ventral vessel that transports blood posteriorly.

6 The nervous system is composed of a pair of dorsal anterior cerebral ganglia (brain), a pair of connectives around the gut, and a pair of longitudinal ventral nerve cords. Paired lateral nerves arise from a ganglionic swelling of the ventral nerve cord in each segment.

7 Marine annelids (polychaetes) have separate sexes, and the gametes usually arise from the peritoneum of most segments, mature in the coelom, and exit by gonoducts, nephridia, or by rupture. Fertilization is external, and a top-shaped trochophore is the larval stage. Freshwater and terrestrial annelids (oligochaetes and leeches) are hermaphroditic, with gonads and gonoducts in a few segments. The eggs are deposited within a cocoon secreted by certain glandular segments called the clitellum.

The phylum **Annelida** includes the familiar earthworms and leeches as well as many marine and freshwater worms. In contrast to the layman's general image of "worms," some marine species are very beautiful animals. More than 11,000 annelids have been described, and many species are widely distributed and occur in great numbers.

Metamerism and Locomotion

The most striking annelid characteristic is the division of the body into similar parts, or **segments,** arranged in a linear series along the anterior-posterior axis. Each segment is termed a **metamere** (Fig. 26.1), and this condition is called **metamerism.** Only the trunk is segmented. Neither the head, or **prostomium,** anterior to the mouth, nor the terminal part of the body, the **pygidium,** which carries the **anus,** is a segment. New segments arise in front of the pygidium; the oldest segments lie just behind the head. Some longitudinal structures, such as the gut and the principal blood vessels and nerves, extend the length of the body, passing through successive segments; other structures are repeated in each segment, reflecting the metameric organization of the body.

Metamerism appears to have evolved twice in the evolutionary history of the Animal Kingdom, once in the evolution of annelids and arthropods and once in the evolution of chordates. In both instances, the condition appears to have evolved as an adaptation for locomotion. In chordates, metamerism probably represented an adaptation for undulatory swimming (p. 644). The ancestors of the annelids were probably elongate coelomate animals that inhabited marine sand and mud; metamerism probably arose as an adaptation for peristaltic burrowing.

The **coelom** is compartmented by transverse **septa,** and the longitudinal and circular muscles are organized as two continuous cylinders outside of the coelomic compartments. One coelomic compartment and its surrounding body wall constitute a segment, which can usually be recognized externally by transverse grooves in the integument encircling the body (Fig. 26.2).

It is important to note that annelidan metamerism is basically a modification of the coelom, making possible a localized hydrostatic skeleton. The nervous, circulatory, and excretory systems are also metameric, i.e. parts are repeated in each segment. But the metamerism of these structures probably reflects an adjustment of these supply systems to the primary metamerism of the coelom (Fig. 26.1).

The typical annelid has a nervous system with a dorsal **brain** and a ventral longitudinal **nerve cord,** a

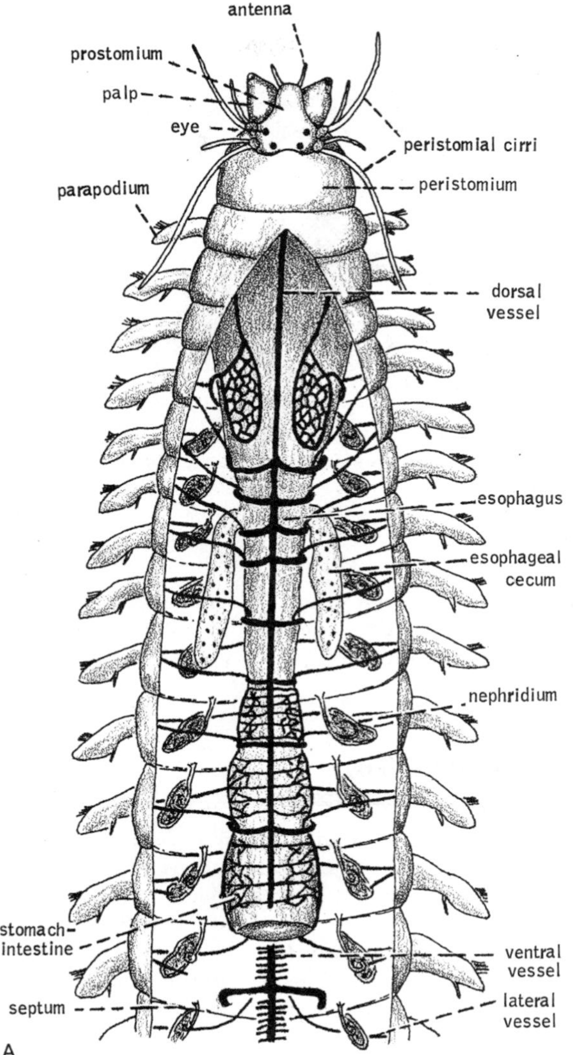

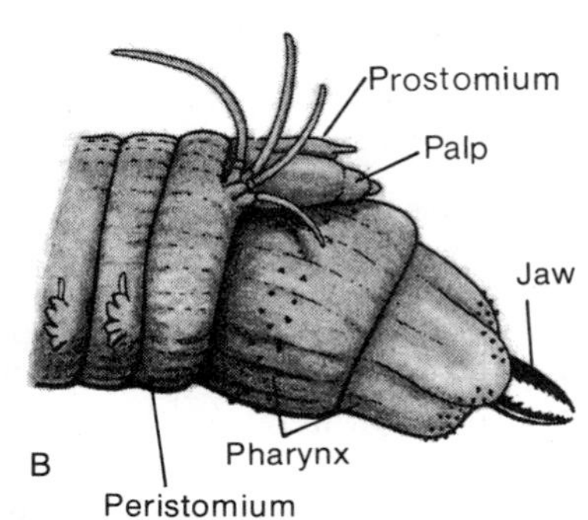

FIGURE 26.1 The polychaete *Nereis. A,* Dorsal dissection of anterior end. *B,* Lateral view of head with pharynx everted.

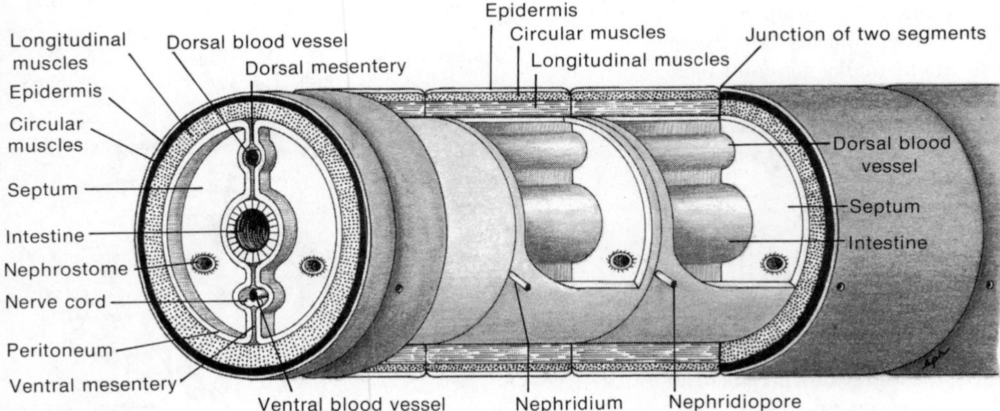

FIGURE 26.2 Diagrammatic lateral view of a series of annelid segments. (Based on a figure by A. Kaestner in *Invertebrate Zoology,* Vol. 1. New York, Interscience Publ., 1967.)

straight tubular **gut**, and a closed circulatory system (Figs. 26.1 and 26.2). The excretory organs are usually **metanephridia** (p. 197), typically one pair per segment.

Classification

The phylum is divided into three classes. The **Polychaeta**, the largest and most diverse of the three classes, contains the marine annelids. The **Oligochaeta** includes the freshwater annelids and earthworms, and the **Hirudinea** comprises the leeches. The latter clearly evolved from the oligochaetes, and the oligochaetes and polychaetes are believed to have evolved separately from some common marine burrowing ancestor.

Class Polychaeta

The members of the class Polychaeta are distinguished from other annelids by the presence of paired segmented appendages, **parapodia** (Gr. *para,* beside + *pous,* foot). The parapodia vary greatly in shape and size, but when well developed, each is composed of an upper **notopodium** and a lower **neuropodium.** Each division contains a large number of setae that provide traction against the substratum (Fig. 26.3). This is the origin of the name Polychaeta (Gr. *polys,* many + *chaite,* hair). In those polychaetes with large movable parapodia, the parapodial muscles are anchored to an internal chitinous rod, called the **aciculum**, with one rod in each division (Fig. 26.3).

The head, or prostomium, of polychaetes is often highly developed and may bear various sensory and feeding structures (Fig. 26.1). The mouth is located beneath the prostomium, just in front of the first one or two modified trunk segments (the **peristomium**) (Gr. *pro,* before + *peri,* around + *stoma,* mouth).

The digestive tract is usually composed of an eversible muscular **pharynx** and a long straight **intestine.** The pharynx is everted (turned inside out) through the mouth by special protractor muscles or by elevated coelomic fluid pressure (Fig. 26.1*B*).

The more than 8000 species of marine polychaetes exhibit great diversity both in form and in habit. An appreciation of their diversity can be gained by examining the adaptive groups that compose the class.

Surface Dwellers. Many polychaetes live on the bottom of the sea beneath stones and shells, in rock and coral crevices, and in algae and sessile animals. The heads of the surface-dwelling polychaetes are well developed and carry several kinds of sensory structures—one or two pairs of **eyes**, up to five **antennae**, and a pair of **palps** (Figs. 26.1 and 26.4).

The parapodia are large and function like legs in crawling (Fig. 26.5). Each makes a small step, and the movements of the parapodium occur in sequence as a result of **waves of contraction** passing along the length of the body (Fig. 26.5). The waves of contraction on one side of the body alternate with those on the opposite side. In rapid movement, alternating contractions of the longitudinal muscles of the body wall throw the body into undulatory waves that generate additional thrust against the substratum.

The crawling mode of locomotion of surface-dwelling polychaetes is reflected in the structure of the body wall

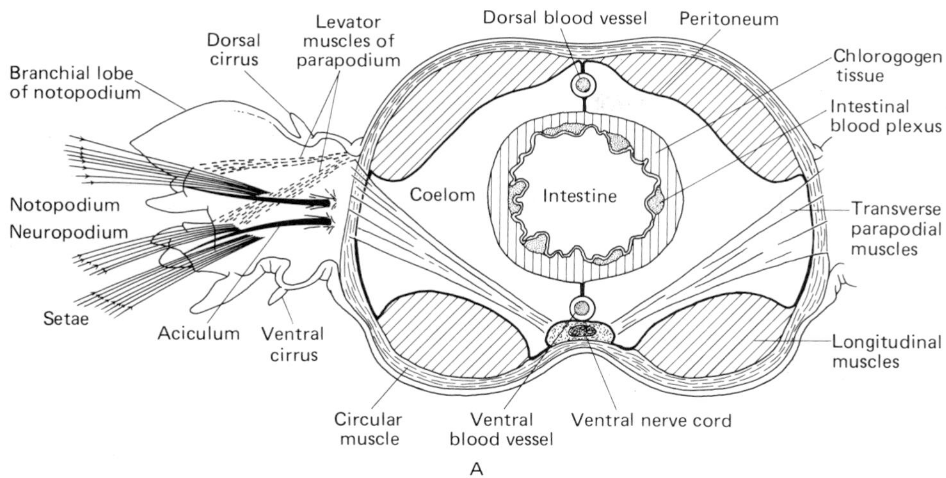

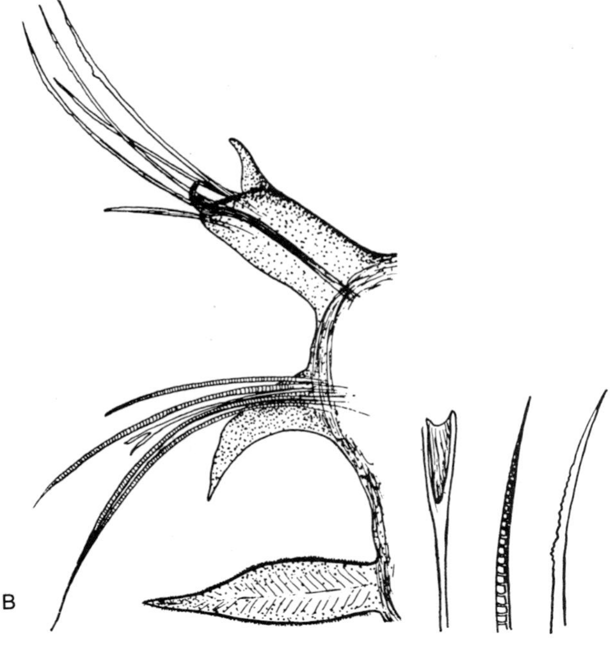

FIGURE 26.3 *A,* Transverse section of *Nereis* at level of intestine. *B,* Parapodium and setae of *Scoloplos rubra.* Note that the setae are not all of the same type. (From Renaud, J. C.: A report on some polychaetous annelids from the Miami-Bimini area. *Am. Mus. Novit., 1812*:1–40, 1956.)

and coelom. The longitudinal muscles are more highly developed than the circular muscles (Fig. 26.3), and the septa are somewhat reduced since the coelom is not as important as a localized hydrostatic skeleton as it is in peristaltic movement.

Many surface dwellers are carnivores and feed on other small invertebrates, including polychaetes. Others are algae eaters, scavengers, or detritus feeders. The pharynx of these surface-dwelling polychaetes is commonly provided with chitinous teeth or jaws, which vary in number from one family to another (Fig. 26.1*B*). The jaws typically swing open when the pharynx is everted and close when the pharynx is retracted.

Pelagic Polychaetes. Several families of polychaetes are adapted for living in the open ocean. They have well-developed heads, and the large parapodia are used as paddles in swimming (Fig. 26.4*B*). They are commonly carnivorous. Like most other small pelagic or planktonic animals, these polychaetes tend to be pale or transparent.

Gallery Dwellers. Many polychaetes are adapted for burrowing in sand or mud. Some excavate extensive burrow systems, or galleries, that open to the surface at numerous points (Fig. 26.6*A*). The wall of the burrow is kept from caving in by its lining of mucus. Their pro-

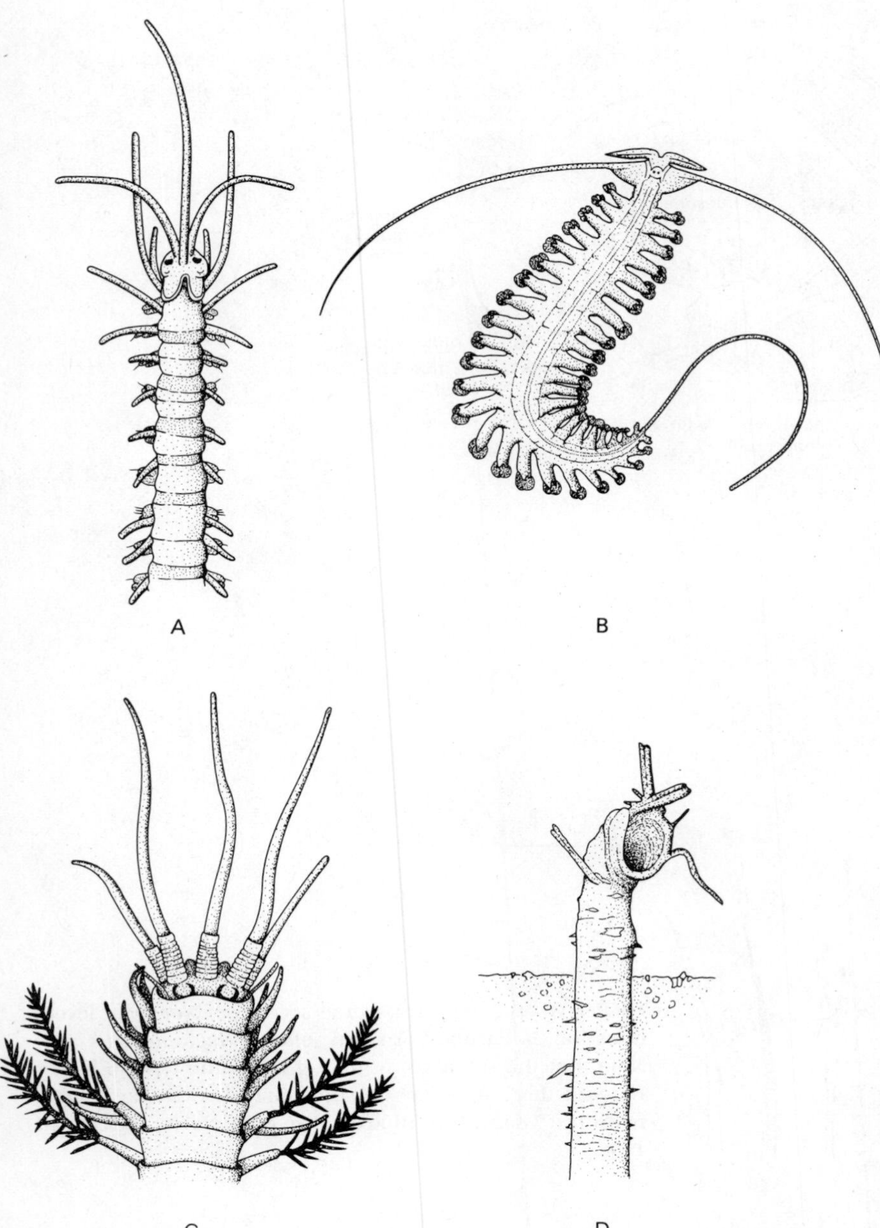

FIGURE 26.4 Polychaete diversity. *A,* Anterior end of *Proceraea fasciata,* a common surface-dwelling syllid on rocks, shells, and pilings along the east coast of the United States. *B, Tomopteris renata,* a pelagic polychaete. *C,* Head and first two gill-bearing segments of *Diopatra cupnea,* a carnivorous tube dweller. *D,* Funnel-like parchment tube of *Diopatra,* to which pieces of shell and algae are attached.

stomium is commonly conical or a simple lobe devoid of eyes and other sensory structures (Fig. 26.6*B* and *C*).

Gallery dwellers may use the parapodia to crawl through the burrow system. However, many crawl by peristaltic contractions; the parapodia of these forms are somewhat reduced and function primarily to anchor the segments against the burrow wall. Septa and circular muscles are well developed. Some gallery dwellers are carnivores. Others are nonselective deposit feeders and consume the substratum through which the galleries penetrated. In many ways they are like earthworms that are gallery dwellers belonging to another class of annelids (Oligochaeta).

One of the best studied gallery dwellers is *Glycera,* the genus containing the bloodworms commonly used as fishing bait. These intertidal to subtidal carnivores lie

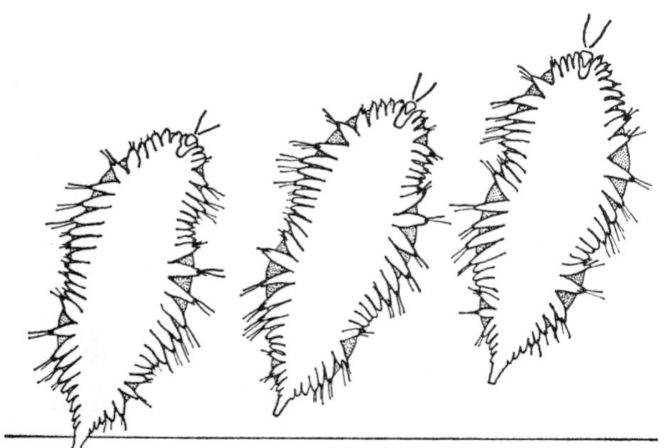

FIGURE 26.5 Ventral view of parapodial movement in polychaete crawling. Waves of movement on one side of body alternate with those on the opposite side. (After Mettam.)

in wait within their gallery systems, and when water pressure waves created by a small crustacean or other prey moving across the surface are detected, the worm moves to a nearby opening. Bloodworms have a very long proboscis-like pharynx that can be shot out with explosive force to seize the prey (Fig. 26.6*B*).

Sedentary Burrowers. Some burrowing polychaetes construct simple vertical burrows with only one or two openings to the surface (Fig. 26.7). These sedentary burrowers, in contrast to gallery dwellers, move about relatively little. Peristaltic crawling is the rule, and part of the parapodium is reduced to a ridge that bears special hooklike setae that aid in gripping the burrow wall. The prostomium lacks sensory structures, although special feeding appendages may be present.

Some sedentary burrowers are nonselective deposit feeders; others are selective deposit feeders. Lugworms (*Arenicola*) are examples of nonselective deposit feeders

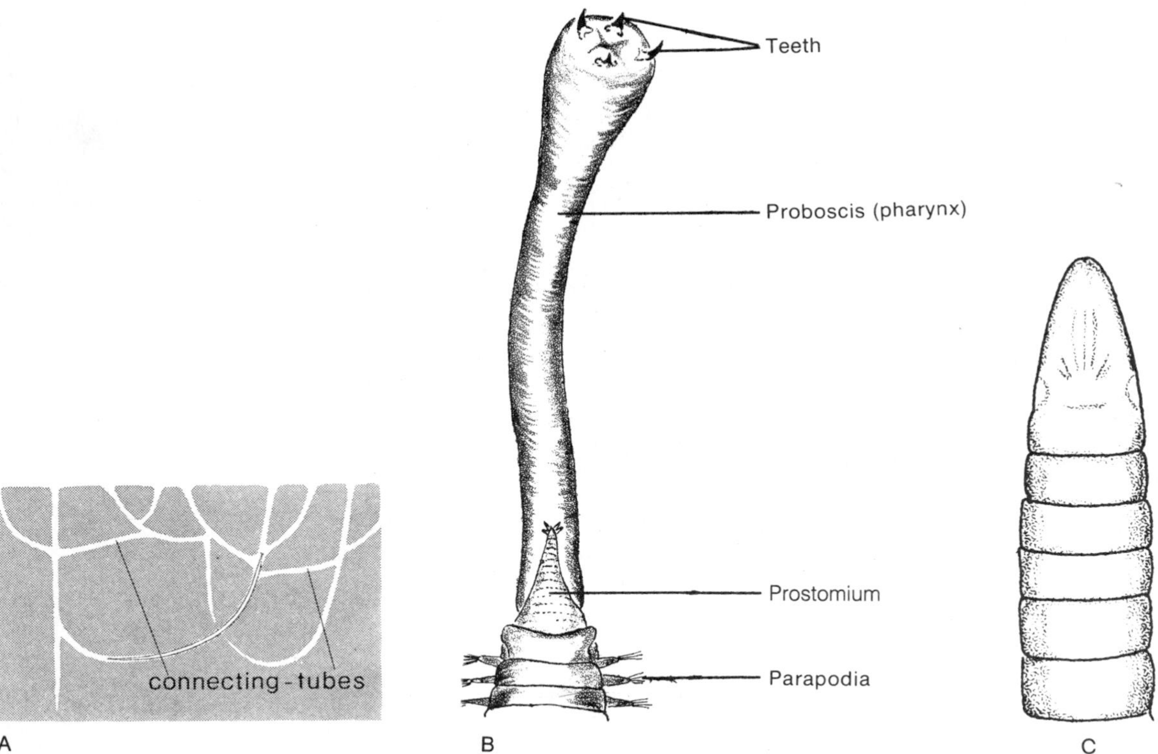

FIGURE 26.6 *A,* Burrow system of *Glycera alba,* showing worm lying in wait for prey. (After Oekelmann and Vahl.) *B,* Anterior end of *Glycera* with proboscis everted. *C,* Anterior end of another gallery dweller, *Drilonereis.* Note that the conical prostomium lacks eyes and sensory appendages. Parapodia are poorly developed on the most anterior segments.

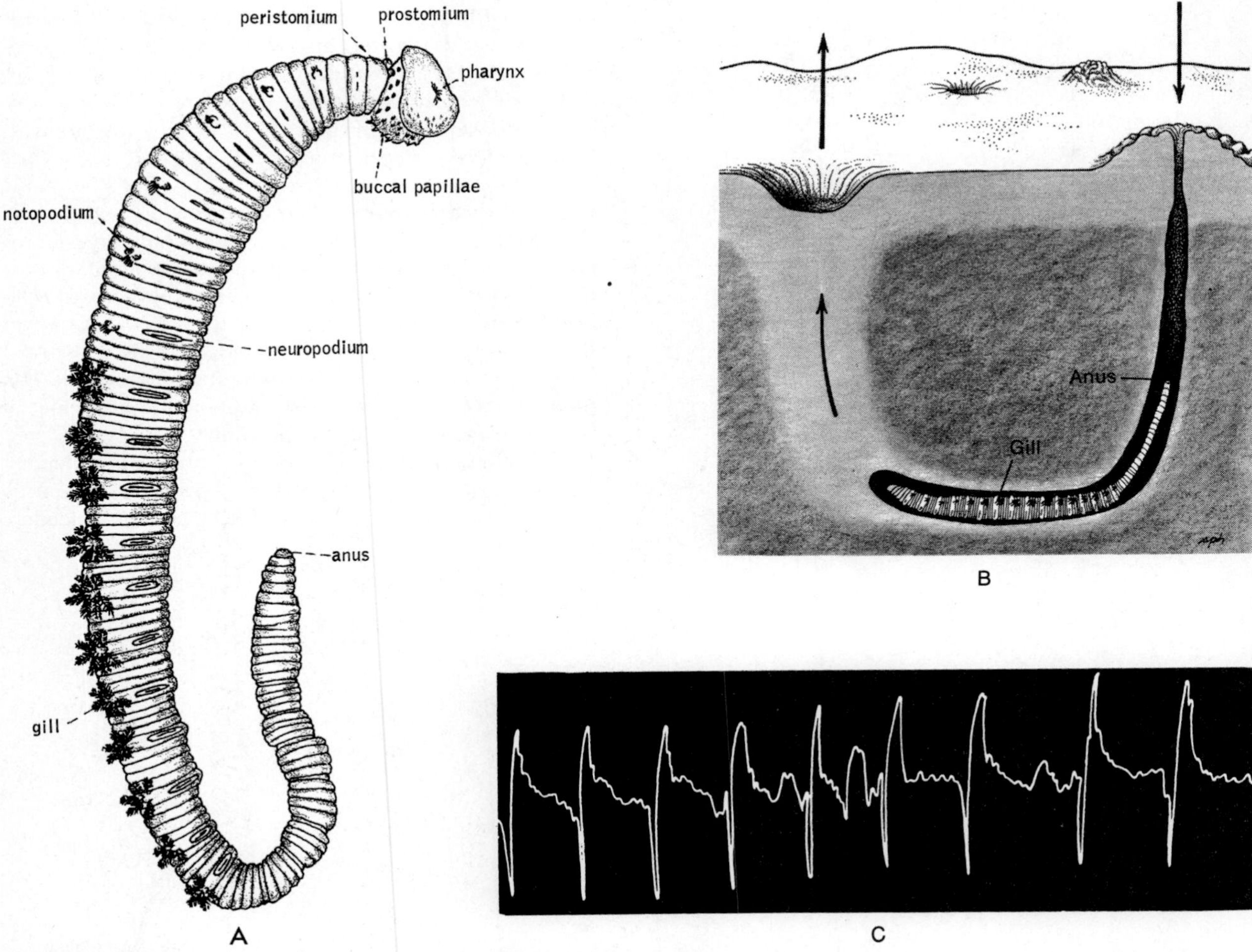

FIGURE 26.7 Lugworm, *Arenicola. A,* Lateral view of worm. Pharynx is everted, obscuring the very small prostomium. *B,* Worm in burrow. Arrows indicate direction of ventilating current produced by worm. *C,* Kymograph tracing of activity cycles of *Arenicola.* The downstroke reflects the worm backing up to the burrow opening to defecate; the sharp upstroke reflects the worm moving back down to the head of the burrow and vigorously resuming ventilation contractions and deposit feeding. Intervals between defecations are about 40 minutes. (*A* modified from Wells: *C* from Wells, G. P.: The behaviour of *Arenicola marina* in sand and the role of spontaneous activity cycles. *Journal of Marine Biology Association (U.K.), 28*:465–478, 1949.)

(Fig. 26.7). These worms live in L-shaped burrowers and ingest the sand at the bottom with an unarmed eversible pharynx. Peristaltic contractions drive a ventilating current of water into the burrow. At rhythmically fixed intervals the worm backs out of its burrow and defecates the mineral material at the surface as conspicuous castings. It then resumes feeding and ventilating. By means of a float attached to a stylus worms in an aquarium can be made to record their activity (Fig. 26.7C).

Selective deposit feeders have special head structures that pick up the organic matter from the surrounding sand grains. The head of *Amphitrite,* for example, is provided with a great mass of long tentacles. These spread over the surface from the opening of the burrow (Fig. 6.4). Detritus material adheres to the mucus on the tentacles and then is conveyed to the mouth in a ciliated tentacular gutter and by tentacular contraction. There is no eversible pharynx.

Tubiculous Polychaetes. Many polychaetes live in protective tubes. In soft bottoms, the opening of the tube projects above the sand surface and thus provides the worm with access to water free of sediment. Tubes may also permit worms to live on firm exposed substrates, such as algae, rock, coral, or shell. The tube may be composed entirely of hardened material secreted by the worm, or it may be composed of foreign material cemented together.

Tube-dwelling polychaetes may be divided into **carnivorous** tube dwellers and **sedentary** tube dwellers. The carnivorous group contains predatory species that feed on small invertebrates. They lie in wait at the mouth of the tube and seize the victim as it passes by. As might be expected, these worms are very agile. Their adaptations are essentially like those of surface dwellers (Fig. 26.4*C* and *D*).

The sedentary tubiculous polychaetes, which include the majority of tubiculous species, move about within the tube less actively. Their head usually lacks special sensory structures, although feeding appendages may be present. The animal moves within the tube by peristaltic contractions, and the parapodia tend to be reduced to ridges with hooked setae.

The structural diversity of tube-dwelling polychaetes is in large part correlated with their different modes of feeding. A few, such as the bamboo worms, are nonselective deposit feeders. The bamboo worm lives head down in sand grain tubes and eats the substratum at the bottom of the tube (Fig. 26.8*B*). Periodically, the

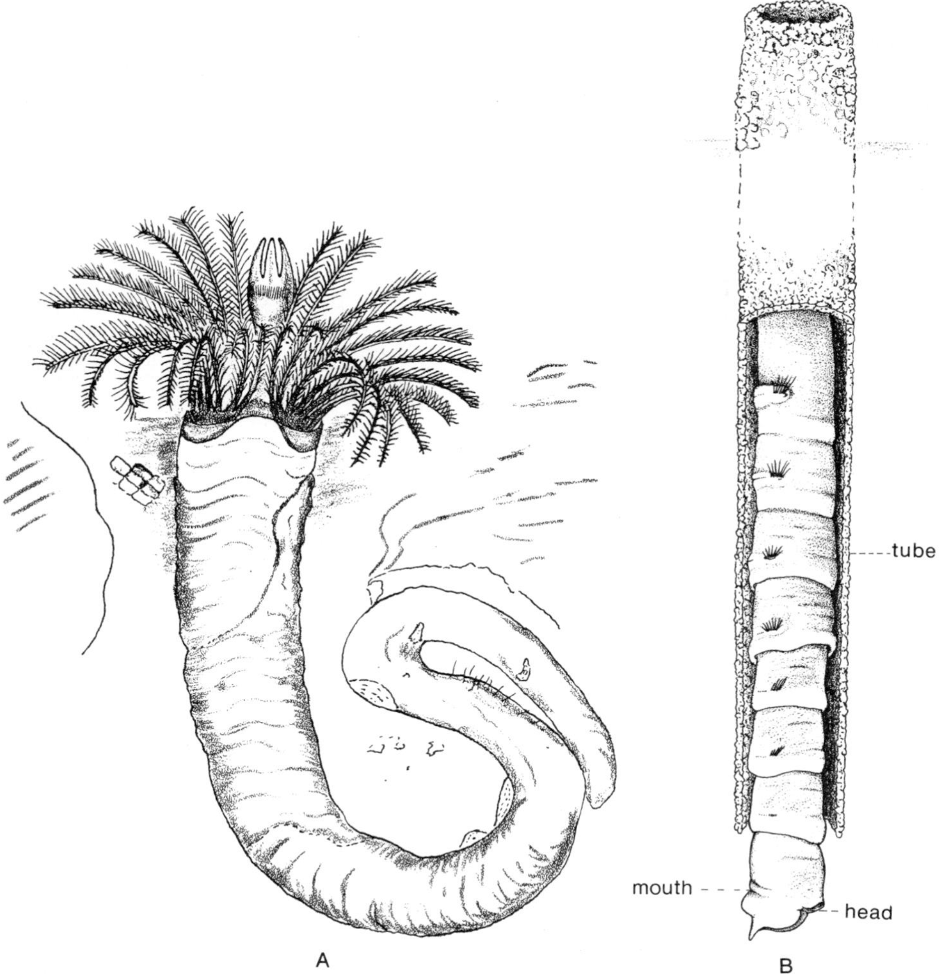

FIGURE 26.8 *A,* A fanworm with radioles extending from end of calcareous tube, which is attached to a shell. *B,* Bamboo worm, a polychaete that lives upside down in a sand grain tube and ingests the substratum at the bottom of the tube. Full extent of the tube is not shown.

BOX 26.1

Members of the family Chaetopteridae feed by filtering water through a mucous bag. Species of *Chaetopterus* live in a large U-shaped tube composed of a secreted parchment-like material. Three piston-like "fan" parapodia in the middle of the body drive water through the tube. A pair of long winglike anterior notopodia secrete a film of mucus, which is collected and rolled up by a ciliated cup. The water current being driven through the tube passes through the mucous film, which is held like a bag. Periodically, mucus secretion is halted, and the rolled up ball of mucus containing trapped food particles is passed forward along a ciliated groove to the mouth. (*B* after MacGinitie.)

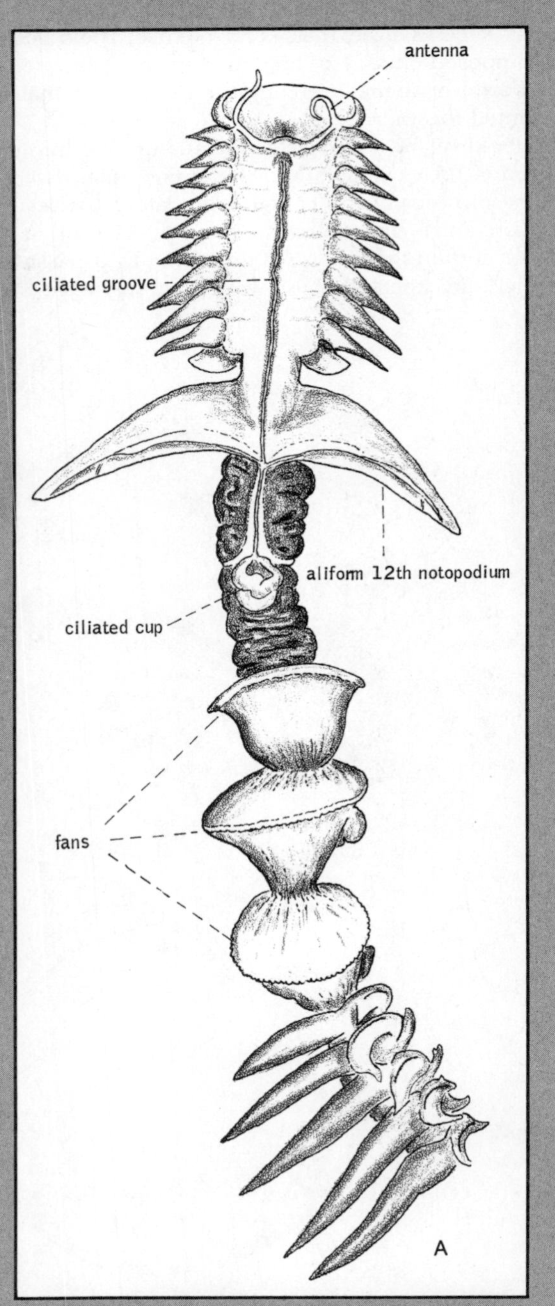

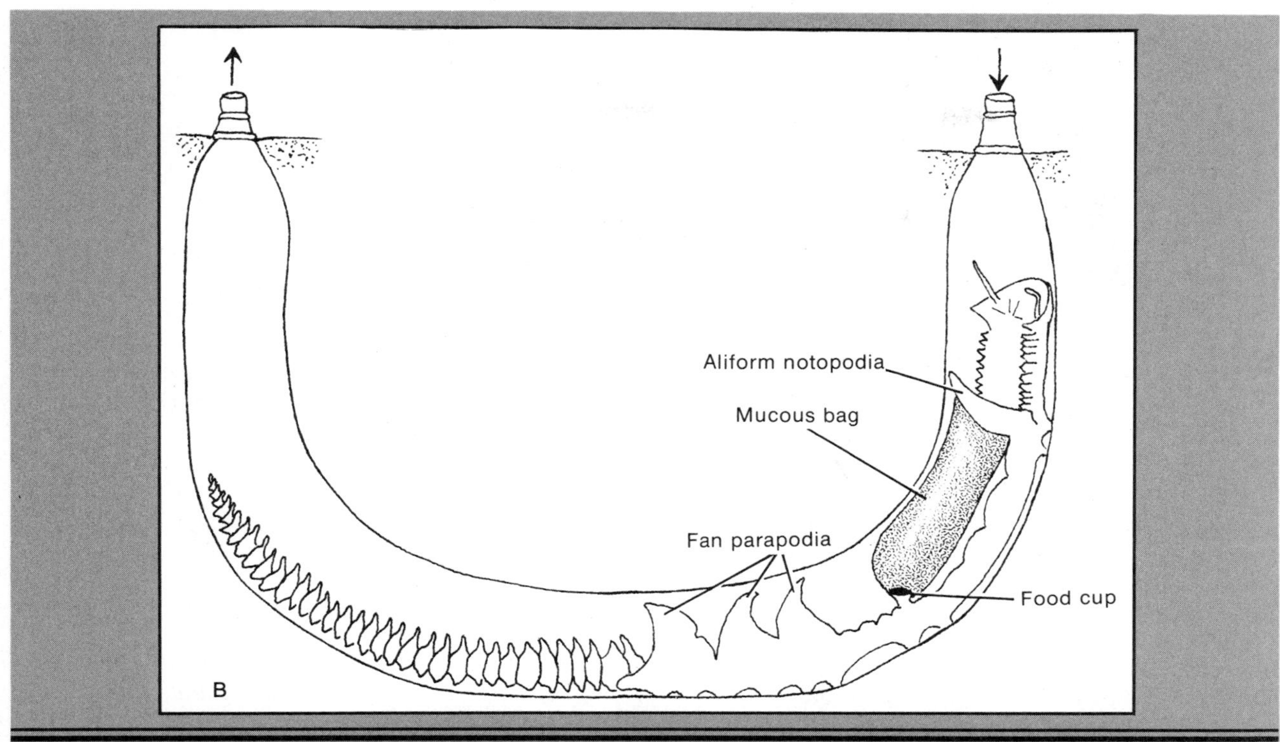

worm backs to the surface and defecates the mineral material. Some sedentary tube dwellers are selective deposit feeders.

Filter feeding has evolved in several families of sedentary, tube-dwelling polychaetes. The beautiful fanworms illustrate one type of filter feeding. They have feather-like head structures called **radioles.** When feeding, the radioles project from the opening of the tube, often in the form of a funnel (Figs. 26.8*A* and 6.3). The cilia on the radioles create a water current, and plankton and other suspended particles passing through the radioles are filtered out. The trapped particles are then transported by cilia down the radiole into the mouth. The tube-dwelling chaetopterid polychaetes are also filter feeders but use a mucous bag as the filter (Box 26.1).

Internal Transport, Gas Exchange, and Excretion in Polychaetes. The closed blood vascular system of polychaetes has a relatively simple basic plan (p. 177). A contractile, longitudinal vessel located above the gut functions as the heart and conveys blood anteriorly. At the anterior end of the body, paired vessels carry the blood around the gut to a ventral longitudinal vessel in which blood passively flows posteriorly. In each segment the ventral vessel gives rise to a vessel supplying the gut and to paired vessels supplying the body wall, parapodia, and nephridia (Fig. 26.9). These segmental vessels eventually return blood to the dorsal vessel.

Coelomic fluid contributes to internal transport in all polychaetes, and in some the blood vascular system is reduced and the coelomic fluid is the principal transporting medium.

Very small polychaetes and those with long, thread-like bodies have no gills. The varying structure and the distribution of the gills present in the larger species indicate that they have evolved independently in different groups. Most commonly, the gills are modifications of the parapodia or outgrowths of some part of the parapodium (Figs. 26.3 and 26.4*C*). However, gills are found in other parts of the body in some polychaetes. The radioles of fanworms serve as gills in addition to functioning as filters (Fig. 26.8).

The ventilating current may be produced by cilia or by contractions of the gills. Gallery dwellers, sedentary burrowers, and tube dwellers drive a ventilating current through the burrow or tube, usually by undulations of the body or by peristaltic contractions.

Many polychaetes, especially those with gills, possess a respiratory pigment, usually **hemoglobin**, either dissolved in the blood plasma or contained within corpuscles in the coelomic fluid. Polychaete excretory organs are protonephridia or metanephridia (p. 197). A pair of

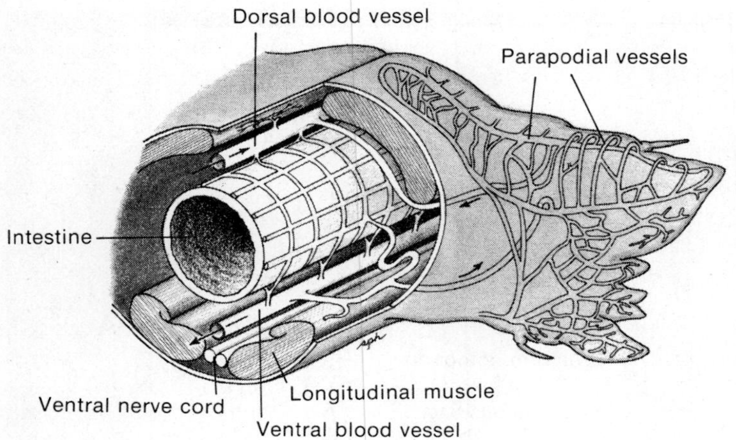

FIGURE 26.9 Vascular system within a segment of *Nereis.* Arrows indicate direction of blood flow; anterior is to the right. (Modified from Nicoll.)

nephridia is present in each segment, but the nephrostome of each nephridium opens through the septum into the segment anterior to the one containing the tubule (Fig. 26.1).

Reproduction and Development. The sexes of polychaetes are separate. The gametes are produced by the coelomic **peritoneum** and mature within the coelomic fluid. Primitively, most segments produce gametes, but in many polychaetes gamete production is restricted to the segments in certain regions of the body. Gametes exit by special gonoducts, by the nephridia, or by rupture of the body wall.

The animals do not copulate. Instead, the gametes are shed into the sea water where fertilization occurs. The likelihood of fertilization in some polychaetes is in-

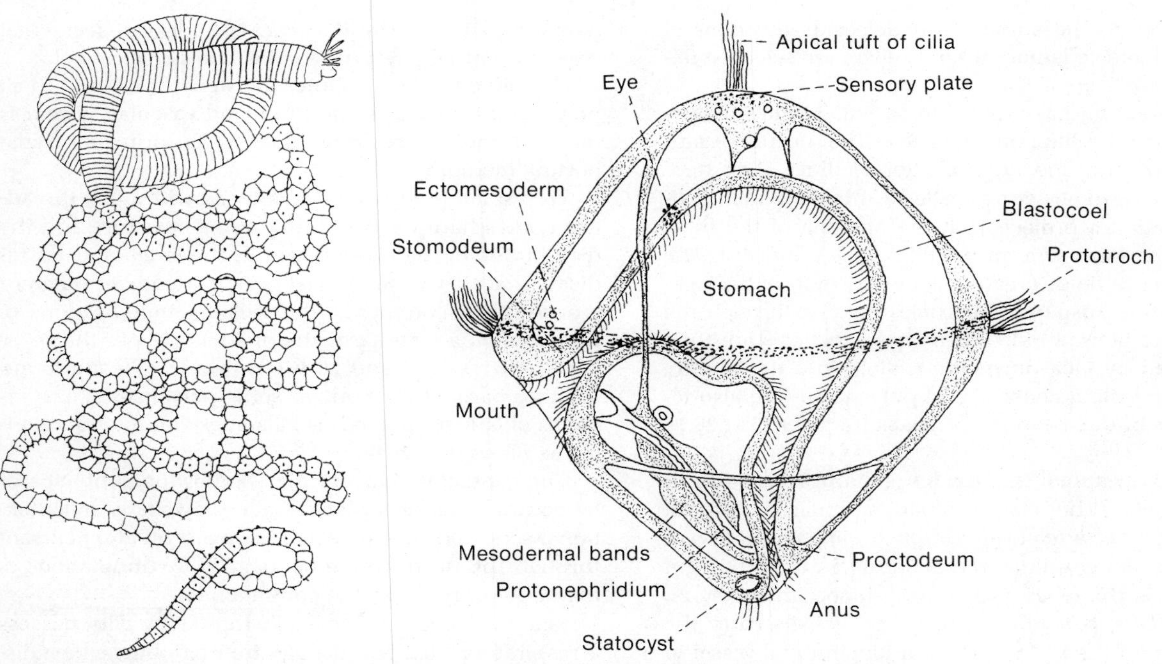

FIGURE 26.10 *Eunice viridis,* the Samoan palolo worm, with posterior region of gamete-bearing segments. (After Woodsworth from Fauvel.) *B,* An annelidan trochophore. (After Shearer from Hyman.)

creased by swarming behavior. Males and females with ripe gametes swim to the surface at the same time in large numbers, shedding their eggs and sperm simultaneously. Swarming usually occurs at specific times of the year, determined by annual, lunar, and tidal periodicities. The times at which swarming will occur in certain species can be predicted. The most notable example is the Samoan palolo worm. At the beginning of the last lunar quarter of October or November the worms release the posterior gamete-bearing regions of the body (Fig. 26.10*A*), which come to the surface in enormous numbers and rupture.

The eggs, especially those of swarming polychaetes, are often planktonic. However, there are species that deposit their eggs within jelly masses attached to the tube or burrow.

Cleavage is typically spiral and leads to a trochophore larva like that of mollusks (Fig. 26.10*B*). During the course of larval life, the trochophore lengthens, and segments with parapodia and setae appear behind the prototroch and mouth.

Class Oligochaeta

In fresh water and on land, the annelids are represented by some 2900 species of oligochaetes. These are less diverse than the polychaetes, and the terrestrial species in many respects parallel the burrowing polychaete gallery dwellers. The simple conical or lobelike **prostomium** lacks sensory appendages (Fig. 26.11). Oligochaetes crawl by peristaltic contractions. Parapodia are completely absent, but **setae** serve to anchor segments in crawling (Figs. 26.11*B* and 26.12*B*). There are fewer setae per segment than in polychaetes, hence the name Oligochaeta (Gr. *oligos*, few + *chaite*,, hair). Correlated with their peristaltic locomotion, the **coelom, septa,** and circular and longitudinal muscles are well developed and reflect a high order of metamerism (Fig. 26.12).

Aquatic oligochaetes are usually less than 3 cm. in length, and a few are almost microscopic. Many species have relatively long setae (Fig. 26.11). They live in algae and in the bottom debris and mud of ponds, lakes, and streams, and there are even a number of species that are found in marine sediments. Some are tubiculous. Most aquatic oligochaetes live in shallow water, but some Tubificidae, a large family of tube-dwellers, occur in enormous numbers in the lower anaerobic regions of deep lakes and have been reported from deep ocean bottoms.

Many oligochaetes live in wet boggy soil and intergrade with the more terrestrial earthworms. The largest annelids are Australian earthworms of the genus *Megascolides*, which may attain a length of 3 m.

Earthworms are gallery dwellers with short heavy setae. They display few structural modifications for a terrestrial existence. Like most other oligochaetes, they

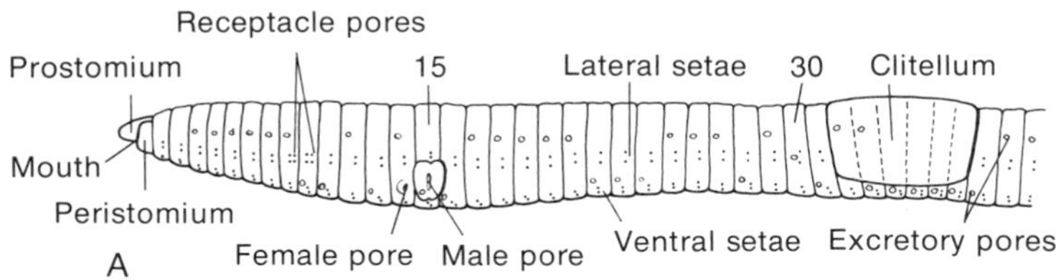

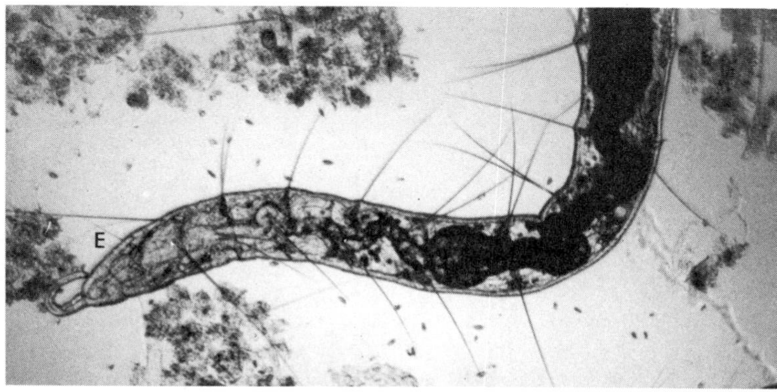

FIGURE 26.11 *A*, Lateral view of the anterior 40 segments of the earthworm *Lumbricus*. Reproductive openings are found on segments 9, 10, 14, and 15. On each segment the excretory pore is either ventral, near the ventral setae, or lateral, above the lateral setae, with much variability between worms. *B*, Anterior end of a freshwater oligochaete, *Stylaria*. In this genus the prostomium is tentacle-like. Note the long setae characteristic of many freshwater oligochaetes. (*B* by Betty M. Barnes.)

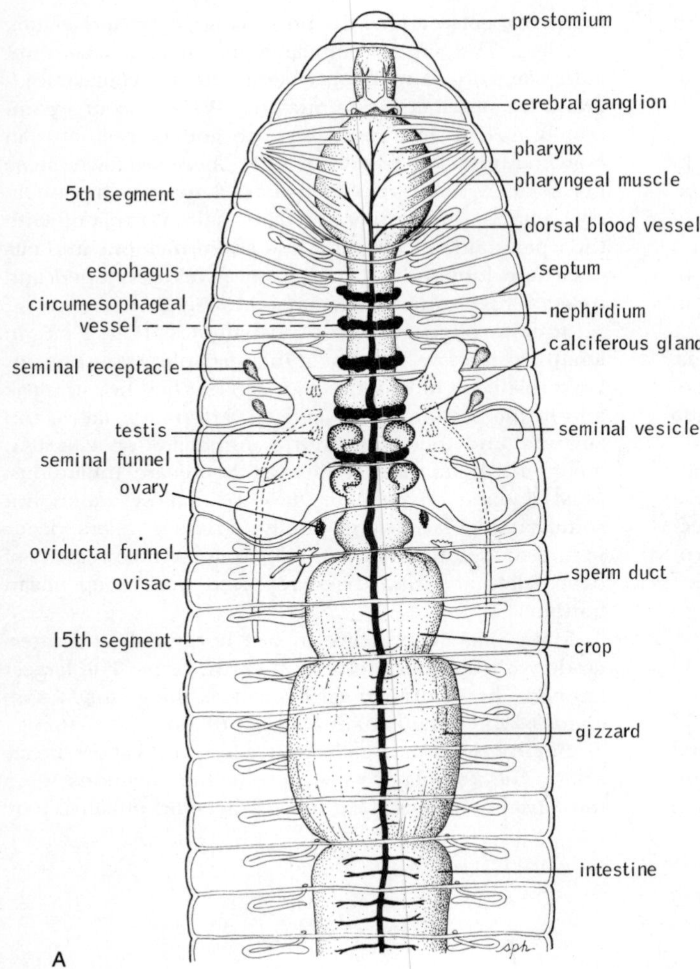

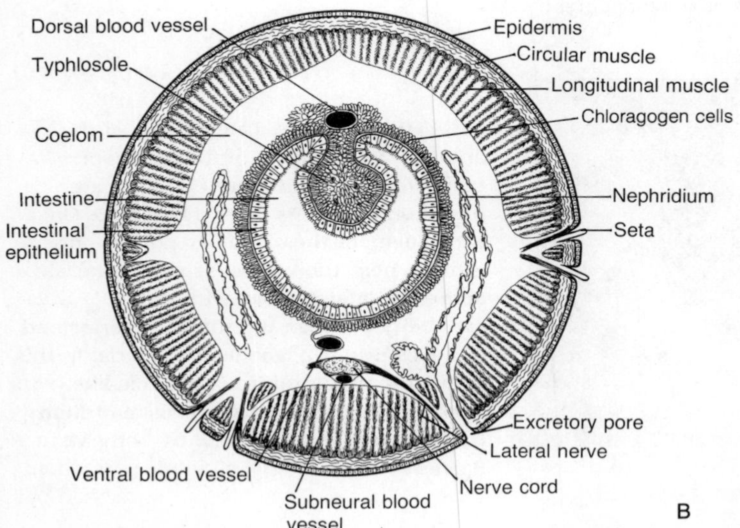

FIGURE 26.12 Structure of the earthworm *Lumbricus. A,* Dorsal view of anterior section of body; *B,* cross section taken at the level of the intestine.

lack special organs for gas exchange. The high concentration of oxygen in air and the highly vascularized body wall permit adequate gas exchange across the general body surface despite the large size that earthworms may attain. The surface is kept moist for gas exchange by epidermal mucous glands and by coelomic fluid, which is released through a dorsal pore between adjacent segments.

Desiccation is minimized by behavioral adaptations. Earthworms come to the surface only at night or during rains. During dry periods or freezing winter weather, earthworms burrow deeper in the soil and become inactive. Water in moist soil can be absorbed across the body surface.

Well-drained soils with a large amount of organic matter provide the best environment for earthworms. Within such soils the population is vertically stratified. Small species and small individuals live in or near the upper humus layer; larger species, such as those of the genus *Lumbricus,* tunnel down to depths of 1 to 2 m.

Nutrition. Decomposing plant tissue is the principal food of most oligochaetes, although some freshwater species feed upon algae, and a few are raptorial on other small invertebrates. Earthworms are selective and nonselective deposit feeders. They consume bits of decaying vegetation at the surface and also cycle soil through the **gut** and digest the organic matter. The castings may be deposited at the surface.

Food is ingested by means of a muscular **pharynx** which, in earthworms, functions as a pump (Fig. 26.12). From the pharynx food passes into the **esophagus,** which is often modified at different levels as a **crop** and one or more **gizzards.** The thin-walled crop functions in storage, and the muscular gizzard grinds the food into smaller particles. **Calciferous glands** over the dorsal esophageal wall excrete excess calcium taken in with food. The calcium is eliminated into the gut as calcite crystals, which have a low solubility.

The remainder of the gut is a long straight **intestine,** the site of extracellular digestion and absorption. The surface area of the earthworm's intestine is nearly doubled by a dorsal longitudinal fold, the **typhlosole** (Fig. 26.12*B*).

Surrounding the intestine is a layer of **chloragogen cells** (Fig. 26.12*B*). These, like the vertebrate liver, are an important site of intermediary metabolism. The synthesis and storage of glycogen and fat, the deamination of amino acids and synthesis of urea (in terrestrial species) occur in chloragogen cells.

The circulatory system, nephridia, and nervous system of oligochaetes are essentially like those of polychaetes. The contractile dorsal blood vessel provides the principal force for blood propulsion. In some oligochaetes, certain anterior commissural vessels (five pairs around the esophagus in *Lumbricus*) are conspicuously contractile (Fig. 26.12*A*). These "hearts" function as accessory pumps.

Hemoglobin is present in solution in the plasma of the larger oligochaetes, including most of the earthworms.

The **nephridia** play a role in salt and water balance, for in freshwater species and in earthworms with a plentiful water supply the urine is hypotonic to the coelomic fluid. Like other freshwater animals, oligochaetes must eliminate the water entering by osmosis.

Giant fibers, present in the ventral nerve cord, transmit impulses controlling rapid contraction of the body. In the earthworm, *Lumbricus,* three conspicuous giant fibers, easily seen in section, are located across the top of the cord.

Sense organs are lacking in most species, but the integument is richly supplied with sensory receptors, including photoreceptors.

Reproduction. In contrast to polychaetes, oligochaetes are **hermaphroditic.** Moreover, there are distinct **gonads,** restricted to a few segments in the anterior third of the body.

The paired **ovaries** are located within a single segment (Fig. 26.13). The eggs mature within the coelomic fluid of that segment and exit through a pair of simple **oviducts.**

Seminal receptacles are present, but in oligochaetes these structures are inpocketings of the ventral body wall and are not connected to the other parts of the female system (Fig. 26.13).

There are one or two testicular segments, each containing a pair of **testes** (Fig. 26.13). In some earthworms, a transverse partition separates the lower part of the coelomic cavity containing the testes, and the sperm mature within this special chamber. The septum of the testicular segment is usually greatly evaginated to form a pouchlike **seminal vesicle.** The paired sperm ducts penetrate the posterior septum and may pass through several segments before opening onto the ventral surface (Figs. 26.11*A* and 26.13). In animals with two pairs of testes, the sperm ducts from the two testes of one side unite.

The epidermis of certain adjacent segments in the anterior third of the body contains many glands that produce secretions for various parts of the reproductive process. These segments, collectively called the **clitellum** (L. *clitellae,* pack-saddle), are characteristic of oligochaetes. The clitellum of many aquatic species is composed of only two segments, those that contain the genital pores. The clitellum may be visible only during the reproductive period. The earthworms of the genus *Lumbricus,* in contrast, have a conspicuous, permanent, girdle-like clitellum of six to seven segments located behind the genital pores (Fig. 26.11).

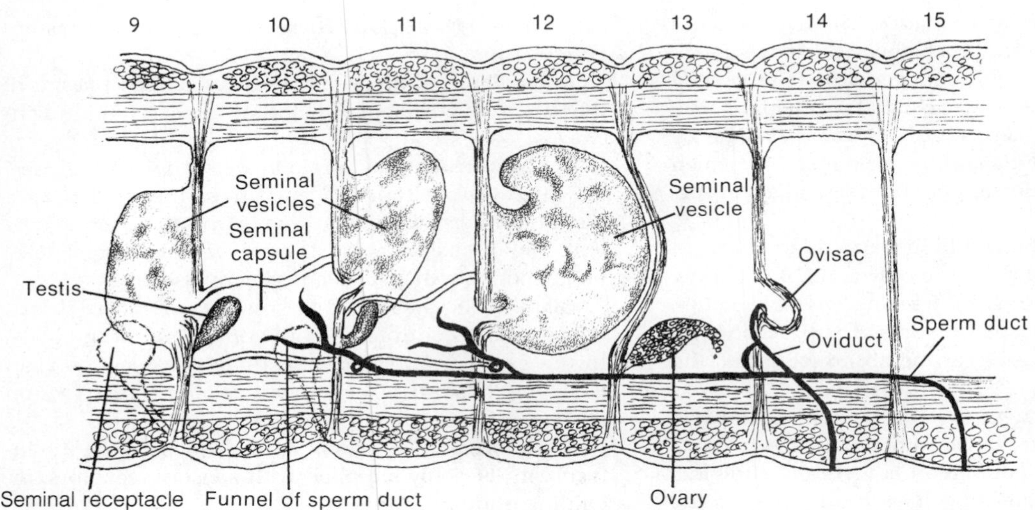

FIGURE 26.13 Reproductive segments of the earthworm *Lumbricus* (lateral view). (After Hesse from Avel.)

FIGURE 26.14 *A,* Two earthworms copulating. Photograph of living animals made at night. (Courtesy of General Biological Supply, Chicago, Illinois.) *B,* Movement of sperm during copulation in lumbricid earthworms. *C,* Earthworm cocoon. (*B* after Grove, A. J., and L. F. Cowley: On the reproductive processes of the brandling worm *Eisenia foetida. Quarterly Journal of Micro. Sci.,* 70:559–581, 1926.)

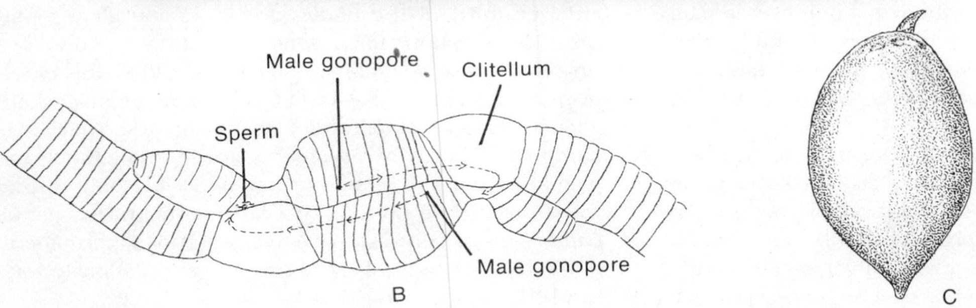

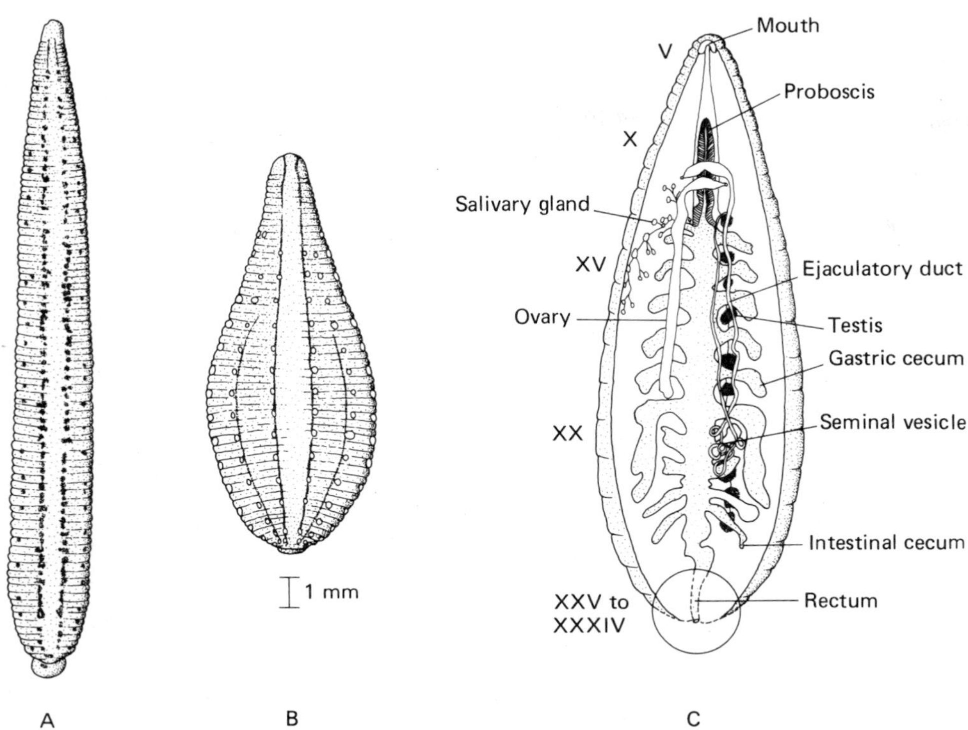

FIGURE 26.15 *A, Erpobdella punctata,* a common North American scavenger and predatory leech; *B, Glossiphonia complanata,* a common European and North American leech that feeds on snails. *C,* Internal structure of *Glossiphonia complanata,* ventral view. (*A* and *B* from Sawyer, R. J.: *North American Freshwater Leeches.* Chicago, University of Illinois Press, 1972; *C* after Harding and Moore from Pennak, R. W.: *Freshwater Invertebrates of the U.S.* 2nd ed. New York, John Wiley and Sons, 1978.)

Copulation with mutual transfer of sperm is characteristic of oligochaetes. Two worms come together with ventral surfaces opposed and with their anterior ends facing in opposite directions. The worms are held together by mucus secreted by the clitellum and sometimes by special genital setae. In most oligochaetes, sperm are passed directly from the male gonopores into the seminal receptacles or spermatheca. However, in *Lumbricus* and other members of the same family, the seminal receptacles of one worm are located some distance anterior to the male gonopores of the copulating partner (Fig. 26.14). The sperm, on being released, pass down a pair of grooves on the ventral surface and then cross over at the level of the seminal receptacles. The grooves are arched over by mucus and thus separated from the grooves of the opposite worm. Copulation in *Lumbricus* is a process requiring two to three hours.

A few days after copulation the clitellum secretes a dense material that will form the **cocoon.** Albumin, serving as a food source for the embryos, is secreted inside the cocoon, especially in terrestrial species with little yolk in the egg. The secretions form an encircling band that slips forward over the body. Eggs from the female gonopores and sperm from the seminal receptacles are collected en route. The cocoon eventually slips off the anterior end of the worm, the two ends sealing in the process (Fig. 26.14). Fertilization takes place within the cocoon that is left in the bottom mud and debris or in the soil. Development is direct, and young worms emerge from the end of the cocoon. Only one egg completes development within the cocoons of *Lumbricus terrestris.*

Class Hirudinea

The members of the class Hirudinea (L. *hirudo,* leech), the **leeches,** constitute the smallest and the most aberrant of the annelid classes. Only some 500 species have been described. The body is dorsoventrally flattened, with the anterior segments modified as a small **sucker** surrounding the mouth and the posterior segments forming a larger sucker behind the anus (Fig. 26.15). The head is greatly reduced, setae are absent, and there is little external evidence of metamerism. The body is

BOX 26.2

Related Phyla

There are four small phyla of marine wormlike animals that are protostomes, and two show some evolutionary relationship to annelids.

A. The phylum Priapulida contains nine wormlike or cucumber-shaped animals, which live buried in marine sands and muds. The cylindrical trunk bears a spiny anterior region (the introvert), which can be invaginated into the trunk. At least some priapulids are predatory on other small marine animals. *B.* The 320 members of the phylum Sipuncula, sometimes called peanut worms, also bear an anterior region that can be invaginated into a cylindrical trunk. They burrow in sand and mud or bore into coralline rock or shells and are deposit feeders with tentacles carried by the end of the introvert. *C.* The phylum Echiura contains about 140 species that possess a single pair of large setae on the underside of a cylindrical trunk. Anteriorly, a large nonretractable ciliated proboscis functions in deposit feeding. Echiurans burrow in sand and mud or live in spaces between shell and rock. *D.* Members of the phylum Pogonophora are deep-water tube-dwelling worms, of which some 80 species have been described to date. A rear section of the body is segmented with setae. At the anterior end of a long middle trunklike region is a forepart bearing long ciliated tentacles. Pogonophorans are gutless, and food is apparently digested and absorbed directly through the tentacles, and, at least in part, is derived from symbiotic chemosynthetic bacteria lodged in certain parts of the body. The largest known pogonophorans are recently discovered species around the deep thermal vents in the Galapagos rift. (*A* after Theel from Hyman; *B* after Fisher; *D* from George, J. D., and Southward, E. C.: A comparative study of the setae of Pogonophora and polychaetous Annelida. *Journal of the Marine Biological Association of the United Kingdom,* 53:403–424, 1973.)

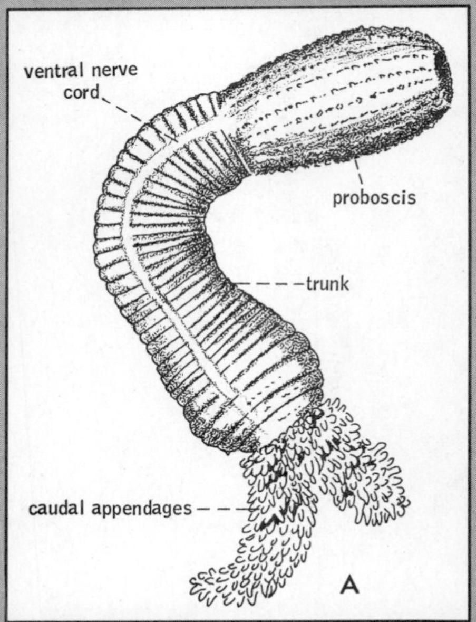

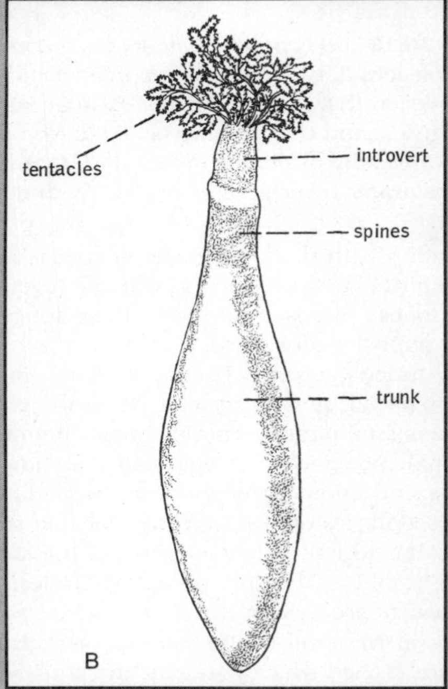

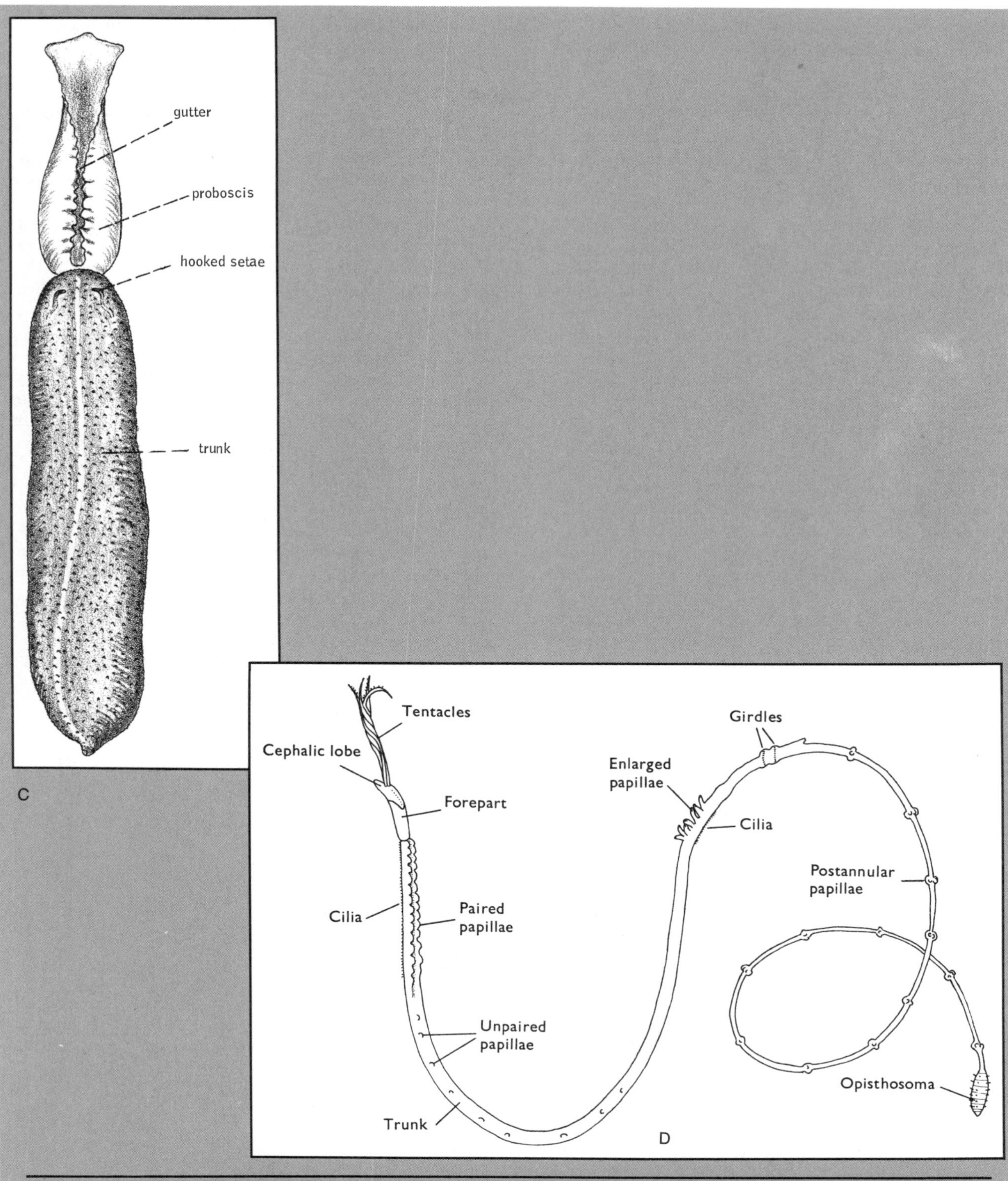
gutter
proboscis
hooked setae
trunk
C
Tentacles
Cephalic lobe
Forepart
Cilia
Paired
papillae
Unpaired
papillae
Trunk
Girdles
Enlarged
papillae
Cilia
Postannular
papillae
Opisthosoma
D

ringed with many annuli, but they do not correspond to the 34 segments that compose the leech body (Fig. 26.15). Only the nephridia and nervous system indicate the original segmentation.

Most leeches are 2 to 5 cm. in length and are never as small as many polychaetes and other annelids. *Hirudo medicinalis,* the medicinal leech, may reach a length of 20 cm.

Locomotion. Leeches crawl in a looping manner. The body is extended and the anterior sucker attached. Then the posterior sucker is released, pulled forward and reattached. Many leeches can swim by dorsoventral undulations of the flattened body. Correlated with these modes of movement, the longitudinal muscles of the body wall are powerfully developed. The septa have been lost, and the coelom is reduced by the invasion of connective tissue to a system of interconnecting spaces, or sinuses (Fig. 26.16).

Nutrition. Contrary to popular notion, not all leeches are bloodsuckers. Many are predaceous and feed upon other small invertebrates, especially oligochaetes, snails, and insect larvae. About three fourths of the known species of leeches are blood-sucking ectoparasites on invertebrates, such as snails and crustaceans, and on vertebrates. All classes of vertebrates may be hosts, but fish and other aquatic forms, including water fowl, are the most common victims. In humid areas of the tropics, leeches attack terrestrial birds and mammals, attaching to the moist thin membranes of the nose and mouth. They climb upon the host when it brushes through vegetation on which the leeches are located. Leeches detect the presence of a host by great sensitivity to water pressure waves, host secretions, or body temperature.

Many leeches, both predatory and parasitic, ingest food through a tubular jawless proboscis that can be extended through the mouth from a proboscis cavity (Fig. 26.15). Predatory species usually swallow the prey whole.

The **pharynx** of other leeches serves as a pump but cannot be protruded. The mouth cavity in many of these forms contains three bladelike **jaws** (Fig. 26.16). When the anterior sucker is attached, the jaws slice through the skin of the host. A salivary secretion, called **hirudin,** prevents the coagulation of blood, which is ingested by the pumping action of the pharynx.

An **esophagus** connects the pharynx with a large crop or stomach (Fig. 26.15). An **intestine** composes the posterior third of the digestive tract and opens through the **anus** just above the posterior sucker. Commonly, pouches (**ceca**) extend laterally from the stomach and intestine. The stomach functions both in digestion and as a storage center. Many leeches may consume 2 to 10 times their weight in blood at a single feeding. Water is removed from the ingested blood, and the remaining material may require as long as six months to be digested. Leeches are unusual in possessing mostly exopeptidases as digestive enzymes. Symbiotic bacteria provide the enzymes for a part of the digestive process and produce some substance that prevents decomposition of the blood cells. Thus a meal may be stored for a

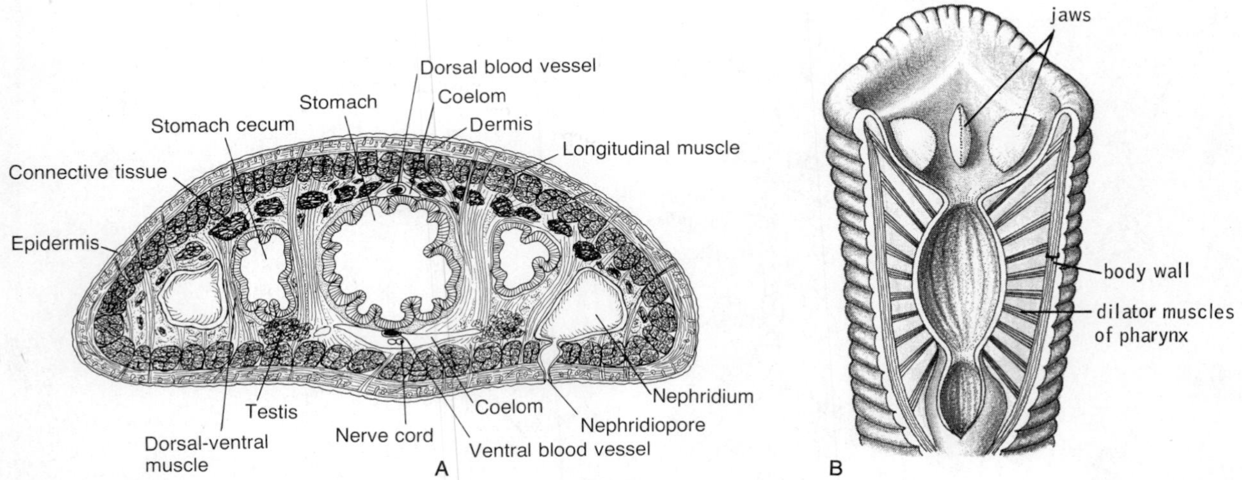

FIGURE 26.16 *A,* Cross section of a leech showing the filling of the coelom by connective tissue. *B,* Anterior end of *Hirudo,* the medicinal leech (ventral dissection). (Modified from Pfurtscheller.)

long time between infrequent feedings. Leeches have survived without feeding for as long as a year and a half.

Some leeches have a blood vascular system like that of their oligochaete ancestors. Others have lost the original blood vessels, and the interconnecting coelomic channels have been converted into a blood vascular system. Hemoglobin, dissolved in the plasma or coelomic fluid, provides for gas transport, but gills are present in only a few leeches. Although the head is reduced, some leeches possess eyes. Special sensory papillae are arranged in rings on the annuli.

Reproduction. The presence of a **clitellum** and a similar **hermaphroditic** reproductive system is taken as evidence that leeches evolved from oligochaetes. Copulation involves a mutual exchange of sperm, which in some leeches is similar to that in oligochaetes. In others **spermatophores** are transferred by hypodermic impregnation. The spermatophores are injected into the ventral surface of the opposite leech, and the injected sperm make their way to the **ovaries** of the female system, where fertilization occurs.

The clitellum produces a **cocoon** that receives fertilized eggs from the female gonopores. Many fish leeches attach their cocoon to the host. Other leeches deposit the cocoon in water, beneath stones and other objects, and in soil. The members of one family of leeches brood their eggs within a modified cocoon attached to the ventral surface of the body.

Classification of the Phylum Annelida

Class Polychaeta. Marine annelids in which setae are carried on lateral segmental parapodia. Metamerism usually well developed. Prostomium (head) variable, but commonly bears sensory or feeding structures. Sexes separate and gametes produced by the peritoneum of numerous segments.

Class Oligochaeta. Freshwater annelids and the terrestrial species known as earthworms. Parapodia absent but setae present. Metamerism well developed. Prostomium a simple lobe without sensory or feeding structures. Hermaphroditic, with gonads present in a few specific segments. Epidermis of certain segments modified as a clitellum for the secretion of a cocoon.

Class Hirudinea. Marine, freshwater, and terrestrial annelids known as leeches. No parapodia or setae. Body more or less dorsoventrally flattened with anterior and posterior segments modified as suckers. Metamerism greatly reduced. Some ectoparasitic. Hermaphroditic, with a clitellum.

Summary

1. Members of the phylum Annelida, called segmented worms, are metameric animals in which the coelom is compartmented by transverse septa to form a localized hydrostatic skeleton. Metamerism is believed to have evolved in annelids as an adaptation for peristaltic burrowing. The segmental features of the excretory system (one pair of nephridia per segment), nervous system (segmental ganglia and lateral segmental nerves along the ventral nerve cord), and internal transport system are probably accommodations to the primary segmentation of the coelom.

2. The class Polychaeta contains mostly marine species that possess lateral segmental appendages called parapodia. An eversible pharynx armed with chitinous jaws is commonly used in carnivorous and herbivorous feeding. Most surface-dwelling polychaetes crawl with the parapodia and possess a well-developed prostomium bearing eyes, antennae, and other structures.

3. Burrowing polychaetes live in galleries that have numerous openings to the surface, or they are more sedentary occupants of simple vertical burrows. Most move by peristaltic body contractions, and the parapodia are reduced. The more active gallery dwellers usually have a simple prostomium with few sense organs and are carnivores or deposit feeders. The sedentary burrowers are selective and nonselective deposit feeders.

4. Tube-dwelling polychaetes construct tubes of secreted material or mineral particles cemented together or both. Active tube dwellers are carnivores and possess adaptations similar to those of surface dwellers. Sedentary tube dwellers are selective or nonselective deposit feeders or filter feeders. They move within the tube by peristaltic contractions and have ridgelike parapodia with hooked setae.

5. Large polychaetes possess gills, and the blood usually contains a respiratory pigment. The sexes are separate; gametes are produced by the peritoneum of many segments, mature in the coelom, and exit by way of gonoducts or nephridia. Fertilization is external, and development usually leads to a trochophore larva.

6. The class Oligochaeta contains mostly freshwater species that live within bottom debris and terrestrial forms (earthworms) that inhabit galleries in soil. They move by peristaltic contractions and possess setae but no parapodia. The prostomium is usually a simple cone without sense organs. Most species are selective or nonselective deposit feeders or both.

7. Oligochaetes are hermaphroditic, with only a few segments involved in reproduction. Two worms copulate and transfer sperm reciprocally. The eggs are fertilized within a cocoon secreted by certain glandular segments called the clitellum. Development is direct.

8. The class Hirudinea contains the leeches. Most are found in fresh water, but some are marine (fish leeches), and some live in tropical jungles. Correlated with their mode of locomotion, leeches have lost setae and developed anterior and posterior suckers. The coelom has been reduced to channels, and septa have disappeared. The body of all leeches is composed of 34 segments.

9. Leeches are predaceous on other invertebrates or are blood suckers, mostly on vertebrates. The feeding organs are jaws and a pumping pharynx or a protrusible proboscis. The gut has a large storage capacity, and the digestive enzymes are largely exopeptidases. Reproduction in leeches is very similar to that in oligochaetes.

References and Selected Readings

Detailed accounts of the annelids may be found in many of the references listed at the end of Chapter 19. The following works are devoted solely to annelids.

Dales, R. P.: *Annelids*. London, Hutchinson University Library, 1963. A brief general account of the annelids.

Edwards, C. A., and J. R. Lofty: *Biology of Earthworms*. 2nd ed. London, Chapman & Hall, Ltd., 1977. This short volume covers the structure, physiology, and ecology of earthworms.

Mann, K. H.: *Leeches (Hirudinea)*. New York, Pergamon Press, 1962. A general biology of leeches.

Mill, P. J. (Ed.): *Physiology of Annelids*. London, Academic Press, 1978.

27 Arthropods

ORIENTATION

The arthropods are a vast asemblage of animals. At least three quarters of a million species have been described, which is more than three times the number of all other animal species combined. They are found in virtually every type of habitat and are perhaps the most successful of all invaders of the terrestrial environment. Arthropods are distinguished from other invertebrates by their hard exoskeleton. This chapter will examine their possible origin from annelidan ancestors and the advantages derived from their external skeleton. The chapter will discuss the four lines of arthropod evolution: the trilobite line (probably the most primitive), the chelicetate line (which gave rise to spiders, scorpions, and ticks), the crustacean line, and the uniramian line (the insects, centipedes, and millipedes). We will explore the diversity of life styles that have evolved within each of these lines and how such diversity might relate to the success of various arthropod groups.

CHARACTERISTICS OF ARTHROPODS

1. The phylum Arthropoda is the largest phylum of the Animal Kingdom and includes such familiar forms as spiders, mites, scorpions, shrimps, crabs, insects, centipedes, and millipedes.
2. The body is covered by an exoskeleton of chitin and protein.
3. Muscles are attached to the inside of the skeleton, and the skeletomuscular system functions as a lever system.
4. Arthropods are metameric, reflecting their annelidan ancestry, but most species exhibit some degree of reduction in metamerism.
5. The segments carry paired jointed appendages, which not only serve in locomotion but also have been adapted for many other kinds of functions.
6. All internal structures derived from invaginations of the body wall have a chitinous lining. The anterior and posterior parts of the gut possess such a lining. The midgut, derived from endoderm, is more restricted than in most animals.
7. The blood vascular system is open, and the dorsal heart is primitively tubular.
8. The plan of the nervous system is like that of annelids.
9. The eggs are generally centrolecithal, and cleavage is commonly superficial.

The phylum **Arthropoda** is the largest group of animals. At least three quarters of a million species have been described. The tremendous adaptive diversity of arthropods has enabled them to survive in virtually every habitat; they are perhaps the most successful of all the invaders of the terrestrial environment.

A number of features indicate that arthropods evolved from annelids, perhaps from some group of surface-dwelling polychaetes. Arthropods are **metameric** and the segments bear lateral appendages, which may be homologous to the parapodia of polychaetes. The nervous system, with its pair of large ventral **nerve cords**, is essentially like that of annelids, and the dorsal tubular **heart** may be the homologue of the dorsal contractile blood vessel of annelids. The process of cleavage in embryonic development is determinate but no longer spiral as in annelids.

Arthropodization. The distinguishing feature of arthropods is the presence of a chitinous exoskeleton. The evolution of this nonliving, protective and supporting covering led in turn to numerous other structural and functional changes that make up the arthropod condition, or arthropodization (Figs. 27.1 and 27.2). Remember from the earlier description (p. 100) that the arthropod exoskeleton is composed of glycoprotein (chitin and protein) organized in two layers, an inner **endocuticle** and an outer **exocuticle**. A thin protein **epicuticle** forms an external layer over the exocuticle. The exocuticle differs from the endocuticle in having the glycoprotein chains cross-linked—a more rigid molecular arrangement called **tanned**. The toughness of shoe leather results from a similar process. Certain parts of the exoskeleton of different arthropods are often especially hard and said to be highly **sclerotized**, which means the exocuticle in such an area is thick and greatly tanned.

Movement is possible because the exoskeleton is divided into plates over the body and numerous cylinders around the appendages. At the junction point, or at joints between plates and cylinders, the exoskeleton is thin and folded (Fig. 27.1C). The name Arthropoda (Gr. *arthron*, joint + *pous*, foot) refers to the jointed cylinders composing the appendages.

Despite the locomotor and supporting advantages of an external skeleton for small animals, it poses problems for growth. The solution to this problem evolved by the arthropods has been the periodic shedding of the

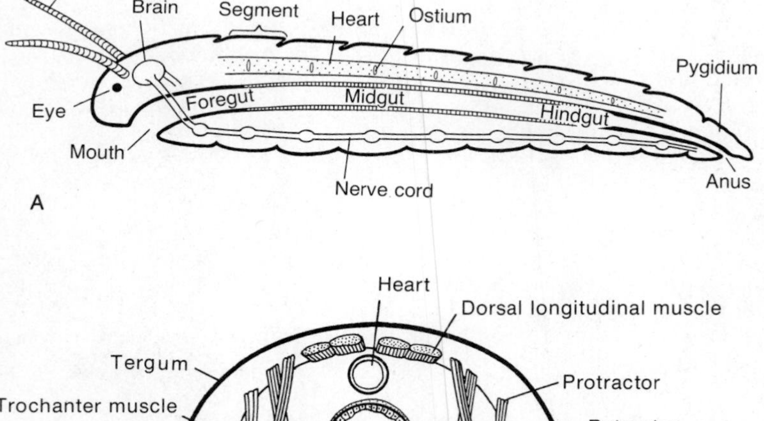

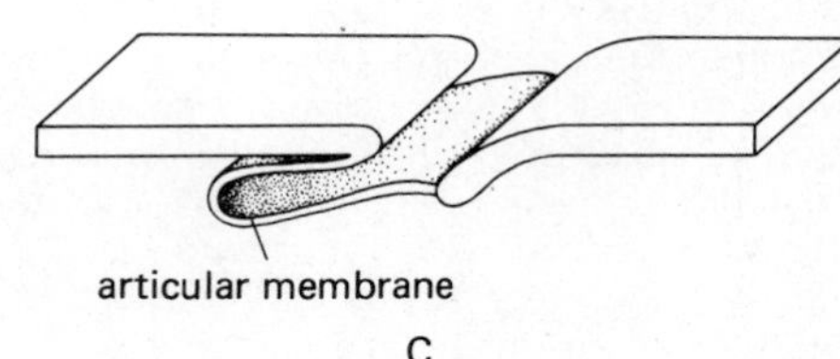

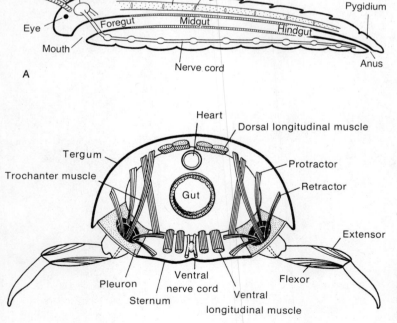

FIGURE 27.1 Structure of a generalized arthropod. *A*, Sagittal section; *B*, cross section. *C*, An arthropod joint, showing thin folded exoskeleton (articular membrane) beneath the junction of two heavier plates. (After Weber.)

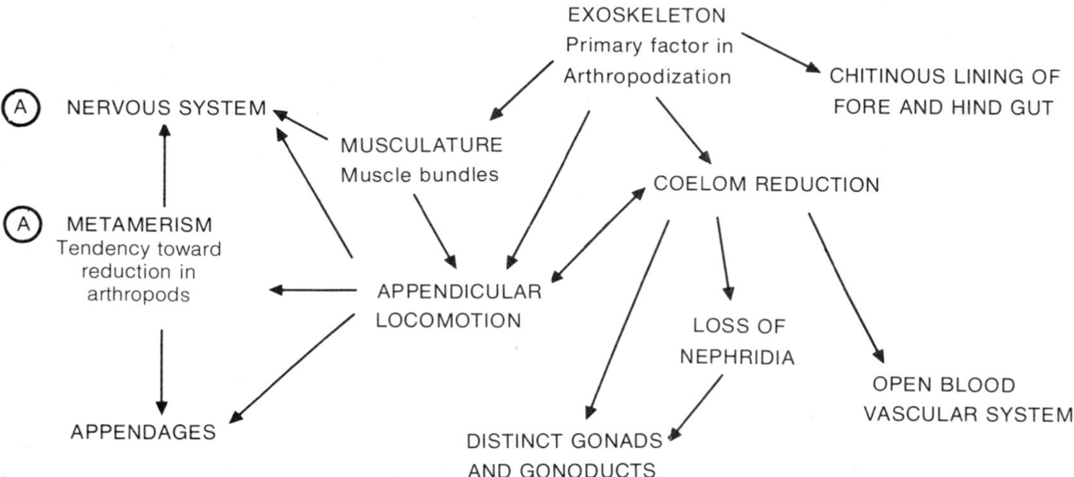

FIGURE 27.2 Interrelationship of changes resulting from the evolution of the arthropod condition (arthropodization). *A* are annelidan features retained by arthropods.

skeleton, a process called **molting** or **ecdysis** described earlier (p. 100).

The stages between molts are known as **instars**, and the length of the instars becomes longer as the animal becomes older. Some arthropods, such as lobsters, continue to molt throughout their life. Other arthropods, such as insects and spiders, have fixed numbers of instars, the last being attained with sexual maturity.

Locomotion. The muscles of arthropods are distinct bundles of muscle fibers attached to the inner side of the exoskeleton (Fig. 27.1*B*). They function with the skeleton as a lever system, much as in vertebrates. Only vestiges of the coelom remain in adult arthropods. The loss of the coelom is probably related to the shift from a fluid internal skeleton to a solid external skeleton.

In arthropods the body tends to be suspended between the legs (Fig. 27.3), rather like in primitive vertebrates, such as salamanders and crocodiles. Each leg of an arthropod performs an effective and a recovery stroke in which the leg is lifted, swung forward, and placed down upon the substratum. The legs on one side of the body carry out these movements in sequence, composing a locomotor wave; the locomotor waves on the two sides of the body alternate with each other, as in crawling polychaetes. The polyneuronal innervation of arthropod muscles is described on page 114.

The primitive arthropod trunk was composed of a large number of segments, each bearing a pair of similar appendages (see Fig. 27.4). The evolution of different modes of existence with different types of locomotion and with different types of feeding behavior has led to a reduction in the number of segments. Segments dropped out, fused together, or became specialized. Appendages became specialized for many functions other than locomotion—prey capture, filter feeding, food handling, gas exchange, ventilation, copulation, egg brooding, and so forth. In the crayfish, for example, the oral appendages are adapted for feeding, the large claws are adapted for grasping, the legs for crawling, the abdominal appendages for egg brooding, and the last appendages form a flipper with the terminal section of the

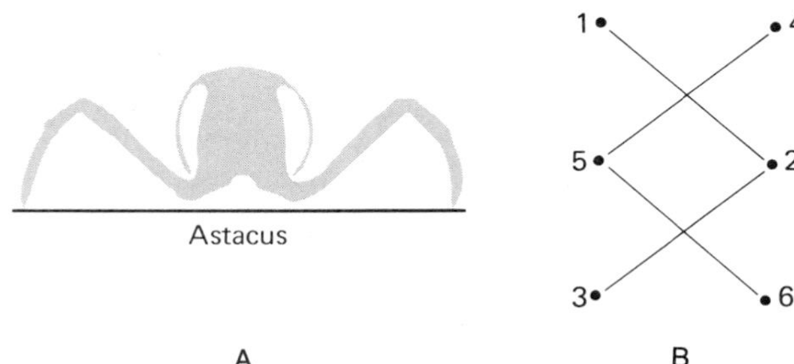

FIGURE 27.3 *A,* Cross section of a lobster, showing suspended position of body in relation to legs. *B,* Alternating sequence of movement of the three pairs of legs of most insects. Numbers indicate the sequence. (Both from Gray, J.: *Animal Locomotion.* New York, W. W. Norton & Co., 1968.)

trunk (Fig. 27.18). Although these appendages are structurally quite different, they are all derived from originally similar segmental appendages. They are therefore said to be **serially homologous.**

Digestive System. Arthropods have become adapted for a great range of diets, and the appendages, especially those just beside and behind the mouth, are utilized in many different modes of feeding. Commonly there is one heavy pair of appendages flanking the mouth that bite or tear, followed by several other pairs that are involved in food handling. The arthropod gut differs from that of most other animals in having large stomodeal and proctodeal regions (Fig. 27.1). The derivatives of these ectodermal portions are lined with chitin and constitute the **foregut** and **hindgut**. The intervening region, derived from endoderm, forms the **midgut**. The foregut is chiefly concerned with ingestion, trituration, and storage of food; its parts are variously modified for these functions depending upon the diet and mode of feeding. The midgut is the site of enzyme production, digestion, and absorption; however, in some arthropods enzymes are passed forward, and digestion begins in the foregut. Very commonly the surface area of the midgut is increased by outpocketings forming pouches or large digestive glands. The hindgut functions in the absorption of water and the formation of feces.

Internal Transport, Gas Exchange, and Excretion. The circulatory system of arthropods is open; a large **pericardial sinus** surrounds the dorsal **heart** (Fig. 27.1). Blood flows from the pericardial sinus into the heart through small lateral openings called **ostia**. In primitive tubular hearts there is one pair of ostia per body segment. When the heart contracts, the ostia close and blood is propelled anteriorly or posteriorly out of the heart through arteries, which deliver the blood to tissue spaces where various exchanges take place. From the tissue spaces (collectively termed the **hemocoel**) blood drains into a system of larger sinuses and eventually returns to the pericardial sinus around the heart. The arterial system varies greatly. Some arthropods, such as insects, have little more than a short vessel leaving the heart; others, such as lobsters and crabs, have an extensive system of vessels delivering blood to tissue spaces in various parts of the body.

Most arthropods have organs for gas exchange. Their **gills** are usually modifications of appendages or outgrowths of the integument associated with an appendage (p. 151).

The gas exchange organs of terrestrial arthropods are typically internal. However, they are derived from invaginations of the integument and thus are lined with chitin. The commonest gas exchange organ in terrestrial arthropods, and one that has evolved independently in many different groups, is a system of air-conducting tubes called **tracheae**, which have been described previously (p. 153).

The arthropods with a respiratory pigment most commonly have **hemocyanin**, a large molecule dissolved in the plasma. Hemoglobin occurs only sporadically among some species. Respiratory pigments are usually absent from the blood of arthropods with tracheal systems; the blood in these animals plays only a small role in gas transport.

The loss of the coelom probably accounts for the disappearance of nephridia and the evolution of the diverse excretory organs of arthropods. Most of the excretory organs are blood-bathed sacs opening by a duct to the body surface. The excretory organs of insects and some arachnids, such as spiders, consist of a few to many blind **malpighian tubules** lying free in the **hemocoel** and opening into the posterior section of the **gut** (Fig. 27.36). Waste materials are removed from the blood and secreted into the lumen of the excretory organs. Aquatic arthropods excrete ammonia. Some terrestrial species, such as insects, excrete uric acid; others, such as spiders, excrete guanine.

Nervous System and Sense Organs. The nervous system of arthropods exhibits the same basic design as that of annelids—a dorsal anterior **brain** and a double ventral **nerve cord** (Figs. 27.1 and 27.36). However, the fusion and loss of segments is reflected in a corresponding forward migration and fusion of the ventral **ganglia** of many arthropods. For example, in a crab all of the ganglia of the ventral nerve cord have fused together anteriorly.

The **sensory receptors** of arthropods are usually associated with some modification of the chitinous exoskeleton, which otherwise would act as a barrier to the detection of external stimuli. An important and very common modification is **bristles**, or **setae**, which contain chemo- or mechanoreceptors. Some arthropods, such as spiders, have hairs so delicate and sensitive that they can respond to airborne vibrations. Other common modifications for receptors are slits, pits, or other openings in the exoskeleton. The opening is covered by a thin membrane, to the underside of which is attached a nerve ending. The dendrite is activated by slight changes in tension of the membrane.

The appendages of arthropods contain many kinds of receptors. A leg, for example, may serve not only for movement but also for touch, taste, and smell. One or more appendages (e.g., the **antennae**) may be given over entirely to monitoring environmental signals.

Most arthropods have **eyes**, which can vary greatly in complexity. Some are simple and have only a few photoreceptors. Others are large, with thousands of retinal cells, and can form a crude image.

Arthropod Classification

Most zoologists agree that there are four lines of arthropod evolution, each constituting at least a different subphylum:

Subphylum Trilobitomorpha—the extinct trilobites
Subphylum Chelicerata—horseshoe crabs, scorpions, spiders, mites, and ticks
Subphylum Crustacea—shrimps, crabs, and allies
Subphylum Uniramia—centipedes, millipedes, and insects

The phylogenetic relationships of the four groups are uncertain, but there is a growing belief that some, if not all, had independent origins from different annelid groups, i.e., arthropodization may have occurred more than once.

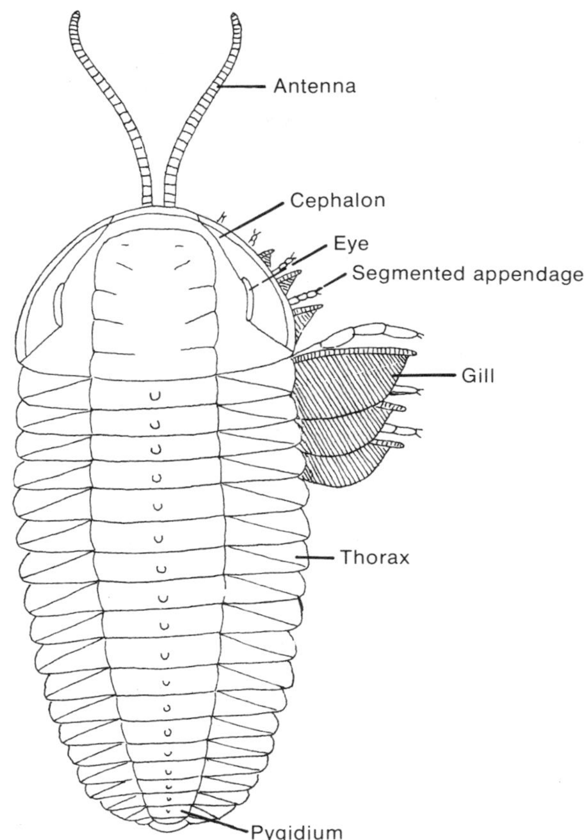

FIGURE 27.4 Dorsal view of the Ordovician trilobite, *Triarthrus eatoni.* (After Walcott and Raymond from Stormer.)

Subphylum Trilobitomorpha

The most primitive known arthropods are the extinct trilobites, which inhabited the Paleozoic oceans from the Cambrian through the Permian. They are usually placed within a separate subphylum, the **Trilobitomorpha**.

Their flattened oval body was about 3 to 10 cm. in length and composed of an anterior **cephalon**, a middle **thorax**, and a posterior **pygidium** (Fig. 27.4). The cephalon, originating from the fusion of the head with four trunk segments, formed a solid shield-shaped **carapace**, which bore a pair of eye clusters on the dorsal surface. The middle section of the body, the thorax, was composed of many unfused segments, and the posterior pygidium contained a number of fused segments. Two dorsal longitudinal furrows extended the length of the body and divided it into three transverse sections, from which the name trilobite is derived.

A pair of antennae was located on the undersurface of the body, just in front of the mouth. Behind the mouth, each segment carried a pair of identical appendages (Fig. 27.4). In all other arthropods, at least some of the segmental appendages are different from others. Each appendage contained two branches—one leglike, on which the animal crawled, and one that contained parallel filaments of uncertain functions—perhaps for swimming, digging, filtering, or gas exchange.

Trilobite fossils are quite common, but most represent only the dorsal part of the body. Our interpretation of appendage function is limited to the very few specimens in which appendages have been preserved. Speculations about life styles are therefore based in large part on body shape and size. Most trilobites were benthic and included some species that burrowed and some capable of rolling up in a ball. But there were also forms that appeared to have been swimmers and some that may have been planktonic. In recent years we have gained some knowledge of trilobite internal anatomy, particularly of the eyes and gut, from X-ray photographs of fossils in which there was some preservation of internal structure.

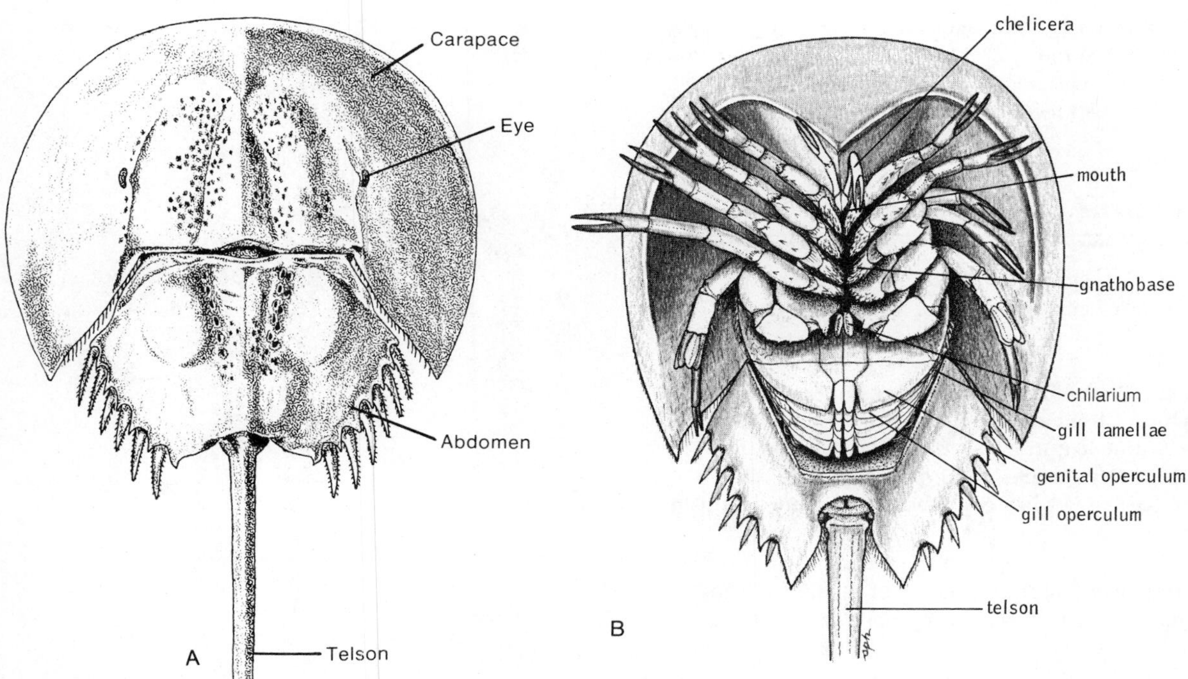

FIGURE 27.5 A horseshoe crab. *A,* Dorsal view; *B,* ventral view showing appendages. (*A* after Van der Hoeven.)

Subphylum Chelicerata

Members of the subphylum Chelicerata are distinguished from all other arthropods in lacking antennae. The first pair of appendages behind the mouth are chelicerae (Gr. *chele*, claw + *keras*, horn), which are modified in various ways for feeding. The body is composed of two regions, a cephalothorax and an abdomen (Fig. 27.5). The cephalothorax contains a number of trunk segments that have fused with the head, and the fused region is usually covered dorsally by a single skeletal piece, the **carapace**. Behind the chelicerae the cephalothorax carries a pair of appendages, called **pedipalps**, and four pairs of legs. There are no abdominal appendages.

Class Merostomata

The members of the class Merostomata, called horseshoe crabs, are the only marine gill-bearing chelicerates (Fig. 27.5). The class has been known from the Ordovician, but there are only five species living today. *Limulus polyphemus* is found along the Atlantic coast of the United States and in the Gulf of Mexico. They are among the largest living chelicerates, reaching a length of 60 cm.

Dorsally, the carapace carries a pair of compound eyes, and the underside of the cephalothorax bears the six pairs of appendages typical of most chelicerates—a pair of chelicerae, a pair of pedipalps, and four pairs of legs. However, the pedipalps are not markedly different from the posterior appendages and should properly be called legs. All except the last pair of appendages are **chelate;** that is, they possess pincers formed by two terminal segments of the appendage. The last pair of appendages is used for sweeping away sand and mud.

The abdominal segments are fused together and terminate in a long spikelike **telson** used to right the animal if flipped over. The underside of the abdomen carries five pairs of **book gills,** each composed of a large number of thin plates (**lamellae**) protected by a flaplike cover.

Horseshoe crabs are harmless animals. They crawl over or push through sand in shallow water, and small individuals swim upside down using the gills as paddles.

Horseshoe crabs are omnivores and scavengers, and their diet includes the softbodied invertebrates and algae they encounter as they plow through the bottom.

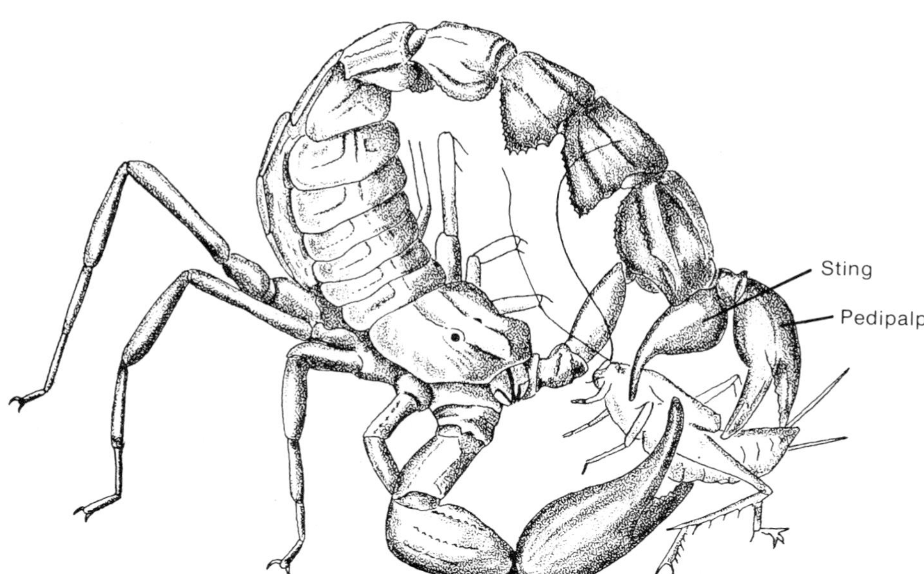

FIGURE 27.6 The North African scorpion, *Androctonus australis*, capturing a grasshopper. (After Vachon from Kaestner, A.: *Invertebrate Zoology*. Vol. II. New York, John Wiley and Sons, 1968.)

During the reproductive period horseshoe crabs migrate into shallow water of sounds and bays to mate. The male clasps the dorsal side of the abdomen of the female with his modified second pair of appendages (the pedipalps of other chelicerates). Eggs are fertilized as they are deposited into shallow depressions in the sand.

Class Arachnida

The great majority of chelicerates, some 60,000 species, are members of the class **Arachnida** (Gr. *arachnē*, spider). In contrast to merostomes, arachnids are terrestrial. They are widely distributed, most living in vegetation, in leaf mold, and beneath bark, logs, and stones. Contributing to the great success of arachnids as land animals has been the evolution of terrestrial gas exchange organs and a waxy epicuticle, which reduces water loss.

The class is divided into ten orders, five of which contain the most common and familiar species. Scorpions (order **Scorpionida**) are large arachnids with big chelate pedipalps and a long segmented abdomen terminating in a **sting** (Fig. 27.6). Paired eyes are mounted on tubercles in the middle of the carapace, and two to five additional pairs of eyes may be present along the anterior lateral margins. Scorpions are secretive, largely nocturnal animals of the tropics and semitropics. In the United States they are common only in the Gulf and southwestern states.

Scorpions are ancient arachnids, known from the Silurian, and probably were among the first terrestrial arthropods.

Pseudoscorpions (order **Pseudoscorpionida**) are only a few millimeters in length and are common inhabitants of leaf mold in both tropical and temperate regions. These tiny arachnids have large chelipeds like scorpions, but the segmented abdomen is short and lacks a terminal sting (Fig. 27.7).

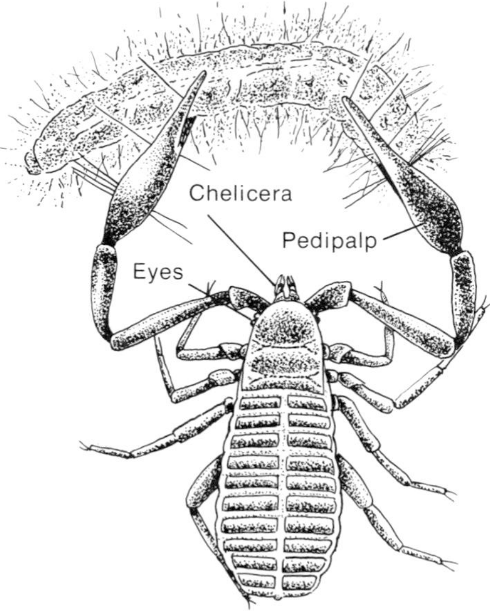

FIGURE 27.7 Dorsal view of a pseudoscorpion catching an insect larva with its pedipalps.

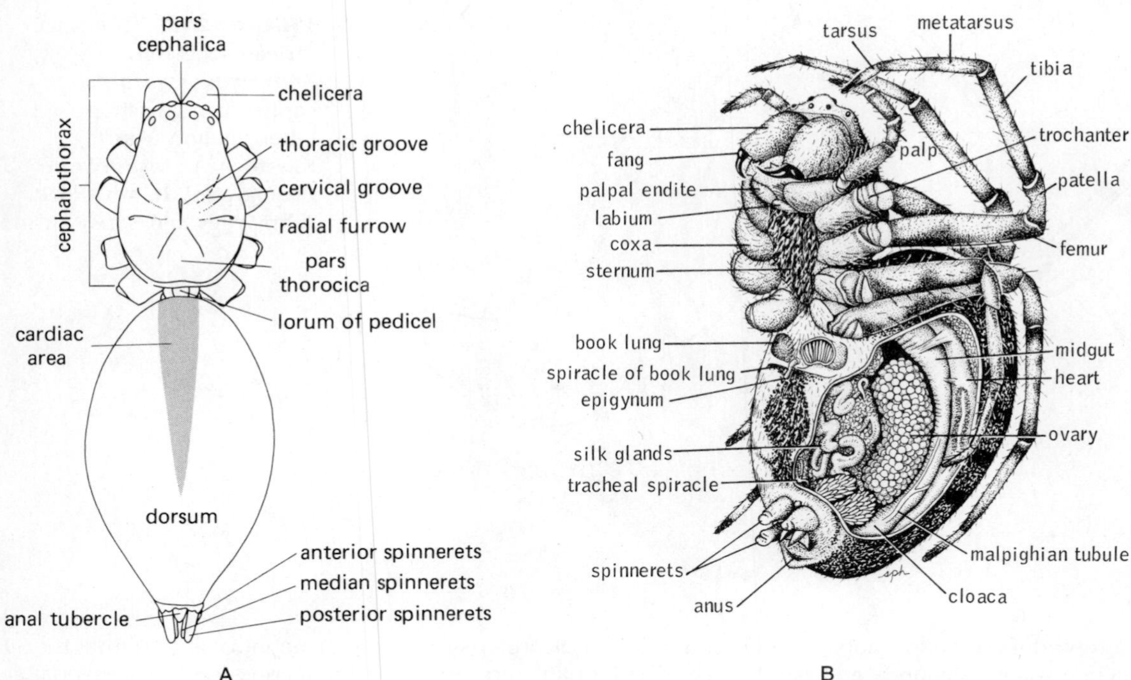

FIGURE 27.8 Structure of spiders. *A,* Dorsal view. *B,* Ventrolateral view. (*A* from Kaston, B. J.: Spiders of Connecticut. *State Geological and Natural History Survey,* Bull. 70:13, 1948; *B* after Pfurtscheller from Kaestner, A.: *Invertebrate Zoology.* Vol. II. New York, John Wiley and Sons, 1968.)

Spiders (order **Araneae**) compose the largest of the arachnid orders. Over 30,000 species have been described, and they frequently occur in far greater numbers than most people are aware of. An ungrazed meadow, for example, may support as many as 2,250,000 spiders per acre. The abdomen is unsegmented and connected to the cephalothorax by a narrow waist (Fig. 27.8*A*). The pedipalps are small and leglike. Usually eight eyes are arranged in two rows of four each across the front of the carapace.

Daddy-long-legs, or harvestmen (order **Opiliones,** or **Phalangida**), are distinguished from other arachnids by their very long legs and segmented abdomen broadly joined to the cephalothorax (Fig. 27.9). A tubercle on the center of the carapace bears a single pair of eyes. The members of this order are common arachnids in both temperate and tropical regions.

The order containing mites and ticks, the **Acarina**, is the second largest and most diverse group of arachnids, and some acarologists believe that the 30,000 known species probably represent considerably less than half of the total number comprising the order. Most of the undescribed species will probably become extinct with the destruction of tropical rain forests and other habitats and will never be known.

Mites are usually less than a millimeter in length, and their adaptive diversity may in part be attributed to their small size, which has enabled them to exploit many types of microhabitats. However, such habitats must contain adequate water vapor, for the small size of mites also makes them vulnerable to desiccation.

The abdomen of mites and ticks is unsegmented and broadly fused with the cephalothorax. The entire body is thus covered dorsally by a single skeletal piece (Figs. 27.10 and 27.12). The pedipalps are small and usually leglike.

Silk. Spiders make a greater use of silk than any other group of animals. The **silk glands** of spiders are located in the abdomen and open through conical **spinnerets** at the end of the abdomen (Fig. 27.13), each spinneret bearing numerous spigots. A particular species may

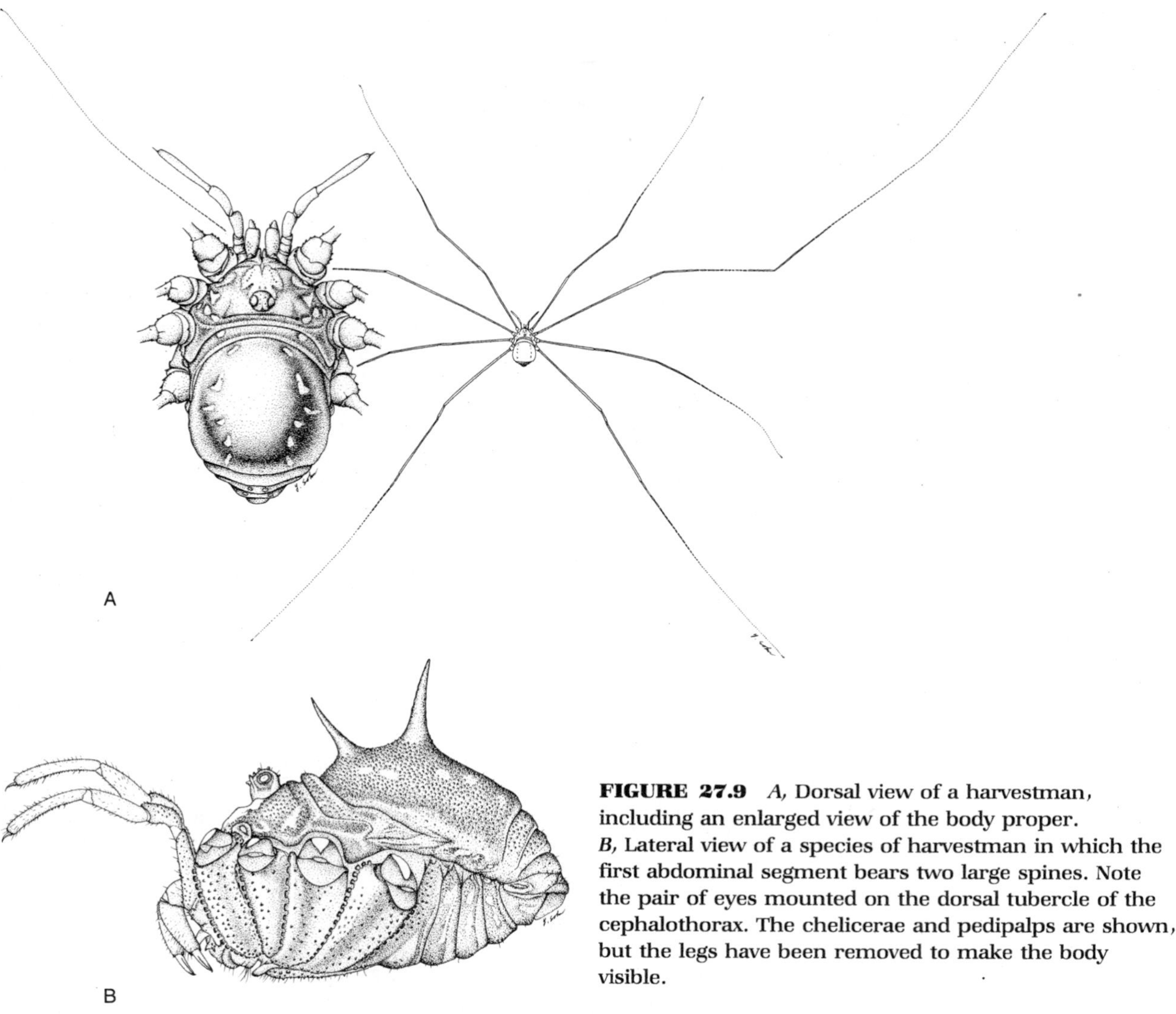

FIGURE 27.9 *A,* Dorsal view of a harvestman, including an enlarged view of the body proper. *B,* Lateral view of a species of harvestman in which the first abdominal segment bears two large spines. Note the pair of eyes mounted on the dorsal tubercle of the cephalothorax. The chelicerae and pedipalps are shown, but the legs have been removed to make the body visible.

possess from two to six different kinds of silk glands. The duct from each gland, of which there may be a large number belonging to any one type, opens at the end of one spigot. Silk is a protein and hardens not on exposure to the air but by polymerization during the process of being drawn out, usually as a result of the spider moving away from the attached end of the line.

Spiders utilize silk in many ways, but contrary to popular notion, only some spiders use silk to construct webs for trapping prey. Silk is used to build nests, which are used as retreats, for reproduction, or for overwintering; in all spiders the eggs are encased within a silken **egg case.**

Most spiders lay down a **dragline** behind them, anchoring it at intervals to the substratum (Fig. 27.13). The dragline not only functions as a safety line for the spider but also is an important means of communication between members of the species. A spider may determine from another dragline whether its owner is male or female and whether it is immature or adult. The dragline is demonstrated when a spider appears suspended in midair after being brushed off clothing or some other object.

Small spiders and newly hatched spiders use the silk as a means of dispersal. They climb to favorable take-off points, tilt the abdomen upward, and release a strand

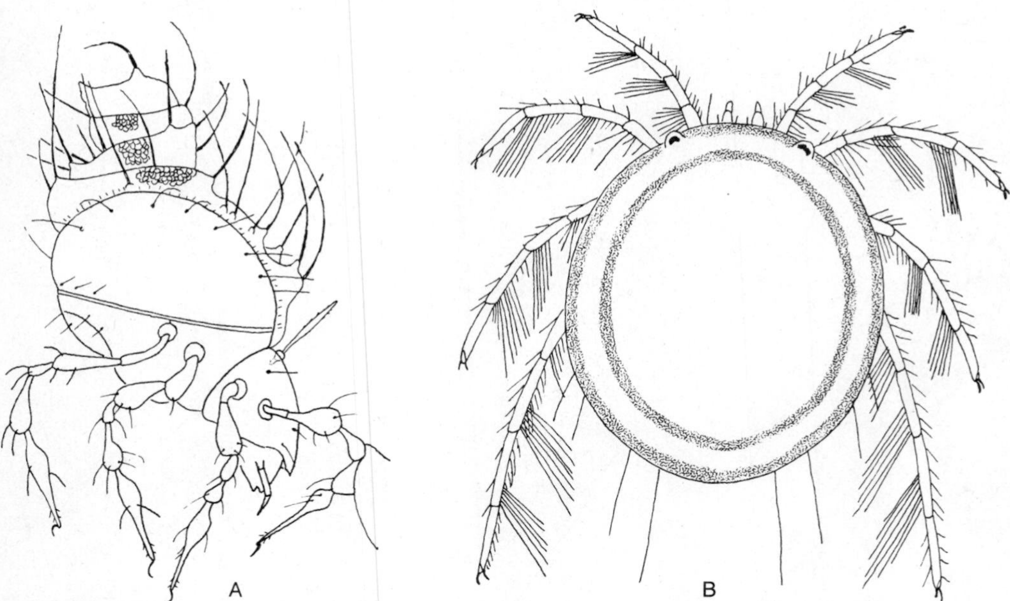

FIGURE 27.10 *A,* An oribatid mite, *Belba jacoti,* carrying five shed nymphal skins. (After Wilson from Baker and Wharton.) *B,* Water mite, *Mideopsis orbicularis.* (After Soar and Williamson from Pennak, R. W.: *Freshwater Invertebrates of the U.S.* 2nd ed. New York, John Wiley and Sons, 1978.)

of silk. When air currents produce sufficient pull, the spider lets go and sails out to whatever new habitat and fate the wind will take it. The wide distribution of many species of spiders is undoubtedly correlated with this ballooning phenomenon. Aside from certain species of mites (spider mites), pseudoscorpions are the only other arachnids that make use of silk. The silk glands of pseudoscorpions open onto the chelicerae and are used in construction of nests, or retreats.

Feeding. Most arachnids are predatory animals, and other arthropods are their usual prey. As an aid to dispatching prey, certain arachnids have independently evolved **poison glands**. The poison glands of scorpions are located in the terminal sting at the end of the abdomen (Fig. 27.6). Their prey is caught with the large pedipalps and then stabbed by the poison barb with a forward thrust of the abdomen. Although the sting of scorpions may be painful, few species have a poison dangerous to human beings. The only dangerous North American scorpions are found in Mexico and the southwestern United States.

Pseudoscorpions catch prey with their large chelate pedipalps (Fig. 27.7), which have a poison gland opening at the end of one or both fingers.

Spiders have a poison gland associated with each chelicera, which consists of a terminal **fang** that folds down against a larger basal piece (Fig. 27.8). The gland

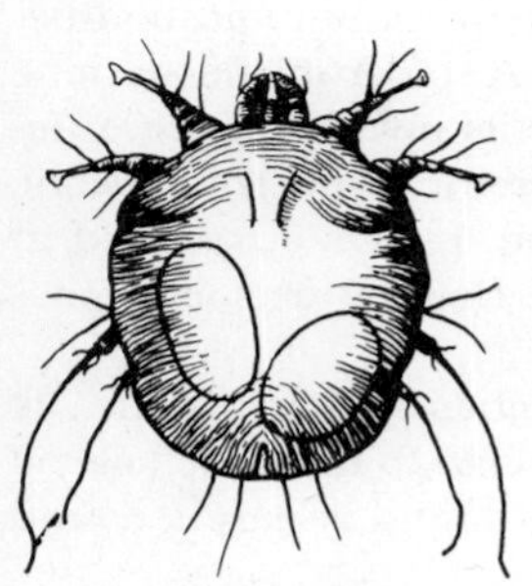

FIGURE 27.11 The mange mite, *Sarcoptes scabiei.* These pass their entire life cycle on the host. Eggs laid in the burrows hatch into young mites that begin burrows of their own. Note the suckers on the anterior legs. (After Craig and Faust.)

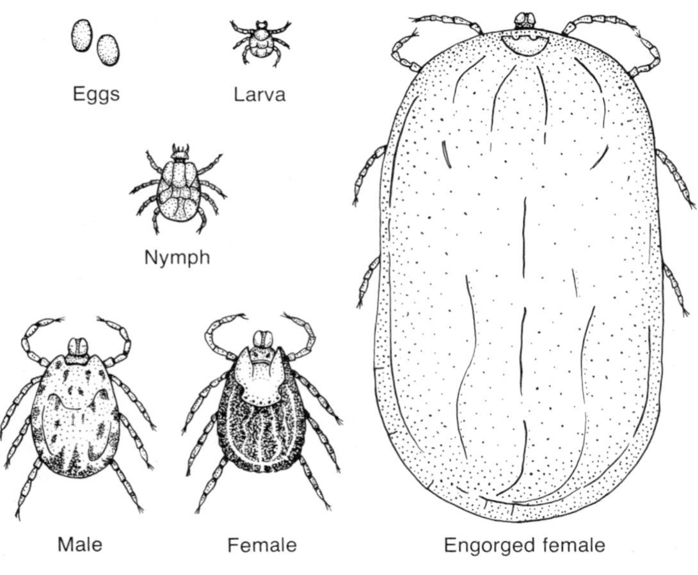

FIGURE 27.12 The common tick, *Dermacentor andersoni.* Eggs laid on the ground hatch into six-legged larvae that feed on small mammals. These drop off and molt into eight-legged nymphs that return to small mammals. After molting on the ground again the adults attack large mammals. The females become enormous after mating and eventually fall to the ground to lay a thousand or more eggs. (After Chandler.)

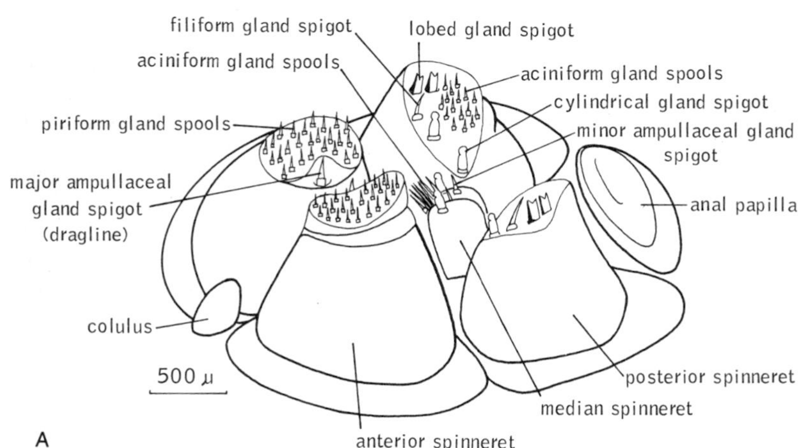

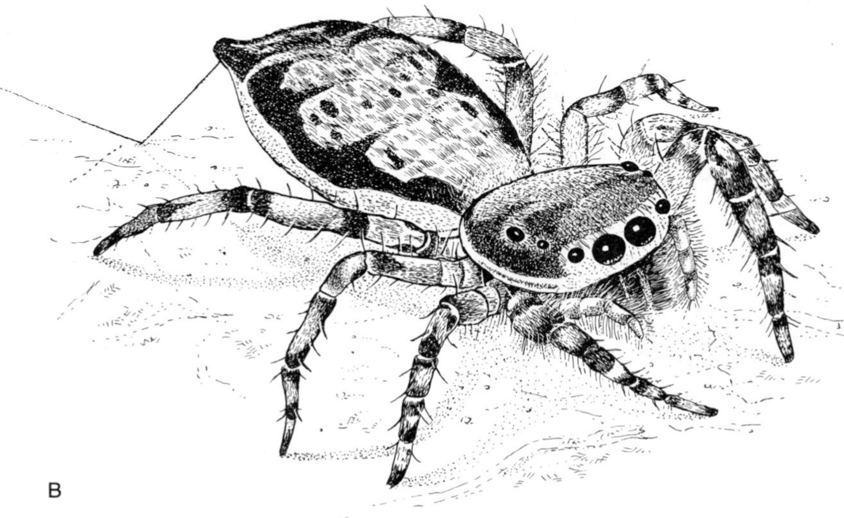

FIGURE 27.13 *A,* Spinnerets of an orb-weaving spider showing distribution of spigots for different silk glands. (From Wilson, R. S.: *American Zoologist, 9*:103–111, 1969.) *B,* Jumping spider laying down dragline. The arrangement of the eight eyes is a characteristic feature of this particular family of hunting spider, which is capable of jumping by rapidly extending the legs. (Based on a figure from Levi, H. W., and L. R. Levi: *A Guide to Spiders and Their Kin.* A Golden Nature Guide. New York, Golden Press, 1968.)

opens by a duct through the end of the fang, and the poison is injected through a bite.

The poison of a very small number of spiders is dangerous to us. Few tarantulas have a toxic bite, despite popular mythology. The members of the cosmopolitan genus *Latrodectus*, which includes the black widows, are perhaps the most notorious of the poisonous species. *Latrodectus mactans* is widely distributed in the United States (Box 27.1). The brown recluse spider, *Loxosceles reclusa*, is another dangerous species in the United States (Box 27.1).

The method of catching prey can be a basis for dividing spiders into two adaptive groups: **cursorial** (hunting) and **web builders.** Cursorial spiders include the tarantulas, wolf spiders, crab spiders, jumping spiders, and others. They spin silk draglines, nests, and cocoons, but they do not use silk to capture prey. Rather, the prey is stalked, pounced upon, and bitten. Hunting spiders generally have heavier legs and more highly developed eyes than do web builders (Fig. 27.13*B*).

Web-building spiders construct various types of webs to trap prey. Orb-web construction is described in Box 36.1. Web-building spiders are aerialists and have long slender legs for climbing about the silken lines (Box 27.1). Eyesight is poor, but web builders are able to detect and interpret the various vibrations of the web with great facility. Web vibrations inform an orb weaver, for example, about the size of the struggling prey and whether it is securely caught. The spider approaches the prey and gives it a fatal bite, sometimes swathing it in silk before or after the bite.

Arachnids are unusual in that most begin digestion of their prey outside their bodies. While the tissues are torn by the chelicerae, enzymes are secreted by the midgut, passed forward through the foregut, and poured out of the mouth into the prey. The partly digested tissues are then sucked in by the pumping action of a part of the foregut. Digestion is completed and absorption occurs in the midgut, which may be greatly evaginated and ramify into various parts of the body.

There are some exceptions to the predatory feeding habit of most arachnids. Harvestmen are omnivores and feed upon vegetable material and dead animal remains in addition to live invertebrates. The greatest diversity in feeding is displayed by mites. Some are predatory; some are herbivorous and have mouthparts adapted for piercing the cells of plants and sucking out the contents. Some feed on plant products, decomposing plant material and fungi. Members of several groups of mites are scavengers. Hair and feather mites spend their lives on the skin of mammals and birds, where they feed upon sloughed skin cells, gland secretions, and fragments of hair and feathers.

Ticks are bloodsucking parasites of reptiles, mammals, and birds, and the chelicerae are adapted for penetrating and anchoring into the skin of the host. A tick feeds until it is engorged. Then it drops off the host and does not feed again until the next instar (Fig. 27.12).

Chiggers, or redbugs, are ectoparasites on terrestrial vertebrates during their posthatching instar. In feeding, the minute mite secretes enzymes that produce a deep well from which it sucks out the digested contents. The mite secretions produce an irritating reaction resembling a mosquito bite but lasting much longer. Following feeding the chigger falls off the host and is predaceous as an adult.

Mange and itch mites (Fig. 27.11) spend their entire lives tunneling through the skin of their bird or mammalian host. Eggs are deposited in the burrows and the young, upon hatching, follow the same existence as the adults.

Gas Exchange. Large arachnids have book-lungs for gas exchange; small arachnids have tracheae (p. 153). Thus scorpions have book-lungs, four pairs with slitlike openings on the ventral surface of the anterior abdominal segments. Pseudoscorpions, harvestmen, and mites possess tracheae. Primitive spiders, such as the large tarantulas, have two pairs of book-lungs, but most spiders have one pair of book-lungs and one pair of tracheae. Some very small spiders have only tracheae. The openings of both types of gas exchange organs are located on the ventral side of the abdomen (Fig. 27.8). Heart and blood vessels are best developed in those arachnids that possess book-lungs, where blood is required for gas transport.

Reproduction and Development. The paired gonads are located in the abdomen, and in both sexes a median **gonopore** opens onto the anterior ventral surface.

Indirect transfer of **spermatophores** appears to be a primitive condition in arachnids and perhaps represents the early arthropod solution to the problem of sperm transmission on land. Spermatophore transfer in scorpions is preceded by a "courtship" dance, during which the large pedipalps of the male are locked with those of the female. In the course of the dance, the male deposits a spermatophore on the ground and then maneuvers the female so that it is taken up into her gonopore (Fig. 27.14).

Pseudoscorpions also utilize spermatophores and exhibit a wide range of behavioral modifications that increase the likelihood of the female finding and picking up the spermatophore. Our knowledge of reproductive behavior in mites is still very poor. Some species transfer sperm indirectly by spermatophores; others transfer sperm directly, utilizing a penis.

The process of sperm transfer in spiders is remarkable and is paralleled in few other animals. The copulatory organs of the male are the ends of the pedipalps,

BOX 27.1

A. A black widow spider, *Latrodectus mactans*. The bite of the larger female is neurotoxic, and symptoms include painful abdominal cramps. It would not ordinarily be fatal except perhaps to a small child, and if correctly diagnosed, it is easily treated. These spiders build tangle-webs in protected places, often around human habitations, a habit they share with other members (nonpoisonous) of the large family to which they belong. A common house spider is a member of this family. (Adapted from figures by Kaston.) *B*. The brown recluse spider, *Loxosceles reclusa*, whose bite produces an ulcerated wound that is difficult to heal. This species is found in the midwestern and southeastern United States.

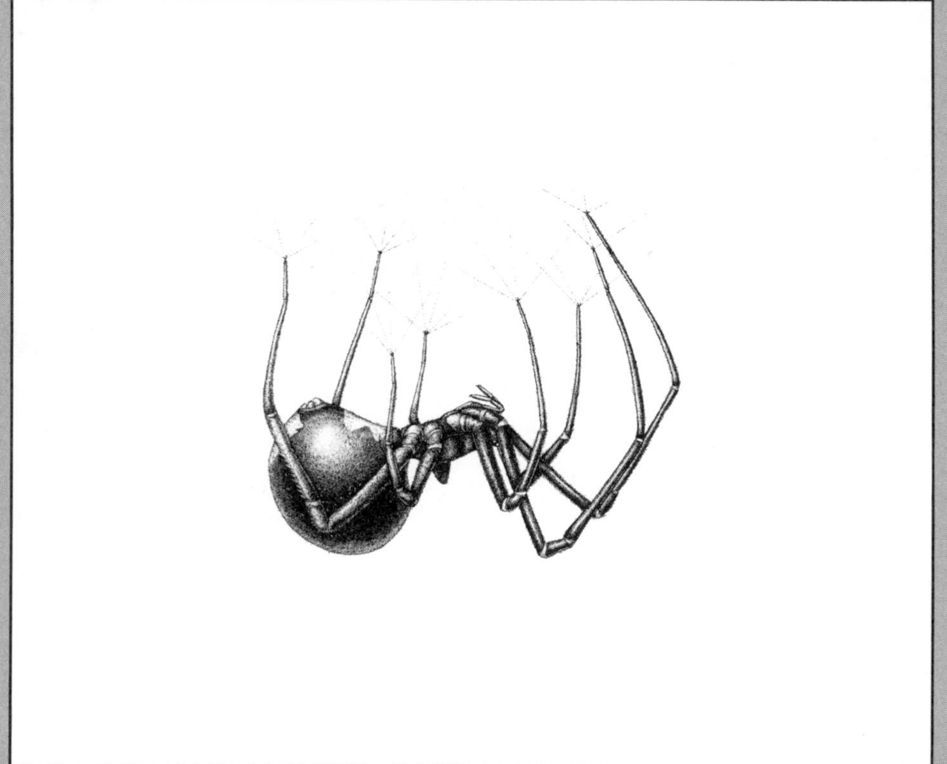

A

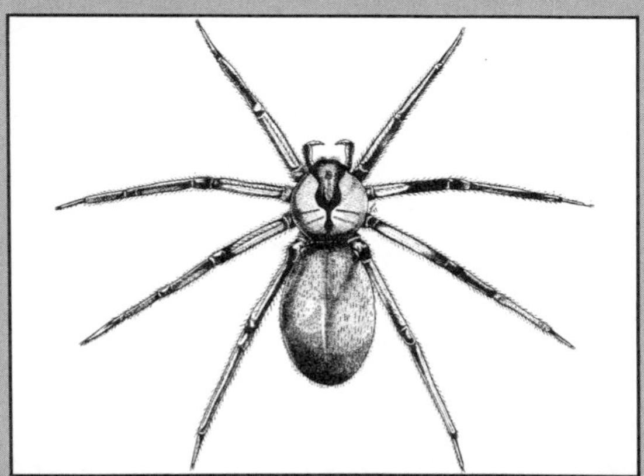

B

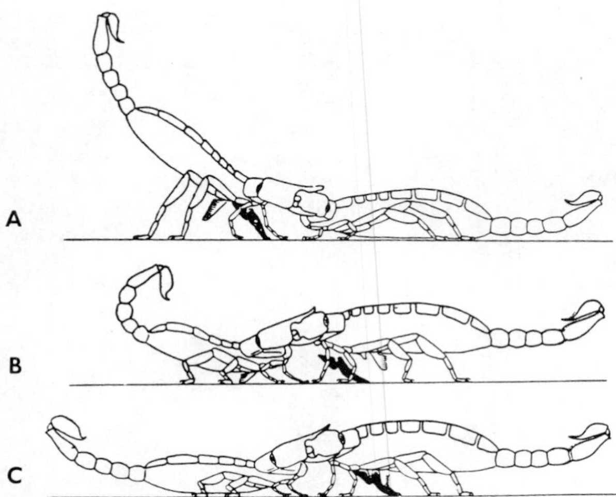

FIGURE 27.14 Sperm transfer in scorpions. *A,* While holding the female's pedipalps with his own, male on left deposits spermatophore on ground. *B,* Female is pulled over spermatophore. *C,* Spermatophore taken up into female gonophore. (After Angermann, H.: *Zeitschrift fur Tierpsychologie (Berlin), 14*:276–302, 1957.)

which resemble a pair of boxing gloves (Fig. 27.15). Prior to mating, the male spins a tiny sperm web on which a droplet of semen is secreted. The two pedipalps are then dipped into the droplet until the semen is taken up within the reservoir of the palpal organ. The male now seeks a female.

The male is frequently smaller than the female, and the predatory nature of spiders makes it important for the male to ensure that the female does not mistake him for potential prey. Complex precopulatory behavior patterns have evolved utilizing visual, tactile, and chemical signals. Significantly, there is considerable difference in the precopulatory behavior of cursorial and web-building spiders, for the sensory cues that are important in prey catching are also the important ones in sex recognition and mating. Thus web vibrations are an important means by which the male signals a female in web-building spiders, and visual signals, such as posturing by the male, are important in many species of cursorial spiders.

Following various forms of precopulatory contact by the male, the palpal organ is locked onto the chitinous plate containing the female reproductive openings, and the ejaculatory process is inserted into the seminal receptacles. Sperm is transferred from one palp at a time. This unusual mode of sperm transfer in spiders probably had its origins in transfer by spermatophore. The male of some arachnid ancestral to the spiders may have used the palp to place a spermatophore into the female gonopore.

Harvestmen and many mites transfer sperm directly without spermatophores. These arachnids deposit their fertilized eggs in soil, in leaf mold, or beneath bark, but spiders place their eggs in silk cases, which are then usually left beneath stones, bark, or leaf mold or are attached to vegetation.

Brooding is common. Wolf spiders and fisher spiders carry their egg cases about with them. After hatching, wolf spiderlings are carried on the back of their mother. The eggs of scorpions develop within the body of the female. Following birth, the young are carried about on the back of the mother. Pseudoscorpions brood their eggs within a membranous sac overlying the gonopore.

Most arachnids have direct development, and the young at hatching or at birth resemble the adult form.

BOX 27.2

Related Class

The Pycnogonida is a small group of long legged marine arthropods called sea spiders. They are like arachnids in lacking antennae and in possessing chelicerae and palps, but the very slender body carries four to six pairs of legs, and there is an additional pair of appendages (ovigerous legs) located behind the palps used in grooming and carrying eggs in the male. Pycnogonids are commonly found slowly crawling over sessile animals, such as hydroids and bryozoans. They feed on the polyps and zooids of these animals or on the organic detritus that accumulates on their surfaces.

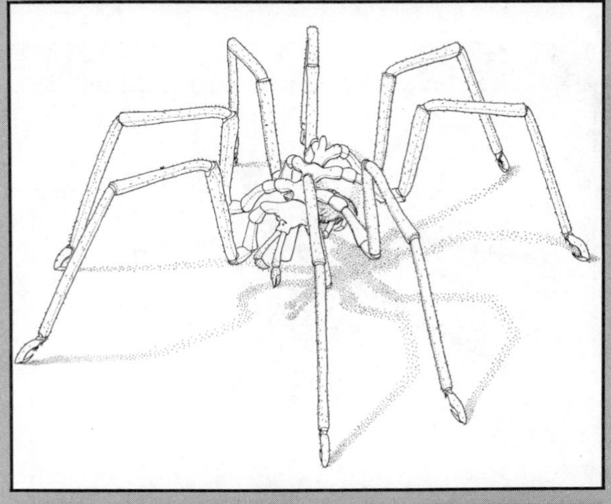

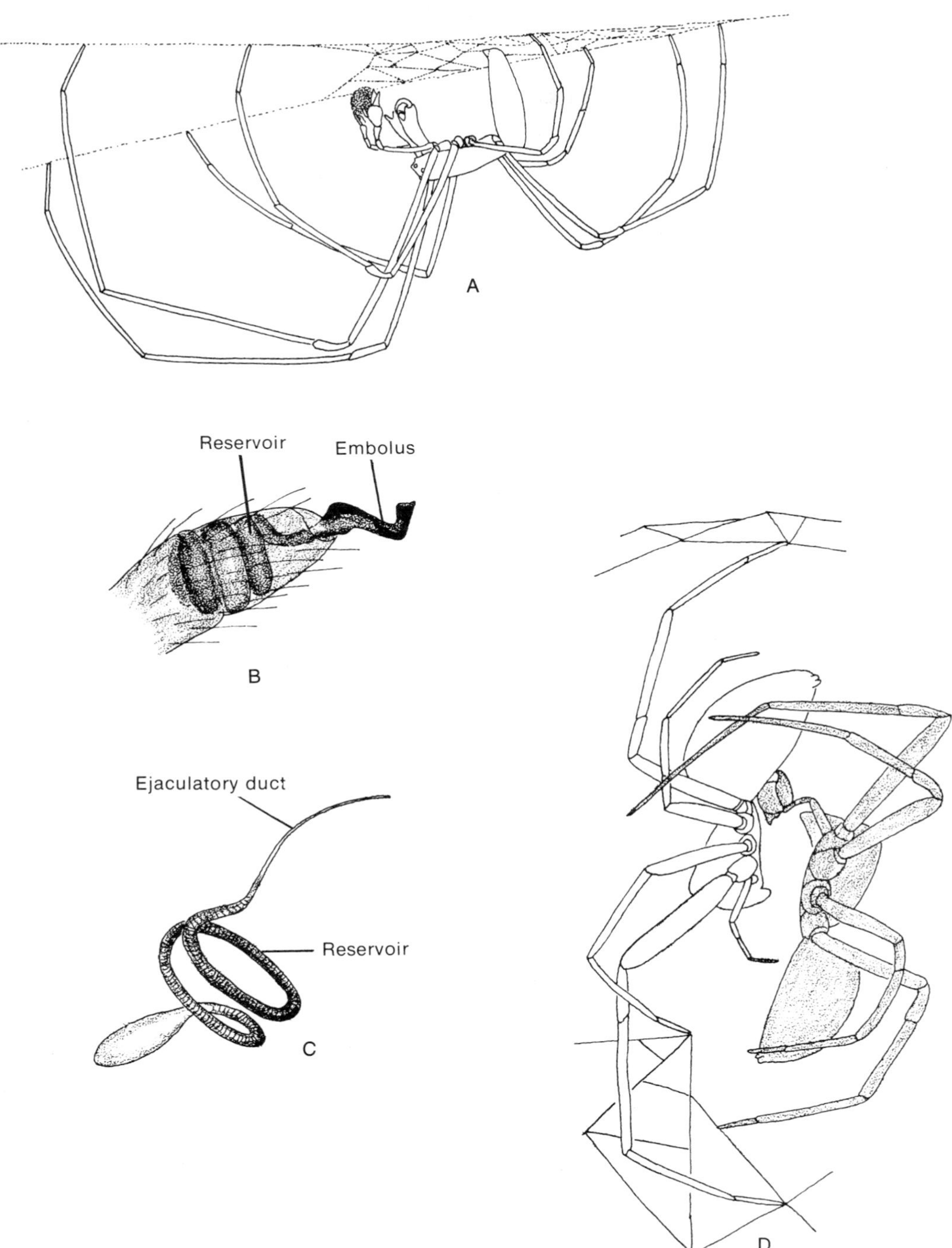

FIGURE 27.15 *A,* Male tetragnath spider in sperm web, filling palps from globule of semen. *B,* Simple palp of *Filistata hibernalis. C,* Receptaculum seminis, the semen-containing part of the male palp. *D,* Mating position of *Chiracanthium* (male shaded). (*A* after Gerhardt; *B* and *C* after Comstock; *D* after Gerhardt.)

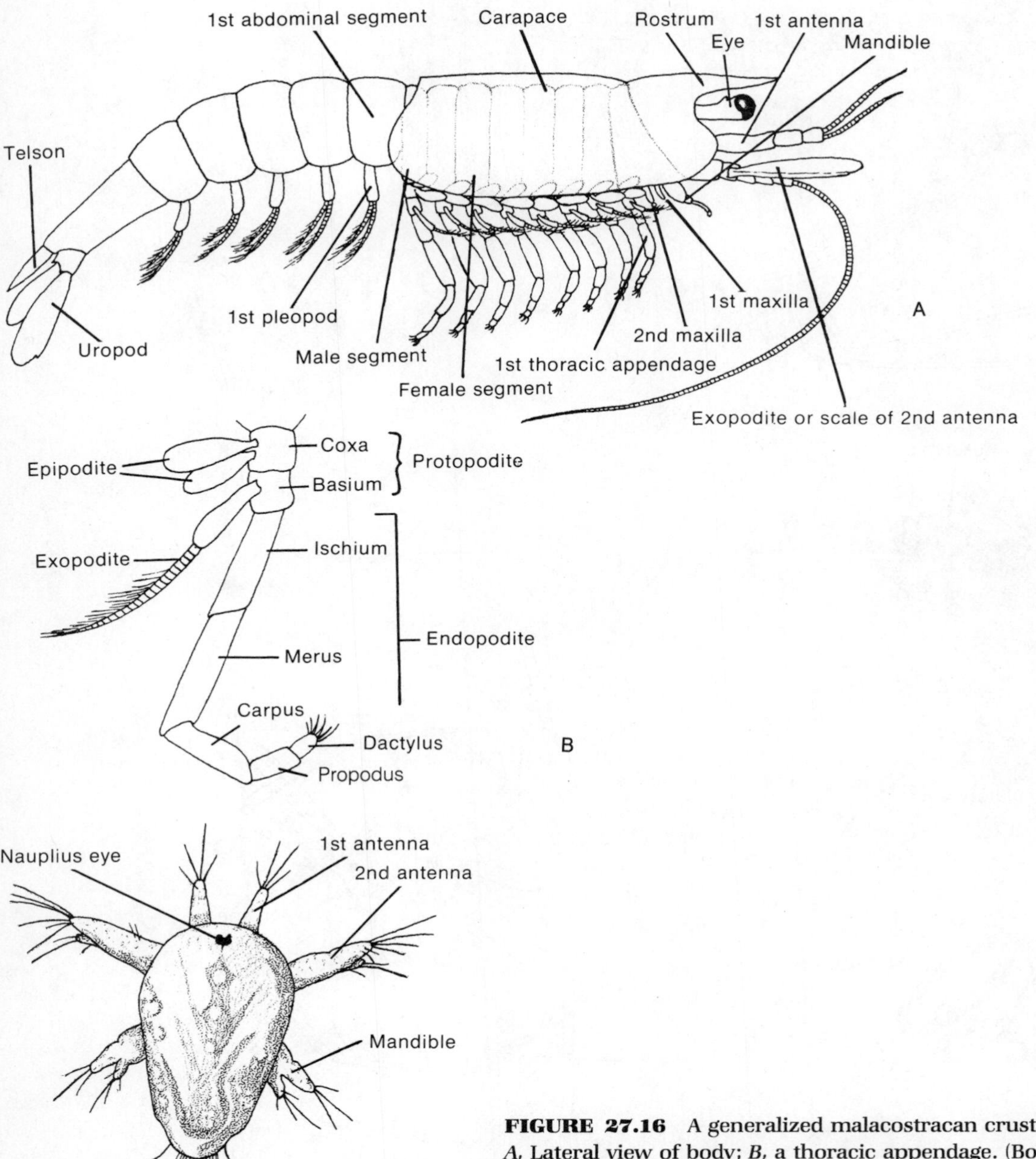

FIGURE 27.16 A generalized malacostracan crustacean. *A,* Lateral view of body; *B,* a thoracic appendage. (Both after Calman.) *C,* Nauplius larva.

Subphylum Crustacea

The subphylum **Crustacea** (L. *crusta,* hard surface), traditionally considered a class of arthropods, contains the shrimp, lobsters, crayfish, and crabs. Most are marine, but many species live in fresh water, and some, principally the sow bugs, are adapted for a terrestrial existence.

Crustaceans differ from all other arthropods in possessing two pairs of antennae. The first pair is probably homologous to the antennae of insects, centipedes, and millipedes, but the second pair is unique to crustaceans. The feeding appendages, located just behind the mouth, are a pair of heavy biting mandibles and two pairs of small maxillae. Thus the head appendages, which are constant for all members of the subphylum, are two pairs of antennae, one pair of mandibles, and two pairs of maxillae (Fig. 27.16).

The trunk of crustaceans varies greatly from one class to another. A **carapace** formed from a posterior fold of the head is commonly present. It may cover only a small part of the dorsal surface of the trunk, or it may greatly overhang the sides of the body (Fig. 27.16). In some crustaceans, the entire body is enclosed within a carapace. The terminology of crustacean appendages is based on the assumption that, primitively, each appendage was a simple two-branched (**biramous**) structure. The outer branch (**exopodite**) and inner branch (**endopodite**) were attached to a basal piece (**protopodite**) (Fig. 27.16). The trunk appendages have been adapted for a wide range of functions.

Crustaceans possess either compound eyes or a simple median eye, but rarely both. The **median,** or **nauplius,** eye is a little cluster of three or four pigment cups containing photoreceptors. It is characteristic of the nauplius larva (Fig. 27.16), but in some groups the larval eye is retained in the adult.

The larger crustaceans are often brightly colored, the pigment being located in the outer layer of the exoskeleton. The red color of boiled lobsters and crabs results from the denaturing of the protein portion of the pigment astaxanthin, which is bluish or green in the living conjugated state. The larger crustaceans also possess **chromatophores**, which are visible through the exoskeleton. These are cells with branching processes, and the pigment granules of various colors flow into and out of the processes. The chromatophores enable the animal to adapt to the color of its background; the most common change is simple darkening or lightening.

Although crustaceans have a wide range of diets and feeding habits, filter feeding is very common, especially among small crustaceans. The **filter** is formed by closely spaced **setae** on certain head or trunk appendages (Fig. 27.17).

A number of generalizations can be made about crustacean reproduction. Copulation is the rule, and certain appendages are adapted for clasping the female and transferring sperm. Fertilization may be internal or it may occur externally at the time of egg deposition, and the eggs are usually brooded on certain parts of the body. The earliest hatching stage, the **nauplius larva**, possesses only the first three pairs of appendages (Fig. 27.16C). The gradual acquisition and differentiation of additional segments and appendages occur in subsequent developmental instars.

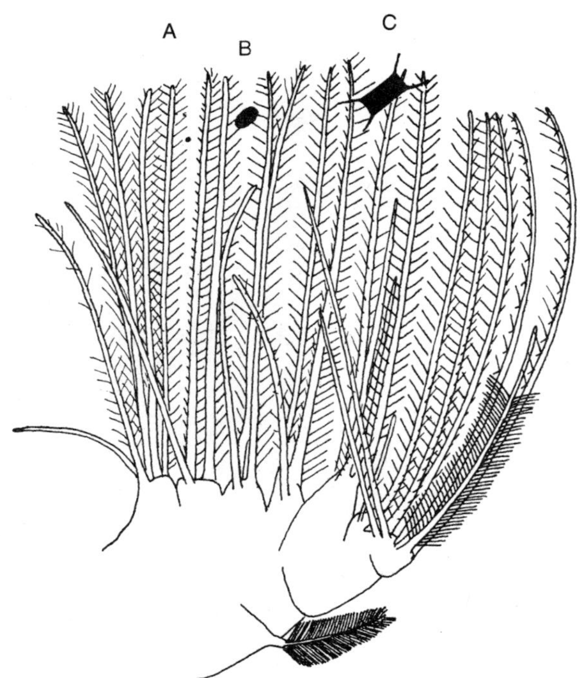

FIGURE 27.17 Filtering setae on the left maxilla of the copepod, *Calanus.* Three diatoms, representing species of different sizes, are drawn to scale to indicate the filtering ability of setae. Diatom *B* is approximately 25 μm. long. (After Dennell from Marshall and Orr.)

Decapods. The **Decapoda**, the largest of the crustacean orders, contains over 8500 species. It also contains the biggest and most familiar crustaceans—shrimp, lobsters, crayfish, and crabs. The name Decapoda refers to the five pairs of legs, including the first pair, which are frequently modified as large claws. In front of the legs are three pairs of smaller appendages called **maxillipeds**, which, like the maxillae, function in food handling (Figs. 27.18 and 27.19). The three pairs of maxillipeds and the five pairs of legs are appendages of the anterior trunk region, called the **thorax**. The thorax is covered by a carapace, and the head-thorax region together is usually called the **cephalothorax** (Fig. 27.18).

Primitively, the **abdomen** is large, as in shrimp, lobsters, and crayfish, and carries six pairs of biramous appendages. The last pair, or uropods, are flattened and form a **tail fan** with the terminal **telson** (Fig. 27.18). The first five pairs are called **swimmerets** or **pleopods** (Fig. 27.17). In a number of different decapod groups the abdomen is reduced and folded beneath the cephalothorax resulting in a short-body form, called a **crab** (Fig.

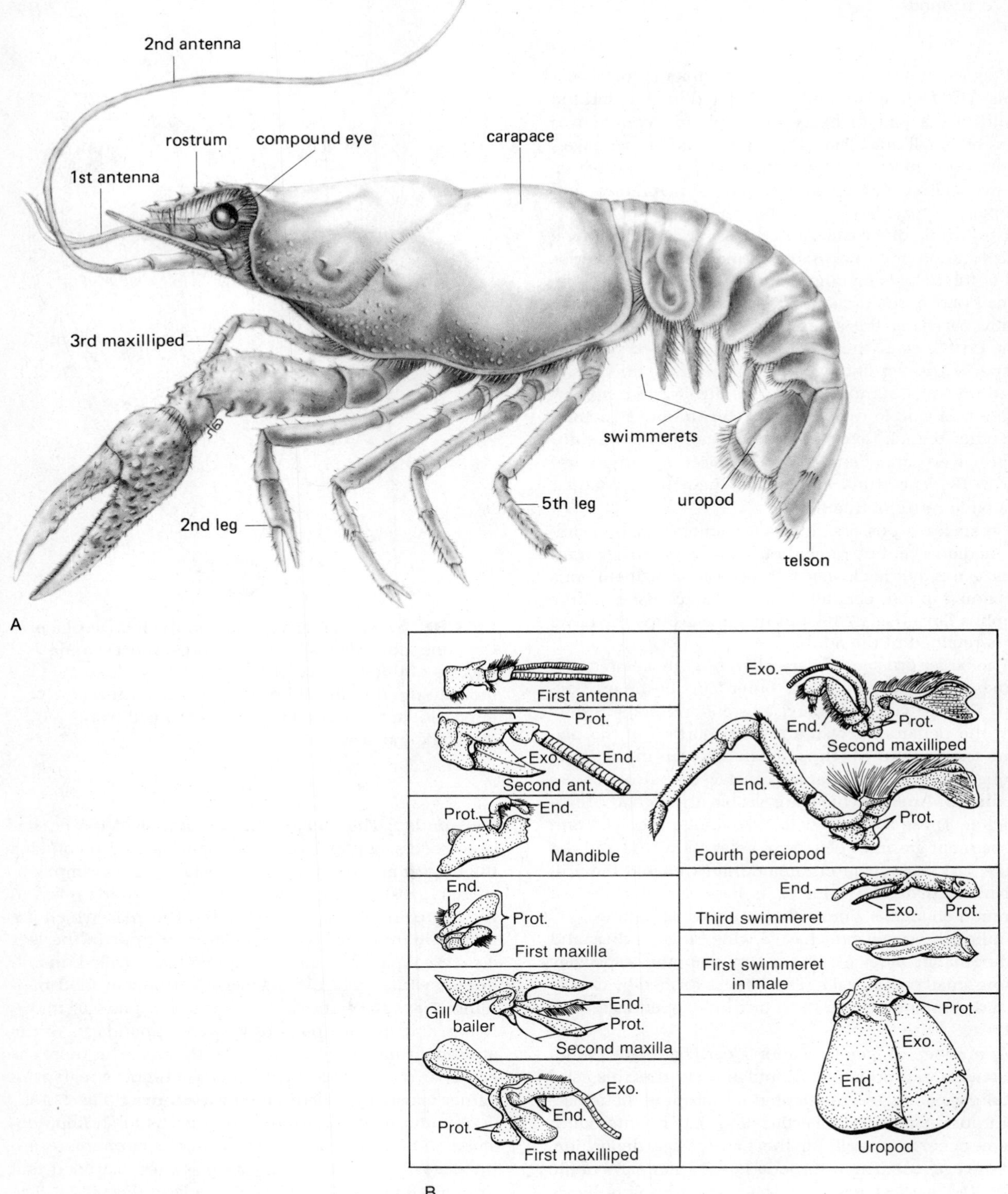

FIGURE 27.18 *A,* Lateral view of a crayfish. *B,* Modifications of the appendages: prot. = protopodite; exo. = exopodite. (*B* after Howes.)

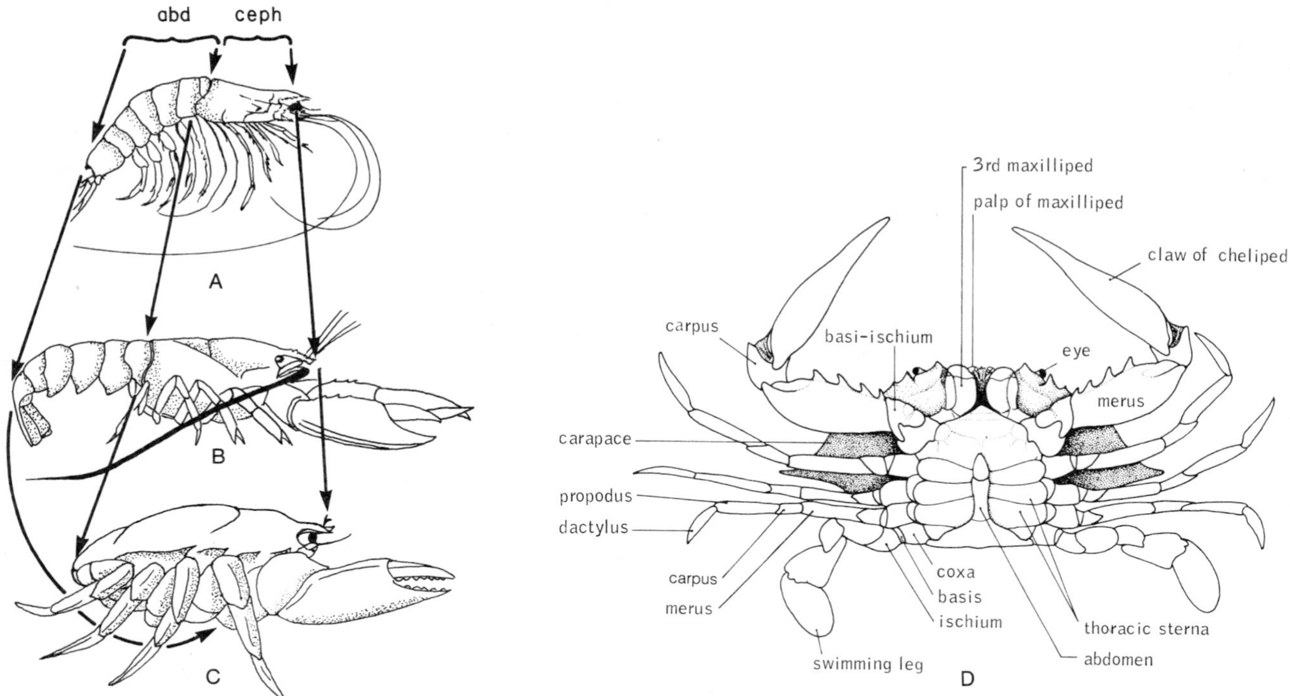

FIGURE 27.19 *A–C,* Evolution of the short body form in decapod crustaceans. In forms with long bodies, such as shrimp (*A*) and lobsters (*B*), a large abdomen extends posterior to the cephalothorax. In forms with short bodies, such as crabs (*A–C*), the abdomen is reduced and folded beneath the cephalothorax. *D,* Ventral view of a brachyuran crab of the family Portunidae. The fifth pair of legs is adapted for swimming. (*A–C* from Glaessner, M. F.: *Treatise on Invertebrate Paleontology.* Part R4(2). Courtesy of the Geological Society of America and the University of Kansas, 1969.)

27.19). Crablike species compose over half of the decapods, but they are not all related.

Most decapods are bottom dwellers, but many shrimp can swim, using the pleopods for propulsion. The best swimmers, however, are the portunid crabs, which have the fifth pair of legs adapted as paddles (Fig. 27.19).

Brachyuran (Gr. *brachys,* short + *oura,* tail) crabs, the largest group of crablike forms, crawl sideways, the legs on one side of the body leading and those on the other side trailing. In these crabs the evolution of the short-body form shifts the center of gravity forward beneath the cephalothorax and probably represents an adaptation for greater motility in their sideways mode of crawling.

The abdomen of hermit crabs is housed within an empty gastropod shell; the soft twisted abdomen bears reduced appendages (Fig. 27.20). When the hermit crab becomes too large, it finds another shell. Only empty shells are used. Hermit crabs probably evolved from forms that backed the body into crevices and other retreats.

As described earlier, decapods possess **gills** that project upward from near the base of the thoracic appendages and are enclosed within a protective branchial chamber (p. 151). The ventilating current is produced by the rapid sculling motion of the **gill bailer**, a semilunar process of the second maxilla.

Several groups of crabs have invaded the land with varying degrees of success. Amphibious fiddler crabs (*Uca*) burrow in the intertidal zone. At high tide the crab remains within its burrow, but at low tide it emerges from the burrow to feed, utilizing water contained within the branchial chambers for gas exchange (Fig. 27.22*B*).

The related ghost crabs (*Ocypode*) are more terrestrial. These crabs live above the high tide mark and in many areas are common inhabitants of dunes. In the tropics, land crabs live in forests and thickets well back from the sea, as much as hundreds of miles inland. Al-

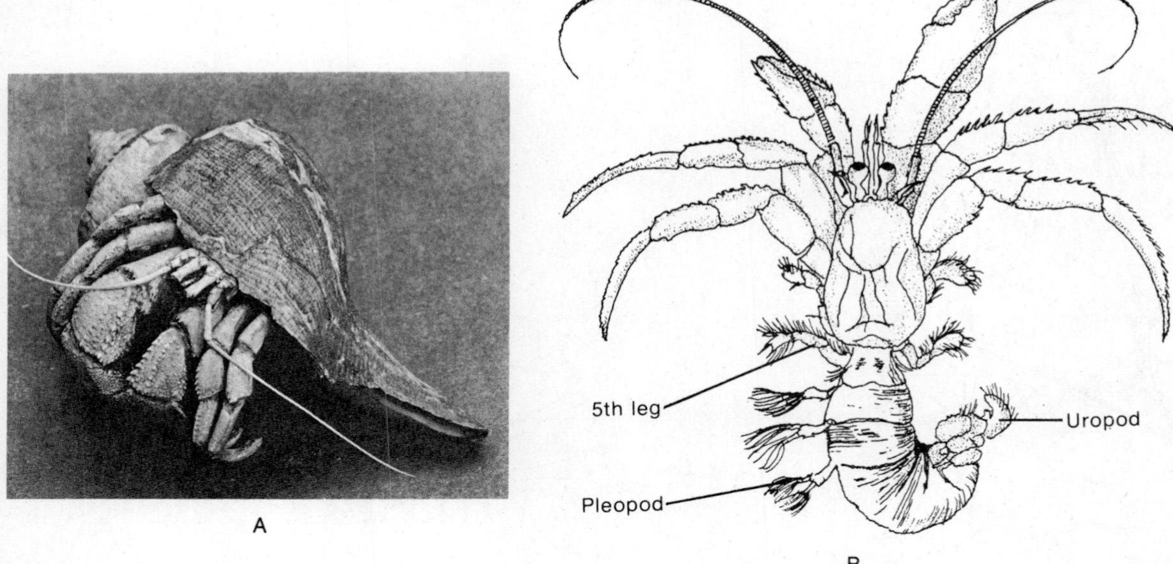

FIGURE 27.20 Hermit crab. *A*, In shell; *B*, out of shell. (*A* courtesy of the American Museum of Natural History; *B* after Calman from Kaestner.)

most all terrestrial species continue to use the gills as organs of gas exchange, which may explain the restriction of land invasions to crabs, for only in these decapods is the branchial chamber sufficiently closed to make possible the retention of the moisture needed for gas exchange. Most land crabs are nocturnal and remain in burrows during the day.

Although the majority of decapods are predators or scavengers, there are many exceptions. Some marine species feed on algae, and recently fallen fruits and leaves are an important food source for many land crabs. The chelipeds are used in collecting and handling food in most decapods. Depending upon the species, these appendages may be adapted for scraping, tearing, cutting or crushing, and some crabs that feed on snails and mussels have one claw adapted for crushing and one for cutting.

Some decapods are filter feeders. An interesting example is the mole crab (Fig. 27.22*A*). Mole crabs burrow backward in sand and project their plumose antennae as filters. Some species, including the Atlantic coast mole crab, live on waveswept beaches and filter the current of the receding waves with their second antennae. Mole crabs do not have the first pair of legs modified as chelipeds, and only the terminal end of the abdomen is flexed beneath the cephalothorax. Mole crabs burrow with the uropods and the fourth pair of legs. The fifth pair of legs is minute, tucked beneath the cephalothorax and used for cleaning the gills.

The amphibious fiddler crabs feed on fine organic matter deposited onto the surface of the intertidal zone of protected beaches, estuaries, and marshes. A mass of sand or mud is placed in the mouthparts, which are then flooded with water pumped forward from the gill chambers. Organic material is floated off and separated by fine setae on the feeding appendages. The remaining mineral material is ejected as a small spitball (Fig. 27.22*B*).

The decapod digestive tract is complex. In lobsters, crayfish, and crabs, the foregut includes a very large chitin-lined cardiac **stomach** containing one dorsal and two lateral **teeth** that form a **gastric mill** (Fig. 27.21). A narrow constriction separates the **cardiac** stomach from the **pyloric** stomach, part of which is derived from the foregut and part from the midgut. Associated with the midgut section and derived from it is a pair of large **digestive glands** also called the **hepatopancreas**, which ramifies out into parts of the cephalothorax. If you have opened steamed crabs, the conspicuous, soft, cream-colored material, sometimes called "mustard," is the digestive gland.

The cardiac stomach functions as both a **crop** and a **gizzard.** The grinding action of the gastric mill plus the action of enzymes passed forward from the digestive

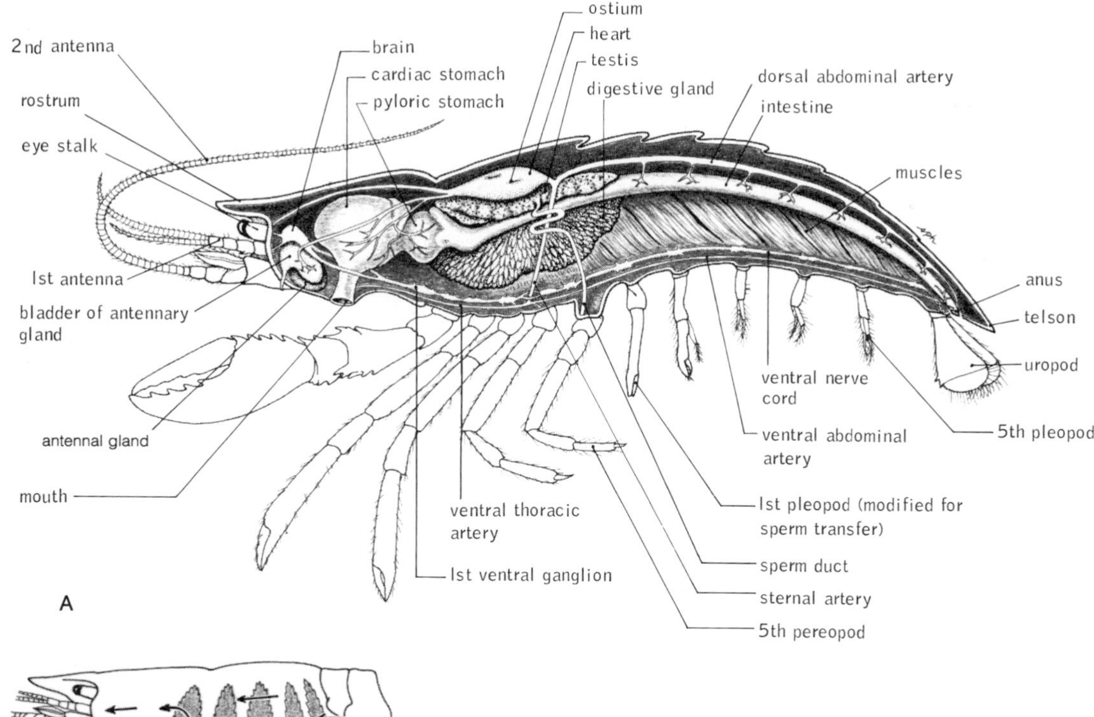

FIGURE 27.21 *A,* Internal anatomy of the crayfish. *B,* Path of ventilating current through the gill chamber of the crayfish.

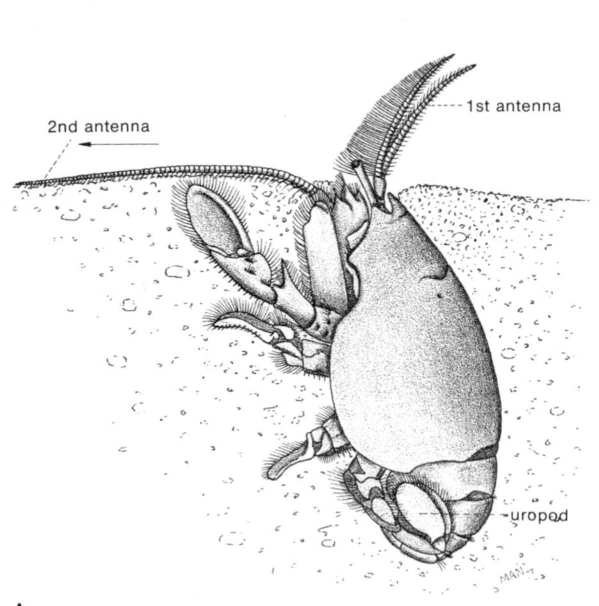

FIGURE 27.22 *A,* Mole crab buried in sand, lateral view. Biramous first pair of antennae form a screening siphon for ventilating current. Setose second antennae are extended into water current (*arrow*) and function in filter feeding. Eyes are on long stalks. *B,* Fiddler crabs (*Uca*). These amphibious crabs are deposit feeders in the intertidal zone at low tide. The male has a large claw which is waved in a species-specific pattern in courting females.

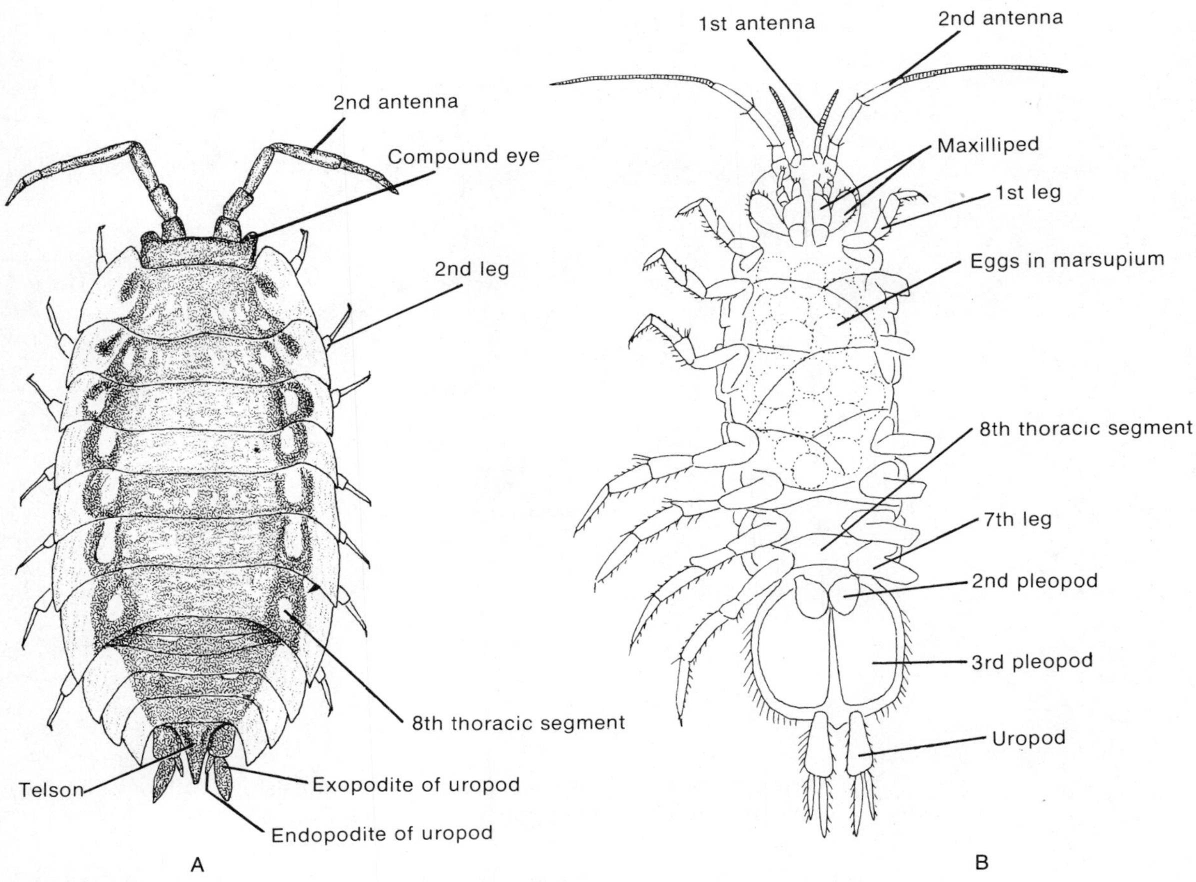

FIGURE 27.23 Isopods. *A,* Dorsal view of *Oniscus asellus,* a terrestrial isopod. *B,* Ventral view of the freshwater isopod *Asellus.* (*A* after Paulmier from Van Name; *B* after Van Name.)

glands reduces the food masses to fluid and fine particulate matter, which is then conducted along channels through the pyloric stomach to the digestive glands. Screens of setae prevent the passage of coarse particles. Absorption takes place in the digestive glands.

The excretory organs of decapods are **antennal glands,** also called green glands. The organ consists of a blood-bathed sac in the head and a duct that opens to the outside by a pore at the base of the second antennae.

Most decapods are marine and are **osmoconformers,** but some shrimp and crabs and the widely distributed crayfish inhabit estuaries and freshwater. The osmoregulatory ability of these species is described on page 206.

Decapods have **compound eyes,** and some species, such as crabs, have a high degree of visual acuity. The eyes are located on **stalks** and have a wide visual arc (Fig. 27.22*A*).

In the base of each first antenna is a **statocyst** that opens by a **pore** to the exterior. The animal uses sand grains as **statoliths** and must replace them following each molt. A newly molted crayfish given iron filings instead of sand grains will place them within the statocyst chamber and then will orient itself to a magnetic field rather than to gravity.

The **hormones** and **neurosecretions** of crustaceans are perhaps better known than those of any other invertebrates. There is experimental evidence of hormones that regulate chromatophores, molting, growth, sexual development, and reproduction. A number of hormones, such as those controlling chromatophores, are released from a gland in the eye stalk.

The single pair of **testes** or **ovaries** is located in the dorsal part of the thorax. The male **sperm duct** descends to the ventral side and opens at the base of the fifth pair of legs (Fig. 27.16). The **oviduct** opens to the exterior at the base of the third pair of legs. During copulation the male assumes various positions astride the female and, using the greatly modified anterior two pairs of pleopods, transfers sperm to the female gonopores or to a median seminal receptacle between the fourth or fifth pair of legs.

The eggs are fertilized internally or on release from the oviduct. The fertilized eggs pass to the ventral surface of the abdomen, where they are attached to the pleopods by an adhesive material on the egg surface. Here they are brooded, often forming a conspicuous mass. Even in crabs the reduced folded abdomen retains its brooding function.

Some shrimp hatch as a nauplius larva; but in most decapods the hatching stage is a later planktonic larval stage, called a **zoea,** in which all of the thoracic appendages are present. The adult form is gradually attained through a series of instars, and the young shrimp, lobster, or crab settles to the bottom and takes up the adult mode of existence.

Copulation and brooding occur on land in terrestrial crabs, but all must go back to the sea (or to fresh water in some tropical families) to permit hatching of the eggs and a planktonic larval life.

Direct development takes place in most freshwater decapods, including the crayfish. In these species, the eggs are brooded on the abdomen throughout development.

The subphylum Crustacea is divided into classes, and the decapods belong to the largest class, the **Malacostraca.** In these crustaceans, the trunk is composed of a thorax of eight segments, which bear the maxillipeds and legs, and an abdomen of six segments, which usually carry five pairs of biramous **pleopods** (Gr. *plein,* to swim + *pous,* foot) and a terminal pair of flattened **uropods** (Fig. 27.16).

Amphipods and Isopods. The two largest orders of malacostracans other than decapods are the **Amphipoda** and **Isopoda**. These malacostracans, most of which are only about a centimeter in length, share a number of features. The **compound eyes** are on the sides of the head and not on stalks, as in decapods. No carapace is present, and the thoracic and abdominal regions are not sharply demarcated on the dorsal side (Fig. 27.23). As in decapods, there are eight pairs of **thoracic appendages**, but only the first pair are food-handling maxillipeds; the other seven are legs. In both orders development is direct, and the young are brooded beneath the thorax in a chamber, the **marsupium**, formed by inward shelflike projections from the bases of the legs (Fig. 27.23).

Isopods tend to be dorsoventrally flattened, and the pleopods are modified for gas exchange (Fig. 27.23). Amphipods, in contrast, are laterally flattened and look somewhat like shrimp (Fig. 27.24). The gills of amphipods are simple processes from the bases of the legs.

The majority of isopods and amphipods are marine, and most are bottom dwellers, swimming only intermit-

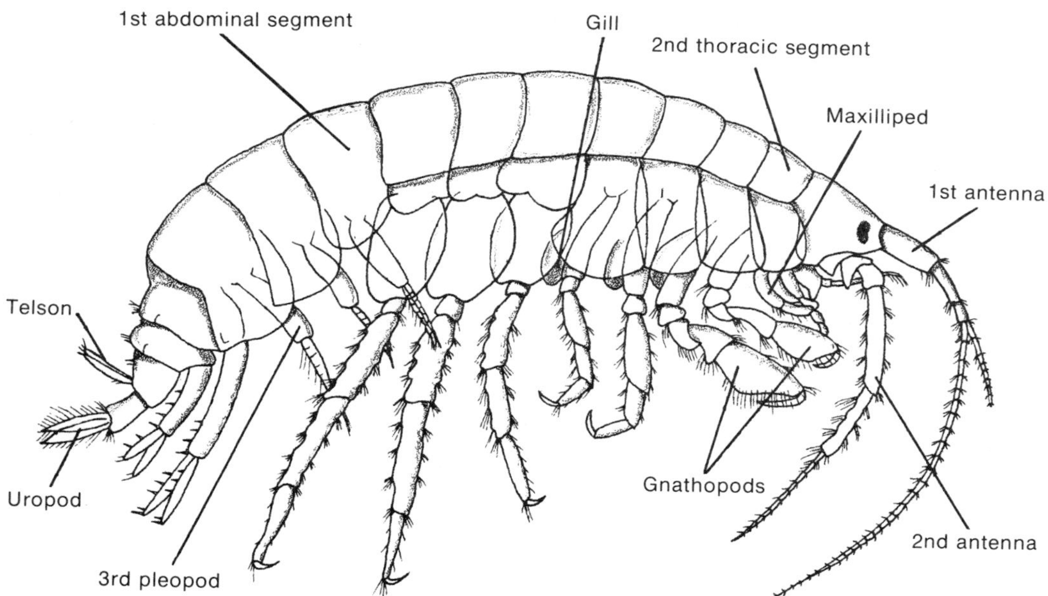

FIGURE 27.24 Male of the amphipod, *Gammarus.* (After Sars from Calman.)

tently with the pleopods. They live in bottom debris, in algae, and among sessile animals. Many amphipods burrow in sand or mud, and some are tube dwellers. Most isopods and amphipods are omnivores and scavengers.

Both orders include freshwater species, and both groups have also invaded land, but most terrestrial amphipods are limited to algae and the undersurfaces of boards washed up on the beach. Such amphipods are called beach fleas because they can jump.

The woodlice, also called sow bugs or pill bugs, are isopods and represent the only truly successful crustacean invasion of land, which we believe they invaded directly from the sea. These isopods are widely distributed in both temperate and tropical regions, where they are commonly found beneath stones and wood and in leaf mold. They reduce desiccation by living in protected habitats, by avoiding light, and by their ability to roll up into a ball, covering the less highly sclerotized ventral surface. They have retained their gills (pleopods) but eliminate nitrogenous wastes as gaseous ammonia. They also brood their eggs in the marsupium, like their aquatic relatives.

Some isopods that burrow at the high tide mark and certain species of amphipod beach fleas are able to return to their habitat when they move down to the intertidal zone to feed or are displaced for other reasons. Orientation is achieved by sensing the slope of the beach and the angle of the sun. The latter ability, however, requires a "map sense," i.e., knowledge of the orientation of the beach with regard to compass points, and an internal clock to compensate for the changing sun angles as the day passes. Animals can be misdirected experimentally by altering the sun's apparent position with a mirror. A comparable orientation ability, using the angle of the sun, has been demonstrated in certain wolf spiders living on the banks of lakes and streams and in bees that return to particular flowers to obtain pollen and nectar.

Branchiopods. Most of the nonmalacostracan crustaceans are very small, averaging only a few millimeters in length. The majority belong to one of four classes. The class **Branchiopoda** contains the water fleas, fairy shrimp, and brine shrimp. All possess flattened **leaflike trunk appendages** bordered by fine setae, but the body form is quite diverse. Water fleas (**cladocerans**) look like plump little birds (Fig. 27.25*A*). The trunk, but not the head, is enclosed within a folded carapace that greatly overhangs the body. The head bears large biramous second antennae and a single median compound eye. Fairy shrimp and brine shrimp have a long trunk with numerous appendages, and there is no carapace (Fig. 27.25*C*).

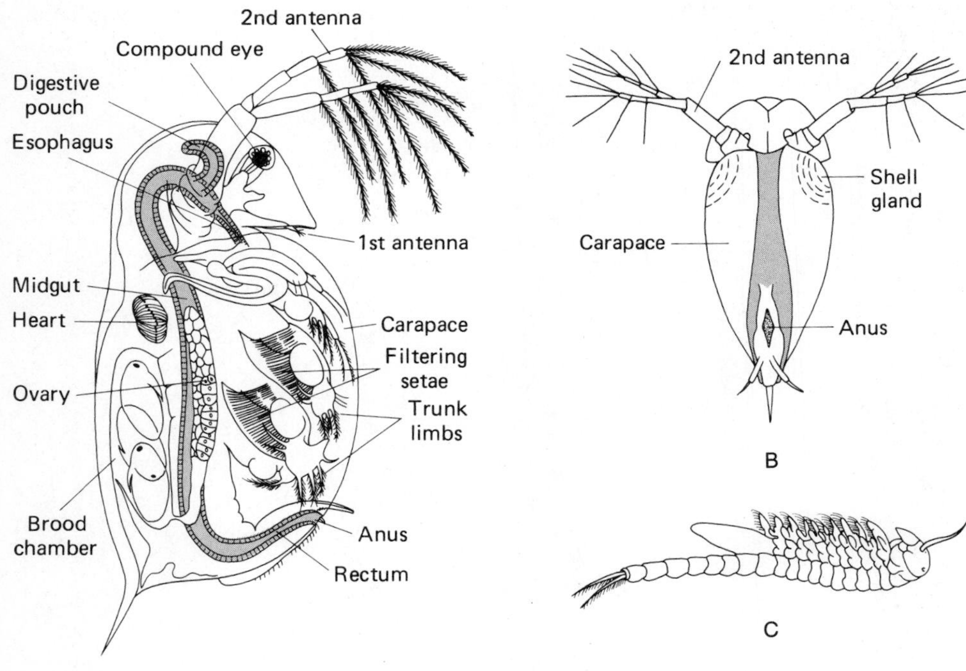

FIGURE 27.25 *Daphnia,* the water flea. *A,* Side view, with one side of the carapace removed to show enclosed body and organs. *B,* Ventral view, with trunk appendages omitted. *C,* A fairy shrimp (they swim upside down). (*C* after Borradaile *et al.*)

The flattened foliaceous appendages of branchiopods function in gas exchange, from which the name Branchiopoda (Gr. *brangchia*, gill + *pous*, foot) is derived. The appendages also serve in filter feeding, the fine setae bordering the appendages functioning as filters. Fairy shrimp also use the appendages for swimming, but water fleas swim with the antennae.

In contrast to most of the other classes, branchiopods are chiefly inhabitants of fresh water, especially ponds and temporary pools. Water fleas are an exception in including planktonic species that live in large lakes and also some that live in the sea. The brine shrimps *Artemia* are adapted for living in salt lakes and can tolerate salinities near the saturation point of salt.

Water fleas brood their eggs in the space beneath the carapace above the back of the trunk (Fig. 27.25). Water fleas have direct development, but in other branchiopods the egg hatches as a nauplius larva. The retention of larval stages in freshwater branchiopods is an exception to the general rule of suppression of larval stages in fresh water.

Branchiopods exhibit a number of reproductive and developmental adaptations to the environmental stresses common in freshwater lakes and ponds. Parthenogenesis is common. Many species produce thin-shelled eggs that hatch rapidly and thick-shelled eggs that remain dormant during periods of drought or freezing. These adaptations are the same as those exhibited by freshwater flatworms and rotifers.

Copepods. The class **Copepoda** is the largest of the nonmalacostracan classes. The 4500 species are mostly marine, but there are also many freshwater forms. The copepod body, usually less than 3 mm. in length, is cylindrical and tapered (Fig. 27.26). The head bears a single **median nauplius eye** but no compound eyes. The first pair of antennae are very large and are typically held at right angles to the body. The trunk is composed of an anterior **thorax** bearing biramous appendages and a narrower posterior **abdomen**, which lacks appendages but usually carries a pair of terminal processes. Thus the body of copepods looks a little like a bomb with a cross-bar (the antennae) at the front end.

Copepods may be planktonic; they may live near the bottom, where they swim about over debris; they may be interstitial (live in the spaces between sand particles); and many are parasitic. The smaller second pair of antennae are the principal swimming appendages of planktonic species; the first antennae function largely to reduce sinking.

Although some copepods are grazers or are predatory, most planktonic forms are filter feeders. Diatoms are the principal source of food, and copepods play an important role in the food chain. Studies on the marine planktonic copepod *Calanus finmarchicus*, which is 5 mm. long, have shown that as many as 373,000 diatoms may be filtered out and digested every 24 hours. About one quarter of the copepod species are parasitic, attacking the skin, fins, and gills of marine and freshwater fish.

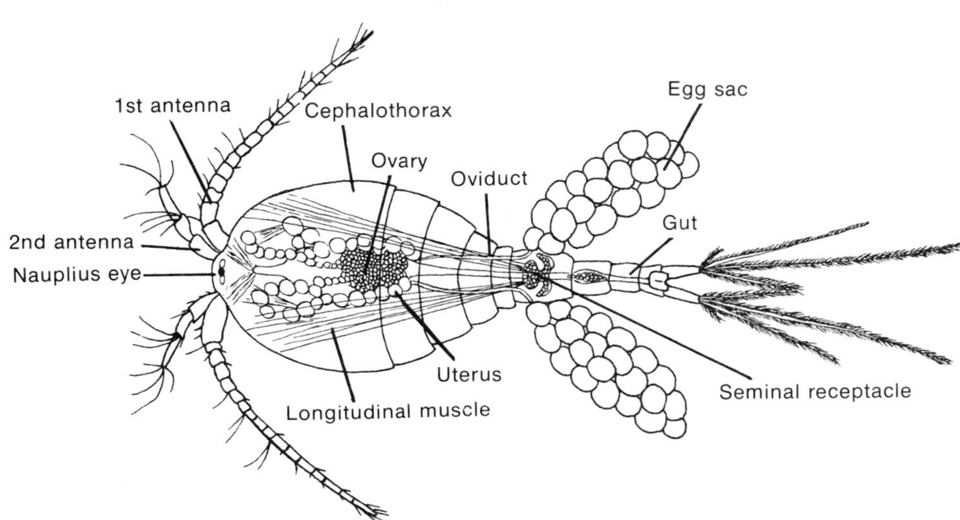

FIGURE 27.26 A cyclopoid copepod. *Macrocyclops albidus* (dorsal view). (After Matthes from Kaestner, A.: *Invertebrate Zoology.* Vol. III. New York, John Wiley and Sons, 1970.)

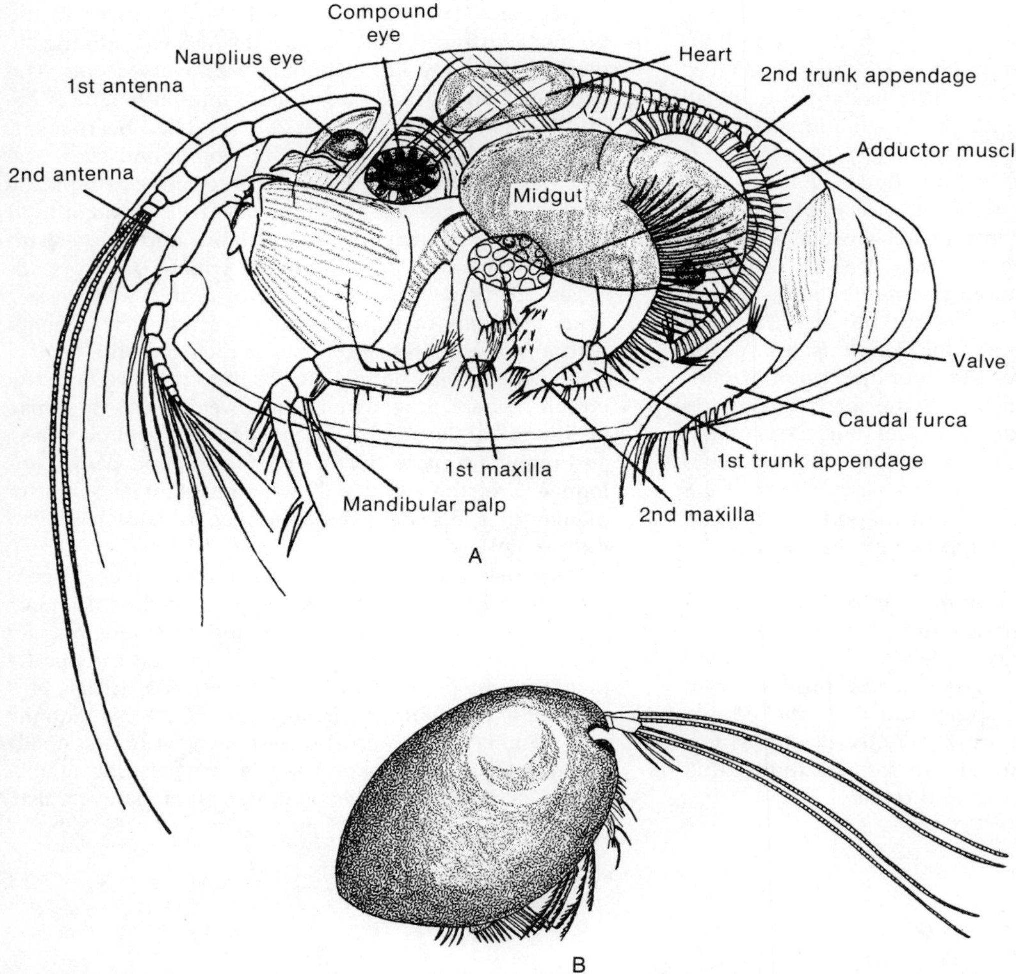

FIGURE 27.27 Ostracods. *A,* Lateral view of a female marine ostracod, *Cypridina,* with left valve removed. *B,* Lateral view of a marine ostracod swimming. Valves contain an antennal notch. (*A* after Claus from Calman; *B* after Muller from Schmitt.)

All degrees of modification from the typical copepod body form exist, and some species are so aberrant they they no longer look like crustaceans.

Some copepods shed their eggs singly into the water, but many brood their eggs within secreted ovisacs attached to the female gonopore. A nauplius larva is the hatching stage in both marine and freshwater forms.

Ostracods. Members of the class **Ostracoda** (Gr. *ostrakōdēs,* shell) are tiny marine and freshwater crustaceans with the body entirely enclosed within a bivalved carapace (Fig. 27.27). Ostracods are sometimes called seed shrimp and parallel the bivalve mollusks in many ways. The two **valves**, usually 2 mm. or less in length, may have interlocking teeth and are held together dorsally by an elastic hinge. The valves are closed by a bundle of **adductor muscle fibers** extending transversely between the two valves near the middle of the body (Fig. 27.27). The chitinous skeleton is impregnated with calcium carbonate. As a consequence, ostracods have an extensive fossil record dating from the Cambrian.

Within the two valves, the ostracod body is mostly head, for the trunk is greatly reduced. The two antennae are large, and the other head appendages are well developed. However, the number of trunk appendages is reduced to two, one, or zero.

There are some planktonic ostracods, but most species are benthic and scurry over the bottom, crawling and swimming with the large antennae. Along with copepods and water fleas, ostracods are very common crustaceans in freshwater ponds and small pools.

Many ostracods are filter feeders utilizing their different postoral appendages as filters.

Eggs are released singly or brooded within the carapace, and a nauplius with a bivalved carapace is the hatching stage.

Barnacles. The class **Cirripedia** (L. *cirrus*, curl + *pes*, foot) contains the barnacles. These marine animals differ from other crustaceans in being attached to the substratum. Correlated with the sessile habit, the body is enclosed within a bivalved carapace covered with protective **calcareous plates** (Fig. 27.28).

If you can imagine an ostracod, which is also enclosed within a bivalve carapace, attached to the substratum by the antennae, and the valves covered by calcareous plates, you have some idea of the structure of barnacles. Indeed, the late larval stage of barnacles, called a **cypris larva**, looks very much like an ostracod (Fig. 27.29). The cypris settles to the bottom and attaches head first by means of cement glands at the base of the first antennae.

All barnacles are attached, but the group may be divided into **stalked** and **sessile** (stalkless) forms (Fig. 27.28). Stalked barnacles appear first in the fossil record and are believed to have given rise to the sessile types. The body of stalked barnacles is composed of a fleshy stalk, or **peduncle**, and a distal **capitulum**. The peduncle represents the preoral part of the body and the capitulum the postoral part. It is the capitulum that is covered by the carapace, and the gape along one margin of the carapace can be opened and closed as in a clam. The number of calcareous plates on the outer surface of the carapace varies, but there are always two pairs, the **terga** and **scuta**, that flank the gape. Although some stalked barnacles are found on rocks, many species live attached to floating objects, such as timbers, or the bodies of larger swimming animals. Whales, porpoises, sea turtles, sea snakes, crabs, and even jellyfish are utilized as attachment surfaces by different species. These commensal species always exhibit some reduction in the size of their plates since some protection is provided by the host.

The body of sessile barnacles, sometimes called **acorn barnacles**, consists mostly of the capitulum, for the preoral part is reduced to a platform on which the capitulum rests and is attached to the substratum (Fig. 27.28*C*). The platform of sessile barnacles and the peduncle of stalked species both contain the antennal **cement gland**, indicating that the two structures are homologous. In sessile barnacles the basal plates of the capitulum form a rigid circular wall surrounding the upper movable lidlike terga and scuta.

Most sessile barnacles have become adapted for a life on rock or other hard substrata, and the heavy, somewhat fused, ring of wall plates probably represents an adaptation for protection against currents, pounding waves, and browsing fish. However, like stalked barnacles, some sessile barnacles have become adapted for living on other animals, such as crabs and whales.

The tendency of barnacles to utilize other animal bodies as substrata probably led to parasitism. Approximately one third of the members of the class are parasitic and are so highly modified that they are recognizable as barnacles only by their larval stages. Other crustaceans are the principal hosts.

Both stalked and sessile barnacles are filter feeders, and the trunk bears six pairs of long, coiled, biramous appendages called **cirri**, from which the name of the class is derived. In feeding, the cirri unroll and project through the gape of the carapace as a large basket (Fig. 27.28*E*). They perform a rhythmic scooping motion or hold the cirri out like a fan, and the many fine setae of the appendages remove planktonic organisms.

Neither eyes, gills, heart, nor blood vessels are present. Correlated with their attached existence is the fact that barnacles, unlike most other arthropods, are **hermaphroditic.** At copulation, a long, extensible, tubular penis is projected from the gape of one individual into the mantle cavity of a neighboring barnacle. Fertilization and brooding occur within the mantle cavity, and a nauplius is the usual hatching stage. Following a number of naupliar instars (Fig. 27.29), the cypris larva develops and is the stage at which settling occurs. Barnacles molt like other crustaceans but they do not shed the outer surface of the carapace bearing the calcareous plates. The plates increase in size by the secretion of additional calcium carbonate to their margins.

Barnacles are major fouling organisms on pilings, sea walls, buoys, and ship bottoms. Much research has been expended to develop antifouling measures, such as antifouling paint, for a badly fouled ship may have its speed reduced by as much as 35 percent.

Subphylum Uniramia

The subphylum Uniramia is the largest division of arthropods, containing the centipedes, millipedes, and insects. In contrast to the marine origins of the chelicerates and crustaceans, the uniramians are believed to have evolved from terrestrial ancestors, i.e., arthropodization occurred on land. We know little about the pre-

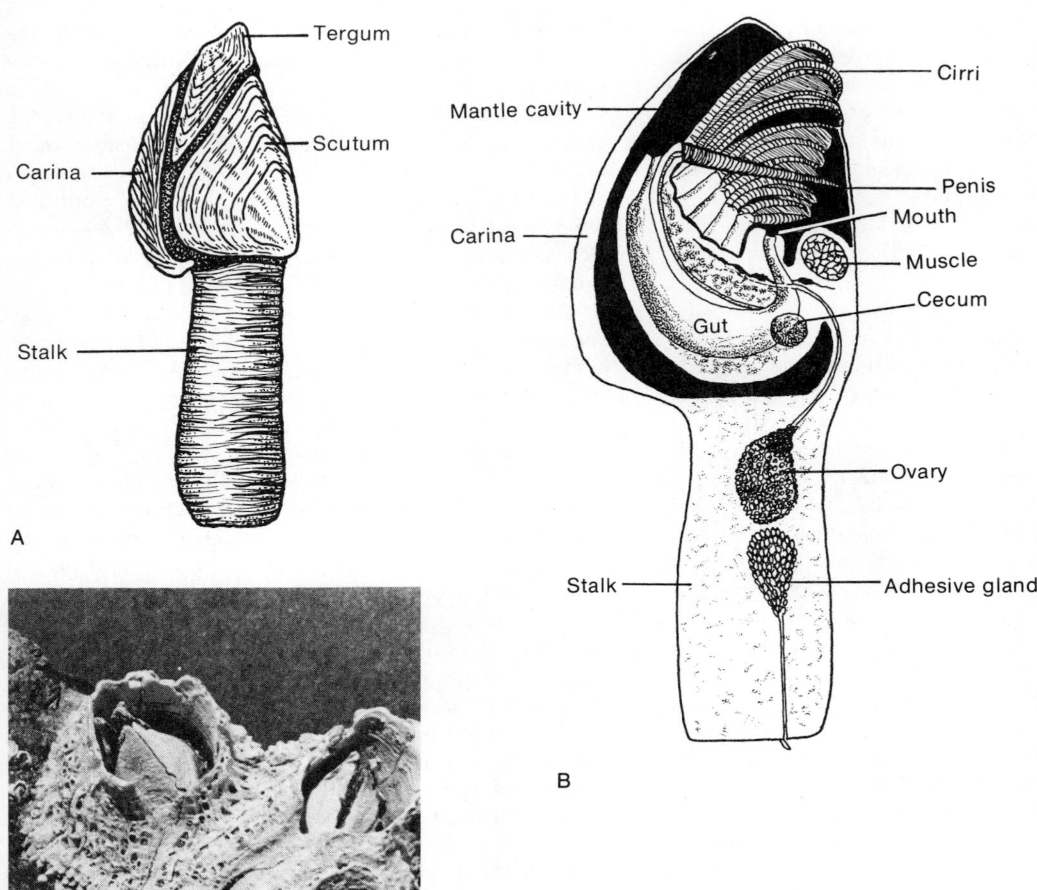

FIGURE 27.28 Barnacles. *A,* External view of *Lepas,* a stalked barnacle. Members of this genus are sometimes called goose barnacles. *B,* Internal structure of a stalked barnacle. *C,* A cluster of sessile (stalkless) barnacles on a rock. *D,* Lateral view of *Balanus,* a sessile barnacle, in section. *E,* A feeding *Balanus,* showing the outstretched cirri. The conspicuous plates at the base of the cirri are the two terga. One scutum can be seen on the right side. (*B* after Broch from Kaestner; *C* after Gruval from Calman.)

arthropod ancestors of uniramians, but they may have been closely related to a little phylum of animals called the Onychophora, which look somewhat like caterpillars (Box 27.3*A*).

All uniramians possess a single pair of antennae, in contrast to the two pairs in crustaceans and the absence of antennae in chelicerates. The first pair of feeding appendages are mandibles, which are followed by one or two pairs of maxillae. The only other conspicuous segmental appendages are legs, which vary from three to many pairs. The name Uniramia refers to the unbranched nature of the appendages. Those appendages that are unbranched in chelicerates and crustaceans are believed to be derived from an original branched condition.

Myriapods: Millipedes and Centipedes

The myriapodous arthropods comprise four classes of uniramians that were once placed within a single

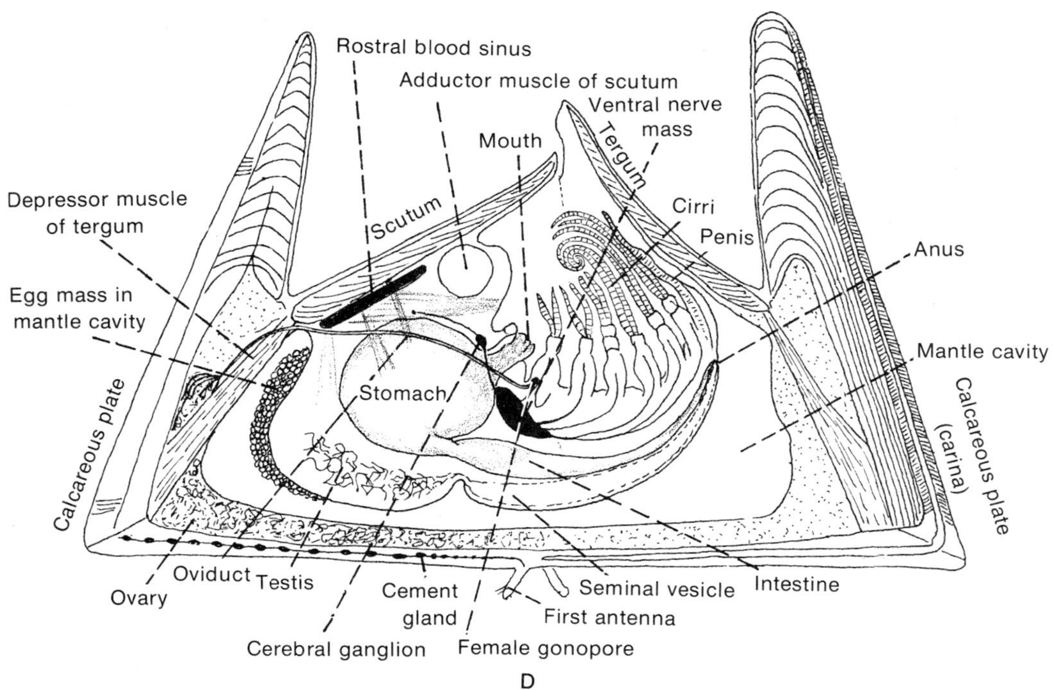

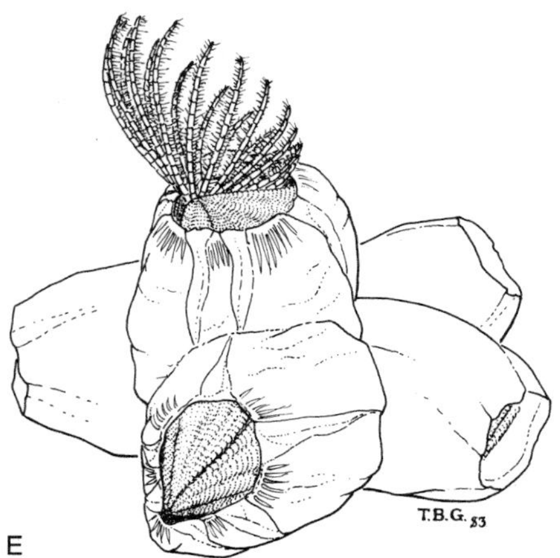

FIGURE 27.28 *Continued.*

class, the **Myriapoda.** They are the **Diplopoda** (which contains the millipedes), the **Chilopoda** (which contains the centipedes), and two small classes, the **Symphyla** and the **Pauropoda.** Although the myriapodous arthropods belong to separate classes, they share a number of characteristics. All are terrestrial and secretive, living in soil and leaf mold and beneath stones, logs, and bark. The body is composed of a **head** and a long **trunk** with many segments and legs (Fig. 27.30). The eyes are not compound, except in a few species.

Gas exchange organs are **tracheae,** and a separate system of tubules and spiracles is present in each segment. Excretory organs are **malpighian tubules,** and the heart is a long dorsal tube with ostia in each segment.

Reproduction parallels the arachnids in that sperm transfer is indirect and in centipedes involves a spermatophore.

Adaptations for locomotion have been a primary theme in the evolutionary history of centipedes and mil-

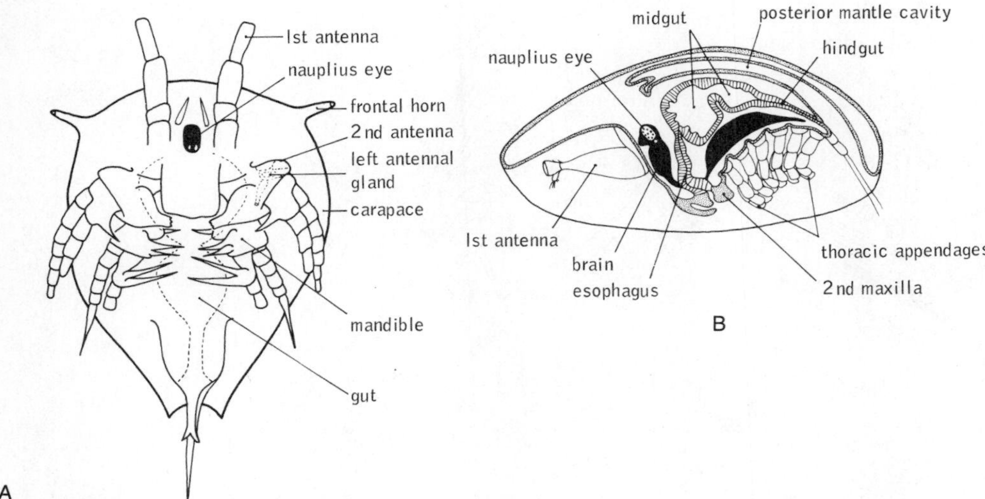

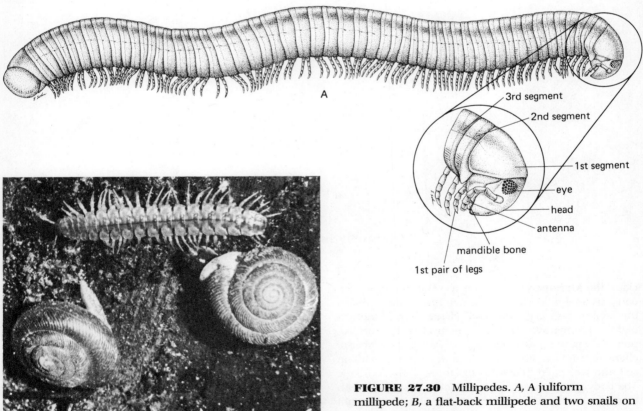

FIGURE 27.29 *A,* Nauplius larva of *Balanus;* setae omitted. *B,* Cypris larva of *Balanus;* side view with left half of carapace removed. (From Walley, L. J.: *Philosophical Transactions of the Royal Society of London (Biol.),* 256,807:237–280, 1970.)

A

3rd segment
2nd segment
1st segment
eye
head
antenna
mandible bone
1st pair of legs

B

FIGURE 27.30 Millipedes. *A,* A juliform millipede; *B,* a flat-back millipede and two snails on a piece of wood.

BOX 27.3

Related Phyla

A. the **Onychophora** is a phylum of ancient wormlike animals that live beneath stones, logs, and leaves in the tropics and south temperate parts of the world. They have a body wall and nephridia like annelids but have a reduced coelom, open internal transport system, and tracheae like arthropods. A pair of antennae and numerous pairs of short peglike legs make them look a little like caterpillars. Most onychophorans are predaceous on other small invertebrates and defend themselves with adhesive secretions squirted from glands in the mouth region.

B. The **Tardigrada** is a phylum of minute animals called water bears. They are found in soil and in fresh water, but most live in the water films of mosses. The name water bear comes from their short body and four pairs of stubby legs with claws. The body is covered by a thin chitinous cuticle that is periodically molted. Tardigrades feed on the contents of algal and moss cells pierced by buccal stylets. Many moss-inhabiting species are able to survive long periods of extreme desiccation. (From Marcus, E., in Bronn, H. G. (Ed.): *Klassen und Ordnugen des erreichs.* Vol. 5, Part 4. Frankfurt, Akademische Verlagsgesellschaft, 1929.)

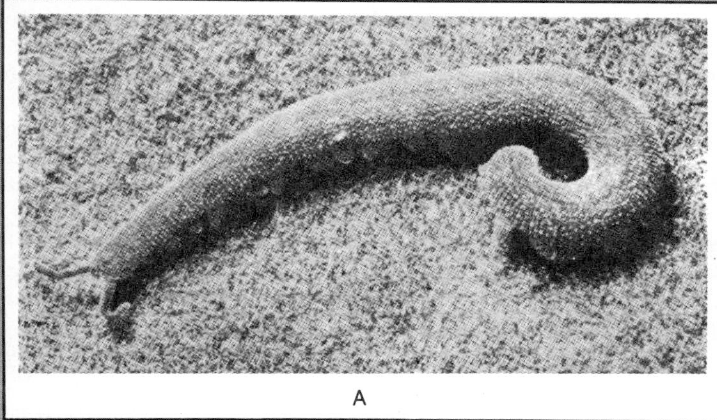

A

B

lipedes. These animals have become adapted for running, climbing, pushing, wedging beneath objects, and burrowing in soil. Yet, in contrast to most other arthropods where increased motility is correlated with a reduction in number of legs and compaction of the body, myriapods have retained a long trunk with many appendages.

Millipedes. Millipedes, members of the class Diplopoda, have the body adapted for pushing, the force being generated by the large number of legs. As many as 12 to 52 legs may be involved in one wave of movement sweeping down the length of the body. The large number of legs requires a large number of segments, but a large number of segments results in a weakened trunk column. This problem was solved by the formation of double segments, or **diplosegments;** i.e., each trunk section of a millipede represents two fused segments. Each double segment contains two pairs of legs, two sets of spiracles, two pairs of heart ostia, and two ventral nerve ganglia (Fig. 27.30).

The common cylindrical (juliform) millipedes are best adapted for pushing through leaf mold and other loose debris (Fig. 27.30). The head is rounded, and the many legs are attached near the midventral line of the cylindrical smooth body. Some species may become longer than pencils in size.

Flatbacked millipedes are adapted for wedging into confining places, such as beneath stones or bark. The body is dorsoventrally flattened, and the laterally projecting terga create a protective working space for the legs (Fig. 27.30*B*).

Both flatbacked and juliform millipedes possess **repugnatorial (stink) glands** on the diplosegments. These glands secrete compounds that contain iodine, quinone, or hydrocyanic acid and are believed to serve a defensive function.

Millipedes are chiefly scavengers, feeding on living and decaying vegetation and on dead animal remains. Millipedes have only one pair of maxillae following the mandibles, in contrast to the two pairs of centipedes and insects.

Centipedes. Centipedes, members of the class Chilopoda, are predaceous and most are adapted for running. Behind the mandibles and the two pairs of maxillae are a pair of large poison claws used for seizing and killing prey (Fig. 27.31). Some species can also pinch with the last pair of legs. Centipedes feed mostly on other arthropods, and although large specimens can inflict a painful bite, only a few tropical species are actually dangerous to humans.

Only one pair of legs is present per trunk segment, for the segments are not doubled as in millipedes. To reduce its tendency to undulate or wobble when running, the trunk of many species is strengthened by having alternately long and short terga (dorsal skeleton) or overlapping terga of different lengths (Fig. 27.31*A* and *C*).

Scutigera, a genus of long-legged centipedes commonly found in bath tubs and sinks, contains the fastest running species. The leg length increases from front to back so that each leg moves to the outside of the preceding leg, reducing interference (Fig. 27.31*C*). Trunk wobble is reduced by overlapping tergal plates. *Scutigera* is the only group of myriapods with compound eyes.

Insects

The Insecta is not only the largest class of arthropods but also the largest group of animals, including more than three quarters of a million species. The great adaptive radiation that the class has undergone has led to the occupation of virtually every type of terrestrial habitat, and some groups have invaded fresh water. The great success of insects over other terrestrial arthropods can be attributed in part to the evolution of flight, which provided advantages for dispersal, escape, access to food or more favorable environmental conditions. The ability to fly evolved in reptiles, birds, and mammals, but the *first* flying animals were insects.

Like arachnids, insects evolved a waxy epicuticle, which greatly reduces evaporative water loss. This more protective outer layer of the exoskeleton enables them to live in a much wider range of terrestrial habitats than is possible for the terrestrial crustacean pill bugs, which lack wax in the epicuticle.

Insects are of great ecological significance in the terrestrial environment. Two thirds of all flowering plants are dependent upon insects for pollination. The principal pollinators are bees, wasps, butterflies, moths, and flies, and the three orders represented by these insects have an evolutionary history that is closely tied to that of the flowering plants, which underwent an explosive evolution in the Cretaceous.

Insects are of enormous importance for humans. Mosquitoes, lice, fleas, bedbugs, and a host of flies can contribute directly to human misery. Some contribute indirectly as vectors of human diseases or diseases of their domesticated animals: mosquitoes (malaria, elephantiasis, and yellow fever); tsetse fly (sleeping sickness); lice (typhus and relapsing fever); fleas (bubonic plague); and the housefly (typhoid fever and dysentery). Our domesticated plants are dependent upon some insects for pollination but are destroyed by others. Vast sums are expended to control insect pests, which can greatly reduce the high agricultural yields necessary to support large human populations. But the overzealous use of pesticides can in turn be hazardous to the environment.

External Structure. Despite the great diversity of insects, the general structure is relatively uniform. The body is composed of **head, thorax,** and **abdomen** (Fig. 27.32). A large pair of **compound eyes** occupies the lateral surface of the head. Between the compound eyes are three small **simple eyes,** or **ocelli,** and a pair of **antennae** (Fig. 27.32). The feeding appendages consist of a pair of **mandibles** and two pairs of **maxillae.** The second pair of maxillae are fused together and called the lower lip, or **labium.** An upper lip, or **labrum,** formed by a shelflike projection of the head, covers the mandibles anteriorly. Near the base of the labium a median process, the **hypopharynx,** projects from the floor of the oral cavity (Fig. 27.32).

The thorax is composed of three segments: a **prothorax, mesothorax,** and **metathorax.** Each segment bears a pair of legs, and the last two segments may each carry a pair of wings (Fig. 27.32).

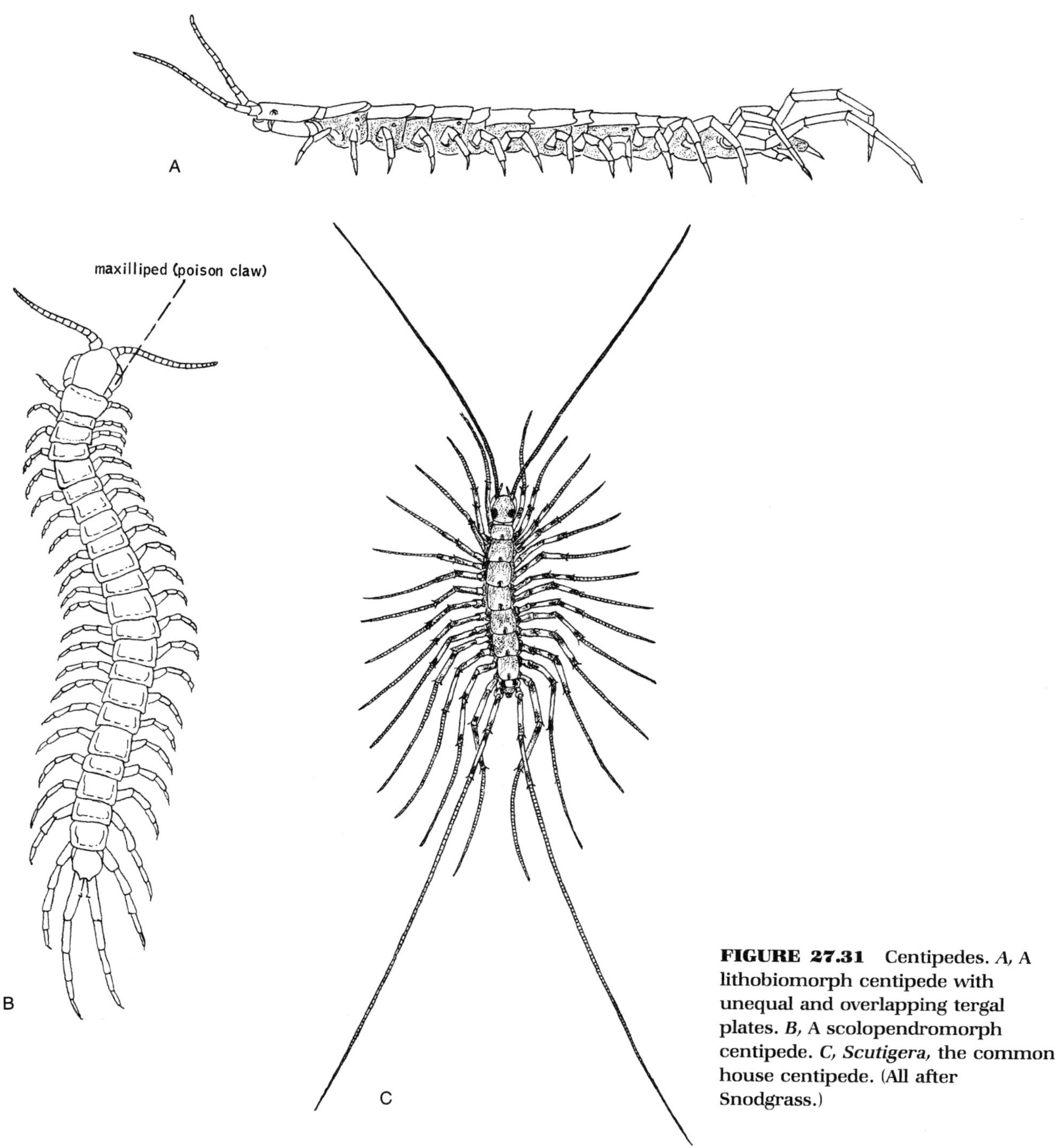

FIGURE 27.31 Centipedes. *A,* A lithobiomorph centipede with unequal and overlapping tergal plates. *B,* A scolopendromorph centipede. *C, Scutigera,* the common house centipede. (All after Snodgrass.)

The abdomen is composed of 9 to 11 segments, which usually lack appendages; however, the terminal reproductive structures are believed by some entomologists to be derived from segmental appendages.

Flight. Most insects have wings and compose the subclass **Pterygota** (Gr. *pteryos,* wing) (Fig. 27.34). Primitive insects, such as proturans, thysanurans, and collembolans (members of the subclass **Apterygota**), are wingless

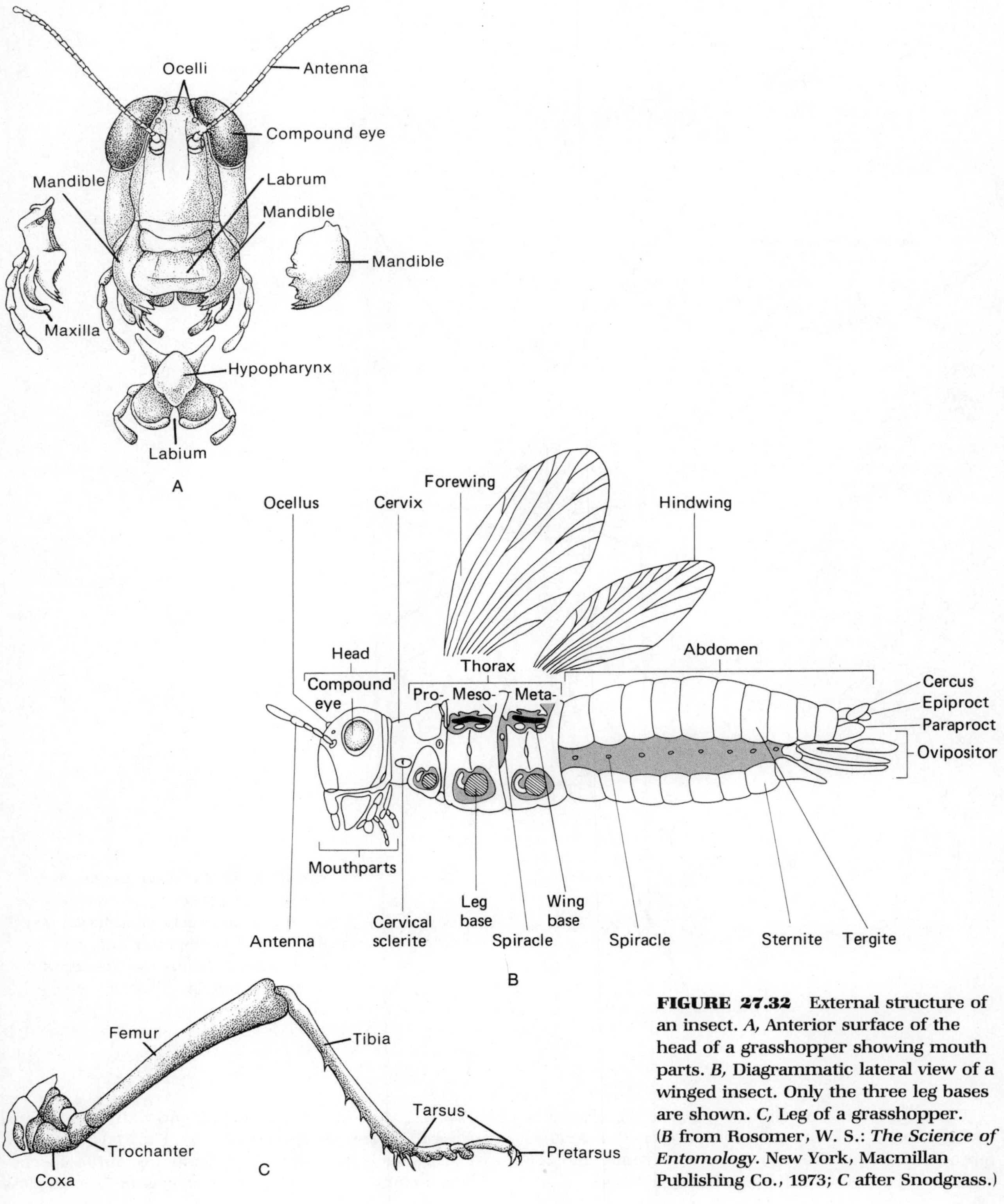

Ocelli
Antenna
Compound eye
Mandible
Labrum
Mandible
Mandible
Maxilla
Hypopharynx
Labium
A
Forewing
Ocellus
Cervix
Hindwing
Head
Thorax
Abdomen
Compound eye
Pro-
Meso-
Meta-
Cercus
Epiproct
Paraproct
Ovipositor
Mouthparts
Leg base
Wing base
Cervical sclerite
Antenna
Spiracle
Spiracle
Sternite
Tergite
B
Femur
Tibia
Tarsus
Pretarsus
Trochanter
Coxa
C

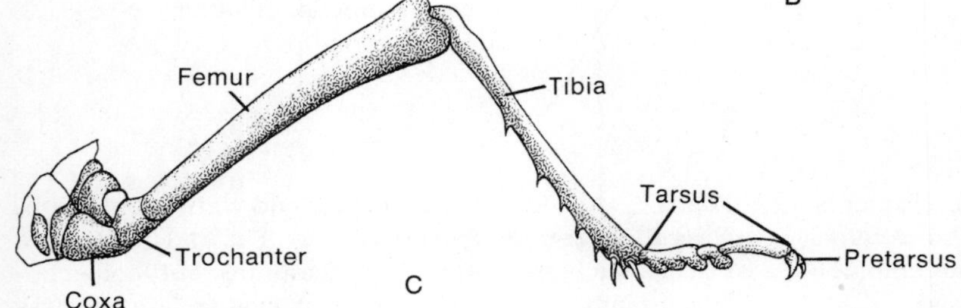

FIGURE 27.32 External structure of an insect. *A,* Anterior surface of the head of a grasshopper showing mouth parts. *B,* Diagrammatic lateral view of a winged insect. Only the three leg bases are shown. *C,* Leg of a grasshopper. (*B* from Rosomer, W. S.: *The Science of Entomology.* New York, Macmillan Publishing Co., 1973; *C* after Snodgrass.)

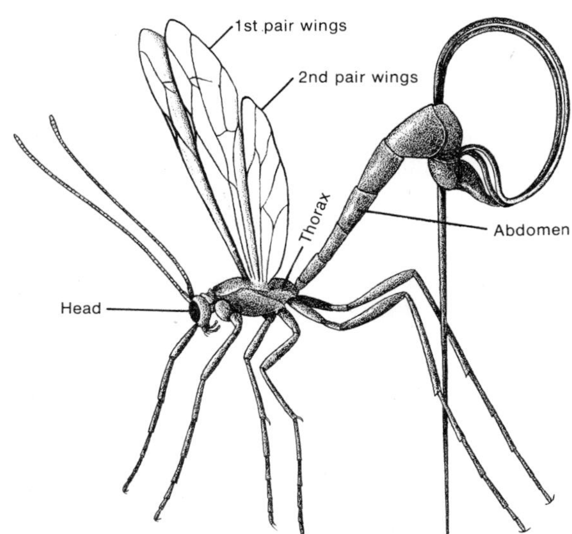

FIGURE 27.33 A parasitic female ichneumon fly (order Hymenoptera). (After Lutz.)

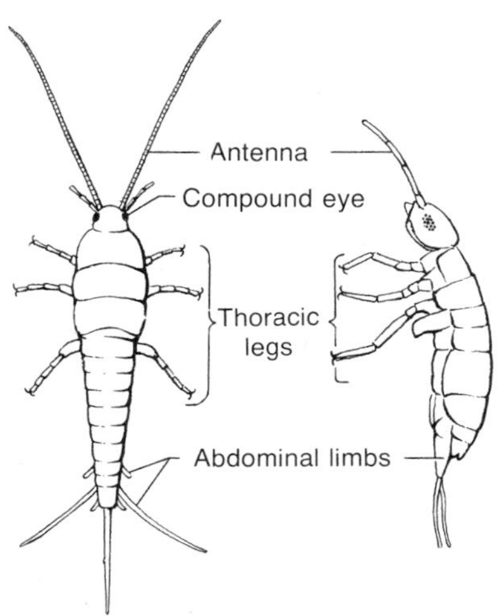

FIGURE 27.34 Primitive wingless insects (*Apterygota*), showing a silverfish (*left*) and a springtail (*right*). (*Left* after Lubbock; *right* after Carpenter and Folsom.)

and apparently diverged from the mainstream of insect evolution before the origin of flight (Fig. 27.34). Some insects—ants, termites, fleas, and lice—are secondarily wingless.

The wings develop as folds of the body wall and are thus composed of two layers of skeletal material applied together. Support is provided by a strutlike arrangement of tubular thickenings called **veins.**

Each wing articulates with the edge of the **tergum,** but its inner end rests on a dorsal pleural (lateral) process, which acts as a fulcrum (Fig. 5.20). The wing is thus somewhat analogous to a seesaw off center. Vertical movements of the wings may be caused directly by muscles attaching onto their bases or indirectly by thoracic muscles, as discussed earlier (p. 115). But up and

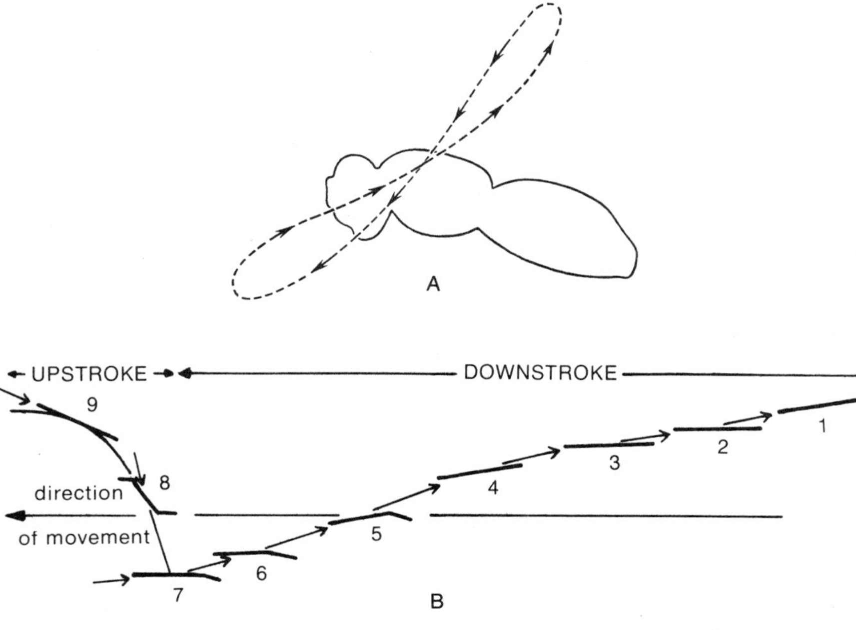

FIGURE 27.35 *A,* An insect in flight, showing the figure eight described by the wing during an upstroke and a downstroke. *B,* Changes in the position of the forewing on a grasshopper during the course of a single beat. Short arrows indicate direction of wind flowing over wing, and numbers indicate consecutive wing positions. (*A* after Magnan from Fox and Fox; *B* after Jensen, M.: *Philosophical Transactions of the Royal Society (B.)* *239*:511–552, 1956.)

down movement alone is not sufficient for flight. The wings must at the same time be moved forward and backward. A complete cycle of a single wing beat describes an ellipse (grasshoppers) or a figure eight (bees and flies), during which the wings are held at different angles to provide both lift and forward thrust (Fig. 27.31). In many insects the front and hind wings are locked together in flight.

The raising or lowering of the wings resulting from the contraction of one set of flight muscles stretches the antagonistic muscles, which then also contract. Insect wing beat thus involves the alternate contraction of the antagonistic elastic systems. The beat frequency varies greatly—4 to 20 beats per second in butterflies and grasshoppers; 190 beats per second in the honeybee and housefly; and 1000 beats per second in certain gnats. At low frequencies (30 beats per second or less) there is usually one nerve impulse to one muscle contraction. At higher frequencies, however, the contraction is **myogenic,** originating from the stretching caused by the contraction of the antagonistic muscles, and there are a number of beats, or oscillations, between each nerve impulse.

Rapid contraction is facilitated by the nature of the muscle insertion. A very slight decrease in muscle length during contraction can bring about a large movement of the wing (as a seesaw with the fulcrum near one end). A large part of the insect flight muscles is occupied by giant mitochondria, which provide for the high rate of respiration in these cells.

Flying ability varies greatly. Many butterflies and damselflies have a relatively slow wing beat and limited maneuverability. At the other extreme, some flies, bees, and moths can hover and dart. The fastest flying insects are hummingbird moths and horseflies, which have been clocked at over 33 miles per hour. Gliding, an important form of flight in birds, occurs in only a few large insects.

There is no flight control center in the insect nervous system, but the eyes and the sensory receptors on the antennae, on the wings, and in the wing muscles themselves provide continual feedback information for flight control. Members of the order **Diptera** (flies, gnats, and mosquitoes) have the second pair of wings reduced to knobs, called **halteres.** The halteres beat with the same frequency as the forewings and function as gyroscopes for the control of flight instability (pitching, rolling, and yawing).

Flight is inhibited by contact of the **tarsi** with the ground. At rest the wings are either held outstretched, directed back over the abdomen, or folded up. The anterior pair of wings of beetles (order **Coleoptera**) are heavy shieldlike covers for the membranous second pair of wings, which are folded over the abdomen at rest.

Nutrition. Primitive insects were herbivores with heavy jawlike **mandibles** for chewing plants (Fig. 27.32), and this feeding habit has been retained by members of a number of major orders, such as grasshoppers, crickets, and beetles. But in the evolution of various groups, the mouthparts have become adapted for many other modes of feeding, and now a wide range of diets is utilized. Sucking, piercing and sucking, and chewing and sucking insects have the mouthparts elongated to form a heavy **beak,** but since this feeding habit evolved independently a number of times, the beak is formed in different ways (Box 27.4).

Salivary secretions play an important role in feeding. Lubrication of food, digestion of sugar and pectins, anticoagulation, and production of venom are provided by the salivary glands of different groups.

The **foregut** of insects is variously modified to suit the diet and mode of feeding, but a **pharynx, crop,** and **proventriculus** are most commonly encountered (Fig. 27.36). The proventriculus may function as a gizzard or as a valve into the midgut.

The insect **midgut** (the **ventriculus,** or **stomach**) is the site of enzyme production, digestion, and absorption, as in other arthropods (Fig. 27.36). In those species that ingest solid foods, the foregut–midgut junction secretes a thin cuticle, the **peritrophic membrane,** which surrounds the food mass as it passes through the midgut. Supposedly, the peritrophic membrane protects the delicate midgut walls from abrasion by the food mass and perhaps more importantly conserves enzymes by dividing the gut lumen into two compartments. The membrane is permeable to some enzymes and the products of digestion. The initial products of digestion pass through the membrane, where they are attacked by a second order of enzymes that are restricted to the space between the gut wall and peritrophic membrane. Outpocketings of the midgut, called **gastric cecae,** are characteristic of many insects, but their function is still uncertain.

The **hindgut,** composed of an **intestine** and **rectum,** opens to the exterior at the end of the abdomen. In many insects, the hindgut is an important site of water absorption, and some species have special rectal structures to facilitate the process.

Excretion, Gas Exchange, and Internal Transport. The excretory organs are **malpighian tubules.** Two to several hundred tubes arise from the hindgut–midgut junction. Sometimes they are long and coiled. They lie more or less freely in the **hemocoel,** and nitrogenous wastes picked up from the surrounding blood are deposited as uric acid in the tubule lumen (p. 198). The sludgelike sediment passes into the intestine where more water may be removed. The fecal matter elimi-

BOX 27.4

The mouth parts of insects are primitively adapted for chewing, but in many groups of insects they have become modified for sucking, piercing, and sponging. *A.* Lateral view of the head of the moth. In moths and butterflies parts of the two maxillae have become modified to form a long sucking tube, which is rolled up when not in use. *B.* Lateral view of the head of a mosquito, showing separated mouth parts. *C.* Cross section of mouthparts of a mosquito in their normal functional position. *D.* Ventral view of anterior half of a hemipteran (a bug), showing piercing beak. *E.* Cross section through a hemipteran beak, showing the food and salivary channels enclosed within the stylet-like maxillae and mandibles. *F.* A hemipteran penetrating plant tissue with its stylets. *G.* Lateral view of the head and mouthparts of a housefly, in which the labium is modified as a sponge. Food is first dissolved with salivary secretions and then sucked back up with a sponge. (*A* and *G* after Snodgrass; *B* and *C* after Waldbauer from Ross H. H.: A Textbook of Entomology, 3rd ed. New York, John Wiley and Sons, 1965; *D* after Hickmann; *E* after Poisson; *F* after Kullenberg.)

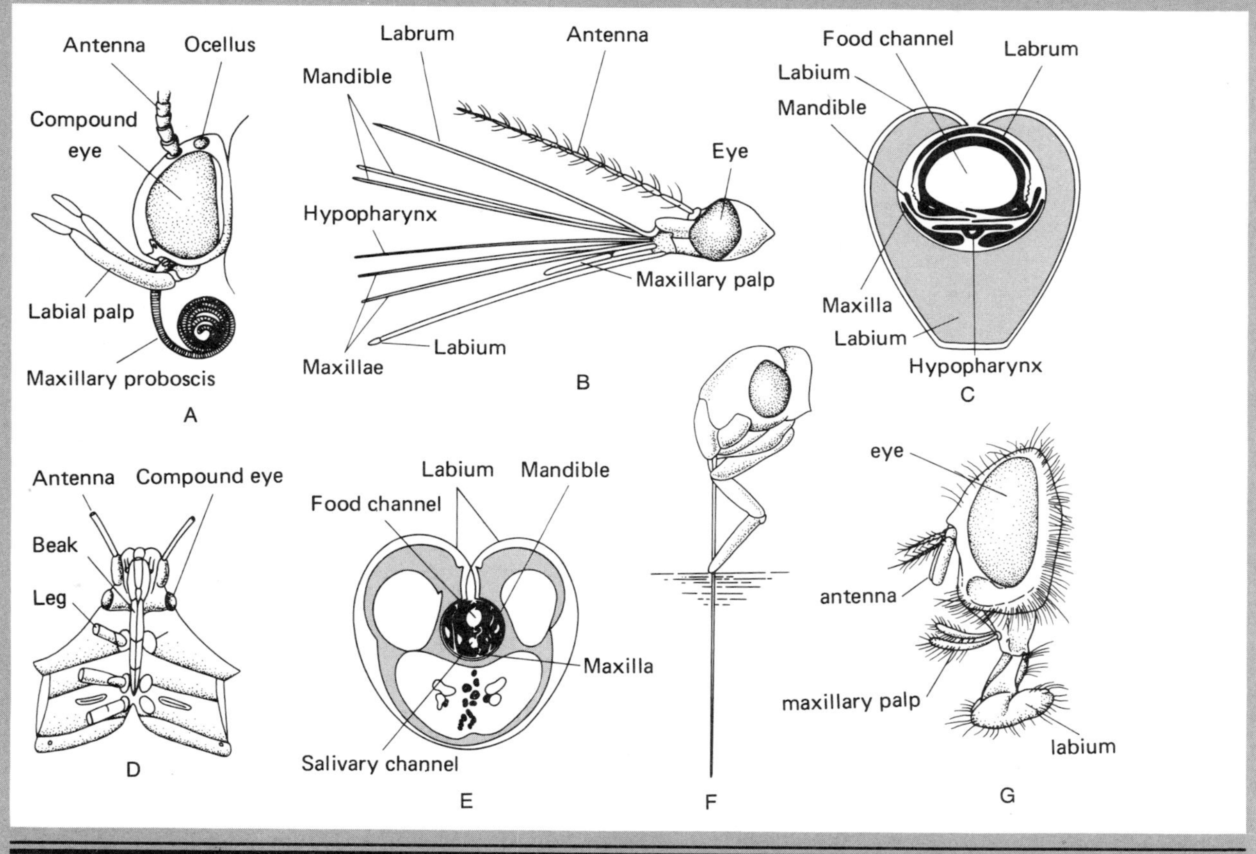

nated from the anus thus consists of undigested residues and excretory wastes.

Tracheae are the gas exchange organs of insects, and the spiracles, or openings, to the tracheal system are located one on each side of the last two thoracic segments and the first seven or eight abdominal segments (Figs. 27.32 and 7.5). The structure and physiology of tracheal systems is described on page 153.

The **heart** is usually a long abdominal tube with nine pairs of ostia, and the only vessel is an anterior **aorta** leading into the thorax and head. Blood flows from posterior to anterior, although reversal of flow is known to occur in some forms.

Reproduction and Development. The single pair of gonads is located in the abdomen, and the gonoducts

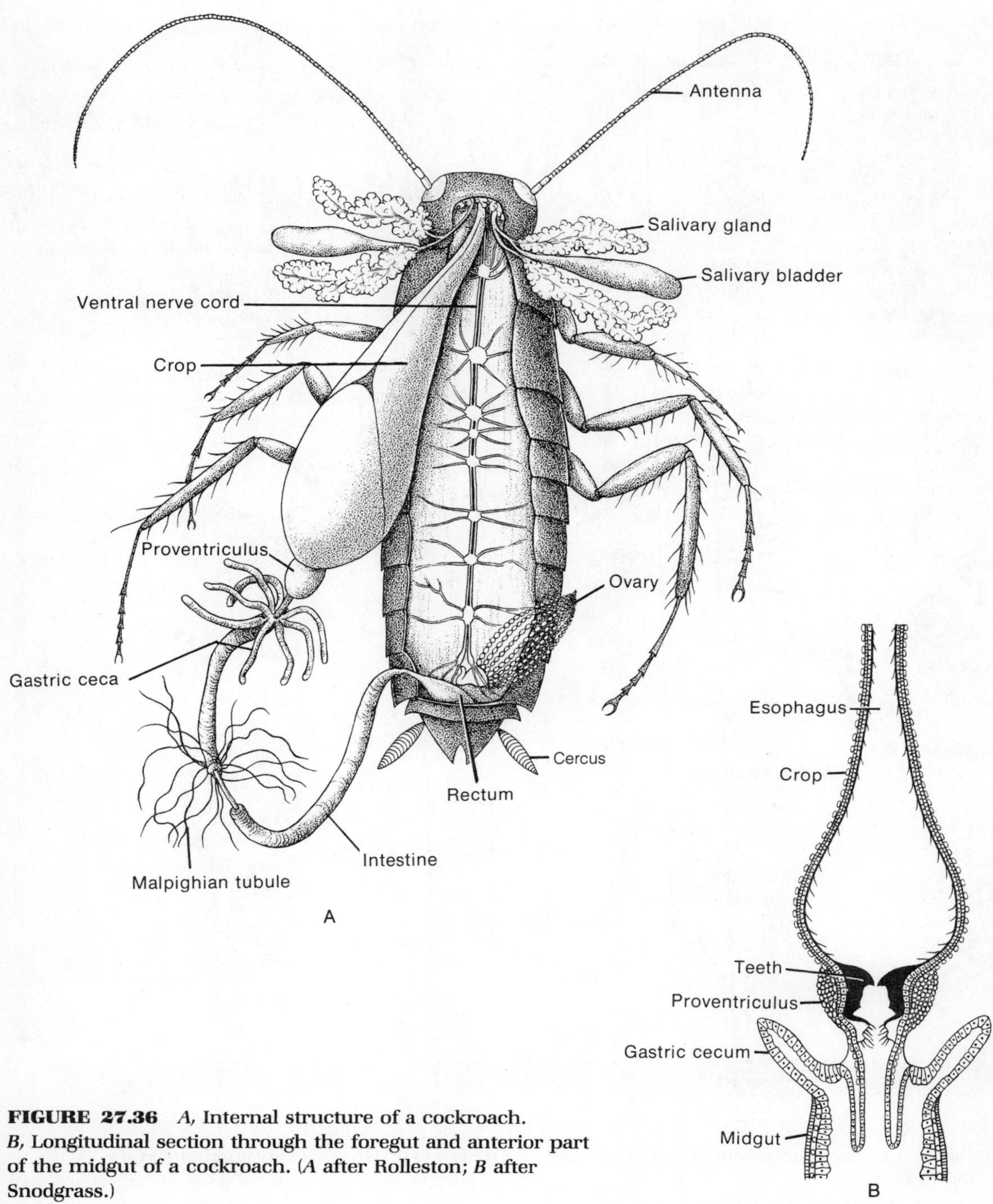

FIGURE 27.36 *A,* Internal structure of a cockroach. *B,* Longitudinal section through the foregut and anterior part of the midgut of a cockroach. (*A* after Rolleston; *B* after Snodgrass.)

from each side unite posteriorly before opening at the end of the abdomen through a short median **vagina** in the female or an **ejaculatory duct** in the male. In addition, the female system usually includes a **seminal receptacle** (spermatheca) and **accessory glands** associated with the vagina, and the male system includes paired **seminal vesicles** and **accessory glands** (Fig. 27.37).

The sperm of many insects are transferred within spermatophores. In a few primitive groups the spermatophores are transferred indirectly. Like arachnids and myriapods the spermatophores are deposited on the ground and then picked up by the female. In most insects, however, transfer is direct, and a tubular **penis** is inserted by the male into the vagina of the female. The posterior abdominal segments of the males of many moths, butterflies, true flies, and other insects bear clasping structures that are used to hold the abdomen of the female during copulation.

Sperm are stored in the seminal receptacle or spermatheca, and fertilization occurs internally as the eggs pass through the oviduct. The egg leaves the ovary encased within a hard shell, but a tiny opening, or **micropyle,** at one end of the egg permits entrance of the sperm.

Certain parts of the terminal segments of the female are modified as an **ovipositor,** through which the eggs pass upon leaving the gonopore. By means of the ovipositor, eggs can be buried in soil, excrement, or carrion, injected into plant tissue or applied to twig, leaf, soil, water, or other surfaces. The eggs are generally deposited in batches, cemented to each other and to the substratum by secretions from the accessory glands.

Parthenogenesis occurs in a number of insect groups. The condition in aphids closely parallels that of the crustacean water fleas, where there are successive generations of parthenogenetic females followed by the appearance of males. In bees, unfertilized eggs produce males; fertilized eggs produce females.

The degree of development at hatching is quite variable. The newly hatched young of most primitive wingless insects are similar to the adults, except in size and sexual maturity. In contrast, the young of the members of winged orders—grasshoppers, crickets, dragonflies, leaf hoppers, bugs, and many others—are similar to the adults but lack the wings and reproductive system, which gradually develop during the course of subsequent instars (Fig. 283). This type of development is

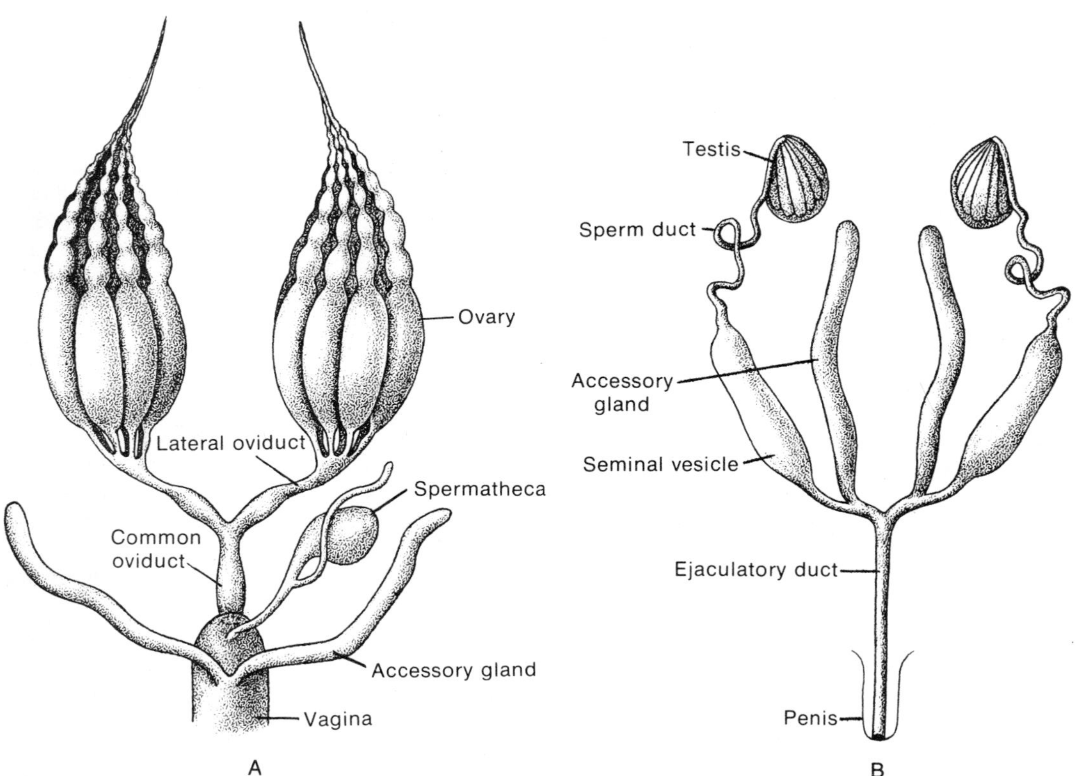

FIGURE 27.37 Dorsal views of the reproductive systems of insects. *A,* Female; *B,* male. (After Snodgrass.)

called **gradual,** or **incomplete (hemimetabolous development),** and the immature stages are called **nymphs,** or when aquatic, **naiads.** Aquatic development with incomplete metamorphosis is characteristic of dragonflies, stoneflies, and mayflies.

The higher orders of insects, such as beetles, butterflies, moths, bees, wasps, and flies, undergo **complete metamorphosis (holometabolous development),** involving more radical changes between the immature and adult forms (Fig. 283). The hatching stage is an active, feeding, wormlike larva called a **caterpillar** (butterflies and moths), **grub** (beetles), or **maggot** (flies). The larva increases in size through successive instars, and at the end of the larval period, passes into a quiescent pupal stage, in which a radical transformation takes place. The transformation of immature insects into reproducing adults is known to be under endocrine control (p. 283).

Insect larvae are a specialized developmental condition and, as we have seen, are absent in primitive groups. Larvae are able to utilize food sources and habitats that would be unavailable to the adults and vice versa. For example, the caterpillars of butterflies are chewing herbivores, in contrast to the nectar-feeding habit of the adults. In some insects, certain moths for instance, the short-lived adults have completely lost the ability to feed.

Parasitism. There are many parasitic insects, and the condition has undoubtedly evolved several times within the class.

Insect parasitism represents an adaptation to meet the habitat-nutrition needs of different stages in the life cycle. For some insects, parasitism provides a new food source and habitat for the adults; for others, a new food source for the larvae.

Adult fleas and lice are blood-sucking ectoparasites on the skin of birds and mammals (there is one group of chewing lice). The eggs and immature stages of fleas may develop on the host or in the host's nest or habitation.

Many species of wasps and flies illustrate larval parasitism. Certain small wasps insert an egg into the leaves and stems of plants. The surrounding plant tissue reacts to form a large mass, or gall. The egg develops within the gall, and the larval stage feeds upon the surrounding plant tissue. Other wasps deposit their eggs in the bodies of insects, especially insect larvae.

The screw-worm fly, a species of blowfly and a pest of domestic animals, lays its eggs in the wounds and nostrils of mammals, and the larvae feed on living tissue. The parasitic condition was probably preceded by the deposition of eggs in carrion, for this is the habit of many nonparasitic species of blowflies.

Communication. Both social and nonsocial insects utilize chemical, tactile, visual, and auditory signals as methods of communication. Many examples of chemical communication by pheromones (p. 777) are now known. For example, the males of some moths can locate females by means of airborne substances detected from a distance as great as 4.5 km., and such pheromones are now used for monitor trapping or in the biological control of some insect pests. Another example of insect pheromones are substances deposited on the ground by ants returning from foraging trips, which serve as trail markers for other individuals in the colony.

Among the more unusual visual signals are the luminescent flashings of fireflies, which play a role in sexual attraction. In species of *Photinus,* for example, flying males flash at definite intervals. Females located on vegetation will flash in response if the male is sufficiently close. The male will then redirect his flight toward her and further flashing will occur.

Sound production is especially notable in grasshoppers, crickets, and cicadas. The chirping sounds of the first two are produced by rasping. The front margin of the forewing or the hind legs acts as a scraper and is rubbed over a file formed by veins of the forewing. Each species of cricket produces a number of songs that differ from the songs of other species. Cricket songs function in sexual attraction and aggression. The staticlike sounds of cicadas, which serve to aggregate individuals, are produced by vibrations of special chitinous abdominal membranes, oscillated by special tymbal muscles (p. 115).

Social Insects. Colonial organization has evolved in a number of animal phyla but reaches its highest degree of development among the social insects.

Social organizations have evolved in two orders of insects: the **Isoptera,** which contains the termites, and the **Hymenoptera,** which includes the ants, bees, and wasps. In all social insects, no individual can exist outside of the colony but the one in which it developed. All social insects exhibit some degree of **polymorphism,** and the different types of individuals of a colony are termed **castes.** Those of termites are shown in Figure 27.38.

Termites have been called social cockroaches, for the two groups are related and in both the gut harbors symbiotic flagellates, which are important in the digestion of cellulose. The contact necessary between individuals

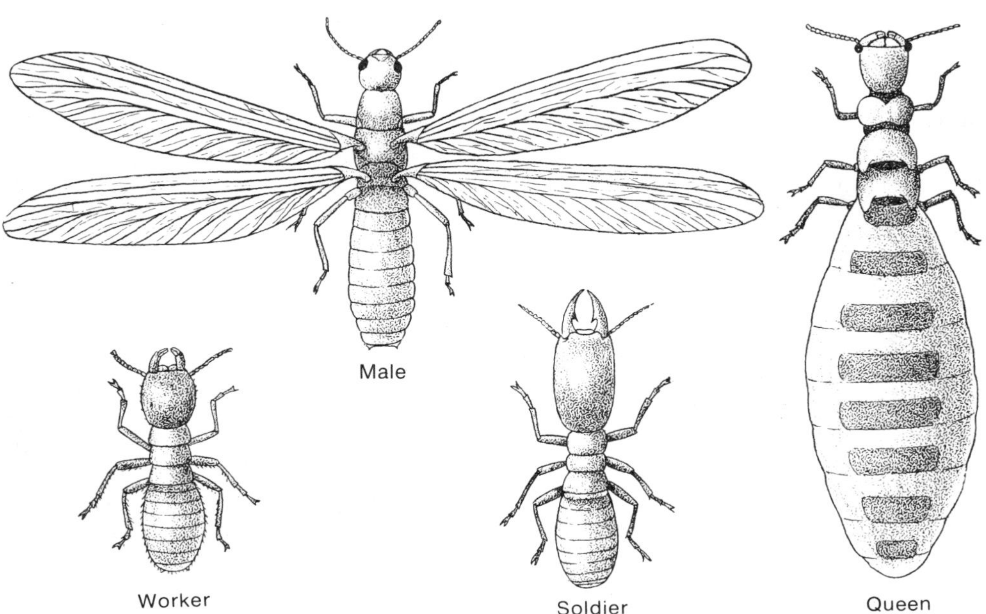

FIGURE 27.38 Castes of the common North American termite, ***Reticulitermes flavipes.*** (After Lutz.)

for the transfer of the symbionts may have been a factor in the evolution of termite social behavior.

Termites live in galleries constructed in wood or soil, and in some species, the colonies may be huge and structurally very complex. The colony is built and maintained by workers and soldiers (Fig. 27.38). The soldiers possess large heads and mandibles and serve for the defense of the colony. Workers and soldiers are sterile and wingless and include both males and females. Wings are present in the fertile males and queens only during a brief nuptial flight. The male, or king, remains with the queen, copulating intermittently and aiding her in the construction of the first nest.

Ant colonies resemble those of termites and are housed within a gallery system in soil, in wood, or beneath stones. There may be a soldier caste in addition to workers. Wings are present only during the nuptial period.

The colonies of ants, bees, and wasps differ from those of termites in being essentially female societies, for all the workers are sterile females. Caste determination of the female is regulated by the presence or absence of certain substances provided in the immature stages by other members of the colony. Males have a brief existence, functioning only in the copulatory nuptial flight. Unlike termite males, they neither contribute to the construction of the first nest nor remain with the queen.

The honeybee, *Apis mellifera,* is the best known social insect. This species is believed to have originated in Africa and to be a recent invader of temperate regions, for unlike other social bees and wasps of temperate regions, the honeybee colony survives the winter, and multiplication occurs by the division of the colony, a process called swarming. Stimulated at least in part by the crowding of workers (20,000 to 80,000 in a single colony), the mother queen leaves the hive along with part of the workers (a **swarm**) to found a new colony. The old colony is left with developing queens. On hatching, a new queen takes several nuptial flights during which copulation with males (**drones**) occurs, and she accumulates enough sperm to last her lifetime. The male dies following copulation, when his reproductive organs are literally exploded into the female. A new queen may also depart with some of the workers as an after-swarm, leaving the remaining workers to yet another developing

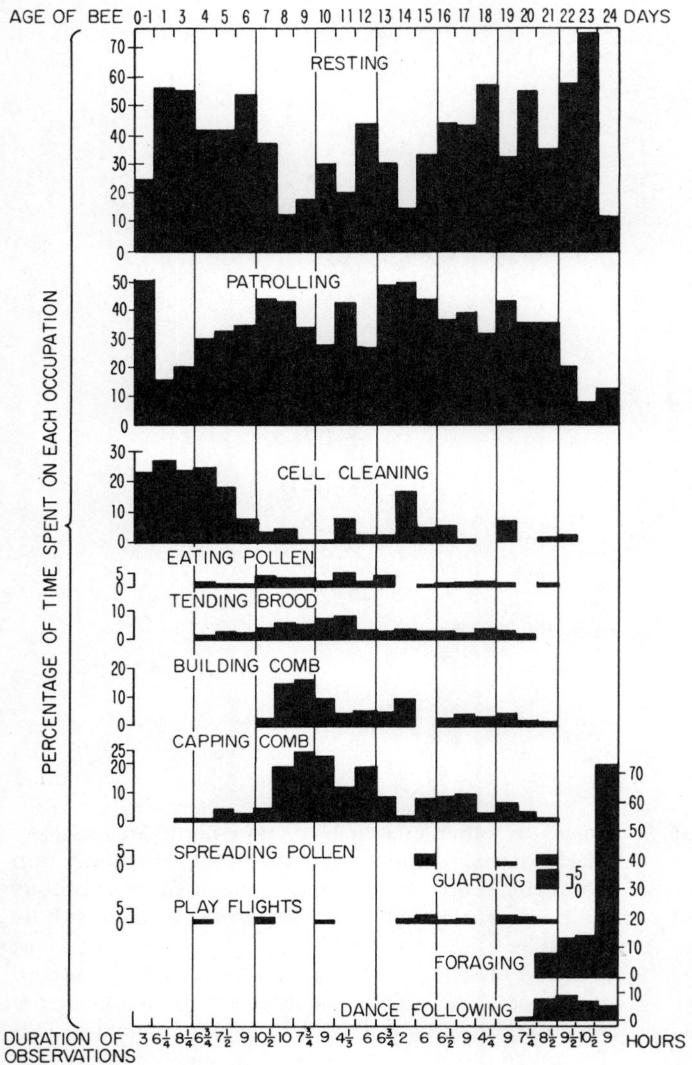

FIGURE 27.39 The activities of a single worker honeybee during the first 24 days of her adult life. (Redrawn from Ribbands, 1953; based on data of Lindauer, 1952.)

queen. Eventually the old colony will consist of about one third of the original number of workers and their new queen.

Honeybee colonies are large. The workers' life span is not long, and a queen may lay 1000 eggs per day. The diet provided these larvae by the nursing workers results in their developing into sterile females, i.e., additional workers. The nursing behavior of the workers is a response to a pheromone ("queen substance") produced by the queen's mandibular glands. At the advent of the swarming or when the vitality of the queen diminishes, the production of this pheromone declines. In the absence of the inhibiting effect of the pheromone, the nursing workers construct royal cells into which eggs and royal jelly are placed. The exact composition of this complex food is still unknown, but those larvae fed upon it develop into queens in about 16 days. At the same time that queens are being produced, unfertilized eggs are deposited in cells similar to those for workers. These haploid eggs develop into drones.

A remarkable feature of honeybee social organization is the temporal division of certain tasks of the workers

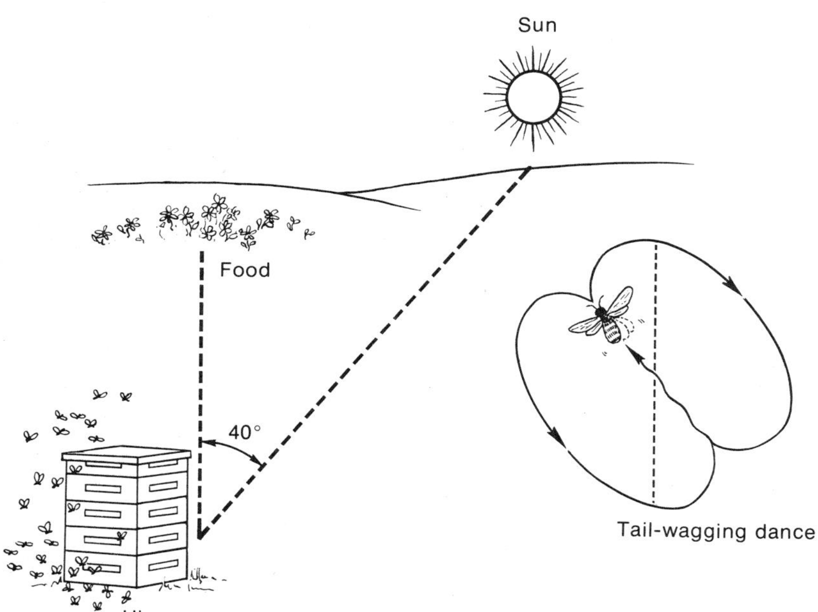

FIGURE 27.40 Diagram illustrating the inclination of the straight tail-wagging run by a scout bee to indicate the location of a food source by reference to the sun. The food source is located at an angle 40 degrees to the left of the sun. The tail-wagging run of the scout bee is therefore upward (indicating that food is toward the sun) and inclined 40 degrees to the left (indicating the angle of the food source to the sun). (After Von Frisch.)

(Fig. 27.39). The first activities of the worker are maintenance tasks within the hive. During this period there is secretion by wax, mandibular, and other glands involved in comb construction, food storage, and larval care. After about three weeks, such glandular activity declines, and the bee begins a period of foraging outside of the hive. A large amount of time is spent by the older worker bees in resting and patrolling.

Communication between members of a honeybee colony is highly evolved; some aspects, such as the tail-wagging dance, set the honeybees apart from all other social insects. A successful foraging scout returns to the hive and communicates to other workers the nature, direction, and distance of a food source. The nectar and pollen and the scout's body provide the information about the kind of food that has been found. The scout bee also executes an excited dance that is a ritualization of the flight path. The dancing bee circles to the right and to the left, with a straight-line run between the two semicircles (Fig. 27.40). During the straight-line run, the bee wags her abdomen and emits audible pulsations. Von Frisch, a pioneer in the study of communication in bees, discovered that the orientation of the circular movements shows the direction of the food, and that the frequency of the tail-wagging runs indicates the distance. The closer the food source is to the hive, the greater the frequency of tail-wagging runs. Bees use the angle of the sun and light polarization as a means of orientation, and the dance of the scout bee indicates the location of the food in reference to the sun's position. If the tail-wagging run is directed upward, the food is located toward the sun; if the tail-wagging run is directed downward, the food is located away from the sun. The inclination of the run to the right or to the left of vertical indicates the angle of the sun to the right or to the left of the food source. An internal "clock" compensates for the passage of time between discovery of the food and the start of the dance so that the information is correct even though the sun has moved during the interval. On cloudy days, the polarization of the light rays and ultraviolet light act as indirect references in the absence of the sun. If the food source is closer than 80 m., the clues provided by chemoreception are sufficient for finding the food, and the tail-wagging dance is not performed by the scout bee.

Although the tail-wagging dance has been decoded, the sensory modality by which it is transmitted to other bees is still uncertain. The hive is dark so that the dance cannot be easily detected visually. The surrounding bees must receive the dancer's vibrations through their antennae or legs. The sound pulsations of the dancer apparently also indicate the distance of the food source from the hive. The average number of vibrations is proportional to the distance of the food from the hive.

Classification of the Phylum Arthropoda*

Subphylum Trilobitomorpha† The extinct trilobites. A single pair of antennae present; the many postoral trunk appendages are all similar.

Subphylum Chelicerata Antennae absent; the first postoral appendages are a pair of chelicerae.

Class Merostomata Body composed of a cephalothorax and abdomen. Marine, with abdominal gills.

Subclass Xiphosura Horseshoe crabs. Abdominal segments fused.

Subclass Eurypterida† Extinct merostomes. Abdominal segments not fused.

Class Arachnida* Body composed of a cephalothorax and abdomen with four pairs of legs. Terrestrial; gills absent. Scorpions, spiders, harvestmen, mites, and ticks.

Order Scorpionida Scorpions. Large chelate pedipalps and long segmented abdomen with terminal sting.

Order Pseudoscorpionida Tiny arachnids with large pedipalps and short segmented abdomens.

Order Araneae Spiders. Abdomen unsegmented and attached to cephalothorax by a narrow waist.

Order Opiliones Harvestmen, or daddy-long-legs. Legs usually very long; the short, segmented abdomen is broadly joined to the cephalothorax.

Order Acarina Mites and ticks. Very small arachnids with unsegmented abdomen broadly fused with cephalothorax.

Class Pycnogonida Sea spiders. Marine chelicerates (?) having a very narrow trunk and four to six pairs of legs.

Subphylum Crustacea Head with two pairs of antennae; first postoral appendages are a pair of mandibles. Appendages primitively branched. Mostly aquatic arthropods.

Class Branchiopoda Fairy shrimp, tadpole shrimp, clam shrimp, and water fleas. Freshwater and marine crustaceans with leaflike setose appendages.

Class Ostracoda Mussel or seed shrimp. Tiny marine and freshwater crustaceans in which the entire body is enclosed within a bivalved carapace.

Class Copepoda Copepods. Very small marine and freshwater crustaceans having a cylindrical tapered body with long first antennae.

Class Cirripedia Barnacles. Marine, sessile crustaceans in which the body is enclosed within a bivalved carapace that is typically covered with calcareous plates.

*Abbreviated
†Extinct taxa
‡Greatly abbreviated

Class Malacostraca‡ Trunk composed of an eight-segmented thorax, on which the legs are located, and a six-segmented abdomen.

Order Amphipoda Amphipods. Small marine and freshwater crustaceans in which the body is laterally compressed.

Order Isopoda Isopods. Marine, freshwater, and terrestrial (wood lice and pill bugs) crustaceans in which the body is dorsoventrally flattened.

Order Decapoda Shrimp, crayfish, lobsters, and crabs. Thorax covered by a carapace; thoracic appendages consist of three pairs of maxillipeds and five pairs of legs.

Subphylum Uniramia Head with one pair of antennae; first postoral appendages are a pair of mandibles. Appendages primitively unbranched. Mostly terrestrial arthropods.

Class Diplopoda Millipedes. Elongate trunk composed of many similar doubled segments, most bearing two pairs of legs.

Class Pauropoda Minute animals that inhabit leaf mold. Elongate trunk of eleven segments, nine or ten of which bear legs.

Class Symphyla Small centipede-like mandibulates that inhabit leaf mold. Trunk contains 12 leg-bearing segments.

Class Chilopoda Centipedes. Elongate trunk of many leg-bearing segments. First trunk segment carries a pair of large poison claws.

Class Insecta Insects. Body composed of head, thorax, and abdomen. Thorax bears three pairs of legs and usually two pairs of wings.

Subclass Apterygota Primary wingless insects. The five orders include the silverfish and springtails.

Subclass Pterygota* Winged insects, or if lacking wings, the wingless condition is secondary.

Order Ephemeroptera Mayflies.
Order Odonata Dragonflies and damselflies.
Order Orthoptera Grasshoppers and crickets.
Order Isoptera Termites.
Order Plecoptera Stoneflies.
Order Dermaptera Earwigs.
Order Mallophaga Chewing lice and bird lice.
Order Anoplura Sucking lice.
Order Thysanoptera Thrips.
Order Hemiptera True bugs.
Order Homoptera Cicadas, leaf hoppers, and aphids.
Order Neuroptera Lacewings, lionflies, snakeflies, and dobsonflies.
Order Coleoptera Beetles and weevils.
Order Trichoptera Caddisflies.
Order Lepidoptera Butterflies and moths.
Order Diptera True flies, midges, gnats, and mosquitoes.
Order Hymenoptera Ants, bees, wasps, and sawflies.
Order Siphonaptera Fleas.

Summary

1. The arthropoda is the largest and most widely distributed of all animal phyla. An external chitinous skeleton, divided into articulating plates and cylinders and periodically molted to permit growth, is the distinguishing feature of arthropods. Metamerism, a paired ventral nerve cord, and perhaps the appendages reflect an annelidan ancestry. The coelom has largely disappeared, nephridia are absent, and distinct muscle bundles are attached to the inner side of the exoskeleton. The circulatory system is open.

2. The head region of the extinct primitive trilobites (Subphylum Trilobitomorpha) bears a pair of eyes, a pair of antennae, and four pairs of postoral appendages like those of the trunk. The remaining three arthropod subphyla, with many living species, exhibit varying degrees of reduction of metamerism through differentiation of appendages and loss and fusion of segments.

3. The subphylum Chelicerata includes the only arthropods without antennae. The body is divided into a cephalothorax and abdomen, and the feeding appendages flanking the mouth are chelicerae. The small class Merostomata, containing the gill-bearing horseshoe crabs, reflects the marine origin of chelicerates, but the large terrestrial class Arachnida contains the scorpions, pseudoscorpions, spiders, harvestmen, mites, and ticks.

4. Arachnids are largely predaceous on other arthropods, and scorpions, pseudoscorpions, and spiders utilize poison in prey-catching. In most species digestion begins externally in the mouth region, with enzymes passed forward from the digestive gland. Spiders make the most extensive use of silk of any group of animals. A waxy epicuticle and gas exchange by book-lungs or tracheae or both are important adaptations of arachnids for life on land. Indirect sperm transmission, often involving a spermatophore, is a common feature of arachnid reproduction.

5. The subphylum Crustacea is a diverse assemblage of mostly aquatic arthropods, distinguished by having two pairs of antennae. All have one pair of mandibles and two pair of maxillae as feeding appendages. Larval stages are typical of many crustaceans, and the earliest stage is a nauplius larva, bearing only the first three pairs of appendages.

6. Members of the class Malacostraca include the largest crustaceans. Their trunk is composed of an eight-segmented thorax and a six-segmented abdomen, each segment bearing appendages. The order Decapoda, the largest in the Malacostraca, is distinguished by possessing three pairs of maxillipeds and five pairs of legs as thoracic appendages. The compound eyes are on stalks. Decapods include shrimp, lobsters, and crabs. Crabs have a short body form resulting from the folding of the abdomen beneath the thorax. Hermit crabs have unfolded abdomens housed within gastropod shells.

7. The gills of decapods arise from the bases of the thoracic appendages and are enclosed by the overhanging carapace. Decapods are mostly scavengers and carnivores. The large midgut digestive gland is the site of absorption. The excretory organs of decapods are green glands opening at the base of each second antenna. At copulation sperm are usually transmitted by anterior copulatory pleopods in the male to a seminal receptacle or oviduct openings in the female. Eggs are brooded on the pleopods, and the hatching stage is usually a zoea larva.

8. The malacostracan orders Amphipoda and Isopoda brood their eggs in a marsupium beneath the thorax. They possess one pair of maxillipeds and seven pairs of legs, but most amphipods are laterally flattened and isopods are dorsoventrally flattened. Both groups have stalkless compound eyes. They are largely marine but are also found in fresh water, and the widespread terrestrial pill bugs (isopods) represent the only successful crustacean invasion of land. Amphipods and isopods are mostly scavengers on animal and plant remains.

9. The class Branchiopoda contains the water fleas, fairy shrimp, and brine shrimp, most of which live in fresh water. They possess foliaceous trunk appendages used in filter feeding. Water fleas have the trunk enclosed within a large carapace and swim with the large second antenna.

10. Members of the class Ostracoda are small, mostly benthic crustaceans. The body is completely enclosed within a hinged bivalve carapace. The trunk is greatly reduced and they swim or scurry over the bottom with the antennae.

11. The class Copepoda includes many species of small planktonic, epibenthic, and interstitial crustaceans. They lack compound eyes but have a median nauplius eye; the large first antennae are held at right angles to the body. The trunk is composed of thorax and abdomen, the latter lacking appendages. In swimming species, the second antennae are the principal locomotor organs, and the first function largely to reduce sinking. Many planktonic species are filter feeders on diatoms and are very important in marine food chains.

12. Barnacles, members of the class Cirripedia, are sessile crustaceans with the body enclosed within a carapace covered with calcareous plates. Both stalked and stalkless barnacles are attached by cement glands that open at the base of the first antennae in the larva. Antennae are lacking in the adult. Barnacles are suspension feeders, utilizing six pairs of long biramous setose trunk appendages as a scoop. Most species are hermaphroditic. Many barnacles are attached to the surface of other animals; about one third of the described species are parasitic.

13. The subphylum Uniramia, which contains the centipedes, millipedes, and insects, is the largest group of arthropods. Most living uniramians are terrestrial and have one pair of antennae and as feeding appendages a pair of mandibles and one or two pairs of maxillae. All of the appendages are primitively unbranched. Gas exchange organs are tracheae, and excretory organs are malpighian tubules. Primitively, sperm transmission is indirect with spermatophores.

14. The uniramian classes Chilopoda (centipedes) and Diplopoda (millipedes) have long trunks of many leg-bearing segments. Centipedes, which live beneath stones and logs, are predaceous, using a pair of large poison claws located behind the second maxillae. Many species run rapidly, and trunk stability is increased by alternating large and small or overlapping tergal plates. Trunk stability in millipedes has been increased by the evolution of diplosegments, each of which bears two pairs of legs. Most species feed on decomposing vegetation.

15. The class Insecta contains over three quarters of a million species. The body usually consists of a head with three ocelli and a pair of compound eyes, a thorax with two pairs of wings and three pairs of legs, and an abdomen without conspicuous segmental appendages. Primitive species are wingless, but most insects are capable of flight, which has been an important factor in the evolutionary success of the class. The mouthparts—one pair of mandibles, one pair of maxillae, a labium (fused second maxillae)—are primitively adapted for chewing vegetation but have become modified for a wide range of other diets and feeding modes. Sperm transmission in primitive insects is indirect by a spermatophore, but most species copulate and have direct sperm transmission. The newly hatched young of many species are more or less similar to the adults but lack wings, which are gradually acquired during the course of subsequent instars (incomplete metamorphosis); other insects hatch as a wormlike larva that undergoes a radical metamorphosis (complete metamorphosis) to attain the adult form.

16. Intraspecific communication by pheromones, sound, and light is utilized by many species. Varying degrees of social organization have evolved in termites, ants, wasps, and bees. Such colonies commonly contain reproductive and worker castes.

References and Selected Readings

Detailed accounts of arthropods may be found in the references listed at the end of Chapter 19. The works listed below deal exclusively with specific arthropod groups.

Bliss, D. E. (Ed.): *The Biology of Crustacea.* Vol. 1–7. New York, Academic Press, 1982–1983. A multivolume work covering all aspects of crustacean biology.

Borror, D. J., D. M. De Long, and C. A. Triplehorn: *An Introduction to the Study of Insects.* 5th ed. Philadelphia, Saunders College Publishing, 1981. A general entomology text emphasizing insect taxonomy.

Borror, D. J., and R. E. White: *A Field Guide to the Insects.* Boston, Houghton Mifflin Co., 1970. A good guide for the identification of common North American insects.

Daly, H. V., and J. T. Doyen: *An Introduction to Insect Biology and Diversity.* New York, McGraw-Hill Book Co., 1979. A general entomology text.

Chapman, R. F.: *The Insects: Structure and Function.* 3rd ed. Cambridge, Massachusetts, Harvard University Press, 1982.

Elzinga, R. J.: *Fundamentals of Entomology.* 2nd ed. New York, Prentice-Hall, 1981. A general entomology text.

Foelix, R. F.: *Biology of Spiders.* Cambridge, Massachusetts, Harvard University Press, 1982.

Gertsch, W. J.: *American Spiders.* 2nd ed. New York, Van Nostrand, 1979. A natural history of spiders by the former curator of spiders at the American Museum of Natural History.

Kaestner, A.: *Invertebrate Zoology.* Vols. II (Chelicerates and Myriapods) and III (Crustaceans). New York, John Wiley & Sons, Inc., 1968 and 1970. The best general accounts in English of the arthropods other than insects.

Romoser, W. S.: *The Science of Entomology,* 2nd ed. New York, Macmillan, 1981. A well-balanced general biology of insects.

Wilson, E. O.: *The Insect Societies.* New York, Academic Press, 1971. A superb general biology of the social insects.

Wicksten, M. K.: Decorator crabs. *Scientific American, 242*(Feb.):146–154, 1980. A study of the habit of certain spider crabs of growing algae and sponges on their backs.

28 Bryozoans

ORIENTATION

This chapter will explore a phylum of minute colonial animals, called the Bryozoa. Although unfamiliar to most laymen, bryozoans are very common marine animals. We will start by using the characteristics of bryozoans to illustrate the predictive value of some basic zoological principles.

The members of this phylum are **sessile, colonial,** and **coelomate,** and individuals are usually less than 0.5 mm. in length. On the basis of these four facts, what other features—structural or functional—would you expect to find in bryozoans? Think a few minutes and make some educated guesses before you continue reading.

CHARACTERISTICS OF BRYOZOANS

The following bryozoan features are correlated with their sessile life, colonial organization, and minute size and are therefore not unexpected:

Correlated with their colonial organization:

1 Bryozoans are **polymorphic.**

Correlated with their sessile condition:

2 Bryozoans are encased within an **exoskeleton.** A protective and supporting skeleton is characteristic of many sessile animals, such as sponges, hydrozoans, corals, and barnacles.

3 Bryozoans are **filter feeders.** Filter feeding, a common adaptation for a slow-moving or immobile life, is illustrated by such animals as sponges, sedentary tubiculous polychaetes, worm shells, slipper shells, bivalves, and barnacles.

4 Bryozoans are **hermaphroditic.** Hermaphroditism is a common adaptation of animals that cannot move about, such as many sponges, some sea anemones, some corals, some bivalves, slipper shells, and barnacles.

5 A **larval stage** is present in the development of bryozoans. Larvae are the principal means of dispersion for most sessile animals.

Correlated with their very small size:

6 Bryozoans have no special internal transport system, for internal distances are short enough to permit transport solely by diffusion.

7 Bryozoans have no gas exchange organs. The ratio of surface area to volume is sufficiently favorable to permit gas exchange across the general body surface.

8 Bryozoans have no excretory system. The excretion of highly soluble ammonia and the large ratio of surface area to volume permit elimination by diffusion across the general body surface.

The members of the phylum Bryozoa or Ectoprocta are common marine organisms, but their small size and atypical form make them virtually unknown to the layman. Sometimes called moss animals (Gr. *bryo,* moss + *zoon,* animal), they live attached to rocks, shells, pilings, jetties, and ship bottoms, and a few species occur in fresh water.

Structure of a Bryozoan Individual

Individuals (**zooids**) of a bryozoan colony are shaped something like a little rectangular box, in which each of the sides represents one of the usual morphological surfaces—anterior, posterior, ventral, and so on (Fig. 28.1). The body is covered with an exoskeleton commonly composed of **chitin.** A layer of calcium carbonate is commonly found just beneath the chitin; both are secreted by a single layer of epidermal cells. In most marine bryozoans a layer of **peritoneum** lies immediately beneath the epidermis; the rigidity of the exoskeleton would make ordinary body wall muscles useless. The interior of the body is occupied by a spacious **coelom.**

The principal organ in the coelom is the **gut,** which is U-shaped to avoid the obstruction resulting from the attachment of the posterior end to other members of the colony. At the anterior end, the skeletal housing is perforated by a circular orifice through which the two ends of the gut project.

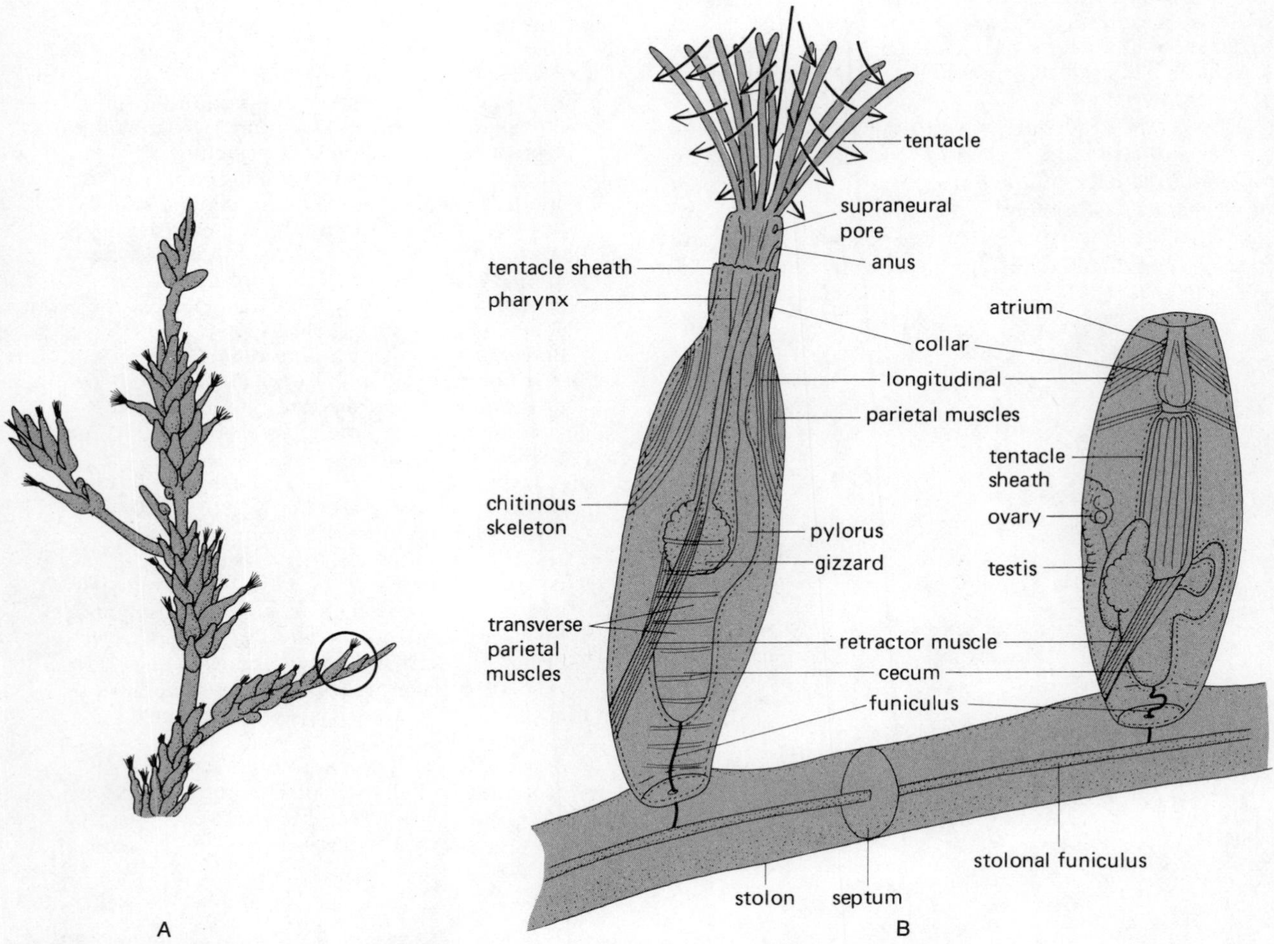

FIGURE 28.1 *Bowerbankia,* a stoloniferous bryozoan. *A,* Section of colony. *B,* Stolon and two zooids of *Bowerbankia,* from circled portion of *A;* one has lophophore protruded, and the other has it retracted. (*A* modified from Gay in Prenant and Bobin; *B* modified from Ryland.)

The feeding organ of bryozoans is a crown of ciliated tentacles, the **lophophore** (Gr. *lophos,* tuft + *pherein,* to bear). When the lophophore is protruded, the outstretched tentacles usually form a funnel with the **mouth** in the center of the base and the **anus** projecting outside of the base (Figs. 28.1 and 28.2). The hollow tentacles contain an extension of the coelom. When the lophophore is protruded, a sheath of body wall extends from the rim of the orifice up to the base of the lophophore. On retraction of the lophophore into the orifice, this tentacular sheath is reversed and surrounds the bunched tentacles within the skeletal housing (Fig. 28.1).

The lophophore is protruded by increased pressure in the coelomic fluid. The pressure is elevated by the inbowing of the body wall on contraction of certain transverse muscle bands. In many bryozoans the muscle bands attach to the inner side of the exposed, or ventral surface, which has become thin and easily depressed. Withdrawal of the lophophore back into the orifice is brought about by two special retractor muscles.

During feeding, the **cilia** on the outstretched tentacles drive water down into and out of the sides of the lophophore funnel, and food particles suspended in the incoming water stream are driven by cilia downward into the mouth (Fig. 28.1).

The nervous system consists of a **ganglion** and **nerve ring** around the anterior end of the gut. Fibers extend from the nerve ring to the tentacles and other parts of the body. No special systems for internal transport, gas exchange, or excretion are present.

Organization of Colonies

In most bryozoans the individuals are so attached to each other that the resulting colonies form erect plantlike growths or encrusting sheets over rocks and shells (Fig. 28.1). To visualize an encrusting colony, imagine a large number of rectangular boxlike individuals lying on their backs on a rock. The dorsal surface is attached to the substratum, the lateral, anterior, and posterior surfaces are attached to surrounding individuals, and the ventral surface is exposed (Fig. 28.2). The orifice of such encrusting species has shifted over to the exposed frontal surface, since the anterior end is connected to adjacent individuals. The colony follows the contours of the rock, shell, or larger algae on which it is growing and may reach many centimeters in width.

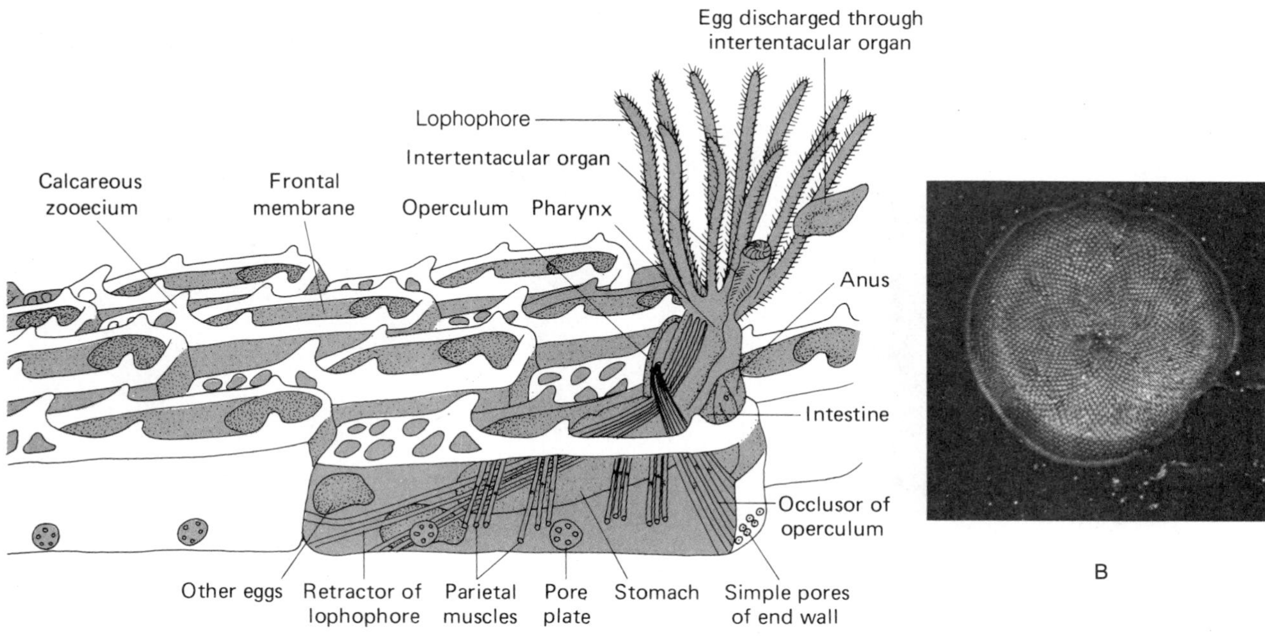

FIGURE 28.2 *A,* Lateral view of a portion of a colony of an encrusting bryozoan, *Electra. B,* Overview of an encrusting colony of *Membranipora* growing on kelp. (*A* modified from Marcus; *B* courtesy of Charles Seaborn.)

Erect colonies may be foliaceous, dendritic, or straplike, but the ventral surface is still the one that is exposed. Individuals are attached together by the dorsal or lateral and posterior surfaces. The anterior end may be free. *Bugula,* one of the most widely distributed bryozoans and a common marine fouling organism, illustrates the erect growth form (Fig. 28.3).

Members of a colony are connected to each other more intimately than by the fusion of their exoskeletons. Pores in the walls permit diffusion of substances from one individual to another, and food material can be distributed to nonfeeding members of the colony (Fig. 28.2).

Bryozoans are **polymorphic,** but the most common members of the colony are the feeding individuals already described. The colonies of some species also include highly modified defensive individuals that protect the colony against other settling organisms (Fig. 28.3C). Some species have vegetative stoloniferous individuals that creep over and anchor to the substratum. Reproductive individuals with special brooding structures are still another type found in many bryozoan colonies (Fig. 28.3C).

Reproduction

Most bryozoans are hermaphroditic, and the gonads shed their gametes into the coelom (Fig. 28.1). Sperm exit through a pore at the end of two or more tentacles, and the eggs exit through a special pore between the tentacles, which is sometimes elevated as an intertentacular organ (Fig. 28.2). Sperm swept in with the feeding current usually fertilize the eggs as they are released.

Most bryozoans brood their few eggs during the early stages of development, commonly in special brooding chambers. For example, *Bugula* and other related forms brood their eggs within a large hoodlike chamber that develops at the anterior end of reproductive individuals (Fig. 28.3).

The developing eggs are released from the brood chambers as larvae that are a little like the trochophore of annelids and mollusks. Following a planktonic existence of varying length, the larva settles to the bottom to become the parent member of a new colony. The

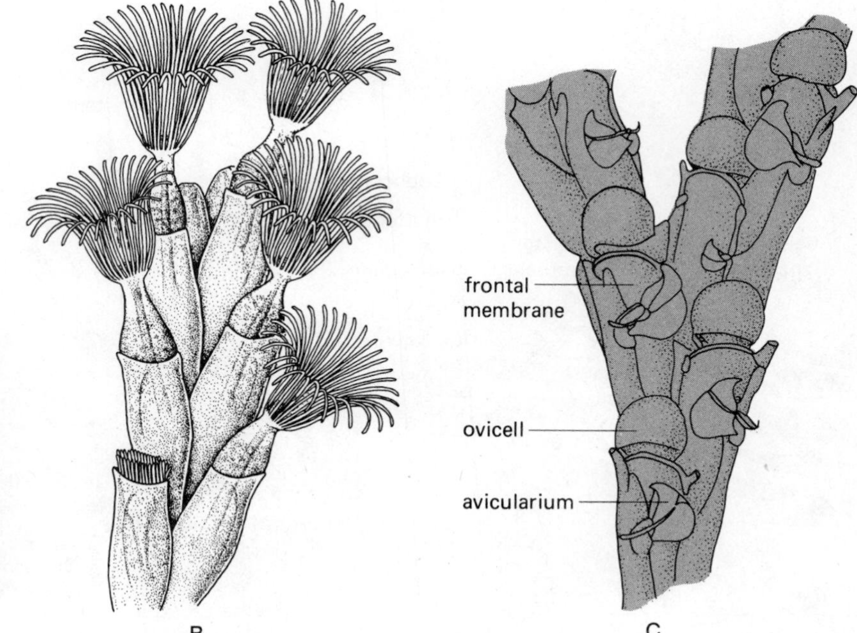

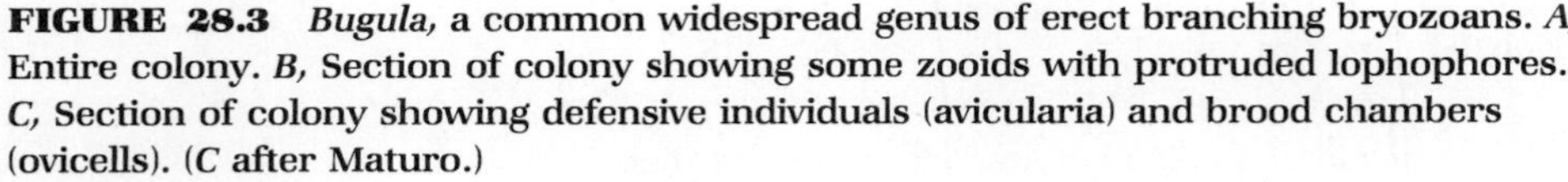

FIGURE 28.3 *Bugula,* a common widespread genus of erect branching bryozoans. *A,* Entire colony. *B,* Section of colony showing some zooids with protruded lophophores. *C,* Section of colony showing defensive individuals (avicularia) and brood chambers (ovicells). (*C* after Maturo.)

body wall evaginates and separates as a new individual, the particular way in which subsequent budding occurs accounting for the growth form of the colony.

References and Selected Readings

Detailed accounts of bryozoans, phoronids, and brachiopods can be found in many of the references listed at the end of Chapter 19. The works listed below deal exclusively with the biology of two of these lophophorate phyla.

Rudwick, M. J. S.: *Living and Fossil Brachiopods.* London, Hutchinson University Library, 1970.

Ryland, J. S.: *Bryozoans.* London, Hutchinson University Library, 1970.

BOX 28.1

Related Phyla

Bryozoans are not the only animals having a lophophore. Two other phyla, the **Phoronida** and the **Brachiopoda,** are lophophorates. The phoronids (*A*) are marine tube-dwelling wormlike animals. Only 15 species are known. The brachiopods (*B* and *C*), although represented today by about 335 species, have a rich fossil record that dates back to the Cambrian. Brachiopods superficially resemble bivalve mollusks in having the body enclosed within two calcareous valves. However, in brachiopods the valves are dorsoventrally oriented, and the ventral valve is larger than the dorsal one. A fleshy stalk, which emerges from a hole in the back of the ventral shell, attaches these animals upside down to the bottom (brachiopod—L. *brachium,* arm + Gr. *pous,* foot). Figure *B* is an external view showing the lophophore between the two valves. Arrows indicate the direction of the water current. Figure *C* is a diagrammatic longitudinal section through a brachiopod. (*B* from Rudwick, M. J. S.: *Living and Fossil Brachiopods.* London, Hutchinson University Library; *C* modified from Williams and Rowell.)

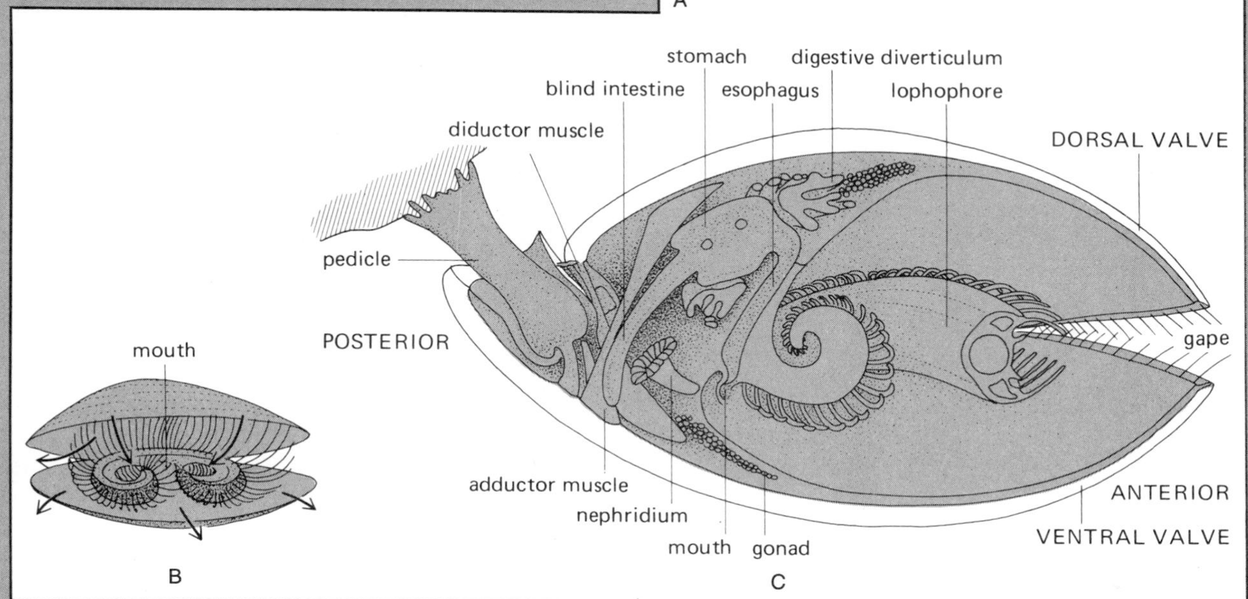

29 Echinoderms

ORIENTATION

The phylum Echinodermata, which includes the familar starfish, brittlestars, sea urchins, and sand dollars, consists of radially symmetrical animals with a calcareous skeleton. Projections of the body wall, called tube feet, part of a unique water vascular system, are used in locomotion, feeding, and gas exchange. This chapter will begin by examining the common features of echinoderms and then will survey each of the five major echinoderm groups. We will see how the basic echinoderm design has become adapted for a variety of life styles. Finally, we will explore the possible evolutionary origin of echinoderms, which might account for their secondary radial symmetry.

CHARACTERISTICS OF ECHINODERMS

1. The Echinodermata is a phylum of marine animals including such familiar forms as starfish, brittle stars, sea urchins, sand dollars, and sea cucumbers.
2. The symmetry is usually radial, typically pentamerous, or five-parted, but is secondarily derived. The larva is bilaterally symmetrical.
3. The body wall contains a skeleton of calcareous ossicles, which are usually movable against one another but may be fused together (sea urchins and sand dollars) or reduced to microscopic size (sea cucumbers). External spines are commonly present and are a part of the skeleton.
4. Echinoderms possess a unique water vascular system of internal canals and external appendages called tube feet. The system functions in locomotion, gas exchange, feeding, and sensory reception.
5. Movement is by means of tube feet or spines or arm movement. Sea lilies are attached, and many fossil echinoderms also lived attached.
6. The mouth is in the center of one side of the radial body, and in most echinoderms this oral surface is directed toward the substratum.
7. Coelomic fluid is the principal means of internal transport, and exchange of gases and wastes between sea water and coelomic fluid takes place across various surface structures.
8. The nervous system is pentamerous and closely associated with the integument on the oral side of the animals.
9. The sexes are separate. Fertilization is external, and early development is usually planktonic and leads to a bilateral larva.

Echinoderms, hemichordates, and chordates constitute the **deuterostomes,** the second great evolutionary line of the Animal Kingdom. In contrast to the protostomes, the mouth arises as a new opening located opposite the blastopore, which forms the anus. Cleavage is radial, not spiral (Fig. 14.1), and the fate of the blastomeres is fixed much later in development than in protostomes. Primitively, the mesoderm and coelom arise as paired outpocketings of the embryonic gut; the coelom is an **enterocoel** (Fig. 14.8), in contrast to the schizocoelous mode of coelom formation in many protostomes (Box 25.1).

The 6000 species of the phylum Echinodermata are entirely marine and include the familiar starfish, brittle

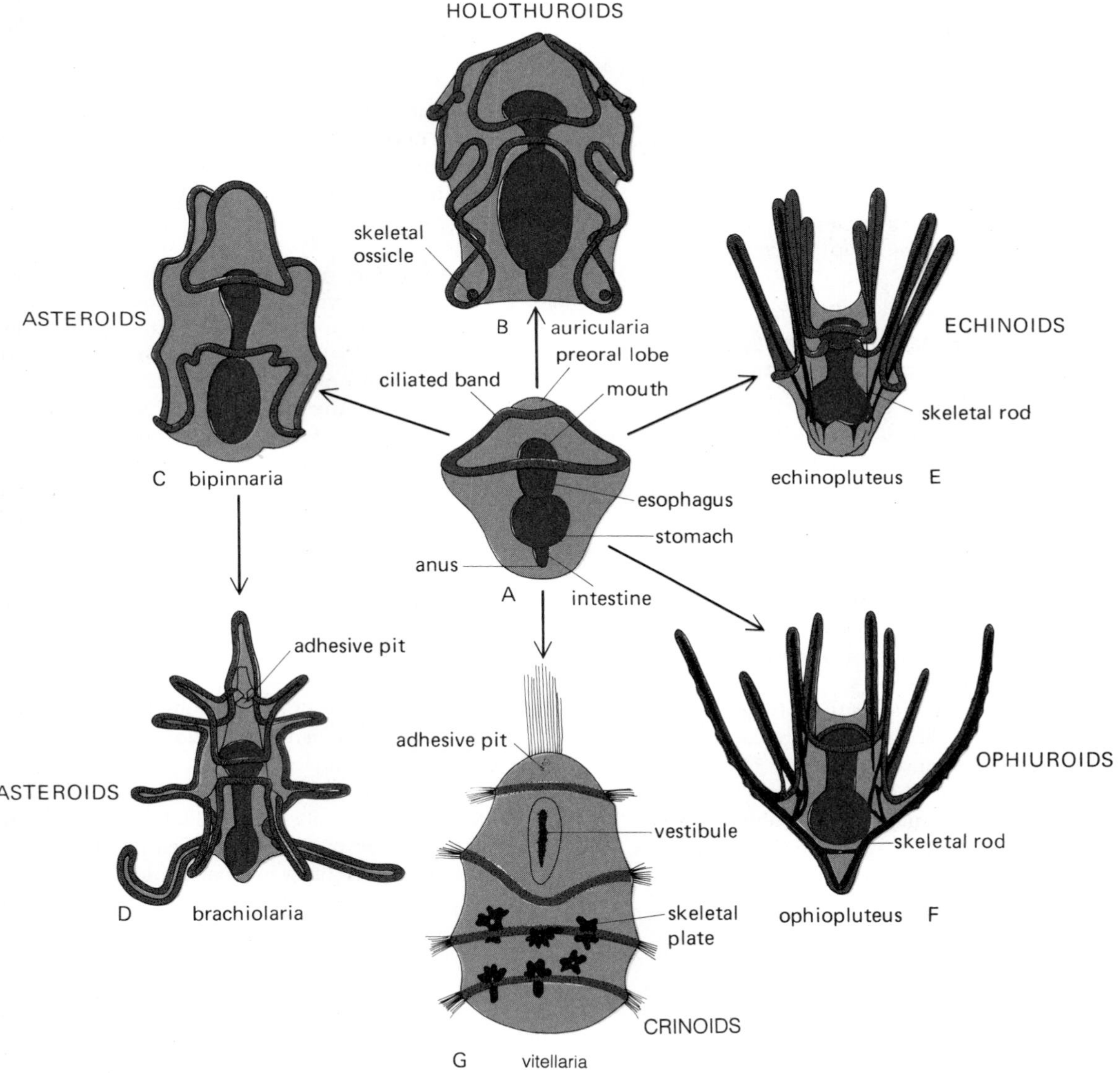

FIGURE 29.1 Comparison of different larval types of echinoderms. Arrows indicate possible evolutionary relationship to a hypothetical ancestral form. Note that all are bilateral; all bear one or more ciliated bands; and the ciliated bands of some run out onto larval arms. (From Ubaghs, G., in Moore, R. C. (Ed.): *Treatise on Invertebrate Paleontology.* Parts S, Vol. 1. Courtesy of the Geological Society of America and the University of Kansas, 1967.)

stars, sea urchins, sand dollars, sea cucumbers, and sea lilies. They are distinguished by possessing a **pentamerous** (five-part) **radial symmetry.** The body wall contains a skeleton of small calcareous pieces, or **ossicles,** which commonly include surface spines, hence the phylum name Echinodermata (Gr. *echinos,* spine + *derma,* skin). There is a unique **water vascular system** of canals and appendages that function in locomotion, feeding, or gas exchange. The larva of echinoderms is bilateral and swims and feeds by means of ciliated bands which wind over the body (Fig. 29.1). At the end of planktonic life the larval symmetry changes from bilateral to radial.

Echinoderms are almost entirely bottom dwellers, and hard substrata of rock, shell, and coral were the habitats of many extinct and contemporary forms. But within each class of echinoderms are species that invaded soft bottoms and became adapted for life in sand.

Class Stelleroidea: Asteroids

The class **Stelleroidea** (L. *stella,* star + Gr. *eidos,* form) contains those echinoderms in which the body is drawn out into arms. The most familiar stelleroids are members of the subclass **Asteroidea** (Gr. *aster,* ray + *eidos,* form), which includes the starfish. In this group the arms are not sharply set off from a central disc. There typically are five arms, but the sun stars have many (Fig. 29.2). The body surface may appear smooth or granular or may bear conspicuous spines. Most starfish are 12 to 24 cm. in diameter, and many are brightly colored in hues of red, orange, or blue.

The mouth is located in the center of the oral surface, which is directed downward. An **ambulacral groove** radiates from the mouth to the tip of each arm and contains the tube feet. The aboral surface may appear smooth or granular or may bear prominent spines. In most starfish a conspicuous button-like body, the **madreporite,** is located to one side of the central disc.

Body Wall. The surface of asteroids is covered by a ciliated epithelium. A thick layer of connective tissue beneath it secretes the skeleton of small calcareous ossicles. The ossicles are somewhat similar to vertebrate bone, for they are perforated by irregular canals filled with cells. A muscle layer below the dermis enables the arms to bend. Ciliated peritoneum forms the innermost layer of the body wall and lines the very large coelom (Fig. 29.4). Circulation of the coelomic fluid is the principal means of internal transport.

All asteroids bear calcareous spines that may be projections of the deeper dermal skeleton or special ossicles resting on top of the deeper skeleton (Figs. 29.2 and 29.3). The spines vary greatly in size and prominence but are always covered by the outer epithelium.

Located between the spines are small fingerlike projections of the body wall called **papulae,** which function in gas exchange and excretion (Figs. 29.3 and 29.4). The thin wall of each papula is composed of an inner peritoneum and outer epidermis, and coelomic fluid circulates within the interior.

Other structures frequently found on the body wall of some starfish are minute jawlike appendages called **pedicellariae** (L. *pediculus,* small foot). A pedicellaria contains two ossicles that form the jaws and muscles for opening and closing them. The pedicellariae are believed to function in the killing of small organisms that might settle on the body surfaces. Such organisms, as well as sediment, are swept clear by the ciliated surface epithelium.

The Water Vascular System. The water vascular system, unique to echinoderms, is composed of tubular outpocketings of the body wall—the **tube feet,** or **podia**—and an internal system of canals derived from the coelom. The system opens to the exterior through the button-like aboral **madreporite,** which is perforated by tiny canals. The canals from the madreporite converge into a vertical **stone canal** that extends orally to a **ring canal** embedded in the ossicles around the mouth (Fig. 29.5). From the ring canal a **radial canal** extends into each arm, passing between the ossicles at the top of the ambulacral groove. At frequent intervals, **lateral canals** leave the radial canals. Each lateral canal terminates in an aborally directed **ampulla** and an orally directed tube foot, which projects into the ambulacral groove (Fig. 29.4).

Muscular contraction of the ampulla forces fluid into the podium and at the same time closes a valve in the lateral canal, preventing backflow. Hydraulic pressure extends the podium, and the sucker at its tip is brought into contact with the substratum. Following adhesion, longitudinal muscles of the podium contract, shortening the tube foot, forcing water back into the ampulla, and pulling the body forward. When a podium is extended, it is also swung forward. Thus each podium performs a little step, and the combined activity of all the podia enables a starfish to grip objects tenaciously and to crawl about. The podia do not move synchronously, but they are coordinated to the extent that they all step in the same direction. One arm acts as the leading arm and exerts a temporary dominance over the other arms. The canal system appears to function in pressure regulation, although blocking the madreporite does not cause any immediate impediment to locomotion.

Podia with suckers represent an adaptation for life on hard substrata such as rock, shell, and coral. Many

FIGURE 29.2 *A,* A species of sun star, *Heliaster,* from the Gulf of California. *B,* Sun stars feeding on intertidal mussels attached to wave-lashed rocks in New Zealand. Large straplike algae (kelps) are also attached to rocks in this part of the intertidal zone. *C,* Two starfish from Fiji—*Fromia monilis* (red and white) and *Fromia indica* (brown and white). *D,* A specimen of the coral-eating *Acanthaster* from Fiji. The tips of the large spines are covered with a poisonous epithelium that can cause a puncture to be very painful. (All by Betty M. Barnes.)

starfish are adapted for living on sandy bottoms and can even burrow to some degree. Soft-bottom starfish have doubled ampullae that provide greater pressure for thrusting the suckerless pointed tube feet into the sand.

Nervous System, Gas Exchange, and Excretion. The nervous system can be regarded as primitive in that it is intimately associated with the epidermal layer. A **nerve ring** encircles the mouth, and a **radial nerve** extends into each arm (Fig. 29.4). Fibers in the nerve ring and the radial nerve make connection with neurons of a general epidermal nerve plexus.

An intact nervous system is necessary for the coordination of the tube feet. If a radial nerve is cut, the tube feet distal to the cut will continue to move but may not step in synchrony with those in other arms.

An **eye spot** at the tip of each arm, composed of a cluster of photoreceptor and pigment cells, constitutes the only sense organ, but individual receptor cells are present in the general body epidermis and are especially concentrated in the epidermis of the podia and the margins of the ambulacral groove.

Gas exchange and excretion in starfish occur through the surface of the papulae and podia, and internal transport is provided by the coelomic fluid. The papulae of soft-bottom starfish are protected from the surrounding sand and sediment by special table-like spines that create a protective cover under which the papulae are

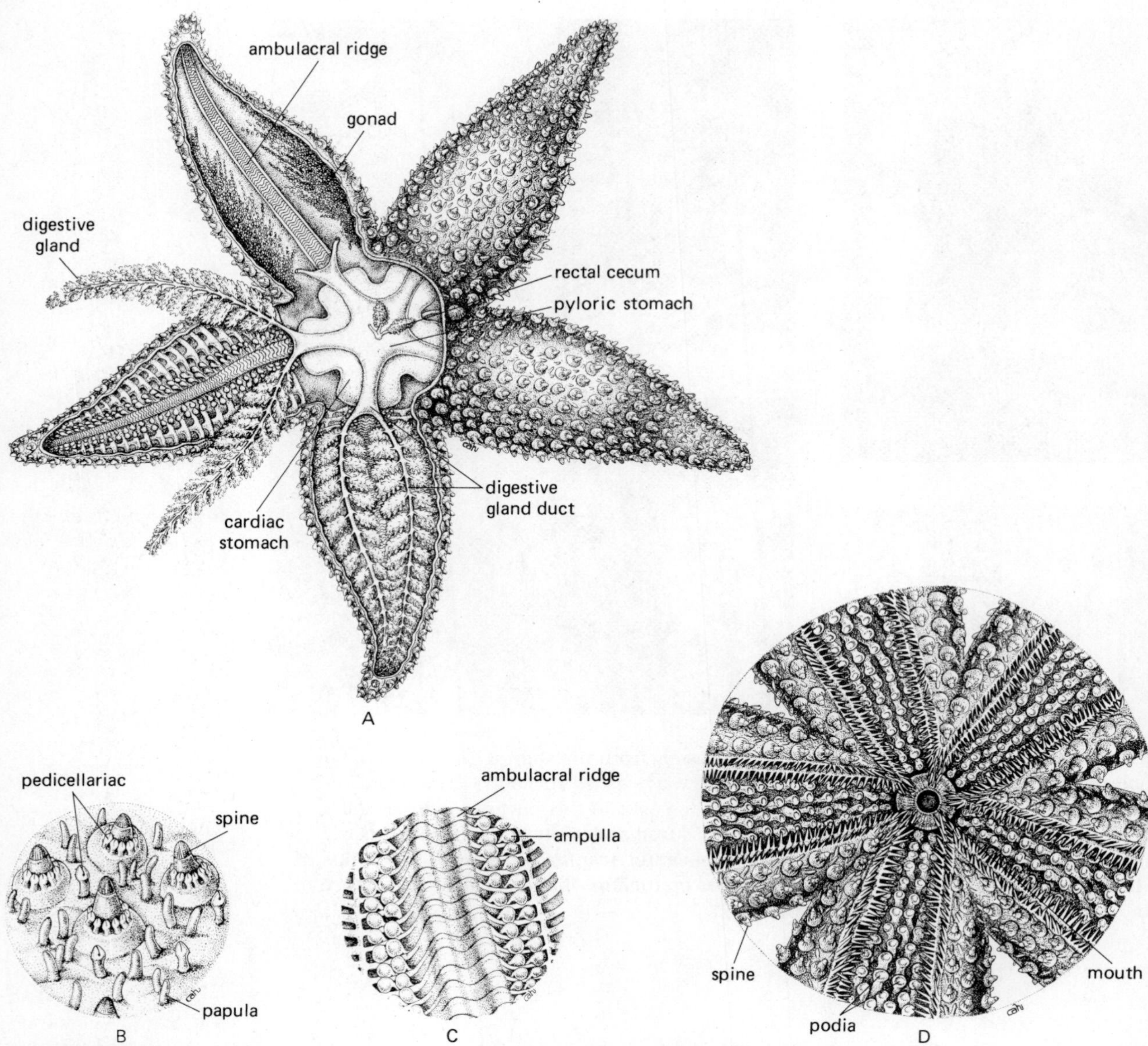

FIGURE 29.3 *A, Asterias* viewed from above with the arms in various stages of dissection. *B,* A small area of the dorsal surface enlarged. *C,* Interior view of a part of one of the ambulacral ridges showing ampullae to either side. *D,* Oral surface of disc.

lodged and a ventilating current flows (Fig. 29.6). It is the table-like spines that account for the smooth appearance of some starfish.

Nutrition. The mouth opens into a large thick-walled **cardiac stomach** that fills most of the central disc. The cardiac stomach opens in turn into a smaller aboral **pyloric stomach** (Fig. 29.3). A pair of **digestive glands** located in each arm discharges into the pyloric stomach. A short intestine extends from the top of the pyloric stomach to the inconspicuous anus at the center of the aboral surface. Associated with the intestine are rectal ceca, outpocketings of unknown function (Fig. 29.3).

Most starfish are carnivores and scavengers. Crustaceans, mollusks, and other echinoderms are common prey. Some species, especially those with short arms,

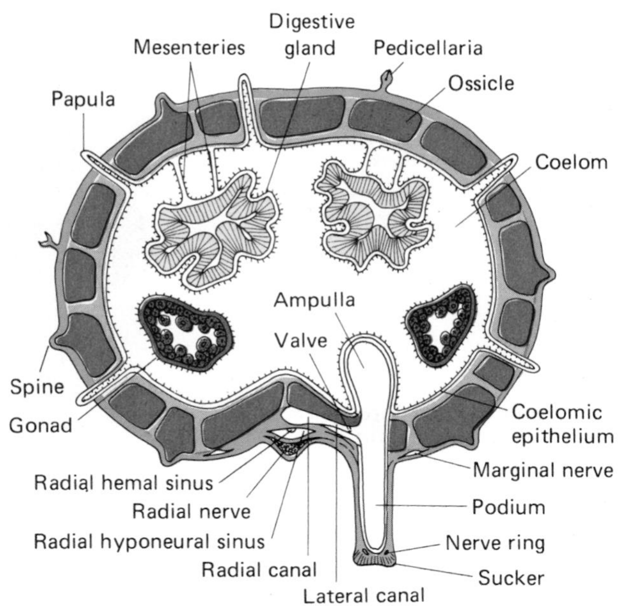

FIGURE 29.4 Diagrammatic cross section through the arm of a starfish. (Based on Cuénot.)

swallow their prey entire; others evert the cardiac stomach over prey. *Asterias* and other species of the same family feed largely on bivalve mollusks, penetrating their prey in a remarkable way. The starfish humps over the clam, the mouth directed over some part of the gape. The pull exerted by the arms produces a very slight opening between the valves, and through this opening the stomach of the starfish is slithered. These animals can pass the stomach through a gape of no more than 0.1 mm. It is this ability, rather than the ability to pull, that enables a starfish to prey upon clams. Many bivalves cannot close the valves tightly enough to prevent entrance of the starfish's stomach. Starfish can be a serious pest of commercial oyster beds. They are removed with a special mop to which the starfish attach by their pedicellariae when the mop is dragged across the oyster bed.

Enzymes produced by the digestive glands are passed to the pyloric and cardiac stomachs. In those species that evert the stomach, the enzymes flow out onto the everted surface and initiate digestion outside the body. Digestion of the clam's adductor muscle causes the valves to gape widely. The stomach is later retracted, bringing with it the partially digested prey. The digestive glands appear to be the principal site of absorption.

Reproduction and Development. Like most echinoderms, starfish have separate sexes; there are usually two gonads to an arm, with a simple gonoduct from each leading to an inconspicuous gonopore at the base of the arm (Fig. 29.3). The eggs are shed freely into the sea water where fertilization takes place. Development occurs in the plankton.

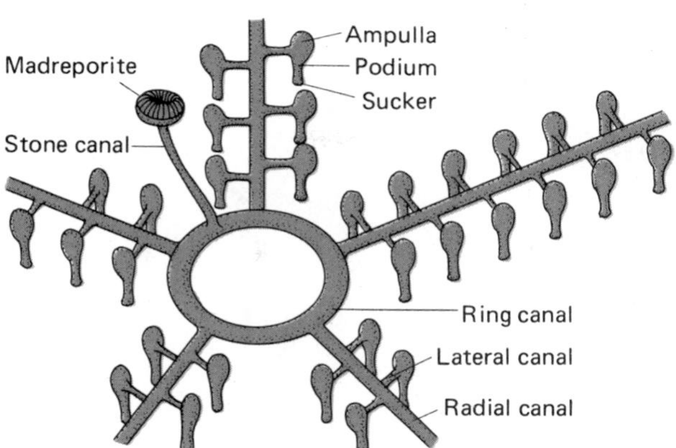

FIGURE 29.5 Diagram of the asteroid water vascular system.

The appearance of ciliated bands on the body surface indicates that development has reached a larval stage, called a **bipinnaria** (Figs. 29.1 and 29.7). The bipinnaria larvae, like larvae of all other echinoderms, are distinctly bilateral and are believed to reflect the symmetry of ancestral echinoderms. The bipinnaria gradually develops body projections, called **larval arms,** over which the ciliated bands extend. The larval arms, which are not the same as the adult arms, are probably an adaptation for increasing the ciliated locomotor and feeding surface. Toward the end of the bipinnaria stage anterior adhesive structures form (three short arms and a sucker), and the larva, now called a **brachiolaria** (Fig. 29.1), settles to the bottom and attaches. A radical and complex metamorphosis ensues. The larval arms and much of the gut degenerate, and the radially symmetrical adult body develops. After metamorphosis the little starfish detaches and crawls away.

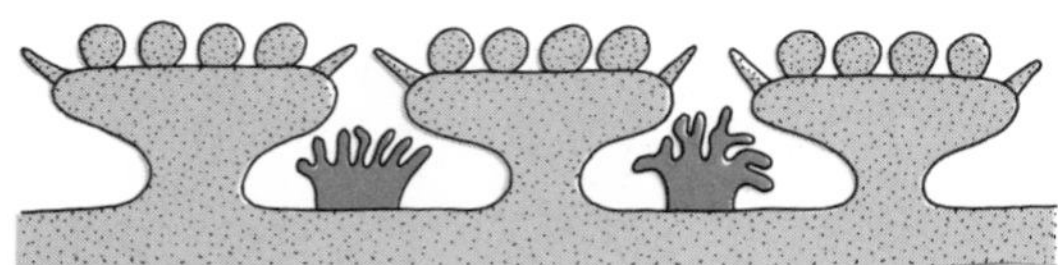

FIGURE 29.6 Diagrammatic section through the surface structures of a soft-bottom asteroid. The large table-like spines bear smaller spines above and protect the papulae (in black) below.

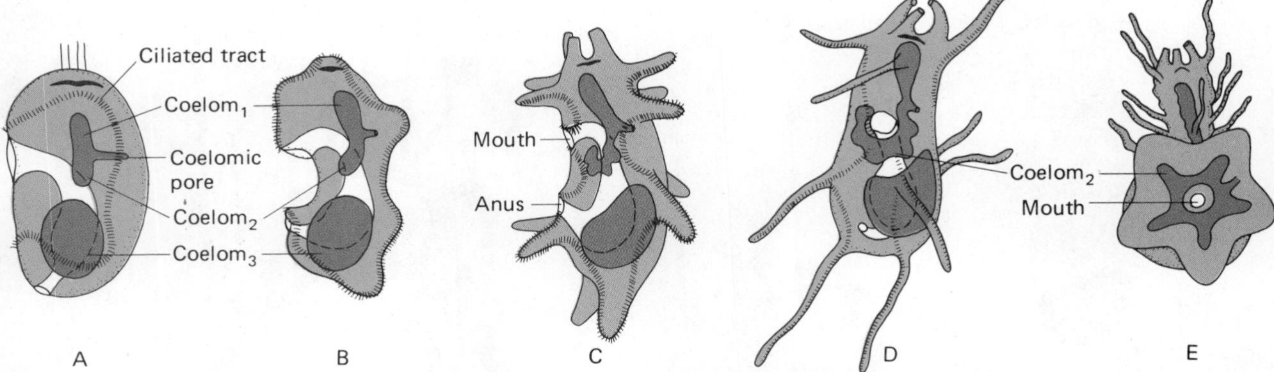

FIGURE 29.7 Diagrammatic lateral views of larval development and metamorphosis of a starfish, showing development of the coelom and water vascular system. *A* and *B*, Early bipinnaria larva. *C*, Brachiolaria larva. *D*, Attached metamorphosing larva. *E*, Young starfish developing from posterior part of old larva.

The regenerative ability of starfish is well known. Any fragment of the body that contains a portion of the central disc is capable of regeneration. But the process is slow and may take as long as a year. Only in some starfish is regeneration coupled with asexual reproduction, when the central disc is cleaved along certain predetermined lines and each part grows into a new starfish.

Class Stelleroidea: Ophiuroids

Closely related to the Asteroidea is the largest group of echinoderms, the subclass **Ophiuroidea** (Gr. *ophis*, serpent + *eidos*, form), which contains the basket stars and brittle stars, or serpent stars. The ophiuroid body,

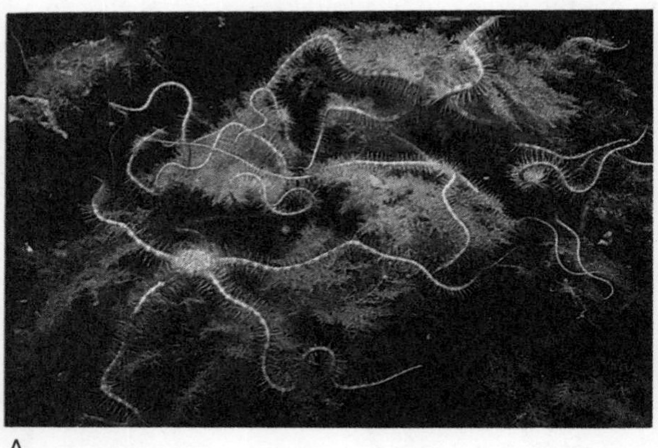
A

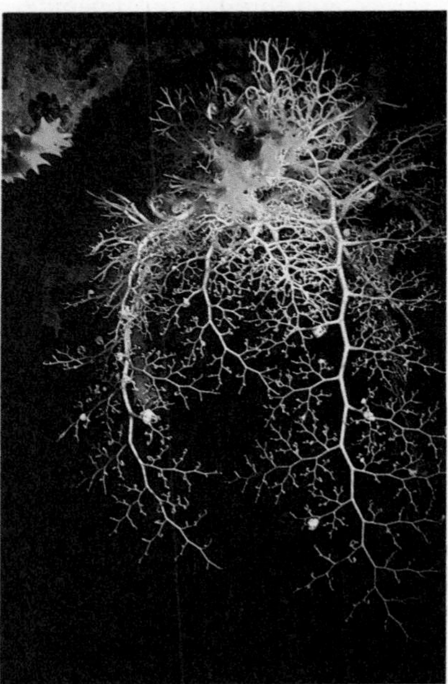
B

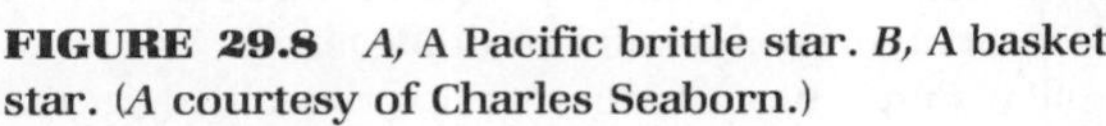
FIGURE 29.8 *A*, A Pacific brittle star. *B*, A basket star. (*A* courtesy of Charles Seaborn.)

like that of the asteroids, is composed of arms and a central disc, but the arms are long, slender and sharply set off from the disc (Fig. 29.8). The arms are highly mobile and easily broken, characteristics that led to the names serpent star and brittle star. In basket stars, the arms are branched (Fig. 29.8). Most brittle stars are small and have a central disc no larger than a penny; basket stars are larger.

The arms of brittle stars are filled by a string of large ossicles, called **vertebrae** because of their superficial similarity to the bones in the vertebrate backbone (Fig. 29.9). There is no ambulacral groove, but podia are located along the length of the oral side of the arm. The articulation of the arm ossicles and their musculature give the arms great mobility; brittle stars move by flexing and pushing with their arms and not by means of the

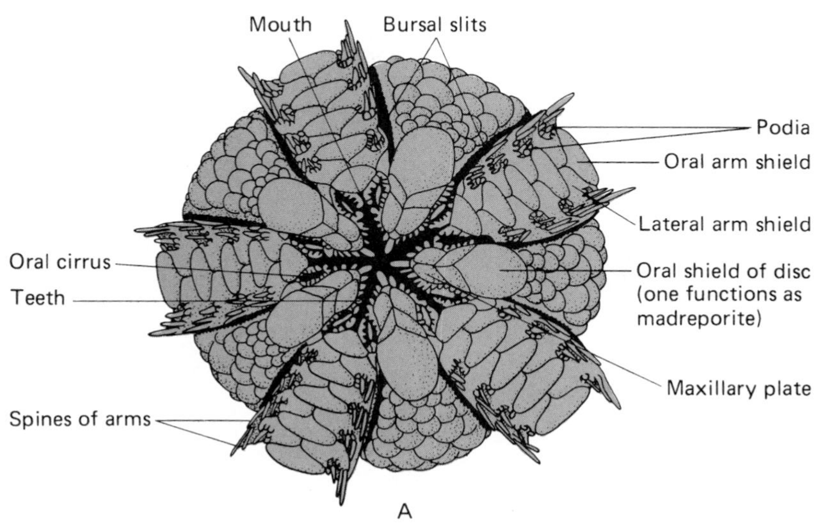

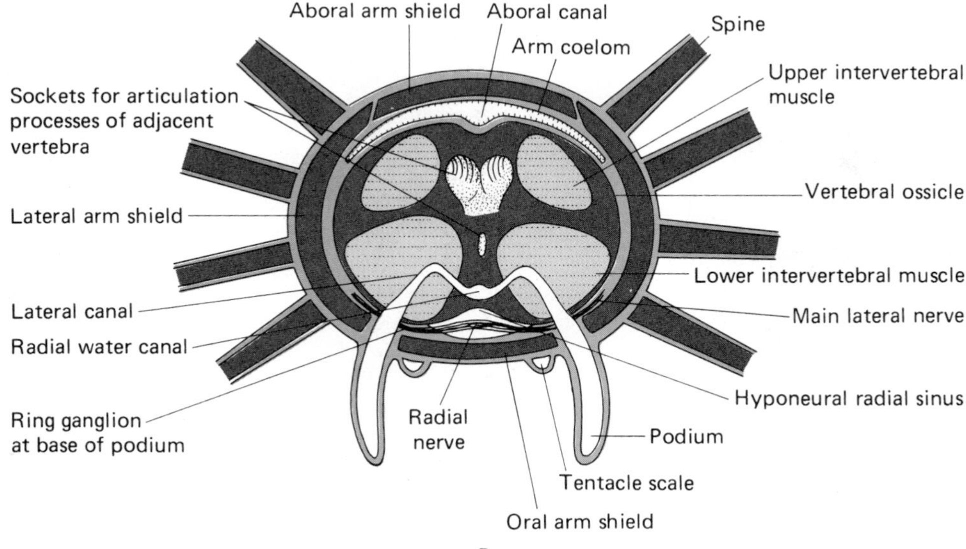

FIGURE 29.9 *A,* Oral view of disc of a brittle star. *B,* Cross section of a brittle star arm. (*A* after Streklov.)

podia. Spines along the sides of the arms increase traction. An arm can be broken at any point if seized by a predator. This ability to undergo selfamputation, or **autotomy,** is an escape adaptation also found in some arthropods, such as crabs.

Ophiuroids are especially common in rock and coral habitats, where they live beneath stones and in crevices and holes. But some even live in burrows on sandy bottoms. Basket stars, which can coil their arms, can climb, especially on branched corals and sea rods.

Brittle stars may be scavengers, deposit feeders, or suspension feeders, and many species utilize more than one mode of feeding. In scavenging the looping motion of the arms and the five jawlike ossicles that frame the mouth are used to ingest the bodies of dead animals. (Fig. 29.9*A*). The podia are used to pick up organic particles in deposit feeding and pass them down the oral side of the arms to the mouth. In suspension feeding the arms are lifted up into the water current, and suspended plankton and detritus are trapped on podia or on strands of mucus draped between spines. The collected material is pushed as balls by the podia along the length of the arms to the mouth.

The great success of ophiuroids is certainly related to the versatility of their feeding habits, their mobility, and their small size, which enable them to exploit habitats unavailable to other echinoderms.

The podia of brittle stars are suckerless and lack ampullae. This may be correlated with their mode of locomotion and the great reduction in the arm coelom. Fluid pressure is derived from contractions of the lateral canals. An ossicle on the oral surface of the disc serves as the madreporite (Fig. 29.9).

Ophiuroids do not have papulae; their gas exchange organs are pouches, called **bursae,** that are infoldings of the body wall of the oral disc to either side of the

A

B

C

D

FIGURE 29.10 Echinoids. *A,* Aboral view of *Tripneustes esculentus. B,* A species of the sea urchin *Eucidaris* with large heavy spines. *C,* Slate pencil urchin. *D,* Heart urchin. (*A* by Betty M. Barnes; *B–D* by Charles Seaborn.)

base of each arm (Fig. 29.9). The ventilating current is produced by cilia or pressure changes within the disc. Coelomic fluid is the medium for internal transport.

Most brittle stars have separate sexes, and the gonads are attached to the coelomic side of the bursal sacs. When mature, the gametes rupture into the bursae and exit out the bursal slits. Development is similar to that of asteroids, but the larva, called a **pluteus** (Fig. 29.1), has a different arrangement of larval arms, and metamorphosis occurs before settling. There is no attachment phase as in asteroids. Many species brood their eggs in the bursae, and a little brittle star crawls out of one of the slits.

Class Echinoidea

The class **Echinoidea** (Gr. *echinos,* sea urchin + *eidos,* form) contains the sea urchins, most of which are adapted for life on hard bottoms, and the sand dollars and heart urchins, which are adapted for burrowing in sand. In both groups the body is not drawn out into arms but is spherical or discoidal (Fig. 29.10). The skeletal ossicles are flattened plates fused together to form a rigid internal shell, or **test,** and the body surface is covered with movable **spines** mounted upon tubercles on the test.

Sea Urchins. The body of a sea urchin is spherical with the oral pole containing the mouth directed downward. Although there are no arms, five ambulacral areas bearing the tube feet radiate out and upward over the test to the aboral pole (Fig. 29.11). Thus the body is divided around the equator into alternating ambulacral and interambulacral meridians. The anus opens at the oral pole within a circular area filled with small plates, called the **periproct.** Around the periproct are five large **genital plates,** each containing a conspicuous gonopore (Fig. 29.11). One of the genital plates also functions as the madreporite.

The long movable spines are usually covered by the surface epithelium and articulate on a tubercle on the test surface. Two sets of muscles enable the spines to be rigidly erected or to be moved about. Stalked pedicellariae are located between the spines.

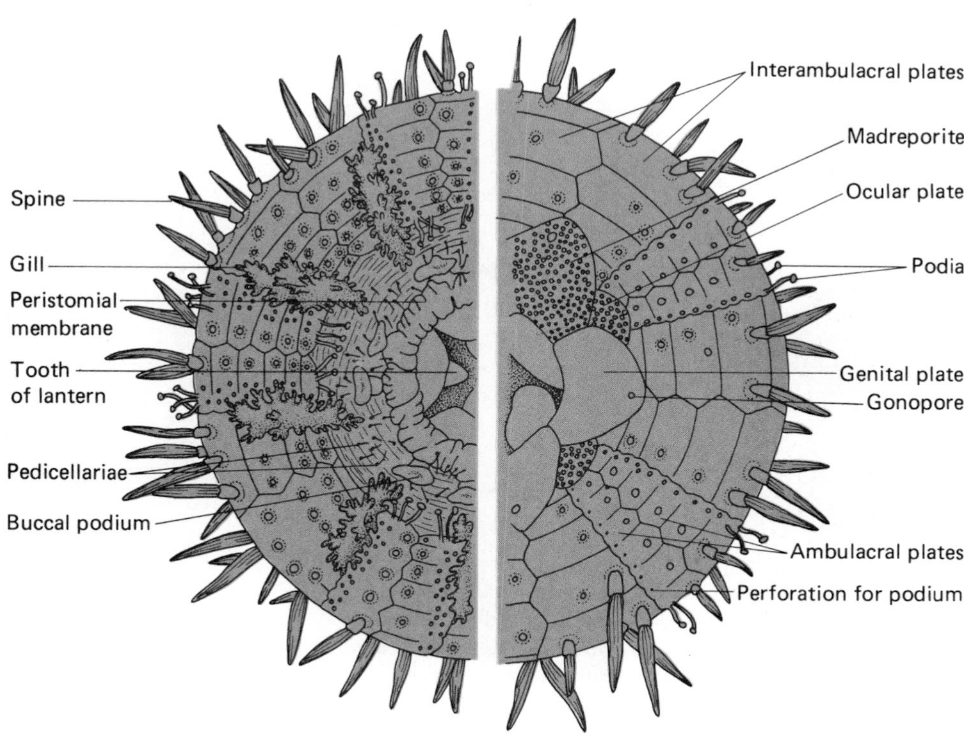

FIGURE 29.11 The common Atlantic coast sea urchin, *Arbacia punctulata. A,* Oral view; *B,* aboral view. (After Reid in Brown, F. A.: *Selected Invertebrate Types.* New York, John Wiley and Sons, 1950.)

Sea urchins use the podia, which can extend beyond the spines, to move about in the same manner as asteroids, and the water vascular system is very similar. However, each ampulla on the inside of the test and the podia on the outside are connected by double perforations through the ossicle (Fig. 29.12). Locomotion may also be aided by the pushing movement of the spines.

Sea urchins feed on algae and encrusting animals, which are scraped up or chewed with a complex movable apparatus (known as **Aristotle's lantern**) containing five projecting teeth (Fig. 29.11*B*). The gut is tubular and loops about inside the test.

Five pairs of gills, which are highly dissected outpocketings of the body wall located to either side of the ambulacral areas at the oral pole, provide for gas exchange (Fig. 29.11). Coelomic fluid is pumped into and out of the gills. As in all echinoderms, the podia, being thin extensions of the body wall, contribute in some degree to gas exchange.

Sand Dollars and Heart Urchins. Sand dollars and heart urchins burrow in sand. These animals, when moving, always keep the same meridian forward and thus have a definitive anterior end. Shifts in the position of the oral center or anus, or both, have led to a degree of bilateral symmetry. In sand dollars, the oral-aboral axis is so depressed that the body is flattened (Figs. 29.10 and 29.13). The mouth is still in the center of the oral surface, but the anus has shifted out of the aboral center and is located eccentrically toward the posterior end. Heart urchins are somewhat egg-shaped and not only have the anus shifted out of the aboral center, but the entire oral center has shifted forward, making these animals even more strikingly bilateral.

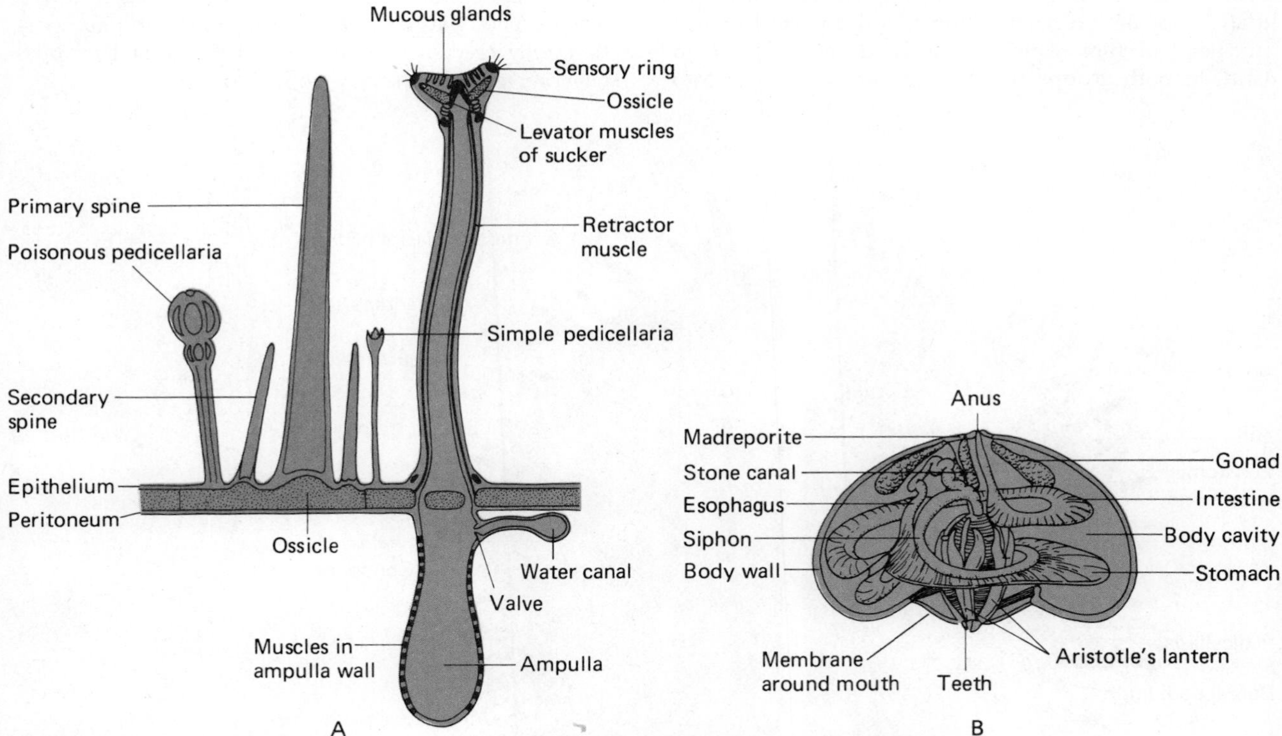

FIGURE 29.12 *A,* Diagrammatic section through the body wall of a sea urchin, showing one ambulacral and one interambulacral ossicle and associated structures. (After Nichols, in part.) *B,* A sea urchin, *Arbacia punctulata,* with one side of the body wall removed and only some of the structures shown. The teeth protrude from Aristotle's lantern, of which only the outer structures are indicated. The digestive tract circles twice around the body, once in each direction. A second tube, the siphon, bypasses the esophagus and stomach. Each of the five gonads opens above, near the anus.

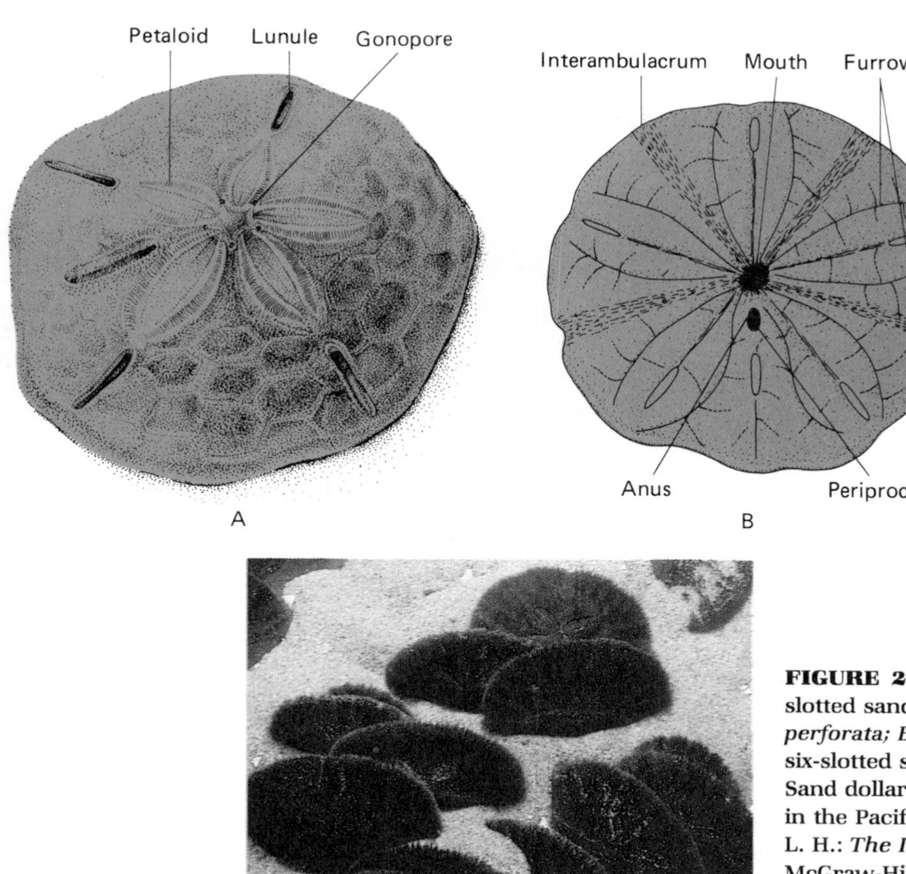

FIGURE 29.13 *A,* Aboral view of the five-slotted sand dollar, *Mellita quinquiesperforata; B,* Oral view of the West Indian six-slotted sand dollar, *M. sexiesperforata. C,* Sand dollars, *Dendraster excentricus,* found in the Pacific Northwest. (*B* after Hyman, L. H.: *The Invertebrates,* Vol. IV. New York, McGraw-Hill Book Co., 1955; *C* Courtesy of Charles Seaborn.)

These animals crawl slowly through the sand by the movement of the tiny spines that cover their surface. Sand dollars burrow into the surface layer of sand; heart urchins construct burrows well below the surface.

On the aboral surface, the podia are very wide and flattened as a modification for gas exchange. The dense covering of spines extends just beyond the podial gills and prevents the gills from being smothered by sand. Each of these five areas of specialized podia is called a **petaloid** since they look like the petal of a flower (Fig. 29.13).

Soft-bottom echinoids are **selective deposit feeders.** In sand dollars, fine particulate organic matter, but not sand grains, can drop down between the spines. The particles are then driven by cilia to the oral surface, where they pass into a branching system of food grooves that converge into five principal grooves corresponding to the ambulacral areas. The food grooves are lined by podia that push the collected food masses to the mouth. The conspicuous slots (**lunules**) in the test of some sand dollars are believed to facilitate the passage of particles from the aboral to the oral side of the body and perhaps to reduce lift created by the flow of water over the flat body surface (Fig. 29.13).

Four or five gonads are suspended radially from the inside of the test, and the short gonoducts open through the gonopores in the genital plates around the periproct. Fertilization is external, and the larva looks much like that of ophiuroids (Fig. 29.1). Metamorphosis occurs very rapidly on settling without an attachment phase.

Class Holothuroidea

The holothuroids,* or **sea cucumbers,** are similar to sea urchins in lacking arms, but in holothuroids the

*Aristotle referred to some marine animal, which was probably not a sea cucumber, as *holothourion* (sea polyp), but the name became associated with this group of animals.

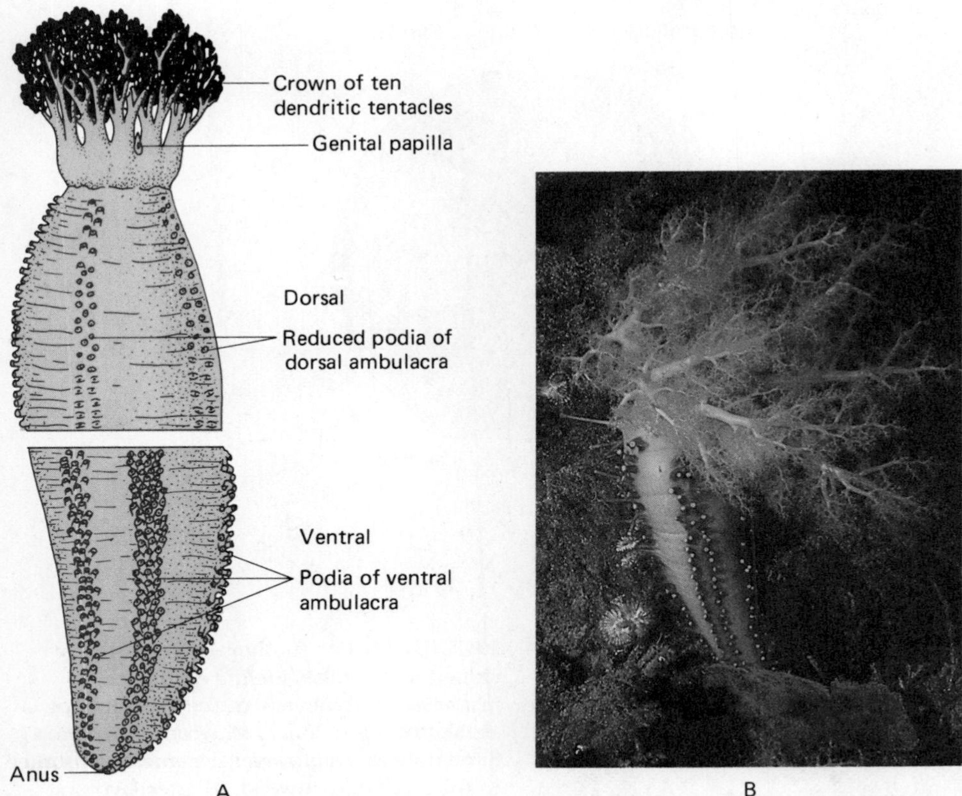

FIGURE 29.14 *A,* The North Atlantic sea cucumber, *Cucumaria frondosa. B,* A Pacific Northwest sea cucumber, *Cucumaria miniata,* with tentacles extended. (*B* courtesy of Charles Seaborn.)

oral-aboral axis is greatly lengthened, so that the animal has a wormlike or cucumber-like shape and lies on its side (Fig. 29.14). Unlike other echinoderms the skeleton is reduced to microscopic ossicles, and the body wall has a leathery texture. Most are from 6 to 30 cm. in length and rather drably colored.

Many sea cucumbers live on hard substrata, lodging themselves beneath and between stones or within coral crevices. These hard-bottom sea cucumbers move by means of tube feet; in some species, three ambulacra are kept against the substratum as a sole and the two upper ambulacra have reduced podia (Fig. 29.14).

Soft-bottom sea cucumbers live in sand or mud burrows with one or two openings to the surface. The very wormlike synaptid sea cucumbers burrow beneath the surface by peristaltic contractions. The podia have completely disappeared.

At the oral end of the body, a circle of **tentacles** representing modified podia surrounds the mouth (Figs. 29.14 and 29.15). The tentacles are outstretched and collect plankton or deposit material from the surrounding sea water or sea bottom. They are retracted and stuffed one at a time into the mouth. The deposit feeders may be selective or nonselective, and some nonselective deposit feeders leave conspicuous castings on the bottom surface.

The gut of sea cucumbers is tubular and terminates before the anus in a muscular **cloaca** that is involved in gas exchange. The gas exchange organs are unusual tubular branching structures called **respiratory trees** that arise as evaginations of the cloacal wall and extend up into the coelom (Fig. 29.15). The pumping action of the cloaca moves a ventilating current of sea water into and out of the respiratory trees.

In the tropics, a little commensal pearlfish (*Carapus*) uses the base of the respiratory tree as a home. To enter its home, the fish nudges the anus of the sea cucumber with its snout and then backs in, tail first, sometimes

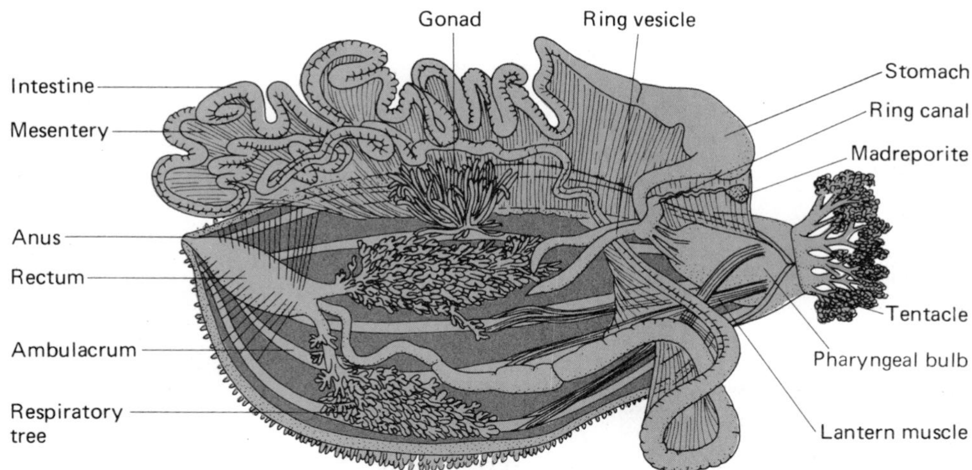

FIGURE 29.15 The sea cucumber, *Thyone briareus,* cut open along one side. The digestive tract has been moved to one side to show the respiratory trees, the retractor muscles of the anterior end, and the internal surface of the body wall with its five ambulacra. In holothurians, the madreporite lies in the body cavity, so that the water vascular system is not filled with sea water but with coelomic fluid.

twisting against the pressure of the closing anus. The fish has reduced scales and no pelvic fins, and the anus has shifted far forward.

Some sea cucumbers possess a cluster of tubular evaginations from the base of the respiratory trees. These tubules, which look like thin spaghetti, are called **tubules of Cuvier** and can be shot out of the anus. They elongate in the process and are adhesive. An intruder or predator that disturbs a sea cucumber enough to evoke the discharge of the tubules can become enmeshed in a death trap of adhesive threads.

There is only one gonad in sea cucumbers, and the gonopore opens between two of the tentacles (Fig. 31.14). The bilateral larva possesses ciliated locomotor bands but never develops the long larval arms characteristic of other echinoderm larvae (Fig. 29.1). Transformation to a young sea cucumber occurs before settling.

Class Crinoidea

The echinoderm classes we have examined thus far have the mouth directed downward and the oral surface placed against the substratum. However, in one living class of echinoderms, the **Crinoidea** (Gr. *krinen,* lily + *eidos,* form), the oral surface is directed upward. Moreover, many members of this class are attached to the substratum by a **stalk.** The attached crinoids are believed to be the most primitive of the living echinoderms and probably illustrate the ancestral mode of existence of the phylum.

Crinoids consist of two groups: the sessile **sea lilies,** in which the aboral surface is connected to the substratum by a stalk (Fig. 29.16), and the free-moving **feather stars,** in which a stalk is absent. Sea lilies are usually less than 70 cm. in length, and most occur in relatively deep water. The stalk, which is composed of ossicles and can bend, is attached to the aboral surface of the body proper, or **crown.** The pentamerous crown bears arms that fork repeatedly in many species. All along the length of the arms there are side branches called **pinnules.** A ciliated ambulacral groove with flanking suckerless podia runs the length of the arms and up onto the pinnules. The grooves from all of the arms converge to the mouth in the center of the oral surface.

The arms are composed mostly of ossicles, but the articulation and musculature permit considerable movement.

Feather stars are similar to sea lilies, from which they clearly have evolved. They are stalked and attached in the last stages of larval development. Then the crown breaks free and swims away as a tiny feather star. Although the animal is unattached, its oral surface is still directed upward (Fig. 29.16). They perch for long periods of time by means of an aboral ring of clawlike pro-

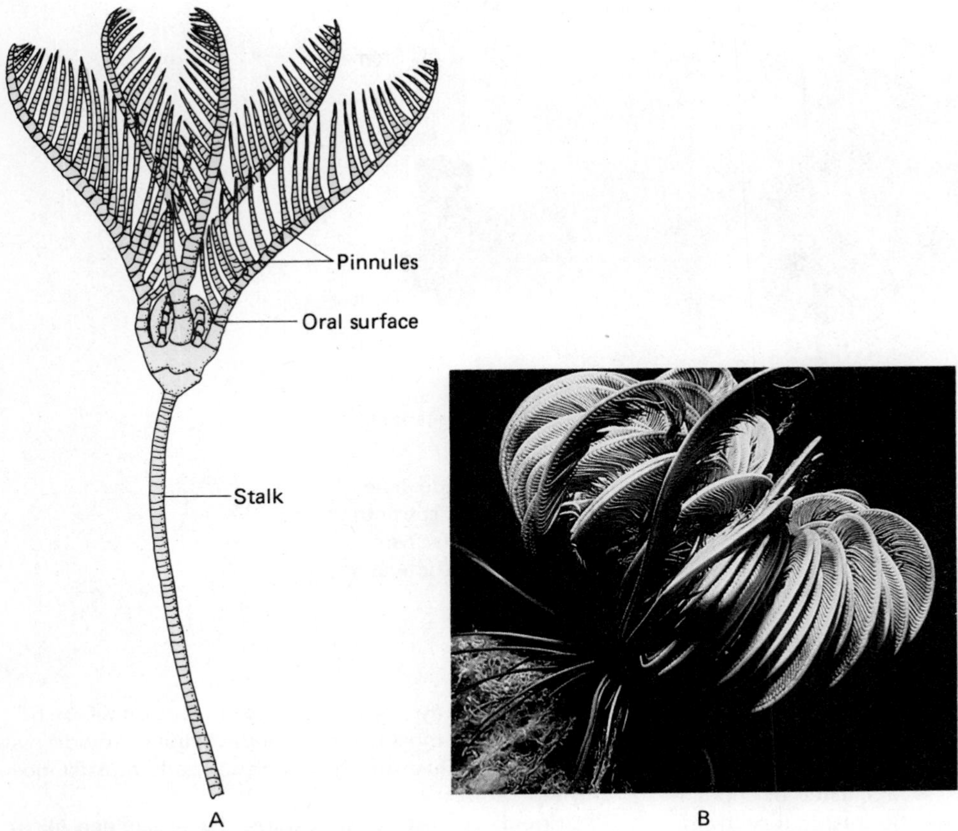

FIGURE 29.16 *A, Ptilocrinus pinnatus,* a stalked crinoid (or sea lily) with five arms. *B,* Philippine 30-armed comatulid (or feather star). (*A* after Clark from Hyman, L. H.: *The Invertebrates,* Vol. IV. New York, McGraw-Hill Book Co., 1955; *B* courtesy of Charles Seaborn.)

jections, called **cirri,** on rocks, on coral, or even on soft substrata. Feather stars swim intermittently by rapidly raising and lowering the arms.

Feather stars are often abundant on coral reefs, especially in the Indo-Pacific Oceans. None occur in very shallow water in the Western Atlantic.

Crinoids are suspension feeders, and this mode of feeding may have been the original function of the water vascular system. On contact, rapid whiplike movements of the podia toss suspended food particles into the ambulacral grooves, where they are entrapped in mucus and conveyed within the ciliated groove down to the mouth. The branching arms and pinnules greatly increase the surface area for collecting food.

The anus of crinoids is located on a short elevation to one side of the central oral surface. Feces can thus be more readily swept away without fouling the ambulacral grooves. Podia are the gas exchange organs.

Gametes develop from coelomic epithelium in the arms and pinnules, and spawning takes place by rupture of the body wall.

Fossil Echinoderms

The phylum made its appearance in the Cambrian. A number of classes, now extinct, flourished during the Paleozic. Most of these earlier echinoderms were attached with the oral surface directed upward, and most had spherical bodies with the ambulacral grooves extending down over the body surface, rather like spine-

BOX 29.1

Related Phyla

The phylum **Chaetognatha** contains a small group of common planktonic animals called **arrowworms.** They appear to be deuterostomes, but their relationship to echinoderms is obscure. Arrowworms have transparent, torpedo-shaped bodies, about 3 cm. long, bearing paired fins. They alternately float and swim with a rapid dartlike motion. Arrowworms feed on other planktonic animals, which they capture with anterior chitinous grasping spines. (*A* after Hertwig; *B* after Ritter-Zahony.)

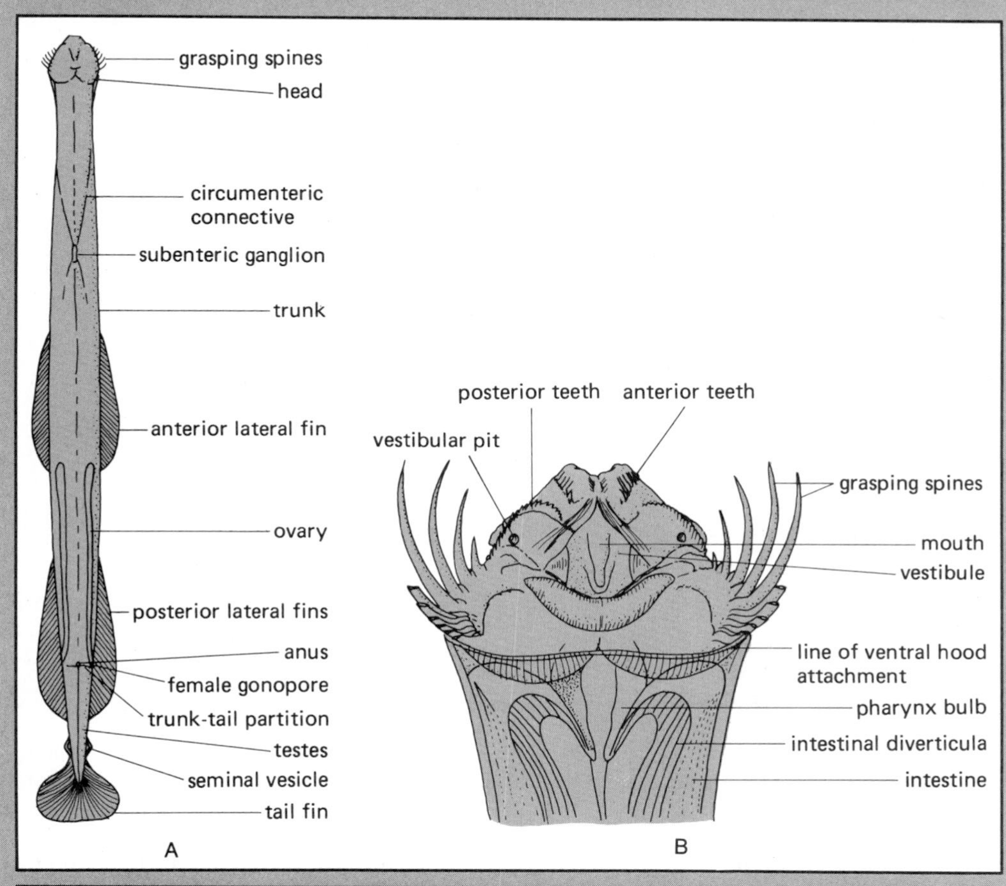

less sea urchins turned upside down. Significantly, the grooves were bordered with brachioles, the equivalent of crinoid pinnules (Fig. 29.17).

All of this seems to suggest that these early fossil echinoderms fed in the same manner as do modern crinoids and that the original function of the water vascular system was feeding. It could scarcely have functioned for locomotion, for the wrong side was directed upward, and most of these animals were attached.

From the fossil record and embryonic development we postulate that the ancestors of echinoderms were some group of motile bilateral coelomates. The ancestral stock took up an attached existence, which resulted in a shift from a bilateral to a more adaptive radial symmetry. Also correlated with an attached mode of existence was the evolution of the calcareous skeleton and of suspension feeding. Modern free-moving echinoderms appeared late in the fossil record, and there are no fossil groups bridging the gap between those species that have the oral surface directed upward and the modern forms that have the oral surface directed downward.

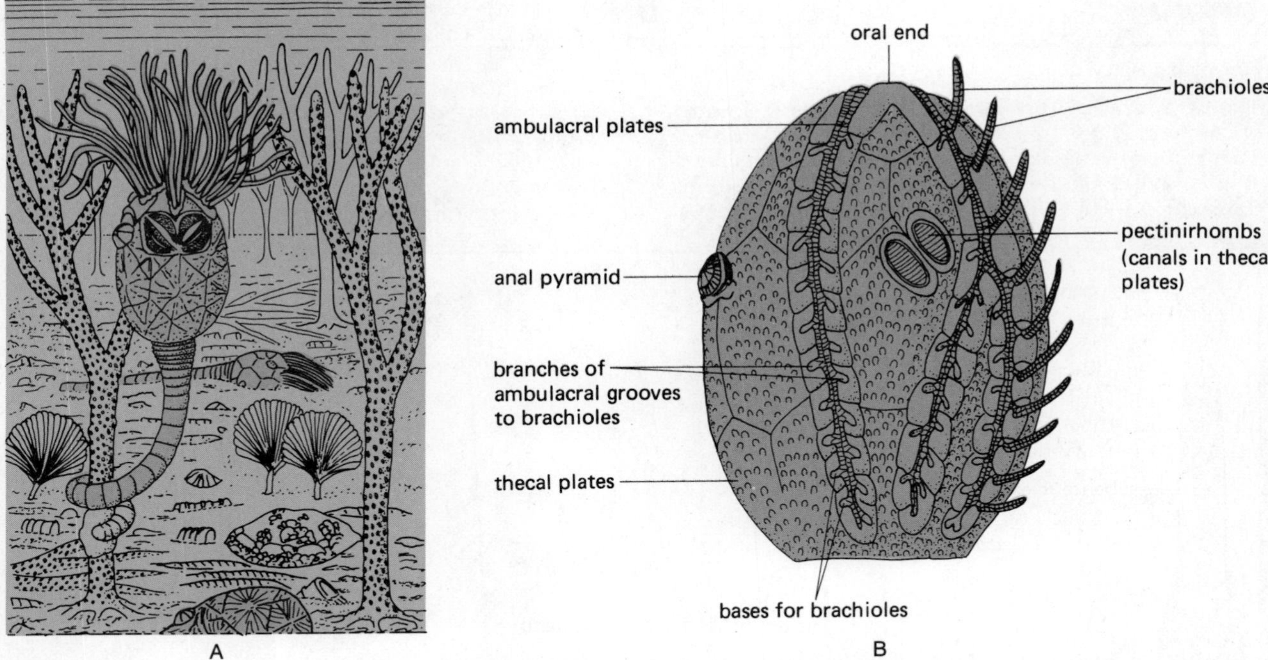

FIGURE 29.17 *A,* Reconstruction of an Ordovician cystoid, *Lepadocytis,* attached to an erect bryozoan. *B,* Lateral view of the cystoid *Callocystites.* (*A* from Kesling, R. V., 1967, in Moore, R. C. (Ed.): *Treatise on Invertebrate Paleontology.* Part S, Vol. 1. Courtesy of the Geological Society of America and the University of Kansas, pp. S85–S286. *B* after Hyman.)

Classification of the Phylum Echinodermata

Subphylum Echinozoa Radially symmetrical, commonly globoid echinoderms without arms or armlike appendages. Includes three extinct and two living classes.

Class Holothuroidea Sea cucumbers. Mouth and anus at opposite ends of cucumber-shaped body. Oral end bearing tentacles derived from podia. Ossicles microscopic.

Class Echinoidea Sea urchins and sand dollars. Body globose or greatly flattened dorsoventrally. Ossicles fused to form a rigid test that is covered with movable spines.

Subphylum Homalozoa† Extinct irregular Paleozoic echinoderms. In some the stalked body was bent over so that one side was directed toward the substratum.

Subphylum Crinozoa Radially symmetrical globoid echinoderms with arms or brachioles. Oral surface directed upward. Includes the extinct cystoids and blastoids, several other extinct classes, and one living class.

Class Crinoidea Sea lilies and feather stars. Living and fossil echinoderms dating from the early Paleozic. Body attached by a stalk (sea lilies) or free (feather stars). Ambulacra located on arms, which are typically branched.

Subphylum Asterozoa Radially symmetrical, unattached star-shaped echinoderms. Oral surface directed downward.

Class Stelleroidea

Subclass Asteroidea Living starfish. Arms grade into central disc. Tube feet are located within a groove on the underside of the arms.

Subclass Ophiuroidea Brittle stars. Arms sharply set off from a central disc and without a groove on the undersurface.

Summary

1. The phylum Echinodermata contains marine animals having a pentamerous radial symmetry, an internal skeleton of calcareous ossicles, which often includes surface spines, and a water vascular system composed of internal canals and surface appendages (podia) used in feeding or locomotion or both. The radial symmetry arises secondarily following the metamorphosis of a bilateral larva, bearing one or more ciliated bands used in locomotion and feeding.

2. The class Stelleroidea contains those echinoderms in which the body is composed of a central disc and arms. In the subclass Asteroidea, which includes the starfish and sun stars, the arms and disc are not sharply set off from each other, and the podia are located within an ambulacral groove on the oral surface of the arms. The body surface bears short spines, many finger-like evaginations of the body wall (papulae) which function in gas exchange and excretion, and in some species minute defensive grasping structures (pedicellariae).

3. The podia of asteroids serve in locomotion. They are extended by hydraulic pressure resulting from contraction of an adjacent ampulla located within the coelom. The many ampullae and podia of an arm are connected by a system of canals, which open to the outside through an aboral madreporite. Most asteroids are carnivores, and the prey, commonly mollusks and other echinoderms, is swallowed entire or digested externally (in part) by an everted stomach. A pair of digestive glands in each arm is the principal source of digestive enzymes and also the site of absorption. An oral nerve ring in the disc and a radial nerve on the oral side of each arm comprises a major part of the nervous system. There are usually two gonads in each arm, fertilization is external, and development is usually planktonic.

4. Brittle stars and basket stars, members of the subclass Ophiuroidea, are stelleroids in which the arms are sharply set off from the disc. They move rapidly by pushing with the arms, and their small size enables many species to live beneath stones, in crevices and holes, and even to burrow in sand. Brittle stars are scavengers, deposit feeders, and suspension feeders, and the podia are used to collect and transport food. The bursae, a pair of pocket-like invaginations of the disc wall at the base of each arm, are sites of gas exchange, excretion, and gamete release.

5. The class Echinoidea includes sea urchins, sand dollars, and heart urchins. The body is not drawn out into arms, and the ossicles are fused to form a test on which are mounted movable spines. The spherical sea urchins move by means of their podia and in some species by their long spines. Protection is provided by spines and pedicellariae. Sea urchins scrape rocks or chew algae with a toothed feeding apparatus. Sand dollars and heart urchins inhabit soft bottoms and display some degree of secondary bilateral symmetry. The many short spines are used in burrowing. They are selective deposit feeders.

6. The class Holothuroidea contains the sea cucumbers. There are no arms, and the oral-aboral axis is greatly elongated. The ossicles are microscopic, and the anterior podia are modified as tentacles around the mouth. Most sea cucumbers creep on the podia of three ambulacra (the sole). The tentacles are used in suspension feeding and selective and nonselective deposit feeding. Gas exchange is provided by water pumped into respiratory trees.

7. Members of the class Crinoidea are the only living echinoderms in which the oral surface is directed upward. Sea lilies are attached to the bottom by a stalk; feather stars are unattached but perch on the substratum with aboral cirri. All crinoids have five to many arms bearing lateral pinnules. Podia on the oral surface of the arms and pinnules collect plankton and other suspended particles, which are transported to the mouth by the cilia of the ambulacral groove. At settling the larva of crinoids attaches to the substratum and metamorphoses into the adult form. Young feather stars detach from a little stalk like that of sea lilies.

8. Most of the many species of Paleozoic echinoderms lived attached with the oral surface directed upward. They are believed to have fed in a manner similar to that of living crinoids.

References and Selected Readings

Detailed accounts of the echinoderms can be found in many of the references listed at the end of Chapter 19. The following works deal with echinoderms alone.

Binyon, J.: *Physiology of Echinoderms.* Oxford, Pergamon Press, 1972. An excellent general physiology of echinoderms.

Clark, A. M.: *Starfishes and Their Relations.* London, British Museum, 1962.

Nichols, D: *Echinoderms.* London, Hutchinson University Library, 1969. A good short biology of the phylum.

Sloan, N.: Coping with stardom: the lives of starfish. *Waters, Journal of the Vancouver Aquarium,* 2(4):1–32, 1977. A brief, beautifully illustrated natural history of starfish that comprises the entire issue of the journal.

30 Proto-chordates

ORIENTATION

The phylum Chordata includes such diverse animals as sea squirts, fish, frogs, birds, and humans. This chapter explains the justification for their being placed within the same phylum. It will then discuss the two groups of invertebrate chordates, the urochordates and cephalochordates. Also discussed are hemichordates, a phylum of animals that shows some similarities to both the chordates and the echinoderms.

CHARACTERISTICS OF THE CHORDATES

1 The phylum Chordata is a large and diverse assemblage of marine, freshwater, and terrestrial animals, which includes sea squirts, fish, amphibians, reptiles, birds, and mammals.

2 All possess a dorsal rodlike notochord, a dorsal hollow nerve cord, and pharyngeal clefts. In many chordates certain of these characteristics are found only in developmental stages.

Subphylum Urochordata

3 Urochordates, containing the marine, sessile sea squirts and a few pelagic forms, are the largest group of invertebrate chordates.

4 The body is covered by an external envelope, called a tunic, which contains cellulose.

5 The large perforated pharynx is adapted for filter feeding.

6 They are hermaphroditic and have a tadpole-like larva, which possesses a notochord and dorsal nerve cord.

Subphylum Cephalochordata

7 Cephalochordates are small, fishlike, marine chordates that burrow in sandy bottoms.

8 They are metameric and possess all of the chordate characteristics in the adult form.

9 The pharyngeal clefts function in filter feeding.

Subphylum Vertebrata

10 Vertebrates are metameric chordates of large body size that possess a backbone of vertebrae around, or replacing, the notochord.

CHARACTERISTICS OF THE HEMICHORDATES

1 Hemichordates are marine animals, most of which (acorn worms) have wormlike bodies and burrow or live beneath stones or algae.

2 The body is divided into an anterior proboscis, a short middle collar region, and a long posterior trunk.

3 The anterior part of the trunk bears lateral gill pores that connect internally with pharyngeal clefts.

4 Most hemichordates are deposit feeders.

The Chordata, the largest of the deuterostome phyla, includes animals with three distinguishing characteristics (see Fig. 30.5): (1) a dorsal hollow **nerve cord,** which arises as an infolding of the surface ectoderm; (2) a dorsal longitudinal skeletal rod, the **notochord,** located beneath the nerve cord; and (3) paired lateral openings, commonly referred to as **gill clefts** or **slits,** through the pharyngeal wall of the gut. Although these three structures are found in the early developmental stages of all chordates and were probably characteristic of the ancestral chordates, not all persist in the adults. The notochord disappears during development in most vertebrates, and only the hollow nerve cord and the pharyngeal clefts, or their derivatives, remain in the adult.

Most of the chordates are vertebrates, but there are two interesting small subphyla of invertebrate chordates, i.e., chordates lacking a vertebral column or backbone.

Subphylum Urochordata

The members of the subphylum **Urochordata,** called **tunicates** or **ascidians,** are the larger group of invertebrate chordates. Most tunicates are sessile, attached to rocks, shells, pilings, and ship bottoms.

The planktonic larva of urochordates (Gr. *oura,* tail + L. *chorda,* string) is about 0.7 mm. in length and looks like a tiny fish or tadpole. The finned tail is the locomotor organ, and it contains longitudinal muscle fibers, the nerve cord, and the notochord (Fig. 30.1). The notochord prevents a shortening, or telescoping, of the tail when the muscle fibers contract and converts muscle contraction into lateral undulations of the tail. The anterior mouth is connected to a large **pharynx** that is perforated by paired lateral clefts. The clefts lead into another chamber, the **atrium.**

At the end of its planktonic existence, the larva settles to the bottom and becomes attached by anterior adhesive papillae. A radical metamorphosis ensues, and the larva develops into the very different adult body form. In the course of the transformation, the tail, including the notochord and neural tube, is resorbed and disappears; the pharynx develops into a large food-collecting chamber; and growth on the anterior ventral side of the body is so much greater than elsewhere that the mouth ends up almost 180° from its original position (Fig. 30.1).

Adult tunicates have more or less spherical bodies, which range in size from a pinhead to a small potato.

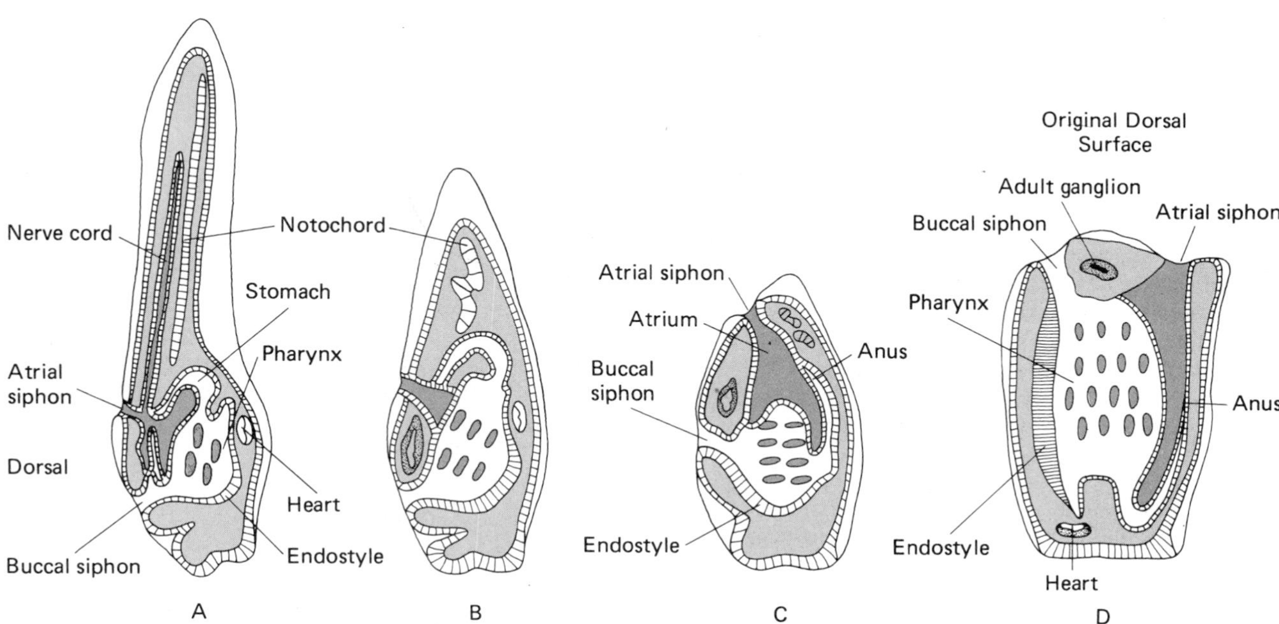

FIGURE 30.1 *A,* Diagrammatic lateral view of a urochordate tadpole larva, which has just attached to the substratum by the anterior end. *B* and *C,* Metamorphosis. *D,* A young individual just after metamorphosis. (Based on a figure by Seeliger.)

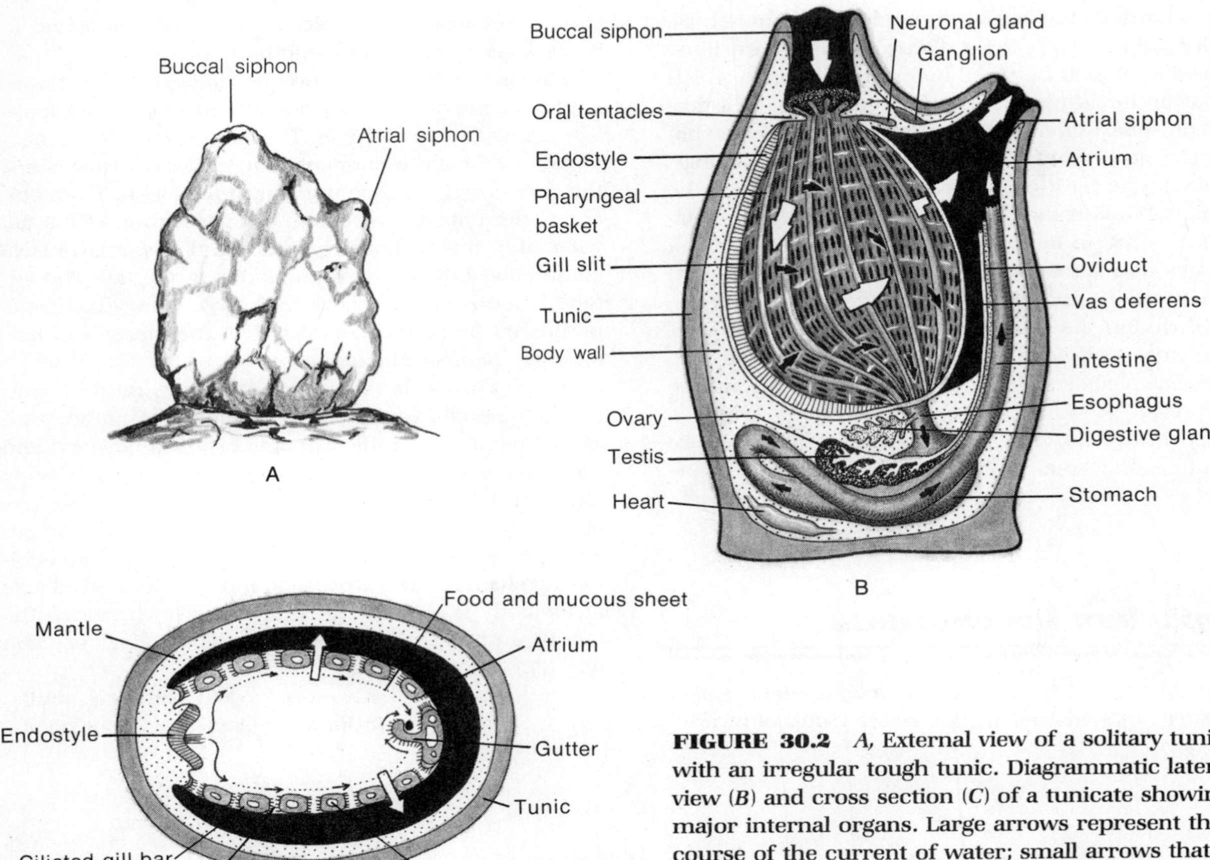

FIGURE 30.2 *A,* External view of a solitary tunicate with an irregular tough tunic. Diagrammatic lateral view (*B*) and cross section (*C*) of a tunicate showing major internal organs. Large arrows represent the course of the current of water; small arrows that of the food and mucous sheet. The stomach, intestine, and other visceral organs are embedded in the mantle.

The body is covered by a thick envelope, or **tunic,** containing **tunicine,** a form of cellulose. Cellulose is rarely encountered in animals. The tunic may be soft and delicate or very tough and rigid.

One end of the urochordate body is attached to the substratum; the opposite end contains two openings: the **buccal** and **atrial siphons** (Fig. 30.2). The buccal siphon opens into a large pharyngeal chamber, which fills the upper half or two thirds of the body of the animal. The wall of the pharynx is perforated by a vast number of slits which open into an outer surrounding chamber, the **atrium,** connected to the exterior through the atrial siphon. Through the pharyngeal and atrial chambers flows a current of water, entering through the buccal siphon and leaving through the atrial siphon. The body below the pharynx, called the **abdomen,** contains the **stomach, intestine, and heart,** and other internal organs.

Nutrition. Tunicates are **filter feeders,** removing plankton from the water stream passing through the pharyngeal chamber. The current is produced by the beating of cilia that border the margins of the pharyngeal perforations. A deep groove, the **endostyle,** extends down the length of the pharyngeal chamber on the side into which the buccal siphon opens (Fig. 30.2). Mucus secreted by the endostyle is driven by cilia out of the groove and over the inner pharyngeal surface as a film. Plankton is trapped in the film and carried across the pharynx by frontal cilia on the gridlike bars that form its walls. The plankton-laden film eventually collects within a gutter on the opposite side of the pharynx. Cilia within the gutter carry the collected material downward to the opening of the esophagus at the bottom of the pharynx (Fig. 30.2).

An enormous amount of water passes through the pharyngeal basket of a tunicate and serves not only for feeding but also for gas exchange. The current can be regulated or halted by opening or closing the siphons. Exposed intertidal species may suddenly contract and eject a spurt of water from the siphons, hence one of their names, sea squirts.

The original function of the chordate pharyngeal slits was probably filter feeding as in urochordates, and the production of mucus by an endostyle was part of the chordate groundplan. The endostyle mucus of urochordates contains iodine bound to tyrosine. In vertebrates the endostyle became internal and acquired a new function as the thyroid gland.

The gut is U-shaped. An intestine leads back upward and opens into the atrium. Fecal matter is swept away by the water current in the atrial siphon.

Excretion and Internal Transport. There are no special organs for excretion of nitrogenous wastes. Most escape by diffusion; some accumulate as inert pigments and crystals, including uric acid, in various parts of the body. Tunicates have an open internal transport system, but the blood follows distinct channels, which course through the pharyngeal basket, abdominal organs, and, in some species, even the tunic. The heart, located near the stomach, periodically reverses its beat and the flow of blood, a most unusual phenomenon.

Colonial Tunicates. Many common tunicates are solitary, but a considerable number are colonial. Some colonial tunicates resemble a vine creeping over the substratum (Fig. 30.3*A*). Others have bodies arranged in clusters; still others are so intimately connected that they are embedded within a common tunic (Fig. 30.3*B* and *C*). Such colonial tunicates have the individuals symmetrically arranged, and the atrial siphons open into a common cloacal chamber (Fig. 30.2). Some colonial and solitary urochordates are unattached and swim using the feeding current as a water jet for locomotion (Box 30.1).

BOX 30.1

Two groups of urochordates are pelagic, and some species are common pelagic animals. Thaliaceans have the atrial opening at the opposite end of the body and use the feeding current as a water jet in swimming. All are transparent gelatinous animals. Most are solitary, called salps (*B*), but there are a few colonial species (*A*) in which the water jet exits from a common cloacal opening at the end of the tubelike colony. Larvaceans retain the larval form into adulthood. They live within or attached to a floating gelatinous "house" that functions as a filter in suspension feeding. (*B* modified after Uljanin and Barrois from Borradaile and others.)

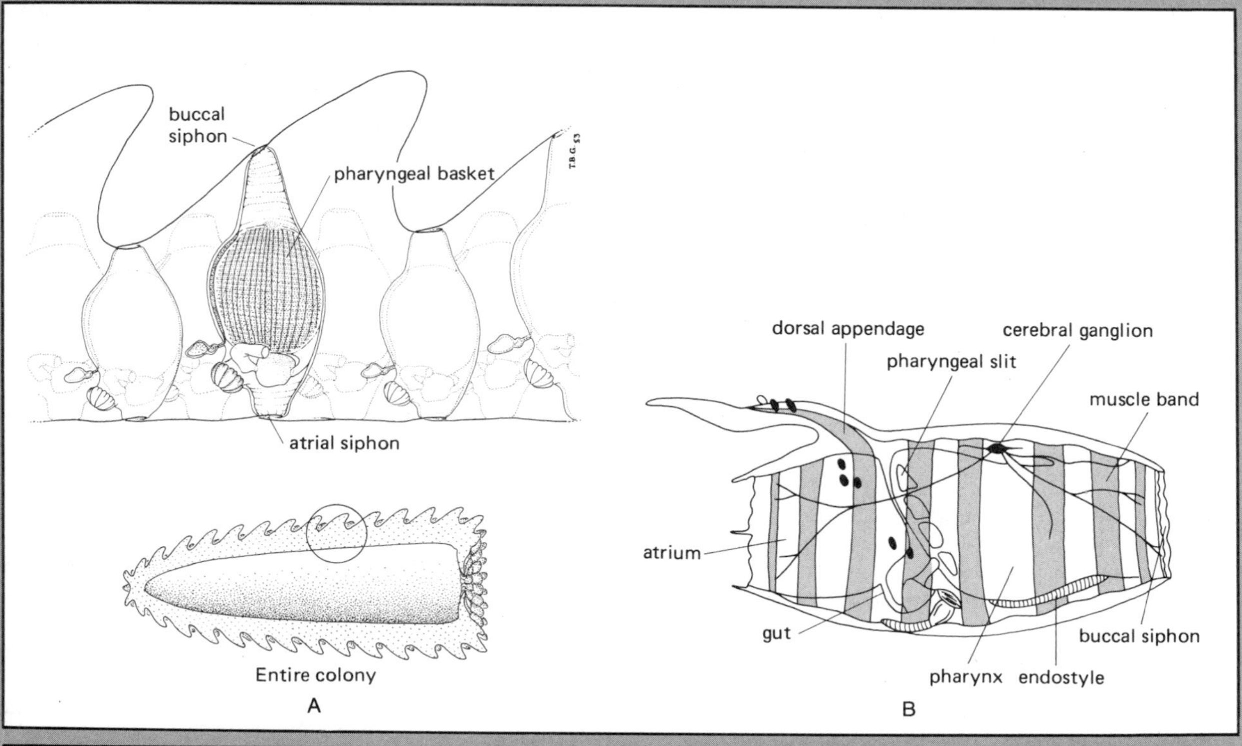

Reproduction. Most tunicates are capable of asexual reproduction by **budding.** All species are **hermaphroditic** with a single abdominal ovary and testis; the oviduct and sperm duct open into the atrium. Fertilization may be external with planktonic development, especially in solitary species. Alternatively, fertilization may take place within the atrium and the eggs may be brooded there through early embryonic stages. Development eventually leads to the tadpole larva already described.

Chordate Metamerism

Although members of another subphylum, the **Cephalochordata,** lack a backbone, they do share with vertebrates the condition of **metamerism.** Metamerism appears to have evolved early in chordate history in the line leading to the cephalochordates and vertebrates but after the divergence of the urochordates. As in annelids, segmentation developed as an adaptation for locomotion—for undulatory swimming or perhaps for rapid burrowing into sand.

In contrast to annelids, it is the musculature rather than the coelom that exhibits the primary segmentation. The segmentation of the nervous system and blood vascular systems represents an adjustment of these systems to supply the muscle blocks. Segmentation in chordates does not involve the coelom. The coelom in chordates is not utilized as a localized hydrostatic skeleton; rather it is the notochord against which the muscle blocks are indirectly pulling.

Subphylum Cephalochordata

Branchiostoma and a related genus of small superficially fish-shaped chordates commonly called amphioxus, constitute the subphylum **Cephalochordata.** Species occur in United States coastal waters south from the Chesapeake and Monterey Bays. They usually lie buried in sand with only their anterior end protruding, but they can also swim very rapidly.

The body of amphioxus (Fig. 30.4) is elongated, tapering at each end, and compressed from side to side. Small median fins and a pair of lateral finlike **metapleural folds** are present. Swimming is accomplished by the contraction of the muscle blocks, or **myomeres.** Shortening of the body is prevented by an unusually long notochord that extends farther anteriorly than in any other chordate, an attribute after which the subphylum is named (Cephalochordata—Gr. *kephalē,* head + L. *chorda,* string).

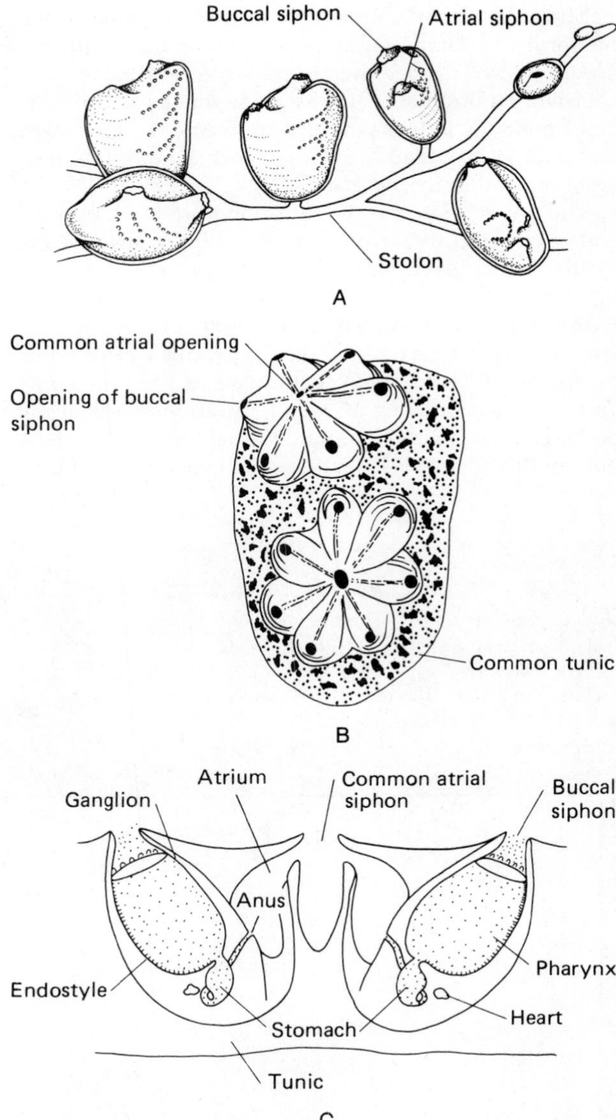

FIGURE 30.3 Colonial tunicates. *A,* A colony of *Perophora viridis; B,* surface view of *Botryllus schlosseri; C,* vertical section through *Botryllus.* (*A* after Miner; *B* after Milne-Edwards from Yonge; *C* after Delage and Herouard.)

Nutrition. Amphioxus is a **filter feeder.** Water and minute food particles are taken in through the **oral hood,** whose edges bear a series of delicate projections, the **cirri,** that act as a strainer to exclude larger particles (Fig. 30.4). The inside of the oral hood is lined with bands of cilia, the **wheel organ,** which, together with

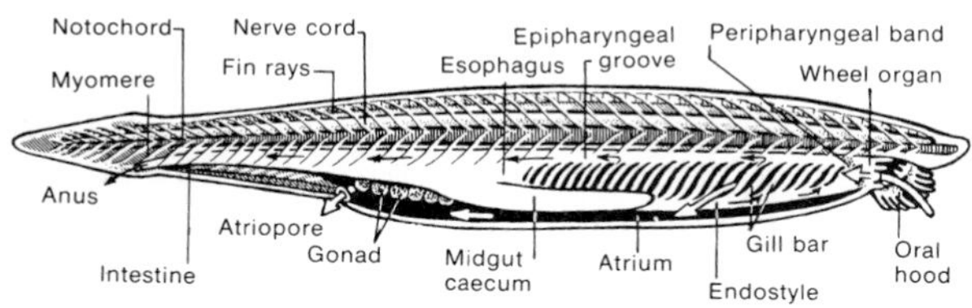

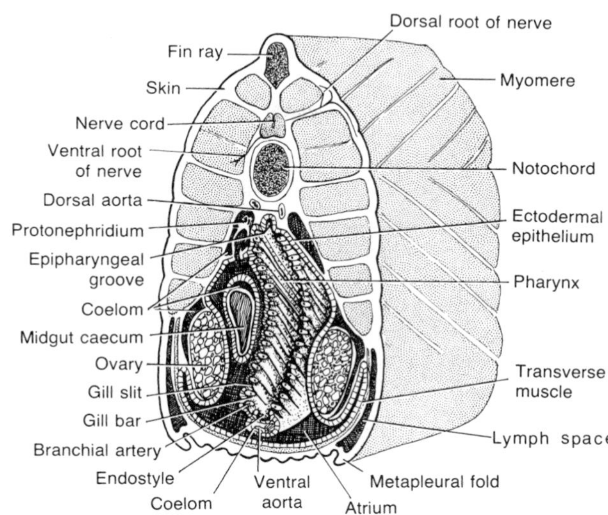

FIGURE 30.4 *Amphioxus. A,* A diagrammatic lateral view of entire animal. White arrows indicate course of water current; black arrows that of food. *B,* A diagrammatic cross section through the posterior part of the pharynx. Branchial arteries extend from the central aorta through the gill bars to the dorsal aortas.

cilia in the pharynx, produce a current of water that enters the mouth.

Food is entrapped within the pharynx in mucus secreted by an endostyle just as it is in urochordates. Water in the pharynx escapes into an atrium through numerous gill slits. Some gas exchange occurs in the pharynx, but the skin is the major respiratory surface. The pharynx is primarily a food-gathering device. Food particles carried back into the intestinal region are sorted out by complex ciliary currents. Large particles continue posteriorly, but small ones are deflected into the **midgut caecum,** a ventral outgrowth from the floor of the gut (Fig. 30.4). Many of these particles are ingested by caecal cells and are digested intracellularly. Undigested material, together with enzymes secreted by certain caecal cells, is carried posteriorly to join the mass of larger food particles. As this mass rotates and is degraded by enzymic action, small particles are carried forward to the caecum to be engulfed and digested. Material not broken down in the midgut is carried posteriorly. Fecal material is discharged through an **anus,** which, as in vertebrates, lies slightly anterior to the posterior tip of the body.

Internal Transport. Absorbed nutrients are distributed by a circulatory system. A series of veins returns blood from the various parts of the body to a sinus that is located ventral to the posterior part of the pharynx and may be comparable to the posterior part of the vertebrate heart. No muscular heart is present; blood is propelled by the contraction of the arteries. A ventral aorta extends from the sinus forward beneath the pharynx and connects with branchial arteries that extend dorsally through the gill bars to a pair of dorsal aortas (Fig. 30.4). The dorsal aortas in turn carry the blood posteriorly to spaces within the tissues. True capillaries are absent, but the general direction of blood flow, i.e., anteriorly in the ventral part of the body and posteriorly in the dorsal part, is similar to that of a vertebrate and different from that of other animals.

The excretory organs are segmentally arranged ciliated **protonephridia** that lie dorsal to certain gill bars and open into the atrium (Fig. 30.4).

Nervous System. The nervous system of amphioxus consists of a tubular nerve cord located dorsal to the notochord (Fig. 30.4). Its anterior end is differentiated slightly but is not expanded to form a brain. Paired segmental nerves, consisting of dorsal and ventral roots as in vertebrates, extend into the tissues. Amphioxus is sensitive to light and to chemical and tactile stimuli, but no elaborate sense organs are present.

Reproduction. The sexes are separate in amphioxus, and numerous testes or ovaries bulge into the atrial cavity (Fig. 30.4). The gametes are discharged into the atrium upon the rupture of the gonad walls. Fertilization and development are external.

Subphylum Vertebrata

The Vertebrata is by far the largest and most important of the chordate subphyla, for all but about 2000 of the approximately 41,500 living species of chordates are vertebrates. Vertebrates share with the lower chordates the three diagnostic characteristics of the phylum. These are clearly represented at some stage in the life history of the various groups. The dorsal tubular nerve cord has differentiated into a brain and spinal cord, present in the embryos and adults of all species (Fig. 30.5). Embryonic vertebrates have a notochord lying ventral to the nerve cord and extending from the middle of the brain nearly to the posterior end of the body, but a vertebral column replaces the notochord in most adults. All embryonic vertebrates have a series of **pharyngeal pouches** that grow laterally from the walls of the pharynx, but these pouches break through the body surface to form gill slits only in fishes and larval amphibians.

Vertebrates evolved as a group of chordates that became more active and developed more aggressive ways of feeding. Most of the distinctive characteristics of vertebrates are related to these changes in mode of life. Their greater activity is reflected in the replacement of the notochord by a vertebral column, in the continued elaboration of the segmented muscular system seen beginning in amphioxus, in an aggregation of nervous tissue (the brain) and elaborate sense organs at the anterior end of the body, and in the protection of these organs by a brain case, or **cranium.**

Early vertebrates probably continued a filter-feeding mode of life but used muscular movements of their pharynx, and not simply ciliary currents, to draw in water and food. This was certainly more efficient and made possible an increase in size. Larger size and the eventual evolution of jaws permitted a yet more active and aggressive mode of life. Respiratory organs (gills or lungs) replaced the general body surface as sites of gas exchange. A muscular **heart** developed that pumps blood effectively through a closed circulatory system. Wastes are excreted by a pair of **kidneys** (Fig. 30.5) composed of numerous tubules quite unlike any invertebrate excretory organ. Vertebrates have become the most successful and dominant group of chordates.

Phylum Hemichordata

We must now examine one last group of deuterostomes, the phylum **Hemichordata,** for the members of this phylum provide some additional evidence linking echinoderms and chordates and must be considered in speculating about chordate origins.

Most hemichordates are marine, wormlike animals, sometimes called acorn worms, that are found beneath stones and shells, in burrows in sand and mud, and a few species occur in tubes. The body is composed of three regions: an anterior **proboscis,** a middle **collar,** and a long posterior **trunk** (Fig. 30.6). The proboscis is simply the anterior end of the body, and the mouth is located at its posterior ventral margin just in front of the short bandlike collar region. The trunk follows the collar and makes up the greater part of the body.

Hemichordates possess **gill clefts.** These are present as a line of many perforations on either side of the anterior end of the trunk (Fig. 30.6). A stream of water enters the mouth, passes into the pharynx, and then exits through the gill pores, driven by cilia. Although hemi-

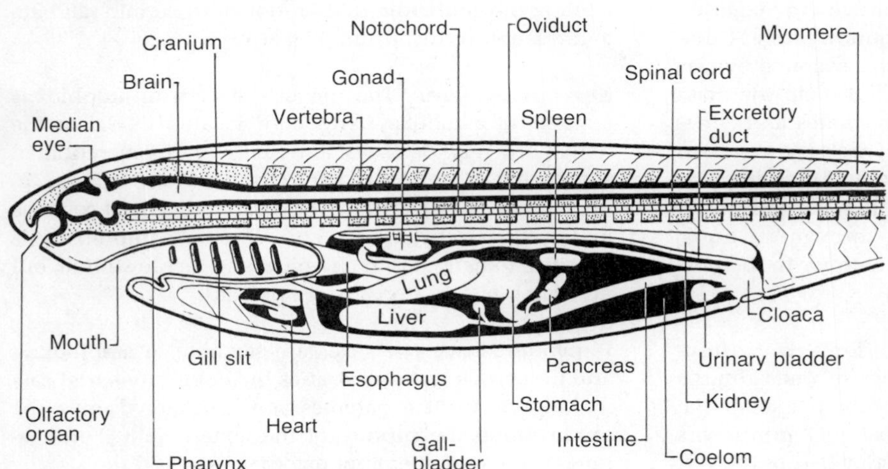

FIGURE 30.5 A diagrammatic sagittal section through a generalized vertebrate to show the characteristics of vertebrates and the arrangement of the major organs.

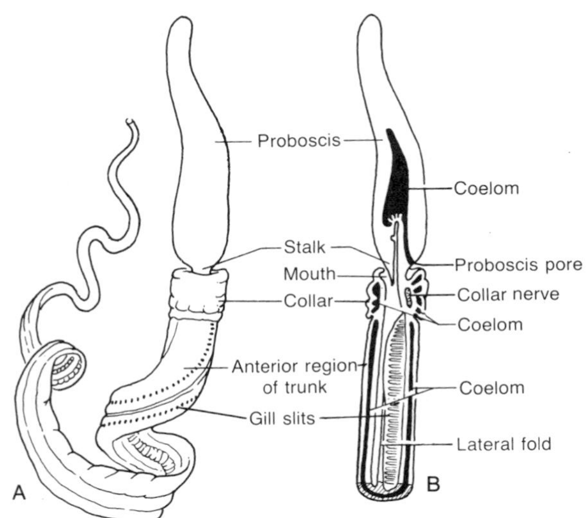

FIGURE 30.6 Phylum Hemichordata (genus *Saccoglossus*). *A*, External view showing external features (after Bateson). *B*, A diagrammatic section through the anterior part of the body showing some of the internal organs. A lateral fold subdivides the pharynx into a ventral channel along which the sand passes and a dorsal channel containing the gill slits.

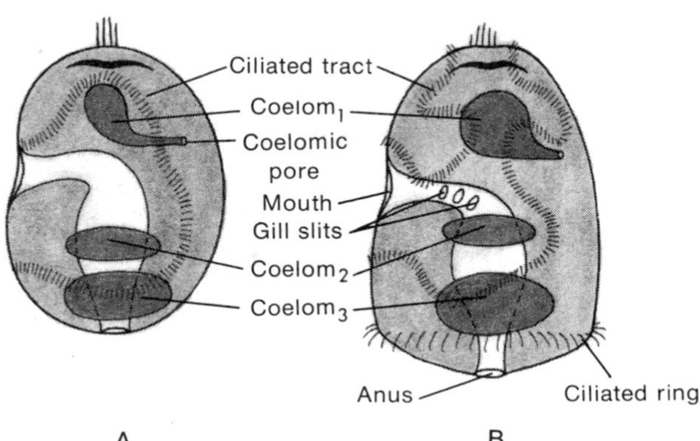

FIGURE 30.7 Diagrammatic side views of the larval development of a hemichordate tornaria larva. Compare with Figure 29.7. *A*, Early larva; *B*, later larva.

chordates do not contain gills, gas exchange occurs as water passes through the pharyngeal clefts; this appears to be their only function.

Nutrition. Hemichordates feed largely on suspended and deposit material that adheres to mucus on the surface of the proboscis and is driven by cilia into the mouth along with the ventilating current. Within the pharynx, the water is separated from the ingested food, which passes into the posterior part of the gut.

Some hemichordates (*Balanoglossus*, for example) that live in burrows ingest large quantities of sand and extract the organic detritus from it. The castings of these animals are often a conspicuous feature on sand flats exposed at low tide.

Nervous System. Despite the presence of pharyngeal clefts, hemichordates are not chordates, as they were once thought to be, for they have no notochord, nor is there a true dorsal nerve cord (Gr. *hemi*, half + L. *chorda*, string). However, within the collar region of some hemichordates, a dense concentration of dorsal epidermal nerve fibers sinks inward and forms a tubelike arrangement (Fig. 30.6).

Reproduction. Hemichordates have separate sexes, and the numerous paired lateral gonads, each with a separate gonopore, are located within either side of the trunk. Fertilization is external. Development follows the typical deuterostome pattern and leads in many species to a **tornaria larva** (Fig. 30.7), which is strikingly like that of echinoderms. After a planktonic existence, the tornaria larva lengthens and settles to the bottom as a young worm.

The preceding discussion has been devoted to the most common hemichordates, members of the class **Enteropneusta** (Gr. *enteron*, intestine + *phesta*, to breathe). A small number of species comprises another class, the **Pterobranchia** (Gr. *pteron*, wing + *branchia*, gills). These are mostly deep-water hemichordates that live attached within secreted tubes. The collar region bears tentacles, and some species lack pharyngeal clefts. Many zoologists consider the pterobranchs to be the most primitive members of the phylum.

Deuterostome Relationships and Chordate Origins

The common pattern in the early embryonic development of echinoderms, hemichordates, and chordates is an important basis for recognizing the deuterostome line of evolution, but there is other evidence of the close evolutionary relationship between these three principal deuterostome phyla. The strikingly similar larval stages clearly indicate a connection between hemichordates and echinoderms. On the other hand, the presence of gill clefts and perhaps the dorsal collar nerve cord relates hemichordates to chordates. Although the precise

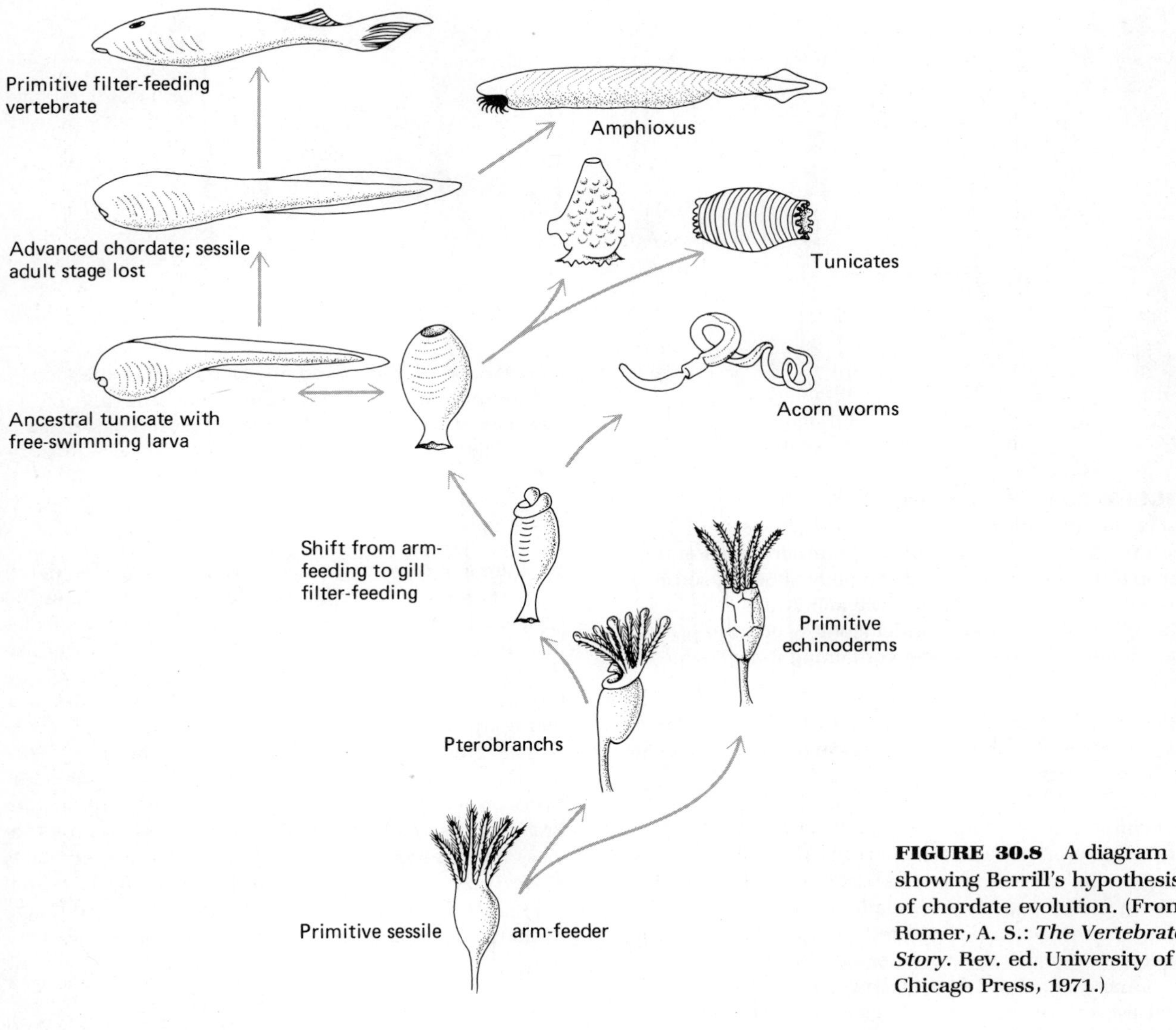

FIGURE 30.8 A diagram showing Berrill's hypothesis of chordate evolution. (From Romer, A. S.: *The Vertebrate Story.* Rev. ed. University of Chicago Press, 1971.)

relationship of the three phyla is unknown, there seems little doubt that they share a common evolutionary history.

The nature of the common ancestral form, as well as the origins of each phylum, is obscure. An idea elaborated by N. J. Berrill and others, and held by many zoologists, postulates that the ancestral chordates were sessile and that urochordates reflect the primitive condition of the phylum (Fig. 30.8). Pharyngeal clefts were originally an adaptation for filter feeding in the adult, and the notochord and finned tail were adaptations for swimming during larval life. According to this theory, the cephalochordates and vertebrates evolved from the urochordate tadpole larva through neoteny and paedomorphosis, i.e., prolongation of larval life, precocious sexual development, and suppression of adult features. Metamerism developed as an adaptation for swimming in the neotenous cephalochordate-vertebrate ancestor (p. 644). The sessile existence of primitive chordates, pterobranch hemichordates, and primitive echinoderms is viewed as a feature resulting from common ancestry.

However, the earliest chordates may have been motile. There is no certainty that the common ancestor of echinoderms and chordates was sessile. Echinoderms appear to have evolved from motile ancestors, and their sessile existence and subsequent radiation may well have been an independent evolutionary event unrelated to the evolution of a sessile existence in urochordates.

Classification of the Phylum Hemichordata

Class Enteropneusta Acorn worms. Wormlike hemichordates having a cylindrical proboscis and collar and a long trunk. Many species burrow in sand or mud; others live beneath stones or within algae.

Class Pterobranchia Minute hemichordates having a shield-shaped proboscis and a collar bearing tentaculate arms. Most are colonial and live in secreted tubes in deep water.

Classification of the Phylum Chordata

Subphylum Urochordata Mostly sessile, nonmetameric invertebrate chordates enclosed within a tunic containing cellulose. Pharynx highly developed and used in filter feeding; notochord and nerve cord present only in the larva.

Class Ascidiacea Sea squirts or tunicates. Sessile; solitary or colonial. This class contains most species of urochordates.

Class Thaliacea Salps. Free-swimming planktonic urochordates; solitary or colonial.

Class Larvacea Small, neotenous, planktonic urochordates living within or attached to a delicate gelatinous "house."

Subphylum Cephalochordata Amphioxus. Small fishlike invertebrate chordates. Metameric but trunk supported by well-developed notochord; no vertebrae present. Jawless suspension feeders; mouth surrounded by an oral hood. *Branchiostoma.*

Subphylum Vertebrata Vertebrates. Metameric chordates with the trunk supported by a linear series of cartilaginous or bony skeletal pieces (vertebrae) surrounding or replacing notochord in adult. Most vertebrates have mouth surrounded by jaws. (Detailed classification of vertebrates provided in subsequent chapters.)

Class Agnatha Lamprey eels and hagfishes.

Class Placodermi Extinct primitive jawed fishes.

Class Chondrichthyes Sharks and rays.

Class Osteichthyes Bony fishes.

Class Amphibia Frogs, toads, and salamanders.

Class Reptilia Alligators, snakes, lizards, and turtles.

Class Aves Birds.

Class Mammalia Mammals.

Summary

1. The phylum Chordata contains a diverse assemblage of animals united in having, at some time in their life history, pharyngeal clefts, a notocord, and a dorsal hollow nerve cord. Although most chordates belong to the subphylum Vertebrata, in which a vertebral backbone surrounds or replaces the notochord, there are two subphyla of invertebrate chordates.

2. The subphylum Urochordata contains mostly sessile marine animals, called sea squirts or tunicates, which have the notochord and dorsal nerve cord only during the tadpole larval stage. Much of the interior of the body is occupied by a large pharynx perforated by many slits. Water enters the pharynx through a buccal siphon (mouth), passes by ciliary propulsion through the slits into a surrounding atrium and then out an exhalant siphon. In the passage through the pharyngeal slits, plankton is filtered from the water current. The body of urochordates is covered by a protective tunic containing cellulose and secreted by the epidermis. The many colonial species of urochordates have the individuals connected together by a stolon or have their tunics fused together.

3. The subphylum Cephalochordata contains a small group of fishlike chordates that burrow in marine sand. They swim or burrow by rapid body undulations produced by the contraction of metameric muscle blocks. Cephalochordates filter feed from the stream of water passing into the mouth and through the pharyngeal clefts. From the dorsal nerve cord emerge paired, segmentally arranged, lateral nerve roots, but there is no anterior brain.

4. The subphylum Vertebrata contains the majority of chordates. Like cephalochordates, they are metameric, but the anterior end of the nerve cord is expanded as a brain. The notochord is surrounded or replaced by vertebrae. Primitively, vertebrates are undulatory swimmers, and the evolution of jaws enabled them to exploit larger food material than did their filter-feeding ancestors.

5. The ancestral chordates are believed by many zoologists to have been sessile animals and the cephalochordates and vertebrates to have evolved through retention and development of the motile larval body form into adulthood.

References and Selected Readings

Accounts of the hemichordates and chordates can be found in many of the references listed at the end of Chapter 19. Those below deal solely with these groups.

Alldredge, A.: Appendicularians. *Scientific American, 235*(1):94–102, 1976. An interesting account of the planktonic urochordates (larvaceans) that secrete gelatinous "houses" for filter feeding.

Barrington, F. J. W.: *The Biology of Hemichordates and Protochordates.* San Francisco, W. H. Freeman & Co., 1965. A general account emphasizing the behavior, physiology, and reproduction of hemichordates, urochordates, and cephalochordates.

Berrill, N. J.: *The Origin of the Vertebrates.* London, Oxford University Press, 1955. This work contains Berrill's theory of chordate and vertebrate origins.

Clark, R. B.: *Dynamics in Metazoan Evolution.* Oxford, Clarendon Press, 1964. A detailed synthesis of the ideas regarding the origins of the coelom and metamerism and their phylogenetic implications.

Goodbody, I.: The physiology of ascidians. *Advances in Marine Biology, 12*:1–149, 1974. A good coverage of all aspects of tunicate physiology.

31 Vertebrates: Fishes

ORIENTATION

Most vertebrates are fishes, and all of them share many features, which we will explore, that adapt them to life in the water. Ancestral fishes were jawless bottom-dwelling species that are assigned to the class Agnatha. Most agnathans are extinct, but the class is represented today by the lampreys and hagfishes. With the evolution of jaws and paired appendages, fishes became more active and capable of feeding in many different ways. Living jawed fishes are grouped into two classes: the sharks and skates of the class Chondrichthyes have a cartilaginous skeleton, whereas minnows, perch, and other familiar fishes of the class Osteichthyes have a skeleton that is at least partly ossified. We will consider the characteristics, evolution, and adaptive diversity of the major fish groups in this chapter. The distinctive characteristics of the living classes are summarized below.

CHARACTERISTICS OF THE CLASS AGNATHA

1 Jaws are absent.
2 Paired fins are absent in most species, but pectoral flaps were present in some extinct ones.
3 Early species had heavy bony scales in their skin, but these have been lost in living species.
4 The deeper parts of the skeleton are cartilaginous in living forms and also appear to have been unossified in extinct species. The embryonic notochord persists in the adult.
5 A median, light-sensitive pineal eye is present.
6 Living and most extinct species have a single, median nostril located in front of the pineal eye.
7 Seven or more gill pouches are present. The pharynx is used in filter feeding in the larvae and in the adults of the extinct species.

CHARACTERISTICS OF THE CLASS CHONDRICHTHYES

1 Jaws and paired fins are present.
2 Bony scales are either reduced to tiny placoid scales or are lost completely.
3 The deeper parts of the skeleton are entirely cartilaginous.
4 The pineal eye is lost.
5 They are heavy-bodied fishes without lungs or a swim bladder. Their body is flattened anteroventrally, and most species retain the primitive heterocercal tail.
6 Their nostrils are paired.
7 The five pairs of gill pouches open independently on the body surface in most species rather than into a chamber covered by an operculum.
8 Their intestine is short, and its surface area is increased by a spiral valve.
9 Males have a clasper on the pelvic fin with which sperm are transferred to the female. Fertilization is internal.

CHARACTERISTICS OF THE CLASS OSTEICHTHYES

1 Bony scales usually are present, but the primitive superficial layers of ganoine and cosmine are lost in most living species.
2 The deeper skeleton always contains some bone; in most species it is nearly completely ossified.
3 The pineal eye is retained in primitive species.
4 Lungs or a swim bladder are present, except in a few bottom-dwelling species in which they have been secondarily lost. As is to be expected in a light-bodied fish, the tail has become homocercal in most living species.
5 The gill pouches open into a common chamber covered by an operculum.
6 The intestinal spiral valve is lost in all but the most primitive species. Surface area is increased by an increase in intestinal length and by pyloric caeca.
7 Most species are oviparous and fertilization is external. In the few viviparous species with internal fertilization the copulatory organ of the male is a modified part of the anal fin.

There are over 20,000 species of fishes living today, more than all terrestrial vertebrates combined. They exhibit tremendous adaptive diversity, having specialized for life in shallow waters, the open ocean, and the deep sea. A few even make brief excursions upon the land. They are usually grouped into three living and one extinct class, each with their own characteristics, but all share common adaptations for life in the water.

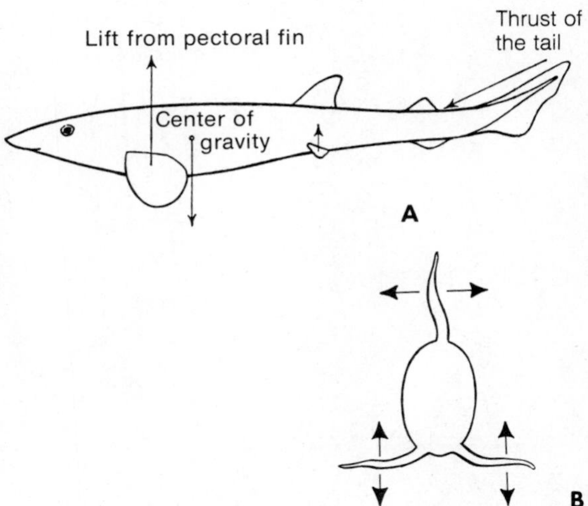

FIGURE 31.1 Role of the tail and fins of a dogfish in generating lift and providing stability. *A,* Lateral view; *B,* composite transverse section through dorsal and pectoral fins. (Modified after Marshall and Hughes.)

Aquatic Adaptations

Water is denser than air, has greater thermal stability, and contains less oxygen. These are facts of life to which all fishes are adapted. The fish integument has the basic structure characteristic of all vertebrates (p. 97), but there is little deposition of horny keratin in the epidermal cells. **Bony scales** develop in the dermis of most groups. **Mucous glands** are abundant. Discharged mucus spreads over the body surface, where it helps to protect the body from ectoparasites, prevents undue exchanges of body water with the environment, and reduces friction between the body and water.

Fishes have a streamlined, fusiform body shape that enables them to move through the water with a minimum of effort. Their skeleton is not so strong as in terrestrial vertebrates, for the buoyancy of the water also provides considerable support. Trunk muscles are segmented **myomeres** whose successive activation causes the lateral undulations of the trunk and tail by which the animal swims (p. 120). Since flesh is denser than water, a fish tends to sink. Primitive fishes lie on the bottom when they rest, but when they swim, their flattened heads and broad pectoral fins have a planing effect that causes the front end of the body to rise. Their **heterocercal** tail (Gr. *heteros,* other + *kerkos,* tail), in which the vertebral and muscular axis of the body extends into an enlarged dorsal lobe of the tail, has a compensating effect (Fig. 31.1*A*). As the tail moves from side to side, the thrust that it generates is directed slightly downward as well as forward so it tends to drive the front of the body down and keeps the fish on an even keel. Primitive fishes require muscular effort to stay afloat. More advanced fishes have evolved a swim bladder (p. 156), a sac of air located dorsal to the body cavity, that gives them a density close to that of water, so they can float with little muscular effort. Their bodies are not flattened anteriorly, and their tails have become superficially symmetrical, or **homocercal.** The medial fins, as well as the paired pectoral and pelvic fins, provide stability against rolling, pitching, and yawing (Fig. 31.1*B*). The paired fins are also used in turning and in changing depth.

The sense organs of fishes are adapted for receiving stimuli from the water. Movable eyelids and tear glands are absent; the eyes are bathed and cleansed by the surrounding water. Light is refracted by a spherical lens. The nose is simply an olfactory organ. Water does not pass through it on the way to the pharynx except in lungfishes. The sense of smell is very important to fishes in finding food, mates, and so on. Salmon are able to detect, using olfactory clues, the stream in which they hatched and to return to it to spawn. Often each nasal sac has two openings on the body surface to facilitate the circulation of water through it. An inner ear is present, but fishes lack the external and middle ear found in terrestrial vertebrates. The pressure waves in the water pass easily through the tissues of the fish to the inner ear. Low-frequency vibrations and certain water movements are detected by the **lateral line system,** a unique aquatic sensory system consisting of groups of receptor cells, called **neuromasts,** that are usually embedded in canals in the skin. The canals open to the surface by pores (Fig. 31.2). Water movements bend hairlike cytoplasmic processes on the receptor cells and affect the cells' rate of discharge. The system has been called one of "distant touch" for it enables a fish to detect water movements and pressures so that it senses objects it is approaching or objects moving toward it. It

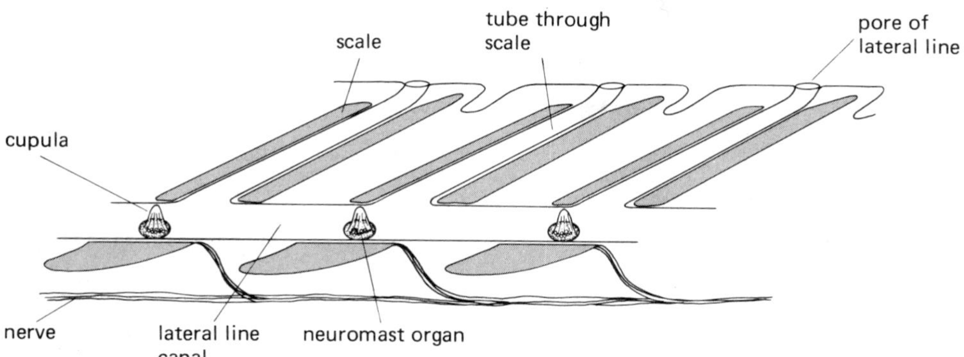

FIGURE 31.2 Diagram of a vertical section through the lateral line of a fish. (From Bond, C. E.: *Biology of Fish.* Philadelphia, W. B. Saunders Co., 1979.)

plays an important role in the schooling behavior of fishes.

Because of the thermal stability of the water, fishes are not subjected to the extreme temperature fluctuations encountered by terrestrial animals. Although some have mechanisms that elevate muscle temperature at high swimming speeds, fishes are basically ectothermic vertebrates with rather low levels of metabolism. Gases are exchanged between water and blood by diffusion across the gills. Water typically enters the pharynx through the mouth and is discharged through the gill pouches (Fig. 7.4, p. 153). The heart receives blood low in oxygen content, and its contraction drives the blood through the gill capillaries. Oxygenated blood is distributed from the gills to the body under relatively low pressure.

The movable muscular tongue characteristic of terrestrial vertebrates is absent in fishes, but the floor of the mouth and pharynx is specialized in some groups to aid in holding and swallowing prey. In many species teeth are present on the roof of the mouth and gill arches, as well as on the margins of the jaws. The respiratory current of water also helps carry food posteriorly into the pharynx.

Nitrogenous metabolic wastes are eliminated by diffusion through the gills and by opisthonephric kidneys, which are drained by archinephric ducts (p. 198). The kidneys also play an important role in water and salt balance. Excess salt swallowed by marine fishes is eliminated by special salt-excreting cells or glands.

Most fishes are oviparous. As the female lays eggs, the male discharges sperm. Fertilization is external, and there is often an aquatic larval stage during the development.

Class Agnatha

The class **Agnatha** (Gr. *a,* without + *gnathos,* jaws) includes the oldest fossil vertebrates, who appear to have been ancestral to other fish groups. Most are extinct, but the class is represented today by about 50 species of lampreys and hagfishes. Although the origins of the Agnatha are obscure, they probably evolved from one of the lower chordate groups by paedomorphosis (p. 425 and p. 648). Scales of primitive agnathous fishes have been found in marine deposits of the late Cambrian, 540 million years ago. Although the first vertebrates, like their lower chordate ancestors, were probably marine, agnathous fishes soon entered fresh water and underwent an extensive adaptive radiation. They flourished during the first part of the Paleozoic era but were replaced by more progressive fishes in the Devonian period (Fig. 31.3). Six orders of agnathous fishes are recognized by some authors. Differences between the five extinct ones need not concern us. All had an extensive development of heavy bony scales and plates in their skins so are collectively called the **ostracoderms** (Gr. *ostrakon,* shell + *derma,* skin). Most ostracoderms were small bottom-feeding fishes that were flattened dorsoventrally (Fig. 31.4). Their scales were very similar to the **cosmoid scales** of some other primitive fishes (Fig. 31.5*A*), for beneath a thin layer of enamel-like **ganoine** there was a thick layer of dentine-like **cosmine.** The rest of the scale consisted of a layer of **spongy bone** containing many vascular spaces and a layer of more compact **lamellar bone.** These heavy scales certainly offered some mechanical protection, possibly against aquatic scorpion-like eurypterids of the period. In ad-

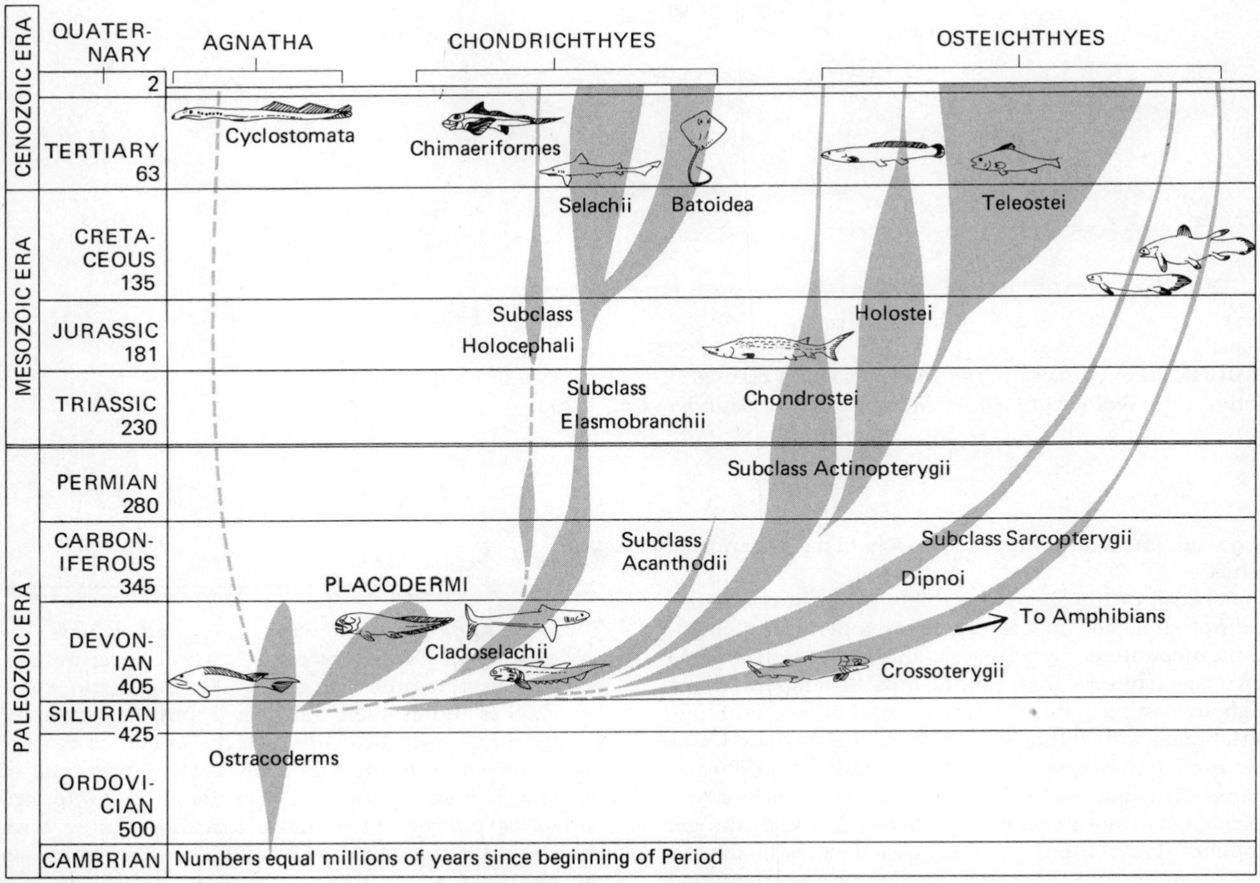

FIGURE 31.3 An evolutionary tree of fishes. The relationships of the various groups, their relative abundance, and their distribution in time are shown.

dition, the layers of ganoine and cosmine may have offered some protection against an excessive inflow of water from the freshwater environment in which most of these fishes lived. The bone, as in all vertebrates, was an important reservoir for calcium and phosphate ions. It is doubtful that fishes could have entered fresh water without such a reservoir.

Ostracoderms had median fins, and some had paired pectoral fins, but paired fins were not well developed in most groups. Most had the heterocercal tail characteristic of primitive fishes.

A single **median nostril** was present on the top of the head in the best known ostracoderms. A pair of lateral eyes and a single median **pineal eye** were on the

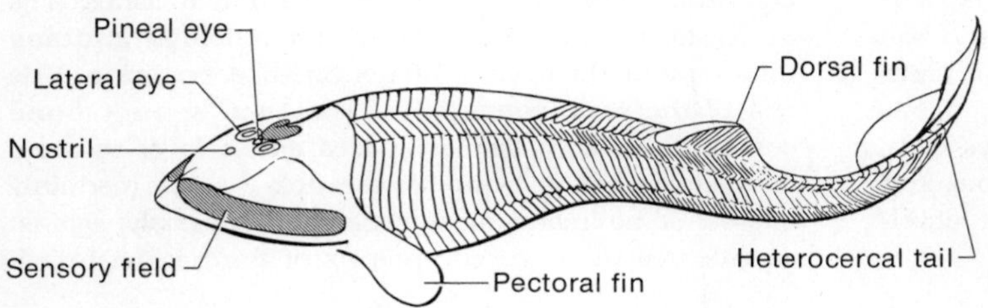

FIGURE 31.4 Laterodorsal view of *Hemicyclaspis*, a representative ostracoderm of the early Devonian period. This fish was about 20 cm. long. (Modified after Stensiö.)

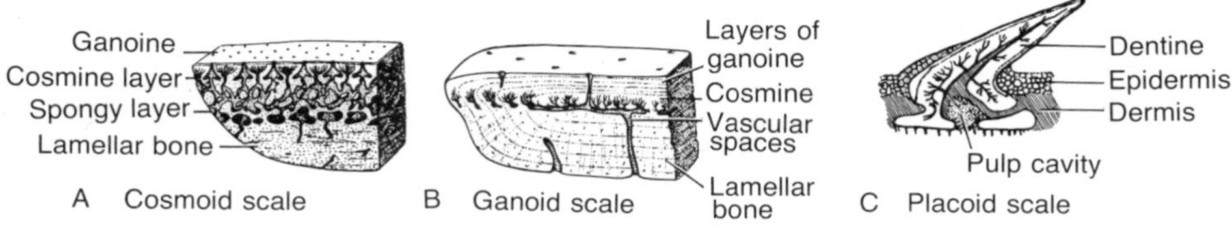

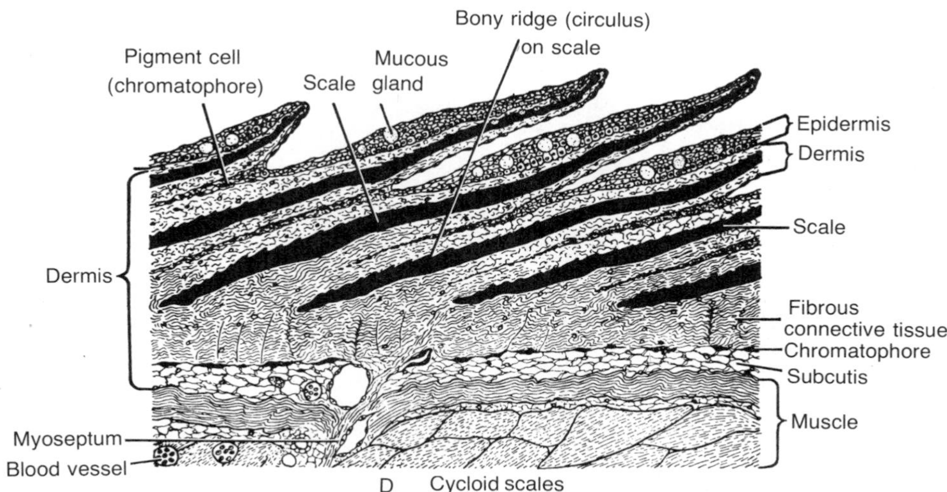

FIGURE 31.5 Types of fish scales. Vertical sections through *(A)* a cosmoid scale of the type seen in certain ostracoderms and sarcopterygians, *(B)* a ganoid scale of the type seen in acanthodians and primitive actinopterygians, *(C)* a placoid scale of a shark, and *(D)* the skin and cycloid scales of teleost. *(A–C* from Romer; *D* courtesy of General Biological Supply House.)

top of the head posterior to the nostril. A median eye is found in many primitive fishes and terrestrial vertebrates, and it is retained in a few living fishes and reptiles. It is a general light-monitoring organ rather than an image-forming eye and receives stimuli that enable the organism to adjust its physiological activity to the diurnal cycle. A dorsomedial and a pair of dorsolateral areas of the head contained small plates beneath which were enlarged cranial nerves. It is believed that these areas were sensory fields, possibly a part of the lateral line system. Much of the ventral surface of the head was covered with small plates forming a flexible floor to the pharynx. Movement of this floor presumably drew water and minute particles of food into the jawless mouth. The water then left the pharynx through as many as nine pairs of small **gill slits,** but the food particles were somehow trapped in the pharynx. It seems probable that the ancestral vertebrates, like the present lower chordates, were filter-feeders.

Living Jawless Vertebrates

Lampreys and hagfishes of the order **Cyclostomata** (Gr. *kyklos,* circle + *stoma,* mouth) are a specialized remnant of the class Agnatha (Figs. 31.6 and 31.7). They are jawless, have more gill pouches than other living fishes, lack paired appendages, retain a pineal eye, and have a single median nostril. Besides leading to an olfactory sac, this nostril opens into a **hypophyseal sac** that passes beneath the front of the brain. Much of the pituitary gland is derived from an embryonic hypophysis. Unlike ostracoderms, cyclostomes have an eel-like shape and slimy, scaleless skin; they are predators or scavengers.

Most lampreys live in fresh water, but some spend their adult life in the ocean and return to fresh water only to reproduce. A familar example of the group is the sea lamprey, *Petromyzon marinus.* The chief axial support for the body, the **notochord,** persists throughout

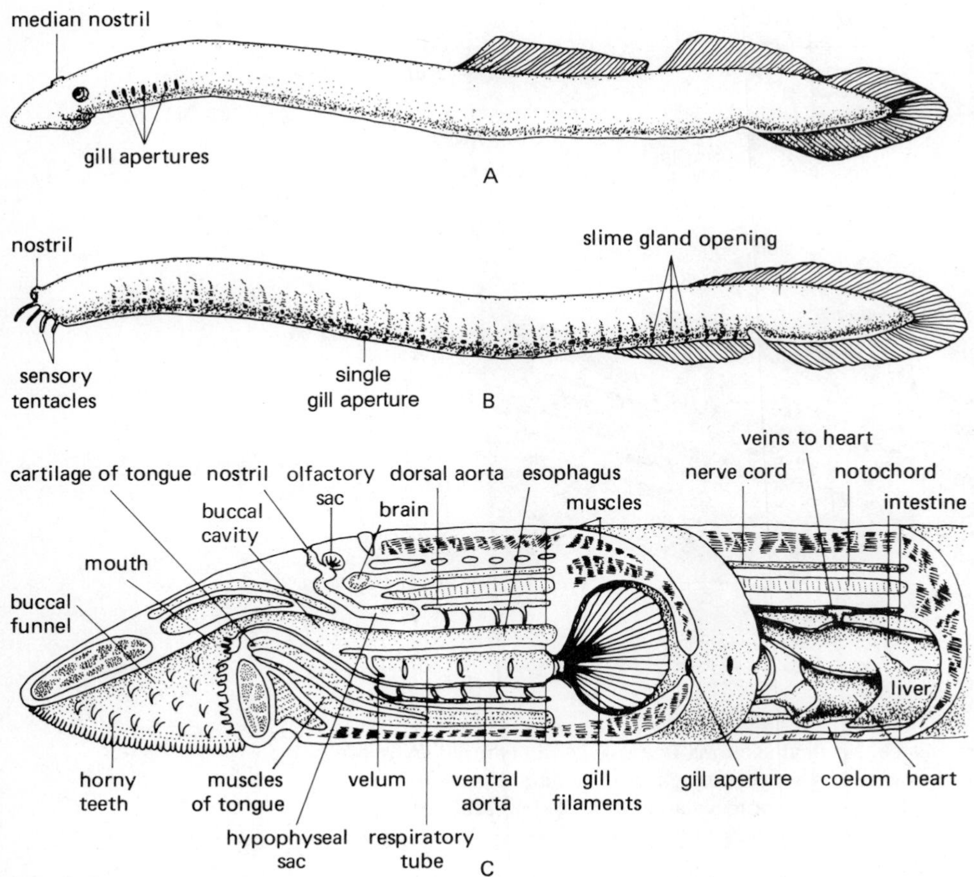

FIGURE 31.6 Living cyclostomes. *A*, A lamprey; *B*, a hagfish; *C*, internal anatomy of the anterior part of a lamprey.

life and is never replaced by vertebrae (Fig. 31.6*C*). Rudimentary vertebrae are present on each side of the notochord and spinal cord. The brain is encased by a cartilaginous **cranium.** The gills are supported by a complex cartilaginous **branchial basket** which may be homologous to the visceral skeleton of other fishes.

The **mouth** lies deep within a **buccal funnel,** a suction-cup mechanism with which the lamprey attaches to other fishes. The mobile **tongue,** armed with horny "teeth," rasps away at the prey's flesh, and the lamprey sucks in the blood and bits of tissue. An anticoagulant secreted by special **oral glands** keeps the blood flowing freely. From the mouth cavity, the food enters a specialized **esophagus** that bypasses the pharynx to lead into a straight **intestine.** There is no stomach or spleen. A **liver** is present, but the pancreas is represented only by cells embedded in the wall of the intestine and in the liver.

The respiratory system consists of seven pairs of gill pouches that connect to a modified pharynx known as the **respiratory tube.** The respiratory tube can be isolated from the food passage by a valvelike arrangement called the **velum** so the animal can pump water in and out of the gill pouches through the external gill slits while it is feeding. The pumping of the pharyngeal region also squeezes upon and changes the pressure in the hypophyseal sac, and water is thereby circulated across the olfactory sac.

The **opisthonephric kidneys** are drained by **archinephric ducts.** These ducts carry only urine, for sperm or eggs pass from the large **testis** or **ovary** into the coelom. A pair of **genital pores** leads from the coelom into a **urogenital sinus** formed by the fused posterior ends of the archinephric ducts and thence to the cloaca and outside. The absence of genital ducts may be a very primitive feature.

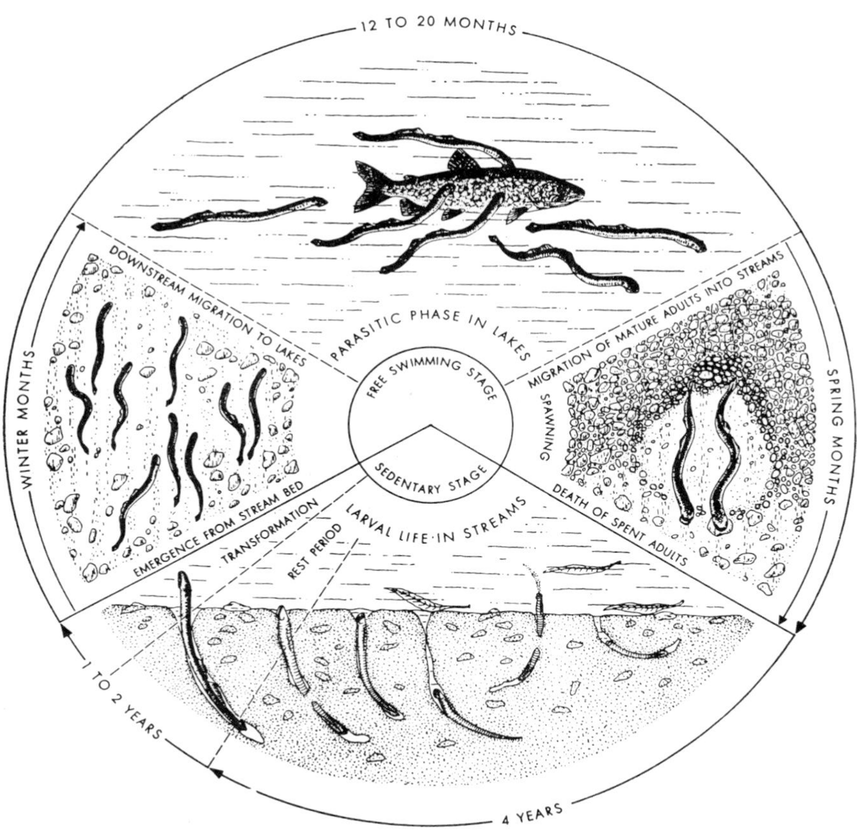

FIGURE 31.7 The life of the sea lamprey in the Great Lakes. The lamprey spends all but a year or two of its six and one half to seven and one half years of life as a larva. (From Applegate and Moffett: *The Sea Lamprey.* Copyright © by Scientific American, Inc., April, 1955. All rights reserved.)

The eggs are laid on the bottom of streams in a shallow nest, which the lampreys make by removing the large stones (Fig. 31.7). Fertilization is external, and the adults die after spawning.

Developing sea lampreys pass through a larval stage that lasts five to six years. The larva is so different in appearance from adult lampreys that originally it was believed to be a different kind of animal and was named **Ammocoetes.** The ammocoetes larva is eel-shaped but lacks the specialized feeding mechanism of the adult. It lies within burrows in the mud at the bottom of streams and sifts minute food particles from water passing through the pharynx. Like the lower chordates, it has a mucus-producing **endostyle** that aids in trapping food.

Sea lampreys reached the Great Lakes in the 1920s, presumably having circumnavigated Niagara Falls through the Welland Canal. They did considerable damage to commercial fisheries in the upper Lakes during the 1940s and 1950s, but control programs using poisons that selectively kill the larvae have greatly reduced the problem in recent years.

Brook lampreys probably evolved from the predaceous sea or lake species. They too spend most of their lives as filter-feeding larvae. During metamorphosis their digestive tract atrophies so they cannot feed. After a few weeks they reproduce and die.

The hagfish are exclusively marine. The best known species, *Myxine glutinosa* of the Atlantic and *Bdelostoma stouti* of the Pacific, are primarily scavengers feeding at the bottom on dead or injured fish and soft-bodied invertebrates. But they can also attack netted fish, entering the body through the anus or gill pouches, and then feed upon the soft internal organs. They are of considerable interest to zoologists for they are unique in several ways. They are the only vertebrates to resemble marine invertebrates in having a blood that is isosmotic to sea water. This suggests that hagfish have always been a marine group. Other vertebrates are osmotically independent of their environment, and this characteristic probably evolved during a freshwater stage of their ancestry. Hagfish are also hermaphroditic, but any individual produces only eggs (from the anterior part of the gonad) or sperm (from the posterior part) in a given season.

Jaws and Paired Appendages

The absence of jaws and paired appendages confines agnathous fishes to relatively inactive lives as filter feeders, blood suckers, and scavengers. Late in the Silurian and early in the Devonian periods, jaws and paired appendages evolved. All jawed vertebrates are frequently called **gnathostomes.** Jaws enabled fishes to feed upon a wider variety of food than could the jawless ostracoderms. Paired appendages, coupled with the evolution of a more streamlined body shape and some reduction in the heavy ancestral armor, permitted greater mobility. Paired appendages are essential to maintain stability. Fishes blossomed in the Devonian, and most lines of evolution became established (Fig. 31.3).

We can most easily visualize the evolution of jaws and paired appendages by examining some of the **acanthodians** (Gr. *akanthodes,* spiny), often called "spiny sharks," although they are probably more closely allied to bony fishes than to sharks (Fig. 31.8). Some ostracoderms had bony plates bordering their mouths that probably helped in picking up bits of food. As time went on, the mouth opening appears to have migrated posteriorly and became associated with the anterior part of the gill or visceral skeleton (p. 103). One or two visceral arches may have been lost in the process, for jawed fishes have fewer gill slits and arches than jawless ones. The most anterior of the remaining visceral arches, the **mandibular arch,** became enlarged and, together with dermal bones developed in the skin adjacent to it, formed the jaws. The second visceral arch, known as the **hyoid arch,** lay close behind the mandibular arch and in some species extended as a prop from the brain case to the posterior end of the upper jaw as it does in many contemporary species (Fig. 5.19, p. 103). The remaining visceral arches are **branchial arches** that supported the gills.

Acanthodians had a series of spines protruding from the side of their trunk (Fig. 31.8*A*). Grooves along the posterior borders of the spines suggested that each supported a web of flesh, so they were in effect paired appendages. Their number varied among different species of acanthodians. In other Devonian fishes, only a pair of pectoral and pelvic fins were present. In all of these early fishes, the paired fin was attached to the trunk by a broad base. Such a fin would have been an effective stabilizing keel but could not have twisted and turned and helped the fish maneuver. Its role in maneuvering evolved later when the fin base narrowed.

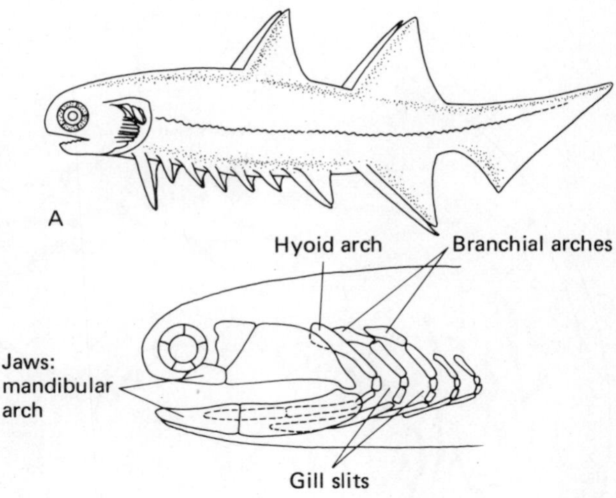

FIGURE 31.8 *A,* The "spiny shark," *Climatius,* was among the first jawed vertebrates. This fish was about 8 cm. long. After removal of the gill covering and superficial bony scales and plates on the head of a related genus *(Acanthodes, B),* it can be seen that the jaws are modified gill arches. (*A* after Watson; *B* modified after Watson.)

There is a bewildering array of primitive jawed fishes from the Devonian, and paleontologists are not in agreement as to how to sort them out. Some are obviously related to contemporary sharks and rays; others, to bony fishes; still others may be unique, extinct groups without living descendants. Among the last are a group of peculiar fishes that are often assigned to the **class Placodermi** (Gr. *plax,* plate + *derma,* skin). Early members of the group, like the ostracoderms, were small freshwater creatures, but the most spectacular genus, *Dunkelosteus* (Fig. 31.9), was a monster of the late Devonian sea living over what is now Cleveland, Ohio. *Dunkelosteus* was 2 to 3 m. long. Its head and anterior trunk were covered by bony plates, but most of the body was naked. Little is known about the structure of its paired fins, but it had formidable jaws. Large bony plates with cleaver-like edges were attached to the underlying mandibular arch.

Class Chondrichthyes

The class Chondrichthyes (Gr. *chondros,* cartilage + *ichthys,* fish) includes about 700 species of contemporary sharks, rays, and similar fishes with a skeleton com-

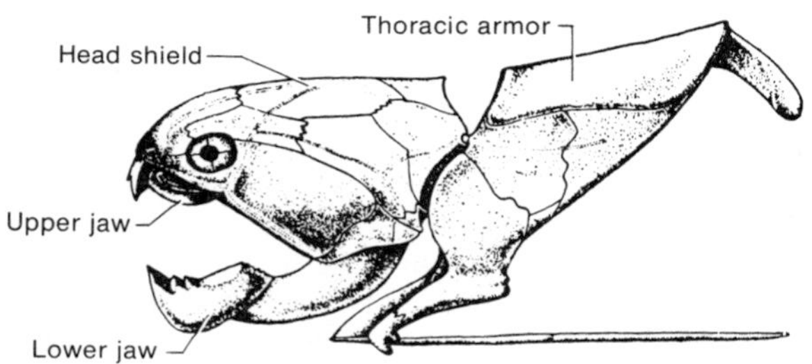

FIGURE 31.9 A lateral view of the anterior end of the placoderm *Dunkleosteus.* The dermal plates composing the upper and lower jaws were attached to the mandibular arch. This fish attained a length of 3 m.

posed only of cartilage. Chondrichthyans are an ancient group extending back to the early Devonian (Fig. 31.3). Whether they evolved from placoderms or directly from ostracoderms is uncertain. The earliest known species were marine, and the group has remained primarily a marine group ever since, although one extinct and several contemporary species have secondarily adapted to fresh water.

The deeper skeleton of all vertebrates is cartilaginous during embryonic life. Cartilaginous fishes retain a cartilaginous skeleton (Fig. 5.9). Although the cartilage is not replaced by bone, calcium salts are sometimes deposited within it and strengthen it. The extensive ancestral dermal armor has been reduced to minute **placoid scales** embedded in the skin (Fig. 31.5*C*). Their enamel-like and dentine-like superficial layers resemble the outer parts of cosmoid scales. The triangular teeth of sharks closely resemble enlarged placoid scales and have doubtless evolved from them.

The **mouth cavity** is continuous posteriorly with a long **pharynx** (Fig. 31.10*A*). A **spiracle,** containing a vestigial gill, and the **gill pouches,** containing functional gills, open from the pharynx to the body surface. A wide **esophagus** leads from the back of the pharynx to a J-shaped **stomach.** A short, straight **valvular intestine,** which receives secretions from the **liver** and **pancreas,** continues back to the **cloaca.** It contains an elaborate spiral fold known as the **spiral valve;** this helical fold serves both to slow the passage of food and to increase the digestive and absorptive surface of the intestine (Fig. 31.10*B*).

The circulatory system is of the primitive type in which the **heart** is undivided and pumps only venous blood forward to the gills where it is aerated (Fig. 8.11). Blood, which is also cooled in the gills, is distributed to the tissues under relatively low pressure.

The primitive **opisthonephric kidneys** (p. 200) are drained in the female by an **archinephric duct.** In males, the archinephric duct carries only sperm that enters the front of the kidney from the testis, and the

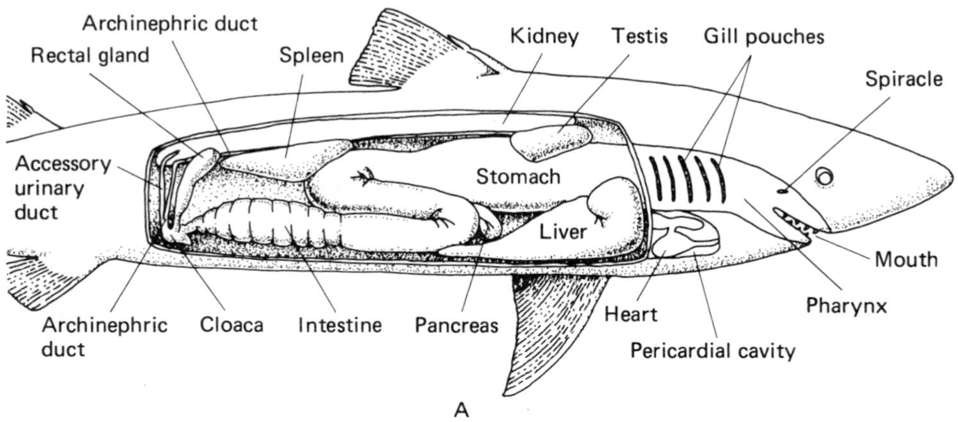

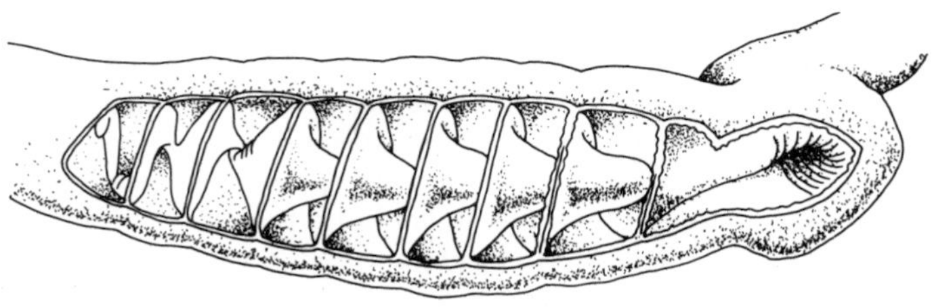

FIGURE 31.10 *A,* The visceral organs of the dogfish, *Squalus,* a representative cartilaginous fish. *B,* The intestine has been cut open to show the spiral valve.

urine-producing part of the kidney is drained by an **accessory urinary duct** (Fig. 31.10*A*). Except in the freshwater species, the kidneys and gills retain much urea so the body fluids are hyperosmotic to sea water (p. 208). Some ions are eliminated by the kidneys, but most excess salt taken in with the food is eliminated by a salt-excreting **rectal gland** that empties into the end of the intestine. Intestine and urogenital ducts discharge into a common **cloaca** that opens on the underside of the body.

A specialized part of the male pelvic fin forms a **clasper** that is used to transfer sperm to the female. The eggs are fertilized in the upper part of the oviduct, and a horny protective capsule is secreted around them by certain oviducal cells. Skates are **oviparous,** but there is no free larval stage as there is in many fishes. The eggs are very heavily laden with yolk, and the embryos develop within the protective capsule. Some sharks are also oviparous, but most brood their young internally. The fertilized eggs develop in a modified portion of the oviduct known as the **uterus;** in some an intimate association is established between each embryo's yolk sac and the uterine lining, forming a **yolk sac placenta;** and the embryos have a greater dependence for their nutrient requirements upon the mother than upon food stored in the yolk. However, in most sharks, including our common dogfish, the eggs develop within a uterus, but there is a greater dependence upon food stored in the yolk. In some cases, a portion of the nutritional requirements is derived by the absorption of materials secreted by the mother into the uterine fluid or by certain embryos consuming other embryos. An intimate placental relationship is not established.

Adaptive Radiation of Cartilaginous Fishes

Cartilaginous fishes are grouped into two subclasses. The **Elasmobranchii** (Gr. *elasmos,* metal plate + *branchia,* gills) include the sharks and skates and are characterized by having separate gill slits for each gill pouch. The **Holocephali** (Gr. *holos,* whole + *kephale,* head) are a side branch that includes the present-day ratfish (*Chimaera,* Fig. 31.11). In this group the gill pouches open into an opercular chamber, scales are absent, the lateral line forms an open groove on the body surface, and the upper jaw is fused to the cranium (whence the name for the group). Chimaeras have crushing tooth plates and feed upon seaweed, mollusks, crustaceans, and similar invertebrates.

The ancestral cartilaginous fish were elasmobranchs belonging to the order **Cladoselachii** (Fig. 31.3). Their broad-based fins and their inability to protrude their jaws suggest that they were not as efficient in swimming and feeding as contemporary species. Contemporary elasmobranchs are grouped into two orders. The Selachii include the sharks, and the Batoidea, the skates and rays.

Sharks. Most selachians are active fishes with protrusible jaws that enable them to feed voraciously with their sharp triangular teeth upon other fishes, crustaceans, and certain mollusks. Whale sharks (*Rhineodon,* Fig. 31.11), which may reach a length of about 12 m., have minute teeth and feed entirely upon small crustaceans and other organisms that form the drifting plankton. They gulp mouthfuls of water, and as the water passes out of the gill slits, the food is kept in their pharynx by a branchial sieve. Whale sharks are the largest living fishes.

Skates and Rays. Most skates and rays are bottom-dwelling fishes that are flattened dorsoventrally. The undulations of their enormous pectoral fins propel the fish along the bottom (Fig. 31.11). Their mouth is often buried in the sand or mud, and water for respiration enters the pharynx via the pair of enlarged spiracles. A spiracular valve in each one is then closed, and the water is forced out the typical gill slits. Most skates and rays have crushing teeth and feed upon shellfish. The sawfish (*Pristis*) has an elongated, blade-shaped snout armed with toothlike scales. By thrashing about in a shoal of small fishes, it can disable many and eat them at leisure. The largest members of the group (the devilfish, *Manta*) have reduced teeth and are plankton-feeders.

Class Osteichthyes

Most living fishes are bony fishes, members of the **class Osteichthyes** (Gr. *osteon,* bone + *ichthys,* fish). The class includes such species as the sturgeon, salmon, minnows, perch and lungfish. Terrestrial vertebrates evolved from crossopterygians, early members of this group.

Basic Features of Bony Fishes

The cartilaginous embryonic skeleton of bony fishes is largely replaced by bone during embryonic development. The thick, primitive bony scales of their ancestors have lost the superficial layers of ganoine and cosmine, and there is little left but a thin disc of bone that develops in overlapping dermal folds of the skin (Fig. 31.5*D*). As a fish grows, increments of bone are added to the scales, and these appear as rings called **circuli.** In some species living in temperate environments, the rate of

FIGURE 31.11 Adaptive radiation among the cartilaginous fishes.

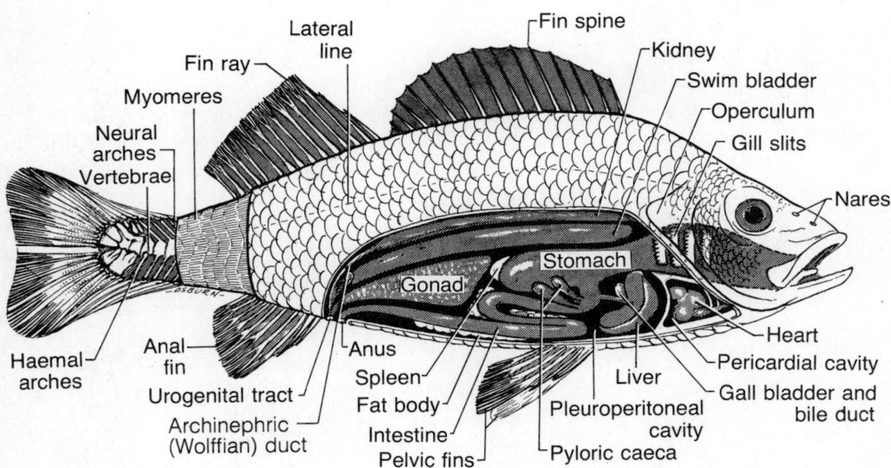

FIGURE 31.12 The visceral organs of the perch.

growth slows in winter, and circuli formed during this period are crowded together forming a **check mark.** Estimates of a fish's age can be made by counting check marks.

The gill pouches, bearing gills, open into a common opercular chamber. Blood and water move through and across the gills in opposite directions so there is a very efficient countercurrent exchange of gases between blood and water (Fig. 7.2, p. 151). A swim bladder, a hydrostatic organ, lies in the dorsal part of the body cavity (Fig. 31.12). Gases, primarily oxygen, can be secreted into it, or be reabsorbed, through specialized capillaries in its wall so the specific gravity of the fish can be adjusted to give the fish neutral buoyancy at different water depths.

The swim bladder of advanced bony fishes and the lungs of more primitive species are among the most distinctive and important characteristics of the group. These organs probably had a common evolutionary origin (Fig. 31.13). The ancestral organ was more like a lung than a swim bladder. The swim bladder loses its connection to the digestive tract only in adult teleost fishes, the most advanced group of bony fishes. We believe that the ancestral bony fishes had lungs similar to those of the living African lungfish (*Protopterus*). In the lungfish, a pair of saclike lungs develops as a ventral outgrowth from the posterior part of the pharynx (Fig. 31.13). The lungs enable the fish to survive conditions of stagnant water and drought. The rivers in which the African lungfish live may completely dry up, but the fish can

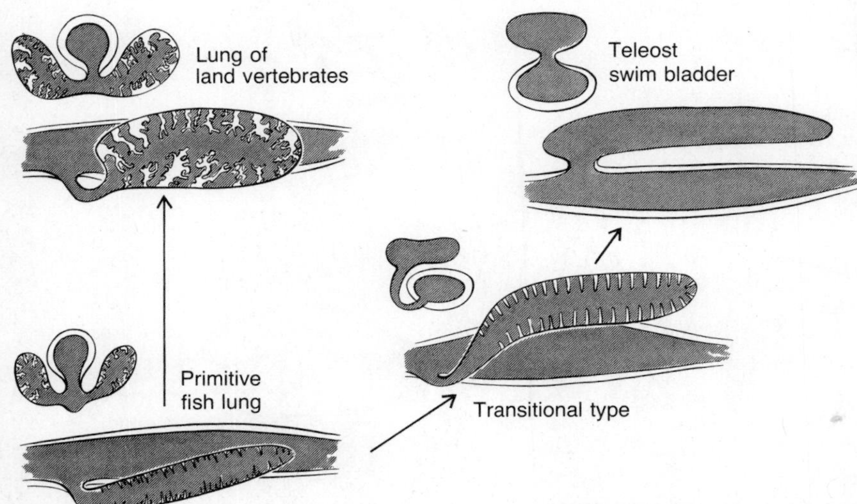

FIGURE 31.13 A diagram to illustrate the evolution of lungs and the swim bladder. The connection to the digestive tract is lost in most adult teleosts. (After Dean.)

survive curled up within a mucous cocoon that it secretes around itself in the dried mud. A small opening from the cocoon to the surface of the mud enables the fish to breathe air during this period.

Air-breathing probably evolved in fishes as a supplement to gill respiration. Presumably, early bony fishes evolved lungs as an adaptation to the unreliable freshwater conditions of the Devonian period. Bodies of fresh water undoubtedly either became stagnant swamps with a low oxygen content or dried up completely. Only fishes with lungs could survive these conditions. The others became extinct or migrated to the sea, as did many placoderms and cartilaginous fishes. Groups of bony fishes that have remained in fresh water throughout their history tended to retain lunglike organs, but those that went to sea no longer needed lungs, for ocean waters usually contain a constant and adequate oxygen supply. Their useless lungs evolved into useful hydrostatic organs. What are presumed to be intermediate stages in this shift can still be seen in certain species. Later, when conditions were more favorable, many salt-water bony fishes reentered fresh water but retained their swim bladders.

The intestine of most bony fishes has lost the primitive spiral valve and is coiled upon itself (Fig. 31.12). Several short **pyloric caeca** extend outward from the anterior end of the intestine just behind the stomach and increase the absorptive surface area.

The body fluid of freshwater bony fishes is hyperosmotic to their environment (p. 208). Excess water that enters the body is excreted by the kidneys. Lost salts are replaced in part by salts in the food and in part by special salt-absorbing cells in the gills. (Similar cells excrete salt in marine species.)

Unlike most vertebrates, the sperm of bony fish do not pass from the testes to the kidneys, or to structures derived from the kidneys (epididymis), but directly to sperm ducts that join the posterior ends of the urinary ducts. The ovaries of the perch and most bony fishes discharge directly into an oviduct rather than into the coelom. Perch are oviparous and fertilization is external.

Evolution of Bony Fishes

Bony fishes can be traced back to the early Devonian and late Silurian (Fig. 31.3), when several lines of evolution within the class were already established. Since all share such features as a partly ossified internal skeleton, well-developed bony scales, and an operculum (a combination of characters not present in other classes of fish), it is assumed that they had a common evolutionary origin, possibly from some ostracoderm group. Members of the subclass **Acanthodii** were characterized by prominent supporting spines at the anterior edge of the dorsal, anal, and paired fins (Fig. 31.8). Acanthodians were the earliest of the bony fishes, and it is possible that other lines of bony fish evolved from them. They flourished during the Devonian period but were replaced by more progressive types by the end of the Paleozoic era. Other bony fishes are usually grouped into the subclasses **Actinopterygii** and **Sarcopterygii**.

Actinopterygians. The actinopterygians (Gr. *aktis*, ray + *pterygion*, fin) are the familiar ray-finned fishes, such as the perch. Their paired fins are fan-shaped (Fig. 31.14). Skeletal elements enter their base, but most of the fin is supported by dermal rays that evolved from rows of bony scales. Their paired olfactory sacs connect only with the outside and not with the mouth cavity.

The infraclass **Chondrostei** has dwindled to a few species, of which the Nile bichir (*Polypterus*) and the sturgeon (*Acipenser*) are examples (Fig. 31.14). The infraclass **Holostei** has also dwindled and is represented to-

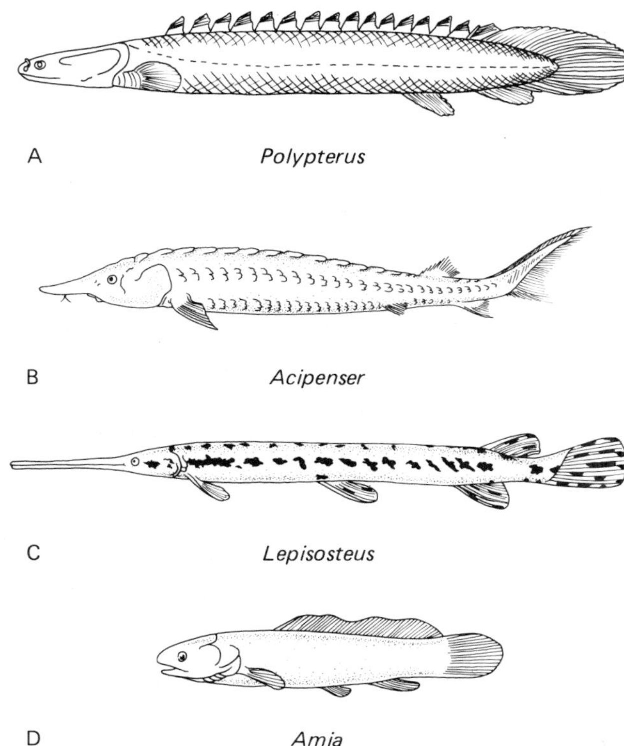

FIGURE 31.14 A group of primitive actinopterygian fishes that have survived to the present day. *A*, the Nile Bichir *(Polypterus)*, and *B*, the sturgeon *(Acipenser)*, are chondrosteans. *C*, The gar *(Lepisosteus)*, and *D*, the bowfin *(Amia)*, are holosteans. (Redrawn from various sources.)

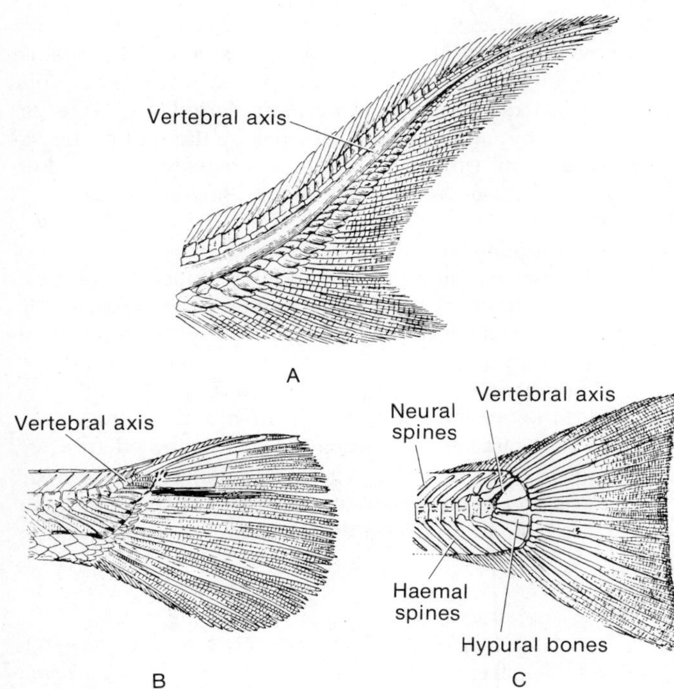

FIGURE 31.15 Caudal fin types of bony fishes. *A,* The primitive heterocercal tail in the sturgeon, *Acipenser; B,* the abbreviated heterocercal tail of the garpike, *Lepisosteus; C,* the homocercal tail of a teleost. (*A* and *B* from Jordan; *C* from Romer.)

day by such relict species as the gar (*Lepisosteus*) and bowfin (*Amia*). The infraclass **Teleostei,** in contrast, has been continuously expanding since its origin in the Mesozoic era. It is to this group that the minnows, perch, and most familiar fishes belong.

In the course of actinopterygian evolution, the vertebral column became more thoroughly ossified and stronger, thereby providing a better attachment for powerful axial muscles. The functional lungs of early actinopterygians became transformed into swim bladders with little respiratory function. In connection with increased buoyancy and better streamlining, the primitive heterocercal tail of most chondrosteans became superficially symmetrical in teleosts, but the caudal skeleton still shows indications of the upward tilt of the vertebral column (Fig. 31.15). Such a tail is said to be homocercal. Increased buoyancy relieved the fins of their primitive hydroplaning function. Their base became narrower, and the point of attachment of the pectoral fins shifted dorsally so that they were closer to a horizontal plane passing through the center of gravity. These changes improved the braking and turning functions of the fins. Early actinopterygians were clothed with thick bony **ganoid** scales (Fig. 31.5*B*). During subsequent evolution, the scales became thinner and lighter. The bone was reduced to a thin disc that develops in the dermis of the skin (Fig. 31.5*D*). Such a scale is termed **cycloid** if its surface is smooth; **ctenoid** if the posterior portion bears minute spiny processes.

Teleosts are an exceedingly large and diverse infraclass of 20,000 or more species. Most living bony fishes are teleosts. They are grouped by most investigators into 30 orders and 9 superorders. The latter are shown in the classification on page 669. Teleosts vary considerably in body shape and other adaptations for their modes of life, as we shall see, but despite this variety certain evolutionary trends can be seen by comparing a primitive teleost such as a herring with a more recently evolved species such as a perch (Fig. 31.16). Primitive teleosts have elongated, streamlined bodies, a single dorsal fin, pelvic fins located near the posterior part of the trunk, fins supported by flexible and branching bony rays rather than by spines, cycloid scales covering the tail and trunk but not extending onto the head, and a duct connecting the swim bladder and digestive tract. The jaws are short but do not extend forward very far when the mouth is opened. The most advanced teleosts, such as the sunfishes and perch, are often rather short and deep-bodied (from dorsal to ventral) fishes. There is a tendency for the dorsal fin to split into two parts, the anterior being supported by spines, the posterior by flexible bony rays. The pelvic fins have shifted forward to a point beneath the pectoral fins, and spines are present in the anterior border of these fins. The scales

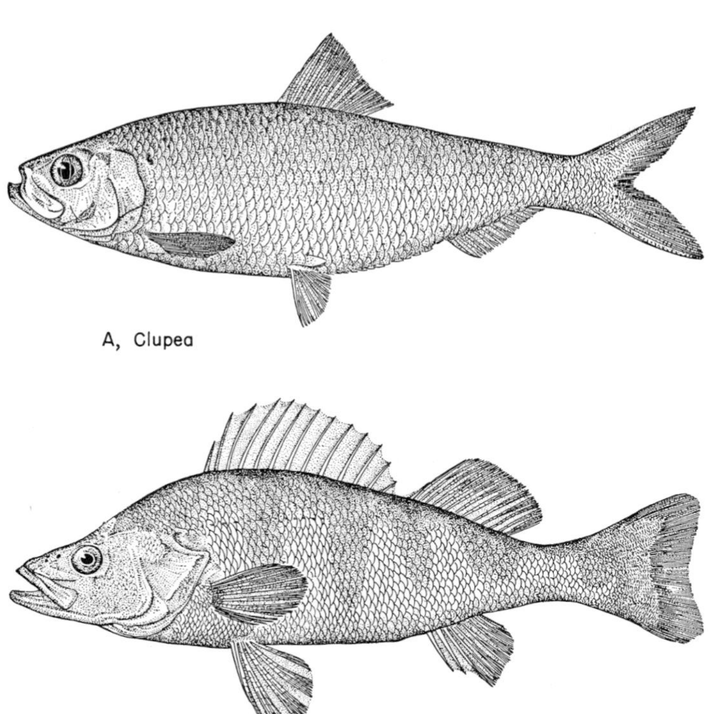

FIGURE 31.16 *A,* A primitive teleost, the herring; *B,* an advanced teleost, the perch. Differences in body proportions and fins are apparent. (From Romer, A. S., and T. S. Parsons: *The Vertebrate Body.* 5th ed. Philadelphia, Saunders College Publishing, 1977.)

have become ctenoid and have extended onto the head and operculum. The duct to the swim bladder is lost. The jaws are highly specialized and protrude rapidly as the mouth opens. This and the concomitant expansion of the buccal and pharyngeal cavities create a suction that helps draw food in.

Teleosts between these two extremes show a mixture of primitive and advanced characteristics and often contain distinctive features of their own. Most of our common freshwater teleosts, such as minnows, suckers, and catfish, resemble primitive teleosts generally but are distinctive in having a set of small bones, the **weberian ossicles,** extending from the anterior end of the swim bladder to the inner ear (Fig. 31.17). This mechanism apparently acts as a hydrophone, for the acoustical sensitivity of these fishes is considerably greater than that of any other group.

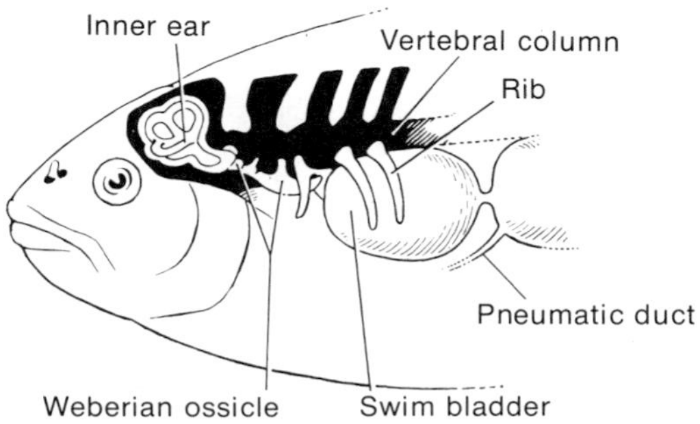

FIGURE 31.17 A lateral view of the head of a carp to show the connection between inner ear and swim bladder made by the weberian ossicles. (From Portman.)

Adaptive Radiation of Teleosts. Few groups of vertebrates have undergone such an extensive adaptive radiation as teleost fishes. They are found in every conceivable aquatic environment. They can tolerate temperatures ranging from the arctic ocean to hot springs. They live in the open ocean, shallow waters, and the deep sea. They are specialized to feed in nearly every possible way. Some are filter feeders; many are carnivores and herbivores. Some have jaw and tooth modifications enabling them to crush coral, and some are specialized to feed on the scales or nibble the eyes of other fishes! We can only examine a few of the fascinating adaptations of this interesting group.

Eels have long, snakelike bodies and have lost their pelvic fins and usually their scales (Fig. 31.18). This body form is particularly well adapted to living in crevices in

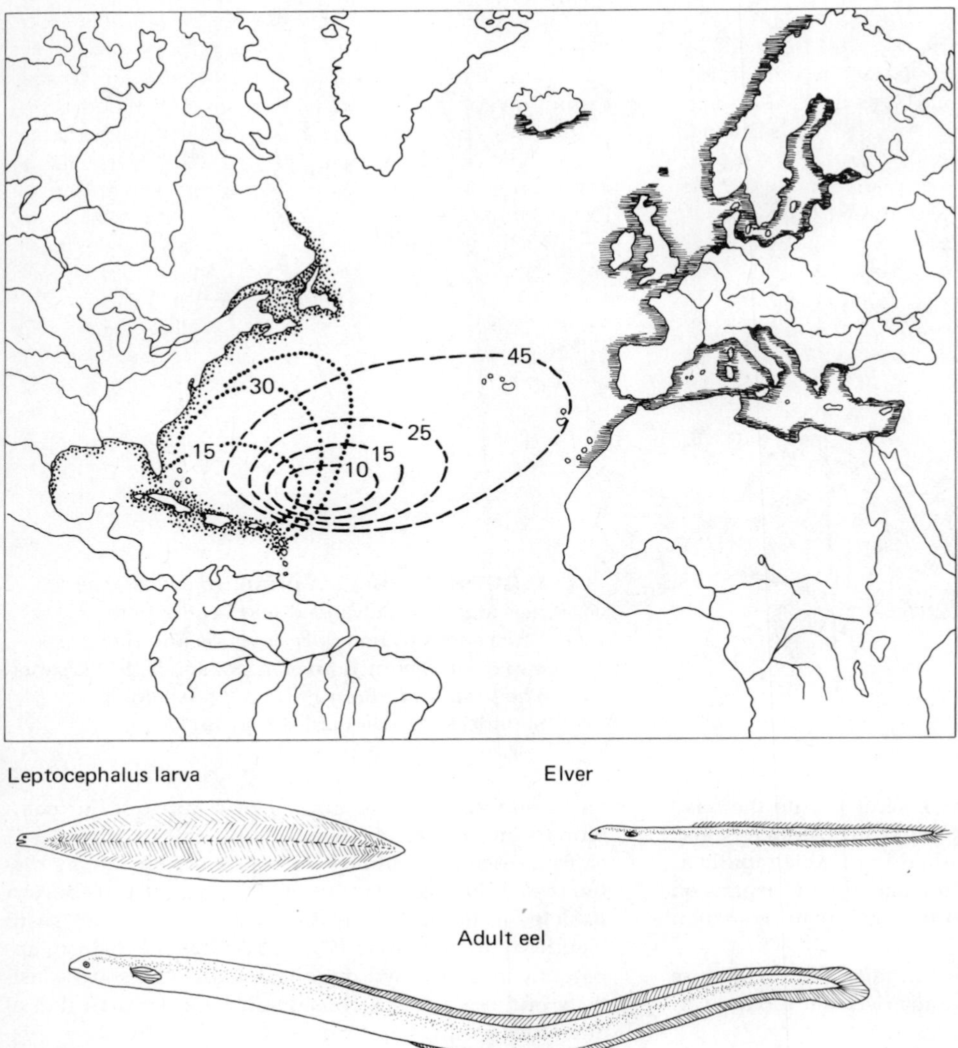

FIGURE 31.18 Life cycle of two species of eels. The North American species *(Anguilla rostrata)* lives in watersheds that enter the coast in areas indicated by stipples; the European species *(Anguilla vulgaris),* in watersheds that enter coastal areas outlined by horizontal lines. Curved lines and numbers in the Atlantic indicate larval distribution and length in millimeters. The leptocephalus larva, elver, and adult eel are shown below the map. (Map after Norman, in Orr R. T.: *Vertebrate Biology.* Philadelphia, Saunders College Publishing, 1982.)

coral reefs, as many marine species do, or in waters with considerable mud and vegetation, as freshwater species do. Freshwater eels have a particularly interesting life cycle, which was determined in the early part of this century by Johann Schmidt, a Danish investigator. With the cooperation of sea captains, he was able to catch thousands of larval eels in plankton nets from various regions of the Atlantic Ocean. He could determine where they bred and their migration by correlating larval size with where they were caught. When they are ready to breed, both the European species, *Anguilla vulgaris,* and the American species, *Anguilla rostrata,* descend the rivers and migrate at considerable depth to the Sargasso Sea, southeast of Bermuda. The young larvae move to the surface and are carried by the Gulf Stream towards the coasts of Europe and North America. The **leptocephalus larvae** are transparent, ribbon-like creatures that do not look at all like eels. Larvae of the two species can be distinguished by having slightly different numbers of vertebrae. The European species takes three years to develop into a small eel-like **elver,** by which time it has reached the coastal rivers and is ready to ascend them. The American species has a shorter distance to travel and reaches the elver stage in one year. The two species breed in different parts of the Sargasso Sea and drift in somewhat different parts of the Gulf Stream, but should the European larvae be carried to North America, they would reach the coast long before they were elvers and would die. Both species mature and live in fresh water, sometimes as long as 20 years, before migrating back to the ocean to spawn and die.

Mackerel and tuna are fish of the open ocean. They are highly streamlined and are very fast and powerful swimmers (Fig. 31.19*A*). They seldom rest; indeed, they depend upon their forward motion to ventilate their gills (ram ventilation, p. 152). If they are caged so they cannot swim, the oxygen tension in their blood drops to low levels. As should be expected, they have a larger proportion of red muscle (p. 113) than other fishes. When temperature probes are inserted into this musculature in freshly caught specimens, it is found that much of this musculature is maintained at a temperature nearly 10° C. above that of the water and body surface. Fish blood is cooled to water temperature as it flows through the gills, so how is a high temperature maintained in these muscles? Blood is distributed to the red muscles not by branches of the dorsal aorta, as in other fishes, but by arteries travelling close to the skin. This blood enters the muscle through many small arteries that parallel, but flow in an opposite direction to, many small veins leaving the muscles. Heat generated by muscular activity is absorbed by the venous blood and transferred back to the cooler arterial blood. Be-

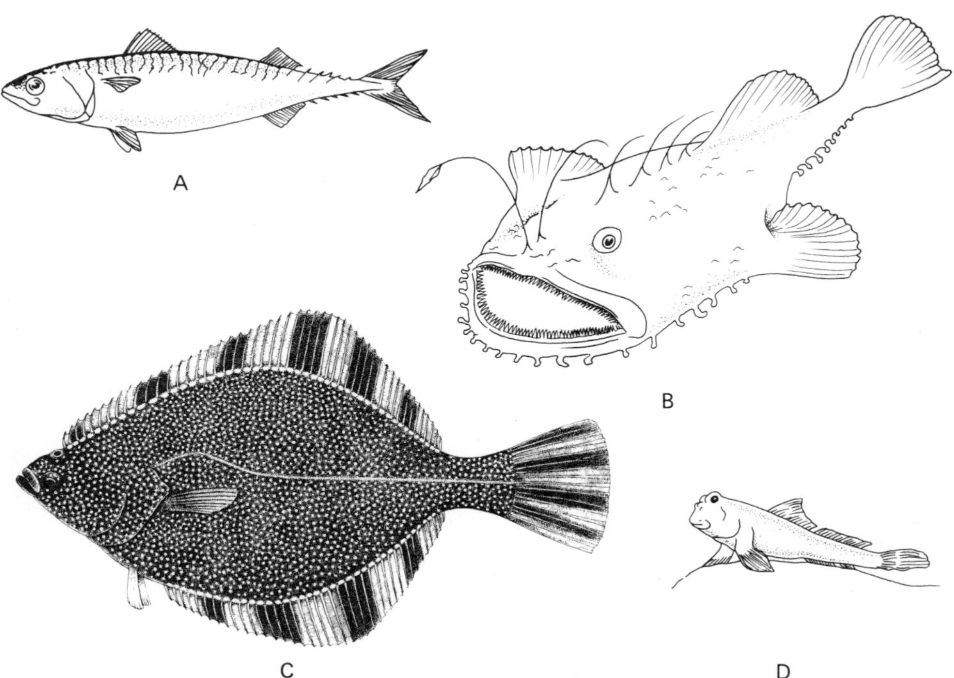

FIGURE 31.19 Some adaptations of teleosts for different modes of life. *A,* the mackerel *(Scomber)* lives in the open ocean; *B,* the angler-fish *(Lopius)* lives on the bottom awaiting prey; *C,* the starry flounder *(Platichthys)* swims slowly over the bottom; *D,* the mud-skipper *(Perophthalmus)* occasionally leaves the water. (*C* from Policansky, D.: The asymmetry of flounders. *Scientific American, 246*(May):116, 1982. Others from Young, J. Z.: *Life of Vertebrates.* Oxford, Clarenden Press, 1981, partly after Norman.)

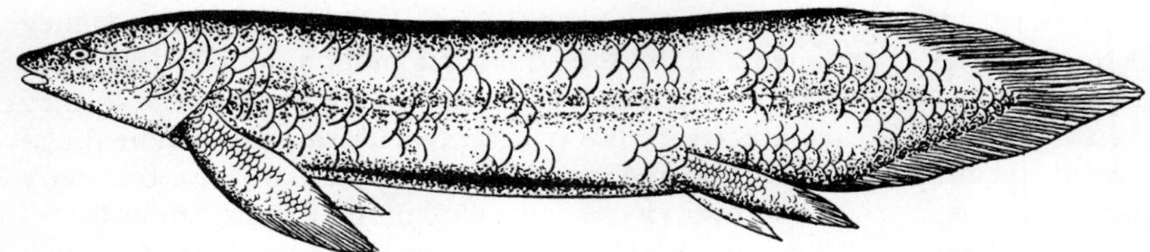

FIGURE 31.20 The Australian lungfish, *Epiceratodus.* (After Norman.)

cause of this countercurrent heat exchange system, high temperatures can be maintained in the red muscles, and this increases their power output.

Fish that live on the bottom tend to reduce or lose their swim bladder, and they are frequently flattened. They can hug the bottom as they feed and hide. Teleosts have evolved a flattened shape in two different ways. Angler fish (Fig. 31.19*B*) are flattened dorsoventrally, have a concealing color pattern, and lie on the bottom with their large mouths directed upward. In some species, the anterior dorsal spine bears a lure that overhangs the mouth. Some other species use a red, wormlike fold in the floor of the mouth as a lure. Flounders, sole, and halibut (Fig. 31.19*C*) are flattened from side to side and lie on or swim slowly over the bottom as they feed on invertebrates. The downward side, which may be either the left or the right side, according to the species, has lost its pigment, but the upward side has a concealing color pattern. During the course of embryonic development, the eye that would be on the downward side migrates to the upward side. Mouth position, however, does not change.

A few teleosts, such as the mud-skipper of Southwest Asia (Fig. 31.19*D*), even leave the water for brief periods and climb up on stones or tree roots with muscular pectoral fins as they chase a crustacean or escape a predator. By closing their operculum they can maintain a high humidity in the opercular chamber and obtain oxygen from the air with their gills.

Sarcopterygians. Although of great interest because of their diversity, teleosts are a side issue in the total picture of vertebrate evolution. The branch toward the higher vertebrates passed through the less spectacular sarcopterygians of ancient Devonian swamps. The sarcopterygians (Gr. *sarx,* flesh + *pterygion,* fin) include the lungfishes and crossopterygians (Figs. 31.20 and 31.21). Their paired appendages are typically elongate, lobe-shaped, and supported internally by an axis of flesh and bone. In many species, each of the olfactory sacs connects to the body surface through an **external nostril** and to the front part of the roof of the mouth cavity through an **internal nostril.** Sarcopterygian evolution diverged at an early time into two lines—the lungfishes (order **Dipnoi**) and the crossopterygians (order **Crossopterygii**). The primitive crossopterygians had a well-ossified internal skeleton, a unique jointed chondrocranium, and small conical teeth suited for seizing prey. It is from this group that the amphibians arose. Throughout their evolutionary history, lungfishes have had a weak skeleton with little ossification of the vertebral column. They developed specialized crushing tooth plates, which enabled them to feed effectively upon shellfish, and showed tendencies toward reduction of the paired appendages. In certain other features, lungfishes and crossopterygians have paralleled actinopterygian evolution. They evolved symmetrical tails, though of a type that is symmetrical internally as well as externally (**diphycercal**), and their primitive, thick

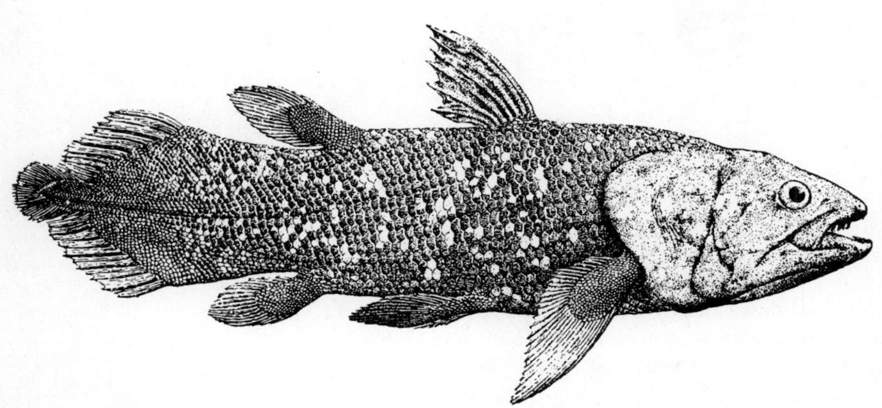

FIGURE 31.21 *Latimeria,* a living coelacanth found off the coast of the Comoro Islands. (From Millot.)

bony scales, which were characterized by having a thick layer of dentine-like **cosmine** (Fig. 31.5), have tended to thin to the cycloid type.

Both crossopterygians and lungfishes were successful in the fresh waters of the Devonian but have dwindled to a few relict species today. Lungfishes retained their lungs and have survived in the unstable freshwater environments of tropical South America, Africa, and Australia. It was long believed that all crossopterygians had become extinct, as indeed the primitive freshwater ones have. However, a few specimens of a somewhat specialized marine side branch (the coelacanths) have been found in recent years near the Comoro Islands between Africa and Madagascar (Fig. 31.21).

Classification of Fishes

Class Agnatha Jawless fishes. The oldest and most primitive class of vertebrates; all members lack jaws and pelvic fins; many lack pectoral fins.

†**Ostracoderms.** Several orders of small Paleozoic fishes are collectively called the ostracoderms; most were heavily armored. *Hemicyclaspis.*

Order Cyclostomata Lampreys and hagfishes. Body eel-shaped; round mouth without jaws; no paired appendages; scaleless slimy skin. The sea lamprey *Petromyzon.*

†**Class Placodermi** Several orders of primitive jawed fish of the Paleozoic; bony head shield movably articulated with thoracic shield; most of trunk without scales. *Dunkleosteus.*

Class Chondrichthyes Extinct and living cartilaginous fishes that have completely lost the primitive covering of dermal plates. Placoid scales usually present; internal skeleton entirely cartilaginous; lungs or swim bladder absent; nearly all are marine.

Subclass Elasmobranchii Gill slits open independently on the body surface; placoid scales present; lateral line system embedded in the skin.

Order Selachii Modern sharks with narrow-based paired fins; well-formed vertebral centra; streamlined bodies; heterocercal tail; short protrusible jaws usually provided with sharp triangular teeth. The dogfish, *Squalus*; whale shark, *Rhineodon.*

Order Batoidea Sawfish, skates, and rays. Body dorsoventrally flattened; pectoral fins enlarged; tail reduced; teeth usually in the form of crushing plates. Common skate, *Raja;* devilfish, *Manta;* sawfish, *Pristis.*

Subclass Holocephali Gill slits covered by an operculum; scales absent; lateral line system in the form of open grooves. The ratfish, *Chimaera.*

Class Osteichthyes Bony fishes. Skeleton includes considerable bone; various types of bony scales, other than placoid, usually present; lungs or swim bladder usually present; abundant in fresh and salt water.

†**Subclass Acanthodii** Several orders of Paleozoic fishes. Oldest jawed fishes; a prominent spine in front of dorsal, anal, and paired fins; usually more than two pairs of paired fins. The "spiny shark," *Climatius.*

Subclass Actinopterygii Ray-finned fishes. Paired fins fan-shaped, supported by radiating bony rays or spines; each nasal cavity has an entrance and exit on snout surface.

Infraclass Chondrostei Two extinct and two living orders of primitive ray-finned fishes. Scales usually ganoid; mouth opening large; tail usually heterocercal. The bichir, *Polypterus,* and the sturgeon, *Acipenser.*

Infraclass Holostei Three extinct and two living orders of intermediate ray-finned fishes. Scales ganoid or cycloid; mouth opening smaller than in chondrosteans; abbreviated heterocercal tail. Garpikes, *Lepisosteus,* and bowfins, *Amia.*

Infraclass Teleostei Specialized ray-finned fishes. Scales cycloid or ctenoid; tail homocercal; a hydrostatic swim bladder usually present; jaws short and usually protrusible.

†**Superorder Leptolepimorpha** One order of primitive herring-like teleosts still retaining some holostean characters, including traces of ganoine on scales.

Superorder Elpomorpha Three orders that include the tarpon and eels. These are primitive teleosts with a "leptocephalous" larva. American eel, *Anguilla.*

Superorder Clupeomorpha One order that includes herring, sardines, shad, and their allies. Primitive teleosts; scales cycloid; head and operculum not scaled; fins without spines; trunk not greatly shortened; single dorsal fin; pelvic fins abdominal; a duct connects swim bladder and digestive tract; most marine. Common herring, *Clupea.*

Superorder Osteoglossomorpha One order of primitive freshwater teleosts of the Old and New World tropics with a unique bony tongue that pushes against palatal teeth. The North American mooneye, *Hiodon;* and the African elephant-snouted fish, *Mormyrus.*

Superorder Protacanthopterygii Four orders that include the salmon, trout, and pike. Primitive tel-

†Extinct groups.

eosts, some of which are beginning to acquire certain of the advanced features of acanthopterygians. The salmon, *Salmo.*

Superorder Ostariophysi Two orders that include the carp, minnows, suckers, and catfish. A large group of primarily freshwater teleosts; primitive in most respects, but have webberian ossicles. The shiner, *Notropis.*

Superorder Atherinomorpha One order that includes the top minnows, flying fishes, and their allies. Teleosts with a mixture of primitive and advanced characters; scales usually cycloid; scales beginning to spread onto head and operculum; some spines present in fins; pelvic fins abdominal in position. Killifish, *Fundulus;* silverside, *Atherina;* flying fish, *Exocoetus.*

Superorder Paracanthopterygii Five small orders of specialized teleost fishes with advanced characteristics, including pirate perch, toadfish, clingfish, anglers, and cod. Scales usually ctenoid; spines usually present in fins; pelvic fins located far forward. The cod, *Gadus.*

Superorder Acanthopterygii Twelve orders of specialized teleosts including sea horses, sculpins, perch, bass, barracuda, halibut, flounders, and puffers. Advanced teleosts; scales ctenoid; head and operculum usually scaled; spines present in fins; trunk often short; often two dorsal fins; pelvic fin thoracic in position; swim bladder without a duct. The perch, *Perca.*

Subclass Sarcopterygii Fleshy-finned fishes. Paired fins lobe-shaped with a central axis of flesh and bone; internal nostrils usually present; primitive members had cosmoid scales.

Order Crossopterygii Crossopterygians. Primitive fleshy-finned fishes, conical teeth present; median eye usually present.

†*Suborder Rhipidistia* Primitive freshwater crossopterygians; includes the ancestor of amphibians. *Osteolepis.*

Suborder Coelacanthini More specialized freshwater and marine crossopterygians; one living genus. *Latimeria.*

Order Dipnoi Lungfishes. Specialized fleshy-finned fishes; teeth forming crushing tooth plates; median eye usually absent; three living genera confined to Australia (*Neoceratodus*), Africa (*Protopterus*), and South America (*Lepidosiren*).

†Extinct groups

Summary

1. Fishes are well suited for an aquatic life. They are streamlined. Their skeleton is not as strong as in terrestrial vertebrates. The segmented muscles and the tail provide the thrust for locomotion, and other fins provide stability and maneuverability. The structure of their sense organs enables them to detect changes beneath the water. Their heart pumps only venous blood through the gills. A muscular tongue is absent.

2. The earliest fishes, which go back to the late Cambrian period, were heavily armored ostracoderms of the class Agnatha. Most were freshwater and fed on the bottom with jawless mouths. They lacked well-developed paired fins and were not very active fishes.

3. The only living agnathous vertebrates are the lampreys and hagfishes of the order Cyclostomata. They too lack jaws and paired appendages.

4. Jaws, which first occur in the acanthodians, evolved from an enlarged visceral arch, the mandibular arch. Dermal bones may attach to the mandibular arch. Acanthodians had more than two sets of paired appendages, which were supported by spines.

5. Placoderms are an extinct class of primitive fishes, many of which had cleaver-like jaws.

6. Cartilaginous fishes of the class Chondrichthyes are characterized by having small placoid scales, no bone in the skeleton, no lungs or swim bladder, a heterocercal tail, a spiral valve in the intestine, and a pelvic clasper in the male. Fertilization is internal. They may be oviparous or brood their young internally, with varying dependence upon yolk or maternal nutrition.

7. In sharks and rays of the subclass Elasmobranchii, each gill pouch opens independently on the body surface. Ratfish of the subclass Holocephali have an opercular fold that covers the gill pouches. Sharks are predaceous; skates and rays are flattened, bottom-dwelling species that feed on mollusks and crustaceans.

8. Most living fishes are bony fishes belonging to the class Osteichthyes. Bony scales are retained in most cases. The deeper skeleton is partly or nearly completely ossified. Lungs or a swim bladder are present. The tail is usually homocercal. The spiral valve has been lost in most species, and pyloric ceca are present. The gills are covered by an operculum. Fertilization is external, and development is oviparous in most.

9. Ancestral bony fishes lived in fresh water subject to seasonal stagnation and drought. Lungs probably evolved as an accessory respiratory organ. Lungfishes, which have remained in fresh water, have retained lungs. Others became marine, and the lungs were transformed into a hydrostatic swim bladder. Many of these fishes reentered fresh water and retained the swim bladder.

10. The class Osteichthyes is divided into three subclasses. The Acanthodii, an extinct group, had broad-based, paired fins supported by single spines. The Actinopterygii (minnows, perch, and similar species) have fan-shaped paired fins supported primarily by soft rays. The Sarcopterygii (lungfish and crossopterygians) have lobe-shaped, paired fins supported by a central axis of flesh and bone.

11. The subclass Actinopterygii is divided into three infraclasses: the Chondrostei, which are represented by a few relict species (bichir and sturgeon); the Holostei, also represented by a few relict species (gar and bowfin); and the Teleostei, which includes most living species. During evolution from the more primitive chondrosteans to the teleosts, the lungs became a swim bladder, the tail shifted from heterocercal to homocercal, and the scales changed from ganoid to cycloid.

12. Teleosts in the course of their evolution have become deeper-bodied, the originally single dorsal fin has divided, pelvic fins have moved far forward, spines have developed in most fins, the scales have changed from cycloid to ctenoid and spread over the operculum and head, the swim bladder has lost its connection with the digestive tract, and the mouth has become highly protrusible. Teleosts have undergone an extensive adaptive radiation.

13. The sarcopterygians are grouped into two orders. The Dipnoi (lungfishes) have poorly ossified skeletons and crushing tooth plates that enable them to feed on crustaceans and mollusks. Three species survive in tropical South America, Africa, and Australia. The crossopterygians have a stronger skeleton and many conical teeth. Most are extinct, but a marine coelacanth has survived. Terrestrial vertebrates evolved from early freshwater crossopterygians.

References and Selected Readings

The following references contain useful information on many aspects of the biology of fishes and other groups of vertebrates.

Alexander, R. McN: *The Chordates.* 2nd ed. Cambridge, Cambridge University Press, 1981. An outstanding discussion of the major vertebrate groups in which biomechanical, physiological, and ecological factors are skillfully integrated with structure.

Fishbein, S. L. (Ed.): *Our Continent, A Natural History of North America.* Washington, National Geographic Society, 1976. Five chapters in this superbly illustrated and authoritative book deal with the evolution of vertebrates.

Olson, E. E.: *Vertebrate Paleozoology.* New York, Wiley-Interscience, 1971. A useful reference on vertebrate paleontology; discusses the adaptive nature of many evolutionary changes.

Orr, R. T.: *Vertebrate Biology.* 5th ed. Philadelphia, Saunders College Publishing, 1982. A valuable text and reference on many aspects of vertebrates; a chapter is devoted to each major group of vertebrates and to such general topics as territory, dormancy, and population dynamics.

Romer, A. S.: *The Vertebrate Story.* Revised ed. Chicago, University of Chicago Press, 1971. A very well-written and nontechnical account of the evolution of vertebrates.

Stahl, B. J.: *Vertebrate History: Problems in Evolution.* New York, McGraw-Hill Book Co., 1974. A fascinating account of vertebrate history with some emphasis upon alternative interpretations of the data and unresolved problems.

Young, J. Z.: *The Life of Vertebrates.* 3rd ed. Oxford, Clarendon Press, 1981. The anatomy, physiology, and evolution of all groups of vertebrates are explored in detail.

Fishes

Bond, C. E.: Biology of Fishes. Philadelphia, Saunders College Publishing, 1979. A general text covering the relationships, structure, physiology, and behavior of fishes.

Carey, F. G.: Fish with warm bodies. *Scientific American, 288*:(Feb.)36, 1973. An account of the countercurrent heat exchange system of tuna and mackerel.

Greenwood, P. H., D. E. Rosen, and S. H. Myers: Phyletic studies of teleostean fishes, with a provisional classification of living forms. *Bulletin of the American Museum of Natural History,* 131:339, 1966.

Hardisty, M. W., and I. C. Potter (Eds.): *The Biology of Lampreys.* New York, Academic Press, 1971. All aspects of lamprey biology are thoroughly explored in chapters written by experts in the field.

Herald, E. S.: *Living Fishes of the World.* Garden City, New York, Doubleday & Co., 1961. A well-written, nontechnical account of the biology of the various groups of fish. As with other books in this series, the photographs are superb; many are in color.

Hoar, W. S., and D. J. Randall (Eds.): *Fish Physiology.* New York, Academic Press, 1969–1979. An eight-volume source book; chapters on various aspects of fish physiology have been written by leading investigators.

McCosker, J. E.: Great white shark. *Science, 81*(Jl.–Aug.):40, 1981. Great white sharks are attracted to the electric fields generated by their prey.

Marshall, N. B.: *Explorations in the Life of Fishes.* Cambridge, Harvard University Press, 1971. Deep-sea fishes and convergent evolution among fishes are emphasized.

Moy-Thomas, J. A.: *Paleozoic Fishes.* 2nd ed. Edited by R. S. Miles. Philadelphia, W. B. Saunders Co., 1971. An extensive revision of the classic book on primitive fishes.

Nelson, J. S.: *Fishes of the World.* New York, John Wiley & Sons, 1976. A systematic treatment of the major groups of fishes.

Norman, J. R.: *A History of Fishes.* 3rd ed. Revised by P. H. Greenwood. London, Benn, 1975. A revision of a classic treatise on all aspects of fish biology.

Partridge, B. L.: The structure and function of fish schools. *Scientific American, 246*:(Jun.)116, 1982. Both the lateral line and eyes play important roles in schooling behavior, which reduces the risk of being eaten.

Policansky, D.: The asymmetry of flounders. *Scientific American, 246*:(May)116, 1982. A study of migration of the eye from one side of the head to the other.

Thompson, K. S.: The biology of lobe-finned fishes. *Biological Reviews, 44*:91, 1969. A review of lung fish and crossopterygian biology.

32 Vertebrates: Amphibians and Reptiles

ORIENTATION

After examining support, water conservation, and other problems associated with terrestrial life, we will consider the extent to which salamanders, frogs, and other amphibians have met these problems. Amphibians do penetrate the terrestrial environment, but they retain certain fishlike features that limit them to moist habitats. Turtles, lizards, snakes, and other reptiles have evolved features that adapt them more completely to life on land, and many can live in very dry environments. Both groups have been very successful, and we will also consider their evolution and adaptive radiation.

CHARACTERISTICS OF AMPHIBIANS

1 Amphibians are ectothermic vertebrates.

2 The bony scales of fishes have been lost except in a few primitive species, and the skin is smooth and moist. Mucous glands are abundant, and there is little cornification of the epidermis.

3 Successive vertebrae interlock to form a strong yet flexible vertebral column. Ribs are very short and frequently fused to the vertebrae in contemporary species. The skull tends to be short, broad, and incompletely ossified.

4 Movable eyelids and tear glands protect and cleanse the eye. Internal nostrils are present.

5 Amphibians have a muscular and protrusible tongue. Their intestine is divided into small and large segments.

6 Larval external gills are lost at metamorphosis, and gas exchange with the environment is by means of moist membranes in the lungs, skin, and buccopharyngeal cavity.

7 Separate right and left atria are present in the heart and receive primarily venous and arterial blood, respectively. These blood streams remain separated to a large extent in their passage through the single ventricle.

8 Most nitrogen is eliminated from the kidneys as urea. Amphibians have a urinary bladder.

9 Many yolk-laden eggs are produced in large ovaries. Jelly layers are secreted around the eggs as they pass down the oviducts. Fertilization usually is external. Most amphibians are oviparous. Aquatic larvae usually undergo metamorphosis into terrestrial adults.

CHARACTERISTICS OF REPTILES

1 Reptiles are ectothermic or heliothermic vertebrates.

2 Their heavily cornified epidermis forms horny scales or plates over the body surface. Cutaneous glands are absent except for a few scent glands.

3 Most of the trunk ribs are long, and the anterior ones articulate with a sternum.

4 A nictitating membrane helps protect the eye. The tympanic membrane frequently is located at the bottom of an external auditory meatus. A lateral line system is never present. No cerebral cortex is present.

5 The internal surface of the lungs is quite large, and the lungs of most groups are ventilated by rib movements.

6 The division of the heart atrium, a partial division of the ventricle, and a complex tripartite division of the conus arteriosus prevent the mixing of arterial and venous blood under most circumstances.

7 Most nitrogenous wastes are eliminated as urea and uric acid through kidneys which are drained by ureters. A large amount of water is reabsorbed in the urinary bladder.

8 The large ovaries produce fewer eggs than those in lower vertebrates, but the eggs contain much more yolk. Males have a copulatory organ, and fertilization is internal. Albuminous materials and a shell are secreted around the fertilized eggs as they pass down the oviduct.

9 Primitive reptiles are oviparous and lay their eggs on land. The cleidoic egg has sufficient yolk, and the embryo develops extraembryonic membranes so that development can proceed directly to a miniature adult before hatching. A few reptiles have become viviparous.

The Transition from Water to Land

The transition from fresh water to land was a momentous step in vertebrate evolution that opened up vast new areas for exploitation, but successful adaptation to the terrestrial environment necessitated structural and functional changes throughout the body because the physical conditions are very different between water and land. Air affords less support and offers less resistance to movement than water. A strong skeleton evolved as did ways of locomotion by thrusting the legs upon the ground rather than by thrusting the tail against the water. Receiving sensory cues is quite different in the air than in the water, and the eyes, ears, and nose were modified. Changes in the nervous system were a natural corollary of the more complex movements, changes in the muscular system, and altered sense organs. Oxygen is more abundant on land than in the water, but gas must be exchanged with the air without an undue loss of body water. Changes in the site of gas exchange from gills to lungs necessitated concomitant changes in the circulatory system which culminated in the evolution of a double circulation through the heart. Water must be conserved in the terrestrial environment, and this was facilitated by changes in nitrogen metabolism. Most nitrogenous wastes are in the form of urea and uric acid which, being less toxic than ammonia, require less water to flush them out of the system. Changes in the excretory system also make it possible for water to be reabsorbed from the urine. Water has a high thermal stability so temperature changes occur slowly. When vertebrates ventured upon the land they had to adapt to a wider range of ambient temperatures. Finally changes in reproduction occurred. The delicate, free-swimming, aquatic larval stage was suppressed, and methods of reproducing upon the land evolved.

In view of the magnitude of these changes, it is not surprising that the transition from water to land took millions of years. The evolution of terrestrial vertebrates, or **tetrapods,** has involved a continual improvement in their adjustment to terrestrial conditions. The crossopterygian fishes unwittingly made the first steps in the transition. Their lungs were probably an adaptation to survive in stagnant water or during temporary drought. Their relatively strong, lobate paired fins may have enabled them to squirm from one drying and overcrowded swamp to another, more favorable one or to push through water choked with aquatic plants. Crossopterygians were not trying to get onto the land, but, in adapting to their own environment, they evolved features that made them viable in a new and different environment. That is, they became **preadapted** to certain terrestrial conditions (see p. 422). The first amphibian fossils are found in strata that were formed nearly 50 million years later than those containing the first crossopterygians.

Class Amphibia

Most of us have seen a toad hopping about a garden in the evening, a frog jumping from a stream bank into the shelter of the water as we approach, and salamanders in a pond or under a log in the woods. These animals, together with the less familiar caecilians of the tropics, are all members of the class Amphibia. There are nearly 3000 living species of frogs, salamanders, and caecilians, all of which must live in the water or in damp habitats on the land. Nearly all return to the water to reproduce, whence their class name (Gr. *amphi,* double + *bios,* life). The living species are descendants of the first vertebrates that left the water and continued the penetration of the terrestrial environment begun by the crossopterygians. Amphibians acquired many terrestrial adaptations, but they also retained many fishlike characteristics that limit them to habitats with a high humidity.

The Adaptations and Evolution of Amphibians

Adaptations of Amphibians. Amphibians continued the transition from water to land and generally became well adapted to damp habitats upon the land. The earliest amphibians were the **labyrinthodonts,** a subclass that diverged from the crossopterygian fishes during the Devonian period (Fig. 32.1) and became extinct early in the Mesozoic Era. They shared with crossopterygians a peculiar labyrinthine infolding of the enamel of their teeth. Their name (Gr. *labyrinthos,* labyrinth + *odontos,* tooth) is derived from this feature.

Labyrinthodonts were rather large (as much as 1 m. long) salamander-shaped creatures with rudimentary necks and heavy muscular tails (Fig. 32.2). Their vertebral column was much stronger than that in fishes. Vertebral centra were more thoroughly ossified, and successive vertebrae interlocked securely by means of overlapping **articular processes** on the vertebral arches (Fig. 5.11, p. 105). The humerus and femur moved primarily in the horizontal plane rather than being nearly vertical and more or less under the body as they are in mammals. The muscles of the trunk were powerful, and lateral undulations of the trunk and tail helped to advance and retract the feet. The footfall pat-

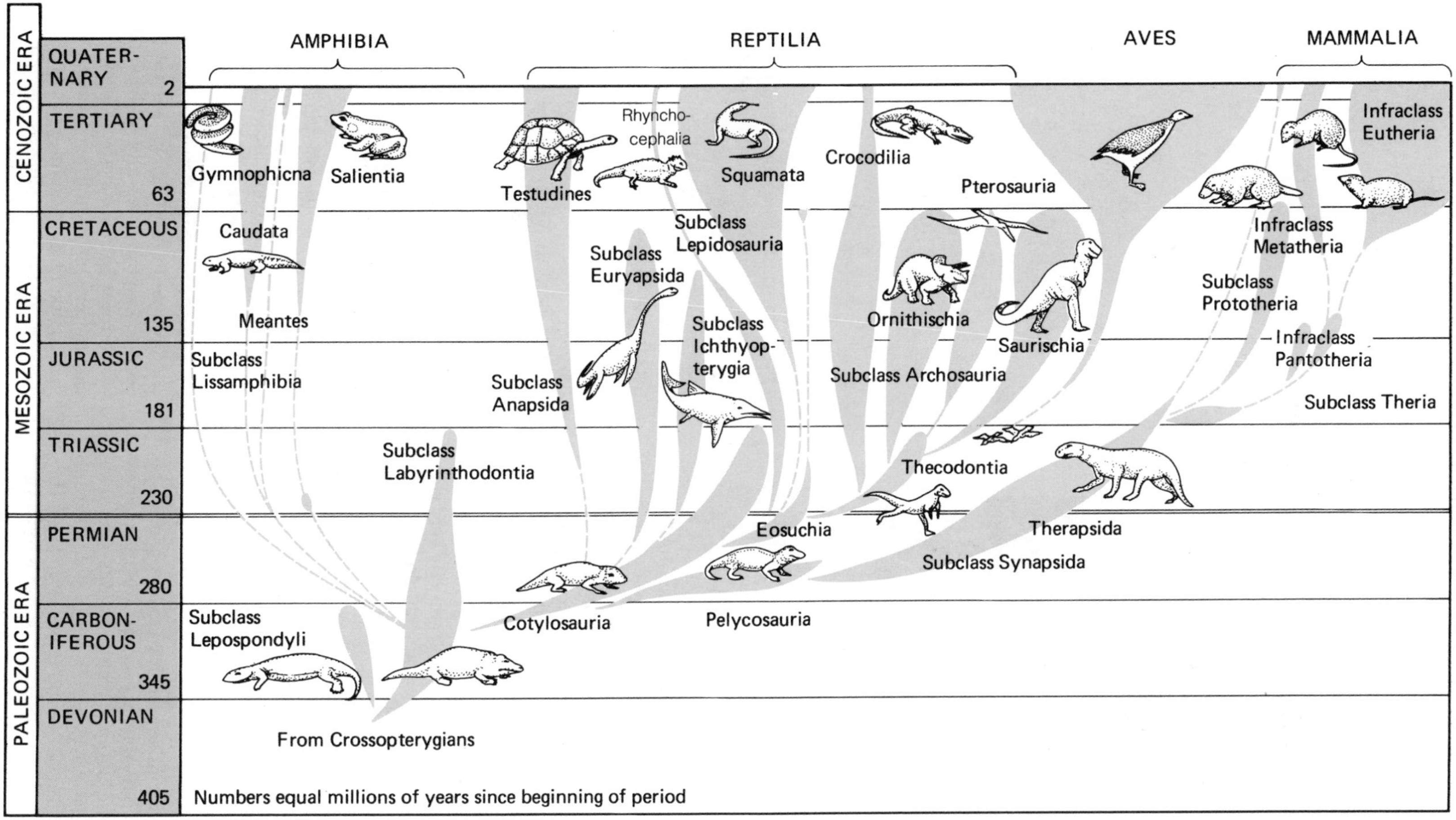

FIGURE 32.1 An evolutionary tree of terrestrial vertebrates. The relationship of the various groups, their relative abundance, and their distribution are shown.

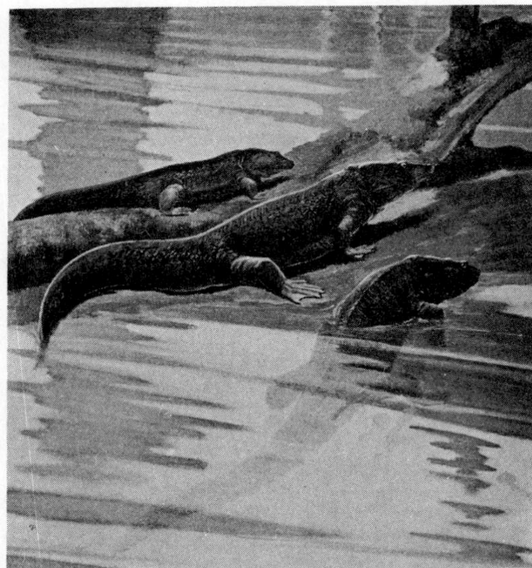

FIGURE 32.2 A restoration of life in a Carboniferous swamp 345 million years ago. The labyrinthodont amphibians were the first terrestrial vertebrates. (Courtesy of the American Museum of Natural History.)

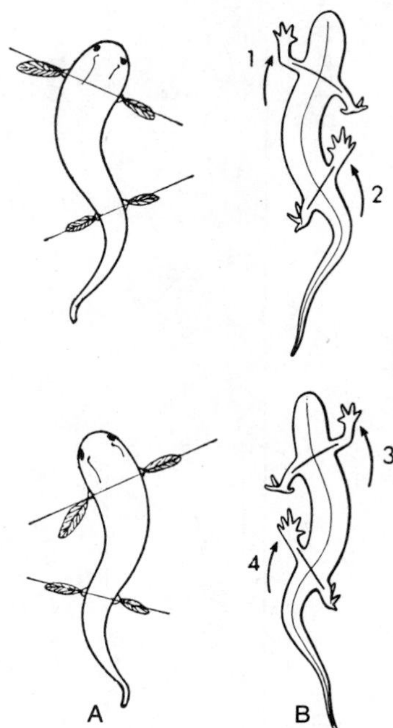

FIGURE 32.3 A comparison of locomotion in *(A)* a fish and *(B)* a labyrinthodont, or salamander. Fishlike undulation of trunk and tail continue in the salamander and help to determine the placement of the limbs upon the ground. Numbers indicate the sequence of limb movements. (*A* after Marshall and Hughes; *B* after Romer.)

tern seen in labyrinthodonts and living amphibians follows naturally from the positions assumed by the fins of a fish as a result of the undulations of the trunk (Fig. 32.3). Advancement of the left front foot is followed in turn by the right rear foot, right front foot, and left rear foot. During most stages of a stride, just one foot is advanced, leaving three feet to support the body.

Early labyrinthodonts probably spent more time in the water than most contemporary salamanders and frogs. Their legs were relatively small, and one very early species had fishlike rays supporting a tail fin.

Many changes occurred in the sensory apparatus during the evolution of the amphibians. Some labyrinthodonts retained the lateral line system, and it is present in the aquatic larvae of amphibians living today, but it is lost in terrestrial adults. In contemporary amphibians, the eye is transformed at metamorphosis. The lens flattens as the cornea assumes more importance in light refraction, and **eyelids** and **tear glands** that protect and cleanse the eye develop.

Frogs have evolved an ear sensitive to air-borne vibrations that includes a **tympanic membrane** and a single auditory ossicle (the **stapes**) that transmits vibrations across the middle ear cavity to the oval window, which is located on the side of the inner ear (Fig. 11.9). They use the tympanic membrane and stapes primarily to detect and transmit the high-frequency vibrations of their mating calls. In addition they have a small **opercular bone,** that fits into the oval window beside the stapes. The opercular bone is connected by a slender muscle to the scapula of the pectoral girdle. Experiments have demonstrated that the opercular apparatus is necessary to detect low-frequency environmental noises, but just how it works is uncertain. Salamanders lack a tympanic membrane, and the opercular apparatus is the more important mechanism.

The brain remains essentially fishlike, although subtle changes occur in relation to changing patterns of sensory perception and locomotion. Lateral line centers are absent in adults.

The digestive system has changed in many ways in comparison with that of primitive fishes. A muscular and frequently protrusible **tongue** has evolved that is used to catch insects and other small invertebrates on which most amphibians feed. Numerous small teeth are usually present on the jaw margins and palate and aid in seizing and holding the prey. Mucus secreted by buccal and pharyngeal glands lubricates the food, and the tongue aids in swallowing it. The pharynx, of course, lacks gill slits in adults. **Chitinase** is among the enzymes secreted by the gastric glands and pancreas and hydrolyzes the arthropod cuticle. The intestine has lost the primitive spiral valve and is longer than in most fishes. Usually it is divided into small and large intestines. A cloaca is still present.

The small bony scales present in the skin of early labyrinthodonts would have limited cutaneous respiration that supplements pulmonary respiration in contemporary species. Air breathing must have depended primarily on lungs, which probably were ventilated by movements of the well-developed ribs along with a pumping action of the floor of the mouth and pharynx. The loss of scales and the decrease in body size in living species, which increases body surface area relative to mass, made cutaneous respiration more efficient. The broadening of the head of later labyrinthodonts, with the concomitant widening of the sheets of muscle in the floor of the mouth and pharynx, suggests the increasing importance of the buccopharyngeal pump in lung ventilation. The ribs of modern amphibians are short and frequently fused to the vertebrae so that they have no function in respiratory movements.

The substitution of lungs and skin for gills as the major sites of gas exchange had important consequences in the circulatory system. Oxygen-depleted blood from the body and oxygen-rich blood from the lungs are kept separate to some extent as they flow through the heart of contemporary amphibians (p. 179). Depleted blood for the most part is sent to the lungs and skin, and oxygen-rich blood is distributed to the body. Since blood leaving the heart on its way to the body does not first go through the gills, blood pressure in the dorsal aorta of a frog is nearly double that in the dorsal aorta of a dogfish. In this respect, terrestrial vertebrates have a more efficient circulatory system than do fishes. The presence of a lymphatic system is a correlate of increased blood pressure.

These features adapt amphibians to terrestrial life, but amphibians also retain many features characteristic of freshwater fishes that restrict them to moist habitats and prevent them from fully exploiting the terrestrial environment. Their ability to conserve body water is rudimentary. Considerable water is lost by evaporation through their thin, moist skin. Their urine is copious and dilute because they must excrete a fairly large volume of water to flush out toxic excretory products through their kidney tubules. Although they excrete less nitrogenous wastes as highly toxic ammonia than do freshwater fishes, some is present. Most nitrogen is eliminated as urea (Table 32.1). Urine is stored temporarily in a **urinary bladder,** and some water is reabsorbed here, especially in the more terrestrial species, such as toads.

Amphibians remain ectotherms in the sense that their body temperature fluctuates with the environmental temperature. They do, however, avoid temperature extremes. They retreat to cool, moist habitats if the temperature becomes too warm. Temperate species move before winter to areas that do not freeze and enter a dormant state known as **hibernation.** Since they cannot maintain their body temperature higher than that of the environment, amphibians bury themselves in the mud at the bottom of ponds or burrow into soft ground below the frost line. During hibernation metabolic activities are at a minimum. The only food utilized is that stored within the body; respiration and circulation are very slow. Some tropical amphibians during the hottest and driest parts of the year go into a comparable dormant state known as **aestivation.**

Amphibians are unable to reproduce under truly terrestrial conditions. Most species must return to the water to lay their eggs. Even the terrestrial toad has no means of internal fertilization, and sperm cannot be sprayed over eggs upon the land. In most species, an egg develops into an aquatic larva that has a powerful muscular tail, external gills, well-developed mouth parts, kidneys, and other organs needed to lead an active and independent life in the water. Larval features are lost and adult terrestrial ones are acquired during **metamorphosis.**

Evolution of Amphibians. Labyrinthodonts included not only the ancestors of other amphibians but also those of reptiles and all higher tetrapods. The subclass **Lepospondylii** (Fig. 32.1), probably diverged very early from labyrinthodonts. Lepospondyls differ from labyrinthodonts in having spool-shaped vertebral centra rather than centra composed of chunks of bone surrounding the notochord. They became extinct at the end of the Carboniferous period without leaving descendants.

Paleontologists believe that the four orders of contemporary amphibians descended from labyrinthodont ancestors, although the structure of their vertebral centra is somewhat similar to that of lepospondyls. They are grouped into a common subclass, the **Lissamphibia,** for all share such features as the peculiar opercular bone of the middle ear, reduced ribs, and, when legs are present, an unossified carpus and tarsus, and never more than four toes in the front foot.

Salamanders retain the primitive labyrinthodont body form with short legs, long trunk, and a well-developed tail. Herpetologists now divide salamanders into two orders: the true salamanders (order **Caudata**), and the aquatic, eel-shaped sirens (order **Meantes**). Biochemical evidence suggests that these two groups diverged late in the Paleozoic era. Frogs and toads (order **Salientia**), with their short trunk, lack of a tail, and powerful hind legs are specialized for a hopping or jumping mode of life (Fig. 32.4). One froglike amphibian of the early Triassic was intermediate between contemporary species and more primitive amphibians (see Fig. 32.9*A*). It has been placed in the order **Proanura.** Caecilians (order **Gymnophiona**) are legless, burrowing species confined to the tropics.

Salamanders

In many ways salamanders are closer to ancestral amphibians than frogs and caecilians. Although much smaller than labyrinthodonts, they retain the primitive body form. True salamanders of the order **Caudata** (L. *cauda,* tail) are found beneath stones and logs in damp woods, and some are aquatic. Salamanders occur primarily in the temperate regions of the world, and most species are found in North America. About 300 species have been described.

Anatomically and physiologically salamanders resemble other amphibians, but some species have unusual developmental and life history characteristics. Salamanders reproduce in the water and their larvae are aquatic. They differ from the larvae of frogs and toads in retaining external gills and in having teeth and other mouth structures adapted for a carnivorous rather than a herbivorous diet. Most adults are terrestrial, but the newt of the eastern United States returns to the water after passing through a terrestrial juvenile stage. The juvenile, known as the red eft (Fig. 36.17), has a striking orange-red coloration. This is an example of an **aposematic,** or warning, color (p. 782) for it is coupled with noxious skin secretions. Bird predators soon learn to recognize this color and avoid eating the eft. Although most salamanders complete their metamorphosis into terrestrial adults, many groups have become **paedomorphic** (p. 425) and retain many larval features throughout their lives. Paedomorphosis among salamanders has evolved by **neoteny,** that is, somatic development has slowed down relative to sexual maturation so the animals become sexually mature before they have completed their metamorphosis. The tiger salamander, *Ambystoma tigrinum,* in the Mexican plateau lives in cool mountain ponds, retains external gills and many other larval features, reproduces as a larva, and never completes its metamorphosis (Fig. 32.5*A*). This permanent larva is called an **axolotl.** At lower elevations in the United States the tiger salamander normally metamor-

FIGURE 32.4 Three stages in the leap of a bullfrog. Observe that the pelvic girdle is flexed when the frog is at rest, and this causes a characteristic bump in the back *(A)*; the hind legs and pelvic girdle extend at take-off *(B)*; and the frog lands on its front legs *(C)*. (Courtesy of Mr. Earle R. Edminston.)

A

B

C

FIGURE 32.5 Neotenic salamanders. *A,* The axolotl, or neotenic, form of the tiger salamander. *B,* The adult tiger salamander, *Ambystoma tigrinum. C,* The mudpuppy, *Necturus maculosus.* (*A,* Courtesy of the Philadelphia Zoological Society; *B* and *C,* courtesy of the Ohio Department of Natural Resources, photographs by A. E. Staffan.)

FIGURE 32.6 The two-lined salamander, *Eurycea bislineata,* is a lungless plethodont. (Courtesy of the Ohio Department of Natural Resources, photograph by A. E. Staffan.)

phoses (Fig. 32.5*B*). The hormone **thyroxine,** secreted by the thyroid gland, is necessary for amphibian metamorphosis. Its secretion appears to be inhibited at the low temperatures on the Mexican plateau but not at lower elevations. The tiger salamander is an example of a **facultative neotenic** species. The mudpuppy, *Necturus,* is an **obligatory neotenic** species and does not metamorphose under any circumstances (Fig. 32.5*C*). Thyroxine is produced in *Necturus* for the thyroid of this species hastens metamorphosis when transplanted to frog tadpoles. The failure of *Necturus* to metamorphose appears to result from the inability of its tissue to respond to thyroxine, possibly because the species lacks the receptors for this hormone.

The most abundant of our American salamanders are woodland types and species that live in and by the sides of streams, such as the two-lined salamander *Eurycea bislineata* (Fig. 32.6). Both belong in the family Plethodontidae. A particularly interesting feature of plethodonts is their complete loss of lungs; gas exchange occurs entirely across the moist membranes lining the mouth and pharynx and the skin. The skin is a more effective respiratory organ than in other salamanders

FIGURE 32.7 The siren, *Siren intermedia.* (Courtesy of G. R. Zug, the Smithsonian Institution, photograph by R. W. McDiarmid, National Museum of Natural History.)

because the epidermis is very thin and capillaries come close to the surface. Loss of lungs may seem to be a curious adaptation for a terrestrial vertebrate, but it has been postulated that early in their evolution plethodonts became adapted for life in rapid mountain streams. The oxygen content of cold water is rather high, and the low temperature reduces the rate of metabolism of the animals. Air in the lungs would be disadvantageous under these conditions, for the animals would float and be washed away. Lungs may have been lost in adapting to this habitat. Many plethodonts still live in streams; others have moved onto the land but never regained the lost lungs.

The sirens, order **Meantes** (L. *meantes,* going; a reference to their reduced legs), are obligatory neotenic species found only in ponds and rivers of the southern United States (Fig. 32.7). They have an eel-like body shape, no hind legs, and only vestiges of front legs. They have many unusual features, including a heart with a nearly completely divided ventricle that is more like a reptile's heart than an amphibian's. If the pond in which they live dries up, they can secrete a slimy mucous cocoon around themselves and aestivate in the manner of an African lungfish.

Frogs and Toads

Frogs and toads of the order **Salientia** (L. *saliens,* leaping) are the most abundant and diverse of all living amphibians. About 2500 species are distributed throughout most of the temperate and tropical regions of the world, although they are absent from oceanic islands and New Zealand. Their adult size ranges from a tiny Cuban species only 1 cm. long to a West African species, *Gigantorana goliath,* over 30 cm. long. Our most common frogs, such as the leopard frog, *Rana pipiens,* which is used in many biology classes, the green frog, *Rana clamitans,* and the bullfrog, *Rana catesbeiana* (Fig. 32.8*A* and *B*) live in water, on the shores of ponds and streams, and in damp meadows and woods. Their skin is thin and moist, and their powerful hind legs, with webbed feet, are very effective in swimming as well as in jumping.

Although they return to the water to reproduce (Fig. 32.8*D*) adult toads live in more terrestrial habitats than frogs. Toads are frequently found in the woods, fields, and gardens. A web may be present on their hind feet, but it frequently is reduced in size. Spurs on the hind feet of many species help them burrow backwards into the soil where they shelter during the hotter and drier parts of the day. Toads are most active in the early morning and in the evening. Their skin is drier and more cornified than in frogs, but toads also lose a lot of water through it. There is less cutaneous gas exchange in toads than in frogs, and the internal surface area of their lungs is increased slightly. The warts on a toad's back are aggregations of poison glands in the skin. The secretions are only mildly irritating in most species, but an experienced snake, raccoon, or other predator tends to avoid eating toads.

Body size and weight in most species of tree frogs and toads are low. This makes it easier for them to cling to small branches and leaves with the **digital pads** on the tips of their toes (Fig. 32.8*C*). The adhesive properties of the pads come not from the secretion of sticky substances but from the presence of numerous minute folds and ridges that catch in irregularities in the substratum. Frequently the skin of tree frogs is dry and warty, as in toads.

Some frogs are entirely aquatic and seldom leave the water. Among these is the African clawed frog, *Xenopus laevis,* which is widely used in experimental studies in this country because it is so easy to breed and raise in the laboratory. *Xenopus* has adapted to an aquatic life by becoming neotenic and retaining many larval characteristics. Among them is the presence of a lateral line sensory system and the absence of many adult features, such as a muscular tongue, eyelids, and a tympanic membrane.

The jumping specializations of frogs and toads are particularly evident in the skeleton (Fig. 32.9). The vertebral column is short and relatively stiff. Short embryonic ribs fuse onto the transverse processes of the vertebrae. The hind legs are long and strong. Elongation of two tarsal bones increases foot length and makes, in effect, another limb segment for jumping. Fusion of the tibia and fibula strengthens the leg. Elongation of the ilium provides an additional lever for the caudal thrust because the pelvic girdle can swivel at the sacroiliac joint. When the frog is at rest, its hind legs are flexed,

FIGURE 32.8 Common North American frogs. *A,* The leopard frog, *Rana pipiens; B,* the bullfrog, *Rana catesbeiana; C,* the gray treefrog, *Hyla versicolor; D,* a male American toad, *Bufo americanus,* singing. (*A* and *D,* photographs by W. F. Walker, Jr.; *B* and *C,* courtesy of the Ohio Department of Natural Resources, photographs by A. E. Staffan.).

and the ilium and the urostyle, which represents several fused caudal vertebrae, are also flexed, thus giving the frog a "humped-back" appearance (Fig. 32.4*A*). The legs are quickly extended during a leap, and additional thrust is given by the extension of the urostyle and ilium (Fig. 32.4*B*). The front legs act as shock absorbers when the animal lands, and their strength has been increased by the fusion of the radius and ulna. Muscles are also modified, and those used in the extension of the hind legs and pelvis are particularly powerful.

Most frogs and toads have the typical amphibian feeding pattern, but the aquatic tadpoles are herbivores. Their horny teeth are well suited for grazing on plant material. Their intestine, which lodges a bacterial colony that helps digest plant material, is very long and coiled like a watch spring. There is an extensive remodeling of the digestive system at metamorphosis for the adults are carnivorous.

Modifications in development have occurred among many tropical frogs, probably as a protection against seasonal drought and numerous aquatic enemies, such as predaceous insect larvae. A Brazilian tree frog (*Hyla faber*) protects its young by laying its eggs in mud craters that the male builds by circling in shallow water and pushing up mud. A more striking means of protection is seen in a small South American frog, *Dendrobates,* in which the male carries the eggs and tadpoles upon his back until they are young froglets able to fend for themselves (Fig. 32.10*A*). This species is also one with very toxic skin secretions. A brilliant coloration warns predators of the danger and they soon learn to avoid this species. Such a warning coloration is known as **aposematic coloration.** In the completely aquatic Surinam toad, *Pipa pipa,* of tropical South America, the male pushes the fertilized eggs into the puffy skin on the back of the female. The eggs gradually sink deeper into the skin, and the outer membrane of each one forms a protective cap over its temporary skin pocket. The young emerge as little froglets. The eggs and larvae of all these frogs are equipped with a large supply of

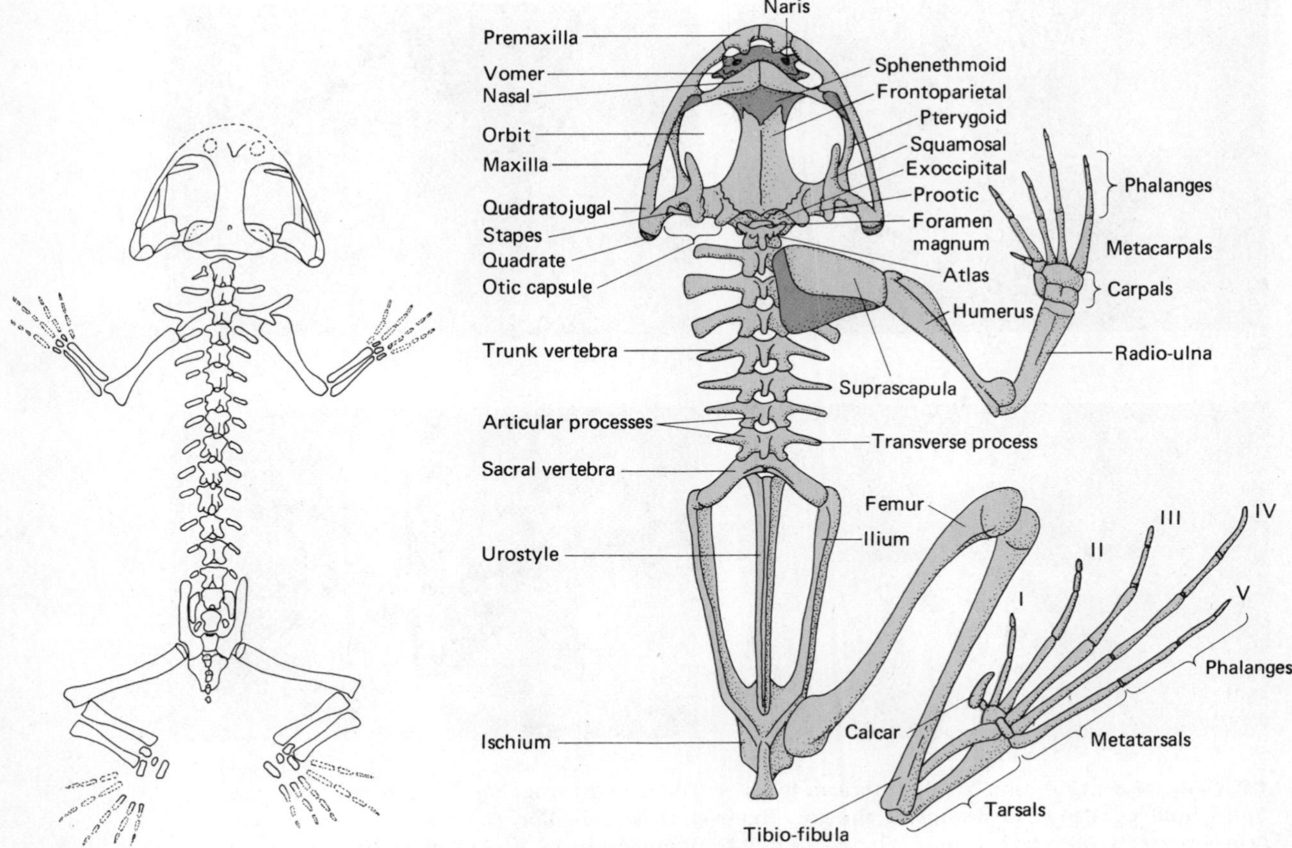

FIGURE 32.9 Comparison of the skeletons of *(A) Triadobatracus,* a possible ancestral frog from the Triassic and (*B*) a modern frog. (*A* after Estes in Vial (Ed.): *Evolutionary Biology of Anurans.* University of Missouri Press, 1973; *B* modified after Parker and Haswell.)

yolk, but in other respects the larvae are fairly typical and simply develop in a sheltered environment.

In certain species, the vulnerable larval stage is omitted, and the embryo develops directly into a miniature adult. Frogs with **direct development** include the marsupial frog, *Gastrotheca,* and *Eleutherodactylus*—both of the New World tropics. The former carries her eggs in a dorsal brood pouch (Fig. 32.10); the latter lays eggs in protected damp places, such as beneath stones or in the axil of leaves. The jelly layers about the egg of *Eleutherodactylus* help to prevent desiccation; sufficient yolk is stored within the egg for the nutritive requirements of the embryo; such larval features as horny teeth, gills, and opercular fold are vestigial or absent; the fins of the larval tail are expanded, become highly vascular, and form an organ for gas exchange; and the period of development is shortened. It is possible that the labyrinthodont amphibians that gave rise to the reptiles developed means of terrestrial reproduction before the adults left the water to live on the land.

Caecilians

Caecilians, which belong to the order **Gymnophiona** (Gr. *gymnos,* naked + *ophioneos,* snakelike), are the least familiar group of living amphibians since they occur only in the tropics and lead a subterranean life burrowing through damp soil in search of small invertebrates on which they feed. About 150 species are known. All have an elongated body form without legs or girdles

A B

FIGURE 32.10 Some frog adaptations. *A,* The male of *Dendrobates* carries eggs and tadpoles upon his back. *B,* The eggs of the marsupial frog, *Gastrotheca,* develop directly into miniature adults within a brood pouch on the back of the female. *Dendrobates,* which has toxic skin secretions, has a reddish aposematic coloration; *Gastrotheca,* a greenish cryptic coloration. (*A,* courtesy of E. S. Ross, California Academy of Sciences; *B,* courtesy of B. R. Zug, Smithsonian Institution, photograph by K. Miyata, National Museum of Natural History.)

(Fig. 32.11). On a quick glance a caecilian could be confused with an earthworm. They are about the same size, and annular folds in their skin give the appearance of segments. However, caecilians have a head with small jaws, nostrils, and a pair of vestigial eyes partly hidden beneath the skin. Their primary sense organ is a pair of protrusible tentacles that lie in a pit between each nostril and eye. Their cloacal aperture is not quite at the end of the body so there is a short tail. Small bony scales are embedded in the skin. Fertilization is internal, an unusual situation among amphibians. The male's cloaca is protrusible and can form a penis-like organ. Most species are oviparous and lay their eggs in damp soil. In some species the eggs develop directly into miniature adults; in others, the eggs hatch and the larvae complete their development in nearby bodies of water. A few species are viviparous, but have no placenta.

Class Reptilia

The familiar snakes, lizards, alligators, and crocodiles, together with dinosaurs and other extinct species, are reptiles (L. *repto,* to crawl). Over 6,000 living species have been described. As a group they have continued the penetration of the land begun by the crossopterygian fishes and the amphibians, and generally they are well adapted to terrestrial life. They can live in drier habitats than amphibians, and many are found in deserts. Some have readapted to an aquatic life, but all must lay their eggs upon the land unless, like the sea snakes, they are viviparous. Even sea turtles leave the water to lay their eggs on beaches. The distribution of reptiles is limited primarily by their inability to maintain a body temperature higher than that of the environment. They are not active in very cold weather, and they are not found in subarctic regions.

FIGURE 32.11 The caecilian *Siphonops annulatus* eating an earthworm. (Courtesy of C. Gans, University of Michigan.)

Major Adaptations of Reptiles

Reptiles are active and agile vertebrates, with adaptations that permit a more complete exploitation of the land. Mucous glands have been lost from their skin, and the epidermis is dry and horny. Considerable **keratin,** a water-insoluble protein, is deposited in the epidermal cells, which form **horny scales** or plates on the surface (Fig. 5.3). Contrary to popular opinion, snakes are not the least bit slimy. The only integumentary glands of reptiles are a few scent glands whose secretions are used in sexual attraction. In some lizards and crocodiles, small plates of bone develop in the dermis beneath the horny scales.

Although reptiles are ectothermic in the sense that they cannot maintain a body temperature much above that of their environment, small reptiles do maintain a high and fairly constant temperature in a warm, sunny environment, primarily by changes in behavior. Spiny lizards of the southwestern deserts of the United States maintain their body temperature at about 34° C. (93° F.) during much of the day. If body temperature falls below the threshold for normal activity, the lizard lies at right angles to the sun's rays, thereby exposing a maximum amount of body surface to the sun. If body temperature rises too far, the lizard seeks shelter or lies parallel to the sun's rays. Large reptiles, of course, cannot always find the needed shade, but their large body size provides some thermal stability simply because a considerable time is needed for a large mass to heat up or cool down. This may have been the primary thermal regulatory mechanism of the dinosaurs. Beyond behavioral regulation, reptiles can either dissipate or reduce the loss of body heat as needed by controlling the amount of blood flow through the skin. The chuckwalla (*Sauromalus*) can also eliminate body heat by panting (Fig. 32.12). Color changes are an additional means of temperature regulation in some lizards. In cold weather, the skin is dark and absorbs a maximum amount of heat; as temperatures rise, pigment migrates to the center of the chromatophores, and the skin becomes lighter in color. At night, body temperature falls to ambient levels, and in prolonged cold weather reptiles become dormant. Body temperature is regulated in reptiles, as it is in mammals, by a center in the hypothalamus of the brain. Since exposure to sunlight by behavioral changes plays such a large role in reptile thermal regulation, they are sometimes called heliotherms (Gr. *helios,* sun).

Reptiles have a stronger skeleton than amphibians. Usually their humerus and femur move back and forth close to the horizontal plane, but some rapidly running lizards pull their hind legs under their bodies and become temporary bipeds. Many of the dinosaurs were also bipeds. Carpal and tarsal bones are ossified. Most of the movement at the ankle comes not at the joint between tibia and tarsus, but at a joint in the middle of the tarsus. The thrust of the hind legs is transferred to the vertebral column through two sacral vertebrae rather than by a single one as in amphibians. Claws on the toes enable the feet to get a better grip upon the

FIGURE 32.12 The chuckwalla (*Sauromalus obesus*) lives in the deserts of the southwestern United States. When unable to shelter, it prevents overheating by panting and placing its body parallel to the sun's rays. (From Bogert, C. M.: ***How Reptiles Regulate Their Body Temperature.*** Copyright © by Scientific American, Inc. All rights reserved.)

ground. The muscular system is more complex than in amphibians, and the central nervous system is better developed. The cerebral hemisphere has enlarged somewhat by the growth of a mass of gray matter, the **corpus striatum,** deep within it. Reptiles do not have a cerebral cortex.

Most reptiles are carnivores, and their digestive system is not very different from that of amphibians, although food catching and swallowing mechanisms are sometimes quite specialized (e.g., in snakes, p. 689). An ileocolic valve separates small and large intestines, and a small cecum frequently is present at the beginning of the large intestine. Tortoises and some lizards are herbivores. Their large intestine usually is longer than the small intestine, is divided by internal folds into compartments, and the passage of food through it is slow. A bacterial colony within the large intestine digests cellulose into fatty acids that are absorbed there.

The dry horny skin of reptiles reduces cutaneous respiration to a negligible amount, but an increase in the respiratory surface of the lungs not only compensates for this but also provides for the increased volume of gas exchange necessitated by a general increase in activity. Mechanisms for moving air into and out of the lungs are more efficient. Instead of pumping air into the lungs by froglike throat movements, reptiles decrease the pressure within their body cavity, and atmospheric pressure drives in air. A subatmospheric pressure is created around the lungs during inspiration by the forward movement of the ribs and the concomitant increase in size of the body cavity. The contraction of abdominal muscles and the elastic recoil of the lungs force out air.

The heart is more completely divided in reptiles than in amphibians for a partial or, in crocodiles, a complete interventricular septum is present. There is an unusual tripartite division of the conus arteriosus leaving the heart (Fig. 8.12). As a consequence of these features, venous blood returning to the heart from the body and going to the lungs is nearly completely separated from the arterial blood returning to the heart from the lungs and going to the body. Certain persistent interconnections either between the two ventricles or between the left and right sides of the conus arteriosus permit the lungs to be bypassed when the animal is beneath water.

The metanephric kidneys of reptiles are similar to those of birds and mammals. Each is drained by a **ureter** rather than by the primitive archinephric duct. Reptiles use less water to remove nitrogenous wastes from the blood than amphibians because a large portion of the wastes is excreted as nontoxic **uric acid** (Table 32.1). Much of the water that is removed is later reabsorbed by other parts of the kidney tubules and by the urinary bladder. The urine of animals excreting uric acid typically has a pastelike consistency.

TABLE 32.1 Types of Nitrogen Excretion in Representative Vertebrates*

Animal	Ammonia	Urea	Uric Acid	Other
Freshwater minnow, *Cyprinus*	60.0	6.2	0.2	33.6
Bullfrog, *Rana*	3.2	84.0	0.4	12.4
European tortoise, *Testudo*	4.1	22.0	51.9	22.0
Hen	3.0	10.0	87.0	0.0
Adult human	3.5	85.0	2.2	9.3

*Data from Prosser and Brown. Figures in percent of total nitrogen excretion.

Male reptiles have evolved **copulatory organs** that introduce the sperm directly into the female reproductive tract. Fertilization is internal, and the delicate sperm are not exposed to the external environment. A large quantity of nutritive **yolk** is stored within the egg while it is still in the ovary. As the eggs pass down the oviduct after ovulation, they are fertilized, and additional substances and a **shell** are secreted around each egg by certain oviducal cells. **Albumin** and similar materials around the egg provide additional food, ions, and water. The leathery or calcareous shell serves for protection against mechanical injury and desiccation, yet it is porous enough to permit gas exchange. Such an egg, which contains, or has the means of providing, all substances necessary for the complete development of the embryo to a miniature adult, has, in a sense, been the key to terrestrial life. It is called a **cleidoic egg** (Gr. *kleidos,* key). Reptiles lay fewer eggs than lower vertebrates, but the eggs are larger, better equipped, and laid in sheltered situations so that the mortality is low. A collared lizard lays only four to 24 eggs; a leopard frog lays 2000 or more.

The developing embryo separates from the yolk, which becomes suspended in a **yolk sac** (Fig. 32.13). A protective **chorion** and **amnion** develop as folds over the embryo (see p. 325). The **allantois** is a saclike outgrowth from the embryo's hindgut. It is homologous to the urinary bladder but extends beyond the body wall, passing between the amnion and the chorion. It functions in gas exchange and excretion. It has been postulated that the evolution of the cleidoic egg was correlated with a shift in nitrogen metabolism. Little water is available to carry off ammonia and urea in the terrestrial environment in which the eggs are laid, and nitrogenous wastes cannot be eliminated in gaseous form; much of it is converted to inert uric acid, which can be stored. These adaptations for terrestrial reproduction are found in the embryos of all reptiles, birds, and mammals. These groups of vertebrates are therefore called **am-**

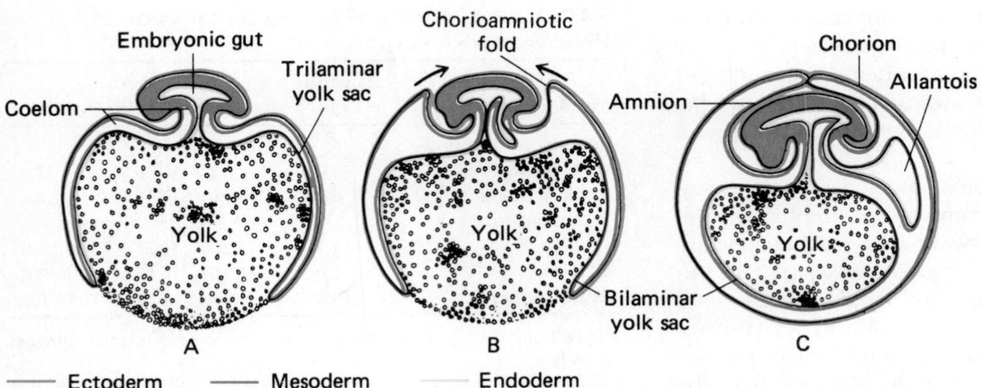

FIGURE 32.13 Sections of vertebrate embryos to show the extraembryonic membranes. *A,* The trilaminar yolk sac of a large yolk fish embryo consists of all three germ layers. *B,* The chorioamniotic folds of an early embryo of a reptile appear to have evolved from the ectoderm and part of the mesoderm of a trilaminar yolk sac. *C,* A later reptile embryo in which the extraembryonic membranes are complete. Notice that the yolk sac is bilaminar. The albumin and shell, which surround the reptile embryo and extraembryonic membranes, have not been shown.

niotes. In contrast, the various fish groups and amphibians lacking an amnion are called **anamniotes.**

Most reptiles are **oviparous** and bury their eggs in soil, sand, or leaf mold where the heat of the sun or of plant decomposition will help to incubate them. Some lizards and snakes give birth to living young. Their eggs are retained in a uterus, and the embryos develop there using primarily food stored in the yolk. Only a few species have developed a placental relationship with the mother's uterine lining.

Evolution of Reptiles

Stem Reptiles. Having "solved" the essential problems of terrestrial life at a time when there were few competitors upon the land, the reptiles multiplied rapidly, spread into all of the habitats available to them, and became specialized accordingly. Much of their adaptive radiation involved different methods of locomotion and feeding. Different feeding patterns have entailed, among other things, modification of the jaw muscles, and this in turn has affected the structure of the temporal region of the skull. Skull morphology, therefore, provides a convenient way to sort out the various lines of reptile evolution. Only a small part of the skull of a primitive tetrapod actually surrounds the brain (Fig. 32.14*A*). A large space lateral to the brain case and posterior to the eyes is largely filled with jaw muscles that extend down to the lower jaw. In ancestral reptiles belonging to the order **Cotylosauria** (Gr. *kotye,* cup-shaped hollow + *sauros,* lizard), a solid roof of dermal bone (the anapsid skull) covered these muscles dorsally and laterally (Fig. 32.14). As a consequence of some brain enlargement and an enlargement of jaw muscles, various types of openings have evolved in the temporal roof of most later reptilian groups. Jaw muscles arise from the brain case and from the periphery of the temporal fenestrae, and they can bulge through the openings when they contract.

Turtles. Turtles (order **Testudines,** L. *testudo,* tortoise) are believed to be direct descendants of cotylosaurs (Fig. 32.1). Most have retained the anapsid skull, but they have many specialized features. Teeth have been lost and the jaws are covered by sharp, horny plates. They are encased in a protective shell composed of bony plates overlaid by horny scales. The bony plates have ossified in the dermis of the skin but have fused with the ribs and some other deeper parts of the skeleton. The portion of the shell covering the back is known as the **carapace;** the ventral portion, the **plastron** (Fig. 32.15*A*).

Ancestral turtles were stiff-necked creatures, unable to retract their heads, but modern species can withdraw theirs into the shell. This is accomplished in most species by bending the neck in an S-shaped loop in the vertical plane (Fig. 32.15*A*), but one group of species in South America, Africa, and Australia has evolved a different method. They retract their neck in the horizontal plane, and, therefore, are known as the side-necked turtles (Fig. 32.15*B*).

Despite their apparent clumsiness, turtles are an ancient and very successful group whose ancestry can be traced back to the Triassic period. They have undergone an extensive adaptive radiation of their own and are

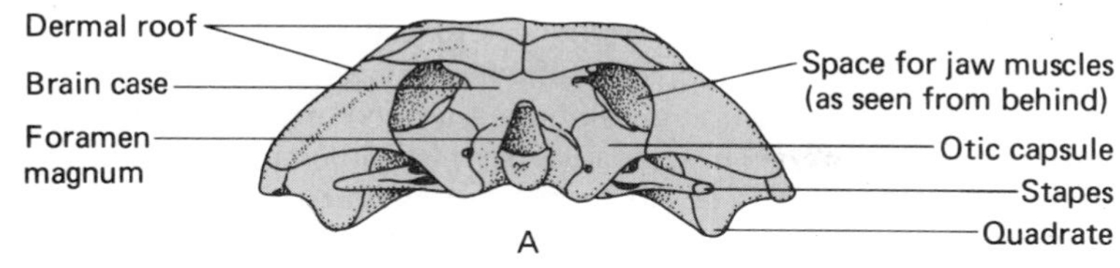

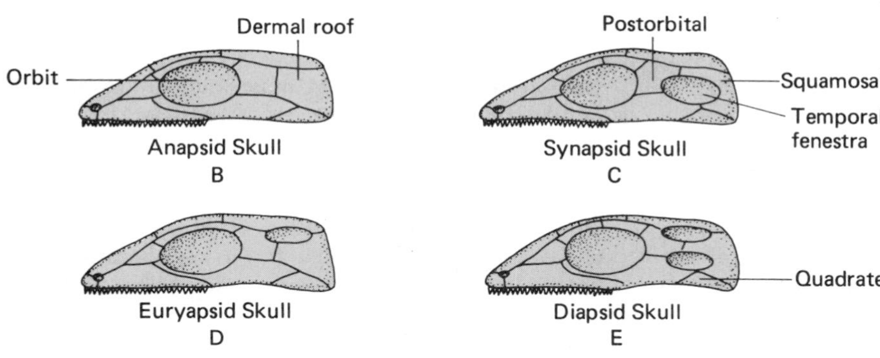

FIGURE 32.14 Diagrams to show the types of temporal roofs found in reptile skulls. *A,* Posterior view of the skull of a cotylosaur to show the relationship between temporal roof and brain case; *B–D,* lateral views of reptile skulls. (*A* from Romer after Price; *B–D* from Romer.)

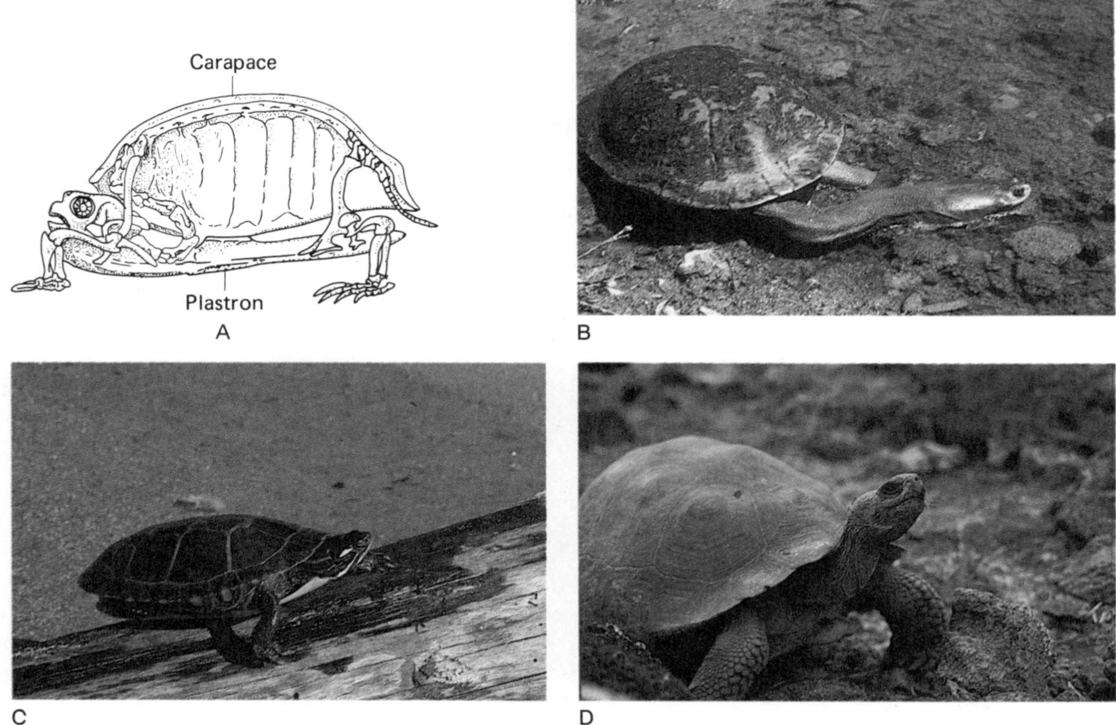

FIGURE 32.15 *A,* Diagram of a turtle skeleton showing the way most species retract their neck in the vertical plane. *B,* An Australian long-necked chelid, *Chelodina,* folds its neck under its shell in the horizontal plane. *C,* The semiaquatic painted turtle of North America, *Chrysemys picta. D,* The terrestrial Galapagos tortoise, *Geochelone elephantopus.* (*A,* from W. K. Gregory, Evolution Emerging, The Macmillan Company, 1951; *B,* courtesy of J. Cann, University of Utah; *C* and *D,* photographs by W. F. Walker, Jr.)

represented today by about 230 species. Although most are semiaquatic, tortoises have adapted to a completely terrestrial life, and several groups are entirely aquatic (Fig. 32.15*C* and *D*). The front legs of sea turtles are modified as flippers, and these turtles spend most of their lives in the water, coming on shore only to lay their eggs.

Marine Blind Alleys. In the Mesozoic, two lines of reptiles evolved with adaptations to marine conditions. Plesiosaurs (order **Sauropterygia,** Gr. *sauros,* lizard + *pterygos,* fin) were superficially turtle-shaped (though they lacked the shell), with squat heavy bodies and long necks (Fig. 32.16). Some species reached a length of 12 m. They propelled themselves by means of large paddle-shaped appendages. Ichthyosaurs (order **Ichthyosauria,** Gr. *ichyths,* fish + *sauros,* lizard), were porpoise-like in size and shape, and probably in habits (Fig. 32.16). They moved with fishlike undulations of the trunk. Both groups had euryapsid skulls with a temporal opening high up on the side (Fig. 32.14).

Plesiosaurs probably came onto the beaches to lay their eggs, but the extreme aquatic adaptation of the ichthyosaurs would have precluded their doing so. How then did they reproduce, for cleidoic eggs cannot develop submerged in water? In an unusual fossil, several intact small ichthyosaur skeletons are lodged in the posterior part of the mother's abdominal cavity, and one is part way out the cloaca. These apparently were offspring about to be born, for the skeletons of individuals that had been eaten would not remain intact during a passage through the digestive tract. These reptiles were viviparous, retaining the eggs in the oviduct until embryonic development was complete.

These marine reptiles flourished during the Mesozoic, competing with the more primitive kinds of fishes. Just why they became extinct near the close of this era is uncertain, but their extinction coincides with the evolution and increase of the teleosts. Possibly they could not compete successfully with these fishes.

Rhynchocephalians. The most abundant of our present-day reptiles are the lizards, snakes, and the closely related tuatara (*Sphenodon,* Fig. 32.17) of New Zealand. The tuatara is the most primitive member of this group and is the sole surviving species of the order **Rhynchocephalia** (Gr. *rhynchos,* beak + *kephale,* head). Rhynchocephalians are lizard-like in general appearance but have a somewhat beak-shaped snout and a **diapsid** skull with two temporal openings, one above and one below the squamosal-postorbital bar (Fig. 32.14). At one time, the group was widespread but now is limited to a few small islands off the coast of New Zealand. *Sphenodon* is a surviving "fossil," for it has not changed greatly from species that were living 150 million years ago.

Lizards. The familiar lizards, the little known worm lizards of the tropics, and snakes, though superficially different, are similar enough in basic structure to be placed in the single order **Squamata** (L. *squamatus,* scaly). Lizards are the oldest and most primitive squamates and probably evolved from some rhynchocephal-

FIGURE 32.16 Aquatic reptiles of the Mesozoic era. Plesiosaurs on the left; ichthyosaurs on the right. Plesiosaurs reached a length of 12 m.; ichthyosaurs, a length of about 3 m. (Courtesy of the Chicago Museum of Natural History.)

FIGURE 32.17 **The New Zealand tuatara, *Sphenodon punctatus,* is the sole surviving species of the order Rhynchocephalia. (Courtesy of R. Goellner, St. Louis Zoological Park.)**

ian-like ancestor early in the Mesozoic era. They are placed in the suborder **Sauria** or **Lacertilia** (Gr. *sauros,* lizard). A distinctive feature of lizards is the reduction of the temporal region of the skull roof. The lower bar of bone extending in the diapsid skull from under the eye to the quadrate bone, which is present in the rhynchocephalian skull (see Fig. 32.14), has been lost. As a result, the quadrate bone, to which the lower jaw attaches, is not held so firmly in place. It can move to some extent. The jaw apparatus is more flexible, the mouth can open wider, and larger prey can be captured and swallowed. In other respects lizards retain many primitive features. Most species have legs, movable eyelids, and a tympanic membrane set in the base of an external ear canal.

Lizards have been a very successful group and have undergone an extensive adaptive radiation. About 3000 species occur throughout the tropical and much of the temperate regions of the world. Most lizards are terrestrial quadrupeds that feed upon insects and other small animals during the daylight hours, but the tropical geckos forage at night. The African chameleons have a prehensile tail and an odd foot structure in which the toes of each foot are fused together into two groups that oppose each other like the jaws of a pair of pliers (Fig. 32.18*A*). They catch insects by a rapid flick of their sticky tongue that can be protruded nearly a full body's length. Many species change color to match their background. Some species are adapted to very arid conditions. The moloch of the deserts of central Australia (Fig. 32.18*B*) has a grotesque shape and coloration that provides a very effective camouflage. Microscopic striations on its head scales trap and carry dew by capillary action to its mouth. The Komodo dragons, now restricted to certain Indonesian islands, are the largest lizards, reaching lengths of 3 m. or more. They feed upon small deer, wild pigs, and goats (Fig. 32.18*C*). Some groups have become herbivores. The marine iguanas of the Galapagos islands feed upon seaweed (Fig. 32.18*D*).

The only poisonous lizards are the beaded lizards, such as the Gila monster of the Southwestern United States (Fig. 32.18*E*). Modified glands in the floor of the mouth discharge a neurotoxic poison, which is injected into the victim by means of grooved teeth. This is a relatively inefficient method, so the bite is not as dangerous as the bite of most poisonous snakes.

A curious adaptation of many lizards is their ability to shed much of their tail when seized by a predator. Often, as in the blue-tailed skink, the tail is very conspicuous (Fig. 32.18*F*). In some species, anaerobic metabolism sustains tail thrashing movements for a considerable period after autotomy, thus distracting the predator and increasing the time it needs to subdue and swallow the tail. In the meantime the lizard moves safely away. Tails break at a special autotomy plane and new ones regenerate from this point.

Snakes. Snakes are placed in the suborder **Serpentes** or **Ophidia** (L. *serpentis,* serpent). The additional loss of the postorbital bone in the dermal skull roof (see Fig. 32.14) frees the squamosal bone. The jaw mechanism is exceptionally flexible, and snakes can swallow prey several times their own diameter. In a sense, snakes have five jaw joints (Fig. 32.19): (1) the usual one between quadrate and lower jaw, (2) one between quadrate and squamosal, (3) one between squamosal and brain case, (4) one about halfway along the lower jaw, and (5) one at the chin, for the two lower jaws are not united at this point. Each side of the lower jaw can be moved independently as prey is pulled into the mouth. Other features that characterize snakes are the absence of movable eyelids, of a tympanic membrane and middle ear cavity, and of legs and girdles. The absence of the pectoral girdle is a necessary correlate of swallowing animals larger than the diameter of the body. There are exceptions to these generalizations, for geckos do not have movable eyelids, glass "snakes" (a lizard) lack legs and some of the more primitive snakes, such as the python, have vestigial hind legs.

The wormlike body form of snakes, their lack of a tympanic membrane, and certain degenerative changes in the eye suggest that snakes evolved from some primitive burrowing lizard group. Their forked tongue, which is often seen darting from the mouth, is an organ concerned with touch and smell. Particles adhere to it, the tongue is withdrawn into the mouth, and the tip is projected into a specialized part of the nasal cavity (**Jacobson's organ**) where the odor of the particle can be detected. Although this organ is present in many other

FIGURE 32.18 Lizard adaptive radiation. *A,* The African chameleon, *Chameleo gracilis,* is adapted for life in the trees. *B,* The moloch, *Moloch horridus,* is a desert species of central Australia. *C,* Several Komodo dragons, *Varanus komodoensis,* of Indonesia feeding upon the carcass of a goat. *D,* The marine iguana, *Amblyrhynchus cristatus,* of the Galapagos islands feeds upon sea weed. *E,* The Gila monster, *Heloderma horridum,* is a poisonous species. *F,* The blue-tailed skink, *Eumeces fasciatus,* has a conspicuous blue tail that can be shed when the animal is attacked. (*A* and *E* courtesy of J. Cadle, Smithsonian Institution; *B* courtesy of C. Gans, University of Michigan; *C,* photograph by C. A. Villee, Jr.; *D,* photograph by W. F. Walker, Jr.; *F* courtesy of E. S. Ross, California Academy of Science.)

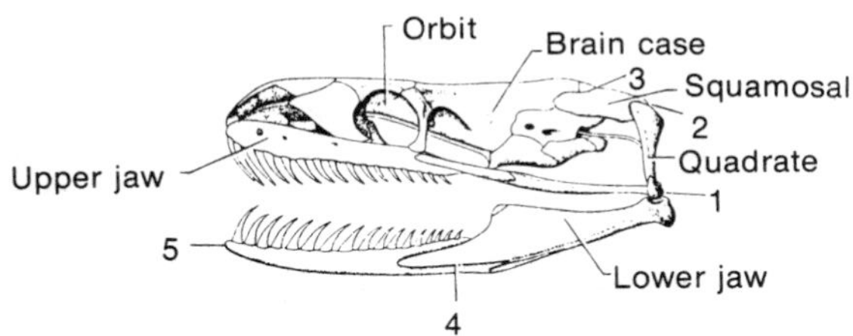

FIGURE 32.19 Lateral view of the skull of a python to show the five major points at which motion can occur when the jaws are opened. (From Romer after M. Smith.)

FIGURE 32.20 A white-lipped python feeding on a mouse. (Photograph by W. F. Walker, Jr.)

terrestrial vertebrates, it is particularly well developed in snakes, and it is needed by snakes to follow prey trails and for sex recognition.

A few primitive snakes still burrow, but during their subsequent evolution most snakes adapted to epigean life and underwent an extensive adaptive radiation. Nearly 2700 species occur in the temperate and tropical regions of the world. A pattern of locomotion evolved that is dependent primarily upon undulatory movements of the long trunk. When a snake moves through grass, for example, loops of the trunk form behind the head and move posteriorly. When these loops meet protuberances from the ground, they push upon them, and the resultants of these forces move the snake forward. Snakes that climb trees or move across shifting sands have developed other patterns of locomotion.

Snakes prey upon fish, frogs, lizards, birds, and small mammals. Those that prey upon very active animals, such as some birds and mammals capable of inflicting a severe wound to the snake with their bills or teeth, have evolved methods of prey immobilization. Boa constrictors, pythons, and many harmless rodent-eating snakes, such as king snakes, quickly entwine their prey in loops of the trunk and suffocate the prey by stopping their respiratory movements (Fig. 32.20). Other groups have evolved poison glands associated with grooved or hollow hypodermic-like teeth—the **fangs.** Old World vipers and New World pit vipers (rattlesnakes, copperheads, cotton-mouths, and water moccasins) have a pair of large hollow fangs at the front of the mouth that are articulated to bones of the upper jaw and palate in such a way that they are folded against the roof of the mouth when the mouth is closed and automatically brought forward when the mouth is opened. After injecting their hemolytic poison, which causes a breakdown of the red blood cells, they back off and wait for the prey to become quiet before attempting to swallow it. Coral snakes and cobras have short, immovable fangs at the front of the upper jaw with which they inject a neurotoxic poison.

Pit vipers have a prominent **sensory pit** on each side of the head between the nostril and eye. These organs enable the snakes to seek out and strike accurately at objects warmer than their surroundings and help the snakes to feed upon small nocturnal mammals.

Many snakes vibrate their tails when they are in danger. Rattlesnakes have increased the efficiency of this habit by developing rattles on the ends of their tails (Fig. 32.21). The rattles are simply remnants of dry, horny skin that have been prevented from being completely shed by catching on an enlargement on the end of the tail when the animals molt. The center of the rattlesnake evolution probably was the western plains of the United States, and the rattles may have developed as a warning mechanism that prevented the snakes from being trampled by bison.

Amphisbaenids. The worm lizards are now placed in the suborder **Amphisbaenia** rather than being included with the lizards for they appear to have been an offshoot from very early squamates. Their body form is wormlike, and they are highly specialized burrowers of the tropics and subtropics (Fig. 32.22). About 140 species are known. One genus has short front legs, but the others lack legs and the girdles are vestigial. There are no external ear openings, but they detect ground vibrations by an extension of the stapes to the lower jaw. Their rudimentary eyes are hidden beneath the skin and they track prey primarily by hearing. The skull is exceptionally strong and spade-shaped, and the short tail is also very strong. The name of the group (G. *amphi,* both + *baino,* to go) refers to their ability to move easily both forward and backward in their burrows.

Dinosaurs and Their Allies. During the Mesozoic era, the land was dominated by the archosaurs—reptiles

FIGURE 32.21 The western diamond rattlesnake, *Crotalus atrox*, from Arizona. (Courtesy of J. Cadle, the Smithsonian Institution.)

that shared features such as a diapsid skull and a tendency to evolve a two-legged gait. Reduced pectoral appendages, enlarged pelvic appendages, and a heavy tail that could act as a counterbalance for the trunk were correlated with this method of locomotion.

Saurischian dinosaurs (order **Saurischia**) are characterized by having a lizard-like pelvis as their name indicates (Gr. *sauros*, lizard + *ischion*, hip or pelvis). They evolved from ancestors that were approximately 1 m. long, but later saurischians became giants of the land and swamps. *Tyrannosaurus* (Fig. 32.23*A*) was the largest terrestrial carnivore that the world has ever seen. It stood about 6 m. high and had large jaws armed with dagger-like teeth 15 cm. long—a truly formidable creature! Other saurischian dinosaurs were herbivores that reverted to a quadruped gait, but their bipedal ancestry is reflected in their retention of a heavy tail and in having hind legs larger than the front ones. *Brontosaurus* (Fig. 32.23*B* and *C*) was an enormous herbivore that attained a length of 25 m. and an estimated weight of 45 metric tons. Its huge size led to the hypothesis that it lived in swamps where it would be supported partly by the buoyancy of the water. A more recent view is that it fed in a giraffe-like fashion upon high vegetation. Its limb structure resembled that of elephants, and its tail was not flattened, as is usually the case in aquatic vertebrates.

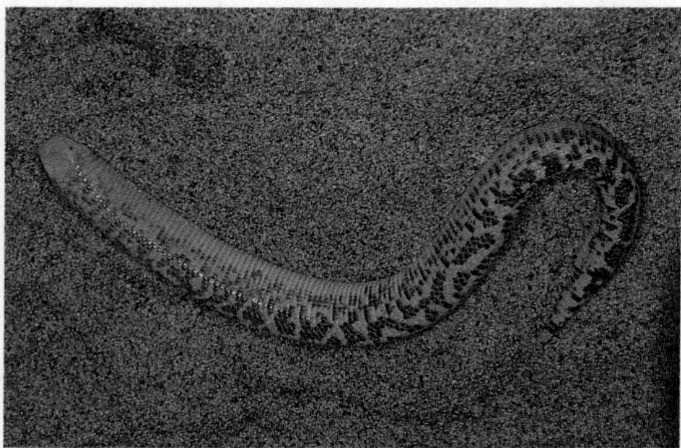

FIGURE 32.22 A desert amphisbaenid from Somali, *Agamodon anguiliceps*. (Courtesy of C. Gans, University of Michigan.)

Dinosaurs belonging to the order **Ornithischia** were characterized by having a birdlike pelvis (Gr. *ornithos*, bird + *ischion*, pelvis) in which part of the pubis was turned posteriorly beside the ischium. All were herbivores. Teeth at the front of the jaws were replaced by horny beaks, and the teeth in the back of the jaws were modified for grinding. Some lived in swamps, others in the uplands. Many reverted to a quadruped gait and increased in size. These animals undoubtedly formed much of the diet of carnivores, such as *Tyrannosaurus*, and many, such as *Stegosaurus* and *Triceratops* (Fig. 32.23*D* and *E*), evolved protective devices, such as spiked tails, bony plates on the body, and horned skulls.

The reasons for the evolution of large size are not entirely clear. Within limits, large size has a protective value, but it may also have been a way of achieving a more nearly constant body temperature, as mentioned earlier (p. 684).

Dinosaurs of both orders carried their limbs in mammal-like fashion, more or less under their bodies. John Ostrom of Yale University believes that this implies that they were active animals capable of rapid locomotion. The greater energy expenditure in turn suggests that they may have been endothermic ("warm-blooded"). Robert Bakker of Johns Hopkins University points out that their bone histology was closer to mammalian than to reptilian bone; a few may have lived in rather cool climates, and analysis of fossil communities suggests that the ratio of carnivorous dinosaurs to their herbivorous prey was closer to the ratio we see for a lion to its prey than that for a carnivorous lizard to its prey. A warm-blooded predator requires about 10 times more food than a cold-blooded one. Many paleontologists do not believe that they were endothermic in the mammalian sense. If they lived in a warm climate, as most certainly did, their large mass would give them a certain thermal stability at a high temperature close to the average environmental temperature.

A bipedal gait freed the front legs from use in terrestrial locomotion. The front legs became reduced in many dinosaurs, but in one group of archosaurs, they

FIGURE 32.23 Representatives of the main groups of dinosaurs that flourished during the late Mesozoic era. *A, Tyrannosaurus,* a carnivorous saurischian; *B* and *C, Brontosaurus* reconstructed as a swamp dweller or browser of high vegetation; *D, Stegosaurus,* an ornithischian; *E, Triceratops,* another ornithischian. (*C* courtesy of Fauna; others courtesy of the Chicago Museum of Natural History.)

evolved into wings. The wings of the flying reptiles (order **Pterosauria**, Gr. *pteron,* wing + *sauros,* lizard) consisted of a membrane of skin supported by a greatly elongated fourth finger (Fig. 32.24). The fifth finger was lost, and the others probably were used for clinging to cliffs. The hind legs were very feeble, and the animal must have been nearly helpless on the ground. Certain pterosaurs became very large. By extrapolating from recently discovered fragments of wing bones, paleontologists estimate that one species had a wing span of 12 m.! The large amount of energy needed for flight suggests that pterosaurs may have been warm-blooded.

FIGURE 32.24 *Pteranodon,* one of the largest of the flying reptiles, or pterosaurs, that lived during the late Mesozoic era. (Courtesy of the American Museum of Natural History.)

Support for this notion has come from A. G. Sharvov's discovery in 1970 in Russia of a specimen in which the fossil shows clear impressions of fur!

Most of the archosaurs became extinct toward the end of the Mesozoic, but the reason for this is not entirely clear. At the end of the Cretaceous period, environmental temperatures became somewhat cooler, and large inland seas that would have buffered temperature fluctuations disappeared. Seasonal and diurnal temperature changes were probably more pronounced than earlier in the Mesozoic Era. The inability of large dinosaurs to retreat to shelter or to hibernate would make them vulnerable, especially if their ability to conserve body heat was rudimentary. Changes in vegetation that accompanied climatic changes may also have been a factor. Inability of the specialized herbivores to adapt to changes in vegetation would have led to their death. Their disappearance, in turn, would deprive the huge carnivores of most of their food supply. The factors mentioned may not have wiped out a population but simply reduced it to a size at which it became very difficult for the species to maintain itself.

Only one group of archosaurs survived this wholesale extinction—the alligators and crocodiles (order **Crocodilia,** L. *crocodilus,* a crocodile). Crocodiles have reverted to a quadruped gait (though their hind legs are much longer than the front) and an amphibious mode of life. Twenty-three species are known, but only two occur in the United States—the American alligator, which can be distinguished by its rounded snout, and the American crocodile, which has a much more pointed snout. Their tail is flattened from side to side so it is a very effective swimming organ. Eye, ear, and nose openings are situated on elevations from the top of the head so they protrude above water when the animal is submerged (Fig. 32.25).

Mammal-like Reptiles. Another line of evolution, destined to lead to mammals, diverged from the cotylosaurs millions of years before the advent of archosaurs. Early mammal-like reptiles are placed in the order **Pelycosauria** (Gr. *pelykos,* an axe + *sauros,* lizard). They were similar to cotylosaurs but had a rather deep, somewhat axe-shaped synapsid skull with a temporal opening ventral to the squamosal and postorbital bones (Fig. 32.14). They were medium-sized, somewhat clumsy, terrestrial quadrupeds with limbs sprawled out at right angles to the body. Some pelycosaurs (*Dimetrodon,* Fig. 32.26*A*) had bizarre sails upon their backs supported by elongated vertebral spines. There has been much speculation as to the sail's function, but it did help to maintain a constant ratio of body surface to mass as members of this evolutionary line became larger. It has been suggested that the sail was a primitive thermoregulator, acting as a heat receptor or radiator.

Later mammal-like reptiles are placed in the order **Therapsida** (Gr. *therion,* mammal + *apsida,* loop, a reference to the arches of the synapsid skull). *Lycaenops* (Fig. 32.26*B*) is quite representative. They came to resemble mammals more closely. Their limbs were more nearly beneath the body, where they could provide better support and move more rapidly back and forth. Their jaws were strong, and their teeth were specialized, like those of a mammal, into ones suited for cropping, stabbing, and cutting. A secondary palate separated the mouth and nasal passages so that breathing could continue as the animal chewed its food. Long ribs were limited to the thorax, as in mammals, and lumbar ribs were very short, indicative of a diaphragm. The existence of

FIGURE 32.25 The broad-nosed caiman *(Caiman latirostris)* from South America. The nose, eye, and ear *(arrow)* of a crocodilian can remain above the surface when the rest of the body is submerged. (From Bellairs: *The Life of Reptiles.* Weidenfield & Nicolson.)

FIGURE 32.26 Two mammal-like reptiles. *A, Dimetrodon,* an early member of the group; *B, Lycaenops,* a later mammal-like reptile similar to those that gave rise to mammals. (Courtesy of the American Museum of Natural History.)

numerous pits for blood vessels and nerves in the snout bones suggests a bare and moist nasal area (rhinarium), seen in mammals, and possibly the presence of modified facial hairs, the whiskers or vibrissae. The late mammal-like reptiles were quite active creatures beginning to show some mammalian attributes. The major osteological characters that separated them from mammals were the nature of the jaw joint and the sound-transmitting apparatus. In all nonmammalian vertebrates, the jaw joint lies between the posterior ends of the mandibular arch, which are usually ossified as a **quadrate bone** in the upper jaw and an **articular bone** in the lower jaw. In mammals, the jaw joint is between two dermal bones (the **dentary** of the lower jaw and **squamosal** of the upper jaw) that lie just anterior to the quadrate and articular. The mammalian homologues of the quadrate and the articular (the **incus** and **malleus**, respectively) are covered by a tympanic membrane and form, with the stapes, a chain of three delicate auditory ossicles that transmit airborne vibrations from the tympanic membrane to the inner ear (Fig. 11.10). These characters had not been achieved by the late therapsids, but changes in feeding mechanisms that resulted in a stronger bite had led to an enlargement of the dentary, which grew posteriorly close to the squamosal. The quadrate and articular were very small, and they may have been involved in transmitting sound vibrations impinging on the tympanic membrane, for the tympanic membrane was located on the posteroventral corner of the lower jaw next to the jaw joint. In a few very late therapsids a dentary-squamosal joint was present beside the quadrate-articular joint.

Classification of Amphibians and Reptiles

Class Amphibia The amphibians. Ectothermic vertebrates usually having an aquatic larval stage; adults typically terrestrial; skin moist and slimy; scales absent in most groups.

†Subclass Labyrinthodontia Several orders of extinct and primitive amphibians collectively called labyrinthodonts; vertebral centra of the "arch" type, each usually consisting of two or three arches of bone encasing the notochord.

†Subclass Lepospondyli Several orders of extinct amphibia with vertebral centra of the "spool" type; each centrum a single structure, spool-shaped, pierced by a longitudinal canal for persistent notochord.

Subclass Lissamphibia Modern amphibians, probably evolved from labyrinthodonts. Skull usually flat and broad; pineal opening lost; brain case poorly ossified; pubis never ossified; carpals and tarsals cartilaginous; never more than four toes on front foot.

†Order Proanura Ancestral frogs of the Triassic. Trunk short, short tail still present; limbs not highly specialized. *Triadobatrachus*.

Order Salientia Modern frogs and toads. Short trunk; tail absent or very small; caudal vertebrae form a urostyle; legs specialized for jumping. The leopard frog, *Rana*; tree frog, *Hyla*; American toad, *Bufo*.

Order Caudata Salamanders. Tail long; legs usually present. The spotted salamander, *Ambystoma*; red-backed salamander, *Plethodon;* mudpuppy, *Necturus*.

Order Meantes Sirens. Neotenic species with a long eel-like trunk and tail; small front legs present; hind legs absent. *Siren*.

Order Gymnophiona Caecilians. Wormlike trunk; limbs absent; tail very short; vestiges of dermal scales in the skin. Confined to the tropics.

Class Reptilia The reptiles. Ectothermic or heliothermic terrestrial vertebrates that reproduce upon the land; body covered with horny scales or plates.

Subclass Anapsida Primitive reptiles characterized by a solid roof in the temporal region of the skull.

†Order Cotylosauria The cotylosaurs. Ancestral reptiles retaining many primitive features.

Order Testudines The turtles.

†Subclass Euryapsida Several orders of ancient marine reptiles that propelled themselves with long paddle-shaped limbs; single temporal opening high on the skull. The plesiosaurs.

†Subclass Ichthyopterygia One order of ancient, marine, fishlike reptiles; single temporal opening high on the skull. The ichthyosaurs.

Subclass Lepidosauria Primitive reptiles with two temporal openings in the skull (diapsid); pineal foramen usually retained; teeth on palate as well as jaw margins, usually not set in sockets; usually quadrupeds.

Order Rhynchocephalia Primitive lizard-like reptiles. *Sphenodon*.

Order Squamata Advanced lepidosaurians. Lizards, snakes, and amphisbaenids.

Subclass Archosauria Advanced diapsid reptiles. Pineal foramen usually absent; teeth usually restricted to jaw margin, set in sockets; an extra fenestra usually present in front of the eye or on the lower jaw, or both; most are bipedal reptiles or

†Extinct.

show signs of bipedal ancestry. Four extinct orders of ruling reptiles, including the dinosaurs and flying pterosaurs.

Order Crocodilia The only living archosaurs; alligators and crocodiles.

†**Subclass Synapsida** Mammal-like reptiles characterized by a single temporal opening on the lateral surface of the skull.

Order Pelycosauria Early mammal-like reptiles retaining many primitive features; limbs in primitive tetrapod position with humerus and femur moving in the horizontal plane. *Dimetrodon*.

Order Therapsida Advanced mammal-like reptiles beginning to show many mammalian characteristics; limbs rotated beneath the body with humerus and femur moving close to the vertical plane. *Lycaenops*.

Summary

1. In adapting from life in water to life on land, vertebrates evolved strong body support, different methods of locomotion, ways of receiving sensory cues from the air, and methods of obtaining oxygen without an undue loss of body water. They had to regulate their body temperature in the face of extreme fluctuations of ambient temperature and to find ways of reproducing upon the land.

2. The extinct labyrinthodonts were the first amphibians, and they probably gave rise to the contemporary amphibians as well as to the reptiles.

3. Amphibians are well adapted to terrestrial life with respect to support, locomotion, and their sensory-nervous systems. They ventilate their lungs with a buccopharyngeal pump. Cutaneous gas exchange supplements pulmonary exchange. The heart atrium is divided into left and right sides, and there is little mixing of the bloodstreams from the body and lungs in the single ventricle.

4. Amphibians are limited to moist habitats by their thin, moist, and scaleless skin; their production of a copious and dilute urine; their inability to regulate their body temperature; and the necessity of laying their eggs in water or in very damp terrestrial locations.

5. Salamanders (order Caudata) retain a long tail, and most species have short legs. Many species have become neotenic. Some terrestrial salamanders are lungless and depend upon cutaneous gas exchange.

6. Sirens (order Meantes) are a neotenic group of salamander-like amphibians with reduced front legs and no hind legs.

7. Frogs and toads (order Salientia) are highly specialized for jumping. Although most species are aquatic, toads are quite terrestrial, and tree frogs and toads are arboreal. Some tropical frogs have evolved interesting reproductive modifications that protect the delicate larval stage.

8. Caecilians (order Gymnophiona) are wormlike amphibians specialized for a burrowing mode of life.

9. Reptiles are better adapted to terrestrial life than are amphibians. Their nearly glandless skin is covered with horny scales that greatly reduce water loss. In warm environments they can maintain a high and nearly constant body temperature primarily by regulating their exposure to the sun, but they cannot maintain a high temperature as ambient temperatures fall. Gas exchange occurs only through the lungs, which are ventilated primarily by rib movements. The heart is nearly completely divided into left and right sides. Nitrogen is excreted primarily as uric acid rather than as ammonia, so little water need be excreted. Reptiles have evolved a cleidoic egg that can, and indeed must, be laid on the land or retained within the uterus. A few species of lizards and snakes are viviparous.

10. As might be expected of a successful group of terrestrial vertebrates, reptiles underwent an extensive adaptive radiation. Dinosaurs and other groups were the predominant terrestrial vertebrates during the Mesozoic era, which is known as the age of reptiles. Plesiosaurs and ichthyosaurs readapted to an aquatic existance. Pterosaurs evolved true flight.

11. Turtles (order Testudines) are encased in a bony shell. Living species can withdraw their neck and limbs.

12. The tuatara of New Zealand is the only surviving species of the order Rhynchocephalia. It retains an unmodified diapsid skull.

13. Lizards, snakes, and the burrowing amphisbaenids of the order Squamata have a modified diapsid skull that increases jaw flexibility. This reaches its highest development among snakes and enables them to swallow prey several times their own diameter.

14. Crocodiles and alligators (order Crocodilia) are closely related to the extinct dinosaurs and pterosaurs. Many features enable them to spend considerable time in the water.

15. Mammals evolved from a group of synapsid or mammal-like reptiles that diverged at least 280 million years ago from the lines of evolution of the other reptilian groups. Late mammal-like reptiles had become quite active and possibly warm-blooded, but they retained the characteristic reptilian jaw joint between the dentary and squamosal bones.

References and Selected Readings

Many of the general references on vertebrates cited at the end of Chapter 31 contain considerable information on the biology of amphibians and reptiles.

Bellairs, A.: *The Life of Reptiles.* London, Weidenfield and Nicolson, 1969. Deals with the anatomy, physiology, development, ecology, and evolution of reptiles. The modern successor to Gadow's famous book (listed here).

Conant, R.: *A Field Guide to Reptiles and Amphibians of Eastern and Central North America.* 2nd ed. Boston, Houghton Mifflin Co., 1975. A very useful guide for the field identification of amphibians and reptiles; similar to the Peterson bird guides.

Desmond, A. J.: *The Hot-Blooded Dinosaurs.* New York, The Dial Press, 1976. A popular account of the history of archosaur discovery and of changing interpretations of their mode of life.

Dunkle, T.: A perfect serpent. *Science, 81*(Oct.):30–35, 1981. The adaptations of the rattlesnake as a four-speed, self-propelled, spring-loaded, heat-seeking hypodermic.

Ernst, C. H., and R. W. Barbour: *Turtles of the United States.* Lexington, University Press of Kentucky, 1972. The most recent comprehensive reference work on the taxonomy and natural history of turtles.

Gadow, H.: *Amphibia and Reptilia.* Weinheim, Germany, Engelmann, 1958. A recent reprint of a book originally published as a part of the "Cambridge Natural History" in 1901; it is still a valuable account of the anatomy and evolution of amphibians and reptiles.

Gans, C.: *Biomechanics: An Approach to Vertebrate Biology.* Philadelphia, J. B. Lippincott Co., 1974. Many examples of the feeding, locomotor, and burrowing adaptations of reptiles are discussed.

Goin, O. J., O. B. Goin, and G. R. Zug: *Introduction to Herpetology.* 3rd ed. San Francisco, W. H. Freeman & Co., 1978. An excellent account of the biology of amphibians and reptiles.

Langstron, W., Jr.: Pterosaurs. *Scientific American, 244*:(Feb.)122, 1981. The diversity of pterosaurs and their probable modes of life are described.

Newman, E. A., and P. H. Hartline: The infrared "vision" of snakes. *Scientific American, 246*:(Mar.)166, 1982. A fascinating analysis of the sensory pits of rattlesnakes and pythons and of how information from the pits and eyes is integrated in the brain to provide a unique picture of the world.

Schmidt, K. P., and R. F. Inger: *Living Reptiles of the World.* New York, Doubleday & Co., 1957. A beautifully illustrated account of the natural history of the families of living reptiles.

33 Vertebrates: Birds

ORIENTATION

Birds are the largest class of vertebrates. Their success can be attributed partly to the evolution of endothermy, which enables them to be very active under a wide range of environmental temperatures, and partly to flight, which gives them great mobility. We will first examine how birds fly and their adaptive features that make endothermy and flight possible. Then we will consider the origin, evolution, and adaptive diversity of birds. Flight enables many birds to travel far from home in their search for food and to migrate seasonally. How they orient and navigate is considered in the last part of the chapter.

CHARACTERISTICS OF BIRDS

1 Birds are endothermic vertebrates.

2 Horny scales are retained on the feet, but feathers cover most of the body. Cutaneous glands are absent except for a uropygeal oil gland.

3 Skeletal bones are exceptionally light and frequently pneumatic. The pectoral appendages are wings, the sternum is broad and usually keeled, and the reduced number of caudal vertebrae form a pygostyle.

4 The eyes and visual centers in the brain are very important and large. The inner ear contains a cochlea, but it is not long and coiled.

5 The narrow jaws form a horn-covered beak in contemporary species. Teeth are absent. Villi are present in the small intestine.

6 The lungs are relatively small, but an unusual pattern of air passages and air sacs produces an exceptionally efficient one-way passage of air across the respiratory surfaces.

7 The flows of venous and arterial blood are completely separated; the atrium and ventricle of the heart are divided.

8 Nitrogenous wastes are eliminated primarily as uric acid. The urinary bladder has been lost.

9 One ovary and oviduct have been lost. The large eggs are heavily laden with yolk. Albuminous materials and a calcareous shell are secreted around the eggs as they pass down the oviduct. Birds are oviparous.

Birds, class **Aves** (L. *avis*, bird), are the largest group of terrestrial vertebrates, comprising about 8700 species. They adapted to flight early in their evolution, and most species are excellent fliers. Even the few species that have reverted to a completely terrestrial life show anatomical and physiological features indicative of their evolution from flying ancestors. Adaptation for flight has imposed a certain uniformity in basic structure and physiology on all birds. In addition to feathers and wings, or vestiges of wings in certain terrestrial species, flight requires a high expenditure of energy. All birds are endothermic, but have developed ways of achieving this in a body of light weight. When not in the air, birds live upon the ground or in the water, or both, and, within the constraints imposed by flight, they are adapted to these habitats as well. Endothermy and their powers of flight have enabled birds to penetrate and to adapt to a wider range of habitats than any other vertebrate group. They range from the polar regions to the equator, and live in mountains, deserts, forests, and jungles. Some spend most of their lives on the ocean and return to land only to nest.

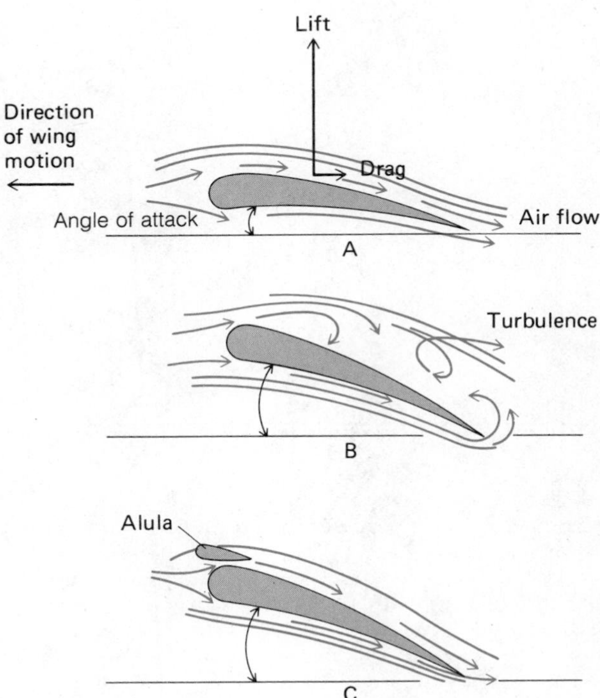

FIGURE 33.1 The effect of a wing on the airstream. *A,* Air moving more rapidly along the longer upper surface of the wing reduces the pressure on the upper surface and generates a force that can be resolved into lift and drag forces; *B,* if the angle of attack of the wing becomes too great, air swirls into the low pressure area, causing turbulence and a reduction in lift; *C,* turbulence can be reduced by raising the alula and forming a slot through which the air moves rapidly and smoothly.

Principles of Flight

Since flight is so central to bird evolution, it is important to understand its mechanisms at the outset. Bird wings are modified pectoral appendages, and the flight surfaces are composed of feathers. A wing is shaped like an airfoil, thick in front and thin and tapering behind, and usually cambered so that it is slightly concave on the undersurface and convex on the upper surface. As the air stream flows across the wing, the stream moves faster along the longer upper surface than the shorter lower surface. Bernoulli's law states that in a fluid stream the pressure is least where the velocity is greatest. The differential air speed decreases the pressure above the wing in relation to the underside. This produces a **lift force** acting perpendicular to the plane of wing motion and a **drag force** parallel to this plane (Fig. 33.1*A*). For the bird to fly, the lift force must equal the force of gravity on the bird, and a propulsive force must overcome the drag force. The somewhat teardrop shape of the wing allows a smooth flow of air across the surface and minimizes lift-reducing eddies. Some of the air flow, however, does roll up as a vortex, which is shed from the trailing margin and tips of the wings as a pair of vortex lines. This often can be seen in high-flying aircraft as a pair of vapor trails because the rapid rotation of air and consequent low pressure in the core of the vortex lines causes condensation.

Many factors affect lift. It increases in direct proportion to the surface area of the wing. Wing areas differ among different species of bird and can be varied in individual birds by the degree to which the wing is stretched out or unfolded. Lift also increases greatly as the speed of airflow across the wing increases, for lift is proportional to the square of the speed. Fast flying birds, such as a swift, have relatively smaller wings than slower flying species. When a bird is flying at low speeds, or during taking off and landing, the wing can be tilted so that its anterior edge is considerably higher, a procedure known as increasing the angle of attack (Fig. 33.1*B*). This increases lift but also tends to create lift-reducing turbulence above the wing. Separating certain feathers to produce slots through which the air moves very rapidly can reduce turbulence and make it possible for the wing to generate a very high lift. A small group of feathers supported by the first digit, the **alula** (L. diminutive of *ala,* wing), can produce a slot at the

FIGURE 33.2 Photograph of a red-shouldered hawk landing. Air speed is low, and lift is increased by several slots: *1*, alulae; *2*, separation of primary wing feathers; *3*, separation of feathers on leading edge of wing; *4*, auxiliary feathers on upper surface of wing. (From Hertel.)

front of the wing (Figs. 33.1*C* and 33.2). Additional slots are often formed along the trailing margin of the wing and at the wing tip. The latter slots reduce the turbulence known as tip vortex. Some birds obtain additional lift on landing by fanning out the tail feathers and bending them down. The tail, then, acts both as a brake and as a high-lift, low-speed airfoil.

The simplest kind of flight is **gliding**, in which the wings provide lift, and the forward motion comes from falling through the air. Altitude is lost in a glide of this type, but altitude can be maintained or even increased if the bird also soars. Land birds, such as the turkey vulture or the osprey (Fig. 33.3*A*), circle and fall within a rising current of warm air or above a bluff where air is deflected upward. Birds that engage in **static soaring** of this type have relatively short, broad wings that enable them to maneuver easily in the capricious air currents. Such wings have a low **aspect ratio** (wing length : wing width). Flight is slow so the wings need a large surface area to provide a lift adequate to support the bird. This is called a low **wing loading ratio** (weight of bird : surface area of wing). Additional lift is generated by considerable slotting of the wing, particularly near the tip. Oceanic birds engage in **dynamic soaring**, which makes use of the increase in air speed with increasing elevation above the ocean surface. Friction with the ocean causes air speed to be slowest at the ocean surface (Fig. 33.3*C*). Starting at a high elevation, these birds glide rapidly downward with the wind. Just above the ocean, they wheel into the wind and use the momentum gained in the glide to start to gain altitude. As they gain altitude, they encounter increasingly fast air speeds that in turn generate additional lift. In this way, the birds regain their original altitude. Dynamic soarers, since they are making use of air speed for much of their lift, have higher wing loading ratios than static soarers. They also have long, narrow wings (high aspect ratios) with little slotting. This reduces the tip vortices and keeps the vortices far apart.

Flapping flight is more complex than gliding or soaring because the wings are not stationary. One way of viewing flapping flight is illustrated in Figure 33.4. The wings are extended and move downwards and forwards during the downstroke. They are also inclined from the horizontal plane with their leading edge lower than the trailing edge. This changes the direction of the local lift and drag forces acting on each part of the wing, and this gives the lift force a forward component and also reduces the retarding component of the drag force. In this way the wing, or at least the part distal to the wrist, which inclines from the horizontal to a greater extent than the rest of the wing, has a propeller effect. Many of the individual flight feathers on the distal part of the wing are also separated, and to some extent they act as individual propellers. On the upstroke the wing is flexed and moves upward and backward. This is simply a recovery stroke and generates no useful aerodynamic forces. Other lifting and propelling forces appear to be generated by the air movements caused by the wings (Box 33.1). In all types of flight, the tail helps to support and balance the body and is used as a rudder.

The Adaptive Features of Birds

Most of the features of the anatomy of birds can be related directly or indirectly to endothermy and flight. They are adapted structurally and functionally to provide a high energy output in a body of low weight.

Thermoregulation

Birds have retained the horny scales of reptiles on parts of their legs, on their feet and, in modified form, as a covering for their beaks, but the scales that cover the rest of the reptilian body have been transformed into feathers. Feathers are an important component of the thermoregulatory apparatus, which maintains a balance between heat production by metabolic processes

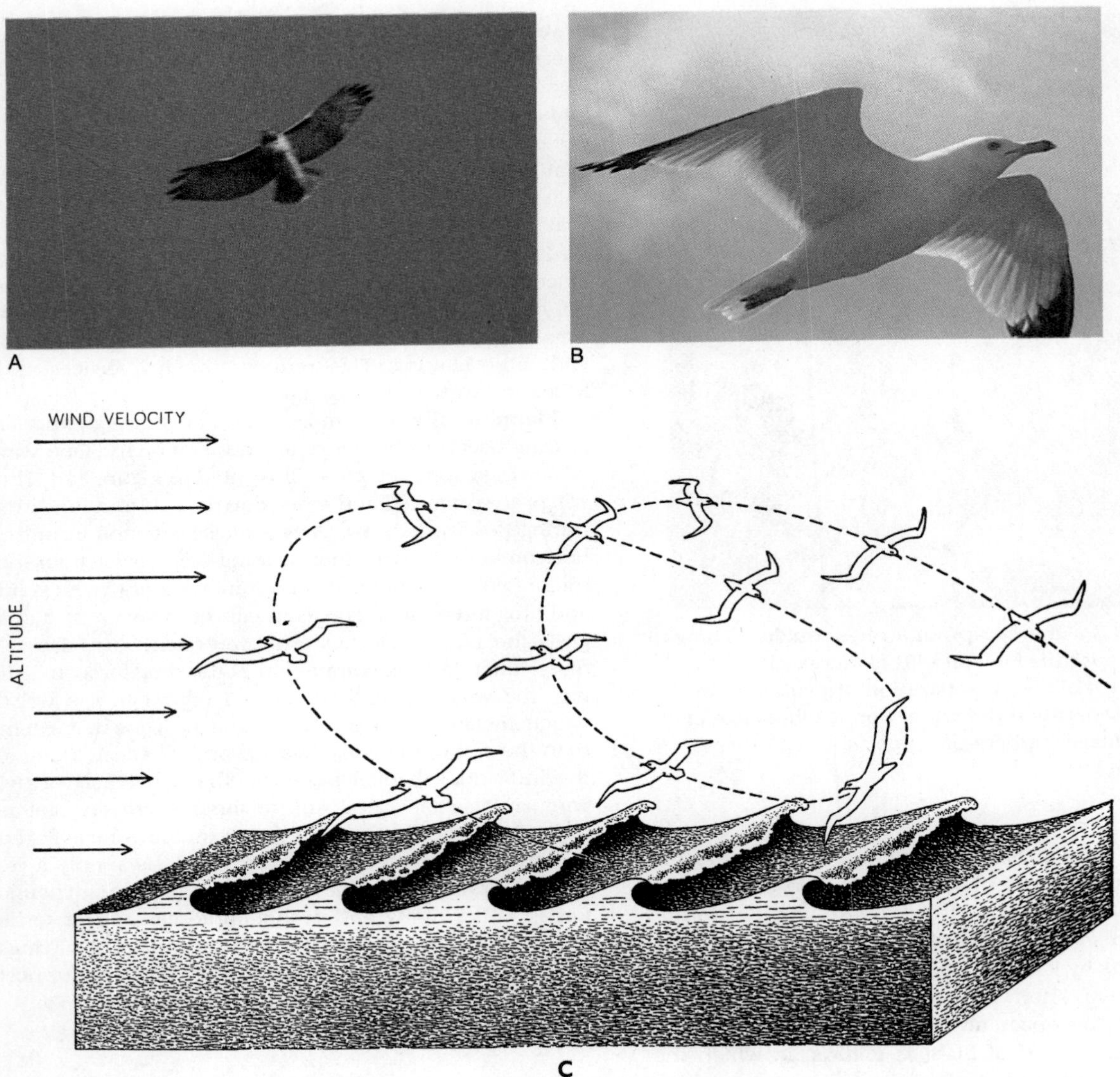

FIGURE 33.3 Soaring flight. *A,* Static soarers like this red-tailed hawk have short, broad wings. *B,* Dynamic soarers like this ring-billed gull have narrow, tapered wings. *C,* Diagram of dynamic soaring. (*A* and *B* courtesy of Louis W. Campbell; *C* from *The Soaring Flight of Birds,* by C. D. Cone, Jr. Copyright © by Scientific American, Inc. All rights reserved.)

and heat loss. Feathers overlap and entrap air between them. This reduces loss of body heat because it prevents air convection currents across the body surface that would carry heat away, and the entrapped air is itself a poor conductor of heat. Water, a very good conductor of heat, is prevented from penetrating the feathers by an oily secretion produced by the **uropygeal gland** (Gr. *oura,* tail + *pugē,* rump) located on the back near the tail base (see Fig. 33.10). When a bird preens, it spreads this secretion over the feathers. Water fowl have very large uropygeal glands. When temperatures fall, the feathers are fluffed out; this increases the thickness of the insulating layer of air. If temperatures fall very low, the bird must produce more heat by raising its meta-

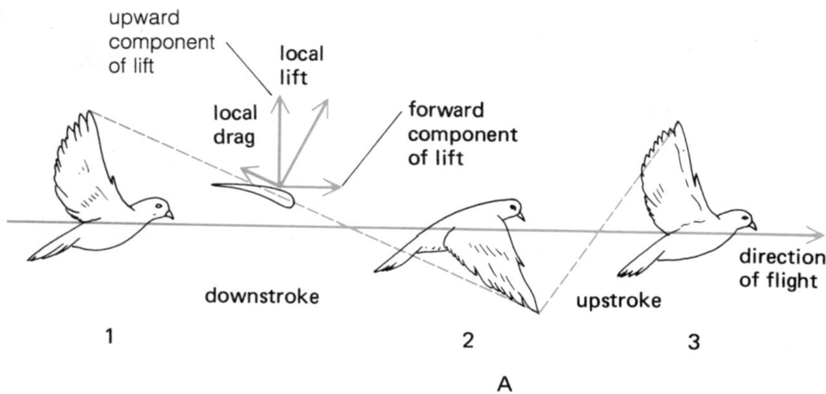

FIGURE 33.4 Flapping flight of birds. *A,* Stages in the wing cycle to show how the forward and downward tilt of the wing on the downstroke generates a local forward thrust. *B,* Photograph of a great egret in the middle of the downstroke showing the separation of the primary flight feathers. (*A* modified from Rayner, J. M. V. in H. Y. Elder and E. R. Trueman (Eds.): *Aspects of Animal Locomotion.* Cambridge University Press, 1980. *B,* Courtesy of Helen Cruickshank/VIREO.)

bolic level, as mammals do (Fig. 34.3, p. 726). When heat is to be lost, the feathers are held closer to the body, more blood is directed through the skin (and especially to uninsulated areas, such as the legs), and panting starts. Birds have no sweat glands. The interaction of these mechanisms enables birds to maintain their body temperature constant at relatively high levels, 40° to 43° C.

Feathers

Feathers, like horny scales, are epidermal outgrowths whose cells have accumulated large amounts of keratin and are no longer living. Pigments deposited in these cells during the development of the feather, together with surface modifications that reflect certain light rays, are responsible for the brilliant colors of birds. Although feathers cover a bird, they fan out in most species from localized **feather tracts** rather than growing out uniformly from all of the body surface. Feathers, more than any other single feature, characterize birds.

The **contour feathers** that cover the body and provide the flying surface consist of a stiff central shaft, whose base, the **calamus** (Gr. *kalamos,* quill), is embedded in a follicle in the skin (Fig. 33.5). The distal part of the shaft, the **rachis** (Gr. *rhakhis,* spine), bears a vane composed of numerous side branches, the **barbs** (L. *barba,* beard). Each barb bears minute hooked branches, **barbules,** along its side that interlock with the barbules of adjacent barbs to hold the barbs together. If the barbs separate, the bird can preen the feather with its bill until they hook together again; thus, the vane is a strong, light, and easily repaired surface ideal for insulation and flight. In birds that have lost the power of flight, such as the ostrich, hooklets are not present upon the barbules and the feather is very fluffy. A small **afterfeather** arises from the distal end of the calamus on the feathers of primitive birds, but it is reduced to a tuft of barbs on the feathers of more specialized species. When present, the afterfeather provides extra insulation.

The contour feathers on the posterior border of the arm, hand, and tail form the flying surfaces and are

BOX 33.1

Dr. J. M. V. Rayner of the University of Bristol recently proposed that the primary lift and propelling forces in flapping flight are generated not so much from the local forces acting on the wings as from the reactions against the bird of the air motions that the wings generate (see the figure below). As the air stream flows across the wings, some of the air leaving from beneath their under surface is drawn into the reduced pressure area above them. This air rolls up as a vortex which is shed from the wing tips. Since the wing tips come close together at the top of the upstroke and at the bottom of the downstroke, the vortices at the wing tips come close together and fuse at these points. The vortices leave the bird at the bottom of the downstroke as a vortex ring. A smoke ring is a familiar example of a vortex ring. Circulation of air in the vortex ring itself induces a downward flow of air through the center of the ring. The ring is drawn downward and backward with a certain momentum (mass times velocity). In accordance with Newton's third law of motion, there is an opposite and equal reaction to the shedding vortex rings that pushes upward and forward against the bird, thereby generating lift and thrust. (Figure modified from Rayner, J. M. V., in H. Y. Elder and E. R. Trueman (Eds.): *Aspects of Animal Locomotion.* Cambridge, Cambridge University Press, 1980.)

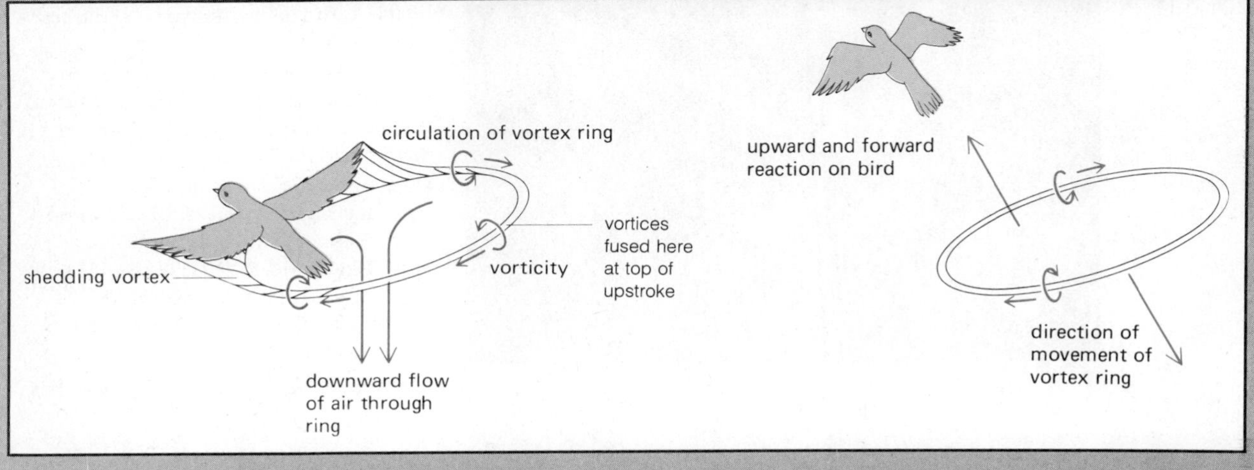

called the **flight feathers**. In birds that are good fliers, the rachis of a wing feather is close to the leading edge, thus thickening this part and giving the feather properties of an airfoil.

Other major types of feathers include down, filoplumes, and bristles. **Down** covers young birds and is found under the contour feathers in the adults of certain species, particularly aquatic ones. It is unusually good insulation, for it has a reduced shaft and long fluffy barbs arising directly from the distal end of the calamus. A **filoplume** consists of a slender, hairlike rachis bearing a few barbs at its tip. Their follicles are richly supplied with nerve endings, which suggests that they may serve as sense organs that help control the movements of other feathers. **Bristles** are stiff, vaneless feathers often found around the eyes and nose where they help keep out dirt. Some insect-catching birds, such as nighthawks, have long bristles around their mouths that act as insect nets.

Most birds **molt** once a year, usually after the breeding season. Those with special breeding plumages also molt before the breeding season. Feathers are lost and replaced in a characteristic sequence for each species. The process is gradual in most species, and the birds can move about normally during molting. Certain male ducks, however, shed the large flight feathers on their wings so rapidly that they are unable to fly for a while.

Skeleton

Many adaptations for flight are apparent in the skeleton of birds. The thin, hollow bones are very light in weight. Extensions from the lungs enter the limb bones in many species. The skeleton of a frigate bird having a

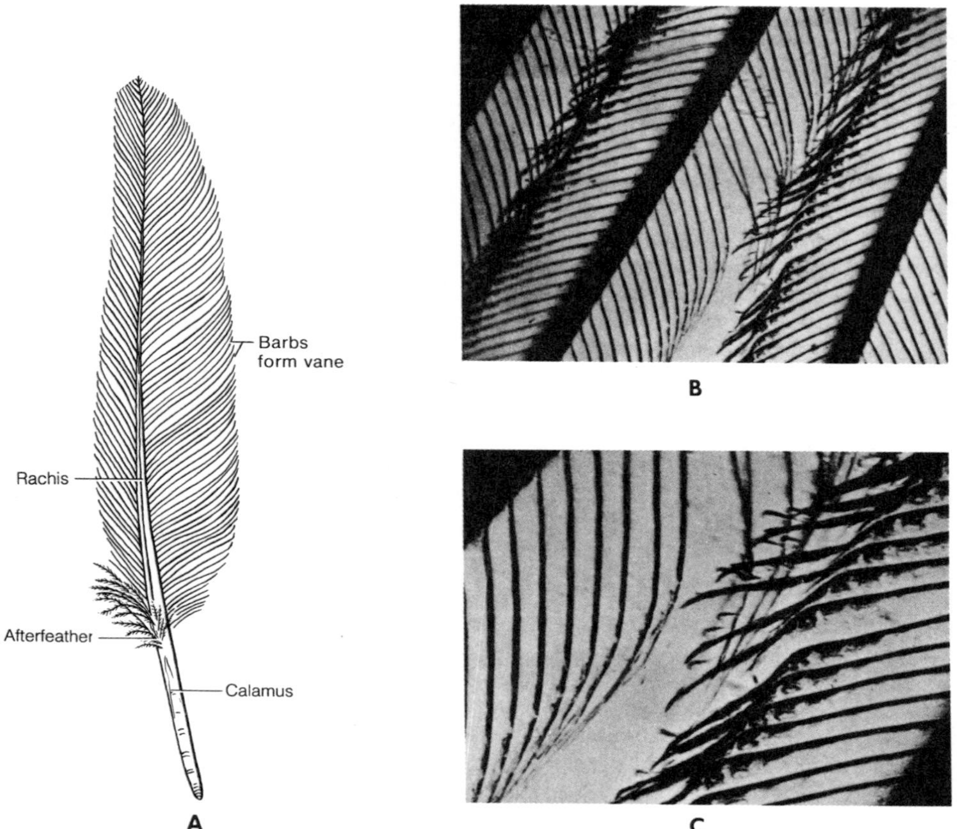

FIGURE 33.5 Contour feathers of a bird. *A,* Entire feather; *B* photomicrograph of barbs and interlocking barbules; *C,* further enlargement to show hooks and spines on barbules. (*A* modified after Young; *B* and *C* from Welty.)

wingspread of over 2 m. weighed only 115 g., which was less than the weight of its feathers! The skeletons of all birds weigh less in relation to their body weight than do the skeletons of mammals. Most of the bone substance is located at the periphery, where it gives better structural support. Many bird bones are further strengthened by internal struts of bone arranged in a manner similar to the trusses inside the wing of an airplane (Fig. 33.6).

The **skull** is notable for the large size of the cranial region, the large orbits, and the toothless beak (Fig. 33.7). The neck region is very long, and the **cervical vertebrae** are articulated in such a way that the head and neck are very mobile. Since the bird's bill is used for feeding, preening, nest building, defense, and the like, freedom of movement of the head is very important. The trunk region, in contrast, is shortened, and the **trunk vertebrae** are firmly united to form a strong fulcrum for the action of the wings and a strong point of attachment for the pelvic girdle and hind legs. In the pigeon, 13 of the posterior trunk, sacral, and caudal vertebrae are fused together to form a **synsacrum,** with which the

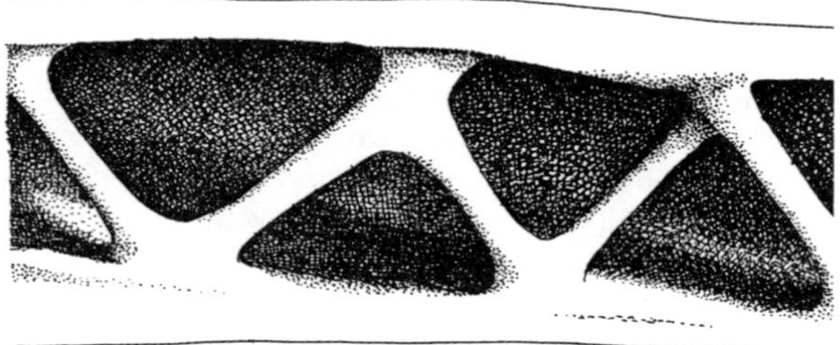

FIGURE 33.6 Longitudinal section of a metacarpal bone from a vulture's wing. Notice the internal trussing similar to that in an airplane's wing. (From D'Arcy Thompson.)

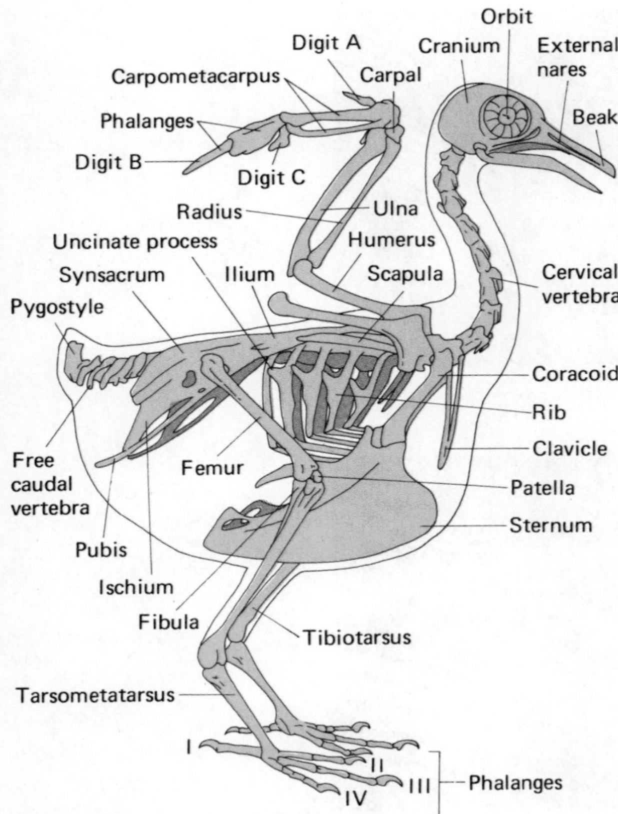

FIGURE 33.7 Skeleton of a pigeon. The distal part of the right wing has been omitted. (Modified after Heilmann.)

pelvic girdle is fused. Several free **caudal vertebrae,** which permit movement of the tail, follow the synsacrum. The terminal caudal vertebrae are fused together as a **pygostyle** (Gr. *pugē*, rump + L. *stylus*, a writing instrument) and support the large tail feathers.

The last two cervical vertebrae of the pigeon and the thoracic vertebrae bear distinct **ribs**. The thoracic basket is firm yet flexible. Extra firmness is provided by the ossification of the ventral portions of the thoracic ribs and by posteriorly projecting **uncinate processes** (L. *uncinus*, a hook) on the dorsal portions of the ribs, which overlap the next posterior ribs. Flexibility, needed in respiratory movements, is made possible by the joints between the dorsal and ventral portions of the ribs. The **sternum,** or breastbone, is greatly expanded and, in all but the flightless birds, has a large midventral **keel** that increases the area available for the attachment of the flight muscles.

The bones of the wing are homologous to those of the pectoral appendage of other tetrapods. The **humerus, radius,** and **ulna** can be recognized easily, but the bones of the hand have been greatly modified. Three short digits arise from a fused **carpometacarpus.** The most anterior of these supports the alula. Paleontological evidence suggests that the fingers are homologous to the first three of reptiles. The pectoral girdle, which supports the wing, consists of a narrow dorsal **scapula**, a stout **coracoid** extending as a prop from the shoulder joint to the sternum, and a delicate **clavicle,** which unites distally with its mate of the opposite side to form the wishbone.

The legs of birds resemble the hind legs of bipedal dinosaurs. The **femur** articulates distally with a reduced **fibula** and a large **tibiotarsus** (fusion of the tibia with certain tarsals). The remaining tarsals and the elongated metatarsals have fused to form a **tarsometatarsus.** The fifth toe has been lost in all birds and the fourth in some species. The first toe is turned posteriorly in the pigeon and many other birds. It serves as a prop and increases the grasping action of the foot when the bird perches. The efficiency of the leg in running on the ground and jumping at take-off is increased by the elongation of the metatarsals and by the elevation of the heel off the ground. The various fusions of the limb bones reduce the chance of dislocation and injury, for birds' legs must act as shock absorbers when they land. The pelvic girdle is equally sturdy; the **ilium, ischium,** and **pubis** of each side are firmly united with each other and with the vertebral column. The pubes and ischia of the two sides do not unite to form a midventral pelvic symphysis as they do in other tetrapods. This permits a more posterior displacement of the viscera, which, together with the shortened trunk, shifts the center of gravity of the body nearer to the hind legs.

The feet of birds have undergone a variety of modifications as birds have adapted to particular modes of life (Fig. 33.8). The foot and toes become particularly sturdy in ground-dwelling species. The woodpeckers have sharp claws, and the fourth toe is turned backward with the first to form a foot ideally suited for clinging onto the sides of trees. Swimming birds have a web stretching between certain of their toes. The marsh-dwelling jacana of the tropics has a foot with exceedingly long toes and claws that enable it to scamper across lily pads and other floating vegetation. Swifts and hummingbirds have very small feet barely strong enough to grasp a perch. These birds spend most of their time on the wing and almost never alight on the ground. The power of grasping is especially well developed in such perching specialists as our songbirds. In perching birds, the tendons of the foot are so arranged that the weight of the body

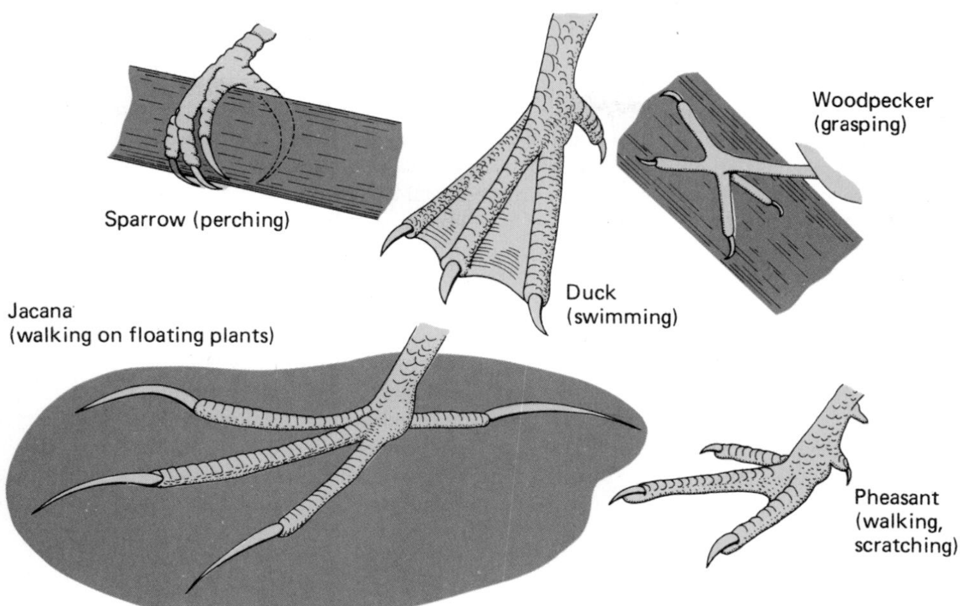

FIGURE 33.8 Adaptations of the feet of birds.

automatically causes the toes to flex and grasp the perch when the bird alights upon a branch.

Muscles

The strong and intricate movements of the wings and the support of the body by a single pair of legs entail numerous modifications of the muscular system. The flight muscles include the **pectoralis**, which originates on the sternum and inserts on the ventral surface of the humerus. It is responsible for the powerful downstroke of the wings (Fig. 33.9). One might expect that dorsally placed muscles would be responsible for the recovery stroke, but instead, another ventral muscle, the **supracoracoideus**, is responsible for the upstroke. The origin of the supracoracoideus is on the sternum dorsal to the pectoralis. Its tendon passes through a canal in the pectoral girdle near the shoulder joint and inserts on the dorsal surface of the humerus. These two muscles are exceptionally large and together make up as much as 25 to 35 percent of body weight in birds that are powerful flyers. In ducks and other birds that fly a great deal the flight muscles consist mostly of dark phasic fibers, rich in myoglobin. In chickens and other birds that beat their wings rapidly and intermittently, the wing muscle fibers are glycolytic and lighter in color.

Muscles within the wing are responsible for its folding and unfolding and the regulation of its shape and angles during flight. Other muscles attach to the follicles of the large flight feathers of the wings and tail and control their positions.

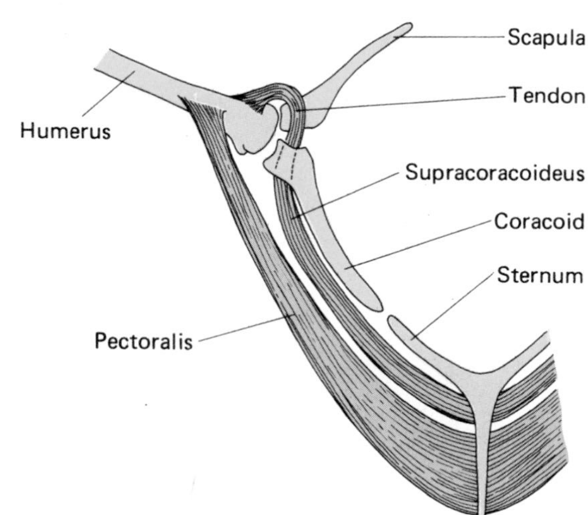

FIGURE 33.9 A diagrammatic cross section through the shoulder region and sternum showing the arrangement of major flight muscles. (From Welty after Storer.)

Special Senses and Nervous System

The sense of smell is less important than many other senses in animals that spend much of their life off the ground, and the olfactory organ and olfactory parts of the brain are reduced in most birds. Nevertheless, many carrion feeders and ground-dwelling species have a well-developed sense of smell. The nocturnal kiwi (see Fig. 33.19*B*) of New Zealand has nostrils at the tip of a long bill and finds earthworms by smelling them as it probes the ground.

Sight is very important and the eyes are large. Rods and cones are packed more closely in the retina than in mammals so the visual acuity of birds, that is, their ability to distinguish objects as they become smaller and closer together, is several times greater than that of human beings. Birds can also accommodate very rapidly for they must change quickly from distant to near vision as they maneuver among the branches of a tree or swoop to the ground from a considerable height. To focus on a near object, the ciliary body contracts and squeezes upon the periphery of the lens, thereby increasing its thickness, rather than by relaxing and permitting the lens to bulge by its own elasticity, as is the case in mammals (p. 253).

The sense of hearing, too, is highly developed in most birds, as one would expect from the importance of songs in their behavior. Although the cochlea of their ear is not as long as it is in mammals, experiments have shown that birds detect nearly as wide a range of frequencies as we do, and their ability to detect rapid changes in frequency is greater than ours.

Birds have large brains in which the cerebrum, optic lobes, and cerebellum are particularly well developed. The large cerebrum results from the enlargement of a deeply situated mass of gray matter, the **corpus striatum** rather than from a large cortex, which makes up much of the brain of mammals. The cerebral cortex is thin, and removal of it has little effect on behavior. Extirpation of parts of the corpus striatum, on the other hand, seriously affects eating, locomotion, vocalization, and reproductive behavior. The large **optic lobes** are the primary visual association area. The large **cerebellum** regulates the intricate coordination of motor activities needed in flight, nest building, and other activities.

Digestive System

The increased activity and high metabolic rate of birds depends upon the intake and processing of an adequate food supply. The digestive system (Fig. 33.10) is compact but so effective that some of the smaller birds process an amount of food equivalent to 30 percent of the body weight each day! Birds eat a variety of insects

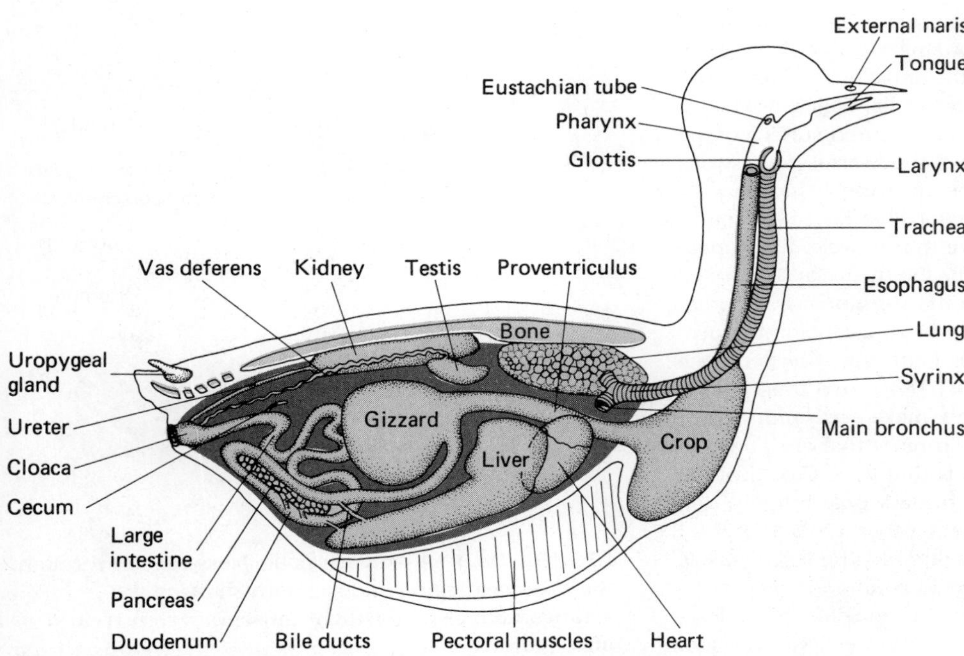

FIGURE 33.10 A lateral dissection of a pigeon to show the major visceral organs.

and other animals and such plant food as fruit and seeds. They do not attempt to eat such bulky, low-caloric foods as leaves and grass. The bills of birds are modified according to the nature of the food they eat. Finches have short heavy bills well suited for picking up and breaking open seeds. The hooked beak of hawks is ideal for tearing apart small animals that they have seized with their powerful talons. Herons use their long sharp bills for spearing fish and frogs, which they deftly flip into their mouths. The length and shape of hummingbirds' bills are correlated with the structure of the flowers from which they extract nectar. Whip-poor-wills fly about in the evening catching insects with their gaping mouths surrounded by bristles. When feeding, the skimmer flies just above the ocean with its elongated lower jaw skimming the surface. Any fish or other organisms that are hit are flicked into its open mouth. The woodcock's long and sensitive bill is adapted for probing for worms in the soft ground. The woodcock can open the tip of its bill slightly to grasp a worm without opening the rest of the mouth!

Food taken into the mouth is mixed with a lubricating saliva and passes through the pharynx and down the esophagus without further treatment, for birds have no teeth. In grain-eating species, such as the pigeon, the lower end of the esophagus is modified to form a **crop** in which the seeds are temporarily stored and softened by the uptake of water. Food is mixed with peptic enzymes in the **proventriculus** (Gr. *pro*, in front of + L. *ventriculus*, stomach), or first part of the stomach, and then passes into the **gizzard** (Old Fr. *gezier*, cooked entrails), the highly modified posterior part of the stomach characterized by thick muscular walls and modified glands that secrete a horny lining. Small stones that have been swallowed are usually found in the gizzard and aid in grinding the food to a pulp and mixing it with the gastric juices. The **intestinal region** is relatively short and lined with **villi** that greatly increase the absorptive surface area.

Respiratory System

The lungs of birds are relatively small, compact organs that connect with an extensive system of **air sacs** that extend into many parts of the body, some entering the bones (Fig. 33.11). Functionally the sacs can be grouped into anterior and posterior sets. They interconnect with the bronchi and lungs as shown in Figure 33.12. Each **main bronchus** extends through the lungs to the posterior air sacs. A branch of the main bronchus also leads to groups of small, parallel passages, the **parabronchi**, that make up much of the lung. Another bronchial branch leads from the parabronchi to the anterior air sacs and back to the main bronchus. Minute branching and anastomosing **air capillaries**, which are surrounded by blood capillaries, extend from the parabronchi (Fig. 33.12*B*).

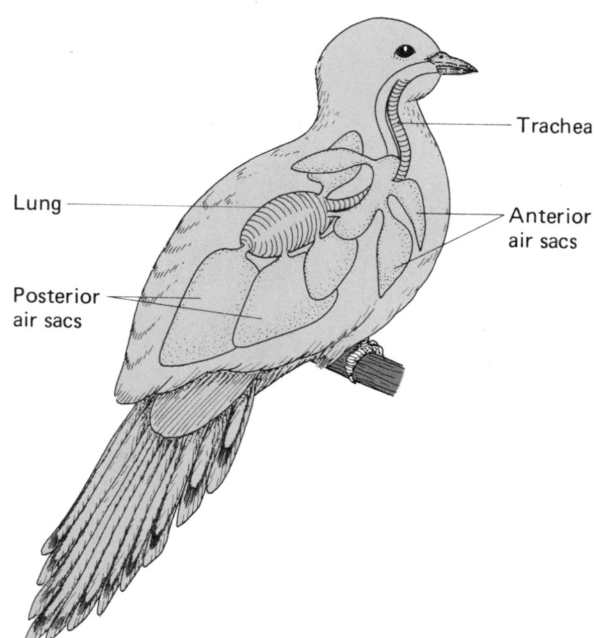

FIGURE 33.11 The lungs and major air sacs of a bird.

By using minute oxygen electrodes and flow meters, Bretz and Schmidt-Nielson determined the course of air flow through the lungs. Two cycles of inspiration and expiration are required for a unit of air to move through the system (Fig. 33.12*A*). During the first inspiration the sternum is lowered and the lungs and air sacs expand, drawing air directly through the main bronchus to the posterior sacs. During the first expiration the sternum is raised, and air in the posterior sacs flows into the parabronchi and air capillaries. On the second inspiration the air in the parabronchi is discharged into the anterior sacs. On the second expiration air in the anterior sacs is expelled to the outside. Notice that during inspiration both sets of air sacs expand, but they receive different types of air: the posterior sacs, oxygen-rich air from the outside; the anterior sacs, oxygen-depleted air from the lungs. During expiration both sets of sacs contract but send air to different places: the posterior sacs, to the parabronchi; the anterior sacs, to the outside.

Although a bird's lung is smaller than a mammal's, the flow is unidirectional through a bird's lung so the air in the air capillaries and adjacent to the blood capillaries has a higher oxygen content than is the case in the blind alveoli of a mammal's lung. The tidal volume

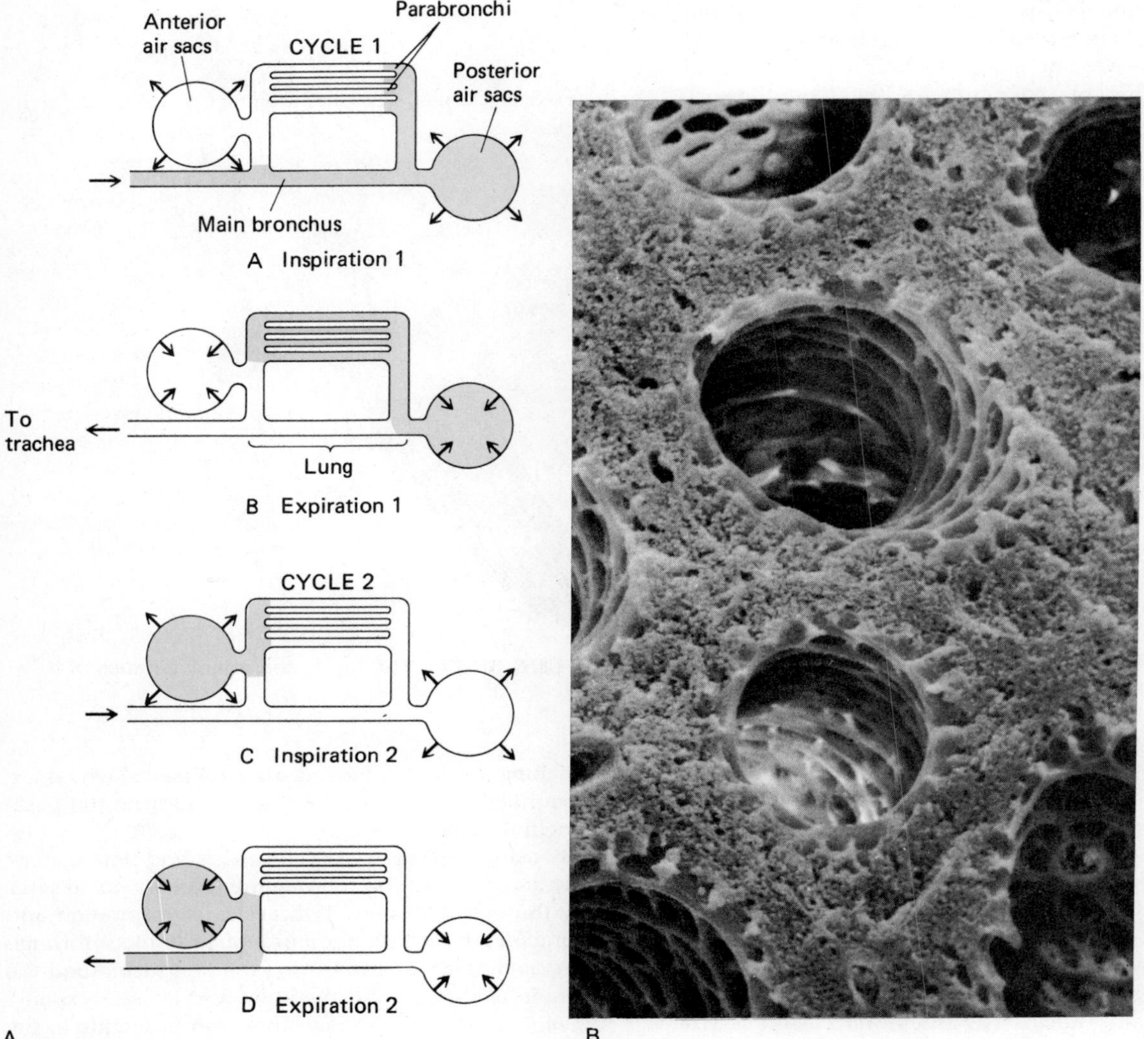

FIGURE 33.12 *A,* Diagrams of the movement of a single volume of air through the lungs and air sacs of a bird in two cycles of inspiration and expiration. *B,* Scanning electron micrograph of a section through the lung of a domestic fowl showing parabronchi in cross section and associated air capillaries. (*A* from Schmidt-Nielson, K.: *Animal Physiology.* 2nd ed. Cambridge, Cambridge University Press. *B* Photograph by H. R. Duncker from Schmidt-Nielson, K.: *How Birds Breathe.* Copyright © by Scientific American, Inc. All rights reserved.)

of air through the whole system is also greater, about twice that of a mammal of comparable size. Birds, therefore, can have a lower ventilation rate than mammals. A 1-kg. bird has a breathing rate of 17 cycles per minute, whereas a 1-kg. mammal has a rate of 54. The relatively high oxygen content of parabronchial air in comparison with alveolar air allows birds to obtain adequate oxygen when flying at high altitudes where the partial pressure of oxygen is low. In an experiment simulating an altitude of 20,000 feet, Tucker found that a sparrow could fly, but a similar sized mouse was barely able to crawl.

The air sacs also have other advantages. To the extent that they enter the bones and replace marrow, they lighten the bird. They provide a large surface area through which water evaporates and are important in cooling the body.

A mechanism for the production of sounds is associated with the air passages. Membranes are set vibrating by the movement of air in a **syrinx** (Gr. *syrinx*, shepherd's pipe) at the posterior end of the trachea (Fig. 33.10). Muscles associated with the syrinx vary the pitch of the notes.

Circulatory System

The circulatory system of birds, as would be expected in extremely active, endothermic animals, is very efficient. As in mammals, the heart is relatively large and completely divided into left and right sides so venous and arterial blood do not mix. Most birds are small animals, and their heart beats very rapidly, 400 to 500 times per minutes in a sparrow. This is comparable to the heart beat of a mammal of similar size. Vessels supplying the flight muscles are very large. In many birds, arteries carrying blood down the legs break up into a network of small vessels that are entwined with veins returning from the feet. This network, or **rete mirabile**, acts as a countercurrent heat exchanger. Heat flowing peripherally in the arterial blood is transferred to the cooler venous blood, and body heat is conserved. Birds do not "try" to keep their feet very warm. At an air temperature of −18°C., the temperature of a pheasant's feet is 2.7°C., but body temperature remains close to 40°C.

Excretory System

Nitrogenous wastes are removed from the blood by a pair of **metanephric kidneys** that are drained by the ureter into the cloaca (Fig. 33.10). There is no urinary bladder. The high rate of metabolism of birds, however, requires a much greater number of kidney tubules. Indeed, most birds have relatively more tubules than do mammals; a cubic millimeter of tissue from the cortex of a bird's kidney contains 100 to 500 renal corpuscles in contrast to 15 or less in a comparable amount of mammalian kidney. Some of the tubules form small loops of Henle. Body water is conserved by tubular reabsorption and the elimination of most of the nitrogenous waste as uric acid. Additional water is reabsorbed in the cloaca, and the uric acid is discharged as a white, crystalline material mixed with the feces.

Body salts are generally conserved by terrestrial vertebrates; any excess can be eliminated by the kidneys. Sea birds, however, take in a great deal of salt with their food and water and must eliminate more salts than can be disposed of by the kidneys. **Salt-excreting glands** in the herring gull are located above the eyes and discharge a concentrated salt solution into the nasal cavities. This solution leaves these cavities through the external nares and drips from the end of the beak (Fig. 33.13). Each gland consists of many lobes, each of which is composed of many vascularized secretory tubules radiating from a central canal.

Reproductive System, Reproduction

The reproductive systems of birds and reptiles are very similar. Sperm produced in the testis is discharged through an epididymis and vas deferens to the cloaca (Fig. 33.10). A few male birds, including the ostriches, ducks, and geese, have a copulatory organ for the transfer of sperm, but sperm is transferred in most species by a brief cloacal apposition by the two sexes. The ovary is small, except during the reproductive season when it enlarges greatly as the eggs accumulate their store of yolk. Ovulated eggs enter the oviduct, are fertilized in the upper part of the duct, and are covered by secretions of albumin, a shell membrane, and a shell as they continue down the oviduct. Possibly as an adaptation for weight reduction in a flying animal, birds have lost the right ovary and oviduct. The absence of a pelvic symphysis facilitates laying large, fragile eggs.

A great deal of the activity of birds is focused on reproduction and the rearing of young. Prior to the breeding season, the males of most species establish a mating and nesting **territory**, that is, an area that the male will defend aggressively against the incursions of other males of the same species (Fig. 33.14). The brilliant plumage of many males and their colorful songs both warn other males to stay away and attract a potential mate. Territorial behavior both spaces the birds and prevents the disadvantages of overcrowding, as discussed later (p. 773). After a territory has been established and a female has taken up residence, the birds engage in **courtship**.

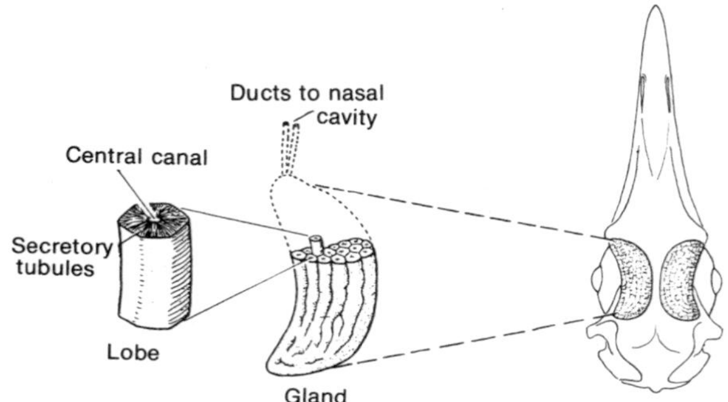

FIGURE 33.13 The salt-excreting glands of the herring gull. (Modified after Fange, Schmidt-Nielson, and Osaki.)

A

B

C

FIGURE 33.14 Reproduction in birds. Courtship display between Chilean flamingos *(A)* and between Magnificent Frigates on the Galapagos Islands *(B)*. *C*, Male feeding young scarlet tanager. (*A* by Juan and Carmecita Munoz/Photo Researchers, Inc.; *B* by George Holton/Photo Researchers, Inc.; and *C* by Frederick Flikinger, courtesy of Louis W. Campbell.)

Some displays are very elaborate. Courtship establishes that the two birds are in fact members of the same species, leads to the establishment of a strong pair bond between them, and prepares the partners physiologically and psychologically for effective copulation. During this period the pair engage in nest building. Nests may be as simple as a depression on the ground used by whip-poor-wills or as elaborate as the woven, colonial nests of African weaver birds. After eggs have been laid, they must be incubated or **brooded.** Many birds loose some feathers on the undersides of their abdomens at this time and form **brood patches** that facilitate heat transfer from the parents to the eggs. The two parents share equally in brooding in gulls and many other sea and aquatic birds, but the female is the chief or sole brooder in most song birds. The male brings her food. Near the end of the embryonic period, the chick develops a thickening, or **egg tooth**, on the end of its bill and powerful dorsal neck muscles. It uses these to break through the shell at hatching. Ducks, shore birds, chickens, and quail are examples of **precocial** (L. *praecox*, ripening before its time) birds. Their eyes are open and the birds are covered with down at hatching. The chicks follow their parent right away, and the parents help them to find food and may feed them for a short time. Song birds are **altricial** (L. *altrix*, nourisher). Their eyes are closed, and they are featherless at hatching. The parents must continue to brood and feed them until they are ready to leave the nest. The nestlings' first attempts at flight often are rudimentary, and they alight on the ground, but the parents are close by to try to scare off potential predators or keep them from being led astray. Raising a brood of young is a very time-consuming activity for most birds. Some cuckoos, cowbirds, and a few other species avoid these chores by practicing **brood parasitism**. They lay their eggs in the nest of other birds, and the foster parents usually accept them and treat them as their own. The European cuckoo helps to ensure this by laying eggs that resemble those of the host. Often the young cuckoo hatches before the host's own eggs and then rolls the unhatched eggs out of the nest!

The Origin and Evolution of Birds

Five specimens of a Jurassic bird are clearly intermediate between archosaurs and modern birds. Some of the fossils from Bavaria are preserved with remarkable detail in a fine-grained, lithographic limestone.

FIGURE 33.15 *A,* Restoration of *Archaeopteryx; B,* Ostrom's hypothesis of the ancestor of birds as a running predator. (*A* from a painting by Rudolf Freund, courtesy of the Carnegie Museum of Natural History. *B* from Ostrom, J. H.: *American Scientist,* (Jan./Feb.) 1979. Copyright 1979 Sigma Xi.)

Archaeopteryx lithographica (Fig. 33.15*A*) was about the size of a crow. Its skeleton is reptilian in having toothed jaws, no fusion of trunk or sacral vertebrae, a long tail, and a poorly developed sternum. Birdlike tendencies are evident in the enlarged orbits, some expansion of the brain case, and particularly in the winglike structure of the hand. As in modern birds, the "hand" is elongated, and only three "fingers" are present; however, there is little fusion of bones and each finger bears a claw. The first fossils to be discovered showed impressions of feathers (Fig. 17.5, p. 394), and this points clearly to a relationship with birds. If the skeleton alone had been discovered, this creature probably would have been regarded as a peculiar archosaurian reptile. It probably could not fly well. It had a large body relative to its wing area and there is no clear indication of a sternum to which flight muscles attached. However, it probably flew to some extent, for the location of the rachis close to the leading edge of the wing feathers is characteristic of the feathers of flying birds rather than of those of flightless birds where the rachis is close to the center of the feather. These most primitive birds are placed in the subclass **Archaeornithes** (Gr. *arkaios*, ancient + *ornis*, bird).

In 1975, John H. Ostrum of Yale University presented compelling evidence that birds evolved from coelurosaurs, a group of early saurischian dinosaurs (Fig. 32.1). Coelurosaurs were small bipedal creatures with a long tail, a long flexible neck, and arms bearing only three clawed fingers; some species resembled *Archaeopteryx* very closely. It is possible that the ancestors of birds were becoming more active and possibly warm-blooded, and feathers may have first been of value in helping to conserve body heat. Professor Ostrum believes that their enlargement, along the posterior edge of the forelimb, enabled bird ancestors to use these limbs as nets to seize insects (Fig. 33.15*B*). Further enlargement of wing and tail feathers may have conferred stability in running rapidly along the ground and in rudimentary gliding from low branches. Recent aerodynamic analysis suggests these feathers would be very important in providing stability in a small running and jumping biped. Feathers and wings may have evolved gradually in this way and may have been adaptive at all stages of their evolution. On reaching a certain threshold of size, they could be used for true flight.

The next fossil birds, found in Cretaceous deposits, had lost the long reptilian tail and evolved a well-developed sternum. A true pygostyle had not yet evolved, and teeth were present. *Hesperornis* (Fig. 33.16) was a large diving species with powerful hind legs and vestigial wings. *Ichthyornis* was a tern-sized flying species. Although they are placed in the subclass **Neornithes** (Gr. *neos*, new + *ornis*, bird) along with modern birds, the more primitive nature of these Cretaceous species is recognized by placing them in a distinct superorder—the **Odontognathae** (Gr. *odontos*, tooth + *gnathos*, jaw).

Remaining birds lack teeth and are placed in the superorder **Neognathae** (Gr. *neos*, new + *gnathos*, jaw). Bird bones are very fragile and do not fossilize well so we cannot be completely certain of the interrelationships of the 30 living and extinct orders. Possible interrelationships are shown in Figure 33.17. Nonetheless, it is clear from fossils that have been found that neognathous birds radiated widely in the late Cretaceous and early Tertiary periods. By mid-Tertiary, 15 to 20 million years ago, most contemporary groups had evolved. Among the late Cretaceous fossils were loons, grebes, pelicans, flamingos, rails, and sandpipers. Clearly neognathous birds adapted to shore dwelling and aquatic modes of life early in their history, but different groups adapted in different ways (Fig. 33.18). Loons have webbed feet that include the front three toes. Grebes swim not with webbed feet but with flattened and paddle-like toes. Pelicans have a webbed foot that includes all four toes. Flamingos are long-legged waders; rails are marsh-dwelling with neither long legs nor webbed feet.

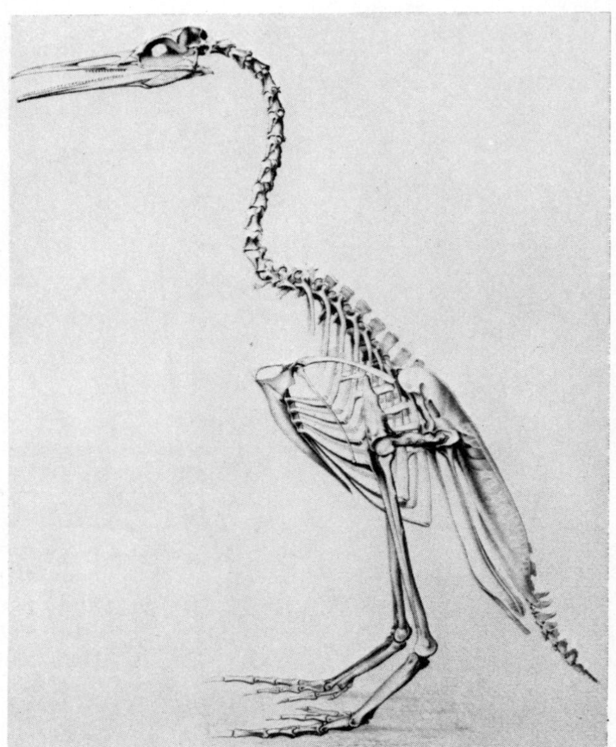

FIGURE 33.16 Marsh's reconstruction of *Hesperornis,* a cretaceous flightless bird. A vestige of one wing can be seen crossing the rib cage. (From Feduccia, A.: *The Age of Birds.* Cambridge, Harvard University Press.)

Penguins (Fig. 33.18*C*) appear in the fossil record in the early Tertiary. Their center of evolution was Antarc-

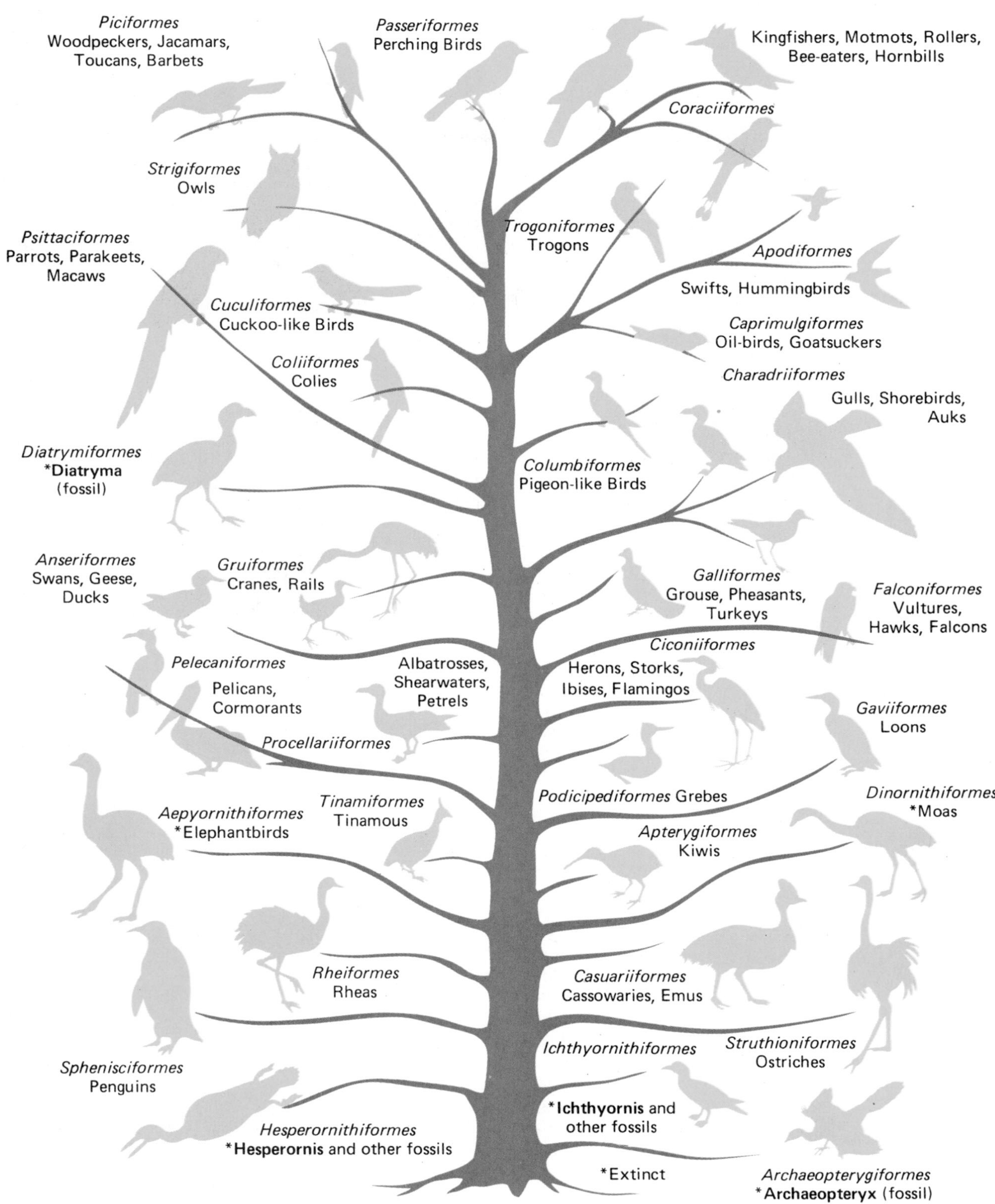

FIGURE 33.17 A hypothetical phylogenetic tree of the extinct and living orders of birds. (Courtesy of Cornell University, Laboratory of Ornithology.)

FIGURE 33.18 Wading and swimming birds. *A,* A great blue heron at its nest; *B,* a great egret; *C,* a king penguin on South Georgia Island; *D,* a wood duck. (*A* and *D* courtesy of Louis W. Campbell; *B* photographed by Warren F. Walker, Jr.; *C* courtesy of Douglas R. Johnson.)

tica, although one species has followed the cold Humbolt current north to the Galapagos Islands located on the equator west of Ecuador. Penguins are flightless for their wings are modified as paddles with which they swim under water. One fossil species was nearly our size, 1.5 m. tall.

Many other birds also lost their power of flight and adapted to a ground-living mode of life. Among these are the ostriches of Africa, the rheas of South America, the cassowaries of Australia, and the kiwi of New Zealand (Fig. 33.19). That these birds evolved from flying ancestors is evident from their broad, but keelless, sternum and the presence of vestigial wings. Their adaptations to terrestrial life include the great enlargement of their hind legs and feet. At one time paleontologists thought that these birds were related so they were grouped together as ratites. It is probable that they evolved independently and that their shared features resulted from convergent evolution. Many other flightless birds have also evolved. The giant moas, New Zealand relatives of the kiwi, were nearly 3 m. tall! They became extinct in historic times, probably about 300 years ago. *Diatryma* (Fig. 33.19C), a relative of the rails, was a flightless predator that stood 2 m. high and had a skull the size of a horse's with a massive hooked beak. The legendary dodo, a giant flightless pigeon from islands in the Indian ocean, was slaughtered by early sailors seeking fresh meat. The presence of many flightless birds early in the Tertiary has led some paleontologists to believe that early mammals and birds were in competition for the terrestrial resources that became available with the extinction of many reptilian groups. Mammals became the dominant terrestrial vertebrates, but flightless birds remained in niches where there were no mammalian competitors, or they became large enough to protect themselves or swift enough to escape.

A

B

C

FIGURE 33.19 Representative flightless birds. *A,* Ostrich; *B,* a kiwi with its relatively huge egg; *C,* a restoration of *Diatryma,* a giant rail of the early Tertiary. (*A* from Grzimek, B., in *Natural History,* Vol. *LXX,* No. 1; *B* and *C* courtesy of the American Museum of Natural History.)

Other adaptive radiations include the hawks and owls, which are rapacious birds with long talons and sharp, hooked beaks adapted to tear small animals apart (Fig. 33.20). Hawks are diurnal predators. Owls feed at night using a very keen sense of hearing to locate their prey in dim light. Grouse, pheasants, and fowl are predominantly terrestrial species, but they can fly well for short distances. The perching and song birds, which comprise 60 percent of all birds, are the most recent group to have evolved (Fig. 33.21). They have diverged widely in feeding habits, eating seeds, insects, small animals, and so forth.

Migration and Navigation

An aspect of bird biology of particular interest is the seasonal migration of many species and their uncanny ability to navigate. Many vertebrates migrate and find their way home if displaced, but birds' power of flight has endowed them with a more spectacular range of movement than in other vertebrates.

The migration of birds from winter quarters in temperate or tropical regions to breeding areas in the north permits them to spread into an area where the days are

A

B

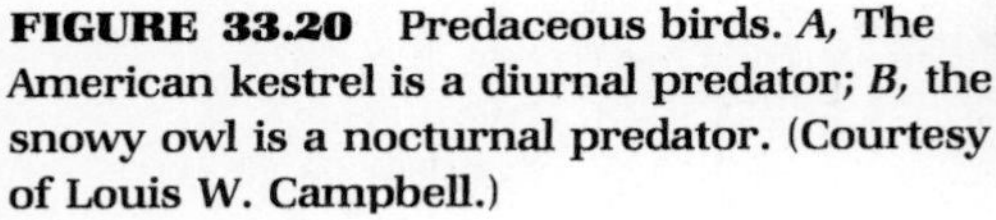

FIGURE 33.20 **Predaceous birds.** ***A,*** **The American kestrel is a diurnal predator;** ***B,*** **the snowy owl is a nocturnal predator. (Courtesy of Louis W. Campbell.)**

long in the summer and a large food supply develops for a few months. Birds can establish territories with a minimum of effort, and they have long hours of daylight to obtain food at a time when their population is increasing greatly. As conditions become inclement, the birds return to winter quarters. Migration prevents predator populations from increasing greatly, for the predators of a particular region do not have a sustained food supply if the birds move out at intervals. But there are hazards as well as advantages to migration, for many migrants are caught in storms and perish.

Many factors interact in disposing certain birds to migrate. In the spring of the year in the northern hemisphere, the hypothalamus stimulates the pituitary gland to secrete more gonadotropic hormones, and the gonads increase in size. There is also a rapid increase in fat deposits and an increased activity, or restlessness, of the birds. In the fall, the gonad decreases in size, but the birds again become restless and accumulate food reserves for the return trip. Hereditary factors, including an innate circa-annual rhythm, also appear to be involved. In northern California certain populations of the Oregon junco accumulate fat and become restless prior to their spring migration, whereas other populations, living under the same environmental conditions, do not show signs of migratory behavior and remain sedentary.

FIGURE 33.21 *A,* A red-headed woodpecker; *B,* a magnolia warbler. (Courtesy of Louis W. Campbell.)

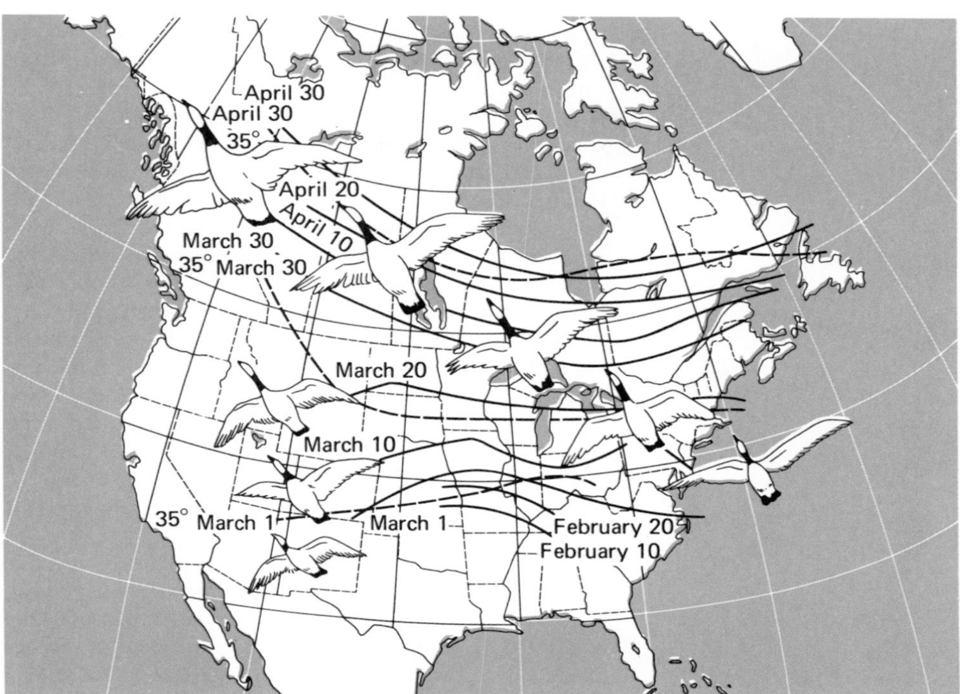

FIGURE 33.22 The northward migration of the Canada goose keeps pace with spring, following the isotherm of 2° C. (35° F.). (Modified after Lincoln.)

Once birds are in a migratory condition, favorable weather and other external factors trigger the onset of migration.

Most passerine migrants travel at night, stopping to feed and rest during the day. Some may fly several hundred kilometers during a single night but then may rest for several days. Many larger birds, including hawks and herons, migrate by day, and ducks and geese may migrate either at night or in the day. Flight ground speeds of migrants, as determined by radar tracking, range from about 30 to 70 km. per hour. Many species tend to follow the advance of certain temperature lines, or isotherms (Fig. 33.22) so their arrival at certain locations is affected by the weather. Other species have a nearly calendar-like regularity. Swallows reach San Juan Capistrano Mission in California on nearly the same date each year, but not with as precise timing as legend holds.

The length of migration and the route taken are consistent for each type of bird but vary with the species. The Canada goose winters in the United States from the Great Lakes south, breeds in Canada as far north as the arctic coast, and migrates along a broad front between the two areas. Other species follow narrower paths, and in some cases these are different for the spring and fall migrations. The golden plover breeds in the Arctic (Fig. 33.23). In the fall it flies south through the Canadian Maritime Provinces and northern New England before crossing the Atlantic to its winter quarters in southeastern South America. It flies north in the spring through Central America and the prairies of the United States and Canada. Availability of food has been a factor in determining this migration route. The birds fatten on late summer berries in the Maritime Provinces and New England prior to their flight across the Atlantic, but this route would be very inclement and devoid of appropriate food in the spring. A route through the grasslands of mid America is better at this time of year. The longest migration is that of the arctic tern. This species breeds in the Arctic, then migrates to its winter quarters in the South Atlantic, about 20,000 km. away.

How birds navigate and find their way remains an intriguing, incompletely solved problem of animal behavior. Obviously, the birds must know where they are going; there must be some feature of the environment that is related to the goal of the bird; and the bird must have some way of perceiving this feature. It is likely that some combination of methods is used in navigating and

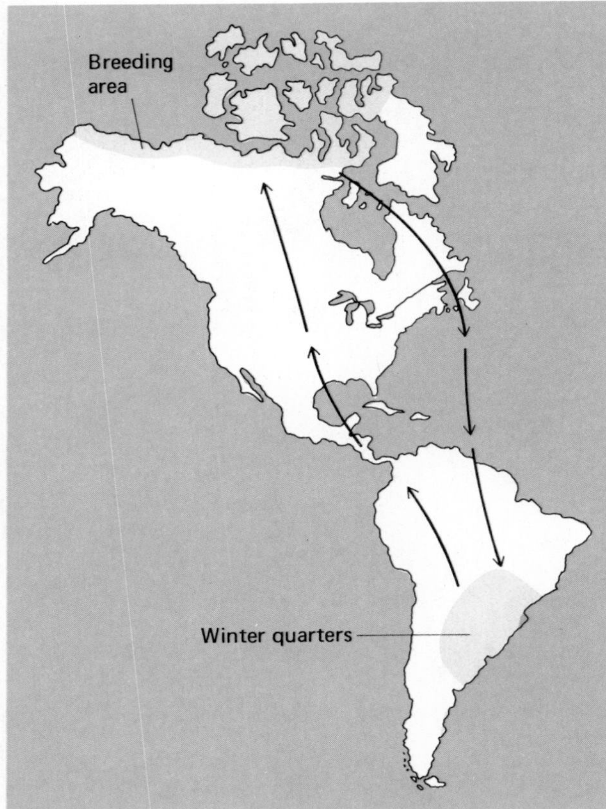

FIGURE 33.23 Migration routes of the Golden Plover. (From Welty, J. C.: *The Life of Birds.* 3rd ed. Saunders College Publishing, after Lincoln.)

that the specific choices vary between bird species or even within one species according to the bird's experience and environmental conditions. Certain species have an innate sense of direction and migrate along a predetermined compass course. During their fall migration, certain populations of European starlings fly from Northwestern Europe to the southeast. In 1958, Perdeck caught and banded thousands of specimens in Holland and released them in Switzerland about 600 km. southeast. Young birds continued in their "innate" direction and ended up far to the east of their normal winter quarters, whereas most of the older birds made an appropriate correction and ended in their normal winter range.

Many species use visual landmarks to some extent in migrating or in finding their way home if they have been displaced. The visual clues used by birds include topographic features, such as coast lines and mountain ranges, and ecological features, such as deserts, prairies, and forest. In 1949, Griffin displaced gannets (a sea bird) in Maine, 340 km. southwest from their nests on an island off the coast. He followed their return in an airplane. They flew in widening circles until they came in sight of the coast and then flew more directly home.

Some birds use a sun compass to find their way. Professor Matthews released lesser black-backed gulls some distance south of home and observed their direction as they vanished from view. Most birds took off in the direction of home on sunny days but became disoriented on cloudy days.

Studies by Sauer suggest that night migrants use the star pattern to navigate. Spring migrants heading north were caught and placed in a cage in a planetarium where they oriented themselves to the north of the artificial sky regardless of true north. To use celestial features, sun or stars, in navigation requires that birds have a keen sense of time, an internal clock of some sort, because celestial clues for direction vary with the time of day. That this is true can be demonstrated by subjecting birds to artificial light–dark cycles out of phase with natural day and night. Their clocks become reset. In one experiment, starlings with their clocks advanced six hours assumed it was dawn when in reality it was midnight. Their flight direction was shifted about 90° counterclockwise from what it should have been.

Other investigators have shown that some species can perceive the earth's or other magnetic fields and that this might be used in navigation. Caged European robins caught during migration oriented appropriately to their migratory direction in the absence of any celestial clues, but their choice of direction was altered by subjecting them to artificial magnetic fields. We do not know how birds sense magnetic fields, but it appears that the vertical component of the field (the angle between the magnetic force and the pull of gravity) is more important than its compass direction.

Classification of Birds

Class Aves The birds. Endothermic, typically flying vertebrates covered with feathers.

†Subclass Archaeornithes Ancestral birds of the upper Jurassic period. They retain many reptilian features including jaws with teeth, long tail, and three unfused fingers, each bearing a claw. *Archaeopteryx.*

Subclass Neornithes Birds with a reduced number of caudal vertebrae; wing composed of three highly modified fingers partly fused together.

[†]Superorder Odontognathae Cretaceous birds retaining teeth. *Hesperornis, Ichthyornis.*

Superorder Neognathae Late Cretaceous and post Cretaceous toothless birds. The first seven orders listed are flightless, or have reduced powers of flight, and are often call ratites; most of the remaining have well-formed sternal keels and are called carinates. It is doubtful that ratites and carinates are natural phylogenetic groups.

Order Struthioniformes Ostriches. Huge flightless birds with small wings; unkeeled sternum; head and neck largely devoid of feathers; large powerful legs with only two toes. Africa and western Asia, *Struthio.*

Order Rheiformes Rheas. Large flightless birds with unkeeled sternum; head and neck feathered; heavy legs with three toes. South America, *Rhea.*

Order Casuariiformes Cassowaries and emus. Large flightless birds with small wings and unkeeled sternum; long hairlike feathers with long after-shaft; heavy legs with three toes. New Guinea and Australia, *Casuarius.*

[†]Order Aepyornithiformes Elephant birds. Turkey to ostrich-sized flightless birds of Africa and Madagascar; four toes; became extinct in historic times; laid largest eggs known, 33 × 24 cm. *Aepyornis.*

[†]Order Dinornithiformes Moas. Largest of the flightless birds, attained a height of 3 m. Became extinct about 300 years ago. New Zealand, *Dinornis.*

Order Apterygiformes Kiwis. Hen-sized flightless birds with unkeeled sternum and vestigial wings; four toes on feet; long bill with nostrils near the tip used in probing soft ground for food, nocturnal in habits. New Zealand, *Apteryx.*

Order Tinamiformes Tinamous. Largely a ground-dwelling group, but with weak powers of flight; sternum retains a keel. Mexico to South America, *Tinamus.*

Order Gaviformes Loons. Legs located far back on the body; webbed feet; reduced tail; long, compressed and sharply pointed bill; very good divers. The common loon, *Gavia.*

Order Sphenisciformes Penguins. Flightless oceanic birds with four anteriorly directed toes with a web between three of them; wings modified as paddles; excellent divers. Confined to the southern hemisphere, chiefly Antarctica. One species occurs on the Galapagos Islands. The emperor penguin, *Aptenodytes.*

Order Podicipediformes Grebes. Legs located far back on the body; lobate toes; reduced tail; very good divers. Eared grebe, *Colymbus.*

Order Procellariiformes Albatrosses, shearwaters, fulmars, petrels, and tropic birds. Webbed feet; fourth toe vestigial; long narrow wings; hooked beak; tubular nostrils. The petrel, *Procellaria.*

Order Pelecaniformes Pelicans, gannets, cormorants, water-turkey, and man-o-war bird. Totipalmate swimmers with four toes included in the webbed foot; tendency for the development of a gular sac. The pelican, *Pelecanus.*

Order Ciconiiformes Herons, bitterns, storks, ibises, and flamingos. Long-legged and long-necked wading birds; feet broad, but usually not webbed; the portion of the head between the eye and nostril (the lores) usually devoid of feathers. Stork, *Ciconia.*

Order Anseriformes Ducks, geese, and swans. Short-legged, web-footed swimming and diving birds; bill usually broad and flat with transverse horny ridges adapted for filtering mud. The mallard, *Anas*; white-fronted goose, *Anser.*

Order Falconiformes Vultures, kites, hawks, falcons, and eagles. Diurnal birds of prey with strong hooked bill; sharp curved talons. Cooper's hawk, *Accipiter;* duck hawk, *Falco.*

Order Galliformes Grouse, quails, partridges, pheasants, turkeys, and chickens. Seed- and plant-eating, largely ground-dwelling birds; short stout bill; heavy feet with short strong claws, adapted for running and scratching in the ground; wings relatively short; often sexually dimorphic. The chicken, *Gallus.*

Order Gruiformes Cranes, rails, gallinules, and coots. Marsh birds; feet not webbed, but toes sometimes lobed; legs elongate in some groups; lores feathered. European crane, *Grus.*

[†]Order Diatrymiformes Large flightless birds of the early Cenozoic. *Diatryma.*

Order Charadriiformes Plovers, woodcock, snipe, sandpipers, stilts, phalaropes, gulls, terns, skimmers, auks, and puffins. A diverse group of shore birds. The killdeer, *Charadrius.*

Order Columbiformes Pigeons and doves. Short slender bill with fleshy pad at its base overhanging the slitlike nostrils; short legs. The domestic pigeon, *Columba.*

Order Psittaciformes Parrots. Feet adapted for grasping, with fourth toe capable of being turned back beside the first toe; bill heavy and hooked; often brilliantly colored plumage. Carolina parakeet, *Conurus.*

Order Cuculiformes Cuckoos and road-runners. Foot with fourth toe capable of being turned back beside the first toe. Tail long. The cuckoo, *Cuculus.*

Order Strigiformes Owls. Nocturnal birds of prey with strong hooked bills; sharp curved talons; feathers arranged as a facial disc around the large, forwardly turned eyes. The barred owl, *Strix.*

Order Caprimulgiformes Nighthawks, goat suckers, and whip-poor-wills. Twilight flying birds with small bills, but large mouths surrounded by bristle-like feathers that help net insects; legs and feet small. The whip-poor-will, *Caprimulgus.*

Order Apodiformes Swifts and hummingbirds. Fast flying birds with long narrow wings; legs and feet very small. The chimney swift, *Chaetura.*

Order Coliiformes The colies of Africa. Small birds with long tails; first and fourth toes can be turned posteriorly. The mousebird, *Colius.*

Order Trogoniformes Trogons. Short stout bill; small feet with first and second toes directed backward; often green iridescent plumage. The copperytailed trogon, *Trogon*.

Order Coraciiformes Kingfishers. Strong sharp bill; foot syndactylous with third and fourth toes fused at their bases; feathers often forming a crest on the head. The belted kingfisher, *Megaceryle*.

Order Piciformes Woodpeckers and toucans. Bill chisel-like (woodpeckers) or very large (toucans); zygodactylous foot with fourth toe permanently turned posteriorly. The flicker, *Colaptes*.

Order Passeriformes The perching birds and songbirds. The largest order of birds, it comprises 60 percent of all birds and includes the flycatchers, larks, swallows, crows, jays, chickadees, nuthatches, creepers, wrens, dippers, thrashers, thrushes, robins, bluebirds, kinglets, pipets, waxwings, shrikes, starlings, vireos, wood warblers, weaver finches, blackbirds, orioles, tanagers, finches, sparrows, and so forth. Foot adapted for perching; three toes in front opposed by one well-developed toe behind. The English sparrow, *Passer*.

Summary

1. Differences in air speed across the top and bottom of the wings generate lift and drag forces. Lift is proportional to the wing surface area and to the square of the air speed across the wings. It can be increased by increasing the angle of attack of the wings and by forming slots between wing feathers.

2. Gliding birds overcome drag and gain forward motion by loss of altitude; static soarers maintain or gain altitude and forward motion by falling in a rising air current; dynamic soarers use increasing air speeds at higher elevations to gain altitude.

3. During flapping flight, the inclination of the wings on the downstroke changes the direction of the lift force in such a way that it has a forward component, or thrust. Additional thrust and lift come from reactions upon the bird of air movements produced by the wings.

4. The birds' high level of metabolism and covering of feathers enable them to maintain a high body temperature and energy output.

5. Feathers are modified horny scales. The most conspicuous types are the contour feathers that cover the bird and form the flying surfaces; other types include down, filoplumes, and bristles.

6. Bird bones are very light and strong. The skull is large, the neck is flexible, and the trunk and sacral vertebrae are firmly united. Tail feathers are attached to a group of terminal, fused caudal vertebrae. The sternum is broad and usually keeled. The pectoral appendages form wings. The pelvic appendages are structured as good levers for taking off and as shock absorbers on landing.

7. The flight muscles of birds are large. Both those responsible for the downstroke (pectoralis) and upstroke (supracoracoideus) arise from the sternum.

8. The sense of smell and olfactory parts of the brain are reduced in most species. Sight and optic centers are important. Although their cochlea is not long, birds have a keen sense of hearing. The corpus striatum is the major cerebral correlation center. The cerebellum is large and important.

9. Birds feed on high-energy foods. They have no teeth; food is broken down mechanically in the gizzard and mixed with enzymes secreted in the proventriculus. In many seed eaters, food may be softened first by the uptake of water in the crop. Intestinal villi increase the absorptive area of the rather short intestine.

10. Bird lungs and air sacs are so arranged that there is a one-way flow of air through the parabronchi and air capillaries where gas exchange occurs. Gas adjacent to the blood capillaries has a higher oxygen content than the air in the alveoli of mammal lungs. Sound is produced in a syrinx at the posterior end of the trachea.

11. Birds have a double circulation through the heart. Vessels supplying flight muscles are large.

12. Some of the kidney tubules have short loops of Henle, but water is conserved primarily by eliminating nitrogen as uric acid. Marine birds have salt-excreting glands in the nasal cavities.

13. Most male birds transfer sperm to the female by means of cloacal apposition. Females have only one ovary and oviduct. Males establish a territory. When a female takes up residence, courtship and nest building begin. Usually the female broods the eggs and young. Chicks break their egg shell with a temporary egg "tooth."

14. *Archaeopteryx*, a Jurassic bird, had teeth, a long tail, distinct, clawed fingers, and many other reptilian features. It probably evolved from small, running dinosaurs. *Hesperornis* and *Ichthyornis* from the Cretaceous retained teeth but were modern in most other respects. Later birds have lost their teeth. They have undergone an extensive adaptive radiation and now occupy most terrestrial, aquatic and aerial habitats.

15. Not all birds migrate, but there are certain advantages to migration: ease of establishing territories, obtaining food, and rearing young. Migration also prevents a great increase in predator populations. Many factors interact to dispose certain birds to migrate: hereditary factors, day length, hormonal changes, fat deposition, and increased restlessness. Some birds migrate by day; other species, by night. Migratory paths are specific for each kind of bird.

16. Birds use some combination of the following clues in navigation: an innate sense of direction, visual landmarks, a sun compass, star patterns, and magnetic fields. Those using celestial clues also need a keen sense of time.

References and Selected Readings

Many of the general references cited at the end of Chapter 31 contain considerable information on the biology of birds.

Bent, A. C.: *Life Histories of North American Birds.* New York, Dover Publications, 1961–1968. A reprinting of Bent's famous multivolume study of the natural history of birds. Originally published between 1919 and 1958 as *Bulletins of the U.S. National Museum.*

Cone, C. D., Jr.: The soaring flight of birds. *Scientific American, 206*(Apr.):130, 1962. The nature of air currents, gliding, and static and dynamic soaring are described.

Dorst, J.: *The Life of Birds.* New York, Columbia University Press, 1974. A thorough account in two volumes of the anatomy, physiology, and natural history of birds. Translated from the French by I. C. J. Galbraith.

Emlen, S. T.: The stellar-orientation system of migratory birds. *Scientific American, 233*(Aug.):102, 1975. The interaction of stellar and other clues used by the indigo bunting.

Farner, D. S., and J. R. King (Eds.): *Avian Biology.* New York, Academic Press, 1971–1975. A five-volume treatise covering most aspects of the biology of birds.

Fedducia, A.: *The Age of Birds.* Cambridge, Harvard University Press, 1980. An excellent account of the evolution and adaptive radiation of birds.

Gillard, E. T.: *The Living Birds of the World.* New York, Doubleday & Co., 1958. The major groups of birds are summarized and superbly illustrated.

Keeton, W. T.: The mystery of pigeon homing. *Scientific American, 231*(Dec.):96, 1974. Pigeons have more than one system for determining direction.

Knudsen, E. I.: The hearing of the barn owl. *Scientific American, 245*(Dec):112, 1981. An analysis of the uncanny ability of the barn owl to find prey in the dark.

Matthews, G. V.: *Bird Navigation.* 2nd ed. Cambridge, Cambridge University Press, 1968. A thorough account is given of the different theories of navigation and homing.

Peterson, R. T.: *A Field Guide to Birds.* 4th ed. Boston, Houghton Mifflin Co., 1980. The standard and widely used guide for the field identification of birds from the Great Plains to the East Coast.

Peterson, R. T.: *Field Guide to Western Birds.* Revised ed. Boston, Houghton Mifflin Co., 1961. A companion to the preceding volume, it covers the birds from the Pacific Coast to the western parts of the Great Plains.

Pettingill, O. S., Jr.: *A Laboratory and Field Manual of Ornithology.* 4th ed. Minneapolis, Burgess Publishing Co., 1970. A manual on the structure, habits, and ecology of birds for the serious student of ornithology.

Rayner, J. M. V.: Vorticity and animal flight. In Elder, H. Y., and Trueman, E. R.: *Aspects of Animal Movement.* Cambridge, Cambridge University Press, 1980, p. 177. A summary of the vortex theory of flight.

Rüppell, G.: *Bird Flight.* New York, Van Nostrand Reinhold Company, 1977. An up-to-date account of the physics of flight, anatomical and physiological adaptations for flight, and the types of flight used by birds. Translated from the German by M. A. Biederman-Thorson.

Schmidt-Nielsen, K.: How birds breathe. *Scientific American, 225*:(Dec.)73, 1971. A summary of new findings on unique lung ventilating mechanism of birds.

Terres, J. K.: *The Audubon Society Encyclopedia of North American Birds.* New York, Alfred A. Knopf, 1980. This is probably the most comprehensive, single-volume treatment of birds in print. It includes complete descriptions of the natural history of all North American birds along with outstanding color photographs of most of them. Excellent accounts of the anatomy, physiology, flight, migration, and many other important topics are also included.

Thomson, Sir A. L. (Ed.): *A New Dictionary of Birds.* New York, McGraw-Hill Book Co., 1964. A very useful one-volume encyclopedia on all aspects of the biology of birds, prepared under the auspices of the British Ornithologist Union. Some entries are only a few lines long; others run to several pages.

Welty, J. C.: *The Life of Birds.* 3rd ed. Philadelphia, Saunders College Publishing, 1982. A comprehensive one-volume work on all aspects of the biology of birds.

34 Vertebrates: Mammals

ORIENTATION

Mammals are able to maintain a constant and relatively high body temperature through an internal regulation of heat gain and loss. That is, they are endothermic. This enables them to be active under a wide range of environmental conditions. In this chapter we will discuss how this capacity evolved and explore the connection between endothermy and modifications of all of the organ systems of the body. We will examine different reproductive strategies of primitive mammals and placental mammals with their relative limitations and advantages. Last we will indicate the wide geographic distribution of placental mammals, the most successful of all mammals, and their adaptations to their varied modes of life.

CHARACTERISTICS OF MAMMALS

1. Mammals are endothermic vertebrates.
2. Hair and subcutaneous fat, or both, form an insulating layer. Cutaneous glands are abundant.
3. The limbs of most mammals are situated more or less under the body. The skull is of the synapsid type and has a relatively large brain case. The jaw joint lies between the dentary and the squamosal (temporal) bones.
4. There are three auditory ossicles in the middle ear and a spiral cochlea in the inner ear. Enlarged nasal cavities are separated from the mouth cavity by a hard palate and contain folded turbinate bones.
5. The cerebrum is large and has a gray cortex. Large cerebellar hemispheres are present.
6. Teeth are heterodont, have a precise occlusion, and their replacement is limited. The small intestine has numerous multicellular intestinal glands and microscopic villi. Most species lack a cloaca.
7. Respiratory and digestive passages are nearly completely separated in the oral and pharyngeal regions. Numerous lung alveoli greatly increase surface area. A muscular diaphragm plays a major role in lung ventilation.
8. Venous and arterial blood are completely separated as they move through the heart in adult mammals.
9. Nitrogenous wastes are eliminated primarily as urea by kidneys drained by ureters. Long loops of Henle in the renal tubules make possible the production of a urine hyperosmotic to the blood.
10. The testes of most mammals either lie permanently within a scrotum or descend into the scrotum during the reproductive season. Males have a copulatory organ, and fertilization is internal.
11. Except for primitive egg-laying mammals, the ovaries are small and produce few eggs; little yolk is deposited in the eggs. The oviducts have differentiated into vaginal, uterine, and uterine tube regions.
12. Monotremes are oviparous; other mammals are viviparous. The uterine lining and extraembryonic membranes unite to form a placenta. Mammary glands are always present in females.

Of all vertebrate groups, mammals (class **Mammalia,** L. *mamma,* breast) are of particular interest to us. We are mammals, as are our domestic animals, which help us in our labors and provide us with wool, leather, and much of our food. Although mammals are not a large class—there are only about 4100 species—they are a very diverse group. The class includes the egg-laying duckbilled platypus and spiny anteater (Monotremes) of the Australian region, the opossum, kangaroo, and other pouched marsupials, and the wide variety of true placental mammals, which range in size from the pigmy shrew weighing only a few grams to the giant whales that exceed 100 tons. As a group mammals are very active and agile vertebrates with a high, sustained level of metabolism. They produce few young but invest considerable time and energy in caring for them. Increased activity and greater care of the young have been the touchstones of mammalian evolution, and most of our characteristics are related to these.

Major Adaptations of Mammals

Temperature Regulation

Although reptiles can maintain a high and fairly constant body temperature in a warm and sunny environment, most have no way of maintaining body temperature when the sun goes down. Reptiles must be diurnal creatures, foraging in the daytime. Mammals can generate heat internally, control its gain and loss, and maintain a relatively high and constant body temperature both day and night. They are endothermic. Many species of mammals are nocturnal; many live in colder parts of the world than any reptile can.

Evolution of Endothermy. Crompton, Taylor, and Jagger have proposed that mammalian endothermy evolved in two steps. First mammals became nocturnal thermoregulators, and second they became diurnal thermoregulators. The earliest mammals of the late Triassic and early Jurassic periods were small, mouse-sized creatures (Fig. 34.1). Many investigators believe that they were nocturnal insect eaters. Their dentition was adapted for feeding upon insects, and their skull structure indicates that there was a great elaboration of the cochlear region of the ear and of the olfactory organ. These are senses of particular importance for nocturnal creatures. To be active at night at ambient temperatures of 25° to 30° C., early mammals must have been able to generate heat internally and to prevent its loss by the presence of insulating layers of subcutaneous fat and fur. They probably did not maintain a body temperature much higher than that of their surroundings. The European hedgehog and the tenrecs of Africa and Madagascar (Fig. 34.2) are contemporary nocturnal insectivores of this type that probably have occupied this niche throughout their evolutionary history. They maintain their body temperature only a few degrees above

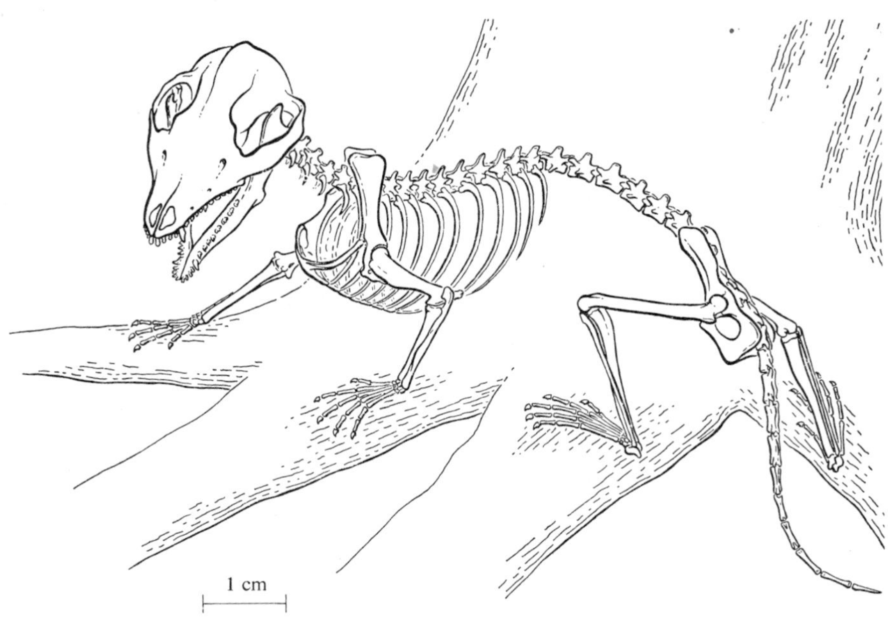

FIGURE 34.1 Skeletal reconstruction of a primitive Triassic triconodont based primarily on *Megazostrodon.* Head-to-body length was about 10 cm. (From Jenkins, F. A., Jr., and F. R. Parrington: The postcranial skeletons of the Triassic mammals *Eozostrodon, Megazostrodon,* and *Erythrotherium. Philosophical Transactions of the Royal Society of London B.* 273:387–431.)

FIGURE 34.2 The tenrec *(Tenrec ecaudatus)* is among the most primitive of living mammals, and it probably has occupied the nocturnal insectivorous niche throughout its history. (From Vaughan, T. A.: *Mammalogy.* 2nd ed. Philadelphia, Saunders College Publishing, 1978. Photograph by J. F. Eisenberg and Edwin Gould.)

the nocturnal ambient temperature and can do so with no more oxygen consumption than a reptile of similar size and activity at the same temperature.

When some early mammals became diurnal, they had to adapt to activity at the higher daytime temperatures. To maintain a body temperature as low as that of their nocturnal ancestors, i.e., 25° to 30° C., would require considerable evaporative cooling and water loss. They "opted" for a higher body temperature of approximately 35° to 40° C. But to maintain this temperature at cooler times of the day requires greater energy expenditure than that of a reptile. Their energetics have changed, and their metabolic rate is three to five times higher than that of reptiles of similar size and under similar conditions.

Mechanisms of Regulation. The mechanisms of temperature regulation in contemporary mammals are well known. Heat is produced internally by their high level of metabolism. Its loss is reduced by a layer of subcutaneous fat and by the hair that entraps an insulating layer of still air next to the skin. Heat can be lost by increasing the amount of blood flowing through the skin and by the evaporative cooling of sweat produced by sweat glands or by panting. Panting, which is the evaporation of water from the respiratory passages, is an important cooling mechanism in heavily furred mammals that have few sweat glands and in some rodents that have none. Mammals can maintain their body temperature over an ambient temperature range of about 10° C. by these mechanisms without any additional consumption of oxygen. This is known as the **thermal neutral zone** (Fig. 34.3), and it is bounded by upper and lower **critical temperatures.** Beyond the critical temperatures, however, body temperature can be maintained only by the expenditure of significantly more metabolic work. The rise in oxygen consumption below the lower critical temperature reflects the extra heat production needed to maintain body temperature. Shivering, an involuntary activity of superficial muscles, is one mechanism of increased heat production, but there is also a general increase in metabolic activity in many parts of the body. The rise in oxygen consumption above the upper critical temperature reflects the added metabolic work needed to dissipate heat. The heart rate and rate of circulation through the skin are increased. Profuse sweating occurs in some mammals and panting in others.

Thermal receptors in the skin signal environmental temperature changes to which mammals adjust partly by changes in behavior. The major thermal control center in the hypothalamus responds to slight changes in blood temperature and initiates the changes needed to adjust heat loss and production to the environmental context.

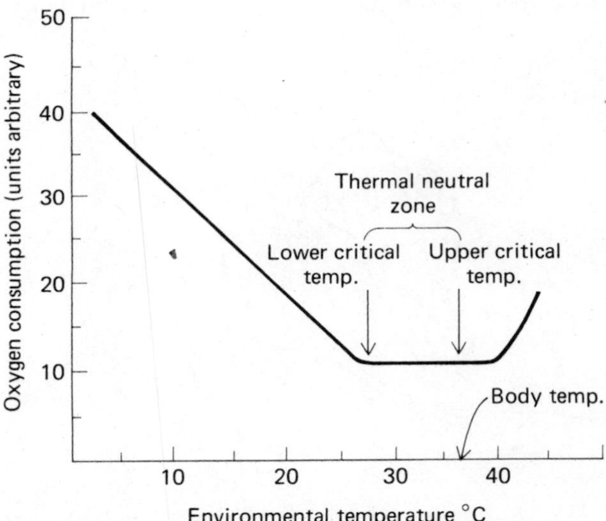

FIGURE 34.3 A graph to show the relationship between oxygen consumption and environmental temperature in a typical mammal with a body temperature of 37° C. (Modified after Gordon: *Animal Function: Principles and Adaptations.* New York, Macmillan.)

Adaptations for Cold Stress. Mammals that live in areas where the environment is rigorous have evolved adaptations that supplement thermal regulation. Many **molt** in early fall. They gradually lose their hair as a much thicker **undercoat** develops for the winter. A second molt occurs in the spring. In addition to an undercoat, the pelage of many mammals includes long, coarse, **guard hairs** that protect the undercoat. A thicker fur for winter both lowers the lower critical temperature and the steepness of the slope of the environmental temperature-oxygen consumption curve. The appendages cannot be insulated as well as the rest of the body, and their temperature is permitted to fall below that of the body core. Arteries carrying blood to the limbs are sometimes closely intermeshed with the veins returning blood so that a countercurrent exchange mechanism is set up whereby much body heat moves from the arteries to the veins and is not lost. Enough heat must be permitted to enter the appendages, however, to keep them from freezing. Nerves and other organs in the distal part of the limbs are adapted to function at lower temperatures.

Some arctic and temperate mammals, notably many insectivores, bats, and rodents, adjust to winter weather by going into a period of dormancy known as **hibernation.** During this period, they lose considerable control over the mechanisms regulating body temperature (the thermostat in the hypothalamus is turned down to conserve energy), and their body temperature approaches the ambient temperature. Metabolism is very low during hibernation yet sufficient to sustain life and to keep the body from freezing. There are certain advantages to hibernation for a small endotherm. Small mammals have a relatively higher rate of metabolism than large ones. They lose a great deal of heat through their surface area and must consume much food just to maintain body temperature. In many regions, insects and certain types of plant food are not available in quantity during the winter. If an animal can permit its body temperature to drop, it can get by on less food or even on the food reserves within its body.

Some other mammals avoid the problem of low temperature by retreating to warmer climates. Many small rodents remain active all winter beneath the snow cover where the microclimate seldom falls far below 0° C. Occasionally they venture forth on the snow surface. A few of the larger mammals undergo extensive seasonal migrations. The caribou of Alaska and Canada summer on the arctic tundra but retreat south to the boreal forests in the winter.

Adaptations for Heat Stress. Mammals living in very hot climates also have special adaptations that help keep them cool without an excessive loss of body water, which often is in short supply. Small desert rodents are nocturnal and burrow so they are living in a cool and moist microhabitat. Their food is rich in fats, which on oxidation yield a considerable amount of metabolic water. Elephants have a large body size that provides some thermal stability, little hair, and large ears that are efficient radiators. Camels have evolved several ways that conserve water. Under very dry conditions, dry mucus and cellular debris in the nasal passages have a hygroscopic effect and absorb moisture from the air being exhaled. As a consequence, less respiratory water is lost than is usually the case. Camels can also tolerate a body temperature as high as 41° C. in the daytime so they need not lose much water in trying to remain cool. Their body cools down at night when ambient temperatures fall. Permitting body temperature to rise is possible in many large mammals living in hot, open habitats because they can keep the critical brain temperature lower by a countercurrent vascular arrangement (Fig. 34.4). Arteries supplying the brain first break up into minute passages that are entwined with veins returning cool venous blood from the nasal passages. Considerable heat passes from the warm arterial blood to the cooler venous blood before the arterial blood reaches the brain.

Locomotion and Coordination

Changes in all of the organ systems are closely correlated with the increased activity made possible by endothermy. Greater activity and agility is reflected in the

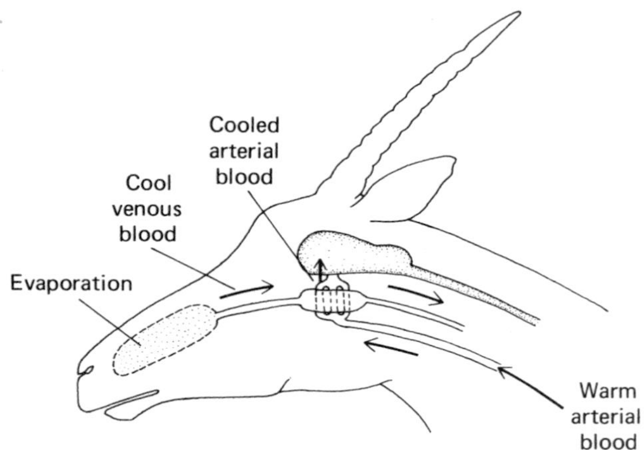

FIGURE 34.4 The countercurrent circulatory mechanism that cools the brain of a gazelle. Many hoofed mammals living in open country have similar mechanisms. (From Schmidt-Nielsen, K.: Animal Physiology, 2nd ed. Cambridge University Press, 1979; after C. R. Taylor.)

skeletal system of even the earliest mammals of the Triassic period (Fig. 34.1). The posterior inclination of the vertebral spines of the thoracic vertebrae and the anterior inclination of the spines of the lumbar vertebrae are typical of quadruped mammals and are correlated with the abandonment of lateral trunk undulations in locomotion. The elbow and knee have moved in close to the trunk so that the legs extend down to the ground more or less under the body. This provides better mechanical support and the potential for a longer swing of the appendage, increased stride length, and greater speed. Primitively the feet were placed flat upon the ground, a posture called **plantigrade.** Most mammals have three sacral vertebrae; this strengthens the articulation between the pelvic girdle and vertebral column. Arboreal species use the tail for balancing, and it plays a major role in the propulsion of aquatic mammals, such as the whales, but in most mammals it has lost its primitive locomotor function and is frequently reduced in size or is absent. Many specialized patterns of locomotion evolved as mammals radiated and adapted to different modes of life.

More complex patterns of locomotion, and probably increased exploratory behavior and agility, require more complex musculature, sensory, and nervous systems. The sense of smell and hearing were probably very acute in primitive mammals. A great expansion of the part of the skull housing the cochlea of the ear occurred as the jaw joint shifted to the dentary and squamosal bones and the former reptilian jaw joint bones became specialized as auditory ossicles, the malleus and incus (p. 106). Sight may have been of less importance in ancestral nocturnal species, but the eyes are well developed in diurnal species. The brain is extraordinarily well developed. The cerebrum is greatly enlarged, with a gray cortex containing centers associated with the sensory input of the sense organs and important motor centers. The cerebellum also enlarges as motor coordination becomes more intricate.

Metabolic Systems

To sustain their high level of metabolism, mammals must obtain large supplies of food and oxygen, eliminate a large volume of waste products, and efficiently transport materials throughout the body. The dentition of mammals is much better adapted than that of reptiles for processing many different kinds of food. Reptiles use their teeth to seize, hold, and sometimes tear their food apart. Spaces between teeth are of little consequence, and new teeth grow in throughout life to replace those that have been lost. Mammals use their teeth in different ways, and the teeth are more differentiated, or **heterodont,** than in any reptile. Chisel-shaped **incisors** at the front of each jaw are used for nipping and cropping. Next is a single **canine** tooth, which is primitively a long, sharp tooth, useful in attacking and stabbing prey or in defense. A series of **premolars** and **molars** follow the canine. These teeth tear, cut, and crush the food (Fig. 34.5*A*). In primitive living mammals, each molar

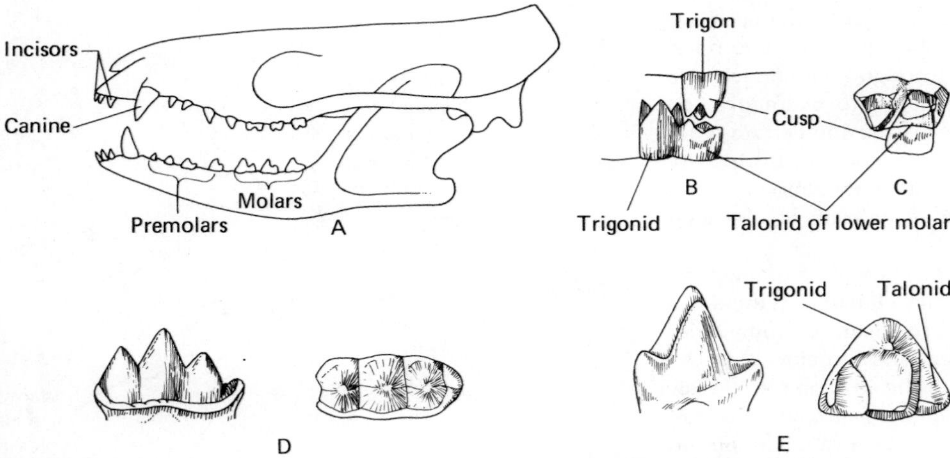

FIGURE 34.5 The teeth of primitive mammals. *A,* The type of teeth found in an insectivore; *B* and *C,* lateral and crown views of the left upper and lower molars of an insectivore to show their occlusion; *D,* the linear cusps of the molar of a triconodont; *E,* the triangular lower molar of a pantother. (*D* and *E* from Stahl, B. J.: *Vertebrate History: Problem in Evolution.* New York, McGraw-Hill Book Co., 1974. After G. G. Simpson.)

tooth has three conical cusps arranged in a triangle, which is called a **trigon** in an upper molar and a **trigonid** in a lower molar (Fig. 34.5*B* and *C*). Trigon and trigonid are mirror images of each other, so good shearing action occurs as parts of the trigon and trigonid slide past each other. Crushing of food occurs when the primary cusp of the trigon falls upon a low heel, or **talonid,** located on the posterior surface of the trigonid. Additional cutting action occurs as the premolars shear across each other. In order to perform these functions, precise occlusion is needed and this would not be possible with continuous tooth replacement. Young mammals are suckled and are born without teeth. A set of milk incisors, canines and premolars develop as the young begin to feed for themselves. As the jaw grows in size, good occlusion is maintained as the **milk teeth** are gradually replaced by larger **permanent teeth.** The molar teeth appear sequentially as a mammal matures and the jaw enlarges further. They are not replaced.

As mammals chew their food, they mix it with saliva that, in addition to lubricating the food, usually contains an amylase that begins the digestion of carbohydrates. Digestion continues in the stomach and intestinal region. Numerous microscopic villi line the small intestine and increase the surface area available for absorption.

A greater exchange of oxygen and carbon dioxide is made possible by the evolution of pulmonary alveoli that greatly increase the respiratory surface of the lungs and by the evolution of a diaphragm that increases the efficiency of ventilation. A **secondary palate,** a horizontal partition of bone and flesh, separates the air and food passages in the mouth cavity and pharynx (Fig. 34.6). The secondary palate permits nearly continuous breathing, which is a necessity for organisms with a high rate of metabolism. Mammals can manipulate food in their mouths, for food and air passages cross only in the laryngeal part of the pharynx. Breathing need be interrupted only momentarily when the food is swallowed.

Mammals, like birds, have evolved an efficient system of internal transport of materials between sites of intake, utilization, and excretion. Their heart is completely divided internally so there is no mixing of venous and arterial blood. Increased blood pressure also contributes to a more rapid and efficient circulation.

Approximately 99 percent of the water that starts down the kidney tubules is later reabsorbed, so that the net loss of water in mammals is minimal. The generally high metabolic rate of mammals results in the formation of a large amount of wastes to be eliminated. An increase in blood pressure, and hence in blood flow through the kidney, and an increase in the number of kidney tubules has enabled mammals to increase the rate of excretion.

Care of the Young

Upon hatching from its egg, or upon birth, a young reptile must feed and fend for itself. Some are successful, many are not. Mammals have evolved a different reproductive strategy. Fewer young are produced, but considerable maternal energy and care are invested in the few that are conceived and raised. The embryos of endothermic mammals must themselves develop in a warm and closely controlled environment. The platypus and spiny anteater resemble reptiles in being oviparous, but the eggs are carefully brooded, and the newborn are fed milk secreted by the **mammary glands.** Milk is a

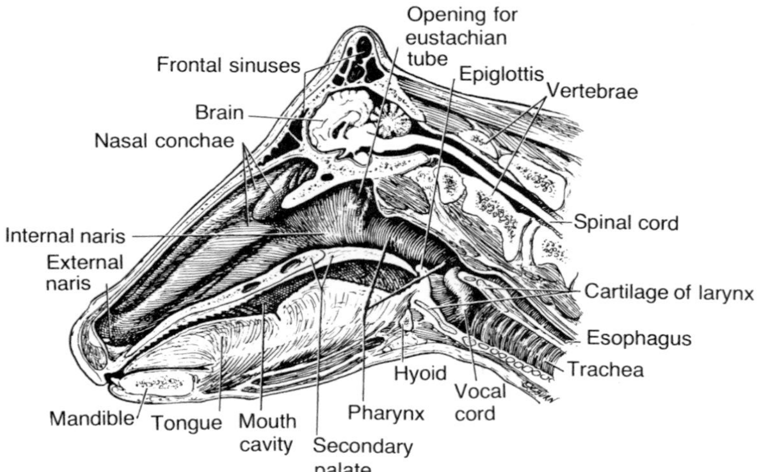

FIGURE 34.6 A sagittal section of the head of a cow showing the relationship between the digestive and respiratory systems. (Modified after Sisson and Grossman.)

very reliable food source because it will be produced if necessary from energy reserves stored in the mother's body. The young are less dependent upon the immediate availability of food to the mother than is the case for other animals.

Other mammals are viviparous and retain their embryos in a uterus. All of the extraembryonic membranes characteristic of reptiles are present, although there is little yolk in the yolk sac. In the opossum, kangaroo, and other marsupials, uterine secretions are absorbed by the yolk sac before a **placenta** is established between the yolk sac and the uterine lining. The placental relationship is very brief, and the young complete their development attached to teats in a pouch on the belly of the mother. Placental mammals establish a placenta very early by the union of the uterine lining and the embryonic chorioallantoic membrane.

Different species of placental mammals are born at different stages of maturity. Certain mice, for example, are extremely **altricial,** being born naked and with closed eyes and plugged ears. Newborn deer and other large herbivores are quite **precocial** and can run about and largely care for themselves.

Primitive Mammals

Early Mammalian Evolution

The line of evolution to mammals separated from the therapsid reptiles late in the Triassic period, about 200 million years ago. The first ones, as we have pointed out, were mouse-sized creatures whose endothermy probably permitted them to exploit a nocturnal insect-eating niche that was not available to the ectothermic reptiles (Fig. 34.1). They had the mammalian pattern of tooth eruption, which indicates that their young fed upon milk. Their molar teeth differed from those of later mammals in having three conical cusps arranged in a linear series, whence their group name of **triconodonts** (Fig. 34.5*D*). Their limb skeleton showed that they had a plantigrade foot, sharp claws, quite flexible joints, and a thumb and great toe that could be abducted and adducted slightly. Probably they scampered about in the ground litter and climbed low branches in their search for food.

Monotremes

The many lines of mammalian evolution during the Mesozoic Era may all have diverged from triconodonts, but the fossil record is not complete enough to be certain. One line probably led to contemporary **monotremes,** the duckbilled platypus (*Ornithorhychus*) and spiny anteater (*Tachyglossus*) of the Australian region (Fig. 37.7*A* and *B*). These are certainly the most primitive mammals living today, but their relationships are somewhat uncertain. Few fossil monotremes are known, and living species lack teeth, which have been used extensively in sorting out lines of mammalian evolution. They lay eggs and retain many other reptilian characteristics, including a cloaca. The ordinal name for the group, **Monotremata,** refers to the presence of a single opening for the discharge of feces, urinary, and genital products (Gr. *monas,* single + *trematos,* hole). In other mammals, the cloaca has become divided, and the opening of the intestine, the **anus,** is separate from that of the urogenital ducts. Because of their primitive nature, monotremes and triconodonts are grouped in the subclass **Prototheria** (Gr. *protos,* first + *therion,* beast).

Monotremes are unusual animals that have survived to the present only because they have been isolated from serious competition. The platypus is a semiaquatic species with webbed feet, short hairs, and a bill like a duck's used in grubbing in the mud for soft-bodied invertebrates. Claws are retained, and they dig long burrows in muddy banks. Spiny anteaters have large claws and a long beak adapted for feeding upon ants and termites. The animal can burrow very effectively. Many of its hairs are modified as quills. Facial muscles are poorly developed in both species, and they do not have fleshy lips. Young monotremes cannot suckle in the usual mammalian way; rather they lap up milk discharged from teatless mammary glands onto tufts of hair.

Therians

In another line of evolution from triconodonts, the three linearly arranged molar cusps shifted to the triangular pattern, and a low heel was added to the posterior edge of each lower molar (Fig. 34.5*E*). These changes, first seen in the extinct **pantothers,** improved the efficiency of the molar teeth in cutting and crushing insects and other small invertebrates. Primitive marsupials and placental mammals living today have similar molar teeth. Because of similarities in their tooth structure, pantothers, marsupials, and placental mammals are grouped together in the subclass **Theria** (Gr. *therion,* beast).

Reproductive Strategies

Marsupials (infraclass **Metatheria,** Gr. *meta,* next to) and placental mammals (infraclass **Eutheria,** Gr. *eu,* true) separated from a common pantother ancestor early in the Cretaceous period and have evolved quite different reproductive strategies. An intimate placental relationship develops in eutherians between the embryonic chorioallantoic membrane and the uterine lining. This is made possible by the **trophoblast** (p. 326), which is the early embryonic precursor of the chorioallantoic membrane. The trophoblast appears to establish an immunological barrier between the mother and the embryo so the embryo, half of whose genes are foreign to the mother, is not rejected. The gestation period is relatively long, and the embryo is provided with a good start in life.

The marsupial egg contains more yolk than that of eutherians and is surrounded by a **shell membrane** of maternal origin. The embryo is nourished for more than half of the gestation period by the stored yolk and by uterine secretions. The shell membrane acts as an immunological barrier. It is absorbed near the end of gestation and a placenta is established, usually between the yolk sac and uterine lining. The placenta is not an intimate union, and it lasts only a few days before birth occurs, probably because of an immunological rejection. The total gestation period of marsupials is relatively short, ranging from 13 days in our opossum to 35 in an Australian wallaby. The young are very small and immature at birth but have special adaptations, including well-developed pectoral appendages, that enable them to crawl up the belly of the mother and attach to teats, which usually are located within a pouch, or **marsupium** (Gr. *marsypion,* a small purse). Pouch life is several times longer than the gestation period, and development is completed there.

As Dr. Pamela Parker has pointed out, there are advantages and disadvantages to each reproductive strategy. One is not inherently superior to the other. Placental mammals are born at a more mature age, but their maturity is derived at the expense of the investment of considerable maternal energy. A placental mother is committed to carrying her embryos until their relatively late birth, even at the expense of her own fitness if food and other environmental resources fail. During the lactation period the young can be abandoned if need be, but then there will have been a loss of the mother's considerable reproductive effort. Marsupial young are less developed and perhaps more vulnerable at the time of their early birth, but few maternal resources have been invested in them. The life of the mother has hardly been affected in any way. If environmental resources fail, the pouch young can be aborted easily and the mother will have lost little reproductive effort. She will have a better chance to survive and reproduce again.

In addition to their short gestation period, kangaroos and wallabies have an additional way of adjusting their reproduction to environmental conditions. Unlike other mammals they can become pregnant again shortly after giving birth and during their lactation period, but the development of the new embryo is arrested in the blastocyst stage. This phenomenon is called **embryonic diapause.** As lactation of the young in the pouch diminishes, development of the blastocyst resumes. Environmental conditions in the "out back" of Australia are often harsh, but when the rains come and the grass grows, these marsupials can reproduce rapidly. They may have a "joey" at heel, who occasionally jumps back in the pouch for some milk, a recently born young in the pouch, who is suckling on a different type of milk, and an arrested blastocyst in the uterus.

Marsupial Adaptive Radiation

Marsupials appear to have originated in the early Cretaceous period in North America and Western Europe. They spread from there through South America, Africa, and across Antarctica to Australia, for these lands were still broadly interconnected as Gondwana (Fig. 17.7). The discovery in 1982 of marsupial fossils in Antarctica supports the notion of their spread into Australia by way of Antarctica. Although a few primitive placental groups reached South America, none spread as far as Antarctica. As Gondwana broke up, Antarctica moved south and became glaciated, Africa became connected to Europe and Asia, and South America separated from North America. South America and Australia were effectively isolated as havens for marsupial evolution. Marsupials radiated widely and occupied most of the ecological niches exploited elsewhere by placental mammals. When a connection was reestablished between North and South America late in the Tertiary period, many placental groups spread into South America, and many primitive species living there became extinct. Opossums survived and reentered North America (Fig. 34.7*C*). The Australian region remained as a haven for marsupial evolution since the only placental mammals to reach there before human beings were bats and a few rodents. Australia, Tasmania, and New Guinea still have a tremendous adaptive diversity of marsupials, such as the carnivorous Tasmanian wolf, anteating species, arboreal phalangers, and koala bears (the original "teddy bear"), plains-dwelling kangaroos, and rabbit-like bandicoots (Fig. 34.7*D* and *E*). There is such a wide variety, and different groups have been evolving independently for so long, that some authorities divide the marsupials into several distinct orders.

Adaptive Radiation of Eutherians

During the Mesozoic era most of the terrestrial resources were exploited by the numerous reptilian groups, and the few contemporary primitive mammals

FIGURE 34.7 Monotremes and marsupials. *A,* The duckbilled platypus; *B,* the spiny anteater; *C,* opossum and young; *D,* koala bear; *E,* kangaroo. The platypus and anteater are monotremes; the others are marsupials. (*A* and *B* courtesy of the New York Zoological Society; *C* and *D* courtesy of American Museum of Natural History; *E* from Australian News and Information Bureau.)

occupied niches not available to the reptiles. Climatic and other changes that occurred in the late Cretaceous and early Tertiary periods led to the extinction of many reptiles, and food and other resources became available to mammals. Placental mammals flourished and radiated widely from their primitive insectivorous niche. They have adapted to nearly every conceivable mode of life upon the land. Others have readapted successfully to an aquatic life, and some have evolved true flight (Fig. 34.8).

Insectivores. The most primitive eutherians, the stem group from which the others evolved, were rather generalized, insect-eating types of the order **Insectivora** (L. *insectum,* segmented animal + *voro,* to eat). Among contemporary species are the tenrecs of Africa and Madagascar (Fig. 34.2), the European hedgehog, the shrews and moles, and the tree shrews of India and Southeastern Asia (Fig. 34.9). Tree shrews have acquired some arboreal specializations, and some authorities regard them as primitive primates. All of these small creatures retain many primitive mammalian features. The limbs of moles are specialized for digging, but limb structure is quite generalized in the other species. Five clawed toes are retained, and the foot posture is plantigrade. Insectivores have a primitive dentition consisting of three incisors, one canine, four premolars, and three molars in each side of the upper and lower jaw (Fig. 34.5*A*). This can be expressed as a dental formula: I^3_3, C^1_1, Pm^2_2, M^3_3. Except for the toothed whales, no placental mammal has more teeth than this, and the number of teeth is reduced in many groups. We, for example, have a dental formula of I^2_2, C^1_1, Pm^2_2, M^3_3.

Toothless Mammals. The South American anteater (Fig. 34.10*A*) and its allies, order **Edentata** (L. *ex,* without + *dens,* tooth), diverged from primitive insectivores in the Cretaceous period and specialized upon a diet of ants and termites, which are abundant in the tropics. They underwent an extensive radiation in South America while this continent was isolated from North America. Many unusual and now extinct types have been dis-

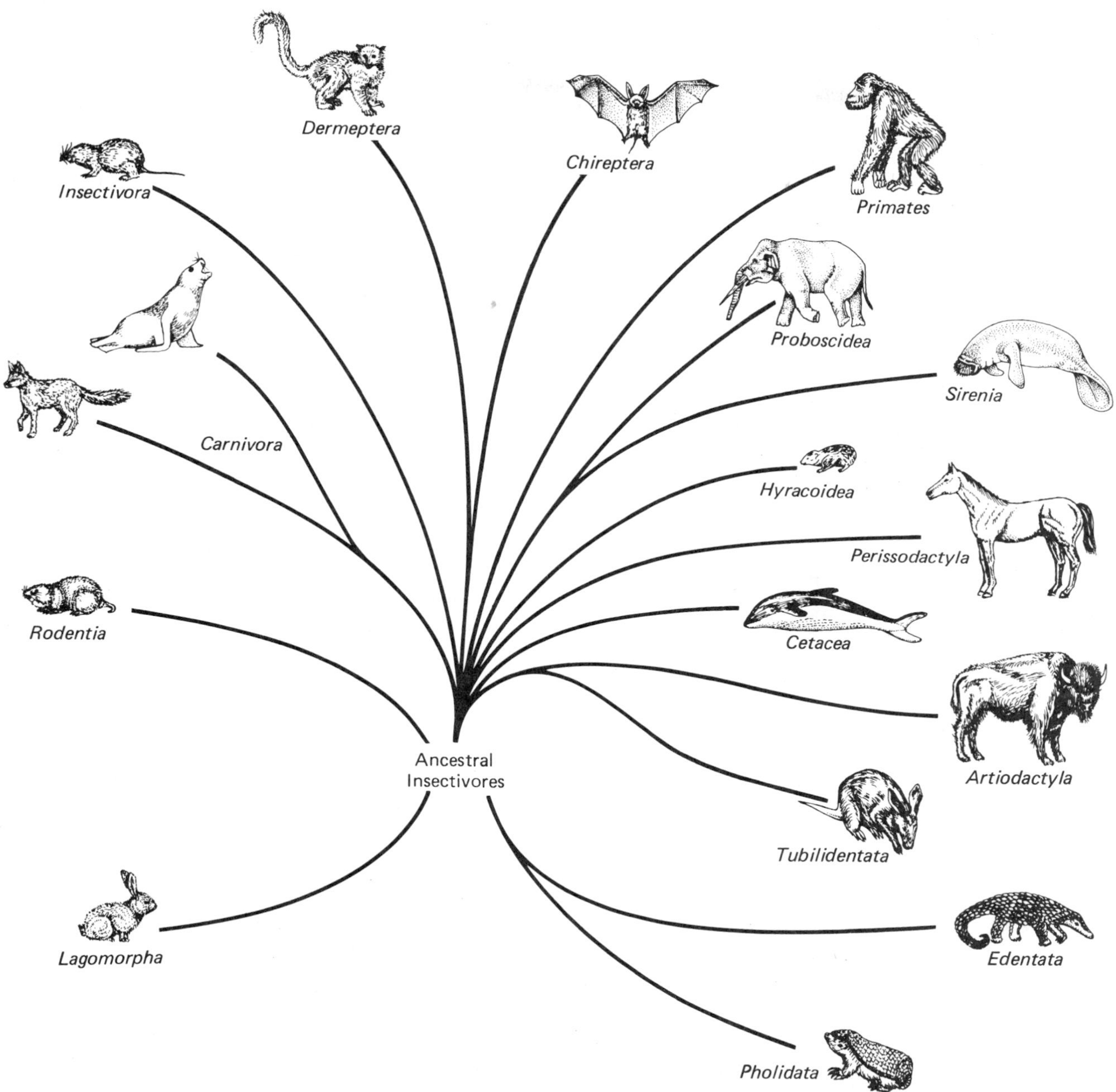

FIGURE 34.8 The radiation of contemporary eutherian mammals. Probable relationships are based on data in Eisenberg, *The Mammalian Radiations.*

covered. An edentate has large claws that enable it to tear open ant and termite nests. Then it laps up the insects with its long tongue, which is covered by a very sticky saliva secreted by enlarged salivary glands. In contrast to a primitive insectivore, which crushes its insect food with its teeth, an anteater swallows whole the insects that it eats. Ants are crushed and ground up by the very muscular pyloric region of the stomach. Its teeth were not needed for survival and have been lost. The tree sloth and armadillo belong to this same order,

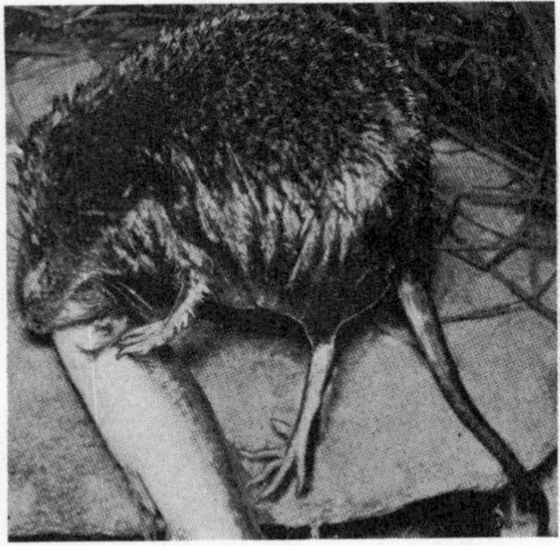

A

B

C

FIGURE 34.9 Insectivores. *A,* A shrew eats more than its own weight every day; *B,* a mole in its burrow; *C,* a tree shrew upon a log. (*A* from Conoway, C. H., in *Natural History, LXVIII,* No. 10; *B,* courtesy of the American Museum of Natural History; *C* from Vaughan, T. A.: *Mammalogy.* 2nd ed. Philadelphia, Saunders College Publishing, 1978. Photograph by M. W. Sorenson.)

A

B

C

FIGURE 34.10 Ant- and termite-eating mammals. *A,* The South American giant anteater; *B,* the African tree pangolin; *C,* the African earth pig, or aardvark, at a termite nest in Kenya. (*A* courtesy of the American Museum of Natural History; *B* from Walker, E. P.: *Mammals of the World.* 3rd ed. Baltimore, The Johns Hopkins Press, 1975. Photograph by Jean-Luc Perret; *C* courtesy of J. Shoshani, Wayne State University.)

though they retain vestiges of teeth. Armadillos entered North America after its reconnection with South America late in the Tertiary period.

Pangolins (order **Pholidota,** Gr. *pholis,* scale) are a closely related group that evolved in Africa and Southeastern Asia. The protective horny scales of the pangolin are composed of hairs cemented together (Fig. 34.10*B*). When disturbed, the animal rolls into a ball.

The aardvark of Africa (order **Tubilidentata,** L. *tubulus,* small tube + *dens,* tooth) independently evolved a somewhat similar mode of life, feeding upon ants and termites (Fig. 34.10*C*). They have small, peg-shaped teeth. They may be related to early hoofed mammals.

Flying Mammals. Bats, order **Chiroptera** (Gr. *cheir,* hand + *pteron,* wing) are closely related to insectivores and are sometimes characterized as flying insectivores. Bat wings (Fig. 34.11) are structurally closer to those of pterosaurs than to birds' wings, for the flying surface is a leathery membrane, but the wing of a bat is supported by four elongated fingers (the second to fifth) rather than by a single one as in the pterosaur. The wing membrane attaches onto the hind leg and, in many bats, the tail is included in the membrane. Bat wings are more cambered (i.e., concave on the underside) than bird wings, and this increases their lift at low speeds. The first finger usually is free of the wing, bears a small claw, and is used for grasping and clinging. The hind legs are small and are of little use upon the ground, but they, too, are effective grasping organs and are used for clinging to a perch from which the bats hang upside down when at rest. While they are active, bats maintain their body temperature between 30° and 35° C., but they permit it to drop to 20° C. or lower when they rest. This economizes on the amount of food they must gather. Most temperate bats hibernate during the winter.

Our familiar bats are insect eaters that fly about at dusk in search of their prey; thus they are utilizing resources not used by diurnal birds. Other bats eat fruit, pollen, and nectar, blood (vampire bats), and many kinds of small vertebrates, including fish. Fish-eating bats catch their prey near the surface of the water by means of hooked claws on their rather powerful feet.

Insectivorous, nocturnally flying bats have evolved a system of echolocation that enables them to avoid obstacles at dusk, find their way in dark caves, and also

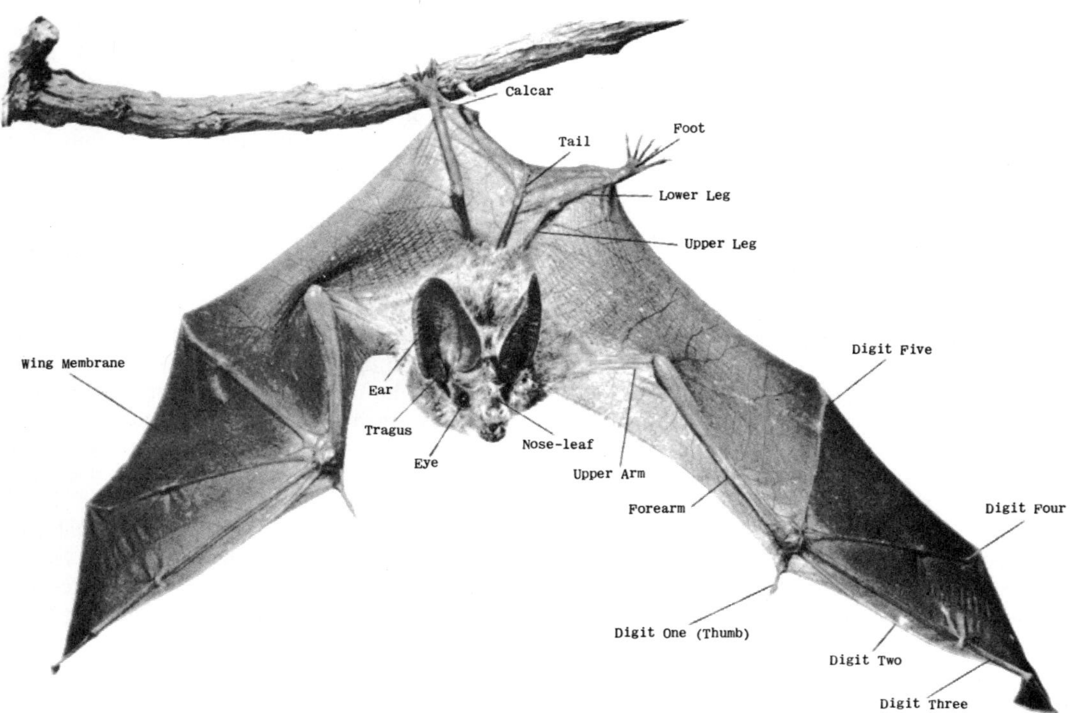

FIGURE 34.11 The Mexican big-eared bat *(Macrotus mexicanus)* about to take off. (From Walker, E. P.: *Mammals of the World.* 3rd ed. Baltimore, The Johns Hopkins Press, 1975. Photograph by E. P. Walker.)

enables many species to find their prey. As early as 1793, Spallanzani observed that a blinded bat could find its way about, but that one in which the ears had been plugged was helpless. However, it was not until the availability of sophisticated electronic apparatus about the time of World War II that Dr. Donald Griffin and others were able to show that bats emitted ultrasonic sounds that bounced off an object and returned as an echo. By analyzing the echoes, many bats can determine the distance, direction, size, and possibly the texture of the object. As bats fly about at dusk searching for insects, they emit ultrasonic pulses that range from 25 to 150 kHz., well above our threshold of hearing. Sounds at these frequencies have short wave lengths and hence can produce sharp echoes from small objects. Different groups of bats have evolved somewhat different sonar systems, but in our most common species the sounds are emitted through their mouths. The pulses are frequency-modulated and drop about an octave (i.e., from 60 to 80 kHz. to 30 or 40 kHz.) during their one- to four-millisecond duration. While the bat is searching, pulses are emitted at the rate of about 10 per second, but the frequency of emission increases tenfold, and the duration of the pulses shortens when a bat detects an insect and homes in on it. The emitted sounds are at very high energy levels, 60 dynes per square centimeter, which is over twice that in a boiler factory. Horseshoe bats emit sounds through their unusual-shaped noses. Tiny muscles in the middle ear contract during sound emission, thereby damping the movement of the auditory ossicles and protecting the inner ear. These muscles relax as the echo returns. Distance appears to be perceived by the time interval between the emitted pulse and the echo, but less well understood mechanisms are involved in some cases. Directionality appears to be determined by a comparison of the differences in intensity of the echo between the two ears. Bats can perceive meaningful signals in the presence of considerable extraneous noise.

Some other mammals stretch a loose skin fold between their front and hind legs and glide from tree to tree. The flying lemurs (order **Dermoptera,** Gr. *derma,* skin + *pteron,* wing) of the East Indies and Philippines have been observed to glide 136 m. while losing no more than 12 m. in elevation (Fig. 34.12).

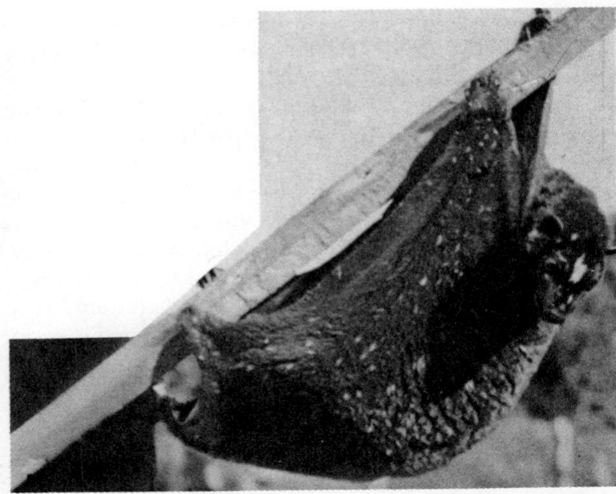

A

FIGURE 34.12 The flying lemur of the East Indies and Philippines. (From Walker, E. P.: *Mammals of the World.* 3rd ed. Baltimore, The Johns Hopkins Press, 1975. Photograph by J. N. Hamlet.)

Carnivores. Certain mammals have evolved specializations for a flesh-eating mode of life. Weasels, dogs, raccoons, bears, and cats (Fig. 34.13) are familiar members of the order **Carnivora** (L. *carno,* flesh + *voro,* to eat). Carnivores have teeth specialized for killing and cutting prey and limb structures adapted to provide the speed needed to run down prey. Canines are large, and in contemporary species the last upper premolar tooth and first lower molar are specialized to form a set of shearing **carnassial** teeth (Fig. 34.14). The jaws are hinged in such a way that they close like a pair of scissors. The posterior teeth come together before the front teeth. One of the most conspicuous adaptations for speed has been a lengthening of the limb, which increases stride length. Most carnivores have shifted from the primitive plantigrade to a **digitigrade** foot posture (Fig. 34.15*B*). They stand upon their toes (though not on their toe tips), with the rest of the foot raised off of the ground in the manner of a sprinter. The foot lengthens to a greater extent than the proximal limb segments. This keeps most of the muscle mass, which is concentrated in the proximal limb segments, close to the body and on a part of the limb that moves through a shorter distance during a limb cycle. The proximal parts of a limb, therefore, have a lower velocity than the distal part. Since the force needed to move a limb segment is equal to its mass times its acceleration, energy is conserved by keeping most of the mass proximal.

Most carnivores are semiarboreal or terrestrial, but one branch of the order, which includes the seals, sea lions, and walruses, early became specialized for exploiting the resources of the sea. In addition to their adaptations as carnivores, which include the large canine tusks of the walrus used in gathering shellfish, these species evolved flippers and other aquatic modifications. When they swim, the large pelvic flippers are turned posteriorly and are moved from side to side in the manner of a fish's tail.

FIGURE 34.13 Representative carnivores and cetaceans. *A*, Raccoon; *B*, walrus; *C*, the birth of a porpoise; *D*, the whalebone plates of a toothless whale hang down from the roof of the mouth; *E*, weasels in summer pelage. The porpoise and whale are cetaceans; the others are carnivores. (*A*, *B*, *D*, and *E* courtesy of the American Museum of Natural History; *C* courtesy of Marine Studios.)

Ungulates. Horses, cows, and similar mammals have become highly specialized for a plant diet. This has entailed a considerable change in their dentition, for plant food must be thoroughly ground by the teeth before it can be acted upon by the digestive enzymes. The molars of primitive plant-eating mammals (and those of omnivorous species, such as human beings) have become square, as seen in a surface view. An extra cusp is added

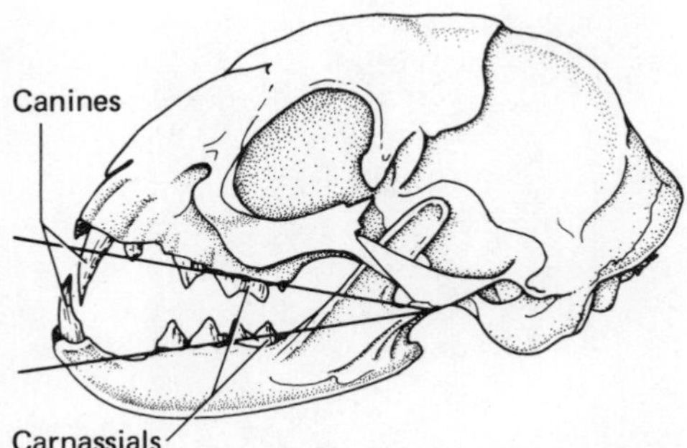

FIGURE 34.14 Teeth and jaw specializations in carnivores as represented by a cat.

to the primitive trigone and trigonid, so that four primary, rounded cusps are present on each molar (*cf.* Figs. 34.5 and 34.16*A*). This is called a **bunodont** tooth (Gr. *bounos,* round + *odous,* tooth). In more advanced species of ungulates the cusps form a pattern of ridges (**lophodont teeth,** Gr. *lophos,* crest) or crescents (**selenodont teeth,** Gr. *selene,* crescent). Ungulates that feed upon leaves and other soft vegetation retain the primitive low-crowned molar teeth, but those that graze upon harder grasses have evolved high-crowned molar teeth (Fig. 34.16*B*). A high-crowned tooth extends a considerable distance above the gum line, and cement (a hard material previously found only on the roots of the teeth) has grown up over the surface of the tooth and into the "valleys" between the elongated cusps. More tooth is provided to wear away, and the tooth is more resistant to wear. Premolar teeth frequently acquire the form of molar teeth. The hinging of the jaw changes so that all of the teeth are brought together at the same time (Fig. 34.16*C*). The configuration of the joint surfaces also permits more fore and aft and sideways movement of the lower jaw than occurs in carnivores.

Herbivores constitute the primary food supply of carnivores and protect themselves primarily by running away. Adaptations for speed have entailed a further lengthening of the legs, especially their distal portions. Their feet are very long, and the animals walk upon their toe tips, a gait termed **unguligrade** (Fig. 34.15*C*). Lateral and medial toes that do not reach the ground became vestigial or disappeared, and the primitive claws on the remaining toes were transformed into hooves—a characteristic that gives the name ungulate to these

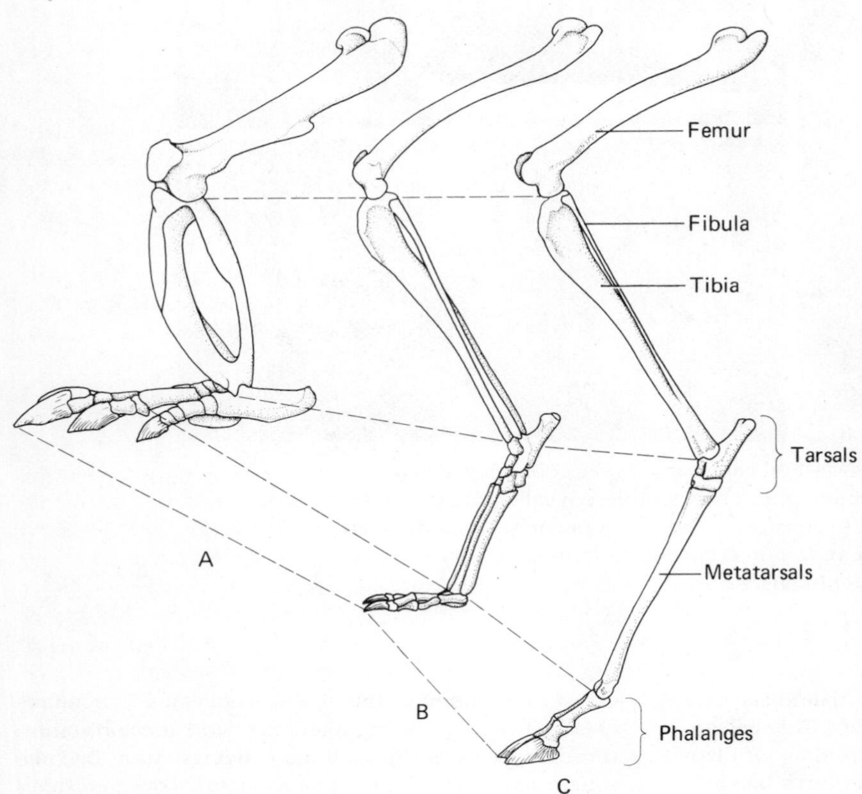

FIGURE 34.15 The hind leg of representative mammals drawn with a constant femur length so that changes in proportion are evident. *A,* The armadillo retains the primitive plantigrade foot although its toes are greatly enlarged for digging. *B,* The foot of the coyote and other carnivores is lengthened, and the animal walks upon its toes with the heel raised off the ground. This is the digitigrade posture. *C,* In the pronghorn and other ungulates, the foot is lengthened further, and the animal walks upon its toe tips, in the unguligrade posture. (From Vaughan, T. A.: *Mammalogy.* 2nd ed. Philadelphia, Saunders College Publishing, 1978.)

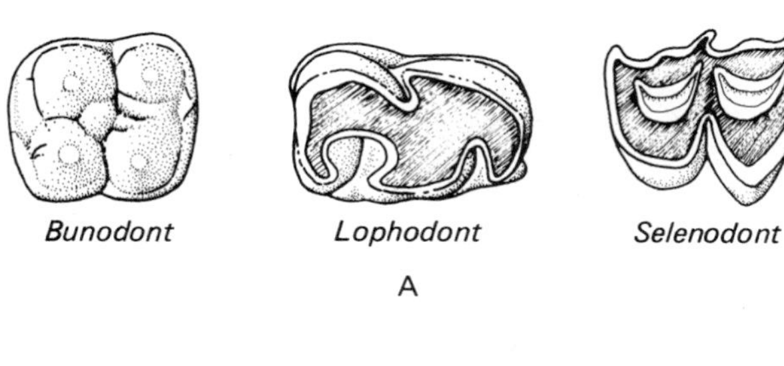

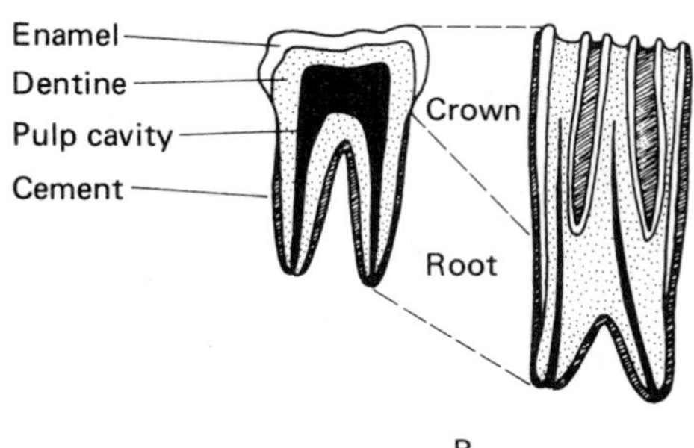

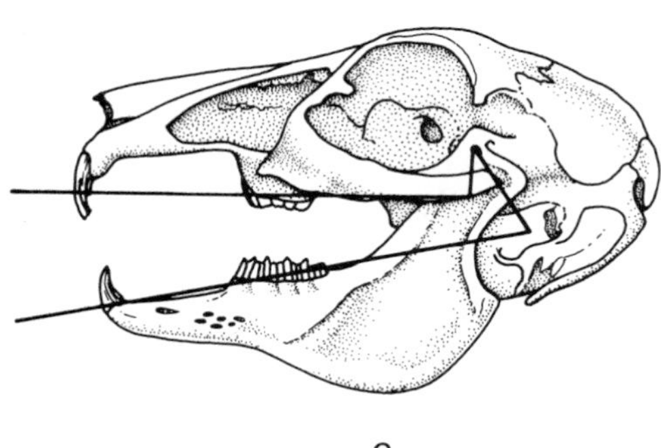

FIGURE 34.16 Modifications of the molar teeth and jaws in plant-eating mammals. *A,* Crown views of cusp patterns; *B,* vertical sections of a low- and high-crowned tooth; *C,* the method of jaw hinging in an herbivore as represented by a rabbit. (*A* and *B* from Vaughan, T. A.: *Mammalogy,* 2nd ed. Philadelphia, Saunders College Publishing, 1978; *C* from Walker, W. F.: *Vertebrate Dissection.* 6th ed. Philadelphia, Saunders College Publishing, 1980.)

mammals (L. *ungula,* hoof + *gradus,* step). Capacity for limb rotation is lost. The limbs are in effect jointed pendulums that can swing rapidly back and forth in the vertical plane.

Contemporary ungulates (Fig. 34.17) are grouped into two orders. In the **Perissodactyla,** the axis of the foot passes through the third toe, and this is always the largest. Ancestral perissodactyls, including the primitive forest-dwelling horses of the early Tertiary, had three well-developed toes (the second, third, and fourth) and sometimes a trace of a fourth toe (the fifth). The tapir and rhinoceros retain the middle three toes as functional toes, but only the third is left in modern, plains-dwelling horses. Perissodactyls are characterized by having an odd number of toes (Gr. *perissos,* odd + *daktylos,* finger). The molar teeth, which are low-crowned and bunodont in primitive perissodactyls, are high-crowned and lophodont.

In the order **Artiodactyla,** the axis of the foot passes between the third and fourth toes, which are equal in size and importance. Ancestral artiodactyls had four toes (the second, third, fourth, and fifth). Pigs and their allies, which move across soft ground, retain these four toes, though the second and fifth are reduced in size. Vestiges of the second and fifth toes, the dew claws, are present in some deer, but camels, giraffes, antelope, sheep, and cattle retain only the third and fourth toes. Artiodactyls are even-toed ungulates (Gr. *artios,* even + *daktylos,* finger). The molar teeth of pigs are low-crowned and bunodont, but those of cattle are high-crowned and selenodont. It is probable that perissodactyls and artiodactyls had a separate evolutionary origin and owe their points of similarity to parallel evolution.

In addition to running away, ungulates protect themselves by kicking with their powerful legs and hooves. Many artiodactyls also have evolved weapons for defense or for combat among males. Wild boars have canine **tusks,** male deer, **antlers,** and sheep and cattle of both sexes, **horns** (Fig. 34.18). Both antlers and horns are bony outgrowths from the skull. Antlers branch, are covered by skin (the velvet) only during their growth, and are shed annually; horns are permanent non-branching structures covered by heavily cornified skin.

Subungulates. Subungulates are a group of plant eaters that have certain ungulate-like characteristics. Elephants (order **Proboscidea,** Gr. *proboskis,* trunk) have five toes, each ending in a hooflike nail (Fig. 34.19*A*). They also walk to some extent upon their toe tips, but a pad of elastic tissue posterior to the digits supports most of the body weight. Elephants are noted for their enormous size. Though large mammals have a relatively lower metabolic rate than small mammals, elephants must obtain large quantities of food. The trunk, which represents the drawn out upper lip and nose, is an effective food-gathering organ. Elephants have a unique dentition in which all of the front teeth are lost except for one pair of incisors, which are modified as tusks. Their premolars, which have come to resemble molars, and their molars are very effective organs for grinding up large quantities of rather coarse plant food. They are high-crowned and so large that there is room for only one in each side of the upper and lower jaws at a time.

FIGURE 34.17 Contemporary ungulates. *A,* Tapir in Rangoon Zoo, Burma; *B,* expedition camel, Kalgan, China; *C,* cattle egret warns the weak-sighted rhinoceros of approaching danger; *D,* hippopotamus. (*A, B,* and *D* courtesy of the American Museum of Natural History; *C* from *Natural History LXVIII,* No. 10, 1959.)

Premolars and molars are replaced sequentially. As one is worn down, a new one moves in. By using up their premolars and molars one at a time, elephants have evolved an interesting way of prolonging total tooth life.

Living elephants are restricted to Africa and tropical Asia and are only a small remnant of a once worldwide and varied proboscidean population. During the Pleistocene Epoch, mastodons, mammoths, and other proboscideans were abundant in North America.

The conies of the Middle East (order **Hyracoidea**), though superficially rabbit-like animals, show an affinity to the elephants in their foot structure and in certain features of their dentition.

The sea cows or manatees (order **Sirenia,** Gr. *seiren,* a mythical creature that lured mariners to destruction) live in warm coastal waters and feed upon seaweed, grinding it up with molars that are replaced from behind in elephant-like fashion. Sea cows have a powerful,

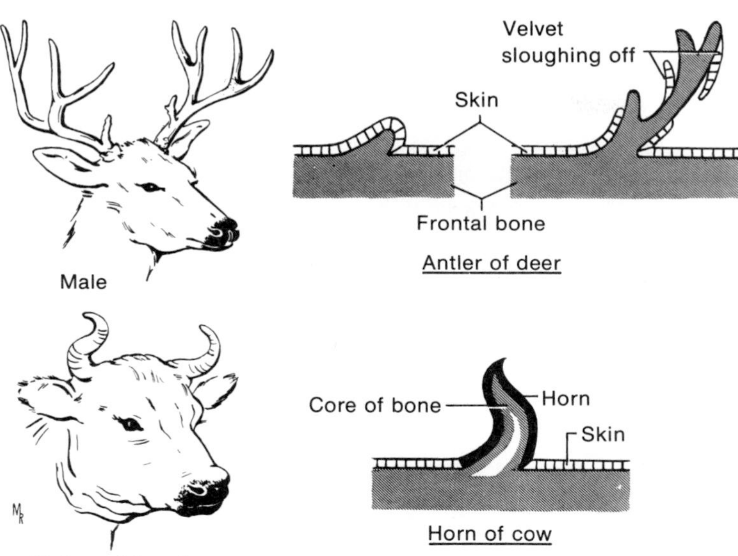

FIGURE 34.18 A diagram to show the differences between antlers (deer) and horns (cow). Antlers are annual growths that are shed in the winter; horns are permanent outgrowths.

horizontally flattened tail and well-developed pectoral flippers (Fig. 34.19*B*). These features, together with a very mobile and expressive snout and a single pair of pectoral mammary glands, led mariners of long ago to regard them as mermaids.

Rodents and Lagomorphs. Other herbivorous mammals gnaw and, in addition to grinding molars, have an upper and lower pair of enlarged, chisel-like incisor teeth that grow out from the base as fast as they wear away at the tip. Gnawing has been a very successful mode of life; in fact, there are more species, and possibly more individuals, of gnawing mammals, or rodents (order **Rodentia,** L. *rodere,* to gnaw) than of all other mammals combined. Rodents have undergone their own adaptive radiation and have evolved specializations for a variety of ecological niches (Fig. 34.20). Rats, mice, and chipmunks live on the ground; gophers and woodchucks burrow; squirrels and porcupines are adept at climbing trees; the flying squirrel can stretch a skin fold

FIGURE 34.19 Subungulates. The elephant *(A)* and the manatee *(B)* are believed to have had a common ancestry. (Courtesy of the American Museum of Natural History.)

FIGURE 34.20 Rodents and lagomorphs. *A,* A flying squirrel; *B,* the pika; *C,* a chipmunk shelling a nut; *D,* a group of beavers. (Courtesy of the American Museum of Natural History.)

located between its front and hind legs and glide to the ground; and muskrats and beavers are semiaquatic.

Rabbits and the related pika of our Western mountains are superficially similar to rodents and were at one time placed in this order. True rodents, however, have only one pair of incisors in each jaw, whereas rabbits have a reduced second pair hidden behind the large pair of upper incisors. Rabbits and the pika are assigned to a separate order, the **Lagomorpha** (Gr. *lagos,* hare + *morphe,* form); their resemblance to rodents is probably a result of parallel evolution.

Whales. Whales, dolphins, and porpoises, of the order **Cetacea** (L. *cetus,* whale), are highly specialized marine mammals that may have diverged from very primitive ungulates. They have a fish-shaped body, pectoral flippers for steering and balancing, no pelvic flippers, and horizontal flukes on a powerful tail that is moved up and down to propel the animal through the water. Some species have even re-evolved a dorsal fin. Despite these fishlike attributes, cetaceans are air-breathing, viviparous, and suckle their young (Fig. 34.13*C*). Some hair is present in the fetus, but it is vestigial or lost in the adult stage, in which its insulating function is performed by a thick layer of blubber.

Certain species have evolved sonar-like systems that help them to avoid obstacles and find their prey even in muddy river waters that a few enter.

Most cetaceans have a good complement of conical teeth well-suited for feeding upon fish, but the largest whales have lost their teeth and feed upon plankton. With fringed horny plates (the whalebone) that hang down from the palate (Fig. 34.13*D*), a toothless whale strains these small organisms from water passing through its mouth. The richness of the plankton together with the buoyancy of the water has enabled these whales to attain enormous size. The blue whale, which reaches a length of 30 m. and a weight of 135 metric tons, is the largest animal that has ever existed.

Classification of Mammals

Class Mammalia The mammals. Endothermic tetrapods, generally covered with hair; mammary glands present; jaw joint between dentary and squamosal bones; three auditory ossicles.

Subclass Prototheria Primitive mammals retaining many reptilian features, including the egg-laying habit and cloaca in living species. Teeth absent or variable, but molars not triangular. Extinct triconodonts and living monotremes. The platypus, *Ornithorhynchus;* spiny anteater, *Tachyglossus.*

Subclass Theria Typical mammals. All living ones are viviparous.

†*Infraclass Pantotheria* Two orders of Mesozoic mammals with molar teeth having three cusps arranged in a triangle; a talonid sometimes present on lower molars. Ancestral to higher therians. Pantotheres.

Infraclass Metatheria Pouched mammals. Young are born at an early stage of development and complete their development attached to teats that are located in a skin pouch; usually three premolar teeth and four molars in each jaw.

Order Marsupialia Marsupials. The opossum, *Didelphis.*

Infraclass Eutheria Placental mammals. Young develop to a relatively mature stage in the uterus; primitive dental formula is I^3_3, C^1_1, Pm^4_4, M^3_3. Living orders are listed here.

Order Insectivora Insectivores, including tenrecs, shrews, moles, and hedgehog. Small mammals, usually with long pointed snouts; sharp cusps on molar teeth adapted for insect eating; feet retain five toes and claws. The common shrew, *Sorex.*

Order Edentata New World edentates, including sloths, anteaters, and armadillos. Teeth reduced or lost; large claws on toes. The armadillo, *Dasypus.*

Order Pholidota The pangolin, *Manis,* of Africa and Southeastern Asia. Teeth lost; long tongue used to feed on insects; body covered with overlapping horny plates.

Order Tubulidentata The aardvark, *Oryceteropus,* of South Africa. Teeth reduced; long tongue used to feed on insects.

Order Chiroptera The bats. Pectoral appendages modified as wings; hind legs small and included in wing membranes. The little brown bat, *Myotis.*

Order Dermoptera The flying lemur of the East Indies and the Philippines. A gliding animal with a lateral fold of skin.

Order Primates The primates. Rather generalized mammals retaining five digits on hands and feet; first digit usually opposable; claws usually replaced by finger- and toenails; eyes typically large and turned forward; often considerable reduction in length of snout. Lemurs, tarsiers, monkeys, apes, and human beings, *Homo.*

Order Carnivora The carnivores. Flesh-eating mammals; large canines; certain premolars and molars modified as shearing teeth; claws well developed. Dogs, raccoons, bears, skunk, mink, cats, hyenas. The domestic cat, *Felis.*

Order Proboscidea Elephants and related extinct mammoths and mastodons. Massive ungulates retaining five toes, each with a small hoof; two upper incisors elongated as tusks; nose and upper lip modified as a proboscis. African elephant, *Loxodonta;* Indian elephant, *Elephas.*

Order Sirenia Sea cows. Marine herbivores; pectoral limbs paddle-like; pelvic limbs lost; large horizontally flattened tail used in propulsion. Florida manatee, *Trichechus.*

Order Hyracoidea Conies. Small, guinea-pig–like herbivores of the Middle East; four toes on front foot, three on hind foot, each with a hoof. *Procavia.*

Order Perissodactyla Odd-toed ungulates. Axis of support passes through third digit; lateral digits reduced or lost. Tapirs, rhinoceroses, and horses, *Equus.*

Order Artiodactyla Even-toed ungulates. Axis of support passes between third and fourth toes; first toe lost; second and fifth toes reduced or lost. Pigs, camels, deer, giraffes, antelopes, cattle, sheep, and goats. American buffalo, *Bison.*

Order Cetacea The whales and their allies. Large marine mammals; pectoral limbs reduced to flippers; pelvic limbs lost; large tail bears horizontal flukes, which are used in propulsion. The bottlenosed dolphin, *Tursiops.*

Order Rodentia The rodents. Gnawing mammals with two pairs of the chisel-like incisor teeth. The largest order of mammals, it includes the squirrels, chipmunks, marmots, gophers, beavers, rats, mice, muskrats, lemmings, voles, porcupines, guinea pigs, capybaras, and chinchillas. The woodchuck, *Marmota.*

Order Lagomorpha Hares, rabbits, and pikas. Gnawing mammals with two pairs of chisel-like incisors and an extra pair of small upper incisors that lie behind the enlarged first pair. The rabbit, *Lepus.*

Summary

1. Endothermy probably evolved in ancestral mammals as they adapted to an insectivorous, nocturnal life style that was not available to the ectothermic reptiles. They probably maintained their body temperature close to the ambient nocturnal range, but those that became diurnal maintained their temperature at a higher level.

2. Mammals can adjust their level of metabolism, the amount of blood flow through the skin, and the amount of evaporative cooling by panting or sweating to maintain a proper balance between heat production and loss.

3. All of the organ systems have undergone changes as mammals became more active and agile. Their arched vertebral column includes three or more sacral vertebrae. Their limbs are usually rotated beneath the body. Their sense organs are highly developed, and the cerebrum and cerebellum of the brain are greatly expanded. Their jaw structure and palate permit the mastication of food, the evolution of a secondary palate separates the food and air passages, and breathing can continue while food is in the mouth. Lung surface area is greatly expanded, and the lungs are ventilated by rib and diaphragm movements. Intestinal villi increase the digestive and absorptive area of the small intestine. Venous and arterial blood are completely separated in their passage through the heart. The kidneys can remove an increased volume of nitrogenous wastes yet conserve salts and water.

4. All female mammals nourish their young on milk secreted by mammary glands. A delay in tooth eruption is correlated with this and is evidence that primitive, extinct species possessed mammary glands.

5. Ancestral mammals had molar teeth with three cusps arranged in a linear series. Living monotremes are oviparous and have no teeth but are placed in the subclass Prototheria along with the ancestral triconodonts. All other mammals (subclass Theria) are viviparous; most have more complex molar teeth that evolved from ones in which the three primary cusps were arranged in a triangle.

6. Two different reproductive strategies have evolved among therian mammals. The trophoblast of eutherian, or placental, embryos appears to act as an immunological barrier and prevents rejection of the embryo. Embryos develop an efficient chorioallantoic placenta. Placental mammals are born at a relatively advanced stage of development, but their maturity requires a considerable maternal investment. A shell membrane apparently protects the early marsupial embryo from immunological rejection. After it has been absorbed, the embryos are born at a very immature stage, attach to teats in the marsupium, and complete their development there. Few maternal resources are invested in the uterine young, and those in the marsupium can be aborted easily if food or other environmental resources fail.

7. With the extinction of many reptiles and the freeing up of environmental resources that occurred in the late Cretaceous period, marsupials and eutherian mammals underwent extensive adaptive radiations. Marsupial radiation was most extensive in South America and Australia because these continents were isolated from other parts of the world for a long period of time.

References and Selected Readings

Many of the general references on vertebrates cited at the end of Chapter 31 contain considerable information on the biology of mammals.

Anderson, H. T. (Ed.): *The Biology of Marine Mammals.* New York, Academic Press, 1969. Chapters deal with the swimming, diving, echolocation, and other aspects of the biology of cetaceans and other marine mammals.

Burt, W. H., and R. P. Grossenheider: *A Field Guide to the Mammals.* Boston, Houghton Mifflin Co., 1952. A useful guide, in the style of Peterson bird guides, for the field identification of mammals.

Dawson, T. J.: Kangaroos. *Scientific American, 237* (Aug.):78, 1977. Their adaptations and way of life.

Eisenberg, J. F.: *The Mammalian Radiations.* Chicago, University of Chicago Press, 1981. A scholarly analysis of evolutionary trends among mammalian groups and of mammalian adaptations and behavior.

Fenton, M. B., and J. H. Fullard: Moth hearing and feeding strategies of bats. *American Scientist, 69*:266, 1981. Variations in the echolocation behavior of bats correlate with the hearing-based defenses that have evolved in their insect prey.

Irving, L.: Adaptations to cold. *Scientific American, 214* (Jan.):94, 1966. Adaptations by which endothermic birds and mammals adjust to cold environments.

Schmidt-Nielsen, K.: Counter current systems in animals. *Scientific American, 244*(May):118, 1981. A discussion of various countercurrent mechanisms, including the pattern of air flow in the camel's nasal passages, which reduces the amount of water lost in breathing.

Schmidt-Nielsen, K.: *Desert Animals.* Oxford, Clarendon Press, 1964. The adaptations of camels, kangaroo rats, human beings, and other animals to desert life are thoroughly analyzed.

Schmidt-Nielsen, K., L. Bolis, and C. R. Taylor (Eds.): *Comparative Physiology: Primitive Mammals.* Cambridge, Cambridge University Press, 1980. Many aspects of the evolution and adaptations of primitive mammals are covered in a series of papers presented at the Fourth International Congress on Comparative Physiology.

Stonehouse, B., and D. Gilmore (Eds.): *The Biology of Marsupials.* Baltimore, University Park Press, 1977. A collection of papers on the chromosomes, evolution, behavior, anatomy and physiology, cell biology, and other aspects of marsupial biology.

Taylor, C. R.: The eland and the oryx. *Scientific American, 220*(Jan.):88, 1969. An examination of how African antelopes adapt to hot, dry environments.

Vaughan, T. A.: *Mammalogy.* 2nd ed. Philadelphia, Saunders College Publishing, 1978. An excellent textbook with chapters on the origins of mammals, the various groups, ecology, zoogeography, behavior, and various aspects of mammalian physiology.

Walker, E. P., et al.: *Mammals of the World.* 3rd ed. Revised by J. L. Paradiso. Baltimore, The Johns Hopkins Press, 1975. Each known genus of mammals is discussed and illustrated in the first two volumes of this treatise. A third volume is devoted to a classified bibliography of the literature regarding mammalian groups and their anatomy, physiology, ecology, and so forth.

Wynsott, W. A. (Ed.): *Biology of Bats.* 2nd ed. New York, Academic Press, 1977. A multivolume treatise that includes chapters on the evolution, anatomy and physiology of the organ systems, thermoregulation and hibernation, development, echolocation, and ecology of bats.

Young, J. Z., and M. J. Hobbs: *The Life of Mammals, Their Anatomy and Physiology.* 2nd ed. Oxford, Clarendon Press, 1975. A very valuable source book emphasizing the anatomy and physiology of mammals.

35 Vertebrates: Primates

ORIENTATION

After examining the nature of the arboreal adaptations that primates acquired early in their evolution, we will review the major groups of primates and study the differences between them. Building on this background, we will next explore the characteristics that distinguish human beings from other primates and consider when, and under what circumstances, these developed in the course of human evolution.

Primate Adaptations

Primates, which include human beings as well as the lemurs, monkeys, and apes, diverged from insectivorous ancestors probably as early as the late Cretaceous. Certain primitive insectivores and primates are difficult to distinguish. The tree shrews, which we have described with insectivores, are considered to be primates by some investigators. As a group, primates have become more adapted to an arboreal life. Baboons, human beings, and a few others have reverted to a terrestrial life, but they too bear the stamp of prior arboreal adaptations. Our flexible limbs and grasping hands are fundamentally adaptations for life in the trees. Claws were transformed into finger- and toenails when grasping hands and feet evolved. Molar teeth became square, each with four low-crowned cusps. This is a configuration well adapted for crushing the variety of soft food encountered in the trees: insects, leaves, shoots, and fruit. The reduction of the olfactory organ and olfactory portion of the brain and the development of stereoscopic, or binocular, vision represent other adaptations of primates to arboreal life. Olfactory trails would be interrupted as an animal jumped from branch to branch and so the sense of smell would be of less value for survival than keen vision and the ability to perceive depth of field. Muscular coordination is also very important, and the cerebellum of primates is unusually well developed. The evolution of stereoscopic vision, increased agility, and the influx of a new sort of sensory information gained by the handling of objects with a grasping hand was accompanied by an extraordinary development of the cerebral hemispheres. To some extent the increased input of sensory information and the growth in size and complexity of the cerebrum are coupled. As more sensory information became available, natural selection favored the evolution of a brain that could process it. As the cerebrum enlarged and became more complex, it, in turn, could deal with a greater sensory input. Higher mental functions, such as symbolization and conceptual thought, that characterize higher primates could not have evolved until the brain had attained a certain threshold of size and complexity under other selective influences. In a very real sense, we and other primates are products of the trees.

The Groups of Primates

Prosimians

Primates are usually divided into two suborders. The **Prosimii** (Gr. *pro,* before + L. *simia,* ape) include five families of small, mostly arboreal, primates of the Old World tropics, extending from Africa to the Philippine Islands. Madagascar has a particularly diverse assemblage. Lemurs, indrids, lorises, and the aye-aye of Madagascar and the tarsier (*Tarsius*) of the East Indies and Philippines (Fig. 35.1) all show the beginning of primate adaptations. Some lemurs retain rather long snouts, for olfaction is still important in their social structure. Their triangular molars are beginning to become square, grasping feet have evolved, and most of the claws have been transformed into nails. The tarsier has large eyes adapted for nocturnal life. Its elongated hind feet enable it to hop through the trees. Elongation has been accomplished through the lengthening of certain tarsal bones. The needed leverage is provided, and the grasping digits are retained. Digital pads are borne upon the ends of the toes.

Anthropoids

Monkeys, apes, and human beings are placed in the suborder **Anthropoidea** (Gr. *anthropos,* man + *oeides,* resembling). Most of them have a relatively flat face that is at least partly devoid of fur, well-developed stereoscopic vision, the capacity to sit on their haunches and to examine objects with their hands, and an unusually large brain and globular brain case. Anthropoids appear in the fossil record early in the Tertiary period. It is believed that they evolved from certain prosimians, but the record is not clear enough to know from which group.

New World monkeys and marmosets constitute the superfamily **Ceboidea.** They differ from other anthropoids in retaining three premolar teeth. All are arboreal. New World monkeys cannot abduct the base of their thumb to the extent that Old World anthropoids can. Some have evolved a prehensile tail that serves as a fifth limb (Fig. 35.2*A*). Marmosets retain claws on most of their digits, their thumb is nonopposable, but they can grasp with their hind feet.

Old World monkeys, apes, and human beings have only two premolar teeth. Contemporary Old World monkeys (family **Cercopithecidae**) constitute a large and diverse group, including the forest-dwelling langurs and green monkeys, the macaques (who are only partially arboreal), and the terrestrial baboons and mandrills. All are quadrupeds, but they tend to sit upright upon **ischial callosities**—hardened skin pads upon their buttocks (Fig. 35.2*B*). The skin around the callosities becomes swollen and brilliantly colored in many females during estrus. The thumb is completely opposable. Cusps of each molar tooth have fused to form two transverse ridges that help in grinding the plant food on which this group largely feeds. Many species have cheek pouches in which food can be stored temporarily. Social structure is highly developed in many species; baboons

A

B

FIGURE 35.1 *A,* Lemur; *B,* tarsier. (*A* courtesy of Mark A. Rosenthal, Lincoln Park Zoo; *B* courtesy of the American Museum of Natural History.)

travel in troops and cooperate in obtaining food and protecting the females and young.

Apes and human beings are placed together in the superfamily **Hominoidea** because they share many features. They are larger than other anthropoids, lack a tail, and show more tendencies toward assuming an upright posture, at least some of the time. The smallest of the contemporary apes are the gibbon and siamang of Malaysia (family **Hylobatidae**). Their weight ranges from 5 to 13 kg. Ischial callosities develop late in life. Other apes are the larger great apes (**Pongidae**): the orangutan (*Pongo*) of Borneo and Sumatra and the chimpanzee (*Pan*) and gorilla (*Gorilla*) of tropical Africa (Fig. 35.3). Some gorillas attain a weight of 270 kg. We are more

A B

FIGURE 35.2 Monkeys. *A,* Woolly monkey from South America; *B,* langur from Indochina. (*A* from Walker, E. P., and J. L. Paradiso: *Mammals of the World.* Baltimore, The Johns Hopkins Press, 1975; *B* courtesy of the Chicago Museum of Natural History.)

FIGURE 35.3 Apes. *A,* Gibbon; *B,* gorilla. (From Campbell, B. G.: *Human Evolution.* Chicago, Aldine Publishing Co., 1974.)

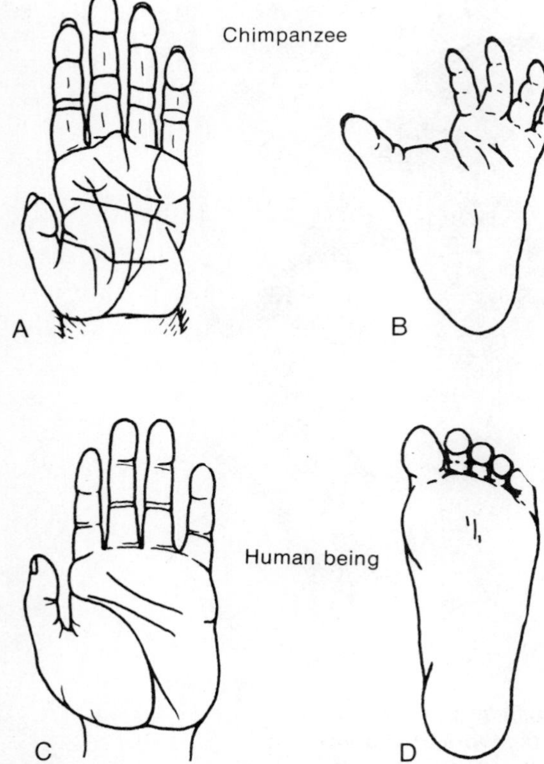

FIGURE 35.4 The hand of a chimpanzee *(A)* is adapted for brachiating and knuckle walking, and its foot *(B)* for grasping; the human hand *(C)* is adapted for grasping, and the foot *(D)* for bipedal locomotion. (*A* and *C* modified from Biegert; *B* and *D* from Morton.)

closely related to the great apes than to the gibbon. Detailed comparisons of chromosome banding patterns show that 18 of the 23 pairs of chromosomes of humans and the great apes are the same. Only slight differences occur in the remaining pairs.

All apes have limb specializations that enable them to **brachiate** (L. *brachium,* arm), that is, to swing from branch to branch using their arms alternately. Their arms are longer than their legs. They retain the grasping hind foot, but the thumb is short in relation to the elongated palm (Fig. 35.4*A*). The hand is not as effective in grasping as in most monkeys and human beings; rather it is used as a hook when brachiating. Gibbons and siamangs are the best brachiators and can clear 3 m. or more with each swing. When walking along a branch or on the ground, they stand nearly erect with arms outstretched as balancers. Orangutans are primarily arboreal climbers, but they can brachiate. The pigmy chimpanzee and young gorillas spend much time in the trees and are good brachiators, but larger chimps and gorillas spend more time on the ground. They can walk bipedally over short distances but prefer a modified quadrupedal gait known as **knuckle walking,** in which they support the front of the body upon the knuckles of their elongated hands (Fig. 35.3*B*).

Apes are primarily herbivores, feeding upon fruit, young leaves, and other soft plant material. Occasionally they will eat meat. Their teeth are large and their jaws powerful. Prominent bony brow ridges above the orbits help resist the stresses set up in the skull by the powerful jaw mechanisms (Fig. 35.5). The gorilla's skull also has large sagittal and nuchal crests that increase the area available for the attachment of jaw muscles. The cusps of ape molars remain distinct and do not fuse to form transverse crests as they do in Old World monkeys.

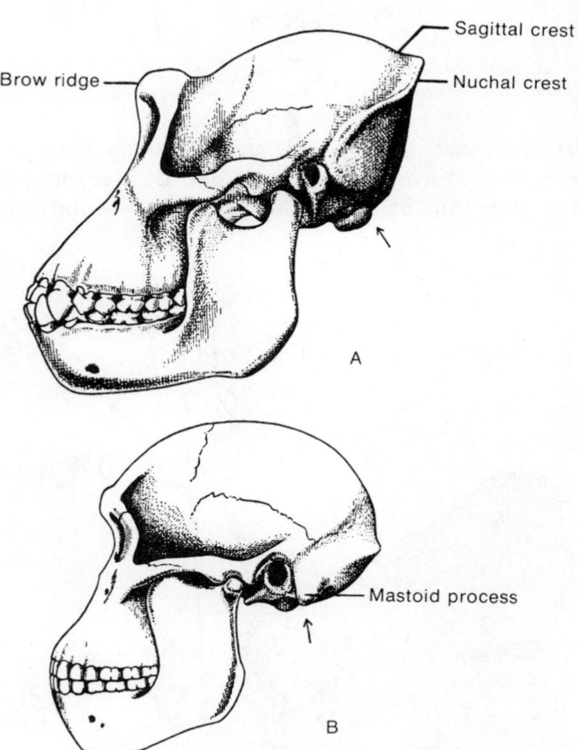

FIGURE 35.5 Lateral view of the skull and lower jaw of *A,* a gorilla; *B, Australopithecus.* The arrow shows the inclination of the foramen magnum, which indicates the relative posture of these two anthropoids. (From LeGros Clark: *The Antecedents of Man.* Quadrangle Books, Inc.)

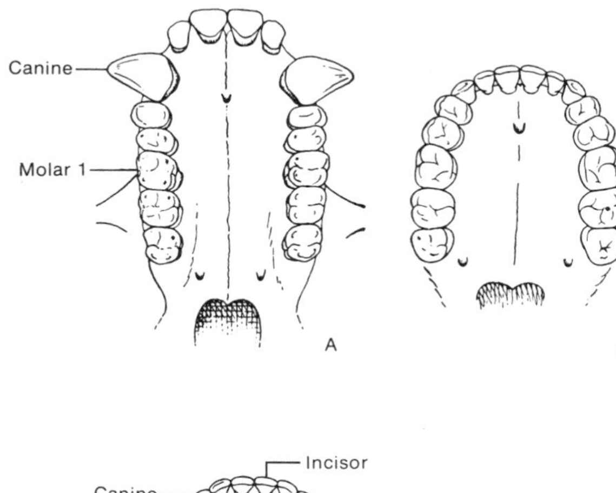

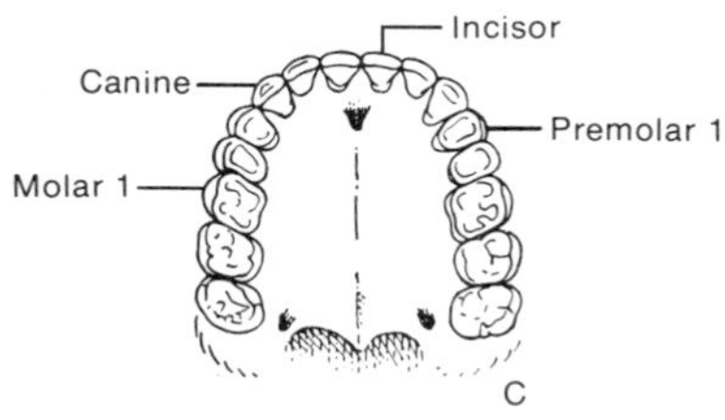

FIGURE 35.6 Palate and upper teeth of *A*, gorilla; *B, Australopithecus; C,* modern human. The large canine and the parallel rows of premolar and molar teeth give the row of teeth in apes a nearly rectangular shape. Australopithecines and humans have a more rounded tooth row.

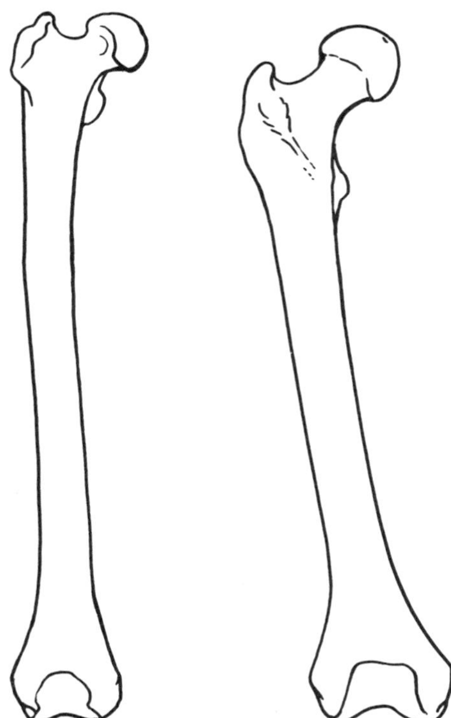

FIGURE 35.7 Femora of *A*, an extinct ape *(Dryopithecus); B,* a modern human being. The axis of the shaft of the human femur inclines medially and is no longer perpendicular to the distal articular surface. This is correlated with our bipedal stance in which the knees are brought close together under the body. (Modified from Campbell, B. G.: *Human Evolution.* Chicago, Aldine Publishing Co., 1974.)

A fifth small cusp is present on the posterior edge of the lower molars. The tooth row has a somewhat squarish or U-shaped appearance, for the molars are in parallel rows, and the canine, which is used in defense, is large (Fig. 35.6). The first lower premolar and the upper canine are sectorial, that is, they form a bladelike cutting surface between them.

Apes are very intelligent creatures. They live in groups, often led by a dominant male, communicate with each other by primitive sounds and facial expressions, and can use sticks and other objects as tools to reach food.

Human Characteristics

Human beings differ enough from apes to be placed in a separate anthropoid family, the **Hominidae.** We differ from contemporary apes in being well adapted to a bipedal gait. We have a lumbar curve in our back that places our center of gravity over the pelvis and hind legs. Our ilium is broad and flaring, providing a large surface for the attachment of gluteal and other muscles that hold us erect. Our legs are longer and stronger than our arms. The distal ends of our femora are brought close to the midline, giving us a knock-kneed appearance but placing the foot under the projection of the body's center of gravity, thereby enabling us to balance easily on one foot when the other is off the ground (Fig. 35.7). Our foot has lost its primitive grasping ability, for the toes are short and parallel each other (Fig. 35.4*D*). The heel bone is large, the tarsals and metatarsals form strong supporting arches, and the great toe, with which we push off when walking, is enlarged. Our head is balanced upon the top of the vertebral column. The foramen magnum is far under the skull; the nuchal area on the back of the skull for the attachment of neck muscles is much reduced; and the mastoid process is enlarged. Body hair is very much reduced in human beings, and a great increase in the number of sweat glands may be correlated with this.

We use our hands not for locomotion but for carrying things and for making and using tools. Our thumb is longer and the metacarpal portion of our hand shorter than in apes (Fig. 35.4*C*).

We include a great deal of meat in our diet as well as a wider variety of plant food. Although early humans retained powerful jaws, fire was used by later people to cook and soften food. Our teeth and jaws are less massive than in apes, and our face does not protrude as much. Our tooth row is more rounded, and the canines are small (Fig. 35.6*C*). Hominids defend themselves with tools and weapons rather than teeth.

Our brain differs from an ape's not only in size but also in organization. The parietal, frontal, and temporal areas of the cortex are enlarged in relation to the ape's. These are regions that are important in sensory and motor integration and in association, memory, and speech.

Human patterns of reproduction and development also differ from those of apes in important ways. Adult female apes seldom accept males except during the few days of estrus, which occurs near the middle of their monthly ovarian cycle. They do not copulate during pregnancy or lactation. Birth is relatively simple because the fetus's head is substantially smaller than the pelvic canal. The single young is at a relatively mature stage at birth and within a few days can cling to its mother's fur with its hands and feet as she moves about. Only one young is raised every few years. Human beings are sexually more active. Mature females copulate throughout their cycles and during pregnancy and lactation. The infant's head is relatively large, and birth sometimes is difficult. Newborn are quite immature and must be carried by one of the parents for months. During her reproductive years a human female can, and frequently does, give birth every year or two. Juvenile and adolescent life is prolonged for years as the children acquire the brain development and knowledge they will need to become integrated into a complex society.

Early Evolution of Apes and Hominids

We differ, then, from contemporary apes in locomotion, the use of our hands, diet, brain size and capacity, and reproduction and development. These differences become blurred as we trace lines of descent back in time. At some point, apes and hominids had a common ancestry, but when was this? *Parapithecus,* a monkey-sized creature of the Oligocene known from a number of mandibles and upper jaw fragments, had the two premolars characteristic of Old World anthropoids. It probably was on the line of ancestry to Old World monkeys, but it may not have been far from the line to apes because it lacked certain dental specializations of later monkeys (Fig. 35.8). The separation between monkeys and apes occurred very early, for the same beds yield fossils (*Propliopithecus, Aegyptopithecus*) that show the beginnings of ape specializations. Gibbons and siamangs appear to have diverged from other apes soon thereafter, for *Pliopithecus* of Miocene age already resembled gibbons in skull characteristics.

Many ape remains have been recovered from Miocene deposits 15 to 20 million years old in Europe, Africa, and South Asia. Originally assigned to different genera, most are now considered to represent different species of *Dryopithecus.* Remains of the limb skeleton indicate that the dryopithecines, although clearly apes, were not as specialized for brachiating as modern species. Limb fragments suggest that they were agile creatures at home in the trees and on the ground.

Until about 10 million years ago, East Africa was heavily forested. Shortly thereafter climatic changes accompanied such geologic disturbances as rifting and volcanic activity. Forests began to break up, and open woodlands interspersed with tropical grasslands, or savannah, became more prevalent. It is likely that these changes provided the context in which natural selection led to the separation of ape and human lines of evolution, but the fossil record is very sparse for this critical period.

The Ape-Men

In 1924, before many of the ape fossils discussed earlier had been found, an endocranial cast and part of the skull of a child were discovered in cave deposits in South Africa. Raymond Dart, Professor of Anatomy at the University of Witwatersrand, described it as *Australopithecus* (L. *australis,* south + *pithecus,* ape) *africanus.* Subsequently, other specimens have been discovered in many parts of the Great Rift valley extending north through east Africa into Ethiopia. The australopithecines range in age from late Pliocene (3.8 million years ago) into the Pleistocene (one million years ago). Many of the fossils were assigned originally to different genera, but most physical anthropologists now agree that they represent varieties of two or three species (Fig. 35.8). The best known, *Australopithecus africanus,* was a lightly built, gracile creature that stood about 1.2 m. tall. Structurally *africanus* is closer to humans than to apes so this species usually is regarded as the first hominid. The configuration of the skull, pelvis, femur, and foot indicate that they were bipeds. The small canine and rounded shape of the tooth row are human (Fig. 35.6*B*), but *africanus* had very large molar teeth with an unusually heavy coating of enamel. The jaws were massive and brow ridges well developed (Fig. 35.5*B*). Their cranial ca-

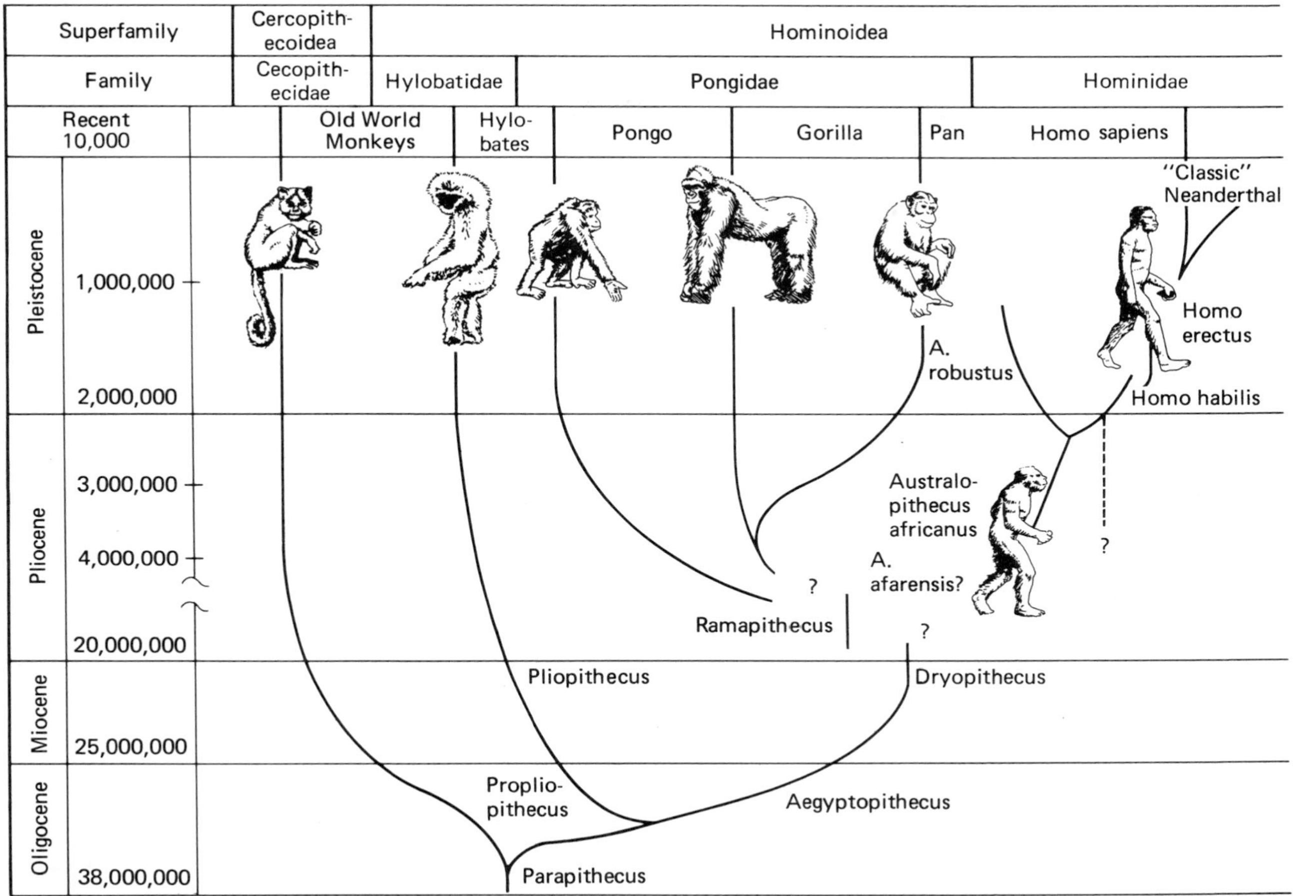

FIGURE 35.8 A possible phylogeny of the major groups of Old World primates. Time, shown on the vertical axis, has been greatly foreshortened preceding the mid-Pliocene. The grouping of species into families is shown on the top of the page.

pacity ranged from about 430 to 550 cc., which is close to the brain size of the great apes.

The oldest specimens, discovered by Donald Johanson at Hadar in Ethiopia (and originally christened "Lucy") and by Mary Leakey at Laetoli in Tanzania, are slightly different from later specimens. Johanson considers them to be a distinct species, which he has named *afarensis*, but many investigators do not agree with him.

The lightly built *africanus* was replaced about two million years ago by a larger species named *Australopithecus robustus*, who stood about 1.5 m. tall. *Robustus* was adapted to eat tougher food, possibly even to crack open bones with its teeth, for the teeth and jaw apparatus are truly massive. A sagittal crest upon the skull provided extra surface for the attachment of large jaw muscles. Although the line of evolution to higher human species may have passed through *africanus*, *robustus* was a side line that became extinct about a million years ago.

The Protohominids

Where did the australopithecines come from and when did they separate from the ape lineage? Remains of *Ramapithecus* from Kenya, southeastern Europe, and Pakistan belong to creatures 9 to 14 million years old, with large, thickly enameled teeth very similar to those of *Australopithecus*. Although fragmentary, these fossils suggest hominid tendencies and a separation of the human-ape lineage going back at least 14 million years. But more recent discoveries, including parts of the facial skeleton, indicate that these creatures may have been closer to apes and not human ancestors.

In the absence of a clear fossil record, we can only speculate on the nature of our protohominid ancestors. We know from the australopithecines that bipedalism and a change in tooth structure were the first human features to evolve. What selective forces would promote these? Dental changes indicate a change in diet to the type of food found in open woodlands and savannahs. Our ancestors probably were leaving the forest some of the time to forage in this new habitat on lizards, eggs, small rodents, occasionally a dead antelope, and on tubers and other tough plant food. As with contemporary apes, they may have picked up sticks to help dig up or catch food. Very likely they carried some of their food back to sheltered areas to eat, for, as Helen Fisher has pointed out, "He who lingers in the grasslands provides someone else with dinner." Many anthropologists believe that bipedalism evolved as an efficient way to carry food and probably also the infants. A foot becoming adapted to bipedalism would also be losing its capability for grasping. Infants could no longer cling to their mothers and would have to be carried.

Sexual changes, of course, leave no fossil record, but it is quite possible that changes in reproductive habits began to occur as well. A more dependent infant would require more maternal attention. Among contemporary apes, males occasionally bring food to the females, especially those in heat. A mother protohominid with an infant in arms would need help, and males may have provided food. More frequent sexual access to the female would promote more male attention and make it easier for a female to raise young.

A change in habitat and diet, acquisition of bipedalism, and changes in reproduction may be linked. But Owen Lovejoy believes that the sexual changes and bipedalism evolved as a strategy that increased infant survival while the protohominids were still primarily forest dwelling. The acquisition of bipedalism in the context of providing the female and young with food then made it possible for the protohominids to move into and exploit open woodlands. Changes in diet and dentition followed.

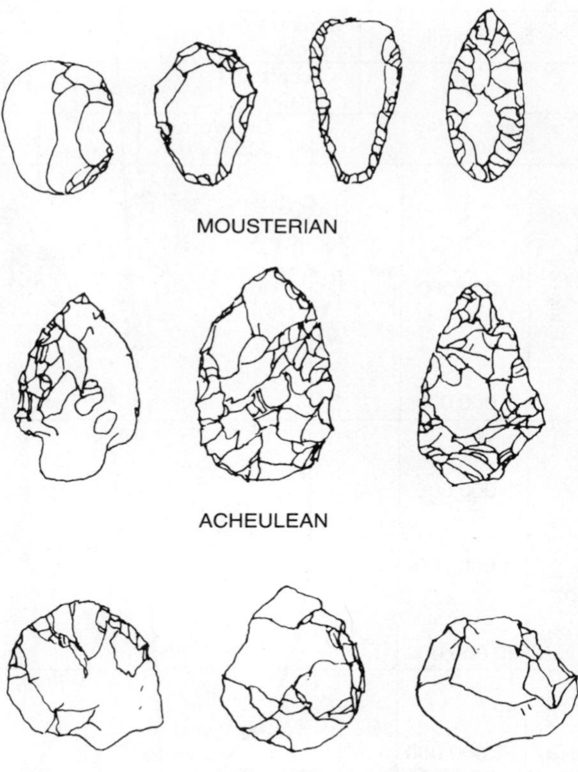

FIGURE 35.9 Paleolithic tool kits. From bottom to top, these three sets of tools illustrate major stages in the early refinement of stoneworking. In the Olduwan culture, cutting tools were made by knocking off large flakes from the surface of a stone. Further shaping and refinements of the cutting edge were done in the Acheulean and Mousterian cultures by chipping away small flakes.

Early Homo

Homo Habilis

One of these scenarios very likely led to the separation of the protohominids from the apes and to the evolution of the australopithecines. We must await the discovery of more fossils before this period of our history becomes clear.

The australopithecines were still small-brained, and there is no evidence that they made stone tools. In 1961 Louis Leakey discovered at Olduvai Gorge in Tanzania the partial skull of a hominid with a cranial capacity of 700 cc., 200 cc. larger than that of most of the australopithecines. Stone tools were associated with the remains, so Leakey named his discovery *Homo habilis* (L. *homo,* man + *habilis,* handy). The tools, primitive choppers of the Olduwan culture, were little more than rounded stones sharpened by breaking off one end (Fig. 35.9), yet cut marks on associated animal bones prove that they were used to butcher. Subsequent discoveries, one made by Louis's son, Richard, show that *Homo habilis* had a cranial capacity ranging from 600 to 800 cc. (Fig. 35.10). Larger brain size and the presence of tools indicate that evolution was beginning to favor an increase in brain complexity and in hunting skills.

Growth in the brain and improvement in tools, trends that dominate later human evolution, also have

FIGURE 35.10 Lateral view of skull KNM-ER 1470 from deposits two million years old at Lake Turkana, Kenya, discovered by Richard E. F. Leakey. This skull is now assigned to *Homo habilis*. (From Leakey, R. E. F.: Evidence for an advanced Plio-Pleistocene hominid from East Rudolf, Kenya. *Nature, 242*:447–450, 1973.)

developmental and reproductive implications. For a brain to enlarge, its growth must accelerate relative to other parts of the body. This could pose problems at birth because head size would approach or exceed the size of the pelvic canal. This is avoided by giving birth at a relatively immature age. Human infants are less developed than those of apes and are far more dependent on parental care. More parental care, in turn, requires greater cooperation between males and females, pair bonding, family structure, and so forth. More sophisticated tools also imply a more complex social structure and more efficient ways of communicating. All of these changes must have been under way at the time of *Homo habilis*.

The oldest remains of *Homo habilis* are dated at two million years, but the species may be older for Olduwan tools have been found in deposits 2.5 million years old. *Homo habilis* lived until 1.5 million years ago, so was a contemporary of *Australopithecus robustus*. *Australopithecus africanus* is old enough and has the appropriate structure to have been the ancestor of *Homo habilis* as well as of *Australopithecus robustus*. Many believe this, but the Leakeys feel that *Homo* has had a far longer independent history (Fig. 35.8). More fossil discoveries are needed to settle this question.

Homo Erectus

Long before the discovery of the australopithecines and *Homo habilis*, Haeckel in Germany had postulated a missing link between apes and modern human beings. This stirred the imagination of the Dutchman, Eugene Dubois, who searched diligently for the missing link in Java, finally discovering in 1894 fossils of a primitive hominid he called *Pithecanthropus erectus*. Shortly afterwards, Davidson Black discovered the remains of similar creatures in caves near Peking, which he named *Sinanthropus pekinensis*. Other fossils of similar hominids have been found in the East Indies, China, Africa, and Europe and have been given a variety of names (Algerian man, Heidelberg man), but physical anthropologists regard all of them as representing a single, widespread species now called *Homo erectus*.

Homo erectus stood upright and was probably about 1.7 m. tall. Brain size ranged from 850 to 1300 cc., which approaches that of *Homo habilis* on one extreme and overlaps that of modern humans on the other. Frontal areas of the brain, however, were poorly developed, for *Homo erectus* had a low sloping forehead (Fig. 35.11*A*). His face was somewhat brutish, protruding slightly, with rather heavy brow ridges and no chin. Teeth, though large, were essentially modern in their configuration.

The oldest specimens of *Homo erectus* have been discovered in Olduvai Gorge in deposits lying above those containing *Homo habilis*. These go back in time 1.5 million years. The youngest remains so far discovered are 400,000 years old. During this period *Homo erectus* developed a sophisticated culture. He made a variety of Acheulean stone tools (named from St. Acheul in France where they were first found). Large, pear-shaped hand axes used for chopping and pounding were fashioned from a piece of flint or other fine-grain stone by chipping away the edges (Fig. 35.9). A variety of smaller scraping and boring tools were made from flakes of stone chipped from the larger pieces. Charred bones indicated the use of fire. Fire, and perhaps crude

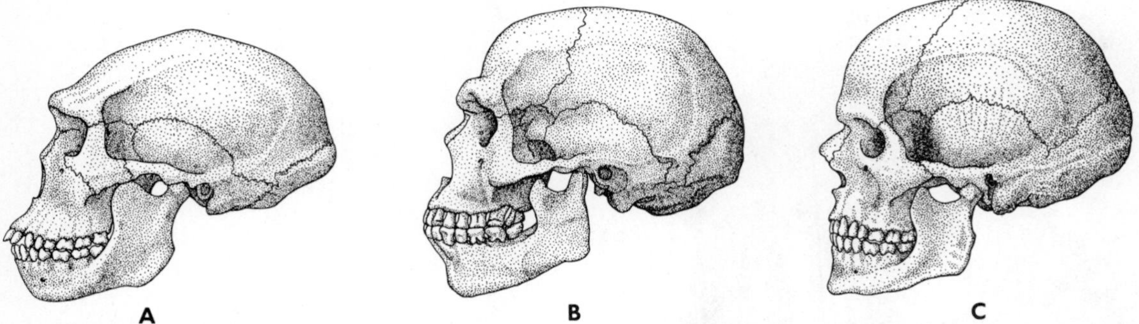

FIGURE 35.11 Lateral view of the skull and lower jaw of *A, Homo erectus* (Java); *B, Homo sapiens* (Neanderthal); *C, Homo sapiens* (Cro-Magnon). (From T. H. Eaton, Jr.: *Evolution.* Copyright © 1970 by W. W. Norton & Co., Inc.)

hide clothing, would have been essential for this hominid to penetrate central Europe and Asia in the Pleistocene, when continental glaciers were advancing. It is clear that *Homo erectus* had become a hunter of large game, for campsites contain the bones of bears, horses, and even elephants. Remains from a hunting site in Spain show that brush fires were set to drive game into an ambush. These were intelligent people, living in groups, with an ability to communicate and to teach the young to make tools and to hunt and with a knowledge of the seasons and habits of game (Fig. 35.12).

Homo Sapiens

Our species, *Homo sapiens* (L. *sapere,* to be wise) evolved from and replaced *Homo erectus.* Specimens 300,000 years old from Swanscombe, England, and Steinheim, Germany, indicated that the first *sapiens* were rather robust people with heavily built faces and jaws. Traces of brow ridges were present, and the forehead was somewhat sloping, but all features are within the range of variation of modern people. Acheulean tools are associated with Swanscombe man.

During most of the last glacial stage, unglaciated parts of western Europe were occupied by a stocky fellow called Neanderthal man (his remains were first discovered in the Neander valley). Originally considered to be a distinct species, Neanderthals are now regarded as an extinct race of *Homo sapiens.* These "classic" Neanderthals were powerful individuals with large brains, strong jaws, heavy brow ridges, and receding chins (Fig. 35.11*B*). They often built their fires in hearths in cave floors, and their Mousterian culture included a large tool kit of stone axes, scrapers, borers, knives, spear

FIGURE 35.12 Reconstruction of the life of *Homo erectus* living in the Pleistocene at Peking. (By permission of the trustees of the British Museum (Natural History).)

points, and saw-edged and notched tools probably used in making spear handles and other simple wooden implements. The stones show considerable secondary chipping to refine the shapes and sharpen the edges (Fig. 35.9). The Neanderthals probably had developed a belief in the supernatural and in an afterlife, for they buried their dead with food and tools.

Some physical anthropologists limit the term Neanderthal to these classic individuals of western Europe, but others use the term in a far broader sense for all humans living in Europe, Africa, and Asia during the last part of the Pleistocene. All used variants of the Mousterian culture.

Cro-Magnons replaced the classic Neanderthals in western Europe about 35,000 years ago. It is likely that Cro-Magnon evolved in other parts of Europe or the Near East and moved into western Europe as the last continental glaciers retreated. Probably Cro-Magnons killed some of the classic Neanderthals and intermarried with others. Physically, Cro-Magnons are indistinguishable from many present-day Europeans (Fig. 35.11C). This hominid developed a very sophisticated Aurignacian culture, using delicate stone, bone, and wooden tools and has left fine examples of painting in caves in France and Spain.

Mankind today can be divided into dozens of geographic races, that is, populations found in particular parts of the world and sharing certain gene frequencies and traits. Examples of races are the Ainu of northern Japan, the Nordics of northern Europe, the Eskimos of arctic America, and the American Indians. Races differ from one another not in single features but in having different frequencies of many alleles and characteristics affecting body proportions, skull shape, degree of skin pigmentation, texture of head hair, abundance of body hair, form of eyelids, thickness of lips, frequency of various blood groups, ability to taste phenylthiocarbamide, and many other anatomical and physiological traits. While certain of these differences, such as skin pigmentation, probably are adaptive, the significance of many is unknown.

On the basis of skeletal remains and the present-day differences in the characteristics and distribution of races, Carleton Coon recognizes five major racial groups: (1) the **Caucasoids,** which include the Nordic, Alpine, and Mediterranean races of Europe; the Armenoids and Dinarics of Eastern Europe, the Near East, and North Africa; and the Hindus of India; (2) the **Mongoloids,** which include the Chinese, Japanese, Ainu, Eskimos, and American Indians; (3) the **Congoids,** or Negroes and Pygmies; (4) the **Capoids,** or Bushmen and Hottentots of Africa; and (5) the **Australoids,** which include the Australian aborigines, Negritos, Tasmanians, and Papuomelanesians.

Fossil evidence indicates that early hominid evolution occurred in Africa, but by midPleistocene, hominids had spread widely through the Old World. Exactly when Mongoloids first crossed the Bering Strait from Asia to the New World is uncertain. Presumably it was during one or more of the glacial periods of the Pleistocene, for at that time much water would have been utilized in forming continental glaciers, the sea level would have been lower, and Siberia and Alaska would have been connected by at least a series of close islands. Geologic evidence indicates that parts of Siberia and Alaska were unglaciated during the ice ages, so ice would not have blocked the route. The oldest sign of people in the New World is a group of primitive tools found in Venezuela and associated with the remains of extinct mammals, such as mastodons and glyptodons (giant anteaters). These have been dated at 16,000 B. C. The earliest Indian sites in North America are in Colorado and Arizona and are dated at 8800 and 9300 B.C., respectively.

Summary

1. Lemurs, tarsiers, monkeys, apes, and human beings all show indications of arboreal adaptations by early primates: grasping hands, finger- and toenails, crushing molar teeth, reduction of the sense of smell and facial length, keen vision, and a well-developed brain.

2. Primate features are just beginning to appear in lemurs and other prosimians but are well advanced in the anthropoids: monkeys, apes, and human beings. Apes and human beings, collectively called hominoids, are larger, lack a tail, and assume more of an upright posture than monkeys.

3. Humans differ from apes in (a) being adapted for bipedalism rather than brachiation, (b) having a hand better adapted for grasping, (c) having dental and jaw features associated with an omnivorous diet, (d) having a larger and more complex brain, and (e) having sexual activity throughout the reproductive cycle, more dependent infants, and longer postnatal development.

4. Apes diverged early from the monkeys, soon after the origin of anthropoids in the Oligocene. Dryopithecines, abundant during the Miocene, may have been the last common ancestors of the great apes and humans.

5. The earliest known humans were the australopithecines of south and east Africa. The oldest species, *Australopithecus africanus,* extends back into the late Pliocene, 3.8 million years ago. These hominoids were bipeds and had a dentition similar to modern humans except for unusually large and thickly enameled molar teeth. The jaws were still massive and the brain apesized.

6. It is clear from the structure of the australopithecines that protohominids evolved bipedalism early and developed a more varied diet than apes. Bipedalism appears to be related to carrying food and, probably, more dependent infants. Changes in dentition indicate a shift from feeding on soft forest food to foraging on coarser and more varied food in open woodlands and savannahs.

7. *Homo habilis,* who lived between 2 and 1.5 million years ago, probably evolved from *Australopithecus africanus* and was a contemporary of *A. robustus.* The cranial capacity was larger (600 to 800 cc.), and this hominid made primitive Olduwan stone tools.

8. The brain continued to evolve in *Homo erectus,* who replaced *Homo habilis* in east Africa 1.5 million years ago. Cranial capacity ranged from 850 to 1300 cc., overlapping that of modern human beings. *Homo erectus* spread out of Africa, and by 400,000 years ago had reached Europe, Asia, and the East Indies. This species developed a sophisticated Acheulean stone culture, used fire, and hunted large game.

9. Our species, *Homo sapiens,* is first known from fossils in Germany and England that are 300,000 years old. During the last part of the glacial period, a variant of *Homo sapiens,* classic Neanderthal, was isolated in western Europe. As continental glaciers retreated, *Neanderthal* was replaced by Cro-Magnon man. Humans reached North America via the Bering Straits 15,000 to 20,000 years ago.

References and Selected Readings

Campbell, B. G.: *Human Evolution.* 2nd ed. Chicago, Aldine Publishing Co., 1974. An excellent account of human evolution with an emphasis on man's unique anatomical adaptations.

Clark, W. E. L.: *The Antecedents of Man.* 3rd ed. Chicago, Quadrangle Books, 1971. An excellent presentation of primates and their evolution.

Coon, C. S.: *The Origin of Races.* New York, Alfred A. Knopf, 1962. Describes contemporary human races and traces their origin back to the Middle Pleistocene; all human fossils known to date of publication are described.

Editors of Time-Life Books: *The Emergence of Man.* New York, Time-Life Books, 1972–1973. A series of authoritative and superbly illustrated books on human evolution, written by the editors of Time-Life Books in consultation with leading anthropologists. Volumes on *Life Before Man, The Missing Link (Australopithecines), The First Men (Homo erectus),* and *The Neanderthals* are particularly interesting.

Fisher, H. E.: *The Sex Contract.* Willian Morrow & Co., New York, 1982. A popular and well-done account of human evolution emphasizing probable behavioral changes.

Hay, R. L., and M. D. Leakey: The fossil footprints of Laetoli. *Scientific American, 246*(Feb.):2, 1982. An account of the remarkable human footprints found in volcanic tuff 3.5 million years old.

Johanson, D. C., and M. E. Edey: *Lucy.* New York, Simon and Schuster, 1981. The discovery of the oldest human fossil and reasons for interpreting it as a new species, *Australopithecus afarensis.*

Lovejoy, C. O.: The origin of man. *Science, 211*(Jan.):341–350, 1981. An authoritative summary of human evolution with emphasis upon the reproductive and behavioral changes that may have led to the evolution of bipedalism.

Pilbeam, D. L.: *Ascent of Man.* New York, Macmillan Publishing Co., 1972. An excellent textbook on physical anthropology.

Rensberger, B.: Facing the past. *Science, 212*(Oct.):8, 1981. New reconstructions and interpretations of Neanderthal suggest he was not the brute previously thought.

Tobias, P. V.: *The Brain in Hominid Evolution.* New York, Columbia University Press, 1971. A thorough review of brain evolution and its relation to cultural development.

PART FIVE Animals and Their Environment

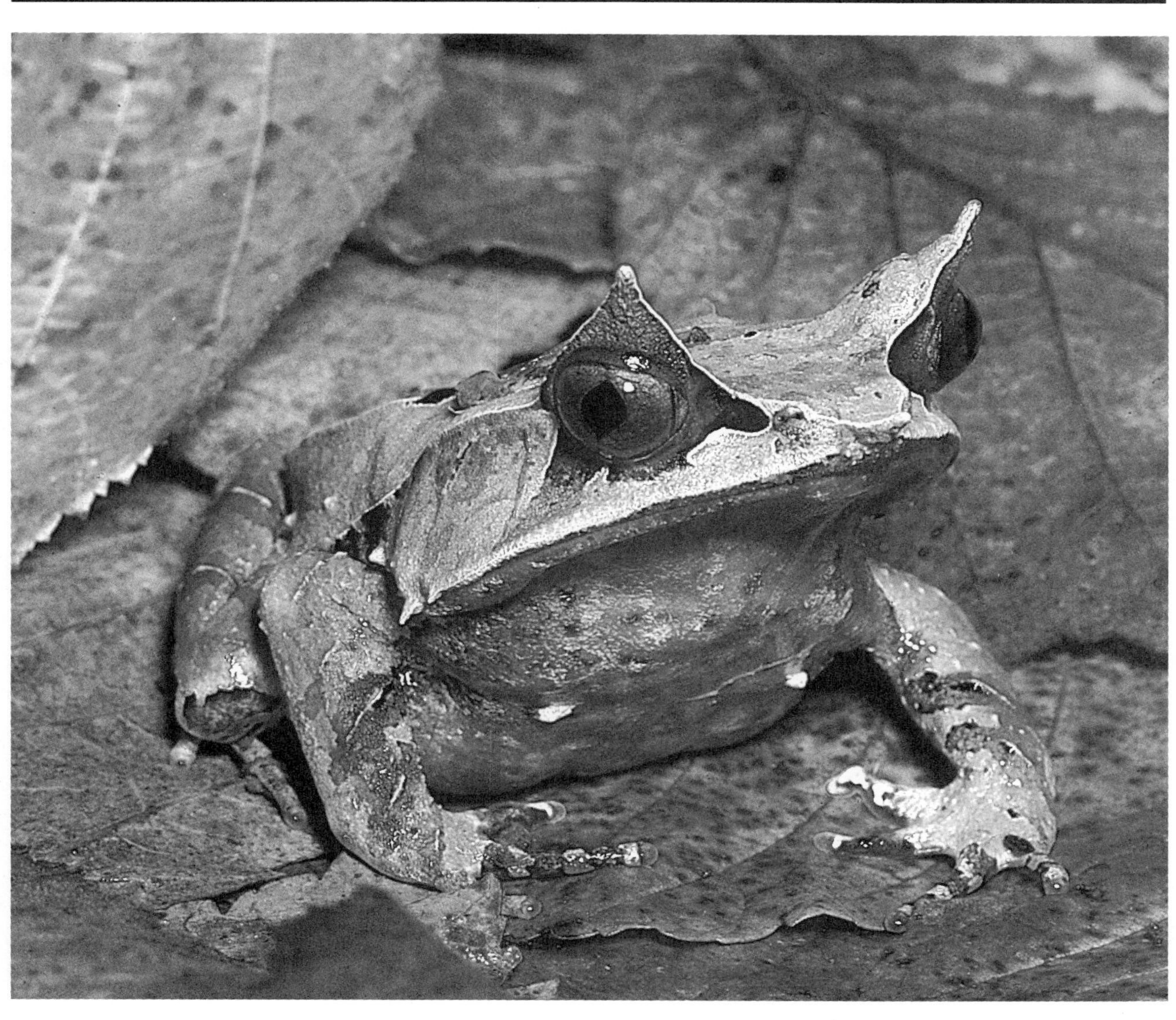

36 Behavior

ORIENTATION

The responses of an animal to its environment are not random but are highly organized and intricate patterns of activity that we call behavior. In this chapter we will examine the role of the nervous and endocrine systems in enabling an animal to recognize signs and in generating the motor responses that are released. We will then explore the way instincts and learned behavior develop and the genetic basis of that development. Finally, we will examine the relationship between behavioral patterns and the ecological contexts in which animals live. Among the topics considered are the adaptive significance of agonistic behavior, territoriality, sexual behavior, communication, social behavior, feeding and defensive behavior, and orientation.

Multicellular plants and animals are the conspicuous members of the world of living organisms, but they reflect their living condition in very different ways. The living state of a plant is most readily seen in daily or seasonal growth, such as the development of new shoots, flowering, and fruiting. The living state of animals is easily recognized by body movements that result from the response of effectors, especially muscle cells.

An animal's effector responses are not random but are highly organized and integrated patterns of activity that enable the species to interact with its environment and to meet the physiological demands of its existence. Such patterns of activity are called behavior, and the study of behavior is termed **ethology** (Gr. *ethos*, custom + *logos*, study). An animal's repertory of behavior patterns may be relatively simple, such as the feeding responses of a sea anemone, or complex, such as the homing of birds. Whether the pattern is simple or complex, it fits the needs and life style of the animal.

The purpose of this chapter is not to describe animal behavior, although we will provide many examples. Rather, our task here is to examine the generalizations that have been established from comparative studies and to explore those aspects of behavior that are most commonly encountered in animals.

Most animal behavior is a sequence of responses related to some need of the species, such as, getting food, acquiring and maintaining living space, protection, and reproduction. For example, certain species of shrimp, called cleaning shrimp, pluck parasites and dead tissue from the bodies of certain fish. The shrimp may even extend its cheliped beneath the fish's operculum to remove parasites from the gills. The behavior benefits both parties of this symbiotic relationship. The shrimp gains food, and the fish is rid of harmful parasites. Some species of cleaning shrimp (*Periclimenes*) have cleaning stations to which "client" fish come. The shrimp signals an approaching fish by waving its long antennae and rocking its body. The fish comes closer to the shrimp and strikes a characteristic pose. The shrimp then strokes the fish with its antennae, climbs aboard, and proceeds to clean it. A precise sequence of behavior is characteristic of the species and is performed in the same manner by all members of the population throughout its geographical range. The steps are depicted in Fig. 36.1 as a diagram called an **ethogram.**

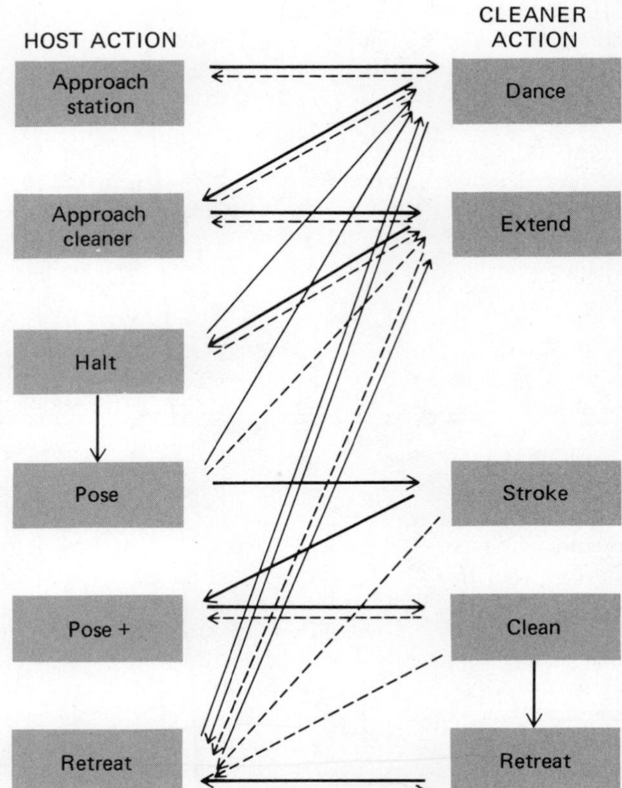

FIGURE 36.1 An ethogram of the sequence of behavior in the interaction between a host fish and the cleaning shrimp *Periclimenes anthophilus.* Arrow from Approach station to Dance, for example, means that the approach of the fish to the cleaning station initiated dancing behavior in the shrimp. The heaviest arrows indicate the most frequently observed patterns. (Modified from Sargent, R. C., and G. E. Wagenbach: Cleaning behavior of the shrimp *Periclimenes anthophilus. Bulletin of Marine Science,* 25(4):466–472.)

The behavioral patterns of either the fish or the shrimp could be studied in a number of ways. They might be analyzed to determine the components that comprise the entire sequence, the specific stimuli that trigger responses, or whether the responses require certain physiological conditions in order to be triggered. An ethologist might ask how fixed or automatic the patterns are: Can they be altered or modified by environmental conditions? How is the pattern determined by the nervous and hormonal integrating systems of the animal? How do they arise in the course of development? How is one part of the pattern related to another? For example, why does the fish pose? All of these questions are sometimes called proximate questions because they deal with causes or are concerned with the way the pattern develops in the life of the animal.

But there are two equally important questions that might be asked: What is the adaptive or ecological significance of the behavior, and how did it evolve? These are sometimes called ultimate questions. For the particular example we have given, the adaptive significance is not difficult to understand, but the steps that might have led to its evolution are more difficult to surmise.

Thus, behavior can be examined from four perspectives: (1) **causation**: the causes of behavior; (2) **development**: the formation of the behavior in the development of the individual; (3) **evolution**: the origin of the behavior in the evolution of the species or group; and (4) **behavioral ecology (function)**: the adaptive significance of the behavior in the life of the animal. The first two are proximate perspectives, and the last two are ultimate ones.

In this chapter we will explore behavior from all four of these perspectives, but in the history of ethology they have not received equal attention. Most of the early studies of animal behavior were carried out during the first half of the twentieth century by physiologists and psychologists and were concerned with causes and development of behavior. There was considerable experimental work with animals, especially laboratory rats. However, such studies treated laboratory animals largely as substitutes for humans, and there was little concern about how the behavior might relate to the existence of that species in the wild. The best known animal psychologists of that period were the Americans J. B. Watson and B. F. Skinner. They believed that most behavior was a consequence of learning, and the school of thought that developed around their ideas became known as Behaviorism and had great influence, especially in psychology. Contemporary comparative psychologists have a broader perspective of behavior and investigate a wider range of animal species, but they continue to focus largely on causes and development of behavior.

The German zoologist Konrad Lorenz laid the foundations for modern ethology and coined the term ethology. Lorenz was interested not only in causes and development of behavior but also in how they function and evolved in the species. It is these two latter concerns that distinguish ethology from comparative psychology. The Dutch zoologist Nikko Tinbergen brought experimental design into ethological research, and many of his studies, as well as those of Lorenz, are now classics in the literature on behavior.

Causation

Elements of Behavior. The waving of the cleaning shrimp's antennae signals the fish to perform a "ready to be cleaned" posture. A male jumping spider responds to the visual presence of the female by "dancing" before her, elevating certain legs bearing colored hairs, and tilting his abdomen. A model cut out of cardboard will elicit the dancing response provided the dimensions are no greater than five times the actual frontal area of the spider and no smaller than one third the area. A shiny band (a foreign object) around the leg of a baby bird will evoke the parents of some species to push the band out of the nest, even though the baby goes with it.

In each of these behavioral patterns, a **sign stimulus**—waving antennae, female frontal area, metal band—acts to release a specific behavioral response, or **motor program.** The sign stimulus, or **releaser,** is the essential stimulus that sets off an **innate releasing mechanism.** The motor programs in these examples are very rigid behavioral patterns—**fixed action patterns.**

A classic example of a releasing mechanism is egg rolling in geese, which was studied by Lorenz and Tinbergen and led to the initial understanding of such behavioral elements. A nesting goose, on observing a nearby displaced egg, will use its beak to roll it back into the nest (Fig. 36.2). Lorenz and Tinbergen discovered that a goose will roll all sorts of inappropriate objects back into the nest, provided that the objects exhibit certain appropriate sizes and shapes. Thus, round objects of a certain size act as a visual stimulus (the sign signal) to trigger an innate releasing mechanism that initiates the egg rolling motor pattern.

Motivation and **drive** are two terms long used to describe behavior. Motivation is a state of responsiveness that is reflected in a drive, or need. The goose's drive to return the wayward eggs back into the nest is only manifested during the nesting period. Only during this period of responsiveness is the goose said to be motivated for egg rolling. Motivation is, therefore, a short-term responsiveness that can appear or disappear. Moreover, the degree of motivation or threshold level of the releas-

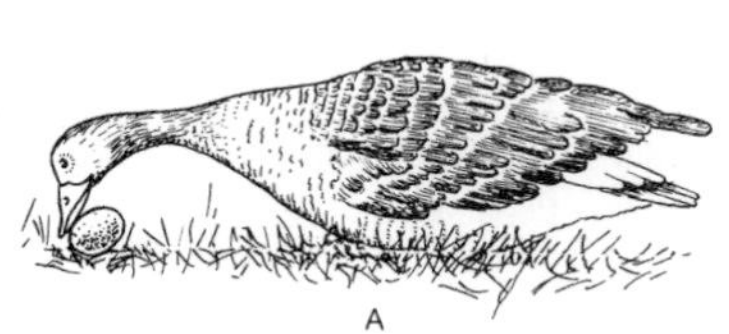

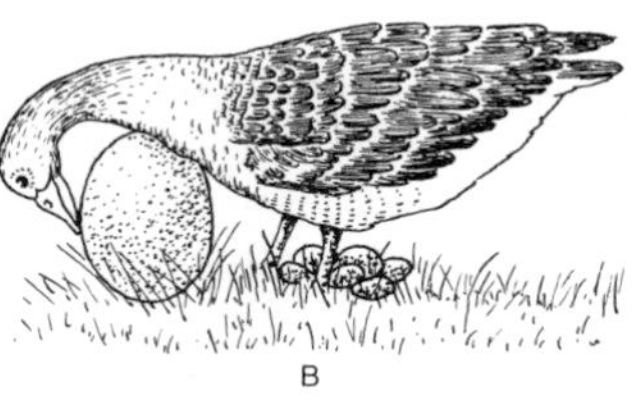

FIGURE 36.2 *A,* Greylag goose retrieving an egg that is outside the nest. This movement is very stereotyped in form and used by many groundnesting birds. *B,* The goose attempts to retrieve a giant egg in precisely the same fashion. (From Lorenz, K. Z., and N. Tinbergen: *Z. Tierpsychol.* 2, 1, 1938.)

ing mechanism can vary over time and as a result of continual stimulation. A goose will rake beer cans into its nest but not an endless number of beer cans.

The concepts of motivation and drive are viewed with less favor by modern ethologists than they were in the past because the concepts are probably not uniformly applicable as explanations of behavior. For example, such widely recognized drives as hunger and thirst are mechanisms that ensure physiological homeostasis. When the animal drinks, the need for water is fulfilled, internal homeostasis is restored, and the thirst motivation and drive to drink are abated or lost. Clearly, egg rolling in a goose is not quite the same sort of behavior. The goose is not fulfilling a need by rolling eggs, nor is the motivation reduced by rolling more eggs.

Physiology of Behavior. The processes of the nervous and hormonal systems of an animal are the physiological basis of behavior. Although this has long been recognized, it has been difficult to make precise correlations between behavioral patterns, i.e., motor programs (not reflexes) and the firing of specific neurons. However, in recent years such correlations have been demonstrated in the nervous systems of leeches, crayfish, sea hares, and sea slugs. These animals have relatively few centrally located neurons that are large enough to be easily dissected and probed with electrodes.

The sea slug *Tritonia diomedia,* a shell-less marine snail on the West Coast of the United States, can escape predators (such as certain sunstars) by a burst of swimming activity (Fig. 36.3). Arthur Willows at the University of Washington discovered the neuronal circuits that govern the swimming behavior of *Tritonia* by careful investigation of individual neurons. He first dissected the central ganglia of the sea slug nervous system and mapped the larger neurons. He then electrically stimulated each of these neurons with fine probes and recorded not only their action potential but the specific motor responses each neuron evoked. He found that certain neurons governed specific turning movements or contractions on one side of the body; others could produce movement simultaneously on both sides of the body; and still others could bring about the swimming behavior in its entirety. Moreover, the swimming lasts some 30 seconds, long after the command neurons have ceased to fire. The command capability is made possible by junctions with other neurons along the length of the command neuron. Thus, an impulse generated in the command neuron is transmitted to all of the motor neurons involved in the swimming response and in the proper sequence. Apparently, reverberating (self-stimulating) circuitry will maintain the swimming behavior after the initial command has ceased. Under natural conditions, contact with a predator sunstar is the stimulus that generates the escape swimming command.

Neuronal circuitry and the physiological regulation of neuronal junctions—threshold requirements, facilitation, spatial and temporal summation, inhibition (see p.

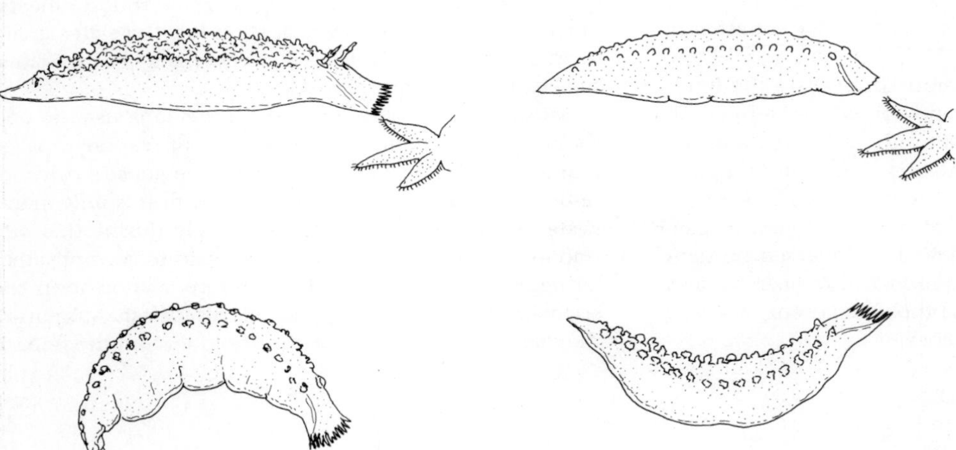

FIGURE 36.3 Escape response of the sea slug *Tritonia diomedia.* On contact with a predatory sunstar, the sea slug withdraws, contracts its dorsal gills and undergoes dorsoventral flexions. The flexions enable the animal to swim some distance away from the predator. (From Willows, A. O. D.: Giant brain cells in mollusks. *Scientific American, 224*(Feb.):69–75, 1971.)

220)—are the underlying causes of behavior, but each animal has a specific system adapted for its own peculiarities of life style. This applies not only to motor circuits, such as those described earlier, but also to sensory neurons delivering data from outlying receptors. Many kinds of stimuli continually bombard an animal; only a few (those important for the species) are selected for internal reporting. To a large extent selection is determined simply by the specific receptors the animal possesses. The specific receptors act as a sort of filter that records some stimuli and ignores others. However, a signal important for one animal is not necessarily important for another.

The female silkworm, *Bombyx mori,* when receptive to mating, emits a very volatile organic substance called bombykol that attracts males of the same species. The male can detect very low concentrations of bombykol and thus can sense the presence of a female from a considerable distance. The receptors are located on the antennae, and as many as 70 percent are specific for just this one compound. Of the many volatile molecules wafted on night breezes, including similar substances released by females of other species of moths, the male silkworm is tuned only to bombykol. Others do not get through the sensory filter. One molecule of bombykol may fire a receptor that may in turn cause the male to fly upwind. As the concentration of bombykol rises, more antennal receptors are activated. Eventually, the male homes in on the target female. She mates only once in her life and with the first male to reach her.

The sensory filter of web-building spiders selects a very different kind of signal. Slit sense organs located in the joint between the last two sections of the legs are very sensitive to vibrations transmitted through the web, and such vibrations are the principal source of information for the spider about its surrounding world. The orb-weaver *Argiope argentata* sits in the hub of its web (Fig. 36.4). It is alerted by vibrations of an ensnared insect and may pluck the web to determine the prey load and position. The spider then runs out to the prey. Large prey is given a long bite and then wrapped; small prey is wrapped and given a short bite. If the prey is very small, it is simply seized and then wrapped at the hub. After wrapping, large prey is cut out and carried back to the hub and rewrapped prior to feeding. When a male orb-web spider enters the web, his vibrations can be distinguished by the female from those of prey.

Hormones affect behavior by controlling or setting the physiological stage required for neuronal processes and motor patterns to take place. Hormones may also have organizational effects, such as the determination of much of the structural and physiological course of sexual development in mammals and other animals.

Hormonal and nervous processes, behavioral responses, and environmental stimuli can all interact to determine the sequence of behavior for a given species. Such interaction is illustrated by the reproductive biology of ring doves (Fig. 36.5). Under the influence of androgens, male ring doves court females; the courting behavior stimulates the female pituitary (via visual stimulation) to secrete follicle stimulating hormone (FSH). FSH causes the development of ovarian follicles that secrete estrogen. The pair soon begin nest building and copulate. Secretion of progesterone, either as a consequence of ovulation or perhaps as a consequence of nest building stimuli, induces incubation behavior when the eggs are laid. Behavioral interactions continue in this manner through care of the young. Note that hormones can evoke behavior, that behavior can stimulate secretion of hormones, and that other kinds of environmental cues, such as the nest or eggs, can evoke behavior or stimulate hormone secretion.

The interaction of hormonal, neural, and environmental cues plays a primary role in controlling the timing of behavioral activity and the shifting of priorities in the life of the species. White-crowned sparrows are subjected to increasing day length after December in their southern winter range (Fig. 36.6). The increasing photoperiod causes the hypothalamus (via the eyes and some neural clock mechanism in the brain) to secrete pituitary releasing hormones. The ovaries and testes have greatly atrophied prior to the migratory trip south, and the pituitary gonadotropins initiate their redevelopment. Feeding now increases, and the birds begin to deposit fat in preparation for the energy cost of the spring migration. The activity level continues to rise, and when a threshold level is reached the birds fly north to the summer range, where the reproductive phase of their life takes place.

Biological Rhythms and Clocks. It is to an organism's advantage that control mechanisms evolve that regulate its metabolic processes and behavior so that there is a synchronization between them and cyclic changes in the external environment. These control mechanisms are cyclic so that metabolic processes and behavior change at appropriate repetitive intervals, which may range from a day or less to a year or more. If an animal's food organisms, for example, are most plentiful in the early morning, its cycle of activity must be regulated so that the organism becomes active shortly before dawn, even though dawn changes slightly from day to day. As the adage says, "The early bird catches the worm." Periods of activity and sleep, feeding and drinking, body temperature, and many other pro-

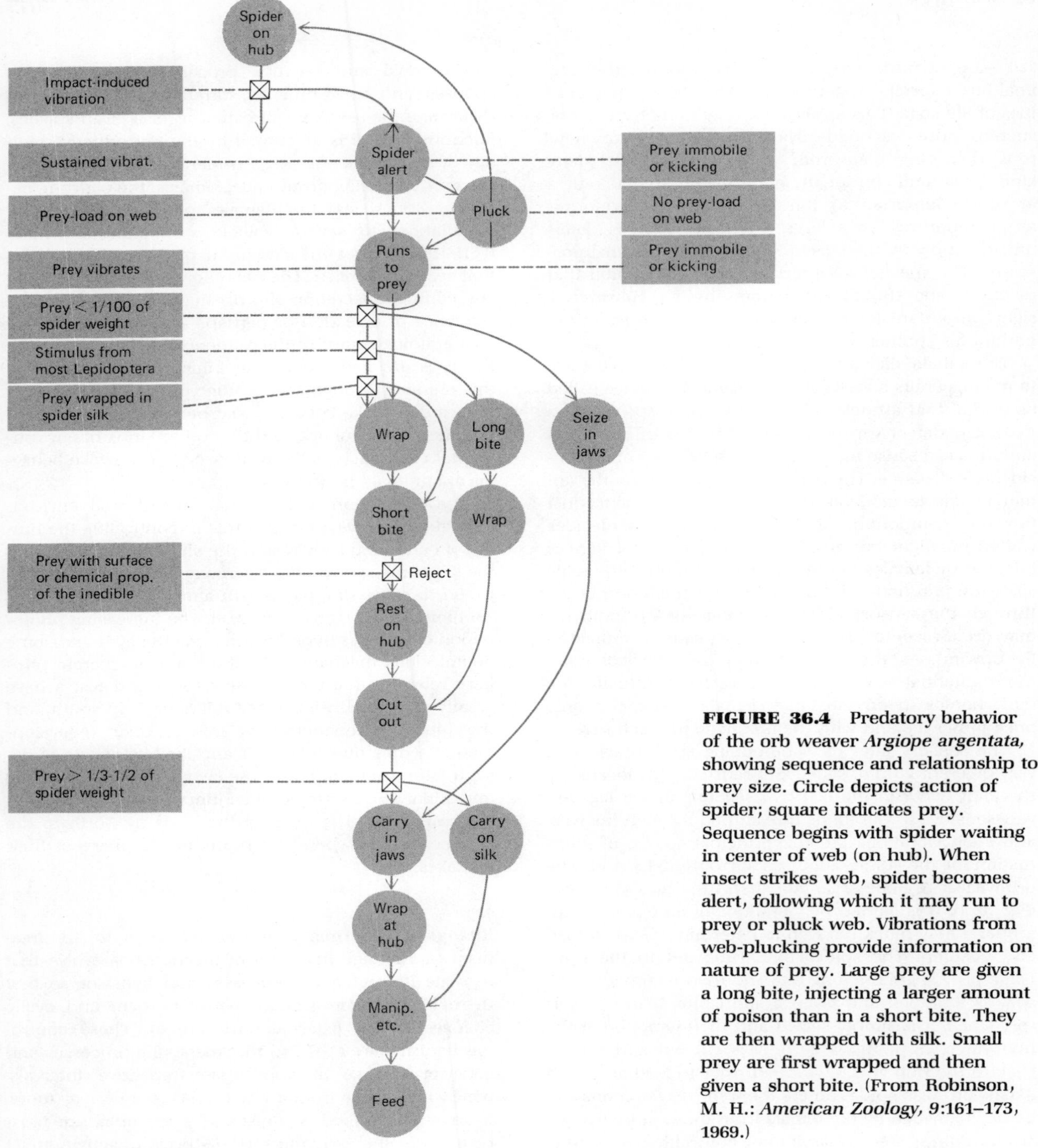

FIGURE 36.4 Predatory behavior of the orb weaver *Argiope argentata,* showing sequence and relationship to prey size. Circle depicts action of spider; square indicates prey. Sequence begins with spider waiting in center of web (on hub). When insect strikes web, spider becomes alert, following which it may run to prey or pluck web. Vibrations from web-plucking provide information on nature of prey. Large prey are given a long bite, injecting a larger amount of poison than in a short bite. They are then wrapped with silk. Small prey are first wrapped and then given a short bite. (From Robinson, M. H.: *American Zoology, 9*:161–173, 1969.)

cesses have a cycle that is approximately 24 hours long, hence they are called **circadian rhythms** (L. *circa,* approximately + *dies,* day). Some other cycles are shorter and many are longer. For example, many marine animals that live along the shore are tied to tidal cycles. Fiddler crabs on the eastern coast of the United States emerge from their burrows to feed at each low tide (twice every 24 hours). Certain intertidal snails, on the other hand, release their eggs only at the very high spring tides, which occur twice each month. The men-

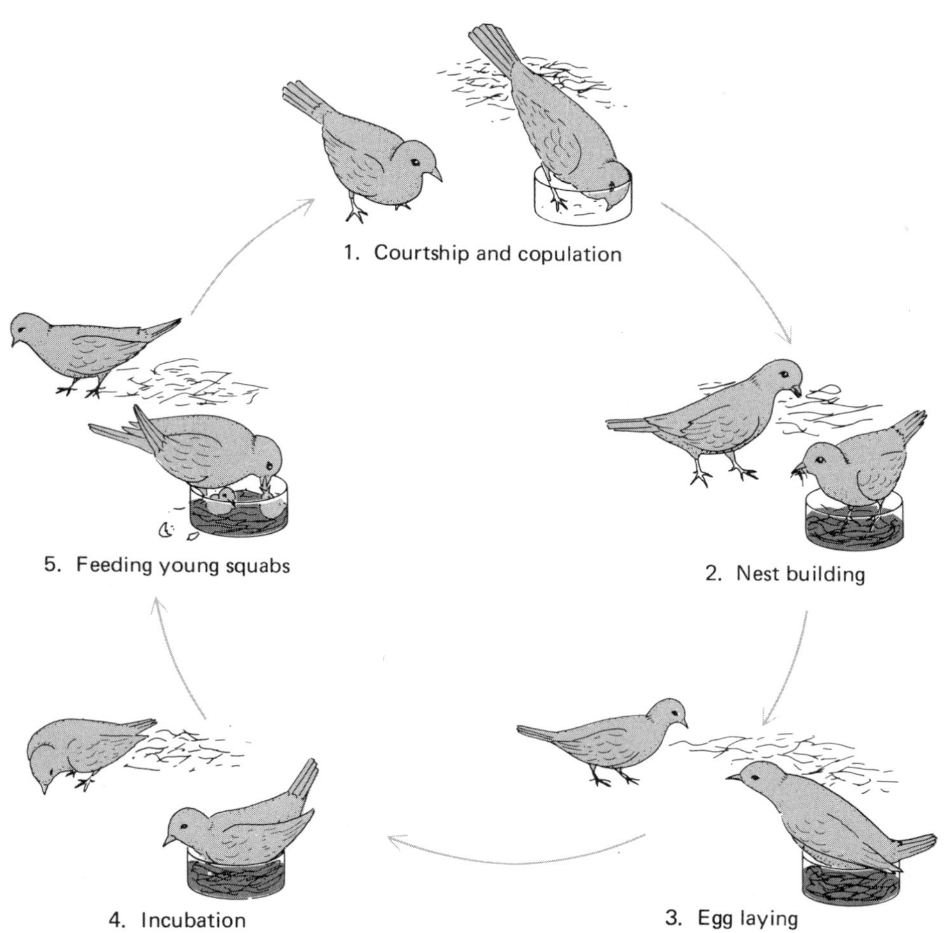

FIGURE 36.5 Cycle of reproductive behavior in the ring dove, in which behavior results from an interaction of hormones and external environmental cues, including the behavior of the mate. See the text for an explanation. (From Drickamer, L. C., and S. H. Vessey: *Animal Behavior,* copyright 1982 by PWS Publishers. All rights reserved. Used by permission of Willard Grant Press.)

strual cycle of women is approximately 28 days, or a lunar month. Many animals migrate to and from breeding grounds twice a year, and many have annual cycles in reproduction, hibernation, and so forth. We will examine circadian rhythms as examples of cyclic phenomena because they have been found in virtually all organisms and many play an important role in behavior.

The existence of circadian rhythms has long been known, but that they were dependent on something more than the obvious cycle of night and day was not recognized until 1729 when the French astronomer, de Mairan, reported that the sensitive plant (*Mimosa*) continued to open and fold up its leaves in a 24-hour cycle even when placed in constant darkness. Many investigators have argued that, although not controlled directly by light cycles, circadian rhythms must be **exogenous** and be controlled by some subtle environmental rhythm that the organisms detect. Others have argued that circadian rhythms are **endogenous** and are regulated by an internal biological clock capable of detecting the passage of time. The endogenous hypothesis received considerable support in the 1950s through the investigations of Pittendrigh, Renner, and others. Renner trained honey bees to collect sugar water in his Paris laboratory between 8:15 and 10:15 in the evening. He then transported his bees on a flight to New York and found that they continued to collect sugar water between 8:15 and 10:15 P.M. Paris time! Environmental influences could not play a role in this case. Further support for the endogenous hypothesis has come in recent years from studies in many organisms that show that the biological clock, whatever it may be, has a genetic basis. Normal fruit flies, *Drosophila,* have a clock that has a *free running period* of 24.2 hours. The free running period is the clock's repetitive cycle when the animals are isolated from environmental cycles and kept under constant

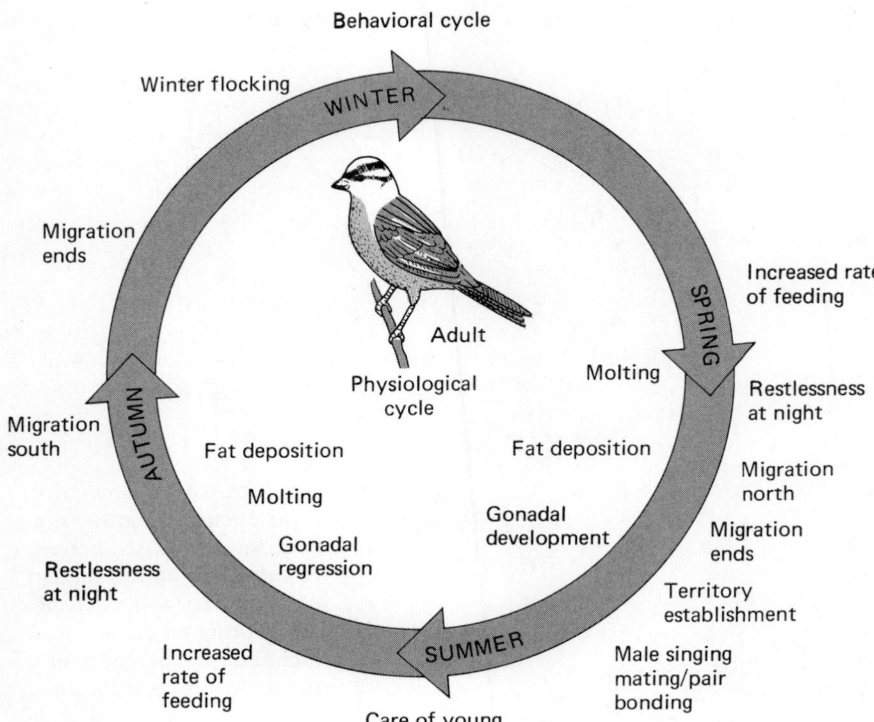

FIGURE 36.6 Seasonal changes in the physiology and behavior of the white-crowned sparrow. (From Alcock, J.: *Animal Behavior: An Evolutionary Approach.* 2nd ed. Sunderland, MA, Sinauer Associates, 1979.)

conditions. Mutants have been discovered with free running periods of 19 and 28 hours. Each has been traced to the same locus on the X chromosome.

Most biological clocks are free running for they continue at a certain period when the organism is isolated from environmental influences. Exposure of an organism to the 24-hour cyclic changes in light and to other environmental factors brings the period of the internal clock into phase with the external clock. The internal clock is said to be set, or **entrained.** An environmental factor capable of entraining the biological clock has been called a **zeitgeber** (German, time giver). Many environmental factors have been shown to be capable of acting as a zeitgeber, but light, temperature, and (for some marine organisms) tides are particularly important ones. Once the clock is set, it will continue to run on the set time for a while, even when environmental conditions are abruptly changed, as in transporting bees from Paris to New York, but as time goes on it gradually becomes reset and comes into phase with the new conditions. Normally the internal clock is adjusted slightly day by day as day length and other factors change. When the organism is isolated from external cycles and kept under constant conditions, the internal clock slowly drifts to its own free running period.

The nature of the internal clock is not entirely clear, but it appears to depend on one or more self-sustained biochemical or physiological oscillators analogous to such oscillators as the respiratory center in the brain or pacemaker in the heart. The complexity of the circadian timing system in mammals suggests that multiple oscillators are involved. They probably have their own periods, and they must be coupled to each other and be connected to certain sense organs so that they can be entrained to external conditions. A group of nerve cell bodies located in the hypothalamus above the optic chiasma appears to be a part of the mammalian biological clock. Lesions in this region will destroy many circadian rhythms. A neuronal pathway from the eye that terminates here has recently been discovered. There are probably other parts to the biological clock as well, because lesions in the suprachiasmatic nucleus of humans, although they stop the activity cycle, do not block circadian rhythms in the body temperature. The other parts have not been well defined, but another nucleus in the hypothalamus appears to be involved. The pineal eye or gland also plays a role in the timing systems of rats, birds, and some other vertebrates.

Development

Behavior is the organized effector responses to nerve impulses and hormonal secretions. Therefore, the development of behavior, even that which is learned, is dependent upon the embryonic development of neural and hormonal systems, which in turn are under genetic direction.

Animal psychologists and ethologists have long recognized two sorts of behavior, instinctive and learned. **Instinctive behavior** is innate, comprised of motor patterns that can be carried out correctly the first time they are evoked; **learned behavior** has been shaped by the environment. The relative prevalence and importance of instinctive behavior versus learned behavior (nature versus nurture) has been the basis of much past and present debate. The argument really centers on the way behavior develops. To what extent does behavior result from the interaction of the preprogrammed neural base with the environment? As knowledge of comparative behavior has grown, ethologists have come to recognize that most animals have some innate motor patterns (fixed action patterns) and some that result from environmental interaction. The latter, learned behavior, develops in different ways and varies in the extent to which the preprogrammed component can be modified.

Various schemes have been formulated to classify the developmental spectrum of behavior; we will follow one proposed by J. Alcock (see the references at the end of the chapter).

Closed Instincts. Closed instincts are preprogrammed, fixed motor patterns that are functional from the moment the neural circuitry is in place and are not modified by the environment. The first web of orb-weaving spiders, for example, is complete in all detail and repeatedly built in the same manner throughout the life of the spider (Box 36.1).

Open Instincts. In many species, behavior that is functional when first performed is capable of modification as a result of interaction with the environment. Herring gull chicks peck the beaks of the parents, which regurgitate partially digested food for the chick. The chick is attracted by a red spot on the beak and by the beak's shape and downward movement (sign stimulus). This "begging behavior" is sufficiently functional to get the chick its first meal, but there is much energy wasted in the pecking. Some pecks fail to reach the parent's beak and some are off target. However, the begging behavior becomes more efficient over time, and in experiments with models of the parent's beak, the chicks become increasingly selective of the shape necessary to evoke the begging response. Thus, the initial functional instinct is perfected by environmental interaction.

Open instincts may also be capable of suppression. The escape response of many animals, for example, can be reduced as the animal learns through repetition that a particular stimulus is not dangerous. The animal is said to have become **habituated** to the stimulus, but the escape response still functions for other signals.

Restricted Learning. Unlike instincts, learned behavior develops only after interaction with the environment. In restricted learning, also called programmed learning, a particular behavior pattern forms as a result of precise environmental stimuli. Two of the best studied examples of restricted learning are imprinting and bird song learning.

For many ground birds and herd mammals—chickens, ducks, geese, horses, and sheep—the ability of the young to follow the parent is critical for survival. The development of this type of behavior was investigated by Konrad Lorenz in what is now considered a classic study of ethology. In chicks and goslings, any moving object of appropriate size will evoke an innate following response. This could be very dangerous if restrictions were not imposed on what the chicks should follow. For a brief critical period (12 to 14 hours, peak time) following hatching, a chick is very sensitive to the characteristics of a moving object about it, usually the mother. After this **imprinting,** only the moving object with those particular characteristics will direct the following response. In the absence of a mother, Lorenz was able to make goslings imprint upon him as a surrogate mother, after which they would follow him. The chick must actually follow to be imprinted, and the imprinting is irreversible. It is as if the characteristics of the parent were indelibly stamped on the nervous system of the chick during a brief receptive period of development. Imprinting for the following response is not restricted to visual stimuli. Depending upon the species, it may involve sounds and odors, again usually those of the parent (Fig. 36.7).

In many species of birds, male recognition of the female of the same species is dependent upon imprinting. The critical period for sex recognition imprinting usually occurs after that for the following response. In geese these periods coincide, leading Lorenz to the mistaken conclusion that one imprinting served two functions. In some birds, such as those in which there is danger of receiving eggs or chicks of another bird in the nest because of crowded nesting conditions, the parents imprint on their own eggs shortly after laying them (or on chicks after hatching). They can then differentiate their eggs from those of an intruder.

The songs of male birds, which advertise their sexual maturity and territorial claims, are innate in some species but learned in others. Sparrows, in which song learning has been most extensively studied, must hear a song in order to reproduce it, and they are only capable of learning the song of their own species. The juvenile bird "records" the adult song during a critical period of about two months (10 to 50 days in white crown spar-

BOX 36.1

The familiar orb-webs are constructed by several families of spiders, although most of those seen in woods, meadows, and gardens are produced by the large, widespread orb-weaving family, the Araneidae. The spider begins the task of web construction by establishing a bridge line. A strand of silk released from the spinnerets is carried by air currents (*1*). When it strikes and attaches to some nearby object, the spider attaches the opposite end and may reinforce it with additional strands (*2*). Next the spider spins a vertical thread from one of the bridge lines, attaches it to a lower branch, and then tightens it to form a Y (*3–4*). The fork of the Y will become the hub of the future web, and the arms and base of the Y, the first radii. When frame threads and additional radii have been added (*5–8*), the spider lays down a scaffolding spiral of dry threads from the inside out (*9–10*). Finally, starting at the outside, the spider removes the temporary spiral scaffold (eating it) and simultaneously replaces it with a permanent spiral of sticky thread (*11–12*). (From Levi, H. W.: Orb-weaving spiders and their webs. *American Scientist, 66*:734–742, 1978.)

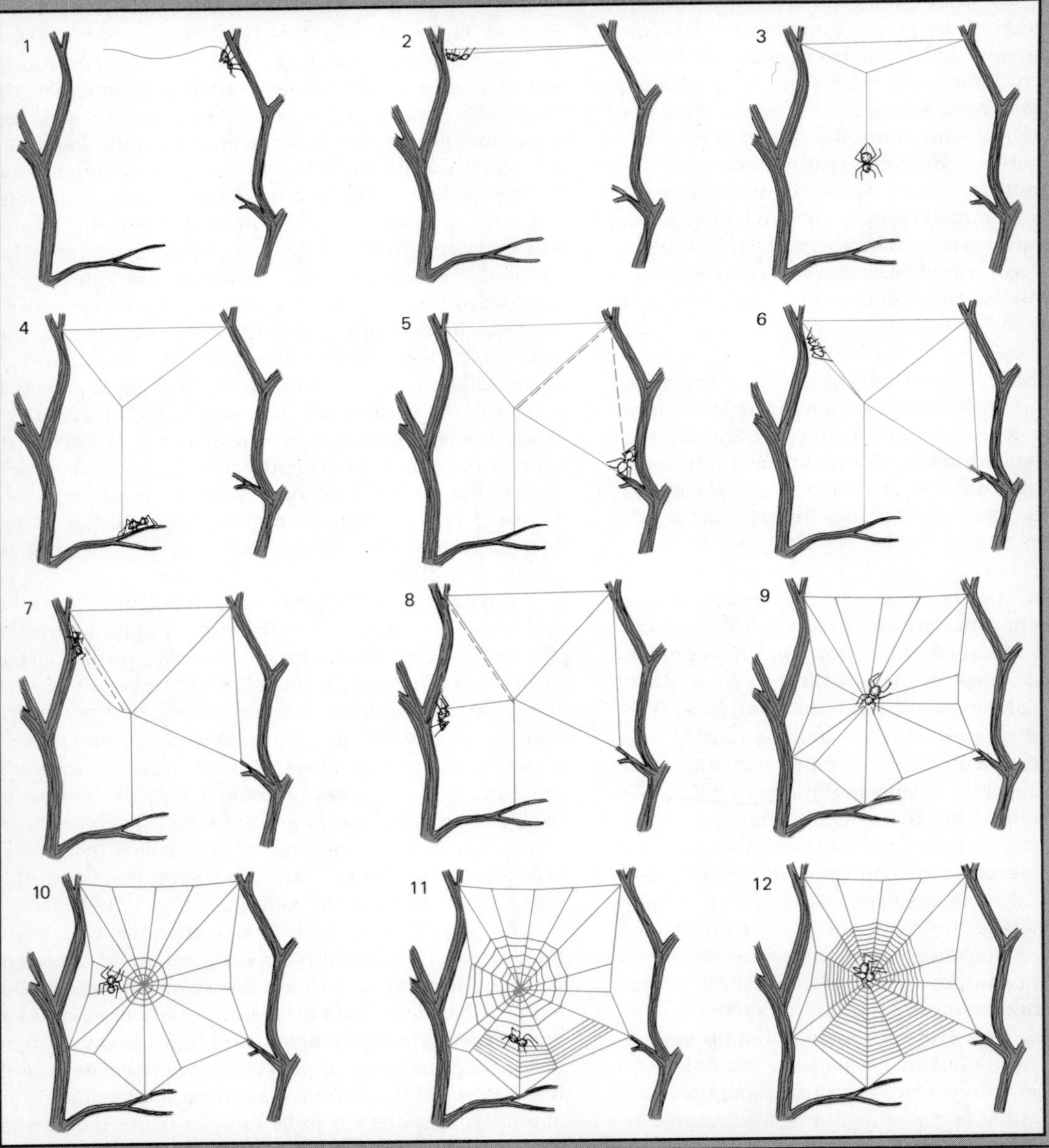

FIGURE 36.7 The young of the European shrew recognize their mother's odor and follow her as a result of early olfactory imprinting. (From Alcock, J.: *Animal Behavior: An Evolutionary Approach.* 2nd ed. Sunderland, MA, Sinauer Associates, 1979.)

rows). During this time, the nervous system of young birds is primed to "memorize" the call. Certain qualities of the call trigger the memorization process, and out of the many surrounding calls the young birds encounter, only the call of its own species is selected for recording.

When the young males begin to sing some four months later, the song is at first just twitterings, called subsong; gradually the song becomes matched with the brain's recording, and the correct song of the species is reproduced (Fig. 36.8). In white crowned sparrows, the perfected song occurs by about seven months. Given the nature of this learning process, the results of experimental studies are predictable. Hand-reared sparrows kept in isolation cannot reproduce the species call (Fig. 36.6), but those isolated after the early critical period can. Birds deafened after the critical period cannot perfect the song (Fig. 36.8), because they cannot hear themselves sing, which seems to be necessary for matching the "memory" of the song.

Flexible Learning. Some behavior of many animals undergoes continual modification as a result of interaction with the environment, and there may be a wide range of environmental data that can be learned. Flexible learning is perhaps best illustrated by the learning associated with exploration around a "home site" and with the development of food preferences.

Many animals that have dens, burrows, or nests learn the location of their home and important landmarks in the surrounding area. Deer mice (*Peromyscus*) have an extensive system of runways leading out from the nest. Under experimental conditions, they are capable of learning every detail of a maze containing the equivalent of one quarter mile of runways in no more than three days. Some of the territorial information the mouse learns is of immediate importance, such as how to return from a foraging trip. Other information may be important at a later time, as in darting beneath a log to escape a predator. This utilization of unrelated stored information for a new experience is called **latent learning.** Deer mice kept in a room with logs and other objects were much more successful in avoiding the clutches of an owl released into the room than were another set of deer mice raised in laboratory cages and then placed in the habitat room.

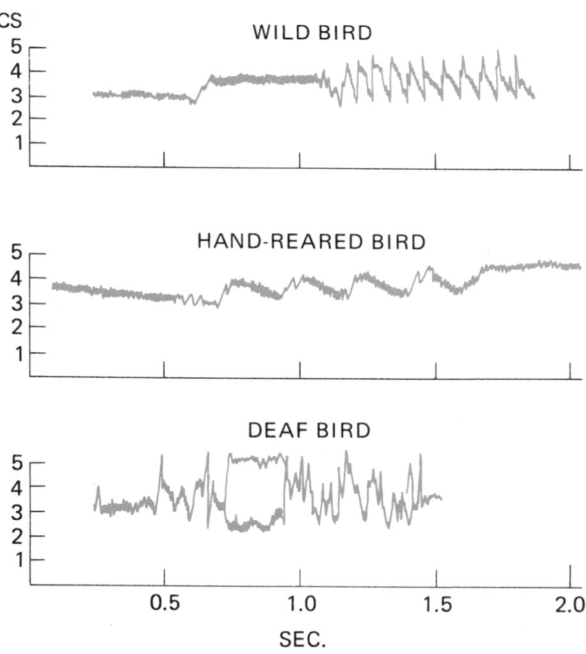

FIGURE 36.8 Songs of the white-crowned sparrow under varying experimental conditions. Top figure shows a spectrogram of the song of a normal wild bird. The middle spectrogram was produced by a bird hatched and reared in isolation from other birds. The bottom spectrogram was produced by a bird that was deafened after hearing other birds but before it began to sing itself. (Spectrogram photographs by M. Konishi from Alcock, J.: *Animal Behavior: An Evolutionary Approach.* 2nd ed. Sunderland, MA, Sinauer Associates, 1979.)

The time and conditions for exploratory learning for some animals can be very restrictive. In bees, for example, the foraging bee learns the position of the hive only on the first trip out in the morning, learns the position of the floral nectar source only as it approaches the flower, and learns the flower color only just as it lands. In fact, exploratory learning in bees should probably be considered restrictive rather than flexible learning.

The development of food preferences is a type of flexible learning that involves a process called **conditioning.** As a result of "reward" or "punishment," a behavioral response becomes associated with a particular stimulus and thus becomes more frequently practiced (or avoided). Juvenile toads initially feed on virtually any moving arthropod but will spit out a noxious species and thereafter avoid it. Many birds and mammals learn acceptable food by its systemic effects. A small amount of new food is sampled, and if there is no ill effect after a period of time, the food item is integrated into the diet. Significantly, the length of the waiting period varies with each species, as well as the food-identifying stimulus—odor, color, shape, movement, and so on. Rats, for example, discriminate food by odor and wait a number of hours before resampling a new food.

Much of exploration and food preference learning involves **trial and error.** A correct pathway or food choice that is randomly selected becomes reinforced by reward.

Genetics and Evolution of Behavior

Since the development of behavior is dependent on the development of the nervous and endocrine systems, it follows that the underlying basis for behavior, like that for the activity of any of the organ systems, must be coded by genes. That behavior has a genetic basis is clear from classic genetic experiments in which two separate species, or races of the same species with differences in mating calls or other behavioral attributes, are crossed. Hybrids typically have an intermediate condition, and there is some segregation of the behavioral characters when hybrids are back crossed to the parents. Inbreeding experiments support this conclusion. Bovet has analyzed the rate at which mice learn to avoid a noxious stimulus. The learning curves for the initial heterogeneous population varied greatly from one individual to another (Fig. 36.9). He then developed many different inbred strains and found that there was little variation in the rate of learning within particular strains. This was expected because inbreeding leads to homozygosity (p. 359). However, different strains, presumably homozygous for different alleles, had quite different learning curves. Although these experiments show that complex behavioral patterns have a genetic basis, the patterns are based on the interaction of so many genes

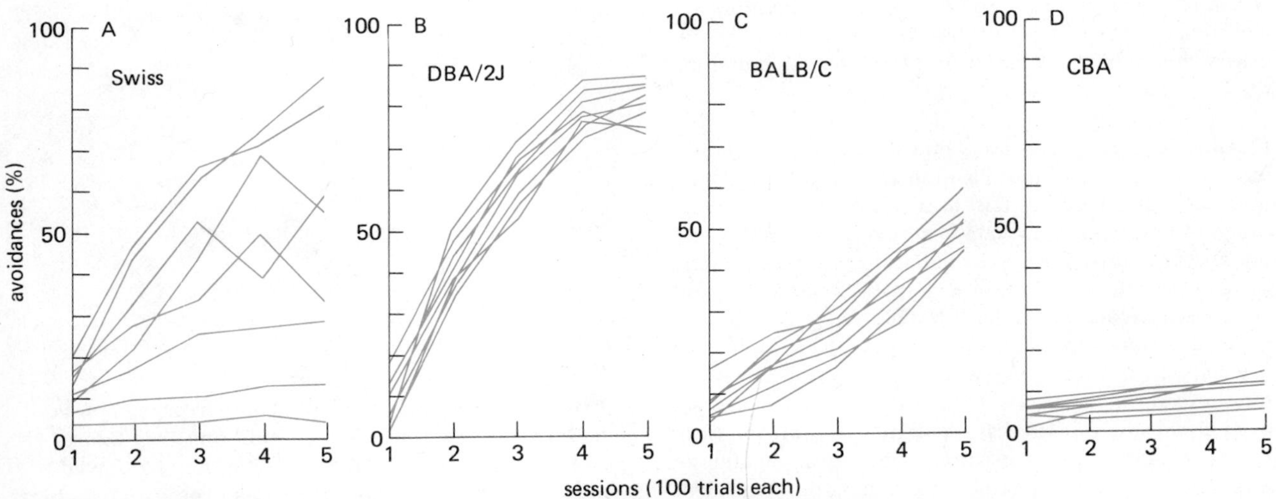

FIGURE 36.9 The rate, expressed in number of sessions, at which mice learn to avoid a certain stimulus. *A,* The learning curves for the initial, genetically heterogeneous population. *B–D,* The learning curves for members of three different inbred strains. (From Bovet, D., F. Bovet-Nitti, and A. Oliverio: *Science, 163*:239–249, 1969. Copyright © 1969 by A. A. A. S.)

that it is difficult to determine the role of individual genes in behavior.

Some idea of how changes in individual genes affect components of behavior can be gained by searching for individual mutants that affect some aspect of behavior. The fruit fly, *Drosophila,* has proven to be a good experimental animal. Seymour Benzer and his associates have discovered mutants that change the flies' normal 24-hour circadian rhythm, their response to light and dark, various aspects of their courtship, and even their learning ability. Males that have a sex-linked mutant called "fruitless" are unable to distinguish between sexes so they will court any fly, male or female, that they come upon. Sex recognition in flies is largely olfactory and depends upon a pheromone secreted by the female. Analysis has shown that the "fruitless" males produce the female pheromone and so smell themselves when they approach another fly! In this case the allele for "fruitless" must be coding for some enzyme involved in the synthesis of the pheromone.

Since behavior has an underlying genetic basis and varies between groups and individuals, it must be subject to natural selection and adaptive change. Behavior appropriate to a particular environmental situation, e.g., genes for moving toward light for an animal that feeds in the daytime, will be favored and increase in frequency in gene pools, whereas genes for inappropriate behavior will be selected against. Behavioral patterns are as adaptive as structural and functional ones.

Behavioral Ecology

Behavioral ecologists are interested in the particular adaptations forged by evolution that enable members of a species to interact favorably with members of their own species and other species with whom they must coexist. All animals must find a place to live, obtain food, avoid being eaten, and reproduce. They must perform these activities in ways that perpetuate themselves yet at the same time do not interfere with other species with which they interact and on which they depend in many ways. A predator cannot exterminate its prey or else it too will cease to exist.

Agonistic Behavior and Dominance. The size of the population of many species, particularly those that live in habitats with relatively uniform conditions, or conditions that change seasonally in a predictable way, is held at a nearly constant size by the resources of the environment (p. 807). Competition for those resources in shortest supply is keen. These may be food, shelter, mates, nesting sites, and so forth. **Agonistic behavior** (Gr. *agonistes,* champion) is the pattern of aggressive behavior by which members of the same species adjust to conflicts that arise from competition for the same limited resource. Agonistic behavior takes many forms. It may be overt **aggression,** in which one individual harms another, the threat of aggression, or it may be submission, in which one individual withdraws. In aggression against members of their own species, the combat usually is ritualized. Postures are taken, threats are made, some combat may ensue, but before one individual is hurt, the one being bested exhibits a submissive signal and withdraws from the dispute (Fig. 36.10). Charles Darwin recognized this in 1872 in his book *Expression of the Emotions in Man and Animals.* When one dog threatens another, the aggressors face their opponents, hairs on the scruff of their necks are raised, fangs are bared. There may be a brief tussle, but usually one dog will show signs of submission by tail wagging or crouching. The combatant who submits lives to fight another day and perhaps with a weaker rival. The victor has established a **dominance** and gains access to perhaps a mate, territory, or other resources. Pursuing the conflict would have little benefit and could result in the victor's injury. Selection apparently has favored genetic mechanisms that lead to submission and its acceptance by the victor.

In some social arthropods and vertebrates, ritualized combats lead to the establishment of a **dominance hierarchy.** This was originally observed in chickens by Schjelderup-Ebbe in 1922. One individual, *A,* is dominant over all others; a second, *B,* dominant over all except *A,* and so on through a **pecking order.** Dominant individuals have better access to food or to mates. This is an example of a linear hierarchy, but there are also triangular ones and other relationships, e.g.,

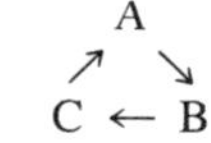

Dominance hierarchies are not necessarily stable. They may vary with the resource in question, with one individual having first access to water and another to mates. They may vary with age. The oldest male elephant seal is dominant and maintains a harem, but in the course of time he is challenged and eventually replaced by a younger male. Dominance hierarchies are a way to ritualize tensions that arise in a social group and to avoid the need for an agonistic encounter every time a resource is in question.

Territoriality. In many cases agonistic behavior is directed toward the defense of a certain physical space. An area that is defended by an individual or pair for their *exclusive use* is known as their **territory.** Territorial behavior is not universal, but it has been observed

FIGURE 36.10 Agonistic behavior. *A,* two male fiddler crabs use their large claws in an agonistic encounter. *B,* Threat display between two male silver gulls. (Courtesy of John Alcock, Arizona State University.)

in such diverse animals as limpets, crustaceans, insects, and all the vertebrate classes. Some species in these groups may exhibit territoriality only during a critical part of the year, such as the breeding season. Territoriality ensures the individuals of nesting sites, food, and other resources they need, but the defense of a territory requires constant vigilance, considerable energy expenditure in advertisement, threats, and other agonistic displays. For territoriality to be effective, the resources must be worth fighting for and be concentrated in a defensible area. Both territorial and nonterritorial harvester ants of the genus *Pogonomyrmex* live in Arizona. The territorial species gathers seeds from clumps of certain species of plants, and workers will defend a favorable cluster against intrusion by workers from another colony. The nonterritorial species forages widely for certain scattered seeds. Defending their large foraging area would not be cost-effective. The males of many song birds establish territories in the spring by their songs and displays. These are small enough to be defensible yet large enough to provide good nesting sites, nesting materials, and the food needed to rear the young. Gulls and other sea birds can hardly defend their vast foraging area but do establish smaller, defensible territories around their nests. After the breeding season, birds abandon their territories.

Mammals that forage widely have **home ranges,** that is, areas in which they can be found 95 percent of the time. Home ranges of different individuals may overlap. In some cases a territory for nesting or some other special purpose may be established within the home range. Home ranges vary greatly in size according to the mobility and needs of the species. African hunting dogs forage over an area of 1500 square miles, whereas a mouse may have a home range of a fraction of an acre.

Forms of space other than territory and home range are recognized. Many species maintain a certain **individual distance** between members. For sparrows roosting on a wire this may be the distance at which one bird can peck another. Often only sexual partners are permitted within the individual distance. Human beings have an individual distance, and we are uncomfortable if a stranger approaches too closely.

Sexual Behavior and Reproduction. Since an individual that reproduces perpetuates its genes, in contrast

FIGURE 36.11 A male damselfly holds and guards his mate with his abdominal claspers while she deposits her fertilized eggs. (Courtesy of John Alcock, Arizona State University.)

to one that does not, it is not surprising that natural selection has favored mechanisms, including behavior, that promote successful reproduction. In bisexual animals, sperm and eggs must be released at about the same time. In many animals as we have seen, environment stimuli induce, often via hormones, the development of the gonads so that gametes are ready at the appropriate breeding season. Among the higher invertebrates and vertebrates, behavioral patterns then ensure the coming together of the gametes. Males and females have evolved different behavioral patterns which, as Robert Trivers proposed in 1972, correlate with differences in the amount of their **parental investment** in gametes and offspring. Females make a greater investment per gamete than males because the eggs are much larger than sperm and include enough stored energy to provide for at least the initial development of the embryos. The yolk stored in some individual bird eggs may be equivalent to 20 percent of the female's weight. Given a finite energy supply, females produce far fewer eggs than males do sperm. A female maximizes her genetic contribution to succeeding generations by having her valuable eggs fertilized by the most superior males, so a female usually makes the choice of a breeding partner. Since sperm are relatively cheap energetically, a male maximizes his genetic contribution by mating with as many females as possible during his reproductive life.

In order to fertilize as many females as possible, many males compete intensely with one another for dominance and choice territories. **Sexual competition** among males of the same species often has contributed to the evolution of large male size, brilliant breeding colors, ornaments, antlers, and other features that give a male an advantage in establishing dominance and attracting females. It also has led to the evolution of strategies by which less successful males may occasionally be able to mate. A low-status male in some fish species may mimic the behavior of a female, gain access to the dominant male's nesting territory, and spawn newly laid eggs of the resident female. Sexual selection has also led to strategies whereby a successful male will protect an inseminated female from copulation with other males. After copulation a male damselfly continues to grasp and fly with the female until she has deposited her eggs (Fig. 36.11). A successful drone honeybee discharges much of his genital apparatus into the virgin queen's genital passages, thereby blocking them against additional insemination.

Since the female usually chooses the mate, **epigamic selection** (Gr. *epi,* upon + *gamos,* marriage) has affected those male attributes that enhance a male's attractiveness. It has also affected those female characters that enable her to ascertain that the quality of the male is worthy of her investment and that the male is conspecific. Success of a male in dominance encounters with other males is an important indicator to the female of male fitness. Some females accept the first male that is able to reach her, but other females test the males by provoking encounters. Female baboons and chimpanzees in estrus have enlarged, brilliantly colored genital swellings that attract all males and incite competition among them. A female frog grasped by a small young male during the breeding season may swim off in search of a larger male before shedding eggs.

FIGURE 36.12 Field photograph of a female praying mantis (*Pyrogomantis,* top) eating the male with whom she is copulating. (Courtesy of E. S. Ross, California Academy of Sciences.)

The victorious male courts the female. A primary function of courtship is to ensure that the male is a member of the same species, but it also provides the female further opportunity to assess the quality of the male. Courtship in some cases is also needed as a release signal to trigger nest building or ovulation. Courtship rituals may be long and elaborate. The first display of the male releases a counter behavior of a conspecific female. This, in turn, releases additional male behavior, and so on until the pair are psychologically and physiologically ready for copulation. An example of courtship is shown in Box 36.2. Certain male spiders make an offering of food to the female during courtship. This inhibits any aggressive tendencies that the female may have on being approached and also provides the female with some of the food needed for egg production. The male praying mantis often offers his own body, and the female happily munches on his head while the rest of his body copulates with her (Fig. 36.12)!

Care of the young is an additional component of successful reproduction in many species, but it too requires a parental investment. The benefit of parental care is the increased likelihood of the survival of the offspring con-

BOX 36.2

Courtship in the Three-spined Stickleback

The appearance of the female in breeding condition elicits a zigzag dance by a territory-holding male (left). The female follows the male, who leads her to the entrance of the nest he has built. When the female is in the nest (upper right), the male's tactile prodding elicits egg laying. (After Tinbergen, N.: *The Study of Instinct.* London, Oxford University Press, 1951.)

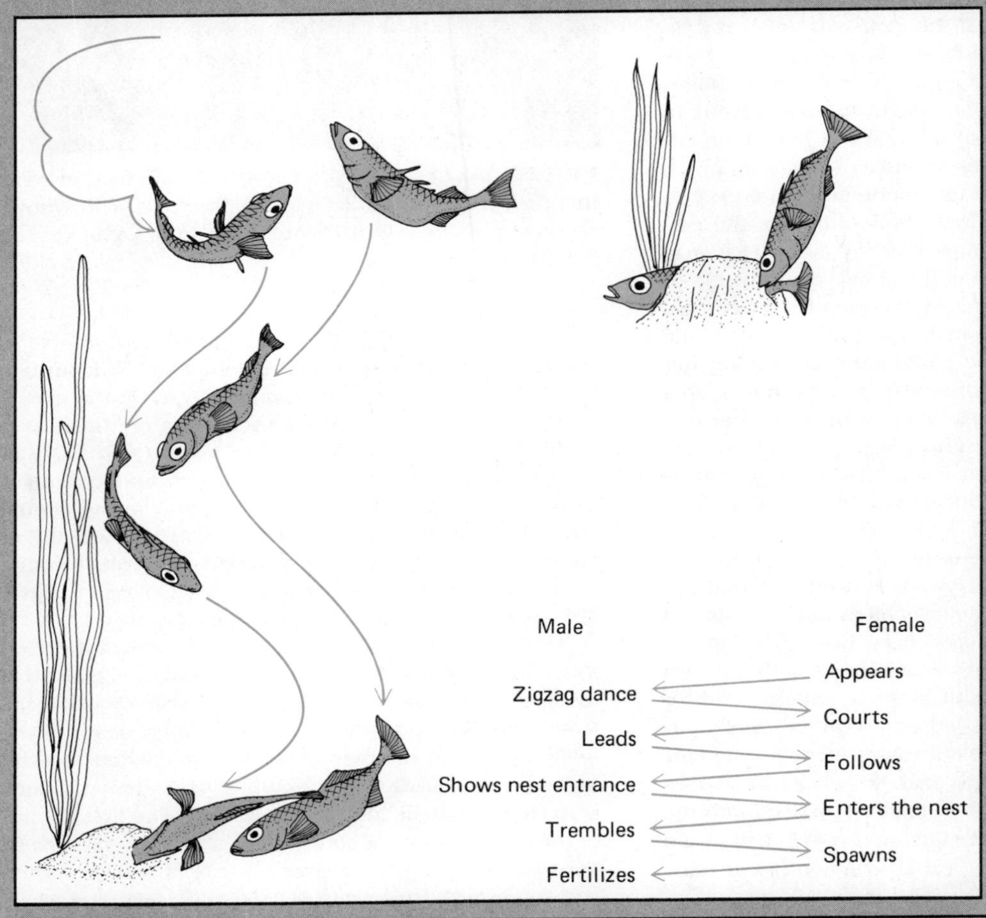

ceived, but the cost is a reduction in the number of offspring that can be produced. As with the gametes, a female has more to lose than a male if the young conceived do not develop. Females are more likely to brood eggs than males, and usually the females invest more in parental care. A high investment in parental care usually is less advantageous to a male for time spent in parenting is time lost in inseminating other females. Moreover, it may not be certain who fathered the offspring. Raising some other male's offspring is to the genetic disadvantage of a male. Under some circumstances, however, it may be to the male's advantage to help rear the young. Receptive females may be few and far between, only a few young of certain paternity may be produced, food and other resources may be scarce and require more than one parent can provide, and the young may need protection against severe predation.

Male and female reproductive strategies enhance the perpetuation of their genes. The interaction of these strategies within the context of particular environments determines the mating systems that have evolved. Male polygamy, known as **polygyny** (Gr. *polys,* many + *gyne,* female), is favored in situations where dominant males can monopolize choice territories with ample resources to raise the young of several females, and the males need not make a large contribution to parenting. Males maximize their genetic contribution, and it is to the female's reproductive advantage to join a mated male in a good territory rather than an unmated one controlling an inferior territory. Polygyny is quite common among mammals, partly because only the females can nurse the young so the males are released from some parenting. Sometimes a male may control a good territory, but the important factor may be simply his dominance over other males. Female elephants usually travel and feed together for protection. During the breeding season they are joined by a dominant male who adds little to their protection and can offer little more than his presumably superior genes.

Monogamy (Gr. *monos,* single + *gamos,* marriage) is favored primarily in situations where there are advantages to both parents in sharing in raising the young. Most bird species are monogamous and have a close pair bonding that lasts for the breeding season, if not for the life of one of the parents. One parent alone could not brood the eggs, protect them against predation, or feed the voracious young even when the male controls a favorable territory.

Female polygamy, or **polyandry** (Gr. *polys,* many + *aner,* male) is quite rare for it seldom is to the advantage of either sex. It is most likely to occur in situations where the male assumes most, if not all, of the parenting duties. Some male fish, such as sticklebacks (Box 36.2), are better equipped to guard the nest than the females, and they need to do so because of a high predation on the protein-rich eggs. The male cannot leave in search of other females or else he would lose his investment in the young, but after laying a clutch of eggs the female is free to go elsewhere and find another male to fertilize her next batch.

Communication. Behavioral interactions of all types require effective communication, for only by exchanging mutually recognizable signals can one animal influence the behavior of another. Communication may facilitate finding food, as in the elaborate dance of the bees that we discussed earlier (p. 613). It may hold a group together, warn a group of danger, indicate social status as in a dominance hierarchy, solicit or indicate the willingness to provide care, identify members of the same species, and indicate sexual state, such as maturity and receptivity. Which sensory channel is used by a species at a given time depends upon a variety of conditions. Will the signal be sent at night or during the daytime? What type of information is to be conveyed? How far is it to go? What are the costs of sending and receiving the information? Is there danger of its interception by a potential predator? In most species natural selection has favored the evolution of one or more communication mechanisms that match the type and amount of information being sent to the circumstances under which it is sent.

The most primitive and widespread means of communication is the secretion by exocrine glands of odoriferous compounds known as **pheromones** (Gr. *pherein,* to bear + *hormaein,* to excite). The secretion of pheromones, which act as signals to other members of the same species, is the only communication mechanism available to unicellular organisms and to many of the simpler invertebrates because other channels require rather complex sending and receiving mechanisms. Pheromone communication has been discovered in nearly all organisms. It has many advantages beyond its simplicity. There is little energetic expense to synthesizing the simple but distinctive organic compounds involved. Conspecific individuals have receptors attuned to the molecular configuration of the pheromone, and it is ignored by other species. Pheromones are effective in the dark, they can pass around obstacles, and they last for several hours or longer. Major disadvantages are a slowness of transmission and a limited information content. The latter disadvantage is compensated for in some animals by secreting different pheromones with different meanings. Black-tailed deer release pheromones in their urine, feces, and from glands located on different parts of the feet and head. Pheromones may be deposited as scent markers on the ground or be carried by water or wind. Wind-blown ones are small enough molecules to be volatile yet are large enough to be distinctive. Note that pheromones are defined behaviorally, not chemically. There are many different kinds of substances that act as pheromones.

Many types of signals can be conveyed by pheromones. Most act as **releasers** and elicit a very specific but transitory type of behavior. Others act as **primers** and elicit slower and longer-lasting responses. Some may act in both ways. In social insects some releaser pheromones supplement visual and tactile clues in leading members of the colony back to a food source, as in the trail pheromones of ants. Some enable members of a colony to recognize each other, and some warn the colony of foreign intruders. When termites from another nest enter a colony, some members of the resident colony release an alarm pheromone that attracts members of the soldier caste, who fend off the invaders. The composition of a beehive is controlled by an antiqueen pheromone, 9-keto-decenoic acid, secreted by the queen. It acts as a releaser for it inhibits the workers from raising a new queen, and it also has a primer effect for it inhibits the development of the ovaries in the workers. If the queen dies, or if the colony becomes so large that the inhibiting effect of the pheromone is dissipated, the workers begin to feed some developing larvae the special food that promotes their development into new queens (p. 753).

Pheromones are important in attracting the opposite sex and in sex recognition in many species. Many female insects produce pheromones that attract males of the appropriate species. An example is bombykol secreted at night by female silkworm moths (p. 765). We have taken advantage of some sex attractant pheromones to help control such pests as gypsy moths by luring the males to traps baited with synthetic analogues of the female pheromone. Male butterflies, who mate in the daytime, are attracted to any flapping object that more or less resembles a female. After approaching the potential mate, the male engages in "hair penciling" behavior. Delicate pencil-like processes covered with pheromones are dangled in front of the prospective mate. If the flapping object is indeed a receptive female of the appropriate species, she will settle down and permit copulation.

Some aspects of the sexual cycle of vertebrates are affected by pheromones. In some species of mice the odor of a strange male will block the successful pregnancy of a recently impregnated female. Nerve impulses from the nose reach the female's hypothalamus and block the prolactin-releasing factor. The lack of prolactin leads to the regression of her corpora lutea and the failure of the young embryos to implant. In some cases, the odor of a male mouse introduced among a group of females will cause their estrous cycles to become synchronized.

It is not known whether or not pheromones affect human behavior, but some curious phenomena suggestive of pheromone influences have been reported. Analysis of the menstrual cycles of the students in an American women's college showed a statistically significant tendency for the increasing synchronization of the menstrual cycles among roommates and close friends. The study rules out many possible explanations for the phenomenon but suggests some pheromonal effect between young women who are together much of the time.

Sounds are frequently used and can, of course, travel faster than pheromones. The sound receptor mechanisms of some species are attuned only to a few frequencies of significance to the animals. Other frequencies are not detected. Certain large moths are particularly sensitive to the high-frequency calls used by bats to find prey, and the moths initiate an erratic avoidance flight upon hearing the bat. Several species of frog may gather in the same pond during the breeding season, and there may be a confusing babel of sounds emitted by different males. The ear of the female frog is structured in such a way that she is sensitive only to the call of the conspecific males. Sound communication is most elaborate in birds and some mammals. Certain birds have a distinct alarm call, territorial songs, and songs that attract a female. Most of the songs are quite stereotyped with the same one being used in the same circumstances. Only a few mammals, including humans, can put a repertory of sounds together in different ways to convey different sorts of information.

Touch is an effective short-range form of communication. Antennae of insects contain touch as well as olfactory receptors, and these are used when other signals would be less effective. Touch is important in the dance of the bees for it is very dark inside a hive. Touch is also very important in the copulatory behavior of many animals.

Sight can be used by those invertebrates and vertebrates with well-developed eyes during periods of sufficient ambient light. Visual signals have the advantage of quick and precise location of the signaler, and they can convey a great deal of information. Since its chromatophores are under neural control, an octopus can change colors quickly and signal its mood, aggression, and receptivity. Some animals combine two or more visual signals. A male peacock spreads its tail and struts before its prospective partner. Zebras (Fig. 36.13) elevate their

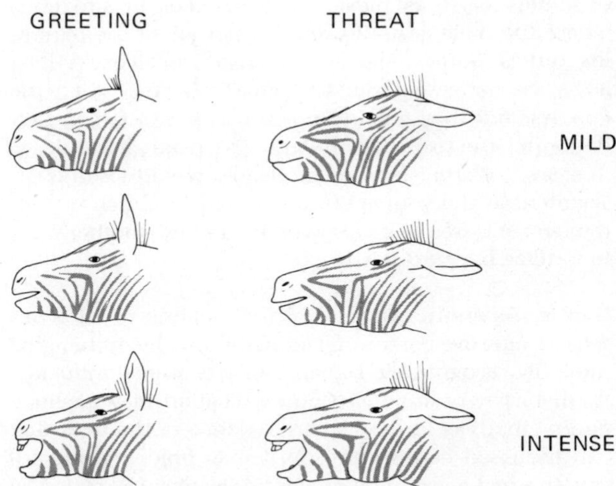

FIGURE 36.13 Visual signals of the zebra. Ear position indicates greeting or threat, and intensity of feeling increases as the mouth is opened wider. (From Drickamer, L. C., and S. H. Vessey: *Animal Behavior,* copyright 1982 by PWS Publishers. All rights reserved. Used by permission of Willard Grant Press.)

ears to signal a greeting, lower them as a threat, and open their mouths to different degrees to indicate the intensity of their feeling.

Social Behavior. A **social group,** as the term is usually used, is an association of members of the same species that have come together because of certain mutual advantages and are held together by an exchange of signals. Common advantages are a reduction in predation or an increase in foraging efficiency. The size and shape of a school of fish may intimidate a predator, but even if it does not, a specific individual in the school is less likely to be caught than if it were solitary. By nesting together in colonies and synchronizing their egg laying, bank swallows reduce predation. A large, vulnerable population of nestlings is present for only a short time, and the adults band together to mob a predator that tries to gain access to the colony. In the season when zebras, wildebeest, and other large herbivores are abundant on the African plains, lions live together as a pride and cooperate in hunting (Fig. 36.14). A solitary lion can catch a zebra only about 15 percent of the time, but a group of five lions can fell one zebra 40 percent of the time, the zebra providing food for the pride for several days. When the larger game migrate, the lions must prey on the much smaller gazelles, which only provide enough food for one lion for a day or two. The prides break up and the lions hunt individually.

FIGURE 36.14 Lions killing a zebra in Kenya. (Courtesy of Animals Animals, Inc./David C. Fritts.)

Social groups also have disadvantages. There often is increased competition within the group for food, mates, and other resources. Bank swallows sometimes steal nesting materials from each other, and neighboring males may copulate with a female whose partner is away. The risk of disease is higher. Hoagland and Sherman have shown that the level of nest infestation with bird lice increases with the size of bank swallow colonies. The advantages must be greater than the disadvantages before social groups evolve.

Social groups range in complexity from simple aggregations of unrelated individuals that may cooperate in chasing off a predator to the complex groups of ants, bees, and other social insects in which there is a close relationship within the group and a precise division of labor. **Altruistic behavior,** in which one individual appears to act in such a way as to benefit others rather than itself, is frequently seen in the more complex social groups. A particularly clear case of altruistic behavior has been observed by Watts and Stokes in the mating of wild turkeys. Several different groups of males, in each of which there is a dominance hierarchy, gather in a special mating territory and go through their displays of tail spreading, wing dragging, and gobbling in front of females who come to the area to copulate. Because of cooperation between the males within a group, one group attains a dominance over other groups, and its dominant member is the one to copulate most frequently with the females. Seemingly the males who helped establish the dominant group but have a low status within the group gain nothing. Close analysis has shown that members of a group are brothers from the same brood. Since they share many genes with the successful male, they are indirectly perpetuating many of their genes. In this case altruism appears to have evolved by kin selection. Kin selection has also led to the evolution of the complex societies of social insects in which some individuals specialize for reproduction and other close relatives do the chores of the colony (p. 93).

Feeding and Defense. Most of the types of behavioral ecology that we have been considering deal with interactions between members of the same species. Different species also interact with each other as they search for favorable habitats, for living space, for food, and for avoiding being eaten. Most animals, as we have seen in considering animal groups, are adapted morphologically and physiologically to feed on a limited range of food and to gather it in specific ways. Herbivores select particular kinds of plants; carnivores, the type and size of prey captured. Even some detritus feeders sort out organic material from sand and other debris. Behavior is important in finding food of a particular type, gathering it in efficient ways, and deciding when and where to move on. A great deal of research is now being done to determine the extent to which **foraging strategies** maximize net energy intake, as we would expect them to do if their evolution has been guided by natural selection.. The **net energy intake** represents the difference between the energy available in the food less that

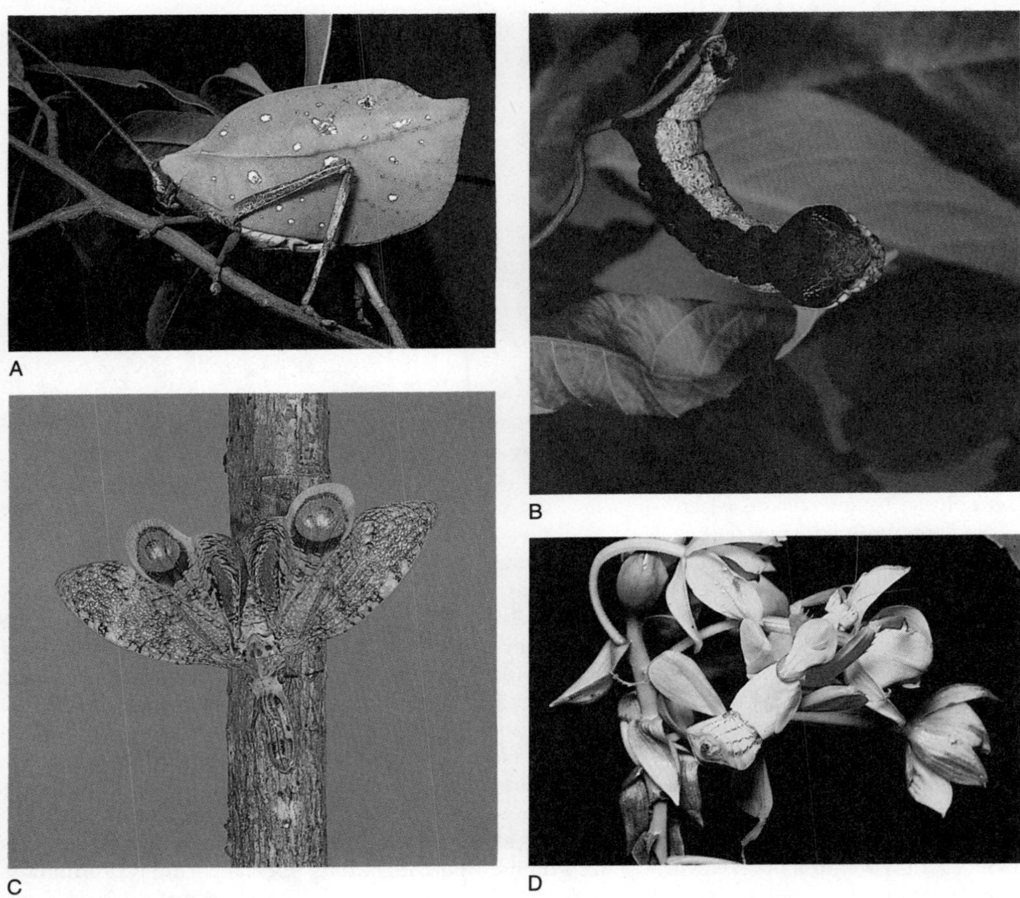

FIGURE 36.15 Concealment and deception. *A,* A Brazilian katydid is protected by its resemblance in shape and color to a mold spotted leaf. *B,* The tropical American sphinx moth caterpillar, *Leucorhampha,* has a cryptic coloration but when disturbed engages in a threat display in which it resembles a snake. *C,* The lantern bug of tropical America, *Lantinaria,* startles a predator that becomes too inquisitive by suddenly uncovering its hind wings and revealing a pair of large eye spots. *D,* The flower mantis of Malaysia, *Hymenopus,* has the shape and color of the orchid flower on which it lies in wait for prey. (*A* and *D* courtesy of E. S. Ross, California Academy of Sciences; B and C, photographs by Lincoln P. Brower, University of Florida.)

used in searching for the food, gathering or catching it, and eating it. Under these circumstances foraging behavior and food choices should vary with ecological conditions. In the relatively unproductive marshes of the northeast, great blue herons must spend a great deal of time finding food, and they eat a fairly wide variety of small vertebrates. But in Florida, where marsh productivity is much higher and food is easier to find, the herons select a more limited range of food types. As animals feed, they frequently develop **search images** and concentrate on finding only one or a few types of prey at a time, passing up other palatable items. This seems to improve the efficiency of foraging behavior. We too develop search images as we look through a box of mixed chocolates for one or two types that we fancy at the time.

Many strategies have evolved that maximize net energy intake. As we have seen (p. 419), natural selection promotes resource partitioning among related competing species. Different species of Darwin's finches maximize net energy intake by feeding on seeds of different sizes. Some herbivores manipulate their plant resources so as to maximize production. Leaf-cutting ants of the American tropics carry bits of appropriate leaves into their subterranean galleries, chew them up, fertilize them with fecal droppings, and "seed" them with bits of mold. As the mold grows, they feed on it. We would call such practices "agriculture." Some carnivorous animals construct traps, use sticks and other tools to gather food, wait in ambush, or engage in **aggressive mimicry.** The orchid mantis (Fig. 36.15*D*) is a striking example as it lies in wait for insects that come to the orchid flower for nectar. In this case, the body shape, color, and behavior that lead the mantis to select flowers that it matches all interact to deceive the prey.

Many prey organisms have evolved interesting counter measures and strategies that reduce the likelihood of their being eaten. Some, such as a mouse, simply retreat to a home refuge when danger threatens. A few remain motionless, or freeze, when approached by a predator. This strategy is usually associated with a concealing or **cryptic coloration** (Gr. *kryptos,* hidden). Some animals combine a cryptic coloration with a deceptive body shape. Many insects are shaped, as well as colored, to resemble twigs, leaves (Fig. 36.15*A*), or even fecal droppings. If concealment fails and predators become too inquisitive, some insects startle the predators by uncovering conspicuous eye spots or assuming the appearance and posture of a predator themselves (Fig. 36.15*B, C,* and *D*).

A great many organisms are dangerous, toxic, or distasteful to predators. The caterpillar of the blue swallowtail butterfly discharges toxic secretions from tentacle-like structures at its anterior end. The bombardier beetle has an unusually effective toxic mechanism. It synthesizes and stores hydrogen peroxide and quinones in separate compartments. When the beetle is threatened, these secretions are released and mixed with enzymes in an outer vestibule, heat quickly to the boiling point, and are discharged as a caustic spray (Fig. 36.16). These species, bees, and many other animals (see Fig. 32.10*A*) that have dangerous toxins, or whose tissues contain distasteful materials, frequently have a conspicuous, **aposematic coloration** (Gr. *apo,* away + *sema,* signal) that advertises their presence. After several encounters predators learn to associate this pattern with a disagreeable experience and avoid these animals. Aposematic coloration provides an opportunity for certain tasteful animals to gain protection by mimicking the distasteful species. **Batesian mimicry,** as this is called, is quite common in insects and some vertebrates (Fig. 36.17). The terrestrial juvenile stage of the newt of the eastern United States has a striking orange-red color and is called the red eft. The red eft is distasteful to bird predators, who soon learn to recognize the eft and avoid eating it. The similarly colored red salamander is perfectly palatable, but it too is avoided by birds that have had encounters with red efts.

In some cases several different distasteful and aposematic species resemble each other. **Mullerian mimicry** of this type is found among several groups of Amazonian butterflies (Fig. 36.18). Since some individuals are

A

B

FIGURE 36.16 Advertisement. *A,* When threatened, the caterpillar of the blue swallowtail butterfly, *Battus philenor,* discharges toxic secretions from the pair of brownish colored osmeteria at its front end. Its conspicuous aposematic coloration advertises its danger. *B,* A bombardier beetle aims a caustic spray between its legs at a predator. (*A* courtesy of E. S. Ross, California Academy of Sciences; *B* courtesy of T. Eisner, Cornell University.)

A B

FIGURE 36.17 Batesian mimicry. *A,* The juvenile, red eft stage of the red-spotted newt, *Notophthalmus viridescens,* is avoided by predators who learn to associate its red aposematic coloration and toxic skin secretions. *B,* The harmless red salamander, *Pseudotriton ruber,* gains protection by its resemblance to the red eft. (*A* photograph by W. F. Walker, Jr.; *B* courtesy of the Ohio Department of Natural Resources, photograph by A. L. Staffan.)

sacrificed in educating the bird predators, sharing the burden of education among several species is advantageous to all.

FIGURE 36.18 Müllerian mimicry. Several different, distasteful species of tropical American butterflies have a similar "tiger pattern" of aposematic coloration. (Photograph by Lincoln P. Brower, University of Florida.)

Some social animals cooperate in group defense. At least one member of a large group is more likely to spot a predator and give an alarm than an individual feeding by itself. Once the group has been alerted, the predator may save its energy and try to sneak up on something else. Some birds may mob a predator, as we have seen. If a predator continues its attack, the members of some species form a circle, with females and young in the center. Musk oxen do this when attacked by wolves.

Orientation. As animals move about in search of habitats, mates, and food they sometimes travel considerable distances. Many vertebrates, ranging from fish to mammals, undertake extensive seasonal migrations. Finding their way from one area to another, or back to a nest after foraging or accidental displacement, is very important. Sometimes homing is as simple as learning visual landmarks in the territory or home range. Often complex navigation is involved, using the sun and stars, a sense of time because the position of these celestial clues changes during the day, and a knowledge of one's destination. We have previously considered two striking examples of homing (in bees, p. 613, and birds, p. 717), but orientation is very widespread among animals. Limpets, for example, can find their way back to resting spots after grazing away on algae. Many animals use sensory clues not available to us. Ants and bees can detect the plane of polarized light and hence the sun's position on cloudy days. Salmon and dogs use a keen sense of smell. Some birds appear to be able to detect the earth's magnetic field.

Summary

1. An animal's responses to its environment are highly organized and intricate patterns of activity that we call behavior, the study of which is termed ethology.

2. In behavioral patterns an animal first recognizes a sign stimulus given by another animal or some object. This releases in the receiver a pattern of motor responses. Frequently the responses are quite stereotyped. A nesting goose observing a displaced egg, or a reasonable facsimile of one, will try to roll it back into the nest. Many behaviors also require a certain state of responsiveness, sometimes called motivation or drive. A goose's drive to retrieve a wayward egg is manifested only in the nesting season.

3. The signs an animal recognizes and the responses it makes depend upon the nature of its receptors and the activation of appropriate effectors by the nervous and endocrine systems. Each animal has receptors attuned to signals of importance to it and a neuronal circuitry that generates responses adaptive to its particular needs. Hormones affect the development of sexual and other responses and set the physiological stage.

4. Many aspects of behavior, as well as metabolism, follow a 24-hour, or circadian, rhythm. Although these rhythms are based on an internal biological clock that has its own period, which is usually somewhat less or more than 24 hours, the internal clock becomes set so that it comes into phase with, and keeps in phase with, daily changes in light and other environmental factors that follow a 24-hour cycle. The biological clock of mammals appears to be based on groups of neurons in the hypothalamus that have an inherent oscillating activity.

5. Instinctive behavior depends on the activation of preprogrammed neuronal circuits. Appropriate fixed action responses are made the first time the animal sees the necessary release sign. Some instinctive behavior, such as a spider's web spinning, are closed and are not modified by experience. Other instincts are open and, although functional when first performed, are modifiable and improve with practice.

6. Learning requires an interaction between an animal and its environment to become functional. The interaction occurs early in life in restricted learning and is not subsequently modified. Many newly hatched birds imprint on the first moving object they see, which is normally their mother, and follow only this object as they continue to develop. Flexible learning continues to be modified as a result of continued interactions with the environment. Some flexible learning is accompanied by conditioning.

7. Evidence that behavior, like other attributes of organisms, is coded by the genes has been discovered by breeding experiments and by an analysis of the effects of individual mutants on behavior. Given an underlying genetic basis, it is to be expected that behavior is subject to evolutionary change and is adaptive. Behavioral ecologists study behavior in its ecological context to ascertain its evolution and adaptive nature.

8. Agonistic behavior is the way members of the same species resolve conflicts for food, mates, and other limited resources. Combats tend to be ritualized and end with the dominance of one individual and the submission of another. Frequently a dominance hierarchy develops within the group. In many cases agonistic behavior secures a certain territory for the exclusive use of an individual or pair, but territories are not established unless the value gained by controlling resources outweighs the costs of defending them.

9. Much behavior increases reproductive success and maximizes an individual's genetic contribution to succeeding generations. Since there is a finite amount of energy available for gamete production, and eggs contain more energy reserves than sperm, females produce fewer eggs than males do sperm. It is to the female's reproductive advantage to select superior males to fertilize her relatively few eggs, while it is to the male's advantage to fertilize as many eggs as possible. Sexual competition among males for access to females, and epigamic selection by females of superior males, have frequently led to the development of large male size, ornaments, and weapons and to elaborate courtship rituals.

10. Parental care of the young increases the probability of their survival but reduces the number of young produced. In general females have a greater investment in the young than the males and are more likely to assume the major care. Intense predation against the young and certain other factors may make it advantageous for the male to help. Whether the mating system is polygyny, monogamy, or polyandry depends to a large extent on the amount of male involvement in parenting.

11. Only by effective communication can one individual influence the activity of others. All sensory channels have been used by one species or another in communicating: odor, sound, touch, and sight.

12. A social group is an aggregation of individuals of the same species that have come together because of mutual advantages, such as a reduction in predation or an increase in foraging efficiency. Disadvantages are increased competition within the group and greater risk of disease. Advantages must outweigh disadvantages before social groups form. Social groups range in complexity from simple aggregations of unrelated individuals to complex societies of closely related individuals in which there is considerable division of labor. Altruistic behavior frequently is seen in the more complex groups.

13. Members of different species interact with each other as they search for habitats, for food, and try to avoid being eaten. Natural selection has favored the evolution of foraging strategies that maximize net energy intake. Defensive strategies include escape, cryptic coloration and behavior, deception of predators, poison mechanisms, the accumulation of distasteful materials in body tissues, aposematic coloration, Batesian mimicry, Mullerian mimicry, and group defense.

14. As animals migrate and travel in their search for food, mates, and nesting sites, they must be able to return to home.

References and Selected Readings

Aidley, D. J. (Ed.): *Animal Migration.* Cambridge, Cambridge University Press, 1981. A collection of papers presented at a symposium of the Society of Experimental Biology dealing with the migration and navigation of insects, fish, birds, and whales.

Alcock, J.: *Animal Behavior, An Evolutionary Approach.* 2nd ed. Sunderland, Massachusetts, Sinauer Associates, Inc., 1979. A general text with an emphasis on the ecology and adaptive nature of behavior.

Bastock, M.: Courtship: *A Zoological Study.* London, Heinemann Educational, 1967. An interesting book describing courtship and reproductive behavior in a variety of animal types.

Benzer, S.: Genetic dissection of behavior. *Scientific American, 229*(Dec.): 24, 1973. A discussion of his research on the genetic components of behavior.

Brower, L. P.: Ecological chemistry. *Scientific American, 220*(Feb.):22, 1969. Some palatable insects mimic others that eat distasteful plants.

Carthy, J. D.: *The Behavior of Arthropods.* New York, Academic Press, 1968. A fascinating summary of studies of mechanisms controlling insect behavior.

Drickamer, L. C., and S. H. Vessey: *Animal Behavior: Concepts, Processes, and Methods.* Boston, Willard Grant Press, 1982. A balanced text dealing with the physiological basis of behavior and its ecological and evolutionary significance. Questions ethologists ask and their methods of inquiry are emphasized.

Gould, J.: *Ethology: The Mechanisms and Evolution of Behavior.* New York, W. W. Norton & Co., 1982. A balanced text dealing with the neural mechanisms of behavior, behavioral genetics, the evolution of behavior, and human ethology.

Guthrie, D. M.: *Neuroethology, An Introduction.* New York, John Wiley & Sons, 1980. A thorough analysis of the neurological basis of behavior.

Lewis, D. B., and D. M. Gower: *Biology of Communication.* New York, John Wiley & Sons, 1980. An analysis of the neurosensory basis of communication and of the types of communicating systems.

Limbaugh, C.: Cleaning symbiosis. *Scientific American, 205*(Aug.):42, 1961. Many marine animals earn a living by removing parasites from other animals

Lore, R., and K. Flannelly: Rat societies. *Scientific American, 236*(May):106, 1977. Interactions in a rat society that increase their chances of survival in a hostile environment are discussed.

Lorenz, K.: *On Aggression.* New York, Harcourt, Brace & World, Inc., 1966. A classic volume discussing aggressive behavior.

Moore-Ede, M. C., F. C. Sulzman, and C. A. Fuller: *The Clocks That Time Us.* Cambridge, Harvard University Press, 1982. A review of the circadian timing system with an emphasis upon the mammalian system and its importance to human beings.

Palmer, J. D.: Biological clocks of the tidal zone. *Scientific American, 232*(Feb.):70–79, 1975. A study of some of the marine animals that have their biological rhythms set to tidal clocks.

Tinbergen, N.: *Social Behavior in Animals.* London, Science Paperbacks, Methuen & Co., Ltd., 1965. A reprint of Tinbergen's classic study.

Wehner, R.: Polarized-light navigation by insects. *Scientific American, 235*(Jul.):106, 1976. A remarkably sophisticated detection system enables ants and bees to utilize polarized light in homing.

Wilson, E. O.: Animal communication. *Scientific American,* (Sept.): 1972. Methods of communication from insects to humans.

Wilson, E. O.: *Sociobiology.* Cambridge, Harvard University Press, 1975. The biological basis of social behavior with a wealth of information about animal studies.

37 Dynamic Processes in Ecology

ORIENTATION

Animals do not live at random in the world but in very specific areas in which they have important trophic and energy relations with plants and other animals. Ecology, the study of organisms in their natural environment, is concerned with the biology of groups of organisms and their relations with the environment. In this chapter we will discuss the hierarchy of populations, communities, and ecosystems. We will explore the various factors that limit the distribution of each species of plant and animal. We will explain the concepts of habitat, ecologic niche, ecosystem, and food chains and describe the roles of producers, consumers, and decomposers in an ecosystem. We will show how climatic and edaphic factors in the physical environment influence the numbers and distribution of plants and animals. Finally we will describe the cyclic use of the elements, such as carbon and nitrogen, how these are exchanged between living and nonliving things in the course of the cycle, and how the one-way flow of energy entering from solar radiation passes through a succession of plants and animals before being radiated back into space.

If a spaceship from another planet reaching this one were to contain a crew of biologists, they would surely be struck by the enormous diversity of species here—several million species of animals and plants. As they circled the earth in their spaceship they would notice that certain species live in certain places and that more than one species may live in the same space. They would realize that only a few places on the planet are uninhabited—the craters of active volcanos, the Dead Sea, and the bottom of the Black Sea. Relatively few organisms are present at the poles, but the number of kinds of organisms and the total number of organisms would increase as they approached the equator. Closer inspection (perhaps from a landing module) would reveal that no species lives alone but that populations are mixed. If they held their spaceship stationary over a single area they could observe that different species are present at different times—night and day, winter and summer, or present and past. When they focused closely on a single species they would find that the abundance of that species is not constant but increases and decreases as the environment changes. Ecologists are biologists who study these phenomena and attempt to understand them so that predictions can be made about future trends and changes.

The word "ecology" is very much in the public consciousness today as we become aware of some of our past and current ecological malpractices. It is important for everyone to know and appreciate the principles of this aspect of biology so that he or she can form an intelligent opinion regarding topics such as insecticides, detergents, mercury pollution, sewage disposal, the different types of power generation, and their effects on mankind, our civilization, and the world we live in.

The Greek *oikos* means "house" or "place to live," and **ecology** *(oikos logos)* is literally the study of organisms "at home," in their native environment. The term was proposed by the German biologist Ernst Haeckel in 1869, but many of the concepts of ecology antedated the term by a century or more. Ecology is concerned with the biology of groups of organisms and their relations to the environment. The term **autecology** refers to studies of individual organisms and their relations to their environment. The contrasting term, **synecology,** refers to studies of groups of organisms that live together and form a functional unit of the environment. Groups of organisms may be associated in three different levels of organization—populations, communities, and ecosystems. In ecological usage, a **population** is a group of individuals of any one species that live in a particular area. A community in the ecological sense, a **biotic community,** includes all of the populations of different species occupying a given defined physical area. The community, together with the physical, nonliving environment, composes an **ecosystem.** Thus synecology is concerned with the many relationships within communities and ecosystems. The ecologist deals, for example, with such questions as how members of a community modify space, light, temperature, and other factors for each other. The ecologist is concerned with the energy source for the community and its flow from one individual to the next in a food chain.

The Concepts of Ranges and Limits

Probably no species of plant or animal is found everywhere in the world; some parts of the earth are too hot, too cold, too wet, too dry, or too something else for the organism to survive there. The environment may not kill the adult animal or plant directly, but it effectively keeps the species from becoming established if it prevents its reproducing or kills off the egg, embryo, or some other stage in the life cycle.

Most species are not even found in all the regions of the world where they could survive. The existence of barriers prevents their further dispersal and enables us to distinguish the major biogeographic realms (p. 400) characterized by certain assemblages of plants and animals.

Biologists early in the nineteenth century were aware that each species requires certain materials for growth and reproduction and can be restricted if the environment does not provide a certain minimal amount of each one of these materials. Justus Liebig stated in 1840 what is now known as his **law of the minimum:** The rate of growth of each organism is limited by whatever essential nutrient is present in a minimal amount. Liebig, who studied the mineral elements affecting the growth of plants, found that the yield of crops was often limited not by a nutrient required in large amounts, such as water or carbon dioxide, but by something needed only in trace amounts, such as boron or manganese. Liebig's law is strictly applicable only under steady-state conditions, when the inflow of energy and materials equals the outflow. In addition, there may be interactions between factors such that a very high concentration of one nutrient may alter the rate of utilization of another (the rate-limiting one) and hence alter the effective minimal amount required. Certain plants, for example, require less zinc when growing in the shade than when growing in the sunlight.

V. E. Shelford pointed out in 1913 that the distribution of each species is determined by its **range of tolerance** to variations in each environmental factor. Much work has been done to define the limits of tolerance, the limits within which species can exist, and this concept, sometimes called Shelford's **law of tolerance,** has been helpful in understanding the distribution of organisms.

Certain stages in the life cycle may be critical in limiting organisms—seedlings and larvae are frequently more sensitive than adult plants and animals. Adult blue crabs, for example, can survive in water with a low salt content and can migrate for some distance up river from the sea, but their larvae cannot, and the species cannot become permanently established there.

Some organisms have very narrow ranges of tolerance to environmental factors; others can survive within much broader limits. Any given organism may have narrow limits of tolerance for one factor and wide limits for another. Ecologists use the prefixes **steno-** and **eury-** to refer to organisms with narrow and wide, respectively, ranges of tolerance to a given factor. A stenothermic organism is one that will tolerate only narrow variations in temperature. The housefly is a eurythermic organism, for it can tolerate temperatures ranging from 5° to 45° C. The adaptation to cold of the antarctic fish *Trematomus bernacchi* is remarkable. It is extremely stenothermic and will tolerate temperatures only between −2° C. and +2° C. At 1.9° C. this fish is immobile from heat prostration!

For most organisms, light, temperature, and water are the principal limiting factors. The supply of oxygen and carbon dioxide is usually not limiting for land organisms, except for animals living deep in the soil, on the tops of mountains, or within the bodies of other animals. But those living in aquatic environments may be limited by the amount of dissolved oxygen present. The amount dissolved in water varies widely. There are many other factors that may limit the distribution of certain animals—water currents, soil types, trace elements, the presence of other kinds of organisms which might provide food or cover, and so forth.

Whether a given species of animal can become established in a given region is the result of a complex interplay of such physical factors as solar radiation, temperature, light (both the total amount of light and the photoperiod, the length of the period of daylight, are important), water, winds, salts, and biotic factors, such as the plants and other animals in that region that serve as food, compete for food or space, or act as predators or parasites.

Habitat and Ecologic Niche

Two basic concepts useful in describing the ecological relations of organisms are the habitat and the ecologic niche. The **habitat** of an organism is the place where it lives, a physical area, some specific part of the earth's surface, air, soil, or water. It may be as large as the ocean or a prairie or as small as the underside of a rotten log or the intestine of a termite, but it is always a tangible, physically demarcated region. More than one animal or plant may live in a particular habitat. Habitats are named for some prominent physical feature or dominant plant group: stream habitat, rotten log habitat, prairie habitat, coniferous forest habitat.

The **ecologic niche** is a more inclusive term that includes not only the physical space occupied by an organism but also its functional role as a member of the community—that is, its trophic position and its position in the gradients of temperature, moisture, pH, and other conditions of the environment. The ecologic niche of an organism depends not only on where it lives but also on what it does—that is, how it transforms energy, how it behaves in response to and modifies its physical and biotic environment, and how it is acted upon by other species.

The term "ecological niche" defines how a species interacts with all of the components or resources (chemical, physical, biological) in its environment. Early ecologists equated habitat with a species "address" and niche with its "profession." In a sense this is acceptable, but contemporary ecologists use the term "niche" in a more precise and quantitative way to define the exact position of a species within its community. One can think of the environment as being made up of a series of resources, each varying along a gradient: degree of moisture, temperature, amount of food eaten, and so on. The range of tolerance for each resource is determined, the limits being the points beyond which members of the species cannot survive. A certain species of bird can feed on seeds of a certain size range, but there will be seeds that are too small for it to pick up with its particular bill structure or too large for it to open. If a niche could be defined by three resource gradients, the niche of a species within its community could be depicted as a three-dimensional volume within a rectangle. Conceptually one can visualize far more than three resources, i.e., n resources, in which case the niche would be an n-dimensional hyperspace.

The ecologic niche is an abstraction that includes all of the physical, chemical, and biotic factors that an organism needs in order to survive. To describe an animal's ecologic niche we must know what it eats and what eats it, what its activities and range of movements are, and what effects it has on other organisms and on the nonliving parts of the surroundings. To define completely the ecologic niche of any species would require detailed knowledge of a large number of biological characteristics and physical properties of the organism and

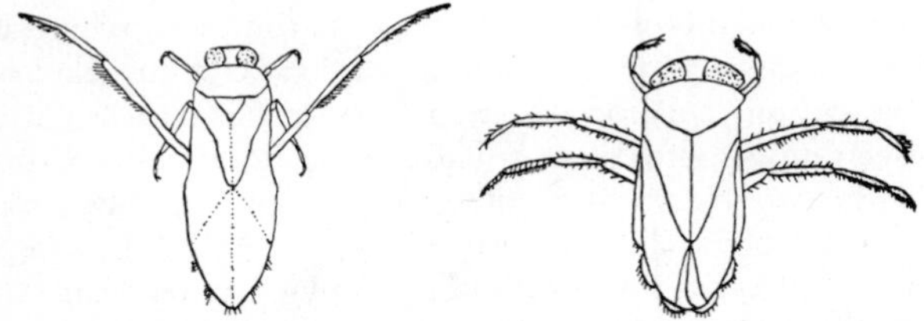

FIGURE 37.1 *Notonecta,* the "back-swimmer" *(left),* and *Corixa,* the "water boatman" *(right),* are two aquatic bugs occupying the same habitat—the shallow, vegetation-choked edges of ponds and lakes—but having different ecological niches.

its environment. Since this is very difficult to obtain, the concept of ecologic niche is used most often to describe *differences* between species with regard to one or a few of their major features.

In the shallow waters at the edge of a lake, you might find many different kinds of water bugs. They all live in the same place and hence have the same habitat. Some of these water bugs, such as the "back-swimmer" *Notonecta,* are predators, catching and eating other animals of about their size (Fig. 37.1). Other water bugs, such as *Corixa,* the water boatman, feed on dead and decaying organisms. Each has quite a different role in the biological economy of the lake, and each occupies an entirely different ecological niche. Thus the lake habitat may be viewed as composed of many ecological niches, each niche filled by a different species and all together dividing up the lake resources, such as food, space, and so on. There are no empty niches. **Resource partitioning** has been a major factor in the evolution of species.

Niche separation and resource partitioning are commonly very subtle, enabling a much larger number of species to occupy a habitat than you would expect. For example, a group of species may all feed on the same kind of food but partition that particular resource in several ways. Each species may be restricted to a certain food size (e.g., seed size or prey size); they may be separated by the time in which they feed; and they may be separated by where they feed (spatial separation). Not all dimensions of two similar niches are necessarily separate. There may be considerable overlap as long as there is resource partitioning at critical points of potential competition.

Penelope Williamson's classic study on niche separation in three species of vireos is a good example of fine-scale resource partitioning. These birds perch at the ends of twigs in trees and remove insects from adjacent leaves. As a group they are thus separated from other insectivorous forest birds which feed by hovering,

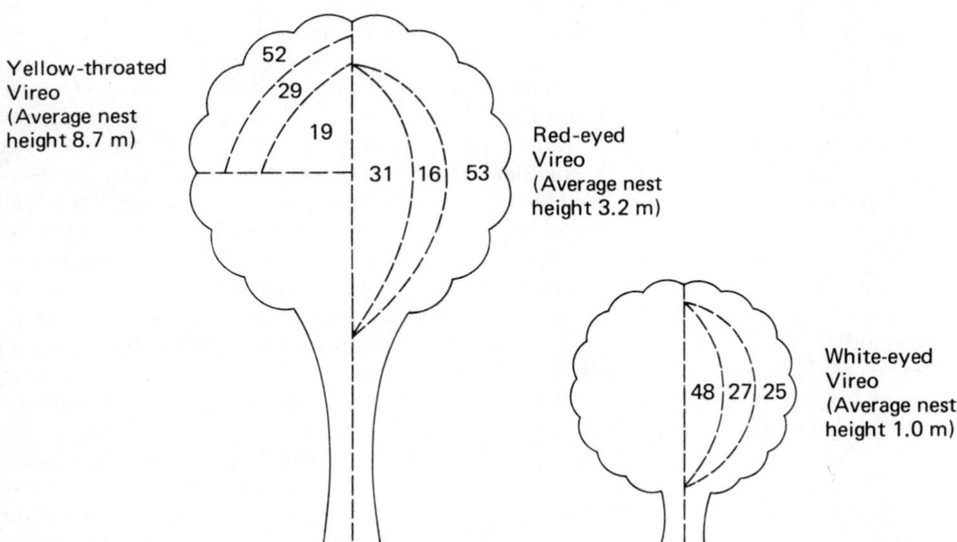

FIGURE 37.2 Summer foraging patterns of three species of vireo in Maryland forests. The numbers represent percentages of foraging time spent in outer, middle, and inner portions of trees. (From Richardson, J. L.: *Dimensions of Ecology.* Baltimore, Williams and Wilkins, 1977.)

by searching beneath bark, and in other ways. Williamson found that the three leaf-gleaning vireos were in turn partitioned by the amount of time they spent feeding in various parts of trees and in the location of the trees they utilized (Fig. 37.2).

Animals with distinctly different stages in their life history may occupy different niches in succession. The frog tadpole is a primary consumer, feeding on plants, but an adult frog is a secondary consumer, feeding on insects and other animals. In contrast, young river turtles are secondary consumers, eating snails, worms, and insects, whereas the adult turtles are primary consumers and eat green plants, such as tape grass.

Competitive Exclusion Principle

Two species may compete for the same space, food, or light, or in escaping from predators or disease; this may be summarized as competition for the same ecologic niche. Competition may result in one species dying off or being forced to change its ecologic niche—to move to a different space or use a different food. Careful ecologic studies usually reveal that there is only one species in an ecologic niche (Gause's rule). One of the clearest examples of ecologic competition was provided by the classic experiments of Gause with populations of paramecia. When either of two closely related species, *Paramecium caudatum* or *Paramecium aurelia,* was cultured separately on a fixed amount of bacteria as food, it multiplied and finally reached a constant level (Fig. 37.3). But when both species were placed in the same culture vessel with a limited amount of food, only *Paramecium aurelia* was left at the end of 16 days. The *Paramecium aurelia* had not attacked the other species or secreted any harmful substance; it simply had been more successful in competing for the limited food supply. Studies in the field generally corroborate Gause's rule. Two fish-eating, cliff-nesting birds, the cormorant and the shag, which seemed at first glance to have survived despite occupying the same ecologic niche, were found upon analysis to have slightly different niches. The cormorant feeds on bottom-dwelling fish and shrimps, whereas the shag hunts fish and eels in the upper levels of the sea. Further study showed that these birds typically have slightly different nesting sites on the cliffs as well.

Experiments similar to Gause's ones with paramecia have been carried out in many laboratories using many kinds of organisms. One species always dies off, show-

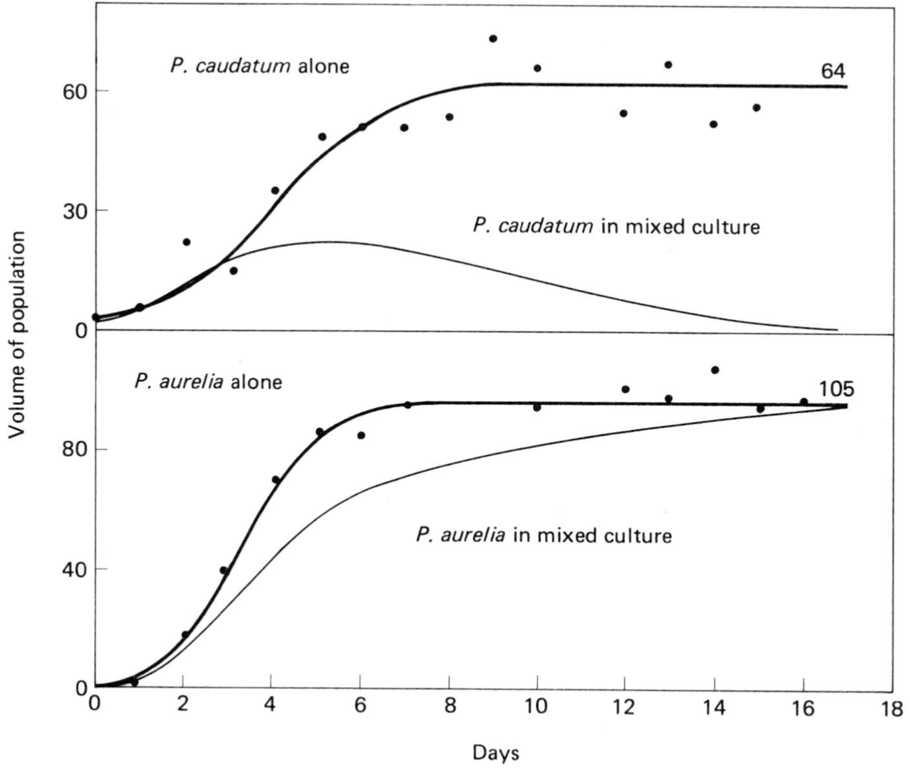

FIGURE 37.3 Competition between two closely related species of protozoa that have identical niches. When separate, *Paramecium caudatum* and *Paramecium aurelia* exhibit normal sigmoid growth in controlled cultures with constant food supply; when together, *P. caudatum* is eliminated. (From Allee, W. C., *et al: Principles of Animal Ecology.* Philadelphia, Saunders College Publishing, 1949; after Gause, G. F.: *The Struggle for Existence.* Baltimore, Maryland, The Williams & Wilkins Co., 1934.)

ing that "competitive exclusion" does occur. Comparable conditions probably do not often occur in nature. Only rarely would two species have exactly the same diet or nest sites. Two competing species can coexist if the effects of crowding are more severe within each species than between the species. In other words, two species can coexist if intraspecific competition is more strict than interspecific competition. If interspecific competition is greater one of the two species will be eliminated. Coexistence can be achieved by small differences in the utilization of resources that are limiting to the populations.

Two species of organisms that occupy the same or similar ecologic niches in different geographical locations are termed **ecological equivalents.** The array of species present in a given type of community in different biogeographic regions may differ widely. However, similar ecosystems tend to develop wherever there are similar physical habitats; the equivalent functional niches are occupied by whatever biological groups happen to be present in the region. Thus, a savanna biome tends to develop wherever the climate permits the development of extensive grasslands, but the species of grass and the species of animals eating the grass may be quite different in different parts of the world. On each of the four continents there are grasslands with large grazing herbivores present. These herbivores are all ecological equivalents. However, in North America the grazing herbivores were bison and prong-horn antelope; in Eurasia, the saga antelope and wild horses; in Africa, other species of antelope and zebra; and in Australia, the large kangaroos. In all four regions, these native herbivores have been replaced to a greater or lesser extent by domesticated sheep and cattle. As examples of ecological equivalents, the species occupying three marine ecologic niches in four different regions of the coast are listed in Table 37.1. The same kinds of ecologic niches are usually present in similar habitats in different parts of the world. Comparisons of such habitats and analyses of the similarities and differences in the species that are ecological equivalents in these different habitats have been helpful in clarifying the interrelations of these different ecologic niches in any given habitat.

TABLE 37.1 Ecological Equivalents: Species Occupying Comparable Ecologic Niches on North American Coasts*

	Grazers on Intertidal Rocks or Marshgrass	Fish Feeding on Plankton
Northeast coast	Periwinkles *Littorina littorea*	Alewife, Atlantic herring
Gulf coast	*Littorina irrorata*	Menhaden, threadfin
Northwest coast	*Littorina danaxis*	Sardine, Pacific herring
Tropical coast	*Littorina scutelata* *Littorina ziczac*	Anchovy

*Adapted from Odum, E. P.: *Fundamentals of Ecology*. 3rd ed. Philadelphia, W. B. Saunders Co., 1971.

The Concept of the Ecosystem

All the living organisms that inhabit a certain area compose the biotic community. A larger unit, termed the **ecosystem,** includes the organisms in a given area and the encompassing physical environment. In the ecosystem, a flow of energy, derived from organism–environment interactions, leads to a clearly defined trophic structure with biotic diversity and to the cyclic exchange of materials between the living and nonliving parts of the system. From the trophic standpoint, an ecosystem has two components: an **autotrophic** part, in which light energy is captured or "fixed" and used to synthesize complex organic compounds from simple inorganic ones, and a **heterotrophic** part, in which the complex molecules undergo rearrangement, utilization, and decomposition. In describing an ecosystem, it is convenient to recognize the following components: (1) the inorganic substances, such as carbon dioxide, water, nitrogen, and phosphate, that are involved in material cycles; (2) the organic compounds, such as proteins, carbohydrates, and lipids, that are synthesized in the biotic phase; (3) the climate, temperature, and other physical factors; (4) the producers, autotrophic organisms (mostly green plants) that can manufacture complex organic materials from simple inorganic substances; (5) the consumers, heterotrophic organisms (mostly animals) that ingest other organisms or chunks of organic matter; and (6) the decomposers, heterotrophic organisms (mostly fungi and bacteria) that break down the complex compounds of dead organisms, absorb some of the decomposition products, and release inorganic nutrients that are made available to the producers to complete the various cycles of elements.

The producers, consumers, and decomposers make up the biomass of the ecosystem—the living weight. In analyzing an ecosystem, the investigator studies the energy circuits present, the food chains, the patterns of biological diversity in time and space, the nutrient cycles, the development and evolution of the ecosystem, and the factors that control the composition of the ecosystem. It is important to appreciate that the ecosystem is the basic functional unit in ecology. It includes both the biotic communities and the abiotic environment in a given region. Each of these influences the properties of the other and both are needed to maintain life on the earth.

A classic example of an ecosystem compact enough to be investigated in quantitative detail is a small lake or pond (Fig. 37.4). The nonliving parts of the lake include the water, dissolved oxygen, carbon dioxide, inorganic salts, such as phosphates, nitrates, and chlorides of sodium, potassium, and calcium, and a multitude of organic compounds. The living part of the lake can be subdivided according to the functions of the organisms, i.e., what they contribute toward keeping the ecosystem operating as a stable, interacting whole. In a lake there are two types of producers: the larger plants growing along the shore or floating in shallow water and the microscopic floating plants, most of which are algae, that are distributed throughout the water as deep as light will penetrate. These tiny plants, collectively known as **phytoplankton,** are usually not visible unless they are present in great abundance and give the water a greenish tinge. They are usually much more important as food producers for the lake than are the more readily visible plants.

The consumers include insects and insect larvae, crustaceans, fish, and perhaps some freshwater clams. Primary consumers are the plant eaters, and secondary

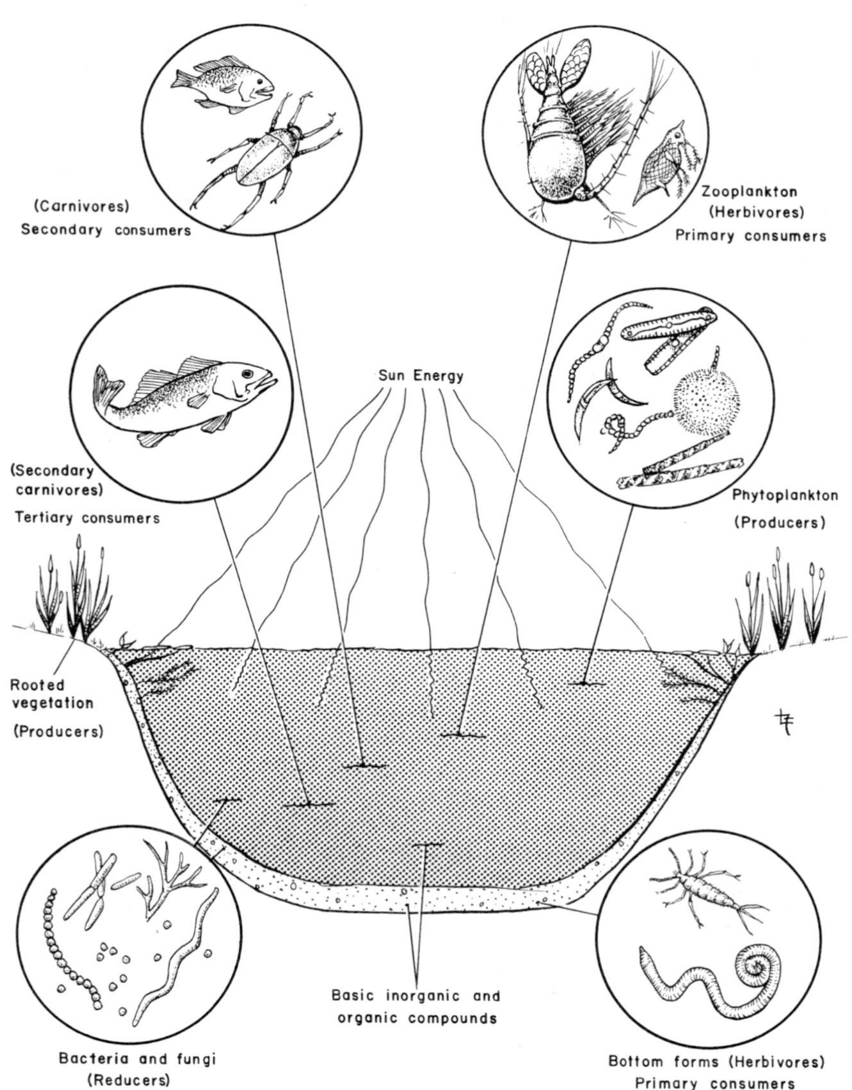

FIGURE 37.4 A small freshwater pond as an example of an ecosystem. The component parts—producer, consumer, and decomposer, or reducer, organisms—plus the nonliving parts are indicated.

consumers are the carnivores that eat the primary consumers. There might be some tertiary consumers that eat the carnivorous secondary consumers. The ecosystem is completed by decomposer organisms, bacteria and fungi that break down the organic compounds of cells from dead producer and consumer organisms either into small organic molecules, which they utilize themselves, or into inorganic substances that can be used as raw materials by green plants. The two major energy circuits in any ecosystem are the **grazing circuit,** in which animals eat living plants or parts of plants, and the contrasting **organic detritus circuit,** in which dead materials accumulate and are decomposed by bacteria and fungi.

The Physical Environment

The numbers and distribution of plants are influenced by climate (**climatic factors**) and by soil (**edaphic factors**). The kinds of plants together with climatic and edaphic factors influence the number and distribution of the various kinds of animals, and this, in turn, may influence plants. The soil on the surface of the earth is not uniform, for weathering, erosion, and sedimentation have produced marked geochemical differentiation of the earth's surface. The distribution of organisms is greatly affected by the kind of soil present. The converse relationship is also true—namely, that the organisms present influence soil formation by contributing to the breakdown of mineral material and to the accumulation of organic debris.

The weathering of the earth's crust has led to the deposition on the surface of the earth of a coat, the characteristics of which depend on the kind of parent rock that was weathered, the kind of weathering processes to which it has been subjected, and its overall age. The term **soil** is applied to this mixture of weathered rock plus organic debris. Most soils remain where they were formed from the parent rock, but some have been carried from their place of origin to another location by the wind (sand dunes or loess), by water (alluvial deposits at the deltas of rivers), or by glaciers. Most of the soils of Canada and the northeastern United States were deposited by the action of glaciers.

After its formation, a soil undergoes development, controlled both directly and indirectly by the climate. The temperature and rainfall determine the rate at which materials in solution and suspension are transported by percolating water out of the soil to places where they accumulate. The nature of the climate determines the kind of vegetation present, and this, in turn, determines what kind of organic materials will be available to be incorporated into the soil. In cold, humid regions where rainfall is greater than evaporation, the conifers produce an acid humus, giving an ash-gray soil termed **podzol.** High temperatures and heavy rainfall in tropical regions result in the development of a red **lateritic** soil with a high iron and aluminum content. In the warm, moist tropical forest the clay minerals rapidly decompose, releasing bases and preventing the soil from becoming acidic. The humus decomposes rapidly and does not accumulate. The silica is leached out, and the oxides of iron, aluminum, and manganese remain, giving the soil a reddish or yellowish color.

In regions with low rainfall and high rates of evaporation, the soil tends to be calcified, rich in calcium carbonate. The different characteristics of the various types of soil determine the kinds of plants that can grow in the region. Soils supply anchorage for the plants, water, mineral nutrients, and aeration of the roots. Some of the important characteristics of a soil are its texture—whether it is gravel, sand, silt, or clay, its organic content, the amount of soil water present, the amount of air entrapped in the soil, and its acidity and salinity.

As soils develop, they tend to become stratified. The uppermost layer is one from which nutrients have been removed by water percolating through it. Below this is a layer of accumulated materials derived from the layer above. The bottom layer is composed of unweathered parent material.

The Cyclic Use of Matter

The total mass of all the organisms that have lived on the earth in the past 3.4 billion years is much greater than the mass of carbon and nitrogen atoms present. According to the law of conservation of matter, matter is neither created nor destroyed; obviously the carbon and nitrogen atoms must have been used over and over again in the course of time. The earth neither receives any great amount of matter from other parts of the universe nor does it lose significant amounts of matter to outer space. The atoms of each element—carbon, hydrogen, oxygen, nitrogen, and the rest—are taken from the abiotic environment, made a part of some cellular component, and finally, perhaps by a quite circuitous route involving several other organisms, are returned to the abiotic environment to be used over again. An appreciation of the roles of green plants, animals, fungi, and bacteria in this cyclic use of the elements can be gained from considering the details of the more important biogeochemical cycles.

The Carbon Cycle. There are about six tons of carbon (as carbon dioxide) in the atmosphere over each acre of the earth's surface. Yet each year an acre of plants, such as sugar cane, will extract as much as twenty tons of carbon from the atmosphere and incorporate it into the plant bodies. If there were no way to renew the supply, the green plants would eventually, perhaps in a few centuries, use up the entire atmospheric supply of carbon. Carbon dioxide fixation by bacteria and animals is another, but quantitatively minor, drain on the supply of carbon dioxide. Carbon dioxide is returned to the air by the decarboxylations that occur in cellular respiration. Plants carry on respiration continuously. Plant tissues are eaten by animals which, by respiration, return more carbon dioxide to the air. But respiration alone would be unable to return enough carbon dioxide to the air to balance that withdrawn by photosynthesis. Vast amounts of carbon would accumulate in the compounds making up the dead bodies of plants and animals. The carbon cycle is balanced by the decay bacteria and fungi that cleave the carbon compounds of dead plants and animals and convert the carbon to carbon dioxide (Fig. 37.5).

When the bodies of plants are compressed under water for long periods of time, they are not decayed by bacteria but undergo a series of chemical changes to form **peat,** later brown coal, or **lignite,** and finally **coal.** The bodies of certain marine plants and animals may undergo somewhat similar changes to form **petroleum.** The peat, coal, and petroleum serve as a **sink** into which the carbon atoms are withdrawn from the cycle for a long period of time. Eventually geologic changes or man's mining and drilling bring the coal and oil to the surface to be burned to carbon dioxide and restored to the cycle.

Other carbon atoms are withdrawn from the cycle into **reservoirs** from which they can readily be retrieved. When carbon dioxide dissolves in water, some reacts with the water to form carbonic acid, bicarbonates, and carbonates. Some carbonates, such as calcium carbonate, are very poorly soluble and tend to precipitate and are deposited as sediments at the bottom of lakes and oceans. Since these reactions are reversible their net effect is to buffer the carbon dioxide content of the air and to keep it relatively constant. If the content of carbon dioxide in the air is decreased in some particular region, carbon dioxide is released from water into the air. If the content of carbon dioxide in the air is increased, more carbon dioxide is dissolved in water and converted to carbonates. The increased burning of fossil fuels, coal, and petroleum in recent decades was

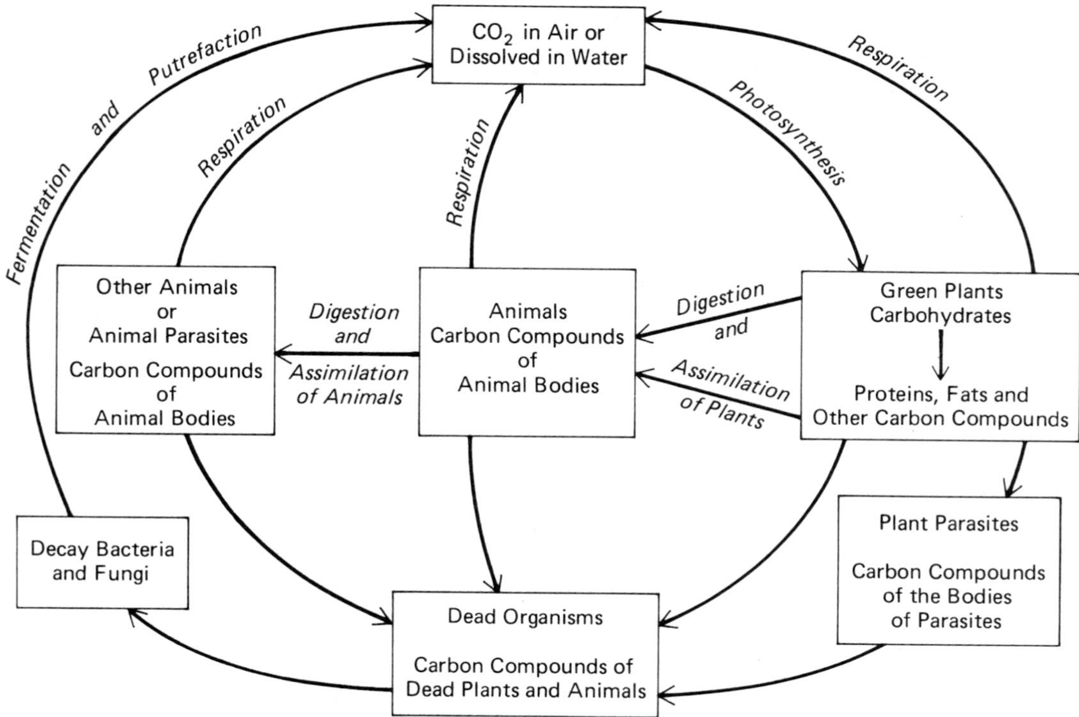

FIGURE 37.5 The carbon cycle in nature. See text for discussion.

enough to raise the content of carbon dioxide in the air by 2.3 parts per million each year. The carbon dioxide content of the air has increased but only by about one part per million each year; the remainder has been absorbed by the buffering system in the oceans.

Most of the earth's carbon atoms are present in limestone and marble as carbonates. The rocks are very gradually worn down, and the carbonates in time are added to the carbon cycle. But other rocks are forming at the bottom of the sea from the calcareous skeletal remains of dead animals and plants, so that the amount of carbon in the carbon cycle remains nearly constant.

The Nitrogen Cycle. The nitrogen for the synthesis of amino acids and proteins is taken up from the soil and water by plants as nitrates (Fig. 37.6). The nitrates are converted to amino groups and used by the plant cells in the synthesis of amino acids and proteins. Animals may eat the plants and utilize the amino acids from the plant proteins in the synthesis of their own proteins and other cellular constituents. When animals and plants die, the decay bacteria convert the nitrogen of their proteins and other compounds into ammonia. Animals excrete several kinds of nitrogen-containing wastes—urea, uric acid, creatinine, and ammonia—and the decay bacteria convert these wastes to ammonia. Most of the ammonia is converted by nitrite bacteria to nitrites, and this in turn is converted by nitrate bacteria to nitrates, thus completing the cycle. Denitrifying bacteria convert some of the ammonia to atmospheric nitrogen. Atmospheric nitrogen can be "fixed," converted to organic nitrogen compounds, such as amino acids, by many species of blue-green algae found in soil, in fresh water, and in the sea, and by the soil bacteria *Azotobacter* and *Clostridium.*

Other bacteria of the genus *Rhizobium,* although unable to fix atmospheric nitrogen themselves, can do this when in combination with cells from the roots of legumes, such as peas and beans. The bacteria invade the roots and stimulate the formation of root nodules, a sort of harmless tumor. The combination of legume cell and bacteria is able to fix atmospheric nitrogen (something neither one can do alone), and for this reason legumes are often planted to restore soil fertility by increasing the content of fixed nitrogen. Nodule bacteria may fix as much as 50 to 100 kg. of nitrogen per acre per year, and free soil bacteria as much as 12 kg. per acre per year.

Atmospheric nitrogen can also be fixed by electrical energy, supplied either by lightning or by the electric company. Although 80 percent of the atmosphere is nitrogen, no animals and only these few plants can utilize it in this form. When the bodies of the nitrogen-fixing bacteria decay, the amino acids are metabolized to ammonia, and this in turn is converted by the nitrite and nitrate bacteria to nitrates to complete the cycle.

The Water Cycle. The great reservoir of water is the ocean. The sun's heat vaporizes water and forms clouds. These, moved by winds, may pass over land, where they are cooled enough to precipitate the water as rain or snow. Some of the precipitated water is evaporated directly back into the atmosphere. Some soaks into the ground; some runs off the surface into streams and goes back to the sea. The ground water is returned to the surface by springs, by pumps, and by transpiration—the movement of water in plants from roots to leaves. In some areas, such as the Amazon basin, most of the rainfall represents recycled transpirational water. Because of this deforestation can result in great changes in climates. This has occurred in the Mediterranean region, which at one time was heavily forested. Water inevitably ends up back in the sea, but it may become incorporated into the bodies of several different organisms, one after another, en route. The energy to run the cycle—the heat needed to evaporate water—comes from sunlight.

Mineral Cycles. As water runs over rocks, it gradually wears away the surface and carries off a variety of minerals, some in solution and some in suspension. Some of these minerals, such as phosphates, sulfates, calcium, magnesium, and others, are necessary for the growth of plants and animals. Phosphorus, an extremely important constituent of all cells, is taken in by plants as inorganic phosphate and converted to a variety of organic phosphates (which are intermediates in the metabolism of carbohydrates, nucleic acids, and fats). Animals get their phosphorus as inorganic phosphate in the water they drink or as inorganic plus organic phosphates in the food they eat.

The phosphorus cycle is not completely balanced, for phosphates are being carried into the sediments at the bottom of the sea faster than they are being returned by the actions of marine birds and fish. Sea birds play an important role in returning phosphorus to the cycle by depositing phosphate-rich **guano** on land. Man and other animals, by catching and eating fish, also recover some phosphorus from the sea. In time, geologic upheavals bring some of the sea bottom back to the surface as new mountains are raised, and in this way minerals are recovered from the sea bottom and made available for use once more.

Solar Radiation

A striking feature of the earth is the great variation in its physical conditions, which range from Arctic tundra to tropical rain forests. Even the oceans are very patchy, nonuniform places. The earth derives nearly all its energy from the sun, but even the sun's energy is not uni-

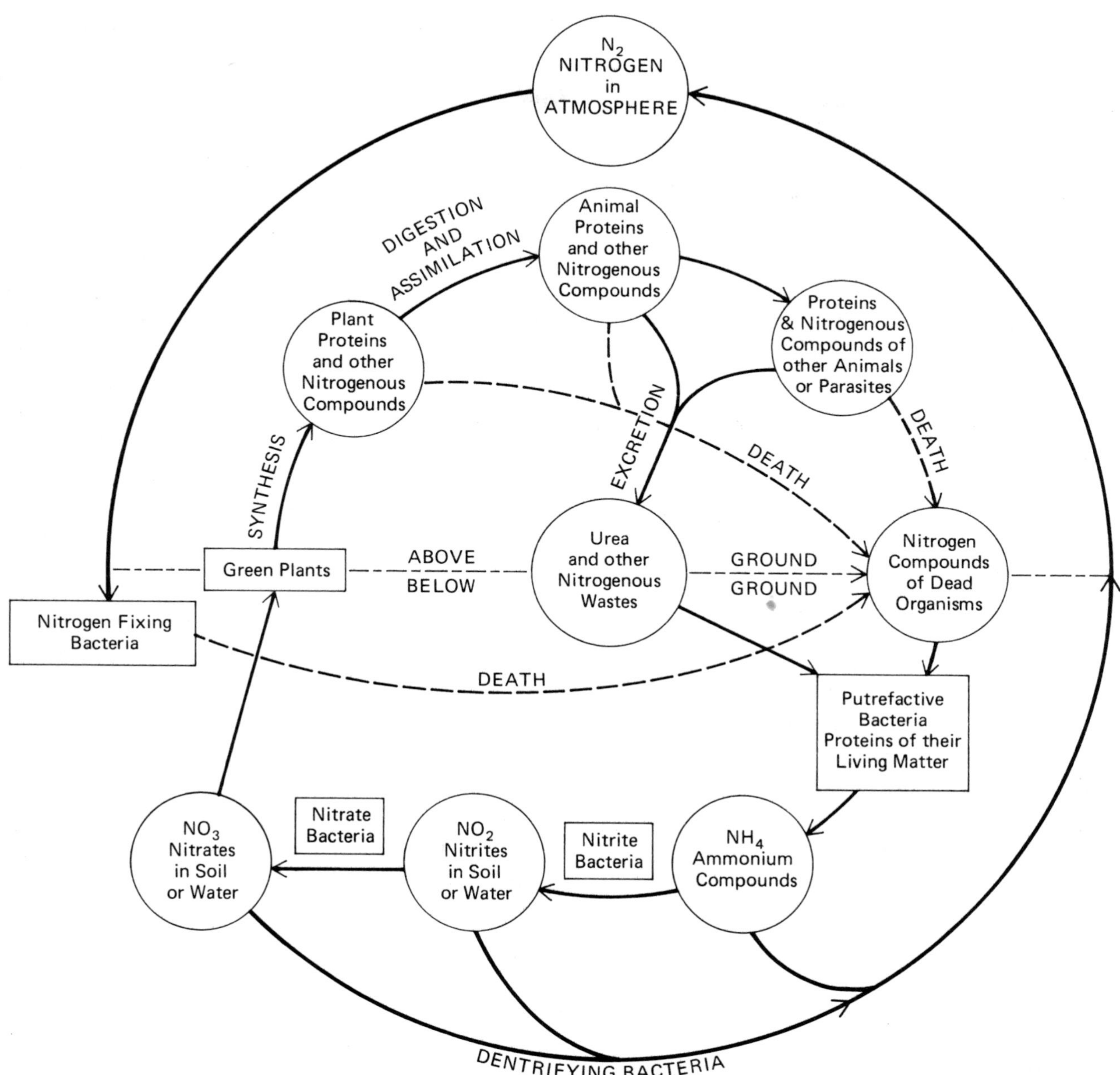

FIGURE 37.6 The nitrogen cycle in nature. See text for discussion.

formly distributed over the face of the globe. The solar radiation reaching the surface of the earth varies with the length of the path that the sun's rays take through the atmosphere (whether it is vertical or at an angle), the area of horizontal surface over which is spread a "bundle" of the sun's rays of a given cross-sectional area, the distance of the earth from the sun (which changes seasonally because of the elliptical orbit of the earth around the sun), the amount of water vapor, dust, and pollutants in the atmosphere, and the total length of the day (the **photoperiod**). At higher latitudes the angle of incidence of the sun rays is less than at middle latitudes, and the energy is spread more thinly. The rays must pass through a thicker layer of atmosphere (Fig.

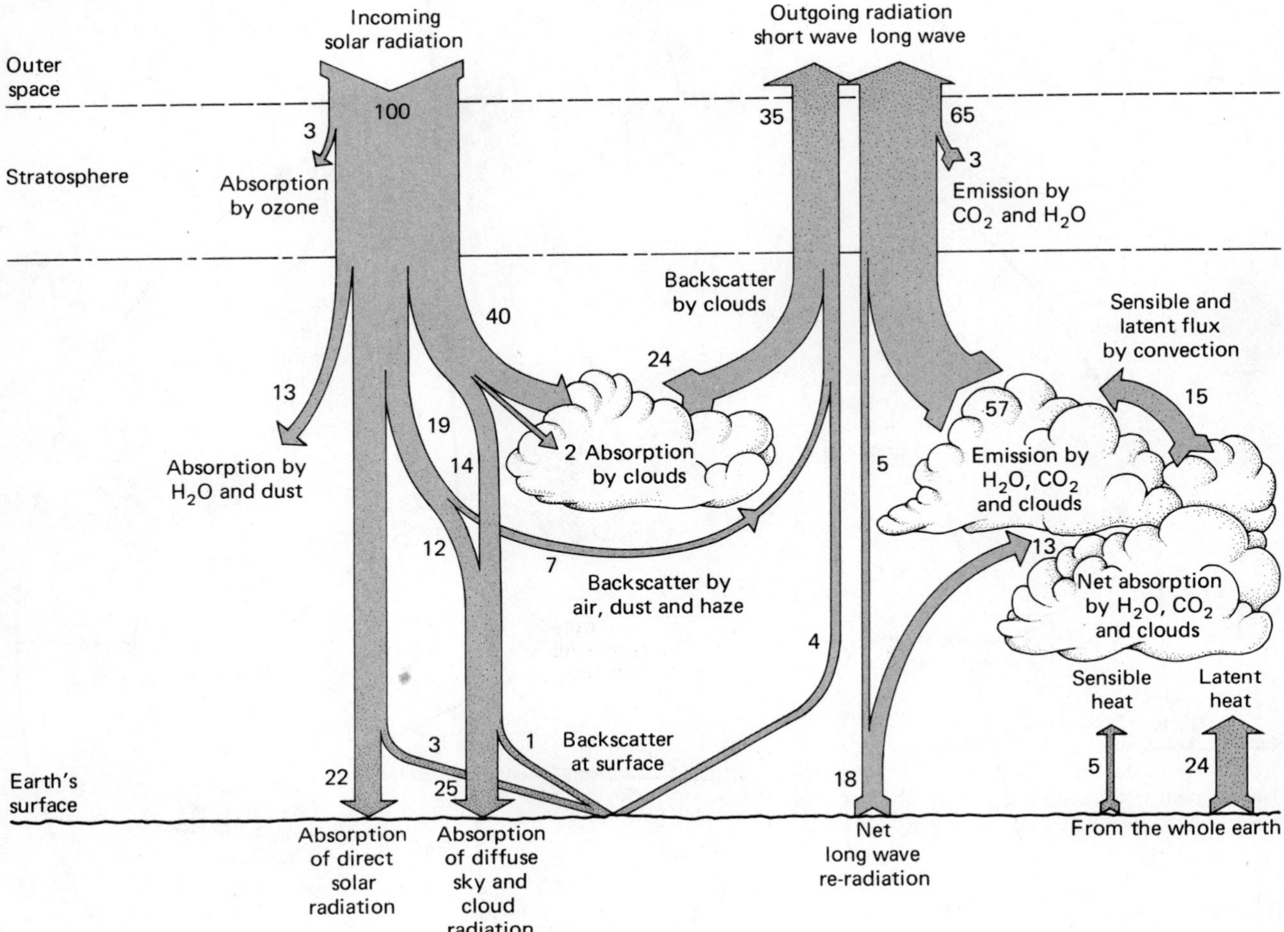

FIGURE 37.7 The annual heat budget of the earth, showing the balance between the heat gained through solar radiation and that lost from the earth. The numbers indicate the average percent loss or gain from various pathways. (From Gross, M. G.: *Oceanography: A View of the Earth.* 2nd ed. Englewood Cliffs, N. J., Prentice-Hall, 1977.)

37.7); consequently, the polar regions receive less radiant energy in the course of a year than do equatorial regions.

The major variations in the amount of incoming solar energy are related to the movements the earth makes with respect to the sun (Fig. 37.8). One complete annual orbit of the earth around the sun requires 365¼ days. The axis of the earth is tilted 23½ degrees in relation to its plane of orbit, and therefore the distribution of energy varies throughout the year. The northern hemisphere receives more radiant energy during the period between March 21 and September 21 than in the other half of the year, not only because there are more hours of daylight but also because the angle of incidence of sunlight is more nearly vertical during that period.

The rotation of the earth on its own axis every 24 hours produces day and night and the energy changes associated with these periods. The changes in temperature lag behind the changes in the amount of light energy received. The hottest days in the northern hemisphere are in July and August, not on June 21. The maximum temperature during the day is usually in

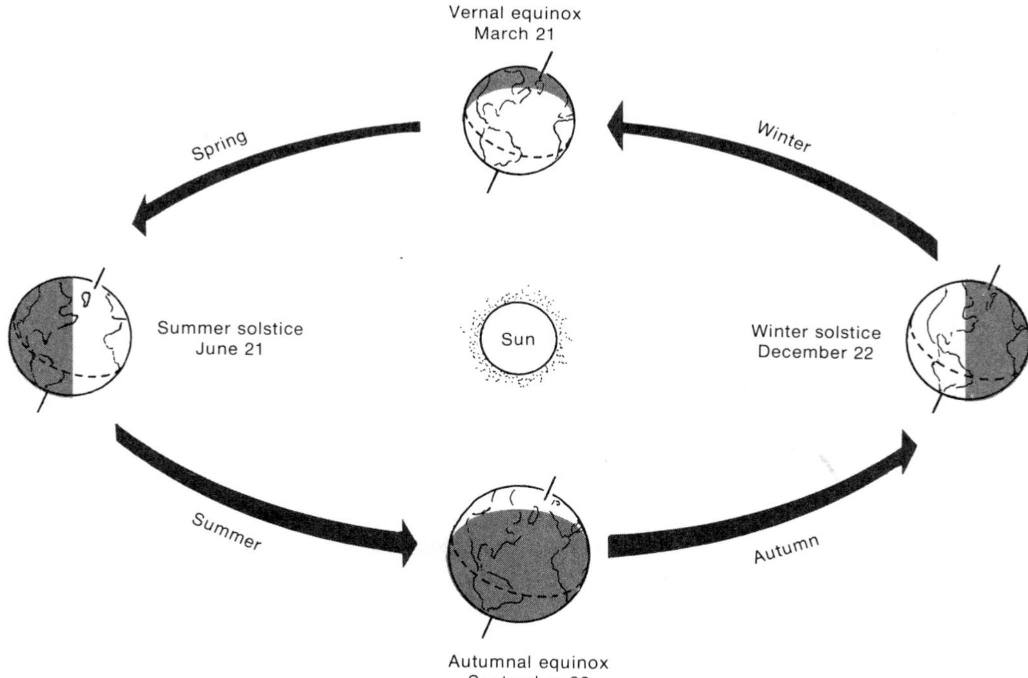

FIGURE 37.8 The sunlit portions of the Northern Hemisphere are seen to vary from greater than one half in summer to less than one half in winter. The proportion of any latitude that is sunlit is also the proportion of the 24-hour day between sunrise and sunset. (From MacArthur, R. H., and J. H. Connell: *The Biology of Populations.* New York, John Wiley & Sons, Inc., 1966.)

midafternoon, and the lowest temperature is just before sunrise. The temperature of the soil tends to lag even more than this. The atmosphere changes the distribution of the energy of different wave lengths so that the nature of sunlight actually reaching the earth is different from the sunlight some 100 miles above the atmosphere. Radiant energy of short wave lengths is not absorbed by water or water vapor; it thus passes through the atmosphere with little diminution. Some of the energy of sunlight is absorbed by the earth; some is reflected back as longer wave lengths or heat. Much of this reflected heat is absorbed by molecules of carbon dioxide and water before being radiated back into space. The earth returns to space as much energy as it receives (Fig. 37.7).

Snow, water, and light-colored soils reflect solar radiation, whereas bare ground with dark soils absorbs it. This situation has been altered further by man's activities in paving large areas of the earth and in building, plowing, removing trees, and causing air pollution. It is estimated that as much as 40 percent of the heat of the atmosphere is derived from the condensation of water vapor produced from the evaporation of water from the surface of the ocean. The moisture-laden air rises, moves to higher latitudes where it is cooled, and gives up its moisture as snow or rain. The heat of condensation is then absorbed by the atmosphere. The carbon dioxide in the atmosphere, heated from below and radiating heat back to the surface of the earth, serves as a heat trap, as does the roof of a greenhouse whose glass substitutes for clouds and water vapor.

Since water is slow to heat up and cool down, the vast oceanic waters store and distribute heat derived from solar radiation. Sea water heated in the tropics is circulated by ocean currents to northern latitudes where heat is released. The warm waters of the Gulf Stream, for example, moderate the climate of northwestern Europe.

The force of trade winds at the ocean's surface causes westerly equatorial currents in the oceans north and south of the equator. When these encounter the east coasts of Asia and America they are reflected north and south. The currents turn eastward at high latitudes, reach the west coasts of the Americas, Europe, and Africa, and then return to the equator. These currents are responsible for the enormous clockwise circulations of water in the north Pacific and Atlantic Oceans and for the counterclockwise circulations in the oceans of the Southern Hemisphere (Fig. 37.9).

Energy Flow and Food Chains

Most of the sun's energy that reaches the earth is eventually lost as heat. A small proportion of the energy of sunlight is absorbed by plants, and a small portion of this is transformed into the potential energy of stored food products. The rest of the energy leaves the plant and becomes part of the earth's general heat loss. All living things, except green plants, obtain their energy by taking in the products of photosynthesis, carried out by green plants, or the products of chemosynthesis, carried out by microorganisms.

As the potential energy of sunlight is transferred from plants and other primary producers, through herbivores and their carnivorous predators and parasites, and ultimately, following their deaths, through the decomposer microorganisms, a large portion of the energy is lost at each step as heat. Because of this progressive loss of energy as heat, the total energy flow at each succeeding level is less and less. When an animal eats food, less than 20 percent of the foodstuff is eventually converted into the flesh of the animal that is doing the eating. The domestic pig is one of the more efficient converters; under the best feeding practices, a pig will convert about 20 percent of the mass of the food that it eats into pork chops and bacon.

The transfer of energy through a biological community begins when the energy of sunlight is fixed in a green plant by photosynthesis. It is estimated that only 8 percent of the energy of the sun reaching the planet strikes green plants and that only 1 percent of this is utilized in photosynthesis. Part of this energy is used by

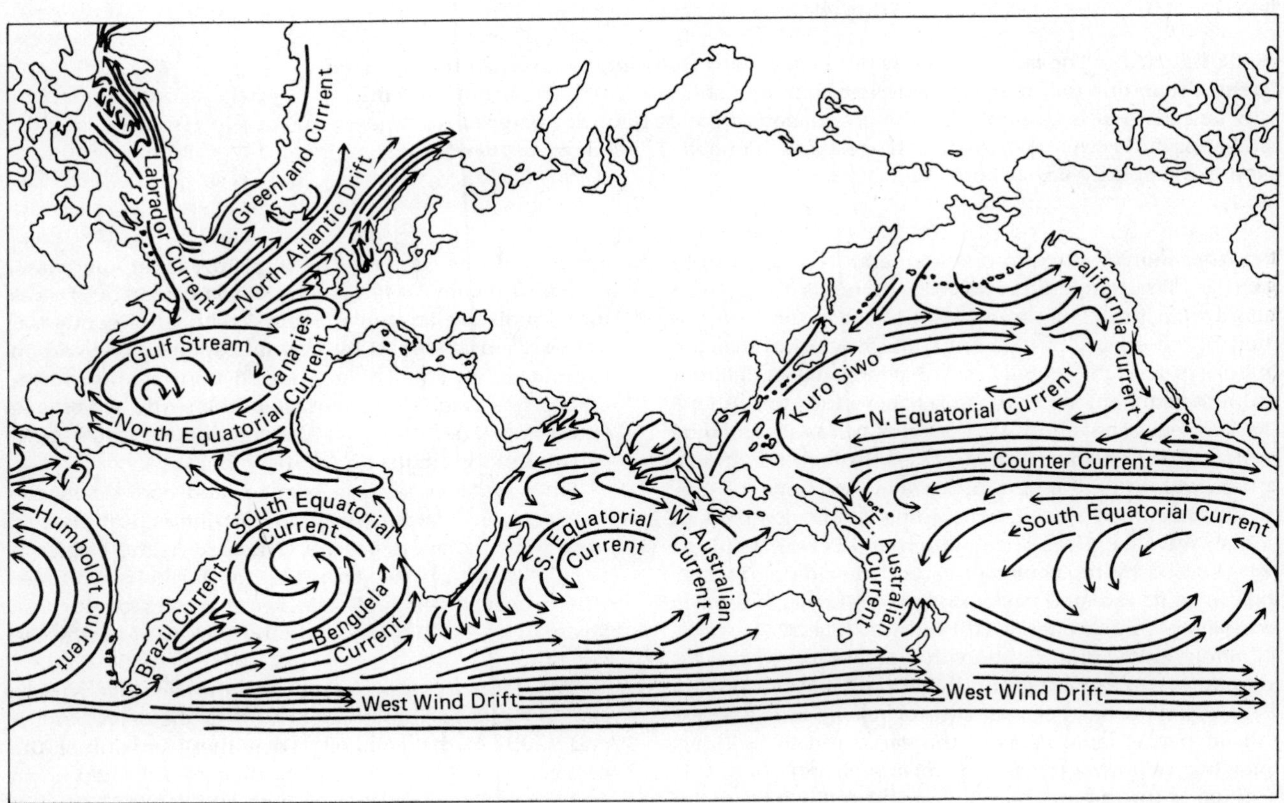

FIGURE 37.9 Pattern of the world's ocean currents. (From MacArthur, R. H., and Connell, J. H.: *The Biology of Populations.* New York, John Wiley & Sons, Inc., 1966.)

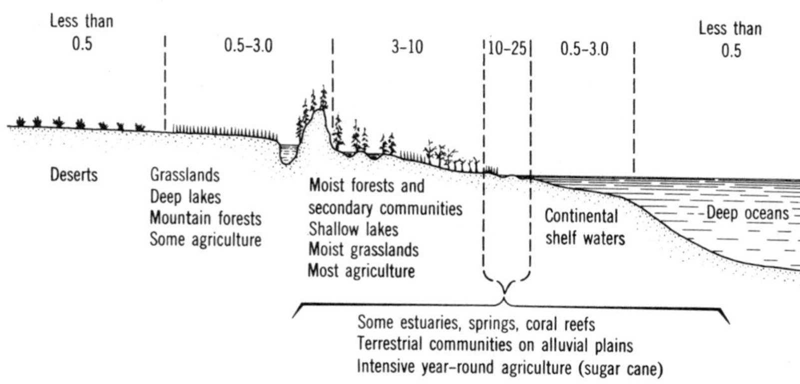

FIGURE 37.10 The world distribution of primary production in grams of dry matter per square meter per day, as indicated by average daily rates of gross production in major ecosystems. (From Odum, E. P.: *Fundamentals of Ecology*. 3rd ed. Philadelphia, Saunders College Publishing, 1971.)

the plant itself to drive the many processes required for maintenance. The amount left over that is stored and expressed as new cytoplasm, or growth, represents the **net primary production.**

The net primary production of a field of sugar cane in Hawaii was 190 kcal. per square meter per day. The average insolation was about 4000 kcal. per square meter per day. From this we can calculate that the net efficiency of the sugar cane is about 4.8 percent. Such values can be achieved only by crops under intensive cultivation during a favorable growing season. On an overall annual basis, sugar cane fields have an efficiency of about 1.9 percent and tropical forests have an efficiency of about 2 percent. The stored energy accumulates as living material or biomass. Part is recycled each season by death and decomposition of the organisms; the part that is alive at any particular moment is called the **standing crop biomass.** This, of course, can vary greatly with the season. In grasslands, there is an annual turnover of the biomass, but in forests, much of the energy is tied up in wood. The most productive ecosystems on an energy basis are coral reefs and the estuaries of rivers (Fig. 37.10). The least productive are deserts and the open ocean. By and large the production of plant material in each area of the earth has reached an optimum level that is limited only by soil and climate.

The transfer of energy from its ultimate source in a plant through a series of organisms, each of which eats the preceding and is eaten by the following, is known as a **food chain,** and each step in the chain is called a **trophic level.** Since a particular species is usually eaten by several different consumers, and a particular consumer usually feeds on a number of different species, the trophic relationships between species occupying the same area are really a **food web.** However, any one path through a food web constitutes a food chain. Some species not only are members of different food chains but also may occupy different positions in different food chains (Fig. 37.11). An animal may be a primary consumer in one chain, eating green plants, but a secondary or tertiary consumer in other chains, eating herbivorous animals or other carnivores.

At each step in the food chain there is loss of energy, for when one animal eats another, (1) some of the ingested organic material may not be usable (e.g., fish scales, chitinous exoskeleton, cellulose), (2) some must be used as fuel for cell respiration, and (3) only some is used for the synthesis of new cytoplasm, the **net production** for this trophic level and that which will be available to the next.

The number of steps in a food chain is limited to perhaps four or five because of the great decrease in available energy at each step. The percentage of food energy consumed that is converted to new cellular material, and thus is available as food energy for the next animal in the food chain, is known as the percentage efficiency of energy transfer.

The flow of energy in ecosystems, from sunlight through photosynthesis in autotrophic producers, through the tissues of herbivorous primary consumers and the tissues of carnivorous secondary consumers, determines the number and total weight (**biomass**) of organisms at each level in the ecosystem. The flow of energy is greatly reduced at each successive level of nutrition because of the heat losses at each transformation of energy. This decreases the total biomass produced in each level.

Humans are at the end of a number of food chains; for example, we eat a fish, such as a black bass, that ate little fish that ate small invertebrates that ate algae. The ultimate size of the human population (or the population of any animal) is limited by the length of the food chain, the percentage efficiency of energy transfer at each step in the chain, and the amount of light energy falling on the earth.

Since humans can do nothing about increasing the amount of incident light energy and very little about the percentage efficiency of energy transfer, they can in-

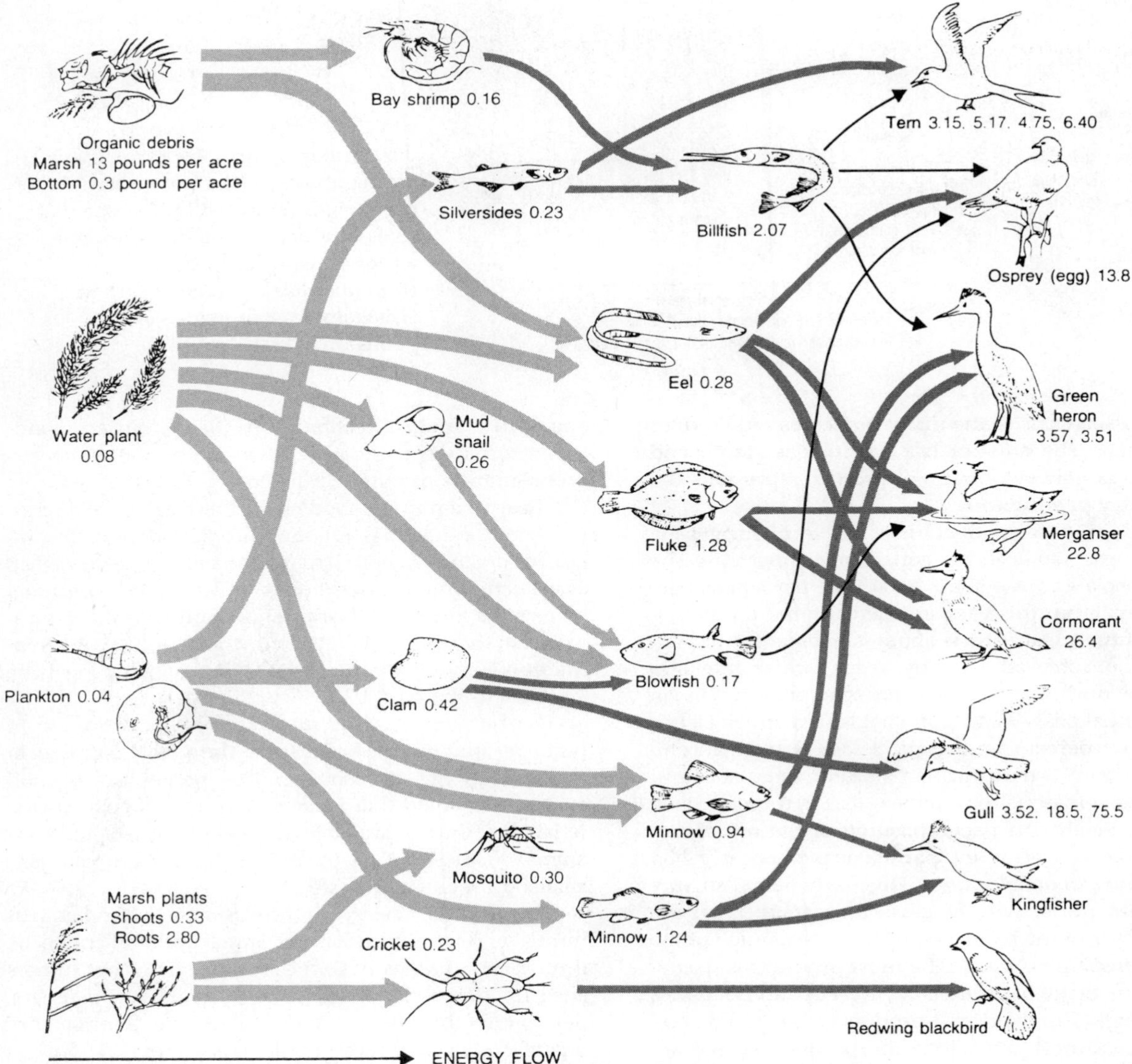

FIGURE 37.11 Part of the food web in the estuaries of Long Island. The thickest arrows are between the first and second trophic levels; the thinnest arrows are between the second and third levels. This figure was taken from a study of DDT accumulation in the food chain, and the numbers after each organism indicate the DDT content in parts per million. The rising numbers thus reflect the energy flow through the food web and the accumulation of pesticides in biomass at higher levels. (After Woodwell, G. M.: "Toxic substances and ecological cycles." Copyright © 1967 by Scientific American, Inc. All rights reserved.)

crease their supply of food energy only by shortening the food chain, i.e., by eating the primary producers, plants, rather than animals. In overcrowded countries such as India and China people are largely vegetarians because this food chain is shortest and a given area of land can in this way support the greatest number of people. Steak is a luxury in both ecological and economic terms, but hamburger is just as much an ecological luxury as steak is.

In addition to predator food chains, such as the human–black bass–minnow–crustacean one, there are parasite food chains. For example, mammals and birds are parasitized by fleas; in the fleas live protozoa that are, in turn, the hosts of bacteria. Since the bacteria might be parasitized by viruses, there could be a five-step parasite food chain.

A third type of food chain is one in which unusable plant and animal materials in feces and dead plant and animal remains are converted into dead organic matter, **detritus,** before being eaten by animals, such as millipedes and earthworms on land, by marine worms and mollusks, or by bacteria and fungi. In a community of organisms in the shallow sea, about 30 percent of the total energy flows via detritus chains, but in a forest community, with a large biomass of plants and a relatively small biomass of animals, as much as 90 percent of energy flow may be via detritus pathways. In an intertidal salt marsh, where most of the animals—shellfish, snails, and crabs—are detritus eaters, 90 percent or more of the energy flow is via detritus chains.

Since, in any food chain, there is a loss of energy at each step, it follows that there is a smaller biomass in each successive step. H. T. Odum has calculated that 8100 kg. of alfalfa plants are required to provide the food for 1000 kg. of calves, which provide enough food to keep one 12-year-old, 48-kg. boy alive for one year. Although boys eat many things other than veal, and calves other things besides alfalfa, these numbers illustrate the principle of a food chain. A food chain may be visualized as a **pyramid;** each step (trophic level) in the pyramid is much smaller than the one on which it feeds. Since the predators are usually larger than the ones on which they prey, the pyramid of numbers of individuals in each step of the chain is even more striking than the pyramid of the mass of individuals in successive steps: one boy requires 4.5 calves, which require 20,000,000 alfalfa plants.

The trophic relationships of populations are always pyramidal if total production of all individuals is included over a number of years but not necessarily if we consider only standing crops. Marine food chains commonly exhibit a pyramid of numbers and an inverse pyramid of biomass. For example, there may be more diatoms in the standing crop than grazing copepods and more copepods than menhaden (fish), but the menhaden biomass is greater than that of copepods, which is greater than that of diatoms. The longer life spans of the members of the higher trophic levels account for this. The biomass of menhaden represents accumulation from several generations of copepods, each of which last about a year. Copepods in turn are grazing on diatoms, which divide mitotically, and produce many generations during a single year. But the standing crop represents the biomass and numbers of only one generation at any particular moment in time.

Standing crops of forest exhibit a pyramid of biomass but an inverted pyramid of numbers (or diamond pattern) because there are fewer trees than herbivores feeding upon them, but the total biomass of the long-lived trees is very great.

Summary

1. Ecology is the study of living organisms in their native environment. The term "autecology" refers to studies of individual organisms and their relations to the environment, whereas "synecology" refers to studies of groups of organisms—populations, communities, and ecosystems—that live together and form a functional unit of the environment.

2. The distribution of each species is limited by barriers to migration and by its range of tolerance to variations in the many environmental factors, such as light, photoperiod, temperature, and water.

3. The habitat of an organism is the place where it lives, a physical area, some specific part of the earth's surface—air, soil, or water. The ecologic niche of an organism includes not only the physical space it occupies but also its functional role in the community, its trophic position. The ecologic niche includes not only where the organism lives but what it does and how it is acted upon by other species. Two species that occupy the same or similar ecologic niches in different geographical locations are termed ecological equivalents.

4. The numbers and distribution of plants are influenced by climatic and edaphic (soil) factors, and the kinds of plants, together with the climatic and edaphic factors, influence the numbers and distribution of animals, which in turn may influence plants.

5. The atoms of each element are taken from the abiotic environment, made a part of some organism, and, perhaps by a circuitous route involving several other organisms, are returned to the abiotic environment to be reused. An appreciation of the roles of green plants, animals, fungi, and bacteria in the cyclic use of matter is gained by considering some of the tangle of interrelationships in the carbon, nitrogen, water, and mineral cycles.

6. An outstanding feature of the earth is the nonuniformity of its physical conditions on both land and sea. The sun's energy in solar radiation is also not uniformly distributed over the earth. The rotation of the earth and the heating from solar radiation sets up the trade winds, and these determine the major currents in the ocean that move clockwise in the Northern Hemisphere and counterclockwise in the Southern Hemisphere. These currents have a marked impact on the climate of the adjacent continents.

7. The potential energy of sunlight is captured by the producers, green plants, in photosynthesis and then transferred through primary consumers, herbivores, and secondary and tertiary consumers and parasites, ultimately, after they die, through the decomposers, the bacteria and fungi. A large fraction of the energy is lost at each step as heat. Thus the total energy flow at each succeeding level is less and less. The efficiency of conversion of foodstuffs into the flesh of the animal eating it is less than 20 percent. The transfer of energy through a series of organisms, each of which eats the preceding and is eaten by the following, is called a "food chain." The flow of energy through the system determines the number and total weight (the biomass) of the organisms at each trophic level.

References and Selected Readings

Benton, A. H., and W. E. Werner: *Field Biology and Ecology.* 3rd ed. New York, McGraw Hill, 1974. A standard text of ecology.

Brewer, R.: *Principles of Ecology.* New York, Holt, Rinehart & Winston, 1979. A standard text of ecology.

Cox, G. W.: *Laboratory Manual of General Ecology.* 4th ed. New York, William C. Brown, 1980. A fine laboratory manual for a course in ecology.

DuBose, R.: *The Wooing of Earth.* New York, Scribners, 1980. A popular presentation of current ecological problems.

Edgerton, F.: *History of American Ecology.* 3rd ed. Salem, New Hampshire, Ayer Co., 1978. A history of the ecology movement in America.

Haynes, R.: *Environmental Science Methods.* New York, Methuen, 1982. A description of field and laboratory methods used in ecological investigation.

Ito, Y.: *Comparative Ecology.* 2nd ed. New York, Cambridge University Press, 1981. A general text of ecology.

Kormondy, E. J.: *Concepts of Ecology.* 2nd ed. Englewood Cliffs, New Jersey, Prentice-Hall, 1976. An excellent paperback survey of the principles of ecology.

Krebs, C.: *Ecology.* New York, Harper & Row, 1979. A general text of ecology; a good reference book.

McNaughton, S. J., and L. L. Wolf: *General Ecology.* 2nd ed. New York, Holt, Rinehart & Winston, 1979. A standard text of ecology.

Milne, L. J., and Milne, M.: *Ecology Out of Joint.* New York, Charles Scribners Sons, 1977. A popular presentation of some of the controversies in ecology.

Mory, R. M.: *Theoretical Ecology.* 2nd ed. Sunderland, Massachusetts, Sinauer Associates, 1981. A text of the biological theories underlying ecology for advanced students.

Owen, D.: *What is Ecology?* 2nd ed. New York, Oxford University Press, 1980. A general text of ecology.

Smith, R. L.: *Elements of Ecology and Field Biology.* New York, Harper & Row, 1977. A general text of ecology, somewhat simplified.

Smith, L. R.: *Ecology and Field Biology.* 3rd ed. New York, Harper & Row, 1980. A very useful general text of ecology and field biology.

Tribe, M. A.: *The Ecology Game.* New York, Cambridge University Press, 1976. A popular book on the economics, politics, and biology of ecological controversies.

38 Ecology of Populations and Communities

ORIENTATION

The many species of plants and animals are organized into groups that interact at many levels and in many ways with each other and with the physical environment. Ecologists distinguish a population, a group of organisms of the same species that occupy a given area, from a community, an assemblage of populations living in a defined area. In this chapter we will examine the characteristics of populations, the nature of a population growth curve, the factors that determine its shape, and the meaning of the terms "biotic potential" and "carrying capacity." We will explore the nature of a survivorship curve and the factors that determine its shape. We will discuss the meaning of the terms "r selection" and "K selection," their roles in evolution, and the factors that limit the size of a given population. We will describe the cyclic increase and decrease in the size of a given population and look into the factors responsible for these population cycles. The tendency of certain populations to disperse or to aggregate in clumps will be examined, together with the factors that govern these phenomena. We will explain the factors that determine the structure of biotic communities and their changes with time, the succession of seral stages of the community to the climax community. Finally, we will describe the large, distinct, easily differentiated community units called biomes, distinguished by the kinds of plants and animals that compose them.

Populations and Their Characteristics

A **population** may be defined as a group of organisms of the same species that occupy a given area. It has characteristics that are a function of the whole group and not of the individual members; these are **population density, birth rate, death rate, age distribution, biotic potential, rate of dispersion, and growth form.** Although individuals are born and die, individuals do not have birth rates or death rates; these are characteristics of the population as a whole.

One important attribute of a population is its **density** —the number of individuals per unit area or volume, e.g., human inhabitants per square mile, trees per acre in a forest, millions of diatoms per cubic meter of sea water. This is a measure of the population's success in a given region. In ecological studies it is important to know not only the population density but also whether it is changing and, if so, what the rate of change is.

A measure of the population density of a large area, which is usually hard to determine precisely, can be made by various methods of sampling, such as counting the number of limpets on one square meter of intertidal rock, the number of insects caught per hour in a standard trap, or the number of birds seen or heard per hour. A method that will give good results when used with the proper precautions is that of capturing, let us say, 100 animals, tagging them in some way, and then releasing them. On some subsequent day, another 100 animals are trapped, and the proportion of tagged animals is determined. This assumes that animals caught once are neither more likely nor less likely to be caught again and that both sets of trapped animals are random samples of the population. If the ratio of untagged to tagged animals in the second sample is 100/20, then the ratio of total population size in the area (x) to the first sample is $x/100$, and $x = 500$, for $x/100 = 100/20$.

For many types of ecological investigations, an estimate of the number of individuals per total area or volume, known as the "crude density," is not sufficiently precise. Only a fraction of that total area may be a suitable habitat for the population, and the size of the individual members of a population may vary tremendously. Ecologists, therefore, calculate an **ecologic density,** defined as the number, or more exactly as the mass, of individuals per area or volume of habitable space. Trapping and tagging experiments might give an estimate of 500 rabbits per square mile, but if only half of that square mile actually consists of areas suitable for rabbits to live in, then the ecologic density will be 1000 rabbits per square mile of rabbit habitat. With species whose members vary greatly in size, such as fish, live weight or some other estimate of the total mass of living fish is a much more satisfactory estimate of density than simply the number of individuals present.

A graph in which the number of organisms is plotted against time is a **population growth curve** (Fig. 38.1). Such curves are characteristic of populations, and are amazingly similar for populations of almost all organisms from bacteria to human beings.

Population growth curves have a characteristic shape. When a few individuals enter a previously unoccupied area, growth is slow at first (the positive acceleration phase), then becomes rapid and increases exponentially (the logarithmic phase). The growth rate eventually slows down as environmental resistance gradually increases (the negative acceleration phase) and finally reaches an equilibrium or saturation level. The upper asymptote of the sigmoid curve is termed the **carrying capacity** of the environment.

Sheep were introduced into Tasmania in 1800. Careful records of their numbers were kept, and by 1850 the sheep population had reached 1,700,000. It remained more or less constant at this carrying capacity for nearly a century.

The birth rate, or natality, of a population is simply the number of new individuals produced per individual per unit time. The **intrinsic reproductive rate (maximum birth rate)** is the largest number of organisms that could be produced per unit time under ideal con-

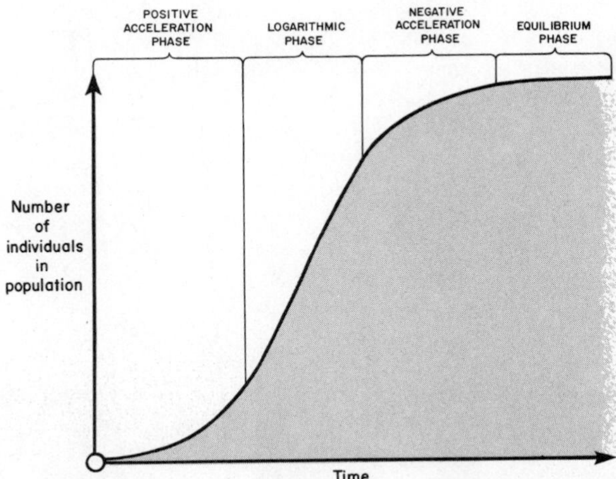

FIGURE 38.1 A typical sigmoid (*S*-shaped) growth curve of a population in which the total number of individuals is plotted against time. The absolute units of time and the total number in the population would vary from one species to another, but the shape of the growth curve would be similar for all populations.

ditions, when there are no limiting factors. This is a constant for a species, determined by physiological factors, such as the number of eggs produced per female per unit time, the proportion of females in the species, and so on. The actual birth rate is usually considerably less than this, for not all the eggs laid are able to hatch, not all the larvae or young survive, and so on. The size and composition of the population and a variety of environmental conditions affect the actual birth rate. It is difficult to determine the intrinsic reproductive rate, for it is difficult to be sure that all limiting factors have been removed. However, under experimental conditions, one can get an estimate of this value that is useful in predicting the rate of increase of the population and in providing a yardstick for comparison with the actual birth rate.

The mortality rate of a population refers to the number of individuals dying per unit time. There is a theoretical **minimum mortality,** somewhat analogous to the maximum birth rate, that is the number of deaths that would occur under ideal conditions, deaths due simply to the physiological changes of old age. This minimum mortality rate is also a constant for a given population. The actual mortality rate will vary depending upon physical factors and on the size, composition, and density of the population.

By plotting the number of survivors in a population against time, one gets a **survivorship curve** (Fig. 38.2). If the units of the time axis are expressed as the percentage of total life span, the survivorship curves for organisms with very different total life spans can be compared. Modern medical and public health practices have greatly increased the average life expectancy in developed countries, and the curve for human survival approaches the curve for minimal mortality. From such curves one can determine at what stage in the life cycle a particular species is most vulnerable. Reducing or increasing mortality in this vulnerable period will have the greatest effect on the future size of the population. Since the death rate is more variable and affected to a greater extent by environmental factors than the birth rate, it has a primary role in controlling the size of a population.

It is obvious that populations that differ in the relative numbers of young and old will have different characteristics, different birth and death rates, and different prospects. Death rates typically vary with age, and birth rates are usually proportional to the number of individuals able to reproduce. Three ages can be distinguished in a population in this respect: prereproductive, reproductive, and postreproductive. A. J. Lotka has shown from theoretical considerations that a population will tend to become stable and have a constant proportion of individuals of these three ages.

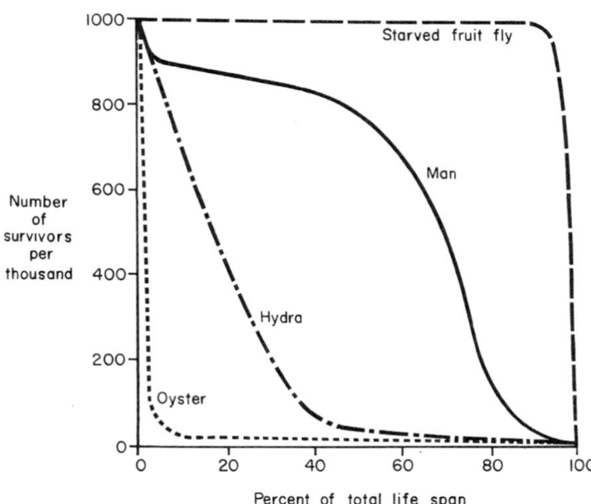

FIGURE 38.2 Survivorship curves of four different animals, plotted as number of survivors left at each fraction of the total life span of the species. The total life span for man is about 100 years; the solid curve indicates that about 10 percent of the babies born die during the first few years of life. Only a small fraction of the human population dies between ages 5 and 45, but after 45 the number of survivors decreases rapidly. Starved fruit flies live only about five days, but almost the entire population lives the same length of time and dies at once. The vast majority of oyster larvae die, but the few that become attached to the proper sort of rock or to an old oyster shell survive. The survivorship curve of hydras is one typical of most animals and plants, in which a relatively constant fraction of the population dies off in each successive time period.

Much is learned about the population biology of a species by systematic sampling over time. For example, monthly samples of marine copepod populations provide information about the age structure of the population at any time in the year, i.e., the relative numbers of individuals that are adult or in various larval stages. Such sampling data indicate the sex ratio, the time or times of egg production, the average life span, as well as the length of the various larval stages. It will show the continually changing population structure over the course of the year. For longer-lived plants and animals, such censuses are of value in predicting population trends. Rapidly growing populations have a high proportion of young forms, and when the age classes are graphed, the population structure is reflected in a pyramid with a very wide base. Declining populations have a smaller base of young individuals than certain of the older age classes. The age of fishes can be determined from the growth rings on their scales, and studies of the age ratios of commercial fish catches are of great use in predicting future catches and in preventing overfishing of a region.

The term **biotic potential,** or reproductive potential, refers to the intrinsic power of a population to increase in numbers when the age ratio is stable and all environmental conditions are optimal. The biotic potential is defined mathematically as the slope of the population growth curve during the logarithmic phase of growth. When environmental conditions are less than optimal, the rate of population growth is less. The difference between the potential ability of a population to increase and the actual change in the size of the population is a measure of environmental resistance.

Even when a population is growing rapidly in number, each *individual* organism of the reproductive age carries on reproduction at the same rate as at any other time; the increase in numbers is due to increased survival. The rates of increase, r, expressed as rate of increase per female per year are human, 0.0055; laboratory rat, 5.4; flour beetle, 36.8. If exponential growth continued for one year, the human population would increase 1.0055 times, the rat population 221 times, and the flour beetle population 1.06×10^{15} times. However, factors in the environment intercede, and this rate is not maintained.

The sum of the physical and biological factors that prevent a species from reproducing at its maximum rate is termed the **environmental resistance.** The difference between the theoretical rate of growth and the actual rate is a measure of environmental resistance. Environmental resistance is often low when a species is first introduced into a new territory so that the species increases in number at a fantastic rate, as when the rabbit was introduced into Australia and the English sparrow and Japanese beetle were brought into the United States. But as a species increases in number the environmental resistance to it also increases, in the form of organisms that prey upon it or parasitize it, and the competition between the members of the species for food and living space.

In an essay in 1798 the Englishman Robert Malthus pointed out this tendency for populations to increase in size until checked by the environment. He realized that these same principles apply to human populations and suggested that wars, famines, and pestilences are inevitable and necessary as brakes on population growth. Since Malthus' time the earth's productive capacity has increased tremendously as has the total human population. But Malthus' basic principle, that there are physical limits to the amount of food that can be produced for any species, remains true. The earth has a finite carrying capacity for human beings just as it does for any other animal. As environmental resistance increases, the rate of increase of the human population will eventually have to decrease. An equilibrium will be reached either by decreasing the birth rate or by increasing the mortality rate.

To deal with the concepts of biotic potential and environmental resistance in mathematical terms, the symbol "r" is used to represent biotic potential or intrinsic rate of natural increase and the symbol "K" is used to represent the carrying capacity; that is, the maximal population size that can be borne by the resources of the area.

Some populations appear to overshoot the carrying capacity and then drop back rather sharply so that the first part of the population growth curve resembles the letter J (Fig. 38.3). There may be a lag between the time at which the population attains a certain size and the time in which the unfavorable effects of that level of crowding become manifest. In the meantime, the population continues to grow.

The size of the population has an effect in evolution. In natural selection the intrinsic rate of natural increase (r) usually becomes maximized. In other words, if a population has two genotypes differing in their rate of reproduction, the one with the greater reproductive rate

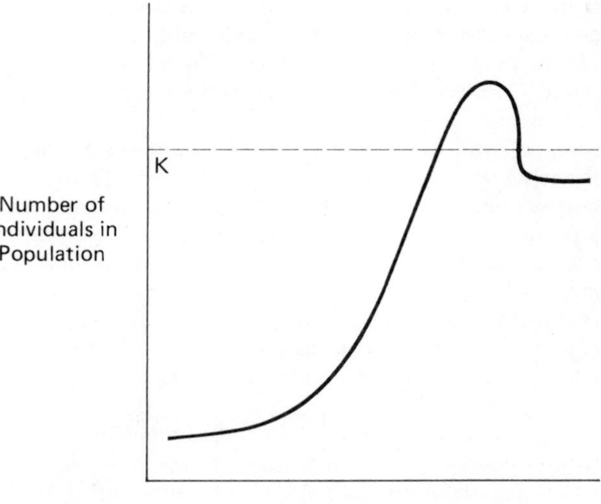

FIGURE 38.3 The J-shaped population growth curve in which population size overshoots the carrying capacity, K, and then declines sharply. This has been termed a "boom or bust" growth curve. The unfavorable effects of overcrowding on population size may lag behind population size itself, permitting the relatively brief large population.

BOX 38.1

Biotic Potential and Carrying Capacity

Exponential population growth may be represented by the formula $dN/dt = rN$. In this differential equation dN/dt is a population growth rate and refers to the change in the numbers: dN, per time interval, dt, when the interval is very small. Biotic potential, r, is the increase in numbers of individuals per time period per individual and thus is a combination of birth rate and death rate. N is the number of individuals in the population. As indicated in the equation, the growth rate of a population is greater when r is large. The growth rate also depends upon the size of N. When the total population, N, is small, the growth rate is slow, and when N is large the growth rate is rapid. If the value of r is 0.5 and the population size, N, is 10, then dN/dt will be 5, but if the population size is 1000, then dN/dt will be 500.

To look at the population size at various times during exponential growth of the population the equivalent integral equation $N_t = N_0e^{rt}$ is used. N_0 is the number in the population at zero time (the beginning of the period), and N_t is the number at time t, the end of the period under study. e is the base of natural logarithms (2.718), r is the biotic potential, and t is the time period under study. The real population growth in the logistic equation is given by $dN/dt = rN\left(\frac{K-N}{K}\right)$. In this the formula for exponential population growth has been modified to include the term $\left(\frac{K-N}{K}\right)$. K represents the carrying capacity, and the expression $\frac{K-N}{K}$ is a measure of environmental resistance or the effect of crowding on population growth. When N is small, $\frac{K-N}{K}$ is very nearly 1 so that the biotic potential, r, is nearly completely realized and population growth is rapid. When N is large growth becomes much slower, for $\frac{K-N}{K}$ becomes a small fraction.

The expression $\frac{K-N}{K}$ is an *inverse* measure of environmental resistance; that is, when environmental resistance is low the numerical value of $\frac{K-N}{K}$ is very nearly 1, but when environmental resistance is high the value of $\frac{K-N}{K}$ approaches 0. The population size at any particular time during logistic growth, that is, growth in which environmental resistance slows the rate of population increase, is given by the equivalent integral equation, $N_t = \frac{K}{1 + e^{a-rt}}$. The symbol a specifies how close the population is to the carrying capacity at the initial time period.

will increase in numbers relative to the slower one. It is generally true that genotypes that tend to produce fewer eggs, later sexual maturation, and so on will lose out in the course of natural selection.

Natural selection may take different courses in species that generally live under uncrowded conditions from those that generally live at or near the carrying capacity, that is, at K. We can differentiate selection under uncrowded conditions as "r selection" and selection near the carrying capacity as "K selection."

r-selected species are those that live in short-lived or unstable ecosystems, such as early successional stages. Because the community is so short-lived or because conditions in it fluctuate widely, the species rarely approaches K. Instead it often lives under uncrowded conditions, and intraspecific competition is rarely important in survival. Traits favored by natural selection under r selection are those that favor a high intrinsic rate of natural increase. This would include a large number of young, eggs, or seeds and reproduction at an early age. Small body size may be favored so that energy is diverted from the synthesis of body tissue to the reproductive process. In contrast, K-selected species are those that generally live under conditions at or near the carrying capacity of the environment. These include species inhabiting stable sites, such as climax forests. Very large species that range over large areas of land also tend to live under fairly stable conditions and at or near K. Under these conditions competition among members of a species, intraspecific competition, is generally important in survival. Selection may favor a lower r, and the energy that is saved may be used for increasing competitive ability. Under K selection the evolutionary tendency is towards the development of larger body size, longer life, and perhaps increased specialization in feeding.

The concept of r and K selection is most useful in comparing related organisms. If an agent such as DDT makes some proportion of an insect population sterile or prevents them from reproducing successfully, the results will be very different in r-selected versus K-selected species. The avian ecologist, Howard Young, compared r- and K-selected species of birds. An r-selected species of bird (e.g., the robin), breeds at age 1, has a clutch of nearly six eggs and a mean annual mortality rate of 74 percent, whereas a K-selected species of bird (e.g., the eagle) breeds at age 6, lays 1.3 eggs per year, and has a 50 percent mortality the first year with a 12.5 percent each succeeding year. If 90 percent of each population fails to reproduce, the population sizes of the two species will decrease. However, the decrease in the r-selected species will become evident very quickly for, by the second year, the population will have decreased to about 25 percent of the initial size. In the K-selected species, however, even after four years of little or no successful reproduction, the population has still not been halved. The K-selected population may become top-heavy with elderly individuals, and hence prospects for recovery of the population are very poor. This can happen even before a decline in population size has become evident.

What does limit the size of populations? Some species are abundant; some are rare. The physical environment is a crucial factor for some species that live under very harsh conditions. This appears to be especially true of small organisms that have short life cycles and high rates of metabolism per gram. In general, numbers may be limited by a shortage of natural resources (e.g., food, places in which to breed), by inaccessibility of these resources relative to the animals' capacities for dispersal and searching (e.g., plants too widely dispersed), and by too little time for reproduction (e.g., limited wet season, limited day length as in the Arctic). Herbivorous insects in a desert are limited by the amount of food available during a short growing season. When drought sets in, the plant crop disappears. The size of the particular insect population depends on its intrinsic rate of increase and the duration of the growing season.

Populations of larger organisms have longer life cycles, and their numbers and biomass more clearly reflect energy flow. These are not limited by the physical environment as much as they are regulated by interactions among their own members or by interactions with competitors, predators, and parasites. This is not to say that these factors do not affect small organisms, nor does it mean that the lack of food or some essential growth substance may not limit the population of larger organisms. White-tailed deer occasionally increase to the point where many die in the winter because the available food supply is inadequate.

Intrinsic factors tending to regulate population size may be physiological or behavioral, or both. **Stress** is one factor that operates physiologically. For example, when some rodent populations become too dense, the animals encounter one another more often. Fights ensue, conditions in general are more stressful, and there is enlargement of the adrenals; the resulting hormonal imbalance adversely affects mating and reproduction. Mortality also increases.

Population Cycles

Once a population becomes established in a certain region and has reached the equilibrium level, the numbers will vary up and down from year to year, depending on variations in environmental resistance or on factors intrinsic to the population. Some of these population variations are completely irregular, but others are regular and cyclic.

One of the best known of these is the regular 9- to 10-year cycle of abundance and scarcity of the snowshoe hare and the lynx in Canada that can be traced from the records of the number of pelts received by the Hudson's Bay Company. The peak of the hare population comes about a year before the peak of the lynx population (Fig. 38.4). Since the lynx feeds on the hare, it is obvious that the lynx cycle is related to the hare cycle. The population cycles may be ultimately caused by periodic declines in the quantity or quality of the plants on which the hares feed. This leads to a decline in the population of hares and secondarily to a decline in the lynx population. This permits the plants to recover from the overharvesting by the hares, and the cycle can begin again.

Lemmings and voles are small mouselike animals living in the northern tundra region. Every three or four years there is a great increase in the number of lemmings; they eat all the available food in the tundra and then migrate in vast numbers looking for food. They may invade villages in hordes, and finally many reach the sea and drown. The numbers of arctic foxes and snowy owls, which feed on lemmings, increase similarly. When the lemming population decreases, the foxes starve and the owls migrate south—thus there is an invasion of snowy owls in the United States every three or four years.

Although some cycles recur with great regularity, others do not. For example, in the carefully managed forests of Germany the numbers of four species of moths whose caterpillars feed on pine needles were estimated from censuses made each year from 1880 to 1940. The numbers varied from less than 1 to more than 10,000 per thousand square meters. The cycles of maxima and minima of the four species were quite indepen-

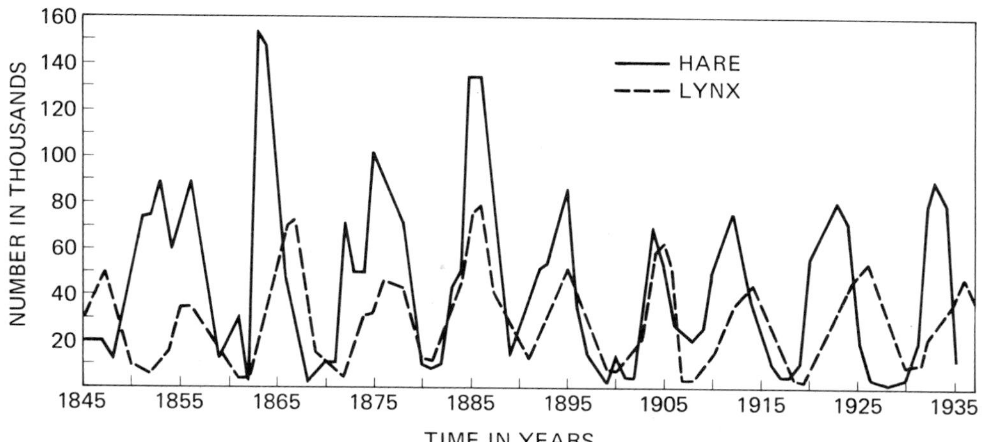

FIGURE 38.4 Changes in the abundance of the lynx and the snowshoe hare, as indicated by the number of pelts received by the Hudson's Bay Company. This is a classic case of cyclic oscillation in population density. (Redrawn from MacLulich, D. A.: Fluctuations in the numbers of the varying hare *(Lepus americanus). University of Toronto Studies, Biol. Series,* no. 43, 1937.)

dent and were irregular in their frequency and duration. Ecologically speaking, each species was marching to its own tune.

Attempts to explain these vast oscillations in numbers on the basis of climatic changes have been unsuccessful. At one time it was believed that these were caused by sunspots, and the sunspot and lynx cycles do appear to correspond during the early part of the nineteenth century. However, the cycles are of slightly different lengths and by 1920 were completely out of phase, sunspot maxima corresponding to lynx minima. Attempts to correlate these cycles with other periodic weather changes and with cycles of disease organisms have been unsuccessful.

Snowshoe hares die off cyclically even in the absence of predators and in the absence of known disease organisms. The animals apparently die of "shock," characterized by low blood sugar, exhaustion, convulsions, and death, symptoms that resemble the "alarm response" induced in laboratory animals subjected to physiologic stress. As population density increases, there is increasing physiologic stress on individual hares, owing to crowding and competition for food. Some individuals are forced into poorer habitats where food is less abundant and predators more abundant. The physiologic stresses stimulate the adrenal medulla to secrete epinephrine, which stimulates the pituitary via the hypothalamus to secrete more ACTH. This, in turn, stimulates the adrenal cortex to produce corticosteroids, an excess or imbalance of which produces the alarm response or physiologic shock.

In the latter part of the winter of a year of peak abundance, the stress of cold weather and lack of food cause the pituitary-adrenal system to fail. It becomes unable to maintain its normal control of carbohydrate metabolism, and low blood sugar, convulsions and death ensue. The onset of the new reproductive season puts additional demands on the pituitary to produce and secrete gonadotropins.

Population Dispersion and Territoriality

Populations have a tendency to disperse, or spread out in all directions, until some barrier is reached. Within the area, the members of the population may occur at random (this is rarely found), they may be distributed more or less uniformly throughout the area (this occurs when there is competition or antagonism to keep them apart), or, most commonly, they may occur in small groups or clumps.

Aggregation in clumps may increase the competition between the members of the group for food or space, but this is more than counterbalanced by the greater survival power of the group during unfavorable periods. A group of animals has much greater resistance than a single individual to adverse conditions, such as desiccation, heat, cold, or poisons. The combined effect of the protective mechanisms of the group is effective in countering the adverse environment, whereas that of a single individual is not.

Aggregation may be caused by local habitat differences, weather changes, reproductive urges, or social attractions. Such aggregations of individuals may have a definite organization involving social hierarchies of dominant and subordinate individuals arranged in a "peck-order."

Other species of animals are regularly found spaced apart; each member tends to occupy a certain area or **territory,** which it defends against intrusion by other members of the same species and sex. Usually a male establishes a territory of his own (perhaps by fighting with other males) and then, by making himself conspicuous, tries to entice a female to share the territory with him.

It has been suggested that territoriality may have survival value for a species in ensuring an adequate amount of food, nesting materials, and cover for the young, in protecting the female and young against other males, and in limiting the population to a density that can be supported by the environment. Many species of birds and some mammals, fish, crabs, and insects establish such territories, either as regions for gathering food or as nesting areas.

Biotic Communities

A biotic community is an assemblage of populations living in a defined area or habitat; it can be either large or small. The interactions of the various kinds of organisms maintain the structure and function of the community and provide the basis for the ecological regulation of community succession. The concept that animals and plants live together in an orderly manner, not strewn haphazardly over the surface of the earth, is an important principle of ecology.

Sometimes adjacent communities are sharply defined and separated from each other; more frequently they blend imperceptibly together. Why certain plants and animals compose a given community, how they affect each other, and how humans can control them to their advantage are some of the major problems of ecological research.

In trying to control some particular species, it has frequently been found more effective to modify the community rather than to attempt direct control of the species itself. For example, the most effective way to increase the quail population is not to raise and release birds, nor even to kill off predators, but to maintain the particular biotic community in which quail are most successful.

Although each community may contain hundreds or thousands of species of plants and animals, most of these are relatively unimportant, and only a few exert a major control of the community owing to their size, numbers, or activities. In land communities these major species are usually plants, for they both produce food and provide shelter for many other species. Many land communities are named for their dominant plants—sagebrush, oak-hickory, pine, and so on. Aquatic communities, containing no conspicuous large plants, are usually named for some physical characteristic—stream rapids community, mud flat community, or sandy beach community.

Biotic communities show marked **vertical stratification,** determined in large part by vertical differences in physical factors, such as temperature, light, and oxygen. The operation of such physical factors in determining vertical stratification in lakes and in the ocean is quite evident. In a forest, there is a vertical stratification of plant life, from mosses and herbs on the ground to shrubs, low trees, and tall trees. Each of these strata has a distinctive animal population. Even such highly motile animals as birds are restricted, more or less, to certain layers. Some species of birds are found only in shrubs, others only in the tops of tall trees. There are daily and seasonal changes in the populations found in each stratum, and some animals are found first in one, then in another layer as they pass through their life histories. These strata are interrelated in many diverse ways, and most ecologists consider them to be subdivisions of one large community rather than separate communities. Vertical stratification, by increasing the number of ecologic niches in a given surface area, reduces competition between species and enables more species to coexist in a given area.

In ecological investigations, it is unnecessary (and indeed usually impossible) to consider all of the species present in a community. Usually a study of the major plants that control the community, the larger populations of animals, and the fundamental energy relations (the food chains) of the system will define the ecological relations within the community. For example, in studying a lake one would first investigate the kinds, distribution, and abundance of the important producer plants and the physical and chemical factors of the environment that might be limiting. Then the reproductive rates, mortality rates, age distributions, and other population characteristics of the important game fish would be determined. A study of the kinds, distribution, and abundance of the primary and secondary consumers of the lake, which constitute the food of the game fish, and the nature of other organisms that compete with these fish for food would elucidate the basic food chains in the lake. Quantitative studies of these would reveal the basic energy relationships of the system and show how efficiently the incident light energy is being converted into the desired end product, the flesh of game fish. On the basis of this knowledge the lake could intelligently be managed to increase the production of game fish.

Detailed studies of simpler biotic communities, such as those of the arctic or desert, where there are fewer organisms and their interrelations are more evident, have provided a basis for studying and understanding the much more varied and complex forest communities.

A thorough ecological investigation of a particular region requires that the region be studied at regular intervals throughout the year for a period of several years. The physical, chemical, climatic, and other factors of the region are carefully evaluated, and an intensive study is made of a number of carefully delimited areas that are large enough to be representative of the region but small enough to be studied quantitatively. The number and kinds of plants and animals in these study areas are estimated by suitable sampling techniques. Estimates are made periodically throughout the year to determine not only the components of the community at any one time but also their seasonal and annual variations. The biological and physical data are correlated, the major and minor communities of the region are identified, and the food chains and other important ecological relationships of the members of the community are analyzed. The particular adaptations of the animals and plants for their respective roles in the community can then be appreciated.

Community Succession

Any given area tends to have an orderly sequence of communities that change together with the physical conditions and lead finally to a stable mature community or **climax community.** The entire sequence of communities characteristic of a given region is termed a **sere,** and the individual transitional communities are called **seral stages** or seral communities. In successive stages there is not only a change in the species of organisms present but also an increase in the number of species and in the total biomass.

Climax communities tend to be more highly organized and more complex than the earlier seral stages. They are more diverse and have more species than immature communities, which tend instead to have large populations of relatively few species. The organisms that make up the climax community tend to be *K*-selected; that is, relatively long-lived, relatively large, and with a low biotic potential. The species present in earlier stages of succession tend to be smaller, shorter-lived, and with a higher biotic potential—they are *r*-selected. Game animals, such as rabbits, are species typically found in early successional stages. They produce a large number of young each year, many of which are eaten by predators.

Succession may be **primary,** with the first stages occurring in water or on bare rock. Water plants, members of the pioneer community, contribute to filling the lake or pond, which changes the environment and enables fewer hydrophytic plants to become established. On rock, the pioneer mosses and lichens aid in the breakdown of rock to form soil, permitting colonization by other plants.

One of the classic studies of ecological succession was made on the shores of Lake Michigan (Fig. 38.5). As the lake has become smaller it has left successively younger sand dunes, and one can study the stages in ecological succession as one goes away from the lake. The youngest dunes, nearest the lake, have only grasses and insects; the next older ones have shrubs, then trees, such as cottonwoods, then evergreens, and finally there is a beech-maple climax community, with deep rich soil.

As the lake retreated it also left a series of ponds. The youngest of these contain little rooted vegetation and lots of bass and bluegills. Later the ponds become choked with vegetation and smaller in size as the basins fill. Finally the ponds become marshes and then dry ground, invaded by shrubs and ending in the beech-maple climax forest. Man-made ponds, such as those behind dams, similarly tend to become filled up.

Secondary succession comes about from disturbance by fire or human cultivation of an established community, such as a forest. Succession starts over but on land with a well-developed soil, not on rock or in water. When farm land is abandoned, a series of communities will eventually lead back to the self-perpetuating climax community characteristic of the region. In the eastern United States, pine in the south and cedar in the north are conspicuous stages in secondary succession of old fields.

Successional series are so regular in many parts of the world that an ecologist, recognizing the particular seral community present in a given area, can predict the sequence of future changes. The ultimate causes of these successions are not clear. Climate and other physical factors play some role, but the succession is directed in part by the nature of the community itself, for the action of each seral community is to make the area less favorable for itself and more favorable for other species until the stable climax community is reached. For example, in old field succession in the southeastern United States, the grass called broomsedge precedes pine. The broomsedge community provides a favorable environment for the germination of pine seeds blown in from adjacent forest by wind, but as the young pines

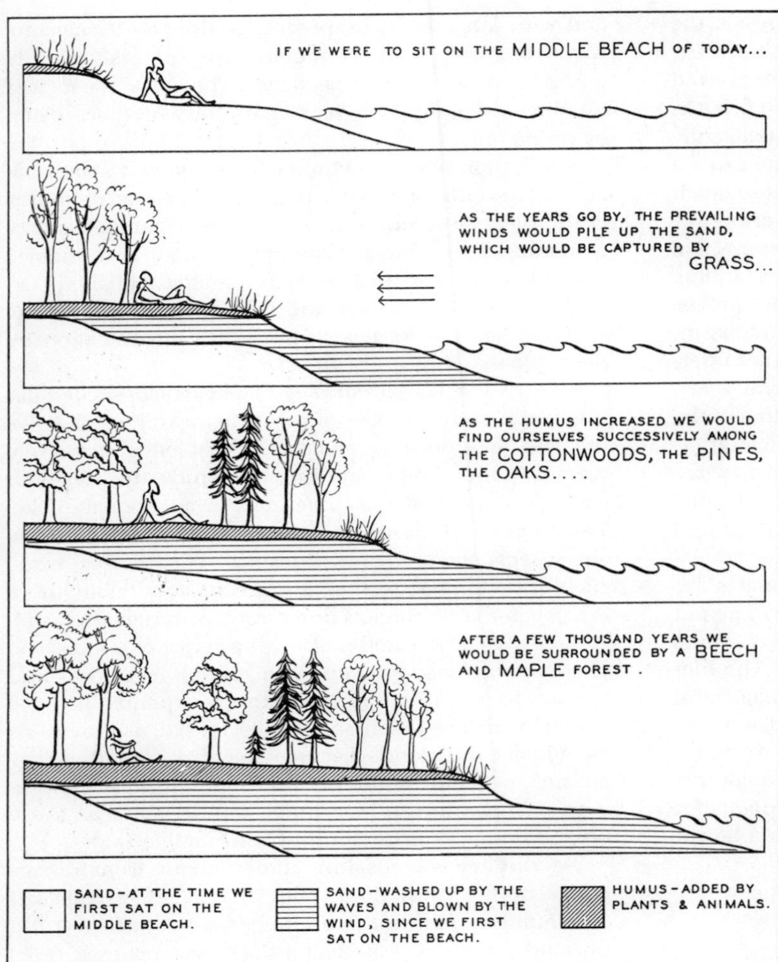

IF WE WERE TO SIT ON THE MIDDLE BEACH OF TODAY...
AS THE YEARS GO BY, THE PREVAILING WINDS WOULD PILE UP THE SAND, WHICH WOULD BE CAPTURED BY GRASS...
AS THE HUMUS INCREASED WE WOULD FIND OURSELVES SUCCESSIVELY AMONG THE COTTONWOODS, THE PINES, THE OAKS....
AFTER A FEW THOUSAND YEARS WE WOULD BE SURROUNDED BY A BEECH AND MAPLE FOREST.
SAND-AT THE TIME WE FIRST SAT ON THE MIDDLE BEACH.
SAND-WASHED UP BY THE WAVES AND BLOWN BY THE WIND, SINCE WE FIRST SAT ON THE BEACH.
HUMUS-ADDED BY PLANTS & ANIMALS.

FIGURE 38.5 Diagram of the succession of communities with time along the shores of Lake Michigan in northern Indiana. (From Allee, W. C., *et al: Principles of Animal Ecology.* Philadelphia, Saunders College Publishing, 1949. After Buchsbaum.)

grow they gradually eliminate the broomsedge by reducing the amount of light that reaches the ground. Physical factors, such as the nature of the soil, the topography, and the amount of water, may cause the succession of communities to stop short of the expected climax community in what is called an **edaphic climax.**

The original **monoclimax hypothesis** stated that within a given area all of the land eventually tends to be occupied by a single kind of community which is the climax community. The climax is determined by the climate of the region and, if the climate is stable, the climax community is also stable. Other ecologists argue that even in areas that have been totally undisturbed it is difficult to find sizeable regions of uniform vegetation. These ecologists argue that several different communities may be the climax community in a given geographical area. This is the **polyclimax hypothesis.** The climax community may be determined by soil factors, so-called edaphic climax, by the elevation or slope of an area, a topographical climax, or by fire or water. Each of these might lead to a different kind of climax community. The debate about the nature of community succession continues.

Geographical Ecology

The biogeographic realms, discussed on page 400, are regions made up of whole continents, or of large parts of a continent, separated by major geographic barriers and characterized by the presence of certain unique animals and plants. Within these biogeographic realms and established by a complex interaction of climate, other physical factors, and biotic factors are large, distinct, easily differentiated community units called **biomes.** Biomes are geographical regions with similar kinds of climax communities. A biome includes all regions of similar climatic conditions that support similar types of ecosystems. In each biome the *kind* of climax vegetation is uniform—grasses, conifers, deciduous trees—but the particular *species* of plant may vary in different parts of the biome. The kind of climax vegetation depends upon the physical environment, and the two together determine the kind of animals present. The definition of biome includes not only the actual climax community of a region but also seral stages of succession.

There is usually no sharp line of demarcation between adjacent biomes; instead each blends with the next through a fairly broad transition region termed an **ecotone.** There is, for example, an extensive region in northern Canada where the tundra and coniferous forests blend in the tundra coniferous forest ecotone. The ecotonal community typically consists of some organisms from each of the adjacent biomes plus some that are characteristic of, and perhaps restricted to, the ecotone. There is a tendency (called the **edge effect**) for the ecotone to contain both a greater number of species and a higher population density than either adjacent biome.

Some of the biomes recognized by ecologists are **tundra, coniferous forest, deciduous forest, broad-leaved evergreen subtropical forest, grassland, desert, chaparral,** and **tropical rain forest.** These biomes are distributed, though somewhat irregularly, as belts around the world (Fig. 38.6), and as one travels from the equator to the pole one may traverse tropical rain for-

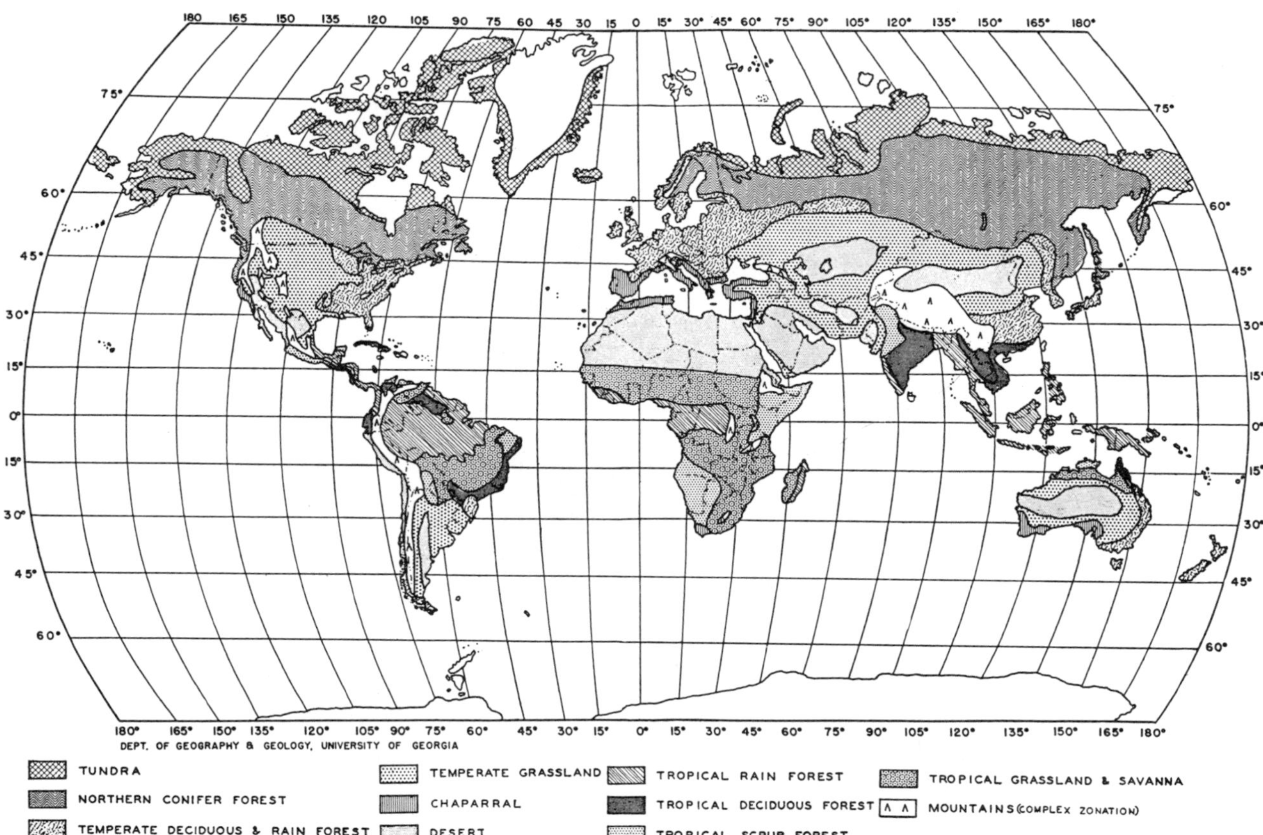

FIGURE 38.6 A map of the biomes of the world. Note that only the tundra and northern coniferous forest are more or less continuous bands around the world. Other biomes are generally isolated in different biogeographic realms and may be expected to have ecologically equivalent but taxonomically unrelated species. (From Odum, E. P.: *Fundamentals of Ecology*. 3rd ed. Philadelphia, Saunders College Publishing, 1971.)

ests, grassland, desert, deciduous forest, coniferous forest, and finally reach the tundra in northern Canada, Alaska, or Siberia.

Since climatic conditions at higher altitudes are in many ways similar to those at higher latitudes, there is a similar succession of biomes on the slopes of high mountains (Fig. 38.7). For example, as one goes from the San Joaquin Valley of California into the Sierras, one passes from desert through grassland and chaparral to deciduous forest and coniferous forest, then, above timberline, to the alpine tundra, a region resembling the tundra of the Arctic.

The Tundra Biome

Between the Arctic Ocean and polar icecaps and the forests to the south lies a band of treeless, wet, arctic sedges called the **tundra** (Fig. 38.8).

Some five million acres of tundra stretch across northern North America, northern Europe, and Siberia. The primary characteristics of this region are the low temperatures and short growing season. The amount of precipitation is rather small, but water is usually not a limiting factor because the rate of evaporation is also very low.

The ground usually remains frozen except for the uppermost 10 or 20 cm., which thaw during the brief summer season. The permanently frozen deeper soil layer is called **permafrost.** The rather thin carpet of vegetation includes lichens, mosses, grasses, sedges, and a few low shrubs. The animals that have adapted to survive in the tundra are caribou or reindeer, the arctic hare, arctic fox, polar bear, wolves, lemmings, snowy owls, ptarmigans, and, during the summer, swarms of flies, mosquitoes, and a host of migratory birds.

The caribou and reindeer are highly migratory because there is not enough vegetation produced in any one local area to support them. Although casual inspec-

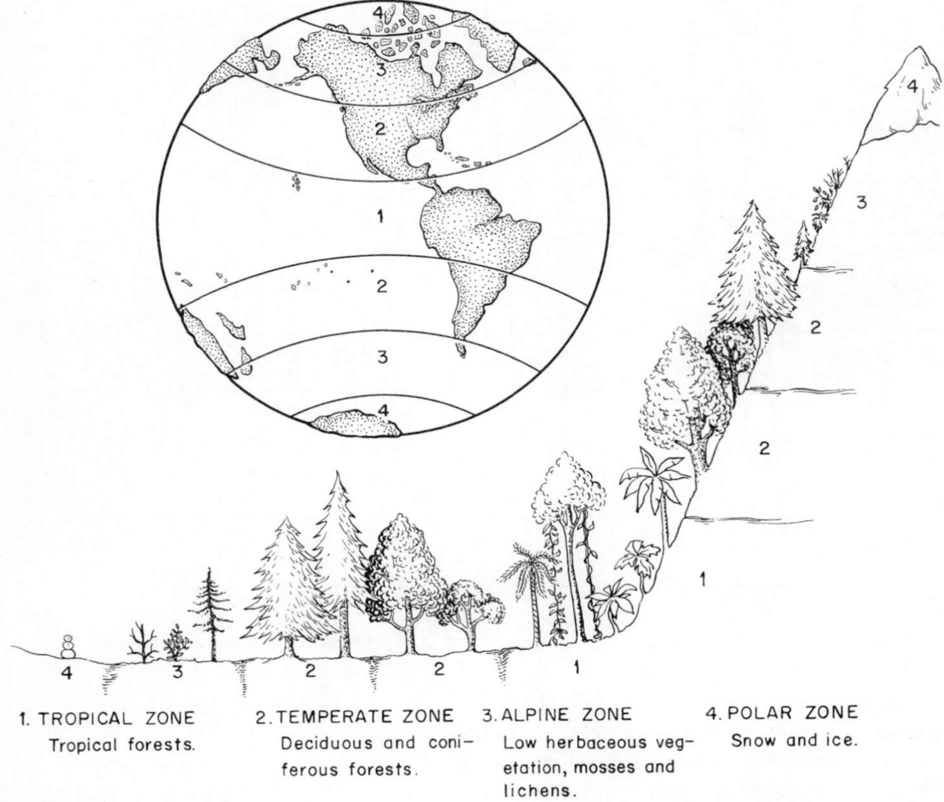

FIGURE 38.7 Diagram showing correspondence of life zones at successively higher altitudes at the same latitude (*1* to *4, right*) and at successively higher latitudes at the same altitude (*1* to *4, left,* and *inset*).

FIGURE 38.8 The tundra biome. *Above*: View of the low tundra near Churchill, Manitoba, in July. Note the numerous ponds. *Below:* View of tundra vegetation showing "lumpy" nature of low tundra and a characteristic tundra bird, the willow ptarmigan. (Lower photo by C. Lynn Haywood.)

tion might suggest that tundras are rather barren areas, a surprisingly large number of organisms have become adapted to survive the cold. During the long daylight hours of the brief summer, the rate of primary production is quite high. The production from the vegetation on the land, from the plants in the many shallow ponds that dot the landscape, and from the phytoplankton in the adjacent Arctic Ocean provides enough food to support a variety of resident mammals and many kinds of migratory birds and insects.

The Forest Biomes

Several different types of forest biomes can be distinguished. These are generally arranged on a gradient from north to south or from high altitude to lower altitude. Adjacent to the tundra region either at high latitude or high altitude is the **northern coniferous forest** (Fig. 38.9), which stretches across both North America and Eurasia just south of the tundra. This is characterized by spruce, fir, and pine trees and by animals such as the snowshoe hare, the lynx, and the wolf.

The evergreen conifers provide dense shade throughout the year; this tends to inhibit development of shrubs and a herbaceous undergrowth. The continuous presence of green leaves permits photosynthesis to occur throughout the year despite the low temperature during the winter and results in a fairly high annual rate of primary production. The northern coniferous forest, like the tundra, shows a marked seasonal periodicity, and the populations of animals undergo striking peaks and depressions in numbers.

The pigmy conifer or **piñon-juniper biome** is found in west central California and in the Great Basin and Colorado River regions of Nevada, Utah, Colorado, New Mexico, and Arizona. This occupies a belt between the desert or grasslands at lower altitudes and the true

FIGURE 38.9 The coniferous forest biome covers part of Canada, northern Europe, and Siberia and extends southward at higher altitudes on the larger mountain ranges. (From Orr, R. T.: *Vertebrate Biology*. 3rd ed. Philadelphia, Saunders College Publishing, 1971.)

FIGURE 38.10 The pinon-juniper biome in Arizona. The small pinon pines and cedars each grow some distance from the neighboring trees, giving an open, parklike appearance to the woodland. (Photograph by U. S. Forest Service.)

northern coniferous forest found at higher altitudes where there is more rainfall. In this region, the annual rainfall of 25 to 50 cm. is irregularly distributed throughout the year. The small piñon pines and cedars tend to be widely spaced, and the biome has an open, parklike appearance (Fig. 38.10).

Along the west coast of North America from Alaska south to central California is a region termed the **moist coniferous forest biome,** characterized by a much greater humidity, somewhat higher temperatures, and smaller seasonal ranges than the classic coniferous forest farther north. There is high rainfall, from 75 to 375 cm. per year, and, in addition, a great deal of moisture is contributed by the frequent fogs. There are forests of Sitka spruce in the northern section, western hemlock, arbor vitae, and Douglas fir in the Puget Sound area, and the coastal redwood, *Sequoia sempervirens,* in California. The potential production of this region is very great, and with careful foresting and replanting, the annual crop of lumber is very high.

The **temperate deciduous forest biome** (Fig. 38.11) is found in areas with abundant, evenly distributed rainfall (75 to 150 cm. annually) and moderate temperatures with distinct summers and winters. Temperate deciduous forest biomes originally covered eastern North America, all of Europe, parts of Japan, China, and Australia, and the southern portion of South America.

The trees present in the North American deciduous forest biome—beech, maple, oak, hickory, and chestnut—lose their leaves during half the year; thus, the contrast between winter and summer is very marked. The undergrowth of shrubs and herbs is generally well

FIGURE 38.11 An example of a temperate deciduous forest, Noble County, Ohio. The dominant trees are white and red oaks with an understory of hickory. (Photograph by U. S. Forest Service.)

developed. The animals originally present in the forest were deer, bears, squirrels, gray foxes, bobcats, wild turkeys, and woodpeckers. Much of this forest region has now been replaced by cultivated fields and cities.

In regions of fairly high rainfall but where temperature differences between winter and summer are less marked, as in Florida, the **broad-leaved evergreen subtropical forest biome** is found. The vegetation includes live oaks, magnolias, tamarinds, and palm trees, with many vines and epiphytes, such as orchids and Spanish moss.

The variety of life reaches its maximum in the **tropical rain forest biome** (Fig. 38.12), which occupies low-lying areas near the equator with annual rainfalls of 200 cm. or more. The thick rain forests, with a tremendous variety of plants and animals, are found in the valleys of the Amazon, Orinoco, Congo, and Zambesi rivers and in parts of Central America, Malaya, Borneo, and New Guinea.

The extremely dense vegetation makes it difficult to study or even photograph the rain forest biome. The vegetation is vertically stratified with tall trees often covered with vines, creepers, lianas, and epiphytes. Under the tall trees is a continuous evergreen carpet, the canopy layer, some 25 to 35 m. tall. The lowest layer is an understory that becomes dense where there is a break in the canopy.

The diversity of species is remarkable, and no single species of animal or plant is present in large enough numbers to be dominant. The trees of the tropical rain forest are usually evergreen and rather tall. Their roots are often shallow and have swollen bases or flying buttresses.

The tropical rain forest is the ultimate of jungles, although the low light intensity at the ground level may result in sparse herbaceous vegetation and actual bare spots in certain areas. Many of the animals live in the upper layers of the vegetation. Among the characteristic animals are monkeys, sloths, termites, ants, anteaters, many reptiles, and many brilliantly colored birds—parakeets, toucans, and birds of paradise.

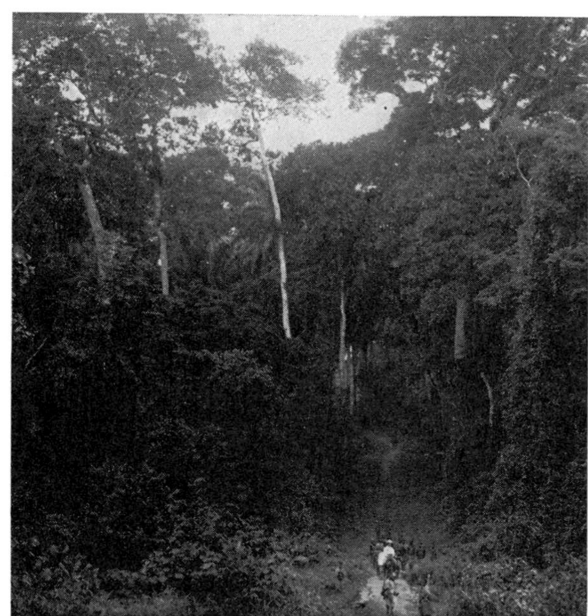

FIGURE 38.12 The rain forest biome: border of a clearing in the Ituri Forest of Nala, The Congo. (Photograph by Herbert Lang, Courtesy of the American Museum of Natural History, New York.)

The Grassland Biome

The **grassland biome** (Fig. 38.13) is found where rainfall is about 25 to 75 cm. per year, not enough to support a forest, yet more than that of a true desert. Grasslands typically occur in the interiors of continents—the prairies of the western United States, and those of Argentina, Australia, southern Russia, and Siberia. Grasslands provide natural pasture for grazing animals, and many of our principal agricultural food plants have been developed by artificial selection from the grasses.

FIGURE 38.13 A region of short-grass grassland with a herd of bison, originally one of the major grazing animals in the grassland biome of the western United States and Canada. The bison in the center is wallowing. (From Odum, E. P.: *Fundamentals of Ecology*. 3rd ed. Philadelphia, Saunders College Publishing, 1971.)

The mammals of the grassland biome are either grazing or burrowing forms—bison, antelope, zebras, wild horses and asses, rabbits, ground squirrels, prairie dogs, and gophers. These characteristically aggregate into herds or colonies; this aggregation probably provides some protection against predators.

Depending upon the amount of rainfall, the species of grasses present in any given grassland may range from tall species, 150 to 250 cm. in height, to short species of grass that do not exceed 15 cm. in height. Trees and shrubs may occur in grasslands either as scattered individuals or in belts along the streams and rivers. The soil of grasslands is very rich in humus because of the rapid growth and decay of the individual plants. The grassland soils are well suited for growing cultivated food plants, such as corn and wheat, that are species of cultivated grasses. The grasslands are also well adapted to serve as natural pastures for cattle, sheep, and goats. However, when grasslands are subjected to consistent overgrazing and overplowing, they can be turned into man-made deserts. In the United States only a few scattered remnants of the original prairie still exist.

There is a broad belt of tropical grassland, or **savanna,** in Africa lying between the Sahara desert and the tropical rain forest of the Congo basin (Fig. 38.14). Other savannas are found in South America and Australia. Although the annual rainfall is high, as much as 125 cm., a distinct, prolonged dry season prevents the development of a forest. During the dry season there may be extensive fires, which play an important role in the ecology of the region. In this region are great numbers and great varieties of grazing animals and predators, such as lions.

How to make best use of these African grasslands is a problem now facing the new nations of Africa as they work to raise the level of nutrition in their human populations. Many ecologists are of the opinion that it would be better to harvest the native herbivores—antelope, hippopotamuses, and wildebeests—on a sustained yield basis rather than try to exterminate them completely and substitute cattle. The diversity of the natural population would mean broader use of all the resources of primary production, and the native species are immune to the many tropical parasites and diseases that plague the cattle that have been introduced.

The Chaparral Biome

In mild temperate regions of the world with relatively abundant rain in the winter but with very dry summers the climax community includes trees and shrubs with hard, thick evergreen leaves. This type of vegetation is called "chaparral" in California and Mexico, "macchie" around the Mediterranean, and "mellee scrub" on Australia's south coast.

The trees and shrubs common in California's chaparral are chamiso and manzanita. Eucalyptus trees introduced from Australia's south coast into California's chaparral region have prospered mightily and have replaced to a considerable extent the native woody vegetation in areas near cities. During the hot dry season, there is an ever present danger of fire that may sweep rapidly over the chaparral slopes. Following a fire, the shrubs sprout vigorously after the first rains and may reach maximum size within 20 years.

The Desert Biome

In regions with less than 25 cm. of rain per year or in certain hot regions where there may be more rainfall but with an uneven distribution in the annual cycle, vegetation is sparse and consists of greasewood, sagebrush, or cactus. The individual plants in the desert are typically widely spaced with large bare areas separating them. In the brief rainy season, the California desert be-

FIGURE 38.14 The savanna biome; characteristic animals of the African grasslands, zebra and wildebeest, Kruger National Park, Transvaal, Republic of South Africa. (From Odum, E. P.: *Fundamentals of Ecology.* 2nd ed. Philadelphia, Saunders College Publishing, 1959. Photograph by Herbert Lang.)

comes carpeted with an amazing variety of wild flowers and grasses, most of which complete their life cycle from seed to seed in a few weeks. The animals present in the desert are reptiles, insects, and burrowing rodents, such as the kangaroo rat and pocket mouse, both of which are able to live without drinking water by extracting the moisture from the seeds and succulent cactus they eat.

The small amount of rainfall may be due to continued high barometric pressure, as in the Sahara and Australian deserts; a geographical position in the rain shadow of a mountain, as in the western North American deserts; or high altitude, as in the deserts in Tibet and Bolivia. The only absolute deserts, where little or no rain ever falls are those of northern Chile and of the central Sahara.

Two types of deserts can be distinguished on the basis of their average temperatures: "hot" deserts, such as that found in Arizona, characterized by the giant saguaro cactus, palo verde trees, and the creosote bush, and "cool" deserts, such as that present in Idaho, dominated by sagebrush (Fig. 38.15).

Certain reptiles and insects are well adapted for survival in deserts because of their thick, impervious integuments and the fact that they excrete dry waste matter. A few species of mammals have become secondarily adapted to the desert by excreting very concentrated urine. They avoid the sun by remaining in their burrows during the day. The camel and the desert birds must have an occasional drink of water but can go for long periods of time using the water stored in the body.

When deserts are irrigated, the large volume of water passing through the irrigation system may lead to the accumulation of salts in the soil as some of the water is evaporated, and this will eventually limit the area's productivity. The water supply itself can fail if the watershed from which it is obtained is not cared for appropriately. The ruins of old irrigation systems and of the civilizations they supported in the deserts of North Africa and the Near East remind us that the irrigated desert will retain its productivity only when the entire system is kept in appropriate balance.

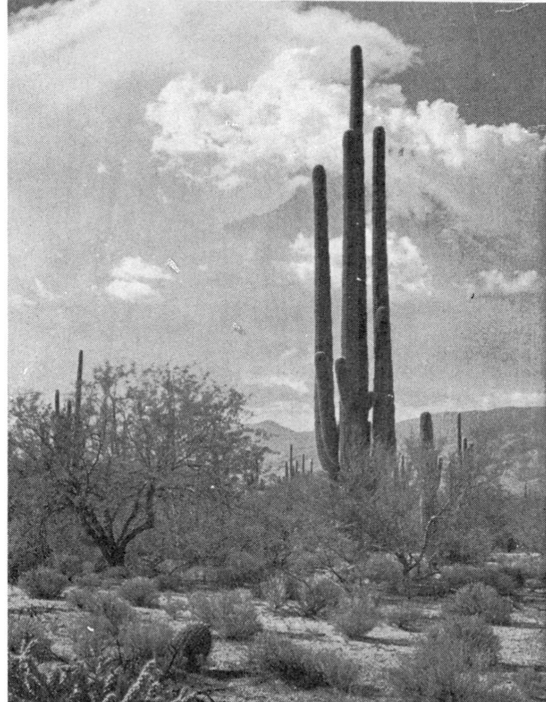

FIGURE 38.15 Two types of desert in western North America: a "cool" desert in Idaho, dominated by sagebrush *(above),* and *(below)* a rather luxuriant "hot" desert in Arizona, with giant cactus (Saguaro) and palo verde trees, in addition to creosote bushes and other desert shrubs. In extensive areas of desert country the desert shrubs alone dot the landscape. (Upper photograph by U. S. Forest Service; lower photograph by U. S. Soil Conservation Service.)

The Edge of the Sea: Marshes and Estuaries

Where the sea meets the land there may be one of several kinds of ecosystems with distinctive characteristics: a rocky shore, a sandy beach, an intertidal mud flat, or a tidal estuary containing salt marshes. The word estuary refers to protected bays and sounds where a river meets the sea. There is tidal flow, but the salinity is less than in the open ocean, intermediate between the sea and fresh water. Most estuaries, particularly those in temperate and arctic regions, undergo marked variations in temperature, salinity, and other physical properties in the course of a year. To survive there, estuarine organisms must have a wide range of tolerance to these changes (they must be euryhaline, eurythermic, and so on).

The waters of estuaries are among the most naturally fertile in the world, frequently having a much greater productivity than the adjacent sea or the fresh water up the river. Great expanses of accumulated sediment in shallow water support rich stands of algae, sea grasses, marsh grasses, and mangroves. Much of this vegetation is converted to detritus and is consumed by the clams, crabs, and other marine detritus eaters. Estuaries are highly productive and important regions for fish, oyster, clam, shrimp, and crab fisheries.

Estuaries and marshes are high on the list of ecological regions of the world that are seriously threatened by man's activities. They were long considered to be worthless regions in which waste materials could be dumped. Many have been irretrievably lost by being drained, filled, and converted to housing developments or industrial sites. We are just beginning to appreciate that the best interests of all are served by maintaining estuaries in their natural state and protecting them from waste material and thermal and oil pollution.

Marine Life Zones

The oceans, which cover 70 percent of the earth's surface, are continuous one with another, but marine organisms are restrained from spreading to all parts of the ocean by factors such as temperature, salinity, and depth. The salinity of the open ocean is about 35 parts per thousand and temperatures range from about −2 C. in the polar seas to 32° C. or more in the tropics; but the annual range of variation in any given region is usually no more than 6° C.

The waters of the seas are continually moving in vast currents, such as the Gulf Stream, the North Pacific Current, and the Humboldt Current, which circle in a clockwise fashion in the Northern Hemisphere and counterclockwise in the Southern Hemisphere. These currents not only influence the distribution of marine forms but also have marked effects on the climates of the adjacent land masses. In addition, there are very slow currents of cold dense water flowing at great depths from the polar regions toward the equator.

Like the land, the ocean consists of regions characterized by different physical conditions and consequently inhabited by specific kinds of plants and animals.

A gently sloping continental shelf usually extends some distance offshore; beyond this, the ocean floor (the **continental slope**) drops steeply to the abyssal region. The region of shallow water over the continental shelf is called the **neritic zone;** it can be subdivided into **supratidal** (above the high tide mark), **intertidal** (between the high and low tide lines, a region also known as the "littoral"), and **subtidal** regions (Fig. 38.16).

The open sea beyond the edge of the continental shelf is the **oceanic zone.** The upper part of the ocean, into which enough light can penetrate to be effective in photosynthesis, is known as the **euphotic zone.** The av-

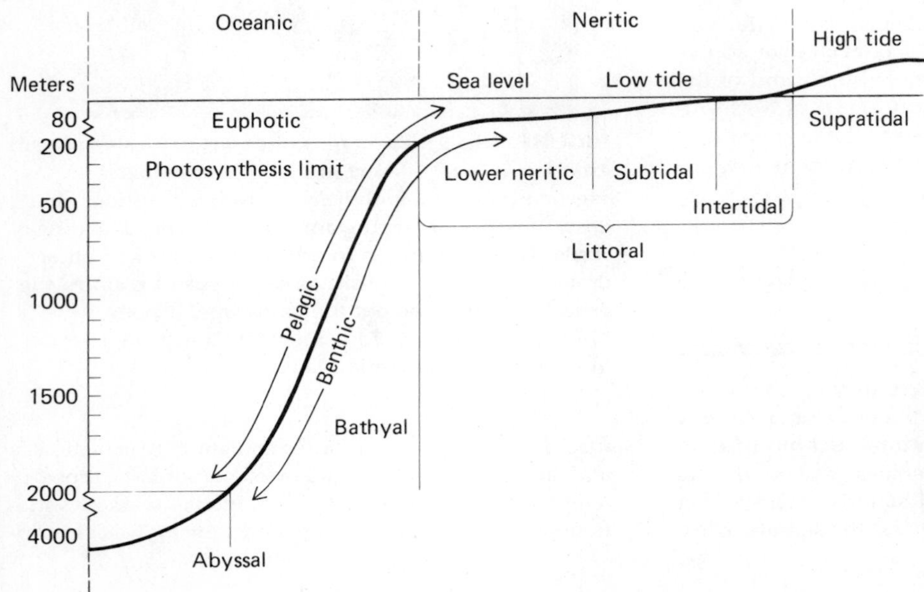

FIGURE 38.16 Zonation in the sea. (Redrawn from Hedgpeth, J. W.: *The Classification of Estuarine and Brackish Waters and the Hydrographic Climate.* Rpt. 11. National Research Council Committee on a Treatise on Marine Ecology and Paleoecology, 1951.)

erage lower limit of this is at about 100 m., but in a few regions of clear tropical water this may extend to twice that depth. The regions of the ocean beneath the euphotic zone are called the **bathyal zone** over the continental slope to a depth of perhaps 2000 m.; the depths of the ocean beyond that compose the **abyssal zone.**

The floor of the ocean is not uniformly flat but is thrown into gigantic ridges and trenches. Some of the ridges rise nearly to the surface (or above it where there are oceanic islands), and some of the trenches lie 10,000 m. below the surface of the sea. Huge underwater avalanches occur from time to time as parts of the slopes and ridges tumble into the lower depths.

Many organisms are bottom dwellers, collectively called **benthos;** they creep or crawl over the bottom or are **sessile** (attached to it). Others are **pelagic,** living in the open water, and are either active swimmers, **nekton,** or organisms that are moved by the current, **plankton.** Plankton may be classified as **phytoplankton,** containing algae, and **zooplankton,** containing animals. The latter may be divided into **holoplankton,** which includes animals like copepods that spend their entire lives as planktonic organisms, and **meroplankton,** which includes the larvae of animals that are benthic as adults.

The edge of the sea is the marine environment most familiar to biologists and laymen alike. The interacting gravitational forces of the sun, moon, and earth produce tidal bulges that once or twice every 24 hours, depending upon the location of the coast, expose or flood the intertidal zone with seawater. Although surf beaches (high-energy beaches) are populated by large numbers of humans, the pounding waves on unstable sand bottoms are too stressful an environment for all but a few specialized burrowers, such as mole crabs and coquina shells.

Rocky coastlines support much richer and more interesting flora and fauna than do sand beaches. Many intertidal animals can tolerate surge and pounding waves if they can anchor firmly to a stable substratum. There is a distinct zonation of algae and various kinds of animals between the low tide mark and the high supratidal spray zone (Fig. 38.17).

The sandy shore is subject to all of the extremes of the rocky shore plus the inconvenience of a constantly shifting substratum. Zonation on a sandy beach (Fig.

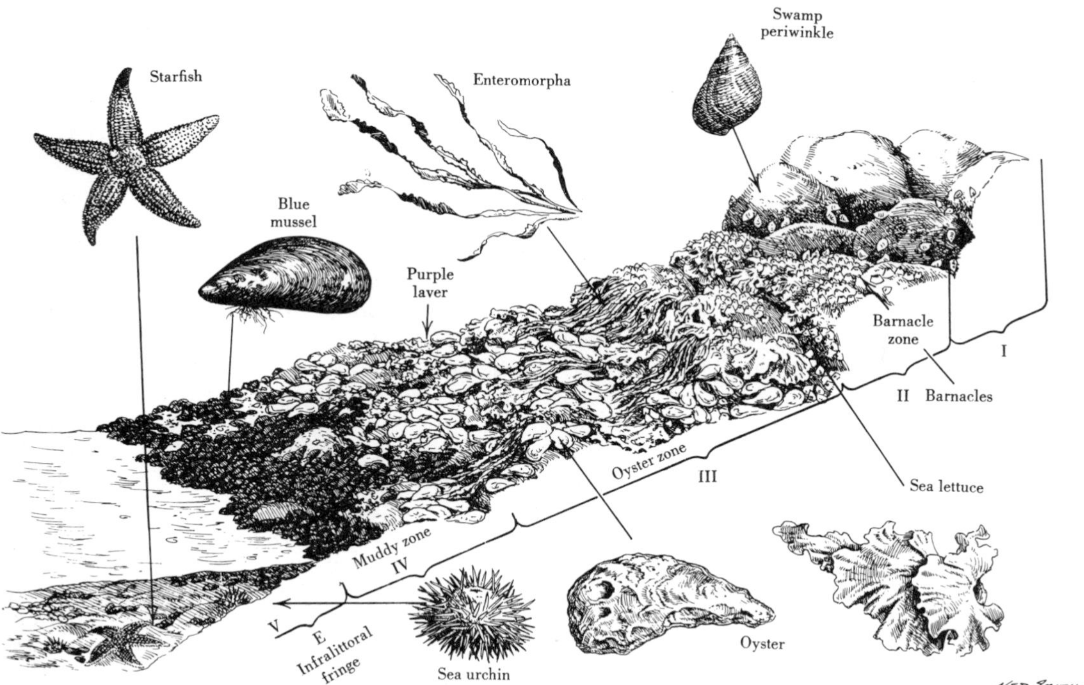

FIGURE 38.17 Zonation along a low-energy, rocky shore—mid-Atlantic coast line. *(I)* Bare rock with some black algae and swamp periwinkle; *(II)* barnacle zone; *(III)* oyster zone: oysters, Enteromorpha, sea lettuce, and purple laver; *(IV)* muddy zone: mussel beds; *(V)* infralittoral fringe: starfish and so on. Note absence of kelps. (Zonation drawing based on Stephenson, 1952; sketches done from life or specimens. From Smith, R. L.: *Ecology and Field Biology.* New York, Harper & Row, Publishers, 1966.)

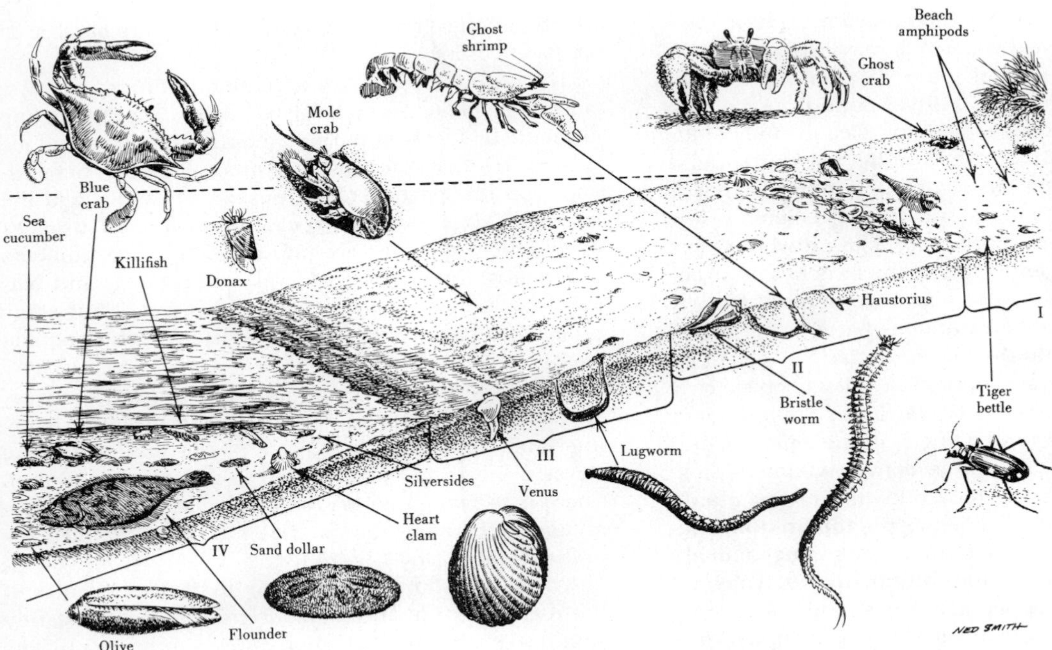

FIGURE 38.18 Life on a sandy ocean beach along the Atlantic coast. The diagram is a composite of some animals typical of a low-energy beach and others typical of a high-energy beach. It is unlikely that you would find mole crabs and lugworms on the same beach. Although strong zonation is absent, organisms still change on a gradient from land to sea. *(I)* Supratidal zone: ghost crabs and sand fleas; *(II)* flat beach zone: ghost shrimp, bristle worms, and clams; *(III)* infratidal zone: clams, lugworms, and mole crabs; *(IV)* subtidal zone: the dotted line indicates high tide. (From Smith, R. L.: *Ecology and Field Biology.* New York, Harper & Row, Publishers, 1966.)

38.18) does not conform to a universal pattern as does that of the rocky shore. Most of the animals burrow into the substratum, above or below the water line.

Although there are large benthic algae and sea grasses in shallow water, the primary producer organisms in the sea equivalent to the flowering plants on land are the **phytoplankton** (Fig. 38.19), consisting principally of diatoms and various unicellular flagellates. It is difficult to appreciate their importance because they are so small. In temperate regions, the phytoplankton typically undergoes two seasonal population explosions or "blooms," one in the spring, the other in late summer or fall. The mechanism is similar to that responsible for "blooms" in lakes. In the wintertime, low temperatures and reduced light restrict photosynthesis to a low level, but when spring brings higher temperatures and more light, photosynthesis accelerates. The nutrient supply is ample because the winter mixing of surface and deep water brings up nutrients that have accumulated at lower levels. Within a fortnight, the diatoms multiply ten-thousand-fold. This prodigious growth accounts for the spring bloom. Soon, however, the nutrients are exhausted. Replacement from lower layers no longer occurs because warming of the surface water keeps it on top (it is less dense) and prevents mixing. Nutrients are now locked in the bodies of animals that have eaten the phytoplankton or are slowly falling to the bottom in dead bodies. Whereas temperature and light were the limiting factors during the winter, nutrient level is the limiting factor during the summer, especially since existing phytoplankton is now being consumed by animals. Now nutrients begin to accumulate again in lower layers. As fall approaches, the upper layers of water begin to cool again. The accompanying density change, together with the autumn equinoctial gales, begins mixing the water again. Water rich in phosphates and nitrates is brought up from below. Other forms of phytoplankton, especially nitrogen-fixing blue-green algae, now bloom until reduced nutrients or temperature again intercedes (Fig. 38.20).

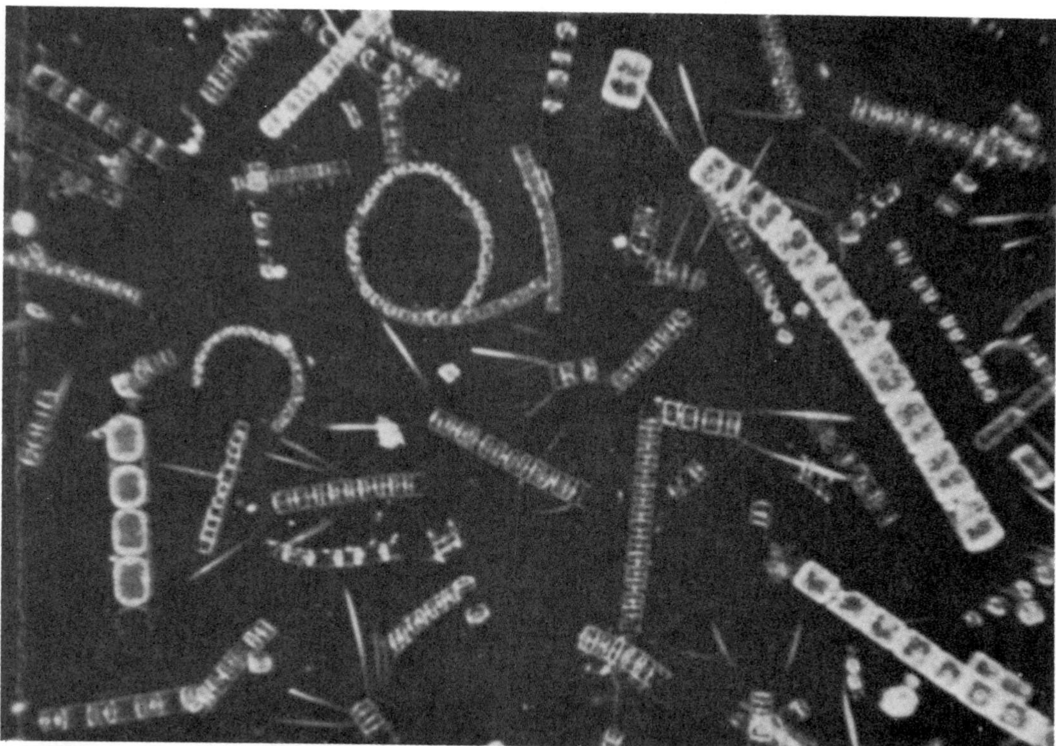

FIGURE 38.19 Living plants of plankton (phytoplankton). Magnification ×10. Chains of cells of several species of *Chaetoceros* (those with spines), a chain of *Thalassiosira condensata* (at and pointing to *bottom right* corner), and a chain of *Lauderia borealis* (above the last named). By electronic flash. (From Hardy, A.: *The Open Sea*. Vol. 1. London, William Collins Sons & Co., Ltd., 1966.)

Tropical oceans, except for coral reefs, coastal waters, and certain areas of upwelling, are less productive than temperate and cold seas. The permanently warm upper layer of water (warm water is lighter than cold water) receives only a slow replacement of nutrients from the colder, heavier layers below. Productive seas are green or gray in color because of the high content of plankton. The blue color of clear water is sometimes called the color of ocean deserts.

Zooplankton contain most of the pelagic marine grazers and are important as an intermediate trophic level in the flow of energy to fish and larger pelagic animals.

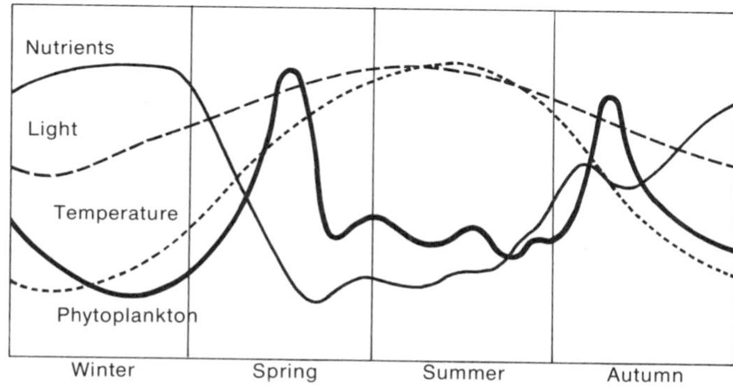

FIGURE 38.20 The probable mechanisms for phytoplankton "blooms." See text for explanation. (From Odum, E. P.: *Fundamentals of Ecology*. 3rd ed. Philadelphia, Saunders College Publishing, 1971.)

FIGURE 38.21 Living animals of the plankton (zooplankton). Magnification ×16. The copepods *Calanus finmarchicus* (the largest animal) and *Pseudocalanus elongatus* (similar in shape, but much smaller than *Calanus* and the one with the cluster of eggs); two small anthomedusae with long tentacles; a fish egg (the circular object); a young arrow-worm *Sagitta* (to the *right* of the fish egg); small nauplius (larval stage) of copepod (close to *left* side of *Calamus*); and the planktonic tunicate *Oikopleura* (curly objects *top right* and *middle bottom*). By electronic flash but partially narcotized. (From Hardy, A.: *The Open Sea*. Vol. 1. London, Williams Collins Sons & Co., Ltd., 1966.)

Copepods are usually the most numerous and ecologically important of the planktonic herbivores (Fig. 38.21).

Eighty-eight percent of the ocean is more than 1 mile deep. It is continuous throughout the world except for the deep water of the Arctic Ocean, which is cut off from the rest by a narrow submerged mountain range connecting Greenland, Iceland, and Europe. This is the area of great pressure and of perpetual night. Since no photosynthesis is possible, the only source of energy is the constant rain of organic debris, the bodies and waste products of organisms in the surface layers, that falls toward the bottom. The other prerequisite for life, oxygen, gets to the bottom by means of the oceanic circulation.

Life in the deep is poorly known because of the enormous difficulty of observation. The knowledge available has been gleaned from studies of net tows, dredging and core samples, and by observation from special undersea craft or via underwater photography and television.

Much of the bottom of the sea is covered by sediments composed of clay particles and the skeletal remains of foraminiferans, diatoms, and other planktonic organisms. Organic detritus is an important food source for many of the invertebrates that live on the ocean floor. Apparently even the greatest "deeps" are inhabited, for tube-dwelling worms have been dredged up from depths of 8000 m., and sea urchins, starfish, bryozoa, and brachiopods have been found at depths of 6000 m. (Fig. 38.22).

Freshwater Life Zones

Freshwater habitats can be divided into **standing water**—lakes, ponds, and swamps—and **running water**—

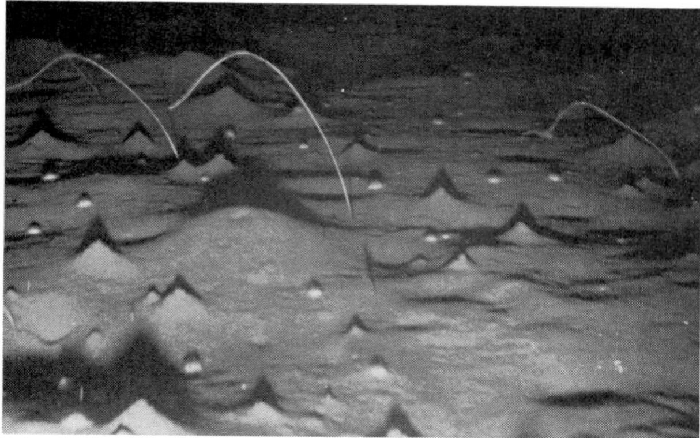
A

B

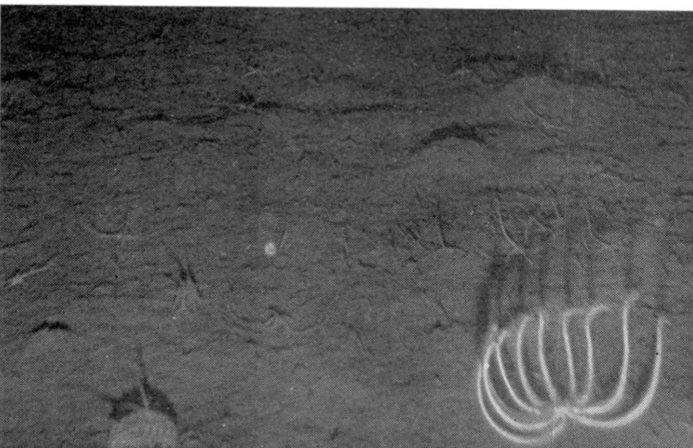
C

FIGURE 38.22 Photographs of the ocean bottom at three different depths off southern California (San Diego Trough). *A,* At 95 m. Note abundant sea urchins (probably *Lytechinus*) appearing as globular, light-colored bodies and the long curved sea whips (probably *Acantoptilum*). Burrowing worms have built the conical piles of sediment at the mouth of their burrows. (Emery, 1952.) *B,* At 1200 m: Vertical photograph of about 4 $m.^2$ of bottom composed of green silty mud having a high organic content. Note the numerous brittle stars *(Ophiuroidea)* and several large sea cucumbers *(Holothuroidea).* (Official Navy photo, courtesy U. S. Navy Electronics Laboratory, San Diego.) *C,* At 1400 m. Note in the right foreground the ten arms of what is probably a crinoid (a relative of the starfish that is attached to the bottom by a stalklike part). Small worm tubes and brittle stars litter the surface, and two sea cucumbers may be seen in the left foreground. Continual activity of burrowing animals keeps the sea bottom "bumpy." The bottom edge of the picture represents a distance of about 2 m. (From Emery, K. O.: Submarine photography with the benthograph. *(Scientific Monthly, 75*:3–11, 1952.)

rivers, creeks, and springs—each of which can be further subdivided.

Standing water, such as a lake, can be divided (much as the zones of the ocean were distinguished) into the shallow water near the shore (the **littoral zone**), the surface waters away from the shore (the **limnetic zone**), and the deep waters under the limnetic zone (the **profundal zone**).

Aquatic life is most conspicuous in the littoral zone. Within this zone the plant communities form concentric rings around the pond or lake as the depth increases (Fig. 38.23). At the shore proper are the cattails, bulrushes, arrowheads, and pickerelweeds—the emergent, firmly rooted vegetation linking water and land environments. Out slightly deeper are the rooted plants with floating leaves, such as the water lilies. Still deeper are the fragile thin-stemmed water weeds, rooted but totally submerged. Here also are found diatoms, blue-green algae, and green algae. Common green pond scum is one of the latter.

The littoral zone is also the scene of the greatest concentration of animals distributed in recognizable communities. In or on the bottom are various dragonfly nymphs, crayfish, isopods, worms, snails, and clams. Other animals live in or on plants and other objects projecting up from the bottom. These include the climbing dragonfly and damselfly nymphs, rotifers, flatworms, bryozoans, hydras, snails, and others. The zooplankton consists of water fleas, such as *Daphnia,* rotifers, and ostracods (Fig. 38.24). The larger freely swimming fauna (**nekton**) includes diving beetles and bugs, dipterous larvae (e.g., mosquitoes), and large numbers of many other insects. Among the vertebrates are frogs, salamanders, snakes, and turtles. Floating members of the community (**neuston**) include whirligig beetles, water stri-

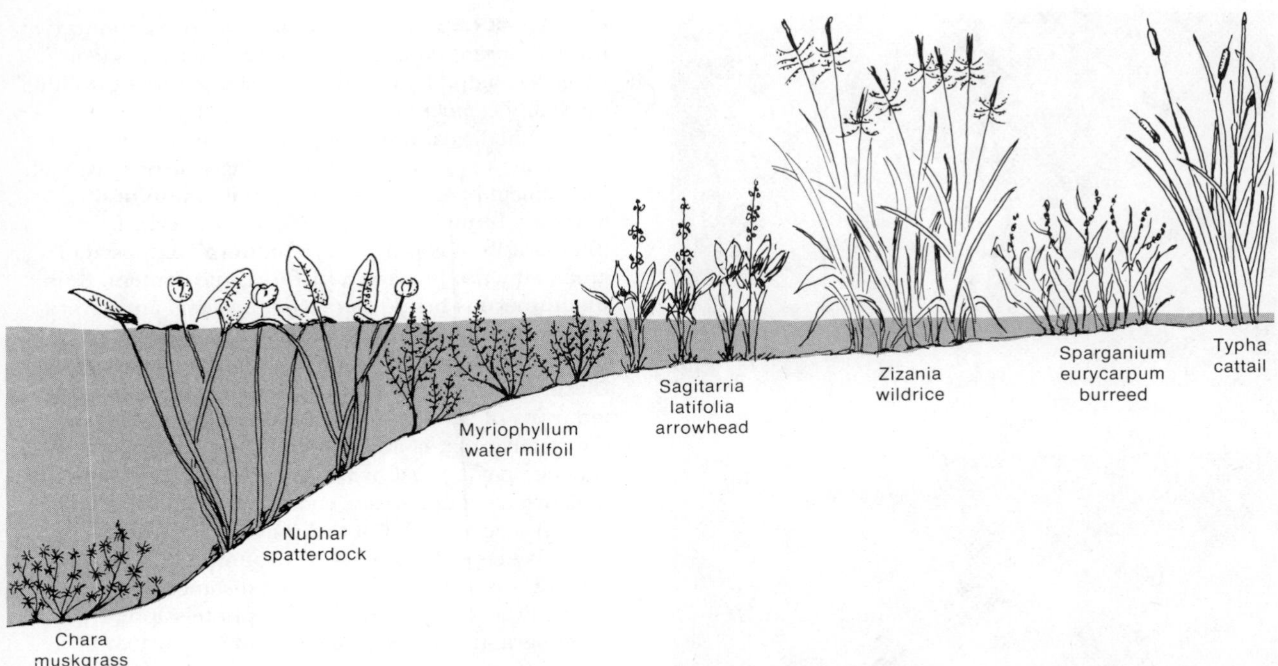

FIGURE 38.23 Zonation of vegetation about ponds and along river banks. Note the changes in vegetation with water depth. (After Dansereau, P.: *Biogeography: An Ecological Perspective.* New York, The Ronald Press Co., 1959; from Smith, R. L.: *Ecology and Field Biology.* New York, Harper & Row, 1966.)

ders, and numerous protozoans. Many pond fish (sunfish, top minnows, bass, pike, and gar) spend much of their time in the littoral zone.

The limnetic or open-water zone is occupied by many phytoplanktonic organisms (dinoflagellates, *Euglena* and *Volvox*), many small crustaceans (copepods, cladocera, and so on), and many fish.

Deep (profundal) life consists of bacteria, fungi, clams, blood worms (larvae of midges), annelids, and other small animals capable of surviving in a region of little light and low oxygen.

As compared to ponds where the littoral zone is large, the water usually shallow, and temperature stratification usually absent, lakes have large limnetic and profundal zones, a marked **thermal stratification,** and a seasonal cycle of heat and oxygen distribution. In the summertime, the surface water (**epilimnion**) of lakes becomes heated, while that below (**hypolimnion**) remains cold. There is no circulatory exchange between upper and lower layers, with the result that the lower layers frequently become deprived of oxygen. Between the two is a region of steep temperature decline (**thermocline**). As the cooler weather of fall approaches, the surface water cools, the temperature is equal at all levels, the water of the whole lake begins to circulate, and the deep is again oxygenated. This is the "**fall overturn.**" In winter, the heaviest water (4° C.) at the bottom is overlaid by lighter, colder water and ice. The bottom is now warmer than the top. Because bacterial decomposition and respiration are less at low temperatures and cold water holds more oxygen, there is usually no great winter stagnation. The formation of ice may, however, cause oxygen depletion and result in a heavy winterkill of fish. The "**spring overturn**" occurs when the ice melts and the surface and deep water are at the same temperature and can mix (Fig. 38.25).

Moving waters differ in a number of respects from lakes and ponds: current is a controlling and limiting factor; land–water interchange is great because of the small size and depth of moving water systems, as compared with lakes; organic detritus from land plants (fallen leaves and branches) is a major source of energy; and oxygen is almost always in abundant supply except when there is pollution. Temperature extremes tend to be greater than in standing water. Plants and animals living in streams are usually attached to surfaces or, in the case of animals, are exceptionally strong swimmers. Characteristic stream organisms are: caddis fly larvae,

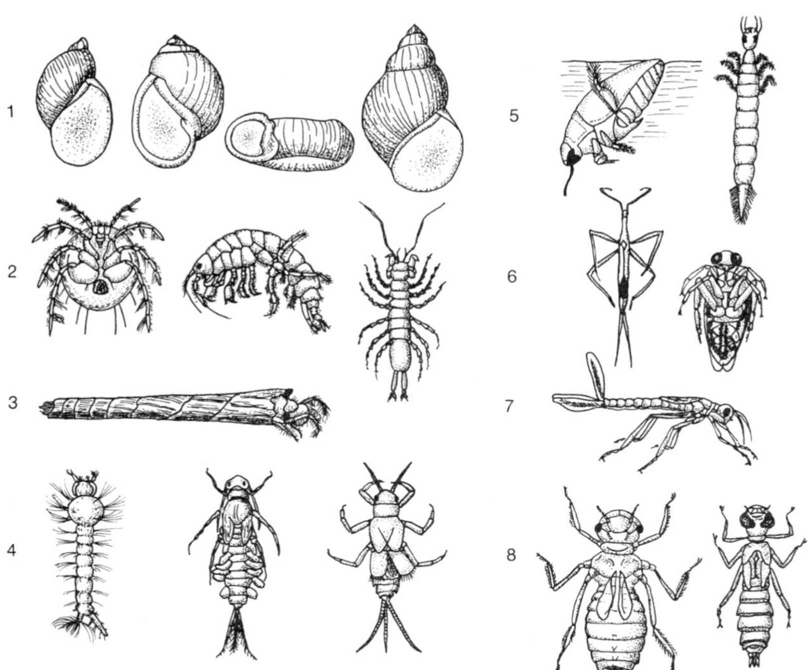

FIGURE 38.24 Some representative animals of the littoral zone of ponds and lakes. Series *1* to *4* are primarily herbivorous forms (primary consumers); series *5* to *8* are predators (secondary consumers). *1,* Pond snails *(left to right): Lymnaea* (pseudosuccinea) *columella; Physa gyrina; Helisoma trivolvis; Campeloma decisum.* *2,* Small arthropods living on or near the bottom or associated with plants or detritus *(left to right):* a water mite, or Hydracarina *(Mideopsis);* an amphipod *(Gammarus);* an isopod *(Asellus). 3,* A pond caddis fly larva *(Triaenodes),* with its thin, light portable case. *4, (Left to right)* A mosquito larva *(Culex pipiens);* a clinging or periphytic mayfly nymph *(Cloeon);* a benthic mayfly nymph *(Caenis)*—note gill covers that protect gills from silt. *5,* A predatory diving beetle, *Dytiscus,* adult and *(right)* larva. *6,* Two predaceous Hemipterans, a water scorpion, *Ranatra* (Nepidae) and *(right)* a backswimmer, *Notonecta. 7,* A damsel fly nymph, *Lestes* (Odonata-Zyoptera); note three caudal gills. *8,* Two dragonfly nymphs (Odonata-Anisoptera), *Helocordulia,* a long-legged sprawling type (benthos) and *(right) Aeschina,* a slender climbing type (periphyton). (After Pennak, R. W.: *Freshwater Invertebrates of the United States.* New York, The Ronald Press Co., 1953.)

blackfly larvae, attached green algae, encrusting diatoms, and aquatic mosses.

Freshwater habitats change much more rapidly than other life zones; ponds become swamps, swamps become filled in and converted to dry land, and streams erode their banks and change their course. The kinds of plants and animals present may change markedly and show ecological successions similar to those on land. The large lakes, such as the Great Lakes, are relatively stable habitats and have more stable populations of plants and animals. Lake Baikal in the Soviet Union is the oldest and deepest lake in the world, formed during the Mesozoic Era and containing many species of fish and other animals found nowhere else.

The Dynamic Balance of Nature

The concept of the dynamic state of the cellular constituents was discussed in Chapter 3, and we learned that the protein, fat, carbohydrate, and other constituents of the cells are constantly being broken down and resynthesized. A biotic community undergoes an analogous constant reshuffling of its constituent parts; the concept of the dynamic state of biotic communities is an important ecological principle. Plant and animal populations are constantly subject to changes in their physical and biotic environment and must adapt or die. In addition,

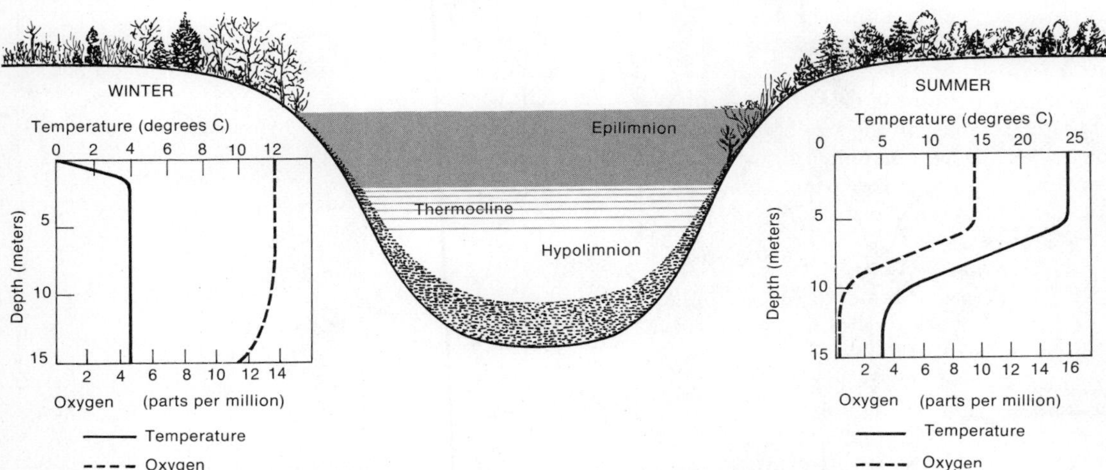

FIGURE 38.25 Thermal stratification in a north temperate lake (Linsley Pond, Connecticut). Summer conditions are shown on the right, winter conditions on the left. Note that in summer the oxygen-rich circulating layer of water, the epilimnion, is separated from the cold oxygen-poor hypolimnion waters by a broad zone, called the thermocline, that is characterized by a rapid change in temperature and oxygen with increasing depth. (From Deevey, E. S., Jr.: *Life in the Depths of a Pond.* Copyright © by Scientific American, Inc. All rights reserved.)

communities undergo a number of rhythmic changes—daily, tidal, lunar, seasonal, and annual—in the activities or movements of their constituent organisms. These result in periodic changes in the composition of the community as a whole. A population may vary in size, but if it outruns its food supply, equilibrium is quickly restored. Communities of organisms are comparable in many ways to a many-celled organism and exhibit growth, specialization, and interdependence of parts, characteristic form, and even development from immaturity to maturity, old age, and death.

Human Ecology

The principles of ecology apply to human populations, as well as to populations of animals and plants. Indeed, one of the most important problems facing us at present is the current explosive growth of the human population. This has had marked effects on our environment, resulting in a decline in the water and mineral resources of the planet, in increased erosion of the soil, and in the increasing pollution of our air, water, and soil with pesticides, industrial wastes, and radioactive wastes. The large and increasing size of the human population and its age composition (in some countries 50 percent of the population is less than 15 years of age) clearly signal that the problem will not readily be solved. The increasing population size has had marked negative effects on our agricultural practices, on our forests, on wildlife management, on our use and control of rivers, lakes, and wetlands, on the mineral resources of the planet, and on public health. The myriad problems of pollution that occupy our attention so strongly at present are due to the demands that the increasing size of the human population have made on our environmental resources. These have forced us to consider the ultimate limitations of the earth, that is, its carrying capacity for humans.

Most of our current problems relating to the supply of energy, food, and mineral resources and the pollution of the air, water, and soil stem directly from the increased population size. The increase in population size is due almost entirely to decreased death rates in both developed and developing countries around the world. Human birth rates have not increased significantly. The programs in many countries to reduce the birth rate by contraceptive measures and abortion are effective, but the age structure of the population in many countries means that there will be a continued impetus towards increasing population size for another 50 to 100 years.

The possibility of widespread famine, though not imminent, becomes more probable as the population exceeds the possible food supply and the amount of land available for farming. The increase in agricultural production, termed the "green revolution," was achieved only by greatly increased use of energy, fertilizer, pesticides, and irrigation, along with the introduction of specially bred strains of protein-rich, high-yield wheat, rice, and other cereals.

The movement of people into the cities has resulted in an exceedingly rapid increase in the size of cities and in all the problems that accompany rapid urbanization—the problem of supplying food, water, hospitals and other health facilities, schools and other educational facilities, together with all the other urban services, such as transportation, fire and police protection, and social services—and has put tremendous strains on the social and political structure of cities.

Governments in many parts of the world have begun to take drastic steps to meet the problem. Mainland China, with a population of more than one billion people, has been successful in attacking the overpopulation problem; its crude birth rate dropped from 32 per 1000 in 1970 to 12 per 1000 in 1980. Over this same period there was a 37 percent decline in the birth rate in the United States.

Earlier predictions that the world's population will double in the next generation and double again in the following generation may prove to be quite unrealistic. In most countries a population doubling would result in unmanageable ecological, economic, and political stresses. In other words, the carrying capacity of the environment for humans would be exceeded, and the increased environmental resistance would limit the population size. However, even today no serious attempts have been made by the governments of some parts of the world to slow down the rate of growth of their populations.

Most biologists and social scientists believe that the danger of human overpopulation is both great and imminent. The growth of the human population in the past 300 years follows an exponential curve (Fig. 38.26). The productivity and carrying capacity of the earth for human beings can be maintained and perhaps increased somewhat, but eventually the human biomass must be brought into equilibrium with the space and food available. There is no agreement as to what factor will become critical in limiting human populations—the amount of food, the amount of drinkable water, the amount of breathable air, the amount of some specific environmental pollutant, or something else.

Application of the principles of systems analysis to human population dynamics, published as *The Limits of Growth* (1972), led to very gloomy predictions of a worldwide decline in resources, increasing global pollution, and the ultimate collapse of the entire system because of depletion of nonrenewable resources. In this book the human population eventually crashes owing to the increasing death rate caused by deficiencies of food and health services. This book evoked a storm of controversy and a reexamination using somewhat different models and assumptions, but they too predicted drastic changes in the early twenty-first century.

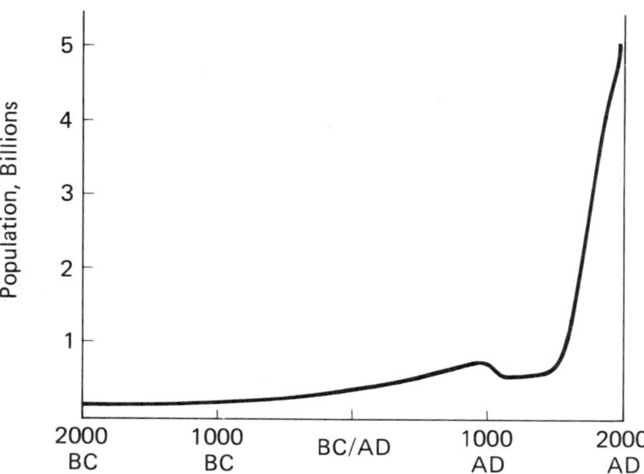

FIGURE 38.26 Growth of the human population of the world, extrapolated to 2000 A.D.

Summary

1. Populations have characteristics that are a function of the whole group, not of the individual members: population density, birth rate, death rate, age distribution, biotic potential, rate of dispersion, and growth form.

2. Plotting the number of organisms against time yields a population growth curve with a characteristic shape defined by the biotic potential of the organism and the carrying capacity of the environment. Plotting the number of survivors in a population against time yields a survivorship curve.

3. The structure of a population is determined by its birth rate and death rate at specific ages. Death rates vary with age, and birth rates are proportional to the number of females in the reproductive years, those able to reproduce.

4. The biotic potential, r, of a population refers to its intrinsic power to increase in number when the age ratio is stable and all environmental conditions are optimal. The sum of the physical and biological factors that prevent a species from reproducing at its maximal rate is termed environmental resistance. The carrying capacity of the environment, K, is the maximal population size that can be borne by the resources of that area.

5. The population size at various times during exponential growth of the population may be calculated from the equation $N_t = N_0e^{rt}$. To determine population size during logistic growth, when the effect of carrying capacity is manifest, the equation $N_t = \frac{K}{1 + e^{a - rt}}$ is used. Natural selection occurring under uncrowded conditions is termed r selection and that under crowded conditions is termed K selection.

6. The size of a population may increase and decrease in regular cycles due to changes in environmental resistance or to factors within the population. Populations have a tendency to disperse, to spread out in all directions until a barrier is reached. Within an area members of a population typically occur in small groups or clumps, which may facilitate their survival during unfavorable periods. When two closely related species compete for the same ecologic niche one of the species usually dies off or is forced to seek a new ecologic niche. This is termed the competitive exclusion principle.

7. A biotic community is an assemblage of populations living in a defined area. These interact in many ways to maintain the structure and function of the community and to direct community succession through a series of seral communities to the climax community or communities.

8. Geographical ecology describes the distinctive features of the major habitats and their subdivisions, how they are organized, and the organisms present in each and identifies the major producers, consumers, and decomposers present.

9. We can distinguish four major habitats: marine, estuarine, freshwater, and terrestrial. Within these major habitats, established by a complex interaction of climate, other physical factors, and biotic factors, are large, distinct community units called biomes. In each biome the kind of climax vegetation—grasses, conifers, deciduous trees—is uniform, but the particular species of plant may vary in different parts of the biome. Between adjacent biomes are fairly broad transition regions called ecotones.

10. The terrestrial biomes include tundra, coniferous forest, deciduous forest, evergreen subtropical forest, tropical rain forest, grassland, desert, and chaparral, distributed in irregular belts around the world and in belts at higher altitudes.

11. The estuarine biomes, found at the mouth of a river or a coastal bay, are among the most naturally fertile in the world; much of the vegetation is converted to detritus and consumed by marine detritus eaters.

12. The marine life zones are distinguished by their depth and their relation to the shore and continental shelf.

13. Freshwater life zones include running water—rivers—and standing water—lakes. The latter is subdivided according to water depth. The water in freshwater ponds is separated into surface and deep layers that do not mix except during the spring and fall when the surface and deep waters are equal in temperature.

14. Each biotic community is in dynamic balance and undergoes a constant reshuffling of its constituents. Some of these changes occur rhythmically, at daily, tidal, seasonal, and annual intervals. Communities, like organisms, exhibit growth, specialization, characteristic form, and development from immaturity to maturity and death.

References and Selected Readings

Altman, I.: *The Environment and Social Behavior.* New York, Irvington Publications, 1981. An interesting description of the effects of environment on the behavior of social animals.

Anderson, J. M.: *Ecology for Environmental Sciences: Biosphere, Ecosystems, and Man.* New York, Halsted, 1981. An excellent discussion of the components of the biosphere and their interactions with humans.

Anderson, R. A.: *Abandon Earth: Last Call.* Mountain View, California, Pacific Press Pub. Assoc., 1982. A doomsday account of the crisis in our handling of the ecology of the earth.

Barrington, E. J.: *Environmental Biology.* New York, Halstead, 1980. A well-written description of biological principles applied to the environment.

Bender, D., and B. Leone: *The Ecology Controversy.* St. Paul, Minnesota, Greenhaven, 1981. A popular account of the major controversies in the plans for managing the environment.

Bourliere, F.: *Tropical Savannas.* New York, Elsevier, 1983. A beautifully illustrated description of one of the important biomes.

Bradshaw, M. J.: *The Earth: The Living Planet.* New York, Halsted, 1977. A popular account of the ecology of the earth.

Brewer, R., and M. McCann: *Field and Laboratory Manual of Ecology.* Philadelphia, Saunders College Publishing, 1982. A text and laboratory manual of ecology.

Ewusie, J. Y.: *Elements of Tropical Ecology.* Exeter, New Hampshire, Henemann, 1980. A presentation of the principles and examples of the ecology of tropical regions.

Gauch, H. G., Jr.: *Multivariate Analysis in Community Ecology.* New York, Cambridge University Press, 1982. An advanced presentation of the analysis of the ecology of community by mathematical modeling.

Gormon, M. L.: *Island Ecology.* New York, Methuen, 1979. A discussion of the special problems of animals and plants growing on islands.

Hazen, W. E. (Ed.): *Readings in Population and Community Ecology.* 3rd ed. Philadelphia, W. B. Saunders Co., 1975. A carefully selected group of papers discussing the ecology of populations and communities.

Hinckley, A. D.: *Applied Ecology.* New York, MacMillan, 1976. A presentation of the application of the principles of ecology to practical problems.

Innis, G. S., and R. V. O'Neill: *Systems Analysis of Ecosystems.* Bentonsville, Maryland, International Publishers, 1979. An advanced account of the application of systems analysis to an understanding of specific ecosystems.

Patten, B.: *Systems Analysis and Simulation in Ecology.* New York, Academic Press, 1976. A further discussion of the use of systems analysis in attempting to predict the outcome of perturbations in the environment.

Pianka, E. R.: *Evolutionary Ecology.* 3rd ed. New York, Harper & Row, 1982. A standard text of the evolution of ecological relations.

Richardson, J. L.: *Dimensions of Ecology.* New York, Oxford University Press, 1977. A description of the application of ecological principles to practical problems.

Ripley, E. A., and R. E. Redmann: *Energy Exchange in Ecosystems.* New York, Elsevier, 1983. A discussion of specific ecosystems and the transfer of energy in successive trophic levels.

Sinclair, G. R., and M. S. Norton-Griffiths: *Dynamics of an Ecosystem.* Chicago, University of Chicago Press, 1979. A detailed study of factors causing changes in a specific ecosystem in the course of a year.

Wagner, R. H.: *Environment and Man.* 3rd ed. New York, Norton, 1978. A popular account of human ecology.

Whittaker, R. H.: *Communities and Ecosystems.* 2nd ed. New York, Macmillan, 1975. Detailed coverage of ecosystems and their components.

Wiegert, R.: *Ecological Energetics.* New York, Academic Press, 1976. A good coverage of ecology that emphasizes the energy relations involved in ecosystems.

Watt, K. E. F.: *Ecology and Resource Management.* New York, McGraw-Hill, 1968. A discussion of the application of ecological principles to the management and conservation of natural resources.

Wilson, E. O.: *Sociobiology.* Cambridge, Belknap Press, 1975. A thought-provoking analysis of the application of ecological principles to human populations.

Index/ Glossary

A

α-helix, 25
Aardvark, 734
Abalone, 527
Abducens nerve, 226
ABO blood groups, 175, 411
  inheritance of, 358
Abortion, 828
Absolute refractory period, 217
Absorption (ab-sorp′shun) The taking up of a substance, as by the skin, mucous surfaces, or lining of the digestive tract.
  of foods, 126
Absorption spectrum A measure of the amount of energy at specific wave lengths that has been absorbed as light passes through a substance. Each type of molecule has a characteristic absorption spectrum.
Abyssal zone, 821
Acantharia, 460
Acanthocephala, 515
Acanthodians, 658
Acanthodii, 663
Acarina, 578
Acetabularia, 34
Acetyl coenzyme A, 70
Acetylcholine (as″ĕ-til-ko′lēn) The acetic acid ester of the organic base choline, normally secreted at the ends of many neurons; responsible for the transmission of a nerve impulse across a synapse.
Aciculum, 552
acid (as′id) A substance whose molecules or ions release hydrogen ions (protons) in water. Acids have a sour taste, turn blue litmus paper red and unite with bases to form salts, 20
Acoela, 503
Acoelomate (a-se′lo-mate) **body plan** The condition of having no body cavity, as in flatworms, 492
Acorn barnacles, 597
Acquired characters, 403
Acromegaly, 274
Acron (ak′ron) The anterior nonsegmental part of the body of a metameric animal, 92
Acrosomal reaction, 297
Acrosome (ak′ro-sōm) A caplike structure covering the head of the spermatozoon, 294
Actin (ak′tin) Minute protein filaments found in most contractile cells. In striated muscle cells, thin filaments of actin overlap and alternate with thicker filaments of myosin to form myofilaments. The interactions between actin and myosin lead to muscle contraction, 48, 108
Actinopterygii (ak″tĭ-nop′tĕr-ij-ĭe) A member of a subclass of Osteichthyes in which the fins are supported by bony rays; includes sturgeons, gars, and teleosts, 663
Actinulae larvae, 477
Action potentials A slight current generated when any tissue becomes active—as when a muscle contracts, a gland secretes, or a nerve conducts an impulse, 109, 216, 240
  temporal pattern of, 241
Active transport The transfer of a substance into or out of a cell across the cell membrane against a concentration gradient by a process that requires the expenditure of energy.
Adaptation, 13
  of eye to light intensity, 254
  sensory, 240
Adaptive radiation The evolution from a single ancestral species of a variety of species that occupy different habitats, 420, 423
Adaptive shifts, 422
Adaptive trends, 422
Adaptive value A measure of the success of one genotype in a population relative to the most successful one, which is given the adaptive value of 1. The complement of selection coefficient (q.v.), 412
Addison's disease, 271
Adenine, 26, 364
Adenosine triphosphate (ATP), 27, 66
Adipose tissue, 23, 56
Adrenal cortical hyperplasia, 272
Adrenal glands, 270
Adrenal medulla, 270
Adrenocorticotropic hormone (ACTH), 265, 271, 274
Aegyptopithecus, 752
Aerobic (a-er-o′bik) Growing or metabolizing only in the presence of molecular oxygen.
Aestivation (es″tĭ-va′shun) The dormant state of decreased metabolism in which certain animals pass hot, dry seasons, 528, 677
Afferent (af′er-ent) Conveying toward a center; designating vessels or neurons that transmit blood or impulses toward a point of reference; afferent neurons are sensory neurons conducting impulses toward the central nervous system.
Age distribution, 804
Agglutination (ah-gloo″tĭ-na′shun) The collection into clumps of cells or particles in a fluid, 175
Agglutinin, 358
Agglutinogen, 358
Aggression, 773
Agnatha (ag′na-thah) The jawless vertebrates. A class of vertebrates including lampreys, hagfish, and the extinct ostracoderms, 653
  characteristics of, 651
Air sacs, 709
Albinism, 344, 380
Albino, 355
Aldolase, 73
Aldosterone, 205, 265, 271
Alkaptonuria, 380
Allantois (ah-lan′to-is) One of the extraembryonic membranes of reptiles, birds, and mammals; a pouch growing out of the posterior part of the digestive system and serving as an embryonic urinary bladder or as a source of blood vessels to and from the chorion or placenta, 325, 685
Allele(s) (ah-lēl′) One of a group of alternative forms of a gene that may occur at a given site (locus) on a chromosome, 339
  multiple, 358
Allergy A hypersensitivity to some substance in the environment, manifested as hay fever, skin rash, or asthma, 175
Alligator, 694
Allopolyploidy (al′o-pol″e-ploi-de) An increase in number of chromosome sets resulting from the hybridization of two species. The hybrid continues

Allopolyploidy *(continued)*
as a third species, 421
Altricial, 712, 730
Alveolus (al-ve'o-lus) A small saclike dilation or cavity; the terminal chamber of air passages in the mammalian lung, 158
Amacrine cells, 255
Ambulacral groove, 624
Ameboid motion (ah-me'boid) The movement of a cell by means of the slow oozing of the cellular contents.
Amino acid(s) (am'ĭ-no) An organic compound containing an amino group (—NH_2) and a carboxyl group (—COOH); amino acids may be linked together to form the peptide chains of protein molecules, 25
activation of for protein synthesis, 373
essential, 26, 141
Amino acid oxidation, 73
Amino group, 25
Aminopeptidases, 138
Ammocoetes larva, 657
Ammonia, 195
Ammonoidea, 547
Amniocentesis, 379
Amnion (am'ne-on) One of the extraembryonic membranes of reptiles, birds, and mammals; a fluid-filled sac around the embryo, 325, 685
Amniote (am'ne-ōt) A vertebrate characterized by having an amnion during its development; a reptile, bird, or mammal, 685
Amoeba(s), 34, 454
Amoeboid movement (ah-me'boid) The movement of a cell by means of the slow oozing of the cellular contents, 117
Amphibian (am-fib'e-an) A member of a class of vertebrates in which the larvae are usually aquatic and the adults terrestrial; includes frogs, salamanders, and related animals.
adaptations of, 674
characteristics of, 670
classification of, 695
evolution of, 677
locomotion of, 676
Amphiblastula larva, 468
Amphioxus, 644
Amphipods, 593–594
Amphisbaenia, 692
Amphitrite, 556
Ampulla, 244
starfish, 624
Amylase (am'ĭ-lās) An enzyme that catalyzes the hydrolysis of starches; it cleaves α-1 $\rightarrow$ 4 glucosidic bonds of polysaccharides, 66, 138
Anabolic, 13
Anabolism (ah-nab'o-lizm) Chemical reactions in which simpler substances are combined to form more complex substances, resulting in the storage of energy, the production of new cellular materials and growth.
Anaerobic (an″a-er-o'bik) Growing or metabolizing only in the absence of molecular oxygen.
Analogy (ah-nal'o-je) A similarity in function among nonhomologous (q.v.) structures. Analogous structures may or may not resemble each other superficially, 389, 435
Anamniotes (an-am'ne-ōt) A vertebrate characterized by the absence of the amnion during embryonic development; a fish or amphibian, 686
Anaphase (an'ah-fāz) Stage in mitosis or meiosis, following the metaphase, in which the chromosomes move apart toward the poles of the spindle, 45
Anatomy, 4
Androgen(s) (an'dro-jen) Any substance that possesses masculinizing activities, such as testosterone or one of the other male sex hormones, 271
Angler fish, 668
Anion(s) (an'i-on) An ion carrying a negative charge, 17
Annelids, 550–570
Anteater, 732
Antelope, 739
Antenna (an-ten'ah) A projecting, usually filamentous organ equipped with sensory receptors, 574
Anterior lobe, hormones of, 274
Anthozoa, 481–485
Anthropoidea (an'thro-poid″eah) A suborder of primates that includes monkeys, apes, and human beings, 743
Antibody (an'ti-bod″e) A protein produced in response to the presence of some foreign substance in the blood or tissues, 171, 173
Anticodon (ăn'ti-kōdŏn) A sequence of three nucleotides in transfer RNA that is complementary to, and combines with, the three nucleotide codon on messenger RNA, thereby binding the amino acid–transfer RNA combination to the mRNA.
Antidiuretic hormone (ADH), 204. See also *Vasopressin.*
Antidromic impulse, 214
Antigen (an'tĭ-jen) A foreign substance, usually a protein or protein-polysaccharide complex, which elicits the formation of specific antibodies within an organism, 171
Antimolting hormone, 286
Antiparallel strands of DNA, 365
Antlers, 739 Size of, evolution of, 422
size of, evolution of, 422
Ants, 610
Aorta (a-or'tah) One of the primary arteries of the body, e.g., the ventral aorta of fishes that distributes venous blood to the gills or the dorsal aorta that distributes blood to the body, 178
Aortic arches, 178
Ape-men, 752
Apes, 748
early evolution of, 752
Aphasia, 236
Apicomplexa, 457
Apis, 611–613
Aplacental viviparity, 326
Aplacophora, 547
Apoenzyme (ap″o-en'zīm) Protein portion of an enzyme; requires the presence of a specific coenzyme to become a complete functional enzyme, 66
Aposematic coloration, 678, 681, 781
Appendix, 137, 172
Apterygota, 603
Aquatic adaptations, 652
Aqueous humor, 251
Arachnida, 577–585
Arachnoid, 229
Araneae, 578
Arbacia, 631
Archaeopteryx, 394, 713
Archaeornithes, 714
Archenteron (ar-ken'ter-on) The central cavity of the gastrula, lined with endoderm, which forms the rudiment of the digestive system, 311
Archinephric duct (ar″ke-nef'ric) The primitive kidney duct; drains the pronephros, mesonephros, and opisthonephros, 198
Archosaurs, 690
Arctic foxes, 808
Arctic tern, 718
Arenicola, 555
Aristotle, 7, 401
Aristotle's lantern, 632
Armadillo, 733
Arrowworms, 637
Artemia, 595
Arteriole (ar-te're-ōl) A minute arterial branch, especially one just proximal to a capillary, 189
Artery A vessel through which the blood passes away from the heart to the various parts of the body; typically has thick elastic walls, 187
Arthropodization, 572
Arthropods, 571–616
compound eye of, 256
hormones of, 283
Artificial selection, 391
Artiodactyla, 739
Ascaroids, 512
Ascending tracts, 228
Aschelminthes, 507
Ascidians, 641–644
Asconoid sponges, 465
Asexual reproduction, 290
Aspartic acid, 73
Association areas of brain, 235
Aster, 37
Asterias, 626
Asteroidea, 624–628
Atolls, 487
Atom(s) The smallest quantity of an element that can retain the chemical properties of the element, composed of an atomic nucleus containing protons and neutrons together with electrons that circle the nucleus in specific orbitals, 14
Atomic orbital Distribution of an electron around the atomic nucleus.
Atomic structure, 14
Bohr planetary model of, 15
Atomic weights, 20
Atrium (a'tre-um) An entrance chamber to another structure or organ; a chamber of the heart receiving blood from a vein and pumping it to a ventricle, 177, 179
of sponges, 463
Auditory ossicle (os'sĭ-k'l) One or more small bones (malleus, incus, and stapes in mammals) that transmit vibrations across the middle ear cavity from the tympanic membrane to the inner ear, 106, 695, 728
Aurelia, 481
Auricle (aw'ri-k'l) The ear-shaped portion

of the mammalian atrium (q.v.); sometimes used as a synonym for atrium. The external ear flap of mammals.
Australopithecus, 752
Autecology, 786
Autoimmune disease, 174
Autonomic nervous system, 225, 227
Autopolyploidy (aw'-to-pol"-e-ploi-de) An increase in the number of chromosome sets within a single species.
Autosome(s) (aw'to-sōm) Any ordinary paired chromosome, as distinguished from a sex chromosome, 348
Autotomy, in brittle stars, 630
Autotrophs (aw'-to-trofs) Organisms that manufacture organic nutrients from inorganic raw materials, 126, 433
Aves (a'vez) A class of flying vertebrates; the birds, 697
Avicularia, 620
Axial filament, 294
Axolotl, 678
Axon(s), 58, 213
  giant, of earthworm, 222
Axon hillock, 240
Axopods, 455
Azaguanine, 376

B

Baboon, 743
Bacon, Roger, 8
Balanced polymorphism (pol"e-mor'fizm) An equilibrium mixture of homozygotes and heterozygotes maintained by separate and opposing forces of natural selection.
Balanoglossus, 647
Baltimore, David, 372
Bamboo worm, 557
Barnacles, 597
Baroreceptor A receptor that detects changes in pressure, 186
Barrier, 400, 419
Barrier reefs, 487
Basal body, 37
Base A compound that releases hydroxyl ions (OH^-) when dissolved in water; turns red litmus paper blue, 21
Base pairing, 365
Basement membrane, 54
Basilar membrane, 247
Basket stars, 628–631
Basophils, 58
Bat, 734
Bathyal zone, 821
Batoidea, 660
Bayliss, William, 9
Beak, in cephalopods, 545
Bear, 736
Bees, 610, 611–613
Behavior, 758
  agonistic, 773
  altruistic, 778
  begging in gulls, 769
  cleaning by shrimp, 762
  egg rolling in goose, 763
  elements of, 763
  embryonic development of, 768
  escape in sea slug, 764
  evolution of, 773
  genetics of, 772
  mating in silkworms, 765
  physiology of, 764
  predatory in spiders, 765
  reproductive, 774
  reproductive in ring doves, 765
  social, 778
  web building in spiders, 769
Behavioral ecology, 773
Benthos (ben'thos) The flora and fauna of the bottom of oceans or lakes, 821
Bergmann's rule, 418
Beriberi, 142
Bernard, Claude, 9
Bilateral symmetry, 88
Bilateria, 439
Bile, 136
Bioenergetics, 13
Biogeographic realms, 400
Biological clock An internal timing mechanism of animals that is involved in controlling circadian rhythms (q.v.) and other cycles, and also plays a role in some types of navigation, 767
Biological oxidation Process in which electrons removed from an atom or molecule are transferred through the electron transmitter system of the mitochondrion, 70
Bioluminescence (bi"-o-loo"mī-nes ens) Emission of light by living cells or by enzyme systems prepared from living cells, 82, 546
  in cephalopods, 546
Biomass The total weight of all the organisms in a particular habitat, 790, 799
Biome (bi'om) Large, easily differentiated community unit arising as a result of complex interactions of climate, other physical factors and biotic factors, 813
Biosynthetic processes, 81
Biotic community(ies), 786, 810
  vertical stratification in, 810
Biotic potential Inherent power of a population to increase in numbers when the age ratio is stable and all environmental conditions are optimal, 804, 806
Bipedalism
  adaptations for, 751
  evolution of, 754
Bipinnaria larvae, 627
Biradial symmetry, 89
Birds
  adaptive features of, 701
  classification of, 720
  evolution of, 712
  origin of, 712
  songs of, development of, 769
Birth, 331
Birth rate, 804
Bivalvia, 531–539
Black widow, 583
Bladder, urinary, 200
Bladder worm, 502
Blastocoel (blas'to-sēl) The fluid-filled cavity of the blastula, the mass of cells produced by cleavage of a fertilized ovum, 309
Blastocyst (blas'to-sist) The modified blastula stage of embryonic mammals; it consists of an inner cell mass, which develops into the embryo, and a peripheral layer of cells, the trophoblast, which contributes to the placenta, 326
Blastomeres, 309
Blastopore (blas'to-pōr) The opening, in the gastrula stage of development, from the archenteron to the surface, 311
Blastula (blas'tu-lah) Usually spherical structure produced by cleavage of a fertilized ovum, consisting of a single layer of cells surrounding a fluid-filled cavity, 309
Blood, 57, 167
  clotting of, 169
  flow of, control of, 8
Blood groups, 175
Bloodworms, 554
Blooms, phytoplankton, 822
Blowfly, 610
Blubber, 23
Blue babies, 183
Body cavity, 91
Body fluids, 194
Body wall, 91
Bohr effect, 161
Bolus, 134
Bombardier beetle, 781
Bonds
  covalent, 17
  energy rich, 27
  ester, 18, 19
  glycoside, 18, 19
  hydrogen, 17
  hydrophobic, 19
  ionic, 17
  peptide, 18, 19
  phosphate ester, 18
  phosphodiester, 26
  thioester, 18
  van der Waal's, 19
Bone, 56, 102. See also *Cartilage replacement bone* and *Dermal bone.*
Bony fishes
  basic features of, 660
  evolution of, 663
Book gills, 576
"Boom or bust" growth curve, 806
Boring sponges, 468
Bottleneck, 411
Bowditch, Henry, 9
Bowfin, 664
Bowman's capsule Double-walled, hollow sac of cells that surrounds the glomerulus at the end of each kidney tubule, 200
Brachiation, 749
Brachiolaria, 627
Brachiopods, 621
Brachyuran crabs, 589
Brain, 229
  changes in human evolution, 754
Branchial (brang'ke-al) Pertaining to the gills or gill region, 151
Branchial basket, 656
Branchiopods, 594
Branchiostoma, 644
Brine shrimp, 595
Bristles, 704
Brittle stars, 628–631
Broad-leaved evergreen subtropical forest biome, 817
Bromouracil, as a mutagen, 377
Bronchi, 158
Brontosaurus, 692
Brood parasitism, 712
Brooding, 325

Brown, Robert, 11
Brownian movement Motion of small particles in solution or suspension resulting from their being bumped by water molecules, 51
Brown recluse spider, 583
Bryozoans, 617–621
Buccal cavity, 127
Budding Asexual reproduction in which a small part of the parent's body separates from the rest and develops into a new individual, eventually either taking up an independent existence or becoming a more or less independent member of a colony, 290
Buffers Substances in a solution that tend to lessen the change in hydrogen ion concentration (*p*H), which otherwise would be produced by adding acids or bases.
Bugula, 620
Bulk flow The transport of materials by the movement of the gas or liquid in which they are contained, 150, 166
Bursae, in ophiuroids, 630
Busycon, 524
Byssal threads, 538

C

Caecilians, 682
Calcarea, 468
Calcareous sponges, 468
Calciferous glands, 563
Calcitonin (kal″sĭ-to′nin) A polypeptide hormone composed of 32 amino acids in a single chain, secreted by parafollicular cells in the thyroid; it counters the effect of parathyroid hormone and causes the deposition of calcium and phosphate in bones, 265, 267
Calcium, 16
Calorie, 65
Camel(s), 739
  regulation of internal body fluids, 210
Capillaries (kap′ĭ-lar″e) Microscopic thin-walled vessels located in the tissues, connecting arteries and veins, and through the walls of which substances pass to the interstitial fluid, 187
  exchange in, 187
  fenestrated, 188
  flow in, 189
Capitulum, 597
Carapace (kar′ă-pās) A bony or chitinous shield covering the back of an animal, 576, 686
  in crustaceans, 587
Carbohydrases, 126
Carbohydrates, 21
Carbon, 15
Carbon cycle, 792
Carbonic anhydrase (kar′bon-ik an-hi′drās) An enzyme that catalyzes the reaction carbon dioxide + water ⇌ carbonic acid; abundant in erythrocytes, 161
Carboxyl group, 25
Carboxypeptidases, 138
Cardiac control, 186
Cardiac muscle, 57
Cardiac output, 185
Cardiac sphincter, 136
Cardiac stomach
  in decapods, 590
  in starfish, 626
Caribou, 814
Carnassial teeth, 736
Carnivore (kar′nĭ-vōr) An animal that eats flesh, 128, 736
Carnivorous diets, 128
Carotid bodies (kar′ŏ-tid) Chemoreceptors on the carotid arteries that monitor the oxygen and carbon dioxide content of the blood, 187
Carrying capacity, 807
Cartilage, 56
Cartilage cells, 56
Cartilage replacement bone Bone that develops in and around a cartilaginous rudiment that it replaces, 102
Cartilaginous fishes, 659
Cassowary, 715
Castes, 610
Castings, 130
Castration, 277
Catabolic, 13
Catalase, 65
Catalysis, 65
Caterpillar, 610
Catfish, 665
Cation(s) (kat′-ion) An ion bearing a positive charge, 17
Cattle, 739
Caudata, 678
Cave paintings, 6
Ceboidea The family of New World monkeys; they often have prehensile tails, 743
Cecum (se′kum), in mammals A blind pouch into which open the ileum, the colon, and the vermiform appendix, 136
Cell(s)
  and environment, exchanges of material between, 52
  as unit of life, 11
  chemical constituents of, 21
  organic compounds in, 28
Cell body of neuron, 213, 214
Cell constancy An extreme example of mosaic development that results in all individuals in a species having the same number of cells in comparable tissues performing similar functions, 506
Cell cycle, 41
  phases of, 42
Cell hybridization, 42
Cell size, factors limiting, 41
Cell theory The generalization that all living things are composed of cells and cell products, that new cells are formed by the division of preexisting cells, that there are fundamental similarities in the chemical constituents and metabolic activities of all cells, and that the activity of an organism as a whole is the sum of the activities and interactions of its independent cell units, 11, 31
Cellular biosynthesis, principles of, 81
Cellular constituents, dynamic state of, 80
Cellular respiration, 63, 69
Cellulases, 129
Cellulose(s), 21, 23
Cement, 134
Center of origin, 400
Centimorgans, 352
Centipedes, 602
Central dogma of biology, 363
Central nervous system, 228, 229
Centriole(s) (sen′trĭ-ōl) Small, dark-staining organelle lying near the nucleus in the cytoplasm of animal cells, and forming the spindle during mitosis and meiosis, 36, 45
Centrolecithal egg (sen′tro-les′ĭ-thal) Type of arthropod egg having the yolk arranged as a large sphere around the central nucleus, 297
Centromere (sen′tro-mēr) The point on a chromosome to which the spindle fiber is attached; during mitosis or meiosis it is the first part of the chromosome to pass toward the pole.
Cephalization (sef″al-i-za′shun) Head formation, concentration of nervous tissue and sense organs at the anterior end of the body, 88
Cephalochordata, 644–646
Cephalon, 575
Cephalopod eye, 256
Cephalopoda, 539–546
Cephalothorax, in crustaceans, 587
Cerata, 524
Cercaria (ser-ka′re-ah) **larva** The final free-swimming larval stage of a trematode parasite, 499
Cercopithecidae The family of Old World monkeys, 743
Cerebellum (ser″e-bel′um) The part of the vertebrate brain that controls muscular coordination, 229, 232
Cerebral aqueduct of Sylvius, 229
Cerebral hemispheres, 234
Cerebrosides, 24
Cerebrospinal fluid, 229
Cerebrum (ser′e-brum) A major portion of the vertebrate brain, occupying the upper part of the cranium; the two cerebral hemispheres, united by the corpus callosum, form the largest part of the central nervous system in humans.
Cervix, 305
Cestoda, 500–502
Cetacea, 742
Chaetognatha, 440, 637
Chaetopteridae, 558
Chagas disease, 453
Chameleons, 689
Chance, Britton, 67
Chaparral biome, 818
Character displacement The morphological divergence that occurs between related species as they compete for limited resources and specialize on the utilization of certain ones, 419
Characteristics, sex-linked, 350
Chelate appendages (ke′lāt) Having the terminal parts of certain appendages in the form of pincers, as in crabs and scorpions, 577
Chelicera (ke-lis′er-a) A pair of pincer-like head appendages found in spiders, scorpions, and other arachnids, 576

Chelicerata, 576–585
Chemical bonds, 17
Chemical compounds, 19
Chemical reactions, 64
Chemicals, mutagenic, 377
Chemiosmotic hypothesis, 76
Chemoreception at synapse, 219
Chemoreceptors, 248
Chiggers, 582
Chimaera, 660
Chimpanzee, 748
Chinese liver fluke, 500
Chiroptera, 734
Chitin (ki′tin) An insoluble horny polysaccharide bound to protein to form a glycoprotein. Abundant in the exoskeleton of arthropods, 23, 100
  in arthropod skeleton, 572
Chitons, 528
Chlamydomonas, 450
Chlorine, 16
Chlorogogen cells, 563
Chlorophyll, 63
Choanae, 157
Choanocyte (ko′ă-no″sīt) A unique cell type with a flagellum surrounded by a thin cytoplasmic collar; characteristic of sponges and one group of protozoa, 463
Choanoflagellates, 452
Cholecystokinin-pancreozymin, 139
Cholesterol, 23, 24, 32
Chonchae, 156
Chondrichthyes (kon″drik′thi-ēz) A class of fishes with cartilaginous skeletons; includes sharks, skates, and related animals, 658
  characteristics of, 651
Chondrocranium (kon″dro-kra′ne-um) Cartilage or cartilage replacement bone that contributes to the vertebrate brain case and encapsules the nose and inner ear, 103
Chondroitin, 56
Chondrostei, 663
Chorion (ko′re-on) An extraembryonic membrane in reptiles, birds and mammals that forms an outer cover around the embryo and in mammals contributes to the formation of the placenta, 325, 685
Chorionic gonadotropin, 265, 281
Choroid coat, 251
Choroid plexuses, 229
Chromatid, 45
Chromatin (kro′mah-tin) The readily stainable portion of the cell nucleus, forming a network of fibrils within the nucleus; composed of DNA and proteins, 36, 363
Chromatin spot An aggregation of chromatin at the periphery of the nucleus, evident in cells of human skin or from the mucosal lining of the mouth; makes possible the determination of the "nuclear sex" of an individual. Most of the cells of a female and none of the cells of a male have a chromatin spot, 348
Chromatophore(s) (kro′mah-to-for) Any pigment-producing and pigment-containing cell such as those in the skin, 99
  in cephalopods, 545
  in crustaceans, 587
Chromomeres, 45
Chromonema, 45
Chromosomes (kro′mo-sōm) Filamentous or rod-shaped bodies in the cell nucleus that contain the hereditary units, the genes, 36, 44
  chemistry of, 363
  giant, 381
  homologous, 339
  puffing of, 285
Chylomicrons, 138
Chyme, 137
Chymotrypsin, 138
Cicada songs, 610
Cilia, 13
Ciliary body, 252
Ciliary movement Movement of a cell, or of material across the surface of a cell, by the beating of microscopic cilia, 117
Ciliature, in ciliates, 445
Ciliophora, 444–449
Cilium (sil′e-um) One of numerous cytoplasmic processes on the exposed surface of many cells. Cilia typically are shorter and more numerous than flagella (q.v.) and have a more oarlike action, 117
Circadian rhythms (sir″kah-de′an) Repeated sequences of events that occur at about 24-hour intervals.
Circulatory system
  closed, 167, 177, 187
  of amphibians, 178
  of annelids, 177
  of arthropods, 177
  of fishes, 177
  of reptiles, birds, mammals, 178
  open, 167, 177, 187
Cirri, 636
Cirripedia, 597
Cisternae, 37
Citric acid, 71
Cladistics (klad′is-tics) The classification of organisms based entirely on the identification of points from which one group diverged from another, 436
Cladocerans, 594
Cladoselachii, 660
Clasper, 660
Class. See *Taxon.*
Cleavage, 309–311
Cleidoic (kli-do′ik) **egg** The eggs of insects, reptiles, and birds that are largely self-contained systems, 324, 685
Climatic factors, 792
Climax community, 811
Clionidae, 468
Clitellum (kli-tel′um) Glandular segments in earthworms and leeches that secrete the cocoon, 563
Cloaca (klo-a′kah) A common chamber receiving the discharge of the digestive, excretory, and reproductive systems in most of the lower vertebrates, 136, 200
Clonal selection theory Theory that lymphocytes that are genetically programmed to respond to a particular antigen are stimulated by the presence of this antigen and produce a clone of responding cells, 174
Clone A population of cells descended by mitotic division from a single ancestral cell.
  ciliate, 449
Cnida, 472
Cnidaria, 470–490
Cnidocil, 472
Cnidocyte (ni′dŏ-site) Cells of cnidarians containing explosive stinging structures, called nematocytes, 471
Coal, 793
Cobalamin (ko-bal′ah-min) Vitamin B_{12}; substance essential to the manufacture of red cells.
Cobalt, 16, 141
Coccidians, 458
Cochlea (kok′le-ah) The coiled portion of the inner ear containing the sound receptive cells, 247, 728
Cochlear duct, 245, 247
Cocoon, oligochaete, 563
Coding relationships, 368
Codominant genes, 345
Codons A sequence of three adjacent nucleotides that code for a single amino acid, 367
Coelom (se′lom) Body cavity of triploblastic animals lying within the mesoderm and lined by it, 91
  annelid, 551
Coenenchyme, 485
Coenzyme(s) (ko-en′zīm) A substance that is required for some particular enzymatic reaction to occur; participates in the reaction by donating or accepting some reactant; loosely bound to enzyme, 27, 66, 71
Coenzyme Q, 70
Cold stress, 727
Colds, prevention of by vitamin C, 6
Coleoidea, 547
Colinearity, 368
Collagen, 56
Collagen cuticle, 506
Collagen fibers, 56
Collar, in hemichordata, 646
Collar cells, 463
Collecting tubule, 200
Colon, 136
Colonial organization, 93
Colony (kol′o-ne) An association of unicellular or multicellular organisms of the same species; each individual is separate or essentially so, but sometimes there are connections among the members of the colony and some division of labor among them, e.g., feeding and reproductive polyps of certain Hydrozoa, 93
Color blindness, 350, 355
Columella, 526
Columnar epithelium, 54
Comb jellies, 472
Comb shape, inheritance of in chickens, 354
Commensalism, 497
Communicating rami, 225
Communication, 777
Community An assemblage of populations that live in a defined area or habitat, which can be either very large or quite small. The organisms constituting the community interact in various ways with one another, 810
Community succession, 811

Competitive exclusion The exclusion by the better adapted species of another related species from its particular ecological niche (q.v.), 419, 789
Complement system A system of plasma enzymes that may be activated by antibodies and lead to the lysis of invading microorganisms, 174
Complementary genes, 353
Complete metamorphosis, 610
Compound eye(s), 256
  characteristics of, 257
Conditioning, 772
Conductile region, 214
Cones, 12
  in retina, 252
Conformation, molecular, 22
Conjugation, 448
Conjunctiva, 251
Connective tissue, types of, 55
Conservation of energy, law of A fundamental law of physics that states that in any given system the amount of energy is constant; energy is neither created nor destroyed, but only transformed from one form to another.
Constipation, 139
"Consumer" organisms Those elements of an ecosystem, plants or animals, that eat other plants or animals.
Consumers, 790
Continental drift, 396
Continental slope, 820
Contraception (kon-trah-sep′shun) Method of birth control that involves the use of mechanical or chemical agents to prevent the sperm from reaching and fertilizing the egg, 828
Contraceptives, oral, 281
Contractile vacuole(s) Osmoregulatory organelle of protozoa and sponges, 53, 208, 448
Control group, 6
Cony, 740
Copepods, 595
Copper, 16
Copulation (kop″u-la′shun) Sexual union; act of physical joining of two animals during which sperm cells are transferred from one to the other, 300
Coral reefs, 485–488
Coral snake, 690
Corals, 484
Corixa, 788
Cornea, 251
Corpora allata, 285
Corpus callosum, 236
Corpus cardiacum, 283
Corpus luteum (kor′pus lu′tĭ-um) A yellow glandular mass in the ovary formed by the cells of an ovarian follicle that has matured and discharged its ovum.
Corpus striatum (stri-a′tum) A large subcortical mass of neuron cell bodies and fibers in the base of each cerebral hemisphere, 235, 685, 708
Correns, 338
Cortex (kor′teks) The outer layer of an organ, 200
Cortical reaction, 297
Corticotropin releasing hormone, 277
Cortisol, 265, 271
Cotylosauria, 686
Countercurrent exchange, 727. See also *Rete mirabile.*
Countercurrent multiplying mechanism, in kidney, 204
"Coupling factors," 79
Courtship, 417, 710, 775
Covalent bonds, 17
Crab, body form of, 587
Cranial nerves, 225
Cranium (kra′ne-um) The part of the skull that surrounds the brain; the brain case, 104
Crayfish, 588
Creatine phosphate, 77
Cretinism, 267
Cricket, song of, 610
Crinoidea, 635–636
Cristae, 37, 77
Crocodilia, 694
Cro-Magnon man, 757
Crop A sac in the digestive tract in which food is stored before digestion begins, 128, 709
Cross
  dihybrid, 346
  monohybrid, 339
Cross-fiber patterning, 241
Crossing over Process during meiosis in which the homologous chromosomes undergo synapsis and exchange segments, 350
Crossopterygian (krŏ-sop′tĕ-rĭ-jĭ-an) A member of an order of sarcopterygian fishes ancestral to terrestrial vertebrates, 668, 674
Crown, crinoids, 635
Crude density, 804
Crustacea, 586–597
Cryptic coloration, 781
Crypts of Lieberkühn, 137
Crystalline style, 535
Ctenophora, 472
Cuboidal epithelium, 54
Cubomedusae, 488
Curare, 219
Cutaneous (ku-ta′ne-us) Pertaining to the integument or skin.
Cutaneous respiration, 156
Cuticle (ku-tĭ′cle) A noncellular protective layer on the surface of an organism, 100, 506
  in nematodes, 510
  in pseudocoelomates, 506
Cuttlefish, 545
Cuvier, Georges, 9
Cyanopsin, 253
Cycle(s)
  of elements, 792
  tricarboxylic acid, 71
Cyclic AMP, 48, 262
Cyclic GMP, 48
Cyclostomata, 655
Cypris larva, 597
Cystocercus, 502
Cystoid, 638
Cysts, ciliates, 449
Cytidine triphosphate (CTP), 27
Cytochrome oxidase, 68
Cytochromes (si′to-krom) The iron-containing heme proteins of the electron transmitter system that are alternately oxidized and reduced in biological oxidation, 70, 393
Cytogenetics, 377, 391
Cytokinesis (si″to-ki-ne′sis) The division of the cytoplasm during mitosis or meiosis, 45
Cytology, 4
Cytopharynx (si″to-far′inks) Gullet-like organelle of ciliate and certain other protozoa, 446
Cytoplasmic organelles, 36
Cytoplast, 40
Cytoproct, 447
Cytosine, 26, 364
Cytostome (si′to-stōm) The mouth opening of ciliate and certain other protozoans, 446

D

Daddy-long-legs, 578
Dalton, 20
Darwin, 388, 400, 401, 403
Darwin's finches, 419, 781
Da Vinci, Leonardo, 8
Deafness, 248
Deamination (de-am′ĭ-na′shun) Removal of an amino group (—NH_2) from an amino acid or other organic compound, 26, 73, 194
Decapods, 587–593
Decomposers, 790
Deer, 739
Dehydration, 73
Dehydroepiandrosterone, 265
Dehydrogenation (de-hi″dro-jen-a′shun) A form of oxidation in which hydrogen atoms are removed from a molecule, 70, 73
Deletion, 376
Deme (dēm) A local interbreeding population occupying a circumscribed area, 408
Demospongiae, 468
Denaturation (de-na-tūr-a′shun) Alteration of physical properties and three-dimensional structure of a protein, nucleic acid, or other macromolecule by mild treatment that does not break the primary structure, 25, 68
Dendrites, 58, 213
Denitrifying bacteria, 794
Dentine, 134
Deoxyribonucleic acid (DNA), 26
  replication of, 366, 369
  transcription of, 371
  Watson-Crick model of, 364
Deoxyribose, 23, 364
Deposit feeding Utilization as a food source of organic detritus that has become mixed with mineral particles in aquatic and terrestrial habitats, 130
Derived character, 435
Dermal bone Bone that develops directly in connective tissue and is not preceded by a cartilaginous rudiment, 102
Dermis (der′mis) The deeper layer of the skin of vertebrates and some invertebrates, 97
Dermoptera, 736
Descartes, Rene, 9
Descending tracts, 228
Desert biome, 818
Deserts, types of, 819

Desmosomes Discontinuous button-like plaques present on the two opposing cell surfaces and separated by the intercellular space; they apparently serve to hold the cells together.
Determinate cleavage, 311
Detorsion, 524
Detritus, 128
Detritus chains, 801
Deuterostome(s) (du'ter-o-stōm") A major branch of the Animal Kingdom containing animals in which the site of the blastopore is posterior—far from the mouth, which forms anew at the anterior end, 518, 623
De Vries, Hugo, 338
Diabetes insipidus, 273
Diabetes mellitus, 22, 204, 269
Dialysis, 52
Diapause, 285
Diaphragm (di'ah-fram) The fibromuscular partition separating the thoracic and abdominal cavities of mammals; its contraction plays the major role in inspiration, 158
Diarrhea, 139
Diastole (di-as'to-le) Relaxation of the heart muscles, especially those of the ventricles, during which the lumen becomes filled with blood, 184
Diatryma, 715
Diencephalon, 229
Diets, 128
Differential gene activity, 381
Differentiation, 314–317
  genetic basis of, 380
Diffusion The movement of molecules from a region of high concentration to one of low concentration, brought about by their kinetic energy, 51, 150, 163
Digestion of foods, 126
Digestive functions, control of, 139
Digestive processes, in vertebrates, 132–140
Digitigrade foot, 736
Dihydroxyacetone phosphate, 73
Dihydroxyphenylalanine (dopa), 380
Dimetrodon, 694
Dinoflagellates, 450
Dinosaurs, 692
  extinction of, 694
Dioecious condition, 301
1-3-diphosphoglyceric acid, 73
Diploid (dip'loid) A chromosome number twice that found in gametes; containing two sets of chromosomes, 36, 291
Diplosegments, 601
Dipnoi, 668
Diptera, 606
Direct development, 324
Dizygotic twins, 329
DNA. See *Deoxyribonucleic acid.*
DNA polymerase, 367
DNA-dependent RNA polymerase, 371
Dogs, 736
Dominance, incomplete, 345
Dominance hierarchy, 773
Dominant gene, 340
Dopa decarboxylase, 285
Dorsal ramus, 225
Dorsal root, 224
Dorsal root ganglion, 225
Double helix, 364
Down, 704
Down's syndrome, 378
Dracunculus, 513
Dragline, 579
Drive, 763
Dryopithecus, 752
Duckbilled platypus, 729, 730
Ductus arteriosus (duk'tus ar-te"-re-o'sus) A short vessel, derived from the dorsal part of the sixth aortic arch, that connects the pulmonary artery and dorsal aorta in the larva or embryo; diverts blood from the developing lungs, 183
Dugesia, 494
Dunkelosteus, 658
Duogland systems, 492
Duplication, 367
Dura mater, 229
Dutrochet, Rene, 11

E

Ear, structure of, 247
Earth
  heat budget of, 796
  rotation of, 796
Earthworm(s), 561
  nerve cord of, 222
  regulation of internal body fluids in, 209
Ecdysis (ek'dĭ-sis) The shedding, or molting, of the arthropod exoskeleton, 100
Ecdysone (ek-di'son) The hormone that induces molting (ecdysis) in arthropods, 284
Echinococcus, 502
Echinoderms, 622–638
Echinoidea, 631–633
Echinostoma, 499
Echiura, 566
Echolocation, 734
Ecological density, 804
Ecological equivalents, 790
Ecological niche The status of an organism within a community or ecosystem; depends on the organism's structural adaptations, physiologic responses and behavior, 787
Ecology (e-kol'o-je) The study of the interrelations between living things and their environment, both physical and biotic, 4, 786
  dynamic processes in, 785
  human, 828
Ecosystem (ek"o-sis'tem) A natural unit of living and nonliving parts that interact to produce a stable system; all of the organisms of a given area and the encompassing physical environment, 786
  concept of, 790
Ecotone A fairly broad transition region between adjacent biomes; contains some organisms from each of the adjacent biomes plus some that are characteristic of, and perhaps restricted to, the ecotone, 813
Ectoderm (ek'to-derm) The outer of the two germ layers of the gastrula; gives rise to the epidermis of the skin and nervous system, 311
Ectoparasite(s), 129, 497
Ectoprocta, 618
Ectothermic animals, 143
Ectothermy Condition in which the internal temperature of an animal is dependent upon the temperature of its environment; cold blooded, 653, 677
Edaphic climax, 812
Edaphic factors, 792
Edema (i-de'ma) An excessive accumulation of liquid in the tissues, 189
Edentata, 732
Eel, 665
Effector (ef-fek'tor) Structures of the body by which an organism acts; the means by which it reacts to stimuli, e.g., muscles, glands, cilia.
Egg cells, 59
Egg envelopes, 302
Egg types, 297
Elasmobranchii (e-las'mo-brank-ĭe) A member of a subclass of cartilaginous fishes in which the gills are platelike and each gill pouch opens independently on the body surface; sharks, skates and related animals, 660
Elastin, 56
Electric field, 216
Electric organs, evolution of, 422
Electrolyte(s) (e-lek'tro-līt) Substance that dissociates in solution into charged particles, ions, and thus permits the conduction of an electric current through the solution, 20, 21
Electron(s), 14
Electron cascade, 69
Electron orbitals, 14
Electron shells, 15
Electron transmitter system, 69, 73
Electroreceptors, 258
Element(s) (el'ĕ-ment) One of the hundred or so types of matter, natural or man-made, composed of atoms all of which have the same number of protons in the atomic nucleus and the same number of electrons circling in the orbits, 16
  periodic chart of, 16
Elephant, 739
Elephantiasis, 513
Elongation of peptide chain, 374
Elongation process, 375
Embryology, 4
Embryonic development, 308—334
Embryonic diapause, 731
Enamel, 133
End bulbs of Krause, 243
Endocrine (en'do-krin) Secreting internally; applied to organs whose function is to secrete into the blood or lymph a substance that has a specific effect on another organ or part.
Endocrine glands, 261
  human, 264
Endocuticle, 572
Endoderm (en'do-derm) The inner germ layer of the gastrula, lining the archenteron; forms the epithelial cells of the digestive tract and its outgrowths—the liver, lungs, and pancreas, 311
Endolymph, 244, 247
Endometrium, 279
Endoparasite(s), 129, 497

Endopeptidase (en″do-pep′tĭ-dās) A proteolytic enzyme that cleaves peptide bonds within a peptide chain, 126
Endoplasmic reticulum
rough, 37
smooth, 37
Endopodite, 587
Endoskeleton (en″do-skel′ĕton) Bony and cartilaginous supporting structures within the body; provide support from within, 101
Endostyle (en′do-stīl) A longitudinal groove in the floor of the pharynx of certain lower chordates and larval cyclostomes; its glandular and ciliated cells secrete mucus which is moved by ciliated cells through the pharynx and entraps minute food particles, 642, 657
Endothermic animals, 143
Endothermy Condition in which the internal temperature of an animal is dependent upon its metabolic processes and is held at a relatively high and constant level; warm blooded.
evolution of, 725
Energy, 14, 50
conservation of, 64
free, 64
transfer of, 798
"Energy currency," 27
Energy transformations, 51, 63
Energy-rich bonds, 27
Energy-rich phosphate bonds, 63
Engram (en′gram) The term is applied to the presumed change that occurs in the brain as a consequence of learning; a memory trace.
Entamoeba, 455
Enterocoel (en′ter-o-sēl″) A body cavity formed by outpocketing from the primitive gut, 314
Enterogastrone, 139
Enterokinase, 138
Enteropneusta, 647
Entoprocta, 440
Entropy, 64
Environment, carrying capacity of, 804
Environmental resistance, 806
Enzyme(s) (en′zīm) A protein catalyst produced within a living organism that accelerates specific chemical reactions, 25, 65
active site of, 67
properties of, 65
turnover number of, 65
Enzyme cascade, 262
Enzyme catalysis, mechanism of, 66
Enzyme induction, 382
Enzyme inhibitors, 68
Enzymic activity, factors affecting, 68
Eosinophils, 58
Ephyra, 481
Epiboly (e-pib′o-le) A method of gastrulation by which the smaller blastomeres at the animal pole of the embryo grow over and enclose the cells of the vegetal hemisphere, 313
Epicuticle, 572
Epidermis (ep′ĭ-der′mis) The outermost layer of cells of an organism, 97
cnidarian, 471
Epidinium, 31
Epigenesis, 380
Epiglottis (ep′ĭ-glot′is) The lidlike structure that deflects food around the glottis (q.v.), 157
Epilimnion (ĕp-i-lĭm′nĭ-ŏn) The uppermost layer, or surface water, of a lake or pond, 826
Epinephrine, 141, 265
Epiphragm, 528
Epithelial tissues, 54
Epitheliomuscle cells, 471
Equatorial plane, 45
Erythroblastosis fetalis, 358
Erythrocytes (e-rith′ro-sīt) Red blood cells; they contain hemoglobin and transport gases, 58, 159, 168
Erythropoietin (e-rith′ro-poy-e-tin) A hormone that promotes red blood cell production, 168
Esophagus, 128
in vertebrates, 134
Ester bonds, 81
Estivation, 528
Estradiol, 265, 278
Estrogen (es′tro-jen) One of the female sex hormones, produced by the ovarian follicle, which promotes the development of the secondary sex characteristics.
Estrous cycle, 279
Estrus (es′trus) The recurrent, restricted period of sexual receptivity in many female mammals, marked by intense sexual urge.
Estuaries, 819
Ethogram, 762
Ethology, 762
Euglenids, 450
Eumetazoa, 439
Eunuch, 277
Euryhaline animals (u-re-ha′līn) Animals that can tolerate a relatively wide range of environmental salt concentrations (salinities), 206
Eurypterida, 614
Eurythermic, 787
Eustachian tube, 246
Eutely, 506
Eutheria (u-ther′ĭ-ah) One of the placental mammals in which a well-formed placenta is present during development and the young are born at a relatively advanced stage of development; includes all living mammals except for the monotremes and marsupials, 730
adaptive radiation of, 731, 733
Evaporative water loss, 209
Evolution, 4, 408. See also *Macroevolution* and *Microevolution.*
convergent, 435
divergent, 435
evidences for, 388
history of, 401
impact on religion, 404
mechanisms of, 406
synthetic theory of, 408
transpecific, 421
Excretion (eks-kre′shun) Removal of metabolic wastes by an organism, 195
Excretory organs, 195–205
Exergonic (ek″ser-gon′ik) A reaction characterized by the release of energy.
Exocuticle, 572
Exons, 372
Exopeptidase (ek″so-pep′tĭ-dās) A proteolytic enzyme that cleaves only the bond joining a terminal amino acid to a peptide chain, 126
Exopodite, 587
Exoskeleton (ek′so-skel′ĕ-ton) Calcareous, chitinous, or other hard material covering the body surface and providing protection or support, 99, 572
Experiments, design of, 6
Expressivity, 384
Extinction, 423
Extracellular digestion, 127
Extraembryonic membranes, 325
Extrafusal fibers, 243
Extrapyramidal pathway, 235
Eye
cephalopod, 256
formation of, 317
muscles of, 252
pineal, 654
vertebrate, structure of, 251
Eye spots, 249

F

Facial nerve, 226
Facilitation (fah-sil″i-ta′shun) The promotion or hastening of any natural process; the reverse of inhibition, 112
at synapse, 221
Fall overturn, 826
Family. See *Taxon.*
Fang(s), 690
in spiders, 580
Fanworms, 559
Fasciola, 500
Fat, 23
Fatty acid(s), 23
essential, 141
unsaturated, 23
Fatty acid oxidation, 72
Feather mites, 582
Feather stars, 635–636
Feathers, 98
contour, 703
flight, 704
Feeding mechanisms, 128, 129
Feet, adaptations of in birds, 707
Fenestra (fē-nes′trah) A moderate-sized opening in a structure, e.g., the fenestra ovalis (oval window) in the middle ear cavity.
Fermentation (fer″men-ta′shun) Anaerobic decomposition of an organic compound by an enzyme system; energy is made available to the cell for other processes.
Fertilization (fer″tĭ-lĭ-za′shun) The fusion of a spermatozoon with an ovum to initiate development of the resulting zygote, 297–298
external, 299
internal, 300
in mammals, 306
Fertilization adaptations, 298–302
Fertilization cone, 297
Fertilization membrane, 297
Fetus (fe′tus) The unborn offspring after it has largely completed its embryonic

development, 330
Fibers
adrenergic, 220
cholinergic, 220
Fibroblasts, 48
Fibrous connective tissue, 55
Fiddler crabs, 589, 590
Filaments, cytoplasmic, 40
Filarioids, 513
File shells, 538
Filoplume, 704
Filter feeding, 129
Filtration (fil-tra'shun) The passage of a liquid through a filter following a pressure gradient; occurs in capillary beds, including the glomeruli of the kidney, 201
kidney, 201
Fireflies, 610
Fischer, Emil, 66
Fish(es)
classification of, 669
evolution of, 654
osmoregulation in, 206, 208
Fission (fish'un) Process of asexual reproduction in which an organism divides into two approximately equal parts, 290
Fitness. See *Adaptive value.*
Fixed action patterns, 763
Flagellar movement. See *Ciliary movement.*
Flagellated canals, 465
Flagellates, 449–454
Flagellum(a) (flă-jel'um) A whiplike cytoplasmic process on the exposed surface of many cells. Structurally similar to a cilium (q.v.), but typically flagella are longer, are fewer in number, and have a more sinuous movement, 13, 117
Flamingo, 714
Flatworms, 491–503
Flavin adenine dinucleotide (FAD), 27
Fleas, 610
Flight
flapping, 701
principles of, 697
Flight muscles, 707
Flounder, 668
Fluke Common name of the parasitic flatworms belonging to the classes Monogenea and Trematoda, 498–500
Flying lemur, 736
Follicle-stimulating hormone (FSH), 265, 274
Food chain(s) A sequence of organisms through which energy is transferred from its ultimate source in a plant; each organism eats the preceding and is eaten by the following member of the sequence, 798
Food supply, regulation of, 140–141
Food web, 799
Foods Organic compounds used in the synthesis of new biomolecules and as fuels in the production of cellular energy, 126
Foot
in mollusks, 518
in rotifers, 507
Foraging strategies, 778
Foramen (fo-ra'men) A small opening in a body structure, 105
Foramen ovale An oval opening between the right and left atria of the fetal heart by which some blood entering the right atrium by-passes the lungs and goes to the aorta, 182
Foraminiferans, 455
Foregut, 606
arthropod, 574
Forest biomes, 815
Formylmethionyl tRNA, 375
Fossil (fos'il) Any remains of an organism or its activity that have been preserved in geologic strata, 393
Founder effect, 412, 419
Fovea, 253
Fowl, 717
Free energy, 64
Free nerve endings, 242
Fringing reefs, 485
Frogs, 680
reproduction of, 681
skeleton of, 680
Fructose, 21, 23
Fructose-1, 6-diphosphate, 73
Fructose-6-phosphate, 73
Fuel utilization, 143–146
Fumaric acid, 71

G

G_1 phase, 41
G_2 phase, 41
Galactosamine, 23
Galen, 7
Gall insect, 610
Gallbladder, 136
Gamete (gam'ēt) A reproductive cell; an egg or sperm whose union, in sexual reproduction, initiates the development of a new individual, 290
Gametogenesis, 290
Gamma globulin, 172, 174
Gamow, George, 367
Ganglion(a) (gang'gle-on) A knotlike mass of the cell bodies of neurons located outside the central nervous system; in invertebrates, includes the swellings of the central nervous system, 59, 214
Gap junction A tight union between cells that contains a minute gap crossed by beadlike processes; permits the passage of ions or action potentials from one cell to the next, 111
Gar, 664
Gas
reactions in blood, 161
transport of, 159
Gastric juices, 136
Gastric mill, 590
Gastrin, 139
Gastrodermis (gas"tro-der'mis) The tissue lining the gut cavity that is responsible for digestion and absorption, 471
Gastropoda, 521–528
Gastrotricha, 509
Gastrovascular, 471
Gastrozooid, 474
Gastrula (gas'troo-lah) Early embryonic stage that follows the blastula; consists initially of two layers, the ectoderm and the endoderm, and of two cavities, the blastocoel between ectoderm and endoderm and the archenteron, formed by invagination, lying within the endoderm, and opening to the exterior through the blastopore, 311
Gastrulation (gas"troo-la'shun) The process by which the young embryo becomes a gastrula and acquires first two then three layers of cells, 311–314
Gause's rule, 789
Gelatin, 56
Gemmule(s), 467
Gene flow. See *Migration.*
Gene pool The totality of all of the genes and their alleles of all of the individuals of a population, 344, 408
integrity of, 416
variability in, 410
Gene-enzyme reactions, 380
Genes (jēn) The biologic units of genetic information; self-reproducing and located in definite positions (loci) on particular chromosomes, 36, 339
changes in, 375
coadaptation of, 417, 419
complementary, 353
dominant, 340
lethal, 383
mutation of, spontaneous, 377
recessive, 340
regulatory, 424
supplementary, 354
Genetic carriers, frequency of, 344
Genetic code
degeneracy of, 368
universality of, 368
Genetic diseases, carriers of, 345
Genetic drift The tendency within small interbreeding populations for heterozygous gene pairs to become homozygous for one allele or the other by chance rather than by selection, 411, 419
Genetic equilibrium The situation in which the distribution of alleles in a population is constant in successive generations (unless altered by selection or mutation), 344
Genetic information, transmission of, 363
Genetic load, 416
Genetics, 4
history of, 338
mathematical basis of, 341
problems in, solutions for, 346
Genic interactions, 352
Genome (je'nom) A complete set of hereditary factors, contained in the haploid assortment of chromosomes.
Genotype (jen'o-tip) The fundamental hereditary constitution, assortment of genes, of any given organism, 340, 344
Genus. See *Taxon.*
Geochronometers, 397. See also *Geological time.*
Geographical distribution, 399
Geographical ecology, 812
Geographical isolates, 419
Geological chronology, 394
Geological eras, 397
Geological time, 395, 398
Germinal vesicle, 59
Germ layers, 311
Ghost crabs, 589
Gibbon, 748
Gigantism, 274
Gila monster, 689

Gill(s) Gas exchange organ whose surface area is increased by filaments, lamellae, or other folds evaginated from a surface, 151, 153
of bivalves, 535
of invertebrates, 151
of vertebrates, 152
Gill bailer, 589
Gill clefts, 641
Gill slit(s) An opening to the outside from the pharynx that arises during development. Water taken in at the mouth passes out through the gill slits, aiding respiration and the filtering of food, 641
embryonic, 319
Giraffe, 739
Gizzard (giz'erd) A portion of the digestive tract, usually a part of the stomach, that mechanically breaks up food, 128, 709
in earthworms, 563
Glass sponges, 468
Glaucoma, 252
Gliding, 701
Globigerina, 455
Globulin (glob'u-lin) One of a class of proteins in blood plasma, some of which (gamma globulins) function as antibodies.
Glochidium, 539
Glomerular filtration, 201
Glomerulus (glo-mer'u-lus) A tuft of minute blood vessels or nerve fibers; specifically, the knot of capillaries at the proximal end of a kidney tubule, 200
Glossopharyngeal nerve, 226
Glottis (glot'is) The opening between the pharynx and larynx; it is bounded by the vocal cords in higher vertebrates, 157
Glucagon, 141, 265, 269
Glucocorticoids, 271
Glucosamine, 23
Glucose, 20, 21
Glucose-6-phosphate, 73
Glutamic acid, 73
Glycera, 554
Glyceraldehyde-3-phosphate, 73
Glycerol, 23
Glycogen, 22
Glycolipids, 23, 32
Glycolysis (gli-kol'ĭ-sis) The metabolic conversion of sugars into simpler compounds, 73
Glycoprotein hormones, subunits of, 275
Glycoproteins, 23
Glycosidic bonds, 81
Glycosuria, 269
Gnathostomes (na'tho-stōmes) The jawed vertebrates, 658
Gnathostomulida, 439
Goblet cells, 137
Goiter, exophthalmic, 267
Golden plover, 718
Golgi bodies, 39
Golgi complex, 38
Golgi organ, 243
Gonad(s) (gon'ad) A gamete-producing gland; an ovary or testis, 293, 302
Gonadotropin releasing hormone, 277
Gonionemus, 478
Gonochoristic (gon-o-ko-ris-tik), or gonochoric, animals Animal species having separate sexes.
Gonoducts General term for reproductive ducts of any animal.
Gonopore External opening of any reproductive system, 293
Gonozooids, 475
Goose, 718
Goose (stalked) barnacles, 597
Gorilla, 748
Gradual development, 610
Granuloreticulosia, 460
Grasshopper songs, 610
Grassland biome, 817
Gray matter, 228
Grebe, 714
Green gland Excretory organ of certain crustaceans, such as shrimp and crabs; also called antennal gland, 198
Green revolution, 828
Growth, 13
Growth curve, population, 804
Growth hormone, 265
Growth hormone releasing hormone, 277
Grub, 610
Guanine, 26, 195, 364
Guanosine triphosphate (GTP), 27
Guinea worm, 513
Gymnophiona, 682

H

Habitat (hab'ĭ-tat) The natural abode of an animal or plant species; the physical area in which it may be found, 787
Habituation A gradual decrease in response to successive stimulation due to changes in the central nervous system.
Hagfish, 657
Hair, 98
Hair mites, 582
Halibut, 668
Halteres, 245, 606
Hämmerling, 34
Hardy, G.H., 343
Hardy-Weinberg principle, 344
Haploid chromosome number, 36, 291
Harrison, Ross, 31
Harvestmen, 578
Harvey, William, 8
Haversian canals, 56
Hawk, 717
Hearing, anatomical basis of, 248
Heart. See also *Cardiac.*
integration of beat, 184
of mammals, 184
types of, 183
Heart embryonic development, 319
Heart murmur Sounds in the heart indicative of the back flow of blood through one or more valves, 184
Heart urchins, 632
Heartworm, 513
Heat stress, 727
Hedgehog, 732
Heliotherms, 683
Heliozoans, 455
Helix, regulation of internal body fluids in, 209
Helix, double, 364
Hemichordata, 646
Hemimetabolous development, 610
Hemipteran mouthparts, 607
Hemocoel (he'mo-sēl) The spaces between the cells and tissues of many invertebrates, which are filled with blood and a definitive epithelial lining. Part of an open blood vascular system, 167
in arthropods, 574
Hemocyanin (he"mo-si'an-in) A blue copper-containing respiratory pigment (q.v.) found in the blood of many invertebrates, 161
in arthropods, 574
Hemoglobin (he"mo-glo'bin) The red, iron-containing respiratory pigment (q.v.) found in the blood of vertebrates and some invertebrates; transports gases and aids in the regulation of pH, 159
Hemolymph (he'mo-limf) The bloodlike fluid of animals with open circulatory systems; combines the properties of blood and lymphlike interstitial fluid, 167
Hemophilia (he"mo-fil'e-ah) Hereditary disease in which prothrombin is not activated because of a deficiency of one of the blood clotting factors; blood does not clot properly; "bleeder's disease," 170, 350
Hemostasis (hē"mo-stā'sis) The stopping of blood flow from an injured vessel; includes the contraction of the vessel and the formation of a blood clot, 168
Heparin (hep'ar-in) A blood clotting inhibitor produced primarily by basophilic leucocytes and mast cells, 170
Hepatic portal, 140
Hepatopancreas, 590
Herbivore (her'bĭ-vōr) A plant-eating animal, 128, 129
Herbivorous diets, 128
Hermaphroditism (her-maf'ro-dit-izm) A state characterized by the presence of both male and female sex organs in the same organism, 301
Hermit crabs, 589
Herring, 664
Hesperornis, 714
Heterocercal tail, 652
Heterogeneous nuclear RNA, 372
Heteropod(s), 525, 527
Heterosis (het"e-rō'sis) Hybrid vigor. The offspring from matings of individuals from unrelated strains or different species frequently are much more vigorous than either parent, but they may not be able to reproduce, 359, 414, 418
Heterotrophs (het'er-o-trofs) Organisms that cannot synthesize their own food from inorganic materials and therefore must live either at the expense of autotrophs or upon decaying matter, 126, 433
Heterozygous (het"er-o-zi'gus) Possessing two different alleles for a given character at the corresponding loci of homologous chromosomes, 340
Hexactinellida, 468
Hexokinase, 73

Hexoseaminidase, A, 345
Hibernation (hi″ber-na′shun) The dormant state of decreased metabolism in which certain animals pass the winter, 145, 677, 727
Hindgut, 606
in arthropods, 574
Hirudin, 568
Hirudinea, 565–570
Hirudo, 568
Histamine (his′tă-mēn) A material released by animal and some plant tissues that stimulates capillary dilation and gastric secretion, 170
Histochemical studies, 49
Histology, 4
Histones, 363
Holocephali, 660
Holometabolous development, 610
Holonephros, 198
Holoplankton, 821
Holostei, 663
Holothuroidea, 633–635
Holozoic nutrition, 126
Home range, 774
Homeostasis (ho″me-o-stā′sis) The tendency to maintain uniformity or stability in the internal environment of the organism, 163
Homeothermic (ho-moi′o-ther″mic) Constant-temperature animals, e.g., birds and mammals that maintain a constant body temperature despite variations in environmental temperature, 143
Homing, 782
Hominidae (hom′ĭ-nĭ-de) The family of human beings, 751
Hominoidea (hom′ĭ-noid-eah) A superfamily that includes both the families of apes and the family of human beings, 748
Homo erectus, 755
Homo habilis, 754
Homo sapiens, 756
Homocercal tail, 652
Homogentisic acid, 380
Homoiotherm (ho-moi′o-therm″) Animals that maintain a constant temperature despite changes in external environment, e.g., birds and mammals. See also *Endotherm.*
Homolecithal eggs, 297
Homologous chromosomes, 291
Homology (ho-mol′o-je) Basic similarity in structure between organs in different animals that results from inheritance from a common ancestor; homologous organs have a common basic plan and mode of development, 388, 435
Homozygous, 340
Honeybees, 611–613
Hooke, Robert, 9
Hookworms, 513
Hormonal integration, 260
Hormones (hor′mōns) Substances produced in cells in one part of the body that diffuse or are transported by the blood stream to cells in other parts of the body where they regulate and coordinate their activities, 261
action of, molecular mechanisms of, 261
role in behavior, 765
Horns, 739
Horsehair worm, 515
Horses, 739
evolution of, 423
Horseshoe crabs, 576
Host, 497
Housefly mouthparts, 607
Human beings, 748
characteristics of, 751
evolution of, 752
races of, 757
skin color of, inheritance of, 355
Human chromosomal abnormalities, 379
Human ecology, 828
Humus (hu′mus) Organic matter in the soil; a dark mold of decayed vegetable tissue that gives soil a brown or black color, 128
Hunter, John, 9
Hyaluronidase, 306
Hybrid vigor (hi′brid) The mating of genetically dissimilar individuals of totally unrelated strains; may yield offspring better adapted to survive than either parental strain, 359
Hydatid cysts, 502
Hydra, 479
Hydras, 471
Hydrocorallina, 488
Hydrogen, 15, 28
Hydrogen bond(s) A weak bond between two molecules formed when a hydrogen atom is shared between two atoms, one of which is usually oxygen; of primary importance in the structure of nucleic acids and proteins, 17
Hydroids, 474
Hydrolysis (hi-drol′ĭ-sis) The splitting of a compound into parts by the addition of water between certain of its bonds, the hydroxyl group being incorporated in one fragment and the hydrogen atom in the other, 126
Hydromedusae, 475
Hydroskeleton (hy′drō-skel″e-tun) A turgid column of liquid within one of the body spaces that provides support or rigidity to an organism or one of its parts, e.g., the coelom and coelomic fluid of an annelid worm, 101
Hydrozoa, 474–479
Hylobatidae, 748
Hymen, 306
Hyperglycemia, 269
Hyperosmotic, 53
Hypertonic (hi″per-ton′ik) Having a greater concentration of solute molecules and a lower concentration of solvent (water) molecules and hence an osmotic pressure greater than that of the solution with which it is compared.
Hypodermic impregnation, 301, 508
Hypoglossal nerve, 226
Hypolimnion, 826
Hypoosmotic, 53
Hypopharynx, 602
Hypophyseal pituitary portal vein, 273
Hypophysis cerebri, 272
Hypothalamic releasing hormones, 276
Hypothalamus (hy″po-thal′ah-mus) A region of the forebrain, the floor of the third ventricle, which contains various centers controlling visceral activities, water balance, temperature, sleep and so on, 233
Hypothesis, 5
Hypotonic (hi″po-ton′ik) Having a lower concentration of solute molecules and a higher concentration of solvent 1(water) molecules and hence an osmotic pressure lower than that of the solution with which it is compared, 53
Hypotricha, 445
Hyracoidea, 740

I

Ichthyornis, 714
Ichthyosaurs, 687
Iguana, 689
Ileum, 136
Immune response (ĭ-mūn′) The production of antibodies in response to antigens, 172
Immunity A resistance to disease resulting from having once produced antibodies in response to the disease or to a vaccination (active immunity), or to having acquired them from another source (passive immunity), 171
Imprinting A form of restricted learning in which a young animal forms a strong attachment to an individual, usually one of the parents, within a few hours of hatching or birth, 769
Inbreeding, 359, 411
Incomplete cleavage, 309
Incomplete development, 610
Incomplete dominance, 345
Incus, 246
Independent assortment, law of, 339
Indeterminate cleavage, 311
Indirect development, 322
Induction (in-duk′shun) The production of a specific morphogenetic effect in one tissue of a developing embryo through the influence of an organizer or another tissue, 316
Inductive reasoning, 7
Inflammation (in″flah-ma′shun) The reaction of tissues to injury: pain, vasodilation, increased temperature, redness, leukocyte accumulation, 170
Infraciliature, 445
Infundibulum, 272
Ingestion (in-jes′chun) The act of taking food into the body by mouth, 128
Inheritance, polygenic, 355
Inhibition at synapse, 221
Initiation complex, 375
Initiation of protein synthesis, 374
Ink gland, in cephalopods, 545
Innate releasing mechanism, 763
Insectivores, 732
Insects, 602–613
sense of taste in, 249
Instar The stage of an insect or arthropod between successive molts, 573
Instincts, 769
Insulin, 25, 141, 202, 265, 269
Integument (in-teg′u-ment) Skin, the covering of the body, 91, 97
Intercerebral gland, 283, 284

Intergradation (in″ter-grā-dā′shun) The mingling of characteristics, resulting from interbreeding, between two subspecies of the same species in the area where they meet (primary intergradation), or between closely related species whose ranges overlap before isolating mechanisms (q.v.) are fully developed (secondary intergradation), 418
Interneurons, 214
Interstitial Referring to the spaces, fluid, or animals that lie between small structures, such as cells or sand grains.
Interstitial cells, cnidarian, 472
Interstitial fauna, 492
Interstitial fluid (in″ter-stish′-al) The liquid that bathes the cells of the body. In vertebrates, it is a filtrate of the plasma and differs from it in lacking most of the large protein molecules, 163
Intertidal zone, 820
Intervening sequences, 372
Interventricular foramen of Monro, 229
Intestine, 128
  in vertebrates, 136–138
Intracellular digestion, 127
Intrafusal fibers, 243
Intrinsic rate of natural increase (r), 806
Introns, 372
Invagination (in-vaj″ĭ-na′shun) The infolding of one part within another, specifically a process of gastrulation in which one region infolds to form a double-layered cup, 311
Inversion, 376
Involution, 313
Iodine, 16
Iodopsin, 255
Ion(s) (i′on) An atom or a group of atoms bearing an electric charge, either positive (cation) or negative (anion), 17
  effects of on enzymes, 21
Iris, 251
Iron, 16
Irritability, 12, 213
Ischial callosity, 743
Islets of Langerhans, 136, 141, 269
Isocitric acid, 71
Isolating mechanisms Mechanisms that normally prevent two closely related species living in the same geographic area from interbreeding.
  evolution of, 419
  types of, 417
Isolecithal eggs, 297
Isomer (i′so-mer) Molecule with the same molecular formula as another but a different structural formula: e.g., glucose and fructose.
Isopods, 593–594
Isoptera, 610
Isosmotic, 53
Isothermal systems, 64
Isotopes (i′so-tōps) Alternate forms of a chemical element having the same atomic number (that is, the same number of nuclear protons and orbital electrons) but possessing different atomic masses (that is, different numbers of neutrons), 17
Isozymes (i′so-zīms) Different molecular forms of proteins with the same enzymatic activity.
Itch mites, 582

J

Jacobson's organ, 689
Jaws
  changes in joint, 106, 695
  evolution of, 658
Jejunum, 136
Jellyfish, 471
Jungles, 817
Juvenile hormone, 285
Juxtaglomerular apparatus, 205
Juxtamedullary nephrons, 201

K

K, 807. See also *Carrying capacity.*
"K selection," 807
Kala-azar, 453
Kangaroo, 731
Kangaroo rats, regulation of internal body fluids in, 210
Keilin, David, 67
Keratin (ker′ah-tin) A horny, water-insoluble protein found in the epidermis of terrestrial vertebrates and in nails, feathers, and horn, 97
α-ketoglutaric acid, 71
Kidney(s), vertebrate, 198–205
Kinetic energy, 50, 63
Kinetochore, 45
Kinetosomes, 445
Kinety, 445
Kinorhyncha, 439
Kiwi, 715
Klinefelter's syndrome, 348, 379
Knee jerk, 223
Knuckle walking, 749
Komodo dragon, 689
Krebs cycle, 71

L

Labia, 306
Labial palps, 534
Labium, 602
Labrum, 602
Labyrinthodont (lab″ĭ-rin′tho-dont) A member of a subclass of extinct amphibians; included the first terrestrial vertebrates and the ancestors of modern amphibians and reptiles, 674
Lactase, 138
Lactation, hormonal control of, 281
Lactic acid, oxidation of, 70
Lactose, 22
  biosynthesis of, 81
"Ladder of nature," 7
Lagena, 245
Lagomorpha, 742
Lamarck, Jean Baptiste de, 10, 403
Lamellibranchs, 535
Lamprey, 655
Lanolin, 24
Larva (lar′vah) A motile and sometimes feeding stage in the early development of certain animals, 322
  insect, 283
Larvaceans, 643
Larynx (lar′inks) The cartilaginous structure located at the entrance of the trachea; houses the vocal cords in many vertebrates, 106, 157
Lateral fissure of Sylvius, 235
Lateral line, 652
Lateritic soil, 792
Latrodectus, 583
Law of the conservation of energy, 50–51
Law of the minimum, 786
Laws of heredity, chromosomal basis of, 339
Lead, 16
Leaf-cutting ants, 781
Learning, 769, 770
Lectins, 43
Leeches, 565–570
Leishmania, 453
Lemmings, 808
Lens, 251
Lepospondyli, 677
Leptocephalus larva, 666
Lethal genes, 383
Leuconoid sponges, 465
Leukocytes (lu′ko-sīts) White blood cells; colorless cells many of which exhibit phagocytosis (q.v.) and amoeboid movement. They are important in inflammatory and immune reactions, 58, 170
Lice, 610
Life zones
  freshwater, 824
  marine, 820
Ligament, 56
Light, refraction of, 253
Light organs, 82
Lignite, 793
Limb buds, 319
Limb formation, 319
Limnetic zone, 825
Limpet(s) 525, 526, 527
Limulus, 576
Linkage The tendency for a group of genes located in the same chromosome to be inherited together in successive generations, 350
Linnaeus, 9, 437
Lipase (lip′ās) An enzyme that catalyzes the hydrolysis of fats; it cleaves the ester bonds joining fatty acids to glycerol, 66, 126, 138
Lipid(s), 21, 23
Lipid bilayer, 32
Lipoic acid, 71
Lipoprotein lipase, 141
Lissamphibia, 677
Liter (le′ter) The unit of volume in the metric system. Defined for most practical purposes as the volume of 1000 grams of pure water under specified conditions. Subdivisions of the liter use the same prefixes, factors and abbreviations as used for subdivisions of the meter (q.v.).
Littoral (lit′o-ral) The region of shallow water near the shore between the high and low tide marks, 825
Liver, 136
Living things

characteristics of, 11
properties of, 28
Lizards, 687
tail autotomy of, 689
Lobes of brain, 234
Lobosa, 460
Lobsters, 587
Local circuit theory of propagation, 218
Locomotion, 118
lever systems in, 122
of earthworm, 120
of fish, 120
of mammals, 727
of primitive tetrapods, 676
Locus (lo'kuss) The particular point on the chromosome at which the gene for a given trait occurs, 339
Loon, 714
Loop of Henle (hen'lē) (Henle was a German anatomist of the 19th century.) The U-shaped loop of a mammalian kidney tubule that dips down into the medulla; lies between the proximal and distal convoluted tubules, 200
Lophophore (lof'o-for) The circular or horseshoe-shaped ridge with a set of ciliated tentacles around the mouth of bryozoans, phoronids, and brachiopods, 619
Loxosceles, 583
Luciferase, 82
Luciferin (lu-sif'er-in) Substrate present in certain organisms capable of bioluminescence, producing light, when acted upon by the enzyme luciferase, 82
Lugworms, 555
Lumbricus, 562
Lumirhodopsin, 254
Lung Gas exchange organ whose surface area is increased by invaginations from a body surface.
book, 154
evolution of, 661
of birds, 709
of frog, 156
of mammals, 158
types of, 154
Lungfish, 661, 668
Lunules, 633
Luteinizing hormone (LH), 265, 274
Lycaenops, 694
Lyell, 403
Lymph (limf) The liquid in the lymphatic system; similar in composition to interstitial fluid (q.v.). It contains many white blood cells, especially lymphocytes, 190
Lymph node One of the nodules of lymphatic tissue that occur in groups along the course of the lymph vessels; produces and contains lymphocytes and phagocytic cells, 181
Lymphatic system, 181
Lymphoblasts, 172
Lymphocytes, 58, 172
Lynx, 808
Lyon hypothesis, 350
Lysis (li'sis) The process of disintegration of a cell or some other structure, 174
Lysosome(s) (lī'so-sōm) Intracellular organelle present in many animal cells; contains a variety of hydrolytic enzymes that are released when the lysosome ruptures, 40

M

M phase, 41
Mackerel, 666
Macroevolution The evolution of taxonomic categories above the species level, 421
Macromere (ma'kro-mēēr) The larger cell resulting from unequal cell division, 309
Macronucleus (makro-nu'kle-us) The large nucleus in ciliates that governs activities not associated with reproduction, 448
Macrophage (mak'rō-fāge) Large phagocytic cells that differentiate from certain leukocytes at the site of an infection, 170
Madreporite, 624
Magendie, Francois, 9
Maggot, 610
Magnesium, 16
Magnus, Albertus, 8
Malacostraca, 593
Malaria, 458
Maleness, inheritance of, 348
Malformations, 331
Malic acid, 71
Malleus, 246
Malpighi, Marcello, 9
Malpighian tubule(s) The excretory organ of many arthropods, named for the seventeenth century Italian anatomist Marcello Malpighi, 198
in insects, 606
Maltase, 126, 138
Malthus, Robert, 806
Maltose, 22
Mammalia (mam'al-iă) The class of vertebrates characterized by having endothermic metabolism, hair, and mammary glands; includes such diverse species as shrews, bats, cats, whales, cattle, and human beings, 725
Mammals
adaptations of, 725
characteristics of, 721
early evolution of, 730
Mammary glands, 729
Mammoth, 740
Manatee, 740
Manganese, 16, 141
Mange mites, 582
Mantle Integument of mollusks that is covered by and secretes the shell, 518
Mantle cavity, 518
Manubrium, 475
Marine biological laboratories, 11
Marine life zones, 820
Marmoset, 743
Marsupials (mar-su'pe-als) A group of mammals characterized by having a pouch in which the young are carried and develop for some time after leaving the uterus in a very premature condition, 730
adaptive radiation of, 731
Marsupium, in isopods and amphipods, 593
Mass spectrometer, 17
Mast cells Tissue cells that become active in inflammatory reactions and release heparin and histamine, 170
Mastax, 507
Mastigophora, 449–454
Mastodon, 740
Mating type In Protozoa, a sex. As many as eight sexes are known in some species, 449
Matrix (ma'triks) Nonliving material secreted by and surrounding the connective tissue cells; frequently contains a thick, interlacing matted network of microscopic fibers.
Matter, 14
cyclic use of, 792
Maxillipeds, 587
McClung, C. E., 338
Meantes, 680
Mechanoreceptors, 242
Medulla (me-dul'lah) The inner part of an organ, e.g., the medulla of the kidney; the most posterior part of the brain, lying next to the spinal cord, 201
Medulla oblongata, 229, 230
Medusoid cnidarians, 471, 475
Megazostrodon, 725
Meiosis (mi-o'sis) Kind of nuclear division, usually two successive cell divisions, which results in daughter cells with the haploid number of chromosomes, one half the number of chromosomes in the original cell, 290–293
Meissner's corpuscles, 243
Melanin (mel'ah-nin) A dark-brown or black pigment common in the integument of many animals and sometimes found in other organs; usually occurs within chromatophores, 99, 344, 380
Melanocyte-stimulating hormone (MSH), 265, 273
Melatonin, 265, 277
Membrane, differentially permeable, 52
Membrane proteins, 32
Membrane theory of nerve conduction, 216
Membranelles, 447
Membranous labyrinth, 244, 247
Mendel, Gregor Johann, 338
Mendel's laws, 338
Meninges, 229
Menstrual cycle, 279
Menstruation, 279
Mercenaria, 532
Mercury, 16
Merkel's discs, 243
Meroplankton, 821
Merostomata, 576
Merozoites (mer″o-zo'īt) One of the young forms derived from splitting up of the schizont in the human cycle of the malarial parasite, *Plasmodium;* it is released into the circulating blood and attacks new erythrocytes, 458
Meselson and Stahl experiment, 369
Mesenchyme (mes'eng-kim) A meshwork of loosely associated, often stellate cells found in the embryos of many animals.
Mesenterial filament, 484
Mesentery(ies) (mes'en-ter″e) One of the membranes in vertebrates that extend

Mesentery(ies) *(continued)*
from the body wall to the visceral organs or from one organ to another; consists of two layers of coelomic epithelium and enclosed connective tissue, vessels, and nerves, 91
in anthozoans, 482
Mesoderm (mes'o-derm) The middle layer of the three primary germ layers of the embryo, lying between the ectoderm and the endoderm, 311
Mesoglea (mes″o-gle'ah) A gelatinous matrix located between the epidermis and gastrodermis of cnidarians, 471
Mesohyl, 463
Mesonephros (mes″o-nef'ros) An embryonic vertebrate kidney that succeeds the pronephros; its tubules develop adjacent to the middle portion of the coelom and drain into the archinephric duct, 199
Mesozoa, 439
Messenger RNA (mRNA) A particular kind of ribonucleic acid that is synthesized in the nucleus and passes to the ribosomes in the cytoplasm; combines with RNA in the ribosomes and provides a template for the synthesis of an enzyme or some other specific protein, 371
Metabolic rates, 143–146
Metabolic wastes, 194
Metabolism (mĕ-tab'o-lizm) The sum of all the physical and chemical processes by which living organized substance is produced and maintained; the transformations by which energy and matter are made available for the uses of the organism, 13, 62
Metamerism (met-am'er-izm) The division of the body into a linear series of similar parts or segments, as in annelids and chordates, 91
in annelids, 551
in chordates, 644
Metamorphosis (met″ah-mor'fo-sis) An abrupt transition from one developmental stage to another; e.g., from a larva to an adult, 283, 677
Metanephridium (met″ah-nĕ-frid'e-um) An excretory tubule of invertebrates in which the inner end opens into the coelom by way of a ciliated funnel, 197
Metanephros (met″ah-nef'ros) The adult kidney of reptiles, birds, and mammals, 200
Metaphase, 45
Metarhodopsin, 254
Metatheria, 730
Metazoa (met″ah-zo'ah) All multicellular animals whose cells become differentiated to form tissues; all animals except the protozoa, 444
synopsis of, 439
Metencephalon, 229
Meter (mē'ter) The unit of length in the metric scale; abbreviated m. Defined as 1,650,763.73 wave lengths of krypton under specified conditions; equal to 39.3701 inches. Common multiples and divisions of the meter and their abbreviations are:
centimeter, 10^{-2}m, cm
kilometer, 10^3m, km
micrometer (micron), 10^{-6}m, µm
millimeter, 10^{-3}m, mm
nanometer, 10^{-9}m, nm
Michaelis, Leonor, 66
Microevolution Evolutionary changes that occur within populations, 408
Micromere (mi'kro-mēer) The smaller cell, following unequal cell division, 309
Micronucleus (mi″kro-nu'kle-us) A small nucleus that governs reproduction in ciliates, 448
Microspora, 457, 459
Microtrabecular lattice, 40
Microtubules, 39
Microvilli, 32
Midgets, 274
Midgut, 606
Migration, 411, 765, 782
of birds, 717
of mammals, 727
Milk, 729
Milk ejection reflex, 282
Millipedes, 601
Mimicry (mim'ik-re') An adaptation in which an animal resembles some other living or nonliving object.
aggressive, 781
Batesian, 781
Mullerian, 781
Mineral cycles, 794
Mineral salts, in cells, 28
Mineralocorticoids, 271
Minimum mortality, 805
Minnow, 665
Miracidium larvae, 498
Mitchell, Peter, 76
Mites, 578
Mitochondria (mīt″-o-kon'dre-ah) Spherical or elongate intracellular organelles that contain the electron transmitter system and certain other enzymes; site of oxidative phosphorylation, 37
molecular organization of, 77
Mitosis (mi-to'sis) A form of cell or nuclear division by means of which each of the two daughter nuclei receives exactly the same complement of chromosomes as the parent nucleus had, 44
regulation of, 48
white fish blastula, 47
Mitotic apparatus, 46
MN blood groups, 344
Modifying factors, 357
Moist coniferous forest biome, 816
Mole (mōl) The amount of a chemical compound whose mass in grams is equivalent to its molecular weight, the sum of the atomic weights of its constituent atoms, 20, 732
Mole crabs, 590
Molecular biology, 4
Molecular motion, 51
Molecular weight, 20
Molecule (mol'ĕ-kul) The smallest particle of a covalently bonded element or compound having the composition and properties of a larger part of the substance, 20
Mollusks, 517–549
Moloch, 689
Molting, 727
hormonal control of, 283
Molybdenum, 141
Mongolism. See *Down's syndrome.*
Monkeys, 743
Monoclimax hypothesis, 812
Monocytes, 58
Monogamy, 777
Monogenea, 498–500
Monoplacophora, 530
Monosomic, 379
Monotreme (mon'o-trēm) A member of the most primitive order of living mammals characterized by retaining a cloaca, into which the digestive and urogenital tracts discharge, and an egg-laying habit; includes the duckbilled platypus and spiny anteater, 730
Monozygotic twins, 329
Morgan, T. H., 338
Morphogenetic movements, 314–317
Morphogenetic substances, 382
Mosquito mouthparts, 607
Moth mouthparts, 607
Motivation, 763
Motor neurons, 214
Motor unit All the skeletal muscle fibers supplied by a single neuron, 111
Mouth cavity, 127
in vertebrates, 132
Movement, 12
Mucosa, 137
Mudpuppy, 679
Mud-skipper, 668
Müller, Johannes, 9
Multiple alleles Three or more alternate conditions of a single locus that produce different phenotypes, 357
Muscle(s), 107
actions of, 107
cardiac, 184
crustacean, 114
flight and tymbal, 115
force of, 113
human, 116
molluscan catch, 114
of fish, 115
phasic, 112
power of, 114
tonic, 113
twitch. See *Muscle, phasic.*
work of, 114
Muscle contraction
biochemistry of, 108
physiology of, 111
Muscle spindle, 243
Muscle tissues, comparison of, 57
Muscular tissues, 56
Mussels, 538
Mutation A stable, inherited change in a gene (point mutation), or in a chromosome, or in the number of sets of chromosomes, 375, 408
chromosomal, 376, 419
effects of, 409
gene, 376
rates of, 408
Mutation pressure, 411
Mutualism, 497
Myelencephalon, 229
Myelin (mi'ĕ-lin) The fatty material that forms a sheath around the axons of nerve cells in the central nervous

system and in certain peripheral nerves, 214
Myelin sheath, 59
Myofibrils (mi″o-fi′brils) The contractile fibrils visible in muscle tissue with light microscopy. Composed of groups of myofilaments of actin and myosin (q.v.), 56, 108
Myofilaments, 108
Myoglobin (mi′o-glo-bin) A type of hemoglobin found in some muscles; facilitates the transfer of oxygen from the blood to the muscles, 113
Myomere (mi′o-mēer) The muscle segment of an adult animal, 115
Myopia (mi-o′pe-ah) Nearsightedness; the eyeball is too long and the retina too far from the lens; light rays converge at a point in front of the retina, and are again diverging when they reach it, resulting in a blurred image.
Myosin (mi′o-sin) Minute protein filaments found in most contractile cells; usually associated with actin (q.v.), 48, 108
Myriapods, 598–602
Myxedema (mik″sĕ-de′mah) A condition that results from a deficiency of thyroxin secretion in an adult; characterized by a low metabolic rate and decreased heat production, 264

N

Naiads, 610
Nares (na′rēz) An external opening of a nasal cavity, 156
Natural selection, 401, 404, 412. See also *Selection*.
  direction of, 410
Nature, dynamic balance of, 827
Nauplius eye, 587
Nauplius larva (no′plĭ-us) A larva with three pairs of appendages—future head limbs—characteristic of the crustaceans, 587
Nautiloidea, 547
Nautilus, 542
Navigation, 782
  of birds, 718
Neanderthal man, 756
Necator, 513
Necturus, 679
Nekton (nek′ton) Collective term for the organisms that are active swimmers, 821, 825
Nematocyst (nem′ah-to-sist) A minute stinging structure found on cnidarians and used for anchorage, for defense, and for the capture of prey, 472
Nematoda, 509–514
Nematomorpha, 515
Nemerteans, 502
Neobenedenia, 499
Neognathae, 714
Neopilina, 531
Neornithes, 714
Neoteny, 425, 678
Nephridium(a) (nĕ-frid′e-um) The excretory tubules of many invertebrate animals. May be a protonephridium or a metanephridium (which see).
  in oligochaetes, 563
Nephron (nef′ron) The anatomical and functional unit of the vertebrate kidney, 198
Nereis, 551
Neritic zone, 820
Nerve cells, in cnidarians, 473
Nerve fiber, cable properties of, 217
Nerve impulse, 216
Nerve net A relatively unorganized, diffuse net of nerve cells with no obvious directionality in the transmission of impulses.
Nerve net systems, 221
Nerves
  cranial, 225
  spinal, 224
Nervous system, 213
  evolution of, 221
  organization of, 221
  peripheral, 224
  vertebrate, 223
Nervous tissues, 58
Net energy intake, 778
Net primary production, 799
Neural crest, 317
Neural integration, 212
Neural tube formation, 317
Neurilemma, 59, 214
Neurofibrils, 214
Neuroglia (nū-rŏg′lĭ-ă) Connecting and supporting cells in the central nervous system surrounding the neurons, 59
Neuromasts, 652
Neuromuscular junction, 219
Neuron(s) (nu′ron) A nerve cell with its processes, collaterals, and terminations; the structural unit of the nervous system, 58, 213
  bipolar, 240
  regeneration of, 216
  regions of, 214
  role in behavior, 764
  types of, 312
Neuronal membrane, differential permeability of, 216
Neurophysin, 222
Neuropodium, 552
Neurosecretion (nu″ro-se-kre′shun) The production of hormones by nerve cells, 222, 276, 277
  at synapse, 219
Neurotransmitters, 220
Neurotubules, 214
Neurula (nu′roo-lah) The early embryonic stage during which the primitive nervous system forms, 317
Neuston, 825
Neutrons, 14
Neutrophils, 58
Niche separation, 788
Nicotinamide, 70
Nicotinamide adenine dinucleotide (NAD), 27
Nicotinamide adenine dinucleotide phosphate (NADP), 27
Nictitating membrane, 252
Nissl substance, 214
Nitrate bacteria, 794
Nitrite bacteria, 794
Nitrogen, 15
Nitrogen cycle, 794
Nitrogen excretion, 685
Nitrogenous wastes, 194–195
Nodes of Ranvier, 214, 218
Nondisjunction (non″dis-junk′shun) The failure of a pair of homologous chromosomes to separate normally during the reduction division at meiosis; both members of the pair are carried to the same daughter nucleus and the other daughter cell is lacking in that particular chromosome, 378
Nonelectrolytes, 21
Norepinephrine, 265, 270
Normal distribution curve, 356
Northern coniferous forest, 815
Notochord (no′to-kord) A rod of turgid cells located along the back of chordate embryos ventral to the nerve cord. Acts as a hydroskeleton (q.v.); usually replaced by the vertebral column in adult vertebrates, 101, 312, 314, 641
Notonecta, 788
Notopodium, 552
Nuclear membrane, 12
Nuclear pores, 34
Nuclear sex, 348
Nucleic acids, 21, 26
Nucleoli, 36
Nucleosome, 363
Nucleotide(s) (nu′kle-o-tīd) A molecule composed of a phosphate group, a 5-carbon sugar—ribose or deoxyribose—and a nitrogenous base—a purine or a pyrimidine; one of the subunits into which nucleic acids are split by the action of nucleases, 26
Nucleus (nu′kle-us) The organelle of a cell containing the hereditary material; a group of nerve cell bodies in the central nervous system.
  atomic, 16
  functions of, 33
  in brain, 231
Nucleus-cytoplasmic mass theory, 48
Nudibranch, 525
Nutrition, 125–147
Nutritive muscle cells, 471
Nymph (nimf) A juvenile insect that often resembles the adult and that will become an adult without an intervening pupal stage, 610

O

Obelia, 477
Occipital lobe, 235
Ocean(s)
  buffering action of, 793
  currents in, 797
  depths of, life in, 824
  productivity of, 823
Ocellus(i) (o-sel′us) A simple light receptor found in many different types of invertebrate animals, 250, 602
Octocorallian corals, 484
Octopods, 545
Oculomotor nerve, 226
Odontognathae, 714
Odontophore, 519
Olfaction, 249
Olfactory bulb, neural circuitry of, 241
Olfactory nerve, 226
Oligodendrocytes, 214
Oligosaccharidases, 126

Ommatidium (om″ah-tid′ĭ-um) One of the elements of a compound eye, itself complete with lens and retina, 256
Oncomiracidium, 499
Oncosphere, 502
Onychophora, 601
Oöcytes, 295
Oögenesis (o″o-jen′e-sis) The origin and development of the ovum, 293, 295–297
Oögonium (o′o-go′ne-um) The primordial cell from which the ovarian egg arises; undergoes growth to become a primary oöcyte, 295
Oötid, 295
Opercular apparatus, 676
Operculum, 526
  of gill chamber, 152
Operon The genes whose codes are transcribed on a single mRNA molecule and are under the control of a single repressor.
Ophiuroidea, 628–631
Opiliones, 578
Opisthaptor, 498
Opisthobranchia, 524
Opisthonephros (o′pis-tho-nef′ros) The adult kidney of most fishes and amphibians; its tubules extend from the mesonephric region to the posterior end of the coelom; drained by the archinephric duct and sometimes also by accessory urinary ducts, 200
Opisthorchis, 500
Optic cup, 317
Optic lobes, 233
Optic nerve, 226, 255
Oral contraceptives, 280
Orangutan, 748
Orbital (or′bĭ-tal) The distribution of an electron around the atomic nucleus, 14
Order. See *Taxon.*
Organ(s), 12, 31
  of Corti, 247
  of taste, 248
Organ systems, 12, 31
Organic compounds, 21
Organogenesis, 317
Origin of life, 426
Ornithine cycle, 196
Ornithischia, 692
Orthodromic impulse, 214
Osculum, 463
Osmoconformers (oz″mo-con-for′mers) Marine animals in which the salt concentration of the extracellular body fluids conforms to, or is the same as, that of the environment, 206
Osmoregulation
  in freshwater organisms, 208
  in terrestrial animals, 208–210
Osmoregulators (oz″mo′rĕg′u-lā-tors) Animals in which the salt content of the body is maintained at a different level from that of the environment, 206
Osmosis (os-mo′sis) The passage of solvent molecules from the lesser to the greater concentration of solute when two solutions are separated by a membrane that selectively prevents the passage of solute molecules but is permeable to the solvent, 52
Osmotic regulation in marine animals, 205–208
Osphradia (oz-fra′di a) Sense organ in the mantle cavity of many mollusks, 521
Osphradium, 528
Osteichthyes (os″te-ik′thĭ-ēz) A class of fishes in which bone is present in the skeleton; includes sturgeons, teleosts, and lungfish, 660
  characteristics of, 651
Ostium(a)
  bivalve, 535
  oviduct, 305
Ostracoderm (os′-tra-ko-derm″) A member of one of several jawless and heavily armored fishes that were abundant during the Devonian period; the ancestral vertebrates, 653
Ostracods, 596
Ostrich, 715
Otoliths, 244
Outbreeding, 359
Ovalbumin, 368
Ovary(ies) (o′vah-re) The female gonad that produces eggs, 293
  endocrine functions of, 278
Overpopulation, human, 828
Ovicells, 620
Oviduct, vertebrate, 305
Oviparity, 326
Ovoviviparity, 326
Ovulation (ōv″u-la′shun) The discharge of a mature ovum from the graafian follicle of the ovary, 279
  reflex, 280
  spontaneous, 279
Ovum (o′vum) The female reproductive cell, which after fertilization by a sperm develops into a new member of the same species, 295
Owen, Richard, 10
Owl, 717
Oxaloacetic acid, 71
Oxalosuccinic acid, 71
Oxidations, 28, 69
Oxidative phosphorylation (ok″sĭ-da′tiv fos″fōr-ĭ-la′shun) The conversion of inorganic phosphate to the energy-rich phosphate of ATP by reactions coupled to the transfer of electrons in the electron transmitter system of the mitochondria, 70
  chemiosmotic theory of, 76
  control of, 76
Oxyconformers, 143
Oxygen, 15
  availability of, 123
Oxygen conformers, 123
Oxygen debt The amount of oxygen required to oxidize completely the lactic acid that accumulates during vigorous muscular exercise, 111
Oxygen dissociation curves, 162
Oxygen regulators, 123
Oxyhemoglobin, 161
Oxytocin, 234, 265, 273
Oyster, 538

P

Pacinian corpuscles, 243
Paedomorphosis, 425, 678
Paired appendages, evolution of, 658
Palate, 134
Palmitic acid, 72
Palmityl coenzyme A, 72
Palolo worm, 561
Pancreas, 136
  β cells of, 269
Pancreozymin, 265
Pangolin, 734
Pantotheres, 730
Papulae, 624
Paramecium, 445, 447
Paramyosin (par-ă-mi′o-sin) A type of protein found in the shell adductor muscle of many bivalved mollusks; characterized by the ability to sustain a contraction for a long period of time with minimal energy expenditure, 114
Parapodia (par″ah-po′de-ah) Paired appendages extending laterally from each segment of polychaete worms and bearing setae, 552
Parasite food chains, 800
Parasitism (par″ah-sit′izm) A type of heterotrophic nutrition found among both plants and animals; a parasite lives in or on the living body of a plant or animal (host) and obtains its nourishment from it, 129, 497
  in insects, 610
Parasitology, 4
Parasympathetic systems, 228
Parathyroid glands, 268
Parathyroid hormone (PTH), 265, 267
Parathyroids (par″ah-thi′roids) Small, pea-sized glands situated in the substance of the thyroid gland; their secretion is concerned chiefly with regulating the metabolism of calcium and phosphorus by the body.
Parazoa, 439
Parenchyma, 492
Parenchymula larva, 468
Parental care, 775
Parthenogenesis (par″thĕ-no-jen′ĕ-sis) The development of an unfertilized egg into an adult organism; common among honeybees, wasps, and certain arthropods, 298
Parturition (par″tu-rish′un) The process of giving birth to a child, 331
Pasteur, Louis, 13
Pauropoda, 614
Peanut worms, 566
Pearlfish, 634
Pearls, 533
Peas, inheritance in, 338
Peat, 793
Pedal creeping, 526
Pedicellariae, 624
Pedipalps, 576
Peduncle, 597
Pelagic (pe-laj′ik) An organism that inhabits open water, as in midocean, 821
Pelecypoda, 531
Pelican, 714
Pellicle, in ciliates, 444
Pelycosauria, 694
Penetrance, 384
Penguin, 714
Penis (pe′nis) The copulatory organ of the male; found in most of those species of animals in which fertilization is internal, 300
  in vertebrates, 305
Pentastomida, 440

Pepsin, 136
Peptide bonds, 25, 81
Peranema, 450
Perch, 664
Pericardial cavity (per"ĭ-kar'de-al) The chamber of the coelom containing the heart, 184
Pericardial sinus An enlarged portion of the hemocoel (q.v.) surrounding the heart in many invertebrates with open circulatory systems, 177
  in arthropods, 574
Perilymph, 244, 247
Periosteum (per"ĭ-os'te-um) The layer of connective tissue that covers bones during life, 56, 107
Periostracum, 518
Peripheral nervous system, 223, 224
Periproct, 631
Perissodactyla, 739
Peristalsis (per"ĭ-stal'sis) Powerful, rhythmic waves of muscular contraction and relaxation in the walls of hollow tubular organs such as the ureter or the parts of the digestive tract; serve to move the contents through the tube, 134, 137
Peristomium, 552
Peritoneum (per"ĭ-to-ne'um) Coelomic epithelium and supporting connective tissue that lines a coelomic cavity, 91
Peritrophic membrane, 606
Permafrost, 814
Permeable, 52
Peroxidase, 66, 67
Petaloid, 633
Petroleum, 793
Petromyzon, 655
pH The negative logarithm of the hydrogen ion concentration, by which the degree of acidity or alkalinity of a fluid may be expressed, 21
Phagocytosis (fag"o-si-to'sis) The engulfing of microorganisms, other cells, and foreign particles by a cell such as a white blood cell, 168
Phalangida, 578
Pharyngeal pouches, 319
Pharynx (far'inks) Anterior region of the gut, generally muscular and adapted for ingestion. In vertebrates, that part of the digestive tract from which the gill pouches or slits develop; in higher vertebrates it is bounded anteriorly by the mouth and nasal cavities and posteriorly by the esophagus and larynx, 128, 157
  in turbellarians, 492
  in urochordates, 641
  in vertebrates, 134
Phase contrast lenses, 48
Phenetics, 436
Phenocopy (fe'no-kop"e) The simulation by an individual of traits characteristic of another genotype; results from physical or chemical influences in the environment that change the course of development and produce a trait that mimics that of an individual with a different genotype, 382
Phenotype, 340
Phenylketonuria, 345
Pheromone (fĕr'o-mōn) A substance secreted by one individual into the external environment that influences the behavior or development of another individual, 777
Pholidota, 734
Phonoreception, 245
  in tetrapods, 246
Phoronida, 621
Phosphate esters, 19
Phosphoanhydride bonds, 68
Phosphodiester bridges, 365
Phospholipids, 23, 32, 138
Phosphorus, 16
Phosphorylase, 262
Phosphorylation, substrate level, 71
Photon (fo'ton) A particle of electromagnetic radiation, one quantum of radiant energy.
Photoperiodism (fo"to-pe're-od-izm) The physiologic response of animals and plants to variations of light and darkness, 795
Photophores, 546
Photoreceptors, 249
Photosynthesis, 63
Phylogenetic tree, 437
Phylogeny (fi-loj'e-ne) The evolutionary history of a group of organisms, 434, 435
Phylum. See *Taxon.*
Physalia, 476
Physiology, 4
Phytoflagellates, 449–452
Phytoplankton (fi"to-plank'ton) Microscopic floating plants, most of which are algae, which are distributed throughout the ocean or a lake, 791
Pia mater, 229
Pig, 739
Pika, 742
Pill bugs, 594
  regulation of internal body fluids in, 210
Pinacocytes, 463
Pineal gland, 277
Pinnules, 635
Pinocytotic vesicles (pi"no-si-to'-tic). Ultramicroscopic vesicles that can engulf materials on one surface of a plasma membrane, pass through the membrane, and release the materials on the other side, 188
Pinon-juniper biome, 815
Pinworm, 513
Pithecanthropus. See *Homo erectus*.
Pituitary (pĭ-tu'ĭ-tār"e) A small gland that lies just below the hypothalamus of the brain, to which it is attached by a narrow stalk; the anterior lobe forms in the embryo as an outgrowth of the roof of the mouth and the posterior lobe grows down from the floor of the brain.
  lobes of, 272
Placenta (plah-sen'tah) A structure formed partly from embryonic tissues and partly from the mother's uterine lining by means of which the embryo receives nutrients and oxygen and eliminates wastes, 326, 730
Placental lactogen, 265, 281
Placental mammals, 730. See also *Eutherians.*
Placental viviparity, 326
Placodermi (plak'o-der"mi) A class of Paleozoic jawed fishes with a bony head shield movably articulated with a thoracic shield, 658
Placozoa, 439
Planarians, 492
Plankton (plank'ton) Minute plants and animals suspended in fresh or salt water, 129, 821
Plantigrade foot, 728
Planula (plan'u-lah) Larval stage of cnidarians (hydras, jellyfish, sea anemones, and corals).
Plasma (plaz'mah) The liquid portion of the blood in which the corpuscles are suspended; differs from serum in containing fibrinogen, 57, 167
Plasma cells (plaz'mah) Antibody-producing cells in the tissues that have differentiated from activated B-lymphocytes, 172
Plasma membrane, 12, 32
  characteristics of, 32
  components of, 32
  lipid components of, 23
Plastron A skeletal plate on the ventral surface of an animal, 686
Plate tectonics, 395
Platelets (plāt'let) A small, colorless blood corpuscle of mammals involved in blood clotting, 58, 169
Plato, 401
Platyhelminthes, 491–503
Pleiotropic gene (plī"ō-trō'pik) A gene that affects a number of different characteristics in an individual organism, 409
Pleopods, 587
Plesiosaurs, 687
Plethodonts, 679
Pliny, 7
Pliopithecus, 752
Pluteus larva, 631
Podia, 624
Podocytes, 201
Podzol, 792
Pogonophora, 566
Poikilothermic animals, 143
Poison glands, in arachnids, 580
Polar body Small cell that consists of practically nothing but a nucleus; formed during oögenesis, maturation of the egg, and appears as a speck at the animal pole of the egg, 295
Polar covalent bond, 19
Polyandry, 777
Polychaeta, 522–561
Polycladida, 503
Polyclimax hypothesis, 812
Polygamy, 777
Polygenes (pol"e-jēns') Two or more pairs of genes that affect the same trait in an additive fashion.
Polygenic inheritance, 355
Polygyny, 777
Polymorphism (pol"e-mor'fizm) Differences in form among the members of a species; occurrence of several distinct phenotypes in a population, 93, 414
  in cnidarians, 474
  of proteins, 410
Polypeptide chain, 25
  synthesis of, 373
Polyplacophora, 528
Polyploid, 376
Polyploidy The condition of having more than two full sets of homologous chromosomes, 421. See also

Polyploidy *(continued)*
Allopolyploidy and *Autopolyploidy*.
Polypoid cnidarians, 471
Polyps (pol'ips) Hydra-like animals; the sessile form of many cnidarians; protruding growths from a mucous membrane, 470
Polypterus, 663
Polysaccharidases, 126
Polysomes, 39
Pond, freshwater, as an ecosystem, 791
Pongidae The family of great apes, 748
Pons (ponz) The ventral portion of the metencephalon; it relays certain impulses from the cerebrum to the cerebellum and interconnects the two sides of the cerebellum.
Population(s) The group of individuals of a given species inhabiting a specified geographic area, 786
characteristics of, 804
trophic relationships of, 801
Population cycles, 808
Population dispersion, 809
Population genetics, 342
Population growth curve, 804
Population size, factors limiting, 808
Porifera, 462–469
Porocytes, 463
Porphyropsin, 253
Porpoise, 742
Portal vein (por'tal) A vein, or group of veins, that drain one region and lead to a capillary bed in another organ rather than directly to the heart, e.g., the hepatic portal vein and renal portal vein, 178
Porter, Keith, 40
Portuguese man-of-war, 475
Postganglionic neuron, 227
Postnatal development, 332
Potential energy, 50, 63
Preadaptation The acquisition by an ancestral group of certain characteristics that usually are adaptive to the ancestral mode of life yet at the same time enable a shift in mode of life, e.g., lungs in fish ancestral to tetrapods, 422, 674
Precocial, 712, 730
Precursor (pre-kur'sor) A substance that precedes another substance in a metabolic pathway; a substance from which another substance is synthesized.
Predator food chains, 800
Preformation theory, 380
Preganglionic neuron, 227
Pregnancy, hormones of, 281
Pre-proinsulin, 269
Priapulida, 439, 566
Primates
adaptations of, 743
groups of, 743
Primitive character, 436
Primitive (prim'ĭ-tiv) **streak** A longitudinal groove that develops on the embryonic disc of the eggs of fishes, reptiles, birds, and mammals as a consequence of the movement of cells and formation of mesoderm; it is homologous to the lips of the blastopore and marks the future longitudinal axis of the embryo, 314
Primordium (pri-mor'de-um) The earliest discernible indication during embryonic development of an organ or part, 317
Probability
laws of, 341
methods of calculating, 342
Proboscidea, 739
Proboscis (pro-bos'is) Any tubular process of the head or snout of an animal, usually used in feeding, 527
Proctodeum, 319
Producers, 790
Product law, 341
Profundal zone, 825
Progeny selection, 341
Progesterone, 265, 278
Proglottids (pro-glot'idz) The body sections of a tapeworm, 500
Prohormone, 269
Proinsulin, 269
Prokaryotes Oorganisms that lack membrane-bound nuclei; the bacteria and blue-green algae, 426
Prolactin (LTH), 265, 274, 281
Prolactin-releasing hormone, 281
Pronephros (pro-nef'ros) The first formed kidney of embryonic or larval vertebrates; its tubules develop adjacent to the cranial end of the coelom and form the archinephric duct, 199
Prophase, 45
Propliopithecus, 752
Proprioception, 243
Prosencephalon, 230
Prosobranchia, 524
Prostaglandins, 265, 282
effects of, 282
Prostate gland, in vertebrates, 305
Prosthetic groups, 71
Prostomium, 551
in oligochaetes, 561
in polychaetes, 552
Protandry, 302
Protease (pro'te-ās) An enzyme that catalyzes the digestion of proteins, 126
Protein(s), 21, 24
metabolism of, 194
primary structure of, 25
quaternary structure of, 28
structure of, 25
Protein kinase, 262
Protein synthesis, stages of, 373
Prothoracicotropic hormone, 284
Protobranch bivalves, 533
Protochordates, 640–650
Protogyny, 302
Protohominids, 753
Proton(s) (pro'ton) A basic physical particle present in the nuclei of all atoms that has a positive electric charge and a mass similar to that of a neutron; a hydrogen ion, 14
Protonephridium (pro"to-nĕ-frid'e-um) The flame-cell excretory organs of certain invertebrates, 198
Protostome(s) (pro'to-stōm) An animal in which the blastopore contributes to the formation of the mouth, 439, 518
Prototheria, 730
Prototroch, 521
Protozoa (pro"to-zo'ah) Unicellular organisms that with flagellates, slime molds, certain algae, and fungi constitute the Kingdom Protista, 31, 438, 443–461
Proventriculus, 606, 709
Pseudocoel (su"do-seal) A body cavity between the mesoderm and endoderm; a persistent blastocoel, 506
Pseudocoelom (su"do-se'lom) A body cavity between the mesoderm and endoderm; a persistent blastocoel, 506
Pseudocoelomates, 505–516
Pseudopodia (su'do-pod-ia) Temporary cytoplasmic protrusions of an amoeboid cell; function in amoeboid locomotion (q.v.) and feeding, 99, 117, 454
Pseudoscorpions, 577
Pterobranchia, 647
Pterosauria, 693
Pterygota, 603
Pulmonata, 524
Pulp cavity, 134
Punctuated equilibria The notion that evolution is characterized by the abrupt appearance of species that continue unchanged for millions of years and then usually become extinct, 423
Punnett, R.C., 340
Punnett square, 340
Pupa (pu'pah) A stage in the development of an insect, between the larva and the imago (adult); a form that neither moves nor feeds, 283, 610
Pupil, 251
Putrefaction (pu"trĕ-fak'shun) The enzymatic anaerobic degradation of proteins and amino acids.
Pycnogonida, 584
Pygidium (pi-jid'i-um) The nonsegmental posterior part of a metameric animal that usually bears the anus, 92
Pygostyle, 706
Pyloric caeca, 663
Pyloric sphincter, 136
Pyloric stomach
in decapods, 590
in starfish, 626
Pyramidal pathway, 235
Pyruvic acid, 70
Python, 690

Q

Quantum (kwon'tum) A unit of radiant energy; has no electric charge and very little mass; the energy of a quantum is an inverse function of the wavelength of the radiation.
Quantum mechanics, 14

R

r, 806. See also *Biotic potential.*
"r selection," 807
Rabbit, 742
Race A division of a species; a population that differs from other populations with respect to the frequency of one or more genes; a subgroup of a species distinguished by a certain

combination of morphologic and physiologic traits.
Radial cleavage Type of cleavage pattern characteristic of echinoderms and vertebrates in which the spindle axes are parallel or at right angles to the polar axis, 309
Radial symmetry, 88
Radiata, 439
Radiolarians, 456
Radioles, 129, 559
Radula (raj'oo-la) A rasplike structure in the alimentary tract of chitons, snails, squids, and certain other mollusks, 519
Ram ventilation, 152
Ramapithecus, 753
Rana, 680
Range The portion of the earth in which a given species is found.
Range of tolerance, 786
Raptorial feeding, 129
Rate of dispersion, 804
Rathke's pouch, 272
Rattlesnake, 690
Ray, John, 9
Rays, 660
Reabsorption Term applied to the selective removal of certain substances from the glomerular filtrate by the cells of the convoluted tubules of the kidney and their secretion into the blood stream, 202
Reactions
- endergonic, 65
- exergonic, 65
- oxidation-reduction, 28

Recapitulation The tendency for embryos in the course of their development to repeat, often in modified form, some of the features seen in the embryos of ancestral species, 391
Receptor(s) A sensory cell, or sometimes a free nerve ending, which responds to a given type of stimulus, 238
- for peptide hormones, 262
- for protein hormones, 262
- for steroid hormones, 262
- functions of, 239
- phasic, 242
- types of, 239

Receptor cells, sensitivity of, 239
Receptor mechanisms, 239
Receptor potential, 240
Recessive genes Genes that do not express their phenotype unless carried by both members of a set of homologous chromosomes, i.e., genes that produce their effect only when homozygous, when present in "double dose," 340
Recombination, in populations, 410
Rectal gland, 660
Rectum, 128, 136
Red blood cells. See *Erythrocytes.*
Red tides, 452
Redbugs, 582
Redi, Francesco, 13
Redia (re'dĭ-ah) The second stage of flukes. It reproduces asexually in snails, 499
Reduction, 28, 70
Reefs, 485–488
Reflex(es) (re'fleks) An inborn, automatic, involuntary response to a given stimulus that is determined by the anatomic relations of the involved neurons; the functional units of the nervous system, 223
- conditioned, 223
- inherited, 223

Reflex arc A sequence of sensory, internuncial and motor neurons that conduct the nerve impulses for a given reflex, 223
- monosynaptic, 223

Refractory period The period of time that elapses after the response of a neuron or muscle fiber to one impulse before it can respond again.
Regeneration Regrowth of a lost or injured tissue or part of an organism, 290
Regulation of internal body fluids, 193–211
Reindeer, 814
Relaxin, 265, 278
Releaser, 763
Releasing factors, 234, 375
Releasing hormones Short peptides synthesized in the hypothalamus, secreted into the hypothalamo-hypophyseal portal system and carried to the pituitary, where they initiate the synthesis and release of specific pituitary hormones.
Renal (re'nal) Pertaining to the kidney, 200
Renal corpuscle The complex formed by a glomerulus and the surrounding Bowman's capsule of a kidney tubule; filtration, the first step in urine formation, occurs here, 200
Renal pyramids, 201
Rennin, 136
Replication, semiconservative, 369
Reproduction, 13
- changes in human evolution, 754
- embryonic development in, 308–334
- gamete formation and fertilization in, 289–307
- of birds, 710
- of mammals, 730

Reproductive rate, intrinsic, 804
Reproductive strategies, 776
Reproductive tissues, 59
Reptiles (rep'tīl) A member of a class of terrestrial vertebrates that are covered with horny scales or plates; living representatives include turtles, lizards, snakes and crocodiles.
- adaptations of, 683
- characteristics of, 670
- classification of, 695
- evolution of, 686
- mammal-like, 694

Repugnatorial glands, 602
Resemblances
- biochemical, 391
- embryonic, 390
- physiological, 391

Resource partitioning, 419, 788
Respiration (res"pĭ-ra'shun) Process by which cells utilize oxygen, produce carbon dioxide, and conserve the energy of foodstuff molecules in biologically useful forms such as ATP; also the act of breathing.
Respiratory assembly, 78
Respiratory membrane, 149
Respiratory organ, 150
Respiratory pigments Metal-containing proteins that bind reversibly with oxygen and sometimes with carbon dioxide. They transport gases in the blood: hemoglobin and hemocyanin (q.v.), 160
Respiratory trees, 634
Resting membrane potential, 217
Rete mirabile, 710
Rete testis, 305
Reticular fibers, 56
Reticular formation, 232
Retina (ret'ĭ-nah) The innermost of the three tunics of the eyeball, surrounding the vitreous body and continuous posteriorly with the optic nerve; contains the light-sensitive receptor cells, rods and cones, 251, 252
- organization of, 255

Retinal isomerase, 254
Retinula, 256
Retroperitoneal organs, 91
Rh factor An agglutinogen, originally discovered in the rhesus monkey, which is found in the erythrocytes of about 85 per cent of the white population, 176, 358
Rhabdome, 256
Rhea, 715
Rhinoceros, 739
Rhizopoda, 460
Rhodopsin (ro-dop'sin) A substance in the retina of the eye (visual purple) made up of retinene, a derivative of vitamin A, and a protein, opsin; undergoes a chemical reaction triggered by light that stimulates the receptor cell to send an impulse to the brain, resulting in the sensation of sight, 253
Rhombencephalon, 230
Rhopalium, 244, 479
Rhynchocephalia, 687
Rhynchocoela, 502
Ribbon worms, 502
Riboflavin, 143
Ribonucleic acid (RNA) (ri"bo-nu'kle-ik as'id) Nucleic acid containing the sugar ribose; present in both nucleus and cytoplasm and of prime importance in the synthesis of proteins, 26
- processing of, 372
- synthesis of, 371

Ribose, 23
Ribosomal functions, 373
Ribosomal RNA (rRNA), 371
Ribosomes (ri'bo-sōms) Minute granules composed of protein and ribonucleic acid either free in the cytoplasm or attached to the membranes of the endoplasmic reticulum of a cell; the site of protein synthesis, 37, 371
- subunits of, 37

RNA. See *Ribonucleic acid.*
Rocky shore, 821
Rod(s) (rod) In zoology, the rod-shaped photoreceptive cells of the retina that are particularly sensitive to dim light and mediate black and white vision, 12, 252

Rodenta, 741
Root nodules, 794
Rotifera, 507–509
Roundworms, 509–514
Ruffini's endings, 243

S

S phase, 41
Saccule (sak'ūl) Small, hollow sac in the inner ear lined with sensitive hair cells and containing small stones made of calcium carbonate; contains receptors for the sense of static balance, 244
Salamanders, 678
Salienta, 680
Salivary glands, 134
Salps, 643
Salt, 21
Salt marshes, 819
Saltatory conduction, 218
Salt-excreting glands, 710
Sand dollar, 632
Sandy shore, 821
Saprozoic nutrition, 126
Sarcodina, 454–456
Sarcomastigophora, 449–456
Sarcomere (sar'ko-mēer) The portion of a myofibril between the Z lines, 108
Sarcopterygian (sar'kop-tĕ-rij-ī-an) A member of a subclass of Osteichthyes in which the fins are supported by a central axis of flesh and bone; includes lungfish and crossopterygians, 668
Saturation deficit, 209
Sauria, 689
Saurischia, 692
Sawfish, 660
Scales
  bony, 97, 655
  cosmoid, 653
  ctenoid, 664
  cycloid, 664
  ganoid, 664
  horny, 97, 683
  placoid, 659
Scallops, 538
Scaphopoda, 547
Schizocoel (skiz'o-cēl) A body cavity formed by the splitting of embryonic mesoderm into two layers, 314
Schleiden, M. J., 11, 31
Schwann, Theodor, 11, 31
Schwann cells, 215
Scientific method, 5
Sclera (skle'rah) The tough, fibrous supporting wall of the eyeball forming approximately the posterior five-sixths of the wall; it is continuous anteriorly with the cornea, 251
Scleractinian corals, 484
Sclerospongiae, 468
Scolex, 500
Scorpions, 577
Screw-worm fly, 610
Scrotum (skro'tum) The pouch that contains the testis in most mammals; its wall is composed of integument and muscular and connective tissue layers of the body wall that are everted during the descent of the testis, 304
Scurvy, 141
Scutigera, 602
Scyphistoma, 481
Scyphomedusae, 479
Scyphozoa, 479–481
Sea anemones, 483
Sea butterflies, 524, 525, 527
Sea cow, 740
Sea cucumbers, 633–635
Sea fans, 484
Sea hares, 524, 525, 527
Sea lilies, 635–636
Sea lion, 736
Sea nettles, 479
Sea pansies, 484
Sea pens, 484
Sea slugs, 524, 525
Sea spiders, 584
Sea urchins, 631
Sea walnuts, 472
Sea wasps, 479, 488
Sea whips, 484
Seal, 736
Search image, 780
Sebaceous glands, 98
Secondary palate, 729
Secretin, 265
Segmentation (seg″men-ta'shun) Division of a body or structure into more or less similar parts, 91
Segregation, law of, 339
Selachii, 660
Selection, 419
  balancing, 414
  directional, 413
  disruptive, 414
  epigamic, 775
  kin, 416
  positive and negative, 416
  sexual, 412, 775
  stabilizing, 412
Selection coefficient A measure on a scale of 1 to 10 of the failure of one genotype relative to others. Lethal genotypes have a selection coefficient of 1; the complement of adaptive value (q.v.), 412
Selective affinities, 316
Selective reabsorption, 197
Semicircular canals The three canals or ducts of the membranous labyrinth that lie at right angles to each other in the vertical and horizontal planes of the body; they detect changes in the angular acceleration of the body.
Semicircular ducts, 244
Semilunar fold, 252
Seminal receptacle A portion of the female reproductive tract in which sperm are stored after mating, 301
Seminal vesicle(s) A portion of the male reproductive tract in which sperm are stored before mating; in mammals produces part of the seminal fluid, 300
  in vertebrates, 305
Seminiferous tubules, 302
Sendai viruses, 42
Sense organs, 238
  structure of, 239
  tonic, 243
Sensory coding, 241
Sensory neurons, 214
Sensory pit, 690
Septa, in annelids, 551
Septibranchia, 547
Seral stages, 811
Sere (sēr) A sequence of communities that replace one another in succession in a given area; the transitory communities are called seral stages. The series ends with a climax community typical of the climate in that part of the world, 811
Serial homology, 574
Serpentes, 689
Sertoli cells, 294
Serum (se'rum) The clear portion of a biological fluid separated from its particular elements; light yellow liquid left after clotting of blood has occurred, 169
Sessile life styles (ses'il) Living attached to the substratum like sea anemones, sponges, and barnacles, 88
Setae (se'tah) Bristle-like projection of the body surface composed of some skeletal material, most commonly chitin.
  in oligochaetes, 561
  in polychaetes, 552
Sex, genetic determination of, 347
Sex chromosomes, 347
Sexual reproduction, 290
Sham operation, 34
Sharks, 660
Sheep, 739
Sheep liver fluke, 500
Shell
  in bivalves, 531
  in cephalopods, 542
  in gastropods, 523, 524
Shell membrane, 731
Shipworms, 538
Shrew, 732
Shrimp, 587
Siamang, 748
Sickle cell anemia, 409, 414
Sign stimulus, 763
Silverfish, 605
Sinanthropus. See *Homo erectus*.
Sinoatrial node (si″no-ā'tre-al nod) A small mass of modified cardiac muscle located at the point where the superior vena cava empties into the right atrium; initiates the heart beat and controls the rate of contraction, 185
Sinus glands, 285
Sinus venosus (si'nus ve-no'sus) The first chamber of the heart of lower vertebrates; it receives the systemic veins and opens into the atrium, 177
Siphonoglyph, 483
Siphonophora, 475
Siphons, in bivalves, 537
Sipuncula, 566
Siren, 680
Sirenia, 740
Silk, spider, 578
Size factor in animals, 92
Skates, 660
Skeletal muscle, 57
Skeleton, 99
  of birds, 704
  of fish, 103
  of frog, 680
  of human, 104
  parts of, 101

Skink, 689
Skull, 106
temporal fenestration of, 686
Sleeping sickness, 453
Slipper shells, 527
Slugs, regulation of internal body fluids in, 209
Smell, sense of, 249
Smooth muscle, 57
Snails, regulation of internal body fluids in, 209
Snakes, 689
feeding mechanisms of, 689
prey immobilization, 690
Snowshoe hare, 808
Snowy owls, 808
Soaring, 701
Social group, 778
Social insects, 610
Sodium, 16
Sodium pump, 217, 240
Solar radiation, 794
Sole, 668
Solute (so'lūt) A substance dissolved in a true solution; a solution consists of a solute and a solvent.
Solvent (sol'vent) The fluid medium in which the solute molecules are dissolved in a true solution; a liquid that dissolves or that is capable of dissolving.
Somatotropin, 274
Somites (so'mīts) Paired, blocklike masses of mesoderm, arranged in a longitudinal series alongside the neural tube of the embryo, forming the vertebral column and dorsal muscles, 314
Song birds, 717
Sow bugs, 594
Speciation The evolutionary process by which one species arises from another, 417
allopatric, 418
geographic, 418
phyletic, 423
sympatric, 420
Species (spe'shēz) A group of similar individuals that are capable of interbreeding with each other but are reproductively isolated in nature from other interbreeding groups. See also *Taxon.*
allopatric, 417
classification of, 434
definition of, 408
endemic, 400, 420
on islands, 400
sympatric, 417
Species flocks, 420
Specific base pairing, 365
Sperm, 290
structure of, 59
Sperm cells, 59
Spermatids, 294
Spermatocytes, 294
Spermatogenesis, 293–295
Spermatogonia, 293
Spermatophore(s) (sper-mat'o-fōr) A package secreted by part of the male reproductive tract, enclosing a number of sperm, 300
in arachnids, 582
in cephalopods, 546
in leeches, 569
Spermatozoa, 294–295
Sphenodon, 687
Spicules (spik'ūls) Microscopic mineral needle-like or rodlike skeletal pieces found in sponges and certain other animals, 463
Spiders, 578
Spinal accessory nerve, 226
Spinal cord, 228
Spinal nerve, 224
Spinal reflex, 223
Spindle, 37
mitotic, 45
Spinnerets, 578
Spiny anteater, 729, 730
Spiny sharks, 658
Spiral cleavage A cleavage pattern characteristic of a number of invertebrate phyla, such as annelids and mollusks, in which the cleavage planes are oriented obliquely to the polar axis of the egg, 310
Spiral valve, 659
Spleen, 168
Sponges, 462–469
Spongin (spun'jin) A flexible skeletal material or protein fiber found in many sponges, 463
Spongocoel, 463
Spontaneous generation, 13, 426
Spontaneous ovulators, 280
Sporocyst, 499
Sporozoans, 457–459
Sporozoite, 458
Spring overturn, 826
Squamata, 687
Squamous epithelium, 54
Squids, 544
Stalked barnacles, 597
Standing crop biomass, 799
Stapes, 246
Starches, 21
Starfish, 624–628
Starling, Ernest, 9
Statocyst(s) (stat'o-sist) A cellular cyst containing one or more granules that is used in a variety of animals to sense the direction of gravity, 244
Statolith, 244
Stegosaurus, 692
Steinheim man, 756
Stelleroidea, 624–631
Stenohaline (sten'o-ha'līn) Descriptive of animals restricted to a narrow range of environmental salt concentrations, 206
Stenothermic, 787
Stentor, 445, 447
Sternum (ster'num) The breastbone of a terrestrial vertebrate, or the ventral skeletal piece of an arthropod segment, 104
Steroids (ste'roids) Complex molecules containing carbon atoms arranged in four interlocking rings, three of which contain six carbon atoms each and the fourth of which contains five; the male and female sex hormones and the adrenal cortical hormones, 24
synthesis of, 271
Stink glands, 602
Stomach, 128
in vertebrates, 134–136
Stomodeum (sto"mo-de'um) An ectodermal pouch invaginated at the front of an embryo; contributes to the mouth cavity, 319
Stone canal, 624
Stony corals, 484
Striations of skeletal muscles, 57
Strobila, 481, 500
Structural formula, 22
Sturgeon, 663
Style sac, 520
Stylet A hard structure shaped like a needle or dagger, usually functioning in feeding or sperm transfer, 510
Sublingual glands
mandibular, 134
parotid, 134
Submucosa, 137
Subspecies A subdivision of a species that can be distinguished from other subdivisions on the basis of a set of characteristics and a distinctive geographic range. Subspecies normally interbreed to some extent, 418
Subtidal zone, 820
Subungulates, 739
Succession
primary, 811
secondary, 811
Succinic acid, oxidation of, 70
Succinyl coenzyme A, 71
Suckers, 565, 665
Sucrase, 65, 126, 138
Sucrose, 22
Sugars, 21
Sulcus of Rolando, 234
Sum law, 341
Summation The adding of one response to another, as in types of muscle contraction in which stimuli follow one another before the effects of previous ones have dissipated (temporal summation), or additional motor units are activated (spacial summation), 113
Sun stars, 624
Sunfish, 664
Superficial cleavage, 309
Supplementary genes, 354
Suprabranchial cavity, 535
Supratidal zone, 820
Surfactants, 154
Survivorship curve, 805
Suspension feeding, 129
Sutton, W.S., 338
Swallow, 718
Swammerdam, Jan, 9
Swanscombe man, 756
Sweat glands, 98, 143
Swim bladder A gas-filled chamber, homologous to a lung, situated in the dorsal part of the abdominal cavity of teleost fishes. By controlling its gas content, a fish can attain neutral buoyancy, 156, 652, 661
Swimmerets, 587
Syconoid sponges, 465
Symbiosis (sim"bi-o'sis) The living together of two dissimilar organisms; association may form mutualism, commensalism, or parasitism, 497
Symmetry, 88
Sympathetic system (sim"pah-thet'ik) A division of the autonomic nervous system in which fibers leave the central nervous system with certain

Sympathetic system *(continued)* thoracic and lumbar nerves and go to the sweat glands and visceral organs; has an effect on most organs that is antagonistic to parasympathetic stimulation, 228
Symphyla, 614
Synapse (sin'aps) The junction between the axon of one neuron and the dendrite of the next, 58
  transmission at, 218
Synapsis (sĭ-nap'sis) The pairing and union side by side of homologous chromosomes from the male and female pronuclei early in meiosis, 291
Synaptic vesicles, 220
Synchrony, reproductive, 298
Syncytium (sin-sit'e-um) A multinucleate mass of cytoplasm produced by the merging of cells.
Synecology, 786
Synsacrum, 705
Syrinx, 710
Systems analysis, 829
Systole (sĭs'to-le) The contraction of the heart, especially its ventricles, 184

T

Tactile hairs of insects, 240
Tactile sense, 242
Tadpole larva, in urochordates, 641
Taenia, 502
Taeniarhynchus, 502
Taiga (ti'ga) Northern coniferous forest biome found particularly in Canada, northern Europe and Siberia, 815
Tapeworms, 500–502
Tapir, 739
Tarantulas, 582
Tardigrada, 601
Tarsi, 606
Taste buds, 12, 249
Taxon (tak'son) One of the units in the classification of organisms. The major taxa from the largest, or most inclusive, to the smallest are kingdom, phylum, class, order, family, genus, and species, 434
Taxonomic nomenclature, 436
Taxonomy (taks-on'o-me) The science of naming, describing, and classifying organisms, 4, 388, 434
Tay-Sachs disease, 345
Tectorial membrane, 247
Teeth
  of herbivores, 737
  of mammals, 728
  structure of, 133
  types of, 738
Tegument, 498
Telencephalon (tel″en-sef'ah-lon) The most anterior of the five major subdivisions of the brain; includes the olfactory bulbs and cerebral hemispheres, 229
Teleost(s) (tel'e-ost) A member of the most advanced group of actinopterygian fishes; includes most of the familiar species of fish, such as herring, salmon, eels, minnows, suckers, catfish, bass, and perch, 664
  adaptive radiation of, 665
Telodendria, 214
Telolecithal eggs, 297
Telophase (tel'o-fāz) The last of the four stages of mitosis, during which the two daughter nuclei appear and the cytoplasm usually divides, 48
Telson, 576
Temin, Howard, 371
Temperate deciduous forest biome, 816
Temporal lobe, 235
Tendon(s), 56, 107
Tenrec, 732
Termination of peptide chain, 374
Terminator codons, 375
Termites, 610
Territoriality (ter'i-tor'ĭ-al'ĭ-te) Behavior pattern in which an animal (usually a male) delineates a territory of his own and defends it against intrusion by other members of the same species and sex, 710, 773, 809
Tesdudines, 686
Test, in echinoids, 631
Testis(es), 293
  endocrine functions of, 277
  in vertebrates, 302
Testosterone, 265, 277
Tetany, 268
Tetrad (tet'rad) A bundle of four homologous chromatids produced at the end of the first meiotic prophase, 291
Tetrapod (tet'ra-pod) The terrestrial vertebrates, most of which have four limbs; amphibians, reptiles, birds, and mammals, 674
  evolution of, 675
Thalamus, 233
Theory, 5
Therapsids (ther-ap'sids) An order of mammal-like reptiles of the Permian period from which mammals evolved, 694
Thermal stratification, 826
Thermocline, 826
Thermodynamics (ther″mo-di-nam'iks) **first law of** Law that states that energy is neither created nor destroyed but only transformed from one kind to another, 64
  second law of, 64
Thermoreceptors, 257
Thermoregulation. See also *Cold Stress* and *Heat Stress.*
  in birds, 701
  in mammals, 726
  in reptiles, 683
Thiamine pyrophosphate, 71
Thioesters, 19
Thorax, in insects, 602
Thrombin (throm'bin) The enzyme derived from prothrombin that converts fibrinogen to fibrin, 169
Thrombus (throm'bus) A clot in a blood vessel or one of the chambers of the heart that remains at the point of its formation, 170
Thymine, 26, 364
Thymus (thī'mus) An organ of pharyngeal pouch origin located in the base of the neck. Site of production of T-lymphocytes in young mammals; atrophies at maturity, 172
Thyroglobulin, 264
Thyroid gland, 264
Thyroid-stimulating hormone (TSH), 265, 266, 274
Thyrotropin releasing hormone (TRH), 267, 277
Thyroxine, 264, 265, 679
Ticks, 578
Tidal cycles, 766
Tissue(s), 12, 31
  epithelial, 54
  types of, 53
Tissue culture, 48
Tissue proteins, half-life of, 80
Toads, 680
Tolerance, range of, 786
Tongue, 134, 676
Tonsils, 172
Tonus, 223
Tools
  Acheulean, 755
  Aurignacian, 757
  Mousterian, 756
  Olduwan, 754
Tornaria larva, 647
Torsion, 521
Trace elements, 16, 66
Trachea(e) (tra'ke-ah) An air-conducting tube. In terrestrial vertebrates it is the main trunk of the system of tubes through which air passes to and from the lungs; in terrestrial arthropods it is one of a system of minute tubules that permeate the body and deliver air to the tissues, 153, 157
  in insects, 607
Tracheal tubules. See *Trachea.*
Tracheoles, 153
Trade winds, 797
Traits
  allelomorphic, 339
  sex-influenced, 350
Transamination, 195
Transcription
  initiation of, 371
  termination of, 371
Transducers (trans-du'sers) Devices receiving energy from one system in one form and supplying it to a second system in a different form; e.g., converting radiant energy to chemical energy.
Transduction, 363
Transfer RNA (tRNA) A form of RNA composed of about 70 nucleotides that serve as adaptor molecules in the synthesis of proteins. An amino acid is bound to a specific kind of transfer RNA and then arranged in order by the complementary nature of the nucleotide triplet (codon) in template or messenger RNA and the triplet anticodon of transfer RNA, 371
Transforming agent, 363
Transition from water to land, 674
Translocation, 376
Transmissional region, 214
Tree shrew, 732
Tree sloth, 733
Trematoda, 498–500
Triacylglycerols, 23
Tricarboxylic acid cycle, 71
Triceratops, 692
Trichinella, 514
Trichinosis, 513
Trichocyst (trik'o-sist) A cellular organelle in the cytoplasm of ciliated

protozoa such as *Paramecium* that can discharge a filament that may aid in trapping and holding prey, 445
Trichoplax, 439
Tricladida, 503
Triconodonts, 730
Trigeminal nerve, 226
Trilobitomorpha, 575
Triplet code The sequences of three nucleotides that compose the codons, the units of genetic information in DNA that specify the order of amino acids in a peptide chain.
Triplet state The state resulting when an electron is activated by absorbing a photon, moves to an outer orbital of higher energy and pairs with an electron of like spin.
Triploid (trip'loid) An individual or cell having three sets of chromosomes.
Trisomic, 379
Trituration, 128
Trochlear nerve, 226
Trochophore (tro'ko-fōr) The top-shaped larva of marine mollusks and polychaete annelids that bears a girdle of cilia, 518, 521
Trochophore larva, in annelids, 561
Trophic level, 799
Trophoblast, 326, 731
Tropical rain forest biome, 817
Trypanosoma, 453
Trypanosomids, 452
Trypsin, 138
Trypsinogen, 138
Tryptophan synthetase, subunits of, 368
Tuatara, 687
Tube feet, 624
Tuber cinereum, 280
Tubilidentata, 734
Tubule(s)
  kidney, 200
  of Cuvier, 635
Tubulin (tub'il-in) The protein filaments that make up the walls of the microtubules of cilia and flagella, 39
Tuna, 666
Tundra A treeless plain between the taiga in the south and the polar ice cap in the north; characterized by low temperatures, a short growing season and ground that is frozen most of the year, 814
Tunic, 642
Tunicates, 641–644
Tunicine, 642
Turbellaria, 492–497
Turner's syndrome, 348, 379
Turnover number, 65
Turtles, 686
Tusks, 739
Tympanic canal, 247
Tympanic membrane, 246
Tyrannosaurus, 692
Tyrosinase, 344, 380
Twinning, 329

U

Ubiquinone Coenzyme Q, a component of the electron transmitter system; consists of a head, a six-membered carbon ring, which can take up and release electrons, and a long tail composed of a chain of carbon atoms, 70
Umbilical cord (um-bil'i-k'l) The stalk, or cord, attached to the navel and connecting the embryo with the placenta; it contains the umbilical arteries and veins and the remnants of the allantois and yolk stalk, 326
Umbilical vessels, 326
Umbrella substance, 34
Undulating membranes, 447
Ungulates (ung'gu-lātes) Four-legged mammals that walk on the tips of their digits and have lost one or more toes. The end of each remaining digit is protected by a hoof, 737
Unguligrade foot, 738
Uniformitarianism, 403
Uniramia, 597–613
Uracil, 26
Uranium, 28
Urbanization, 829
Urea (u-re'ah) One of the end products of protein metabolism; the diamide of carbonic acid, NH_2CONH_2; soluble in water, 195
  synthesis of, 196
Urease, 66
Ureter (u-re'ter) The fibromuscular tube that conveys urine from a metanephric kidney to the cloaca or exterior of the body, 200
Urethra (u-re'thrah) The membranous canal conveying urine from the bladder to the exterior of the body, 200
Uric acid (u'rik) An end product of nucleic acid and protein metabolism with a low solubility in water; particularly abundant in certain terrestrial animals; $C_5H_4N_4O_3$, 195
Uridine triphosphate (UTP), 27
Urine, 197
  formation of, 201
Urochordata, 641–644
Urochrome, 201
Uropygeal gland, 702
Uterus (u'ter-us) in vertebrates The womb; the hollow, muscular organ of the female reproductive tract in which the fetus undergoes development, 305
Utricle, 244

V

Vaccination (vak'sēn-a-shun) Treatment of a patient with a commercially prepared antigen that is strong enough to stimulate the patient to make antibodies but not sufficiently strong to cause the disease's harmful effects, 172
Vacuoles, 40, 208
Vagina (vah-ji'nah) In many kinds of animals, the terminal portion of the female reproductive tract; receives the male copulatory organ, 300, 305
Vagus nerve, 226
Valence (va'lens) An expression of the number of atoms of hydrogen (or its equivalent) that one atom of a chemical element can hold in combination, if negative, or displace in a reaction, if positive; the number of electrons gained, lost or shared by the atom in forming bonds with one or more other atoms.
Valence electrons, 19
Van der Waal's bonds, 19
Van Leeuwenhoek, Antony, 9
Vanadium, 53
Variation
  forces acting on, 411
  geographic, 418
  in populations, 408
Vas deferens, 305
Vasa efferentia, 304
Vascular tissues, 57
Vasopressin, 204, 234, 265, 273
Veins (vān) A vessel through which blood passes from the tissues toward the heart; typically has thin walls and valves that prevent a reverse flow of blood, 190
  in insect wings, 605
Veliger (vel'e-jer) Larval stage of many marine clams and snails, 528
Velum, 475
Ventral ramus, 225
Ventral root, 225
Ventricle(s) (ven'trĭ-k'l) A cavity of an organ such as one of the chambers of the heart that receive blood from the atria, or one of the several chambers of the brain, 177, 179, 182, 184
  of brain, 229
Venus's flower-basket, 464
Vertebrae, 105
  in brittle stars, 629
Vertebrata, 646
Vertebrates (ver'tĕ-brāt) The subphylum of chordates characterized by the presence of a vertebral column; includes fishes, amphibians, reptiles, birds, and mammals.
  origin of, 425
Vesalius, Andreas, 8
Vestibular canal, 247
Vestibulocochlear nerve, 226
Vestigial organ An organ present in one animal that is a remnant of a well-developed homologous organ in another animal; vestigial organs are without functions, or their original functions are greatly reduced, 390
Villus(i) (vil'lus) A small, finger-like vascular process or protrusion, especially a protrusion from the free surface of a membrane such as the lining of the intestine, 137
  placental, 326
Virchow, 31
Visceral arches (vis'er-al) An arch of cartilage or bone that develops in the wall of the pharynx between the gill slits; they support the gills in fishes, but certain ones become incorporated into the skull, hyoid apparatus, and larynx of higher vertebrates, 103, 106
Vision, chemistry of, 253
Visual acuity, 257
Visual images, processing of, 255
Visual purple, 253
Vitamin(s) (vi-tah-min) An organic substance necessary in small amounts for the normal metabolic functioning of a given organism; must be present in the diet because the organism

Vitamin(s) *(continued)*
cannot synthesize an adequate amount of it, 141–144
Vitamin C, in treatment of colds, 6
Vitellaria, 623
Vitelline membrane, 297
Vitreous humor, 252
Viviparity, 326
Vocal cords, 157
Voles, 808
Volvocida, 450
Von Baer, Karl Ernst, 9
Von Tschermak, 338
Vorticella, 445, 447

W

Wallabies, 731
Wallace, 400, 401, 403
Walrus, 736
Warning coloration. See *Aposematic coloration.*
Wasps, 610
Water, 28
functions of, 20
hydrogen bonding of, 18
properties of, 20
Water cycle, 794
Water expulsion vesicle Osmoregulatory organelle of protozoa and sponges, 208, 448
Water fleas, 594
Watson-Crick model, 366
Waxes, 24
Web-building spiders, 582
Weberian ossicles, 665
Weinberg, G., 343
Weismann, August, 31
Whale, 742
Whalebone, 742
Wheel organ, 644
White blood cells. See *Leukocytes.*
White matter, 228
Wing, 697, 706
Wolff, Kaspar, 9
Woodlice, 594
Worm lizards, 690
Worm shells, 525, 527
Wuchereria, 513

X

X chromosomes, 348
X organ Organ present in crustacea that produces hormones that regulate molting, metabolism, reproduction, the distribution of pigment in the compound eyes, and the control of pigmentation of the body, 285
Xenopus, 680
Xiphosura, 614

Y

Y chromosome, 348
Y organ, 285
Yolk, 297
Yolk plug, 314
Yolk sac A pouchlike outgrowth of the digestive tract of certain vertebrate embryos that grows around the yolk, digests it, and makes it available to the rest of the organism, 325, 685
Yolk sac placenta, 660

Z

Zinc, 16, 141
Zoantharia, 488
Zoea larva, 593
Zooflagellates, 452–454
Zooids, in bryozoans, 618
Zoology
applications of, 11
history of, 6
Zooplankton, 821
Zooxanthellae (zo″ahz-an-thel′-e) Dinoflagellates living symbiotically with certain marine animals, especially corals, 484
Zygote (zi′got) The cell formed by the union of two gametes; a fertilized egg, 298

LENGTH

Centimeters Inches

A centimetre is less than half an inch.

A pen cap is about 1 centimetre in diameter at its widest part

A metre is a little longer than a yard

A kilometre is about 3/5 of a mile

Speed limit in town is 50 kph (31 mph)

150 kph (93 mph) is usually a speeding ticket

Length Conversion

1 in. = 2.5 cm	1 mm = 0.04 in.
1 ft. = 30 cm	1 cm = 0.4 in.
1 yd = 0.9 m	1m = 40 in.
1 mi = 1.6 km	1 m = 1.1 yd
	1 km = 0.6 mi

WEIGHT

A plastic pen cap weighs about 1 gram

A recipe for lasagna calls for 700 grams of ground beef and 113 grams of shredded cheese

This book weighs about 1.5 kilograms

A football player can weigh as much as 126 kilograms

Weight Conversion

1 oz = 28 g	1 g = 0.035 oz
1 lb = 0.45 kg	1 kg = 2.2 lb